Organic Chemistry

Seventh Edition

Janice Gorzynski Smith
University of Hawai'i at Mānoa

ORGANIC CHEMISTRY

Some ancillaries, including electronic and print components, may not be available to customers outside the United States.

This book is printed on acid-free paper.

1 2 3 4 5 6 7 8 9 LWI 27 26 25 24 23

ISBN 978-1-266-22393-8
MHID 1-266-22393-2

Cover Image: *Ethan Daniels/Stocktrek Images, Inc./Alamy Stock Photo*

mheducation.com/highered

About the Author

Daniel C. Smith

Janice Gorzynski Smith was born in Schenectady, New York. She received an A.B. degree *summa cum laude* in chemistry from Cornell University, and a Ph.D. in Organic Chemistry from Harvard University under the direction of Nobel Laureate E. J. Corey. During her tenure with the Corey group, she completed the total synthesis of the plant growth hormone gibberellic acid.

Following her postdoctoral work, Jan joined the faculty of Mount Holyoke College, where she was employed for 21 years. During this time she was active in teaching organic chemistry lecture and lab courses, conducting a research program in organic synthesis, and serving as department chair. Her organic chemistry class was named one of Mount Holyoke's "Don't-miss courses" in a survey by *Boston* magazine. Jan and her family moved to Hawai'i in 2000, and she became a faculty member at the University of Hawai'i at Mānoa. In 2003, she received the Chancellor's Citation for Meritorious Teaching.

Jan resides in Hawai'i with her husband Dan, an emergency medicine physician, pictured with her in the Galápagos Islands in 2021. She has four children and nine grandchildren. When not teaching, writing, or enjoying her family, Jan bikes, hikes, snorkels, and scuba dives in sunny Hawai'i, and time permitting, enjoys travel and Hawaiian quilting.

Contents in Brief

Prologue 1
1 Structure and Bonding 5
2 Acids and Bases 57
3 Introduction to Organic Molecules and Functional Groups 92
4 Alkanes 130
5 Stereochemistry 178
6 Understanding Organic Reactions 220
7 Alkyl Halides and Nucleophilic Substitution 253
8 Alkyl Halides and Elimination Reactions 305
9 Alcohols, Ethers, and Related Compounds 342
10 Alkenes and Addition Reactions 393
11 Alkynes and Synthesis 436
12 Oxidation and Reduction 466
Spectroscopy A Mass Spectrometry 507
Spectroscopy B Infrared Spectroscopy 529
Spectroscopy C Nuclear Magnetic Resonance Spectroscopy 557
13 Radical Reactions 606
14 Conjugation, Resonance, and Dienes 645
15 Benzene and Aromatic Compounds 686
16 Reactions of Aromatic Compounds 722
17 Introduction to Carbonyl Chemistry; Organometallic Reagents; Oxidation and Reduction 777
18 Aldehydes and Ketones—Nucleophilic Addition 832
19 Carboxylic Acids and Nitriles 888
20 Carboxylic Acids and Their Derivatives—Nucleophilic Acyl Substitution 928
21 Substitution Reactions of Carbonyl Compounds at the α Carbon 976
22 Carbonyl Condensation Reactions 1015
23 Amines 1053
24 Carbon–Carbon Bond-Forming Reactions in Organic Synthesis 1106
25 Pericyclic Reactions 1141
26 Carbohydrates 1174
27 Amino Acids and Proteins 1222
28 Nucleic Acids and Protein Synthesis 1268
29 Lipids 1295 (Available online)
30 Metabolism 1328 (Available online)
31 Synthetic Polymers 1358 (Available online)
Appendices A-1
Glossary G-1
Index I-1

Contents

Preface xx

Acknowledgments xxvii

Prologue 1

What Is Organic Chemistry? 1

Some Representative Molecules 2

The Marine Environment 4

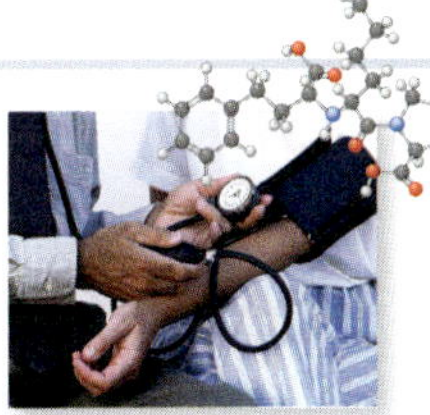

Russell Illig/Photodisc/Getty Images

1 Structure and Bonding 5

1.1 The Periodic Table 6
1.2 Bonding 8
1.3 Lewis Structures 10
1.4 Isomers 15
1.5 Exceptions to the Octet Rule 16
1.6 Resonance 16
1.7 Determining Molecular Shape 23
1.8 Drawing Organic Structures 27
1.9 Hybridization 33
1.10 Ethane, Ethylene, and Acetylene 36
1.11 Bond Length and Bond Strength 41
1.12 Electronegativity and Bond Polarity 42
1.13 Polarity of Molecules 44
1.14 Oxybenzone—A Representative Organic Molecule 45

Chapter 1 Review 47
Key Concepts 47
Key Skills 48
Multiple-Choice Self-Test 50
Problems 51

Comstock/Getty Images

2 Acids and Bases 57

2.1 Brønsted–Lowry Acids and Bases 58
2.2 Reactions of Brønsted–Lowry Acids and Bases 59
2.3 Acid Strength and pK_a 62
2.4 Predicting the Outcome of Acid–Base Reactions 65
2.5 Factors That Determine Acid Strength 66
2.6 Common Acids and Bases 76
2.7 Aspirin 77
2.8 Lewis Acids and Bases 78

Chapter 2 Review 81
Key Concepts 81
Key Skills 82
Multiple-Choice Self-Test 84
Problems 85

Pixtal/age fotostock

3 Introduction to Organic Molecules and Functional Groups 92

3.1 Functional Groups 93
3.2 An Overview of Functional Groups 94
3.3 Intermolecular Forces 101
3.4 Physical Properties 104
3.5 Application: Vitamins 111
3.6 Application of Solubility: Soap 112
3.7 Application: The Cell Membrane 113
3.8 Functional Groups and Reactivity 116
3.9 Biomolecules 118
Chapter 3 Review 119
Key Concepts 119
Key Skills 121
Multiple-Choice Self-Test 122
Problems 123

Daniel C. Smith

4 Alkanes 130

4.1 Alkanes—An Introduction 131
4.2 Cycloalkanes 133
4.3 An Introduction to Nomenclature 134
4.4 Naming Alkanes 135
4.5 Naming Cycloalkanes 141
4.6 Natural Occurrence of Alkanes 142
4.7 Properties of Alkanes 144
4.8 Conformations of Acyclic Alkanes—Ethane 145
4.9 Conformations of Butane 148
4.10 An Introduction to Cycloalkanes 152
4.11 Cyclohexane 153
4.12 Substituted Cycloalkanes 156
4.13 Oxidation of Alkanes 162
4.14 Lipids—Part 1 165
Chapter 4 Review 167
Key Concepts 167
Key Skills 169
Multiple-Choice Self-Test 171
Problems 172

George Ostertag/
Alamy Stock Photo

5 Stereochemistry 178

5.1 Starch and Cellulose 179
5.2 The Two Major Classes of Isomers 181
5.3 Looking Glass Chemistry—Chiral and Achiral Molecules 181
5.4 Stereogenic Centers 184
5.5 Stereogenic Centers in Cyclic Compounds 188
5.6 Labeling Stereogenic Centers with *R* or *S* 190
5.7 Disastereomers 195
5.8 Meso Compounds 197

5.9 *R* and *S* Assignments in Compounds with Two or More Stereogenic Centers 199
5.10 Disubstituted Cycloalkanes 199
5.11 Isomers—A Summary 201
5.12 Physical Properties of Stereoisomers 202
5.13 Chemical Properties of Enantiomers 206
Chapter 5 Review 209
Key Concepts 209
Key Skills 210
Multiple-Choice Self-Test 212
Problems 213

wasanajai/Shutterstock

6 Understanding Organic Reactions 220

6.1 Writing Equations for Organic Reactions 221
6.2 Kinds of Organic Reactions 222
6.3 Bond Breaking and Bond Making 224
6.4 Bond Dissociation Energy 228
6.5 Thermodynamics 231
6.6 Enthalpy and Entropy 233
6.7 Energy Diagrams 235
6.8 Energy Diagram for a Two-Step Reaction Mechanism 237
6.9 Kinetics 239
6.10 Catalysts 242
6.11 Enzymes 243
Chapter 6 Review 244
Key Concepts 244
Key Equations 245
Key Skills 245
Multiple-Choice Self-Test 246
Problems 247

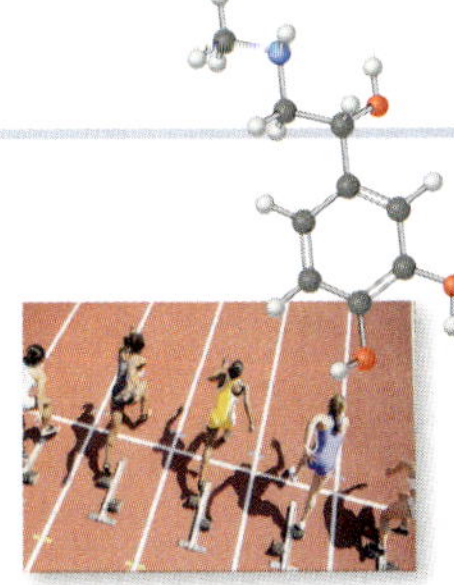

Randy Faris/Corbis/Getty Images

7 Alkyl Halides and Nucleophilic Substitution 253

7.1 Introduction to Alkyl Halides 254
7.2 Nomenclature 255
7.3 Properties of Alkyl Halides 256
7.4 Interesting Alkyl Halides 256
7.5 The Polar Carbon–Halogen Bond 258
7.6 General Features of Nucleophilic Substitution 259
7.7 The Leaving Group 261
7.8 The Nucleophile 263
7.9 The Mechanisms for Nucleophilic Substitution 266
7.10 The S_N2 Mechanism 268
7.11 The S_N1 Mechanism 274
7.12 Carbocation Stability 278
7.13 The Hammond Postulate 279
7.14 When Is the Mechanism S_N1 or S_N2? 282
7.15 Biological Nucleophilic Substitution 288

7.16 Vinyl Halides and Aryl Halides 291
7.17 Organic Synthesis 291
Chapter 7 Review 293
Key Concepts 293
Key Skills 295
Key Mechanism Concepts 296
Multiple-Choice Self-Test 297
Problems 298

Forest & Kim Starr

8 Alkyl Halides and Elimination Reactions 305

8.1 General Features of Elimination 306
8.2 Alkenes—The Products of Elimination Reactions 307
8.3 The Mechanisms of Elimination 311
8.4 The E2 Mechanism 311
8.5 The Zaitsev Rule 314
8.6 Stereochemistry of the E2 Reaction 316
8.7 The E1 Mechanism 320
8.8 S_N1 and E1 Reactions 323
8.9 When Is the Mechanism E1 or E2? 324
8.10 E2 Reactions and Alkyne Synthesis 324
8.11 When Is the Reaction S_N1, S_N2, E1, or E2? 326
Chapter 8 Review 330
Key Concepts 330
Key Skills 331
Key Mechanism Concepts 333
Multiple-Choice Self-Test 334
Problems 335

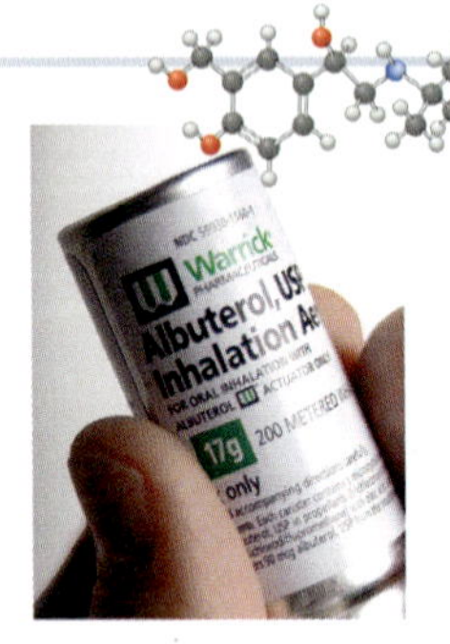

Jill Braaten/McGraw Hill

9 Alcohols, Ethers, and Related Compounds 342

9.1 Introduction 343
9.2 Structure and Bonding 344
9.3 Nomenclature 344
9.4 Properties of Alcohols, Ethers, and Epoxides 348
9.5 Interesting Alcohols, Ethers, and Epoxides 349
9.6 Preparation of Alcohols, Ethers, and Epoxides 350
9.7 General Features—Reactions of Alcohols, Ethers, and Epoxides 353
9.8 Dehydration of Alcohols to Alkenes 354
9.9 Carbocation Rearrangements 357
9.10 Dehydration Using $POCl_3$ and Pyridine 359
9.11 Conversion of Alcohols to Alkyl Halides with HX 361
9.12 Conversion of Alcohols to Alkyl Halides with $SOCl_2$ and PBr_3 364
9.13 Tosylate—Another Good Leaving Group 367
9.14 Reaction of Ethers with Strong Acid 370
9.15 Thiols and Sulfides 371
9.16 Reactions of Epoxides 374
9.17 Application: Epoxides, Leukotrienes, and Asthma 378

9.18 Application: Benzo[*a*]pyrene, Epoxides, and Cancer 380

Chapter 9 Review 380

Key Concepts 380

Key Reactions 381

Key Skills 382

Key Mechanism Concepts in Reactions of Alcohols 384

Multiple-Choice Self-Test 385

Problems 386

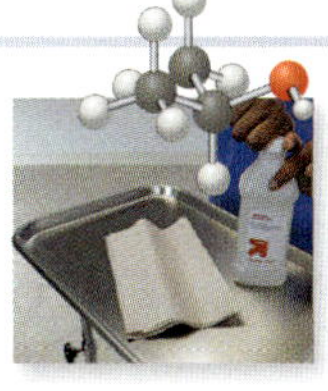

McGraw Hill

10 Alkenes and Addition Reactions 393

10.1 Introduction 394

10.2 Calculating Degrees of Unsaturation 395

10.3 Nomenclature 397

10.4 Properties of Alkenes 401

10.5 Interesting Alkenes 402

10.6 Lipids—Part 2 403

10.7 Preparation of Alkenes 405

10.8 Introduction to Addition Reactions 405

10.9 Hydrohalogenation—Electrophilic Addition of HX 407

10.10 Markovnikov's Rule 408

10.11 Stereochemistry of Electrophilic Addition of HX 410

10.12 Hydration—Electrophilic Addition of Water 412

10.13 Halogenation—Addition of Halogen 413

10.14 Stereochemistry of Halogenation 414

10.15 Halohydrin Formation 416

10.16 Hydroboration–Oxidation 419

10.17 Keeping Track of Reactions 423

10.18 Alkenes in Organic Synthesis 425

Chapter 10 Review 427

Key Concepts 427

Key Reactions 428

Key Skills 428

Key Mechanism Concepts 429

Multiple-Choice Self-Test 430

Problems 431

McGraw Hill

11 Alkynes and Synthesis 436

11.1 Introduction 437

11.2 Nomenclature 438

11.3 Properties of Alkynes 439

11.4 Interesting Alkynes 439

11.5 Preparation of Alkynes 441

11.6 Introduction to Alkyne Reactions 441

11.7 Addition of Hydrogen Halides 443

11.8 Addition of Halogen 445

11.9 Addition of Water 446

11.10 Hydroboration–Oxidation 448

11.11 Reaction of Acetylide Anions 450

11.12 Synthesis 454
Chapter 11 Review 457
Key Reactions 457
Key Skills 457
Multiple-Choice Self-Test 459
Problems 461

Amarita/Shutterstock

12 Oxidation and Reduction 466

12.1 Introduction 467
12.2 Reducing Agents 468
12.3 Reduction of Alkenes 469
12.4 Application: Hydrogenation of Oils 472
12.5 Reduction of Alkynes 474
12.6 The Reduction of Polar C—X σ Bonds 477
12.7 Oxidizing Agents 478
12.8 Epoxidation 479
12.9 Dihydroxylation 482
12.10 Oxidative Cleavage of Alkenes 485
12.11 Oxidative Cleavage of Alkynes 487
12.12 Oxidation of Alcohols 488
12.13 Green Chemistry 491
12.14 Biological Oxidation 492
12.15 Sharpless Epoxidation 493
Chapter 12 Review 496
Key Concepts 496
Key Reactions 497
Key Skills 498
Multiple-Choice Self-Test 499
Problems 501

MizC/Getty Images

Spectroscopy A Mass Spectrometry 507

A.1 Mass Spectrometry and the Molecular Ion 508
A.2 Alkyl Halides and the M + 2 Peak 513
A.3 Fragmentation 515
A.4 Fragmentation Patterns of Some Common Functional Groups 518
A.5 Other Types of Mass Spectrometry 520
Chapter Review 523
Key Concepts 523
Key Skills 523
Multiple-Choice Self-Test 524
Problems 524

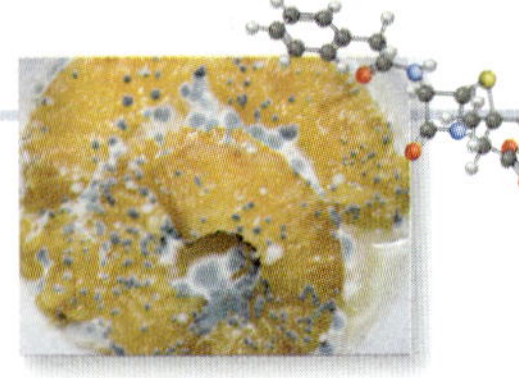

T.Daly/Alamy Stock Photo

Spectroscopy B Infrared Spectroscopy 529

B.1 Electromagnetic Radiation 530
B.2 The General Features of Infrared Spectroscopy 532
B.3 IR Absorptions 534
B.4 Infrared Spectra of Common Functional Groups 540

B.5 IR and Structure Determination 547

Chapter Review 549
Key Concepts 549
Key Skills 550
Multiple-Choice Self-Test 551
Problems 552

Daniel C. Smith

Spectroscopy C Nuclear Magnetic Resonance Spectroscopy 557

C.1 An Introduction to NMR Spectroscopy 558
C.2 ^{1}H NMR: Number of Signals 561
C.3 ^{1}H NMR: Position of Signals 566
C.4 The Chemical Shifts of Protons on sp^2 and *sp* Hybridized Carbons 570
C.5 ^{1}H NMR: Intensity of Signals 572
C.6 ^{1}H NMR: Spin–Spin Splitting 573
C.7 More-Complex Examples of Splitting 577
C.8 Spin–Spin Splitting in Alkenes 580
C.9 Other Facts About ^{1}H NMR Spectroscopy 582
C.10 Using ^{1}H NMR to Identify an Unknown 585
C.11 ^{13}C NMR Spectroscopy 588
C.12 Magnetic Resonance Imaging (MRI) 591

Chapter Review 592
Key Concepts 592
Key Skills 593
Multiple-Choice Self-Test 596
Problems 597

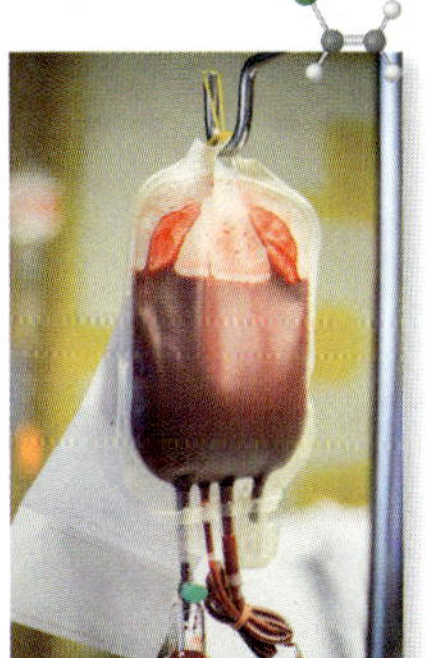

Hin255/iStock/Getty Images

13 Radical Reactions 606

13.1 Introduction 607
13.2 General Features of Radical Reactions 608
13.3 Halogenation of Alkanes 610
13.4 The Mechanism of Halogenation 611
13.5 Chlorination of Other Alkanes 614
13.6 Chlorination Versus Bromination 614
13.7 Halogenation as a Tool in Organic Synthesis 617
13.8 The Stereochemistry of Halogenation Reactions 618
13.9 Application: The Ozone Layer and CFCs 620
13.10 Radical Halogenation at an Allylic Carbon 622
13.11 Application: Oxidation of Unsaturated Lipids 625
13.12 Application: Antioxidants 626
13.13 Radical Addition Reactions to Double Bonds 627
13.14 Polymers and Polymerization 630

Chapter 13 Review 633
Key Concepts 633
Key Reactions 633
Key Skills 634
Multiple-Choice Self-Test 636
Problems 637

Pixtal/age fotostock

14 Conjugation, Resonance, and Dienes 645

14.1 Conjugation 646
14.2 Resonance and Allylic Carbocations 648
14.3 Common Examples of Resonance 649
14.4 The Resonance Hybrid 651
14.5 Electron Delocalization, Hybridization, and Geometry 653
14.6 Conjugated Dienes 654
14.7 Interesting Dienes and Polyenes 655
14.8 The Carbon–Carbon σ Bond Length in Buta-1,3-diene 656
14.9 Stability of Conjugated Dienes 657
14.10 Electrophilic Addition: 1,2- Versus 1,4-Addition 658
14.11 Kinetic Versus Thermodynamic Products 660
14.12 The Diels–Alder Reaction 663
14.13 Specific Rules Governing the Diels–Alder Reaction 664
14.14 Other Facts About the Diels–Alder Reaction 669
14.15 Conjugated Dienes and Ultraviolet Light 672
Chapter 14 Review 674
Key Concepts 674
Key Reactions 675
Key Skills 676
Multiple-Choice Self-Test 678
Problems 679

Schafer & Hill/Photolibrary/
Getty Images

15 Benzene and Aromatic Compounds 686

15.1 Background 687
15.2 The Structure of Benzene 688
15.3 Nomenclature of Benzene Derivatives 690
15.4 Spectroscopic Properties 692
15.5 Interesting Aromatic Compounds 693
15.6 Benzene's Unusual Stability 694
15.7 The Criteria for Aromaticity—Hückel's Rule 696
15.8 Examples of Aromatic Compounds 698
15.9 Aromatic Heterocycles 701
15.10 What Is the Basis of Hückel's Rule? 706
15.11 The Inscribed Polygon Method for Predicting Aromaticity 708
Chapter 15 Review 711
Key Concepts 711
Key Skills 712
Multiple-Choice Self-Test 713
Problems 714

Comstock/Getty Images

16 Reactions of Aromatic Compounds 722

16.1 Electrophilic Aromatic Substitution 723
16.2 The General Mechanism 724
16.3 Halogenation 725
16.4 Nitration and Sulfonation 727

16.5 Friedel–Crafts Alkylation and Friedel–Crafts Acylation 729
16.6 Substituted Benzenes 736
16.7 Electrophilic Aromatic Substitution of Substituted Benzenes 739
16.8 Why Substituents Activate or Deactivate a Benzene Ring 743
16.9 Orientation Effects in Substituted Benzenes 744
16.10 Limitations on Electrophilic Substitution Reactions with Substituted Benzenes 748
16.11 Disubstituted Benzenes 750
16.12 Synthesis of Benzene Derivatives 752
16.13 Nucleophilic Aromatic Substitution 753
16.14 Reactions of Substituted Benzenes 757
16.15 Multistep Synthesis 762

Chapter 16 Review 764
Key Concepts 764
Key Reactions 765
Key Skills 766
Key Mechanism Concepts 767
Multiple-Choice Self-Test 768
Problems 769

Folio Images/Alamy Stock Photo

17 Introduction to Carbonyl Chemistry; Organometallic Reagents; Oxidation and Reduction 777

17.1 Introduction 778
17.2 General Reactions of Carbonyl Compounds 779
17.3 A Preview of Oxidation and Reduction 782
17.4 Reduction of Aldehydes and Ketones 784
17.5 The Stereochemistry of Carbonyl Reduction 786
17.6 Enantioselective Carbonyl Reductions 787
17.7 Reduction of Carboxylic Acids and Their Derivatives 790
17.8 Oxidation of Aldehydes 795
17.9 Organometallic Reagents 796
17.10 Reaction of Organometallic Reagents with Aldehydes and Ketones 799
17.11 Retrosynthetic Analysis of Grignard Products 803
17.12 Protecting Groups 805
17.13 Reaction of Organometallic Reagents with Carboxylic Acid Derivatives 807
17.14 Reaction of Organometallic Reagents with Other Compounds 811
17.15 α,β-Unsaturated Carbonyl Compounds 812
17.16 Summary—The Reactions of Organometallic Reagents 815
17.17 Synthesis 815

Chapter 17 Review 818
Key Reactions 818
Key Skills 821
Key Mechanism Concepts 822
Multiple-Choice Self-Test 823
Problems 824

Brett Hondow/Alamy Stock Photo

18 Aldehydes and Ketones—Nucleophilic Addition 832

18.1 Introduction 833
18.2 Nomenclature 834
18.3 Properties of Aldehydes and Ketones 838
18.4 Interesting Aldehydes and Ketones 840
18.5 Preparation of Aldehydes and Ketones 840
18.6 Reactions of Aldehydes and Ketones—General Considerations 842
18.7 Nucleophilic Addition of H^- and R^-—A Review 845
18.8 Nucleophilic Addition of ^-CN 846
18.9 The Wittig Reaction 848
18.10 Addition of 1° Amines 854
18.11 Addition of 2° Amines 856
18.12 Addition of H_2O—Hydration 860
18.13 Addition of Alcohols—Acetal Formation 862
18.14 Acetals as Protecting Groups 867
18.15 Cyclic Hemiacetals 868
18.16 An Introduction to Carbohydrates 871

Chapter 18 Review 873
Key Concepts 873
Key Reactions 873
Key Skills 874
Multiple-Choice Self-Test 876
Problems 877

Daniel C. Smith

19 Carboxylic Acids and Nitriles 888

19.1 Structure and Bonding 889
19.2 Nomenclature 890
19.3 Physical and Spectroscopic Properties 893
19.4 Interesting Carboxylic Acids and Nitriles 894
19.5 Aspirin, Arachidonic Acid, and Prostaglandins 895
19.6 Preparation of Carboxylic Acids 897
19.7 Carboxylic Acids—Strong Organic Brønsted–Lowry Acids 898
19.8 Inductive Effects in Aliphatic Carboxylic Acids 901
19.9 Substituted Benzoic Acids 903
19.10 Extraction 905
19.11 Amino Acids 908
19.12 Nitriles 911

Chapter 19 Review 916
Key Concepts 916
Key Reactions 917
Key Skills 917
Multiple-Choice Self-Test 918
Problems 920

Likit Supasai/Shutterstock

20 Carboxylic Acids and Their Derivatives—Nucleophilic Acyl Substitution 928

20.1 Introduction 929
20.2 Structure and Bonding 931
20.3 Nomenclature 932
20.4 Physical and Spectroscopic Properties 936
20.5 Interesting Esters and Amides 938
20.6 Introduction to Nucleophilic Acyl Substitution 939
20.7 Reactions of Acid Chlorides 943
20.8 Reactions of Anhydrides 945
20.9 Reactions of Carboxylic Acids 946
20.10 Reactions of Esters 952
20.11 Application: Lipid Hydrolysis 954
20.12 Reactions of Amides 956
20.13 Application: The Mechanism of Action of β-Lactam Antibiotics 957
20.14 Summary of Nucleophilic Acyl Substitution Reactions 958
20.15 Natural and Synthetic Fibers 959
20.16 Biological Acylation Reactions 961

Chapter 20 Review 963
Key Concepts 963
Key Reactions 963
Key Skills 965
Multiple-Choice Self-Test 966
Problems 967

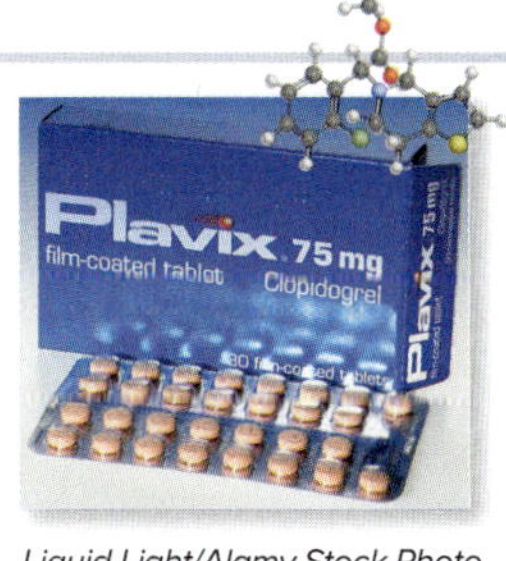

Liquid Light/Alamy Stock Photo

21 Substitution Reactions of Carbonyl Compounds at the α Carbon 976

21.1 Introduction 977
21.2 Enols 977
21.3 Enolates 981
21.4 Enolates of Unsymmmetrical Carbonyl Compounds 985
21.5 Racemization at the α Carbon 987
21.6 A Preview of Reactions at the α Carbon 988
21.7 Halogenation at the α Carbon 989
21.8 Direct Enolate Alkylation 993
21.9 Malonic Ester Synthesis 996
21.10 Acetoacetic Ester Synthesis 1001

Chapter 21 Review 1004
Key Reactions 1004
Key Skills 1005
Multiple-Choice Self-Test 1006
Problems 1008

Corbis/SuperStock

22 Carbonyl Condensation Reactions 1015

22.1 The Aldol Reaction 1016
22.2 Crossed Aldol Reactions 1021
22.3 Directed Aldol Reactions 1025
22.4 Intramolecular Aldol Reactions 1027
22.5 The Claisen Reaction 1030
22.6 The Crossed Claisen and Related Reactions 1032
22.7 The Dieckmann Reaction 1034
22.8 The Michael Reaction 1035
22.9 The Robinson Annulation 1037
Chapter 22 Review 1041
Key Reactions 1041
Key Skills 1042
Multiple-Choice Self-Test 1043
Problems 1045

kai4107/Shutterstock

23 Amines 1053

23.1 Introduction 1054
23.2 Structure and Bonding 1054
23.3 Nomenclature 1055
23.4 Physical and Spectroscopic Properties 1058
23.5 Interesting and Useful Amines 1059
23.6 Preparation of Amines 1062
23.7 Reactions of Amines—General Features 1068
23.8 Amines as Bases 1069
23.9 Relative Basicity of Amines and Other Compounds 1071
23.10 Amines as Nucleophiles 1077
23.11 Hofmann Elimination 1079
23.12 Reactions of Amines with Nitrous Acid 1082
23.13 Substitution Reactions of Aryl Diazonium Salts 1083
23.14 Coupling Reactions of Aryl Diazonium Salts 1088
23.15 Application: Synthetic Dyes and Sulfa Drugs 1090
Chapter 23 Review 1093
Key Concepts 1093
Key Reactions 1093
Key Skills 1095
Multiple-Choice Self-Test 1096
Problems 1098

DEA/M. GIOVANOLI/De Agostini Picture Library/ Getty Images

24 Carbon–Carbon Bond-Forming Reactions in Organic Synthesis 1106

24.1 Coupling Reactions of Organocuprate Reagents 1107
24.2 Suzuki Reaction 1109
24.3 Heck Reaction 1114
24.4 Stille Coupling 1116
24.5 Carbenes and Cyclopropane Synthesis 1119
24.6 Simmons–Smith Reaction 1122
24.7 Metathesis 1123

Chapter 24 Review *1128*
Key Reactions *1128*
Key Skills *1129*
Multiple-Choice Self-Test *1131*
Problems *1132*

Ron Nichols/USDA Natural Resources Conservation Service

25 Pericyclic Reactions 1141

25.1 Types of Pericyclic Reactions 1142
25.2 Molecular Orbitals 1143
25.3 Electrocyclic Reactions 1146
25.4 Cycloaddition Reactions 1152
25.5 Sigmatropic Rearrangements 1156
25.6 Summary of Rules for Pericyclic Reactions 1162
Chapter 25 Review *1163*
Key Concepts *1163*
Key Reactions *1163*
Key Skills *1164*
Multiple-Choice Self-Test *1165*
Problems *1166*

Daniel C. Smith

26 Carbohydrates 1174

26.1 Introduction 1175
26.2 Monosaccharides 1176
26.3 The Family of D-Aldoses 1182
26.4 The Family of D-Ketoses 1183
26.5 The Cyclic Forms of Monosaccharides 1184
26.6 Glycosides 1192
26.7 Reactions of Monosaccharides at the OH Groups 1195
26.8 Reactions at the Carbonyl Group—Oxidation and Reduction 1196
26.9 Reactions at the Carbonyl Group—Adding or Removing One Carbon Atom 1199
26.10 Disaccharides 1202
26.11 Polysaccharides 1205
26.12 Other Important Sugars and Their Derivatives 1208
Chapter 26 Review *1210*
Key Reactions *1210*
Key Skills *1211*
Multiple-Choice Self-Test *1214*
Problems *1216*

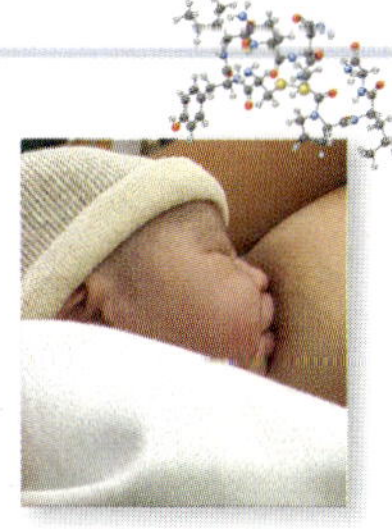

Daniel C. Smith

27 Amino Acids and Proteins 1222

27.1 Amino Acids 1223
27.2 Separation of Amino Acids 1227
27.3 Enantioselective Synthesis of Amino Acids 1230
27.4 Peptides 1232
27.5 Peptide Sequencing 1236
27.6 Peptide Synthesis 1240

27.7 Automated Peptide Synthesis 1245
27.8 Protein Structure 1247
27.9 Important Proteins 1253
27.10 Enzymes 1255
Chapter 27 Review 1258
Key Reactions 1258
Key Skills 1259
Multiple-Choice Self-Test 1261
Problems 1262

Daniel C. Smith

28 Nucleic Acids and Protein Synthesis 1268
28.1 Nucleosides and Nucleotides 1269
28.2 Nucleic Acids 1273
28.3 The DNA Double Helix 1274
28.4 Replication 1277
28.5 Ribonucleic Acids and Transcription 1278
28.6 The Genetic Code, Translation, and Protein Synthesis 1280
28.7 DNA Sequencing 1283
28.8 The Polymerase Chain Reaction 1285
28.9 Viruses 1286
Chapter 28 Review 1288
Key Concepts 1288
Key Skills 1288
Multiple-Choice Self-Test 1290
Problems 1291

Fcafotodigital/iStock/Getty Images

29 Lipids 1295 (Available online)
29.1 Introduction 1296
29.2 Waxes 1297
29.3 Triacylglycerols 1297
29.4 Phospholipids 1302
29.5 Fat-Soluble Vitamins 1304
29.6 Eicosanoids 1304
29.7 Terpenes 1308
29.8 Steroids 1314
Chapter 29 Review 1319
Key Concepts 1319
Key Reactions 1320
Key Skills 1321
Multiple-Choice Self-Test 1321
Problems 1322

Samuel Borges Photography/ Shutterstock

30 Metabolism 1328 (Available online)

30.1 ATP and Coupled Reactions 1329
30.2 Overview of Metabolism 1330
30.3 Key Oxidizing and Reducing Agents in Metabolism 1332
30.4 The Catabolism of Triacylglycerols by β-Oxidation 1336
30.5 The Catabolism of Carbohydrates—Glycolysis 1339
30.6 The Fate of Pyruvate 1344
30.7 The Citric Acid Cycle and ATP Production 1346
Chapter 30 Review 1351
Key Concepts 1351
Key Skills 1353
Multiple-Choice Self-Test 1353
Problems 1354

Stuar/Shutterstock

31 Synthetic Polymers 1358 (Available online)

31.1 Introduction 1359
31.2 Chain-Growth Polymers—Addition Polymers 1360
31.3 Anionic Polymerization of Epoxides 1366
31.4 Ziegler–Natta Catalysts and Polymer Stereochemistry 1367
31.5 Natural and Synthetic Rubbers 1368
31.6 Step-Growth Polymers—Condensation Polymers 1370
31.7 Polymer Structure and Properties 1375
31.8 Green Polymer Synthesis 1377
31.9 Polymer Recycling and Disposal 1379
Chapter 31 Review 1382
Key Concepts 1382
Key Reactions 1383
Key Skills 1383
Multiple-Choice Self-Test 1384
Problems 1385

Appendix A Periodic Table of the Elements A-1
Appendix B Common Abbreviations, Arrows, and Symbols A-2
Appendix C pK_a Values for Selected Compounds A-4
Appendix D Nomenclature A-6
Appendix E Bond Dissociation Energies for Some Common Bonds [A–B → A· + ·B] A-10
Appendix F Reactions That Form Carbon–Carbon Bonds A-11
Appendix G Characteristic IR Absorption Frequencies A-12
Appendix H Characteristic NMR Absorptions A-13
Appendix I General Types of Organic Reactions A-15
Appendix J How to Synthesize Particular Functional Groups A-17

Glossary G-1
Index I-1

Preface

My goal in writing *Organic Chemistry* was to create a text that showed students the beauty and logic of organic chemistry by giving them a book that they would *use. Organic Chemistry* is my attempt to simplify and clarify a course that intimidates many students—to make organic chemistry interesting, relevant, and accessible to *all* students, both chemistry majors and those interested in pursuing careers in biology, medicine, and other disciplines, without sacrificing the rigor they need to be successful in the future.

The Basic Features

Organic Chemistry, seventh edition, continues the successful student-oriented approach used in prior editions. This text uses less prose and more diagrams and bulleted summaries for today's students, who rely more heavily on visual imagery to learn than ever before. Each topic is broken down into small chunks of information that are more manageable and easily learned. Sample Problems illustrate stepwise problem solving, and relevant examples from everyday life are used to illustrate topics. New concepts are introduced one at a time so that the basic themes are kept in focus.

The organization of this text provides the student with a logical and accessible approach to an intense and fascinating subject. The text begins with a healthy dose of review material in Chapters 1 and 2 to ensure that students have a firm grasp of the fundamentals. Stereochemistry, the three-dimensional structure of molecules, is introduced early (Chapter 5) and reinforced often. Certain reaction types with unique characteristics and terminology are grouped together. These include acid–base reactions (Chapter 2), oxidation and reduction (Chapters 12 and 17), reactions of organometallic reagents (Chapters 17 and 24), and radical reactions (Chapter 13). Each chapter ends with a Chapter Review, end-of-chapter summaries that succinctly organize the main concepts and reactions.

New to this Edition

Students sometimes ask me if the facts of organic chemistry have significantly changed since the last edition. While the basic principles remain the same—carbon forms four bonds in stable compounds and oppositely charged species attract each other—organic chemistry is a dynamic subject that is continually refined as new facts are determined, and new editions reflect current understanding. Each year, novel compounds are discovered and new drugs are marketed, and these compounds replace older examples to illustrate particular concepts. In this edition, for example, every effort has been made to include content on COVID-19, the devastating disease responsible for the worldwide pandemic that began in late 2019. This material is included in the Prologue, Problems 3.60 and 29.25, and Sections 19.4, 20.5, and 28.9.

General

Problems In response to reviewer feedback, the level of difficulty in some Sample Problems has been increased, so that students can more readily tackle related, more challenging end-of-chapter problems. There are now more in-chapter problems, either new to this edition or moved from the end-of-chapter material, to give students more immediate practice on the topics they have just learned. In all, this edition contains more than 300 new problems.

Multiple-Choice Self-Test Each chapter now contains a Multiple-Choice Self-Test following the Chapter Review, which gives students added practice on key principles prior to solving the end-of-chapter problems. Answers to the self-test are given at the end of each chapter.

Chapters Two new chapters are added to this edition and three chapters are now available in the eBook and for customizable versions.

- **Chapter 28** provides an in-depth discussion of the structure and properties of the nucleic acids DNA and RNA. Three key processes are presented: replication—how DNA makes copies of itself; transcription—how the genetic information in DNA is passed onto RNA; and translation—how the coded genetic information in RNA is used to synthesize proteins. The chapter concludes with discussions of manipulating DNA in the laboratory and how viruses act.
- **Chapter 30** focuses on the biochemical reactions involved in metabolism. The discussion centers on three components: the breakdown of fats, the metabolism of the carbohydrate glucose to the three-carbon unit pyruvate by glycolysis, and the citric acid cycle, a key cyclic metabolic pathway used for amino acids, carbohydrates, and fats.

Chapters 1–28 are available in the print edition of *Organic Chemistry,* and Chapters 29 (Lipids), 30 (Metabolism), and 31 (Synthetic Polymers) are available in the eBook and for customizable versions.

Other New Coverage

Several sections include new material.

- Section 7.4 contains material on the prevalence of fluoro and trifluoromethyl groups in several new drugs.
- Three examples of recently approved drugs that contain epoxides or ethers, along with other functional groups, have been added to Section 9.5B.
- Section B.4 has been completely reorganized so that infrared spectra of related compounds can be more easily compared.
- In NMR spectroscopy, the spin–spin splitting discussion in Section C.8 has been revised and a new example is used in Figure C.8 to better illustrate this concept.
- Heterocyclic aromatic compounds are moved into their own section (15.9), and material on the aromatic bases in DNA has been added.
- Figures 17.2–17.5 have all been updated with new molecules and examples using drugs and marine natural products.
- A new Sample Problem 18.4 on distinguishing ethers and acetals in a complex molecule has been added.
- New material on biological enols is now in Section 21.2B.
- In Chapter 22, two sections have been added: Section 22.1B on the retro-aldol reaction and Section 22.2C on enantioselective aldol reactions and the 2021 Nobel Prize in Chemistry.
- Chapter 24 now contains the Stille coupling reaction in Section 24.4.
- Section 27.10 on enzymes has been added to Chapter 27.
- New material has been added to Chapter 31 to reflect recent advances on biodegradable polymers and the use of polymers in controlled drug release.

Features of the Text to Make Learning Organic Chemistry Easier

Illustrations *Organic Chemistry* is supported by a well-developed illustration program. Besides traditional skeletal (line) structures and condensed formulas, there are numerous ball-and-stick molecular models and electrostatic potential maps to help students grasp the three-dimensional structure of molecules (including stereochemistry) and to better understand the distribution of electronic charge.

Micro-to-Macro Illustrations Unique to *Organic Chemistry* are micro-to-macro illustrations, where line art and photos combine with chemical structures to reveal the underlying molecular structures giving rise to macroscopic properties of common phenomena. Examples include starch and cellulose (Chapter 5), adrenaline (Chapter 7), partial hydrogenation of vegetable oil (Chapter 12), and dopamine (Chapter 23).

Spectra Over 100 spectra created specifically for *Organic Chemistry* are presented mainly in the spectroscopy chapters. The spectra are color-coded by type and generously labeled. Mass spectra are green; infrared spectra are red; and proton and carbon nuclear magnetic resonance spectra are blue.

Mechanisms Curved arrow notation is used extensively to help students follow the movement of electrons in reactions.

Problem Solving

Sample Problems Sample Problems show students how to solve organic chemistry problems in a logical, stepwise manner. More than 800 follow-up problems are located throughout the chapters to test whether students understand concepts covered in the Sample Problems.

How To's *How To*'s provide students with detailed instructions on how to work through key processes.

Applications and Summaries

Chapter Reviews Succinct summary tables reinforcing important principles and concepts are provided at the end of each chapter.

Margin Notes Margin notes are placed carefully throughout the chapters, providing interesting information relating to topics covered in the text. Some margin notes are illustrated with photos to make the chemistry more relevant.

Learning Resources for Instructors and Students

ALEKS (Assessment and Learning in Knowledge Spaces) is a web-based system for individualized assessment and learning available 24/7 over the Internet. ALEKS uses artificial intelligence to accurately determine a student's knowledge and then guides them to the material that they are most ready to learn. ALEKS offers immediate feedback and access to ALEKSPedia—an interactive text that contains concise entries on chemistry topics. ALEKS is also a full-featured course management system with rich reporting features that allow instructors to monitor individual and class performance, set student goals, assign/grade online quizzes, and more. ALEKS allows instructors to spend more time on concepts while ALEKS teaches students practical problem-solving skills. And with ALEKS 360, your student also has access to this text's eBook. Learn more at **www.aleks.com/highered/science**

McGraw Hill Virtual Labs is a must-see, outcomes-based lab simulation. It assesses a student's knowledge and adaptively corrects deficiencies, allowing the student to learn faster and retain more knowledge with greater success. First, a student's knowledge is adaptively leveled on core learning outcomes: Questioning reveals knowledge deficiencies that are corrected by the delivery of content that is conditional on a student's response. Then, a simulated lab experience requires the student to think and act like a scientist: recording, interpreting, and analyzing data using simulated equipment found in labs and clinics. The student is allowed to make mistakes—a powerful part of the learning experience! A virtual coach provides subtle hints when needed, asks questions about the student's choices, and allows the student to reflect on and correct those mistakes. Whether your need is to overcome the logistical challenges of a traditional lab, provide better lab prep, improve student performance, or make your online experience one that rivals the real world, McGraw Hill Virtual Labs accomplishes it all.

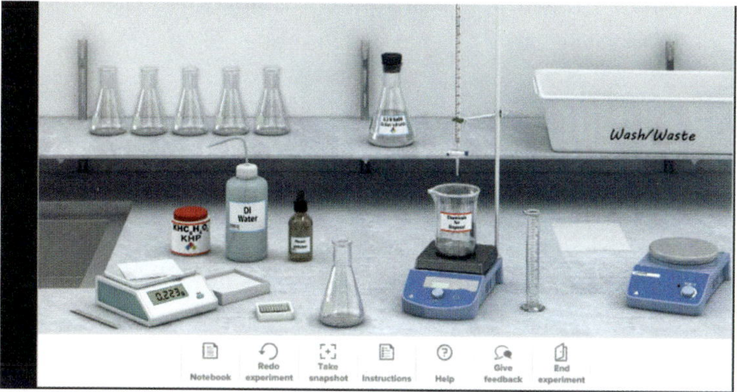

The following items may accompany this text. Please consult your McGraw Hill representative for policies, prices, and availability as some restrictions may apply.

Presentation Tools

Within the Instructor's Presentation Tools, instructors have access to editable Accessible PowerPoint lecture outlines, which appear as ready-made presentations that combine art and lecture notes for each chapter of the text. For instructors who prefer to create their lecture notes from scratch, all illustrations and photos are pre-inserted by chapter into a separate set of PowerPoint slides. They are also available as individual .jpg files. All assets are copyrighted by McGraw Hill Higher Education, but can be used by instructors for classroom purposes.

Classroom Response System Questions (Clicker Questions)

Nearly 600 questions covering the content of the *Organic Chemistry* text are available on the *Organic Chemistry* site in PowerPoint format for use with any classroom response system.

Test Bank

A test bank with over 1500 questions is available with the seventh edition. The test bank is available in the TestGen test-generating software to quickly create customized exams.

Captioned Videos

Closed caption videos covering the most important topics for *Organic Chemistry* are provided.

Break down barriers and build student knowledge

Students start your course with varying levels of preparedness. Some will get it quickly. Some won't. ALEKS is a course assistant that helps you meet each student where they are and provide the necessary building blocks to get them where they need to go. You determine the assignments and the content, and ALEKS will deliver customized practice until your students truly get it.

Experience the ALEKS Difference

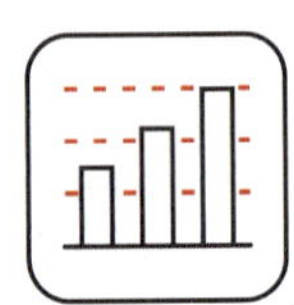

Easily Identify Knowledge Gaps

Gain greater visibility into student performance so you know immediately if your lessons clicked.

- **ALEKS's "Initial Knowledge Check"** helps accurately evaluate student levels and gaps on day one, so you know precisely where students are at and where they need to go when they start your course.
- **You know when students are at risk of falling behind** through ALEKS Insights so you can remediate—be it through prep modules, practice questions, or written explanations of video tutorials.
- **Students always know where they are,** know how they are doing, and can track their own progress easily.

Gain More Flexibility and Engagement

Teach your course your way, with best-in-class content and tools to immerse students and keep them on track.

- **ALEKS gives you flexibility** to assign homework, share a vast library of curated content including videos, review progress, and provide student support, anytime anywhere.
- **You save time** otherwise spent performing tedious tasks while having more control over and impact on your students' learning process.
- **Students gain a deeper level of understanding** through interactive and hands-on assignments that go beyond multiple-choice questions.

with ALEKS® Constructive Learning Paths.

Narrow the Equity Gap

Efficiently and effectively create individual pathways for students without leaving anyone behind.

- **ALEKS creates an equitable experience for all students,** making sure they get the support they need to successfully finish the courses they start.
- **You help reduce attrition,** falling enrollment, and further widening of the learning gap.
- **Student success rates improve**—not just better grades, but better learning.

Count on Hands-on Support

A dedicated Implementation Manager will work with you to build your course exactly the way you want it and your students need it.

- **An ALEKS Implementation Manager** is with you every step of the way through the life of your course.
- **You never have to figure it out on your own** or be your student's customer service. We believe in a consultative approach and take care of all of that for you, so you can focus on your class.
- **Your students benefit** from more meaningful moments with you, while ALEKS—directed by you—does the rest.

Student Study Guide/Solutions Manual

Written by Janice Gorzynski Smith and Erin R. Smith, the Student Study Guide/Solutions Manual provides step-by-step solutions to all in-chapter and end-of-chapter problems. Each chapter begins with an overview of key concepts and includes a short-answer practice test on the fundamental principles and new reactions.

Create

Your Book, Your Way

McGraw Hill's Content Collections Powered by Create® is a self-service website that enables instructors to create custom course materials—print and eBooks—by drawing upon McGraw Hill's comprehensive, cross-disciplinary content. Choose what you want from our high-quality textbooks, articles, and cases. Combine it with your own content quickly and easily, and tap into other rights-secured, third-party content such as readings, cases, and articles. Content can be arranged in a way that makes the most sense for your course and you can include the course name and information as well. Choose the best format for your course: color print, black-and-white print, or eBook. When you are finished customizing, you will receive a free digital copy to review in just minutes! Visit McGraw Hill Create®—www.mcgrawhillcreate.com—today and begin building!

Acknowledgments

Although many years have passed since I first began writing textbooks with McGraw Hill, I continue to appreciate the hard work and insight of the publishing professionals with whom I work.

I must specifically single out three individuals, whose expertise has been invaluable in the publishing of this seventh edition: Mary Hurley, senior Product Developer; Shirley Hino, Director of Digital Content; and John Murdzek, freelance Developmental Editor. I can never say enough about their contributions, as well as the skill of the production, design, and marketing teams, in bringing my projects to successful publication.

My immediate family has experienced the day-to-day demands of living with a busy author. Thanks go to my husband Dan and my children and grandchildren, all of whom keep me grounded during the time-consuming process of writing and publishing a textbook.

Among the many others that go unnamed but who have profoundly affected this work are the thousands of students I have been lucky to teach over the last 30 years. I have learned so much from my daily interactions with them, and I hope that the wider chemistry community can benefit from this experience by the way I have presented the material in this text.

This seventh edition has evolved based on the helpful feedback of many people who reviewed the sixth edition text and digital products, class-tested the book, and attended focus groups or symposiums. These many individuals have collectively provided constructive improvements to the project.

Reviewers of the sixth edition of *Organic Chemistry* include:

David Baker, *Delta College*
Miranda Beam, *Campbellsville University*
Kerry Breno, *Whitworth University*
Michael Carta, *Three Rivers Community College*
Patrick Coppock, *Georgia Gwinnett College*
David Forbes, *University of South Alabama*
Bruce N. Hietbrink, *University of Delaware*
Ara Kahyaoglu, *Bergen Community College*
Adam Keller, *Columbus State Community College*
Oleg Larionov, *The University of Texas at San Antonio*
Siva Murru, *University of Louisiana Monroe*
Phuong-Truc Pham, *Penn State Scranton*
Jacob Plummer, *Wingate University*
Carson Prevatte, *Middle Tennessee State University*
Michael Rennekamp, *Columbus State Community College*
Chatney Spencer, *Motlow State Community College*
Sharon Hann Stickley, *Columbus State Community College*
Brandon Tutkowski, *Wingate University*
Sheela Venkitachalam, *Delaware Valley University*
Omar Villanueva, *Georgia Gwinnett College*
Elizabeth Walters, *University of North Carolina Wilmington*

Listed below are the reviewers of previous editions of this text:

Steven Castle, *Brigham Young University*
Ihsan Erden, *San Francisco State University*
Andrew Frazer, *University of Central Florida, Orlando*
Tiffany Gierasch, *University of Maryland, Baltimore County*
Anne Gorden, *Auburn University*
Michael Lewis, *Saint Louis University*
Eugene A. Mash Jr., *University of Arizona*
Mark McMills, *Ohio University*
Joan Mutanyatta–Comar, *Georgia State University*
Felix Ngassa, *Grand Valley State University*
Michael Rathke, *Michigan State University*
Jacob Schroeder, *Clemson University*
Keith Schwartz, *Portland State University*
John Selegue, *University of Kentucky*
Paul J. Toscano, *University at Albany, SUNY*
Jane E. Wissinger, *University of Minnesota*

The following contributed to the editorial direction of the text by responding to our survey on the MCAT and the organic chemistry course student population:

Chris Abelt, *College of William and Mary*
Orlando Acevedo, *Auburn University*
Kim Albizati, *University of California, San Diego*
Merritt Andrus, *Brigham Young University*
Ardeshir Azadnia, *Michigan State University*
Susan Bane, *Binghamton University*
Russell Barrows, *Metropolitan State University of Denver*
Peter Beak, *University of Illinois, Urbana Champaign*
Phil Beauchamp, *Cal Poly, Pomona*
Michael Berg, *Virginia Tech*
K. Darrell Berlin, *Oklahoma State University*
Thomas Bertolini, *University of South Carolina*
Ned Bowden, *University of Iowa*
David W. Brown, *Florida Gulf Coast University*
Rebecca Broyer, *University of Southern California*
Arthur Bull, *Oakland University*
K. Nolan Carter, *University of Central Arkansas*
Steven Castle, *Brigham Young University*
Victor Cesare, *St. John's University*
Manashi Chatterjee, *University of Nebraska, Lincoln*
Melissa Cichowicz, *West Chester University*
Jeff Corkill, *Eastern Washington University, Cheney*

Sulekha Coticone, *Florida Gulf Coast University*
Michael Crimmins, *University of North Carolina at Chapel Hill*
Eric Crumpler, *Valencia College*
David Dalton, *Temple University*
Rick Danheiser, *Massachusetts Institute of Technology*
Tammy Davidson, *University of Florida*
Brenton DeBoef, *University of Rhode Island*
Amy Deveau, *University of New England*
Kenneth M. Doxsee, *University of Oregon*
Larissa D'Souza, *Johns Hopkins University*
Philip Egan, *Texas A&M University, Corpus Christi*
Seth Elsheimer, *University of Central Florida*
John Esteb, *Butler University*
Steve Fleming, *Temple University*
Marion Franks, *North Carolina A&T State University*
Andy Frazer, *University of Central Florida*
Brian Ganley, *University of Missouri, Columbia*
Robert Giuliano, *Villanova University*
Anne Gorden, *Auburn University*
Carlos G. Gutierrez, *California State University, Los Angeles*
Scott Handy, *Middle Tennessee State University*
Rick Heldrich, *College of Charleston*
James Herndon, *New Mexico State University*
Kathleen Hess, *Brown University*
Sean Hickey, *University of New Orleans*
Carl Hoeger, *University of California, San Diego*
Javier Horta, *University of Massachusetts, Lowell*
Bob A. Howell, *Central Michigan University*
Jennifer Irvin, *Texas State University*
Phil Janowicz, *Cal State, Fullerton*
Mohamad Karim, *Tennessee State University*
Mark L. Kearley, *Florida State University*
Amy Keirstead, *University of New England*
Margaret Kerr, *Worcester State University*
James Kiddle, *Western Michigan University*
Jisook Kim, *University of Tennessee at Chattanooga*
Angela King, *Wake Forest University*
Margaret Kline, *Santa Monica College*
Dalila G. Kovacs, *Grand Valley State University*
Deborah Lieberman, *University of Cincinnati*
Carl Lovely, *University of Texas, Arlington*
Kristina Mack, *Grand Valley State University*
Daniel Macks, *Towson University*
Vivian Mativo, *Georgia Perimeter College, Clarkston*
Mark McMills, *Ohio University*
Stephen Mills, *Xavier University*
Robert Minto, *Indiana University–Purdue University, Indianapolis*
Debbie Mohler, *James Madison University*
Kathleen Morgan, *Xavier University of Louisiana*
Paul Morgan, *Butler University*
James C. Morris, *Georgia Institute of Technology*
Linda Munchausen, *Southeastern Louisiana University*
Toby Nelson, *Oklahoma State University*
Felix Ngassa, *Grand Valley State University*
George A. O'Doherty, *Northeastern University*
Anne Padias, *University of Arizona*
Dan Paschal, *Georgia Perimeter College*
Richard Pennington, *Georgia Gwinnett College*
John Pollard, *University of Arizona*
Gloria Proni, *John Jay College*
Khalilah Reddie, *University of Massachusetts, Lowell*
Joel M. Ressner, *West Chester University of Pennsylvania*
Christine Rich, *University of Louisville*
Carmelo Rizzo, *Vanderbilt University*
Harold R. Rogers, *California State University, Fullerton*
Paul B. Savage, *Brigham Young University*
Deborah Schwyter, *Santa Monica College*
Holly Sebahar, *University of Utah*
Laura Serbulea, *University of Virginia*
Abid Shaikh, *Georgia Southern University*
Kevin Shaughnessy, *The University of Alabama*
Joel Shulman, *University of Cincinnati*
Joseph M. Simard, *University of New England*
Rhett Smith, *Clemson University*
Priyantha Sugathapala, *University at Albany, SUNY*
Claudia Taenzler, *University of Texas at Dallas*
Robin Tanke, *University of Wisconsin, Stevens Point*
Richard T. Taylor, *Miami University, Oxford*
Edward Turos, *University of South Florida*
Ted Wood, *Pierce College*
Kana Yamamoto, *University of Toledo*

Andrea Leonard and Ryan Simon of the University of Louisiana at Lafayette, revised the PowerPoint Lectures and Test Bank for the seventh edition.

Although every effort has been made to make this text and its accompanying Student Study Guide/Solutions Manual as error-free as possible, some errors undoubtedly remain. Please feel free to email me about any inaccuracies, so that subsequent editions may be further improved.

With much aloha,

Janice Gorzynski Smith
jgsmith@hawaii.edu

Prologue

What is Organic Chemistry?
Some Representative Organic Molecules
The Marine Environment

Organic chemistry. You might wonder how a discipline that conjures up images of eccentric old scientists working in basement laboratories is relevant to you, a student in the twenty-first century.

Consider for a moment the activities that occupied your past 24 hours. You likely showered with soap, drank a caffeinated beverage, ate at least one form of starch, took some medication, and traveled in a vehicle that had rubber tires and was powered at least partly by fossil fuels. If you did any *one* of these, your life was touched by organic chemistry.

What Is Organic Chemistry?

- Organic chemistry is the chemistry of compounds that contain the element carbon.

Some compounds that contain the element carbon are *not* organic compounds. Examples include carbon dioxide (CO_2), sodium carbonate (Na_2CO_3), and sodium bicarbonate ($NaHCO_3$).

Organic chemistry was singled out as a separate discipline for historical reasons. Originally, it was thought that compounds in living things, termed ***organic compounds,*** were fundamentally different from those in nonliving things, called ***inorganic compounds.*** Although we have known for more than 150 years that this distinction is artificial, the name *organic* persists. Today the term refers to the study of the compounds that contain carbon, many of which, incidentally, are found in living organisms.

It may seem odd that a whole discipline is devoted to the study of a single element in the periodic table, when more than 100 elements exist. It turns out, though, that there are far more organic compounds than any other type. Clothes, foods, medicines, gasoline, refrigerants, and soaps are composed almost solely of organic compounds. Some, like cotton, wool, or silk, are *naturally occurring;* that is, they can be isolated directly from natural sources. Others, such as nylon and polyester, are *synthetic,* meaning they are produced by chemists in the laboratory. **By studying the principles and concepts of organic chemistry, you can learn more about compounds such as these and how they affect the world around you.**

Realize, too, what organic chemistry has done for us. Organic chemistry has made available both comforts and necessities that were previously nonexistent, or reserved for only the wealthy. We have seen an enormous increase in life span, from 47 years in 1900 to over 70 years currently. To a large extent this is due to the isolation and synthesis of new drugs to fight infections and the availability of vaccines for diseases. Chemistry has also given us the tools to control insect populations that spread disease, and there is more food for all because of fertilizers, pesticides, and herbicides. Our lives would be vastly different today without the many products that result from organic chemistry (Figure 1).

Figure 1 Products of organic chemistry used in medicine

a. Oral contraceptives

Comstock Images/PictureQuest/Getty Images

b. Plastic syringes

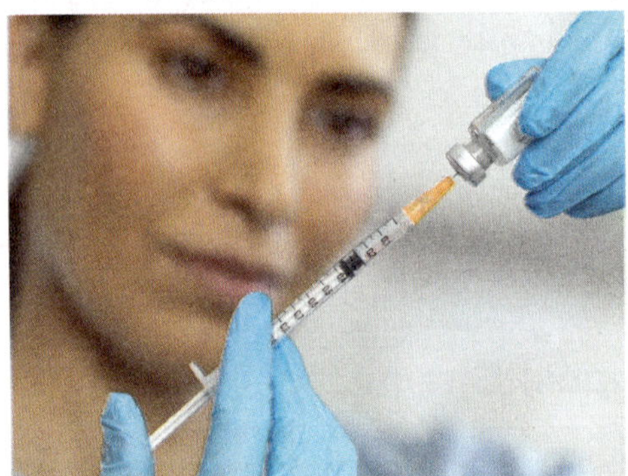

Andrew Brookes/Getty Images

c. Antibiotics

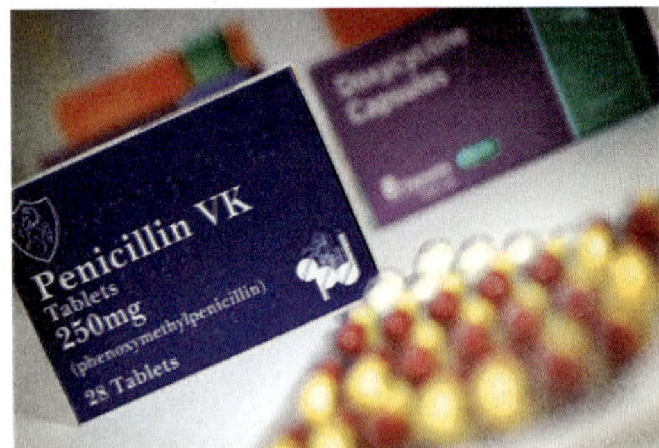

Julian Claxton/Alamy Stock Photo

d. Synthetic heart valves

Layne Kennedy/Corbis NX/Getty Images

- Organic chemistry has given us contraceptives, plastics, antibiotics, and the knitted material used in synthetic heart valves.

Ball-and-stick models are described in Section 1.7.

Some Representative Organic Molecules

Perhaps the best way to appreciate the variety of organic molecules is to look at a few. Three simple organic compounds are **methane, ethanol,** and **trichlorofluoromethane.**

CH_4
methane

- **Methane,** the simplest of all organic compounds, contains one carbon atom. Methane—the main component of natural gas—occurs widely in nature. Like other **hydrocarbons**—organic compounds that contain only carbon and hydrogen—methane is combustible; that is, it burns in the presence of oxygen. Methane is the product of the anaerobic (without air) decomposition of organic matter by bacteria. The natural gas we use today was formed by the decomposition of organic material millions of years ago. Hydrocarbons such as methane are discussed in Chapter 4.

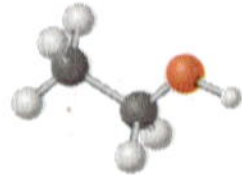

CH_3CH_2OH
ethanol

- **Ethanol,** the alcohol present in beer, wine, and other alcoholic beverages, is formed by the fermentation of sugar, possibly the oldest example of organic synthesis. Ethanol can also be made in the lab by a totally different process, but **the ethanol produced in the lab is *identical* to the ethanol produced by fermentation.** Alcohols including ethanol are discussed in Chapter 9.

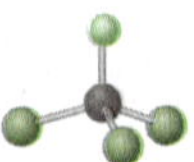

$CFCl_3$
trichlorofluoromethane

- **Trichlorofluoromethane** is a member of a class of molecules called **c**hloro**f**luoro**c**arbons, or **CFCs,** which contain one or two carbon atoms and several halogens. Trichlorofluoromethane is an unusual organic molecule in that **it contains no hydrogen atoms.** Because it has a low molecular weight and is easily vaporized, trichlorofluoromethane has been used as an aerosol propellant and refrigerant. It and other CFCs have been implicated in the destruction of the stratospheric ozone layer, a topic discussed in Chapter 13.

Three complex organic molecules that are important medications are **amoxicillin, fluoxetine,** and **remdesivir.**

- **Amoxicillin** is one of the most widely used antibiotics in the penicillin family. The discovery and synthesis of such antibiotics in the twentieth century made routine the treatment of infections that were formerly fatal. You were likely given some amoxicillin to treat an ear infection when you were a child. The penicillin antibiotics are discussed in Chapter 20.

Complex organic structures are drawn with shorthand conventions described in Chapter 1.

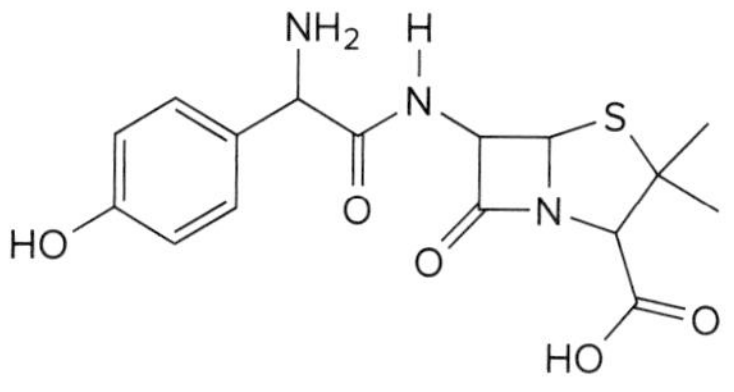

amoxicillin

- **Fluoxetine** is the generic name for the antidepressant **Prozac.** Prozac was designed and synthesized by chemists in the laboratory, and is now produced on a large scale in chemical factories. Because it is safe and highly effective in treating depression, over 25 million prescriptions for fluoxetine were written in the United States in 2018.

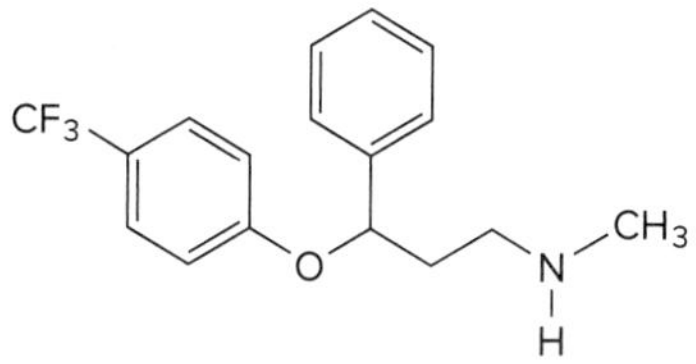

fluoxetine

- **Remdesivir,** an antiviral drug (trade name Veklury) first developed to treat hepatitis C, was given emergency use authorization to treat COVID-19 in 2020, and is recommended for some patients suffering from moderate disease. Remdesivir represents a chemical success to a different challenge: synthesizing agents that combat viral infections.

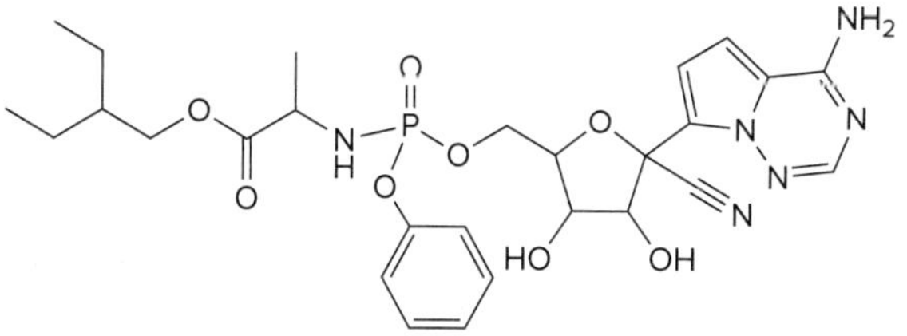

Generic name: remdesivir
Trade name: Veklury

What are the common features of these organic compounds?

- All organic compounds contain carbon atoms and most contain hydrogen atoms.
- All the carbon atoms have four bonds. A stable carbon atom is said to be *tetravalent*.
- Other elements may also be present. Any atom that is not carbon or hydrogen is called a *heteroatom*. Common heteroatoms include N, O, S, P, and the halogens.
- Some compounds have chains of atoms and some compounds have rings.

These features explain why there are so many organic compounds: **Carbon forms four strong bonds with itself and other elements. Carbon atoms combine together to form rings and chains.**

The Marine Environment

Coral reefs play a vital role in maintaining the health of ocean ecosystems (Figure 2a). Sea life use the reef for shelter from predators, and the reef provides the habitat for numerous species to breed and establish nurseries. Moreover, many coastal societies around the world depend on the reef for food or income. Coral reefs also protect shorelines by reducing wave impact and damage from flooding and storms.

Bleaching is a phenomenon that occurs when corals expel symbiotic algae from their tissues in response to an external stress, causing the coral to turn white (Figure 2b). Because these single-celled organisms provide the coral with nutrients, prolonged bleaching causes the corals to starve, often resulting in death.

Figure 2
Healthy coral compared to bleached coral

a. Healthy corals

Daniel C. Smith

b. Bleached corals

Buttchi 3 Sha Life/Shutterstock

Although coral bleaching is most often associated with an increase in water temperature, many other factors—salinity changes, increased acidity, and pollutants—also trigger bleaching. Research at the University of Hawai‘i suggests that minute amounts of compounds such as oxybenzone contribute to bleaching. **Oxybenzone** effectively filters a broad spectrum of harmful ultraviolet light, so it is a common sunscreen component, but it can be washed off while swimming, leading to a low but potentially harmful concentration in the water. As a result, the state of Hawai‘i has now banned the sale of sunscreens that contain oxybenzone and related organic compounds. Our understanding of the chemistry and biology of the coral reef and the effect of humankind on this habitat is crucial to maintaining the health of the marine environment.

OH O

O

oxybenzone

In this introduction, we have seen a variety of molecules that have diverse structures. They represent a miniscule fraction of the organic compounds currently known and the many thousands that are newly discovered or synthesized each year. The principles you learn in organic chemistry will apply to all of these molecules, from simple ones like methane and ethanol, to more complex ones like remdesivir and oxybenzone. It is these beautiful molecules, their properties, and their reactions that we will study in organic chemistry.

WELCOME TO THE WORLD OF ORGANIC CHEMISTRY!

1 Structure and Bonding

Russell Illig/Photodisc/Getty Images

1.1 The periodic table
1.2 Bonding
1.3 Lewis structures
1.4 Isomers
1.5 Exceptions to the octet rule
1.6 Resonance
1.7 Determining molecular shape
1.8 Drawing organic structures
1.9 Hybridization
1.10 Ethane, ethylene, and acetylene
1.11 Bond length and bond strength
1.12 Electronegativity and bond polarity
1.13 Polarity of molecules
1.14 Oxybenzone—A representative organic molecule

Hypertension, or high blood pressure, is a condition that affects millions of Americans and increases the risk of heart attack and stroke. **Lisinopril,** an oral medication used to treat high blood pressure, belongs to a class of drugs called ACE (angiotensin-converting enzyme) inhibitors. Lisinopril decreases blood pressure by blocking the synthesis of angiotensin, a protein that narrows blood vessels and increases blood pressure. In Chapter 1, we learn about the structure, bonding, and properties of organic compounds like lisinopril.

Structure and Bonding?

Before examining organic molecules in detail, we must review topics about structure and bonding learned in previous chemistry courses. We will discuss these concepts primarily from an organic chemist's perspective, and spend time on only the particulars needed to understand organic compounds.

Important topics in Chapter 1 include drawing Lewis structures, predicting the shape of molecules, determining what orbitals are used to form bonds, and how electronegativity affects bond polarity. Equally important is Section 1.8 on drawing organic molecules, both shorthand methods routinely used for simple and complex compounds, and three-dimensional representations that allow us to more clearly visualize them.

1.1 The Periodic Table

All matter is composed of the same building blocks called **atoms.** There are two main components of an atom.

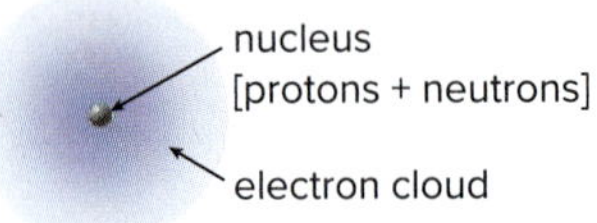

- **The nucleus contains positively charged protons and uncharged neutrons. Most of the mass of the atom is contained in the nucleus.**
- **The electron cloud is composed of negatively charged electrons. The electron cloud comprises most of the volume of the atom.**

The charge on a proton is equal in magnitude but opposite in sign to the charge on an electron. In a neutral atom, the **number of protons in the nucleus equals the number of electrons.** This quantity, called the **atomic number,** is unique to a particular element. For example, every neutral carbon atom has an atomic number of six, meaning it has six protons in its nucleus and six electrons surrounding the nucleus.

In addition to neutral atoms, we will encounter **charged ions.**

- A *cation* is positively charged and has fewer electrons than protons.
- An *anion* is negatively charged and has more electrons than protons.

The number of neutrons in the nucleus of a particular element can vary. **Isotopes** are two atoms of the same element having a different number of neutrons. The **mass number** of an atom is the total number of protons and neutrons in the nucleus. **Isotopes have *different* mass numbers.** The **atomic weight** of a particular element is the weighted average of the mass of all its isotopes, reported in atomic mass units (amu).

Isotopes of carbon and hydrogen are sometimes used in organic chemistry. The most common isotope of hydrogen has one proton and no neutrons in the nucleus, but 0.02% of hydrogen atoms have one proton and one neutron. This isotope of hydrogen is called **deuterium** and is sometimes symbolized by the letter **D.**

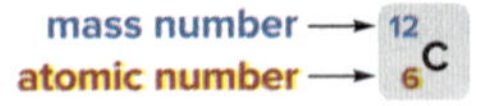

Each atom is identified by a one- or two-letter abbreviation that is the characteristic symbol for that element. Carbon is identified by the single letter **C.** Sometimes the atomic number is indicated as a subscript to the left of the element symbol, and the mass number is indicated as a superscript. Using this convention, the most common isotope of carbon, which contains six protons and six neutrons, is designated as $^{12}_{6}\mathbf{C}$.

A **row** in the periodic table is called a ***period,*** and a **column** is called a ***group.*** A periodic table is located in Appendix A for your reference.

The **periodic table** is a schematic arrangement of the more than 100 known elements, arranged in order of increasing atomic number. The periodic table is composed of rows and columns. Each column in the periodic table is identified by a **group number,** an Arabic (1 to 8) or Roman (I to VIII) numeral followed by the letter A or B. Carbon is located in group **4A** in the periodic table in this text.

- **Elements in the same row are similar in *size.***
- **Elements in the same column have similar *electronic and chemical properties.***

Although more than 100 elements exist, most are not common in organic compounds. Figure 1.1 contains a truncated periodic table, indicating the handful of elements that are routinely seen in this text. **Most elements in organic compounds are located in the first and second rows of the periodic table.**

Figure 1.1
A periodic table of the common elements seen in organic chemistry

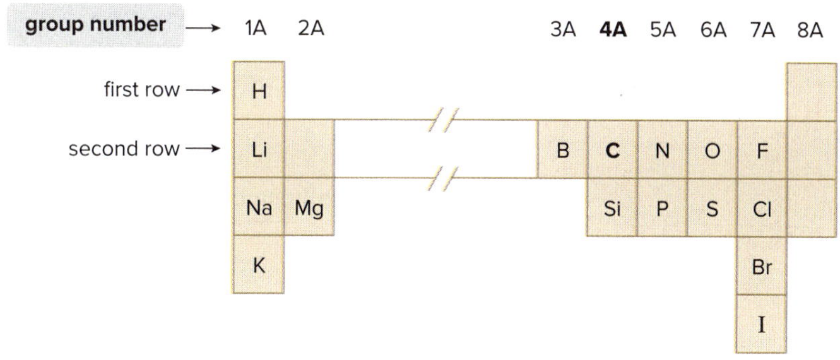

- Carbon is located in the second row, group **4A.**

Carbon's entry in the periodic table:

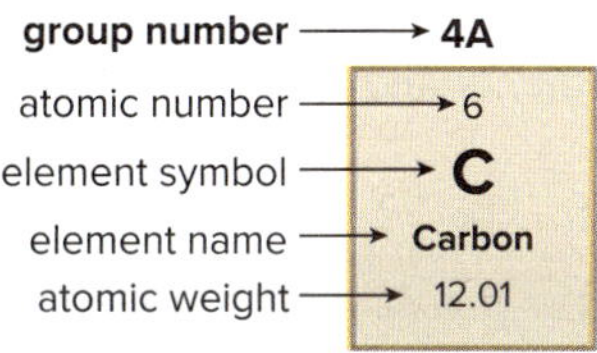

Across each row of the periodic table, electrons are added to a particular shell of orbitals around the nucleus. The shells are numbered 1, 2, 3, and so on. Adding electrons to the first shell forms the first row. Adding electrons to the second shell forms the second row. **Electrons are first added to the shells closest to the nucleus.**

Each shell contains a certain number of **orbitals.** An orbital is a region of space that is high in electron density. There are four different kinds of orbitals, called ***s, p, d,*** and ***f.*** The first shell has only one orbital, an *s* orbital. The second shell has two kinds of orbitals, ***s*** and ***p,*** and so on. Each type of orbital has a particular shape.

For the first- and second-row elements, we must consider only ***s*** **orbitals** and ***p*** **orbitals.**

- An ***s*** **orbital** has a **sphere of electron density.** It is ***lower in energy*** than other orbitals of the same shell, because electrons are kept closer to the positively charged nucleus.
- A ***p*** **orbital** has a **dumbbell shape.** It contains a **node of electron density** at the nucleus. A node means there is *no* electron density in this region. A *p* orbital is ***higher in energy*** than an *s* orbital (in the same shell) because its electron density is farther away from the nucleus.

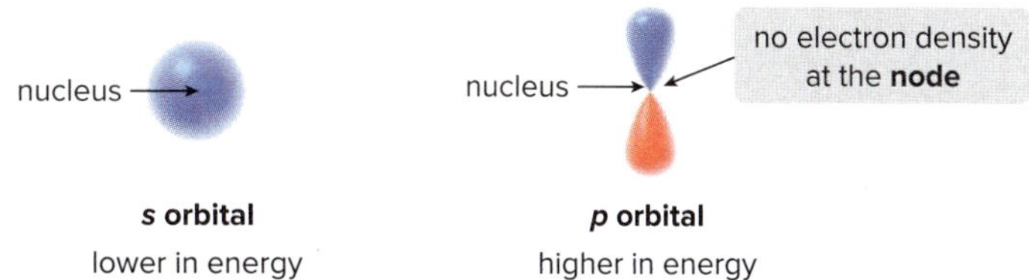

An *s* orbital is filled with electrons before a *p* orbital in the same shell.

1.1A The First Row

The first row of the periodic table is formed by adding electrons to the first shell of orbitals around the nucleus. There is only one orbital in the first shell, called the **1*s* orbital.**

- **Each orbital can have a maximum of two electrons.**

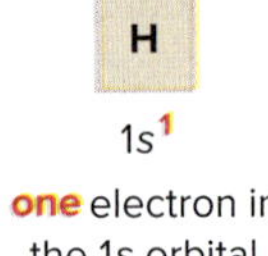

$1s^1$
one electron in the 1s orbital

As a result, there are **two elements in the first row,** one having one electron added to the 1*s* orbital, and one having two. The element **hydrogen (H)** has what is called a $1s^1$ configuration with one electron in the 1*s* orbital, and **helium (He)** has a $1s^2$ configuration with two electrons in the 1*s* orbital.

1.1B The Second Row

Every element in the second row has a filled first shell of electrons. Thus, all second-row elements have a $1s^2$ configuration. Each element in the second row of the periodic table also has four orbitals available to accept additional electrons:

- **one 2*s* orbital,** the *s* orbital in the second shell
- **three 2*p* orbitals,** all dumbbell-shaped and perpendicular to each other along the *x*, *y*, and *z* axes

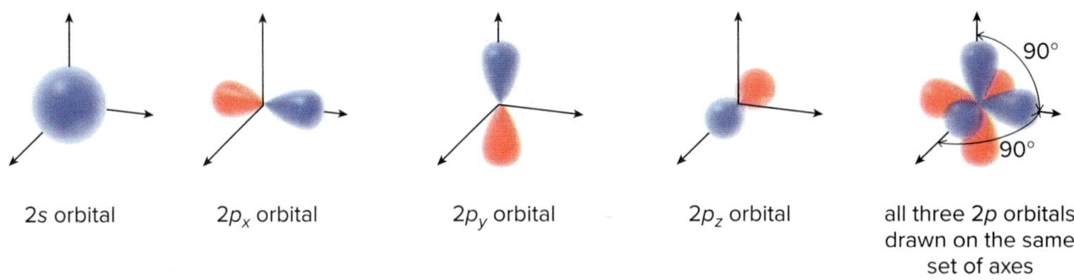

2*s* orbital | $2p_x$ orbital | $2p_y$ orbital | $2p_z$ orbital | all three 2*p* orbitals drawn on the same set of axes

Because each of the four orbitals in the second shell can hold two electrons, there is a **maximum capacity of *eight* electrons** for elements in the second row. The second row of the periodic table consists of eight elements, obtained by adding electrons to the 2*s* and three 2*p* orbitals.

group number	1A	2A	3A	4A	5A	6A	7A	8A
second row	Li	Be	B	C	N	O	F	Ne
number of valence electrons	1	2	3	4	5	6	7	8

Worked-out solutions to all in-chapter and end-of-chapter problems can be found in the *Student Study Guide/Solutions Manual to Accompany Organic Chemistry, 7e.*

The outermost electrons are called **valence electrons.** The valence electrons are more loosely held than the electrons closer to the nucleus, and as such, they participate in chemical reactions. **The group number of a second-row element reveals its number of valence electrons.** For example, carbon in group 4A has **four** valence electrons, and oxygen in group 6A has **six.**

Problem 1.1 While the most common isotope of nitrogen has a mass number of 14 (nitrogen-14), a radioactive isotope of nitrogen has a mass number of 13 (nitrogen-13). Nitrogen-13 is used in PET (positron emission tomography) scans by physicians to monitor brain activity and diagnose dementia. For each isotope, give the following information: (a) the number of protons; (b) the number of neutrons; (c) the number of electrons in the neutral atom; (d) the group number; and (e) the number of valence electrons.

1.2 Bonding

Until now our discussion has centered on individual atoms, but it is more common in nature to find two or more atoms joined together.

- ***Bonding*** **is the joining of two atoms in a stable arrangement.**

Joining two or more elements forms **compounds.** Although only about 100 elements exist, more than 50 million compounds are known. Examples of compounds include hydrogen gas (H_2), formed by joining two hydrogen atoms, and methane (CH_4), the simplest organic compound, formed by joining a carbon atom with four hydrogen atoms.

One general rule governs the bonding process.

- **Through bonding, atoms attain a complete outer shell of valence electrons.**

Because the noble gases in group 8A of the periodic table are especially stable as atoms having a filled shell of valence electrons, the general rule can be restated.

Jill Braaten

NaCl

In the ionic compound NaCl, each positively charged Na^+ ion is surrounded by six negatively charged Cl^- ions, and each Cl^- ion is surrounded by six Na^+ ions.

- **Through bonding, atoms gain, lose, or share electrons to attain the electronic configuration of the noble gas closest to them in the periodic table.**

What does this mean for first- and second-row elements? **A first-row element like hydrogen can accommodate *two electrons* around it.** This would make it like the noble gas helium at the end of the same row. **A second-row element is generally most stable with *eight valence electrons* around it** like neon. Elements that behave in this manner are said to follow the **octet rule.**

There are two different kinds of bonding: **ionic bonding** and **covalent bonding.**

- ***Ionic bonds* result from the *transfer* of electrons from one element to another.**
- ***Covalent bonds* result from the *sharing* of electrons between two nuclei.**

The type of bonding is determined by the location of an element in the periodic table. An ionic bond generally occurs when elements on the **far left** side of the periodic table combine with elements on the **far right** side, ignoring the noble gases, which form bonds only rarely. **The resulting ions are held together by extremely strong electrostatic interactions.** A positively charged **cation** formed from the element on the left side attracts a negatively charged **anion** formed from the element on the right side. Ionic compounds form extended crystal lattices that maximize the positive and negative electrostatic interactions. Examples of ionic inorganic compounds include sodium chloride (NaCl), common table salt, and potassium iodide (KI), an essential nutrient added to make iodized salt.

- **The transfer of electrons forms stable salts composed of cations and anions.**

A **compound** may have either ionic or covalent bonds. A **molecule** has only covalent bonds.

The second type of bonding, **covalent bonding,** occurs with elements like carbon in the middle of the periodic table, which would otherwise have to gain or lose several electrons to form an ion with a complete valence shell. **A covalent bond is a two-electron bond,** and a compound with covalent bonds is called a **molecule.** Covalent bonds also form between two elements from the same side of the table, such as two hydrogen atoms or two chlorine atoms. H_2, Cl_2, and CH_4 are all examples of covalent molecules.

Problem 1.2 Label each bond in the following compounds as ionic or covalent.

a. F_2 b. LiBr c. CH_3CH_3 d. $NaNH_2$ e. $NaOCH_3$

How many covalent bonds will a particular atom typically form? As you might expect, it depends on the location of the atom in the periodic table. In the first row, **hydrogen forms one covalent bond** using its one valence electron. When two hydrogen atoms are joined in a bond, each has a filled valence shell of two electrons. **A *solid line* indicates a two-electron bond.**

Second-row elements can have no more than eight valence electrons around them. For neutral molecules, two consequences result.

- **Atoms with one, two, three, or four valence electrons form one, two, three, or four bonds, respectively, in neutral molecules.**
- **Atoms with five or more valence electrons form enough bonds to give an octet. In this case, the predicted number of bonds = 8 – the number of valence electrons.**

For example, B has **three** valence electrons, so it forms **three** bonds, as in BF_3. N has **five** valence electrons, so it also forms **three** bonds ($8 - 5 =$ **3** bonds), as in NH_3.

Nonbonded pair of electrons = unshared pair of electrons = lone pair

These guidelines are used in Figure 1.2 to summarize the usual number of bonds formed by the common atoms in organic compounds. When second-row elements form fewer than four bonds their octets consist of both **bonding (shared) electrons** and **nonbonding (unshared) electrons.** Unshared electrons are also called **lone pairs.**

Problem 1.3 How many covalent bonds are predicted for each atom?

a. O b. Al c. Br d. Si e. Se

Figure 1.2 The usual number of bonds of common neutral atoms

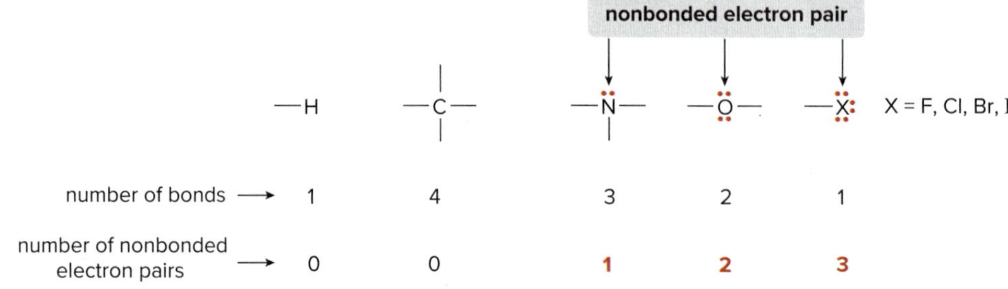

1.3 Lewis Structures

Lewis structures **are electron dot representations for molecules.** Three rules are used for drawing Lewis structures.

1. Draw only the valence electrons.
2. Give every second-row element no more than eight electrons.
3. Give each hydrogen two electrons.

1.3A A Procedure for Drawing Lewis Structures

Follow a stepwise procedure to draw a Lewis structure.

How To Draw a Lewis Structure

Step [1] **Arrange atoms next to each other that you think are bonded together.**

- Always place hydrogen atoms and halogen atoms on the periphery because H and X (X = F, Cl, Br, and I) form only one bond each.

The letter **X** is often used to represent one of the halogens in group 7A: F, Cl, Br, or I.

For **CH_3Cl:** H C H (Cl above C, H below C) ***not*** H C H H (Cl above C)

This H cannot form two bonds.

- As a first approximation, use the common bonding patterns in Figure 1.2 to arrange the atoms.

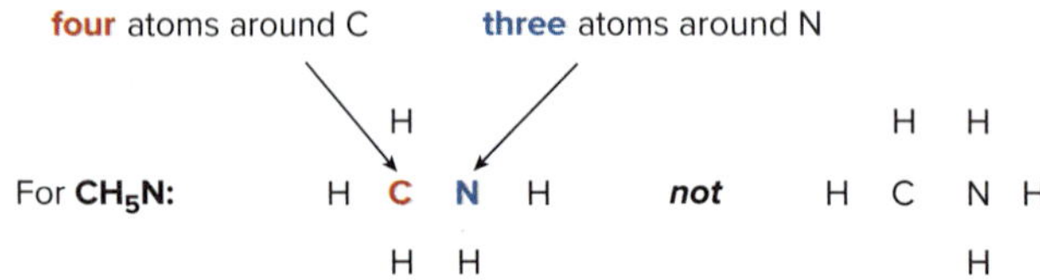

- In truth, the proper arrangement of atoms may not be obvious, or more than one arrangement may be possible (Section 1.4). Even in many simple molecules, the connectivity between atoms must be determined experimentally.

—Continued

Step [2] **Count the electrons.**

- Count the number of valence electrons from all atoms.
- Add one electron for each negative charge.
- Subtract one electron for each positive charge.
- This sum gives the total number of electrons that must be used in drawing the Lewis structure.

Step [3] **Arrange the electrons around the atoms.**

- Place a bond between every two atoms, giving **two electrons to each H** and **no more than eight to any second-row atom.**
- Use all remaining electrons to **fill octets with lone pairs.**
- If all valence electrons are used and an atom does not have an octet, form multiple bonds, as shown in Sample Problem 1.2.

Step [4] **Assign formal charges to all atoms.**

- Formal charges are discussed in Section 1.3C.

Sample Problem 1.1 illustrates how to draw the Lewis structure of a simple organic molecule.

Sample Problem 1.1 Drawing a Lewis Structure for a Simple Molecule

Draw a Lewis structure for methanol, a compound with molecular formula CH_4O.

Solution

Step [1] **Arrange the atoms.**

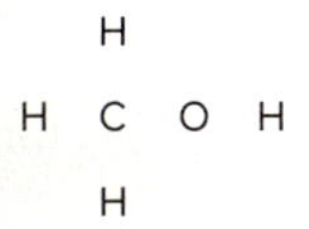

```
   H
H  C  O  H
   H
```

- Place the second-row elements, C and O, in the middle.
- Place three H's around C to surround C by four atoms.
- Place one H next to O to surround O by two atoms.

Step [2] **Count the electrons.**

$$\begin{aligned} 1\ C \times 4\ e^- &= 4\ e^- \\ 1\ O \times 6\ e^- &= 6\ e^- \\ 4\ H \times 1\ e^- &= 4\ e^- \\ &\mathbf{14\ e^-\ total} \end{aligned}$$

Step [3] **Add the bonds and lone pairs.**

- Add five two-electron bonds to form the C–H, C–O, and O–H bonds, using 10 of the 14 electrons.
- Place two lone pairs on the O atom to use the remaining four electrons and give the O atom an octet.

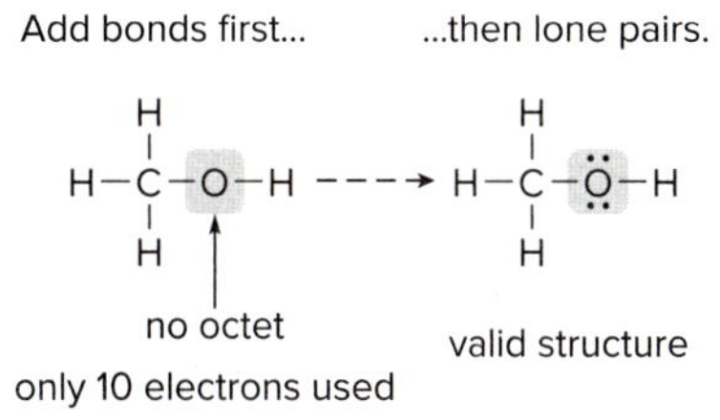

Add bonds first... ...then lone pairs.

```
   H                  H
   |                  |
H—C—O—H  - - - ->  H—C—Ö—H
   |  ↑               |  ..
   H  |               H
```

no octet
only 10 electrons used

valid structure

This Lewis structure is valid because it uses all 14 electrons, each H is surrounded by two electrons, and each second-row element is surrounded by no more than eight electrons.

Problem 1.4 Draw a valid Lewis structure for each species.

a. CH_3CH_3 b. CH_5N c. C_2H_5Br

1.3B Multiple Bonds

Sample Problem 1.2 illustrates an example of a Lewis structure with a double bond.

Sample Problem 1.2 Drawing a Lewis Structure with a Multiple Bond

Draw a Lewis structure for ethylene, a compound of molecular formula C_2H_4, in which each carbon is bonded to two hydrogens.

Solution

Follow Steps [1] to [3] to draw a Lewis structure.

Step [1] **Arrange the atoms.**

H C C H
H H

• Each C gets 2 H's.

Step [2] **Count the electrons.**

$$2\ \text{C} \times 4\ e^- = 8\ e^-$$
$$4\ \text{H} \times 1\ e^- = 4\ e^-$$
$$\mathbf{12\ e^-\ total}$$

Step [3] **Add the bonds and lone pairs.**

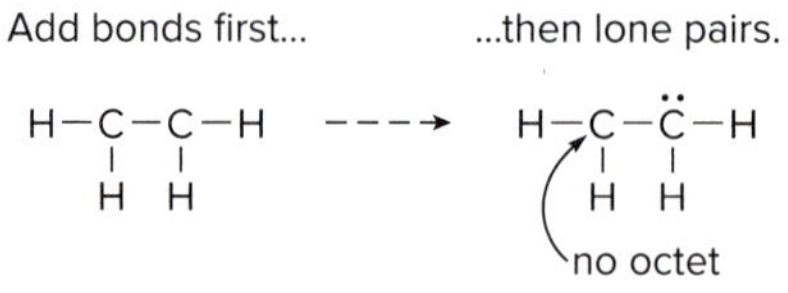

Add bonds first... ...then lone pairs.

H–C–C–H (each C bonded to H below) ---→ H–C–C̈–H (each C bonded to H below) — no octet

After placing five bonds between the atoms and adding the two remaining electrons as a lone pair, one C still has no octet.

To give both C's an octet, **change *one* lone pair into *one* bonding pair of electrons between the two C's, forming a double bond.**

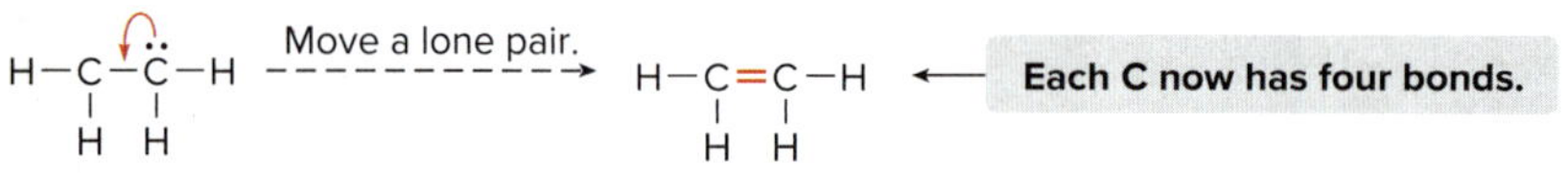

H–C–C̈–H —Move a lone pair.→ H–C=C–H ← **Each C now has four bonds.**

ethylene
a valid Lewis structure

This uses all 12 electrons, each C has an octet, and each H has two electrons. The Lewis structure is valid. **Ethylene contains a carbon–carbon double bond.**

Problem 1.5 Draw an acceptable Lewis structure for each compound, assuming the atoms are connected as arranged. Formaldehyde (H_2CO) is a preservative, glycolic acid ($HOCH_2CO_2H$) is used to make dissolving sutures, and acetyl chloride (CH_3COCl) is a reagent used in chemical synthesis.

a. H_2CO

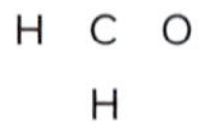

```
H C O
  H
```

b. $HOCH_2CO_2H$

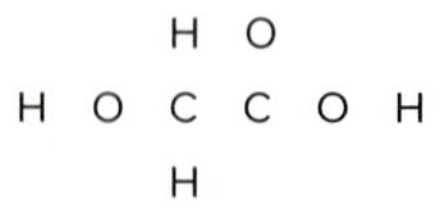

```
    H O
H O C C O H
    H
```

c.

```
  H O
H C C Cl
  H
```

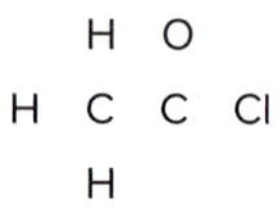

More Practice: Try Problems 1.48, 1.49.

• **After placing all electrons in bonds and lone pairs, use a lone pair to form a multiple bond if an atom does not have an octet.**

Carbon always forms four bonds in stable organic molecules. Carbon forms single, double, and triple bonds to itself and other elements.

You must change *one* lone pair into *one* new bond for each *two* electrons needed to complete an octet. In acetylene, a compound with molecular formula C_2H_2, placing the 10 valence electrons gives a Lewis structure in which one or both of the C's lack an octet.

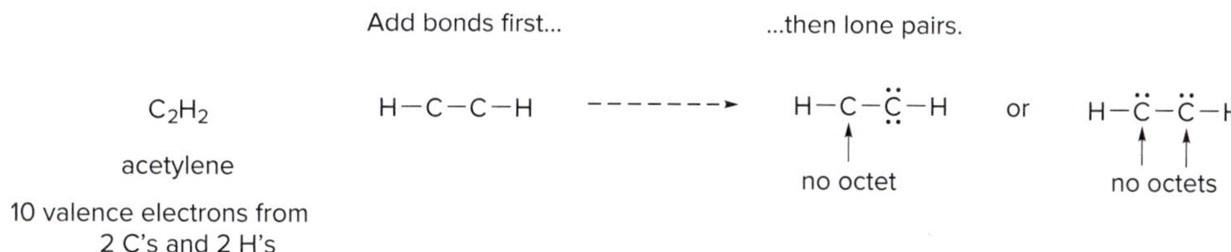

In this case, **change *two* lone pairs into *two* bonding pairs of electrons, forming a triple bond.**

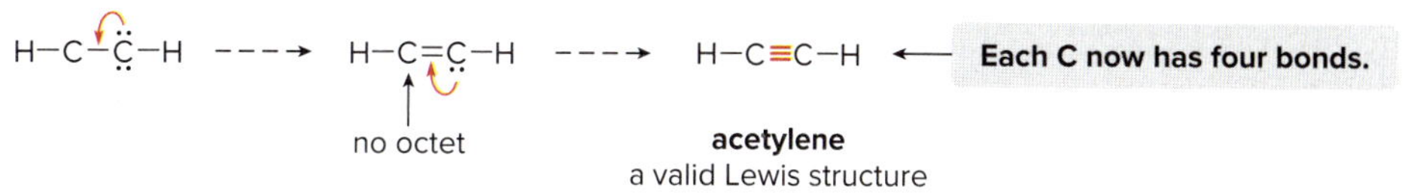

H—C—C—H ---→ H—C=C—H ---→ H—C≡C—H ←— **Each C now has four bonds.**

no octet

acetylene
a valid Lewis structure

Problem 1.6 Draw an acceptable Lewis structure for each compound, assuming the atoms are connected as arranged.

a. HCN H C N

b. C_3H_4 H C C C H (with H above and H below the first C)

1.3C Formal Charge

To manage electron bookkeeping in a Lewis structure, chemists use **formal charge.**

- ***Formal charge* is the charge assigned to individual atoms in a Lewis structure.**

By calculating formal charge, we determine how the number of electrons around a particular atom compares to its number of valence electrons. Formal charge is calculated as follows:

formal charge = **number of valence electrons** − **number of electrons an atom "owns"**

The number of electrons "owned" by an atom is determined by its number of bonds and lone pairs.

- **An atom "owns" *all* of its unshared electrons and *half* of its shared electrons.**

number of electrons owned = **number of unshared electrons** + $\frac{1}{2}$ **[number of shared electrons]**

The number of electrons "owned" by different carbon atoms is indicated in the following examples:

—C— (four single bonds)
- C shares eight electrons.
- C "owns" **four** electrons.

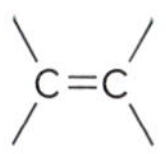

C=C
- Each C shares eight electrons.
- Each C "owns" **four** electrons.

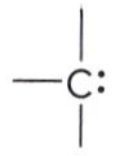

—C: (three bonds, one lone pair)
- C shares six electrons.
- C has two unshared electrons.
- C "owns" **five** electrons.

Sample Problem 1.3 illustrates how formal charge is calculated on the atoms of a polyatomic ion. **The sum of the formal charges on the individual atoms equals the net charge on the molecule or ion.**

Sample Problem 1.3 Determining the Formal Charge on an Atom

Determine the formal charge on each atom in the ion H_3O^+.

$$\left[\text{H}-\ddot{\text{O}}-\text{H} \atop \quad | \atop \quad \text{H} \right]^+$$

Solution

To calculate the formal charge on each atom:

- Determine the number of valence electrons from the group number.
- Determine the number of electrons an atom "owns" from the number of bonding and nonbonding electrons it has.
- Subtract the second quantity from the first to give the formal charge.

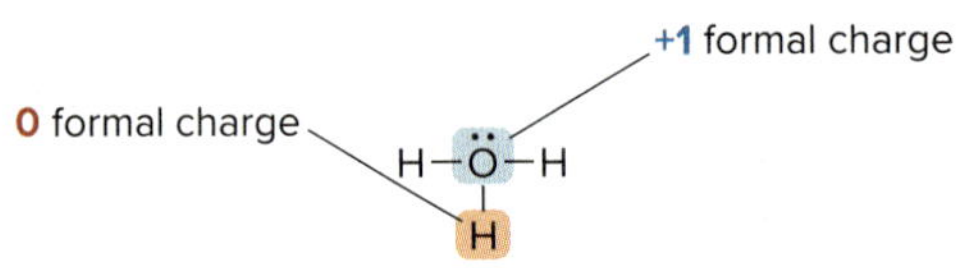

For the O atom (group 6A):

- number of valence electrons = 6
- number of bonding electrons = 6
- number of nonbonding electrons = 2

$$\text{formal charge} = 6 - \left[2 + \frac{1}{2}(6)\right] = +1$$

For each H atom (group 1A):

- number of valence electrons = 1
- number of bonding electrons = 2
- number of nonbonding electrons = 0

$$\text{formal charge} = 1 - \left[0 + \frac{1}{2}(2)\right] = 0$$

The formal charge on the O atom is +1 and the formal charge on each H is 0. The overall charge on the ion H_3O^+ is the sum of all of the formal charges on the atoms: 1 + 0 + 0 + 0 = +1.

Problem 1.7 Calculate the formal charge on each second-row atom.

a. $[NH_4]^+$ (H—N—H with H above and below) b. $CH_3—N{\equiv}C:$ c. $:\ddot{O}{=}\ddot{O}{-}\ddot{\underset{..}{O}}:$ d. $CH_3—N{=}\ddot{O}:$ with $:\ddot{\underset{..}{O}}:$ bonded to N

More Practice: Try Problem 1.47.

Problem 1.8 Draw a Lewis structure for each ion.

a. CH_3O^- b. HC_2^- c. $(CH_3NH_3)^+$ d. $(CH_3NH)^-$

Problem 1.9 Draw an acceptable Lewis structure such that all atoms have zero formal charge.

a. diethyl ether, $CH_3CH_2OCH_2CH_3$, the first general anesthetic used in medical procedures
b. acrylonitrile, CH_2CHCN, starting material used to manufacture synthetic Orlon fibers

When you first add formal charges to Lewis structures, use the procedure in Sample Problem 1.3. With practice, you will notice that certain bonding patterns always result in the same formal charge. For example, any N atom with four bonds (and thus no lone pairs) has a +1 formal charge. Table 1.1 lists the bonding patterns and resulting formal charges for carbon, nitrogen, and oxygen.

Table 1.1 Formal Charge Observed with Common Bonding Patterns for C, N, and O

Atom	Number of valence electrons	Formal charge +1	Formal charge 0	Formal charge −1
C	4	 three bonds, +	 four bonds	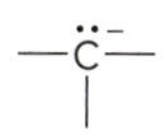 three bonds, one lone pair, −
N	5	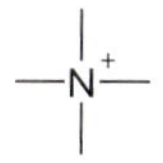 four bonds, +	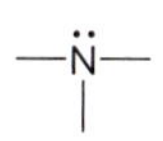 three bonds, one lone pair	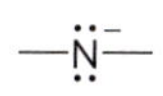 two bonds, two lone pairs, −
O	6	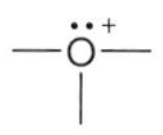 three bonds, one lone pair, +	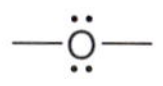 two bonds, two lone pairs	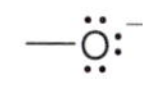 one bond, three lone pairs, −

Problem 1.10 What is the formal charge on the O atom in each of the following species that contains a multiple bond to O?

a. 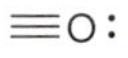 ≡O: b. 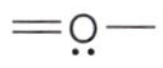 =Ö— c. 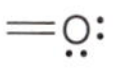 =Ö:

Problem 1.11 Assign formal charges to each N and O atom in the given molecules. All lone pairs have been drawn in. The five- and six-membered rings are explained in Section 1.8B.

a. 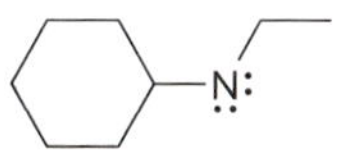 b. 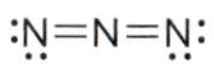 :N̤=N=N̤: c. 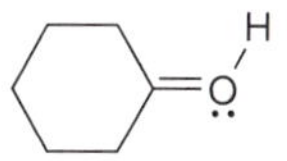 d.

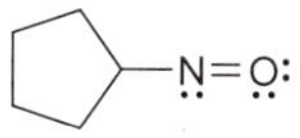

1.4 Isomers

Sometimes in drawing a Lewis structure, more than one arrangement of atoms is possible for a given molecular formula. For example, there are two acceptable arrangements of atoms for the molecular formula C_2H_6O.

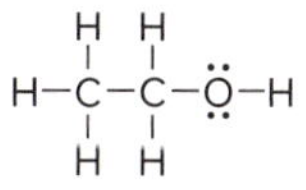

ethanol 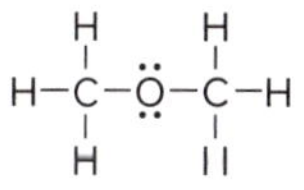 dimethyl ether

same molecular formula
C_2H_6O
isomers

Both are valid Lewis structures, and both molecules exist. One is called ethanol, and the other, dimethyl ether. These two compounds are called **isomers.**

- ***Isomers* are different molecules having the same molecular formula.**

Ethanol and dimethyl ether are **constitutional isomers** because they have the same molecular formula, but the ***connectivity of their atoms is different.*** Ethanol has one C–C bond and one O–H bond, whereas dimethyl ether has two C–O bonds. A second class of isomers, called **stereoisomers,** is introduced in Section 4.12B.

Problem 1.12 Draw Lewis structures for each molecular formula.

a. $C_2H_4Cl_2$ (two isomers)
b. C_3H_8O (three isomers)
c. C_3H_6 (two isomers)
d. C_3H_7Cl (two isomers)
e. C_2H_4O (three isomers)
f. C_3H_9N (four isomers)

1.5 Exceptions to the Octet Rule

Most of the common elements in organic compounds—**C, N, O, and the halogens**—follow the octet rule. Hydrogen is a notable exception, because it accommodates only two electrons in bonding. Additional exceptions include boron and beryllium (second-row elements in groups 3A and 2A, respectively), and elements in the third row (particularly phosphorus and sulfur).

Elements in groups 2A and 3A of the periodic table, such as beryllium and boron, do not have enough valence electrons to form an octet in a neutral molecule. Lewis structures for BeH_2 and BF_3 show that these atoms have only four and six electrons, respectively, around the central atom. There simply aren't enough electrons to form an octet. Because the Be and B atoms each have less than an octet of electrons, these molecules are highly reactive.

H—Be—H

four electrons around Be

six electrons around B

A second exception to the octet rule occurs with some elements located in the third row and later in the periodic table. These elements have empty *d* orbitals available to accept electrons, and thus they may have ***more than eight* electrons** around them. For organic chemists, the two most common elements in this category are **phosphorus** and **sulfur,** which can have 10 or even 12 electrons around them, as shown in dimethyl sulfoxide, sulfuric acid, and alendronic acid.

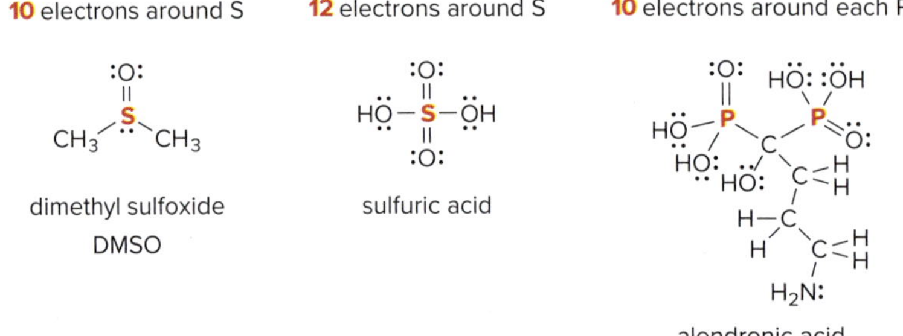

dimethyl sulfoxide DMSO

sulfuric acid

alendronic acid

Alendronic acid, sold as a sodium salt under the trade name **Fosamax,** is used to prevent osteoporosis, a medical condition resulting in bone loss, in women.

1.6 Resonance

Some molecules can't be adequately represented by a single Lewis structure. For example, two valid Lewis structures can be drawn for the anion $(HCONH)^-$. One structure has a negatively charged N atom and a C–O double bond; the other has a negatively charged O atom and a C–N double bond. These structures are called **resonance structures** or **resonance forms.** A **double-headed arrow** is used to separate two resonance structures.

double-headed arrow

- ***Resonance structures*** **are two Lewis structures having the *same* placement of atoms but a *different* arrangement of electrons.**

Which resonance structure is an accurate representation for $(HCONH)^-$? **The answer is *neither* of them.** The true structure is a composite of both resonance forms, and is called a **resonance hybrid.** The hybrid shows characteristics of ***both*** resonance structures.

Each resonance structure implies that electron pairs are localized in bonds or on atoms. In actuality, resonance allows certain electron pairs to be ***delocalized*** over two or more atoms, and this delocalization of electron density adds stability. **A molecule with two or more resonance structures is said to be *resonance stabilized.***

1.6A An Introduction to Resonance Theory

Keep in mind the following basic principles of resonance theory.

- **Resonance structures are *not* real. An individual resonance structure does not accurately represent the structure of a molecule or ion.**
- **Resonance structures are *not* in equilibrium with each other. There is no movement of electrons from one form to another.**
- **Resonance structures are *not* isomers. Two isomers differ in the arrangement of *both* atoms and electrons, whereas resonance structures differ *only* in the *arrangement of electrons.***

For example, ions **A** and **B** are resonance structures because the atom position is the same in both compounds, but the location of an electron pair is different. In contrast, compounds **C** and **D** are isomers because the atom placement is different; **C** has an O–H bond, and **D** has an additional C–H bond.

A B C D

- **A** and **B** are **resonance** structures.
- The position of one electron pair (in red) is different.

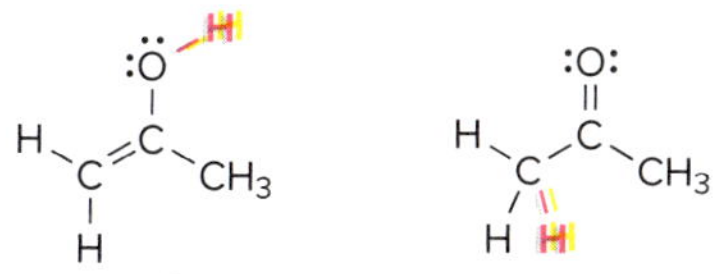

- **C** and **D** are **isomers.**
- The position of a H atom (in red) is different.

Problem 1.13 Classify each pair of compounds as isomers or resonance structures.

a. :N=C=O: and :C≡N–O: b. H–O–C(=O)–O: and H–O–C(–O:)=O:

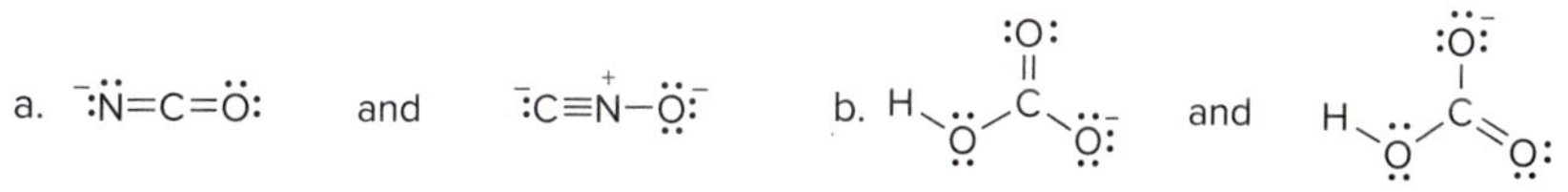

Problem 1.14 Considering structures **A–D**, classify each pair of compounds as isomers, resonance structures, or neither: (a) **A** and **B;** (b) **A** and **C;** (c) **A** and **D;** (d) **B** and **D.**

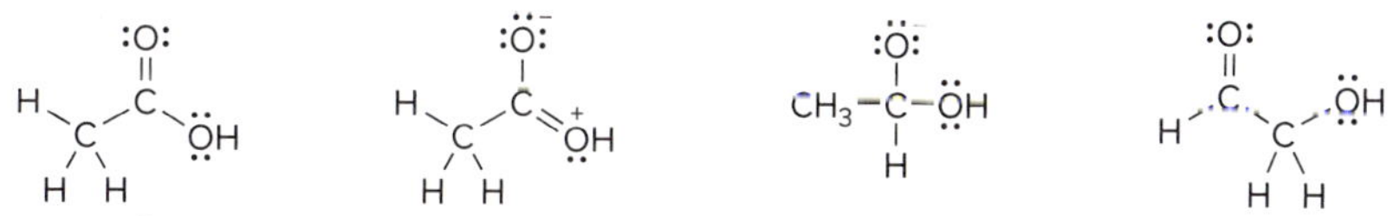

A B C D

1.6B Drawing Resonance Structures

To draw resonance structures, use three criteria.

Rule [1] **Two resonance structures differ in the position of multiple bonds and nonbonded electrons. The placement of atoms and single bonds always stays the same.**

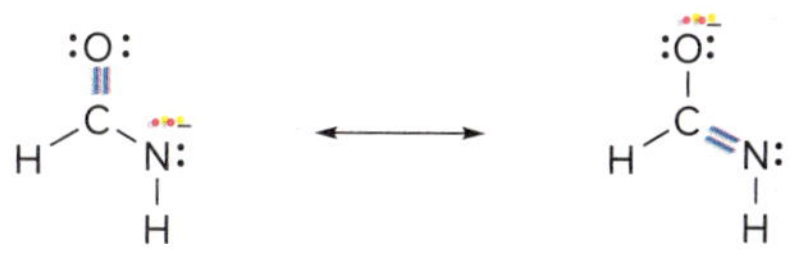

A ⟷ B

- The position of a double bond (in blue) is different.
- The position of a lone pair (in red) is different.

Rule [2] **Two resonance structures must have the same number of unpaired electrons.**

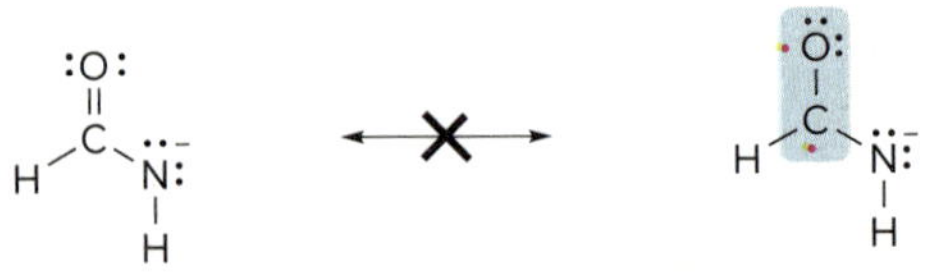

A — no unpaired electrons

C — **two** unpaired electrons
not a resonance structure of **A** (or **B**)

Rule [3] **Resonance structures must be valid Lewis structures. Hydrogen must have two electrons, and a second-row element can have no more than *eight* electrons.**

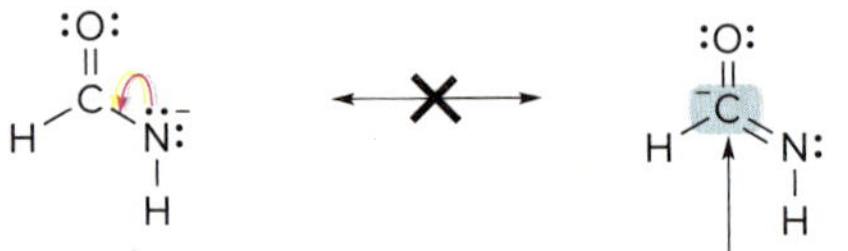

A

10 electrons around C
not a valid resonance structure of **A**

Curved arrow notation is a convention that shows how electron position differs between the two resonance forms.

- ***Curved arrow notation shows the movement of an electron pair.*** **The tail of the arrow always begins at an electron pair, in either a bond or lone pair. The head points to where the electron pair "moves."**

A curved arrow always begins at an electron pair. It ends at an atom or a bond.

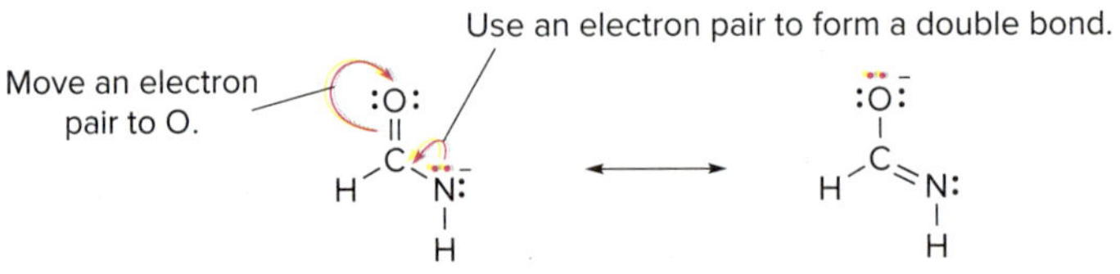

Move an electron pair to O.

Use an electron pair to form a double bond.

A

B

Resonance structures **A** and **B** differ in the location of two electron pairs, so two curved arrows are needed. To convert **A** to **B,** take the lone pair on N and form a double bond between C and N. Then, move an electron pair in the C–O double bond to form a lone pair on O. Curved arrows thus show how to reposition the electrons in converting one resonance form to another. **The electrons themselves do not actually move.** Sample Problem 1.4 illustrates the use of curved arrows to convert one resonance structure to another.

Sample Problem 1.4 Using Curved Arrows

Follow the curved arrows to draw a second resonance structure for each ion.

a.

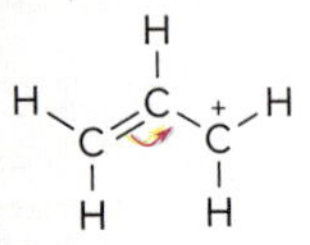

b.

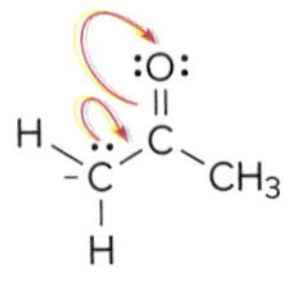

Solution

a. The curved arrow tells us to move **one** electron pair in the double bond to the adjacent C–C bond. Then determine the formal charge on any atom whose bonding is different.

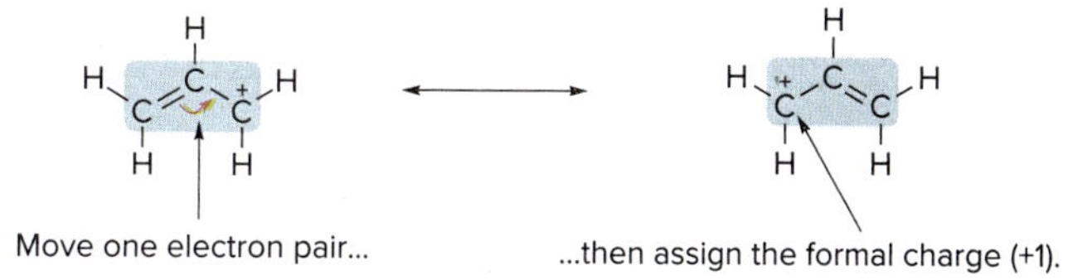

Move one electron pair...

...then assign the formal charge (+1).

Positively charged carbon atoms are called **carbocations.** Carbocations are unstable intermediates because they contain a carbon atom that is lacking an octet of electrons.

b. **Two** curved arrows tell us to move **two** electron pairs. The second resonance structure has a formal charge of (–1) on O.

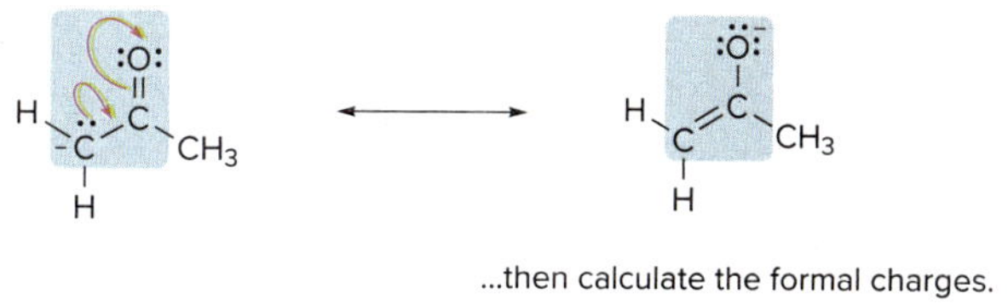

Move the electron pairs...

...then calculate the formal charges.

This type of resonance-stabilized anion is called an **enolate anion.** Enolates are important intermediates in many organic reactions, and all of Chapters 21 and 22 is devoted to their preparation and reactions.

Problem 1.15 Follow the curved arrows to draw a second resonance structure for each species.

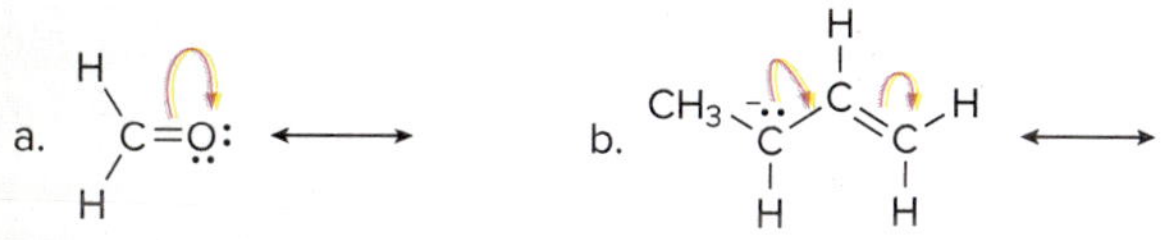

a.

b.

More Practice: Try Problems 1.54, 1.55.

Problem 1.16 Use curved arrow notation to show how the first resonance structure can be converted to the second.

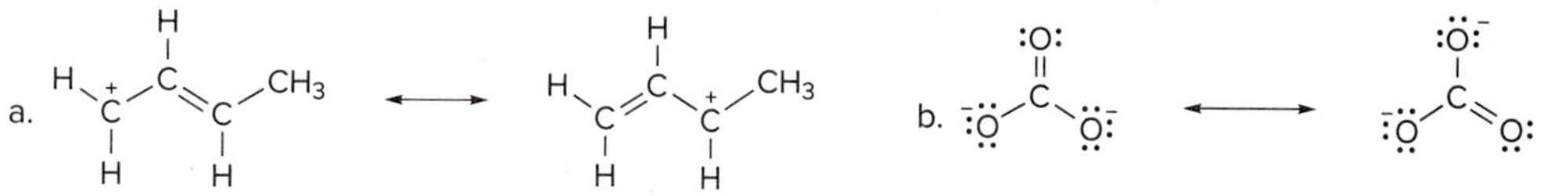

a.

b.

Two resonance structures can have exactly the same kinds of bonds, as they do in the carbocation in Sample Problem 1.4a, or they may have different types of bonds, as they do in the enolate in Sample Problem 1.4b. Either possibility is fine as long as the individual resonance structures are valid Lewis structures.

A resonance structure can have an atom with *fewer* than eight electrons around it. **B** is a resonance structure of **A** even though the carbon atom is surrounded by only six electrons.

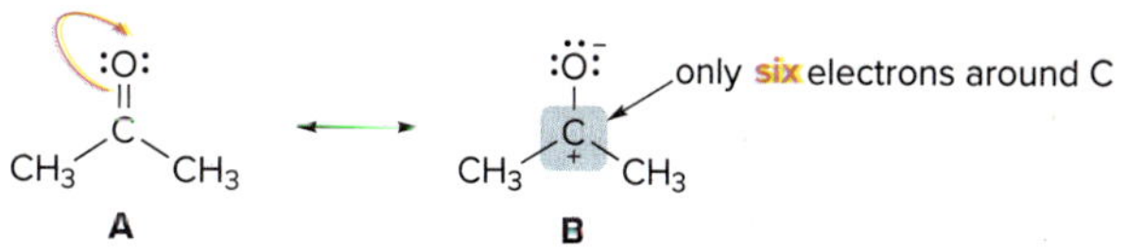

only six electrons around C

A **B**

valid resonance structure

In contrast, a resonance structure can *never* have a second-row element with more than eight electrons. **C** is *not* a resonance structure of **A** because the carbon atom is now surrounded by 10 electrons.

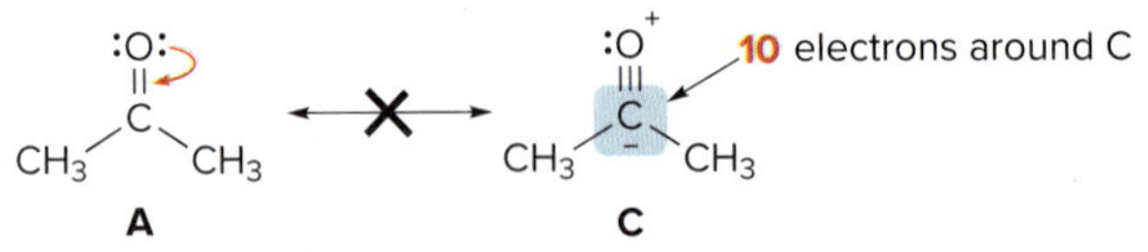

10 electrons around C

A C

not a valid resonance structure

We will learn much more about resonance in Chapter 14.

The ability to draw and manipulate resonance structures is a necessary skill that will be used throughout your study of organic chemistry. With practice, you will begin to recognize certain common bonding patterns for which more than one Lewis structure can be drawn. For instance, both the carbocation in Sample Problem 1.4a and the enolate anion in Sample Problem 1.4b are specific examples of one general type of resonance observed in certain three-atom systems.

- **In a group of three atoms having a multiple bond X=Y joined to an atom Z having a *p* orbital with zero, one, or two electrons, two resonance structures can be drawn.**

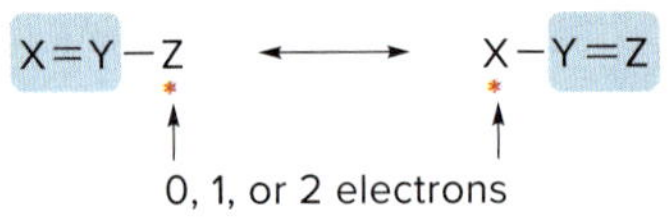

X=Y–Z* ⟷ X–Y=Z*

0, 1, or 2 electrons

The * corresponds to a charge, a lone pair, or a single electron.

* = +, –, ·, or :

Recall from the Prologue that a ***heteroatom*** **is an atom other than carbon or hydrogen.**

X, Y, and Z may all be carbon atoms or they may be **heteroatoms** such as nitrogen or oxygen. The atom Z can be charged (positive or negative) or neutral (with a lone pair or a single electron), corresponding to the [*] in the general structure X=Y–Z*. The two resonance structures differ in the location of the multiple bond and the [*].

In the enolate anion in Sample Problem 1.4b, X corresponds to oxygen and [*] is a lone pair, which gives carbon a net negative charge. Moving the double bond and the lone pair and readjusting charges gives the second resonance structure.

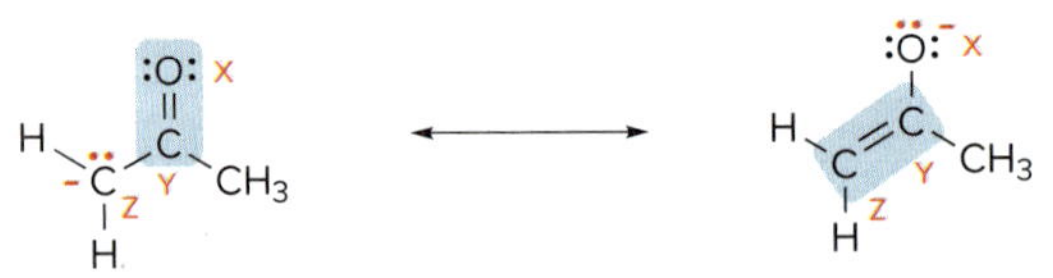

- The position of the double bond changes.
- The location of a lone pair changes.

In Chapter 14, we will learn more about the orbitals involved in this type of resonance.

Sample Problem 1.5 Drawing Resonance Structures

Draw a second resonance structure for acetamide.

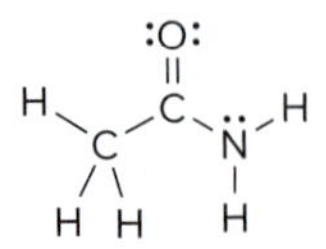

acetamide

Solution

Always look for a three-atom system that contains a multiple bond joined to an atom Z with zero, one, or two nonbonded electrons. **Move the double bond (from X=Y to Y=Z)** and **move the [*] from Z to X.** Recalculate formal charges on X and Z.

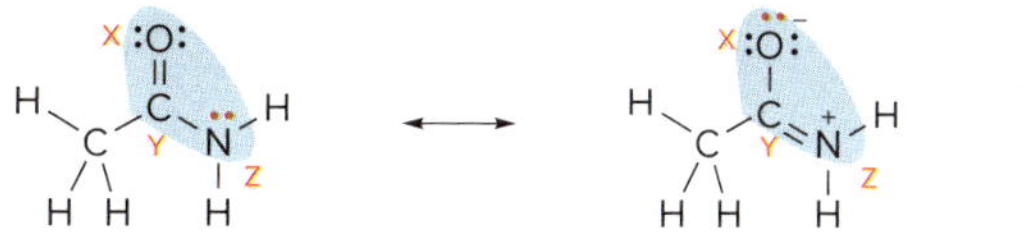

acetamide
- C=O
- lone pair on N

second resonance structure
- C=N
- additional lone pair on O

In this example, the three-atom system for resonance (X=Y–Z*) is O=C–N with a lone pair on N. After moving the double bond and the lone pair, the formal charges on O and N are –1 and +1, respectively, calculated using the procedure for determining formal charges.

Problem 1.17 Draw a second resonance structure for each species in parts (a), (b), and (c). Draw two additional resonance structures for the ion in part (d).

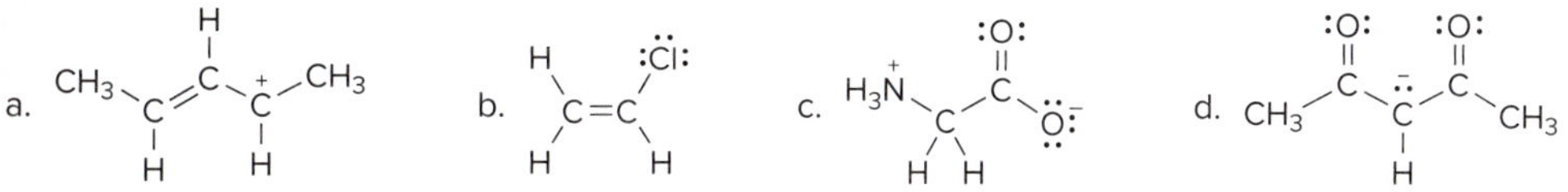

a. b. c. d.

More Practice: Try Problems 1.52, 1.56, 1.57.

Problem 1.18 (a) Draw a second resonance structure for **A.** (b) Why can't a second resonance structure be drawn for **B?**

A **B**

Problem 1.19 Creatine is a dietary supplement used by some athletes to boost their athletic performance. (a) Draw in all lone pairs in creatine. (b) Draw two additional resonance structures showing all lone pairs and formal charges.

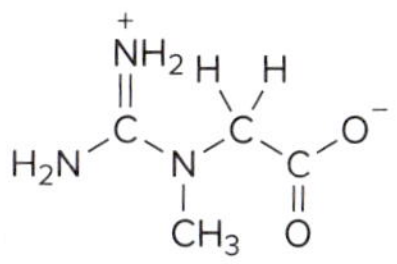

creatine

1.6C The Resonance Hybrid

The **resonance hybrid** is the composite of all possible resonance structures. In the resonance hybrid, the electron pairs drawn in different locations in individual resonance structures are ***delocalized.***

- **The resonance hybrid is more stable than any resonance structure because it delocalizes electron density over a larger volume.**

What does the hybrid look like? When all resonance forms are identical, as they were in the carbocation in Sample Problem 1.4a, each resonance form contributes **equally** to the hybrid.

When two resonance structures are different, the hybrid looks more like the "better" resonance structure. The "better" resonance structure is called the **major contributor** to the hybrid, and all others are **minor contributors.** The hybrid is the weighted average of the contributing resonance structures. What makes one resonance structure "better" than another? There are many factors, but for now, we will learn one fact.

- **A "better" resonance structure is one that has *more bonds* and *fewer charges.***

X
more bonds
fewer charges

major contributor

Y
minor contributor

Comparing resonance structures **X** and **Y, X** is the major contributor because it has more bonds and fewer charges. Thus, the hybrid looks more like **X** than **Y.**

How can we draw a hybrid, which has delocalized electron density? First, we must determine what is different in the resonance structures. Two differences commonly seen are the **position of a multiple bond** and the **site of a charge.** The anion $(HCONH)^-$ illustrates two conventions for drawing resonance hybrids.

A **B**

individual resonance structures

resonance hybrid

- The (–) charge is delocalized on N and O.
- The double bond is delocalized between O, C, and N.

Common symbols and conventions used in organic chemistry are summarized in Appendix B.

- **Double bond position.** Use a dashed line for a bond that is single in one resonance structure and double in another.
- **Location of charge.** Use a δ– (partial negative charge) or δ+ (partial positive charge) for an atom that is neutral in one resonance structure and charged in another.

The hybrid for $(HCONH)^-$ shows two dashed bonds, indicating that both the C–O and C–N bonds have partial double bond character. Both the O and N atoms bear a partial negative charge (δ–) because these atoms are neutral in one resonance structure and negatively charged in the other.

This discussion of resonance is meant to serve as an introduction only. You will learn many more facets of resonance theory in later chapters. In Chapter 2, for example, the enormous effect of resonance on acidity is discussed.

Problem 1.20 Label the resonance structures in each pair as major, minor, or equal contributors to the hybrid. Then draw the hybrid.

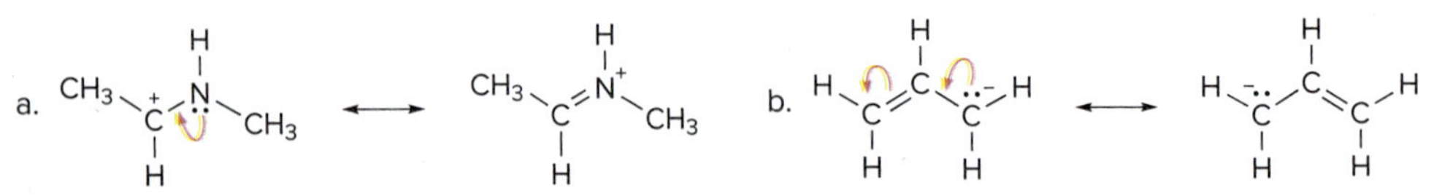

1.7 Determining Molecular Shape

Consider the H_2O molecule. The Lewis structure tells us which atoms are connected to each other, but it implies nothing about the geometry. What does the overall molecule look like? Is H_2O a bent or linear molecule? Two variables define a molecule's structure: **bond length** and **bond angle.**

1.7A Bond Length

Although the SI unit for bond length is the picometer (pm), the angstrom (Å) is still widely used in the chemical literature; 1 Å = 10^{-10} m. As a result, 1 pm = 10^{-2} Å, and 95.8 pm = 0.958 Å.

***Bond length* is the average distance between the centers of two bonded nuclei.** Bond lengths are typically reported in picometers (pm), where 1 pm = 10^{-12} m. For example, the O–H bond length in H_2O is 95.8 pm. Average bond lengths for common bonds are listed in Table 1.2.

- **Bond length *decreases* across a row of the periodic table as the size of the atom *decreases*.**

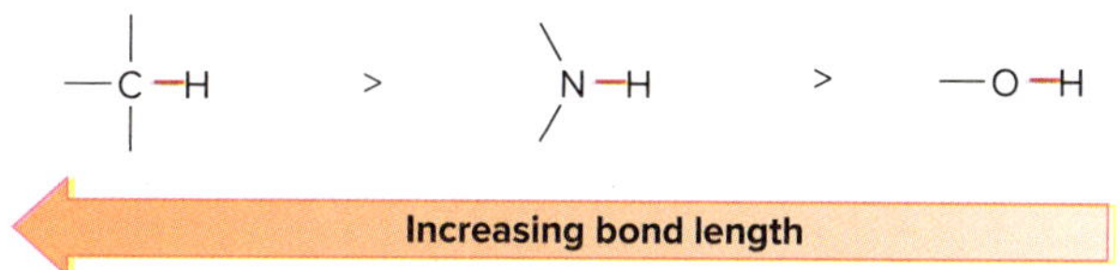

- **Bond length *increases* down a column of the periodic table as the size of an atom *increases*.**

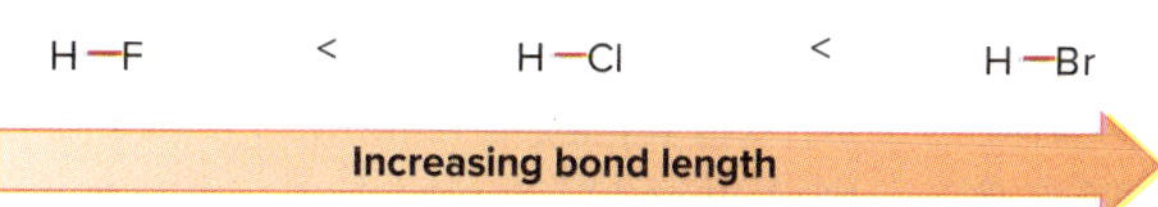

Table 1.2 Average Bond Lengths

Bond	Length (pm)	Bond	Length (pm)	Bond	Length (pm)
H–H	74	H–F	92	C–F	133
C–H	109	H–Cl	127	C–Cl	177
N–H	101	H–Br	141	C–Br	194
O–H	96	H–I	161	C–I	213

1.7B Bond Angle

Bond angle determines the shape around any atom bonded to two other atoms. To determine the bond angle and shape around a given atom, first count how many groups surround the atom. **A group is either an atom or a lone pair of electrons.** Then use the **valence shell electron pair repulsion (VSEPR) theory** to determine the shape. VSEPR is based on the fact that electron pairs repel each other; thus:

- **The most stable arrangement keeps the groups around an atom as far away from each other as possible.**

A second-row element has only three possible arrangements, defined by the number of groups surrounding it.

To determine geometry: [1] Draw a valid Lewis structure; [2] count groups around a given atom.

Number of groups	Geometry	Bond angle
• **two** groups	**linear**	180°
• **three** groups	**trigonal planar**	120°
• **four** groups	**tetrahedral**	109.5°

Let's examine several molecules to illustrate this phenomenon. We first need a valid Lewis structure, and then we count groups around a given atom to predict its geometry.

Two Groups Around an Atom

Any atom surrounded by only two groups is linear and has a bond angle of 180°. For example, each carbon atom in **HC≡CH** (acetylene) is surrounded by two atoms and no lone pairs, so each H–C–C bond angle in acetylene is 180°. Therefore all four atoms in HC≡CH are linear.

180°
H–C≡C–H =
180°
two atoms around each C

ball-and-stick model

two groups
linear carbons

Most students in organic chemistry find that building models helps them visualize the shape of molecules. Invest in a set of models *now*.

Acetylene illustrates an important feature: ***ignore multiple bonds in predicting geometry.* Count only atoms and lone pairs.**

We will represent molecules with models having balls for atoms and sticks for bonds, as in the ball-and-stick model of acetylene just shown. These representations are analogous to a set of molecular models. Balls are color-coded using accepted conventions: carbon (black), hydrogen (white or gray), oxygen (red), and so forth, as shown.

Common element colors are also shown in Appendix B.

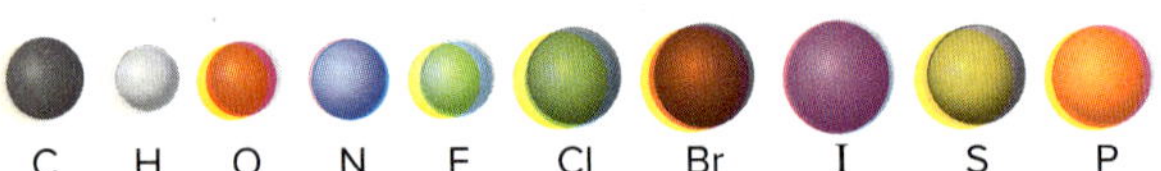

Three Groups Around an Atom

Any atom surrounded by three groups is trigonal planar and has bond angles of 120°. For example, each carbon atom in **$CH_2{=}CH_2$** (ethylene) is surrounded by three atoms and no lone pairs, making *each* H–C–C bond angle 120°. All six atoms of $CH_2{=}CH_2$ lie in one plane.

H H
C=C 120° =
H H
three atoms around each C

ethylene

three groups
trigonal planar carbons

Four Groups Around an Atom

Any atom surrounded by four groups is tetrahedral and has bond angles of approximately 109.5°. The simple organic compound methane, **CH_4,** has a central carbon atom with bonds to four hydrogen atoms, each pointing to a corner of a tetrahedron. This arrangement keeps four groups farther apart than a square planar arrangement in which all bond angles would be only 90°.

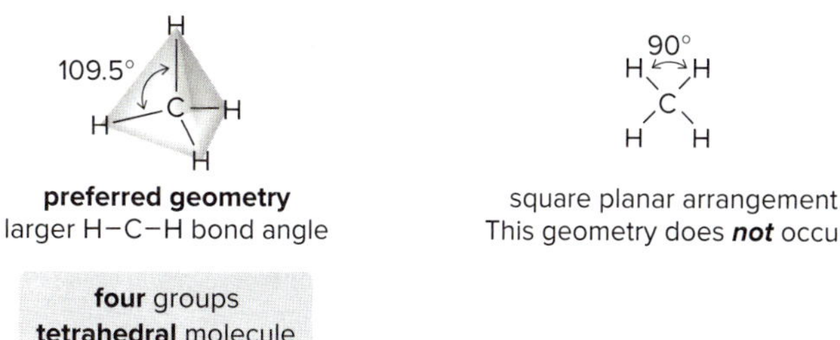

How can we represent the three-dimensional geometry of a tetrahedron on a two-dimensional piece of paper? **Place two of the bonds in the plane of the paper, one bond in front and one bond behind,** using the following conventions:

- A *solid line* is used for a bond *in* the plane.
- A *wedge* is used for a bond *in front* of the plane.
- A *dashed wedge* is used for a bond *behind* the plane.

bonds **in** the plane

bond **behind**

bond **in front**

ball-and-stick model of CH_4

All carbons in stable molecules are ***tetravalent,*** but the geometry varies with the number of groups around the particular carbon.

This is just one way to draw a tetrahedron for CH_4. We can turn the molecule in many different ways, generating many equivalent representations. All of the following are acceptable drawings for CH_4, because each drawing has two solid lines, one wedge, and one dashed wedge.

- These representations are equivalent.
- The **wedge** can be skewed to the left or the right of the **dashed wedge**.

The two H atoms are really aligned.

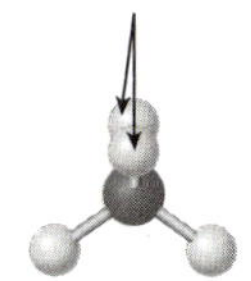

Finally, **wedges and dashed wedges are used for groups that are really *aligned one behind another.*** It does not matter in the two margin drawings whether the wedge or dashed wedge is skewed to the left or right, because the two H atoms are really aligned as shown in the three-dimensional model.

Ammonia (NH_3) and water (H_2O) both have atoms surrounded by four groups, some of which are lone pairs. In $\mathbf{NH_3}$, the three H atoms and one lone pair around N point to the corners of a tetrahedron. The H—N—H bond angle of 107° is close to the theoretical tetrahedral bond angle of 109.5°. This molecular shape is referred to as **trigonal pyramidal,** because one of the groups around the N is a nonbonded electron pair, not another atom.

One corner of the tetrahedron has an **electron pair,** not a bond.

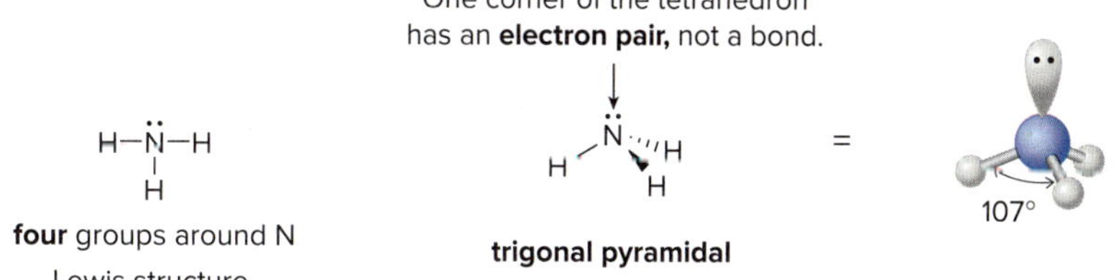

In H_2O, the two H atoms and two lone pairs around O point to the corners of a tetrahedron. The H—O—H bond angle of 105° is close to the theoretical tetrahedral bond angle of 109.5°. Water has a **bent** molecular shape, because two of the groups around oxygen are lone pairs of electrons.

H—Ö—H

four groups around O
Lewis structure

Two corners of the tetrahedron have **electron pairs,** not bonds.

a bent molecule

105°

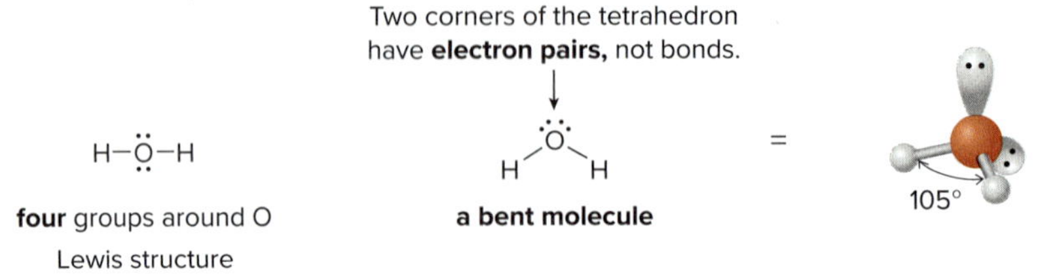

In both NH_3 and H_2O, the bond angle is somewhat smaller than the theoretical tetrahedral bond angle because of repulsion of the lone pairs of electrons. The bonded atoms are compressed into a smaller space with a smaller bond angle.

Predicting geometry based on counting groups is summarized in Table 1.3.

Table 1.3 Summary: Determining Geometry Based on the Number of Groups

Number of groups around an atom	Geometry	Bond angle	Examples
2	linear	180°	HC≡CH
3	trigonal planar	120°	CH_2=CH_2
4	tetrahedral	109.5°	CH_4, NH_3, H_2O

Sample Problem 1.6 Determining the Geometry Around a Second-Row Atom

Determine the geometry around the highlighted atom in each species.

a. :Ö=C=Ö: b. $H-\overset{+}{N}H_2-H$ (H—N⁺—H with H above and below)

Solution

a.

:Ö=C=Ö:
180°

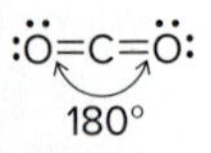

two atoms around C
no lone pairs

two groups

linear

b.

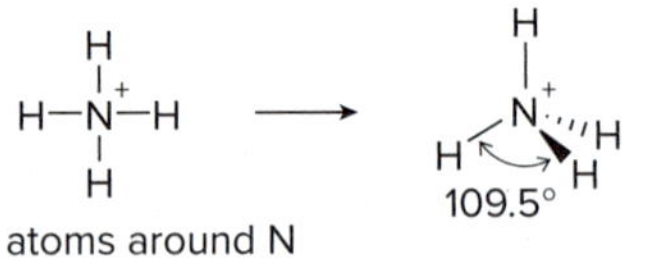

H—N⁺—H (with H above and below) ⟶ tetrahedral N⁺ with 109.5°

four atoms around N
no lone pairs

four groups ⟶ **tetrahedral**

Problem 1.21 Determine the geometry around all second-row elements in each compound drawn as a Lewis structure with no implied geometry.

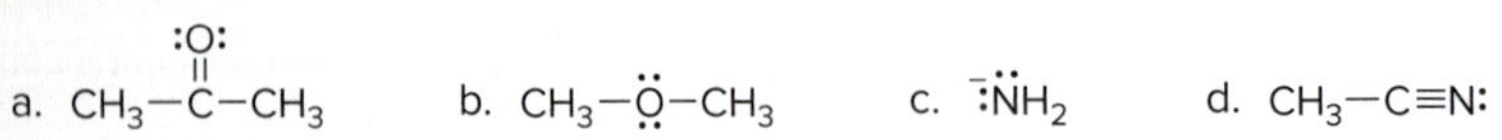

a. CH_3—C(=O:)—CH_3 b. CH_3—Ö—CH_3 c. $^-$:NH$_2$ d. CH_3—C≡N:

More Practice: Try Problems 1.60, 1.61, 1.76c.

Problem 1.22 Predict the indicated bond angles in each compound drawn as a Lewis structure with no implied geometry.

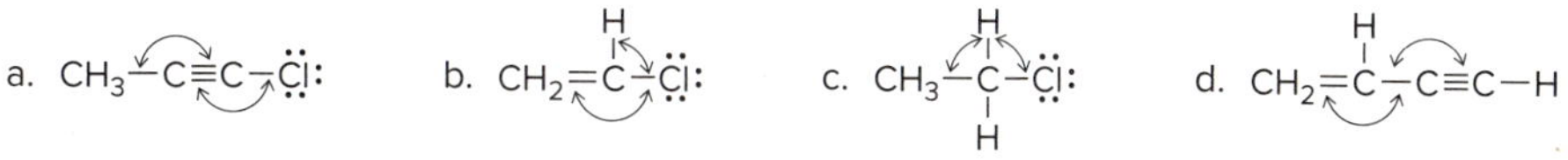

a. $CH_3-C\equiv C-\ddot{\underset{..}{Cl}}:$ b. $CH_2=C(H)-\ddot{\underset{..}{Cl}}:$ c. $CH_3-C(H)(H)-\ddot{\underset{..}{Cl}}:$ d. $CH_2=C(H)-C\equiv C-H$

Water hemlock, which grows in wet marshy areas in the western part of North America, is the source of cicutoxin (Problem 1.23), a convulsant toxic to both livestock and humans. *calyptra/123RF*

Problem 1.23 Using the principles of VSEPR theory, you can predict the geometry around any atom in any molecule, no matter how complex. Cicutoxin is a poisonous compound isolated from water hemlock, a highly toxic plant that grows in temperate regions in North America. Predict the geometry around the highlighted atoms in cicutoxin.

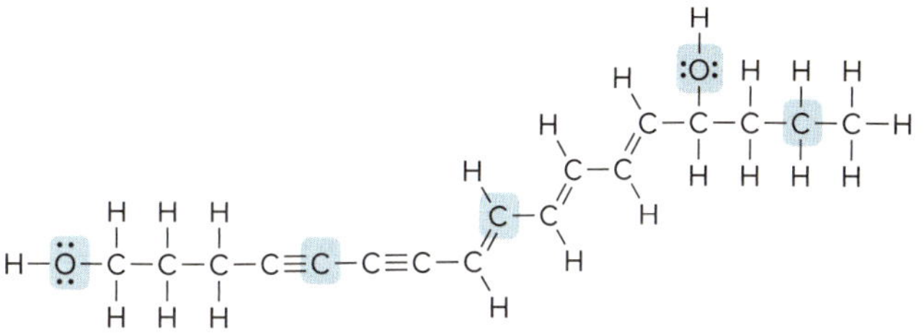

cicutoxin

1.8 Drawing Organic Structures

Drawing organic molecules presents a special challenge. Because they often contain many atoms, we need shorthand methods to simplify their structures. The two main types of shorthand representations used for organic compounds are **condensed structures** and **skeletal structures.**

1.8A Condensed Structures

Condensed structures can be used for compounds having a chain of atoms bonded together. The following conventions are used:

- All of the atoms are drawn in, but the two-electron bond lines are generally omitted.
- Atoms are usually drawn next to the atoms to which they are bonded.
- Parentheses are used around similar groups bonded to the same atom.
- Lone pairs are omitted.

To interpret a condensed formula, it is usually best to start at the *left side* of the molecule and remember that the ***carbon atoms must be tetravalent.*** A carbon bonded to three H atoms becomes $\mathbf{CH_3}$; a carbon bonded to two H atoms becomes $\mathbf{CH_2}$; and a carbon bonded to one H atom becomes **CH.**

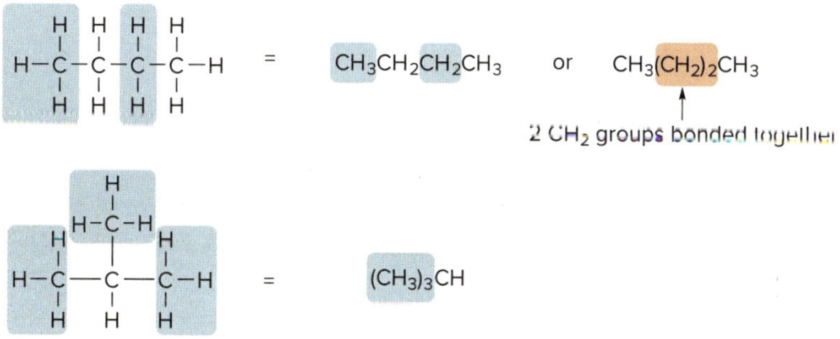

= $CH_3CH_2CH_2CH_3$ or $CH_3(CH_2)_2CH_3$

2 CH_2 groups bonded together

= $(CH_3)_3CH$

Other examples of condensed structures with heteroatoms and carbon–carbon multiple bonds are given in Figure 1.3.

Figure 1.3 Examples of condensed structures

[1] = $CH_3CH_2CHCH_2CH_3$ (with CH_3 bonded to the CH) or $CH_3CH_2CH(CH_3)CH_2CH_3$

[2] = $CH_3CH{=}CHCH_3$

[3] = $(CH_3)_2CHCH_2OH$

[4] = $Cl_2CH(CH_2)_2OC(CH_3)_3$

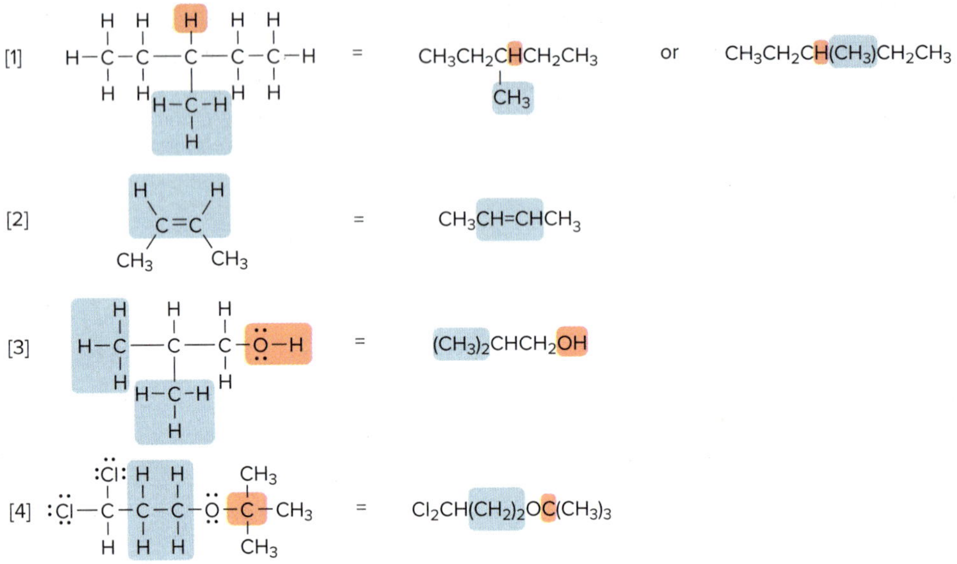

- Entry [1]: Draw the H atom next to the C to which it is bonded, and use parentheses around CH_3 to show it is bonded to the carbon chain.
- Entry [2]: Keep the carbon–carbon double bond and draw the H atoms after each C to which they are bonded.
- Entry [3]: Omit the lone pairs on the O atom in the condensed structure.
- Entry [4]: Omit the lone pairs on Cl and O and draw the two CH_2 groups as $(CH_2)_2$.

Translating some condensed formulas is not obvious, and it will come only with practice. This is especially true for compounds containing a carbon–oxygen double bond. Some noteworthy examples in this category are given in Figure 1.4. While carbon–carbon double bonds are generally drawn in condensed structures, carbon–oxygen double bonds are usually omitted.

Figure 1.4 Condensed structures containing a C–O double bond

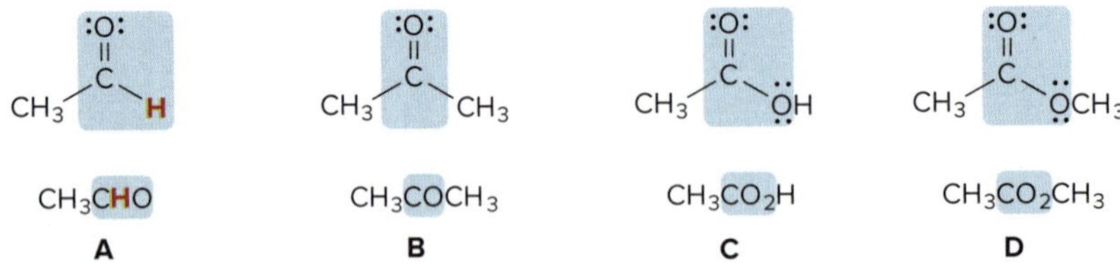

- In **A,** the H atom is bonded to C, *not* O.
- In **B,** each CH_3 group is bonded to C, *not* O.
- In **C** and **D,** the C atom is doubly bonded to one O and singly bonded to the other O.

Sample Problem 1.7 Converting a Condensed Structure to a Lewis Structure

Convert each condensed formula to a Lewis structure.

a. $(CH_3)_2CHOCH_2CH_2CH_2OH$ b. $CH_3(CH_2)_2CO_2C(CH_3)_3$

Solution

Start at the left and proceed to the right, making sure that each carbon has four bonds. Give each O atom two lone pairs to have an octet.

a. $(CH_3)_2CHOCH_2CH_2CH_2OH$

One C atom (labeled in blue) is bonded to 2 CH_3's, 1 H, and 1 O.

b. $CH_3(CH_2)_2CO_2C(CH_3)_3$

One C atom (labeled in blue) is bonded to both O's.

In part (a), the O atom is singly bonded to two C's, whereas in part (b), a C=O is needed to give each C and O an octet.

Problem 1.24 Convert each condensed formula to a Lewis structure.

a. $CH_3(CH_2)_4CH(CH_3)_2$

b. $(CH_3)_3CCH(OH)CH_2CH_3$

c. $(CH_3)_2CHCHO$

d. $(HOCH_2)_2CH(CH_2)_3C(CH_3)_2CH_2CH_3$

More Practice: Try Problem 1.49.

The tartness of rhubarb is primarily due to malic acid (Problem 1.26).
Mikeledray/Shutterstock

Problem 1.25 During periods of strenuous exercise, the buildup of lactic acid [$CH_3CH(OH)CO_2H$] causes the aching feeling in sore muscles. Convert this condensed structure to a Lewis structure of lactic acid.

Problem 1.26 Draw an acceptable Lewis structure for malic acid [$HO_2CCH(OH)CH_2CO_2H$], assuming the atoms are connected as arranged.

1.8B Skeletal Structures

Skeletal structures are used for organic compounds containing both rings and chains of atoms. Three rules are used to draw them.

- **Assume a carbon atom is located at the junction of any two lines or at the end of any line.**
- **Assume each carbon has enough hydrogens to make it tetravalent.**
- **Draw in all heteroatoms and the hydrogens directly bonded to them.**

Carbon chains are drawn in a **zigzag** fashion, and rings are drawn as **polygons,** as shown for hexane and cyclohexane.

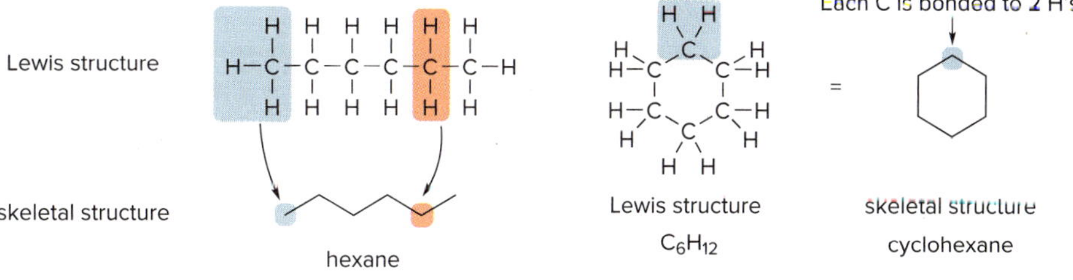

How To Interpret a Skeletal Structure

Example Draw in all C atoms, H atoms, and lone pairs in the following molecule:

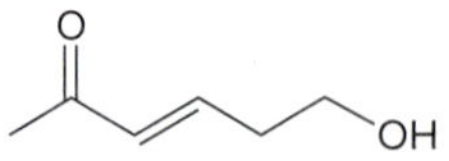

Step [1] **Place a C atom at the intersection of any two lines and at the end of any line.**

- This molecule has six carbons, including the C labeled in red at the left end of the chain.
- There are two C's (labeled in green) between the C=C and the OH group.

Step [2] **Add enough H's to make each C tetravalent.**

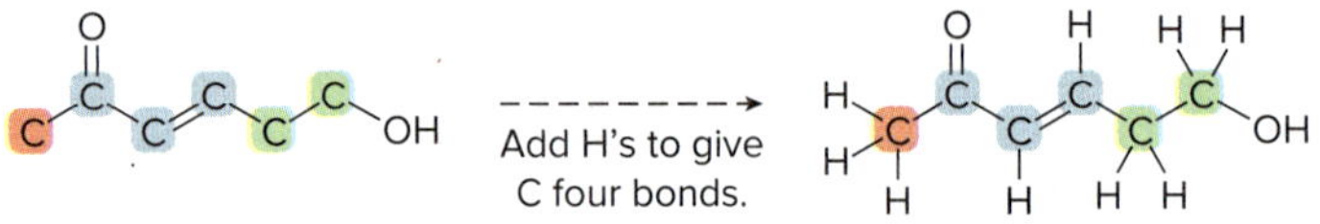

Add H's to give C four bonds.

- The end C labeled in red needs three H's to be tetravalent.
- Each C on the C=C has three bonds already, so only one H must be drawn.
- There are two CH_2 groups between the C=C and the OH group.

Step [3] **Add lone pairs to give each heteroatom an octet.**

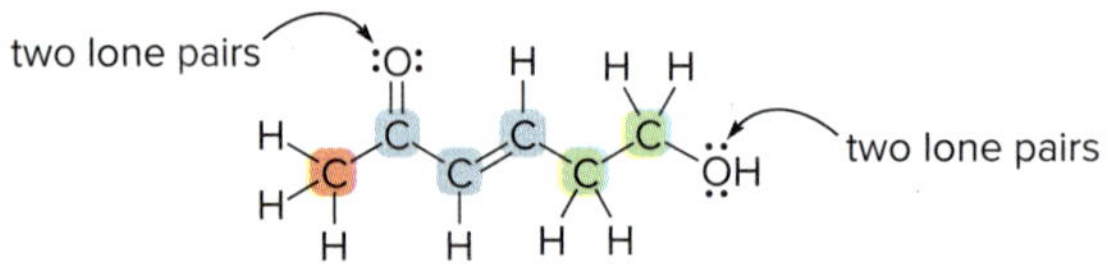

two lone pairs

two lone pairs

Each O needs **two** lone pairs for an octet.

Figure 1.5 shows other examples of skeletal structures, and Sample Problem 1.8 illustrates how to interpret the skeletal structure for a more complex cyclic compound.

Figure 1.5 Interpreting skeletal structures

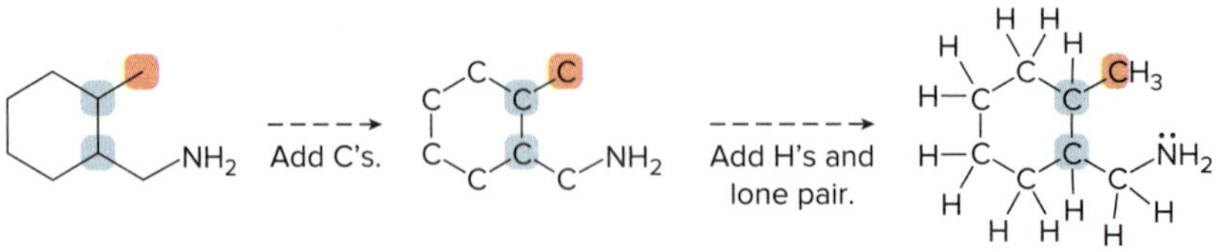

Add C's. Add H's and lone pair.

- C labeled in red needs 3 H's.
- C's labeled in blue need 1 H.
- All other C's need 2 H's.
- N needs one lone pair.

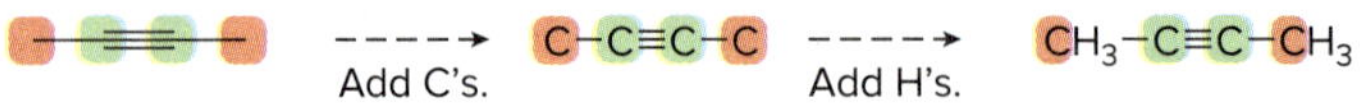

C–C≡C–C Add C's. Add H's. CH_3–C≡C–CH_3

- C's labeled in red need 3 H's.
- C's labeled in green have four bonds to C.

Sample Problem 1.8 Converting a Skeletal Structure to a Lewis Structure

To write a molecular formula for an organic compound, write the elements of C and H first, and then other elements in alphabetical order.

Draw a complete structure for vanillin showing all C atoms, H atoms, and lone pairs, and give the molecular formula. Vanillin is the principal component of the extract of the vanilla bean.

vanilla bean

Vast natalia/Alamy Stock Photo

Solution

- Skeletal structures have a C atom at the junction of any two lines and at the end of any line.
- Each C must have enough H's to make it tetravalent.
- Each O atom needs two lone pairs to have a complete octet.

- C in blue has three H's.
- C in green has four bonds to other C's.
- C in red is doubly bonded to O.

Elite Images/McGraw Hill

Problem 1.27 How many hydrogen atoms are present around each highlighted carbon atom in the following molecules? What is the molecular formula for each molecule? Both compounds are active ingredients in some common sunscreens.

a. octinoxate (2-ethylhexyl 4-methoxycinnamate)

b. avobenzone

More Practice: Try Problems 1.62, 1.63, 1.74a.

Problem 1.28 Convert each skeletal structure to a complete structure with all C's, H's, and lone pairs drawn in.

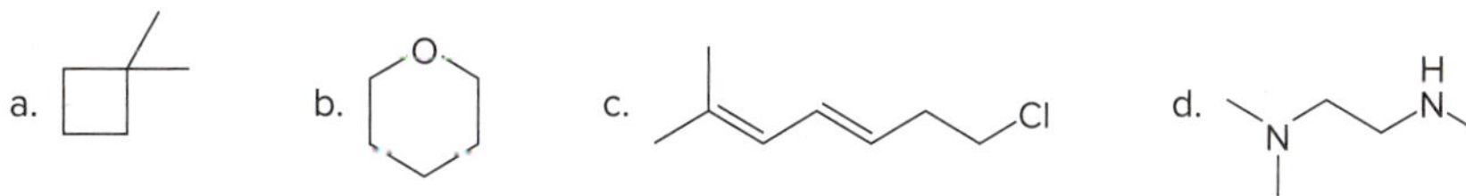

Problem 1.29 How many hydrogen atoms are present around each highlighted carbon atom in the following molecules? What is the molecular formula for each molecule? 2'-Deoxyguanosine is a component of DNA, and riboflavin (vitamin B_2) is a yellow, water-soluble vitamin obtained in the diet from leafy greens, soybeans, almonds, and liver.

a. 2'-deoxyguanosine
DNA component

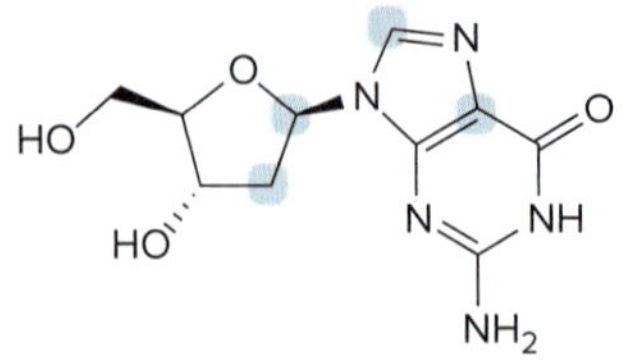

b. riboflavin
vitamin B_2

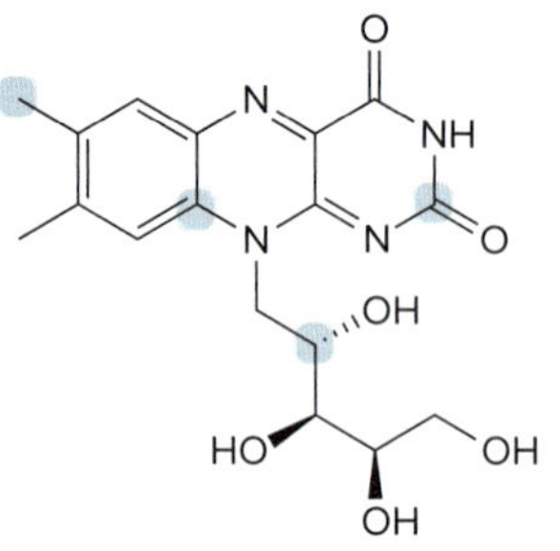

When heteroatoms are bonded to a carbon skeleton, the **heteroatom is joined *directly* to the carbon to which it is bonded,** with no H atoms in between. Thus, an OH group is drawn as OH or HO depending on where the OH is located. In contrast, when carbon appendages are bonded to a carbon skeleton, the **H atoms will be drawn to the *right* of the carbon to which they are bonded regardless of the location.**

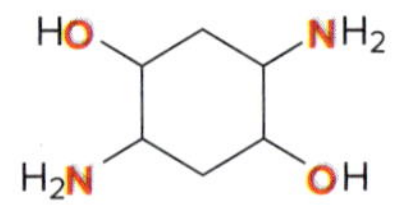

Place the O and N atoms *directly* joined to the ring.

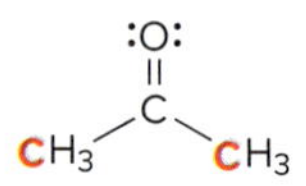

Two C atoms in red are bonded to the middle C.

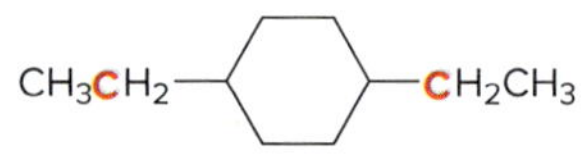

Two C atoms in red are bonded to the ring.

1.8C Skeletal Structures with Charged Atoms

Take care in interpreting skeletal structures for positively and negatively charged carbon atoms, because *both* the hydrogen atoms *and* the lone pairs are omitted. Keep in mind the following:

- **A charge on a carbon atom takes the place of one hydrogen atom.**
- **The charge determines the number of lone pairs. Negatively charged carbon atoms have one lone pair and positively charged carbon atoms have none.**

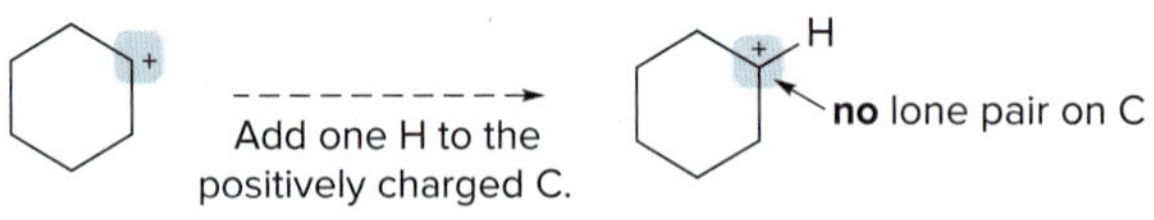

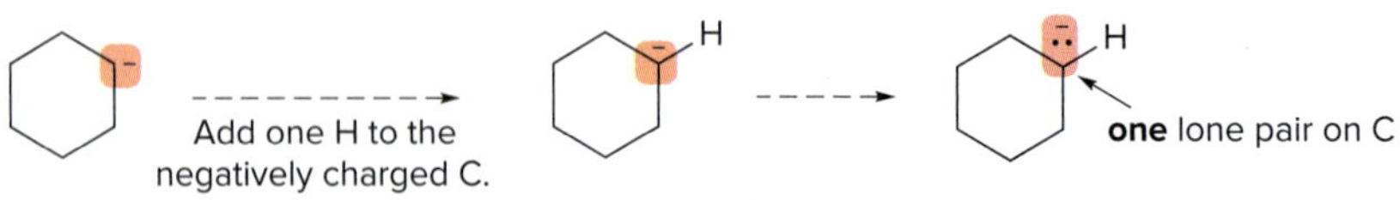

Add one H to the positively charged C. **no** lone pair on C

Add one H to the negatively charged C. **one** lone pair on C

Skeletal structures often leave out lone pairs on heteroatoms, but *don't forget about them.* Use the formal charge on an atom to determine the number of lone pairs. For example, a neutral O atom with two bonds needs two additional lone pairs, and a positively charged O atom with three bonds needs only one lone pair.

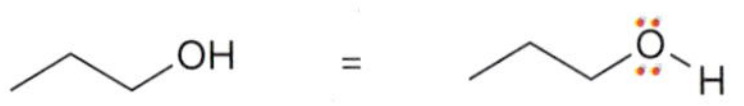

neutral O atom
two lone pairs

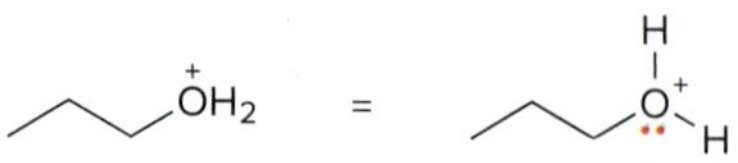

positively charged O atom
one lone pair

Problem 1.30 Draw in all hydrogens and lone pairs on the charged carbons in each ion.

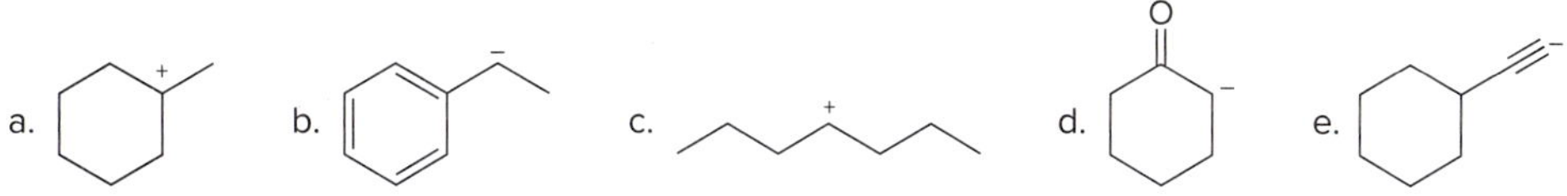

Problem 1.31 Use the formal charge to draw in the lone pairs on each N or O atom in the following compounds.

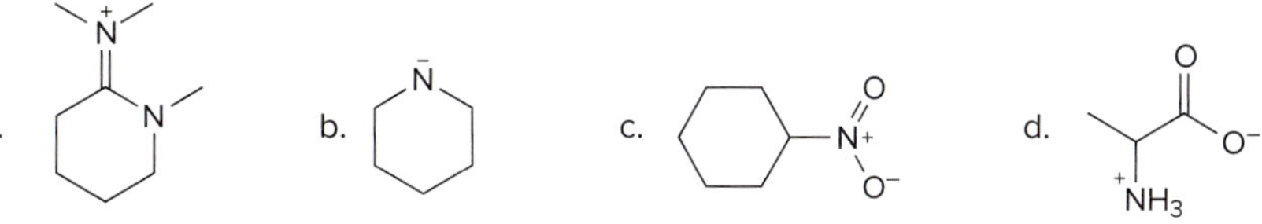

Problem 1.32 Draw in all the hydrogen atoms and nonbonded electron pairs in each ion.

a. b. c. d. $\equiv\overset{+}{N}H$ e.

Problem 1.33 Draw a skeletal structure for the molecules in parts (a) and (b), and a condensed structure for the molecules in parts (c) and (d).

a. $CH_3O(CH_2)_2COCH{=}C(CH_3)_2$

b. CH_3–C(H)–C(H)–CH_2CH_2Cl (four-membered ring), H_2N–C(H)–C(H)–H

c.

d. HO

1.9 Hybridization

What orbitals do the first- and second-row atoms use to form bonds?

1.9A Hydrogen

Recall from Section 1.2 that two hydrogen atoms share each of their electrons to form H_2. Thus, the 1*s* orbital on one H overlaps with the 1*s* orbital on the other H to form a bond that concentrates electron density between the two nuclei. This type of bond, called a **σ (sigma) bond,** is cylindrically symmetrical because the electrons forming the bond are distributed symmetrically about an imaginary line connecting the two nuclei.

H_{1s} + H_{1s} ⟶ H—H

σ bond

- **A σ bond concentrates electron density on the axis that joins two nuclei. All single bonds are σ bonds.**

1.9B Bonding in Methane

To account for the bonding patterns observed in more complex molecules, we must take a closer look at the 2*s* and 2*p* orbitals of atoms of the second row. Let's illustrate this with methane, CH_4.

Carbon has **four valence electrons.** To fill atomic orbitals in the most stable arrangement, electrons are placed in the orbitals of lowest energy. For carbon, this places two electrons in the 2*s* orbital and one each in two 2*p* orbitals.

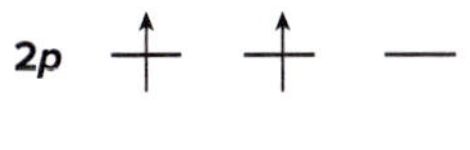

ground state for carbon's four valence electrons

- **This lowest-energy arrangement of electrons for an atom is called its ground state.**

In this description, **carbon should form *only two bonds*** because it has only two unpaired valence electrons, and CH_2 should be a stable molecule. In reality, however, CH_2 is a highly reactive species because carbon does not have an octet of electrons.

Because the carbon atom in CH_4 forms four bonds to hydrogen and all C–H bonds are *identical,* chemists have proposed that atoms like carbon do *not* use pure *s* and pure *p* orbitals in forming bonds. Instead, atoms use a set of new orbitals called **hybrid orbitals.** The mathematical process by which these orbitals are formed is called **hybridization.**

- ***Hybridization* is the combination of two or more atomic orbitals to form the same number of hybrid orbitals, each having the same shape and energy.**

Hybridization of one 2*s* orbital and three 2*p* orbitals for carbon forms four hybrid orbitals, each with one electron. These new hybrid orbitals are intermediate in energy between the 2*s* and 2*p* orbitals.

2*p* 2*p* 2*p*

2*s*

four atomic orbitals

hybridize

sp^3 sp^3 sp^3 sp^3

four hybrid orbitals

- **These hybrid orbitals are called sp^3 *hybrids* because they are formed from *one s* orbital and *three p* orbitals.**

p orbital sp^3 hybrid orbital

What do these new hybrid orbitals look like? Mixing a spherical 2*s* orbital and three dumbbell-shaped 2*p* orbitals together produces four orbitals having one large lobe and one small lobe, oriented toward the corners of a tetrahedron. Each large lobe concentrates electron density in the bonding direction between two nuclei. **Bonds formed from hybrid orbitals are *stronger* than bonds formed from pure *p* orbitals.**

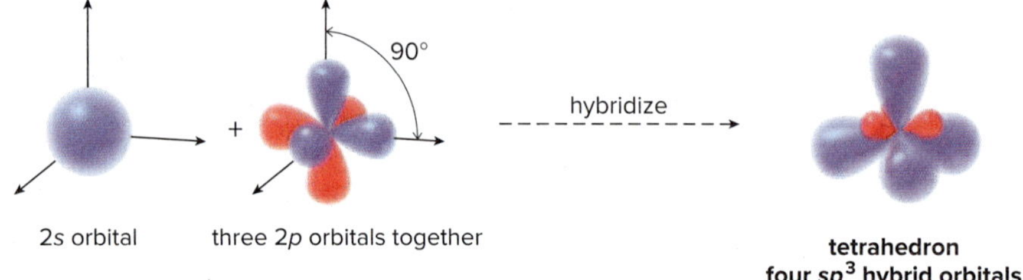

The four hybrid orbitals form four equivalent bonds. We can now explain the observed bonding in CH_4.

- **Each bond in CH_4 is formed by overlap of an sp^3 hybrid orbital of carbon with a 1s orbital of hydrogen. These four bonds point to the corners of a tetrahedron.**

All four C–H bonds in methane are **σ bonds,** because the electron density is concentrated on the axis joining C and H. An orbital picture of the bonding in CH_4 is given in Figure 1.6.

Figure 1.6
Bonding in CH_4 using sp^3 hybrid orbitals

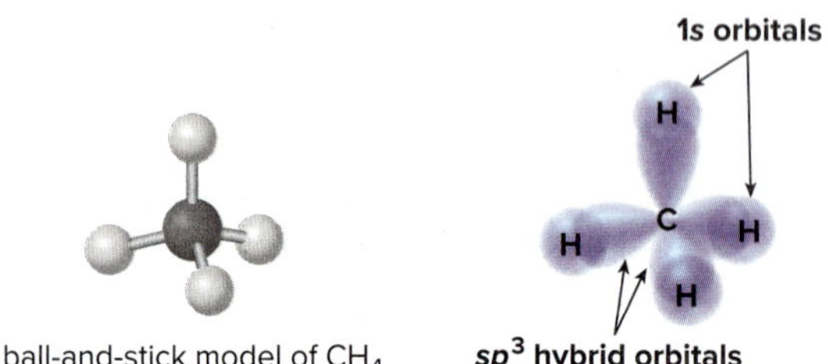

- All four C–H bonds are σ bonds. Each is formed by overlap of an sp^3 hybrid orbital on carbon and a 1s orbital on hydrogen.

Problem 1.34 What orbitals are used to form each of the C–C and C–H bonds in $CH_3CH_2CH_3$ (propane)? How many σ bonds are present in this molecule?

- **Any atom surrounded by four groups (atoms and lone pairs) is sp^3 hybridized.**

The N atom in $\mathbf{NH_3}$ and the O atom in $\mathbf{H_2O}$ are both surrounded by four groups, making them sp^3 hybridized. Each N–H and O–H bond in these molecules is formed by overlap of an sp^3 hybrid orbital with a 1s orbital from H. The lone pairs of electrons on N and O also occupy sp^3 hybrid orbitals, as shown in Figure 1.7.

Figure 1.7
Hybrid orbitals of NH_3 and H_2O

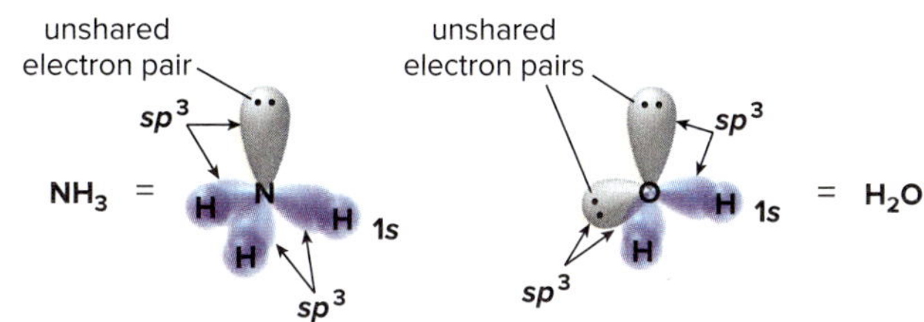

1.9C Other Hybridization Patterns—sp and sp^2 Hybrid Orbitals

Forming sp^3 hybrid orbitals is just one way that 2s and 2p orbitals can hybridize. Three common modes of hybridization are seen in organic molecules. The number of orbitals is always conserved in hybridization; that is, a **given number of atomic orbitals hybridizes to form an equivalent number of hybrid orbitals.**

- ***One* 2s orbital and *three* 2p orbitals form *four* sp^3 hybrid orbitals.**
- ***One* 2s orbital and *two* 2p orbitals form *three* sp^2 hybrid orbitals.**
- ***One* 2s orbital and *one* 2p orbital form *two* sp hybrid orbitals.**

We have already seen pictorially how four sp^3 hybrid orbitals are formed from one 2s and three 2p orbitals. Figures 1.8 and 1.9 illustrate the same process for sp and sp^2 hybrids. Each sp and sp^2 hybrid orbital has one large and one small lobe, much like an sp^3 hybrid orbital. Note, however, that both sp^2 and sp hybridization **leave one and two 2*p* orbitals *unhybridized*,** respectively, on each atom.

Figure 1.8
Forming two sp hybrid orbitals

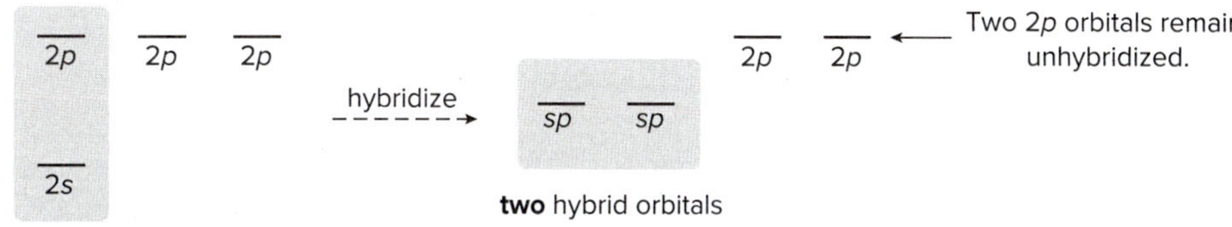

- Forming **two *sp* hybrid orbitals** uses **one 2*s*** and **one 2*p* orbital,** leaving **two 2*p* orbitals *unhybridized.***

Figure 1.9
Forming three sp^2 hybrid orbitals

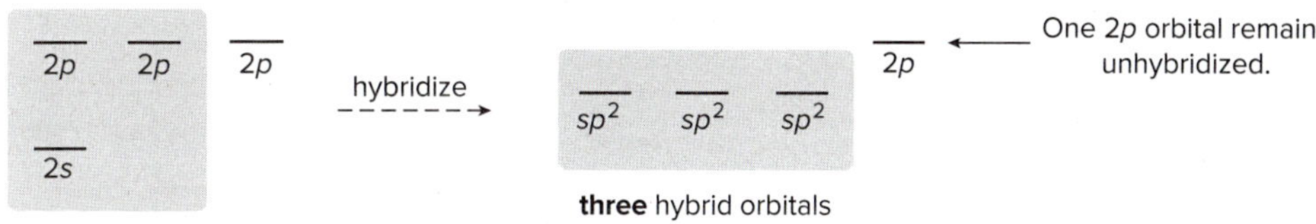

- Forming **three sp^2 hybrid orbitals** uses **one 2*s*** and **two 2*p* orbitals,** leaving **one 2*p* orbital *unhybridized.***

The **superscripts** for hybrid orbitals correspond to the **number of atomic orbitals** used to form them. The number "1" is understood.

For example: $sp^3 = s^1p^3$

one 2*s* + three 2*p* orbitals used to make each hybrid orbital

To determine the hybridization of an atom in a molecule, we count groups (atoms and lone pairs) around the atom, just as we did in determining geometry.

- **The number of groups around an atom equals the number of atomic orbitals that are hybridized to form hybrid orbitals (Table 1.4).**

Table 1.4 Three Types of Hybrid Orbitals

Number of groups	Number of orbitals used	Type of hybrid orbital
2	2	**two** *sp* hybrid orbitals
3	3	**three** sp^2 hybrid orbitals
4	4	**four** sp^3 hybrid orbitals

Hybridization in various carbon compounds is presented in Section 1.10.

Sample Problem 1.9 Determining the Hybridization of an Atom

What orbitals are used to form each bond in methanol, CH_3OH?

Solution

To solve this problem, **draw a valid Lewis structure** and **count groups around each atom.** Then, use the rule to determine hybridization: **two groups = *sp*, three groups = sp^2, and four groups = sp^3.**

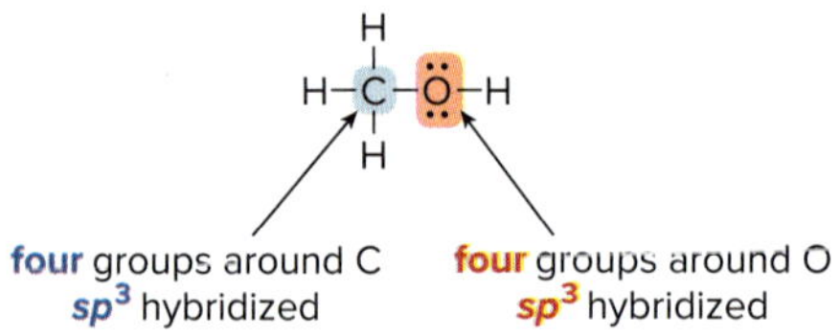

- All C–H bonds are formed from $C_{sp^3}–H_{1s}$.
- The C–O bond is formed from $C_{sp^3}–O_{sp^3}$.
- The O–H bond is formed from $O_{sp^3}–H_{1s}$.

Problem 1.35 What orbitals are used to form each highlighted bond in the following molecule? In what type of orbital do the lone pairs on each O and N reside?

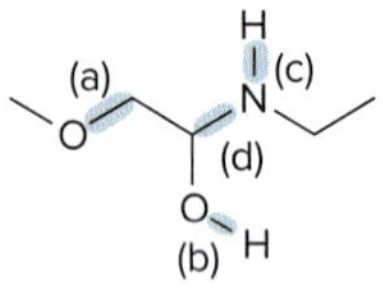

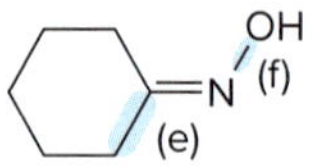

More Practice: Try Problems 1.45e, 1.46e, 1.67, 1.68, 1.75c, 1.78d.

1.10 Ethane, Ethylene, and Acetylene

The principles of hybridization determine the type of bonds in **ethane, ethylene,** and **acetylene.**

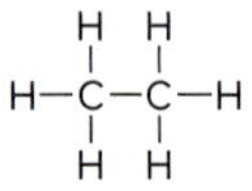

ethane 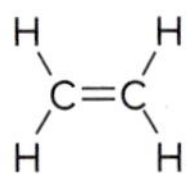 ethylene  acetylene

1.10A Ethane—CH_3CH_3

Based on the Lewis structure for **ethane, CH_3CH_3,** each carbon atom is singly bonded to four other atoms. As a result:

- **Each carbon is tetrahedral.**
- **Each carbon is sp^3 hybridized.**

CH_3CH_3 = [tetrahedral structure] = [ball-and-stick model]
ethane
tetrahedral C atoms

All of the bonds in ethane are σ bonds. The C–H bonds are formed from the overlap of one of the three sp^3 hybrid orbitals on each carbon atom with the 1*s* orbital on hydrogen. The C–C bond is formed from the overlap of an sp^3 hybrid orbital on each carbon atom.

Ethane is a constituent of natural gas. *Steve Allen/Stockbyte/Getty Images*

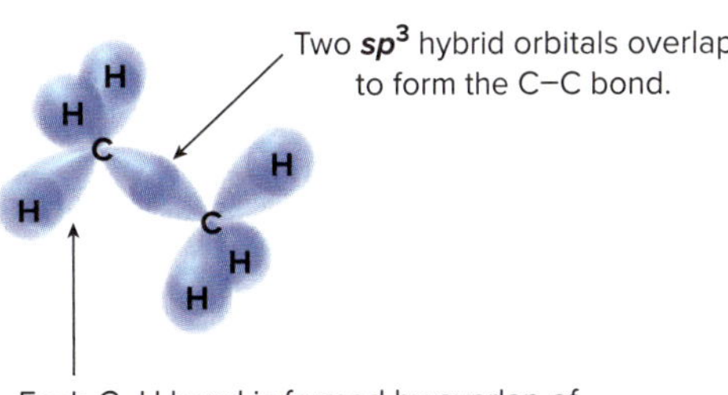

A model of ethane shows that **rotation can occur around the central C–C σ bond.** The relative position of the H atoms on the adjacent CH_3 groups changes with bond rotation, as seen in the location of the labeled red H atom before and after rotation. This process is discussed in greater detail in Chapter 4.

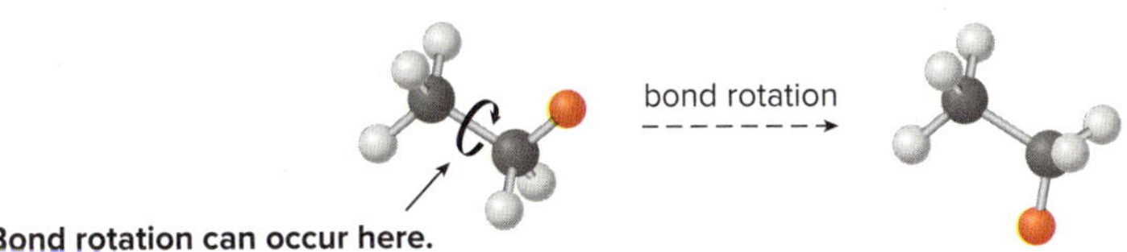

1.10B Ethylene—C_2H_4

Ethylene is an important starting material in the preparation of the plastic polyethylene. *Nextdoor Images/Creatas/PunchStock*

Based on the Lewis structure of **ethylene, $CH_2{=}CH_2$,** each carbon atom is singly bonded to two H atoms and doubly bonded to the other C atom, so each C is surrounded by three groups. As a result:

- **Each carbon is trigonal planar (Section 1.7B).**
- **Each carbon is sp^2 hybridized.**

$CH_2{=}CH_2$ = [structure] 120°
ethylene
three groups around each C

What orbitals are used to form the two bonds of the C–C double bond? Recall from Section 1.9 that **sp^2 hybrid orbitals** are formed from **one 2*s* and two 2*p* orbitals,** leaving one **2*p* orbital unhybridized.** Because carbon has four valence electrons, **each of these orbitals has one electron** that can be used to form a bond.

Each C–H bond results from the end-on overlap of an sp^2 hybrid orbital on carbon and the 1*s* orbital on hydrogen. Similarly, one of the C–C bonds results from the end-on overlap of

An sp^2 hybridized C in $CH_2{=}CH_2$ has three sp^2 hybrid orbitals and one higher-energy, unhybridized *p* orbital:

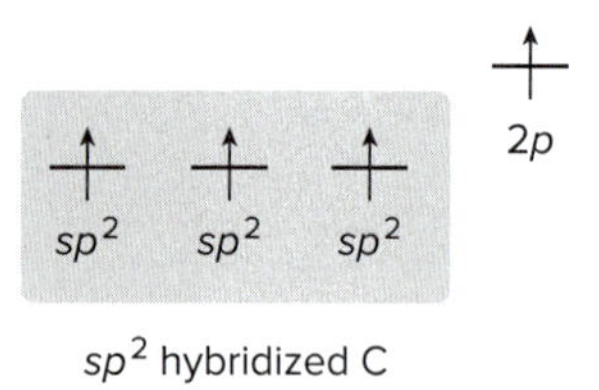

sp^2 hybridized C

an sp^2 hybrid orbital on each carbon atom. Each of these bonds is a **σ bond.** All five σ bonds lie in the same plane, viewed from above in the following representation, and from the side in Figure 1.10a.

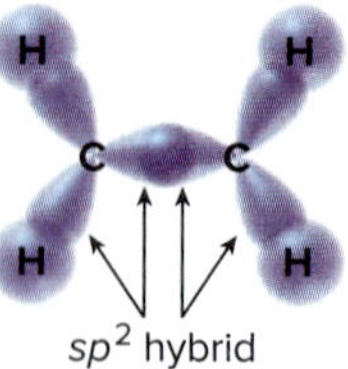

- Each C has three sp^2 hybrid orbitals.
- The C–H bonds and the C–C bond are σ bonds.

The second C–C bond results from the side-by-side overlap of the 2*p* orbitals on each carbon. Because the unhybridized 2*p* orbitals are located perpendicular to the plane of the molecule, side-by-side overlap creates an area of electron density above and below the plane containing the sp^2 hybrid orbitals (that is, the plane containing the six atoms in the σ bonding system), as shown in Figure 1.10b.

Figure 1.10 The σ and π bonds in ethylene

a.

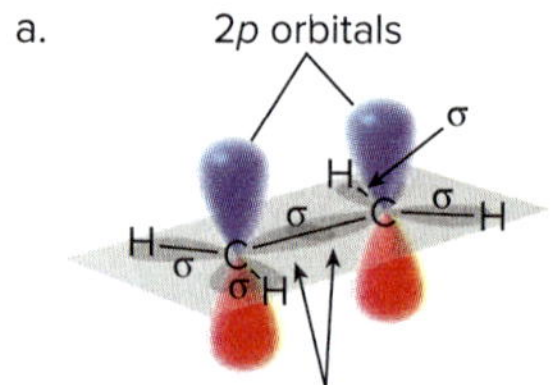

- Overlap of the two sp^2 hybrid orbitals forms the C–C σ bond.

b.

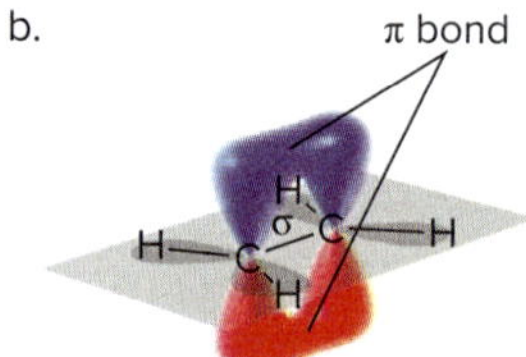

- Overlap of the two 2*p* orbitals forms the C–C π bond.
- The π bond extends **above and below the plane** of the molecule.

In this second bond, the electron density is ***not*** concentrated on the axis joining the two nuclei. This new type of bond is called a **π bond.** Because the electron density in a π bond is farther from the two nuclei, **π bonds are usually weaker and therefore more easily broken than σ bonds.**

Thus, a carbon–carbon double bond has two components:

- **a σ bond, formed by end-on overlap of two sp^2 hybrid orbitals;**
- **a π bond, formed by side-by-side overlap of two 2*p* orbitals.**

Unlike the C–C single bond in ethane, rotation about the C–C double bond in ethylene is **restricted.** It can occur only if the π bond first breaks and then re-forms, a process that requires considerable energy.

All double bonds are composed of one σ and one π bond.

Rotation around a C=C bond does *not* occur.

1.10C Acetylene—C_2H_2

Because acetylene produces a very hot flame on burning, it is often used in welding torches. The fire is very bright, too, so it was once used in the lamps worn by spelunkers—people who study and explore caves. *Phillip Spears/Getty Images*

Based on the Lewis structure of **acetylene, HC≡CH,** each carbon atom is singly bonded to one hydrogen atom and triply bonded to the other carbon atom, so each carbon atom is surrounded by two groups. As a result:

- **Each carbon is linear (Section 1.7B).**
- **Each carbon is *sp* hybridized.**

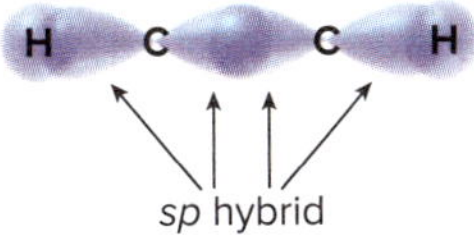

acetylene

What orbitals are used to form the bonds of the C−C triple bond? Recall from Section 1.9 that ***sp* hybrid orbitals** are formed from **one 2*s* and one 2*p* orbital,** leaving **two 2*p* orbitals unhybridized.** Because carbon has four valence electrons, **each of these orbitals has one electron** that can be used to form a bond.

An *sp* hybridized C in HC≡CH has two *sp* hybrid orbitals and two higher-energy, unhybridized *p* orbitals:

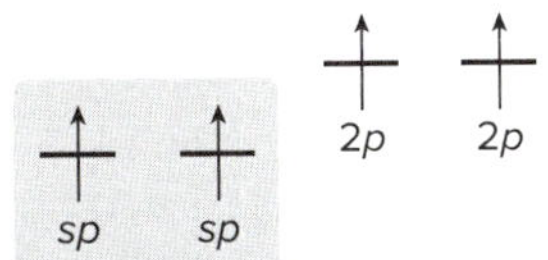

sp hybridized C

Each C−H bond results from the end-on overlap of an *sp* hybrid orbital on carbon and the 1*s* orbital on hydrogen. Similarly, one of the C−C bonds results from the end-on overlap of an *sp* hybrid orbital on each carbon atom. Each of these bonds is a **σ bond.**

H C C H

sp hybrid

- Each C has two *sp* hybrid orbitals.
- The C−H bonds and C−C bond are σ bonds.

Each carbon atom also has two **unhybridized 2*p* orbitals** that are perpendicular to each other and to the *sp* hybrid orbitals (Figure 1.11a). Side-by-side overlap between the two 2*p* orbitals on one carbon with the two 2*p* orbitals on the other carbon creates the second and third bonds of the C−C triple bond (Figure 1.11b). The electron density from one of these two bonds is above and below the axis joining the two nuclei, and the electron density from the second of these two bonds is in front of and behind the axis, so both of these bonds are **π bonds.**

Figure 1.11 The σ and π bonds in acetylene

a.

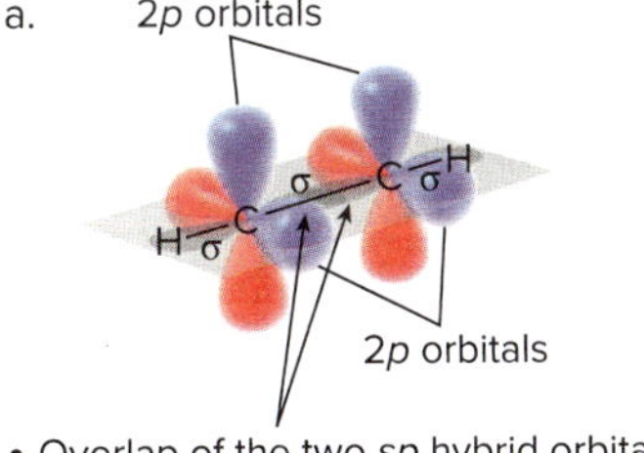

- Overlap of the two *sp* hybrid orbitals forms the C−C σ bond.

b.

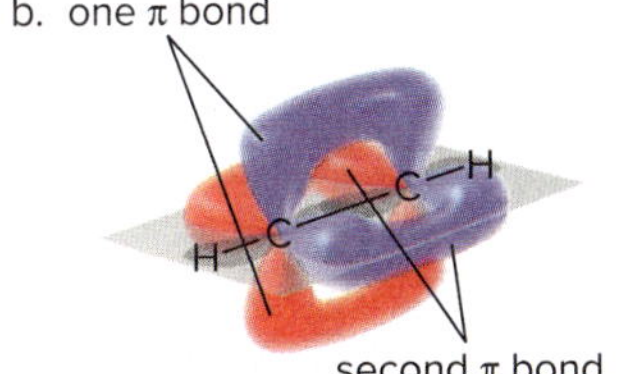

- Overlap of two sets of two 2*p* orbitals forms two C−C π bonds.
- Two π bonds extend out from the axis of the linear molecule.

The side-by-side overlap of two *p* orbitals always forms a π bond.

All triple bonds are composed of one σ and two π bonds.

Thus, a carbon–carbon triple bond has three components:

- **a σ bond, formed by end-on overlap of two *sp* hybrid orbitals;**
- **two π bonds, formed by side-by-side overlap of two sets of 2*p* orbitals.**

Table 1.5 summarizes the three possible types of bonding in carbon compounds.

Table 1.5 A Summary of Covalent Bonding Seen in Carbon Compounds

Number of groups bonded to C	Hybridization	Bond angle	Example	Observed bonding
4	sp^3	109.5°	CH_3CH_3 ethane	one σ bond $C_{sp^3}–C_{sp^3}$
3	sp^2	120°	$CH_2{=}CH_2$ ethylene	one σ bond $C_{sp^2}–C_{sp^2}$ + one π bond $C_{2p}–C_{2p}$
2	sp	180°	$HC{\equiv}CH$ acetylene	one σ bond $C_{sp}–C_{sp}$ + two π bonds $C_{2p}–C_{2p}$ $C_{2p}–C_{2p}$

Sample Problem 1.10 Determining Hybridization

Answer each question for cyclohexanone.

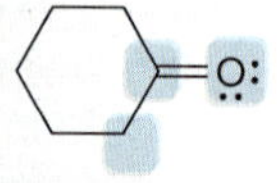

cyclohexanone

a. Determine the hybridization of the highlighted atoms.
b. What orbitals are used to form the C–O double bond?
c. In what type of orbital does each lone pair reside?

Solution

a.

three groups around C
sp^2 hybridized

three groups around O
sp^2 hybridized

four groups around C
sp^3 hybridized

b. • The σ bond is formed from the end-on overlap of $C_{sp^2}–O_{sp^2}$.
• The π bond is formed from the side-by-side overlap of $C_{2p}–O_{2p}$.

c. The O atom has three sp^2 hybrid orbitals.
• One is used for the σ bond of the double bond.
• The remaining two sp^2 hybrids are occupied by the lone pairs.

Problem 1.36 Determine the hybridization around the highlighted atoms in each molecule.

a. $CH_3–C{\equiv}CH$ b. (cyclohexylidene)=N–CH_3 c. $CH_2{=}C{=}CH_2$ d. nicotine

More Practice: Try Problems 1.45d, 1.46d, 1.66, 1.74c, 1.75a, 1.76a, 1.78c.

Problem 1.37 Acalabrutinib, approved for use in the United States in 2017 and sold under the trade name Calquence, is a drug used to treat mantle cell lymphoma. (a) How many *sp*, sp^2 and sp^3 hybridized carbons does acalabrutinib contain? (b) What is the hybridization of the O atom labeled in blue? (c) What orbitals are used to form the carbon–nitrogen double bond labeled in red?

Generic name: acalabrutinib
Trade name: Calquence

1.11 Bond Length and Bond Strength

Let's now examine the relative bond length and bond strength of the C–C and C–H bonds in ethane, ethylene, and acetylene.

1.11A A Comparison of Carbon–Carbon Bonds

While the SI unit of energy is the **joule** (J), organic chemists often report energy values in **calories** (cal). For this reason, energy values in the tables in this text are reported in joules, followed by the number of calories in parentheses.
1 cal = 4.18 J

An inverse relationship exists between bond length and bond strength. The shorter the bond, the closer the electron density is kept to the nucleus, and the harder the bond is to break. ***Shorter* bonds are *stronger* bonds.**

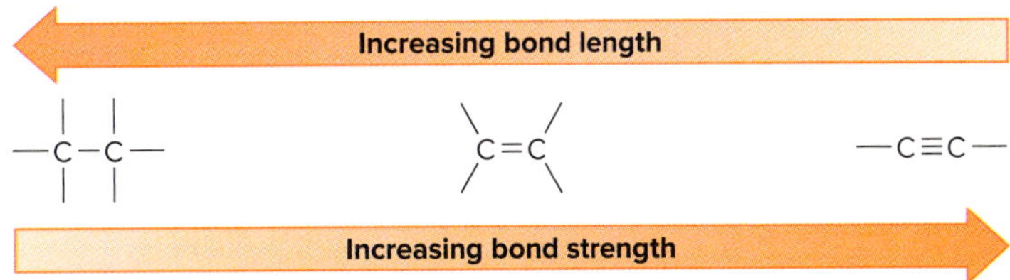

- **As the number of electrons between two nuclei *increases*, bonds become shorter and stronger.**
- **Triple bonds are shorter and stronger than double bonds, which are shorter and stronger than single bonds.**

Values for bond lengths and bond strengths for CH_3CH_3, $CH_2{=}CH_2$, and $HC{\equiv}CH$ are listed in Table 1.6. Be careful not to confuse two related but different principles regarding multiple bonds such as C–C double bonds. **Double bonds, consisting of both a σ and a π bond, are *strong*.** The **π component** of the double bond, however, is usually much ***weaker* than the σ component.** This is a particularly important consideration when studying alkenes in Chapter 10.

Table 1.6 Bond Lengths and Bond Strengths for Ethane, Ethylene, and Acetylene

Compound	C–C bond length (pm)	Bond strength kJ/mol (kcal/mol)
$CH_3{-}CH_3$	153	368 (88)
$CH_2{=}CH_2$	134	635 (152)
$HC{\equiv}CH$	121	837 (200)
	Increasing bond length (↑)	Increasing bond strength (↓)

Compound	C–H bond length (pm)	Bond strength kJ/mol (kcal/mol)
$CH_3CH_2{-}H$	111	410 (98)
$CH_2{=}CH{-}H$	110	435 (104)
$HC{\equiv}C{-}H$	109	523 (125)
	Increasing bond length (↑)	Increasing bond strength (↓)

1.11B A Comparison of Carbon–Hydrogen Bonds

The length and strength of a C–H bond vary slightly depending on the hybridization of the carbon atom.

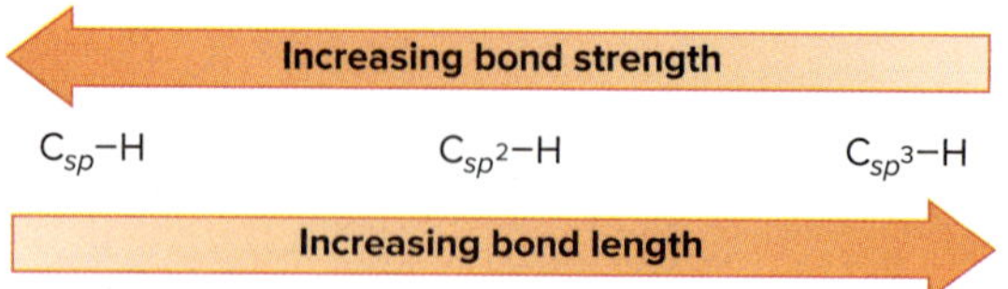

To understand why this is so, we must look at the atomic orbitals used to form each type of hybrid orbital. A single 2*s* orbital is always used, but the number of 2*p* orbitals varies with the type of hybridization. The **percent *s*-character** indicates the fraction of a hybrid orbital due to the 2*s* orbital used to form it.

***sp* hybrid**	one 2*s* orbital / **two** hybrid orbitals	**= 50% *s*-character**
***sp*² hybrid**	one 2*s* orbital / **three** hybrid orbitals	**= 33% *s*-character**
***sp*³ hybrid**	one 2*s* orbital / **four** hybrid orbitals	**= 25% *s*-character**

Why should the percent *s*-character of a hybrid orbital affect the length of a C–H bond? A 2*s* orbital keeps electron density closer to a nucleus compared to a 2*p* orbital. As the **percent *s*-character *increases,*** a hybrid orbital holds its electrons closer to the nucleus, and the **bond becomes *shorter* and *stronger.***

- **Increased percent *s*-character → Increased bond strength → Decreased bond length**

Problem 1.38 Which of the bonds shown in red in each compound or pair of compounds is shorter?

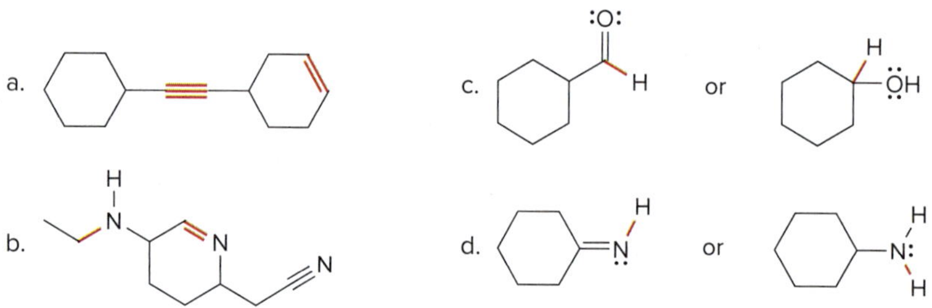

The seeds of some types of sandalwood are rich in santalbic acid (Problem 1.39), an unusual fatty acid that contains a carbon–carbon triple bond.
Bijayakumar/Shutterstock

Problem 1.39 Rank the labeled bonds in santalbic acid, a fatty acid obtained from the seeds of the sandalwood tree used in cosmetics, in order of increasing bond length.

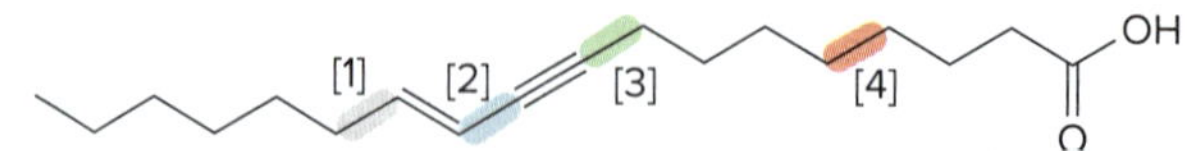

1.12 Electronegativity and Bond Polarity

***Electronegativity* is a measure of an atom's attraction for electrons in a bond.** Electronegativity indicates how much a particular atom "*wants*" electrons.

- Electronegativity *increases* across a row of the periodic table as the nuclear charge increases (excluding the noble gases).
- Electronegativity *decreases* down a column of the periodic table as the atomic radius increases, pushing the valence electrons farther from the nucleus.

As a result, the *most* electronegative elements are located at the **upper right-hand corner** of the periodic table, and the *least* electronegative elements in the **lower left-hand corner.** A scale has been established to represent electronegativity values arbitrarily, from 0 to 4, as shown in Figure 1.12.

Figure 1.12 Electronegativity values for some common elements

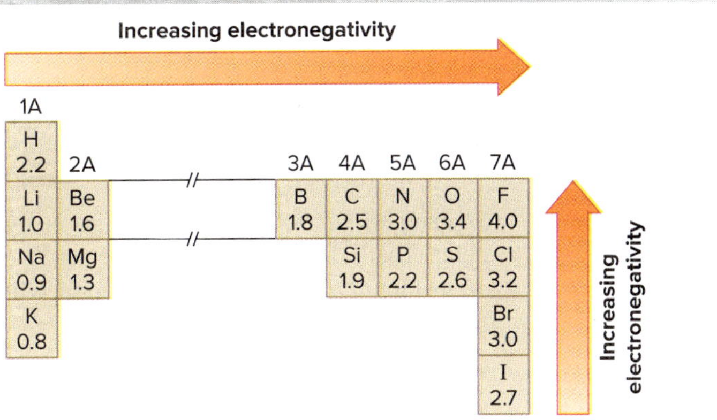

Electronegativity values are relative, so they can be used for comparison purposes only. When comparing two different elements, one is **more electronegative** than the other if it attracts electron density toward itself. One is less electronegative—**more electro*positive***—if it gives up electron density to the other element.

Problem 1.40 Rank the following atoms in order of increasing electronegativity. Label the most electronegative and most electropositive atom in each group.

a. Se, O, S b. P, Na, Cl c. Cl, S, F d. C, O, Si

Electronegativity values are used as a guideline to indicate whether the electrons in a bond are **equally shared** or **unequally shared** between two atoms. Whenever two identical atoms are bonded together, each atom attracts the electrons in the bond to the same extent. The electrons are equally shared, and the **bond is *nonpolar.*** Thus, a **carbon–carbon bond is nonpolar.** Whenever two different atoms having similar electronegativities are bonded together, the bond is also **nonpolar. C–H bonds are considered to be nonpolar,** because the electronegativity difference between C (2.5) and H (2.2) is small.

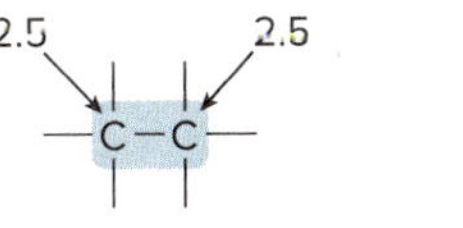

nonpolar bond

2.5 2.2 C–H

nonpolar bond

The small electronegativity difference between C and H is ignored.

Bonding between atoms of different electronegativity values results in the **unequal sharing** of electrons. In a C–O bond, the electrons are pulled away from C (2.5) toward O (3.4), the element of higher electronegativity. **The bond is *polar,*** or ***polar covalent.*** The bond is said to have a **dipole;** that is, **a partial separation of charge.**

A C–O bond is a **polar** bond.

C is electron deficient. O is electron rich.

$\delta+$ $\delta-$ C–O

a bond dipole

The direction of polarity in a bond is often indicated by an arrow, with the head of the arrow pointing toward the more electronegative element. The tail of the arrow, with a perpendicular line drawn through it, is positioned at the less electronegative element. Alternatively, the symbols δ+ and δ– indicate this unequal sharing of electron density.

- **δ+ means an atom is electron deficient (has a partial positive charge).**
- **δ– means an atom is electron rich (has a partial negative charge).**

Problem 1.41 Show the direction of the dipole in each bond. Label the atoms with δ+ and δ–.

a. H–F b. B–C c. –C–Li d. –C–Cl

Students often wonder how large an electronegativity difference must be to consider a bond polar. That's hard to say. We will set an arbitrary value for this difference and use it as an *approximation.* **Usually, a polar bond will be one in which the electronegativity difference between two atoms is $\geq$ 0.5 unit.**

The distribution of electron density in a molecule can be shown using an **electrostatic potential map.** These maps are color coded to illustrate areas of high and low electron density. Electron-rich regions are indicated in red, and electron-deficient sites are indicated in blue. Regions of intermediate electron density are shown in orange, yellow, and green.

An electrostatic potential map of CH_3Cl indicates the polar nature of the C–Cl bond (Figure 1.13). The more electronegative Cl atom pulls electron density toward it, making it electron rich. This is indicated by the red around the Cl in the plot. The carbon is electron deficient, and this is shown with blue.

Figure 1.13 Electrostatic potential plot of CH_3Cl

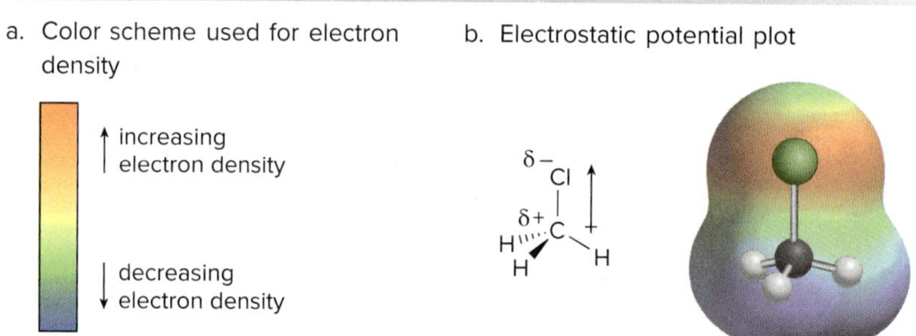

1.13 Polarity of Molecules

A **polar molecule** has either one polar bond, or two or more bond dipoles that reinforce. A **nonpolar molecule** has either no polar bonds, or two or more bond dipoles that cancel.

Thus far, we have been concerned with the polarity of one bond. To determine whether a molecule has a net dipole, use the following two-step procedure:

[1] Use electronegativity differences to **identify all of the polar bonds and the directions of the bond dipoles.**

[2] **Determine the geometry** around individual atoms by counting groups, and decide if individual dipoles **cancel** or **reinforce each other in space.**

Whenever C or H is bonded to N, O, and all halogens, the bond is ***polar.*** Thus, the C–I bond is considered polar even though the electronegativity difference between C and I is small. Remember, electronegativity is just an approximation.

The two molecules $\mathbf{H_2O}$ and $\mathbf{CO_2}$ illustrate different outcomes of this process. In H_2O, each O–H bond is polar because the electronegativity difference between O (3.4) and H (2.2) is large. Because H_2O is a **bent** molecule, the two dipoles reinforce (both point *up*). Thus, **$\mathbf{H_2O}$ has a net dipole, making it a polar molecule.** CO_2 also has polar C–O bonds because the electronegativity difference between O (3.4) and C (2.5) is large. However, CO_2 is a **linear** molecule, so the two dipoles, which are equal and opposite in direction, **cancel.** Thus, CO_2 is a **nonpolar molecule** with **no net dipole.**

H–O–H: δ– (O), δ+ (H), δ+ (H); net dipole (δ+ → δ–)

net dipole

O=C=O: δ– δ+ δ–

no net dipole

Electrostatic potential plots for H_2O and CO_2 appear in Figure 1.14. Additional examples of polar and nonpolar molecules are given in Figure 1.15.

Problem 1.42 Indicate which of the following molecules is polar because it possesses a net dipole. Show the direction of the net dipole if one exists.

a. CH_3Br
b. CH_2Br_2
c. CF_4
d. (Cl)(H)C=C(Cl)(H) — both Cl on the same side
e. (Cl)(H)C=C(H)(Cl) — Cl atoms on opposite sides
f. CF_2Cl_2
g. (Cl)(Cl)C=C(H)(H)
h. CH_3CH_2OH
i. $(CH_3CH_2)_2C{=}O$

Figure 1.14 Electrostatic potential plots for H_2O and CO_2

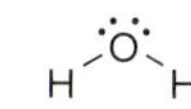

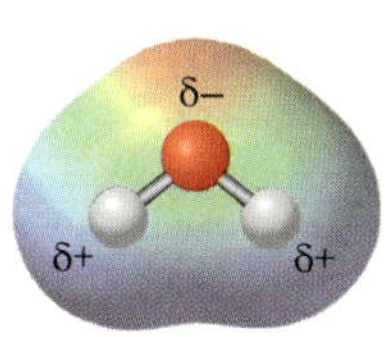

- The electron-rich (red) region is concentrated on the more electronegative O atom. Both H atoms are electron deficient (blue-green).

:Ö=C=Ö:

δ+
δ–
δ–

- Both electronegative O atoms are electron rich (red), and the central C atom is electron deficient (blue).

Figure 1.15 Examples of polar and nonpolar molecules

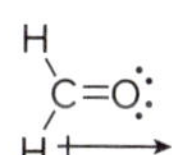

one polar bond
a **polar** molecule

three polar bonds
All dipoles cancel.
no net dipole
a **nonpolar** molecule

net dipole
three polar bonds
All dipoles reinforce.
a **polar** molecule

net dipole
two polar bonds
Both dipoles reinforce.
a **polar** molecule

four polar bonds
All dipoles cancel.
no net dipole
a **nonpolar** molecule

1.14 Oxybenzone—A Representative Organic Molecule

The principles learned in this chapter apply to all organic molecules, regardless of size or complexity. We now know a great deal about the structure of oxybenzone, the molecule that appears on the cover of this text.

Sample Problem 1.11 Applying the Principles of Bonding, Geometry, and Polarity to a Representative Organic Molecule

Answer each question about oxybenzone, the popular sunscreen component described in the Prologue.

OH O

oxybenzone

a. How many lone pairs does oxybenzone contain?
b. What is the molecular formula of oxybenzone?
c. What is the hybridization and geometry around each atom labeled in blue?
d. In what type of orbital(s) are any lone pairs on the O atom in red located?
e. Label all polar bonds.

Solution

a, b. Each O atom needs two lone pairs for an octet, so oxybenzone has six lone pairs. In determining the molecular formula from the skeletal structure, assume there is a C atom at the end of any line and at the intersection of two lines, and that each C has enough H's to make it tetravalent; molecular formula = $C_{14}H_{12}O_3$.

sp^2 hybridized and trigonal planar

sp^3 hybridized and tetrahedral

sp^2 hybridized and trigonal planar

$C_{14}H_{12}O_3$

c, d. Count groups to determine hybridization and geometry; with four groups an atom is sp^3 hybridized and tetrahedral; with three groups an atom is sp^2 hybridized and trigonal planar. The O atom in red is surrounded by three groups—one atom and two lone pairs—so it is sp^2 hybridized and its lone pairs occupy sp^2 hybrid orbitals.

e. All C–O and O–H bonds are polar because of the large electronegativity difference between the atoms.

Lisinopril (trade name Zestril, Problem 1.43) is an ACE inhibitor, a drug that lowers blood pressure by decreasing the amount of angiotensin in the blood. Angiotensin narrows blood vessels, thus increasing blood pressure. *alon harel/Alamy Stock Photo*

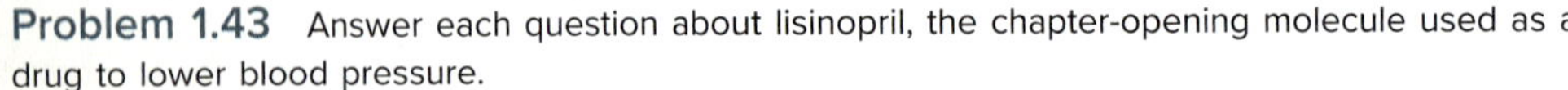

Problem 1.43 Answer each question about lisinopril, the chapter-opening molecule used as a drug to lower blood pressure.

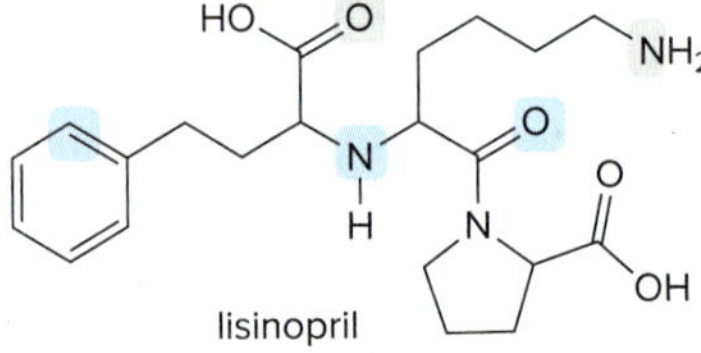

lisinopril

a. What is the molecular formula of lisinopril?
b. How many carbon atoms are sp^2 hybridized?
c. What is the hybridization and shape around each atom labeled in blue?
d. In what type of orbital do the lone pairs on each atom labeled in gray reside?

More Practice: Try Problems 1.45, 1.46, 1.74, 1.76, 1.78.

Problem 1.44 Use the ball-and-stick model of vitamin B_6 to answer each question.

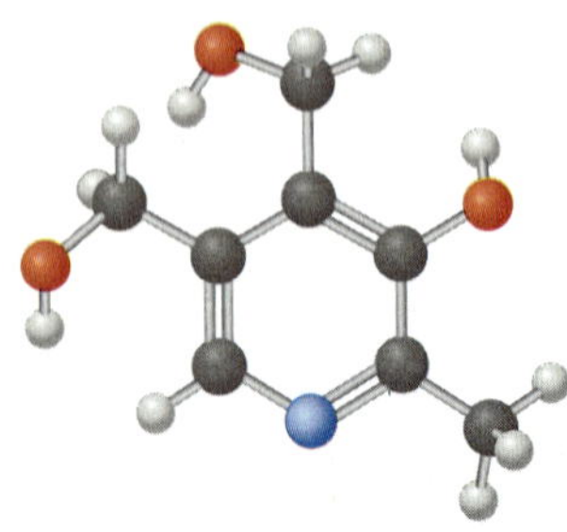
vitamin B_6

a. Draw a skeletal structure of vitamin B_6.
b. How many sp^2 hybridized carbons are present?
c. What is the hybridization of the N atom in the ring?

Chapter 1 REVIEW

KEY CONCEPTS

Resonance (1.6)

1 Drawing resonance structures

- Look for lone pairs and multiple bonds.
- Atoms and σ bonds do not change location.

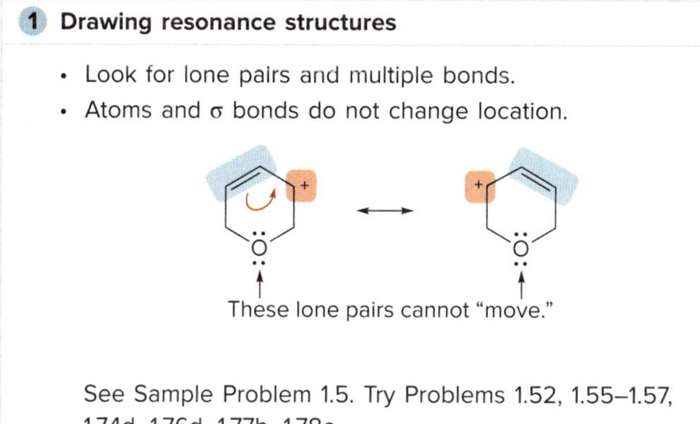

See Sample Problem 1.5. Try Problems 1.52, 1.55–1.57, 1.74d, 1.76d, 1.77b, 1.78e.

2 Drawing the resonance hybrid

- Draw σ bonds and lone pairs that do not move.
- Use a dashed line for a bond that is single in one resonance structure and multiple in another.
- Use a δ+ (or δ–) for an atom that is neutral in one structure and charged in another.

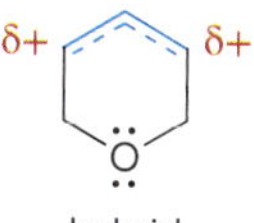

hybrid

Try Problems 1.56, 1.77c.

Periodic Trends

1 Bond length of H–Z bonds (1.7A)

2 Electronegativity (1.12)

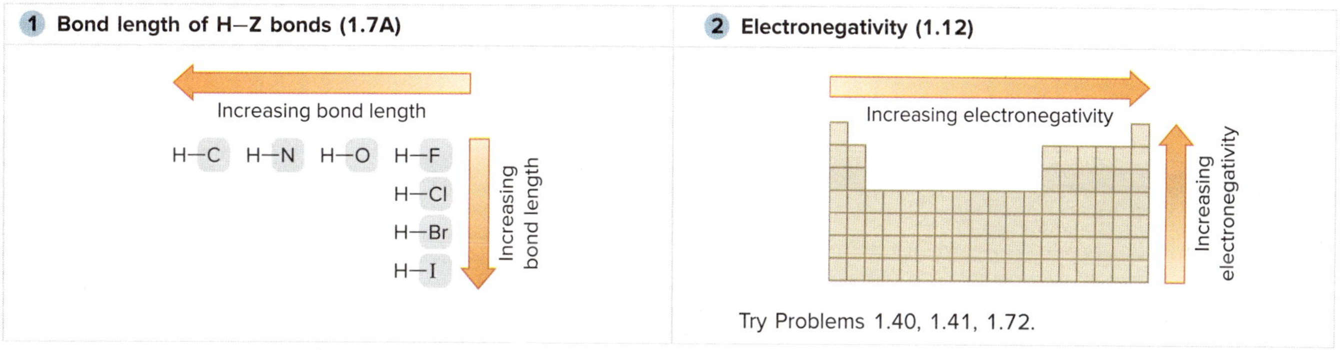

Try Problems 1.40, 1.41, 1.72.

Bond Length and Bond Strength (1.11)

1 Bond length and bond strength of C–C bonds (1.11A)

- Bonds become shorter and stronger as the number of electrons between two nuclei increases.

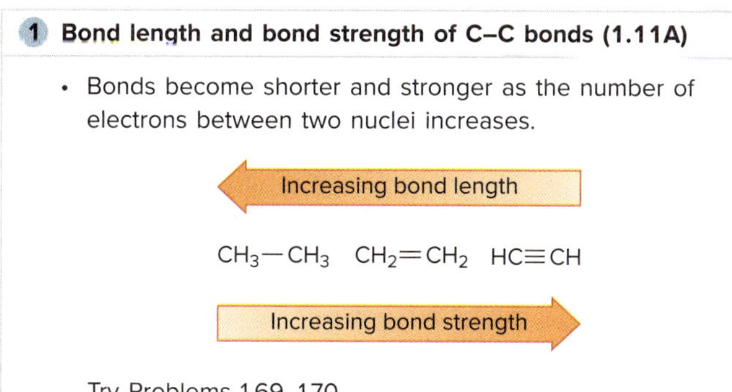

Try Problems 1.69, 1.70.

2 Bond length and bond strength of C–H bonds (1.11B)

- Bonds become shorter and stronger as the percent *s*-character increases.

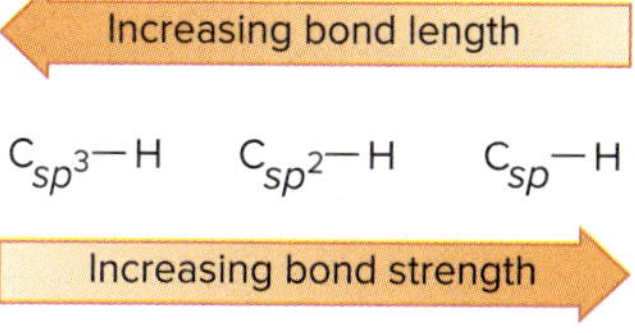

See Table 1.6.

Drawing Organic Structures (1.8)

Abbreviate the structure of complex molecules with skeletal structures or condensed structures.

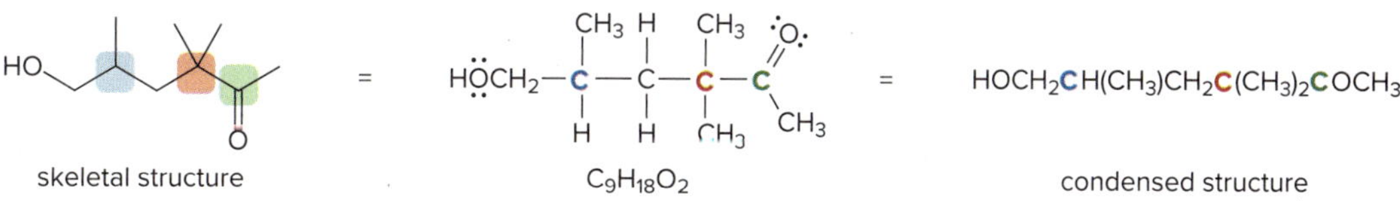

See Figures 1.3, 1.4, 1.5, Sample Problems 1.7, 1.8. Try Problems 1.62–1.65.

KEY SKILLS

[1] Drawing a valid Lewis structure (1.3); example: CH_3CHO

1 Arrange the atoms with H's on the periphery.	2 Count valence electrons.	3 Add single bonds.	4 Complete octets with multiple bonds and lone pairs.
H H H C C O H	2 C's x 4 e^- = 8 4 H's x 1 e^- = 4 1 O x 6 e^- = 6 total e^- = 18	H–C–C–O (with three H's on C) 12 e^- used.	H–C–C=Ö (with three H's on C, one H on the second C) Add one double bond and two lone pairs to complete O and C octets.

See Sample Problems 1.1, 1.2. Try Problems 1.48, 1.49.

[2] Calculating formal charge (1.3C)

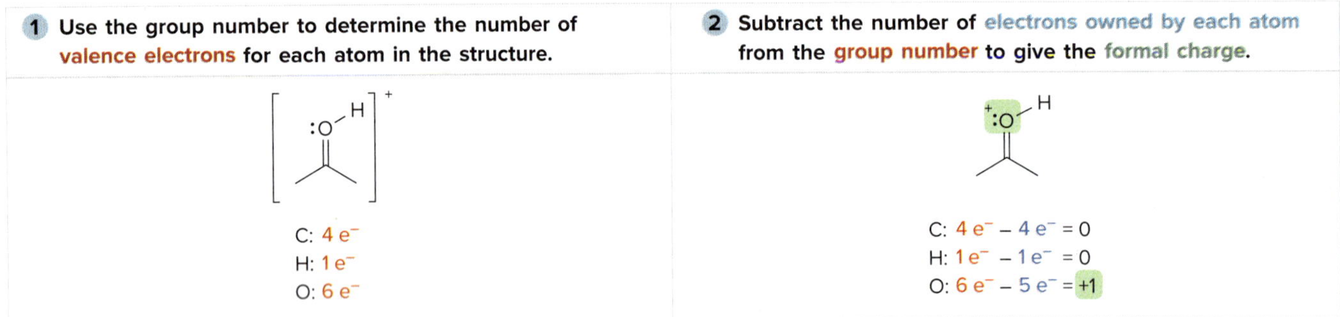

See Sample Problem 1.3. Try Problem 1.47.

[3] Predicting geometry from a valid Lewis structure (1.7)

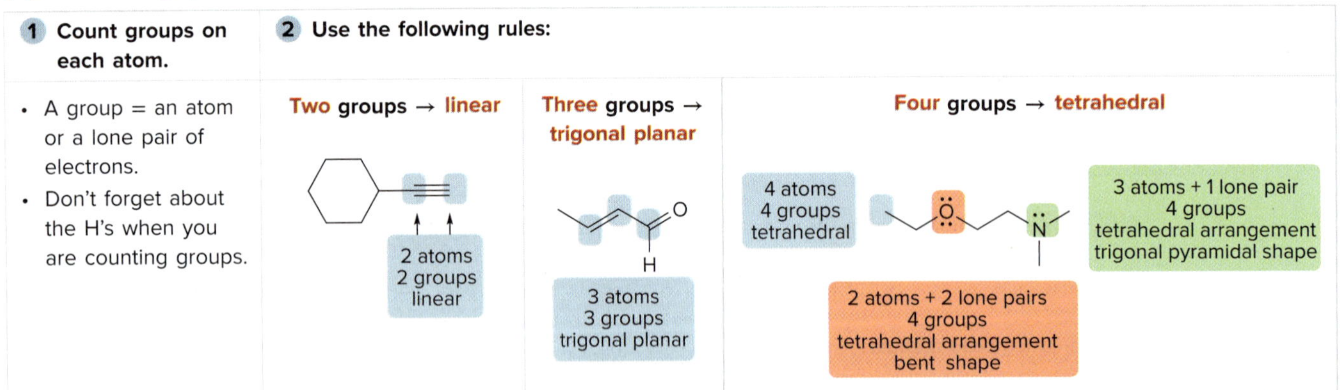

See Sample Problem 1.6. Try Problems 1.60, 1.61, 1.76c.

[4] Identifying isomers and resonance structures (1.4, 1.6)

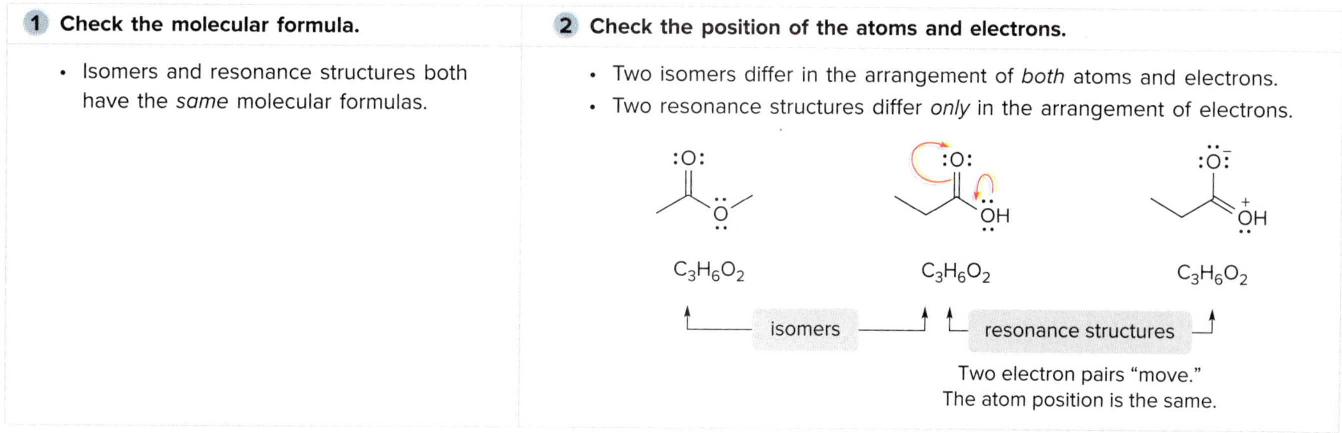

Try Problems 1.51–1.53.

[5] Using curved arrows (1.6B)

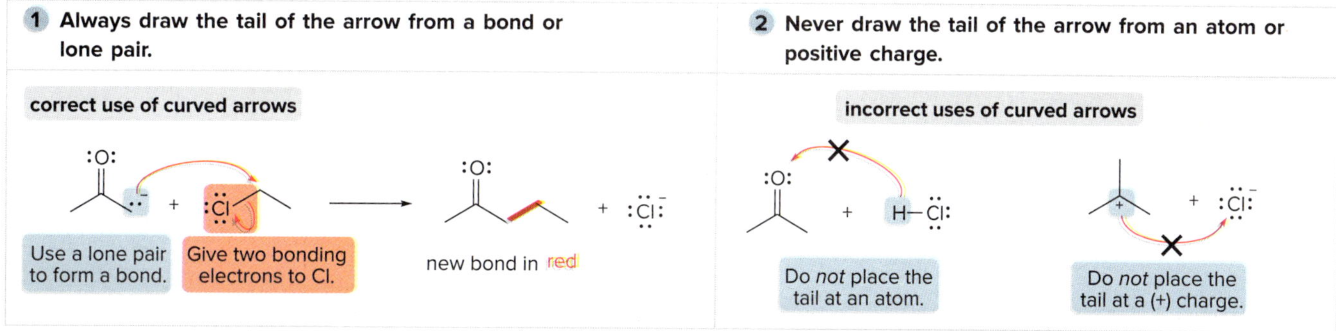

See Sample Problem 1.4. Try Problems 1.54, 1.55.

[6] Predicting hybridization from a valid Lewis structure (1.9)

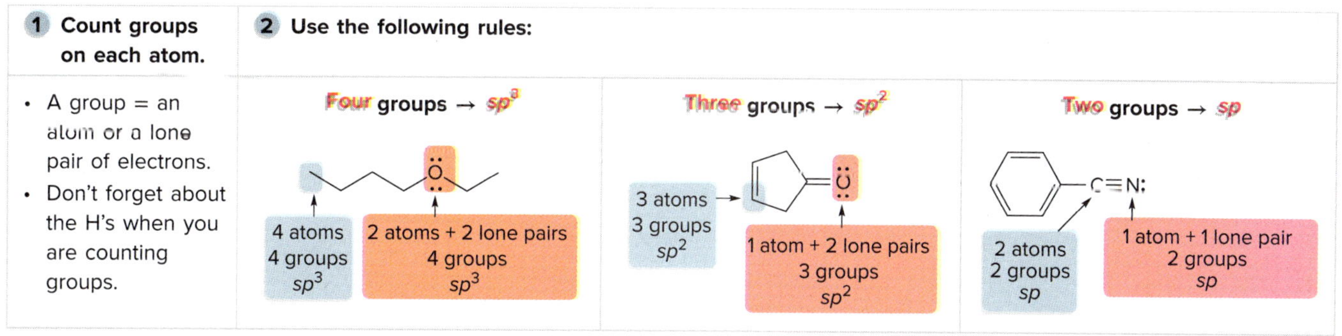

See Sample Problem 1.10. Try Problems 1.45d, 1.46d, 1.66, 1.74c, 1.75a, 1.76a, 1.78c.

[7] Determining if a molecule has a net dipole from a valid Lewis structure (1.13); example: CH_3OH

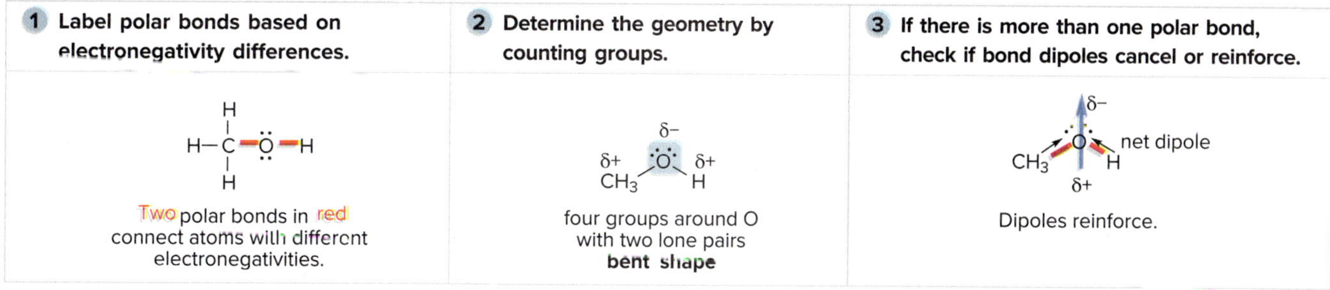

Try Problem 1.73.

CHAPTER 1 MULTIPLE-CHOICE SELF-TEST

The Self-Test consists of multiple-choice questions similar to those found on the American Chemical Society organic chemistry exam. Answers are given at the end of the chapter.

1. Which of the following cations is *not* a resonance structure of **A?**

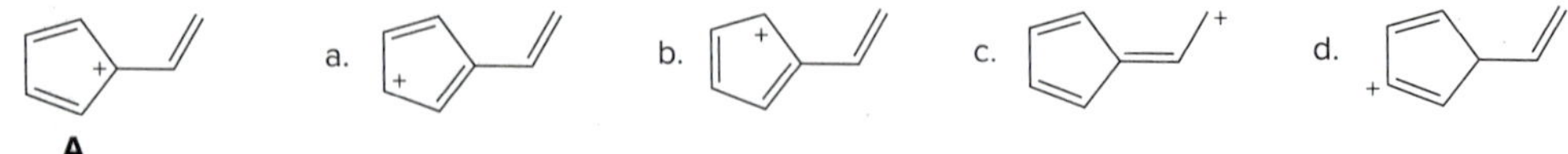

2. Classify **A, B,** and **C** as resonance structures or isomers of each other.

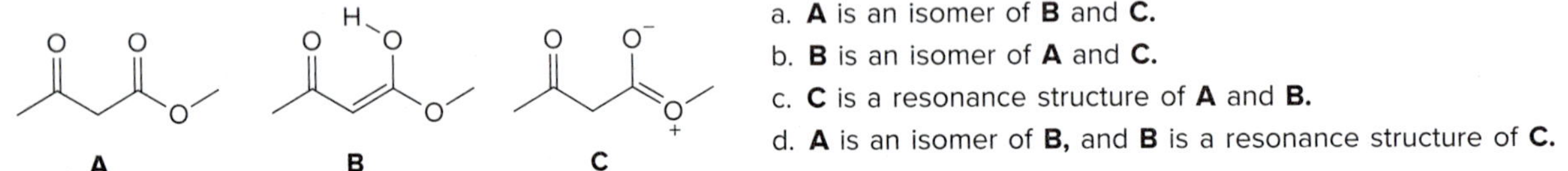

a. **A** is an isomer of **B** and **C.**
b. **B** is an isomer of **A** and **C.**
c. **C** is a resonance structure of **A** and **B.**
d. **A** is an isomer of **B,** and **B** is a resonance structure of **C.**

3. Assign formal charge to each labeled atom. All nonbonded electron pairs are drawn in.

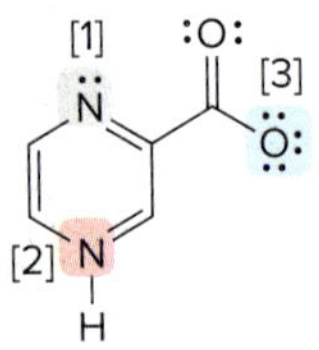

a. [1] 0; [2] +1; [3] −2
b. [1] −1; [2] 0; [3] −1
c. [1] 0; [2] 0; [3] +1
d. [1] 0; [2] +1; [3] −1

4. What is the molecular formula for entacapone, a drug used to treat Parkinson's disease?

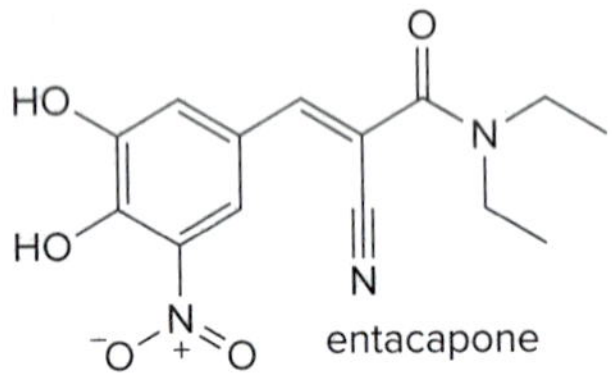

entacapone

a. $C_{13}H_{15}N_3O_5$
b. $C_{14}H_{15}N_3O_5$
c. $C_{14}H_9N_3O_5$
d. $C_{13}H_{17}N_3O_5$

5. What is the hybridization of the N, C, and O atoms in the given molecule?

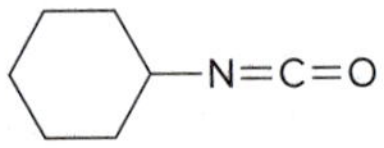

a. N sp^2, C sp, O sp^2
b. N sp, C sp^2, O sp
c. N sp^3, C sp^2, O sp^2
d. N sp^2, C sp^2, O sp^2

6. Which of the following compounds does *not* have a net dipole: (a) $BrCH_2Br$; (b) CH_3OCH_3; (c) $(CH_3)_2CO$; (d) Cl_3CCCl_3?

7. Label the most electronegative and most electropositive elements in the following group: nitrogen, oxygen, sulfur, and phosphorus.
   a. N is the most electronegative element, and S is the most electropositive element.
   b. N is the most electronegative element, and P is the most electropositive element.
   c. O is the most electronegative element, and S is the most electropositive element.
   d. O is the most electronegative element, and P is the most electropositive element.

8. How many lone pairs are contained in the amino acid arginine: (a) 6; (b) 7; (c) 8; (d) 9?

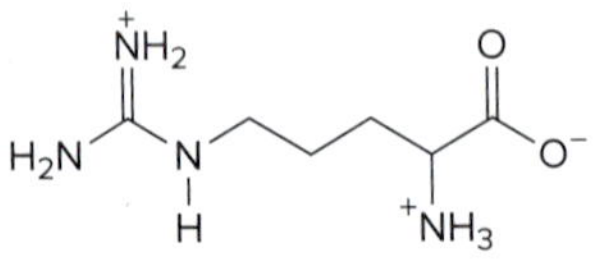

arginine

9. What is the shape around the labeled atoms in the following ion?

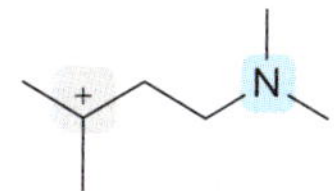

a. C and N are both trigonal planar.
b. C is tetrahedral and N is trigonal planar.
c. C is trigonal planar and N is trigonal pyramidal.
d. C is tetrahedral and N is trigonal pyramidal.

10. Which structure is another way of drawing compound **A?**

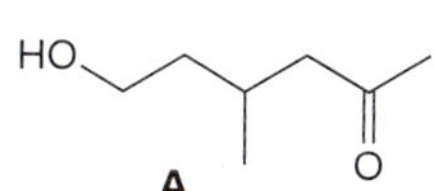

a. $HOCH_2CH_2$—C(CH$_3$)(H)—C(H)(H)—C(=O)H
b. $CH_3COCH_2CH(CH_3)CH_2CH_2OH$
c. $CH_3COCH_2CH_2CH(CH_3)CH_2OH$
d. $CH_3OCH_2CH_2CH(CH_3)CH_2CH_2OH$

PROBLEMS

Problems Using Three-Dimensional Models

1.45 Citric acid is responsible for the tartness of citrus fruits, especially lemons and limes.

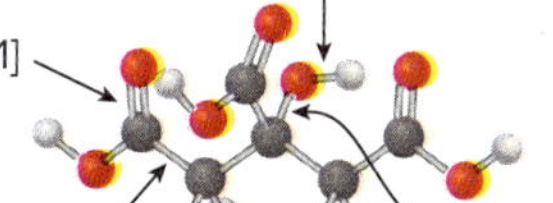

citric acid

a. What is the molecular formula for citric acid?
b. How many lone pairs are present?
c. Draw a skeletal structure.
d. How many sp^2 hybridized carbons are present?
e. What orbitals are used to form each indicated bond ([1]–[4])?

1.46 2'-Deoxyadenosine is a component of DNA.

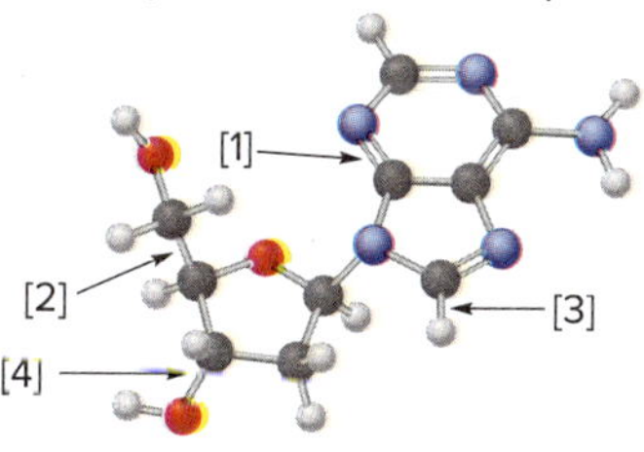

2'-deoxyadenosine

a. What is the molecular formula for 2'-deoxyadenosine?
b. How many lone pairs are present?
c. Draw a skeletal structure.
d. How many sp^2 hybridized carbons are present?
e. What orbitals are used to form each indicated bond ([1]–[4])?

Lewis Structures and Formal Charge

1.47 Give the formal charge on the highlighted carbon in each species. All H's and electrons on the highlighted carbon are drawn in.

a. 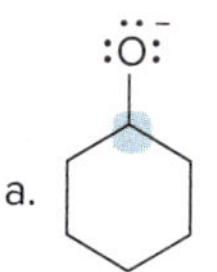

b. 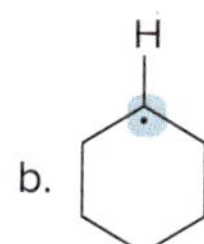

c. 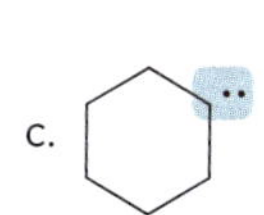

d. 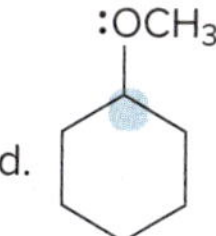

e. 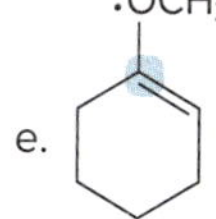

f. 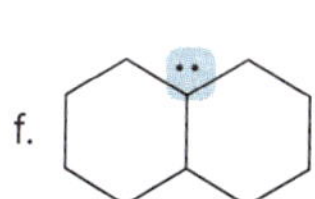

1.48 Draw one valid Lewis structure for each compound. Assume the atoms are arranged as drawn.

a. CH_2N_2 H C N N (H below C)
b. CH_3NO_2 H C N O (H above C; H O below)
c. CH_3CNO H C C N O (H above and below first C)
d. $(CH_2CN)^-$ H C C N (H below first C)

1.49 Draw an acceptable Lewis structure from each condensed structure, such that all atoms have zero formal charge.
a. dihydroxyacetone, $(HOCH_2)_2CO$, an ingredient in sunless tanning products
b. acetic anhydride, $(CH_3CO)_2O$, a reagent used to synthesize aspirin

Isomers and Resonance Structures

1.50 Draw Lewis structures for the nine isomers having molecular formula C_3H_6O, with all atoms having a zero formal charge.

1.51 With reference to anion **A,** label compounds **B–E** as an isomer or resonance structure of **A.** For each isomer, indicate what bonds differ from **A.**

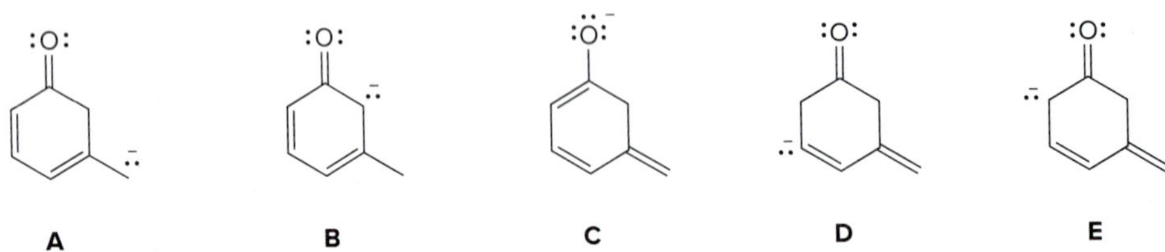

1.52 Which of the following species is a valid resonance structure of **A?** Use curved arrows to show how **A** is converted to any valid resonance structure. When a compound is not a valid resonance structure of **A,** explain why not.

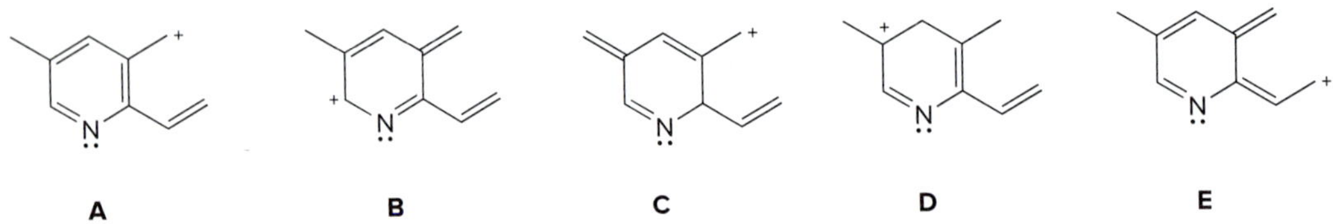

1.53 How are the molecules or ions in each pair related? Classify them as resonance structures, isomers, or neither.

a. 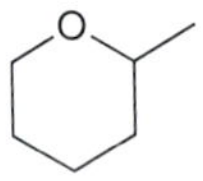 and OH

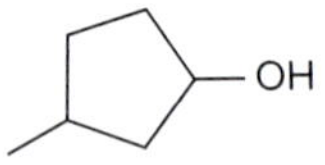

c. 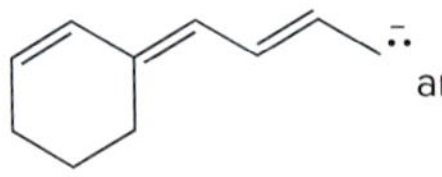 and

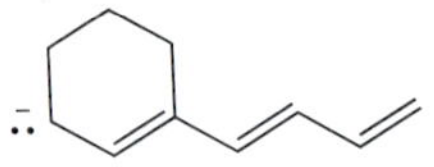

b. 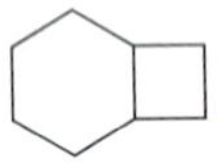 and

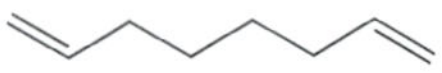

d. 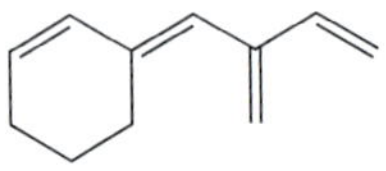 and

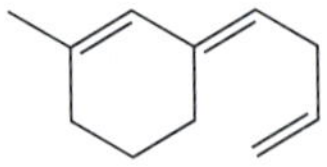

1.54 Add curved arrows to show how the first resonance structure can be converted to the second.

a. 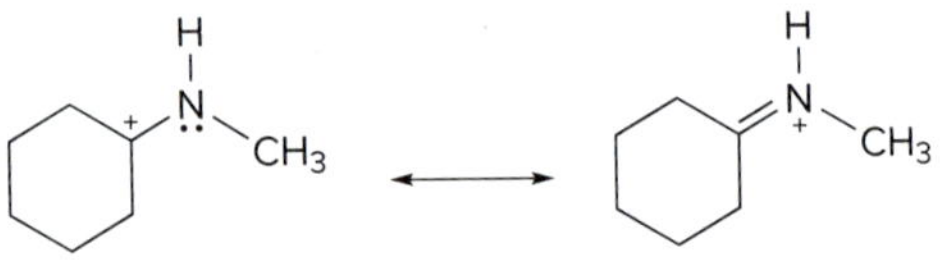 H N CH_3 ⟷ H N CH_3

b.  ⟷

1.55 Follow the curved arrows to draw a second resonance structure for each species.

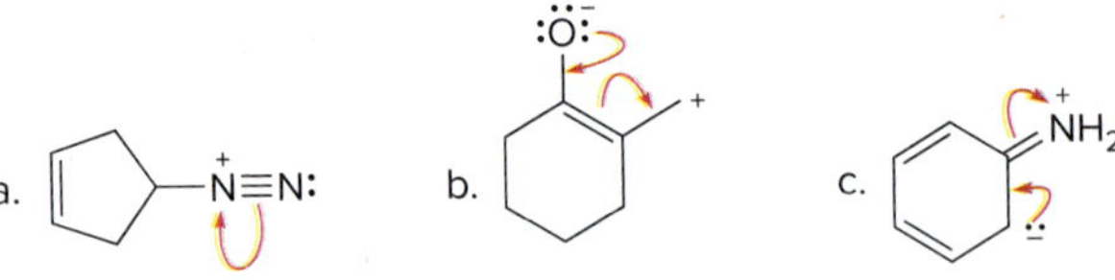

a. $\overset{+}{N}\equiv N$:

b.

c. $\overset{+}{N}H_2$

1.56 Draw a second resonance structure for each ion. Then, draw the resonance hybrid.

a. 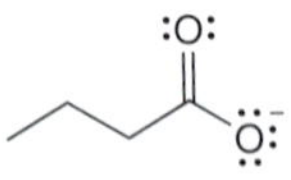 :O: :Ö:

b. 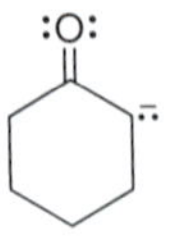 :O:

c. 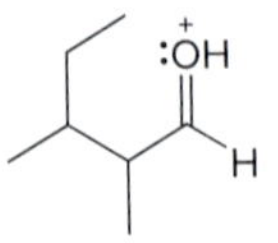 $\overset{+}{:OH}$ H

1.57 Draw all reasonable resonance structures for each species.

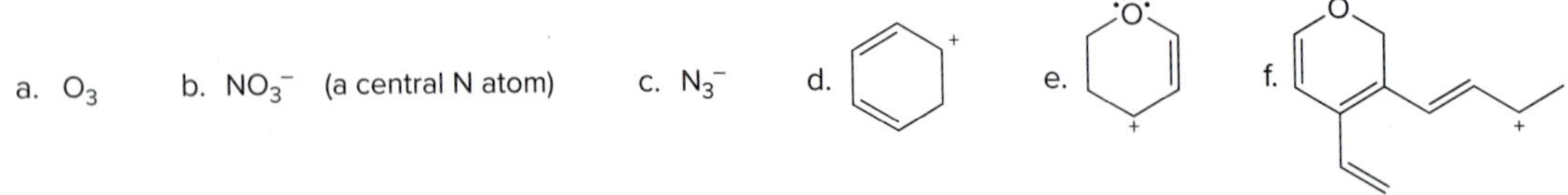

a. O_3 b. NO_3^- (a central N atom) c. N_3^- d. e. f.

1.58 Which of the given resonance structures (**A, B,** or **C**) contributes most to the resonance hybrid? Which contributes least?

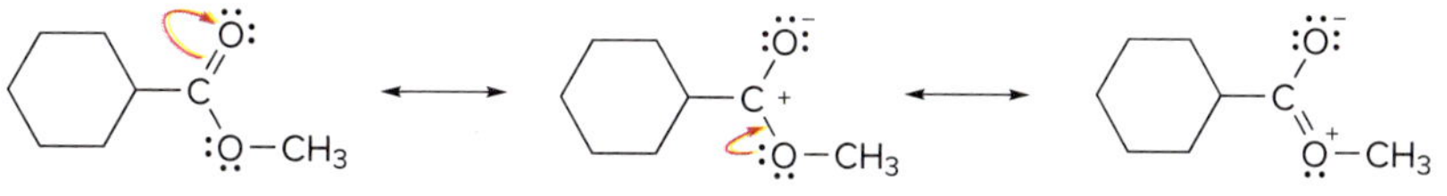

A B C

1.59 Consider the compounds and ions with curved arrows drawn below. When the curved arrows give a second valid resonance structure, draw the resonance structure. When the curved arrows generate an invalid Lewis structure, explain why the structure is unacceptable.

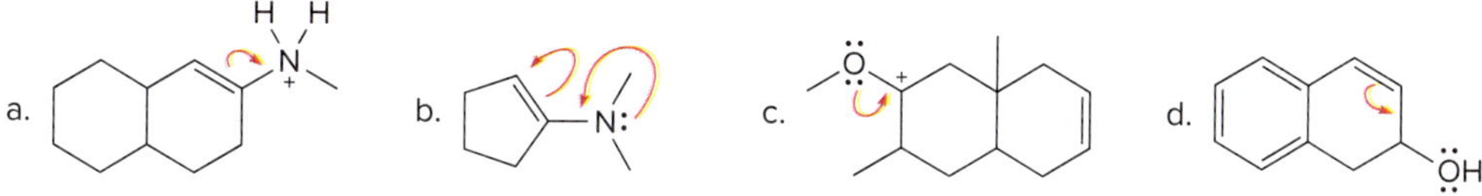

Geometry

1.60 Predict all bond angles in each compound.

a. CH_3Cl b. NH_2OH c. $CH_2{=}NCH_3$ d. $HC{\equiv}CCH_2OH$ e. 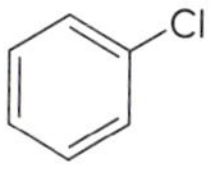 (chlorobenzene structure, Cl)

1.61 Predict the geometry around each highlighted atom.

a.  (structure) b. $(CH_3)_2N^-$ c.  (structure) d. 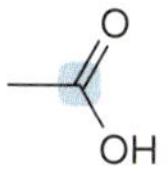 (structure, O, OH) e. $(CH_3)_3N$

Drawing Organic Molecules

1.62 How many hydrogens are present around each carbon atom in the following molecules?

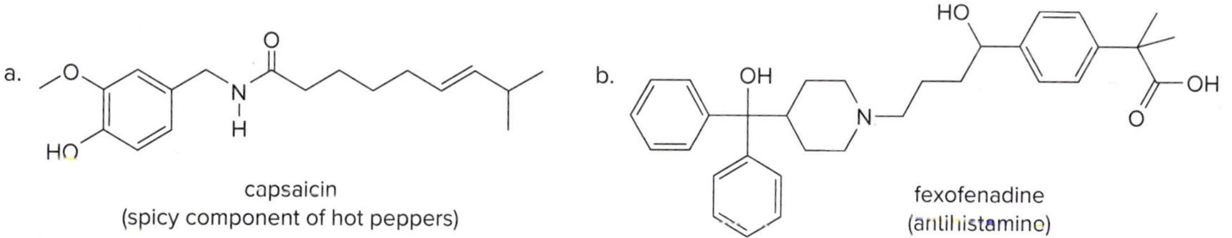

capsaicin
(spicy component of hot peppers)

fexofenadine
(antihistamine)

1.63 Draw in all the carbon and hydrogen atoms in each molecule.

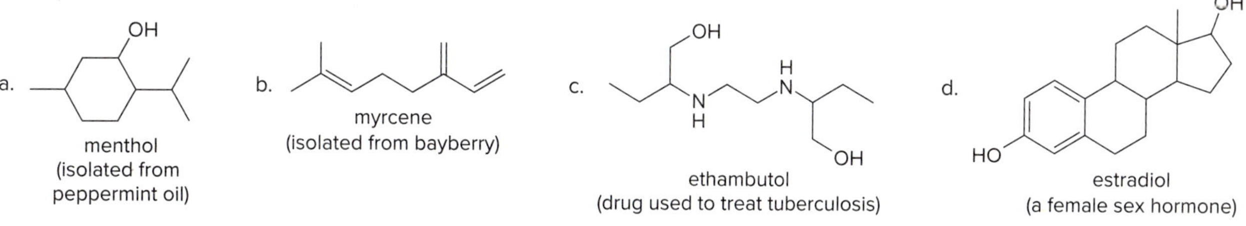

menthol
(isolated from peppermint oil)

myrcene
(isolated from bayberry)

ethambutol
(drug used to treat tuberculosis)

estradiol
(a female sex hormone)

1.64 Convert each molecule to a skeletal structure.

a. $(CH_3)_2CHCH_2CH_2CH(CH_3)_2$

b. $CH_3CH(Cl)CH(OH)CH_3$

c. $CH_3(CH_2)_2C(CH_3)_2CH(CH_3)CH(CH_3)CH(Br)CH_3$

d. 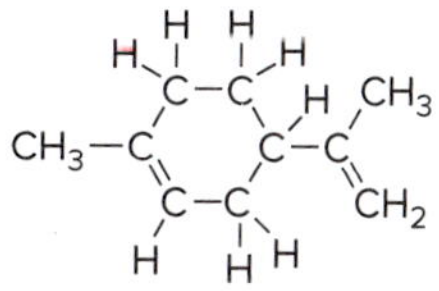 (structure)

limonene
(oil of lemon)

1.65 Convert the following condensed formulas into skeletal structures.

a. $CH_3CONHCH_3$ b. CH_3COCH_2Br c. $(CH_3)_3COH$ d. CH_3COCl e. $CH_3COCH_2CO_2H$ f. $HO_2CCH(OH)CO_2H$

Hybridization

1.66 Predict the hybridization and geometry around each highlighted atom.

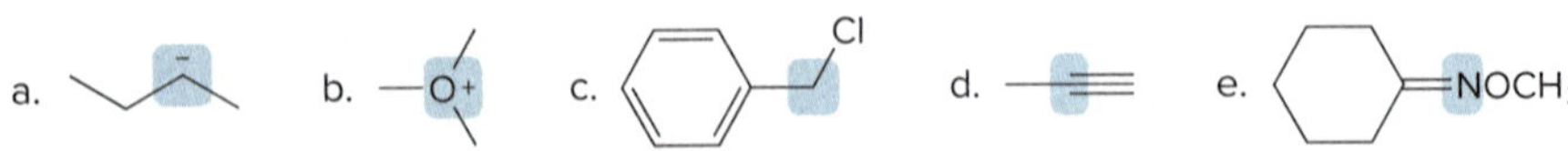

1.67 What orbitals are used to form each highlighted bond? For multiple bonds, indicate the orbitals used in individual bonds.

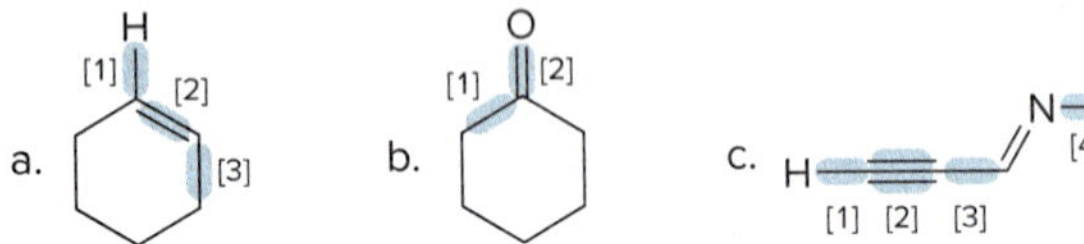

1.68 Ketene, $CH_2{=}C{=}O$, is an unusual organic molecule that has a single carbon atom doubly bonded to two different atoms. Determine the hybridization of both C atoms and the O in ketene. Then, draw a diagram showing what orbitals are used to form each bond (similar to Figures 1.10 and 1.11).

Bond Length and Strength

1.69 Rank the labeled bonds in safinamide, a drug used to treat Parkinson's disease approved in the United States in 2017, in order of increasing bond length.

[2] O [3] F

O [1] H N

H_2N

safinamide

1.70 Answer the following questions about compound **A.**

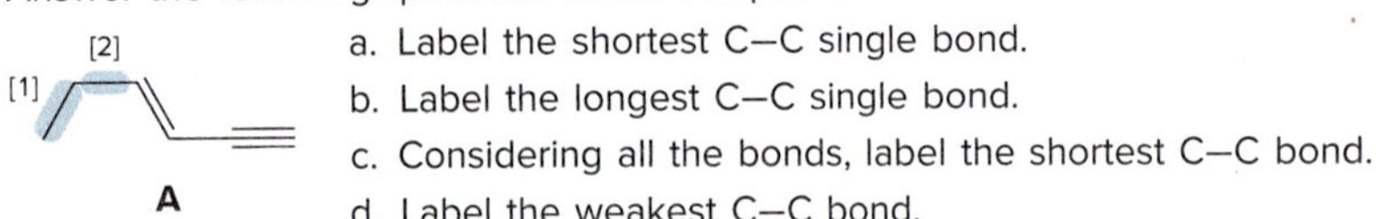

a. Label the shortest C–C single bond.
b. Label the longest C–C single bond.
c. Considering all the bonds, label the shortest C–C bond.
d. Label the weakest C–C bond.
e. Label the strongest C–H bond.
f. Explain why bond [1] and bond [2] are different in length, even though they are both C–C single bonds.

1.71 Two useful organic compounds that contain Cl atoms are vinyl chloride ($CH_2{=}CHCl$) and chloroethane (CH_3CH_2Cl). Vinyl chloride is the starting material used to prepare poly(vinyl chloride), a plastic in insulation, pipes, and bottles. Chloroethane (ethyl chloride) is a local anesthetic. Why is the C–Cl bond in vinyl chloride stronger than the C–Cl bond in chloroethane?

Bond Polarity

1.72 Use the symbols δ+ and δ– to indicate the polarity of the highlighted bonds.

a. $NH_2{-}OH$ b. NH_2 c. Li

1.73 Label the polar bonds in each molecule. Indicate the direction of the net dipole (if there is one).

a. $CHBr_3$
b. $CH_3CH_2OCH_2CH_3$
c. Br Br
d. Cl Cl

Problems That Combine Concepts

1.74 Anacin is an over-the-counter pain reliever that contains aspirin and caffeine. Answer the following questions about each compound.

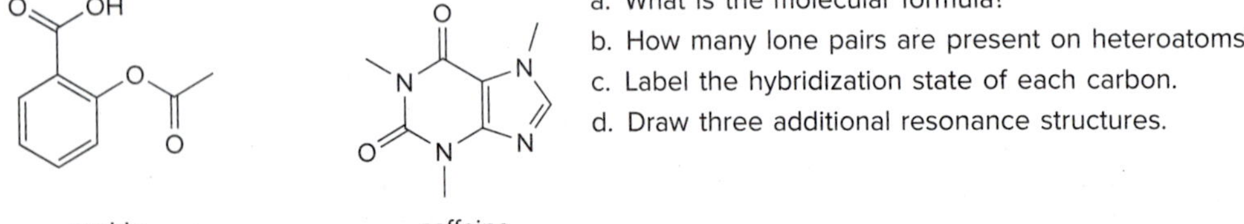

aspirin (acetylsalicylic acid)

caffeine

a. What is the molecular formula?
b. How many lone pairs are present on heteroatoms?
c. Label the hybridization state of each carbon.
d. Draw three additional resonance structures.

1.75 Answer the following questions about acetonitrile ($CH_3C{\equiv}N{:}$).

a. Determine the hybridization of both C atoms and the N atom.

b. Label all bonds as σ or π.

c. In what type of orbital does the lone pair on N reside?

d. Label all bonds as polar or nonpolar.

1.76 As mentioned in the Prologue, remdesivir (trade name Veklury) is a drug approved to treat COVID-19 patients with moderate symptoms. Answer each question on remdesivir.

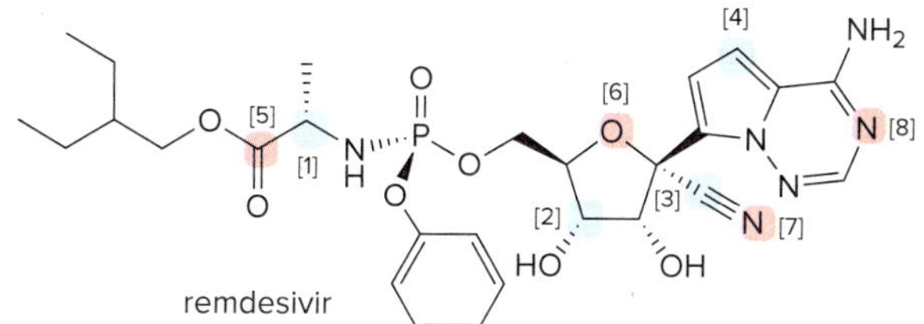

a. What is the hybridization of the C atoms labeled in blue ([1]–[4])?

b. How many lone pairs are present in remdesivir?

c. What is the geometry around each atom labeled in red ([5]–[8])?

d. Draw two additional resonance structures and label them as major, minor, or equal contributors to the hybrid.

1.77 (a) Add curved arrows to show how the starting material **A** is converted to the product **B.** (b) Draw all reasonable resonance structures for **B.** (c) Draw the resonance hybrid for **B.**

A → **B** + $:\ddot{Cl}:^-$

1.78 As we will learn in Chapter 30, acetyl coenzyme A (acetyl CoA) is a key organic reactant in many biochemical transformations in cells.

a. How many lone pairs does acetyl CoA contain?

b. Which atoms in the structure of acetyl CoA do not follow the octet rule?

c. Give the hybridization of each atom highlighted in blue.

d. In what type of orbital do the lone pairs on each atom shown in red reside?

e. Draw two additional resonance structures.

acetyl coenzyme A

1.79 CH_3^+ and CH_3^- are two highly reactive carbon species.

a. What is the predicted hybridization and geometry around each carbon atom?

b. Two electrostatic potential plots are drawn for these species. Which ion corresponds to which diagram and why?

Challenge Problems

1.80 The N atom in CH_3CONH_2 (acetamide) is sp^2 hybridized, even though it is surrounded by four groups. Using this information, draw a diagram that shows the orbitals used by the atoms in the $-CONH_2$ portion of acetamide, and offer an explanation as to the observed hybridization.

1.81 Use the observed bond lengths to answer each question. (a) Why is bond [1] longer than bond [2] (143 pm versus 136 pm)? (b) Why are bonds [3] and [4] equal in length (127 pm), and shorter than bond [2]?

$CH_3{-}\ddot{O}H$ [1] $\qquad$ $CH_3C(=\ddot{O})\ddot{O}H$ [2] $\qquad$ $CH_3C(=\ddot{O})\ddot{O}:^-$ [3] [4]

1.82 Secnidazole is a medication used to treat infections of the lower gastrointestinal tract caused by a single-celled parasite.

a. Draw a resonance structure that contributes equally to the resonance hybrid.

b. Draw at least five resonance structures that are minor contributors to the resonance hybrid.

secnidazole

1.83 When two carbons having different hybridization are bonded together, the C–C bond contains a slight dipole. In a C_{sp^2}–C_{sp^3} bond, what is the direction of the dipole? Which carbon is considered more electronegative?

1.84 Draw all possible isomers having molecular formula C_4H_8 that contain one π bond.

1.85 The curved arrow notation introduced in Section 1.6B is a powerful method used by organic chemists to show the movement of electrons not only in resonance structures, but also in chemical reactions. Because each curved arrow shows the movement of two electrons, following the curved arrows illustrates what bonds are broken and formed in a reaction. Consider the following three-step process. (a) Add curved arrows in Step [1] to show the movement of electrons. (b) Use the curved arrows drawn in Step [2] to identify the structure of **X. X** is converted in Step [3] to phenol and HCl.

[1]

[2] X [3]

phenol

1.86 The curved arrow notation in Section 1.6B is also used to illustrate biological reactions that occur in the presence of enzymes. These reactions are often drawn with two or more bonds broken and formed at the same time, and an acid, symbolized by HA, or a base, symbolized by B: may be required. Follow the curved arrows and draw the products of each reaction involved in metabolism. In part (b), "SCoA" is derived from coenzyme A, a key reactant in many biological transformations (Chapter 30).

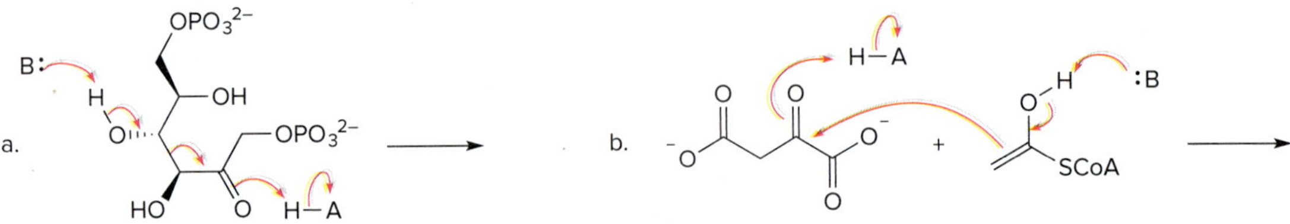

SELF-TEST ANSWERS

1. d 2. b 3. d 4. b 5. a 6. d 7. d 8. b 9. c 10. b

Acids and Bases

2

Comstock/Getty Images

2.1 Brønsted–Lowry acids and bases
2.2 Reactions of Brønsted–Lowry acids and bases
2.3 Acid strength and pK_a
2.4 Predicting the outcome of acid–base reactions
2.5 Factors that determine acid strength
2.6 Common acids and bases
2.7 Aspirin
2.8 Lewis acids and bases

The rich flavor and aroma of a freshly brewed cup of coffee results from a myriad of organic compounds. The mild acidity of coffee made from the beans of plants grown at higher altitudes or in volcanic soil is in part due to **quinic acid,** an organic acid present in low concentration in green coffee beans. Quinic acid concentration increases during processing, as more-complex compounds are degraded by the heat of roasting, and it contributes to the increase in the perceived acidity of coffee that has been warmed for a long time on a hot surface. In Chapter 2, we learn about acidity and acid–base reactions.

Why Study. . . Acids and Bases?

Chemical terms such as *anion* and *cation* may be unfamiliar to most nonscientists, but *acid* has found a place in everyday language. Commercials advertise the latest remedy for the heartburn caused by excess stomach *acid.* The news may report the latest environmental impact of *acid* rain. Wine lovers know that wine sours because its alcohol has turned to *acid. Acid* comes from the Latin word *acidus,* meaning "sour," because when tasting compounds was a routine method of identification, these compounds were sour.

In Chapter 2, we concentrate on two definitions of acids and bases: the **Brønsted–Lowry** definition, which describes acids as **proton donors** and bases as **proton acceptors,** and the **Lewis** definition, which describes acids as **electron pair acceptors** and bases as **electron pair donors.**

2.1 Brønsted–Lowry Acids and Bases

The general words "acid" and "base" usually mean a *Brønsted–Lowry* acid and *Brønsted–Lowry* base.

H^+ = proton.

HA = Brønsted–Lowry acid.
B: = Brønsted–Lowry base.

The Brønsted–Lowry definition describes acidity in terms of protons: positively charged **hydrogen ions,** H^+**.**

- A Brønsted–Lowry acid is a *proton donor.*
- A Brønsted–Lowry base is a *proton acceptor.*

A Brønsted–Lowry acid must contain a *hydrogen* atom. This definition of an acid is often familiar to students, because many inorganic acids in general chemistry are Brønsted–Lowry acids. The symbol **HA** is used for a general Brønsted–Lowry acid.

A Brønsted–Lowry base must be able to form a bond to a proton. Because a proton has no electrons, **a base must contain an "available" electron pair** that can be easily donated to form a new bond. These include **lone pairs** or electron pairs in **π bonds.** The symbol **B:** is used for a general Brønsted–Lowry base. Examples of Brønsted–Lowry acids and bases are given in Figure 2.1.

Charged species such as ^-OH and $^-NH_2$ are used as **salts,** with cations such as Li^+, Na^+, or K^+ to balance the negative charge. These cations are called **counterions** or **spectator ions,** and their **identity is usually inconsequential.** For this reason, the counterion is often omitted.

salt		counterion	base
NaOH	=	Na^+	^-OH
KOH	=	K^+	^-OH

Compounds like H_2O and CH_3OH that contain both hydrogen atoms and lone pairs may be either an acid or a base, depending on the particular reaction. These fundamental principles

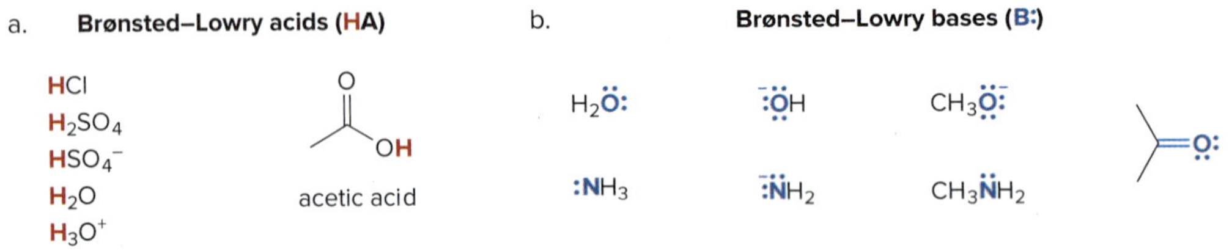

Figure 2.1 Examples of Brønsted–Lowry acids and bases

a. **Brønsted–Lowry acids (HA)**

HCl
H_2SO_4
HSO_4^-
H_2O
H_3O^+
acetic acid

- **All Brønsted–Lowry acids contain a proton.**
- The net charge may be zero, (+), or (–).

b. **Brønsted–Lowry bases (B:)**

$H_2\ddot{O}:$
$:\ddot{O}H^-$
$CH_3\ddot{O}:^-$
$:NH_3$
$:\ddot{N}H_2^-$
$CH_3\ddot{N}H_2$

- **All Brønsted–Lowry bases contain a lone pair of electrons or a π bond.**
- The net charge may be zero or (–).

Morphine is obtained from the opium poppy. *mafoto/Getty Images*

are true no matter how complex the compound. For example, the addictive pain reliever **morphine** is a Brønsted–Lowry acid because it contains many hydrogen atoms. It is also a Brønsted–Lowry base because it has lone pairs on O and N, and four π bonds.

morphine

- **H** atoms on O make morphine an acid.
- Lone pairs and π bonds (in **blue**) make morphine a base.

Problem 2.1
a. Which compounds are Brønsted–Lowry acids: HBr, NH_3, CCl_4?
b. Which compounds are Brønsted–Lowry bases: CH_3CH_3, $(CH_3)_3CO^-$, $HC{\equiv}CH$?
c. Classify each compound as an acid, a base, or both: CH_3CH_2OH, $CH_3CH_2CH_2CH_3$, $CH_3CO_2CH_3$.

2.2 Reactions of Brønsted–Lowry Acids and Bases

A Brønsted–Lowry acid–base reaction results in transfer of a proton from an acid to a base. These acid–base reactions, also called ***proton transfer reactions,*** are fundamental to the study of organic chemistry.

Consider, for example, the reaction of the acid HA with the base :B. **In an acid–base reaction, one bond is broken and one is formed.**

- **The electron pair of the base B: forms a new bond to the proton of the acid.**
- **The acid HA loses a proton, leaving the electron pair in the HA bond on A.**

Recall from Section 1.6 that a curved arrow shows the movement of an **electron pair. The tail of the arrow always begins at an electron pair,** and the head points to where that electron pair "moves."

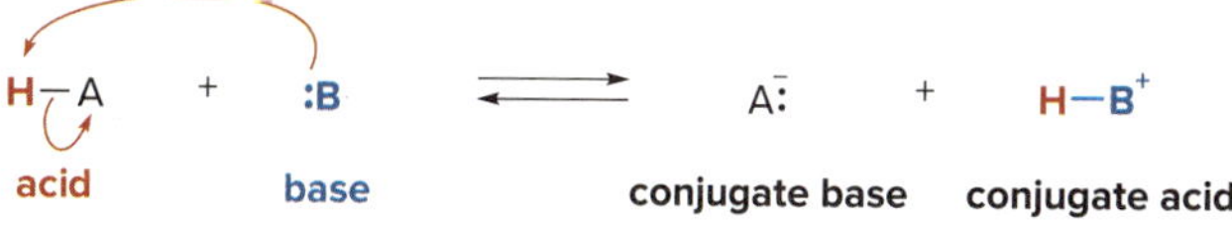

This "movement" of electrons in reactions can be illustrated using curved arrow notation. Because **two electron pairs** are involved in this reaction, **two curved arrows** are needed. Two products are formed.

- Loss of a proton from an acid forms its *conjugate base.*
- Gain of a proton by a base forms its *conjugate acid.*

The **net charge must be the same** on both sides of any equation. In this example, the net charge on each side is zero. Individual charges can be calculated using formal charges. **A double reaction arrow** is used between starting materials and products to indicate that the reaction can proceed in the forward and reverse directions. These are **equilibrium arrows.**

Two examples of proton transfer reactions are drawn here with curved arrow notation.

A double reaction arrow indicates equilibrium.

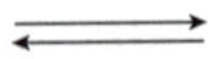

equilibrium arrows

Remove H^+ from an acid to form its conjugate base. **Add H^+** to a base to form its conjugate acid.

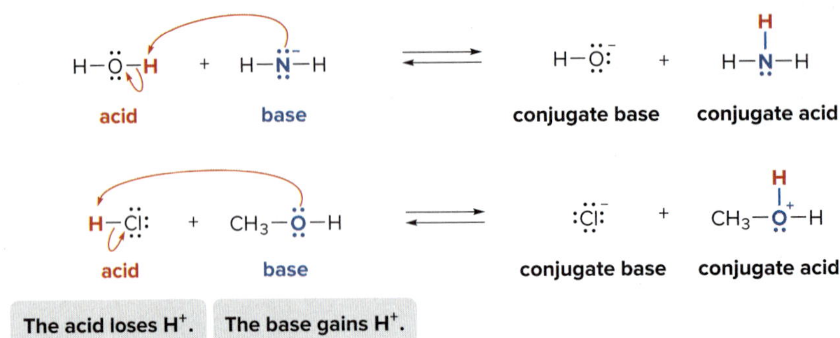

- **Brønsted–Lowry acid–base reactions always result in the transfer of a proton from an acid to a base.**

The ability to identify and draw a conjugate acid or base from a given starting material is illustrated in Sample Problems 2.1 and 2.2.

Sample Problem 2.1 Drawing a Conjugate Acid and a Conjugate Base

a. What is the conjugate acid of CH_3O^-?
b. What is the conjugate base of NH_3?

Solution

a. **Add H^+ to CH_3O^- to form its conjugate acid.**

$CH_3-\ddot{O}:^-$ (base) → Add H^+ to a lone pair. → $CH_3-\ddot{O}-H$ (conjugate acid)

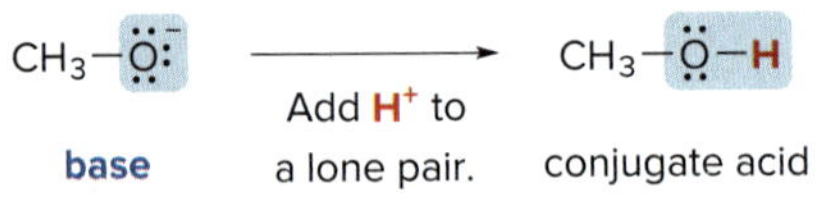

b. **Remove H^+ from NH_3 to form its conjugate base.**

NH_3 (acid) → Remove H^+. → $H-\ddot{N}:^-$ with H (conjugate base) ← The electron pair stays on N.

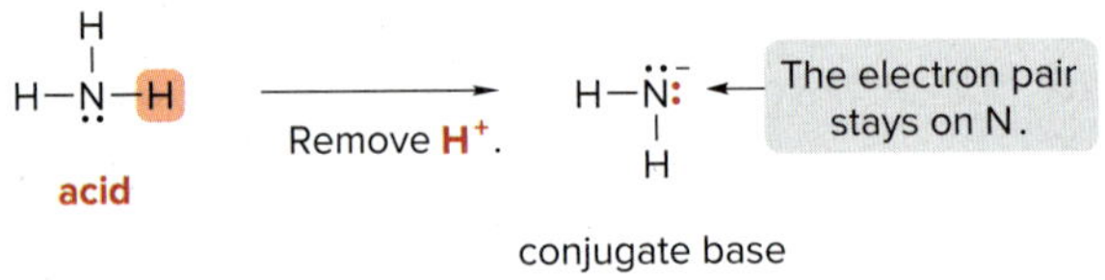

Problem 2.2 a. Draw the conjugate acid of each base: NH_3, Cl^-, $(CH_3)_2C{=}O$.
b. Draw the conjugate base of each acid: HBr, HSO_4^-, CH_3OH.

More Practice: Try Problems 2.45, 2.46.

Problem 2.3 (a) What is the conjugate base of $CH_2{=}CH_2$? (b) What is the conjugate acid of $CH_3CH_2^-$?

Sample Problem 2.2 Determining the Acid, Base, Conjugate Acid, and Conjugate Base in a Reaction

Label the acid and base, and the conjugate acid and base, in the following reaction. Use curved arrow notation to show the movement of electron pairs.

A + B ⇌ C + D

Solution

A is the base because it accepts a proton, forming its conjugate acid, **C. B** is the acid because it donates a proton, forming its conjugate base, **D.** Two curved arrows are needed because two electron pairs are involved. One shows that the lone pair on **A** bonds to a proton of **B,** and the second shows that the electron pair in the O–H bond remains on O.

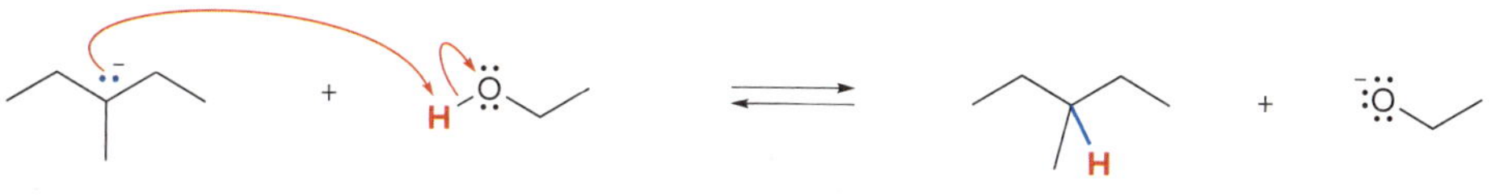

A + B ⇌ C + D

A	B	C	D
base	acid	conjugate acid	conjugate base
The base gains H^+.	The acid loses H^+.		

Problem 2.4 Label the acid and base, and the conjugate acid and base, in the following reactions. Use curved arrows to show the movement of electron pairs.

a. 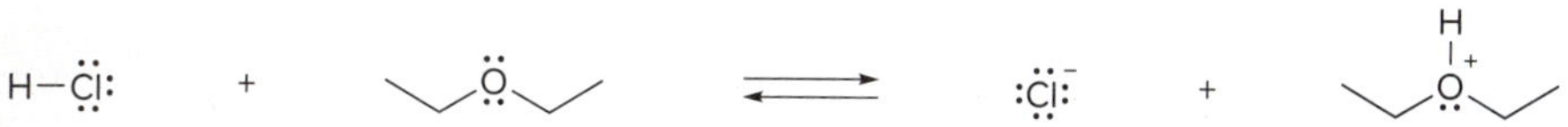 H–Cl: + ⇌ :Cl:⁻ +

b.  + ⇌ +

c. 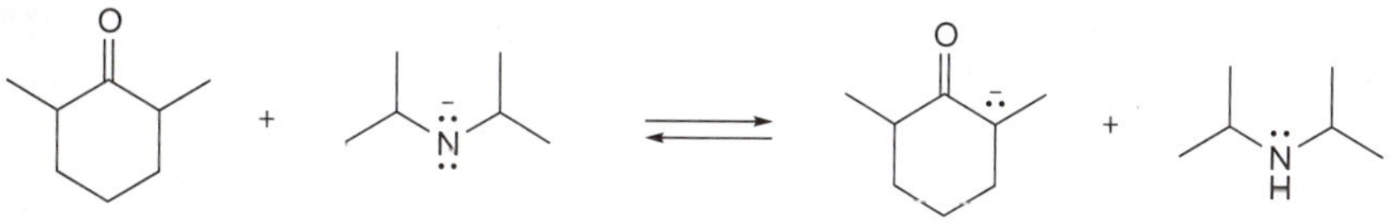 + ⇌ +

More Practice: Try Problem 2.47.

In all proton transfer reactions, the **electron-rich base** donates an electron pair to the acid, which usually has a polar HA bond. The H of the acid bears a partial positive charge, making it **electron deficient.** This is the first example of a general pattern of reactivity.

- **Electron-rich species react with electron-deficient ones.**

Given two starting materials, how do you know which is the acid and which is the base in a proton transfer reaction? Use the following generalizations:

[1] **Common acids and bases introduced in general chemistry are used in the same way in organic reactions. HCl and H_2SO_4 are strong acids, and ^-OH is a strong base.**

[2] **When only one starting material contains a hydrogen, it must be the acid. If only one starting material has a lone pair or a π bond, it must be the base.**

[3] **A starting material with a net positive charge is usually the acid. A starting material with a negative charge is usually the base.**

Figure 2.2 shows how to use these generalizations to identify the acid and base with pairs of compounds.

Figure 2.2 Identifying the acid and the base in a proton transfer reaction

a. When one reactant has a net (+) or (–) charge:

A + $:\ddot{N}(CH_3)_2^-$ base **B**

B has a (–) charge, so **B** is the base.

b. When one reactant is an inorganic acid or base:

C + $H\ddot{Cl}:$ acid

HCl is a strong inorganic acid, so HCl is the acid.

c. When only one reactant has a H or lone pair:

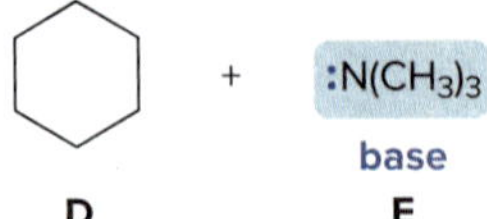

E has a lone pair and **D** does not, so **E** is the base.

Problem 2.5 Identify the acid and base in each pair of compounds.

a. cyclohexyl–$\ddot{O}H$ + NaH b. $(CH_3CH_2)_2\ddot{N}H$ + $(CH_3)_2C=\overset{+}{O}H$ c. quinoline + HCl

Problem 2.6 Decide which compound is the acid and which is the base, and draw the products of each proton transfer reaction.

a. $Cl_3C-COOH$ + $^-:\ddot{O}CH_3$ ⇌ c. cyclopentyl–NH_2 + HCl ⇌

b. $Cl_3C-C\equiv C-H$ + $H:^-$ ⇌ d. $CH_3CH_2CH_2OH$ + H_2SO_4 ⇌

Problem 2.7 Draw the products formed from the acid–base reaction of HCl with each compound.

a. OH b. O c. N d. NH

Problem 2.8 Rimantadine is an oral medication used to treat influenza. What product is formed when rimantadine reacts with the HCl present in the stomach?

NH_2

rimantadine

2.3 Acid Strength and pK_a

Acid strength is the tendency of an acid to donate a proton.

- **The more readily a compound donates a proton, the *stronger* the acid.**

Acidity is measured by an equilibrium constant. When a Brønsted–Lowry acid HA is dissolved in water, an acid–base reaction occurs, and an equilibrium constant K_{eq} can be written for the reaction.

$$H-A + H-\ddot{O}-H \rightleftharpoons A:^- + H-\overset{H}{\underset{+}{\ddot{O}}}-H$$

acid base solvent

$$K_{eq} = \frac{[\text{products}]}{[\text{starting materials}]} = \frac{[H_3O^+][A:^-]}{[HA][H_2O]}$$

Because the concentration of the solvent H_2O is essentially constant, the equation can be rearranged and a new equilibrium constant, called the **acidity constant, K_a,** can be defined.

$$K_a = [H_2O]K_{eq} = \frac{[H_3O^+][A:^-]}{[HA]}$$

How is the magnitude of K_a related to acid strength?

- The *stronger the acid,* the further the equilibrium lies to the right and the *larger the* K_a.

For most organic compounds, K_a is small, typically 10^{-5} to 10^{-50}. This contrasts with the K_a values for many inorganic acids, which range from 10^0 to 10^{10}. Because using exponents can be cumbersome, it is often more convenient to use pK_a values instead of K_a values.

$$pK_a = -\log K_a$$

How does pK_a relate to acid strength?

Recall that a **log** is an **exponent;** for example, log $10^{-5} = -5$.

K_a values of typical organic acids

10^{-5} → 10^{-50}

larger number stronger acid | smaller number weaker acid

pK_a values of typical organic acids

+5  +50

smaller number stronger acid | larger number weaker acid

- The *smaller* the pK_a, the *stronger* the acid.

Problem 2.9 Which compound in each pair is the stronger acid?

a. $CH_3CH_2CH_3$ (pK_a = 50) or CH_3CH_2OH (pK_a = 16)

b. phenol (OH on benzene ring), $K_a = 10^{-10}$, or toluene (CH_3 on benzene ring), $K_a = 10^{-41}$

Problem 2.10 Use a calculator when necessary to answer the following questions.

a. What is the pK_a for each K_a: 10^{-10}, 10^{-21}, and 5.2×10^{-5}?

b. What is the K_a for each pK_a: 7, 11, and 3.2?

An inverse relationship exists between acidity and basicity.

- A *strong acid* readily donates a proton, forming a *weak conjugate base.*
- A *strong base* readily accepts a proton, forming a *weak conjugate acid.*

Table 2.1 is a brief list of pK_a values for some common compounds, ranked in order of ***increasing* pK_a** and therefore ***decreasing* acidity.** Because strong acids form weak conjugate bases,

Table 2.1 Selected pK_a Values

Acid	pK_a	Conjugate base
H–Cl	−6	Cl^-
CH_3CO_2–H	4.8	$CH_3CO_2^-$
HO–H	11	HO^-
CH_3CH_2O–H	16	$CH_3CH_2O^-$
HC≡CH	25	$HC≡C^-$
H–H	35	H^-
H_2N–H	38	H_2N^-
CH_2=CH_2	44	CH_2=$\bar{C}H$
CH_3–H	50	CH_3^-

Increasing acidity (upward, acid column); Increasing basicity (downward, conjugate base column)

this list also ranks their conjugate bases, in order of ***increasing*** **basicity.** CH_4 is the weakest acid in the list, because it has the highest pK_a (50). Its conjugate base, CH_3^-, is therefore the strongest conjugate base. An extensive pK_a table is located in Appendix C.

Comparing pK_a values tells us the **relative acidity of two acids,** and the **relative basicity of their conjugate bases,** as shown in Sample Problem 2.3.

Sample Problem 2.3 Using pK_a Values to Determine Relative Acidity and Basicity

Rank the following compounds in order of increasing acidity, and then rank their conjugate bases in order of increasing basicity.

$CH_2{=}CH_2$ H–Cl CH_3CO_2H

Solution

Use the pK_a values in Table 2.1 and the rule: **the lower the pK_a, the stronger the acid.**

$CH_2{=}CH_2$	CH_3CO_2H	H–Cl
pK_a = 44	pK_a = 4.8	pK_a = –6

Increasing acidity →

Remove a proton to draw the conjugate bases. Because strong acids form weak conjugate bases, the **basicity of conjugate bases increases with increasing pK_a** of their acids.

$CH_2{=}CH^-$ $CH_3CO_2^-$ Cl^-

← Increasing basicity

Problem 2.11 Rank the conjugate bases of each group of acids in order of increasing basicity.

a. NH_3, H_2O, CH_4 b. $CH_2{=}CH_2$, $HC{\equiv}CH$, CH_4

Problem 2.12 Consider two acids: HCO_2H (formic acid, pK_a = 3.8) and pivalic acid [$(CH_3)_3CCO_2H$, pK_a = 5.0]. (a) Which acid has the larger K_a? (b) Which acid is the stronger acid? (c) Which acid forms the stronger conjugate base? (d) When each acid is dissolved in water, for which acid does the equilibrium lie further to the right?

The pK_a values in Table 2.1 span a large range (–6 to 50). The pK_a scale is logarithmic, so a small difference in pK_a translates into a large numerical difference. The difference between the pK_a values of NH_3 (38) and $CH_2{=}CH_2$ (44) is six pK_a units, so NH_3 is 10^6 or *one million times more acidic* than $CH_2{=}CH_2$.

Although Table 2.1 is abbreviated, it is a useful tool for *estimating* the pK_a of a compound similar though not identical to one in the table. Suppose you are asked to estimate the pK_a of the N–H bond of CH_3NH_2. Although CH_3NH_2 is not listed in the table, we have enough information to *approximate* its pK_a. Because the pK_a of the N–H bond of NH_3 is 38, we can estimate the pK_a of the N–H bond of CH_3NH_2 to be 38. Its actual pK_a is 40, so this is a good first approximation.

Problem 2.13 Estimate the pK_a of each of the indicated bonds.

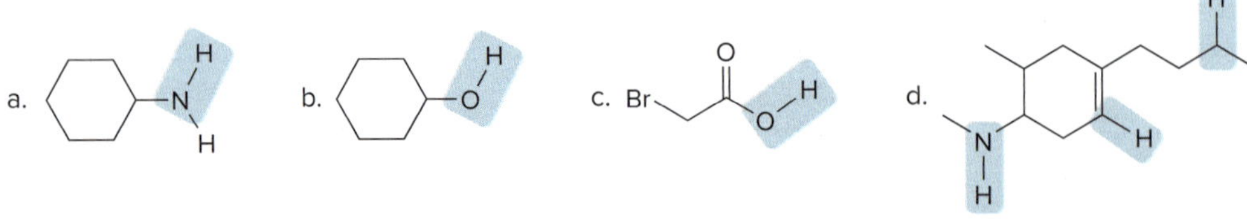

2.4 Predicting the Outcome of Acid–Base Reactions

In a proton transfer reaction, the **stronger acid reacts with the stronger base** to form the weaker acid and the weaker base.

A proton transfer reaction represents an equilibrium. Because an acid donates a proton to a base, forming a conjugate acid and conjugate base, there are always two acids and two bases in the reaction mixture. Which pair of acids and bases is favored at equilibrium? **The position of the equilibrium depends on the relative strengths of the acids and bases.**

- **Equilibrium always favors formation of the *weaker* acid and base.**

Because a strong acid readily donates a proton and a strong base readily accepts one, these two species react to form a weaker conjugate acid and base that do not donate or accept a proton as readily. Comparing pK_a values allows us to determine the position of equilibrium, as illustrated in Sample Problem 2.4.

Sample Problem 2.4 Using pK_a Values to Predict the Direction of Equilibrium

Determine the direction of equilibrium when acetylene (HC≡CH) reacts with $^-NH_2$ in a proton transfer reaction.

Solution

Follow three steps to determine the position of equilibrium:

Step [1] Identify the acid and base in the starting materials.

- $^-NH_2$ is the base because it bears a net negative charge, so HC≡CH is the acid.

Step [2] Draw the products of proton transfer and identify the conjugate acid and base in the products.

H–C≡C–H (acid, $pK_a = 25$) + $:\ddot{N}H_2^-$ (base) ⇌ H–C≡C:$^-$ (conjugate base) + $:NH_3$ (conjugate acid, $pK_a = 38$)

Step [3] Compare the pK_a values of the acid and the conjugate acid. Equilibrium favors formation of the *weaker* acid with the *higher* pK_a.

Use unequal equilibrium arrows (⇌ with longer reverse arrow or ⇌ with longer forward arrow) for a reversible reaction in which products or reactants are favored at equilibrium.

H–C≡C–H ($pK_a = 25$, stronger acid) + $:\ddot{N}H_2^-$ (base) ⇀ H–C≡C:$^-$ (conjugate base) + $:NH_3$ ($pK_a = 38$, **weaker acid**)

Equilibrium favors the products, forming the weaker acid.

- Because the pK_a of the starting acid (25) is ***lower*** than the pK_a of the conjugate acid (38), HC≡CH is a ***stronger*** acid and equilibrium favors the products.

Problem 2.14 Draw the products of each reaction and determine the direction of equilibrium.

a. $H_2C=CH_2$ + $H:^-$ ⇌

b. CH_4 + $:\ddot{O}H^-$ ⇌

c. CH_3COOH + $^-:\ddot{O}CH_2CH_3$ ⇌

d. $:\ddot{C}l:^-$ + $H\ddot{O}CH_2CH_3$ ⇌

More Practice: Try Problems 2.55, 2.56.

How can we know if a particular base is strong enough to deprotonate a given acid, so that the equilibrium lies to the right? The pK_a table readily gives us this information, as shown in Sample Problem 2.5.

Sample Problem 2.5 Determining if a Base Is Strong Enough to Deprotonate an Acid

Which of the following bases is strong enough to deprotonate *N,N*-dimethylacetamide [$CH_3CON(CH_3)_2$, pK_a = 30], so that equilibrium favors the products: (a) $NaNH_2$; (b) NaOH?

Solution

- **Draw the structure of the conjugate acid of each base,** and determine its pK_a from Table 2.1 or Appendix C. **Equilibrium favors the side with the weaker acid that has the higher pK_a.**
- **Compare the pK_a values of the starting acid and the conjugate acid.** If the conjugate acid has a *higher* pK_a than the starting acid, the conjugate acid is the *weaker* acid and equilibrium favors the *products.* **The base is strong enough** to deprotonate the acid.
- If the conjugate acid has a *lower* pK_a than the starting acid, the conjugate acid is the *stronger* acid and equilibrium favors the *starting materials.* **The base is *not* strong enough** to deprotonate the acid.

a. Na^+ is a counterion and $^-NH_2$ is the base in $NaNH_2$.

$^-$:NH$_2$ (base) —Add H$^+$.→ :NH$_3$ (conjugate acid, pK_a = 38, **weaker acid**)

The conjugate acid (NH_3) of the base is a *weaker* acid than $CH_3CON(CH_3)_2$ (pK_a = 30), so the base *is* strong enough to deprotonate the acid, and **equilibrium favors the products.**

b. Na^+ is a counterion and ^-OH is the base in NaOH.

$^-$:ÖH (base) —Add H$^+$.→ H_2Ö: (conjugate acid, pK_a = 14, **stronger acid**)

The conjugate acid (H_2O) of the base is a *stronger* acid than $CH_3CON(CH_3)_2$ (pK_a = 30), so the base is *not* strong enough to deprotonate the acid, and **equilibrium favors the starting materials.**

Problem 2.15 Using the data in Appendix C, determine which of the following bases is strong enough to deprotonate acetonitrile (CH_3CN), so that equilibrium favors the products: (a) NaH; (b) Na_2CO_3; (c) NaOH; (d) $NaNH_2$; (e) $NaHCO_3$.

More Practice: Try Problem 2.54.

Problem 2.16 Which of the following bases are strong enough to deprotonate C_6H_5OH (pK_a = 10) so that equilibrium favors the products: (a) H_2O; (b) NaOH; (c) $NaNH_2$; (d) CH_3NH_2; (e) $NaHCO_3$; (f) NaSH; (g) NaH?

Problem 2.17 Which of the labeled protons can be removed with NH_3 so that equilibrium favors the products?

a. C_6H_5–**SH** b. C_6H_5–**OH** c. C_6H_5–C(=O)–**OH** d. C_6H_5–**H**

Because Table 2.1 is arranged from low to high pK_a, **an acid can be deprotonated by the conjugate base of any acid *below* it in the table.**

Sample Problem 2.5 illustrates a fundamental principle in acid–base reactions.

- **An acid can be deprotonated by the conjugate base of any acid having a *higher* pK_a.**

2.5 Factors That Determine Acid Strength

The wide range of pK_a values in Table 2.1 illustrates that a tremendous difference in acidity exists among compounds. HCl (pK_a < 0) is an extremely strong acid, water (pK_a = 14) is moderate in acidity, and CH_4 (pK_a = 50) is an extremely weak acid. How are these differences explained? One general rule governs acid strength.

- **Anything that *stabilizes* a conjugate base A:$^-$ makes the starting acid HA *more* acidic.**

Four factors affect the acidity of HA:

[1] Element effects
[2] Inductive effects
[3] Resonance effects
[4] Hybridization effects

No matter which factor is discussed, follow the same procedure. To compare the acidity of any two acids:

- **Draw the conjugate bases.**
- **Determine which conjugate base is more stable.**
- **The *more stable* the conjugate base, the *more acidic* the acid.**

2.5A Element Effects—Trends in the Periodic Table

The most important factor determining the acidity of HA is the location of A in the periodic table.

Comparing Elements in the Same Row of the Periodic Table

To examine acidity trends **across a row** of the periodic table, we compare CH_4 and H_2O, two compounds having H atoms bonded to a second-row element. We know from Table 2.1 that **H_2O has a much *lower* pK_a and therefore is much *more acidic* than CH_4,** but why is this the case?

To answer this question, first draw both conjugate bases and then determine which is more stable. Each conjugate base has a net negative charge, but the negative charge in ^-OH is on oxygen and in CH_3^- it is on carbon.

H–ÖH —loss of H^+→ :ÖH⁻
conjugate base
negative charge on **O**
***more* stable**

CH_4 —loss of H^+→ CH_3^-
conjugate base
negative charge on **C**
***less* stable**

Because the oxygen atom is much **more electronegative** than carbon, oxygen more readily accepts a negative charge, making ^-OH much more stable than CH_3^-. **H_2O is a stronger acid than CH_4 because ^-OH is a more stable conjugate base than CH_3^-.** This is a specific example of a general trend.

- **Across a row of the periodic table, the acidity of HA *increases* as the electronegativity of A increases.**

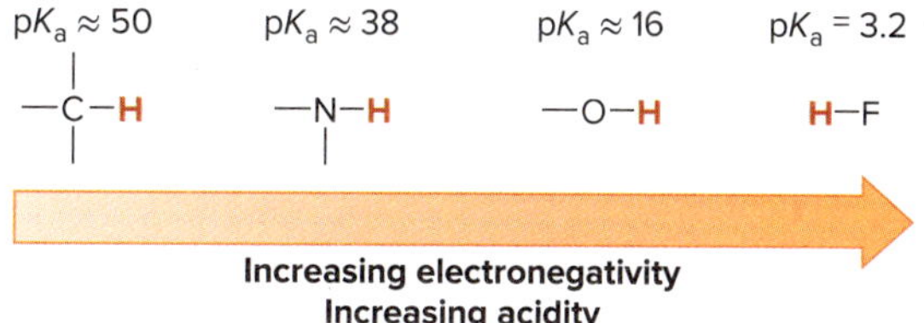

The enormity of this effect is evident by comparing the pK_a values for these bonds. **A C–H bond is approximately 10^{47} times *less acidic* than H–F.**

Comparing Elements Down a Column of the Periodic Table

To examine acidity trends down a column of the periodic table, we compare H–F and H–Br. Draw both conjugate bases and then determine which is more stable. In this case, removal of a proton forms F^- and Br^-.

H–F: —loss of H^+→ :F:⁻
conjugate base

H–Br: —loss of H^+→ :Br:⁻
conjugate base

There are two important differences between F^- and Br^-—electronegativity and size. In this case, **size is more important than electronegativity.** The size of an atom or ion *increases* down a column of the periodic table, so Br^- is much larger than F^-, and this stabilizes the negative charge.

F⁻

smaller anion
less stable conjugate base

- **Positive or negative charge is stabilized when it is spread over a larger volume.**

Because Br^- is larger than F^-, Br^- is more stable than F^-, and H–Br is a stronger acid than H–F.

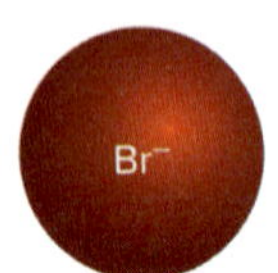

larger anion
more stable conjugate base

H–F	H–Br
$pK_a = 3.2$	$pK_a = -9$
less acidic	**more acidic**

This again is a specific example of a general trend.

- **Down a column of the periodic table, the acidity of HA *increases* as the size of A increases.**

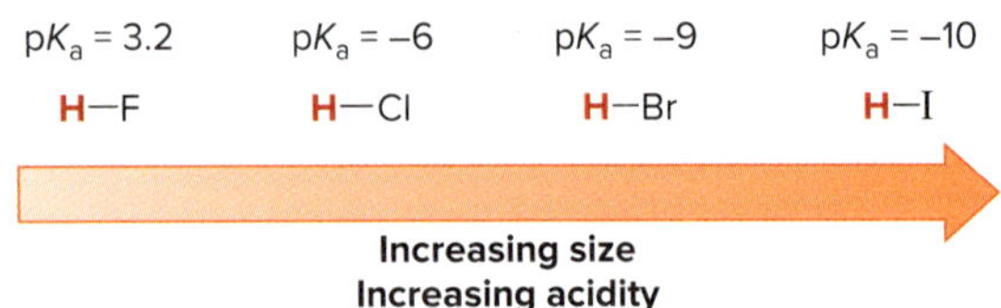

Because of carbon's position in the periodic table (in the second row and to the left of O, N, and the halogens), **C–H bonds are usually the *least acidic* bonds in a molecule.**

This is ***opposite*** to what would be expected on the basis of electronegativity differences between F and Br, because F is more electronegative than Br. **Size and *not* electronegativity determines acidity down a column.** Combining both trends together:

- **The acidity of HA *increases* both left-to-right across a row and down a column of the periodic table.**

Sample Problem 2.6 Using the Identity of X in HX to Determine Relative Acidity

Without reference to a pK_a table, decide which compound in each pair is the stronger acid:

a. H_2O or HF b. H_2S or H_2O

Solution

a. H_2O and HF both have H atoms bonded to a second-row element. Because the acidity of HA *increases across a row* of the periodic table, the H–F bond is more acidic than the H–O bond. **HF is a stronger acid than H_2O.**

b. H_2O and H_2S both have H atoms bonded to elements in the same column. Because the acidity of HA *increases down a column* of the periodic table, the H–S bond is more acidic than the H–O bond. **H_2S is a stronger acid than H_2O.**

Problem 2.18 Without reference to a pK_a table, decide which compound in each pair is the stronger acid.

a. [zigzag alkane structure] or H_2O b. [cyclopentane structure] or H_2S

More Practice: Try Problems 2.43a, 2.44a, 2.61, 2.64a, 2.67, 2.76a.

Problem 2.19 Rank the labeled H atoms in the following compound in order of increasing acidity.

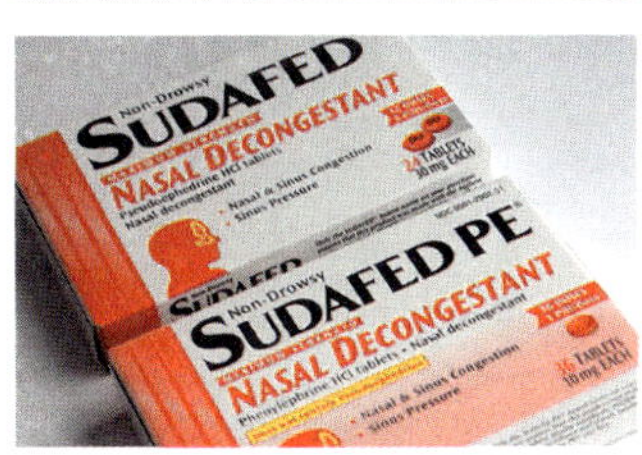

Because the pseudoephedrine (Problem 2.21) in Sudafed can be readily converted to the illegal, addictive drug methamphetamine, products that contain pseudoephedrine are now stocked behind the pharmacy counter so that their sale can be more closely monitored. Sudafed PE is a related product that contains a decongestant less easily converted to methamphetamine.
Jill Braaten/McGraw Hill

When discussing acidity, the most acidic proton in a compound is the one removed first by a base. Although four factors determine the overall acidity of a particular hydrogen atom, **the element effect—the identity of A—is the single most important factor in determining the acidity of the HA bond.**

To decide which hydrogen is most acidic, **first determine what element each hydrogen is bonded to and then decide its acidity based on periodic trends.** For example, $CH_3NHCH_2CH_2CH_2CH_3$ contains only C–H and N–H bonds. Because the acidity of HA increases across a row of the periodic table, the single H on N is the most acidic H in this compound.

most acidic H shown in red

Problem 2.20 Which hydrogen in each molecule is most acidic?

Problem 2.21 Which hydrogen in pseudoephedrine, the nasal decongestant in the commercial medication Sudafed, is most acidic?

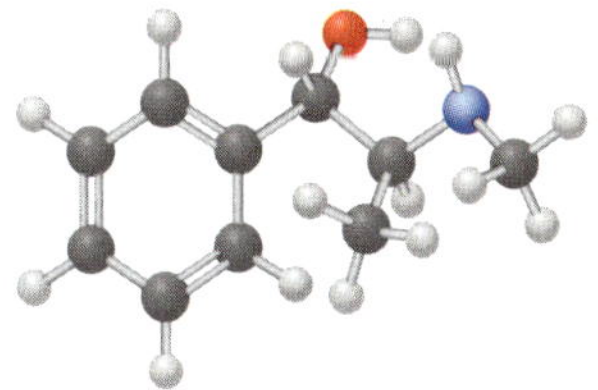

pseudoephedrine

Problem 2.22 Rank the compounds in each group in order of increasing acidity.

a.

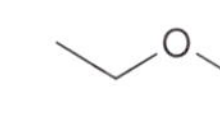

b.

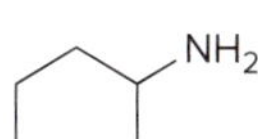

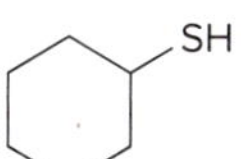

2.5B Inductive Effects

A second factor affecting the acidity of HA is the presence of atoms more electronegative than carbon. To illustrate this phenomenon, compare ethanol (CH_3CH_2OH) and 2,2,2-trifluoroethanol (CF_3CH_2OH), two compounds containing O–H bonds. The pK_a table in Appendix C indicates that CF_3CH_2OH is a stronger acid than CH_3CH_2OH. We are comparing the acidity of the O–H bond in both compounds, so what causes the difference?

ethanol
pK_a = 16

2,2,2-trifluoroethanol
pK_a = 12.4
stronger acid

Draw both conjugate bases and then determine which is more stable. Both bases have a negative charge on an electronegative oxygen, but the second anion has three very electronegative fluorine atoms. These fluorine atoms withdraw electron density from the carbon to which they are bonded, making it electron deficient. Furthermore, this electron-deficient carbon pulls electron density through σ bonds from the negatively charged oxygen atom, stabilizing the negative charge. This is called an **inductive effect.**

No additional electronegative atoms stabilize the conjugate base.

CF_3 withdraws electron density, stabilizing the conjugate base.

- **An *inductive effect* is the pull of electron density through σ bonds caused by electronegativity differences of atoms.**

In this case, the electron density is pulled away from the negative charge through σ bonds by the very electronegative fluorine atoms, so it is called an **electron-*withdrawing* inductive effect. Thus, the three very electronegative fluorine atoms stabilize the negatively charged conjugate base $CF_3CH_2O^-$, making CF_3CH_2OH a stronger acid than CH_3CH_2OH.** We have learned two important principles from this discussion:

- **More electronegative atoms stabilize regions of high electron density by an *electron-withdrawing* inductive effect.**
- **The acidity of HA *increases* with the presence of electron-withdrawing groups in A.**

Inductive effects result because an electronegative atom stabilizes the negative charge of the conjugate base. **The *more electronegative* the atom and the *closer* it is to the site of the negative charge, the greater the effect.** This effect is discussed in greater detail in Chapter 19.

Problem 2.23 Which compound in each pair is the stronger acid?

a. or

b. or

c. or

d. CH_3O ... NH_2 or ... NH_2

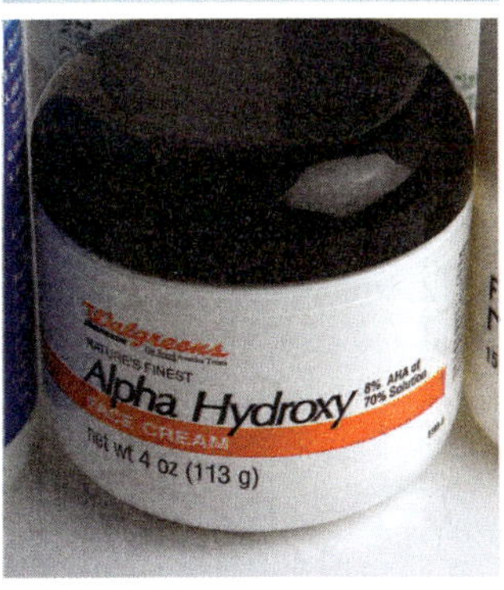

α-Hydroxy acids (Problem 2.24) are used in skin care products that purportedly smooth fine lines and improve skin texture by reacting with the outer layer of skin cells, causing them to loosen and flake off.
Jill Braaten/McGraw Hill

Problem 2.24 Glycolic acid, $HOCH_2CO_2H$, is the simplest member of a group of compounds called α-hydroxy acids, ingredients in skin care products that have an OH group on the carbon adjacent to a CO_2H group. Would you expect $HOCH_2CO_2H$ to be a stronger or weaker acid than acetic acid, CH_3CO_2H?

Problem 2.25 Explain the apparent paradox: HBr is a stronger acid than HCl, but HOCl is a stronger acid than HOBr.

2.5C Resonance Effects

A third factor that determines acidity is resonance. Recall from Section 1.6 that resonance occurs whenever two or more different Lewis structures can be drawn for the same arrangement of atoms. To illustrate this phenomenon, compare ethanol (CH_3CH_2OH) and acetic acid (CH_3CO_2H), two compounds containing O–H bonds. Based on Table 2.1, CH_3CO_2H is a stronger acid than CH_3CH_2OH.

Resonance structures are two Lewis structures having the same placement of atoms but a different arrangement of electrons.

ethanol
$pK_a = 16$

acetic acid
$pK_a = 4.8$
stronger acid

Draw the conjugate bases of these acids to illustrate the importance of resonance. For ethoxide ($CH_3CH_2O^-$), the conjugate base of ethanol, only one Lewis structure can be drawn. The negative charge of this conjugate base is *localized* on the O atom.

ethanol
acid

ethoxide
conjugate base
only **one** Lewis structure

With acetate ($CH_3CO_2^-$), however, two resonance structures can be drawn.

acetic acid

acetate
conjugate base
two resonance structures

hybrid
resonance-stabilized conjugate base

Resonance delocalization often produces a larger effect on pK_a than the inductive effects discussed in Section 2.5B. Resonance makes CH_3CO_2H ($pK_a = 4.8$) a much stronger acid than CH_3CH_2OH ($pK_a = 16$), whereas the inductive effects due to three electronegative F atoms make CF_3CH_2OH ($pK_a = 12.4$) a somewhat stronger acid than CH_3CH_2OH.

These two resonance structures differ in the **position of a π bond** and a **lone pair.** Although each resonance structure of acetate implies that the negative charge is localized on an O atom, in actuality, charge is ***delocalized*** over both O atoms. **Delocalization of electron density stabilizes acetate, making it a weaker base.**

Remember that neither resonance form adequately represents acetate. The true structure is a **hybrid** of both structures. In the hybrid, the electron pairs drawn in different locations in individual resonance structures are ***delocalized.*** With acetate, a dashed line is used to show that each C–O bond has partial double bond character. The symbol δ– (partial negative) indicates that the charge is delocalized on both O atoms in the hybrid.

Thus, **resonance delocalization makes $CH_3CO_2^-$ more stable than $CH_3CH_2O^-$, so CH_3CO_2H is a stronger acid than CH_3CH_2OH.** This is another example of a general rule.

- **The acidity of HA *increases* when the conjugate base A:⁻ is resonance stabilized.**

Electrostatic potential plots of $CH_3CH_2O^-$ and $CH_3CO_2^-$ in Figure 2.3 indicate that the negative charge is concentrated on a single O in $CH_3CH_2O^-$, but delocalized over the O atoms in $CH_3CO_2^-$.

Figure 2.3 Electrostatic potential plots of $CH_3CH_2O^-$ and $CH_3CO_2^-$

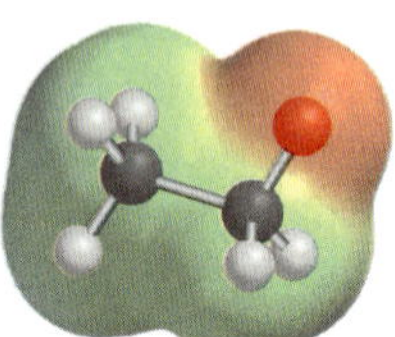

$CH_3CH_2O^-$
The negative charge is concentrated on the single oxygen atom, making this anion *less stable*.

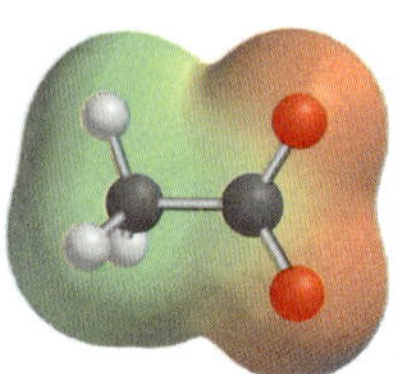

$CH_3CO_2^-$
The negative charge is delocalized over both oxygen atoms, making this anion *more stable*.

Problem 2.26 The C–H bond in acetone, $(CH_3)_2C{=}O$, has a pK_a of 19.2. Draw two resonance structures for its conjugate base. Then, explain why acetone is much more acidic than propane, $CH_3CH_2CH_3$ (pK_a = 50).

Problem 2.27 Rank the labeled protons in the following molecule in order of increasing pK_a.

H_aO, H_b, OH_c

2.5D Hybridization Effects

The final factor affecting the acidity of HA is the hybridization of A. To illustrate this phenomenon, compare ethane (CH_3CH_3), ethylene ($CH_2{=}CH_2$), and acetylene ($HC{\equiv}CH$). Appendix C indicates that there is a considerable difference in the pK_a values of these compounds.

ethane	ethylene	acetylene
pK_a = 50	pK_a = 44	pK_a = 25

Increasing acidity →

The conjugate bases formed by removing a proton from ethane, ethylene, and acetylene are **carbanions—species with a negative charge on carbon.**

sp^3 hybridized C	sp^2 hybridized C	sp hybridized C
25% s-character	**33%** s-character	**50%** s-character

Increasing percent s-character
Increasing stability →

The hybridization of the carbon bearing the negative charge is different in each anion, so the lone pair of electrons occupies an orbital with a different percent *s*-character in each case. A higher percent *s*-character means a hybrid orbital has a larger fraction of the lower-energy *s* orbital.

- **The *higher* the percent *s*-character of the hybrid orbital, the more stable the conjugate base.**

Thus, **acidity increases from CH_3CH_3 to $CH_2{=}CH_2$ to $HC{\equiv}CH$ as the negative charge of the conjugate base is stabilized by increasing percent *s*-character.** Once again this is a specific example of a general trend.

- **The acidity of HA *increases* as the percent *s*-character of A:⁻ increases.**

Problem 2.28 For each pair of compounds: [1] Which indicated H is more acidic? [2] Draw the conjugate base of each acid. [3] Which conjugate base is stronger?

a.  (cyclopentyl)–C≡C–**H** or (cyclopentyl)–CH=CH–C(H)(**H**)H   b. 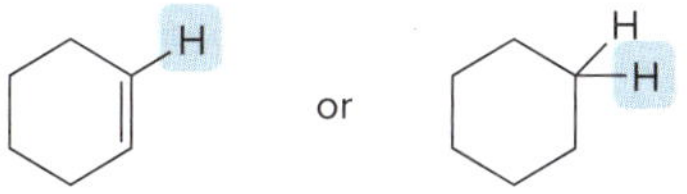 (cyclohexene with CH–**H**) or (cyclohexane with CH(H)–**H**)

Problem 2.29 Rank the following ions in order of increasing basicity.

a. $CH_3CH_2^-$, CH_3O^-, CH_3NH^-

b. CH_3^-, HO^-, Br^-

c. 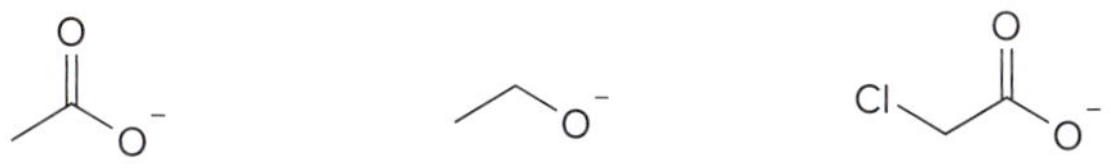 CH_3COO^- (acetate), $CH_3CH_2O^-$, $ClCH_2COO^-$

d. 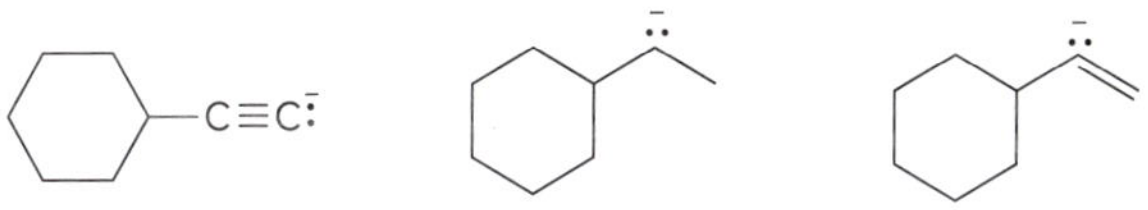 (cyclohexyl)–C≡C:⁻, (cyclohexyl)–CH₂–C̈H⁻–CH₃, (cyclohexyl)–CH₂–C̈⁻=CH₂

2.5E Summary of Factors Determining Acid Strength

The ability to recognize the most acidic site in a molecule will be important throughout the study of organic chemistry. All the factors that determine acidity are therefore summarized in Figure 2.4. The following two-step procedure shows how these four factors can be used to determine the relative acidity of protons.

Figure 2.4 Summary of the factors that determine acidity

Factor	Example		
1. **Element effects:** The acidity of HA increases both left-to-right across a row and down a column of the periodic table.	CH_4	and	H_2O **more acidic**
2. **Inductive effects:** The acidity of HA increases with the presence of electron-withdrawing groups in A.	$CH_3CH_2O{-}H$	and	$CF_3CH_2O{-}H$ **more acidic**
3. **Resonance effects:** The acidity of HA increases when the conjugate base A:⁻ is resonance stabilized.	$CH_3CH_2O{-}H$	and	$CH_3CO_2{-}H$ **more acidic**
4. **Hybridization effects:** The acidity of HA increases as the percent *s*-character of A:⁻ increases.	$CH_2{=}CH_2$	and	$H{-}C{\equiv}C{-}H$ **more acidic**

How To Determine the Relative Acidity of Protons

Step [1] **Identify the atoms bonded to hydrogen, and use periodic trends to assign relative acidity.**

- The most common HA bonds in organic compounds are C–H, N–H, and O–H. Because acidity increases left-to-right across a row, the relative acidity of these bonds is **C–H < N–H < O–H.** Therefore, H atoms bonded to C atoms are usually *less acidic* than H atoms bonded to any heteroatom.

Step [2] **If the two H atoms in question are bonded to the same element, draw the conjugate bases and look for other points of difference. Ask three questions:**

- Do electron-withdrawing groups stabilize the conjugate base?
- Is the conjugate base resonance stabilized?
- How is the conjugate base hybridized?

Sample Problem 2.7 shows how to apply this procedure to actual compounds.

Sample Problem 2.7 Determining the Relative Acidity of Compounds

Rank the following compounds in order of increasing acidity of their most acidic hydrogen atom.

OH, Cl — **A** | OH — **B** | NH_2 — **C**

Solution

[1] Compounds **A, B,** and **C** contain C–H, N–H, and O–H bonds. Because acidity increases left-to-right across a row of the periodic table, the **O–H bonds are most acidic.** Compound **C** is thus the least acidic because it has *no* O–H bonds.

[2] The only difference between compounds **A** and **B** is the presence of an electronegative Cl in **A.** The Cl atom stabilizes the conjugate base of **A,** making it more acidic than **B.** Thus,

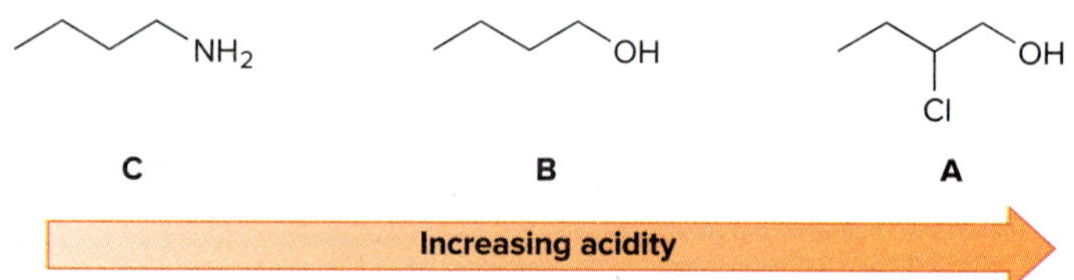

Problem 2.30 Rank the compounds in each group in order of increasing acidity.

a. (alkane) | OH | NH_2

b. Br, O, OH | O, OH | OH

c. NH_2 | N | OH

More Practice: Try Problems 2.57, 2.58, 2.78a.

Problem 2.31 Rank the following compounds in order of increasing acidity.

a. OH | NH_2 | (alkane)

b. Cl | OH | SH

c. (alkane) | Cl, OH | OH

d. (alkyne) | (alkane) | (alkene)

Problem 2.32 Which anion (**A** or **B**) is the stronger base?

A B

Sample Problem 2.8 Determining the Most Acidic Proton in a More Complex Molecule

Which proton in quinic acid, the chapter-opening compound present in a brewed cup of coffee, is most acidic?

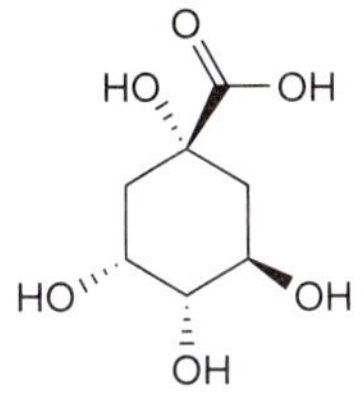

quinic acid

Solution

Because acidity increases left-to-right across a row of the periodic table, **concentrate on the O–H bonds,** which are more acidic than the C–H bonds. By drawing the conjugate bases formed by removal of the O–H protons, we can see that the OH groups bonded to the six-membered ring are different from the OH group bonded to the C=O.

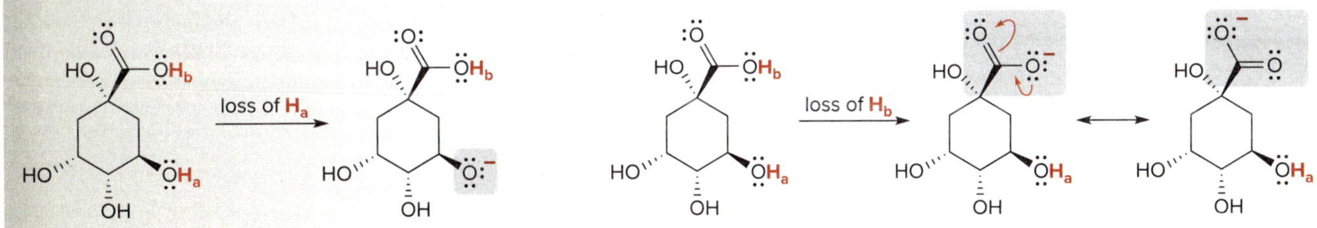

loss of H_a — conjugate base, only **one** Lewis structure

loss of H_b — **two** resonance structures, resonance-stabilized conjugate base

Removal of H_a (or any of the protons on the OH groups bonded to the six-membered ring) forms a conjugate base for which only *one* Lewis structure can be drawn, so the negative charge is *localized* on one O atom. Removal of H_b forms a conjugate base for which *two* resonance structures can be drawn, so the negative charge is *delocalized* on two O atoms. Delocalization makes this conjugate base more stable, so **H_b is the most acidic proton in quinic acid.**

Problem 2.33 Which proton in each of the following drugs is most acidic? THC is the active component in marijuana, ketoprofen is an anti-inflammatory agent, and chloroquine is an antimalarial drug.

a. 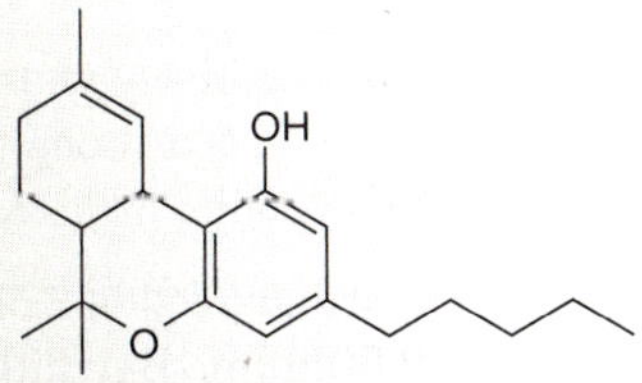 THC tetrahydrocannabinol

b. 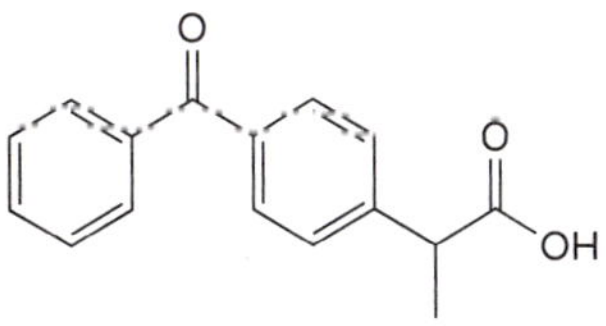 ketoprofen

c. 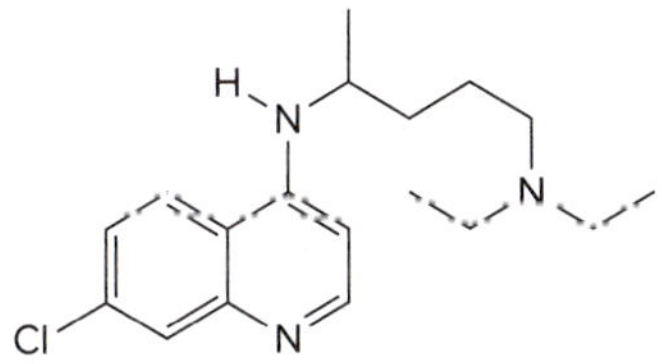 chloroquine

More Practice: Try Problems 2.43a, 2.44a, 2.50, 2.51, 2.64a, 2.67, 2.76a, 2.78a.

2.6 Common Acids and Bases

Sulfuric acid is formed when sulfur oxides, emitted into the atmosphere by burning fossil fuels high in sulfur content, dissolve in water. This makes rainwater acidic, forming acid rain, which has destroyed acres of forests worldwide.
Mary Terriberry/Shutterstock

Many strong or moderately strong acids and bases are used as reagents in organic reactions.

2.6A Common Acids

Several organic reactions are carried out in the presence of strong inorganic acids, most commonly **HCl** and **H_2SO_4.** These strong acids, with **pK_a values ≤ 0,** should be familiar from previous chemistry courses.

Two organic acids are also commonly used, namely **acetic acid** and ***p*-toluenesulfonic acid** (usually abbreviated as **TsOH**). Although acetic acid has a higher pK_a than the inorganic acids, making it a weaker acid, it is more acidic than most organic compounds. *p*-Toluenesulfonic acid is similar in acidity to the strong inorganic acids. Because it is a solid, small quantities can be easily weighed on a balance and then added to a reaction mixture.

acetic acid
pK_a = 4.8

p-toluenesulfonic acid = TsOH
pK_a = −1.3

2.6B Common Bases

Three common kinds of strong bases include:

[1] Negatively charged oxygen bases: **$^-$OH** (hydroxide) and its organic derivatives
[2] Negatively charged nitrogen bases: **$^-NH_2$** (amide) and its organic derivatives
[3] Hydride **(H^-)**

Figure 2.5 gives examples of these strong bases. Each negatively charged base is used as a salt with a spectator ion (usually Li^+, Na^+, or K^+) that serves to balance charge.

Figure 2.5 Some common negatively charged bases

Oxygen bases

Na^+	$^-$:ÖH	sodium hydroxide
Na^+	$^-$:ÖCH$_3$	sodium methoxide
Na^+	$^-$:ÖCH$_2$CH$_3$	sodium ethoxide
K^+	$^-$:ÖC(CH$_3$)$_3$	potassium *tert*-butoxide

Nitrogen bases

Na^+	$^-$:NH$_2$	sodium amide
Li^+	$^-$:N[CH(CH$_3$)$_2$]$_2$	lithium diisopropylamide

Hydride

Na^+	H:$^-$	sodium hydride

- Strong bases have weak conjugate acids with high pK_a values, usually > 12.

Strong bases have a net negative charge, but not all negatively charged species are strong bases. For example, none of the halides, F^-, Cl^-, Br^-, or I^-, is a strong base. These anions have very strong conjugate acids and have little affinity for donating their electron pairs to a proton.

Carbanions, negatively charged carbon atoms discussed in Section 2.5D, are especially strong bases. Perhaps the most common example is **butyllithium.** Butyllithium and related compounds are discussed in greater detail in Chapter 17.

$CH_3CH_2CH_2\ddot{C}H_2^-$ Li^+
butyllithium

Two other weaker organic bases are **triethylamine** and **pyridine.** These compounds have a lone pair on nitrogen, making them basic, but they are considerably weaker than the amide bases because they are neutral, not negatively charged.

Problem 2.34 Draw the products formed when propan-2-ol [$(CH_3)_2CHOH$], the main ingredient in rubbing alcohol, is treated with each acid or base: (a) NaH; (b) H_2SO_4; (c) $Li^{+-}N[CH(CH_3)_2]_2$; (d) CH_3CO_2H.

2.7 Aspirin

Aspirin is one of the most widely used over-the-counter drugs. Whether you purchase Anacin, Bufferin, Bayer; or a generic, the active ingredient is the same—**acetylsalicylic acid.**

Aspirin is the most well known member of a group of compounds called **salicylates.** Although aspirin was first used in medicine for its analgesic (pain-relieving), antipyretic (fever-reducing), and anti-inflammatory properties, today it is commonly used as an antiplatelet agent in the treatment and prevention of heart attacks and strokes. **Aspirin is a synthetic compound;** it does not occur in nature, although some related salicylates are found in willow bark and meadowsweet blossoms (Figure 2.6).

Figure 2.6 Salicin, an analgesic in willow bark

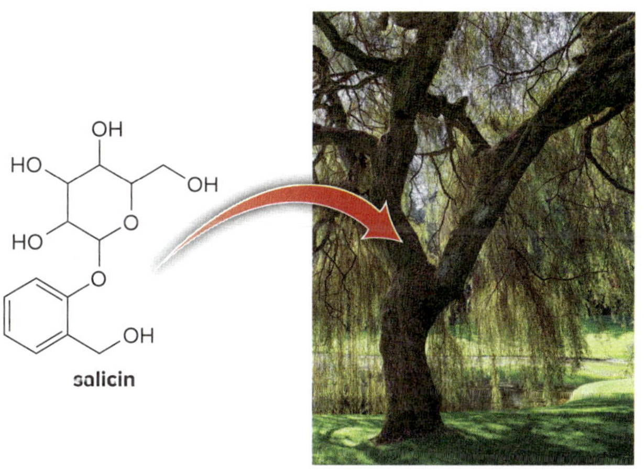

willow tree

july7th/iStock/Getty Images

- The modern history of aspirin dates back to 1763 when Reverend Edmund Stone reported on the analgesic effect of chewing on the bark of the willow tree. Willow bark is now known to contain *salicin,* which is structurally related to aspirin.

Like many drugs, aspirin undergoes a proton transfer reaction. Its most acidic proton is the H bonded to O, and in the presence of base, this H is readily removed.

:B + H—B$^+$

aspirin
acetylsalicylic acid
neutral form

This form exists in the **stomach.**

conjugate base
ionic form

This form exists in the **intestine.**

Why is this acid–base reaction important? After ingestion, aspirin first travels into the stomach and then the intestine. In the acidic environment of the stomach, aspirin remains in its neutral form, but in the basic environment of the small intestine, aspirin is deprotonated to form its conjugate base, an ion. Likewise, in the slightly basic environment of the blood, aspirin exists primarily as its ionic conjugate base.

We will learn more about solubility and the cell membrane in Section 3.7.

Whether aspirin is a neutral acid or an ionic conjugate base affects its transport throughout the body and its ability to pass through a cell membrane. In its ionic form, aspirin is readily soluble in the aqueous environment of the blood, so it is transported in the bloodstream to tissues. Once aspirin has reached its target location, however, its conjugate base must be re-protonated to form the neutral acid that can pass through the nonpolar interior of a cell membrane where it inhibits prostaglandin synthesis, as we will learn in Chapter 19. Thus, in the body, aspirin undergoes acid–base reactions and these reactions are crucial in determining its properties and action.

Problem 2.35

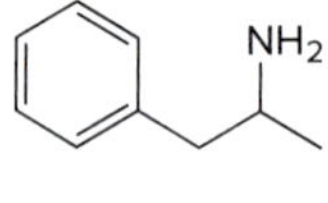

amphetamine

Compounds like amphetamine that contain nitrogen atoms are protonated by the HCl in the gastric juices of the stomach, and the resulting salt is then deprotonated in the basic environment of the intestine to regenerate the neutral form. Write proton transfer reactions for both of these processes. In which form will amphetamine pass through a cell membrane?

2.8 Lewis Acids and Bases

The Lewis definition of acids and bases is more general than the Brønsted–Lowry definition.

All Brønsted–Lowry bases are Lewis bases.

- **A Lewis acid is an *electron pair acceptor.***
- **A Lewis base is an *electron pair donor.***

Lewis bases are structurally the same as Brønsted–Lowry bases. Both have an **available electron pair**—a lone pair or an electron pair in a π bond. A Brønsted–Lowry base always donates this electron pair to a proton, but a Lewis base donates this electron pair to anything that is electron deficient. Simple Lewis bases are shown in Figure 2.7.

Figure 2.7 Simple Lewis acids and bases

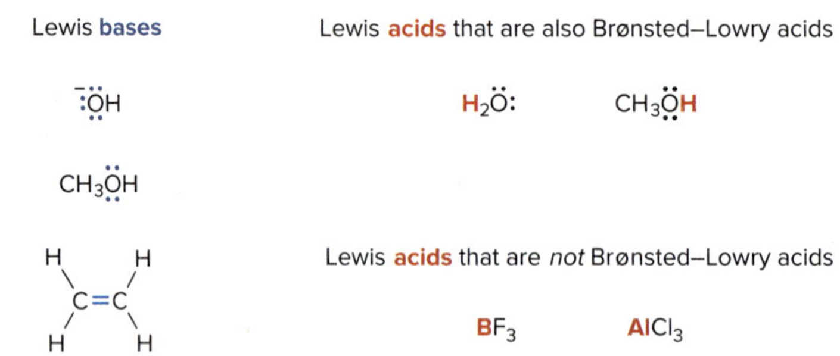

The vacant unhybridized *p* orbital extends above and below the plane of BF_3.

120° F B F F

sp^2 hybridized trigonal planar

A Lewis acid must be able to accept an electron pair, but there are many ways for this to occur. **All Brønsted–Lowry acids are also Lewis acids, but the reverse is not necessarily true.** Any species that is electron deficient and capable of accepting an electron pair is also a Lewis acid, as shown in Figure 2.7.

Common examples of Lewis acids (which are not Brønsted–Lowry acids) include $\mathbf{BF_3}$ and $\mathbf{AlCl_3}$. These compounds contain elements in group 3A of the periodic table that can accept an electron pair because they do not have filled valence shells of electrons. For example, BF_3 contains an sp^2 hybridized, trigonal planar B atom with a vacant unhybridized *p* orbital that can accept two electrons.

Problem 2.36 Which species are Lewis bases?

a. NH_3 b. $CH_3CH_2CH_3$ c. H^- d. H–C≡C–H

Problem 2.37 Which species are Lewis acids?

a. BBr_3 b. CH_3CH_2OH c. $(CH_3)_3C^+$ d. Br^-

Any reaction in which one species donates an electron pair to another species is a Lewis acid–base reaction.

In a Lewis acid–base reaction, a Lewis base donates an electron pair to a Lewis acid. Most reactions in organic chemistry involving movement of electron pairs can be classified as Lewis acid–base reactions. Lewis acid–base reactions illustrate a general pattern of reactivity.

- **Electron-rich species react with electron-poor species.**

In the simplest Lewis acid–base reaction, one bond is formed and no bonds are broken. This is illustrated with the reaction of BF_3 with H_2O. BF_3 has only six electrons around B, so it is the electron-deficient Lewis acid. H_2O has two lone pairs on O, so it is the electron-rich Lewis base.

Lewis **acid** + Lewis **base** → one new bond formed

H_2O donates an electron pair to BF_3 to form one new bond. The electron pair in the new B–O bond comes from the oxygen atom, and a single product, a Lewis acid–base complex, is formed. Both B and O bear formal charges in the product, but the overall product is neutral.

Nucleophile = nucleus loving.
Electrophile = electron loving.

- **A Lewis acid is called an *electrophile*.**
- **When a Lewis base reacts with an electrophile other than a proton, the Lewis base is called a *nucleophile*.**

In this Lewis acid–base reaction, **BF_3 is the electrophile** and **H_2O is the nucleophile.**

In a Lewis acid–base reaction, the **electron pair is not removed from the Lewis base;** instead, the electron pair is donated to an atom of the Lewis acid, and one new covalent bond is formed.

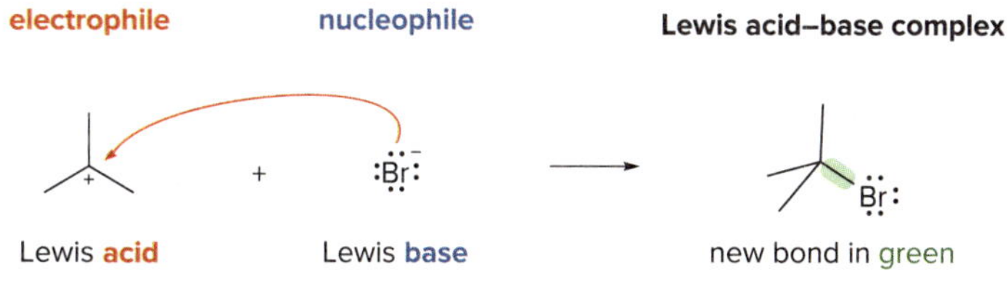

Problem 2.38 For each reaction, label the Lewis acid and base. Use curved arrow notation to show the movement of electron pairs.

a. BF_3 + CH_3OCH_3 → F_3B^-–$O^+(CH_3)_2$

b. $(CH_3)_2CH^+$ + $^-$:OH → $(CH_3)_2CHOH$

Problem 2.39 Draw the products of each reaction, and label the nucleophile and electrophile.

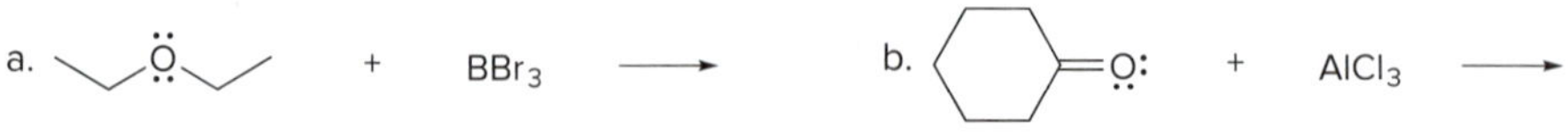

Problem 2.40 Draw the product formed when $(CH_3CH_2)_3N$:, a Lewis base, reacts with each Lewis acid: (a) $B(CH_3)_3$; (b) $(CH_3)_3C^+$; (c) $AlCl_3$.

In some Lewis acid–base reactions, one bond is formed and one bond is broken. To draw the products of these reactions, keep the following steps in mind.

> **[1] Always identify the Lewis acid and base first.**
>
> **[2] Draw a curved arrow from the electron pair of the base to the electron-deficient atom of the acid.**
>
> **[3] Count electron pairs and break a bond when needed to keep the correct number of valence electrons.**

For example, draw the Lewis acid–base reaction between cyclohexene and H—Cl. The Brønsted–Lowry acid HCl is also a Lewis acid, and cyclohexene, having a π bond, is the Lewis base.

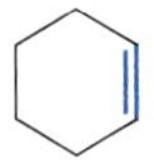

H—Cl
δ+ δ−

cyclohexene
Lewis **base**

Lewis **acid**

Recall from Section 1.6B that a positively charged carbon atom is called a **carbocation.**

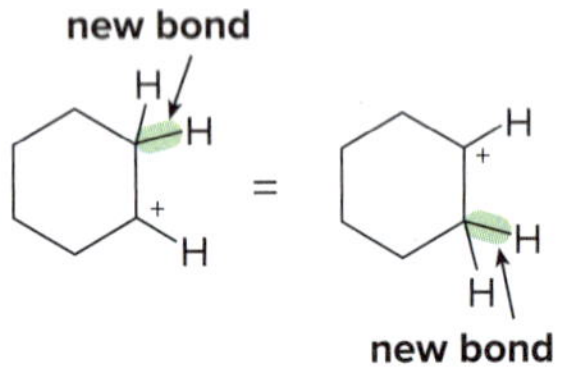

In the reaction of cyclohexene with HCl, the new bond to H could form at **either carbon of the double bond,** because the same carbocation results.

To draw the product of this reaction, the electron pair in the π bond of the Lewis base forms a new bond to the proton of the Lewis acid, forming a carbocation. The H—Cl bond must break, giving its two electrons to Cl, forming Cl^-. Because two electron pairs are involved, two curved arrows are needed.

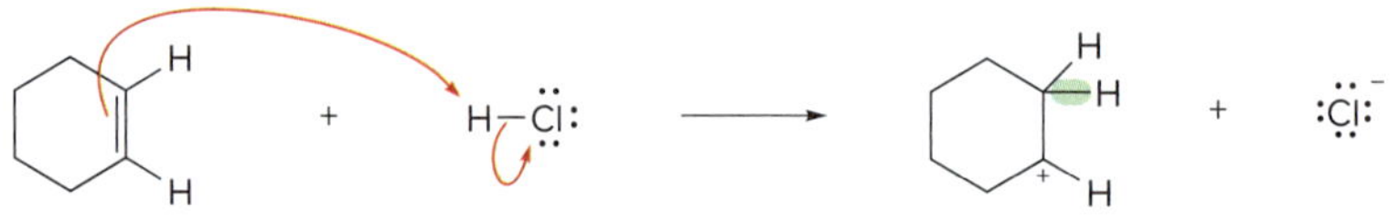

Lewis **base**

Lewis **acid**

carbocation
new bond in green

The Lewis acid–base reaction of cyclohexene with HCl is a specific example of a fundamental reaction of compounds containing C—C double bonds, as discussed in Chapter 10.

Problem 2.41 Label the Lewis acid and base. Use curved arrow notation to show the movement of electron pairs.

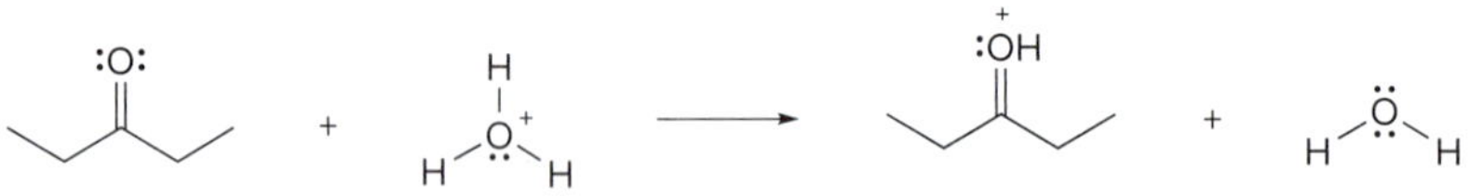

Problem 2.42 Draw the product formed when the Lewis acid $(CH_3CH_2)_3C^+$ reacts with each Lewis base: (a) CH_3OH; (b) $(CH_3)_2O$; (c) $(CH_3)_2NH$.

Chapter 2 REVIEW

KEY CONCEPTS

Brønsted–Lowry and Lewis Acids and Bases (2.1, 2.8)

1 Brønsted–Lowry acids

- A **Brønsted–Lowry acid** is a **proton donor.**

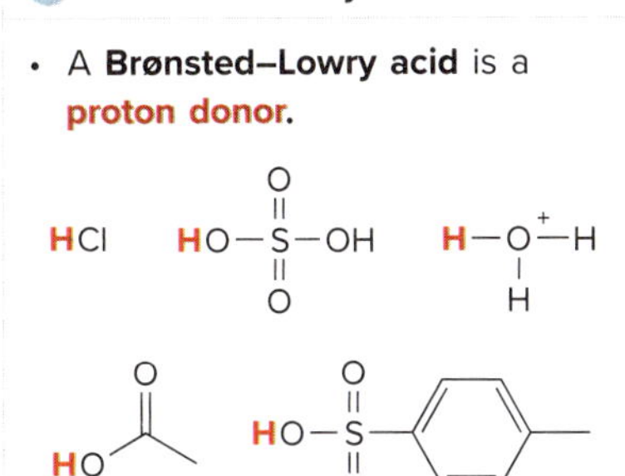

2 Brønsted–Lowry bases and Lewis bases

- A **Brønsted–Lowry base** is a **proton acceptor.**
- A **Lewis base** is an **electron pair donor.**

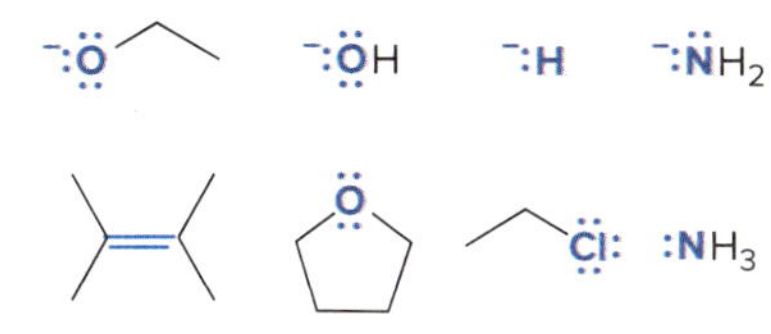

3 Lewis acids

- A **Lewis acid** is an **electron pair acceptor.**

$AlCl_3$ BF_3 $FeBr_3$

HNO_3 HCl HO

See Figures 2.1, 2.7. Try Problems 2.69, 2.70.

Acid–Base Reactions

1 Drawing the products of a Brønsted–Lowry acid–base reaction (2.2)

- A **Brønsted–Lowry acid donates a proton** to a **Brønsted–Lowry base.**

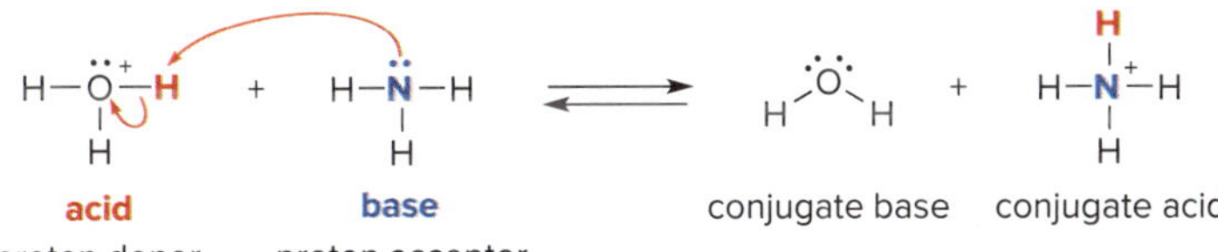

acid proton donor | **base** proton acceptor | conjugate base | conjugate acid

See Sample Problems 2.1, 2.2, Figure 2.2.
Try Problems 2.48, 2.49, 2.55, 2.56.

2 Drawing the products of a Lewis acid–base reaction (2.8)

- A **Lewis base donates an electron pair** to a **Lewis acid.**

Lewis **acid** **electrophile** | Lewis **base** **nucleophile**

- Electron-rich species react with electron poor ones.
- Nucleophiles react with electrophiles.

Try Problems 2.71, 2.72.

Periodic Trends (2.5A)

1 Trends in acidity **2 Trends in basicity**

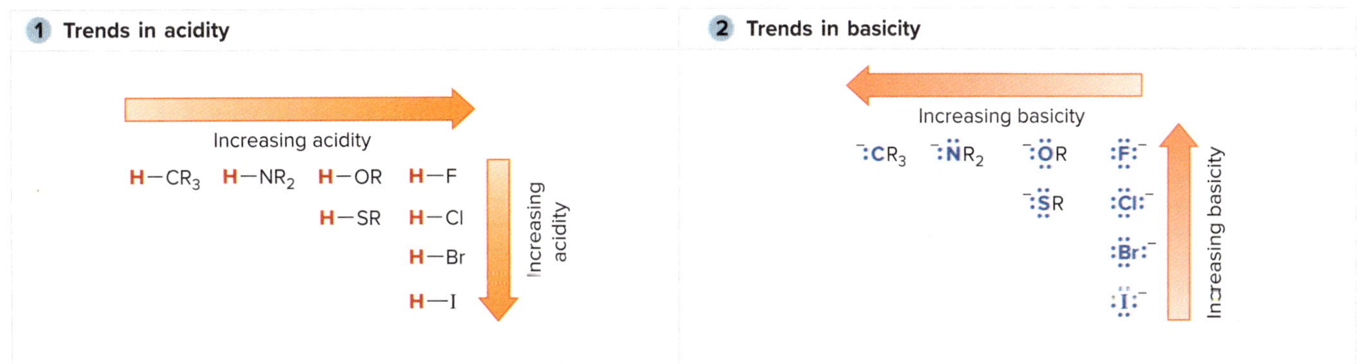

Acid and Base Strength and pK_a (2.3)

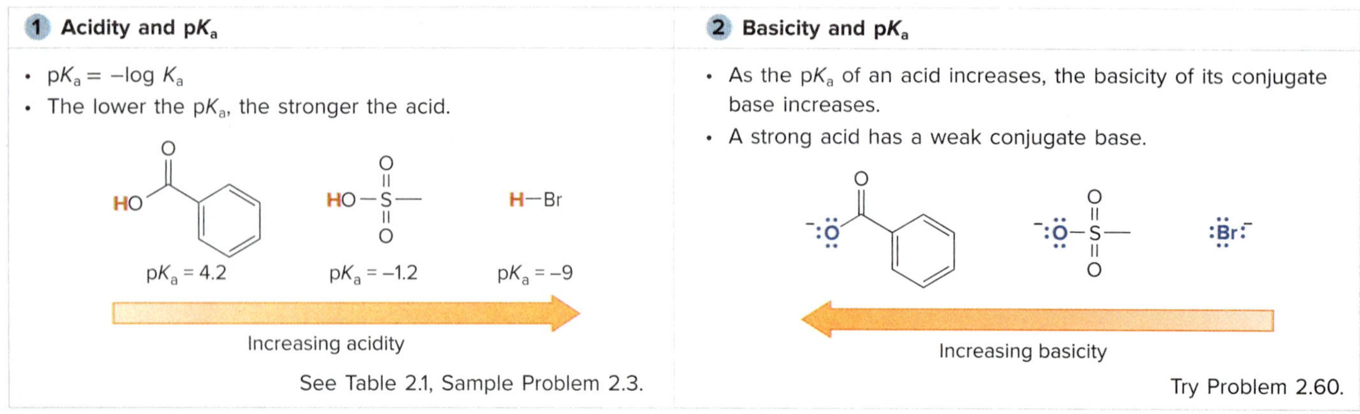

KEY SKILLS

[1] Drawing the products of a Brønsted–Lowry acid–base reaction (2.2)

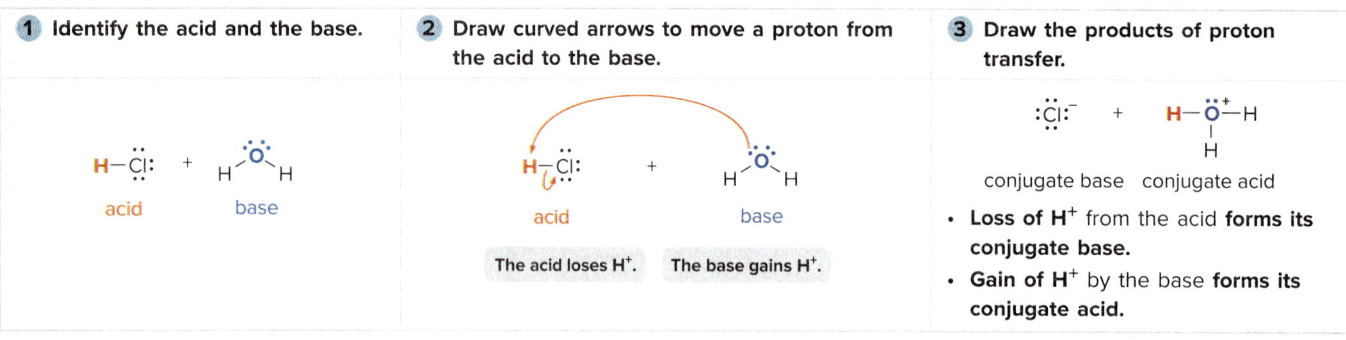

Try Problems 2.48–2.50.

[2] Determining the direction of equilibrium using pK_a (2.4)

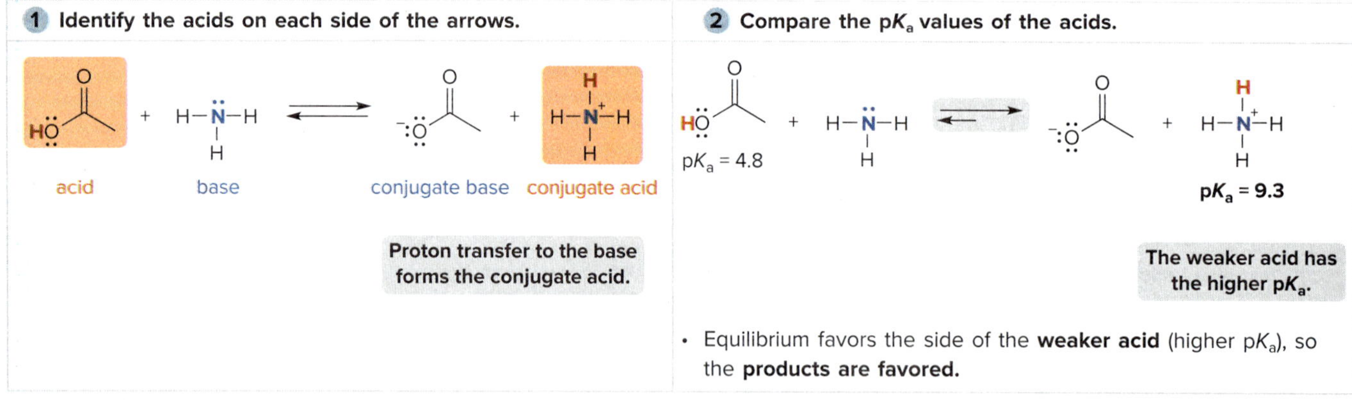

See Sample Problems 2.4, 2.5. Try Problems 2.55, 2.56.

[3] Determining acidity using trends in the periodic table (2.5A)

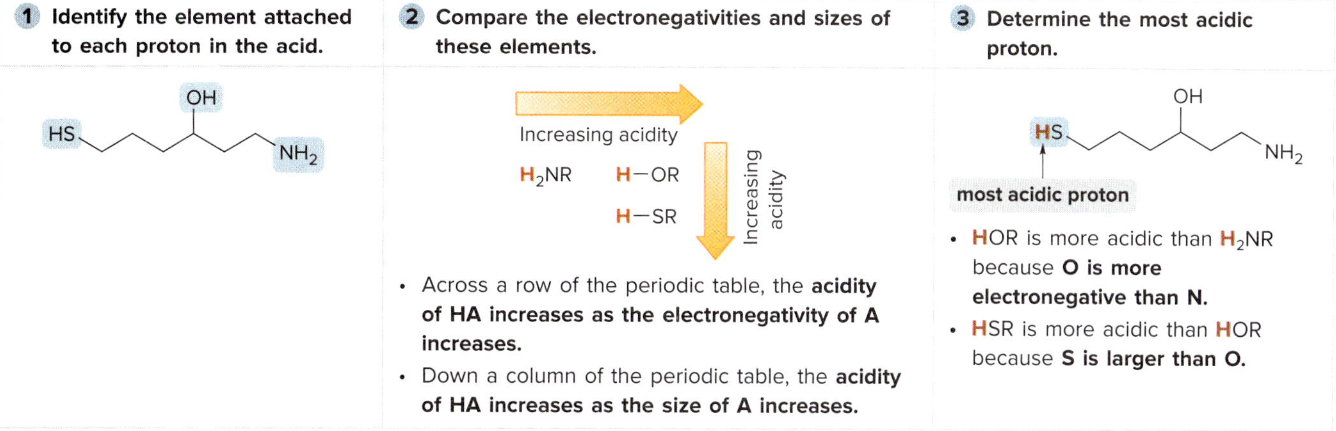

See Sample Problem 2.6. Try Problems 2.43a, 2.44a, 2.61, 2.64a, 2.67, 2.76a.

[4] Determining acidity using inductive effects (2.5B)

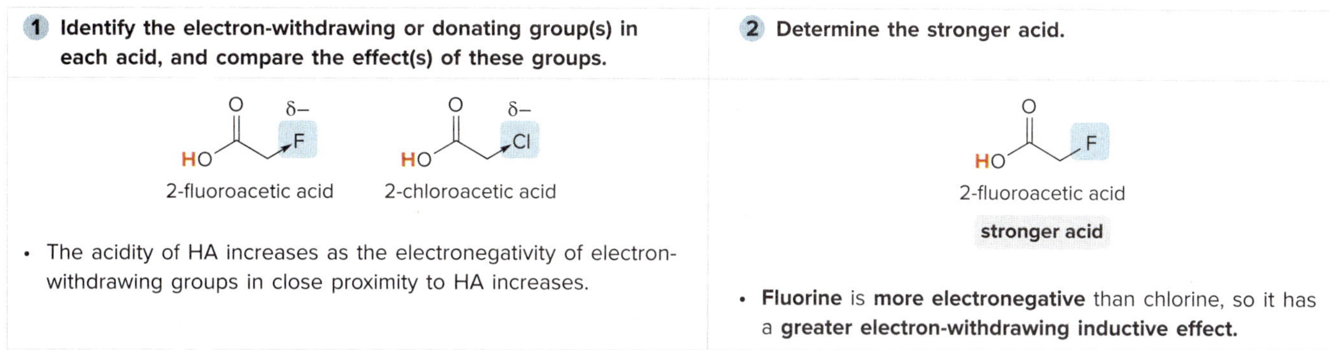

Try Problem 2.58.

[5] Determining acidity using resonance effects (2.5C)

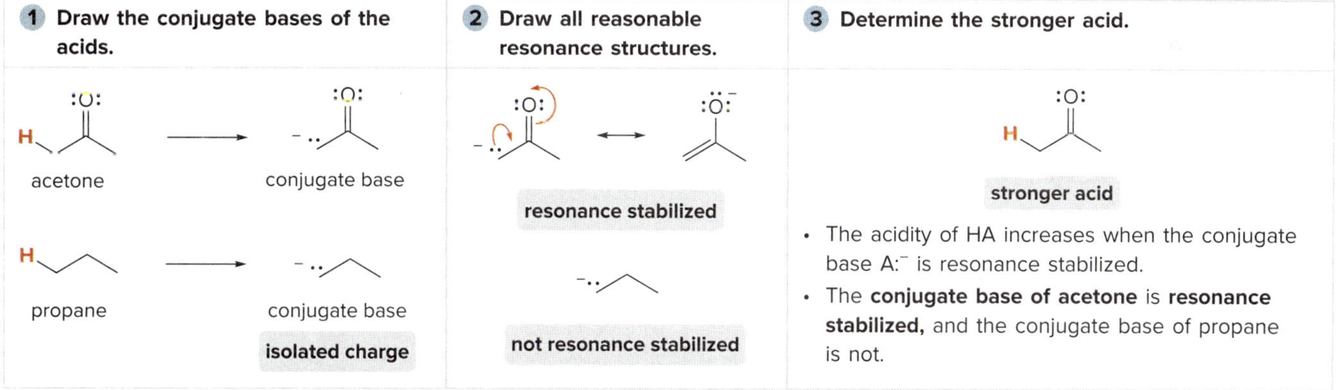

See Figure 2.3. Try Problems 2.60, 2.63, 2.66.

[6] Determining acidity using hybridization effects (2.5D)

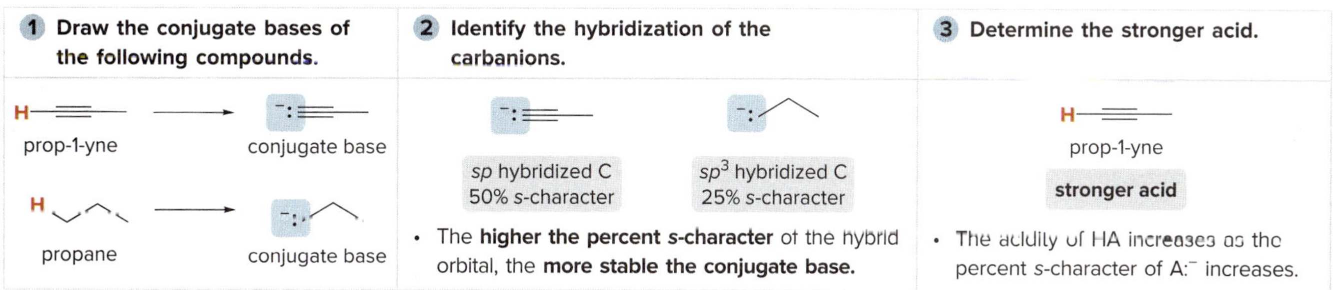

Try Problem 2.78a.

[7] Determining the most acidic proton (2.5E)

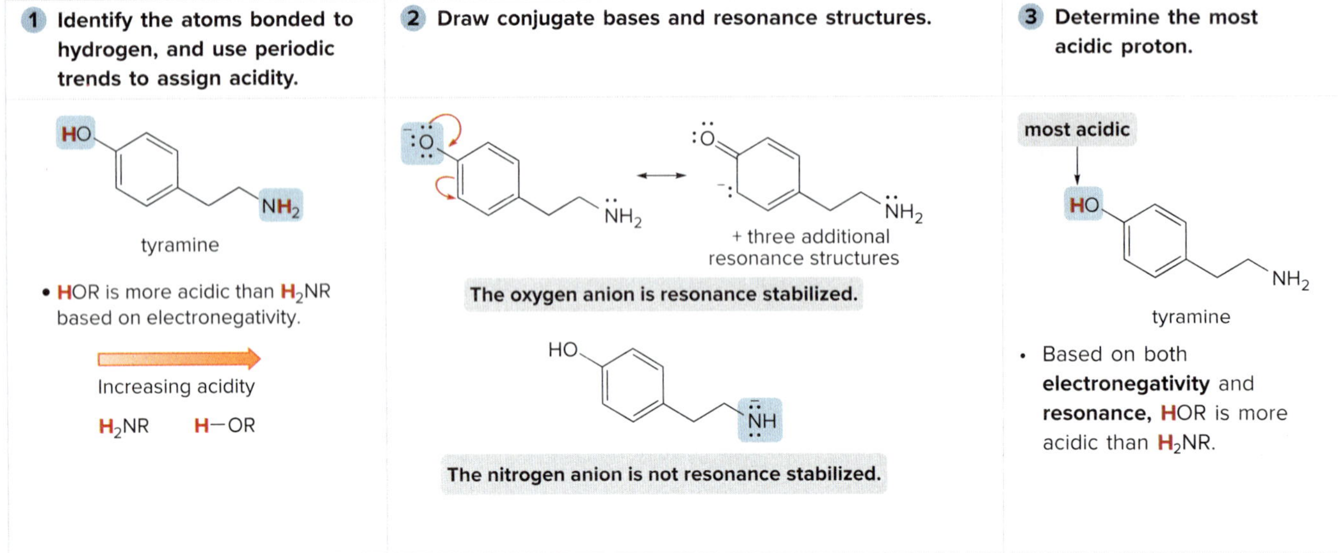

See *How To* (p. 74), Figure 2.6, Sample Problems 2.7, 2.8.
Try Problems 2.43a, 2.44a, 2.51, 2.57, 2.64a, 2.66, 2.67.

[8] Identifying the most basic electron pair (2.6B)

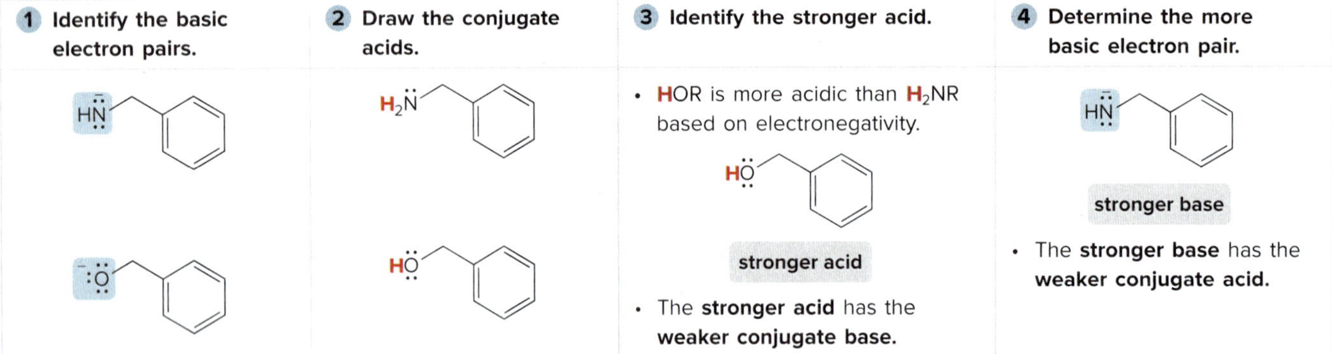

Try Problems 2.43c, 2.59, 2.64b, 2.76c, 2.78a.

CHAPTER 2 MULTIPLE-CHOICE SELF-TEST

The Self-Test consists of multiple-choice questions similar to those found on the American Chemical Society organic chemistry exam. Answers are given at the end of the chapter.

1. Which of the following compounds is *not* a Lewis base?

a. (cyclopentyl)–OCH_3 b. (cyclopentyl)–Cl c. (cyclopentyl)–NH_2 d. (cyclopentyl)–CH_3

2. Which of the following compounds has the lowest pK_a?

a. CH_3CONH_2 b. CH_3COOH c. $(CH_3)_3COH$ d. $(CH_3)_3CNH_2$

3. Which of the following anions is the strongest base: (a) CH_3O^-; (b) F^-; (c) $CH_3CH_2^-$; (d) CH_3NH^-?

4. Which statement describes the relationship between pK_a, acidity, and basicity?

a. A higher pK_a value means the acid is less acidic.

b. In an acid–base reaction, the equilibrium lies on the side of the acid with the lower pK_a.

c. If the pK_a of an acid is 10.2, it can be deprotonated by a base whose conjugate acid has a pK_a < 10.2.

d. A lower pK_a value for the acid means the conjugate base is more basic.

5. Rank the labeled protons in the given molecule in order of increasing acidity.

H_c, H_a, H_b, H_d

a. $H_b < H_d < H_a < H_c$

b. $H_d < H_b < H_a < H_c$

c. $H_d < H_b < H_c < H_a$

d. $H_a < H_d < H_b < H_c$

6. Which of the following species is amphoteric?

a. (cyclohexyl)–OH b. (cyclohexane) c. (cyclohexyl)–$\overset{+}{N}H_3$ d. $Cl_3C–CCl_3$

7. Which of the following anions is a strong enough base to deprotonate 1-butyne ($CH_3CH_2C{\equiv}CH$, pK_a = 25) so that equilibrium lies to the right: (a) NH_3; (b) CH_3^-; (c) ^-OH; (d) Cl^-?

8. Which statement describes a possible outcome of the reaction of $(CH_3)_3C^+$ with H_2O?

a. In a Lewis acid–base reaction, $(CH_3)_3COH$ and H^+ are formed.

b. In a Brønsted–Lowry acid–base reaction, the conjugate acid H_3O^+ is formed.

c. In a Lewis acid–base reaction, the conjugate base ^-OH is formed.

d. In a Brønsted–Lowry acid–base reaction, the conjugate base ^-OH is formed.

9. Which statement is true about the reaction of **A** with H_2SO_4?

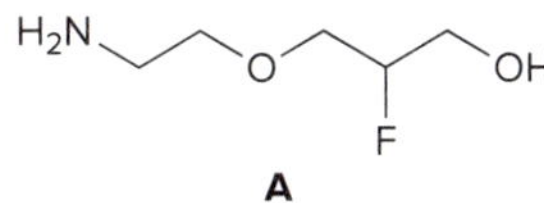

A

B **C** **D**

a. The conjugate base **F**$^-$ is formed.

b. The conjugate acid **B** is formed.

c. The conjugate acid **C** is formed.

d. The conjugate acid **D** is formed.

10. Label **A, B,** and **C** as a Lewis acid, a Brønsted–Lowry acid, both, or neither: **A** = BBr_3; **B** = CH_3CO_2H; **C** = CBr_4.

a. **A** Lewis acid, **B** both, **C** neither

b. **A** Lewis acid, **B** Brønsted–Lowry acid, **C** neither

c. **A** both, **B** Brønsted–Lowry acid, **C** neither

d. **A, B,** and **C** are all Lewis acids.

PROBLEMS

Problems Using Three-Dimensional Models

2.43 Propranolol is an antihypertensive agent—that is, it lowers blood pressure.

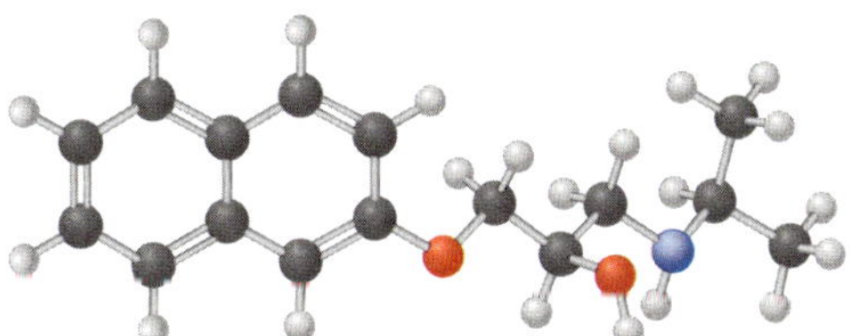
propranolol

a. Which proton in propranolol is most acidic?

b. What products are formed when propranolol is treated with NaH?

c. Which atom is most basic?

d What products are formed when propranolol is treated with HCl?

2.44 Amphetamine is a powerful stimulant of the central nervous system.

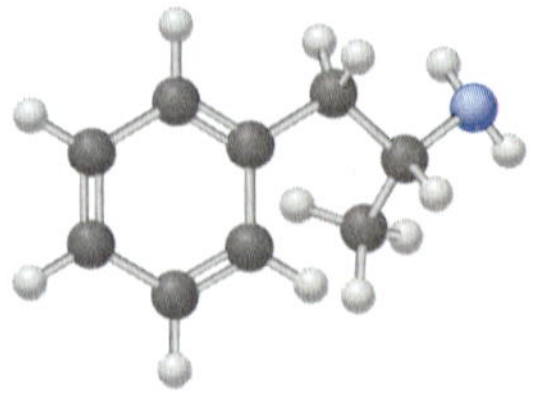
amphetamine

a. Which proton in amphetamine is most acidic?

b. What products are formed when amphetamine is treated with NaH?

c. What products are formed when amphetamine is treated with HCl?

Brønsted–Lowry Acids and Bases

2.45 What is the conjugate acid of each base?

a. HCO_3^- b. 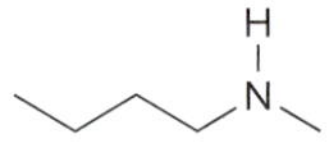 c. 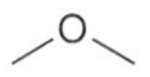 d.

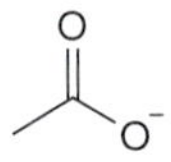

2.46 What is the conjugate base of each acid?

a. HCO_3^- b. 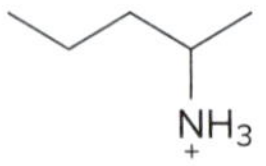 $\overset{+}{N}H_3$ c. 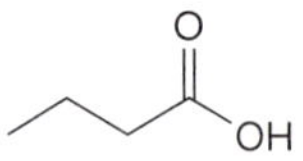 OH d.

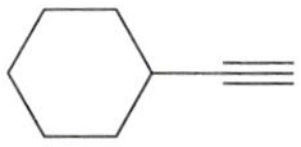

Reactions of Brønsted–Lowry Acids and Bases

2.47 As we will see in later chapters, many steps in key reaction sequences involve acid–base reactions. (a) Draw curved arrows to illustrate the flow of electrons in Steps [1]–[3]. (b) Identify the base and its conjugate acid in Step [1]. (c) Identify the acid and its conjugate base in Step [3].

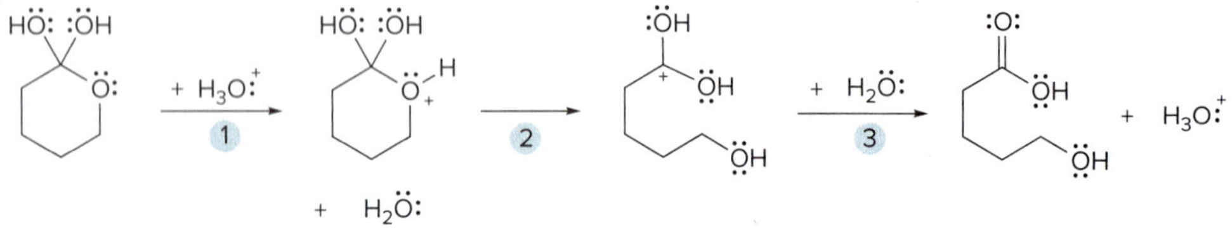

HÖ: :ÖH Ö: + H_3O^+ [1] → HÖ: :ÖH O^+–H + H_2O: [2] → :ÖH ÖH ÖH + H_2O: [3] → :O: ÖH ÖH + H_3O^+

2.48 Draw the products formed from the acid–base reaction of H_2SO_4 with each compound.

a. 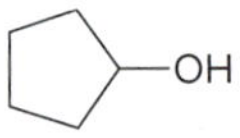 –OH b. 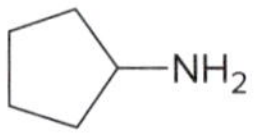 –NH_2 c. 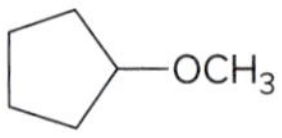 –OCH_3 d. 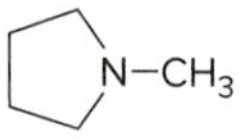 N–CH_3

2.49 Draw the products of each proton transfer reaction. Label the acid and base in the starting materials, and the conjugate acid and base in the products.

a. 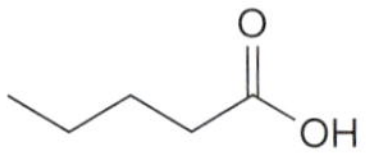 OH + $CH_3\ddot{O}:^-$ ⇌

b. 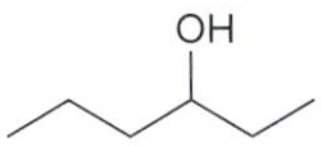 OH + HBr ⇌

c. 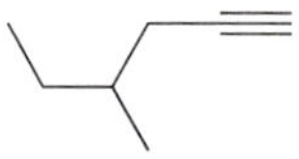 + $NaNH_2$ ⇌

d. 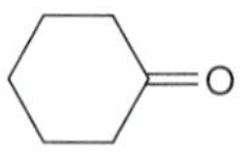 =O + H_2SO_4 ⇌

2.50 Draw the products of each acid–base reaction.

a. 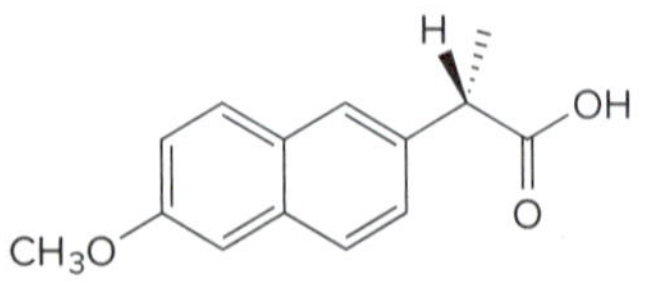 H OH O CH_3O + NaOH ⇌

naproxen
anti-inflammatory agent

b. 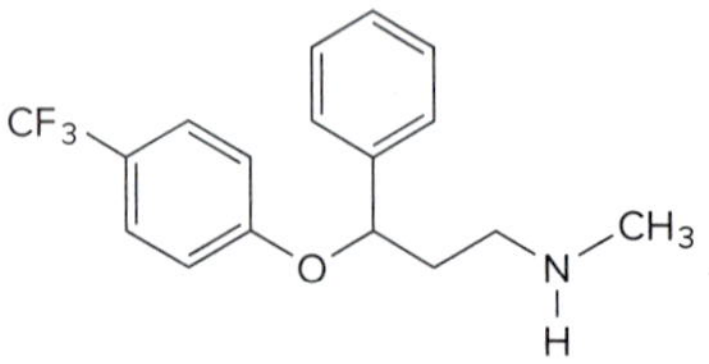 CF_3 O N–H CH_3 + HCl ⇌

fluoxetine
antidepressant

2.51 What product is formed when each compound is treated with NaH? Each of these acid–base reactions was a step in a synthesis of a commercially available drug.

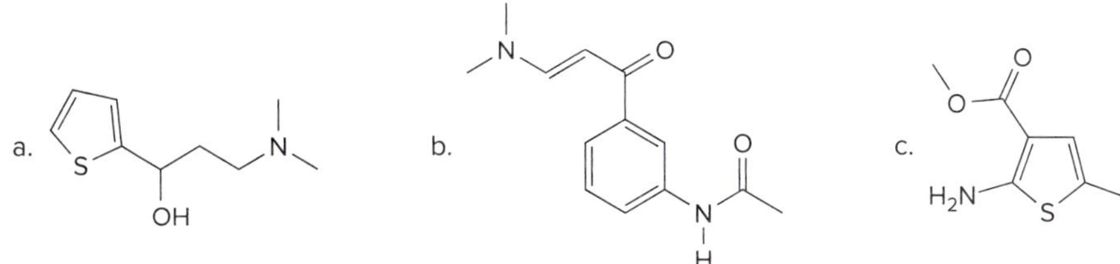

pK_a, K_a, and the Direction of Equilibrium

2.52 What is the K_a for each compound?

a. H_2S pK_a = 7.1

b. $ClCH_2COOH$ pK_a = 2.8

c. HCN pK_a = 9.2

2.53 What is the pK_a for each compound?

a. (benzyl–$\overset{+}{N}H_3$) $K_a = 4.7 \times 10^{-10}$

b. (phenyl–$\overset{+}{N}H_3$) $K_a = 1.3 \times 10^{-5}$

c. CF_3COOH $K_a = 3.02 \times 10^{-1}$

2.54 Which of the following bases are strong enough to deprotonate $CH_3CH_2CH_2C{\equiv}CH$ (pK_a = 25) so that equilibrium favors the products: (a) H_2O; (b) NaOH; (c) $NaNH_2$; (d) NH_3; (e) NaH; (f) CH_3Li?

2.55 Draw the products of each reaction. Use the pK_a table in Appendix C to decide if the equilibrium favors the starting materials or products.

a. CH_3NH_2 + H_2SO_4 $\rightleftharpoons$

b. CH_3CH_2COOH + NaCl $\rightleftharpoons$

c. (phenol, OH) + $NaHCO_3$ $\rightleftharpoons$

d. H–C≡C–H + $CH_3CH_2^-\ Li^+$ $\rightleftharpoons$

2.56 Draw the products of each reaction and decide if equilibrium favors the starting materials or the products.

a. (cyclohexyl)–$\ddot{N}H_2$ + (benzoate, $C_6H_5CO_2^-$) $\rightleftharpoons$

b. (cyclohexyl)–$\ddot{N}H^-$ + (benzoic acid, C_6H_5COOH) $\rightleftharpoons$

c. (cyclohexyl)–$\overset{+}{N}H_3$ + (benzoate, $C_6H_5CO_2^-$) $\rightleftharpoons$

d. (cyclohexyl)–$\ddot{N}H_2$ + (benzoic acid, C_6H_5COOH) $\rightleftharpoons$

Relative Acid Strength

2.57 Rank the labeled protons in the following molecule in order of increasing pK_a.

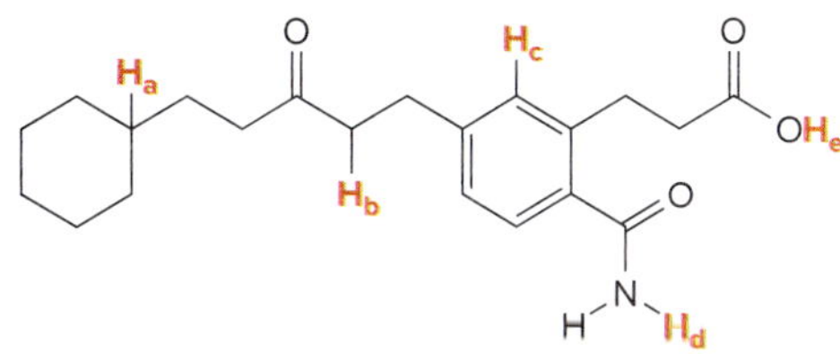

2.58 Rank the following Brønsted–Lowry acids in order of increasing acidity. Which compound forms the strongest conjugate base?

A (Br, OH) B (O, F) C (NH) D (OH)

2.59 Which of the following anions is the stronger base? Explain your choice.

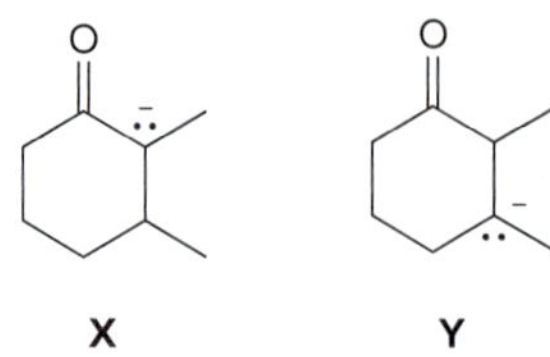

2.60 The pK_a of three CH bonds is given below.

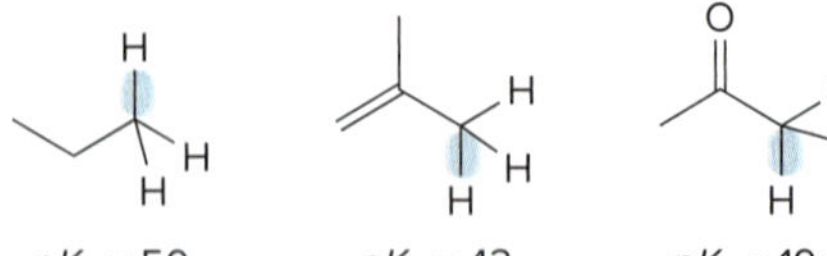

a. For each compound, draw the conjugate base, including all possible resonance structures.

b. Explain the observed trend in pK_a.

2.61 a. What is the conjugate acid of **A**?

b. What is the conjugate base of **A**?

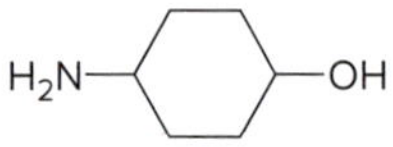

H_2N–(cyclohexane)–OH

A

2.62 Explain why the N–H proton in **X** is more acidic than the O–H proton. **X** was a key intermediate in the synthesis of the antibiotic levofloxacin.

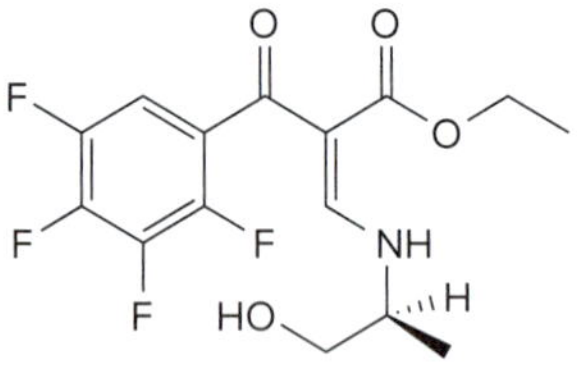

X

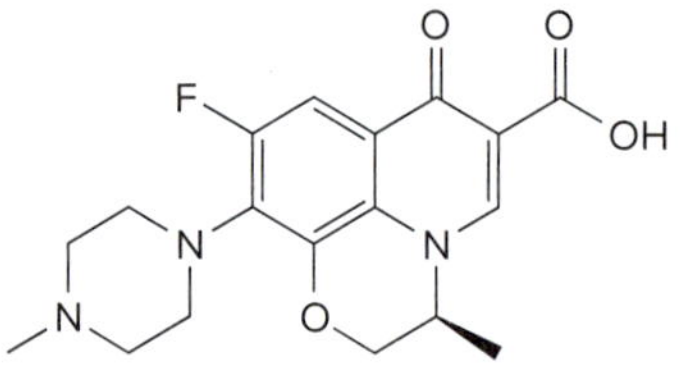

levofloxacin

2.63 RNA is a biological molecule that translates the genetic information in DNA into protein synthesis.

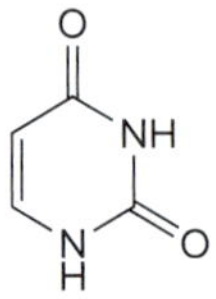

uracil

a. Identify the most acidic proton in uracil, one of the components of RNA, and explain your choice.

b. If uracil is treated with two equivalents of very strong base, what dianion is formed?

2.64 Many drugs are Brønsted–Lowry acids or bases.

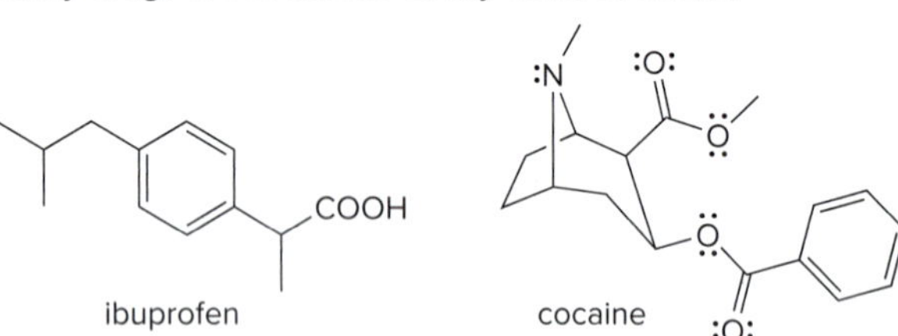

a. What is the most acidic proton in the analgesic ibuprofen? Draw the conjugate base.

b. What is the most basic electron pair in cocaine? Draw the conjugate acid.

2.65 Dimethyl ether (CH_3OCH_3) and ethanol (CH_3CH_2OH) are isomers, but CH_3OCH_3 has a pK_a of 40 and CH_3CH_2OH has a pK_a of 16. Why are these pK_a values so different?

2.66 Atenolol is a β (beta) blocker, a drug used to treat high blood pressure. Which of the indicated N–H bonds is more acidic? Explain your reasoning.

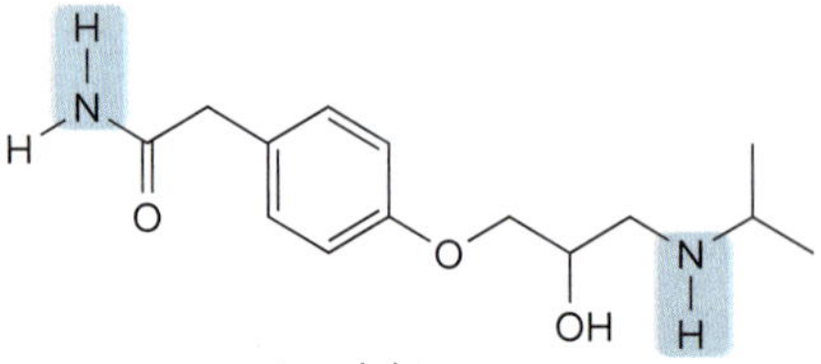

atenolol

2.67 Use the principles in Section 2.5 to label the most acidic hydrogen in each drug. Explain your choice.

a. valproic acid (used to treat epilepsy)

b. paroxetine Trade name: Paxil (used to treat depression)

c. metoprolol (used to treat high blood pressure)

2.68 The pK_a of the two most acidic protons of the amino acid leucine are shown.

leucine: pK_a = 9.74 ($H_3\overset{+}{N}$); pK_a = 2.33 (OH)

a. Draw the product formed, including all reasonable resonance structures, when leucine is treated with one equivalent of base.

b. Draw the product formed, including all reasonable resonance structures, when leucine is treated with two equivalents of base.

c. Is NaOH a strong enough base to remove both protons? Why or why not?

Lewis Acids and Bases

2.69 Classify each compound as a Lewis base, a Brønsted–Lowry base, both, or neither.

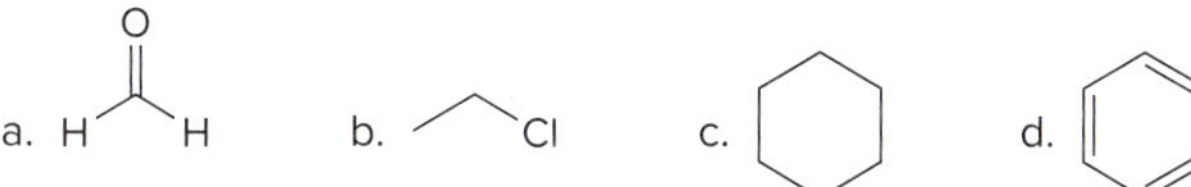

2.70 Classify each species as a Lewis acid, a Brønsted–Lowry acid, both, or neither.

a. H_3O^+ b. Cl_3C^+ c. BCl_3 d. BF_4^-

Lewis Acid–Base Reactions

2.71 Label the Lewis acid and Lewis base in each reaction. Use curved arrows to show the movement of electron pairs.

a. Cl^- + BCl_3 ⟶ BCl_4^-

b. acyl chloride + ^-OH ⟶ tetrahedral product

2.72 Draw the products of each Lewis acid–base reaction. Label the electrophile and nucleophile.

a. diethyl sulfide + $AlCl_3$ ⟶

b. acetone + BF_3 ⟶

c. carbocation + H_2O ⟶

Problems That Combine Concepts

2.73 Answer the following questions about the four species **A–D**.

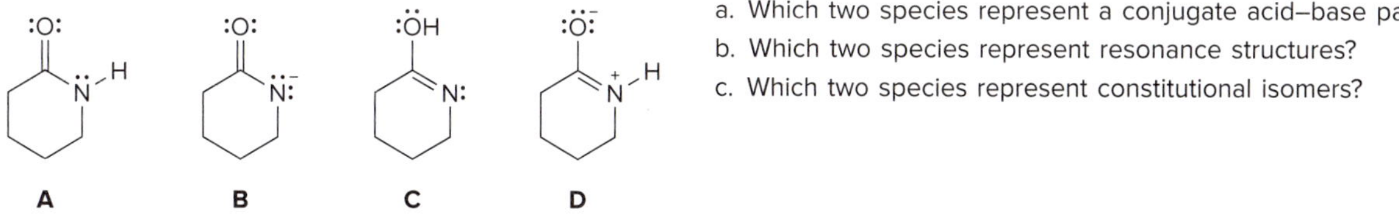

a. Which two species represent a conjugate acid–base pair?

b. Which two species represent resonance structures?

c. Which two species represent constitutional isomers?

2.74 Classify each reaction as either a proton transfer reaction, or a reaction of a nucleophile with an electrophile. Use curved arrows to show how the electron pairs move.

a. 2-propanol + HBr ⟶ $\overset{+}{O}H_2$ + Br^- ⟶ 2-bromopropane + H_2O

b. benzene + Br^+ ⟶ carbocation $\xrightarrow{Br^-}$ bromobenzene + HBr

2.75 Hydroxide (^{-}OH) can react as a Brønsted–Lowry base (and remove a proton) or as a Lewis base (and attack a carbon atom). (a) What organic product is formed when ^{-}OH reacts with the carbocation $(CH_3)_3C^+$ as a Brønsted–Lowry base? (b) What organic product is formed when ^{-}OH reacts with $(CH_3)_3C^+$ as a Lewis base?

2.76 Answer the following questions about esmolol, a drug used to treat high blood pressure sold under the trade name Brevibloc.

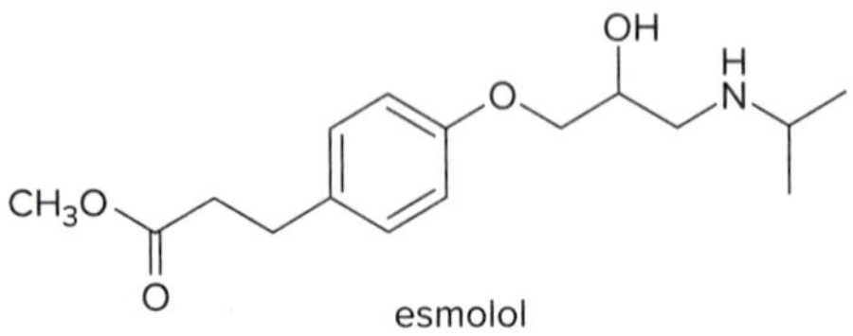

esmolol

a. Label the most acidic hydrogen atom in esmolol.
b. What products are formed when esmolol is treated with NaH?
c. What products are formed when esmolol is treated with HCl?
d. Label all sp^2 hybridized C atoms.
e. Label the only trigonal pyramidal atom.
f. Label all C's that bear a δ+ charge.

Challenge Problems

2.77 Caffeic acid is an organic acid isolated from coffee beans. Predict which labeled hydrogen (H_a or H_b) is more acidic and explain your choice.

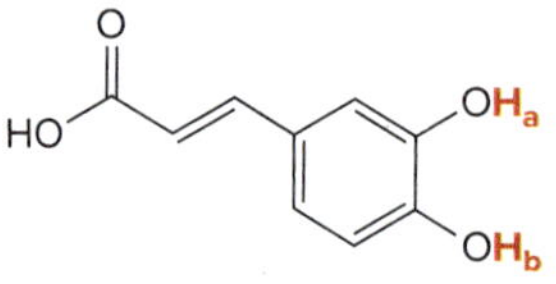

caffeic acid

2.78 Bosutinib is a drug approved by the Food and Drug Administration in 2012 for the treatment of a form of leukemia.

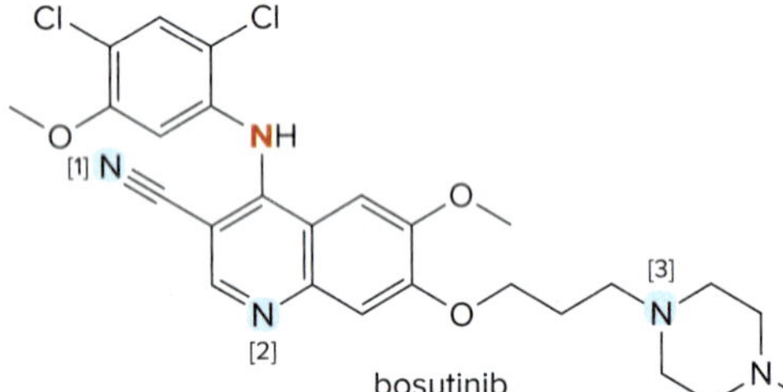

bosutinib

a. Rank the labeled N atoms ([1]–[3]) of bosutinib in order of increasing basicity.
b. Explain why the N atom in red is less basic than N[3].

2.79 Molecules like acetamide (CH_3CONH_2) can be protonated on either their O or N atoms when treated with a strong acid like HCl. Which site is more readily protonated and why?

2.80 Two pK_a values are reported for malonic acid, a compound with two COOH groups. Explain why one pK_a is lower and one pK_a is higher than the pK_a of acetic acid (CH_3COOH, pK_a = 4.8).

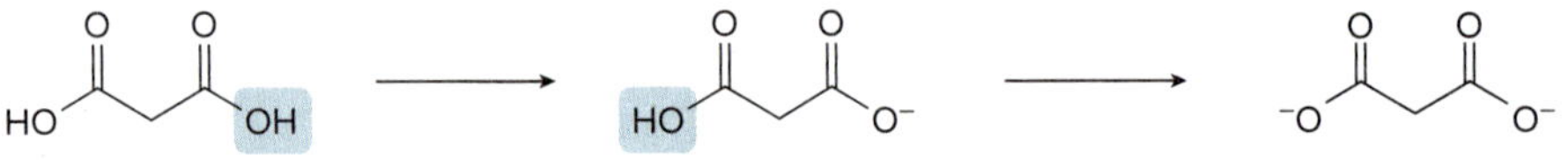

malonic acid
pK_a = 2.86

pK_a = 5.70

2.81 Write a stepwise reaction sequence using proton transfer reactions to show how the following reaction occurs. (Hint: As a first step, use ^{-}OH to remove a proton from the CH_2 group between the C=O and C=C.)

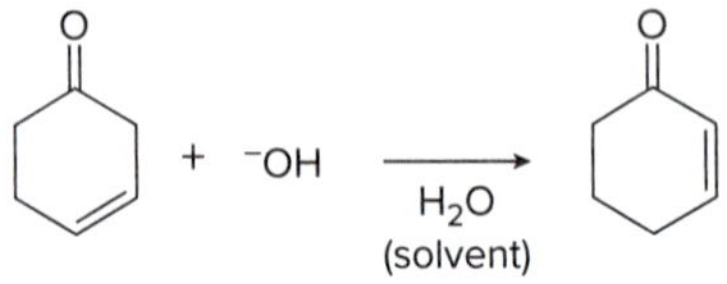

+ ^{-}OH → H_2O (solvent)

2.82 Elenolic acid is a component of olive oil. Use the pK_a values in Appendix C to predict which H atoms in elenolic acid might have a $pK_a < 25$.

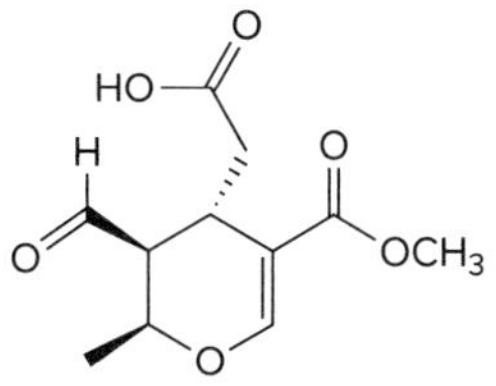

elenolic acid

2.83 Which H atom in vitamin C (ascorbic acid) is most acidic?

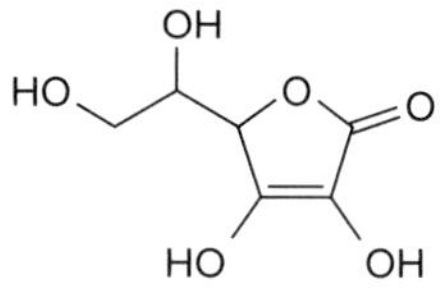

vitamin C
ascorbic acid

SELF-TEST ANSWERS

1. d 2. b 3. c 4. a 5. b 6. a 7. b 8. b 9. d 10. a

3 Introduction to Organic Molecules and Functional Groups

3.1 Functional groups
3.2 An overview of functional groups
3.3 Intermolecular forces
3.4 Physical properties
3.5 Application: Vitamins
3.6 Application of solubility: Soap
3.7 Application: The cell membrane
3.8 Functional groups and reactivity
3.9 Biomolecules

Pixtal/age fotostock

Vitamin B_5, pantothenic acid, is a water-soluble vitamin that is converted to coenzyme A, a biological reactant needed in the metabolism of carbohydrates, lipids, and amino acids. In addition to dairy foods and eggs, avocados are an excellent source of vitamin B_5. Avocados, which are grown in tropical and subtropical climates, contain a buttery green flesh that surrounds a large seed. In Chapter 3, we learn why some vitamins like vitamin A can be stored in the fat cells in the body, whereas others like pantothenic acid are excreted in urine.

Why Study . . .

Functional Groups?

Having learned some basic concepts about structure, bonding, and acid–base chemistry in Chapters 1 and 2, we will now concentrate on organic molecules.

- What are the characteristic features of an organic compound?
- What determines the properties of an organic compound?

After these questions are answered, we can understand some common phenomena. For example, why do we store some vitamins in the body and readily excrete others? How does soap clean away dirt? We will also use the properties of organic molecules to explain some basic biological phenomena, such as the structure of cell membranes and the transport of species across these membranes.

3.1 Functional Groups

What are the characteristic features of an organic compound? Most organic molecules have C–C and C–H σ bonds. These bonds are strong, nonpolar, and not readily broken. Organic molecules may have these structural features as well:

- **Heteroatoms—atoms other than carbon or hydrogen.** Common heteroatoms are nitrogen, oxygen, sulfur, phosphorus, and the halogens.
- **π Bonds.** The most common π bonds occur in C–C and C–O double bonds.

These structural features distinguish one organic molecule from another. They determine a molecule's geometry, physical properties, and reactivity, and comprise what is called a **functional group.**

- A *functional group* is an atom or a group of atoms with characteristic chemical and physical properties. It is the *reactive part* of the molecule.

Why do heteroatoms and π bonds confer reactivity on a particular molecule?

- Heteroatoms have lone pairs and create electron-deficient sites on carbon.
- π Bonds are easily broken in chemical reactions. A π bond makes a molecule a base and a nucleophile.

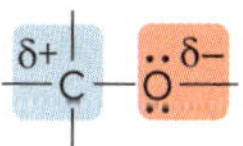

- Lone pairs make O a base and a nucleophile.
- The C atom is electron deficient, making it an electrophile.

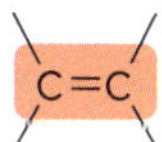

- The π bond is easily broken.
- The π bond makes a compound a base and a nucleophile.

The C–C and C–H σ bonds form the **carbon backbone** or **skeleton** to which the functional groups are bonded. A functional group usually behaves the same whether it is bonded to a carbon skeleton having as few as two or as many as 20 carbons. For this reason, we often abbreviate the carbon and hydrogen portion of the molecule by a capital letter **R,** and draw the **R** bonded to a particular functional group.

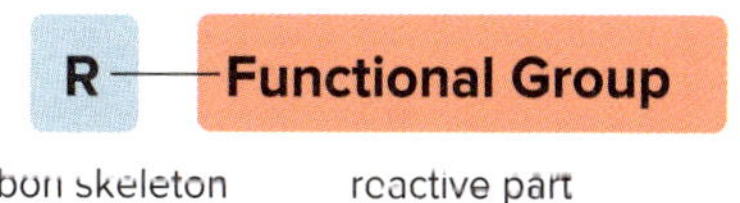

carbon skeleton reactive part

Ethane, for example, has only C–C and C–H σ bonds, so it has *no* functional group. Ethane has no polar bonds, no lone pairs, and no π bonds, so it has **no reactive sites.** Because of this, ethane and molecules like it are very unreactive.

Ethanol, on the other hand, has two carbons and five hydrogens in its carbon backbone, as well as an OH group, a functional group called a **hydroxy** group. Ethanol has lone pairs and polar bonds that make it reactive with a variety of reagents, including the acids and bases discussed in Chapter 2. The hydroxy group makes the properties of ethanol very different from the properties of ethane. Moreover, any organic molecule containing a hydroxy group has properties similar to ethanol.

ethane

- all C–C and C–H σ bonds
- no functional group

hydroxy group

ethanol

- polar C–O and O–H bonds
- two lone pairs

Most organic compounds can be grouped into a relatively small number of categories, based on the structure of their functional group. Ethane, for example, is an **alkane,** whereas ethanol is a simple **alcohol.**

Problem 3.1 What reaction occurs when CH_3CH_2OH is treated with (a) H_2SO_4? (b) NaH? What happens when CH_3CH_3 is treated with these same reagents?

3.2 An Overview of Functional Groups

We can subdivide the most common functional groups into three types.

- **Hydrocarbons**
- **Compounds containing a C–Z σ bond** where Z = an electronegative element
- **Compounds containing a C=O group**

3.2A Hydrocarbons

To review the structure and bonding of the simple aliphatic hydrocarbons, return to Section 1.10.

The word *aliphatic* is derived from the Greek word *aleiphas* meaning "fat." Aliphatic compounds have physical properties similar to fats.

***Hydrocarbons* are compounds made up of only the elements carbon and hydrogen. They may be aliphatic or aromatic.**

[1] Aliphatic hydrocarbons. Aliphatic hydrocarbons can be divided into three subgroups.

- **Alkanes have only C–C σ bonds and no functional group. Ethane, CH_3CH_3, is a simple alkane.**
- **Alkenes have a C–C double bond as a functional group. Ethylene, $CH_2{=}CH_2$, is a simple alkene.**
- **Alkynes have a C–C triple bond as a functional group. Acetylene, HC≡CH, is a simple alkyne.**

[2] Aromatic hydrocarbons. This class of hydrocarbons was so named because many of the earliest known aromatic compounds had strong, characteristic odors.

The simplest aromatic hydrocarbon is **benzene.** The six-membered ring and three π bonds of benzene comprise a *single* functional group. Benzene is a component of the **BTX** mixture (**B** for **b**enzene) added to gasoline to boost octane ratings.

benzene
molecular formula C_6H_6

phenyl group
C_6H_5–
phenylcyclohexane

When a benzene ring is bonded to another group, it is called a **phenyl group.** In phenylcyclohexane, for example, a phenyl group is bonded to the six-membered cyclohexane ring. Table 3.1 summarizes the four different types of hydrocarbons.

Table 3.1 Hydrocarbons

Type of compound	General structure	Example	Functional group
Alkane	R—H	CH_3CH_3	—
Alkene	>C=C<	$H_2C=CH_2$	double bond
Alkyne	—C≡C—	H—C≡C—H	triple bond
Aromatic compound	benzene ring	benzene ring	phenyl group

Polyethylene is a synthetic plastic first produced in the 1930s, and initially used as insulating material for radar during World War II. It is now a plastic used in milk containers, sandwich bags, and plastic wrapping. Over 100 billion pounds of polyethylene are manufactured each year.

Alkanes, which have no functional groups, are notoriously unreactive except under very drastic conditions. For example, **polyethylene** is a synthetic plastic and high-molecular-weight alkane, consisting of chains of $-CH_2-$ groups bonded together, hundreds or even thousands of atoms long. Because it is an alkane with no reactive sites, it is a very stable compound that does not readily degrade and thus persists for years in landfills.

polyethylene

The chain continues in both directions.

Carbon atoms in alkanes and other organic compounds are classified by the number of other carbons directly bonded to them.

- A *primary carbon* (1° carbon) is bonded to *one* other C atom.
- A *secondary carbon* (2° carbon) is bonded to *two* other C atoms.
- A *tertiary carbon* (3° carbon) is bonded to *three* other C atoms.
- A *quaternary carbon* (4° carbon) is bonded to *four* other C atoms.

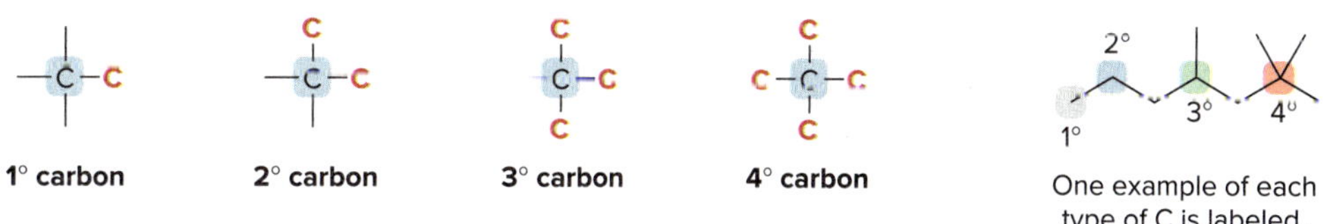

Hydrogen atoms are classified as **primary (1°), secondary (2°),** or **tertiary (3°)** depending on the **type of carbon atom** to which they are bonded.

- A *primary hydrogen* (1° H) is on a C bonded to one other C atom.
- A *secondary hydrogen* (2° H) is on a C bonded to two other C atoms.
- A *tertiary hydrogen* (3° H) is on a C bonded to three other C atoms.

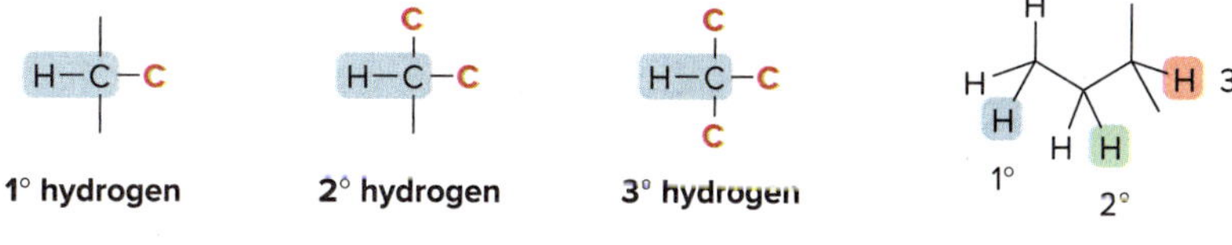

Sample Problem 3.1 Classifying the Carbons and Hydrogens in a Molecule

Classify the designated carbon atoms in **A** as 1°, 2°, 3°, or 4°. Classify the designated hydrogen atoms in **B** as 1°, 2°, or 3°.

A

B

Solution

- Classify C's by the number of other C's bonded to them.
- Classify H's by the type of C to which they are bonded; a 1° H is bonded to a 1° C, etc.

Problem 3.2 (a) Classify the carbon atoms in each compound as 1°, 2°, 3°, or 4°. (b) Classify the hydrogen atoms in each compound as 1°, 2°, or 3°.

More Practice: Try Problem 3.35.

Bilobalide (Problem 3.3) is obtained from *Ginkgo biloba*, the oldest seed-producing plant that currently lives on earth. Extracts from the leaves, roots, bark, and seeds of the ginkgo tree have been used in traditional Chinese medicine and currently comprise a widely used herbal supplement. *Michael Pettigrew/Getty Images*

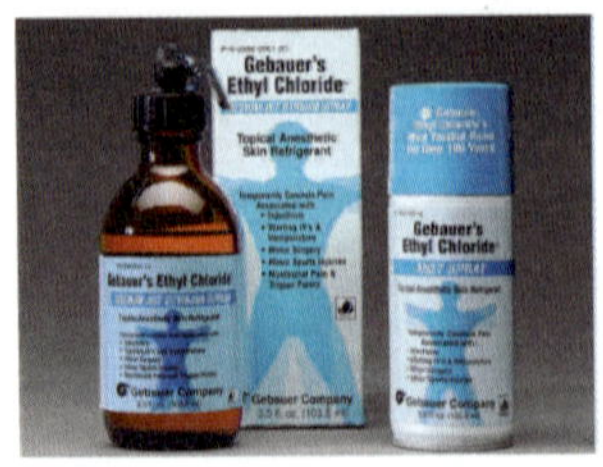

Chloroethane, CH_3CH_2Cl, is a local anesthetic. *Courtesy of Gebauer Company, Cleveland, Ohio*

Problem 3.3 Classifying a carbon atom by the number of carbons to which it is bonded can also be done in more-complex molecules that contain heteroatoms. Classify each sp^3 hybridized carbon atom in bilobalide, a compound isolated from *Ginkgo biloba* extracts, as 1°, 2°, 3°, or 4°.

HO OH O O O O O

bilobalide

3.2B Compounds Containing C–Z σ Bonds

Functional groups that contain C–Z σ bonds include **alkyl halides, alcohols, ethers, amines, thiols, and sulfides** (Table 3.2). The electronegative heteroatom Z creates a polar bond, making carbon electron deficient. The lone pairs on Z are available for reaction with protons and other electrophiles, especially when Z = N or O.

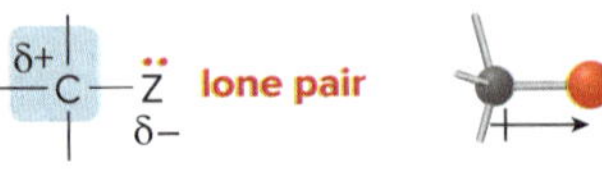

Several simple compounds in this category are widely used. As an example, chloroethane (CH_3CH_2Cl, commonly called ethyl chloride) is an alkyl halide used as a local anesthetic. Chloroethane quickly evaporates when sprayed on a wound, causing a cooling sensation that numbs the site of an injury.

Table 3.2 Compounds Containing C–Z σ Bonds

Type of compound	General structure	Example	3-D structure	Functional group
Alkyl halide	R–X: (X = F, Cl, Br, I)	CH_3–Br:	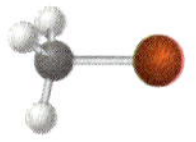	**–X** halo group
Alcohol	R–OH	CH_3–O–H		**–OH** hydroxy group
Ether	R–O–R	CH_3–O–CH_3		**–OR** alkoxy group
Amine	R–NH_2 or R_2NH or R_3N	CH_3–N(H)–H		**–NH_2** amino group
Thiol	R–SH	CH_3–S–H	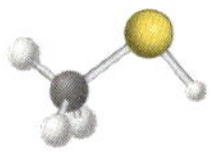	**–SH** mercapto group
Sulfide	R–S–R	CH_3–S–CH_3	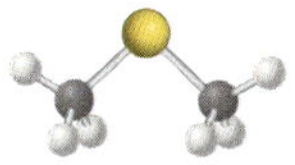	**–SR** alkylthio group

Molecules containing these functional groups may be simple or very complex. Diethyl ether, the first common general anesthetic, is a simple ether because it contains a single O atom, depicted in red, bonded to two C atoms. Hemibrevetoxin B, on the other hand, contains four ether groups, in addition to other functional groups.

Hemibrevetoxin B is a neurotoxin produced by algal blooms referred to as "red tides," because of the color often seen in shallow ocean waters when these algae proliferate.

diethyl ether

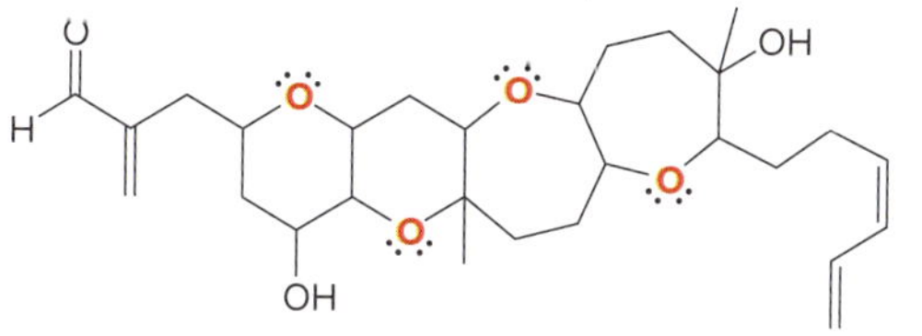

hemibrevetoxin B

Alkyl halides and alcohols are classified as **primary (1°), secondary (2°),** or **tertiary (3°)** based on the number of carbon atoms bonded to the carbon bearing the halogen or OH group. The classification of four functional groups in sucralose, the synthetic sweetener sold as Splenda, is shown.

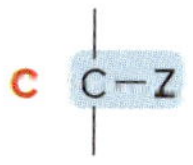

	C–C–Z	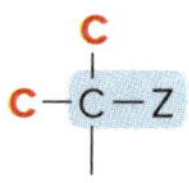 C–C(C)–Z	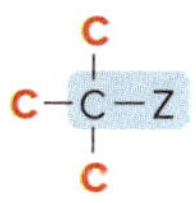 C–C(C)(C)–Z
Z = X	1° alkyl halide	2° alkyl halide	3° alkyl halide
Z = OH	1° alcohol	2° alcohol	3° alcohol

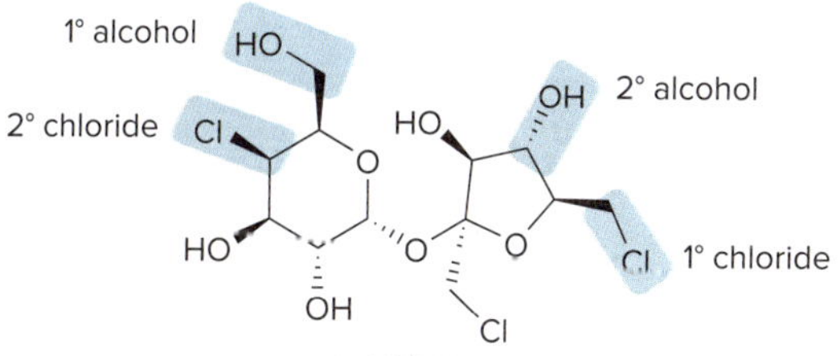

1° alcohol, 2° chloride, 2° alcohol, 1° chloride

sucralose

Problem 3.4 Classify each alkyl halide and alcohol as 1°, 2°, or 3°.

a. b. c. d.

Problem 3.5 Classify each OH group and halogen in dexamethasone, a synthetic steroid, as 1°, 2°, or 3°.

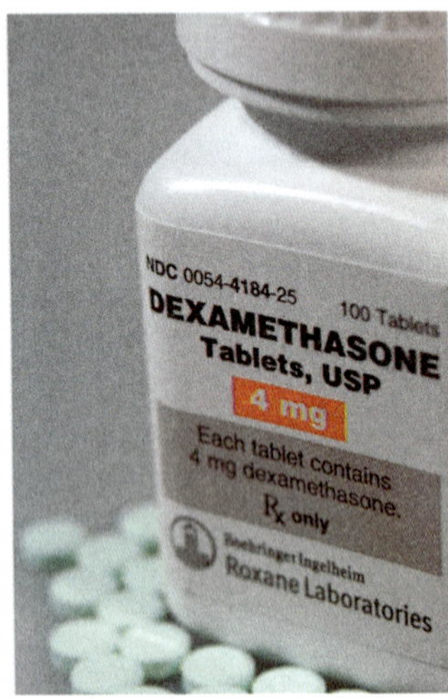

Dexamethasone (Problem 3.5) relieves inflammation and is used to treat some forms of arthritis, skin conditions, and asthma. Dexamethasone has been effectively used to treat patients who are critically ill with COVID-19. *Jill Braaten*

dexamethasone

Amines are classified as **primary (1°), secondary (2°),** or **tertiary (3°)** based on the number of carbon atoms bonded to the *nitrogen* atom.

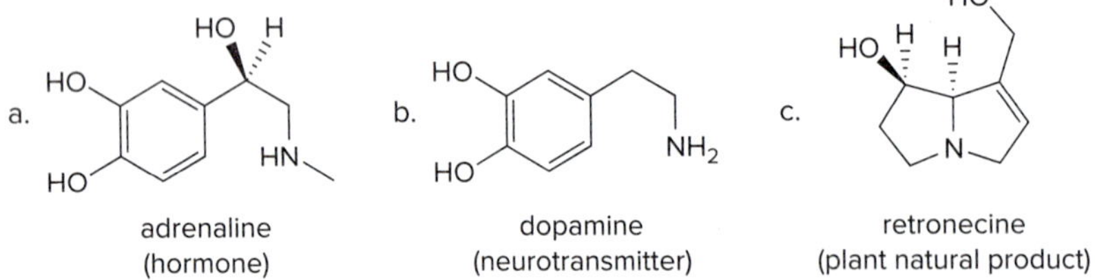

Classifying amines is different from classifying alcohols and alkyl halides as primary (1°), secondary (2°), or tertiary (3°). Amines are classified by the number of carbon–*nitrogen* bonds, whereas alkyl halides and alcohols are classified by the type of *carbon* bonded to the halogen or hydroxy group.

Problem 3.6 Classify each amine in the following compounds as 1°, 2°, or 3°.

a. adrenaline (hormone)

b. dopamine (neurotransmitter)

c. retronecine (plant natural product)

Problem 3.7 Draw the structure of a compound of molecular formula $C_4H_{11}NO$ that fits each description: (a) a compound that contains a 1° amine and a 3° alcohol; (b) a compound that contains a 3° amine and a 1° alcohol.

3.2C Compounds Containing a C=O Group

Many different types of functional groups possess a C–O double bond (a **carbonyl group**), including **aldehydes, ketones, carboxylic acids, esters, amides, and acid chlorides** (Table 3.3). The polar C–O bond makes the carbonyl carbon an **electrophile,** while the lone pairs on O allow it to react as a **nucleophile** and **base.** The carbonyl group also contains a π bond that is more easily broken than a C–O σ bond.

δ– δ+

carbonyl group

Table 3.3 Compounds Containing a C=O Group

Type of compound	General structure	Example	Condensed structure	3-D structure	Functional group
Aldehyde	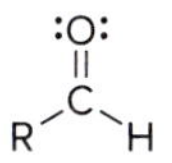 :O: ‖ R–C–H	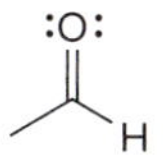 :O: ‖ –C–H	CH_3CHO	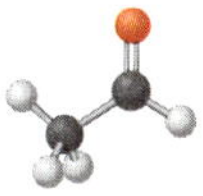	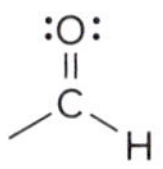 :O: ‖ –C–H
Ketone	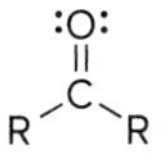 :O: ‖ R–C–R	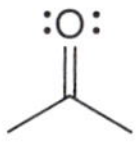 :O: ‖ –C–	$(CH_3)_2CO$	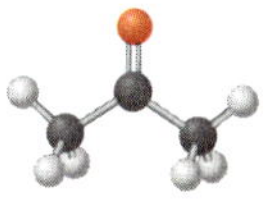	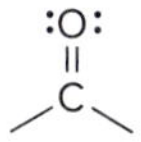 :O: ‖ –C– carbonyl group
Carboxylic acid	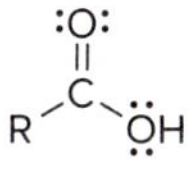 :O: ‖ R–C–ÖH	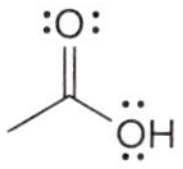 :O: ‖ –C–ÖH	CH_3CO_2H	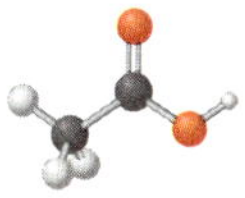	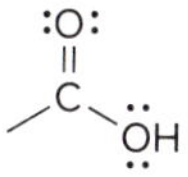 :O: ‖ –C–ÖH carboxy group
Ester	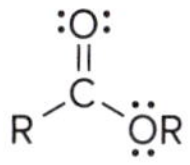 :O: ‖ R–C–ÖR	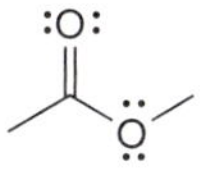 :O: ‖ –C–Ö–	$CH_3CO_2CH_3$	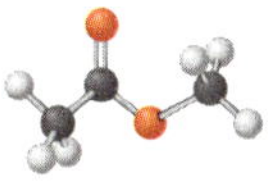	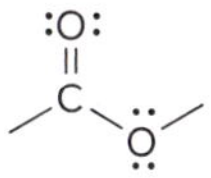 :O: ‖ –C–Ö–
Amide	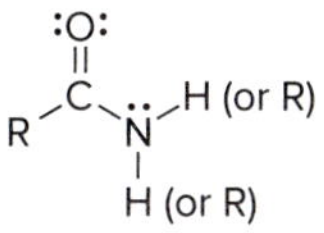 :O: ‖ R–C–N̈–H (or R), H (or R)	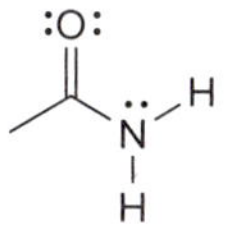 :O: ‖ –C–N̈–H, H	CH_3CONH_2	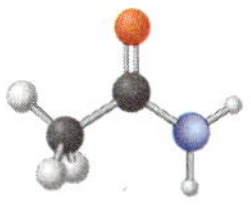	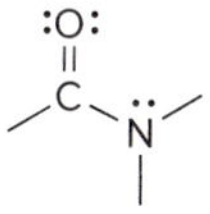 :O: ‖ –C–N̈–
Acid chloride	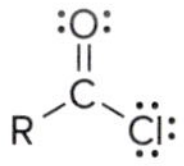 :O: ‖ R–C–C̈l:	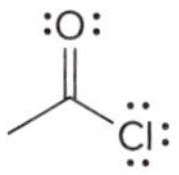 :O: ‖ –C–C̈l:	CH_3COCl	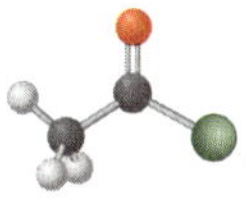	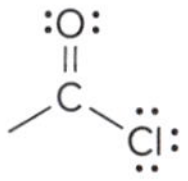 :O: ‖ –C–C̈l:

Amides, compounds that contain a nitrogen atom bonded directly to the carbonyl carbon, are classified as **primary (1°), secondary (2°),** or **tertiary (3°)** based on the number of carbon atoms bonded to the nitrogen atom.

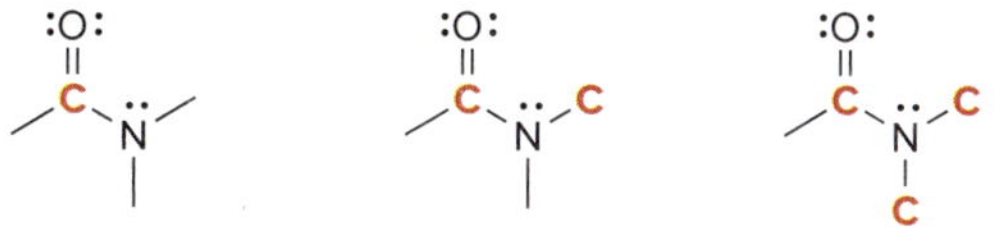

1° amide 2° amide 3° amide

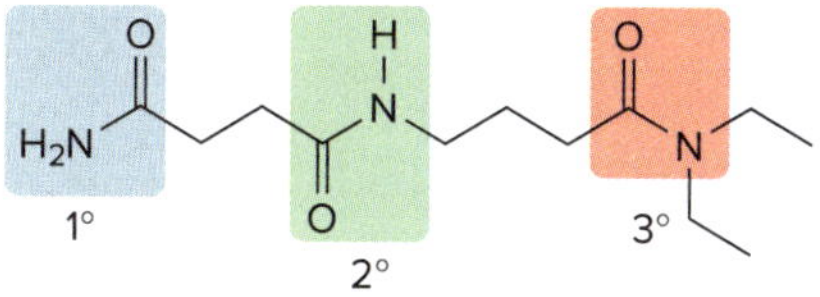

H₂N 1° 2° 3°

Problem 3.8 Classify the amides in thyrotropin-releasing hormone (TRH), a hormone produced by the hypothalamus, as 1°, 2°, or 3°.

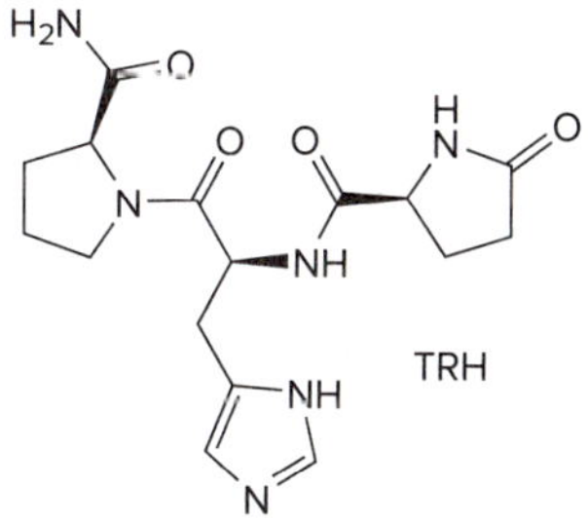

TRH

Problem 3.9 Classify each compound as an ether, ketone, or neither: (a) $(CH_3)_2CO$; (b) $CH_3CH_2CH_2COCH_2CH_3$; (c) $(CH_3)_3COCOCH_3$; (d) $CH_3CH_2CH(CH_3)CH_2OCH(CH_3)_2$.

The importance of a functional group cannot be overstated. A functional group determines a molecule's bonding and shape, type and strength of intermolecular forces, physical properties, nomenclature, and chemical reactivity.

Sample Problem 3.2 Identifying Functional Groups in a Complex Molecule

Identify the functional groups in two drugs, atenolol and donepezil. Atenolol is a β (beta) blocker, a drug used to treat hypertension (high blood pressure), and donepezil (trade name Aricept) is used to treat mild to moderate dementia associated with Alzheimer's disease.

atenolol
(used to treat high blood pressure)

donepezil
(used to treat Alzheimer's disease)

Solution

Concentrate on the heteroatoms and π bonds. With carbonyl groups, pay attention to what is bonded to the carbonyl carbon—hydrogen, carbon, or a heteroatom.

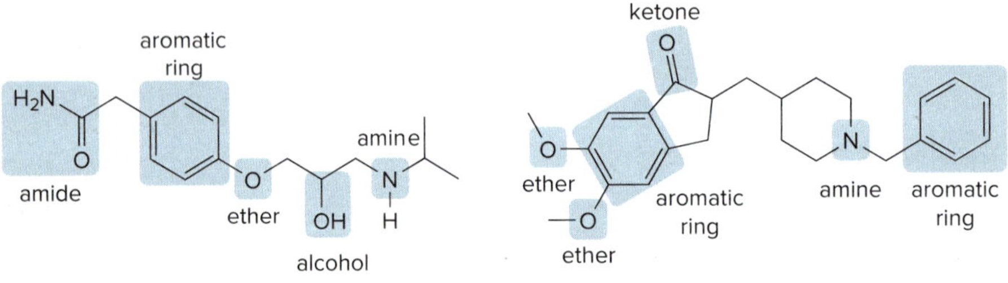

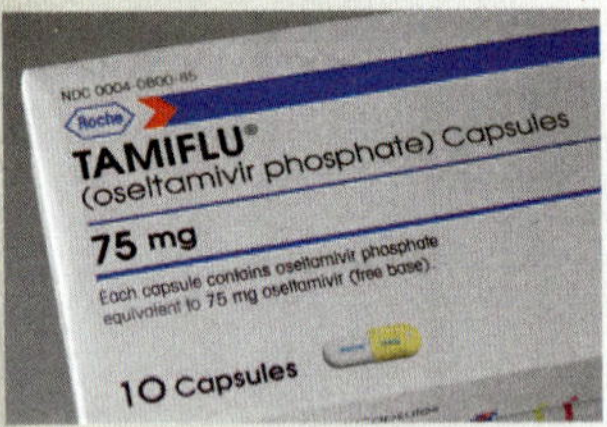

Tamiflu (Problem 3.10) is the trade name for oseltamivir, an antiviral drug used to treat influenza. *Jill Braaten*

Problem 3.10 Oseltamivir can be prepared in 10 steps from shikimic acid. Identify the functional groups in oseltamivir and shikimic acid.

shikimic acid → (10 steps) → oseltamivir

More Practice: Try Problems 3.33a; 3.34a; 3.36; 3.37; 3.59a; 3.60a; 3.61a, b; 3.62a, b; 3.64a.

Problem 3.11 Draw the structure of a compound fitting each description:

a. an aldehyde with molecular formula C_4H_8O
b. a ketone with molecular formula C_4H_8O
c. a carboxylic acid with molecular formula $C_4H_8O_2$
d. an ester with molecular formula $C_4H_8O_2$

Problem 3.12 Identify the functional groups in leukotriene C_4, a major contributor to the inflammation associated with asthma.

leukotriene C_4

3.3 Intermolecular Forces

Intermolecular forces are also referred to as **noncovalent interactions** or **nonbonded interactions.**

Intermolecular forces are the interactions that exist *between* molecules. A functional group determines the type and strength of these interactions.

3.3A Ionic Compounds

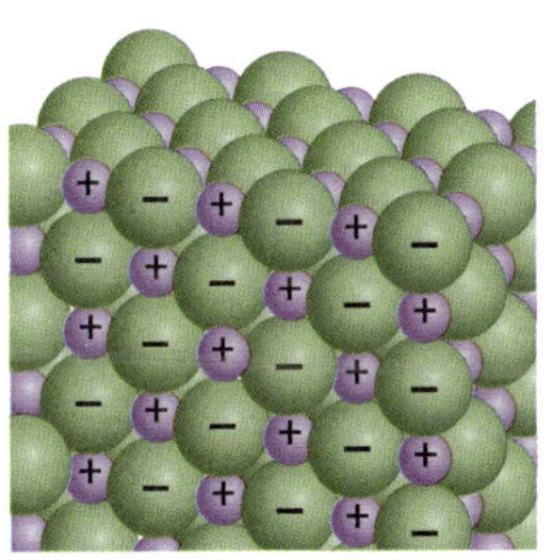

strong electrostatic interaction between Na^+ and Cl^-

Ionic compounds, such as NaCl, contain oppositely charged particles held together by **extremely strong electrostatic interactions.** These ionic interactions are much *stronger* than the intermolecular forces present between covalent molecules, so it takes a great deal of energy to separate oppositely charged ions from each other.

3.3B Covalent Compounds

Covalent compounds are composed of discrete molecules. The nature of the forces between the molecules depends on the functional group present. There are three different types of interactions, presented here in order of *increasing strength:*

- **van der Waals forces**
- **dipole–dipole interactions**
- **hydrogen bonding**

Van der Waals Forces

Van der Waals forces, also called **London forces,** are very weak interactions caused by the **momentary changes in electron density in a molecule.** Van der Waals forces are the only attractive forces present in nonpolar compounds.

For example, although a nonpolar CH_4 molecule has no net dipole, at any one instant its electron density may not be completely symmetrical, creating a *temporary* dipole. This can induce a temporary dipole in another CH_4 molecule, with the partial positive and negative charges arranged close to each other. **The weak interaction of these temporary dipoles constitutes van der Waals forces.** All compounds exhibit van der Waals forces.

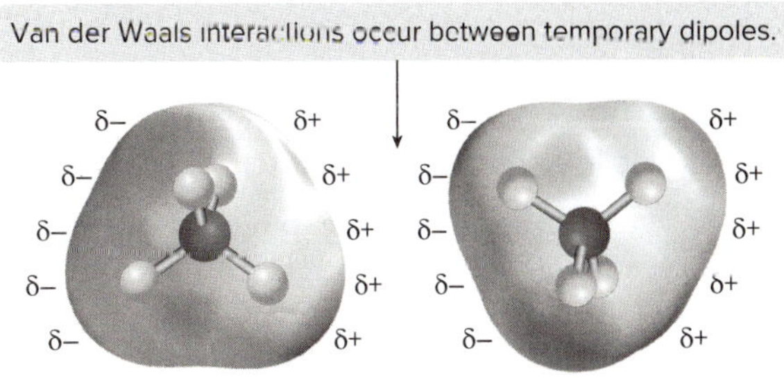

Although any single van der Waals interaction is weak, a large number of van der Waals interactions creates a strong force. For example, geckos stick to walls and ceilings by van der Waals interactions of the surfaces with the 500,000 tiny hairs on each foot.
(top): Don Mennig/Alamy Stock Photo; (bottom): Matthew D. H. Smith

The surface area of a molecule determines the strength of the van der Waals interactions. **The *larger* the surface area, the *larger* the attractive force between two molecules, and the *stronger* the intermolecular forces.** Long, sausage-shaped molecules such as $CH_3CH_2CH_2CH_2CH_3$ (pentane) have stronger van der Waals interactions than compact, spherical ones like $C(CH_3)_4$ (2,2-dimethylpropane).

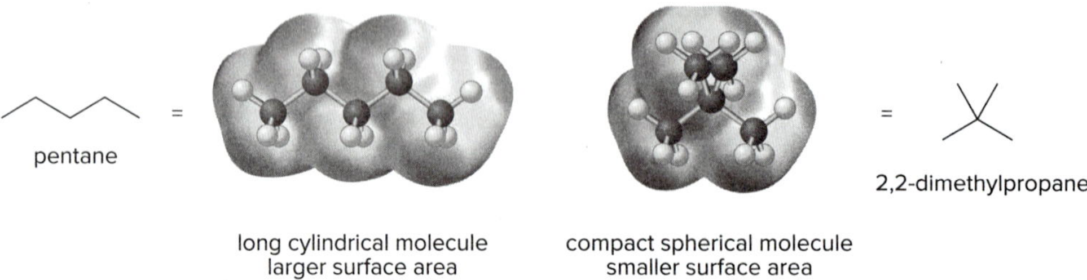

Another factor affecting the strength of van der Waals forces is **polarizability.**

- ***Polarizability* is a measure of how the electron cloud around an atom responds to changes in its electronic environment.**

Larger atoms like iodine, which have more loosely held valence electrons, are more polarizable than smaller atoms like fluorine, which have more tightly held electrons. Because larger atoms have more easily induced dipoles, compounds containing them possess stronger intermolecular interactions.

Two F_2 molecules have little force of attraction between them, because the electrons are held very tightly and temporary dipoles are difficult to induce. On the other hand, **two I_2 molecules exhibit a much stronger force of attraction,** because the electrons are held much more loosely and temporary dipoles are easily induced.

- **Compounds with large, polarizable atoms have stronger intermolecular forces than compounds with small, less polarizable atoms.**

Dipole–Dipole Interactions

***Dipole–dipole interactions* are the attractive forces between the permanent dipoles of two polar molecules.** In acetone, $(CH_3)_2C{=}O$, for example, the dipoles in adjacent molecules align so that the partial positive and partial negative charges are in close proximity. These attractive forces caused by permanent dipoles are much stronger than weak van der Waals forces.

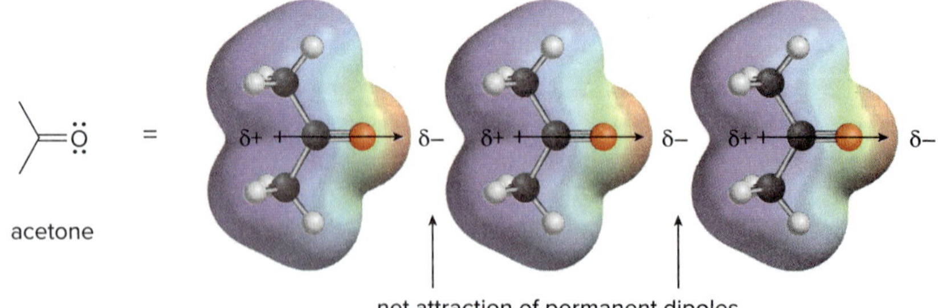

Hydrogen bonding helps determine the three-dimensional shape of large biomolecules such as carbohydrates, proteins, and nucleic acids. See Chapters 26–28 for details.

Hydrogen Bonding

***Hydrogen bonding* typically occurs when a hydrogen atom bonded to O, N, or F is electrostatically attracted to a lone pair of electrons on an O, N, or F atom in another molecule.** Thus, H_2O molecules can hydrogen bond to each other. When they do, a H atom covalently bonded to O in one water molecule is attracted to a lone pair of electrons on the O in another water molecule. Hydrogen bonds are the *strongest* of the three types of intermolecular forces, though they are still much weaker than any covalent bond.

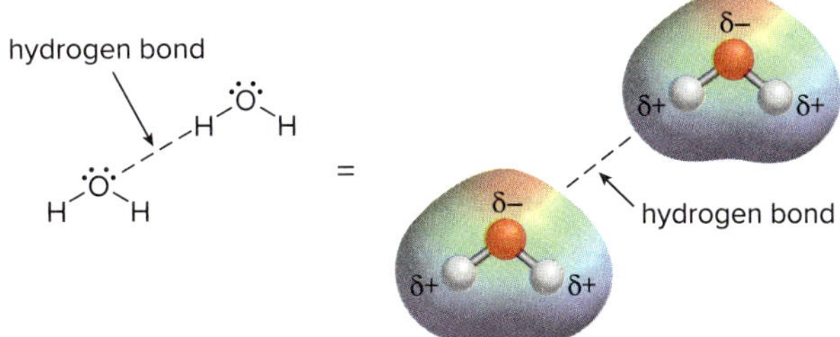

Sample Problem 3.3 illustrates how to determine the relative strength of intermolecular forces for a group of compounds. Table 3.4 summarizes the four types of interactions that affect the properties of all compounds.

Table 3.4 Summary of Types of Intermolecular Forces

Type of force	Relative strength	Exhibited by	Example
van der Waals	weak	all molecules	$CH_3CH_2CH_2CH_2CH_3$ $CH_3CH_2CH_2CHO$ $CH_3CH_2CH_2CH_2OH$
dipole–dipole	moderate	molecules with a net dipole	$CH_3CH_2CH_2CHO$ $CH_3CH_2CH_2CH_2OH$
hydrogen bonding	strong	molecules with an O–H, N–H, or H–F bond	$CH_3CH_2CH_2CH_2OH$
ion–ion	very strong	ionic compounds	NaCl, LiF

Sample Problem 3.3 Determining Intermolecular Forces in Organic Compounds

Rank the following compounds in order of increasing strength of intermolecular forces: $CH_3CH_2CH_2CH_2CH_3$ (pentane), $CH_3CH_2CH_2CH_2OH$ (butan-1-ol), and $CH_3CH_2CH_2CHO$ (butanal).

Solution

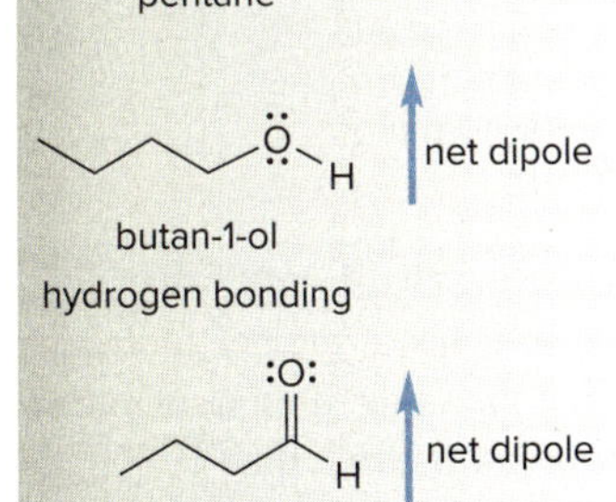

- Pentane has only nonpolar C–C and C–H bonds, so its molecules are held together by only **van der Waals** forces.
- Butan-1-ol is a polar bent molecule, so it can have **dipole–dipole** interactions in addition to **van der Waals** forces. Because it has an O–H bond, butan-1-ol molecules are held together by intermolecular **hydrogen bonds** as well.
- Butanal has a trigonal planar carbon with a polar C=O bond, so it exhibits **dipole–dipole** interactions in addition to **van der Waals** forces. There is *no* H atom bonded to O, so two butanal molecules *cannot* hydrogen bond to each other.

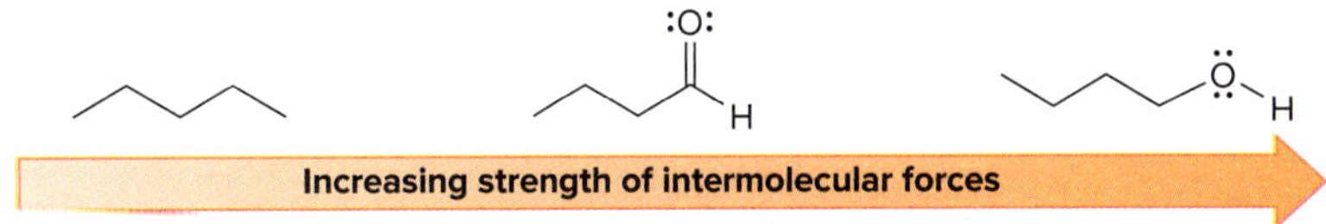

Problem 3.13 What types of intermolecular forces are present in each compound?

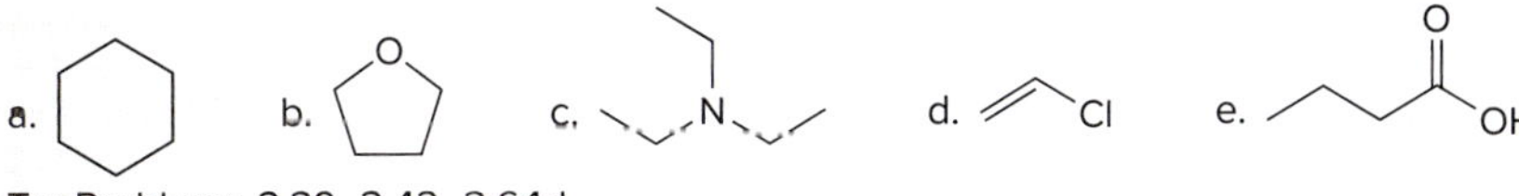

More Practice: Try Problems 3.39, 3.42, 3.64d.

3.4 Physical Properties

The strength of a compound's intermolecular forces determines many of its physical properties, including its boiling point, melting point, and solubility.

3.4A Boiling Point (bp)

The *boiling point* of a compound is the temperature at which a liquid is converted to a gas. In boiling, energy is needed to overcome the attractive forces in the more ordered liquid state.

- The *stronger* the intermolecular forces, the *higher* the boiling point.

Because **ionic compounds** are held together by extremely strong interactions, they have **very high boiling points.** The boiling point of NaCl, for example, is 1413 °C. **With covalent molecules, the boiling point depends on the identity of the functional group.** For compounds of approximately the same molecular weight:

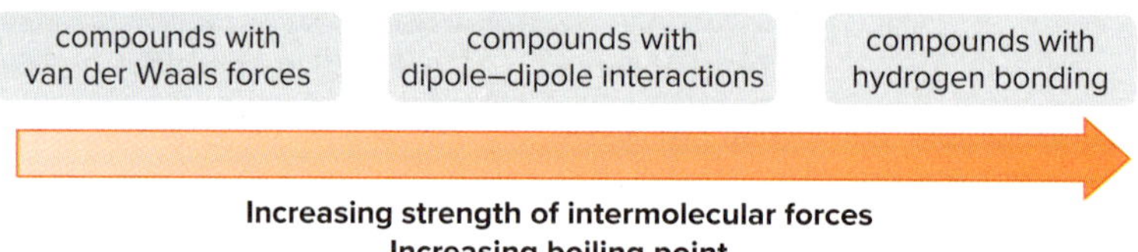

Recall from Sample Problem 3.3, for example, that the relative strength of the intermolecular forces increases from pentane to butanal to butan-1-ol. The boiling points of these compounds increase in the same order.

pentane bp = 36 °C | butanal bp = 76 °C | butan-1-ol bp = 118 °C

Increasing strength of intermolecular forces
Increasing boiling point

Because surface area and polarizability affect the strength of intermolecular forces, they also affect the boiling point. For two compounds with similar functional groups:

- The *larger* the surface area, the *higher* the boiling point.
- The *more polarizable* the atoms, the *higher* the boiling point.

Examples of each phenomenon are illustrated in Figure 3.1. In comparing two ketones that differ in size, pentan-3-one has a higher boiling point than acetone because it has a greater molecular weight and *larger surface area.* In comparing two alkyl halides having the same number of carbon atoms, CH_3I has a higher boiling point than CH_3F because I is *more polarizable* than F.

- In comparing two compounds, the lower-boiling compound is said to be *more volatile* and the higher-boiling compound is said to be *less volatile.*

Figure 3.1 Effect of surface area and polarizability on boiling point

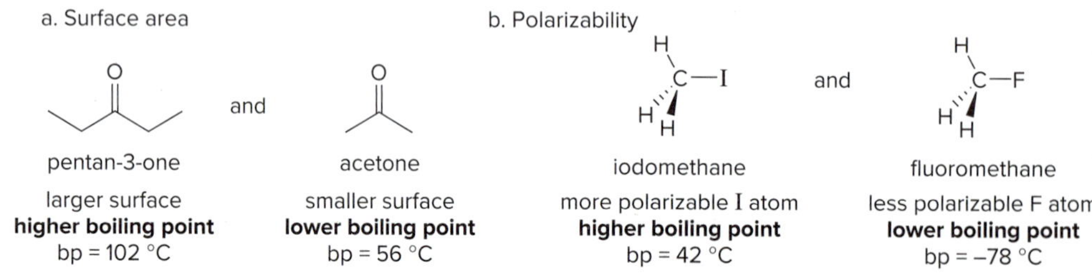

Sample Problem 3.4 Determining Relative Boiling Points

Which compound in each pair has the higher boiling point? Which compound in each pair is more volatile?

a.  A or B b. C or D

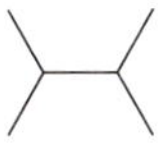

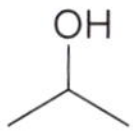

Solution

a. Isomers **A** and **B** have only nonpolar C–C and C–H bonds, so they exhibit only van der Waals forces. Because **B** is more compact, it has less surface area and a lower boiling point. The lower-boiling compound is more volatile, so **B** is more volatile than **A.**

b. Compounds **C** and **D** have approximately the same molecular weight but different functional groups. **C** is a nonpolar alkane, exhibiting only van der Waals forces. **D** is an alcohol with an O–H group available for hydrogen bonding, so it has stronger intermolecular forces and a higher boiling point. **C** is more volatile than **D.**

Problem 3.14 Which compound in each pair has the higher boiling point?

a. 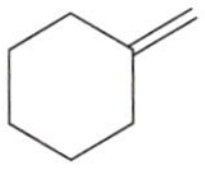 or 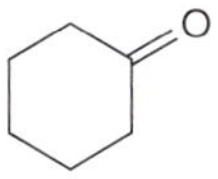 c. 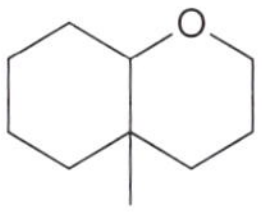 or

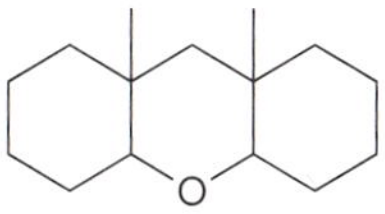

b. 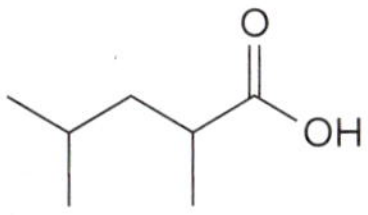 or 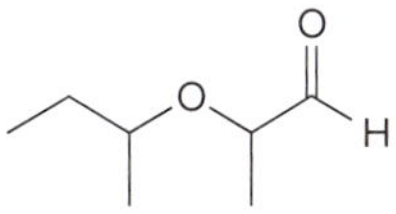 d. 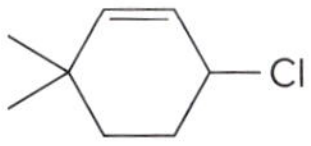 or

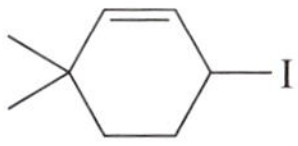

More Practice: Try Problems 3.43, 3.44.

Problem 3.15 Explain why the boiling point of propanamide, $CH_3CH_2CONH_2$, is considerably higher than the boiling point of *N,N*-dimethylformamide, $HCON(CH_3)_2$ (213 °C vs. 153 °C), even though both compounds are isomeric amides.

Problem 3.16 Rank the following compounds in order of increasing boiling point.

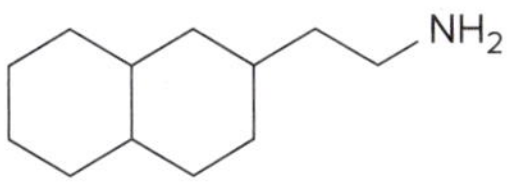

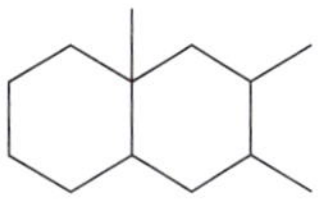

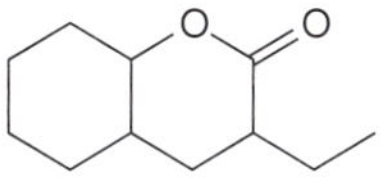

A B C

3.4B Melting Point (mp)

The *melting point* is the temperature at which a solid is converted to a liquid. In melting, energy is needed to overcome the attractive forces in the more ordered crystalline solid. Two factors determine the melting point of a compound.

- **The *stronger* the intermolecular forces, the *higher* the melting point.**
- **Given the same functional group, the *more symmetrical* the compound, the *higher* the melting point.**

Because **ionic compounds** are held together by extremely strong interactions, they have **very high melting points.** For example, the melting point of NaCl is 801 °C. With covalent molecules, the melting point once again depends on the identity of the functional group. For compounds of approximately the same molecular weight:

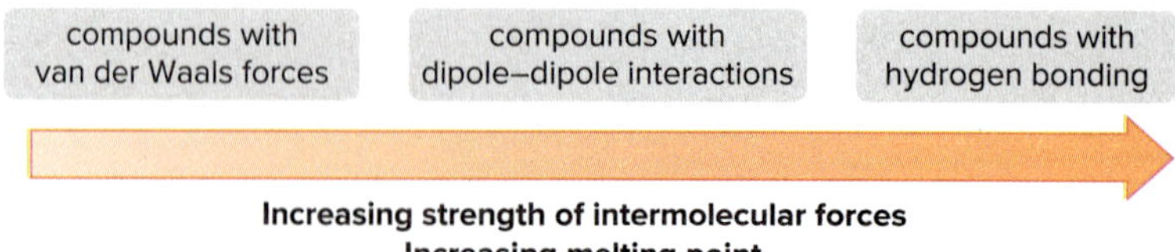

Increasing strength of intermolecular forces
Increasing melting point

The trend in the melting points of pentane, butanal, and butan-1-ol parallels the trend observed in their boiling points.

:O:
H
H
pentane
mp = –130 °C
butanal
mp = –96 °C
butan-1-ol
mp = –90 °C

Increasing strength of intermolecular forces
Increasing melting point

Symmetry also plays a role in determining the melting points of compounds having the same functional group and similar molecular weights, but very different shapes. A compact symmetrical molecule like 2,2-dimethylpropane packs well into a crystalline lattice whereas 2-methylbutane, which has a CH_3 group dangling from a four-carbon chain, does not. Thus, 2,2-dimethylpropane has a much higher melting point.

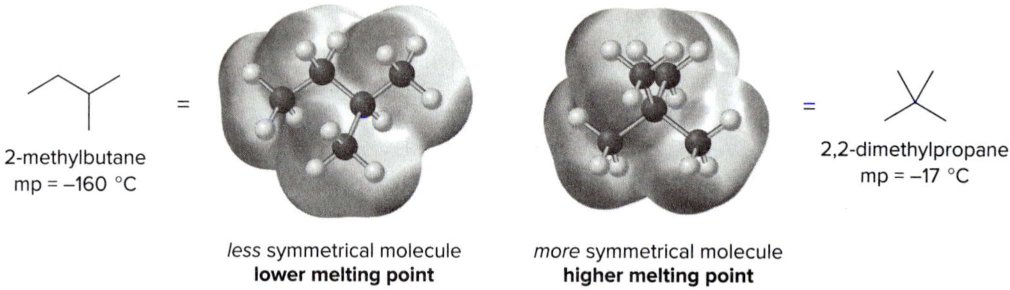

less symmetrical molecule
lower melting point

more symmetrical molecule
higher melting point

Problem 3.17 Predict which compound in each pair has the higher melting point.

a. or NH_2 b. or

Problem 3.18 Consider acetic acid (CH_3CO_2H) and its conjugate base, sodium acetate (CH_3CO_2Na). (a) What intermolecular forces are present in each compound? (b) Explain why the melting point of sodium acetate (324 °C) is considerably higher than the melting point of acetic acid (17 °C).

Problem 3.19 Rank **A–C** in order of increasing melting point.

OH O

A **B** **C**

3.4C Solubility

Quantitatively, a compound may be considered soluble when 3 g of solute dissolves in 100 mL of solvent.

Solubility* is the extent to which a compound, called the *solute,* dissolves in a liquid, called the *solvent. In dissolving a compound, the energy needed to break up the interactions between the molecules or ions of the solute comes from new interactions between the solute and the solvent.

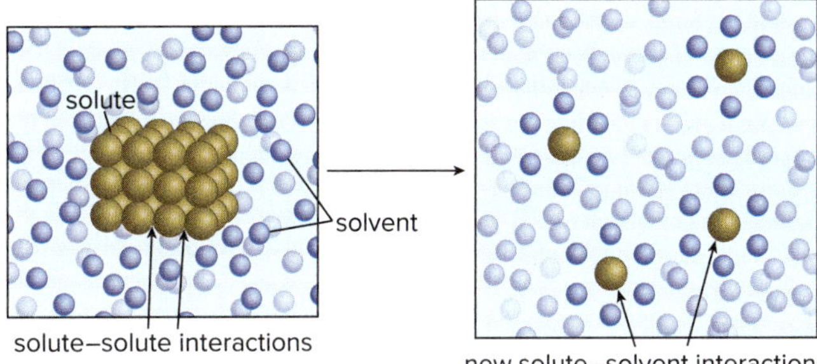

Compounds dissolve in solvents having similar kinds of intermolecular forces.

- "Like dissolves like."
- Polar compounds dissolve in polar solvents. Nonpolar or weakly polar compounds dissolve in nonpolar or weakly polar solvents.

Water and **organic liquids** are two different kinds of solvents. Water is very polar because it is capable of hydrogen bonding with a solute. Many organic solvents are either nonpolar, like carbon tetrachloride (CCl_4) and hexane [$CH_3(CH_2)_4CH_3$], or weakly polar like diethyl ether ($CH_3CH_2OCH_2CH_3$).

Ionic compounds are held together by strong electrostatic forces, so they need very polar solvents to dissolve. **Most ionic compounds are soluble in water, but are insoluble in organic solvents.** To dissolve an ionic compound, the strong ion–ion interactions must be replaced by many weaker **ion–dipole interactions,** as illustrated in Figure 3.2.

Figure 3.2 Dissolving an ionic compound in H_2O

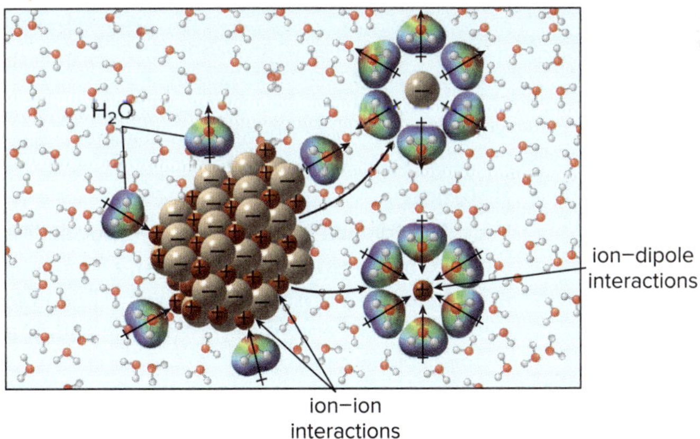

- When an ionic solid is dissolved in H_2O, the ion–ion interactions are replaced by ion–dipole interactions. Though these forces are weaker, there are so many of them that they compensate for the stronger ionic bonds.

Most organic compounds are soluble in organic solvents (remember, *like dissolves like*). **An organic compound is water soluble only if it contains one polar functional group capable of hydrogen bonding with the solvent for every five C atoms it contains.** In other words, a water-soluble organic compound has an O- or N-containing functional group that solubilizes its nonpolar carbon backbone.

Compare, for example, the solubility of butane and acetone in H_2O and CCl_4.

butane: CCl_4 → **soluble**; H_2O → **insoluble**

acetone: CCl_4 → **soluble**; H_2O → **soluble**

Because butane and acetone are both organic compounds having a C–C and C–H backbone, they are soluble in the organic solvent CCl_4. Butane, a nonpolar molecule, is insoluble in the polar solvent H_2O. Acetone, however, is H_2O soluble because it contains only three C atoms and its O atom can hydrogen bond with one H atom of H_2O. In fact, acetone is so soluble in water that acetone and water are **miscible**—they form solutions in all proportions with each other.

$(CH_3)_2C{=}O$ molecules cannot hydrogen bond to each other because they have no OH group. However, $(CH_3)_2C{=}O$ can hydrogen bond to H_2O because its O atom can hydrogen bond to one of the H atoms of H_2O.

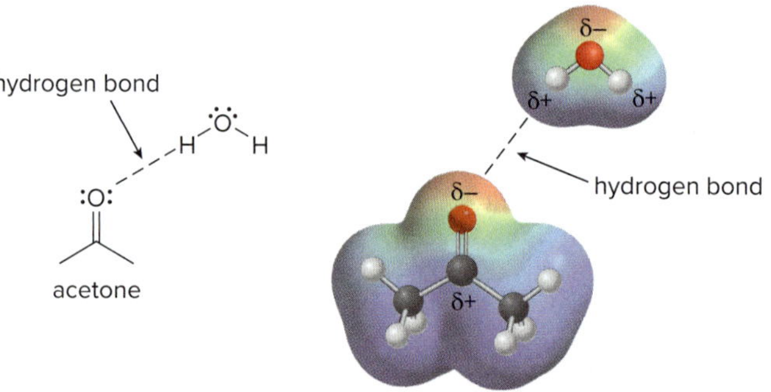

Sample Problem 3.5 Determining Hydrogen Bonding

(a) Which of the following compounds can hydrogen bond to another molecule like itself? (b) Which of the following compounds can hydrogen bond to water?

meperidine (a narcotic) Trade name: Demerol

acetylsalicylic acid (aspirin)

Solution

- **To hydrogen bond to another molecule like itself, a compound needs an O–H or N–H bond.**
- **To hydrogen bond with water, a compound needs an O or N atom.**

a. Only acetylsalicylic acid has an O–H bond for intermolecular hydrogen bonding, so two molecules of acetylsalicylic acid can hydrogen bond to each other, but two molecules of meperidine cannot.

b. Both meperidine and acetylsalicylic acid have electronegative O atoms and meperidine has an electronegative N atom, so both compounds can hydrogen bond to water. One possibility for each compound:

hydrogen bond **hydrogen bond**

Problem 3.20 (a) At which sites can **C** hydrogen bond to another molecule like itself? (b) At which sites can **D** hydrogen bond to water?

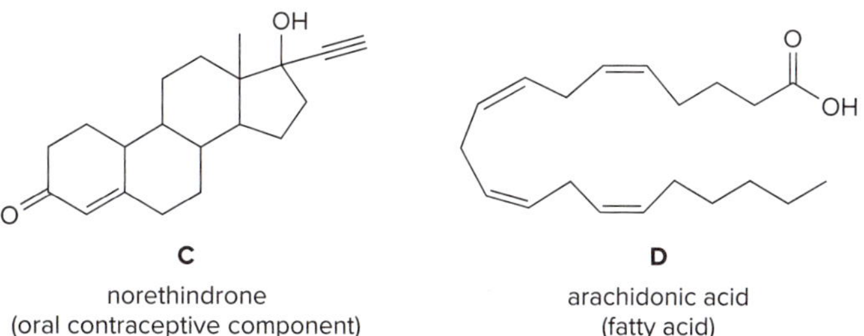

C
norethindrone
(oral contraceptive component)

D
arachidonic acid
(fatty acid)

More Practice: Try Problems 3.40; 3.41; 3.59c; 3.60b; 3.61c, d.

Problem 3.21 Which of the following molecules can hydrogen bond to another molecule like itself? Which can hydrogen bond to water?

a. Br, Br, Br b. N c. NH_2, O d. O, O

For an organic compound with one functional group, **a compound is water soluble only if it has ≤ five C atoms and contains an O or N atom.**

The size of an organic molecule with a polar functional group determines its water solubility. A low-molecular-weight alcohol like **ethanol is water soluble** because it has a small carbon skeleton (≤ five C atoms) compared to the size of its polar OH group. Cholesterol, on the other hand, has 27 carbon atoms and only one OH group. Its carbon skeleton is too large for the OH group to solubilize by hydrogen bonding, so **cholesterol is insoluble in water.**

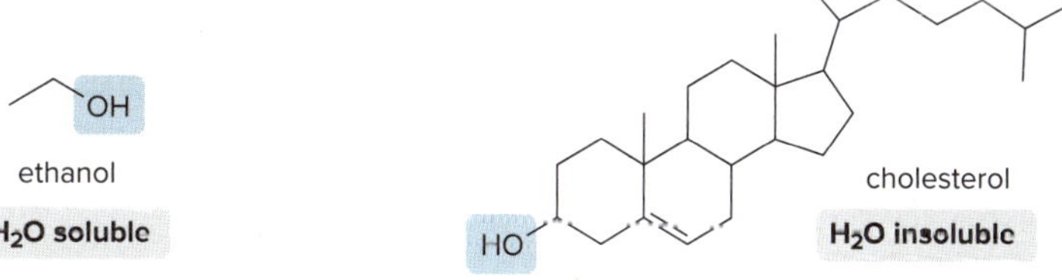

Hydrophobic = afraid of H_2O.
Hydrophilic = H_2O loving.

- **The nonpolar part of a molecule that is not attracted to H_2O is said to be *hydrophobic*.**
- **The polar part of a molecule that can hydrogen bond to H_2O is said to be *hydrophilic*.**

In cholesterol, for example, the **hydroxy group is hydrophilic,** whereas the **carbon skeleton is hydrophobic.**

MTBE
tert-butyl methyl ether

Cl Cl
4,4'-dichlorobiphenyl
(a polychlorinated biphenyl, PCB)

MTBE (*tert*-butyl methyl ether) and 4,4'-dichlorobiphenyl (a polychlorinated biphenyl, abbreviated as PCB) demonstrate that solubility properties can help determine the fate of organic compounds in the environment.

MTBE, a high-octane additive in unleaded gasoline, is **water soluble.** Small amounts of MTBE have contaminated the drinking water in several communities, making it unfit for consumption. For this reason, the use of MTBE as a gasoline additive has steadily declined in the United States since 1999.

4,4'-Dichlorobiphenyl is a polychlorinated biphenyl **(PCB),** a compound that contains two benzene rings and two or more chlorine atoms. PCBs have been used as plasticizers in polystyrene coffee cups and coolants in transformers. They have been released into the environment during production, use, storage, and disposal, making them one of the most widespread organic pollutants. **PCBs are insoluble in H_2O, but very soluble in organic media,** so they are soluble in fatty tissue, including that found in all types of fish and birds around the world. Although PCBs are not acutely toxic, frequently ingesting large quantities of fish contaminated with PCBs has been shown to retard growth and memory retention in children.

Solubility properties of some representative compounds are summarized in Table 3.5.

Table 3.5 Summary of Solubility

Type of compound	Solubility in H_2O	Solubility in organic solvents (such as CCl_4)
Ionic		
NaCl	**soluble**	**insoluble**
Covalent		
$CH_3CH_2CH_2CH_3$	**insoluble** (no N or O atom to hydrogen bond to H_2O)	**soluble**
$CH_3CH_2CH_2OH$	**soluble** ($\leq$ 5 C's and an O atom for hydrogen bonding to H_2O)	**soluble**
$CH_3(CH_2)_{10}OH$	**insoluble** (> 5 C's; too large to be soluble even though it has an O atom for hydrogen bonding to H_2O)	**soluble**

Sample Problem 3.6 Predicting the Water Solubility of a Compound

Which compounds are water soluble?

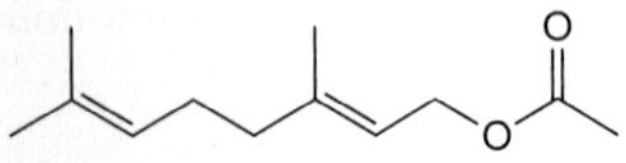

geranyl acetate
(from lily of the valley)
A

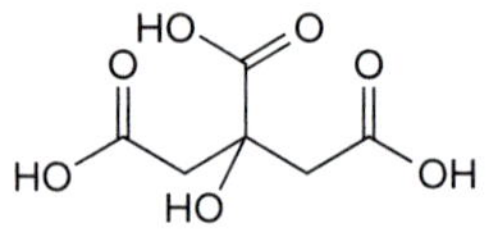

citric acid
(tart acid from citrus fruits)
B

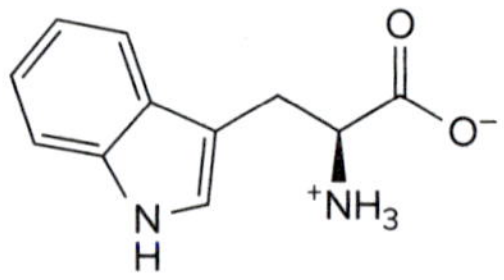

tryptophan
(an amino acid)
C

Solution

Water-soluble compounds are ionic or contain a functional group with an O or N atom that can hydrogen bond to water for every 5 C's.

- **A** has 12 C's and only one oxygen-containing functional group (an ester), so **A** is water insoluble.
- **B** has 6 C's and many OH groups and C=O's that can hydrogen bond to water, so **B** is water soluble.
- **C** is ionic, so **C** is water soluble.

Problem 3.22 Which compounds are water soluble?

a.

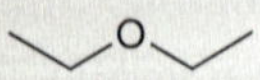

b. 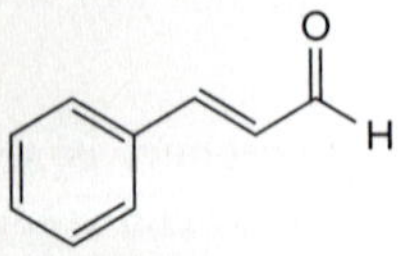 cinnamaldehyde
(odor of cinnamon)

c. 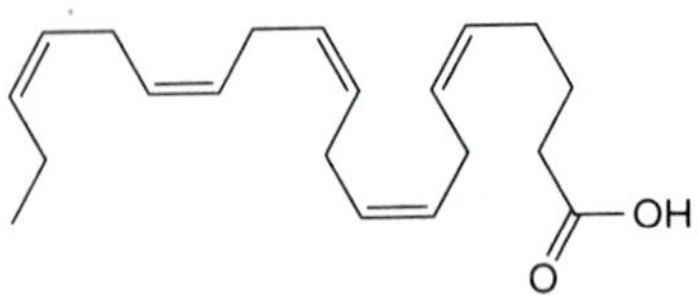 eicosapentaenoic acid
(fatty acid from fish oil)

d. 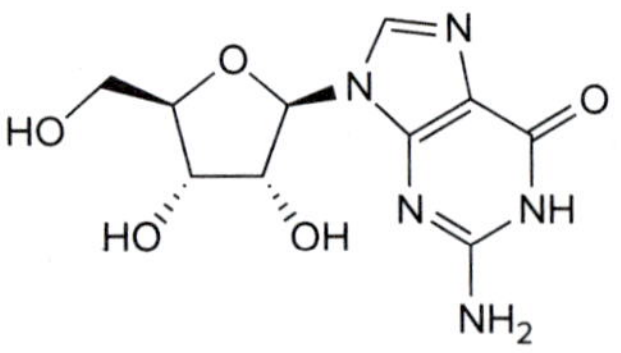 guanosine
(DNA component)

More Practice: Try Problems 3.47; 3.48b, c; 3.49–3.52.

Problem 3.23 Rank the following compounds in order of increasing water solubility.

A B C D

3.5 Application: Vitamins

***Vitamins* are organic compounds needed in small amounts for normal cell function.** Our bodies cannot synthesize these compounds, so they must be obtained in the diet. Most vitamins are identified by a letter, such as A, C, D, E, and K. There are several different B vitamins, though, so a subscript is added to distinguish them: for example, B_1, B_2, and B_{12}.

Whether a vitamin is **fat soluble** (it dissolves in organic media) or **water soluble** can be determined by applying the solubility principles discussed in Section 3.4C. Vitamins A and C illustrate the differences between fat-soluble and water-soluble vitamins.

3.5A Vitamin A

Vitamin A, or **retinol,** is an essential component of the vision receptors in the eyes. It also helps to maintain the health of mucous membranes and the skin, so many anti-aging creams contain vitamin A. A deficiency of this vitamin leads to a loss of night vision.

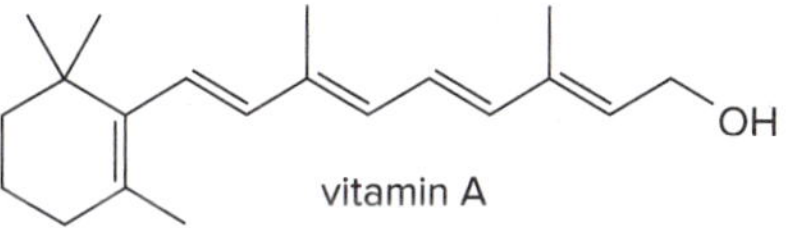

Vitamin A is synthesized from β-carotene, the orange pigment in carrots. *Purestock/SuperStock*

Vitamin A contains 20 carbons and a single OH group, making it **water insoluble.** Because it is organic, it is **soluble in any organic medium.** To understand the consequences of these solubility characteristics, we must learn about the chemical environment of the body.

About 70% of the body is composed of water. Fluids such as blood, gastric juices in the stomach, and urine are largely water with dissolved ions such as Na^+ and K^+. Vitamin A is insoluble in these fluids. There are also fat cells composed of organic compounds having C–C and C–H bonds. Vitamin A is soluble in this organic environment, and thus it is readily stored in these fat cells, particularly in the liver.

Vitamin A may be obtained directly from the diet. In addition, β-carotene, the orange pigment found in many plants including carrots, is readily converted to vitamin A in our bodies.

β-carotene

Eating too many carrots does not result in an excess of stored vitamin A. If you consume more β-carotene than you need, your body stores this precursor until it needs more vitamin A. Some β-carotene reaches the surface tissues of the skin and eyes, giving them an orange color. This phenomenon may look odd, but it is harmless and reversible. When stored β-carotene is converted to vitamin A and is no longer in excess, these tissues will return to their normal hue.

Vitamin C is obtained by eating citrus fruits and a wide variety of other fruits and vegetables. Individuals can also obtain the recommended daily dose of vitamin C by taking tablets that contain vitamin C prepared in the laboratory. Both the "natural" vitamin C in oranges and the "synthetic" vitamin C in vitamin supplements are identical. *Mary Reeg/McGraw Hill*

3.5B Vitamin C

Although most animal species can synthesize vitamin C, humans, guinea pigs, the Indian fruit bat, and the bulbul bird must obtain this vitamin from dietary sources. Citrus fruits, strawberries, kiwi, tomatoes, and sweet potatoes are all excellent sources of vitamin C.

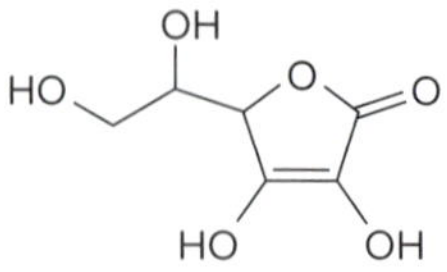

vitamin C
(ascorbic acid)

Vitamin C has six carbon atoms, each bonded to an oxygen atom that is capable of hydrogen bonding, making it **water soluble.** Vitamin C thus dissolves in urine. Although it has been acclaimed as a deterrent for all kinds of diseases, from the common cold to cancer, the consequences of taking large amounts of vitamin C are not really known, because any excess of the minimum daily requirement is excreted in the urine.

Problem 3.24 Predict the water solubility of each vitamin.

a. 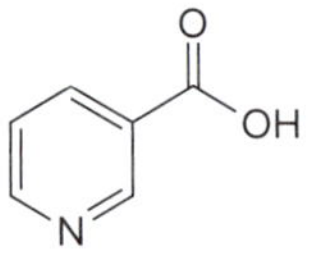 vitamin B_3
(niacin)

b. 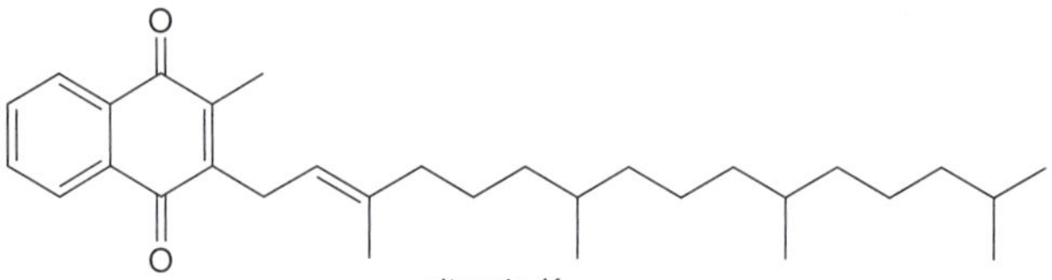 vitamin K_1
(phylloquinone)

Problem 3.25 (a) Identify the functional groups in pantothenic acid (vitamin B_5), the chapter-opening molecule. (b) At which sites can pantothenic acid hydrogen bond to water? (c) Predict the water solubility of pantothenic acid.

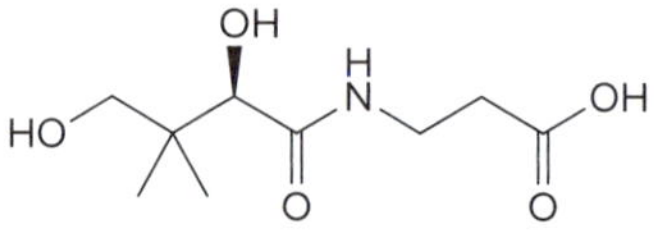

pantothenic acid
vitamin B_5

3.6 Application of Solubility: Soap

Soap has been used by humankind for some 2000 years. Historical records describe its manufacture in the first century and document the presence of a soap factory in Pompeii. Before this time clothes were cleaned by rubbing them on rocks in water, or by forming soapy lathers from the roots, bark, and leaves of certain plants. These plants produced natural materials called *saponins,* which act in much the same way as modern soaps.

On a molecular level, soap has two distinct parts:

- **a hydrophilic portion composed of ions, called the *polar head***
- **a hydrophobic carbon chain of nonpolar C–C and C–H bonds, called the *nonpolar tail***

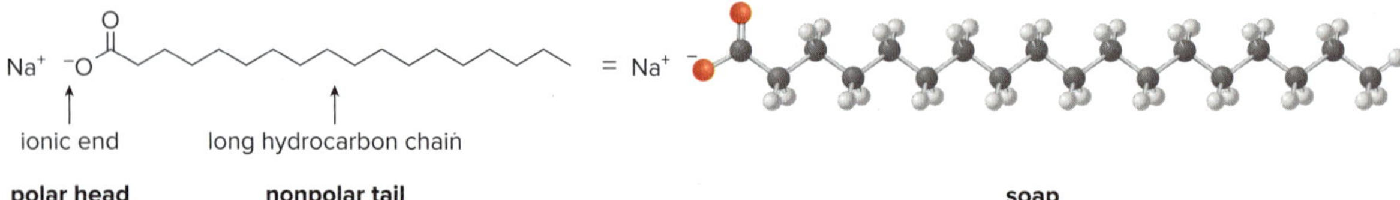

Dissolving soap in water forms ***micelles,*** **spherical droplets having the ionic heads on the surface and the nonpolar tails packed together in the interior,** as shown in Figure 3.3.

In this arrangement, the ionic heads are solvated by the polar solvent water, thus solubilizing the nonpolar, "greasy" hydrocarbon portion of the soap.

Figure 3.3
Dissolving soap in water

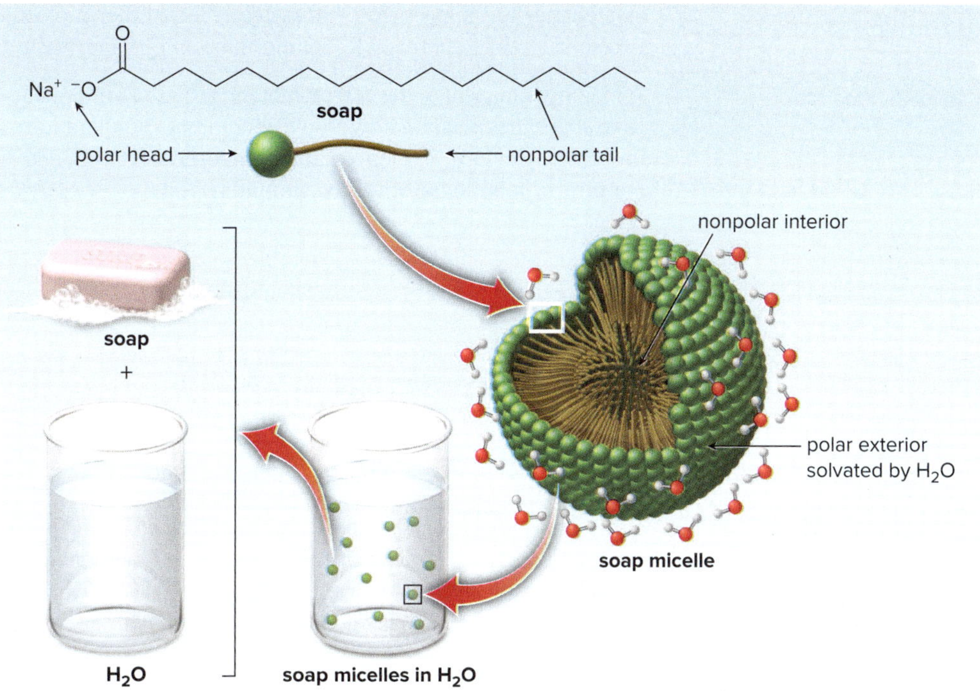

- When soap is dissolved in H_2O, it forms micelles with the nonpolar tails in the interior and the polar heads on the surface. The polar heads are solvated by ion–dipole interactions with H_2O molecules.

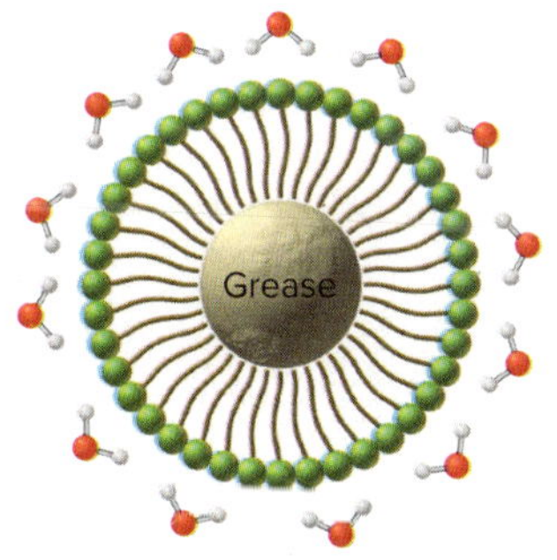

Cross-section of a soap micelle with a grease particle dissolved in the interior

How does soap dissolve grease and oil? Water alone cannot dissolve dirt, which is composed largely of nonpolar hydrocarbons. When soap is mixed with water, however, the nonpolar hydrocarbon tails dissolve the dirt in the interior of the micelle. The polar head of the soap remains on the surface of the micelle to interact with water. The nonpolar tails of the soap are so well sealed off from the water by the polar head groups that the micelles are water soluble, allowing them to separate from the fibers of our clothes and be washed down the drain with water. In this way, soaps do a seemingly impossible task: they remove nonpolar hydrocarbon material from skin and clothes, by solubilizing it in the polar solvent water.

Problem 3.26 Which of the following structures represent soaps? Explain your answers.

a. O, O^- Na^+

b. O, O^- Na^+

c. O, OH

Problem 3.27 Some commercial shower gels contain sodium lauryl sulfate, not soap, as a cleaning agent. Explain how sodium lauryl sulfate cleans away dirt.

O O S O O^- Na^+

sodium lauryl sulfate

3.7 Application: The Cell Membrane

The cell membrane is a beautifully complex example of how the principles of organic chemistry come into play in a biological system.

3.7A Structure of the Cell Membrane

The basic unit of living organisms is the **cell.** The cytoplasm is the aqueous medium inside the cell, separated from water outside the cell by the **cell membrane.** The cell membrane acts as a barrier to the passage of ions, water, and other molecules into and out of the cell, and it is also selectively permeable, letting nutrients in and waste out.

A major component of the cell membrane is a group of organic compounds called **phospholipids.** Like soap, they contain a hydrophilic ionic portion and a hydrophobic hydrocarbon portion, in this case two long carbon chains composed of C–C and C–H bonds. **Phospholipids thus contain a polar head and *two* nonpolar tails.**

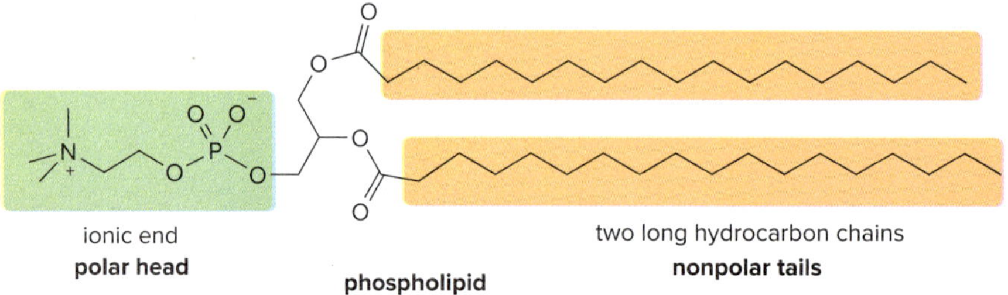

When phospholipids are mixed with water, they assemble in an arrangement called a **lipid bilayer,** with the ionic heads oriented on the outside and the nonpolar tails on the inside. The polar heads electrostatically interact with the polar solvent H_2O, while the nonpolar tails are held in close proximity by numerous van der Waals interactions. This is schematically illustrated in Figure 3.4.

Figure 3.4
The cell membrane

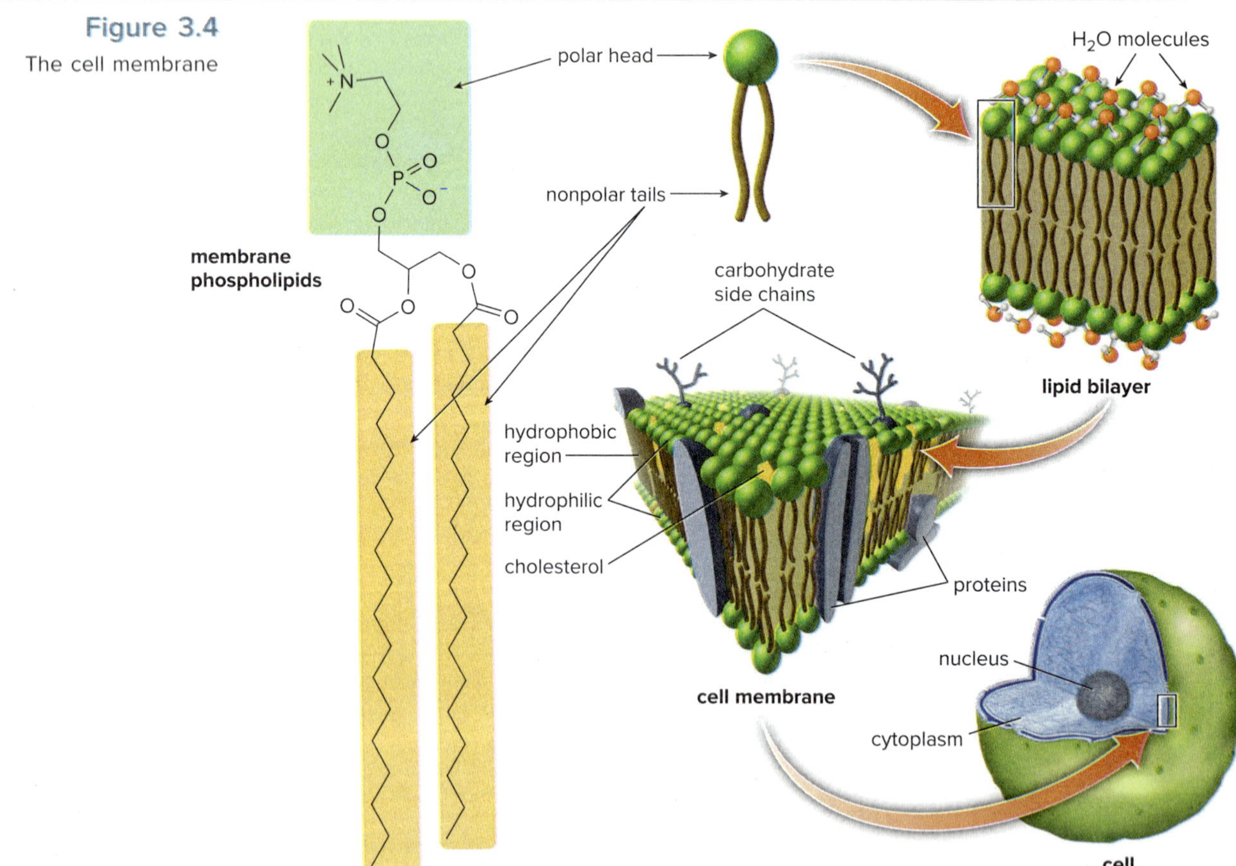

- Phospholipids contain an ionic or polar head, and two long nonpolar hydrocarbon tails. In an aqueous environment, phospholipids form a lipid bilayer, with the polar heads oriented toward the aqueous exterior and the nonpolar tails forming a hydrophobic interior. Cell membranes are composed largely of this lipid bilayer.

Cell membranes are composed of these lipid bilayers. The charged heads of the phospholipids are oriented toward the aqueous interior and exterior of the cell. The nonpolar tails form the hydrophobic interior of the membrane, thus serving as an insoluble barrier that protects the cell from the outside.

The nonpolar interior of the cell membrane is especially important in protecting the human brain from fluctuation in the concentration of compounds in the blood, as well as the passage of unwanted substances into the brain. The blood–brain barrier consists of a tight layer of cells in the blood capillaries of the brain, and all substances must pass through the cell membrane of these capillaries to enter the brain. Because ions are not soluble in the nonpolar interior of the cell membrane, the blood–brain barrier is only slightly permeable to ions. On the other hand, uncharged organic molecules like nicotine, caffeine, and heroin are very soluble in the interior of the cell membrane, so they readily pass into the brain.

nicotine

caffeine

heroin

sevoflurane

General anesthetics such as sevoflurane are also weakly polar compounds that can penetrate the blood–brain barrier because they are soluble in the lipid bilayer of the blood capillaries.

Problem 3.28 (a) What types of intermolecular forces do morphine and heroin each possess? (b) Which compound can cross the blood–brain barrier more readily, and therefore serve as the more potent pain reliever?

morphine

heroin

3.7B Transport Across a Cell Membrane

How does a polar molecule or ion in the water outside a cell pass through the nonpolar interior of the cell membrane and enter the cell? Some nonpolar molecules like O_2 are small enough to enter and exit the cell by diffusion. Polar molecules and ions, on the other hand, may be too large or too polar to diffuse efficiently. Some ions are transported across the membrane with the help of molecules called **ionophores.**

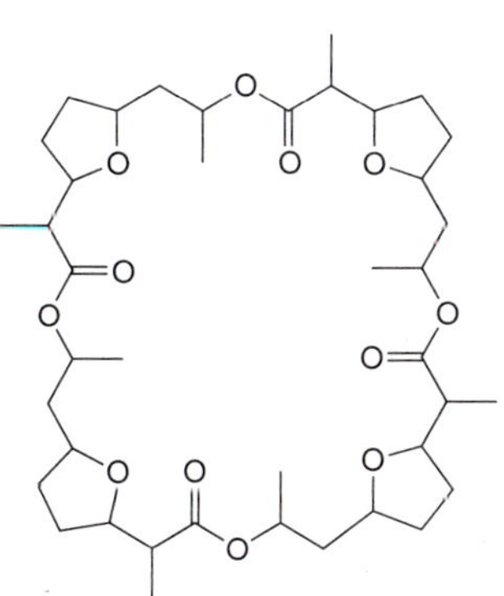

nonactin

***Ionophores* are organic molecules that complex cations.** They have a hydrophobic exterior that makes them soluble in the nonpolar interior of the cell membrane, and a central cavity with several oxygen atoms whose lone pairs complex with a given ion. The size of the cavity determines the identity of the cation with which the ionophore complexes. **Nonactin** is a naturally occurring antibiotic that acts as an ionophore.

Several synthetic ionophores have also been prepared, including one group called **crown ethers. *Crown ethers* are cyclic ethers containing several oxygen atoms that bind**

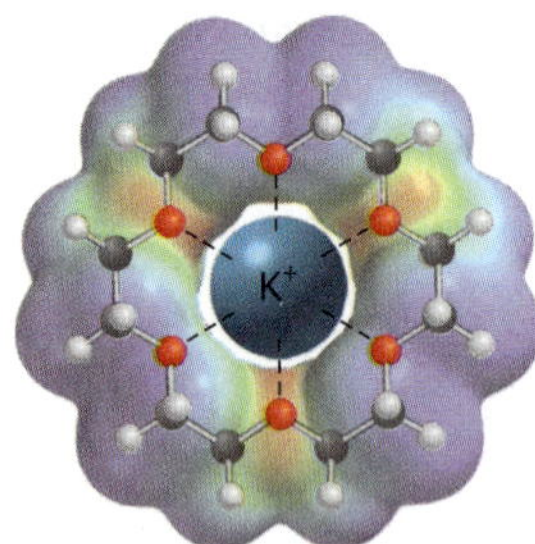

complex with K^+

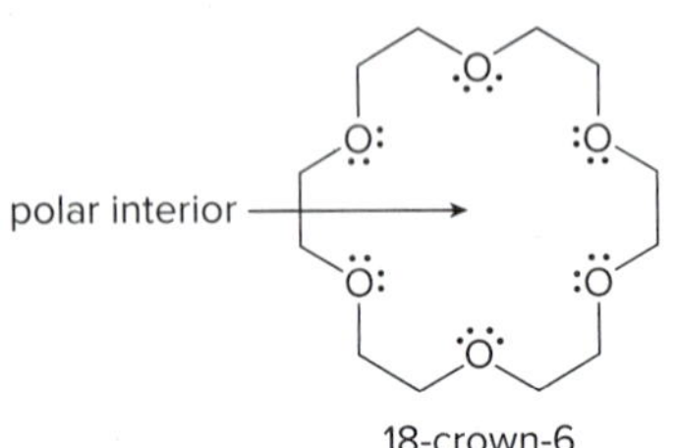

18-crown-6

specific cations depending on the size of their cavity. Crown ethers are named according to the general format ***x*-crown-*y*,** where *x* is the total number of atoms in the ring and *y* is the number of oxygen atoms. For example, 18-crown-6 contains 18 atoms in the ring, including 6 O atoms. This crown ether binds potassium ions. Sodium ions are too small to form a tight complex with the O atoms, and larger cations do not fit in the cavity.

How does an ionophore transfer an ion across a membrane? The ionophore binds the ion on one side of the membrane in its polar interior. It can then move across the membrane because its hydrophobic exterior interacts with the hydrophobic tails of the phospholipid. The ionophore then releases the ion on the other side of the membrane. This ion-transfer role is essential for normal cell function. This process is illustrated in Figure 3.5.

In this manner, antibiotic ionophores like nonactin transport ions across a cell membrane of bacteria. This disrupts the normal ionic balance in the cell, thus interfering with cell function and causing the bacteria to die.

Problem 3.29 Now that you have learned about solubility, explain why aspirin (Section 2.7) crosses a cell membrane as a neutral carboxylic acid rather than an ionic conjugate base.

Figure 3.5 Transport of ions across a cell membrane

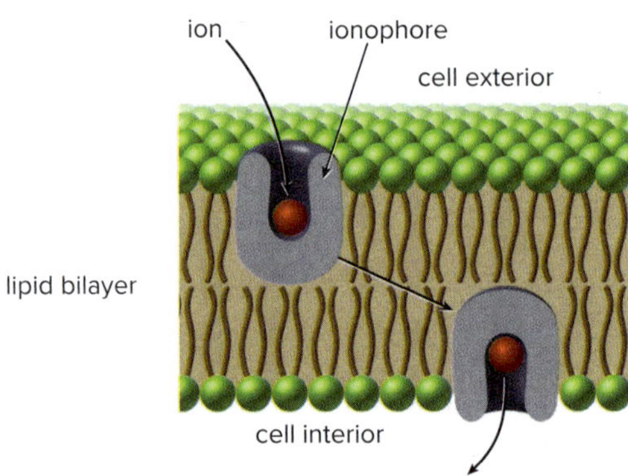

- By binding an ion on one side of a lipid bilayer (where the concentration of the ion is high) and releasing it on the other side of the bilayer (where the concentration of the ion is low), an ionophore transports an ion across a cell membrane.

3.8 Functional Groups and Reactivity

Much of Chapter 3 has been devoted to how a functional group determines the strength of intermolecular forces and, consequently, the physical properties of molecules. A functional group also determines reactivity. What type of reaction does a particular kind of organic compound undergo? Begin by recalling two fundamental concepts:

- Functional groups create reactive sites in molecules.
- Electron-rich sites react with electron-poor sites.

All functional groups contain a heteroatom, a π bond, or both, and these features make electron-deficient (or electrophilic) sites and electron-rich (or nucleophilic) sites in a

molecule. To predict reactivity, first locate the functional group and then determine the resulting electron-rich or electron-deficient sites it creates. Keep three guidelines in mind:

- **An electronegative heteroatom like N, O, or X makes a carbon atom *electrophilic*.**

- **A lone pair on a heteroatom makes it *basic* and *nucleophilic*.**

- **π Bonds create *nucleophilic* sites and are more easily broken than σ bonds.**

C=C

one easily broken π bond

—C≡C—

two easily broken π bonds

Problem 3.30 Label the electrophilic and nucleophilic sites in each molecule.

a. Br

b. S

c. O, O, N, H, H

By identifying the nucleophilic and electrophilic sites in a compound you can begin to understand how it will react. In general, electron-rich sites react with electron-deficient sites:

- **An electron-deficient carbon atom reacts with a nucleophile, symbolized as :Nu⁻.**
- **An electron-rich carbon reacts with an electrophile, symbolized as E^+.**

:Nu⁻ = a nucleophile;
E^+ = an electrophile.

At this point we don't know enough organic chemistry to draw the products of many reactions with confidence. We do know enough, however, to begin to predict if two compounds might react together based solely on electron density arguments, and at what atoms that reaction is most likely to occur.

For example, alkenes contain an electron-rich C–C double bond, so they react with electrophiles, E^+. On the other hand, alkyl halides possess an electrophilic carbon atom, so they react with electron-rich nucleophiles.

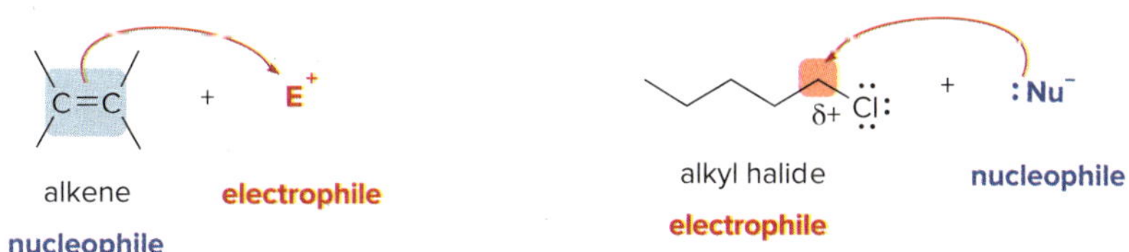

You don't need to worry about the products of these reactions. At this point you should only be able to find reactive sites in molecules and begin to understand why a reaction might occur at these sites. After you learn more about the structure of organic molecules in Chapters 4 and 5, we will begin a detailed discussion of organic reactions in Chapter 6.

Problem 3.31 Considering only electron density, state whether the following reactions will occur.

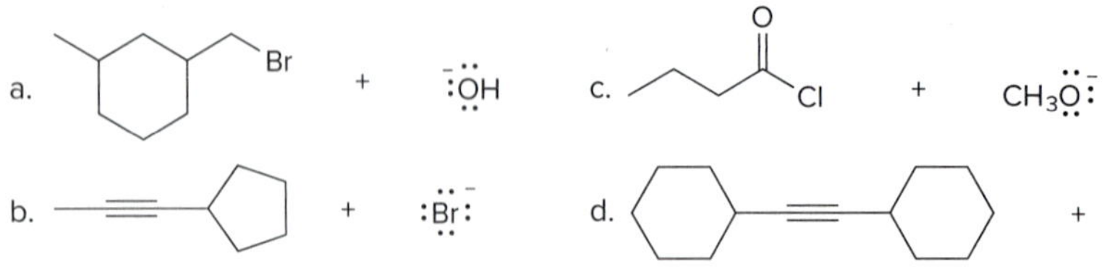

a. Br + $:\ddot{O}H^-$

b. + $:\ddot{Br}:^-$

c. O, Cl + $CH_3\ddot{O}:^-$

d. + $:\ddot{Br}:^+$

3.9 Biomolecules

***Biomolecules* are organic compounds found in biological systems.** Many are relatively small, with molecular weights of less than 1000 g/mol. There are four main families of these small molecules—simple sugars, amino acids, lipids, and nucleotides. Many simple biomolecules are used to synthesize larger compounds that have important cellular functions.

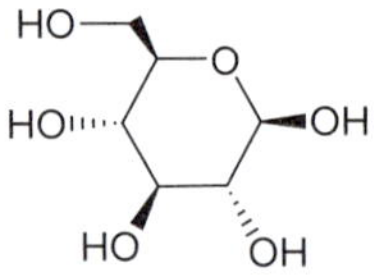

glucose
a simple sugar

oleic acid
a fatty acid

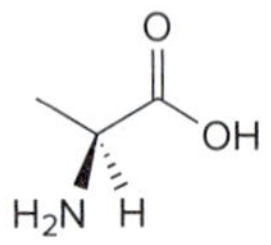

alanine
an amino acid

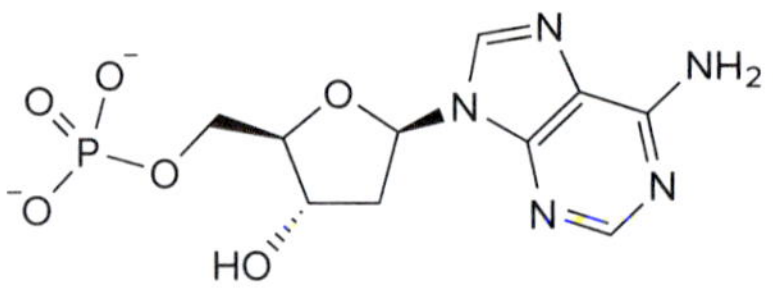

2'-deoxyadenosine 5'-monophosphate
a nucleotide

Simple sugars such as glucose combine to form the complex carbohydrates starch and cellulose, as described in Chapter 26. Alanine is an amino acid used to synthesize proteins, the subject of Chapter 27. Fatty acids such as oleic acid react with alcohols to form triacylglycerols, the most prevalent lipids, first mentioned in Chapter 10, and discussed in more detail in Chapters 18 and 29. Finally, 2'-deoxyadenosine 5'-monophosphate is a nucleotide that combines with thousands of other nucleotides to form DNA, deoxyribonucleic acid, the high-molecular-weight polynucleotide that stores the genetic information of an organism, as described in Chapter 28. The properties and reactions of these biomolecules are explained by the principles of basic organic chemistry.

Problem 3.32 The fact that sweet-tasting carbohydrates like table sugar are also high in calories has prompted the development of sweet, low-calorie alternatives.

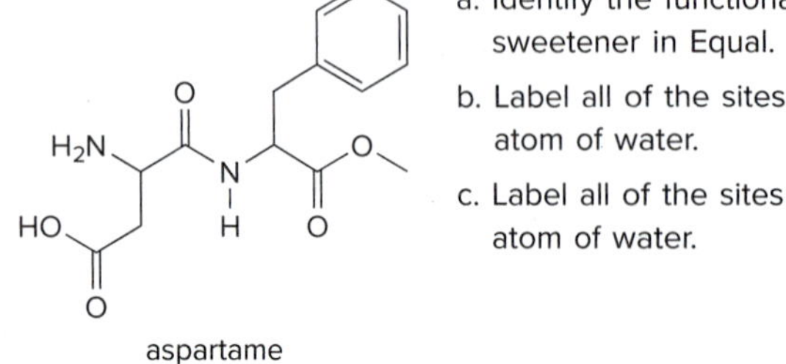

aspartame

a. Identify the functional groups in aspartame, the artificial sweetener in Equal.

b. Label all of the sites that can hydrogen bond to the oxygen atom of water.

c. Label all of the sites that can hydrogen bond to a hydrogen atom of water.

Chapter 3 REVIEW

KEY CONCEPTS

[1] Classifying atoms and functional groups (3.2)

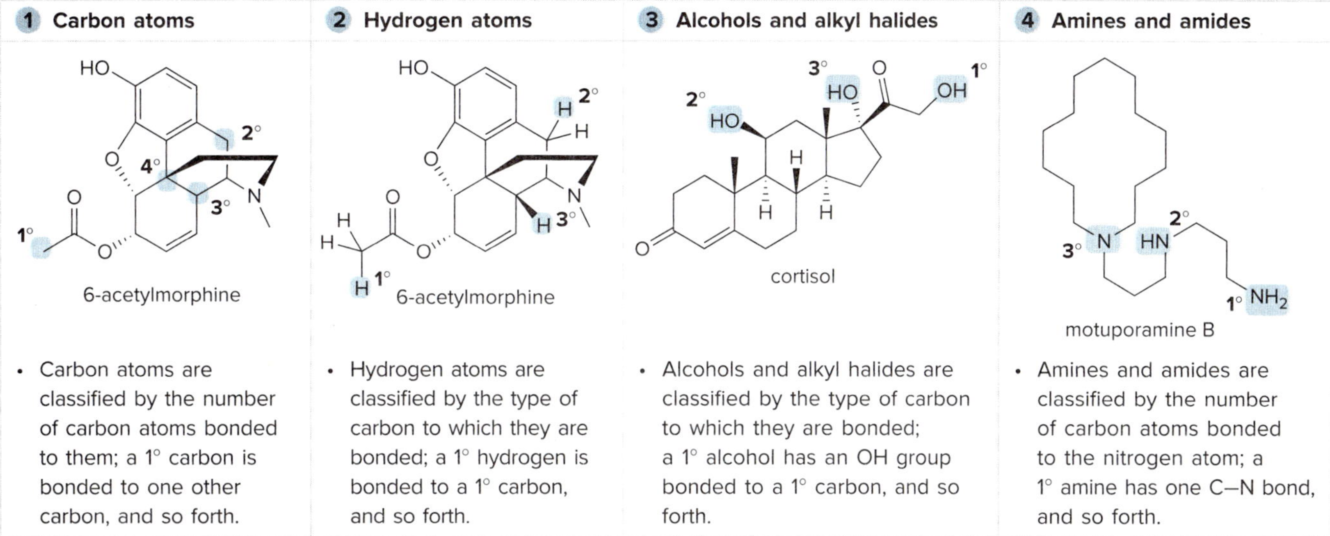

1 Carbon atoms	2 Hydrogen atoms	3 Alcohols and alkyl halides	4 Amines and amides
• Carbon atoms are classified by the number of carbon atoms bonded to them; a 1° carbon is bonded to one other carbon, and so forth.	• Hydrogen atoms are classified by the type of carbon to which they are bonded; a 1° hydrogen is bonded to a 1° carbon, and so forth.	• Alcohols and alkyl halides are classified by the type of carbon to which they are bonded; a 1° alcohol has an OH group bonded to a 1° carbon, and so forth.	• Amines and amides are classified by the number of carbon atoms bonded to the nitrogen atom; a 1° amine has one C–N bond, and so forth.

See Sample Problem 3.1. Try Problem 3.35.

[2] Types of intermolecular forces (3.3)

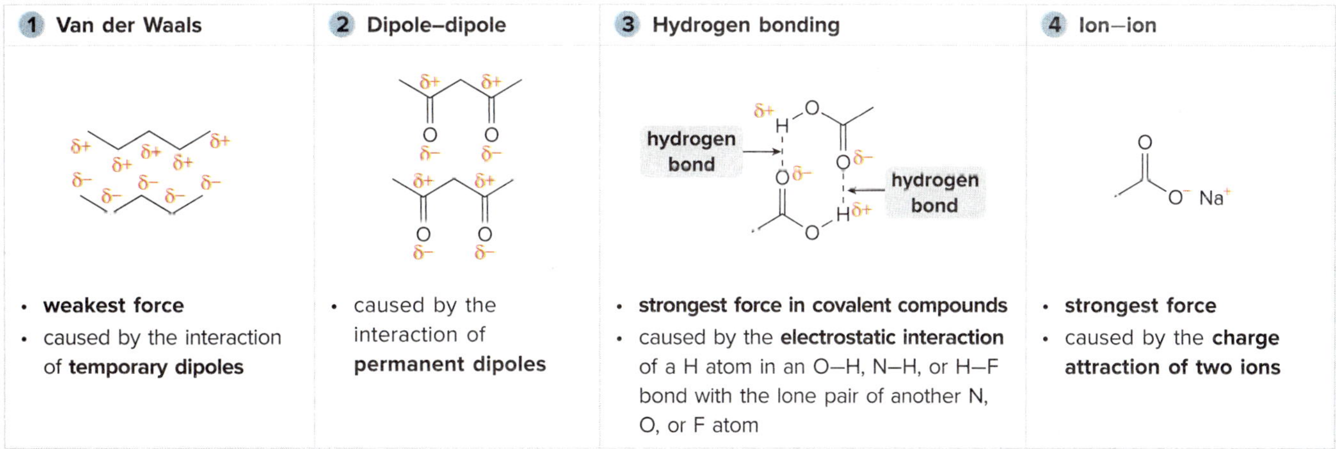

1 Van der Waals	2 Dipole–dipole	3 Hydrogen bonding	4 Ion–ion
• **weakest force** • caused by the interaction of **temporary dipoles**	• caused by the interaction of **permanent dipoles**	• **strongest force in covalent compounds** • caused by the **electrostatic interaction** of a H atom in an O–H, N–H, or H–F bond with the lone pair of another N, O, or F atom	• **strongest force** • caused by the **charge attraction of two ions**

See Table 3.4, Sample Problem 3.3. Try Problems 3.39, 3.42, 3.64d.

[3] Factors that determine boiling point (3.4A)

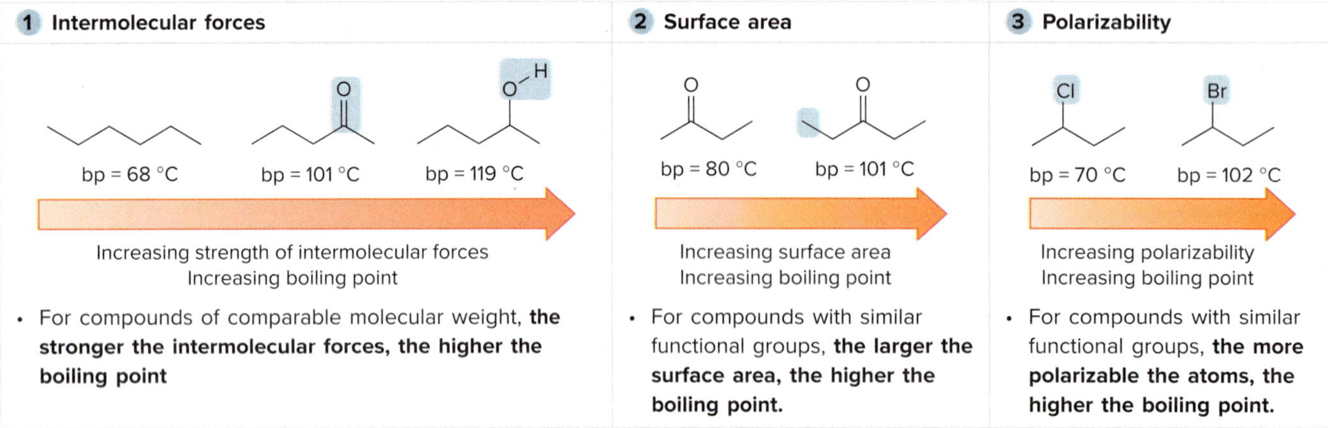

- For compounds of comparable molecular weight, **the stronger the intermolecular forces, the higher the boiling point**
- For compounds with similar functional groups, **the larger the surface area, the higher the boiling point.**
- For compounds with similar functional groups, **the more polarizable the atoms, the higher the boiling point.**

See Figure 3.1, Sample Problem 3.4. Try Problems 3.43, 3.44.

[4] Factors that determine melting point (3.4B)

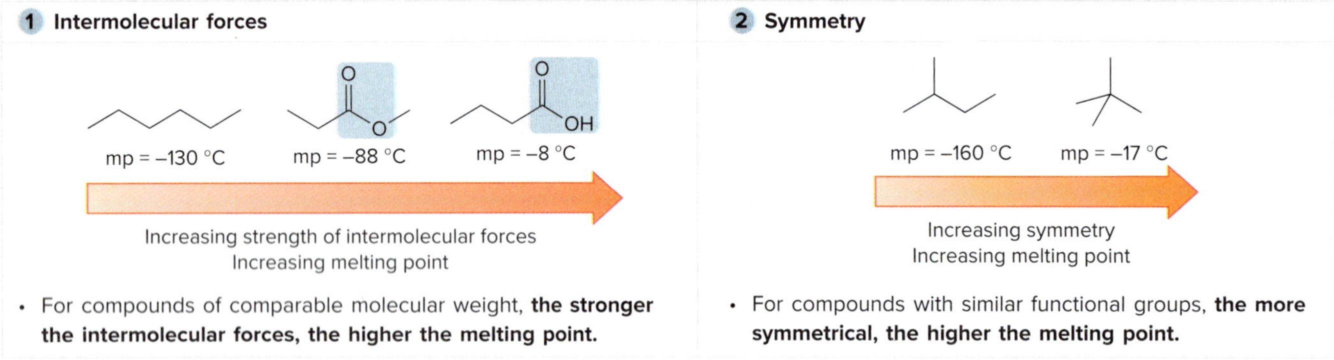

- For compounds of comparable molecular weight, **the stronger the intermolecular forces, the higher the melting point.**
- For compounds with similar functional groups, **the more symmetrical, the higher the melting point.**

Try Problems 3.45, 3.48a.

[5] Factors that determine solubility (3.4C, 3.5)

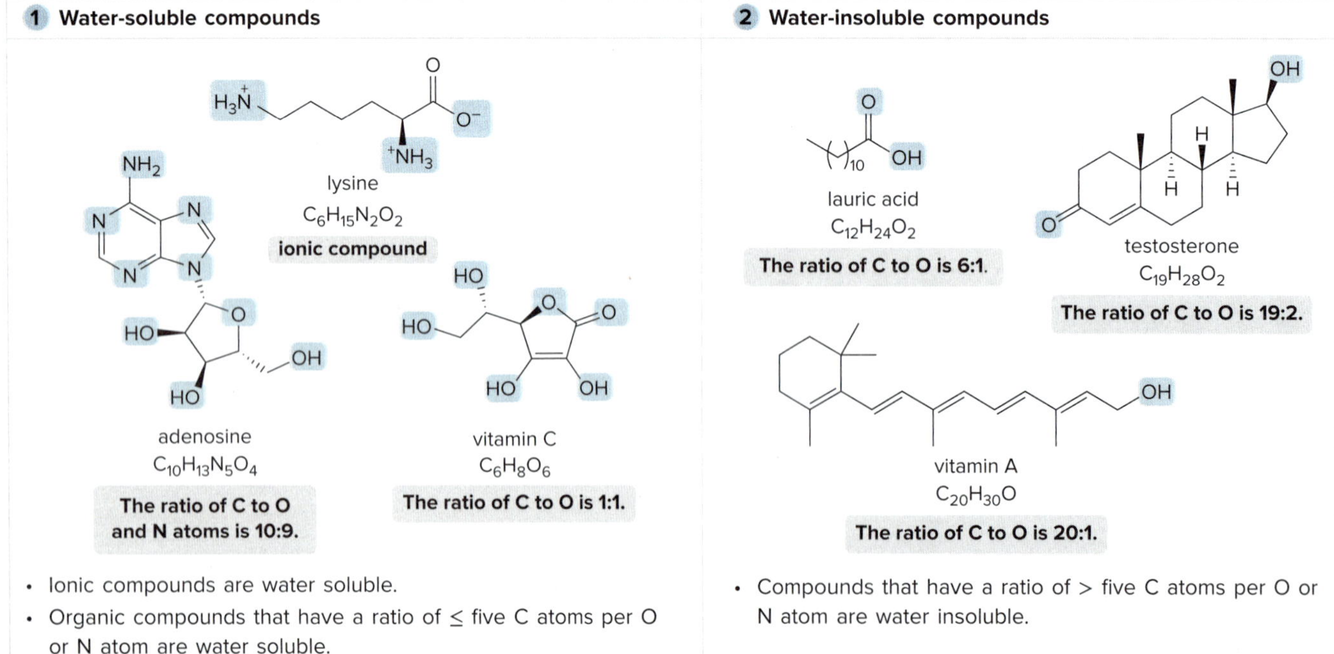

- Ionic compounds are water soluble.
- Organic compounds that have a ratio of ≤ five C atoms per O or N atom are water soluble.
- Compounds that have a ratio of > five C atoms per O or N atom are water insoluble.

See Figure 3.2, Table 3.5. Try Problems 3.47; 3.48b, c; 3.49–3.52.

[6] Reactivity of functional groups (3.8)

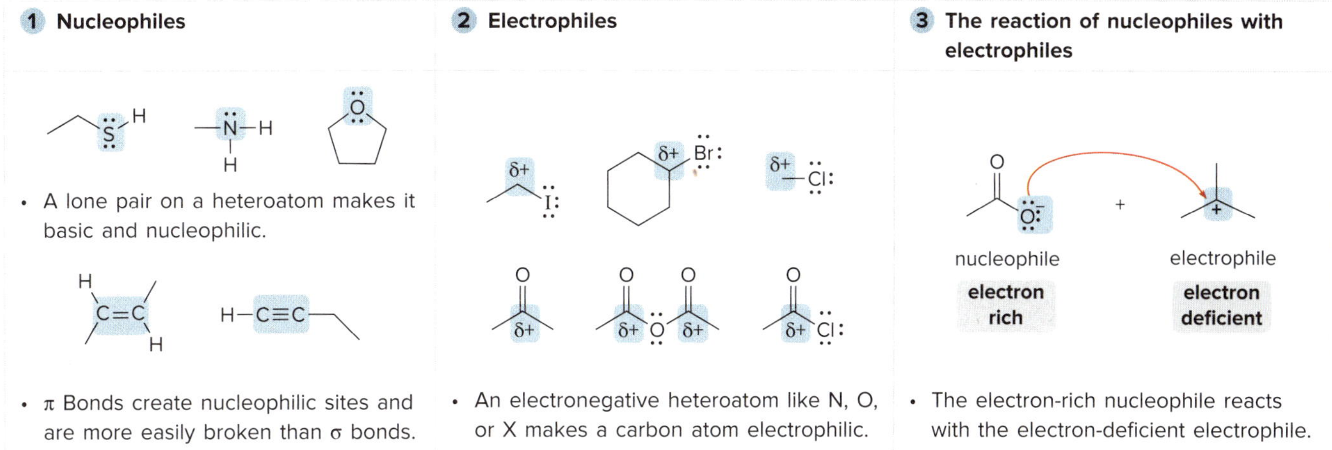

Try Problems 3.33c; 3.34c; 3.55; 3.56; 3.59d, e; 3.60c; 3.64g.

KEY SKILLS

[1] Predicting boiling points (3.4A)

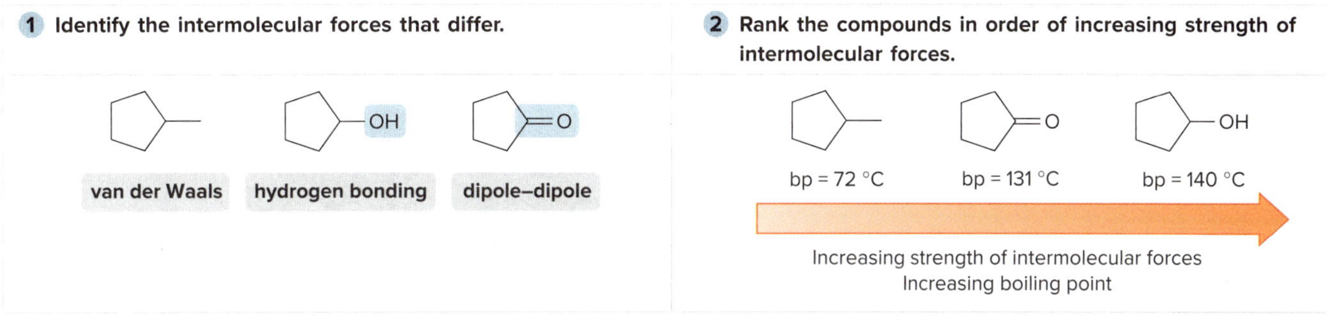

See Sample Problem 3.3. Try Problems 3.43, 3.44.

[2] Determining sites of hydrogen bonding between two identical molecules (3.4C)

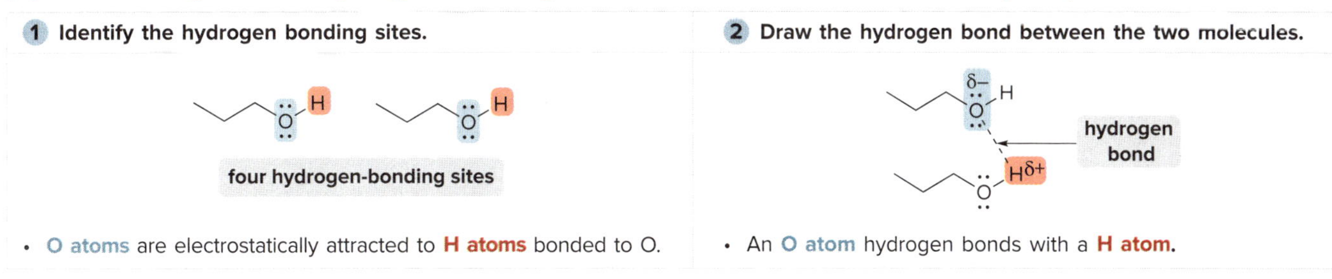

See Sample Problem 3.5. Try Problems 3.40a, 3.41a, 3.60b.

[3] Determining sites of hydrogen bonding between an organic molecule and H_2O (3.4C)

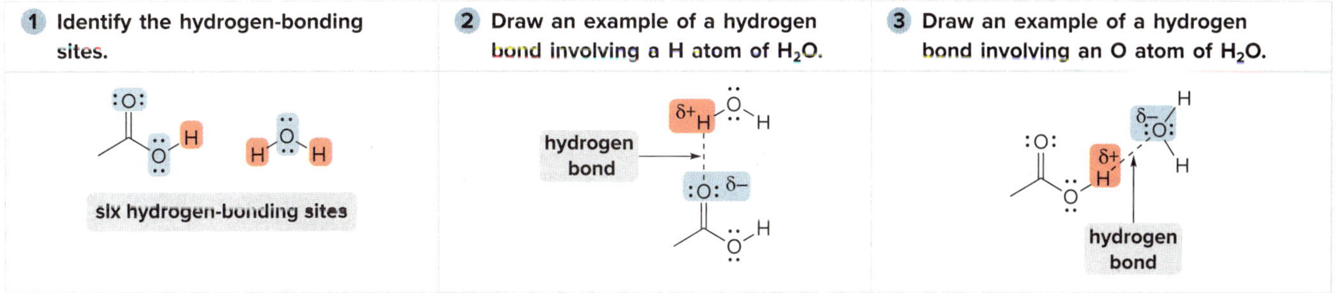

Try Problems 3.40b, 3.41b, 3.59c, 3.61c, 3.64f.

[4] Drawing curved arrows to show the reaction between a nucleophile and an electrophile (3.8)

1 Identify the nucleophile and electrophile.

nucleophile — **electron rich**

electrophile — **electron deficient**

2 Draw a curved arrow from the nucleophile to the electrophile.

nucleophile — electrophile

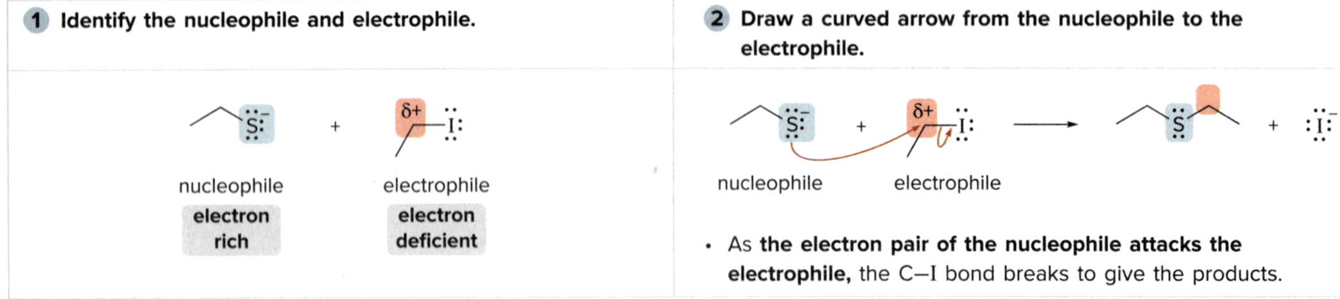

- As **the electron pair of the nucleophile attacks the electrophile,** the C–I bond breaks to give the products.

Try Problem 3.56.

CHAPTER 3 MULTIPLE-CHOICE SELF-TEST

The Self-Test consists of multiple-choice questions similar to those found on the American Chemical Society organic chemistry exam. Answers are given at the end of the chapter.

1. Which of the following compounds can hydrogen bond both to another molecule like itself and to water?

a.
b.
c.
d.

2. Rank **A, B,** and **C** in order of increasing boiling point.

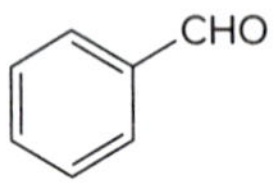

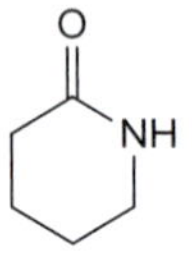

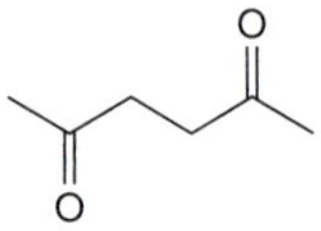

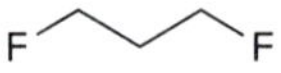

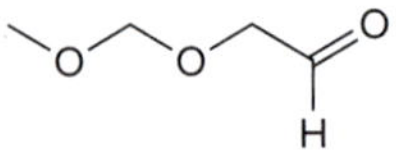

A

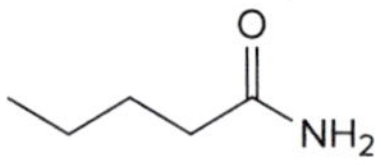

B

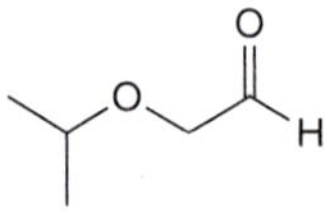

C

a. **B** < **A** < **C**
b. **A** < **C** < **B**
c. **C** < **B** < **A**
d. **C** < **A** < **B**

3. Which compound exhibits the weakest intermolecular forces?

a.
b.
c.
d.

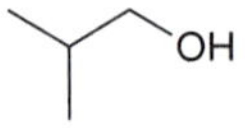

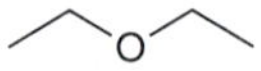

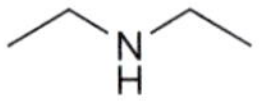

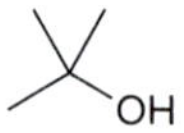

4. Which compound is *not* a ketone: (a) $(CH_3CH_2)_2CO$; (b) $CH_3CH_2COCH_3$; (c) $(CH_3)_2CHOCH_2CH_3$; (d) $CH_3COCH(CH_3)_2$?

5. In addition to an aromatic ring, what functional groups are present in metoprolol?

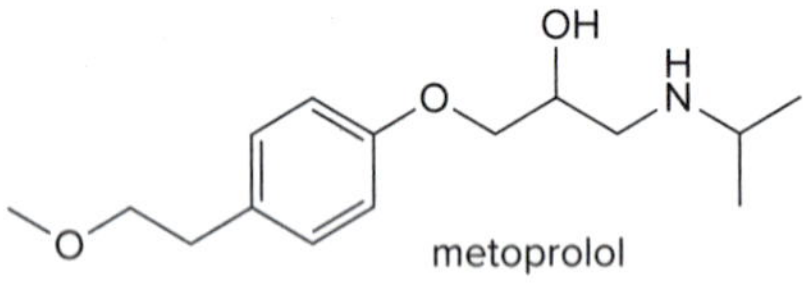

metoprolol

a. two ethers, 1° alcohol, 2° amine
b. ether, ester, 1° alcohol, 1° amine
c. two ethers, 2° alcohol, 2° amine
d. two ethers, 2° alcohol, 2° amide

6. How many 3° C's and 2° H's are present in the given compound?

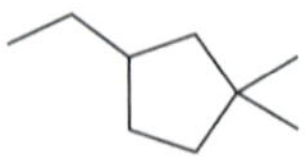

a. one 3° C, eight 2° H's
b. two 3° C's, eight 2° H's
c. two 3° C's, six 2° H's
d. one 3° C, nine 2° H's

7. What functional group is *not* present in penicillin G?

penicillin G

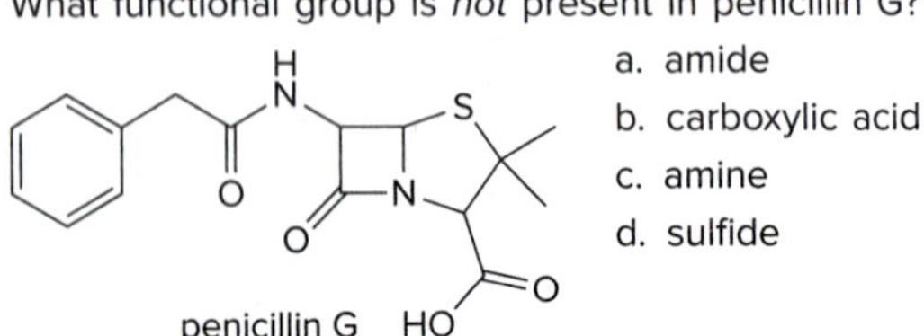

a. amide
b. carboxylic acid
c. amine
d. sulfide

8. Which compound is *not* water soluble?

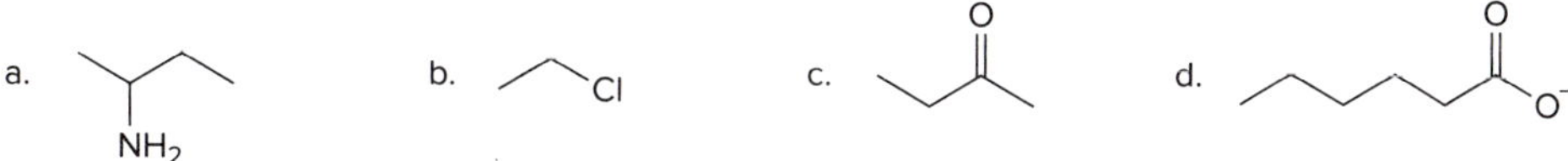

9. Which statement describes the physical properties of **A, B,** and **C?**

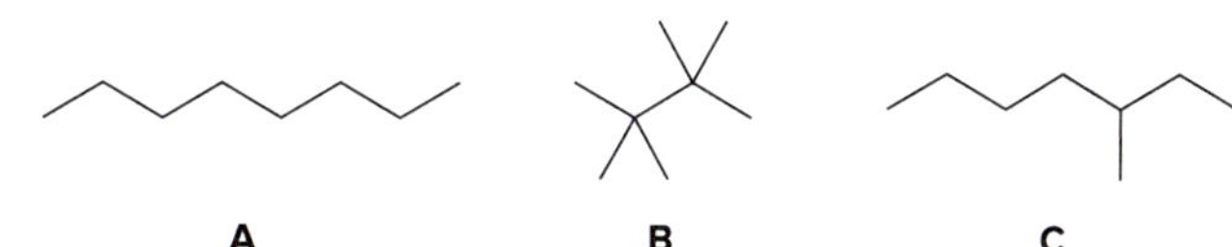

a. The bp of **A** is higher than the bp of **B,** and the mp of **B** is higher than the mp of **A.**
b. The bp and mp of **A** are higher than the bp and mp of **B.**
c. The bp and mp of **B** are higher than the bp and mp of **C.**
d. The bp of **A** is higher than the bp of **C,** and the mp of **C** is higher than the mp of **A.**

10. Which labeled site is unlikely to react with a nucleophile?

OH a. b. Cl c. d. O O

PROBLEMS

Problems with Three-Dimensional Models

3.33

elemicin

a. Identify the functional groups in the ball-and-stick model of elemicin, a compound partly responsible for the flavor and fragrance of nutmeg.
b. Draw a skeletal structure of a constitutional isomer of elemicin that should have a higher boiling point and melting point.
c. Label all electrophilic carbon atoms.

3.34

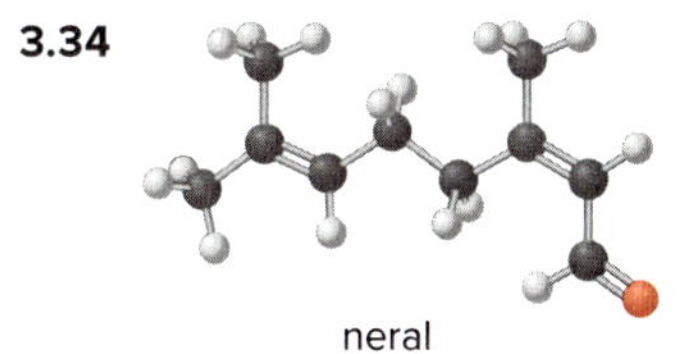

neral

a. Identify the functional groups in the ball-and-stick model of neral, a compound with a lemony odor isolated from lemongrass.
b. Draw a skeletal structure of a constitutional isomer of neral that should be more water soluble.
c. Label the most electrophilic carbon atom.

Functional Groups

3.35 For each alkane: (a) classify each carbon atom as 1°, 2°, 3°, or 4°; (b) classify each hydrogen atom as 1°, 2°, or 3°.

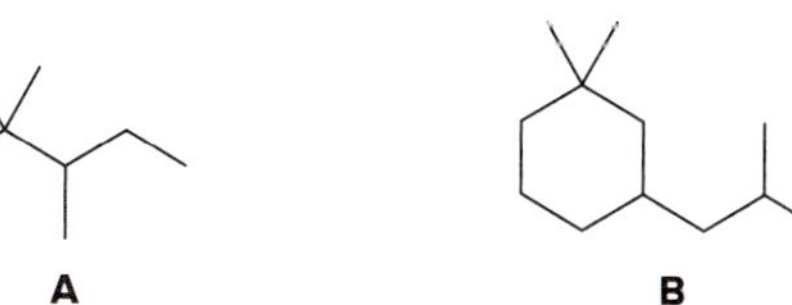

3.36 Identify the functional groups in each molecule. Classify each alcohol, alkyl halide, amide, and amine as 1°, 2°, or 3°.

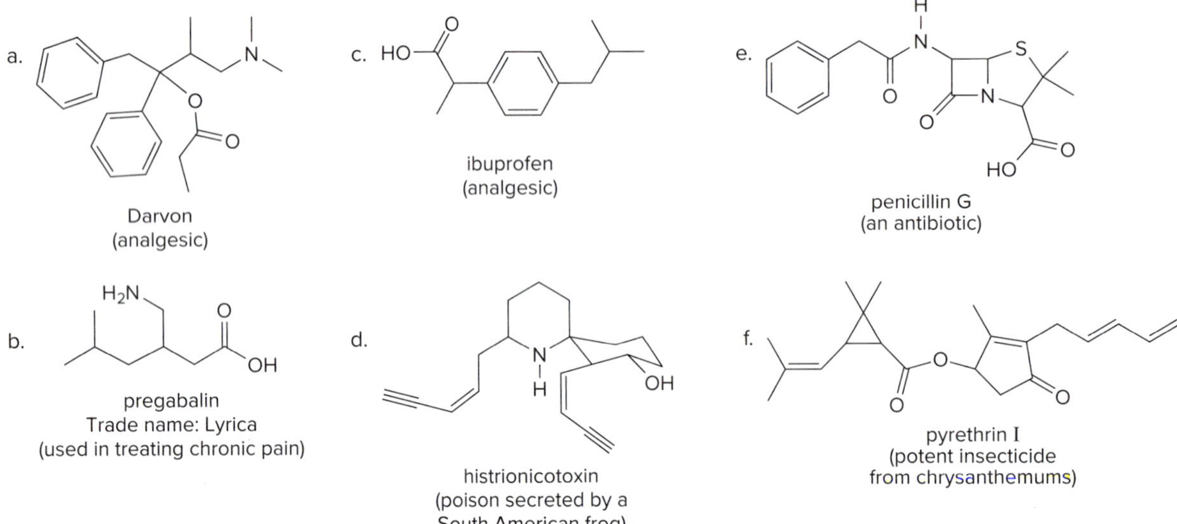

3.37 Salinosporamide A is an anticancer agent isolated from marine sediment.

a. Identify the functional groups in salinosporamide A.

b. Classify each alcohol, alkyl halide, amide, and amine as 1°, 2°, or 3°.

salinosporamide A

3.38 Draw seven constitutional isomers with molecular formula $C_3H_6O_2$ that contain a carbonyl group. Identify the functional group(s) in each isomer.

Intermolecular Forces

3.39 What types of intermolecular forces are exhibited by each compound?

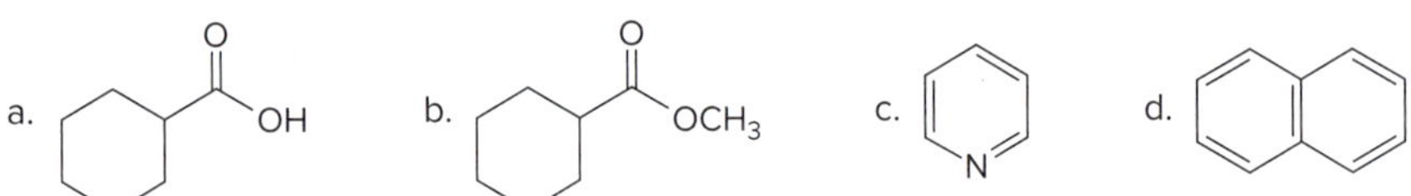

3.40 (a) Which of the following molecules can hydrogen bond to another molecule like itself? (b) Which of the following molecules can hydrogen bond to water?

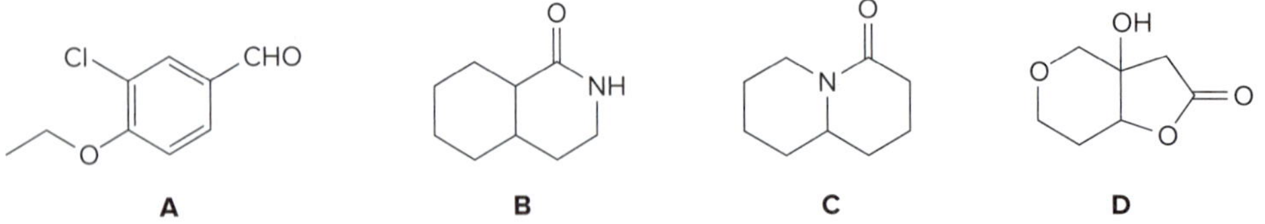

3.41 Indinavir (trade name Crixivan) is a drug used to treat HIV. (a) At which sites can indinavir hydrogen bond to another molecule like itself? (b) At which sites can indinavir hydrogen bond to water?

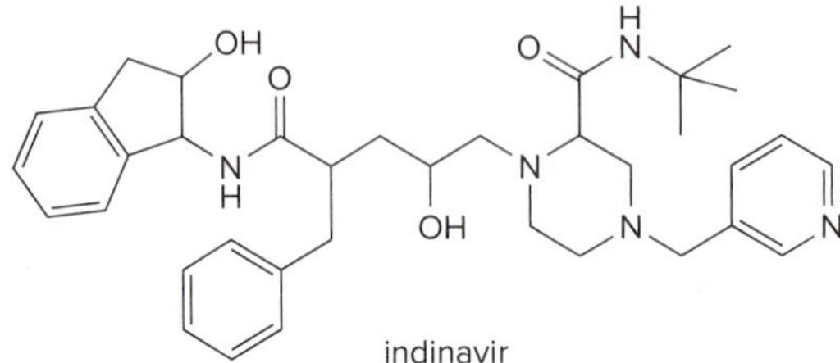

3.42 Intramolecular forces of attraction are often important in holding large molecules together. For example, some proteins fold into compact shapes, held together by attractive forces between nearby functional groups. A schematic of a folded protein is drawn here, with the protein backbone indicated by a blue ribbon, and various appendages drawn dangling from the chain. What types of intramolecular forces occur at each labeled site (**A–F**)?

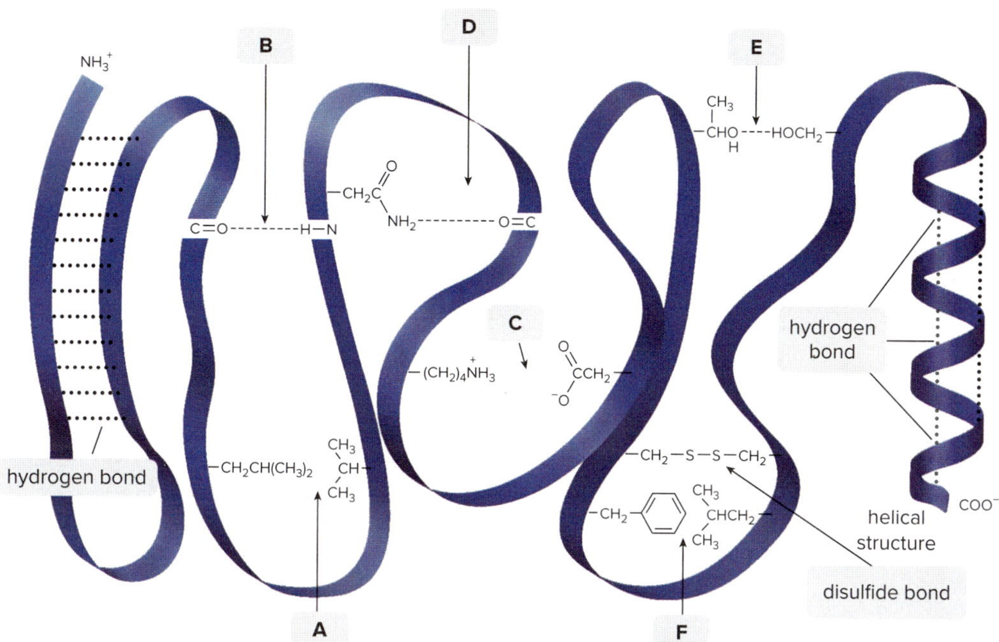

Physical Properties

3.43 (a) Draw four compounds with molecular formula $C_6H_{12}O$, each containing at least one different functional group. (b) Predict which compound has the highest boiling point, and explain your reasoning.

3.44 Rank the compounds in each group in order of increasing boiling point.

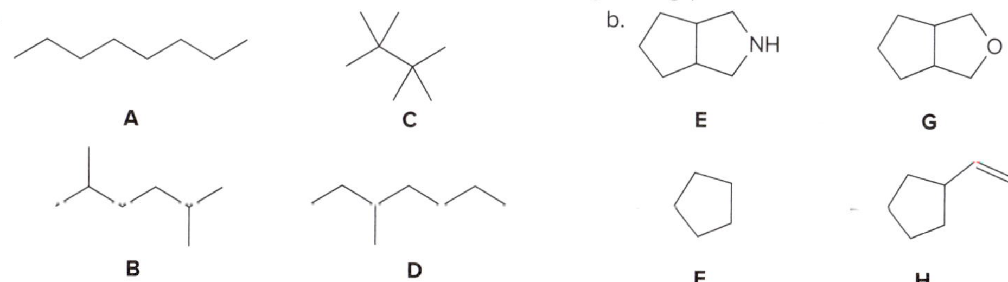

3.45 Menthone and menthol are both isolated from mint. Explain why menthol is a solid at room temperature but menthone is a liquid.

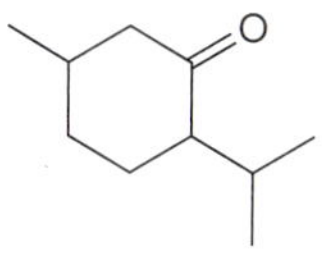

O	OH
menthone 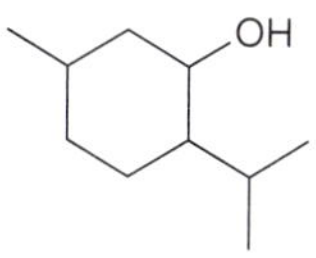	menthol

3.46 Explain why benzene has a lower boiling point but much higher melting point than toluene.

benzene
bp = 80 °C
mp = 5 °C

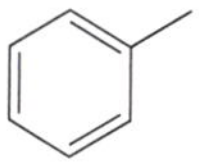

toluene
bp = 111 °C
mp = −93 °C

3.47 Predict the water solubility of each of the following organic molecules.

a. 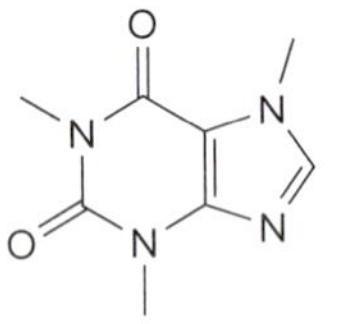 caffeine (stimulant in coffee, tea, and many soft drinks)

b. 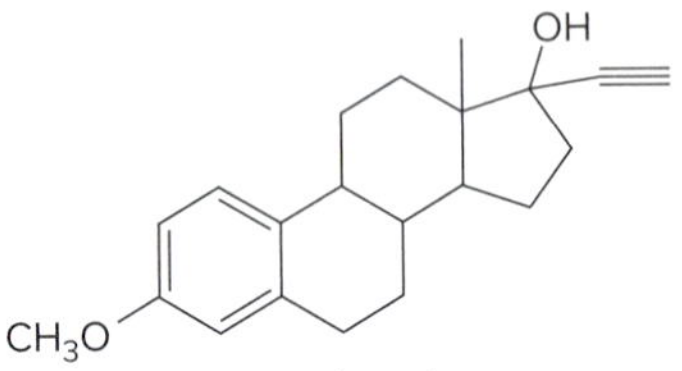 mestranol (component in oral contraceptives)

c. acetylcholine (neurotransmitter)

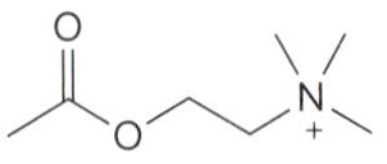

d. 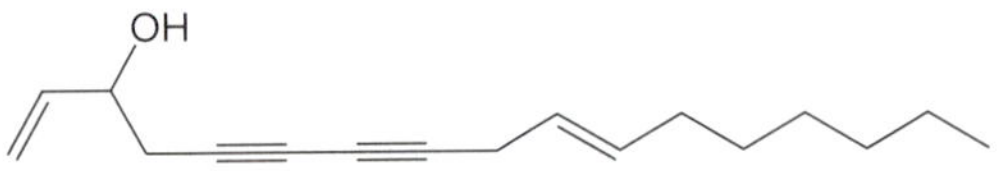 carotatoxin (neurotoxin isolated from carrots)

3.48 Use the structures of keto acid **A** and amino acid **B** to answer each question.

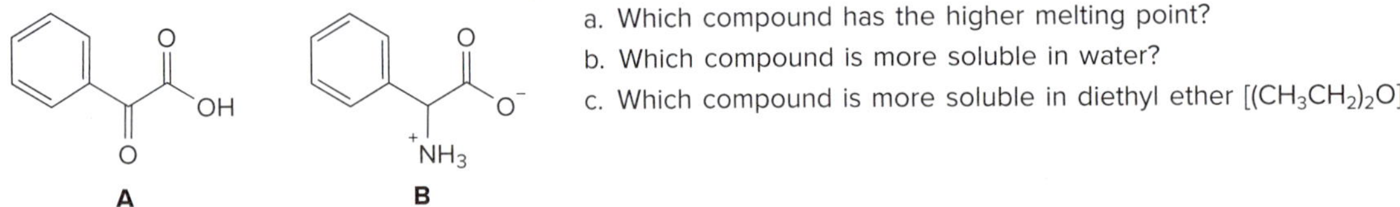

a. Which compound has the higher melting point?

b. Which compound is more soluble in water?

c. Which compound is more soluble in diethyl ether [$(CH_3CH_2)_2O$]?

Applications

3.49 Predict the solubility of each of the following vitamins in water and in organic solvents.

a. 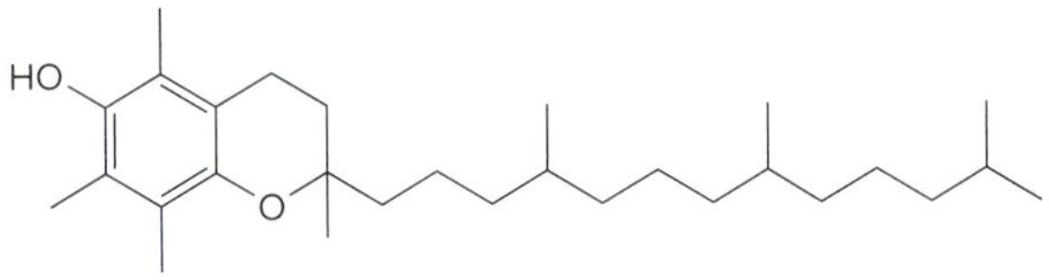 vitamin E

b. 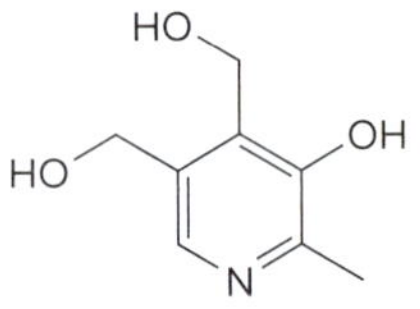 pyridoxine vitamin B_6

3.50 Avobenzone and dioxybenzone are two commercial sunscreens. Using the principles of solubility, predict which sunscreen is more readily washed off when an individual goes swimming. Explain your choice.

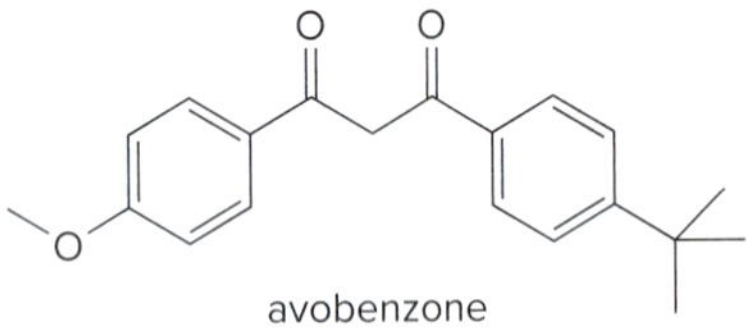

avobenzone

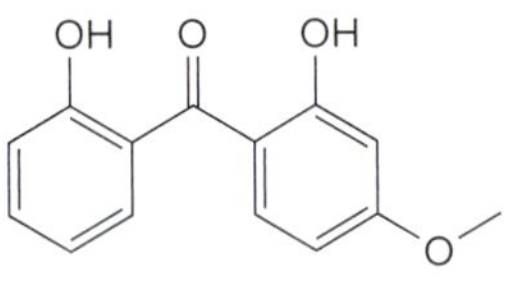

dioxybenzone

3.51 Poly(ethylene glycol) (PEG) and poly(vinyl chloride) (PVC) are examples of polymers, large organic molecules composed of repeating smaller units covalently bonded together. Polymers have very different properties depending (in part) on their functional groups. Discuss the water solubility of each polymer and suggest why PEG is used in shampoos, whereas PVC is used to make garden hoses and pipes. Synthetic polymers are discussed in detail in Chapters 13 and 31.

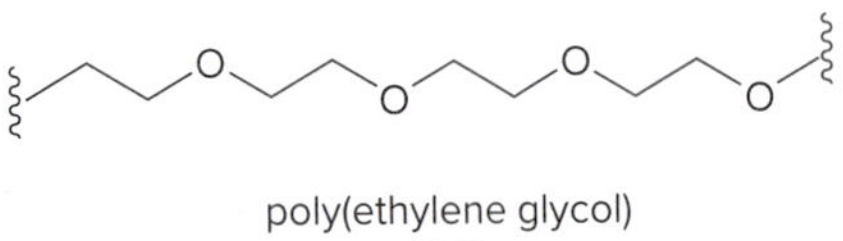

poly(ethylene glycol) PEG

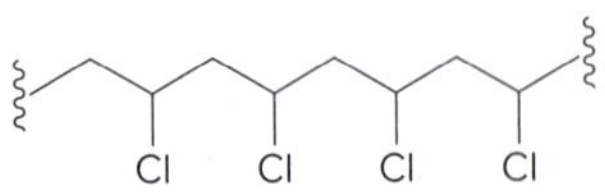

poly(vinyl chloride) PVC

3.52 THC is the active component in marijuana, and ethanol is the alcohol in alcoholic beverages. Explain why drug screenings are able to detect the presence of THC but not ethanol weeks after these substances have been introduced into the body.

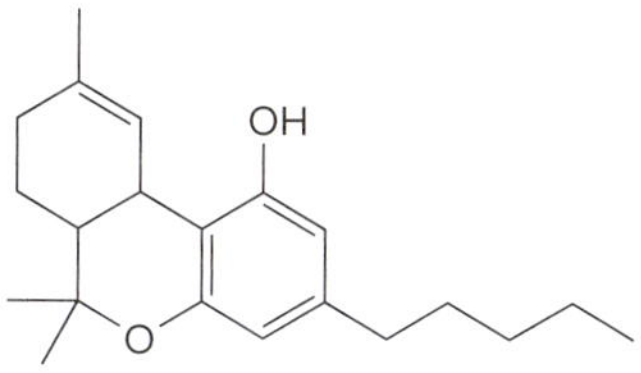

tetrahydrocannabinol
THC

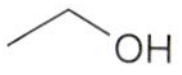

ethanol

3.53 Cocaine is a widely abused, addicting drug. Cocaine is usually obtained as its hydrochloride salt (cocaine hydrochloride) but can be converted to crack (the neutral organic molecule) by treatment with base. Which of the two compounds here has a higher boiling point? Which is more soluble in water? How does the relative solubility explain why crack is usually smoked but cocaine hydrochloride is injected directly into the bloodstream?

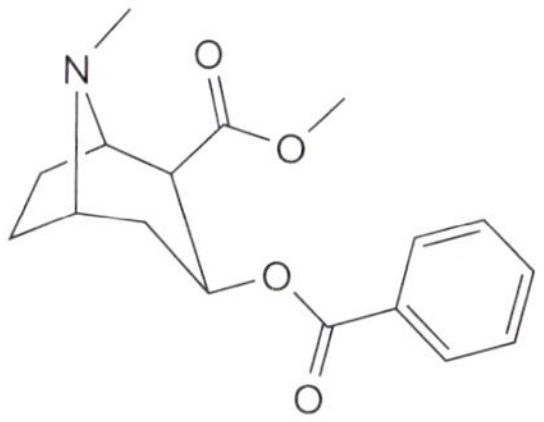

cocaine (crack)
neutral organic molecule

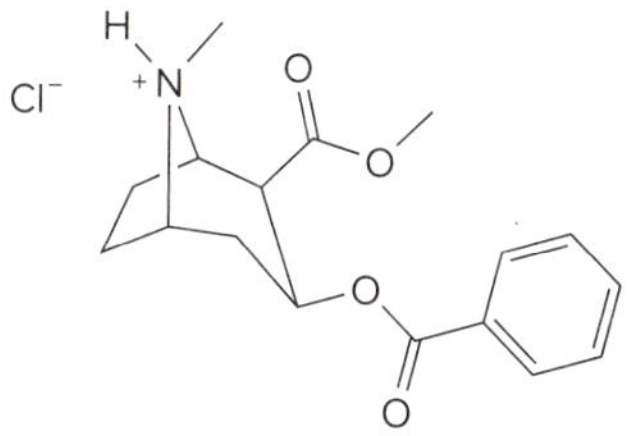

cocaine hydrochloride
a salt

3.54 Many drugs are sold as their hydrochloride salts ($R_2NH_2^+$ Cl^-), formed by reaction of an amine (R_2NH) with HCl.

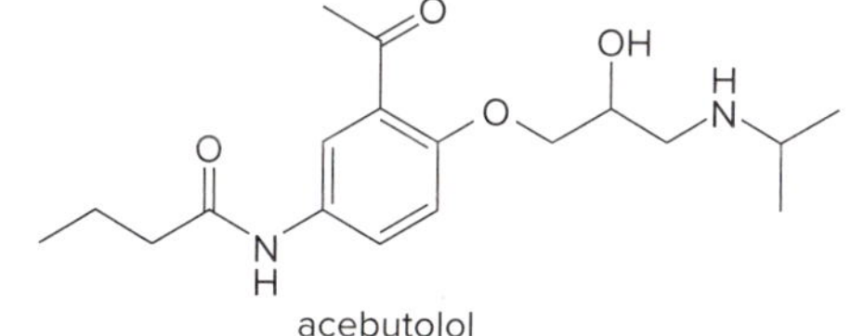

acebutolol

a. Draw the product (a hydrochloride salt) formed by reaction of acebutolol with HCl. Acebutolol is a β blocker used to treat high blood pressure.

b. Discuss the solubility of acebutolol and its hydrochloride salt in water.

c. Offer a reason as to why the drug is marketed as a hydrochloride salt rather than a neutral amine.

Reactivity of Organic Molecules

3.55 Label the electrophilic and nucleophilic sites in each molecule.

a. 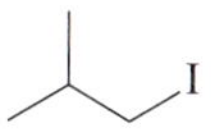 b. 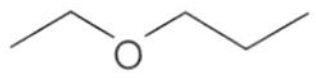 c. 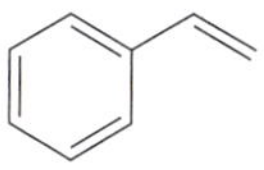 d.

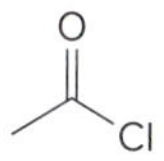

3.56 By using only electron density arguments, determine whether the following reactions will occur.

a.  + Br^- ⟶

b. 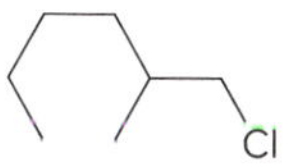 + ^-CN ⟶

c. 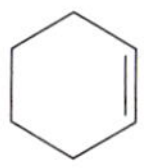 + ^-OH ⟶

d. 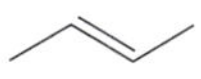 + H_3O^+ ⟶

Cell Membrane

3.57 The composition of a cell membrane is not uniform for all types of cells. Some cell membranes are more rigid than others. Rigidity is determined by a variety of factors, one of which is the structure of the carbon chains in the phospholipids that comprise the membrane. One example of a phospholipid was drawn in Section 3.7A, and another, having C–C double bonds in its carbon chains, is drawn here. Which phospholipid would be present in the more rigid cell membrane and why?

phospholipid

3.58 Which compound is more likely to be a general anesthetic? Explain your choice.

A B

Problems That Combine Concepts

3.59 Naloxone is a drug used to treat overdoses of heroin and prescription opioids.

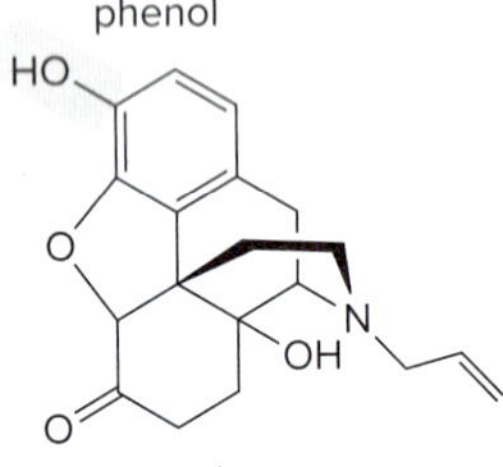

naloxone

a. In addition to the labeled phenol (a functional group that consists of an OH group bonded to a benzene ring), what other functional groups are present in naloxone?

b. Classify any amine or alcohol as 1°, 2°, or 3°.

c. At which sites can naloxone hydrogen bond to water?

d. Label two nucleophilic sites.

e. Label two electrophilic sites.

3.60 In March of 2021 the drugmaker Pfizer began clinical trials of PF-07321332 (generic name nirmatrelvir), an oral antiviral medicine, for the treatment of COVID-19. PF-07321332 binds to a viral enzyme in SARS-CoV-2, the virus that causes COVID-19, thus preventing the virus from replicating (making copies of itself) in the cell.

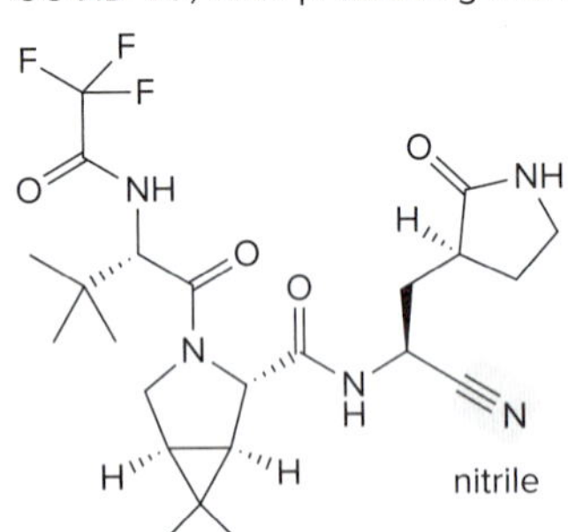

PF-07321332

a. In addition to the labeled nitrile, a functional group we will learn about in later chapters, what other functional groups are present in PF-07321332? Classify any amide or amine as 1°, 2°, or 3°.

b. At what sites can PF-07321332 hydrogen bond to another molecule like itself?

c. Label three electrophilic sites.

d. How many *sp*, sp^2, and sp^3 hybridized C's are present?

3.61 Quinapril (trade name Accupril) is a drug used to treat hypertension and congestive heart failure.

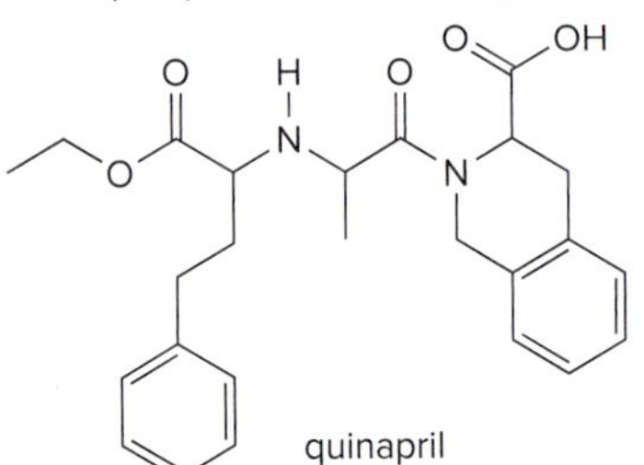
quinapril

a. Identify the functional groups in quinapril.

b. Classify any alcohol, amide, or amine as 1°, 2°, or 3°.

c. At which sites can quinapril hydrogen bond to water?

d. At which sites can quinapril hydrogen bond to acetone $[(CH_3)_2CO]$?

e. Label the most acidic hydrogen atom.

f. Which site is most basic?

3.62 Answer each question about oxycodone, a narcotic analgesic used for severe pain.

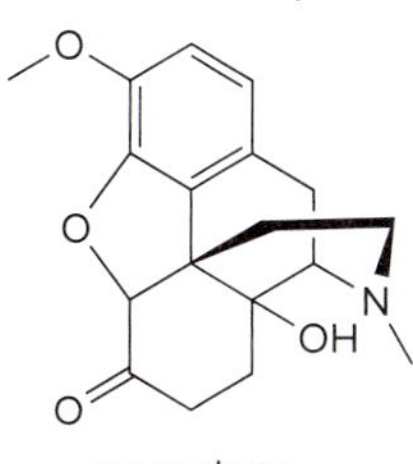

oxycodone

a. Identify the functional groups in oxycodone.
b. Classify any alcohol, amide, or amine as 1°, 2°, or 3°.
c. Which proton is most acidic?
d. Which site is most basic?
e. What is the hybridization of the N atom?
f. How many sp^2 hybridized C atoms does oxycodone contain?

Challenge Problems

3.63 Although diethyl ether and tetrahydrofuran are both four-carbon ethers, one compound is much more water soluble than the other. Predict which compound has higher water solubility and offer an explanation.

diethyl ether   tetrahydrofuran

3.64 Answer the following questions by referring to the ball-and-stick model of fentanyl, a potent narcotic analgesic used in surgical procedures.

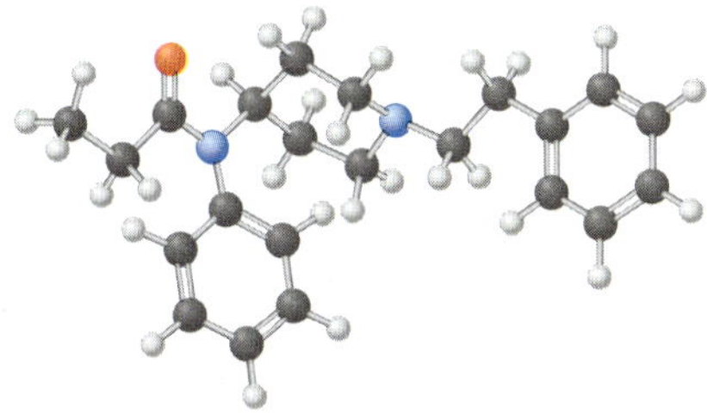
fentanyl

a. Identify the functional groups.
b. Label the most acidic proton.
c. Label the most basic atom.
d. What types of intermolecular forces are present between two molecules of fentanyl?
e. Draw an isomer predicted to have a higher boiling point.
f. Which sites in the molecule can hydrogen bond to water?
g. Label all electrophilic carbons.

3.65 Hexafluoroethane (CF_3CF_3, bp = –78 °C) has a lower boiling point than fluoroethane (CH_3CH_2F, bp = –32 °C), even though its molecular weight is significantly higher. Suggest a reason for the observed difference.

3.66 Explain why **A** is less water soluble than **B**, even though both compounds have the same functional groups.

3.67 Recall from Section 1.10B that there is restricted rotation around carbon–carbon double bonds. Maleic acid and fumaric acid are two isomers with vastly different physical properties and pK_a values for loss of both protons. Explain why each of these differences occurs.

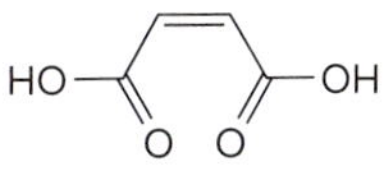

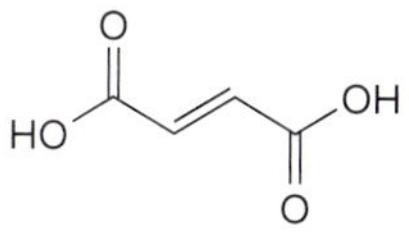

	maleic acid	fumaric acid
mp (°C)	130	286
solubility (g/L) in H_2O at 25 °C	788	7
pK_{a1}	1.9	3.0
pK_{a2}	6.5	4.5

SELF-TEST ANSWERS

1. b 2. d 3. b 4. c 5. c 6. a 7. c 8. b 9. a 10. c

4 Alkanes

Daniel C. Smith

4.1 Alkanes—An introduction
4.2 Cycloalkanes
4.3 An introduction to nomenclature
4.4 Naming alkanes
4.5 Naming cycloalkanes
4.6 Natural occurrence of alkanes
4.7 Properties of alkanes
4.8 Conformations of acyclic alkanes—Ethane
4.9 Conformations of butane
4.10 An introduction to cycloalkanes
4.11 Cyclohexane
4.12 Substituted cycloalkanes
4.13 Oxidation of alkanes
4.14 Lipids—Part 1

Many alkanes occur in nature. Methane, the simplest alkane, is a gas in the atmosphere that comes from natural and humanmade sources. A significant amount of atmospheric methane comes from the decomposition of organic material by microorganisms in wetlands and flooded rice fields when no oxygen is present. Methane is a greenhouse gas with significant global warming potential. In Chapter 4, we learn about methane and other alkanes.

Why Study . . .

Alkanes?

In Chapter 4, we apply the principles of bonding, shape, and reactivity discussed in Chapters 1–3 to our first family of organic compounds, the **alkanes.** Because alkanes have no functional group, they are much less reactive than other organic compounds, and for this reason, much of Chapter 4 is devoted to learning how to name and draw them, as well as to understanding what happens when rotation occurs about their carbon–carbon single bonds.

Studying alkanes also provides an opportunity to learn about **lipids,** a group of biomolecules similar to alkanes, in that they are composed mainly of nonpolar carbon–carbon and carbon–hydrogen σ bonds. Section 4.14 serves as a brief introduction only, so we will return to lipids in Chapters 10 and 29 (online).

4.1 Alkanes—An Introduction

Secretion of **undecane** by a cockroach causes other members of the species to aggregate. Undecane is a ***pheromone,*** **a chemical substance used for communication** in an animal species, most commonly an insect population. *God of Insects*

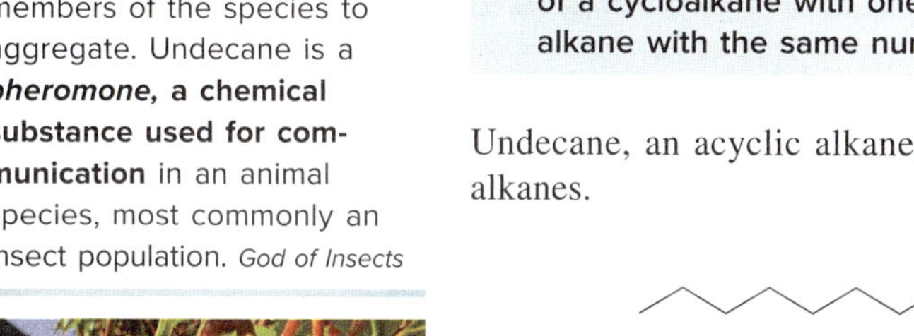

Cyclohexane is one component of the mango, the most widely consumed fruit in the world. *Daniel C. Smith*

Recall from Section 3.2 that **alkanes are aliphatic hydrocarbons having only C–C and C–H σ bonds.** Because their carbon atoms can be joined together in chains or rings, they can be categorized as acyclic or cyclic.

- **Acyclic alkanes** have the molecular formula C_nH_{2n+2} (where n = an integer) and contain only linear and branched chains of carbon atoms. Acyclic alkanes are also called **saturated hydrocarbons** because they have the maximum number of hydrogen atoms per carbon.
- **Cycloalkanes** contain carbons joined in one or more rings. Because the general formula of a cycloalkane with one ring is C_nH_{2n}, they have two fewer H atoms than an acyclic alkane with the same number of carbons.

Undecane, an acyclic alkane, and cyclohexane, a cycloalkane, are two naturally occurring alkanes.

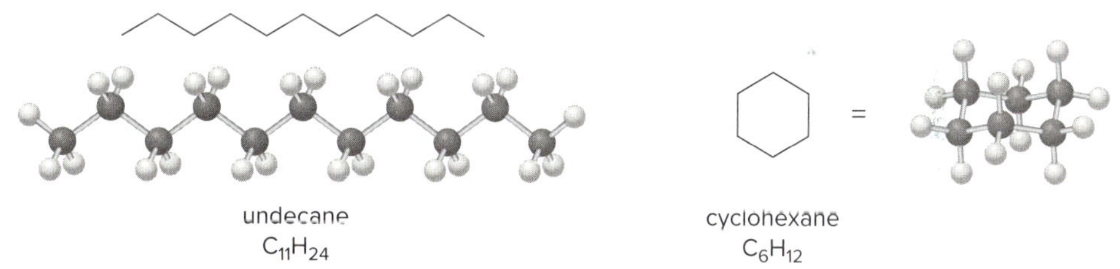

4.1A Acyclic Alkanes Having One to Five Carbons

Structures for the two simplest acyclic alkanes were given in Chapter 1. **Methane, CH_4,** has a single carbon atom, and **ethane, CH_3CH_3,** has two. All C atoms in an alkane are surrounded by four groups, making them ***sp*³ hybridized** and **tetrahedral,** and all bond angles are 109.5°.

CH_4 methane = H–C(H)(H)H — 109 pm, 109.5°

CH_3CH_3 ethane = H₃C–CH₃ — 153 pm

To draw the structure of an alkane, join the carbon atoms together with single bonds, and add enough H atoms to make each C tetravalent.

The three-carbon alkane $\mathbf{CH_3CH_2CH_3}$, **propane,** has molecular formula C_3H_8. Each carbon in the three-dimensional drawing has two bonds in the plane (solid lines), one bond in front (on a wedge), and one bond behind the plane (on a dashed wedge).

$CH_3CH_2CH_3$ =
propane

Problem 4.1 Both olives and the leaves of olive trees contain alkanes with long carbon chains. A predominant alkane in olives has 27 carbons, whereas a major alkane component in olive leaves has 31 carbons. What is the molecular formula of each of these alkanes?

The alkane content of olives and olive leaves is somewhat different (Problem 4.1), so it is possible to use alkane identity to determine the presence of leaf material in olive oil.
Patricia Fenn/Flickr Open/Moment Open/Getty Images

There are two different ways to arrange four carbons, giving two compounds with molecular formula C_4H_{10}, named **butane** and **2-methylpropane** (or isobutane).

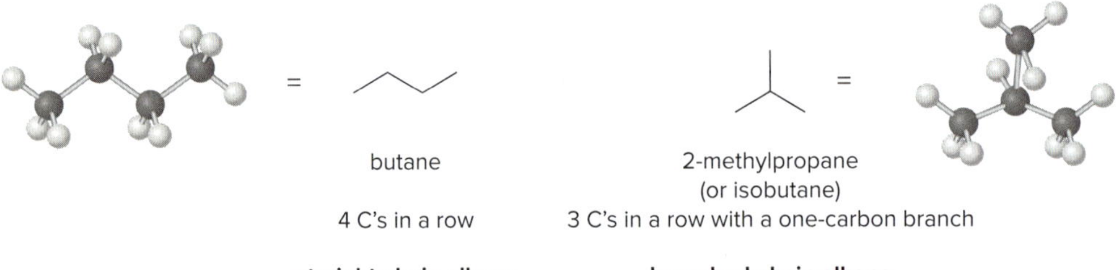

Butane and 2-methylpropane are ***isomers,*** **two different compounds with the same molecular formula** (Section 1.4). They belong to one of the two major classes of isomers called **constitutional** or **structural isomers.** We will learn about the second major class of isomers, called **stereoisomers,** in Section 4.12B.

- ***Constitutional isomers*** **differ in the way the atoms are connected to each other.**

The molecular formulas for methane, ethane, and propane fit into the general molecular formula for an alkane, $\mathbf{C_nH_{2n+2}}$.
- Methane = CH_4 = $C_1H_{2(1)+2}$
- Ethane = C_2H_6 = $C_2H_{2(2)+2}$
- Propane = C_3H_8 = $C_3H_{2(3)+2}$

Butane, which has four carbons in a row, is a **straight-chain** or **normal alkane** (an ***n*-alkane**). 2-Methylpropane, on the other hand, is a **branched-chain alkane.**

With alkanes having more than four carbons, the names of the straight-chain isomers are systematic and derive from Greek roots: ***pent***ane for five C atoms, ***hex***ane for six, and so on. There are three constitutional isomers for the five-carbon alkane, each having molecular formula C_5H_{12}: **pentane, 2-methylbutane** (or isopentane), and **2,2-dimethylpropane** (or neopentane).

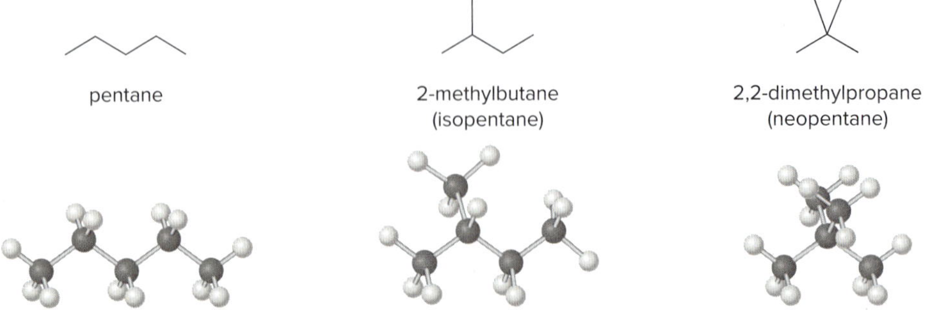

Take care in interpreting skeletal structures. Although pentane is typically drawn using a zigzag structure, the carbon skeleton can be drawn in a variety of ways, and still represent the same compound. Each of the following representations has five carbon atoms in a row, so each represents pentane, not an isomer of pentane.

Problem 4.2 (a) Which compounds (**B–F**) are identical to **A?** (b) Which compounds (**B–F**) represent a constitutional isomer of **A?**

A **B** **C** **D** **E** **F**

4.1B Acyclic Alkanes Having More Than Five Carbons

The maximum number of possible constitutional isomers increases dramatically as the number of carbon atoms in the alkane increases, as shown in Table 4.1. For example, there are 75 possible isomers for an alkane having 10 carbon atoms, and 366,319 possible isomers for one having 20 carbons.

Each entry in Table 4.1 is formed from the preceding entry by adding a CH_2 group. **A CH_2 group is called a *methylene group*. A group of compounds that differ by only a CH_2 group is called a *homologous series*.** The names of all alkanes end in the suffix ***-ane,*** and the syllables preceding the suffix identify the number of carbon atoms in the chain.

Table 4.1 Summary: Straight-Chain Alkanes

Number of C atoms	Molecular formula	Name (*n*-alkane)	Number of constitutional isomers	Number of C atoms	Molecular formula	Name (*n*-alkane)	Number of constitutional isomers
1	CH_4	methane	—	9	C_9H_{20}	nonane	35
2	C_2H_6	ethane	—	10	$C_{10}H_{22}$	decane	75
3	C_3H_8	propane	—	11	$C_{11}H_{24}$	undecane	159
4	C_4H_{10}	butane	2	12	$C_{12}H_{26}$	dodecane	355
5	C_5H_{12}	pentane	3	13	$C_{13}H_{28}$	tridecane	802
6	C_6H_{14}	hexane	5	14	$C_{14}H_{30}$	tetradecane	1858
7	C_7H_{16}	heptane	9	15	$C_{15}H_{32}$	pentadecane	4347
8	C_8H_{18}	octane	18	20	$C_{20}H_{42}$	icosane	366,319

Problem 4.3 Draw the five constitutional isomers having molecular formula C_6H_{14}.

Problem 4.4 Review classifying carbons and hydrogens in Section 3.2, and draw the structure of an alkane with molecular formula C_7H_{16} that contains (a) one 4° carbon; (b) only 1° and 2° carbons; (c) 1°, 2°, and 3° hydrogens.

4.2 Cycloalkanes

Cycloalkanes have molecular formula C_nH_{2n} and contain carbon atoms arranged in a ring. Think of a cycloalkane as being formed by removing two H atoms from the end carbons of a chain, and then bonding the two carbons together. Simple cycloalkanes are named by adding the prefix ***cyclo-*** to the name of the acyclic alkane having the same number of carbons.

Cycloalkanes with three to six carbon atoms are shown.

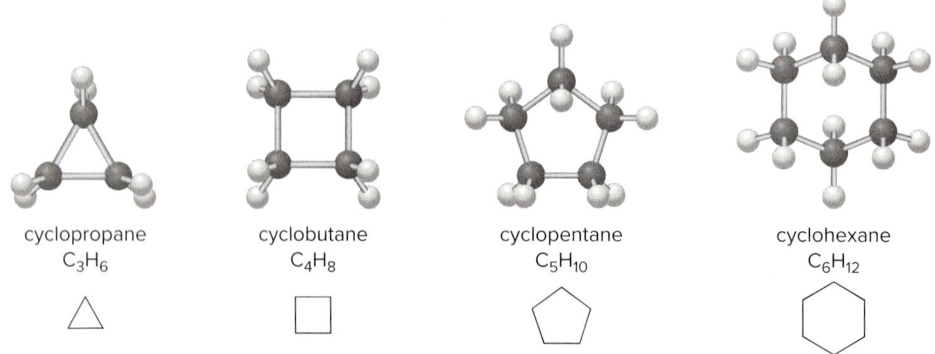

Problem 4.5 Draw the five constitutional isomers that have molecular formula C_5H_{10} and contain one ring.

4.3 An Introduction to Nomenclature

Garlic has been used in Chinese herbal medicine for more than 4000 years. **Allicin,** the molecule largely responsible for garlic's odor, is not stored in the garlic bulb, but instead is produced by the action of enzymes when the bulb is crushed or bruised.
Pixtal/age fotostock

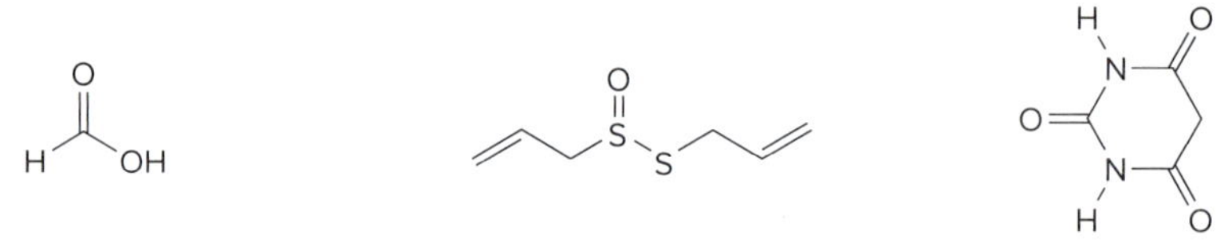

How are organic compounds named? Long ago, the name of a compound was often based on the plant or animal source from which it was obtained. For example, the name for **formic acid,** a caustic compound isolated from certain ants, comes from the Latin word *formica,* meaning "ant;" and **allicin,** the pungent principle of garlic, is derived from the botanical name for garlic, *Allium sativum.* Other compounds were named by their discoverer for personal reasons. Adolf von Baeyer supposedly named barbituric acid after a woman named Barbara, although speculation continues on Barbara's identity.

formic acid (obtained from certain ants)

allicin (odor of garlic)

barbituric acid (named for Barbara?)

With the isolation and preparation of thousands of new organic compounds it became clear that each organic compound must have an unambiguous name, derived from a set of easily remembered rules. A systematic method of naming compounds was developed by the ***I****nternational* ***U****nion of* ***P****ure and* ***A****pplied* ***C****hemistry*. It is referred to as the **IUPAC system of nomenclature;** how it can be used to name alkanes and cycloalkanes is explained in Sections 4.4 and 4.5.

The IUPAC system of nomenclature has been regularly revised since it was first adopted in 1892. Revisions in 1979, 1993, and 2013 have given chemists a variety of acceptable names for compounds. Many changes are minor. For example, the 1979 nomenclature rules assign the name 1-butene to $CH_2{=}CHCH_2CH_3$, whereas the 1993 rules assign the name but-1-ene; that is, only the position of the number differs. The 2013 revision assigns a preferred IUPAC name (PIN) to each compound, but considers other unambiguous IUPAC names as acceptable as well. In this text we will use a recent, generally accepted IUPAC name for each compound.

Naming organic compounds has become big business for drug companies. The IUPAC name of an organic compound can be long and complex, and may be comprehensible only to a chemist. As a result, most drugs have three names:

- **Systematic:** The systematic name follows the accepted rules of nomenclature and indicates the compound's chemical structure; this is the IUPAC name.
- **Generic:** The generic name is the official, internationally approved name for the drug.
- **Trade:** The trade name for a drug is assigned by the company that manufactures it. Trade names are often "catchy" and easy to remember. Companies hope that the public will continue to purchase a drug with an easily recalled trade name long after a cheaper generic version becomes available.

In the world of over-the-counter anti-inflammatory agents, the compound a chemist calls 2-[4-(2-methylpropyl)phenyl]propanoic acid has the generic name ibuprofen. It is marketed under a variety of trade names including Motrin and Advil.

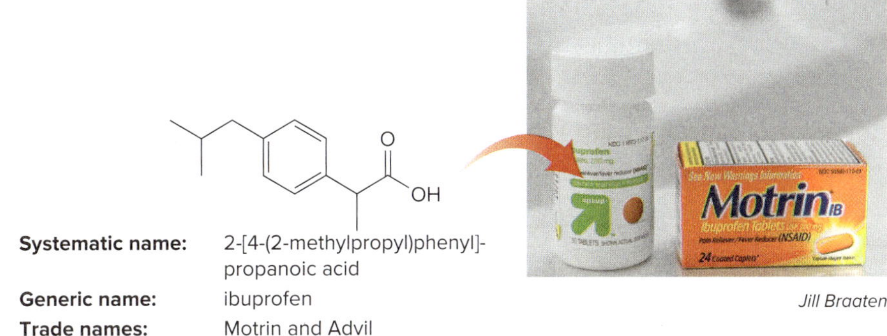

Systematic name: 2-[4-(2-methylpropyl)phenyl]-propanoic acid
Generic name: ibuprofen
Trade names: Motrin and Advil

Jill Braaten

4.4 Naming Alkanes

The name of every organic molecule has three parts:

- The **parent name** indicates the number of carbons in the longest continuous carbon chain in the molecule.
- The **suffix** indicates what functional group is present.
- The **prefix** reveals the identity, location, and number of substituents attached to the carbon chain.

prefix	+	**parent**	+	**suffix**
What and where are the substituents?		What is the longest carbon chain?		What is the functional group?

The names listed in Table 4.1 of Section 4.1B for the simple *n*-alkanes consist of the parent name, which indicates the number of carbon atoms in the longest carbon chain, and the suffix ***-ane,*** which indicates that the compounds are alkanes. The parent name for **one carbon is *meth-,*** for **two carbons is *eth-,*** and so on. Thus, we are already familiar with two parts of the name of an organic compound.

To determine the third part of a name, the prefix, we must learn how to name the carbon groups or *substituents* that are bonded to the longest carbon chain.

4.4A Naming Substituents

Carbon substituents bonded to a long carbon chain are called **alkyl groups.**

- **An *alkyl group* is formed by removing one hydrogen from an alkane.**

An alkyl group is a part of a molecule that is now able to bond to another atom or a functional group. **To name an alkyl group, change the *-ane* ending of the parent alkane to *-yl.*** Thus, **methane** (CH_4) becomes **methyl** (CH_3–) and **ethane** (CH_3CH_3) becomes **ethyl** (CH_3CH_2–). As we learned in Section 3.1, **R** denotes a general carbon group bonded to a functional group. **R** thus denotes any alkyl group.

Naming three- and four-carbon alkyl groups is more complicated because the parent hydrocarbons have more than one type of hydrogen atom. Propane has both 1° and 2° H atoms, and removal of each of these H atoms forms a different alkyl group, **propyl** or **isopropyl.**

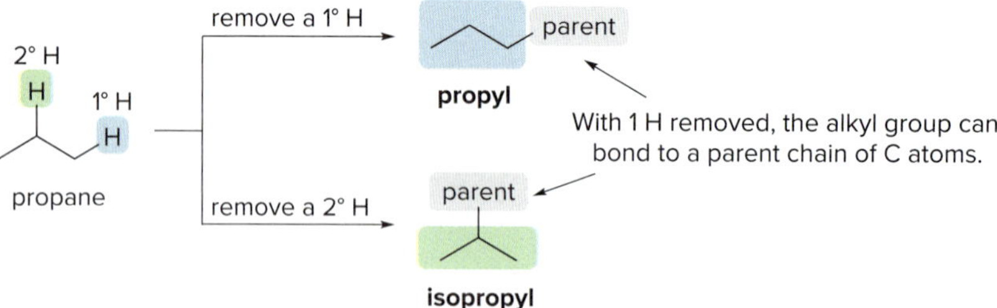

The prefix ***iso-*** is part of the words *propyl* and *butyl,* forming a single word: **isopropyl** and **isobutyl.** The prefixes ***sec-*** and ***tert-*** are separated from the word *butyl* by a hyphen: ***sec-*****butyl** and ***tert*****-butyl.**

The prefix *sec-* is short for *secondary.* A *sec*-butyl group is formed by removal of a **2° H.** The prefix *tert-* is short for *tertiary.* A *tert*-butyl group is formed by removal of a **3° H.**

Because there are two different butane isomers to begin with, each with two different kinds of H atoms, there are four possible alkyl groups containing four carbon atoms: **butyl, *sec*-butyl, isobutyl,** and ***tert*-butyl.**

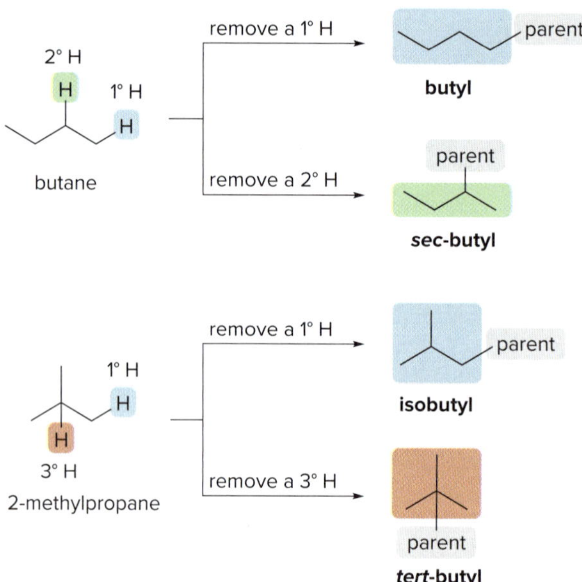

Abbreviations are sometimes used for certain common alkyl groups.

- methyl **(Me)**
- ethyl **(Et)**
- butyl **(Bu)**
- *tert*-butyl **(*t*-Bu)**

The names isopropyl, *sec*-butyl, isobutyl, and *tert*-butyl are recognized as acceptable substituent names in IUPAC nomenclature. A general method to name these substituents, as well as alkyl groups that contain five or more carbon atoms, is described in Appendix D.

4.4B Naming an Acyclic Alkane

Four steps are needed to name an alkane.

How To Name an Alkane Using the IUPAC System

Step [1] **Find the parent carbon chain and add the suffix.**

- Find the *longest continuous* carbon chain, and name the molecule by using the parent name for that number of carbons, given in Table 4.1. To the name of the parent, add the suffix ***-ane*** for an alkane. Each functional group has its own characteristic suffix.

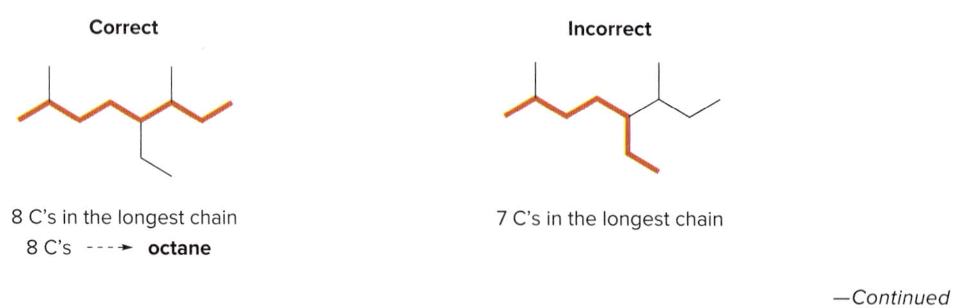

—*Continued*

- Finding the longest chain is a matter of trial and error. Place your pencil on one end of the chain, go to the other end without picking it up, and count carbons. Repeat this procedure until you have found the chain with the largest number of carbons.
- **It does not matter if the chain is *straight* or has *bends*.** All of the following representations are equivalent, and each longest chain has eight carbons.

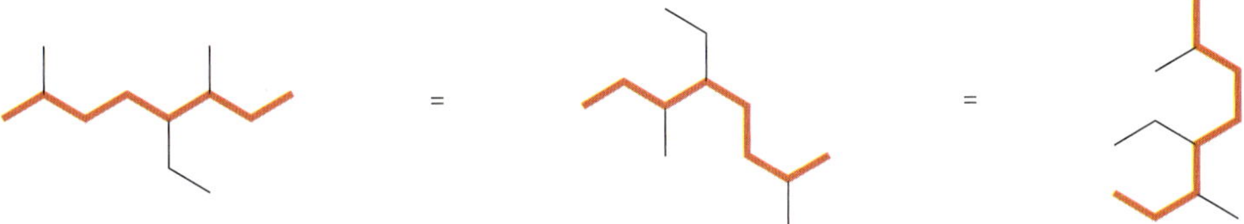

- **If there are two chains of equal length, pick the chain with *more* substituents**. In the following example, two different chains in the same alkane contain 7 C's, but the compound on the left has two alkyl groups attached to its long chain, whereas the compound to the right has only one.

Correct	Incorrect
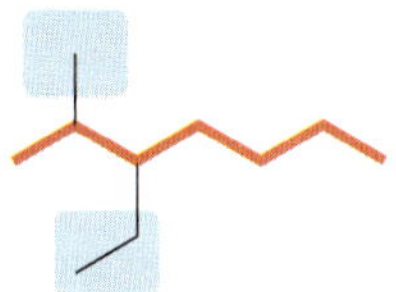	
7 atoms in the longest chain **2** substituents	**7** atoms in the longest chain **only 1** substituent
***more* substituents**	***fewer* substituents**

Step [2] **Number the atoms in the carbon chain.**

- Number the longest chain to give the *first* substituent the lower number.

Correct	Incorrect
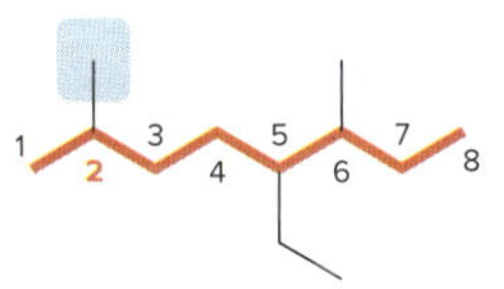	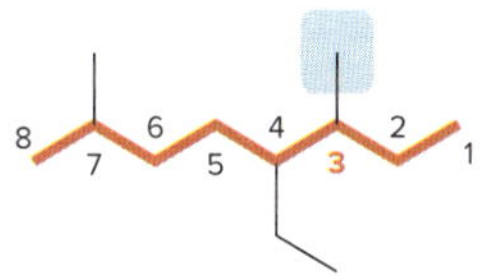
first substituent at C2	**first substituent at C3**

- If the first substituent is the same distance from both ends, number the chain to give the *second* substituent the lower number. **Always look for the first point of difference** in numbering from each end of the longest chain.

Correct	Incorrect
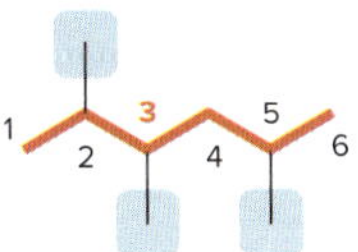	
CH_3 groups at C2, C3, and C5 The second CH_3 group has the *lower* number (C3).	CH_3 groups at C2, C4, and C5 The second CH_3 group has the *higher* number (C4).

- When numbering a carbon chain results in the *same* numbers from either end of the chain, **assign the lower number *alphabetically*** to the first substituent.

Correct	Incorrect
ethyl	methyl
• ethyl at **C3** • methyl at **C5**	• methyl at **C3** • ethyl at **C5**

***Earlier* letter ⟶ *lower* number**

—*Continued*

How To, continued . . .

Step [3] **Name and number the substituents.**

methyl at C2 methyl at C6

1 2 5 6 8

ethyl at C5

8 C's in the longest chain

- Name the substituents as alkyl groups, and use the numbers from Step [2] to designate their location.
- Every carbon belongs to *either* the longest chain or a substituent, but *not both.*
- **Each substituent needs its *own* number.**
- If two or more identical substituents are bonded to the longest chain, use prefixes to indicate how many: ***di-*** for two groups, ***tri-*** for three groups, ***tetra-*** for four groups, and so forth. This molecule has two methyl substituents, so its name contains the prefix *di-* before the word methyl → *di*methyl.

Step [4] **Combine substituent names and numbers + parent + suffix.**

- Precede the name of the parent by the names of the substituents.
- Alphabetize the names of the substituents, **ignoring all prefixes except *iso-*,** as in isopropyl and isobutyl.
- Precede the name of each substituent by the number that indicates its location. There must be **one number for each substituent.**
- Separate numbers by commas and separate numbers from letters by hyphens. The name of an alkane is a single word, with no spaces after hyphens or commas.

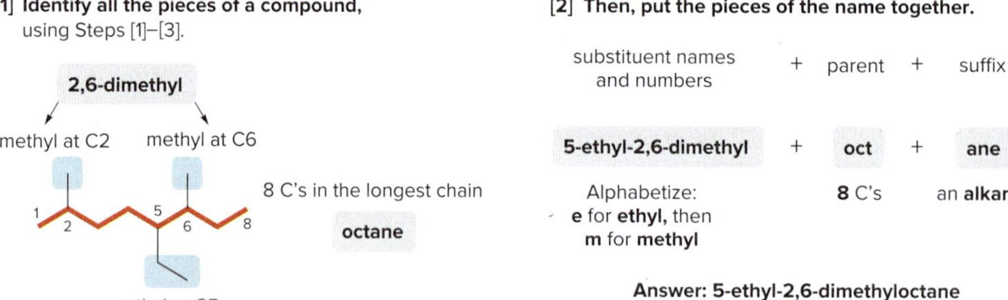

Problem 4.6 Using the information in the previous *How To* and alkanes **A–D,** answer each question. (a) Which two compounds have the same longest carbon chain? (b) Which two compounds have the same number of substituents?

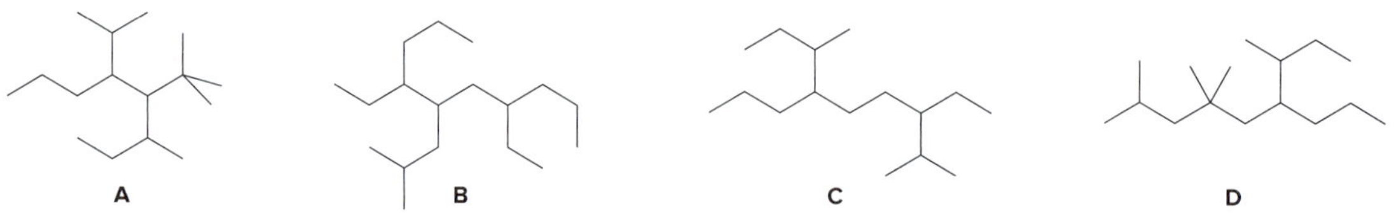

Several additional examples of alkane nomenclature are given in Figure 4.1.

Figure 4.1 Examples of alkane nomenclature

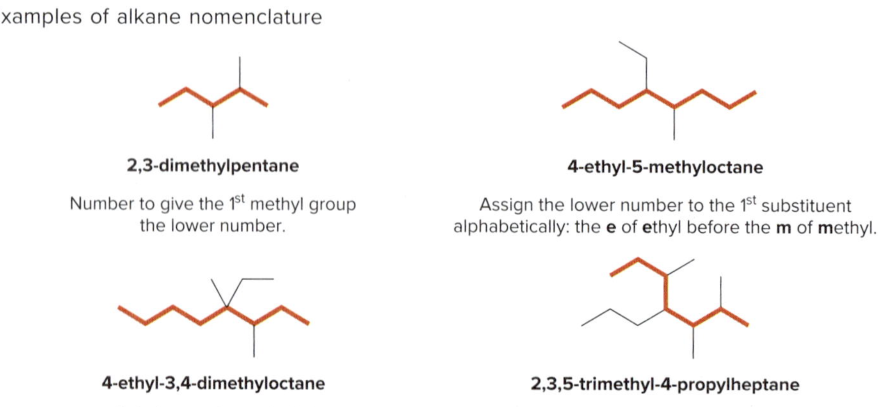

- The carbon atoms of each long chain are drawn in **red.**

Sample Problem 4.1 Naming an Alkane

Give the IUPAC name for the following compound.

Solution

To help identify which carbons belong to the longest chain and which are substituents, **box in or highlight the atoms of the long chain.** Every other carbon atom then becomes a substituent that needs its own name as an alkyl group.

Step 1: Name the parent.

9 C's in the longest chain

nonane

Step 2: Number the chain.

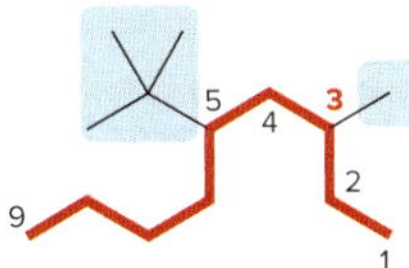

first substituent at C**3**

Step 3: Name and number the substituents.

tert-butyl at C5

methyl at C3

Step 4: Combine the parts.

- Alphabetize: the **b** of **butyl** before the **m** of **methyl**

Answer: 5-*tert*-butyl-3-methylnonane

Problem 4.7 Give the IUPAC name for each compound.

a.

b.

c.

d.

e.

f.

More Practice: Try Problems 4.38; 4.42a–d, h, j; 4.69a.

Problem 4.8 Give the IUPAC name for each compound.

a. $(CH_3)_3CCH_2CH(CH_2CH_3)_2$

c. $CH_3(CH_2)_3CH(CH_2CH_2CH_3)CH(CH_3)_2$

b.

d.

You must also know how to derive a structure from a given name. Sample Problem 4.2 illustrates a stepwise method.

Sample Problem 4.2 Deriving a Structure from a Name

Give the structure corresponding to the following IUPAC name: 6-isopropyl-3,3,7-trimethyldecane.

Solution

Follow three steps to derive a structure from a name.

Step [1] **Identify the parent name and functional group found at the *end* of the name.**

decane ---→ **10** C's ---→

Step [2] **Number the carbon skeleton in *either* direction.**

1 2 3 4 5 6 7 8 9 10

Step [3] **Add the substituents at the appropriate carbons.**

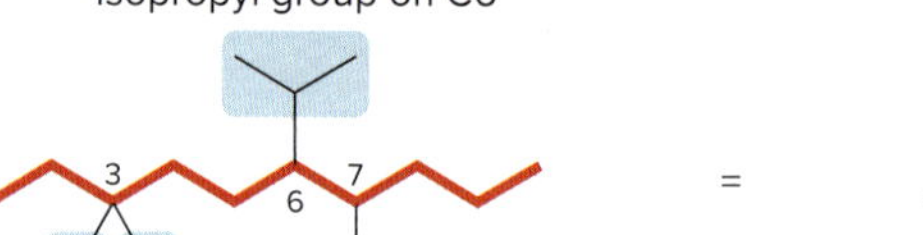

=

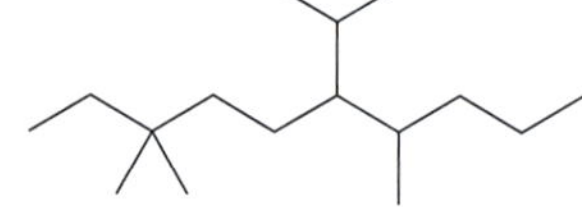

Answer

Problem 4.9 Give the structure corresponding to each IUPAC name.

a. 3-methylhexane
b. 3,3-dimethylpentane
c. 3,5,5-trimethyloctane
d. 3-ethyl-4-methylhexane
e. 3-ethyl-5-isobutylnonane
f. 3-ethyl-6-isopropyl-9,9-dimethyldodecane

More Practice: Try Problem 4.43a, c, g, h.

Problem 4.10 Give the IUPAC name for each of the five constitutional isomers of molecular formula C_6H_{14} in Problem 4.3.

4.4C Common Names

Some organic compounds are identified using **common names** that do not follow the IUPAC system of nomenclature. Many of these names were given to molecules long ago, before the IUPAC system was adopted. These names are still widely used. For example, isopentane, an older name for 2-methylbutane, is still allowed by IUPAC rules. We will follow the IUPAC system except in cases in which a common name is widely accepted.

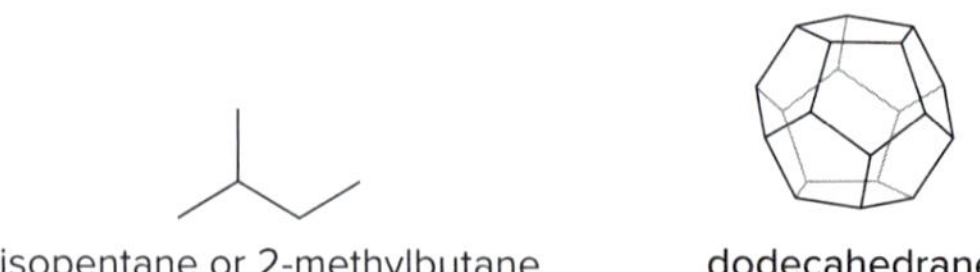

isopentane or 2-methylbutane dodecahedrane

In the past several years, organic chemists have attempted to synthesize some unusual cycloalkanes not found in nature. **Dodecahedrane** is one such molecule. The IUPAC name for dodecahedrane is undecacyclo[9.9.0.0^{2,9}.0^{3,7}.0^{4,20}.0^{5,18}.0^{6,16}.0^{8,15}.0^{10,14}.0^{12,19}.0^{13,17}]icosane, a name so complex that few trained organic chemists would be able to identify its structure. Because systematic names for polycyclic compounds are so unwieldy, a common name that is more descriptive of its shape and structure is often assigned. Dodecahedrane is named because its 12 five-membered rings resemble a dodecahedron.

4.5 Naming Cycloalkanes

Cycloalkanes are named by using similar rules, but the prefix ***cyclo-*** immediately precedes the name of the parent.

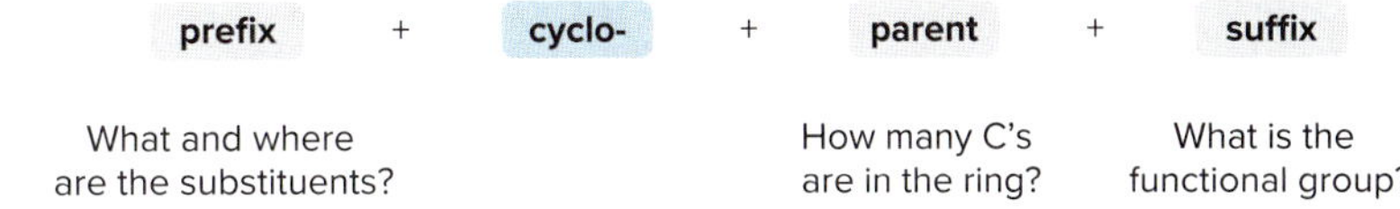

How To Name a Cycloalkane Using the IUPAC System

Step [1] **Find the parent cycloalkane.**

- Count the number of carbon atoms in the ring and use the parent name for that number of carbons. Add the prefix ***cyclo-*** and the suffix ***-ane*** to the parent name.

6 C's in the ring
cyclohexane

Step [2] **Name and number the substituents.**

- No number is needed to indicate the location of a single substituent.

methylcyclohexane *tert*-butylcyclopentane

- For rings with more than one substituent, **begin numbering at one substituent** and proceed around the ring clockwise or counterclockwise to **give the second substituent the *lower* number.**

CH_3 groups at C1 and C**3**
The 2nd substituent has a lower number.
Correct: 1,3-dimethylcyclohexane

CH_3 groups at C1 and C**5**
Incorrect: 1,5-dimethylcyclohexane

- **With two different substituents,** number the ring to **assign the *lower* number to the substituents *alphabetically.***

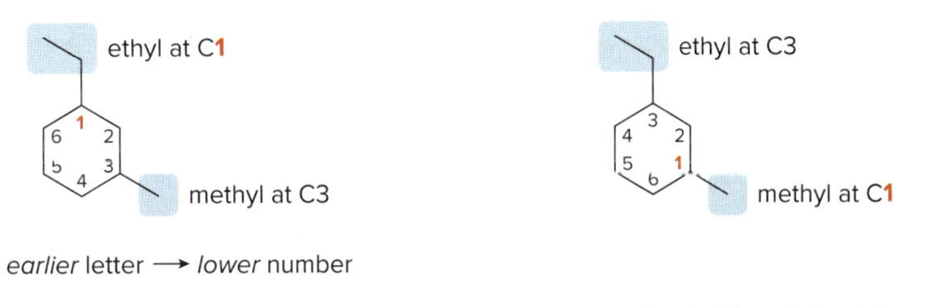

Correct: 1-ethyl-3-methylcyclohexane

Incorrect: 3-ethyl-1-methylcyclohexane

Several examples of cycloalkane nomenclature are given in Figure 4.2.

Figure 4.2 Examples of cycloalkane nomenclature

ethylcyclobutane

No number is needed with only one substituent.

1-*sec*-butyl-3-methylcyclohexane

Assign the lower number to the 1st substituent alphabetically: the **b** of **b**utyl before the **m** of **m**ethyl.

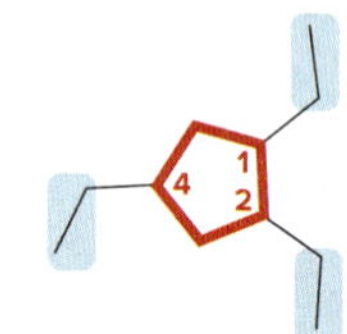

1,2,4-triethylcyclopentane

Number to give the 2nd CH_3CH_2 group the lower number: 1,2,4- not 1,3,4- or 1,3,5-.

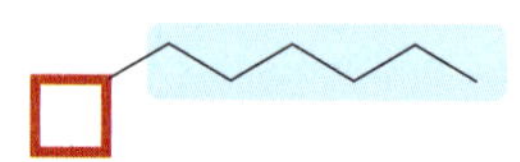

hexylcyclobutane

Name as a cycloalkane with an alkyl substituent even when the alkyl group has more C's than the ring.

Problem 4.11 Give the IUPAC name for each compound.

a. 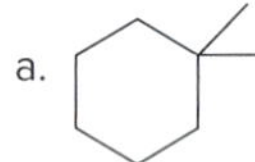
c. 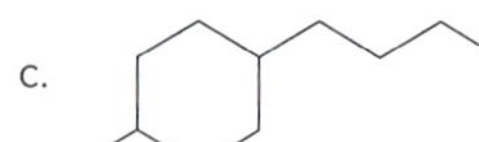
e. 

b. 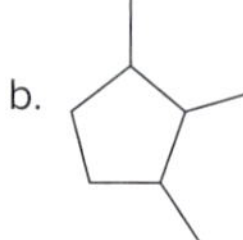
d. 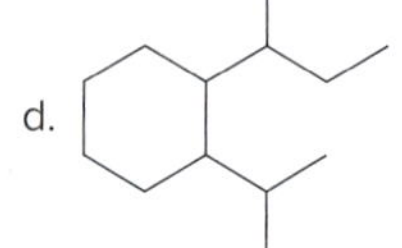
f. 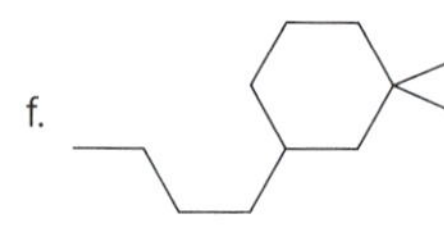

Problem 4.12 Give the structure corresponding to each IUPAC name.

a. 1,2-dimethylcyclobutane
b. 1,1,2-trimethylcyclopropane
c. 4-ethyl-1,2-dimethylcyclohexane
d. 1-*sec*-butyl-3-isopropylcyclopentane
e. 1,1,2,3,4-pentamethylcycloheptane

4.6 Natural Occurrence of Alkanes

Many alkanes occur in nature, primarily in natural gas and petroleum. Both of these fossil fuels serve as energy sources, formed from the degradation of organic material long ago.

Natural gas is composed largely of **methane** (60% to 80% depending on its source), with lesser amounts of ethane, propane, and butane. These organic compounds burn in the presence of oxygen, releasing energy for cooking and heating.

As mentioned in the chapter opener, methane in the atmosphere comes from natural and human-made sources. As global temperatures increase, methane trapped in permafrost and glaciers is released with melting. Microorganisms in the gut of ruminant animals produce methane that is released during defecation and belching. The microorganisms in wetlands and flooded rice fields decompose organic material to form methane under anaerobic conditions. Although methane does not persist in the atmosphere as long as carbon dioxide (Section 4.13), methane is a greenhouse gas with significant global warming potential, and its concentration has increased significantly in the last 200 years.

Petroleum is a complex mixture of compounds, most of which are hydrocarbons containing 1–40 carbon atoms. Distilling crude petroleum, a process called **refining,** separates it into usable fractions that differ in boiling point (Figure 4.3). Most products of petroleum refining provide fuel for home heating, automobiles, diesel engines, and airplanes. Each fuel type has a different composition of hydrocarbons: gasoline (C_5H_{12}–$C_{12}H_{26}$), kerosene ($C_{12}H_{26}$–$C_{16}H_{34}$), and diesel fuel ($C_{15}H_{32}$–$C_{18}H_{38}$).

Figure 4.3 Refining crude petroleum into usable fuel and other petroleum products

Glow Images

- **An oil refinery.** At an oil refinery, crude petroleum is separated into fractions of similar boiling point by the process of **distillation.**

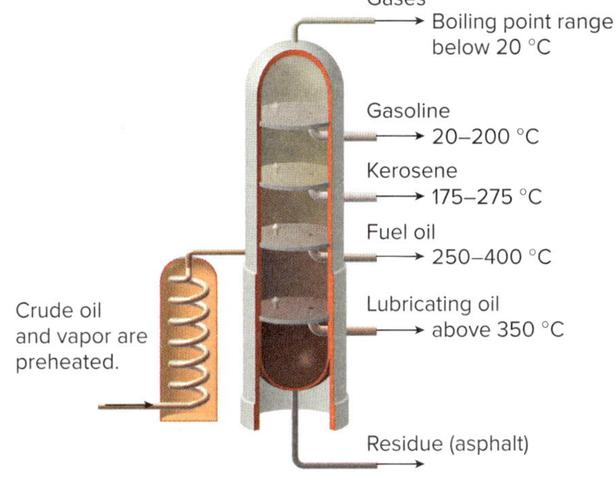

- **Schematic of a refinery tower.** As crude petroleum is heated, the lower boiling, more volatile components distill first, followed by fractions of progressively higher boiling point.

Petroleum provides more than fuel. About 3% of crude oil is used to make plastics and other synthetic compounds including drugs, fabrics, dyes, and pesticides. These products are responsible for many of the comforts we now take for granted in industrialized countries. Imagine what life would be like without air conditioning, refrigeration, anesthetics, and pain relievers, all products of the petroleum industry.

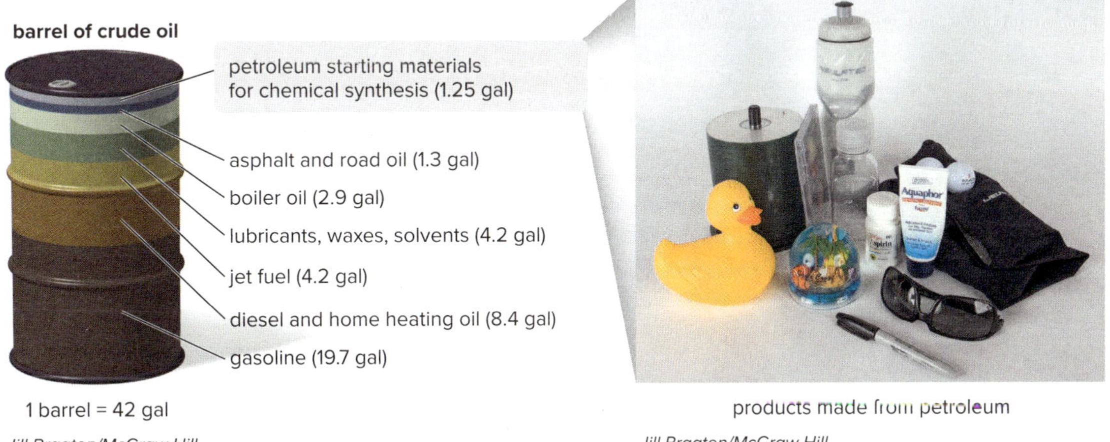

Jill Braaten/McGraw Hill

Jill Braaten/McGraw Hill

Energy from petroleum is ***nonrenewable,*** and the remaining known oil reserves are limited. Given our dependence on petroleum, not only for fuel, but also for the many necessities of modern society, it becomes clear that we must both conserve what we have and find alternate energy sources.

4.7 Properties of Alkanes

4.7A Physical Properties

Crude oil spills are difficult to clean up and can damage an ecosystem for years. *Narongsak Nagadhana/Shutterstock*

The mutual insolubility of nonpolar oil and very polar water leads to the common expression "Oil and water don't mix."

Alkanes contain only nonpolar C–C and C–H bonds, and as a result they exhibit only **weak van der Waals forces.** Table 4.2 summarizes how these intermolecular forces affect the physical properties of alkanes.

The gasoline industry exploits the dependence of boiling point and melting point on alkane size by seasonally changing the composition of gasoline in locations where it gets very hot in the summer and very cold in the winter. Gasoline is refined to contain a larger fraction of higher-boiling hydrocarbons in warmer weather, so it evaporates less readily. In colder weather, it is refined to contain more lower-boiling hydrocarbons, so it freezes less readily.

Because nonpolar alkanes are not water soluble, crude petroleum that leaks into the sea from an oil tanker or offshore oil well creates an insoluble oil slick on the surface. The insoluble hydrocarbon oil poses a special threat to birds whose feathers are coated with natural nonpolar oils for insulation. Because these hydrophobic oils dissolve in the crude petroleum, birds lose their layer of natural protection and many die.

Problem 4.13 Arrange the following compounds in order of increasing boiling point.

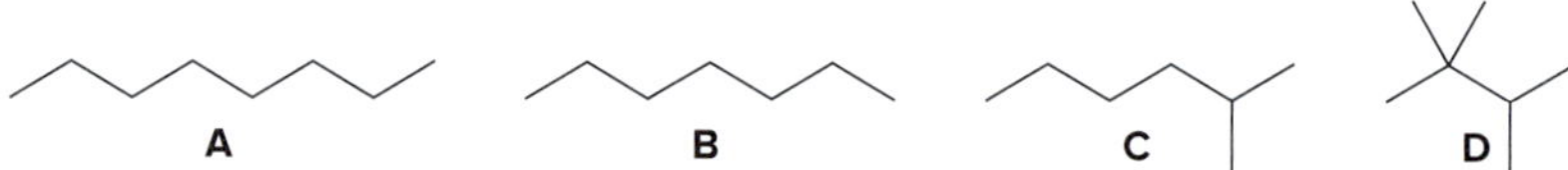

Table 4.2 Physical Properties of Alkanes

Property	Observation
Boiling point and melting point	• Alkanes have low bp's and mp's compared to more polar compounds of comparable size. • Bp and mp increase as the number of carbons increases because of increased surface area. bp = 0 °C, mp = –138 °C; bp = 69 °C, mp = –95 °C; OH: bp = 139 °C, mp = –78 °C **Increasing strength of intermolecular forces** **Increasing boiling point and melting point**
	• The bp of isomers decreases with branching because of decreased surface area. • Mp increases with increased symmetry. bp = 10 °C, mp = –17 °C; bp = 30 °C, mp = –160 °C **more branching—lower boiling point** **more symmetry—higher melting point**
Solubility	• Alkanes are soluble in organic solvents. • Alkanes are insoluble in water.

Key: bp = boiling point; mp = melting point

4.7B Spectroscopic Properties

Students who would like to learn about the spectroscopic properties of alkanes are referred to the following sections in later chapters:

- **Mass spectrometry:** Sections A.1A and A.3, especially Figure A.5 and Sample Problem A.6
- **Infrared spectroscopy:** Section B.4A and Table B.2

4.8 Conformations of Acyclic Alkanes—Ethane

Let's now take a closer look at the three-dimensional structure of alkanes. The three-dimensional structure of molecules is called **stereochemistry.** In Chapter 4, we examine the effect of rotation around single bonds. In Chapter 5, we will learn about other aspects of stereochemistry.

Recall from Section 1.10A that **rotation occurs around carbon–carbon σ bonds.** Thus, the two CH_3 groups of ethane rotate, allowing the hydrogens on one carbon to adopt different orientations relative to the hydrogens on the other carbon. These arrangements are called **conformations.**

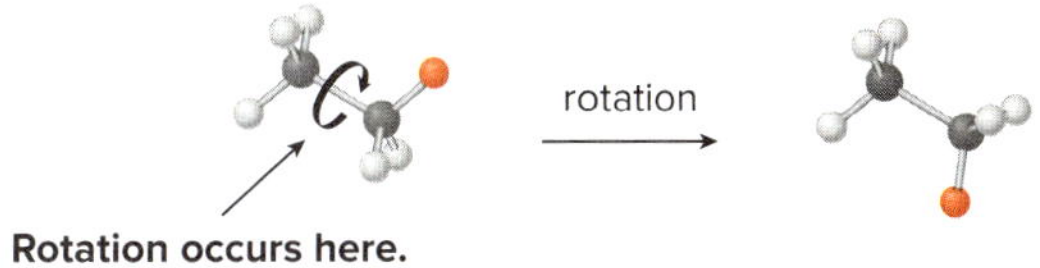

- *Conformations* are different arrangements of atoms that are interconverted by rotation around single bonds.

Two different arrangements are the **eclipsed conformation** and the **staggered conformation.**

- In the *eclipsed conformation,* the C–H bonds on one carbon are directly aligned with the C–H bonds on the adjacent carbon.
- In the *staggered conformation,* the C–H bonds on one carbon bisect the H–C–H bond angle on the adjacent carbon.

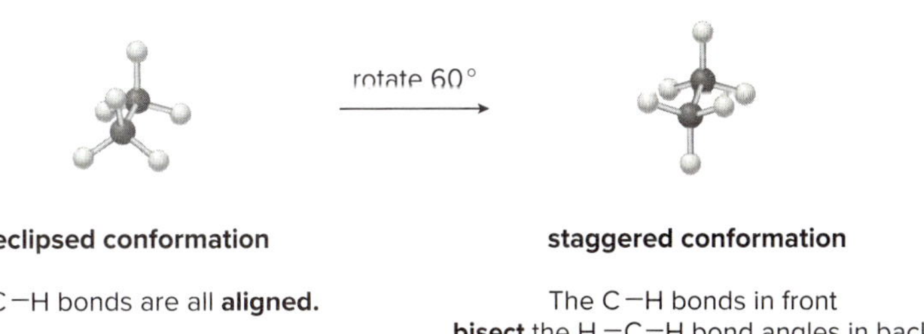

Rotating the atoms on one carbon by 60° converts an eclipsed conformation into a staggered conformation, and vice versa. These conformations are often viewed end-on—that is, looking directly down the carbon–carbon bond. The angle that separates a bond on one atom from a bond on an adjacent atom is called a **dihedral angle.** For ethane in the staggered conformation, the dihedral angle for the C–H bonds is **60°.** For eclipsed ethane, it is **0°.**

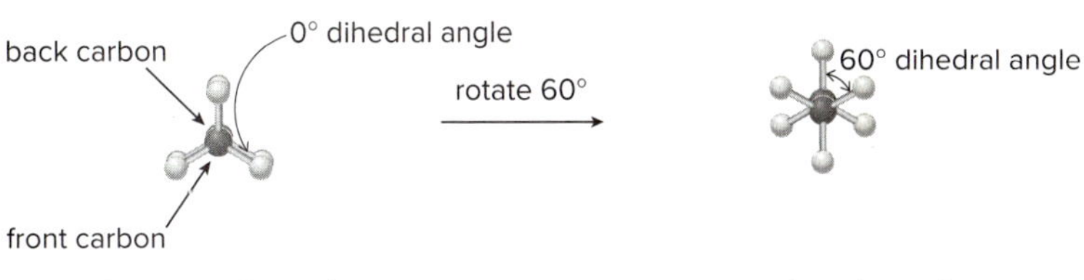

End-on representations for conformations are commonly drawn using a convention called a **Newman projection.** A Newman projection is a graphic that shows the three groups bonded to each carbon atom in a particular C–C bond, as well as the dihedral angle that separates them.

How To Draw a Newman Projection

Step [1] **Look directly down the C–C bond (end-on), and draw a circle with a dot in the center to represent the carbons of the C–C bond.**

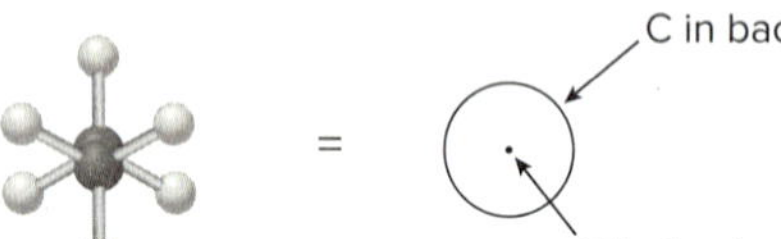

- The circle represents the back carbon and the dot represents the front carbon.

Step [2] **Draw in the bonds.**

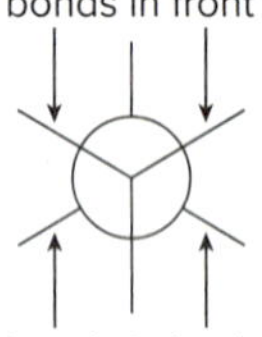

- Draw the bonds on the **front** C as three lines **meeting at the center** of the circle.
- Draw the bonds on the **back** C as three lines coming **out of the edge** of the circle.

Step [3] **Add the atoms on each bond.**

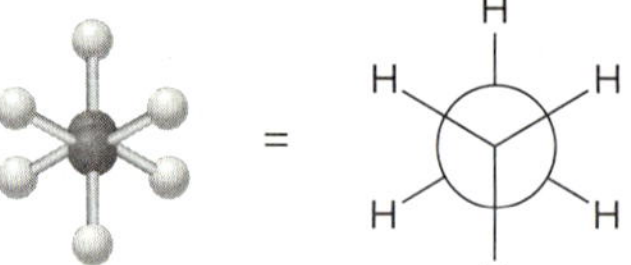

- Each C has 3 H's in ethane.

Figure 4.4 illustrates the Newman projections for both the staggered and eclipsed conformations for ethane.

Figure 4.4 Newman projections for the staggered and eclipsed conformations of ethane

dihedral angle 60°

staggered conformation

dihedral angle 0°

eclipsed conformation

Follow this procedure for any C–C bond. With a Newman projection, **always consider *one* C–C bond only and draw the atoms bonded to the carbon atoms, *not* the carbon atoms in the bond itself.** Newman projections for the staggered and eclipsed conformations of propane are drawn in Figure 4.5.

Figure 4.5 Newman projections for the staggered and eclipsed conformations of propane

Consider one C–C bond only.

propane

- Arbitrarily pick one C to be in front and one C to be in back.
- 3 H's on one C
- 2 H's and 1 CH_3 on the other C

staggered conformation

eclipsed conformation

Problem 4.14 Convert each representation to a Newman projection around the indicated bond.

a. b. c. d.

Problem 4.15 Which of the following is (are) possible Newman projections for 2-methylpentane?

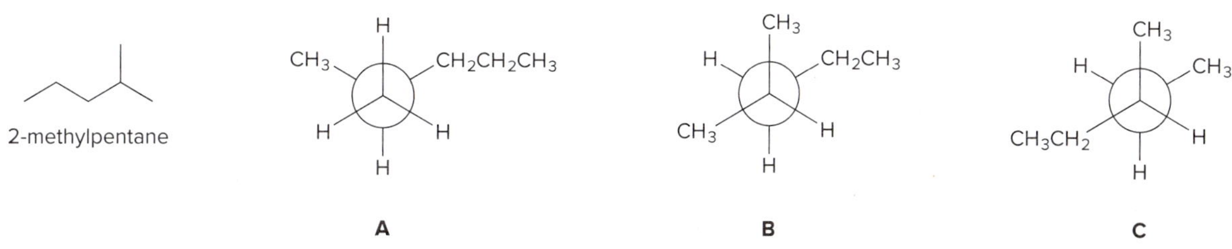

The staggered and eclipsed conformations of ethane interconvert at room temperature, but **each conformation is *not* equally stable.**

- **The staggered conformations are more stable (lower in energy) than the eclipsed conformations.**

The cause of this stability difference has been the subject of some debate in the chemical literature. A contributing factor may be increased electron–electron repulsion between the bonds in the eclipsed conformation compared to the staggered conformation, where the bonding electrons are farther apart.

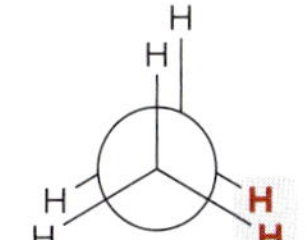

Each H,H eclipsing increases energy by 4.0 kJ/mol.

The difference in energy between the staggered and eclipsed conformations is 12 kJ/mol (2.9 kcal/mol), a small enough difference that the rotation is still very rapid at room temperature, and the conformations cannot be separated. Because three eclipsed C–H bonds increase the energy of a conformation by 12 kJ/mol, **each eclipsed C–H bond results in an increase in energy of 4.0 kJ/mol (1.0 kcal/mol).** The energy difference between the staggered and eclipsed conformations is called **torsional energy.** Thus, eclipsing introduces **torsional strain** into a molecule.

- ***Torsional strain* is an increase in energy caused by eclipsing interactions.**

Strain results in an **increase in energy.** Torsional strain is the first of three types of strain discussed in this text. The other two are **steric strain** (Section 4.9) and **angle strain** (Section 4.10).

The graph in Figure 4.6 shows how the potential energy of ethane changes with dihedral angle as one CH_3 group rotates relative to the other. **The staggered conformation is the most stable arrangement, so it is at an *energy minimum.*** As the C–H bonds on one carbon are rotated relative to the C–H bonds on the other carbon, the energy increases as the C–H bonds get closer until a **maximum is reached after 60° rotation to the eclipsed conformation.** As rotation continues, the energy decreases until after 60° rotation, when the staggered conformation is reached once again.

- **An energy minimum and maximum occur every 60° as the conformation changes from staggered to eclipsed. Conformations that are neither staggered nor eclipsed are intermediate in energy.**

Problem 4.16 The torsional energy in propane is 14 kJ/mol (3.4 kcal/mol). Because each H,H eclipsing interaction is worth 4.0 kJ/mol (1.0 kcal/mol) of destabilization, how much is one H,CH_3 eclipsing interaction worth in destabilization? (See Section 4.9 for an alternate way to arrive at this value.)

Figure 4.6

Graph: Energy versus dihedral angle for ethane

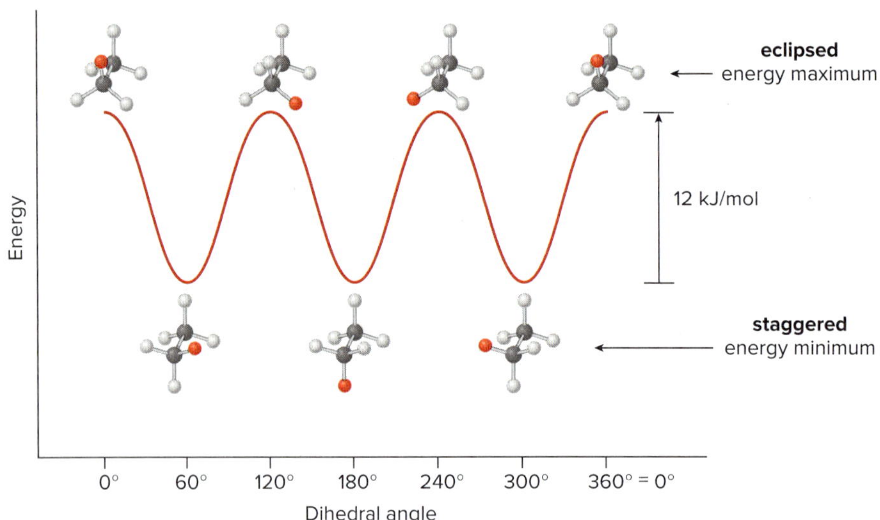

- Note the position of the labeled H atom after each 60° rotation. All three staggered conformations are identical (except for the position of the label), and the same is true for all three eclipsed conformations.

4.9 Conformations of Butane

Butane and higher-molecular-weight alkanes have several carbon–carbon bonds, all capable of rotation.

butane

Consider rotation at C2–C3.

Each C is bonded to 2 H's and 1 CH_3 group.

To analyze the different conformations that result from rotation around the C2–C3 bond, begin arbitrarily with one—for example, the staggered conformation that places two CH_3 groups 180° from each other—then,

It takes six 60° rotations to return to the original conformation.

- **Rotate one carbon atom in 60° increments either clockwise or counterclockwise, while keeping the other carbon fixed. Continue until you return to the original conformation.**

Figure 4.7 illustrates the six possible conformations that result from this process.

Figure 4.7 Six different conformations of butane

1 staggered, anti — rotate 60° → 2 eclipsed — rotate 60° → 3 staggered, gauche — rotate 60° → 4 eclipsed — rotate 60° → 5 staggered, gauche — rotate 60° → 6 eclipsed — rotate 60° → 1

Although each 60° bond rotation converts a staggered conformation into an eclipsed conformation (or vice versa), neither all the staggered conformations nor all the eclipsed conformations are the same. For example, the dihedral angle between the methyl groups in staggered conformations **3** and **5** are both 60°, whereas it is 180° in staggered conformation **1.**

- **A staggered conformation with two larger groups 180° from each other is called *anti*.**
- **A staggered conformation with two larger groups 60° from each other is called *gauche*.**

Similarly, the methyl groups in conformations **2** and **6** both eclipse hydrogen atoms, whereas they eclipse each other in conformation **4.**

The staggered conformations (**1, 3,** and **5**) are lower in energy than the eclipsed conformations (**2, 4,** and **6**), but how do the energies of the individual staggered and eclipsed conformations compare to each other? The relative energies of the individual staggered conformations (or the individual eclipsed conformations) depend on their **steric strain.**

- ***Steric* strain is an increase in energy resulting when atoms are forced too close to one another.**

The methyl groups are farther apart in the anti conformation (**1**) than in the gauche conformations (**3** and **5**), so among the staggered conformations, **1** is lower in energy (more stable) than **3** and **5.** In fact, the anti conformation is 3.8 kJ/mol (0.9 kcal/mol) lower in energy than either gauche conformation because of the steric strain that results from the proximity of the methyl groups in **3** and **5.**

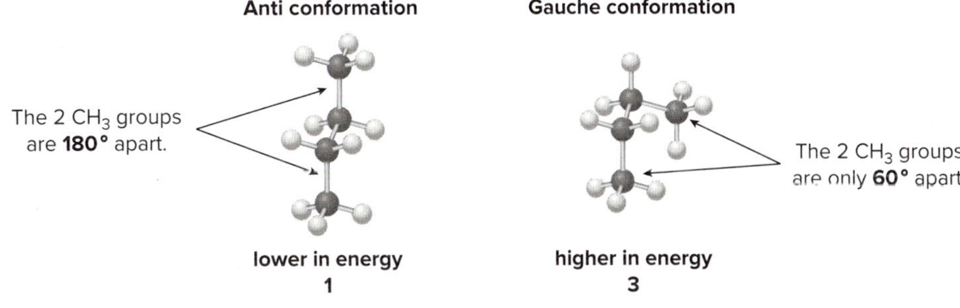

- **Gauche conformations are generally *higher* in energy than anti conformations because of steric strain.**

Steric strain caused by two eclipsed CH_3 groups

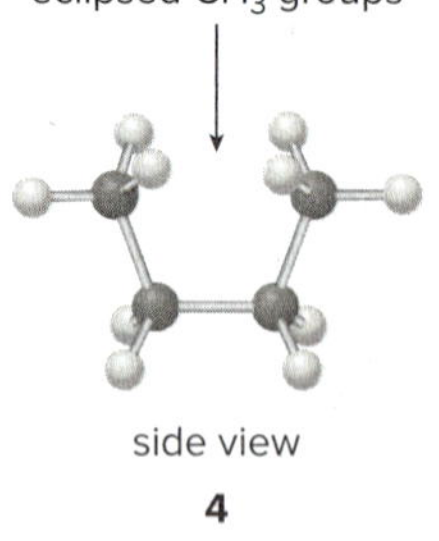

side view

4

Steric strain also affects the relative energies of eclipsed conformations. Conformation **4** is higher in energy than **2** or **6,** because the two larger CH_3 groups are forced close to each other, introducing considerable steric strain.

Problem 4.17 Label the sites of torsional and steric strain in each conformation.

a. H, CH_3, H, CH_3, H, CH_3

b. CH_3, H, H, CH_3, H, CH_3

c. H H, H H, CH_2CH_3, CH_3CH_2

To graph energy versus dihedral angle, keep in mind two considerations:

- **Staggered conformations are at energy minima and eclipsed conformations are at energy maxima.**
- **Unfavorable steric interactions increase energy.**

For butane, this means that anti conformation **1** is lowest in energy, and conformation **4** with two eclipsed CH_3 groups is the highest in energy. The relative energy of other conformations is depicted in the energy versus rotation diagram for butane in Figure 4.8.

CH_3,CH_3 eclipsing
11 kJ/mol destabilization

We can now use the values in Figure 4.8 to estimate the destabilization caused by other eclipsed groups. For example, conformation **4** is 19 kJ/mol less stable than the anti conformation **1.** Conformation **4** possesses two H,H eclipsing interactions, worth 4.0 kJ/mol each in destabilization (Section 4.8), and one CH_3,CH_3 eclipsing interaction. Thus, the **CH_3,CH_3 interaction** is worth $19 - 2(4.0) =$ **11 kJ/mol** of destabilization.

Figure 4.8 Graph: Energy versus dihedral angle for butane

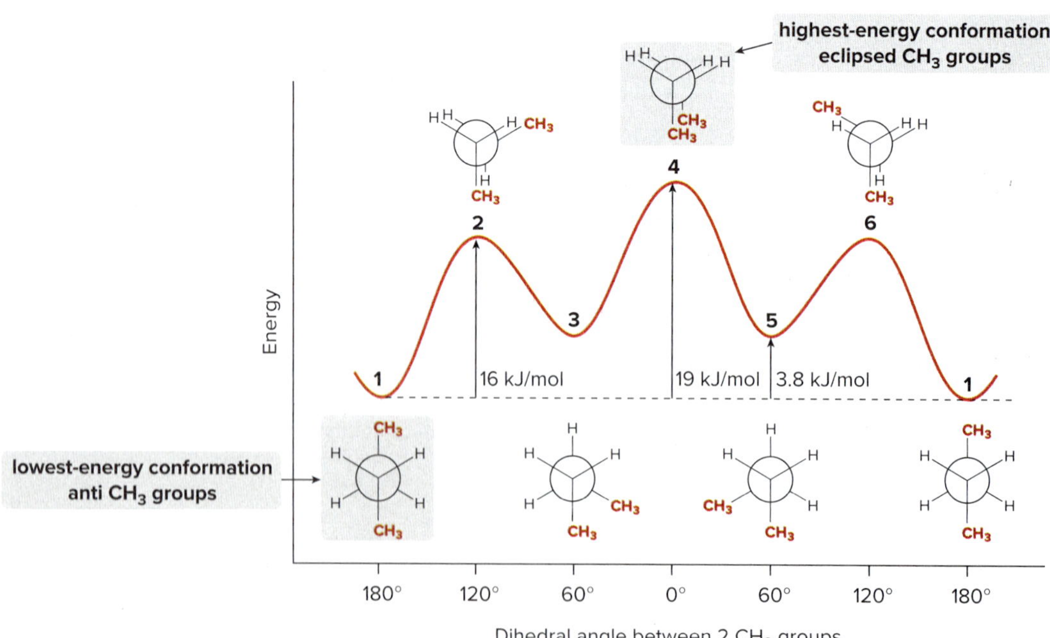

- Staggered conformations **1, 3,** and **5** are at energy minima.
- Anti conformation **1** is lower in energy than gauche conformations **3** and **5,** which possess steric strain.
- Eclipsed conformations **2, 4,** and **6** are at energy maxima.
- Eclipsed conformation **4,** which has additional steric strain due to two eclipsed CH_3 groups, is highest in energy.

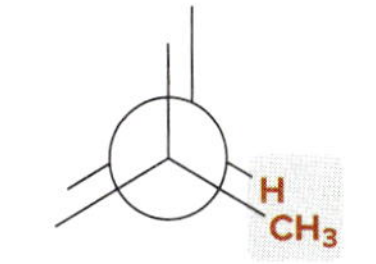

H,CH_3 eclipsing
6.0 kJ/mol destabilization

Similarly, conformation **2** is 16 kJ/mol less stable than the anti conformation **1,** and possesses one H,H eclipsing interaction (worth 4.0 kJ/mol of destabilization) and two H,CH_3 interactions. Thus, **each H,CH_3 interaction** is worth 1/2(16 − 4.0) = **6.0 kJ/mol** of destabilization. These values are summarized in Table 4.3.

- **The energy difference between the lowest- and highest-energy conformations is called the *barrier to rotation.***

Table 4.3 Torsional and Steric Strain Energies in Acyclic Alkanes

	Energy increase	
Type of interaction	kJ/mol	kcal/mol
H,H eclipsing	4.0	1.0
H,CH_3 eclipsing	6.0	1.4
CH_3,CH_3 eclipsing	11	2.6
gauche CH_3 groups	3.8	0.9

We can use these same principles to determine conformations and relative energies for any acyclic alkane. Because the **lowest-energy conformation has all bonds staggered and all large groups anti,** alkanes are often drawn in zigzag skeletal structures to indicate this.

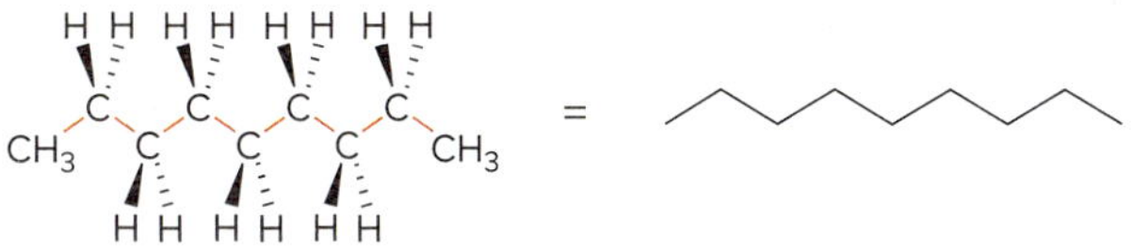

Problem 4.18

a. Draw the three staggered and three eclipsed conformations that result from rotation around the bond labeled in red using Newman projections.

b. Label the most stable and least stable conformation.

Problem 4.19 The eclipsed conformation of CH_3CH_2Cl is 15 kJ/mol less stable than the staggered conformation. How much is the H,Cl eclipsing interaction worth in destabilization?

Problem 4.20 Rank the following conformations in order of increasing energy.

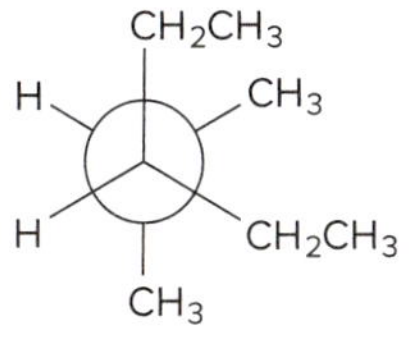

A 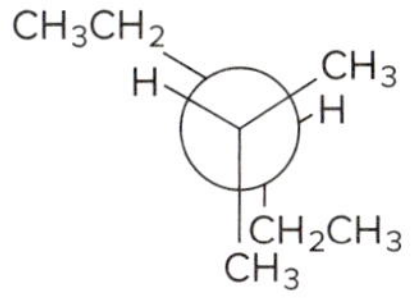 B 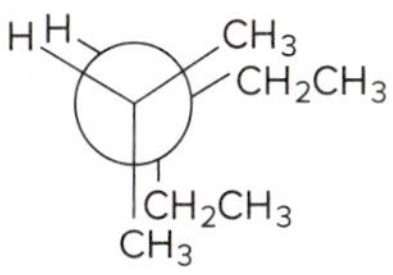 C 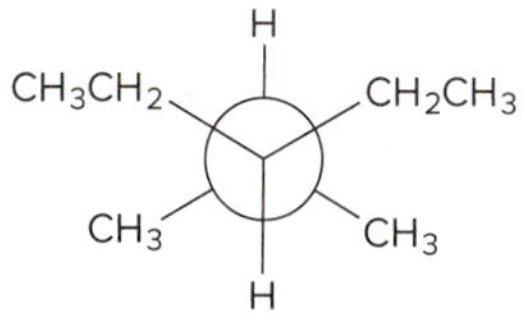 D

Problem 4.21 Consider rotation around the carbon–carbon bond in 1,2-dichloroethane ($ClCH_2CH_2Cl$).

a. Using Newman projections, draw all of the staggered and eclipsed conformations that result from rotation around this bond.

b. Graph energy versus dihedral angle for rotation around this bond.

Problem 4.22 Calculate the destabilization present in each eclipsed conformation.

a. b.

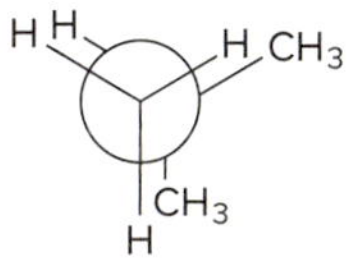

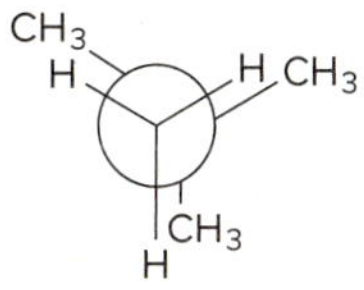

4.10 An Introduction to Cycloalkanes

Besides torsional strain and steric strain, the conformations of cycloalkanes are affected by **angle strain.**

- ***Angle* strain is an increase in energy when tetrahedral bond angles deviate from the optimum angle of 109.5°.**

Originally cycloalkanes were thought to be flat rings, with the bond angles between carbon atoms determined by the size of the ring. For example, a flat cyclopropane ring would have 60° internal bond angles, a flat cyclobutane ring would have 90° angles, and large flat rings would have very large angles. It was assumed that rings with bond angles so different from the tetrahedral bond angle would be very strained and highly reactive. This is called the **Baeyer strain theory.**

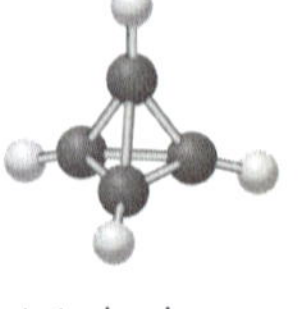

tetrahedrane

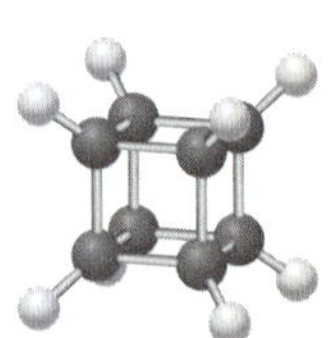

cubane

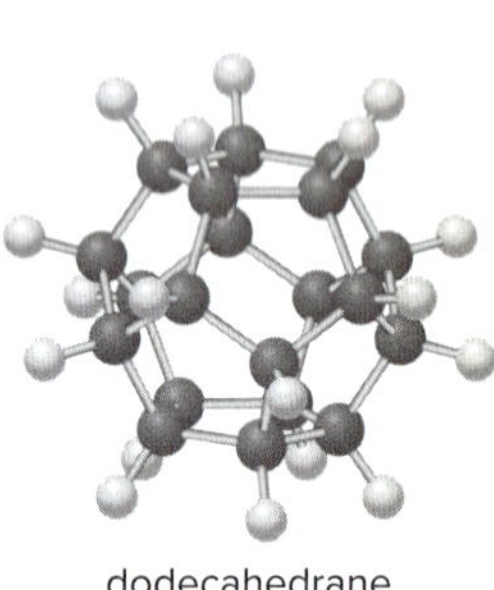

dodecahedrane

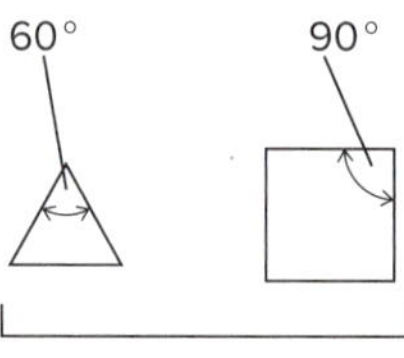

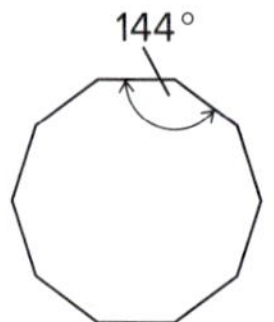

It turns out, though, that **cycloalkanes with more than three C atoms in the ring are not flat molecules.** They are puckered to **reduce strain,** both angle strain and torsional strain. The three-dimensional structures of some simple cycloalkanes are shown in Figure 4.9. Three- and four-membered rings still possess considerable angle strain, but puckering reduces the internal bond angles in larger rings, thus reducing angle strain.

Many polycyclic hydrocarbons are of interest to chemists. For example, **dodecahedrane,** containing 12 five-membered rings bonded together (Section 4.4C), is one member of a family of three hydrocarbons that contain several rings of one size joined together. The two other members of this family are **tetrahedrane,** consisting of four three-membered rings, and **cubane,** consisting of six four-membered rings. These compounds are the simplest regular polyhedra whose structures resemble three of the highly symmetrical Platonic solids: the tetrahedron, the cube, and the dodecahedron.

Figure 4.9 Three-dimensional structure of some cycloalkanes

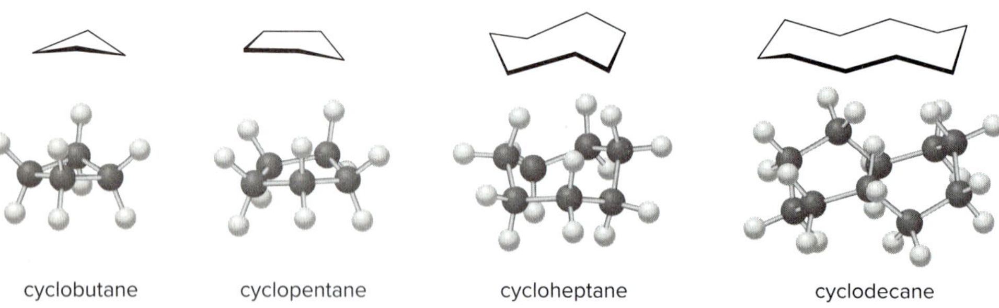

How stable are these compounds? Tetrahedrane (with internal 60° bond angles) is so strained that all attempts to prepare it have been thus far unsuccessful. Although cubane is also highly strained because of its 90° bond angles, it was first synthesized in 1964 and is a stable molecule at room temperature. Finally, dodecahedrane is very stable because it has bond angles very close to the tetrahedral bond angle (108° versus 109.5°). Its synthesis eluded chemists for years not because of its strain or inherent instability, but because of the enormous challenge of joining 12 five-membered rings together to form a sphere.

4.11 Cyclohexane

Let's now examine the conformation of **cyclohexane,** the most common ring size in naturally occurring compounds.

4.11A The Chair Conformation

A planar cyclohexane ring would experience angle strain, because the internal bond angle between the carbon atoms would be 120°, and torsional strain, because all of the hydrogens on adjacent carbon atoms would be eclipsed.

If a cyclohexane ring were flat...

120°

The internal bond angle is > 109.5°.
angle strain

All H's are aligned.
torsional strain

In reality, cyclohexane adopts a puckered conformation, called the **chair** form, which is more stable than any other possible conformation.

chair form = carbon skeleton of chair cyclohexane

Visualizing the chair. If the cyclohexane chair conformation is tipped downward, we can more easily view it as a chair with a back, seat, and foot support.

The chair conformation is so stable because it eliminates angle strain **(all C–C–C bond angles are 109.5°)** and torsional strain (all hydrogens on adjacent carbon atoms are **staggered,** not eclipsed).

109.5°

All H's are **staggered.**

- **In cyclohexane, three C atoms pucker up and three C atoms pucker down, alternating around the ring. These C atoms are called *up* C's and *down* C's.**

Each carbon in cyclohexane has two different kinds of hydrogens.

Each cyclohexane carbon atom has one axial and one equatorial hydrogen.

- ***Axial* hydrogens are located above and below the ring (along a perpendicular axis).**
- ***Equatorial* hydrogens are located in the plane of the ring (around the equator).**

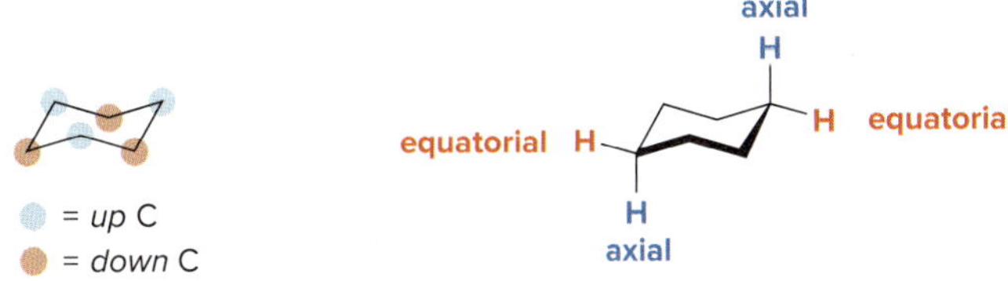

A three-dimensional representation of the chair form is shown in Figure 4.10.

Figure 4.10
A three-dimensional model of the chair form of cyclohexane with all H atoms drawn

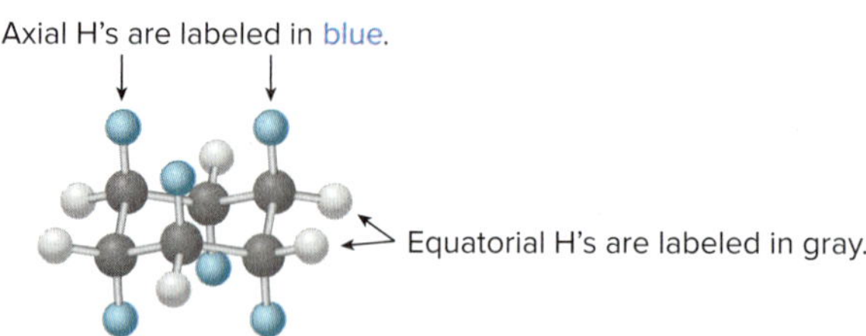

- Cyclohexane has **six axial H's** and **six equatorial H's.**

How To Draw the Chair Form of Cyclohexane

Step [1] **Draw the carbon skeleton.**

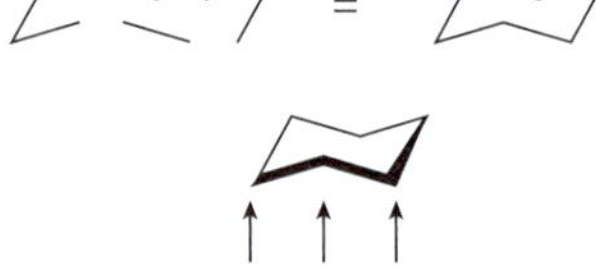

- Draw three parts of the chair: **a wedge, a set of parallel lines,** and **another wedge.**
- Then, join them together.
- The bottom 3 C's come out of the page, and for this reason, bonds to them are sometimes highlighted in bold.

Step [2] **Label the *up* C's and *down* C's on the ring.**

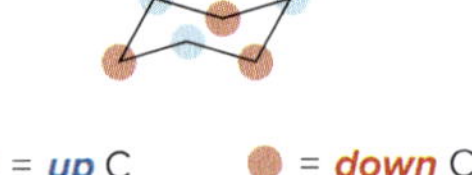

= ***up*** C = ***down*** C

- There are 3 *up* and 3 *down* C's, and they *alternate* around the ring.

Step [3] **Draw in the axial H atoms.**

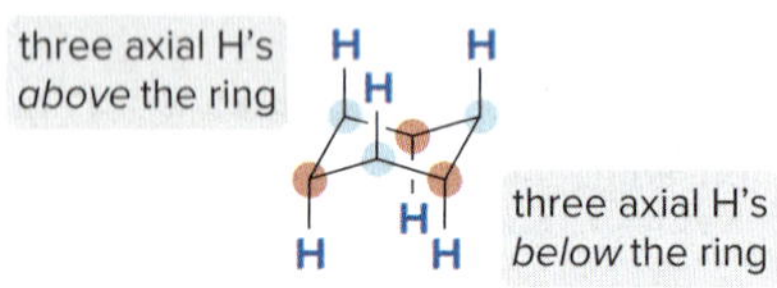

- On an ***up* C** the axial H is ***up.***
- On a ***down* C** the axial H is ***down.***

Step [4] **Draw in the equatorial H atoms.**

- **The axial H is down on a down C,** so the equatorial H must be up.
- **The axial H is up on an up C,** so the equatorial H must be down.

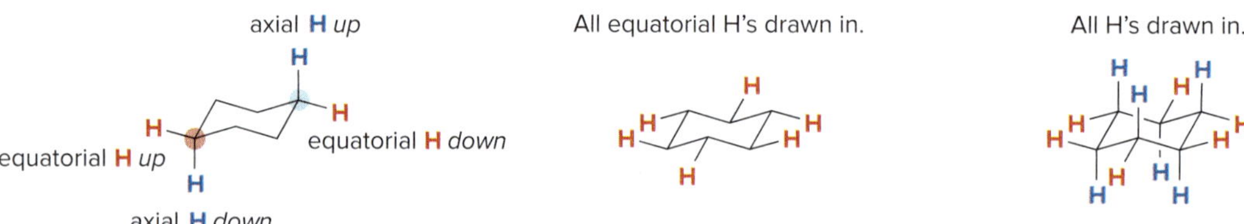

Problem 4.23 Classify the ring carbons as *up* C's or *down* C's. Identify the bonds highlighted in bold as axial or equatorial.

Problem 4.24 Using the cyclohexane with the C's numbered as shown, draw a chair form that fits each description.

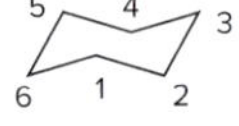

a. The ring has an axial CH_3 group at C1 and an equatorial OH on C2.
b. The ring has an equatorial CH_3 group on C6 and an axial OH group on C4.
c. The ring has equatorial OH groups on C1, C2, and C5.

4.11B Ring-Flipping

Like acyclic alkanes, **cyclohexane does not remain in a single conformation.** The bonds twist and bend, resulting in new arrangements, but the movement is more restricted. One conformational change involves **ring-flipping,** which can be viewed as a stepwise process.

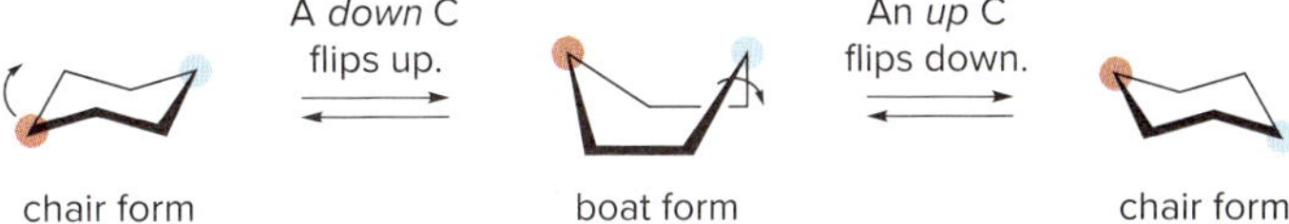

- **A *down* carbon flips up.** This forms a new conformation of cyclohexane called a **boat.** The boat form has two carbons oriented above a plane containing the other four carbons.
- The boat form can flip in two possible ways. The carbon labeled with a red circle can flip down, re-forming the initial conformation; or the **second *up* carbon,** labeled with a blue circle, **can flip down. This forms a second chair conformation.**

Because of ring-flipping, the ***up* carbons become *down* carbons and the *down* carbons become *up* carbons.** Thus, cyclohexane exists as two different chair conformations of equal stability, which rapidly interconvert at room temperature.

The process of ring-flipping also affects the orientation of cyclohexane's hydrogen atoms.

- **Axial and equatorial H atoms are interconverted during a ring flip. Axial H atoms become equatorial H atoms, and equatorial H atoms become axial H atoms (Figure 4.11).**

Figure 4.11

Ring-flipping interconverts axial and equatorial hydrogens in cyclohexane.

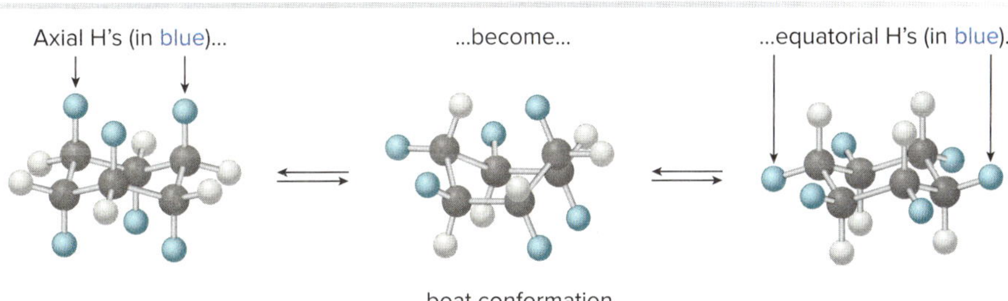

The chair forms of cyclohexane are 30 kJ/mol more stable than the boat forms. The boat conformation is destabilized by torsional strain because the hydrogens on the four carbon atoms in the plane are eclipsed. Additionally, there is steric strain because two hydrogens at either end of the boat—the **flagpole hydrogens**—are forced close to each other, as shown in Figure 4.12.

Figure 4.12

Two views of the boat conformation of cyclohexane

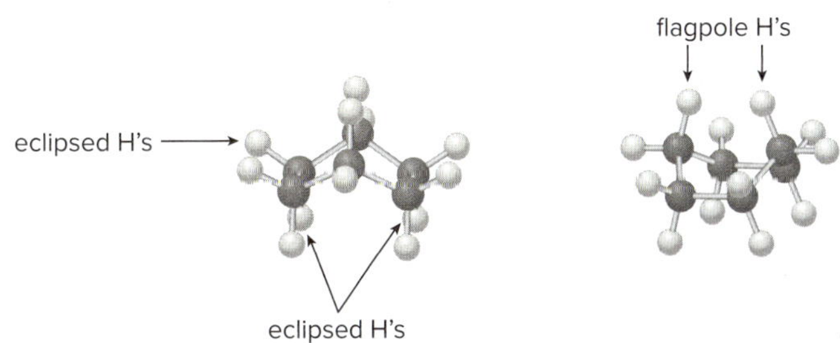

The boat form of cyclohexane is less stable than the chair forms for two reasons:

- Eclipsing interactions between H's cause **torsional strain.**
- The proximity of the flagpole H's causes **steric strain.**

4.12 Substituted Cycloalkanes

What happens when one hydrogen on cyclohexane is replaced by a larger substituent? Is there a difference in the stability of the two cyclohexane conformations?

- **The equatorial position has more room than the axial position, so *larger* substituents are more stable in the *equatorial* position.**

4.12A Cyclohexane with One Substituent

There are two possible chair conformations of a monosubstituted cyclohexane, such as methylcyclohexane, as shown in the following *How To.*

How To Draw the Two Conformations for a Substituted Cyclohexane

Step [1] **Draw one chair form and add the substituents.**

- Arbitrarily pick a ring carbon, classify it as an *up* or *down* carbon, and draw the bonds. **Each C has one axial and one equatorial bond.**
- Add the substituents, in this case H and CH_3, arbitrarily placing one axial and one equatorial. In this example, the CH_3 group is drawn equatorial.
- This forms one of the two possible chair conformations, labeled **A.**

up C
Add bonds.
axial bond
up
equatorial bond
down
Add substituents.
axial
H
CH_3
equatorial
A

Step [2] **Ring-flip the cyclohexane ring.**

- Convert *up* C's to *down* C's and vice versa. The chosen *up* C now puckers down.

***up* C**
ring-flip
***down* C**

Step [3] **Add the substituents to the second conformation.**

- Draw axial and equatorial bonds. **On a *down* C the axial bond is *down.***
- Ring-flipping converts axial bonds to equatorial bonds, and vice versa. The equatorial methyl becomes axial.
- This forms the other possible chair conformation, labeled **B.**

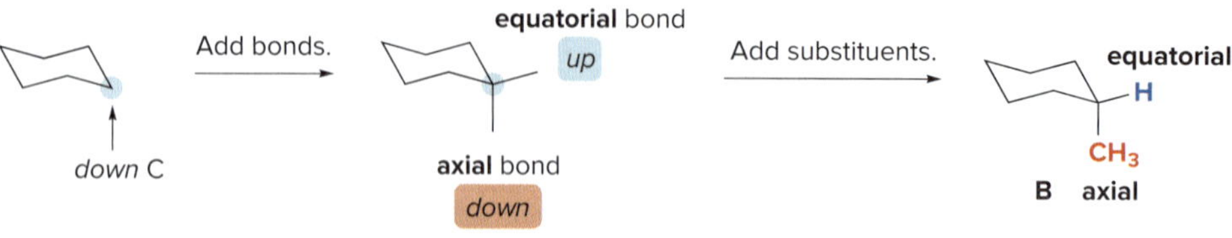

Although the CH_3 group flips from equatorial to axial, it starts on a down bond and stays on a down bond. **It *never* flips from below the ring to above the ring.**

- **A substituent always stays on the *same side* of the ring—either below or above—during the process of ring-flipping.**

Each carbon atom has one *up* and one *down* bond. An ***up*** bond can be either axial or equatorial, depending on the carbon to which it is attached. **On an *up* C, the axial bond is *up*,** but on a *down* C, the equatorial bond is *up*.

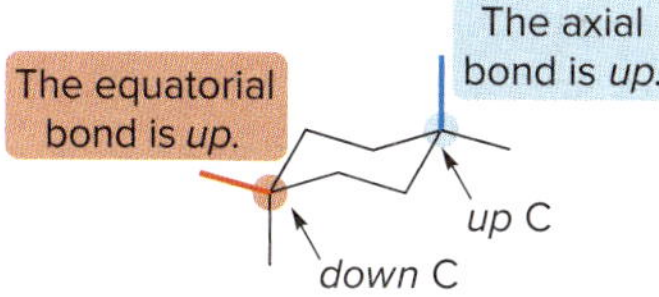

The two conformations of methylcyclohexane are different, so they are not equally stable. In fact, **A,** which places the larger methyl group in the roomier equatorial position, is considerably more stable than **B,** which places it axial.

H
CH_3
equatorial
A
The larger CH_3 group is equatorial.
more stable
95%

H
CH_3
B axial
less stable
5%

Why is a substituted cyclohexane ring more stable with a larger group in the equatorial position? Figure 4.13 shows that with an equatorial CH_3 group, steric interactions with nearby groups are minimized. An axial CH_3 group, however, is close to two other axial H atoms, creating two destabilizing steric interactions called **1,3-diaxial interactions.** Each unfavorable H,CH_3 interaction destabilizes the conformation by 3.8 kJ/mol, so **B** is 7.6 kJ/mol less stable than **A.**

Figure 4.13 Three-dimensional representations for the two conformations of methylcyclohexane

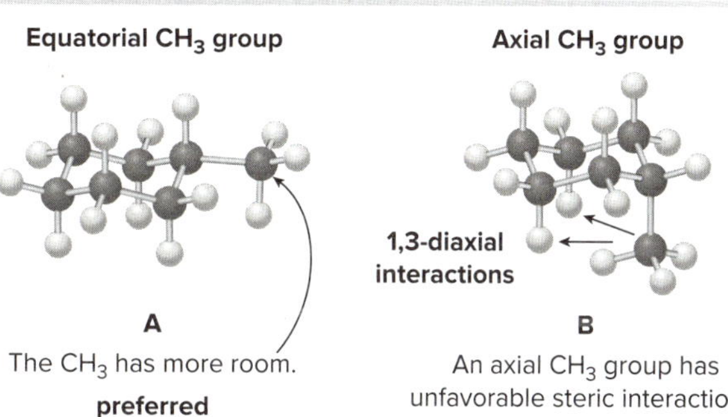

- **Larger axial substituents create unfavorable 1,3-diaxial interactions, destabilizing a cyclohexane conformation.**

The *larger* the substituent on the six-membered ring, the *higher* the percentage of the conformation containing the equatorial substituent at equilibrium. With a very large substituent like *tert*-butyl [$(CH_3)_3C$–], essentially none of the conformation containing an axial *tert*-butyl group is present at room temperature, so **the ring is essentially anchored in a single conformation having an equatorial *tert*-butyl group.** This is illustrated in Figure 4.14.

Figure 4.14 The two conformations of *tert*-butylcyclohexane

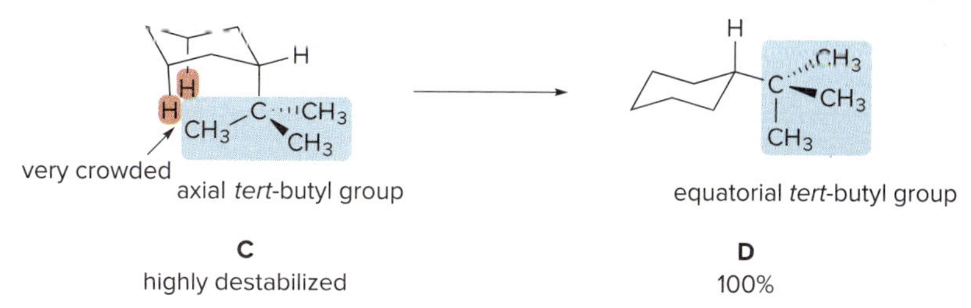

- The large *tert*-butyl group anchors the cyclohexane ring in conformation **D.**

Problem 4.25 Draw a second chair conformation for each cyclohexane. Then decide which conformation is present in higher concentration at equilibrium.

a. Br b. Cl c. CH_2CH_3

Problem 4.26 Draw both conformations for 1-ethyl-1-methylcyclohexane and decide which conformation (if any) is more stable.

4.12B A Disubstituted Cycloalkane

Rotation around the C–C bonds in the ring of a cycloalkane is restricted, so **a group on one side of the ring can *never* rotate to the other side of the ring.** As a result, there are two different 1,2-dimethylcyclopentanes—one having two CH_3 groups on the **same side** of the ring and one having them on **opposite sides** of the ring.

Wedges indicate bonds in front of the plane of the ring, and dashed wedges indicate bonds behind. For a review of this convention, see Section 1.7B. If a ring carbon is bonded to a CH_3 group in **front** of the ring (on a wedge), it is *assumed* that the other atom bonded to this carbon is hydrogen, located **behind** the ring (on a dashed wedge).

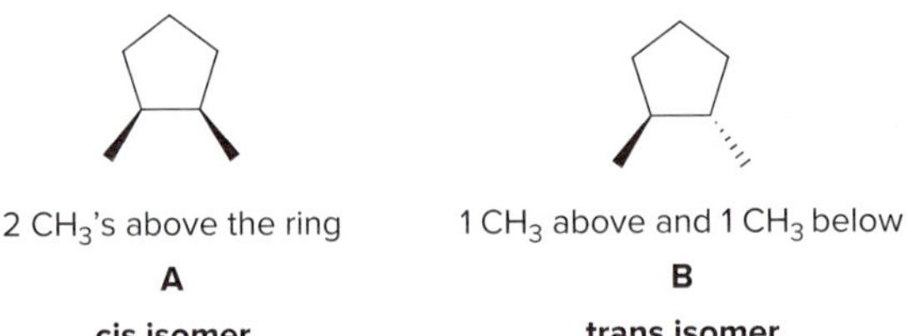

2 CH_3's above the ring | 1 CH_3 above and 1 CH_3 below
A | **B**
cis isomer | **trans isomer**

A and **B** are **isomers,** because they are different compounds with the same molecular formula, but they represent the second major class of isomers called **stereoisomers.**

- ***Stereoisomers*** **are isomers that differ *only* in the way the atoms are oriented in space.**

Cis and **trans** isomers are named by adding the prefixes *cis* and *trans* to the name of the cycloalkane. Thus, **A** is *cis*-1,2-dimethylcyclopentane, and **B** is *trans*-1,2-dimethylcyclopentane.

The prefixes **cis** and **trans** are used to distinguish these stereoisomers.

- **The cis isomer has two groups on the *same side* of the ring.**
- **The trans isomer has two groups on *opposite sides* of the ring.**

Problem 4.27 Draw the structure for each compound using wedges and dashed wedges.

a. *cis*-1,2-dimethylcyclopropane b. *trans*-1-ethyl-2-methylcyclopentane

Problem 4.28 For *cis*-1,3-diethylcyclobutane, draw (a) a stereoisomer; (b) a constitutional isomer.

4.12C A Disubstituted Cyclohexane

A disubstituted cyclohexane like 1,4-dimethylcyclohexane also has cis and trans stereoisomers. In addition, each of these stereoisomers has two possible chair conformations.

***trans*-1,4-dimethylcyclohexane** | ***cis*-1,4-dimethylcyclohexane**

All disubstituted cycloalkanes with two groups bonded to *different* atoms have cis and trans isomers.

To draw both conformations for each stereoisomer, follow the procedure in Section 4.12A for a monosubstituted cyclohexane, keeping in mind that two substituents must now be added to the ring.

How To Draw Two Conformations for a Disubstituted Cyclohexane

Step [1] **Draw one chair form and add the substituents.**

- For *trans*-1,4-dimethylcyclohexane, arbitrarily pick two C's located 1,4- to each other, classify them as *up* or *down* C's, and draw in the substituents.
- **The trans isomer must have one group *above* the ring (on an *up* bond) and one group *below* the ring (on a *down* bond).** The substituents can be either axial or equatorial, as long as one is up and one is down. The easiest trans isomer to visualize has two axial CH_3 groups. This arrangement is said to be **diaxial.**
- This forms one of the two possible chair conformations, labeled **A.**

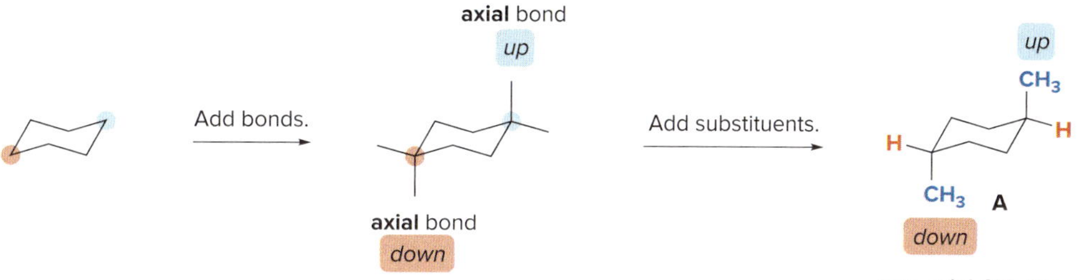

Step [2] **Ring-flip the cyclohexane ring.**

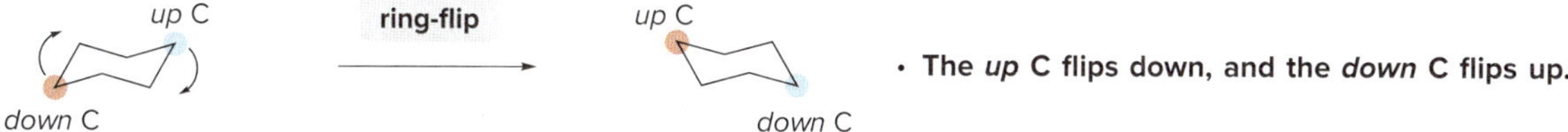

- **The *up* C flips down, and the *down* C flips up.**

Step [3] **Add the substituents to the second conformation.**

H
up
CH_3
CH_3
down
H
two equatorial CH_3 groups
B

- **Ring-flipping converts axial bonds to equatorial bonds, and vice versa.** The diaxial CH_3 groups become **diequatorial.** This trans conformation is less obvious to visualize. It is still trans, because one CH_3 group is above the ring (on an *up* bond), and one is below (on a *down* bond).

Conformations **A** and **B** are not equally stable. **Because B has both larger CH_3 groups in the roomier equatorial position, B is *lower* in energy.**

2 CH_3 groups in the more crowded **axial** position

2 CH_3 groups in the more roomy **equatorial** position

A
diaxial conformation

B
diequatorial conformation
more stable

The cis isomer of 1,4-dimethylcyclohexane also has two conformations, as shown in Figure 4.15. Because each conformation has one CH_3 group axial and one equatorial, they are **identical in energy.** At room temperature, therefore, the two conformations exist in a 50:50 mixture at equilibrium.

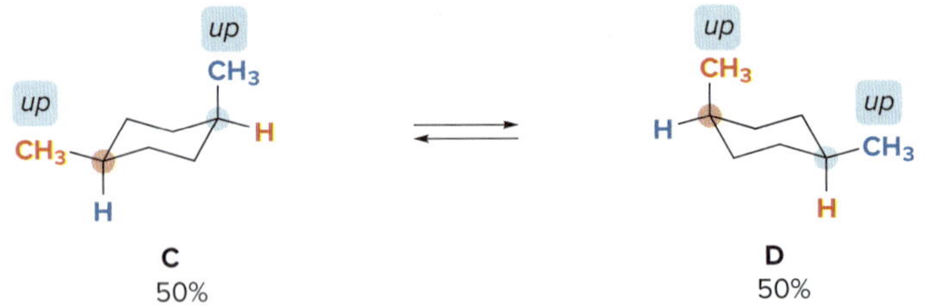

Figure 4.15 The two conformations of *cis*-1,4-dimethylcyclohexane

up
CH_3
H
H
C
50%

up
CH_3
H
up
CH_3
H
D
50%

- **A cis isomer has two groups on the same side of the ring,** either both *up* or both *down*. In this example, Conformations **C** and **D** have two CH_3 groups drawn *up*.
- Both conformations have one CH_3 group axial and one equatorial, making them equally stable.

Problem 4.29 For each compound drawn below: (a) label each OH, Br, and CH_3 group as axial or equatorial; (b) classify each conformation as cis or trans.

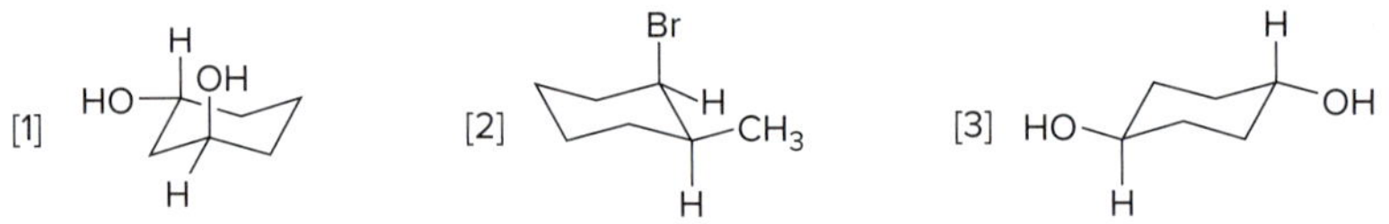

[1] HO, OH, H, H
[2] Br, H, CH_3, H
[3] HO, H, H, OH

The relative stability of the two conformations of any disubstituted cyclohexane can be analyzed using this procedure.

- **A *cis* isomer has two substituents on the *same side*, either both on *up* bonds or both on *down* bonds.**
- **A *trans* isomer has two substituents on *opposite sides*, one *up* and one *down*.**
- **Whether substituents are axial or equatorial depends on the relative location of the two substituents (on carbons 1,2-, 1,3-, or 1,4-).**

Sample Problem 4.3 Drawing Two Conformations for a Disubstituted Cycloalkane

Draw both chair conformations for *trans*-1,3-dimethylcyclohexane.

Solution

Step [1] **Draw one chair form and add substituents.**

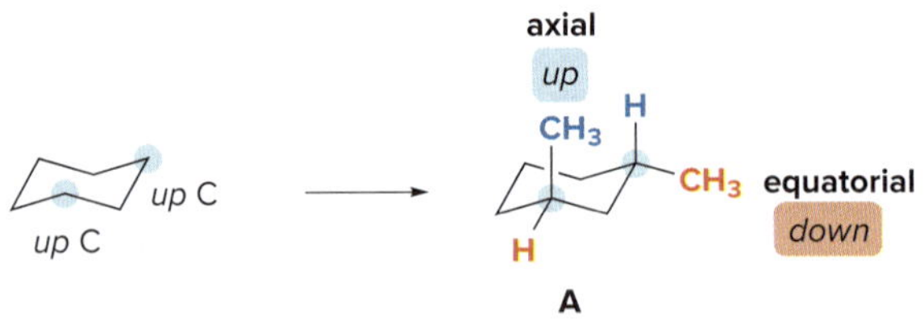

up C
up C

axial
up
CH_3
H
CH_3 equatorial
down
H
A

- Pick two C's 1,3- to each other.
- **The trans isomer has two groups on opposite sides.** In Conformation **A,** one CH_3 is axial (on an *up* bond), and one group is equatorial (on a *down* bond).

Steps [2–3] **Ring-flip and add substituents.**

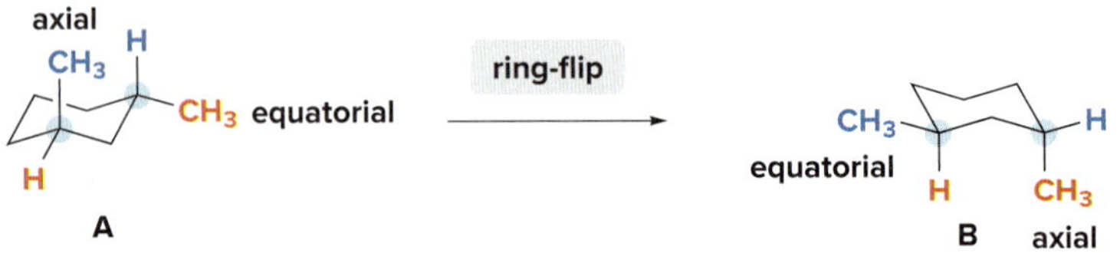

axial
CH_3
H
CH_3 equatorial
H
A

ring-flip

CH_3
equatorial
H
H
CH_3
B
axial

- The two *up* C's flip down.
- The axial CH_3 flips equatorial (still an *up* bond) and the equatorial CH_3 flips axial (still a *down* bond). Conformation **B** is *trans* because the two CH_3's are still on *opposite* sides.
- **Conformations A and B are equally stable** because each has one CH_3 equatorial and one axial.

Problem 4.30 Consider 1,2-dimethylcyclohexane.

a. Draw structures for the cis and trans isomers using a hexagon for the six-membered ring.
b. Draw the two possible chair conformations for the cis isomer. Which conformation, if either, is more stable?
c. Draw the two possible chair conformations for the trans isomer. Which conformation, if either, is more stable?
d. Which isomer, cis or trans, is more stable and why?

More Practice: Try Problems 4.39, 4.55, 4.56.

Problem 4.31 Label each compound as cis or trans. Then draw the second chair conformation.

a. b. c.

Problem 4.32 Draw a chair conformation of cyclohexane with one CH_3CH_2 group and one CH_3 group that fits each description.

a. a 1,1-disubstituted cyclohexane with an axial CH_3CH_2 group
b. a cis-1,2-disubstituted cyclohexane with an axial CH_3 group
c. a trans-1,3-disubstituted cyclohexane with an equatorial CH_3 group
d. a trans-1,4-disubstituted cyclohexane with an equatorial CH_3CH_2 group

Sample Problem 4.4 Converting a Hexagon with Substituents to a Chair Form

Draw the two chair forms for the following trisubstituted cyclohexane, and label the more stable conformation.

Solution

Use the wedges and dashed wedges to determine what groups are above and below the ring, respectively. Start at a substituent and proceed in the *same* direction around the ring—clockwise or counterclockwise—to convert the hexagon to both chair forms.

Step [1] Draw one chair form and add substituents.

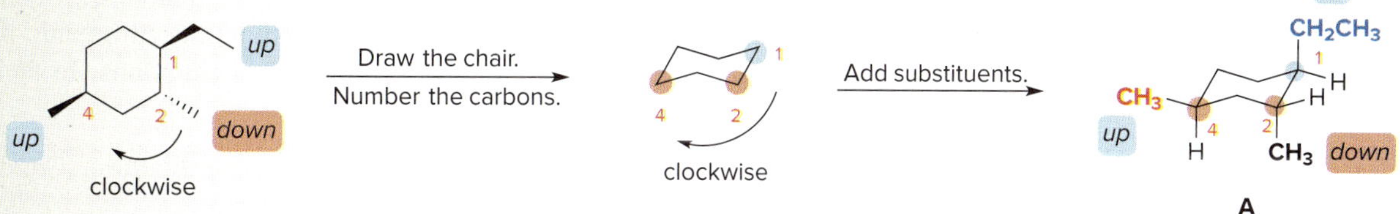

- If the ring C's are numbered 1 → 2 → 4 in the *clockwise* direction in the flat hexagon, number the C's in the chair in the *clockwise* direction. Any C in the chair can be assigned C1.
- The ethyl group on an *up* bond at C1 is axial, the methyl group on a *down* bond at C2 is axial, and the methyl on an *up* bond at C4 is equatorial.
- Conformation **A** has two axial and one equatorial substituents.

Step [2] **Ring-flip and add substituents.**

axial CH_2CH_3; CH_3 equatorial; H; H; H; CH_3 axial

A

two axial substituents

ring-flip

Add substituents.

axial CH_3; H; H; CH_2CH_3 equatorial; CH_3; H; equatorial

B

only **one** axial substituent
more stable conformation

- Ring-flipping the cyclohexane generates the second chair form **B,** which now has one axial and two equatorial substituents, making it the **more stable** conformation.

Problem 4.33 (a) Draw **C** in its more stable chair conformation. (b) Convert **D** to a hexagon with substituents on wedges and dashed wedges.

C

H H CH_3 CH_3 $CH_2CH_2CH_3$ H

D

More Practice: Try Problems 4.56a; 4.60a, b; 4.62c.

Problem 4.34 Label each pair of compounds as identical or stereoisomers: (a) **A** and **B;** (b) **A** and **C;** (c) **A** and **D;** (d) **B** and **C.**

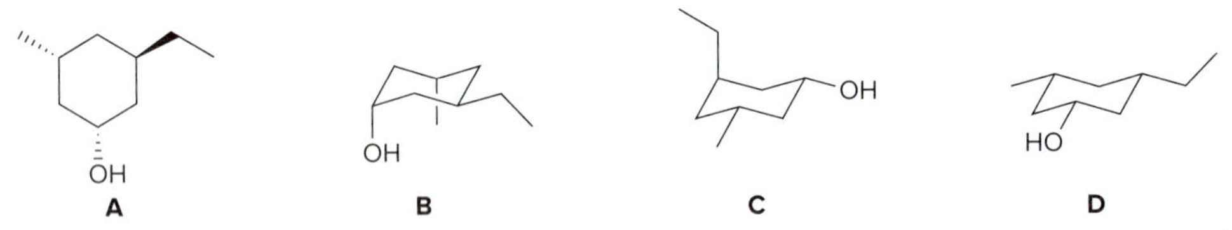

4.13 Oxidation of Alkanes

In Chapter 3, we learned that a functional group contains a heteroatom or π bond and constitutes **the reactive part of a molecule.** Alkanes are the only family of organic molecules that has no functional group, and therefore, **alkanes undergo few reactions.** In fact, alkanes are inert to reaction unless forcing conditions are used.

In Chapter 4, we consider only one reaction of alkanes—**combustion.** Combustion is an **oxidation–reduction** reaction.

4.13A Oxidation and Reduction Reactions

Compounds that contain many C–H bonds and few C–Z bonds are said to be in a ***reduced state,*** whereas those that contain few C–H bonds and more C–Z bonds are in a ***more oxidized state.*** CH_4 is highly reduced, whereas CO_2 is highly oxidized.

- ***Oxidation*** **is the** ***loss*** **of electrons.**
- ***Reduction*** **is the** ***gain*** **of electrons.**

Oxidation and reduction are opposite processes. As in acid–base reactions, there are always two components in these reactions. **One component is oxidized and one is reduced.**

To determine if an organic compound undergoes oxidation or reduction, we concentrate on the carbon atoms of the starting material and product, and **compare the relative number of C–H and C–Z bonds,** where Z = an element *more electronegative* than carbon (usually O, N, or X). Oxidation and reduction are then defined in two complementary ways.

- ***Oxidation*** **results in an** ***increase*** **in the number of C–Z bonds;** ***or***
- ***Oxidation*** **results in a** ***decrease*** **in the number of C–H bonds.**

- ***Reduction*** **results in a** ***decrease*** **in the number of C–Z bonds;** ***or***
- ***Reduction*** **results in an** ***increase*** **in the number of C–H bonds.**

Because Z is more electronegative than C, replacing C–H bonds with C–Z bonds decreases the electron density around C. Loss of electron density = oxidation.

Figure 4.16 illustrates the oxidation of CH_4 by replacing C–H bonds with C–O bonds (from left to right). The symbol **[O]** indicates oxidation. Because reduction is the reverse of oxidation, the molecules in Figure 4.16 are progressively reduced moving from right to left, from CO_2 to CH_4. The symbol **[H]** indicates reduction.

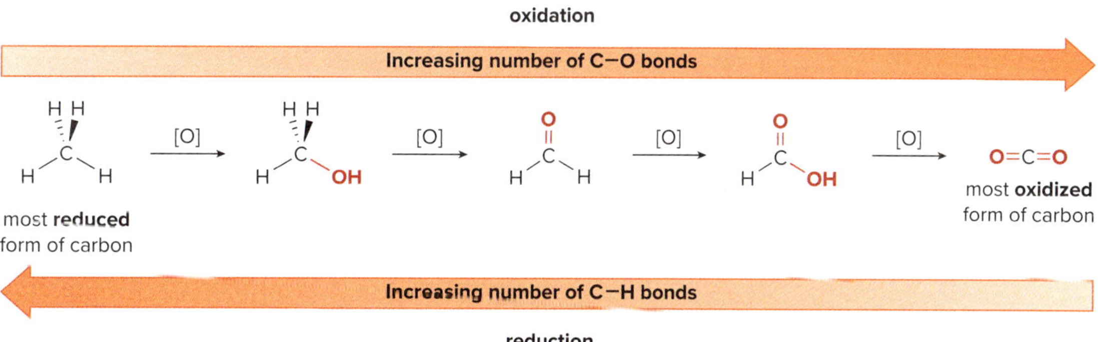

Figure 4.16 The oxidation and reduction of a carbon compound

Sample Problem 4.5 Determining Whether a Compound Is Oxidized or Reduced

Determine whether the organic compound is oxidized or reduced in each transformation.

a. ethanol ⟶ acetic acid b. cyclohexene ⟶ cyclohexane

Solution

a. The conversion of ethanol to acetic acid is an **oxidation** because the number of C–O bonds increases: CH_3CH_2OH has one C–O bond and CH_3COOH has three C–O bonds.

b. The conversion of cyclohexene (C_6H_{10}) to cyclohexane (C_6H_{12}) is a **reduction** because the number of C–H bonds increases: cyclohexane has two more C–H bonds than cyclohexene.

Problem 4.35 Classify each transformation as an oxidation, reduction, or neither.

a.  [benzaldehyde] → [benzoic acid]

b.  [bicyclic ketone] → [bicyclic alkane]

c. [acetone] → [2,2-propanediol]

d. 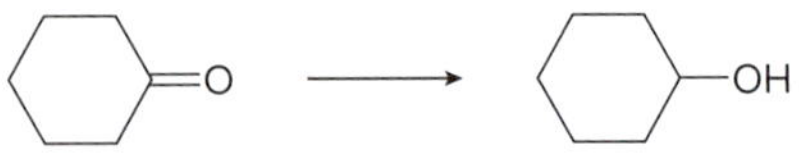 [cyclohexanone] → [cyclohexanol]

More Practice: Try Problems 4.64, 4.68a.

4.13B Combustion of Alkanes

When an organic compound is *oxidized* by a reagent, the reagent itself is *reduced.* Similarly, when an organic compound is *reduced* by a reagent, the reagent is *oxidized.* **Organic chemists identify a reaction as an oxidation or reduction by what happens to the *organic* component of the reaction.**

Alkanes undergo **combustion**—that is, **they burn in the presence of oxygen to form carbon dioxide and water.** This is a practical example of oxidation. Every C–H and C–C bond in the starting material is converted to a C–O bond in the product. The products, **$CO_2 + H_2O$,** are the same, regardless of the identity of the starting material. Combustion of alkanes in the form of natural gas, gasoline, or heating oil releases energy for heating homes, powering vehicles, and cooking food.

$$\underset{\text{methane}}{CH_4} + 2\,O_2 \xrightarrow{\text{flame}} CO_2 + 2\,H_2O + \text{(heat) energy}$$

$$2\ \underset{\text{2,2,4-trimethylpentane (isooctane)}}{C_8H_{18}} + 25\,O_2 \xrightarrow{\text{flame}} 16\,CO_2 + 18\,H_2O + \text{(heat) energy}$$

2,2,4-trimethylpentane (isooctane): **reduced** starting material

$16\,CO_2$: **oxidized** product

Combustion requires a spark or a flame to initiate the reaction. Gasoline, therefore, which is composed largely of alkanes, can be safely handled and stored in the air, but the presence of a spark or match causes immediate and violent combustion.

Driving an automobile 10,000 miles at 25 miles per gallon releases ~10,000 lb of CO_2 into the atmosphere.

The combustion of alkanes and other hydrocarbons obtained from fossil fuels adds a tremendous amount of CO_2 to the atmosphere each year. Quantitatively, data show over a 30% increase in the atmospheric concentration of CO_2 in the last 63 years (from 315 parts per million in 1958 to 414 parts per million in 2021; Figure 4.17). Although the composition of the atmosphere has changed over the lifetime of the earth, this may be the first time that the actions of humankind have altered that composition significantly and so quickly.

An increased CO_2 concentration in the atmosphere will have long-range and far-reaching effects. CO_2 absorbs thermal energy that normally radiates from the earth's surface, and redirects it back to the surface. Higher levels of CO_2 contribute to an increase in the average temperature of the earth's atmosphere. The global climate change resulting from these effects will lead to melting of the polar ice caps, a rise in sea level, and many more unforeseen consequences.

Problem 4.36 Draw the products of each combustion reaction.

a.  [propane] + O_2 $\xrightarrow{\text{flame}}$

b. 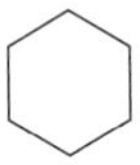 [cyclohexane] + O_2 $\xrightarrow{\text{flame}}$

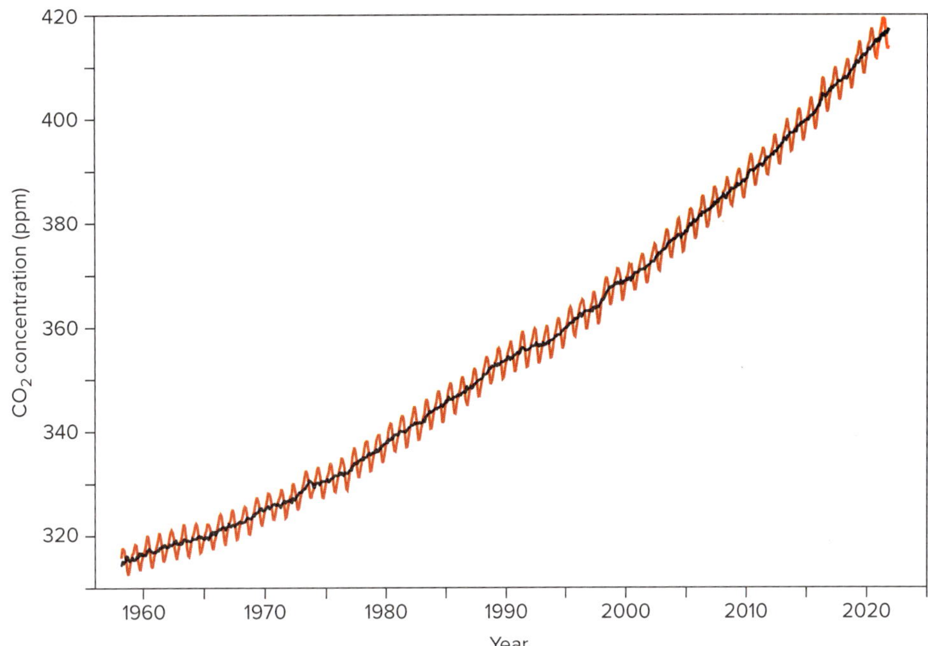

Figure 4.17 The changing concentration of CO_2 in the atmosphere since 1958

Source: US Department of Commerce, National Oceanic & Atmospheric Administration, NOAA Monitoring Laboratory.

- The increasing level of atmospheric CO_2 is clearly evident on the graph. Two data points are recorded each year. The sawtooth nature of the graph is due to seasonal variation of CO_2 level with the seasonal variation in photosynthesis. (Data recorded at Mauna Loa, Hawai'i)

4.14 Lipids—Part 1

Lipids that contain carbon–carbon double bonds are discussed in Section 10.6.

Lipids are biomolecules whose properties resemble those of alkanes and other hydrocarbons. They are unlike any other class of biomolecules, because they are defined by a **physical property,** not by the presence of a particular functional group.

- **Lipids are biomolecules that are soluble in organic solvents and insoluble in water.**

Lipids have varied sizes and shapes, and a diverse number of functional groups. Fat-soluble vitamins like vitamin A and the phospholipids that comprise cell membranes are two examples of lipids that were presented in Sections 3.5A and 3.7A. Other examples are shown in Figure 4.18. One unifying feature accounts for their solubility.

Figure 4.18 Two representative lipid molecules

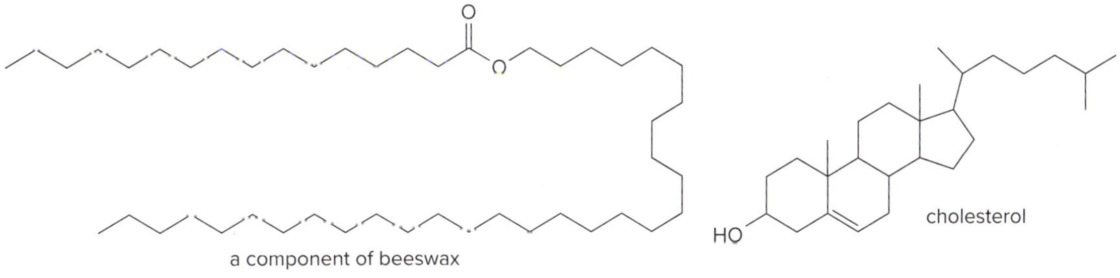

O
O
a component of beeswax
HO
cholesterol

- **Lipids are composed of many nonpolar C–H and C–C bonds, and have few polar functional groups.**

Riorita/E+/Getty Images

Waxes are lipids having two long alkyl chains joined by a single oxygen-containing functional group. Because of their many C–C and C–H bonds, **waxes are hydrophobic.** They form a protective coating on the feathers of birds to make them water repellent, and on leaves to prevent water evaporation. Bees secrete $\mathbf{CH_3(CH_2)_{14}COO(CH_2)_{29}CH_3}$, a wax that forms the honeycomb in which they lay eggs.

Cholesterol is a member of the steroid family, a group of lipids having four rings joined together. Because it has just one polar OH group, cholesterol is insoluble in the aqueous medium of the blood. It is synthesized in the liver and transported to other cells bound to water-soluble organic molecules. Elevated cholesterol levels can lead to coronary artery disease.

Cholesterol is a vital component of the cell membrane. Its hydrophobic carbon chain is embedded in the interior of the lipid bilayer, and its hydrophilic hydroxy group is oriented toward the aqueous exterior (Figure 4.19). Because its tetracyclic carbon skeleton is quite rigid compared to the long floppy side chains of a phospholipid, cholesterol stiffens the cell membrane somewhat, giving it more strength.

Figure 4.19 Cholesterol embedded in a lipid bilayer of a cell membrane

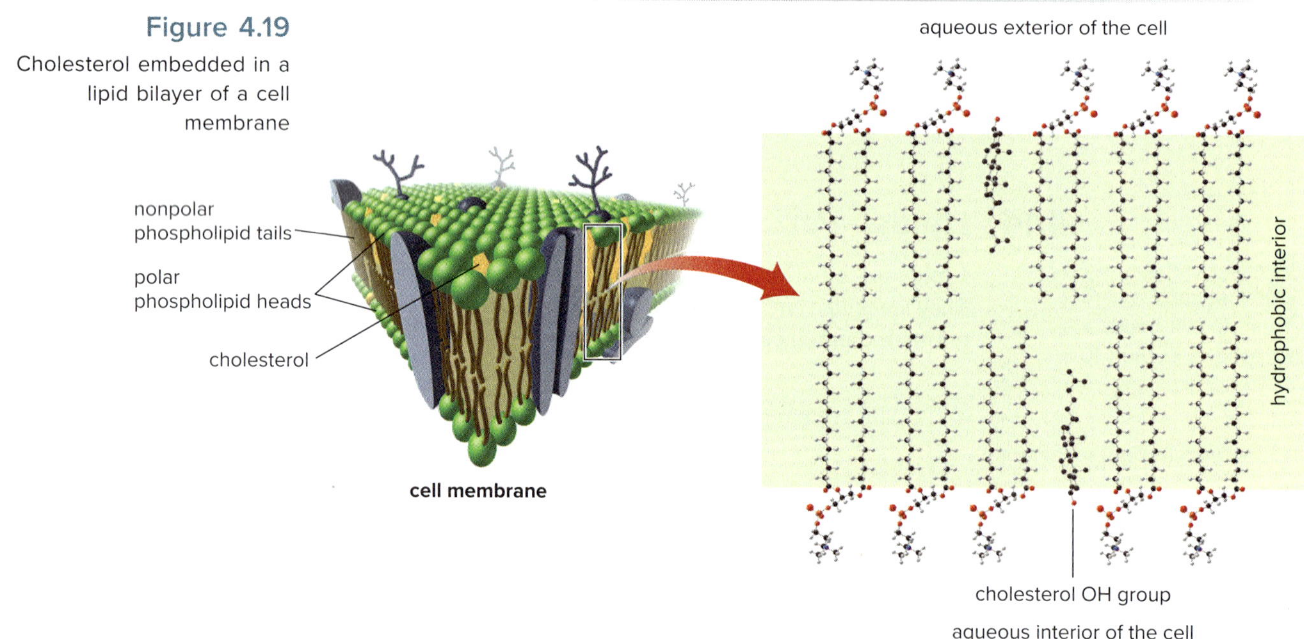

- The nonpolar hydrocarbon skeleton of cholesterol is embedded in the nonpolar interior of the cell membrane. Its rigid carbon skeleton stiffens the fluid lipid bilayer, giving it strength.
- Cholesterol's polar OH group is oriented toward the aqueous media inside and outside the cell.

Lipids have a high energy content, meaning that much energy is released on their metabolism. Because lipids are composed mainly of C–C and C–H bonds, they are oxidized with the release of energy, just like alkanes are. In fact, lipids are the most efficient biomolecules for the storage of energy. **The combustion of alkanes provides heat for our homes, and the metabolism of lipids provides energy for our bodies.**

Problem 4.37 Explain why beeswax is insoluble in H_2O, slightly soluble in ethanol (CH_3CH_2OH), and soluble in chloroform ($CHCl_3$).

Chapter 4 REVIEW

KEY CONCEPTS

[1] General facts about alkanes

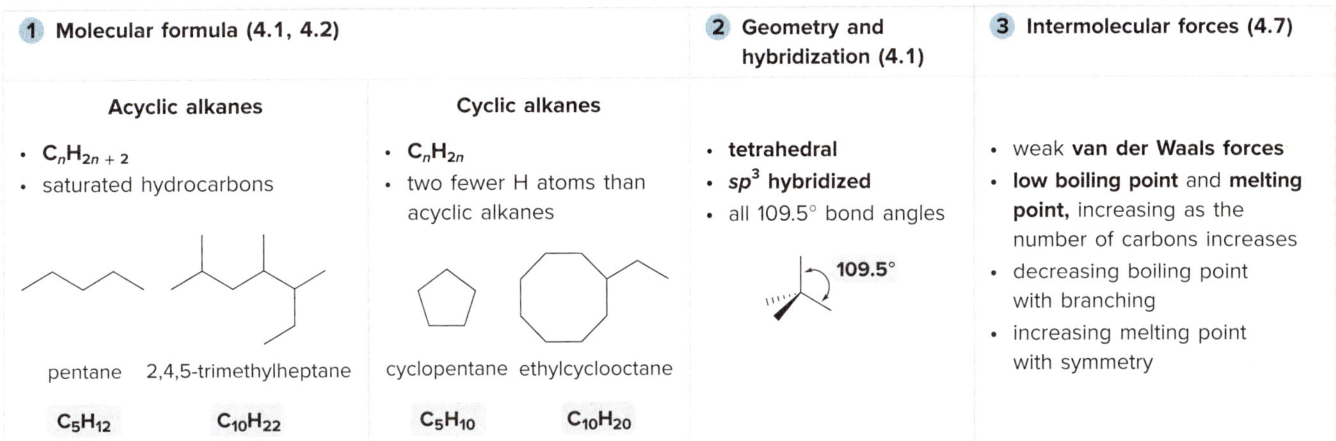

1 Molecular formula (4.1, 4.2)		2 Geometry and hybridization (4.1)	3 Intermolecular forces (4.7)
Acyclic alkanes • C_nH_{2n+2} • saturated hydrocarbons pentane C_5H_{12} 2,4,5-trimethylheptane $C_{10}H_{22}$	**Cyclic alkanes** • C_nH_{2n} • two fewer H atoms than acyclic alkanes cyclopentane C_5H_{10} ethylcyclooctane $C_{10}H_{20}$	• **tetrahedral** • ***sp*^{3} hybridized** • all 109.5° bond angles 109.5°	• weak **van der Waals forces** • **low boiling point** and **melting point,** increasing as the number of carbons increases • decreasing boiling point with branching • increasing melting point with symmetry

See Table 4.2. Try Problems 4.46, 4.47.

[2] Names of alkyl groups (4.4A)

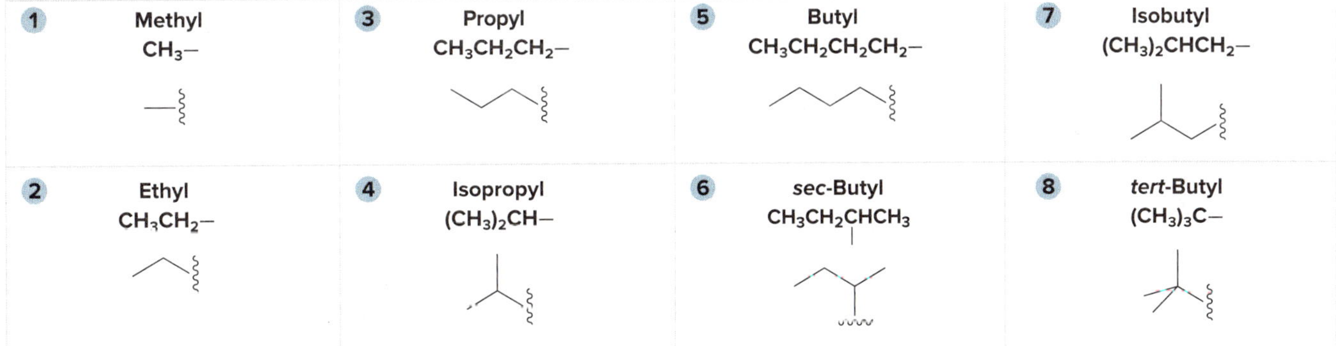

1 **Methyl** $CH_3—$	3 **Propyl** $CH_3CH_2CH_2—$	5 **Butyl** $CH_3CH_2CH_2CH_2—$	7 **Isobutyl** $(CH_3)_2CHCH_2—$
2 **Ethyl** $CH_3CH_2—$	4 **Isopropyl** $(CH_3)_2CH—$	6 ***sec*-Butyl** $CH_3CH_2CHCH_3$	8 ***tert*-Butyl** $(CH_3)_3C—$

[3] Conformations of acyclic alkanes (4.8, 4.9)

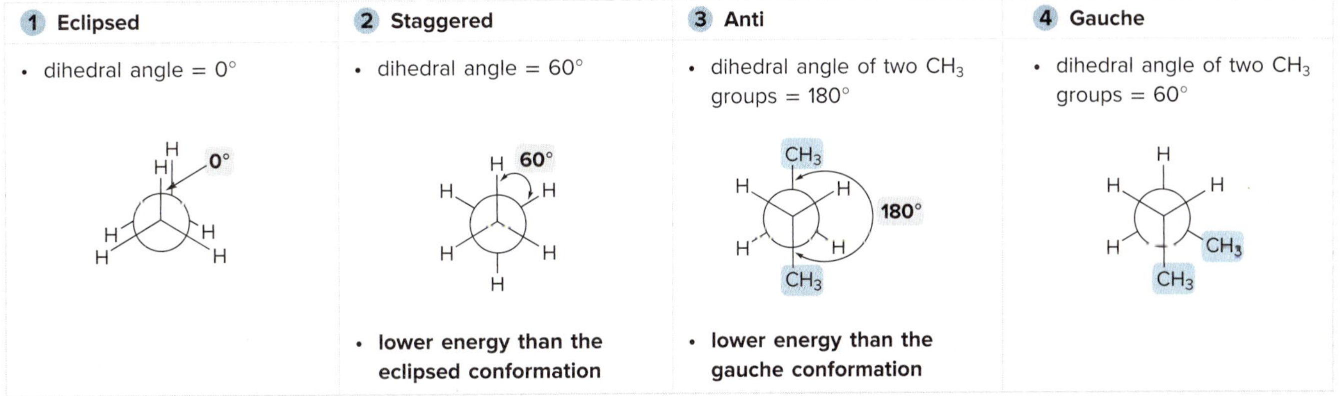

1 Eclipsed	2 Staggered	3 Anti	4 Gauche
• dihedral angle = 0°	• dihedral angle = 60°	• dihedral angle of two CH_3 groups = 180°	• dihedral angle of two CH_3 groups = 60°
0°	60°	180°	
	• **lower energy than the eclipsed conformation**	• **lower energy than the gauche conformation**	

See Figure 4.7. Try Problems 4.40, 4.48–4.51, 4.52a, 4.54a.

[4] Types of strain

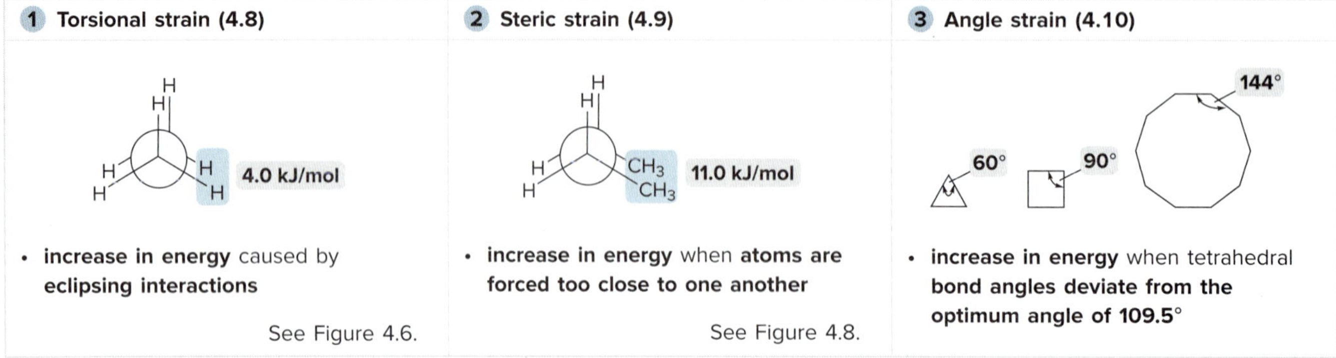

[5] Chair cyclohexane and monosubstituted cyclohexanes

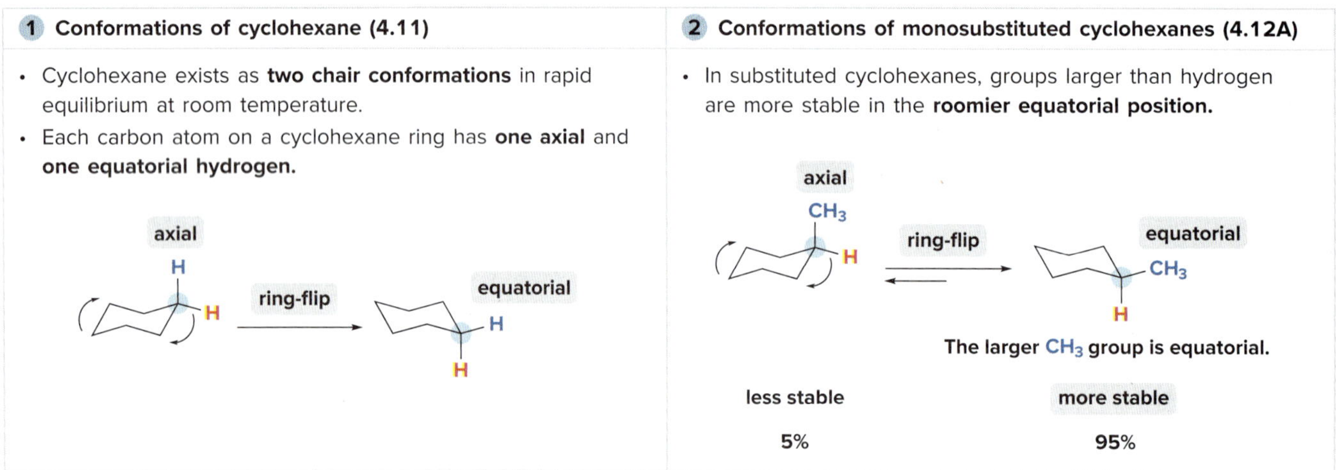

See *How To*'s p. 154, p. 156, Figures 4.11, 4.13, 4.14.

[6] Disubstituted cyclohexanes (4.12C)

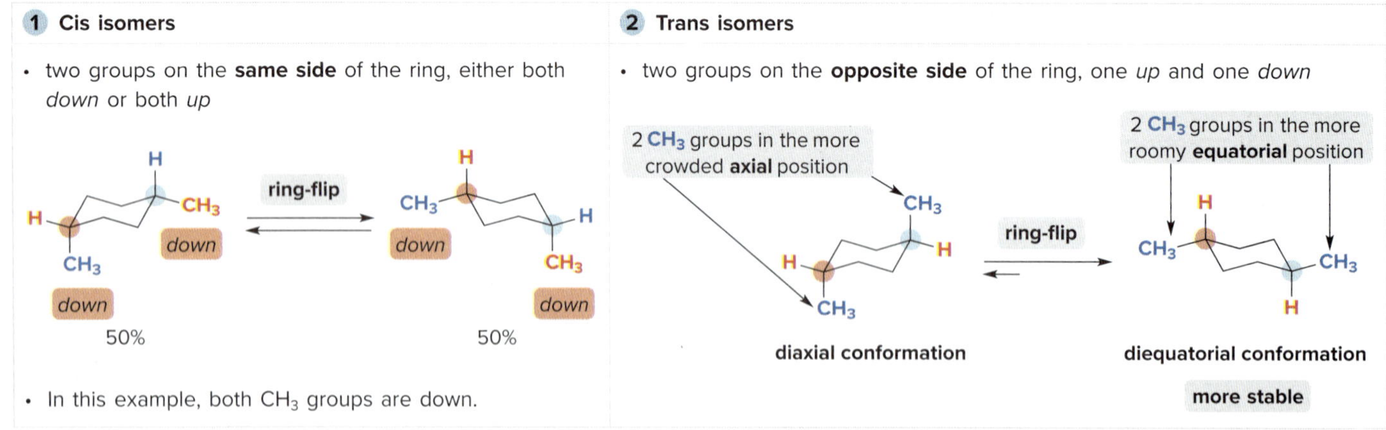

See *How To* p. 159, Figure 4.15, Sample Problem 4.3. Try Problems 4.55–4.58, 4.60a, 4.62b.

[7] Two types of isomers

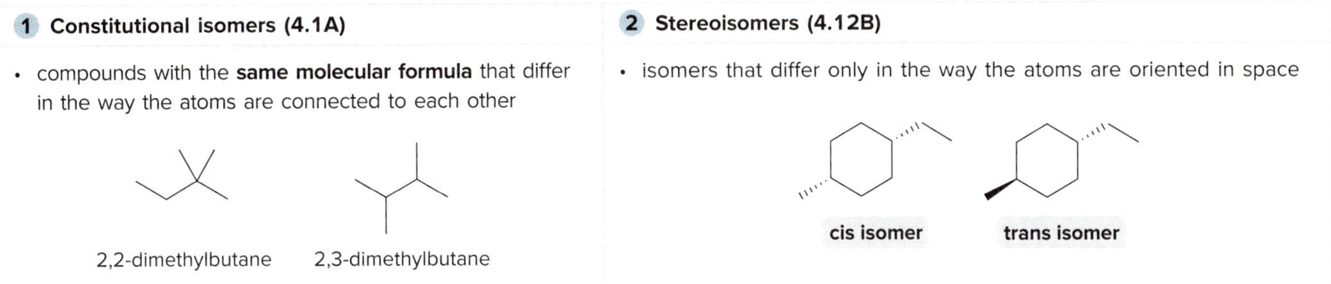

Try Problems 4.41, 4.61, 4.62a, 4.63.

KEY SKILLS

[1] Naming an alkane using the IUPAC system (4.4)

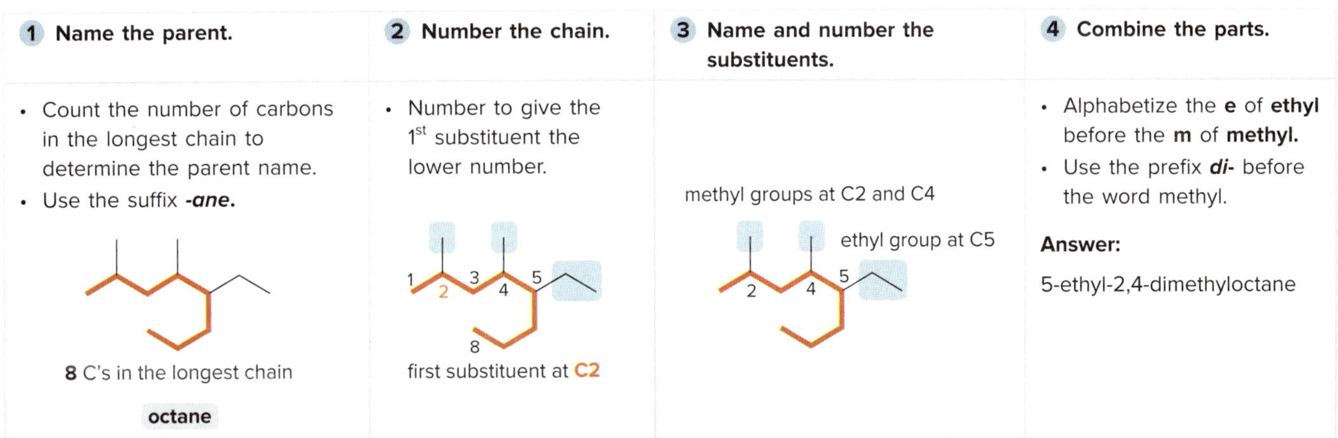

See Table 4.1, *How To* p. 136, Sample Problem 4.1, Figure 4.1. Try Problems 4.38; 4.42a–d, h, j; 4.43a, c, g, h; 4.45; 4.69a.

[2] Naming a cycloalkane using the IUPAC system (4.5)

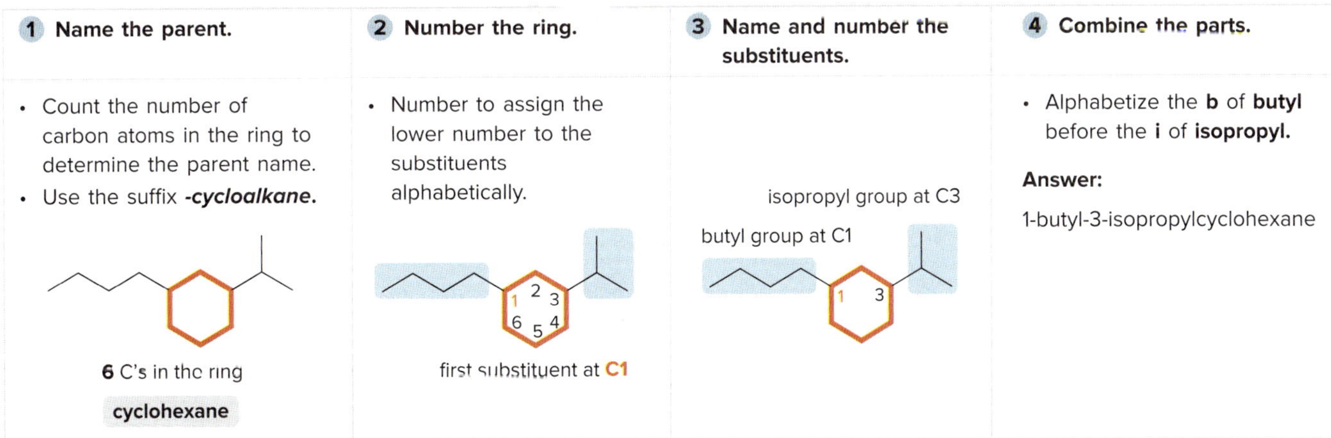

See *How To* p. 141, Figure 4.2. Try Problems 4.42e–g, i; 4.43b, d–f, i, j.

[3] Determining the highest- and lowest-energy conformations using Newman projections (4.9); example: $ClCH_2CH_2Cl$

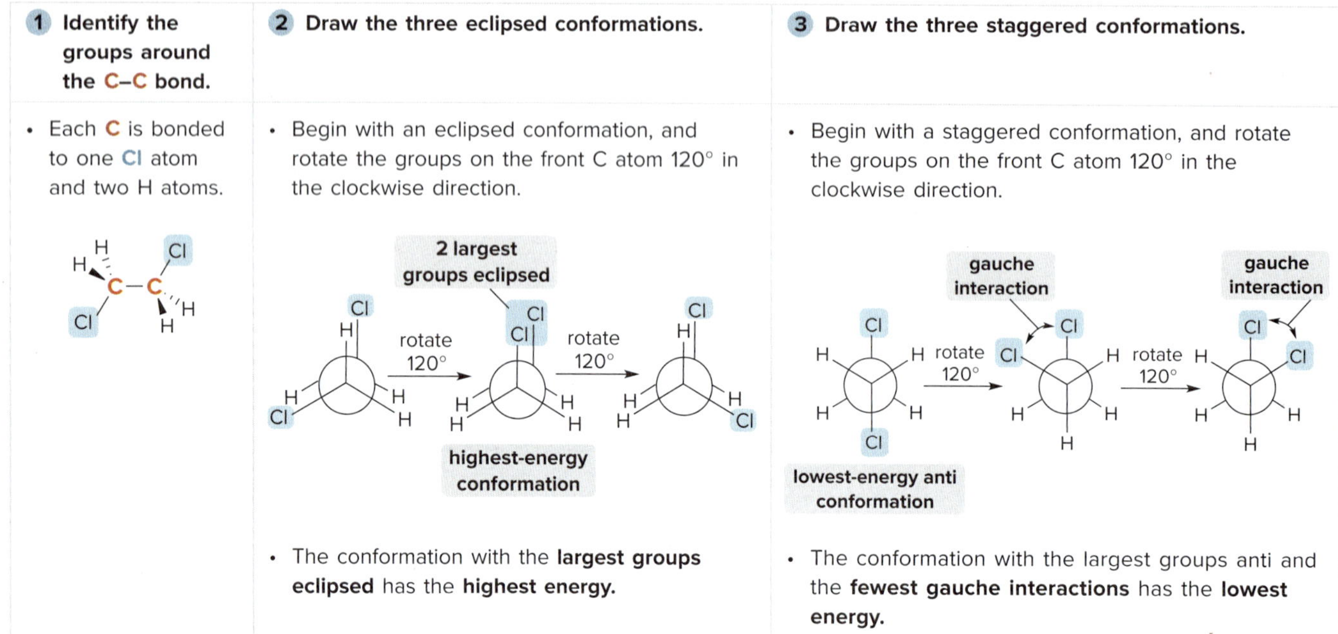

See *How To* p. 146, Figures 4.4, 4.6, 4.7, 4.8. Try Problems 4.48–4.51, 4.52a.

[4] Drawing two conformations for a disubstituted cyclohexane (4.12C); example: *cis*-1-(*tert*-butyl)-2-methylcyclohexane

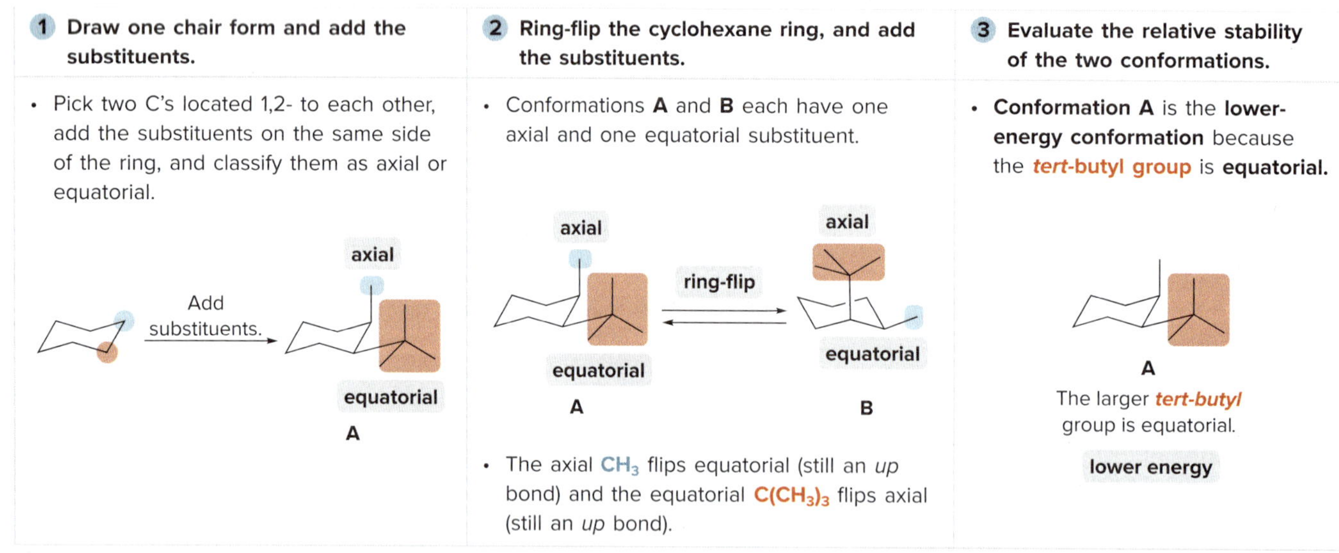

See *How To* p. 159, Figure 4.15, Sample Problem 4.3. Try Problems 4.39, 4.55, 4.56b–d, 4.57, 4.58, 4.60a.

[5] Determining whether a compound is oxidized or reduced (4.13A)

1 **Count the number of C–O bonds in the starting material and compare to the product.**

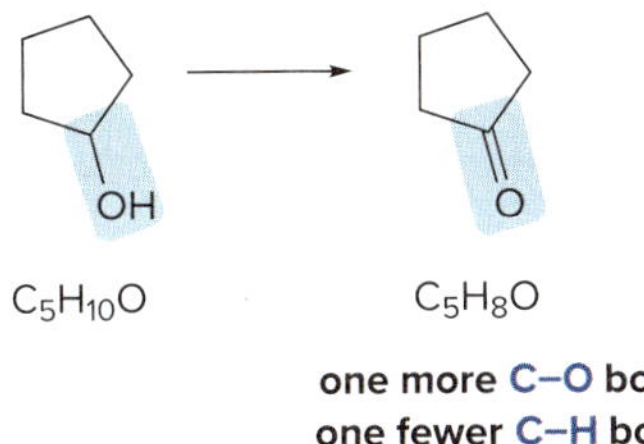

$C_5H_{10}O$ C_5H_8O

one more C–O bond
one fewer C–H bond

oxidation

- **Oxidation** results in an ***increase*** **in the number of C–Z bonds** or a ***decrease*** **in the number of C–H bonds.**

2 **Count the number of C–H bonds in the starting material and compare to the product.**

C_6H_{10} C_6H_{12}

two more C–H bonds

reduction

- **Reduction** results in a ***decrease*** **in the number of C–Z bonds** or an ***increase*** **in the number of C–H bonds.**

See Sample Problem 4.5. Try Problems 4.64, 4.68a.

CHAPTER 4 MULTIPLE-CHOICE SELF-TEST

The Self-Test consists of multiple-choice questions similar to those found on the American Chemical Society organic chemistry exam. Answers are given at the end of the chapter.

1. Give the IUPAC name for the given alkane.

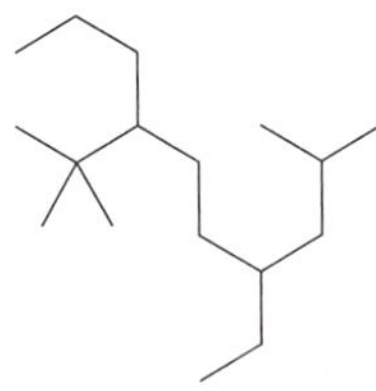

a. 4-ethyl-2-methyl-7-*tert*-butyldecane
b. 3-isobutyl-6-*tert*-butylnonane
c. 6-ethyl-3-propyl-2,2,8-trimethylnonane
d. 7-*tert*-butyl-4-ethyl-2-methyldecane

2. Give the IUPAC name for the given cycloalkane.

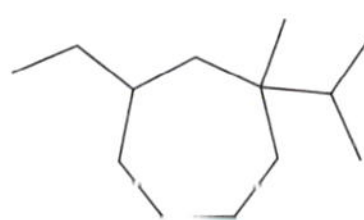

a. 1-ethyl-3-isopropyl-3-methylcycloheptane
b. 3-ethyl-1-methyl-1-isopropylcycloheptane
c. 3-ethyl-1-isopropyl-1-methylcycloheptane
d. 6-ethyl-1-methyl-1-isopropylcycloheptane

3. Which compound does *not* contain a trans disubstituted cyclohexane?

a.  b. 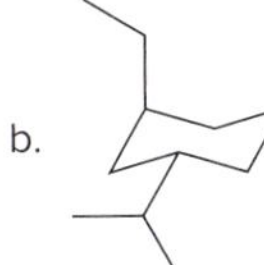 c.  d. 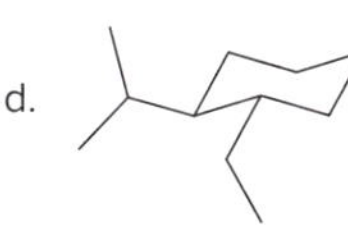

4. Classify each reaction as an oxidation, reduction, or neither.

[1] cyclohexanone → 2-chlorocyclohexanone (=O, Cl) [2] cyclohexanone → cyclohexenone (=O) [3] cyclohexene → cyclohexanol (OH)

a. [1], [2], and [3] oxidation
b. [1] and [2] oxidation, [3] neither
c. [1] and [3] neither, [2] oxidation
d. [1] neither, [2] reduction, [3] oxidation

5. How are **A** and **B** related to each other?

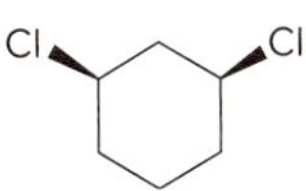

A

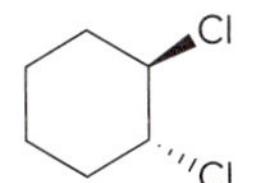

B

a. constitutional isomers
b. stereoisomers
c. identical
d. not isomers of each other

6. How are **C** and **D** related to each other?

C: CH_3, CH_3CH_2, H, H, CH_3, CH_2CH_3

D: $CH(CH_3)_2$, H, H, H, CH_2CH_3, CH_3

a. constitutional isomers
b. stereoisomers
c. identical
d. not isomers of each other

7. Which statement describes the substituents in the more stable chair conformation of **E?**

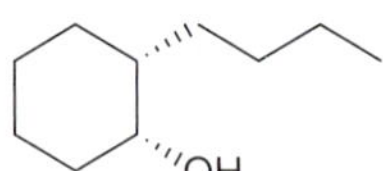

E

a. The butyl and OH groups are both equatorial.
b. The butyl group is equatorial and the OH is axial.
c. The OH group is equatorial and the butyl group is axial.
d. The butyl and OH groups are both axial.

8. Considering rotation around the C2–C3 bond of $(CH_3)_2CHCH(CH_3)_2$, which Newman projection depicts the lowest-energy conformation?

a. 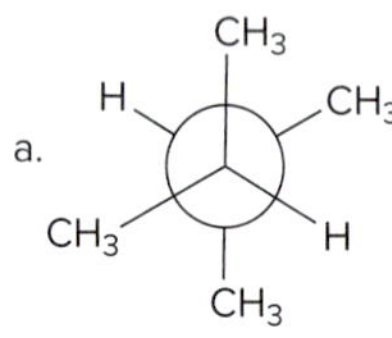

b. CH_3, H, CH_3, H, CH_3, CH_3

c. H, CH_3, CH_3, CH_3, H, CH_3

d. CH_3, H, CH_3, CH_3, H, CH_3

9. Which compound has the highest boiling point?

a. H, CH_2CH_3, H, H, H, CH_2CH_3

b. H, H, $CH_2CH_2CH_2CH_3$, H, H, H

c. H, H, CH_2CH_3, CH_3, H, CH_3

d. H, H, H, H, H, $CH_2CH_2CH_3$

10. Which compound is different from the other three?

a. 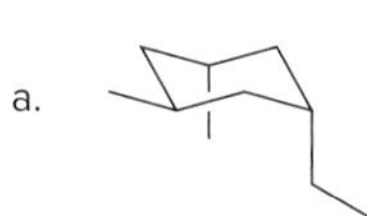

b.

c.

d.

PROBLEMS

Problems Using Three-Dimensional Models

4.38 Name each alkane using the ball-and-stick model, and classify each carbon as 1°, 2°, 3°, or 4°.

a. 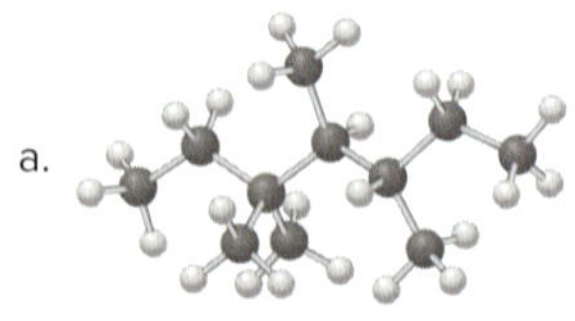

b. 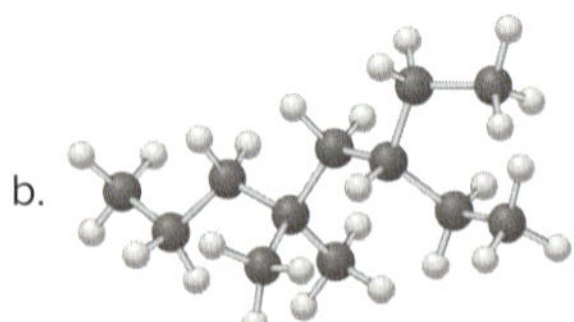

4.39 Consider the substituted cyclohexane shown in the ball-and-stick model.

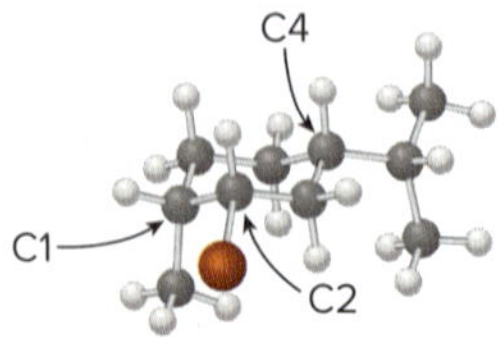

a. Label the substituents on C1, C2, and C4 as axial or equatorial.
b. Are the substituents on C1 and C2 cis or trans to each other?
c. Are the substituents on C2 and C4 cis or trans to each other?
d. Draw the second possible conformation in the chair form, and classify it as more stable or less stable than the conformation shown in the three-dimensional model.

4.40 Convert each three-dimensional model to a Newman projection around the indicated bond.

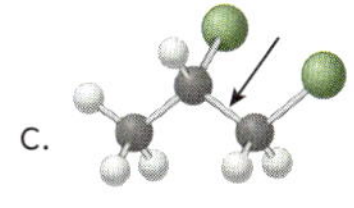

Constitutional Isomers

4.41 Draw the structure of all compounds that fit the following descriptions.

a. five constitutional isomers having the molecular formula C_4H_8

b. nine constitutional isomers having the molecular formula C_7H_{16}

c. twelve constitutional isomers having the molecular formula C_6H_{12} and containing one ring

IUPAC Nomenclature

4.42 Give the IUPAC name for each compound.

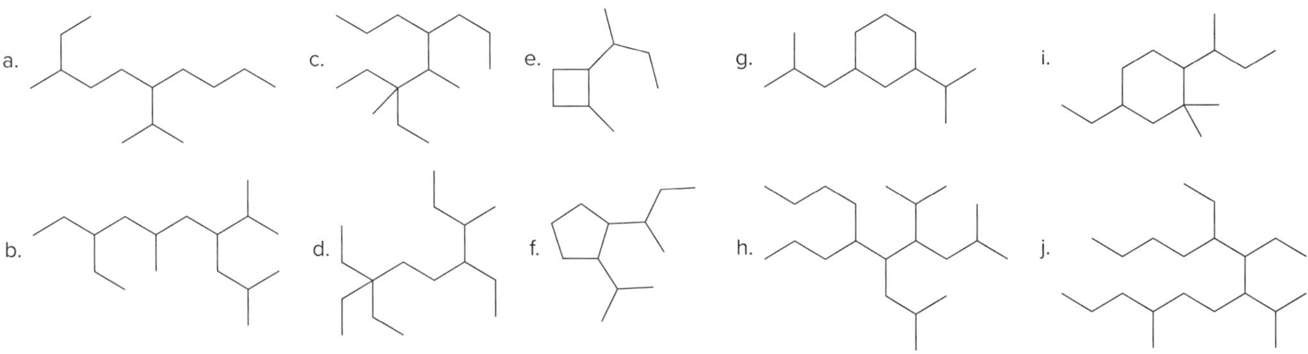

4.43 Draw the structure corresponding to each IUPAC name.

a. 3-ethyl-2-methylhexane

b. *sec*-butylcyclopentane

c. 4-isopropyl-2,4,5-trimethylundecane

d. cyclobutylcycloheptane

e. 3-ethyl-1,1-dimethylcyclohexane

f. 4-butyl-1,1-diethylcyclooctane

g. 6-isopropyl-2,3-dimethyldodecane

h. 2,2,6,6,7-pentamethyloctane

i. *cis*-1-ethyl-3-methylcyclopentane

j. *trans*-1-*tert*-butyl-4-ethylcyclohexane

4.44 Draw the structure of each alkane and cycloalkane from the given incorrect name. Then, give the IUPAC name for each compound.

a. 7-ethyl-3,6-dimethylnonane

b. 4-ethyl-3-isopropylheptane

c. 3-ethyl-1,4-dimethylcycloheptane

d. 1-ethyl-3-methyl-5-isopropylcyclohexane

4.45 Give the IUPAC name for each compound.

a. CH_3 (front, top); H (front, upper left); $CH_2CH_2CH_3$ (back, upper right); CH_3 (back, lower left); H (front, lower right); $CH_2CH_2CH_3$ (back, bottom)

b. CH_3 (top); CH_3 (upper left); CH_2CH_3 (upper right); H (lower left); H (lower right); CH_2CH_3 (bottom)

c. CH_3 (top); CH_3CH_2 (upper left); $CH_2CH_2CH_3$ (upper right); $CH_3CH_2CH_2$ (lower left); H (lower right); CH_2CH_3 (bottom)

Properties of Alkanes

Students who have already learned about mass spectrometry can try Problems A.2b, A.5b, A.10, A.11, A.17, A.22, and A.30. Students who have already learned about infrared spectroscopy can try Problem B.15a.

4.46 Rank the following alkanes in order of increasing boiling point.

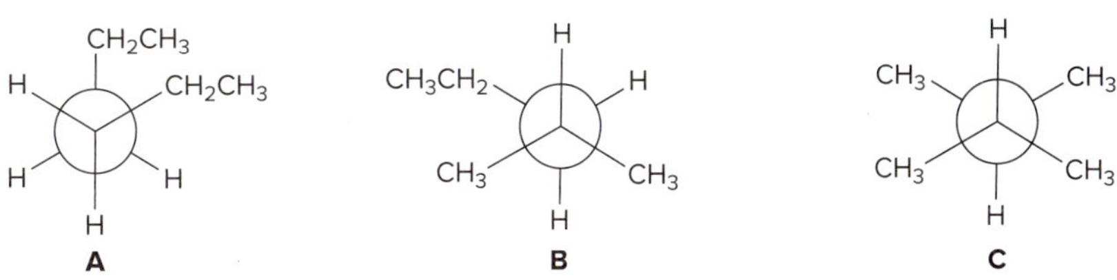

4.47 The melting points and boiling points of two isomeric alkanes are as follows: $CH_3(CH_2)_6CH_3$, mp = −57 °C and bp = 126 °C; $(CH_3)_3CC(CH_3)_3$, mp = 102 °C and bp = 106 °C. (a) Explain why one isomer has a lower melting point but higher boiling point. (b) Explain why there is a small difference in the boiling points of the two compounds, but a huge difference in their melting points.

Conformation of Acyclic Alkanes

4.48 Which conformation in each pair is *higher* in energy? Calculate the energy difference between the two conformations using the values given in Table 4.3.

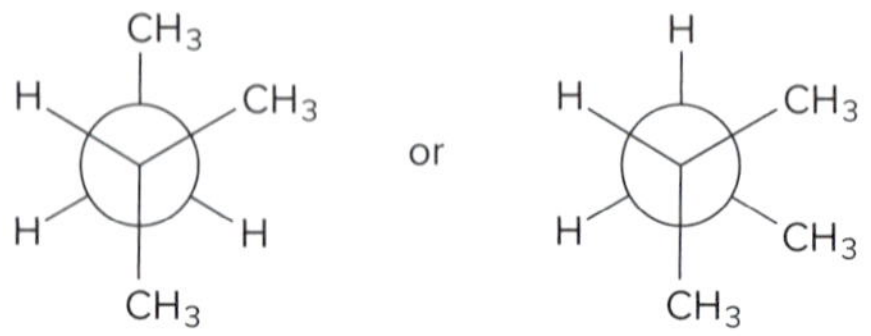

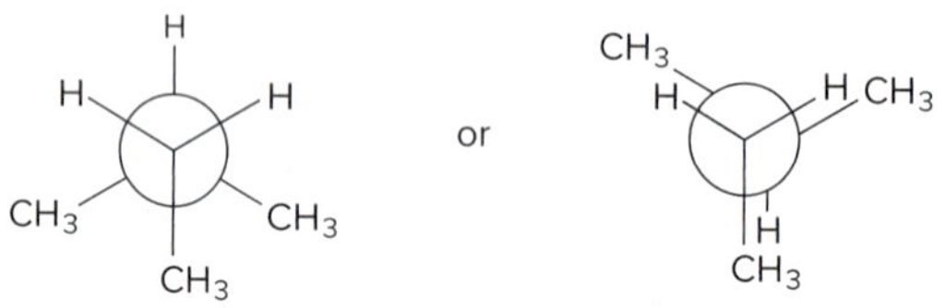

4.49 Considering rotation around the C3–C4 bond in hexane, draw Newman projections for the most stable and least stable conformations.

4.50 Rank the following Newman projections in order of increasing energy.

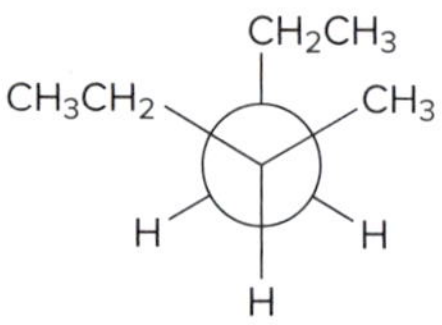

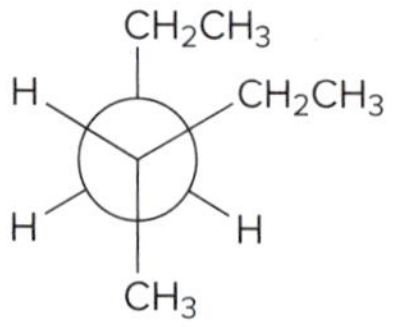

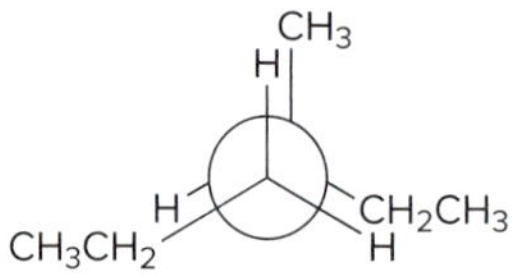

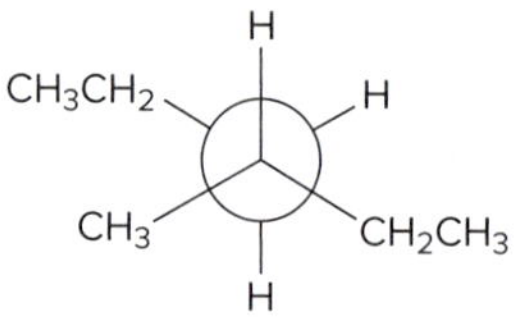

4.51 Classify each conformation as staggered or eclipsed around the indicated bond, and rank the conformations in order of increasing stability.

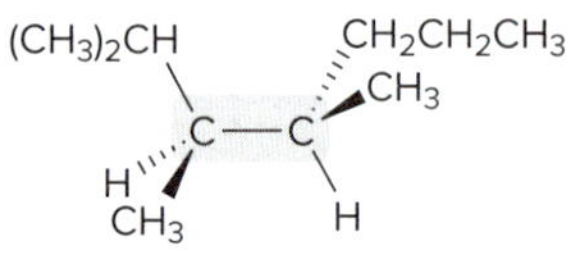

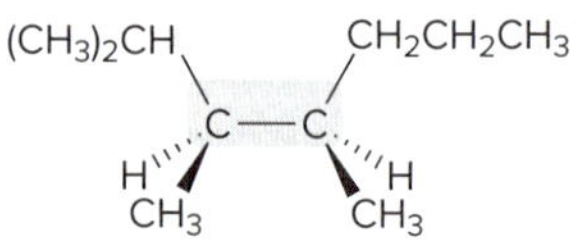

4.52 (a) Using Newman projections, draw all staggered and eclipsed conformations that result from rotation around the bond highlighted in red in each molecule; (b) draw a graph of energy versus dihedral angle for rotation around this bond.

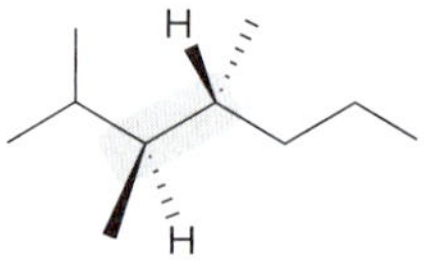

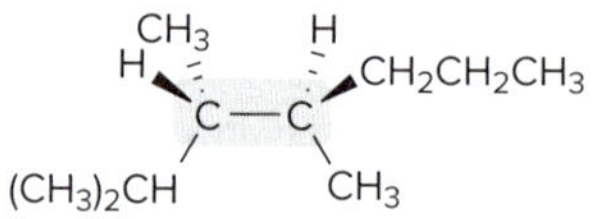

4.53 Calculate the barrier to rotation for each bond highlighted in red.

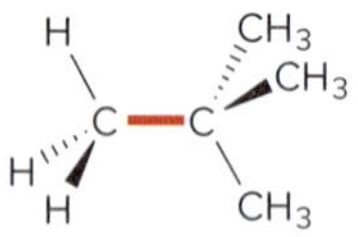

4.54 (a) Draw the anti and gauche conformations for ethylene glycol ($HOCH_2CH_2OH$). (b) Ethylene glycol is unusual in that the gauche conformation is more stable than the anti conformation. Offer an explanation.

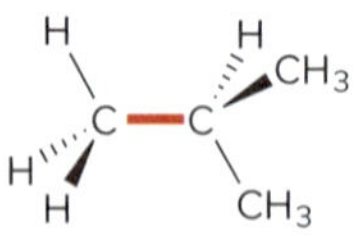

Conformations and Stereoisomers in Cycloalkanes

4.55 Draw the more stable chair conformation for each compound.

a. *trans*-1-isopropyl-3-methylcyclohexane

b. *cis*-1-*sec*-butyl-4-ethylcyclohexane

c. *cis*-1-ethyl-2-isobutylcyclohexane

d. *trans*-1,2-dibutylcyclohexane

4.56 For each compound drawn below:

a. Draw representations for the cis and trans isomers using a hexagon for the six-membered ring, and wedges and dashed wedges for substituents.

b. Draw the two possible chair conformations for the cis isomer. Which conformation, if either, is more stable?

c. Draw the two possible chair conformations for the trans isomer. Which conformation, if either, is more stable?

d. Which isomer, cis or trans, is more stable and why?

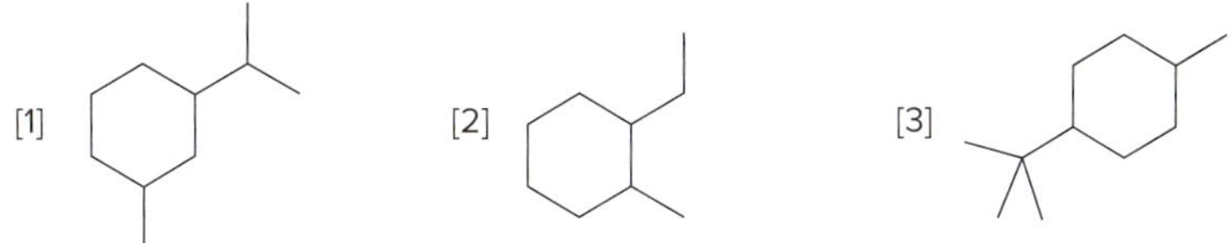

4.57 Convert each of the following structures to its more stable chair form. One structure represents menthol and one represents isomenthol. Menthol, the more stable isomer, is used in lip balms and mouthwash. Which structure corresponds to menthol?

4.58 Which isomer in each pair of trisubstituted cyclohexanes is more stable?

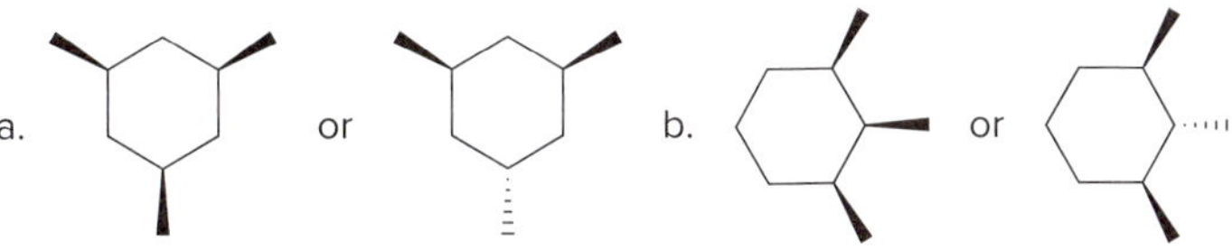

4.59 Answer the following questions about compound **A,** which contains a CH_3 group and OH group bonded to the carbon skeleton that consists of three six-membered rings in the conformation shown.

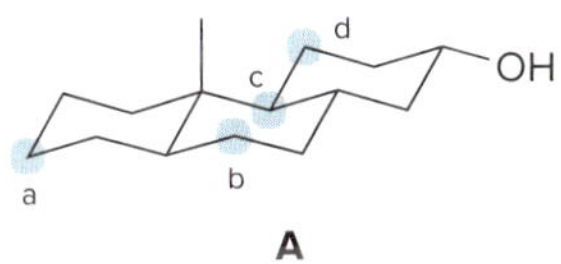

a. Are the CH_3 and OH groups oriented cis or trans to each other?

b. Is a substituent on C_a that is cis to the CH_3 group located in the axial or equatorial position?

c. Is an equatorial Br at C_b oriented cis or trans to the OH group?

d. Is the H atom on C_c located cis or trans to the OH group?

e. Is a substituent on C_d that is trans to the OH group located in the axial or equatorial position?

4.60 Glucose is a simple sugar with five substituents bonded to a six-membered ring.

HO— O HO— OH HO OH glucose

a. Using a chair representation, draw the most stable arrangement of these substituents on the six-membered ring.

b. Convert this representation to one that uses a hexagon with wedges and dashed wedges.

c. Draw a constitutional isomer of glucose.

d. Draw a stereoisomer that has an axial OH group on one carbon.

Constitutional Isomers and Stereoisomers

4.61 Classify each pair of compounds as constitutional isomers, stereoisomers, identical molecules, or not isomers of each other.

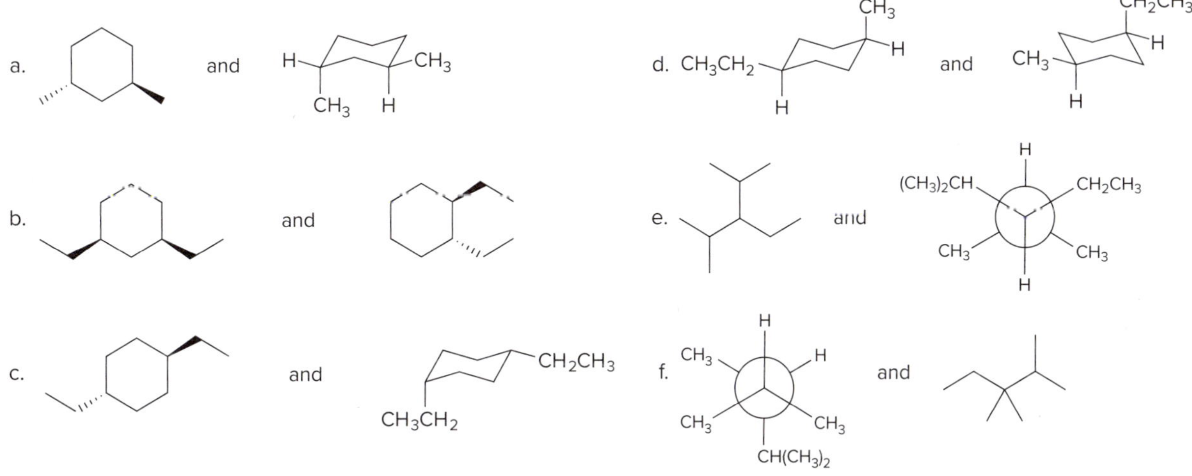

4.62 Answer the following questions about compounds **A–D.**

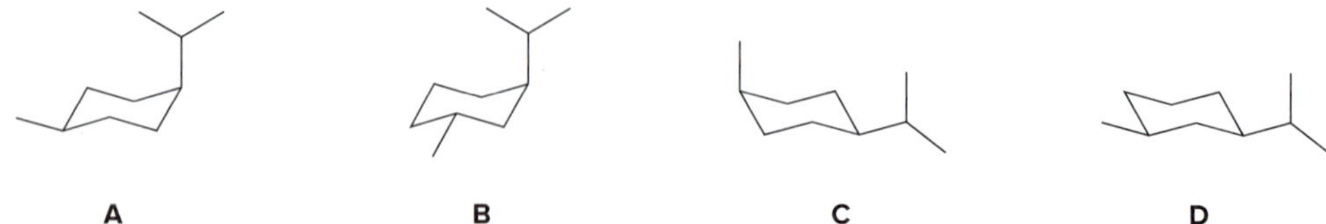

a. How are the compounds in each pair related? Choose from constitutional isomers, stereoisomers, or identical molecules: **A** and **B; A** and **C; B** and **D.**

b. Label each compound as a cis or trans isomer.

c. Draw **B** as a hexagon with wedges and dashed wedges to show the stereochemistry of substituents.

d. Draw a stereoisomer of **A** as a hexagon using wedges and dashed wedges to show the orientation of substituents.

4.63 Draw the three constitutional isomers having molecular formula C_7H_{14} that contain a five-membered ring and two methyl groups as substituents. For each constitutional isomer that can have cis and trans isomers, draw the two stereoisomers.

Oxidation and Reduction

4.64 Classify each reaction as oxidation, reduction, or neither.

a.

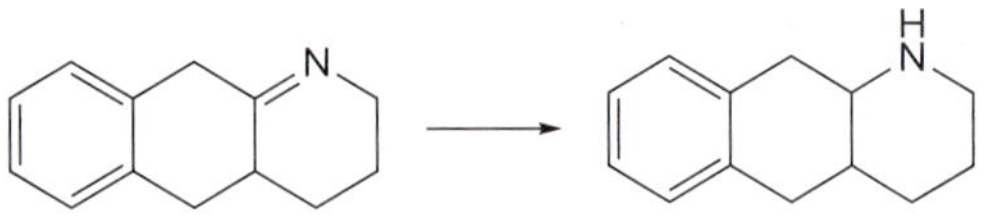

b.

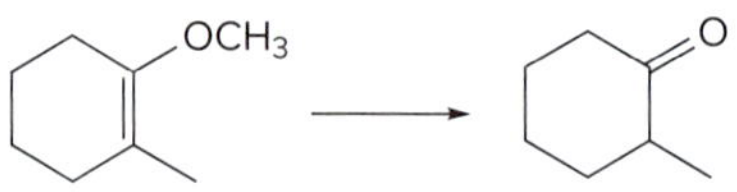

4.65 Draw the products of combustion of each alkane.

a.

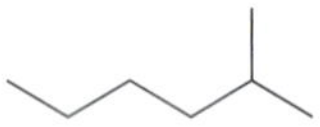

b.

Lipids

4.66 Cholic acid, a compound called a **bile acid,** is converted to a **bile salt** in the body. Bile salts have properties similar to soaps, and they help transport lipids through aqueous solutions. Explain why this is so.

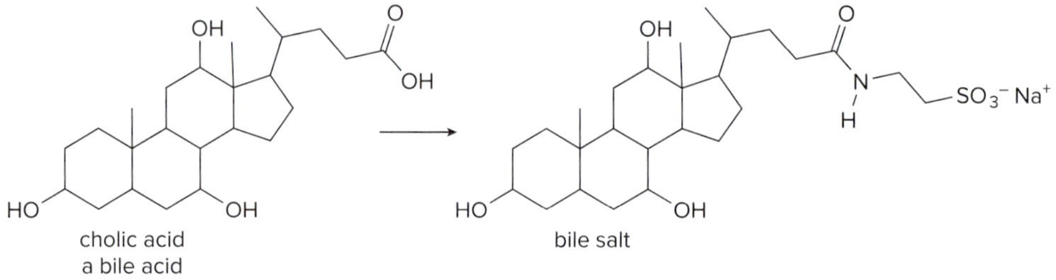

cholic acid
a bile acid

bile salt

4.67 Mineral oil, a mixture of high-molecular-weight alkanes, is sometimes used as a laxative. Why are individuals who use mineral oil for this purpose advised to avoid taking it at the same time they consume foods rich in fat-soluble vitamins such as vitamin A?

Problems That Combine Concepts

4.68 Hydrocarbons like benzene are metabolized in the body to arene oxides, which rearrange to form phenols. This is an example of a general process in the body, in which an unwanted compound (benzene) is converted to a more water-soluble derivative called a *metabolite,* so that it can be excreted more readily from the body.

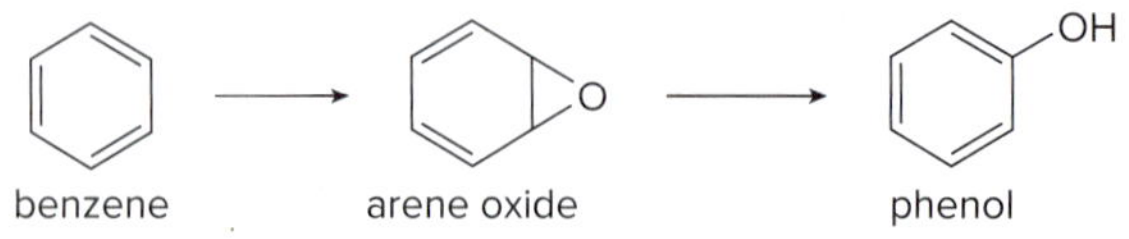

benzene

arene oxide

phenol

a. Classify each of these reactions as oxidation, reduction, or neither.

b. Explain why phenol is more water soluble than benzene. This means that phenol dissolves in urine, which is largely water, to a greater extent than benzene.

4.69 In an effort to develop more sustainable fuels for aircraft, aviation biofuels, which are blended with petroleum-based fuels, can be produced from plant sources and municipal solid waste.

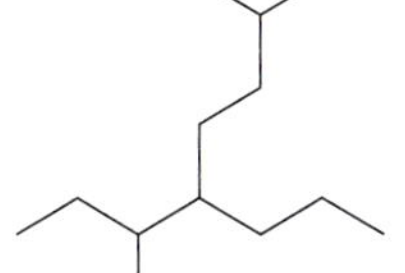

a. Name the given alkane, one component of jet fuels derived from plant and waste sources.
b. Write a balanced equation for the combustion of this alkane with O_2.
c. Explain why using plant-based fuels helps to reduce the carbon footprint of jet fuels.

Challenge Problems

4.70 Although penicillin G has two amide functional groups, one is much more reactive than the other. Which amide is more reactive and why?

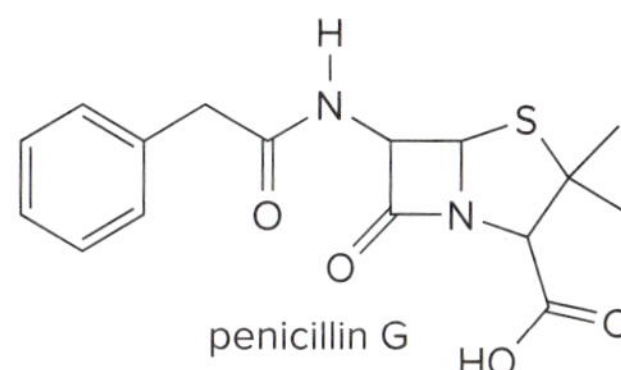

penicillin G

4.71 Haloethanes (CH_3CH_2X, X = Cl, Br, I) have similar barriers to rotation (13.4–15.5 kJ/mol) despite the fact that the size of the halogen increases, Cl → Br → I. Offer an explanation.

4.72 When two six-membered rings share a C–C bond, this bicyclic system is called a **decalin.** There are two possible arrangements: *trans*-decalin having two hydrogen atoms at the ring fusion on opposite sides of the rings, and *cis*-decalin having the two hydrogens at the ring fusion on the same side.

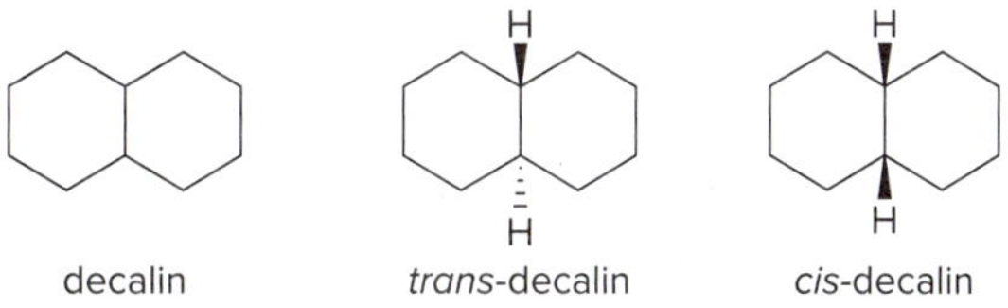

a. Draw *trans*- and *cis*-decalin using the chair form for the cyclohexane rings.
b. The trans isomer is more stable. Explain why.

4.73 Consider the tricyclic structures **A** and **B.** (a) Label each substituent on the rings as axial or equatorial. (b) Draw a skeletal structure using wedges and dashed wedges to show whether the substituents are located above or below the rings.

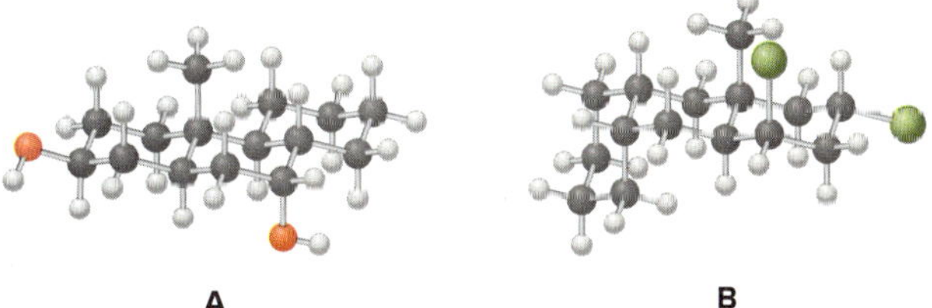

4.74 Read Appendix D on naming bicyclic compounds. Then give the IUPAC name for each of the following compounds.

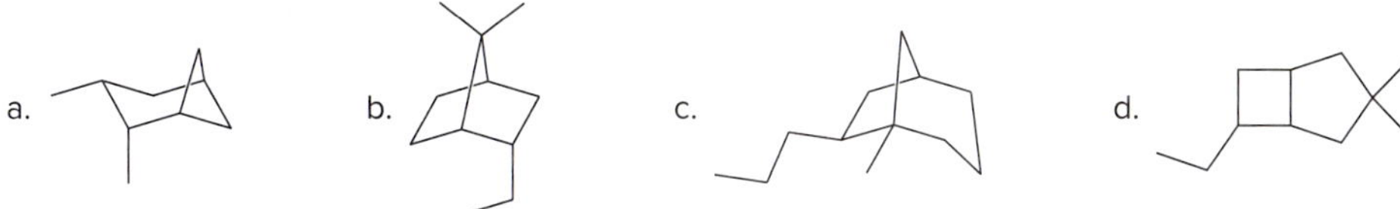

4.75 Using the method of naming bicyclic compounds described in Appendix D, decalin (Problem 4.72) is called bicyclo[4.4.0]decane.
a. Draw a constitutional isomer of bicyclo[4.4.0]decane that contains a bridged ring system with at least one six-membered ring, and name the isomer you have drawn.
b. Draw a constitutional isomer of bicyclo[4.4.0]decane that contains a spiro ring system with at least one six-membered ring, and name the isomer you have drawn.

SELF-TEST ANSWERS

1. d 2. c 3. c 4. b 5. a 6. c 7. b 8. a 9. b 10. b

5 Stereochemistry

5.1 Starch and cellulose
5.2 The two major classes of isomers
5.3 Looking glass chemistry—Chiral and achiral molecules
5.4 Stereogenic centers
5.5 Stereogenic centers in cyclic compounds
5.6 Labeling stereogenic centers with *R* or *S*
5.7 Diastereomers
5.8 Meso compounds
5.9 *R* and *S* assignments in compounds with two or more stereogenic centers
5.10 Disubstituted cycloalkanes
5.11 Isomers—A summary
5.12 Physical properties of stereoisomers
5.13 Chemical properties of enantiomers

George Ostertag/Alamy Stock Photo

Paclitaxel (trade name Taxol), a potent anticancer agent active against ovarian, breast, and several other cancers, was discovered in 1962 and approved for use by the Food and Drug Administration in 1992. Initial studies with paclitaxel were carried out with material isolated from the bark of the Pacific yew tree, but stripping the bark killed these magnificent trees. Paclitaxel was synthesized in the laboratory in 1994, and is now produced by a plant cell fermentation process. Like other widely used drugs, paclitaxel is biologically active because of its complex structure and the particular three-dimensional arrangement of its functional groups. In Chapter 5, we learn about the stereochemistry of molecules like paclitaxel.

Why Study . . . Stereochemistry?

Are you left-handed or right-handed? If you're right-handed, you've probably spent little time thinking about your hand preference. If you're left-handed, though, you probably learned at an early age that many objects—like scissors and baseball gloves—"fit" for righties, but are "backwards" for lefties. **Hands, like many objects in the world around us, are mirror images that are *not* identical.**

In Chapter 5, we examine the "handedness" of molecules, and learn about the importance of the three-dimensional shape of a molecule.

5.1 Starch and Cellulose

Recall from Chapter 4 that ***stereochemistry* is the three-dimensional structure of a molecule.** How important is stereochemistry? Two biomolecules—starch and cellulose—illustrate how apparently minor differences in structure can result in vastly different properties.

Starch and **cellulose** are two polymers that belong to the family of biomolecules called **carbohydrates** (Figure 5.1). **A *polymer* is a large molecule composed of repeating smaller units—called monomers—that are covalently bonded together.**

Figure 5.1 Starch and cellulose—Two common carbohydrates

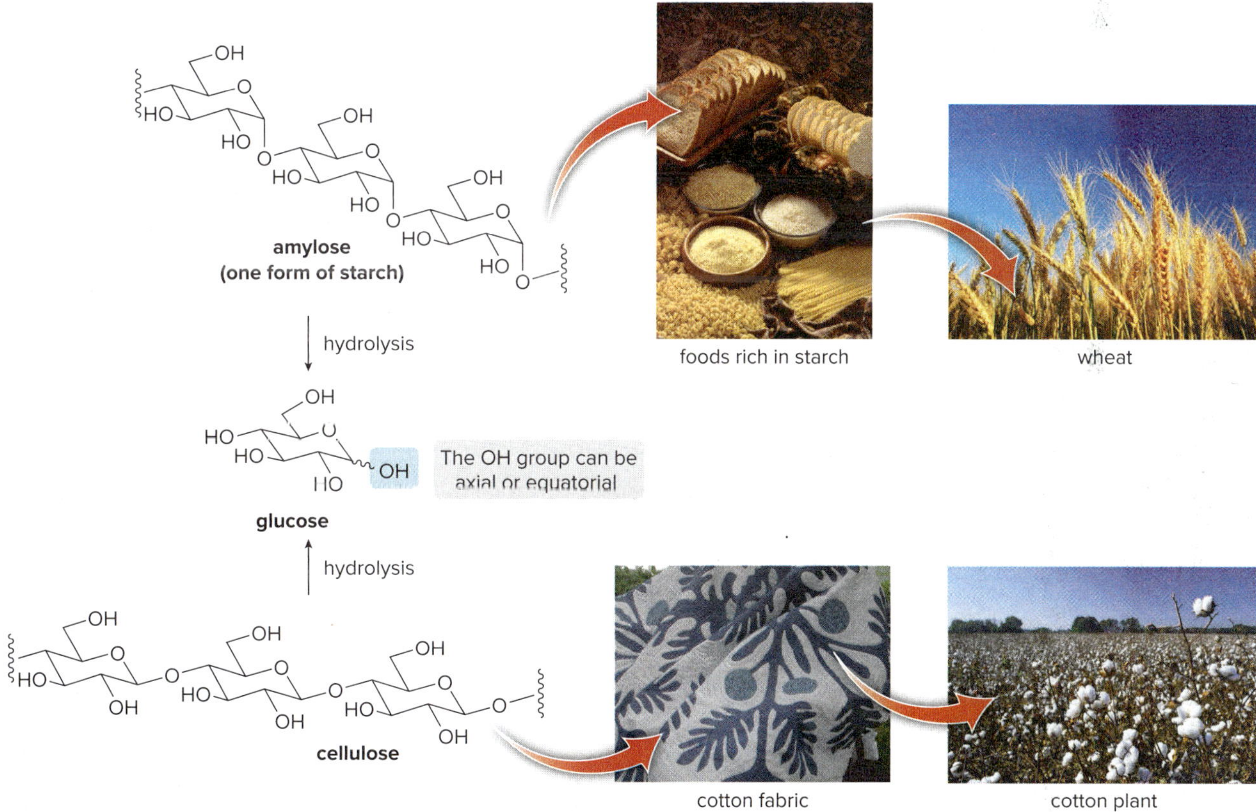

Source: (Top left): Scott Bauer/Agricultural Research Service/USDA; (top right): Bryan Mullennix/Pixtal/age fotostock; (bottom left): Daniel C. Smith; (bottom right): David Frazier/Corbis/Getty Images

Starch is the main carbohydrate in the seeds and roots of plants. When we humans ingest wheat, rice, or potatoes, we consume starch, which is then hydrolyzed to the simple sugar **glucose,** one of the compounds our bodies use for energy. **Cellulose,** nature's most abundant organic material, gives rigidity to tree trunks and plant stems. Wood, cotton, and flax are composed largely of cellulose. Complete hydrolysis of cellulose also forms glucose, but unlike starch, humans cannot hydrolyze cellulose to glucose. In other words, we can digest starch but not cellulose.

Cellulose and starch are both composed of the same repeating unit—a six-membered ring containing an oxygen atom and three OH groups—joined by an oxygen atom. They differ in the position of the O atom joining the rings together.

repeating unit

In cellulose, the O occupies the **equatorial** position.

In starch, the O occupies the **axial** position.

- **In cellulose, the O atom joins two rings using two equatorial bonds.**
- **In starch, the O atom joins two rings using one equatorial and one axial bond.**

cellulose
two equatorial bonds (in red)

starch
one axial bond (in blue) and one equatorial bond (in red)

Starch and cellulose are **isomers** because they are different compounds with the same molecular formula $(C_6H_{10}O_5)_n$. They are **stereoisomers** because only the three-dimensional arrangement of atoms is different.

How the six-membered rings are joined together has an enormous effect on the shape and properties of these carbohydrate molecules. Cellulose is composed of long chains held together by intermolecular hydrogen bonds, forming sheets that stack in an extensive three-dimensional network. The axial–equatorial ring junction in starch creates chains that fold into a helix (Figure 5.2). Moreover, the human digestive system contains the enzyme necessary to hydrolyze starch by cleaving its axial C–O bond, but not an enzyme to hydrolyze the equatorial C–O bond in cellulose.

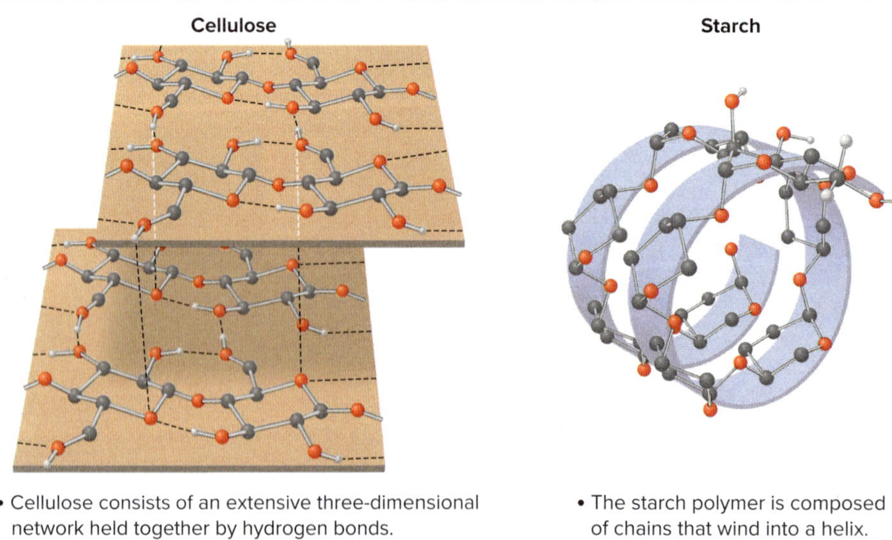

Figure 5.2 Three-dimensional structure of cellulose and starch

- Cellulose consists of an extensive three-dimensional network held together by hydrogen bonds.
- The starch polymer is composed of chains that wind into a helix.

Thus, a **change in the three-dimensional arrangement of groups—axial or equatorial—confers very different properties on starch and cellulose.**

5.2 The Two Major Classes of Isomers

Because an understanding of isomers is integral to the discussion of stereochemistry, let's begin with an overview of isomers.

- **Isomers are different compounds with the same molecular formula.**

There are two major classes of isomers: **constitutional isomers** and **stereoisomers.** ***Constitutional (or structural) isomers*** **differ in the way the atoms are connected to each other.** Constitutional isomers have

- different IUPAC names;
- the same or different functional groups;
- different physical properties, so they are separable by physical techniques such as distillation; and
- different chemical properties. They behave differently or give different products in chemical reactions.

Stereoisomers **differ *only* in the way atoms are oriented in space.** Stereoisomers have identical IUPAC names (except for a prefix like cis or trans). Because they differ only in the three-dimensional arrangement of atoms, stereoisomers always have the same functional group(s).

A particular three-dimensional arrangement is called a *configuration.* Thus, stereoisomers differ in configuration. The cis and trans isomers in Section 4.12B and the biomolecules starch and cellulose in Section 5.1 are two examples of stereoisomers.

Figure 5.3 illustrates examples of both types of isomers. Chapter 5 concentrates on the types and properties of stereoisomers.

Problem 5.1 Classify each pair of compounds as constitutional isomers or stereoisomers.

a. 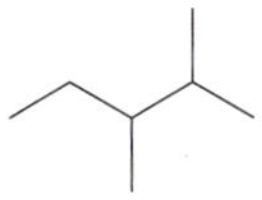 and

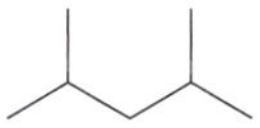

b.  and OH

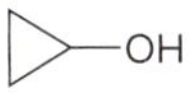

c.  and

d.  and

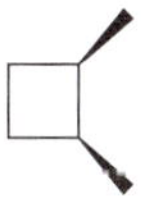

Figure 5.3 A comparison of constitutional isomers and stereoisomers

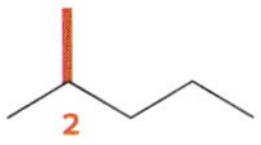

2-methylpentane C_6H_{14} and 3-methylpentane C_6H_{14}

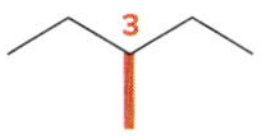

same molecular formula
different names

constitutional isomers

cis-1,2-dimethylcyclopentane and *trans*-1,2-dimethylcyclopentane

same molecular formula
same name except for the prefix

stereoisomers

5.3 Looking Glass Chemistry—Chiral and Achiral Molecules

Everything has a mirror image. What's significant is **whether a molecule is *identical* to or *different* from its mirror image.**

Some molecules are like hands. **Left and right hands are mirror images of each other, but they are *not* identical.** If you try to mentally place one hand inside the other hand, you can never superimpose either all the fingers, or the tops and palms. To *superimpose* an object on

Despite the dominance of right-handedness over left-handedness, even identical twins can exhibit differences in hand preference. Pictured are Matthew (right-handed) and Zachary (left-handed), identical twin sons of the author.
Daniel C. Smith

its mirror image means to align *all* parts of the object with its mirror image. With molecules, this means aligning all atoms and all bonds.

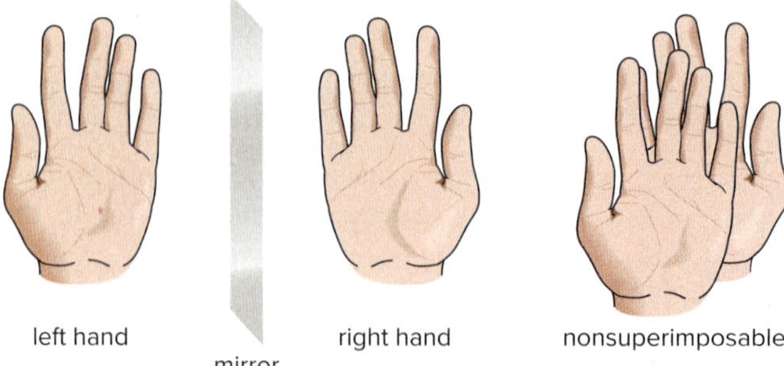

- **A molecule (or object) that is *not* superimposable on its mirror image is said to be *chiral.***

Other molecules are like socks. **Two socks from a pair are mirror images that *are* superimposable.** One sock can fit inside another, aligning toes and heels, and tops and bottoms. A sock and its mirror image are *identical*.

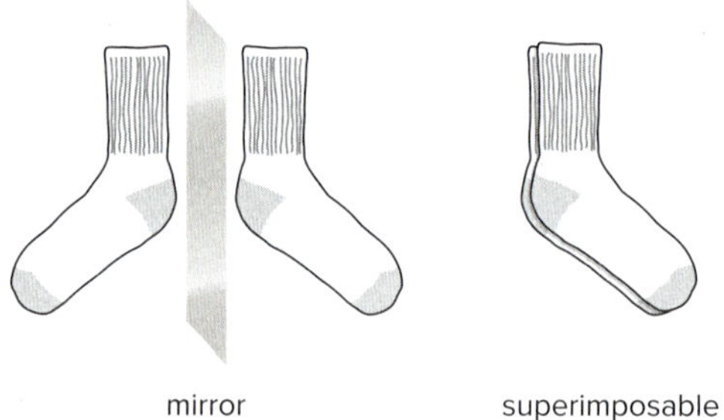

- **A molecule (or object) that *is* superimposable on its mirror image is said to be *achiral.***

Let's determine whether three molecules—H_2O, CH_2BrCl, and CHBrClF—are superimposable on their mirror images; that is, **are H_2O, CH_2BrCl, and CHBrClF chiral or achiral?**

To test chirality:

- Draw the molecule in three dimensions.
- Draw its mirror image.
- Try to align all bonds and atoms. To superimpose a molecule and its mirror image, you can perform any rotation but **you cannot break bonds.**

The adjective ***chiral*** comes from the Greek word *cheir*, meaning "hand." Left and right hands are **chiral:** they are mirror images that do *not* superimpose on each other.

Few beginning students of organic chemistry can readily visualize whether a compound and its mirror image are superimposable by looking at drawings on a two-dimensional page. Molecular models can help a great deal in this process.

Following this procedure, H_2O and CH_2BrCl are both **achiral** molecules because each molecule is superimposable on its mirror image.

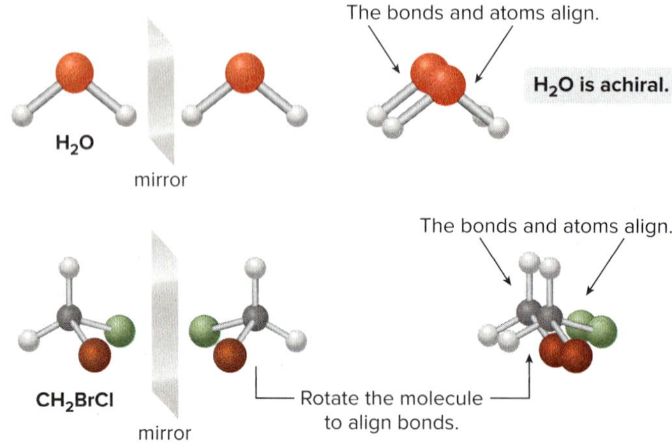

With CHBrClF, the result is different. The molecule (labeled **A**) and its mirror image (labeled **B**) are *not* superimposable. No matter how you rotate **A** and **B,** all the atoms never align. **CHBrClF is thus a chiral molecule,** and **A** and **B** are different compounds.

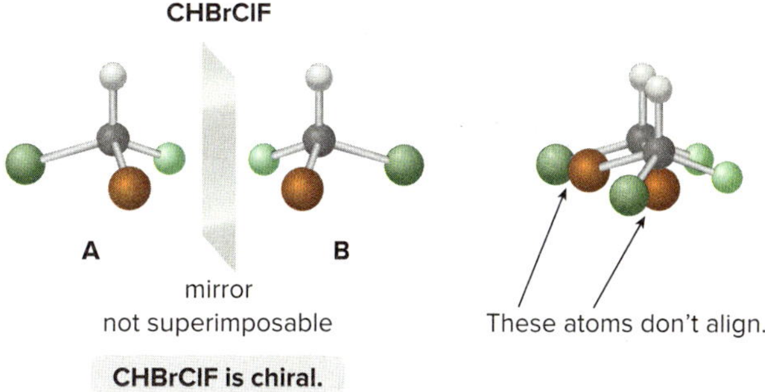

A and **B** are **stereoisomers** because they are isomers differing only in the three-dimensional arrangement of substituents. These stereoisomers are called **enantiomers.**

- ***Enantiomers* are mirror images that are not superimposable.**

CHBrClF contains a carbon atom bonded to four different groups. **A carbon atom bonded to four different groups is called a tetrahedral *stereogenic center.*** Most chiral molecules contain one or more stereogenic centers.

The general term *stereogenic center* refers to any site in a molecule at which the interchange of two groups forms a stereoisomer. A **carbon atom with four different groups is a *tetrahedral* stereogenic center,** because the interchange of two groups converts one enantiomer into another. We will learn about another type of stereogenic center in Section 8.2B.

Naming a carbon atom with four different groups is a topic that currently has no firm agreement among organic chemists. The IUPAC recommends the term *chirality center,* but the term has not gained wide acceptance among organic chemists since it was first suggested in 1996. Other terms in common use are chiral center, chiral carbon, asymmetric carbon, stereocenter, and stereogenic center, the term used in this text.

Molecules can contain zero, one, or more stereogenic centers.

- **With no stereogenic centers, a molecule generally is not chiral.** H_2O and CH_2BrCl have *no* stereogenic centers and are *achiral* molecules. (There are a few exceptions to this generalization, as we will learn in Section 15.5.)
- **With one tetrahedral stereogenic center, a molecule is *always* chiral.** CHBrClF is a *chiral* molecule containing *one* stereogenic center.
- **With two or more stereogenic centers, a molecule *may* or *may not* be chiral,** as we will learn in Section 5.8.

Problem 5.2 Draw the mirror image of each compound. Label each molecule as chiral or achiral.

a. (Cl, Br) b. (Br, Cl, H) c. (O) d. (Br, H, F)

When trying to distinguish between chiral and achiral compounds, keep in mind:

- A *plane of symmetry* is a mirror plane that cuts a molecule in half, so that one half of the molecule is a reflection of the other half.
- Achiral molecules usually contain a plane of symmetry, but chiral molecules do not.

CH_2BrCl
plane of symmetry

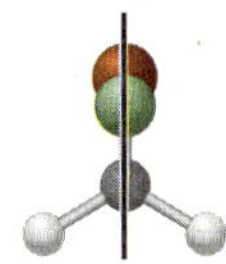

This molecule has **two identical halves.**

CH_2BrCl is achiral.

The achiral molecule CH_2BrCl has a plane of symmetry, but the chiral molecule CHBrClF does not.

Figure 5.4 summarizes the main facts about chirality we have learned thus far.

Figure 5.4 The basic principles of chirality

- Everything has a mirror image. The fundamental question is whether a molecule and its mirror image are superimposable.
- If a molecule and its mirror image are *not* superimposable, the molecule and its mirror image are ***chiral.***
- The terms ***stereogenic center*** and ***chiral molecule*** are related but distinct. In general, a chiral molecule must have one or more stereogenic centers.
- The presence of a ***plane of symmetry*** makes a molecule achiral.

Problem 5.3 Draw in a plane of symmetry for each molecule.

a. b. c. d. Cl Cl

Problem 5.4 A molecule is achiral if it has a plane of symmetry in *any* conformation. Each of the following conformations does not have a plane of symmetry, but rotation around a carbon–carbon bond forms a conformation that does have a plane of symmetry. Draw this conformation for each molecule.

a. CH_3, H, Br, C—C, Br, H, CH_3 b. OH, OH

When a right-handed shell is held in the right hand with the thumb pointing toward the wider end, the opening is on the right side. *Jill Braaten/McGraw Hill*

Stereochemistry may seem esoteric, but chirality pervades our very existence. On a molecular level, many biomolecules fundamental to life are chiral. On a macroscopic level, many naturally occurring objects possess handedness. Examples include chiral helical seashells shaped like right-handed screws, and plants such as honeysuckle that wind in a chiral left-handed helix. The human body is chiral, and hands, feet, and ears are not superimposable.

5.4 Stereogenic Centers

A necessary skill in the study of stereochemistry is the ability to locate and draw tetrahedral stereogenic centers.

5.4A Stereogenic Centers on Carbon Atoms That Are Not Part of a Ring

Ephedrine is isolated from ma huang, an herb used to treat respiratory ailments in traditional Chinese medicine. Once a popular drug to promote weight loss and enhance athletic performance, ephedrine has now been linked to episodes of sudden death, heart attack, and stroke. *Mark W. Skinner*

Recall from Section 5.3 that any carbon atom bonded to four different groups is a tetrahedral stereogenic center. To locate a stereogenic center, examine each *tetrahedral* carbon atom in a molecule, and look at the four ***groups***—not the four *atoms*—bonded to it. CBrClFI has one stereogenic center because its central carbon atom is bonded to four different elements. 3-Bromohexane also has one stereogenic center because one carbon is bonded to H, Br, CH_2CH_3, and $CH_2CH_2CH_3$. We consider all atoms in a group as a *whole unit,* not just the atom bonded directly to the carbon in question. Although C3 of 3-bromohexane is bonded to two carbon atoms, one is part of an ethyl group and one is part of a propyl group.

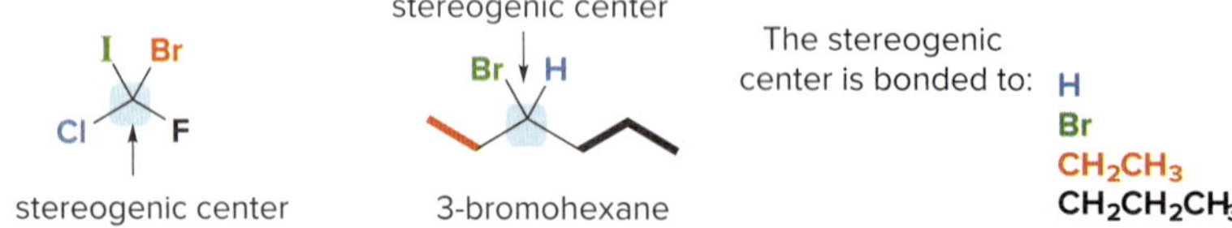

Always omit from consideration all C atoms that can't be tetrahedral stereogenic centers. These include

- $\mathbf{CH_2}$ and $\mathbf{CH_3}$ groups (more than one H bonded to C); and
- any ***sp*** or ***sp***^{2} hybridized C (less than four groups around C).

Larger organic molecules can have two, three, or even hundreds of stereogenic centers. **Propoxyphene** and **ephedrine** each contain two stereogenic centers, and **fructose,** a simple carbohydrate, has three.

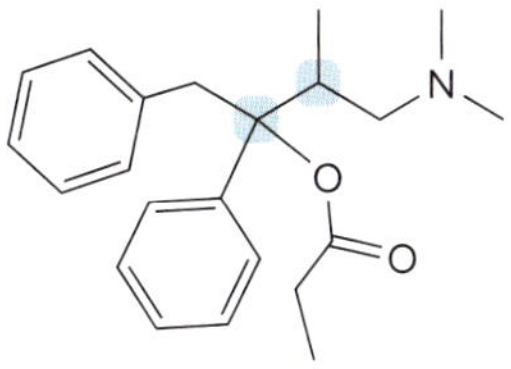

propoxyphene
Trade name: Darvon
(analgesic)

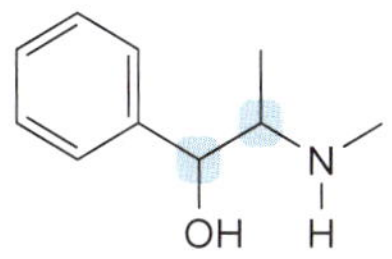

ephedrine
(bronchodilator, decongestant)

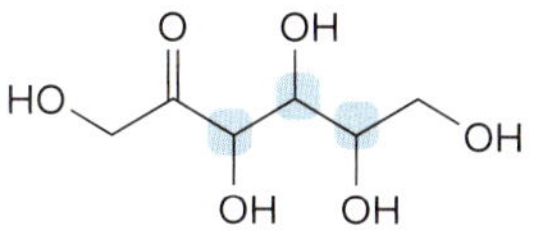

fructose
(a simple sugar)

= stereogenic center

Sample Problem 5.1 Locating Stereogenic Centers

Locate the stereogenic centers in each drug. Albuterol is a bronchodilator—that is, it widens airways—so it is used to treat asthma. Chloramphenicol is an antibiotic used extensively in developing countries because of its low cost.

a. albuterol

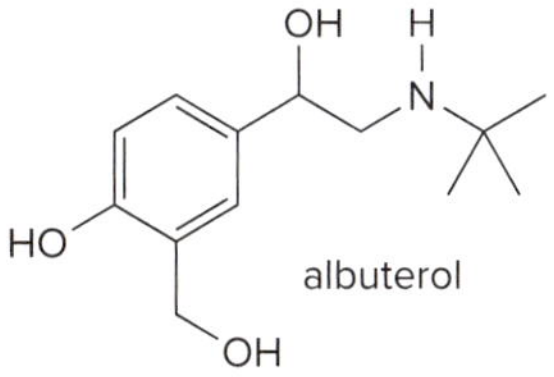

b. chloramphenicol

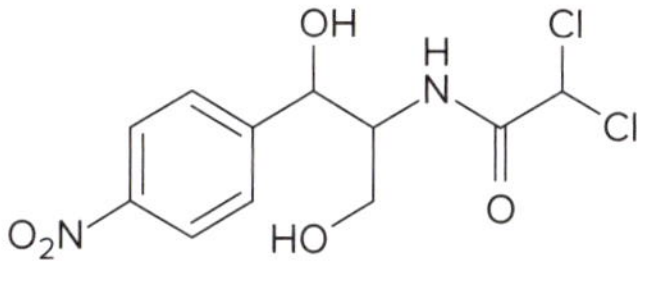

Solution

Heteroatoms surrounded by four different groups are also stereogenic centers. Stereogenic N atoms are discussed in Chapter 23.

Omit all CH_2 and CH_3 groups and all doubly bonded (sp^2 hybridized) C's. In albuterol, one C has three CH_3 groups bonded to it, so it can be eliminated as well. Draw in H atoms on tetrahedral C's in skeletal structures to more clearly see the groups. This leaves one C in albuterol and two C's in chloramphenicol surrounded by four different groups, making them stereogenic centers.

a. one stereogenic center

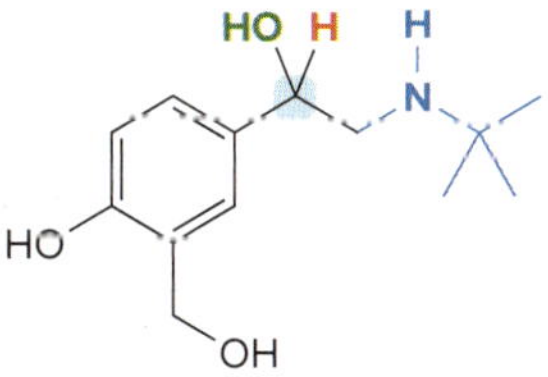

b. two stereogenic centers

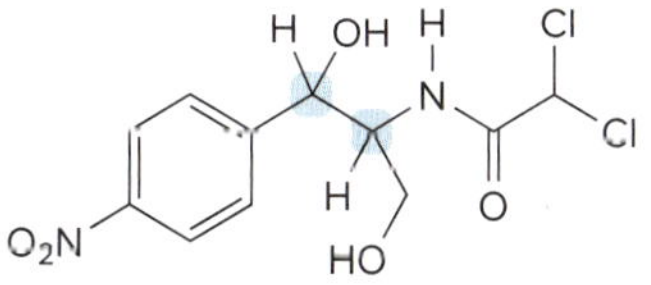

Problem 5.5 Locate the stereogenic centers in each molecule. Compounds may have one or more stereogenic centers.

a.

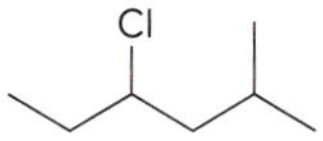

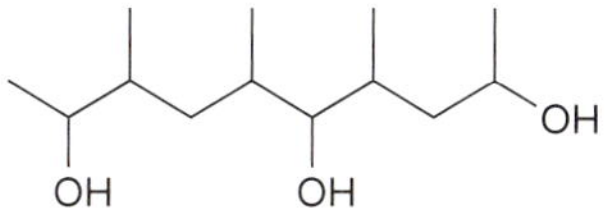

b.

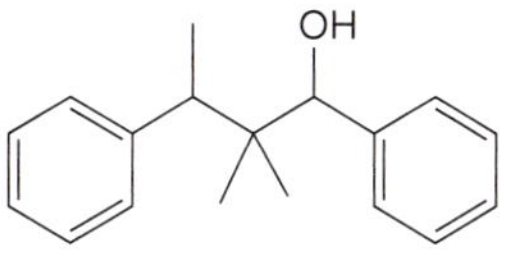

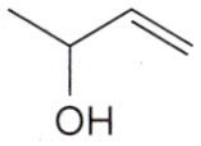

c.

d.

e.

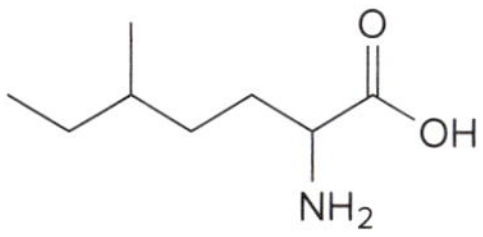

f.

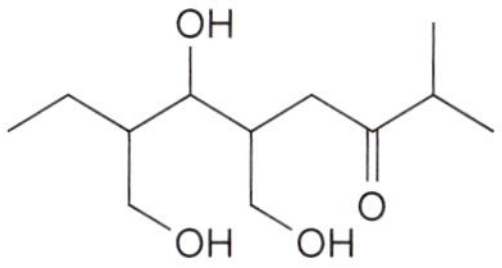

More Practice: Try Problem 5.41b–d.

Problem 5.6 The principles in Section 5.4A can be used to locate stereogenic centers in any molecule, no matter how complicated. Always look for carbons surrounded by four different groups. With this in mind, locate the four stereogenic centers in aliskiren, a drug introduced in 2007 for the treatment of hypertension.

aliskiren

In Section 26.2, we will learn about Fischer projection formulas, an older convention used for drawing stereogenic centers utilized mainly in carbohydrate chemistry.

5.4B Drawing a Pair of Enantiomers

- **Any molecule with one tetrahedral stereogenic center is a chiral compound and exists as a pair of enantiomers.**

butan-2-ol
one stereogenic center

Butan-2-ol, for example, has one stereogenic center. To draw both enantiomers, use the typical convention for depicting a tetrahedron: **place two bonds in the plane, one in front of the plane on a wedge, and one behind the plane on a dashed wedge.** Then, to form the first enantiomer **A,** arbitrarily place the four groups—H, OH, CH_3, and CH_2CH_3—on any bond to the stereogenic center.

In drawing a tetrahedron using solid lines, wedges, and dashed wedges, always **draw the two solid lines *first;*** then draw the wedge and the dashed wedge on the ***opposite side*** of the solid lines.

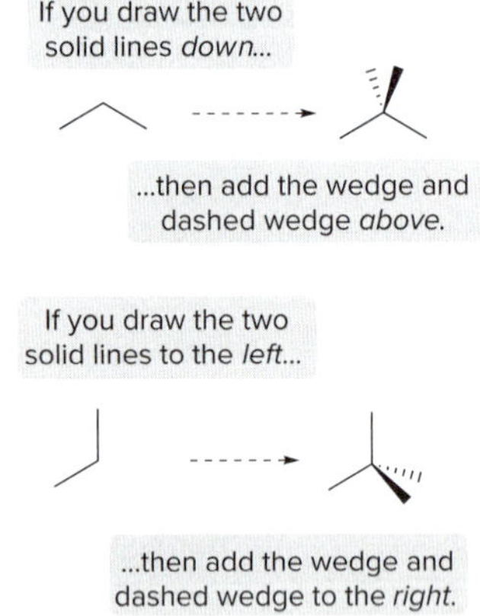

Draw the molecule...then the mirror image.

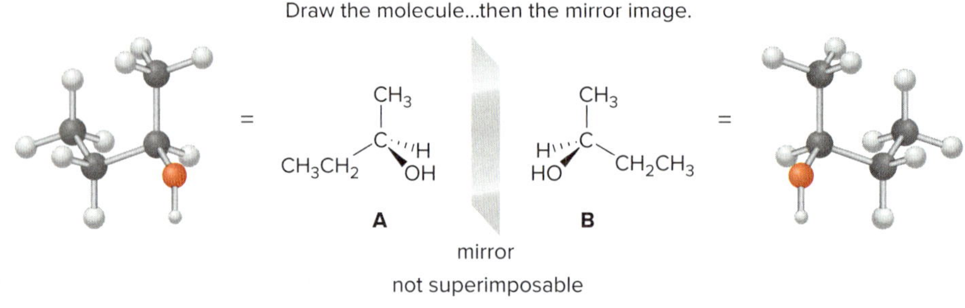

mirror
not superimposable
enantiomers

Then, draw a mirror plane and arrange the substituents in the mirror image so that they are a reflection of the groups in the first molecule, forming **B.** No matter how **A** and **B** are rotated, it is impossible to align all of their atoms. Because **A** and **B** are mirror images and not superimposable, **A** and **B** are a pair of **enantiomers.**

This is just one way to draw an enantiomer, as shown in Figure 5.5a for 3-bromohexane. Another way to draw an enantiomer (Figure 5.5b), especially for compounds with more than one stereogenic center, is to keep the carbon skeleton in the *same* position, but *invert* the configuration at all stereogenic centers by converting bonds in front (on wedges) to bonds in back (on dashed wedges), and vice versa.

Figure 5.5 Different ways of drawing an enantiomer

a. Drawing an enantiomer as a reflection.

Draw all groups as a reflection:

3-bromohexane — enantiomer

- Groups on wedges and dashed wedges stay the *same*.
- The position of the C's in the long chain is *different*.
- Remember that H and Br are directly aligned.

b. Drawing an enantiomer by inverting the configuration of a stereogenic center

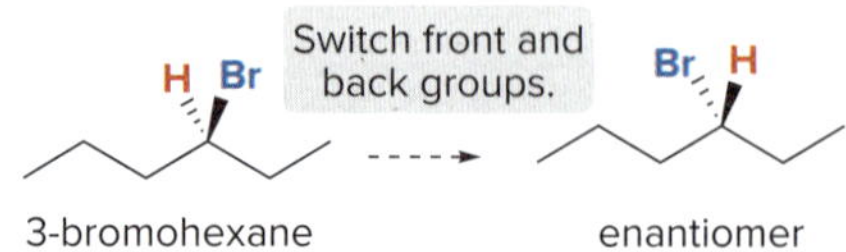

3-bromohexane — enantiomer

- Groups on wedges and dashed wedges *interchange*.
- The position of the C's in the long chain stays the *same*.

The two representations labeled "enantiomer" are *identical,* just drawn in different ways.

Sample Problem 5.2 Different Ways of Drawing an Enantiomer

Locate the stereogenic center in the amino acid alanine, and draw the enantiomer using the two methods shown in Figure 5.5.

alanine

Solution

The stereogenic center is the carbon with four different groups, labeled in blue.

[1] Project a mirror plane and draw all groups on the stereogenic center as a reflection of the groups in alanine.

Draw all groups as a reflection:

alanine enantiomer

- Groups in front and behind stay in the same position.

[2] Keep the carbon skeleton the same, and switch the position of groups that lie in front of and behind the plane.

alanine enantiomer

- The H *behind* the plane becomes an H in *front* on a wedge.
- The NH_2 in *front* of the plane becomes an NH_2 in *back* on a dashed wedge.

Problem 5.7 Draw the enantiomer of each compound.

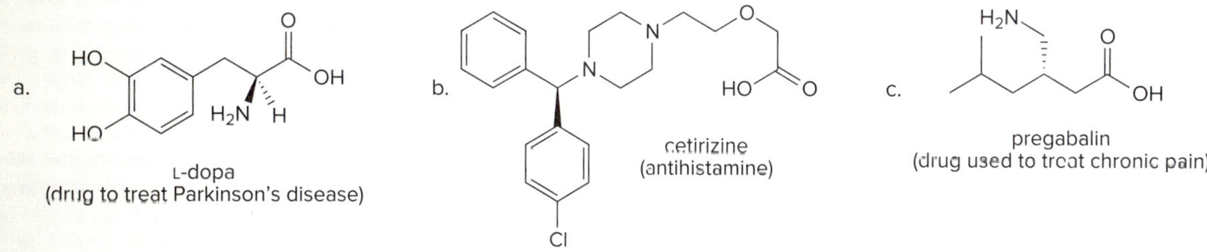

a. L-dopa (drug to treat Parkinson's disease)

b. cetirizine (antihistamine)

c. pregabalin (drug used to treat chronic pain)

More Practice: Try Problems 5.42, 5.61b, 5.62c, 5.63b.

Problem 5.8 Locate the stereogenic center in each compound and draw both enantiomers.

a. b. c.

Problem 5.9 Determine if each compound is identical to or an enantiomer of **A**.

A a. b. c.

5.5 Stereogenic Centers in Cyclic Compounds

Stereogenic centers may also occur at carbon atoms that are part of a ring. To find stereogenic centers on ring carbons, always draw the rings as flat polygons, and look for tetrahedral carbons that are bonded to four different groups, as usual. Each ring carbon is bonded to two other atoms in the ring, as well as two substituents attached to the ring.

Does methylcyclopentane have a stereogenic center? All of the carbon atoms are bonded to two or three hydrogen atoms except for C1, the ring carbon bonded to the methyl group. Next, compare the ring atoms and bonds on both sides equidistant from C1, and **continue until a point of difference is reached, or until both sides meet,** either at an atom or in the middle of a bond. In this case, there is no point of difference on either side, so C1 is bonded to identical alkyl groups that happen to be part of a ring. **C1, therefore, is *not* a stereogenic center.**

two identical groups, equidistant from C1

methylcyclopentane

C1 is *not* a stereogenic center.

With 3-methylcyclohexene, the result is different. All carbon atoms are bonded to two or three hydrogen atoms or are sp^2 hybridized except for C3, the ring carbon bonded to the methyl group. In this case, the atoms equidistant from C3 are different, so C3 is bonded to *different* alkyl groups in the ring. **C3 is therefore bonded to four different groups, making it a stereogenic center.**

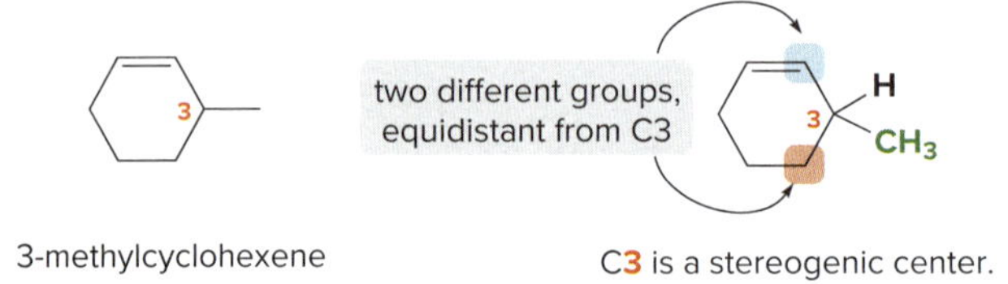

3-methylcyclohexene

C3 is a stereogenic center.

Two enantiomers are *different* compounds. To convert one enantiomer to another, you must **switch the position of two atoms.** This amounts to breaking bonds.

Because 3-methylcyclohexene has one tetrahedral stereogenic center, it is a chiral compound and exists as a pair of enantiomers.

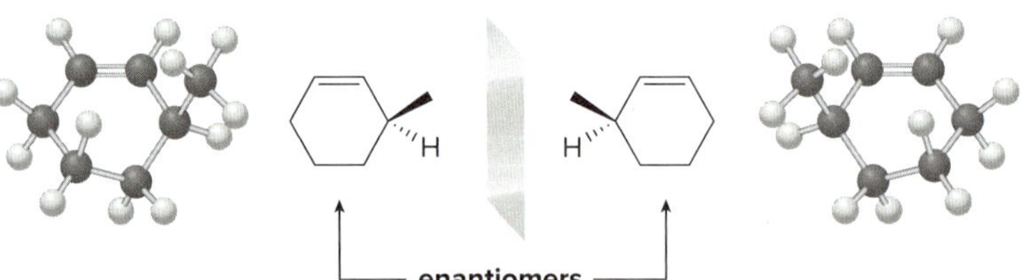

Many biologically active compounds contain one or more stereogenic centers on ring carbons. For example, **thalidomide,** a drug once prescribed as a sedative and anti-nausea agent for pregnant women in Great Britain and Europe, contains one stereogenic center, so it exists as a pair of enantiomers.

anti-nausea drug

teratogen

thalidomide **enantiomers**

Today, thalidomide is prescribed under strict controls for the treatment of Hansen's disease (leprosy). Because it was once thought to be highly contagious, individuals in Hawai'i with Hansen's disease were sent to Kalaupapa, a remote and inaccessible peninsula on the north shore of the Hawaiian island of Moloka'i. Hansen's disease is now known to be a curable bacterial infection, which is treated by the sulfa drugs discussed in Section 23.15.
Lokuttara/Shutterstock

Unfortunately thalidomide was sold as a mixture of its two enantiomers, and each of these stereoisomers has a different biological activity. This is a property not uncommon in chiral drugs, as we will see in Section 5.13A. Although one enantiomer was an effective sedative and anti-nausea drug, the other enantiomer was responsible for thousands of catastrophic birth defects in children born to women who took the drug during pregnancy. Thalidomide was never approved for use in the United States due to the diligence of Frances Oldham Kelsey, a medical reviewing officer for the Food and Drug Administration, who insisted that the safety data on thalidomide were inadequate.

Sucrose and **paclitaxel** (the chapter-opening molecule) are two useful compounds with several stereogenic centers at ring carbons. Identify the stereogenic centers in these more complicated compounds in exactly the same way, **looking at one carbon at a time. Sucrose,** with nine stereogenic centers on two rings, is the carbohydrate used as table sugar. **Paclitaxel,** with 11 stereogenic centers, is an anticancer agent active against ovarian, breast, and some lung tumors.

sucrose
(table sugar)

paclitaxel
Trade name: Taxol
(anticancer agent)

To draw the enantiomer of a complex compound with many stereogenic centers, change all groups above the plane on wedges to dashed wedges, and all groups behind the plane on dashed wedges to wedges as shown in Figure 5.5b for a compound with one stereogenic center. For example, the antibiotic cephalexin has three stereogenic centers, whose configurations are inverted in its enantiomer.

cephalexin
Trade name: Keflex

enantiomer

Problem 5.10 Locate the stereogenic centers in each compound. A molecule may have one or more stereogenic centers. Gabapentin enacarbil [part (d)] is used to treat seizures and certain types of chronic pain.

a.

b.

c.

d.

gabapentin enacarbil

e.

f.

cholesterol

Problem 5.11 Locate the stereogenic centers in each compound and draw the enantiomer. Exemestane (trade name Aromasin) is used to treat breast cancer, and zanamivir is used to treat and prevent influenza.

a. exemestane

b. zanamivir

5.6 Labeling Stereogenic Centers with *R* or *S*

Naming enantiomers with the prefix *R* or *S* is called the Cahn–Ingold–Prelog system after the three chemists who devised it.

Because enantiomers are two different compounds, we need a method to distinguish them by name. This is done by adding the prefix ***R*** or ***S*** to the IUPAC name of the enantiomer. To designate an enantiomer as *R* or *S,* first **assign a priority** (1, 2, 3, or 4) to each group bonded to the stereogenic center, and then use these priorities to label one enantiomer *R* and one *S.*

Rules Needed to Assign Priority

Rule 1 **Assign priorities (1, 2, 3, or 4) to the atoms directly bonded to the stereogenic center in order of *decreasing* atomic number. The atom of *highest* atomic number gets the *highest* priority (1).**

- In CHBrClF, priorities are assigned as follows: Br (1, highest) → Cl (2) → F (3) → H (4, lowest). In many molecules the lowest-priority group will be H.

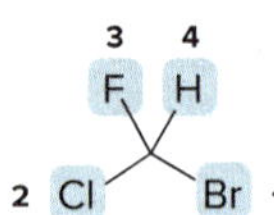

Rule 2 **If two atoms on a stereogenic center are the *same,* assign priority based on the atomic number of the atoms bonded to these atoms. *One* atom of higher atomic number determines a higher priority.**

- With butan-2-ol, the O atom gets highest priority (1) and H gets lowest priority (4) using Rule 1. Butan-2-ol also has two carbon atoms bonded to the stereogenic center, one that is part of a CH_3 group and one that is part of a CH_2CH_3 group. To assign priority (either 2 or 3) to the two C atoms, look at what atoms (other than the stereogenic center) are bonded to each C.

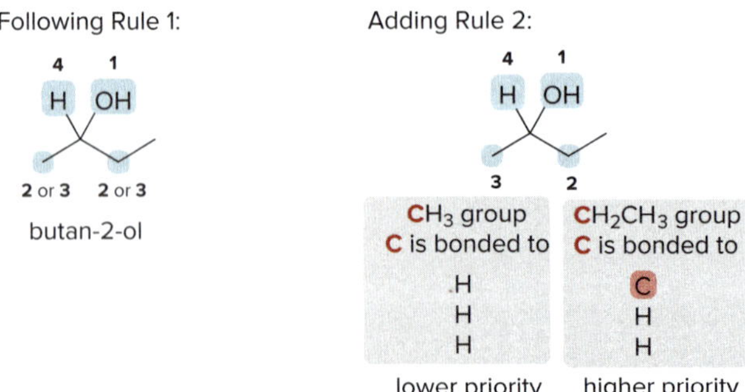

- The CH_2CH_3 gets higher priority (2) than the CH_3 group (priority 3) because the carbon of the ethyl group is bonded to another carbon.
- The order of priority of groups in butan-2-ol is –OH **(1),** –CH_2CH_3 **(2),** –CH_3 **(3),** and –H **(4).**
- If priority still cannot be assigned, continue along a chain until a point of difference is reached.

Rule 3 **If two isotopes are bonded to the stereogenic center, assign priorities in order of *decreasing mass* number.**

- In comparing two isotopes of the element hydrogen, deuterium, which has a mass number of two (one proton and one neutron), has a higher priority than hydrogen, which has a mass number of one (one proton only).

Rule 4 **To assign a priority to an atom that is part of a multiple bond, treat a multiply bonded atom as an equivalent number of singly bonded atoms.**

- The C of a C=O is considered to be bonded to two O atoms.

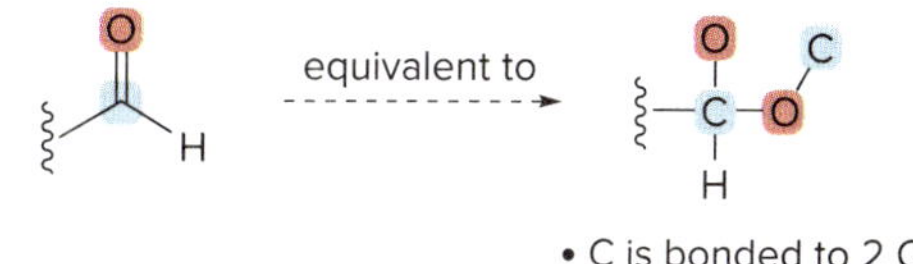

- C is bonded to 2 O's.
- O is bonded to 2 C's.

- Other common multiple bonds are drawn below.

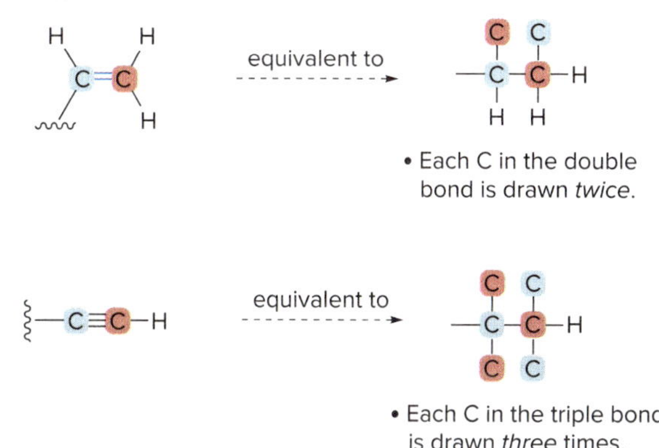

Figure 5.6 gives examples of priorities assigned to stereogenic centers.

Figure 5.6 Examples of assigning priorities to stereogenic centers

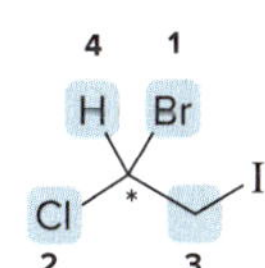

- The stereogenic center is bonded to Br, Cl, C, and H.
- The stereogenic center is *not* bonded directly to I.

3 2 4 1

- **C**H(CH_3)$_2$ gets the highest priority because the **C** is bonded to 2 other C's.

3 2 HO O OH H OH 4 1

- OH gets the highest priority because O has the highest atomic number.
- CO_2H (three bonds to O) gets higher priority than CH_2OH (one bond to O).

[* = stereogenic center]

Problem 5.12 Which group in each pair is assigned the *higher* priority?

a. $-CH_3$, $-CH_2CH_3$
b. $-I$, $-Br$
c. $-H$, $-D$
d. $-CH_2Br$, $-CH_2CH_2Br$
e. $-CH_2CH_2Cl$, $-CH_2CH(CH_3)_2$
f. $-CH_2OH$, $-CHO$

Problem 5.13 Rank the following groups in order of *decreasing* priority.

a. $-COOH$, $-H$, $-NH_2$, $-OH$
b. $-H$, $-CH_3$, $-Cl$, $-CH_2Cl$
c. $-CH_2CH_3$, $-CH_3$, $-H$, $-CH(CH_3)_2$
d. $-CH{=}CH_2$, $-CH_3$, $-C{\equiv}CH$, $-H$

R is derived from the Latin word *rectus* meaning "right," and *S* is from the Latin word *sinister* meaning "left."

Once priorities are assigned to the four groups around a stereogenic center, we can use three steps to designate the center as either *R* or *S*.

How To Assign *R* or *S* to a Stereogenic Center

Example Label each enantiomer as *R* or *S*.

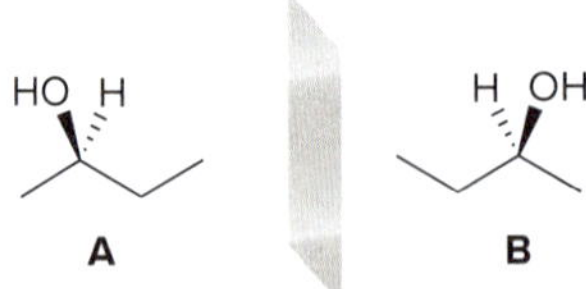

two enantiomers of butan-2-ol

Step [1] **Assign priorities from 1 to 4 to each group bonded to the stereogenic center.**

- The priorities for the four groups around the stereogenic center in butan-2-ol were given in Rule 2, on page 190.

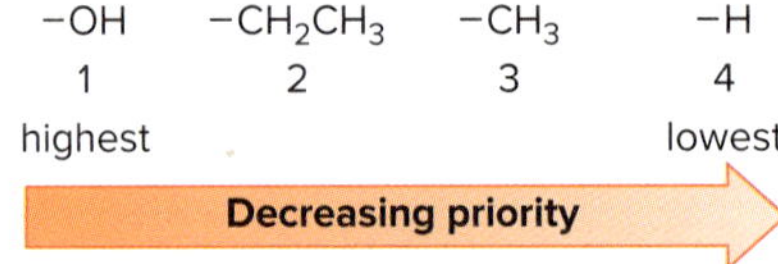

Step [2] **Orient the molecule with the lowest-priority group (4) *back* (on a *dashed wedge*), and visualize the relative positions of the remaining three groups (priorities 1, 2, and 3).**

- For each enantiomer of butan-2-ol, **look toward the lowest-priority group,** drawn behind the plane, down the C–H bond.

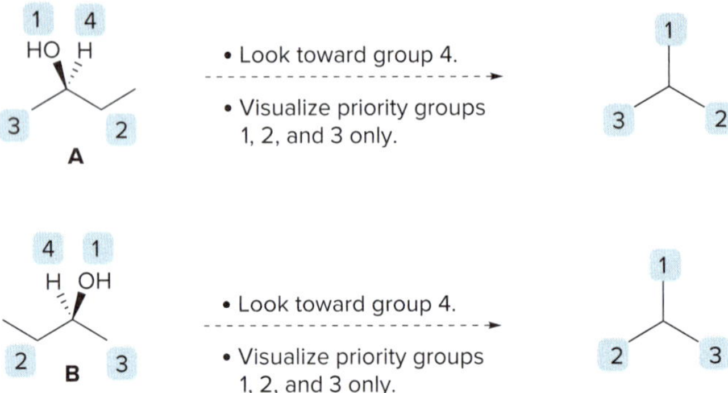

Step [3] **Trace a circle from priority group 1 → 2 → 3.**

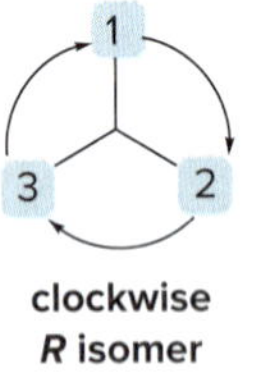

clockwise
***R* isomer**

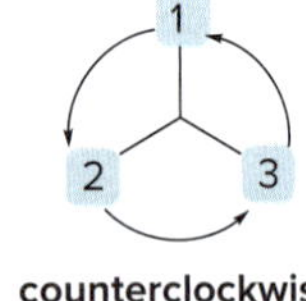

counterclockwise
***S* isomer**

- If tracing the circle goes in the **clockwise** direction—to the right from the noon position—the isomer is named ***R*.**
- If tracing the circle goes in the **counterclockwise** direction—to the left from the noon position—the isomer is named ***S*.**

- The letter *R* or *S* precedes the IUPAC name of the molecule. For the enantiomers of butan-2-ol:

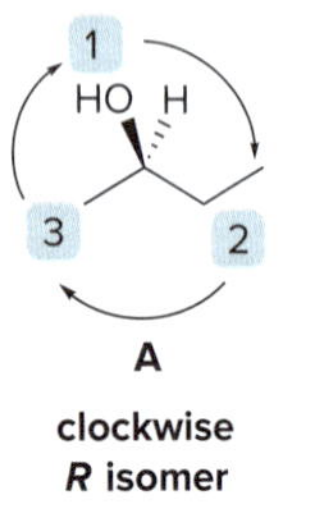

A

clockwise
***R* isomer**
(*R*)-butan-2-ol

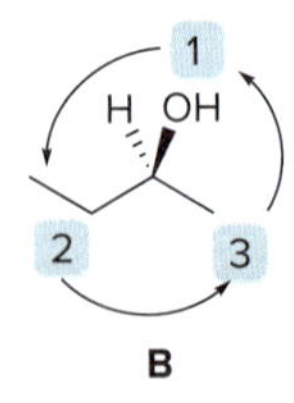

B

counterclockwise
***S* isomer**
(*S*)-butan-2-ol

Sample Problem 5.3 Labeling a Stereogenic Center as *R* or *S*

Label the stereogenic center in each compound as *R* or *S*.

a. 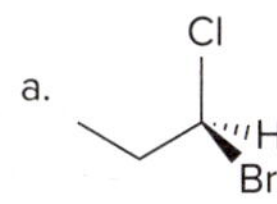

b. Cl

Solution

a. Assign priorities...

2 Cl, 4 H, 3, Br 1

...then look toward the lowest-priority group (H), and trace a circle, 1 → 2 → 3.

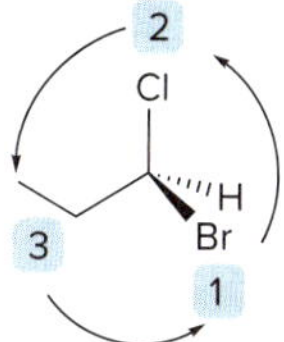

counterclockwise

Answer: ***S*** isomer

b. Assign priorities...

3, 1 Cl, H 4, 2

...then look toward the lowest-priority group (H), and trace a circle, 1 → 2 → 3.

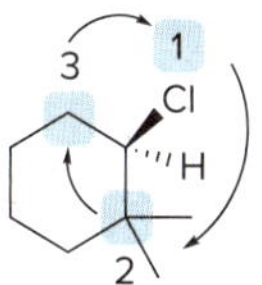

clockwise

Answer: ***R*** isomer

Problem 5.14 Label each stereogenic center in the following compounds as *R* or *S*.

a. HO H b. F, Br c. HO H, H, O

More Practice: Try Problem 5.45a, e.

How do you assign *R* or *S* to a molecule when the lowest-priority group is not oriented toward the back, on a dashed wedge? You could rotate and flip the molecule until the lowest-priority group is in the back, as shown in Figure 5.7; then follow the stepwise procedure for assigning the configuration. Or, if manipulating and visualizing molecules in three dimensions is difficult for you, try the procedure suggested in Sample Problem 5.4.

Figure 5.7 Orienting the lowest-priority group in back

2, 1, 3, 4 —rotate→ 2, 1, 4, 3 = 2, 1, 3

clockwise
R isomer

4, 1, 3, 2 —flip 180°→ 4, 1, 2, 3 = 1, 2, 3

counterclockwise
S isomer

- In rotating a molecule about a single bond, the position of three groups changes.
- In flipping a molecule 180°, the position of all four groups changes.

Sample Problem 5.4 Designating a Stereogenic Center as *R* or *S* When the Lowest-Priority Group Is Not Drawn Back

Label each stereogenic center as *R* or *S*.

a. b.

Solution

In both parts, the lowest-priority group is not oriented behind the page. To assign *R* or *S* in this case:

- **Switch** the position of the lowest-priority group with the group located ***behind*** the page.
- Determine *R* or *S* in the usual manner.
- **Reverse the answer.** Because we switched the position of two groups on the stereogenic center to begin with, and there are only two possibilities, the answer is **opposite** to the correct answer.

a. [1] Assign priorities. [2] Switch groups 4 and 1. [3] Trace a circle, 1 → 2 → 3, and reverse the answer.

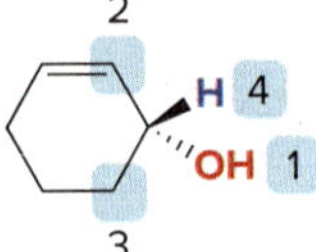

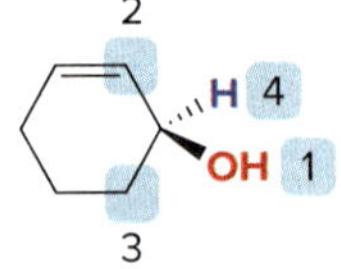

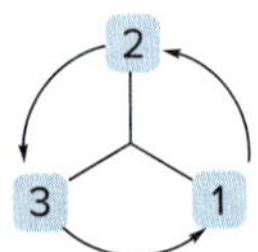

counterclockwise
It looks like an *S* isomer, but we must reverse the answer, because we switched groups 1 and 4, *S* → *R*.

Answer: ***R*** isomer

b. [1] Assign priorities. [2] Switch groups 4 and 1. [3] Trace a circle, 1 → 2 → 3, and reverse the answer.

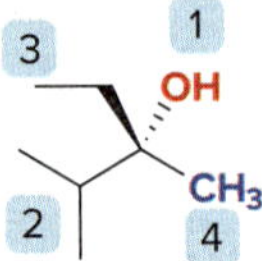

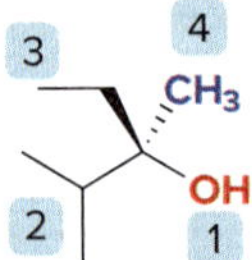

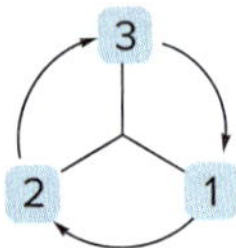

clockwise
It looks like an *R* isomer, but we must reverse the answer, because we switched groups 1 and 4, *R* → *S*.

Answer: ***S*** isomer

Problem 5.15 Label each stereogenic center as *R* or *S*.

a. Cl H Br
b. HO O H OH
c.
d. Br Cl OH
e. H N OH
f. O H OH

More Practice: Try Problems 5.45, 5.46, 5.50, 5.61a, 5.63a.

Problem 5.16 Draw both enantiomers of pretomanid, an antibiotic developed by the TB Alliance, a not-for-profit organization that seeks to produce new and affordable tuberculosis medications. If pretomanid is sold as a single enantiomer with the *S* configuration, which enantiomer corresponds to pretomanid?

pretomanid

Problem 5.17 Locate the stereogenic centers in ritonavir, a drug used to treat HIV, and label each stereogenic center as *R* or *S*. Ritonavir was approved for use in the United States as an oral generic drug in 2020.

ritonavir

5.7 Diastereomers

We have now seen many examples of compounds containing one tetrahedral stereogenic center. The situation is more complex for compounds with two stereogenic centers, because more stereoisomers are possible. Moreover, a molecule with two or more stereogenic centers ***may* or *may not* be chiral.**

- **For n stereogenic centers, the maximum number of stereoisomers is 2^n.**

- **When $n = 1$, $2^1 = 2$.** With one stereogenic center, there are always two stereoisomers and they are **enantiomers.**
- **When $n = 2$, $2^2 = 4$.** With two stereogenic centers, the maximum number of stereoisomers is four, although sometimes there are *fewer* than four.

Problem 5.18 Locate the stereogenic centers and determine the maximum number of stereoisomers possible for each compound.

a. b. c.

Let's illustrate a stepwise procedure for finding all possible stereoisomers using 2,3-dibromopentane. Because 2,3-dibromopentane has two stereogenic centers, the maximum number of stereoisomers is four.

In testing to see if one compound is superimposable on another, rotate atoms and flip the entire molecule, but **do not break any bonds.**

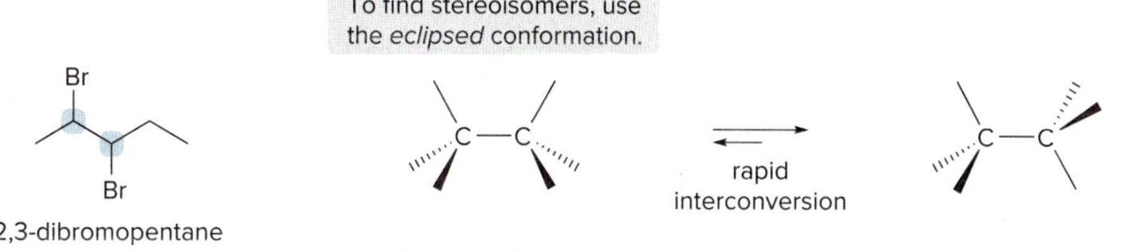

How To Find and Draw All Possible Stereoisomers for a Compound with Two Stereogenic Centers

Step [1] **Draw one stereoisomer by arbitrarily arranging substituents around the stereogenic centers. Then draw its mirror image.**

- Arbitrarily add the H, Br, CH_3, and CH_2CH_3 groups to the stereogenic centers, forming **A.** Then draw the mirror image **B** so that substituents in **B** are a reflection of the substituents in **A.**
- Determine whether **A** and **B** are superimposable by flipping or rotating one molecule to see if all the atoms align.
- In this case, the atoms of **A** and **B** do not align, making **A** and **B** nonsuperimposable mirror images—**enantiomers. A** and **B** are two of the four possible stereoisomers for 2,3-dibromopentane.

- H and Br do *not* align.
- **A** and **B** are *different* compounds.

Step [2] **Draw a third possible stereoisomer by switching the positions of any two groups on only *one* stereogenic center. Then draw its mirror image.**

- Switching the positions of H and Br (or any two groups) on one stereogenic center of either **A** or **B** forms a new stereoisomer (labeled **C** in this example), which is different from both **A** and **B.** Then draw the mirror image of **C,** labeled **D. C** and **D** are nonsuperimposable mirror images—**enantiomers.** We have now drawn four stereoisomers for 2,3-dibromopentane, the maximum number possible.

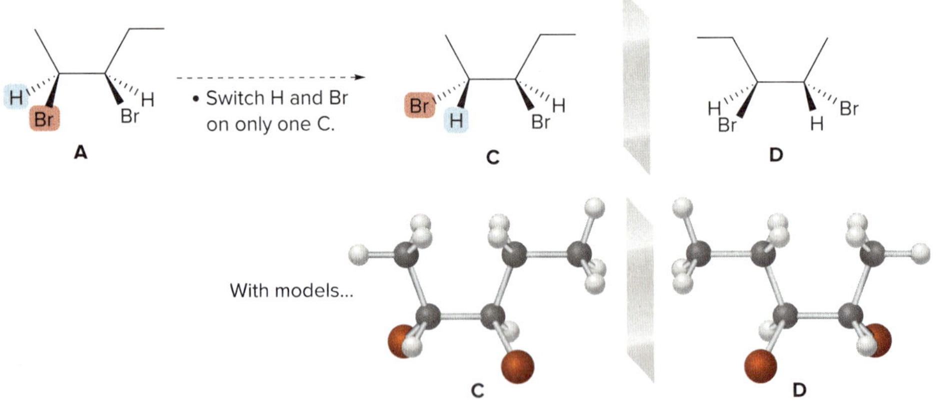

There are only two types of stereoisomers: ***Enantiomers* are stereoisomers that *are* mirror images. *Diastereomers* are stereoisomers that are *not* mirror images.**

There are four stereoisomers for 2,3-dibromopentane: enantiomers **A** and **B,** and enantiomers **C** and **D.** What is the relationship between two stereoisomers like **A** and **C? A** and **C** represent the second class of stereoisomers, called **diastereomers.** ***Diastereomers* are stereoisomers that are *not* mirror images of each other.** **A** and **B** are diastereomers of **C** and **D,** and vice versa. Figure 5.8 summarizes the relationships between the stereoisomers of 2,3-dibromopentane.

Figure 5.8 The four stereoisomers of 2,3-dibromopentane

A and B are enantiomers; C and D are enantiomers.

A and **B** are diastereomers of **C** and **D.**

- Pairs of enantiomers: **A** and **B; C** and **D.**
- Pairs of diastereomers: **A** and **C; A** and **D; B** and **C; B** and **D.**

Problem 5.19 Label the two stereogenic centers in each compound and draw all possible stereoisomers. Glutathione [part (c)] is an antioxidant that destroys harmful oxidizing agents in cells.

a. b. c. (glutathione)

Problem 5.20 Compounds **E** and **F** are two isomers of 2,3-dibromopentane drawn in staggered conformations. Which compounds (**A–D**) in Figure 5.8 are identical to **E** and **F?**

E **F**

5.8 Meso Compounds

Whereas 2,3-dibromopentane has two stereogenic centers and the maximum of four stereoisomers, **2,3-dibromobutane** has two stereogenic centers but fewer than the maximum number of stereoisomers.

2,3-dibromobutane
= stereogenic center

To find and draw all the stereoisomers of 2,3-dibromobutane, follow the same stepwise procedure outlined in Section 5.7. Arbitrarily add the H, Br, and CH_3 groups to the stereogenic centers, forming one stereoisomer **A,** and then draw its mirror image **B. A** and **B** are nonsuperimposable mirror images—**enantiomers.**

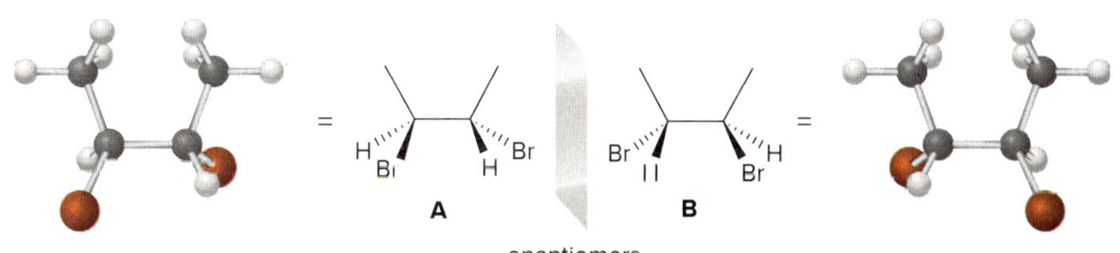

To find the other two stereoisomers (if they exist), switch the position of two groups on *one* stereogenic center of only *one* enantiomer. In this case, switching the positions of H and Br on one stereogenic center of **A** forms **C,** which is different from both **A** and **B** and is thus a new stereoisomer.

A

• Switch H and Br on only one C.

C

D

identical
C = D

With models...

C

D

However, the mirror image of **C,** labeled **D,** is superimposable on **C,** so **C** and **D** are *identical.* Thus, **C** is **achiral,** even though it has two stereogenic centers. **C** is a **meso compound.**

- **A *meso compound* is an achiral compound that contains tetrahedral stereogenic centers.**

C contains a **plane of symmetry. Meso compounds generally have a plane of symmetry,** so they possess two identical halves.

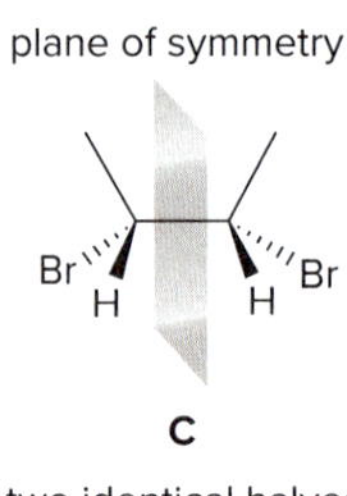

Because one stereoisomer of 2,3-dibromobutane is superimposable on its mirror image, there are only three stereoisomers and not four, as summarized in Figure 5.9.

Figure 5.9
The three stereoisomers of 2,3-dibromobutane

A

B

C

enantiomers

meso compound

A and **B** are diastereomers of **C.**

- Pair of enantiomers: **A** and **B.**
- Pairs of diastereomers: **A** and **C; B** and **C.**

Problem 5.21 Draw all the possible stereoisomers for each compound, and label pairs of enantiomers and diastereomers.

a. OH OH

b. Cl OH

c. O O

Problem 5.22 Which compounds are meso compounds?

a. b. c.

5.9 *R* and *S* Assignments in Compounds with Two or More Stereogenic Centers

(2*S*,3*R*)-2,3-dibromopentane

When a compound has more than one stereogenic center, the *R* or *S* configuration must be assigned to each of them. In the stereoisomer of 2,3-dibromopentane drawn here, C2 has the *S* configuration and C3 has the *R*, so the complete name of the compound is (2*S*,3*R*)-2,3-dibromopentane.

R,*S* configurations can be used to determine whether two compounds are identical, enantiomers, or diastereomers.

- **Identical compounds have the *same R,S* designations at every tetrahedral stereogenic center.**
- **Enantiomers have exactly *opposite R,S* designations.**
- **Diastereomers have the *same R,S* designation for at least one stereogenic center and the *opposite* for at least one of the other stereogenic centers.**

For example, if a compound has two stereogenic centers, both with the *R* configuration, then its enantiomer is *S*,*S* and the diastereomers are either *R*,*S* or *S*,*R*.

Problem 5.23 If the two stereogenic centers of a compound are *R*,*S* in configuration, what are the *R*,*S* assignments for its enantiomer and two diastereomers?

Problem 5.24 Without drawing out the structures, label each pair of compounds as enantiomers or diastereomers.

a. (2*R*,3*S*)-hexane-2,3-diol and (2*R*,3*R*)-hexane-2,3-diol
b. (2*R*,3*R*)-hexane-2,3-diol and (2*S*,3*S*)-hexane-2,3-diol
c. (2*R*,3*S*,4*R*)-hexane-2,3,4-triol and (2*S*,3*R*,4*R*)-hexane-2,3,4-triol

Lychee (Problem 5.25), which is grown extensively in Asia and South Africa, contains a tough inedible rind that encloses a sweet edible fruit. *Pixtal/age fotostock*

Problem 5.25 Hypoglycin A, an amino acid derivative found in unripened lychee, is a compound that is acutely toxic and can lead to death when ingested in large amounts by undernourished children. (a) Draw all possible stereoisomers for hypoglycin A, and give the *R*,*S* designation for each stereogenic center. (b) Use the *R*,*S* designations to label pairs of enantiomers and diastereomers.

hypoglycin A

5.10 Disubstituted Cycloalkanes

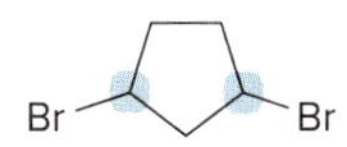

1,3-dibromocyclopentane

Let us now turn our attention to disubstituted cycloalkanes, and draw all possible stereoisomers for **1,3-dibromocyclopentane.** Because 1,3-dibromocyclopentane has two stereogenic centers (labeled in blue), it has a maximum of four stereoisomers.

To draw all possible stereoisomers, remember that a disubstituted cycloalkane can have two substituents on the *same* side of the ring (**cis isomer,** labeled **A**) or on *opposite* sides of the ring (**trans isomer,** labeled **B**). These compounds are **stereoisomers but not mirror**

images of each other, making them **diastereomers. A** and **B** are two of the four possible stereoisomers.

Br Br
A
cis isomer

Br Br
B
trans isomer

A and **B** are diastereomers.

In determining chirality in substituted cycloalkanes, always draw the rings as **flat polygons.** This is especially true for cyclohexane derivatives, where having two chair forms that interconvert can make analysis especially difficult.

To find the other two stereoisomers (if they exist), draw the mirror image of each compound and determine whether the compound and its mirror image are superimposable.

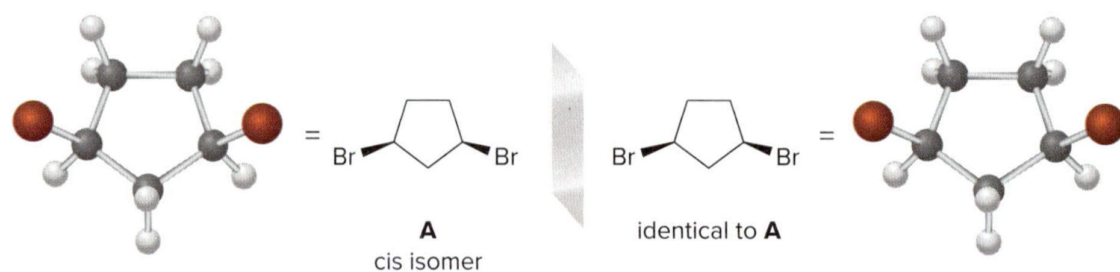

A
cis isomer

identical to **A**

cis-1,3-Dibromocyclopentane contains a plane of symmetry.

plane of symmetry

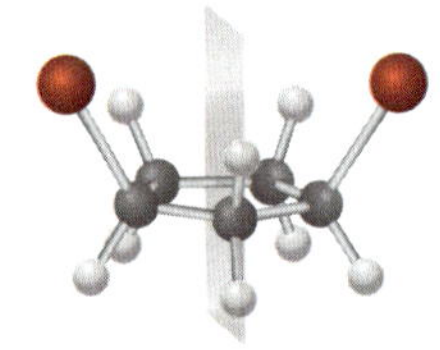

two identical halves

- The cis isomer is superimposable on its mirror image, making them *identical.* Thus, **A** is an **achiral meso compound.**

= Br Br Br Br =

B
trans isomer

C
trans isomer

B and **C** are enantiomers.

- The trans isomer **B** is *not* superimposable on its mirror image, labeled **C,** making **B** and **C** different compounds. Thus, **B** and **C** are **enantiomers.**

Because one stereoisomer of 1,3-dibromocyclopentane is superimposable on its mirror image, there are only three stereoisomers, not four. **A** is an achiral meso compound, and **B** and **C** are a pair of chiral enantiomers. **A** and **B** are diastereomers, as are **A** and **C.**

Problem 5.26 Which of the following cyclic molecules are meso compounds?

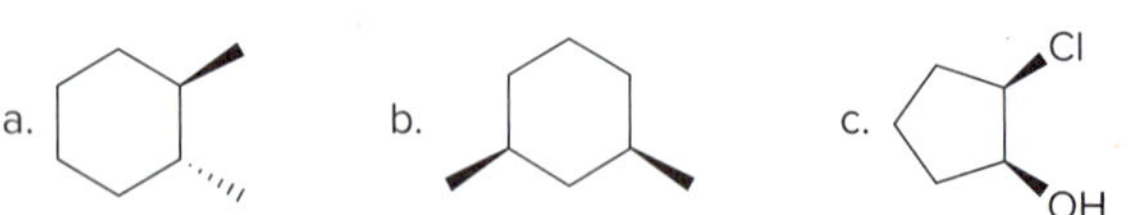

Problem 5.27 Draw all possible stereoisomers for each compound. Label pairs of enantiomers and diastereomers. Dolutegravir [part (c)] was approved for use in the United States in 2013 for the treatment of HIV.

a. b. Cl Cl c. O OH O N N O F F H N O

dolutegravir
Trade name: Tivicay

5.11 Isomers—A Summary

Before moving on to other aspects of stereochemistry, take the time to review Figures 5.10 and 5.11. Keep in mind the following facts, and use Figure 5.10 to summarize the types of isomers.

- **There are two major classes of isomers: constitutional isomers and stereoisomers.**
- **There are only two kinds of stereoisomers: enantiomers and diastereomers.**

Figure 5.10
Summary—Types of isomers

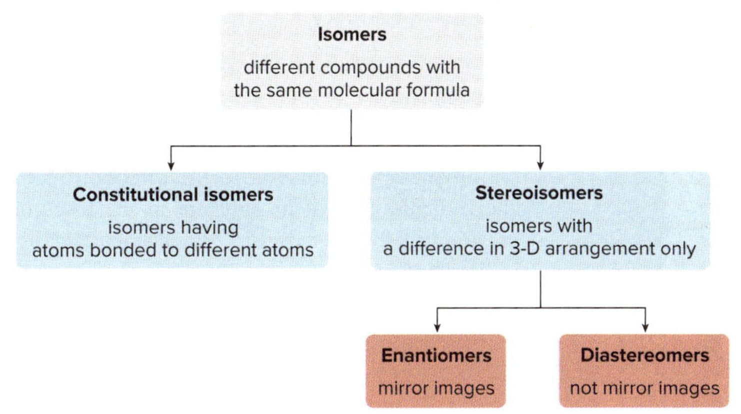

Then, to determine the relationship between two nonidentical molecules, refer to the flowchart in Figure 5.11.

Figure 5.11
Determining the relationship between two nonidentical molecules

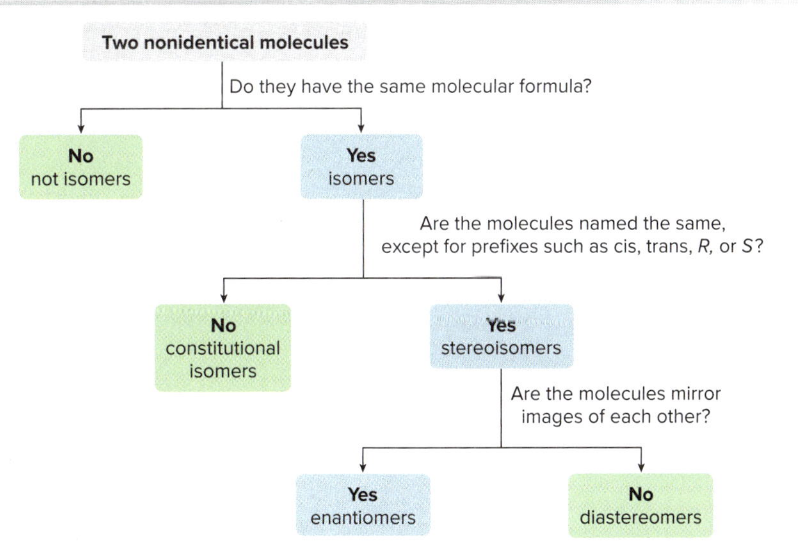

Problem 5.28 State how each pair of compounds is related. Are they enantiomers, diastereomers, constitutional isomers, or identical?

a. 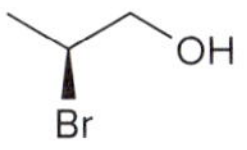 and

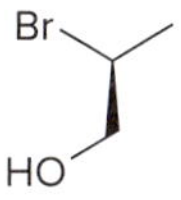

b. 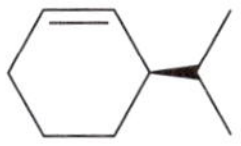 and

c. 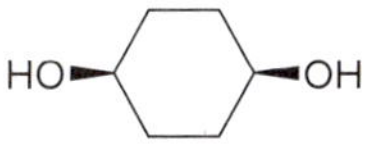 and

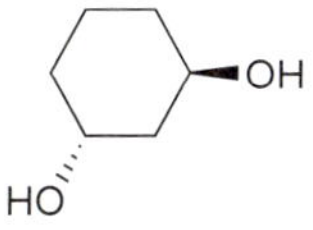

d. 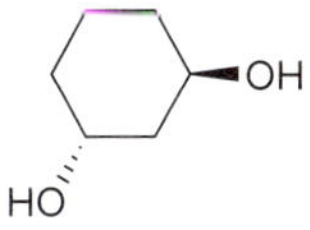 and

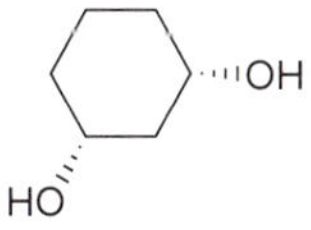

5.12 Physical Properties of Stereoisomers

Recall from Section 5.2 that constitutional isomers have different physical and chemical properties. How, then, do the physical and chemical properties of enantiomers compare?

- **The chemical and physical properties of two enantiomers are *identical* except in their interaction with *chiral* substances.**

5.12A Optical Activity

Two enantiomers have identical physical properties—melting point, boiling point, solubility—except for how they interact with plane-polarized light.

What is plane-polarized light? Ordinary light consists of electromagnetic waves that oscillate in all planes perpendicular to the direction in which the light travels. Passing light through a polarizer allows light in only one plane to come through, resulting in **plane-polarized light** (or simply **polarized light**). Plane-polarized light has an electric vector that oscillates in a single plane.

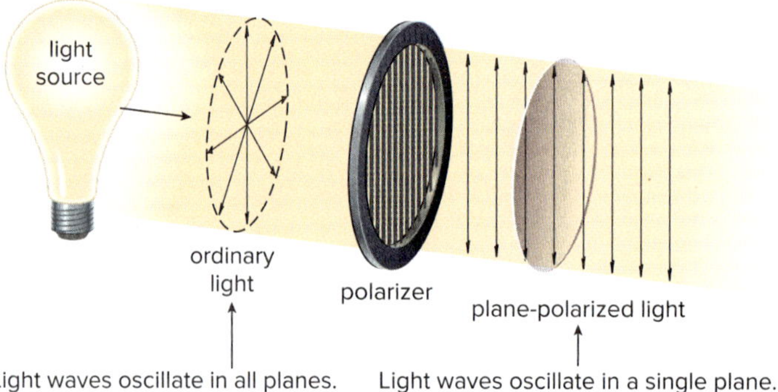

A **polarimeter** is an instrument that allows plane-polarized light to travel through a sample tube containing an organic compound. After the light exits the sample tube, an analyzer slit is rotated to determine the direction of the plane of the exiting polarized light. With **achiral compounds,** the light exits the sample tube *unchanged,* and the plane of the polarized light is in the same position it was before entering the sample tube.

- **A compound that does not change the plane of polarized light is said to be *optically inactive*.**

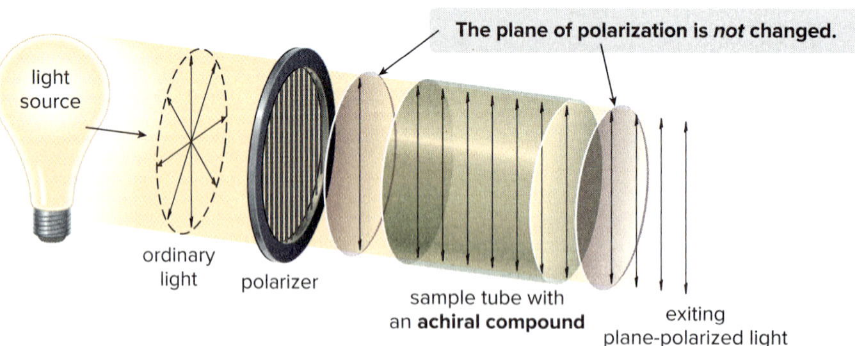

With **chiral compounds,** the plane of the polarized light is rotated through an angle α. The angle α, measured in degrees (°), is called the **observed rotation.**

- **A compound that rotates the plane of polarized light is said to be *optically active*.**

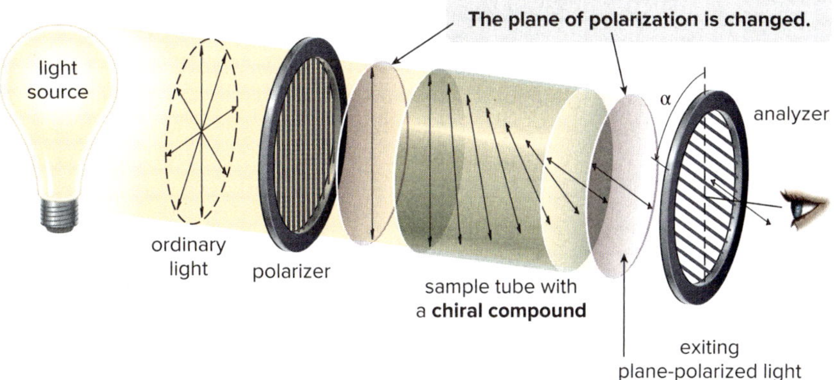

The achiral compound CH_2BrCl is optically *inactive,* whereas a single enantiomer of CHBrClF, a chiral compound, is optically *active.*

The rotation of polarized light can be in the **clockwise** or **counterclockwise** direction.

- **If the rotation is *clockwise* (to the right from the noon position), the compound is called *dextrorotatory.* The rotation is labeled *d* or (+).**
- **If the rotation is *counterclockwise* (to the left from noon), the compound is called *levorotatory.* The rotation is labeled *l* or (–).**

(*S*)-(–)-glyceraldehyde

(*S*)-(+)-lactic acid

No relationship exists between the *R* and *S* prefixes that designate configuration and the (+) and (–) designations indicating optical rotation. For example, the *S* enantiomer of lactic acid is dextrorotatory (+), whereas the *S* enantiomer of glyceraldehyde is levorotatory (–).

How does the rotation of two enantiomers compare?

- **Two enantiomers rotate plane-polarized light *to an equal extent* but in the *opposite* direction.**

Thus, if enantiomer **A** rotates polarized light +5°, then the same concentration of enantiomer **B** rotates it –5°.

5.12B Racemic Mixtures

What is the observed rotation of an equal amount of two enantiomers? Because **two enantiomers rotate plane-polarized light to an equal extent but in opposite directions, the rotations cancel,** and no rotation is observed.

- **An equal amount of two enantiomers is called a *racemic mixture* or a *racemate.* A racemic mixture is optically *inactive.***

Besides optical rotation, other physical properties of a racemate are not readily predicted. The melting point and boiling point of a racemic mixture are not necessarily the same as either pure enantiomer, and this fact is not easily explained. The physical properties of two enantiomers and their racemic mixture are summarized in Table 5.1.

Table 5.1 The Physical Properties of Enantiomers **A** and **B** Compared

Property	A alone	B alone	Racemic A + B
Melting point	identical to **B**	identical to **A**	may be different from **A** and **B**
Boiling point	identical to **B**	identical to **A**	may be different from **A** and **B**
Optical rotation	equal in magnitude but opposite in sign to **B**	equal in magnitude but opposite in sign to **A**	0°

5.12C Specific Rotation

The observed rotation depends on the number of chiral molecules that interact with polarized light. This in turn depends on the concentration of the sample and the length of the sample tube. To standardize optical rotation data, the quantity **specific rotation** ([α]) is defined using a specific sample tube length (usually 1 dm), concentration, temperature (25 °C), and wavelength (589 nm, the D line emitted by a sodium lamp).

$$\text{specific rotation} = [\alpha] = \frac{\alpha}{l \times c}$$

α = observed rotation (°)
l = length of sample tube (dm)
c = concentration (g/mL)

[dm = decimeter
1 dm = 10 cm]

Specific rotations are physical constants just like melting points or boiling points, and are reported in chemical reference books for a wide variety of compounds.

Problem 5.29 The amino acid (*S*)-alanine has the physical characteristics listed under the structure.

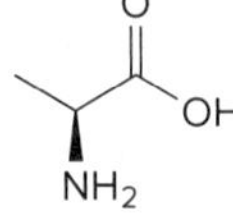

(*S*)-alanine
[α] = +8.5
mp = 297 °C

a. What is the melting point of (*R*)-alanine?
b. How does the melting point of a racemic mixture of (*R*)- and (*S*)-alanine compare to the melting point of (*S*)-alanine?
c. What is the specific rotation of (*R*)-alanine, recorded under the same conditions as the reported rotation of (*S*)-alanine?
d. What is the optical rotation of a racemic mixture of (*R*)- and (*S*)-alanine?
e. Label each of the following as optically active or inactive: a solution of pure (*S*)-alanine; an equal mixture of (*R*)- and (*S*)-alanine; a solution that contains 75% (*S*)- and 25% (*R*)-alanine.

Problem 5.30 A natural product was isolated in the laboratory, and its observed rotation was +10° when measured in a 1 dm sample tube containing 1.0 g of compound in 10 mL of water. What is the specific rotation of this compound?

5.12D Enantiomeric Excess

Sometimes in the laboratory we have neither a pure enantiomer nor a racemic mixture, but rather a mixture of two enantiomers in which one enantiomer is present in excess of the other. The **enantiomeric excess (*ee*),** also called the **optical purity,** tells how much more there is of one enantiomer.

- **Enantiomeric excess = ee = % of one enantiomer − % of the other enantiomer.**

Enantiomeric excess tells how much one enantiomer is present in excess of the racemic mixture. For example, if a mixture contains 75% of one enantiomer and 25% of the other, the enantiomeric excess is 75% − 25% = 50%. There is a 50% excess of one enantiomer over the racemic mixture.

Problem 5.31 What is the ee for each of the following mixtures of enantiomers **A** and **B?**

a. 95% **A** and 5% **B** b. 85% **A** and 15% **B**

Knowing the *ee* of a mixture makes it possible to calculate the amount of each enantiomer present, as shown in Sample Problem 5.5.

Sample Problem 5.5 Using Enantiomeric Excess to Calculate the Amount of Each Enantiomer

If the enantiomeric excess is 95%, how much of each enantiomer is present?

Solution

Label the two enantiomers **A** and **B** and assume that **A** is in excess. A 95% *ee* means that the solution contains an excess of 95% of **A,** and 5% of the racemic mixture of **A** and **B.** Because a racemic mixture is an equal amount of both enantiomers, it has 2.5% of **A** and 2.5% of **B.**

- Total amount of **A** = 95% + 2.5% = 97.5%
- Total amount of **B** = 2.5% (or 100% − 97.5%)

Problem 5.32 For the given *ee* values, calculate the percentage of each enantiomer present.

a. 90% *ee* b. 99% *ee* c. 60% *ee*

More Practice: Try Problem 5.60d.

The enantiomeric excess can also be calculated if two quantities are known—the specific rotation [α] of a mixture and the specific rotation [α] of a pure enantiomer.

$$ee = \frac{[\alpha]\text{ mixture}}{[\alpha]\text{ pure enantiomer}} \times 100\%$$

Sample Problem 5.6 Calculating Enantiomeric Excess

Camphor (Sample Problem 5.6), a fragrant compound obtained from the wood of camphor laurel, has been utilized for centuries in cooking, ointments, and medicine. Much of the camphor used today is synthesized from petroleum products. *Blew1/123RF*

Pure camphor has a specific rotation of +44. If a sample prepared in the lab has a specific rotation of +12, what is the enantiomeric excess of this sample of camphor?

Solution

Calculate the *ee* of the mixture using the given formula.

$$ee = \frac{[\alpha]\text{ mixture}}{[\alpha]\text{ pure enantiomer}} \times 100\% = \frac{+12}{+44} \times 100\% = 27\%\ ee$$

Answer

Problem 5.33 Pure MSG, a common flavor enhancer, exhibits a specific rotation of +24. (a) Calculate the *ee* of a solution whose [α] is +10. (b) If the *ee* of a solution of MSG is 80%, what is [α] for this solution?

^{-}O ... $O^{-}\ Na^{+}$, $\overset{+}{N}H_3$

MSG
monosodium glutamate

More Practice: Try Problems 5.60a, b, e; 5.62e, f.

Problem 5.34 (*S*)-Lactic acid has a specific rotation of +3.8. (a) If the *ee* of a solution of lactic acid is 60%, what is [α] for this solution? (b) How much of the dextrorotatory and levorotatory isomers does the solution contain?

5.12E The Physical Properties of Diastereomers

Diastereomers are not mirror images of each other, and as such, **their physical properties are different, including optical rotation.** Figure 5.12 compares the physical properties of the three stereoisomers of tartaric acid, consisting of a meso compound that is a diastereomer of a pair of enantiomers.

Figure 5.12 The physical properties of the three stereoisomers of tartaric acid

- **A** and **B** are enantiomers.
- **A** and **B** are diastereomers of **C.**

Property	A	B	C	A + B (1:1)
melting point (°C)	171	171	146	206
solubility (g/100 mL H_2O)	139	139	125	139
[α]	+13	−13	0	0
R,S designation	*R,R*	*S,S*	*R,S*	—
d,l designation	*d*	*l*	none	*d,l*

- The physical properties of **A** and **B** differ from their diastereomer **C.**
- The physical properties of a racemic mixture of **A** and **B** (last column) can also differ from either enantiomer and diastereomer **C.**
- **C** is an achiral meso compound, so it is optically inactive; [α] = 0.

Whether the physical properties of a set of compounds are the same or different has practical applications in the lab. Physical properties characterize a compound's physical state, and two compounds can usually be separated only if their physical properties are different.

Two enantiomers can be separated by the process of **resolution,** as described in Section 27.2.

- **Because two enantiomers have identical physical properties, they cannot be separated by common physical techniques like distillation.**
- **Diastereomers and constitutional isomers have different physical properties, and therefore they can be separated by common physical techniques.**

Problem 5.35 Compare the physical properties of the three stereoisomers of 1,3-dimethylcyclopentane.

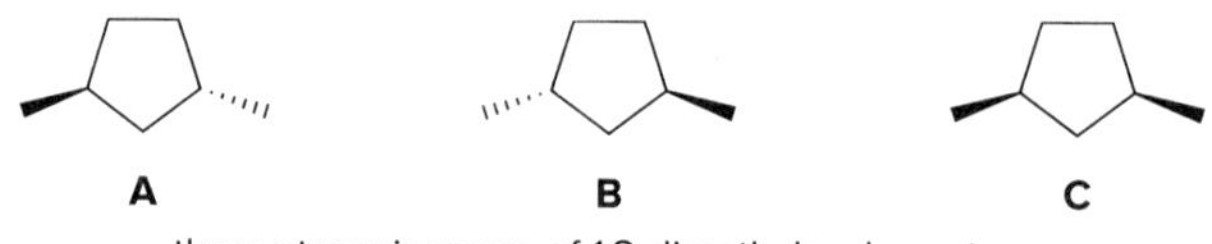

three stereoisomers of 1,3-dimethylcyclopentane

a. How do the boiling points of **A** and **B** compare? What about those of **A** and **C?**

b. Characterize a solution of each of the following as optically active or optically inactive: pure **A;** pure **B;** pure **C;** an equal mixture of **A** and **B;** an equal mixture of **A** and **C.**

c. A reaction forms a 1:1:1 mixture of **A, B,** and **C.** If this mixture is distilled, how many fractions would be obtained? Which fractions would be optically active and which would be optically inactive?

5.13 Chemical Properties of Enantiomers

When two enantiomers react with an achiral reagent, they react at the same rate, but when they react with a chiral, non-racemic reagent, they react at different rates.

- **Two enantiomers have exactly the same chemical properties except for their reaction with chiral, non-racemic reagents.**

For an everyday analogy, consider what happens when you are handed an achiral object like a pen and a chiral object like a right-handed glove. Your left and right hands are enantiomers, and they can both hold the achiral pen in the same way. With the glove, however, only your right hand can fit inside it, not your left.

We will examine specific reactions of chiral molecules with both chiral and achiral reagents later in this text. Here, we examine two more general applications.

5.13A Chiral Drugs

A living organism is a sea of chiral molecules. Many drugs are chiral, and often they must interact with a chiral receptor or a chiral enzyme to be effective. One enantiomer of a drug may treat a disease whereas its mirror image may be ineffective. Alternatively, one enantiomer may trigger one biochemical response and its mirror image may elicit a totally different response.

Although (*R*)-ibuprofen shows no anti-inflammatory activity itself, it is slowly converted to the *S* enantiomer in vivo.

The drugs ibuprofen and fluoxetine each contain one stereogenic center, and thus exist as a pair of enantiomers, only one of which exhibits biological activity. **(*S*)-Ibuprofen** is the active component of the anti-inflammatory agents Motrin and Advil, and **(*R*)-fluoxetine** is the active component in the antidepressant Prozac.

(*S*)-ibuprofen
anti-inflammatory agent

(*R*)-fluoxetine
antidepressant

Changing the orientation of two substituents to form a mirror image can also alter biological activity to produce an undesirable side effect in the other enantiomer. The *S* enantiomer of **naproxen** is an active anti-inflammatory agent, but the *R* enantiomer is a harmful liver toxin.

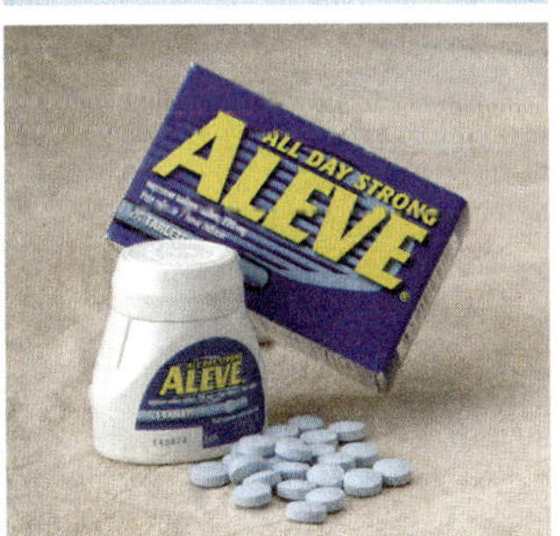

(*S*)-Naproxen is the active drug in the widely used pain relievers Naprosyn and Aleve.
Elite Images/McGraw Hill

(*S*)-naproxen
anti-inflammatory agent

(*R*)-naproxen
liver toxin

If a chiral drug could be sold as a single active enantiomer, it should be possible to use smaller doses with fewer side effects. Many established chiral drugs were first sold as racemic mixtures, and because it is more difficult and costly to modify an existing method to produce a single enantiomer, they are still sold as racemates. In contrast, newer technology has made single enantiomers more readily available, so recently approved chiral drugs are often marketed as a single enantiomer.

5.13B Enantiomers and the Sense of Smell

Research suggests that the odor of a particular molecule is determined more by its shape than by the presence of a particular functional group. For example, hexachloroethane (Cl_3CCCl_3) and cyclooctane have no obvious structural similarities, but they both have a camphor-like odor, a fact attributed to their similar spherical shape. Each molecule binds to spherically shaped olfactory receptors present on the nerve endings in the nasal passage, resulting in similar odors (Figure 5.13).

Figure 5.13 The shape of molecules and the sense of smell

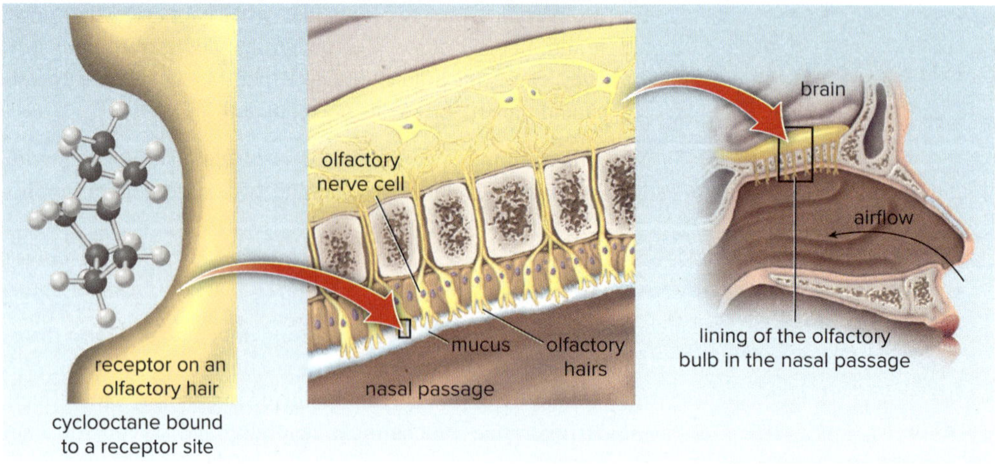

- Cyclooctane and other molecules similar in shape bind to a particular olfactory receptor on the nerve cells that lie at the top of the nasal passage. Binding results in a nerve impulse that travels to the brain, which interprets impulses from particular receptors as specific odors.

Because enantiomers interact with chiral smell receptors, some enantiomers have different odors. There are a few well-characterized examples of this phenomenon in nature. For example, (*S*)-carvone is responsible for the odor of caraway, whereas (*R*)-carvone is responsible for the odor of spearmint.

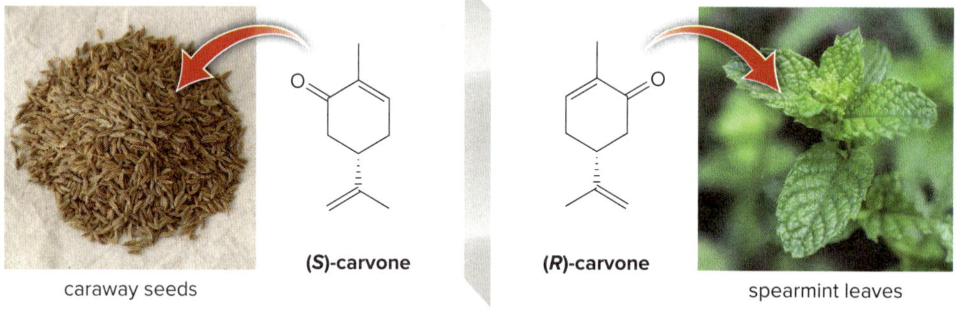

Elite Images/McGraw Hill *DAJ/amana images/Getty Images*

These examples demonstrate that understanding the three-dimensional structure of a molecule is very important in organic chemistry.

The foul odor of human sweat is due in part to the presence of unequal amounts of both enantiomers of 3-methyl-3-sulfanylhexan-1-ol (Problem 5.36). *Daniel C. Smith*

Problem 5.36 Although fresh human sweat has no odor, enzymes in skin bacteria form compounds with distinctive, foul odors, including 3-methyl-3-sulfanylhexan-1-ol (**A**). The two enantiomers of **A** smell different and are present in sweat in unequal amounts. The *S* enantiomer, the major isomer, has a strong onion-like odor, whereas the *R* isomer smells like grapefruit. Draw both enantiomers and label each compound as *R* or *S*.

SH

OH

3-methyl-3-sulfanylhexan-1-ol

A

Chapter 5 REVIEW

KEY CONCEPTS

[1] Two types of isomers (5.2, 5.11); example: $C_5H_{10}BrCl$

1 **Constitutional isomers**—same molecular formula, but different connectivity of atoms

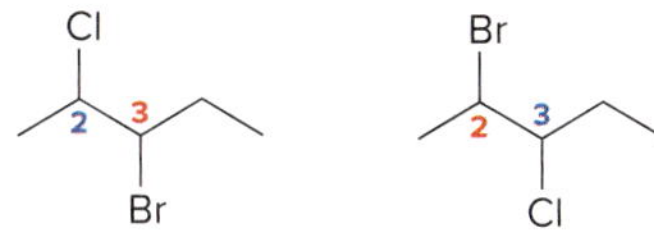

3-bromo-2-chloro-pentane

2-bromo-3-chloro-pentane

- Constitutional isomers have different IUPAC names.

See Figure 5.3.

2 **Stereoisomers**—same molecular formula and connectivity of atoms, but different spatial orientation of atoms

Enantiomers (5.4, 5.5)

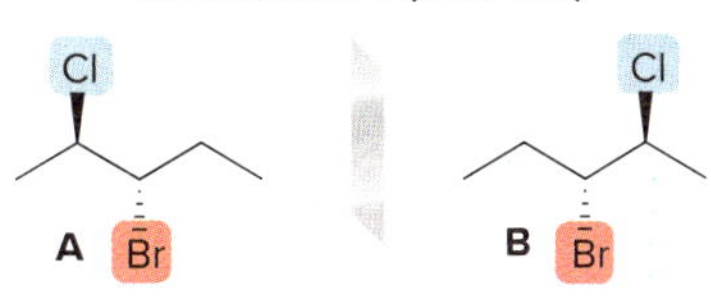

nonsuperimposable mirror images

A and B are enantiomers.

See Figure 5.5.

Diastereomers (5.7)

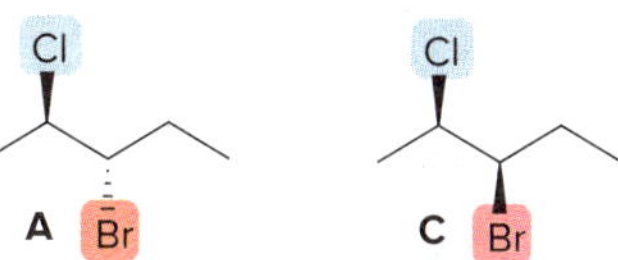

stereoisomers but not mirror images

A and C are diastereomers.

See Figures 5.8, 5.10, 5.11. Try Problems 5.37, 5.38, 5.54–5.57, 5.59a.

[2] Stereochemical terms

1 **Chiral compounds (5.3–5.5)**

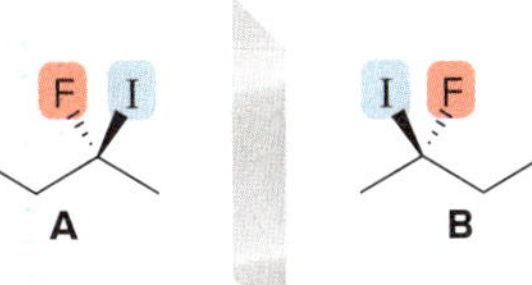

nonsuperimposable mirror images
no plane of symmetry
tetrahedral stereogenic center

A and B are chiral.

2 **Achiral compound (5.3)**

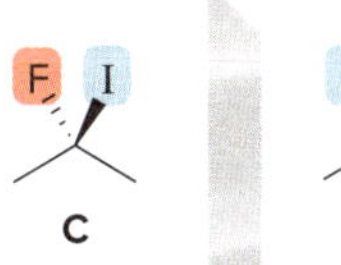

superimposable mirror images
plane of symmetry
no tetrahedral stereogenic center

C and D are identical and achiral.

- An **achiral compound** is **superimposable on its mirror image.**

3 **Meso compound (5.8)**

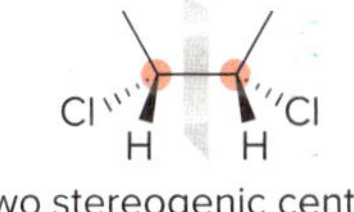

two stereogenic centers
plane of symmetry
achiral

4 **Racemic mixture (5.12B)**

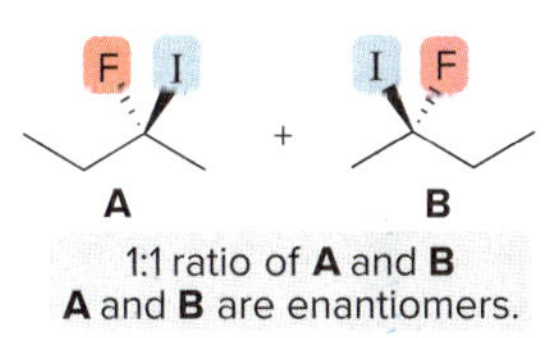

1:1 ratio of **A** and **B**
A and **B** are enantiomers.

Try Problems 5.39, 5.40, 5.59b.

[3] Optical activity (5.12)

1 **An optically active solution contains:**

- **a chiral compound**

2 **An optically inactive solution contains one of the following:**

- **an achiral compound** with no stereogenic centers
- **a meso compound**
- **a racemic mixture** of two enantiomers

Try Problems 5.58; 5.59c, g.

[4] The prefixes *R* and *S* compared with *d* (+) and *l* (–) (5.6, 5.12)

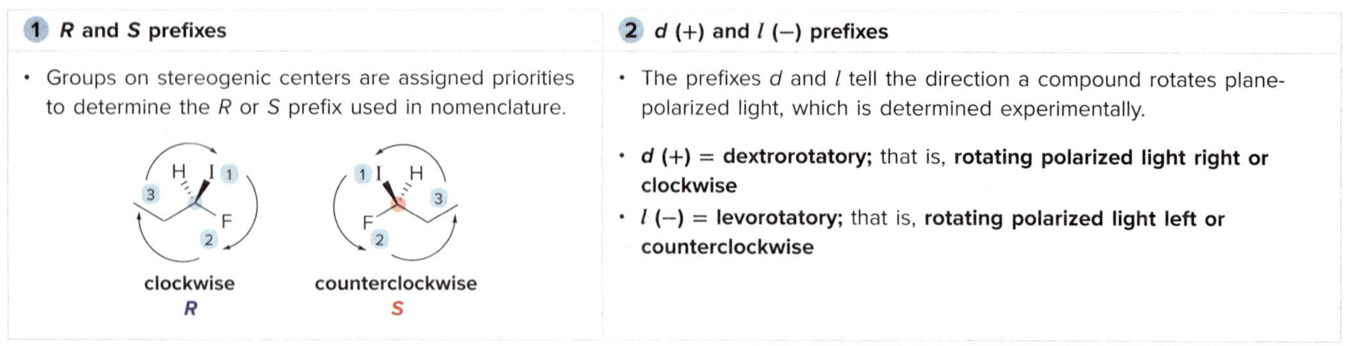

[5] *R* and *S* assignments in compounds with two or more stereogenic centers (5.9); example: 3-bromo-2-chloropentane

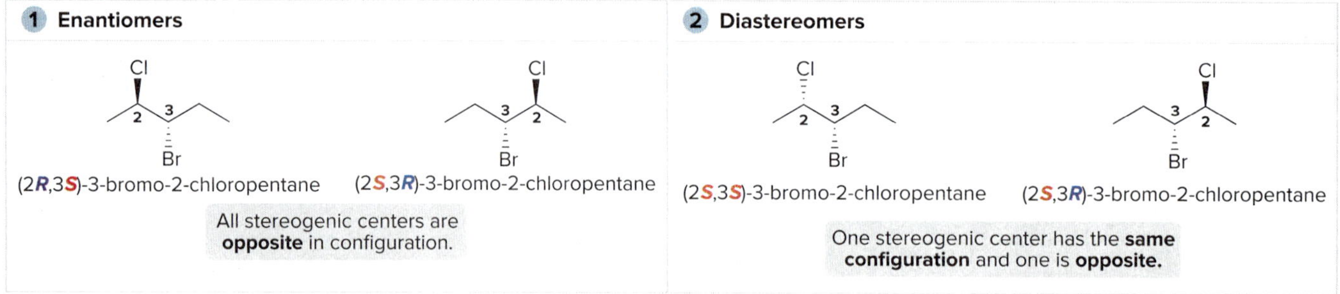

Try Problem 5.51.

[6] Physical and chemical properties of isomers (5.12, 5.13)

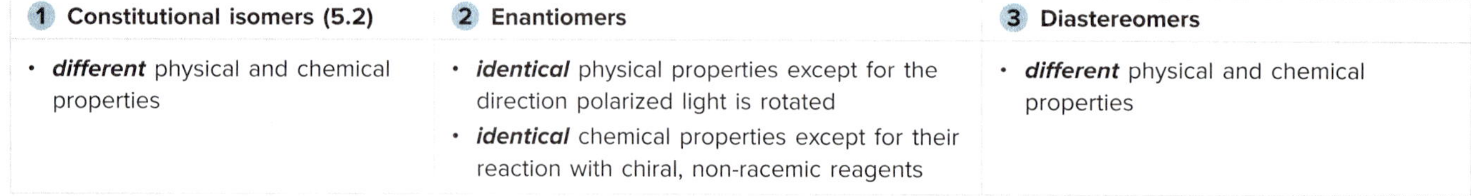

1 Constitutional isomers (5.2)	2 Enantiomers	3 Diastereomers
• ***different*** physical and chemical properties	• ***identical*** physical properties except for the direction polarized light is rotated • ***identical*** chemical properties except for their reaction with chiral, non-racemic reagents	• ***different*** physical and chemical properties

See Figure 5.12. Try Problems 5.58b, 5.59e.

KEY SKILLS

[1] Locating stereogenic centers (5.4, 5.5); example: adenosine

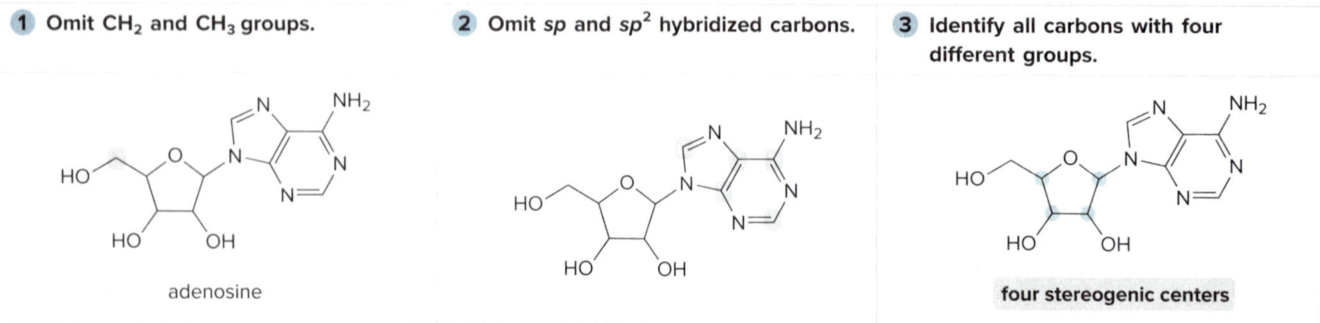

See Sample Problem 5.1. Try Problems 5.41, 5.62a.

[2] Labeling stereogenic centers with *R* or *S* (5.6); example: 1-aminoethan-1-ol

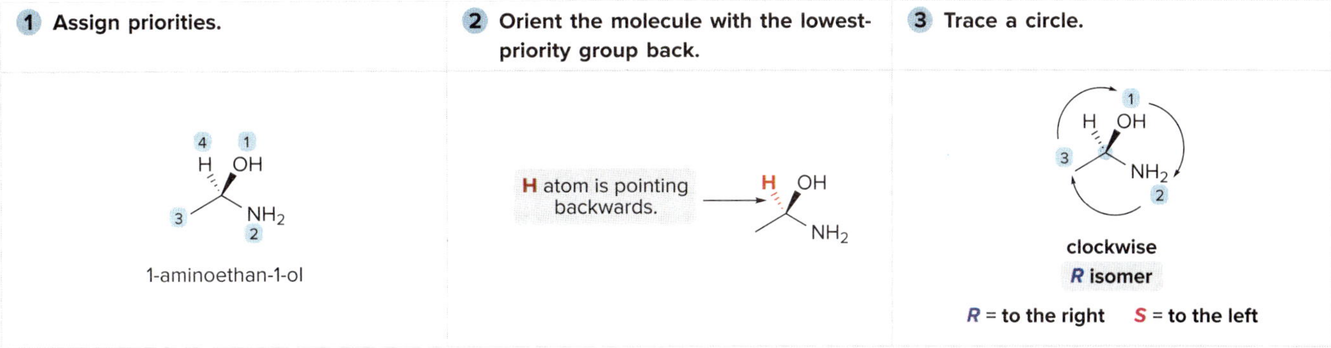

See *How To* p. 192, Sample Problem 5.3, Figure 5.6. Try Problem 5.45a, e.

[3] Assigning *R* or *S* when the lowest-priority group is not oriented toward the back (5.6); example: propranolol

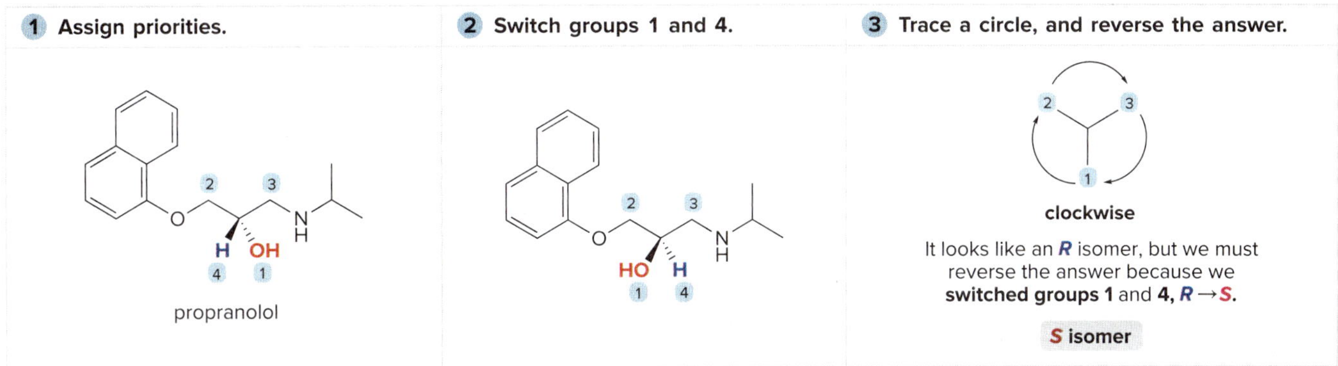

See Figure 5.7, Sample Problem 5.4. Try Problems 5.45, 5.46, 5.61a, 5.63a.

[4] Finding and drawing all stereoisomers for a compound with two stereogenic centers (5.7, 5.8)

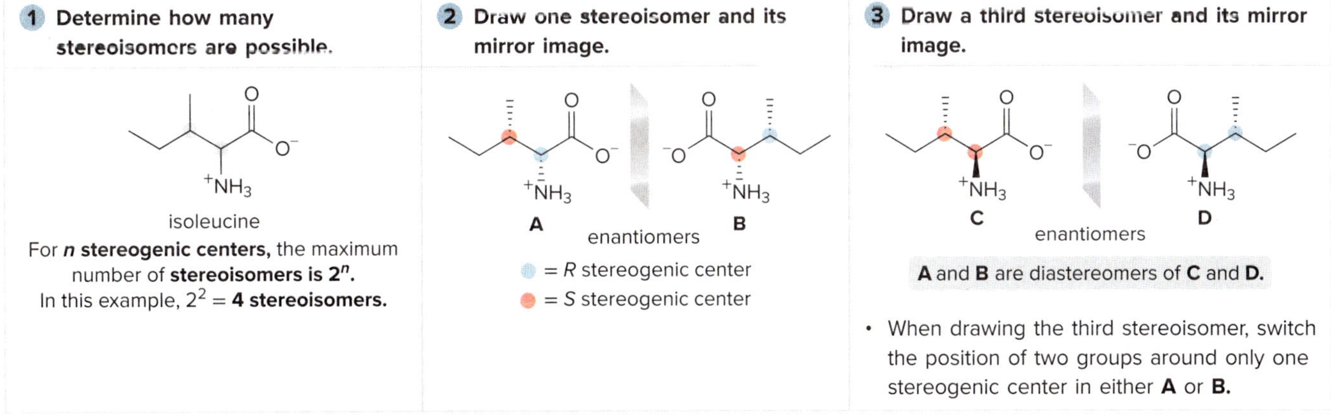

See *How To* p. 196, Figures 5.8, 5.9. Try Problems 5.52, 5.53.

[5] Determining if two nonidentical compounds are constitutional isomers, enantiomers, or diastereomers (5.11); example: menthol and isomers

1 Assess the connectivity of atoms, and assign the *R* or *S* configuration to each stereogenic center.

- Menthol and isomers **A** and **B** are **stereoisomers** because they have the same connectivity of atoms, but differ only in the spatial orientation of groups.

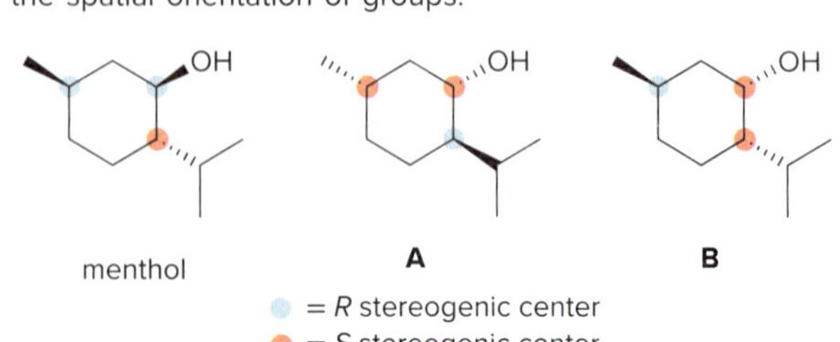

menthol A B

= *R* stereogenic center
= *S* stereogenic center

2 Use configurations to determine whether compounds are enantiomers or diastereomers.

- Menthol and isomer **A** are **enantiomers** because they have exactly ***opposite R,S*** designations at all stereogenic centers.
- Menthol and isomer **B** are **diastereomers** because they have the ***same R,S*** designations for two stereogenic centers and the ***opposite R,S*** designation for one stereogenic center.

See Figure 5.10. Try Problems 5.54–5.57, 5.59a.

[6] Calculations involving enantiomeric excess (ee) (5.12D)

1 Determine the % of each enantiomer given the ee.

97% *ee* of enantiomer **A** (97% excess **A** over the racemic mixture)

3% racemic mixture of **A + B** (1.5% **A** + 1.5% **B**)

***ee* = % of one enantiomer −% of the other enantiomer**

- Total amount of **A** = 97% + 1.5% = **98.5%**
- Total amount of **B** = 100% − 98.5% = **1.5%**

2 Determine *ee* given the observed rotation of a mixture.

OH H N

(1*S*,2*S*)-pseudoephedrine
[α] pure = +51

[α] of **mixture** of enantiomers = **+20**

$$ee = \frac{[\alpha]\ \text{mixture}}{[\alpha]\ \text{pure enantiomer}} \times 100\%$$

$$= \frac{+20}{+51} \times 100\%$$

= **39% *ee*** of (1*S*,2*S*)-pseudoephedrine

See Sample Problems 5.5, 5.6. Try Problems 5.60; 5.62e, f.

CHAPTER 5 MULTIPLE-CHOICE SELF-TEST

The Self-Test consists of multiple-choice questions similar to those found on the American Chemical Society organic chemistry exam. Answers are given at the end of the chapter.

1. Label each stereogenic center as *R* or *S*.

OH O 2 3 H OH

a. C2 *R*, C3 *R*
b. C2 *R*, C3 *S*
c. C2 *S*, C3 *R*
d. C2 *S*, C3 *S*

2. Rank the following four groups in order of decreasing priority around a stereogenic center: $-Cl$, $-CH_2Br$, $-C{\equiv}CH$, $-CH_2OH$.

a. $-Cl > -CH_2Br > -CH_2OH > -C{\equiv}CH$
b. $-CH_2Br > -Cl > -C{\equiv}CH > -CH_2OH$
c. $-CH_2Br > -Cl > -CH_2OH > -C{\equiv}CH$
d. $-CH_2Br > -CH_2OH > -C{\equiv}CH > -Cl$

3. How are compounds **A** and **B** related to each other?

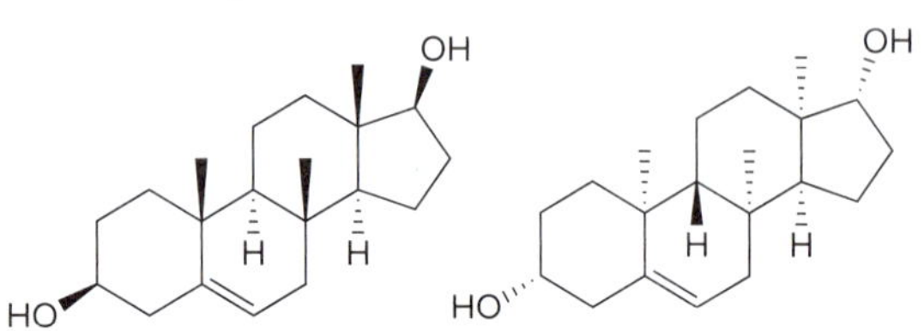

a. enantiomers
b. diastereomers
c. constitutional isomers
d. identical

4. Which compound is optically active?

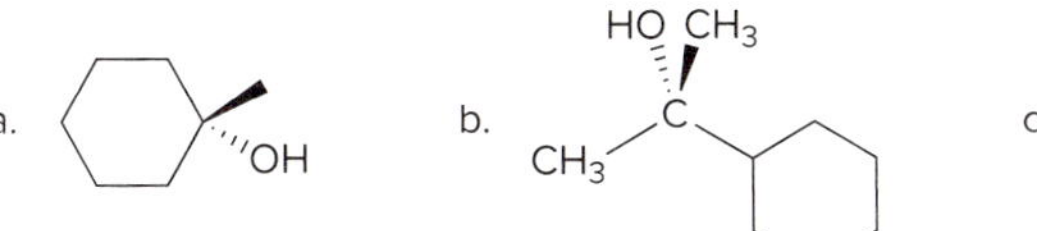

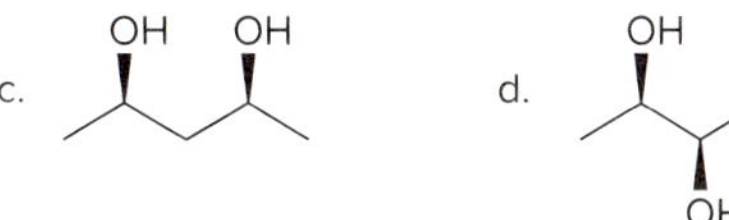

5. The specific rotation of pure **A** is +25. What is [α] of a mixture that contains 30% **A** and 70% of its enantiomer **B:** (a) −10; (b) +10; (c) +7.5; (d) −17.5?

6. If an equal mixture of **A–D** was distilled using an efficient fractional distillation, how many total fractions and how many optically active fractions would be obtained?

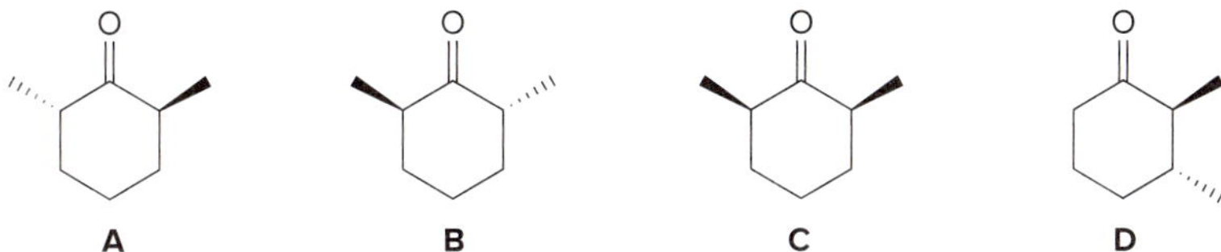

a. four total, three optically active

b. three total, three optically active

c. three total, one optically active

d. three total, two optically active

7. How many stereogenic centers are present in ingenol?

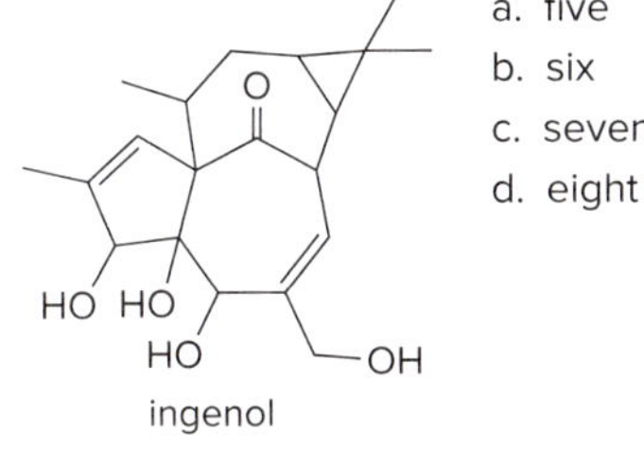

a. five

b. six

c. seven

d. eight

8. What is the maximum number of possible stereoisomers for **E?**

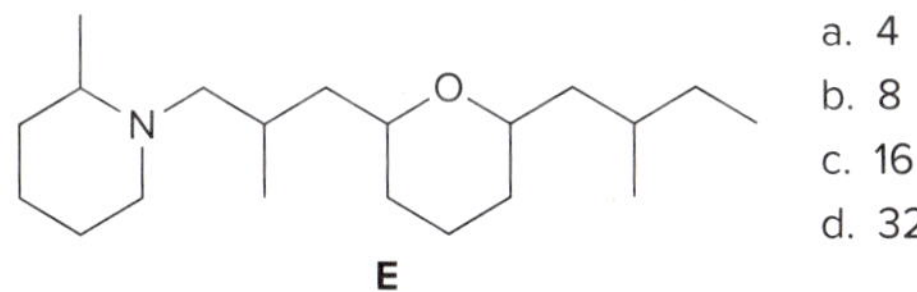

a. 4

b. 8

c. 16

d. 32

9. Which compound is an enantiomer of **F?**

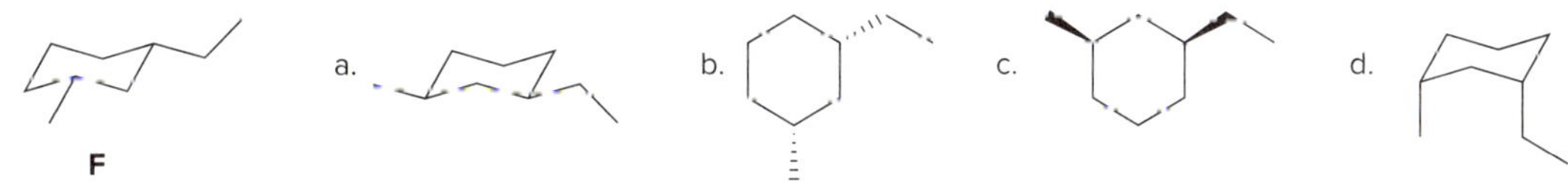

10. Which of the following statements about **A, B,** and **C** is *not* true?

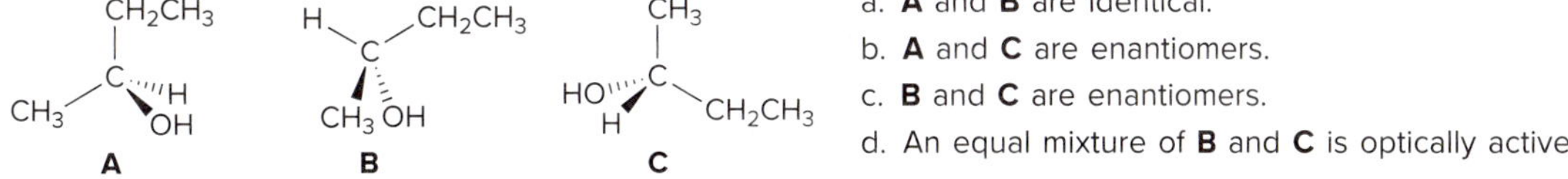

a. **A** and **B** are identical.

b. **A** and **C** are enantiomers.

c. **B** and **C** are enantiomers.

d. An equal mixture of **B** and **C** is optically active.

PROBLEMS

Problem Using Three-Dimensional Models

5.37 Consider the ball-and-stick models **A–D.** How is each pair of compounds related: (a) **A** and **B;** (b) **A** and **C;** (c) **A** and **D;** (d) **C** and **D?** Choose from identical molecules, enantiomers, or diastereomers.

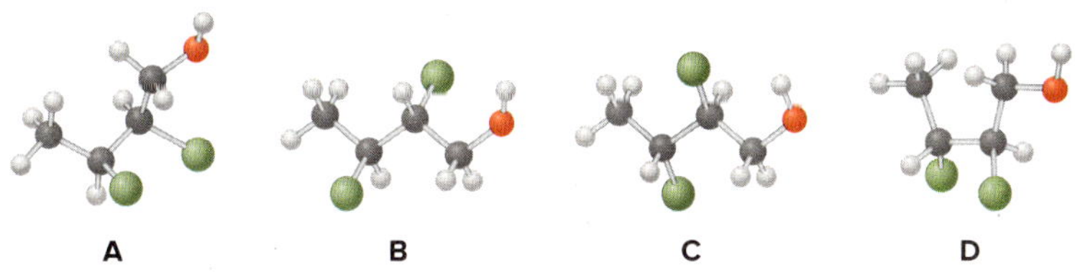

Constitutional Isomers Versus Stereoisomers

5.38 Label each pair of compounds as constitutional isomers, stereoisomers, or not isomers of each other.

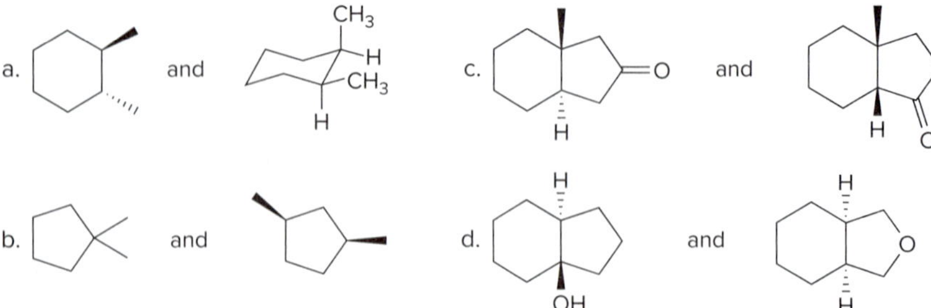

Mirror Images and Chirality

5.39 Label each compound as chiral or achiral.

a. 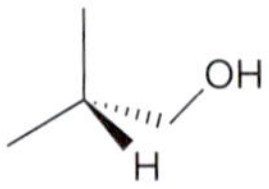 b. 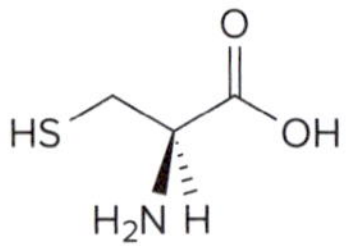 c. 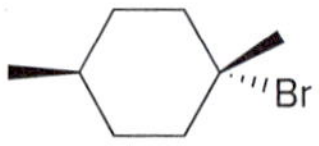 d.

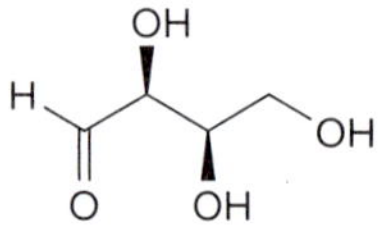

5.40 Indicate a plane of symmetry for each molecule that contains one. A molecule may require rotation around a carbon–carbon bond to see the plane of symmetry.

a. 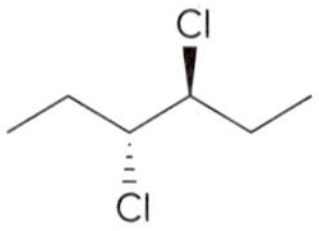 b. 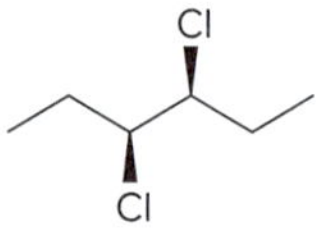 c.  d.

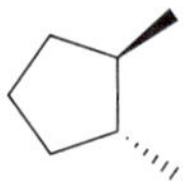

Finding and Drawing Stereogenic Centers

5.41 Locate the tetrahedral stereogenic center(s) in each compound. A molecule may have one or more stereogenic centers.

a. 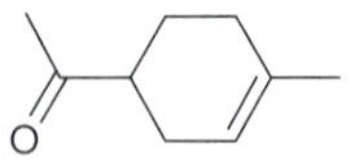 c. 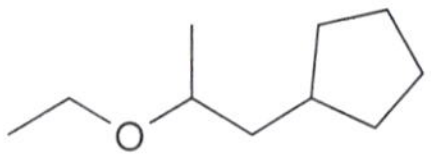 e. 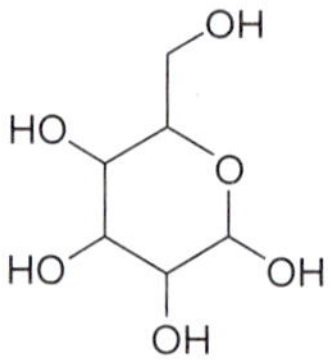 g.

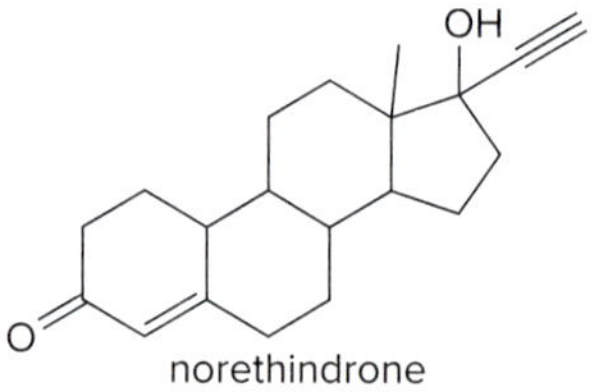

norethindrone
(oral contraceptive component)

b. 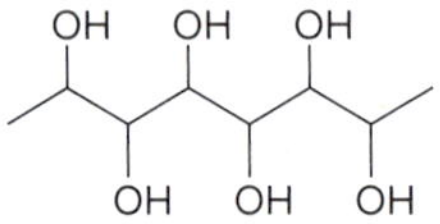 d. 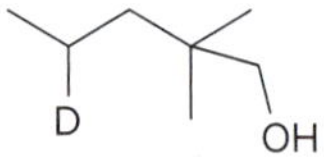 f. 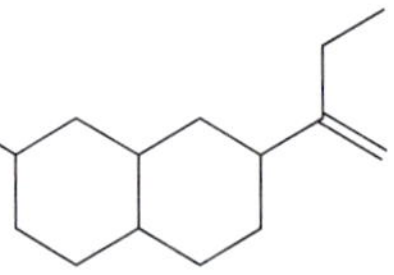 h.

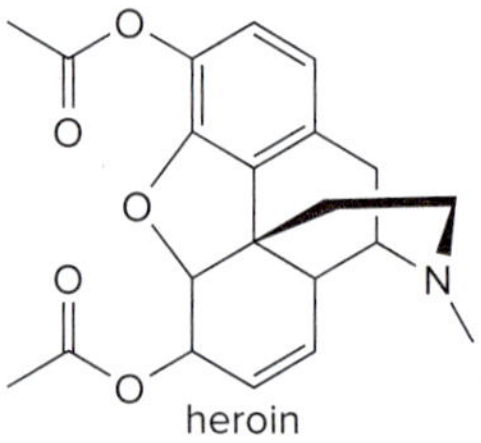

heroin
(an opiate)

5.42 Draw both enantiomers for each biologically active compound.

a.

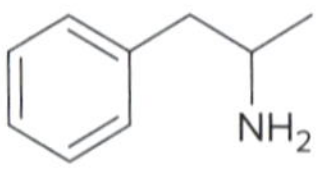

amphetamine
(a powerful central nervous stimulant)

b.

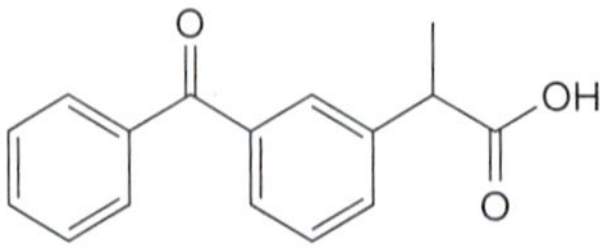

ketoprofen
(analgesic and anti-inflammatory agent)

Nomenclature

5.43 Which group in each pair is assigned the higher priority in *R,S* nomenclature?

a. $-CD_3$, $-CH_3$

b. $-CH(CH_3)_2$, $-CH_2OH$

c. $-CH_2Cl$, $-CH_2CH_2CH_2Br$

d. $-CH_2NH_2$, $-NHCH_3$

5.44 Rank the following groups in order of decreasing priority.

a. $-F$, $-NH_2$, $-CH_3$, $-OH$

b. $-CH_3$, $-CH_2CH_3$, $-CH_2CH_2CH_3$, $-(CH_2)_3CH_3$

c. $-NH_2$, $-CH_2NH_2$, $-CH_3$, $-CH_2NHCH_3$

d. $-COOH$, $-CH_2OH$, $-H$, $-CHO$

e. $-Cl$, $-CH_3$, $-SH$, $-OH$

f. $-C{\equiv}CH$, $-CH(CH_3)_2$, $-CH_2CH_3$, $-CH{=}CH_2$

5.45 Label each stereogenic center as *R* or *S*.

a. 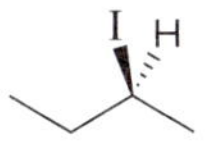 b. 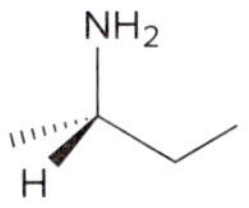 c. d. 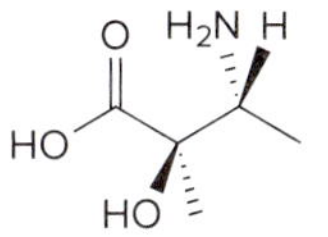 e. f. 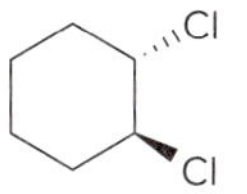 g. h.

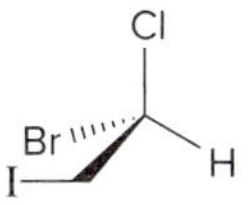

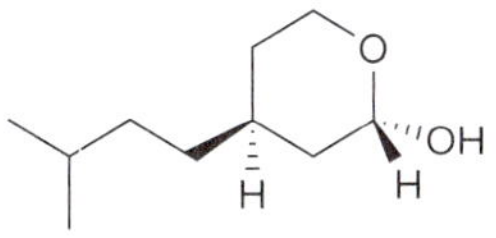

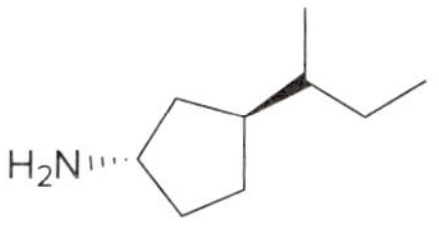

5.46 Locate the stereogenic centers in each Newman projection and label each center as *R* or *S*.

a. 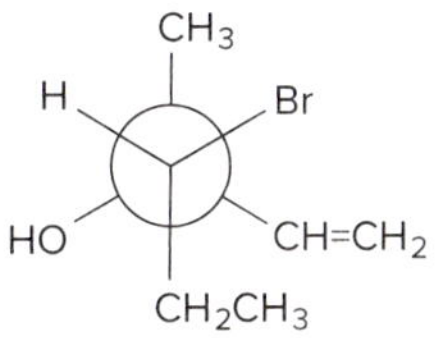 b.

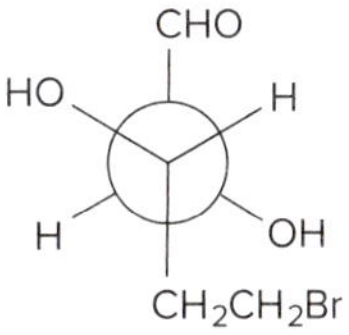

5.47 Draw the structure of (*S*,*S*)-ethambutol, a drug used to treat tuberculosis that is 10 times more potent than any of its other stereoisomers.

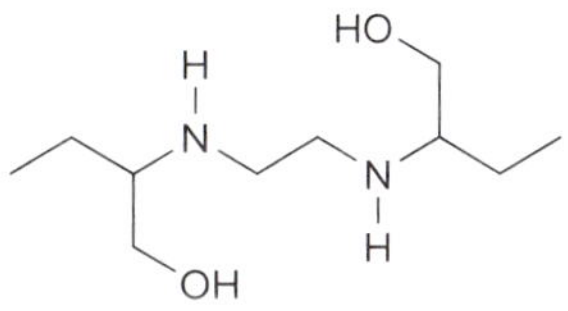

ethambutol

5.48 Draw the structure for each compound.

a. (*R*)-3-methylhexane

b. (4*R*,5*S*)-4,5-diethyloctane

c. (3*R*,5*S*,6*R*)-5-ethyl-3,6-dimethylnonane

d. (3*S*,6*S*)-6-isopropyl-3-methyldecane

5.49 Give the IUPAC name for each compound, including the *R*,*S* designation for each stereogenic center.

a. 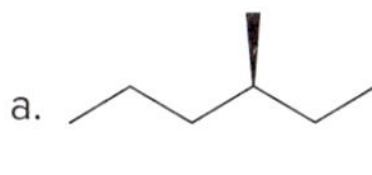

b. 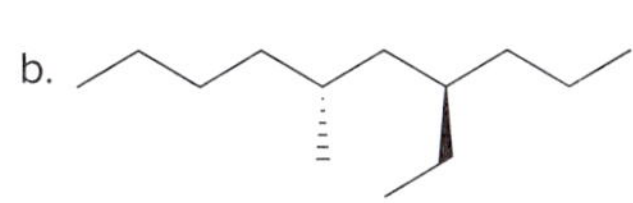

c.

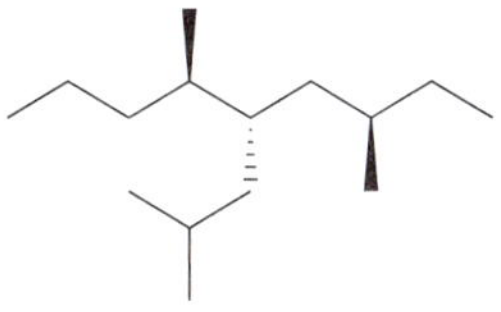

5.50 Ubrogepant (trade name Ubrelvy) is a drug approved by the Food and Drug Administration in 2019 for the treatment of acute migraines. Using wedges and dashed wedges and the given *R*,*S* designations, draw the stereochemistry at each stereogenic center of ubrogepant.

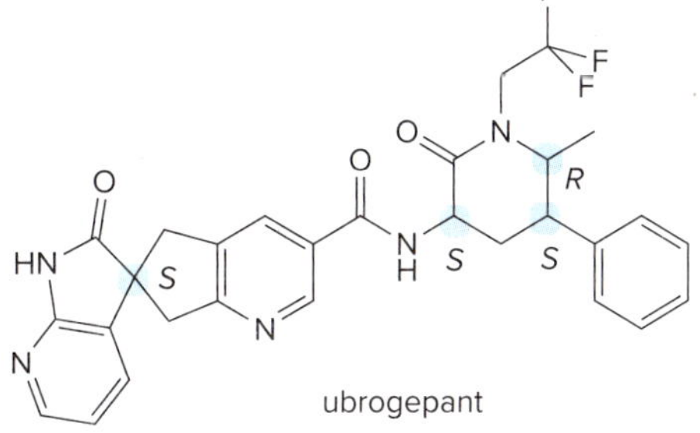

ubrogepant

Compounds with More Than One Stereogenic Center

5.51 The shrub ma huang (Section 5.4A) contains two biologically active stereoisomers—ephedrine and pseudoephedrine—with two stereogenic centers as shown in the given structure. Ephedrine is one component of a once-popular combination drug used by body builders to increase energy and alertness, whereas pseudoephedrine is a nasal decongestant.

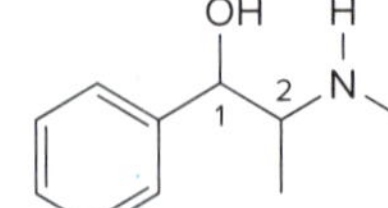

isolated from ma huang

a. Draw the structure of naturally occurring (–)-ephedrine, which has the 1*R*,2*S* configuration.

b. Draw the structure of naturally occurring (+)-pseudoephedrine, which has the 1*S*,2*S* configuration.

c. How are ephedrine and pseudoephedrine related?

d. Draw all other stereoisomers of (–)-ephedrine and (+)-pseudoephedrine, and give the *R*,*S* designation for all stereogenic centers.

e. How is each compound drawn in part (d) related to (–)-ephedrine?

5.52 Draw all possible stereoisomers for each compound. Label pairs of enantiomers and diastereomers. Label any meso compound.

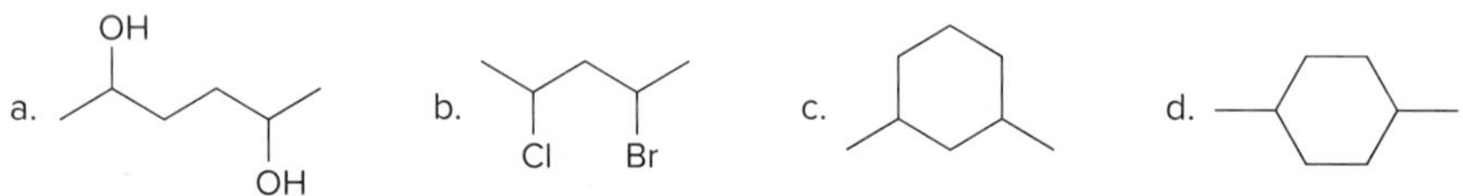

5.53 Draw all possible constitutional isomers and stereoisomers for a compound of molecular formula C_6H_{12} having a cyclobutane ring and two methyl groups as substituents. Label each compound as chiral or achiral.

Comparing Compounds: Enantiomers, Diastereomers, and Constitutional Isomers

5.54 Consider Newman projections (**A–D**) for four-carbon carbohydrates. How is each pair of compounds related: (a) **A** and **B;** (b) **A** and **C;** (c) **A** and **D;** (d) **C** and **D?** Choose from identical molecules, enantiomers, or diastereomers.

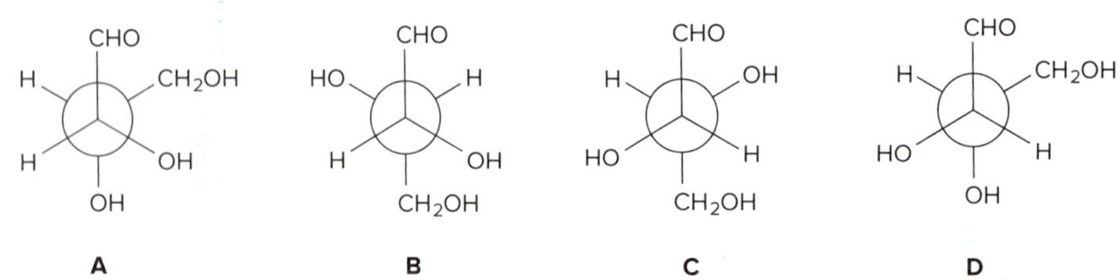

5.55 How is compound **A** related to compounds **B–E?** Choose from enantiomers, diastereomers, constitutional isomers, or identical molecules.

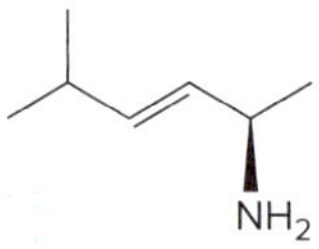

NH_2 NH_2 NH_2 NH_2 NH_2

A 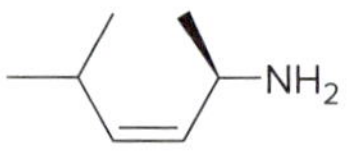 **B** 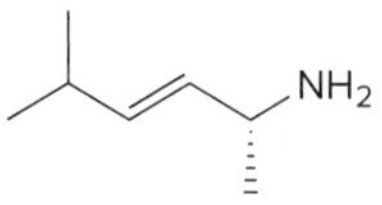 **C** 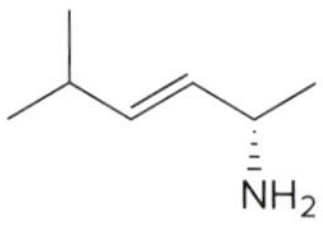 **D** 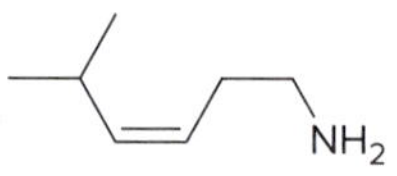 **E**

5.56 How is each compound (**B–D**) related to **A?** Choose from enantiomers, diastereomers, identical molecules, constitutional isomers, or not isomers of each other.

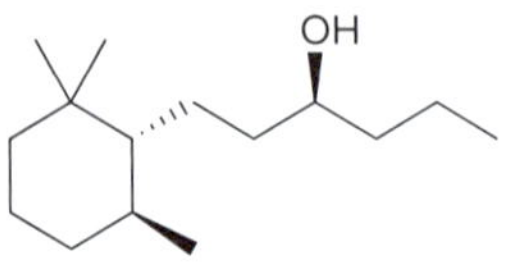

OH OH OH OH

A 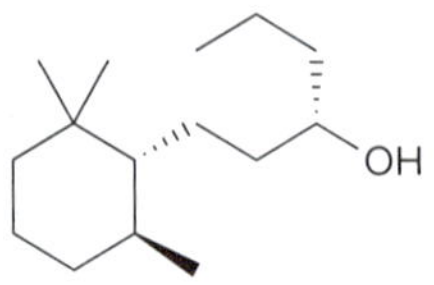 **B**  **C** 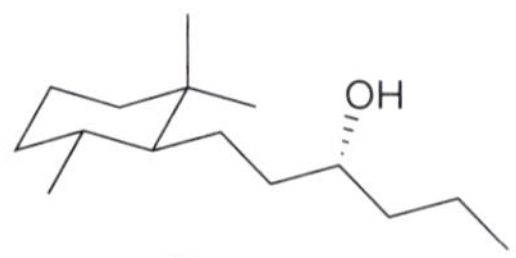 **D**

5.57 How are the compounds in each pair related to each other? Are they identical, enantiomers, diastereomers, constitutional isomers, or not isomers of each other?

a. and

b. and

c. and

d. and

e. and

f. and

Physical Properties of Isomers

5.58 A mixture contains equal amounts of compounds **A–D.**

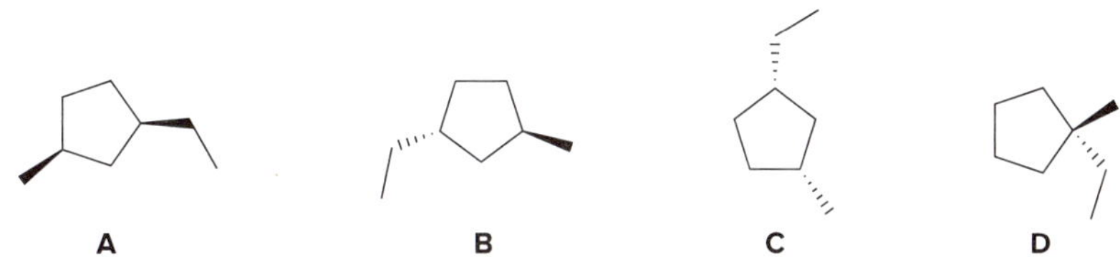

a. Which compounds alone are optically active?

b. If the mixture was subjected to fractional distillation, how many fractions would be obtained?

c. How many of these fractions would be optically active?

5.59 Drawn are four isomeric dimethylcyclopropanes.

a. How are the compounds in each pair related (enantiomers, diastereomers, constitutional isomers): **A** and **B; A** and **C; B** and **C; C** and **D?**

b. Label each compound as chiral or achiral.

c. Which compounds alone would be optically active?

d. Which compounds have a plane of symmetry?

e. How do the boiling points of the compounds in each pair compare: **A** and **B; B** and **C; C** and **D?**

f. Which of the compounds are meso compounds?

g. Would an equal mixture of compounds **C** and **D** be optically active? What about an equal mixture of **B** and **C?**

5.60 The [α] of pure quinine, an antimalarial drug, is −165.

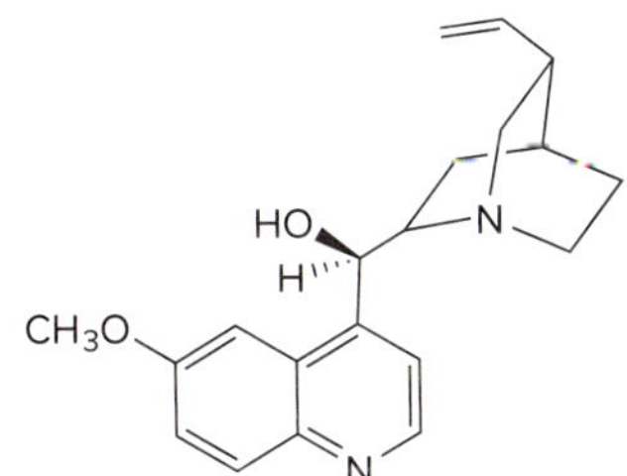

quinine
(antimalarial drug)

a. Calculate the *ee* of a solution with the following [α] values: −50, −83, and −120.

b. For each *ee*, calculate the percent of each enantiomer present.

c. What is [α] for the enantiomer of quinine?

d. If a solution contains 80% quinine and 20% of its enantiomer, what is the *ee* of the solution?

e. What is [α] for the solution described in part (d)?

Problems That Combine Concepts

5.61 Captopril is a drug used to treat high blood pressure and congestive heart failure.

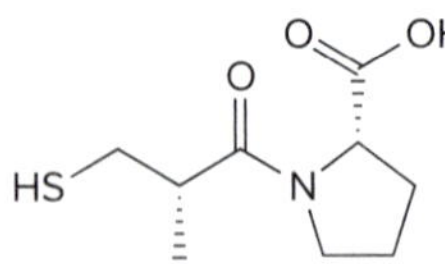

captopril

a. Designate each stereogenic center as *R* or *S*.
b. Draw the enantiomer of captopril.
c. What product is formed when captopril is treated with one equivalent of NaH?
d. What product is formed when captopril is treated with two equivalents of NaH?

5.62 Trabectedin is an anticancer drug sold under the trade name Yondelis.

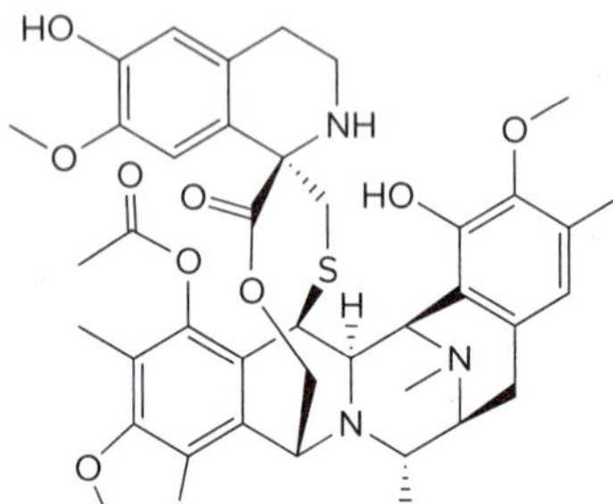

trabectedin

a. Locate the stereogenic centers in trabectedin.
b. What is the maximum number of stereoisomers possible for trabectedin?
c. Draw the enantiomer.
d. Draw a diastereomer.
e. If the specific rotation of trabectedin is +41.5, what is the [α] of a solution that contains 75% trabectedin and 25% of its enantiomer?
f. What is the *ee* of a solution with [α] = +10.5?

5.63 Saquinavir (trade name Invirase) is a protease inhibitor, used to treat HIV (human immunodeficiency virus).

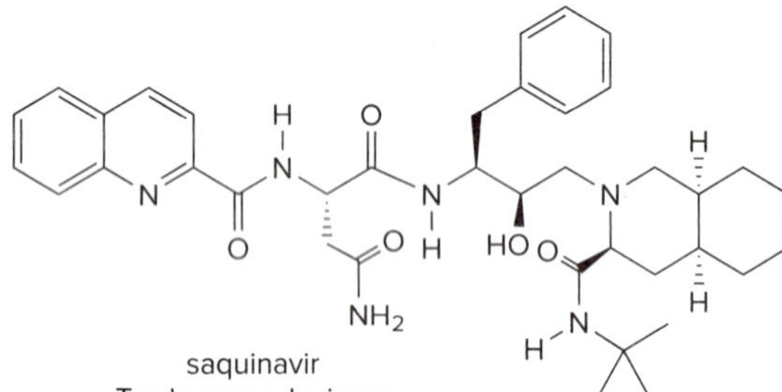

saquinavir
Trade name: Invirase

a. Locate all stereogenic centers in saquinavir, and label each stereogenic center as *R* or *S*.
b. Draw the enantiomer of saquinavir.
c. Draw a diastereomer of saquinavir.
d. Draw a constitutional isomer that contains at least one different functional group.

Challenge Problems

5.64 A limited number of chiral compounds having no stereogenic centers exist. For example, although **A** is achiral, constitutional isomer **B** is chiral. Make models and explain this observation. Compounds containing two double bonds that share a single carbon atom are called *allenes*. Locate the allene in the antibiotic mycomycin and decide whether mycomycin is chiral or achiral.

$(CH_3)_2C{=}C{=}CH_2$

achiral
A

$CH_3(H)C{=}C{=}C(CH_3)H$

chiral
B

$HC{\equiv}C{-}C{\equiv}C{-}CH{=}C{=}CH{-}CH{=}CH{-}CH{=}CH{-}CH_2CO_2H$

mycomycin

5.65 A preliminary study has demonstrated that the antifungal drug amphotericin B may help to restore lung function in individuals with cystic fibrosis, a genetic disease in which the normal pH and viscosity of the lining of lung airways is altered.

amphotericin B

a. Locate all the tetrahedral stereogenic centers in amphotericin B.
b. Certain carbon–carbon double bonds can also be stereogenic centers. With reference to the definition in Section 5.3, explain how this can occur and then locate additional stereogenic centers in amphotericin B.
c. Considering all stereogenic centers, what is the maximum number of stereoisomers possible for amphotericin B?

5.66 Label each compound as chiral or achiral. Compounds that contain a single carbon common to two rings are called spiro compounds. Because carbon is tetrahedral, the two rings are perpendicular to each other.

5.67 An acid–base reaction of (*R*)-*sec*-butylamine with a racemic mixture of 2-phenylpropanoic acid forms two products having different melting points and somewhat different solubilities. Draw the structure of these two products. Assign *R* and *S* to any stereogenic centers in the products. How are the two products related? Choose from enantiomers, diastereomers, constitutional isomers, or not isomers of each other.

2-phenylpropanoic acid (racemic mixture) + (*R*)-*sec*-butylamine ⟶

SELF-TEST ANSWERS

1. c 2. a 3. b 4. d 5. a 6. c 7. d 8. d 9. a 10. d

6 Understanding Organic Reactions

6.1 Writing equations for organic reactions
6.2 Kinds of organic reactions
6.3 Bond breaking and bond making
6.4 Bond dissociation energy
6.5 Thermodynamics
6.6 Enthalpy and entropy
6.7 Energy diagrams
6.8 Energy diagram for a two-step reaction mechanism
6.9 Kinetics
6.10 Catalysts
6.11 Enzymes

wasanajai/Shutterstock

The seeds of *Jatropha curcas,* a shrub that readily grows in a variety of soil types in tropical and subtropical regions, are rich in oils derived from linoleic acid ($C_{18}H_{32}O_2$) and other long-chain carboxylic acids. Like other oils, the combustion of jatropha oil produces CO_2 and H_2O and a great deal of energy, making this oil attractive for use in biodiesel and aviation jet fuels. Fuel yield from jatropha plants far exceeds that produced from soybeans and corn, but the long-term effects of extensive jatropha cultivation are as yet unknown. In Chapter 6, we will learn about energy changes that accompany chemical reactions like combustion.

Why Study . . . Organic Reactions?

Reactions are at the heart of organic chemistry. An understanding of chemical processes has made possible the conversion of natural substances into new compounds with different, and sometimes superior, properties. Aspirin, ibuprofen, nylon, and polyethylene are all products of chemical reactions between substances derived from petroleum.

Reactions are difficult to learn when each reaction is considered a unique and isolated event. *Avoid this tendency.* **Virtually all chemical reactions are woven together by a few basic themes.** After we learn the general principles, specific reactions then fit neatly into a general pattern.

In our study of organic reactions we will begin with the functional groups, looking for electron-rich and electron-deficient sites, and bonds that might be broken easily. These reactive sites give us a clue as to the general type of reaction a particular class of compound undergoes. Finally, we will learn about how a reaction occurs. Does it occur in one step or in a series of steps? Understanding the details of an organic reaction allows us to determine when it might be used in preparing interesting and useful organic compounds.

6.1 Writing Equations for Organic Reactions

Often the solvent and temperature of a reaction are omitted from chemical equations, to further focus attention on the main substances involved in the reaction.

Most organic reactions take place in a **liquid solvent.** Solvents solubilize key reaction components and serve as heat reservoirs to maintain a given temperature. Chapter 7 presents the two major types of reaction solvents and how they affect substitution reactions.

Like other reactions, equations for organic reactions are usually drawn with a single reaction arrow (→) between the starting material and product, but other conventions make these equations look different from those encountered in general chemistry.

The **reagent,** the chemical substance with which an organic compound reacts, is sometimes drawn on the left side of the equation with the other reactants. At other times, the reagent is drawn above or below the reaction arrow itself, to focus attention on the organic starting material by itself on the left side. The solvent and temperature of a reaction may be added above or below the arrow. **The symbols "*hν*" and "Δ" are used for reactions that require *light* or *heat,* respectively.** Figure 6.1 presents an organic reaction in different ways.

When two sequential reactions are carried out without drawing any intermediate compound, the steps are usually numbered above or below the reaction arrow. This convention signifies that the first step occurs *before* the second, and the reagents are added *in sequence,* not at the same time.

The first reaction...

[1] CH_3MgBr

[2] H_2O

...then the second

OH

(HOMgBr)

inorganic by-product (often omitted)

In this equation only the organic product is drawn on the right side of the arrow. Although the reagent CH_3MgBr contains both Mg and Br, these elements do not appear in the organic product, and they are often omitted on the product side of the equation. These elements have not disappeared. They are part of an inorganic by-product (HOMgBr in this case), and are often of little interest to an organic chemist.

Figure 6.1
Different ways of writing organic reactions

a. + Br_2 → Br Br

Br_2 → Br Br

• The reagent (Br_2) can be on the left side or above the arrow.

b. Br_2 / *hν* or Δ / CCl_4 → Br

CCl_4 is the solvent.

hν—Indicates **light** is needed.

Δ—Indicates **heat** is added.

• Other reaction parameters can be indicated.

6.2 Kinds of Organic Reactions

Like other compounds, organic molecules undergo acid–base and oxidation–reduction reactions, as discussed in Chapters 2 and 4. Organic molecules also undergo **substitution, elimination,** and **addition** reactions.

6.2A Substitution Reactions

- ***Substitution*** **is a reaction in which an atom or a group of atoms is *replaced* by another atom or group of atoms.**

Z = H or a heteroatom

In a general substitution reaction, Y *replaces* Z on a carbon atom. **Substitution reactions involve σ bonds: one σ bond breaks and another forms at the same carbon atom.** The most common examples of substitution occur when Z is hydrogen or a heteroatom that is more electronegative than carbon.

O, Cl + $^{-}$OH → O, OH + Cl^{-}

6.2B Elimination Reactions

- ***Elimination*** **is a reaction in which elements of the starting material are "lost" and a π bond is formed.**

Two σ bonds are broken. A π bond is formed.

In an elimination reaction, two groups X and Y are removed from a starting material. **Two σ bonds are broken, and a π bond is formed between adjacent atoms.** The most common examples of elimination occur when X = H and Y is a heteroatom more electronegative than carbon.

HO, H — H_2SO_4 → new π bond + H–OH

loss of H_2O

6.2C Addition Reactions

- ***Addition*** **is a reaction in which elements are added to a starting material.**

+ X–Y → X, Y

A π bond is broken.

Two σ bonds are formed.

In an addition reaction, new groups X and Y are added to a starting material. **A π bond is broken and two σ bonds are formed.**

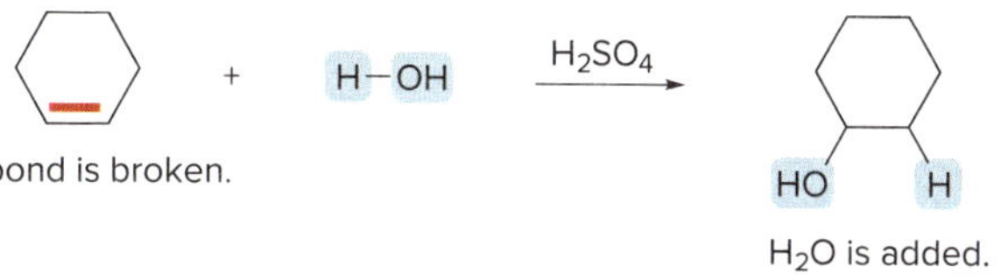

A π bond is broken. + H–OH $\xrightarrow{H_2SO_4}$ HO H

H_2O is added.

A summary of the general types of organic reactions is given in Appendix I.

Addition and elimination reactions are exactly opposite. A π bond is *formed* in elimination reactions, whereas a π bond is *broken* in addition reactions.

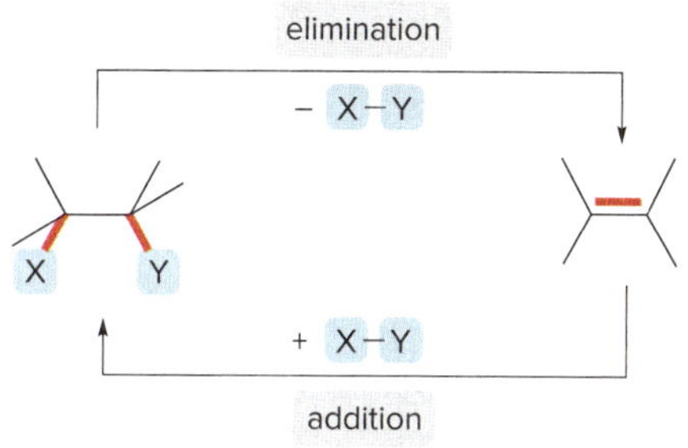

elimination

– X–Y

X Y

+ X–Y

addition

Problem 6.1 Classify each transformation as substitution, elimination, or addition.

a. OH → Br

b. O → OH

c. O OH → O

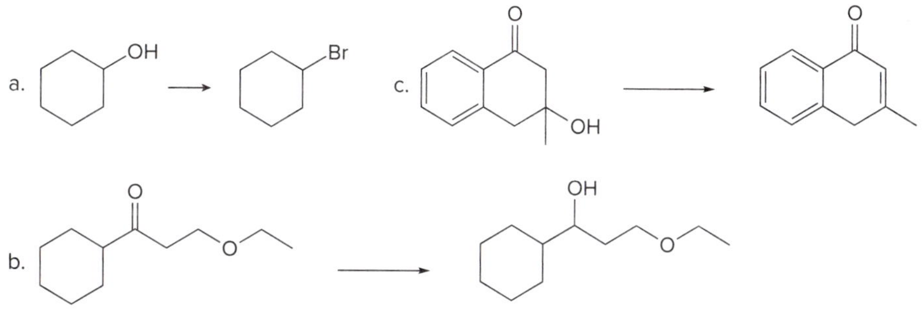

To determine whether a reaction is a substitution, elimination, or addition with a complex starting material, **concentrate on the functional groups that *change*.** The conversion of amine **X** and acid chloride **Y** to the naturally occurring compound capsaicin is a **substitution** reaction, because the N atom of the amine *replaces* the Cl of the acid chloride.

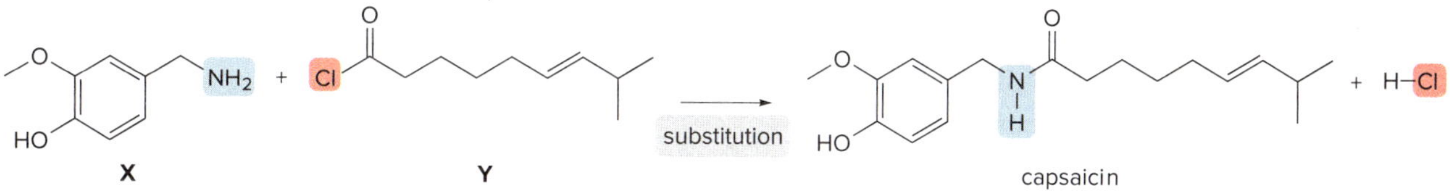

X + Y → substitution → capsaicin + H–Cl

Capsaicin is responsible for the characteristic spicy flavor of jalapeño and habañero peppers. *DNY59/Getty Images*

Problem 6.2 The following enzyme-catalyzed reactions illustrate the last three steps in the citric acid cycle, a critical part of metabolism discussed in Chapter 30 (online). Classify each reaction as a substitution, elimination, or addition.

succinate [1] → fumarate [2] → malate [3] → oxaloacetate

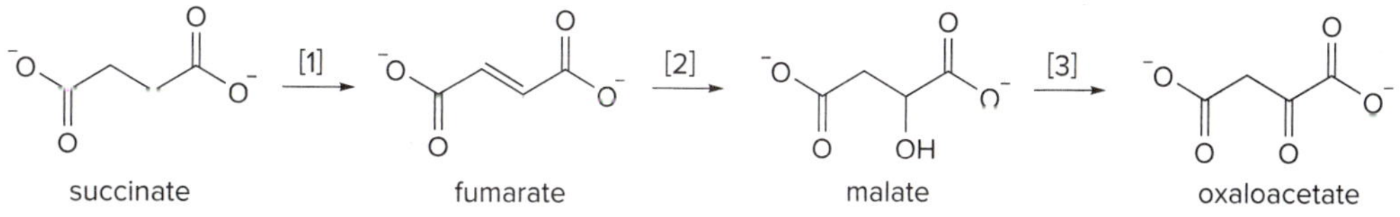

6.3 Bond Breaking and Bond Making

Having now learned how to write and identify some common kinds of organic reactions, we can turn to a discussion of **reaction mechanism.**

- A *reaction mechanism* is a detailed description of how bonds are broken and formed as a starting material is converted to a product.

A reaction mechanism describes the relative order and rate of bond cleavage and formation. It explains all the known facts about a reaction and accounts for all products formed, and it is subject to modification or refinement as new details are discovered.

A reaction can occur either in one step or in a series of steps.

- **A one-step reaction is called a *concerted reaction.*** No matter how many bonds are broken or formed, a starting material is converted *directly* to a product.

- **A stepwise reaction** involves more than one step. A starting material is first converted to an unstable intermediate, called a **reactive intermediate,** which then goes on to form the product.

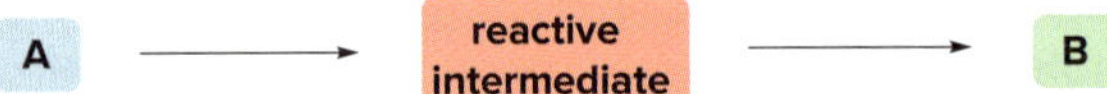

6.3A Bond Cleavage

Bonds are broken and formed in all chemical reactions. When a bond is broken, the electrons in the bond can be divided **equally** or **unequally** between the two atoms of the bond.

- Breaking a bond by ***equally dividing*** **the electrons between the two atoms in the bond is called homolysis or homolytic cleavage.**

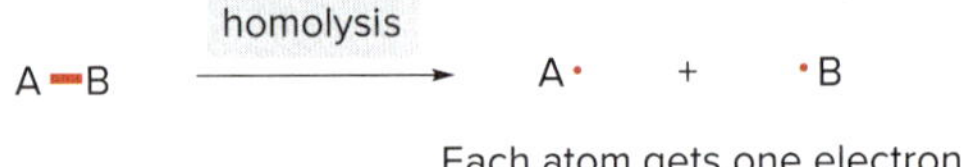

- Breaking a bond by ***unequally dividing*** **the electrons between the two atoms in the bond is called heterolysis or heterolytic cleavage.**

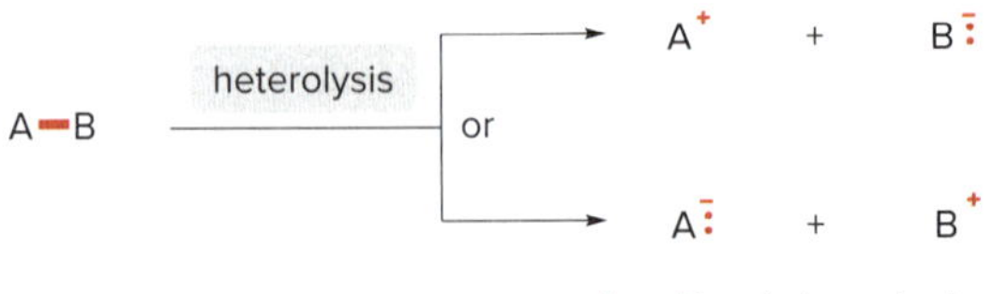

Heterolysis of a bond between **A** and **B** can give either **A** or **B** the two electrons in the bond. When **A** and **B** have different electronegativities, the ***electrons normally end up on the more electronegative atom.***

Homolysis and heterolysis require energy. Both processes generate reactive intermediates, but the products are different in each case.

- Homolysis generates uncharged reactive intermediates with *unpaired* electrons.
- Heterolysis generates *charged* intermediates.

Each of these reactive intermediates has a very short lifetime and reacts quickly to form a stable organic product.

6.3B Radicals, Carbocations, and Carbanions

The curved arrow notation first discussed in Section 1.6B works fine for heterolytic bond cleavage because it illustrates the movement of an **electron pair.** For homolytic cleavage, however, one electron moves to one atom in the bond and one electron moves to the other, so a different kind of curved arrow is needed.

- **To illustrate the movement of a single electron, use a half-headed curved arrow, sometimes called a *fishhook*.**

- Two **half-headed** curved arrows are needed for two **single** electrons.
- One **full-headed** curved arrows is needed for one electron **pair.**

A full-headed curved arrow ( ) shows the movement of an electron *pair.* **A half-headed curved arrow** ( ) shows the movement of a *single* electron.

Figure 6.2 illustrates homolysis and two different heterolysis reactions for a carbon compound using curved arrows. Three different reactive intermediates are formed.

Figure 6.2 Three reactive intermediates resulting from homolysis and heterolysis of a C–Z bond

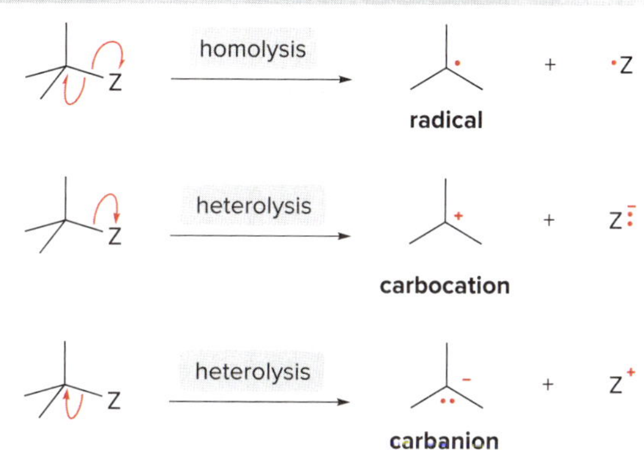

- Radicals are intermediates in **radical** reactions.
- **Ionic intermediates** are seen in **polar** reactions.

Homolysis of the C–Z bond generates two uncharged products with unpaired electrons.

- **A reactive intermediate with a single unpaired electron is called a *radical*.**

Most radicals are highly unstable because they contain an atom that does not have an octet of electrons. Radicals typically have **no charge. They are intermediates in a group of reactions called *radical reactions*,** which are discussed in detail in Chapter 13.

Heterolysis of the C–Z bond can generate a **carbocation** or a **carbanion.**

- **Giving two electrons to Z and none to carbon generates a positively charged carbon intermediate called a *carbocation*.**
- **Giving two electrons to C and none to Z generates a negatively charged carbon species called a *carbanion*.**

Both carbocations and carbanions are unstable reactive intermediates: A carbocation contains a carbon atom surrounded by only six electrons. A carbanion has a negative charge on carbon, which is not a very electronegative atom. **Carbocations (electrophiles)** and **carbanions (nucleophiles)** can be intermediates in ***polar reactions*—reactions in which a nucleophile reacts with an electrophile.**

The chemistry of **carbenes,** another type of organic reactive intermediate, is discussed in Section 24.5.

Thus, homolysis and heterolysis generate radicals, carbocations, and carbanions, the three most common reactive intermediates in organic chemistry.

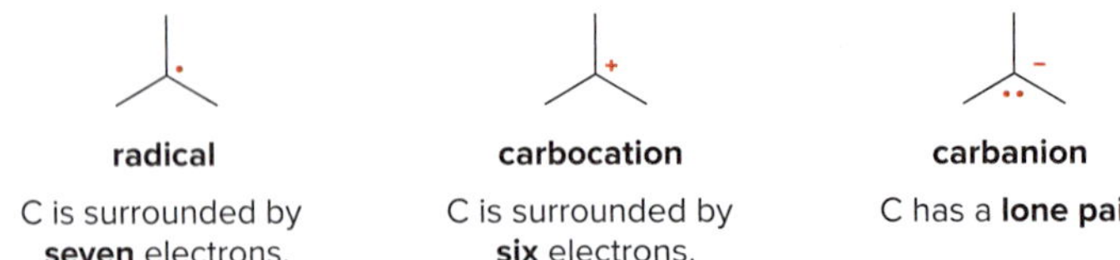

Problem 6.3 Draw the homolysis products of nonpolar bonds and heterolysis products of polar bonds indicated with an arrow. Use electronegativity differences to decide the identity of any charges. Classify the organic reactive intermediate as a carbocation, carbanion, or radical.

a. OH b. Li c.

6.3C Bond Formation

Like bond cleavage, bond formation occurs in two different ways. Two radicals can each donate **one electron** to form a two-electron bond. Alternatively, two ions with unlike charges can come together, with the negatively charged ion donating **both electrons** to form the resulting two-electron bond. **Bond formation always releases energy.**

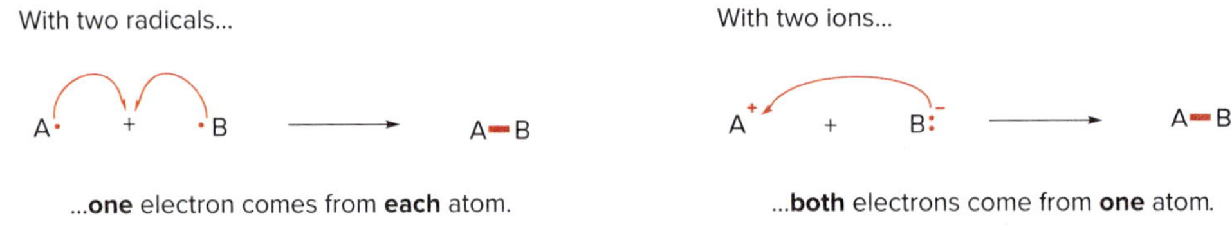

6.3D All Kinds of Arrows

Table 6.1 summarizes the many kinds of arrows used in describing organic reactions. Curved arrows are especially important because they explicitly show what electrons are involved in a reaction, how these electrons move in forming and breaking bonds, and if a reaction proceeds via a radical or polar pathway.

A more complete summary of the arrows used in organic chemistry is given in Appendix B, Common Abbreviations, Arrows, and Symbols.

Table 6.1 A Summary of Arrow Types in Chemical Reactions

Arrow	Name	Use
→	Reaction arrow	Drawn between the starting materials and products in an equation (6.1)
⇄	Double reaction arrows (equilibrium arrows)	Drawn between the starting materials and products in an equilibrium equation (2.2)
↔	Double-headed arrow	Drawn between resonance structures (1.6B)
	Full-headed curved arrow	Shows movement of an electron pair (1.6B, 2.2)
	Half-headed curved arrow (fishhook)	Shows movement of a single electron (6.3B)

Sample Problem 6.1 Using Curved Arrows in an Equation

Use curved arrows to show the movement of electron pairs in each reaction.

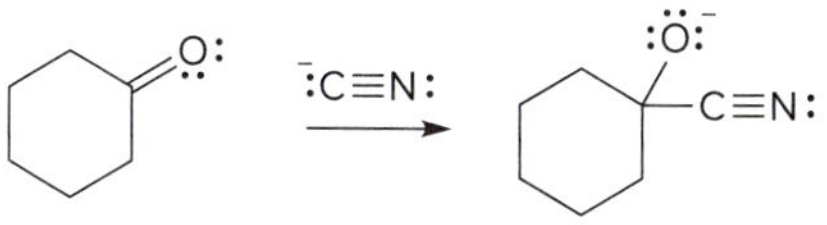

Solution

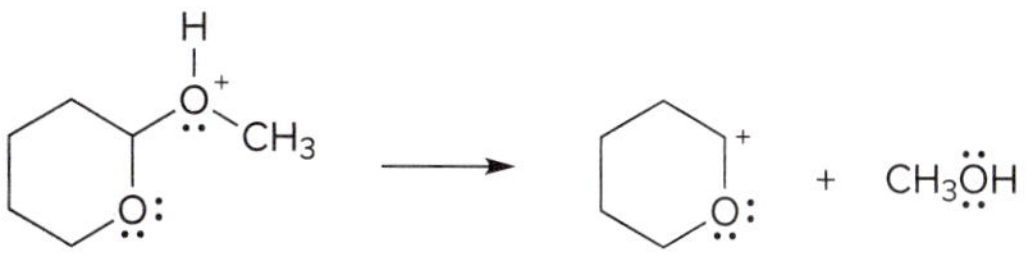

Concentrate on bonds that are broken or formed, and pay attention to atoms that have different charges in the reactants and products.

a. Only *one* C–O bond is broken, so only *one* curved arrow is needed. The electron pair in the C–O bond (in red) ends up on O.

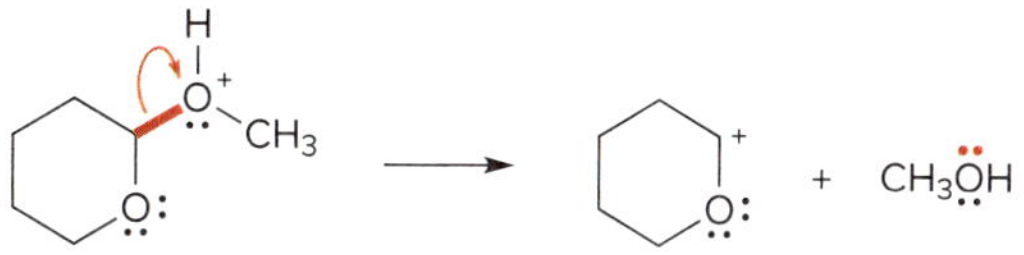

b. *Two* curved arrows are needed because *two* electron pairs are involved. The lone pair on C in ^{-}CN forms a new bond to the carbonyl carbon, and an electron pair in the C=O moves onto O.

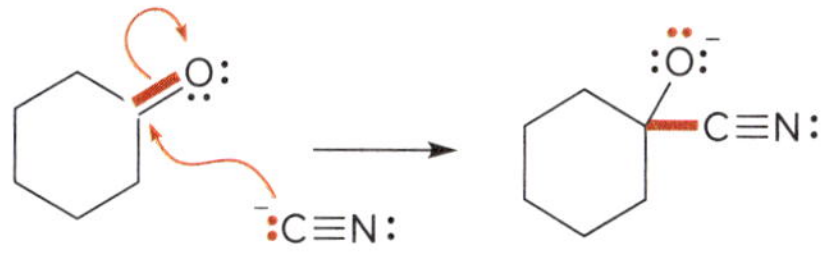

Problem 6.4 Use curved arrows to show the movement of electrons in each equation.

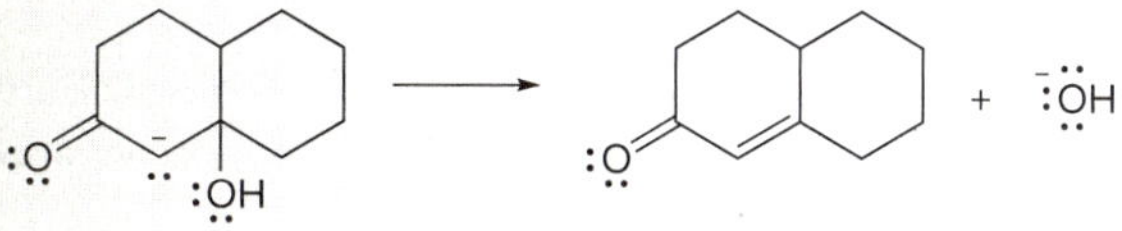

c. $CH_3CH_2\ddot{B}r:$ + $^{-}:\ddot{O}H$ → $CH_3CH_2\ddot{O}H$ + $:\ddot{B}r:^{-}$

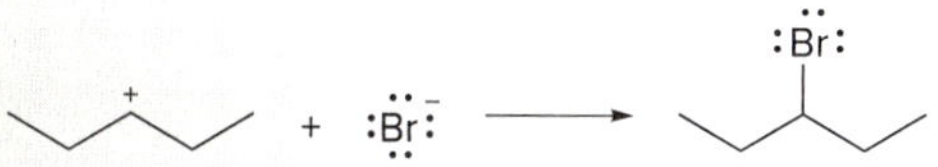

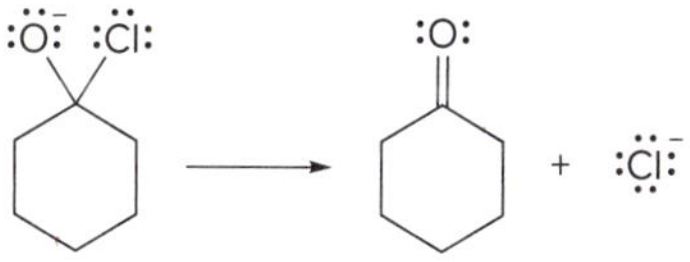

More Practice: Try Problems 6.28, 6.30a, 6.31a, 6.32, 6.33a, 6.42a, 6.47a, 6.49a, 6.50a, 6.53a.

Sample Problem 6.2 Following Curved Arrows to Draw a Reaction Product

Follow the curved arrows and draw the products of the following reaction.

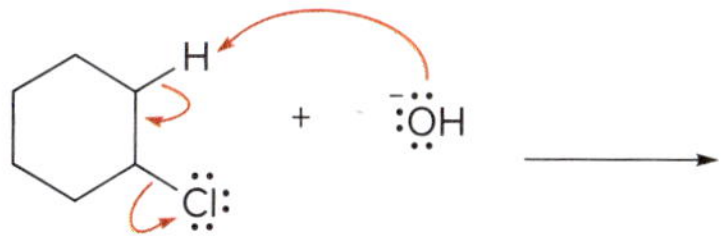

Solution

Three full-headed curved arrows are drawn, so three electron *pairs* take part in the reaction. Arrow **1** shows that a lone pair on ^{-}OH forms a new bond to H, forming H_2O. Arrow **2** indicates that the electron pair in the C–H bond forms a carbon–carbon double bond. Arrow **3** shows that the electron pair in the C–Cl bond ends up on Cl, forming Cl^{-}. After breaking and making bonds, formal charges on the atoms involved in the reaction are adjusted when necessary.

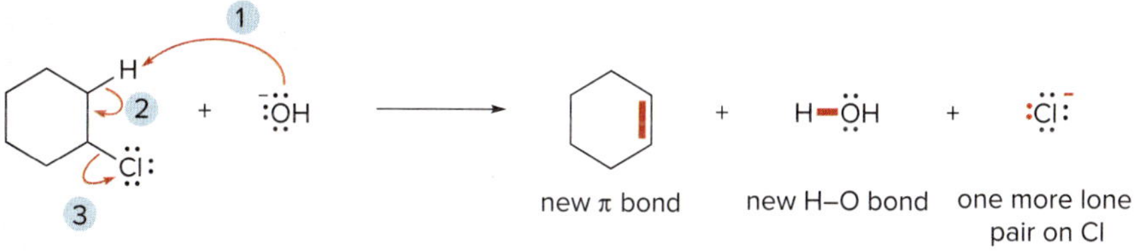

new π bond

new H–O bond

one more lone pair on Cl

Problem 6.5 Follow the curved arrows and draw the products of each reaction.

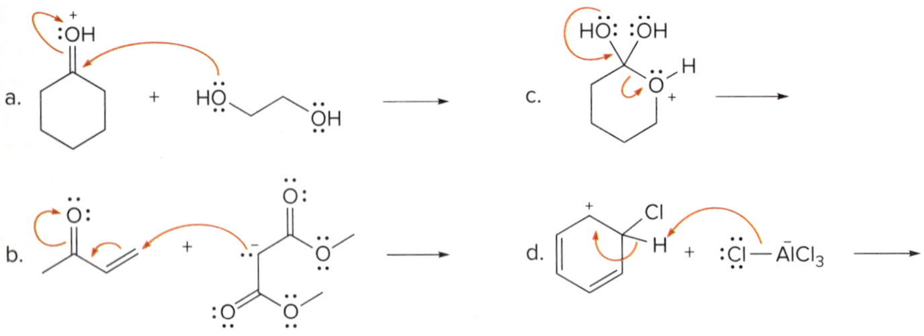

a. [cyclohexanone protonated, $:\overset{+}{O}H$] + $HO\!\!\diagup\!\!\diagdown OH$ ⟶

c. [$HO:$ $:OH$ on cyclic oxonium with $\overset{+}{O}$–H] ⟶

b. [enone] + [enolate diester] ⟶

d. [cyclohexadienyl cation with Cl and H] + $:\ddot{Cl}—\bar{A}lCl_3$ ⟶

More Practice: Try Problems 6.29, 6.31b, 6.33b, 6.53b, 6.55a.

6.4 Bond Dissociation Energy

Bond breaking can be quantified using the bond dissociation energy.

- **The *bond dissociation energy* is the energy needed to homolytically cleave a covalent bond.**

$$A—B \longrightarrow A\cdot + \cdot B \qquad \Delta H^\circ = \text{bond dissociation energy}$$

Homolysis requires energy.

The superscript (°) means that values are determined under standard conditions (pure compounds in their most stable state at 25 °C and 1 atm pressure).

The energy absorbed or released in any reaction, symbolized by **$\Delta H°$**, is called the **enthalpy change** or **heat of reaction.**

- **When $\Delta H°$ is positive (+), energy is absorbed and the reaction is *endothermic.***
- **When $\Delta H°$ is negative (–), energy is released and the reaction is *exothermic.***

Additional bond dissociation energies for C–C multiple bonds are given in Table 1.6.

A table of bond dissociation energies also appears in Appendix E.

A bond dissociation energy is the $\Delta H°$ for a specific kind of reaction—the homolysis of a covalent bond to form two radicals. Because bond breaking requires energy, **bond dissociation energies are always *positive* numbers,** and homolysis is always **endothermic.** Conversely, **bond formation always *releases* energy,** so this reaction is always **exothermic.** The H–H bond requires +435 kJ/mol to cleave and releases –435 kJ/mol when formed. Table 6.2 contains a representative list of bond dissociation energies for many common bonds.

$$H—H \xrightarrow[\text{endothermic reaction}]{\Delta H = +435 \text{ kJ/mol}} H\cdot + \cdot H$$

$$H\cdot + \cdot H \xrightarrow[\text{exothermic reaction}]{\Delta H = -435 \text{ kJ/mol}} H—H$$

Comparing bond dissociation energies is equivalent to comparing **bond strength.**

- **The *stronger* the bond, the *higher* its bond dissociation energy.**

For example, the H–H bond is stronger than the Cl–Cl bond because its bond dissociation energy is higher [Table 6.2: 435 kJ/mol (H_2) versus 242 kJ/mol (Cl_2)]. The data in Table 6.2 demonstrate that **bond dissociation energies *decrease* down a column of the periodic table**

Table 6.2 Bond Dissociation Energies for Some Common Bonds [A–B → A• + •B]

Bond	$\Delta H°$ kJ/mol	(kcal/mol)
H–Z bonds		
$H{-}F$	569	(136)
$H{-}Cl$	431	(103)
$H{-}Br$	368	(88)
$H{-}I$	297	(71)
$H{-}OH$	498	(119)
Z–Z bonds		
$H{-}H$	435	(104)
$F{-}F$	159	(38)
$Cl{-}Cl$	242	(58)
$Br{-}Br$	192	(46)
$I{-}I$	151	(36)
$HO{-}OH$	213	(51)
R–H bonds		
$CH_3{-}H$	435	(104)
$CH_3CH_2{-}H$	410	(98)
$CH_3CH_2CH_2{-}H$	410	(98)
$(CH_3)_2CH{-}H$	397	(95)
$(CH_3)_3C{-}H$	381	(91)
$CH_2{=}CH{-}H$	435	(104)
$HC{\equiv}C{-}H$	523	(125)
$CH_2{=}CHCH_2{-}H$	364	(87)
$C_6H_5{-}H$	460	(110)
$C_6H_5CH_2{-}H$	356	(85)
R–R bonds		
$CH_3{-}CH_3$	368	(88)
$CH_3{-}CH_2CH_3$	356	(85)
$CH_3{-}CH{=}CH_2$	385	(92)
$CH_3{-}C{\equiv}CH$	489	(117)
R–X bonds		
$CH_3{-}F$	456	(109)
$CH_3{-}Cl$	351	(84)
$CH_3{-}Br$	293	(70)
$CH_3{-}I$	234	(56)
$CH_3CH_2{-}F$	448	(107)
$CH_3CH_2{-}Cl$	339	(81)
$CH_3CH_2{-}Br$	285	(68)
$CH_3CH_2{-}I$	222	(53)
$(CH_3)_2CH{-}F$	444	(106)
$(CH_3)_2CH{-}Cl$	335	(80)
$(CH_3)_2CH{-}Br$	285	(68)
$(CH_3)_2CH{-}I$	222	(53)
$(CH_3)_3C{-}F$	444	(106)
$(CH_3)_3C{-}Cl$	331	(79)
$(CH_3)_3C{-}Br$	272	(65)
$(CH_3)_3C{-}I$	209	(50)
R–OH bonds		
$CH_3{-}OH$	389	(93)
$CH_3CH_2{-}OH$	393	(94)
$CH_3CH_2CH_2{-}OH$	385	(92)
$(CH_3)_2CH{-}OH$	401	(96)
$(CH_3)_3C{-}OH$	401	(96)

as the valence electrons used in bonding are farther from the nucleus. Bond dissociation energies for a group of methyl–halogen bonds exemplify this trend.

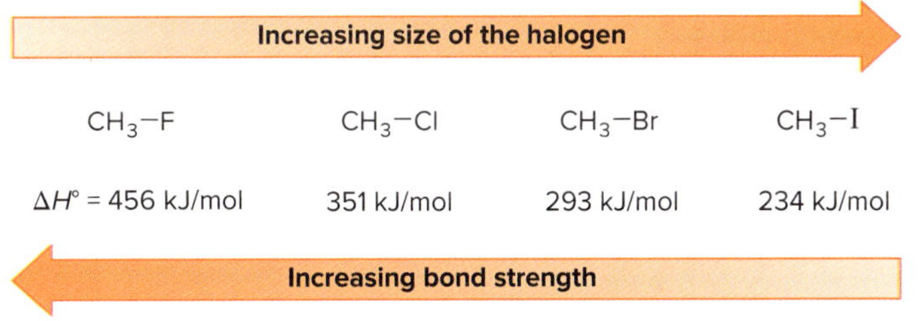

Because bond length increases down a column of the periodic table, bond dissociation energies are a quantitative measure of the general phenomenon noted in Chapter 1—***shorter*** **bonds are** ***stronger*** **bonds.**

Problem 6.6 Rank the indicated bonds in order of increasing bond dissociation energy.

a. F NH_2 OH b.

Bond dissociation energies are also used to calculate the enthalpy change ($\Delta H°$) in a reaction in which several bonds are broken and formed. **$\Delta H°$ indicates the relative strength of bonds broken and formed in a reaction.**

- When $\Delta H°$ is *positive*, more energy is needed to break bonds than is released in forming bonds. The bonds broken in the starting material are *stronger* than the bonds formed in the product.

- When $\Delta H°$ is *negative*, more energy is released in forming bonds than is needed to break bonds. The bonds formed in the product are *stronger* than the bonds broken in the starting material.

To determine the overall $\Delta H°$ for a reaction:

[1] Beginning with a *balanced* equation, add the bond dissociation energies for all bonds broken in the starting materials. This (+) value represents the **energy needed** to break bonds.

[2] Add the bond dissociation energies for all bonds formed in the products. This (−) value represents the **energy released** in forming bonds.

[3] **The overall $\Delta H°$ is the sum in Step [1] *plus* the sum in Step [2].**

$\Delta H°$ overall enthalpy change	=	sum of $\Delta H°$ of bonds broken	+	(−) sum of $\Delta H°$ of bonds formed

Sample Problem 6.3 Using Bond Dissociation Energies to Calculate $\Delta H°$

Use the values in Table 6.2 to determine $\Delta H°$ for the following reaction.

$(CH_3)_3C{-}Cl + H{-}O{-}H \longrightarrow (CH_3)_3C{-}OH + H{-}Cl$

Solution

[1] Bonds broken

	$\Delta H°$ (kJ/mol)
$(CH_3)_3C{-}Cl$	+331
H−OH	+498
Total	+829 kJ/mol

Energy needed to break bonds.

[2] Bonds formed

	$\Delta H°$ (kJ/mol)
$(CH_3)_3C{-}OH$	−401
H−Cl	−431
Total	−832 kJ/mol

Energy released in forming bonds.

[3] Overall $\Delta H°$=

sum in Step [1] + sum in Step [2]

+829 kJ/mol
−832 kJ/mol

Answer: −3 kJ/mol

Because $\Delta H°$ is a negative value, this reaction is **exothermic** and energy is released. **The bonds broken in the starting material are *weaker* than the bonds formed in the product.**

Problem 6.7 Use the values in Table 6.2 to calculate $\Delta H°$ for each reaction. Classify each reaction as endothermic or exothermic.

a. $CH_3CH_2Br + H_2O \longrightarrow CH_3CH_2OH + HBr$ b. $CH_4 + Cl_2 \longrightarrow CH_3Cl + HCl$

More Practice: Try Problems 6.34, 6.42b.

The oxidation of both glucose and linoleic acid, the molecule that introduced Chapter 6, forms CO_2 and H_2O.

linoleic acid + $25\ O_2$ ⟶ $18\ CO_2$ + $16\ H_2O$ $\Delta H° = (-)$

glucose + $6\ O_2$ ⟶ $6\ CO_2$ + $6\ H_2O$ $\Delta H° = (-)$

The combustion of gasoline and the metabolism of glucose during exercise are exothermic reactions that release energy.
S-F/Shutterstock

$\Delta H°$ is *negative* for both oxidations, so both reactions are *exothermic*. **Both linoleic acid and glucose release energy on oxidation because the bonds in the products are *stronger* than the bonds in the reactants.**

Bond dissociation energies have two important limitations. They present only *overall* energy changes. They reveal nothing about the reaction mechanism or how fast a reaction proceeds. Moreover, bond dissociation energies are determined for reactions in the gas phase, whereas most organic reactions are carried out in a liquid solvent where solvation energy contributes to the overall enthalpy of a reaction. As such, bond dissociation energies are imperfect indicators of energy changes in a reaction. Despite these limitations, using bond dissociation energies to calculate $\Delta H°$ gives a useful approximation of the energy changes that occur when bonds are broken and formed in a reaction.

Problem 6.8 Calculate $\Delta H°$ for each combustion reaction. Each equation is balanced as written; remember to take into account the coefficients in determining the number of bonds broken or formed. [$\Delta H°$ for O_2 = 497 kJ/mol; $\Delta H°$ for one C=O in CO_2 = 535 kJ/mol]

a. $CH_4 + 2\ O_2 \longrightarrow CO_2 + 2\ H_2O$ b. $2\ CH_3CH_3 + 7\ O_2 \longrightarrow 4\ CO_2 + 6\ H_2O$

6.5 Thermodynamics

For a reaction to be practical, the equilibrium must favor the products, *and* the reaction rate must be fast enough to form them in a reasonable time. These two conditions depend on the **thermodynamics** and the **kinetics** of a reaction, respectively.

- ***Thermodynamics*** **describes energy and equilibrium. How do the *energies* of the reactants and the products compare? What are the relative *amounts* of reactants and products at equilibrium?**

Reaction kinetics are discussed in Section 6.9.

- ***Kinetics*** **describes reaction rates. How *fast* are reactants converted to products?**

K_{eq} was first defined in Section 2.3 for acid–base reactions.

The **equilibrium constant, K_{eq},** is a mathematical expression that relates the amount of starting material and product at equilibrium. For example, when starting materials **A** and **B** react to form products **C** and **D,** the equilibrium constant is given by the following expression:

$$\mathbf{A} + \mathbf{B} \rightleftharpoons \mathbf{C} + \mathbf{D}$$

$$K_{eq} = \frac{[\text{products}]}{[\text{starting materials}]} = \frac{[\mathbf{C}][\mathbf{D}]}{[\mathbf{A}][\mathbf{B}]}$$

The size of K_{eq} tells about the position of equilibrium; that is, it expresses whether the starting materials or products predominate once equilibrium has been reached.

- When $K_{eq} > 1$, **equilibrium favors the *products*** (**C** and **D**) and the equilibrium lies to the *right* as the equation is written.
- When $K_{eq} < 1$, **equilibrium favors the *starting materials*** (**A** and **B**) and the equilibrium lies to the *left* as the equation is written.

- **For a reaction to be useful, the equilibrium must favor the products, and $K_{eq} > 1$.**

What determines whether equilibrium favors the products in a given reaction? **The position of equilibrium is determined by the relative energies of the reactants and products.** The free energy of a molecule, also called its **Gibbs free energy,** is symbolized by $G°$. The **change in free energy** between reactants and products, symbolized by $\Delta G°$, determines whether the starting materials or products are favored at equilibrium.

- **$\Delta G°$ is the overall energy difference between reactants and products.**

$$\Delta G° = G°_{\text{products}} - G°_{\text{reactants}}$$

($G°_{\text{products}}$ = free energy of the products; $G°_{\text{reactants}}$ = free energy of the reactants)

$\Delta G°$ is related to the equilibrium constant K_{eq} by the following equation:

$$\Delta G° = -2.303RT \log K_{eq}$$

[R = 8.314 J/(K•mol), the gas constant; T = Kelvin temperature (K)]

At 25 °C, $2.303RT = 5.7$ kJ/mol; thus, $\Delta G° = -5.7\log K_{eq}$.

$K_{eq} > 1$ **when** $\Delta G° < 0$, and equilibrium favors the *products*. $K_{eq} < 1$ **when** $\Delta G° > 0$, and equilibrium favors the *starting materials*.

Using this expression, we can determine the relationship between the equilibrium constant and the free energy change between reactants and products.

- **When $K_{eq} > 1$, log K_{eq} is positive, making $\Delta G°$ negative, and energy is *released*. Thus, equilibrium favors the products when the energy of the products is *lower* than the energy of the reactants.**
- **When $K_{eq} < 1$, log K_{eq} is negative, making $\Delta G°$ positive, and energy is *absorbed*. Thus, equilibrium favors the reactants when the energy of the products is *higher* than the energy of the reactants.**

Compounds that are lower in energy have increased stability. Thus, **equilibrium favors the products when they are *more stable* (lower in energy) than the starting materials of a reaction.** This is summarized in Figure 6.3.

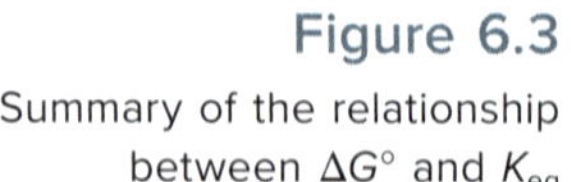

Figure 6.3 Summary of the relationship between $\Delta G°$ and K_{eq}

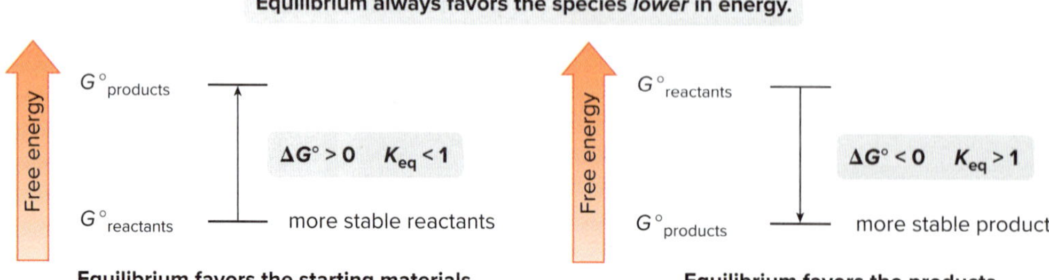

Table 6.3 Representative Values for $\Delta G°$ and K_{eq} at 25 °C, for a Reaction A → B

$\Delta G°$ (kJ/mol)	K_{eq}	Relative amount of A and B at equilibrium	
+18	$\mathbf{10^{-3}}$	**Essentially all A (99.9%)**	Increasing [product] ↓
+12	10^{-2}	100 times as much **A** as **B**	
+6	10^{-1}	10 times as much **A** as **B**	
0	1	Equal amounts of **A** and **B**	
−6	10^{1}	10 times as much **B** as **A**	
−12	10^{2}	100 times as much **B** as **A**	
−18	$\mathbf{10^{3}}$	**Essentially all B (99.9%)**	

Because $\Delta G°$ depends on the logarithm of K_{eq}, a **small change in energy corresponds to a large difference in the relative amount of starting material and product at equilibrium.** Several values of $\Delta G°$ and K_{eq} are given in Table 6.3.

Problem 6.9 (a) Which K_{eq} corresponds to a negative value of $\Delta G°$, $K_{eq} = 1000$ or $K_{eq} = .001$? (b) Which K_{eq} corresponds to a lower value of $\Delta G°$, $K_{eq} = 10^{-2}$ or $K_{eq} = 10^{-5}$?

Problem 6.10 Given each of the following values, is the starting material or product favored at equilibrium?

a. $K_{eq} = 5.5$ b. $\Delta G° = 40$ kJ/mol

The symbol ~ means "approximately."

Problem 6.11 Given each of the following values, is the starting material or product lower in energy?

a. $\Delta G° = 8.0$ kJ/mol b. $K_{eq} = 10$ c. $\Delta G° = -12$ kJ/mol d. $K_{eq} = 10^{-3}$

6.6 Enthalpy and Entropy

The **free energy change ($\Delta G°$)** depends on the **enthalpy change ($\Delta H°$)** and the **entropy change ($\Delta S°$).** $\Delta H°$ indicates relative bond strength, but what does $\Delta S°$ measure?

Entropy is a rather intangible concept that comes up again and again in chemistry courses. One way to remember the relation between entropy and disorder is to consider a handful of chopsticks. Dropped on the floor, they are arranged randomly (a state of high entropy). Placed end to end in a straight line, they are arranged intentionally (a state of low entropy). The more disordered, random arrangement is favored and easier to achieve.

***Entropy* ($S°$) is a measure of the randomness in a system.** The more freedom of motion or the more disorder present, the higher the entropy. Gas molecules move more freely than liquid molecules and are higher in entropy. Cyclic molecules have more restricted bond rotation than similar acyclic molecules and are lower in entropy.

The *entropy change* ($\Delta S°$) is the change in the amount of disorder between reactants and products. $\Delta S°$ is positive (+) when the products are more disordered than the reactants. $\Delta S°$ is negative (−) when the products are less disordered (more ordered) than the reactants.

- **Reactions resulting in an *increase in entropy* are favored.**

$\Delta G°$ is related to $\Delta H°$ and $\Delta S°$ by the following equation:

$\Delta G°$	=	$\Delta H°$	−	$T\Delta S°$
total energy change		change in **bonding energy**		change in **disorder**

[T = Kelvin temperature]

This equation tells us that the total energy change in a reaction is due to two factors: the change in the **bonding energy** and the change in **disorder.** The change in bonding energy can be calculated from bond dissociation energies (Section 6.4). Entropy changes, on the other hand, are more difficult to assess, but they are important when the number of molecules of starting material *differs* from the number of molecules of product in the balanced chemical equation. The entropy of a system also changes when an acyclic molecule is *cyclized* to a cyclic one, or a cyclic molecule is converted to an acyclic one.

For example, **when a single starting material forms two products,** as in the homolytic cleavage of a bond to form two radicals, **entropy increases** and favors formation of the products. In contrast, **entropy decreases when an acyclic compound forms a ring,** because a ring has fewer degrees of freedom. In this case, therefore, entropy does *not* favor formation of the product.

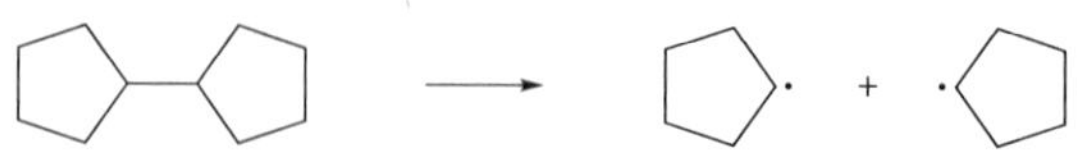

single reactant → two products
Entropy *increases*.
Entropy favors the **products.**

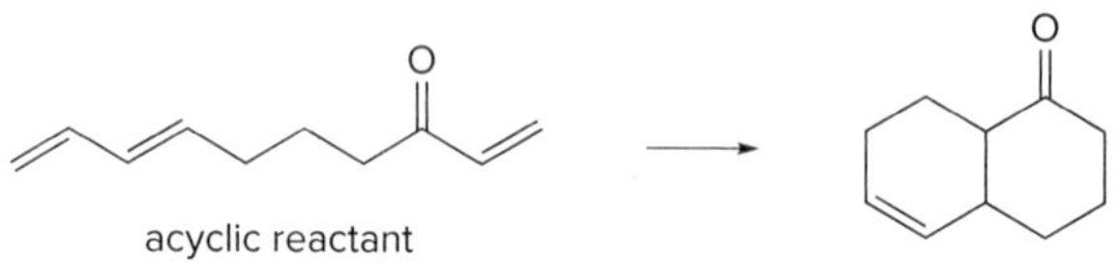

acyclic reactant → cyclic product
Entropy favors the **reactants.**
Entropy *decreases*.

The metabolism of glucose (Section 6.4) is favored by entropy because the number of molecules of products formed (6 CO_2 and 6 H_2O) is greater than the number of molecules of reactants ($C_6H_{12}O_6$ and 6 O_2). Moreover, a cyclic reactant is cleaved to form 12 acyclic product molecules.

Problem 6.12 For which reactions does entropy favor the products?

a. 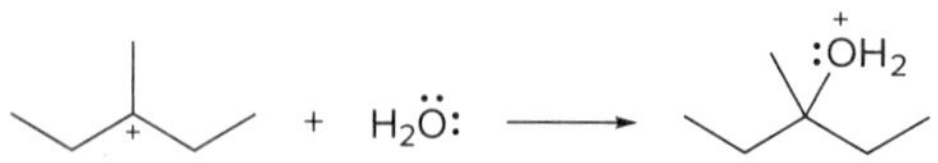 + $H_2\ddot{O}$: → $:\overset{+}{O}H_2$ b. 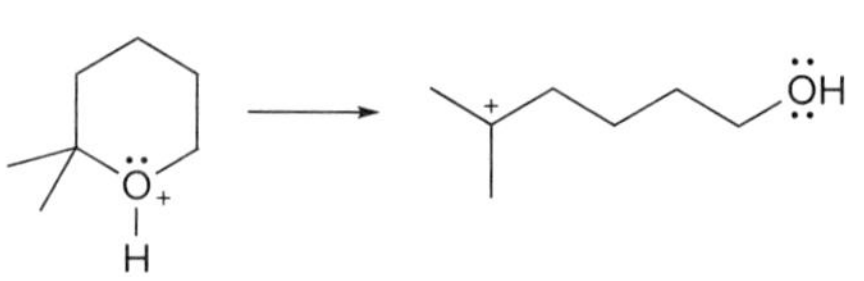 → $\ddot{O}H$

In most reactions that are not carried out at high temperature, the entropy term ($T\Delta S°$) is small compared to the enthalpy term ($\Delta H°$) and it can be neglected. Thus, **we will often approximate the overall free energy change of a reaction by the change in the bonding energy only.** Keep in mind that this is an approximation, but it gives us a starting point from which to decide if the reaction is energetically favorable.

Recall from Section 6.4 that a reaction is endothermic when $\Delta H°$ is positive and exothermic when $\Delta H°$ is negative. A reaction is **endergonic when $\Delta G°$ is positive** and **exergonic when $\Delta G°$ is negative.** $\Delta G°$ is usually approximated by $\Delta H°$ in this text, so the terms endergonic and exergonic are rarely used.

$$\Delta G° \approx \Delta H°$$

According to this approximation:

- **The product is favored when $\Delta H°$ is a *negative* value; that is, the bonds in the product are *stronger* than the bonds in the starting material.**
- **The starting material is favored when $\Delta H°$ is a *positive* value; that is, the bonds in the starting material are *stronger* than the bonds in the product.**

Problem 6.13 Considering each of the following values and neglecting entropy, tell whether the starting material or product is favored at equilibrium: (a) $\Delta H°$ = 80 kJ/mol; (b) $\Delta H°$ = −40 kJ/mol.

Problem 6.14 For a reaction with $\Delta H°$ = 40 kJ/mol, decide which of the following statements is (are) true. Correct any false statement to make it true. (a) The reaction is exothermic; (b) $\Delta G°$ for the reaction is positive; (c) K_{eq} is greater than 1; (d) the bonds in the starting materials are stronger than the bonds in the product; and (e) the product is favored at equilibrium.

6.7 Energy Diagrams

An **energy diagram** is a schematic representation of the energy changes that take place as reactants are converted to products. An energy diagram indicates how readily a reaction proceeds, how many steps are involved, and how the energies of the reactants, products, and intermediates compare.

Consider a concerted reaction between molecule **A–B** with anion **C:⁻** to form products **A:⁻** and **B–C.** If the reaction occurs in a single step, the bond between **A** and **B** is broken *as* the bond between **B** and **C** is formed. Let's assume that the products are lower in energy than the reactants in this hypothetical reaction.

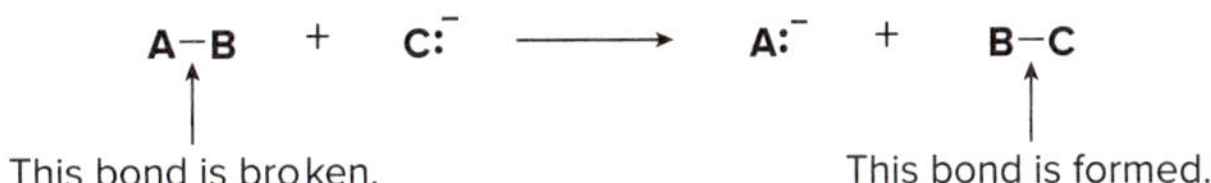

An energy diagram plots **energy on the *y* axis** versus the progress of reaction, often labeled the **reaction coordinate,** on the ***x* axis.** As the starting materials **A–B** and **C:⁻** approach one another, their electron clouds feel some repulsion, causing an increase in energy, until a maximum value is reached. This unstable energy maximum is called the **transition state.** In the transition state the bond between **A** and **B** is partially broken, and the bond between **B** and **C** is partially formed. Because it is at the top of an energy "hill," **a transition state can never be isolated.**

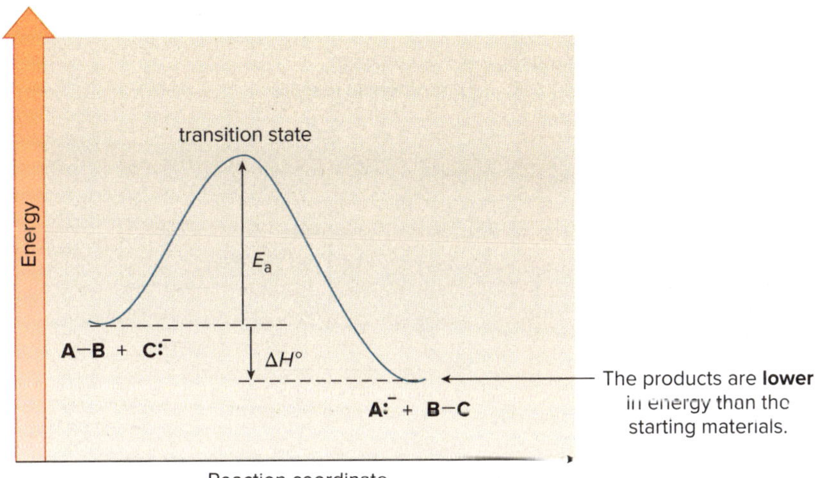

At the transition state, the bond between **A** and **B** can re-form to regenerate starting material, *or* the bond between **B** and **C** can form to generate product. As the bond forms between **B** and **C,** the energy decreases until some stable energy minimum of the products is reached.

- The energy difference between the reactants and products is $\Delta H°$. Because the products are at lower energy than the reactants, this reaction is *exothermic* and energy is *released.*

- The energy difference between the transition state and the starting material is called the *energy of activation,* symbolized by E_a.

The *energy of activation* is the minimum amount of energy needed to break bonds in the reactants. It represents an **energy barrier** that must be overcome for a reaction to occur. The size of E_a tells us about the reaction rate.

A slow reaction has a large E_a. A fast reaction has a low E_a.

- The *larger* the E_a, the *greater* the amount of energy that is needed to break bonds, and the *slower* the reaction rate.

How can we draw the structure of the unstable transition state? The structure of the transition state is somewhere in between the structures of the starting material and product. Any bond that is partially broken or formed is drawn with a *dashed* line. Any atom that gains or loses a charge contains a *partial charge* in the transition state. Transition states are drawn in brackets, with a superscript double dagger (‡).

In the hypothetical reaction between **A–B** and **C:⁻** to form **A:⁻** and **B–C,** the bond between **A** and **B** is partially broken, and the bond between **B** and **C** is partially formed. Because **A** gains a negative charge and **C** loses a charge in the course of the reaction, each atom bears a partial negative charge in the transition state.

$$\left[\overset{\delta-}{A}\cdots B\cdots\overset{\delta-}{C}\right]^{\ddagger}$$

This bond is partially broken. This bond is partially formed.

Several energy diagrams are drawn in Figure 6.4. For any energy diagram:

- **E_a determines the height of the energy barrier.**
- **$\Delta H°$ determines the relative position of the reactants and products.**

Figure 6.4 Some representative energy diagrams

Example [1]

- **Large E_a ⟶ slow reaction**
- **(+) $\Delta H°$ ⟶ endothermic reaction**

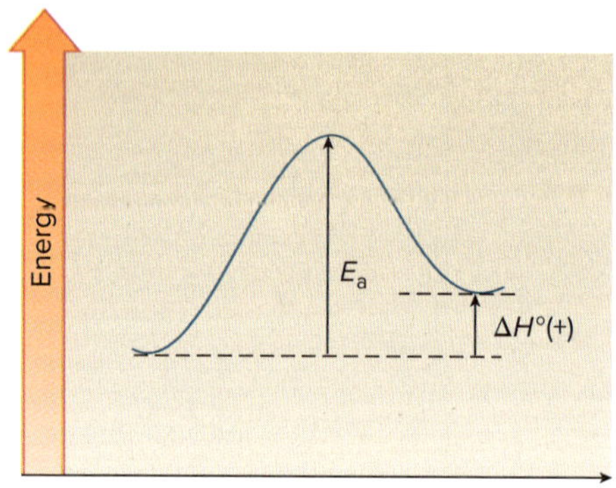

Example [2]

- **Large E_a ⟶ slow reaction**
- **(–) $\Delta H°$ ⟶ exothermic reaction**

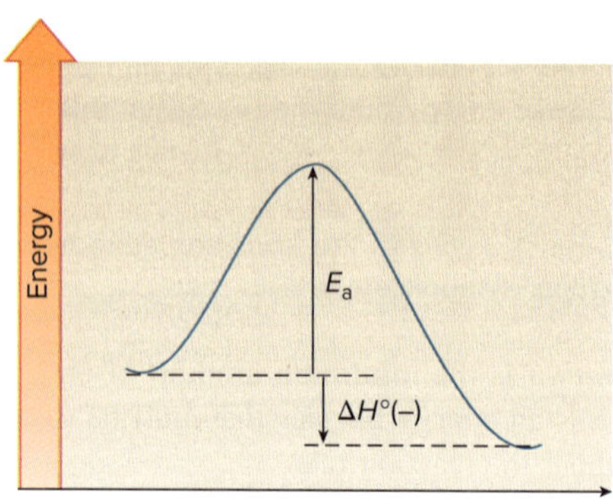

Example [3]

- **Low E_a ⟶ fast reaction**
- **(+) $\Delta H°$ ⟶ endothermic reaction**

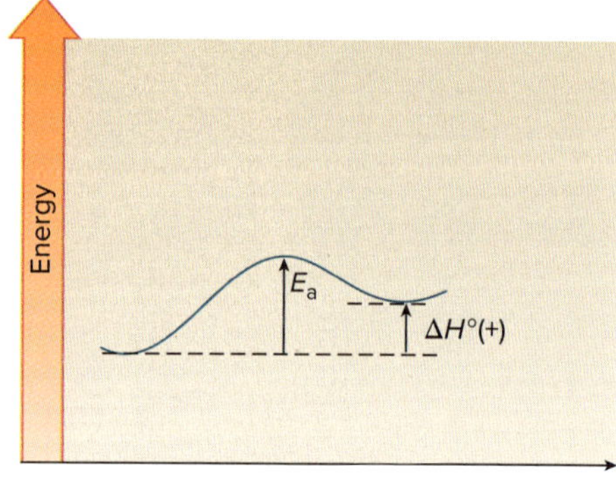

Example [4]

- **Low E_a ⟶ fast reaction**
- **(–) $\Delta H°$ ⟶ exothermic reaction**

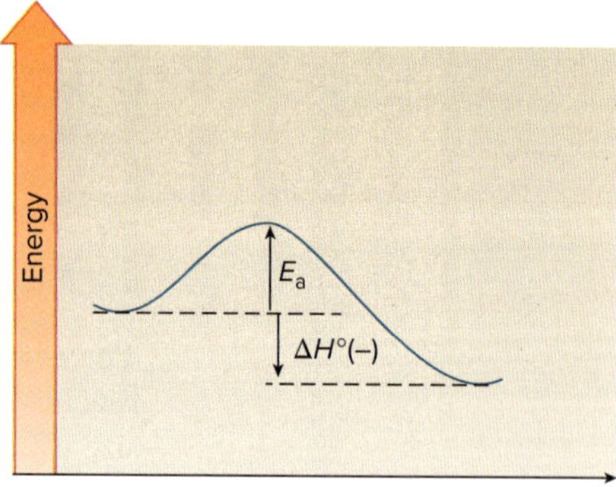

The two variables, **E_a and $\Delta H°$, are independent** of each other. Two reactions can have identical values for $\Delta H°$ but very different E_a values. For two exothermic reactions with the same negative value of $\Delta H°$ but different E_a values, the reaction with the lower E_a is faster.

Problem 6.15 Draw an energy diagram for a reaction in which the products are higher in energy than the starting materials and E_a is large. Clearly label all of the following on the diagram: the axes, the starting materials, the products, the transition state, $\Delta H°$, and E_a.

Problem 6.16 Draw the structure for the transition state in each reaction.

a. $(CH_3)_3C{-}OH_2^+$ ⟶ $(CH_3)_3C^+$ + H_2O

b. $CH_3CH_2CH_2OH$ + ^-OH ⟶ $CH_3CH_2CH_2O^-$ + H_2O

Problem 6.17 Compound **A** can be converted to either **B** or **C.** The energy diagrams for both processes are drawn on the graph below.

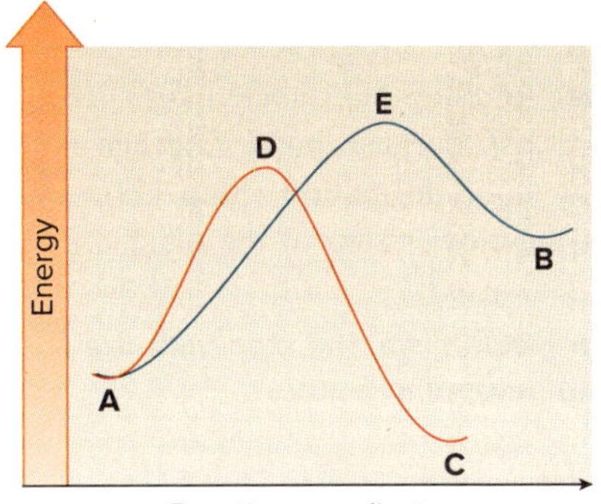

a. Label each reaction as endothermic or exothermic.
b. Which reaction is faster?
c. Which reaction generates the product lower in energy?
d. Which points on the graph correspond to transition states?
e. Label the energy of activation for each reaction.
f. Label the $\Delta H°$ for each reaction.

6.8 Energy Diagram for a Two-Step Reaction Mechanism

Although the hypothetical reaction in Section 6.7 is concerted, many reactions involve more than one step with formation of a reactive intermediate. Consider the same overall reaction, **A–B + C:$^-$** to form products **A:$^-$ + B–C,** but in this case begin with the assumption that the reaction occurs by a *stepwise* pathway—that is, bond breaking occurs *before* bond making. Once again, assume that the overall process is exothermic.

A–B + **C:$^-$** ⟶ **A:$^-$** + **B–C**

This bond is broken... ***before*** ...this bond is formed.

One possible stepwise mechanism involves heterolysis of the **A–B** bond to form two ions **A:$^-$** and **B$^+$**, followed by reaction of **B$^+$** with anion **C:$^-$** to form product **B–C,** as outlined in the accompanying equations. Species **B$^+$** is a **reactive intermediate. B$^+$** is a product in Step [1] that reacts with **C:$^-$** in Step [2].

Step [1]: Heterolysis of the **A–B** bond

A–B ⟶ **A:$^-$** + **B$^+$**

Break one bond.

Step [2]: Formation of the **B–C** bond

B$^+$ + **C:$^-$** ⟶ **B–C**

Form one bond.

B$^+$ is an intermediate.

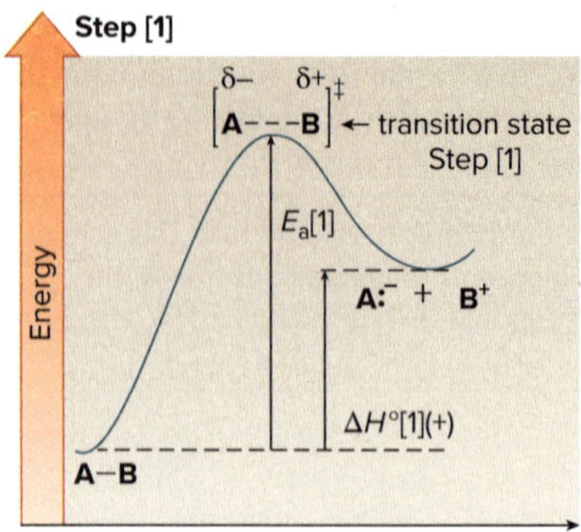

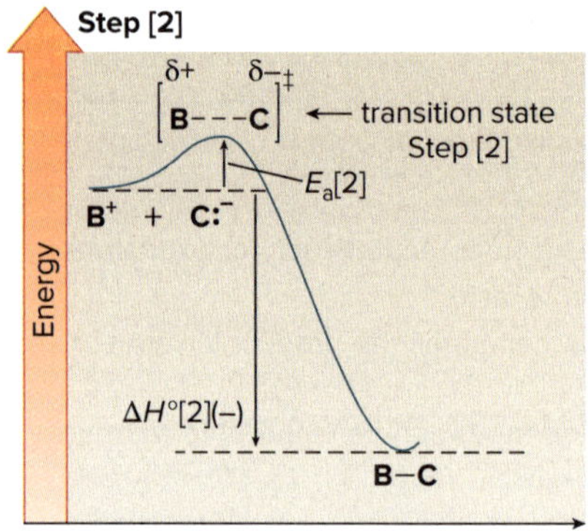

To draw an energy diagram for a two-step mechanism, we must draw an energy diagram for each step, and then combine them together. Each step has its own energy barrier, with a transition state at the energy maximum.

Step [1] is endothermic because energy is needed to cleave the **A–B** bond, making $\Delta H°$ a positive value and placing the products of Step [1] at higher energy than the starting materials. In the transition state, the **A–B** bond is partially broken.

Step [2] is exothermic because energy is released in forming the **B–C** bond, making $\Delta H°$ a negative value and placing the products of Step [2] at lower energy than the starting materials of Step [2]. In the transition state, the **B–C** bond is partially formed.

The overall process is shown in Figure 6.5 as a single energy diagram that combines both steps. Because the reaction has two steps, there are two transition states, each corresponding to an energy barrier. The transition states are separated by an energy minimum, at which the reactive intermediate $\mathbf{B^+}$ is located. Because we made the assumption that the overall two-step process is exothermic, the overall energy difference between the reactants and products, labeled $\Delta H°_{overall}$, has a negative value, and the final products are at a lower energy than the starting materials.

The energy barrier for Step [1], labeled $E_a[1]$, is higher than the energy barrier for Step [2], labeled $E_a[2]$, because bond cleavage (Step [1]) is more difficult (requires more energy) than bond formation (Step [2]). A higher-energy transition state for Step [1] makes it the slower step of the mechanism.

- **In a multistep mechanism, the step with the highest-energy transition state is called the rate-determining step.**

In this reaction, the rate-determining step is Step [1].

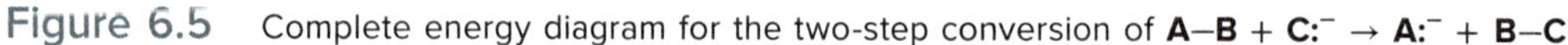

Figure 6.5 Complete energy diagram for the two-step conversion of **A–B** + **C:⁻** → **A:⁻** + **B–C**

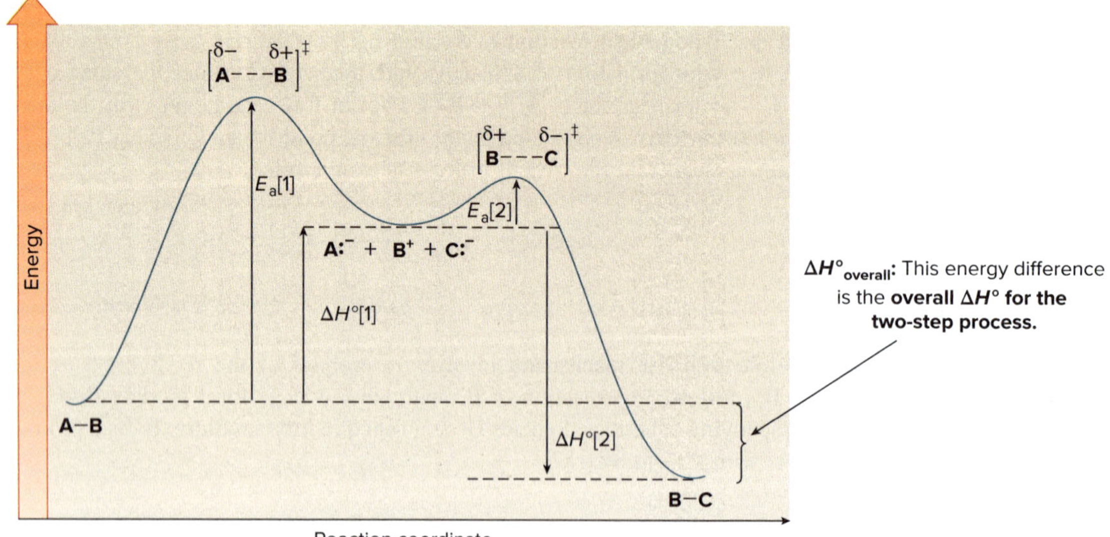

- **The transition states are located at energy maxima, whereas the reactive intermediate B⁺ is located at an energy minimum.**
- Each step has its own value of $\Delta H°$ and E_a.
- The overall energy difference between starting material and products is called $\Delta H°_{overall}$. In this example, the products of the two-step sequence are at lower energy than the starting materials.
- Because Step [1] has the higher-energy transition state, it is the **rate-determining step.**

Problem 6.18 Consider the following energy diagram.

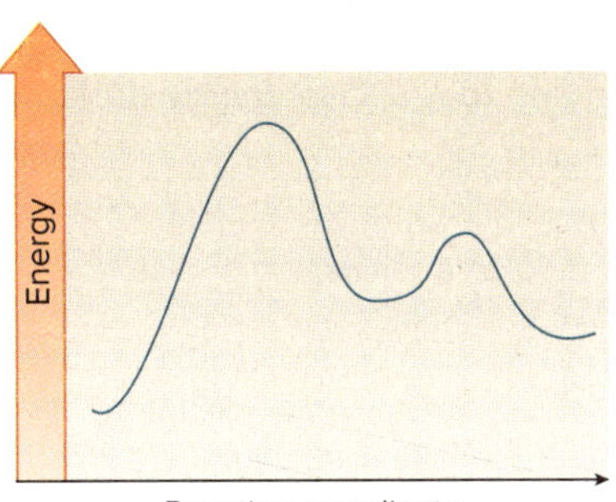

a. How many steps are involved in this reaction?
b. Label $\Delta H°$ and E_a for each step, and label $\Delta H°_{overall}$.
c. Label each transition state.
d. Which point on the graph corresponds to a reactive intermediate?
e. Which step is rate-determining?
f. Is the overall reaction endothermic or exothermic?

Problem 6.19 Draw an energy diagram for a two-step reaction, **A** → **B** → **C,** where the relative energy of these compounds is **C** < **A** < **B,** and the conversion of **B** → **C** is rate-determining.

6.9 Kinetics

Some reactions have a very favorable equilibrium constant (K_{eq} >> 1), but the rate is very slow. Gasoline can be safely handled in the air because its reaction with O_2 is slow unless there is a spark to provide energy to initiate the reaction. *Moodboard/Getty Images*

We now turn to a more detailed discussion of **reaction rate**—that is, how fast a particular reaction proceeds. **The study of reaction rates is called *kinetics*.**

The rate of chemical processes affects many facets of our lives. Aspirin is an effective anti-inflammatory agent because it rapidly inhibits the synthesis of prostaglandins (Section 19.5). DDT (Section 7.4) is a persistent environmental pollutant because it does not react appreciably with water, oxygen, or any other chemical with which it comes into contact. These processes occur at different rates, resulting in beneficial or harmful effects.

6.9A Energy of Activation

As we learned in Section 6.7, the energy of activation, E_a, is the energy difference between the reactants and the transition state. It is the **energy barrier** that must be exceeded for reactants to be converted to products.

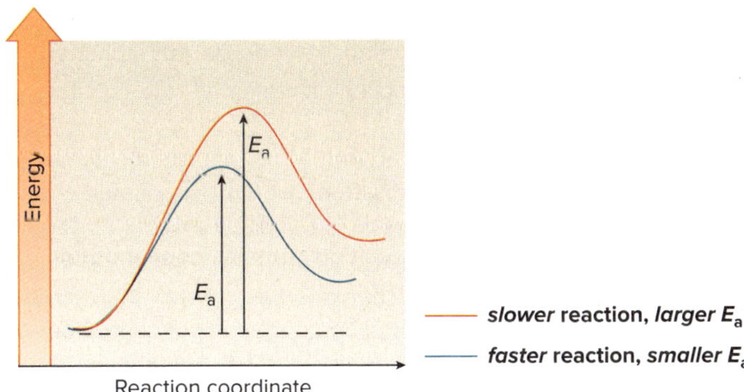

- **The *larger* the E_a, the *slower* the reaction.**

Concentration and temperature also affect reaction rate.

- **The *higher* the concentration, the *faster* the rate. Increasing concentration increases the number of collisions between reacting molecules, which in turn increases the rate.**

- **The *higher* the temperature, the *faster* the rate. Increasing temperature increases the average kinetic energy of the reacting molecules. Because the kinetic energy of colliding molecules is used for bond cleavage, increasing the average kinetic energy increases the rate.**

Practically, the effect of temperature on reaction rate is used to an advantage in the kitchen. Food is stored in a cold refrigerator to slow the reactions that cause spoilage. *Jill Braaten/McGraw Hill*

As a rule of thumb, increasing the temperature by 10 °C doubles the reaction rate. Thus, reactions in the lab are often heated to increase their rates so that they occur in a reasonable amount of time.

Keep in mind that certain **reaction quantities have *no effect* on reaction rate.**

- **$\Delta G°$, $\Delta H°$, and K_{eq} do *not* determine the rate of a reaction. These quantities indicate the direction of equilibrium and the relative energy of reactants and products.**

Problem 6.20 Which value (if any) corresponds to a faster reaction: (a) E_a = 40 kJ/mol or E_a = 4 kJ/mol; (b) a reaction temperature of 0 °C or a reaction temperature of 25 °C; (c) K_{eq} = 10 or K_{eq} = 100; (d) $\Delta H°$ = –10 kJ/mol or $\Delta H°$ = 10 kJ/mol?

Problem 6.21 For a reaction with K_{eq} = 0.8 and E_a = 80 kJ/mol, decide which of the following statements is (are) true. Correct any false statement to make it true. Ignore entropy considerations. (a) The reaction is faster than a reaction with K_{eq} = 8 and E_a = 80 kJ/mol. (b) The reaction is faster than a reaction with K_{eq} = 0.8 and E_a = 40 kJ/mol. (c) $\Delta G°$ for the reaction is a positive value. (d) The starting materials are lower in energy than the products of the reaction. (e) The reaction is exothermic.

6.9B Rate Equations

The rate of a chemical reaction is determined by measuring the decrease in the concentration of the reactants over time, or the increase in the concentration of the products over time. A **rate law** (or **rate equation**) is an equation that shows the relationship between the rate of a reaction and the concentration of the reactants. A rate law is determined *experimentally,* and it depends on the mechanism of the reaction.

A rate law has two important terms: the **rate constant symbolized by *k*** and the **concentration of the reactants.** Not all reactant concentrations may appear in the rate equation, as we shall soon see.

$$\text{rate} = k[\text{reactants}]$$

k = the rate constant

A rate constant k and the energy of activation E_a are inversely related. **A *high* E_a corresponds to a *small* k.**

A rate constant k is a fundamental characteristic of a reaction. It is a complex mathematical term that takes into account the dependence of a reaction rate on temperature and the energy of activation.

- *Fast* reactions have *large* rate constants.
- *Slow* reactions have *small* rate constants.

What concentration terms appear in the rate equation? That depends on the mechanism. For the organic reactions we will encounter:

- **A rate equation contains concentration terms for *all* reactants involved in a *one-step* mechanism.**
- **A rate equation contains concentration terms for *only* the reactants involved in the *rate-determining step* in a multistep reaction.**

In the one-step reaction of **A–B + C:$^-$** to form **A:$^-$ + B–C,** *both* reactants appear in the transition state of the only step of the mechanism. The **concentration of *both* reactants affects the reaction rate,** and *both* terms appear in the rate equation. This type of reaction involving two reactants is said to be **bimolecular.**

$$\text{A–B} + \text{C:}^- \longrightarrow \text{A:}^- + \text{B–C} \qquad \text{rate} = k[\text{AB}][\text{C:}^-]$$

Both reactants are involved in the only step.
Both reactants determine the rate.

sum of the exponents = **2**
Second-order rate equation

The ***order* of a rate equation equals the sum of the exponents of the concentration terms** in the rate equation. In the rate equation for the concerted reaction of **A–B + C:$^-$,** there are two concentration terms, each with an exponent of one. Thus, the sum of the exponents is two and the **rate equation is *second order*** (the reaction follows second-order kinetics).

Because the rate of the reaction depends on the concentration of both reactants, doubling the concentration of *either* **A—B** or **C:$^-$** doubles the rate of the reaction. Doubling the concentration of *both* **A—B** and **C:$^-$** increases the reaction rate by a factor of *four.*

The situation is different in the stepwise conversion of **A—B** + **C:$^-$** to form **A:$^-$** + **B—C.** The mechanism shown in Section 6.8 has two steps: a slow step (the **rate-determining** step) in which the **A—B** bond is broken, and a fast step in which the **B—C** bond is formed.

A—B ⟶ A:$^-$ + B$^+$ —(C:$^-$)⟶ B—C

Step [1] **rate-determining** Step [2]

rate = k[AB]

Only **AB** is involved in the rate-determining step. **Only [AB] determines the rate.**

only one concentration term

First-order rate equation

In a multistep mechanism, a reaction can occur no faster than its rate-determining step. **Only the concentrations of the reactants in the rate-determining step appear in the rate equation.** In this example, the rate depends on the concentration of **A—B** *only,* because only **A—B** appears in the rate-determining step. A reaction involving only one reactant is said to be **unimolecular.** Because there is only one concentration term (raised to the first power), the **rate equation is *first order*** (the reaction follows first-order kinetics).

Because the rate of the reaction depends on the concentration of only *one* reactant, doubling the concentration of **A—B** doubles the rate of the reaction, but **doubling the concentration of C:$^-$ has *no effect* on the reaction rate.**

This might seem like a puzzling result. If **C:$^-$** is involved in the reaction, why doesn't it affect the overall rate of the reaction?

The following analogy is useful. Let's say three students must make 20 peanut butter and jelly sandwiches for a class field trip. Student (**1**) spreads the peanut butter on the bread. Student (**2**) spreads on the jelly, and student (**3**) cuts the sandwiches in half. Suppose student (**2**) is very slow in spreading the jelly. It doesn't matter how fast students (**1**) and (**3**) are; they can't finish making sandwiches any faster than student (**2**) can add the jelly. Five more students can spread on the peanut butter, or an entirely different individual can replace student (**3**), and this doesn't speed up the process. How fast the sandwiches are made is determined entirely by the rate-determining step—that is, spreading the jelly.

Rate equations provide very important information about the mechanism of a reaction. Rate laws for new reactions with unknown mechanisms are determined by a set of experiments that measure how a reaction's rate changes with concentration. Then, a mechanism is suggested based on which reactants affect the rate.

Problem 6.22 The rate equation for the reaction of CH_3CH_2Br with ^-OH is: rate = $k[CH_3CH_2Br][^-OH]$. What effect does the indicated concentration change have on the overall rate of the reaction?

a. tripling the concentration of CH_3CH_2Br only

b. tripling the concentration of ^-OH only

c. tripling the concentration of both CH_3CH_2Br and ^-OH

Problem 6.23 Write a rate equation for each reaction, given the indicated mechanism.

a. (bromopropane) Br + ^-OH ⟶ (propene) + H_2O + Br^-

b. (tert-butyl bromide) Br —slow⟶ (tert-butyl cation) + Br^- —(^-OH), fast⟶ (2-methylpropene) + H_2O

Problem 6.24 Label each statement as true or false. Correct any false statement to make it true.

a. If a reaction is fast, it has a large rate constant.
b. A fast reaction has a large negative $\Delta G°$ value.
c. When E_a is large, the rate constant k is also large.
d. Fast reactions have equilibrium constants > 1.
e. Increasing the concentration of a reactant always increases the rate of a reaction.

6.10 Catalysts

Some reactions do not occur in a reasonable time unless a **catalyst** is added.

- **A *catalyst* is a substance that speeds up the rate of a reaction. A catalyst is recovered unchanged in a reaction, and it does not appear in the product.**

Common catalysts in organic reactions are **acids** and **metals.** Two examples are shown with the catalyst drawn in red.

acetic acid + ethanol $\xrightarrow{H_2SO_4}$ ethyl acetate + H_2O

cyclohexene + H_2 $\xrightarrow{Pd}$ cyclohexane

The reaction of acetic acid with ethanol to yield ethyl acetate and water occurs in the presence of an acid catalyst. The acid catalyst is written over or under the arrow to emphasize that it is not part of the starting materials or the products. The details of this reaction are discussed in Chapter 20.

The reaction of cyclohexene with hydrogen to form cyclohexane occurs only in the presence of a metal catalyst such as palladium, platinum, or nickel. The metal provides a surface that binds both the cyclohexene and the hydrogen, and in doing so, facilitates the reaction. We return to this mechanism in Chapter 12.

Catalysts accelerate a reaction by lowering the energy of activation (Figure 6.6). They have no effect on the equilibrium constant, so they do not change the amount of reactant and product at equilibrium. Thus, catalysts affect how *quickly* equilibrium is achieved, but not the relative amounts of reactants and products at equilibrium. If a catalyst is somehow used up in one step of a reaction sequence, it must be regenerated in another step.

Figure 6.6 The effect of a catalyst on a reaction

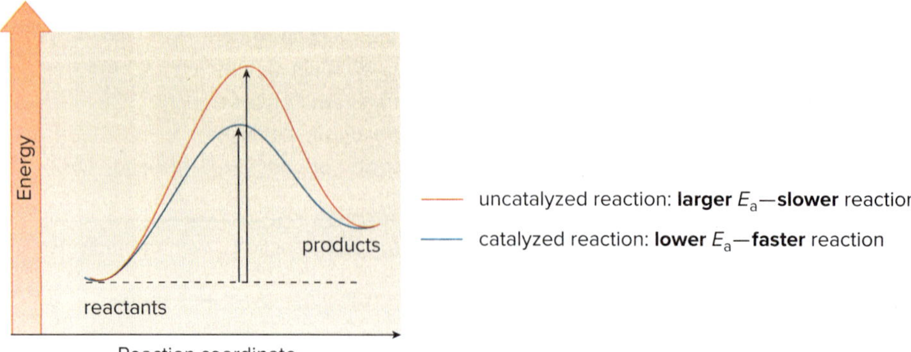

- The catalyst *lowers* the energy of activation, thus ***increasing the rate*** **of the catalyzed reaction.**
- The energy of the reactants and products is the same in both the uncatalyzed and catalyzed reactions, so the **position of equilibrium is unaffected.**

Problem 6.25 Identify the catalyst in each equation.

a. $CH_2{=}CH_2 \xrightarrow[H_2SO_4]{H_2O} CH_3CH_2OH$ b. $CH_3Cl \xrightarrow[^-OH]{I^-} CH_3OH + Cl^-$

6.11 Enzymes

The catalysts that synthesize and break down biomolecules in living organisms are governed by the same principles as the acids and metals in organic reactions. The catalysts in living organisms, however, are usually protein molecules called **enzymes.**

- ***Enzymes*** **are biochemical catalysts composed of amino acids held together in a very specific three-dimensional shape.**

An enzyme contains a region called its **active site,** which binds an organic reactant, called a **substrate.** When bound, this unit is called the **enzyme–substrate complex,** as shown schematically in Figure 6.7 for the enzyme lactase, the enzyme that binds lactose, the principal carbohydrate in milk. Once bound, the organic substrate undergoes a very specific reaction at an enhanced rate. In this example, lactose is converted into two simpler sugars, glucose and galactose. When individuals lack adequate amounts of lactase, they are unable to digest lactose, causing abdominal cramping and diarrhea.

Figure 6.7 Lactase, an example of a biological catalyst

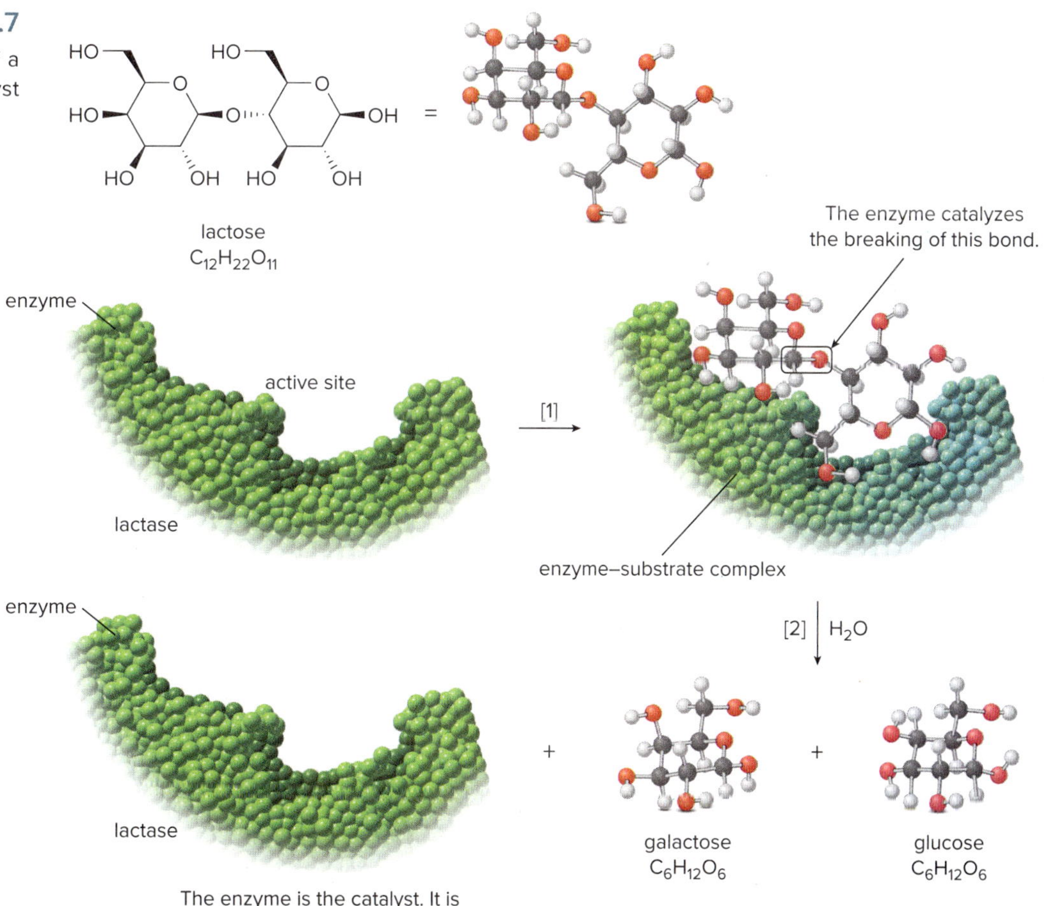

- The enzyme lactase binds the carbohydrate lactose ($C_{12}H_{22}O_{11}$) in its active site in Step [1]. Lactose then reacts with water to break a bond and form two simpler sugars, galactose and glucose, in Step [2]. This process is the first step in digesting lactose, the principal carbohydrate in milk.

An enzyme speeds up a biological reaction in a variety of ways. It may hold reactants in the proper conformation to facilitate reaction, or it may provide an acidic site needed for a particular transformation. Once the reaction is completed, the enzyme releases the substrate and it is then able to catalyze another reaction.

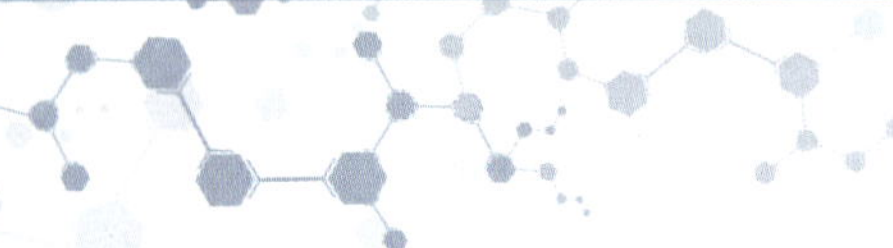

Chapter 6 REVIEW

KEY CONCEPTS

[1] Types of reactions (6.2)

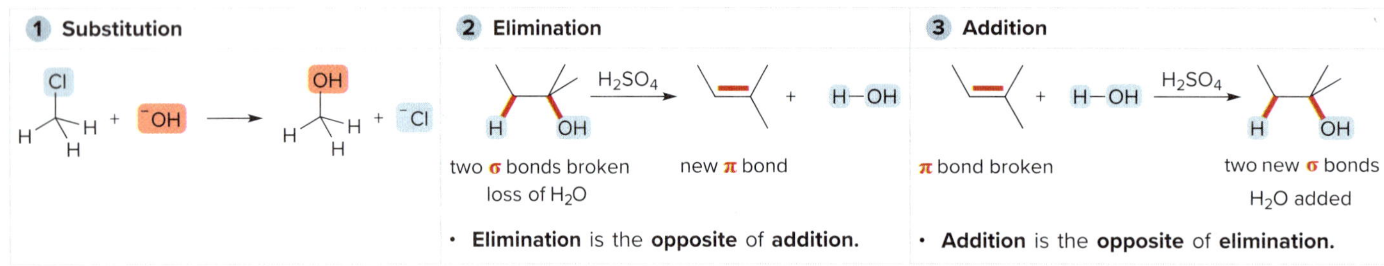

- **Elimination** is the **opposite** of **addition.**
- **Addition** is the **opposite** of **elimination.**

Try Problems 6.27, 6.30b, 6.47e, 6.50e.

[2] Energy trends

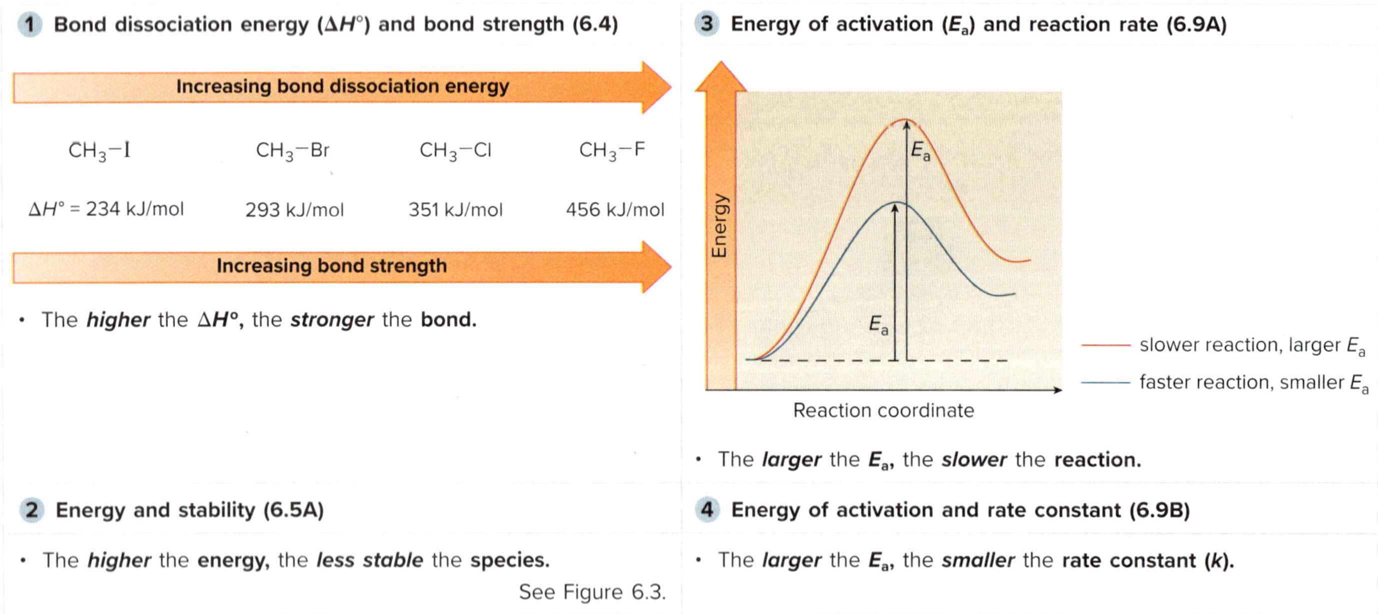

- The ***higher*** the **ΔH°,** the ***stronger*** the **bond.**
- The ***larger*** the $\mathbf{E_a}$, the ***slower*** the **reaction.**

2 Energy and stability (6.5A)

- The ***higher*** the **energy,** the ***less stable*** the **species.**

See Figure 6.3.

4 Energy of activation and rate constant (6.9B)

- The ***larger*** the $\mathbf{E_a}$, the ***smaller*** the **rate constant (*k*).**

[3] Reactive intermediates (6.3)

1 Homolysis generates radicals with unpaired electrons.

2 Heterolysis generates ions.

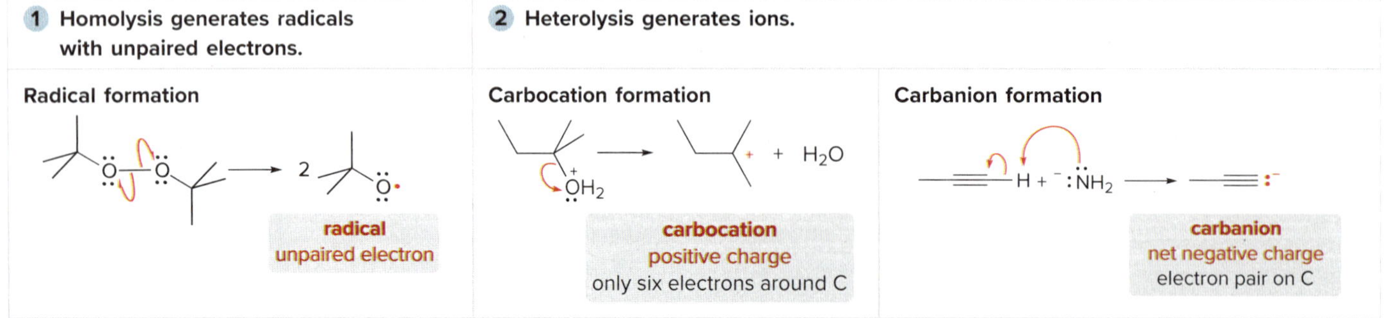

[4] Energy diagrams

1 One-step reaction mechanism (6.7)

transition state
Energy
E_a
reactants
$\Delta H°$
products
Reaction coordinate

See Figure 6.4.

- E_a determines the **rate; larger E_a --→ slower reaction (6.9).**

2 Two-step reaction mechanism (6.8)

transition state [1]
transition state [2]
Energy
E_a [1]
E_a [2]
intermediates
$\Delta H°$ [1]
reactants
$\Delta H°$ [2]
products
Reaction coordinate

See Figure 6.5.

- **$\Delta H°$** is the **difference in bonding energy** between the reactants and products.

Try Problems 6.41; 6.42c; 6.43; 6.44e; 6.57e, f.

[5] Conditions favoring product formation (6.5, 6.6)

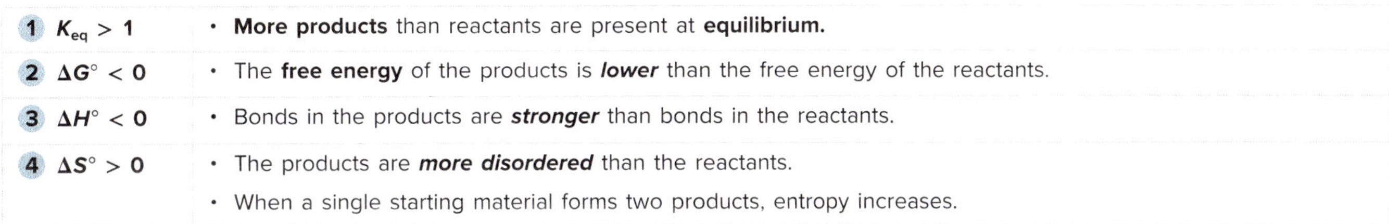

1 $K_{eq} > 1$	• **More products** than reactants are present at **equilibrium.**
2 $\Delta G° < 0$	• The **free energy** of the products is ***lower*** than the free energy of the reactants.
3 $\Delta H° < 0$	• Bonds in the products are ***stronger*** than bonds in the reactants.
4 $\Delta S° > 0$	• The products are ***more disordered*** than the reactants. • When a single starting material forms two products, entropy increases.

Try Problems 6.36, 6.37, 6.39.

KEY EQUATIONS

1

$$\Delta G° = -2.303RT \log K_{eq}$$

K_{eq} depends on the energy difference between reactants and products.

R = 8.314 J/(K•mol), the gas constant
T = Kelvin temperature (K)

2

$$\Delta G° = \Delta H° - T\Delta S°$$

free energy change = change in bonding energy − change in disorder

T = Kelvin temperature (K)

Try Problems 6.37, 6.38.

KEY SKILLS

[1] Using full-headed curved arrows to show the movement of electron pairs (6.3D)

1 Look for all bonds that are broken or formed.

2 Use full-headed curved arrows for electron pairs.

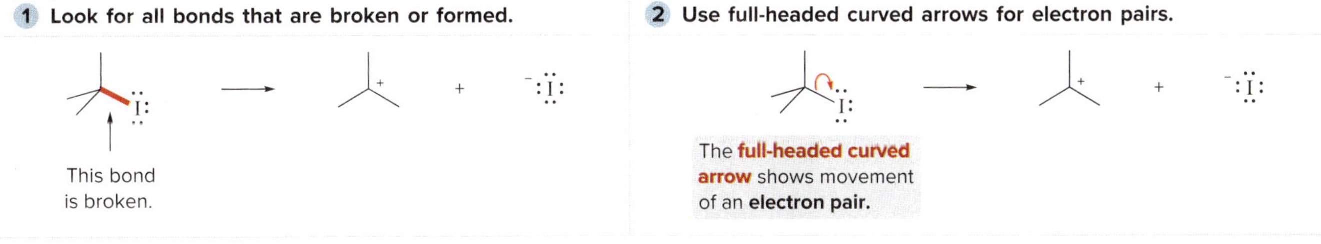

See Figure 6.2, Sample Problems 6.1, 6.2. Try Problems 6.28b, 6.29, 6.30a, 6.31, 6.32, 6.47a, 6.49a, 6.50a, 6.53a, 6.55a.

[2] Using half-headed curved arrows to show the movement of single electrons (6.3B)

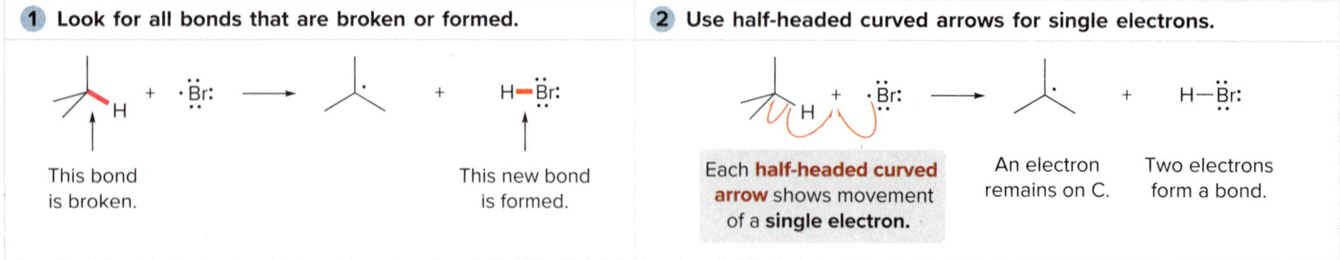

See Figure 6.2. Try Problems 6.28a, 6.33, 6.42a.

[3] Calculating $\Delta H°$ of a reaction (6.4)

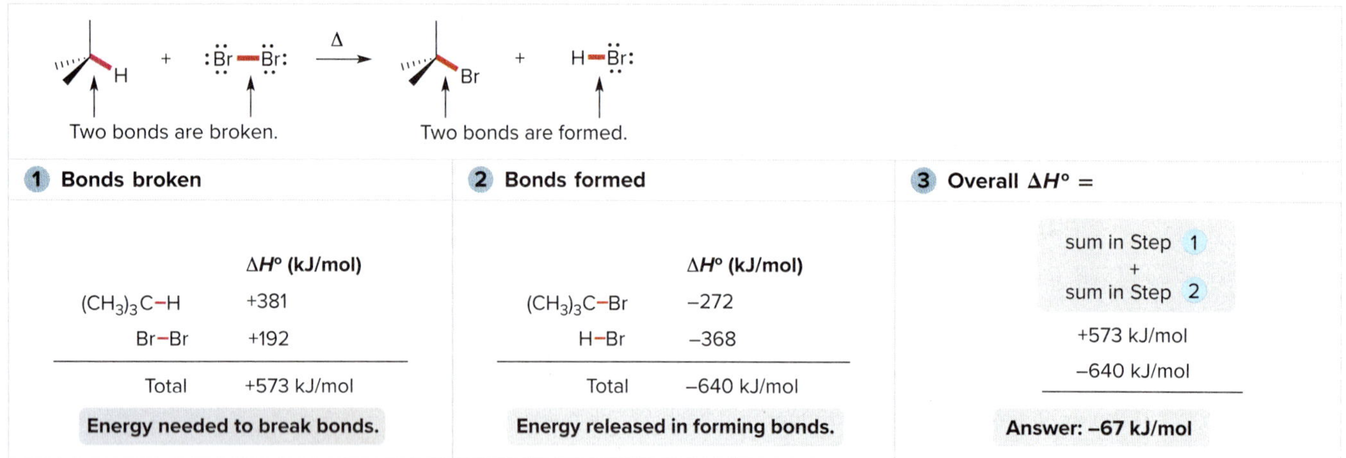

1 Bonds broken	$\Delta H°$ (kJ/mol)	2 Bonds formed	$\Delta H°$ (kJ/mol)	3 Overall $\Delta H°$ =
$(CH_3)_3C–H$	+381	$(CH_3)_3C–Br$	–272	sum in Step 1 + sum in Step 2
Br–Br	+192	H–Br	–368	+573 kJ/mol
Total	+573 kJ/mol	Total	–640 kJ/mol	–640 kJ/mol
Energy needed to break bonds.		**Energy released in forming bonds.**		**Answer: –67 kJ/mol**

See Table 6.2, Sample Problem 6.3. Try Problems 6.34, 6.42b.

CHAPTER 6 MULTIPLE-CHOICE SELF-TEST

The Self-Test consists of multiple-choice questions similar to those found on the American Chemical Society organic chemistry exam. Answers are given at the end of the chapter.

1. Classify the given transformation as (a) an addition; (b) an elimination; (c) a substitution; or (d) a Brønsted–Lowry acid–base reaction.

Br

2. Use the curved arrows to identify the product of the reaction.

HO: OH H a. :OH O+ H b. $H_2\overset{+}{O}$:O: H c. :OH O: d. :O: O+ H

3. Which labeled bond has the highest bond dissociation energy?

a. b. c. d.

4. Ignoring entropy, what can be said about a reaction in which $\Delta H° = +25$ kJ/mol?

a. The bonds in the starting material are stronger than the bonds in the product.
b. $K_{eq} > 1$.
c. The reaction is exothermic.
d. $\Delta G°$ is a negative value.

5. How many bonds are broken or formed in the transition state of the given reaction?

Br: + $^-$:OH → + :Br:$^-$ + H_2O:

a. one c. three
b. two d. four

6. What can be said about a reaction in which $K_{eq} = 0.15$ and $E_a = +25$ kJ/mol?

a. The starting materials are higher in energy than the products.
b. The reaction is slower than a reaction with $K_{eq} = 1.3$ and $E_a = +25$ kJ/mol.
c. $\Delta G°$ is a negative value.
d. The reaction is slower than a reaction with $K_{eq} = 0.15$ and $E_a = +10$ kJ/mol.

7. If the rate equation for a reaction is rate = k[**X**][**Y**], which statement is incorrect?

a. The kinetics are second order.
b. Both **X** and **Y** appear in the only step or the rate-determining step of the reaction.
c. Doubling [**X**] and [**Y**] doubles the rate.
d. Changing the concentration of either **X** or **Y** changes the rate.

8. What products are formed from heterolysis of the indicated bond in **A**?

A: $\ddot{N}H_2$ (on cyclohexane)

a. cyclohexyl radical + $\cdot\ddot{N}H_2$
c. cyclohexane + $:NH_3$
b. cyclohexyl cation + $:\ddot{N}H_2^-$
d. cyclohexyl anion + $:\ddot{N}H_2^-$

9. In which reactions does entropy favor the products?

[1] Br → (alkene) + HBr
[2] H, O, OH → OH, O (cyclic)
[3] 2 (radical) → (alkane)

a. Reaction [1] only
b. Reactions [1] and [3]
c. Reactions [2] and [3]
d. Entropy favors the reactants in [1], [2], and [3].

10. What anion results by following the curved arrows in **A**?

A: $:\ddot{Cl}:$, $:\ddot{O}H^-$, NO_2

a. $\ddot{O}H$, NO_2
b. $:\ddot{Cl}:$, $\ddot{O}H$, NO_2
c. $:\ddot{Cl}:$, $\ddot{O}H$, NO_2
d. $:\ddot{Cl}:$, $\ddot{O}H$, NO_2

PROBLEMS

Problem Using Three-Dimensional Models

6.26 Explain why the bond dissociation energy for bond (a) is lower than the bond dissociation energy for bond (b).

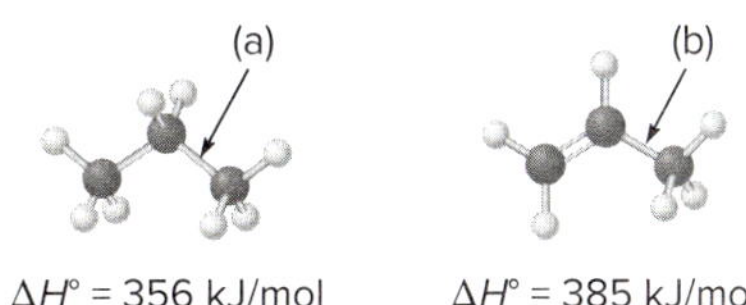

Types of Reactions

6.27 Classify each transformation as substitution, elimination, or addition.

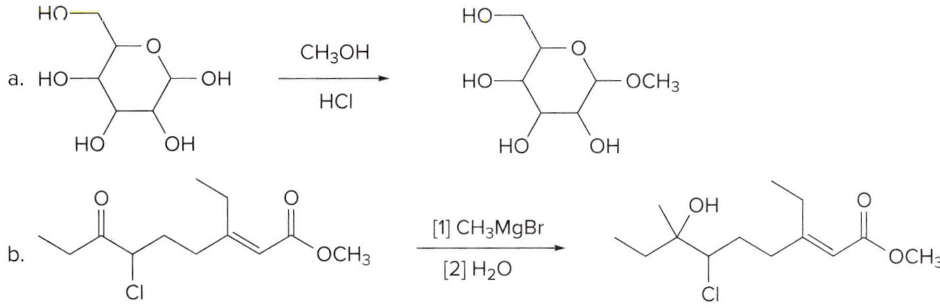

Curved Arrows

6.28 Use full-headed or half-headed curved arrows to show the movement of electrons in each reaction.

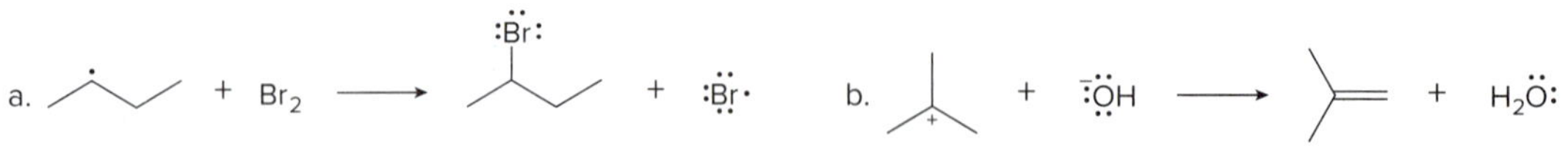

a. + Br_2 ⟶ :Br: + :Br· b. + :OH ⟶ + H_2O:

6.29 Draw the products of each reaction by following the curved arrows.

a. O + :Cl: ⟶ b. :OH_2 N ⟶

6.30 (a) Add curved arrows for each step to show how **A** is converted to the epoxy ketone **C.** (b) Classify the conversion of **A** to **C** as a substitution, elimination, or addition. (c) Draw one additional resonance structure for **B.**

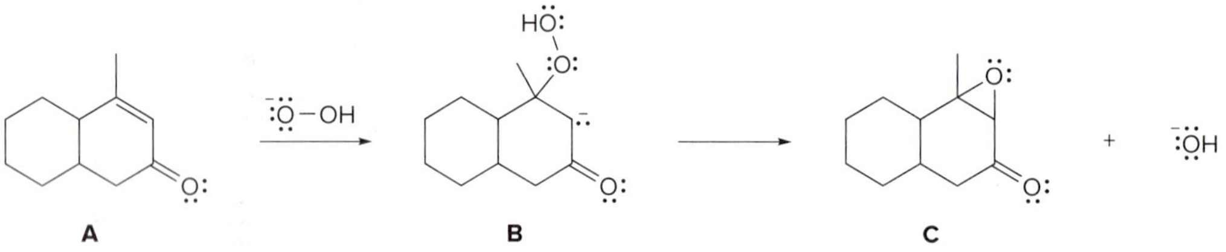

:O: :Ö–OH ⟶ HÖ: :O: :O: ⟶ :O: :O: + :ÖH

A B C

6.31 (a) Draw in the curved arrows to show how **A** is converted to **B** in Step [1]. (b) Identify **X,** using the curved arrows drawn for Step [2].

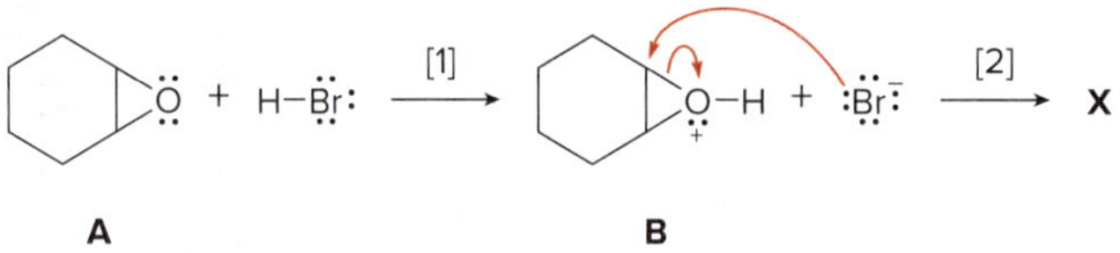

Ö + H–Br: —[1]⟶ O–H + :Br: —[2]⟶ X

A B

6.32 Add curved arrows to each step in the following reaction sequence.

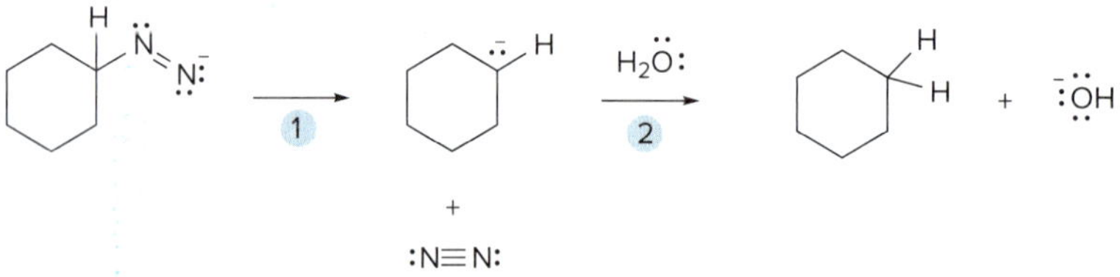

H N N: ⟶(1) H + :N≡N: —H_2O:, (2)⟶ H H + :ÖH

6.33 $PGF_{2\alpha}$ (a lipid discussed in Section 19.5) is synthesized in cells using a cyclooxygenase enzyme that catalyzes a multistep radical pathway. Two steps in the pathway are depicted in the accompanying equations. (a) Draw in curved arrows to illustrate how **C** is converted to **D** in Step [1]. (b) Identify **Y,** the product of Step [2], using the curved arrows that are drawn on compound **D.**

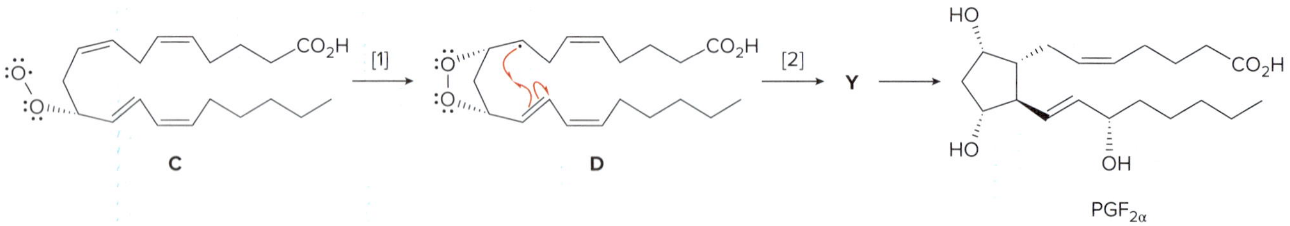

:O· :O CO_2H —[1]⟶ :O :O CO_2H —[2]⟶ Y ⟶ HO HO CO_2H OH

C D $PGF_{2\alpha}$

Bond Dissociation Energy and Calculating $\Delta H°$

6.34 Calculate $\Delta H°$ for each reaction.

a. $HO\cdot + CH_4 \longrightarrow \cdot CH_3 + H_2O$ b. $CH_3OH + HBr \longrightarrow CH_3Br + H_2O$

6.35 Homolysis of the indicated C–H bond in propene forms a resonance-stabilized radical.

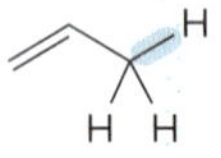

H H H propene

a. Draw the two possible resonance structures for this radical.

b. Use half-headed curved arrows to illustrate how one resonance structure can be converted to the other.

c. Draw a structure for the resonance hybrid.

Thermodynamics, $\Delta G°$, $\Delta H°$, $\Delta S°$, and K_{eq}

6.36 Given each value, determine whether the starting material or product is favored at equilibrium.

a. $K_{eq} = 0.5$
b. $\Delta G° = -100$ kJ/mol
c. $\Delta H° = 8.0$ kJ/mol
d. $K_{eq} = 16$
e. $\Delta G° = 2.0$ kJ/mol
f. $\Delta H° = 200$ kJ/mol
g. $\Delta S° = 8$ J/(K•mol)
h. $\Delta S° = -8$ J/(K•mol)

6.37 a. Which value corresponds to a negative value of $\Delta G°$: $K_{eq} = 10^{-2}$ or $K_{eq} = 10^{2}$?

b. In a unimolecular reaction with five times as much starting material as product at equilibrium, what is the value of K_{eq}? Is $\Delta G°$ positive or negative?

c. Which value corresponds to a larger K_{eq}: $\Delta G° = -8$ kJ/mol or $\Delta G° = 20$ kJ/mol?

6.38 As we learned in Chapter 4, monosubstituted cyclohexanes exist as an equilibrium mixture of two conformations having either an axial or equatorial substituent. When R = CH_2CH_3, K_{eq} for this process is 23. When R = $C(CH_3)_3$, K_{eq} for this process is 4000.

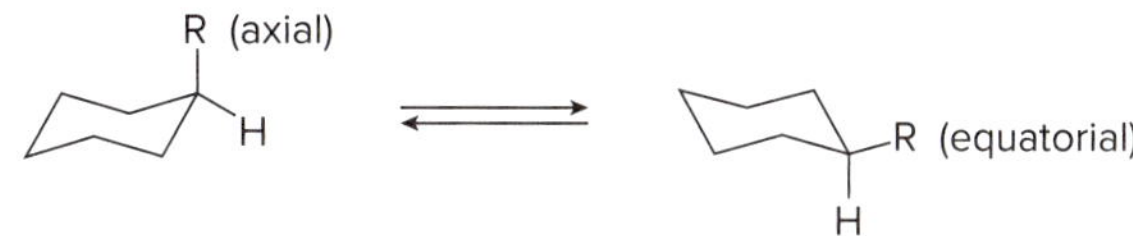

a. When R = CH_2CH_3, which conformation is present in higher concentration?

b. Which R shows the higher percentage of equatorial conformation at equilibrium?

c. Which R shows the higher percentage of axial conformation at equilibrium?

d. For which R is $\Delta G°$ more negative?

e. How is the size of R related to the amount of axial and equatorial conformations at equilibrium?

6.39 For which of the following reactions is $\Delta S°$ a positive value?

a.  [structures] ⟶ [structures] +

b. 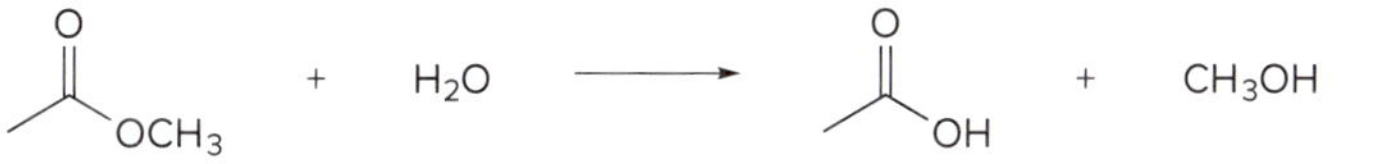 O, OCH_3 + H_2O ⟶ O, OH + CH_3OH

Energy Diagrams and Transition States

6.40 Draw the transition state for each reaction.

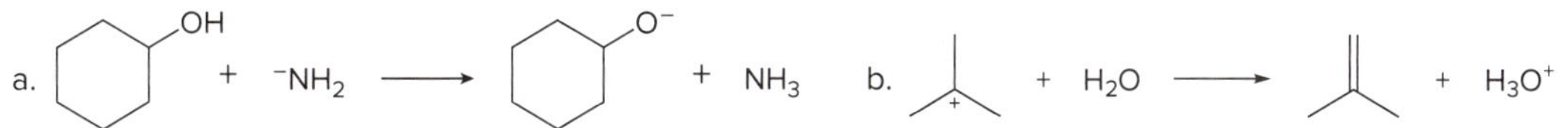

6.41 Draw an energy diagram for each reaction. Label the axes, the starting material, product, transition state, $\Delta H°$, and E_a.

a. a concerted reaction with $\Delta H° = -80$ kJ/mol and $E_a = 16$ kJ/mol

b. a two-step reaction, **A** → **B** → **C**, in which the relative energy of the compounds is **A** < **C** < **B**, and the step **A** → **B** is rate-determining

6.42 Consider the following reaction: CH_4 + Cl• → •CH_3 + HCl.

a. Use curved arrows to show the movement of electrons in this radical reaction.

b. Calculate $\Delta H°$ using the bond dissociation energies in Table 6.2.

c. Draw an energy diagram assuming that $E_a = 16$ kJ/mol.

d. What is E_a for the reverse reaction (•CH_3 + HCl → CH_4 + Cl•)?

6.43 Consider the following energy diagram for the conversion of **A** → **G.**

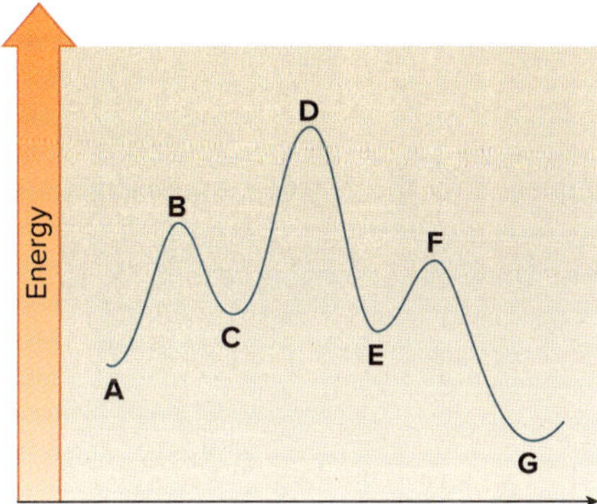

a. Which points on the graph correspond to transition states?

b. Which points on the graph correspond to reactive intermediates?

c. How many steps are present in the reaction mechanism?

d. Label each step of the mechanism as endothermic or exothermic.

e. Label the overall reaction as endothermic or exothermic.

6.44 Consider the following two-step reaction:

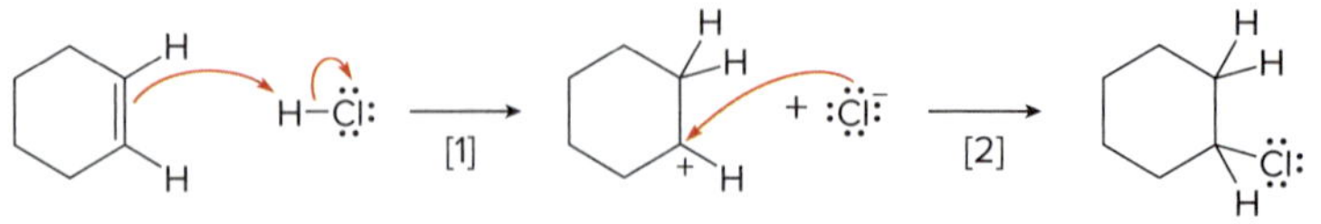

[1] [2]

a. How many bonds are broken and formed in Step [1]? Would you predict the $\Delta H°$ of Step [1] to be positive or negative?
b. How many bonds are broken and formed in Step [2]? Would you predict the $\Delta H°$ of Step [2] to be positive or negative?
c. Which step is rate-determining?
d. Draw the structure for the transition state in both steps of the mechanism.
e. If $\Delta H°_{overall}$ is negative for this two-step reaction, draw an energy diagram illustrating all of the information in parts (a)–(d).

Kinetics and Rate Laws

6.45 Indicate which factors affect the rate of a reaction.

a. $\Delta G°$
b. $\Delta H°$
c. E_a
d. temperature
e. concentration
f. K_{eq}
g. k
h. catalysts

6.46 The following is a concerted, bimolecular reaction: $CH_3Br + NaCN \rightarrow CH_3CN + NaBr$.

a. What is the rate equation for this reaction?
b. What happens to the rate of the reaction if $[CH_3Br]$ is doubled?
c. What happens to the rate of the reaction if [NaCN] is halved?
d. What happens to the rate of the reaction if $[CH_3Br]$ and [NaCN] are both increased by a factor of five?

6.47 The conversion of acetyl chloride to methyl acetate occurs via the following two-step mechanism:

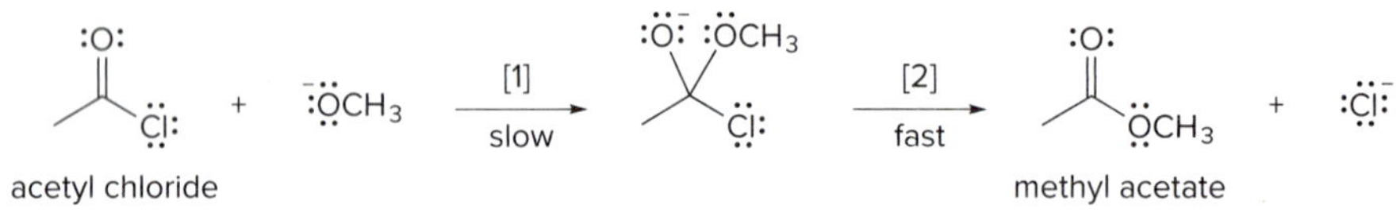

acetyl chloride + $^-OCH_3$ —[1] slow→ intermediate —[2] fast→ methyl acetate + Cl^-

a. Add curved arrows to show the movement of the electrons in each step.
b. Write the rate equation for this reaction, assuming the first step is rate-determining.
c. If the concentration of $^-OCH_3$ were increased 10 times, what would happen to the rate of the reaction?
d. If the concentrations of both CH_3COCl and $^-OCH_3$ were increased 10 times, what would happen to the rate of the reaction?
e. Classify the conversion of acetyl chloride to methyl acetate as an addition, elimination, or substitution.

Problems That Combine Concepts

6.48 Consider the conversion of alkyl halide **A** to ether **B.**

A + CH_3OH → **B** + HBr

a. Classify the conversion of **A** to **B** as substitution, elimination, or addition.
b. The reaction rate depends on the concentration of **A** only. Write the rate equation for the reaction, and explain why the reaction mechanism must involve more than one step.
c. Heterolysis of the polar bond in **A** forms a resonance-stabilized intermediate. Draw all reasonable resonance structures for this intermediate.

6.49 In Chapter 18, we will learn about the hydrolysis of acetals to aldehydes and ketones. Four of the seven steps in the mechanism for this process are shown in the conversion of acetal **A** to hemiacetal **E.**

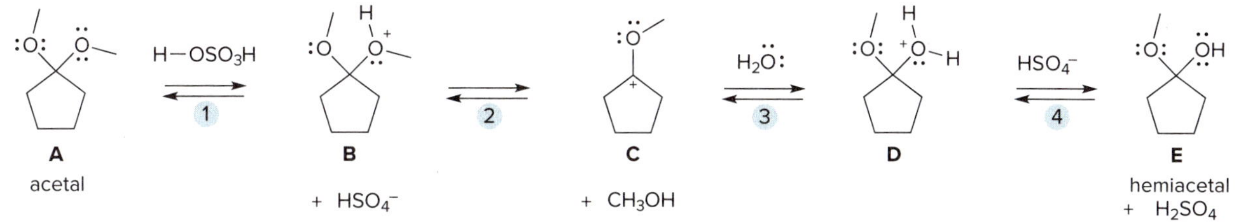

A acetal; $H-OSO_3H$ (1); B $+ HSO_4^-$; (2); C $+ CH_3OH$; $H_2\ddot{O}$: (3); D; HSO_4^- (4); E hemiacetal $+ H_2SO_4$

a. Add curved arrows for each step.
b. Draw another resonance structure for **C.**
c. Identify the nucleophile and electrophile in Step [3].
d. Which steps are Brønsted–Lowry acid–base reactions?

6.50 The Diels–Alder reaction, a powerful reaction discussed in Chapter 14, occurs when a 1,3-diene such as **A** reacts with an alkene such as **B** to form the six-membered ring in **C.**

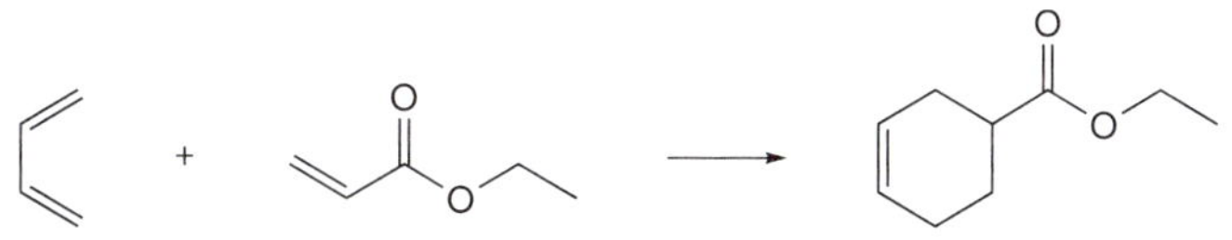

A + B → C

a. Draw curved arrows to show how **A** and **B** react to form **C.**
b. What bonds are broken and formed in this reaction?
c. Would you expect this reaction to be endothermic or exothermic?
d. Does entropy favor the reactants or products?
e. Is the Diels–Alder reaction a substitution, elimination, or addition?

6.51 The conversion of $(CH_3)_3CI$ to $(CH_3)_2C{=}CH_2$ can occur by either a one-step or a two-step mechanism, as shown in Equations [1] and [2].

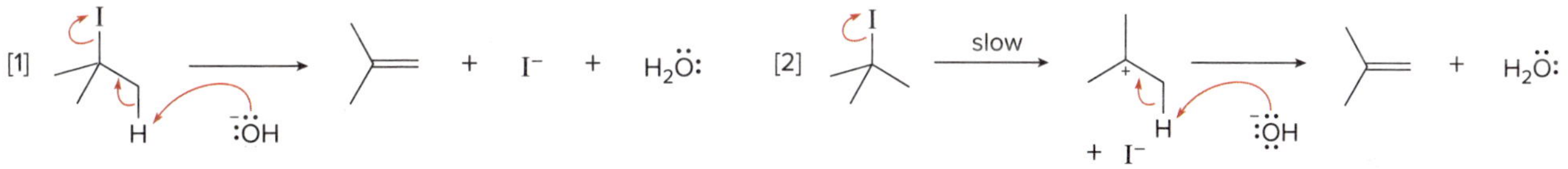

[1] (I, H, :ÖH) → $+ I^- + H_2\ddot{O}$:   [2] (I) —slow→ (+, H, :ÖH) $+ I^-$ → $+ H_2\ddot{O}$:

a. What rate equation would be observed for the mechanism in Equation [1]?
b. What rate equation would be observed for the mechanism in Equation [2]?
c. What is the order of each rate equation (i.e., first, second, and so forth)?
d. How can these rate equations be used to show which mechanism is the right one for this reaction?
e. Assume Equation [1] represents an endothermic reaction and draw an energy diagram for the reaction. Label the axes, reactants, products, E_a, and $\Delta H°$. Draw the structure for the transition state.
f. Assume Equation [2] represents an endothermic reaction and that the product of the rate-determining step is higher in energy than the reactants or products. Draw an energy diagram for this two-step reaction. Label the axes, reactants and products for each step, and the E_a and $\Delta H°$ for each step. Label $\Delta H°_{overall}$. Draw the structure for both transition states.

Challenge Problems

6.52 Explain why HC≡CH is more acidic than CH_3CH_3, even though the C–H bond in HC≡CH has a higher bond dissociation energy than the C–H bond in CH_3CH_3.

6.53 The use of curved arrows is a powerful tool that illustrates even complex reactions.

a. Add curved arrows to show how carbocation **A** is converted to carbocation **B.** Label each new σ bond formed. Similar reactions have been used in elegant syntheses of steroids.

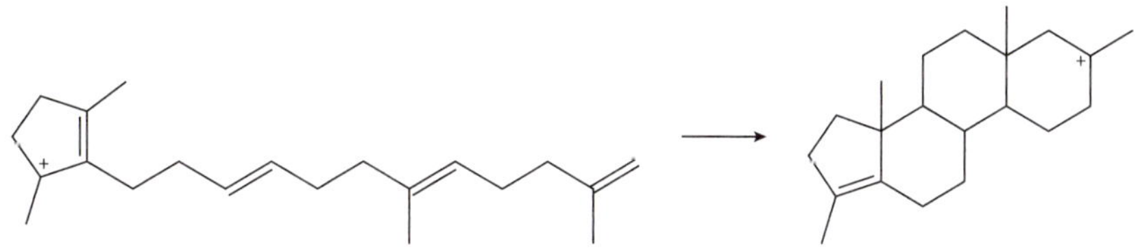

A → B

b. Draw the product by following the curved arrows. This reaction is an example of a [3,3] sigmatropic rearrangement, as we will learn in Chapter 25.

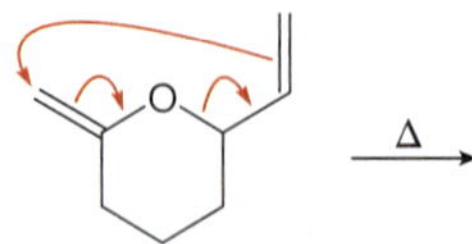

6.54 As we will learn in Section 13.12, many antioxidants—compounds that prevent unwanted radical oxidation reactions from occurring—are phenols, compounds that contain an OH group bonded directly to a benzene ring.

a. Explain why homolysis of the O–H bond in phenol requires considerably less energy than homolysis of the O–H bond in ethanol (362 kJ/mol vs. 438 kJ/mol).

b. Why is the C–O bond in phenol shorter than the C–O bond in ethanol?

O H phenol

O H ethanol

6.55 In Chapter 9 we will learn that carbocations (Section 6.3B) sometimes undergo rearrangement reactions by the migration of a hydrogen atom or an alkyl group from one carbon atom to an adjacent carbon atom. In these 1,2-shifts, the migrating group moves with the two electrons that bonded it to the carbon skeleton, as in the conversion of carbocation **A** to carbocation **B.**

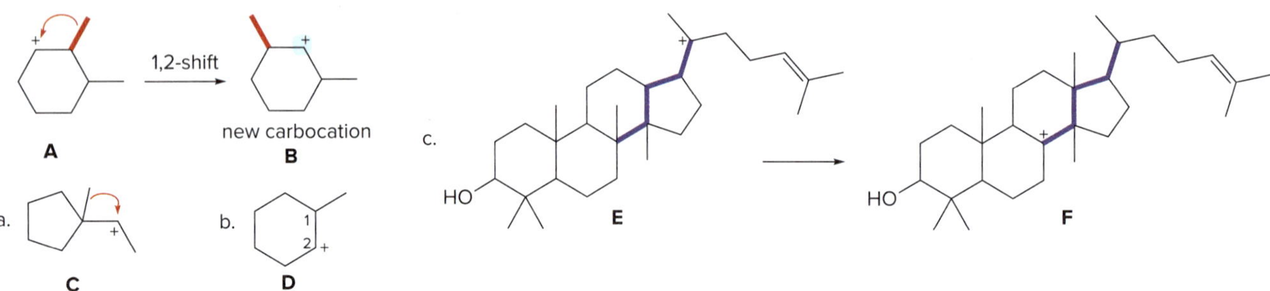

a. Follow the curved arrow to draw the product of a 1,2-shift of a methyl group in carbocation **C.**

b. Draw the product formed when a hydrogen migrates from C1 to C2 in carbocation **D.**

c. When cholesterol is synthesized in the body, carbocation **E** undergoes a series of four 1,2-shifts of H atoms or methyl groups to form carbocation **F.** Draw out the four 1,2-shifts that occur. (Hint: The atoms of the carbon skeleton to which the H or CH_3 groups are bonded are highlighted in blue.)

SELF-TEST ANSWERS

1. c 2. a 3. d 4. a 5. d 6. d 7. c 8. b 9. a 10. b

Alkyl Halides and Nucleophilic Substitution

Randy Faris/Corbis/Getty Images

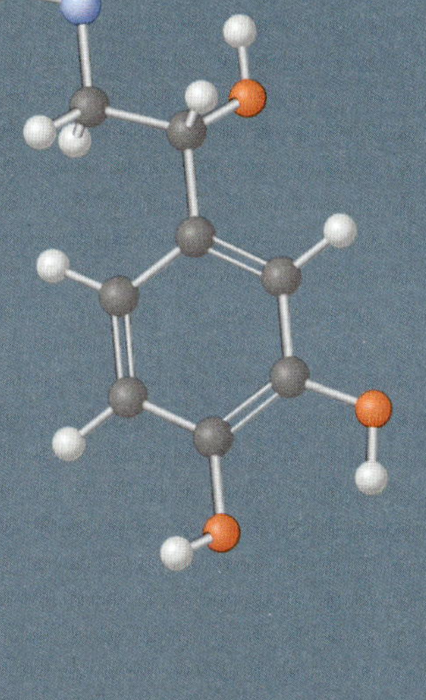

7.1 Introduction to alkyl halides
7.2 Nomenclature
7.3 Properties of alkyl halides
7.4 Interesting alkyl halides
7.5 The polar carbon–halogen bond
7.6 General features of nucleophilic substitution
7.7 The leaving group
7.8 The nucleophile
7.9 The mechanisms for nucleophilic substitution
7.10 The S_N2 mechanism
7.11 The S_N1 mechanism
7.12 Carbocation stability
7.13 The Hammond postulate
7.14 When is the mechanism S_N1 or S_N2?
7.15 Biological nucleophilic substitution
7.16 Vinyl halides and aryl halides
7.17 Organic synthesis

Adrenaline (or **epinephrine**), a hormone secreted by the adrenal gland, increases blood pressure and heart rate, and dilates lung passages. Individuals often speak of the "rush of adrenaline" when undertaking a particularly strenuous or challenging activity. Adrenaline is made in the body by a simple organic reaction called **nucleophilic substitution**. In Chapter 7 we learn about the mechanism of nucleophilic substitution and how adrenaline is synthesized in organisms.

Why Study . . . Alkyl Halides?

This is the first of three chapters dealing with an in-depth study of the organic reactions of compounds containing C–Z σ bonds, where Z is an element more electronegative than carbon. In Chapter 7, we learn about **alkyl halides** and one of their characteristic reactions, **nucleophilic substitution,** a key step in the synthesis of several useful drugs and natural products. In Chapter 8, we look at **elimination**, a second general reaction of alkyl halides. We conclude this discussion in Chapter 9 by examining other molecules that also undergo nucleophilic substitution and elimination reactions. In these chapters, we will learn about many specific details that explain how and why key reactions take place.

7.1 Introduction to Alkyl Halides

Alkyl halides are organic molecules containing a halogen atom X bonded to an sp^3 hybridized carbon atom. As we learned in Section 3.2, alkyl halides are classified as **primary (1°), secondary (2°),** or **tertiary (3°)** depending on the number of carbons bonded to the carbon with the halogen. Whether an alkyl halide is 1°, 2°, or 3° is the *most important factor* in determining the course of its chemical reactions.

Alkyl halides have the general molecular formula $C_nH_{2n+1}X$, and are formally derived from an alkane by replacing a hydrogen atom with a halogen.

C—X

alkyl halide

X = F, Cl, Br, I

C is sp^3 hybridized.

2° chloride

Cl

Br

F

3° bromide 1° fluoride

Four types of organic halides having the halogen atom in close proximity to a π bond are illustrated in Figure 7.1. **Vinyl halides** have a halogen atom bonded to a carbon–carbon double bond, and **aryl halides** have a halogen atom bonded to a benzene ring. These two types of organic halides with X bonded directly to an sp^2 hybridized carbon atom do *not* undergo the reactions presented in Chapter 7, as discussed in Section 7.16.

Figure 7.1 Four types of organic halides (RX) having X near a π bond

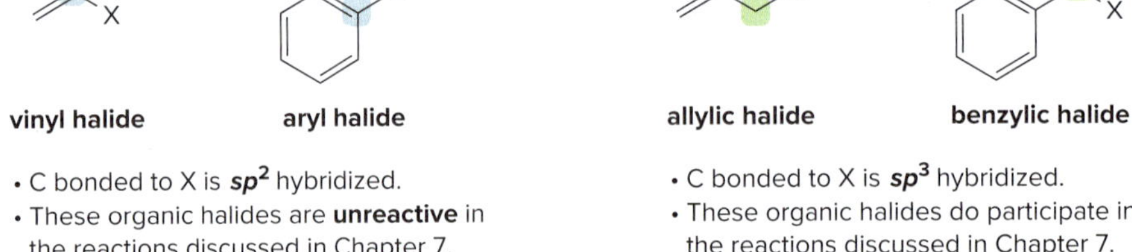

- C bonded to X is sp^2 hybridized.
- These organic halides are **unreactive** in the reactions discussed in Chapter 7.

- C bonded to X is sp^3 hybridized.
- These organic halides do participate in the reactions discussed in Chapter 7.

Allylic halides and benzylic halides have halogen atoms bonded to sp^3 hybridized carbon atoms and *do* undergo the reactions described in Chapter 7. **Allylic halides** have X bonded to the carbon atom *adjacent* to a carbon–carbon double bond, and **benzylic halides** have X bonded to the carbon atom *adjacent* to a benzene ring. The synthesis of allylic and benzylic halides is discussed in Sections 13.10 and 16.14, respectively.

Problem 7.1 Telfairine, a naturally occurring insecticide, and halomon, an antitumor agent, are two polyhalogenated compounds isolated from red algae. (a) Classify each halide bonded to an sp^3 hybridized carbon as 1°, 2°, or 3°. (b) Label each halide as vinyl, allylic, or neither.

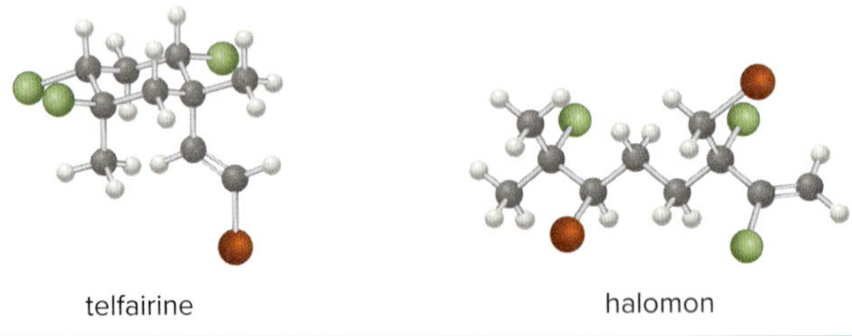

7.2 Nomenclature

The systematic (IUPAC) method for naming alkyl halides follows from the basic rules described in Chapter 4.

7.2A IUPAC System

An alkyl halide is named as an alkane with a halogen substituent—that is, as a ***halo alkane.*** To name a halogen substituent, change the ***-ine*** ending of the name of the halogen to the suffix ***-o*** (chlor*ine* → chlor*o*).

How To Name an Alkyl Halide Using the IUPAC System

Example Give the IUPAC name of the following alkyl halide:

Step [1] **Find the parent carbon chain and name it as an alkane.**

- Name the parent chain as an ***alkane,*** with the halogen as a substituent bonded to the longest chain.

7 C's in the longest chain

7 C's ----→ heptane

Step [2] **Apply all other rules of nomenclature.**

a. **Number** the chain.

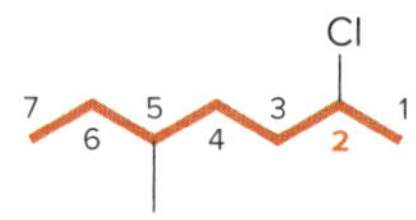

- Begin at the end nearest the first substituent, either alkyl or halogen.

b. **Name** and **number** the substituents.

Cl chloro at C2

methyl at C5

c. **Alphabetize: c** for **c**hloro, then **m** for **m**ethyl.

Answer: 2-chloro-5-methylheptane

7.2B Common Names

Common names for alkyl halides are used only for simple alkyl halides. To assign a common name:

- **Name all the carbon atoms of the molecule as a single alkyl group.**
- **Name the halogen bonded to the alkyl group. To name the halogen, change the *-ine* ending of the halogen name to the suffix *-ide;* for example, brom*ine* → brom*ide*.**
- **Combine the names of the alkyl group and halide, separating the words with a space.**

Other examples of alkyl halide nomenclature are given in Figure 7.2.

Figure 7.2 Examples: Nomenclature of alkyl halides

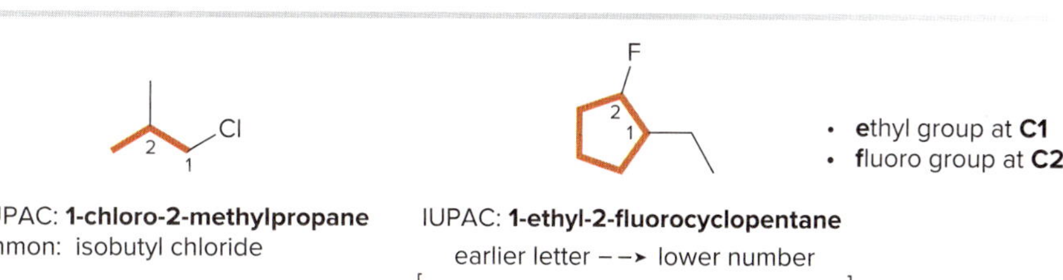

Problem 7.2 Give the IUPAC name for each compound.

Problem 7.3 Give the structure corresponding to each name.

a. 3-chloro-2-methylhexane
b. 4-ethyl-5-iodo-2,2-dimethyloctane
c. *cis*-1,3-dichlorocyclopentane
d. 1,1,3-tribromocyclohexane
e. 6-ethyl-3-iodo-3,5-dimethylnonane
f. (*R*)-1-fluoro-2,6,6-trimethylnonane

7.3 Properties of Alkyl Halides

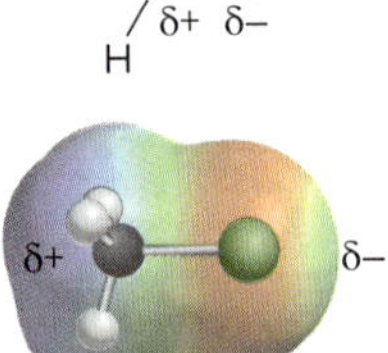

Alkyl halides are weakly polar molecules. They exhibit **dipole–dipole** interactions because of their polar C–X bond, but because the rest of the molecule contains only C–C and C–H bonds they are incapable of intermolecular hydrogen bonding. As a result, alkyl halides are insoluble in water regardless of size. The boiling points and melting points of alkyl halides are higher than alkanes with the same number of carbons, and they increase as the size of R and X increases.

The spectroscopic properties of alkyl halides are discussed in Chapters A–C. Of particular note are the characteristic features of the mass spectra of alkyl chlorides and alkyl bromides, which are discussed in Section A.2.

Problem 7.4 An sp^3 hybridized C–Cl bond is more polar than an sp^2 hybridized C–Cl bond. (a) Explain why this phenomenon arises. (b) Rank the following compounds in order of increasing boiling point.

7.4 Interesting Alkyl Halides

Many simple alkyl halides make excellent solvents because they are not flammable and dissolve a wide variety of organic compounds. Compounds in this category include **$CHCl_3$** (chloroform or trichloromethane) and **CCl_4** (carbon tetrachloride or tetrachloromethane). Large quantities of these solvents are produced industrially each year, but like many chlorinated organic compounds, both chloroform and carbon tetrachloride are toxic if inhaled or ingested. Other simple alkyl halides are shown in Figure 7.3.

Figure 7.3 Two simple alkyl halides

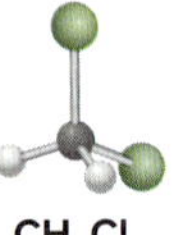

CH_2Cl_2

- **Dichloromethane (or methylene chloride, CH_2Cl_2)** is an important solvent, once used to decaffeinate coffee. Coffee is now decaffeinated by using supercritical CO_2 due to concerns over the possible ill effects of trace amounts of residual CH_2Cl_2 in the coffee. Subsequent studies on rats have shown, however, that no cancers occurred when animals ingested the equivalent of over 100,000 cups of decaffeinated coffee per day.

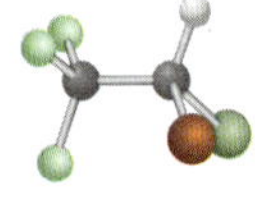

$CF_3CHClBr$

- **Halothane ($CF_3CHClBr$)** is a safe general anesthetic compared to other organic anesthetics such as $CHCl_3$, which causes liver and kidney damage, and $CH_3CH_2OCH_2CH_3$ (diethyl ether), which is very flammable.

Synthetic organic halides are also used in insulating materials, plastic wrap, and coatings. Two such compounds are **Teflon** and **poly(vinyl chloride) (PVC).**

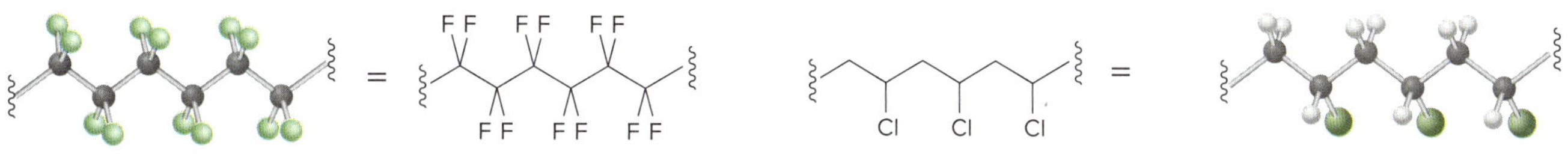

Teflon
(nonstick coating)

poly(vinyl chloride) (PVC)
(plastic used in films, pipes, and insulation)

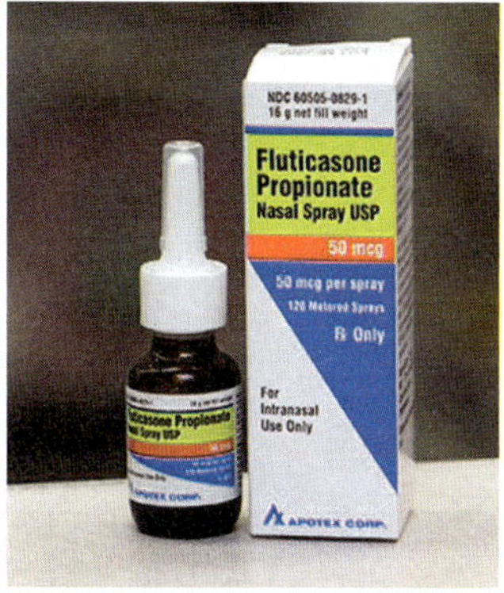

Fluticasone is a synthetic steroid used to treat the chronic inflammation of asthma.
Mark Dierker/McGraw Hill

Several useful drugs contain one or more fluorine atoms. Fluorine is a unique atom: It is the most electronegative element; it forms strong bonds to carbon; and it has very low polarizability. Substitution of one or more fluorine atoms can make a drug more readily dissolve in fats and decrease the rate at which the drug is metabolized. This in turn can affect the pharmacological profile of the compound.

Examples include fluticasone, roflumilast, and travoprost. Fluticasone is an aerosol inhalant used for the treatment of seasonal nasal allergies and asthma, and roflumilast, which was approved by the FDA in 2015, is used for the treatment of severe cases of chronic obstructive pulmonary disease (COPD). Travoprost is a drug that decreases the pressure inside the eye, making it useful in the treatment of glaucoma.

fluticasone
Trade names: Flonase, Flovent

roflumilast
Trade name: Daxas

travoprost
Trade name: Travatan

Although the beneficial effects of many organic halides are undisputed, certain synthetic chlorinated organics such as the **chlorofluorocarbons** and the pesticide **DDT** have caused lasting harm to the environment.

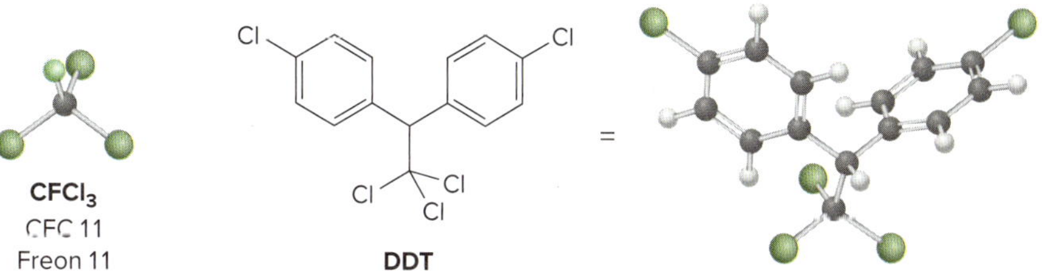

$CFCl_3$
CFC 11
Freon 11

DDT

Chlorofluorocarbons (CFCs) have the general molecular structure $\mathbf{CF_xCl_{4-x}}$. Trichlorofluoromethane [$CFCl_3$, CFC 11, or Freon 11 (trade name)] is an example of these easily vaporized compounds, having been extensively used as a refrigerant and an aerosol propellant. CFCs slowly rise to the stratosphere, where sunlight catalyzes their decomposition, a process that contributes to the destruction of the ozone layer, the thin layer of atmosphere that shields the earth's surface from harmful ultraviolet radiation (Section 13.9). Although it is now easy

DDT, a nonbiodegradable pesticide, has been labeled both a "miraculous" discovery by Winston Churchill in 1945 and the "elixir of death" by Rachel Carson in her 1962 book *Silent Spring*. DDT use was banned in the United States in 1973, but because of its effectiveness and low cost, it is still widely used to control inspect populations in developing countries.

Hundreds of organic halides with diverse structures and biological activities have been isolated from red algae of the genus *Laurencia*, seaweed that grows in shallow water at the edges of reefs (Problem 7.5).
Michael Guiry

to second-guess the extensive use of CFCs, it is also easy to see why they were used so widely. **CFCs made refrigeration available to the general public.** Would you call your refrigerator a comfort or a necessity?

The story of the insecticide **DDT** (**d**ichloro**d**iphenyl**t**richloroethane) follows the same theme: DDT is an organic molecule with valuable short-term effects that has caused long-term problems. DDT kills insects that spread diseases such as malaria and typhus, and in controlling insect populations, DDT has saved millions of lives worldwide. DDT is a weakly polar organic compound that persists in the environment for years. Because DDT is soluble in organic media, it accumulates in fatty tissues. Most adults in the United States have low concentrations of DDT (or a degradation product of DDT) in their bodies. DDT is acutely toxic to many types of marine life (crayfish, sea shrimp, and some fish), but the long-term effect on humans is not known.

Problem 7.5 Ma'ilione is a marine natural product isolated from the Hawaiian red seaweed *Laurencia cartilaginea*. (a) Predict the solubility of ma'ilione in water and CH_2Cl_2. (b) Locate the stereogenic centers. (c) Draw a stereoisomer and a constitutional isomer of ma'ilione.

HO
Br
O
ma'ilione

7.5 The Polar Carbon–Halogen Bond

The properties of alkyl halides dictate their reactivity. The electrostatic potential maps of four simple alkyl halides in Figure 7.4 illustrate that the electronegative halogen X creates a polar C–X bond, making the carbon atom electron deficient. **The chemistry of alkyl halides is determined by this polar C–X bond.**

Figure 7.4 Electrostatic potential maps of four halomethanes (CH_3X)

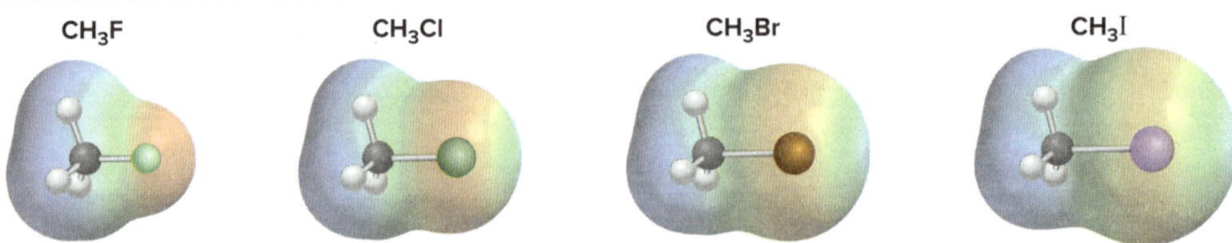

- The polar C–X bond makes the carbon atom *electron deficient* in each CH_3X molecule.

What kind of reactions do alkyl halides undergo? **The characteristic reactions of alkyl halides are substitution and elimination.** Because alkyl halides contain an electrophilic carbon, they react with electron-rich reagents—Lewis bases (nucleophiles) and Brønsted–Lowry bases.

- **Alkyl halides undergo substitution reactions with nucleophiles.**

X + :Nu$^-$ (nucleophile) → (substitution of X by Nu) Nu + :X$^-$ (leaving group)

In a substitution reaction of an alkyl halide, **the halogen X is replaced by an electron-rich nucleophile :Nu$^-$.** The C–X σ bond is broken and the C–Nu σ bond is formed.

- **Alkyl halides undergo elimination reactions with Brønsted–Lowry bases.**

H, X (alkyl halide) + :B (base) $\xrightarrow{\text{elimination of HX}}$ alkene (new π bond) + $H{-}B^+$ + $:X^-$

In an elimination reaction of an alkyl halide, the **elements of HX are removed by a Brønsted–Lowry base :B.**

The remainder of Chapter 7 is devoted to a discussion of the substitution reactions of alkyl halides. Elimination reactions are discussed in Chapter 8.

7.6 General Features of Nucleophilic Substitution

Three components are necessary in any substitution reaction.

[1] An alkyl group containing an sp^3 hybridized carbon bonded to X.

[2] **X**—An atom X (or a group of atoms) called **a leaving group,** which is able to accept the electron density in the C–X bond. The most common leaving groups are halide anions (X^-), but H_2O (from ROH_2^+) and N_2 (from RN_2^+) are also encountered.

[3] **:Nu⁻**—A **nucleophile.** Nucleophiles contain a **lone pair** or a **π bond** but not necessarily a negative charge.

Because these substitution reactions involve electron-rich nucleophiles, they are called ***nucleophilic*** **substitution reactions.** Examples are shown in Equations [1] and [2]. **Nucleophilic substitutions are Lewis acid–base reactions.** The nucleophile donates its electron pair, the alkyl halide (Lewis acid) accepts it, and the C–X bond is heterolytically cleaved. Curved arrow notation can be used to show the movement of electron pairs, as shown in Equation [2].

[1] CH_3CH_2Cl + $:\!OH^-$ ⟶ CH_3CH_2OH + $:Cl^-$

[2] $(CH_3)_2CHCH_2Br$ + $:\!OCH_3^-$ (nucleophile) ⟶ $(CH_3)_2CHCH_2OCH_3$ + $:Br^-$ (leaving group)

Negatively charged nucleophiles like ^-OH and ^-SH are used as **salts** with Li^+, Na^+, or K^+ counterions to balance charge. The identity of the cation is usually inconsequential, and therefore it is often omitted from the chemical equation.

$CH_3CH_2CH_2Br$ + $Na^+\ :\!OH^-$ ⟶ $CH_3CH_2CH_2OH$ + $Na^+\ :Br^-$

When a neutral nucleophile is used, the substitution product bears a positive charge. **All atoms originally bonded to the nucleophile stay bonded to it after substitution occurs.** All three CH_3 groups stay bonded to the N atom in the given example.

neutral nucleophile

All CH_3's remain in the product.

The reaction of alkyl halides with NH_3 to form amines (RNH_2) is discussed in Chapter 23.

Furthermore, when the substitution product bears a positive charge and also contains a *proton* bonded to O or N, the initial substitution product readily loses a proton in a Brønsted–Lowry acid–base reaction, forming a neutral product.

$:NH_3$ excess — nucleophilic substitution — $:NH_3$ — proton transfer — NH_2 + NH_4^+

All of these reactions are nucleophilic substitutions and have the same overall result—**replacement of the leaving group by the nucleophile,** regardless of the identity or charge of the nucleophile. To draw any nucleophilic substitution product:

- **Find the sp^3 hybridized carbon with the leaving group.**
- **Identify the nucleophile, the species with a lone pair or π bond.**
- **Substitute the nucleophile for the leaving group and assign charges (if necessary) to any atom that is involved in bond breaking or bond formation.**

Problem 7.6 Identify the nucleophile and leaving group and draw the products of each substitution reaction.

a. + $^-OCH_2CH_3$ ⟶
b. + NaOH ⟶
c. + N_3^- ⟶
d. + NaCN ⟶
e. + $CH_3CO_2^-$ ⟶
f. + $NaOCH_3$ ⟶

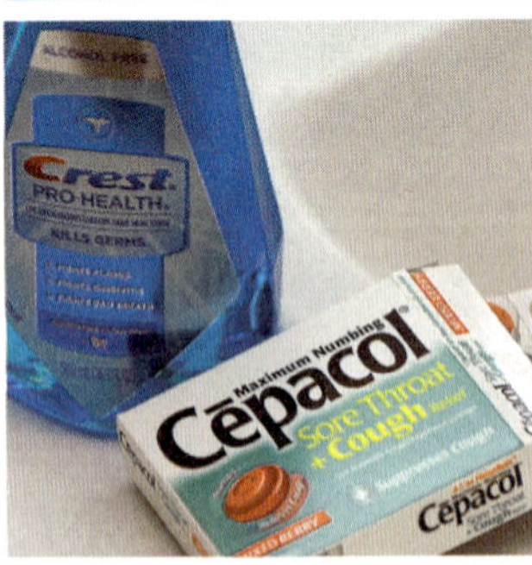

Cepacol throat lozenges and Crest Pro-Health mouthwash contain the antiseptic CPC, which is prepared by nucleophilic substitution (Problem 7.8). *Jill Braaten*

Problem 7.7 Draw the product of nucleophilic substitution with each neutral nucleophile. When the initial substitution product can lose a proton to form a neutral product, draw the product after proton transfer.

a. + $:N(CH_2CH_3)_3$ ⟶
b. + $H_2\ddot{O}:$ ⟶

Problem 7.8 CPC (cetylpyridinium chloride), an antiseptic found in throat lozenges and mouthwash, is synthesized by the following reaction. Draw the structure of CPC.

pyridine + $CH_3(CH_2)_{15}Cl$ ⟶ CPC

Problem 7.9 What neutral nucleophile is needed to convert dihalide **A** to ticlopidine, an antiplatelet drug used to reduce the risk of strokes?

A → ticlopidine

7.7 The Leaving Group

Nucleophilic substitution is a general reaction of organic compounds. Why, then, are alkyl halides the most common substrates, and halide anions the most common leaving groups? To answer this question, we must understand leaving group ability. **What makes a good leaving group?**

In a nucleophilic substitution reaction of R–X, the C–X bond is heterolytically cleaved, and the leaving group departs with the electron pair in that bond, forming X:$^-$. **The more stable the leaving group X:$^-$, the better able it is to accept an electron pair,** giving rise to the following generalization:

- In comparing two leaving groups, the *better* leaving group is the *weaker* base.

Good leaving groups are weak bases.

For example, H_2O is a better leaving group than ^-OH because H_2O is a weaker base. Moreover, the periodic trends in basicity can now be used to identify **periodic trends in leaving group ability:**

- Left-to-right across a row of the periodic table, basicity *decreases* so leaving group ability *increases.*

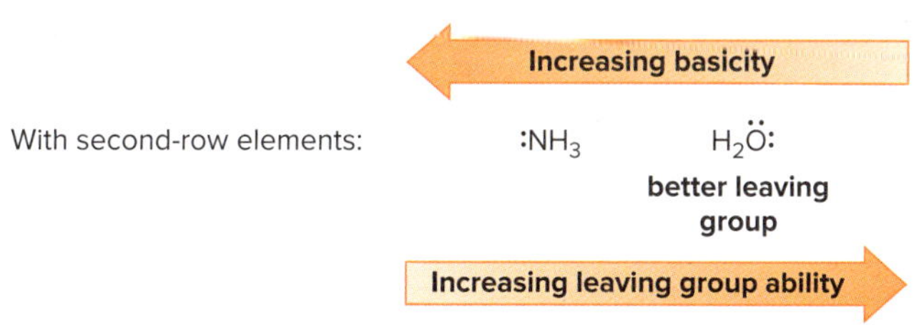

- Down a column of the periodic table, basicity *decreases* so leaving group ability *increases.*

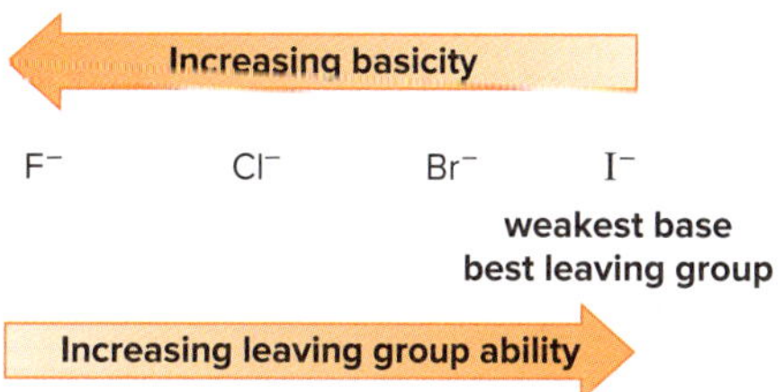

All good leaving groups are weak bases with strong conjugate acids having low pK_a values. Thus, all halide anions except F$^-$ are good leaving groups because their conjugate acids (HCl,

HBr, and HI) have low pK_a values. Tables 7.1 and 7.2 list good and poor leaving groups for nucleophilic substitution reactions, respectively. Nucleophilic substitution does not occur with any of the leaving groups in Table 7.2 because these leaving groups are strong bases.

Table 7.1 Good Leaving Groups for Nucleophilic Substitution

Starting material	Leaving group	Conjugate acid	pK_a
R–Cl	$:\ddot{Cl}:^-$	HCl	–6
R–Br	$:\ddot{Br}:^-$	HBr	–9
R–I	$:\ddot{I}:^-$	HI	–10
R–OH_2^+	$H_2\ddot{O}:$	H_3O^+	0

Table 7.2 Poor Leaving Groups for Nucleophilic Substitution

Starting material	Leaving group	Conjugate acid	pK_a
R–F	$:\ddot{F}:^-$	HF	3.2
R–OH	$^-:\ddot{O}H$	H_2O	14
R–NH_2	$^-:\ddot{N}H_2$	NH_3	38
R–H	$H:^-$	H_2	35
R–R	$R:^-$	RH	50

Problem 7.10 Which molecules contain good leaving groups?

a. 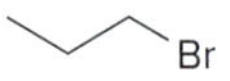 (propyl bromide) Br b.  (propanol) OH c. 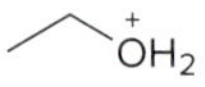 $\overset{+}{O}H_2$ d.  (butane)

Problem 7.11 (a) Which of the labeled atoms in each molecule is the best leaving group? (b) Which of the labeled atoms in each molecule is the worst leaving group?

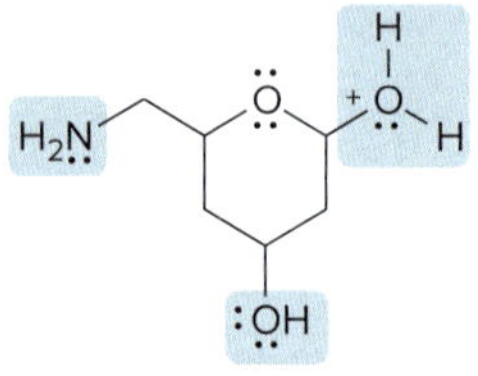

A: ring bearing $H_2\ddot{N}$, $\ddot{O}$, $\overset{+}{O}H_2$, and $:\ddot{O}H$ groups
B: ring bearing HO, Cl, I, and Br groups

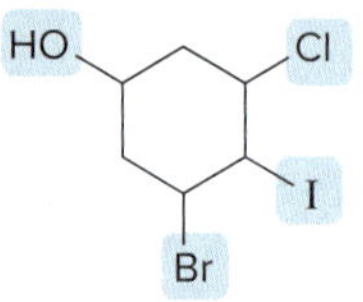

Given a particular nucleophile and leaving group, how can we determine whether the equilibrium will favor products in a nucleophilic substitution? We can often correctly predict the direction of equilibrium by comparing the basicity of the nucleophile and the leaving group.

- **Equilibrium favors the products of nucleophilic substitution when the leaving group is a *weaker base* than the nucleophile.**

Sample Problem 7.1 illustrates how to apply this general rule.

Sample Problem 7.1 Using Basicity to Determine If a Substitution Is Likely to Occur

Will the following substitution reaction favor formation of the products?

$$CH_3CH_2Cl + {}^-:\ddot{O}H \longrightarrow CH_3CH_2\ddot{O}H + :\ddot{Cl}:^-$$

Solution

Determine the basicity of the nucleophile (^-OH) and the leaving group (Cl^-) by comparing the pK_a values of their conjugate acids. **The stronger the conjugate acid, the weaker the base, and the better the leaving group.**

			conjugate acids	
nucleophile	^{-}OH	⟶	H_2O	$pK_a = 14$
leaving group	Cl^-	⟶	HCl	$pK_a = -6$
	weaker base		**stronger acid**	

Because Cl^-, the leaving group, is a weaker base than ^{-}OH, the nucleophile, **the reaction favors the products.**

Problem 7.12 Does the equilibrium favor the reactants or the products in each substitution reaction?

a. $CH_3CH_2CH_2NH_2$ + Br^- ⟶ $CH_3CH_2CH_2Br$ + $^{-}NH_2$

b. $CH_3CH_2CH_2CH_2CH_2I$ + ^{-}CN ⟶ $CH_3CH_2CH_2CH_2CH_2CN$ + I^-

More Practice: Try Problem 7.49.

7.8 The Nucleophile

We use the word *base* to mean *Brønsted–Lowry* base and the word *nucleophile* to mean a *Lewis base* that reacts with electrophiles *other than protons*.

Nucleophiles and bases are structurally similar: both have a lone pair or a π bond. They differ in what they attack.

- **Bases attack protons. Nucleophiles attack other electron-deficient atoms (usually carbons).**

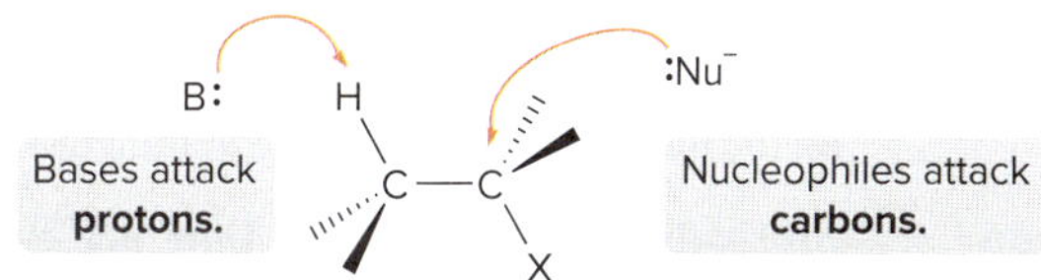

7.8A Nucleophilicity Versus Basicity

How is **nucleophilicity** (nucleophile strength) related to basicity? Although it is generally true that **a strong base is a strong nucleophile,** nucleophile size and steric factors can sometimes change this relationship.

Nucleophilicity parallels basicity in three instances.

[1] For two nucleophiles with the same nucleophilic atom, the *stronger* base is the *stronger* nucleophile.

- The relative nucleophilicity of ^{-}OH and $CH_3CO_2^-$, two oxygen nucleophiles, is determined by comparing the pK_a values of their conjugate acids (H_2O and CH_3CO_2H). CH_3CO_2H ($pK_a = 4.8$) is a stronger acid than H_2O ($pK_a = 14$), so **^{-}OH is a stronger base and stronger nucleophile than $CH_3CO_2^-$.**

[2] A negatively charged nucleophile is always *stronger* than its conjugate acid.

- ^{-}OH is a stronger base and stronger nucleophile than H_2O, its conjugate acid.

[3] Right-to-left across a row of the periodic table, nucleophilicity *increases* as basicity *increases*.

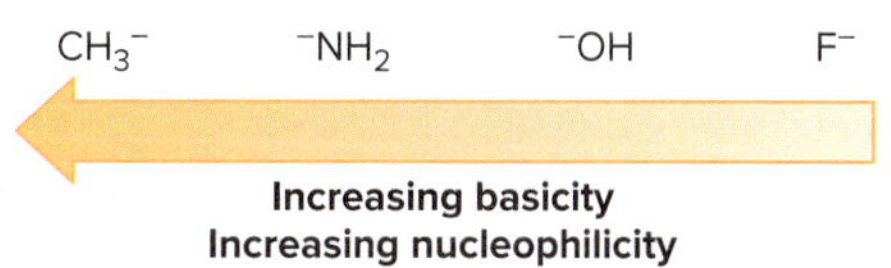

Problem 7.13 Identify the stronger nucleophile in each pair.

a. NH_3, $^-NH_2$ b. CH_3NH_2, CH_3OH c. $CH_3CO_2^-$, $CH_3CH_2O^-$

7.8B Steric Effects and Nucleophilicity

All steric effects arise because two atoms cannot occupy the same space. In Chapter 4, for example, we learned that **steric strain** is an increase in energy when big groups (occupying a large volume) are forced close to each other.

Nucleophilicity does not parallel basicity when **steric hindrance** becomes important. ***Steric hindrance* is a decrease in reactivity resulting from the presence of bulky groups at the site of a reaction.**

For example, although pK_a tables indicate that *tert*-butoxide [$(CH_3)_3CO^-$] is a stronger base than ethoxide ($CH_3CH_2O^-$), **ethoxide is the *stronger* nucleophile.** The three CH_3 groups around the O atom of *tert*-butoxide create steric hindrance, making it more difficult for this big, bulky base to attack a tetravalent carbon atom.

ethoxide
stronger nucleophile

tert-butoxide
stronger base
Three CH_3 groups *crowd* the O.
weaker nucleophile

Steric hindrance decreases nucleophilicity but *not* basicity. Because bases pull off small, easily accessible protons, they are unaffected by steric hindrance. Nucleophiles, on the other hand, must attack a crowded tetrahedral carbon, so bulky groups decrease reactivity.

Sterically hindered bases that are poor nucleophiles are called *nonnucleophilic bases.* Potassium *tert*-butoxide [K^+ $^-OC(CH_3)_3$] is a strong, nonnucleophilic base.

7.8C Comparing Nucleophiles of Different Size—Solvent Effects

Atoms vary greatly in size down a column of the periodic table, and in this case, **nucleophilicity depends on the solvent used in a substitution reaction.** Although solvent has thus far been ignored, most organic reactions take place in a liquid solvent that dissolves all reactants to some extent. Because substitution reactions involve polar starting materials, polar solvents are used to dissolve them. There are two main kinds of polar solvents: **polar *protic* solvents** and **polar *aprotic* solvents.**

Polar Protic Solvents

In addition to dipole–dipole interactions, **polar *protic* solvents are capable of intermolecular hydrogen bonding,** because they contain an O–H or N–H bond. The most common polar protic solvents are water and alcohols (ROH) (Figure 7.5). **Polar protic solvents solvate *both* cations and anions well.**

- **Cations are solvated by ion–dipole interactions.**
- **Anions are solvated by hydrogen bonding.**

Figure 7.5 Polar protic solvents

H_2O	CH_3OH methanol	CH_3CH_2OH ethanol	$(CH_3)_3COH$ *tert*-butanol	CH_3CO_2H acetic acid

For example, if the salt NaBr is used as a source of the nucleophile Br^- in H_2O, the Na^+ cations are solvated by ion–dipole interactions with H_2O molecules, and the Br^- anions are solvated by strong hydrogen-bonding interactions.

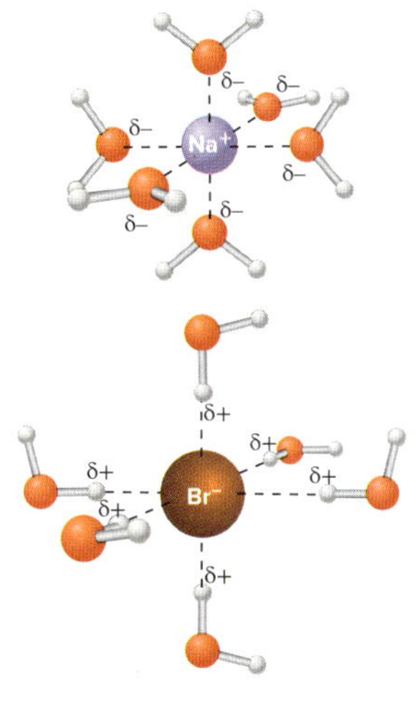

How do polar protic solvents affect nucleophilicity? **In polar protic solvents, nucleophilicity *increases* down a column of the periodic table as the size of the anion increases. This is *opposite* to basicity.** A small electronegative anion like F^- is very well solvated by hydrogen bonding, effectively *shielding* it from reaction. On the other hand, a large, less electronegative anion like I^- does not hold onto solvent molecules as tightly. The *solvent does not "hide" a large nucleophile* as well, and the nucleophile is much more able to donate its electron pairs in a reaction. Thus, **nucleophilicity increases down a column** even though basicity decreases, giving rise to the following trend in polar protic solvents:

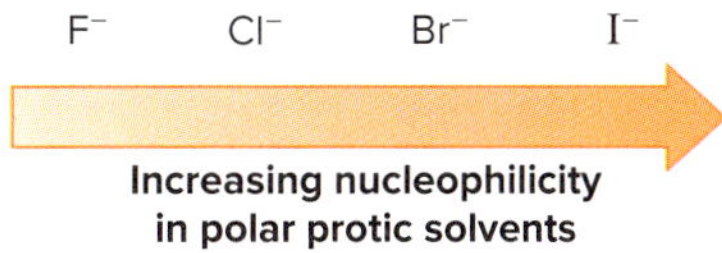

I^- is a *weak base* but a *strong nucleophile* in polar protic solvents.

Polar Aprotic Solvents

Polar *aprotic* solvents also exhibit dipole–dipole interactions, but they have no O–H or N–H bond so they are **incapable of hydrogen bonding.** Examples of polar aprotic solvents are shown in Figure 7.6. **Polar aprotic solvents solvate only cations well.**

- Cations are solvated by ion–dipole interactions.
- Anions are not well solvated because the solvent cannot hydrogen bond to them.

Figure 7.6 Polar aprotic solvents

acetone

$CH_3-C\equiv N$
acetonitrile

tetrahydrofuran
THF

dimethyl sulfoxide
DMSO

dimethylformamide
DMF

$[(CH_3)_2N]_3PO$
hexamethylphosphoramide
HMPA

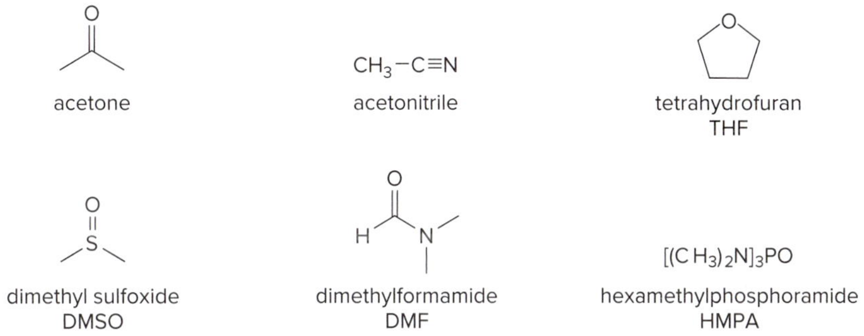

Abbreviations are often used in organic chemistry, instead of a compound's complete name. A list of common abbreviations is given in Appendix B.

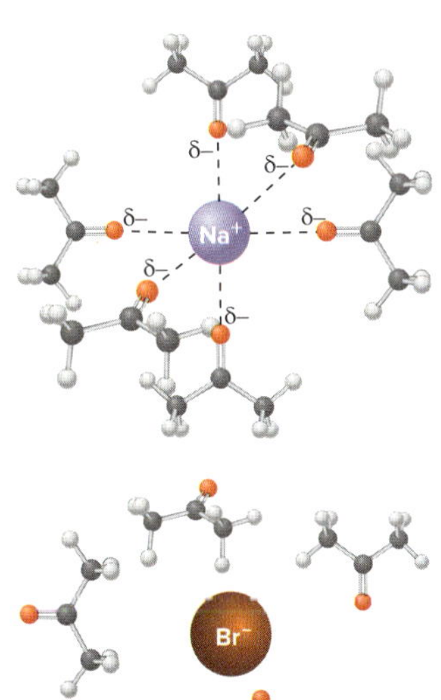

Br⁻ anions are surrounded by solvent but not well solvated by the $(CH_3)_2C{=}O$ molecules.

When the salt NaBr is dissolved in acetone, $(CH_3)_2C{=}O$, the Na^+ cations are solvated by ion dipole interactions with the acetone molecules, but, with no possibility for hydrogen bonding, the **Br^- anions are not well solvated.** Often these anions are called **naked anions** because they are not bound by tight interactions with solvent.

How do polar aprotic solvents affect nucleophilicity? Because anions are not well solvated in polar aprotic solvents, there is no need to consider whether solvent molecules more effectively hide one anion than another. **Nucleophilicity parallels basicity and the stronger base is the stronger nucleophile.** Because basicity decreases with size down a column, nucleophilicity decreases as well:

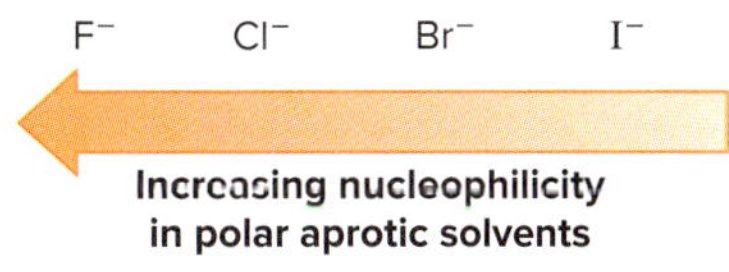

Problem 7.14 Classify each solvent as protic or aprotic.

a. 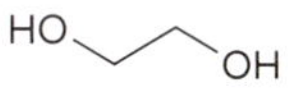 HO OH b. 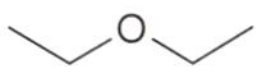 c. 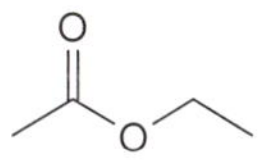 d. $N(CH_3)_3$ e. $HCONH_2$

Problem 7.15 Identify the stronger nucleophile in each pair of anions.

a. Br^- or Cl^- in a polar protic solvent
b. HO^- or Cl^- in a polar aprotic solvent
c. HS^- or F^- in a polar protic solvent
d. CH_3S^- or I^- in a polar aprotic solvent

7.8D Summary

Keep in mind the central relationship between nucleophilicity and basicity in comparing two nucleophiles.

- **It is generally true that the *stronger* base is the *stronger* nucleophile.**
- **In polar *protic* solvents, however, nucleophilicity *increases* with increasing size of an anion (opposite to basicity).**
- **Steric hindrance *decreases* nucleophilicity without decreasing basicity, making $(CH_3)_3CO^-$ a stronger base but a weaker nucleophile than $CH_3CH_2O^-$.**

Table 7.3 lists some common nucleophiles used in nucleophilic substitution reactions.

Problem 7.16 Rank the nucleophiles in each group in order of increasing nucleophilicity.

a. ^-OH, $^-NH_2$, H_2O b. ^-OH, Br^-, F^- (polar aprotic solvent) c. H_2O, ^-OH, $CH_3CO_2^-$

Problem 7.17 What nucleophile is needed to convert $(CH_3)_2CHCH_2CH_2Br$ to each product?

a. [structure: $(CH_3)_2CHCH_2CH_2SH$] b. [structure: $(CH_3)_2CHCH_2CH_2OCH_2CH_3$] c. [structure: $(CH_3)_2CHCH_2CH_2OC(=O)CH(CH_3)$...] d. [structure: $(CH_3)_2CHCH_2CH_2C\equiv CH$]

Table 7.3 Common Nucleophiles in Organic Chemistry

	Negatively charged nucleophiles			Neutral nucleophiles	
Oxygen	^-OH	^-OR	$CH_3CO_2^-$	H_2O	ROH
Nitrogen	N_3^-			NH_3	RNH_2
Carbon	^-CN	$HC\equiv C^-$			
Halogen	Cl^-	Br^-	I^-		
Sulfur	HS^-	RS^-		H_2S	RSH

7.9 The Mechanisms for Nucleophilic Substitution

Now that you know something about the general features of nucleophilic substitution, you can begin to understand the mechanism.

$$R{-}X \quad + \quad :Nu^- \longrightarrow R{-}Nu \quad + \quad :X^-$$

The σ bond is broken. The σ bond is formed.

7.9A Possible Mechanisms

Nucleophilic substitution at an sp^3 hybridized carbon involves two σ bonds: the bond to the leaving group is broken and the bond to the nucleophile is formed. To understand the mechanism of this reaction, though, we must know the timing of these two events; that is, **what is**

the order of bond breaking and bond making? Do they happen at the same time, or does one event precede the other? Consider two possibilities:

[1] **The mechanism has one step, and bond breaking and bond making occur at the *same* time.**

The C–X σ bond is broken... *as* ...the C–Nu σ bond is formed.

- If the C–X bond is broken *as* the C–Nu bond is formed, the mechanism has **one step.** As we learned in Section 6.9, the rate of such a bimolecular reaction depends on the concentration of *both* reactants; that is, the rate equation is **second order,** and **rate = k[RX][:Nu⁻].**

[2] **The mechanism has two steps, and bond breaking occurs *before* bond making.**

:X: 1 carbocation + :X:⁻ :Nu⁻ 2 Nu

The C–X σ bond is broken... **before** ...the C–Nu σ bond is formed.

- If the C–X bond is broken ***first*** and then the C–Nu bond is formed, the mechanism has **two steps** and a **carbocation** is formed as an intermediate. Because the first step is rate-determining, the rate depends on the concentration of RX *only;* that is, the rate equation is **first order,** and **rate = k[RX].**

Two specific nucleophilic substitution reactions illustrate these mechanistic possibilities.

7.9B Two Mechanisms for Nucleophilic Substitution

Rate equations for two different alkyl halides, CH_3Br and $(CH_3)_3CBr$, give us insight into the possible mechanism for nucleophilic substitution.

Reaction of bromomethane (CH_3Br) with acetate ($CH_3CO_2^-$) affords the substitution product methyl acetate with loss of Br^- as the leaving group (Equation [1]). Kinetic data show that the reaction rate depends on the concentration of *both* reactants; that is, the rate equation is **second order.** This suggests a **bimolecular reaction with a one-step mechanism** in which the C–X bond is broken *as* the C–Nu bond is formed.

[1] CH_3–Br: (bromomethane) + acetate → **second-order kinetics** → CH_3–O (methyl acetate) + :Br:⁻

Equation [2] illustrates a similar nucleophilic substitution reaction with a different alkyl halide, $(CH_3)_3CBr$, which also leads to substitution of Br^- by $CH_3CO_2^-$. Kinetic data show that this reaction rate depends on the concentration of only *one* reactant, the alkyl halide; that is, the rate equation is **first order.** This suggests a **two-step mechanism in which the rate-determining step involves the alkyl halide only.**

[2] Br: + acetate → **first-order kinetics** → + :Br:⁻

The numbers **1** and **2** in the names S_N1 and S_N2 refer to the kinetic order of the reactions. For example, S_N**2** means that the kinetics are **second** order. The number 2 does *not* refer to the number of steps in the mechanism.

How can these two different results be explained? Although these two reactions have the same nucleophile and leaving group, **there must be two different mechanisms** because there are two different rate equations. These equations are specific examples of two well-known mechanisms for nucleophilic substitution at an sp^3 hybridized carbon:

- **S_N2 mechanism (substitution nucleophilic bimolecular).**
- **S_N1 mechanism (substitution nucleophilic unimolecular).**

The reaction in Equation [1] illustrates an S_N2 mechanism, whereas the reaction in Equation [2] illustrates an S_N1 mechanism.

7.10 The S_N2 Mechanism

The reaction of CH_3Br with $CH_3CO_2^-$ is an example of an **S_N2 reaction.** What are the general features of this mechanism?

CH_3-Br + acetate → (S_N2 reaction) $CH_3-O-C(=O)CH_3$ + Br^-

acetate

S_N2 reaction

7.10A Kinetics

An S_N2 reaction exhibits **second-order kinetics;** that is, the reaction is **bimolecular** and both the alkyl halide and the nucleophile appear in the rate equation.

- rate = $k[CH_3Br][CH_3CO_2^-]$

Changing the concentration of *either* reactant affects the rate. For example, doubling the concentration of *either* the nucleophile or the alkyl halide doubles the rate. Doubling the concentration of *both* reactants increases the rate by a factor of *four.*

Problem 7.18 What happens to the rate of an S_N2 reaction under each of the following conditions?

a. [RX] is tripled, and [:Nu⁻] stays the same.
b. Both [RX] and [:Nu⁻] are tripled.
c. [RX] is halved, and [:Nu⁻] stays the same.
d. [RX] is halved, and [:Nu⁻] is doubled.

7.10B A One-Step Mechanism

The most straightforward explanation for the observed second-order kinetics is a **concerted reaction—bond breaking and bond making occur at the *same* time,** as shown in Mechanism 7.1.

Mechanism 7.1 The S_N2 Mechanism

One step The C–Br bond breaks as the C–O bond forms.

$CH_3CO_2^-$ + CH_3-Br → $CH_3CO-O-CH_3$ + Br^-

new C–O bond

In the transition state, the C–Br bond is partially broken, the C–O bond is partially formed, and both the attacking nucleophile and the departing leaving group bear a partial negative charge.

7.10C Stereochemistry of the S_N2 Reaction

From what direction does the nucleophile approach the substrate in an S_N2 reaction? There are two possibilities.

- **Frontside attack:** The nucleophile approaches from the ***same*** side as the leaving group.
- **Backside attack:** The nucleophile approaches from the side ***opposite*** the leaving group.

The results of frontside and backside attack of a nucleophile are illustrated with $CH_3CH(D)Br$ as substrate and the general nucleophile $:Nu^-$. This substrate has the leaving group bonded to a stereogenic center, thus allowing us to see the structural difference that results when the nucleophile attacks from two different directions.

In frontside attack, the nucleophile approaches from the same side as the leaving group, forming A. In this example, the leaving group was drawn on the *right*, so the nucleophile attacks from the *right*, and all other groups remain in their original positions. Because the nucleophile and leaving group are in the same position relative to the other three groups on carbon, frontside attack results in **retention of configuration** around the stereogenic center.

Recall from Section 1.1 that D stands for the isotope deuterium (^2H).

frontside attack

A

Nu replaces Br on the *same* side.

In backside attack, the nucleophile approaches from the opposite side to the leaving group, forming B. In this example, the leaving group was drawn on the *right*, so the nucleophile attacks from the *left*. Because the nucleophile and leaving group are in the opposite position relative to the other three groups on carbon, backside attack results in **inversion of configuration** around the stereogenic center.

backside attack

B

Nu replaces Br on the *opposite* side.

The products of frontside and backside attack are *different* compounds. **A** and **B** are stereoisomers that are nonsuperimposable—they are **enantiomers.**

Inversion of configuration in an S_N2 reaction is often called **Walden inversion,** after Latvian chemist Dr. Paul Walden, who first observed this process in 1896.

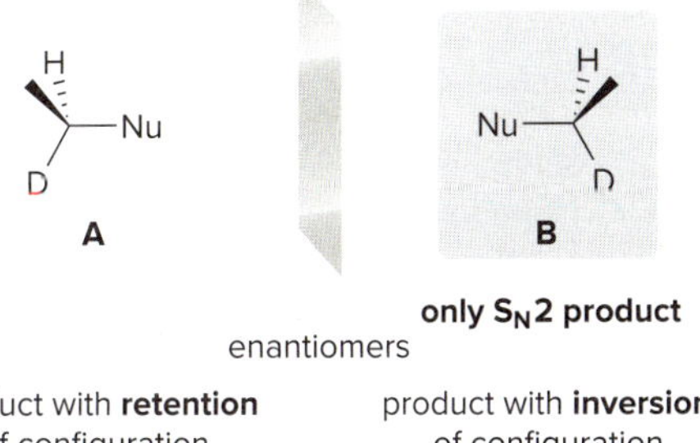

product with **retention** of configuration

product with **inversion** of configuration

Which product is formed in an S_N2 reaction? When the stereochemistry of the product is determined, **only B, the product of backside attack, is formed.**

Backside attack occurs in all S_N2 reactions, but we can observe this change only when the leaving group is bonded to a stereogenic center.

- **All S_N2 reactions proceed with *backside attack* of the nucleophile, resulting in *inversion* of configuration at a stereogenic center.**

One explanation for backside attack is based on an electronic argument. Both the nucleophile and leaving group are electron rich, and these like charges *repel* each other. Backside attack keeps these two groups as far away from each other as possible. In the transition state, the nucleophile and leaving group are 180° away from each other, and the other three groups around carbon occupy a plane, as illustrated in Figure 7.7.

Figure 7.7 Stereochemistry of the S_N2 reaction

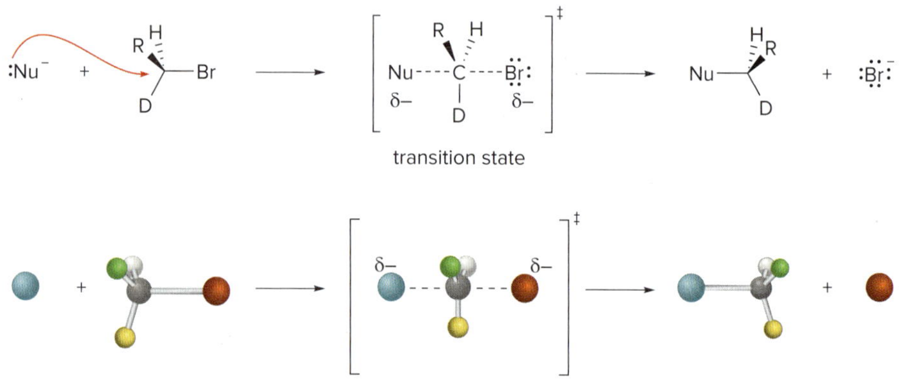

- :Nu^- and Br^- are 180° away from each other, on either side of a plane containing R, H, and D.

Two additional examples of inversion of configuration in S_N2 reactions are given in Figure 7.8.

Figure 7.8 Two examples of inversion of configuration in the S_N2 reaction

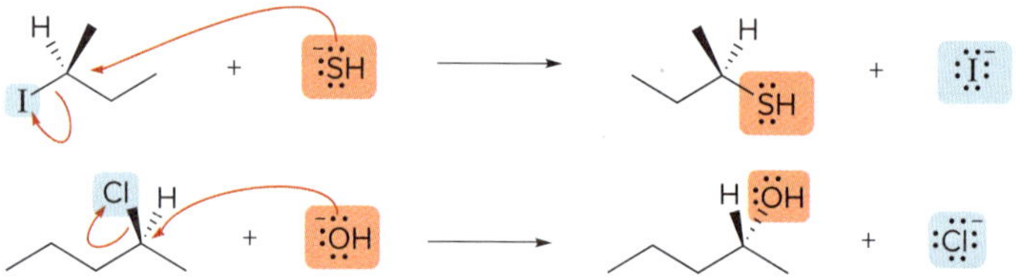

- The bond to the nucleophile in the product is always on the **opposite side** compared to the bond to the leaving group in the starting material. If the leaving group is drawn to the *left,* the nucleophile approaches from the *right.* If the leaving group is drawn in *front* of the plane (on a wedge), the nucleophile approaches from the *back* and ends up on a dashed wedge.

Sample Problem 7.2 Drawing the Product of Inversion in an S_N2 Reaction

Label the nucleophile and leaving group, and draw the product (including stereochemistry) of the following S_N2 reaction.

Solution

Br^- is the leaving group and ^-CN is the nucleophile. Because S_N2 reactions proceed with **inversion** of configuration and the leaving group is drawn *above* the ring (on a wedge), the nucleophile must come in from *below* (ending up on a dashed wedge).

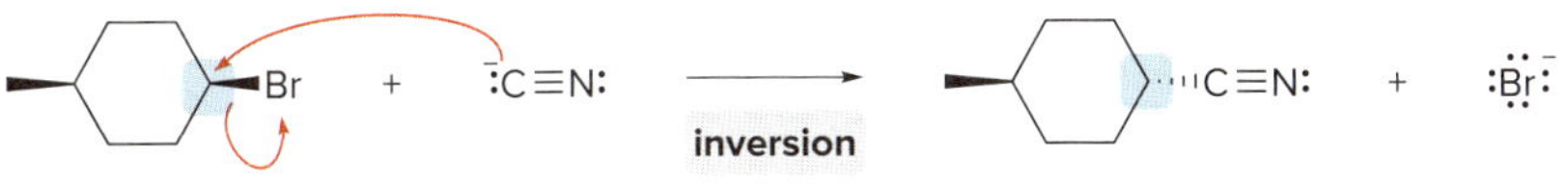

inversion

- **Inversion** of configuration occurs at the C–Br bond.
- **Backside attack** converts the **cis** starting material to a **trans** product because the nucleophile (^-CN) attacks from *below* the plane of the ring.

Problem 7.19 Draw the product of each S_N2 reaction and indicate stereochemistry.

a. 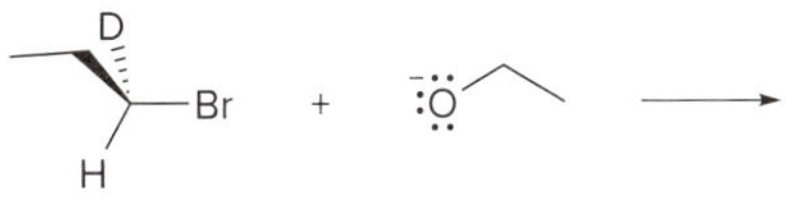 D, H, Br + $^-$:O: (ethoxide) ⟶

b.  I + $^-$:C≡N: ⟶

c. 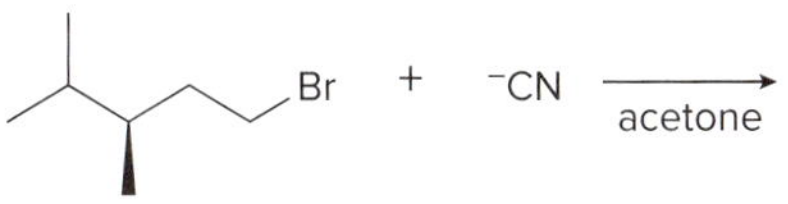 Br + ^-CN ⟶ acetone

d. Br + $^-OCH_2CH_3$ ⟶ DMF

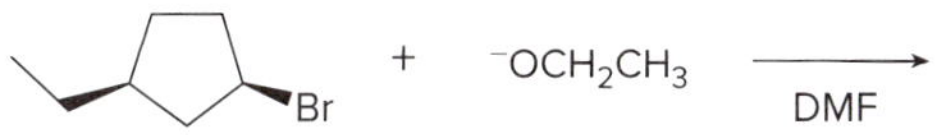

More Problems: Try Problems 7.41, 7.53.

7.10D The Identity of the R Group

How does the rate of an S_N2 reaction change as the alkyl group in the substrate alkyl halide changes from CH_3 ⟶ 1° ⟶ 2° ⟶ 3°?

- **As the number of R groups on the carbon with the leaving group *increases,* the rate of an S_N2 reaction *decreases.***

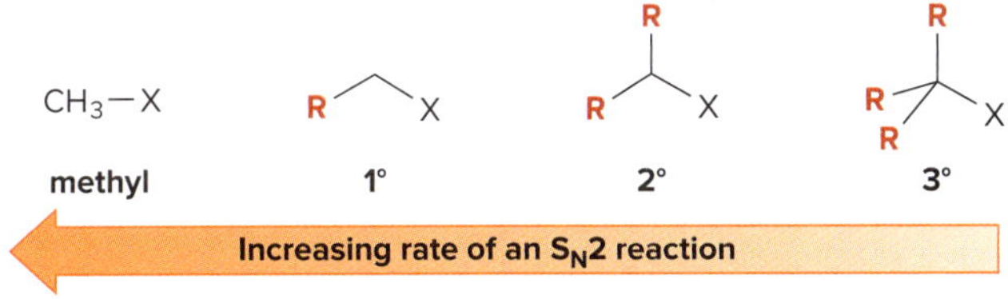

- **Methyl and 1° alkyl halides undergo S_N2 reactions with ease.**
- **2° Alkyl halides react more slowly.**
- **3° Alkyl halides *do not* undergo S_N2 reactions.**

This order of reactivity can be explained by steric effects. As small H atoms are replaced by larger alkyl groups, **steric hindrance caused by bulky R groups makes nucleophilic attack from the back side more difficult,** slowing the reaction rate. Figure 7.9 illustrates the effect of increasing steric hindrance in a series of alkyl halides.

Figure 7.9 Steric effects in the S_N2 reaction

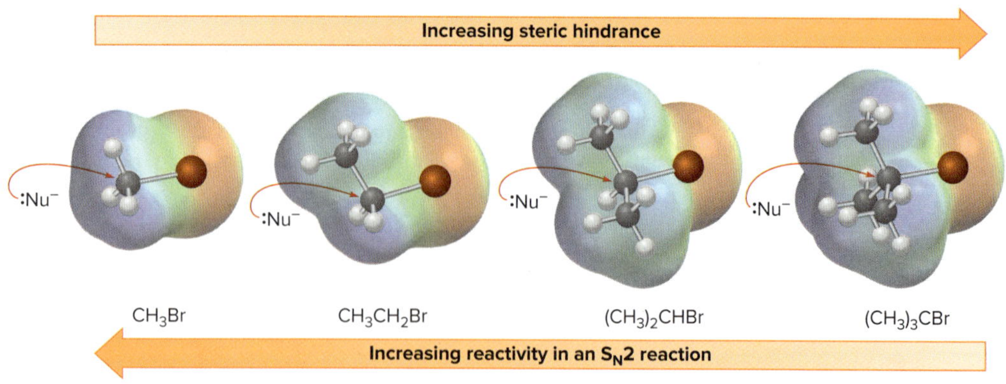

- **The S_N2 reaction is fastest with unhindered halides.**

Table 7.4 summarizes what we have learned thus far about the S_N2 mechanism.

Table 7.4 Characteristics of the S_N2 Mechanism

Characteristic	Result
Kinetics	• **Second-order kinetics;** rate = $k[RX][:Nu^-]$
Mechanism	• **One step**
Stereochemistry	• **Backside attack** of the nucleophile • **Inversion** of configuration at a stereogenic center
Identity of R	• **Unhindered halides react fastest.** • Rate: $CH_3X > RCH_2X > R_2CHX > R_3CX$

Problem 7.20 Which compound in each pair undergoes a faster S_N2 reaction?

a. Cl or Cl   b. Br or Br

The S_N2 reaction is a key step in the laboratory synthesis of many drugs including **ethambutol** (trade name Myambutol), used in the treatment of tuberculosis, and **fluoxetine** (trade name Prozac), an antidepressant, as illustrated in Figure 7.10. Often an S_N2 reaction is preceded by an acid–base reaction that generates a stronger nucleophile, as shown in Sample Problem 7.3.

Figure 7.10 Nucleophilic substitution in the synthesis of two useful drugs

[1] HO, $\ddot{N}H_2$ + Cl, Cl + $H_2\ddot{N}$, OH → HO, H N, N H, OH

ethambutol
(Trade name: Myambutol)

[2] OH, I —$CH_3\ddot{N}H_2$→ OH, N H, CH_3 —one step→ CF_3, O, N H, CH_3

fluoxetine
(Trade name: Prozac)

- In both examples, the initial substitution product bears a positive charge and goes on to lose a proton to form the product drawn.
- The NH_2 group serves as a neutral nucleophile to displace halogen in each synthesis. The new bonds formed by nucleophilic substitution are drawn in red in the products.

Sample Problem 7.3 Drawing an S_N2 Product with More Complex Reactants

Identify **C**, the product of an S_N2 reaction in the synthesis of raloxifene, a drug used to reduce the risk of invasive breast cancer in postmenopausal women.

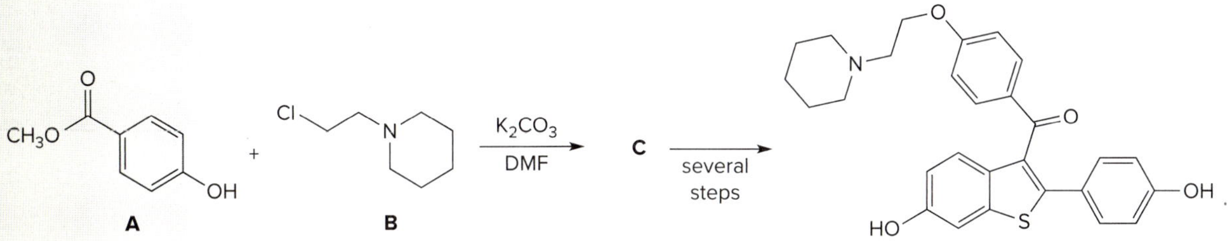

A + B $\xrightarrow[\text{DMF}]{K_2CO_3}$ C $\xrightarrow{\text{several steps}}$ raloxifene

Solution

Even though both starting materials have two functional groups, follow the same strategy used in simpler reactions.

[1] Identify the nucleophile and the leaving group. There are only a limited number of leaving groups (Table 7.1). Because **B** contains a Cl bonded to an sp^3 hybridized C, **B** contains the leaving group, so **A** contains the nucleophile.

Because this reaction is carried out in the presence of base (K_2CO_3) the **most acidic proton in either reactant is removed,** and that is the OH proton in **A.** Removal of the OH proton in **A** forms the negatively charged conjugate base, the **nucleophile.**

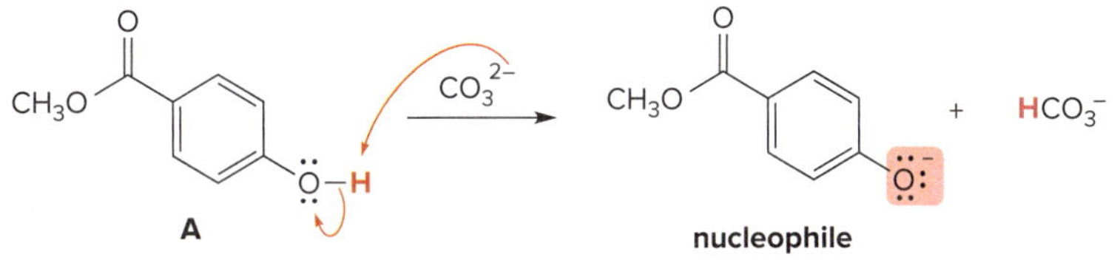

A $\xrightarrow{CO_3^{2-}}$ nucleophile + HCO_3^-

[2] Substitute the nucleophile for the leaving group.

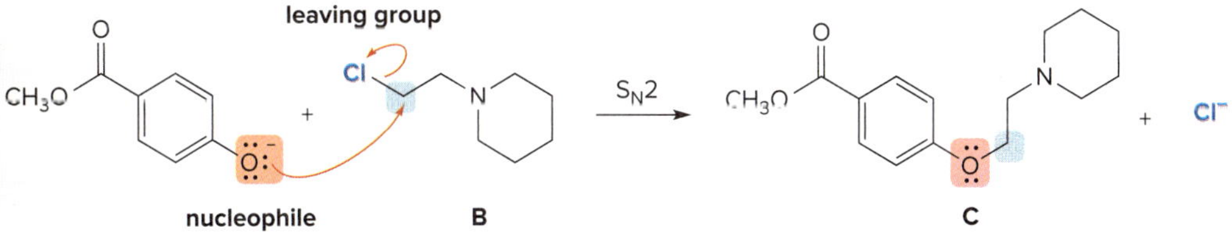

nucleophile (leaving group) + B $\xrightarrow{S_N2}$ C + Cl^-

Nucleophilic substitution forms **C** with a new C–O bond.

Problem 7.21 Draw the product **X** of the following S_N2 reaction. **X** was a key intermediate in the synthesis of rizatriptan, a drug introduced in 1998 for the treatment of migraines.

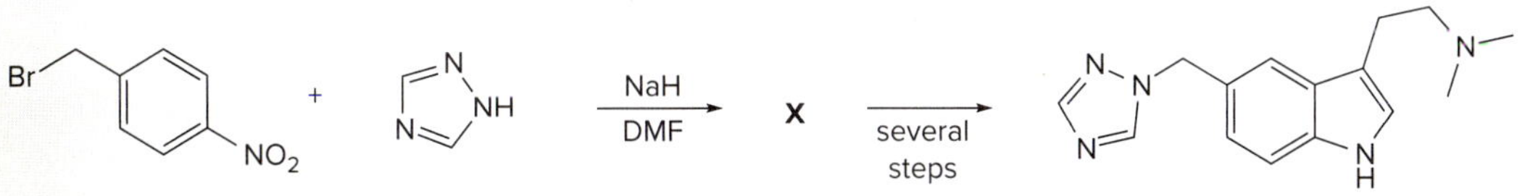

$\xrightarrow[\text{DMF}]{\text{NaH}}$ **X** $\xrightarrow{\text{several steps}}$ rizatriptan

More Practice: Try Problems 7.54–7.57.

7.11 The S_N1 Mechanism

The reaction of $(CH_3)_3CBr$ with $CH_3CO_2^-$ is an example of the second mechanism for nucleophilic substitution, the **S_N1 mechanism.** What are the general features of this mechanism?

acetate

S_N1 reaction

7.11A Kinetics

The S_N1 reaction exhibits **first-order kinetics.**

- rate = $k[(CH_3)_3CBr]$

As we learned in Section 7, the kinetics suggest that the S_N1 mechanism involves **more than one step,** and that the slow step is **unimolecular,** involving *only* the alkyl halide. **The identity and concentration of the nucleophile have *no effect* on the reaction rate.** Doubling the concentration of $(CH_3)_3CBr$ doubles the rate, but doubling the concentration of the nucleophile has *no effect*.

Problem 7.22 What happens to the rate of an S_N1 reaction under each of the following conditions?

a. [RX] is tripled, and [:Nu⁻] stays the same.
b. Both [RX] and [:Nu⁻] are tripled.
c. [RX] is halved, and [:Nu⁻] stays the same.
d. [RX] is halved, and [:Nu⁻] is doubled.

7.11B A Two-Step Mechanism

The most straightforward explanation for the observed first-order kinetics is a **two-step mechanism** in which **bond breaking occurs *before* bond making,** as shown in Mechanism 7.2.

Mechanism 7.2 The S_N1 Mechanism

1 slow

carbocation

+ :Br:⁻

2

1 Heterolysis of the C–Br bond forms a **carbocation** in the rate-determining step.

2 **Nucleophilic attack** of acetate (a Lewis base) on the carbocation (a Lewis acid) forms the new C–O bond.

The key features of the S_N1 mechanism are:

- The mechanism has two steps.
- Carbocations are formed as reactive intermediates.

An energy diagram for the reaction of $(CH_3)_3CBr + CH_3CO_2^-$ is shown in Figure 7.11. Each step has its own energy barrier, with a transition state at each energy maximum. Because the transition state for Step [1] is at higher energy, **Step [1] is rate-determining.** $\Delta H°$ for Step [1] has a positive value because only bond breaking occurs, whereas $\Delta H°$ of Step [2] has a negative value because only bond making occurs. The overall reaction is assumed to be exothermic, so the final product is drawn at lower energy than the initial starting material.

Figure 7.11

An energy diagram for the S_N1 reaction: $(CH_3)_3CBr + CH_3CO_2^- \rightarrow (CH_3)_3COCOCH_3 + Br^-$

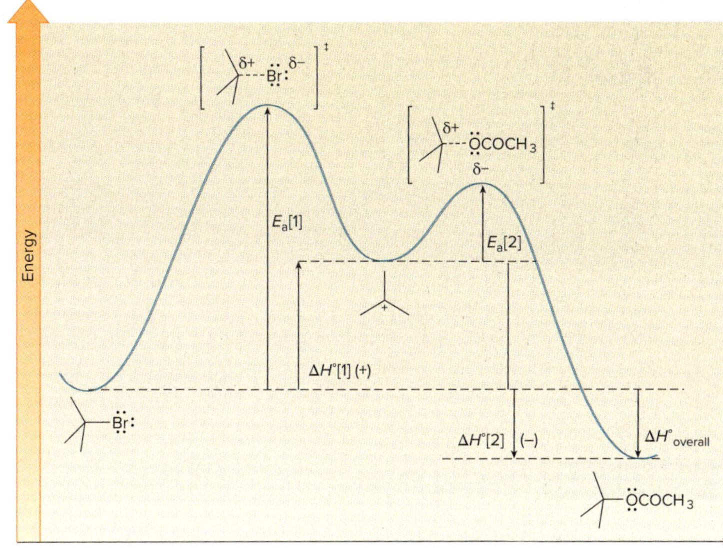

- The S_N1 mechanism has **two steps,** so there are **two energy barriers.**
- $\boldsymbol{E_a[1] > E_a[2]}$ because Step [1] involves bond breaking and Step [2] involves bond formation.
- In each step only one bond is broken or formed, so the transition state for each step has one partial bond.

7.11C Stereochemistry of the S_N1 Reaction

vacant *p* orbital
120°
sp^2 hybridized

To understand the stereochemistry of the S_N1 reaction, we must examine the geometry of the carbocation intermediate.

- **A carbocation (with three groups around C) is** $\boldsymbol{sp^2}$ **hybridized and trigonal planar, and contains a vacant** ***p*** **orbital extending above and below the plane.**

To illustrate the consequences of having a trigonal planar carbocation formed as a reactive intermediate, we examine the S_N1 reaction of a 3° alkyl halide **A** having the leaving group bonded to a stereogenic center.

Br
D
A
[1]
D
planar carbocation
+ :Br:⁻
:Nu⁻
[2]
attack from the **front**
Nu
D
B
attack from the **back**
Nu
D
C

Loss of the leaving group in Step [1] generates a **planar carbocation** that is now *achiral.* Attack of the nucleophile in Step [2] can occur from either the front or the back to afford two products, **B** and **C.** These two products are *different* compounds containing one stereogenic center. **B** and **C** are stereoisomers that are not superimposable—they are **enantiomers.** Because there is no preference for nucleophilic attack from either direction, an equal amount of the two enantiomers is formed—a **racemic mixture.** We say that ***racemization*** has occurred.

Nucleophilic attack from both sides of a planar carbocation occurs in S_N1 reactions, but we see the result of this phenomenon only when the leaving group is bonded to a stereogenic center.

- ***Racemization*** **is the formation of equal amounts of two enantiomeric products from a single starting material.**
- $\boldsymbol{S_N1}$ **reactions proceed with** ***racemization*** **at a single stereogenic center.**

Two additional examples of racemization in S_N1 reactions are given in Figure 7.12.

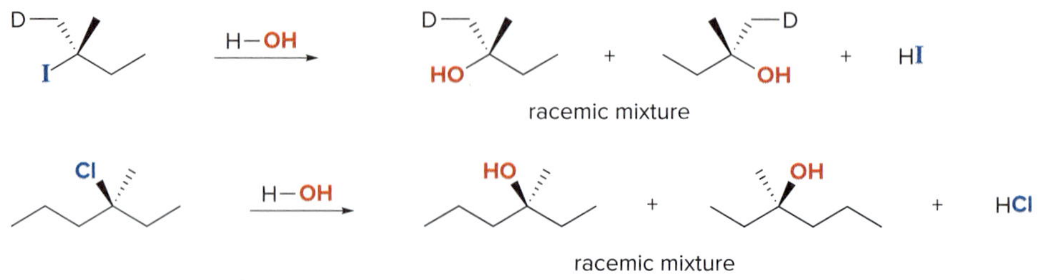

Figure 7.12
Two examples of racemization in the S_N1 reaction

- Nucleophilic substitution of each starting material by an S_N1 mechanism forms a **racemic mixture** of two products.
- With H_2O, a neutral nucleophile, the initial product of nucleophilic substitution (ROH_2^+) loses a proton to form the final neutral product, ROH (Section 7.6).

Sample Problem 7.4 Drawing the Products of an S_N1 Reaction

Label the nucleophile and leaving group, and draw the products (including stereochemistry) of the following S_N1 reaction.

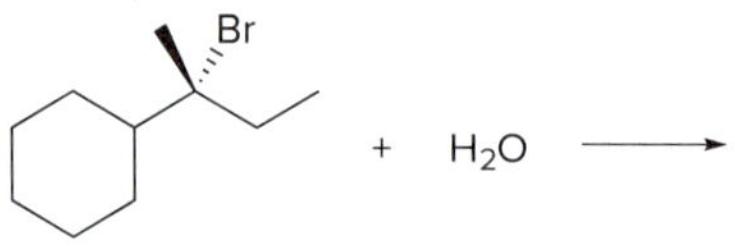

Solution

Br^- is the leaving group and H_2O is the nucleophile. Loss of the leaving group generates **a trigonal planar carbocation,** which can react with the nucleophile from either direction to form two products.

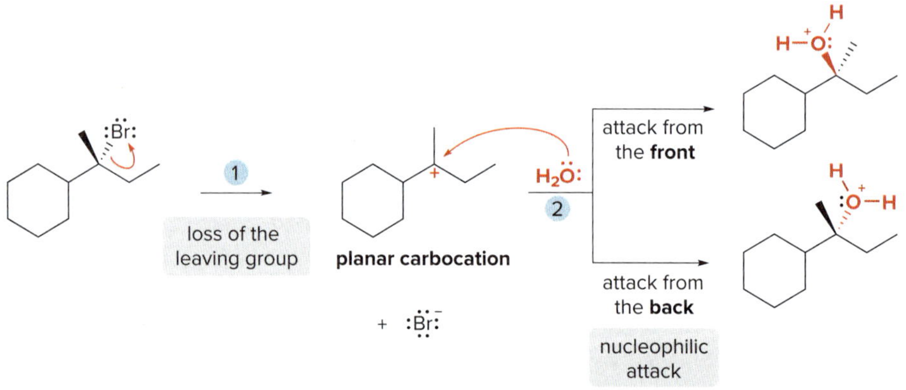

In this example, the initial products of nucleophilic substitution bear a positive charge. They readily lose a proton to form neutral products. The overall process with a neutral nucleophile thus has **three steps:** the first two constitute the **two-step S_N1 mechanism** (loss of the leaving group and attack of the nucleophile), and the third is a **Brønsted–Lowry acid–base reaction** leading to a neutral organic product.

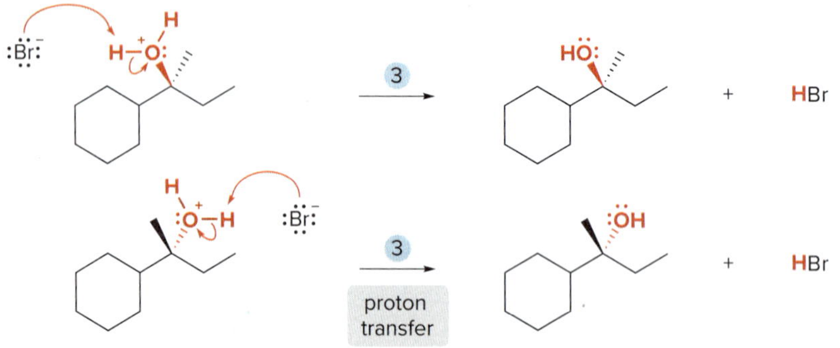

The two products in this reaction are nonsuperimposable mirror images—**enantiomers.** Because nucleophilic attack on the trigonal planar carbocation occurs with equal frequency from both directions, a **racemic mixture is formed.**

Problem 7.23 Draw the products of each S_N1 reaction and indicate the stereochemistry of any stereogenic centers.

a. Br, $\xrightarrow{H_2O}$ b. Cl, $\xrightarrow{CH_3CO_2^-}$ c. Br, $\xrightarrow{CH_3OH}$

More Practice: Try Problem 7.60.

7.11D The Identity of the R Group

How does the rate of an S_N1 reaction change as the alkyl group in the substrate alkyl halide changes from CH_3 ⟶ 1° ⟶ 2° ⟶ 3°?

- As the number of R groups on the carbon with the leaving group *increases,* the rate of an S_N1 reaction *increases.*

CH_3-X **methyl** | R–CH₂–X **1°** | R₂CH–X **2°** | R₃C–X **3°**

Increasing rate of an S_N1 reaction ⟶

- 3° Alkyl halides undergo S_N1 reactions rapidly.
- 2° Alkyl halides react more slowly.
- Methyl and 1° alkyl halides do *not* undergo S_N1 reactions.

This trend is exactly opposite to that observed for the S_N2 mechanism. To explain this result, we must examine the rate-determining step, the formation of the carbocation, and learn about the effect of alkyl groups on **carbocation stability.** Table 7.5 summarizes the characteristics of the S_N1 mechanism.

Table 7.5 Characteristics of the S_N1 Mechanism

Characteristic	Result
Kinetics	• **First-order kinetics;** rate = k[RX]
Mechanism	• **Two steps**
Stereochemistry	• **Trigonal planar carbocation** intermediate • **Racemization** at a single stereogenic center
Identity of R	• **More-substituted halides react fastest.** • Rate: $R_3CX > R_2CHX > RCH_2X > CH_3X$

7.12 Carbocation Stability

Carbocations are classified as **primary (1°), secondary (2°),** or **tertiary (3°)** by the number of R groups bonded to the charged carbon atom. As the number of R groups on the positively charged carbon atom increases, the stability of the carbocation **increases.**

methyl 1° 2° 3°

Increasing carbocation stability

We will examine the reason for this order of stability by invoking two different principles: **inductive effects** and **hyperconjugation.**

Problem 7.24 Classify each carbocation as 1°, 2°, or 3°.

a. b. c. d.

7.12A Inductive Effects

Inductive effects are electronic effects that occur through σ bonds. In Section 2.5B, for example, we learned that more-electronegative atoms stabilize a negative charge by an **electron-withdrawing inductive effect.**

Electron-donor groups (Z) stabilize a (+) charge; Z→Y$^+$. Electron-withdrawing groups (W) stabilize a (–) charge; W←Y$^-$.

To stabilize a positive charge, **electron-donating groups** are needed. **Alkyl groups are electron-donor groups that stabilize a positive charge.** An alkyl group with several σ bonds is more polarizable than a hydrogen atom, and more able to donate electron density. Thus, as R groups successively replace the H atoms in CH_3^+, **the positive charge is more dispersed on the electron-donor R groups, and the carbocation is more stabilized.**

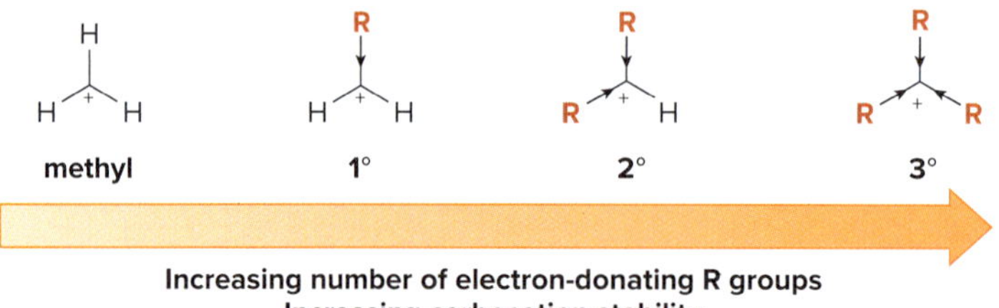

Problem 7.25 Classify the carbocations as 1°, 2°, or 3°, and rank the carbocations in each group in order of increasing stability.

a. b.

7.12B Hyperconjugation

A second explanation for the observed trend in carbocation stability is based on orbital overlap. A 3° carbocation is more stable than a 2°, 1°, or methyl carbocation because the positive charge is *delocalized* over more than one atom.

- Spreading out charge by the overlap of an empty *p* orbital with an adjacent σ bond is called *hyperconjugation.*

For example, CH_3^+ cannot be stabilized by hyperconjugation, but $(CH_3)_2CH^+$ can:

no opportunity for hyperconjugation

Hyperconjugation is possible.

Both carbocations contain an sp^2 hybridized carbon, so both are trigonal planar with a vacant *p* orbital extending above and below the plane. There are no adjacent C–H σ bonds with which the *p* orbital can overlap in CH_3^+, but there *are* adjacent C–H σ bonds in $(CH_3)_2CH^+$. This overlap (the **hyperconjugation**) delocalizes the positive charge on the carbocation, spreading it over a larger volume, and this stabilizes the carbocation.

The *larger* the number of alkyl groups on the adjacent carbons, the *greater* the possibility for hyperconjugation, and the *larger* the stabilization. Hyperconjugation thus provides an alternate way of explaining why **carbocations with a larger number of R groups are more stabilized.**

7.13 The Hammond Postulate

The rate of an S_N1 reaction depends on the rate of formation of the carbocation (the product of the rate-determining step) via heterolysis of the C–X bond.

- **The rate of an S_N1 reaction *increases* as the number of R groups on the carbon with the leaving group *increases*.**
- **The stability of a carbocation *increases* as the number of R groups on the positively charged carbon *increases*.**

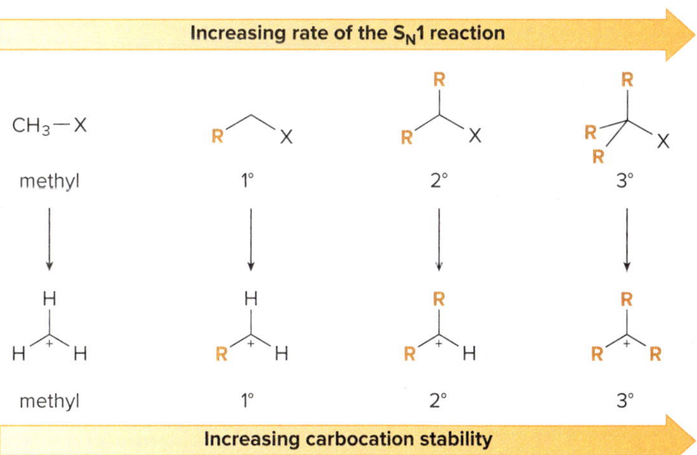

- **Thus, the rate of an S_N1 reaction *increases* as the stability of the carbocation *increases*.**

rate-determining step

The reaction is faster with a more stable carbocation.

The rate of a reaction depends on the magnitude of E_a, and the stability of a product depends on $\Delta G°$. The **Hammond postulate,** first proposed in 1955, **relates rate to stability.**

7.13A The General Features of the Hammond Postulate

The Hammond postulate provides a qualitative estimate of the energy of a transition state. Because the energy of the transition state determines the energy of activation and therefore the reaction rate, predicting the relative energy of two transition states allows us to determine the relative rates of two reactions.

According to the Hammond postulate, the transition state of a reaction resembles the structure of the species (reactant or product) to which it is closer in energy. In **endothermic reactions,** the transition state is closer in energy to the **products.** In **exothermic reactions,** the transition state is closer in energy to the **reactants.**

[1] An endothermic reaction

The transition state resembles the *products* more.

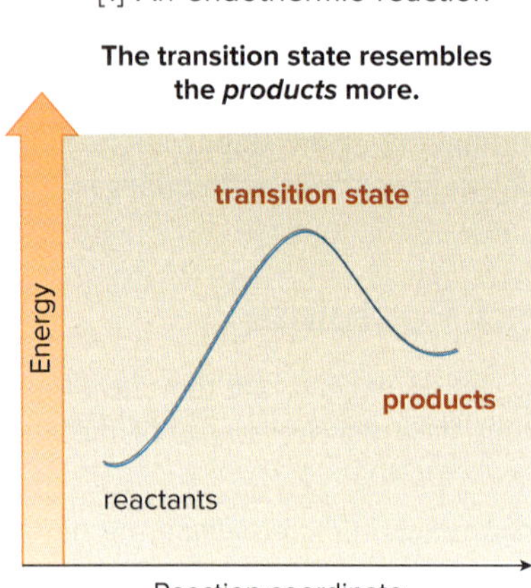

[2] An exothermic reaction

The transition state resembles the *reactants* more.

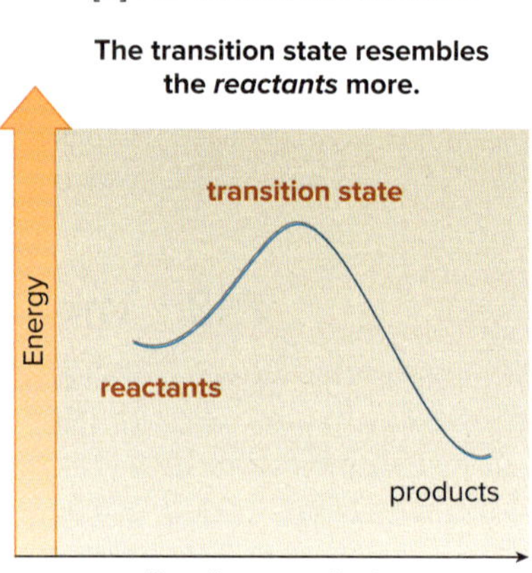

- Transition states in *endothermic* reactions resemble the *products*.
- Transition states in *exothermic* reactions resemble the *reactants*.

What happens to the reaction rate if the energy of the product is lowered? In an **endothermic reaction,** the transition state resembles the products, so anything that stabilizes the product stabilizes the transition state, too. **Lowering the energy of the transition state *decreases* the energy of activation (E_a), which *increases* the reaction rate.**

Suppose there are two possible products of an endothermic reaction, but one is more stable (lower in energy) than the other (Figure 7.13a). According to the Hammond postulate, **the transition state to form the more stable product is lower in energy, so this reaction should occur faster.**

- In an endothermic reaction, the *more stable* product forms *faster*.

What happens to the reaction rate of an **exothermic reaction** if the energy of the product is lowered? The transition state resembles the reactants, so **lowering the energy of the products has little or no effect on the energy of the transition state.** If E_a is unaffected, then the reaction rate is unaffected, too, as shown in Figure 7.13b.

- In an exothermic reaction, the more stable product may or may not form faster because E_a is *similar* for both products.

Figure 7.13 How the energies of the transition state and products are related

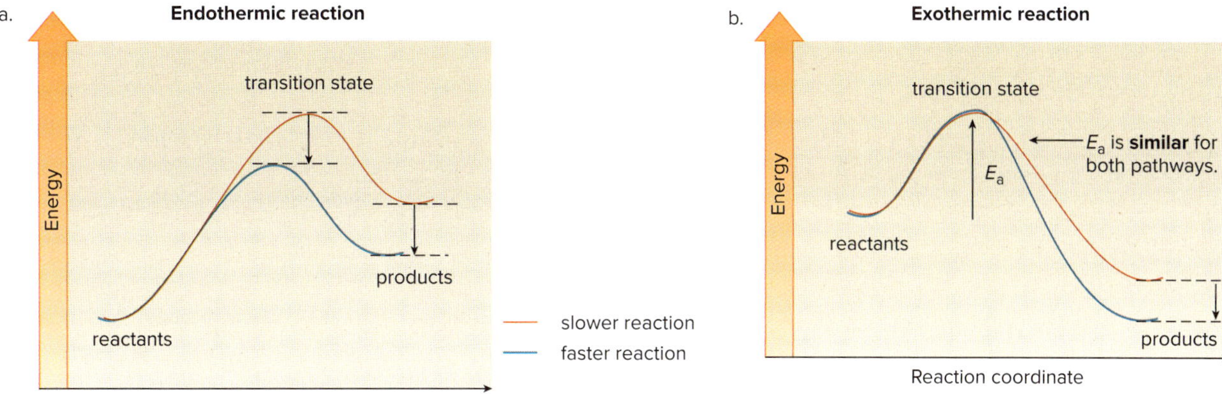

- The *lower*-energy transition state leads to the *lower*-energy product.
- Decreasing the energy of the product often has *little effect* on the energy of the transition state.

7.13B The Hammond Postulate and the S_N1 Reaction

In the S_N1 reaction, the rate-determining step is the formation of the carbocation, an *endothermic* reaction. According to the Hammond postulate, the **stability of the carbocation determines the rate of its formation.**

For example, heterolysis of the C–Cl bond in $(CH_3)_2CHCl$ affords a less stable 2° carbocation, $(CH_3)_2CH^+$ (Equation [1]), whereas heterolysis of the C–Cl bond in $(CH_3)_3CCl$ affords a more stable 3° carbocation, $(CH_3)_3C^+$ (Equation [2]). The Hammond postulate states that Reaction [2] is faster than Reaction [1], because the transition state to form the more stable 3° carbocation is lower in energy. Figure 7.14 depicts an energy diagram comparing these two endothermic reactions.

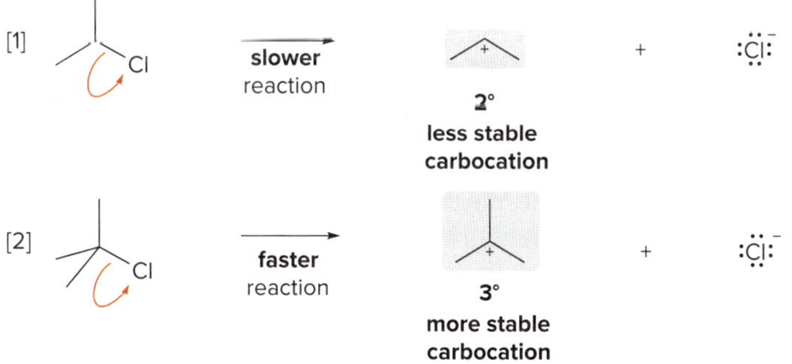

In conclusion, the Hammond postulate can be used to predict the relative rates of two reactions. **In the S_N1 reaction the rate-determining step is endothermic, so the more stable carbocation is formed faster.**

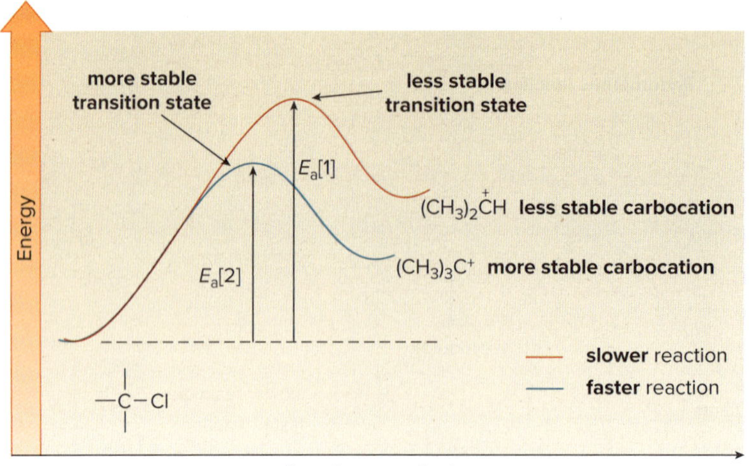

Figure 7.14 Energy diagram for carbocation formation in two different S_N1 reactions

- $(CH_3)_2CH^+$ is less stable than $(CH_3)_3C^+$, so $E_a[1] > E_a[2]$, and Reaction [1] is slower.

Problem 7.26 Rank the labeled halides in order of increasing rate of an S_N1 reaction.

a. b. c.

7.14 When Is the Mechanism S_N1 or S_N2?

Given a particular starting material and nucleophile, how do we know whether a reaction occurs by the S_N1 or S_N2 mechanism? Four factors are examined:

- **The alkyl halide—CH_3X, RCH_2X, R_2CHX, or R_3CX**
- **The nucleophile—strong or weak**
- **The leaving group—good or poor**
- **The solvent—protic or aprotic**

7.14A The Alkyl Halide—The Most Important Factor

The most important factor in determining whether a reaction follows the S_N1 or S_N2 mechanism is the *identity of the alkyl halide.*

- *Increasing* alkyl substitution favors S_N1.
- *Decreasing* alkyl substitution favors S_N2.

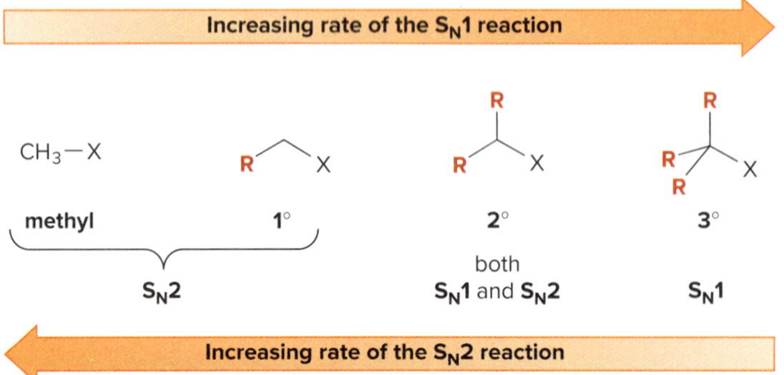

- **Methyl and 1° halides (CH_3X and RCH_2X) undergo only S_N2 reactions.**
- **3° Alkyl halides (R_3CX) undergo only S_N1 reactions.**
- **2° Alkyl halides (R_2CHX) undergo both S_N1 and S_N2 reactions. Other factors determine the mechanism.**

Examples are given in Figure 7.15.

Figure 7.15 The identity of RX and the mechanism of nucleophilic substitution

Br	Br	Br
1° halide	**2°** halide	**3°** halide
S_N2	Both S_N2 and S_N1 are possible.	S_N1

Problem 7.27 What is the likely mechanism of nucleophilic substitution for each alkyl halide?

a. Br b. Br c. Br d. Br

Problem 7.28 Mometasone is an anti-inflammatory steroid used as a cream to treat eczema, a nasal spray to treat hay fever, and in oral inhalation form to treat asthma. It is available worldwide under more than 100 trade names.

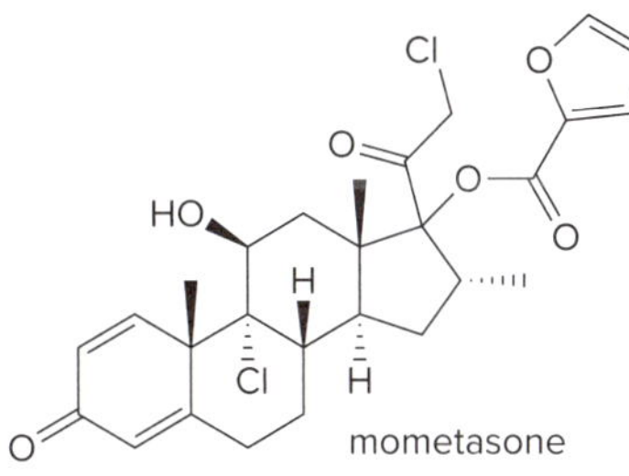

mometasone

a. Which alkyl chloride in mometasone reacts faster in an S_N1 reaction?
b. Which alkyl chloride reacts faster in an S_N2 reaction?

7.14B The Nucleophile

How does the strength of the nucleophile affect an S_N1 or S_N2 mechanism? The rate of the S_N1 reaction is unaffected by the identity of the nucleophile because the nucleophile does not appear in the rate equation (rate = k[RX]). The identity of the nucleophile *is* important for the S_N2 reaction, however, because the nucleophile does appear in the rate equation for this mechanism (rate = $k[RX][:Nu^-]$).

- ***Strong*** **nucleophiles present in high concentration favor S_N2 reactions.**
- ***Weak*** **nucleophiles favor S_N1 reactions by decreasing the rate of any competing S_N2 reaction.**

The most common nucleophiles in S_N2 reactions bear a net negative charge. The most common nucleophiles in S_N1 reactions are weak nucleophiles such as H_2O and ROH. The identity of the nucleophile is especially important in determining the mechanism and therefore the stereochemistry of nucleophilic substitution when 2° alkyl halides are starting materials.

Let's compare the substitution products formed when the 2° alkyl halide **A** (*cis*-1-bromo-4-methylcyclohexane) is treated with either the strong nucleophile ^-OH or the weak nucleophile

H_2O. Because a 2° alkyl halide can react by either mechanism, the strength of the nucleophile determines which mechanism takes place.

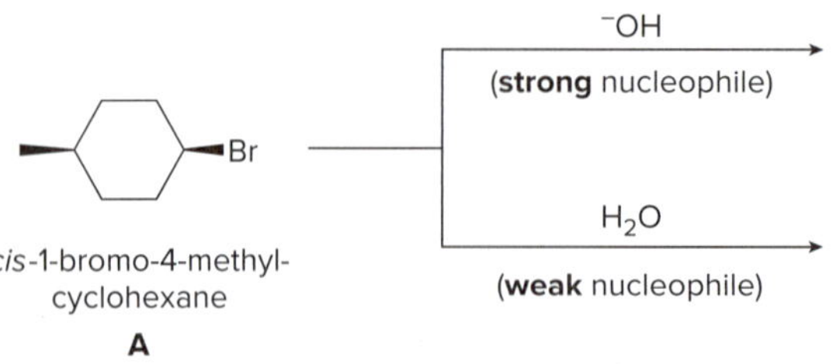

cis-1-bromo-4-methylcyclohexane
A

^-OH (**strong** nucleophile)

H_2O (**weak** nucleophile)

The **strong nucleophile $^-$OH favors an S_N2 reaction,** which occurs with **backside** attack of the nucleophile, resulting in **inversion of configuration.** Because the leaving group Br^- is *above* the plane of the ring, the nucleophile attacks from *below,* and a single product **B** is formed.

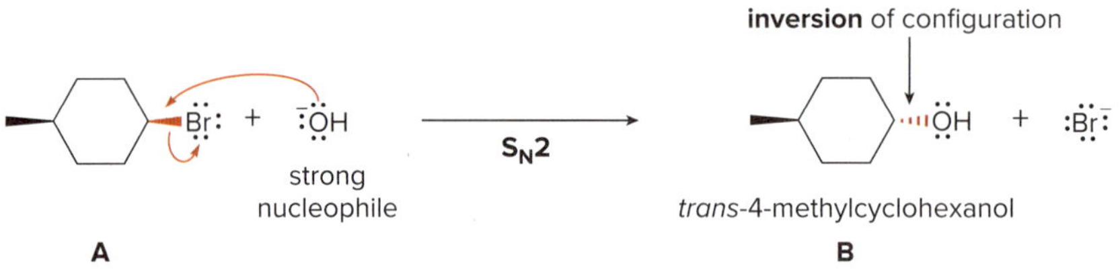

A + :ÖH (strong nucleophile) —S_N2→ trans-4-methylcyclohexanol **B** + :Br:$^-$

inversion of configuration

The **weak nucleophile H_2O favors an S_N1 reaction,** which occurs by way of an intermediate carbocation. Loss of the leaving group in **A** forms the carbocation, which undergoes nucleophilic attack from both above and below the plane of the ring to afford two products, **C** and **D.** Loss of a proton by proton transfer forms the final products, **B** and **E. B** and **E** are diastereomers of each other (**B** is a trans isomer and **E** is a cis isomer).

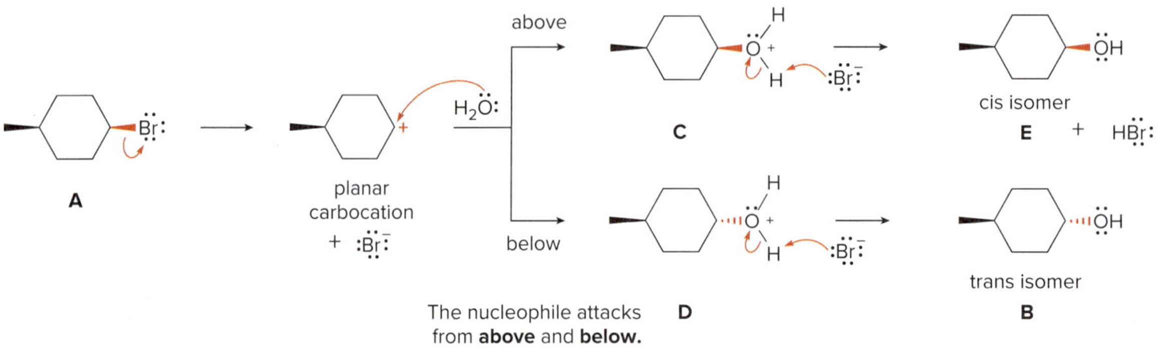

A → planar carbocation + :Br:$^-$; $H_2\ddot{O}$: above → **C** → cis isomer **E** + HBr: ; below → **D** → trans isomer **B**

The nucleophile attacks from **above** and **below.**

Thus, the mechanism of nucleophilic substitution determines the stereochemistry of the products formed.

Problem 7.29 For each alkyl halide and nucleophile: [1] Draw the product of nucleophilic substitution; [2] determine the likely mechanism (S_N1 or S_N2) for each reaction.

a. 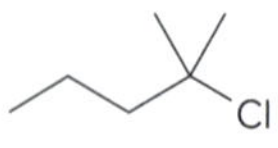 [alkyl chloride] + CH_3OH

b. 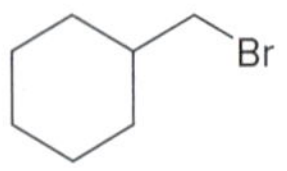 [cyclohexyl alkyl bromide] + ^-SH

c. 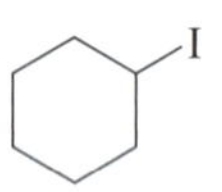 [iodocyclohexane] + $CH_3CH_2O^-$

d. 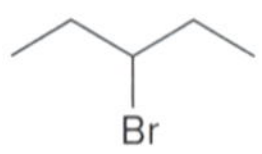 [alkyl bromide] + CH_3OH

Problem 7.30 Draw the products (including stereochemistry) for each reaction.

a. 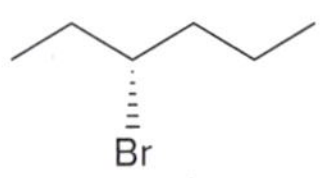 [alkyl bromide] + H_2O →

b. 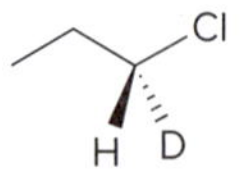 [alkyl chloride, H, D] + $^-$:C≡C–H →

7.14C The Leaving Group

How does the identity of the leaving group affect an S_N1 or S_N2 reaction?

- A *better* leaving group increases the rate of *both* S_N1 and S_N2 reactions.

Because the bond to the leaving group is partially broken in the transition state of the only step of the S_N2 mechanism and the slow step of the S_N1 mechanism, **a better leaving group increases the rate of both reactions.** The better the leaving group, the more willing it is to accept the electron pair in the C–X bond, and the faster the reaction.

For alkyl halides, the following order of reactivity is observed for the S_N1 and the S_N2 mechanisms:

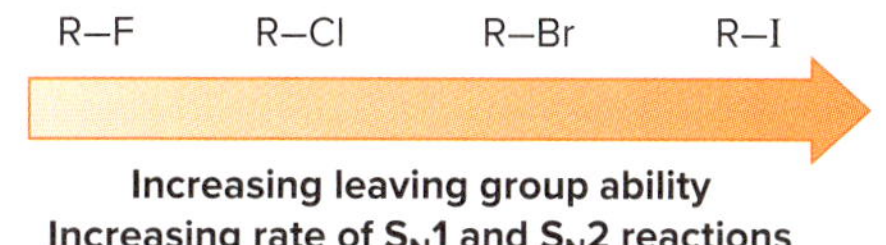

Problem 7.31 Rank the alkyl halides (labeled in red) in the following marine natural product in order of increasing reactivity in the S_N1 reaction.

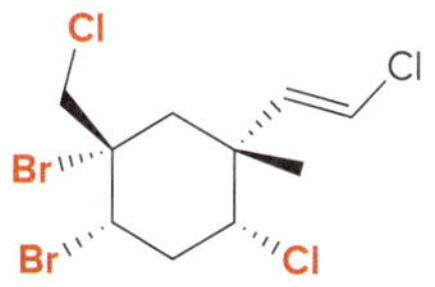

7.14D The Solvent

Polar protic solvents and polar aprotic solvents affect the rates of S_N1 and S_N2 reactions differently.

- Polar *protic* solvents are especially good for S_N1 reactions.
- Polar *aprotic* solvents are especially good for S_N2 reactions.

Summary of solvent effects:
- **Polar protic solvents favor S_N1 reactions** because the ionic intermediates are stabilized by solvation.
- **Polar aprotic solvents favor S_N2 reactions** because nucleophiles are not well solvated, and therefore are more nucleophilic.

Polar protic solvents like H_2O and ROH solvate both cations and anions well, and this characteristic is important for the S_N1 mechanism, in which two ions (a carbocation and a leaving group) are formed by heterolysis of the C–X bond. The carbocation is solvated by ion–dipole interactions with the polar solvent, and the leaving group is solvated by hydrogen bonding, in much the same way that Na^+ and Br^- are solvated in Section 7.8C. These interactions stabilize the reactive intermediate. In fact, a polar protic solvent is generally needed for an S_N1 reaction.

Polar aprotic solvents exhibit dipole–dipole interactions but not hydrogen bonding, and as a result, they do not solvate anions well. This has a pronounced effect on the nucleophilicity of anionic nucleophiles. Because these nucleophiles are not "hidden" by strong interactions with the solvent, they are **more nucleophilic.** Because stronger nucleophiles favor S_N2 reactions, **polar aprotic solvents are especially good for S_N2 reactions.**

Problem 7.32 Which solvents favor S_N1 reactions and which favor S_N2 reactions?

a. 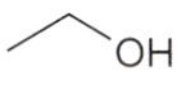 OH b. CH_3CN c. 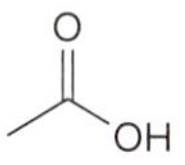 O OH d. 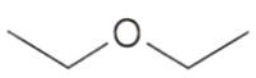 O

Problem 7.33 Decide on the mechanism for each substitution, and then pick the solvent that affords the faster reaction.

a. $(CH_3CH_2)_2CClCH_3$ + CH_3OH in CH_3OH or DMSO
b. $CH_3CH_2CH_2Br$ + ^-OH in H_2O or DMF
c. $(CH_3CH_2)_2CHCl$ + CH_3O^- in CH_3OH or HMPA

7.14E Summary of Factors That Determine Whether the S_N1 or S_N2 Mechanism Occurs

Table 7.6 summarizes the factors that determine whether a reaction occurs by the S_N1 or S_N2 mechanism. Sample Problems 7.5 and 7.6 illustrate how these factors are used to determine the mechanism of a given reaction.

Table 7.6 Summary of Factors That Determine the S_N1 or S_N2 Mechanism

Alkyl halide	Mechanism	Other factors
CH_3X **RCH_2X (1°)**	**S_N2**	Favored by • **strong nucleophiles** (usually a net negative charge) • polar **aprotic** solvents
R_3CX (3°)	**S_N1**	Favored by • **weak nucleophiles** (usually neutral) • polar **protic** solvents
R_2CHX (2°)	**S_N1 or S_N2**	The mechanism depends on the conditions. • **Strong nucleophiles favor the S_N2 mechanism over the S_N1 mechanism.** RO^- is a stronger nucleophile than ROH, so RO^- favors the S_N2 reaction and ROH favors the S_N1 reaction. • **Protic solvents favor the S_N1 mechanism and aprotic solvents favor the S_N2 mechanism.** H_2O and CH_3OH are polar protic solvents that favor the S_N1 mechanism, whereas acetone $[(CH_3)_2C{=}O]$ and DMSO $[(CH_3)_2S{=}O]$ are polar aprotic solvents that favor the S_N2 mechanism.

Sample Problem 7.5 Determining the Mechanism of Nucleophilic Substitution

Determine the mechanism of nucleophilic substitution for each reaction and draw the products.

a. [propyl bromide] Br + $^-$:C≡CH ⟶ b. [cyclopentyl]–Br + $^-$CN ⟶

Solution

a. The alkyl halide is 1°, so it must react by an S_N2 mechanism with the nucleophile $^-$:C≡CH.

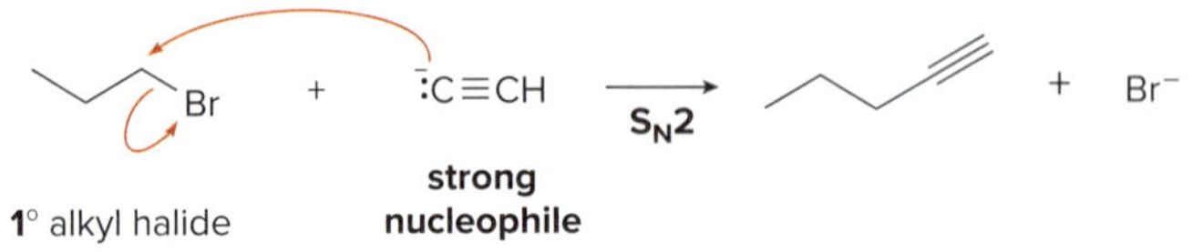

Br + $^-$:C≡CH $\xrightarrow{S_N2}$ [pent-1-yne] + Br^-

1° alkyl halide; **strong nucleophile**

b. The alkyl halide is 2°, so it can react by either the S_N1 or S_N2 mechanism. The strong nucleophile (^-CN) favors the S_N2 mechanism.

Br + $^-$:CN —S_N2→ CN + Br^-

2° alkyl halide | strong nucleophile

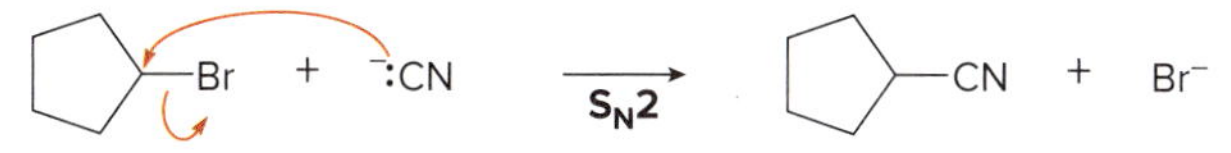

Sample Problem 7.6 Determining the Mechanism and Stereochemistry in Nucleophilic Substitution

Determine the mechanism of nucleophilic substitution for each reaction and draw the products, including stereochemistry.

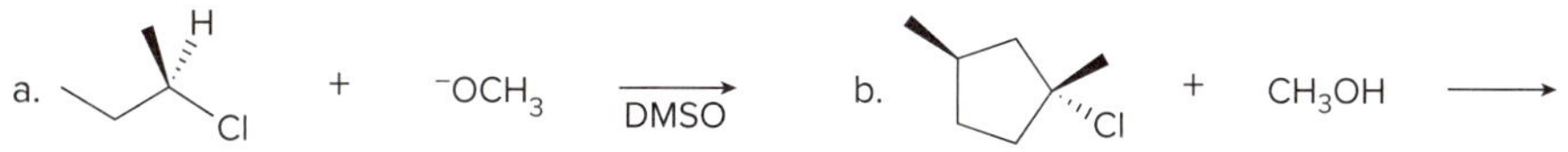

a. + $^-OCH_3$ —DMSO→ b. + CH_3OH ⟶

Solution

a. The 2° alkyl halide can react by either the S_N1 or S_N2 mechanism. **The strong nucleophile ($^-OCH_3$) favors the S_N2 mechanism,** as does the **polar aprotic solvent** (DMSO). S_N2 reactions proceed with **inversion** of configuration.

$CH_3\ddot{O}$:$^-$ + —DMSO, S_N2→ $CH_3\ddot{O}$ + :Cl:$^-$

strong nucleophile | **2°** alkyl halide | **inversion** of configuration

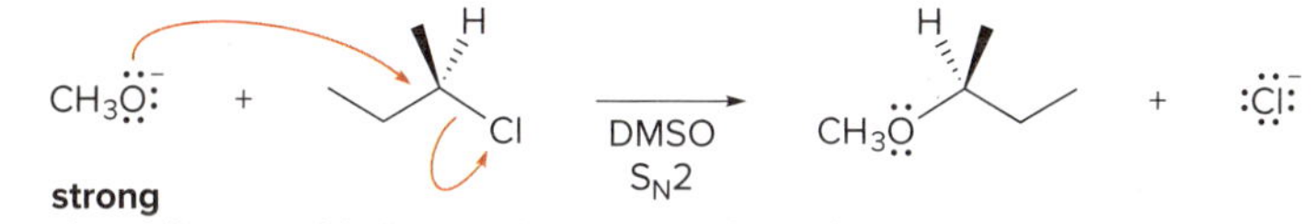

b. The alkyl halide is 3°, so it reacts by an $\mathbf{S_N1}$ mechanism with the weak nucleophile CH_3OH. S_N1 reactions proceed with **racemization** at a single stereogenic center, so two products are formed.

+ $CH_3\ddot{O}H$ —S_N1→ $\ddot{O}CH_3$ + $\ddot{O}CH_3$ + HCl

3° alkyl halide | **weak nucleophile** | two products of nucleophilic substitution

Problem 7.34 Determine the mechanism and draw the products of each reaction. Include the stereochemistry at all stereogenic centers.

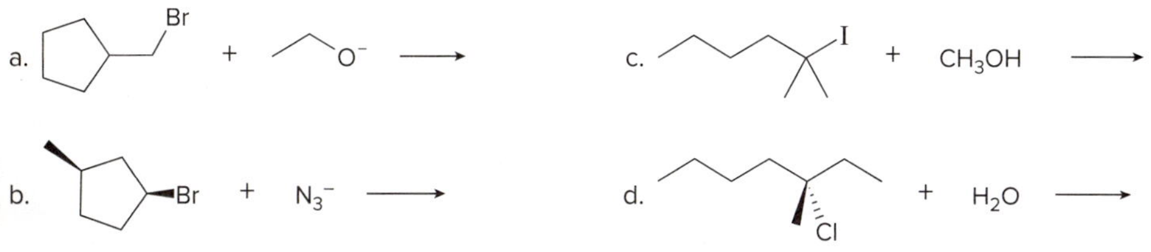

a. Br + O^- ⟶

b. Br + N_3^- ⟶

c. I + CH_3OH ⟶

d. Cl + H_2O ⟶

More Practice: Try Problem 7.63.

7.15 Biological Nucleophilic Substitution

Nucleophilic substitution occurs in a wide variety of biological reactions.

7.15A Leaving Groups Derived from Phosphorus

In contrast to nucleophilic substitutions run in the laboratory that use alkyl halides as substrates and halide anions as leaving groups, biological substitutions often occur with phosphorus leaving groups, such as phosphate (**PO_4^{3-}**, abbreviated as **P_i** for inorganic phosphate), diphosphate (**$P_2O_7^{4-}$**, abbreviated as **PP_i**), and triphosphate (**$P_3O_{10}^{5-}$**, abbreviated as **PPP_i**). These anions are excellent leaving groups because they are **weak, resonance-stabilized bases.**

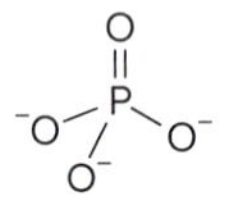

phosphate
P_i

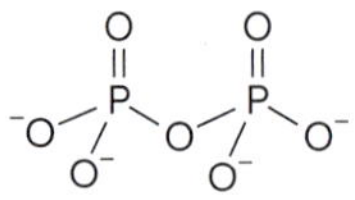

diphosphate
PP_i

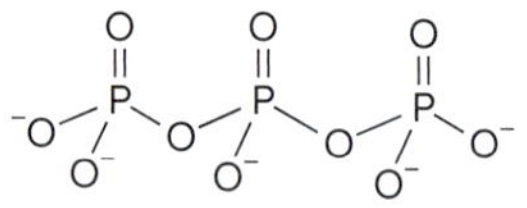

triphosphate
PPP_i

When an organic compound contains a carbon bonded to one of these leaving groups, the compound is called an organic monophosphate, diphosphate, or triphosphate.

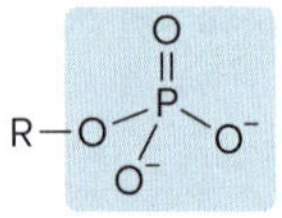

organic monophosphate

R−OP

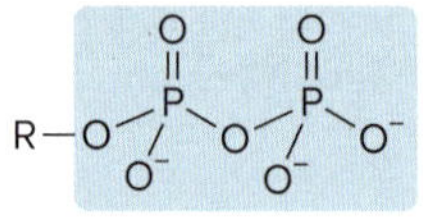

organic diphosphate

R−OPP

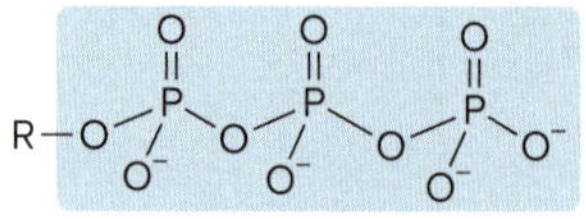

organic triphosphate

R−OPPP

Adenosine triphosphate (ATP) is an organic triphosphate.

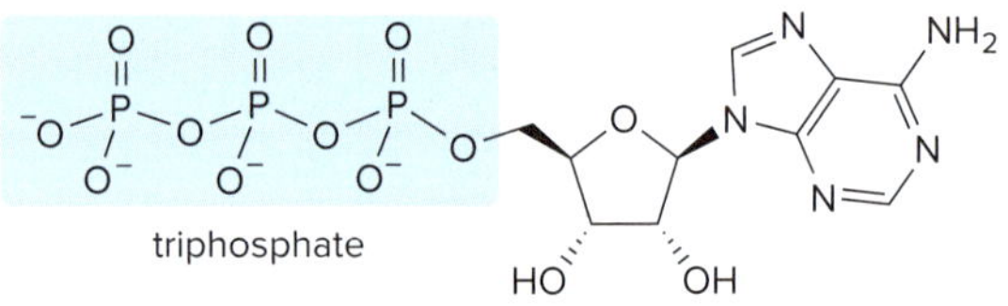

triphosphate

adenosine triphosphate
ATP

Nucleophilic substitutions with these substrates may proceed by either an S_N2 or S_N1 pathway, as shown with the general diphosphate R_2CHOPP.

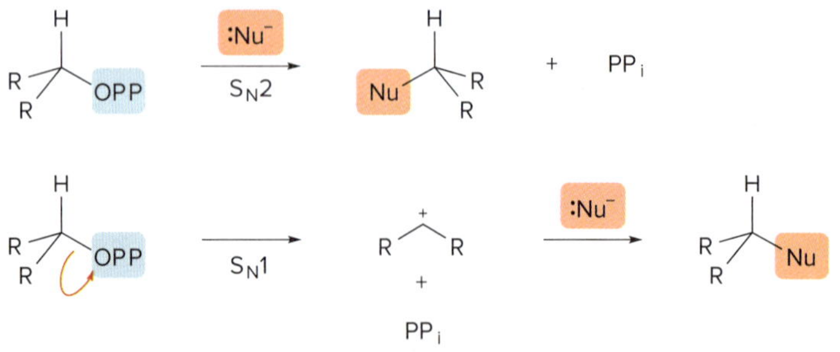

$R_2CHOPP \xrightarrow[S_N2]{:Nu^-} R_2CHNu + PP_i$

$R_2CHOPP \xrightarrow{S_N1} R_2CH^+ + PP_i \xrightarrow{:Nu^-} R_2CHNu$

The final step in the biosynthesis of geraniol, a component of rose oil used in perfumery, is an S_N1 reaction of geranyl diphosphate with water. This reaction occurs by way of a resonance-stabilized carbocation. We will learn more about reactions of diphosphates in Chapter 14.

geranyl diphosphate —S_N1→ [resonance-stabilized carbocation] + PP_i —$H_2\ddot{O}$:→ [$\overset{+}{O}H_2$] —(– H^+)→ geraniol

7.15B *S*-Adenosylmethionine

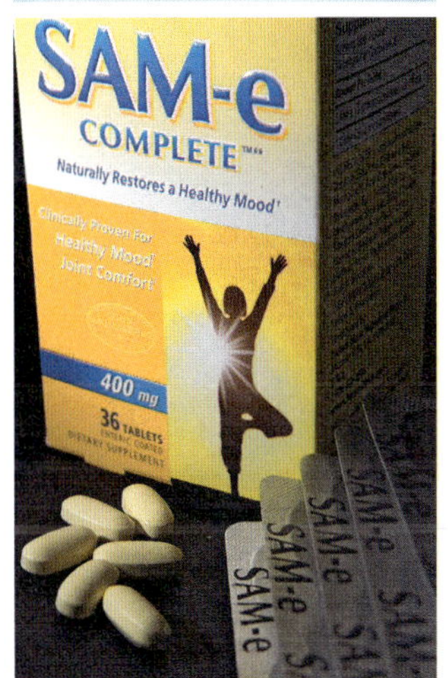

SAM, a nutritional supplement sold under the name SAM-e (pronounced sammy), has been used in Europe to treat depression and arthritis for over 20 years. In cells, SAM is used in nucleophilic substitutions that synthesize key amino acids, hormones, and neurotransmitters.
Jill Braaten

A common nucleophilic substitution occurs with *S*-adenosylmethionine, or **SAM.** SAM is the cell's equivalent of CH_3I. The many polar functional groups in SAM make it soluble in the aqueous environment in the cell.

Nucleophiles attack here. (→ CH_3)

The rest of the molecule is simply a **leaving group.**

S-adenosylmethionine
SAM

simplified as $CH_3-\overset{+}{S}R_2$
a sulfonium salt

The CH_3 group in SAM [abbreviated as $(\mathbf{CH_3SR_2})^+$] is part of a **sulfonium salt,** a positively charged sulfur species that contains a good leaving group. Nucleophilic attack at the CH_3 group of SAM displaces R_2S, a good neutral leaving group. This reaction is called **methylation,** because a CH_3 group is transferred from one compound (SAM) to another (:Nu^-).

$$:Nu^- + CH_3-\overset{+}{S}R_2 \longrightarrow CH_3-Nu + SR_2$$

SAM; S_N2 product; leaving group

For example, **adrenaline** (epinephrine), the molecule described in the chapter opener, is a hormone synthesized in the adrenal glands from noradrenaline (norepinephrine) by nucleophilic substitution using SAM (Figure 7.16). When an individual senses danger or is confronted by stress, the hypothalamus region of the brain signals the adrenal glands to synthesize and release adrenaline, which enters the bloodstream and then stimulates a response in many organs. Stored carbohydrates are metabolized in the liver to form glucose, which is further metabolized to provide an energy boost. Heart rate and blood pressure increase, and lung passages are dilated. These physiological changes result from the "rush of adrenaline," and prepare an individual for "fight or flight."

Figure 7.16
Adrenaline synthesis from noradrenaline in response to stress

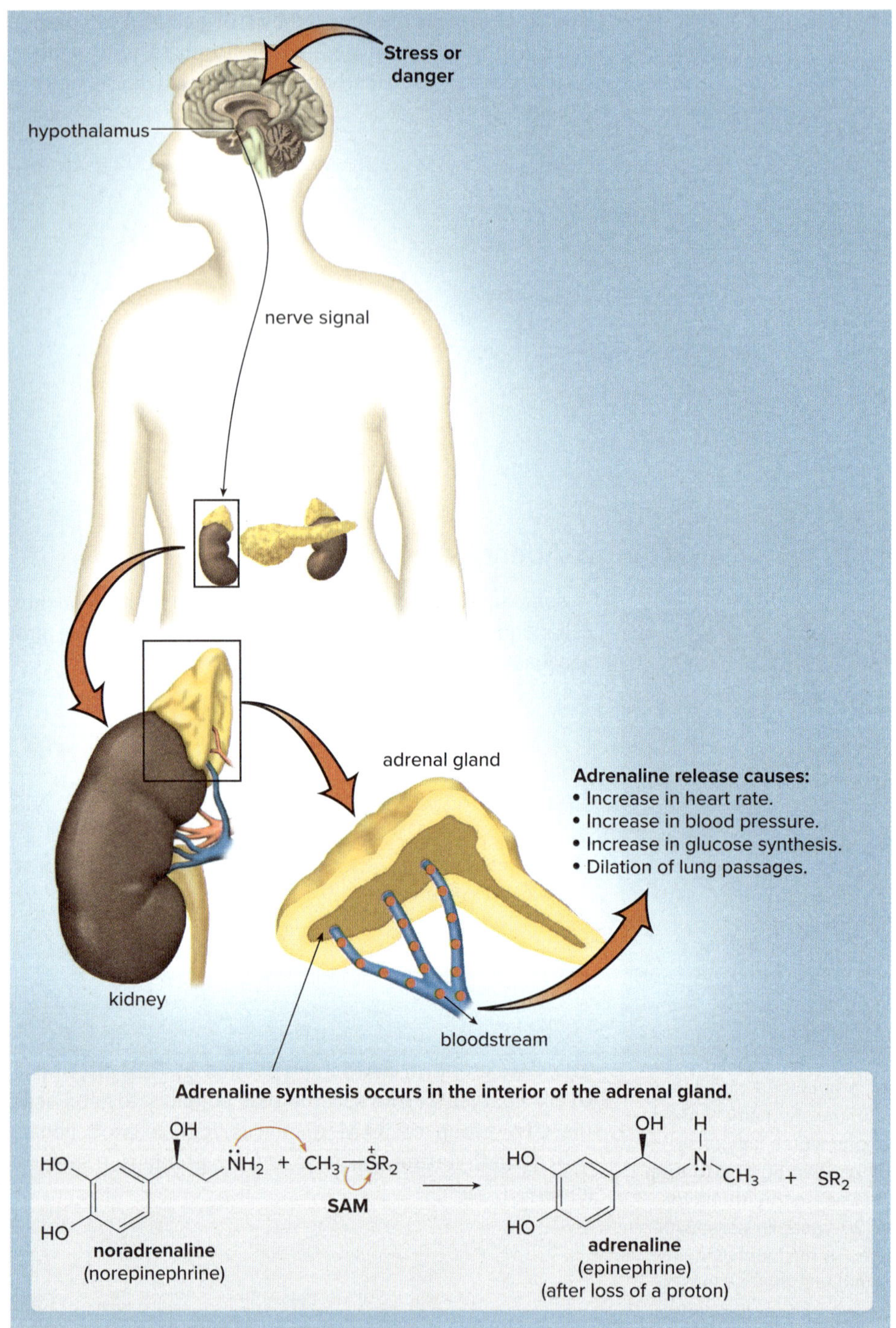

Problem 7.35 Nicotine, a toxic and addictive component of tobacco, is synthesized from **A** using SAM. Write out the reaction that converts **A** into nicotine.

N
H
N
N
N
CH_3
A
nicotine

7.16 Vinyl Halides and Aryl Halides

S_N1 and S_N2 reactions occur only at *sp*^{3} hybridized carbon atoms. Now that we have learned about the mechanisms for nucleophilic substitution, we can understand why **vinyl halides** and **aryl halides,** which have a halogen atom bonded to an sp^2 hybridized C, do not undergo nucleophilic substitution by either the S_N1 or S_N2 mechanism. The discussion here centers on vinyl halides, but similar arguments hold for aryl halides as well.

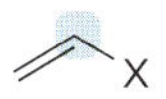

vinyl halide

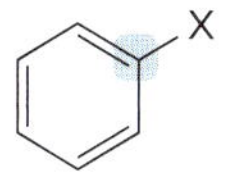

aryl halide

C bonded to X is ***sp*^{2}** hybridized.

Vinyl halides do not undergo S_N2 reactions in part because of the percent *s*-character in the hybrid orbital of the carbon atom in the C–X bond. The higher percent *s*-character in the sp^2 hybrid orbital of the vinyl halide compared to the sp^3 hybrid orbital of the alkyl halide (33% vs. 25%) makes the bond shorter and stronger.

Vinyl halides do not undergo S_N1 reactions because heterolysis of the C–X bond would form a **highly unstable vinyl carbocation.** Because this carbocation has only two groups around the positively charged carbon, it is ***sp*** hybridized. These carbocations are even less stable than 1° carbocations, so the S_N1 reaction does not take place.

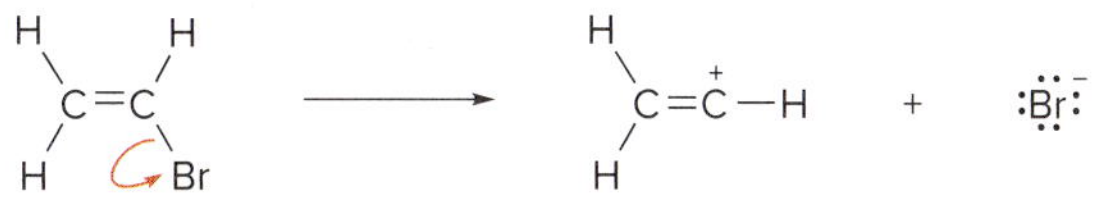

vinyl carbocation
highly unstable

Problem 7.36 Rank the alkyl halides in the following marine natural product in order of increasing reactivity in the S_N2 reaction.

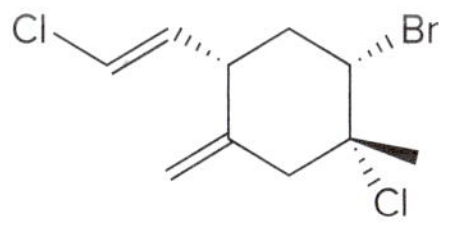

Problem 7.37 (a) Rank **A, B,** and **C** in order of increasing S_N2 reactivity. (b) Rank **A, B,** and **C** in order of increasing S_N1 reactivity.

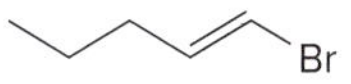

A 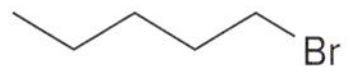 **B** 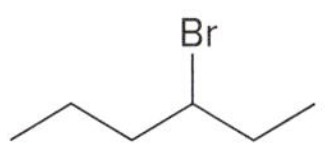 **C**

7.17 Organic Synthesis

Thus far we have concentrated on the starting material in nucleophilic substitution—the alkyl halide—and have not paid much attention to the product formed. Nucleophilic substitution reactions, and in particular S_N2 reactions, introduce a wide variety of different functional groups in molecules, depending on the nucleophile. For example, when ^-OH, ^-OR, and ^-CN are used as nucleophiles, the products are alcohols (ROH), ethers (ROR), and nitriles (RCN), respectively. Table 7.7 lists some functional groups readily introduced using nucleophilic substitution.

By thinking of **nucleophilic substitution as a reaction that *makes* a particular kind of organic compound,** we begin to think about ***synthesis.***

- **Organic synthesis is the systematic preparation of a compound from a readily available starting material by one or many steps.**

Table 7.7 Molecules Synthesized from R–X by the S_N2 Reaction

	Nucleophile (:Nu⁻)	Product	Name
Oxygen compounds	^-OH	R–OH	alcohol
	$^-OR'$	R–OR'	ether
	$^-O-C(=O)-R'$	$R-O-C(=O)-R'$	ester
Carbon compounds	^-CN	R–CN	nitrile
	$^-:C\equiv C-H$	$R-C\equiv C-H$	alkyne
Nitrogen compounds	N_3^-	$R-N_3$	azide
	$:NH_3$	$R-NH_2$	amine
Sulfur compounds	^-SH	R–SH	thiol
	$^-SR'$	R–SR'	sulfide

7.17A Background on Organic Synthesis

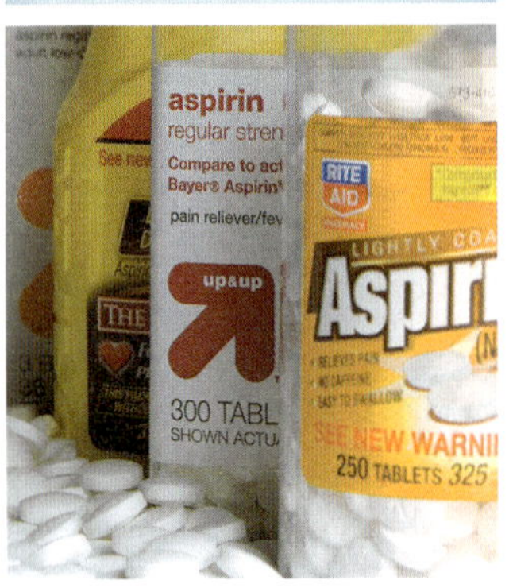

Aspirin is synthesized by a two-step procedure from simple, cheap starting materials (Figure 7.17). *Jill Braaten*

Phenol, the starting material for the aspirin synthesis, is a petroleum product, like most of the starting materials used in large quantities in industrial syntheses. A shortage of petroleum reserves thus affects the availability not only of fuels for transportation, but also of raw materials needed for most chemical synthesis.

Chemists synthesize molecules for many reasons. Sometimes a **natural product,** a compound isolated from natural sources, has useful medicinal properties, but is produced by an organism in only minute quantities. Synthetic chemists then prepare this molecule from simpler starting materials, so that it can be made available to a large number of people.

Sometimes, chemists prepare molecules that do not occur in nature (although they may be similar to those in nature), because these molecules have superior properties to their naturally occurring relatives. **Aspirin, or acetylsalicylic acid** (Section 2.7), is a well-known example. Acetylsalicylic acid is prepared from phenol, a product of the petroleum industry, by a two-step procedure (Figure 7.17). Aspirin has become one of the most popular and widely used drugs in the world because it has excellent analgesic and anti-inflammatory properties, *and* it is inexpensive and readily available.

Figure 7.17 Synthesis of aspirin

phenol → [1] NaOH; [2] CO_2; [3] H_3O^+ → (salicylic acid: OH, COOH) → $(CH_3CO)_2O$, acid → aspirin

7.17B Nucleophilic Substitution and Organic Synthesis

To carry out synthesis we must think backwards. We examine a compound and ask: **What starting material and reagent are needed to make it?** If we are using nucleophilic substitution, we must determine what alkyl halide and what nucleophile can be used to form a specific product. This is the simplest type of synthesis because it involves only one step. In Chapter 11, we will learn about multistep syntheses.

Suppose, for example, that we are asked to prepare $(CH_3)_2CHCH_2OH$ (2-methylpropan-1-ol) from an alkyl halide and any required reagents. To accomplish this synthesis, we must "fill in the boxes" for the starting material and reagent in the accompanying equation.

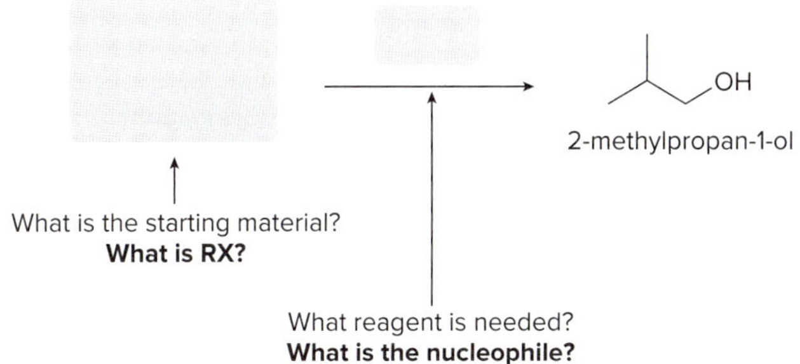

To determine the two components needed for the synthesis, remember that the carbon atoms come from the organic starting material, in this case a 1° alkyl halide $[(CH_3)_2CHCH_2Br]$. The **functional group comes from the nucleophile,** ^-OH in this case. With these two components, we can "fill in the boxes" to complete the synthesis.

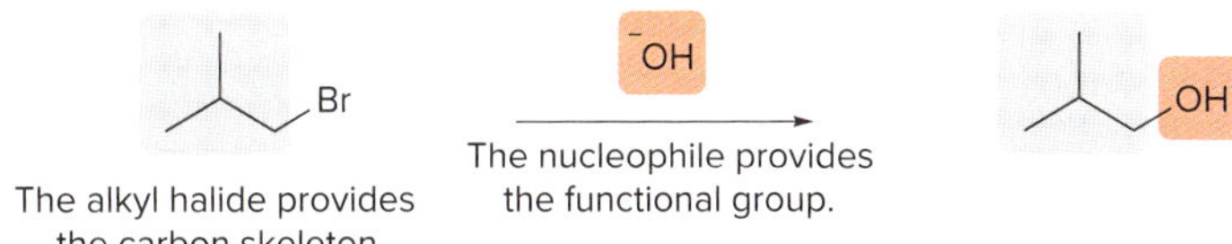

After any synthesis is proposed, check to see if it is reasonable, given what we know about reactions. Will the reaction written give a high yield of product? The synthesis of $(CH_3)_2CHCH_2OH$ is reasonable, because the starting material is a 1° alkyl halide and the nucleophile (^-OH) is strong, and both facts contribute to a successful S_N2 reaction.

Problem 7.38 What alkyl halide and nucleophile are needed to prepare each compound?

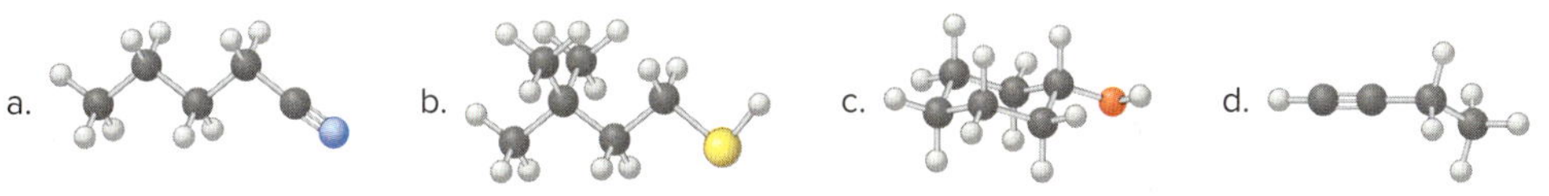

Problem 7.39 The ether, $CH_3OCH_2CH_3$, can be prepared by two different nucleophilic substitution reactions, one using CH_3O^- as nucleophile and the other using $CH_3CH_2O^-$ as nucleophile. Draw both routes.

Chapter 7 REVIEW

KEY CONCEPTS

[1] General facts about the reactions of alkyl halides (RX)

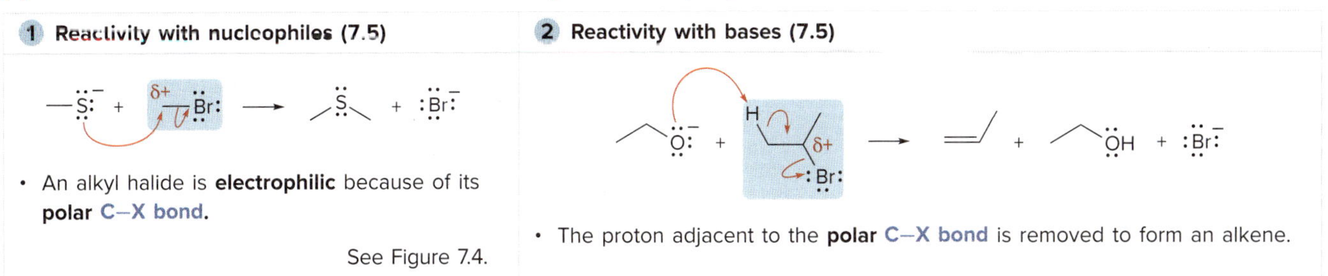

[2] Nucleophilic substitution (7.6)

- A nucleophile replaces a leaving group on an sp^3 hybridized carbon.
- One σ bond is broken and one σ bond is formed. There are two possible mechanisms: S_N1 and S_N2.

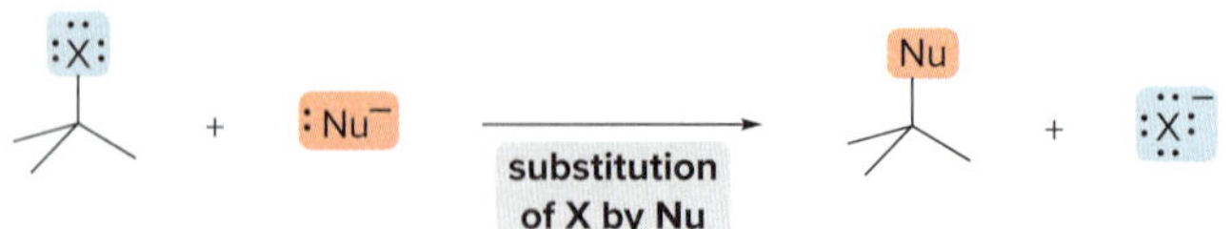

Try Problem 7.46.

[3] Periodic Trends

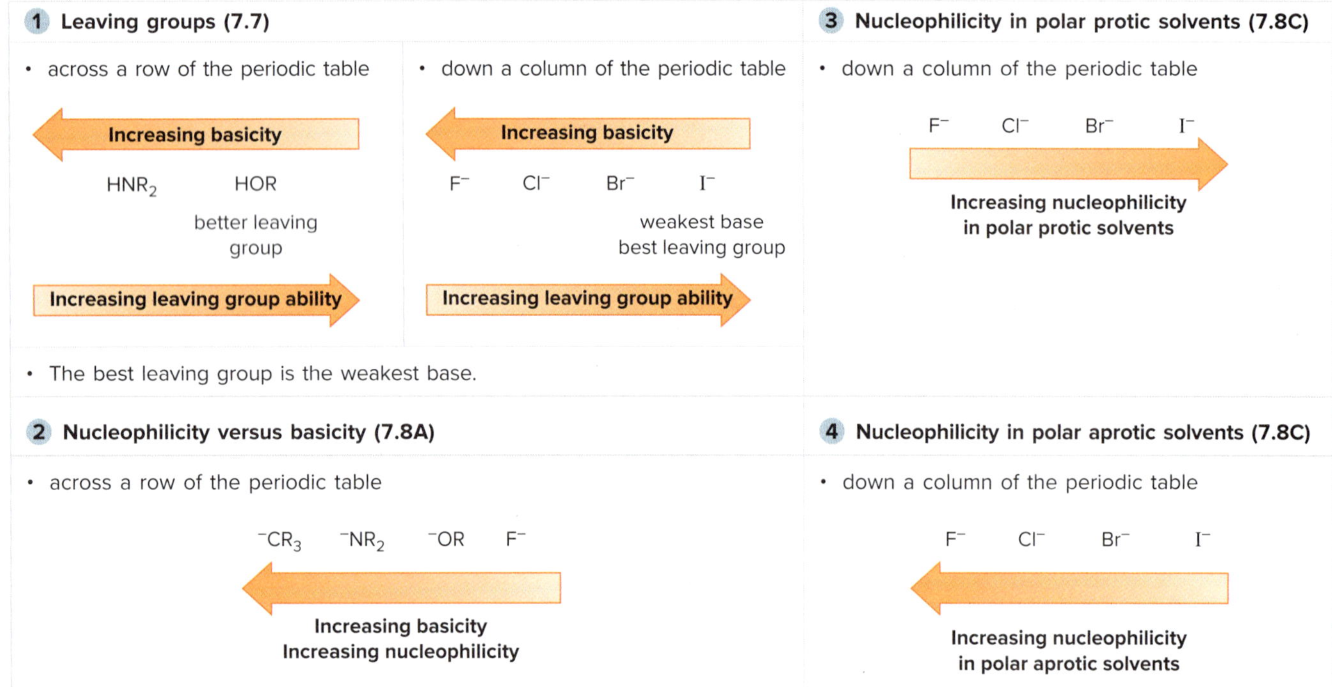

Try Problems 7.47, 7.48, 7.50.

[4] Carbocation stability (7.12)

- The stability of a carbocation increases as the number of electron-donating groups, such as **alkyl** groups, bonded to the **positively charged carbon** increases.

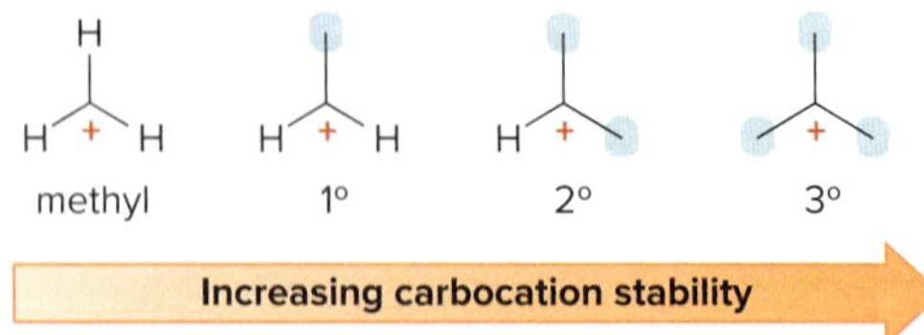

KEY SKILLS

[1] Comparing the nucleophile and leaving group to determine if products are favored (7.7)

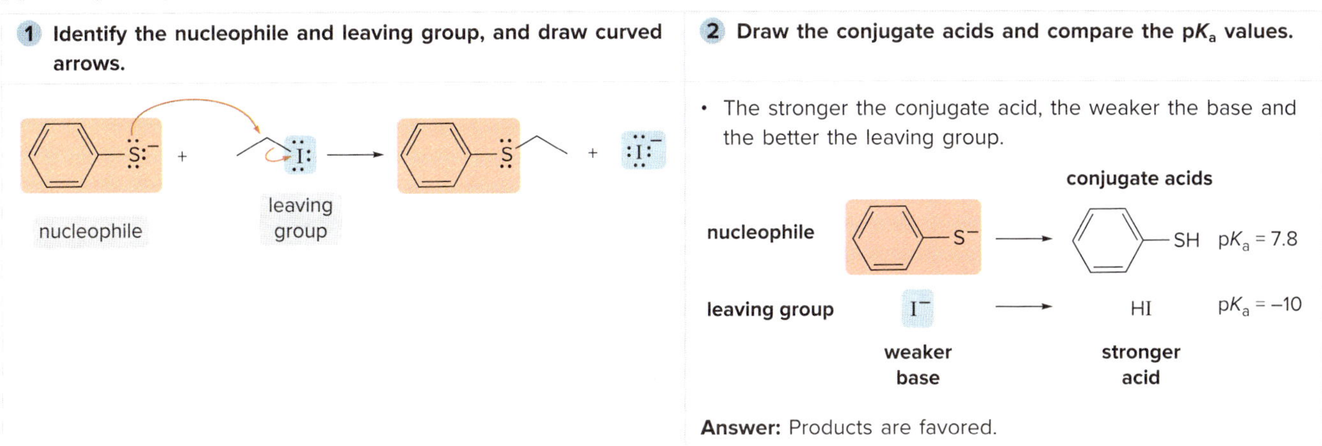

See Sample Problem 7.1. Try Problem 7.49.

[2] Drawing the product(s) of an S_N2 reaction (7.10)

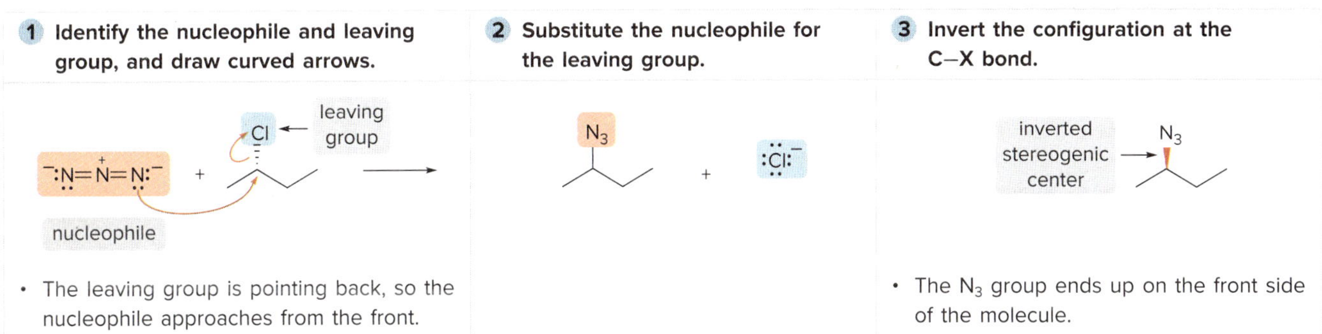

See Sample Problem 7.2, Figures 7.7, 7.8. Try Problems 7.41, 7.53–7.57.

[3] Drawing the product(s) of an S_N1 reaction (7.11)

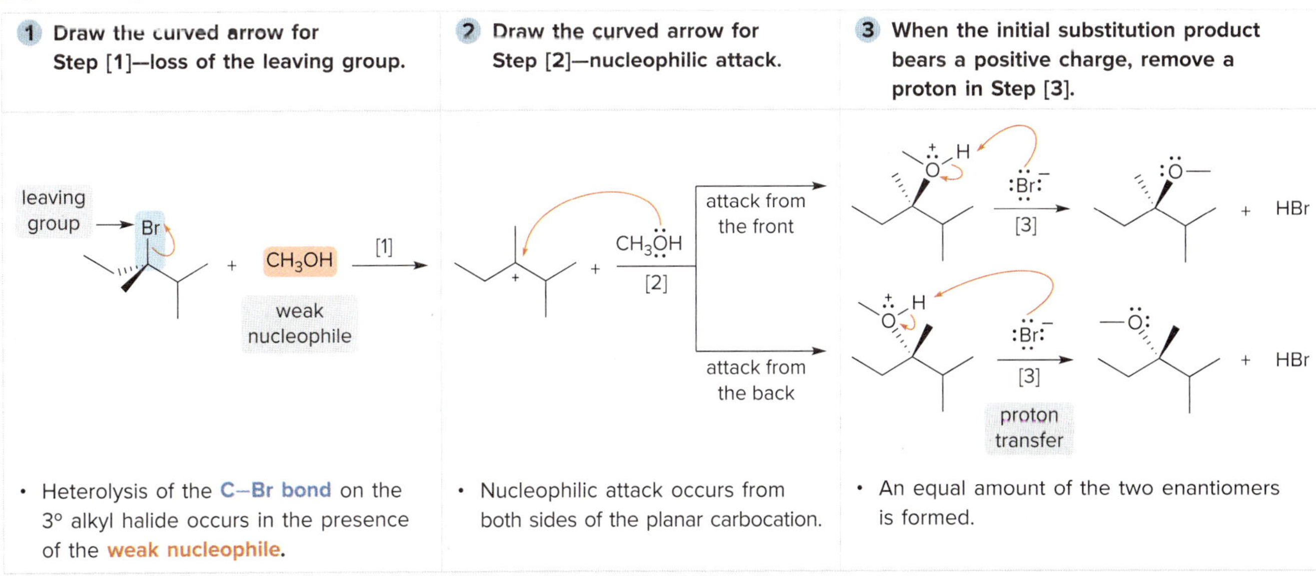

See Sample Problem 7.4, Figure 7.12. Try Problem 7.60.

[4] Deciding if a reaction proceeds by S_N1 or S_N2 (7.14E)

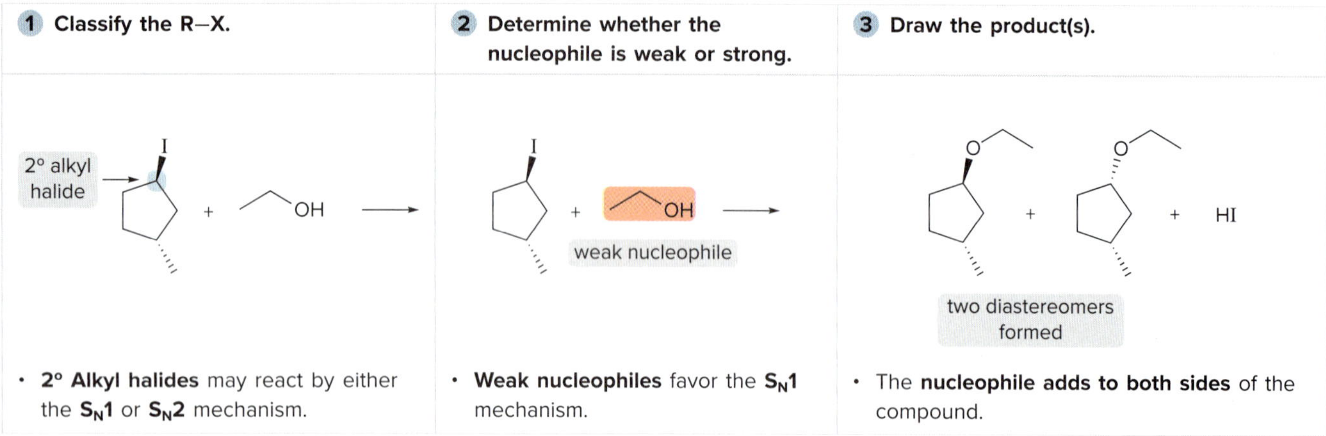

See Sample Problems 7.5, 7.6, Table 7.6. Try Problem 7.63.

KEY MECHANISM CONCEPTS

Comparison of S_N1 and S_N2 reactions

	S_N2 mechanism	S_N1 mechanism
1 Mechanism	• one step (7.10B) 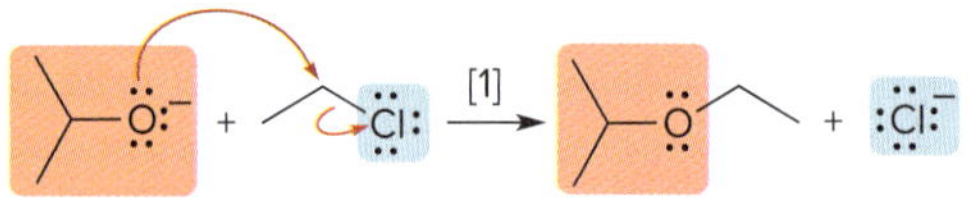 [1]	• two steps (7.11B) 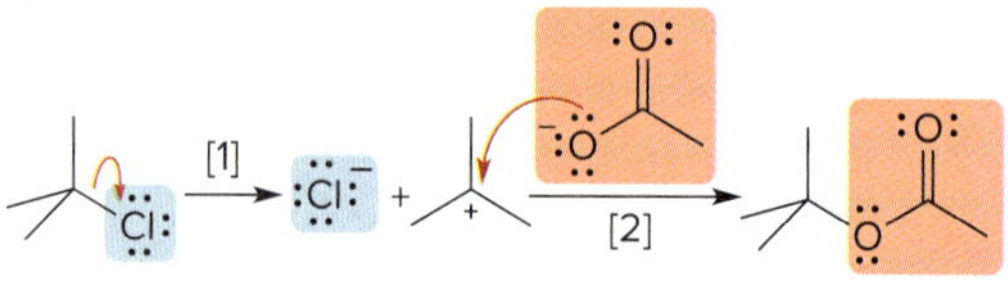 [1] [2]
2 Rate equation	• rate = k[RCl][:Nu$^-$] • second-order kinetics (7.10A)	• rate = k[RCl] • first-order kinetics (7.11A)
3 Alkyl halide	• order of reactivity (7.10D) 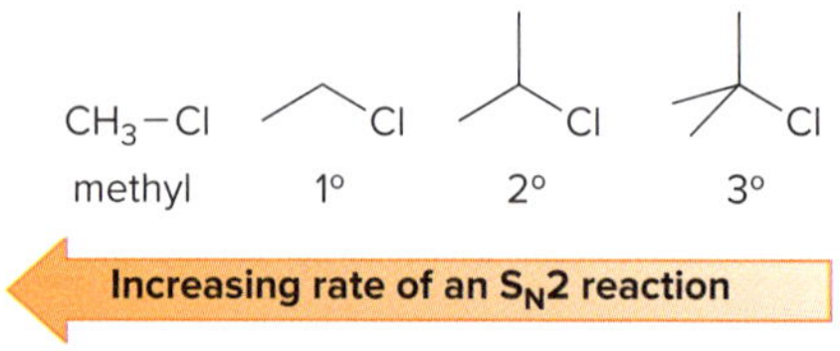 CH_3–Cl methyl, 1°, 2°, 3° Increasing rate of an S_N2 reaction • faster with less steric hindrance around the C–X bond	• order of reactivity (7.11D) 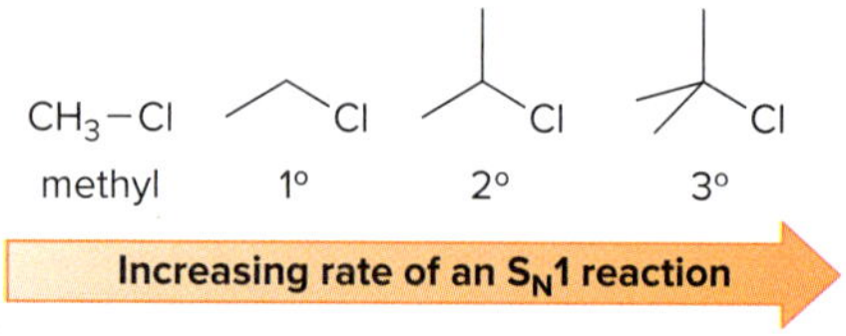 CH_3–Cl methyl, 1°, 2°, 3° Increasing rate of an S_N1 reaction • faster when more stable (more substituted) carbocations are formed (7.13)
4 Stereochemistry	• backside attack by the nucleophile (7.10C) 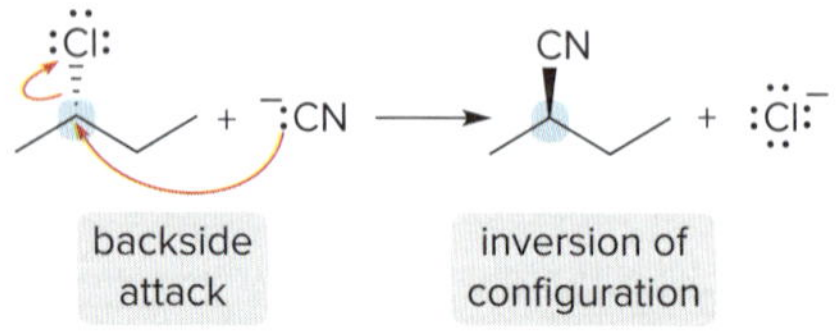 :Cl: + :CN → CN + :Cl:$^-$ backside attack; inversion of configuration	• trigonal planar carbocation intermediate (7.11C) 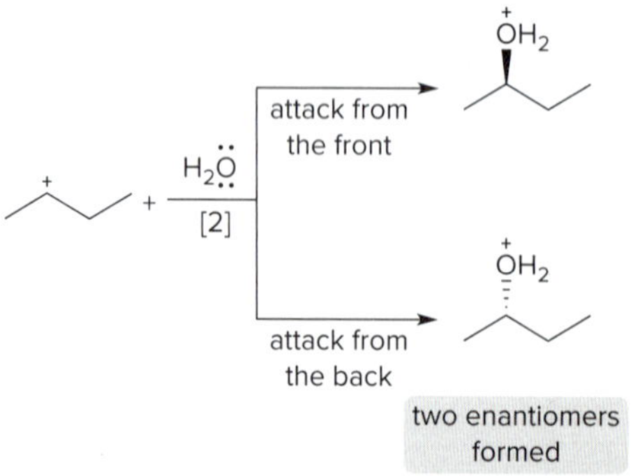 $H_2\ddot{O}$ [2] attack from the front → $\overset{+}{O}H_2$ attack from the back → $\overset{+}{O}H_2$ two enantiomers formed

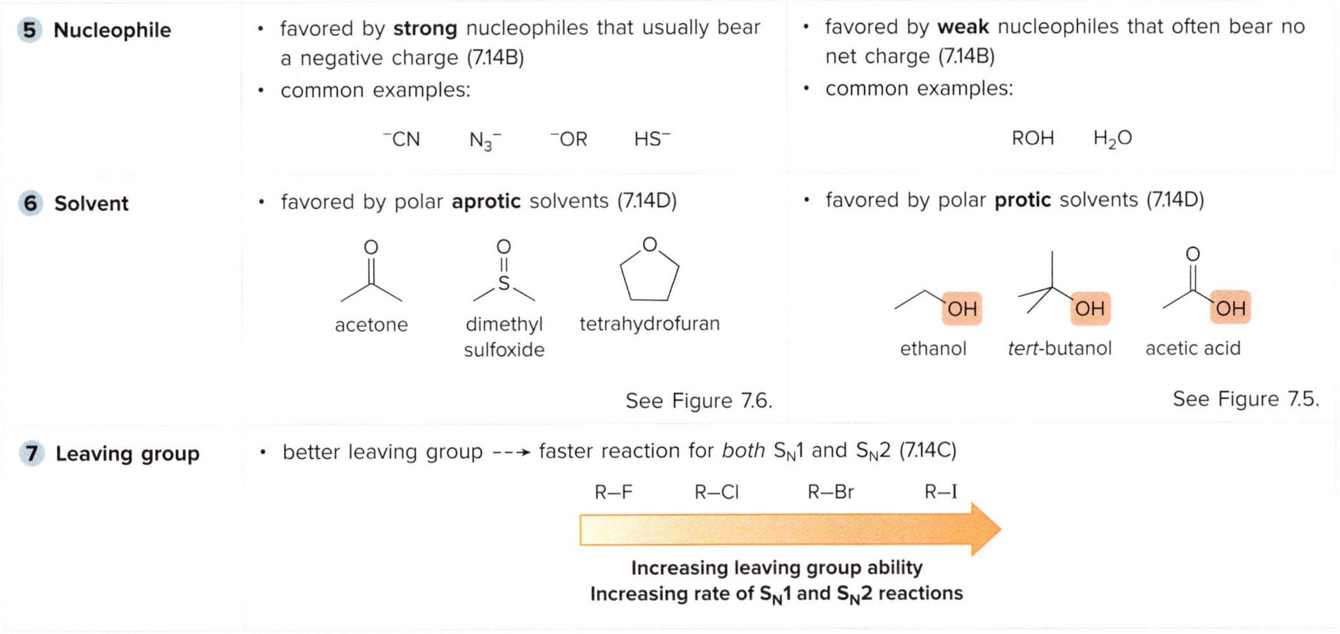

See Tables 7.4, 7.5, 7.6. Try Problems 7.48, 7.50, 7.52, 7.59, 7.63.

CHAPTER 7 MULTIPLE-CHOICE SELF-TEST

The Self-Test consists of multiple-choice questions similar to those found on the American Chemical Society organic chemistry exam. Answers are given at the end of the chapter.

1. Which carbocation is most stable?

a. b. c. d.

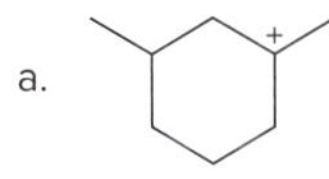

2. Which anion is the best leaving group: (a) H^-; (b) H_2N^-; (c) Cl^-; (d) Br^-?

3. Rank **A–D** in order of increasing reactivity in an S_N2 reaction.

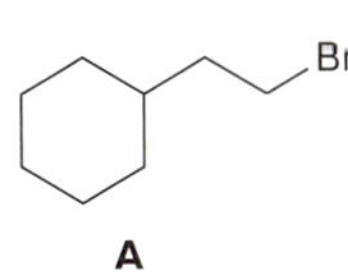

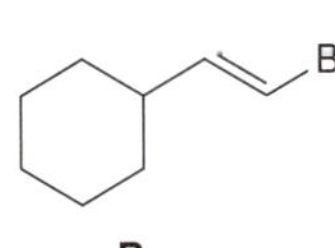

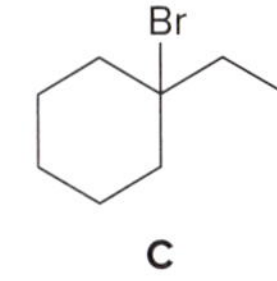

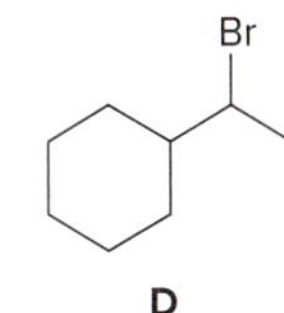

a. **A** < **D** < **C** < **B** b. **C** < **D** < **B** < **A** c. **C** < **B** < **D** < **A** d. **B** < **C** < **D** < **A**

4. Give the IUPAC name of the following halide.

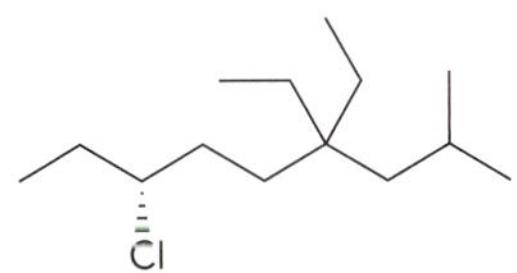

a. (*R*)-7-chloro-4,4-diethyl-2-methylnonane
b. (*R*)-3-chloro-6,6-diethyl-8-methylnonane
c. (*S*)-3-chloro-6,6-diethyl-8-methylnonane
d. (*S*)-7-chloro-4,4-diethyl-2-methylnonane

5. Which statement is *not* true?
a. Electron-donating groups stabilize a nearby positive charge.
b. Steric hindrance decreases nucleophilicity and basicity.
c. In an endothermic reaction, the more stable product is formed faster.
d. Polar aprotic solvents increase nucleophilicity.

6. What product(s) are formed when **A** is treated with ^{-}SH?

Cl SH SH SH

A 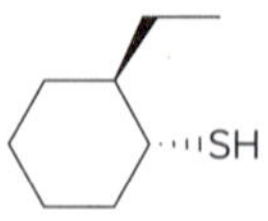 **B** 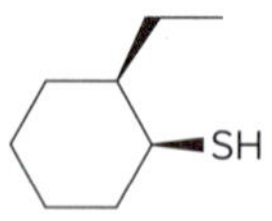 **C** 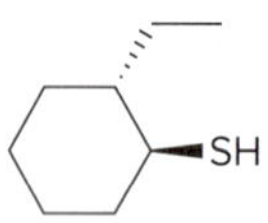 **D**

a. **B** only
b. **C** only
c. **D** only
d. an equal mixture of **B** and **C**

7. Which change causes an increase in the rate of the nucleophilic substitution reaction of **E** with H_2O?

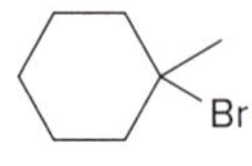

E

a. The leaving group is changed from Br^- to Cl^-.
b. The amount of H_2O is increased by a factor of three.
c. The concentration of **E** is increased.
d. The solvent is changed from H_2O to $(CH_3)_2CO$.

8. Which reaction gives the best yield of $C_6H_5OCH_2CH_2CH_3$?

a.  (benzene)–Br + $^{-}$O(propyl)
b. 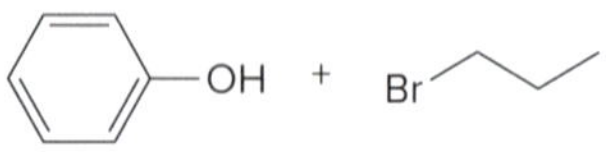 (benzene)–OH + Br(propyl)
c. (benzene)–O^{-} + Br(propyl)
d.  (benzene)–O^{-} + Cl(propyl)

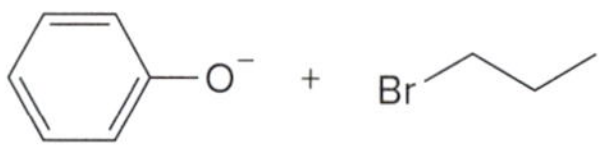

9. Which anion is the strongest nucleophile in polar aprotic solvents: (a) $CH_3CH_2S^-$; (b) $CH_3CH_2O^-$; (c) Cl^-; (d) $CH_3CO_2^-$?

10. What statement best describes the product mixture formed when **F** reacts with CH_3OH?

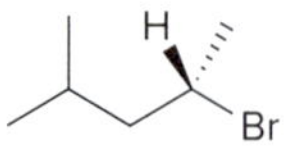

H Br H OCH_3 H OCH_3

F 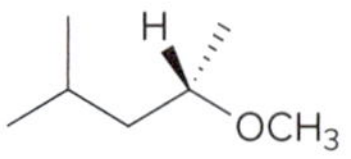 **G** 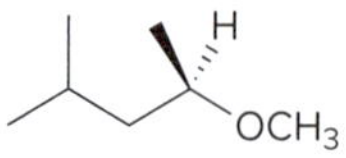 **H**

a. The product is optically active and contains only **G.**
b. The product is optically active and contains only **H.**
c. An optically inactive mixture of **G** and **H** is formed.
d. An optically active mixture of **G** and **H** is formed.

PROBLEMS

Problems Using Three-Dimensional Models

7.40 Give the IUPAC name for each compound, including any *R,S* designation.

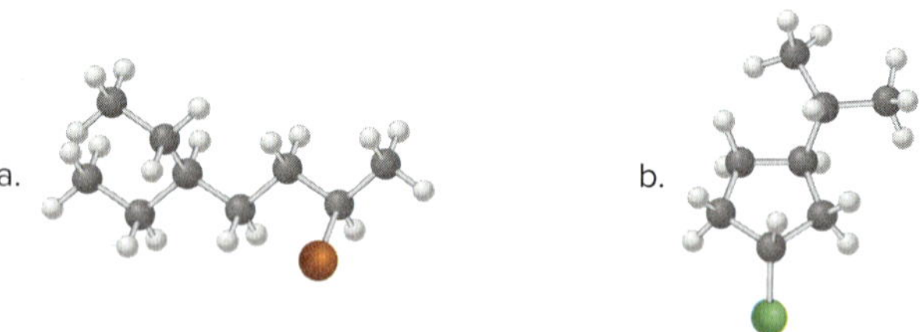

7.41 Draw the products formed when each alkyl halide is treated with NaCN.

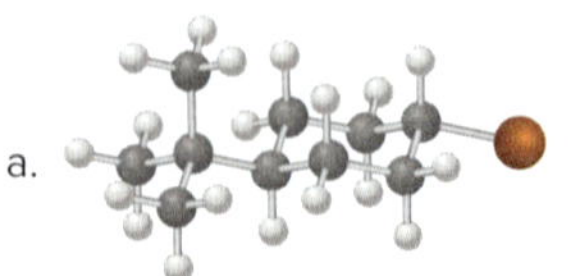

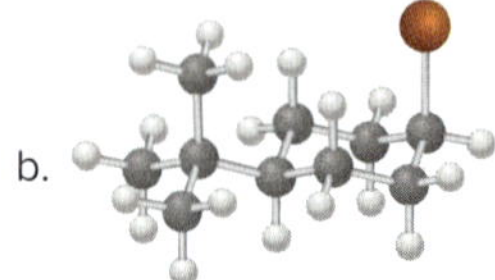

Nomenclature

7.42 Give the IUPAC name for each compound.

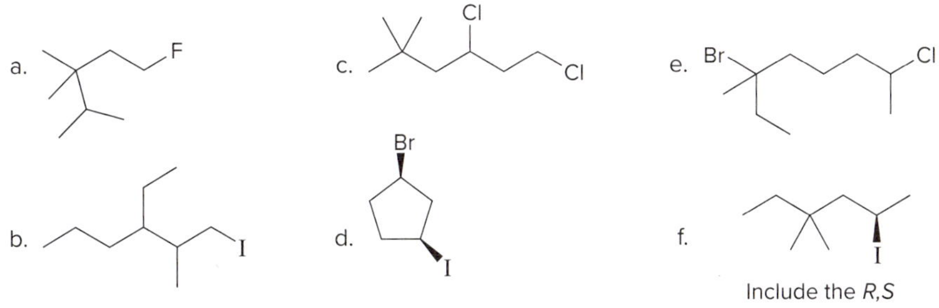

Include the *R,S* designation in the name.

7.43 Give the structure corresponding to each name.

a. 3-bromo-4-ethylheptane
b. 1,1-dichloro-2-methylcyclohexane
c. 1-bromo-4-ethyl-3-fluorooctane
d. (*S*)-3-iodo-2-methylnonane
e. (1*R*,2*R*)-*trans*-1-bromo-2-chlorocyclohexane
f. (*R*)-4,4,5-trichloro-3,3-dimethyldecane

7.44 Draw the eight constitutional isomers having the molecular formula $C_5H_{11}Cl$.

a. Give the IUPAC name for each compound (ignoring *R* and *S* designations).
b. Classify each alkyl halide as 1°, 2°, or 3°.
c. Label any stereogenic centers.
d. For each constitutional isomer that contains a stereogenic center, draw all possible stereoisomers, and label each stereogenic center as *R* or *S*.

Properties

Students who have already learned about mass spectrometry can try Problems A.6; A.7a, b; A.9; A.16d, e; A.19; and A.24(**A**), (**B**). Students who have learned about nuclear magnetic resonance spectroscopy can try Problems C.32; C.62a, b.

7.45 Match the boiling points to each of the given compounds; boiling points: 73 °C, 102 °C, 246 °C, 51 °C, and 130 °C; compounds: (a) $(CH_3)_3CBr$; (b) $CH_3CH_2CH_2CH_2Br$; (c) $ICH_2CH_2CH_2CH_2I$; (d) $CH_3CH_2CH_2CH_2I$; (e) $(CH_3)_3CCl$.

General Nucleophilic Substitution, Leaving Groups, and Nucleophiles

7.46 Draw the products of each nucleophilic substitution reaction.

a. [alkyl iodide] I + NaCN →
b. [cyclohexyl iodide] I + H_2O →
c. [cyclopentyl chloride] Cl + OH →
d. Cl + S →

7.47 Which of the following molecules contain a good leaving group?

a. OH b. Cl c. d. $\overset{+}{O}H_2$

7.48 Rank the following compounds in order of increasing reactivity in a substitution reaction with ^{-}CN as nucleophile.

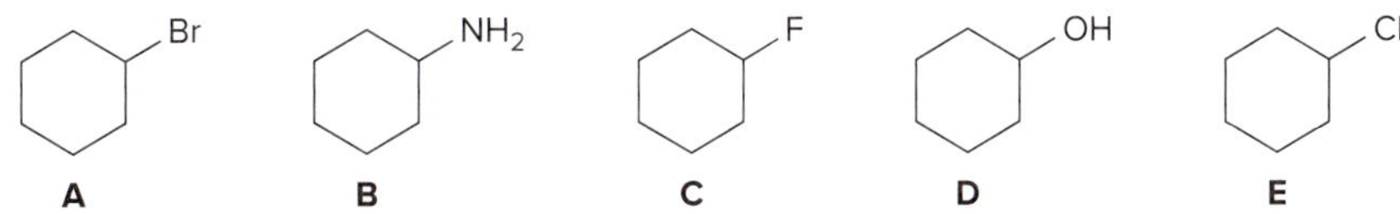

A **B** **C** **D** **E**

7.49 Which of the following nucleophilic substitution reactions will take place?

a. NH_2 + I^- → I + $^{-}NH_2$

b. I + CH_3O^- → O + I^-

7.50 Rank the anions in order of increasing nucleophilicity in acetone: CH_3S^-, CH_3NH^-, I^-, Br^-, and CH_3O^-.

7.51 Classify each solvent as protic or aprotic: (a) $(CH_3)_2CHOH$; (b) CH_3NO_2; (c) CH_2Cl_2; (d) NH_3.

The S_N2 Reaction

7.52 Consider the following S_N2 reaction.

Br + $^-$CN —(acetone)→ CN + Br^-

a. Draw a mechanism using curved arrows.
b. Draw an energy diagram. Label the axes, the reactants, products, E_a, and $\Delta H°$. Assume that the reaction is exothermic.
c. Draw the structure of the transition state.
d. What is the rate equation?
e. What happens to the reaction rate in each of the following instances? [1] The leaving group is changed from Br^- to I^-; [2] The solvent is changed from acetone to CH_3CH_2OH; [3] The alkyl halide is changed from $CH_3(CH_2)_4Br$ to $CH_3CH_2CH_2CH(Br)CH_3$; [4] The concentration of ^-CN is increased by a factor of five; and [5] The concentrations of both the alkyl halide and ^-CN are increased by a factor of five.

7.53 Draw the products of each S_N2 reaction and indicate the stereochemistry where appropriate.

a. (H, D, Cl) + $^-OCH_3$ →

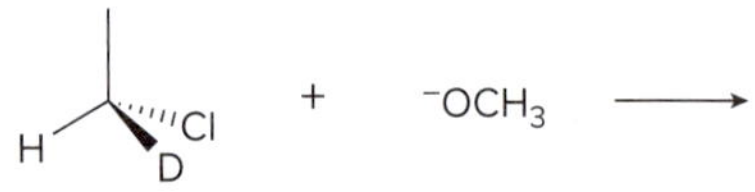

b. Cl + $^-OCH_2CH_3$ →

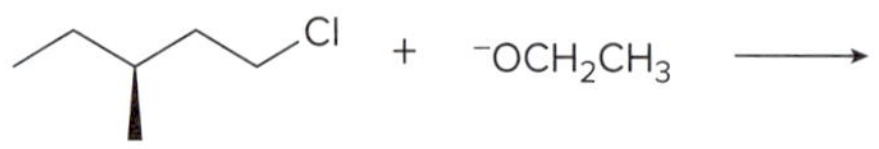

c. Br + $^-$CN →

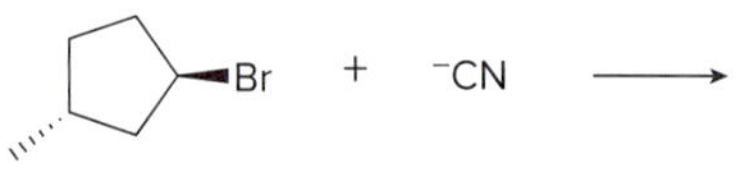

d. Cl + NaSH →

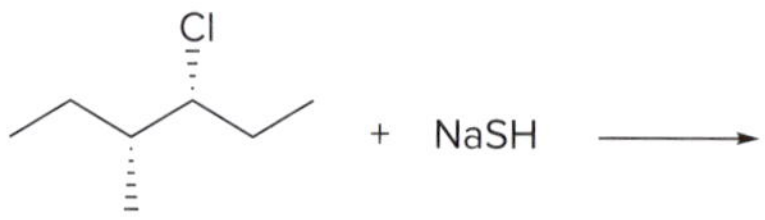

7.54 Salmeterol (trade name Serevent) is a long-acting bronchodilator used to treat asthma and chronic obstructive pulmonary disease (COPD). Salmeterol can be prepared by a stepwise reaction sequence that utilizes two S_N2 reactions. Draw the structure of intermediate **X** and the product salmeterol.

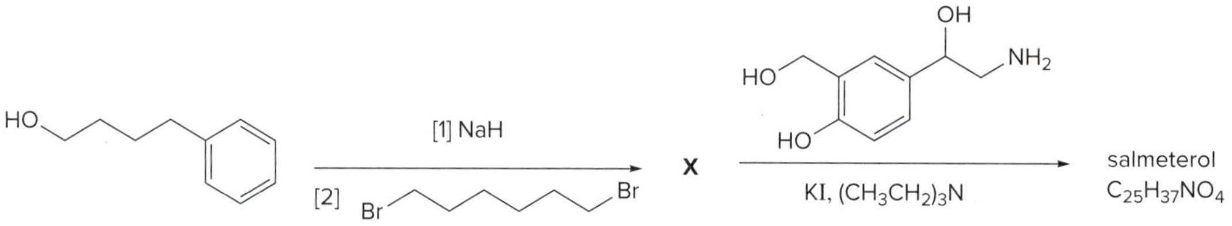

HO —([1] NaH; [2] Br~~~Br)→ **X** —(HO, HO, OH, NH_2; KI, $(CH_3CH_2)_3N$)→ salmeterol $C_{25}H_{37}NO_4$

7.55 Draw the product of the following S_N2 reaction, including the stereochemistry at all stereogenic centers. The product of this reaction is aprepitant, a drug used to treat nausea and emesis (vomiting) in chemotherapy patients.

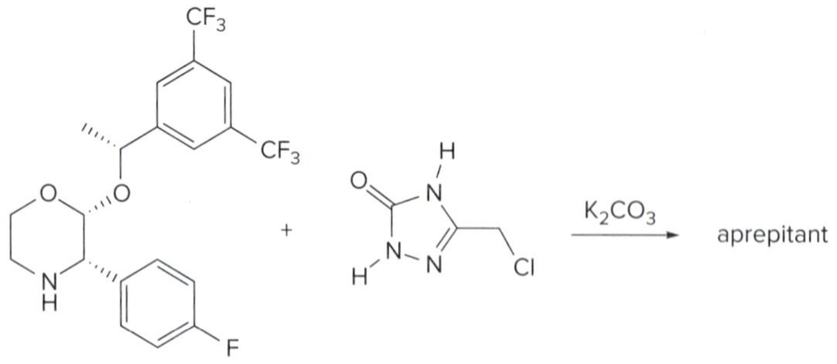

(CF_3, CF_3, O, O, N–H, F) + (O, H–N, N, N, H, Cl) —(K_2CO_3)→ aprepitant

7.56 Identify **M** in the following reaction sequence used to prepare the antiulcer drug omeprazole (trade name Prilosec).

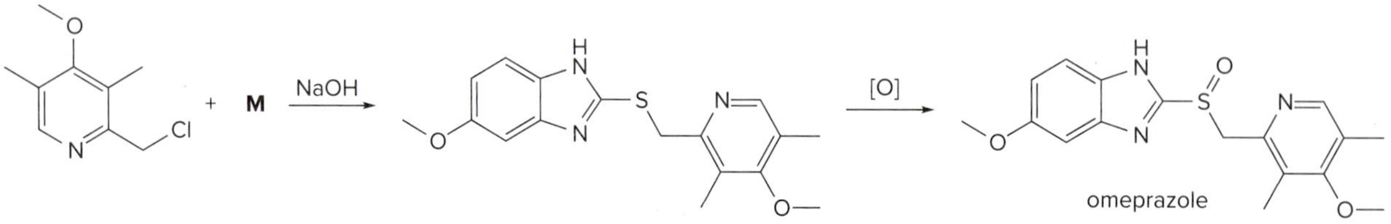

(O, N, Cl) + **M** —(NaOH)→ (O, H–N, N, S, N, O) —([O])→ omeprazole

7.57 The non-sedating antihistamine cetirizine (trade name Zyrtec) is prepared by a reaction sequence that involves two consecutive substitution reactions. Identify **N** and **O** in the following reaction sequence.

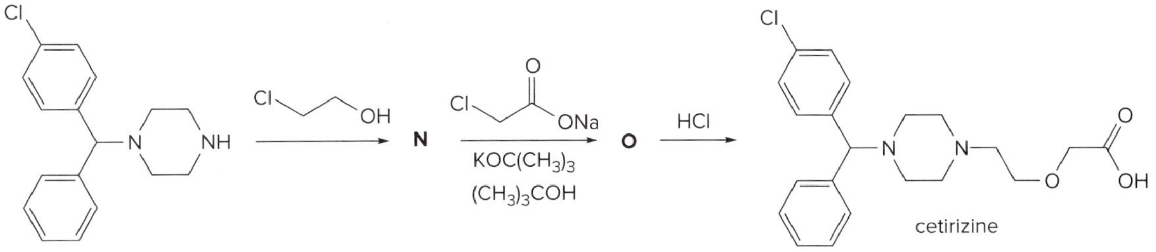

Cl, NH, Cl, OH, **N**, Cl, O, ONa, $KOC(CH_3)_3$, $(CH_3)_3COH$, **O**, HCl, Cl, N, N, O, O, OH

cetirizine

Carbocations

7.58 Which of the following carbocations (**A** or **B**) is more stable? Explain your choice.

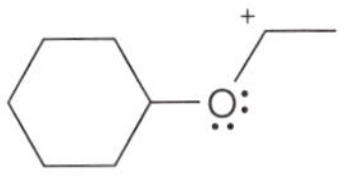

A 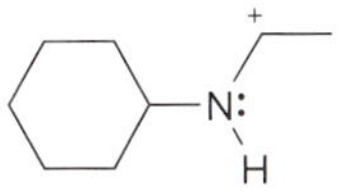 **B**

The S_N1 Reaction

7.59 Consider the following S_N1 reaction.

I + H_2O ⟶ $\overset{+}{O}H_2$ + I^-

a. Draw a mechanism for this reaction using curved arrows.

b. Draw an energy diagram. Label the axes, starting material, product, E_a, and $\Delta H°$. Assume that the starting material and product are equal in energy.

c. Draw the structure of any transition states.

d. What is the rate equation for this reaction?

e. What happens to the reaction rate in each of the following instances? [1] The leaving group is changed from I^- to Cl^-; [2] The solvent is changed from H_2O to DMF; [3] The alkyl halide is changed from $(CH_3)_2C(I)CH_2CH_3$ to $(CH_3)_2CHCH(I)CH_3$; and [4] The concentrations of both the alkyl halide and H_2O are increased by a factor of five.

7.60 Draw the products of each S_N1 reaction and indicate the stereochemistry when necessary.

a.  Br + CH_3CH_2OH ⟶ b. 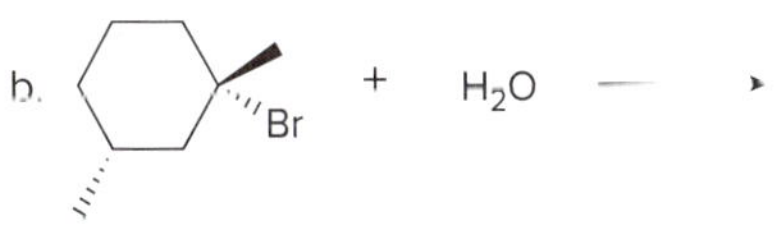 Br + H_2O ⟶

7.61 Draw a stepwise mechanism for the following reaction that illustrates how two substitution products are formed. Explain why 1-bromohex-2-ene reacts rapidly with a weak nucleophile (CH_3OH) under S_N1 reaction conditions, even though it is a 1° alkyl halide.

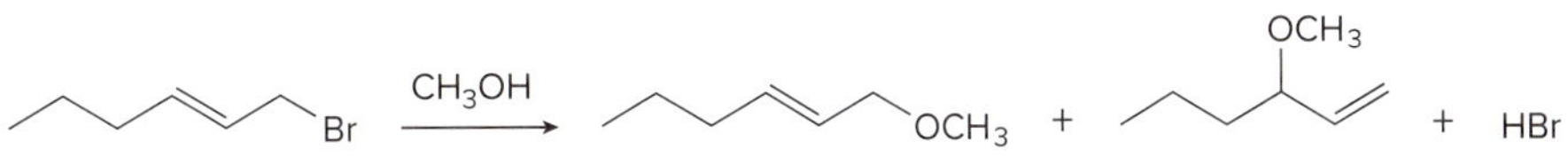

Br $\xrightarrow{CH_3OH}$ OCH_3 + OCH_3 + HBr

1-bromohex-2-ene

S_N1 and S_N2 Reactions

7.62 (a) Which halide in the following marine natural product reacts fastest in the S_N2 reaction? (b) Which halide in the following marine natural product reacts fastest in the S_N1 reaction?

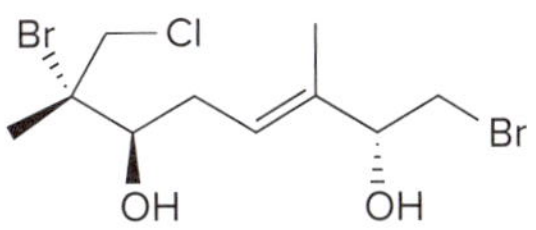

Br, Cl, Br, OH, OH

plocamenol A

7.63 Determine the mechanism of nucleophilic substitution of each reaction and draw the products, including stereochemistry.

a. 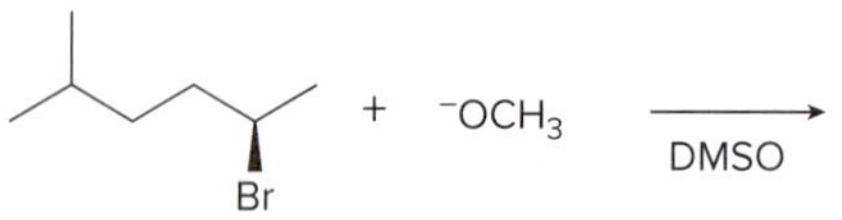 Br + $^{-}OCH_3$ $\xrightarrow{\text{DMSO}}$

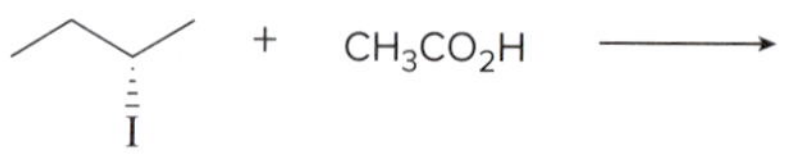

b. 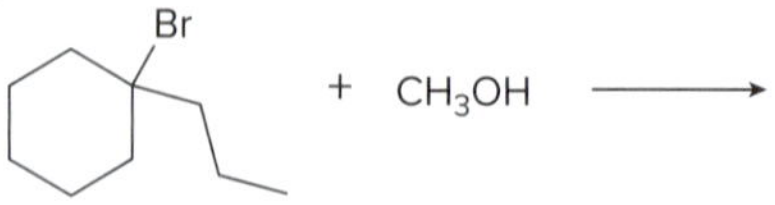 Br + CH_3OH ⟶

c. I + CH_3CO_2H ⟶

d. 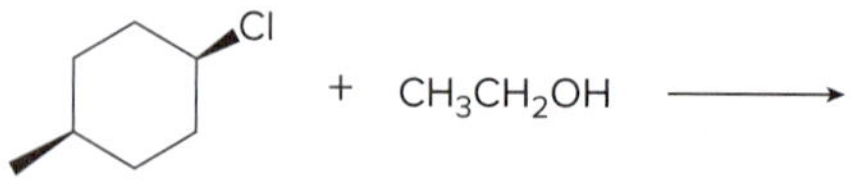 Cl + CH_3CH_2OH ⟶

7.64 Uridine monophosphate (UMP) is one of the four nucleotides that compose RNA, the nucleic acid that translates the genetic information of DNA into proteins needed by cells for proper function and development. A key step in the synthesis of UMP is the S_N1 reaction of **A** with **B** to form **C,** which is then converted to UMP in one step. Draw a stepwise mechanism for this S_N1 reaction.

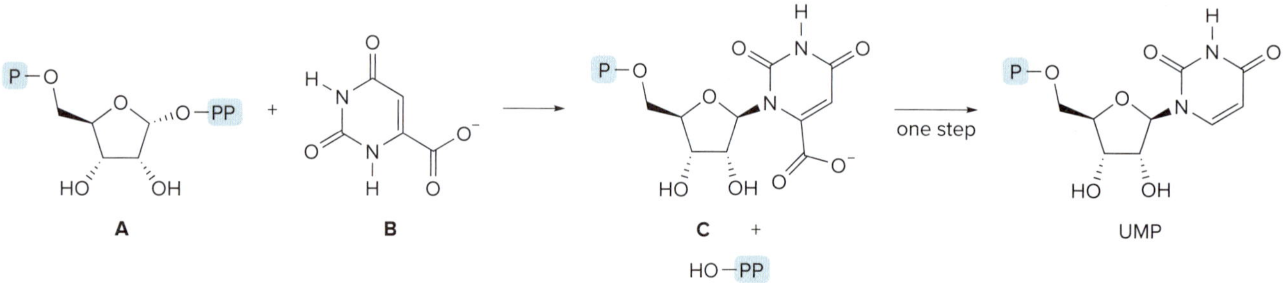

P–O, O–PP, HO, OH — **A** + **B** ⟶ **C** + HO–PP $\xrightarrow{\text{one step}}$ **UMP**

7.65 Diphenhydramine, the antihistamine in Benadryl, can be prepared by the following two-step sequence. What is the structure of diphenhydramine?

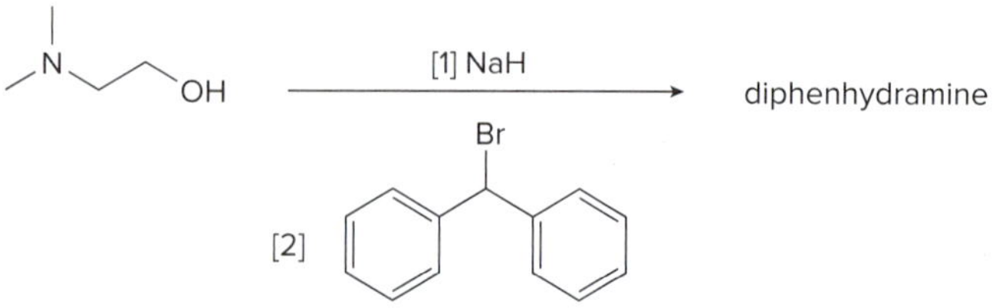

OH $\xrightarrow{[1]\ \text{NaH};\ [2]\ \text{Br}}$ diphenhydramine

7.66 Draw a stepwise, detailed mechanism for the following reaction. Use curved arrows to show the movement of electrons.

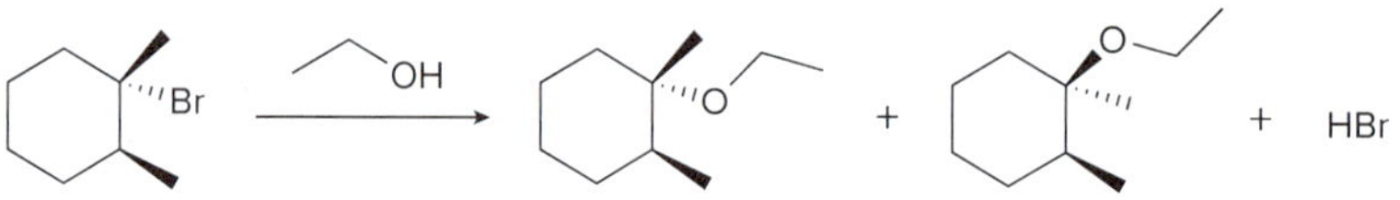

Br $\xrightarrow{\text{OH}}$ O + O + HBr

7.67 When a single compound contains both a nucleophile and a leaving group, an **intramolecular** reaction may occur. With this in mind, draw the product of the following reaction.

OH, Br $\xrightarrow{^{-}OH}$ O⁻, Br + H_2O ⟶ $C_7H_{10}O_2$ + Br^{-}

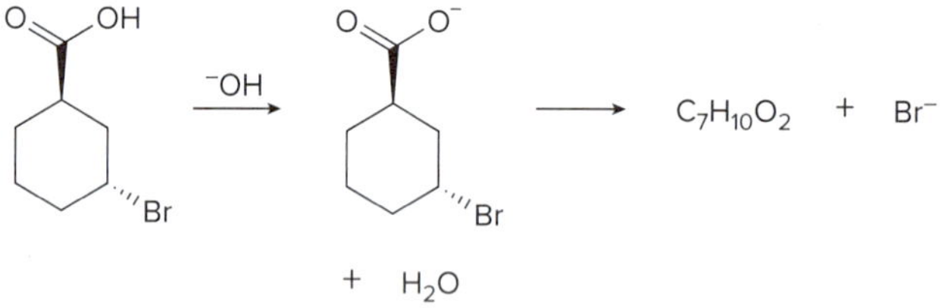

7.68 Quinapril (trade name Accupril) is used to treat high blood pressure and congestive heart failure. One step in the synthesis of quinapril involves reaction of the racemic alkyl bromide **A** with a single enantiomer of the amino ester **B.** (a) What two products are formed in this reaction? (b) Given the structure of quinapril, which one of these two products is needed to synthesize the drug?

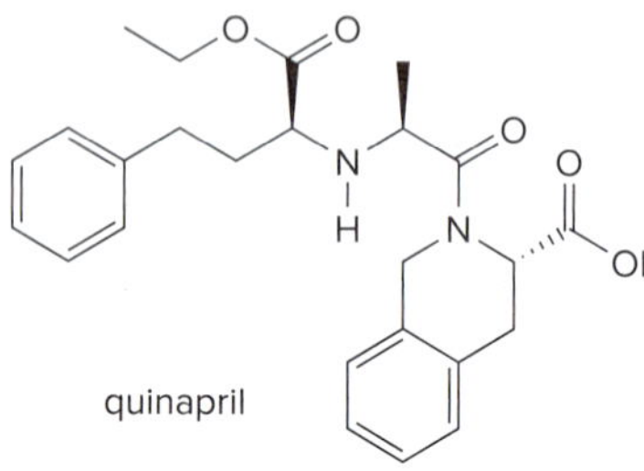

quinapril

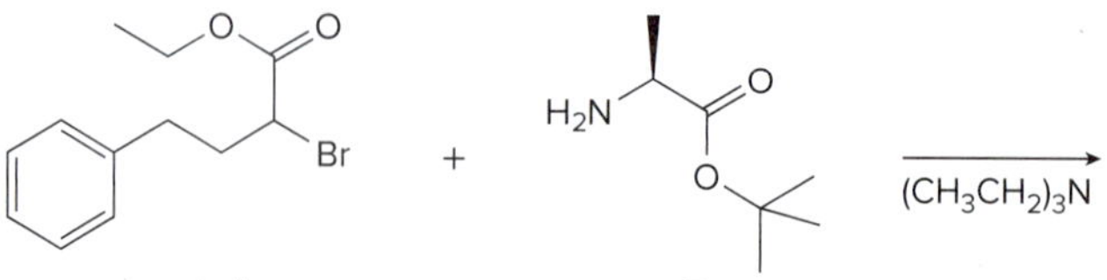

racemic **A** (Br) + H_2N **B** $\xrightarrow{(CH_3CH_2)_3N}$

7.69 Draw a stepwise, detailed mechanism for the following reaction.

Cl ... N ... Cl + CH_3NH_2 (excess) ⟶ N–CH_3 + $CH_3\overset{+}{N}H_3$ Cl^-

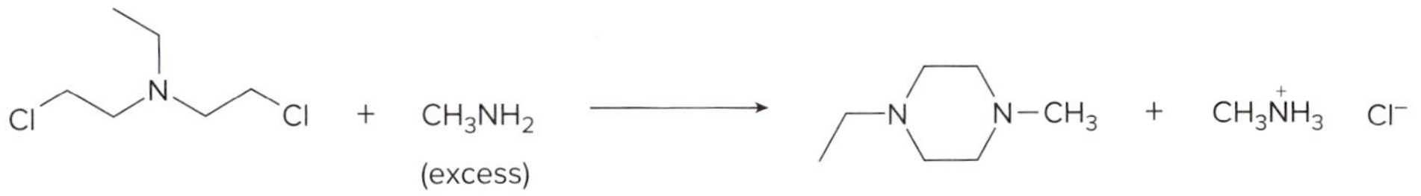

7.70 When (*R*)-6-bromo-2,6-dimethylnonane is dissolved in CH_3OH, nucleophilic substitution yields an optically inactive solution. When the isomeric halide (*R*)-2-bromo-2,5-dimethylnonane is dissolved in CH_3OH under the same conditions, nucleophilic substitution forms an optically active solution. Draw the products formed in each reaction, and explain why the difference in optical activity is observed.

Synthesis

7.71 Fill in the appropriate reagent or starting material in each of the following reactions.

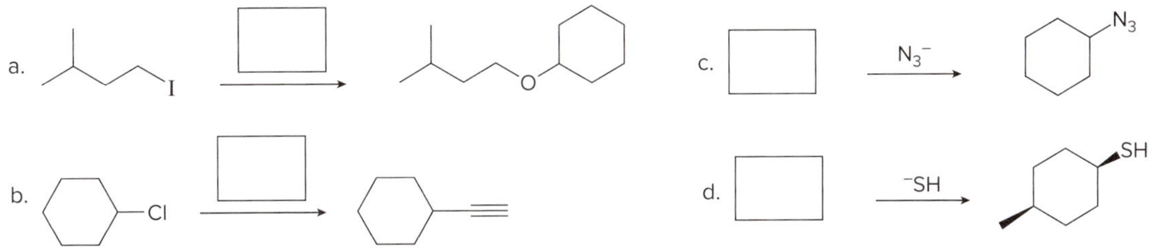

a. I ⟶ [ ] ⟶ O

b. Cl ⟶ [ ] ⟶

c. [ ] $\xrightarrow{N_3^-}$ N_3

d. [ ] $\xrightarrow{^-SH}$ SH

7.72 Devise a synthesis of each compound from an alkyl halide using any other organic or inorganic reagents.

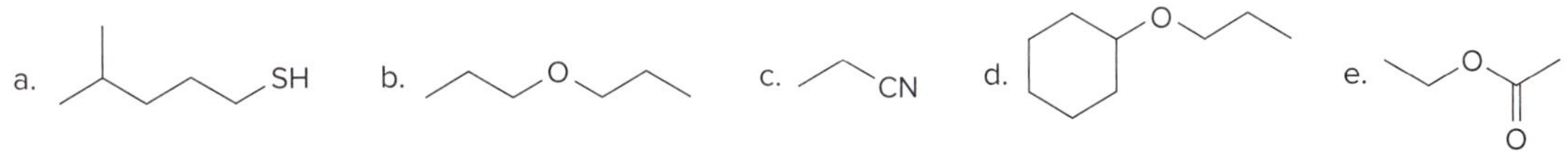

a. SH b. O c. CN d. O e. O, O

7.73 Suppose you have compounds **A–D** at your disposal. Using these compounds, devise two different ways to make **E.** Which one of these methods is preferred, and why?

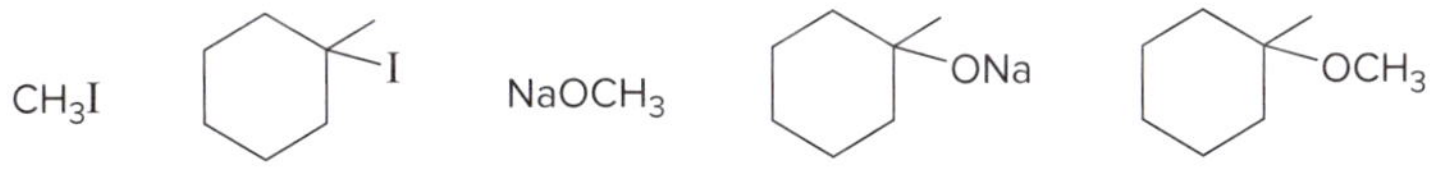

CH_3I (**A**) I (**B**) $NaOCH_3$ (**C**) ONa (**D**) OCH_3 (**E**)

7.74 Muscalure, the sex pheromone of the common housefly, can be prepared by a reaction sequence that uses two nucleophilic substitutions. Identify compounds **A–D** in the following synthesis of muscalure.

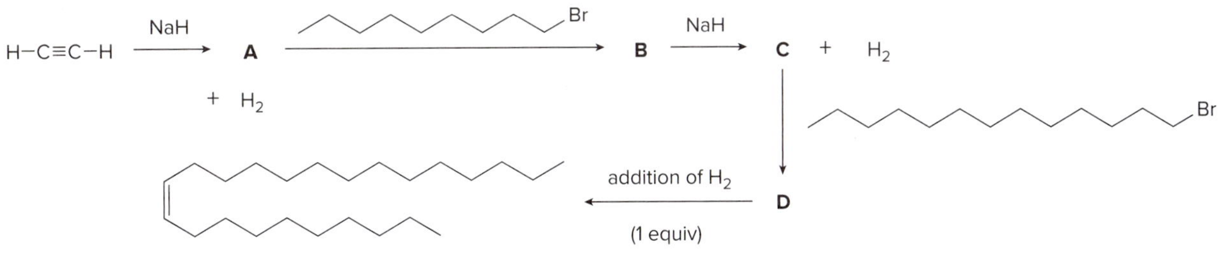

H–C≡C–H $\xrightarrow{NaH}$ **A** + H_2 $\xrightarrow{\text{Br}}$ **B** $\xrightarrow{NaH}$ **C** + H_2 $\xrightarrow{\text{Br}}$ **D** $\xrightarrow[\text{(1 equiv)}]{\text{addition of } H_2}$ muscalure

Challenge Problems

7.75 Explain why quinuclidine is a much more reactive nucleophile than triethylamine, even though both compounds have N atoms surrounded by three R groups.

quinuclidine triethylamine

7.76 Draw a stepwise mechanism for the following reaction sequence.

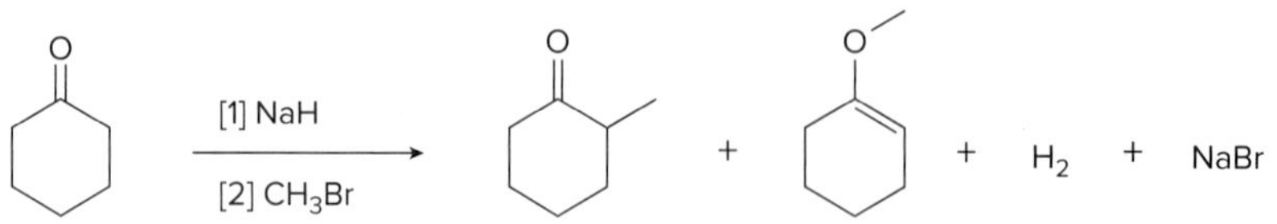

[1] NaH
[2] CH_3Br

\+ H_2 + NaBr

major product minor product

7.77 Which N atom in fentanyl, an opioid pain reliever and anesthetic, is more nucleophilic? Explain your choice.

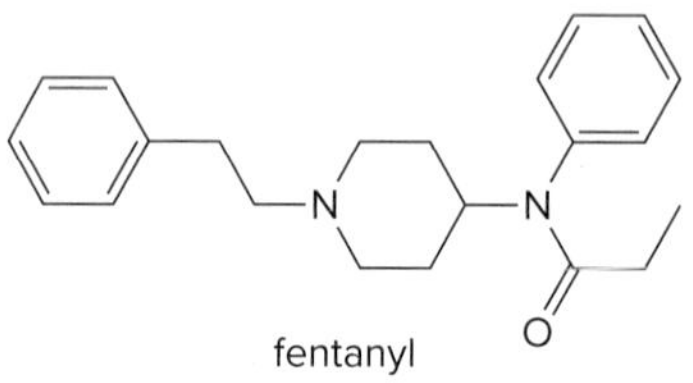

fentanyl

7.78 When trichloride **J** is treated with CH_3OH, nucleophilic substitution forms the dihalide **K.** Draw a mechanism for this reaction and explain why one Cl is much more reactive than the other two Cl's so that a single substitution product is formed.

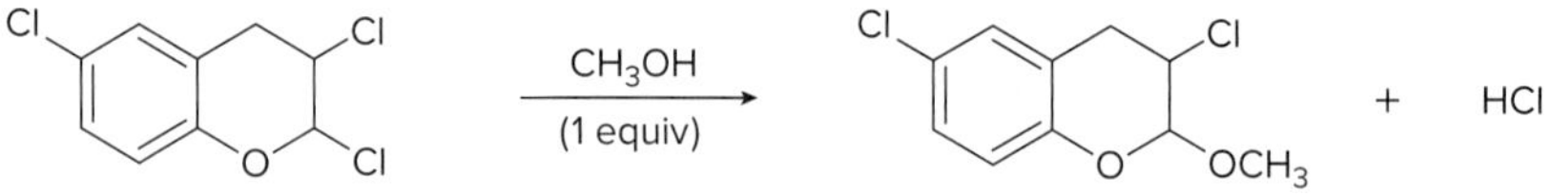

CH_3OH (1 equiv)

Cl Cl Cl — **J**

Cl Cl OCH_3 — **K** + HCl

7.79 In some nucleophilic substitutions under S_N1 conditions, complete racemization does not occur and a small excess of one enantiomer is present. For example, treatment of optically pure 1-bromo-1-phenylpropane with water forms 1-phenylpropan-1-ol. (a) Calculate how much of each enantiomer is present using the given optical rotation data. (b) Which product predominates—the product of inversion or the product of retention of configuration? (c) Suggest an explanation for this phenomenon.

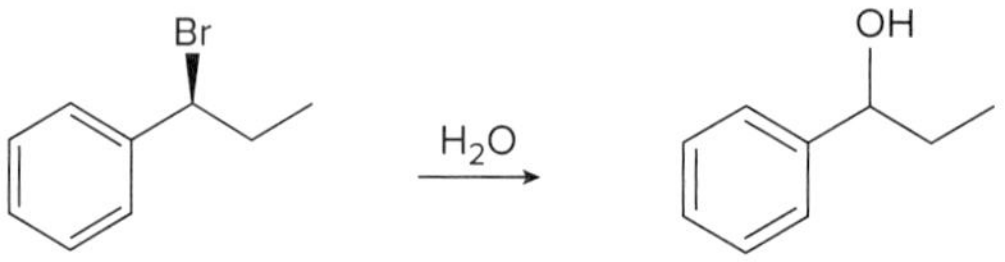

H_2O

1-bromo-1-phenylpropane

1-phenylpropan-1-ol
observed $[\alpha] = +5.0$
optically pure *S* isomer, $[\alpha] = -48$

7.80 A key reaction in the synthesis of some lipids involves the rearrangement of geranyl diphosphate to linalyl diphosphate. Use the information in Section 7.15A to draw a stepwise mechanism for this process.

geranyl diphosphate (OPP) → linalyl diphosphate (OPP)

SELF-TEST ANSWERS

1. a 2. d 3. d 4. a 5. b 6. b 7. c 8. c 9. b 10. c

Alkyl Halides and Elimination Reactions

8.1 General features of elimination

8.2 Alkenes—The products of elimination reactions

8.3 The mechanisms of elimination

8.4 The E2 mechanism

8.5 The Zaitsev rule

8.6 Stereochemistry of the E2 reaction

8.7 The E1 mechanism

8.8 S_N1 and E1 reactions

8.9 When is the mechanism E1 or E2?

8.10 E2 reactions and alkyne synthesis

8.11 When is the reaction S_N1, S_N2, E1, or E2?

Forest & Kim Starr

The elegant synthesis of **quinine** in 1944 is considered by many scientists to be the beginning of modern-day organic synthesis. Quinine, a natural product isolated from the bark of the cinchona tree native to the Andes Mountains, is a powerful antipyretic—that is, it reduces fever—and for centuries, it was the only effective treatment for malaria. Its bitter taste gives tonic water its characteristic flavor. One of the steps in a lengthy synthesis of quinine involves elimination, a characteristic reaction of alkyl halides and the subject of Chapter 8.

Why Study . . . Elimination Reactions?

Elimination reactions introduce π bonds into organic compounds, so they can be used to synthesize **alkenes** and **alkynes**—hydrocarbons that contain one and two π bonds, respectively. Elimination reactions are valuable in organic synthesis because they form functional groups that span two carbons. Like nucleophilic substitution, elimination reactions can occur by two different pathways, depending on the conditions. By the end of Chapter 8, therefore, you will have learned four different reaction mechanisms, two for nucleophilic substitution (S_N1 and S_N2) and two for elimination (E1 and E2).

The biggest challenge with this material is learning how to sort out two different reactions that follow four different mechanisms. **Will a particular alkyl halide undergo substitution or elimination with a given reagent, and by which of the four possible mechanisms?** To answer this question, we conclude Chapter 8 with a summary that allows you to predict which reaction and mechanism are likely for a given substrate.

8.1 General Features of Elimination

All **elimination reactions** involve loss of elements from the starting material to form a new π bond in the product.

- **Alkyl halides undergo elimination reactions with Brønsted–Lowry bases. The elements of HX are lost and an alkene is formed.**

H, X + :B (base) → (elimination of HX) new π bond, alkene + $H{-}B^+$ + $:X^-$

Equations [1] and [2] illustrate examples of elimination reactions. In both reactions a base removes the elements of an acid, HBr or HCl, from the organic starting material.

[1] (alkyl bromide, H, Br) → ($K^+\ ^-OC(CH_3)_3$; [−HBr]) alkene + $HOC(CH_3)_3$ + $K^+\ Br^-$

[2] (chlorocyclohexane, H, Cl) → ($Na^+\ ^-OCH_2CH_3$; [−HCl]) cyclohexene + $HOCH_2CH_3$ + $Na^+\ Cl^-$

Removal of the elements of HX, called **dehydrohalogenation,** is one of the most common methods to introduce a π bond and prepare an alkene. Dehydrohalogenation is an example of **β elimination,** because it involves loss of elements from two adjacent atoms: the **α carbon** bonded to the leaving group X, and the **β carbon** adjacent to it. Three curved arrows illustrate how four bonds are broken or formed in the process.

B: + H–Cβ–Cα–X → C=C (α, β; new π bond) + $H{-}B^+$ + $:X^-$ (leaving group)

- The base (B:) removes a proton on the β carbon, thus forming $H{-}B^+$.
- The electron pair in the β C–H bond forms the new π bond between the α and β carbons.
- The electron pair in the C–X bond ends up on halogen, forming the leaving group $:X^-$.

The most common bases used in elimination reactions are negatively charged oxygen compounds such as **^-OH** and its alkyl derivatives, **^-OR,** called **alkoxides,** listed in Table 8.1. **Potassium *tert*-butoxide, $K^+\ ^-OC(CH_3)_3$, a bulky nonnucleophilic base, is especially useful** (Section 7.8B).

Table 8.1 Common Bases Used in Dehydrohalogenation

Structure	Name
$Na^+\ ^-OH$	Sodium hydroxide
$K^+\ ^-OH$	Potassium hydroxide
$Na^+\ ^-OCH_3$	Sodium methoxide
$Na^+\ ^-OCH_2CH_3$	Sodium ethoxide
$K^+\ ^-OC(CH_3)_3$	Potassium *tert*-butoxide

To draw any product of dehydrohalogenation:

- **Find the α carbon—the sp^3 hybridized carbon bonded to the leaving group.**
- **Identify all β carbons with H atoms.**
- **Remove the elements of H and X from the α and β carbons and form a π bond.**

For example, 2-bromo-2-methylpropane has three β carbons (three CH_3 groups), but because all three are *identical,* only *one* alkene is formed upon elimination of HBr. In contrast, 2-bromobutane has two *different* β carbons (labeled $β_1$ and $β_2$), so elimination affords *two* constitutional isomers by loss of HBr across either the α and $β_1$ carbons, or the α and $β_2$ carbons. We learn about which product predominates and why in Section 8.5.

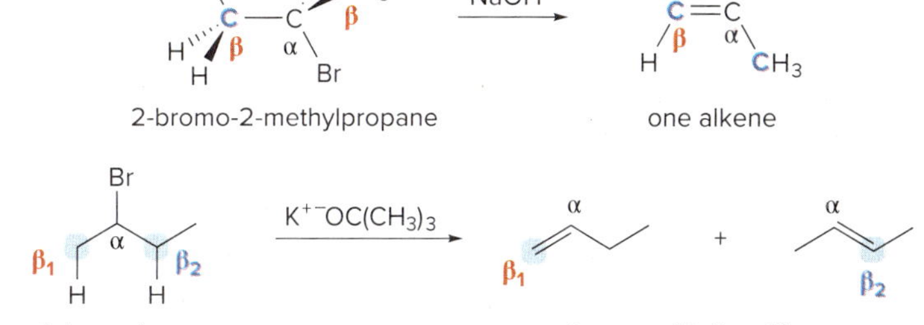

DDE and DDT accumulate in the fatty tissues of predator birds such as osprey. When DDE and DDT concentration is high, female osprey produce eggs with thin shells that are easily crushed, so fewer osprey chicks hatch.
Comstock/Getty Images

An elimination reaction is the first step in the slow degradation of the **pesticide DDT** (Section 7.4). Elimination of HCl from DDT forms the degradation product **DDE** (dichlorodiphenyldichloroethylene). This stable alkene is found in minute concentration in the fatty tissues of most adults in the United States.

–HCl

DDT → DDE

Problem 8.1 Label the α and β carbons in each alkyl halide. Draw all possible elimination products formed when each alkyl halide is treated with $K^{+-}OC(CH_3)_3$.

a. b. c. d.

8.2 Alkenes—The Products of Elimination Reactions

Because elimination reactions of alkyl halides form alkenes, let's review earlier material on alkene structure and learn some additional facts as well.

8.2A Bonding in a Carbon–Carbon Double Bond

Recall from Section 1.10B that alkenes are hydrocarbons containing a carbon–carbon double bond. Each carbon of the double bond is sp^2 hybridized and trigonal planar, and all bond angles are 120°.

ethylene = 120°, sp^2 hybridized

Ethylene, the simplest alkene, is a hormone that regulates plant growth and fruit ripening. A ripe banana placed next to unripe tomatoes speeds up their ripening because the banana gives off ethylene. *Jill Braaten/McGraw Hill*

The double bond of an alkene consists of a σ bond and a π bond.

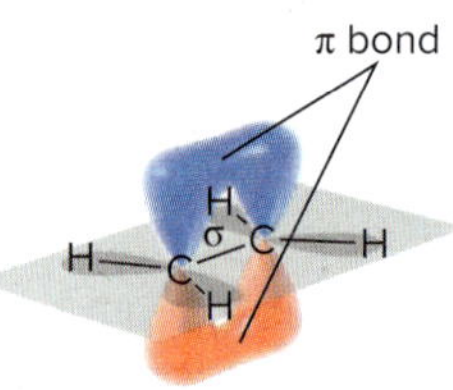

- **The σ bond, formed by end-on overlap of the two sp^2 hybrid orbitals, lies in the plane of the molecule.**
- **The π bond, formed by side-by-side overlap of two 2*p* orbitals, lies perpendicular to the plane of the molecule. The π bond is formed during elimination.**

Alkenes are classified according to the number of carbon atoms bonded to the carbons of the double bond. A **monosubstituted alkene** has *one* carbon atom bonded to the carbons of the double bond. A **disubstituted alkene** has *two* carbon atoms bonded to the carbons of the double bond, and so forth.

$RHC=CH_2$	$RHC=CHR$	$R_2C=CHR$	$R_2C=CR_2$
monosubstituted (***one*** R group)	**di**substituted (***two*** R groups)	**tri**substituted (***three*** R groups)	**tetra**substituted (***four*** R groups)

Figure 8.1 shows several alkenes and how they are classified. You must be able to classify alkenes in this way to determine the major and minor products of elimination reactions, when a mixture of alkenes is formed.

Figure 8.1 Classifying alkenes by the number of R groups bonded to the double bond

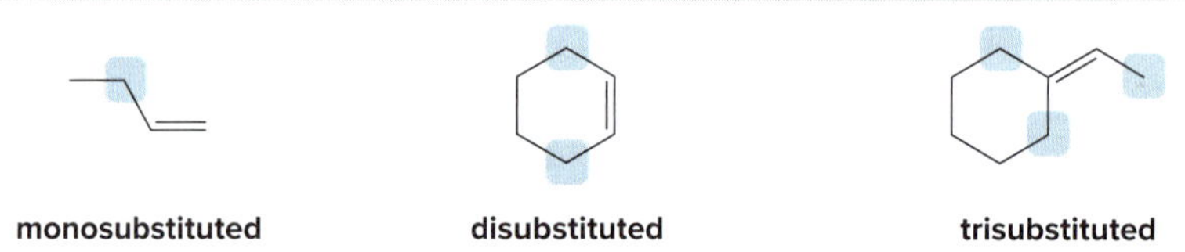

monosubstituted **disubstituted** **trisubstituted**

- Carbon atoms bonded to the double bond are screened in blue.

Problem 8.2 Classify each alkene in the following vitamins by the number of carbon substituents bonded to the double bond.

a. vitamin A (OH)

b. vitamin D_3 (H, HO)

8.2B Restricted Rotation

Figure 8.2 shows that there is free rotation about the carbon–carbon single bonds of butane, but *not* about the carbon–carbon double bond of but-2-ene. Because of restricted rotation, two stereoisomers of but-2-ene are possible.

Figure 8.2 Rotation around C–C and C=C compared

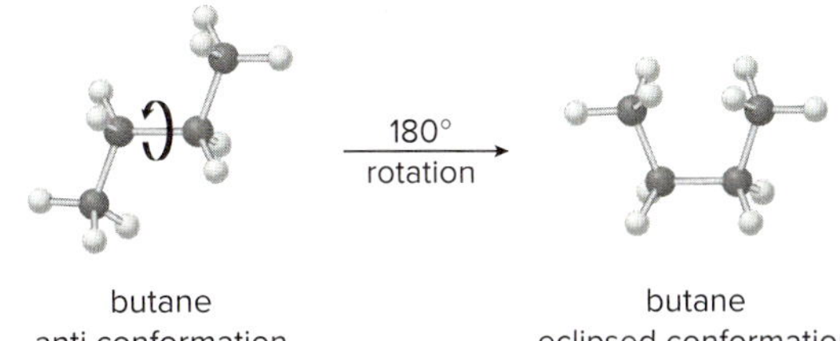

These conformations **interconvert** by rotation.
They represent the **same** molecule.

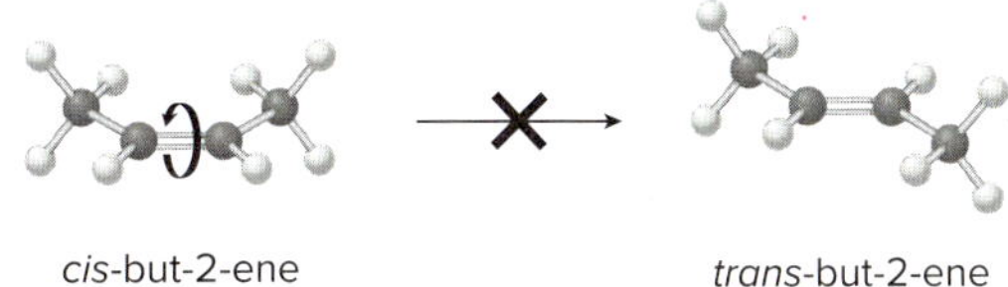

These molecules **do *not* interconvert** by rotation.
They are **different** molecules.

The concept of cis and trans isomers was first introduced for disubstituted cycloalkanes in Chapter 4. In both cases, a ring or a double bond restricts motion, preventing the rotation of a group from one side of the ring or double bond to the other.

- **The cis isomer has two groups on the *same side* of the double bond.**
- **The trans isomer has two groups on *opposite sides* of the double bond.**

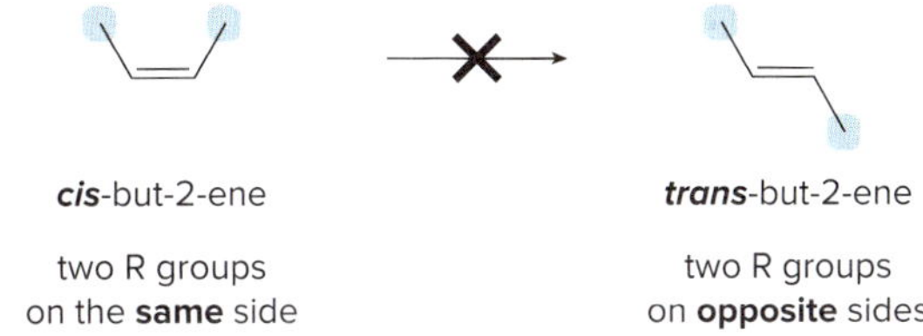

cis-But-2-ene and *trans*-but-2-ene are stereoisomers, but not mirror images of each other, so they are **diastereomers.**

The cis and trans isomers of but-2-ene are a specific example of a general type of stereoisomer occurring at carbon–carbon double bonds. **Whenever the two groups on *each* end of a carbon–carbon double bond are *different from each other,* two diastereomers are possible.**

X and X' must be **different** from *each other*... X X' Y Y' ...and Y and Y' must be **different** from *each other.*

Problem 8.3 For which double bonds are stereoisomers possible?

a. OH

b.

c. OH
nerolidol
(from the angel's trumpet plant)

d.
humulene
(from hops)

The characteristic fragrance of lilac flowers (Problem 8.4) is a mixture of (*E*)-ocimene and other volatile ethers, aldehydes, and alcohols. *Steven P. Lynch*

Problem 8.4 (a) Which double bonds in (*E*)-ocimene, a major component of the odor of lilac flowers, can exhibit stereoisomerism? (b) Draw a diastereomer of (*E*)-ocimene.

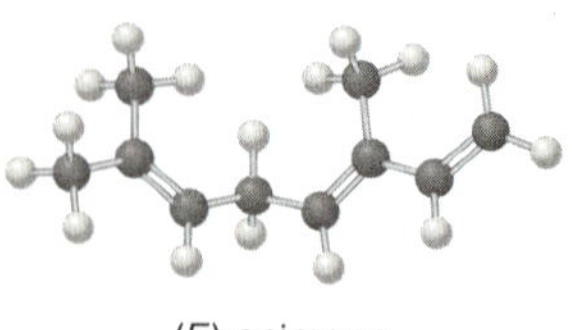

(*E*)-ocimene

Problem 8.5 Label each pair of alkenes as constitutional isomers, stereoisomers, or identical.

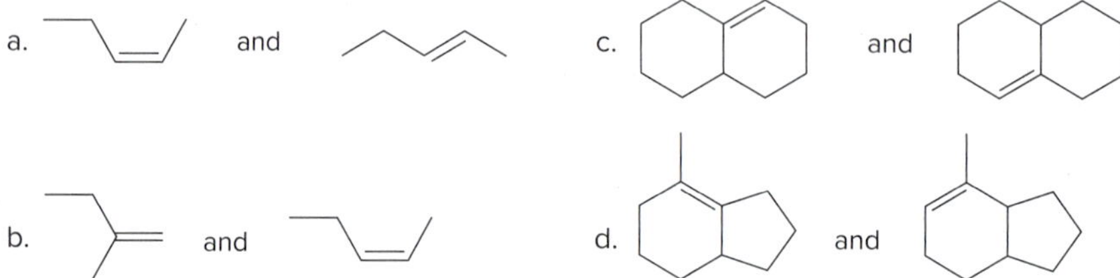

8.2C Stability of Alkenes

Some alkenes are more stable than others. For example, **trans alkenes are generally more stable than cis alkenes** because the larger groups bonded to the double bond carbons are farther apart, reducing steric interactions.

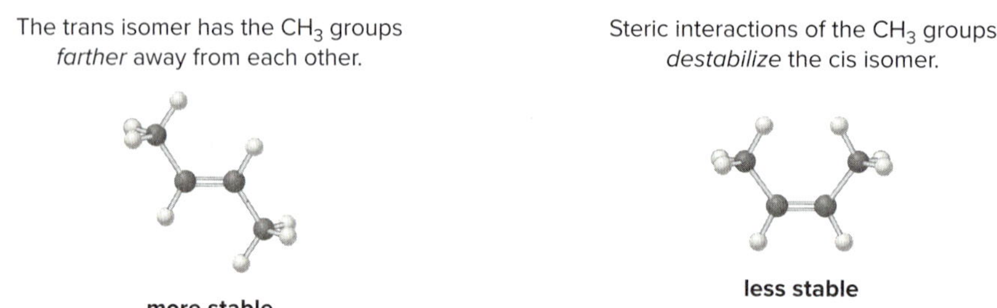

The stability of an alkene *increases,* moreover, as the **number of R groups bonded to the double bond carbons *increases.***

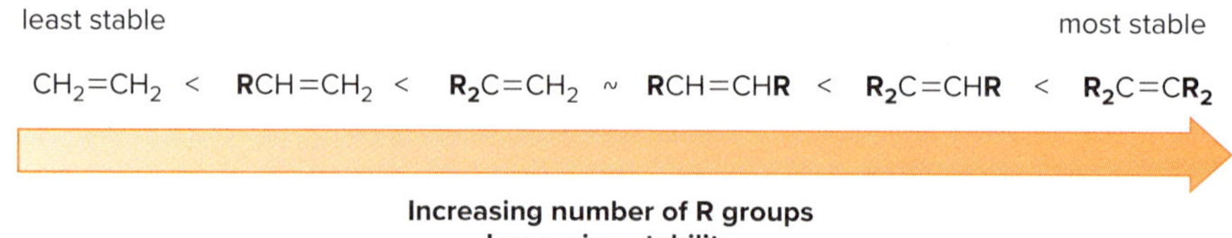

R groups increase the stability of an alkene because R groups are sp^3 hybridized, whereas the carbon atoms of the double bond are sp^2 hybridized. Recall from Sections 1.11B and 2.5D that the percent *s*-character of a hybrid orbital increases from 25% to 33% in going from sp^3 to sp^2. The higher the percent *s*-character, the more readily an atom accepts electron density. Thus, **sp^2 hybridized carbon atoms are more able to *accept* electron density,** and **sp^3 hybridized carbon atoms are more able to *donate* electron density.**

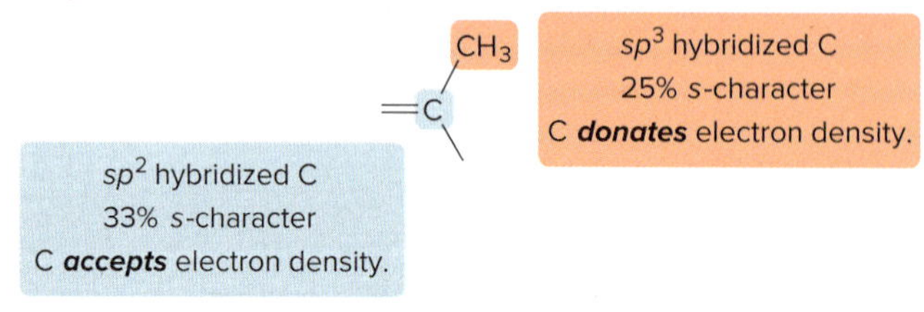

- As a result, *increasing* the number of electron-donating R groups on a carbon atom able to accept electron density makes the alkene *more stable*.

Thus, *trans*-but-2-ene (a disubstituted alkene) is more stable than *cis*-but-2-ene (another disubstituted alkene), but both are more stable than but-1-ene (a monosubstituted alkene).

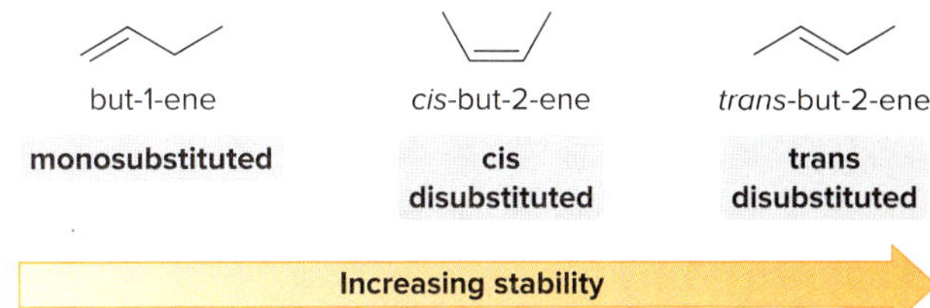

In summary:

- Trans alkenes are *more stable* than cis alkenes because they have fewer steric interactions.
- *Increasing* alkyl substitution *stabilizes* an alkene by an electron-donating inductive effect.

Problem 8.6 Which alkene in each pair is more stable?

a. or b. or

Problem 8.7 Rank the following alkenes in order of increasing stability.

A B C D

8.3 The Mechanisms of Elimination

What is the mechanism for elimination? What is the order of bond breaking and bond making? Is the reaction a one-step process or does it occur in many steps?

There are two mechanisms for elimination—**E2** and **E1**—just as there are two mechanisms for nucleophilic substitution—S_N2 and S_N1.

- The **E2 mechanism (bimolecular elimination)**
- The **E1 mechanism (unimolecular elimination)**

The E2 and E1 mechanisms differ in the timing of bond cleavage and bond formation, analogous to the S_N2 and S_N1 mechanisms. In fact, E2 and S_N2 reactions have some features in common, as do E1 and S_N1 reactions.

8.4 The E2 Mechanism

The most common mechanism for dehydrohalogenation is the E2 mechanism. For example, $(CH_3)_3CBr$ reacts with ^{-}OH to form $(CH_3)_2C{=}CH_2$ via an E2 mechanism.

$(CH_3)_3C{-}CH_2Br$ ... E2 reaction

8.4A Kinetics

An E2 reaction exhibits **second-order kinetics;** that is, the reaction is **bimolecular,** and both the alkyl halide and the base appear in the rate equation.

- rate = $k[(CH_3)_3CBr][^{-}OH]$

8.4B A One-Step Mechanism

The most straightforward explanation for the second-order kinetics is a **concerted reaction: all bonds are broken and formed in a single step,** as shown in Mechanism 8.1.

Mechanism 8.1 The E2 Mechanism

new π bond

- The base $^{-}$**OH removes a proton** from the β carbon, forming H_2O (a by-product).
- The electron pair in the β C–H bond forms the **new π bond.**
- The **leaving group Br^{-} comes off** with the electron pair in the C–Br bond.

Two bonds are broken (C–H and C–Br) and two bonds are formed (H–OH and the π bond) in the single step of an E2 mechanism, so the transition state contains **four partial bonds,** with the negative charge distributed over the base and the leaving group. **Entropy favors the products of an E2 reaction** because two molecules of starting material form three molecules of product.

Problem 8.8 Use curved arrows to show the movement of electrons in the following E2 mechanism. Draw the structure of the transition state.

There are close parallels between the E2 and S_N2 mechanisms in how the identity of the base, the leaving group, and the solvent affect the rate.

The Base

- **The base appears in the rate equation, so the rate of the E2 reaction *increases* as the strength of the base *increases*.**

E2 reactions are generally run with strong, negatively charged bases like $^{-}$OH and $^{-}$OR. Two strong, sterically hindered nitrogen bases, called **DBN** and **DBU,** are also sometimes used.

The IUPAC names for **DBN** and **DBU** are rarely used because the names are complex. **DBN** stands for 1,5-**d**iaza**b**icyclo[4.3.0]-**n**on-5-ene, and **DBU** stands for 1,8-**d**iaza**b**icyclo[5.4.0]-**u**ndec-7-ene.

DBN DBU

The Leaving Group

- **Because the bond to the leaving group is partially broken in the transition state, the *better* the leaving group the *faster* the E2 reaction.**

Increasing leaving group ability
Increasing rate of the E2 reaction

The Solvent

- **Polar aprotic solvents *increase* the rate of E2 reactions.**

Because **polar aprotic solvents** like $(CH_3)_2C{=}O$ do not solvate anions well, a negatively charged base is not "hidden" by strong interactions with the solvent (Section 7.14D), and the base is stronger. **A stronger base increases the reaction rate.**

8.4C The Identity of the Alkyl Halide

The S_N2 and E2 mechanisms differ in how the R group affects the reaction rate.

- **As the number of R groups on the carbon with the leaving group *increases,* the rate of the E2 reaction *increases.***

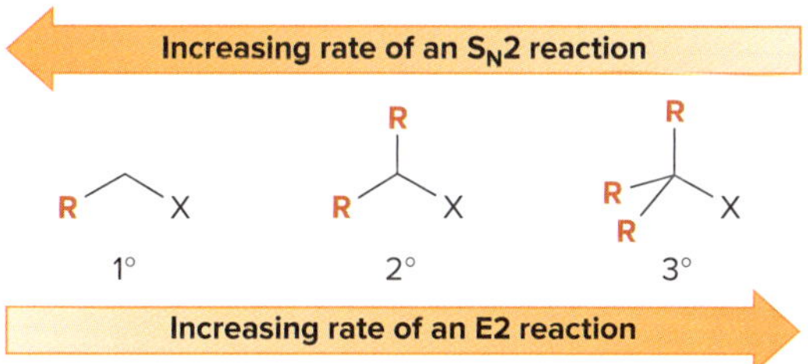

This trend is exactly *opposite* to the reactivity of alkyl halides in S_N2 reactions, where increasing alkyl substitution decreases the rate of reaction (Section 7.10D).

Why does increasing alkyl substitution increase the rate of an E2 reaction? In the transition state, the double bond is partially formed, so *increasing the stability* of the double bond with alkyl substituents *stabilizes* the transition state (i.e., it lowers E_a), which *increases* the rate of the reaction.

The double bond is partially formed.

- **Increasing the number of R groups on the carbon with the leaving group forms more highly substituted, *more stable* alkenes in E2 reactions.**

For example, the E2 reaction of a 1° alkyl halide (1-bromobutane) forms a monosubstituted alkene, whereas the E2 reaction of a 3° alkyl halide (2-bromo-2-methylpropane) forms a disubstituted alkene. The disubstituted alkene is more stable, so the **3° alkyl halide reacts faster than the 1° alkyl halide.**

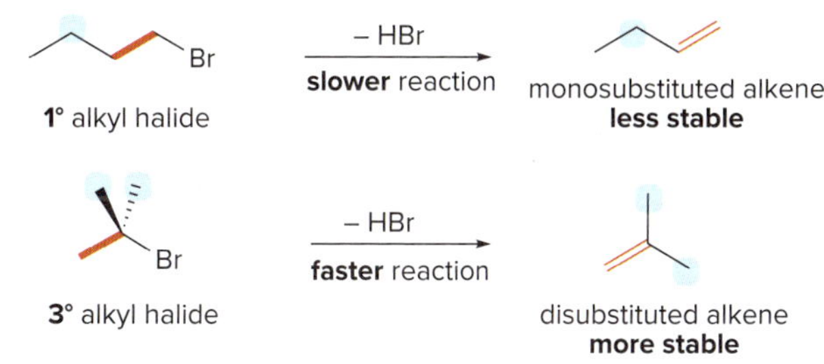

Elimination reactions are often steps in the synthesis of complex natural products. For example, elimination of HCl from compound **A** forms alkene **B,** which was converted to the antimalarial drug quinine, the chapter-opening molecule.

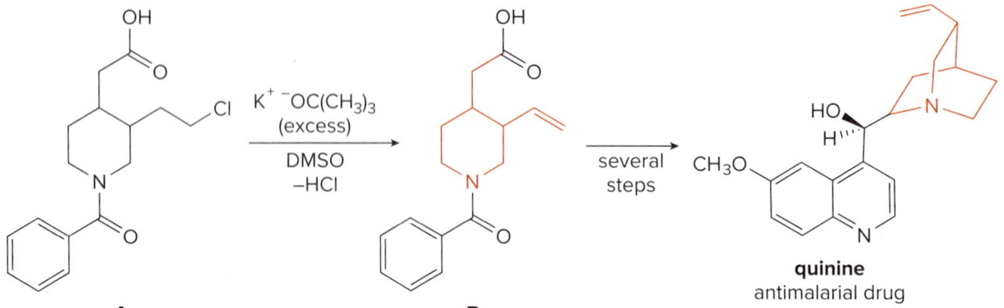

$K^+ \ ^-OC(CH_3)_3$ (excess)

DMSO

$-HCl$

several steps

A

B

quinine
antimalarial drug

Table 8.2 summarizes the characteristics of the E2 mechanism.

Table 8.2 Characteristics of the E2 Mechanism

Characteristic	Result
Kinetics	• **Second order**
Mechanism	• **One step**
Identity of R	• **More substituted halides react faster.** • Rate: $R_3CX > R_2CHX > RCH_2X$
Base	• Favored by **strong bases**
Leaving group	• **Better leaving group** ⇢ faster reaction
Solvent	• Favored by **polar aprotic solvents**

Problem 8.9 Rank the alkyl halides in each group in order of increasing reactivity in an E2 reaction.

a.

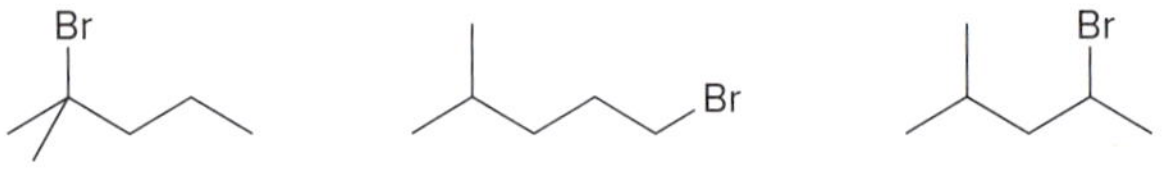

b.

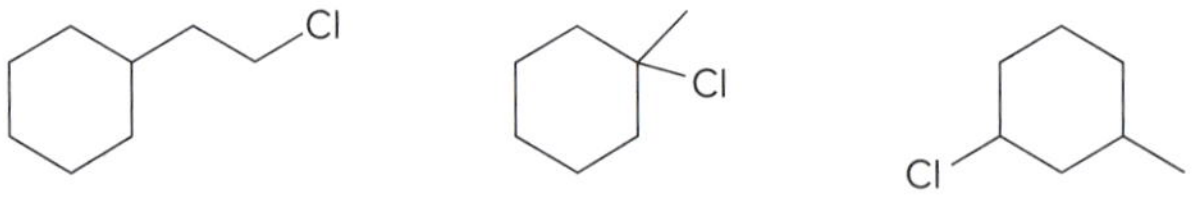

Problem 8.10 How does each of the following changes affect the rate of an E2 reaction?

a. tripling [RX]

b. halving [B:]

c. changing the solvent from CH_3OH to DMSO

d. changing the leaving group from I^- to Br^-

e. changing the base from ^-OH to H_2O

f. changing the alkyl halide from CH_3CH_2Br to $(CH_3)_2CHBr$

8.5 The Zaitsev Rule

Recall from Section 8.1 that a mixture of alkenes can form from the dehydrohalogenation of alkyl halides having two or more different β carbon atoms. When this occurs, one of the products usually predominates. The **major product is the more stable product—the one with the more substituted double bond.** For example, elimination of the elements of H and I from

1-iodo-1-methylcyclohexane yields two constitutional isomers: the trisubstituted alkene **A** (the major product) and the disubstituted alkene **B** (the minor product).

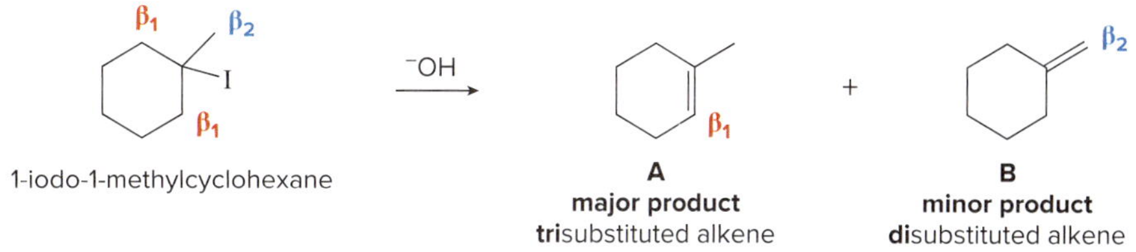

This phenomenon is called the **Zaitsev rule** (also called the **Saytzeff rule,** depending on the translation) for the Russian chemist who first noted this trend.

- **The Zaitsev rule: The major product in β elimination has the more substituted double bond.**

A reaction is *regioselective* when it yields predominantly or exclusively one constitutional isomer when more than one is possible. The E2 reaction is **regioselective** because the more substituted alkene predominates.

The Zaitsev rule results because the double bond is partially formed in the transition state for the E2 reaction. Thus, increasing the stability of the double bond by adding R groups lowers the energy of the transition state, which increases the reaction rate. E2 elimination of HBr from 1-bromo-1-methylcyclopentane yields alkenes **C** and **D. D, having the more substituted double bond, is the major product,** because the transition state leading to its formation is lower in energy.

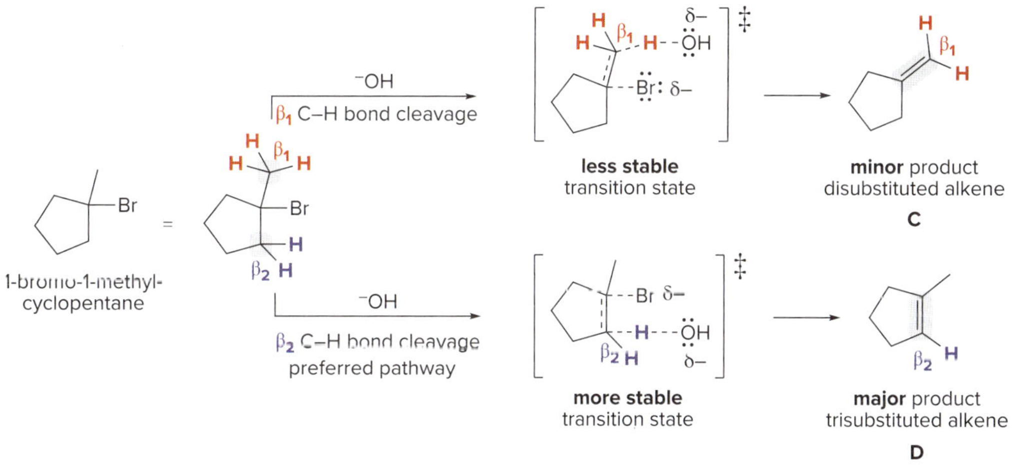

When a mixture of stereoisomers is possible from dehydrohalogenation, the **major product is the more stable stereoisomer.** Dehydrohalogenation of alkyl halide **X** forms a mixture of trans and cis alkenes, **Y** and **Z.** The trans alkene **Y** is the major product because it is more stable.

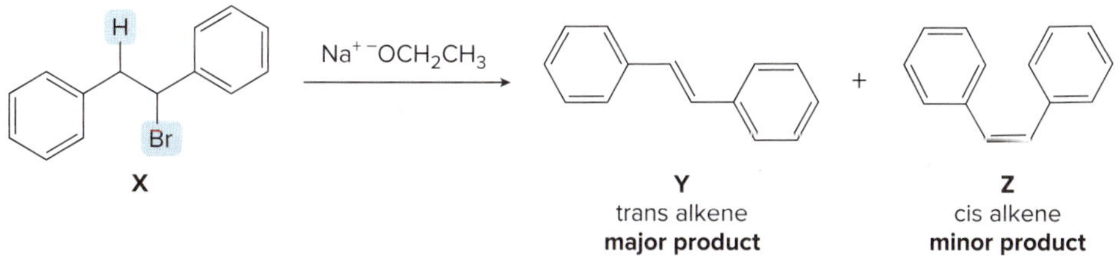

A reaction is *stereoselective* when it forms predominantly or exclusively one stereoisomer when two or more are possible. The **E2 reaction is stereoselective** because one stereoisomer is formed preferentially.

Sample Problem 8.1 Determining the Major Product of an E2 Reaction

Predict the major product in the following E2 reaction.

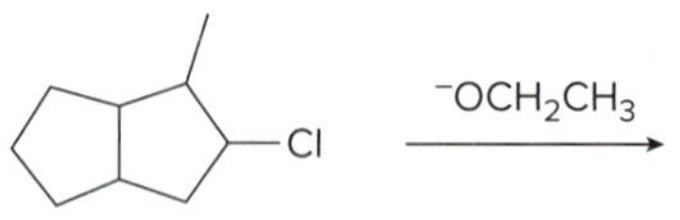

$^{-}OCH_2CH_3$

Solution

The alkyl halide has two different β C atoms (labeled β_1 and β_2), so two different alkenes are possible: one formed by removal of HCl across the α and β_1 carbons, and one formed by removal of HCl across the α and β_2 carbons. Using the Zaitsev rule, the major product should be **A,** because it has the **more substituted double bond.**

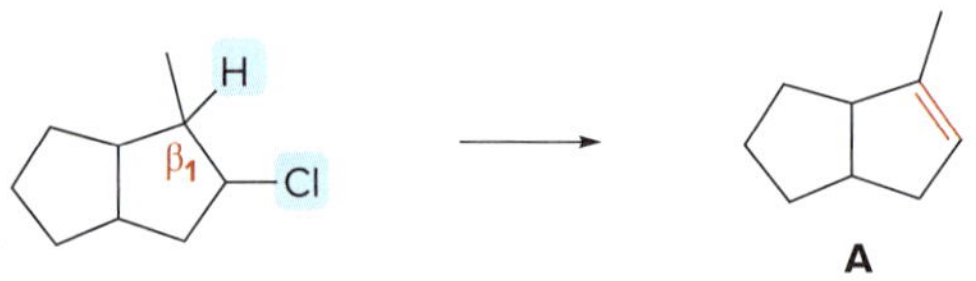

A
trisubstituted alkene
major product

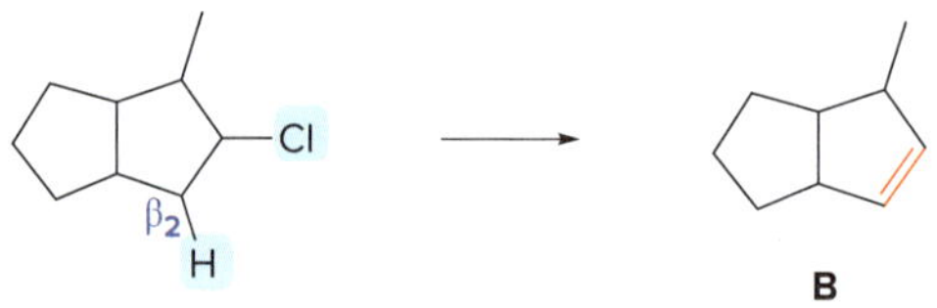

B
disubstituted alkene
minor product

Problem 8.11 What alkenes are formed from each alkyl halide by an E2 reaction? Use the Zaitsev rule to predict the major product.

a. b. c. d. e. f.

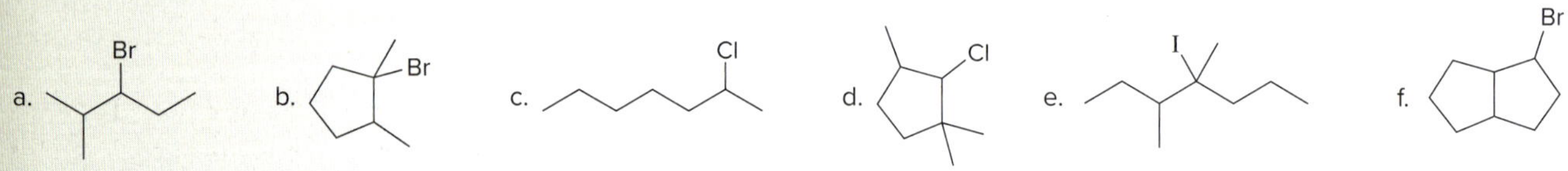

More Practice: Try Problems 8.29, 8.30, 8.32.

8.6 Stereochemistry of the E2 Reaction

The transition state of the E2 reaction consists of four atoms that react at the same time, and they react only if they possess a particular stereochemical arrangement.

8.6A General Stereochemical Features

The transition state of an E2 reaction consists of **four atoms** from the alkyl halide—one hydrogen atom, two carbon atoms, and the leaving group (X)—**all aligned in a plane.** There are two ways for the C–H and C–X bonds to be coplanar:

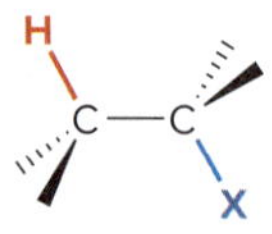

The dihedral angle for the C–H and C–X bonds equals **0°** for the syn periplanar arrangement and **180°** for the anti periplanar arrangement.

- **The H and X atoms can be oriented on the same side of the molecule. This geometry is called *syn periplanar.***
- **The H and X atoms can be oriented on opposite sides of the molecule. This geometry is called *anti periplanar.***

All evidence suggests that **E2 elimination occurs most often in the anti periplanar geometry.** This arrangement allows the molecule to react in the lower-energy *staggered* conformation. It also allows two electron-rich species, the incoming base and the departing leaving group, to be farther away from each other, as illustrated in Figure 8.3.

Anti periplanar geometry is the preferred arrangement for any alkyl halide undergoing E2 elimination, regardless of whether it is cyclic or acyclic. This stereochemical requirement has important consequences for compounds containing six-membered rings.

Figure 8.3 Two possible geometries for the E2 reaction

- An **anti periplanar** arrangement has a **staggered** conformation.
- The two electron-rich groups are far apart.

- A **syn periplanar** arrangement has an **eclipsed** conformation.
- The two electron-rich groups are close.

Problem 8.12 Given that an E2 reaction proceeds with anti periplanar stereochemistry, draw the products of each elimination. The alkyl halides in (a) and (b) are diastereomers of each other. How are the products of these two reactions related? Recall from Section 3.2A that $C_6H_5–$ is a phenyl group, a benzene ring bonded to another group.

a. H, CH_3, C_6H_5 on C—C bearing C_6H_5, H, Br $\xrightarrow{^-OCH_2CH_3}$

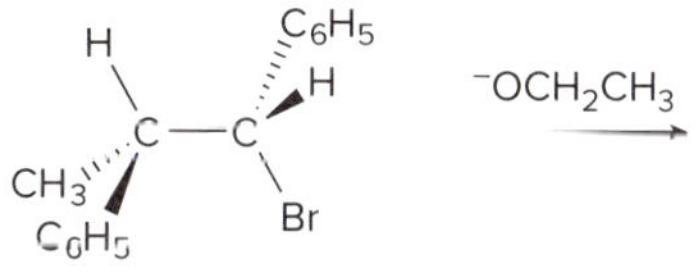

b. H, C_6H_5, CH_3 on C—C bearing C_6H_5, H, Br $\xrightarrow{^-OCH_2CH_3}$

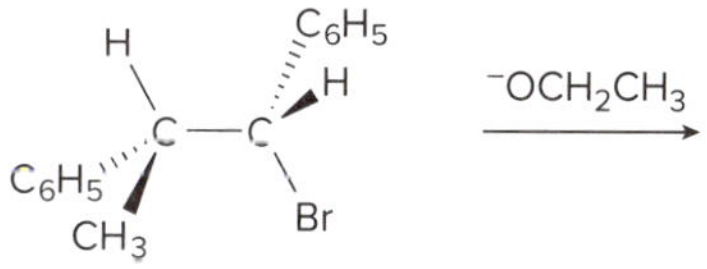

8.6B Anti Periplanar Geometry and Halocyclohexanes

Recall from Section 4.12 that cyclohexane exists as two chair conformations that rapidly interconvert, and that substituted cyclohexanes are more stable with substituents in the roomier **equatorial position.** Chlorocyclohexane exists as two chair conformations, but **X** is preferred because the Cl group is equatorial.

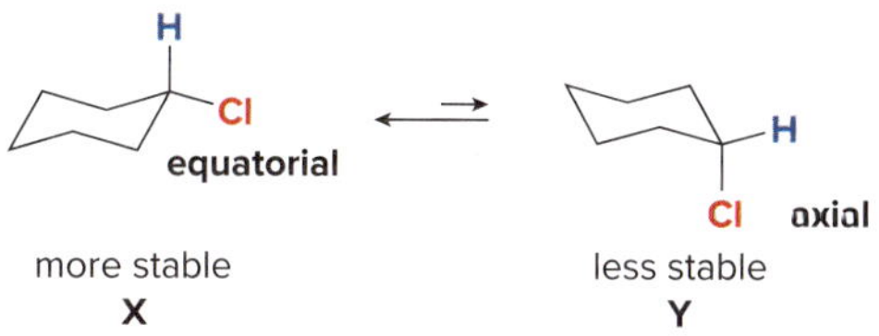

H, Cl equatorial ⇌ H, Cl axial

more stable **X** — less stable **Y**

chlorocyclohexane

For E2 elimination, **the C–Cl bond must be anti periplanar to a C–H bond on a β carbon,** and this occurs only when the H and Cl atoms are both in the **axial** position. This requirement for **trans diaxial geometry** means that E2 elimination must occur from the less stable conformation **Y,** as shown in Figure 8.4.

Figure 8.4

The trans diaxial geometry for the E2 elimination in chlorocyclohexane

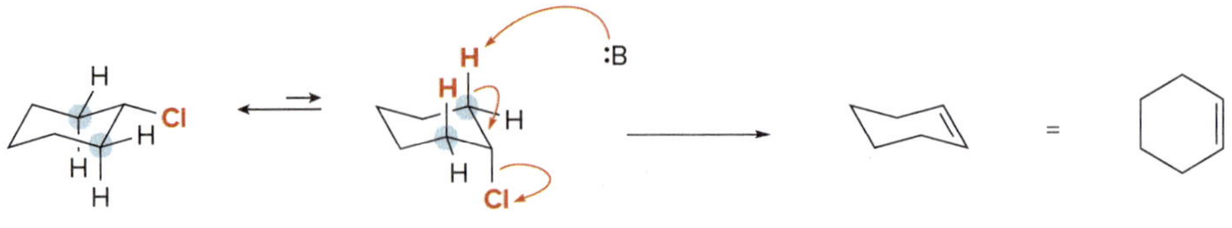

X This conformation does *not* react.

Y This conformation reacts.

- In conformation **X** (**equatorial** Cl group), a β C–H bond and a C–Cl bond are *never* anti periplanar; therefore, **no E2** elimination can occur. β Carbons are highlighted in blue.
- In conformation **Y** (**axial** Cl group), two β C–H bonds and the C–Cl bond are **trans diaxial;** therefore, **E2 elimination occurs**. Axial H's on β carbons that can react are shown in red.

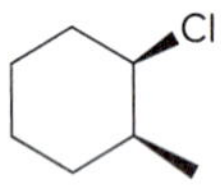

cis-1-chloro-2-methylcyclohexane

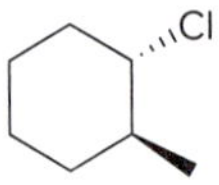

trans-1-chloro-2-methylcyclohexane

Sometimes this rigid stereochemical requirement affects the regioselectivity of the E2 reaction of substituted cyclohexanes. Dehydrohalogenation of *cis*- and *trans*-1-chloro-2-methylcyclohexane via an E2 mechanism illustrates this phenomenon.

The **cis isomer** exists as two conformations (**A** and **B**), each of which has one group axial and one group equatorial. E2 reaction must occur from conformation **B,** which contains an **axial** Cl atom.

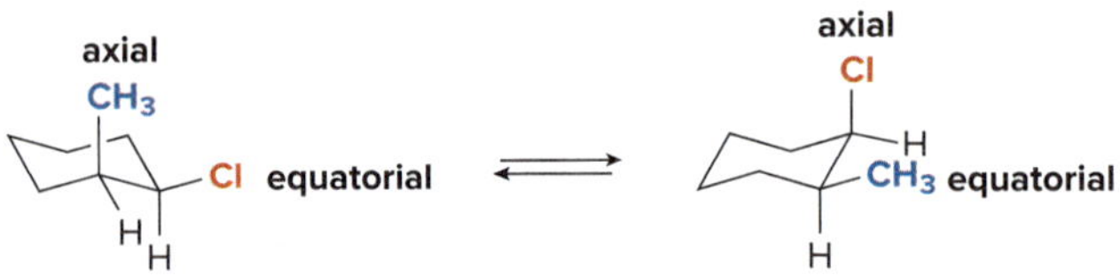

cis isomer **A**

cis isomer **B**

This conformation reacts.

Because conformation **B** has two different axial β H atoms, labeled H_a and H_b, E2 reaction occurs in two different directions to afford two alkenes. **The major product contains the more stable trisubstituted double bond, as predicted by the Zaitsev rule.**

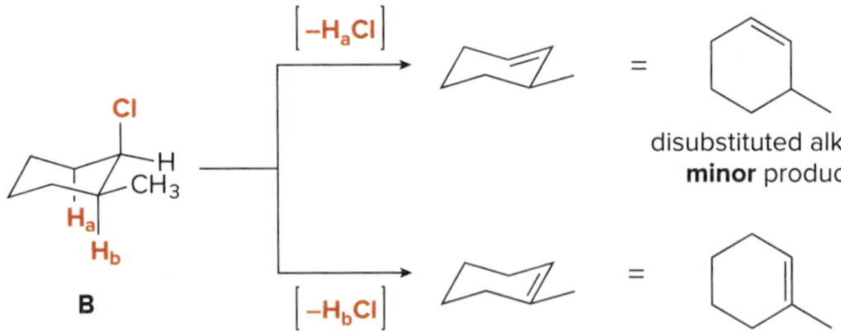

disubstituted alkene
minor product

trisubstituted alkene
major product

The **trans isomer** exists as two conformations, **C,** having two equatorial substituents, and **D,** having two axial substituents. E2 reaction must occur from conformation **D,** which contains an **axial** Cl atom.

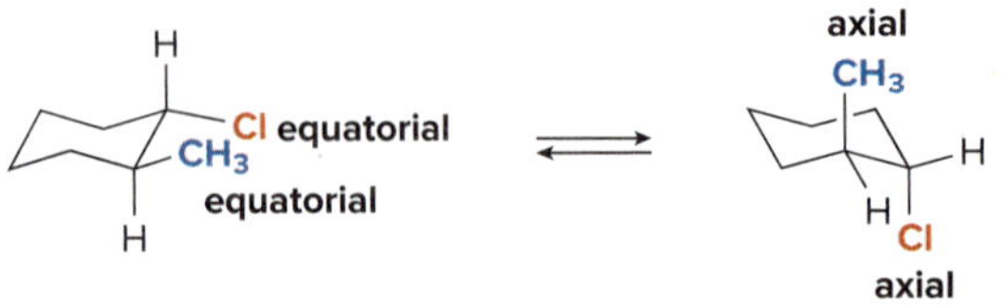

trans isomer **C**

trans isomer **D**

This conformation reacts.

Because conformation **D** has **only one axial β H,** E2 reaction occurs in only *one* direction to afford a **single product,** having the disubstituted double bond. This is *not* predicted by the Zaitsev rule. **E2 reaction requires H and Cl to be trans and diaxial,** and with the trans isomer, this is possible only when the less stable alkene is formed as product.

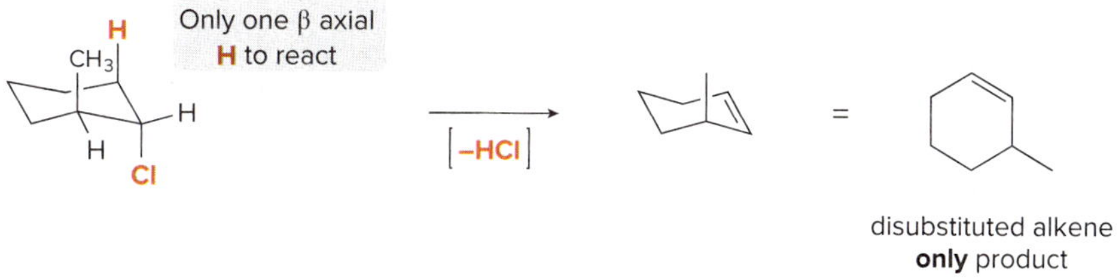

Only one β axial H to react

[–HCl]

disubstituted alkene **only** product

- **With substituted cyclohexanes, E2 elimination must occur with a trans diaxial arrangement of H and X, and as a result of this requirement, the more substituted alkene is not necessarily the major product.**

Sample Problem 8.2 Drawing an E2 Product from a Halocyclohexane

Draw the major E2 elimination product formed from the following alkyl halide.

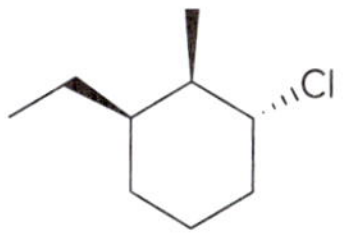

Solution

To draw the elimination products, locate the β carbons and **look for H atoms that are trans to the leaving group.** The given alkyl chloride has two different β carbons, labeled β_1 and β_2. **Elimination can occur only when the leaving group (Cl) and a H atom on the β carbon are *trans*.**

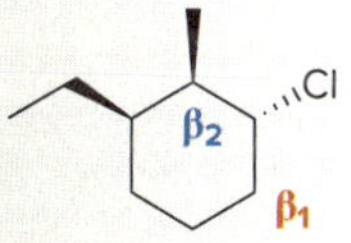

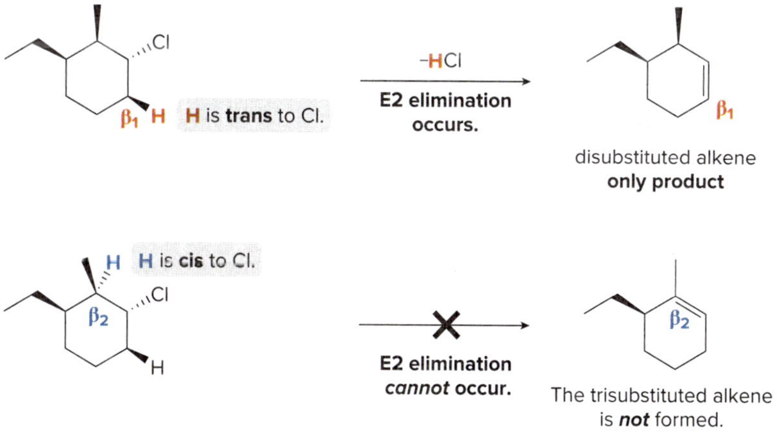

β_1 H **H is trans to Cl.**

–HCl

E2 elimination occurs.

disubstituted alkene **only product**

H **H is cis to Cl.**

E2 elimination *cannot* occur.

The trisubstituted alkene is ***not*** formed.

The β_1 C has a H atom ***trans*** to Cl, so E2 elimination occurs to form a disubstituted alkene. Because there is no trans H on the β_2 C, E2 elimination ***cannot*** occur in this direction, and the more stable trisubstituted alkene is *not* formed. Although this result is not predicted by the Zaitsev rule, it is consistent with the requirement that the **H and X atoms in an E2 elimination must be located trans to each other.**

Problem 8.13 Draw the major E2 elimination product from each of the following alkyl halides.

a. $\xrightarrow{^-OH}$ b. $\xrightarrow{^-OH}$ c. $\xrightarrow{^-OH}$

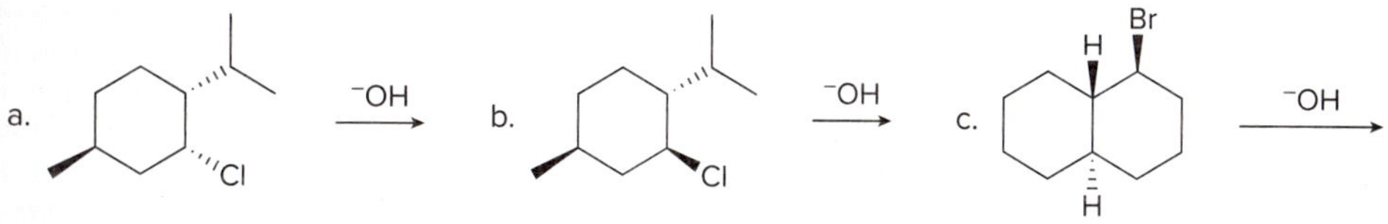

More Practice: Try Problems 8.33, 8.34, 8.36, 8.37, 8.57.

Problem 8.14 Explain why *cis*-1-chloro-2-methylcyclohexane undergoes E2 elimination much faster than its trans isomer.

8.7 The E1 Mechanism

The dehydrohalogenation of $(CH_3)_3CI$ with H_2O to form $(CH_3)_2C{=}CH_2$ can be used to illustrate the second general mechanism of elimination, the **E1 mechanism.**

$(CH_3)_3CI + H_2\ddot{O}: \xrightarrow{\text{E1 reaction}} (CH_3)_2C{=}CH_2 + H{-}\overset{+}{O}H_2 + :\ddot{I}:^-$

8.7A Kinetics

An E1 reaction exhibits **first-order kinetics.**

- rate = $k[(CH_3)_3CI]$

Like the S_N1 mechanism, the kinetics suggest that the reaction mechanism has more than one step, and that the slow step is **unimolecular,** involving *only* the alkyl halide.

8.7B A Two-Step Mechanism

The most straightforward explanation for the observed first-order kinetics is a **two-step reaction: the bond to the leaving group breaks first *before* the π bond is formed,** as shown in Mechanism 8.2.

Mechanism 8.2 The E1 Mechanism

[1] slow: $(CH_3)_3C{-}I \rightarrow$ carbocation + $:\ddot{I}:^-$

[2]: carbocation + $H_2\ddot{O}: \rightarrow (CH_3)_2C{=}CH_2 + H_3\overset{+}{O}:$

1. Heterolysis of the C−I bond forms a **carbocation** in the rate-determining step.
2. A base (either H_2O or I^-) removes a proton from a carbon adjacent to the carbocation, and the electron pair in the C−H bond forms the π bond.

The E1 and E2 mechanisms both involve the same number of bonds broken and formed. **The only difference is the timing.**

- **In an E1 reaction, the leaving group comes off *before* the β proton is removed, and the reaction occurs in *two* steps.**
- **In an E2 reaction, the leaving group comes off *as* the β proton is removed, and the reaction occurs in *one* step.**

An energy diagram for the reaction of $(CH_3)_3CI + H_2O$ is shown in Figure 8.5. Each step has its own energy barrier, with a transition state at each energy maximum. Because its transition state is higher in energy, **Step [1] is rate-determining.** $\Delta H°$ for Step [1] is positive because only bond breaking occurs, whereas $\Delta H°$ of Step [2] is negative because two bonds are formed and only one is broken.

Figure 8.5
Energy diagram for an E1 reaction:
$(CH_3)_3CI + H_2O \rightarrow (CH_3)_2C{=}CH_2 + H_3O^+ + I^-$

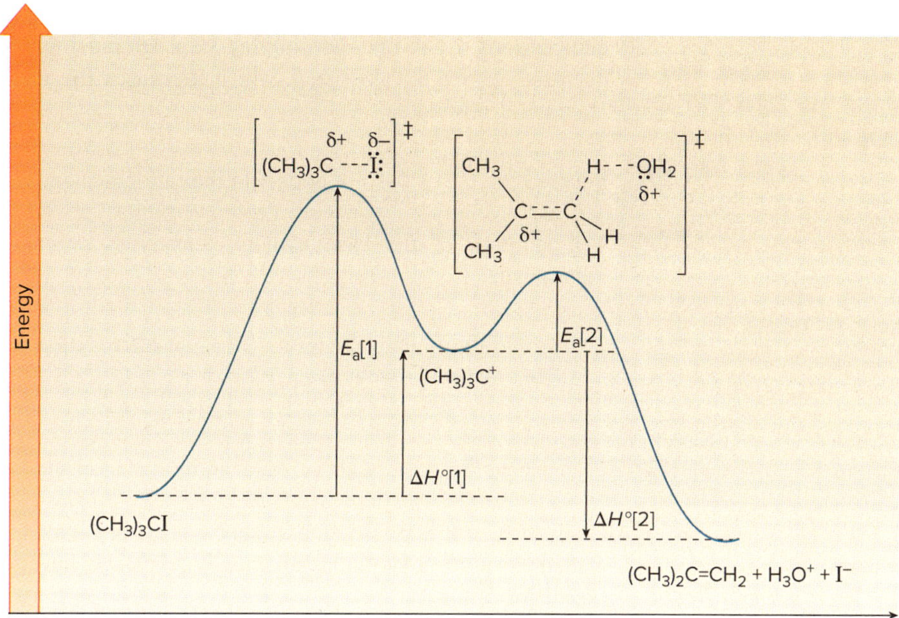

- The E1 mechanism has **two steps,** so there are two energy barriers.
- **Step [1] is rate-determining**.

Problem 8.15 Draw an E1 mechanism for the following reaction. Draw the structure of the transition state for each step.

(alkyl chloride, Cl) + CH_3OH ⟶ (alkene) + $CH_3\overset{+}{O}H_2$ + Cl^-

---

8.7C Other Characteristics of E1 Reactions

Three other features of E1 reactions are worthy of note.

[1] **The rate of an E1 reaction *increases* as the number of R groups on the carbon with the leaving group *increases*.**

Increasing alkyl substitution has the same effect on the rate of *both* an E1 and E2 reaction; increasing rate of the E1 and E2 reactions: RCH_2X (1°) < R_2CHX (2°) < R_3CX (3°).

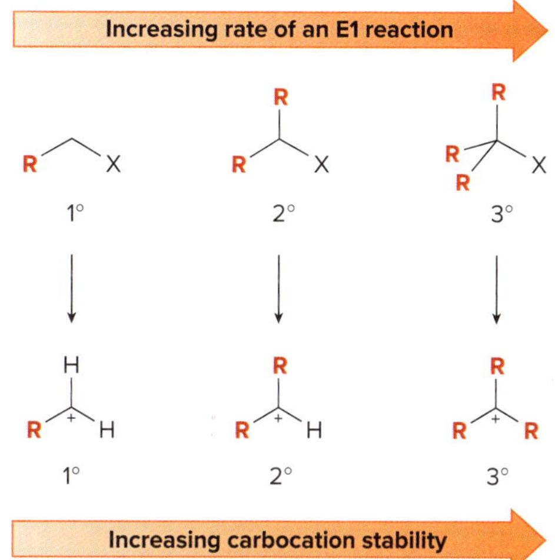

Like an S_N1 reaction, more substituted alkyl halides yield more substituted (and more stable) carbocations in the rate-determining step. **Increasing the stability of a carbocation,** in turn, decreases E_a for the slow step, which **increases the rate of the E1 reaction** according to the Hammond postulate.

[2] Because the base does not appear in the rate equation, *weak bases* favor E1 reactions.

The strength of the base usually determines whether a reaction follows the E1 or E2 mechanism.

- ***Strong* bases like ^{-}OH and ^{-}OR favor E2 reactions, whereas *weaker* bases like H_2O and ROH favor E1 reactions.**

[3] E1 reactions are regioselective, favoring formation of the more substituted, more stable alkene.

The Zaitsev rule applies to E1 reactions, too. For example, E1 elimination of HBr from 1-bromo-1-methylcyclopentane yields alkenes **A** and **B. A,** having the more substituted double bond, is the major product.

β_1 β_2 Br α β_1 — H_2O → A (β_1) + B (β_2)

1-bromo-1-methyl-cyclopentane

A trisubstituted alkene **major product**

B disubstituted alkene **minor product**

Table 8.3 summarizes the characteristics of E1 reactions.

Table 8.3 Characteristics of the E1 Mechanism

Characteristic	Result
Kinetics	• **First order**
Mechanism	• **Two steps**
Identity of R	• **More substituted halides react faster.** • Rate: $R_3CX > R_2CHX > RCH_2X$
Base	• Favored by **weaker bases** such as H_2O and ROH
Leaving group	• A **better leaving group** makes the reaction faster because the bond to the leaving group is partially broken in the rate-determining step.
Solvent	• **Polar protic solvents** that solvate the ionic intermediates are needed.

Problem 8.16 What alkenes are formed from each alkyl halide by an E1 reaction? Use the Zaitsev rule to predict the major product.

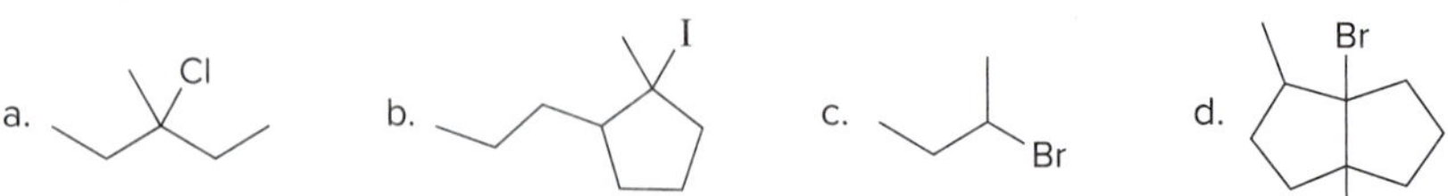

Problem 8.17 How does each of the following changes affect the rate of an E1 reaction?

a. doubling [RX]
b. doubling [B:]
c. changing the halide from $(CH_3)_3CBr$ to $CH_3CH_2CH_2Br$
d. changing the leaving group from Cl^- to Br^-
e. changing the solvent from DMSO to CH_3OH

8.8 S_N1 and E1 Reactions

S_N1 and E1 reactions have exactly the same first step—formation of a carbocation. They differ in what happens to the carbocation.

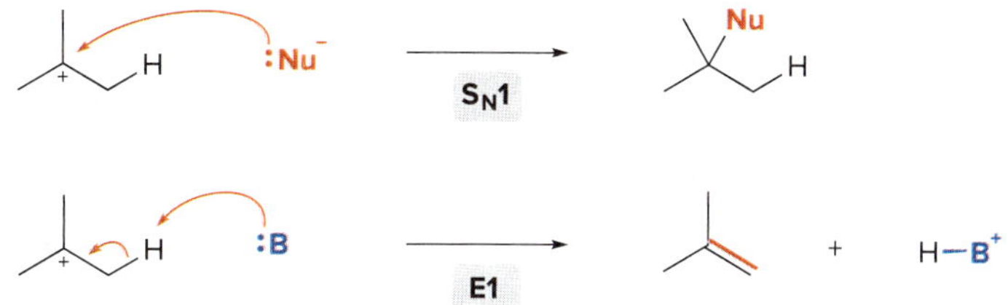

- **In an S_N1 reaction, a nucleophile attacks the carbocation, forming a substitution product.**
- **In an E1 reaction, a base removes a proton, forming a new π bond.**

The same conditions that favor substitution by an S_N1 mechanism also favor elimination by an E1 mechanism: **a 3° alkyl halide as substrate, a weak nucleophile or base as reagent, and a polar protic solvent.** As a result, both reactions usually occur in the same reaction mixture to afford a mixture of products, as illustrated in Sample Problem 8.3.

Sample Problem 8.3 Drawing the S_N1 and E1 Products in a Reaction

Draw the S_N1 and E1 products formed in the reaction of $(CH_3)_3CBr$ with H_2O.

Solution

The first step in both reactions is heterolysis of the C–Br bond to form a **carbocation.**

Br: → (slow) carbocation + :Br:⁻

Reaction of the carbocation with H_2O as a nucleophile affords the substitution product (Reaction [1]). Alternatively, H_2O acts as a base to remove a proton, affording the elimination product (Reaction [2]). **Two products are formed.**

[1] + $H_2\ddot{O}$: (**nucleophile**) → $H_2\ddot{O}$: → (proton transfer) **S_N1 product** + $H_3\ddot{O}^+$:

[2] $H_2\ddot{O}$: → **E1 product** + $H_3\ddot{O}^+$:

Problem 8.18 Draw both the S_N1 and E1 products of each reaction.

a. (cyclohexyl)–Br + H_2O →

b. (alkyl)–Cl + ⁄⁀OH →

c. (dicyclopentyl)–Br + H_2O

d. (alkyl)–Br + CH_3OH

More Practice: Try Problems 8.46d, f; 8.48b, c, f; 8.49a; 8.51; 8.52; 8.54.

Because E1 reactions often occur with a competing S_N1 reaction, **E1 reactions of alkyl halides are *much less useful* than E2 reactions.**

8.9 When Is the Mechanism E1 or E2?

Given a particular starting material and base, how do we know whether a reaction occurs by the E1 or E2 mechanism?

Because the rate of *both* the E1 and E2 reactions increases as the number of R groups on the carbon with the leaving group increases, **you cannot use the identity of the alkyl halide to decide which elimination mechanism occurs.**

- **The strength of the base is the most important factor in determining the mechanism for elimination. Strong bases favor the E2 mechanism. Weak bases favor the E1 mechanism.**

Table 8.4 compares the E1 and E2 mechanisms.

Table 8.4 A Comparison of the E1 and E2 Mechanisms

Mechanism	Comment
E2 mechanism	• Much more common and useful • Favored by **strong, negatively charged bases,** especially ^-OH and ^-OR • The reaction occurs with 1°, 2°, and 3° alkyl halides. Order of reactivity: $\mathbf{R_3CX > R_2CHX > RCH_2X}$.
E1 mechanism	• Much less useful because a mixture of S_N1 and E1 products usually results • Favored by **weaker, neutral bases,** such as H_2O and ROH • This mechanism does not occur with 1° RX because they form highly unstable 1° carbocations.

Problem 8.19 Which mechanism, E1 or E2, will occur in each reaction?

a. 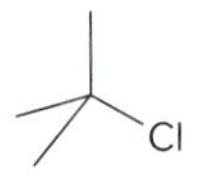 (CH3)3CCl + $^-OCH_3$ ⟶

b. 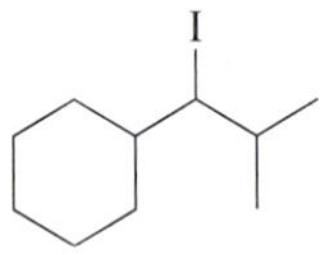 [cyclohexyl–CH(I)–CH(CH3)2] + H_2O ⟶

c. [1-chloro-1-tert-butylcyclohexane] + CH_3OH ⟶

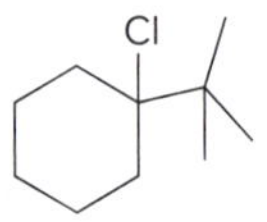

d. 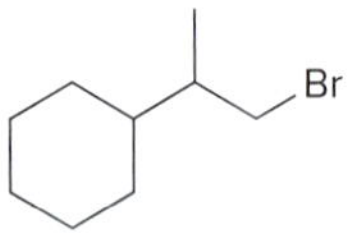 [cyclohexyl–CH2–CH(CH3)–CH2Br] + $^-OC(CH_3)_3$ ⟶

8.10 E2 Reactions and Alkyne Synthesis

Recall from Section 1.10C that the carbon–carbon triple bond of alkynes consists of one σ and two π bonds.

A single elimination reaction produces the π bond of an alkene. **Two consecutive elimination reactions produce the two π bonds of an alkyne.**

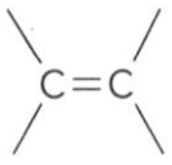

C=C	—C≡C—
alkene	alkyne
one π bond	**two** π bonds
One elimination reaction is needed.	**Two** elimination reactions are needed.

- **Alkynes are prepared by two successive dehydrohalogenation reactions.**

Two elimination reactions are needed to remove two moles of HX from a **dihalide** as substrate. Two different starting materials can be used.

vicinal dihalide

geminal dihalide

- **A vicinal dihalide** has two X atoms on *adjacent* carbon atoms.
- **A geminal dihalide** has two X atoms on the *same* carbon atom.

The word *geminal* comes from the Latin *geminus,* meaning "twin."

Equations [1] and [2] illustrate how two moles of HX can be removed from these dihalides with base. Two equivalents of strong base are used and each step follows an **E2 mechanism.**

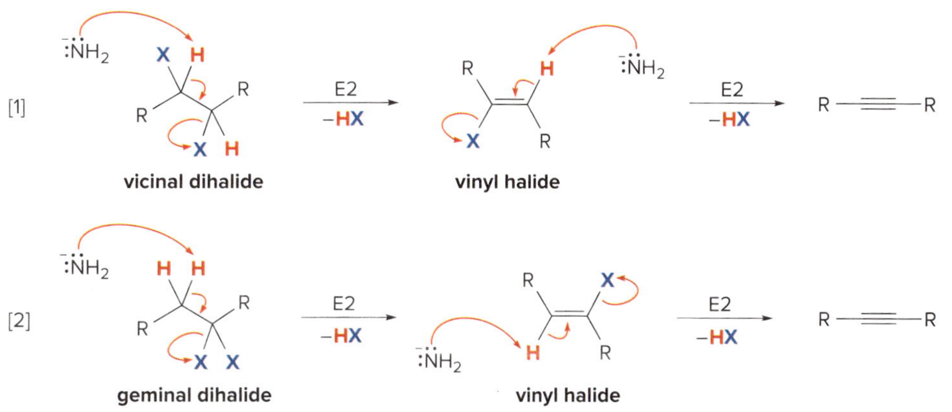

[1] vicinal dihalide —E2, −HX→ vinyl halide —E2, −HX→ R—≡—R

[2] geminal dihalide —E2, −HX→ vinyl halide —E2, −HX→ R—≡—R

The relative strength of C–H bonds depends on the hybridization of the carbon atom: $sp > sp^2 > sp^3$. For more information, review Section 1.11B.

Stronger bases are needed to synthesize alkynes by dehydrohalogenation than are needed to synthesize alkenes. The typical base is **amide ($^-NH_2$),** used as the sodium salt **$NaNH_2$** (sodium amide). $KOC(CH_3)_3$ can also be used with DMSO as solvent. Because DMSO is a polar aprotic solvent, the anionic base is not well solvated, thus **increasing its basicity** and making it strong enough to remove two equivalents of HX. Examples are given in Figure 8.6.

The strongly basic conditions needed for alkyne synthesis result from the difficulty of removing the second equivalent of HX from the intermediate vinyl halide, RCH=C(R)X. Because H and X are both bonded to sp^2 hybridized carbons, these bonds are shorter and stronger than the sp^3 hybridized C–H and C–X bonds of an alkyl halide, necessitating the use of a stronger base.

Figure 8.6 Examples of dehydrohalogenation of dihalides to afford alkynes

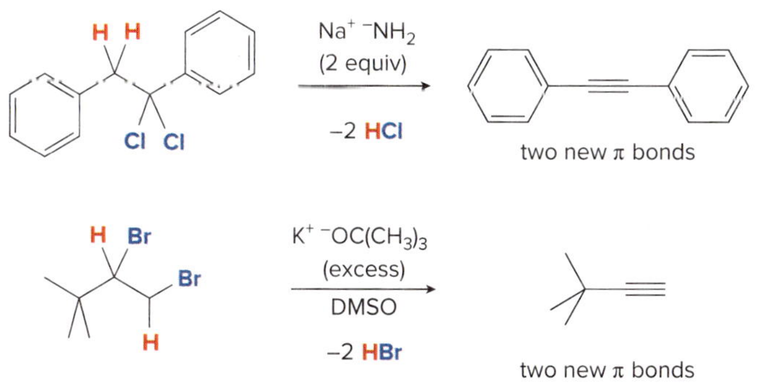

$Na^+ \ ^-NH_2$ (2 equiv), −2 HCl → two new π bonds

$K^+ \ ^-OC(CH_3)_3$ (excess), DMSO, −2 HBr → two new π bonds

Problem 8.20 Draw the alkynes formed when each dihalide is treated with excess base.

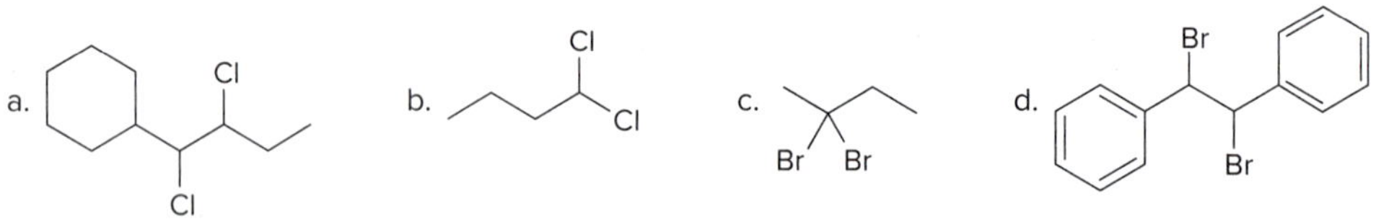

a. b. c. d.

8.11 When Is the Reaction S_N1, S_N2, E1, or E2?

We have now considered two different kinds of reactions (substitution and elimination) and four different mechanisms (S_N1, S_N2, E1, and E2) that begin with one class of compounds (alkyl halides). How do we know if a given alkyl halide will undergo substitution or elimination with a given base or nucleophile, and by what mechanism?

Unfortunately, there is no easy answer, and often mixtures of products result. Two generalizations help to determine whether substitution or elimination occurs.

[1] *Good* nucleophiles that are *weak* bases favor *substitution* over elimination.

Certain anions generally give products of substitution because they are good nucleophiles but weak bases. These include **I^-, Br^-, HS^-, ^-CN,** and **$CH_3CO_2^-$.**

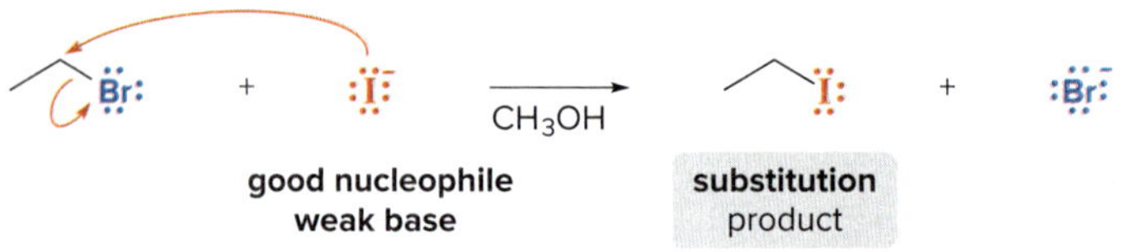

CH_3OH

good nucleophile
weak base

substitution product

[2] *Bulky,* nonnucleophilic bases favor *elimination* over substitution.

$KOC(CH_3)_3$, DBU, and **DBN** are too sterically hindered to attack a tetravalent carbon, but are able to remove a small proton, favoring elimination over substitution.

$K^+ \ ^-:\ddot{O}C(CH_3)_3$ + $H\ddot{O}C(CH_3)_3$ + KBr

strong,
nonnucleophilic base

elimination product

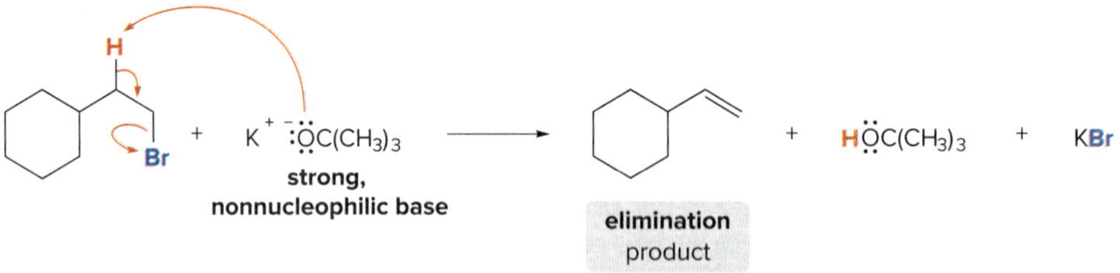

Most often, however, we will have to rely on other criteria to predict the outcome of these reactions. To determine the product of a reaction with an alkyl halide:

[1] Classify the alkyl halide as 1°, 2°, or 3°.

[2] Classify the base or nucleophile as strong, weak, or bulky.

Predicting the substitution and elimination products of a reaction can then be organized by the type of alkyl halide, as summarized in Table 8.5. The explanation that follows the table is organized with 2° alkyl halides last, because their reactions can follow any of the four mechanisms and product mixtures often result.

Table 8.5 Summary of Alkyl Halides and S_N1, S_N2, E1, and E2 Mechanisms

Alkyl halide type	Reaction with		Mechanism
1° RCH_2X	• Strong nucleophile	→	S_N2
	• Strong bulky base	→	E2
2° R_2CHX	• Strong base and nucleophile	→	S_N2 and E2
	• Strong bulky base	→	E2
	• Weak base and nucleophile	→	S_N1 and E1
3° R_3CX	• Weak base and nucleophile	→	S_N1 and E1
	• Strong base	→	E2

8.11A Tertiary Alkyl Halides

Tertiary alkyl halides react by all mechanisms *except* S_N2.

- **With strong bases, elimination occurs by an E2 mechanism.**

Br (3°) + $:\ddot{O}H^-$ (strong base) → E2 product

A strong base or nucleophile favors an S_N2 or E2 mechanism, but 3° halides are too sterically hindered to undergo an S_N2 reaction, so only E2 elimination occurs.

- **With weak nucleophiles or bases, a mixture of S_N1 and E1 products results.**

Br (3°) + $H_2\ddot{O}:$ (weak nucleophile and base) → OH (S_N1 product) + E1 product

A weak base or nucleophile favors S_N1 and E1 mechanisms and both occur.

8.11B Primary Alkyl Halides

Primary alkyl halides react by S_N2 and E2 mechanisms.

- **With strong nucleophiles, substitution occurs by an S_N2 mechanism.**

Br (1°) + $:\ddot{O}H^-$ (strong nucleophile) → OH (S_N2 product)

A strong base or nucleophile favors S_N2 or E2, but 1° halides are the *least* reactive halide type in elimination, so only S_N2 reaction occurs.

- **With strong, bulky bases, elimination occurs by an E2 mechanism.**

Br (1°) + $K^+ \ ^-OC(CH_3)_3$ (strong, bulky base) → E2 product

A strong, bulky base cannot act as a nucleophile, so elimination occurs and the mechanism is E2.

8.11C Secondary Alkyl Halides

Secondary alkyl halides react by *all* mechanisms.

- **With strong bases and nucleophiles, a mixture of S_N2 and E2 products results.**

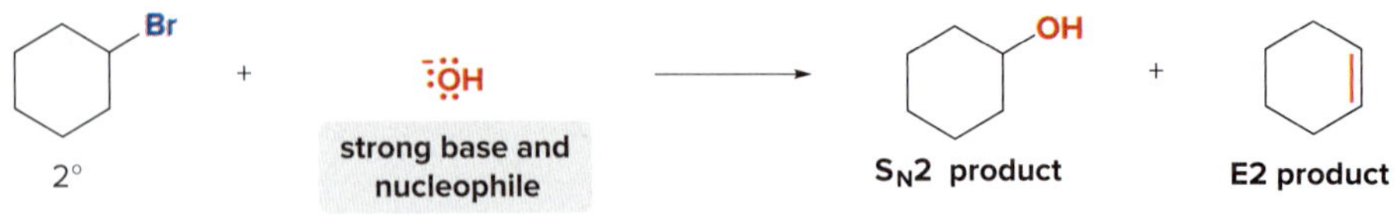

Br, 2° + :ÖH (strong base and nucleophile) → OH (S_N2 product) + E2 product

A strong base that is also a strong nucleophile gives a mixture of S_N2 and E2 products.

- **With strong, bulky bases, elimination occurs by an E2 mechanism.**

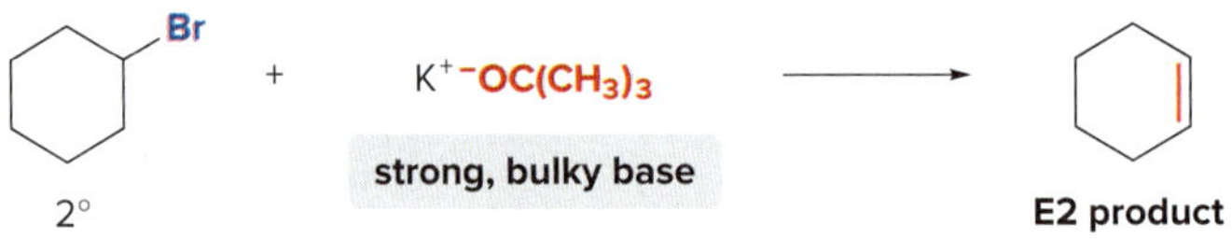

Br, 2° + $K^{+}\,^{-}OC(CH_3)_3$ (strong, bulky base) → E2 product

A strong, bulky base cannot act as a nucleophile, so elimination occurs and the mechanism is E2.

- **With weak nucleophiles or bases, a mixture of S_N1 and E1 products results.**

Br, 2° + $H_2\ddot{O}$: (weak nucleophile and base) → OH (S_N1 product) + E1 product

A weak base or nucleophile favors S_N1 and E1 mechanisms and both occur.

Sample Problems 8.4–8.6 illustrate how to apply the information in Table 8.5 to specific alkyl halides.

Sample Problem 8.4 Determining the Substitution and Elimination Products from an Alkyl Halide

Draw the products of the following reaction.

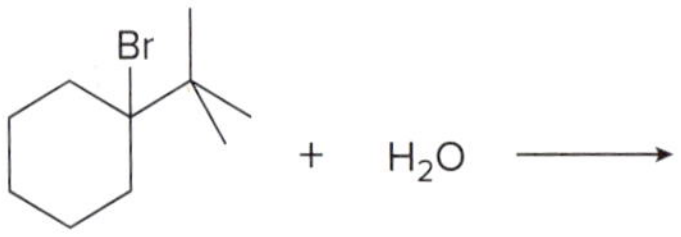

Br + H_2O →

Solution

- **Classify the halide as 1°, 2°, or 3° and the reagent as a strong or weak base (and nucleophile)** to determine the mechanism. In this case, the alkyl halide is 3° and the reagent (H_2O) is a weak base and nucleophile, so products of both $\mathbf{S_N1}$ and **E1** mechanisms are formed.
- To draw the S_N1 product, **substitute the nucleophile (H_2O) for the leaving group (Br^-),** and draw the neutral product after loss of a proton.
- To draw the E1 product, **remove the elements of H and Br** from the α and β carbons. There are two identical β C atoms with H atoms, so only one elimination product is possible.

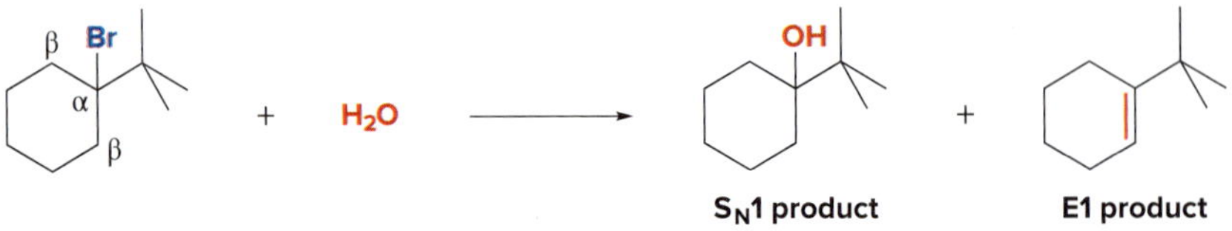

β Br α β + H_2O → OH (S_N1 product) + E1 product

Sample Problem 8.5 Drawing Substitution and Elimination Products from a 2° Alkyl Halide

Draw the products of the following reaction.

(cyclopentyl)–Br + CH_3O^- $\xrightarrow{CH_3OH}$

Solution

- **Classify the halide as 1°, 2°, or 3° and the reagent as a strong or weak base (and nucleophile)** to determine the mechanism. In this case, the alkyl halide is 2° and the reagent (CH_3O^-) is a strong base and nucleophile, so products of both $\mathbf{S_N2}$ and **E2** mechanisms are formed.
- To draw the S_N2 product, **substitute the nucleophile (CH_3O^-) for the leaving group (Br^-).**
- To draw the E2 product, **remove the elements of H and Br** from the α and β carbons. There are two identical β C atoms with H atoms, so only one elimination product is possible.

Br + $CH_3\ddot{O}:^-$ (nucleophile) ⟶ $\ddot{O}CH_3$ — **S_N2 product**

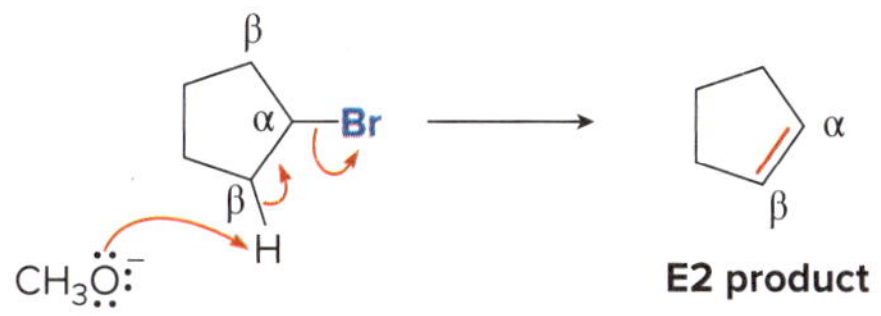

β, α, Br, β, H, $CH_3\ddot{O}:^-$ ⟶ α, β — **E2 product**

Problem 8.21

Draw the products in each reaction.

a. 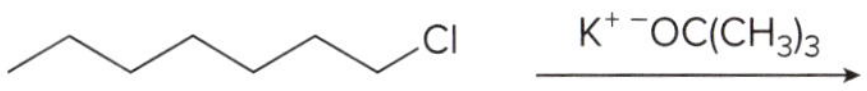 Cl, $K^+\ {}^-OC(CH_3)_3$ ⟶

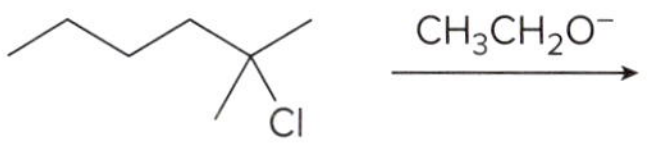

b. 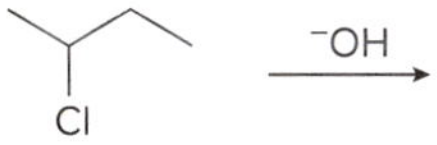 Cl, ^-OH ⟶

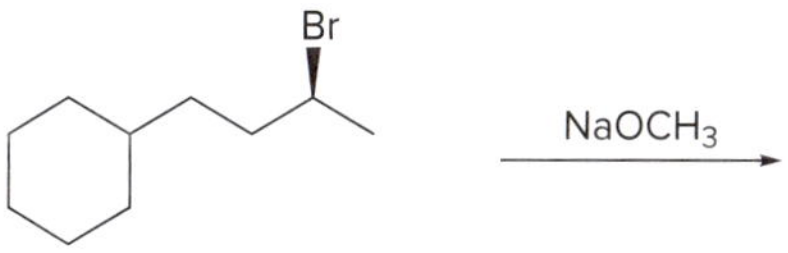

c. 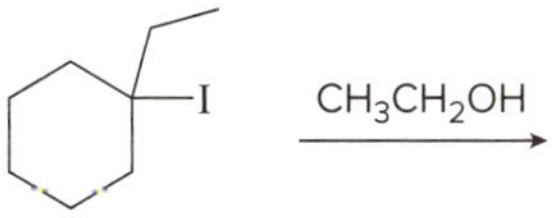 I, CH_3CH_2OH ⟶

d. Cl, $CH_3CH_2O^-$ ⟶

e. Br, $NaOCH_3$ ⟶

f. 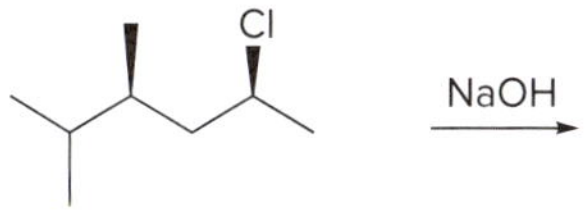 Cl, NaOH ⟶

More Practice: Try Problems 8.46, 8.48, 8.49, 8.53.

Sample Problem 8.6 Drawing the Mechanism When a Reaction Involves Both Substitution and Elimination

Draw the products of the following reaction, and include the mechanism showing how each product is formed.

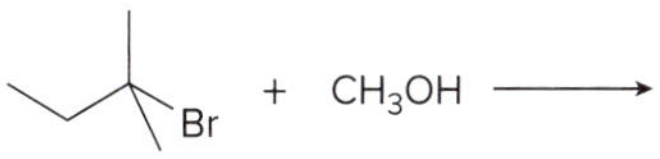

Br + CH_3OH ⟶

Solution

[1] **Classify the halide as 1°, 2°, or 3° and the reagent as a strong or weak base (and nucleophile)** to determine the mechanism. In this case, the alkyl halide is 3° and the reagent (CH_3OH) is a weak base and nucleophile, so products of both $\mathbf{S_N1}$ and **E1** mechanisms are formed.

[2] Draw the steps of the mechanisms to give the products. Both mechanisms begin with the same first step: loss of the leaving group to form a **carbocation.**

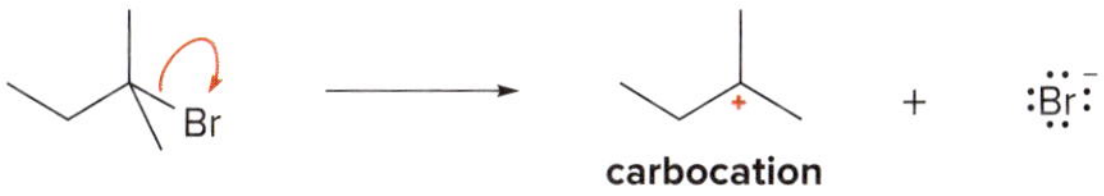

Br ⟶ **carbocation** + $:\ddot{Br}:^-$

- **For S_N1: The carbocation reacts with a nucleophile.** Nucleophilic attack of CH_3OH on the carbocation generates a positively charged intermediate that loses a proton to afford the neutral S_N1 product.

$CH_3\ddot{O}H$ —nucleophilic attack→ [oxonium intermediate with CH_3] $CH_3\ddot{O}H$ —proton transfer→ $\ddot{O}CH_3$ + $CH_3\overset{+}{O}H_2$

S_N1 product

- **For E1: A base (CH_3OH or Br^-) removes a proton from the carbocation.** Two different products of elimination can form because the carbocation has two different β carbons.

β1 — H — $CH_3\ddot{O}H$ → **E1 product** + $CH_3\overset{+}{O}H_2$

H — $CH_3\ddot{O}H$ — β2 → **E1 product** + $CH_3\overset{+}{O}H_2$

In this problem, three products are formed: one from an S_N1 reaction and two from E1 reactions.

Problem 8.22 Draw a stepwise mechanism for the following reaction.

Br —CH_3OH→ OCH_3 + [alkene] + HBr

More Practice: Try Problems 8.51, 8.52, 8.54.

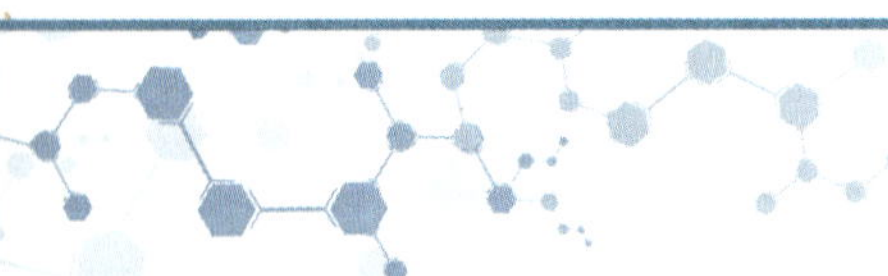

Chapter 8 REVIEW

KEY CONCEPTS

Nucleophiles and bases in S_N1, S_N2, E1, and E2 reactions (8.11)

1 Nucleophiles that are weak bases	2 Strong, bulky bases	3 Strong nucleophiles and strong bases	4 Weak nucleophiles and weak bases
^-SH Br^-	$^-OC(CH_3)_3$	^-OH	H_2O
^-CN I^-	DBU	^-OR	ROH
$CH_3CO_2^-$	DBN		
• **Substitution** is favored over elimination.	• **E2** elimination is favored over substitution.	• **S_N2** and **E2** mechanisms are favored.	• **S_N1** and **E1** mechanisms are favored.

Try Problem 8.53.

KEY SKILLS

[1] Comparing the stability of alkenes (8.2)

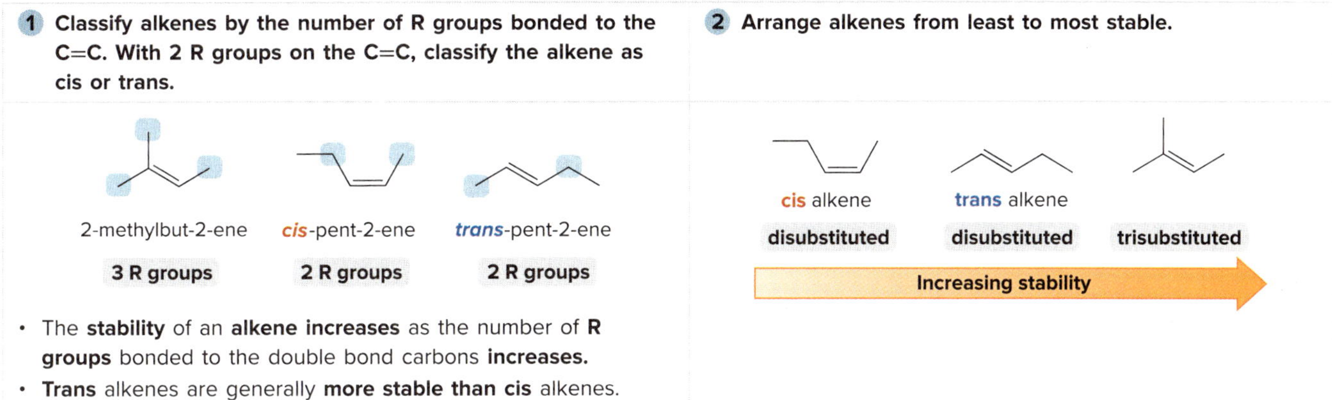

Try Problem 8.27.

[2] Drawing all products and predicting the major product of an elimination reaction (8.5)

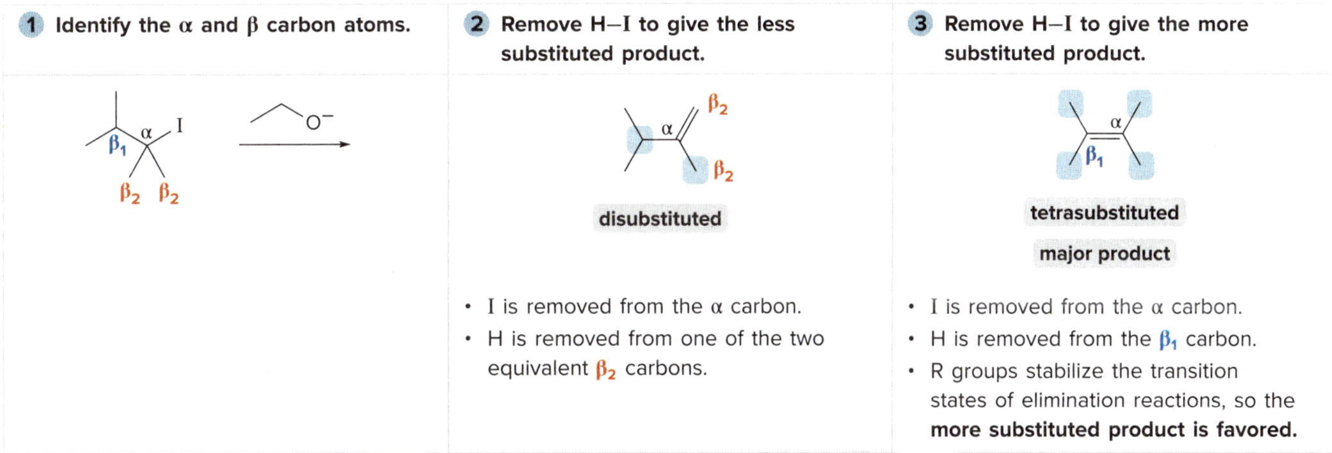

See Sample Problem 8.1. Try Problems 8.29, 8.38.

[3] Drawing the product of an E2 reaction of a halocyclohexane when loss of HX must be anti periplanar (8.6B)

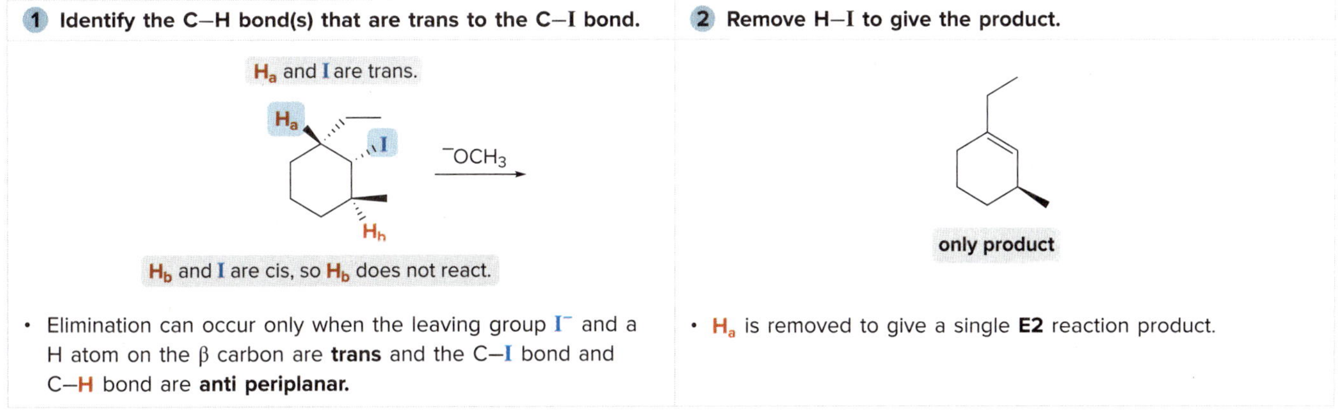

See Figure 8.4, Sample Problem 8.2. Try Problems 8.34–8.36.

[4] Deciding if a β elimination reaction proceeds by an E1 or E2 mechanism (8.9)

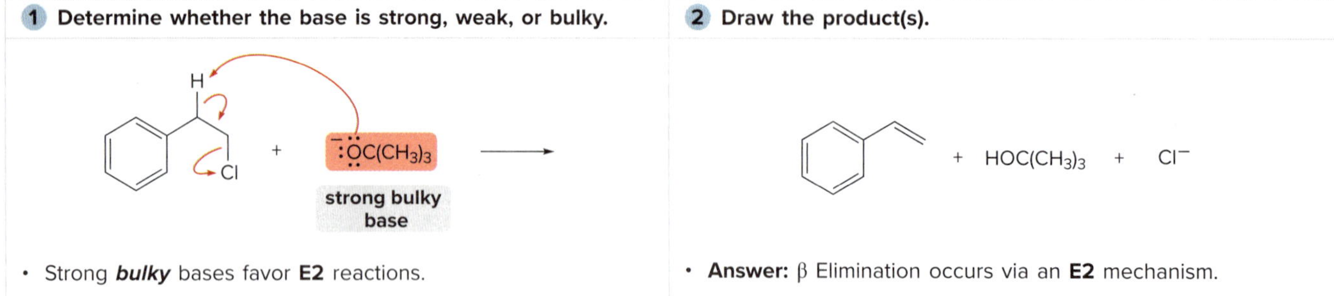

See Table 8.4. Try Problem 8.40.

[5] Deciding if a reaction proceeds by S_N1, S_N2, E1, or E2 (8.11)

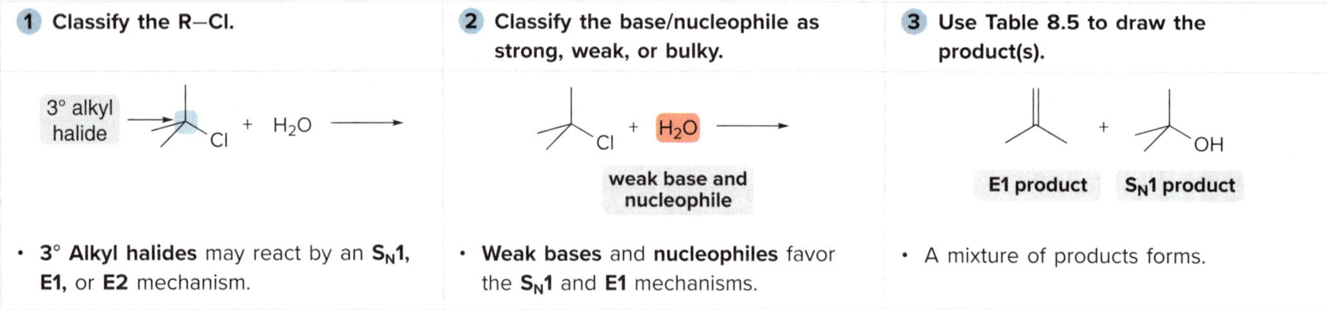

Try Problems 8.46, 8.48, 8.49, 8.53.

[6] Drawing the product(s) of a reaction with a 1° alkyl halide (8.11B)

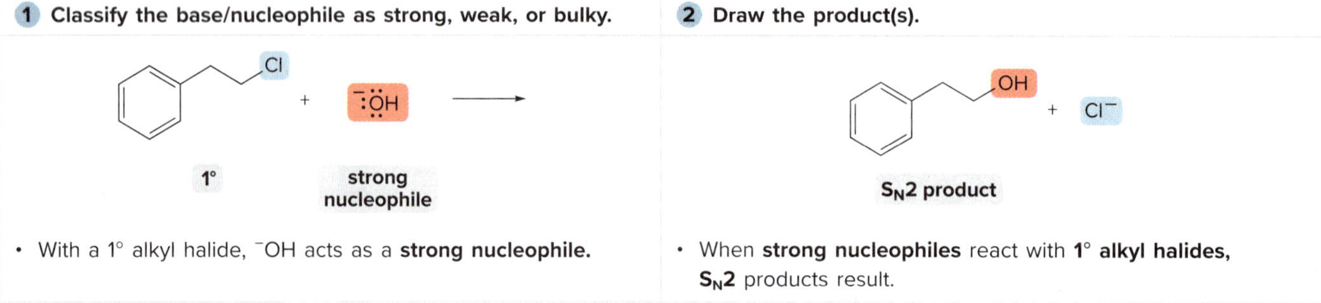

Try Problem 8.46a, c.

[7] Drawing the product(s) of a reaction with a 2° alkyl halide (8.11C)

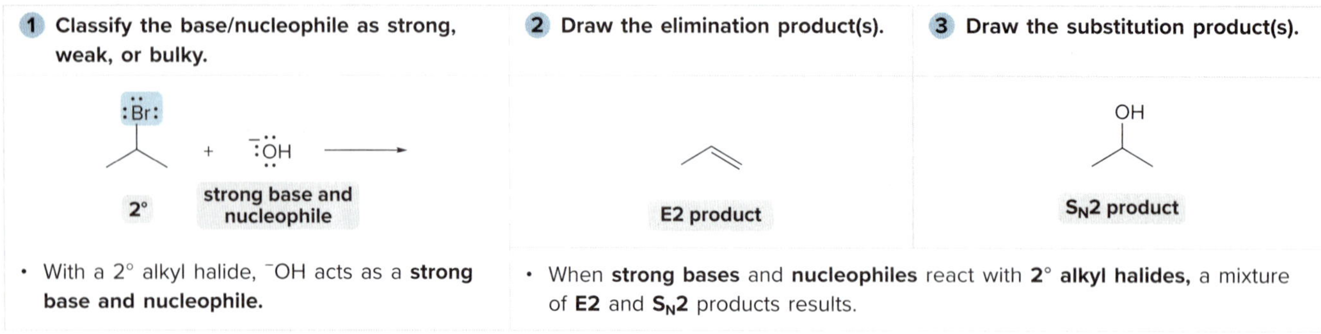

See Sample Problem 8.5. Try Problem 8.48a, b, d, f.

KEY MECHANISM CONCEPTS

[1] Comparison of E1 and E2 reactions

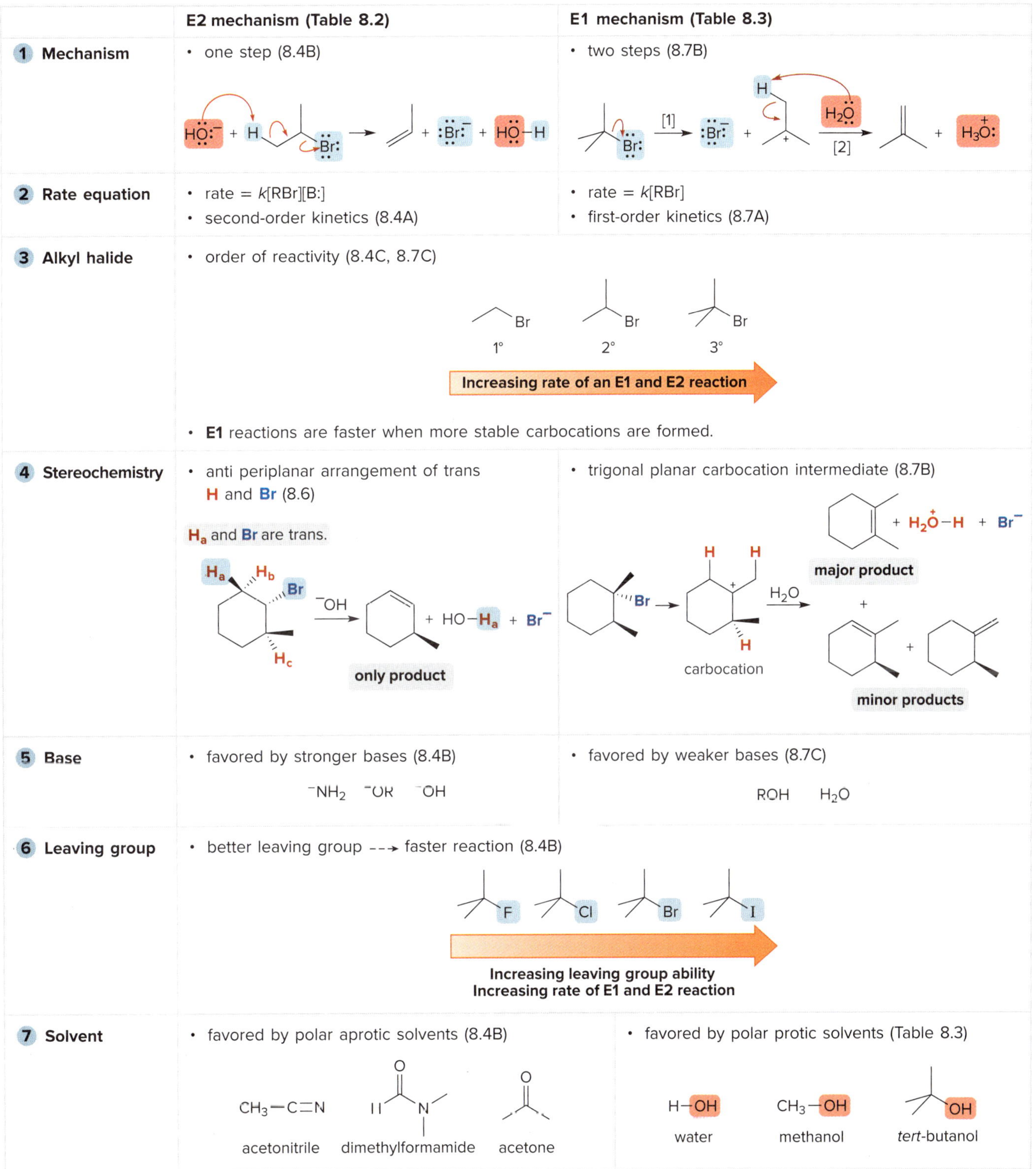

	E2 mechanism (Table 8.2)	E1 mechanism (Table 8.3)
1 Mechanism	• one step (8.4B)	• two steps (8.7B)
2 Rate equation	• rate = k[RBr][B:] • second-order kinetics (8.4A)	• rate = k[RBr] • first-order kinetics (8.7A)
3 Alkyl halide	• order of reactivity (8.4C, 8.7C) 1° 2° 3° Increasing rate of an E1 and E2 reaction • **E1** reactions are faster when more stable carbocations are formed.	
4 Stereochemistry	• anti periplanar arrangement of trans **H** and **Br** (8.6) H_a and **Br** are trans. only product	• trigonal planar carbocation intermediate (8.7B) carbocation major product minor products
5 Base	• favored by stronger bases (8.4B) $^-NH_2$ ^-OR ^-OH	• favored by weaker bases (8.7C) ROH H_2O
6 Leaving group	• better leaving group ---> faster reaction (8.4B) Increasing leaving group ability Increasing rate of E1 and E2 reaction	
7 Solvent	• favored by polar aprotic solvents (8.4B) acetonitrile dimethylformamide acetone	• favored by polar protic solvents (Table 8.3) water methanol *tert*-butanol

8 Product	• More substituted alkene favored in E2 and E1 reactions (Zaitsev rule, 8.5, 8.7C) 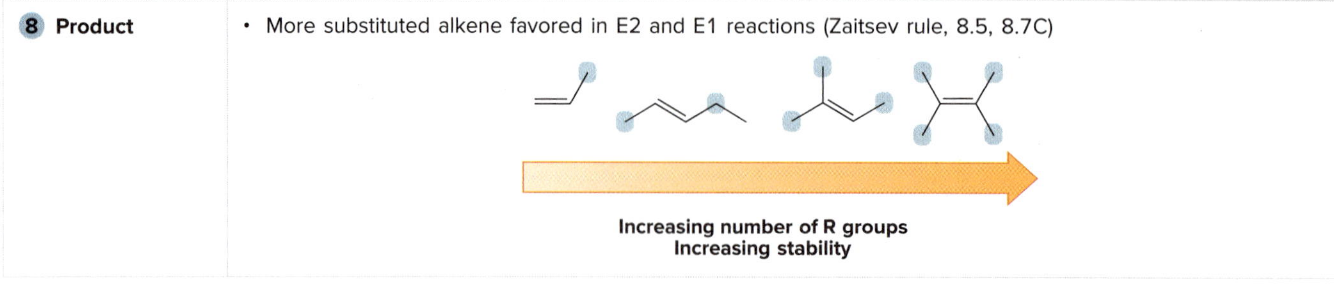

Try Problems 8.31, 8.45, 8.49, 8.51, 8.54.

[2] Summary of S_N1, S_N2, E1, and E2 reactions (8.11)

Alkyl halide type	Reaction with		Mechanism
1 1° RCH_2X	• strong nucleophile	--->	S_N2
	• strong ***bulky*** base	--->	E2
2 2° R_2CHX	• strong base and nucleophile	--->	S_N2 + E2
	• strong ***bulky*** base	--->	E2
	• **weak** base and nucleophile	--->	S_N1 + E1
3 3° R_3CX	• **weak** base and nucleophile	--->	S_N1 + E1
	• strong base	--->	E2

CHAPTER 8 MULTIPLE-CHOICE SELF-TEST

The Self-Test consists of multiple-choice questions similar to those found on the American Chemical Society organic chemistry exam. Answers are given at the end of the chapter.

1. What product is *not* formed when **A** reacts with $NaOCH_2CH_3$?

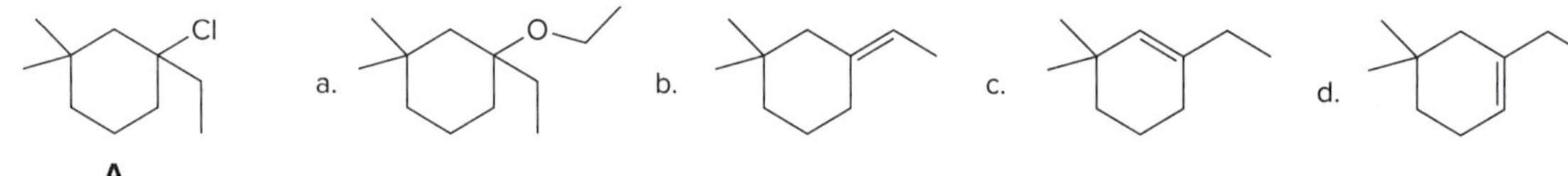

2. Label the most stable and least stable alkene.

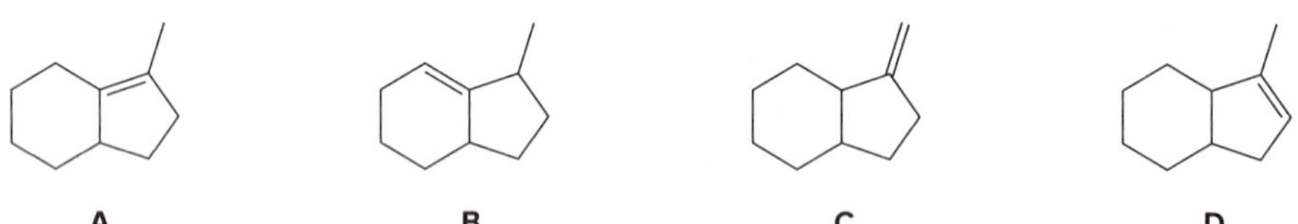

a. **A** is most stable and **B** is least stable.
b. **A** is most stable and **C** is least stable.
c. **D** is most stable and **C** is least stable.
d. **B** is most stable and **A** is least stable.

3. What is the major product formed when **B** undergoes an E2 elimination reaction?

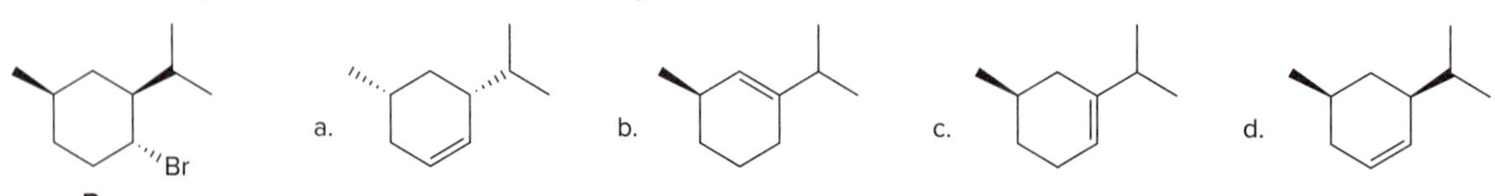

4. Which elimination mechanism is likely observed under each set of reaction conditions?

A 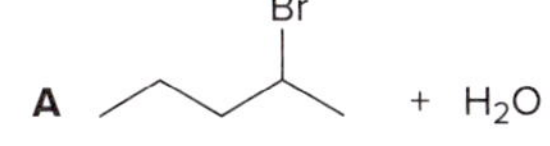

+ H_2O

B 

+ ^{-}OH

C + CH_3O^-

Cl

a. The mechanism is E1 for reaction **C** and E2 for reactions **A** and **B.**
b. The mechanism is E1 for reactions **A** and **C** and E2 for reaction **B.**
c. The mechanism is E1 for reaction **A** and E2 for reactions **B** and **C.**
d. The mechanism is E2 for reactions **A** and **B** and E1 for reaction **C.**

5. Which compound reacts most rapidly in an E2 reaction?

a. 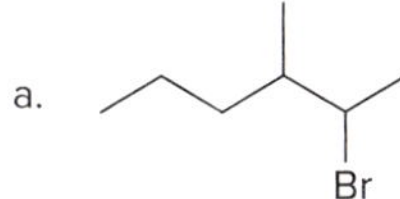

b. 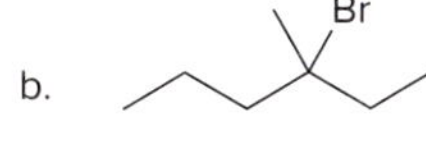

c. 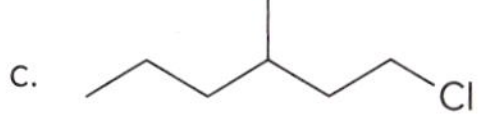

d. 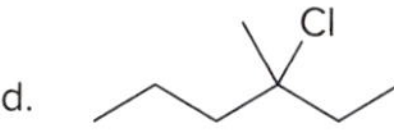

6. From which dichlorides can 2-hexyne be prepared cleanly when treated with strong base?

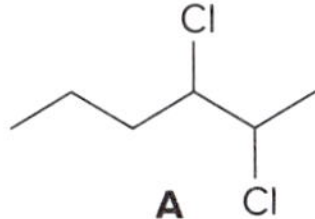

A

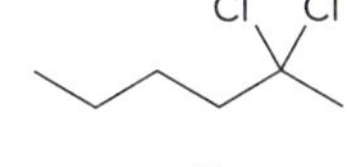

B

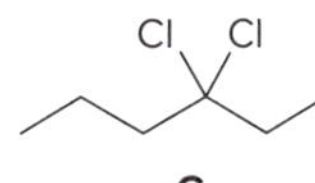

C

a. **A** only b. **B** only c. **C** only d. **A, B,** and **C** all yield 2-hexyne cleanly.

7. Which alkyl halide gives a single constitutional isomer in an E2 elimination reaction?

a. 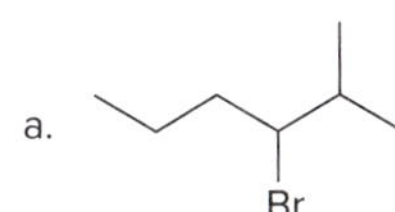

b. 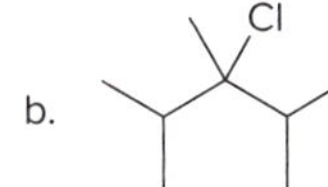

c. 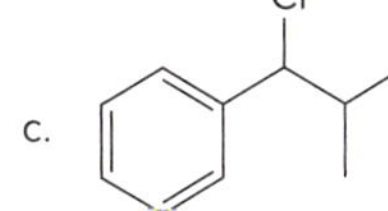

d. 

8. How are isomers **A, B,** and **C** related to each other?

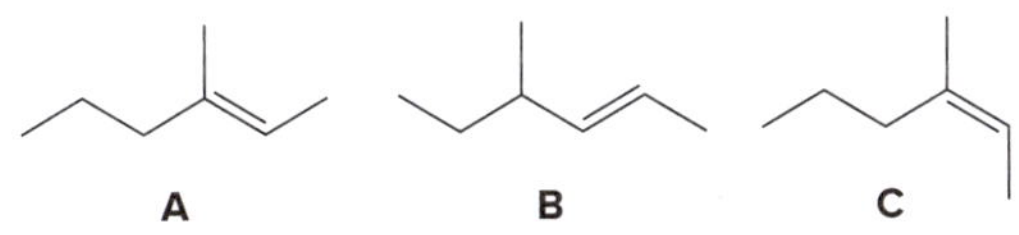

a. **A** is a constitutional isomer of **B** and **C.**
b. **B** is a constitutional isomer of **A** and **C.**
c. **A** is a stereoisomer of **B** and **C.**
d. **C** is a constitutional isomer of **A** and **B.**

9. Which statement is true about the reaction of **D** with CH_3O^-?

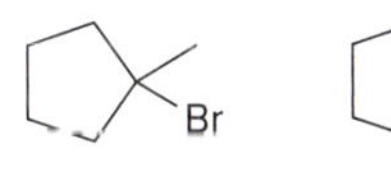

a. The major product is ether **E** by an S_N1 reaction.
b. The reaction exhibits second-order kinetics.
c. A mixture of substitution and elimination products is formed.
d. Changing the leaving group from Br to Cl increases the rate of reaction.

10. When alkyl chloride **F** reacts with CH_3CH_2OH, which product is *not* formed?

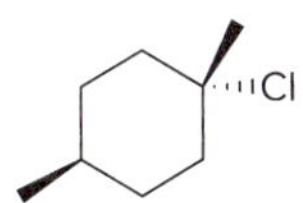

F

a. 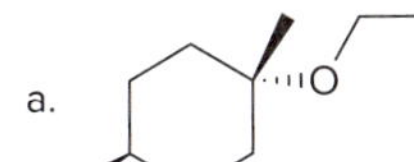

b. 

c. 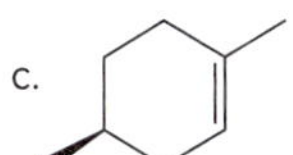
d. 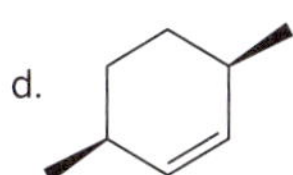

PROBLEMS

Problem Using Three-Dimensional Models

8.23 Name each compound and decide which stereoisomer will react faster in an E2 elimination reaction. Explain your choice.

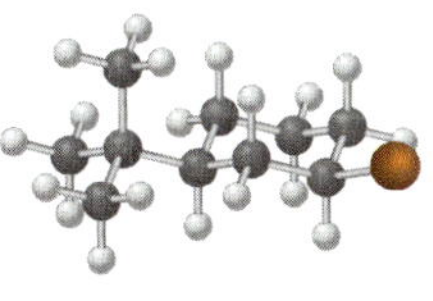
D

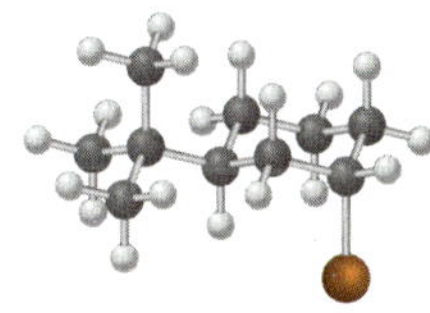
E

General Elimination

8.24 Draw all possible constitutional isomers formed by dehydrohalogenation of each alkyl halide.

a. Br b. Br c. Cl d. I

8.25 What alkyl halide forms each of the following alkenes as the *only* product in an elimination reaction?

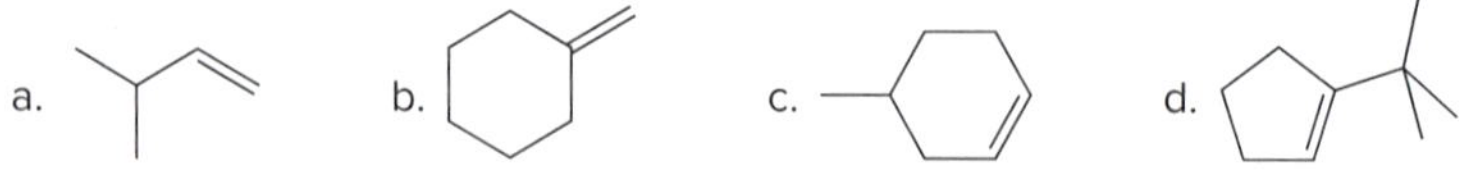

Alkenes

8.26 Label each pair of alkenes as constitutional isomers, stereoisomers, or identical.

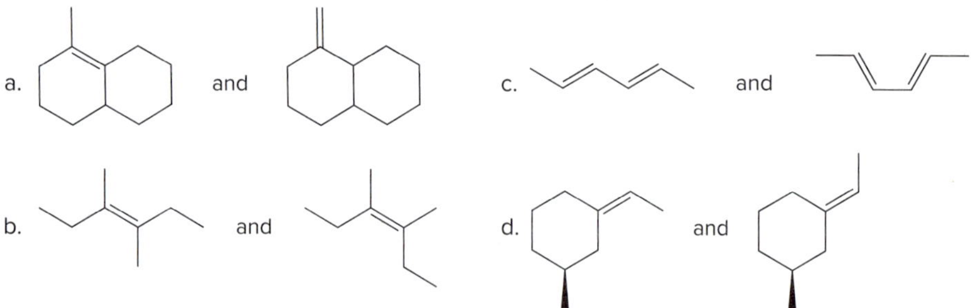

8.27 Rank the following alkenes in order of increasing stability.

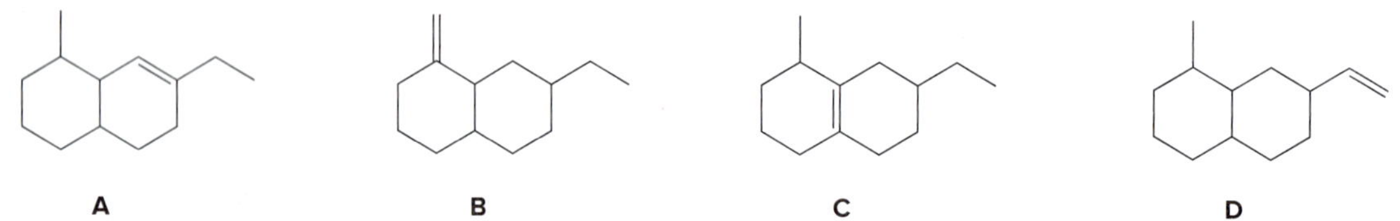

8.28 $\Delta H°$ values obtained for a series of similar reactions are one set of experimental data used to determine the relative stability of alkenes. Explain how the following data suggest that *cis*-but-2-ene is more stable than but-1-ene (Section 12.3A).

but-1-ene + H_2 ⟶ $\Delta H° = -127$ kJ/mol

cis-but-2-ene + H_2 ⟶ $\Delta H° = -120$ kJ/mol

E2 Reaction

8.29 Draw all constitutional isomers formed in each E2 reaction, and predict the major product using the Zaitsev rule.

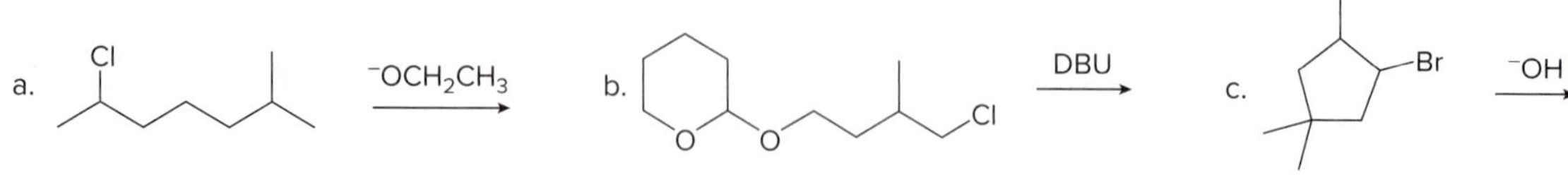

8.30 For each of the following alkenes, draw the structure of two different alkyl halides that yield the given alkene as the only product of dehydrohalogenation.

a. b. O c. O

8.31 Consider the following E2 reaction.

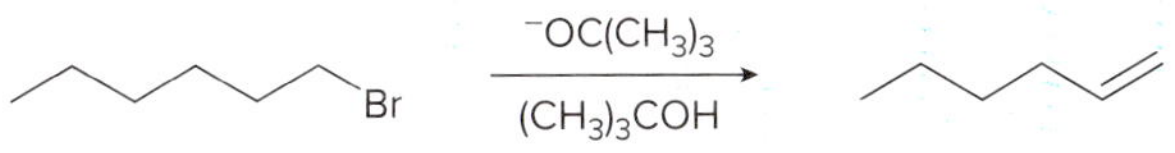

$$\text{CH}_3\text{CH}_2\text{CH}_2\text{CH}_2\text{CH}_2\text{Br} \xrightarrow[(CH_3)_3COH]{^-OC(CH_3)_3} \text{CH}_3\text{CH}_2\text{CH}_2\text{CH}{=}\text{CH}_2$$

a. Draw the by-products of the reaction and use curved arrows to show the movement of electrons.

b. What happens to the reaction rate with each of the following changes? [1] The solvent is changed to DMF. [2] The concentration of $^-OC(CH_3)_3$ is decreased. [3] The base is changed to ^-OH. [4] The halide is changed to $CH_3CH_2CH_2CH_2CH(Br)CH_3$. [5] The leaving group is changed to I^-.

8.32 What is the major stereoisomer formed when each alkyl halide is treated with $KOC(CH_3)_3$?

a. 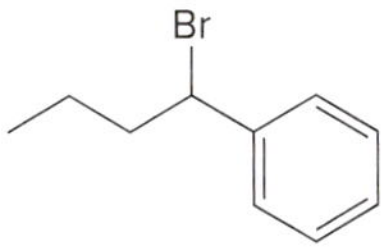 b.

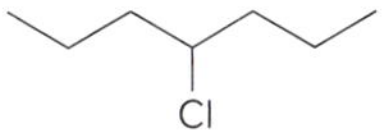

Stereochemistry and the E2 Reaction

8.33 What is the major E2 elimination product formed from each halide?

a. 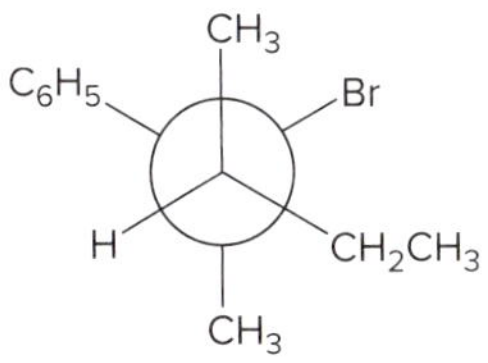 b. 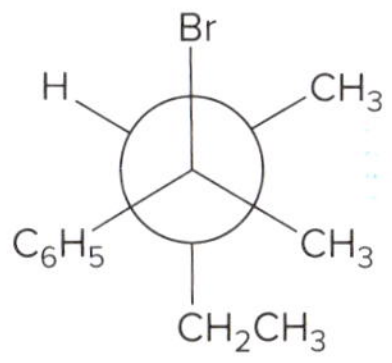 c.

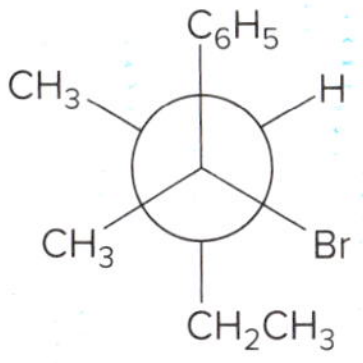

8.34 Taking into account anti periplanar geometry, predict the major E2 product formed from each starting material.

a. 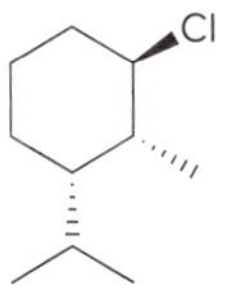 b. 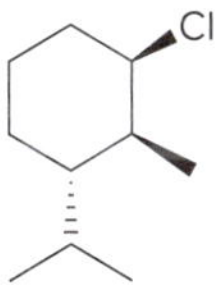 c. 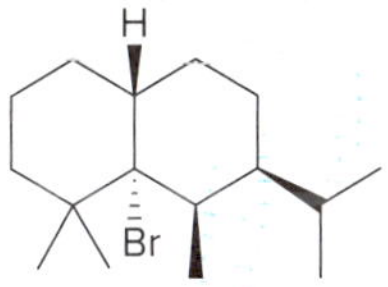 d.

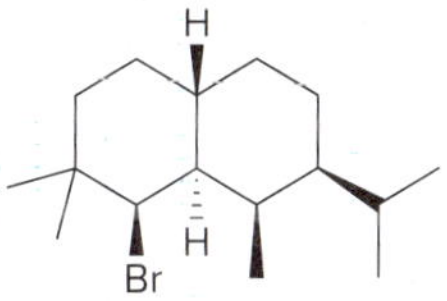

8.35 Does *cis*- or *trans*-1-bromo-4-*tert*-butylcyclohexane react faster in an E2 reaction?

8.36 Draw the structure (including stereochemistry) of an alkyl chloride that forms each alkene as the exclusive E2 elimination product.

a. 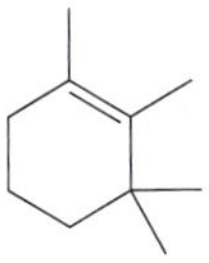 b. 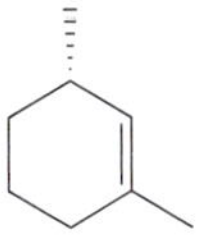 c.

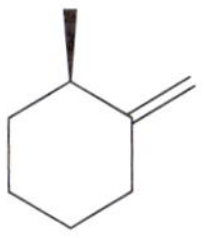

8.37 Draw the major stereoisomer formed when each compound undergoes an elimination reaction with NaOH.

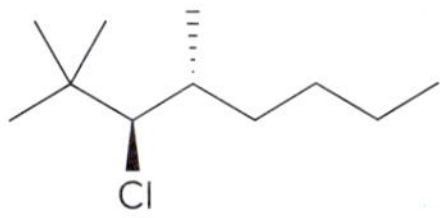

A 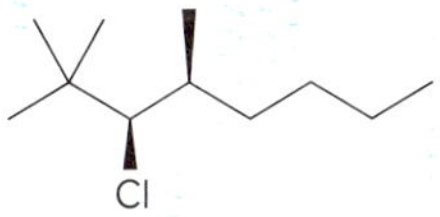 **B**

E1 Reaction

8.38 What alkene is the major product formed from each alkyl halide in an E1 reaction?

a. 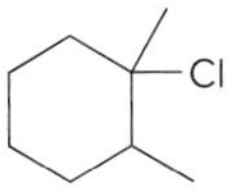 b.  c.

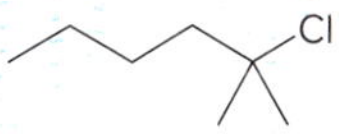

E1 and E2

8.39 Rank the following alkyl halides in order of increasing reactivity in E2 elimination. Then do the same for E1 elimination.

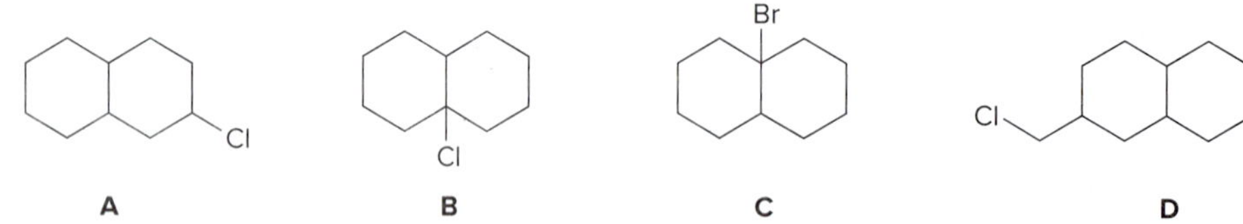

8.40 Draw all constitutional isomers formed in each elimination reaction. Label the mechanism as E2 or E1.

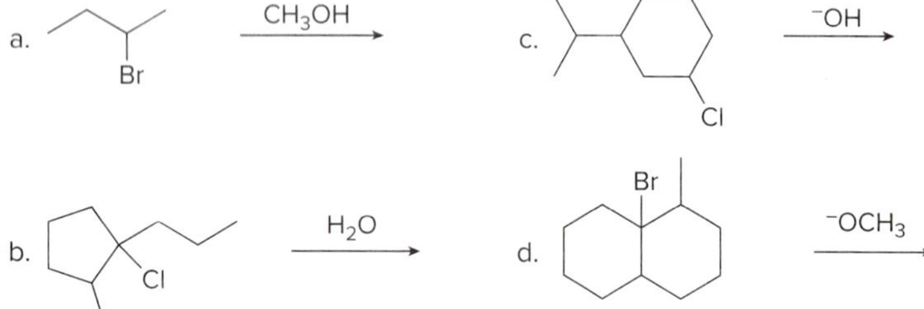

8.41 In the dehydrohalogenation of bromocyclodecane, the major product is *cis*-cyclodecene rather than *trans*-cyclodecene. Offer an explanation.

Alkynes

8.42 Draw the products formed when each dihalide is treated with excess $NaNH_2$.

8.43 Draw the structure of a dihalide that could be used to prepare each alkyne. There may be more than one possible dihalide.

8.44 Under certain reaction conditions, 2,3-dibromobutane reacts with two equivalents of base to give three products, each of which contains two new π bonds. Product **A** has two *sp* hybridized carbon atoms, product **B** has one *sp* hybridized carbon atom, and product **C** has none. What are the structures of **A, B,** and **C?**

S_N1, S_N2, E1, and E2 Mechanisms

8.45 For which reaction mechanisms—S_N1, S_N2, E1, or E2—are each of the following statements true? A statement may be true for one or more mechanisms.

a. The mechanism involves carbocation intermediates.
b. The mechanism has two steps.
c. The reaction rate increases with better leaving groups.
d. The reaction rate increases when the solvent is changed from CH_3OH to $(CH_3)_2SO$.
e. The reaction rate depends on the concentration of only the alkyl halide.
f. The mechanism is concerted.
g. The reaction of CH_3CH_2Br with NaOH occurs by this mechanism.
h. Racemization at a stereogenic center occurs.
i. Tertiary (3°) alkyl halides react faster than 2° or 1° alkyl halides.
j. The reaction follows a second-order rate equation.

8.46 Draw the organic products formed in each reaction.

a. 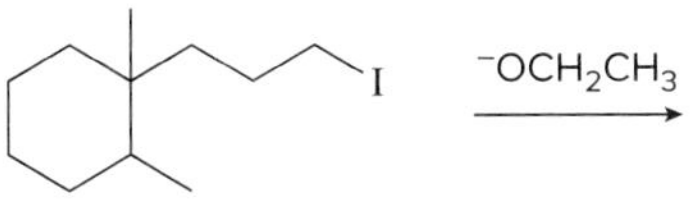 (structure) $\xrightarrow{^{-}OCH_2CH_3}$
b. 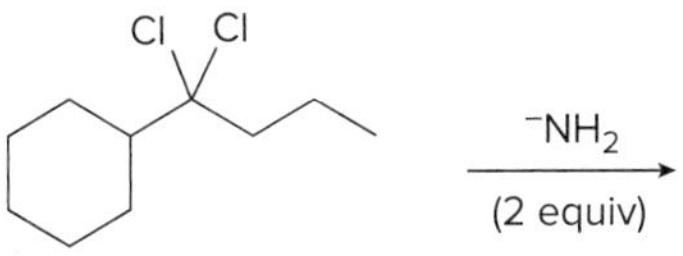 (structure with Cl, Cl) $\xrightarrow[\text{(2 equiv)}]{^{-}NH_2}$
c. 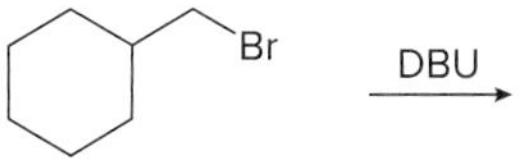 (structure with Br) $\xrightarrow{\text{DBU}}$
d. 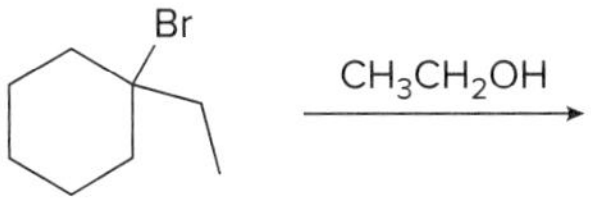 (structure with Br) $\xrightarrow{CH_3CH_2OH}$
e. 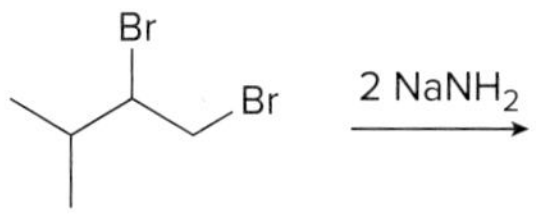 (structure with Br, Br) $\xrightarrow{2\ NaNH_2}$
f. 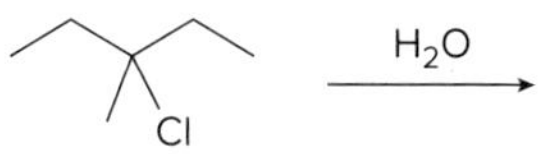 (structure with Cl) $\xrightarrow{H_2O}$

8.47 What reagents and reaction conditions are needed for each of the following conversions?

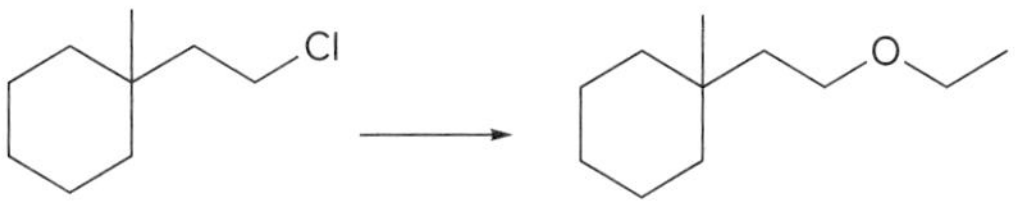

a. 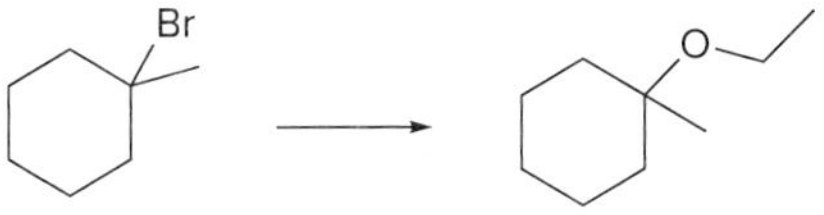 (structure with Br) →
b.  (structure with Br) →
c. (structure with Cl) →
d. 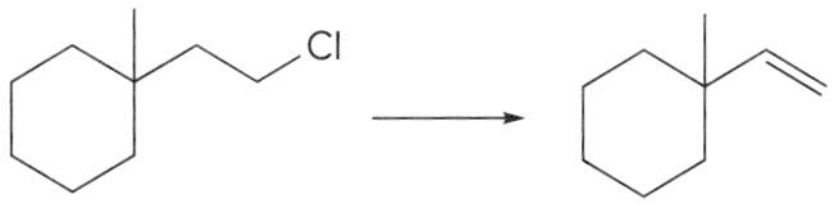 (structure with Cl) →

8.48 Draw all products, including stereoisomers, in each reaction.

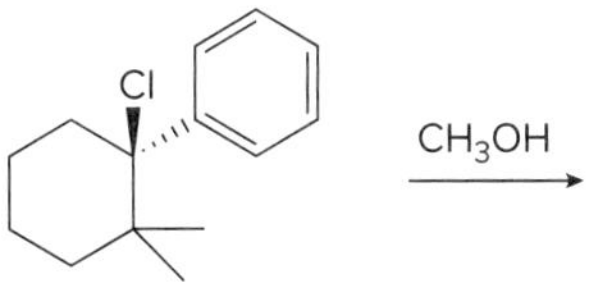

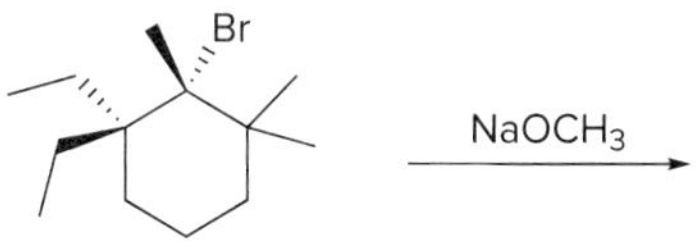

a. 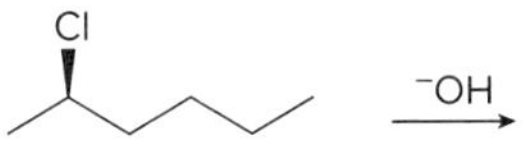 (structure with Cl) $\xrightarrow{^{-}OH}$
b. 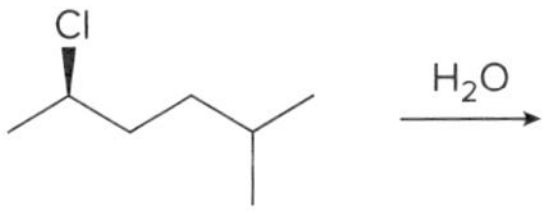 (structure with Cl) $\xrightarrow{H_2O}$
c. (structure with Cl) $\xrightarrow{CH_3OH}$
d. 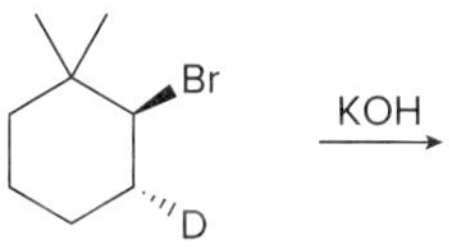 (structure with Br, D) $\xrightarrow{\text{KOH}}$
e. (structure with Br) $\xrightarrow{NaOCH_3}$
f. 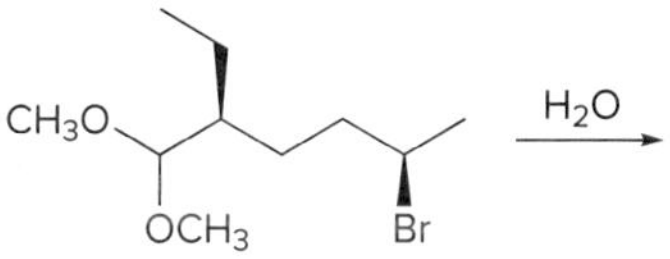 CH_3O (structure with OCH_3, Br) $\xrightarrow{H_2O}$

8.49 (structure with Cl, H) Draw all of the substitution and elimination products formed from the given alkyl halide with each reagent: (a) CH_3OH; (b) KOH. Indicate the stereochemistry around the stereogenic centers present in the products, as well as the mechanism by which each product is formed.

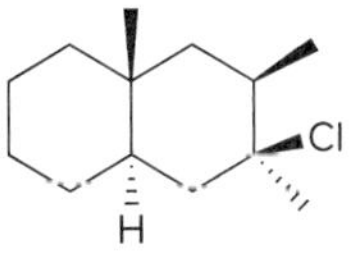

8.50 The following reactions do not afford the major product that is given. Explain why this is so, and draw the structure of the major product actually formed.

a. 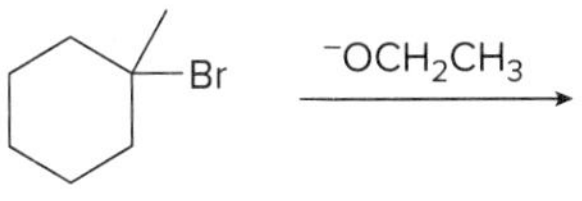 (structure with Br) $\xrightarrow{^{-}OCH_2CH_3}$ (structure with O)
b. 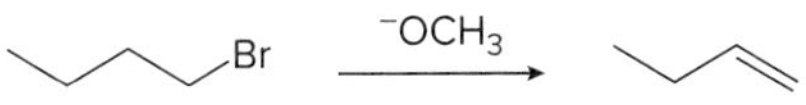 (structure with Br) $\xrightarrow{^{-}OCH_3}$ (structure)
c. 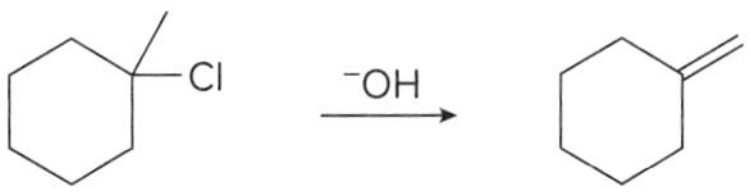 (structure with Cl) $\xrightarrow{^{-}OH}$ (structure)
d. 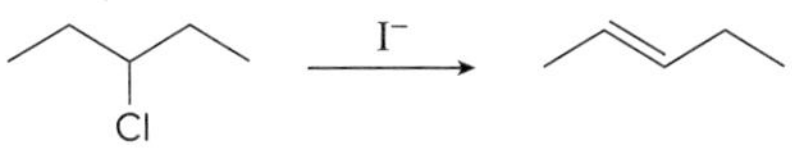 (structure with Cl) $\xrightarrow{I^{-}}$ (structure)

8.51 Draw a stepwise, detailed mechanism for the following reaction.

(structure with Cl) $\xrightarrow{CH_3CH_2OH}$ (structure with O) + (structure) + (structure) + HCl

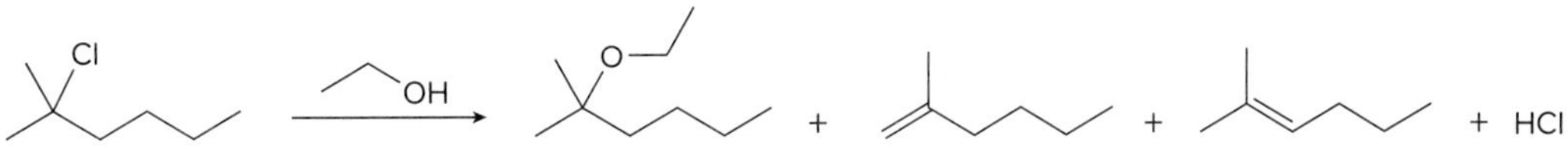

8.52 In Section 7.15A we learned that phosphate (PO_4^{3-}, abbreviated as P_i) is a good leaving group in biological systems. With this in mind, draw a stepwise mechanism for the following E1 reaction, which synthesizes chorismate, an intermediate in the biosynthesis of amino acids.

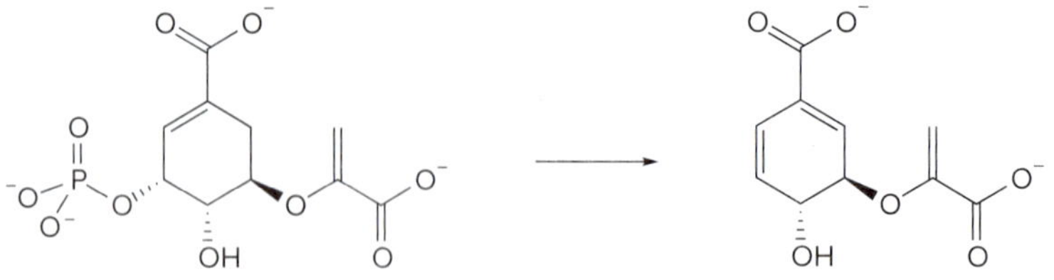

chorismate

8.53 Draw the major product formed when (*R*)-1-chloro-3-methylpentane is treated with each reagent: (a) $NaOCH_2CH_3$; (b) KCN; (c) DBU.

8.54 Draw a stepwise, detailed mechanism for the following reaction.

H_2O + + + HBr

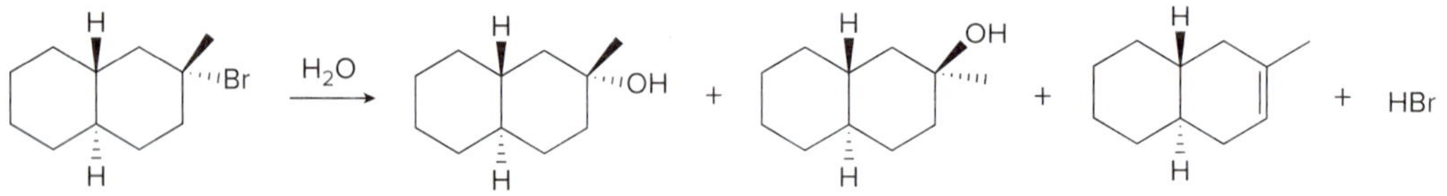

Challenge Problems

8.55 Explain why alkene **A** is more stable than alkene **B,** even though **B** contains more carbon atoms bonded to the double bond. Would you expect **C** to be more or less stable than **A** and **B?**

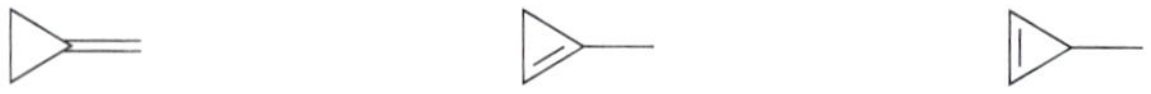

A **B** **C**

8.56 Draw a stepwise detailed mechanism that illustrates how four organic products are formed in the following reaction.

CH_3OH + + + + HCl

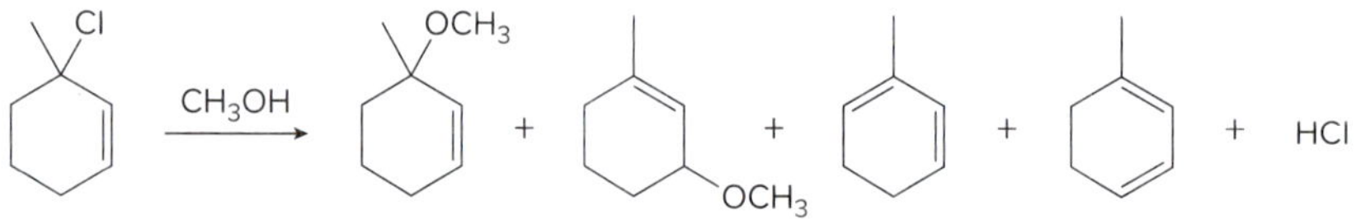

8.57 From what you have learned about the Zaitsev rule and the stereochemistry of dehydrohalogenation in an E2 mechanism, draw the single stereoisomer formed when each compound is treated with strong base.

a. 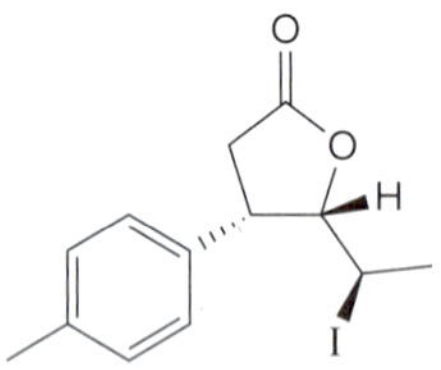 b.

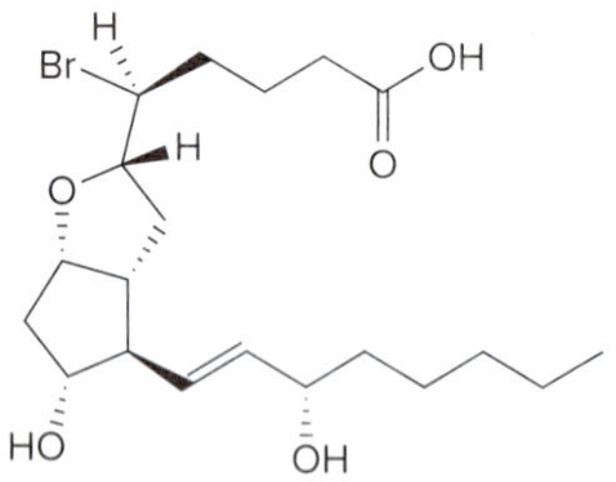

8.58 Draw a stepwise mechanism for the following reaction. The four-membered ring in the starting material and product is called a β-lactam. This functional group confers biological activity on penicillin and many related antibiotics, as is discussed in Chapter 20. (Hint: The mechanism begins with β elimination and involves only two steps.)

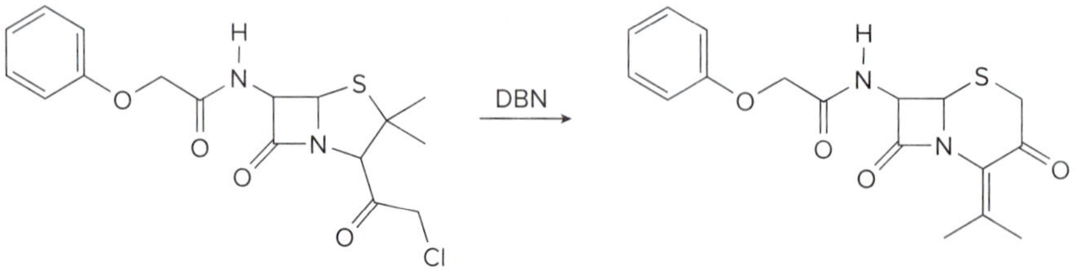

DBN

8.59 Although dehydrohalogenation occurs with anti periplanar geometry, some eliminations have syn periplanar geometry. Examine the starting material and product of each elimination, and state whether the elimination occurs with syn or anti periplanar geometry.

a. 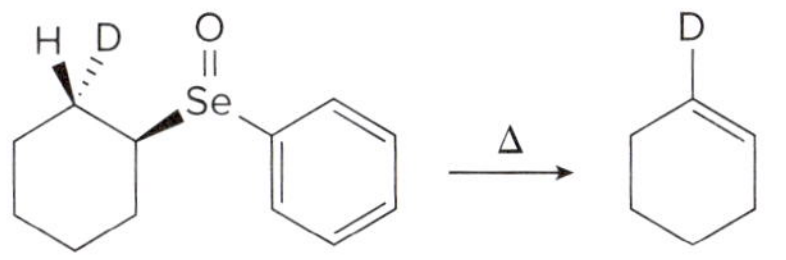 H D O Se Δ D

b. 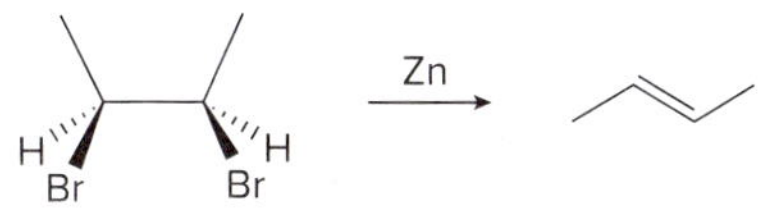 H Br H Br Zn

8.60 Treatment of **X** with base forms a product with molecular formula $C_{38}H_{26}$. Suggest a structure for the product and a mechanism that explains its formation.

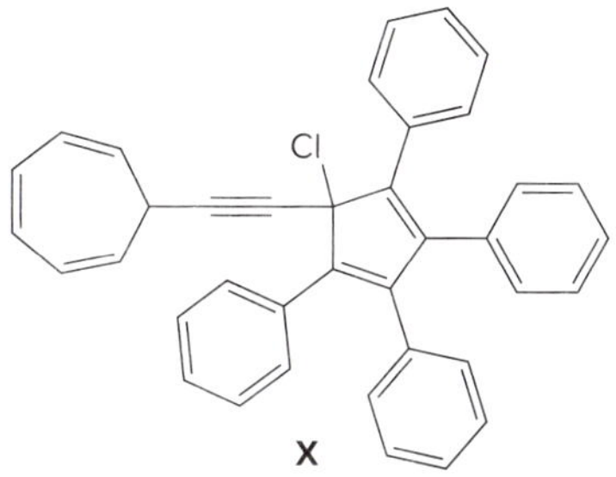

Cl

X

SELF-TEST ANSWERS

1. a 2. b 3. d 4. c 5. b 6. a 7. c 8. b 9. b 10. d

9 Alcohols, Ethers, and Related Compounds

9.1 Introduction
9.2 Structure and bonding
9.3 Nomenclature
9.4 Properties of alcohols, ethers, and epoxides
9.5 Interesting alcohols, ethers, and epoxides
9.6 Preparation of alcohols, ethers, and epoxides
9.7 General features—Reactions of alcohols, ethers, and epoxides
9.8 Dehydration of alcohols to alkenes
9.9 Carbocation rearrangements
9.10 Dehydration using $POCl_3$ and pyridine
9.11 Conversion of alcohols to alkyl halides with HX
9.12 Conversion of alcohols to alkyl halides with $SOCl_2$ and PBr_3
9.13 Tosylate—Another good leaving group
9.14 Reaction of ethers with strong acid
9.15 Thiols and sulfides
9.16 Reactions of epoxides
9.17 Application: Epoxides, leukotrienes, and asthma
9.18 Application: Benzo[*a*]-pyrene, epoxides, and cancer

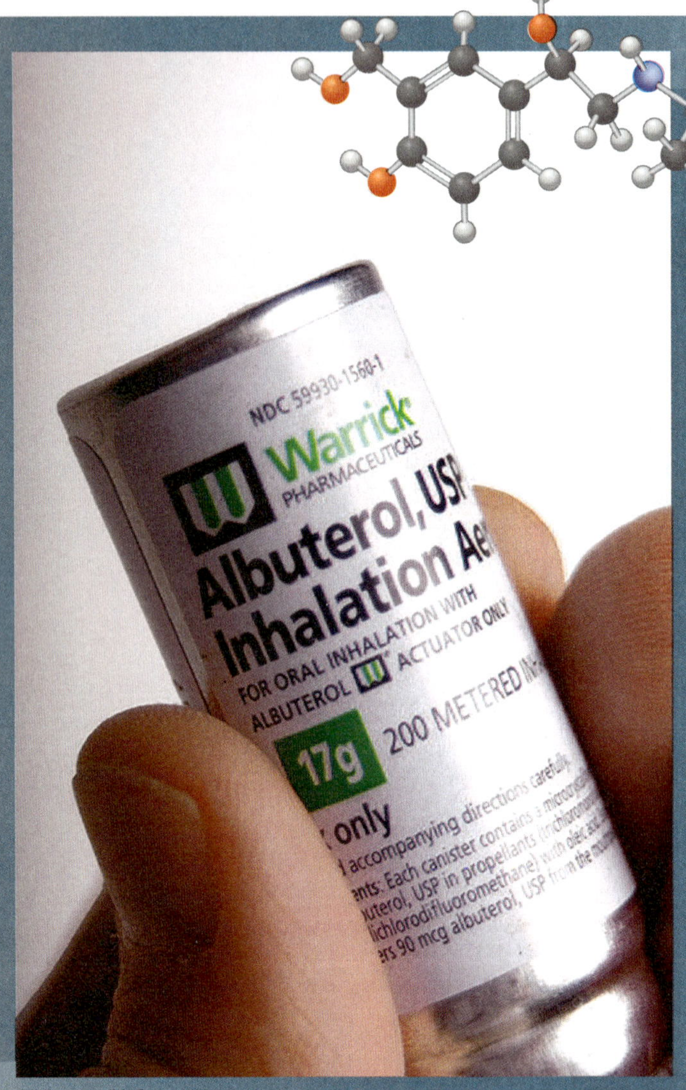

Jill Braaten/McGraw Hill

Albuterol (trade names Proventil, Ventolin, and others) is a bronchodilator, a drug that widens airways, making it an effective treatment for asthma and chronic obstructive pulmonary disease (COPD). It was approved for use in the United States in 1982 and is one of the most common medications in rescue inhalers to open up airways during an asthma attack. Although albuterol contains one stereogenic center and only one enantiomer is responsible for its biological activity, it is sold as a racemic mixture. Albuterol contains two alcohol units and is synthesized from an epoxide, two of the functional groups that are the subject of Chapter 9.

Why Study . . .

Alcohols, Ethers, Epoxides, Thiols, and Sulfides?

In Chapter 9, we take the principles learned in Chapters 7 and 8 about leaving groups, nucleophiles, and bases, and apply them to **alcohols, ethers,** and **epoxides,** three new functional groups that contain polar C–O bonds. The hydroxy group (OH) of an alcohol is especially common in many natural products, and the reactions of alcohols are widely used in organic synthesis. In Chapter 9, you will discover that all of the reactions follow one of the four mechanisms introduced in Chapters 7 and 8—S_N1, S_N2, E1, or E2—so there are **no new general mechanisms to learn.**

Later in the chapter, we will also examine **thiols (RSH)** and **sulfides (R_2S),** sulfur analogues of alcohols and ethers, respectively. These functional groups play a key role in the chemistry of biomolecules, especially the proteins discussed in Chapter 27.

9.1 Introduction

Alcohols, ethers, and **epoxides** are three functional groups that contain carbon–oxygen σ bonds.

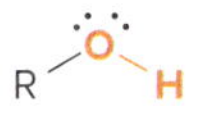

alcohol  ether  epoxide

***Alcohols* contain a hydroxy group** (**OH** group) bonded to an sp^3 hybridized carbon atom. As we learned in Section 3.2, alcohols are classified as **primary (1°), secondary (2°),** or **tertiary (3°)** based on the number of carbon atoms bonded to the carbon with the OH group.

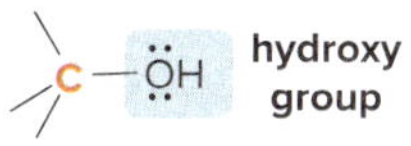

hydroxy group

alcohol

C is sp^3 hybridized.

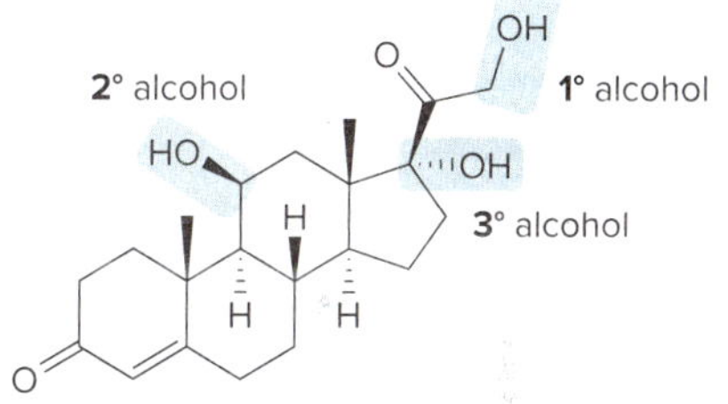

2° alcohol 1° alcohol 3° alcohol

cortisol

anti-inflammatory steroid

Compounds having a hydroxy group on an sp^2 hybridized carbon atom—**enols** and **phenols**—undergo different reactions than alcohols and are discussed in Chapters 11 and 19, respectively. **Enols** have an OH group on a carbon of a C–C double bond. **Phenols** have an OH group on a benzene ring.

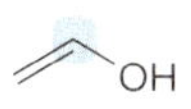

enol 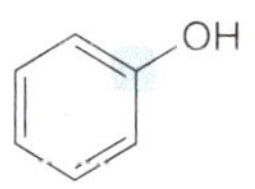 phenol

• C bonded to OH is sp^2 hybridized.

***Ethers* have two alkyl groups bonded to an oxygen atom.** An ether is **symmetrical** if the two alkyl groups are the same, and **unsymmetrical** if they are different. Both alcohols and ethers are organic derivatives of H_2O, formed by replacing one or both of the hydrogens on the oxygen atom by R groups, respectively.

ether

symmetrical ether
identical R groups

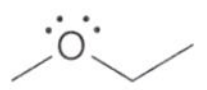

unsymmetrical ether
different R groups

***Epoxides* are ethers having the oxygen atom in a three-membered ring.** Epoxides are also called **oxiranes.**

epoxide or **oxirane**

Problem 9.1 Label each ether and alcohol in brevenal, a marine natural product. Classify each alcohol as 1°, 2°, or 3°.

brevenal

Brevenal (Problem 9.1) is a nontoxic marine polyether produced by *Karenia brevis,* a single-celled organism that proliferates during red tides, vast algal blooms that turn the ocean water red, brown, or green. *Don Paulson Photography/Purestock/Alamy Stock Photo*

9.2 Structure and Bonding

Alcohols, ethers, and epoxides each contain an oxygen atom surrounded by two atoms and two nonbonded electron pairs, making the O atom **tetrahedral** and ***sp*^{3}** hybridized. Only two of the four groups around O are atoms, so alcohols and ethers have a **bent** shape like H_2O. Because oxygen is much more electronegative than carbon or hydrogen, the C–O and O–H bonds are all polar, with the O atom electron rich and the C and H atoms electron poor.

109°

111°

The bond angle around the O atom in an alcohol or ether is similar to the tetrahedral bond angle of **109.5°.** In contrast, the C–O–C bond angle of an epoxide must be **60°,** a considerable deviation from the tetrahedral bond angle. For this reason, **epoxides have angle strain,** making them much more reactive than other ethers.

60°

a strained, three-membered ring

9.3 Nomenclature

To name an alcohol, ether, or epoxide using the IUPAC system, we must learn how to name the functional group either as a substituent or by using a suffix added to the parent name.

9.3A Naming Alcohols

- **In the IUPAC system, alcohols are identified by the suffix *-ol.***

How To Name an Alcohol Using the IUPAC System

Example Give the IUPAC name of the following alcohol:

Step [1] **Find the longest carbon chain containing the carbon bonded to the OH group.**

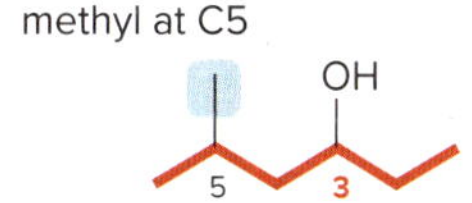

6 C's in the longest chain

6 C's ⟶ **hexane** ⟶ **hexano*l***

- Change the **-*e*** ending of the parent alkane to the suffix ***-ol.***

Step [2] **Number the carbon chain to give the OH group the lower number, and apply all other rules of nomenclature.**

a. **Number** the chain.

- Number the chain to put the OH group at C**3**, not C4.

hexan-3-ol

b. **Name** and **number** the substituents.

methyl at C5

Answer: 5-methylhexan-3-ol

$CH_3CH_2CH_2CH_2OH$ is named as 1-butanol using the 1979 IUPAC recommendations and butan-1-ol using the 1993 IUPAC recommendations.

When an OH group is bonded to a ring, the **ring is numbered beginning with the OH group.** Because the functional group is always at C1, the "1" is usually omitted from the name. The ring is then numbered in a clockwise or counterclockwise fashion to give the next substituent the lower number.

Common names are often used for simple alcohols. To assign a common name:

- **Name all the carbon atoms of the molecule as a single alkyl group.**
- **Add the word *alcohol*, separating the words with a space.**

isopropyl group

isopropyl alcohol

Compounds with two hydroxy groups are called **diols** (using the IUPAC system) or **glycols.** Compounds with three hydroxy groups are called **triols,** and so forth. To name a diol, for example, the suffix ***-diol*** is added to the name of the parent alkane, and numbers are used to indicate the location of the two OH groups.

ethylene glycol (ethane-1,2-diol)

glycerol (propane-1,2,3-triol)

***trans*-cyclopentane-1,2-diol**

Common names are usually used for these simple compounds.

Numbers are needed to show the location of **two** OH groups.

Sample Problem 9.1 Naming an Alcohol Using IUPAC Nomenclature

Give the IUPAC name for each alcohol.

a. 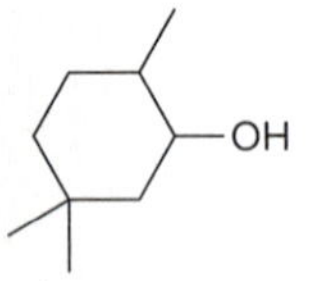 OH

b. 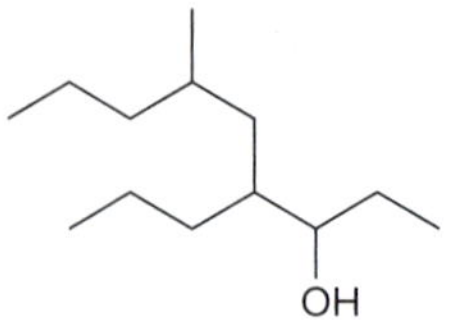 OH

Solution

Part (a):

- Name the molecule as a cyclohexanol with the OH group at C1.
- Number to give the second substituent (CH_3) the lower number.
- Name and number the substituents and combine the parts.

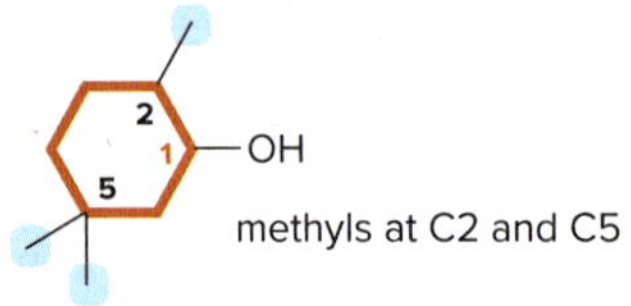

methyls at C2 and C5

- **Answer: 2,5,5-trimethylcyclohexanol**

Part (b):

- Find the longest chain that contains the C to which the OH is bonded; 9 C's → **nonanol.**
- Number to give the OH the lower number and apply all other rules of nomenclature.

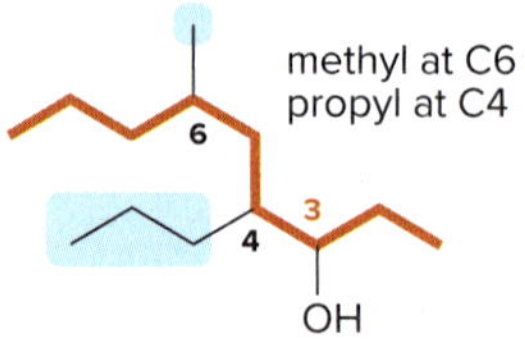

methyl at C6
propyl at C4

- **Answer: 6-methyl-4-propylnonan-3-ol**

Problem 9.2 Give the IUPAC name for each compound.

a. 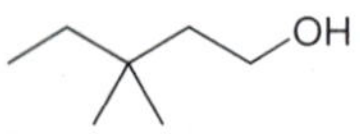 OH

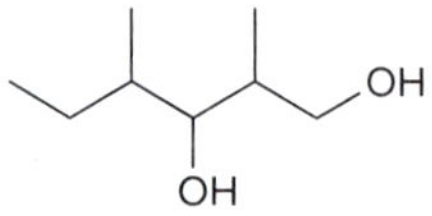

b. 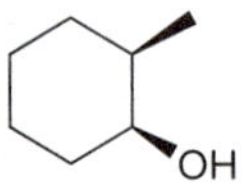 OH

c. 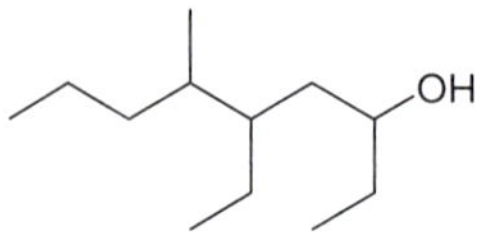 OH

d. 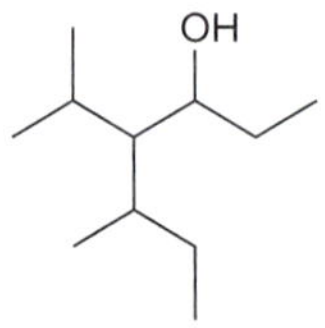 OH

e. OH OH

f.

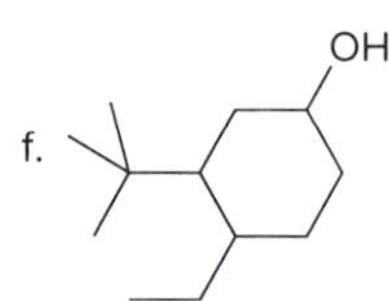

More Practice: Try Problems 9.37a; 9.39; 9.41a–d, g.

Problem 9.3 Give the structure corresponding to each name.

a. 7,7-dimethyloctan-4-ol

b. 5-methyl-4-propylheptan-3-ol

c. 2-*tert*-butyl-3-methylcyclohexanol

d. *trans*-cyclohexane-1,2-diol

9.3B Naming Ethers

Simple ethers are usually assigned common names. To do so, **name both alkyl groups** bonded to the oxygen, arrange these names alphabetically, and add the word ***ether.*** For symmetrical ethers, name the alkyl group and add the prefix ***di-*.**

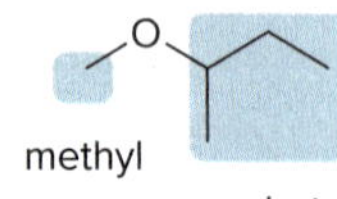

methyl

sec-butyl

***sec*-butyl methyl ether**

Alphabetize the **b** of **b**utyl before the **m** of **m**ethyl.

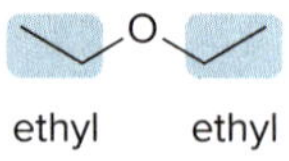

ethyl ethyl

diethyl ether

More complex ethers are named using the IUPAC system. One alkyl group is named as a hydrocarbon chain, and the other is named as part of a substituent bonded to that chain.

- Name the simpler alkyl group + O atom as an **alkoxy** substituent by changing the *-yl* ending of the alkyl group to *-oxy.*
- Name the remaining alkyl group as an alkane, with the alkoxy group as a substituent bonded to this chain.

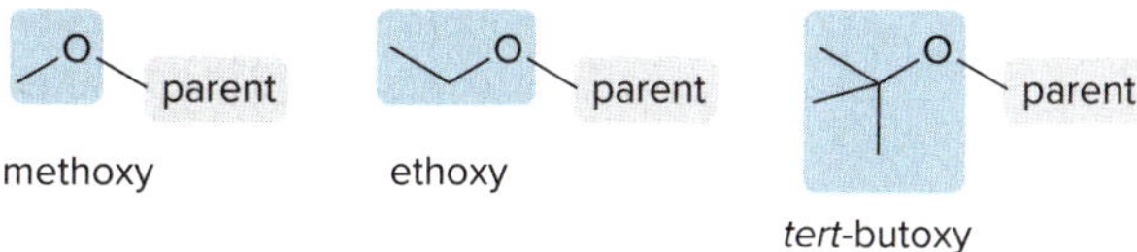

Sample Problem 9.2 Naming an Ether Using IUPAC Nomenclature

Give the IUPAC name for the following ether.

Solution

[1] Name the longer chain as an alkane and the shorter chain as an alkoxy group.

ethoxy group

8 C's ----→ octane

[2] Apply the other nomenclature rules to complete the name.

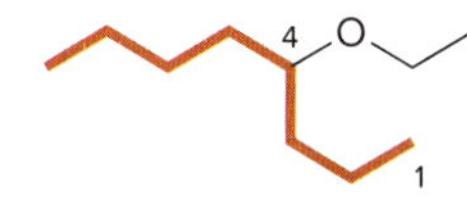

Answer: 4-ethoxyoctane

Problem 9.4 Name each of the following ethers.

a. b. c.

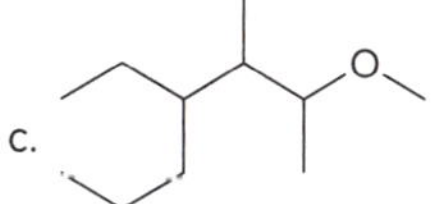

d.

More Practice: Try Problems 9.40a; 9.41f, h.

tetrahydrofuran
THF

Cyclic ethers have an O atom in a ring. A common cyclic ether is **tetrahydrofuran (THF),** a polar aprotic solvent used in nucleophilic substitution (Section 7.8C) and many other organic reactions.

9.3C Naming Epoxides

Any cyclic compound containing a heteroatom is called a *heterocycle.*

Epoxides are named in three different ways—**epoxyalkanes, oxiranes,** or **alkene oxides.**

To name an epoxide as an **epoxyalkane,** first name the alkane chain or ring to which the oxygen is attached, and use the prefix ***epoxy*** to name the epoxide as a substituent. Use two numbers to designate the location of the atoms to which the O's are bonded.

1,2-epoxycyclohexane

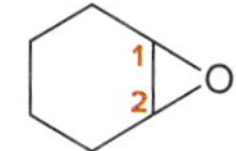

1,2-epoxy-2-methylpropane

***cis*-2,3-epoxypentane**

Epoxides bonded to a chain of carbon atoms can also be named as derivatives of **oxirane,** the simplest epoxide having two carbons and one oxygen atom in a ring. The oxirane ring is numbered to **put the O atom at position "1" and the first substituent at position "2."** No number is used for a substituent in a monosubstituted oxirane.

oxirane   2,2-dimethyloxirane

Epoxides are also named as **alkene oxides,** because they are often prepared by adding an O atom to an alkene (Chapter 12). To name an epoxide this way, mentally replace the epoxide oxygen by a double bond, name the alkene (Section 10.3), and then add the word ***oxide.*** For example, the common name for oxirane is ethylene oxide, because it is an epoxide derived from the alkene ethylene. We will use this method of naming epoxides after the details of alkene nomenclature are presented in Chapter 10.

ethylene → **ethylene oxide** oxirane

Problem 9.5 Name each epoxide.

9.4 Properties of Alcohols, Ethers, and Epoxides

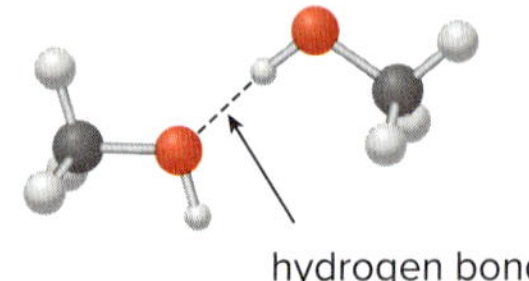

hydrogen bond

Alcohols, ethers, and epoxides exhibit dipole–dipole interactions because they have a bent structure with two polar bonds. **Alcohols are also capable of intermolecular hydrogen bonding** because they possess a hydrogen atom on an oxygen, making alcohols much more polar than ethers and epoxides.

Steric factors affect the extent of hydrogen bonding. Although all alcohols can hydrogen bond, **increasing the number of R groups around the carbon atom bearing the OH group decreases the extent of hydrogen bonding**. Thus, 3° alcohols are least able to hydrogen bond, whereas 1° alcohols are most able to.

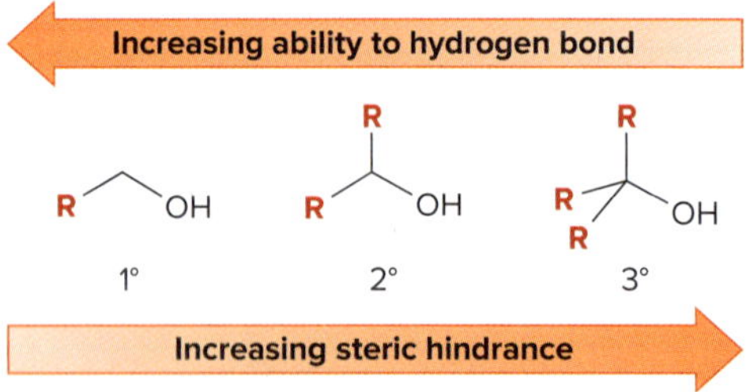

Alcohols, ethers, and epoxides with ≤ 5 C's are water soluble because they each have an oxygen atom that can hydrogen bond to water. The boiling points and melting points of these functional groups increase with size and the extent of hydrogen bonding. As a result, ethers have lower boiling points than alcohols of comparable molecular weight, and the order of boiling points of alcohols follows the trend 3° ROH < 2° ROH < 1° ROH.

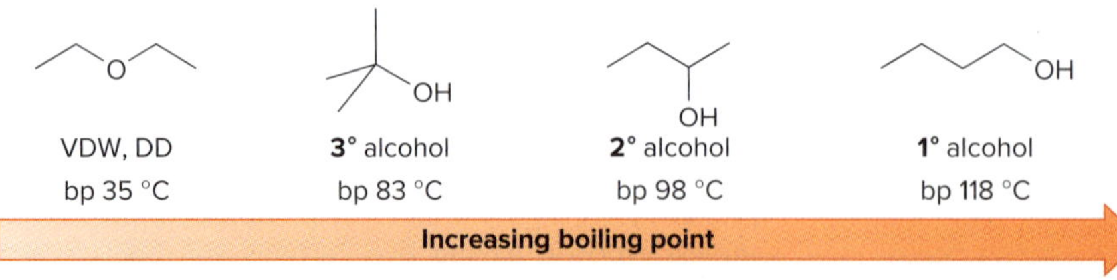

Problem 9.6 Rank the following compounds in order of increasing boiling point.

a.

b.

Students who have already been exposed to spectroscopy or who would like to learn about the spectroscopic properties of alcohols and ethers are referred to the following sections:

- Mass spectrometry: Section A.4B and Figure A.6
- Infrared spectroscopy: Section B.4B, Sample Problem B.3
- Nuclear magnetic resonance spectroscopy: Section C.9A, Figures C.12 and C.14a, Sample Problems C.3 and C.6

9.5 Interesting Alcohols, Ethers, and Epoxides

A large number of alcohols, ethers, and epoxides have interesting and useful properties.

9.5A Ethanol

The most widely known simple alcohol, **ethanol** (CH_3CH_2OH), is formed by the fermentation of the carbohydrates in grains, grapes, and potatoes to yield a variety of alcoholic beverages. It is perhaps the first organic compound synthesized by humans, because alcohol production has been known for at least 4000 years. Ethanol depresses the central nervous system, increases the production of stomach acid, and dilates blood vessels, producing a flushed appearance. Ethanol is also a common laboratory solvent, which is sometimes made unfit to ingest by adding small amounts of benzene or methanol (both of which are toxic).

Ethanol is a common gasoline additive, widely touted as an environmentally friendly fuel source. Two common gasoline–ethanol fuels are gasohol, which contains 10% ethanol, and E-85, which contains 85% ethanol. Ethanol is now routinely prepared from the carbohydrates in corn (Figure 9.1). Starch, a complex carbohydrate polymer, can be hydrolyzed to the simple sugar glucose, which forms ethanol by the process of fermentation. Combining ethanol with gasoline forms a usable fuel, which combusts to form CO_2, H_2O, and a great deal of energy.

Figure 9.1 Ethanol from corn, a renewable fuel source

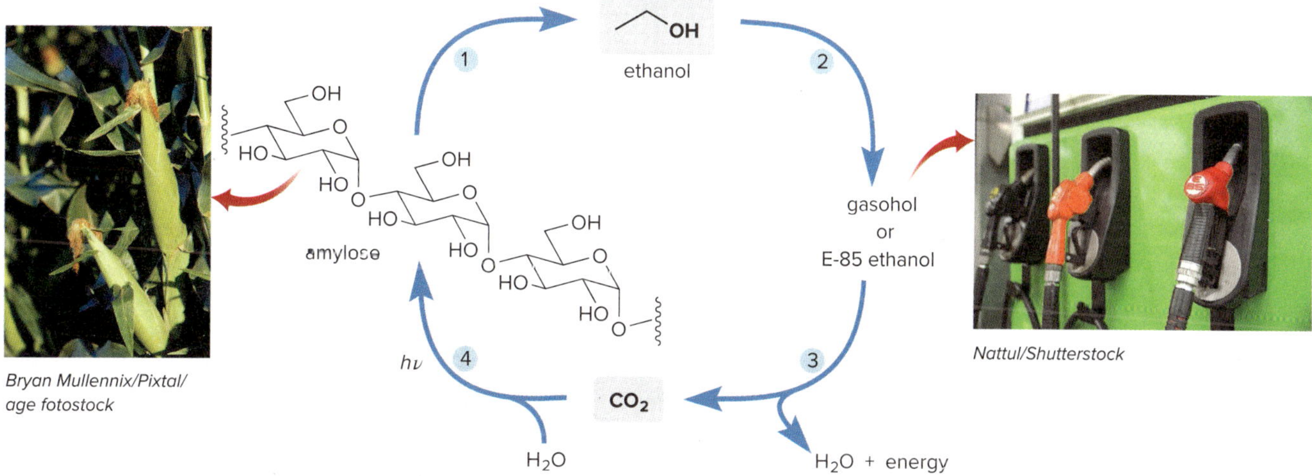

1 Hydrolysis of amylose (one form of starch) and **fermentation** of the resulting simple sugars yield ethanol.
2 Blending ethanol with hydrocarbons from petroleum refining forms usable fuels.
3 **Combustion** of this ethanol–hydrocarbon fuel forms CO_2 and releases a great deal of energy.
4 **Photosynthesis** converts atmospheric CO_2 back to plant carbohydrates, and the cycle continues.

Because green plants use sunlight to convert CO_2 and H_2O to carbohydrates during photosynthesis, next year's corn crop removes CO_2 from the atmosphere to make new molecules of starch as the corn grows. While in this way ethanol is a *renewable* fuel source, the need for large-scale farm equipment and the heavy reliance on fertilizers and herbicides make ethanol expensive to produce. Moreover, many criticize the use of valuable farmland for an energy-producing crop rather than for food production. As a result, discussion continues on ethanol as an alternative to fossil fuels.

9.5B Interesting Ethers and Epoxides

The discovery that **diethyl ether** ($CH_3CH_2OCH_2CH_3$) is a general anesthetic revolutionized surgery in the nineteenth century. Diethyl ether is an imperfect anesthetic, but given the alternatives in the nineteenth century, it was considered a miracle drug that allowed patients to tolerate the excruciating pain of surgery. It is safe, easy to administer, and causes little patient mortality, but it is highly flammable and causes nausea in many patients. For these reasons, it has largely been replaced by sevoflurane and other halogenated ethers, which are nonflammable and cause little patient discomfort.

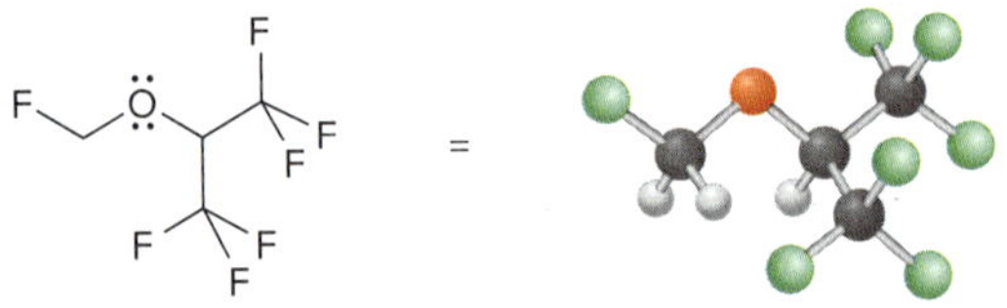

sevoflurane

Several drugs with complex structures contain ethers or epoxides along with other functional groups. These include eplerenone, cabazitaxel, and eribulin mesylate. Eplerenone is an epoxide-containing drug prescribed to reduce the cardiovascular risk in patients who have already had a heart attack. Cabazitaxel, which is related to the anti-cancer drug Taxol described in the Chapter 5 opener, is used to treat a form of prostate cancer. Eribulin mesylate, a synthetic analogue of a more complex compound isolated from a marine sponge, was approved in the United States in 2010 for the treatment of metastatic breast cancer. Ethers and epoxides are shown in red.

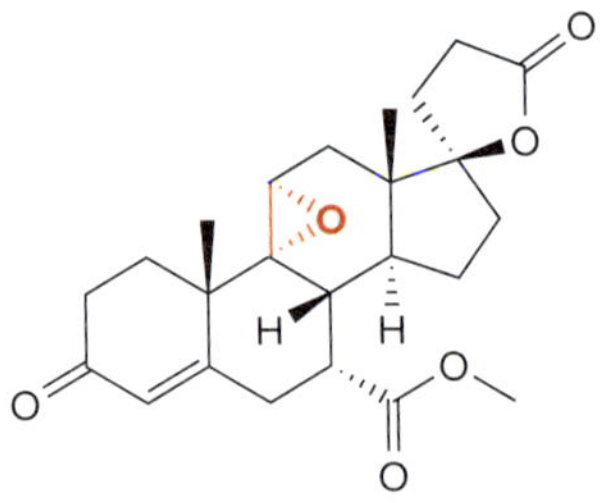

eplerenone
Trade name: Inspra

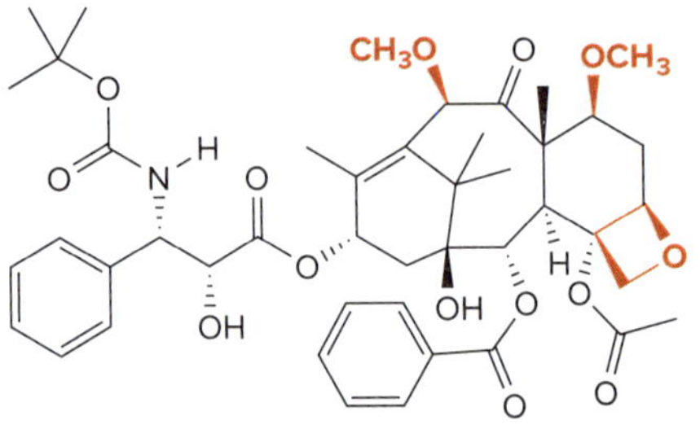

cabazitaxel
Trade name: Jevtana

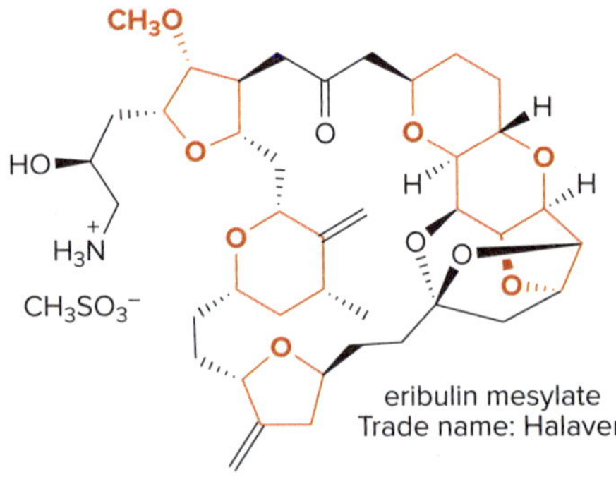

eribulin mesylate
Trade name: Halaven

9.6 Preparation of Alcohols, Ethers, and Epoxides

Alcohols and ethers are both common products of nucleophilic substitution. They are synthesized from alkyl halides by S_N2 reactions using strong nucleophiles. As in all S_N2 reactions, highest yields of products are obtained with unhindered methyl and 1° alkyl halides.

alcohol

ether

The preparation of ethers by this method is called the **Williamson ether synthesis,** and, although it was first reported in the 1800s, it is still the most general method to prepare an ether. Unsymmetrical ethers can be synthesized in two different ways, but often one path is preferred.

For example, ethyl isopropyl ether can be prepared from $CH_3CH_2O^-$ and 2-bromopropane (Path [a]), or from $(CH_3)_2CHO^-$ and bromoethane (Path [b]). Because the mechanism is S_N2, **the preferred path uses the less sterically hindered halide,** CH_3CH_2Br—Path [b].

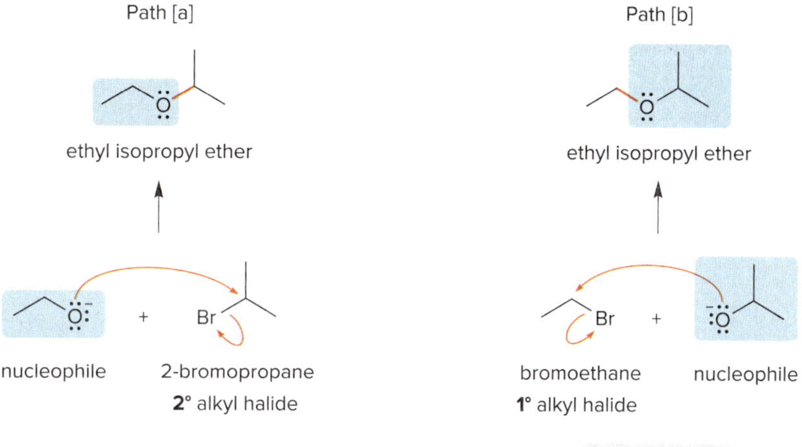

Problem 9.7 Draw the organic product of each reaction.

a. Br + ^-OH ⟶

b. Cl + $^-OCH_3$ ⟶

c. I + $^-OCH(CH_3)_2$ ⟶

d. Br + $^-OCH_2CH_3$ ⟶

Problem 9.8 A key step in the synthesis of the antidepressant paroxetine (trade name Paxil) involves a Williamson ether synthesis of the acyclic ether in **X.** Draw two different routes to this ether and state which is preferred.

paroxetine

X

A **hydroxide** nucleophile is needed to synthesize an alcohol, and salts such as **NaOH** and **KOH** are inexpensive and commercially available. An **alkoxide** salt is needed to make an ether. Simple alkoxides such as sodium methoxide ($NaOCH_3$) can be purchased, but others are prepared from alcohols by a Brønsted–Lowry acid–base reaction. For example, **sodium ethoxide** (**$NaOCH_2CH_3$**) is prepared by treating ethanol with NaH.

NaH is an especially good base for forming an alkoxide, because the by-product of the reaction, H_2, is a gas that just bubbles out of the reaction mixture.

CH_3CH_2OH + $Na^+H:^-$ ⟶ $CH_3CH_2O^- Na^+$ + H_2

alkoxide nucleophile

sodium ethoxide

When an organic compound contains both a hydroxy group and a halogen atom on adjacent carbon atoms, an *intramolecular* version of this reaction forms an epoxide. The starting material for this two-step sequence, a **halohydrin,** is prepared from an alkene, as we will learn in Chapter 10.

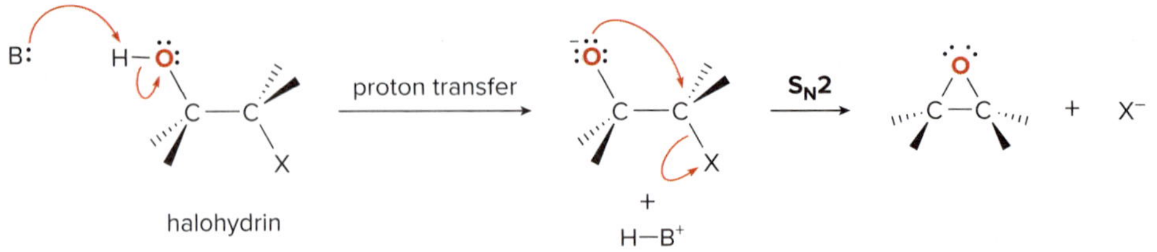

halohydrin — proton transfer → + H–B$^+$ — S_N2 → epoxide + X^-

Sample Problem 9.3 Synthesizing an Ether by a Two-Step Reaction Sequence

Draw the product of the following two-step reaction sequence.

Cyclohexanol (OH) → [1] NaH; [2] ethyl bromide (Br)

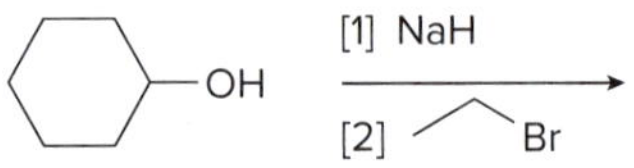

Solution

[1] The base removes a proton from the OH group, forming an alkoxide.

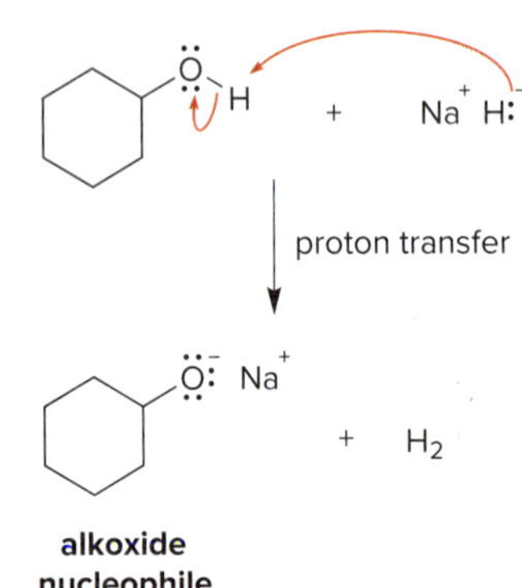

alkoxide nucleophile

[2] The alkoxide acts as a nucleophile in an S_N2 reaction, forming an ether.

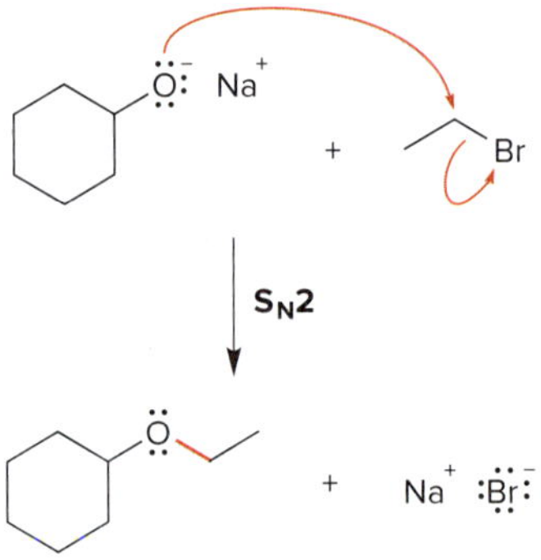

Na$^+$ + Br — S_N2 → ether + Na$^+$:Br:$^-$

- This two-step sequence converts an alcohol to an ether. The new C–O bond of the ether is shown in red.

Problem 9.9 Draw the products of each reaction.

a.  OH + NaH →

b. 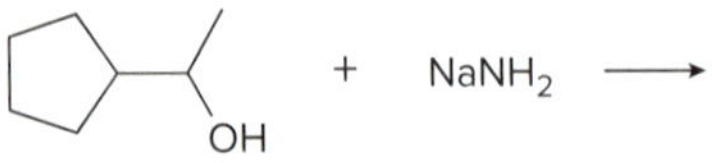 OH + $NaNH_2$ →

c. 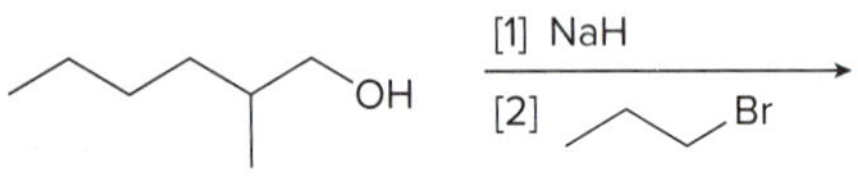 OH → [1] NaH; [2] Br

d. 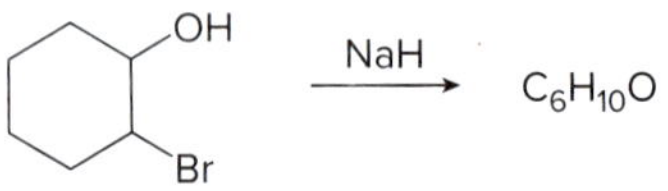 OH, Br → NaH → $C_6H_{10}O$

e. 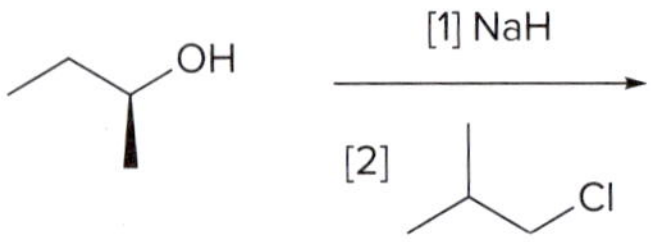 OH → [1] NaH; [2] Cl

f. 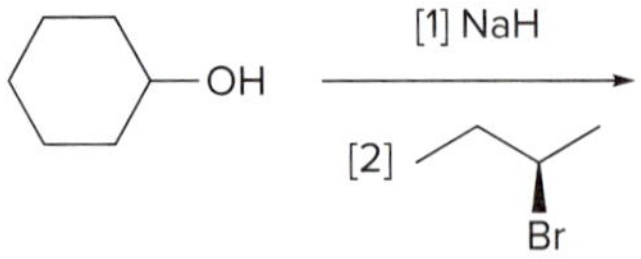 OH → [1] NaH; [2] Br

More Practice: Try Problems 9.42h, 9.45a, 9.46, 9.63i.

9.7 General Features—Reactions of Alcohols, Ethers, and Epoxides

We begin our discussion of the chemical reactions of alcohols, ethers, and epoxides with a look at the general reactive features of each functional group.

9.7A Alcohols

Unlike many families of molecules, the reactions of alcohols do *not* fit neatly into a single reaction class. In Chapter 9, we discuss only the substitution and β elimination reactions of alcohols. Alcohols are also key starting materials in oxidation reactions (Chapter 12), and their polar O–H bond makes them more acidic than many other organic compounds, a feature we will explore in Chapter 19.

Alcohols are similar to alkyl halides in that both contain an electronegative element bonded to an sp^3 hybridized carbon atom. **Alkyl halides contain a good leaving group (X^-), however, whereas alcohols do *not*.** Nucleophilic substitution with ROH as starting material would displace $^-$**OH, a strong base and therefore a poor leaving group.**

R–X + :Nu⁻ ⟶ R–Nu + X⁻ **good** leaving group

R–OH + :Nu⁻ ⟶✕ R–Nu + ⁻OH **poor** leaving group

For an alcohol to undergo a nucleophilic substitution or elimination reaction, the **OH group must be converted into a *better* leaving group.** This can be done by reaction with acid. Treatment of an alcohol with a strong acid like HCl or H_2SO_4 protonates the O atom via an acid–base reaction. This transforms the $^-$OH leaving group into **H_2O, a weak base and therefore a *good* leaving group.**

$$R{-}\ddot{O}H + H{-}Cl \rightleftharpoons R{-}\overset{+}{O}H_2 + Cl^-$$

strong acid (H–Cl); weak base, **good leaving group** ($R{-}\overset{+}{O}H_2$)

If the OH group of an alcohol is made into a good leaving group, alcohols *can* undergo β elimination and nucleophilic substitution, as described in Sections 9.8–9.12.

Because the pK_a of $(ROH_2)^+$ is ~−2, protonation of an alcohol occurs only with very strong acids.

Cyclohexanol (OH) → β elimination, $-H_2O$ → cyclohexene: **Alkenes are formed by β elimination.** (Sections 9.8–9.10)

Cyclohexanol (OH) → nucleophilic substitution → cyclohexyl–X: **Alkyl halides are formed by nucleophilic substitution.** (Sections 9.11–9.12)

9.7B Ethers and Epoxides

Like alcohols, **ethers do *not* contain a good leaving group,** which means that nucleophilic substitution and β elimination do not occur directly. Ethers undergo fewer useful reactions than alcohols.

$R{-}\ddot{O}R$ **poor leaving group**

Epoxides don't have a good leaving group either, but they have one characteristic that neither alcohols nor ethers have: **the "leaving group" is contained in a strained three-membered**

ring. Nucleophilic attack opens the three-membered ring and relieves angle strain, making nucleophilic attack a favorable process that occurs even with the poor leaving group. Specific examples are presented in Section 9.16.

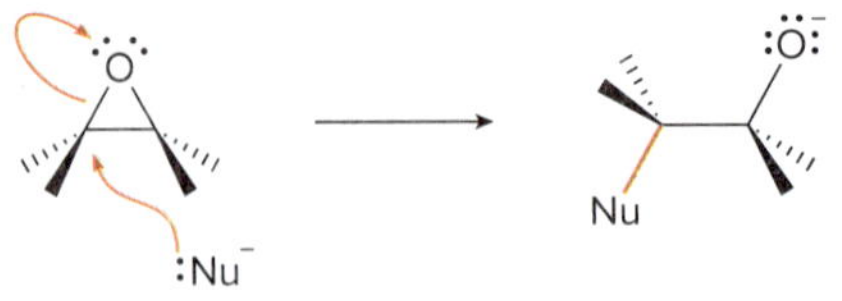

:Nu⁻ Nu :Ö:⁻

9.8 Dehydration of Alcohols to Alkenes

The dehydrohalogenation of alkyl halides, discussed in Chapter 8, is one way to introduce a π bond into a molecule. Another way is to eliminate water from an alcohol in a **dehydration** reaction.

- **Dehydration is a β elimination reaction in which the elements of OH and H are removed from the α and β carbon atoms, respectively.**

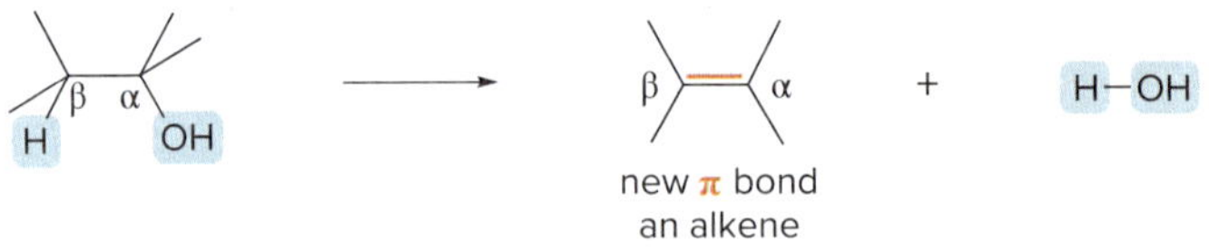

H β α OH → β α + H–OH

new π bond
an alkene

Dehydration is typically carried out using $\mathbf{H_2SO_4}$ and other strong acids, or phosphorus oxychloride (**$\mathbf{POCl_3}$**) in the presence of an amine base. We consider dehydration in acid first, followed by dehydration with $POCl_3$ in Section 9.10.

9.8A General Features of Dehydration in Acid

Alcohols undergo dehydration in the presence of strong acid to afford alkenes, as illustrated in Equations [1] and [2]. Typical acids used for this conversion are H_2SO_4 or *p*-toluenesulfonic acid (abbreviated as TsOH).

Recall from Section 2.6 that *p*-toluenesulfonic acid is a strong organic acid ($pK_a = -1.3$).

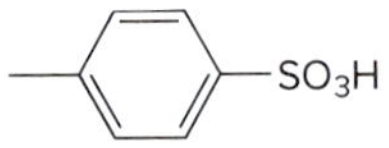

SO_3H

p-toluenesulfonic acid
TsOH

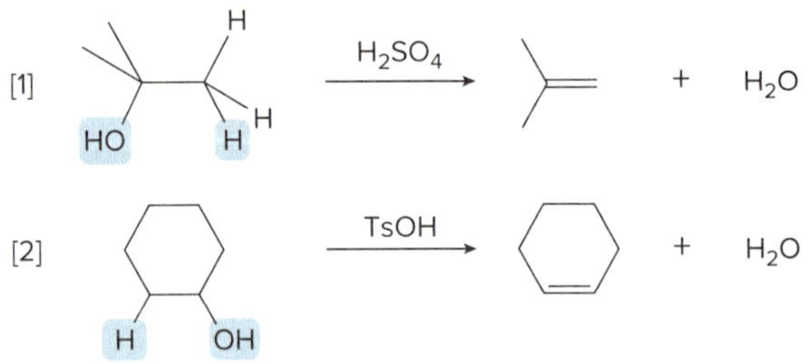

[1] HO H H H $\xrightarrow{H_2SO_4}$ + H_2O

[2] H OH $\xrightarrow{TsOH}$ + H_2O

More substituted alcohols dehydrate more readily, giving rise to the following order of reactivity:

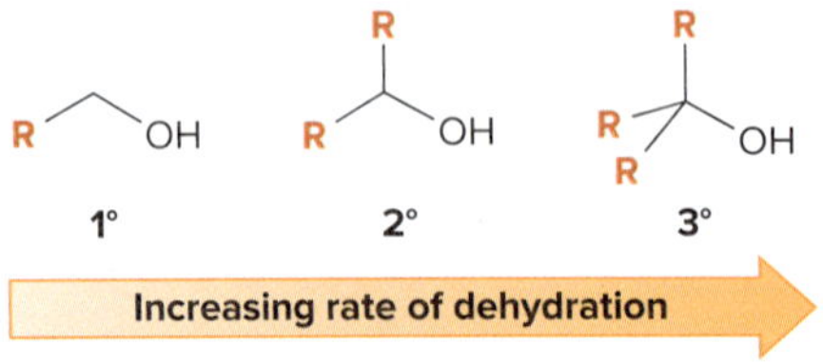

R–CH₂OH (1°) R₂CHOH (2°) R₃COH (3°)

Increasing rate of dehydration

When an alcohol has two or three different β carbons, dehydration is regioselective and follows the Zaitsev rule. **The more substituted alkene is the major product when a mixture of constitutional isomers is possible.** For example, elimination of H and OH from

2-methylbutan-2-ol yields two constitutional isomers: the trisubstituted alkene **A** as *major* product and the disubstituted alkene **B** as *minor* product.

2-methylbutan-2-ol $\xrightarrow{H_2SO_4}$ **A** major product, trisubstituted alkene + **B** minor product, disubstituted alkene

Problem 9.10 Draw the products formed when each alcohol undergoes dehydration with TsOH, and label the major product when a mixture results.

a. b. c.

Problem 9.11 Rank the alcohols in order of increasing reactivity when dehydrated with H_2SO_4.

9.8B The E1 Mechanism for the Dehydration of 2° and 3° Alcohols

The mechanism of dehydration depends on the structure of the alcohol: **2° and 3° alcohols react by an E1 mechanism, whereas 1° alcohols react by an E2 mechanism.** Regardless of the type of alcohol, however, strong acid is *always* needed to protonate the O atom to form a good leaving group.

The E1 dehydration of 2° and 3° alcohols is illustrated with $(CH_3)_3COH$ (a 3° alcohol) as starting material to form $(CH_3)_2C{=}CH_2$ as product (Mechanism 9.1). The mechanism consists of **three steps.**

Mechanism 9.1 Dehydration of 2° and 3° ROH—An E1 Mechanism

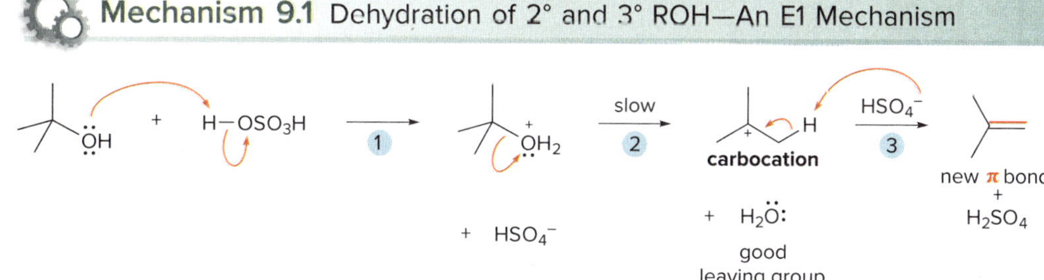

1. **Protonation** of the oxygen atom converts the poor leaving group ($^-$OH) into a **good leaving group** (H_2O).
2. Heterolysis of the C–O bond forms a **carbocation** in the rate-determining step.
3. A base (such as HSO_4^- or H_2O) removes a proton from a carbon adjacent to the carbocation to form the new **π bond**.

Thus, **dehydration of 2° and 3° alcohols occurs via an E1 mechanism with an added first step.** Step [1] protonates the OH group to make a good leaving group. Steps [2] and [3] are the two steps of an E1 mechanism: loss of a leaving group (H_2O in this case) to form a carbocation, followed by removal of a β proton to form a π bond. The acid used to protonate the alcohol in Step [1] is regenerated upon removal of the proton in Step [3], so dehydration is **acid-catalyzed.**

The E1 dehydration of 2° and 3° alcohols with acid gives clean elimination products without by-products formed from an S_N1 reaction. This makes the E1 dehydration of alcohols much more synthetically useful than the E1 dehydrohalogenation of alkyl halides (Section 8.7). Clean elimination takes place because the reaction mixture contains no good nucleophile to react with the intermediate carbocation, so **no competing S_N1 reaction occurs.**

The dehydration of alcohols by an E1 mechanism occurs in biological systems as well. E1 dehydration of diol **A** forms **enol B** with an alkene bonded to an OH group. **B** is an intermediate in the biosynthesis of the amino acid histidine.

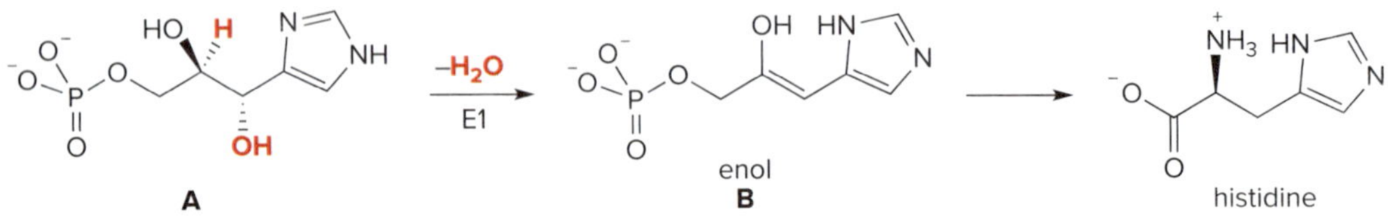

9.8C The E2 Mechanism for the Dehydration of 1° Alcohols

Because 1° carbocations are highly unstable, the dehydration of 1° alcohols cannot occur by an E1 mechanism involving a carbocation intermediate. With 1° alcohols, therefore, **dehydration follows an E2 mechanism.** The two-step process for the conversion of $CH_3CH_2CH_2OH$ (a 1° alcohol) to $CH_3CH{=}CH_2$ with H_2SO_4 as acid catalyst is shown in Mechanism 9.2.

Mechanism 9.2 Dehydration of a 1° ROH—An E2 Mechanism

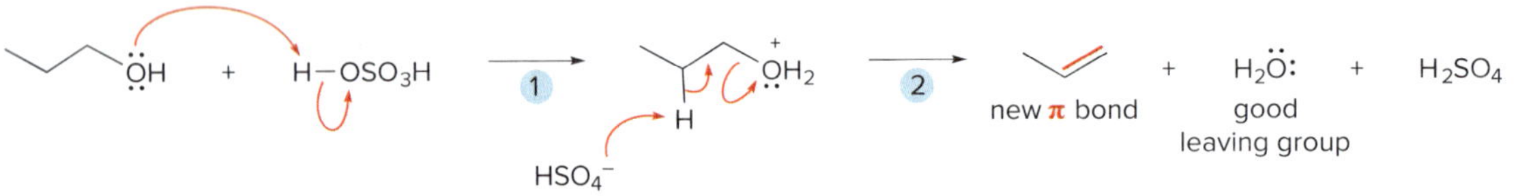

1 **Protonation** of the oxygen atom converts the poor leaving group (^-OH) into a **good leaving group** (H_2O).

2 Two bonds are broken and two bonds are formed. The base (HSO_4^- or H_2O) removes a proton from the β carbon; the electron pair in the β C–H bond forms the new π **bond** and the leaving group (H_2O) departs.

The dehydration of a 1° alcohol begins with the protonation of the OH group to form a good leaving group, just as in the dehydration of a 2° or 3° alcohol. With 1° alcohols, however, loss of the leaving group and removal of a β proton occur at the *same* time, so that **no highly unstable 1° carbocation is generated.**

9.8D Le Châtelier's Principle

Although **entropy favors product formation** in dehydration (one molecule of reactant forms two molecules of products), **enthalpy does *not*,** because the two σ bonds broken in the reactant are *stronger* than the σ and π bonds formed in the products. Product formation can be favored by utilizing **Le Châtelier's principle, which states that a system at equilibrium will react to counteract any disturbance to the equilibrium.** Thus, removing a product from a reaction mixture as it is formed drives the equilibrium to the *right*, forming more product.

In dehydration, the alkene product generally has a lower boiling point than the alcohol reactant. Thus, the alkene can be distilled from the reaction mixture as it is formed, leaving the alcohol and acid to react further, forming more product.

Problem 9.12 Which alcohols will undergo dehydration with the formation of intermediate carbocations?

a. OH b. OH c. OH d. OH

9.9 Carbocation Rearrangements

Sometimes "unexpected" products are formed in dehydration; that is, the carbon skeletons of the starting material and product might be different, or the double bond might be in an unexpected location. For example, the dehydration of 3,3-dimethylbutan-2-ol yields two alkenes, whose carbon skeletons do not match the carbon framework of the starting material.

OH

3,3-dimethylbutan-2-ol $\xrightarrow{H_2SO_4}$ + + H_2O

This phenomenon sometimes occurs when carbocations are reactive intermediates. **A less stable carbocation can rearrange to a more stable carbocation by shift of a hydrogen atom or an alkyl group.** These **1,2-shifts** involve migration of an alkyl group or hydrogen atom from one carbon to an adjacent carbon atom. The migrating group moves with the two electrons that bonded it to the carbon skeleton.

Because the migrating group in a 1,2-shift moves with two bonding electrons, the carbon it leaves behind now has only three bonds (six electrons), giving it a net positive (+) charge.

R (or H) —1,2-shift→ (carbocation rearrangement) R (or H)

- **Movement of a hydrogen atom is called a 1,2-hydride shift.**
- **Movement of an alkyl group is called a 1,2-alkyl shift.**

For example, 2° carbocation **A** rearranges to the more stable 3° carbocation by a 1,2 hydride shift, whereas carbocation **B** does not rearrange because it is 3° to begin with.

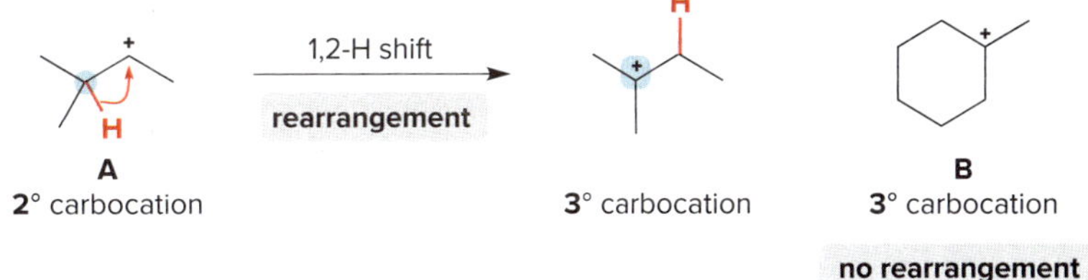

Problem 9.13 Show how a 1,2-shift forms a more stable carbocation from each intermediate.

a. b. c. d. (two different ways)

The dehydration of 3,3-dimethylbutan-2-ol illustrates the rearrangement of a 2° to a 3° carbocation by a **1,2-methyl shift,** as shown in Mechanism 9.3. The carbocation rearrangement occurs in Step [3] of the four-step mechanism.

Mechanism 9.3 A 1,2-Methyl Shift—Carbocation Rearrangement During Dehydration

Part [1] Formation of a 2° carbocation and rearrangement

1 **Protonation** of the oxygen atom converts the poor leaving group (^-OH) into a **good leaving group** (H_2O).

2 Heterolysis of the C–O bond forms a **2° carbocation.**

3 1,2-Shift of a CH_3 group converts a 2° carbocation to a **more stable 3° carbocation.**

Part [2] Loss of a proton to form the π bond

4 **Loss of a proton** from a β carbon (β_1 or β_2) forms two different alkenes.

Steps [1], [2], and [4] in the mechanism for the dehydration of 3,3-dimethylbutan-2-ol are exactly the same steps previously seen in dehydration: protonation, loss of H_2O, and loss of a proton. Only Step [3], rearrangement of the less stable 2° carbocation to the more stable 3° carbocation, is new.

- **1,2-Shifts convert a less stable carbocation to a more stable carbocation.**

Sample Problem 9.4 illustrates a dehydration reaction that occurs with a **1,2-hydride** shift.

Sample Problem 9.4 Drawing a Dehydration Reaction with a Rearrangement

Show how the dehydration of alcohol **X** forms alkene **Y** using a 1,2-hydride shift.

Solution

Steps [1] and [2] Protonation of **X** and loss of H_2O form a 2° carbocation.

Steps [3] and [4] Rearrangement of the 2° carbocation by a **1,2-hydride shift** forms a more stable 3° carbocation. Loss of a proton from a β carbon forms alkene **Y.**

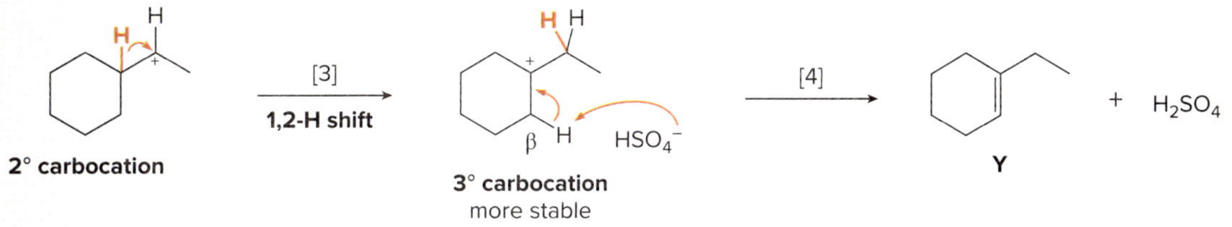

[3] **1,2-H shift** [4] + H_2SO_4

2° carbocation **3° carbocation** more stable **Y**

Problem 9.14 What other alkene is also formed along with **Y** in Sample Problem 9.4? What alkenes would form from **X** if no carbocation rearrangement occurred?

More Practice: Try Problems 9.43d, f; 9.47a; 9.49; 9.50.

Problem 9.15 Draw all constitutional isomers formed when each alcohol undergoes dehydration in the presence of acid. Watch out for rearrangements!

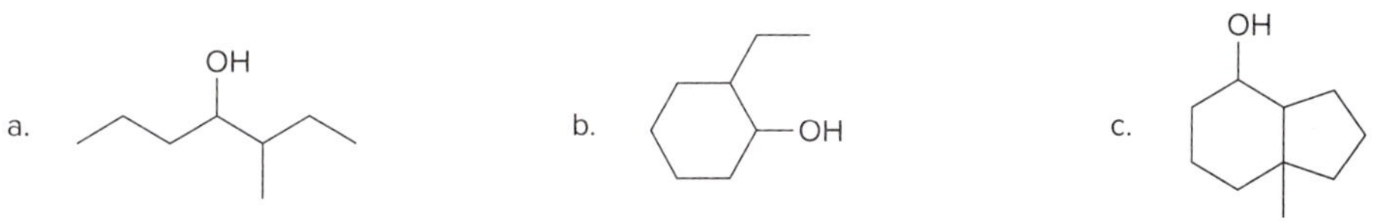

a. b. c.

Rearrangements are not unique to dehydration reactions. **Rearrangements can occur whenever a carbocation is formed as reactive intermediate,** meaning any S_N1 or E1 reaction. In fact, the formation of rearranged products often indicates the presence of a carbocation intermediate.

9.10 Dehydration Using $POCl_3$ and Pyridine

Because some organic compounds decompose in the presence of strong acid, other methods that avoid strong acid have been developed to convert alcohols to alkenes. A common method uses **phosphorus oxychloride ($POCl_3$)** and pyridine (an amine base) in place of H_2SO_4 or TsOH. For example, the treatment of cyclohexanol with $POCl_3$ and pyridine forms cyclohexene in good yield.

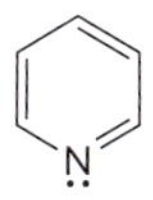

pyridine

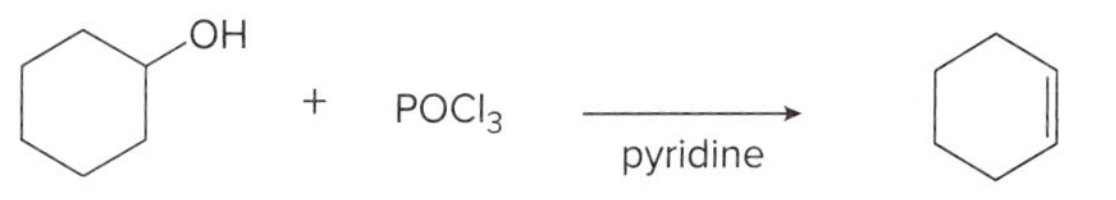

cyclohexanol + $POCl_3$ —pyridine→ cyclohexene

$POCl_3$ serves much the same role as strong acid does in acid-catalyzed dehydration. **It converts a poor leaving group (^-OH) into a good leaving group.** Dehydration then proceeds by an **E2 mechanism,** as shown in Mechanism 9.4. Pyridine is the base that removes a β proton during elimination.

No rearrangements occur during dehydration with $POCl_3$, suggesting that carbocations are *not* formed as intermediates in this reaction. Steps [1] and [2] of the mechanism convert the OH group into a good leaving group. In Step [3], the C–H and C–O bonds are broken and the π bond is formed.

Mechanism 9.4 Dehydration Using $POCl_3$ + Pyridine—An E2 Mechanism

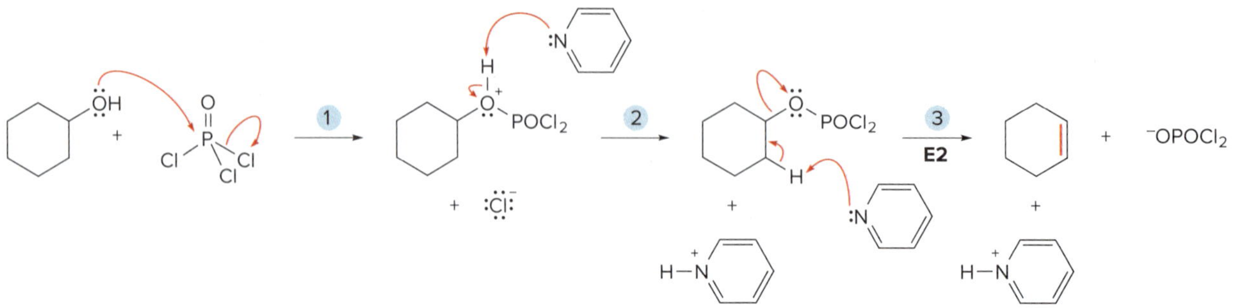

1 – 2 Reaction of the OH with $POCl_3$ followed by loss of a proton converts a poor leaving group (^-OH) into a **good leaving group** ($^-OPOCl_2$).

3 Two bonds are broken and two bonds are formed. The base (pyridine) removes a proton; the electron pair in the β C–H bond forms the **π bond,** and the leaving group ($^-OPOCl_2$) departs.

We have now learned about two different reagents for alcohol dehydration—strong acid (H_2SO_4 or TsOH) and $POCl_3$ + pyridine. The best dehydration method for a given alcohol is often hard to know ahead of time, and this is why organic chemists develop more than one method for a given type of transformation. Two examples of dehydration reactions used in the synthesis of natural products are given in Figure 9.2.

Problem 9.16 If dehydration with $POCl_3$ and pyridine follows the Zaitsev rule and an E2 mechanism, draw the major product formed from each alcohol under these conditions.

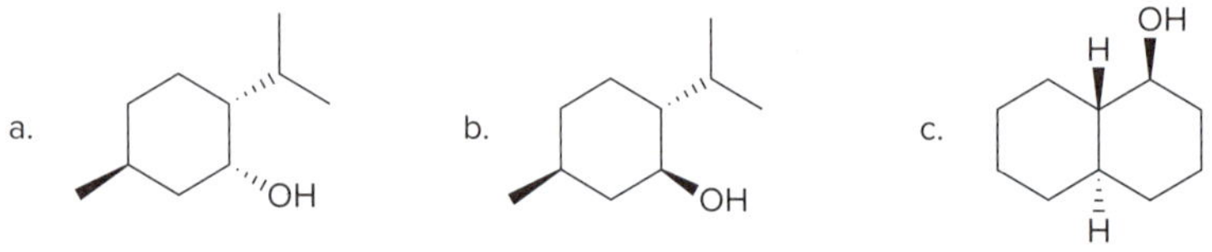

a. b. c.

Figure 9.2 Dehydration reactions in the synthesis of two natural products

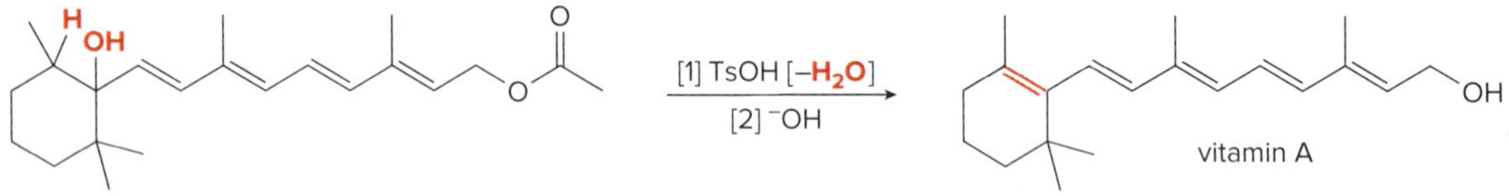

[1] TsOH [$-H_2O$]
[2] ^-OH
vitamin A

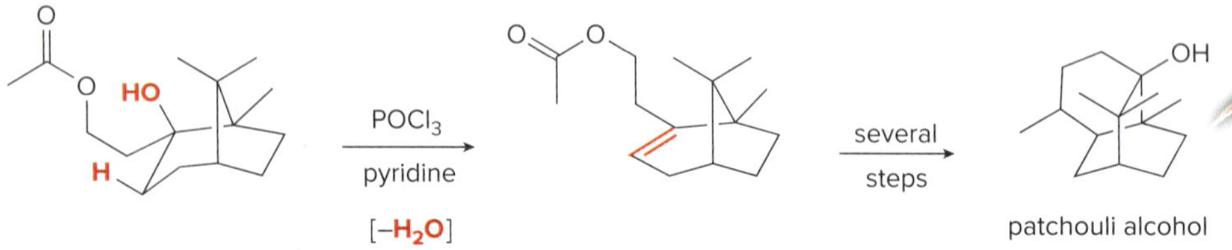

$POCl_3$
pyridine
[$-H_2O$]
several steps
patchouli alcohol

patchouli plant
Stephen Orsillo/Shutterstock

- New double bonds formed by dehydration are shown in red.
- **Patchouli alcohol,** obtained from the patchouli plant native to Malaysia, has been used in perfumery because of its exotic fragrance. In the 1800s, shawls imported from India were often packed with patchouli leaves to ward off insects, thus permeating the clothing with the distinctive odor.

9.11 Conversion of Alcohols to Alkyl Halides with HX

Alcohols undergo nucleophilic substitution reactions only if the OH group is converted to a better leaving group before nucleophilic attack. Thus, substitution does *not* occur when an alcohol is treated with X^- because **$^-$OH is a poor leaving group** (Reaction [1]), but substitution *does* occur on treatment of an alcohol with HX because H_2O is now the leaving group (Reaction [2]).

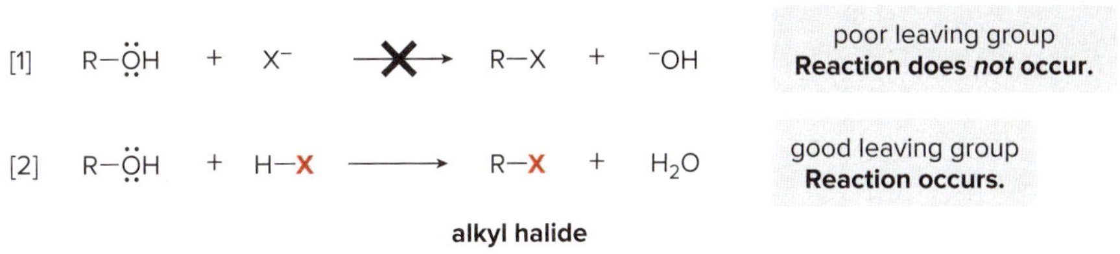

[1] R–ÖH + X^- ⟶✕ R–X + $^-$OH — poor leaving group **Reaction does *not* occur.**

[2] R–ÖH + H–**X** ⟶ R–**X** + H_2O — good leaving group **Reaction occurs.**

alkyl halide

- **The reaction of alcohols with HX (X = Cl, Br, I) is a general method to prepare 1°, 2°, and 3° alkyl halides.**

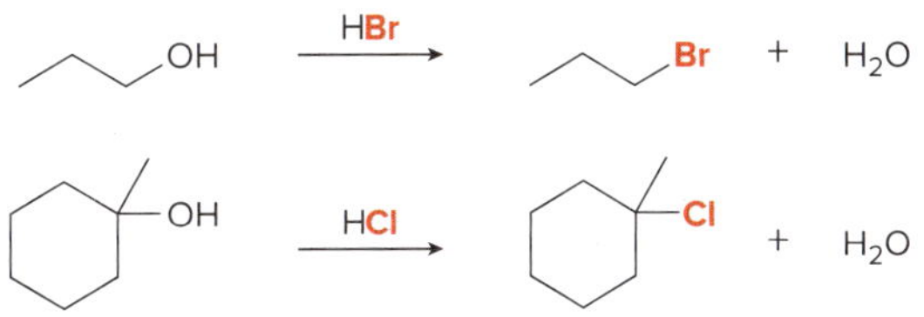

CH₃CH₂CH₂OH —HBr→ CH₃CH₂CH₂**Br** + H_2O

1-methylcyclohexanol —HCl→ 1-chloro-1-methylcyclohexane + H_2O

More substituted alcohols usually react more rapidly with HX:

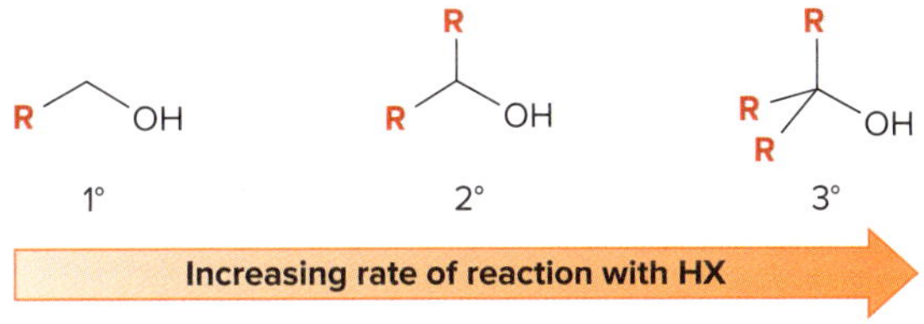

RCH₂OH (1°) — R₂CHOH (2°) — R₃COH (3°)

Increasing rate of reaction with HX

Problem 9.17 Draw the products of each reaction.

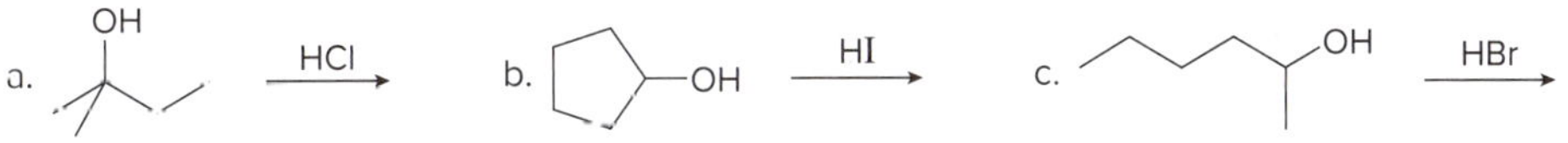

a. (2-methyl-2-butanol type tertiary alcohol) —HCl→ b. cyclopentanol derivative —OH —HI→ c. (secondary alcohol chain) —HBr→

9.11A Two Mechanisms for the Reaction of ROH with HX

How does the reaction of ROH with HX occur? Acid–base reactions are very fast, so the strong acid HX protonates the OH group of the alcohol, forming a **good leaving group** (H_2O) and a **good nucleophile** (the conjugate base, X^-). Both components are needed for nucleophilic substitution. The mechanism of substitution of X^- for H_2O then depends on the structure of the R group.

When there is an oxygen-containing reactant and a strong acid, generally the first step in the mechanism is to **protonate the oxygen atom.**

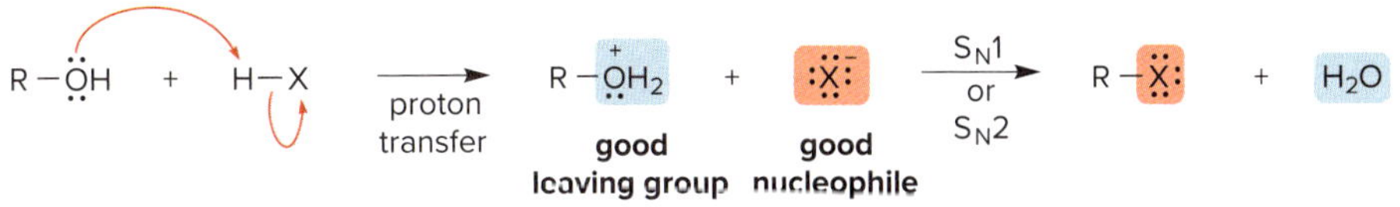

R–ÖH + H–X —proton transfer→ R–$\overset{+}{O}H_2$ (**good leaving group**) + :X:$^-$ (**good nucleophile**) —S_N1 or S_N2→ R–X: + H_2O

- **Methyl and 1° ROH form RX by an S_N2 mechanism.**
- **Secondary (2°) and 3° ROH form RX by an S_N1 mechanism.**

The reaction of CH_3CH_2OH with HBr illustrates the **S_N2** mechanism of a 1° alcohol (Mechanism 9.5). Nucleophilic attack on the protonated alcohol occurs in one step: **the bond to the nucleophile X^- is formed *as* the bond to the leaving group (H_2O) is broken.**

Mechanism 9.5 Reaction of a 1° ROH with HX—An S_N2 Mechanism

1 **Protonation** of the OH group forms a **good leaving group** (H_2O).

2 The bond to the nucleophile forms *as* the leaving group departs.

The reaction of $(CH_3)_3COH$ with HBr illustrates the $\mathbf{S_N1}$ mechanism of a 3° alcohol (Mechanism 9.6). Nucleophilic attack on the protonated alcohol occurs in two steps: **the bond to the leaving group (H_2O) is broken *before* the bond to the nucleophile X^- is formed.**

Mechanism 9.6 Reaction of 2° and 3° ROH with HX—An S_N1 Mechanism

1 **Protonation** of the OH group forms a good leaving group (H_2O).

2 Loss of the leaving group forms a **carbocation.**

3 **Nucleophilic attack** of Br^- forms the substitution product.

Both mechanisms begin with the same first step—protonation of the O atom to form a good leaving group—and both mechanisms give an alkyl halide (RX) as product. The mechanisms differ only in the *timing* of bond breaking and bond making.

The reactivity of hydrogen halides increases with increasing acidity:

H–Cl H–Br H–I

Increasing reactivity toward ROH

Because Cl^- is a poorer nucleophile than Br^- or I^-, the reaction of 1° alcohols with HCl occurs only when an additional Lewis acid catalyst, usually $\mathbf{ZnCl_2}$, is added. $ZnCl_2$ complexes with the O atom of the alcohol in a Lewis acid–base reaction, making a leaving group and facilitating the S_N2 reaction.

R–OH (1° alcohol, Lewis base) + $ZnCl_2$ (Lewis acid) → S_N2 → R–Cl + $ZnCl_2(OH)^-$ (leaving group)

Knowing the mechanism allows us to predict the stereochemistry of the products when reaction occurs at a stereogenic center.

- Primary (1°) alcohols react by an S_N2 mechanism, so *inversion* occurs at a stereogenic center.
- Secondary (2°) and 3° alcohols react by an S_N1 mechanism, so *racemization* occurs at a stereogenic center.

Sample Problem 9.5 Predicting the Stereochemistry When an Alcohol Reacts with a Hydrogen Halide

Draw the products and stereochemistry for each reaction.

a. 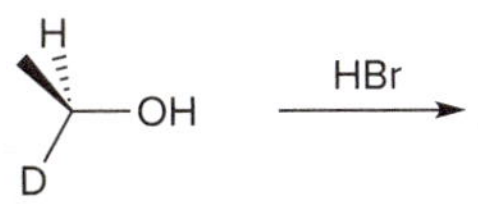 (H, D, OH) —HBr→ b. 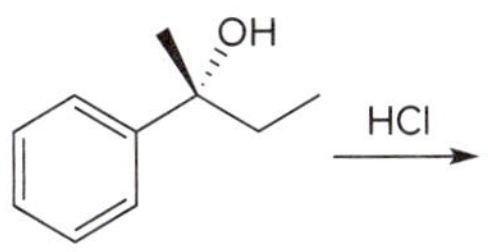 (benzene ring, OH) —HCl→

Solution

a. The alcohol is **1°**, so the mechanism of substitution is **S_N2.** Because the leaving group OH (which is protonated to form H_2O) is drawn on the *right* and S_N2 reactions proceed with **inversion of stereochemistry at a stereogenic center,** the nucleophile approaches from the *left* and a single product is formed.

H, OH, D —HBr, S_N2→ Br, H, D + H–OH

1° alcohol **product of inversion**

b. The alcohol is **3°,** so the mechanism of substitution is **S_N1.** Because S_N1 reactions form a **trigonal planar carbocation,** nucleophilic attack of Cl^- occurs from in front and behind to afford a **racemic mixture** of two enantiomers.

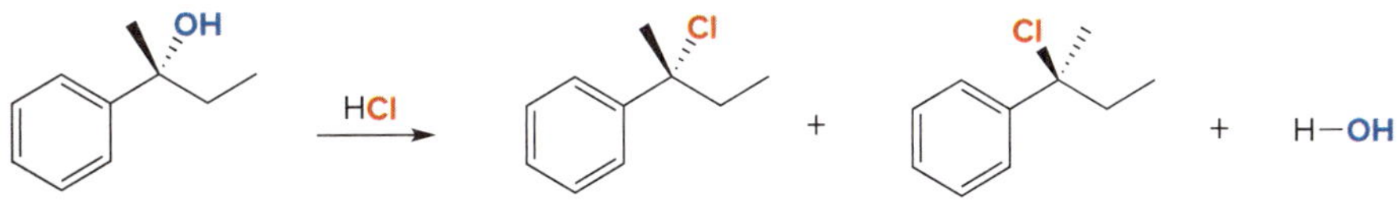

OH —HCl→ Cl + Cl + H–OH

3° alcohol **two enantiomers racemic mixture**

Problem 9.18 Draw the products of each reaction, indicating the stereochemistry around any stereogenic centers.

a. 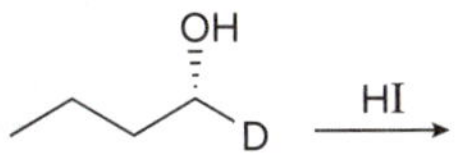 (OH, D) —HI→ b.  (OH) —HBr→ c. 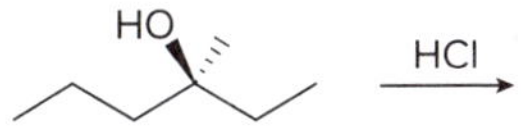 (HO) —HCl→

More Practice: Try Problems 9.38a, c; 9.44a–c, e, f; 9.63b.

9.11B Carbocation Rearrangement in the S_N1 Reaction

Because carbocations are formed in the S_N1 reaction of 2° and 3° alcohols with HX, **carbocation rearrangements are possible,** as illustrated in Sample Problem 9.6.

Sample Problem 9.6 Drawing an S_N1 Mechanism That Involves a Rearrangement

Draw a stepwise mechanism for the following reaction.

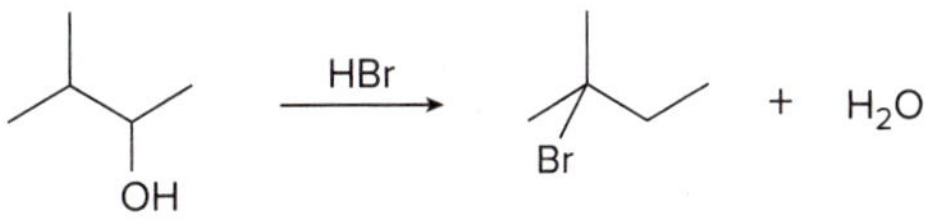

OH —HBr→ Br + H_2O

Solution

A 2° alcohol reacts with HBr by an S_N1 mechanism. Because substitution converts a 2° alcohol to a 3° alkyl halide in this example, a **carbocation rearrangement** must occur.

Steps [1] and [2] Protonation of the O atom and then loss of H_2O form a **2° carbocation.**

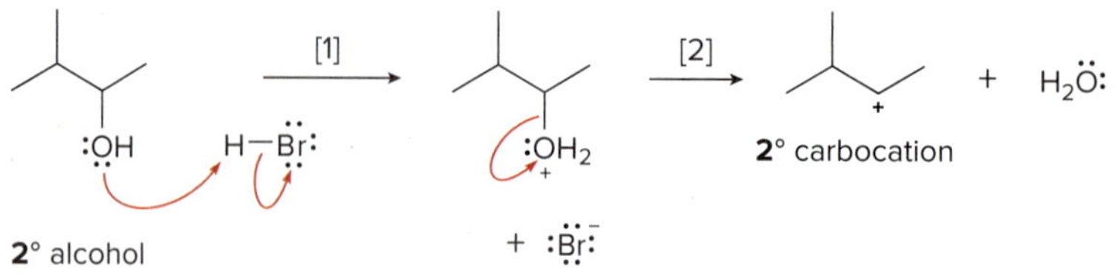

2° alcohol

[1]

[2]

2° carbocation

$+ H_2O$

Steps [3] and [4] **Rearrangement of the 2° carbocation by a 1,2-hydride shift forms a more stable 3° carbocation.** Nucleophilic attack forms the substitution product.

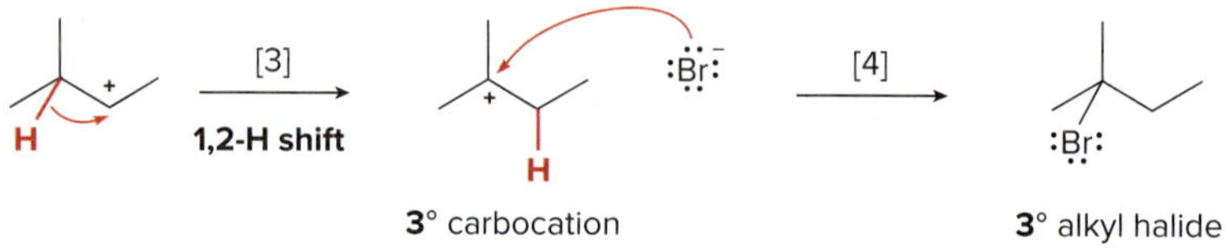

[3]

1,2-H shift

3° carbocation

[4]

3° alkyl halide

Problem 9.19 What is the major product formed when each alcohol is treated with HCl?

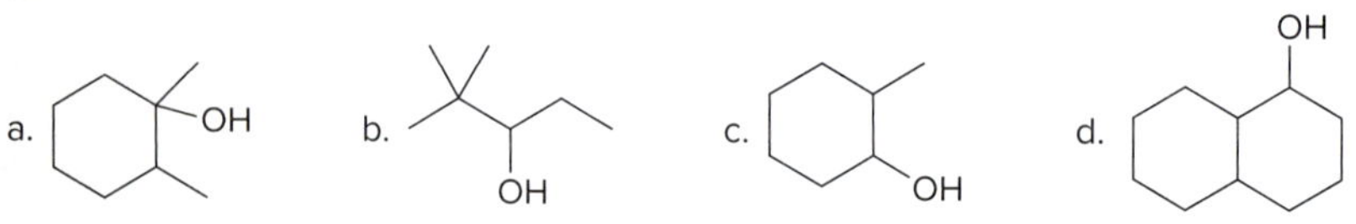

a. b. c. d.

More Practice: Try Problem 9.48.

9.12 Conversion of Alcohols to Alkyl Halides with $SOCl_2$ and PBr_3

Primary (1°) and 2° alcohols can be converted to alkyl halides using $\mathbf{SOCl_2}$ and $\mathbf{PBr_3}$.

- **$SOCl_2$ (thionyl chloride) converts alcohols into alkyl chlorides.**
- **PBr_3 (phosphorus tribromide) converts alcohols into alkyl bromides.**

Both reagents convert $^-$OH into a good leaving group *in situ*—that is, directly in the reaction mixture—as well as provide the **nucleophile,** either Cl^- or Br^-, to displace the leaving group.

9.12A Reaction of ROH with $SOCl_2$

The treatment of a 1° or 2° alcohol with thionyl chloride, $SOCl_2$, and pyridine forms an **alkyl chloride,** with SO_2 and HCl as by-products.

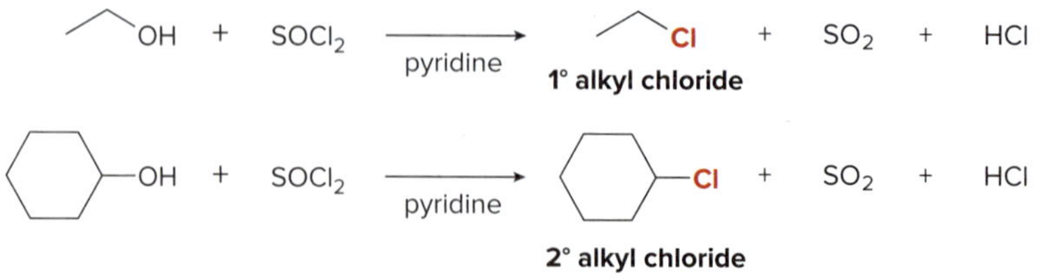

$$\text{CH}_3\text{CH}_2\text{OH} + SOCl_2 \xrightarrow{\text{pyridine}} \text{CH}_3\text{CH}_2\text{Cl} + SO_2 + HCl$$

1° alkyl chloride

$$\text{C}_6\text{H}_{11}\text{OH} + SOCl_2 \xrightarrow{\text{pyridine}} \text{C}_6\text{H}_{11}\text{Cl} + SO_2 + HCl$$

2° alkyl chloride

The mechanism for this reaction consists of two parts: **conversion of the OH group into a better leaving group, and nucleophilic attack by Cl^- via an S_N2 reaction,** as shown in Mechanism 9.7.

Mechanism 9.7 Reaction of ROH with $SOCl_2$ + Pyridine—An S_N2 Mechanism

1 – 2 Reaction of the alcohol with $SOCl_2$ and loss of a proton convert the OH group to OSOCl, a **good leaving group.**

3 **Nucleophilic attack** of chloride and loss of the leaving group (SO_2 and Cl^-) form RCl in a single step.

Problem 9.20 If the reaction of an alcohol with $SOCl_2$ and pyridine follows an S_N2 mechanism, what is the stereochemistry of the alkyl chloride formed from (*R*)-butan-2-ol?

9.12B Reaction of ROH with PBr_3

In a similar fashion, the treatment of a 1° or 2° alcohol with phosphorus tribromide, PBr_3, forms an alkyl bromide.

The mechanism for this reaction also consists of two parts: **conversion of the OH group into a better leaving group, and nucleophilic attack by Br^- via an S_N2 reaction,** as shown in Mechanism 9.8.

Mechanism 9.8 Reaction of ROH with PBr_3—An S_N2 Mechanism

1 Reaction of the alcohol with PBr_3 converts the OH group to $OPBr_2$, a **good leaving group,** and generates the nucleophile, Br^-.

2 **Nucleophilic attack** of bromide and loss of the leaving group form RBr in a single step.

Problem 9.21 If the reaction of an alcohol with PBr_3 follows an S_N2 mechanism, what is the stereochemistry of the alkyl bromide formed from (*R*)-butan-2-ol?

Table 9.1 summarizes the methods for converting an alcohol to an alkyl halide presented in Sections 9.11 and 9.12. Sample Problem 9.7 illustrates that converting a poor ^-OH leaving group into a better X^- leaving group is a useful first step that then allows other functional groups to be introduced.

Table 9.1 Summary of Methods for ROH → RX

Overall reaction	Reagent	Comment
ROH → RCl	HCl	• Useful for all ROH • An S_N1 mechanism for 2° and 3° ROH; an S_N2 mechanism for CH_3OH and 1° ROH
	$SOCl_2$	• Best for CH_3OH, and 1° and 2° ROH • An S_N2 mechanism
ROH → RBr	HBr	• Useful for all ROH • An S_N1 mechanism for 2° and 3° ROH; an S_N2 mechanism for CH_3OH and 1° ROH
	PBr_3	• Best for CH_3OH, and 1° and 2° ROH • An S_N2 mechanism
ROH → RI	HI	• Useful for all ROH • An S_N1 mechanism for 2° and 3° ROH; an S_N2 mechanism for CH_3OH and 1° ROH

Sample Problem 9.7 Converting an Alcohol to an S_N2 Product in Two Steps

Convert propan-1-ol to butanenitrile (**A**).

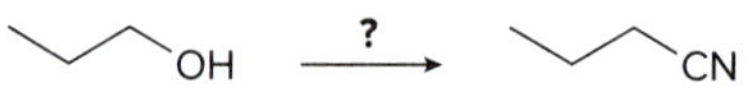

OH —?→ CN

propan-1-ol   butanenitrile **A**

Solution

Direct conversion of propan-1-ol to **A** using ^-CN as a nucleophile is not possible because **^-OH is a poor leaving group.** However, conversion of the OH group to a Br atom forms a good leaving group, which can then readily undergo an S_N2 reaction with ^-CN to yield **A.** The overall result of this two-step sequence is the substitution of ^-OH by ^-CN.

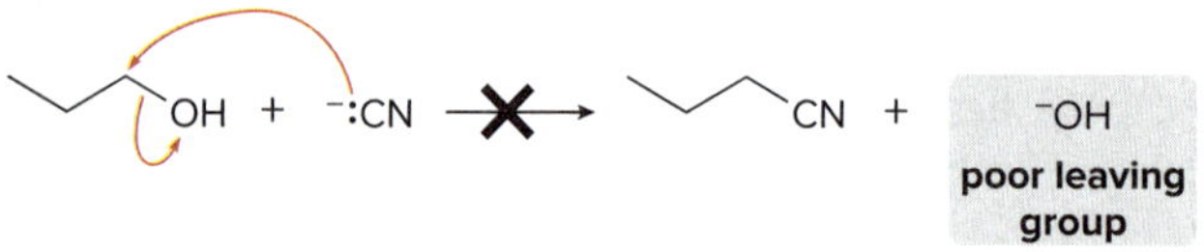

OH + $^-$:CN —✕→ CN + $^-$OH **poor leaving group**

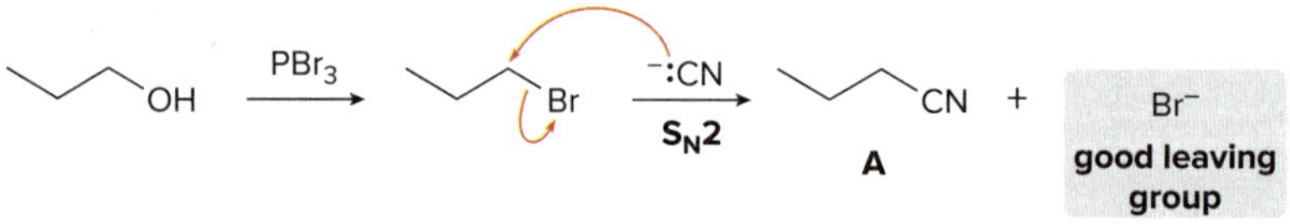

OH —PBr_3→ Br —$^-$:CN, S_N2→ CN (**A**) + Br^- **good leaving group**

Problem 9.22 Convert **A** to each compound (**B** and **C**).

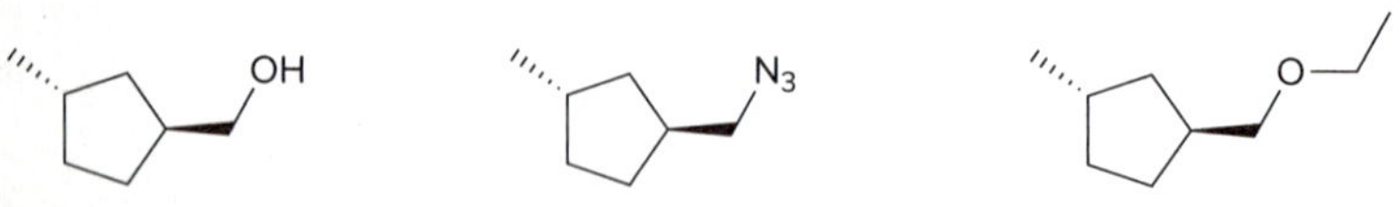

OH N$_3$ O

A **B** **C**

More Practice: Try Problems 9.45c, 9.63f, 9.64.

Problem 9.23 Draw the organic products formed in each reaction, and indicate the stereochemistry of products that contain stereogenic centers.

a. 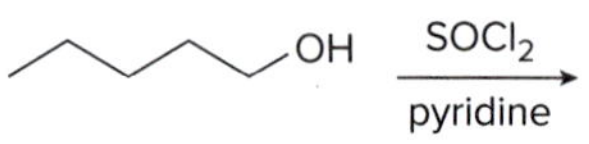 OH —$SOCl_2$, pyridine→ b. 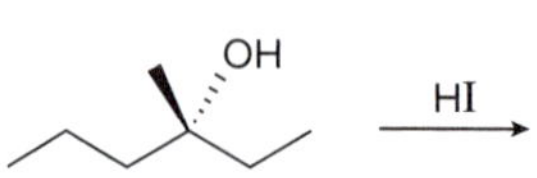 OH —HI→ c. 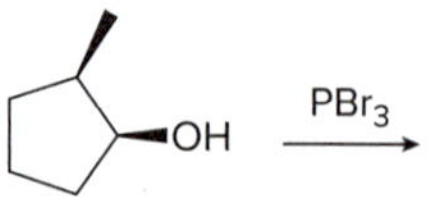 OH —PBr_3→

9.13 Tosylate—Another Good Leaving Group

We have now learned two methods to convert the OH group of an alcohol to a better leaving group: treatment with strong acids (Section 9.8A), and conversion to an alkyl halide (Sections 9.11–9.12). Alcohols can also be converted to **alkyl tosylates.**

Recall from Section 1.5 that a third-row element like sulfur can have 10 or 12 electrons around it in a valid Lewis structure.

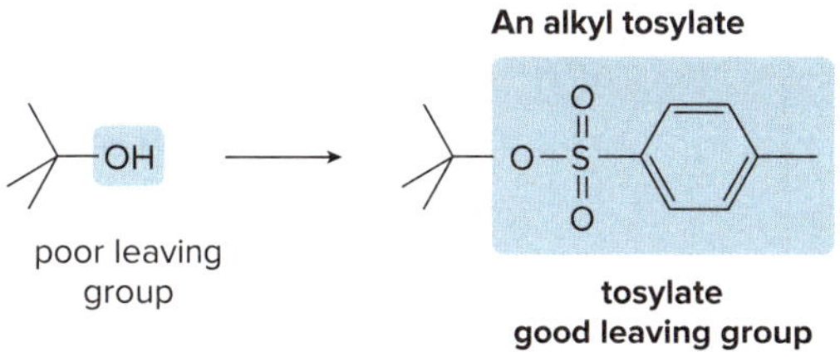

An alkyl tosylate

poor leaving group

tosylate
good leaving group

An alkyl tosylate is often called simply a **tosylate.**

An **alkyl tosylate** is composed of two parts: the **alkyl group R,** derived from an alcohol; and the **tosylate** (short for ***p*-toluenesulfonate**), which is a good leaving group. A tosyl group, $CH_3C_6H_4SO_2$–, is abbreviated as **Ts,** so an alkyl tosylate becomes **ROTs.**

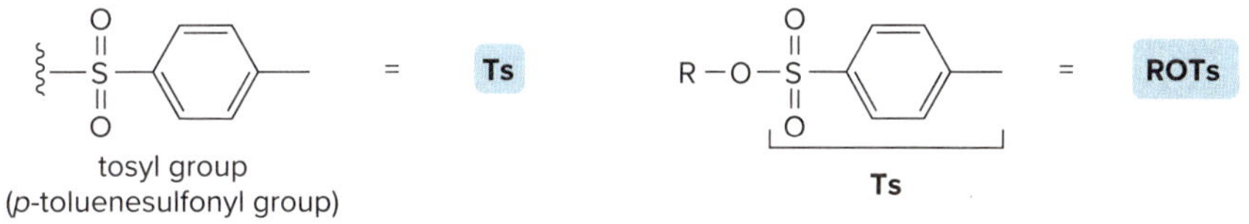

= **Ts**

tosyl group
(*p*-toluenesulfonyl group)

R–O–Ts = **ROTs**

Ts

9.13A Conversion of Alcohols to Alkyl Tosylates

A tosylate (TsO⁻) is similar to I⁻ in leaving group ability.

Alcohols are converted to alkyl tosylates by treatment with *p*-toluenesulfonyl chloride (TsCl) in the presence of pyridine. This overall process converts a poor leaving group ($^{-}$OH) into a good one ($^{-}$OTs). A tosylate is a good leaving group because its conjugate acid, *p*-toluenesulfonic acid ($CH_3C_6H_4SO_3H$, TsOH), is a strong acid (pK_a = –1.3, Section 2.6).

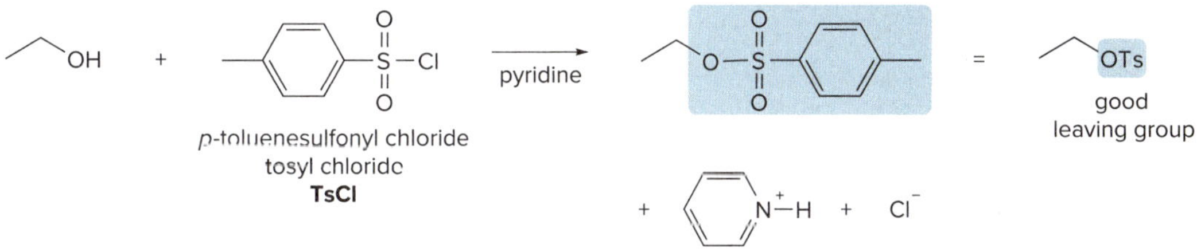

pyridine

p-toluenesulfonyl chloride
tosyl chloride
TsCl

OTs
good
leaving group

$+$ pyridinium (N^+–H) $+$ Cl^-

(*S*)-Butan-2-ol is converted to its tosylate with **retention of configuration** at the stereogenic center. Thus, the C–O bond of the alcohol must *not* be broken when the tosylate is formed.

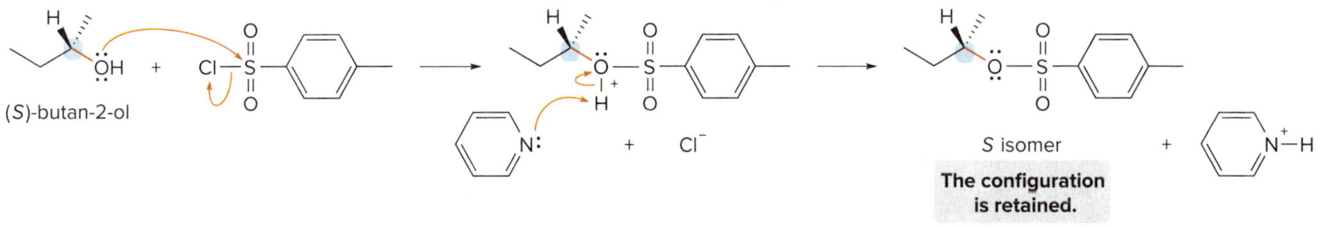

(*S*)-butan-2-ol

S isomer

The configuration is retained.

Problem 9.24 Draw the products of each reaction, and indicate the stereochemistry at any stereogenic center.

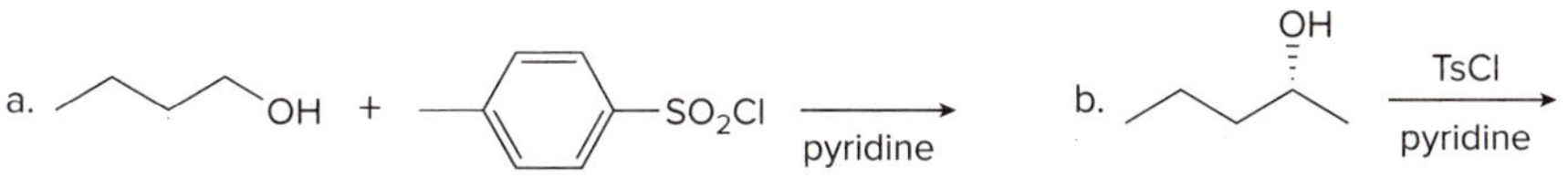

a. butan-1-ol (OH) + $CH_3C_6H_4SO_2Cl$ $\xrightarrow{\text{pyridine}}$

b. (OH) $\xrightarrow[\text{pyridine}]{\text{TsCl}}$

9.13B Reactions of Alkyl Tosylates

Because alkyl tosylates have good leaving groups, **they undergo both nucleophilic substitution and β elimination,** exactly as alkyl halides do. Generally, alkyl tosylates are treated with strong nucleophiles and bases, so that the mechanism of substitution is **S_N2** and the mechanism of elimination is **E2.**

For example, propyl tosylate, which has the leaving group on a 1° carbon, reacts with $NaOCH_3$ to yield methyl propyl ether, the product of nucleophilic substitution by an S_N2 mechanism. Propyl tosylate reacts with $KOC(CH_3)_3$, a strong bulky base, to yield propene by an E2 mechanism.

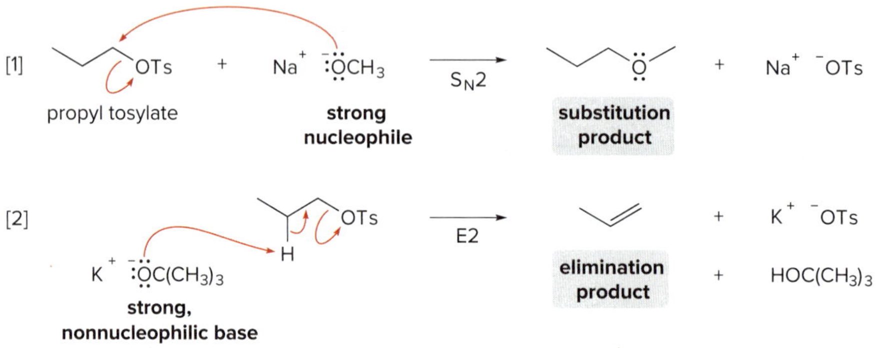

[1] OTs + Na^+ $^-$:ÖCH$_3$ —S_N2→ Ö + Na^+ $^-$OTs

propyl tosylate; **strong nucleophile**; **substitution product**

[2] OTs, H + K^+ $^-$:ÖC(CH$_3$)$_3$ —E2→ + K^+ $^-$OTs + HOC(CH$_3$)$_3$

strong, nonnucleophilic base; **elimination product**

Because substitution occurs via an S_N2 mechanism, **inversion of configuration** results when the leaving group is bonded to a stereogenic center.

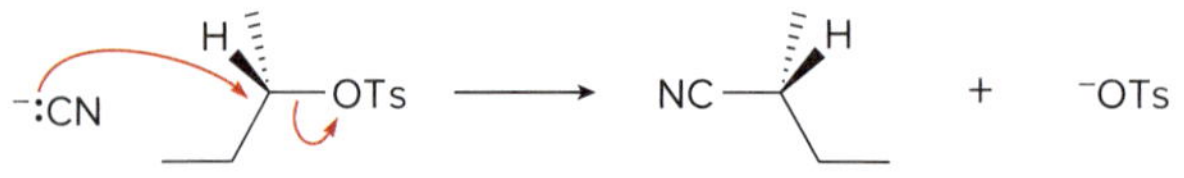

$^-$:CN + H OTs ⟶ NC H + $^-$OTs

Inversion of configuration

Sample Problem 9.8 Drawing the Substitution Product from an Alkyl Tosylate

Draw the product of the following reaction, including stereochemistry.

OTs, D —Na^+ $^-OCH_2CH_3$→

Solution

The 1° alkyl tosylate and the strong nucleophile both favor substitution by an **S_N2** mechanism, which proceeds by backside attack, resulting in **inversion** of configuration at the stereogenic center. The leaving group is drawn in front (on a wedge), so the nucleophile approaches from behind, ending up on a dashed wedge.

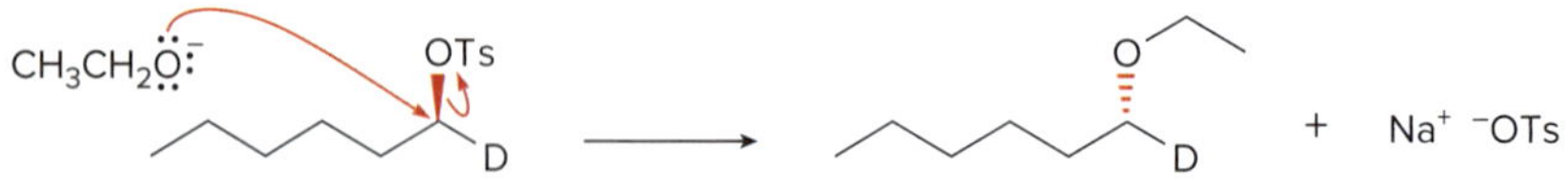

$CH_3CH_2\ddot{O}$:$^-$ + OTs, D ⟶ O, D + Na^+ $^-$OTs

1° tosylate

Problem 9.25 Draw the products of each reaction, and include the stereochemistry at any stereogenic center in the products.

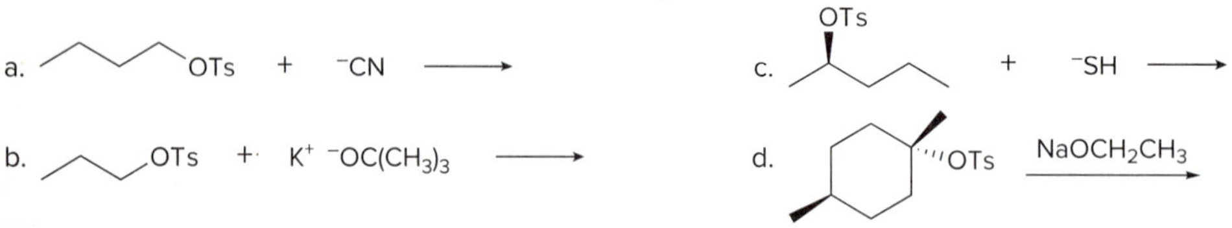

a. OTs + $^-$CN ⟶

b. OTs + K^+ $^-OC(CH_3)_3$ ⟶

c. OTs + $^-$SH ⟶

d. OTs —$NaOCH_2CH_3$→

More Practice: Try Problems 9.44d; 9.45b; 9.63d, f.

9.13C The Two-Step Conversion of an Alcohol to a Substitution Product

We now have another **two-step method to convert an alcohol to a substitution product:** reaction of an alcohol with TsCl and pyridine to form an alkyl tosylate (Step [1]), followed by nucleophilic attack on the tosylate (Step [2]).

$$R-OH \xrightarrow[\text{pyridine [1]}]{TsCl} R-OTs \xrightarrow[{[2]}]{:Nu^-} R-Nu + {}^-OTs$$

Let's look at the stereochemistry of this two-step process.

- Step [1], formation of the tosylate, proceeds with **retention** of configuration at a stereogenic center because the C–O bond remains intact.
- Step [2] is an S_N2 reaction, so it proceeds with **inversion of configuration** because the nucleophile attacks from the back side.
- Overall there is a **net inversion of configuration** at a stereogenic center.

For example, the treatment of *cis*-3-methylcyclohexanol with *p*-toluenesulfonyl chloride and pyridine forms a cis tosylate **A,** which undergoes backside attack by the nucleophile $^-OCH_3$ to yield the trans ether **B.**

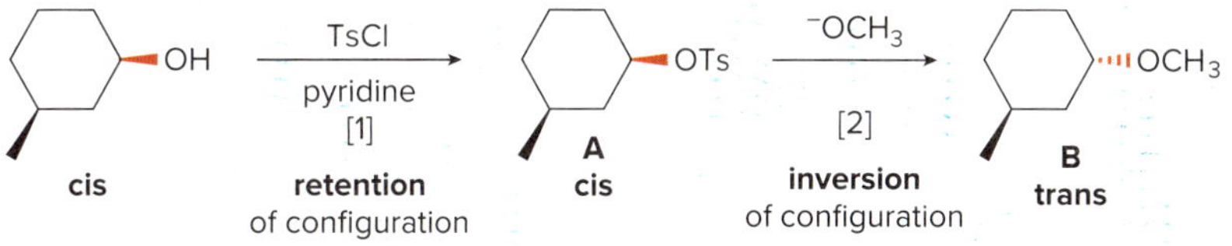

Problem 9.26 Draw the products formed when (*S*)-butan-2-ol is treated with TsCl and pyridine, followed by NaOH. Label the stereogenic center in each compound as *R* or *S*. What is the stereochemical relationship between the starting alcohol and the final product?

9.13D A Summary of Substitution and Elimination Reactions of Alcohols

The reactions of alcohols in Sections 9.8–9.13C share two similarities:

- **The OH group is converted into a better leaving group by treatment with acid or another reagent.**
- **The resulting product undergoes either elimination or substitution, depending on the reaction conditions.**

Figure 9.3 summarizes these reactions with cyclohexanol as starting material.

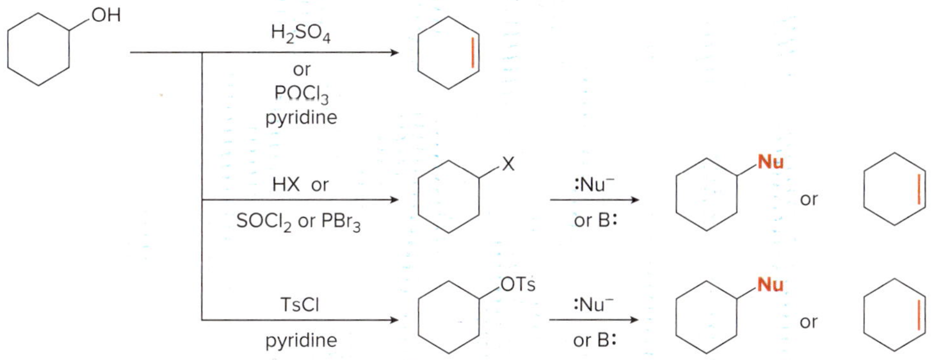

Figure 9.3 Summary: Nucleophilic substitution and β elimination reactions of alcohols

Problem 9.27 Draw the product formed when $(CH_3)_2CHOH$ is treated with each reagent.

a. $SOCl_2$, pyridine
b. TsCl, pyridine
c. H_2SO_4
d. HBr
e. PBr_3, then NaCN
f. $POCl_3$, pyridine

9.14 Reaction of Ethers with Strong Acid

Because ethers are so unreactive, diethyl ether and tetrahydrofuran (THF) are often used as solvents for organic reactions.

Recall from Section 9.7B that ethers have a poor leaving group, so they cannot undergo nucleophilic substitution or β elimination reactions directly. Instead, they must first be converted into a good leaving group by reaction with strong acids. Only **HBr** and **HI** can be used, though, because they are strong acids that are also sources of good nucleophiles (Br^- and I^-, respectively). **When ethers react with HBr or HI, both C–O bonds are cleaved and two alkyl halides are formed as products.**

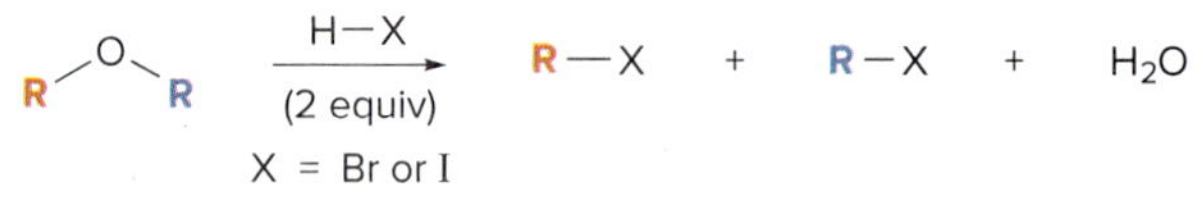

R–O–R + H–X (2 equiv) → R–X + R–X + H_2O

X = Br or I

HBr or HI serves as a strong acid that both protonates the O atom of the ether and is the source of a good nucleophile (Br^- or I^-). Because both C–O bonds in the ether are broken, **two successive nucleophilic substitution reactions occur.**

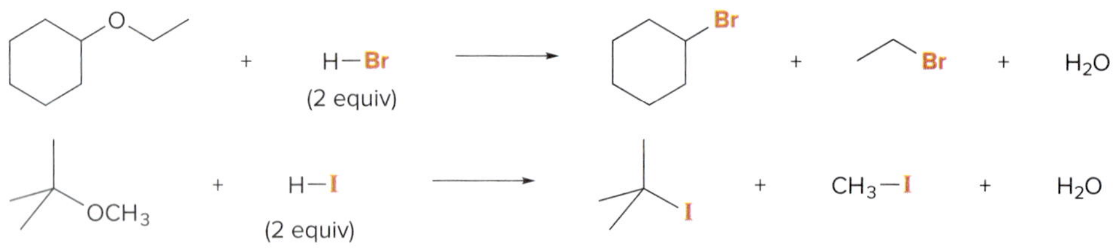

(cyclohexyl ethyl ether) + H–Br (2 equiv) → (bromocyclohexane) + (bromoethane) Br + H_2O

OCH_3 (tert-butyl methyl ether) + H–I (2 equiv) → (tert-butyl iodide) I + CH_3–I + H_2O

- The mechanism of ether cleavage is S_N1 or S_N2, depending on the identity of R.
- With 2° or 3° alkyl groups bonded to the ether oxygen, the C–O bond is cleaved by an S_N1 mechanism involving a carbocation; with methyl or 1° R groups, the C–O bond is cleaved by an S_N2 mechanism.

For example, cleavage of $(CH_3)_3COCH_3$ with HI occurs at two bonds, as shown in Mechanism 9.9. The 3° alkyl group undergoes nucleophilic substitution by an $\mathbf{S_N1}$ mechanism, resulting in the cleavage of one C–O bond. The methyl group undergoes nucleophilic substitution by an $\mathbf{S_N2}$ mechanism, resulting in the cleavage of the second C–O bond.

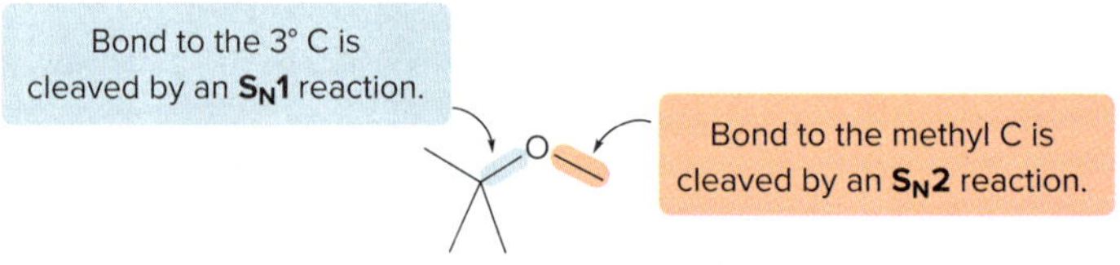

Bond to the 3° C is cleaved by an S_N1 reaction.

Bond to the methyl C is cleaved by an S_N2 reaction.

The mechanism illustrates the central role of HX in the reaction:

- HX protonates the ether oxygen, thus making a good leaving group.
- HX provides a source of X^- for nucleophilic attack.

Problem 9.28 What alkyl halides are formed when each ether is treated with HBr?

a. 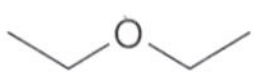 (structure) b. 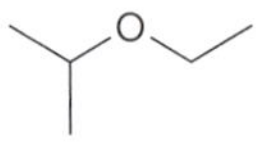 (structure) c. 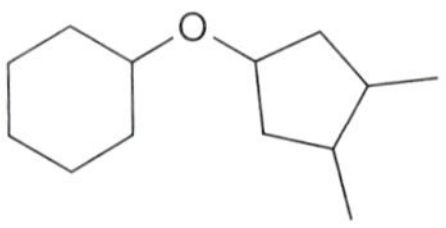 (structure)

Mechanism 9.9 Mechanism of Ether Cleavage in Strong Acid—$(CH_3)_3COCH_3 + HI \rightarrow (CH_3)_3CI + CH_3I + H_2O$

Part [1] Cleavage of the 3° C–O bond by an S_N1 mechanism

1. **Protonation** of the ether O atom forms a **good leaving group.**
2. Cleavage of the C–O bond to the 3° carbon forms a **3° carbocation** and CH_3OH.
3. **Nucleophilic attack of I^-** forms the substitution product.

Part [2] Cleavage of the CH_3–O bond by an S_N2 mechanism

$CH_3-\ddot{O}H + H-I \xrightarrow{[4]} CH_3-\overset{+}{O}H_2 + I^- \xrightarrow[{[5]}]{S_N2} CH_3-I + H_2\ddot{O}$:

4. **Protonation** of the OH group forms a **good leaving group** (H_2O).
5. **Nucleophilic attack** of iodide forms the second alkyl halide, CH_3I. Because the mechanism is S_N2, the C–O bond is broken as the C–I bond (in red) is formed.

Problem 9.29 Explain why the treatment of anisole with HBr yields phenol and CH_3Br, but not bromobenzene.

anisole —HBr→ phenol + CH_3Br [bromobenzene]

9.15 Thiols and Sulfides

Thiols and **sulfides** are sulfur analogues of alcohols and ethers, respectively.

R–S–H thiol   R–S–R sulfide

9.15A Thiols

Thiols, also called mercaptans, contain a mercapto group (SH) bonded to a carbon atom. Because sulfur is below oxygen in the periodic table, the sulfur atom is surrounded by two atoms and two lone pairs, giving thiols a **bent shape.** Unlike alcohols, however, thiols are incapable of intermolecular hydrogen bonds, so thiols have *lower* boiling points and melting points than alcohols with a similar number of carbons.

thiol (mercapto group)   ethanethiol bp 35 °C   ethanol bp 78 °C

Many simple thiols have pungent and disagreeable odors. Skunks, onions, and human sweat all contain thiols (see also Problem 5.36).

propane-1-thiol	3-methylbutane-1-thiol	(*S*)-3-methyl-3-sulfanylhexan-1-ol
onion odor	skunk odor	onion-like odor in human sweat

Thiols are named in a similar method to alcohols, using the suffix ***-thiol*** instead of the suffix ***-ol***. To name a thiol in the IUPAC system:

- **Name the parent carbon chain and add the suffix *-thiol*.**
- **Number the carbon chain to give the SH group the lower number and apply the other rules of nomenclature.**

Examples of thiol nomenclature are given in Figure 9.4.

Figure 9.4 Naming thiols

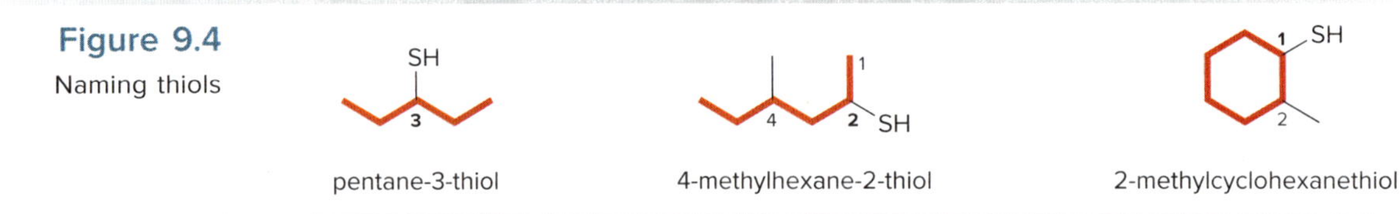

Problem 9.30 Name each thiol.

a. SH

b. SH

c. SH

Thiols are prepared by S_N2 reactions of alkyl halides with ^-SH, a good nucleophile.

Br + :SH$^-$ → SH + :Br:$^-$

Thiols are easily oxidized with Br_2 or I_2 to **disulfides (RSSR),** compounds that contain a sulfur–sulfur bond. This reaction is an oxidation (Section 4.13) because H atoms are removed from the thiol in forming the disulfide. Disulfides are reduced to thiols with Zn and acid.

SH —(Br_2 or I_2)→ S–S

S–S —(Zn, HCl)→ SH + HS

Disulfide formation is especially important in determining the shape and properties of some proteins that contain the amino acid cysteine, as we will learn in Chapter 27.

Problem 9.31 Draw the product of each reaction.

a. 1-bromo-4-methylpentane —NaSH→

b. (D, Cl) —NaSH→

c. —Br_2→ (SH) grapefruit mercaptan

d. (dicyclohexyl disulfide, S–S) —Zn, HCl→

The potent odor of grapefruit mercaptan (Problem 9.31c) contributes to the characteristic aroma of grapefruit.
Purestock/SuperStock

9.15B Sulfides

Sulfides contain two alkyl groups bonded to a sulfur atom. Sulfides are named with the same rules used to name ethers. The suffix ***sulfide*** is used instead of *ether* for simple compounds.

R–S–R **sulfide** · *sec*-butyl ethyl sulfide · diethyl sulfide

To name more complex sulfides using the IUPAC system, one alkyl group is named as a parent chain and the other is named as part of a substituent bonded to that chain.

- **Name the simpler alkyl group + S atom as an *alkylthio* substituent.**
- **Name the remaining alkyl group as an alkane with an alkylthio substituent using the usual rules of nomenclature.**

methyl**thio**cyclohexane · 3-ethyl**thio**-5-methyloctane (5, 3)

Problem 9.32 Give the IUPAC name for each sulfide.

a. (S) b. (S) c. (S)

Sulfides are prepared from thiols by an S_N2 reaction that is analogous to the Williamson ether synthesis.

propane-1-thiol (S–H) —Na^+ $H:^-$→ sulfur nucleophile + H_2 —(Br), S_N2→ ethyl propyl sulfide + $:Br:^-$

Sulfides contain a nucleophilic sulfur atom that reacts readily with unhindered alkyl halides to form **sulfonium ions.**

(S) + (Br) —S_N2→ **sulfonium ion** (S^+) + $:Br:^-$

S-Adenosylmethionine (SAM), a biological sulfonium ion that was introduced in Section 7.15, is synthesized from the amino acid methionine, which contains a nucleophilic sulfide, and adenosine triphosphate (ATP), which contains a triphosphate leaving group (Section 7.15).

methionine + adenosine triphosphate ATP (PPPO: leaving group) —S_N2→ *S*-adenosylmethionine + PPP_i triphosphate **leaving group**

Problem 9.33 Draw the product of each reaction.

a. cyclohexanethiol (SH) —[1] NaH, [2] CH_3Br→

b. sulfide + alkyl chloride (Cl) →

9.16 Reactions of Epoxides

Although epoxides do not contain a good leaving group, they contain a strained three-membered ring with two polar bonds. **Nucleophilic attack opens the strained three-membered ring,** making it a favorable process even with the poor leaving group.

δ− O, δ+, δ+, :Nu⁻ → Nu, :Ö:⁻ leaving group; new C–Nu bond

This reaction occurs readily with strong nucleophiles like ⁻CN, and with acids like HZ, where Z is a nucleophilic atom.

ethylene oxide —[1] ⁻CN, [2] H_2O→ NC–CH_2–CH_2–OH

ethylene oxide —HCl→ HO–CH_2CH_2–Cl

9.16A Opening of Epoxide Rings with Strong Nucleophiles

Virtually all strong nucleophiles open an epoxide ring by a two-step reaction sequence.

epoxide + :Nu⁻ —[1]→ Nu, :Ö:⁻ —H–OH, H_2O [2]→ Nu, :ÖH + ⁻OH

- **Step [1]:** The nucleophile attacks an electron-deficient carbon of the epoxide, cleaving a C–O bond and relieving the strain of the three-membered ring.
- **Step [2]:** Protonation of the alkoxide with water generates a neutral product with two functional groups on adjacent atoms.

Common nucleophiles that open epoxide rings include ^{-}OH, ^{-}OR, ^{-}CN, ^{-}SR, and NH_3. With these strong nucleophiles, the reaction occurs via an **S_N2 mechanism,** resulting in two consequences:

- **The nucleophile opens the epoxide ring from the back side.**

[1] S_N2 backside attack; H_2O [2]

CH_3O and OH are **anti** in the product.

Other examples of the nucleophilic opening of epoxide rings are presented in Sections 12.6 and 17.14.

- **In an unsymmetrical epoxide, the nucleophile attacks at the *less* substituted carbon atom.**

CH_3S^- [1] attack at the less substituted C; H_2O [2]

Problem 9.34 Draw the product of each reaction, and indicate the stereochemistry at any stereogenic center.

a. [1] $CH_3CH_2O^-$ [2] H_2O

b. [1] $H-C\equiv C^-$ [2] H_2O

1,2-Epoxycyclohexane, an achiral epoxide with a plane of symmetry, reacts with $^{-}OCH_3$ to yield two *trans*-1,2-disubstituted cyclohexanes, **A** and **B,** which are **enantiomers;** each has two stereogenic centers.

1,2-epoxycyclohexane — **achiral starting material**; [1] $^{-}OCH_3$ [2] H_2O → **A** + **B** — **enantiomers**

[* denotes a stereogenic center]

Nucleophilic attack of $^{-}OCH_3$ occurs from the back side at *either* C–O bond, because both ends are equally substituted. Because attack at either side occurs with equal probability, an equal amount of the two enantiomers is formed—**a racemic mixture.** This is a specific example of a general rule concerning the stereochemistry of products obtained from an achiral reactant.

The epoxide is above the plane. The nucleophile attacks from below. → trans products → H_2O → **A**, **B** enantiomers

- **Whenever an achiral reactant yields a product with stereogenic centers, the product must be achiral (meso) or racemic.**

This general rule can be restated in terms of optical activity. Recall from Section 5.12 that achiral compounds and racemic mixtures are optically inactive.

- **Optically *inactive* starting materials give optically *inactive* products.**

Problem 9.35 The cis and trans isomers of 2,3-dimethyloxirane both react with ^{-}OH to give butane-2,3-diol. One stereoisomer gives a single achiral product, and one gives two chiral enantiomers. Which epoxide gives one product and which gives two?

cis-2,3-dimethyloxirane

one enantiomer of *trans*-2,3-dimethyloxirane

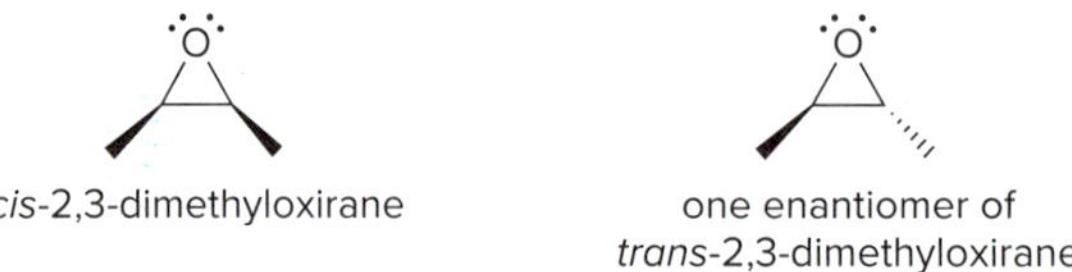

9.16B Reaction with Acids HZ

Acids **HZ** that contain a nucleophile **Z** also open epoxide rings by a two-step reaction sequence.

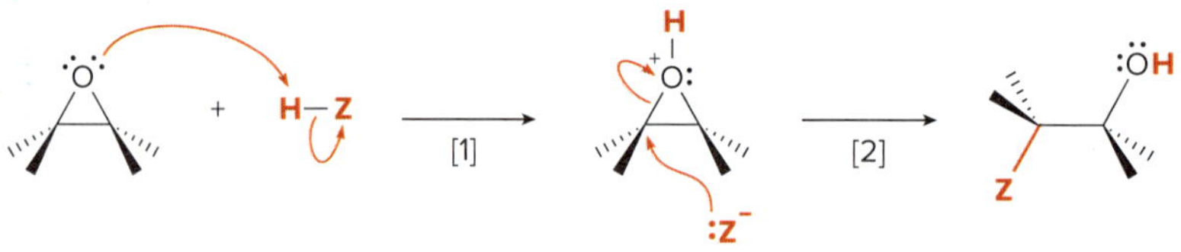

- **Step [1]:** Protonation of the epoxide oxygen with HZ makes the epoxide oxygen into a good leaving group (OH). It also provides a source of a good nucleophile (Z^-) to open the epoxide ring.
- **Step [2]:** The nucleophile Z^- then opens the protonated epoxide ring by **backside** attack.

These two steps—**protonation followed by nucleophilic attack**—are the exact reverse of the opening of epoxide rings with strong nucleophiles, where nucleophilic attack precedes protonation.

HCl, HBr, and **HI** all open an epoxide ring in this manner. $\mathbf{H_2O}$ and **ROH** can, too, but acid must also be added. Regardless of the reaction, the product has an OH group from the epoxide on one carbon and a new functional group Z from the nucleophile on the adjacent carbon. With epoxides fused to rings, ***trans*-1,2-disubstituted cycloalkanes** are formed.

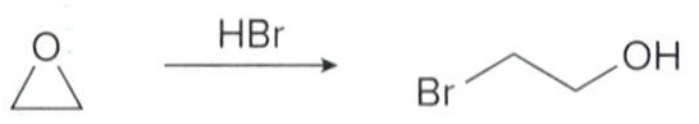

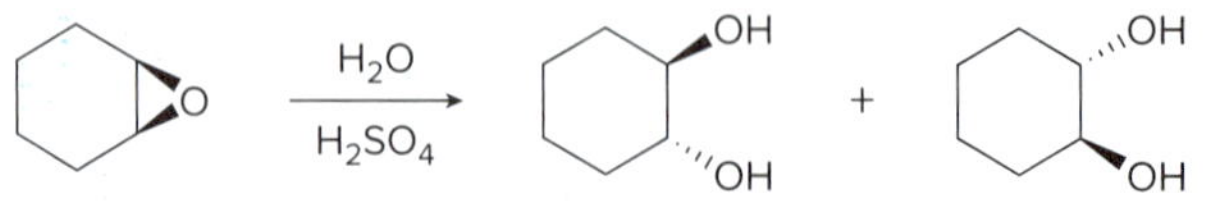

enantiomers

Although backside attack of the nucleophile suggests that this reaction follows an S_N2 mechanism, the regioselectivity of the reaction with unsymmetrical epoxides does not.

- **With unsymmetrical epoxides, nucleophilic attack occurs at the *more* substituted carbon atom.**

For example, the treatment of 2,2-dimethyloxirane with HCl results in nucleophilic attack at the carbon with two methyl groups.

2,2-dimethyloxirane —HCl→ (attack at the *more* substituted C)

Backside attack of the nucleophile suggests an S_N2 mechanism, but attack at the more substituted carbon suggests an S_N1 mechanism. To explain these results, the **mechanism of nucleophilic attack is thought to be somewhere in between S_N1 and S_N2.**

Figure 9.5 illustrates two possible pathways for the reaction of 2,2-dimethyloxirane with HCl. Backside attack of Cl^- at the more substituted carbon proceeds via transition state **A,** whereas backside attack of Cl^- at the less substituted carbon proceeds via transition state **B. Transition state A has a partial positive charge on a more substituted carbon, making it more stable.** Thus, the preferred reaction path takes place by way of the lower-energy transition state **A.**

Figure 9.5
Opening of an unsymmetrical epoxide ring with HCl

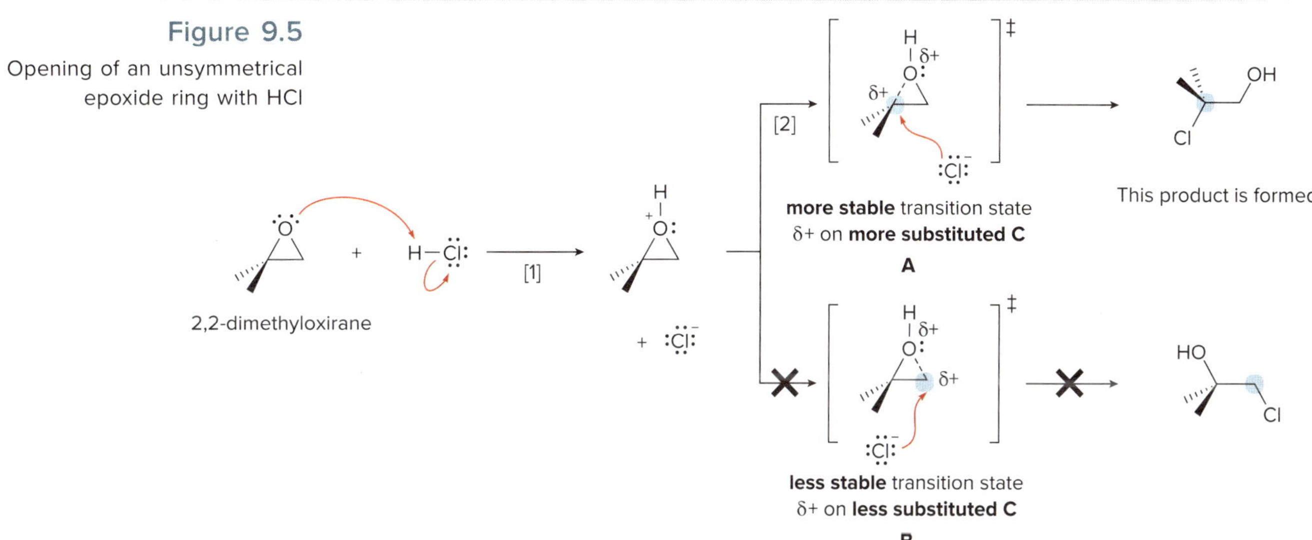

- Transition state **A** is lower in energy because the partial positive charge (δ+) is located on the *more* substituted carbon. In this case, therefore, nucleophilic attack occurs from the back side (an S_N2 characteristic) at the *more* substituted carbon (an S_N1 characteristic).

Opening of an epoxide ring with either a strong nucleophile $:Nu^-$ or an acid HZ is **regioselective,** because one constitutional isomer is the major or exclusive product. The **site selectivity of these two reactions, however, is *exactly the opposite.***

- **With a strong nucleophile, $:Nu^-$ attacks at the *less* substituted carbon.**
- **With an acid HZ, the nucleophile attacks at the *more* substituted carbon.**

Sample Problem 9.9 Determining the Regioselectivity of Opening an Epoxide Ring

What product is formed when 2,2-dimethyloxirane is treated with each set of reagents: $^-OCH_3$ followed by H_2O, or CH_3OH and H_2SO_4?

Solution

All nucleophiles open an epoxide ring from the back side. Classify the nucleophile to determine if nucleophilic attack occurs at the *more* or *less* substituted carbon. With **strong,** negatively charged nucleophiles, attack occurs at the *less* substituted carbon, whereas with **acids HZ,** nucleophilic attack occurs at the *more* substituted carbon.

2,2-dimethyloxirane — [1] $^{-}OCH_3$ / [2] H_2O (attack at the **less** substituted C) → HO ... OCH_3

With a strong nucleophile, **CH_3O ends up on the less substituted C.**

2,2-dimethyloxirane — CH_3OH / H_2SO_4 (attack at the **more** substituted C) → CH_3O ... OH

With acid, **CH_3O** ends up on the **more substituted C.**

Problem 9.36 Draw the product of each reaction.

a. HBr →

b. [1] ^{-}CN / [2] H_2O →

c. CH_3CH_2OH / H_2SO_4 →

d. [1] CH_3O^{-} / [2] CH_3OH →

More Practice: Try Problems 9.58; 9.59; 9.61; 9.63g, h.

The reaction of epoxide rings with nucleophiles is important for the synthesis of many biologically active compounds, including **albuterol,** the bronchodilator used in the treatment of asthma mentioned in the chapter opener (Figure 9.6).

Figure 9.6 The synthesis of albuterol using the opening of an epoxide ring

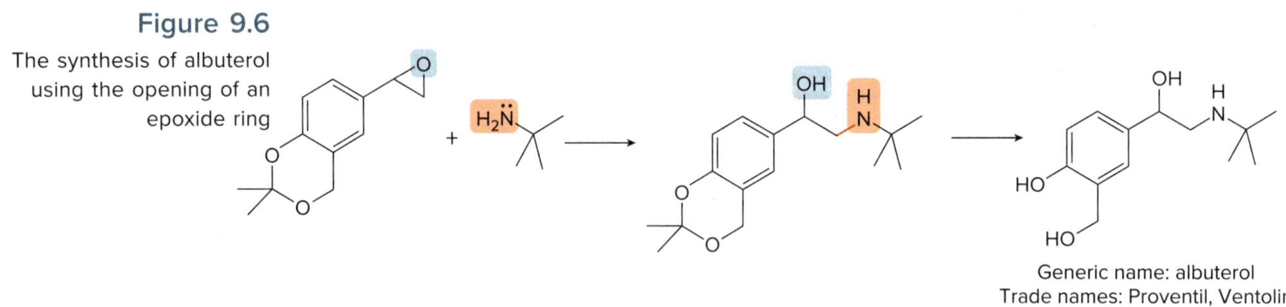

- A key step in the synthesis is the opening of an epoxide ring with a nitrogen nucleophile to form a new C–N bond, shown in red.

9.17 Application: Epoxides, Leukotrienes, and Asthma

The opening of epoxide rings with nucleophiles is a key step in some important biological processes.

9.17A Asthma and Leukotrienes

Asthma is an obstructive lung disease that affects millions of Americans. Because it involves episodic constriction of small airways, bronchodilators such as albuterol (Figure 9.6) are used to treat symptoms by widening airways. Because asthma is also characterized by chronic inflammation, inhaled steroids that reduce inflammation are also commonly used.

Leukotrienes were first synthesized in 1980 in the laboratory of Professor E. J. Corey, the 1990 recipient of the Nobel Prize in Chemistry.

Leukotrienes are molecules that contribute to the asthmatic response. A typical example, **leukotriene C_4,** is shown. Although its biological activity was first observed in the 1930s, the chemical structure of leukotriene C_4 was not determined until 1979. Structure determination and chemical synthesis were difficult because leukotrienes are highly unstable and extremely potent, and are therefore present in tissues in exceedingly small amounts.

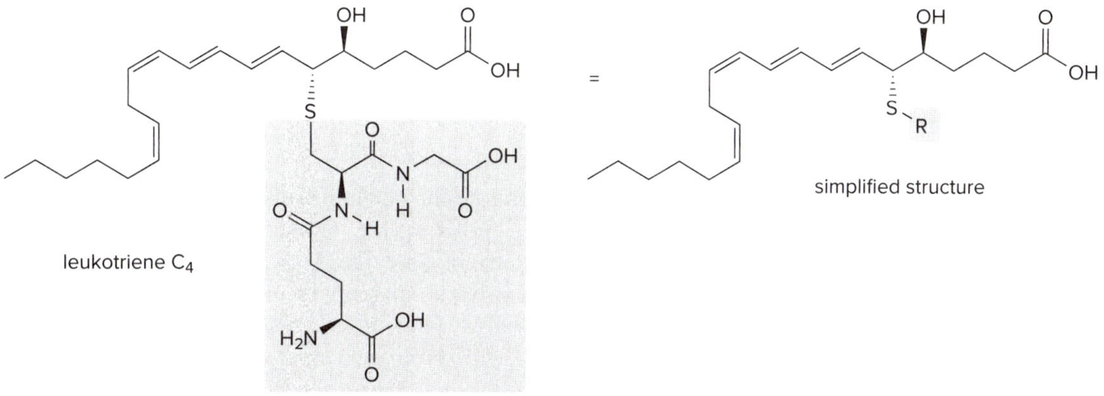

leukotriene C_4

simplified structure

9.17B Leukotriene Synthesis and Asthma Drugs

Leukotrienes are synthesized in cells by the oxidation of **arachidonic acid** to 5-HPETE, which is then converted to an epoxide, **leukotriene A_4.** Opening of the epoxide ring with a sulfur nucleophile **RSH** yields leukotriene C_4.

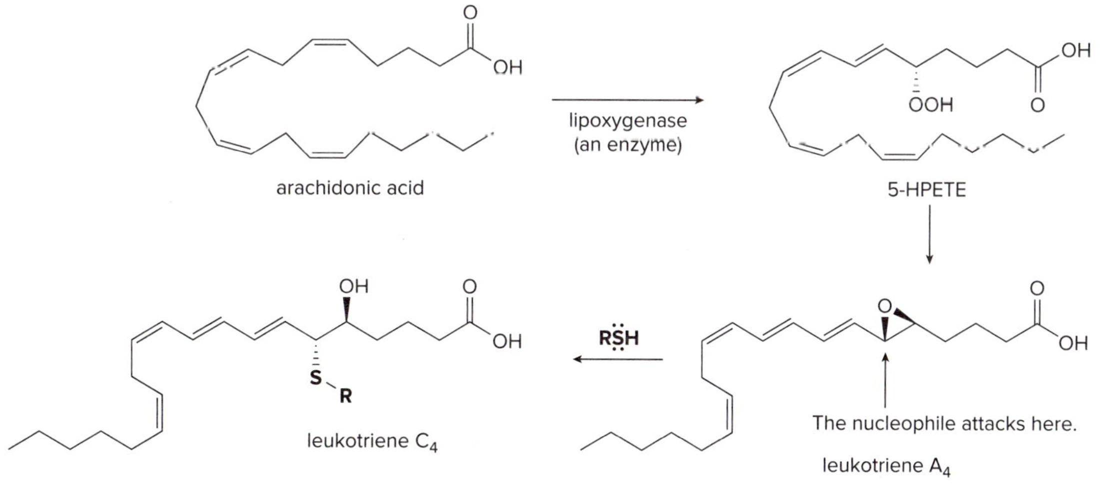

arachidonic acid

lipoxygenase (an enzyme)

5-HPETE

The nucleophile attacks here.

leukotriene A_4

leukotriene C_4

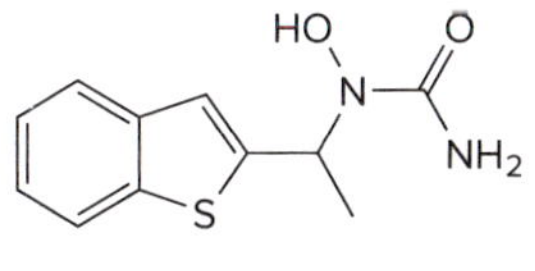

Generic name: zileuton
Trade name: Zyflo CR
anti-asthma drug

New asthma drugs act by blocking the synthesis of leukotriene C_4 from arachidonic acid. For example, **zileuton** (trade name Zyflo CR) inhibits the enzyme (called a lipoxygenase) needed for the first step of this process. By blocking the synthesis of leukotriene C_4, a compound responsible for the disease, zileuton treats the **cause of asthma,** not just its symptoms.

9.18 Application: Benzo[*a*]pyrene, Epoxides, and Cancer

The sooty exhaust from trucks and buses contains PAHs such as benzo[*a*]pyrene. *John Thoeming/McGraw Hill*

Benzo[*a*]pyrene is a widespread environmental pollutant, produced during the combustion of all types of organic material—gasoline, fuel oil, wood, garbage, and cigarettes. It is a **polycyclic aromatic hydrocarbon (PAH),** a class of compounds that is discussed further in Chapter 15.

oxidation →
several steps

:Nu⁻

benzo[*a*]pyrene
water insoluble

a diol epoxide
carcinogen
more water soluble

After this nonpolar and water-insoluble hydrocarbon is inhaled or ingested, it is oxidized in the liver to a diol epoxide. Oxidation is a common fate of foreign substances that are not useful nutrients for the body. The oxidation product has three oxygen-containing functional groups, making it much more water soluble, and more readily excreted in urine. It is also a potent carcinogen. The strained three-membered ring of the epoxide reacts readily with biological nucleophiles :Nu⁻ (such as DNA or an enzyme), leading to ring-opened products that often disrupt normal cell function, causing cancer or cell death.

These examples illustrate the central role of the nucleophilic opening of epoxide rings in two well-defined cellular processes.

Chapter 9 REVIEW

KEY CONCEPTS

Carbocation rearrangements (9.9)

1 Hydride shift

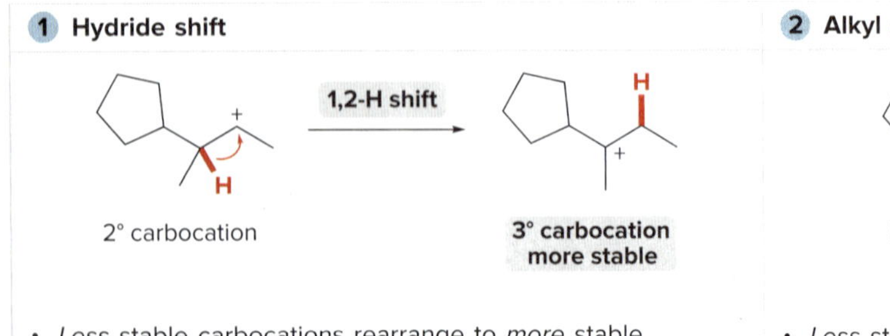

- *Less* stable carbocations rearrange to *more* stable carbocations by the shift of a hydrogen atom.

2 Alkyl shift

1,2-methyl shift

2° carbocation

3° carbocation more stable

- *Less* stable carbocations rearrange to *more* stable carbocations by the shift of an alkyl group.

KEY REACTIONS

[1] Preparation of alcohols, alkoxides, ethers, epoxides, thiols, and sulfides

1. CH_3-X + ^-OH —S_N2 (9.6)→ CH_3-OH (alcohol) + X^-
   X = Cl, Br, I

2. CH_3O-H + Na^+H^- —proton transfer (9.6)→ $CH_3O^-Na^+$ (alkoxide) + H−H

3. ethyl−X + $^-OCH_3$ —S_N2, Williamson ether synthesis (9.6)→ ethyl−OCH_3 (ether) + X^-
   X = Cl, Br, I

4. $Na^+H:^-$ + halohydrin (H−Ö:, X) —[1] (9.6)→ alkoxide + H−H —[2] S_N2→ epoxide + X^-
   X = Cl, Br, I

5. CH_3-X + ^-SH —S_N2 (9.15)→ CH_3-SH (thiol) + X^-
   X = Cl, Br, I

6. ethyl−X + $^-SCH_3$ —S_N2 (9.15)→ ethyl−SCH_3 (sulfide) + X^-
   X = Cl, Br, I

Try Problems 9.42h, i; 9.45; 9.46; 9.53; 9.60; 9.63d, i, k.

[2] Reactions of alcohols

1. cyclohexanol (OH, H) —H_2SO_4 or TsOH, dehydration (9.8)→ alkene + H−OH

2. substituted cyclohexanol (OH, H) —$POCl_3$, pyridine, dehydration (9.10)→ alkene + H−OH

3. tertiary cyclohexanol (OH) —H−X, X = Cl, Br, I (9.11)→ alkyl halide + H−OH

4. cyclohexylmethanol (OH) —$SOCl_2$, pyridine, S_N2 (9.12)→ alkyl chloride + pyridinium ($\overset{+}{N}$−H) Cl^- + SO_2

5. cyclohexylmethanol (OH) —PBr_3, S_N2 (9.12)→ alkyl bromide + $HO-PBr_2$

6. cyclohexylmethanol (OH) —Ts−Cl, pyridine (9.13A)→ alkyl tosylate (O−Ts) + pyridinium ($\overset{+}{N}$−H) Cl^-

See Table 9.1, Figure 9.3. Try Problems 9.37f, 9.38, 9.42–9.46.

[3] Reactions of alkyl tosylates

1. cyclohexyl−OTs + $^-SCH_3$ —S_N2 (9.13B)→ cyclohexyl−SCH_3 + ^-OTs

2. cyclohexyl−OTs (H) + $^-OCH_3$ —E2 (9.13B)→ cyclohexene + $H-OCH_3$ + ^-OTs

See Sample Problem 9.8. Try Problems 9.44d; 9.45b; 9.63d, f.

[4] Reactions of ethers

CH3CH2–O–CH3 → (H–X (2 equiv); X = Br, I; (9.14)) → CH3CH2–X + CH_3–X + H–OH

Try Problems 9.55, 9.63j.

[5] Reactions involving thiols and sulfides

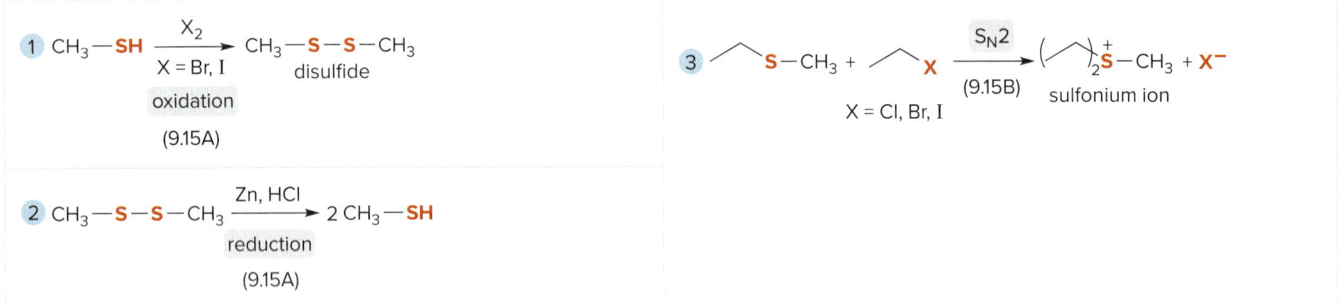

Try Problem 9.63c, k, l.

[6] Reactions of epoxides

See Sample Problem 9.9, Figure 9.5. Try Problems 9.58; 9.59; 9.61; 9.63g, h.

KEY SKILLS

[1] Naming an acyclic alcohol using the IUPAC system (9.3A)

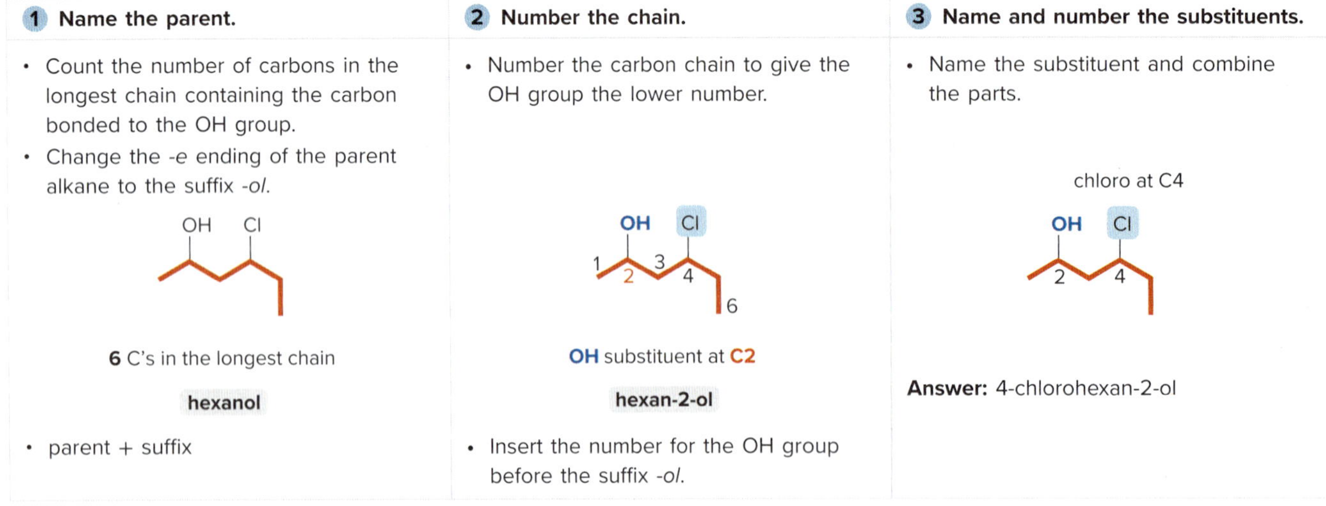

See Sample Problem 9.1 *How To*, p. 345. Try Problems 9.37a; 9.39; 9.41a–d, g.

[2] Using the Williamson ether synthesis to convert an alcohol to an ether (9.6)

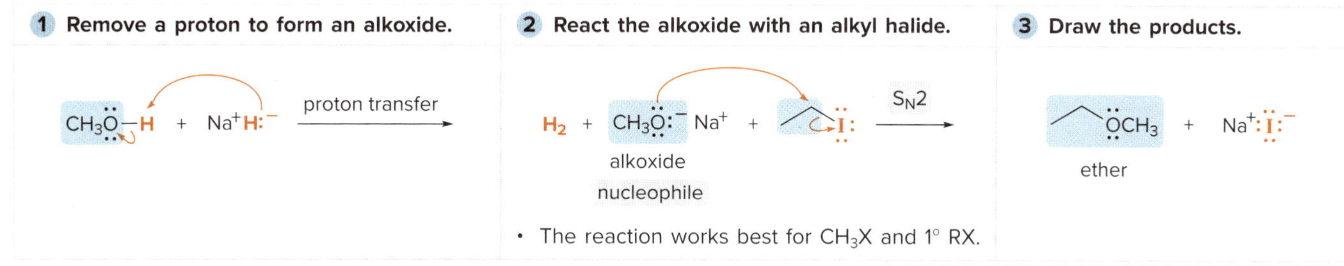

See Sample Problem 9.3. Try Problems 9.42h; 9.45a, c; 9.46; 9.53; 9.63i.

[3] Determining when a carbocation rearrangement might occur (9.9)

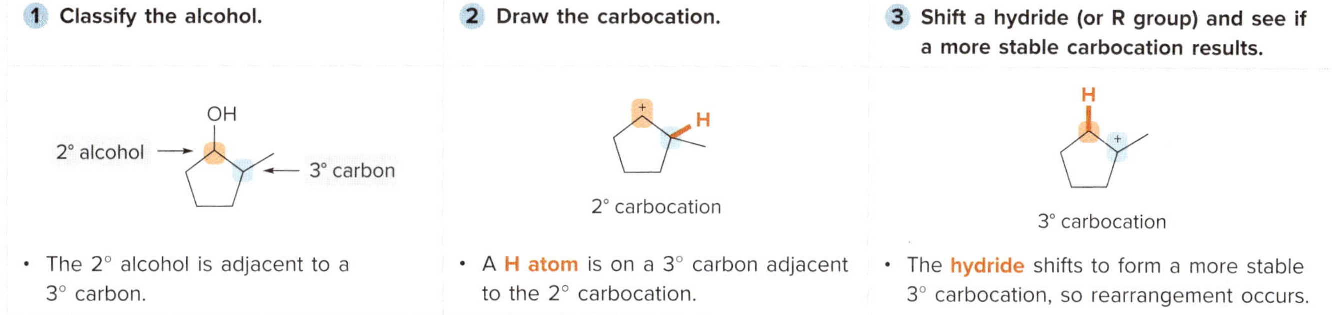

Try Problems 9.43d, f; 9.47a; 9.48; 9.49.

[4] Drawing the products of a dehydration reaction when a 1,2-shift occurs (9.9)

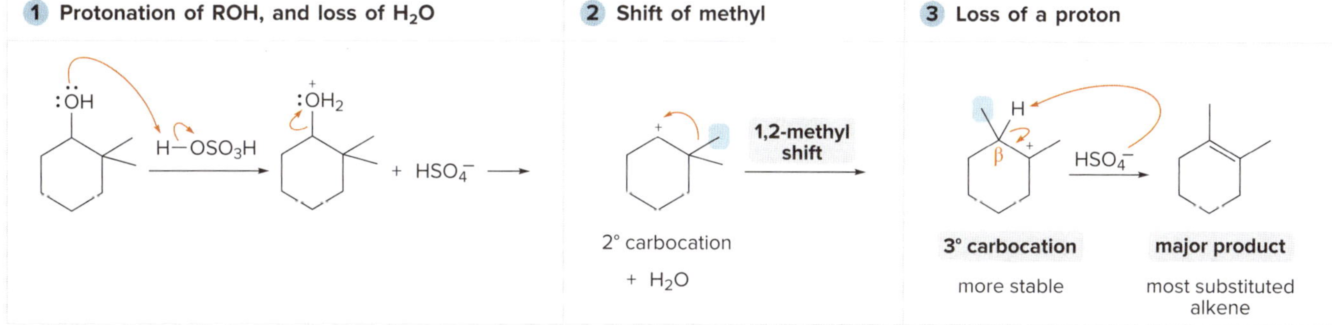

See Sample Problem 9.4. Try Problems 9.43d, f; 9.47a; 9.49.

[5] Drawing the products of an S_N1 reaction when a 1,2-shift occurs (9.11B)

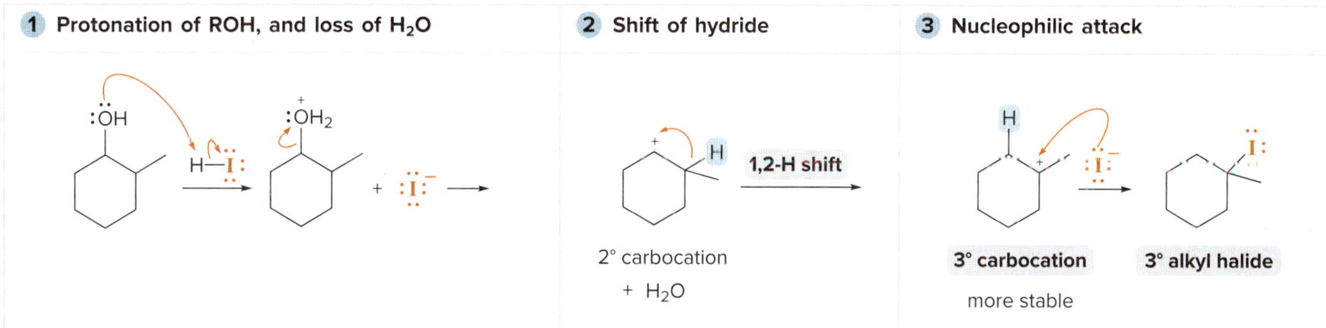

See Sample Problem 9.6. Try Problems 9.37f[4], 9.48.

[6] Determining the stereochemistry in the conversion of ROH to RCl (9.12)

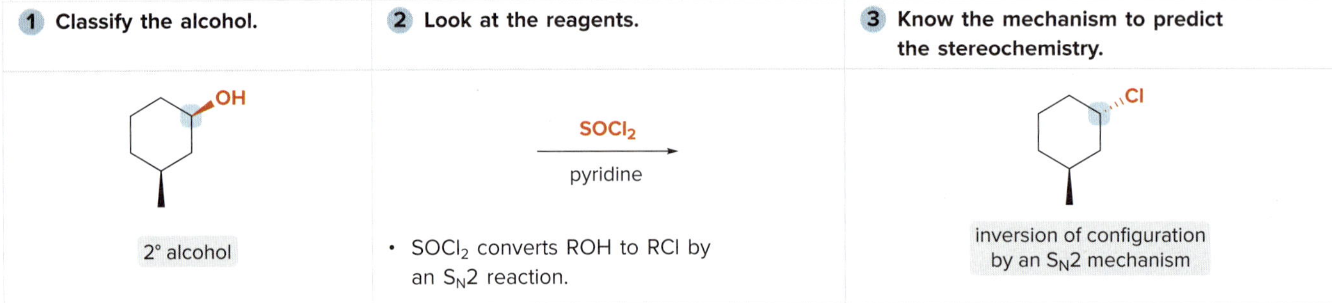

See Sample Problem 9.5. Try Problems 9.37f; 9.38; 9.44a–c, e, f.

[7] Drawing the product of an epoxide ring opening (9.16)

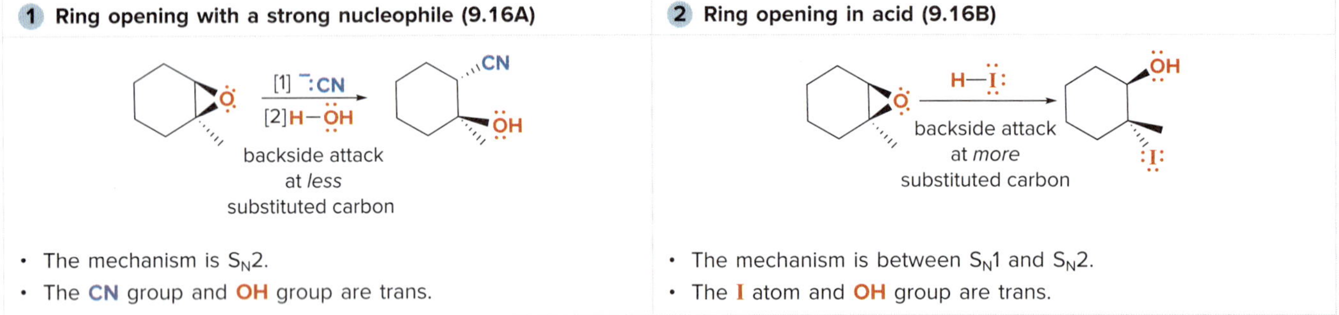

See Sample Problem 9.9, Figure 9.5. Try Problems 9.58; 9.59; 9.61; 9.63g, h.

KEY MECHANISM CONCEPTS IN REACTIONS OF ALCOHOLS

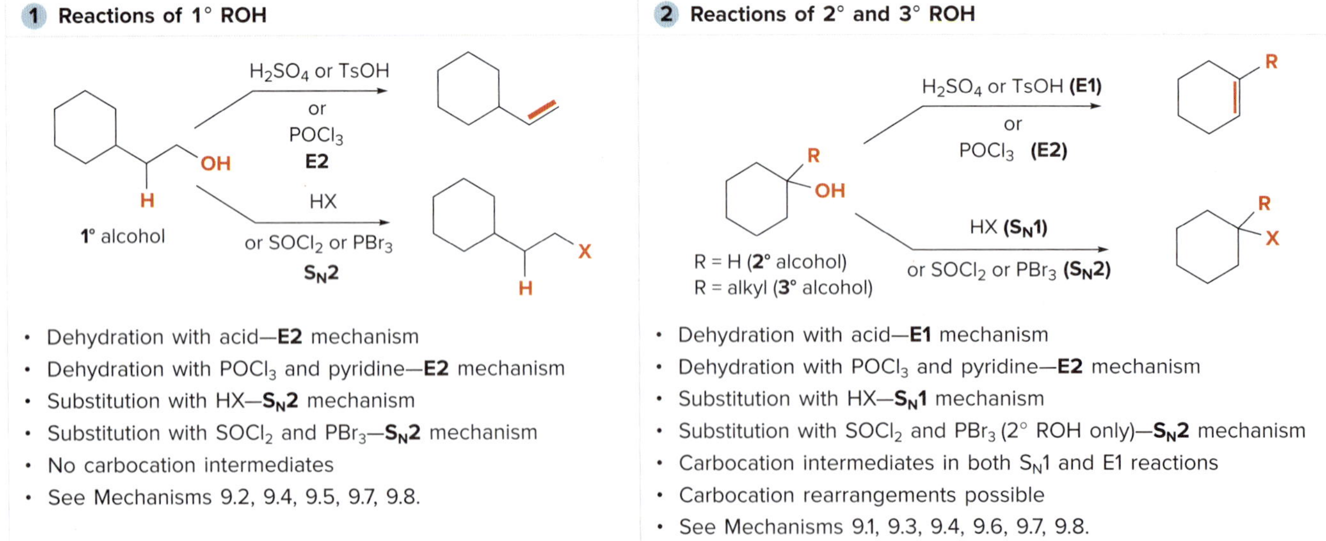

1 Reactions of 1° ROH

- Dehydration with acid—**E2** mechanism
- Dehydration with $POCl_3$ and pyridine—**E2** mechanism
- Substitution with HX—**S_N2** mechanism
- Substitution with $SOCl_2$ and PBr_3—**S_N2** mechanism
- No carbocation intermediates
- See Mechanisms 9.2, 9.4, 9.5, 9.7, 9.8.

2 Reactions of 2° and 3° ROH

- Dehydration with acid—**E1** mechanism
- Dehydration with $POCl_3$ and pyridine—**E2** mechanism
- Substitution with HX—**S_N1** mechanism
- Substitution with $SOCl_2$ and PBr_3 (2° ROH only)—**S_N2** mechanism
- Carbocation intermediates in both S_N1 and E1 reactions
- Carbocation rearrangements possible
- See Mechanisms 9.1, 9.3, 9.4, 9.6, 9.7, 9.8.

CHAPTER 9 MULTIPLE-CHOICE SELF-TEST

The Self-Test consists of multiple-choice questions similar to those found on the American Chemical Society organic chemistry exam. Answers are given at the end of the chapter.

1. Which alcohol reacts fastest in a dehydration reaction?

a. 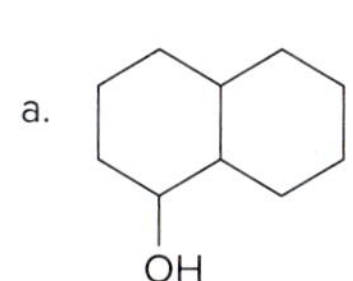

b. 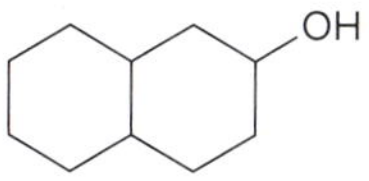 c. 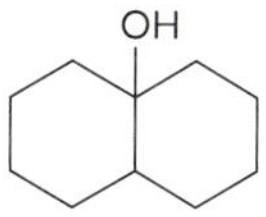 d.

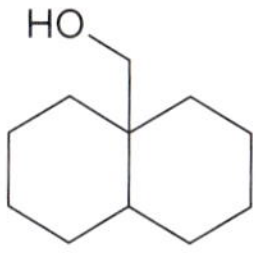

2. What sequence of reactions converts **A** to **B?**

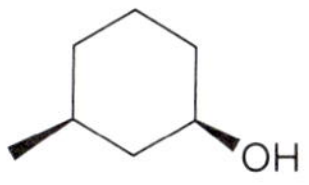

A 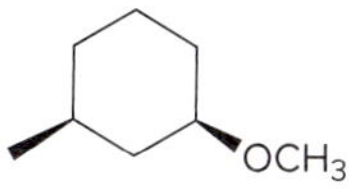 **B**

a. [1] TsCl, pyridine; [2] $NaOCH_3$
b. [1] HBr; [2] $NaOCH_3$
c. [1] HCl; [2] H_2SO_4; [3] $NaOCH_3$
d. [1] NaH; [2] CH_3I

3. Which carbocation is likely to rearrange?

a. 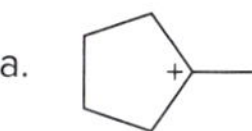
b. 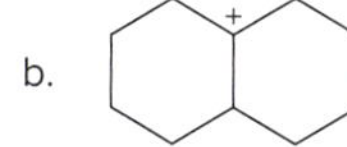
c. 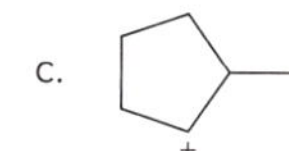
d. 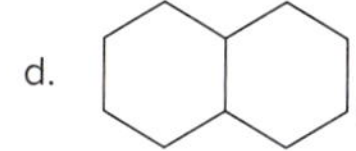

4. Rank **A–D** in order of increasing boiling point.

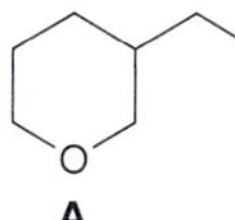

A

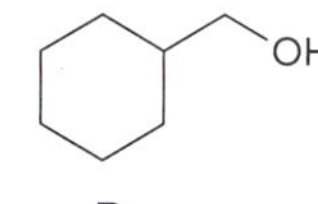

B

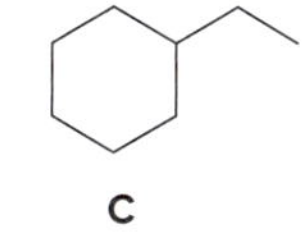
C

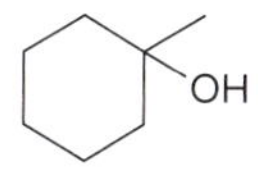

D

a. **C** < **A** < **D** < **B** b. **C** < **A** < **B** < **D** c. **A** < **C** < **D** < **B** d. **B** < **D** < **A** < **C**

5. Give the IUPAC name for the following alcohol.

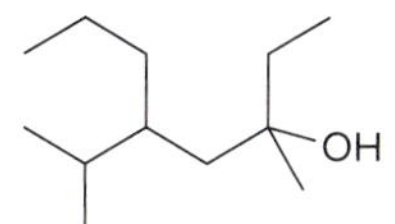

a. 3,6-dimethyl-5-propyloctan-3-ol
b. 5-*sec*-butyl-3-methyloctan-3-ol
c. 3-methyl-5-*sec*-butyloctan-3-ol
d. 3,6-dimethyl-4-propyloctan-6-ol

6. What is the major product formed when **C** undergoes an acid-catalyzed dehydration reaction?

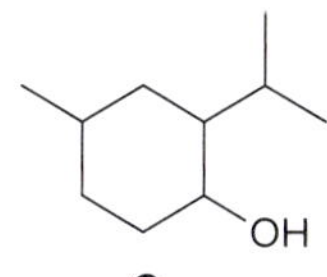

a. 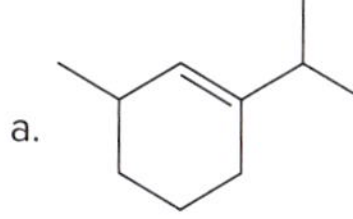
b. 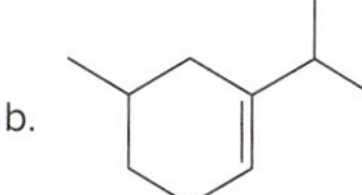
c. 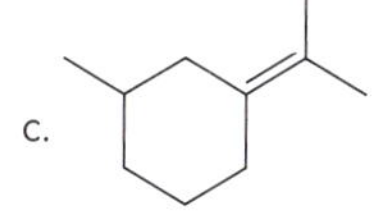
d. 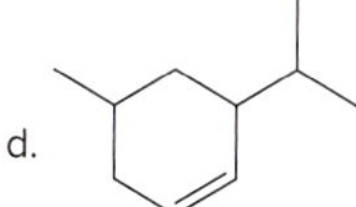

7. What product is formed when epoxide **D** is treated with ^{-}SH, followed by H_2O?

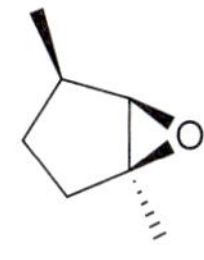

D

a. 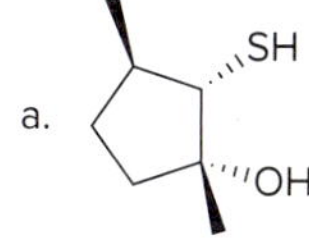

b. 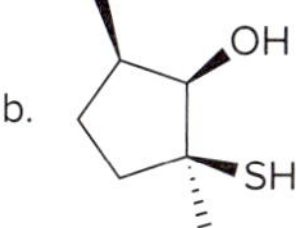

c. 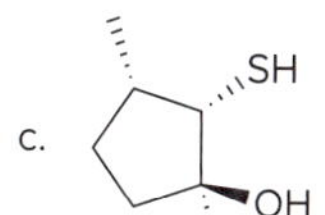

d. 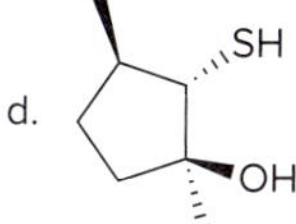

8. What is an acceptable name for epoxide **E?**

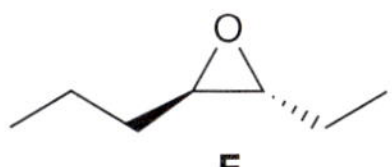

E

a. *trans*-3,4-epoxyhept-3-ene
b. *trans*-4,5-epoxyheptane
c. *trans*-2-ethyl-3-propyloxirane
d. *trans*-1-ethyl-2-propyloxirane

9. What product is formed in the following two-step reaction sequence?

OH [1] PBr_3 [2] NaCN → a. CN b. CN c. CN d. CN

10. Which hydroxy group reacts fastest with HBr?

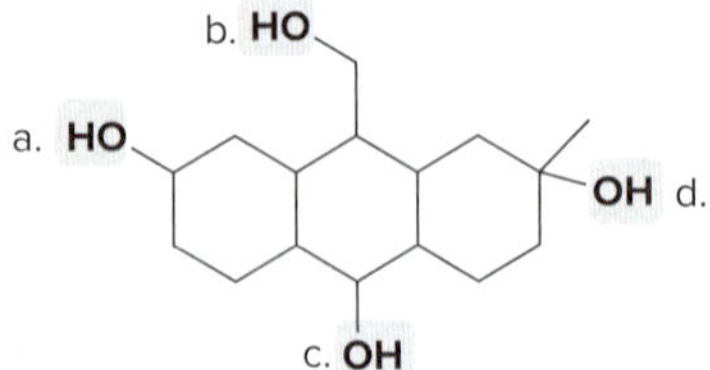

PROBLEMS

Problems Using Three-Dimensional Models

9.37 Answer each question using the ball-and-stick model of compound **A.**

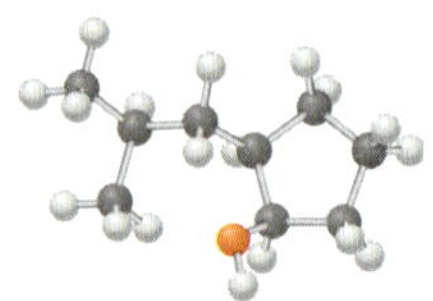

A

a. Give the IUPAC name for **A,** including *R,S* designations for stereogenic centers.
b. Classify **A** as a 1°, 2°, or 3° alcohol.
c. Draw a stereoisomer for **A** and give its IUPAC name.
d. Draw a constitutional isomer that contains an OH group and give its IUPAC name.
e. Draw a constitutional isomer that contains an ether and give its IUPAC name.
f. Draw the products formed (including stereochemistry) when **A** is treated with each reagent: [1] NaH; [2] H_2SO_4; [3] $POCl_3$, pyridine; [4] HCl; [5] $SOCl_2$, pyridine; [6] TsCl, pyridine.

9.38 Draw the product and indicate the stereochemistry when the given alcohol is treated with each reagent: (a) HBr; (b) PBr_3; (c) HCl; (d) $SOCl_2$ and pyridine.

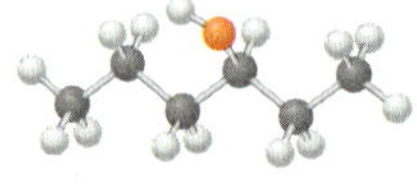

Nomenclature

9.39 Give the IUPAC name for each alcohol.

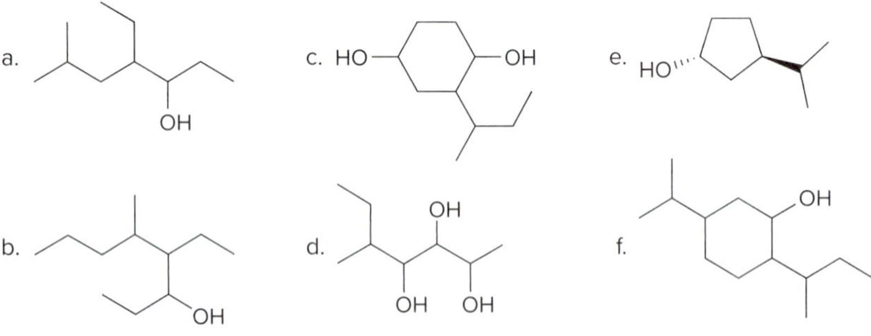

9.40 Name each ether, epoxide, thiol, and sulfide.

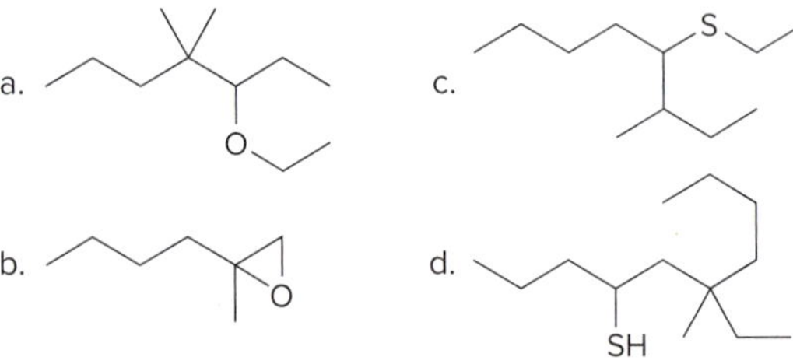

9.41 Give the structure corresponding to each name.

a. *trans*-2-methylcyclohexanol
b. 2,3,3-trimethylbutan-2-ol
c. 6-*sec*-butyl-7,7-diethyldecan-4-ol
d. 3-chloropropane-1,2-diol
e. 1,2-epoxy-1,3,3-trimethylcyclohexane
f. 1-ethoxy-3-ethylheptane
g. (2*R*,3*S*)-3-isopropylhexan-2-ol
h. (*S*)-2-ethoxy-1,1-dimethylcyclopentane
i. 4-ethylheptane-3-thiol
j. 1-isopropylthio-2-methylcyclohexane

Alcohols

9.42 Draw the organic product(s) formed when $CH_3CH_2CH_2OH$ is treated with each reagent.

a. H_2SO_4
b. NaH
c. HCl + $ZnCl_2$
d. HBr
e. $SOCl_2$, pyridine
f. PBr_3
g. TsCl, pyridine
h. [1] NaH; [2] CH_3CH_2Br
i. [1] TsCl, pyridine; [2] NaSH
j. $POCl_3$, pyridine

9.43 What alkenes are formed when each alcohol is dehydrated with TsOH? Label the major product when a mixture results.

a. 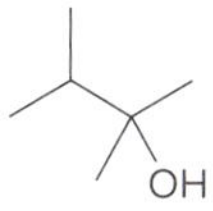 OH b. 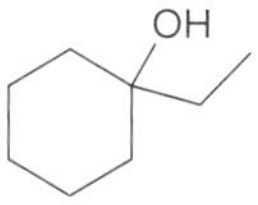 OH c. 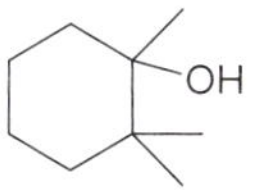 OH d. 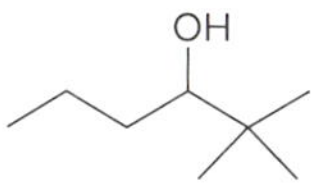 OH e. 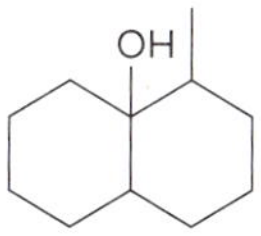 OH f. 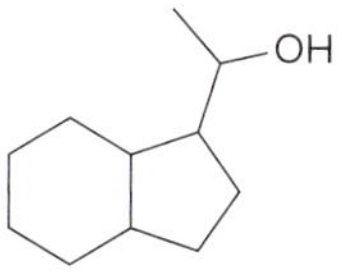 OH

9.44 Draw the products of each reaction and indicate stereochemistry around stereogenic centers.

a. 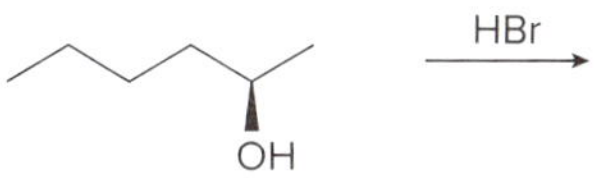 OH → HBr
b. 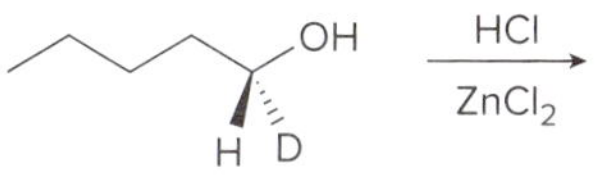 OH, H D → HCl, $ZnCl_2$
c. OH → $SOCl_2$, pyridine
d. 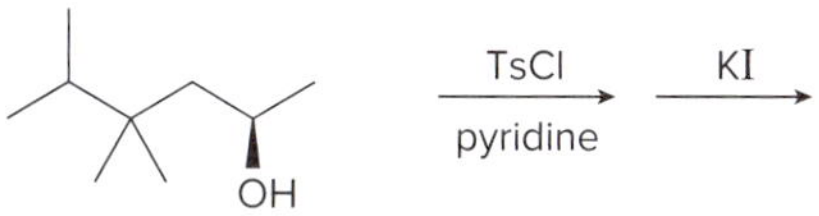 OH → TsCl, pyridine → KI
e. OH → HBr
f. 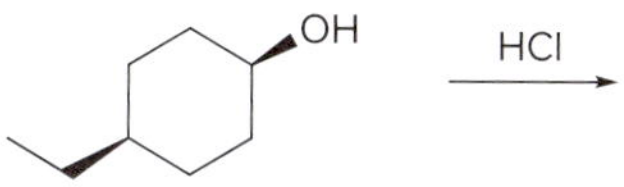 OH → HCl

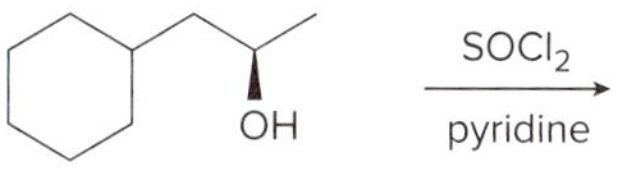

9.45 Draw the substitution product formed (including stereochemistry) when (*R*)-hexan-2-ol is treated with each series of reagents: (a) NaH, followed by CH_3I; (b) TsCl and pyridine, followed by $NaOCH_3$; (c) PBr_3, followed by $NaOCH_3$. Which two routes produce identical products?

9.46 Draw the product (including stereochemistry) formed in each two-step reaction sequence.

a. 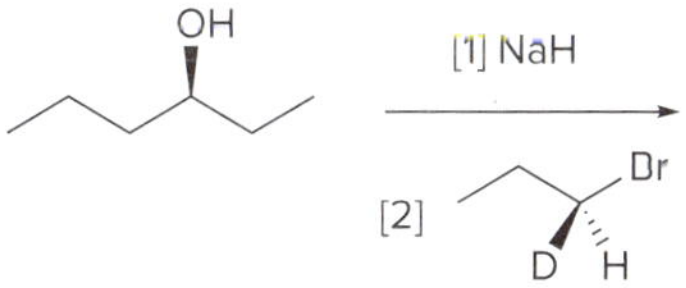 OH → [1] NaH; [2] Br, D H
b. 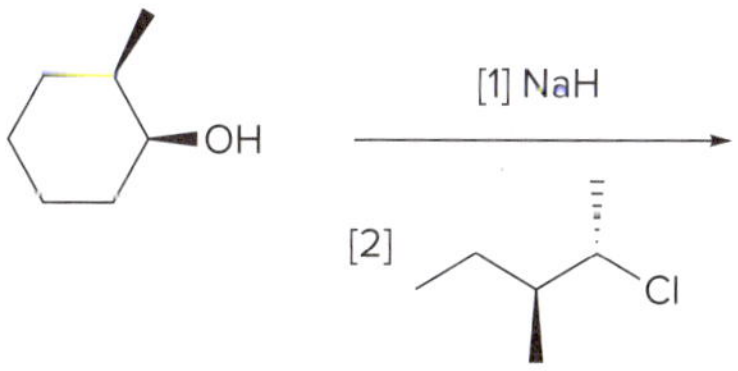 OH → [1] NaH; [2] Cl

9.47 (a) What is the major alkene formed when **A** is dehydrated with H_2SO_4? (b) What is the major alkene formed when **A** is treated with $POCl_3$ and pyridine? Explain why the major product is different in these reactions.

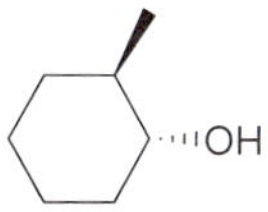

OH

A

9.48 Reaction of 2° alcohol **A** with HCl forms three alkyl chlorides, all of which result from rearrangement of the 2° carbocation initially formed. Draw the structures of these products and a mechanism that illustrates how each is formed.

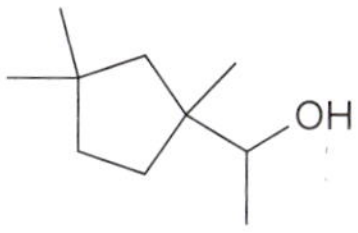

OH

A

9.49 Draw a stepwise mechanism for the following reaction.

OH $\xrightarrow{H_2SO_4}$ + + + H_2O

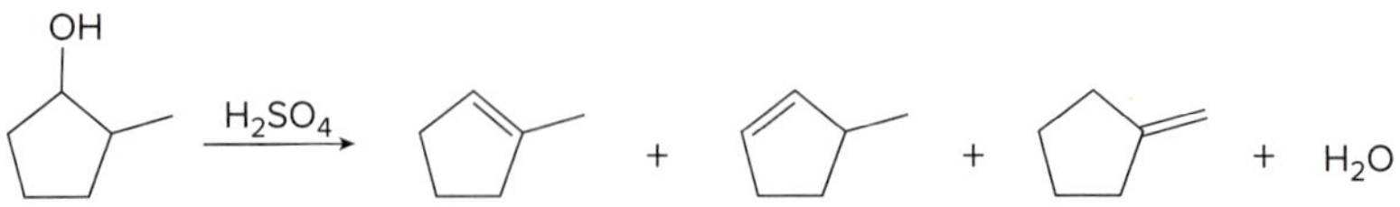

9.50 Sometimes carbocation rearrangements can change the size of a ring. Draw a stepwise, detailed mechanism for the following reaction.

OH $\xrightarrow{H_2SO_4}$ + H_2O

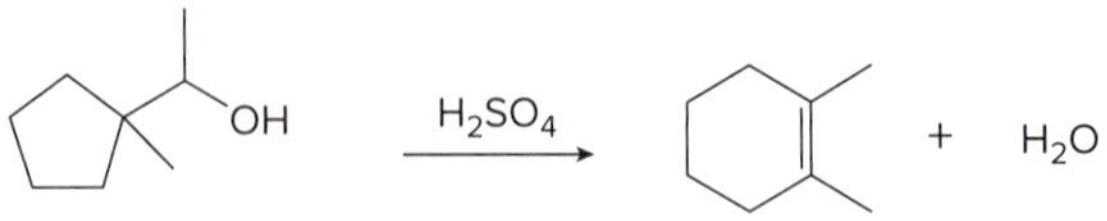

9.51 An allylic alcohol contains an OH group on a carbon atom adjacent to a C–C double bond. Treatment of allylic alcohol **A** with HCl forms a mixture of two allylic chlorides, **B** and **C.** Draw a stepwise mechanism that illustrates how both products are formed.

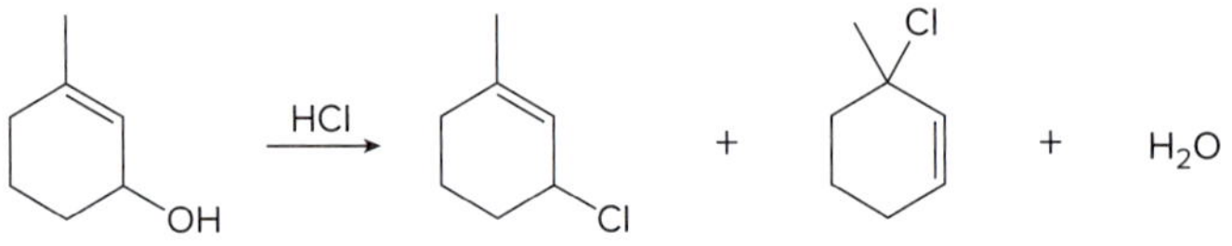

OH $\xrightarrow{HCl}$ Cl + Cl + H_2O

A **B** **C**

9.52 Draw a stepwise, detailed mechanism for the following reaction.

O OH $\xrightarrow{H_2SO_4}$ O + H_2O

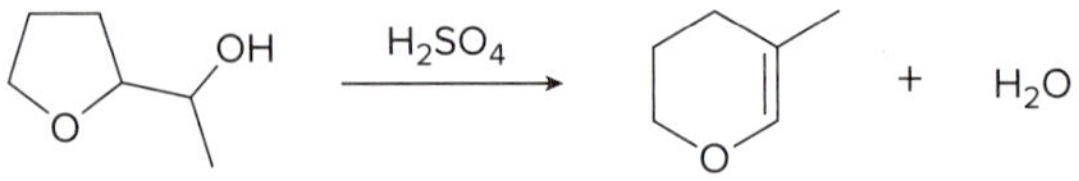

Ethers

9.53 Draw two different routes to each of the following ethers using a Williamson ether synthesis. Indicate the preferred route (if there is one).

a. 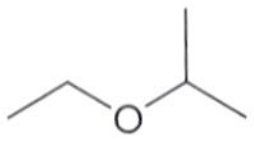 O b. 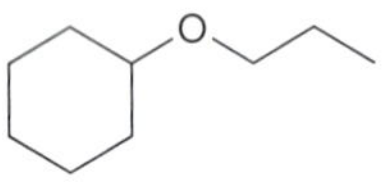 O c. 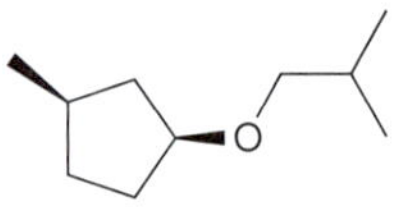 O

9.54 Explain why it is not possible to prepare *tert*-butyl phenyl ether using a Williamson ether synthesis.

9.55 Draw the products formed when each ether is treated with two equivalents of HBr.

a. 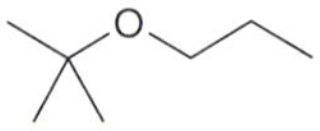 O b. 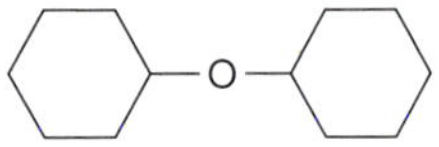 O c. 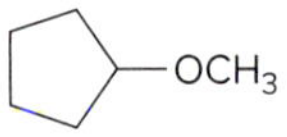 OCH_3

9.56 Draw a stepwise mechanism for each reaction.

a. 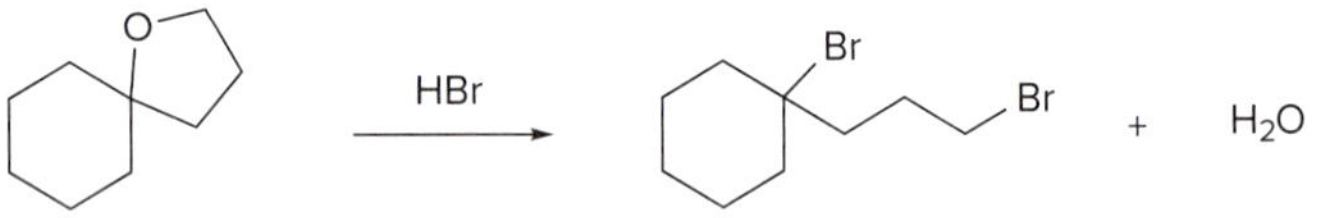 O $\xrightarrow{HBr}$ Br Br + H_2O

b. 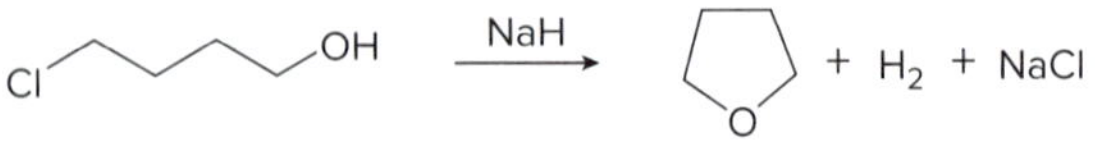 Cl OH $\xrightarrow{NaH}$ O + H_2 + NaCl

9.57 Draw a stepwise mechanism for the following reaction.

O $\xrightarrow{CF_3CO_2H}$ OH +

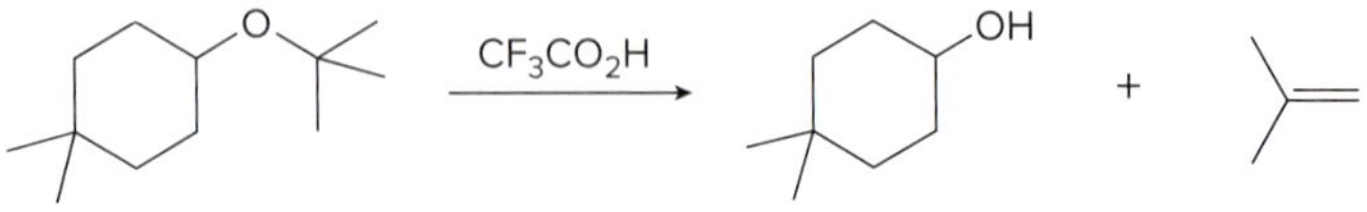

Epoxides

9.58 Draw the products formed when ethylene oxide is treated with each reagent.

a. HBr
b. H_2O (H_2SO_4)
c. [1] $CH_3CH_2O^-$; [2] H_2O
d. [1] $HC\equiv C^-$; [2] H_2O
e. [1] ^-OH; [2] H_2O
f. [1] CH_3S^-; [2] H_2O

9.59 Draw the products of each reaction.

a. $\xrightarrow[H_2SO_4]{CH_3CH_2OH}$

b. $\xrightarrow[\text{[2] } H_2O]{\text{[1] } CH_3CH_2O^- \; Na^+}$

c. $\xrightarrow{HBr}$

d. $\xrightarrow[\text{[2] } H_2O]{\text{[1] NaCN}}$

9.60 When each halohydrin is treated with NaH, a product of molecular formula C_4H_8O is formed. Draw the structure of the product and indicate its stereochemistry.

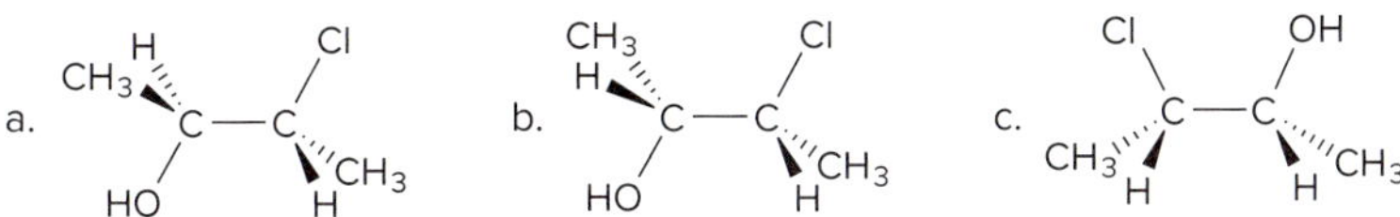

9.61 (a) What reaction conditions are needed to convert (*R*)-2-ethyl-2-methyloxirane to (*R*)-2-methylbutane-1,2-diol? (b) What reaction conditions are needed to convert (*R*)-2-ethyl-2-methyloxirane to (*S*)-2-methylbutane-1,2-diol?

9.62 Tezacaftor is one component of a three-drug therapy (trade name Trikafta) approved in the United States in 2019 for patients with cystic fibrosis. One reaction in a multistep synthesis of tezacaftor involves the reaction of **A** with epoxide **B** to form intermediate **C**. Draw the structure of **C,** including the stereochemistry at any newly formed stereogenic centers.

O_2N, Br, F, NH_2 — **A** + **B** $\xrightarrow{\text{Lewis acid}}$ **C** $\xrightarrow{\text{several steps}}$ tezacaftor

Problems That Combine Concepts

9.63 Draw the products of each reaction, and indicate the stereochemistry where appropriate.

a. OTs $\xrightarrow{KOC(CH_3)_3}$

b. OH $\xrightarrow{HBr}$

c. SH $\xrightarrow{Br_2}$

d. OTs $\xrightarrow{KSH}$

e. OH $\xrightarrow{PBr_3}$

f. OH, D $\xrightarrow[\text{[2] } CH_3CO_2^-]{\text{[1] TsCl, pyridine}}$

g. $\xrightarrow{HBr}$

h. $\xrightarrow[\text{[2] } H_2O]{\text{[1] } NaOCH_3}$

i. OH $\xrightarrow[\text{[2] } \text{CH}_2\text{=CHCH}_2\text{I}]{\text{[1] NaH}}$

j. $\xrightarrow[\text{(2 equiv)}]{HI}$

k. D, Br + $CH_3CH_2\ddot{S}:^-\ Na^+$

l. S $\xrightarrow{CH_3CH_2I}$

9.64 Prepare each compound from cyclopentanol. More than one step may be needed.

a. Cl b. c. OCH_3 d. CN

9.65 Identify **Y** in the following reaction, one step in the synthesis of methylphenidate, a drug used to treat attention deficit hyperactivity disorder (ADHD).

X + Y ⟶ Z ⟶ methylphenidate

9.66 Propranolol, an antihypertensive agent used in the treatment of high blood pressure, can be prepared from 1-naphthol, epichlorohydrin, and isopropylamine using two successive nucleophilic substitution reactions. Devise a stepwise synthesis of propranolol from these starting materials.

propranolol 1-naphthol epichlorohydrin isopropylamine

Spectroscopy

Problems 9.67–9.70 are intended for students who have already learned about spectroscopy in Chapters A–C.

9.67 Propose a structure consistent with each set of spectral data:

a. $C_6H_{14}O$: IR peak at 3600–3200 cm^{-1}; NMR (ppm):
0.8 (triplet, 6 H) 1.5 (quartet, 4 H)
1.0 (singlet, 3 H) 1.6 (singlet, 1 H)

b. $C_6H_{14}O$: IR peak at 3000–2850 cm^{-1}; NMR (ppm):
1.10 (doublet, relative area = 6)
3.60 (septet, relative area = 1)

9.68 As we will learn in Chapter 17, reaction of $(CH_3)_2CO$ with LiC≡CH followed by H_2O affords compound **D,** which has a molecular ion in its mass spectrum at 84 and prominent absorptions in its IR spectrum at 3600–3200, 3303, 2938, and 2120 cm^{-1}. **D** shows the following ^1H NMR spectral data: 1.53 (singlet, 6 H), 2.37 (singlet, 1 H), and 2.43 (singlet, 1 H) ppm. What is the structure of **D?**

9.69 Treatment of $(CH_3)_2CHCH(OH)CH_2CH_3$ with TsOH affords two products (**M** and **N**) with molecular formula C_6H_{12}. The ^1H NMR spectra of **M** and **N** are given below. Propose structures for **M** and **N,** and draw a mechanism to explain their formation.

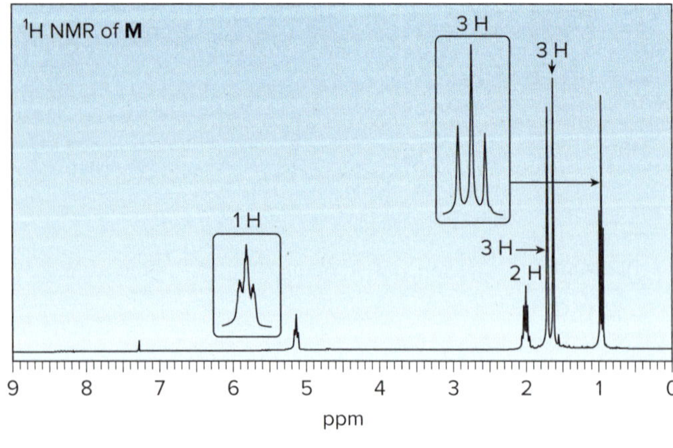

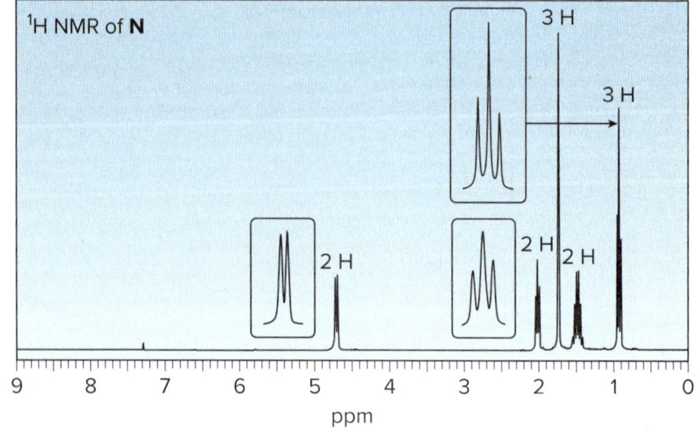

9.70 Use the ^{1}H NMR and IR spectra given below to identify compound **A,** having molecular formula $C_4H_8O_2$.

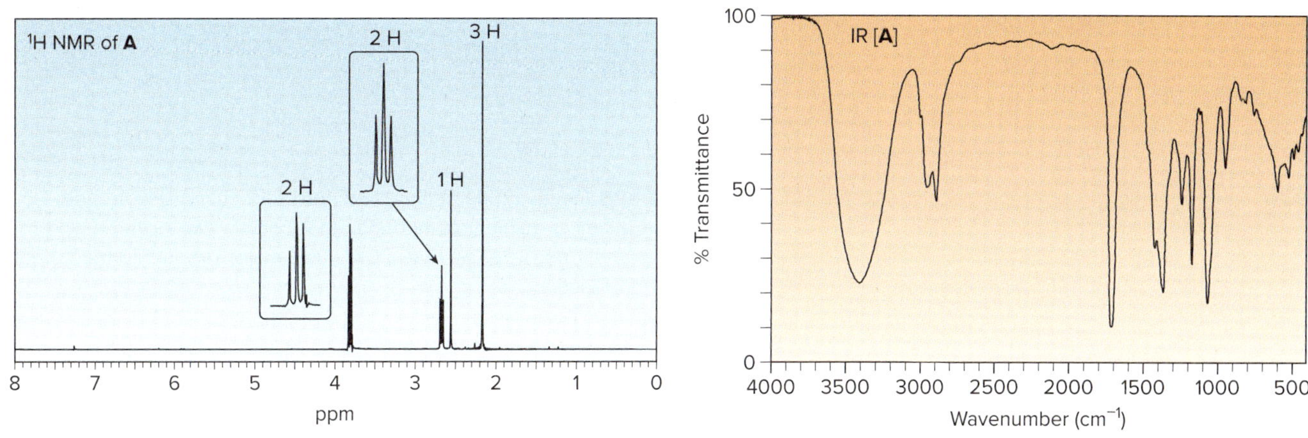

Additional spectroscopy problems involving alcohols, ethers, and epoxides are given in Chapters A–C:

- Mass spectrometry: A.4a, A.13, A.28, A.29a, A.31, A.32, A.34, A.36, A.37
- Infrared spectroscopy: B.10; B.14; B.15b; B.20b; B.21b; B.22b; B.24; B.30(**B**), (**D**); B.31c
- Nuclear magnetic resonance spectroscopy: C.20; C.23a, b; C.25; C.27b; C.29; C.31c; C.36a; C.38b, e, g, h; C.44b, c; C.45c, e, f; C.52b, e; C.54a, c; C.72; C.73b; C.80; C.83

Challenge Problems

9.71 Draw a stepwise mechanism for the following reaction, one step in a published synthesis of a prostaglandin (Section 19.5).

NaH, (epichlorohydrin) → H_2O → (product with OH)

9.72 Epoxides are converted to allylic alcohols with nonnucleophilic bases such as lithium diethylamide [$LiN(CH_2CH_3)_2$]. Draw a stepwise mechanism for the conversion of 1,2-epoxycyclohexane to cyclohex-2-en-1-ol with this base. Explain why a strong bulky base must be used in this reaction.

[1] $LiN(CH_2CH_3)_2$ [2] H_2O → cyclohex-2-en-1-ol + $HN(CH_2CH_3)_2$ + LiOH

9.73 Rearrangements can occur during the dehydration of 1° alcohols even though no 1° carbocation is formed—that is, a 1,2-shift occurs as the $C{-}OH_2^+$ bond is broken, forming a more stable 2° or 3° carbocation, as shown in Equation [1]. Using this information, draw a stepwise mechanism for the reaction shown in Equation [2]. We will see another example of this type of rearrangement in Section 16.5C.

[1] 1° alcohol → no 1° carbocation at this step —1,2-shift→ 2° carbocation + $H_2\ddot{O}$:

[2] H_2SO_4

9.74 Dehydration of 1,2,2-trimethylcyclohexanol with H_2SO_4 affords 1-*tert*-butylcyclopentene as a minor product. (a) Draw a stepwise mechanism that shows how this alkene is formed. (b) Draw other alkenes formed in this dehydration. At least one must contain a five-membered ring.

1,2,2-trimethylcyclohexanol —H_2SO_4→ 1-*tert*-butylcyclopentene + other alkenes

9.75 Draw a stepwise mechanism for the following reaction.

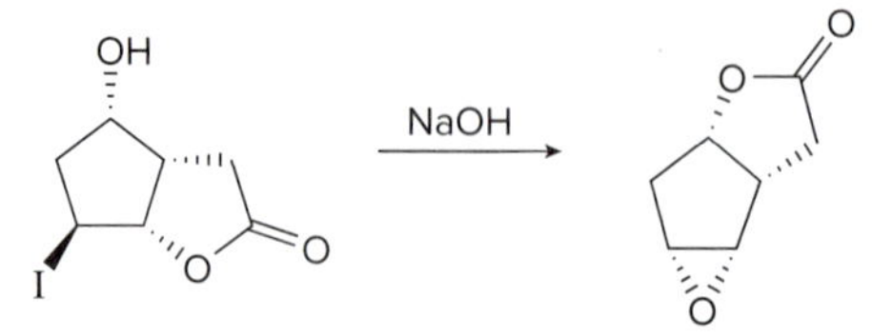

NaOH

9.76 Ring opening of epoxide **A** occurs in water to form a mixture of **B** (major product) and **C** (minor product). (a) Use curved arrows to illustrate how cyclization yields both products. (b) In a similar fashion, compounds that contain two or three epoxides can undergo an epoxide-opening cascade to form two or three new rings, respectively. Draw the major product formed when diepoxide **D** and triepoxide **E** each undergo a similar cyclization reaction.

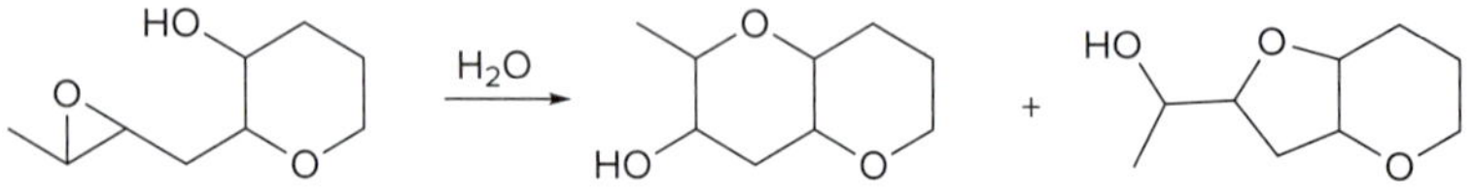

H_2O

A **B** **C**

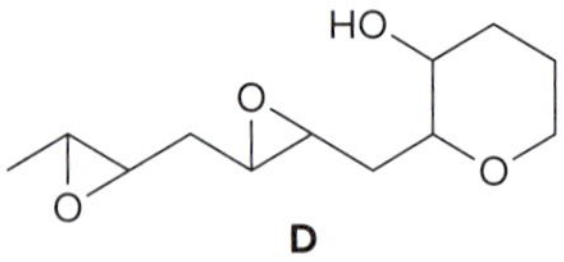

D 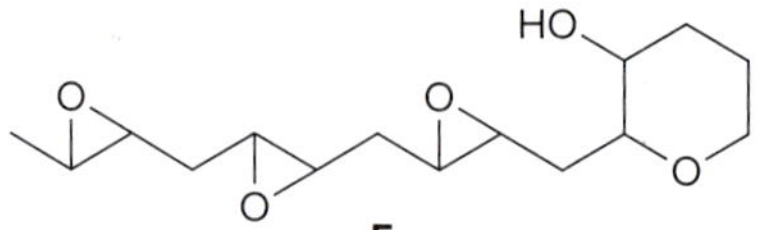 **E**

SELF-TEST ANSWERS

1. c 2. d 3. c 4. a 5. a 6. c 7. d 8. c 9. b 10. d

Alkenes and Addition Reactions

10

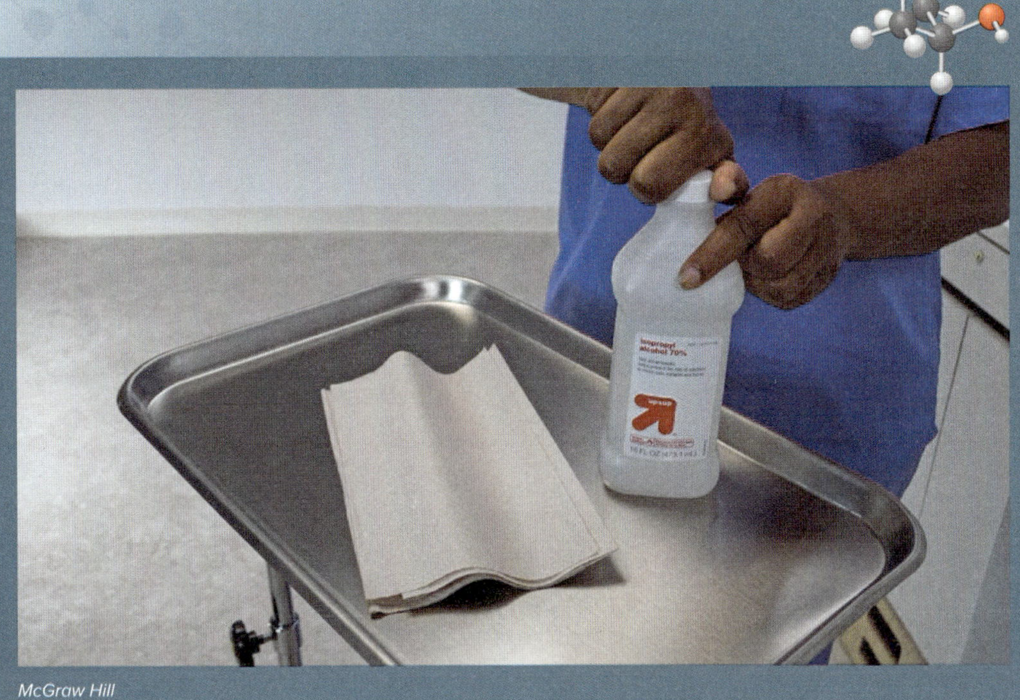

McGraw Hill

10.1 Introduction
10.2 Calculating degrees of unsaturation
10.3 Nomenclature
10.4 Properties of alkenes
10.5 Interesting alkenes
10.6 Lipids—Part 2
10.7 Preparation of alkenes
10.8 Introduction to addition reactions
10.9 Hydrohalogenation—Electrophilic addition of HX
10.10 Markovnikov's rule
10.11 Stereochemistry of electrophilic addition of HX
10.12 Hydration—Electrophilic addition of water
10.13 Halogenation—Addition of halogen
10.14 Stereochemistry of halogenation
10.15 Halohydrin formation
10.16 Hydroboration–oxidation
10.17 Keeping track of reactions
10.18 Alkenes in organic synthesis

2-Propanol, commonly called isopropyl alcohol, is the active ingredient in both rubbing alcohol that disinfects skin and some hand sanitizers that kill germs. Because it dissolves many polar and nonpolar compounds, 2-propanol is used as a solvent in industrial processes and laboratory reactions. Hand sanitizers that contain 70% 2-propanol in water are effective at killing the Sars-CoV-2 virus, the virus responsible for the COVID-19 pandemic, by dissolving the lipid membrane that surrounds the virus core. Much of the 2-propanol used in consumer products is synthesized by the hydration of propene, an addition reaction of an alkene that is the subject of Chapter 10.

Why Study . . . Alkenes?

In Chapters 10 and 11, we turn our attention to **alkenes** and **alkynes,** compounds that contain one and two π bonds, respectively. Because π bonds are easily broken, alkenes and alkynes undergo **addition,** the third general type of organic reaction. These multiple bonds also make carbon atoms electron rich, so alkenes and alkynes react with a wide variety of electrophilic reagents in addition reactions that are very versatile in organic synthesis.

In Chapter 10, we review the properties and synthesis of alkenes first, and then concentrate on reactions. **Every new reaction in Chapter 10 is an *addition reaction.*** The most challenging part is learning the reagents, mechanism, and stereochemistry that characterize each individual reaction.

10.1 Introduction

Alkenes are also called **olefins.**

Alkenes are compounds that contain a carbon–carbon double bond. **Terminal alkenes** have the double bond at the end of the carbon chain, whereas **internal alkenes** have at least one carbon atom bonded to each end of the double bond. **Cycloalkenes** contain a double bond in a ring.

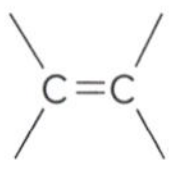

alkene

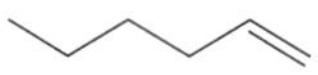

terminal alkene

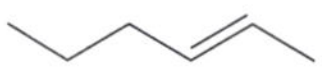

internal alkene

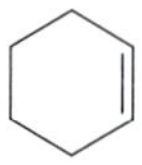

cycloalkene

The double bond of an alkene consists of one σ bond and one π bond. Each carbon is sp^2 hybridized and trigonal planar, and all bond angles are approximately 120° (Section 8.2A).

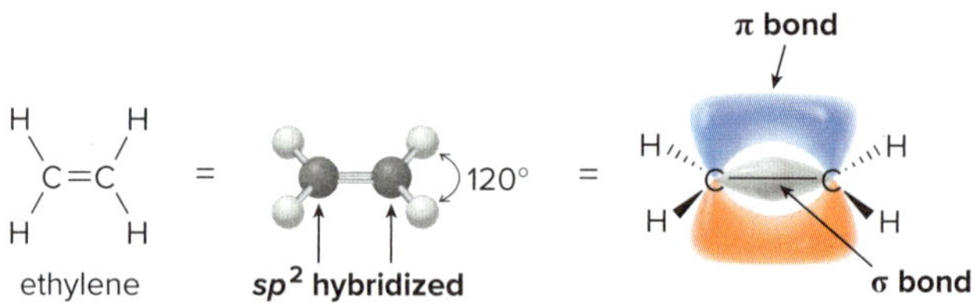

Bond dissociation energies of the C–C bonds in ethane (a σ bond only) and ethylene (one σ and one π bond) can be used to estimate the strength of the π component of the double bond. If we assume that the σ bond in ethylene is similar in strength to the σ bond in ethane (368 kJ/mol), then the π bond is worth 267 kJ/mol.

$CH_2{=}CH_2$		$CH_3{-}CH_3$		
635 kJ/mol (σ + π bond)	−	368 kJ/mol (σ bond)	=	267 kJ/mol **π bond only**

- **The π bond is much *weaker* than the σ bond of a C–C double bond, making it much more easily broken. As a result, alkenes undergo many reactions that alkanes do not.**

Other features of the carbon–carbon double bond, which were presented in Chapter 8, are summarized in Table 10.1.

Cycloalkenes having fewer than eight carbon atoms have a cis geometry. A trans cycloalkene must have a carbon chain long enough to connect the ends of the double bond without introducing too much strain. *trans*-Cyclooctene is the smallest, isolable trans cycloalkene, but it is

Table 10.1 Properties of the Carbon–Carbon Double Bond

Property	Result
Restricted rotation	• **The rotation around the C–C double bond is restricted.** Rotation can occur only if the π bond breaks and then re-forms, a process that is unfavorable (Section 8.2B).
Stereoisomerism	• Whenever the two groups on each end of a C=C are different from each other, two diastereomers are possible. *Cis-* and *trans-*but-2-ene (drawn at the bottom of Table 10.1) are diastereomers (Section 8.2B).
Stability	• **Trans** alkenes are generally more stable than **cis** alkenes. • **The stability of an alkene increases as the number of R groups on the C=C increases** (Section 8.2C). but-1-ene — *cis*-but-2-ene — *trans*-but-2-ene Increasing stability →

considerably less stable than *cis*-cyclooctene, making it one of the few alkenes having a higher-energy trans isomer.

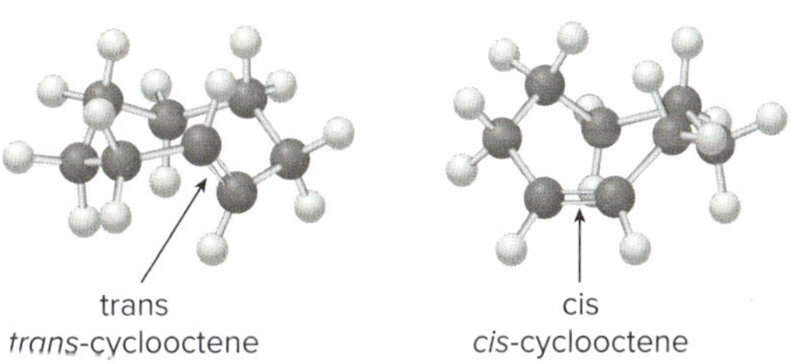

Problem 10.1 Draw the six alkenes of molecular formula C_5H_{10}. Label one pair of diastereomers.

10.2 Calculating Degrees of Unsaturation

An acyclic alkene has the general molecular formula C_nH_{2n}, giving it two fewer hydrogens than an acyclic alkane with the same number of carbons.

• **Alkenes are *unsaturated hydrocarbons* because they have fewer than the maximum number of hydrogen atoms per carbon.**

In Chapter 12, we will learn how to use the hydrogenation of π bonds to determine how many degrees of unsaturation result from π bonds and how many result from rings.

Cycloalkanes also have the general molecular formula C_nH_{2n}. Thus, **each π bond or ring removes two hydrogen atoms from a molecule, and this introduces one *degree of unsaturation.*** The number of degrees of unsaturation for a given molecular formula can be calculated by comparing the *actual* number of H atoms in a compound and the *maximum* number of H atoms possible. Remember that for *n* carbons, the **maximum number of H atoms is 2*n* + 2** (Section 4.1). This procedure gives the total number of rings and π bonds in a molecule.

Sample Problem 10.1 Calculating the Number of Degrees of Unsaturation in a Hydrocarbon

Calculate the number of degrees of unsaturation in a compound of molecular formula C_4H_6, and propose possible structures.

Solution

[1] Calculate the maximum number of H's possible.

- For n carbons, the maximum number of H's is $\mathbf{2n + 2}$; in this example, $2n + 2 = 2(4) + 2 = 10$.

[2] Subtract the actual number of H's from the maximum number and divide by two.

- 10 H's (maximum) − 6 H's (actual) = 4 H's fewer than the maximum number.

$$\frac{\text{4 H's fewer than the maximum}}{\text{2 H's removed for each degree of unsaturation}} =$$

Answer: two degrees of unsaturation

A compound with two degrees of unsaturation has:

	two rings	*or*	**two π bonds**	*or*	**one ring and one π bond**

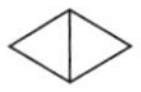

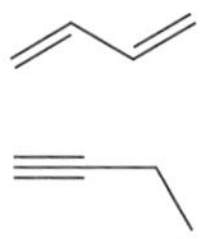

Possible structures for C_4H_6:

Problem 10.2 Calculate the number of degrees of unsaturation for each molecular formula, and propose two possible structures: (a) C_8H_{12}; (b) $C_{10}H_{10}$.

More Practice: Try Problems 10.34a, b; 10.35.

This procedure can be extended to compounds that contain heteroatoms such as oxygen, nitrogen, and halogen, as illustrated in Sample Problem 10.2.

Sample Problem 10.2 Calculating the Number of Degrees of Unsaturation in Compounds with O, X, or N

Calculate the number of degrees of unsaturation for each molecular formula: (a) C_5H_8O; (b) $C_6H_{11}Cl$; (c) C_8H_9N. Propose one possible structure for each compound.

Solution

a. When a compound contains an oxygen atom, **use the given number of C's and H's and *ignore the O atom*** in the calculation; that is, C_5H_8O is equivalent to C_5H_8 when calculating degrees of unsaturation.

[1] For 5 C's, the maximum number of H's = $2n + 2 = 2(5) + 2 = 12$.
[2] Because the compound contains only 8 H's, it has 12 − 8 = 4 H's fewer than the maximum number.
[3] Each degree of unsaturation removes 2 H's, so the answer in Step [2] must be divided by 2.
Answer: two degrees of unsaturation

b. **A compound with a halogen atom is equivalent to a hydrocarbon having one *more* H;** that is, $C_6H_{11}Cl$ is equivalent to C_6H_{12} when calculating degrees of unsaturation.

[1] For 6 C's, the maximum number of H's = $2n + 2 = 2(6) + 2 = 14$.
[2] Because the compound contains only 12 H's, it has 14 − 12 = 2 H's fewer than the maximum number.
[3] Each degree of unsaturation removes 2 H's, so the answer in Step [2] must be divided by 2.
Answer: one degree of unsaturation

c. **A compound with a nitrogen atom is equivalent to a hydrocarbon having one *fewer* H;** that is, C_8H_9N is equivalent to C_8H_8 when calculating degrees of unsaturation.

[1] For 8 C's, the maximum number of H's = $2n + 2 = 2(8) + 2 = 18$.
[2] Because the compound contains only 8 H's, it has 18 − 8 = 10 H's fewer than the maximum number.
[3] Each degree of unsaturation removes 2 H's, so the answer in Step [2] must be divided by 2.
Answer: five degrees of unsaturation

Possible structures:

a. b. Cl c. N H

Problem 10.3 How many degrees of unsaturation are present in each compound?

a. C_6H_6 b. C_8H_{18} c. C_7H_8O d. $C_7H_{11}Br$ e. C_5H_9N

More Practice: Try Problem 10.34c–h.

Problem 10.4 How many degrees of unsaturation does each of the following drugs contain?

a. zolpidem (sleep aid sold as Ambien), $C_{19}H_{21}N_3O$

b. mefloquine (antimalarial drug), $C_{17}H_{16}F_6N_2O$

10.3 Nomenclature

- **In the IUPAC system, an alkene is identified by the suffix *-ene*.**

10.3A General IUPAC Rules

How To Name an Alkene

Example Give the IUPAC name of the following alkene:

Step [1] **Find the longest carbon chain and change the *-ane* ending of the parent alkane to *-ene*.**

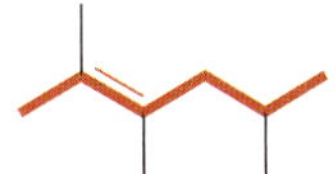

6 C's in the longest chain

hex*ane* ----→ hex*ene*

Step [2] **Number the carbon chain to give the double bond the lower number, and apply all other rules of nomenclature.**

a. **Number** the chain, and name using the ***first* number** assigned to the C=C.

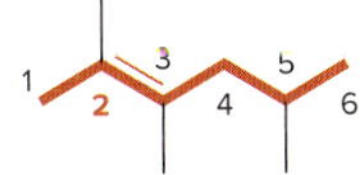

- Number the chain to put the C=C at C**2**, not C4.

hex-**2**-ene

b. **Name** and **number** the substituents.

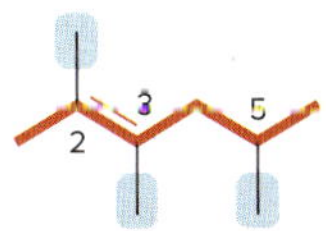

three methyl groups at C2, C3, and C5

Answer: 2,3,5-trimethylhex-2-ene

Compounds with two double bonds are named as **dienes** by changing the ***-ane*** ending of the parent alkane to the suffix ***-adiene.*** Compounds with three double bonds are named as **trienes,** and so forth. In naming cycloalkenes, the **double bond is located between C1 and C2,** and the "1" is usually omitted in the name. The ring is numbered clockwise or counterclockwise to give the first substituent the lower number. Representative examples are given in Figure 10.1.

Figure 10.1 Examples of cycloalkene nomenclature

1-methylcyclopentene

3-methylcycloheptene

[Number clockwise beginning at the C=C and place the CH_3 at C3.]

1,6-dimethylcyclohexene

[Number counterclockwise beginning at the C=C and place the first CH_3 at C1.]

Problem 10.5 Give the IUPAC name for each alkene.

a. b. c. d. e.

$CH_3CH_2CH{=}CH_2$ is named as 1-butene using the 1979 IUPAC recommendations and but-1-ene using the 1993 IUPAC recommendations.

Compounds that contain both a double bond and a hydroxy group are named as **alkenols,** and the chain (or ring) is numbered to **give the OH group the lower number.**

prop-2-en-1-ol

6-methylhept-6-en-2-ol

Problem 10.6 Give the IUPAC name for each polyfunctional compound.

a. b. c.

10.3B Naming Stereoisomers

A prefix is needed to distinguish two alkenes when diastereomers are possible.

Using Cis and Trans as Prefixes

An alkene having one alkyl group bonded to each carbon atom can be named using the prefixes **cis** and **trans** to designate the relative location of the two alkyl groups. For example, *cis*-hex-3-ene has two ethyl groups on the **same side** of the double bond, whereas *trans*-hex-3-ene has two ethyl groups on **opposite sides** of the double bond.

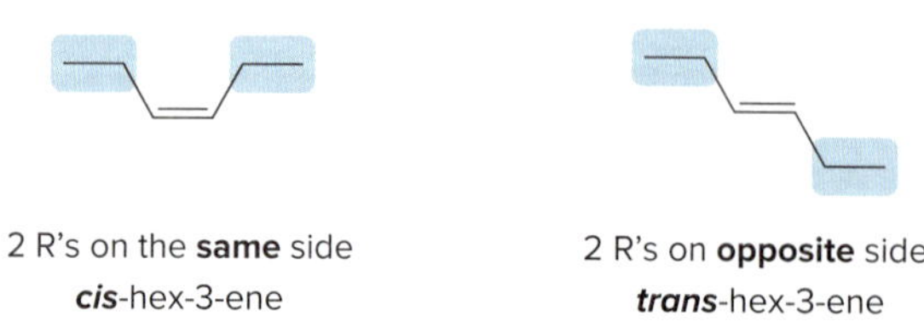

2 R's on the **same** side
***cis*-hex-3-ene**

2 R's on **opposite** sides
***trans*-hex-3-ene**

Using the Prefixes *E* and *Z*

E stands for the German word ***e**ntgegen* meaning "opposite." *Z* stands for the German word ***z**usammen,* meaning "together." Using *E,Z* nomenclature, a cis isomer has the *Z* configuration and a trans isomer has the *E* configuration.

Although the prefixes cis and trans can be used to distinguish diastereomers when two alkyl groups are bonded to the C=C, they cannot be used when there are three or four alkyl groups bonded to the C=C.

A 3-methylpent-2-ene

B 3-methylpent-2-ene

For example, alkenes **A** and **B** are two *different* compounds that are both called 3-methylpent-2-ene. In **A** the two CH_3 groups are cis, whereas in **B** the CH_3 and CH_2CH_3 groups are cis. The ***E,Z*** **system of nomenclature** has been devised to unambiguously name these kinds of alkenes.

How To Assign the Prefixes *E* and *Z* to an Alkene

Step [1] **Assign priorities to the two substituents on each end of the C=C by using the priority rules for *R,S* nomenclature (Section 5.6).**

- **Divide the double bond in half,** and assign the numbers **1** and **2** to indicate the relative priority of the two groups on each end—the higher-priority group is labeled **1,** and the lower-priority group is labeled **2.**

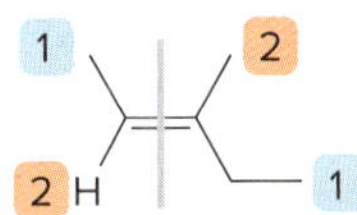

Assign priorities to each side separately.

Step [2] **Assign *E* or *Z* based on the location of the two higher-priority groups (1).**

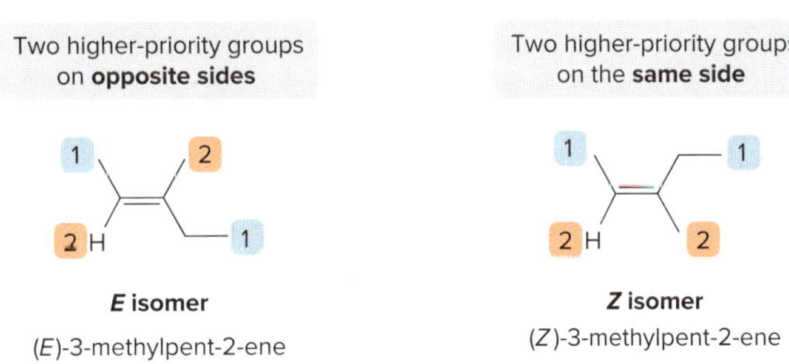

***E* isomer** (*E*)-3-methylpent-2-ene

***Z* isomer** (*Z*)-3-methylpent-2-ene

- The ***E*** isomer has the two higher-priority groups on the **opposite sides.**
- The ***Z*** isomer has the two higher-priority groups on the **same side.**

Problem 10.7 Label each C–C double bond as *E* or *Z*. Clavulanic acid [part (d)] is sold in combination with the antibiotic amoxicillin under the trade name Augmentin.

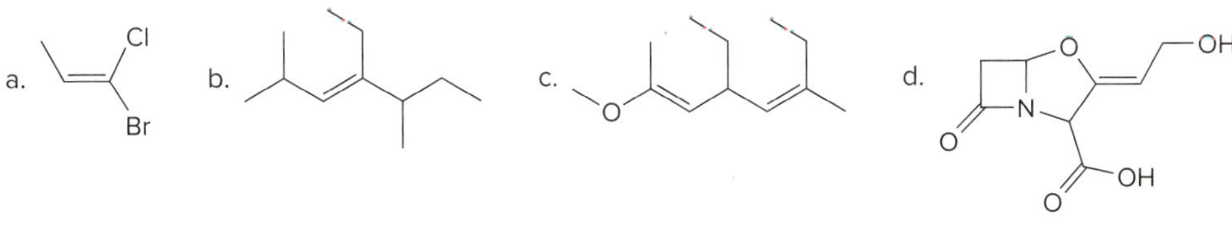

clavulanic acid

Sample Problem 10.3 Naming an Alkene Using the IUPAC System

Give the IUPAC of the following alkene, including an *E,Z* designation for the double bond.

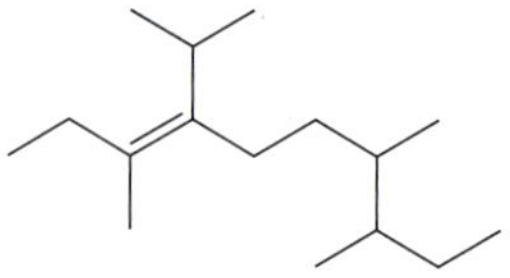

Solution

[1] Name the carbon skeleton and substituents.

- Name the longest carbon chain; 10 C's → **decene.**
- Number to give the C=C the lower number; **dec-3-ene.**
- Name and number the substituents and combine the parts.

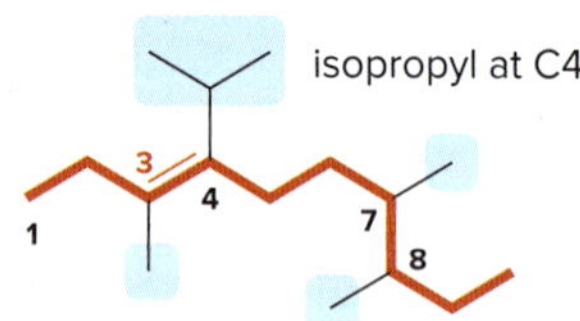

4-isopropyl-3,7,8-trimethyldec-3-ene

[2] Assign the prefix *E* or *Z*.

- Assign priorities to the two substituents on each end of the C=C.
- The *E* isomer has the two higher-priority groups on *opposite* sides.
- The *Z* isomer has the two higher-priority groups on the *same* side.

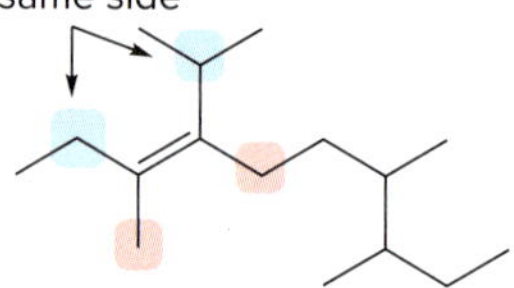

Answer:
(*Z*)-4-isopropyl-3,7,8-trimethyldec-3-ene

Problem 10.8 Give the IUPAC name for each alkene.

a. 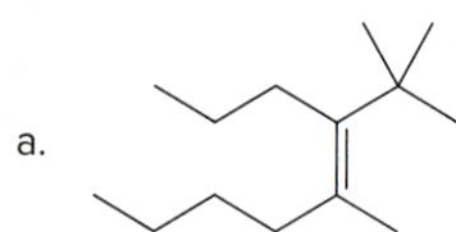

b.

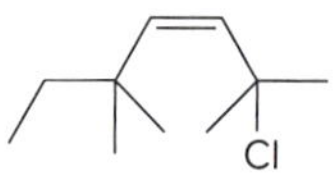

Cl

c.

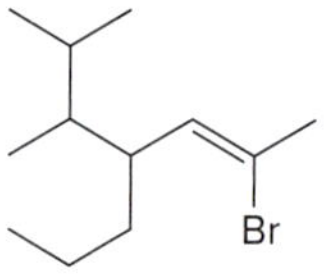

Br

More Practice: Try Problems 10.32a; 10.38a, e; 10.39a, d; 10.40a.

Problem 10.9 Draw the structure corresponding to each IUPAC name.

a. (*Z*)-4-ethylhept-3-ene b. (*E*)-3,5,6-trimethyloct-2-ene c. (*Z*)-2-bromo-1-iodohex-1-ene

Problem 10.10 (a) Name the following compound including *E,Z* and *R,S* designations. (b) Draw a stereoisomer at the carbon–carbon double bond. (c) Draw a stereoisomer at the stereogenic center.

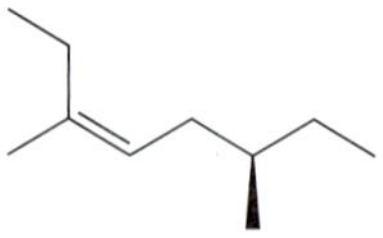

10.3C Common Names

The simplest alkene, $CH_2{=}CH_2$, named in the IUPAC system as **ethene,** is often called **ethylene,** its common name. The common names for three **alkyl groups** derived from alkenes are also used. Three examples of naming organic molecules using these common names are shown in Figure 10.2.

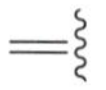

methylene group 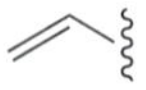 **vinyl** group 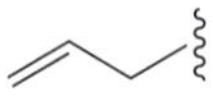 **allyl** group

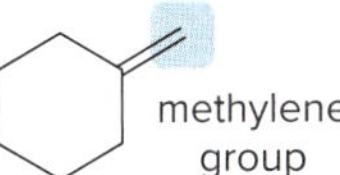

methylenecyclohexane

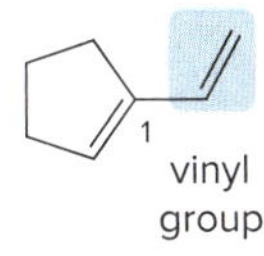

1-vinylcyclopentene

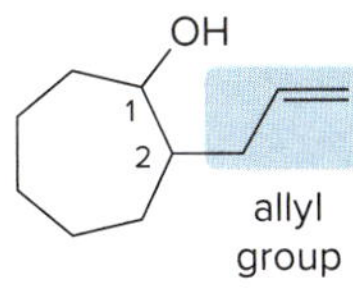

2-allylcycloheptanol

Figure 10.2 Naming alkenes with common substituent names

10.4 Properties of Alkenes

Most alkenes exhibit only weak van der Waals interactions, so their physical properties are similar to those of alkanes of comparable molecular weight.

- **Alkenes have low melting points and boiling points.**
- **Melting points and boiling points increase as the number of carbons increases because of increased surface area.**
- **Alkenes are soluble in organic solvents and insoluble in water.**

Cis and trans alkenes often have somewhat different physical properties. For example, *cis*-but-2-ene has a higher boiling point (4 °C) than *trans*-but-2-ene (1 °C). This difference arises because the C–C single bond between an alkyl group and one of the double bond carbons of an alkene is slightly polar. **The sp^3 hybridized alkyl carbon donates electron density to the sp^2 hybridized alkenyl carbon.**

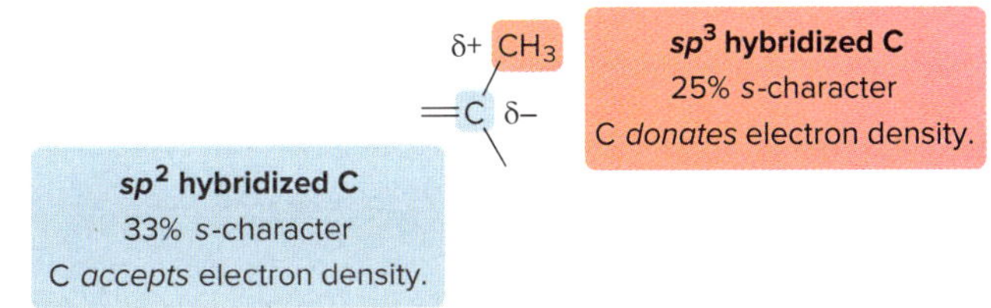

Related arguments involving C_{sp^3}–C_{sp^2} bonds were used in Section 8.2C to explain why the stability of an alkene increases with increasing alkyl substitution.

The bond dipole places a partial negative charge on the alkenyl carbon (sp^2) relative to the alkyl carbon (sp^3) because an sp^2 hybridized orbital has greater percent *s*-character (33%) than an sp^3 hybridized orbital (25%). **In a cis isomer, the two C_{sp^3}–C_{sp^2} bond dipoles reinforce each other, yielding a small net molecular dipole. In a trans isomer, the two bond dipoles cancel.**

CH3 CH3 C=C H H	CH3 H C=C H CH3
a small net dipole more polar isomer	**no net dipole less polar isomer**
cis-but-2-ene	*trans*-but-2-ene
higher bp	**lower bp**

- **A cis alkene is more polar than a trans alkene, giving it a slightly higher boiling point and making it more soluble in polar solvents.**

Problem 10.11 Rank the following isomers in order of increasing boiling point.

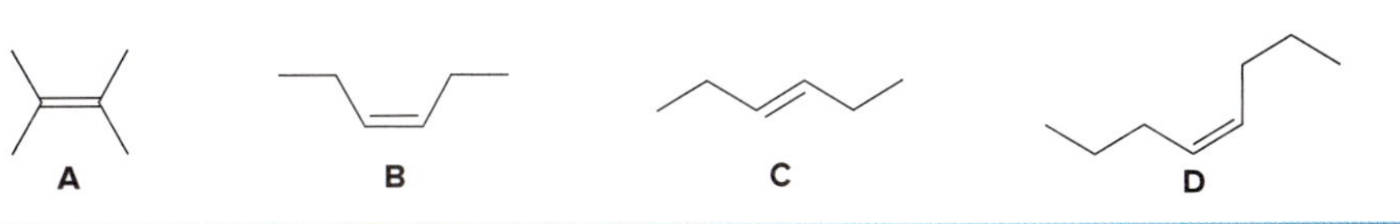

Students who have already been exposed to spectroscopy or who would like to learn about the spectroscopic properties of alkenes are referred to the following sections:

- Infrared spectroscopy: Sections B.3A, B.3D, B.4A; Tables B.1, B.2
- Nuclear magnetic resonance spectroscopy: Section C.8; Tables C.1, C.2, C.4, C.5; Sample Problems C.7c, C.9c

10.5 Interesting Alkenes

Ethylene is prepared from petroleum by a process called **cracking.** Ethylene is the most widely produced organic chemical, serving as the starting material not only for the polymer **polyethylene,** a widely used plastic, but also for many other useful organic compounds, as shown in Figure 10.3.

Figure 10.3

Ethylene, an industrial starting material for many useful products

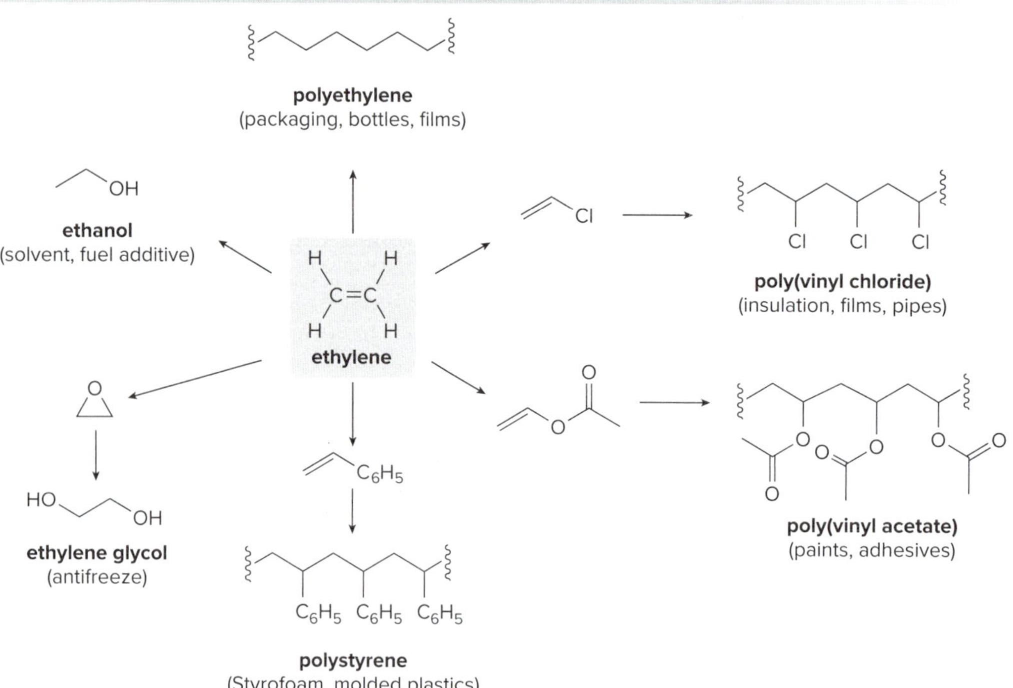

Numerous organic compounds containing carbon–carbon double bonds have been isolated from natural sources, including β-carotene, the orange pigment in carrots (Section 3.5A), and zingiberene, a triene in the oil of ginger.

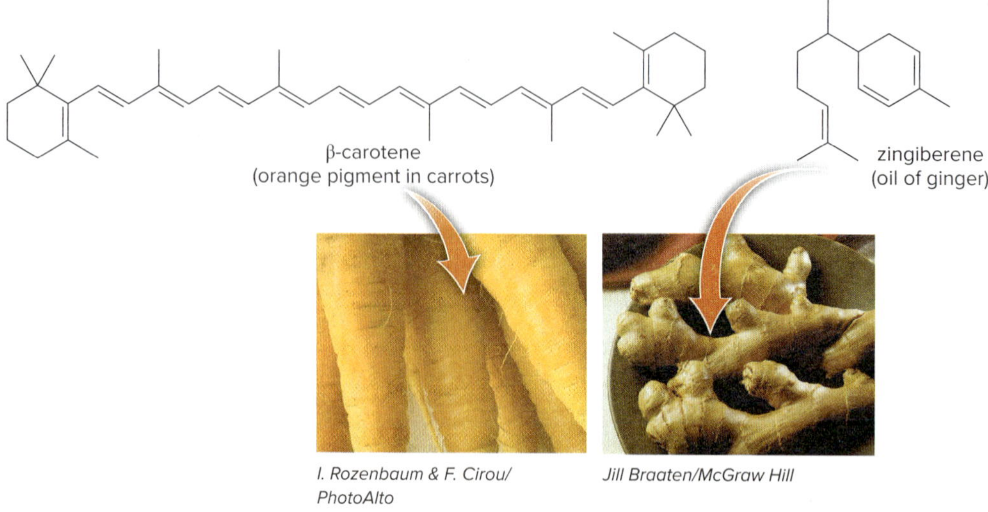

I. Rozenbaum & F. Cirou/PhotoAlto

Jill Braaten/McGraw Hill

10.6 Lipids—Part 2

Lipids are water-insoluble biomolecules composed largely of nonpolar C–C and C–H bonds (Section 4.14).

General structure of an ester: RC(=O)OR'

Understanding the geometry of C–C double bonds provides an insight into the properties of **triacylglycerols,** the most abundant lipids. Triacylglycerols contain three ester groups, each having a long carbon chain (abbreviated as R, R', and R") bonded to a carbonyl group (C=O).

R groups have 11–19 C's.

Three ester groups are labeled in red.

triacylglycerol

10.6A Fatty Acids

Candlenuts, known as kukui nuts in Hawai'i, are rich in linoleic and linolenic acids, two essential fatty acids that cannot be synthesized in the body and must therefore be obtained in the diet. *inga spence/Alamy Stock Photo*

Triacylglycerols are hydrolyzed to glycerol (a triol) and three **fatty acids** of general structure RCO_2H. Naturally occurring fatty acids contain 12–20 carbon atoms, with a carboxy group (CO_2H) at one end.

triacylglycerol —H_2O (H^+ or ^-OH) or enzymes→ glycerol + **fatty acids** ($HOC(=O)R$, $HOC(=O)R'$, $HOC(=O)R''$)

These fatty acids have **12–20 C's.**

- **Saturated fatty acids have no double bonds in their long hydrocarbon chains, and unsaturated fatty acids have one or more double bonds in their hydrocarbon chains.**
- **Double bonds in naturally occurring fatty acids have the *Z* configuration.**

Table 10.2 lists the structure and melting point of four fatty acids containing 18 carbon atoms. Stearic acid is one of the two most common saturated fatty acids, and oleic and linoleic acids are the most common unsaturated ones. The data show the effect of *Z* double bonds on the melting point of fatty acids.

- **As the number of double bonds in the fatty acid *increases,* the melting point *decreases.***

Table 10.2 The Effect of Double Bonds on the Melting Point of Fatty Acids

Name	Structure	Mp (°C)
Stearic acid (**0** C=C)	[structure]	69
Oleic acid (**1** C=C)	[structure]	4
Linoleic acid (**2** C=C)	[structure]	–5
Linolenic acid (**3** C=C)	[structure]	–11

Increasing number of double bonds

The three-dimensional structures of the fatty acids in Figure 10.4 illustrate how *Z* double bonds introduce kinks in the long hydrocarbon chain, decreasing the ability of the fatty acid to pack well in a crystalline lattice. **The *larger* the number of *Z* double bonds, the more kinks in the hydrocarbon chain, and the *lower* the melting point.**

Figure 10.4 Three-dimensional structure of four C_{18} fatty acids

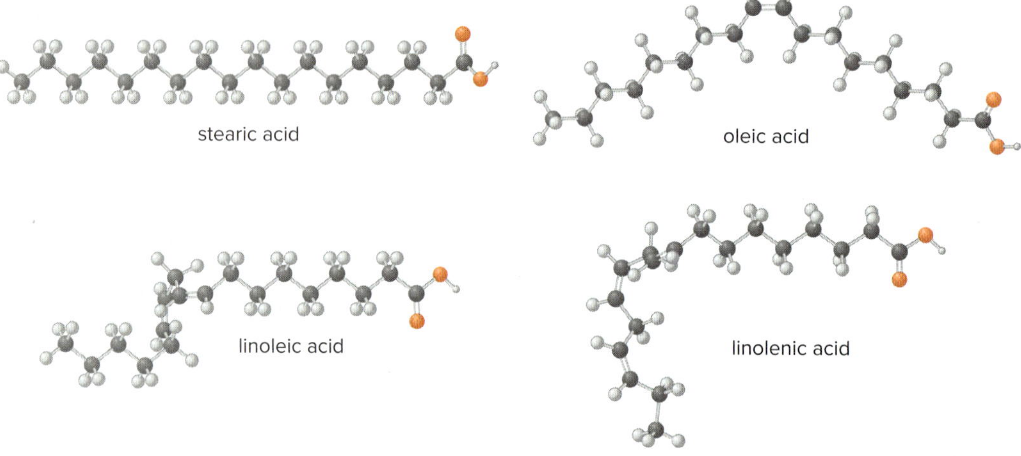

Problem 10.12 Linolenic acid (Table 10.2) and stearidonic acid are omega-3 fatty acids, unsaturated fatty acids that contain the first double bond located at C3, when numbering begins at the methyl end of the chain. Predict how the melting point of stearidonic acid compares with the melting points of linolenic and stearic acids. Soybean oil enriched in stearidonic acid has been approved by the European Food Safety Authority as a healthier alternative to vegetable oils that contain fewer degrees of unsaturation.

C3 →
O
OH
stearidonic acid

Canola, soybeans, and flaxseed are excellent dietary sources of linolenic acid, an essential fatty acid. Oils derived from omega-3 fatty acids (Problem 10.12) are currently thought to be especially beneficial for individuals at risk of developing coronary artery disease. *Jill Braaten/McGraw Hill*

10.6B Fats and Oils

Fats and **oils** are triacylglycerols with different physical properties.

- **Fats have higher melting points—they are *solids* at room temperature.**
- **Oils have lower melting points—they are *liquids* at room temperature.**

The identity of the three fatty acids in the triacylglycerol determines whether it is a fat or an oil. **Increasing the number of double bonds in the fatty acid side chains decreases the melting point of the triacylglycerol.**

- **Fats are derived from fatty acids having few double bonds.**
- **Oils are derived from fatty acids having a larger number of double bonds.**

Saturated fats are typically obtained from animal sources, whereas unsaturated oils are common in vegetable sources. Thus, butter and lard are high in saturated triacylglycerols, and olive oil and safflower oil are high in unsaturated triacylglycerols. An exception to this generalization is coconut oil, which is composed largely of saturated alkyl side chains.

Considerable evidence suggests that an elevated cholesterol level is linked to an increased risk of heart disease. Saturated fats stimulate cholesterol synthesis in the liver, thus increasing the cholesterol concentration in the blood.

10.7 Preparation of Alkenes

Recall from Chapters 8 and 9 that alkenes can be prepared from alkyl halides and alcohols via elimination reactions. For example, **dehydrohalogenation of alkyl halides with strong base yields alkenes via an E2 mechanism** (Sections 8.4 and 8.5).

- Typical bases include **$^-$OH** and **$^-$OR** [especially $^-OC(CH_3)_3$], and nonnucleophilic bases such as **DBU** and **DBN.**
- **Alkyl tosylates** can also be used as starting materials under similar reaction conditions (Section 9.13).

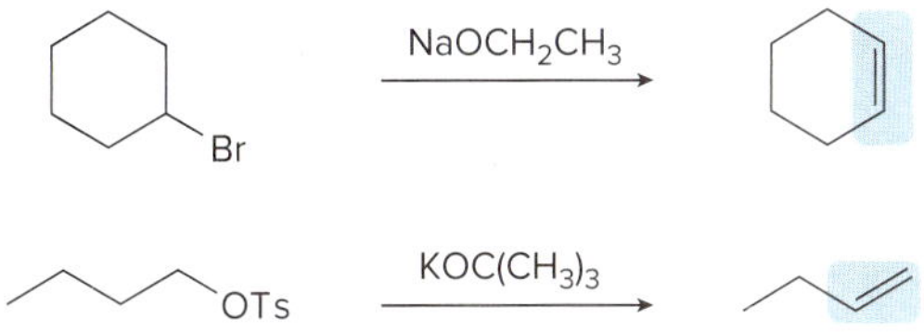

Br — $NaOCH_2CH_3$ →

OTs — $KOC(CH_3)_3$ →

The acid-catalyzed dehydration of alcohols with H_2SO_4 or TsOH yields alkenes, too (Sections 9.8 and 9.9). The reaction occurs via an E1 mechanism for 2° and 3° alcohols, and an E2 mechanism for 1° alcohols. E1 reactions involve carbocation intermediates, so rearrangements are possible. **Dehydration can also be carried out with $POCl_3$ and pyridine** by an E2 mechanism (Section 9.10).

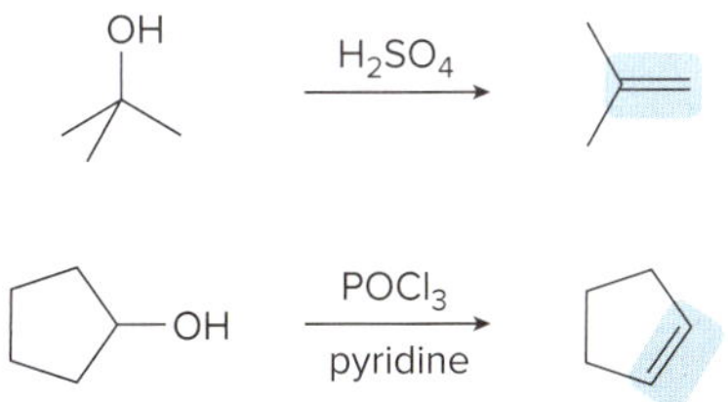

OH — H_2SO_4 →

OH — $POCl_3$, pyridine →

These elimination reactions are **stereoselective** and **regioselective,** so the **most stable alkene is usually formed as the major product.**

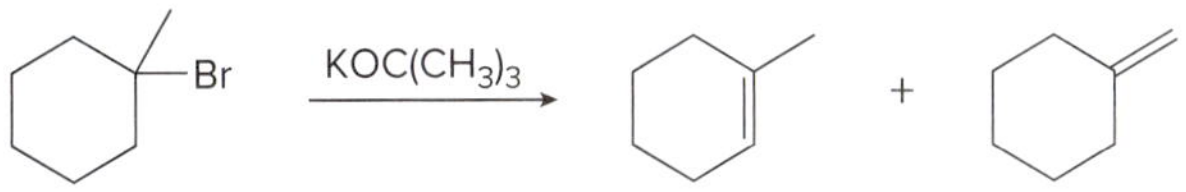

Br — $KOC(CH_3)_3$ → +

major product
trisubstituted alkene

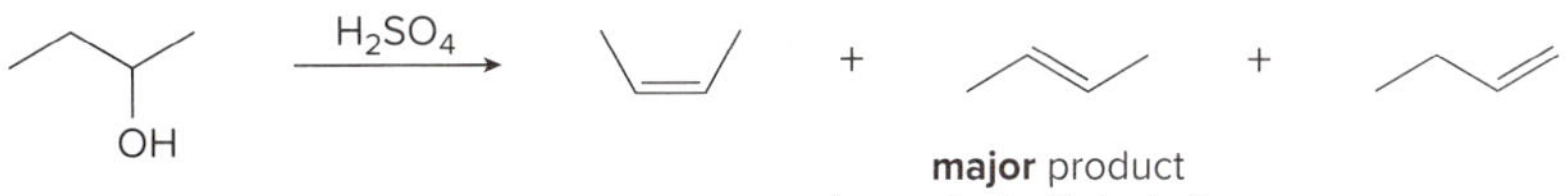

OH — H_2SO_4 → + +

major product
trans disubstituted alkene

Problem 10.13 Draw the products of each elimination reaction.

a. 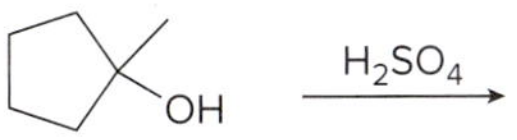 OH — H_2SO_4 →

b. 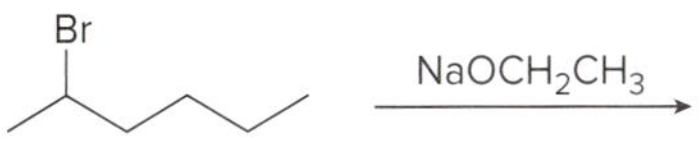 Br — $NaOCH_2CH_3$ →

10.8 Introduction to Addition Reactions

Because the C–C π bond of an alkene is much *weaker* than a C–C σ bond, the characteristic reaction of alkenes is **addition: the π bond is broken and two new σ bonds are formed.**

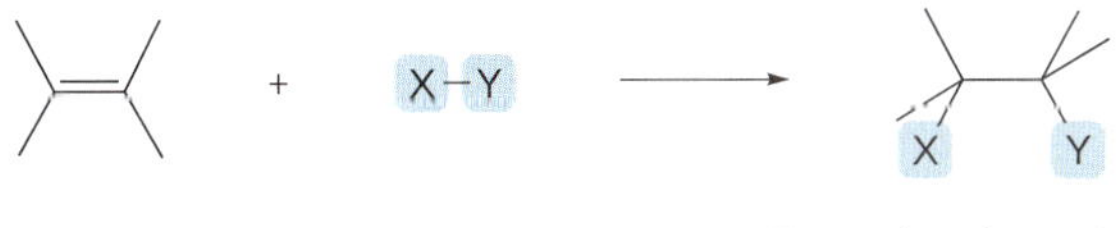

\+ X–Y → X Y

A π bond is broken. Two σ bonds are formed.

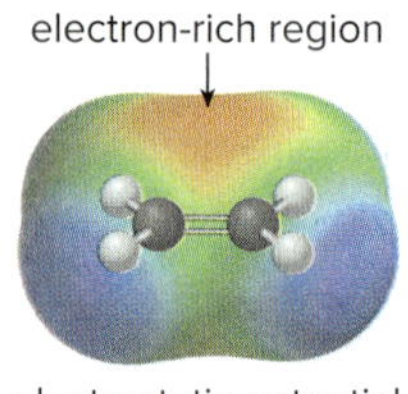

electrostatic potential plot of ethylene

The addition reactions of alkenes are discussed in Sections 10.9–10.16 and in Chapter 12 (Oxidation and Reduction).

Alkenes are electron rich, as seen in the electrostatic potential plot. The electron density of the π bond is concentrated above and below the plane of the molecule, making the π bond more exposed than the σ bond.

What kinds of reagents add to the weak, electron-rich π bond of alkenes? There are many of them, and that can make alkene chemistry challenging. To help you organize this information, keep in mind the following:

- Every reaction of alkenes involves *addition:* the π bond is always broken.
- Because alkenes are electron rich, simple alkenes do *not* react with nucleophiles or bases, reagents that are themselves electron rich. Alkenes react with *electrophiles.*

The stereochemistry of addition is often important in delineating a reaction's mechanism. Because the carbon atoms of a double bond are both trigonal planar, the elements of X and Y can be added to them from the **same side** or from **opposite sides.**

syn addition

anti addition

X–Y

or

X and Y added from the **same side**

X and Y added from **opposite sides**

or

- *Syn addition* takes place when both X and Y are added from the *same* side.
- *Anti addition* takes place when X and Y are added from *opposite* sides.

Five reactions of alkenes are discussed in Chapter 10 and each is illustrated in Figure 10.5, using cyclohexene as the starting material.

Figure 10.5

Five addition reactions of cyclohexene

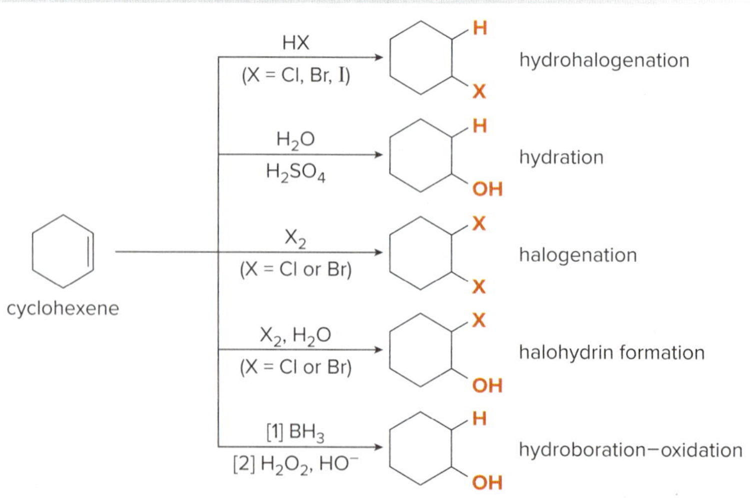

- In each reaction, the π bond is broken and two new σ bonds are formed.

10.9 Hydrohalogenation—Electrophilic Addition of HX

Hydrohalogenation of an alkene to form an alkyl halide is the reverse of the dehydrohalogenation of an alkyl halide to form an alkene, a reaction discussed in detail in Sections 8.4 and 8.5.

Hydrohalogenation results in the addition of hydrogen halides HX (X = Cl, Br, and I) to alkenes to form alkyl halides.

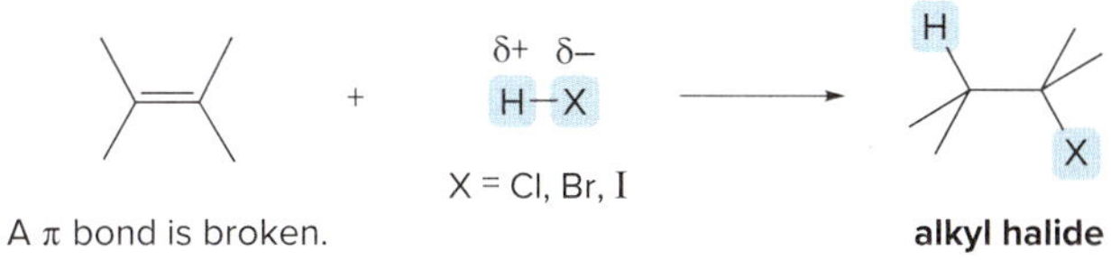

δ+ δ– H–X, X = Cl, Br, I

A π bond is broken. **alkyl halide**

Two bonds are broken in this reaction—the weak π bond of the alkene and the HX bond—and two new σ bonds are formed—one to H and one to X. Because X is more electronegative than H, the H–X bond is polarized, with a partial positive charge on H. Because the electrophilic (H) end of HX is attracted to the electron-rich double bond, these reactions are called **electrophilic additions.**

To draw the products of an addition reaction:

- **Locate the C–C double bond.**
- **Identify the σ bond of the reagent that breaks—namely, the H–X bond in hydrohalogenation.**
- **Break the π bond of the alkene and the σ bond of the reagent, and form two new σ bonds to the C atoms of the double bond.**

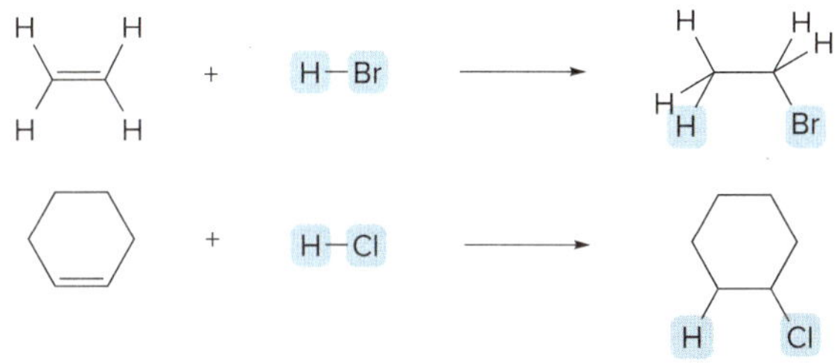

$CH_2{=}CH_2$ + H–Br ⟶ CH_3CH_2Br

cyclohexene + H–Cl ⟶ chlorocyclohexane

Addition reactions are exothermic because the two σ bonds formed in the product are *stronger* than the σ and π bonds broken in the reactants. For example, $\Delta H°$ for the addition of HBr to ethylene is –60 kJ/mol, as illustrated in Figure 10.6.

Figure 10.6 The addition of HBr to $CH_2{=}CH_2$, an exothermic reaction

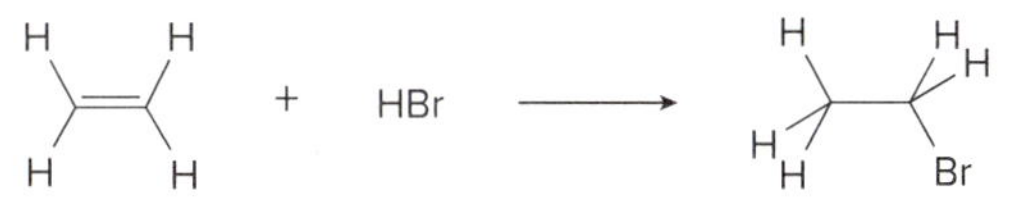

$CH_2{=}CH_2$ + HBr ⟶ CH_3CH_2Br

$\Delta H°$ calculation:

[1] Bonds broken

	$\Delta H°$ (kJ/mol)
$CH_2{=}CH_2$ π bond	+267
H–Br	+368
Total	+635 kJ/mol

Energy needed to break bonds.

[2] Bonds formed

	$\Delta H°$ (kJ/mol)
$BrCH_2CH_2$–H	–410
CH_3CH_2–Br	–285
Total	–695 kJ/mol

Energy released in forming bonds.

[3] Overall $\Delta H°$ =

sum in Step [1] + sum in Step [2]

+635 kJ/mol
–695 kJ/mol

$\Delta H°$ = –60 kJ/mol

The reaction is exothermic.

[Values taken from Appendix E.]

The mechanism of electrophilic addition of HX consists of **two steps:** addition of H^+ to form a carbocation, followed by nucleophilic attack of X^-. The mechanism is illustrated for the reaction of *cis*-but-2-ene with HBr in Mechanism 10.1.

Mechanism 10.1 Electrophilic Addition of HX to an Alkene

H—Br: → [1] slow → H carbocation + :Br:⁻ → [2] → :Br:

cis-but-2-ene

carbocation

new bond shown in red

1 The π bond of the alkene attacks the H of HBr to form a new C–H bond and a **carbocation** in the rate-determining in step.

2 **Nucleophilic attack of Br^-** on the carbocation forms the new C–Br bond.

The mechanism of electrophilic addition consists of two successive Lewis acid–base reactions. In Step [1], the **alkene is the Lewis base** that donates an electron pair to **H–Br, the Lewis acid,** whereas in Step [2], **Br^- is the Lewis base** that donates an electron pair to the **carbocation, the Lewis acid.**

Problem 10.14 What product is formed when each alkene is treated with HCl?

a. b. c.

10.10 Markovnikov's Rule

With an unsymmetrical alkene, HX can add to the double bond to give two constitutional isomers.

propene (C2, C1) —H–Cl→ 1-chloropropane or **2**-chloropropane

only product

For example, HCl addition to propene could in theory form 1-chloropropane by addition of H and Cl to C2 and C1, respectively, and 2-chloropropane by addition of H and Cl to C1 and C2, respectively. In fact, **electrophilic addition forms *only* 2-chloropropane.** This is a specific example of a general trend called **Markovnikov's rule,** named for the Russian chemist who first determined the regioselectivity of electrophilic addition of HX.

- **Markovnikov's rule: In the addition of HX to an unsymmetrical alkene, the H atom bonds to the *less substituted* carbon atom—that is, the carbon that has more H atoms to begin with.**

The basis of Markovnikov's rule is the formation of a carbocation in the rate-determining step of the mechanism. With propene, there are two possible paths for this first step, depending on which carbon atom of the double bond forms the new bond to hydrogen.

[1] H—Cl ✕→ 1° carbocation + Cl^-

[2] H—Cl → (preferred path) 2° carbocation + Cl^- → 2-chloropropane

The Hammond postulate was first introduced in Section 7.13 to explain the relative rate of S_N1 reactions with 1°, 2°, and 3° RX.

Path [1] forms a highly unstable 1° carbocation, whereas Path [2] forms a **more stable 2° carbocation.** According to the Hammond postulate, Path [2] is faster because formation of the carbocation is an endothermic process, so **the transition state to form the more stable 2° carbocation is lower in energy** (Figure 10.7).

- **In the addition of HX to an unsymmetrical alkene, the H atom is added to the *less* substituted carbon to form the *more stable, more substituted* carbocation.**

Figure 10.7 Electrophilic addition and the Hammond postulate

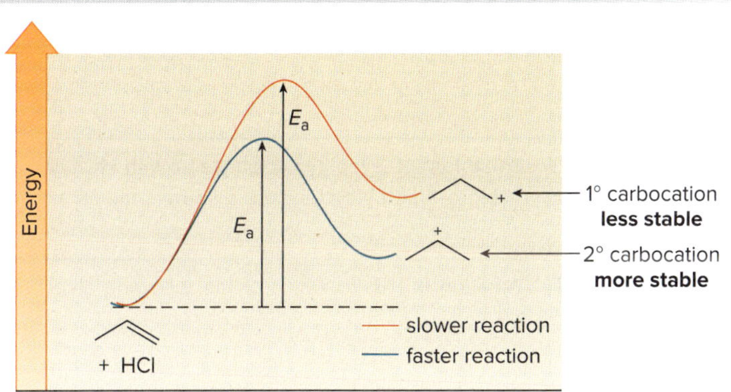

- The E_a for formation of the more stable 2° carbocation is *lower* than the E_a for formation of the 1° carbocation. The 2° carbocation is formed *faster*.

Similar results are seen in any electrophilic addition involving an intermediate carbocation: **the more stable, *more* substituted carbocation is formed by addition of the electrophile to the *less* substituted carbon.**

Problem 10.15 Draw the products formed when each alkene is treated with HCl.

a. b. c. d.

Problem 10.16 Use the Hammond postulate to explain why $(CH_3)_2C{=}CH_2$ reacts faster than $CH_3CH{=}CH_2$ in electrophilic addition of HX.

Because carbocations are formed as intermediates in hydrohalogenation, carbocation rearrangements can occur, as illustrated in Sample Problem 10.4.

Sample Problem 10.4 Drawing Hydrohalogenation with a Carbocation Rearrangement

Draw a stepwise mechanism for the following reaction.

HBr

Br

Solution

Because the carbon skeletons of the starting material and product are *different*—the alkene reactant has a 4° carbon and the product alkyl halide does not—a carbocation rearrangement must have occurred.

Markovnikov addition of HBr adds H^+ to the less substituted end of the double bond, forming a 2° carbocation in Step [1].

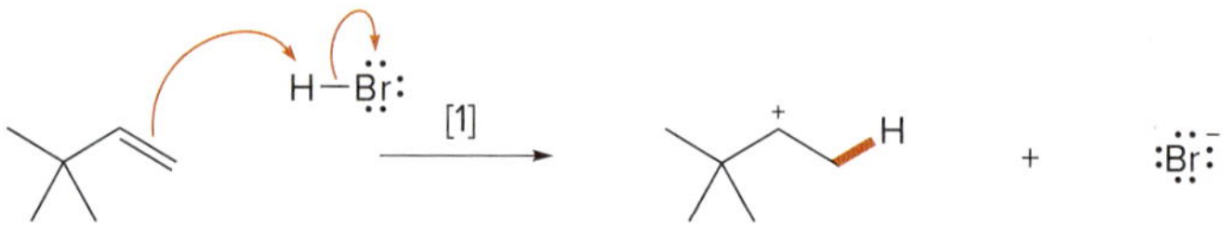

H—Br: [1] H + :Br:$^-$

2° carbocation

new bond shown in red

Rearrangement of the 2° carbocation by a 1,2-methyl shift forms a more stable 3° carbocation in Step [2]. Nucleophilic attack of Br^- forms the product, a 3° alkyl halide, in Step [3].

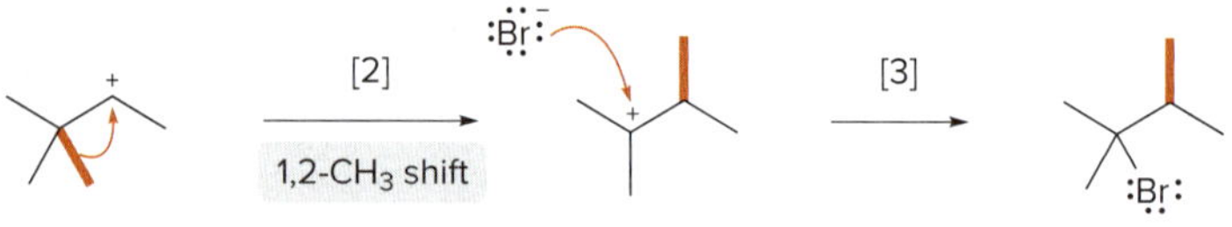

[2] 1,2-CH_3 shift :Br:$^-$ [3] :Br:

3° carbocation

Problem 10.17 Treatment of 3-methylcyclohexene with HCl yields two products, 1-chloro-3-methylcyclohexane and 1-chloro-1-methylcyclohexane. Draw a mechanism to explain this result.

Problem 10.18 Addition of HBr to which of the following alkenes will lead to a rearrangement?

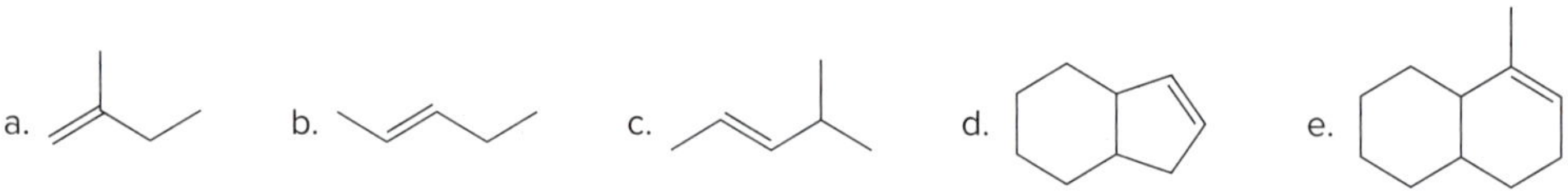

a. b. c. d. e.

10.11 Stereochemistry of Electrophilic Addition of HX

To understand the stereochemistry of electrophilic addition, recall two stereochemical principles learned in Chapters 7 and 9.

- **Trigonal planar atoms react with reagents from two directions with equal probability (Section 7.11C).**
- **Achiral starting materials yield achiral or racemic products (Section 9.16).**

Many hydrohalogenation reactions begin with an **achiral reactant** and form an **achiral product.** For example, the addition of HBr to cyclohexene, an achiral alkene, forms bromocyclohexane, an achiral alkyl halide.

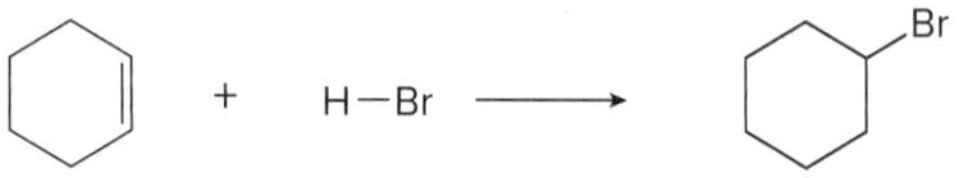

+ H—Br → Br

cyclohexene
achiral starting material

bromocyclohexane
achiral product

Because addition converts sp^2 hybridized carbons to sp^3 hybridized carbons, sometimes new stereogenic centers are formed from hydrohalogenation. Markovnikov addition of HCl to 1,3,3-trimethylcyclohexene, an achiral alkene, forms one constitutional isomer, 1-chloro-1,3,3-trimethylcyclohexane. Because this product now has a stereogenic center at one of the newly formed sp^3 hybridized carbons (labeled in blue), **an equal amount of two enantiomers—a racemic mixture—**must form.

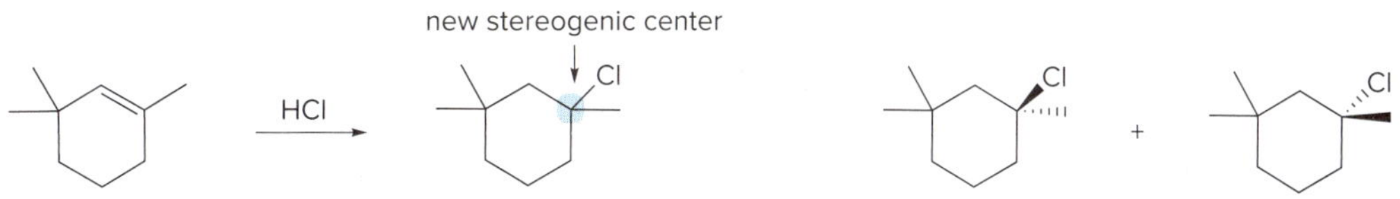

new stereogenic center

HCl

1,3,3-trimethylcyclohexene
achiral starting material

1-chloro-1,3,3-trimethylcyclohexane

A + B

The product has one new stereogenic center, so **two enantiomers are formed.**

The mechanism of hydrohalogenation illustrates why two enantiomers are formed. Initial addition of the electrophile H^+ (from HCl) occurs from **either side of the planar double bond** to form a carbocation. Both modes of addition (from above and below) generate the same **achiral carbocation.** Either representation of this carbocation can then be used to draw the second step of the mechanism.

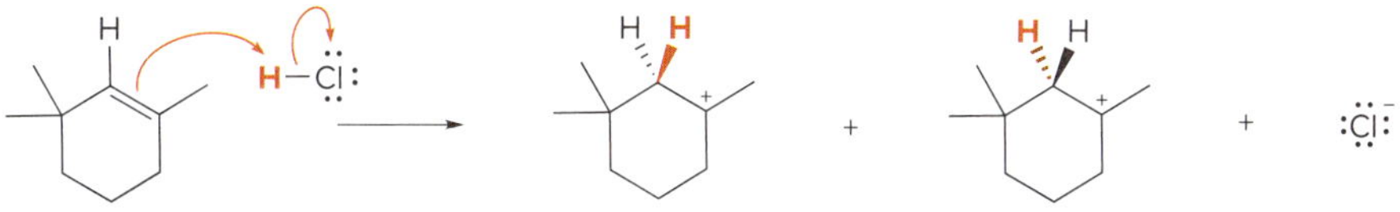

H^+ from **above**

H^+ from **below**

identical carbocations

Nucleophilic attack of Cl^- on the trigonal planar carbocation also occurs from two different directions, forming two products, **A** and **B,** having a new stereogenic center. **A** and **B** are not superimposable, so they are **enantiomers.** Because attack from either direction occurs with equal probability, a **racemic mixture** of **A** and **B** is formed.

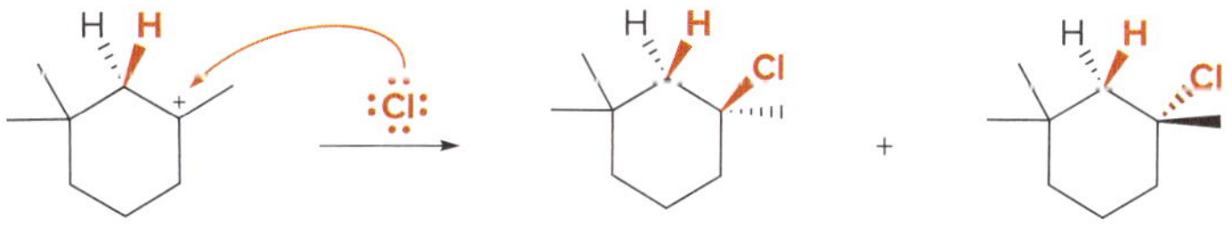

Cl^- from **above**
A
syn addition
H and Cl are added from the **same** side.

Cl^- from **below**
B
anti addition
H and Cl are added from **opposite** sides.

The terms **cis** and **trans** refer to the arrangement of groups in a particular compound, usually an alkene or a disubstituted cycloalkane. The terms **syn** and **anti** describe the stereochemistry of a process—for example, how two groups are added to a double bond.

Because hydrohalogenation begins with a **planar** double bond and forms a **planar** carbocation, addition of H and Cl occurs in two different ways. The elements of H and Cl can both be added from the same side of the double bond—that is, **syn addition**—or they can be added from opposite sides—that is, **anti addition.** *Both* modes of addition occur in this two-step reaction mechanism.

- **Hydrohalogenation occurs with syn and anti addition of HX.**

Table 10.3 summarizes the characteristics of electrophilic addition of HX to alkenes.

Table 10.3 Summary: Electrophilic Addition of HX to Alkenes

	Observation
Mechanism	• The mechanism involves **two steps.** • The rate-determining step forms a **carbocation.** • **Rearrangements** can occur.
Regioselectivity	• **Markovnikov's rule is followed.** In unsymmetrical alkenes, H bonds to the less substituted C to form the more stable carbocation.
Stereochemistry	• **Syn** and **anti** addition occur.

Problem 10.19 Draw the products, including stereochemistry, of each reaction.

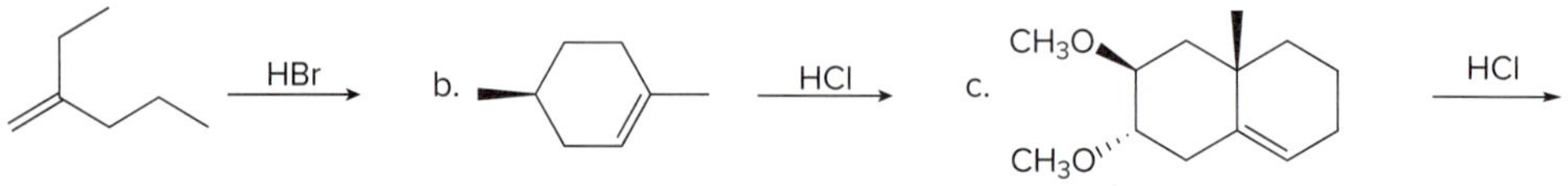

a. $\xrightarrow{HBr}$ b. $\xrightarrow{HCl}$ c. CH_3O, CH_3O $\xrightarrow{HCl}$

Problem 10.20 Draw all stereoisomers formed when 1,2-dimethylcyclohexene is treated with HCl. Label pairs of enantiomers.

10.12 Hydration—Electrophilic Addition of Water

Hydration results in the addition of water to an alkene to form an alcohol. H_2O itself is too weak an acid to protonate an alkene, but with added H_2SO_4, H_3O^+ is formed and addition readily occurs.

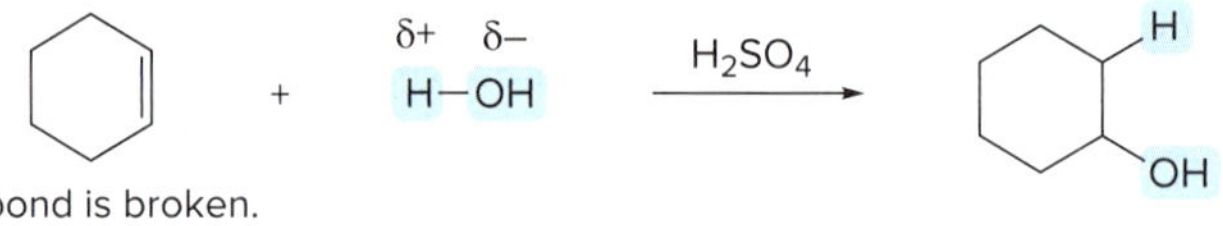

+ $\overset{\delta+}{H}-\overset{\delta-}{OH}$ $\xrightarrow{H_2SO_4}$ H, OH

A π bond is broken.

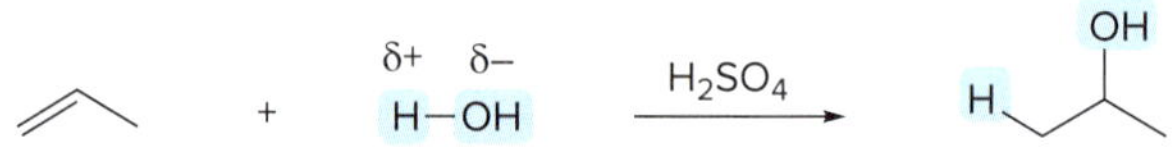

propene + $\overset{\delta+}{H}-\overset{\delta-}{OH}$ $\xrightarrow{H_2SO_4}$ H, OH

2-propanol
present in rubbing alcohol
and hand sanitizer

Hydration is simply another example of **electrophilic addition.** The first two steps of the mechanism are similar to those of electrophilic addition of HX—that is, addition of H^+ (from H_3O^+) to generate a carbocation, followed by nucleophilic attack of H_2O. Mechanism 10.2 illustrates the addition of H_2O to cyclohexene to form cyclohexanol.

Mechanism 10.2 Electrophilic Addition of H_2O to an Alkene—Hydration

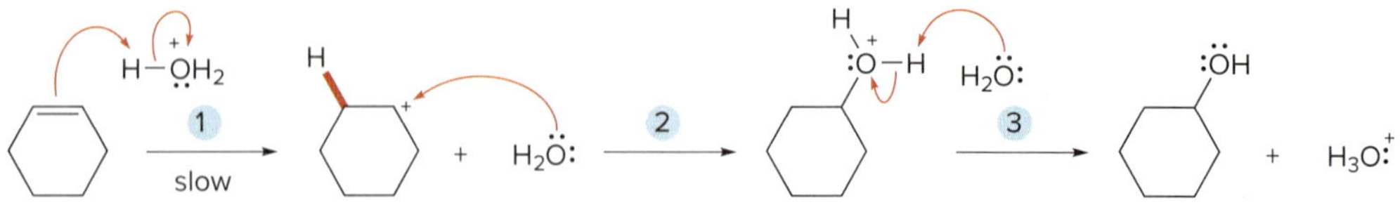

$H-\overset{+}{O}H_2$ $\xrightarrow[\text{slow}]{1}$ H (carbocation) + $H_2\ddot{O}$: $\xrightarrow{2}$ H, $:\overset{+}{O}-H$ $H_2\ddot{O}$: $\xrightarrow{3}$ $:\ddot{O}H$ + H_3O^+

carbocation
new bond shown in red

cyclohexanol

1 The π bond of the alkene attacks the H of H_3O^+ to form a new C–H bond and a **carbocation** in the rate-determining step.

2 **Nucleophilic attack of H_2O** on the carbocation forms the new C–O bond.

3 **Removal of a proton** with H_2O forms a neutral alcohol. Because the acid used in Step [1] is regenerated in Step [3], the reaction is acid-catalyzed.

Hydration of an alkene to form an alcohol is the reverse of the dehydration of an alcohol to form an alkene, a reaction discussed in detail in Section 9.8.

There are three consequences to the formation of carbocation intermediates:

- **In unsymmetrical alkenes, H adds to the *less* substituted carbon to form the *more* stable carbocation; that is, Markovnikov's rule holds.**
- **Addition of H and OH occurs in both a syn and anti fashion.**
- **Carbocation *rearrangements* can occur.**

Alcohols add to alkenes, forming ethers, using the same mechanism. Addition of CH_3OH to 2-methylpropene, for example, forms *tert*-butyl methyl ether (**MTBE**), a high octane fuel additive described in Section 3.4C.

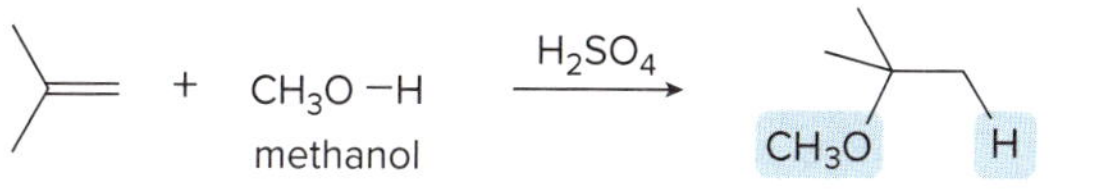

+ CH_3O-H methanol $\xrightarrow{H_2SO_4}$ CH_3O H

tert-butyl methyl ether
MTBE

Problem 10.21 What two alkenes give rise to each alcohol as the major product of acid-catalyzed hydration?

a. 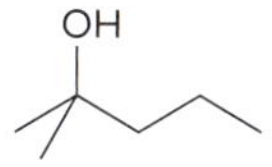 OH b. 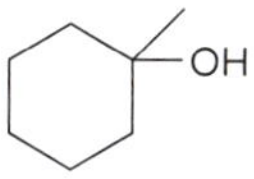 OH c. 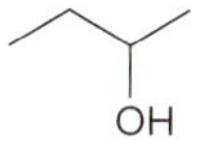 OH

Problem 10.22 Draw all stereoisomers formed in each reaction.

a. 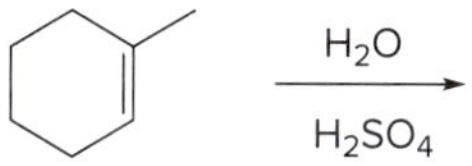 $\xrightarrow[H_2SO_4]{H_2O}$ b. 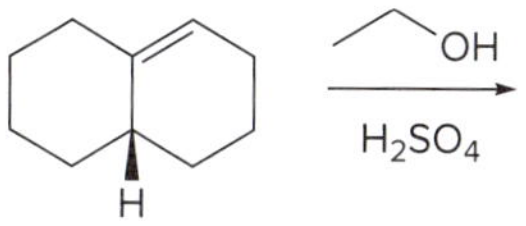 H $\xrightarrow[H_2SO_4]{\text{OH}}$ c. 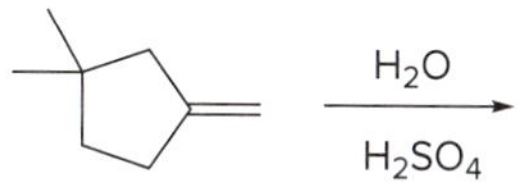 $\xrightarrow[H_2SO_4]{H_2O}$

10.13 Halogenation—Addition of Halogen

Halogenation results in the addition of halogen X_2 (X = Cl or Br) **to an alkene, forming a vicinal dihalide.**

+ X–X ⟶ X X

A π bond is broken. **vicinal dihalide**

Halogenation is synthetically useful only with Cl_2 and Br_2. The dichlorides and dibromides formed in this reaction serve as starting materials for the synthesis of alkynes, as we learned in Section 8.10.

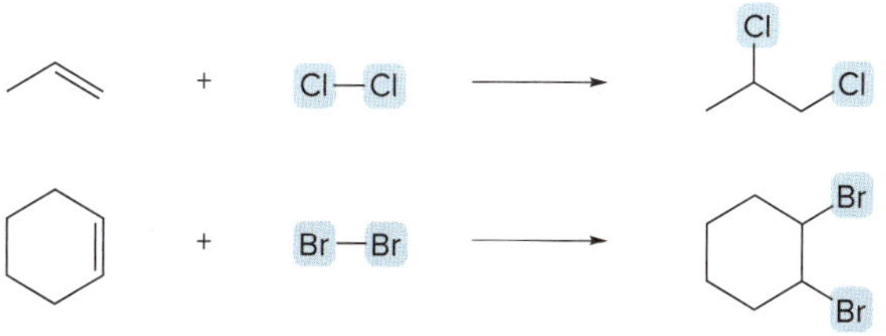

+ Cl–Cl ⟶ Cl Cl

+ Br–Br ⟶ Br Br

Halogens add to π bonds because halogens are **polarizable.** The electron-rich double bond induces a dipole in an approaching halogen molecule, making one halogen atom electron deficient and the other electron rich ($X^{\delta+}-X^{\delta-}$). **The electrophilic halogen atom is then attracted to the nucleophilic double bond,** making addition possible.

Two facts demonstrate that halogenation follows a different mechanism from that of hydrohalogenation or hydration. First, **no rearrangements** occur, and second, only **anti addition of X_2** is observed. For example, treatment of cyclohexene with Br_2 yields two **trans** enantiomers formed by **anti addition.**

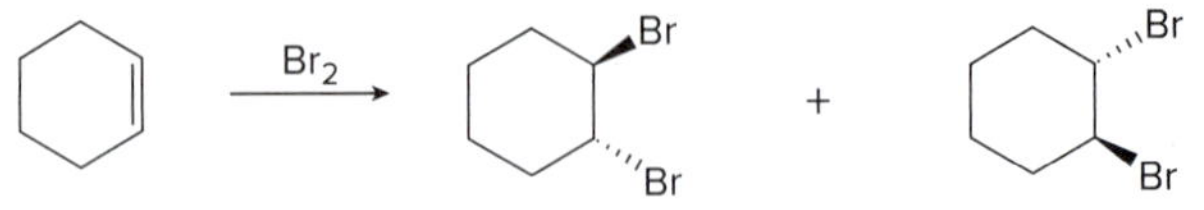

Br_2

Br Br + Br Br

enantiomers

These facts suggest that **carbocations are *not* intermediates in halogenation.** Unstable carbocations rearrange, and both syn and anti addition is possible with carbocation intermediates. The accepted mechanism for halogenation comprises **two steps,** but it does *not* proceed with formation of a carbocation, as shown in Mechanism 10.3.

Mechanism 10.3 Addition of X_2 to an Alkene—Halogenation

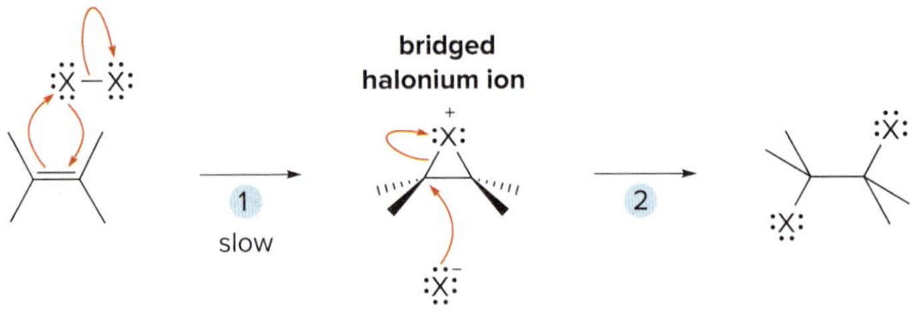

:X–X:

bridged halonium ion

1 slow

2

1 Four bonds are broken or formed to generate an unstable **bridged halonium ion** that contains a three-membered ring. The electron pair in the π bond and a lone pair on a halogen are used to form two new C–X bonds, and the X–X bond is cleaved.

2 **Nucleophilic attack of X^-** ring opens the bridged halonium ion and forms a new C–X bond.

Bridged halonium ions resemble carbocations in that they are short-lived intermediates that react readily with nucleophiles. Carbocations are inherently unstable because only six electrons surround carbon, whereas **halonium ions are unstable because they contain a strained three-membered ring** with a positively charged halogen atom.

carbocation

C has no octet.

bridged halonium ion

The ring has angle strain.

Problem 10.23 Draw the constitutional isomer formed in each reaction.

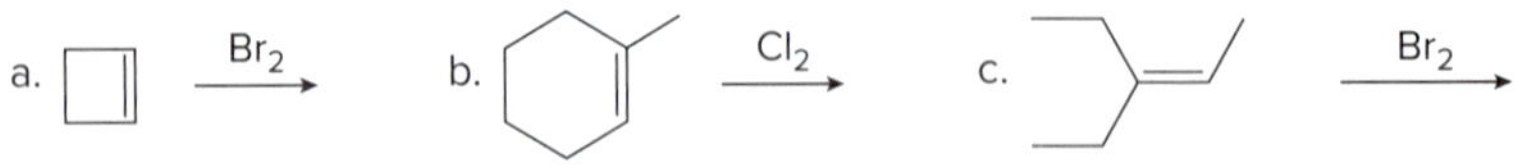

a. Br_2 b. Cl_2 c. Br_2

10.14 Stereochemistry of Halogenation

How does the proposed mechanism invoking a bridged halonium ion intermediate explain the observed **trans products of halogenation?** For example, chlorination of cyclopentene affords both enantiomers of *trans*-1,2-dichlorocyclopentane, with *no* cis products.

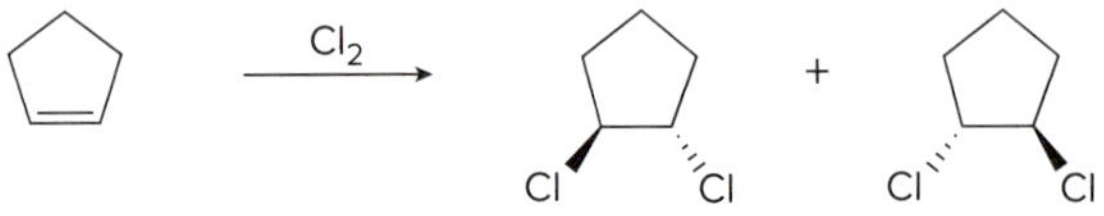

Cl_2

Cl Cl + Cl Cl

trans enantiomers

Initial addition of the electrophile Cl^+ (from Cl_2) occurs from either side of the planar double bond to form the bridged chloronium ion. In this example, both modes of addition (from above and below) generate the same **achiral** intermediate, so either representation can be used to draw the second step.

from **above** or from **below**

identical achiral chloronium ions

The opening of bridged halonium ion intermediates resembles the opening of epoxide rings with nucleophiles discussed in Section 9.16.

In the second step, **nucleophilic attack of Cl^- must occur from the back side**—that is, from the side of the five-membered ring *opposite* to the side having the bridged chloronium ion. Because the nucleophile attacks from below in this example and the leaving group departs from above, the two Cl atoms in the product are oriented **trans** to each other. Backside attack occurs with equal probability at either carbon of the three-membered ring to yield an equal amount of two enantiomers—**a racemic mixture.**

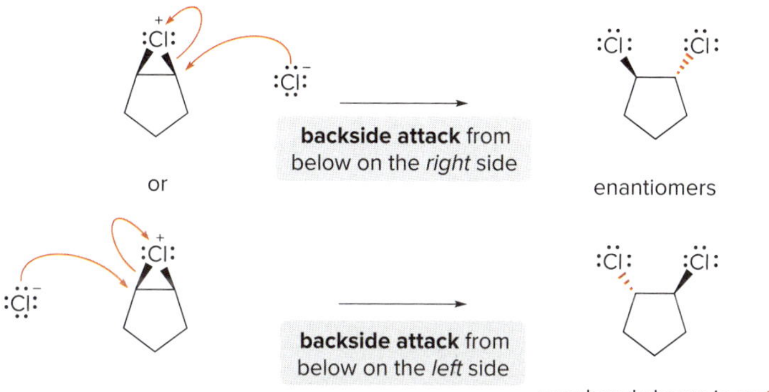

In summary, the mechanism for halogenation of alkenes occurs in two steps.

- **Addition of X^+ forms an unstable bridged halonium ion in the rate-determining step.**
- **Nucleophilic attack of X^- occurs from the *back side* to form trans products. The overall result is *anti addition* of X_2 across the double bond.**

Because halogenation occurs exclusively in an anti fashion, cis and trans alkenes yield different stereoisomers. Halogenation of alkenes is a **stereospecific reaction.**

- **A reaction is *stereospecific* when each of two specific stereoisomers of a starting material yields a particular stereoisomer of a product.**

cis-But-2-ene yields two enantiomers, whereas *trans*-but-2-ene yields a single achiral meso compound, as shown in Figure 10.8.

Problem 10.24 Draw all stereoisomers formed in each reaction.

a. $\xrightarrow{Cl_2}$ b. $\xrightarrow{Br_2}$ c. $\xrightarrow{Br_2}$ d. $\xrightarrow{Cl_2}$

Figure 10.8
Halogenation of *cis*- and *trans*-but-2-ene

cis-but-2-ene $\xrightarrow{Br_2}$ **enantiomers**

trans-but-2-ene $\xrightarrow{Br_2}$ **identical** an achiral meso compound

To draw the products of halogenation:

- Add Br_2 in an **anti** fashion across the double bond, leaving all other groups in their original orientations. With the alkene drawn in the plane of the page, **one Br adds from the front (ending up on a wedge), and one Br adds from the back (ending up on a dashed wedge).**
- Sometimes this reaction produces two stereoisomers, as in the case of *cis*-but-2-ene, which forms an equal amount of **two enantiomers.** Sometimes it produces a single compound, as in the case of *trans*-but-2-ene, where a **meso** compound is formed.

10.15 Halohydrin Formation

Treatment of an alkene with a halogen $\mathbf{X_2}$ and $\mathbf{H_2O}$ forms a **halohydrin** by addition of the elements of **X** and **OH** to the double bond.

A π bond is broken. + X–X $\xrightarrow{H–OH}$ **halohydrin**

$\xrightarrow[H_2O]{Cl_2}$ chlorohydrin

The mechanism for halohydrin formation is similar to the mechanism for halogenation: addition of the electrophile X^+ (from X_2) to form a **bridged halonium ion,** followed by nucleophilic attack by H_2O from the back side on the three-membered ring (Mechanism 10.4). Even though X^- is formed in Step [1] of the mechanism, its concentration is small compared to H_2O (often the solvent), so H_2O and *not* X^- is the nucleophile.

Mechanism 10.4 Addition of X and OH—Halohydrin Formation

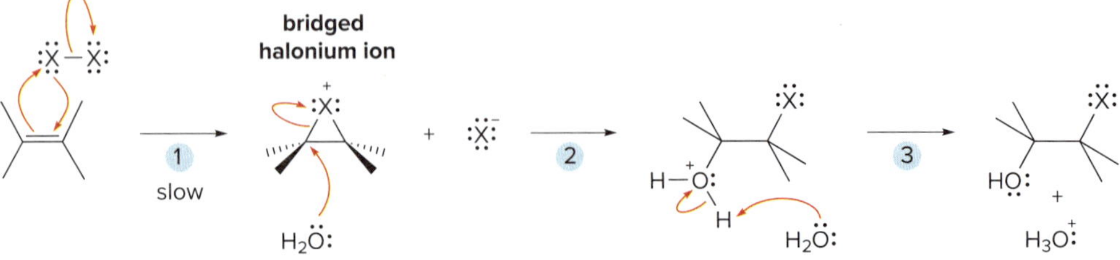

1 Four bonds are broken or formed to generate an unstable **bridged halonium ion** that contains a three-membered ring. The electron pair in the π bond and a lone pair on a halogen are used to form two new C–X bonds, and the X–X bond is cleaved.

2 **Nucleophilic attack of H_2O** ring opens the bridged halonium ion and forms a new C–O bond.

3 Loss of a proton forms the halohydrin.

Recall from Section 7.8C that DMSO (dimethyl sulfoxide) is a polar aprotic solvent.

Although the combination of Br_2 and H_2O effectively forms **bromohydrins** from alkenes, other reagents can also be used. Bromohydrins are also formed with ***N*-bromosuccinimide** (abbreviated as **NBS**) in **aqueous DMSO** [$(CH_3)_2S{=}O$]. NBS serves as a source of Br_2, which then goes on to form a bromohydrin by the same reaction mechanism.

N—Br → Br_2

N-bromosuccinimide
NBS

NBS / DMSO, H_2O → bromohydrin

10.15A Stereochemistry and Regioselectivity of Halohydrin Formation

Because the bridged halonium ion ring is opened by backside attack of H_2O, addition of X and OH occurs in an **anti** fashion and **trans** products are formed.

Br_2 / H_2O →

trans enantiomers

With unsymmetrical alkenes, two constitutional isomers are possible from addition of X and OH, but only one is formed. **The preferred product has the electrophile X^+ bonded to the *less* substituted carbon atom**—that is, the carbon that has more H atoms to begin with in the reacting alkene. Thus, the **nucleophile (H_2O) bonds to the more substituted carbon.**

Br_2 / H_2O →

only product

Br ends up on the *less* substituted C.

not formed

This result is reminiscent of the opening of epoxide rings with acids HZ (Z = a nucleophile), which we encountered in Section 9.16B. As in the opening of an epoxide ring, **nucleophilic attack occurs at the *more* substituted carbon end of the bridged halonium ion** because that carbon is better able to accommodate a partial positive charge in the transition state.

nucleophilic attack at the **more** substituted C

Table 10.4 summarizes the characteristics of halohydrin formation.

Table 10.4 Summary: Conversion of Alkenes to Halohydrins

	Observation
Mechanism	• The mechanism involves **three steps.** • The rate-determining step forms a **bridged halonium ion.** • **No rearrangements** can occur.
Regioselectivity	• The electrophile X^+ bonds to the **less substituted carbon.**
Stereochemistry	• **Anti** addition occurs.

Sample Problem 10.5 Drawing the Halohydrin Formed in an Alkene Addition

Draw the products of the following reaction, including stereochemistry.

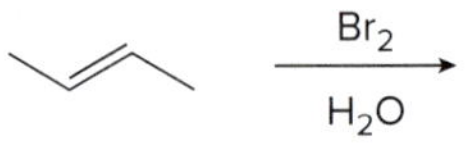

Br_2 / H_2O

trans-but-2-ene

Solution

The reagent ($Br_2 + H_2O$) adds the elements of **Br** and **OH** to a double bond in an **anti** fashion—that is, from **opposite** sides. To draw two products of anti addition: add **Br from above** and **OH from below** in one product; then add **Br from below** and **OH from above** in the other product. In this example, the two products are nonsuperimposable mirror images—**enantiomers.**

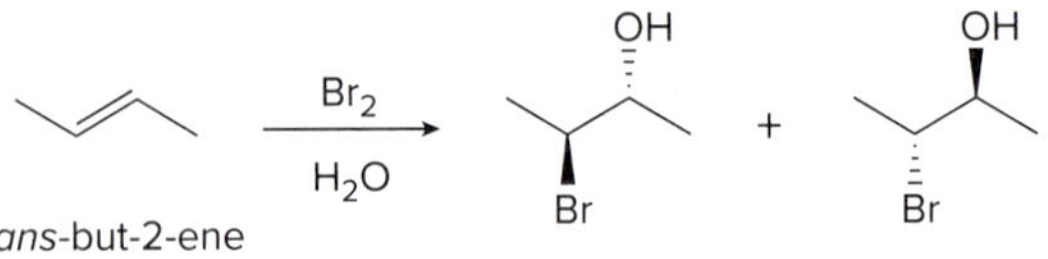

trans-but-2-ene $\xrightarrow[H_2O]{Br_2}$ OH, Br + OH, Br

enantiomers

Problem 10.25 Draw the products of each reaction and indicate their stereochemistry.

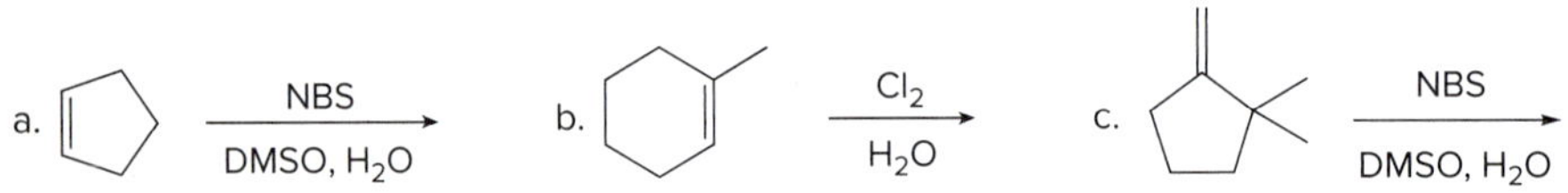

a. $\xrightarrow[DMSO,\ H_2O]{NBS}$ b. $\xrightarrow[H_2O]{Cl_2}$ c. $\xrightarrow[DMSO,\ H_2O]{NBS}$

More Practice: Try Problems 10.45e, f; 10.48c; 10.50b, c.

10.15B Halohydrins: Useful Compounds in Organic Synthesis

Because halohydrins are easily converted to epoxides by intramolecular S_N2 reaction (Section 9.6), they have been used in the synthesis of many naturally occurring compounds. Key steps in the synthesis of estrone, a female sex hormone, are illustrated in Figure 10.9.

Figure 10.9 The synthesis of estrone from a chlorohydrin

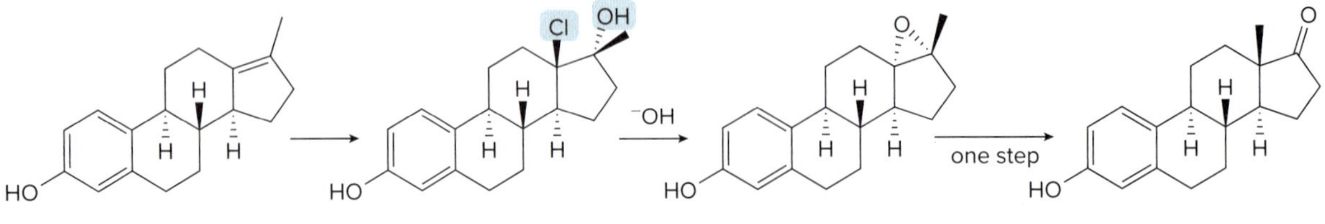

A → B $\xrightarrow{^-OH}$ C $\xrightarrow{\text{one step}}$ estrone

- Chlorohydrin **B,** prepared from alkene **A** by addition of Cl and OH, is converted to epoxide **C** with base. **C** is converted to estrone in one step.

10.16 Hydroboration–Oxidation

Hydroboration–oxidation is a two-step reaction sequence that converts an alkene to an alcohol.

BH_3 (hydroboration) → alkylborane (H, BH_2) → H_2O_2, HO^- (oxidation) → alcohol (H, OH)

- *Hydroboration* is the addition of borane (BH_3) to an alkene, forming an alkylborane.
- *Oxidation* converts the C–B bond of the alkylborane to a C–O bond.

Hydroboration–oxidation results in **addition of H_2O** to an alkene.

BH_3 → H, BH_2 → H_2O_2, HO^- → H, OH

BH_3 → H, BH_2 → H_2O_2, HO^- → H, OH

Borane (BH_3) is a reactive gas that exists mostly as the dimer, diborane (B_2H_6). Borane is a strong **Lewis acid** that reacts readily with Lewis bases. For ease in handling in the laboratory, it is commonly used as a complex with tetrahydrofuran (THF).

BH_3 + THF → $BH_3 \cdot THF$

borane, **Lewis acid**; tetrahydrofuran, **THF**, **Lewis base**; **$BH_3 \cdot THF$**

10.16A Hydroboration

The first step in hydroboration–oxidation is **addition of the elements of H and BH_2** to the π bond of the alkene, forming an intermediate alkylborane.

BH_3 → H, BH_2 **alkylborane**

Because **syn addition** to the double bond occurs and **no carbocation rearrangements** are observed, carbocations are *not* formed during hydroboration, as shown in Mechanism 10.5.

Mechanism 10.5 Addition of H and BH_2—Hydroboration

One step The π bond and H–BH_2 bonds break as the C–H and C–B bonds form.

H–BH_2 → [H---BH_2]‡ transition state → H, BH_2 **syn addition**

The proposed mechanism involves a **concerted addition of H and BH_2 from the same side of the planar double bond:** the π bond and H–BH_2 bond are broken as two new σ bonds are formed. Because four atoms are involved, the transition state is said to be **four-centered.**

Because the alkylborane formed by reaction with one equivalent of alkene still has two B–H bonds, it can react with two more equivalents of alkene to form a trialkylborane. This is illustrated in Figure 10.10 for the reaction of $CH_2{=}CH_2$ with BH_3.

Figure 10.10
Conversion of BH_3 to a trialkylborane with three equivalents of $CH_2{=}CH_2$

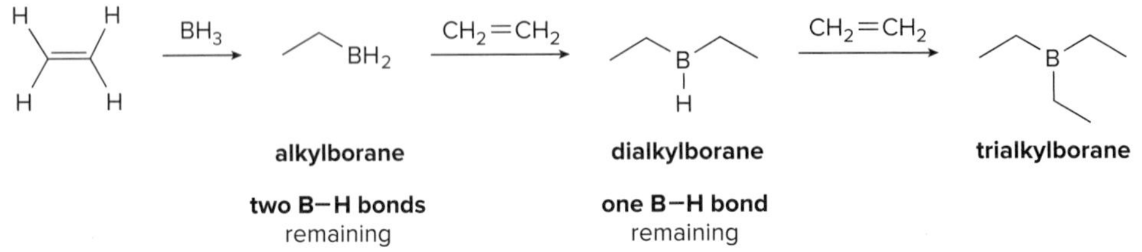

- We often draw hydroboration as if addition stopped after one equivalent of alkene reacts with BH_3. Instead, all three B–H bonds actually react with three equivalents of an alkene to form a trialkylborane. The term **organoborane** is used for any compound with a carbon–boron bond.

Because only one B–H bond is needed for hydroboration, commercially available dialkylboranes having the general structure **R_2BH** are sometimes used instead of BH_3. A common example is 9-borabicyclo[3.3.1]nonane **(9-BBN).** 9-BBN undergoes hydroboration in the same manner as BH_3.

9-borabicyclo[3.3.1]nonane
9-BBN

Hydroboration is regioselective. **With unsymmetrical alkenes, the boron atom bonds to the *less* substituted carbon atom.** For example, addition of BH_3 to propene forms an alkylborane with the B bonded to the terminal carbon atom.

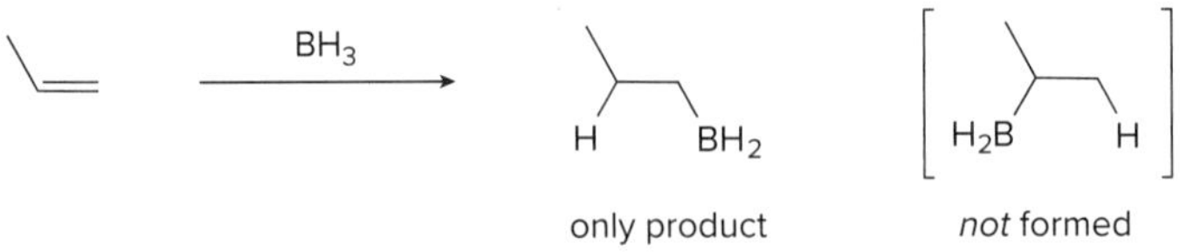

Because H is more electronegative than B, the B–H bond is polarized to give boron a partial positive charge ($H^{\delta-}$–$B^{\delta+}$), making BH_2 the electrophile in hydroboration.

Steric factors explain this regioselectivity. The larger boron atom bonds to the less sterically hindered, more accessible carbon atom.

Electronic factors are also used to explain this regioselectivity. If bond breaking and bond making are not completely symmetrical, boron bears a partial negative charge in the transition state and carbon bears a partial positive charge. Because alkyl groups stabilize a positive charge, the more stable transition state has the partial positive charge on the more substituted carbon, as illustrated in Figure 10.11.

- **In hydroboration, the boron atom bonds to the *less* substituted carbon.**

Figure 10.11 Hydroboration of an unsymmetrical alkene

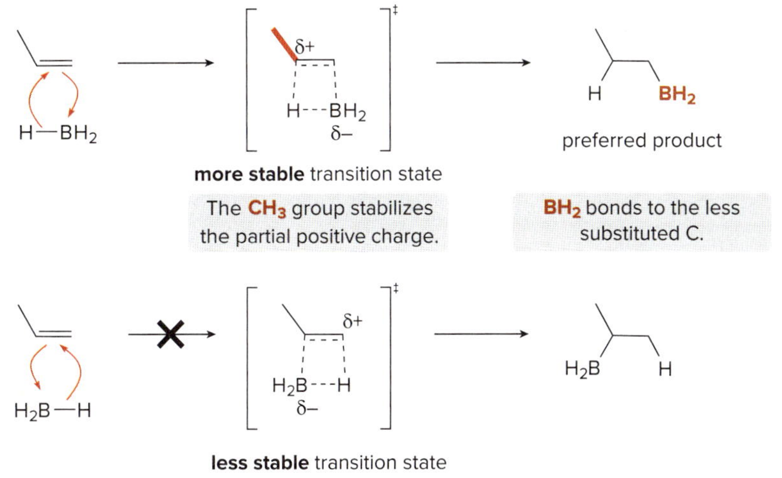

Problem 10.26 What alkylborane is formed from hydroboration of each alkene?

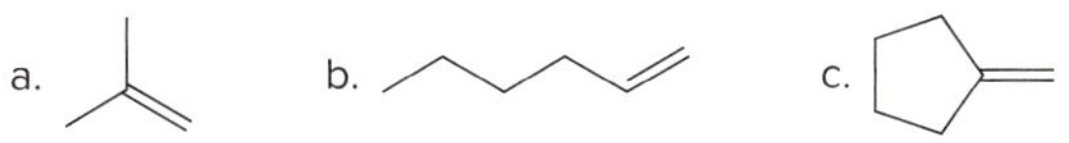

a. b. c.

10.16B Oxidation of the Alkylborane

Because alkylboranes react rapidly with water and spontaneously burn when exposed to the air, they are oxidized, without isolation, with basic hydrogen peroxide (H_2O_2, HO^-). **Oxidation replaces the C–B bond with a C–O bond, forming a new OH group with retention of configuration;** that is, the **OH group replaces the BH_2 group in the same position** relative to the other three groups on carbon.

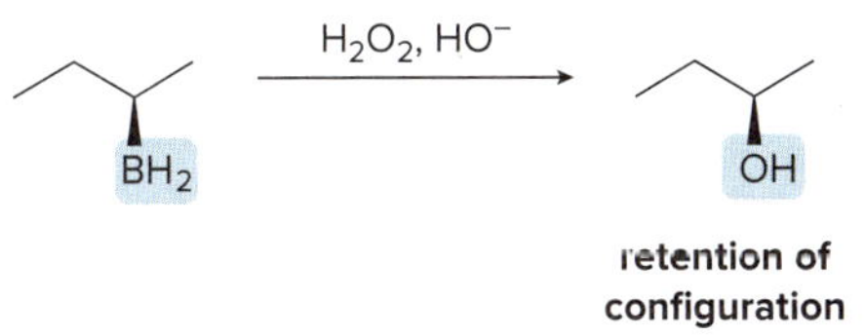

Thus, to draw the product of a hydroboration–oxidation reaction, keep in mind two stereochemical facts:

- **Hydroboration occurs with syn addition.**
- **Oxidation occurs with retention of configuration.**

The overall result of this two-step sequence is **syn addition of the elements of H and OH** to a double bond, as illustrated in Sample Problem 10.6. **The OH group bonds to the *less* substituted carbon.**

Sample Problem 10.6 Drawing the Products of Hydroboration–Oxidation

Draw the product of the following reaction sequence, including stereochemistry.

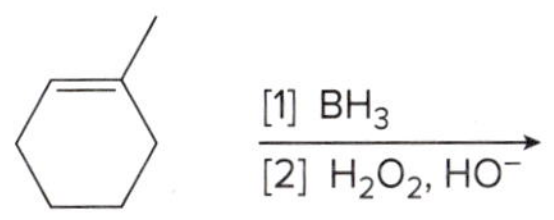

Solution

In Step [1], **syn addition of BH_3 to the unsymmetrical alkene adds the BH_2 group to the *less* substituted carbon from above and below the planar double bond.** Two enantiomeric alkylboranes are formed. In Step [2], oxidation replaces the BH_2 group with OH in each enantiomer with **retention of configuration** to yield two alcohols that are also enantiomers.

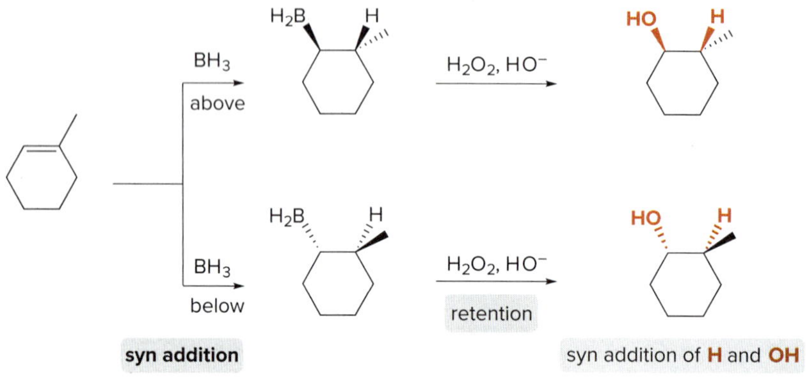

H_2B H
BH_3 above
H_2O_2, HO^-
HO H
H_2B H
BH_3 below
H_2O_2, HO^-
retention
HO H
syn addition
syn addition of **H** and **OH**

Hydroboration–oxidation results in the **addition of H and OH in a syn fashion** across the double bond. The achiral alkene is converted to an equal mixture of two enantiomers—that is, a **racemic mixture of alcohols.**

Problem 10.27 Draw the products formed when each alkene is treated with BH_3 followed by H_2O_2, HO^-. Include the stereochemistry at all stereogenic centers.

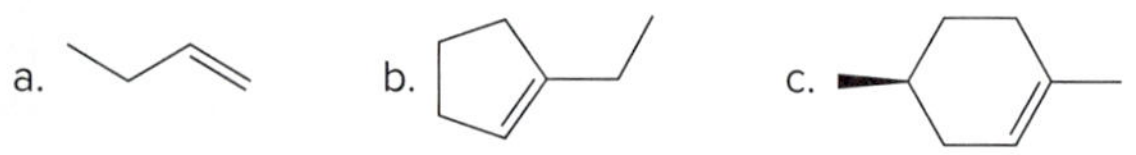

a. b. c.

More Practice: Try Problem 10.45g; 10.48b, d; 10.50g.

Problem 10.28 What alkene can be used to prepare each alcohol as the exclusive product of a two-step hydroboration–oxidation sequence?

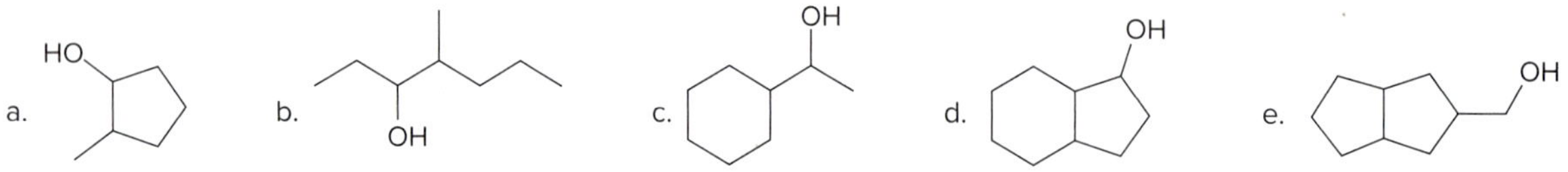

a. HO
b. OH
c. OH
d. OH
e. OH

Table 10.5 summarizes the features of hydroboration–oxidation.

Table 10.5 Summary: Hydroboration–Oxidation of Alkenes

	Observation
Mechanism	• The addition of H and BH_2 occurs in **one step.** • **No rearrangements** can occur.
Regioselectivity	• The **OH group bonds to the less substituted** carbon atom.
Stereochemistry	• **Syn addition** occurs. • OH replaces BH_2 with **retention** of configuration.

Hydroboration–oxidation is a very common method for adding H_2O across a double bond. One example is shown in the synthesis of **artemisinin** (or **qinghaosu**), the active component of **qing-hao,** a Chinese herbal remedy used for the treatment of malaria (Figure 10.12).

Figure 10.12
An example of hydroboration–oxidation in synthesis

- The carbon atoms of artemisinin that come from alcohol **A** are indicated in red.

Artemisia annua, a plant used for hundreds of years in traditional Chinese medicine, is the source of artemisinin, a compound used in combination with other drugs to treat malaria. Tu Youyou received the 2015 Nobel Prize in Physiology or Medicine for her discovery of the antimalarial effects of artemisinin. *Adam Gault/Getty Images*

10.16C A Comparison of Hydration Methods

Hydration (H_2O, H^+) and hydroboration–oxidation (BH_3 followed by H_2O_2, HO^-) both add the elements of H_2O across a double bond. Despite their similarities, these reactions often form different constitutional isomers, as shown in Sample Problem 10.7.

Sample Problem 10.7 Comparing Two Different Methods of Hydration of an Alkene

Draw the product formed when $CH_3CH_2CH_2CH_2CH{=}CH_2$ is treated with either (a) H_2O, H_2SO_4; or (b) BH_3 followed by H_2O_2, HO^-.

Solution

With $H_2O + H_2SO_4$, electrophilic addition of H and OH places the **H atom on the *less* substituted carbon** of the alkene to yield a **2° alcohol.** In contrast, addition of BH_3 gives an alkylborane with the **BH_2 group on the *less* substituted terminal carbon** of the alkene. Oxidation replaces BH_2 by OH to yield a **1° alcohol.**

Problem 10.29 Draw the constitutional isomer formed when the following alkenes are treated with each set of reagents: [1] H_2O, H_2SO_4; or [2] BH_3 followed by H_2O_2, ^-OH.

More Practice: Try Problem 10.47.

10.17 Keeping Track of Reactions

Chapters 7–10 have introduced three basic kinds of organic reactions: **nucleophilic substitution, β elimination,** and **addition.** In the process, many specific reagents have been discussed and the stereochemistry that results from many different mechanisms has been examined. **How can we keep track of all the reactions?**

To make the process easier, **remember that most organic molecules undergo only one or two different kinds of reactions.** For example:

- **Alkyl halides undergo substitution and elimination because they have good leaving groups.**
- **Alcohols also undergo substitution and elimination, but can do so only when OH is made into a good leaving group.**
- **Alkenes undergo addition because they have easily broken π bonds.**

You must still learn many reaction details, and in truth, there is no one method to learn them. ***You must practice these reactions over and over again, not by merely looking at them, but by writing them.*** Some students do this by making a list of specific reactions for each functional group, and then rewriting them with different starting materials. Others make flash cards: index cards that have the starting material and reagent on one side and the product on the other. Whatever method you choose, **the details must become second nature,** much like the answers to simple addition problems, such as, what is the sum of 2 + 2?

Learning reactions is really a two-step process:

- **First, learn the basic type of reaction for a functional group. This provides an overall organization to the reactions.**
- **Then, learn the specific reagents for each reaction. It helps to classify the reagent according to its properties. Is it an acid or a base? Is it a nucleophile or an electrophile? Is it an oxidizing agent or a reducing agent?**

Sample Problem 10.8 illustrates this process.

Sample Problem 10.8 Using the Functional Group and Reagent to Identity the Type of Reaction

Draw the product of each reaction.

a. 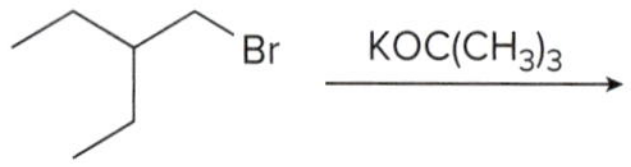 [alkyl bromide] Br $\xrightarrow{KOC(CH_3)_3}$ b. 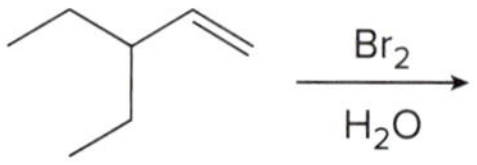 [alkene] $\xrightarrow[H_2O]{Br_2}$

Solution

In each problem, **identify the functional group** to determine the general reaction type—substitution, elimination, or addition. Then, **determine if the reagent is an electrophile, nucleophile, acid, base,** and so forth.

a. The reactant is a **1° alkyl halide,** which can undergo substitution and elimination. The reagent [$KOC(CH_3)_3$] is a **strong nonnucleophilic base,** favoring elimination by an E2 mechanism.

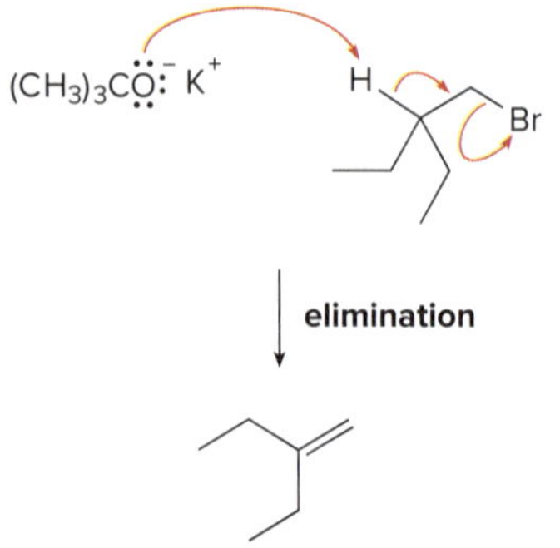

$(CH_3)_3C\ddot{O}$:$^-$ K^+ H Br

↓ elimination

E2 product

b. The reactant is an **alkene,** which undergoes addition reactions to its π bond. The reagent (Br_2 + H_2O) serves as the source of the **electrophile Br^+,** resulting in **addition** of Br and OH to the double bond (Section 10.15).

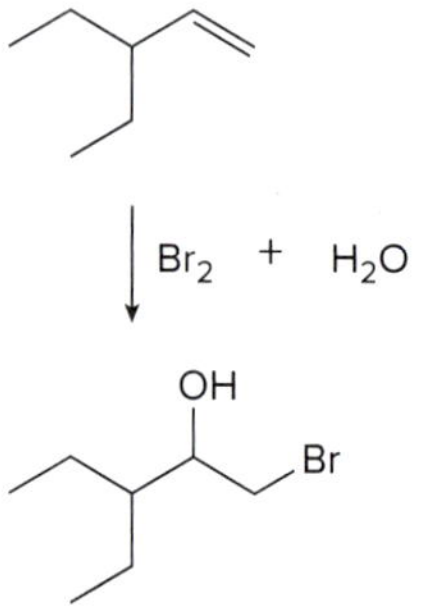

↓ Br_2 + H_2O

OH Br

addition product

Problem 10.30 Draw the products of each reaction using the two-part strategy from Sample Problem 10.8.

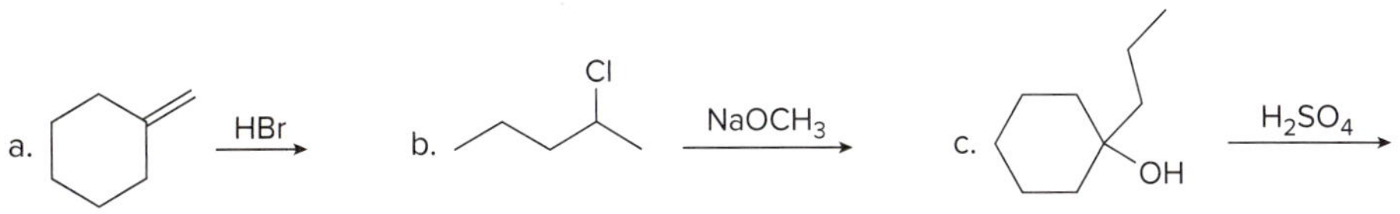

a. [methylenecyclohexane] $\xrightarrow{HBr}$ b. [2-chloropentane] $\xrightarrow{NaOCH_3}$ c. [1-propylcyclohexanol] $\xrightarrow{H_2SO_4}$

More Practice: Try Problems 10.33b, 10.45, 10.48.

10.18 Alkenes in Organic Synthesis

Alkenes are a central functional group in organic chemistry. **Alkenes are easily prepared by elimination reactions** such as dehydrohalogenation and dehydration. **Because their π bond is easily broken, they undergo many addition reactions** to prepare a variety of useful compounds.

Suppose, for example, that we must synthesize 1,2-dibromocyclohexane from cyclohexanol, a cheap and readily available starting material. Because there is no way to accomplish this transformation in one step, this synthesis must have at least two steps.

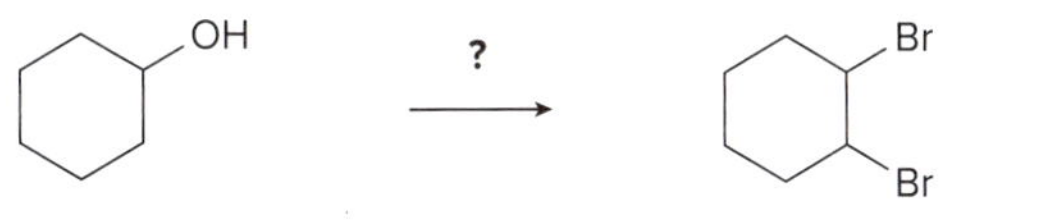

cyclohexanol **starting material** $\xrightarrow{?}$ 1,2-dibromocyclohexane **product**

To solve this problem we must:

- **Work backwards from the product by asking: What type of reactions introduce the functional groups in the product?**
- **Work forwards from the starting material by asking: What type of reactions does the starting material undergo?**

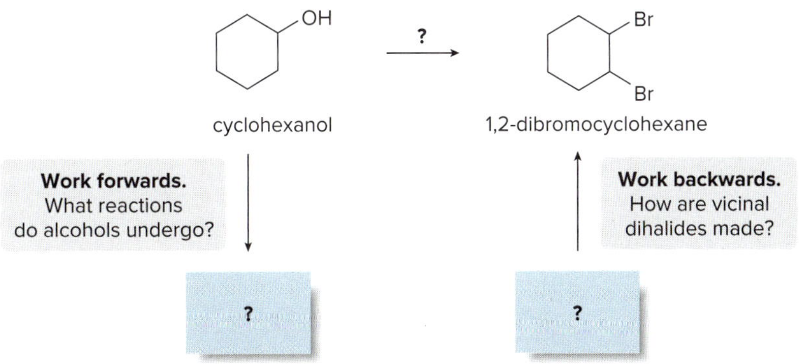

cyclohexanol $\xrightarrow{?}$ 1,2-dibromocyclohexane

Work forwards. What reactions do alcohols undergo? → ?

Work backwards. How are vicinal dihalides made? ← ?

In Chapter 11, we will learn about retrosynthetic analysis in more detail.

Working backwards from the product to determine the starting material from which it is made is called *retrosynthetic analysis*.

We know reactions that answer each of these questions.

Working backwards:

[1] 1,2-Dibromocyclohexane, a vicinal dibromide, can be prepared by the addition of Br_2 to **cyclohexene.**

Working forwards:

[2] Cyclohexanol can undergo acid-catalyzed dehydration to form **cyclohexene.**

cyclohexene —Br_2→ 1,2-dibromocyclohexane

cyclohexanol —H_2SO_4→ cyclohexene

A **reactive intermediate** is an unstable intermediate like a carbocation, which is formed during the conversion of a stable starting material to a stable product. A **synthetic intermediate** is a stable compound that is the product of one step and the starting material of another in a multistep synthesis.

Cyclohexene is called a **synthetic intermediate,** or simply an **intermediate,** because it is the **product of one step and the starting material of another.** We now have a two-step sequence to convert cyclohexanol to 1,2-dibromocyclohexane, and the synthesis is complete. Take note of the central role of the alkene in this synthesis.

cyclohexanol —H_2SO_4→ cyclohexene —Br_2→ 1,2-dibromocyclohexane

a synthetic intermediate

Sample Problem 10.9 Devising a Synthesis from an Alkene

Devise a synthesis of 1-ethoxy-2-methylcyclohexane from 1-methylcyclohexene.

1-methylcyclohexene —?→ 1-ethoxy-2-methylcyclohexane

Solution

Work backwards from the product and forwards from the starting material until a common intermediate is reached.

Working backwards:

[1] Williamson ether synthesis converts 2-methylcyclohexanol to 1-ethoxy-2-methylcyclohexanol by a two-step process: reaction with the base NaH to generate alkoxide **A,** followed by S_N2 reaction with CH_3CH_2Br to generate the ether.

2-methylcyclohexanol —NaH→ **A** + H_2 —CH_3CH_2Br, S_N2→ 1-ethoxy-2-methylcyclohexane

Working forwards:

Hydroboration–oxidation of 1-methylcyclohexene forms 2-methylcyclohexanol, with the OH group on the *less* substituted carbon.

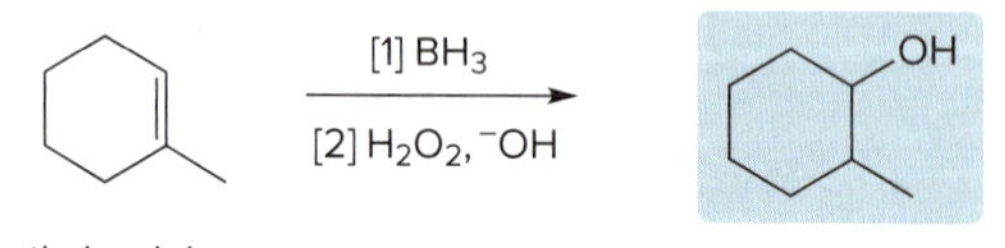

1-methylcyclohexene 2-methylcyclohexanol

2-Methylcyclohexanol is a **synthetic intermediate,** and the synthesis is complete by combining both operations.

1-methylcyclohexene → [1] BH_3; [2] H_2O_2, ^-OH → 2-methylcyclohexanol → [1] NaH; [2] ethyl bromide → 1-ethoxy-2-methylcyclohexane

Problem 10.31 Devise a synthesis of each compound from the indicated starting material.

a. 1-bromobutane → ? → 1-chloro-2-butanol (OH, Cl)

b. 1-methylcyclohexanol → ? → 2-methylcyclohexanol (OH)

More Practice: Try Problems 10.60, 10.61.

Chapter 10 REVIEW

KEY CONCEPTS

Markovnikov's Rule (10.10, 10.12)

In the addition of HX to an unsymmetrical alkene, the H atom bonds to the *less* substituted carbon.

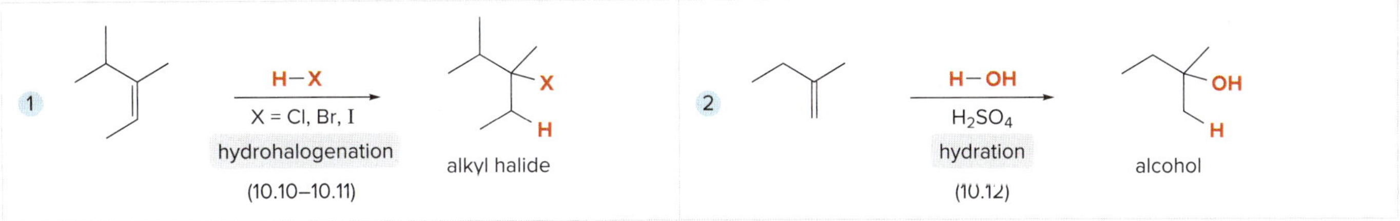

Try Problems 10.33b; 10.45a–c; 10.48a

Stereochemistry of Alkene Addition (10.8)

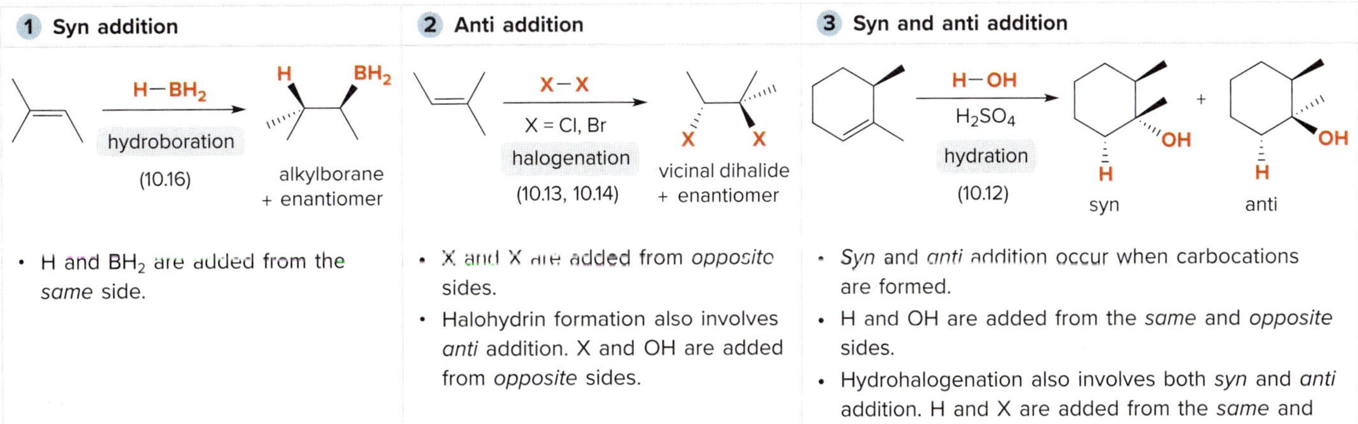

Try Problem 10.50.

KEY REACTIONS

All reactions of alkenes involve **addition**—the weak π bond is broken and two new σ bonds are formed.

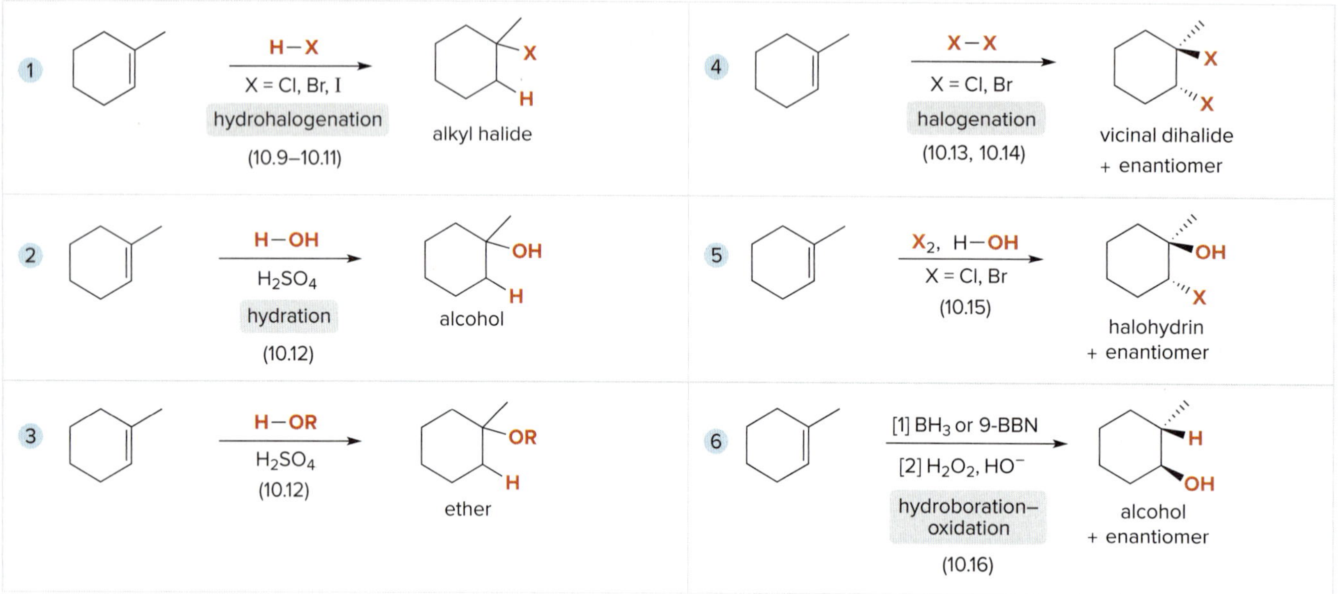

See Sample Problems 10.4, 10.5, 10.6. Try Problems 10.33b, 10.45, 10.48, 10.50.

KEY SKILLS

[1] Calculating degrees of unsaturation (10.2)

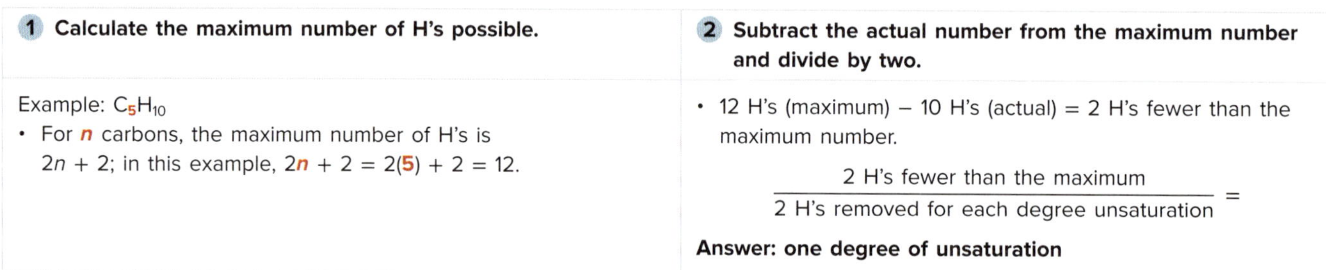

1 Calculate the maximum number of H's possible.	**2** Subtract the actual number from the maximum number and divide by two.
Example: C_5H_{10} • For ***n*** carbons, the maximum number of H's is $2n + 2$; in this example, $2n + 2 = 2(5) + 2 = 12$.	• 12 H's (maximum) − 10 H's (actual) = 2 H's fewer than the maximum number. $\frac{\text{2 H's fewer than the maximum}}{\text{2 H's removed for each degree unsaturation}} =$ **Answer: one degree of unsaturation**

See Sample Problem 10.1. Try Problems 10.34, 10.35.

[2] Assigning *E,Z* in naming an alkene (10.3)

1 Assign priorities to the substituents on each end of the C=C.	**2** Assign *E* or *Z* based on the location of the two higher-priority groups.
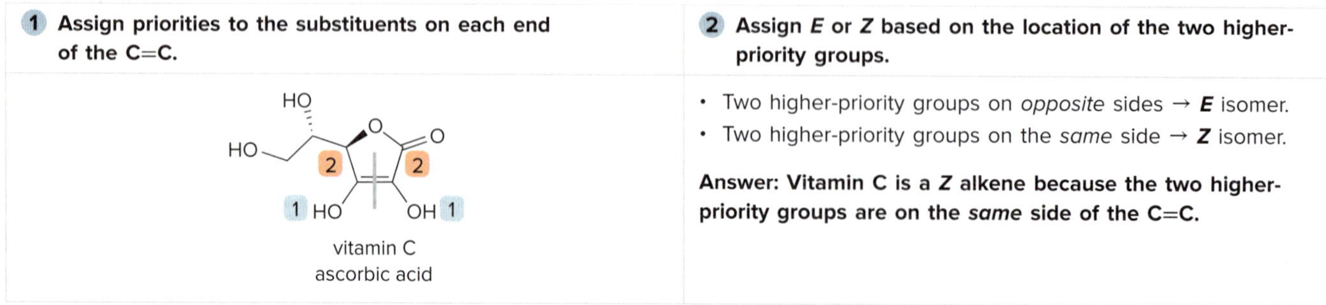	• Two higher-priority groups on *opposite* sides → ***E*** isomer. • Two higher-priority groups on the *same* side → ***Z*** isomer. **Answer: Vitamin C is a *Z* alkene because the two higher-priority groups are on the *same* side of the C=C.**

vitamin C
ascorbic acid

See *How To*, p. 399. Try Problems 10.33a, 10.36, 10.37a.

[3] Drawing the products of an addition reaction (10.8–10.16)

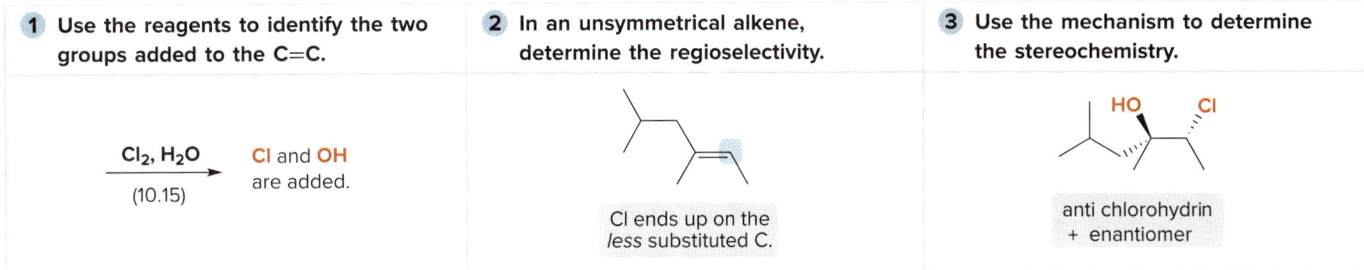

Try Problems 10.33b, 10.45, 10.48, 10.50.

[4] Comparing the products of hydration of an alkene (10.12, 10.16)

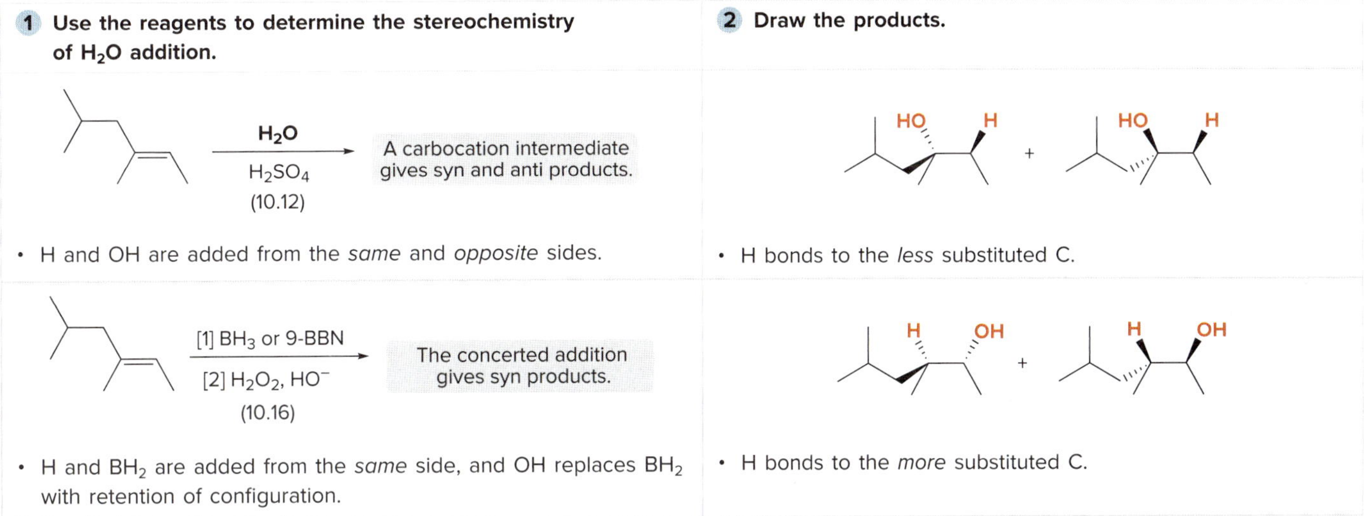

See Sample Problem 10.7. Try Problem 10.47.

KEY MECHANISM CONCEPTS

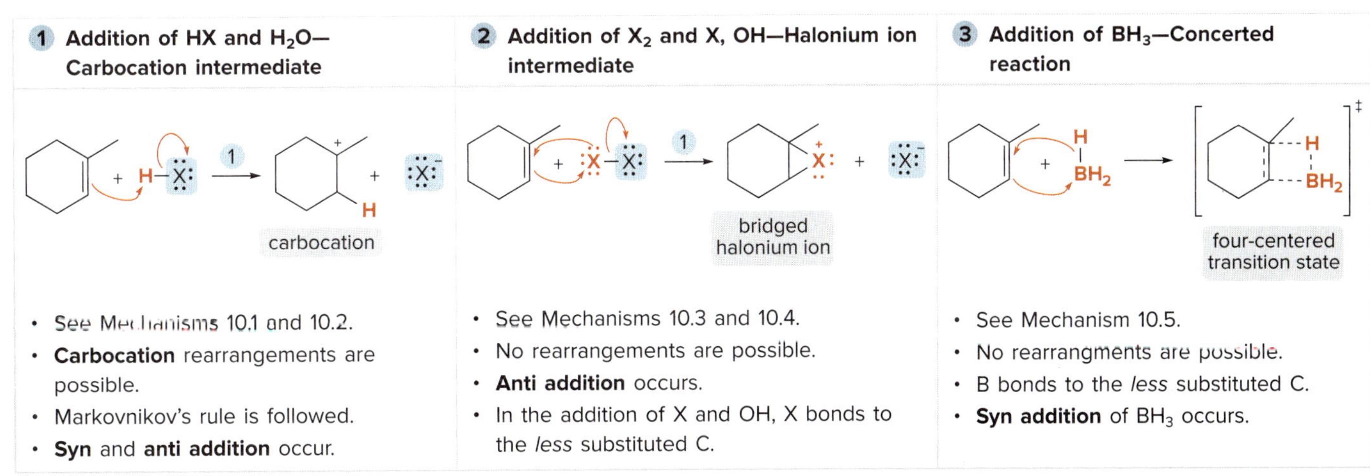

See Tables 10.3, 10.4, 10.5. Try Problems 10.53–10.55.

CHAPTER 10 MULTIPLE-CHOICE SELF-TEST

The Self-Test consists of multiple-choice questions similar to those found on the American Chemical Society organic chemistry exam. Answers are given at the end of the chapter.

1. Label each highlighted double bond in kavain, a natural product isolated from kava, as *E* or *Z*.

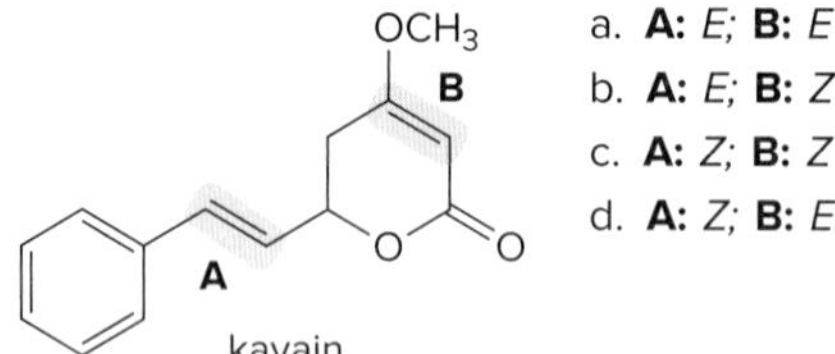

a. **A:** *E*; **B:** *E*
b. **A:** *E*; **B:** *Z*
c. **A:** *Z*; **B:** *Z*
d. **A:** *Z*; **B:** *E*

2. How many degrees of unsaturation are present in a compound of molecular formula $C_8H_{12}O_3$: (a) one; (b) two; (c) three; (d) four?

3. Give the IUPAC name for the following alkene.

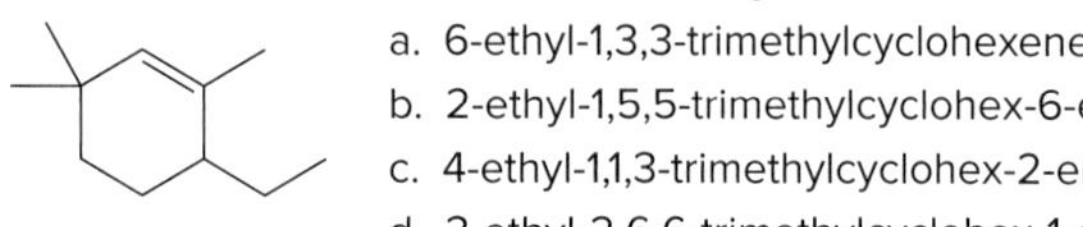

a. 6-ethyl-1,3,3-trimethylcyclohexene
b. 2-ethyl-1,5,5-trimethylcyclohex-6-ene
c. 4-ethyl-1,1,3-trimethylcyclohex-2-ene
d. 3-ethyl-2,6,6-trimethylcyclohex-1-ene

4. Label the most stable and the least stable alkene.

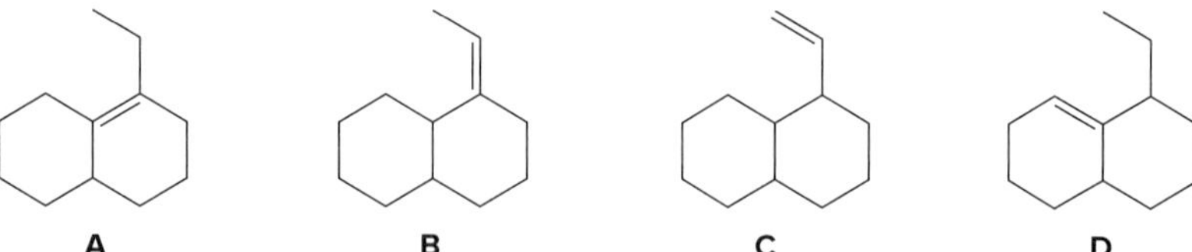

a. **A** is most stable and **B** is least stable.
b. **D** is most stable and **C** is least stable.
c. **A** is most stable and **C** is least stable.
d. **D** is most stable and **B** is least stable.

5. In which reactions might carbocation rearrangements occur?

a. hydroboration–oxidation
b. chlorination with Cl_2
c. halohydrin formation
d. hydration with H_2O, H_2SO_4

6. What product(s) is (are) formed when **A** reacts with HCl?

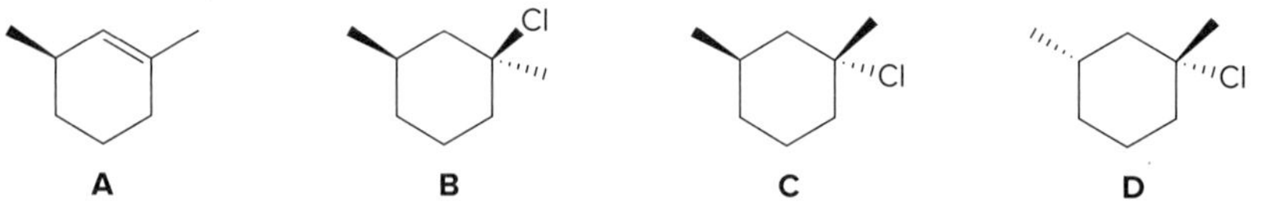

a. **B** only b. **C** only c. a mixture of **B** and **C** d. a mixture of **B, C,** and **D**

7. Which alkene reacts with the given reagents to form alcohol **A** cleanly?

OH
A

a. + H_2O, H_2SO_4
b. + [1] BH_3 [2] H_2O_2, HO^-
c. + H_2O, H_2SO_4
d. + H_2O, NBS, DMSO

8. What halohydrin is formed by addition of Br and OH to 1-methylcyclohexene?

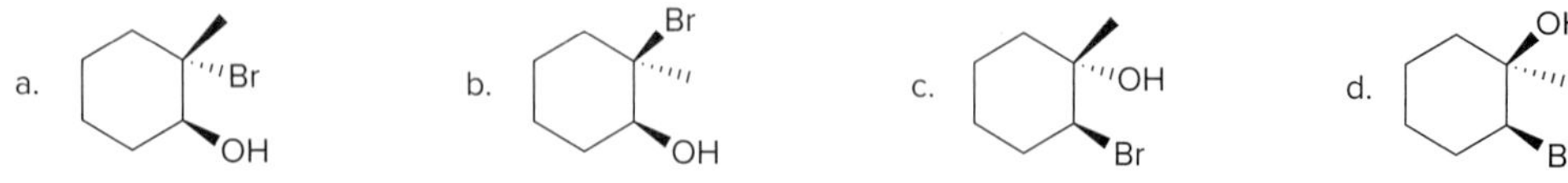

9. Which reagents give both syn and anti addition to a carbon–carbon double bond?

a. BH_3 b. Cl_2, H_2O c. Br_2 d. H_2O, H_2SO_4

10. What reagents are needed to convert **B** to **C**?

Br
B

Br Br
C

a. [1] $KOC(CH_3)_3$; [2] Br_2
b. HBr
c. [1] $KOC(CH_3)_3$; [2] HBr
d. Br_2, H_2O

PROBLEMS

Problems Using Three-Dimensional Models

10.32 Give the IUPAC name for each compound.

a. 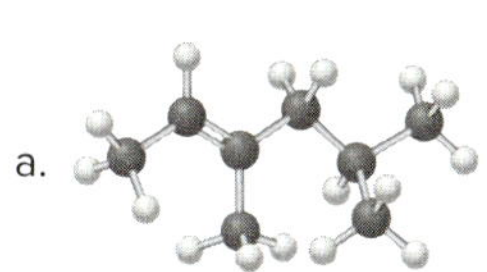

b. 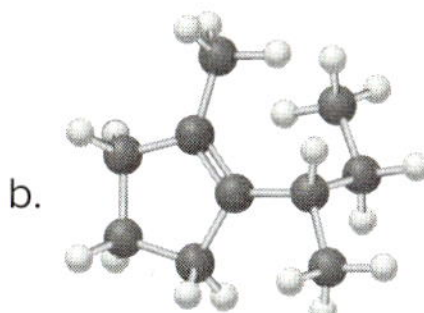

10.33 (a) Label the carbon–carbon double bond in **A** as *E* or *Z*. (b) Draw the products (including stereoisomers) formed when **A** is treated with H_2O in the presence of H_2SO_4.

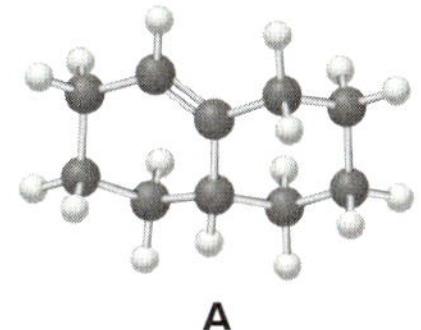

A

Degrees of Unsaturation

10.34 Calculate the number of degrees of unsaturation for each molecular formula.

a. C_6H_8
b. $C_{40}H_{56}$
c. $C_{10}H_{16}O_2$
d. C_8H_9Br
e. C_8H_9ClO
f. $C_7H_{11}N$
g. C_4H_8BrN
h. $C_{10}H_{18}ClNO$

10.35 How many rings and π bonds does a compound with molecular formula $C_{10}H_{14}$ possess? List all possibilities.

Nomenclature and Stereochemistry

10.36 Label the alkene in each drug as *E* or *Z*. Enclomiphene is one component of the fertility drug Clomid. Tamoxifen is an anticancer drug. Cilastatin is one component of a broad-spectrum two-drug antibiotic.

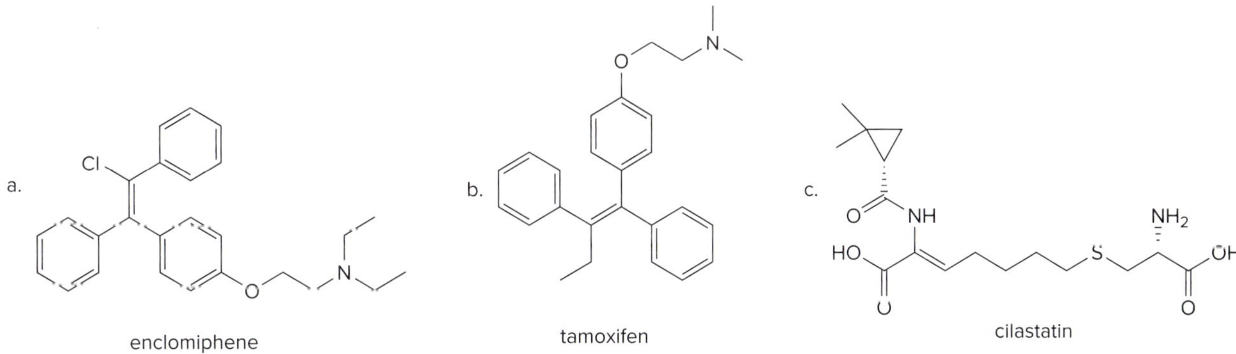

a. enclomiphene b. tamoxifen c. cilastatin

10.37 Brassicadiene, a compound present in cauliflower and broccoli plants, is an attractant for stink bugs that ultimately cause extensive damage to these crops. It is hoped that elucidation of the structure of brassicadiene, which was reported in 2018, will ultimately help farmers protect their crops from this unwanted pest.

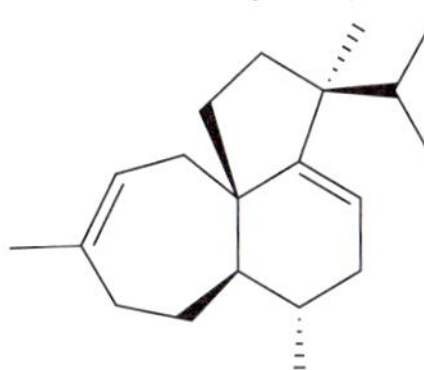

brassicadiene

a. Label each double bond as *E* or *Z*.
b. Label each tetrahedral stereogenic center as *R* or *S*.
c. Given that both double bonds are present in rings with fewer than eight carbons, are stereoisomers possible at either C=C?
d. Keeping in mind your answer in part (c), how many stereoisomers are possible for brassicadiene?

10.38 Give the IUPAC name for each compound.

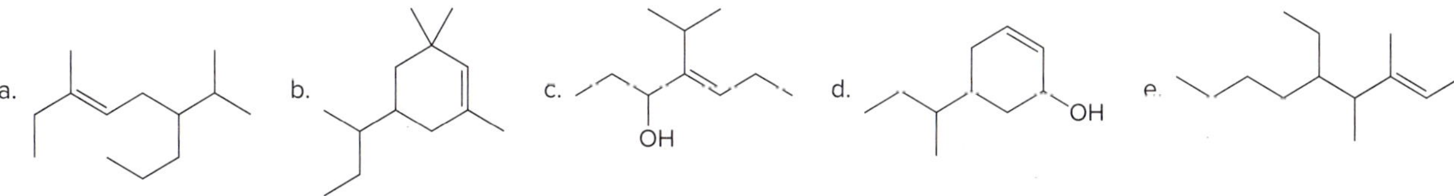

10.39 Give the structure corresponding to each name.

a. (*E*)-4-ethylhept-3-ene
b. 3,3-dimethylcyclopentene
c. 4-vinylcyclopentene
d. (*Z*)-3-isopropylhept-2-ene
e. *cis*-3,4-dimethylcyclopentene
f. 1-isopropyl-4-propylcyclohexene
g. 3,4-dimethylcyclohex-2-enol
h. 3,5-diethylhex-5-en-3-ol

10.40 (a) Draw all possible stereoisomers of 4-methylnon-2-ene, and name each isomer, including its *E,Z* and *R,S* prefixes. (b) Label two pairs of enantiomers. (c) Label four pairs of diastereomers.

10.41 (a) Draw the structure of (1*E*,4*R*)-1,4-dimethylcyclodecene. (b) Draw the enantiomer and name it, including its *E,Z* and *R,S* prefixes. (c) Draw two diastereomers and name them, including the *E,Z* and *R,S* prefixes.

10.42 Now that you have learned how to name alkenes in Section 10.3, name each of the following epoxides as an alkene oxide, as described in Section 9.3.

a. b. c. d.

10.43 Iejimalide B, an anticancer agent with a 24-membered ring, is isolated from a tunicate found off Ie Island in Okinawa. (a) Label each double bond in iejimalide B as *E* or *Z*. (b) Label each tetrahedral stereogenic center as *R* or *S*. (c) How many stereoisomers are possible for iejimalide B?

iejimalide B

Lipids

10.44 Eleostearic acid is an unsaturated fatty acid obtained from the seeds of the tung oil tree (*Aleurites fordii*), a deciduous tree native to China. (a) Draw the structure of a stereoisomer that has a higher melting point than eleostearic acid. (b) Draw the structure of a stereoisomer that has a lower melting point.

eleostearic acid

Reactions of Alkenes

10.45 Draw the products formed when $(CH_3)_2C{=}CH_2$ is treated with each reagent.

a. HBr
b. H_2O, H_2SO_4
c. CH_3CH_2OH, H_2SO_4
d. Cl_2
e. Br_2, H_2O
f. NBS (aqueous DMSO)
g. [1] BH_3; [2] H_2O_2, HO^-

10.46 What alkene can be used to prepare each alkyl halide or dihalide as the exclusive or major product of an addition reaction?

a.

b. 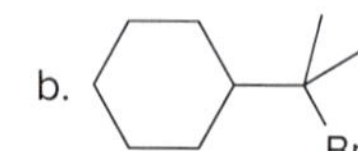

c. 

d.

10.47 Which alcohols can be prepared as a single product by hydroboration–oxidation of an alkene? Which alcohols can be prepared as a single product by the acid-catalyzed addition of H_2O to an alkene?

a. b. c. d.

10.48 Draw the constitutional isomer formed in each reaction.

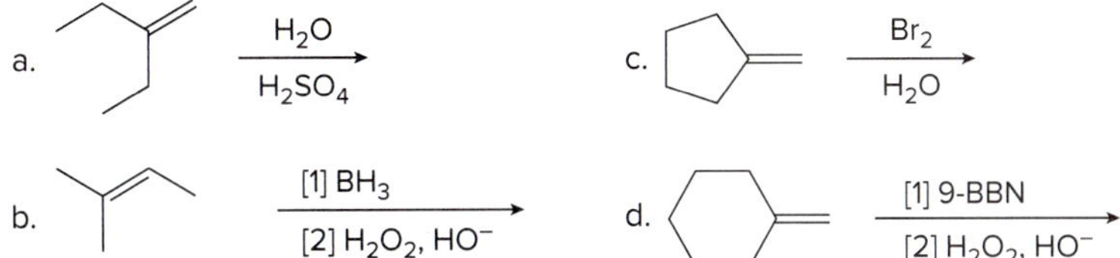

10.49 What three alkenes (excluding stereoisomers) can be used to prepare 3-chloro-3-methylhexane by addition of HCl?

10.50 Draw the products of each reaction, including stereoisomers.

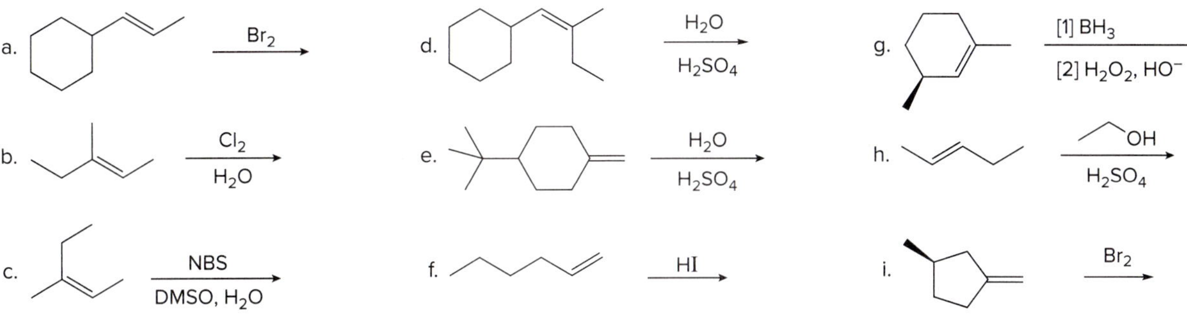

10.51 Which alkene reacts faster with HBr? Explain your choice.

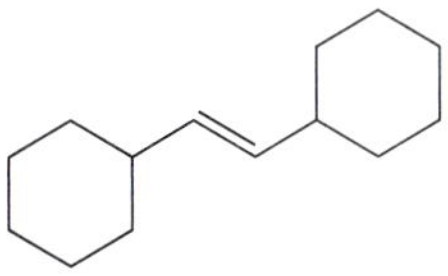

A 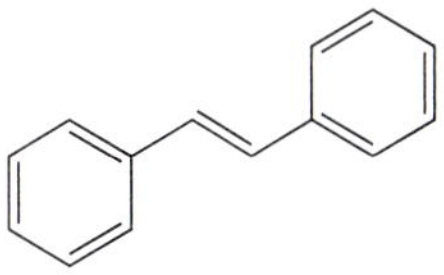 B

10.52 (a) What alkene yields **A** and **B** when it is treated with Br_2 in CCl_4? (b) What alkene yields **C** and **D** under the same conditions?

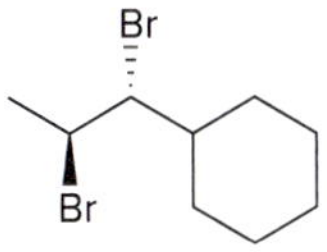

Br Br A · Br Br B · Br Br C · Br Br D

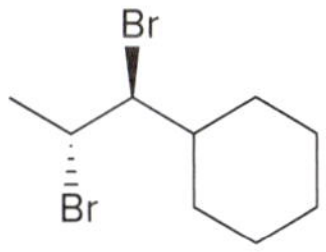

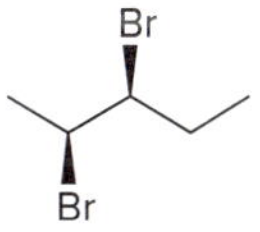

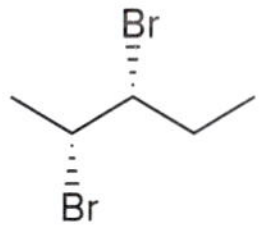

Mechanisms

10.53 Draw a stepwise mechanism for the following reaction, which results in ring expansion of a six-membered ring to a seven-membered ring.

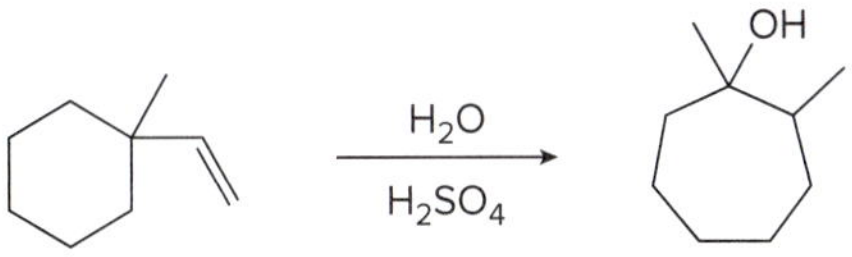

H_2O, H_2SO_4 → OH

10.54 Draw a stepwise mechanism for the conversion of hex-5-en-1-ol to the cyclic ether **A**.

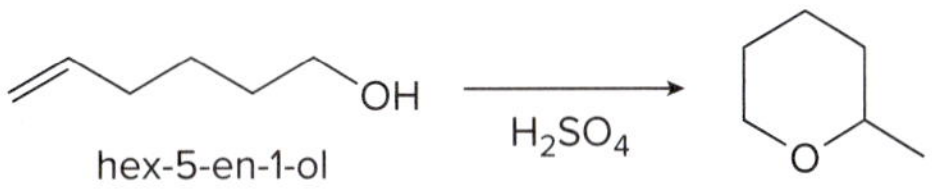

hex-5-en-1-ol (OH) → H_2SO_4 → O, **A**

10.55 Draw a stepwise mechanism that shows how all three alcohols are formed from the bicyclic alkene.

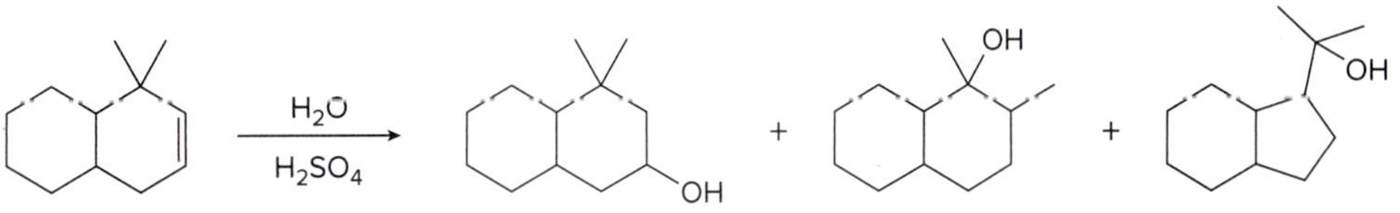

H_2O, H_2SO_4 → OH + OH + OH

10.56 Less stable alkenes can be isomerized to more stable alkenes by treatment with strong acid. For example, 2,3-dimethylbut-1-ene is converted to 2,3-dimethylbut-2-ene when treated with H_2SO_4. Draw a stepwise mechanism for this isomerization process.

10.57 When buta-1,3-diene ($CH_2{=}CH{-}CH{=}CH_2$) is treated with HBr, two constitutional isomers are formed, $CH_3CHBrCH{=}CH_2$ and $BrCH_2CH{=}CHCH_3$. Draw a stepwise mechanism that accounts for the formation of both products.

10.58 Explain why the addition of HBr to alkenes **A** and **C** is regioselective, forming addition products **B** and **D,** respectively.

OCH_3 **A** —HBr→ Br, OCH_3 **B**   O, OCH_3 **C** —HBr→ Br, O, OCH_3 **D**

10.59 Bromoetherification, the addition of the elements of Br and OR to a double bond, is a common method for constructing rings containing oxygen atoms. This reaction has been used in the synthesis of the polyether antibiotic monensin (Problem 18.35). Draw a stepwise mechanism for the following intramolecular bromoetherification reaction.

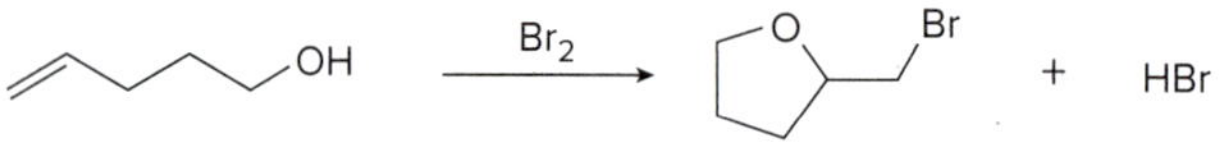

OH —Br_2→ O, Br + HBr

Synthesis

10.60 Devise a synthesis of each product from the given starting material. More than one step is required.

a. 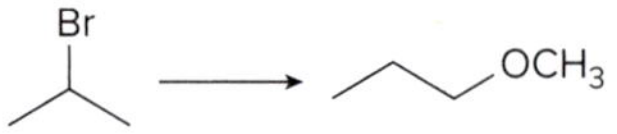 Br → OCH_3

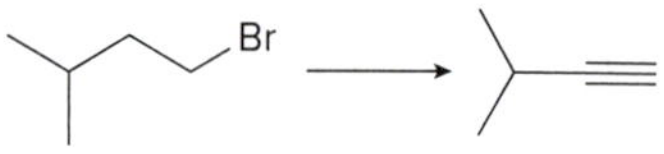

b. 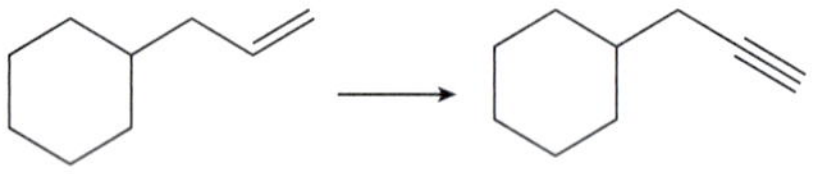 →

c. Br →

d.  Cl → O

10.61 Devise a synthesis of each compound from cyclohexene as the starting material. More than one step is needed.

a. 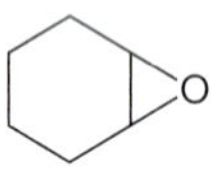 O

b. 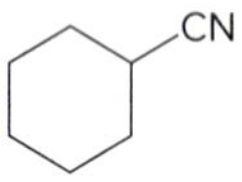 CN

c. 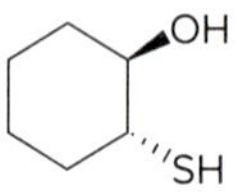 OH, SH

+ enantiomer

Spectroscopy

Problem 10.62 is intended for students who have already learned about spectroscopy in Chapters A–C.

10.62 When 2-bromo-3,3-dimethylbutane is treated with $K^{+}\ {}^{-}OC(CH_3)_3$, a single product **T** having molecular formula C_6H_{12} is formed. When 3,3-dimethylbutan-2-ol is treated with H_2SO_4, the major product **U** has the same molecular formula. Given the following ^1H NMR data, what are the structures of **T** and **U?** Explain in detail the splitting patterns observed for the three split signals in **T.**

^1H NMR of **T:** 1.01 (singlet, 9 H), 4.82 (doublet of doublets, 1 H, J = 10, 1.7 Hz), 4.93 (doublet of doublets, 1 H, J = 18, 1.7 Hz), and 5.83 (doublet of doublets, 1 H, J = 18, 10 Hz) ppm

^1H NMR of **U:** 1.60 (singlet) ppm

Additional problems on the spectroscopy of alkenes are given in Chapters A–C:

- Mass spectrometry: A.18b, A.22, A.25
- Infrared spectroscopy: B.5, B.8(**A**), B.9, B.15c, B.17a
- Nuclear magnetic resonance spectroscopy: C.12a; C.15d, e; C.18c, d; C.21; C.22; C.31d; C.34d; C.40d, f; C.45i, j; C.46; C.47; C.49d, f, h; C.53b; C.54c

Challenge Problems

10.63 Explain why **A** is a stable compound but **B** is not.

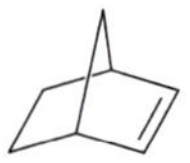

A 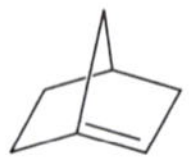   **B**

10.64 Alkene **A** can be isomerized to isocomene, a natural product isolated from goldenrod, by treatment with TsOH. Draw a stepwise mechanism for this conversion. (Hint: Look for a carbocation rearrangement.)

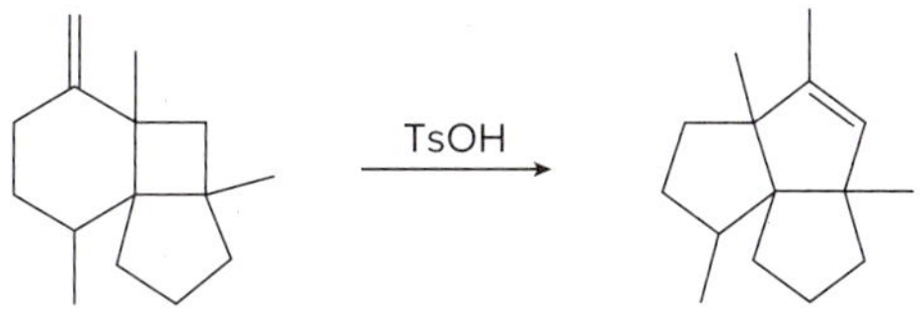

TsOH

A

isocomene

10.65 Lactones, cyclic esters such as compound **A,** are prepared by **halolactonization,** an addition reaction to an alkene. For example, iodolactonization of **B** forms lactone **C,** a key intermediate in the synthesis of prostaglandin $PGF_{2\alpha}$ (Section 19.5). Draw a stepwise mechanism for this addition reaction.

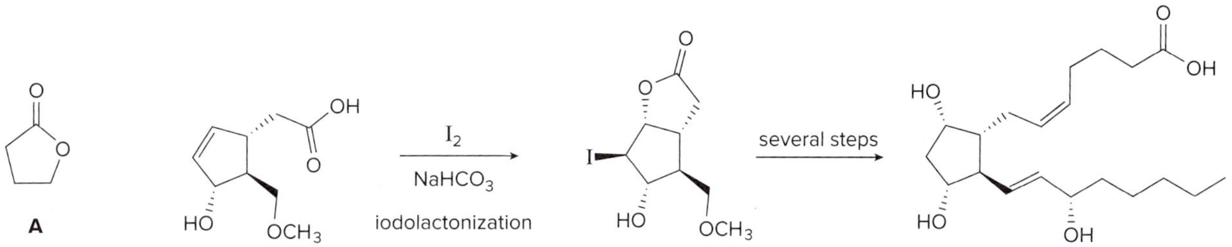

I_2, $NaHCO_3$ — iodolactonization

several steps

A B C $PGF_{2\alpha}$

10.66 Draw a stepwise mechanism for the following reaction.

H_2SO_4 + H_2O

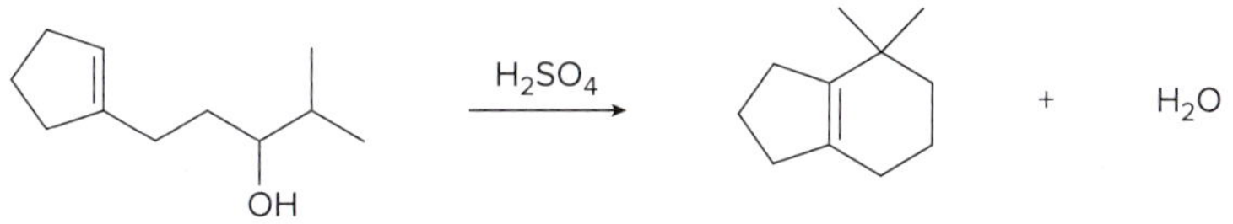

10.67 Like other electrophiles, carbocations add to alkenes to form new carbocations, which can then undergo substitution or elimination reactions depending on the reaction conditions. With this in mind, draw a stepwise mechanism for the following reaction, which involves the addition of an electrophile—a carbocation—to a double bond.

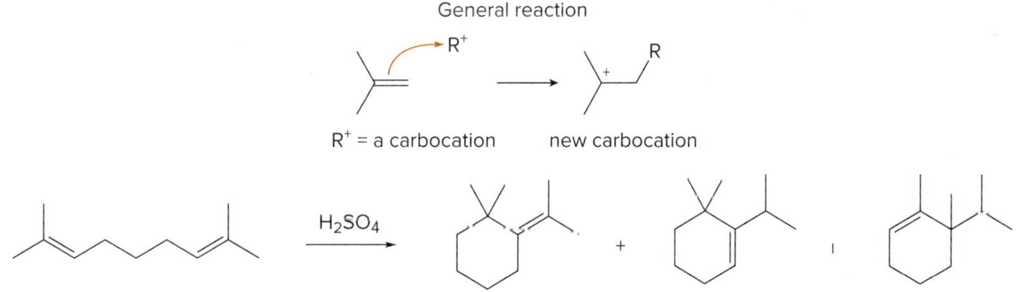

10.68 Draw a stepwise mechanism for the following reaction. This reaction combines two processes: the opening of an epoxide ring with a nucleophile and the addition of an electrophile to a carbon–carbon double bond. (Hint: Begin the mechanism by protonating the epoxide ring.)

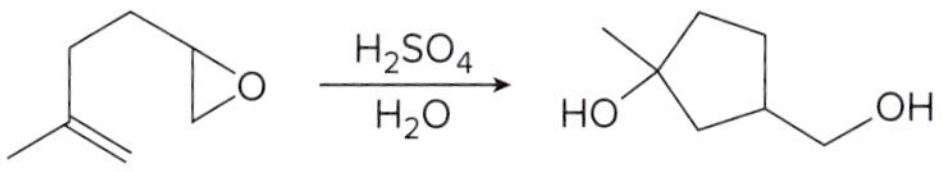

H_2SO_4, H_2O

SELF-TEST ANSWERS

1. a 2. c 3. a 4. c 5. d 6. c 7. b 8. c 9. d 10. a

11 Alkynes and Synthesis

11.1 Introduction
11.2 Nomenclature
11.3 Properties of alkynes
11.4 Interesting alkynes
11.5 Preparation of alkynes
11.6 Introduction to alkyne reactions
11.7 Addition of hydrogen halides
11.8 Addition of halogen
11.9 Addition of water
11.10 Hydroboration–oxidation
11.11 Reaction of acetylide anions
11.12 Synthesis

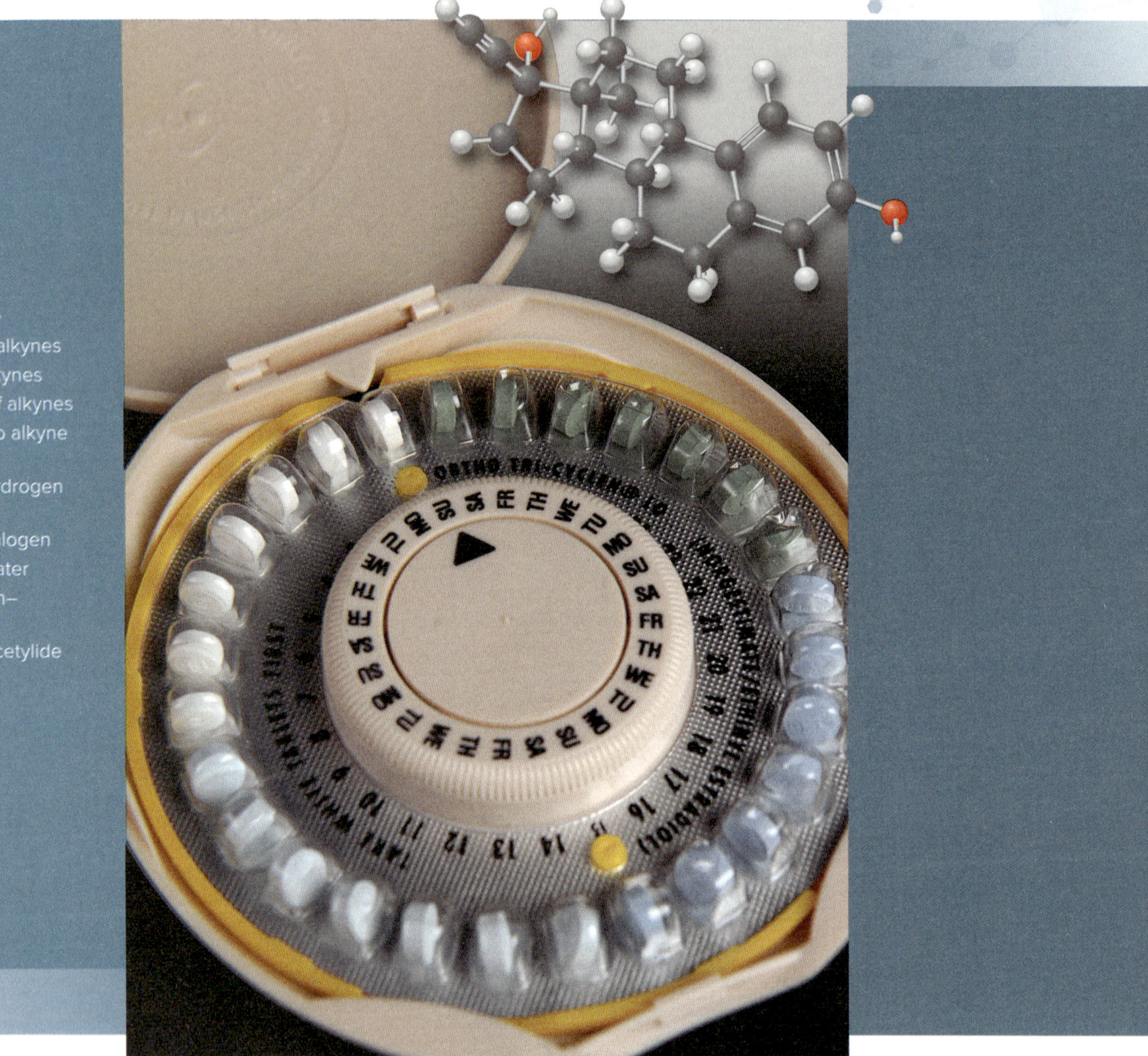

McGraw Hill

Ethynylestradiol is a synthetic compound whose structure closely resembles the carbon skeleton of female estrogen hormones. Because it is more potent than its naturally occurring analogues, it is a component of several widely used oral contraceptives. Ethynylestradiol and related compounds with similar biological activity contain a carbon–carbon triple bond. In Chapter 11, we learn about alkynes, hydrocarbons that contain triple bonds.

Why Study . . . Alkynes?

In Chapter 11, we continue our focus on organic molecules with electron-rich functional groups by examining ***alkynes*****, compounds that contain a carbon–carbon triple bond.** Like alkenes, **alkynes are nucleophiles with easily broken π bonds,** and as such, they undergo **addition** reactions with electrophilic reagents.

Alkynes also undergo a reaction that has no analogy in alkene chemistry. Because a C–H bond of an alkyne is more acidic than a C–H bond of an alkene or an alkane, alkynes are readily deprotonated with strong base. The resulting nucleophiles react with electrophiles to form new carbon–carbon σ bonds, so that complex molecules can be prepared from simple starting materials. The study of alkynes thus affords an opportunity to learn more about organic synthesis.

11.1 Introduction

Alkynes contain a carbon–carbon triple bond. A **terminal alkyne** has the triple bond at the end of the carbon chain, so that a hydrogen atom is bonded directly to a carbon atom of the triple bond. An **internal alkyne** has a carbon atom bonded to each carbon atom of the triple bond.

—C≡C—

alkyne | **terminal** alkyne | **internal** alkyne

An alkyne has the general molecular formula $\mathbf{C_nH_{2n-2}}$, giving it *four* fewer hydrogens than the maximum number possible. Because every degree of unsaturation removes two hydrogens, a **triple bond introduces two degrees of unsaturation.**

Each carbon of a triple bond is ***sp*** hybridized and **linear,** and all bond angles are **180°** (Section 1.10C). The triple bond of an alkyne consists of **one σ bond** and **two π bonds.**

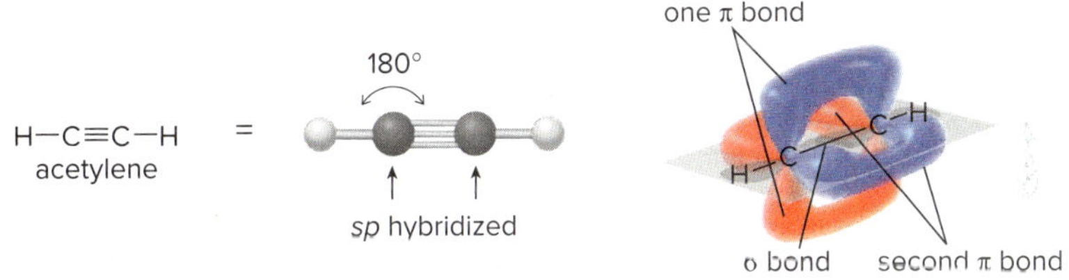

- **The σ bond is formed by end-on overlap of the two *sp* hybrid orbitals.**
- **Each π bond is formed by side-by-side overlap of two 2*p* orbitals.**

Bond dissociation energies of the C–C bonds in ethylene (one σ and one π bond) and acetylene (one σ and two π bonds) can be used to estimate the strength of the second π bond of the triple bond. If we assume that the σ bond and first π bond in acetylene are similar in strength to the σ and π bonds in ethylene (368 and 267 kJ/mol, respectively), then the second π bond is worth 202 kJ/mol.

HC≡CH		$CH_2=CH_2$		
837 kJ/mol	–	635 kJ/mol	=	202 kJ/mol
(σ + two π bonds)		(σ + π bond)		↑ **second π bond**

Skeletal structures for alkynes may look somewhat unusual, but they follow the customary convention: a carbon atom is located at the intersection of any two lines and at the end of any line; thus,

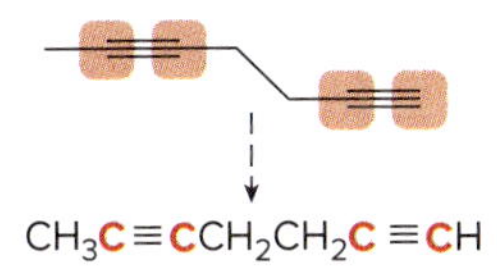

- **Both π bonds of a C–C triple bond are *weaker* than a C–C σ bond, making them much more easily broken. As a result, alkynes undergo many addition reactions.**
- **Alkynes are more polarizable than alkenes because the electrons in their π bonds are more loosely held.**

Problem 11.1 Corticatic acid A is an antifungal acetylenic fatty acid isolated from a marine sponge.

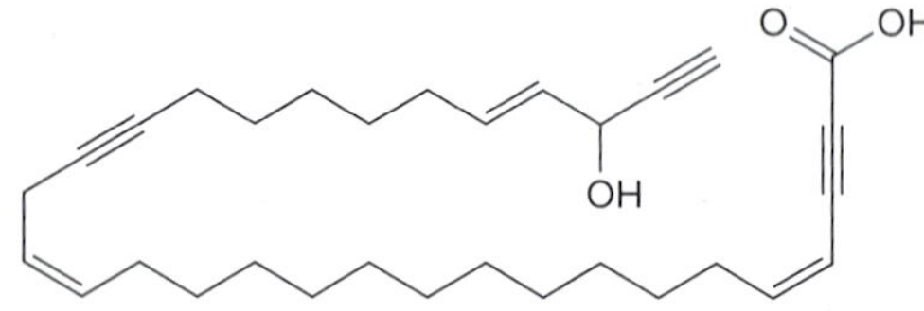

corticatic acid A

a. Label the most acidic H atom.

b. How many degrees of unsaturation does corticatic acid A contain?

c. Label each triple bond as internal or terminal.

d. How many bonds are formed from $C_{sp}–C_{sp^2}$?

11.2 Nomenclature

Alkynes are named in the same way that alkenes were named in Section 10.3.

- **In the IUPAC system, change the *-ane* ending of the parent alkane to the suffix *-yne*.**
- **Choose the longest carbon chain that contains both atoms of the triple bond and number the chain to give the triple bond the lower number.**
- **Compounds with two triple bonds are named as *diynes,* those with three are named as *triynes,* and so forth.**
- **Compounds with both a double and a triple bond are named as *enynes*. The chain is numbered to give the first site of unsaturation (either C=C or C≡C) the lower number.**

Sample Problem 11.1 Naming an Alkyne

Give the IUPAC name for the following alkyne.

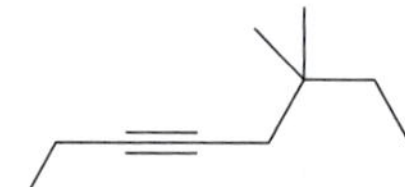

Solution

[1] Find the longest chain.

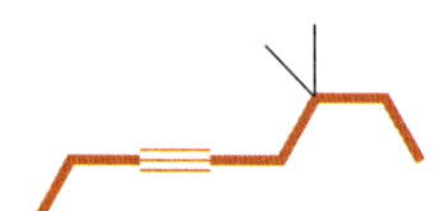

8 C's in the longest chain

oct*ane* ---→ oct*yne*

[2] Number the long chain; then name and number the substituents.

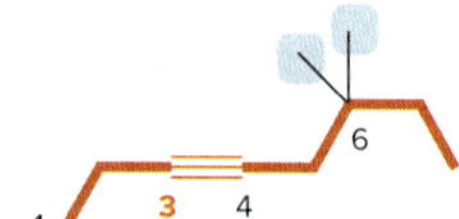

two methyl groups at C6

Answer: 6,6-dimethyloct-3-yne

Problem 11.2 Give the IUPAC name for each compound.

a. 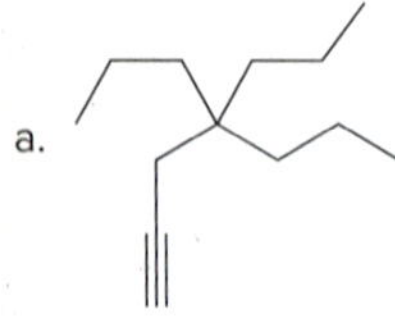

c.

e.

b.

Cl

d.

f.

More Practice: Try Problems 11.27, 11.29, 11.30.

The simplest alkyne, HC≡CH, named in the IUPAC system as **ethyne,** is more often called **acetylene,** its common name. The two-carbon alkyl group derived from acetylene is called an **ethynyl group** (HC≡C–). Examples of alkyne nomenclature are shown in Figure 11.1.

Problem 11.3 Give the structure corresponding to each of the following names.

a. *trans*-2-ethynylcyclopentanol b. 4-*tert*-butyldec-5-yne c. 3,3,5-trimethylcyclononyne

Figure 11.1 Examples of alkyne nomenclature

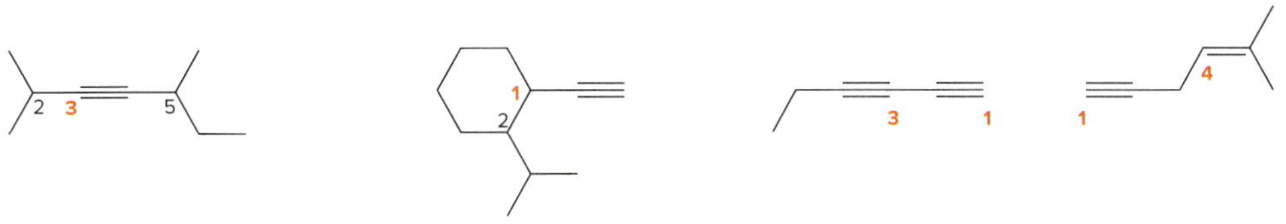

2,5-dimethylhept-3-yne 1-ethynyl-2-isopropylcyclohexane hexa-1,3-diyne 5-methylhex-4-en-1-yne

11.3 Properties of Alkynes

The physical properties of alkynes resemble those of hydrocarbons having a similar shape and molecular weight.

- **Alkynes have low melting points and boiling points.**
- **Melting points and boiling points increase as the number of carbons increases.**
- **Alkynes are soluble in organic solvents and insoluble in water.**

Problem 11.4 Explain why an alkyne often has a slightly higher boiling point than an alkene of similar molecular weight. For example, the bp of pent-1-yne is 39°C, and the bp of pent-1-ene is 30°C.

Students who have already been exposed to spectroscopy or who would like to learn about the spectroscopic properties of alkynes are referred to the following sections:

- Infrared spectroscopy: Sections B.3A, B.4A; Tables B.1, B.2; Sample Problem B.2b
- Nuclear magnetic resonance spectroscopy: Section C.4; Tables C.1, C.2, C.5

11.4 Interesting Alkynes

Acetylene, HC≡CH, is a colorless gas with an ethereal odor that burns in oxygen to form CO_2 and H_2O. Because the combustion of acetylene releases more energy per mole of product formed than other hydrocarbons, it burns with a very hot flame, making it an excellent fuel for welding torches.

Ethynylestradiol, the molecule that opened Chapter 11, and **norethindrone** are two components of oral contraceptives that contain a carbon–carbon triple bond (Figure 11.2). Both molecules are synthetic analogues of the naturally occurring female hormones estradiol and progesterone, but are more potent so they can be administered in lower doses. Most oral contraceptives contain two of these synthetic hormones. They act by artificially elevating hormone levels in a woman, thereby preventing pregnancy.

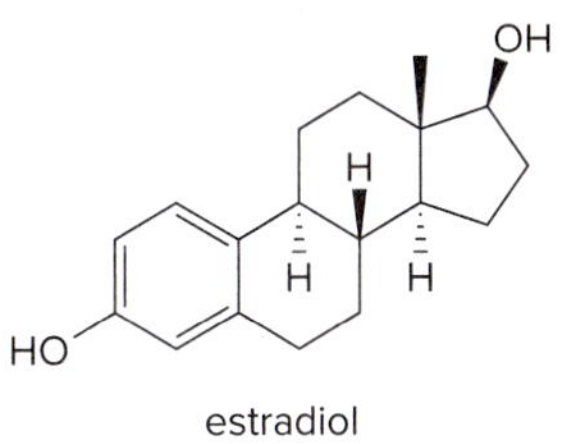

estradiol 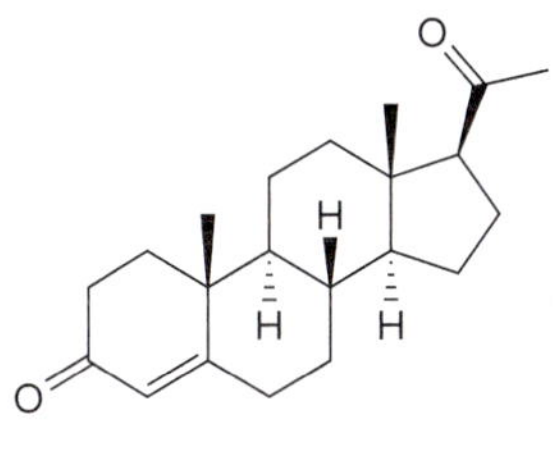 progesterone

Figure 11.2 How oral contraceptives work

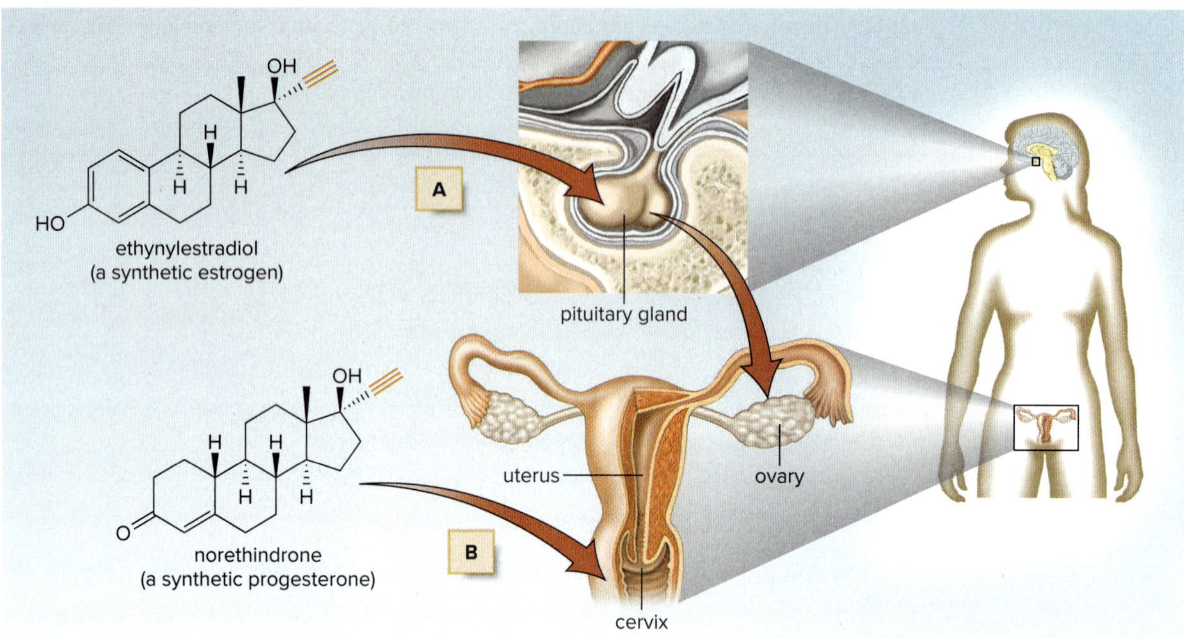

- Monthly cycles of hormones from the pituitary gland cause ovulation, the release of an egg from an ovary. To prevent pregnancy, the two synthetic hormones in many oral contraceptives have different effects on the female reproductive system.
  - **A:** The elevated level of **ethynylestradiol,** a synthetic estrogen, "fools" the pituitary gland into thinking a woman is pregnant, so ovulation does not occur.
  - **B:** The elevated level of **norethindrone,** a synthetic progesterone, stimulates the formation of a thick layer of mucus in the cervix, making it difficult for sperm to reach the uterus.

Two other synthetic hormones with alkynyl appendages are **desogestrel** and **levonorgestrel.** Desogestrel is a synthetic progesterone approved for use in the United States in 1992 as an oral contraceptive and as a treatment for menopausal symptoms. Levonorgestrel interferes with ovulation, so it prevents pregnancy if taken within a few days of unprotected sex.

desogestrel

levonorgestrel
(Trade name: Plan B)

A carbon–carbon triple bond, along with other functional groups, is also present in the drugs ponatinib and efavirenz. Ponatinib is used to treat chronic myeloid leukemia, and efavirenz, sold as a stand-alone medication or in combination with other drugs, is an oral antiretroviral agent used to treat HIV.

ponatinib

efavirenz

11.5 Preparation of Alkynes

Alkynes are prepared by elimination reactions, as discussed in Section 8.10. **A strong base removes two equivalents of HX from a vicinal or geminal dihalide to yield an alkyne** by two successive E2 eliminations.

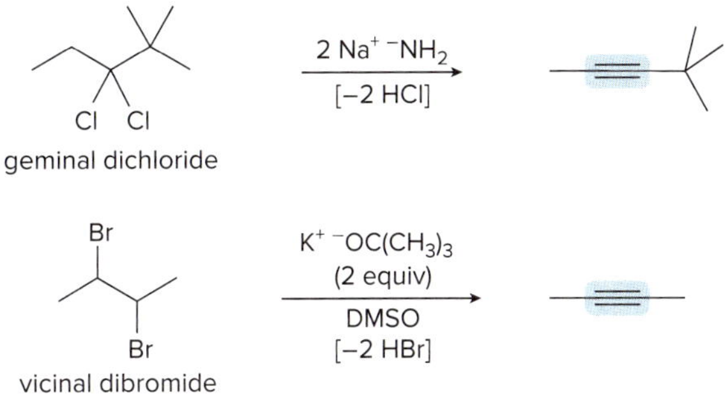

Cl Cl
geminal dichloride

$\xrightarrow[\text{[−2 HCl]}]{2\ Na^+\ {}^-NH_2}$

Br
Br
vicinal dibromide

$K^+\ {}^-OC(CH_3)_3$ (2 equiv), DMSO [−2 HBr]

Because vicinal dihalides are synthesized by adding halogens to alkenes, an alkene can be converted to an alkyne by the two-step process illustrated in Sample Problem 11.2.

Sample Problem 11.2 Converting an Alkene to an Alkyne

Convert alkene **A** into alkyne **B** by a stepwise method.

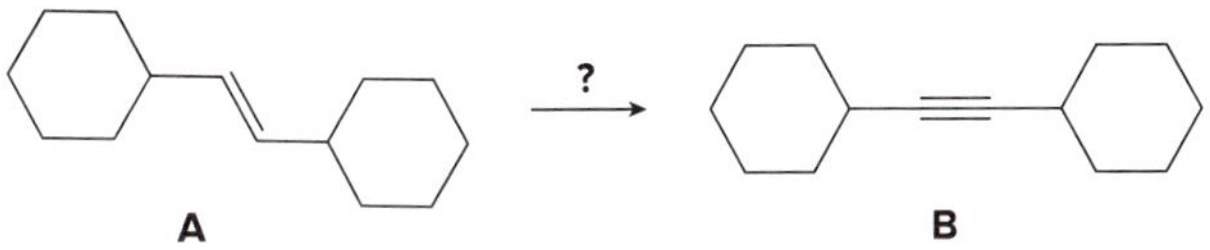

A → ? → **B**

Solution

A two-step method is needed:

- **Addition of X_2** forms a vicinal dihalide.
- **Elimination** of two equivalents of HX forms two π bonds.

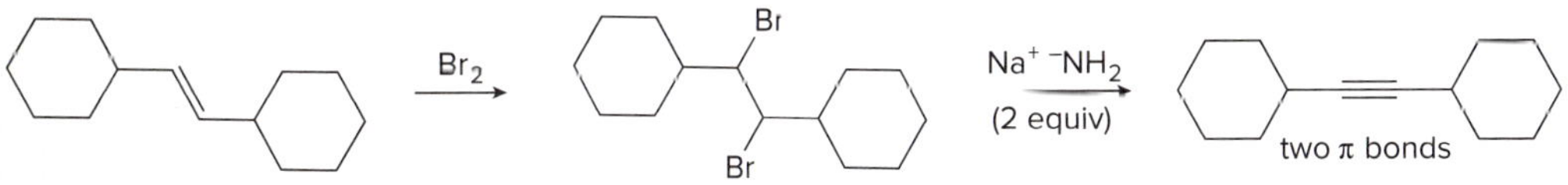

A $\xrightarrow{Br_2}$ vicinal dibromide (Br, Br) $\xrightarrow[\text{(2 equiv)}]{Na^+\ {}^-NH_2}$ **B** (two π bonds)

- **This two-step process introduces one degree of unsaturation:** an alkene with one π bond is converted to an alkyne with two π bonds.

Problem 11.5 Convert each compound to hex-1-yne, $HC{\equiv}CCH_2CH_2CH_2CH_3$.

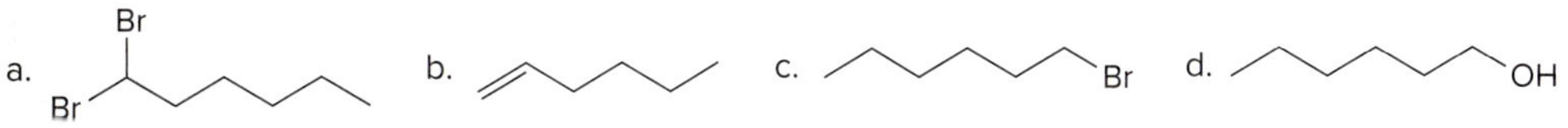

a. Br, Br
b.
c. Br
d. OH

More Practice: Try Problems 11.39c, 11.50a.

11.6 Introduction to Alkyne Reactions

All reactions of alkynes occur because they contain **easily broken π bonds** or, in the case of terminal alkynes, an **acidic, *sp* hybridized C–H bond.**

11.6A Addition Reactions

Like alkenes, **alkynes undergo addition reactions because they contain weak π bonds.** Two sequential reactions take place: addition of one equivalent of reagent forms an alkene, which then adds a second equivalent of reagent to yield a product having **four new bonds.**

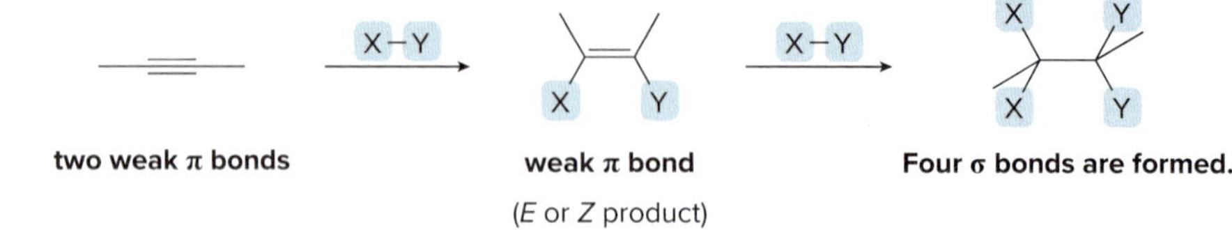

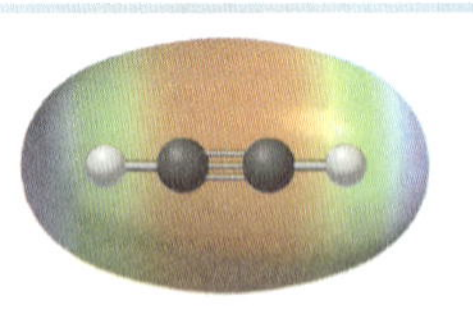

The red electron-rich region between the two carbon atoms in the electrostatic potential plot of acetylene forms a cylinder of electron density.

Alkynes are electron rich, as shown in the electrostatic potential map of acetylene. The two π bonds form a cylinder of electron density between the two *sp* hybridized carbon atoms, and this exposed electron density makes a triple bond nucleophilic. As a result, **alkynes react with electrophiles.** Four addition reactions are discussed in Chapter 11 and illustrated in Figure 11.3 with but-1-yne as the starting material. The oxidation and reduction of alkynes, reactions that also involve addition, are discussed in Chapter 12.

Figure 11.3 Four addition reactions of but-1-yne

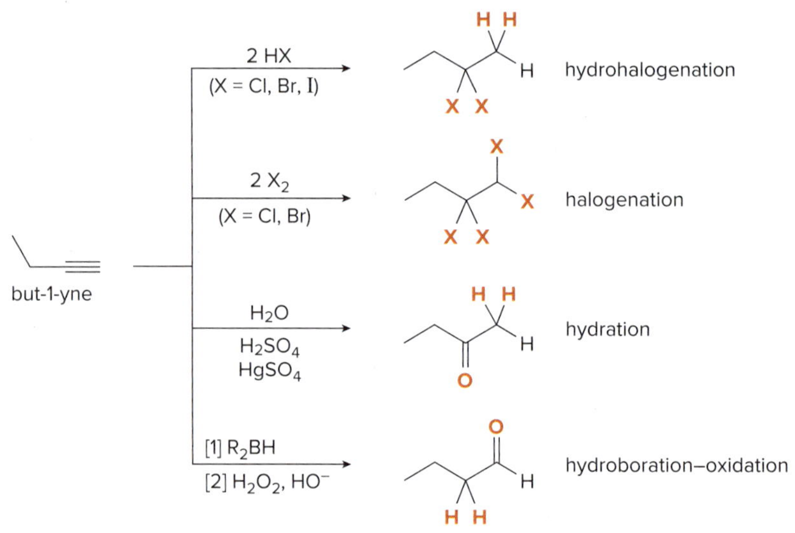

- In each addition, both π bonds of the triple bond are broken, and four new bonds are formed.

11.6B Terminal Alkynes—Reaction as an Acid

Because ***sp* hybridized C–H bonds are more acidic than *sp*^{2} and *sp*^{3} hybridized C–H bonds,** terminal alkynes are readily deprotonated with strong base in a Brønsted–Lowry acid–base reaction. The resulting anion is called an **acetylide anion.**

Recall from Section 2.5D that the acidity of a C–H bond increases as the percent *s*-character of C increases. Thus, the following order of relative acidity results: C_{sp^3}–H < C_{sp^2}–H < C_{sp}–H.

$$\text{R–C}\equiv\text{C–H} \;+\; :\text{B} \;\rightleftharpoons\; \text{R–C}\equiv\text{C}:^- \;+\; \text{H–B}^+$$

terminal alkyne, pK_a ≈ 25 — **acetylide anion**

What bases can be used for this reaction? Because an acid–base equilibrium favors the weaker acid and base, only **bases having conjugate acids with pK_a values *higher* than the terminal alkyne—that is, pK_a values > 25—are strong enough** to form a significant concentration of acetylide anion. As shown in Table 11.1, $^{-}NH_2$ and H^- are strong enough to deprotonate a terminal alkyne, but ^{-}OH and ^{-}OR are not.

Table 11.1 A Comparison of Bases for Alkyne Deprotonation

	Base	pK_a of the conjugate acid
These bases are **strong** enough to deprotonate an alkyne.	$^{-}NH_2$	38
	H^-	35
These bases are ***not*** strong enough to deprotonate an alkyne.	^{-}OH	14
	^{-}OR	15.5–18

Why is this reaction useful? The acetylide anions formed by deprotonating terminal alkynes are **strong nucleophiles** that can react with a variety of electrophiles, as shown in Section 11.11.

$$R-C\equiv C:^- \;+\; E^+ \longrightarrow R-C\equiv C-E$$

nucleophile electrophile new bond

Poison dart frogs, *Dendrobates histrionicus,* inhabit the moist humid floor of tropical rainforests in western Ecuador and Colombia. Histrionicotoxin (Problem 11.7), the defensive toxin secreted by the frogs, acts by interfering with nerve transmission in mammals, resulting in prolonged muscle contraction. *Dirk Ercken/ Shutterstock.com*

Problem 11.6 Which bases can deprotonate acetylene? The pK_a values of the conjugate acids are given in parentheses.

a. CH_3NH^- (pK_a = 40) b. CO_3^{2-} (pK_a = 10.2) c. $CH_2{=}CH^-$ (pK_a = 44) d. $(CH_3)_3CO^-$ (pK_a = 18)

Problem 11.7 Histrionicotoxin is a defensive toxin that protects the poison dart frog *Dendrobates histrionicus* from predators. Use the pK_a of a terminal alkyne and the information in Table 11.1 to rank the labeled protons of histrionicotoxin in order of increasing acidity.

OH_b H_c N H_a

histrionicotoxin

11.7 Addition of Hydrogen Halides

Alkynes undergo **hydrohalogenation, the addition of hydrogen halides, HX** (X = Cl, Br, I). Two equivalents of HX are usually used: addition of one mole forms a **vinyl halide,** which then reacts with a second mole of HX to form a **geminal dihalide.**

R—C≡C—R (two weak π bonds) —[H–X (X = Cl, Br, I)]→ RHC=CXR ((*E* or *Z* product) vinyl halide) —[H–X]→ RCH_2CX_2R (geminal dihalide)

Addition of HX to an alkyne is another example of **electrophilic addition,** because the electrophilic (H) end of the reagent is attracted to the electron-rich triple bond.

- With two equivalents of HX, both H atoms bond to the *same* carbon.
- With a terminal alkyne, both H atoms bond to the *terminal* carbon; that is, the hydrohalogenation of alkynes follows Markovnikov's rule.

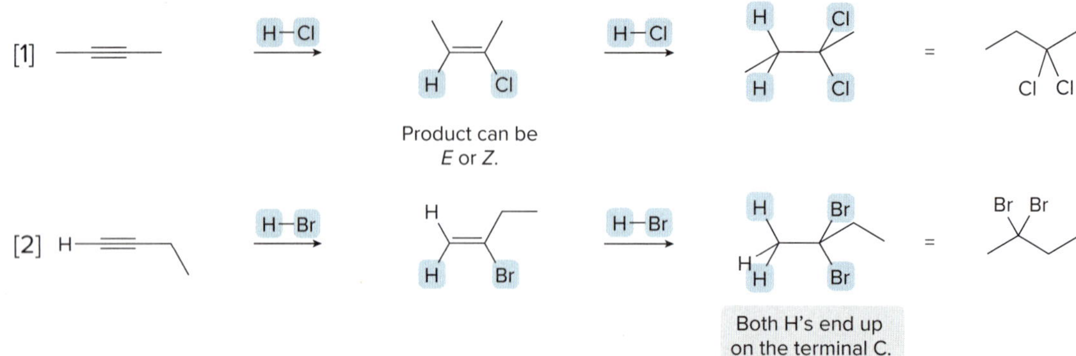

- With only one equivalent of HX, the reaction stops with formation of the vinyl halide.

H–≡– $\xrightarrow[\text{(1 equiv)}]{\text{H–Cl}}$ (H)(H)C=C(Cl)–

a vinyl chloride
(2-chloropropene)

One proposed mechanism for the addition of two equivalents of HX to an alkyne involves **two steps for each addition of HX:** addition of H^+ (from HX) to form a carbocation, followed by nucleophilic attack of X^-. Mechanism 11.1 illustrates the addition of HBr to but-1-yne to yield 2,2-dibromobutane. Each two-step mechanism is similar to the two-step addition of HBr to *cis*-but-2-ene discussed in Section 10.9.

Mechanism 11.1 Electrophilic Addition of HX to an Alkyne

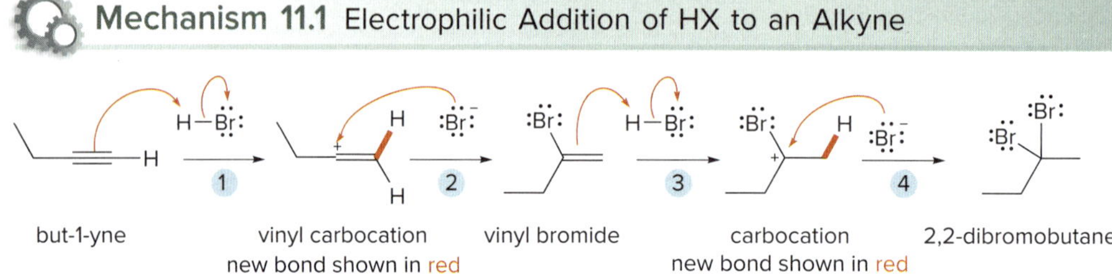

1. **Addition of H^+ forms a vinyl carbocation and follows Markovnikov's rule.** The H atom bonds to the terminal C to form the more substituted carbocation.
2. Nucleophilic attack of Br^- forms a vinyl bromide. One equivalent of HBr adds in two steps.
3. The addition of a second equivalent of HBr follows in the same two-step manner. Addition of H^+ to the vinyl bromide forms a **carbocation.**
4. Nucleophilic attack of Br^- forms a geminal dibromide, 2,2-dibromobutane.

Because of the instability of a vinyl carbocation, other mechanisms for HX addition that avoid formation of a discrete carbocation have been proposed. It is likely that more than one mechanism occurs, depending in part on the identity of the alkyne substrate.

The formation of both carbocations (in Steps [1] and [3]) deserves additional scrutiny. **The vinyl carbocation formed in Step [1] is *sp* hybridized and therefore less stable than a 2° sp^2 hybridized carbocation** (Section 7.17). This makes electrophilic addition of HX to an alkyne ***slower*** than electrophilic addition of HX to an alkene, even though alkynes are more polarizable and have more loosely held π electrons than alkenes.

but-1-yne + H–Br: $\xrightarrow{[1]}$ vinyl carbocation (H, H) + :Br:⁻

***sp* hybridized**
vinyl carbocation

In Step [3], two carbocations are possible but only one is formed. Markovnikov addition in Step [3] places the H on the terminal carbon (C1) to form the more substituted carbocation **A,** rather than the less substituted carbocation **B.** Because the more stable carbocation is formed faster—another example of the Hammond postulate—carbocation **A** must be more stable than carbocation **B.**

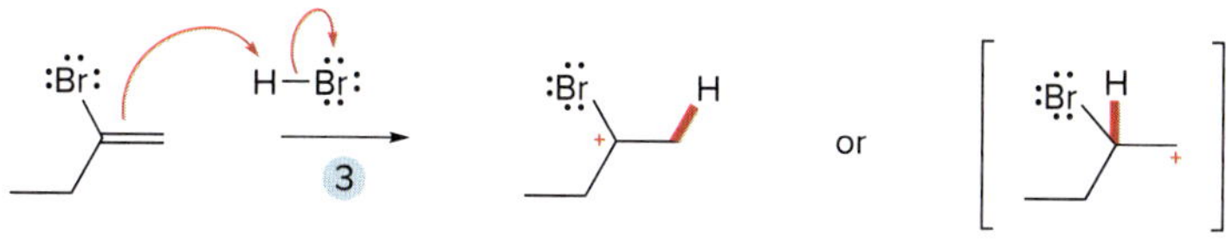

A
more stable carbocation
new bond shown in red

B
not formed
new bond shown in red

Why is carbocation **A,** having a positive charge on a carbon that also has a Br atom, more stable? Shouldn't the electronegative Br atom withdraw electron density from the positive charge, and thus destabilize it? It turns out that **A is stabilized by resonance** but **B** is not. Two resonance structures can be drawn for carbocation **A,** but only one Lewis structure can be drawn for carbocation **B.**

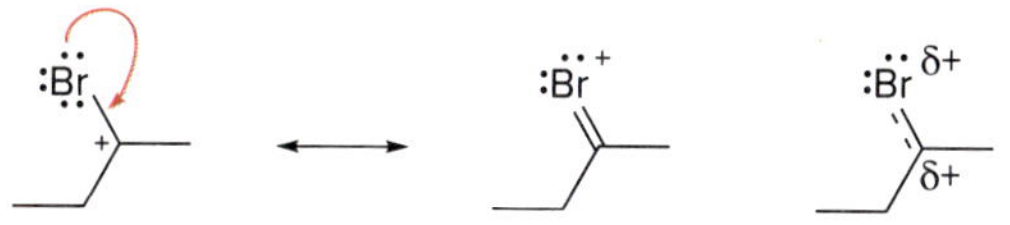

two resonance structures for **A**

hybrid
The positive charge is delocalized.

- **Resonance stabilizes a molecule by delocalizing charge and electron density.**
- **Thus, halogens stabilize an adjacent positive charge by resonance.**

Markovnikov's rule applies to the addition of HX to vinyl halides because **addition of H^+ forms a resonance-stabilized carbocation.** As a result, addition of each equivalent of HX to a triple bond forms the more stable carbocation, so that both H atoms bond to the less substituted C.

Problem 11.8 Draw the organic products formed when each alkyne is treated with two equivalents of HBr.

a. 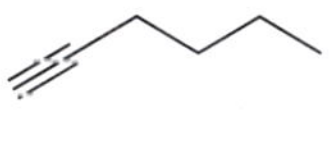 b. 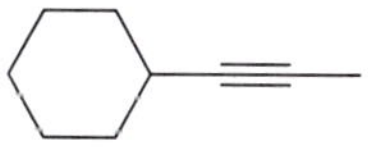 c.

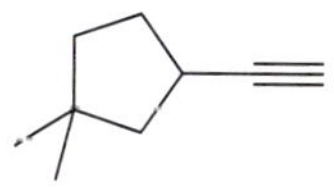

Problem 11.9 Draw additional resonance structures for each cation.

a. 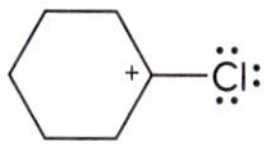 b. 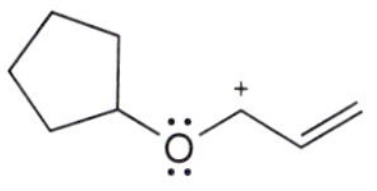 c.

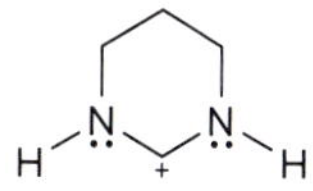

11.8 Addition of Halogen

Halogens, X_2 (X = Cl or Br), add to alkynes in much the same way they add to alkenes (Section 10.13). Addition of one mole of X_2 forms a **trans dihalide,** which can then react with a second mole of X_2 to yield a **tetrahalide.**

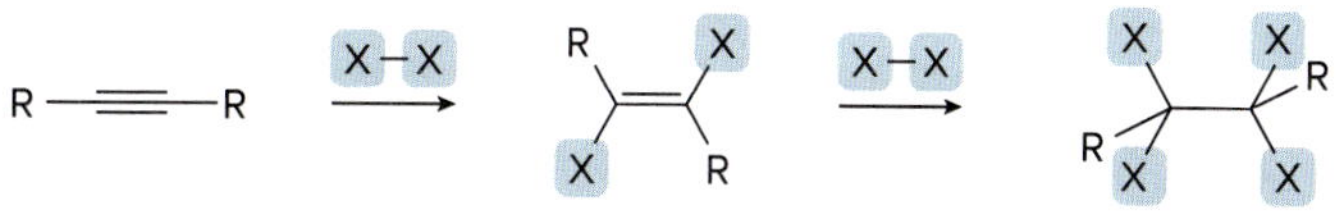

trans dihalide

tetrahalide

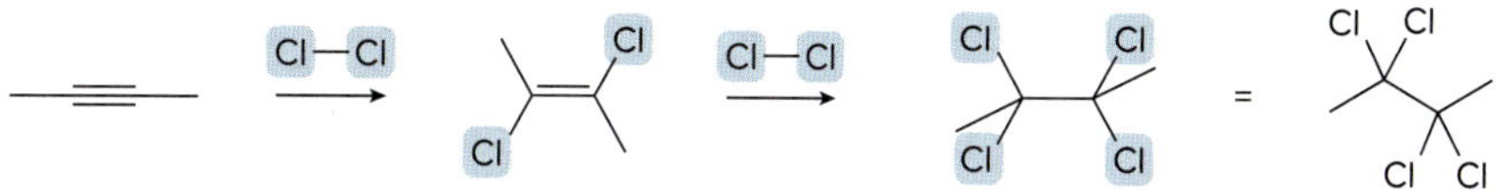

Problem 11.10 Draw the products formed when $CH_3CH_2C{\equiv}CCH_2CH_3$ is treated with each reagent: (a) Br_2 (2 equiv); (b) Cl_2 (1 equiv).

Problem 11.11 Explain the following result. Although alkenes are generally more reactive than alkynes toward electrophiles, the reaction of Cl_2 with but-2-yne can be stopped after one equivalent of Cl_2 has been added.

11.9 Addition of Water

Although the addition of H_2O to an alkyne resembles the acid-catalyzed addition of H_2O to an alkene in some ways, an important difference exists. In the presence of strong acid or Hg^{2+} catalyst, the **elements of H_2O add to the triple bond,** but the initial addition product, an **enol,** is unstable and rearranges to a product containing a **carbonyl group**—that is, a **C=O.** A carbonyl compound having two alkyl groups bonded to the C=O carbon is called a **ketone.**

R—≡—R → (H–OH; H_2SO_4, $HgSO_4$) [less stable **enol**] ⇌ **ketone**

H_2O is added to form a **carbonyl** group.

Internal alkynes undergo hydration with concentrated acid, whereas terminal alkynes require the presence of an additional Hg^{2+} catalyst—usually $HgSO_4$—to yield methyl ketones by **Markovnikov addition of H_2O.**

Because an enol contains both a C=C and a hydroxy group, the name **enol** comes from alk**ene** + alcoh**ol.**

$HgSO_4$ is often used in the hydration of internal alkynes as well, because hydration can be carried out under milder reaction conditions.

(internal alkyne) → (H_2O, H_2SO_4) [**enol**] ⇌ ketone

H—≡— → (H_2O, H_2SO_4, $HgSO_4$) [**enol**] ⇌ methyl ketone

Markovnikov addition of H_2O
H adds to the terminal C.

Let's first examine the conversion of a general enol **A** to the carbonyl compound **B. A** and **B** are called **tautomers: A is the *enol form* and B is the *keto form* of the tautomer.**

- ***Tautomers*** **are constitutional isomers that differ in the location of a double bond and a hydrogen atom. Two tautomers are in equilibrium with each other.**

enol form **A** ⇌ **keto** form **B**

Tautomers differ in the position of a double bond and a hydrogen atom. In Chapter 21 an in-depth discussion of keto–enol tautomers is presented.

- **An enol tautomer has an O–H group bonded to a C=C.**
- **A keto tautomer has a C=O and an additional C–H bond.**

Equilibrium favors the keto form largely because a C=O is much stronger than a C=C. Tautomerization, the process of converting one tautomer into another, is catalyzed by both acid and base. Under the strongly acidic conditions of hydration, tautomerization of the enol to the keto form occurs rapidly by a two-step process: **protonation,** followed by **deprotonation** as shown in Mechanism 11.2.

Mechanism 11.2 Tautomerization in Acid

1 Protonation of the double bond forms a **resonance-stabilized carbocation.**

2 Loss of a proton, which can be drawn with either resonance structure, forms the **carbonyl group.** Because acid is re-formed in this step, tautomerization is acid-catalyzed.

Problem 11.12 Label each pair of compounds as keto–enol tautomers or constitutional isomers, but not tautomers.

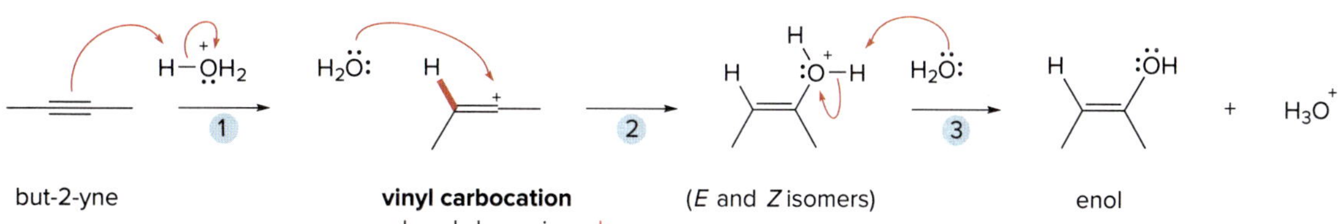

Problem 11.13 (a) Draw two different enol tautomers of 2-methylcyclohexanone. (b) Draw two constitutional isomers that are not tautomers, but contain a C=C and an OH group.

2-methylcyclohexanone

Hydration of an internal alkyne with strong acid forms an enol by a mechanism similar to that of the acid-catalyzed hydration of an alkene (Section 10.12). Mechanism 11.3 illustrates the hydration of but-2-yne with H_2O and H_2SO_4. Once formed, the enol then tautomerizes to the more stable keto form by protonation followed by deprotonation.

Mechanism 11.3 Hydration of an Alkyne

Part [1] Addition of H_2O to form an enol

1 Addition of H^+ forms a **vinyl carbocation.**

2 – 3 **Nucleophilic attack** followed by loss of a proton forms the enol.

Part [2] Tautomerization

4 Tautomerization of the enol to the keto form begins with protonation of the double bond to form a **carbocation.**

5 Loss of a proton, which can be drawn with either resonance structure, forms the **ketone.**

Sample Problem 11.3 Drawing an Enol and a Ketone Formed by Hydration of an Alkyne

Draw the enol intermediate and the ketone product formed in the following reaction.

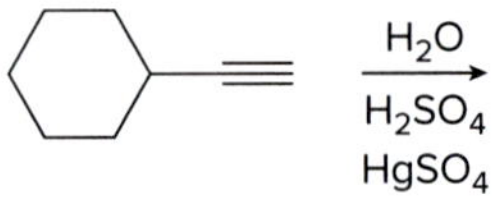

H_2O, H_2SO_4, $HgSO_4$

Solution

First, form the enol by adding H_2O to the triple bond with the **H bonded to the less substituted terminal carbon,** according to Markovnikov's rule.

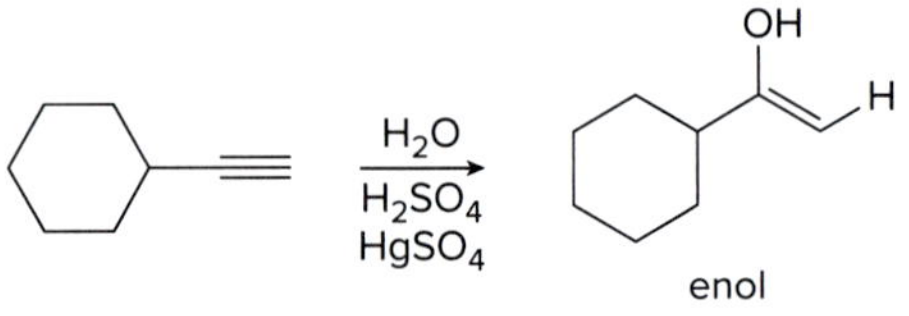

H_2O, H_2SO_4, $HgSO_4$

OH H enol

To convert the enol to the keto tautomer, add a proton to the C=C and remove a proton from the OH group. In tautomerization, the C–OH bond is converted to a C=O, and a new C–H bond is formed on the other enol carbon.

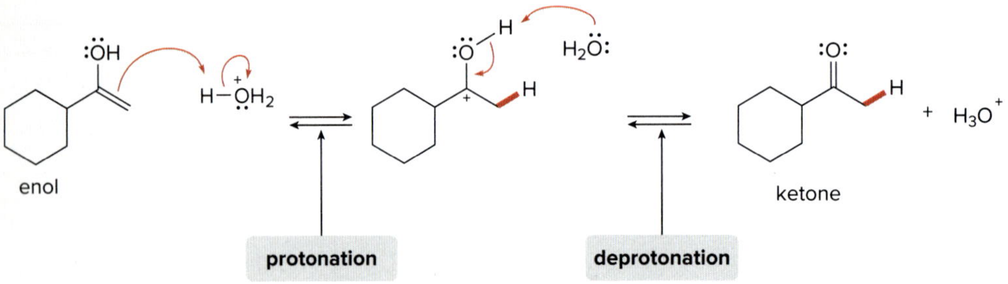

enol $H-\overset{+}{O}H_2$ protonation $H_2\ddot{O}$: deprotonation ketone + H_3O^+

- **The overall result is the addition of H_2O to a triple bond to form a ketone.**

Problem 11.14 Draw the keto tautomer of each enol.

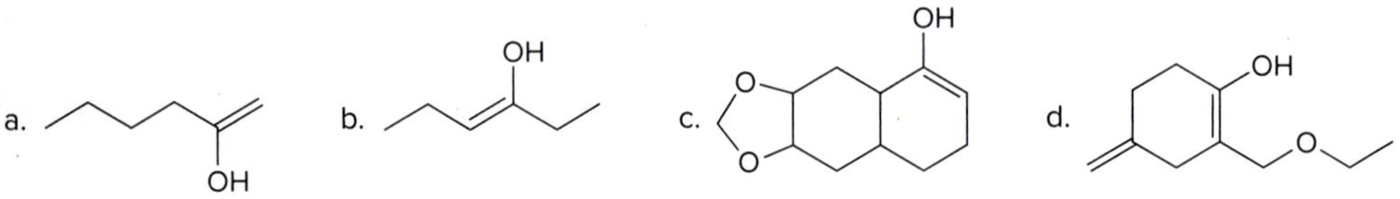

a. b. c. d.

More Practice: Try Problems 11.31–11.33, 11.39f.

Problem 11.15 Draw the enols formed when each alkyne is treated with H_2O, H_2SO_4, and $HgSO_4$. Then draw the ketones formed from these enols after tautomerization.

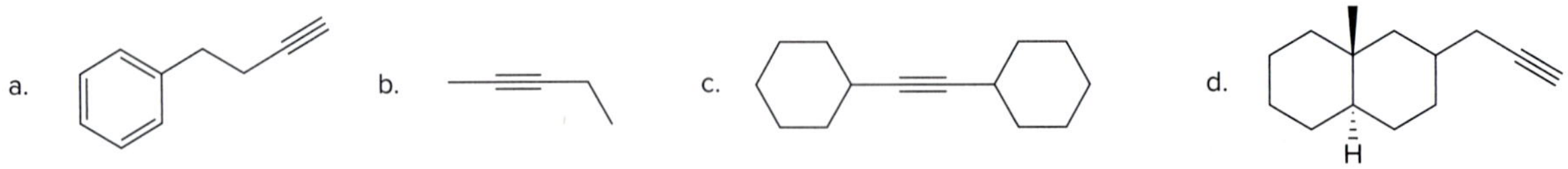

a. b. c. d.

11.10 Hydroboration–Oxidation

Hydroboration–oxidation is a two-step reaction sequence that converts an alkyne to a carbonyl compound.

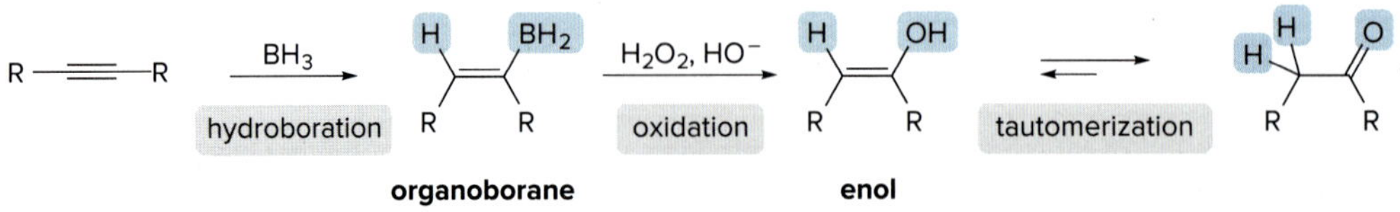

R—≡—R $\xrightarrow{BH_3}$ (hydroboration) organoborane $\xrightarrow{H_2O_2,\ HO^-}$ (oxidation) enol ⇌ (tautomerization)

organoborane **enol**

- **Addition of borane forms an organoborane.**
- **Oxidation with basic H_2O_2 forms an enol.**
- **Tautomerization of the enol forms a carbonyl compound.**
- **The overall result is addition of H_2O to a triple bond.**

Hydroboration–oxidation of an *internal* alkyne forms a **ketone. Hydroboration of a *terminal* alkyne adds boron to the less substituted, terminal carbon.** After oxidation to the enol, tautomerization yields an **aldehyde,** a carbonyl compound having a hydrogen atom bonded to the carbonyl carbon. Hydroboration of a terminal alkyne is generally carried out with a dialkylborane (R_2BH), which has been prepared from BH_3 (Section 10.16).

internal alkyne → [1] BH_3; [2] H_2O_2, HO^- → **enol** ⇌ **ketone**

terminal alkyne → [1] R_2BH; [2] H_2O_2, HO^- → **enol** ⇌ **aldehyde**

OH bonds to the *less* substituted C.

Hydration (H_2O, H_2SO_4, and $HgSO_4$) and **hydroboration–oxidation** (BH_3 or R_2BH followed by H_2O_2, HO^-) both **add the elements of H_2O across a triple bond.** Sample Problem 11.4 shows that different constitutional isomers are formed from terminal alkynes in these two reactions.

- **Addition of H_2O using H_2O, H_2SO_4, and $HgSO_4$ forms methyl ketones from terminal alkynes.**
- **Addition of H_2O using an organoborane, then H_2O_2, HO^- forms aldehydes from terminal alkynes.**

Sample Problem 11.4 Comparing Hydration Products Using Two Different Methods

Draw the product formed when $CH_3CH_2C{\equiv}CH$ is treated with each of the following sets of reagents: (a) H_2O, H_2SO_4, $HgSO_4$; and (b) R_2BH, followed by H_2O_2, HO^-.

Solution

(a) With $H_2O + H_2SO_4 + HgSO_4$, electrophilic addition of H and OH places the **H atom on the *less* substituted carbon** of the alkyne to form a **ketone** after tautomerization. (b) In contrast, addition of R_2BH places the **R_2B group on the *less* substituted terminal carbon** of the alkyne. Oxidation and tautomerization yield an **aldehyde.** The ketone and aldehyde formed in these reactions are constitutional isomers.

→ H_2O, H_2SO_4, $HgSO_4$ → H on the terminal C ⇌ ketone

↓ R_2BH

→ H_2O_2, HO^- → OH on the terminal C ⇌ aldehyde

Problem 11.16 Draw the products formed when the following alkynes are treated with each set of reagents: [1] H_2O, H_2SO_4, $HgSO_4$; or [2] R_2BH followed by H_2O_2, ^-OH.

a. 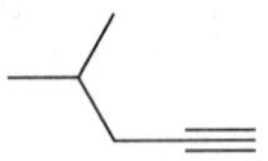 b. 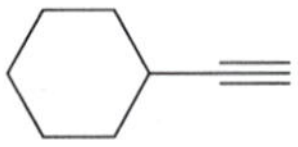 c. 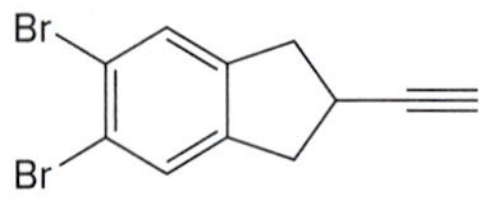 Br Br d.

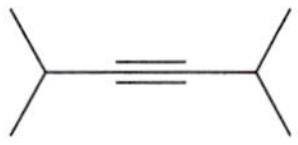

More Practice: Try Problems 11.35d, e; 11.39d, f.

Problem 11.17 What alkyne yields each ketone as the only product both with acid-catalyzed hydration and after hydroboration–oxidation?

a. 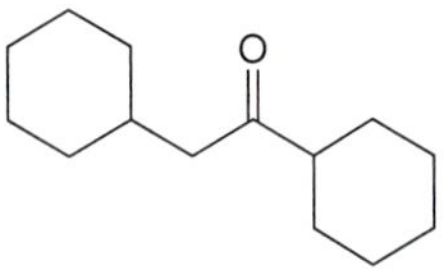 O b. 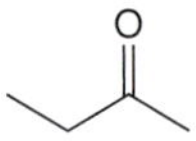 O c. 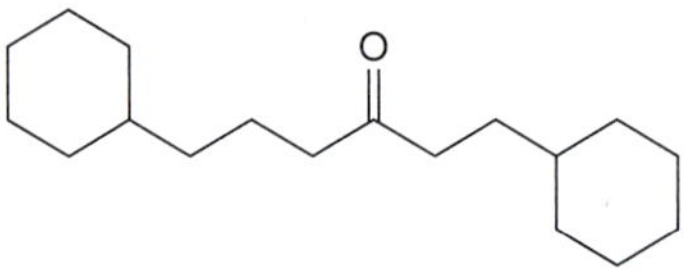 O

11.11 Reaction of Acetylide Anions

Terminal alkynes are readily converted to acetylide anions with strong bases such as $NaNH_2$ and NaH. These anions are strong nucleophiles, capable of reacting with electrophiles such as alkyl halides and epoxides.

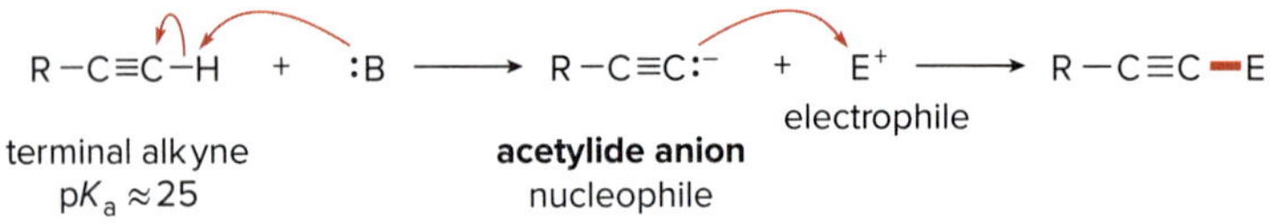

$$R-C\equiv C-H + :B \longrightarrow R-C\equiv C:^- + E^+ \longrightarrow R-C\equiv C-E$$

terminal alkyne $pK_a \approx 25$; **acetylide anion** nucleophile; electrophile

11.11A Reaction of Acetylide Anions with Alkyl Halides

Acetylide anions react with unhindered alkyl halides to yield products of nucleophilic substitution.

$$R-X + {}^-:C\equiv C-R' \xrightarrow{S_N2} R-C\equiv C-R' + X:^-$$

nucleophile; leaving group

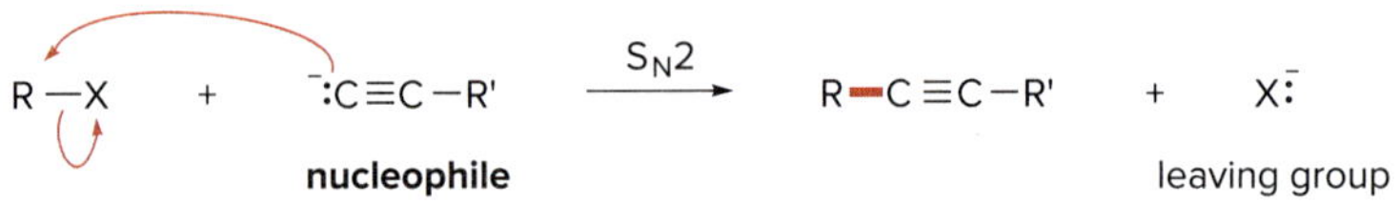

Because acetylide anions are strong nucleophiles, the mechanism of nucleophilic substitution is **S_N2,** and thus the **reaction is fastest with CH_3X and 1° alkyl halides.** Terminal alkynes (Reaction [1]) or internal alkynes (Reaction [2]) can be prepared depending on the identity of the acetylide anion.

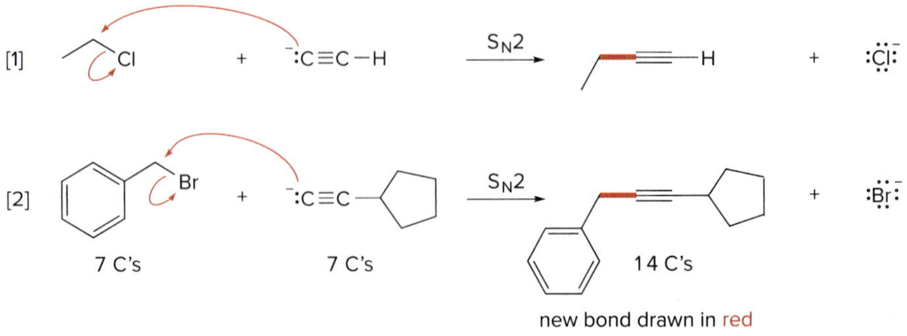

[1] $CH_3CH_2Cl + {}^-:C\equiv C-H \xrightarrow{S_N2} CH_3CH_2C\equiv C-H + :Cl:^-$

[2] $C_6H_5CH_2Br + {}^-:C\equiv C-C_5H_9 \xrightarrow{S_N2} C_6H_5CH_2C\equiv C-C_5H_9 + :Br:^-$

7 C's 7 C's 14 C's

new bond drawn in red

- **Nucleophilic substitution with acetylide anions forms new carbon–carbon bonds.**

Because organic compounds consist of a carbon framework, reactions that form carbon–carbon bonds are especially useful. In Reaction [2], for example, nucleophilic attack of a seven-carbon acetylide anion on a seven-carbon alkyl halide yields a 14-carbon alkyne as product.

Although nucleophilic substitution with acetylide anions is a very valuable carbon–carbon bond-forming reaction, it has the same limitations as any S_N2 reaction. **Steric hindrance around the leaving group causes 2° and 3° alkyl halides to undergo elimination by an E2 mechanism,** as shown with 2-bromo-2-methylpropane. Thus, nucleophilic substitution with acetylide anions forms new carbon–carbon bonds in high yield only with unhindered CH_3X and 1° alkyl halides.

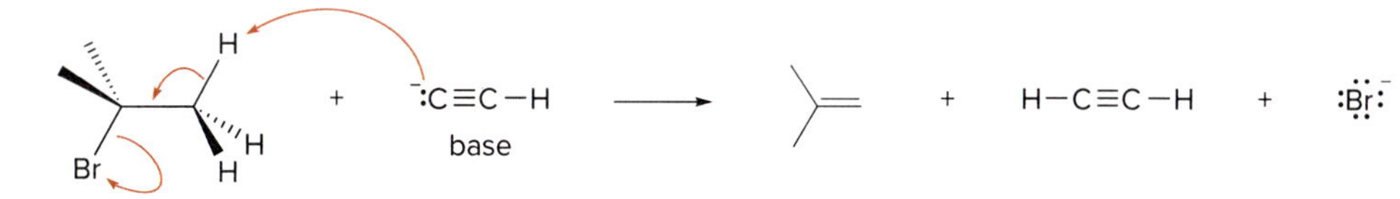

2-bromo-2-methylpropane + $^{-}$:C≡C–H (base) → E2 product + H–C≡C–H + :Br:$^{-}$

2-bromo-2-methylpropane
3° alkyl halide

Steric hindrance prevents an S_N2 reaction.

E2 product

Sample Problem 11.5 Drawing the Products When Acetylide Anions React with Alkyl Halides

Draw the organic products formed in each reaction.

a. 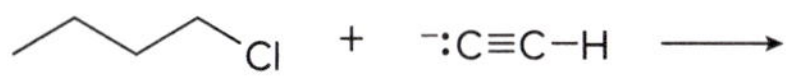 1-chlorobutane + $^{-}$:C≡C–H →

b. 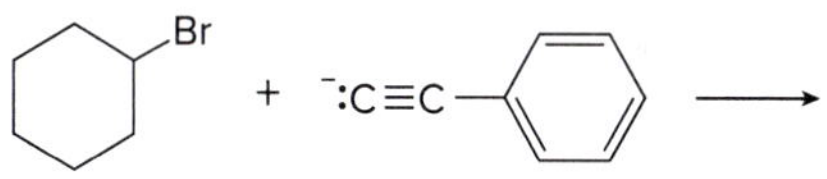 bromocyclohexane + $^{-}$:C≡C–C$_6$H$_5$ →

Solution

a. Because the alkyl halide is **1°** and the acetylide anion is a strong nucleophile, substitution occurs by an $\mathbf{S_N2}$ mechanism, resulting in a new C–C bond.

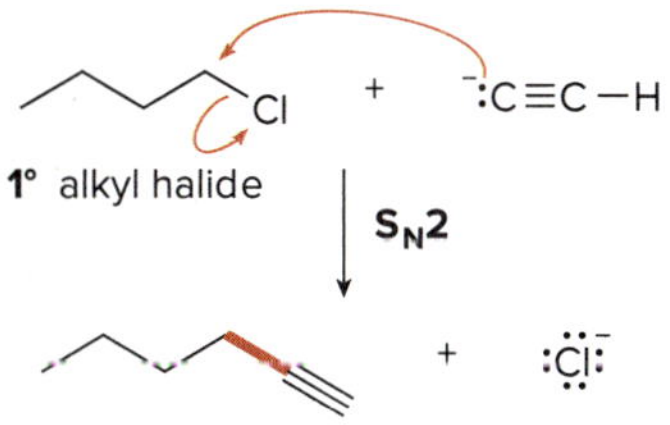

1° alkyl halide + $^{-}$:C≡C–H → (S_N2) hex-1-yne + :Cl:$^{-}$

b. Because the alkyl halide is **2°**, the major product is formed from elimination by an **E2** mechanism.

2° alkyl halide + $^{-}$:C≡C–C$_6$H$_5$ → (E2) cyclohexene (major product) + H–C≡C–C$_6$H$_5$ + Br^{-}

major product

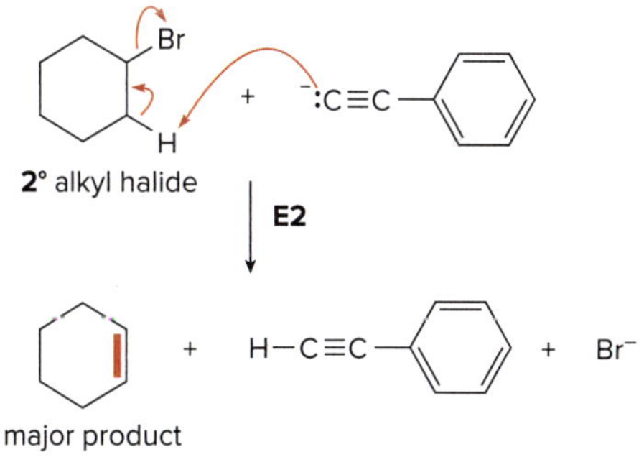

Problem 11.18 Draw the organic products formed in each reaction.

a. 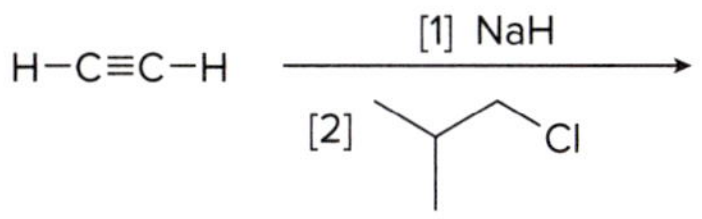 H–C≡C–H → [1] NaH [2] 1-chloro-2-methylpropane

b. 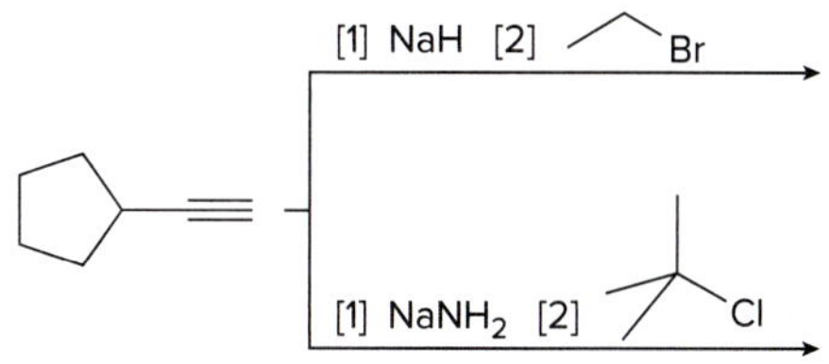 ethynylcyclopentane → [1] NaH [2] bromoethane

ethynylcyclopentane → [1] $NaNH_2$ [2] 2-chloro-2-methylpropane

More Practice: Try Problems 11.35g; 11.39g, i; 11.41a, b; 11.47.

Problem 11.19 What acetylide anion and alkyl halide can be used to prepare each alkyne? Indicate all possibilities when more than one route will work.

a. 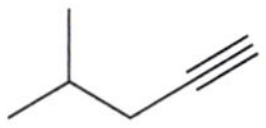 b.  c.

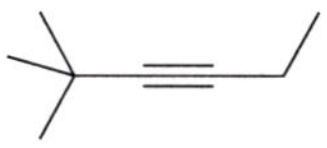

Because acetylene has two *sp* hybridized C–H bonds, two sequential reactions can occur to form **two new carbon–carbon bonds,** as shown in Sample Problem 11.6.

Sample Problem 11.6 Forming an Internal Alkyne by Two Sequential S_N2 Reactions

Identify the terminal alkyne **A** and the internal alkyne **B** in the following reaction sequence.

$$H-C\equiv C-H \xrightarrow[\text{[2] } CH_3CH_2CH_2Br]{\text{[1] } NaNH_2} \textbf{A} \xrightarrow[\text{[2] } CH_3CH_2Cl]{\text{[1] } NaNH_2} \textbf{B}$$

Solution

In each step, the base $^-NH_2$ removes a proton on an *sp* hybridized carbon, and the resulting acetylide anion reacts as a nucleophile with an alkyl halide to yield an S_N2 product. The first two-step reaction sequence forms the **terminal alkyne A** by nucleophilic attack of the acetylide anion on $CH_3CH_2CH_2Br$.

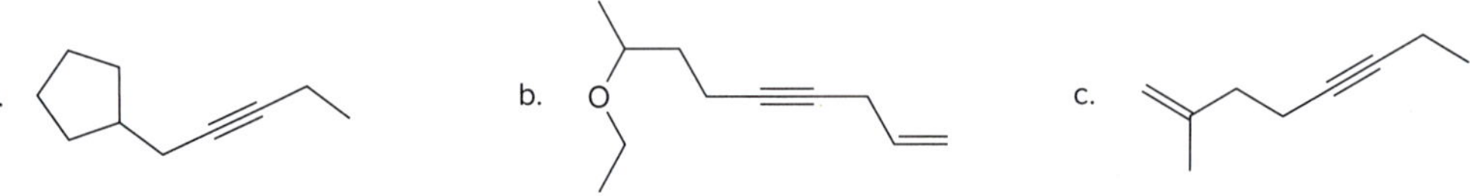

The second two-step reaction sequence forms the **internal alkyne B** by nucleophilic attack of the acetylide anion on CH_3CH_2Cl.

$:NH_2^- + H-C\equiv C-CH_2CH_2CH_3 \longrightarrow {}^-:C\equiv C-CH_2CH_2CH_3$ (acetylide anion) $+ :NH_3$; then with $CH_3CH_2Cl \longrightarrow$ **internal** alkyne **B**

Problem 11.20 Show how HC≡CH, CH_3CH_2Br, and $(CH_3)_2CHCH_2CH_2Br$ can be used to prepare $CH_3CH_2C\equiv CCH_2CH_2CH(CH_3)_2$. Show all reagents, and use curved arrows to show movement of electron pairs.

More Practice: Try Problems 11.35g; 11.39g, i; 11.41a, b; 11.49a; 11.52a.

Problem 11.21 In addition to acetylene, what alkyl halides are needed to synthesize each internal alkyne by nucleophilic substitution reactions?

a. b. c.

The soft coral *Capnella imbricata* is the source of the natural product capnellene (Figure 11.4). *Michael G. Moye*

Sample Problem 11.6 illustrates how a seven-carbon product can be prepared from three smaller molecules by forming two new carbon–carbon bonds.

H–C≡C–H, Cl, Br

new bonds shown in red

Carbon–carbon bond formation with acetylide anions is a valuable reaction used in the synthesis of numerous natural products. Two examples include **capnellene,** isolated from the soft coral *Capnella imbricata,* and **niphatoxin B,** isolated from a red sea sponge, as shown in Figure 11.4.

Figure 11.4

Use of acetylide anion reactions in the synthesis of two marine natural products

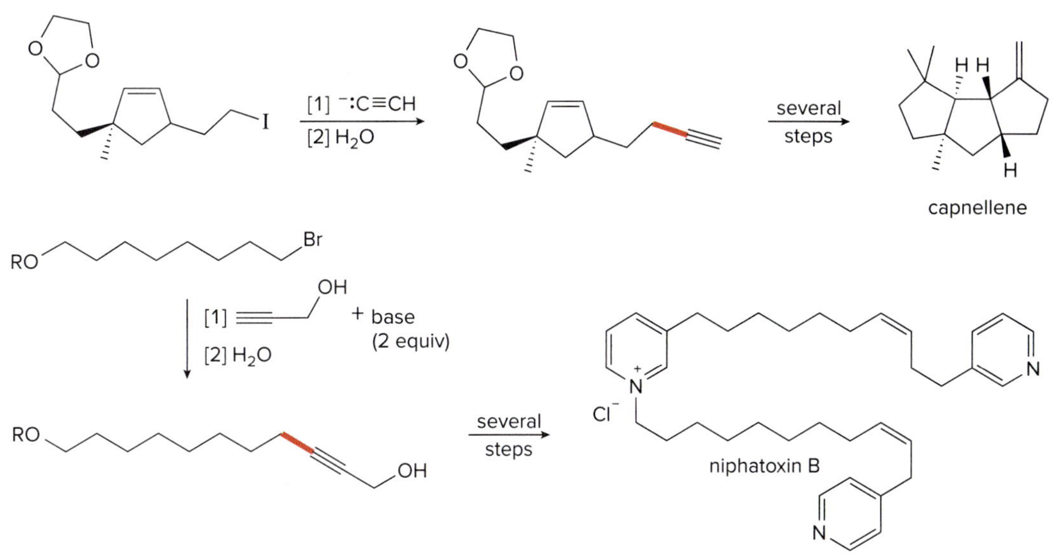
[1] ⁻:C≡CH
[2] H_2O
several steps
capnellene
RO
Br
OH
[1]
+ base
(2 equiv)
[2] H_2O
RO
OH
several steps
Cl^-
N
niphatoxin B

- New carbon–carbon bonds formed from acetylide anions are shown in red.

11.11B Reaction of Acetylide Anions with Epoxides

Acetylide anions are strong nucleophiles that open epoxide rings by an S_N2 mechanism. This reaction also results in the formation of a **new carbon–carbon bond.** Backside attack occurs at the **less substituted** end of the epoxide.

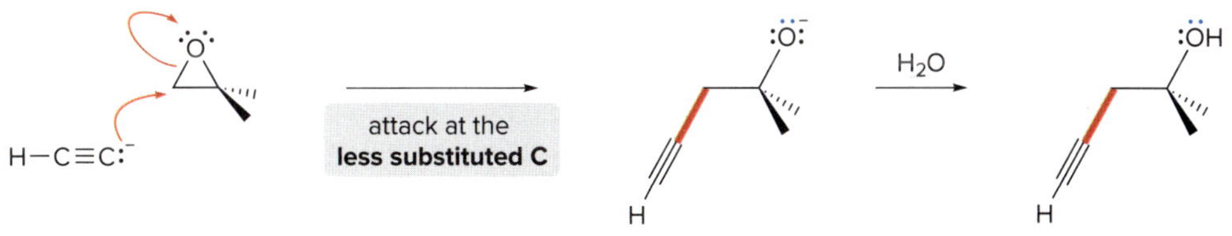
H−C≡C:⁻
attack at the
less substituted C
H_2O
:OH

Opening of epoxide rings with strong nucleophiles was first discussed in Section 9.16A.

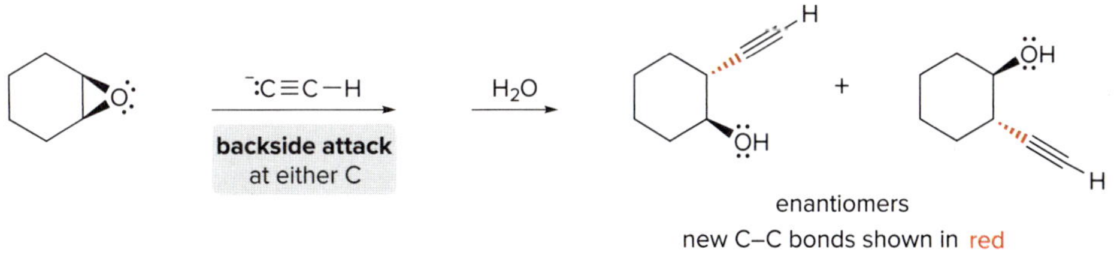
⁻:C≡C−H
backside attack
at either C
H_2O
OH
+
H
enantiomers
new C–C bonds shown in red

Problem 11.22 Draw the products of each reaction.

a. 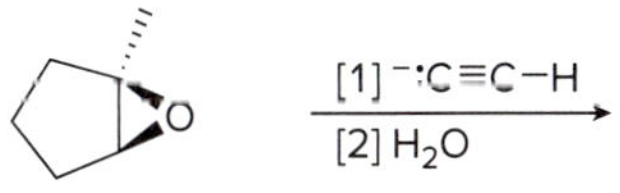
[1] ⁻:C≡C−H
[2] H_2O

b. 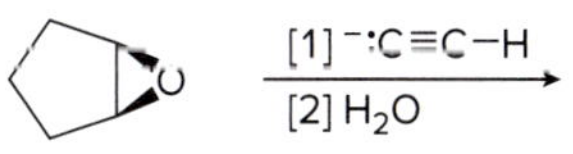
[1] ⁻:C≡C−H
[2] H_2O

Problem 11.23 Draw the products formed when $CH_3CH_2C{\equiv}C^-Na^+$ reacts with each compound.

a. $CH_3CH_2CH_2Br$
b. $(CH_3)_2CHCH_2CH_2Cl$
c. $(CH_3CH_2)_3CCl$
d. $BrCH_2CH_2CH_2CH_2OH$
e. ethylene oxide followed by H_2O
f. propene oxide followed by H_2O

Sample Problem 11.7 Identifying the Acetylide Anion and Epoxide Needed to Synthesize an Alcohol

What acetylide anion and epoxide are needed to synthesize **C?**

C

Solution

To identify the bond formed when the acetylide anion opens an epoxide ring, **locate the carbon bonded to the OH and the adjacent carbon bonded to a C≡C.** The new C–C bond results from nucleophilic attack of the acetylide anion with the less substituted end of the epoxide to form an alkoxide that is protonated with water to yield **C.**

Form this bond.

C

acetylide anion + epoxide → alkoxide with the new C–C bond in red $\xrightarrow{H_2O}$ **C**

Problem 11.24 What acetylide anion and epoxide are needed to synthesize each compound?

a. b. c.

More Practice: Try Problem 11.48.

11.12 Synthesis

The reactions of acetylide anions give us an opportunity to examine organic synthesis more systematically. Performing a multistep synthesis can be difficult. Not only must you know the reactions for a particular functional group, but you must also put these reactions in a logical order, a process that takes much practice to master.

11.12A General Terminology and Conventions

To plan a synthesis of more than one step, we use the process of **retrosynthetic analysis**—that is, working backwards from the desired product to determine the starting materials from which it is made (Section 10.18). To write a synthesis working backwards from the product to the starting material, an **open arrow** (⇒) is used to indicate that the product is drawn on the left and the starting material on the right.

Carefully read the directions for each synthesis problem. Sometimes a starting material is specified, whereas at other times you must begin with a compound that meets a particular criterion; for example, you may be asked to synthesize a compound from alcohols having five or fewer carbon atoms. These limitations are meant to give you some direction in planning a multistep synthesis.

The product of a synthesis is often called the **target compound.** Using retrosynthetic analysis, we must determine what compound can be converted to the target compound by a single reaction. That is, **what is the immediate precursor of the target compound?** After an appropriate precursor is identified, this process is continued until we reach a specified starting material. Sometimes multiple retrosynthetic pathways are examined before a particular route is decided upon.

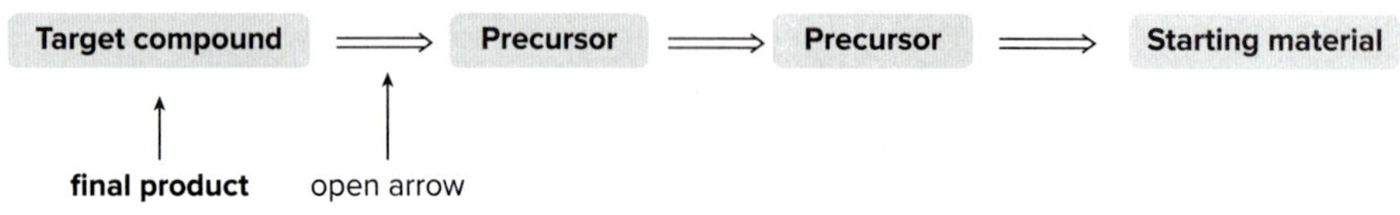

In designing a synthesis, reactions are often divided into two categories:

- **Reactions that form new carbon–carbon bonds.**
- **Reactions that convert one functional group to another—that is, functional group interconversions.**

Appendix F lists the carbon–carbon bond-forming reactions encountered in this text.

Carbon–carbon bond-forming reactions are central to organic synthesis because simpler and less valuable starting materials can be converted to more complex products. Keep in mind that whenever the product of a synthesis has more carbon–carbon bonds than the starting material, the synthesis must contain at least one of these reactions.

How To Develop a Retrosynthetic Analysis

Step [1] **Compare the carbon skeletons of the starting material and product.**

- If the product has more carbon–carbon σ bonds than the starting material, the synthesis must form one or more C–C bonds. If not, only functional group interconversion occurs.
- **Match the carbons in the starting material with those in the product** to see where new C–C bonds must be added or where functional groups must be changed.

Step [2] **Concentrate on the functional groups in the starting material and product and ask:**

- What methods introduce the functional groups in the product?
- What kind of reactions does the starting material undergo?

Step [3] **Work backwards from the product and forwards from the starting material.**

- Ask: **What is the immediate precursor of the product?**
- Compare each precursor to the starting material to determine if there is a one-step reaction that converts one to the other. Continue this process until the starting material is reached.
- Always generate *simpler* precursors when working backwards.
- Use *fewer* steps when multiple routes are possible.
- Keep in mind that you may need to evaluate several different precursors for a given compound.

Step [4] **Check the synthesis by writing it in the synthetic direction.**

- To check a retrosynthetic analysis, write out the steps beginning with the starting material, indicating all necessary reagents.

11.12B Examples of Multistep Synthesis

Retrosynthetic analysis with acetylide anions is illustrated in Sample Problems 11.8 and 11.9.

Sample Problem 11.8 Devising a Short Synthesis

Devise a synthesis of $HC{\equiv}CCH_2CH_2CH_3$ from $HC{\equiv}CH$ and any other organic or inorganic reagents.

Retrosynthetic Analysis

The two C's in the starting material match up with the two *sp* hybridized C's in the product, so a three-carbon unit must be added.

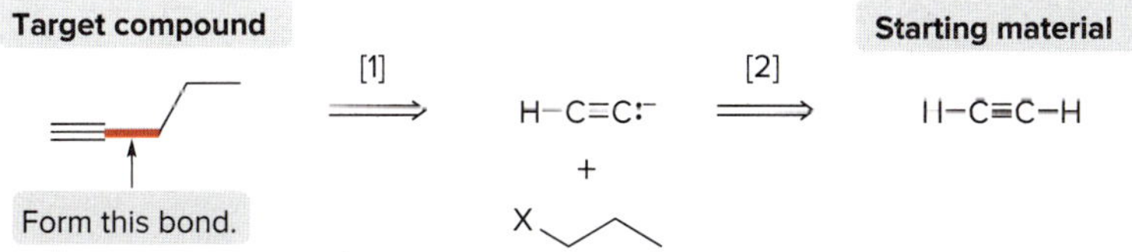

Thinking backwards . . .

[1] Form a new C–C bond using an acetylide anion and a 1° alkyl halide.
[2] Prepare the acetylide anion from acetylene by treatment with base.

Synthesis

Deprotonation of HC≡CH with NaH forms the acetylide anion, which undergoes S_N2 reaction with an alkyl halide to form the target compound, a five-carbon alkyne.

A two-step process:

H−C≡C−H $\xrightarrow{Na^+H^-}$ H−C≡C:⁻ + Cl–CH₂CH₂CH₃ ⟶ (target compound) + Cl⁻
+ H_2

target compound

Problem 11.25 Use retrosynthetic analysis to show how hex-3-yne can be prepared from acetylene and any other organic and inorganic compounds. Then draw the synthesis in the synthetic direction, showing all needed reagents.

More Practice: Try Problem 11.47a.

Sample Problem 11.9 Devising a Synthesis with More Than Two Steps

Devise a synthesis of the following compound from starting materials having two or fewer carbons.

(butan-2-one) ⟹ compounds having ≤ 2 C's

Retrosynthetic Analysis

A carbon–carbon bond-forming reaction must be used to convert the two-carbon starting materials to the four-carbon product.

Target compound — Starting material

(butan-2-one) $\xRightarrow{[1]}$ (but-1-yne) $\xRightarrow{[2]}$ ⁻:C≡C−H + CH₃CH₂X $\xRightarrow{[3]}$ H−C≡C−H

Form this bond.

Thinking backwards . . .

[1] Form the carbonyl group by hydration of a triple bond.
[2] Form a new C–C bond using an acetylide anion and a 1° alkyl halide.
[3] Prepare the acetylide anion from acetylene by treatment with base.

Synthesis

Three steps are needed to complete the synthesis. Treatment of HC≡CH with NaH forms the acetylide anion, which undergoes an S_N2 reaction with an alkyl halide to form a four-carbon terminal alkyne. Hydration of the alkyne with H_2O, H_2SO_4, and $HgSO_4$ yields the target compound.

H−C≡C−H $\xrightarrow{Na^+H^-}$ H−C≡C:⁻ + Cl–CH₂CH₃ ⟶ (but-1-yne) + Cl⁻ $\xrightarrow[H_2SO_4,\ HgSO_4]{H_2O}$ (target compound)
+ H_2

target compound

Problem 11.26 Devise a synthesis of each compound from two-carbon starting materials.

a. (butanal: CH₃CH₂CH₂CHO) b. (2,2-dibromobutane)

More Practice: Try Problems 11.47, 11.49–11.56.

These examples illustrate the synthesis of organic compounds by multistep routes. In Chapter 12, we will learn other useful reactions that expand our capability to do synthesis.

Chapter 11 REVIEW

KEY REACTIONS

[1] Addition reactions

In each addition, both π bonds of the triple bond are broken, and four new bonds are formed.

1. H–X (2 equiv); X = Cl, Br, I — **hydrohalogenation** (11.7) → geminal dihalide
2. X–X (2 equiv); X = Cl, Br — **halogenation** (11.8) → tetrahalide
3. H–OH, H_2SO_4, $HgSO_4$ — **hydration** (11.9) → [enol] ⇌ ketone
4. [1] R_2BH; [2] H_2O_2, HO^- — **hydroboration–oxidation** (11.10) → [enol (*E* and *Z* isomers)] ⇌ aldehyde

See Figure 11.3. Try Problems 11.35a–e; 11.39a, b, d, f.

[2] Reactions involving acetylide anions

1. $HC{\equiv}C{-}H$ + $H:^-$ ⇌ $HC{\equiv}C:^-$ + H_2 (11.6B) acetylide ion
2. $HC{\equiv}C:^-$ + $CH_3{-}X$ → $HC{\equiv}C{-}CH_3$ + X^-; X = Cl, Br, I (11.11A)
3. Epoxide; [1] $HC{\equiv}C:^-$; [2] H–OH (11.11B) → O–H product + ^-OH

See Table 11.1. Try Problems 11.35f–h; 11.39g–j; 11.41; 11.47; 11.48.

KEY SKILLS

[1] Converting an alkene to an alkyne (11.5)

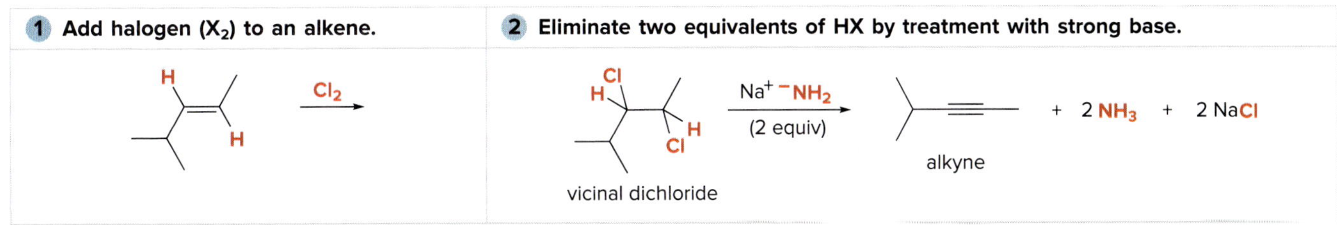

See Sample Problem 11.2. Try Problems 11.39c, 11.50a.

[2] Drawing the product of an addition reaction (11.6–11.10)

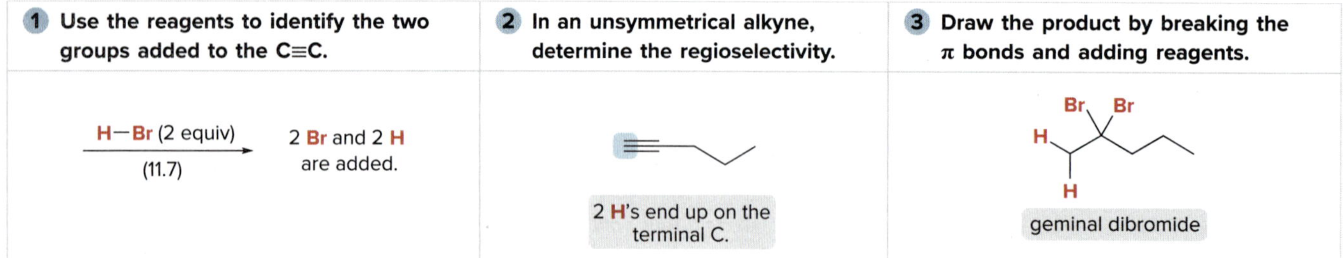

Try Problems 11.35a–e; 11.39a, b, d, f.

[3] Converting an enol to a keto tautomer in acid (11.9)

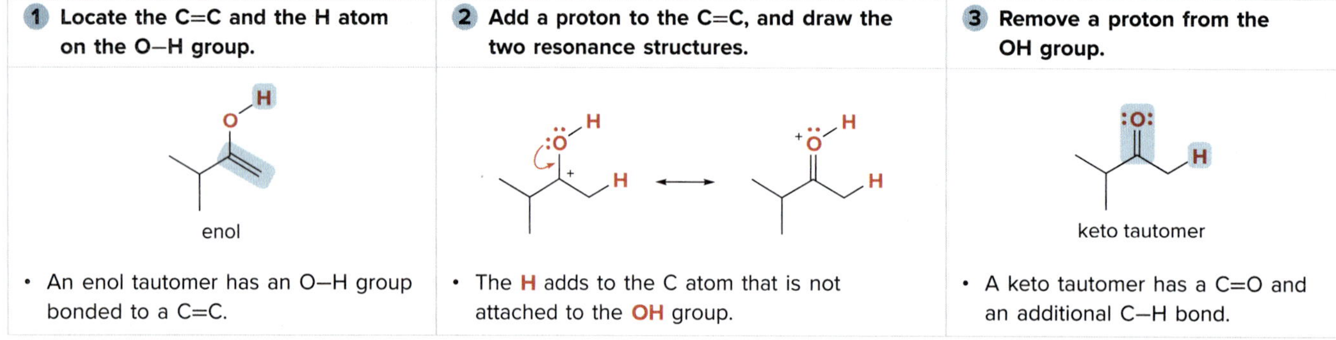

See Mechanism 11.2. Try Problems 11.31c, d; 11.33a.

[4] Converting a keto tautomer to an enol in acid (11.9)

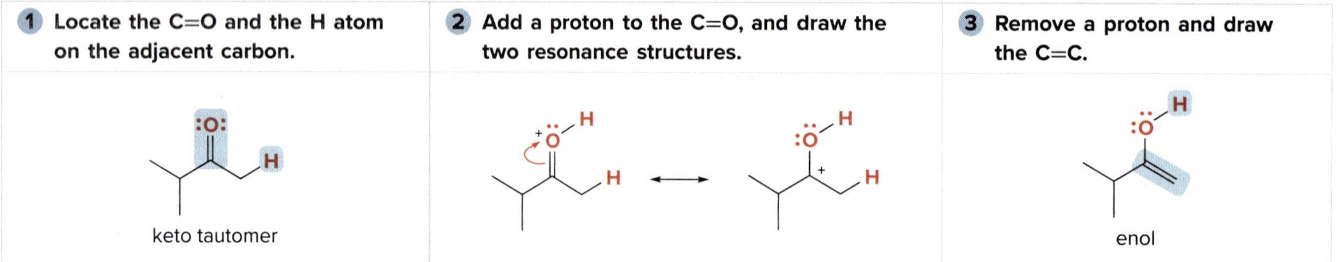

Try Problems 11.31a, b; 11.33b.

[5] Comparing the products of hydration of an alkyne (11.9, 11.10)

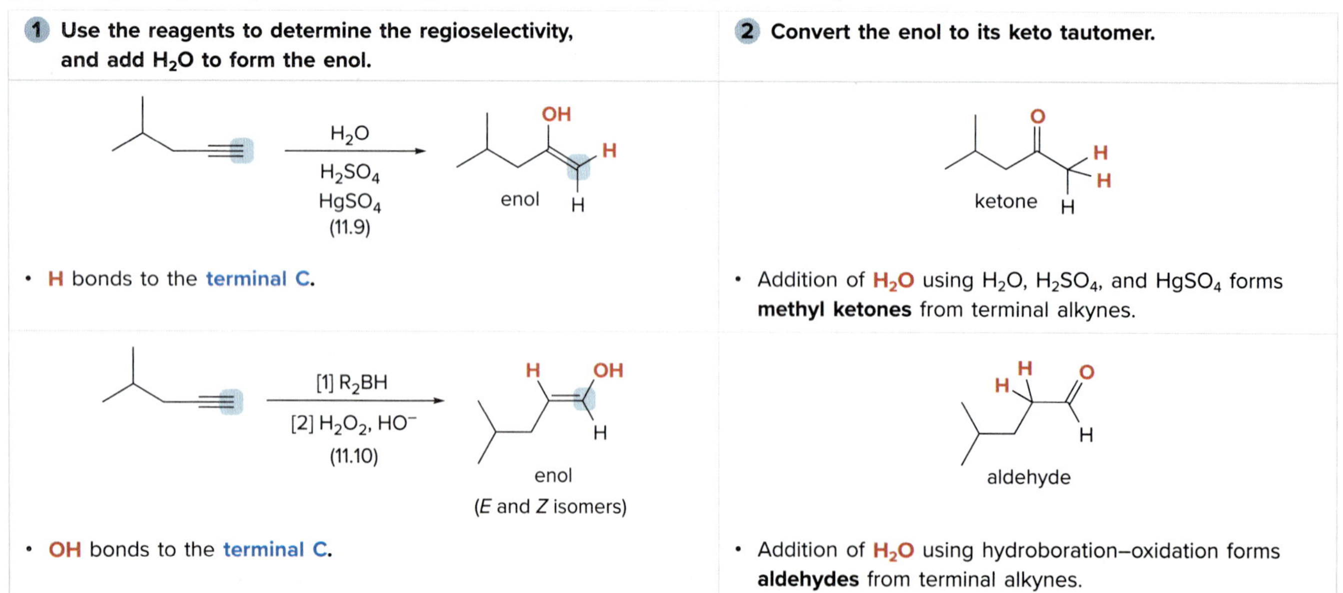

See Sample Problems 11.3, 11.4. Try Problems 11.35d, e; 11.39d, f.

[6] Comparing reactions of acetylide anions (11.11)

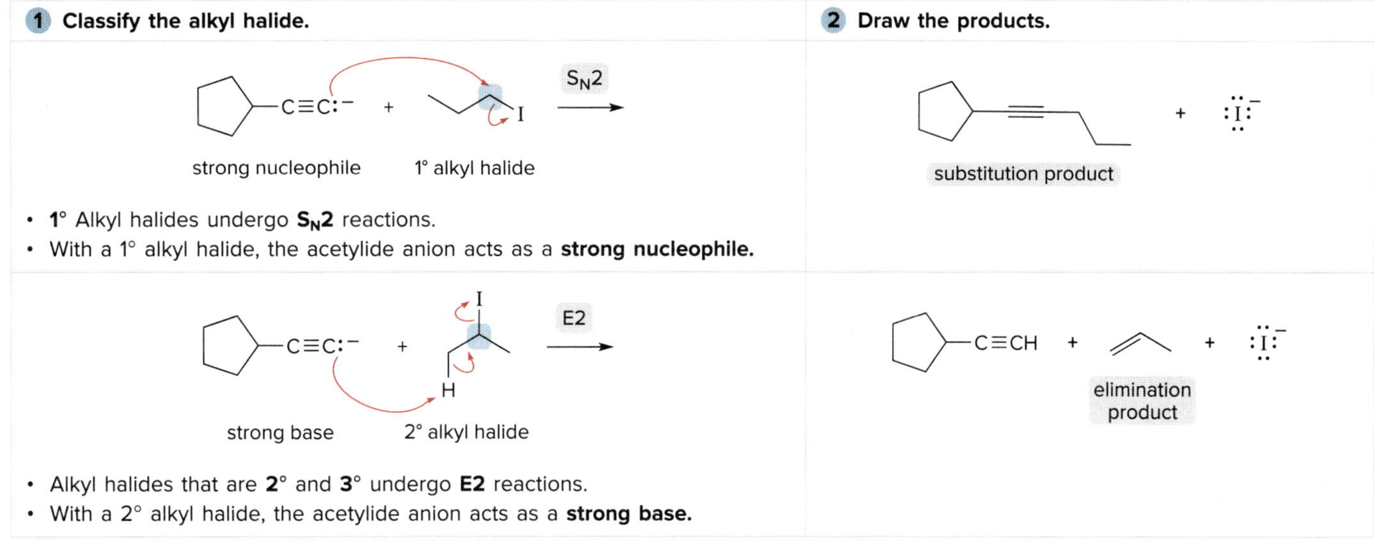

See Sample Problem 11.5. Try Problems 11.35g; 11.39g, i; 11.41a, b.

[7] Devising a synthesis (11.12); example: $(CH_3)_2CHCH_2CHO$ from $(CH_3)_2CHCH{=}CH_2$

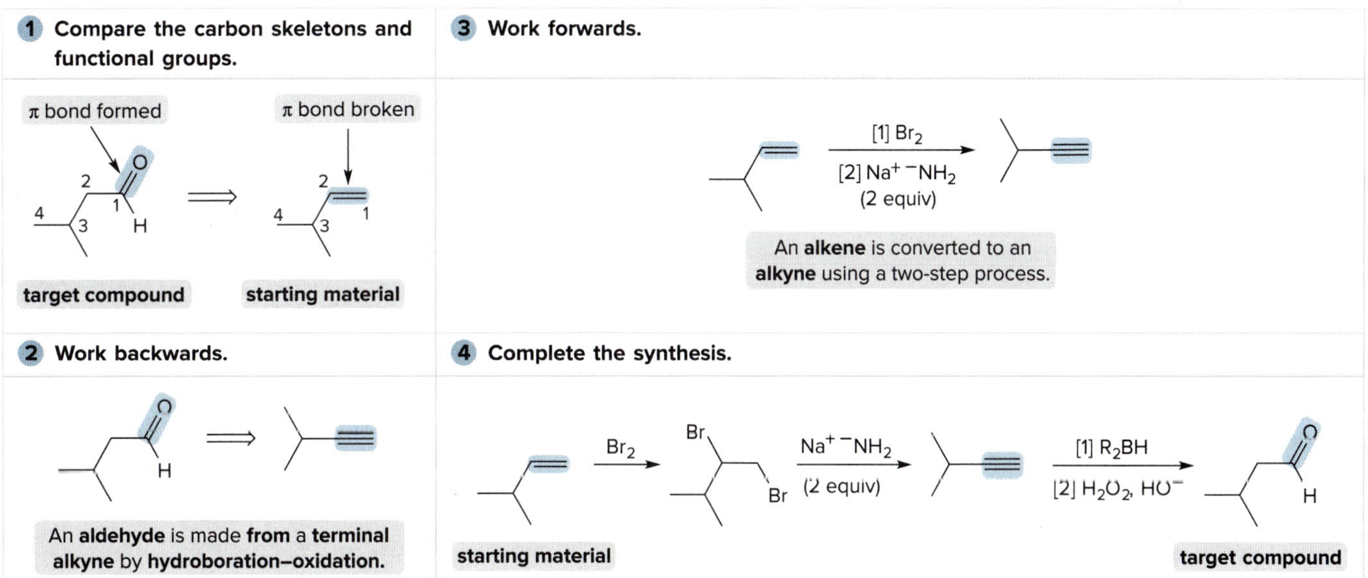

See *How To*, p. 455; Sample Problems 11.8, 11.9. Try Problems 11.49–11.56.

CHAPTER 11 MULTIPLE-CHOICE SELF-TEST

The Self-Test consists of multiple-choice questions similar to those found on the American Chemical Society organic chemistry exam. Answers are given at the end of the chapter.

1. Which base is strong enough to deprotonate hex-1-yne (pK_a = 25) so that equilibrium favors the products? The reported pK_a values are for the conjugate acids of the given bases.

a. $NaOCOCH_3$ (pK_a = 4.8)
b. $NaOCH_2CH_3$ (pK_a = 16)
c. NH_3 (pK_a = 9.3)
d. $CH_3CH_2CH_2CH_2Li$ (pK_a = 50)

2. Give the IUPAC name for the following alkyne.

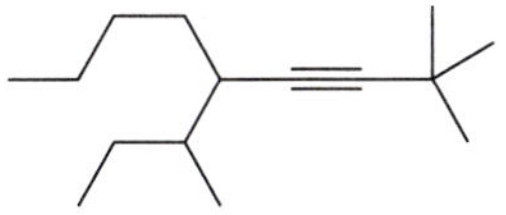

a. 5-*sec*-butyl-2,2-dimethylnon-3-yne
b. 5-isobutyl-2,2-dimethylnon-3-yne
c. 3-*sec*-butyl-1-*tert*-butylhept-1-yne
d. 2,2-dimethyl-5-*sec*-butylnon-3-yne

3. Which compounds are tautomers of **A?**

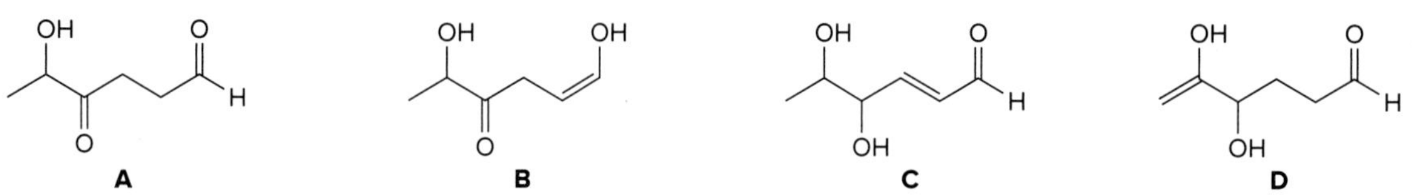

a. **B** only b. **C** only c. **D** only d. both **B** and **D**

4. What product is formed in the following reaction sequence?

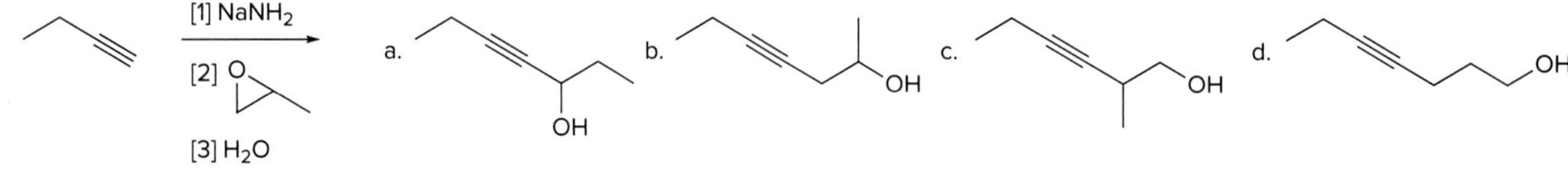

5. What reagent is needed for the following reaction?

Br
Br

a. Br_2
b. [1] $NaNH_2$; [2] Br_2
c. HBr (2 equiv)
d. Br_2, H_2O

6. Which method to prepare compound **A** is preferred?

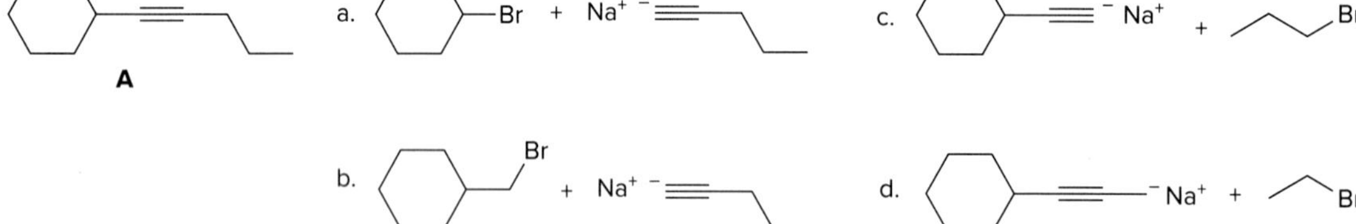

7. What product is formed in the following reaction?

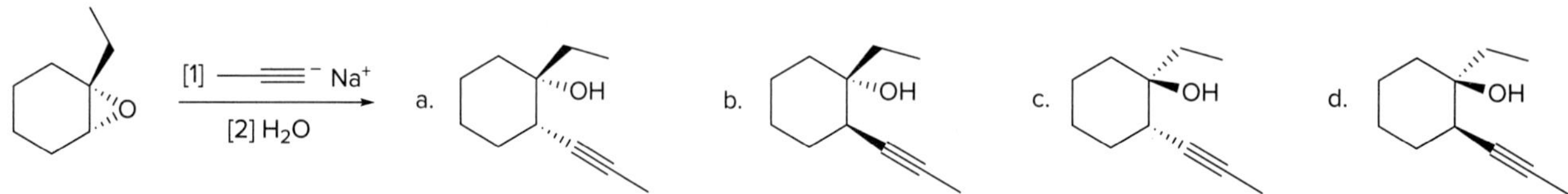

8. Which reaction (or reaction sequence) does *not* yield ketone **C?**

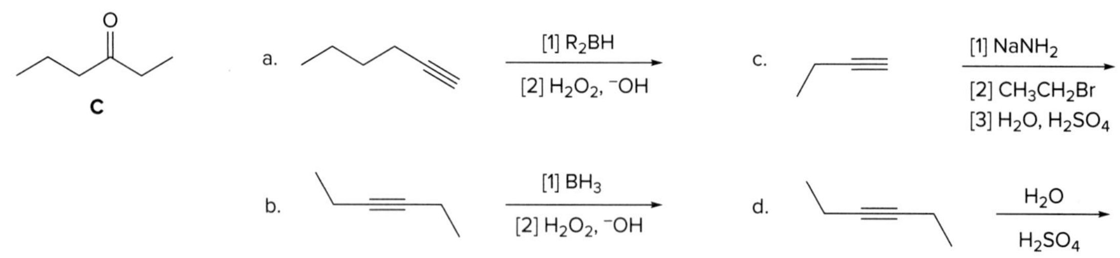

9. What compound is *not* formed in the following reaction?

[1] $NaNH_2$
[2] $(CH_3)_3CBr$
[3] H_2O

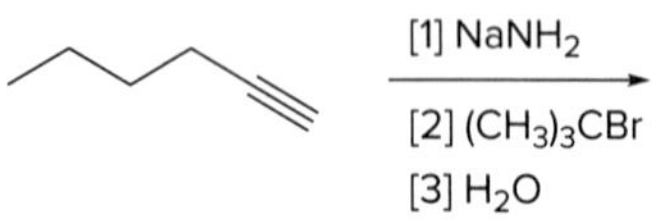

a. 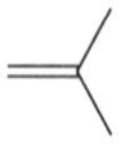 b. 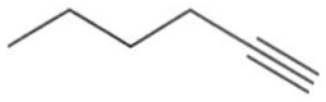 c. 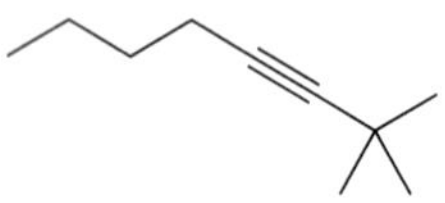 d. NaBr

10. What reagents are needed to convert $CH_3CH{=}CHCH_3$ to $CH_3C{\equiv}CCH_3$?

a. [1] Br_2; [2] $NaNH_2$ (2 equiv)
b. [1] HBr; [2] $KOC(CH_3)_3$
c. [1] Cl_2; [2] NaOH (2 equiv)
d. [1] HCl (2 equiv); [2] $NaNH_2$ (2 equiv)

PROBLEMS

Problem Using Three-Dimensional Models

11.27 Give the IUPAC name for each compound.

a. 

b. 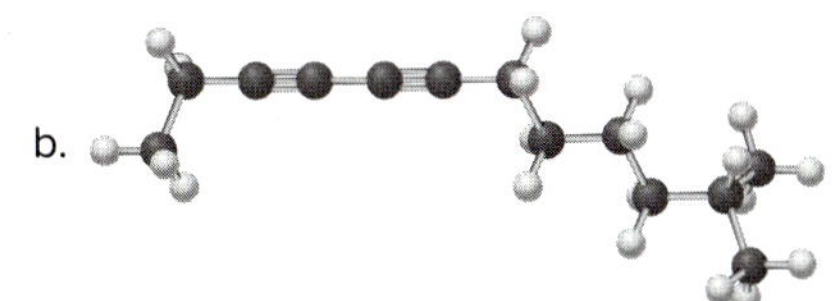

Structure and Nomenclature

11.28 Answer the following questions about erlotinib and terbinafine. Erlotinib, sold under the trade name Tarceva, is used to treat lung cancer and pancreatic cancer. Terbinafine (trade name Lamisil) is an antifungal medication used to treat ringworm and fungal nail infections.

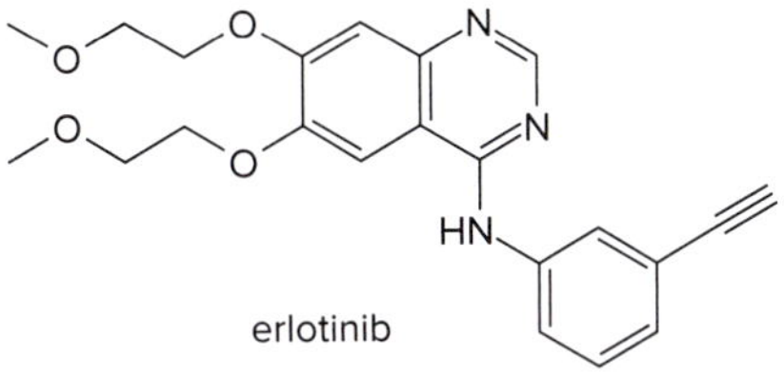

erlotinib

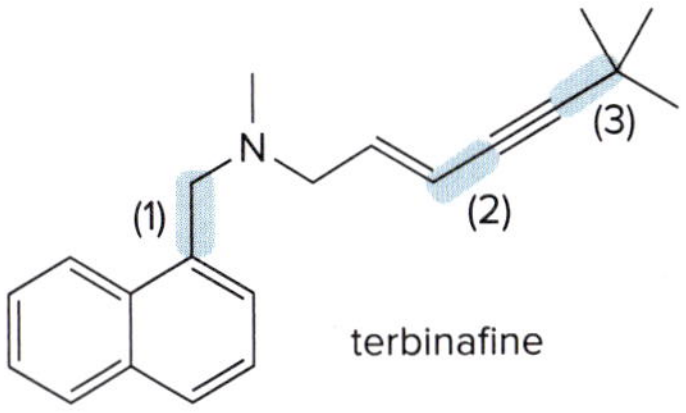

terbinafine

a. Which C–H bond in erlotinib is most acidic?

b. What orbitals are used to form the shortest C–C single bond in erlotinib?

c. Rank the labeled bonds in terbinafine in order of increasing bond strength.

d. Draw two additional resonance structures for terbinafine that contain all uncharged atoms.

11.29 Give the IUPAC name for each alkyne.

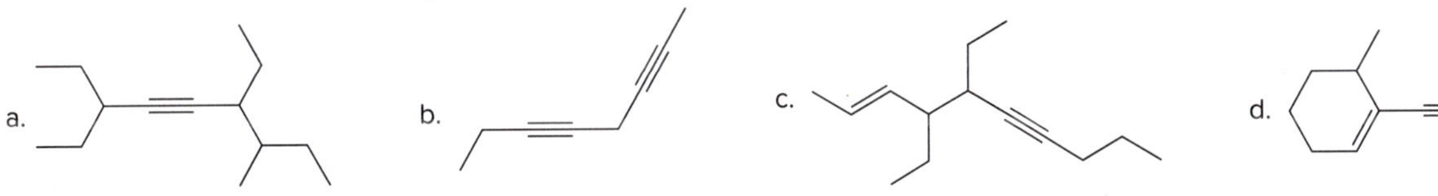

11.30 Give the structure corresponding to each name.

a. 5,6-dimethylhept-2-yne

b. 5-*tert*-butyl-6,6-dimethylnon-3-yne

c. (*S*)-4-chloropent-2-yne

d. *cis*-1-ethynyl-2-methylcyclopentane

e. 3,4-dimethylocta-1,5-diyne

f. (*Z*)-6-methyloct-6-en-1-yne

Tautomers

11.31 Draw the enol form of each keto tautomer in parts (a) and (b), and the keto form of each enol tautomer in parts (c) and (d).

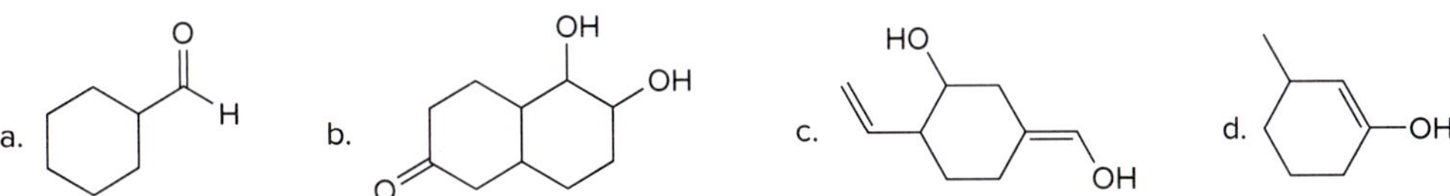

11.32 How is each compound related to **A?** Choose from tautomers, constitutional isomers but not tautomers, or neither.

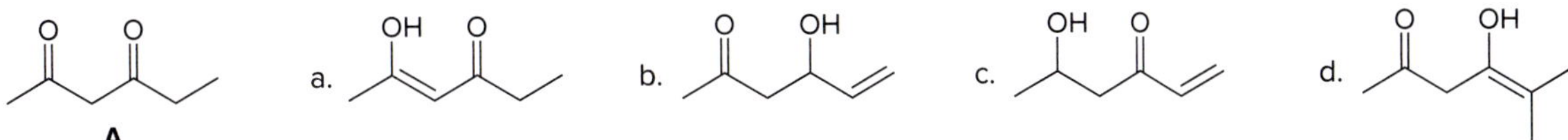

11.33 Warfarin, typically drawn as an enol tautomer, is an anticoagulant used to prevent blood clots.

a. Draw the all-keto tautomer of warfarin.

b. Draw three additional mono enol tautomers of warfarin.

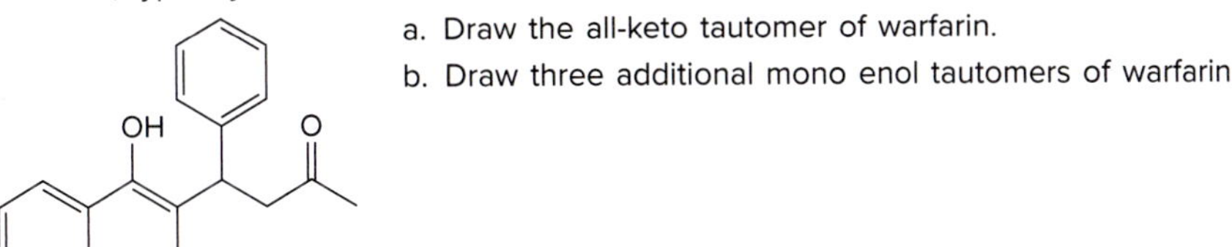

warfarin

11.34 Enamines and imines are tautomers that contain N atoms. Draw a stepwise mechanism for the acid-catalyzed conversion of enamine **X** to imine **Y**.

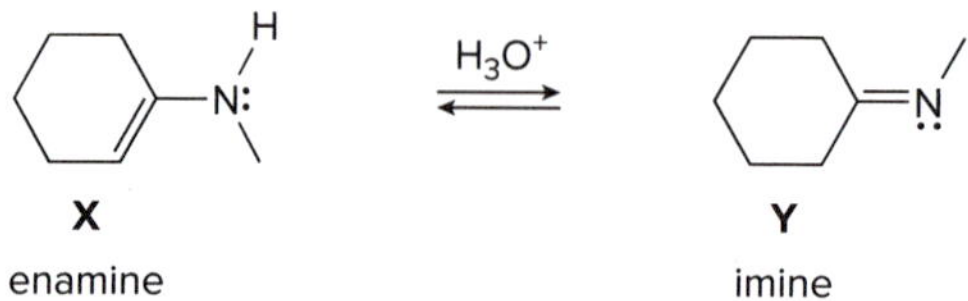

Reactions

11.35 Draw the products formed when hex-1-yne is treated with each reagent.

a. HCl (2 equiv)
b. HBr (2 equiv)
c. Cl_2 (2 equiv)
d. $H_2O + H_2SO_4 + HgSO_4$
e. [1] R_2BH; [2] H_2O_2, HO^-
f. NaH
g. [1] $^-NH_2$; [2] CH_3CH_2Br
h. [1] $^-NH_2$; [2] (ethylene oxide); [3] H_2O

11.36 Explain the apparent paradox: Although the addition of one equivalent of HX to an alkyne is more exothermic than the addition of HX to an alkene, an alkene reacts faster with HX.

11.37 What alkynes give each of the following ketones as the only product after hydration with H_2O, H_2SO_4, and $HgSO_4$?

a. b. c.

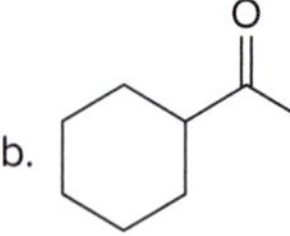

11.38 What alkyne gives each compound as the only product after hydroboration–oxidation?

a. b.

11.39 Draw the organic products formed in each reaction.

a. $\xrightarrow{\text{2 HBr}}$

b. $\xrightarrow{2\ Cl_2}$

c. $\xrightarrow{[1]\ Cl_2;\ [2]\ NaNH_2\ (2\ equiv)}$

d. $\xrightarrow{[1]\ R_2BH;\ [2]\ H_2O_2,\ HO^-}$

e. $HC{\equiv}C^- + D_2O \longrightarrow$

f. $\xrightarrow{H_2O,\ H_2SO_4}$

g. $\xrightarrow{[1]\ NaNH_2;\ [2]}$ OTs

h. $\xrightarrow{[1]\ HC{\equiv}C^-;\ [2]\ H_2O}$

i. $\xrightarrow{[1]\ NaNH_2;\ [2]}$ I

j. $\xrightarrow{[1]\ NaH;\ [2]}$ (ethylene oxide) [3] H_2O

11.40 When alkyne **A** is treated with $NaNH_2$ followed by CH_3I, a product having molecular formula $C_6H_{10}O$ is formed, but it is *not* compound **B**. What is the structure of the product, and why is it formed?

A OH $\xrightarrow{[1]\ NaNH_2;\ [2]\ CH_3I}$ (✗) B OH

11.41 Draw the products formed in each reaction and indicate stereochemistry.

a. 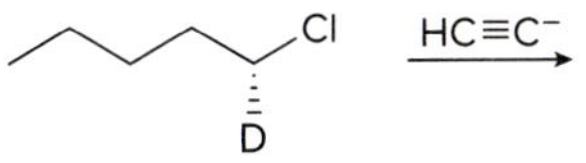 Cl, D; $HC{\equiv}C^-$ b. 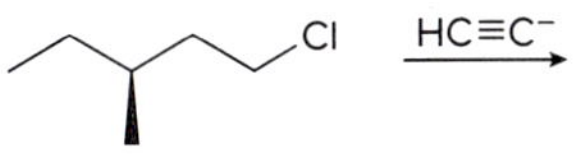 Cl; $HC{\equiv}C^-$ c. O; [1] $HC{\equiv}C^-$ [2] H_2O

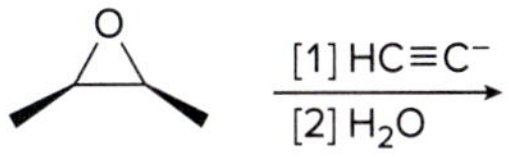

11.42 What reactions are needed to convert alcohol **A** to either alkyne **B** or alkyne **C?**

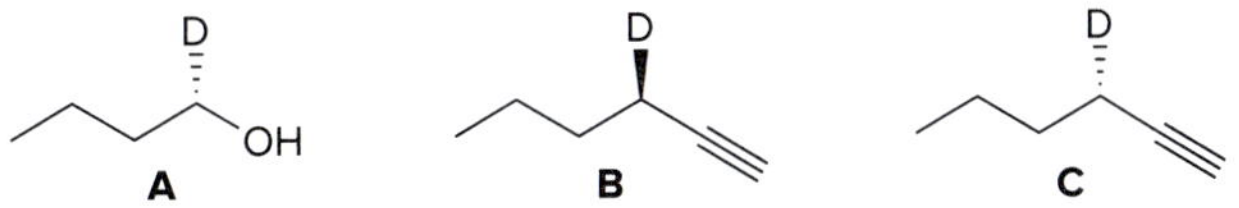

D, OH **A** D **B** D **C**

11.43 Identify the lettered compounds in the following reaction scheme.

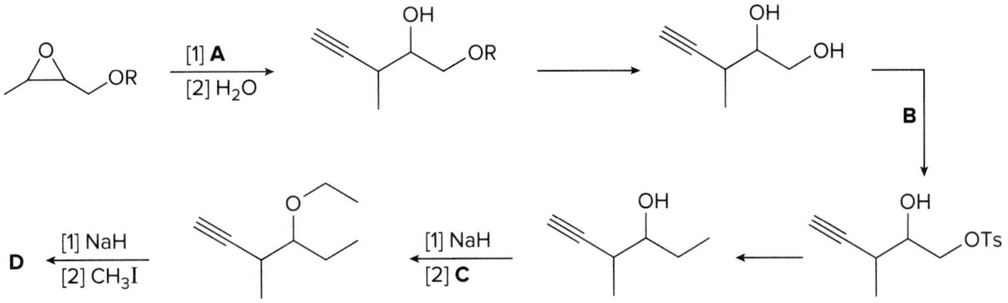

O, OR; [1] **A** [2] H_2O → OH, OR → OH, OH → **B** → OH, OTs → OH → [1] NaH [2] **C** → O → [1] NaH [2] CH_3I → **D**

Mechanisms

11.44 One step in the synthesis of the antihistamine fexofenadine (Section 23.5) involves acid-catalyzed hydration of the triple bond in **A.** Draw a stepwise mechanism for this reaction and explain why only ketone **B** is formed.

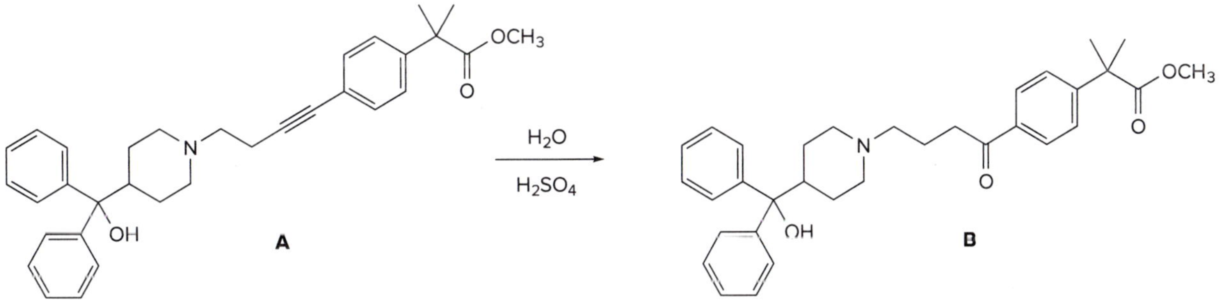

OCH_3, O, N, OH **A** $\xrightarrow[H_2SO_4]{H_2O}$ OCH_3, O, N, O, OH **B**

11.45 Tautomerization in base resembles tautomerization in acid, but deprotonation precedes protonation in the two-step mechanism. (a) Draw a stepwise mechanism for the following tautomerization. (b) Then draw a stepwise mechanism for the reverse reaction, the conversion of the keto form to the enol.

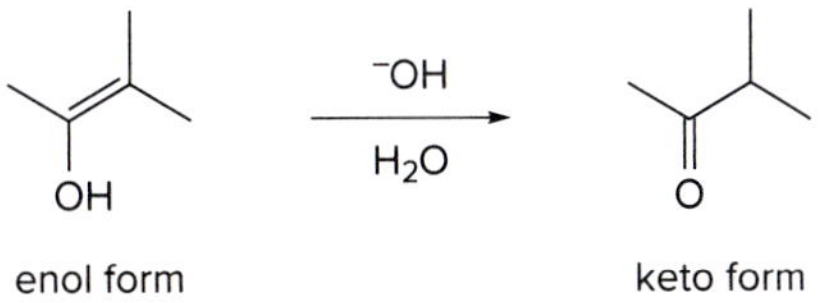

OH enol form $\xrightarrow[H_2O]{^-OH}$ O keto form

11.46 Draw a stepwise mechanism for each reaction.

a. 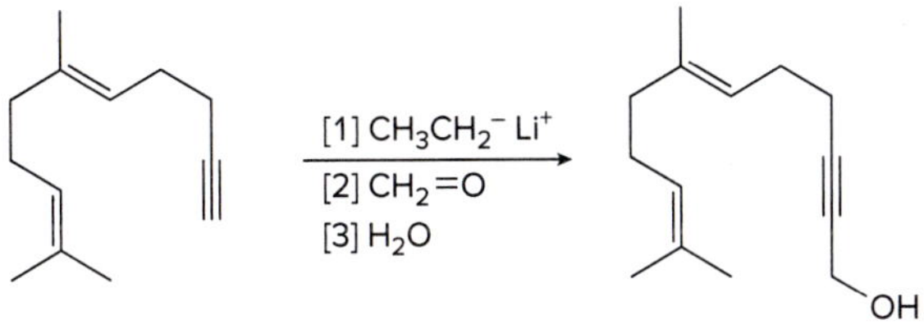 [1] $CH_3CH_2^-\ Li^+$ [2] $CH_2{=}O$ [3] H_2O → OH

b. 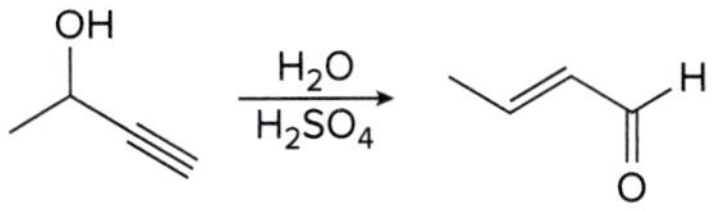 OH $\xrightarrow[H_2SO_4]{H_2O}$ H, O

Synthesis

11.47 What acetylide anion and alkyl halide are needed to synthesize each alkyne?

a. 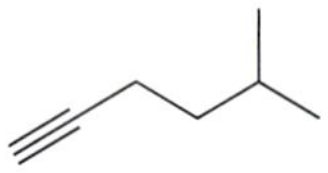 b. 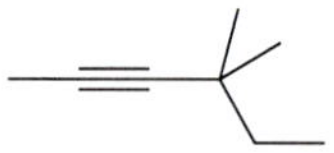 c.

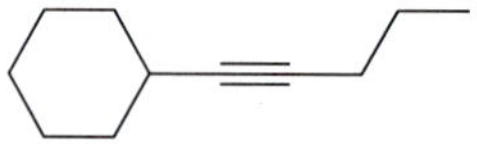

11.48 What acetylide anion and epoxide are needed to synthesize each compound?

a. 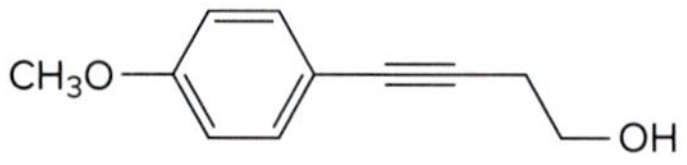 CH_3O —OH b. 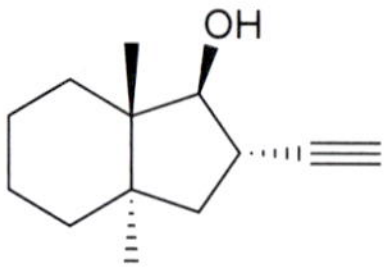 OH c. 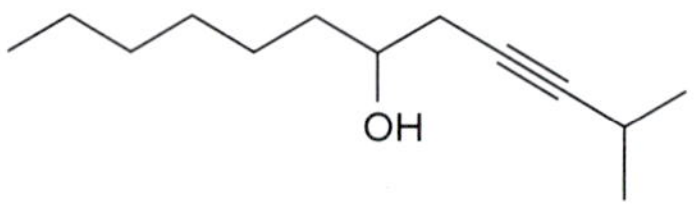 OH

11.49 Synthesize each compound from acetylene. You may use any other organic or inorganic reagents.

a.  b. 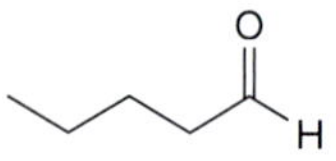 O H c. 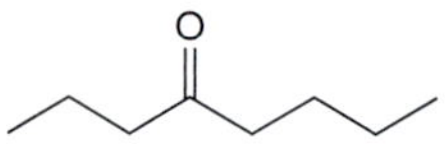 O

11.50 Devise a synthesis of each compound using $CH_3CH_2CH{=}CH_2$ as the starting material. You may use any other organic compounds or inorganic reagents.

a. 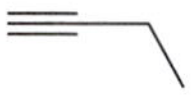 b.  OH c. 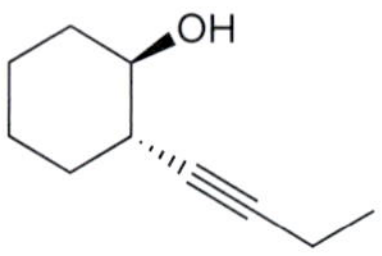 OH

(+ enantiomer)

11.51 Devise a synthesis of alkyne **Y** from alkyl bromide **X,** CH_3I, and any needed reagents.

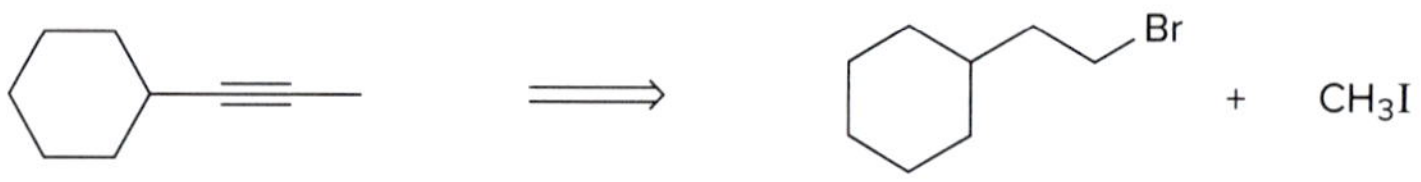

Y ⟹ **X** Br + CH_3I

11.52 Devise a synthesis of each compound. You may use HC≡CH, ethylene oxide, and alkyl halides as organic starting materials and any inorganic reagents.

a. 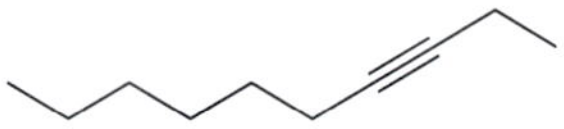 b.  O

11.53 Devise a synthesis of the ketone hexan-3-one, $CH_3CH_2COCH_2CH_2CH_3$, from CH_3CH_2Br as the only organic starting material; that is, all the carbon atoms in hexan-3-one must come from CH_3CH_2Br. You may use any other needed reagents.

11.54 Devise a synthesis of dodec-7-yn-5-ol from hex-1-ene ($CH_3CH_2CH_2CH_2CH{=}CH_2$) as the only organic starting material. You may use any other needed reagents.

OH

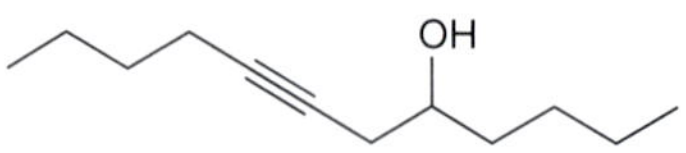

dodec-7-yn-5-ol

11.55 Devise a synthesis of each compound using $CH_3CH_2CH_2OH$ as the only organic starting material: (a) $CH_3C{\equiv}CCH_2CH_2CH_3$; (b) $CH_3C{\equiv}CCH_2CH(OH)CH_3$. You may use any other needed inorganic reagents.

11.56 Devise a synthesis of $CH_3CH_2C{\equiv}CCH_2CH_2OH$ from CH_3CH_2OH as the only organic starting material. You may use any other needed reagents.

Spectroscopy

Problem 11.57 is intended for students who have already learned about spectroscopy in Chapters A–C.

11.57 Compound **Y** (molecular formula C_6H_{10}) gives four lines in its ^{13}C NMR spectrum (27, 30, 67, and 93 ppm) and the IR spectrum given here. Propose a structure for **Y.**

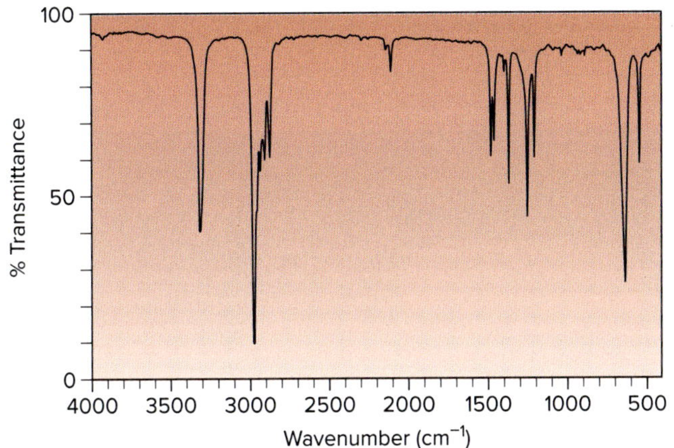

Additional spectroscopy problems on alkynes are given in Chapters B and C:

- Infrared spectroscopy: B.4a, B.5, B.9, B.13, B.20a, B.22a, B.29, B.33
- Nuclear magnetic resonance spectroscopy: C.12a

Challenge Problems

11.58 Explain why the C=C of an enol is more nucleophilic than the C=C of an alkene, despite the fact that the electronegative oxygen atom of the enol inductively withdraws electron density from the carbon–carbon double bond.

11.59 *N*-Chlorosuccinimide (NCS) serves as a source of Cl^+ in electrophilic addition reactions to alkenes and alkynes. Keeping this in mind, draw a stepwise mechanism for the following addition to but-2-yne.

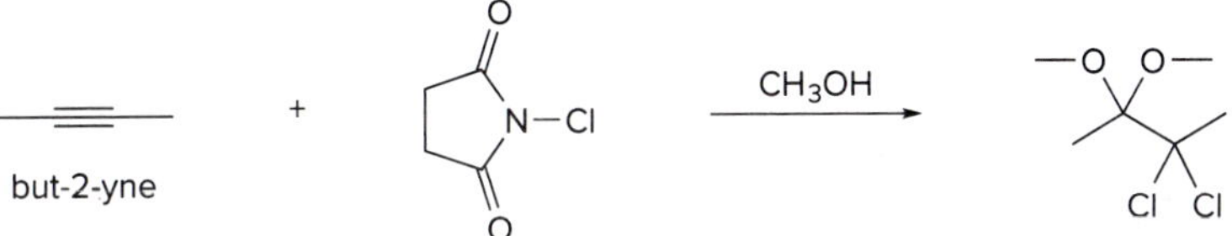

N-chlorosuccinimide

11.60 Draw a stepwise mechanism for the following reaction.

OH $\xrightarrow{H_3O^+}$ O

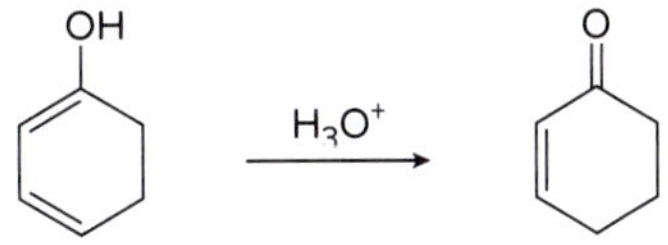

11.61 Draw a stepwise mechanism for the following reaction.

$\xrightarrow[H_2O]{Br_2}$ O Br

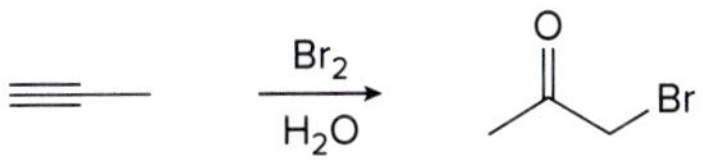

11.62 Draw a stepwise mechanism for the following intramolecular reaction.

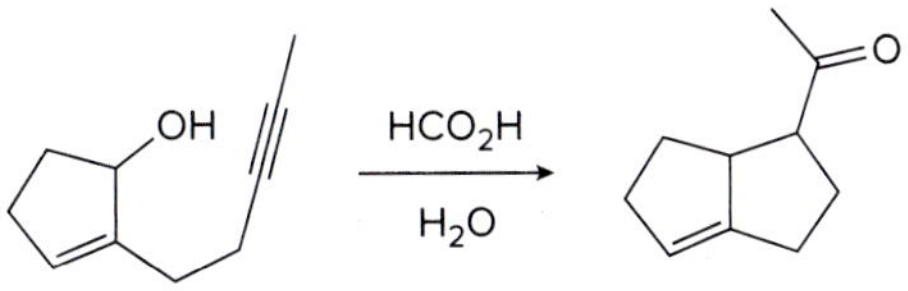

OH $\xrightarrow[H_2O]{HCO_2H}$ O

11.63 Explain why an optically active solution of (*R*)-α-methylbutyrophenone loses its optical activity when dilute acid is added to the solution.

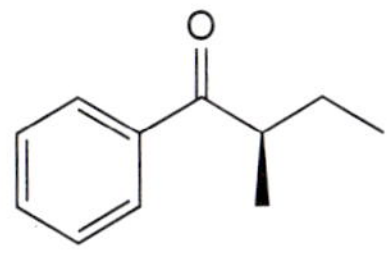

O

(*R*)-α-methylbutyrophenone

SELF-TEST ANSWERS

1. d 2. a 3. a 4. b 5. c 6. c 7. b 8. a 9. c 10. a

12 Oxidation and Reduction

Amarita/Shutterstock

12.1 Introduction
12.2 Reducing agents
12.3 Reduction of alkenes
12.4 Application: Hydrogenation of oils
12.5 Reduction of alkynes
12.6 The reduction of polar C–X σ bonds
12.7 Oxidizing agents
12.8 Epoxidation
12.9 Dihydroxylation
12.10 Oxidative cleavage of alkenes
12.11 Oxidative cleavage of alkynes
12.12 Oxidation of alcohols
12.13 Green chemistry
12.14 Biological oxidation
12.15 Sharpless epoxidation

Soybean oil is rich in **oleic** and **linoleic acids,** two unsaturated fatty acids. When a vegetable oil containing unsaturated fatty acids is treated with hydrogen, some or all of the π bonds add hydrogen, decreasing the number of degrees of unsaturation and increasing the melting point. Adding hydrogen to an alkene is a reduction reaction that increases the number of carbon–hydrogen bonds in the product. In Chapter 12, we learn about oxidation and reduction reactions of alkenes and several other functional groups.

Why Study . . .

Oxidation and Reduction?

In Chapter 12, we discuss the oxidation and reduction of **alkenes** and **alkynes,** as well as compounds with **polar C–X σ bonds**—alcohols, alkyl halides, and epoxides. Although there will be many different reagents and mechanisms, discussing these reactions as a group allows us to more easily compare and contrast them.

The word *mechanism* will often be used loosely here. In contrast to the S_N1 reaction of alkyl halides or the electrophilic addition reactions of alkenes, the details of some of the mechanisms presented in Chapter 12 are known with less certainty. For example, although the identity of a particular intermediate might be confirmed by experiment, other details of the mechanism are suggested by the structure or stereochemistry of the final product.

Oxidation and reduction reactions are very versatile, and knowing them allows us to design many more complex organic syntheses.

12.1 Introduction

Recall from Section 4.13 that the way to determine whether an organic compound has been oxidized or reduced is to compare the **relative number of C–H and C–Z bonds** (Z = an element *more electronegative* than carbon) in the starting material and product.

- ***Oxidation*** **results in an *increase* in the number of C–Z bonds (usually C–O bonds) *or* a *decrease* in the number of C–H bonds.**
- ***Reduction*** **results in a *decrease* in the number of C–Z bonds (usually C–O bonds) *or* an *increase* in the number of C–H bonds.**

Two components are always present in an oxidation or reduction reaction—**one component is oxidized and one is reduced.** When an organic compound is *oxidized* by a reagent, the reagent itself must be *reduced.* Similarly, when an organic compound is *reduced* by a reagent, the reagent becomes *oxidized.*

Thus, an organic compound such as CH_4 can be oxidized by replacing C–H bonds with C–O bonds, as shown in Figure 12.1. Reduction is the opposite of oxidation, so Figure 12.1 also shows how a compound can be reduced by replacing C–O bonds with C–H bonds. The symbols **[O]** and **[H]** indicate oxidation and reduction, respectively.

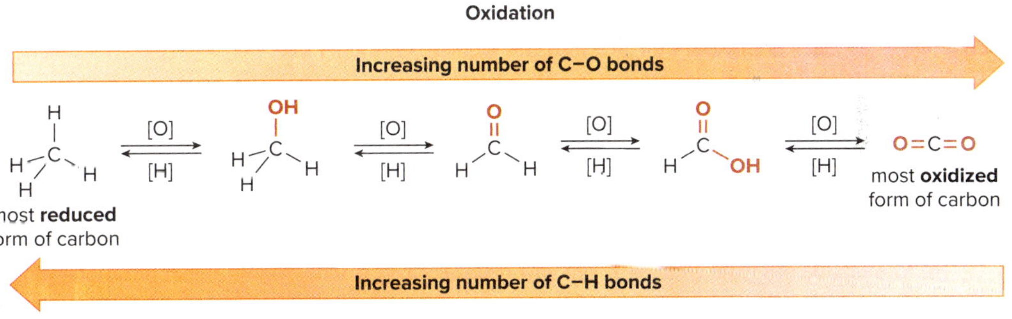

Figure 12.1 A general scheme for the oxidation and reduction of a carbon compound

Sometimes two carbon atoms are involved in a single oxidation or reduction reaction, and the net change in the number of C–H or C–Z bonds at *both* atoms must be taken into account. The conversion of an **alkyne to an alkene** and an **alkene to an alkane** are examples of **reduction,** because each process adds two new C–H bonds to the starting material, as shown in Figure 12.2.

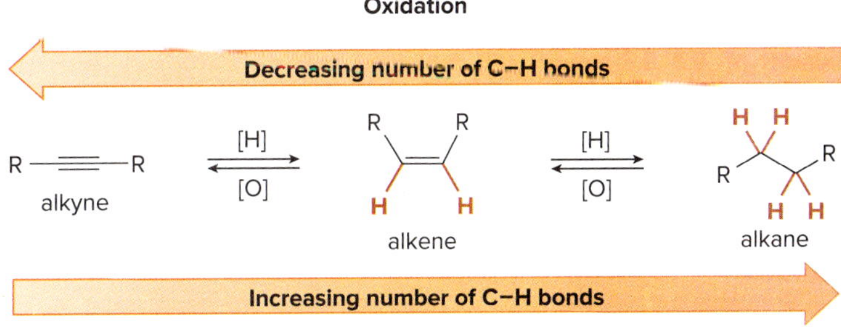

Figure 12.2 Oxidation and reduction of hydrocarbons

Problem 12.1 Classify each reaction as oxidation, reduction, or neither.

a. 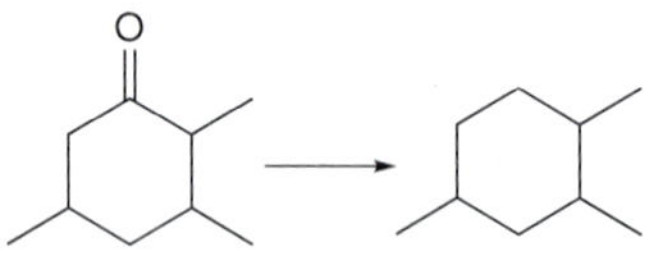 b. 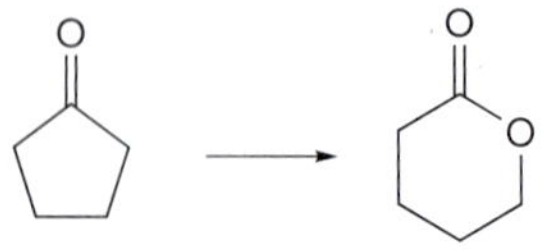 c. d. 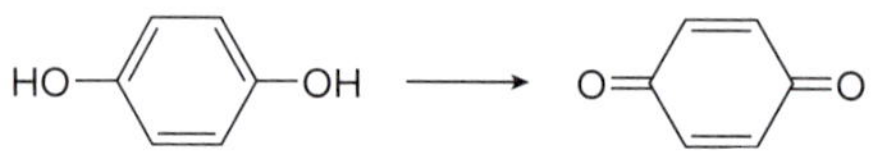 HO—⟨⟩—OH → O=⟨⟩=O e. f.

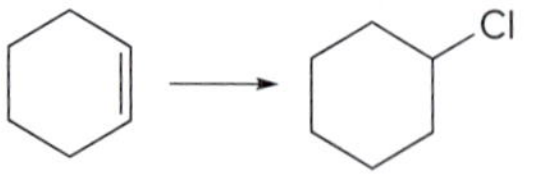

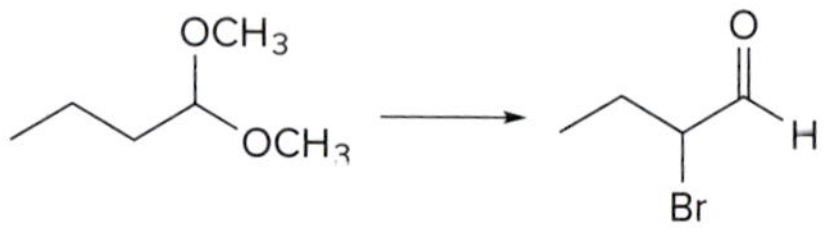

OCH3 OCH3 Cl O H Br OCH3 OCH3 OCH3 O O

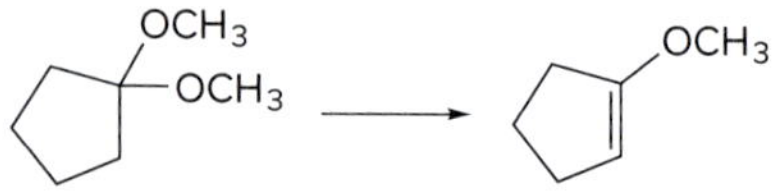

12.2 Reducing Agents

Reducing agents provide the equivalent of two hydrogen atoms, but **there are three types of reductions,** differing in how H_2 is added. The simplest reducing agent is molecular $\mathbf{H_2}$. Reductions of this sort are carried out in the presence of a metal catalyst that acts as a surface on which the reaction occurs.

The second way to deliver H_2 in a reduction is to add two protons and two electrons to a substrate—that is, $\mathbf{H_2 = 2\ H^+ + 2\ e^-}$. Reducing agents of this sort use alkali metals as a source of electrons and liquid ammonia (NH_3) as a source of protons. Reductions with **Na in** $\mathbf{NH_3}$ are called **dissolving metal reductions.**

$$2\ Na \longrightarrow 2\ Na^+ + 2\ e^-$$

$$2\ NH_3 \longrightarrow 2\ {}^-NH_2 + 2\ H^+$$

an equivalent of H_2 for reduction

The third way to deliver the equivalent of two hydrogen atoms is to add **hydride** ($\mathbf{H^-}$) and a **proton** ($\mathbf{H^+}$). The most common hydride reducing agents contain a hydrogen atom bonded to boron or aluminum. Simple examples include **sodium borohydride** ($\mathbf{NaBH_4}$) and **lithium aluminum hydride** ($\mathbf{LiAlH_4}$). These reagents deliver H^- to a substrate, and then a proton is added from H_2O or an alcohol.

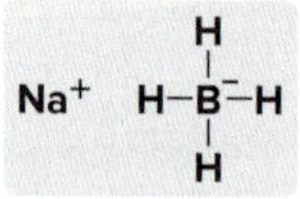

Na^+ $H-\overset{H}{\underset{H}{B^-}}-H$ Li^+ $H-\overset{H}{\underset{H}{Al^-}}-H$

sodium borohydride **$NaBH_4$** lithium aluminum hydride **$LiAlH_4$**

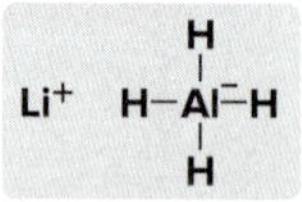

- **Metal hydride reagents act as a source of H^- because they contain polar metal–hydrogen bonds that place a partial negative charge on hydrogen.**

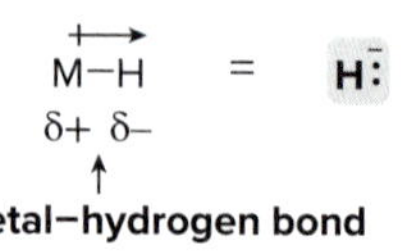

$$\overset{\delta+}{M}-\overset{\delta-}{H} = H:^-$$

a polar metal–hydrogen bond

M = B or Al

12.3 Reduction of Alkenes

Reduction of an alkene forms an alkane by addition of H_2. Two bonds are broken—the **weak π bond** of the alkene and the H_2 σ bond—and two new C–H σ bonds are formed.

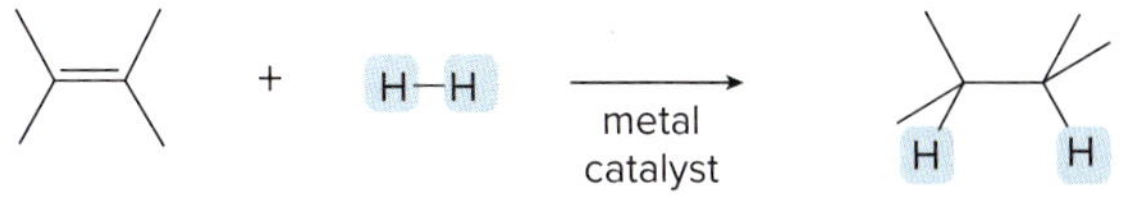

+ H–H → metal catalyst

A π **bond** is broken.

Two C–H σ bonds are formed.

Hydrogenation catalysts are insoluble in common solvents, thus creating a **heterogeneous** reaction mixture. This insolubility has a practical advantage. These catalysts contain expensive metals, but they can be filtered away from the other reactants after the reaction is complete, and then reused.

The addition of H_2 occurs only in the presence of a **metal catalyst,** and thus, the reaction is called **catalytic hydrogenation.** The catalyst consists of a metal—usually Pd, Pt, or Ni—adsorbed onto a finely divided inert solid, such as charcoal. For example, the catalyst 10% Pd on carbon is composed of 10% Pd and 90% carbon, by weight. H_2 adds in a **syn** fashion, as shown in Equation [2].

[1] + H–H → Pd-C

[2] + H–H → Pd-C

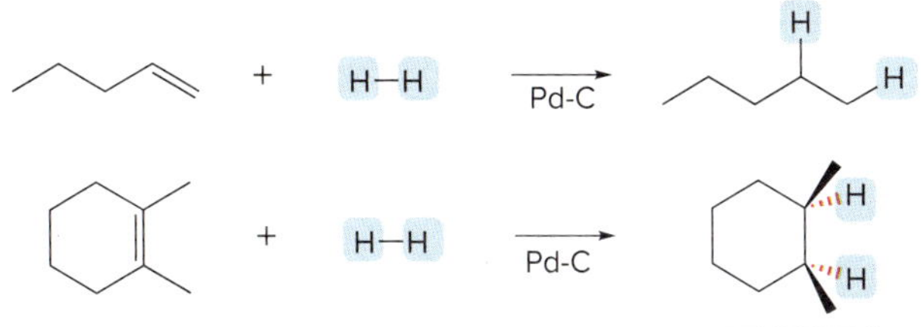

syn addition

Problem 12.2 What alkane is formed when each alkene is treated with H_2 and a Pd catalyst?

a. 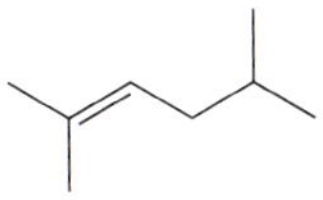 b. 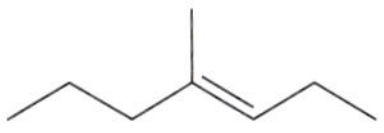 c.

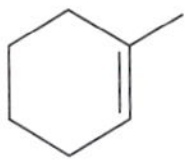

Problem 12.3 Draw all alkenes that react with one equivalent of H_2 in the presence of a palladium catalyst to form each alkane. Consider constitutional isomers only.

a. 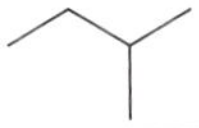 b. 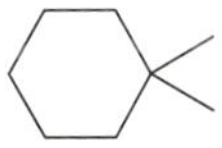 c.

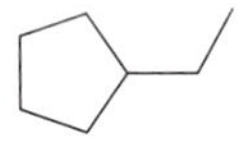

12.3A Hydrogenation and Alkene Stability

Hydrogenation reactions are **exothermic** because the bonds in the product are stronger than the bonds in the starting materials, making them similar to other alkene addition reactions. The $\Delta H°$ for hydrogenation, called the **heat of hydrogenation,** can be used as a measure of the relative stability of two different alkenes that are hydrogenated to the same alkane.

Recall from Chapter 8 that **trans alkenes are generally more stable than cis alkenes.**

For example, both *cis*- and *trans*-but-2-ene are hydrogenated to butane, and the heat of hydrogenation for the trans isomer is less than that for the cis isomer. **Because less energy is released in converting the trans alkene to butane, it must be *lower* in energy (more stable) to begin with.** The relative energies of the butene isomers are illustrated in Figure 12.3.

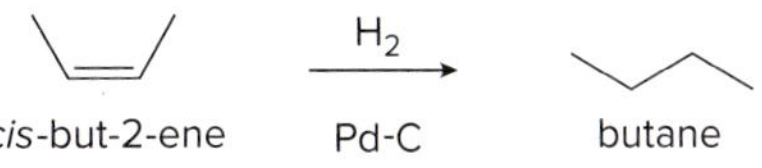

cis-but-2-ene —H_2, Pd-C→ butane $\Delta H° = -120$ kJ/mol

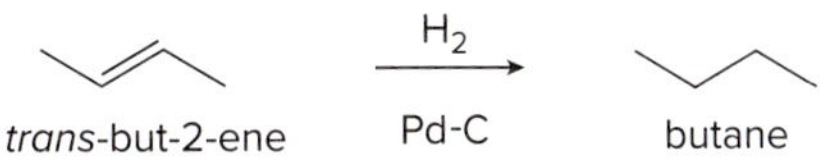

trans-but-2-ene —H_2, Pd-C→ butane $\Delta H° = -115$ kJ/mol

more stable starting material

Less energy is released.

Figure 12.3
Relative energies of *cis*- and *trans*-but-2-ene

Energy

cis isomer — **less stable alkene** — More energy is released. — $\Delta H° = -120$ kJ/mol

trans isomer — **more stable alkene** — Less energy is released. — $\Delta H° = -115$ kJ/mol

- **When hydrogenation of two alkenes gives the same alkane, the more stable alkene has the *smaller* heat of hydrogenation.**

Problem 12.4 Which alkene in each pair has the larger heat of hydrogenation?

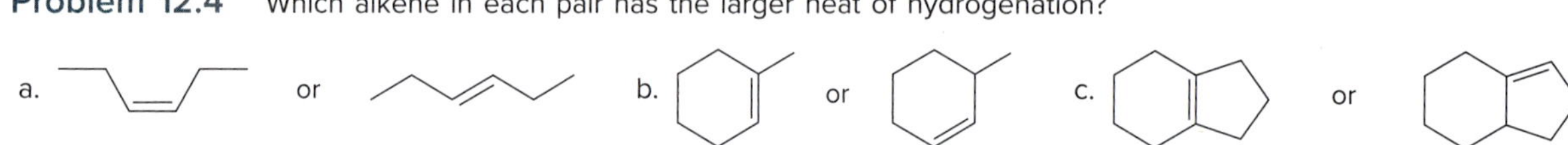

Problem 12.5 Explain why heats of hydrogenation cannot be used to determine the relative stability of 2-methylpent-2-ene and 3-methylpent-1-ene.

12.3B The Mechanism of Catalytic Hydrogenation

In the generally accepted mechanism for catalytic hydrogenation, the surface of the metal catalyst binds both H_2 and the alkene, and H_2 is transferred to the π bond in a rapid but stepwise process (Mechanism 12.1).

Mechanism 12.1 Addition of H_2 to an Alkene—Hydrogenation

1 **H_2 adsorbs to the catalyst surface** with partial or complete cleavage of the H—H bond.

2 The π bond of the alkene complexes with the metal.

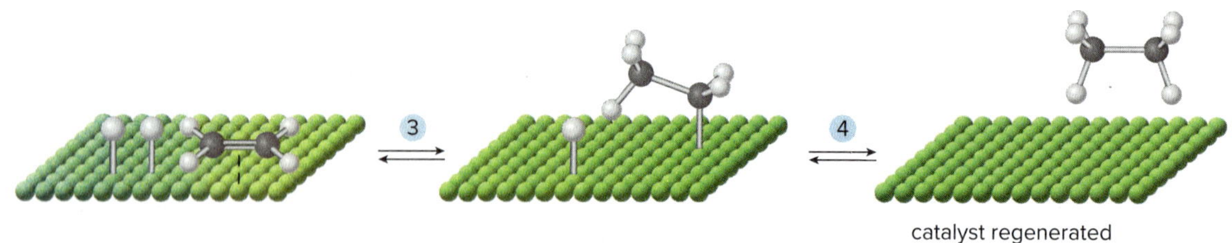

3 – 4 **Two H atoms are transferred sequentially** to the π bond in Steps [3] and [4], forming the alkane. Because the product alkane no longer has a π bond with which to complex to the metal, it is released from the catalyst surface.

The mechanism explains two facts about hydrogenation:

- **Rapid, sequential addition of H_2 occurs from the side of the alkene complexed to the metal surface, resulting in syn addition.**
- **Less crowded double bonds complex more readily to the catalyst surface, resulting in *faster* reaction.**

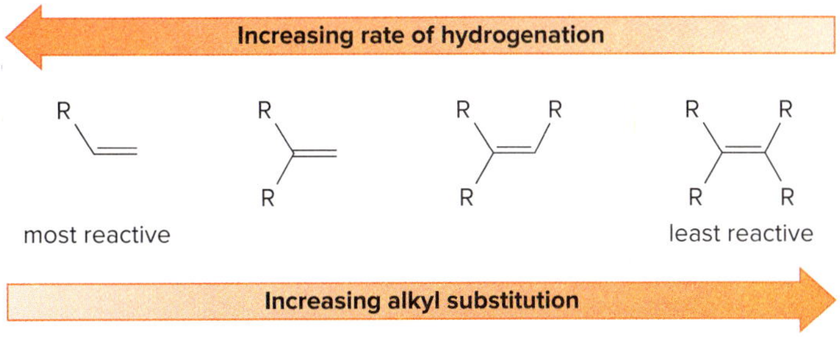

Increasing rate of hydrogenation

most reactive

least reactive

Increasing alkyl substitution

Problem 12.6 Given that syn addition of H_2 occurs from both sides of a trigonal planar double bond, draw all stereoisomers formed when each compound is treated with H_2.

a. 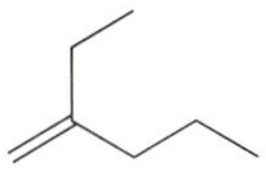 b. 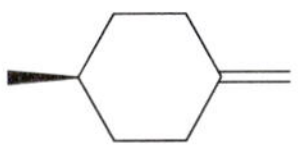 c. 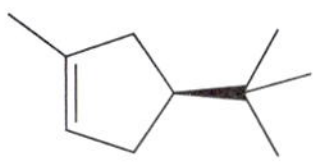 d.

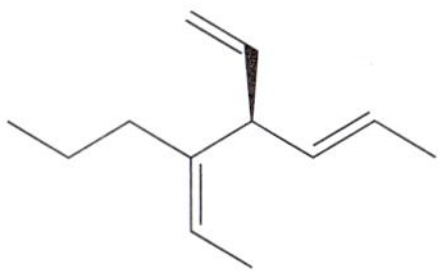

Problem 12.7 (a) Rank the alkenes in each group in order of increasing rate of hydrogenation. (b) Rank the alkenes in each group in order of increasing heat of hydrogenation.

[1] 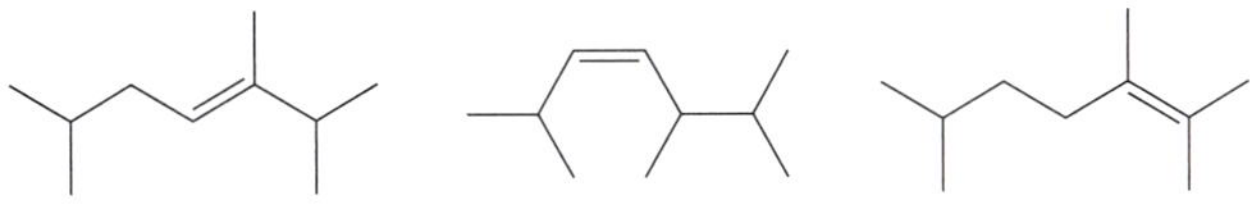 A B C [2] 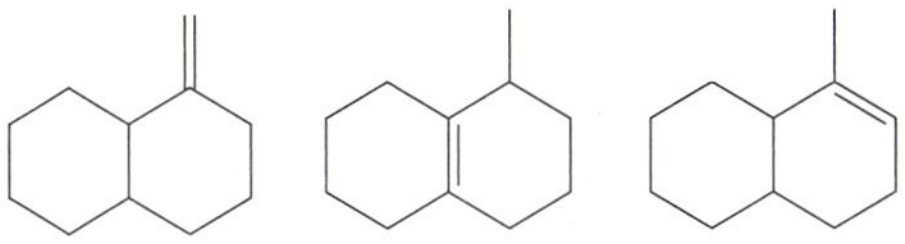 D E F

12.3C Hydrogenation Data and Degrees of Unsaturation

Recall from Section 10.2 that the **number of degrees of unsaturation gives the *total* number of rings and π bonds in a molecule.** Because H_2 adds to π bonds but does *not* add to the C–C σ bonds of rings, hydrogenation allows us to determine how many degrees of unsaturation are due to π bonds and how many are due to rings. This is done by comparing the number of degrees of unsaturation before and after a molecule is treated with H_2, as illustrated in Sample Problem 12.1.

Sample Problem 12.1 Using Hydrogenation Data to Determine the Number of Rings and π Bonds in a Molecule

How many rings and π bonds are contained in a compound of molecular formula C_8H_{12} that is hydrogenated to a compound of molecular formula C_8H_{14}?

Solution

[1] Determine the number of degrees of unsaturation in the compounds before and after hydrogenation.

Before H_2 addition—C_8H_{12}

- The maximum number of H's possible for n C's is $\mathbf{2n + 2}$; in this example, $2n + 2 = 2(8) + 2 = 18$.
- 18 H's (maximum) – 12 H's (actual) = 6 H's fewer than the maximum number.

$$\frac{\text{6 H's fewer than the maximum}}{\text{2 H's removed for each degree of unsaturation}} =$$

three degrees of unsaturation

After H_2 addition—C_8H_{14}

- The maximum number of H's possible for n C's is $\mathbf{2n + 2}$; in this example, $2n + 2 = 2(8) + 2 = 18$.
- 18 H's (maximum) – 14 H's (actual) = 4 H's fewer than the maximum number.

$$\frac{\text{4 H's fewer than the maximum}}{\text{2 H's removed for each degree of unsaturation}} =$$

two degrees of unsaturation

[2] Assign the number of degrees of unsaturation to rings or π bonds as follows:

- The number of degrees of unsaturation that remain in the product after H_2 addition = the **number of rings** in the starting material.
- The number of degrees of unsaturation that react with H_2 = the **number of π bonds.**

In this example, **two** degrees of unsaturation remain after hydrogenation, so the starting material has **two** rings. Thus:

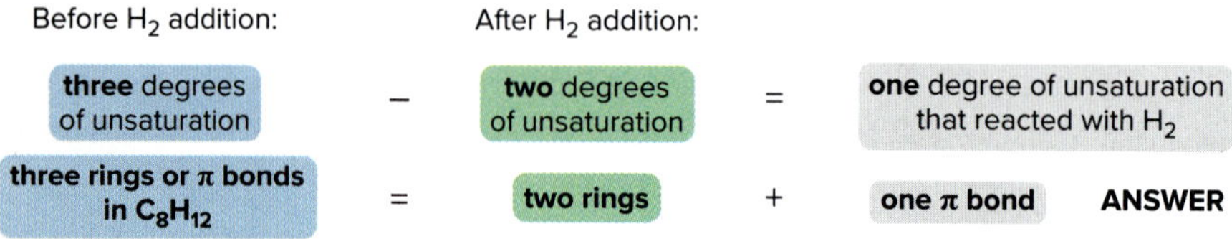

Problem 12.8 Complete the missing information for compounds **A**, **B**, and **C**, each subjected to hydrogenation. The number of rings and π bonds refers to the reactant (**A**, **B**, or **C**) prior to hydrogenation.

Compound	Molecular formula before hydrogenation	Molecular formula after hydrogenation	Number of rings	Number of π bonds
A	$C_{10}H_{12}$	$C_{10}H_{16}$	?	?
B	?	C_4H_{10}	0	1
C	C_6H_8	?	1	?

More Practice: Try Problem 12.36.

12.3D Hydrogenation of Aldehydes and Ketones

Compounds that contain a carbonyl group also react with H_2 and a metal catalyst. For example, **aldehydes and ketones are reduced to 1° and 2° alcohols,** respectively. We return to this reaction in Chapter 17.

aldehyde —(H_2, Pd-C)→ 1° alcohol   ketone —(H_2, Pd-C)→ 2° alcohol

12.4 Application: Hydrogenation of Oils

Many processed foods, such as peanut butter, margarine, and some brands of crackers, contain *partially hydrogenated* vegetable oils. These oils are produced by hydrogenating the long hydrocarbon chains of triacylglycerols.

In Section 10.6 we learned that **fats and oils are triacylglycerols that differ in the number of degrees of unsaturation** in their long alkyl side chains.

- **Fats—usually animal in origin—are solids with triacylglycerols having few degrees of unsaturation.**
- **Oils—usually vegetable in origin—are liquids with triacylglycerols having a larger number of degrees of unsaturation.**

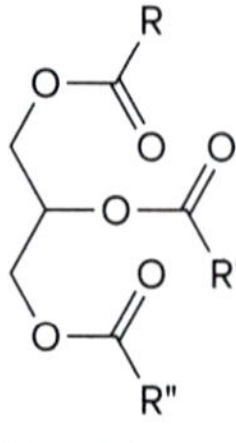

triacylglycerol

The number of double bonds in the R groups of the triacylglycerol determines whether it is a fat or an oil.

When an unsaturated vegetable oil is treated with hydrogen, some (or all) of the π bonds add H_2, decreasing the number of degrees of unsaturation (Figure 12.4). This increases the melting point of the oil. For example, margarine is prepared by partially hydrogenating vegetable oil to give a product having a semi-solid consistency that more closely resembles butter. This process is sometimes called ***hardening.***

Figure 12.4 Partial hydrogenation of the double bonds in a vegetable oil

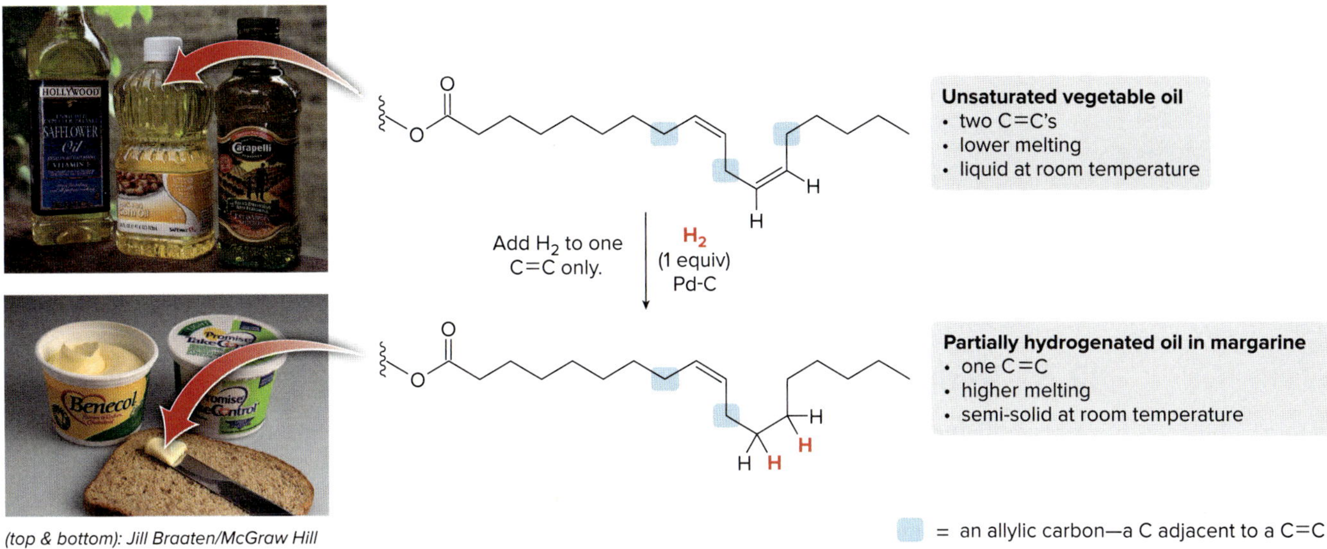

(top & bottom): Jill Braaten/McGraw Hill

- **Decreasing** the number of degrees of unsaturation **increases** the melting point. Only one long chain of the triacylglycerol is drawn.
- When an oil is *partially* hydrogenated, some double bonds react with H_2, whereas some double bonds remain in the product.
- Partial hydrogenation **decreases** the number of allylic sites (shown in blue), making a triacylglycerol **less** susceptible to oxidation, thereby increasing its shelf life.

Peanut butter is a common consumer product that contains partially hydrogenated vegetable oil. *Elite Images/ McGraw Hill*

If unsaturated oils are healthier than saturated fats, why does the food industry hydrogenate oils? There are two reasons—aesthetics and shelf life. Consumers prefer the semi-solid consistency of margarine to a liquid oil. Imagine pouring vegetable oil on a piece of toast or pancakes.

Furthermore, unsaturated oils are more susceptible than saturated fats to oxidation at the **allylic carbon atoms**—the carbons adjacent to the double bond carbons—a process discussed in Chapter 13. Oxidation makes the oil rancid and inedible. Hydrogenating the double bonds reduces the number of allylic carbons (also illustrated in Figure 12.4), thus reducing the likelihood of oxidation and increasing the shelf life of the food product. This process reflects a delicate balance between providing consumers with healthier food products, while maximizing shelf life to prevent spoilage.

One other fact is worthy of note. Because the steps in hydrogenation are reversible and H atoms are added in a sequential rather than concerted fashion, a **cis double bond can be isomerized to a trans double bond.** After addition of one H atom (Step [3] in Mechanism 12.1), an intermediate can lose a hydrogen atom to re-form a double bond with either the cis or trans configuration.

As a result, some of the cis double bonds in vegetable oils are converted to trans double bonds during hydrogenation, forming so-called **"trans fats."** The shape of the resulting fatty acid chain is very different, closely resembling the shape of a *saturated* fatty acid chain. Consequently, trans fats are thought to have the same negative effects on blood cholesterol levels as saturated fats; that is, trans fats stimulate cholesterol synthesis in the liver, thus increasing blood cholesterol levels, a factor linked to increased risk of heart disease.

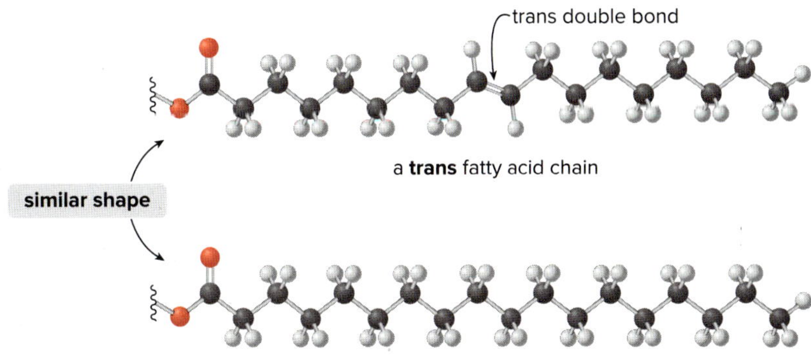

Problem 12.9 Draw the products formed when triacylglycerol **A** is treated with each reagent, forming compounds **B** and **C.** Rank **A, B,** and **C** in order of increasing melting point.

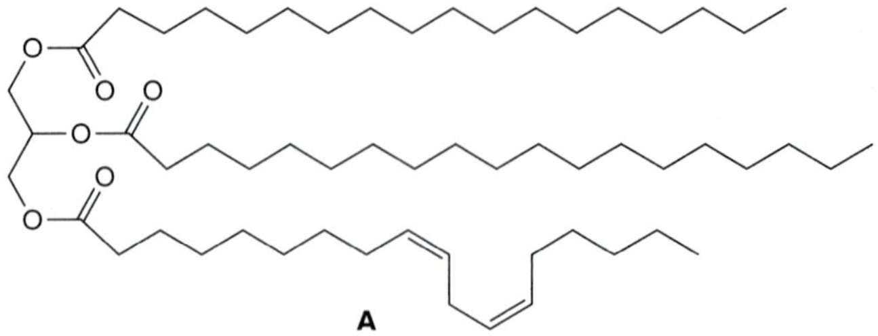

a. H_2 (excess), Pd-C (Compound **B**)
b. H_2 (1 equiv), Pd-C (Compound **C**)

A

12.5 Reduction of Alkynes

Reduction of an alkyne adds H_2 to one or both of the π bonds. There are three different ways by which the elements of H_2 can be added to a triple bond.

- **Adding two equivalents of H_2 forms an alkane.**

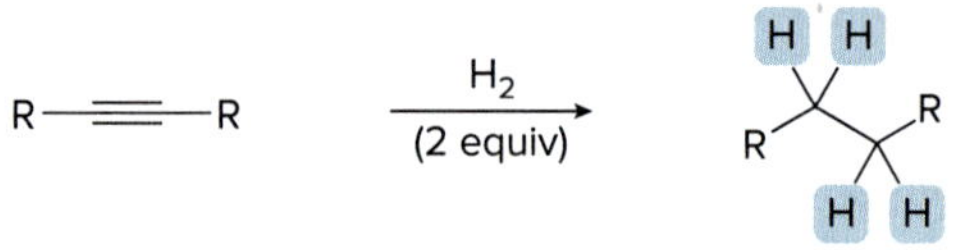

R—≡—R $\xrightarrow[\text{(2 equiv)}]{H_2}$ alkane

- **Adding one equivalent of H_2 in a syn fashion forms a cis alkene.**

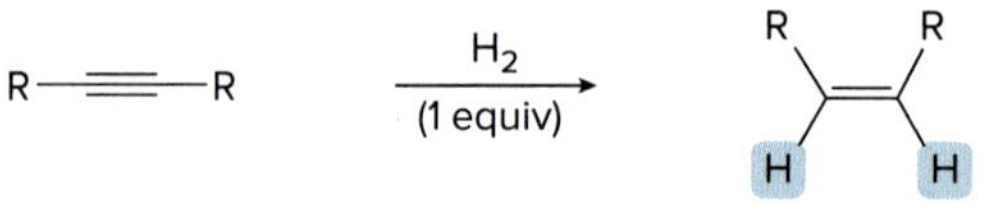

R—≡—R $\xrightarrow[\text{(1 equiv)}]{H_2}$ **cis** alkene — **syn** addition

- **Adding one equivalent of H_2 in an anti fashion forms a trans alkene.**

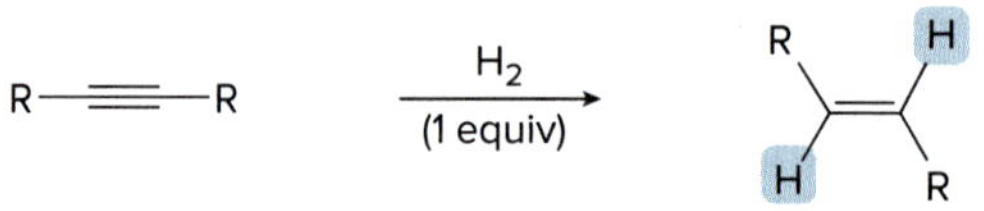

R—≡—R $\xrightarrow[\text{(1 equiv)}]{H_2}$ **trans** alkene — **anti** addition

12.5A Reduction of an Alkyne to an Alkane

When an alkyne is treated with two or more equivalents of H_2 and a Pd catalyst, reduction of *both* π bonds occurs. **Syn addition** of one equivalent of H_2 forms a cis alkene, which adds a second equivalent of H_2 to form an **alkane. Four new C–H bonds are formed.** By using a Pd-C catalyst, it is not possible to stop the reaction after addition of only one equivalent of H_2.

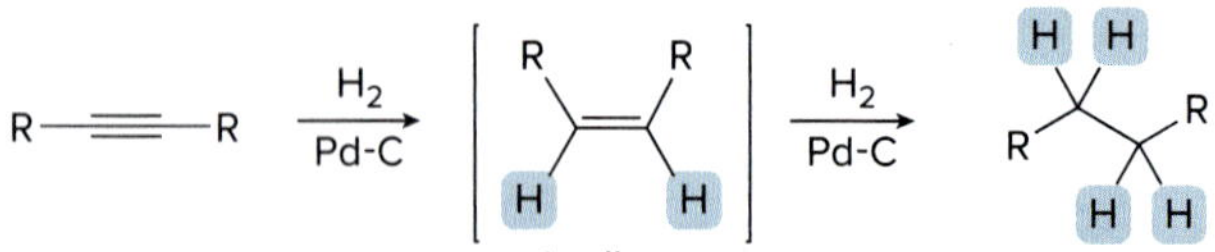

R—≡—R $\xrightarrow[\text{Pd-C}]{H_2}$ [cis alkene (not isolated)] $\xrightarrow[\text{Pd-C}]{H_2}$ **alkane**

$\xrightarrow[\text{Pd-C}]{H_2}$ [ ] $\xrightarrow[\text{Pd-C}]{H_2}$

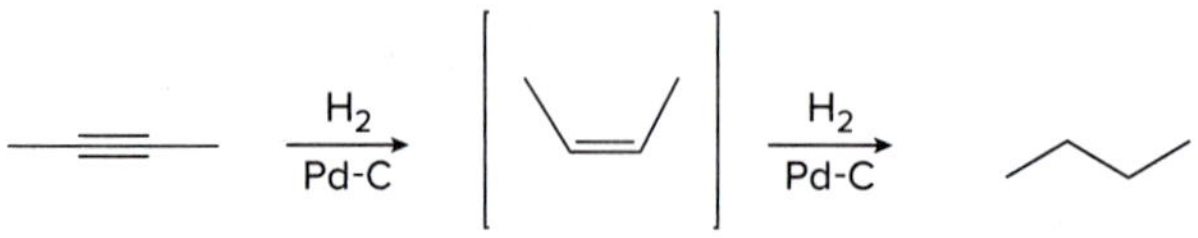

Problem 12.10 Which alkyne has the smaller heat of hydrogenation, $HC{\equiv}CCH_2CH_2CH_3$ or $CH_3C{\equiv}CCH_2CH_3$? Explain your choice.

12.5B Reduction of an Alkyne to a Cis Alkene

Palladium metal is too active a catalyst to allow the hydrogenation of an alkyne to stop after one equivalent of H_2. To prepare a cis alkene from an alkyne and H_2, a less active Pd catalyst is used—Pd adsorbed onto $CaCO_3$ with added lead(II) acetate and quinoline. This catalyst is called the **Lindlar catalyst** after the chemist who first prepared it. Compared to Pd metal, the **Lindlar catalyst is deactivated or "poisoned."**

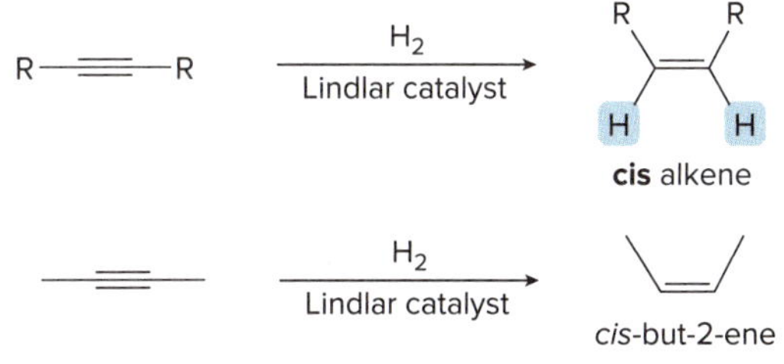

Reduction of an alkyne to a cis alkene is a **stereoselective reaction,** because only one stereoisomer is formed.

With the Lindlar catalyst, one equivalent of H_2 adds to an alkyne, and the cis alkene product is unreactive to further reduction.

R—≡—R → (H_2, Lindlar catalyst) → **cis** alkene (R, R on the same side; H, H)

but-2-yne → (H_2, Lindlar catalyst) → *cis*-but-2-ene

Jasmine flowers are the source of *cis*-jasmone, a perfume component (Problem 12.11).
Charlotte Björnström/EyeEm/ Getty Images

Problem 12.11 What is the structure of *cis*-jasmone, a natural product isolated from jasmine flowers, formed by treatment of alkyne **A** with H_2 in the presence of the Lindlar catalyst?

A → (H_2, Lindlar catalyst) → *cis*-jasmone

Problem 12.12 (a) Draw the structure of a compound of molecular formula C_6H_{10} that reacts with H_2 in the presence of Pd-C but does not react with H_2 in the presence of Lindlar catalyst. (b) Draw the structure of a compound of molecular formula C_6H_{10} that reacts with H_2 when either catalyst is present.

12.5C Reduction of an Alkyne to a Trans Alkene

Although catalytic hydrogenation is a convenient method for preparing cis alkenes from alkynes, it cannot be used to prepare trans alkenes. With a **dissolving metal reduction** (such as Na in NH_3), however, the elements of H_2 are added in an **anti** fashion to the triple bond, thus forming a **trans alkene.** For example, but-2-yne reacts with Na in NH_3 to form *trans*-but-2-ene.

R—≡—R → (Na, NH_3) → **trans** alkene (R, H / H, R)

but-2-yne → (Na, NH_3) → *trans*-but-2-ene

The **mechanism** for the dissolving metal reduction using Na in NH_3 features sequential addition of electrons and protons to the triple bond. Half-headed arrows denoting the movement of a single electron must be used in two steps when Na donates *one* electron. The mechanism can be divided conceptually into two parts, each of which consists of two steps: **addition of an electron followed by protonation of the resulting negative charge,** as shown in Mechanism 12.2.

Mechanism 12.2 Dissolving Metal Reduction of an Alkyne to a Trans Alkene

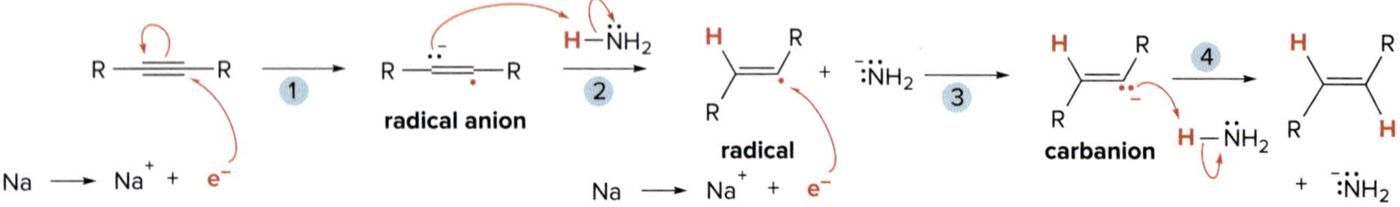

1. Addition of an electron to the triple bond forms a **radical anion,** a species that contains *both* a negative charge *and* an unpaired electron.
2. Protonation of the anion with the solvent NH_3 yields a **radical.** The net result of the first two steps is the addition of a H atom.
3. Addition of a second electron forms a **carbanion.**
4. Protonation of the carbanion forms the **trans alkene.** Steps [3] and [4] add the second H atom to the triple bond.

Although the vinyl carbanion formed in Step [3] could have two different arrangements of its R groups, only the trans alkene is formed from the more stable vinyl carbanion; this carbanion has the larger R groups farther away from each other to avoid steric interactions. Protonation of this anion leads to the more stable trans product.

The larger R groups are farther away from each other.

This **more stable vinyl carbanion** forms the trans alkene.

Steric interactions between closer R groups **destabilize** this carbanion.

Dissolving metal reduction of a triple bond with Na in NH_3 is a **stereoselective reaction** because it forms a trans product exclusively.

- **Dissolving metal reductions always form the more stable trans product preferentially.**

The three methods to reduce a triple bond are summarized in Figure 12.5 using hex-3-yne as starting material.

Figure 12.5 Summary: Three methods to reduce a triple bond

hex-3-yne

2 H_2, Pd-C → hexane

H_2, Lindlar catalyst → *cis*-hex-3-ene

Na, NH_3 → *trans*-hex-3-ene

Finally, terminal alkynes can be reduced to terminal alkenes with H_2 and Lindlar catalyst but *not* with Na and NH_3. The reagent Na in NH_3 removes the acidic proton of a terminal alkyne to form an acetylide anion, which is not reduced to an alkene.

H_2, Lindlar catalyst → terminal alkene

terminal alkyne

Na, NH_3 → acetylide anion ✗→ terminal alkene

Problem 12.13 What product is formed when $CH_3OCH_2CH_2C{\equiv}CCH_2CH(CH_3)_2$ is treated with each reagent: (a) H_2 (excess), Pd-C; (b) H_2 (1 equiv), Lindlar catalyst; (c) H_2 (excess), Lindlar catalyst; (d) Na, NH_3?

Problem 12.14 A chiral alkyne **A** with molecular formula C_6H_{10} is reduced with H_2 and Lindlar catalyst to **B** having the *R* configuration at its stereogenic center. What are the structures of **A** and **B?**

12.6 The Reduction of Polar C–X σ Bonds

Compounds containing polar C–X σ bonds that react with strong nucleophiles are reduced with metal hydride reagents, most commonly lithium aluminum hydride. Two functional groups possessing both of these characteristics are **alkyl halides** and **epoxides.** Alkyl halides are reduced to alkanes with loss of X^- as the leaving group. Epoxide rings are opened to form alcohols.

R–X (alkyl halide) → [1] $LiAlH_4$ [2] H_2O → R–H (alkane)

epoxide → [1] $LiAlH_4$ [2] H_2O → alcohol

Reduction of these C–X σ bonds is another example of nucleophilic substitution, in which $LiAlH_4$ serves as a source of a hydride nucleophile (H^-). Because **H^- is a strong nucleophile,** the reaction follows an **S_N2 mechanism,** illustrated for the one-step reduction of an alkyl halide in Mechanism 12.3.

Mechanism 12.3 Reduction of RX with $LiAlH_4$

RCH_2X + $H–\bar{A}lH_3$ Li^+ → RCH_2H + Li^+X^- + AlH_3

$LiAlH_4$ donates H^-.

- The nucleophile H^- replaces the leaving group X^- in a single step.

Because the reaction follows an S_N2 mechanism:

- **Unhindered CH_3X and 1° alkyl halides are more easily reduced than more substituted 2° and 3° halides.**
- **In unsymmetrical epoxides, nucleophilic attack of H^- (from $LiAlH_4$) occurs at the *less* substituted carbon atom.**

Examples are shown in Figure 12.6.

Figure 12.6
Examples of reduction of C–X σ bonds with $LiAlH_4$

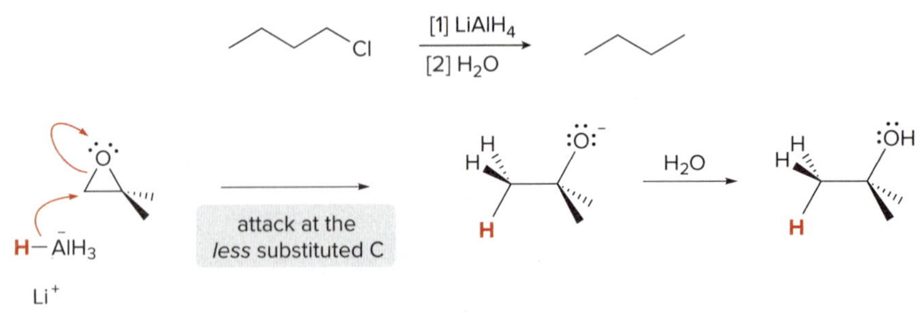

Problem 12.15 Draw the products of each reaction.

a. 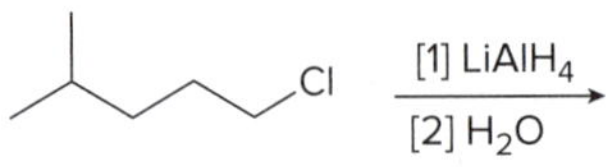 Cl [1] $LiAlH_4$ [2] H_2O

b. 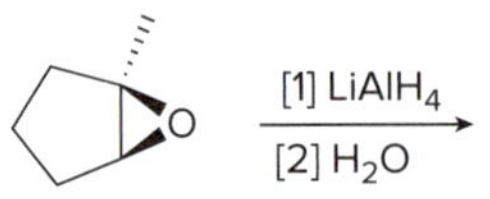 O [1] $LiAlH_4$ [2] H_2O

12.7 Oxidizing Agents

Oxidizing agents fall into two main categories:

- **Reagents that contain an oxygen–oxygen bond**
- **Reagents that contain metal–oxygen bonds**

Oxidizing agents containing an O–O bond include $\mathbf{O_2}$, $\mathbf{O_3}$ (ozone), $\mathbf{H_2O_2}$ (hydrogen peroxide), $\mathbf{(CH_3)_3COOH}$ (*tert*-butyl hydroperoxide), and peroxyacids. **Peroxyacids,** a group of reagents with the general structure $\mathbf{RCO_3H}$, have one more O atom than carboxylic acids (RCO_2H). Some peroxyacids are commercially available whereas others are prepared and used without isolation. Two common peroxyacids are peroxyacetic acid and *meta*-chloroperoxybenzoic acid, abbreviated as **mCPBA.**

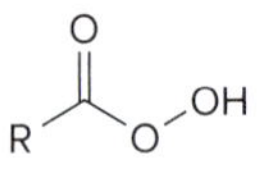

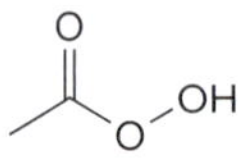

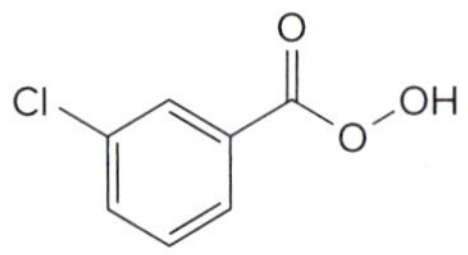

The most common oxidizing agents with metal–oxygen bonds contain either chromium in the +6 oxidation state (six Cr–O bonds) or manganese in the +7 oxidation state (seven Mn–O bonds). Common Cr^{6+} reagents include chromium(VI) oxide **($\mathbf{CrO_3}$)** and sodium or potassium dichromate **($\mathbf{Na_2Cr_2O_7}$ and $\mathbf{K_2Cr_2O_7}$). These reagents are strong oxidants** used in the presence of a strong aqueous acid such as H_2SO_4. **Pyridinium chlorochromate (PCC),** a Cr^{6+} reagent that is soluble in halogenated organic solvents, can be used without strong acid present. This makes it a **more selective $\mathbf{Cr^{6+}}$ oxidant,** as described in Section 12.12.

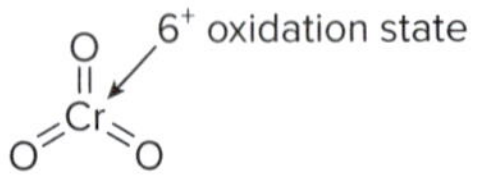

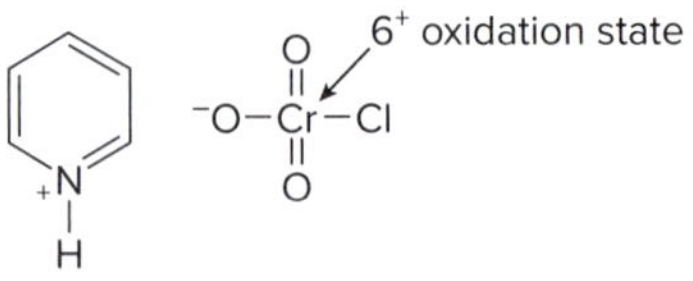

The most common Mn^{7+} reagent is **$KMnO_4$** (potassium permanganate), a strong, water-soluble oxidant. Other oxidizing agents that contain metals include **OsO_4** (osmium tetroxide) and **Ag_2O** [silver(I) oxide].

In the remainder of Chapter 12, the oxidation of alkenes, alkynes, and alcohols—three functional groups already introduced in this text—is presented (Figure 12.7). Addition reactions to alkenes and alkynes that increase the number of C–O bonds are described in Sections 12.8–12.11. Oxidation of alcohols to carbonyl compounds appears in Sections 12.12–12.14.

Figure 12.7
Oxidation reactions of alkenes, alkynes, and alcohols

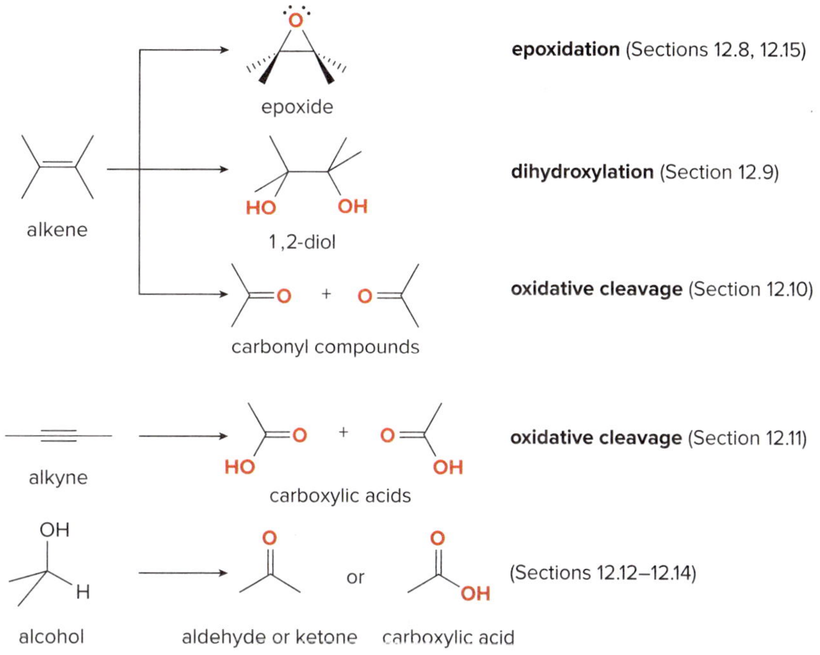

12.8 Epoxidation

Epoxidation is the addition of a single oxygen atom to an alkene to form an **epoxide.**

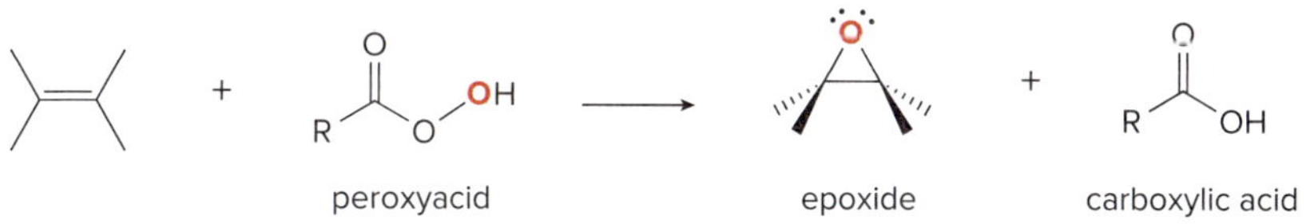

The weak π bond of the alkene is broken and two new C–O σ bonds are formed. Epoxidation is typically carried out with a peroxyacid, resulting in cleavage of the weak O–O bond of the reagent.

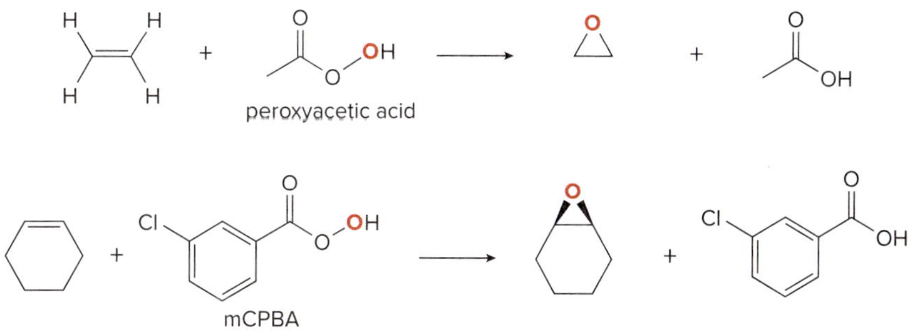

Epoxidation occurs via the concerted addition of one oxygen atom of the peroxyacid to the π bond as shown in Mechanism 12.4. Epoxidation resembles the formation of the bridged halonium ion in Section 10.13, in that two bonds in a three-membered ring are formed in one step.

Mechanism 12.4 Epoxidation of an Alkene with a Peroxyacid

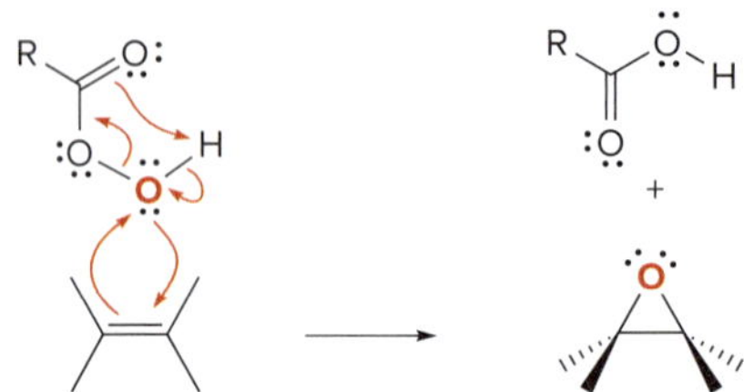

- **All bonds are broken and formed in a single step.** The two epoxide C–O bonds are formed from one electron pair of the π bond and one lone pair of the peroxyacid. The **weak O–O bond is broken.**

Problem 12.16 What epoxide is formed when each alkene is treated with mCPBA?

a. b. c.

12.8A The Stereochemistry of Epoxidation

Epoxidation occurs via **syn addition** of an O atom from either side of the planar double bond, so that both C–O bonds are formed on the same side. The relative position of substituents in the alkene reactant is **retained** in the epoxide product.

- **A cis alkene gives an epoxide with cis substituents. A trans alkene gives an epoxide with trans substituents.**

Epoxidation is a **stereospecific** reaction because cis and trans alkenes yield different stereoisomers as products, as illustrated in Sample Problem 12.2.

Sample Problem 12.2 Drawing the Stereoisomers Formed in Epoxidation

Draw the stereoisomers formed when *cis*- and *trans*-but-2-ene are epoxidized with mCPBA.

Solution

To draw each product of epoxidation, add an O atom from either side of the alkene, and keep all substituents in their *original* orientations. The **cis** methyl groups in *cis*-but-2-ene become **cis** substituents in the epoxide. Addition of an O atom from either side of the trigonal planar alkene leads to the same compound—an **achiral meso compound that contains two stereogenic centers,** labeled in blue.

H C=C H, CH_3 CH_3 —mCPBA→ O + O

cis CH_3 groups
cis-but-2-ene

cis CH_3 groups
O added from above

cis CH_3 groups
O added from below

Products are identical, an **achiral meso compound.**

Epoxidation of *cis*- and *trans*-but-2-ene illustrates the general rule about the stereochemistry of reactions: **an achiral starting material gives achiral or racemic products.**

The **trans** methyl groups in *trans*-but-2-ene become **trans** substituents in the epoxide. Addition of an O atom from either side of the trigonal planar alkene yields an equal mixture of two enantiomers—a **racemic mixture**—with two stereogenic centers labeled in blue.

CH_3, H, H, CH_3 — trans CH_3 groups — *trans*-but-2-ene

mCPBA →

trans CH_3 groups O added from above + **trans CH_3 groups** O added from below

Products are **enantiomers.**

Problem 12.17 Draw all stereoisomers formed when each alkene is treated with mCPBA.

a. b. c. d. e.

More Practice: Try Problems 12.32b; 12.37d, k; 12.38b; 12.41e, f; 12.44a.

12.8B The Synthesis of Disparlure

Disparlure has been used to control the spread of the gypsy moth caterpillar, a pest that has periodically devastated forests in the northeastern United States by defoliating many shade and fruit-bearing trees. The active pheromone is placed in a trap containing a poison or sticky substance, and the male moth is lured to the trap by the pheromone. Alternatively, thousands of disparlure-baited traps are placed along the edges of infestation. When the pheromone permeates the air, males are confused and can't locate individual females, so that mating is disrupted.
Safwan Abd Rahman/Shutterstock

Disparlure, the sex pheromone of the female gypsy moth, is synthesized by a stepwise reaction sequence that uses an epoxidation reaction as the final step. Retrosynthetic analysis of disparlure illustrates three key operations:

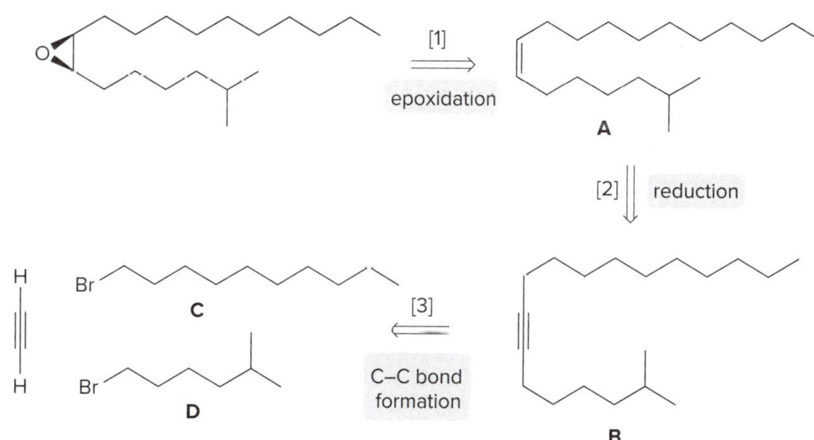

- **Step [1]** The cis epoxide in disparlure is prepared from a cis alkene **A** by epoxidation.
- **Step [2]** **A** is prepared from an internal alkyne **B** by reduction.
- **Step [3]** **B** is prepared from acetylene and two 1° alkyl halides (**C** and **D**) by using S_N2 reactions with acetylide anions.

Figure 12.8 illustrates the synthesis of disparlure beginning with acetylene.

Epoxidation of the cis alkene **A** (Part [3], Figure 12.8) from two different sides of the double bond affords two cis epoxides in the last step—a racemic mixture of two enantiomers. Thus, half of the product is the desired pheromone disparlure, but the other half is its biologically inactive enantiomer. Separating the desired from the undesired enantiomer is difficult and expensive, because both compounds have identical physical properties. A reaction that affords a chiral epoxide from an achiral precursor without forming a racemic mixture is discussed in Section 12.15.

How to separate a racemic mixture into its component enantiomers is discussed in Section 27.2.

Figure 12.8 The synthesis of disparlure

Part [1] Formation of two C–C bonds using acetylide anions (Section 11.11A)

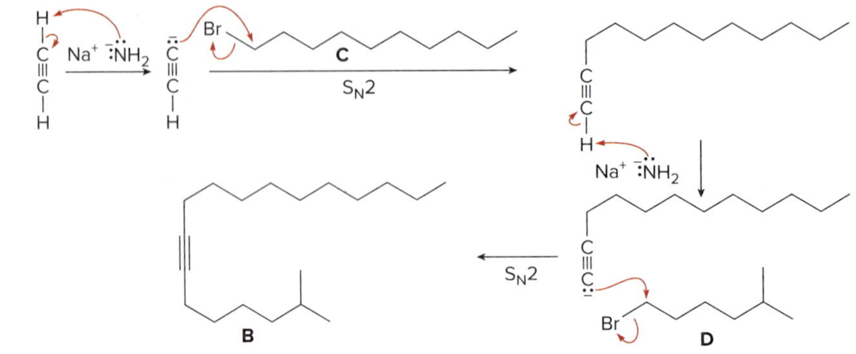

Part [2] Reduction of alkyne **B** to form cis alkene **A** (Section 12.5B)

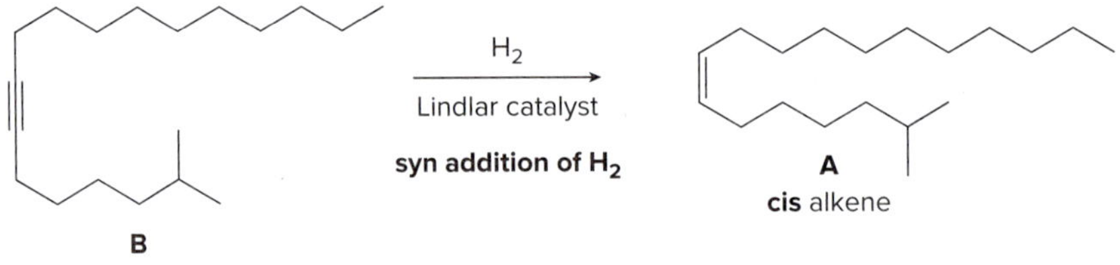

Part [3] Epoxidation of **A** to form disparlure (Section 12.8)

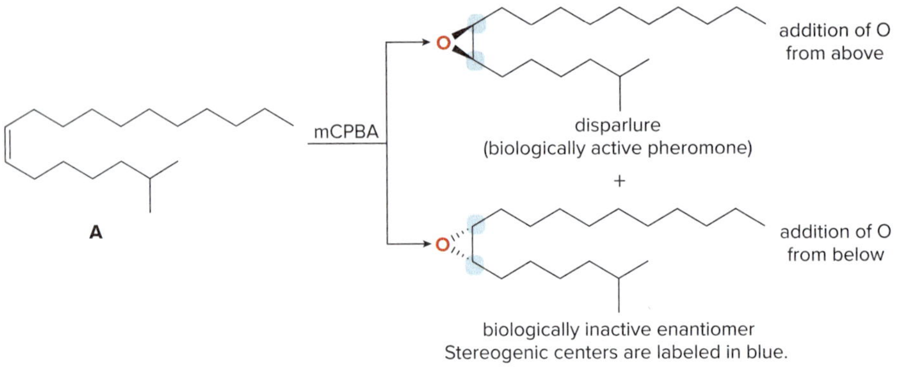

Stereogenic centers are labeled in blue.

The synthesis is conceptually divided into three parts:

- **Part [1]** Acetylene is converted to an internal alkyne **B** by forming two C—C bonds. Each bond is formed by treating an alkyne with base ($NaNH_2$) to form an acetylide anion, which reacts with an alkyl halide (**C** or **D**) in an S_N2 reaction (Section 11.11A).
- **Part [2]** The internal alkyne **B** is reduced to a cis alkene **A** by syn addition of H_2 using the Lindlar catalyst (Section 12.5B).
- **Part [3]** The cis alkene **A** is epoxidized to disparlure using a peroxyacid such as mCPBA.

12.9 Dihydroxylation

Dihydroxylation is the addition of two hydroxy groups to a double bond, forming a **1,2-diol** or **glycol.** Depending on the reagent, the two new OH groups can be added to the opposite sides (**anti** addition) or the same side (**syn** addition) of the double bond.

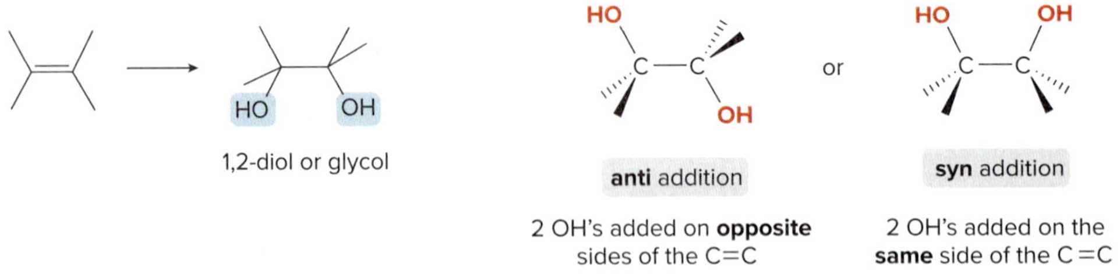

12.9A Anti Dihydroxylation

Anti dihydroxylation is achieved in two steps—epoxidation followed by opening of the ring with ^-OH or H_2O. Cyclohexene, for example, is converted to a racemic mixture of two *trans*-cyclohexane-1,2-diols by anti addition of two OH groups.

[1] RCO_3H
[2] H_2O (H^+ or ^-OH)

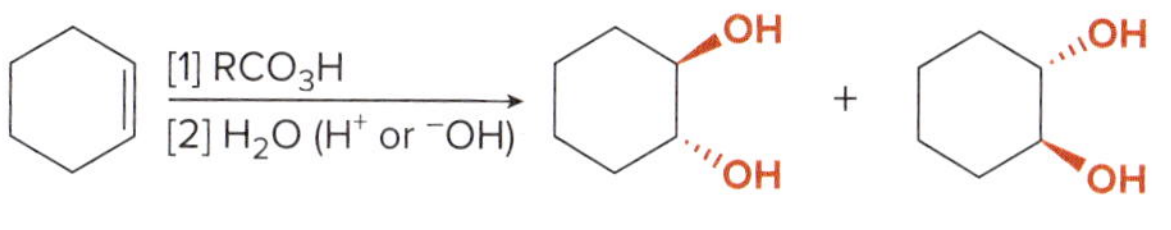

trans-1,2-diols
enantiomers

The stereochemistry of the products can be understood by examining the stereochemistry of each step.

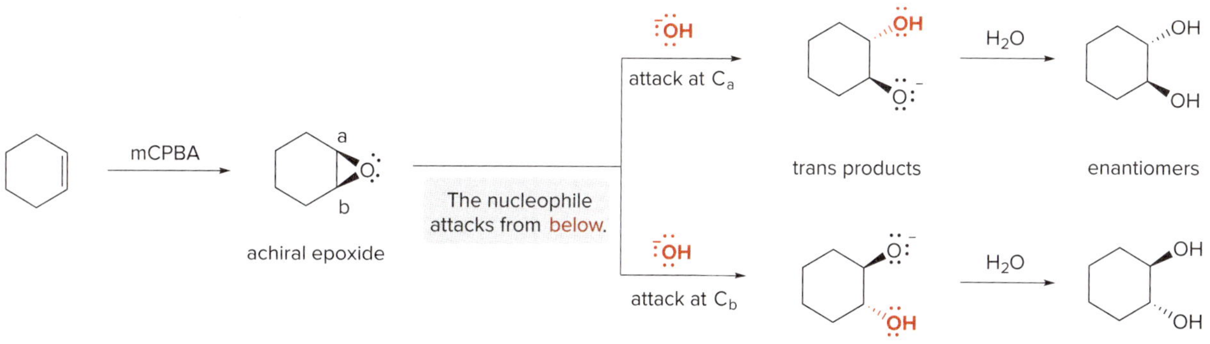

mCPBA
achiral epoxide
The nucleophile attacks from below.
attack at C_a
attack at C_b
trans products
H_2O
enantiomers

Epoxidation of cyclohexene adds an O atom from either above or below the plane of the double bond to form a single **achiral epoxide,** so only one representation is shown. Opening of the epoxide ring then occurs with **backside attack at either C–O bond.** Because the epoxide is drawn above the plane of the six-membered ring, nucleophilic attack occurs from **below** the plane. This reaction is a specific example of the opening of epoxide rings with strong nucleophiles, first presented in Section 9.16A.

Because one OH group of the 1,2-diol comes from the epoxide and one OH group comes from the nucleophile (^-OH), the overall result is **anti addition of two OH groups** to an alkene.

Problem 12.18 Draw the products formed when both *cis*- and *trans*-but-2-ene are treated with a peroxyacid followed by ^-OH (in H_2O). Explain how these reactions illustrate that anti dihydroxylation is stereospecific.

12.9B Syn Dihydroxylation

Syn dihydroxylation results when an alkene is treated with either $\mathbf{KMnO_4}$ or $\mathbf{OsO_4}$.

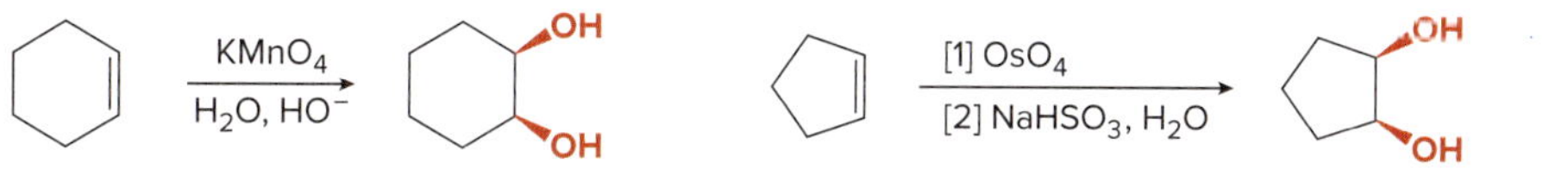

$KMnO_4$
H_2O, HO^-
cis-cyclohexane-1,2-diol

[1] OsO_4
[2] $NaHSO_3$, H_2O
cis-cyclopentane-1,2-diol

Each reagent adds two oxygen atoms to the same side of the double bond—that is, in a **syn** fashion—to yield a cyclic intermediate. Hydrolysis of the cyclic intermediate cleaves the metal–oxygen bonds, forming the *cis*-1,2-diol. With OsO_4, sodium bisulfite ($NaHSO_3$) is also added in the hydrolysis step.

Two O atoms are added to the **same** side of the C=C.

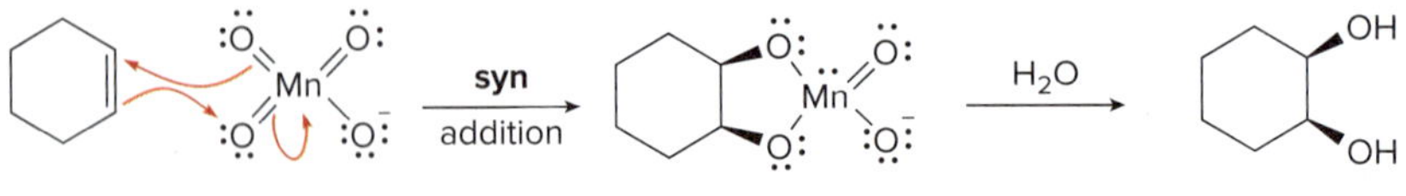

syn addition

H_2O

***cis*-1,2-diol**

syn addition

$NaHSO_3$

H_2O

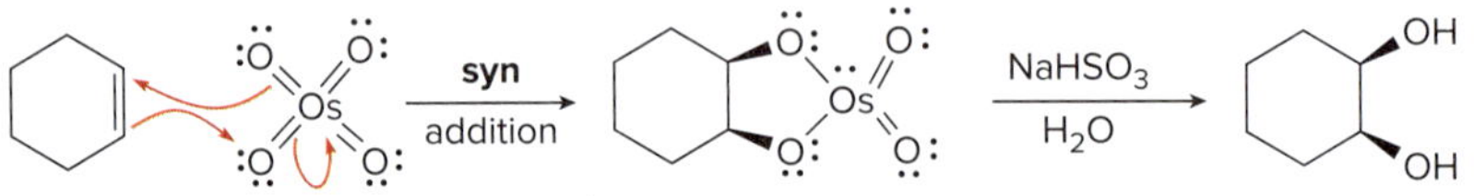

Although $KMnO_4$ is inexpensive and readily available, its use is limited by its insolubility in organic solvents. To prevent further oxidation of the product 1,2-diol, the reaction mixture must be kept basic with added ^-OH.

Although OsO_4 is a more selective oxidant than $KMnO_4$ and is soluble in organic solvents, it is toxic and expensive. To overcome these limitations, dihydroxylation can be carried out by using a *catalytic* amount of OsO_4, if the oxidant ***N*-methylmorpholine *N*-oxide (NMO)** is also added.

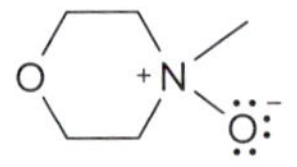

NMO is an **amine oxide.** It is not possible to draw a Lewis structure of an amine oxide having only neutral atoms.

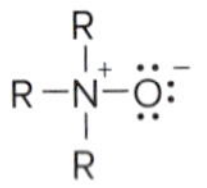

amine oxide

N-methylmorpholine *N*-oxide

NMO

In the catalytic process, dihydroxylation of the double bond converts the Os^{8+} oxidant into an Os^{6+} product, which is then re-oxidized by NMO to Os^{8+}. This Os^{8+} reagent can then be used for dihydroxylation once again, and the catalytic cycle continues.

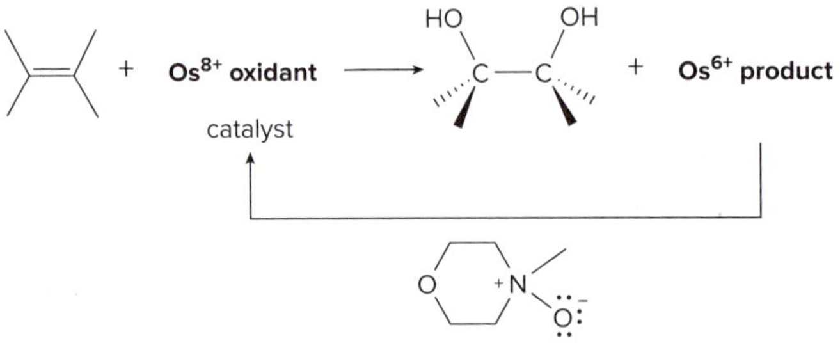

+ **Os^{8+} oxidant** → + **Os^{6+} product**

catalyst

NMO oxidizes the **Os^{6+} product** back to **Os^{8+}** to begin the cycle again.

Problem 12.19 Draw the products formed when both *cis*- and *trans*-but-2-ene are treated with OsO_4, followed by hydrolysis with $NaHSO_3 + H_2O$. Explain how these reactions illustrate that syn dihydroxylation is stereospecific.

Problem 12.20 Draw all stereoisomers formed in each reaction.

a. 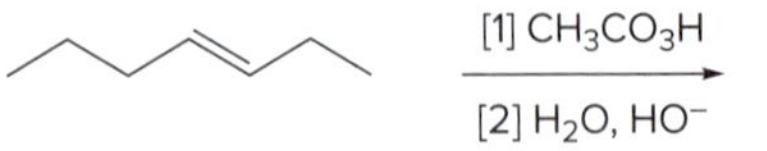 [1] CH_3CO_3H [2] H_2O, HO^-

b. 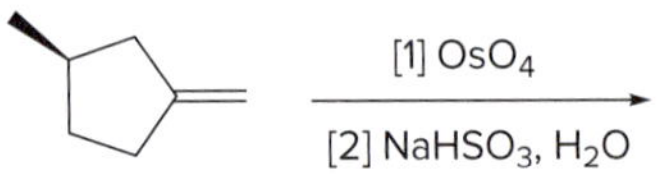 [1] OsO_4 [2] $NaHSO_3$, H_2O

c. 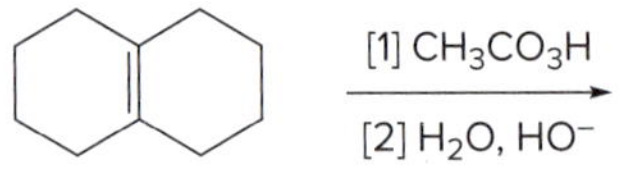 [1] CH_3CO_3H [2] H_2O, HO^-

d. 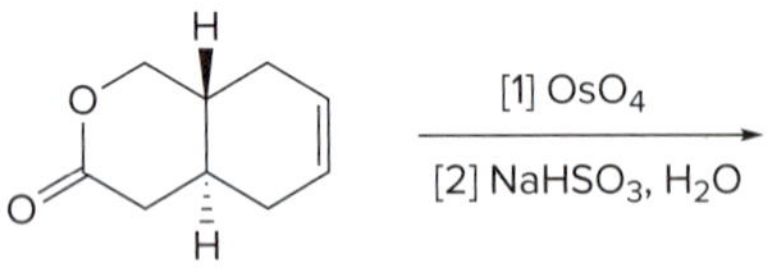 [1] OsO_4 [2] $NaHSO_3$, H_2O

12.10 Oxidative Cleavage of Alkenes

Lightning produces O_3 from O_2 during an electrical storm. Moreover, the pungent odor around a heavily used photocopy machine is O_3 produced from O_2 during the process. O_3 at ground level is an unwanted atmospheric pollutant. In the stratosphere, however, it protects us from harmful ultraviolet radiation, as discussed in Chapter 13.
Balazs Kovacs/Getty Images

Oxidative cleavage of an alkene breaks both the σ and π bonds of the double bond to form two carbonyl groups. Depending on the number of R groups bonded to the double bond, oxidative cleavage yields either **ketones** or **aldehydes.**

[O]

The σ and π bonds are broken. — ketone — aldehyde

One method of oxidative cleavage relies on a two-step procedure using **ozone (O_3) as the oxidant** in the first step. Cleavage with ozone is called **ozonolysis.**

[1] O_3 / [2] Zn, H_2O — ketones — aldehydes

[1] O_3 / [2] CH_3SCH_3

Addition of ozone to the π bond of the alkene forms an unstable intermediate called a **molozonide,** which then rearranges to an **ozonide** by a stepwise process. The unstable ozonide is then reduced without isolation to afford carbonyl compounds. **Zn (in H_2O)** and **dimethyl sulfide (CH_3SCH_3)** are two common reagents used to convert the ozonide to carbonyl compounds.

molozonide — ozonide — Zn, H_2O or CH_3SCH_3 — by-product: $Zn(OH)_2$ or $(CH_3)_2S{=}O$

To draw the product of any oxidative cleavage:

- **Locate all π bonds in the molecule.**
- **Replace each C=C by two C=O bonds.**

Sample Problem 12.3 Drawing the Oxidative Cleavage Products from an Alkene

Draw the products when each alkene is treated with O_3 followed by CH_3SCH_3.

a. b.

Solution

a. Cleave the double bond and replace it with two carbonyl groups.

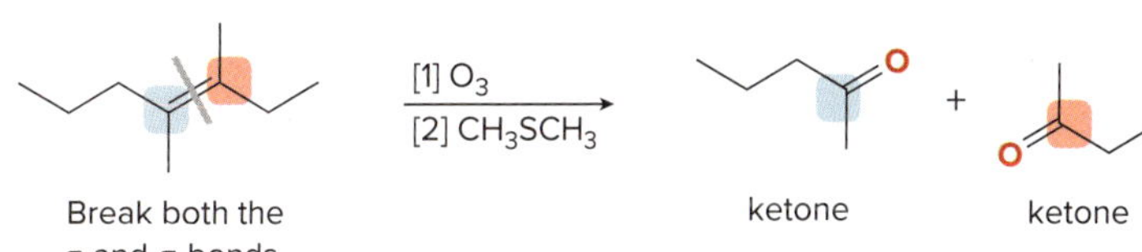

b. For a cycloalkene, oxidative cleavage results in a **single molecule with two carbonyl groups—a dicarbonyl compound.**

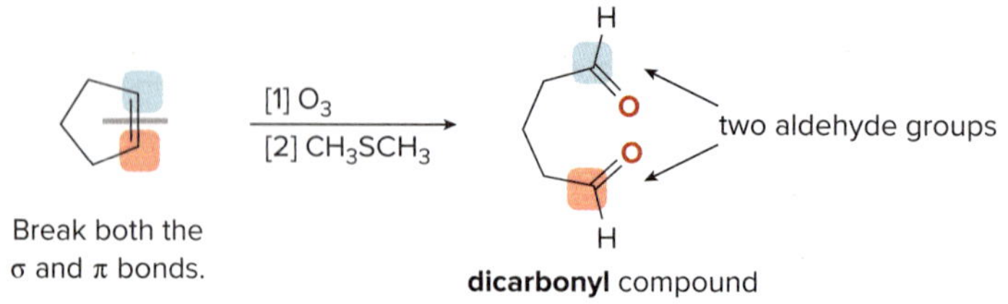

[1] O_3
[2] CH_3SCH_3

two aldehyde groups

Break both the σ and π bonds.

dicarbonyl compound

Problem 12.21 Draw the products formed when each compound is treated with O_3 followed by Zn, H_2O.

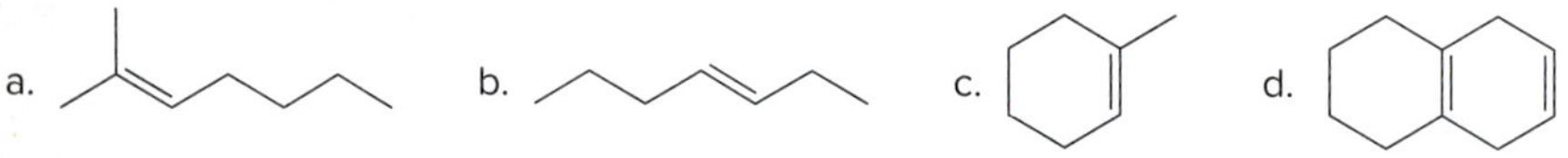

a. b. c. d.

More Practice: Try Problems 12.37i; 12.39c; 12.40a, b; 12.48a, b; 12.51.

Ozonolysis of dienes (and other polyenes) results in oxidative cleavage of all C=C bonds. The number of carbonyl groups formed in the products is *twice* the number of double bonds in the starting material. The *two* double bonds in limonene are converted to products containing *four* carbonyl groups.

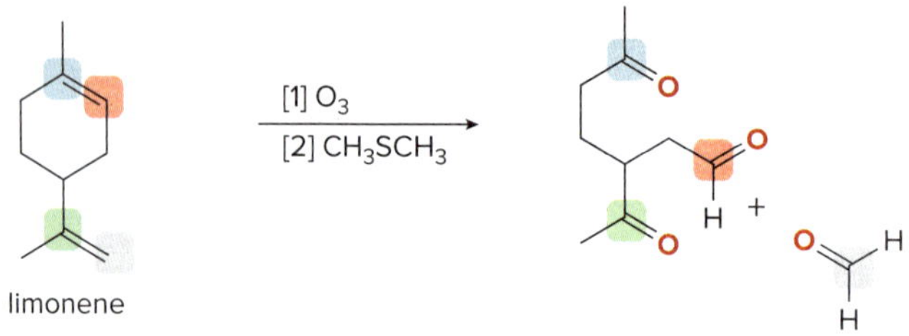

[1] O_3
[2] CH_3SCH_3

limonene

Oxidative cleavage is a valuable tool for structure determination of unknown compounds. The ability to determine what alkene gives rise to a particular set of oxidative cleavage products is thus a useful skill, illustrated in Sample Problem 12.4.

Sample Problem 12.4 Determining the Alkene That Forms a Set of Oxidative Cleavage Products

What alkene forms the following products after reaction with O_3 followed by CH_3SCH_3?

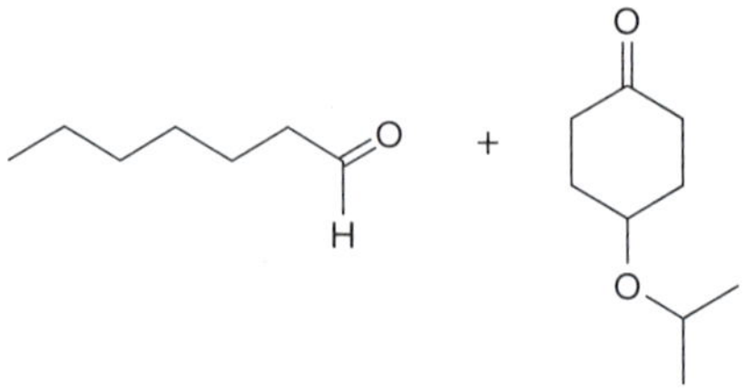

Solution

To draw the starting material, **ignore the O atoms** in the carbonyl groups and **join the carbonyl carbons together by a C=C.**

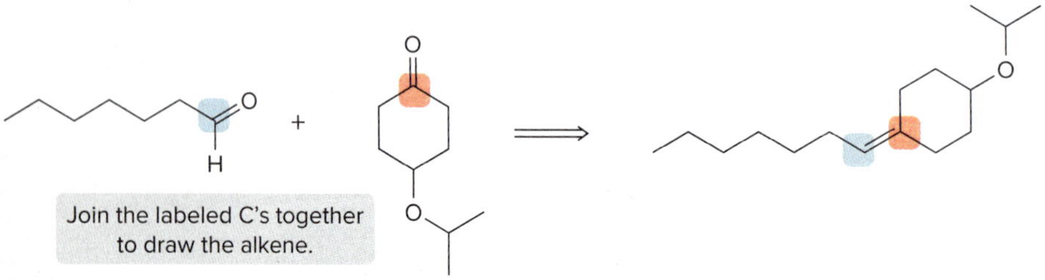

Join the labeled C's together to draw the alkene.

Problem 12.22 What alkene yields each set of oxidative cleavage products?

a. + b. + c. only d. (from $C_{10}H_{18}$)

More Practice: Try Problems 12.45b; 12.49a, b; 12.50.

Problem 12.23 Draw the products formed when cembrene A, which is isolated from soft corals of the genus *Naphthea,* is treated with O_3 followed by CH_3SCH_3. Label each product as chiral or achiral.

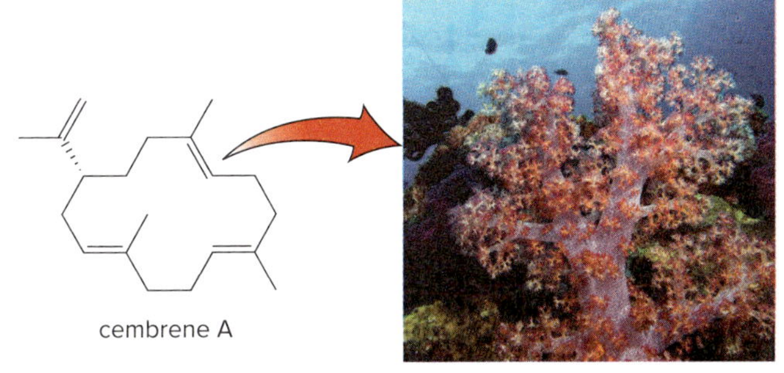

cembrene A

soft coral
magnusdeepbelow/Shutterstock

Problem 12.24 (a) Translate the ball-and-stick model of biformene, a lipid isolated from amber, into a skeletal structure, using wedges and dashed wedges for any stereogenic centers. (b) Draw the products formed when biformene is treated with O_3 followed by CH_3SCH_3.

biformene

amber
Hjochen/Shutterstock

12.11 Oxidative Cleavage of Alkynes

Alkynes also undergo oxidative cleavage of the σ bond and both π bonds of the triple bond. Internal alkynes are oxidized to **carboxylic acids (RCOOH),** whereas terminal alkynes afford carboxylic acids and $\mathbf{CO_2}$ from the *sp* hybridized C–H bond.

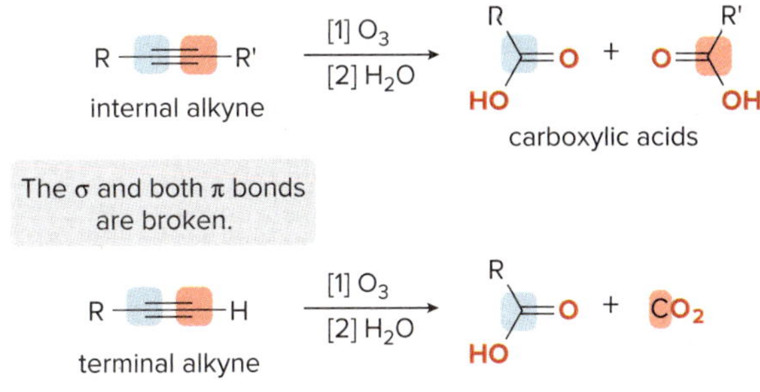

Oxidative cleavage is commonly carried out with O_3, followed by cleavage of the intermediate ozonide with H_2O.

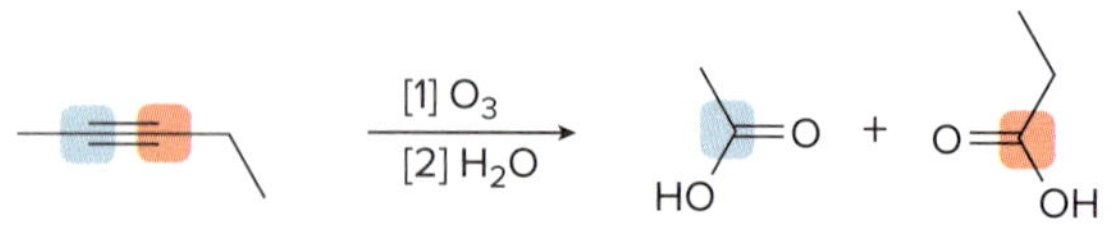

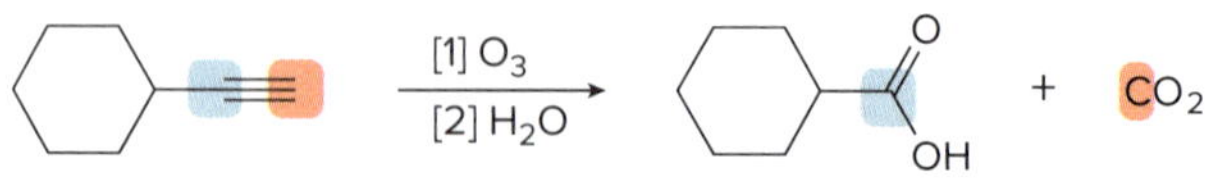

[1] O_3
[2] H_2O
HO + O= OH

[1] O_3
[2] H_2O
O OH + CO_2

Problem 12.25 Draw the products formed when each alkyne or diyne is treated with O_3 followed by H_2O.

a. 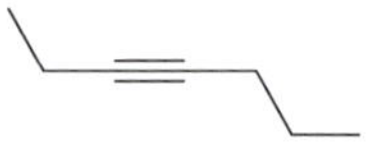 b. 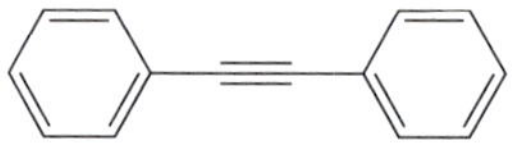 c.

Problem 12.26 What alkyne (or diyne) yields each set of oxidative cleavage products?

a. 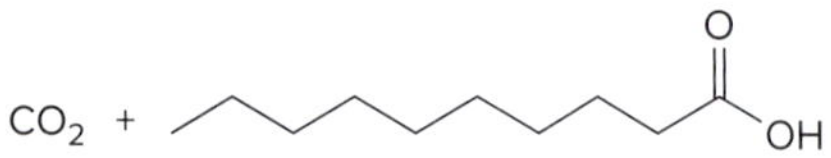 CO_2 + O OH

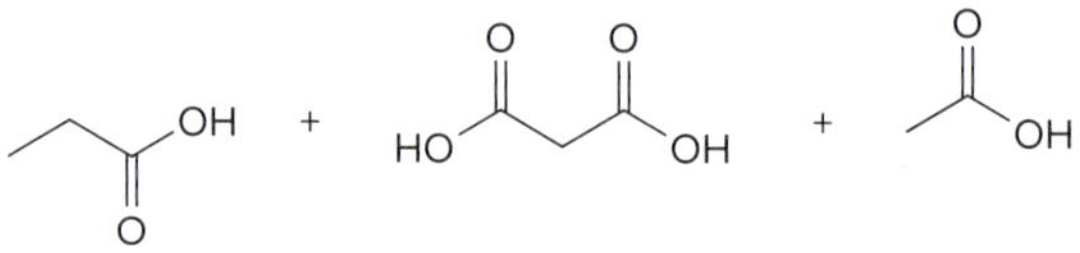

b. 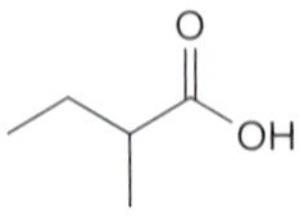 O OH only

c. OH O + HO O O OH + O OH

d. 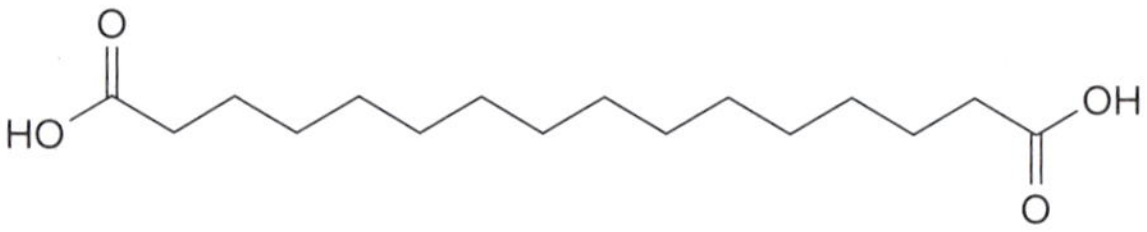 O HO OH O

12.12 Oxidation of Alcohols

Alcohols are oxidized to a variety of carbonyl compounds, depending on the type of alcohol and reagent. Oxidation occurs by replacing the C–H bonds *on the carbon bearing the OH group* by C–O bonds.

- **1° Alcohols** are oxidized to either **aldehydes** or **carboxylic acids** by replacing either one or two C–H bonds by C–O bonds.

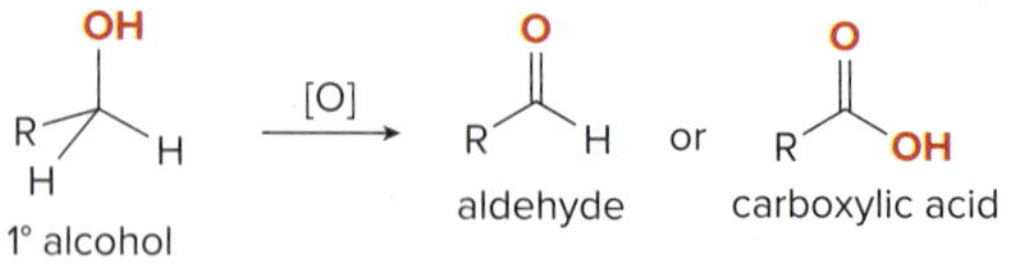

OH R H H → [O] → O R H (aldehyde) or O R OH (carboxylic acid)

1° alcohol
2 C–H bonds

- **2° Alcohols** are oxidized to **ketones** by replacing the one C–H bond by a C–O bond.

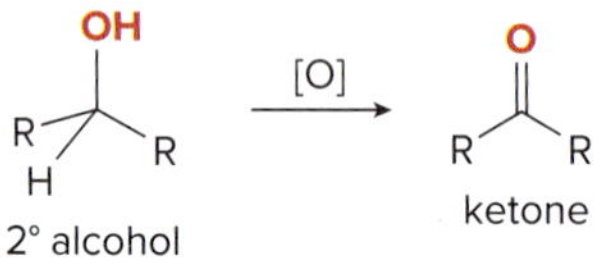

OH R R H → [O] → O R R (ketone)

2° alcohol
1 C–H bond

- **3° Alcohols have no H atoms on the carbon with the OH group,** so they are *not* easily **oxidized.**

R, R, R, OH —[O]→ NO REACTION

3° alcohol

Alcohol oxidations often occur by a pathway that involves bonding a leaving group Z to the oxygen, where Z is typically a metal in a high oxidation state. Elimination with a base then forms a C=O and a metal in a lower oxidation state.

H O–H → B: H O–Z → O + Z:⁻ + HB⁺

alcohol — carbonyl compound

The oxidation of alcohols to carbonyl compounds is typically carried out with Cr^{6+} oxidants. Cr^{6+} oxidations are characterized by a color change, as the **red-orange Cr^{6+} reagent is reduced to Cr^{3+}.**

- **CrO_3, $Na_2Cr_2O_7$, and $K_2Cr_2O_7$** are **strong, nonselective oxidants** used in aqueous acid (H_2SO_4 + H_2O).
- **PCC** (Section 12.7) is soluble in CH_2Cl_2 (dichloromethane), and can be used without strong acid present, making it a **more selective, milder oxidant.**

12.12A Oxidation of 2° Alcohols

Any of the Cr^{6+} oxidants effectively oxidizes 2° alcohols to ketones.

OH —$K_2Cr_2O_7$, H_2SO_4, H_2O→ O

OH —PCC→ O

2° alcohol — ketone

The mechanism for alcohol oxidation has two key parts: **formation of a chromate ester** and **loss of a proton.** Mechanism 12.5 is drawn for the oxidation of a general 2° alcohol with CrO_3.

Mechanism 12.5 Oxidation of an Alcohol with CrO_3

2° alcohol + CrO_3 —1→ —2 proton transfer→ chromate ester + H_2O: —3→ ketone + $HCrO_3^-$ + H_3O^+

1 – 2 Nucleophilic attack of the alcohol on the electrophilic metal (Cr^{6+} oxidation state) followed by proton transfer forms a **chromate ester.**

3 A base removes a proton and the electron pair in the C–H bond forms the **new π bond** of the C=O. Carbon is oxidized because the **number of C–O bonds increases,** and **Cr^{6+} is reduced to Cr^{4+}.**

These three steps convert the Cr^{6+} oxidant to a Cr^{4+} product, which is then further reduced to a Cr^{3+} product by a series of steps.

12.12B Oxidation of 1° Alcohols

1° Alcohols are oxidized to either aldehydes or carboxylic acids, depending on the reagent.

- **1° Alcohols are oxidized to aldehydes (RCHO) under mild reaction conditions—using PCC in CH_2Cl_2.**
- **1° Alcohols are oxidized to carboxylic acids (RCOOH) under harsher reaction conditions: $Na_2Cr_2O_7$, $K_2Cr_2O_7$, or CrO_3 in the presence of H_2O and H_2SO_4.**

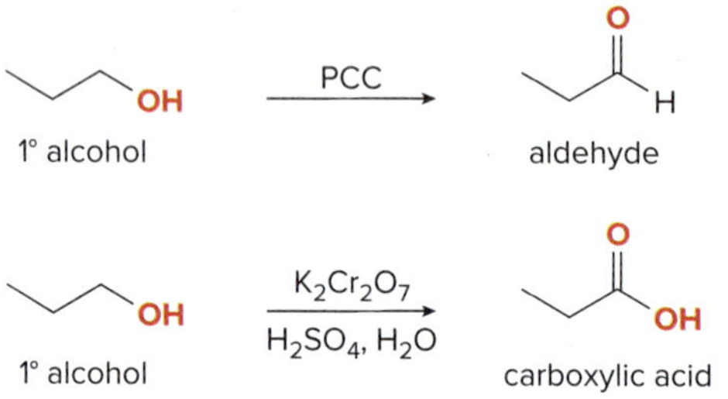

1° alcohol —PCC→ aldehyde

1° alcohol —$K_2Cr_2O_7$, H_2SO_4, H_2O→ carboxylic acid

The mechanism for the oxidation of 1° alcohols to aldehydes parallels the oxidation of 2° alcohols to ketones detailed in Section 12.12A. Oxidation of a 1° alcohol to a carboxylic acid requires three operations: **oxidation first to the aldehyde, reaction with water,** and then further **oxidation to the carboxylic acid,** as shown in Mechanism 12.6.

Mechanism 12.6 Oxidation of a 1° Alcohol to a Carboxylic Acid

1° alcohol —Part 1 [Mechanism 12.5]→ aldehyde + $HCrO_3^-$ —Part 2 [Mechanism 18.8]→ hydrate —Part 3 two steps→ chromate ester + H_2O: → carboxylic acid + $HCrO_3^-$ + H_3O^+

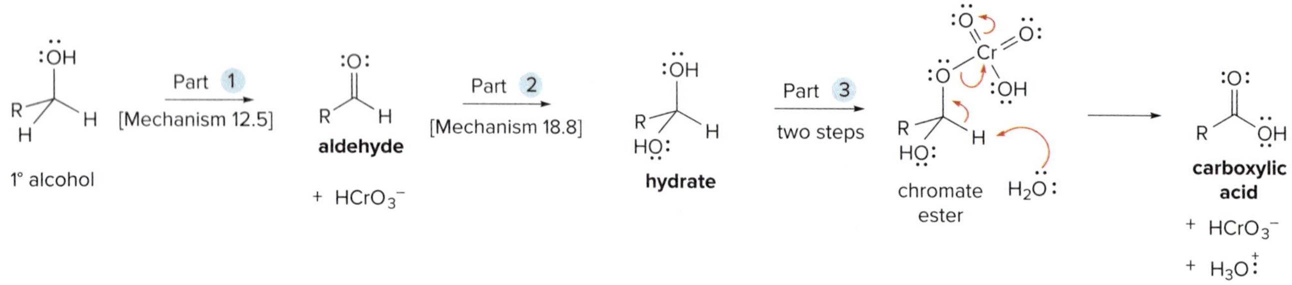

Part 1 The 1° alcohol is oxidized to an aldehyde by the three-step sequence in Mechanism 12.5.

Part 2 Water adds to the C=O to form a **hydrate,** a compound with two OH groups bonded to the same carbon, by a mechanism discussed in Section 18.12.

Part 3 Oxidation of the C–H bond of the hydrate follows Mechanism 12.5—formation of a chromate ester and loss of a proton.

Problem 12.27 Draw the organic products in each of the following reactions.

a. 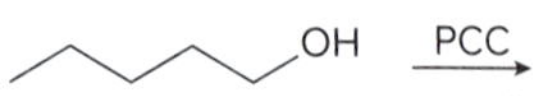 (alcohol) —PCC→

b. 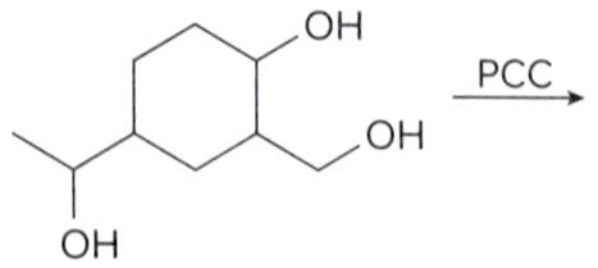 (triol) —PCC→

c. 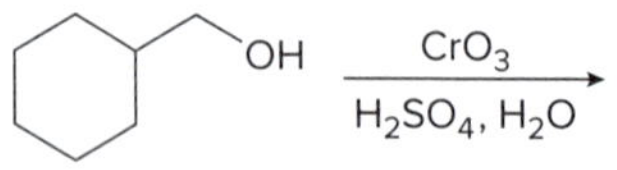 (cyclohexyl alcohol) —CrO_3, H_2SO_4, H_2O→

d. 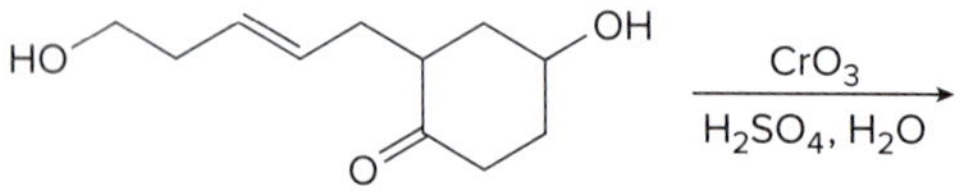 (diol ketone) —CrO_3, H_2SO_4, H_2O→

Problem 12.28 What reagent is needed to convert **A** to each compound?

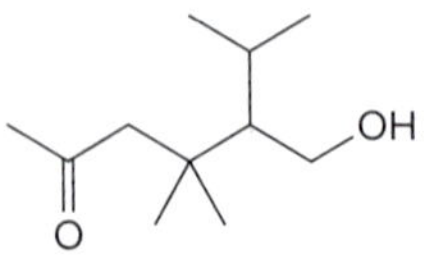

A a. 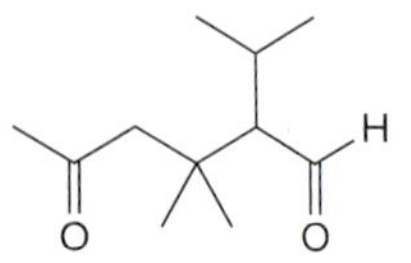 b. 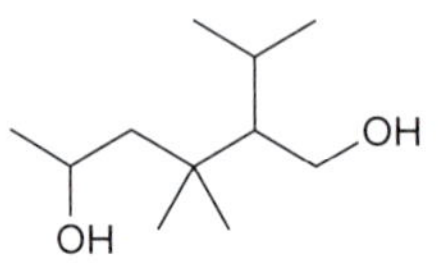 c.

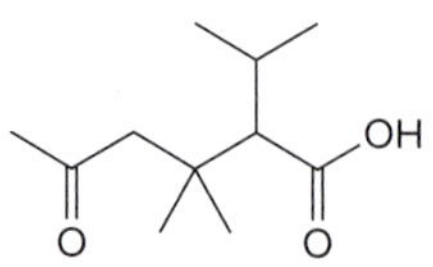

12.13 Green Chemistry

Several new methods of oxidation are based on green chemistry. ***Green chemistry* is the use of environmentally benign methods to synthesize compounds.** Its purpose is to use safer reagents and less solvent, and develop reactions that form fewer by-products and generate less waste.

Since many oxidation methods use toxic reagents (such as OsO_4 and O_3) and corrosive acids (such as H_2SO_4), or they generate carcinogenic by-products (such as Cr^{3+}), alternative reactions have been developed. One method uses a polymer-supported Cr^{6+} reagent—$HCrO_4^-$–Amberlyst A-26 resin—that avoids the use of strong acid, and forms a Cr^{3+} by-product that can be easily removed from the product by filtration.

The Amberlyst A-26 resin consists of a complex hydrocarbon network with cationic ammonium ion appendages that serve as counterions to the anionic chromium oxidant, $HCrO_4^-$. Heating the insoluble polymeric reagent with an alcohol results in oxidation to a carbonyl compound, with formation of an insoluble Cr^{3+} by-product. Not only can the metal by-product be removed by filtration without added solvent, it can also be regenerated and reused in a subsequent reaction.

Polymer —$\overset{+}{N}(CH_3)_3$ $^-O-CrO_2-OH$

Amberlyst A-26 resin — Cr^{6+} oxidant

With $HCrO_4^-$–Amberlyst A-26 resin, **1° alcohols are oxidized to aldehydes and 2° alcohols are oxidized to ketones.**

1° alcohol $\xrightarrow[\text{Amberlyst A-26 resin}]{HCrO_4^-}$ aldehyde + polymer-supported Cr^{3+} by-product

2° alcohol $\xrightarrow[\text{Amberlyst A-26 resin}]{HCrO_4^-}$ ketone + polymer-supported Cr^{3+} by-product

K^+ $^-O-SO_2-O-OH$

potassium peroxymonosulfate

Many other green approaches to oxidation that avoid the generation of metal by-products entirely are also under active investigation. For example, **potassium peroxymonosulfate, $KHSO_5$,** is a sulfate derivative of hydrogen peroxide, sold as a triple salt ($2\,KHSO_5 \cdot KHSO_4 \cdot K_2SO_4$) under the trade name of Oxone. Oxone oxidizes a variety of substrates without the presence of a heavy metal like chromium or manganese, and in some cases, oxidation reactions can be carried out in water or aqueous solutions. The **weak oxygen–oxygen bond of the reagent is cleaved** during oxidation, and a sulfate salt (K_2SO_4) is formed as by-product. Two examples of oxidations of alcohols are shown.

Oxone, NaBr, CH_3CN, H_2O

Oxone, NaCl, ethyl acetate, H_2O

Problem 12.29 Sodium hypochlorite (NaOCl, the oxidant in household bleach) in acetic acid is touted as a "green" oxidizing agent—there are no metals used in the reaction and the by-products of oxidation are NaCl and H_2O. Moreover, NaOCl in CH_3CO_2H selectively oxidizes 2° alcohols in the presence of 1° alcohols. With this fact in mind, draw the product when **A** is treated with each oxidant.

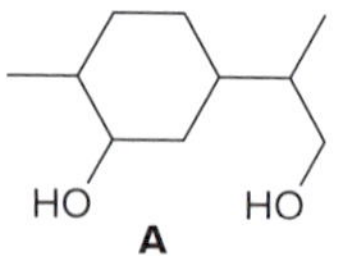

HO HO
A

a. NaOCl, CH_3CO_2H
b. PCC
c. CrO_3, H_2SO_4
d. $HCrO_4^-$–Amberlyst A-26 resin

12.14 Biological Oxidation

Many reactions in biological systems involve oxidation or reduction. Instead of using Cr^{6+} reagents for oxidation, cells use two organic compounds—a high-molecular-weight **enzyme** and a simpler **coenzyme** that serves as the oxidizing agent.

The coenzyme often used to oxidize alcohols in biological systems is **nicotinamide adenine dinucleotide,** abbreviated as **NAD^+.** Although the structure is complex, only a portion of the molecule, drawn in red, participates in redox reactions.

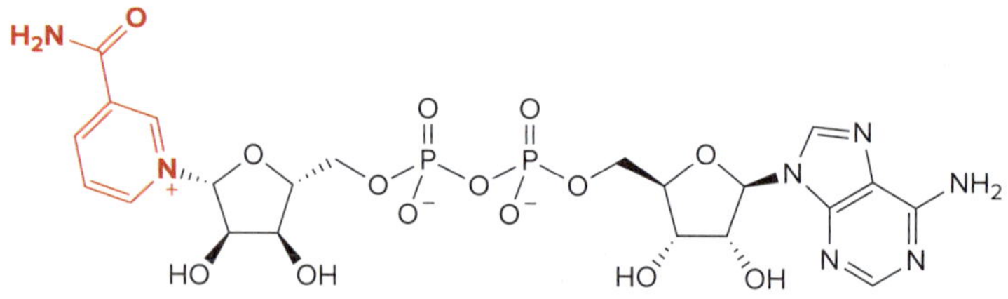

nicotinamide adenine dinucleotide
NAD^+

Biological oxidation of an alcohol occurs by transferring a hydride, a hydrogen atom with two electrons, from the alcohol to NAD^+ to form a carbonyl group. In the process, NAD^+ is reduced to nicotinamide adenine dinucleotide (reduced form), abbreviated as **NADH.** NADH is a biological reducing agent that converts carbonyl compounds to alcohols, as discussed in Section 17.6.

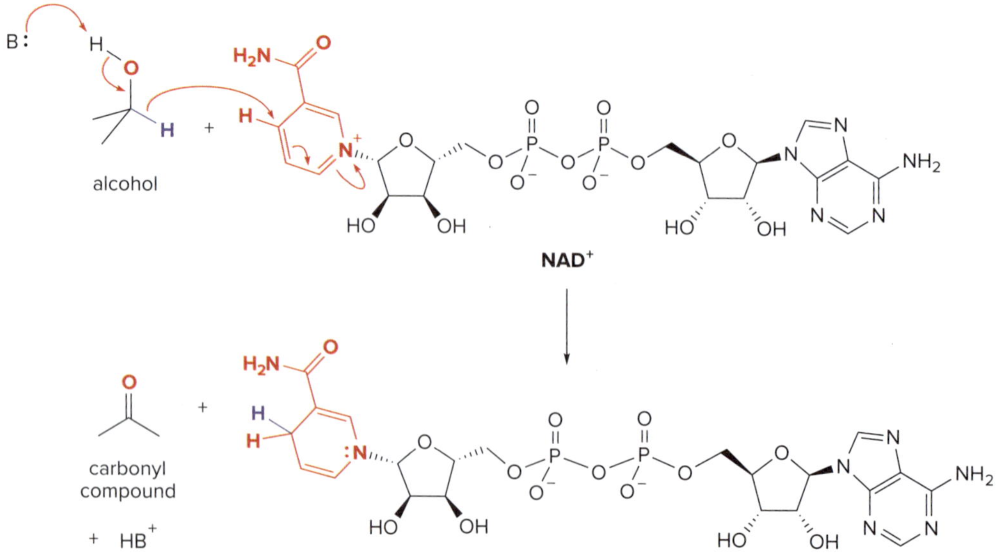

B: alcohol + NAD^+

↓

carbonyl compound + nicotinamide adenine dinucleotide (reduced form) **NADH**

+ HB^+

For example, when CH_3CH_2OH (ethanol) is ingested, it is oxidized in the liver by NAD^+ to CH_3CHO (acetaldehyde), and then to CH_3COO^- (acetate anion, the conjugate base of acetic acid). Acetate is the starting material for the synthesis of fatty acids and cholesterol. Both oxidations are catalyzed by a dehydrogenase enzyme.

ethanol —(NAD^+, alcohol dehydrogenase)→ acetaldehyde —(NAD^+, aldehyde dehydrogenase)→ acetate anion

If more ethanol is ingested than can be metabolized in a given time, the concentration of acetaldehyde builds up. This toxic compound is responsible for the feelings associated with a hangover.

Antabuse

Antabuse, a drug given to individuals with alcoholism to prevent them from consuming alcoholic beverages, acts by interfering with the normal oxidation of ethanol. Antabuse inhibits the oxidation of acetaldehyde to the acetate anion. Because the first step in ethanol metabolism occurs but the second does not, the concentration of acetaldehyde rises, causing an individual to become violently ill.

Like ethanol, methanol is oxidized by the same enzymes to give an aldehyde and an acid: formaldehyde and formic acid. These oxidation products are extremely toxic because they cannot be used by the body. As a result, the pH of the blood decreases, and blindness and death can follow.

CH_3-OH (methanol) —(NAD^+, alcohol dehydrogenase)→ formaldehyde —(NAD^+, aldehyde dehydrogenase)→ formic acid

Because the enzymes have a higher affinity for ethanol than methanol, methanol poisoning is treated by giving ethanol to the afflicted individual. With both methanol and ethanol in the patient's system, the enzymes react more readily with ethanol, allowing the methanol to be excreted unchanged without the formation of methanol's toxic oxidation products.

12.15 Sharpless Epoxidation

In all of the reactions discussed so far, an **achiral starting material has reacted with an achiral reagent to give either an achiral product or a racemic mixture of two enantiomers.** If you are trying to make a chiral product, this means that only half of the product mixture is the desired enantiomer and the other half is the undesired one. The synthesis of disparlure, outlined in Figure 12.8, exemplifies this dilemma.

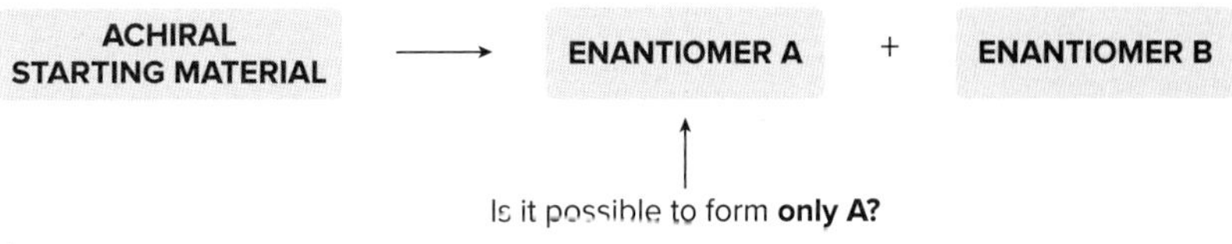

K. Barry Sharpless shared the 2001 Nobel Prize in Chemistry for his work on chiral oxidation reactions.

K. Barry Sharpless, of The Scripps Research Institute, reasoned that using a chiral reagent might make it possible to favor the formation of one enantiomer over the other.

- **An *enantioselective* reaction affords predominantly or exclusively one enantiomer.**
- **A reaction that converts an achiral starting material into predominantly one enantiomer is also called an *asymmetric reaction.***

The Sharpless asymmetric epoxidation is an enantioselective reaction that oxidizes alkenes to epoxides. Only the double bonds of **allylic alcohols**—that is, alcohols having a hydroxy group on the carbon adjacent to a C=C—are oxidized in this reaction.

allylic alcohol → (Sharpless reagent) → O added from **above** or O added from **below**

Sharpless reagent
$(CH_3)_3C{-}OOH$
$Ti[OCH(CH_3)_2]_4$
(+)- or (–)-diethyl tartrate

- With Sharpless reagent, one enantiomer is favored.
- The new stereogenic center is labeled in blue.

The **Sharpless reagent** consists of three components: *tert*-butyl hydroperoxide, $\mathbf{(CH_3)_3COOH}$; a titanium catalyst—usually titanium(IV) isopropoxide, $\mathbf{Ti[OCH(CH_3)_2]_4}$; and **diethyl tartrate (DET).** There are two different chiral diethyl tartrate isomers, labeled as (+)-DET or (–)-DET to indicate the direction in which they rotate polarized light.

(+)-(*R*,*R*)-diethyl tartrate
(+)-DET

(–)-(*S*,*S*)-diethyl tartrate
(–)-DET

(+)-DET is prepared from (+)-(*R*,*R*)-tartaric acid [$HO_2CCH(OH)CH(OH)CO_2H$], a naturally occurring carboxylic acid found in grapes and sold as a by-product of the wine industry. *Jenny Cundy/Image Source*

The identity of the DET isomer determines which enantiomer is the major product obtained in the epoxidation of an allylic alcohol with the Sharpless reagent.

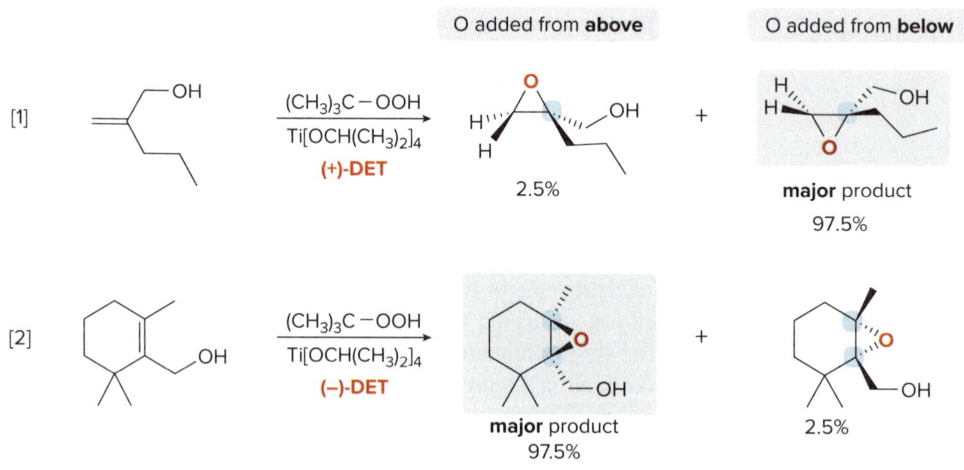

Stereogenic centers are labeled in blue.

The degree of enantioselectivity of a reaction is measured by its enantiomeric excess (***ee***) (Section 5.12D). Reactions [1] and [2] are highly enantioselective because each has an enantiomeric excess of 95% (97.5% of the major enantiomer – 2.5% of the minor enantiomer).

Enantiomeric excess = **ee** = % of one enantiomer – % of the other enantiomer.

To determine which enantiomer is formed for a given isomer of DET, draw the allylic alcohol in a plane, with the **C=C horizontal and the OH group in the upper right corner;** then:

- Epoxidation with (–)-DET adds an oxygen atom from *above* the plane.
- Epoxidation with (+)-DET adds an oxygen atom from *below* the plane.

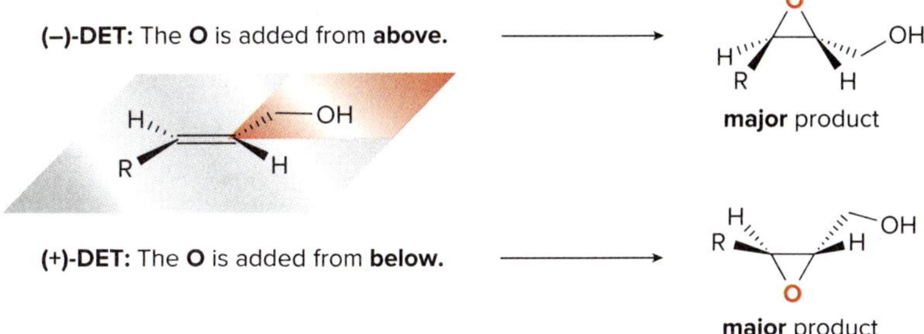

Sample Problem 12.5 Drawing the Product of a Sharpless Epoxidation

Predict the major product in each epoxidation.

a. $\xrightarrow[\text{(+)-DET}]{(CH_3)_3C-OOH,\ Ti[OCH(CH_3)_2]_4}$ b. $\xrightarrow[\text{(–)-DET}]{(CH_3)_3C-OOH,\ Ti[OCH(CH_3)_2]_4}$

Solution

To draw an epoxidation product:

- Draw the allylic alcohol with the **C=C horizontal and the OH group in the upper right corner of the alkene.** Re-draw the alkene if necessary.
- **(+)-DET** adds the O atom from **below,** and **(–)-DET** adds the O atom from **above.**

a. Because the C=C is drawn horizontal with the OH group in the upper right corner, it is not necessary to re-draw the alkene. With **(+)-DET,** the O atom is added from **below.**

$(CH_3)_3C-OOH$, $Ti[OCH(CH_3)_2]_4$, **(+)-DET**

OH in the upper right corner

The O atom is added from **below** the plane.

b. The allylic alcohol must be re-drawn with the C=C horizontal and the OH group in the **upper right corner.** Because **(–)-DET** is used, the O atom is then added from **above.**

Flip the molecule and re-draw.

OH in the upper right corner

$(CH_3)_3C-OOH$, $Ti[OCH(CH_3)_2]_4$, **(–)-DET**

The O atom is added from **above** the plane.

Problem 12.30 Draw the products of each Sharpless epoxidation.

a. $\xrightarrow[\text{(+)-DET}]{(CH_3)_3C-OOH,\ Ti[OCH(CH_3)_2]_4}$ b. $\xrightarrow[\text{(–)-DET}]{(CH_3)_3C-OOH,\ Ti[OCH(CH_3)_2]_4}$

More Practice: Try Problems 12.32e; 12.37j; 12.38e, f; 12.42a; 12.56–12.58.

The Sharpless epoxidation has been used to synthesize many chiral natural products, including two insect pheromones—(+)-α-multistriatin and (–)-frontalin, as shown in Figure 12.9.

Figure 12.9 The synthesis of chiral insect pheromones using asymmetric epoxidation

OH → $(CH_3)_3C{-}OOH$, $Ti[OCH(CH_3)_2]_4$, **(+)-DET** → OH → several steps → (+)-α-multistriatin, pheromone of the European elm bark beetle

OH → $(CH_3)_3C{-}OOH$, $Ti[OCH(CH_3)_2]_4$, **(−)-DET** → OH → two steps → (−)-frontalin, pheromone of the western pine beetle

- The bonds in the products that originate from the epoxide intermediate are indicated in red.

Problem 12.31 Explain why only one C=C of geraniol is epoxidized with the Sharpless reagent.

OH

geraniol

Chapter 12 REVIEW

KEY CONCEPTS

Reaction Selectivity

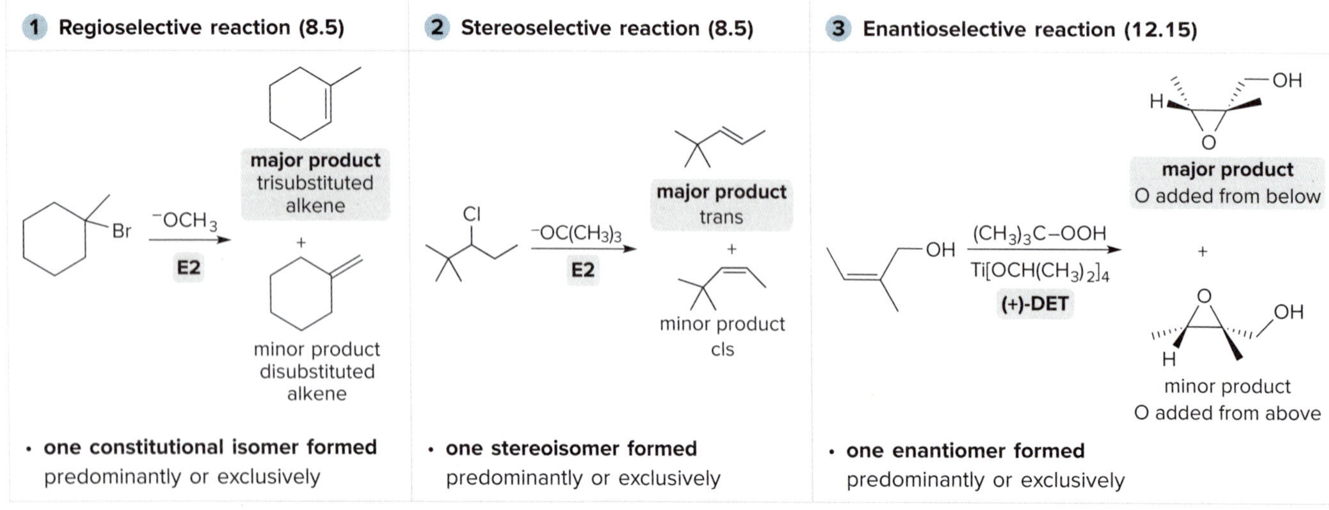

- **one constitutional isomer formed** predominantly or exclusively
- **one stereoisomer formed** predominantly or exclusively
- **one enantiomer formed** predominantly or exclusively

KEY REACTIONS

Reduction Reactions

[1] Reduction of alkenes

H—H
Pd, Pt, or Ni
hydrogenation
(12.3)
alkane

Try Problems 12.32a, 12.34, 12.37a, 12.38a, 12.40c, 12.42b.

[2] Reduction of alkynes

1. H—H (2 equiv), Pd, Pt, or Ni — hydrogenation (12.5A) — alkane

2. H—H, Lindlar catalyst — hydrogenation, syn addition of H_2 (12.5B) — cis alkene

3. Na, NH_3 — hydrogenation, anti addition of H_2 (12.5C) — trans alkene

See Figure 12.5. Try Problem 12.41d.

[3] Reduction of alkyl halides and epoxides

1. [1] $Li^+ H—\bar{A}lH_3$; [2] H_2O — S_N2 (12.6) — X = Cl, Br, I — alkane + Li^+X^- + AlH_3

2. [1] $Li^+ H—\bar{A}lH_3$; [2] H_2O — S_N2 (12.6) — alcohol + $Li^+ \ ^-OH$ + AlH_3

See Figure 12.6. Try Problems 12.37l; 12.38g; 12.41a, c.

Oxidation Reactions

[1] Oxidation of alkenes

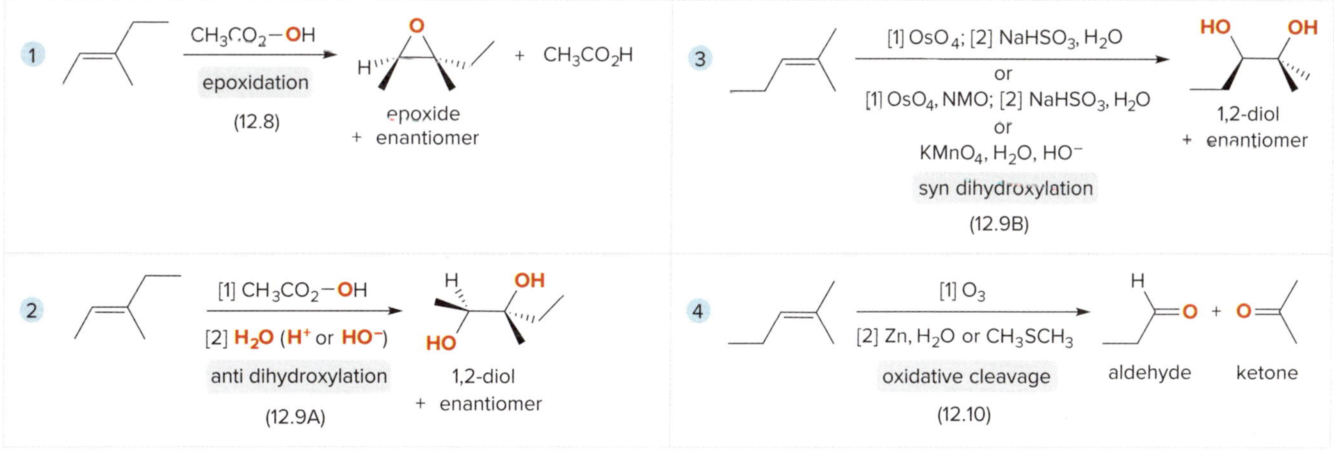

See Sample Problems 12.2, 12.3, Figure 12.7. Try Problems 12.32b; 12.37d–g, i, k; 12.38b; 12.41b, e, f; 12.44a; 12.48a, b; 12.51.

[2] Oxidative cleavage of alkynes

1. internal alkyne — [1] O_3; [2] H_2O — ozonolysis (12.11) — carboxylic acids

2. terminal alkyne — [1] O_3; [2] H_2O — ozonolysis (12.11) — carboxylic acid + CO_2

Try Problem 12.48c, d.

[3] Oxidation of alcohols

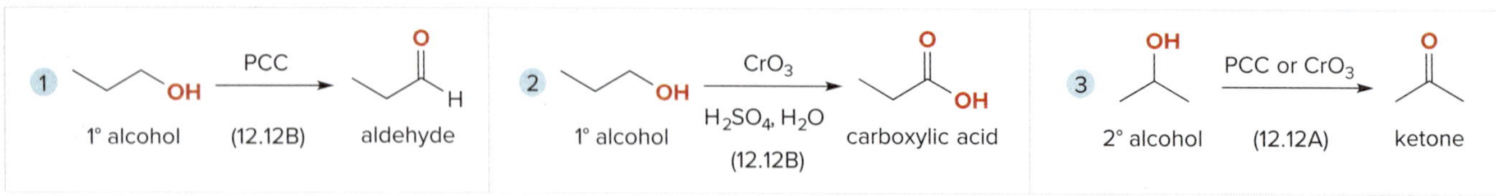

Try Problems 12.32c, d; 12.38c, d, h; 12.42c, d.

[4] Asymmetric epoxidation of allylic alcohols (12.15)

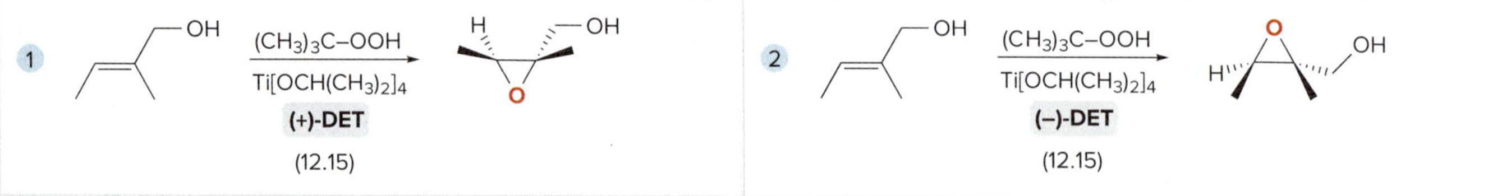

See Sample Problem 12.5, Figure 12.9.
Try Problems 12.32e; 12.37j; 12.38e, f; 12.42a; 12.56–12.58.

KEY SKILLS

[1] Determining the number of rings and π bonds in a compound ($C_{14}H_{20}$) hydrogenated to a compound of molecular formula $C_{14}H_{26}$ (12.3C)

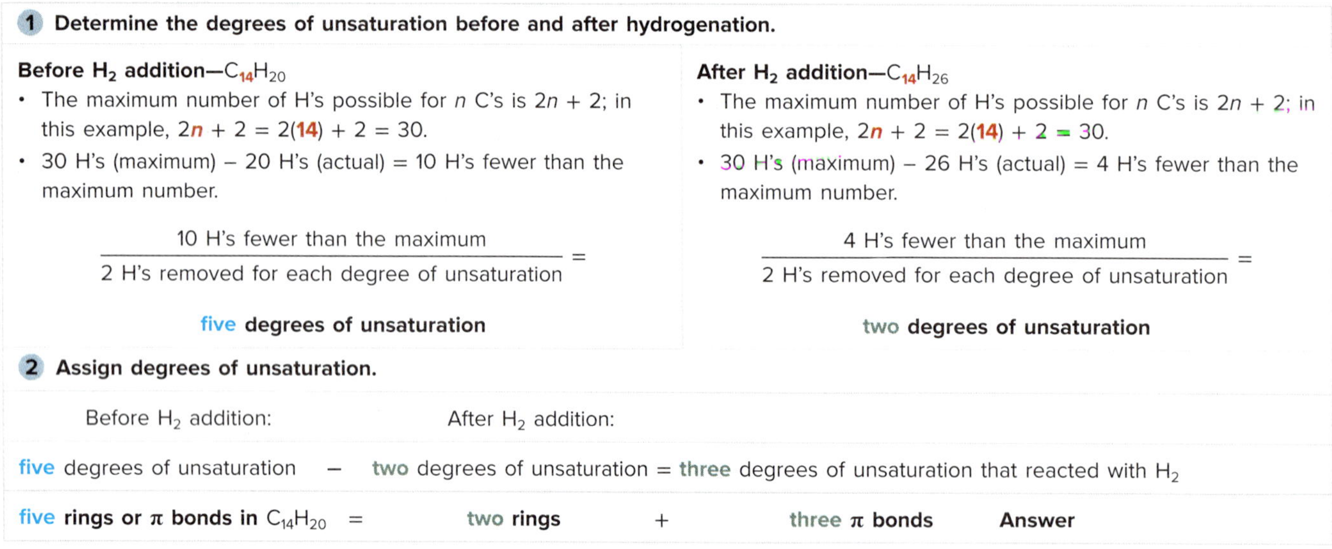

1 Determine the degrees of unsaturation before and after hydrogenation.

Before H_2 addition—$C_{14}H_{20}$

- The maximum number of H's possible for *n* C's is $2n + 2$; in this example, $2n + 2 = 2(14) + 2 = 30$.
- 30 H's (maximum) – 20 H's (actual) = 10 H's fewer than the maximum number.

$$\frac{\text{10 H's fewer than the maximum}}{\text{2 H's removed for each degree of unsaturation}} =$$

five **degrees of unsaturation**

After H_2 addition—$C_{14}H_{26}$

- The maximum number of H's possible for *n* C's is $2n + 2$; in this example, $2n + 2 = 2(14) + 2 = 30$.
- 30 H's (maximum) – 26 H's (actual) = 4 H's fewer than the maximum number.

$$\frac{\text{4 H's fewer than the maximum}}{\text{2 H's removed for each degree of unsaturation}} =$$

two **degrees of unsaturation**

2 Assign degrees of unsaturation.

Before H_2 addition:		After H_2 addition:		
five degrees of unsaturation	–	two degrees of unsaturation = three degrees of unsaturation that reacted with H_2		
five **rings or π bonds in $C_{14}H_{20}$**	=	two **rings**	+ three **π bonds**	**Answer**

See Sample Problem 12.1. Try Problem 12.36.

[2] Drawing the stereoisomers from alkene epoxidation with mCPBA (12.8A); example: (*Z*)-3-methylpent-2-ene

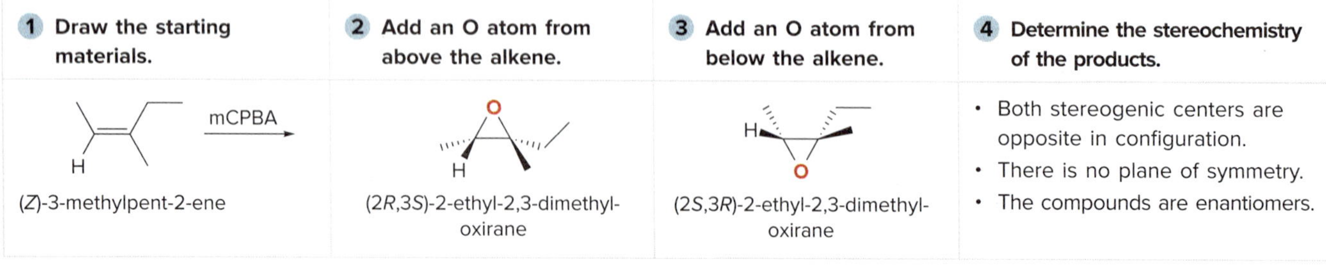

1 Draw the starting materials.	**2 Add an O atom from above the alkene.**	**3 Add an O atom from below the alkene.**	**4 Determine the stereochemistry of the products.**
mCPBA (*Z*)-3-methylpent-2-ene	(2*R*,3*S*)-2-ethyl-2,3-dimethyloxirane	(2*S*,3*R*)-2-ethyl-2,3-dimethyloxirane	• Both stereogenic centers are opposite in configuration. • There is no plane of symmetry. • The compounds are enantiomers.

See Sample Problem 12.2. Try Problems 12.32b; 12.37d, k; 12.38b.

[3] Drawing the products of dihydroxylation of an alkene (12.9)

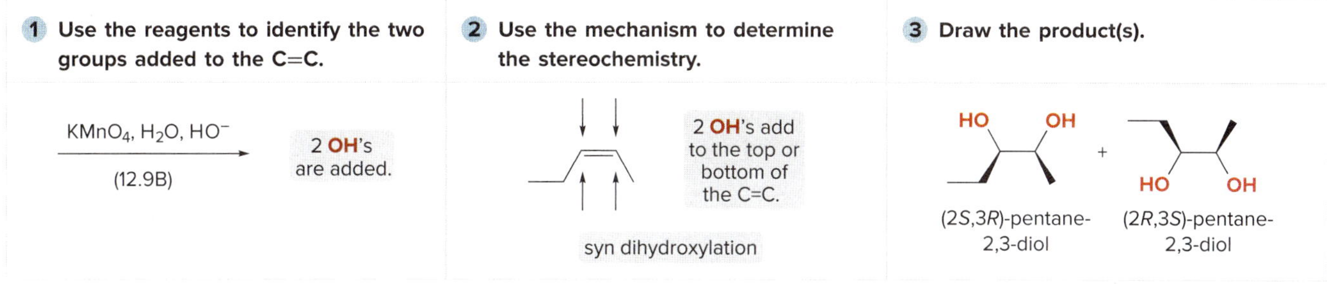

Try Problems 12.37e–g, 12.41b, 12.46.

[4] Drawing the products of an ozonolysis reaction (12.10)

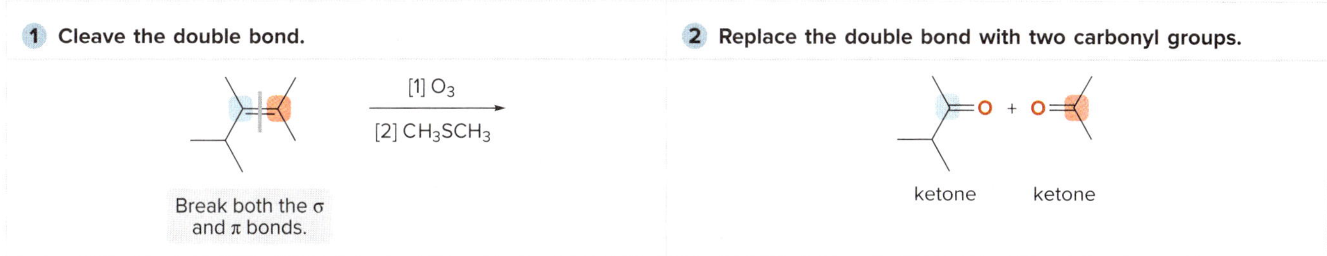

See Sample Problem 12.3. Try Problems 12.37i; 12.39c; 12.40a, b; 12.48a, b; 12.51.

[5] Identifying an alkene from ozonolysis products (12.10)

See Sample Problem 12.4. Try Problems 12.49a, b; 12.50.

[6] Predicting the product of a Sharpless reaction (12.15)

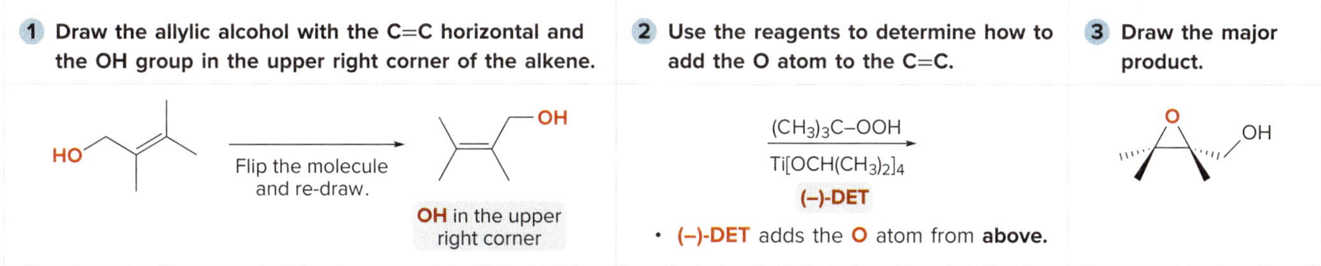

See Sample Problem 12.5. Try Problems 12.32e; 12.37j; 12.38e, f; 12.42a; 12.56–12.58.

CHAPTER 12 MULTIPLE-CHOICE SELF-TEST

The Self-Test consists of multiple-choice questions similar to those found on the American Chemical Society organic chemistry exam. Answers are given at the end of the chapter.

1. How many rings and π bonds are contained in a compound of molecular formula $C_{15}H_{26}$ that is hydrogenated to a compound of molecular formula $C_{15}H_{28}$?

a. three rings

b. two rings and one π bond

c. one ring and two π bonds

d. three π bonds

2. Classify each reaction as oxidation, reduction, or neither.

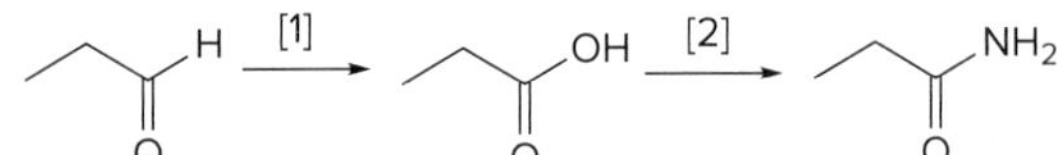

a. both oxidations
b. [1] oxidation, [2] reduction
c. [1] oxidation, [2] neither
d. [1] neither, [2] reduction

3. Rank alkenes **A–D** in order of increasing heat of hydrogenation.

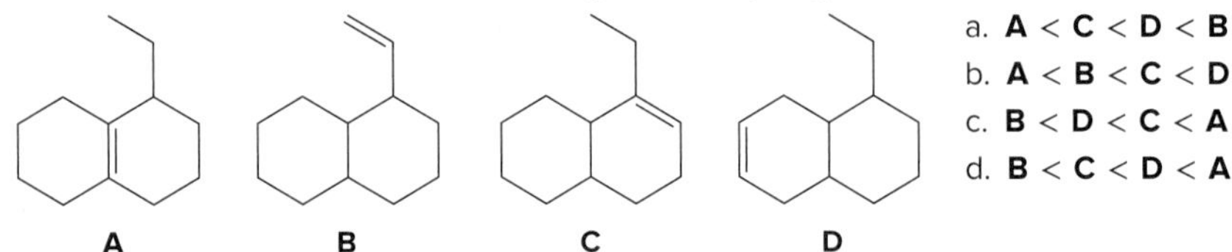

a. **A** < **C** < **D** < **B**
b. **A** < **B** < **C** < **D**
c. **B** < **D** < **C** < **A**
d. **B** < **C** < **D** < **A**

4. What product is formed when **A** reacts with PCC?

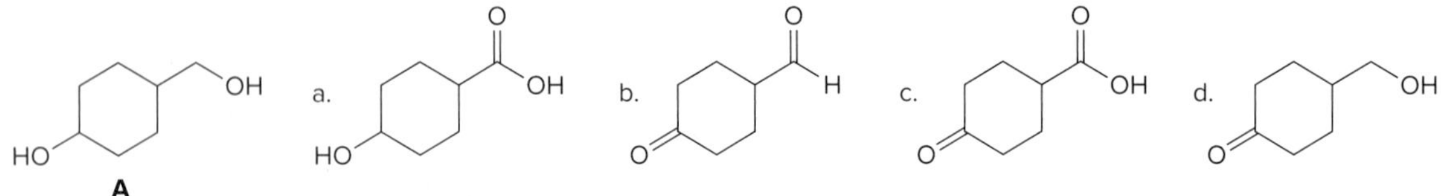

5. What product is *not* formed when **B** undergoes an ozonolysis reaction?

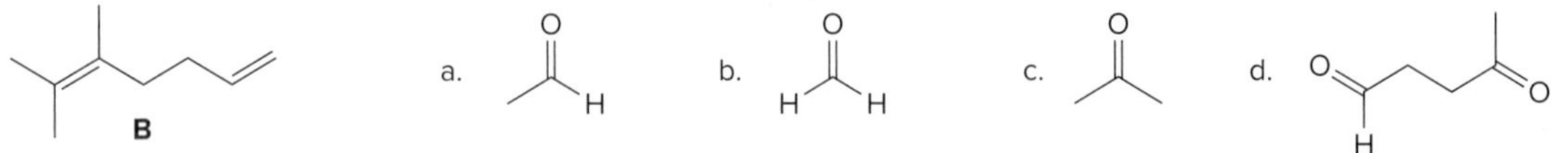

6. Which route does *not* yield the product hexan-2-ol?

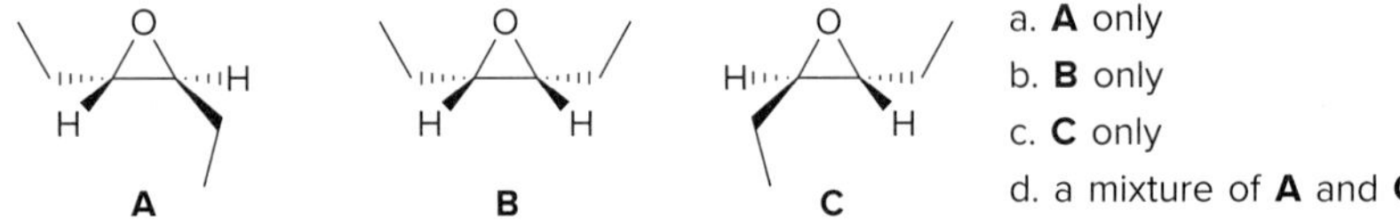

7. Treatment of $CH_3CH_2C{\equiv}CCH_2CH_3$ with Na in NH_3 followed by mCPBA forms what product(s)?

a. **A** only
b. **B** only
c. **C** only
d. a mixture of **A** and **C**

8. Which statement is false?
a. CrO_3 oxidation of a 2° alcohol forms a ketone.
b. Treatment of cyclopentene with OsO_4 followed by aqueous $NaHSO_3$ forms an equal mixture of two enantiomers.
c. $CH_3CH_2C{\equiv}CCH_2CH_3$ reacts with H_2 in the presence of either Pd-C or the Lindlar catalyst.
d. Treatment of *trans*-but-2-ene with RCO_3H followed by ^-OH yields a meso compound.

9. What product is formed when **C** is treated with Sharpless reagent and (–)-DET?

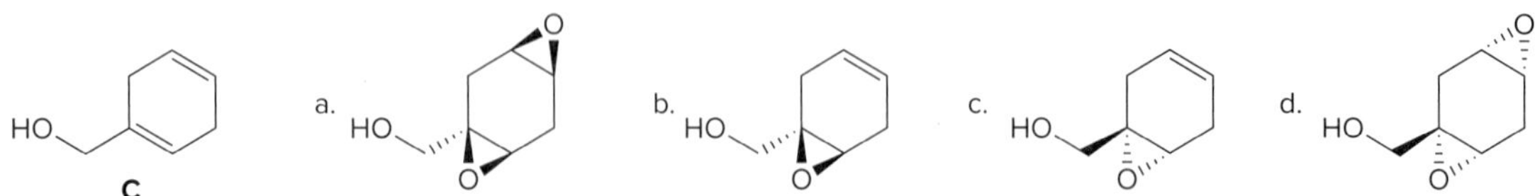

10. Which compound does *not* react with H_2 in the presence of a Pd catalyst?

PROBLEMS

Problems Using Three-Dimensional Models

12.32 Draw the products formed when **A** is treated with each reagent: (a) H_2 + Pd-C; (b) mCPBA; (c) PCC; (d) CrO_3, H_2SO_4, H_2O; (e) Sharpless reagent with (+)-DET.

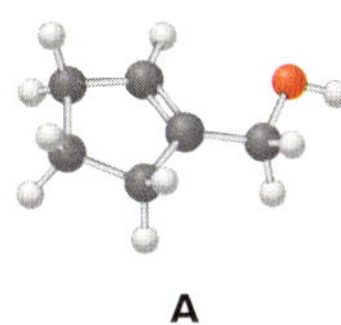

A

12.33 Devise a synthesis of the following compound from acetylene and organic compounds containing two or fewer carbons. You may use any other required reagents.

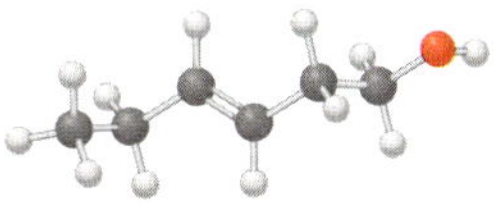

Hydrogenation

12.34 Draw the organic products formed when each compound is treated with H_2, Pd-C. Indicate the three-dimensional structure of all stereoisomers formed.

a. 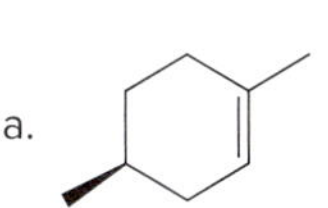 b. 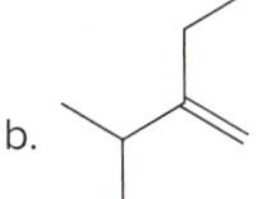 c. 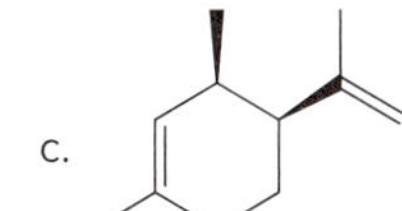 d. 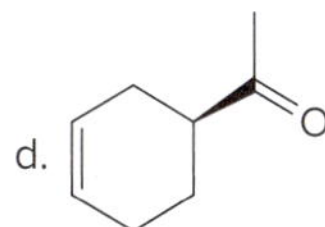

12.35 Match each alkene to its heat of hydrogenation.
Alkenes: 3-methylbut-1-ene, 2-methylbut-1-ene, 2-methylbut-2-ene
$\Delta H°$ (hydrogenation) kJ/mol: –119, –127, –112

12.36 How many rings and π bonds are contained in compounds **A–C?** Draw one possible structure for each compound.

a. Compound **A** has molecular formula C_5H_8 and is hydrogenated to a compound having molecular formula C_5H_{10}.
b. Compound **B** has molecular formula $C_{10}H_{16}$ and is hydrogenated to a compound having molecular formula $C_{10}H_{18}$.
c. Compound **C** has molecular formula C_8H_8 and is hydrogenated to a compound having molecular formula C_8H_{16}.

Reactions—General

12.37 Draw the organic products formed when cyclopentene is treated with each reagent. With some reagents, no reaction occurs.

a. H_2 + Pd-C
b. H_2 + Lindlar catalyst
c. Na, NH_3
d. CH_3CO_3H
e. [1] CH_3CO_3H; [2] H_2O, HO^-
f. [1] OsO_4 + NMO; [2] $NaHSO_3$, H_2O
g. $KMnO_4$, H_2O, HO^-
h. [1] $LiAlH_4$; [2] H_2O
i. [1] O_3; [2] CH_3SCH_3
j. $(CH_3)_3COOH$, $Ti[OCH(CH_3)_2]_4$, (–)-DET
k. mCPBA
l. Product in (k); then [1] $LiAlH_4$; [2] H_2O

12.38 Draw the organic products formed when allylic alcohol **A** is treated with each reagent.

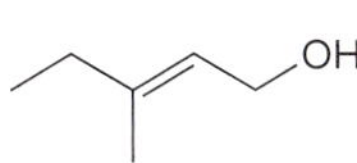

A

a. H_2 + Pd-C
b. mCPBA
c. PCC
d. CrO_3, H_2SO_4, H_2O
e. $(CH_3)_3COOH$, $Ti[OCH(CH_3)_2]_4$, (+)-DET
f. $(CH_3)_3COOH$, $Ti[OCH(CH_3)_2]_4$, (–)-DET
g. [1] PBr_3; [2] $LiAlH_4$; [3] H_2O
h. $HCrO_4^-$–Amberlyst A-26 resin

12.39 For alkenes **A, B, C,** and **D:** (a) Rank **A–D** in order of increasing heat of hydrogenation; (b) rank **A–D** in order of increasing rate of reaction with H_2, Pd-C; (c) draw the products formed when each alkene is treated with ozone, followed by Zn, H_2O.

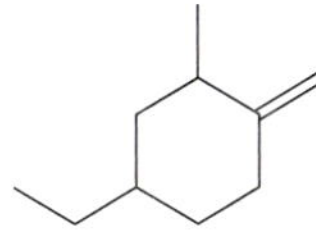 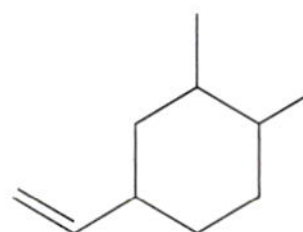 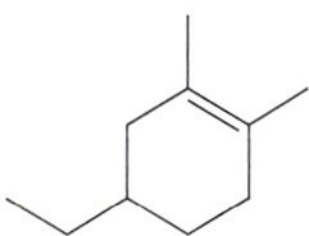 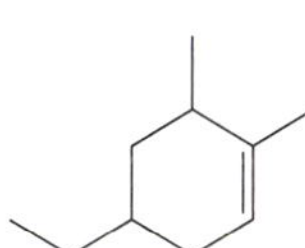

A **B** **C** **D**

12.40 The cadinenes are a group of hydrocarbons isolated from the wood of the Cade juniper shrub native to the Mediterranean region.

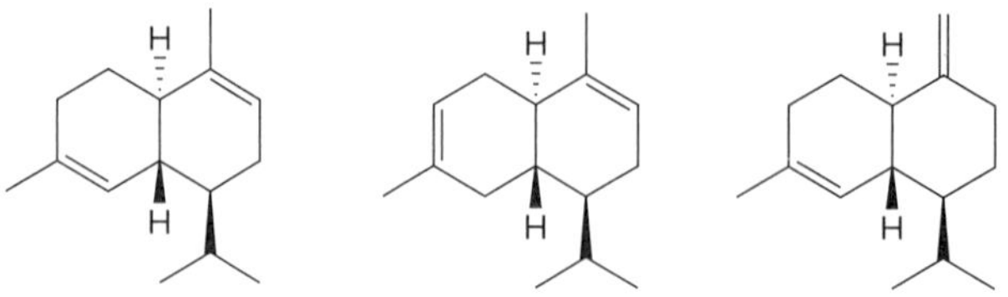

α-cadinene β-cadinene γ-cadinene

a. Which isomer(s) afford an acyclic compound after ozonolysis? Draw the structures of the products formed.
b. Which isomer(s) afford a monocyclic compound after ozonolysis? Draw the structures of the products formed.
c. Draw the structure of all products formed when α-cadinene is treated with excess H_2 in the presence of Pd-C.

12.41 Draw the organic products formed in each reaction.

a. 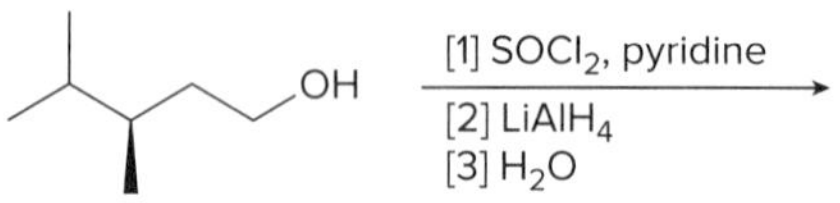 [1] $SOCl_2$, pyridine; [2] $LiAlH_4$; [3] H_2O

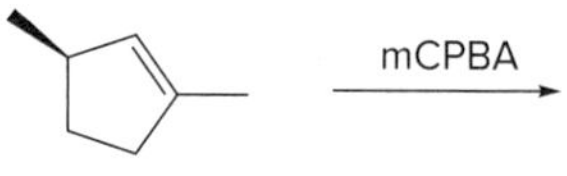

b. 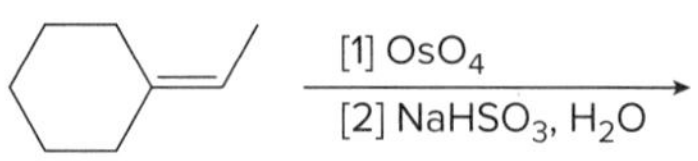 [1] OsO_4; [2] $NaHSO_3$, H_2O

c. 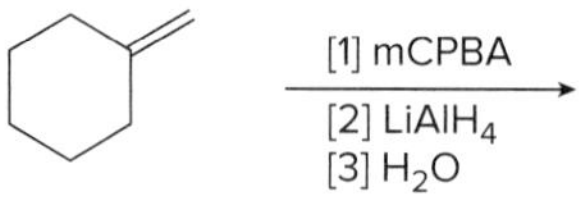 [1] mCPBA; [2] $LiAlH_4$; [3] H_2O

d. 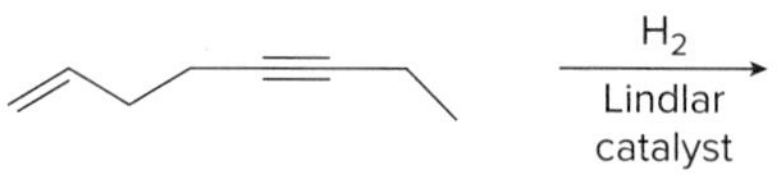 H_2, Lindlar catalyst

e. mCPBA

f. 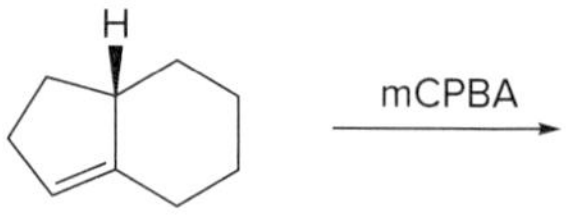 mCPBA

12.42 Draw the product formed in each reaction. Each reaction is one step in the synthesis of a drug or natural product: (a) antidepressant fluoxetine (Prologue); (b) antidepressant sertraline (Section 15.5); (c) anticancer agent exemestane (Section 5.5); (d) antibiotic monensin (Section 18.15).

a. 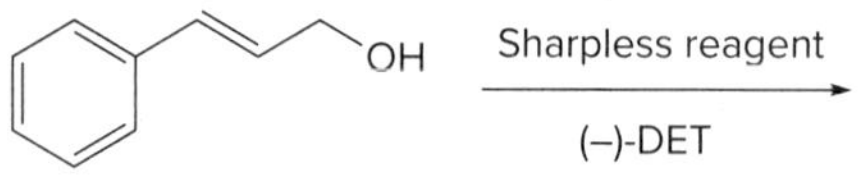 Sharpless reagent, (−)-DET

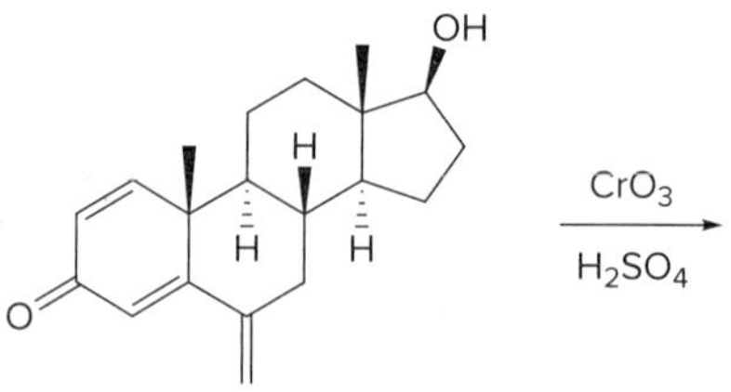

b. 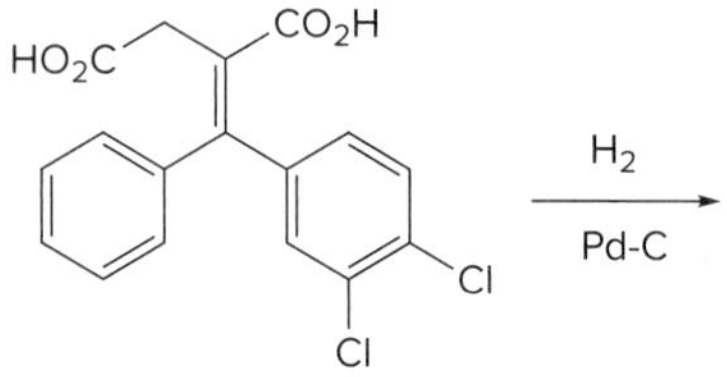 H_2, Pd-C

c. CrO_3, H_2SO_4

d. 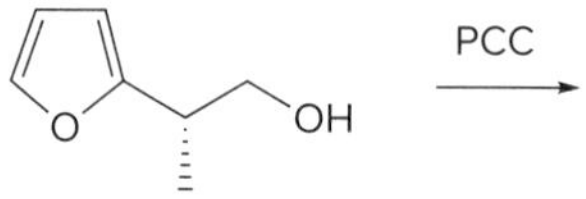 PCC

12.43 Draw the structure of two different epoxides that would yield 2-methylpentan-2-ol [$(CH_3)_2C(OH)CH_2CH_2CH_3$] when reduced with $LiAlH_4$.

12.44 Hydrogenation of alkene **A** with D_2 in the presence of Pd-C affords a single product **B.** Keeping this result in mind, what compound is formed when **A** is treated with each reagent: (a) mCPBA; (b) Br_2, H_2O followed by base? Explain these results.

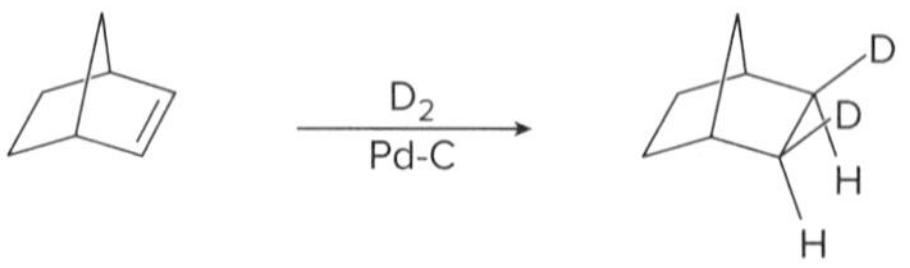

A $\xrightarrow{D_2,\ \text{Pd-C}}$ **B**

12.45 Compound **X,** a fatty acid that is abundant in the brain and retina, has molecular formula $C_{22}H_{32}O_2$. (a) How many degrees of unsaturation does **X** contain? (b) Treatment of **X** with O_3 followed by $(CH_3)_2S$ affords the following mixture of products: $HO_2CCH_2CH_2CHO$, $CH_2(CHO)_2$ (5 equivalents), and CH_3CH_2CHO. What is the structure of **X** assuming any C=C has the *Z* configuration? (c) Draw a possible product **Y** formed when **X** is treated with two equivalents of H_2. (d) Draw the product **W** formed when **X** is treated with excess H_2. (e) Rank **X, Y,** and **W** in order of increasing melting point.

12.46 What alkene is needed to synthesize each 1,2-diol using [1] OsO_4 followed by $NaHSO_3$ in H_2O; or [2] CH_3CO_3H followed by ^-OH in H_2O?

a. 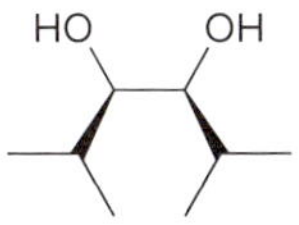 HO OH

b. 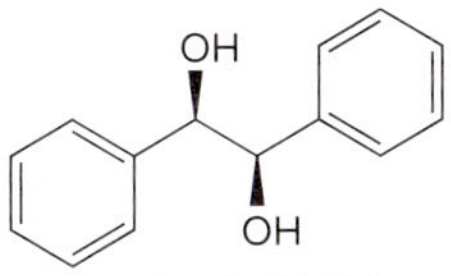 OH OH + enantiomer

c. 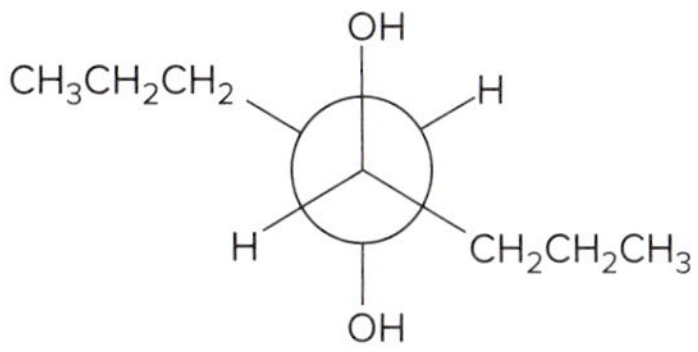 OH $CH_3CH_2CH_2$ H H $CH_2CH_2CH_3$ OH

12.47 (a) What product is formed in Step [1] of the following reaction sequence? (b) Draw a mechanism for Step [2] that accounts for the observed stereochemistry. (c) What reaction conditions are necessary to form chiral **A** from prop-2-en-1-ol (CH_2=$CHCH_2OH$)?

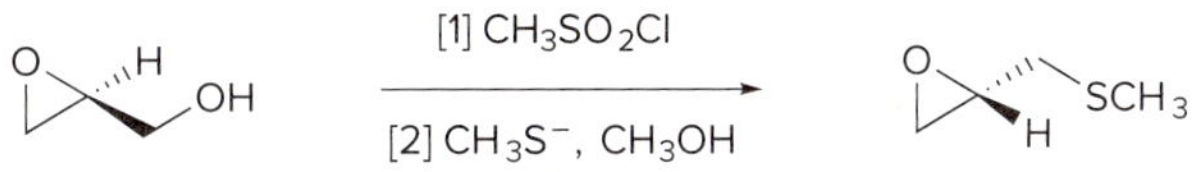

A —[1] CH_3SO_2Cl, [2] CH_3S^-, CH_3OH→ **B**

Oxidative Cleavage

12.48 Draw the products formed in each oxidative cleavage.

a. 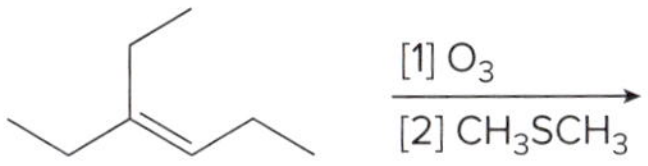 [1] O_3 [2] CH_3SCH_3

b. 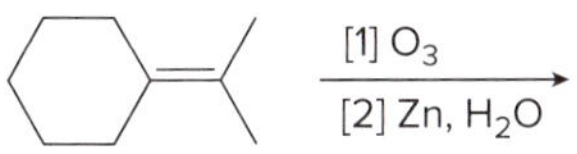 [1] O_3 [2] Zn, H_2O

c. [1] O_3 [2] H_2O

d. 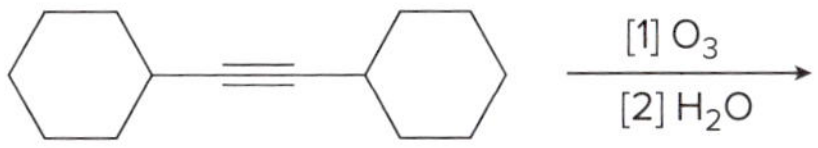 [1] O_3 [2] H_2O

12.49 What alkene or alkyne yields each set of products after oxidative cleavage with ozone?

a. 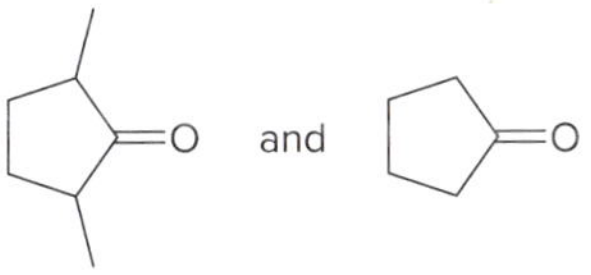 and

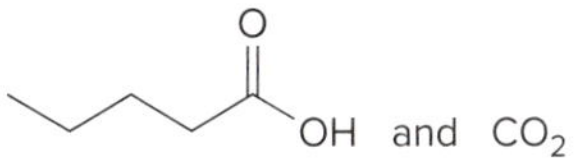

b. 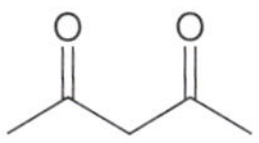 and two equivalents of CH_2=O

c. OH and CO_2

d. 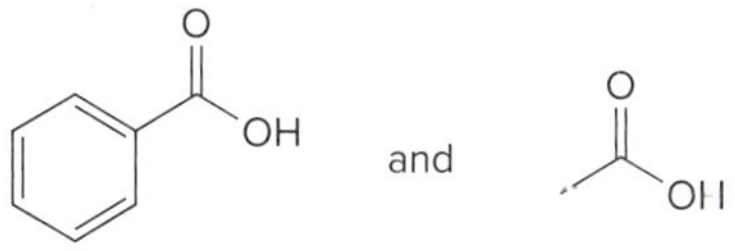 OH and OH

12.50 Identify the starting material in each reaction.

a. $C_{10}H_{18}$ —[1] O_3 [2] CH_3SCH_3→ H

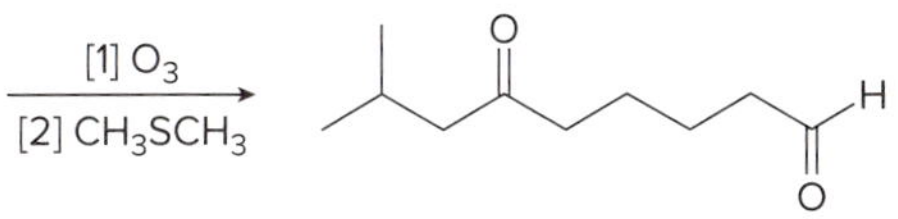

b. $C_{10}H_{16}$ —[1] O_3 [2] CH_3SCH_3→

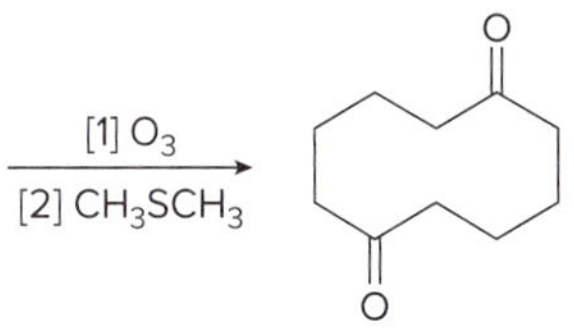

12.51 Draw the products formed when each naturally occurring compound is treated with O_3 followed by Zn, H_2O.

a. 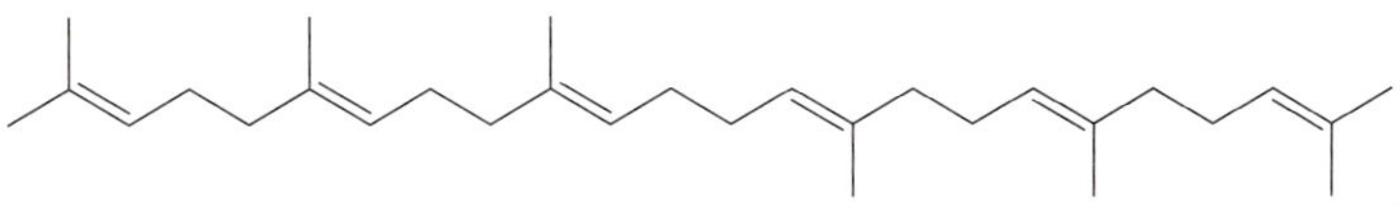 squalene

b. 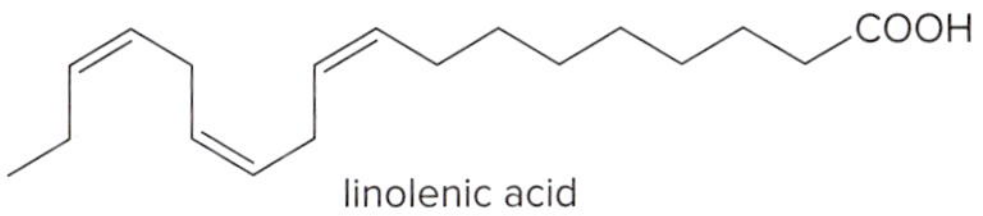 COOH linolenic acid

c. 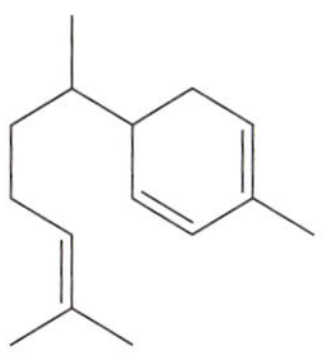 zingiberene

Identifying Compounds from Reactions

12.52 Identify compounds **A** and **B.** Compound **A** has molecular formula C_6H_{10} and gives $(CH_3)_2CHCH_2CH_2CH_3$ when treated with excess H_2 in the presence of Pd. **A** reacts with $NaNH_2$ and CH_3I to form compound **B** (molecular formula C_7H_{12}).

12.53 Oximene and myrcene, two hydrocarbons isolated from alfalfa that have the molecular formula $C_{10}H_{16}$, both yield 2,6-dimethyloctane when treated with H_2 and a Pd catalyst. Ozonolysis of oximene forms $(CH_3)_2C{=}O$, $CH_2{=}O$, $CH_2(CHO)_2$, and CH_3COCHO. Ozonolysis of myrcene yields $(CH_3)_2C{=}O$, $CH_2{=}O$ (two equiv), and $HCOCH_2CH_2COCHO$. Identify the structures of oximene and myrcene.

12.54 One compound that contributes to the "seashore smell" at beaches in Hawai'i is dictyopterene D', a component of a brown edible seaweed called limu lipoa. Hydrogenation of dictyopterene D' with excess H_2 in the presence of a Pd catalyst forms butylcycloheptane. Ozonolysis with O_3 followed by $(CH_3)_2S$ forms $CH_2(CHO)_2$, $HCOCH_2CH(CHO)_2$, and CH_3CH_2CHO. What are possible structures of dictyopterene D'?

12.55 Compound **W** has molecular formula $C_{14}H_{18}$ and reacts with H_2 to form **X.** Oxidative cleavage of **W** with O_3 followed by CH_3SCH_3 affords **Y.** What is the structure of **W?**

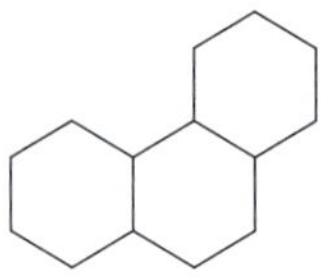

X

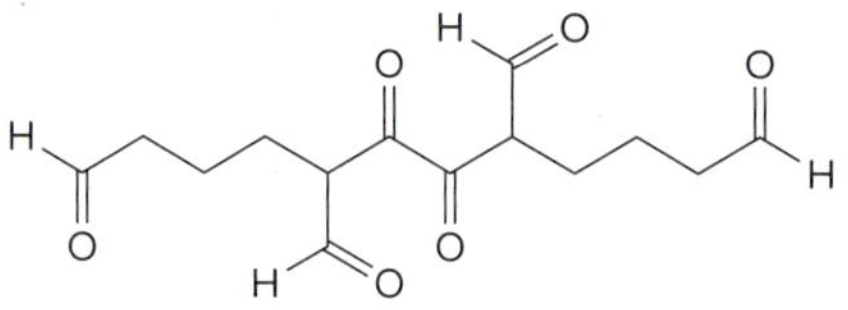

Y

Sharpless Asymmetric Epoxidation

12.56 Draw the product of each asymmetric epoxidation reaction.

a. 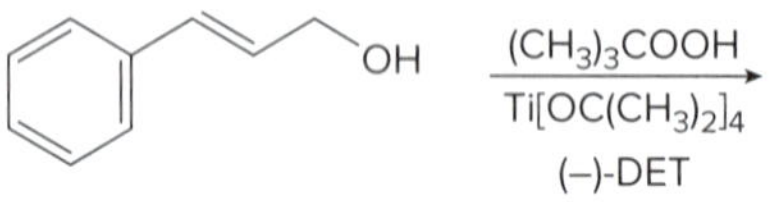 $(CH_3)_3COOH$, $Ti[OC(CH_3)_2]_4$, (–)-DET

b. 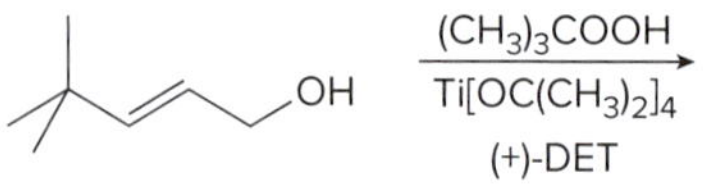 $(CH_3)_3COOH$, $Ti[OC(CH_3)_2]_4$, (+)-DET

12.57 Epoxidation of the following allylic alcohol using the Sharpless reagent with (–)-DET gives two epoxy alcohols in a ratio of 87:13.

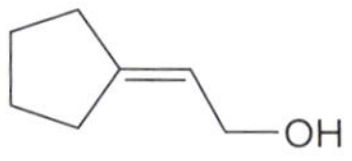

a. Assign structures to the major and minor product.

b. What is the enantiomeric excess in this reaction?

12.58 What allylic alcohol and DET isomer are needed to make each chiral epoxide using a Sharpless asymmetric epoxidation reaction?

a. 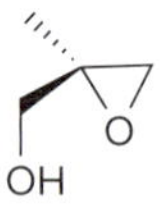 b. 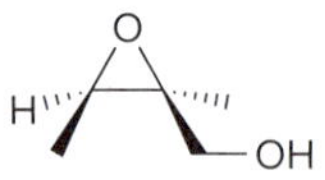 c.

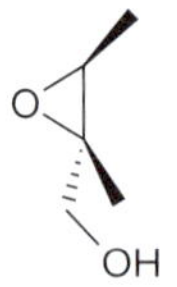

12.59 Identify **A** in the following reaction sequence, and draw a mechanism for the conversion of **A** to **B. B** has been converted to (*S*,*S*)-reboxetine, an antidepressant marketed outside the United States.

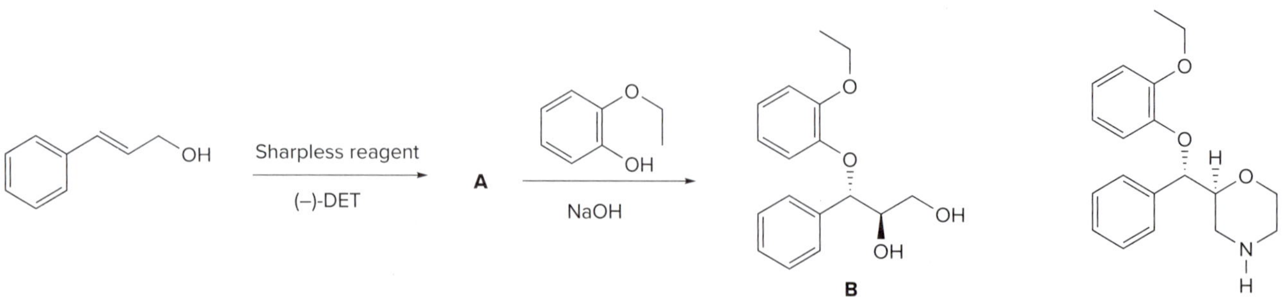

Sharpless reagent, (–)-DET → **A** → NaOH → **B**

(*S*,*S*)-reboxetine

Synthesis

12.60 Orchids release odors that are similar to the sex hormones of female bees in order to attract male bees to their flowers for pollination. Devise a synthesis of (*Z*)-tricosene, one pheromone component secreted by *Ophrys exaltata* orchids, from acetylene and alkyl halides.

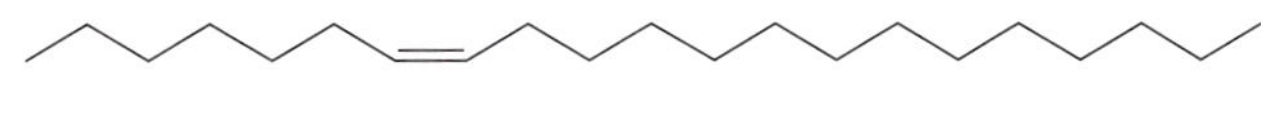

(*Z*)-tricosene

12.61 It is sometimes necessary to isomerize a cis alkene to a trans alkene in a synthesis, a process that cannot be accomplished in a single step. Using the reactions you have learned in Chapters 8–12, devise a stepwise method to convert *cis*-but-2-ene to *trans*-but-2-ene.

12.62 Devise a synthesis of each compound from acetylene and any other required reagents.

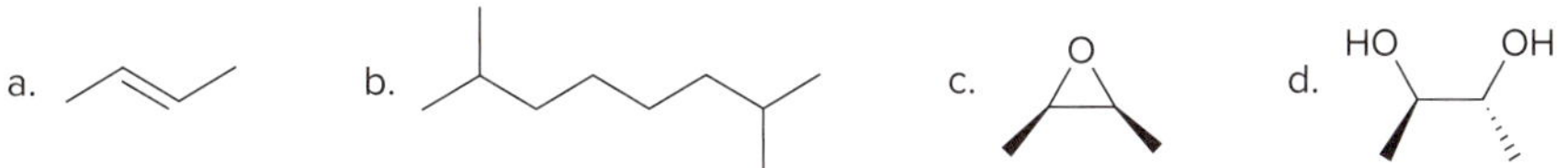

12.63 Devise a synthesis of compound **A** from the given starting materials. You may use any other inorganic reagents or organic alcohols. **A** was used to prepare aliskiren, a drug used to treat hypertension (see also Problem 5.6).

CH_3O ... Br, CH_3O — **A** ⟹ HO, CHO, CH_3O + Br⌒Br

aliskiren

12.64 Devise a synthesis of each compound from the indicated starting material, organic compounds containing one or two carbons, and any other required reagents.

a. → b. → (+ enantiomer) c. H−C≡C−H →

12.65 Devise a synthesis of each compound from the indicated starting material. You may use any other needed organic or inorganic reagents.

a. HO⌒OH ⟹ Bı b. ⟹ OH

12.66 Devise a synthesis of (3*R*,4*S*)-3,4-dichlorohexane from acetylene and any needed organic compounds or inorganic reagents.

12.67 Devise a synthesis of each compound from CH_3CH_2OH as the only organic starting material; that is, every carbon in the product must come from a molecule of ethanol. You may use any other needed inorganic reagents.

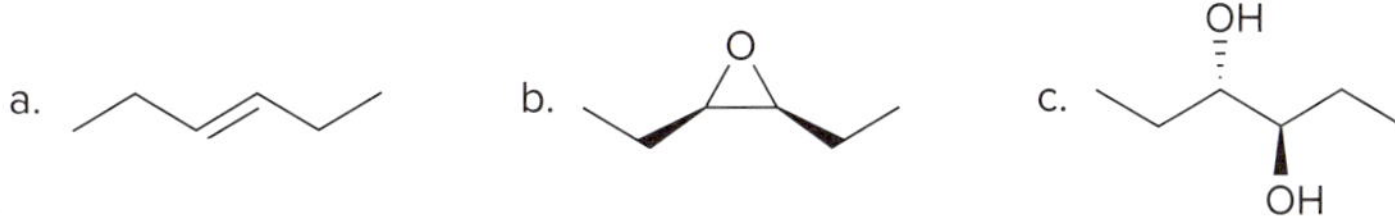

12.68 Devise a synthesis of **A** from the three starting materials given. You may use any other needed organic or inorganic reagents.

A ⟹ H−≡−H + (epoxide) + ⌒OH

Spectroscopy

Problems 12.69 and 12.70 are intended for students who have already learned about spectroscopy in Chapters A–C.

12.69 Treatment of alcohol **A** (molecular formula $C_5H_{12}O$) with CrO_3, H_2SO_4, and H_2O affords **B** with molecular formula $C_5H_{10}O$, which gives an IR absorption at 1718 cm^{-1}. The ^1H NMR spectrum of **B** contains the following signals: 1.10 (doublet, 6 H), 2.14 (singlet, 3 H), and 2.58 (septet, 1 H) ppm. What are the structures of **A** and **B**?

12.70 Treatment of compound **C** (molecular formula $C_9H_{12}O$) with PCC affords **D** (molecular formula $C_9H_{10}O$). Use the ^{1}H NMR and IR spectra of **D** to determine the structures of both **C** and **D**.

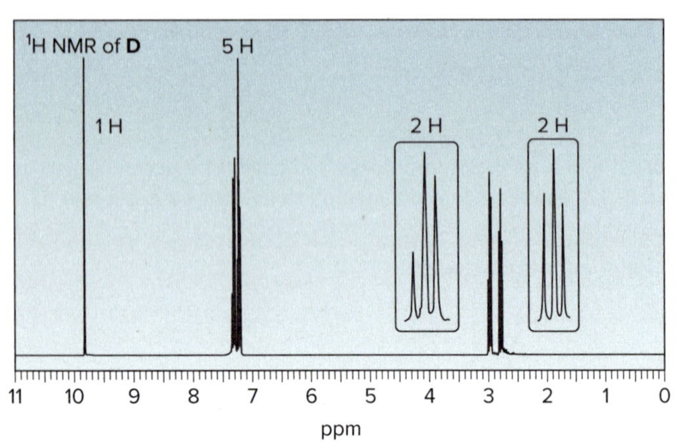

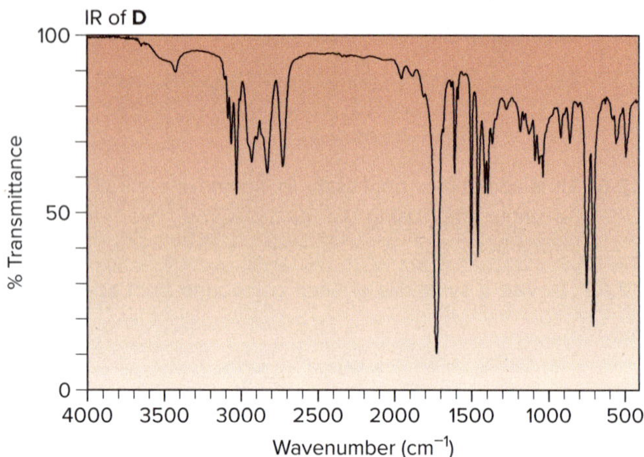

Challenge Problems

12.71 The Birch reduction is a dissolving metal reaction that converts substituted benzenes to cyclohexa-1,4-dienes using Li and liquid ammonia in the presence of an alcohol. Draw a stepwise mechanism for the following Birch reduction.

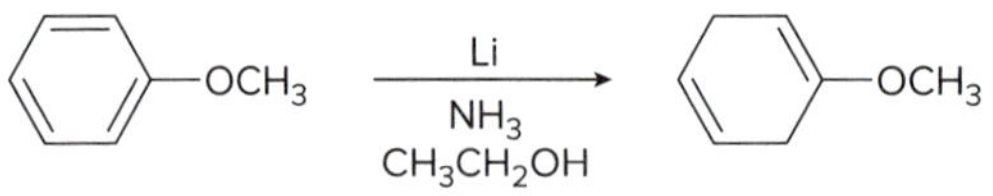

OCH_3 —(Li; NH_3, CH_3CH_2OH)→ OCH_3

12.72 In the Cr^{6+} oxidation of cyclohexanols, it is generally true that sterically hindered alcohols react faster than unhindered alcohols. Which of the following alcohols should be oxidized more rapidly?

OH OH

12.73 Dihydroxylation of an alkene can be carried out with H_2O_2 in HCO_2H. In this reaction, *trans*-but-2-ene affords (2*R*,3*S*)-butane-2,3-diol, whereas *cis*-but-2-ene affords a mixture of (2*R*,3*R*)-butane-2,3-diol and (2*S*,3*S*)-butane-2,3-diol. Does dihydroxylation by this method occur with syn or anti addition?

12.74 Draw a stepwise mechanism for the following reaction.

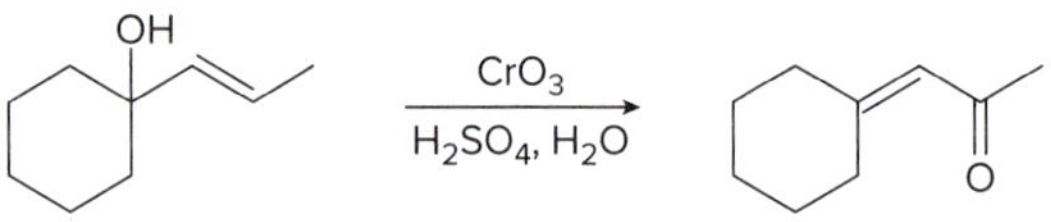

OH —(CrO_3; H_2SO_4, H_2O)→ O

12.75 Sharpless epoxidation of allylic alcohol **X** forms compound **Y**. Treatment of **Y** with NaOH and C_6H_5SH in an alcohol–water mixture forms **Z**. Identify the structure of **Y** and draw a mechanism for the conversion of **Y** to **Z**. Account for the stereochemistry of the stereogenic centers in **Z**. **Z** has been used as an intermediate in the synthesis of chiral carbohydrates.

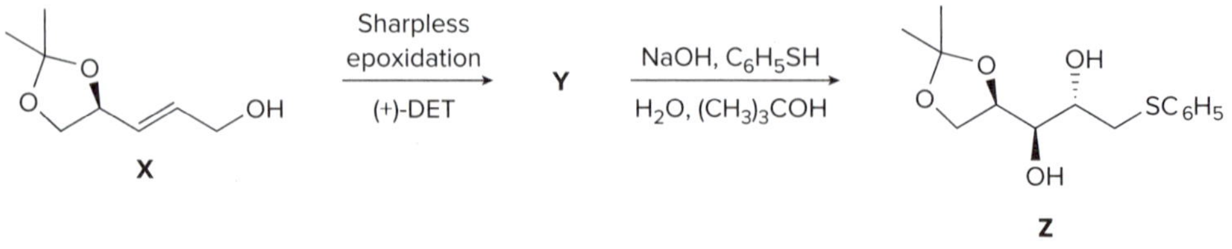

X (O, O, OH) —(Sharpless epoxidation; (+)-DET)→ **Y** —(NaOH, C_6H_5SH; H_2O, $(CH_3)_3COH$)→ **Z** (O, O, OH, SC_6H_5, OH)

SELF-TEST ANSWERS

1. c 2. c 3. a 4. b 5. a 6. b 7. d 8. b 9. b 10. d

SPECTROSCOPY

A

Mass Spectrometry

MizC/Getty Images

A.1 Mass spectrometry and the molecular ion
A.2 Alkyl halides and the M + 2 peak
A.3 Fragmentation
A.4 Fragmentation patterns of some common functional groups
A.5 Other types of mass spectrometry

Nootkatone is a fragrant lipid that is a key component of the aroma of grapefruit. Research from the Centers for Disease Control and Prevention has determined that nootkatone acts as an insecticide and repellent for ticks, mosquitoes, and other insects. Because its effectiveness may last significantly longer than plant-derived repellents like citronella and it is nontoxic, nootkatone is now licensed by the Food and Drug Administration as a component of products that control ticks and mosquitoes. To determine the structure of nootkatone and other complex organic compounds, chemists use a variety of instrumental methods. In Spectroscopy Part A, we examine mass spectrometry, a method to determine the molecular weight of an organic compound.

Why Study . . . Spectroscopy?

Whether a compound is prepared in the laboratory or isolated from a natural source, a chemist must determine its identity. Seventy years ago, determining the structure of an organic compound involved a series of time-consuming operations: measuring physical properties (melting point, boiling point, solubility, and density), identifying the functional groups using a series of chemical tests, and converting an unknown compound into another compound whose physical and chemical properties were then characterized as well.

Although still a challenging task, structure determination has been greatly simplified by modern instrumental methods. These techniques have both decreased the time needed for compound characterization, and increased the complexity of compounds whose structures can be completely determined.

In Spectroscopy A, we are introduced to **mass spectrometry (MS),** which is used to determine the molecular weight and molecular formula of a compound. In Spectroscopy B, we learn how **infrared (IR) spectroscopy** is used to identify a compound's functional groups. Spectroscopy C is devoted to **nuclear magnetic resonance (NMR) spectroscopy,** which is used to identify the carbon–hydrogen framework in a compound, making it the most powerful spectroscopic tool for organic structure analysis. Each method provides valuable information for determining the structure of an organic compound. We examine three methods that rely on the interaction of an energy source with a molecule to produce a change that is recorded in a spectrum.

A.1 Mass Spectrometry and the Molecular Ion

***Mass spectrometry* is a technique used for measuring the molecular weight and determining the molecular formula of an organic molecule.**

A.1A General Features

In the most common type of **mass spectrometer,** a molecule is vaporized and ionized, usually by bombardment with a beam of high-energy electrons, as shown in Figure A.1. The energy

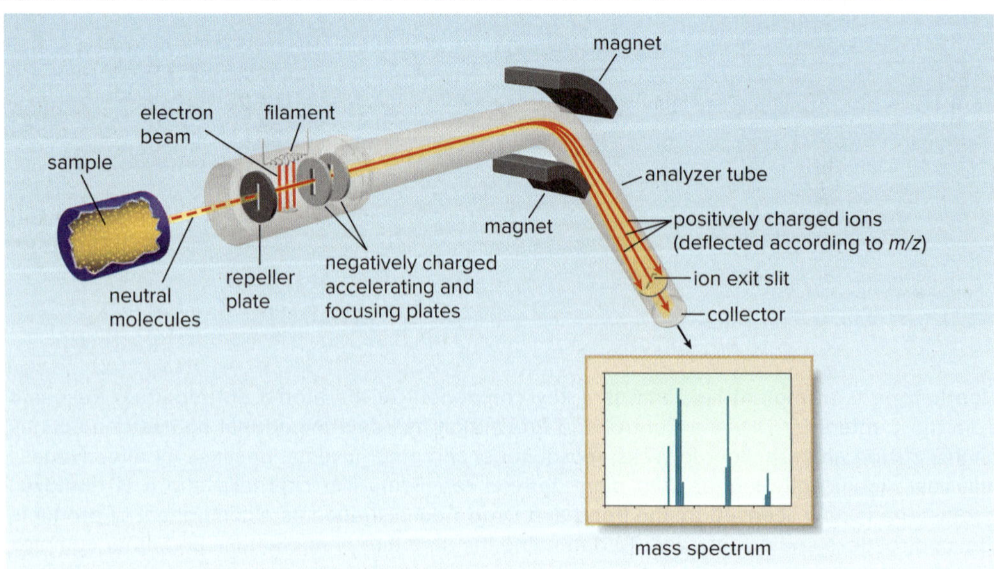

Figure A.1 Schematic of a mass spectrometer

- In a mass spectrometer, a sample is vaporized and bombarded by a beam of electrons to form an unstable radical cation, which then decomposes to smaller fragments. The positively charged ions are accelerated toward a negatively charged plate, and then passed through a curved analyzer tube in a magnetic field, where they are deflected by different amounts depending on their ratio of mass to charge (m/z). A mass spectrum plots the intensity of each ion versus its m/z ratio.

The term **spectroscopy** is usually used for techniques that use electromagnetic radiation as an energy source. Because the energy source in MS is a beam of electrons, the term **mass spectrometry** is used instead.

of these electrons is typically about 6400 kJ, or 70 electron volts (eV). This electron beam ionizes a molecule by causing it to eject an electron.

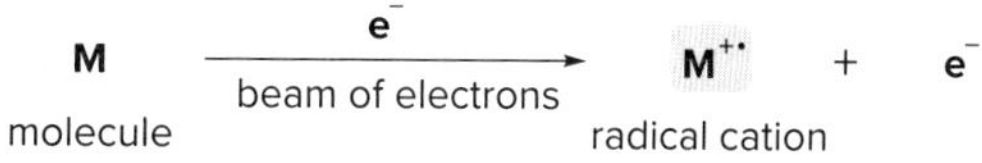

The species formed is a **radical cation,** symbolized $\mathbf{M^{+\bullet}}$**.** It is a radical because it has an unpaired electron, and it is a cation because it has one fewer electron than it started with.

- **The radical cation $M^{+\bullet}$ is called the *molecular ion* or the *parent ion.***

A single electron has a negligible mass, so the **mass of $M^{+\bullet}$ represents the molecular weight** of M. Because the molecular ion $M^{+\bullet}$ is inherently unstable, it decomposes. Single bonds break to form ***fragments,*** **radicals and cations having a lower molecular weight than the molecular ion.** A mass spectrometer analyzes the masses of cations only. The cations are accelerated in an electric field and deflected in a curved path in a magnetic field, thus sorting the molecular ion and its fragments by their **mass-to-charge (*m/z*) ratio.** Because *z* is almost always +1, *m/z* actually measures the mass (*m*) of the individual ions.

$$\underset{\text{molecule}}{\mathbf{M}} \xrightarrow{e^-} \underset{\text{unstable radical cation}}{\mathbf{M^{+\bullet}}} \longrightarrow \textbf{radicals} + \underset{\text{These fragments are analyzed.}}{\textbf{cations}}$$

- **A *mass spectrum* plots the amount of each cation (its relative abundance) versus its mass.**

The whole-number mass of CH_4 is (1 C × 12 amu) + (4 H × 1 amu) = 16 amu; amu = atomic mass unit.

A mass spectrometer analyzes the masses of *individual* molecules, not the weighted average mass of a group of molecules, so the whole-number masses of the most common individual isotopes must be used to calculate the mass of the molecular ion. Thus, the mass of the molecular ion for CH_4 should be 16. As a result, the mass spectrum of CH_4 shows a line for the molecular ion—the parent peak or **M** peak—at $m/z = 16$.

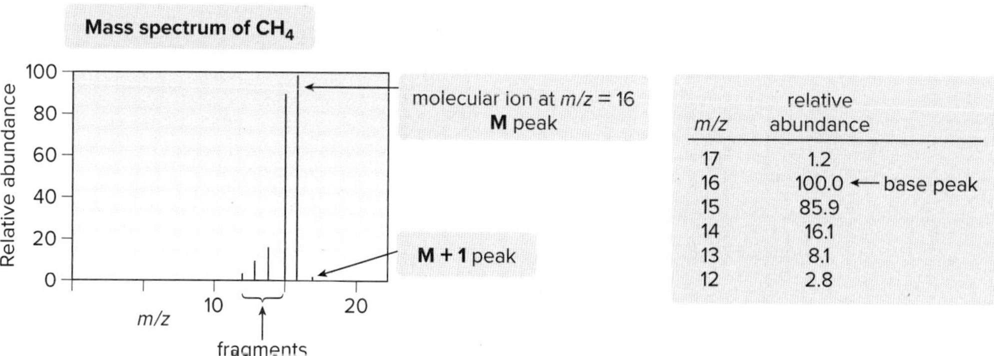

m/z	relative abundance
17	1.2
16	100.0 ← base peak
15	85.9
14	16.1
13	8.1
12	2.8

The tallest peak in a mass spectrum is called the **base peak.** For CH_4, the base peak is also the M peak, although this may *not* always be the case for all organic compounds.

The mass spectrum of CH_4 consists of more peaks than just the M peak. What is responsible for the peaks at $m/z < 16$? Because the molecular ion is unstable, it fragments into other cations and radical cations containing one, two, three, or four fewer hydrogen atoms than methane itself. Thus, the peaks at m/z = 15, 14, 13, and 12, are due to these lower-molecular-weight fragments. The decomposition of a molecular ion into lower-molecular-weight fragments is called **fragmentation.**

$$CH_4 \xrightarrow{e^-} (CH_4)^{+\bullet} \xrightarrow{-H\bullet} CH_3^{+} \xrightarrow{-H\bullet} CH_2^{+\bullet} \xrightarrow{-H\bullet} CH^{+} \xrightarrow{-H\bullet} C^{+\bullet}$$

mass 16 (**molecular ion**); mass 15, mass 14, mass 13, mass 12 (**fragments**)

What is responsible for the small peak at m/z = 17 in the mass spectrum of CH_4? Although most carbon atoms have an atomic mass of 12, 1.1% of them have an additional neutron in the nucleus, giving them an atomic mass of 13. When one of these carbon-13 isotopes forms methane, it gives a molecular ion peak at m/z = 17 in the mass spectrum. This peak is called the **M + 1** peak.

These key features—the molecular ion, the base peak, and the M + 1 peak—are illustrated in the mass spectrum of hexane in Figure A.2.

Figure A.2
Mass spectrum of hexane ($CH_3CH_2CH_2CH_2CH_2CH_3$)

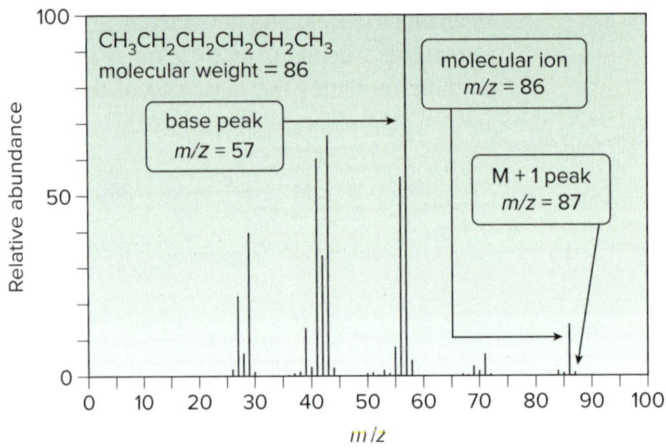

- The molecular ion for hexane (molecular formula C_6H_{14}) is at m/z = 86.
- The base peak (relative abundance = 100) occurs at m/z = 57.
- A small M + 1 peak occurs at m/z = 87.

Problem A.1 Label the molecular ion, the base peak, and the M + 1 peak in the mass spectrum of pentane (C_5H_{12}).

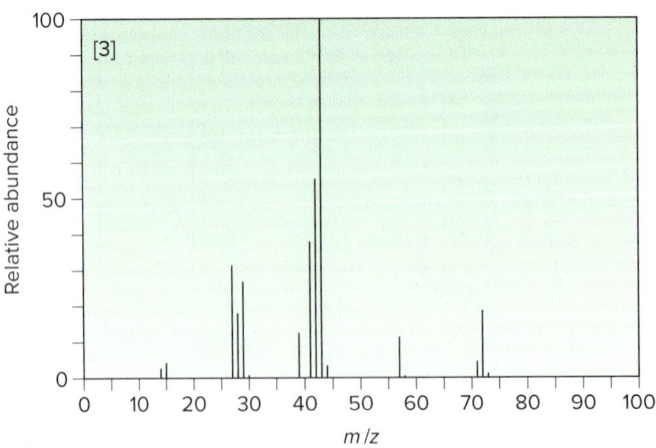

A.1B Analyzing Unknowns Using the Molecular Ion

Because the **mass of the molecular ion equals the molecular weight of a compound,** a mass spectrum can be used to distinguish between compounds that have similar physical properties but different molecular weights, as illustrated in Sample Problem A.1.

Sample Problem A.1 Using the Molecular Ion to Identify a Compound

Pent-1-ene and pent-1-yne are low-boiling hydrocarbons that have different molecular ions in their mass spectra. Match each hydrocarbon to its mass spectrum.

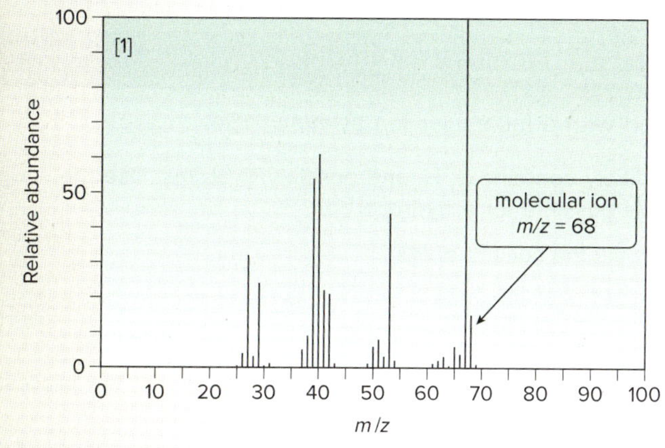

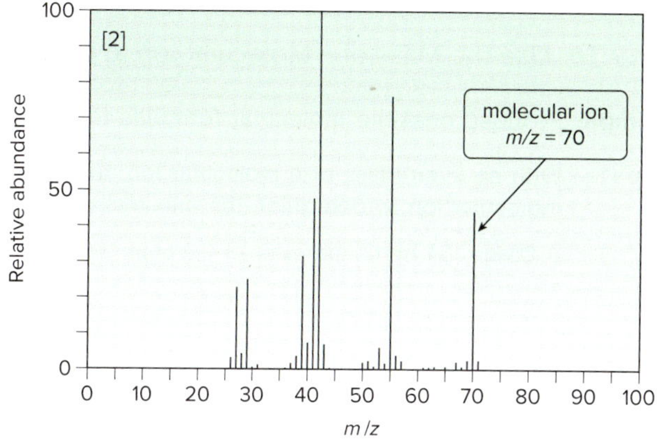

Solution

To solve this problem, first determine the molecular formula and molecular weight of each compound. Then, because the molecular weight of the compound equals the mass of the molecular ion, match the molecular weight to *m/z* for the molecular ion:

Compound	Molecular formula	Molecular weight = *m/z* of molecular ion	Spectrum
pent-1-ene	C_5H_{10}	70	[2]
pent-1-yne	C_5H_8	68	[1]

Problem A.2 What is the mass of the molecular ion formed from compounds having each molecular formula: (a) C_3H_6O; (b) $C_{10}H_{20}$; (c) $C_8H_8O_2$; (d) methamphetamine ($C_{10}H_{15}N$)?

More Practice: Try Problem A.18.

Problem A.3 Which compound gives a molecular ion at m/z = 122: $C_6H_5CH_2CH_2CH_3$, $C_6H_5COCH_2CH_3$, or $C_6H_5OCH_2CH_3$?

Hydrocarbons like methane (CH_4) and hexane (C_6H_{14}), as well as compounds that contain only C, H, and O atoms, always have a molecular ion with an *even* mass. An odd molecular ion generally indicates that a compound contains nitrogen.

The effect of N atoms on the mass of the molecular ion in a mass spectrum is called the **nitrogen rule: A compound that contains an *odd* number of N atoms gives an odd molecular ion.** Conversely, a compound that contains an *even* number of N atoms (including *zero*) gives an *even* molecular ion. Two "street" drugs that mimic the effects of heroin illustrate this principle: 3-methylfentanyl (two N atoms, even molecular weight) and MPPP (one N atom, odd molecular weight).

3-methylfentanyl
$C_{23}H_{30}N_2O$
molecular weight = 350

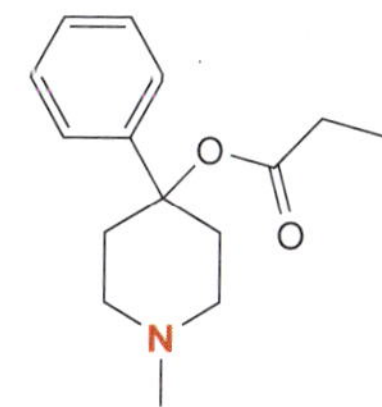

MPPP
(1-methyl-4-phenyl-4-propionoxypiperidine)
$C_{15}H_{21}NO_2$
molecular weight = 247

A.1C Using the Molecular Ion to Propose Molecular Formulas

How to use the molecular ion to propose molecular formulas for an unknown is shown in the stepwise procedure and Sample Problem A.2.

How To Use the Mass of a Molecular Ion to Propose Molecular Formulas for an Unknown

Example Propose possible molecular formulas for a compound with a molecular ion at m/z = 154.

Step [1] **With an even mass of a molecular ion, the compound likely contains C, H, and possibly O atoms. Use the molecular ion to determine the maximum number of C's possible for a hydrocarbon.**

- Divide 154 by 12, the mass of 1 C atom. The remainder gives the number of H's.

$$\frac{154}{12} = \text{12 C's maximum (remainder = 10)} \longrightarrow C_{12}H_{10}$$

Step [2] **To determine another possible molecular formula for a hydrocarbon, replace 1 C by 12 H's. Repeat the process until the formula has more than the maximum number of H's possible.**

$$C_{12}H_{10} \xrightarrow[+12\text{ H's}]{-1\text{ C}} C_{11}H_{22}$$

$$C_{11}H_{22} \xrightarrow[+12\text{ H's}]{-1\text{ C}} C_{10}H_{34}$$

too many H's
not a possible formula

- Because the maximum number of H's for a compound with 11 C's is 24 ($C_{11}H_{2(11)+2}$), $C_{10}H_{34}$ is not a possible formula.

Step [3] **To determine possible molecular formulas for compounds with O atoms, replace CH_4 (mass 16) by O (mass 16) in each formula. Repeat the process to give possible molecular formulas for compounds with two or more O atoms.**

- Four possibilities are shown.

$$C_{12}H_{10} \xrightarrow[+1\text{ O}]{-1\text{ CH}_4} C_{11}H_{6}O$$

$$C_{11}H_{22} \xrightarrow[+1\text{ O}]{-1\text{ CH}_4} C_{10}H_{18}O \xrightarrow[+1\text{ O}]{-1\text{ CH}_4} C_9H_{14}O_2 \xrightarrow[+1\text{ O}]{-1\text{ CH}_4} C_8H_{10}O_3$$

Sample Problem A.2 Using the Molecular Ion to Propose a Molecular Formula

Propose possible molecular formulas for a compound with a molecular ion at m/z = 86.

Solution

Because the molecular ion has an **even** mass, the compound likely contains C, H, and possibly O atoms. Begin by determining the molecular formula for a hydrocarbon having a molecular ion at 86. Then, because the mass of an O atom is 16 (the mass of CH_4), replace CH_4 by O to give a molecular formula containing one O atom. Repeat this last step to give possible molecular formulas for compounds with two or more O atoms.

For a molecular ion at m/z = 86:

Possible hydrocarbons:

- Divide 86 by 12 (mass of 1 C atom). This gives the maximum number of C's possible.

$$\frac{86}{12} = \text{7 C's maximum (remainder = 2)} \longrightarrow C_7H_2$$

- Replace 1 C by 12 H's for another possible molecular formula.

$$C_7H_2 \xrightarrow[+12\text{ H's}]{-1\text{ C}} C_6H_{14}$$

Possible compounds with C, H, and O:

- Substitute 1 O for CH_4. (This can't be done for C_7H_2.)

$$C_6H_{14} \xrightarrow[+1\text{ O}]{-\text{ CH}_4} C_5H_{10}O$$

- Repeat the process.

$$C_5H_{10}O \xrightarrow[+1\text{ O}]{-\text{ CH}_4} C_4H_6O_2$$

Problem A.4 Propose two molecular formulas for each of the following molecular ions: (a) 72; (b) 100; (c) 73.

More Practice: Try Problems A.20, A.21.

Sample Problem A.3 Using the Molecular Ion and Degrees of Unsaturation to Propose a Molecular Formula

Propose a molecular formula for nootkatone, the chapter-opening compound that contains the elements C, H, and O, has five degrees of unsaturation, and has a molecular ion in its mass spectrum at $m/z = 218$.

Solution

Determine possible molecular formulas using the procedure in Sample Problem A.2. Because each degree of unsaturation removes 2 H's, the correct molecular formula has 10 fewer H's than the maximum number.

For a molecular ion at $m/z = 218$: $\frac{218}{12}$ = 18 C's maximum (remainder = 2) ⟶ $C_{18}H_2$ — not enough H's

$$C_{18}H_2 \xrightarrow[+12\text{ H's}]{-1\text{ C}} C_{17}H_{14} \xrightarrow[+1\text{ O}]{-1\text{ CH}_4} C_{16}H_{10}O$$

not enough H's

$$C_{17}H_{14} \xrightarrow[+12\text{ H's}]{-1\text{ C}} C_{16}H_{26} \xrightarrow[+1\text{ O}]{-1\text{ CH}_4} C_{15}H_{22}O$$

five degrees of unsaturation
Answer

The maximum number of H's for a compound with 15 C's is $2n + 2 = 2(15) + 2 = 32$. A compound with 22 H's has 10 fewer H's than the maximum number and thus five degrees of unsaturation.

Problem A.5 Propose a molecular formula for each compound.

a. Cedrol, an alcohol in cedar oil, which has a molecular ion at $m/z = 222$ and three degrees of unsaturation
b. Caryophyllene, a hydrocarbon in hops and rosemary, which has a molecular ion at $m/z = 204$ and four degrees of unsaturation
c. β-Ionone, a ketone in violets, which has a molecular ion at $m/z = 192$ and four degrees of unsaturation

More Practice: Try Problems A.22, A.23.

A.2 Alkyl Halides and the M + 2 Peak

Most of the elements found in organic compounds, such as carbon, hydrogen, oxygen, nitrogen, sulfur, phosphorus, fluorine, and iodine, have one major isotope. **Chlorine** and **bromine,** on the other hand, have two, giving characteristic patterns to the mass spectra of their compounds.

Chlorine has two common isotopes, ^{35}**Cl** and ^{37}**Cl,** which occur naturally in a 3:1 ratio. Thus, **there are two peaks in a 3:1 ratio for the molecular ion of an alkyl chloride.** The larger peak—the **M** peak—corresponds to the compound containing ^{35}Cl, and the smaller peak—the **M + 2** peak—corresponds to the compound containing ^{37}Cl.

- **When the molecular ion consists of two peaks (M and M + 2) in a 3:1 ratio, a Cl atom is present.**

Sample Problem A.4 Determining the Molecular Ions for an Alkyl Chloride

What molecular ions will be present in a mass spectrum of 2-chloropropane, $(CH_3)_2CHCl$?

Solution

Calculate the molecular weight using each of the common isotopes of Cl.

Molecular formula	Mass of molecular ion (*m*/*z*)
$C_3H_7{}^{35}Cl$	78 (M peak)
$C_3H_7{}^{37}Cl$	80 (M + 2 peak)

There should be two peaks in a ratio of 3:1, at m/z = 78 and 80, as illustrated in the mass spectrum of 2-chloropropane in Figure A.3.

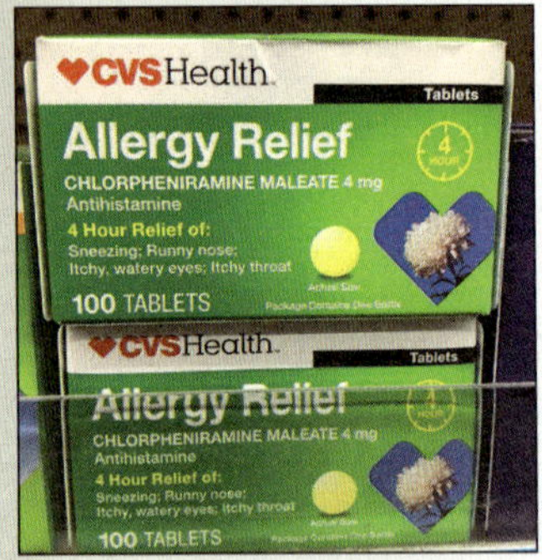

Daniel C. Smith

Problem A.6 What molecular ions will be present in the mass spectrum of the antihistamine chlorpheniramine?

Cl N N

chlorpheniramine

More Practice: Try Problems A.19, A.24.

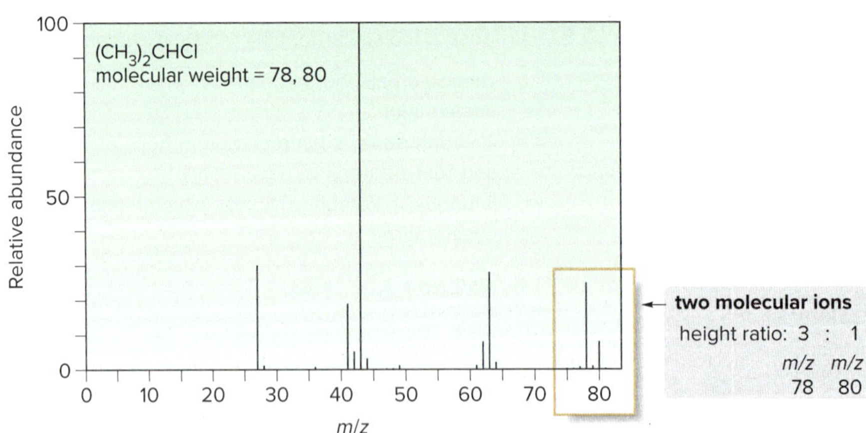

Figure A.3 Mass spectrum of 2-chloropropane $[(CH_3)_2CHCl]$

Bromine has two common isotopes, ^{79}**Br** and ^{81}**Br,** which occur naturally in a 1:1 ratio. Thus, **there are two peaks in a 1:1 ratio for the molecular ion of an alkyl bromide.** In the mass spectrum of 2-bromopropane (Figure A.4), for example, there is an M peak at m/z = 122 and an M + 2 peak at m/z = 124.

- **When the molecular ion consists of two peaks (M and M + 2) in a 1:1 ratio, a Br atom is present in the molecule.**

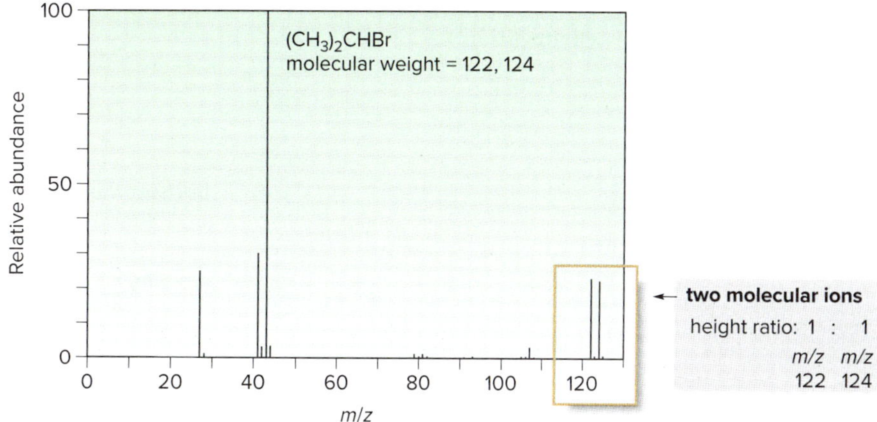

Figure A.4 Mass spectrum of 2-bromopropane [$(CH_3)_2CHBr$]

Problem A.7 What molecular ions would you expect for compounds having each of the following molecular formulas: (a) C_4H_9Cl; (b) C_3H_7F; (c) $C_4H_{11}N$; (d) $C_4H_4N_2$?

Problem A.8 What molecular ion(s) would you expect for each of the following drugs?

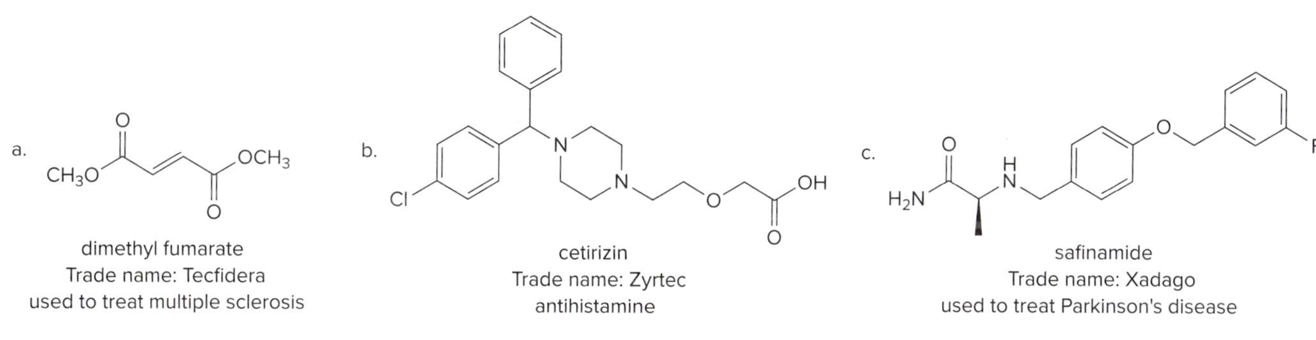

a. dimethyl fumarate
Trade name: Tecfidera
used to treat multiple sclerosis

b. cetirizin
Trade name: Zyrtec
antihistamine

c. safinamide
Trade name: Xadago
used to treat Parkinson's disease

Problem A.9 Assign one or more of the following phrases to compounds **A–D:** (a) odd molecular ion; (b) even molecular ion; (c) prominent M and M + 2 peaks in a 1:1 ratio; (d) prominent M and M + 2 peaks in a 3:1 ratio. Phrases may apply to one or more compounds.

Cl N **A** O OH **B** NH2 N **C** O O Br **D**

A.3 Fragmentation

While many chemists use a mass spectrum to determine only a compound's molecular weight and molecular formula, additional useful structural information can be obtained from fragmentation patterns. Although each organic compound fragments in a unique way, a particular functional group exhibits common fragmentation patterns.

As an example, consider hexane, whose mass spectrum was shown in Figure A.2. When hexane is bombarded by an electron beam, it forms a highly unstable radical cation (m/z = 86) that can decompose by cleavage of any of the C–C bonds. Thus, cleavage of the terminal C–C bond forms $CH_3CH_2CH_2CH_2CH_2^+$ and $CH_3\cdot$. Fragmentation generates a cation and a radical, and **cleavage generally yields the more stable, more substituted carbocation.**

e^-

Cleave the bond shown in red.

radical cation $m/z = 86$

cation $m/z = 71$

+ $\cdot CH_3$ radical

- **Loss of a CH_3 group always forms a fragment with a mass 15 units less than that of the molecular ion.**

As a result, the mass spectrum of hexane shows a peak at $m/z = 71$ due to $CH_3CH_2CH_2CH_2CH_2^+$. Figure A.5 illustrates how cleavage of other C–C bonds in hexane gives rise to other fragments that correspond to peaks in its mass spectrum.

Figure A.5

Identifying fragments in the mass spectrum of hexane

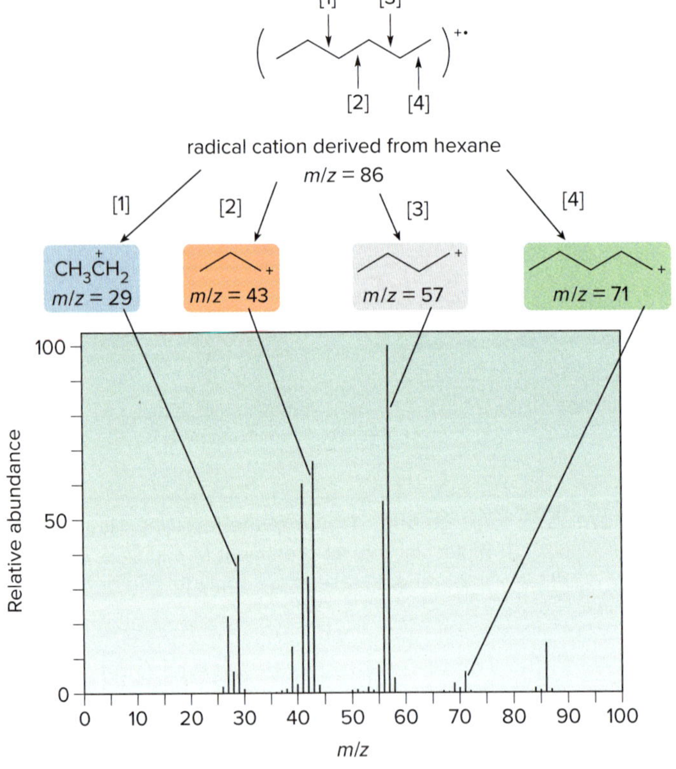

- Cleavage of C–C bonds (labeled [1]–[4]) in hexane forms lower-molecular-weight fragments that correspond to lines in the mass spectrum. Although the mass spectrum is complex, possible structures can be assigned to some of the fragments, as shown.

Sample Problem A.5 Assigning Possible Structures to Fragments in a Mass Spectrum

The mass spectrum of 2,3-dimethylpentane [$(CH_3)_2CHCH(CH_3)CH_2CH_3$] shows fragments at $m/z = 85$ and 71. Propose possible structures for the ions that give rise to these peaks.

Solution

To solve a problem of this sort, first calculate the mass of the molecular ion. Draw out the structure of the compound, break a C–C bond, and calculate the mass of the resulting fragments. Repeat this process on different C–C bonds until fragments of the desired mass-to-charge ratio are formed.

Cleave bond [1].

$m/z = 85$ + $\cdot CH_3$

[1] [2] e^-

$m/z = 100$

Cleave bond [2].

$m/z = 71$ + $\cdot CH_2CH_3$

In this example, 2,3-dimethylpentane has a molecular ion at $m/z = 100$. Cleavage of bond [1] forms a 2° carbocation with $m/z = 85$ and $CH_3\cdot$. Cleavage of bond [2] forms another 2° carbocation with $m/z = 71$ and $CH_3CH_2\cdot$. Thus, the fragments at $m/z = 85$ and 71 are possibly due to the two carbocations drawn.

Problem A.10 The mass spectrum of 2,3-dimethylpentane also shows peaks at $m/z = 57$ and 43. Propose possible structures for the ions that give rise to these peaks.

More Practice: Try Problem A.17.

Sample Problem A.6 Identifying Isomers from Fragmentation Patterns

Which mass spectrum (**A** or **B**) corresponds to each isomer: 2-methylpentane $[(CH_3)_2CHCH_2CH_2CH_3]$ or 2,2-dimethylbutane $[(CH_3)_3CCH_2CH_3]$?

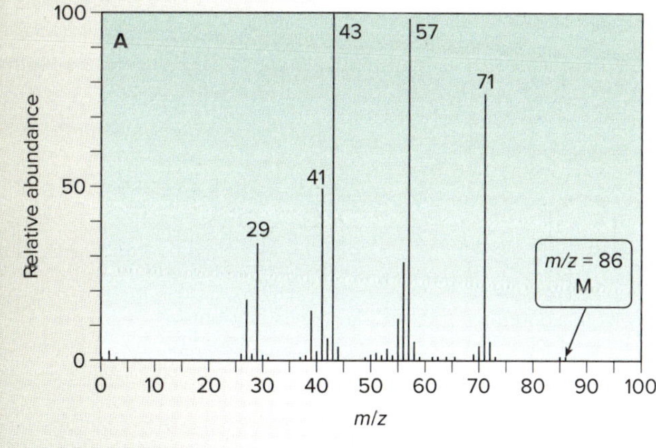

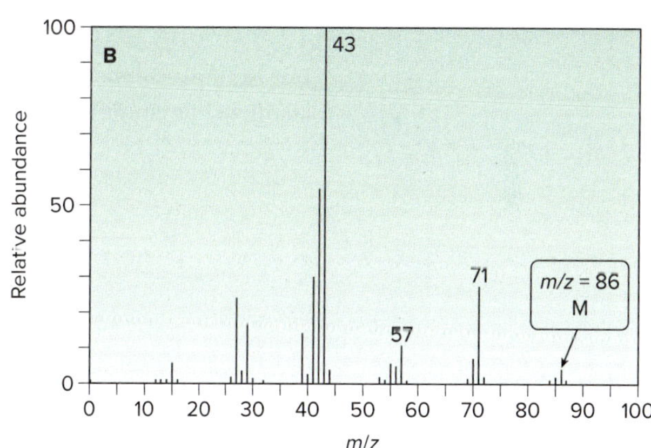

Solution

Because both compounds have the same molecular formula (C_6H_{14}) resulting in the same molecular ion at $m/z = 86$, fragmentation patterns must be used to distinguish the isomers. Look at bond cleavage in each compound that yields fragments that are more abundant in one spectrum than the other.

Break this bond.

e^-

2-methylpentane

$m/z = 43$

Break this bond.

e^-

+ $\cdot CH_2CH_3$

2,2-dimethylbutane

$m/z = 57$

For example, in 2-methylpentane, cleavage of the bond shown in red forms a fragment at $m/z = 43$ (due to a 2° carbocation), the base peak in spectrum **B.** In contrast, in 2,2-dimethylbutane, cleavage of the bond shown in blue forms a fragment at $m/z = 57$ (due to a 3° carbocation), a prominent fragment in spectrum **A.** Using this information, we can assign spectrum **B** to 2-methylpentane, and spectrum **A** to 2,2-dimethylbutane.

Problem A.11 The base peak in the mass spectrum of 2,2,4-trimethylpentane [$(CH_3)_3CCH_2CH(CH_3)_2$] occurs at $m/z = 57$. What ion is responsible for this peak and why is this ion the most abundant fragment?

More Practice: Try Problem A.30.

A.4 Fragmentation Patterns of Some Common Functional Groups

Each functional group exhibits characteristic fragmentation patterns that help to analyze a mass spectrum.

A.4A Aldehydes and Ketones

Aldehydes and ketones often undergo the process of **α cleavage, breaking the bond between the carbonyl carbon and the carbon adjacent to it.** Cleavage yields a neutral radical and a resonance-stabilized acylium ion.

:O:
R R'
Break this bond.
R = H or alkyl
α cleavage
$R-\overset{+}{C}=\ddot{O}:$ ⟷ $R-C\equiv\overset{+}{O}:$ + R'·
resonance-stabilized acylium ion

For example, α cleavage of benzophenone forms a fragment at $m/z = 105$ due to a resonance-stabilized acylium ion.

benzophenone
e⁻
Break this bond.
α cleavage
acylium ion
$m/z = 105$
(base peak)

Sample Problem A.7 Drawing the Fragments Formed from α Cleavage

What mass spectral fragments are formed from α cleavage of pentan-2-one, $CH_3COCH_2CH_2CH_3$?

Solution

Alpha (α) cleavage breaks the bond between the carbonyl carbon and the carbon adjacent to it, yielding a neutral radical and a resonance-stabilized acylium ion. A ketone like pentan-2-one with two different alkyl groups bonded to the carbonyl carbon has two different pathways for α cleavage.

:O:
[1] [2]
Cleave bond [1].
$:O=\overset{+}{C}-$ + $CH_3\cdot$
$m/z = 71$
Cleave bond [2].
$CH_3-\overset{+}{C}=O:$ + ·
$m/z = 43$

As a result, two fragments are formed by α cleavage of pentan-2-one, giving peaks at $m/z = 71$ and 43.

Problem A.12 What cations are formed in the mass spectrometer by α cleavage of each of the following compounds?

a. b. c. d.

More Practice: Try Problems A.29b, c; A.35.

A.4B Alcohols

Alcohols undergo fragmentation in two different ways—α cleavage and dehydration. Alpha (α) cleavage occurs by breaking a bond between an alkyl group and the carbon that bears the OH group, resulting in an alkyl radical and a resonance-stabilized carbocation.

:ÖH — α cleavage → :ÖH ⟷ $^+$ÖH + R·

Break this bond.

resonance-stabilized carbocation

Likewise, alcohols undergo dehydration, the elimination of H_2O, from two adjacent atoms. Unlike fragmentations discussed thus far, dehydration results in the cleavage of two bonds and forms H_2O and the radical cation derived from an alkene.

H OH — dehydration → + H_2O

For example, dehydration of cyclohexanol forms the radical cation of cyclohexene, a fragment with a mass 18 units less than that of the molecular ion, as shown in Figure A.6.

OH H — dehydration → + H_2O

cyclohexanol $m/z = 100$ cyclohexene $m/z = 82$

- **Loss of H_2O from an alcohol always forms a fragment with a mass 18 units less than the molecular ion.**

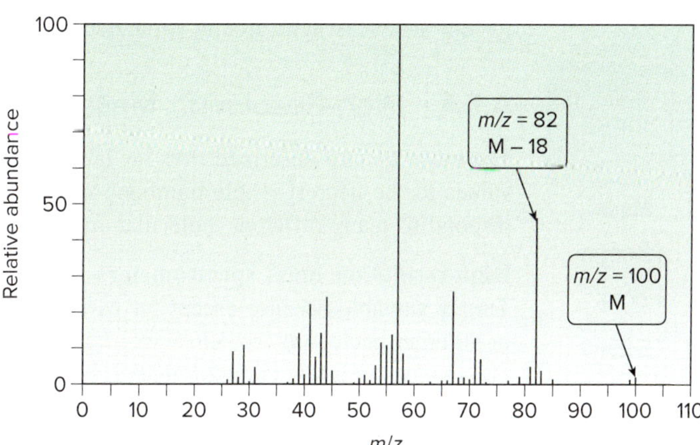

Figure A.6 The mass spectrum of cyclohexanol

Problem A.13 (a) What mass spectral fragments are formed by α cleavage of butan-2-ol, $CH_3CH(OH)CH_2CH_3$?

(b) What fragments are formed by dehydration of butan-2-ol?

A.4C Amines

Like alcohols, amines undergo fragmentation by α cleavage. Alpha (α) cleavage occurs by breaking the bond between an alkyl group and the carbon that bears the amine nitrogen, forming an alkyl radical and a resonance-stabilized carbocation.

:NR'_2 ⁺· — α cleavage → [:NR'_2 ⟷ $^+NR'_2$] + R·

Break this bond.

resonance-stabilized carbocation

For example, α cleavage of triethylamine (molecular ion at m/z = 101) forms CH_3· and a resonance-stabilized cation at m/z = 86.

Break this bond.

α cleavage

triethylamine
m/z = 101

resonance-stabilized carbocation
m/z = 86

+ ·CH_3

Problem A.14 Propose structures for the two fragments of highest abundance in the mass spectrum of hexan-3-amine.

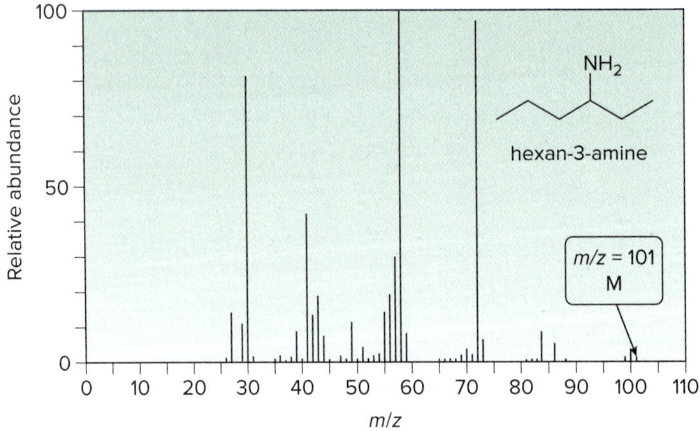

A.5 Other Types of Mass Spectrometry

Recent advances have greatly expanded the information obtained from mass spectrometry.

A.5A High-Resolution Mass Spectrometry

The mass spectra described thus far have been low-resolution spectra; that is, they report m/z values to the nearest whole number. As a result, the mass of a given molecular ion can correspond to many different molecular formulas, as shown in Sample Problem A.2.

High-resolution mass spectrometers measure m/z ratios to four (or more) decimal places. This is valuable because except for carbon-12, whose mass is defined as 12.0000, the masses of all other nuclei are very close to—but not exactly—whole numbers. Table A.1 lists the exact mass values of a few common nuclei. Using these values, it is possible to determine the single molecular formula that gives rise to a molecular ion.

Table A.1 Exact Masses of Some Common Isotopes

Isotope	Mass
^{12}C	12.0000
^{1}H	1.00783
^{16}O	15.9949
^{14}N	14.0031

For example, a compound having a molecular ion at m/z = 60 using a low-resolution mass spectrometer could have the following molecular formulas:

Formula	Exact mass
C_3H_8O	60.0575
$C_2H_4O_2$	60.0211
$C_2H_8N_2$	60.0688

If the molecular ion had an exact mass of 60.0578, the compound's molecular formula is C_3H_8O, because its mass is closest to the observed value.

Problem A.15 The low-resolution mass spectrum of an unknown analgesic **X** had a molecular ion of 151. Possible molecular formulas include $C_7H_5NO_3$, $C_8H_9NO_2$, and $C_{10}H_{17}N$. High-resolution mass spectrometry gave an exact mass of 151.0640. What is the molecular formula of **X?**

A.5B Gas Chromatography–Mass Spectrometry (GC–MS)

Two analytical tools—**gas chromatography (GC)** and **mass spectrometry (MS)**—can be combined into a single instrument **(GC–MS)** to analyze mixtures of compounds (Figure A.7a). The gas chromatograph separates the mixture, and then the mass spectrometer records a spectrum of the individual components.

A gas chromatograph consists of a thin capillary column containing a viscous, high-boiling liquid, all housed in an oven. When a sample is injected into the GC, it is vaporized and swept by an inert gas through the column. The components of the mixture travel through the column at different rates, often separated by boiling point, with lower-boiling compounds exiting the column before higher-boiling compounds. Each compound then enters the mass spectrometer, where it is ionized to form its molecular ion and lower-molecular-weight fragments. The GC–MS records a gas chromatogram for the mixture, which plots the amount of each component versus its **retention time**—that is, the time required to travel through the column. Each component of a mixture is characterized by its retention time in the gas chromatogram and its molecular ion in the mass spectrum (Figure A.7b).

Figure A.7 Compound analysis using GC–MS

a. Schematic of a GC–MS instrument

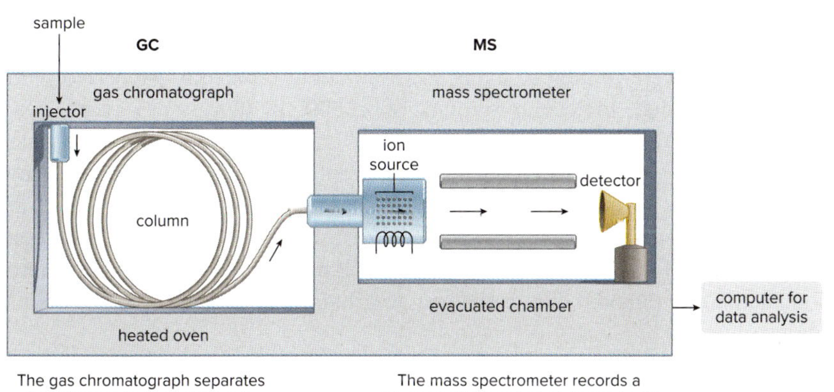

The gas chromatograph separates the mixture into its components.

The mass spectrometer records a spectrum of the individual components.

b. GC trace of a three-component mixture. The mass spectrometer gives a spectrum for each component.

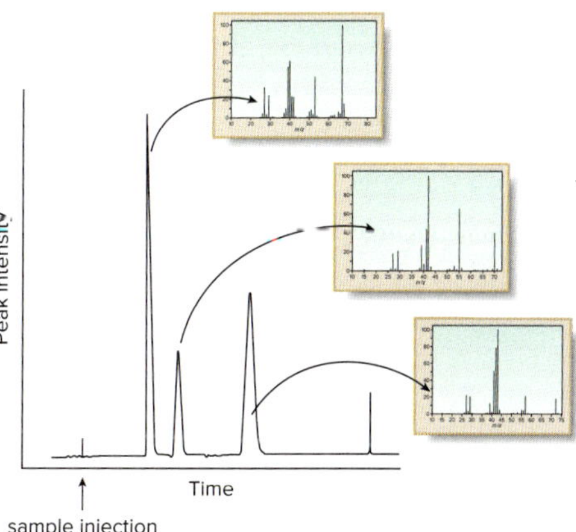

GC–MS is widely used for characterizing mixtures containing environmental pollutants. It is also used to analyze urine and hair samples for the presence of illegal drugs or banned substances thought to improve athletic performance.

To analyze a urine sample for THC (tetrahydrocannabinol), the principal psychoactive component of marijuana, the organic compounds are extracted from urine, purified, concentrated, and injected into the GC–MS. THC appears as a GC peak with a characteristic retention time (for a given set of experimental parameters), and gives a molecular ion at 314, its molecular weight, as shown in Figure A.8.

The recreational use of cannabis, which contains the psychoactive compound tetrahydrocannabinol (THC), is now legal in many parts of the United States. *Adam Bailleaux/Atomazul/123RF*

Figure A.8 Mass spectrum of tetrahydrocannabinol (THC)

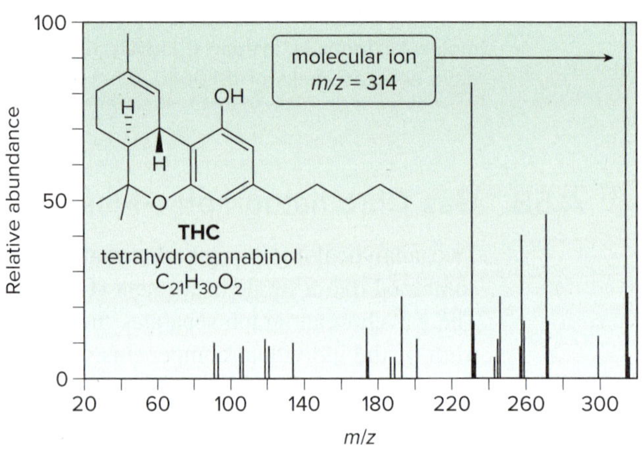

Problem A.16 Benzene, toluene, and *p*-xylene (BTX) are often added to gasoline to boost octane ratings. What would be observed if a mixture of these three compounds were subjected to GC–MS analysis? How many peaks would be present in the gas chromatogram? What would be the relative order of the peaks? What molecular ions would be observed in the mass spectra?

benzene toluene *p*-xylene

A.5C Mass Spectra of High-Molecular-Weight Biomolecules

Dr. John Fenn shared the 2002 Nobel Prize in Chemistry for his development of ESI mass spectrometry.

Until the 1980s mass spectra were limited to molecules that could be readily vaporized with heat under vacuum, and thus had molecular weights of < 800. In the last 35 years, new methods have been developed to generate gas phase ions of large molecules, allowing mass spectra to be recorded for large biomolecules such as proteins and carbohydrates. **Electrospray ionization (ESI),** for example, forms ions by creating a fine spray of charged droplets in an electric field. Evaporation of the charged droplets forms gaseous ions that are then analyzed by their *m*/*z* ratio. ESI and related techniques have extended mass spectrometry into the analysis of nonvolatile compounds with molecular weights greater than 100,000 daltons (atomic mass units).

Spectroscopy A CHAPTER REVIEW

KEY CONCEPTS

Molecular ion (M) in mass spectrometry (A.1, A.2)

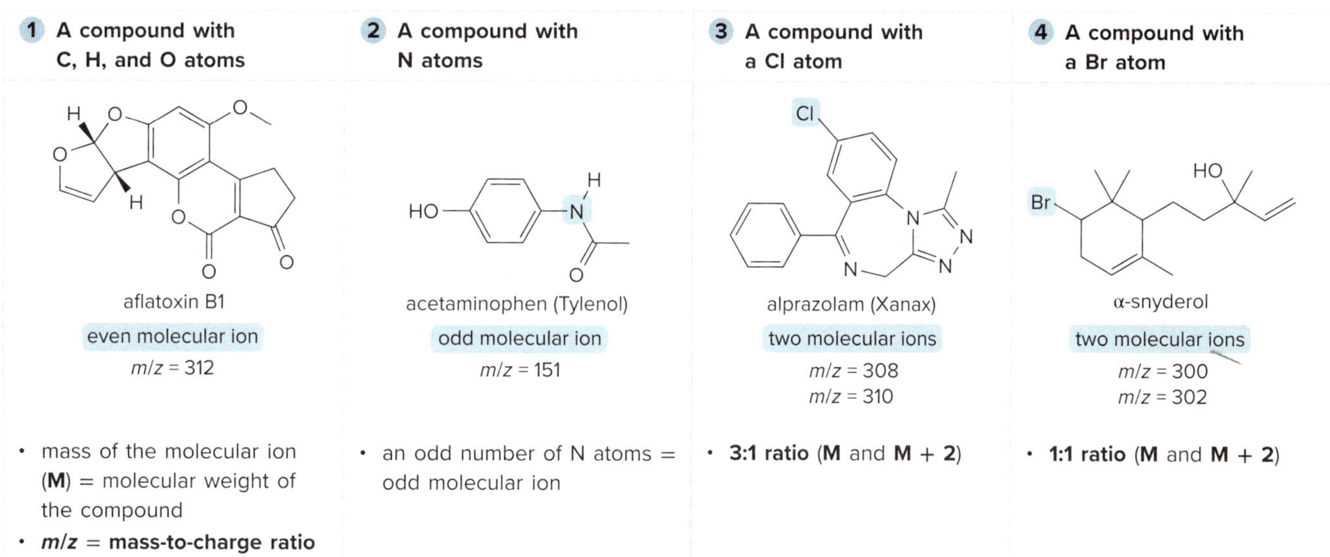

1 A compound with C, H, and O atoms	2 A compound with N atoms	3 A compound with a Cl atom	4 A compound with a Br atom
aflatoxin B1	acetaminophen (Tylenol)	alprazolam (Xanax)	α-snyderol
even molecular ion	odd molecular ion	two molecular ions	two molecular ions
m/z = 312	m/z = 151	m/z = 308 m/z = 310	m/z = 300 m/z = 302
• mass of the molecular ion (**M**) = molecular weight of the compound • ***m/z* = mass-to-charge ratio**	• an odd number of N atoms = odd molecular ion	• **3:1 ratio** (**M** and **M + 2**)	• **1:1 ratio** (**M** and **M + 2**)

Try Problems A.18, A.19, A.24, A.26a–c.

KEY SKILLS

[1] Proposing possible molecular formulas for a compound that contains C, H, and perhaps O with a given molecular ion (A.1); example: m/z = 100

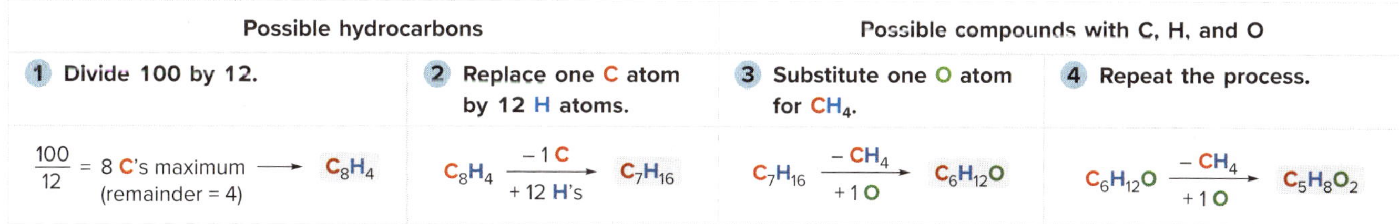

Possible hydrocarbons		Possible compounds with C, H, and O	
1 Divide 100 by 12.	**2 Replace one C atom by 12 H atoms.**	**3 Substitute one O atom for CH_4.**	**4 Repeat the process.**
$\frac{100}{12}$ = 8 C's maximum (remainder = 4) ⟶ C_8H_4	C_8H_4 $\xrightarrow[+12\ H\text{'s}]{-1\ C}$ C_7H_{16}	C_7H_{16} $\xrightarrow[+1\ O]{-CH_4}$ $C_6H_{12}O$	$C_6H_{12}O$ $\xrightarrow[+1\ O]{-CH_4}$ $C_5H_8O_2$

See *How To* (A.1C), Sample Problems A.2, A.3. Try Problems A.20–A.23.

[2] Determining the molecular ions for a compound with Cl or Br (A.2); example: bromocyclohexane ($C_6H_{11}Br$)

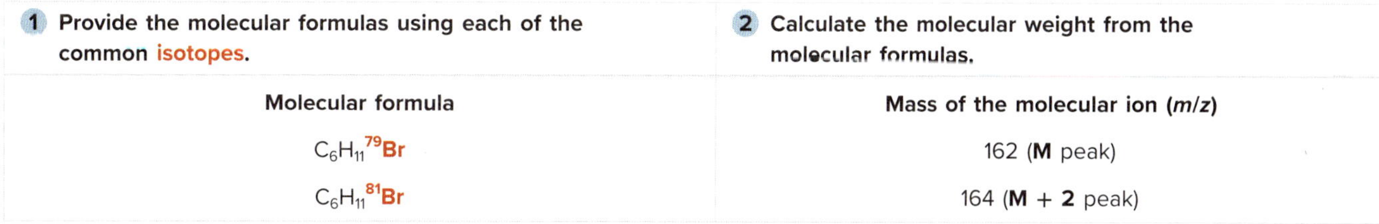

1 Provide the molecular formulas using each of the common isotopes.	2 Calculate the molecular weight from the molecular formulas.
Molecular formula	**Mass of the molecular ion (m/z)**
$C_6H_{11}{}^{79}Br$	162 (**M** peak)
$C_6H_{11}{}^{81}Br$	164 (**M + 2** peak)

See Sample Problem A.4, Figures A.3, A.4. Try Problems A.19, A.24.

[3] Proposing possible structures for fragmentation by α cleavage (A.3, A.4)

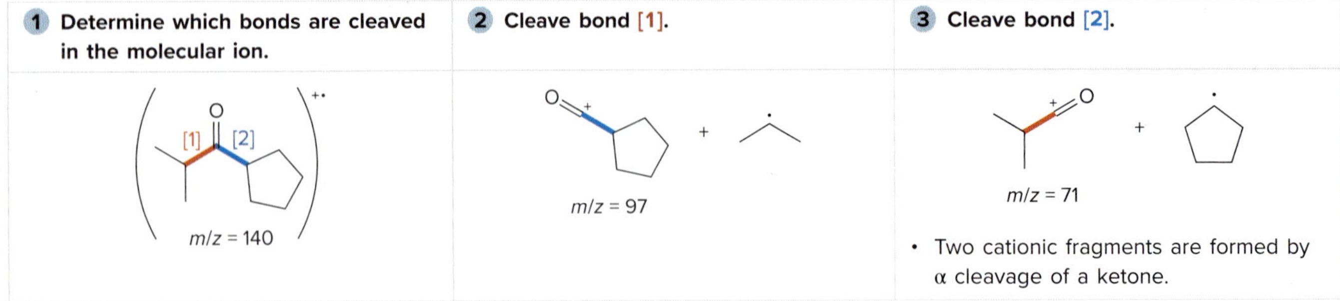

① Determine which bonds are cleaved in the molecular ion.	② Cleave bond [1].	③ Cleave bond [2].
m/z = 140	m/z = 97	m/z = 71 • Two cationic fragments are formed by α cleavage of a ketone.

See Sample Problems A.6, A.7, Figure A.5. Try Problems A.28–A.35.

SPECTROSCOPY CHAPTER A MULTIPLE-CHOICE SELF-TEST

The Self-Test consists of multiple-choice questions similar to those found on the American Chemical Society organic chemistry exam. Answers are given at the end of the chapter.

1. Which compound will show a prominent M + 2 peak in its mass spectrum?

a. b. c. d.

2. Which compound will show an odd molecular ion in its mass spectrum?

a. b. c. d.

3. Which fragment is *not* formed in the mass spectrum of **A?**

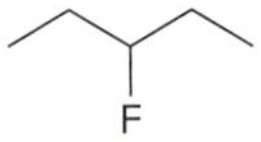

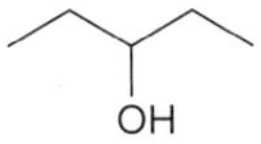

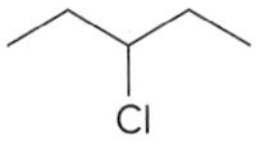

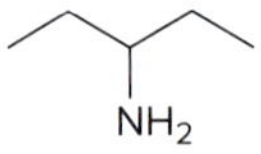

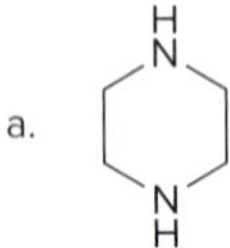

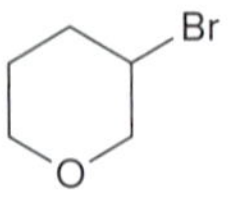

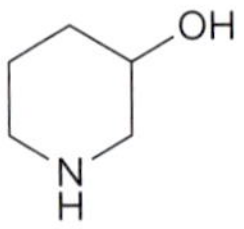

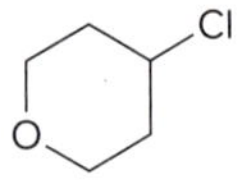

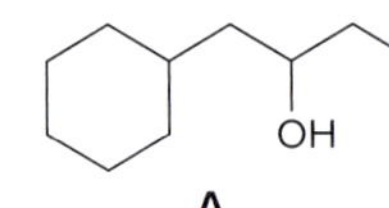

A

a. b. c. d.

4. Which compound has its molecular ion at the highest m/z value?

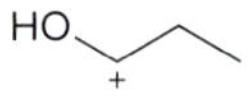

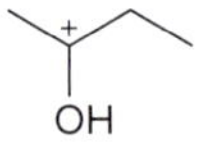

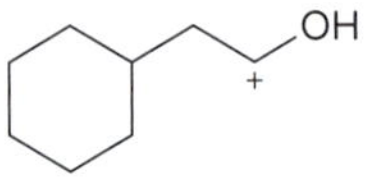

a. b. c. d.

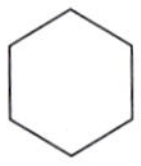

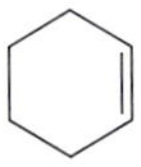

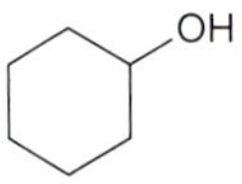

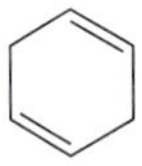

5. Which is *not* a possible molecular formula for a compound with a molecular ion at m/z = 102?

a. C_8H_6 b. C_7H_{18} c. $C_6H_{14}O$ d. $C_5H_{10}O_2$

6. Which fragment likely corresponds to the base peak in the mass spectrum of **B?**

B

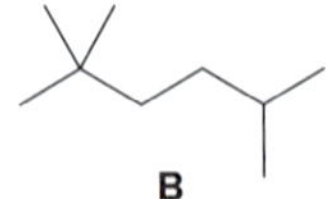

a. b. c. d.

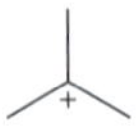

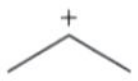

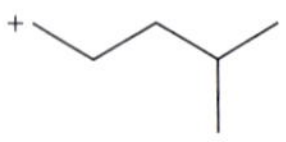

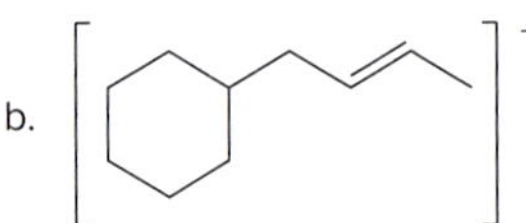

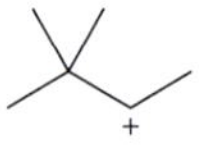

PROBLEMS

Problems that combine mass spectrometry and infrared spectroscopy are located at the end of Spectroscopy B. Problems that combine mass spectrometry, infrared spectroscopy, and nuclear magnetic resonance spectroscopy are found at the end of Spectroscopy C.

Problem Using a Three-Dimensional Model

A.17 The mass spectrum of the following compound shows fragments at m/z = 127, 113, and 85. Propose structures for the ions that give rise to these peaks.

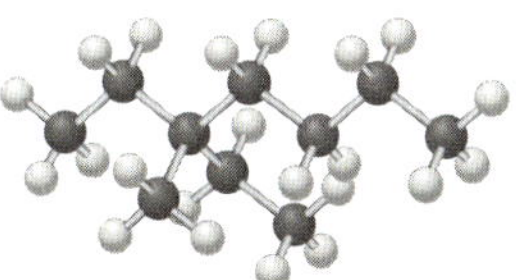

Molecular Ions and Molecular Formulas

A.18 What molecular ion is expected for each compound?

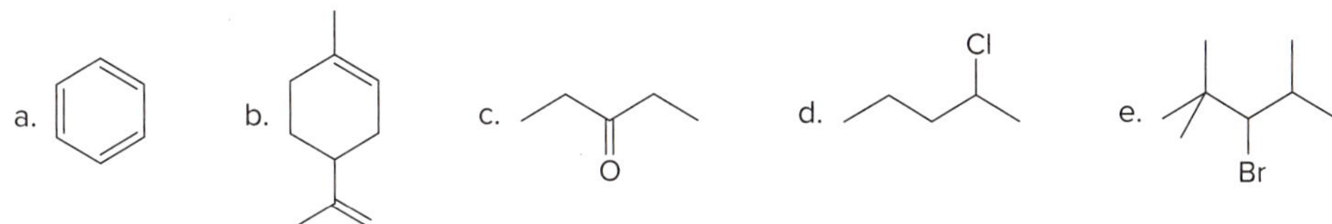

A.19 Diphenhydramine ($C_{17}H_{21}NO$), brompheniramine ($C_{16}H_{19}BrN_2$), and loratadine ($C_{22}H_{23}ClN_2O_2$) are three antihistamines in common over-the-counter products. Besides the difference in their m/z values, how can the characteristics of the molecular ion of each drug make it possible to differentiate one from another?

A.20 Propose two molecular formulas for each molecular ion: (a) 102; (b) 98; (c) 119; (d) 74.

A.21 Propose four possible structures for a hydrocarbon with a molecular ion at m/z = 112.

A.22 What is the molecular formula for α-himachalene, a hydrocarbon obtained from cedar wood, which has four degrees of unsaturation and has a molecular ion in its mass spectrum at m/z = 204?

A.23 Propose a molecular formula for rose oxide, a rose-scented compound isolated from roses and geraniums, which contains the elements of C, H, and O, has two degrees of unsaturation, and has a molecular ion in its mass spectrum at m/z = 154.

A.24 Match each structure to its mass spectrum.

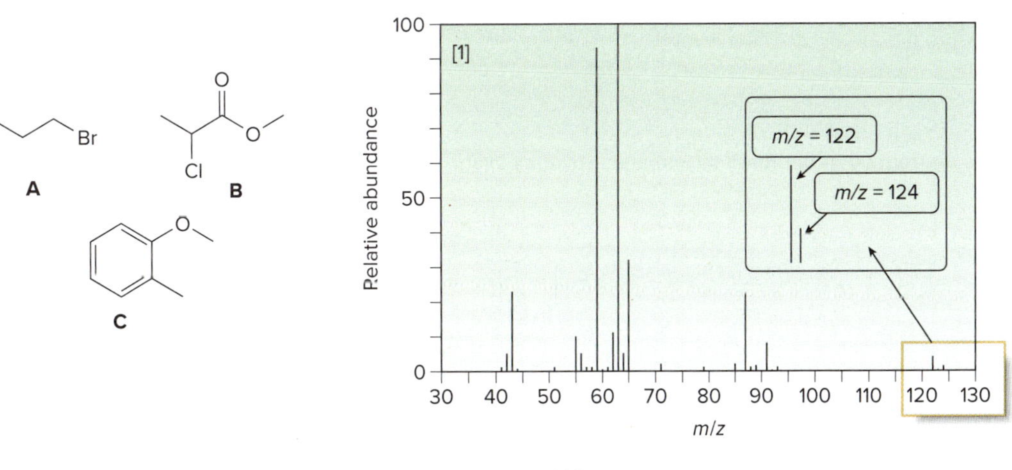

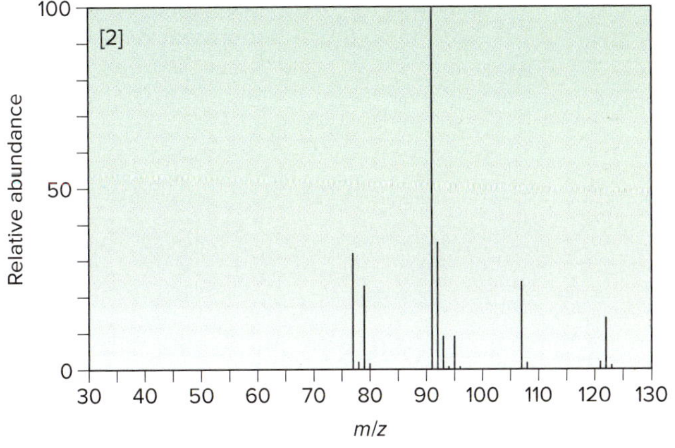

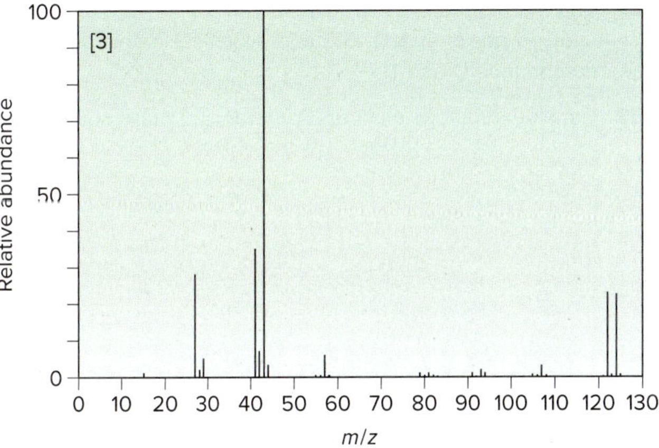

A.25 Propose two possible structures for a hydrocarbon having an exact mass of 96.0939 that forms ethylcyclopentane upon hydrogenation with H_2 and Pd-C.

A.26 Propose a structure consistent with each set of data.

a. a compound that contains a benzene ring and has a molecular ion at m/z = 107

b. a hydrocarbon that contains only sp^3 hybridized carbons and a molecular ion at m/z = 84

c. a compound that contains a carbonyl group and gives a molecular ion at m/z = 114

d. a compound that contains C, H, N, and O and has an exact mass for the molecular ion at 101.0841

A.27 A low-resolution mass spectrum of the neurotransmitter dopamine gave a molecular ion at m/z = 153. Two possible molecular formulas for this molecular ion are $C_8H_{11}NO_2$ and $C_7H_{11}N_3O$. A high-resolution mass spectrum provided an exact mass at 153.0680. Which of the possible molecular formulas is the correct one?

Fragmentation

A.28 Label each of the following in the mass spectrum of hexan-2-ol [$CH_3CH(OH)CH_2CH_2CH_2CH_3$]: the molecular ion, the base peak, the fragment resulting from the loss of H_2O, and α cleavage fragments.

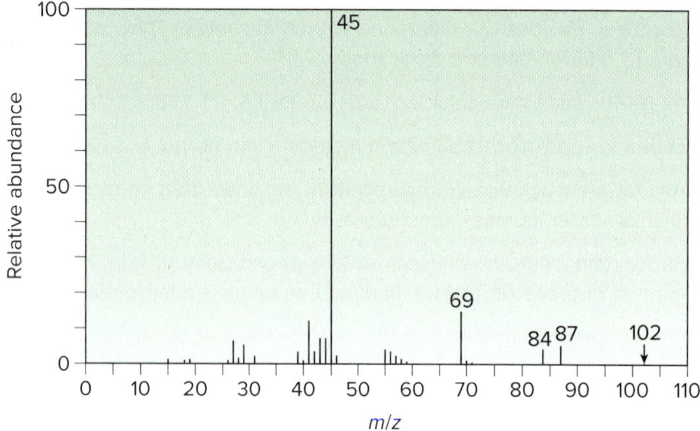

A.29 What cations are formed in the mass spectrometer by α cleavage of each of the following compounds?

a. OH b. O c. O d. NH_2

A.30 The mass spectra of two isomeric hydrocarbons—2,2-dimethylhexane (**A**) and 3-ethyl-2-methylpentane (**B**)—were recorded. Spectrum [1] showed peaks at 99 and 57 (base). Spectrum [2] showed peaks at 85, 71, and 43 (base). Assign spectrum [1] or [2] to compounds **A** and **B.**

A.31 The mass spectra of two isomeric alcohols—3,3-dimethylpentan-2-ol (**C**) and heptan-2-ol (**D**)—were recorded. Spectrum [1] showed peaks at 101, 87, and 71 (base). Spectrum [2] showed peaks at 101, 98, and 45 (base). Assign spectrum [1] or [2] to compounds **C** and **D.**

A.32 Consider isomeric alcohols **A** and **B** and mass spectra [1] and [2].

OH OH

A **B**

(a) Label the molecular ion and base peak in each spectrum. (b) Use the fragmentation patterns to determine which mass spectrum corresponds to isomer **A** and which corresponds to isomer **B.**

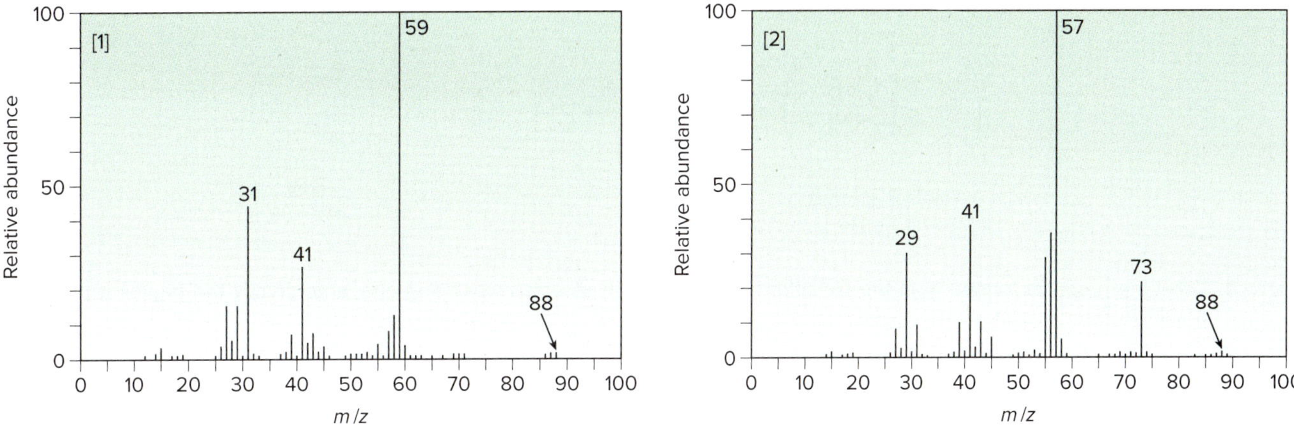

A.33 Consider the mass spectrum of hexan-2-amine. Label the molecular ion and base peak and propose a structure for the fragment that corresponds to the base peak.

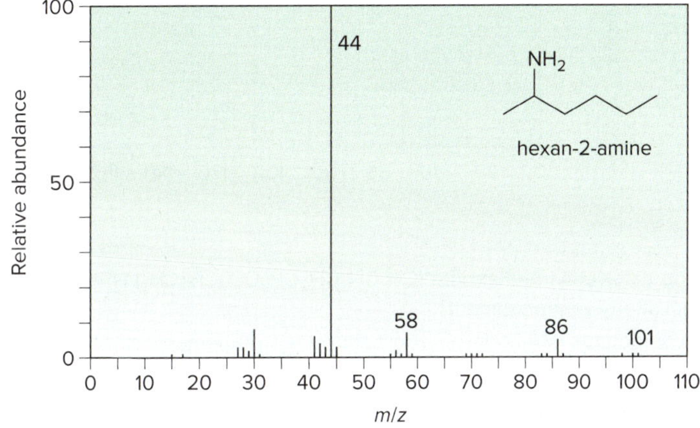

A.34 For each compound, assign likely structures to the fragments at each m/z value, and explain how each fragment is formed.

a. $C_6H_5CH_2CH_2OH$: peaks at $m/z = 104, 91$

b. $CH_2{=}C(CH_3)CH_2CH_2OH$: peaks at $m/z = 71, 68, 41, 31$

A.35 Suppose you have two bottles, labeled ketone **A** and ketone **B.** You know that one bottle contains $CH_3CO(CH_2)_5CH_3$ and one contains $CH_3CH_2CO(CH_2)_4CH_3$, but you do not know which ketone is in which bottle. Ketone **A** gives a fragment at $m/z = 99$ and ketone **B** gives a fragment at $m/z = 113$. What are the likely structures of ketones **A** and **B** from these fragmentation data?

A.36 Primary (1°) alcohols often show a peak in their mass spectra at $m/z = 31$. Suggest a structure for this fragment.

A.37 Like alcohols, ethers undergo α cleavage by breaking a carbon–carbon bond between an alkyl group and the carbon bonded to the ether oxygen atom; that is, the red C–C bond in R–CH_2OR' is broken. With this in mind, propose structures for the fragments formed by α cleavage of $(CH_3)_2CHCH_2OCH_2CH_3$. Suggest a reason why an ether fragments by α cleavage.

Challenge Problems

A.38 What molecular ions would be present in the mass spectrum of a compound that contains C, H, and (a) 1 Br and 1 Cl; (b) 3 Br's? Give the relative peak intensities of the molecular ions in each case.

A.39 In addition to α cleavage, some aldehydes and ketones undergo the McLafferty rearrangement. In the McLafferty rearrangement, a hydrogen on a carbon three atoms from the C=O is transferred to the carbonyl oxygen and a carbon–carbon bond is broken. This process forms an alkene and the radical cation derived from an enol, which appears as a fragment in the mass spectrum.

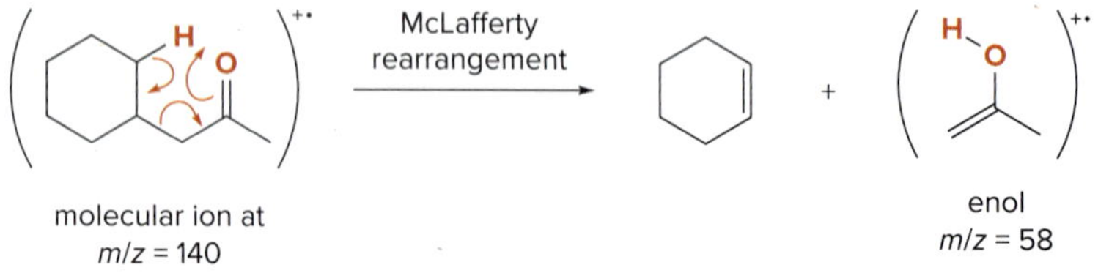

McLafferty rearrangement

molecular ion at $m/z = 140$

enol $m/z = 58$

a. Draw the products formed from the McLafferty rearrangement of 1-phenylpentan-1-one, and identify the fragment that results in the given mass spectrum.

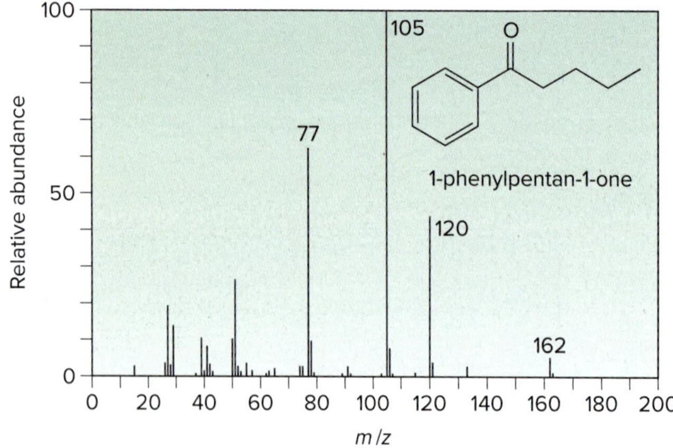

b. If a mass spectrum of the ester ethyl pentanoate ($CH_3CH_2CH_2CH_2CO_2CH_2CH_3$) is recorded, what is the mass of the radical cation formed by the McLafferty rearrangement?

c. Which of the following compounds can undergo a McLafferty rearrangement?

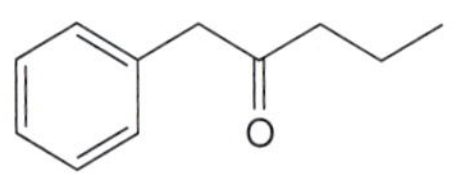

A 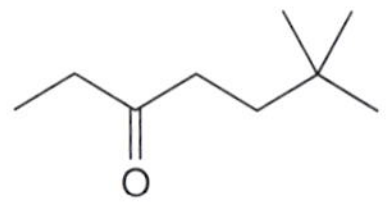 B 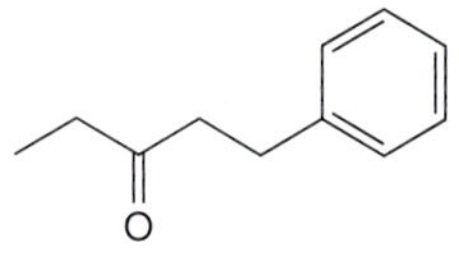 C

SELF-TEST ANSWERS

1. c 2. c 3. d 4. c 5. b 6. a

SPECTROSCOPY

B

Infrared Spectroscopy

T.Daly/Alamy Stock Photo

B.1 Electromagnetic radiation

B.2 The general features of infrared spectroscopy

B.3 IR absorptions

B.4 Infrared spectra of common functional groups

B.5 IR and structure determination

The serendipitous discovery of **penicillin** from a mold of the genus *Penicillium* by Scottish bacteriologist Sir Alexander Fleming in 1928 is considered one of the single most important events in the history of medicine. Penicillin G and related compounds are members of the β-lactam family of antibiotics, all of which contain a strained four membered amide ring that is responsible for their biological activity. Penicillin was first used to cure a streptococcal infection in 1942, and by 1944 penicillin production was given high priority by the U.S. government, because it was needed to treat the many injured soldiers in World War II. The unusual structure of penicillin was elucidated by modern instrumental methods in the 1940s. In Spectroscopy Part B, we learn about infrared spectroscopy, which is used to determine the functional groups in organic compounds like penicillin.

Why Study . . .
Infrared Spectroscopy?

Although mass spectrometry tells us the molecular weight and molecular formula for an organic compound, other forms of spectroscopy must be used to completely delineate the structure of a complex compound. **Infrared spectroscopy** is a technique that uses infrared light to interact with compounds, causing bonds to bend and vibrate, and giving a **spectrum with characteristic absorptions for particular functional groups.** Because the properties and reactions of an organic compound are determined in large part by what functional groups it contains, infrared spectroscopy is a valuable method for determining the structure of compounds isolated from natural sources, and for monitoring the progress of reactions that result in the addition or removal of functional groups.

We begin this chapter by learning about infrared light, the energy source used in infrared spectroscopy.

B.1 Electromagnetic Radiation

Infrared (IR) spectroscopy and **nuclear magnetic resonance (NMR)** spectroscopy (Part C) both use a form of electromagnetic radiation as their energy source. To understand IR and NMR, therefore, you need to understand some of the properties of **electromagnetic radiation**—radiant energy having dual properties of both waves and particles.

The particles of electromagnetic radiation are called **photons,** each having a discrete amount of energy called a **quantum.** Because electromagnetic radiation also has wave properties, it can be characterized by its **wavelength** and **frequency.**

- Wavelength (λ) is the distance from one point on a wave (e.g., the peak or trough) to the same point on the adjacent wave. A variety of different length units are used for λ, depending on the type of radiation.
- Frequency (ν) is the number of waves passing a point per unit time. Frequency is reported in cycles per second (s^{-1}), which is also called hertz (Hz).

Length units used to report wavelength include:

Unit	Length
meter (m)	1 m
centimeter (cm)	10^{-2} m
micrometer (μm)	10^{-6} m
nanometer (nm)	10^{-9} m
Angstrom (Å)	10^{-10} m

You come into contact with many different kinds of electromagnetic radiation in your daily life. You use visible light to see the words on this page, you may cook with microwaves, and you should use sunscreen to protect your skin from the harmful effects of ultraviolet radiation.

The different forms of electromagnetic radiation make up the **electromagnetic spectrum.** The spectrum is arbitrarily divided into different regions, as shown in Figure B.1. All electromagnetic radiation travels at the speed of light (c), 3.0×10^8 m/s.

Figure B.1
The electromagnetic spectrum

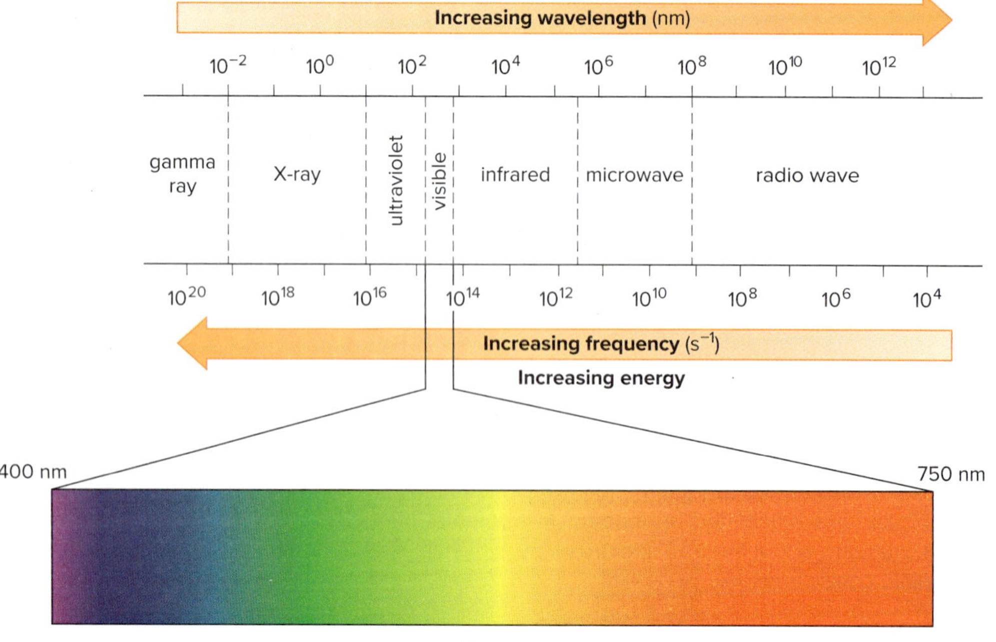

- Visible light occupies only a small region of the electromagnetic spectrum.

The speed of electromagnetic radiation (c) is directly proportional to its wavelength and frequency:

$$c = \lambda\nu$$

The speed of light (c) is a constant, so wavelength and frequency are *inversely* related:

- **$\lambda = c/\nu$: Wavelength increases as frequency decreases.**
- **$\nu = c/\lambda$: Frequency increases as wavelength decreases.**

The energy (E) of a photon is directly proportional to its frequency where h = Planck's constant (6.63×10^{-34} J · s).

$$E = h\nu$$

Frequency and wavelength are *inversely* proportional ($\nu = c/\lambda$), however, so energy and wavelength are *inversely* proportional:

$$E = h\nu = \frac{hc}{\lambda}$$

- **The energy of electromagnetic radiation increases as frequency increases and wavelength decreases.**

When electromagnetic radiation strikes a molecule, some wavelengths—but not all—are absorbed. Only some wavelengths are absorbed because molecules have discrete energy levels. The energies of their electronic, vibrational, and nuclear spin states are *quantized,* not *continuous.*

- **For absorption to occur, the energy of the photon must match the difference between two energy states in a molecule.**

higher-energy state

ΔE

For absorption to occur, the energy of the incident electromagnetic radiation must match ΔE.

lower-energy state

ΔE = the energy difference between two states in a molecule

- **The *larger* the energy difference between two states, the *higher* the energy of radiation needed for absorption, the *higher* the frequency, and the *shorter* the wavelength.**

Problem B.1 Which of the following has the higher frequency: (a) light having a wavelength of 10^2 or 10^4 nm; (b) light having a wavelength of 100 nm or 100 μm; (c) red light or blue light?

Problem B.2 Which of the following has the higher energy: (a) light having a ν of 10^4 Hz or 10^8 Hz; (b) light having a λ of 10 nm or 1000 nm; (c) red light or blue light?

B.2 The General Features of Infrared Spectroscopy

Organic chemists use infrared (IR) spectroscopy to identify the functional groups in a compound.

B.2A Background

Using the wavenumber scale results in IR values in a numerical range that is easier to report than the corresponding frequencies given in hertz (4000–400 cm^{-1} compared to 1.2×10^{14}–1.2×10^{15} Hz).

Infrared radiation (λ = 2.5–25 μm) is the energy source in infrared spectroscopy. Infrared light has somewhat longer wavelengths than visible light, making infrared light lower in frequency and lower in energy than visible light. Frequencies in IR spectroscopy are reported using a unit called the **wavenumber** ($\tilde{\nu}$):

$$\tilde{\nu} = \frac{1}{\lambda}$$

Wavenumber is *inversely* proportional to wavelength and reported in **reciprocal centimeters (cm^{-1}).** Wavenumber ($\tilde{\nu}$) is *proportional* to frequency (ν). **Frequency (and therefore energy) increases as the wavenumber increases.** Using the wavenumber scale, IR absorptions occur from **4000 cm^{-1}** to **400 cm^{-1}.**

- **Absorption of IR light causes changes in the vibrational motions of a molecule.**

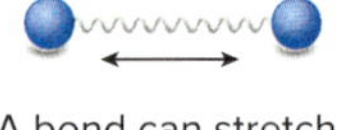

A bond can stretch.

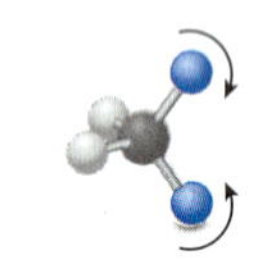

Two bonds can bend.

Covalent bonds are not static. They are more like springs with weights on each end. When two atoms are bonded to each other, the bond stretches back and forth. When three or more atoms are joined together, bonds can also bend. These bond stretching and bending vibrations represent the different vibrational modes available to a molecule.

These vibrations are quantized, so they occur only at specific frequencies, which correspond to the frequency of IR light. **When the frequency of IR light matches the frequency of a particular vibrational mode, the IR light is absorbed,** causing the amplitude of the particular bond stretch or bond bend to increase.

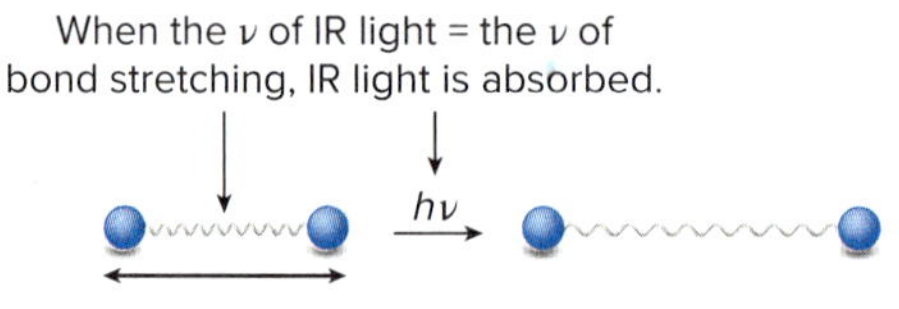

- **Different kinds of bonds vibrate at different frequencies, so they absorb different frequencies of IR light.**
- **IR spectroscopy distinguishes between the different kinds of bonds in a molecule, so it is possible to determine the functional groups present.**

Problem B.3 Which of the following has higher energy: (a) IR light of 3000 cm^{-1} or 1500 cm^{-1} in wavenumber; (b) IR light having a wavelength of 10 μm or 20 μm?

B.2B Characteristics of an IR Spectrum

In an IR spectrometer, light passes through a sample. Frequencies that match vibrational frequencies are absorbed, and the remaining light is transmitted to a detector. A spectrum plots the amount of transmitted light versus its wavenumber. The IR spectrum of propan-1-ol, $CH_3CH_2CH_2OH$, illustrates several important features of IR spectroscopy.

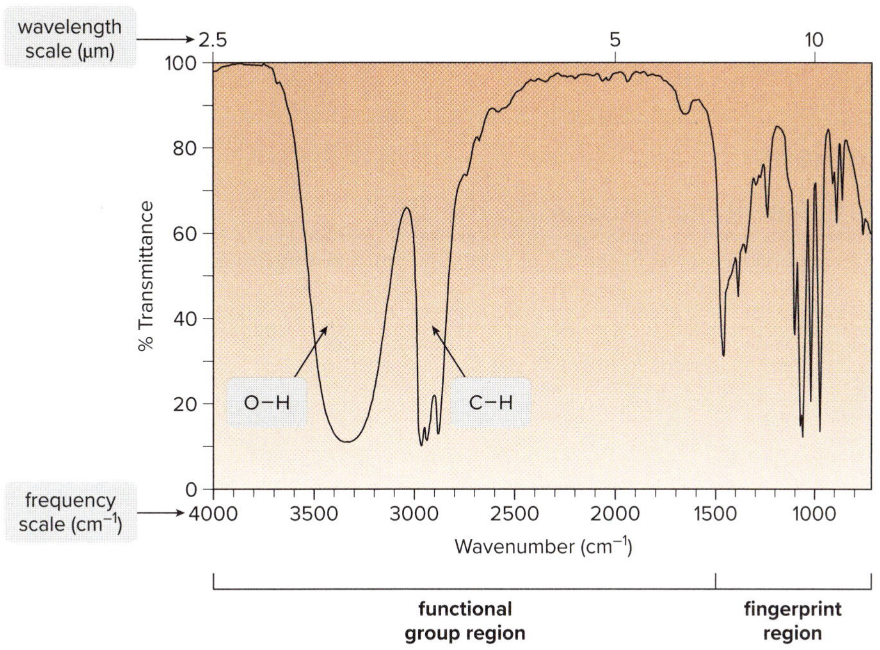

- The absorption peaks go *down* on a page. The *y* axis measures **percent transmittance:** 100% transmittance means that all the light shone on a sample is transmitted and none is absorbed; 0% transmittance means that none of the light shone on a sample is transmitted and all is absorbed. **A strong absorption has a low % transmittance because much light is absorbed.**
- **Each peak corresponds to a particular kind of bond, and each bond type (such as O–H and C–H) occurs at a characteristic frequency.**
- IR spectra have both a wavelength and a wavenumber scale on the *x* axis. Wavelengths are recorded in μm (2.5–25). Wavenumber, frequency, and energy *decrease* from left to right. Where a peak occurs is reported in reciprocal centimeters (cm^{-1}).

Conceptually, the IR spectrum is divided into two regions:

- **The functional group region occurs at ≥ 1500 cm^{-1}. Common functional groups give one or two peaks in this region, at a characteristic frequency.**
- **The fingerprint region occurs at < 1500 cm^{-1}. This region often contains a complex set of peaks and is unique for every compound.**

Compare, for example, the IR spectra of 5-methylhexan-2-one (**A**) and ethyl propanoate (**B**) in Figure B.2. The IR spectra look similar in their functional group regions because both

Figure B.2 Comparing the functional group region and fingerprint region of two compounds

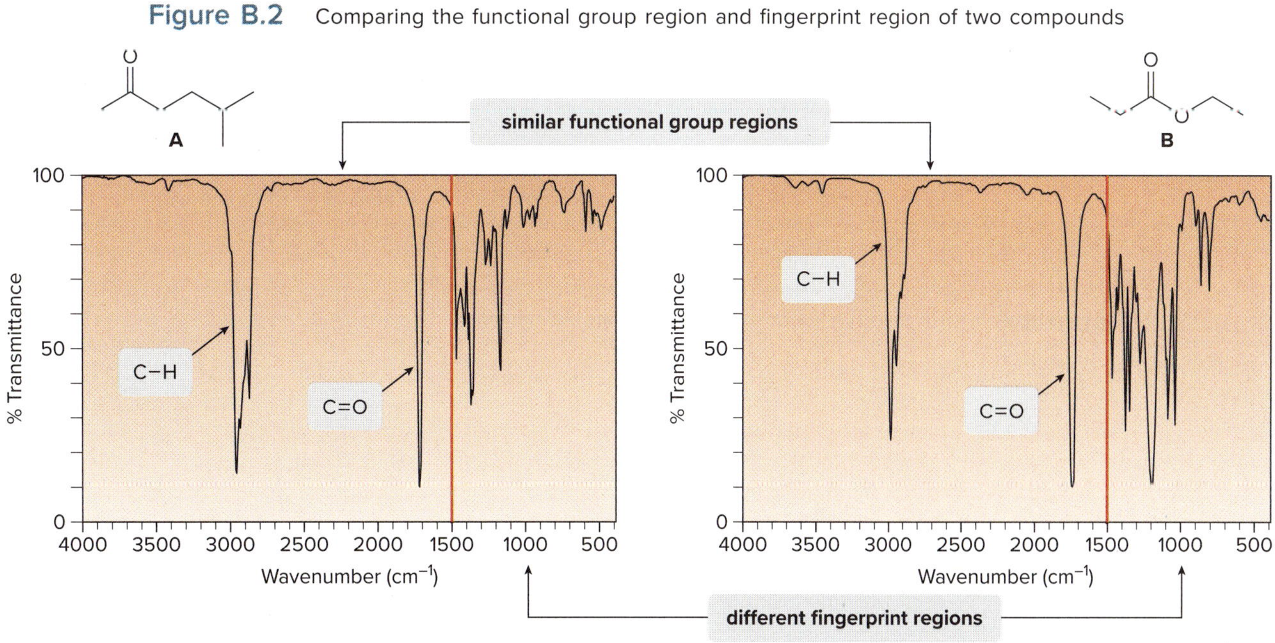

- **A** and **B** show similar absorptions for their C=O group and *sp*^{3} hybridized C–H bonds in the functional group region.
- **A** and **B** are different compounds, so their fingerprint regions are quite different.

compounds contain a carbonyl group (C=O) and several sp^3 hybridized C–H bonds. Because **A** and **B** are different compounds, however, their fingerprint regions look very different.

B.3 IR Absorptions

B.3A Where Particular Bonds Absorb in the IR

Where a particular bond absorbs in the IR depends on **bond strength** and **atom mass.**

- Bond strength: stronger bonds vibrate at higher frequency, so they absorb at higher $\tilde{\nu}$.
- Atom mass: bonds with lighter atoms vibrate at higher frequency, so they absorb at higher $\tilde{\nu}$.

Thinking of bonds as springs with weights on each end illustrates these trends. The strength of the spring is analogous to bond strength, and the mass of the weights is analogous to atomic mass. For two springs with the same weights on each end, the **stronger spring vibrates at a higher frequency.** For two springs of the same strength, **springs with lighter weights vibrate at higher frequency** than those with heavier weights. Hooke's law, as shown in Figure B.3, describes the relationship of frequency to mass and bond strength.

Figure B.3 Hooke's law: How the frequency of bond vibration depends on atom mass and bond strength

The frequency of bond vibration can be derived from Hooke's law, which describes the motion of a vibrating spring:

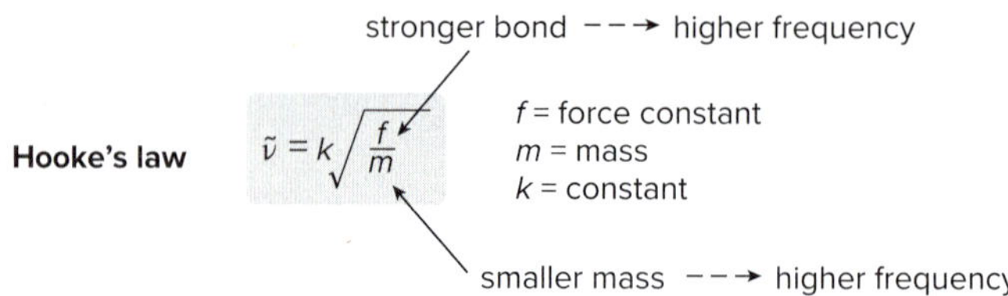

- The force constant (f) is the strength of the bond (or spring). The *larger* the value of f, the *stronger* the bond, and the *higher* the $\tilde{\nu}$ of vibration.
- The mass (m) is the mass of atoms (or weights). The *smaller* the value of m, the *higher* the $\tilde{\nu}$ of vibration.

As a result, **bonds absorb in four predictable regions in an IR spectrum.** These four regions, and the bonds that absorb there, are summarized in Figure B.4. Remembering the information

Figure B.4 Summary: The four regions of the IR spectrum

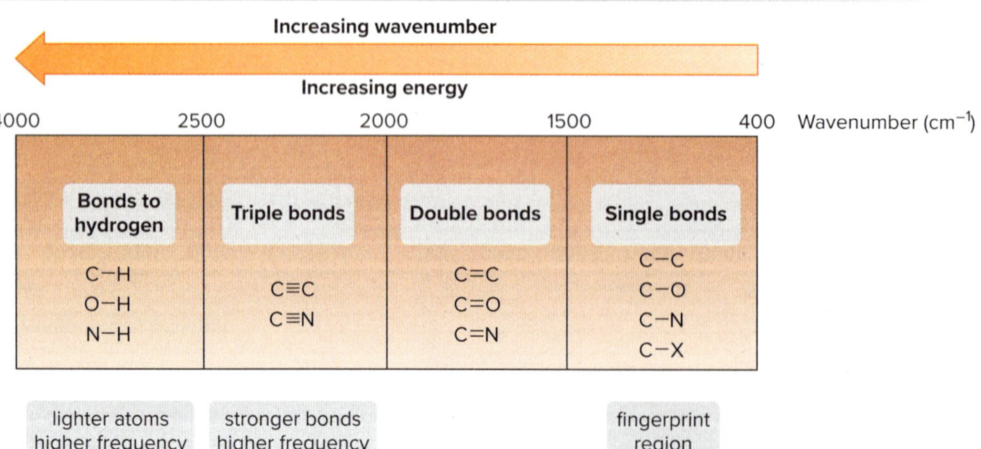

in this figure will help you analyze the spectra of unknown compounds. To help you remember it, keep in mind these two points:

- **Absorptions for bonds to hydrogen always occur on the *left* side of the spectrum (the high wavenumber region). H has so little mass that H–Z bonds (where Z = C, O, and N) vibrate at *high* frequencies.**
- **Bond strength decreases in going from C≡C → C=C → C–C, so the frequency of vibration *decreases*—that is, the absorptions for these bonds move farther to the right side of the spectrum.**

The functional group region consists of absorptions for single bonds to hydrogen (all H–Z bonds), as well as absorptions for all multiple bonds. Most absorptions in the functional group region are due to bond stretching (rather than bond bending). The fingerprint region consists of absorptions due to all other single bonds (except H–Z bonds), often making it a complex region that is very difficult to analyze.

Besides learning the general regions of the IR spectrum, it is useful to learn the specific absorption values for common bonds. Table B.1 lists the most important IR absorptions in the functional group region. Other details of IR absorptions will be presented in later chapters when new functional groups are introduced. Appendix G contains a detailed list of the characteristic IR absorption frequencies for common bonds.

Table B.1 Important IR Absorptions

Bond type	Approximate $\tilde{\nu}$ (cm^{-1})	Intensity
O–H	3600–3200	strong, broad
N–H	3500–3200	medium
C–H	~3000	
• C_{sp^3}–H	3000–2850	strong
• C_{sp^2}–H	3150–3000	medium
• C_{sp}–H	3300	medium
C≡C	2250	medium
C≡N	2250	medium
C=O	1800–1650 (often ~1700)	strong
C=C	1650	medium
(benzene ring)	1600, 1500	medium

Almost all bonds in a molecule give rise to an absorption peak in an IR spectrum, but a few do not. **For a bond to absorb in the IR, there must be a change in dipole moment during the vibration.** Thus, symmetrical, nonpolar bonds do *not* absorb in the IR. The carbon–carbon triple bond of but-2-yne, for example, does not have an IR stretching absorption at 2250 cm^{-1} because the C≡C bond is nonpolar and there is no change in dipole moment when the bond stretches along its axis. This type of vibration is said to be **IR inactive.**

Stretching along the bond axis does not change the dipole moment.

CH_3–C≡C–CH_3

nonpolar bond
IR inactive

Problem B.4 Which highlighted bond in each pair absorbs at higher wavenumber?

a. [alkyne structure] or [alkene structure] b. [C–H structure] or [C–D structure] c. [C=N structure] or [C–N structure]

B.3B The Effect of Percent *s*-Character on C–H Absorptions

Any factor that affects bond strength affects the location of an IR absorption. Recall from Section 1.11 that **the strength of a C–H bond *increases* as the percent *s*-character of the hybrid orbital on carbon *increases;*** thus:

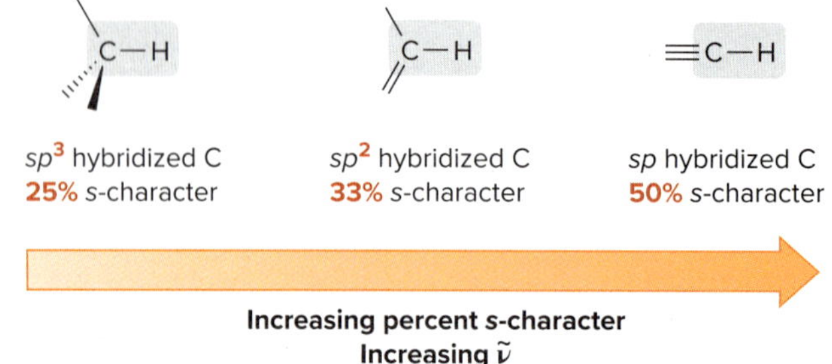

- **The *higher* the percent *s*-character, the *stronger* the bond and the *higher* the wavenumber of the absorption.**

Problem B.5 Rank the indicated bonds in compounds **A** and **B** in order of increasing (a) strength; (b) bond length; (c) percent *s*-character; (d) wavenumber of absorption. Mestranol (**A**) was a component of the first available oral contraceptive, and petroformymic acid (**B**) was isolated from a marine sponge.

mestranol
A

petroformymic acid
B

B.3C The Effect of Resonance on IR Absorptions

When a compound contains a carbonyl group (C=O), often the carbonyl absorption is the most intense peak in the IR spectrum. The exact location of that absorption depends on what groups are bonded directly to the carbonyl carbon.

When a carbonyl group is bonded to a carbon–carbon double bond or a benzene ring, the **two sites of unsaturation are separated by one σ bond and the system is said to be *conjugated.*** As we will learn in Chapter 14, **conjugation** results in many unusual properties of a compound, one of which is reflected in the location of its carbonyl absorption.

1 σ bond — conjugated system

1 σ bond — conjugated system

2 σ bonds — not conjugated

- **Conjugation of the carbonyl group with a C=C or a benzene ring shifts the absorption to lower wavenumber by ~30 cm^{-1}.**

The effect of conjugation on the frequency of the C=O absorption is explained by **resonance.** An α,β-unsaturated carbonyl compound can be written as three resonance structures, two of which place a single bond between the carbon and oxygen atoms of the carbonyl group. Thus, the π bond of the carbonyl group is delocalized, giving the conjugated carbonyl group some single bond character, and making it somewhat **weaker** than an unconjugated C=O. **Weaker bonds absorb at lower frequency (lower wavenumber) in an IR spectrum.**

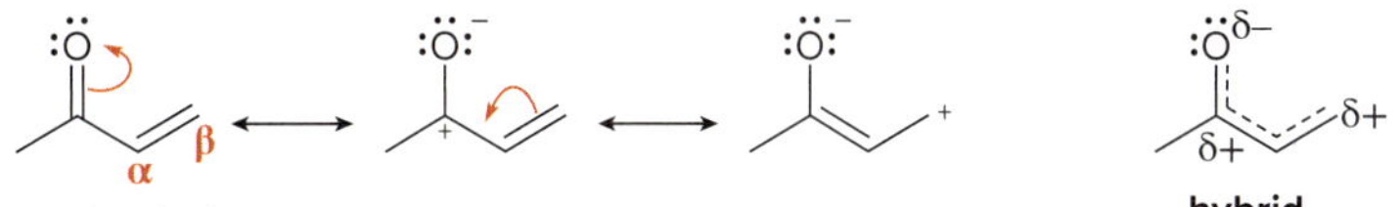

α,β-unsaturated carbonyl group

Two resonance contributors have **a C–O single bond.**

hybrid

The π bonds are delocalized.

Figure B.5 illustrates the effects of conjugation on the location of the carbonyl absorption in some representative compounds.

Figure B.5

The effect of conjugation on the carbonyl absorption in an IR spectrum

1709 cm^{-1} 1685 cm^{-1} 1715 cm^{-1} 1685 cm^{-1}

conjugated C=O lower wavenumber

conjugated C=O lower wavenumber

Resonance also affects the relative position of the carbonyl absorptions of compounds RCOZ, when Z contains a nonbonded electron pair. Three resonance structures can be drawn for RCOZ.

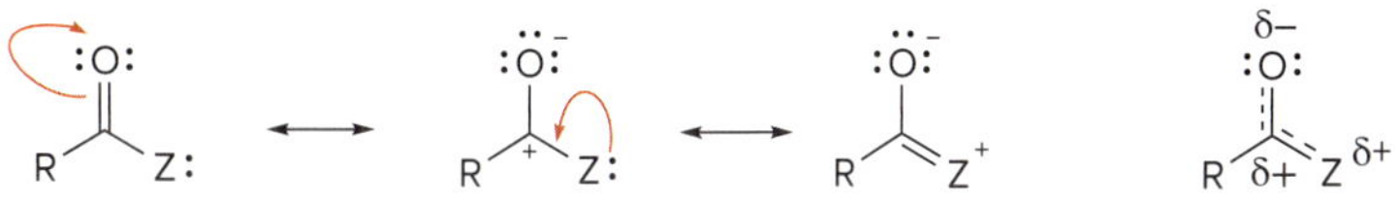

1 **2** **3** **hybrid**

Because resonance structures **2** and **3** contain a carbon–oxygen single bond, the more these structures contribute to the resonance hybrid, the more single bond character the carbonyl group possesses, and the *lower* the frequency of the carbonyl absorption.

- **The more basic Z is, the more it donates its electron pair and the more resonance structure 3 contributes to the hybrid.**
- **As a result, as the basicity of Z *increases*, the frequency of the carbonyl absorption *decreases*.**

To compare the carbonyl absorptions of an ester (RCO_2R') and an amide ($RCONR'_2$), we look at the relative basicity of an –OR' group and an $–NR'_2$ group. Basicity decreases across a row of the periodic table, so an $–NR'_2$ group is more basic than an –OR' group. Thus, **an amide carbonyl has more single bond character than an ester carbonyl, and the carbonyl absorption occurs at *lower* wavenumber.**

For example, the carbonyl absorptions of the ester ethyl acetate and the amide *N,N*-dimethylacetamide occur at 1743 and 1662 cm^{-1}, respectively.

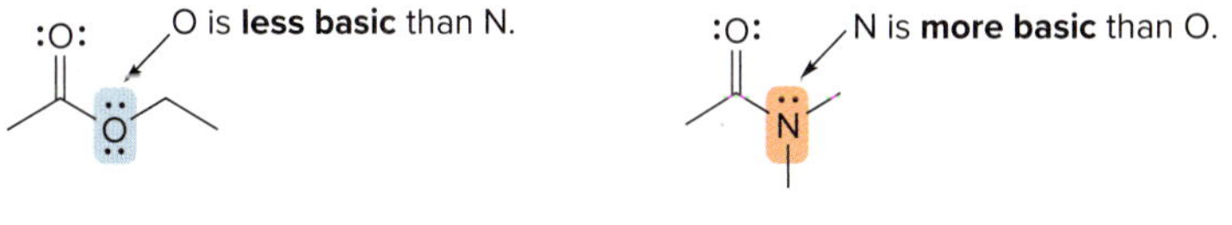

O is **less basic** than N.

N is **more basic** than O.

ethyl acetate
1743 cm^{-1}
higher wavenumber

N,N-dimethylacetamide
1662 cm^{-1}
lower wavenumber

Sample Problem B.1 illustrates how resonance affects the position of the carbonyl absorption in three compounds.

Sample Problem B.1 Predicting the Relative Position of Carbonyl Absorptions

Rank compounds **A, B,** and **C** in order of increasing frequency of the carbonyl absorption in their IR spectra.

A B C

Solution

Compare pairs of compounds. **A** and **B** are both amides, but **B** contains a carbonyl that is conjugated with a benzene ring. Because conjugation shifts the carbonyl absorption to lower wavenumber, **B** absorbs at lower wavenumber than **A. C** is an ester that is not conjugated. In comparing carbonyl absorptions in **A** and **C,** the amide **A** contains a more basic N atom, so the carbonyl absorption of **A** occurs at lower wavenumber than that of ester **C.** Thus, in order of increasing frequency (wavenumber): **B** < **A** < **C.**

Problem B.6 (a) Considering compounds **A, B, C,** and **D,** which compound has a C=O that absorbs at the *highest* wavenumber? (b) Which compound has a C=O that absorbs at the *lowest* wavenumber?

A B C D

More Practice: Try Problems B.25, B.26.

Problem B.7 Rank compounds **A–C** in order of increasing frequency of the carbonyl absorption in their IR spectra.

A B C

B.3D Analyzing an IR Spectrum

The principles learned in this section can be used to determine what types of bonds are present in a compound, as shown in the stepwise *How To.*

How To Analyze an IR Spectrum

Example Which compound—**A, B,** or **C**—gives rise to the given IR spectrum?

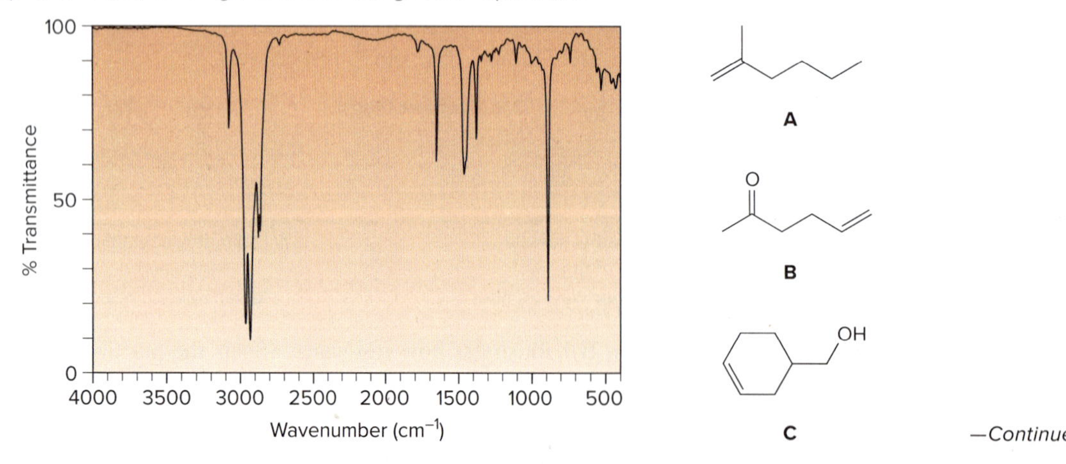

—Continued

Step [1] **Concentrate on the functional group region above 1500 cm^{-1}, and examine the two sections where double and triple bonds absorb, using the values in Table B.1.**

- A C=C absorbs at ~1650 cm^{-1}.
- A C=O absorbs between 1650 and 1800 cm^{-1}, often around 1700 cm^{-1}.
- A C≡C or C≡N absorbs at ~2250 cm^{-1}.

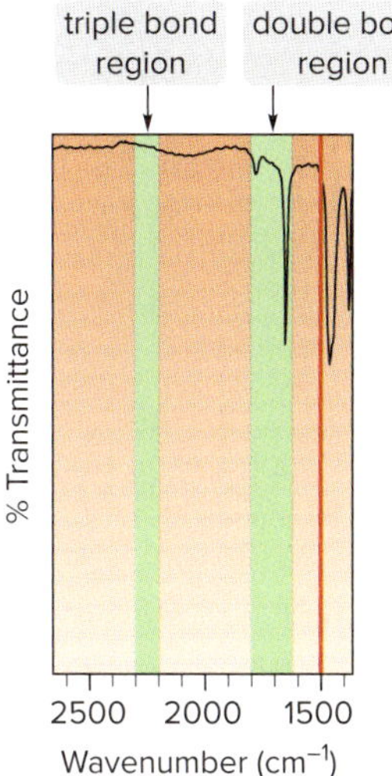

- In this example, there are no absorptions in the triple bond region and an absorption of medium intensity at 1650 cm^{-1}, suggesting that the compound contains a C=C.
- Because there is no strong absorption at ~1700 cm^{-1}, the compound does *not* contain a C=O.
- Using these data, we can eliminate **B** as a possibility because **B** contains a C=O.

Step [2] **Examine the C–H region around 3000 cm^{-1}.**

- C_{sp^3}–H bonds absorb at 3000–2850 cm^{-1}.
- C_{sp^2}–H bonds absorb at 3150–3000 cm^{-1}.
- C_{sp}–H bonds absorb at 3300 cm^{-1}.

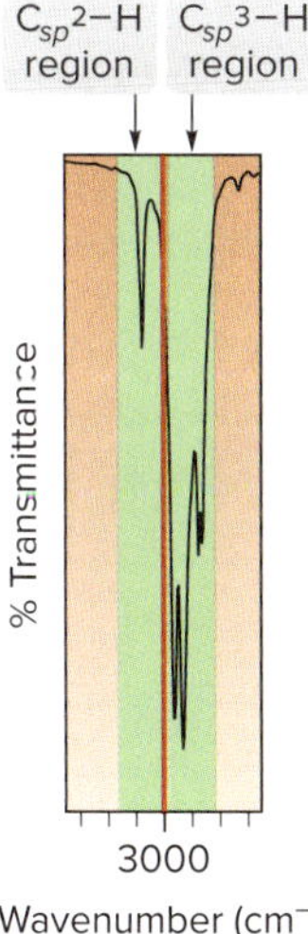

- In addition to C_{sp^3}–H bonds, which are present in almost all organic compounds, the compound contains an absorption in the C_{sp^2}–H region.
- Both **A** and **C** contain C_{sp^2}–H, so both compounds are still possibilities.

Step [3] **Examine the region above 3000 cm^{-1} for O–H and N–H bonds.**

- O–H bonds appear as strong, broad peaks at 3600–3200 cm^{-1}.
- N–H bonds of amines and amides absorb in the 3500–3200 cm^{-1} region, and are of medium intensity.

Because the IR shows no absorption at 3600–3200 cm^{-1}, the compound does not contain an OH group. This eliminates **C** as a possibility, so the IR spectrum is due to **A.**

Sample Problem B.2 Using IR Spectroscopy to Determine the Types of Bonds in a Compound

What types of bonds are responsible for the absorptions above 1500 cm^{-1} in compounds **A** and **B?**

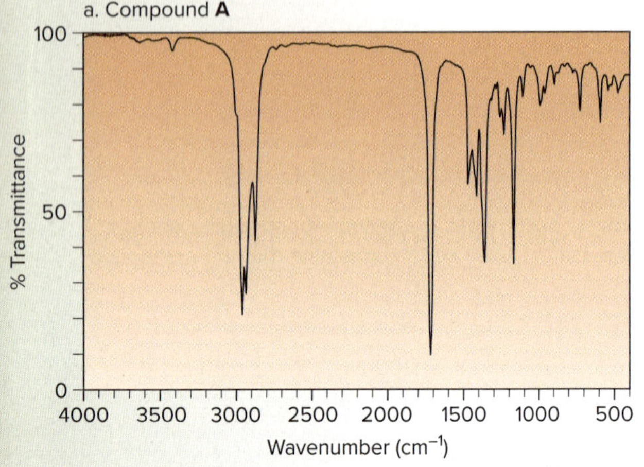

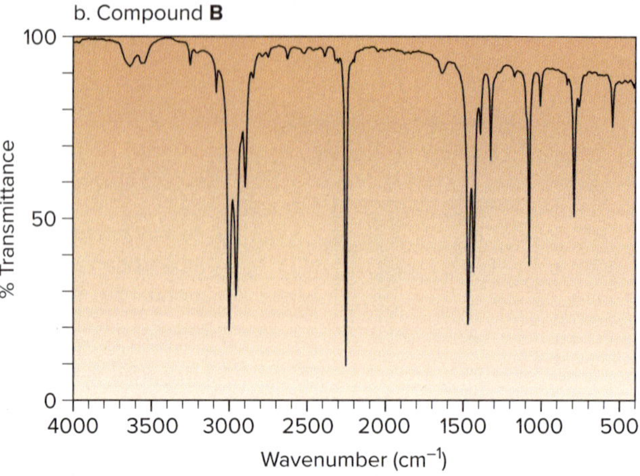

Solution

a. Compound **A** has two major absorptions above 1500 cm^{-1}: The absorption at ~3000 cm^{-1} is due to C–H bonds and the absorption at ~1700 cm^{-1} is due to a C=O group. Because the C–H absorption occurs at < 3000 cm^{-1}, all C–H bonds contain sp^3 hybridized C atoms.

b. Compound **B** has two major absorptions above 1500 cm^{-1}: The absorption at ~3000 cm^{-1} is due to C_{sp^3}–H bonds and the absorption at ~2250 cm^{-1} is due to a triple bond, either a C≡C or a C≡N.

Problem B.8 What types of bonds are responsible for the absorptions above 1500 cm^{-1} in the IR spectra for compounds **A** and **B?**

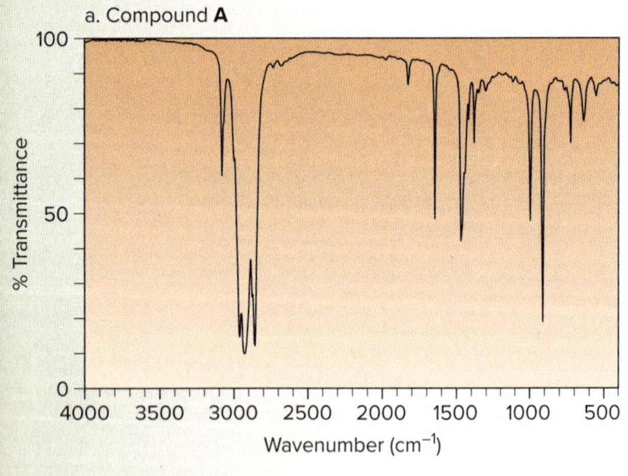

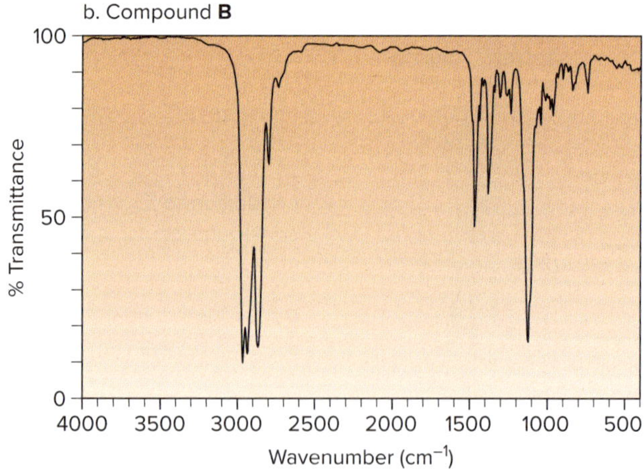

More Practice: Try Problems B.21, B.22, B.23a, B.30.

B.4 Infrared Spectra of Common Functional Groups

Each class of compounds exhibits characteristic absorptions in the infrared.

B.4A IR Absorptions in Hydrocarbons

The IR spectra of an alkane, an alkene, an alkyne, and an aromatic compound with a benzene ring illustrate characteristic differences (Figure B.6):

- An **alkane** like hexane (Figure B.6a) has only C–C single bonds and sp^3 hybridized C atoms. Therefore, it has only one major absorption above 1500 cm^{-1}: C_{sp^3}–H absorption at 3000–2850 cm^{-1}.
- An **alkene** like hex-1-ene (Figure B.6b) has a C=C and C_{sp^2}–H, in addition to its sp^3 hybridized C atoms. Therefore, there are three major absorptions above 1500 cm^{-1}: C_{sp^2}–H at 3150–3000 cm^{-1}, C_{sp^3}–H at 3000–2850 cm^{-1}, and C=C at 1650 cm^{-1}.
- An **alkyne** like hex-1-yne (Figure B.6c) has a C≡C and C_{sp}–H, in addition to its sp^3 hybridized C atoms. Therefore, there are three major absorptions: C_{sp}–H at 3300 cm^{-1}, C_{sp^3}–H at 3000–2850 cm^{-1}, and C≡C at ~2250 cm^{-1}.
- An **aromatic compound** like isopropylbenzene (Figure B.6d) contains a benzene ring and C_{sp^2}–H, in addition to its sp^3 hybridized C atoms. Thus, there are three major absorptions: C_{sp^2}–H at 3150–3000 cm^{-1}, C_{sp^3}–H at 3000–2850 cm^{-1}, and the benzene ring at 1600, 1500 cm^{-1}.

Figure B.6 IR absorptions of hydrocarbons

a. **Alkanes (hexane)**

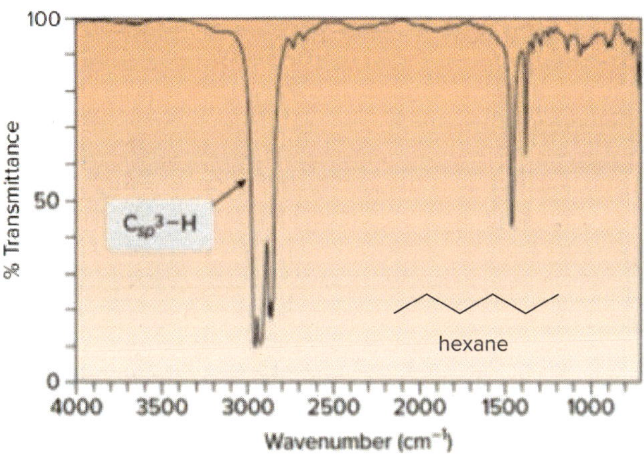

- C_{sp^3}–H absorption at 3000–2850 cm^{-1}

c. **Alkynes (hex-1-yne)**

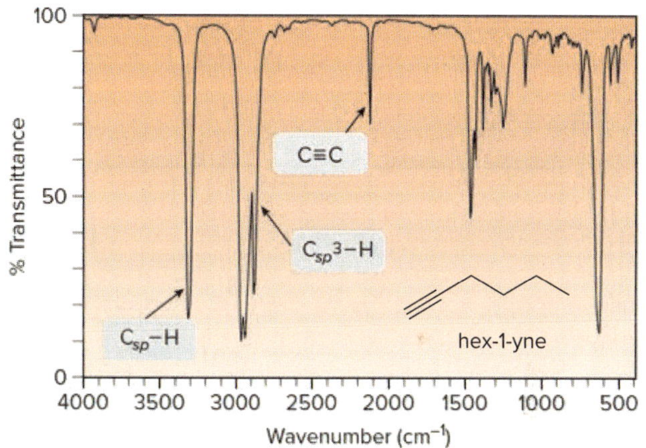

- C_{sp}–H at 3300 cm^{-1}
- C_{sp^3}–H at 3000–2850 cm^{-1}
- C≡C at ~2250 cm^{-1}

b. **Alkenes (hex-1-ene)**

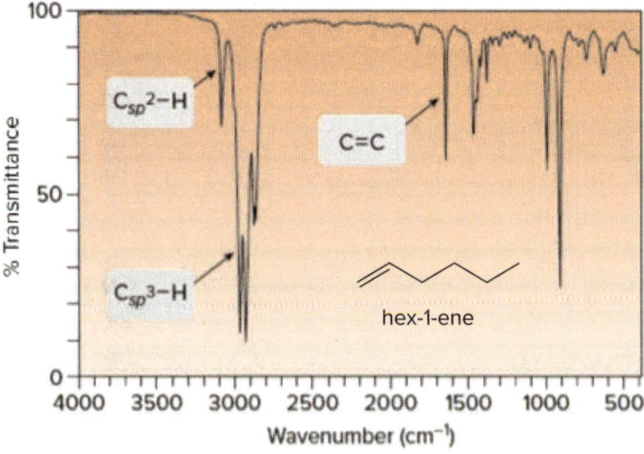

- C_{sp^2}–H at 3150–3000 cm^{-1}
- C_{sp^3}–H at 3000–2850 cm^{-1}
- C=C at 1650 cm^{-1}

d. **Aromatic compounds with benzene rings (isopropylbenzene)**

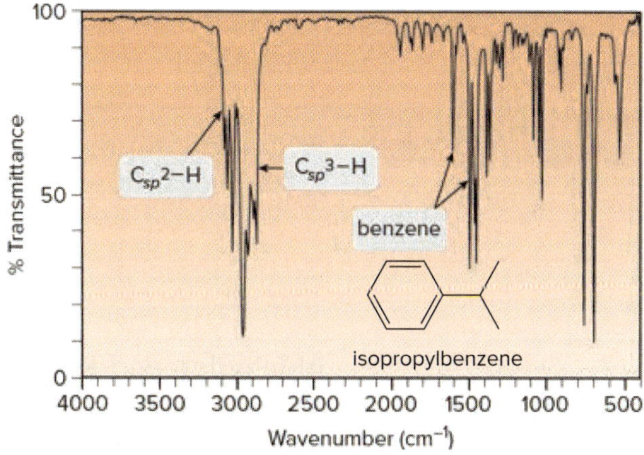

- C_{sp^2}–H at 3150–3000 cm^{-1}
- C_{sp^3}–H at 3000–2850 cm^{-1}
- Benzene ring at 1600, 1500 cm^{-1}

Problem B.9 How do the IR spectra of each pair of isomers differ?

a. cyclopentane and pent-1-ene
b. cyclopentene and pent-1-yne
c. benzene and hexa-1,3-diyne

B.4B IR Absorptions in Oxygen-Containing Compounds

The most important IR absorptions for oxygen-containing compounds occur at **3600–3200 cm^{-1} for an OH group** and at approximately **1700 cm^{-1} for a C=O.**

Alcohols and Ethers (Figure B.7)

- The most prominent absorption for an **alcohol** like butan-2-ol (Figure B.7a) is the broad, strong absorption at **3600–3200 cm^{-1}** due to the **OH** group.
- An **ether** like diethyl ether (Figure B.7b) has neither an OH group nor a C=O, so its only absorption above 1500 cm^{-1} occurs at ~3000 cm^{-1}, due to sp^3 hybridized C–H bonds. **Compounds that contain an oxygen atom but do not show an OH or C=O absorption are ethers.**

Figure B.7 IR absorptions of alcohols and ethers

a. **Alcohols (butan-2-ol)**

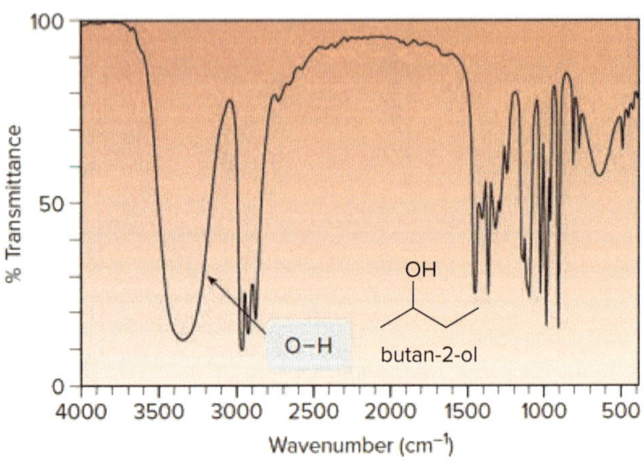

- O–H at 3600–3200 cm^{-1}

b. **Ethers (diethyl ether)**

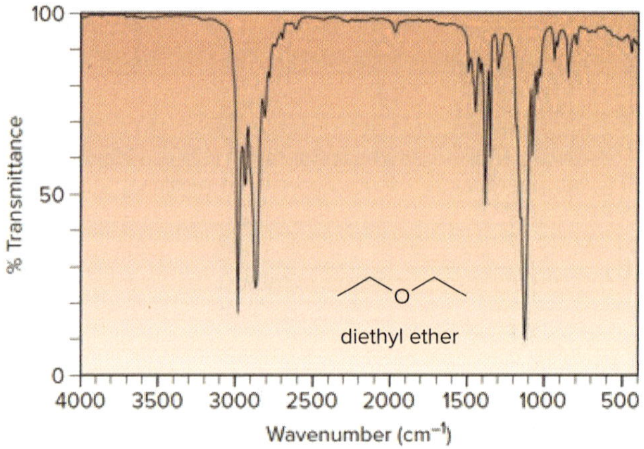

- No absorptions above 1500 cm^{-1} except C–H at ~3000 cm^{-1}

Problem B.10 Propose a structure for each compound from the IR data.

a. a compound of molecular formula $C_6H_{12}O$ that has an absorption at 3000–2850 cm^{-1} and no other peaks in the functional group region
b. a compound of molecular formula $C_7H_{16}O$ that has peaks at 3600–3200 and 3000–2850 cm^{-1}
c. a compound of molecular formula $C_8H_{16}O_2$ that has peaks at 3600–3200 and 3000–2850 cm^{-1}

Aldehydes and Ketones (Figure B.8)

- An **aldehyde** like propanal (Figure B.8a) has a C=O and C_{sp^2}–H. In addition to the absorption of its C_{sp^3}–H, there are two major absorptions: C_{sp^2}–H of the aldehyde C–H at 2830–2700 cm^{-1} (one or two peaks) and C=O at ~1700 cm^{-1}.

- In addition to the absorption of its C_{sp^3}–H, a **ketone** like butan-2-one (Figure B.8b) has one major absorption: C=O at ~1700 cm^{-1}.

Figure B.8 IR absorptions of aldehydes and ketones

a. **Aldehydes (propanal)**

b. **Ketones (butan-2-one)**

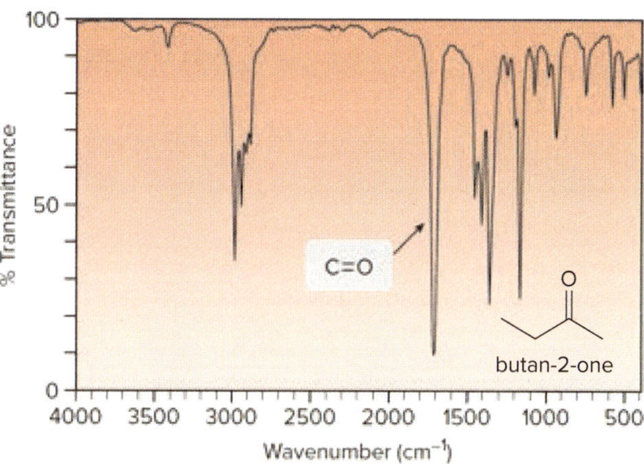

- C_{sp^2}–H of the aldehyde C–H at 2830–2700 cm^{-1} (one or two peaks)
- C=O at ~1700 cm^{-1}

- C=O at ~1700 cm^{-1}

The exact location of the carbonyl absorption provides additional information about a compound. In Section B.3C, we learned that conjugation of the C=O with a C=C or a benzene ring shifts the absorption to lower wavenumber.

When the carbonyl carbon is located in a ring, **ring size** also affects the location of the carbonyl absorption.

- **The carbonyl absorption of cyclic ketones shifts to higher wavenumber as the size of the ring decreases and the ring strain increases.**

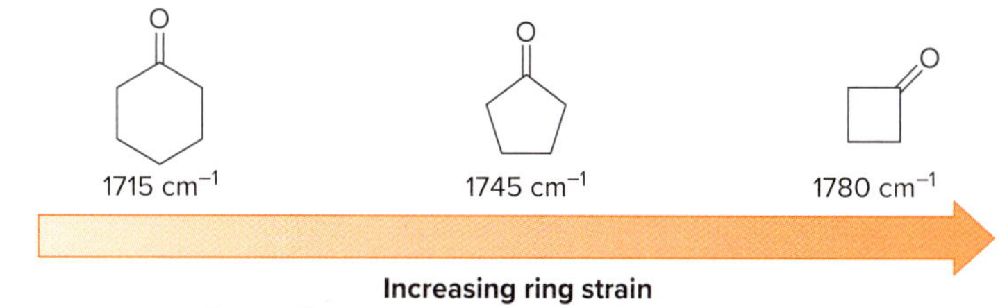

Problem B.11 Rank the following compounds in order of increasing frequency of the carbonyl absorption.

A B C

Carboxylic Acids, Esters, and Amides (Figure B.9)

- A carboxylic acid like butanoic acid (Figure B.9a) has two characteristic IR absorptions: O–H at 3500–2500 cm^{-1}, a broad, strong absorption that almost obscures the C–H peak at ~3000 cm^{-1}, and C=O at ~1710 cm^{-1}.
- In addition to the absorption of its C_{sp^3}–H, an **ester** like methyl propanoate (Figure B.9b) has one major absorption: C=O at ~1745–1735 cm^{-1}.
- An **amide** like propanamide (Figure B.9c) has three characteristic absorptions: N–H stretching peaks at 3400–3200 cm^{-1} (one or two peaks), C=O at 1680–1630 cm^{-1}, and an N–H bending absorption at ~1640 cm^{-1}.

Figure B.9 IR absorptions of carboxylic acids, esters, and amides

a. **Carboxylic acids (butanoic acid)**

- O–H at 3500–2500 cm^{-1}, a broad, strong absorption that almost obscures the C–H peak at ~3000 cm^{-1}.
- C=O at ~1710 cm^{-1}

c. **Amides (propanamide)**

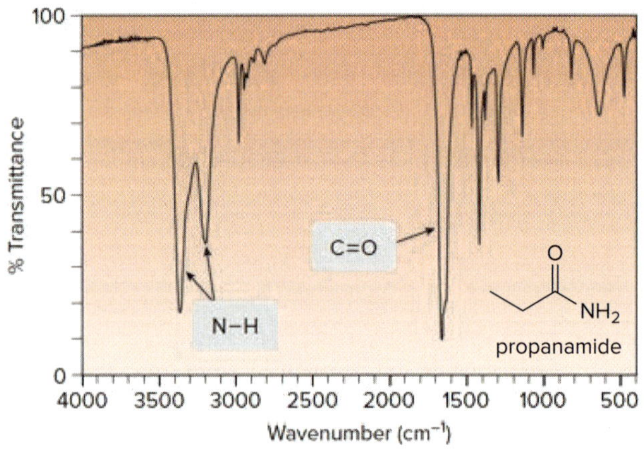

- N–H stretching peaks at 3400–3200 cm^{-1} (one or two peaks)
- C=O at 1680–1630 cm^{-1}
- N–H bending absorption at ~1640 cm^{-1}

b. **Esters (methyl propanoate)**

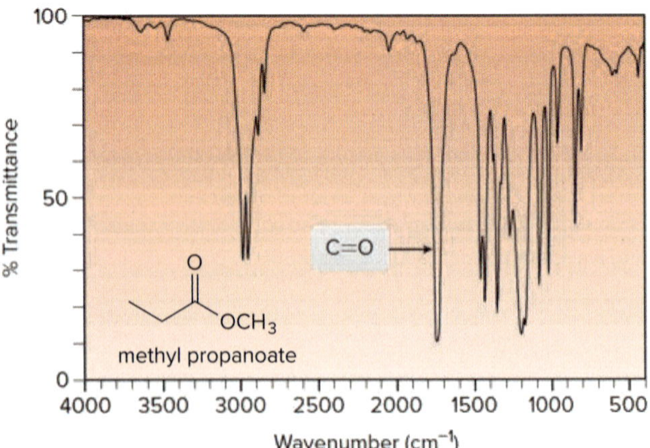

- C=O at ~1745–1735 cm^{-1}

Problem B.12 How would the compounds in each pair differ in their IR spectra?

a. **A** and **B** c. **E** and **F**

b. **C** and **D** d. **G** and **H**

B.4C IR Absorptions in Amines and Nitriles

Common functional groups that contain nitrogen atoms are also distinguishable by their IR absorptions above 1500 cm^{-1} (Figure B.10):

- The **N–H** bonds in an **amine** like octylamine (Figure B.10a) give rise to two weak absorptions at 3300 and 3400 cm^{-1}.
- The **C≡N** group of a **nitrile** like octanenitrile (Figure B.10b) absorbs in the triple bond region at ~2250 cm^{-1}.

Figure B.10 IR absorptions of amines and nitriles

a. **Amines (octylamine)**

N–H

NH_2

octylamine

% Transmittance

Wavenumber (cm^{-1})

- N–H (one or two peaks) at 3500–3300 cm^{-1}

b. **Nitriles (octanenitrile)**

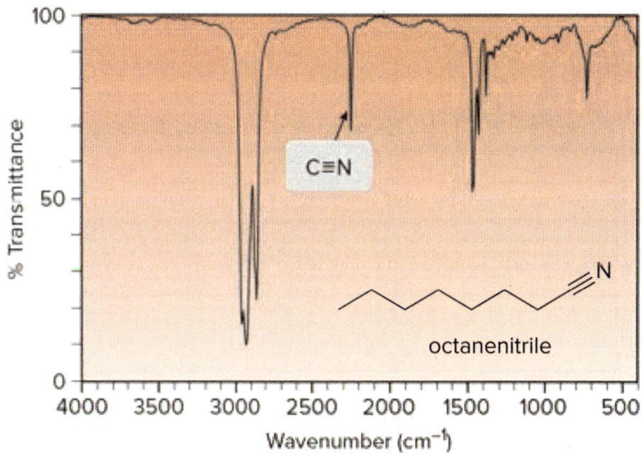

- C≡N at ~2250 cm^{-1}

Problem B.13 How would the following compounds differ in their IR spectra?

N and ≡N

B.4D Summary of IR Absorptions for Common Functional Groups

Table B.2 summarizes the typical IR peaks for common functional groups.

Table B.2 Characteristic IR Absorptions in the Functional Group Region

Compound type	Absorption (cm^{-1})	Intensity
Alkane		
C_{sp^3}–H	3000–2850	strong
Alkene		
C_{sp^2}–H	3150–3000	medium
C=C	1650	medium
Alkyne		
C_{sp}–H	3300	medium
C≡C	2250	medium
Benzene	1600, 1500	medium
Alcohol		
O–H	3600–3200	strong, broad
Amine		
N–H	3500–3300	medium
Carbonyl compounds		
Aldehyde C_{sp^2}–H	2830–2700	medium
Aldehyde C=O	1730	strong
Ketone C=O	1715	strong
Ester C=O	1745–1735	strong
Amide C=O	1680–1630	strong
Amide N–H	3400–3200	medium
Carboxylic acid		
O–H	3500–2500	strong, very broad
C=O	1710	strong
Nitrile		
C≡N	2250	medium

Sample Problem B.3 Using IR Spectroscopy to Distinguish Isomers

How can the two isomers having molecular formula C_2H_6O be distinguished by IR spectroscopy?

Solution

First, draw the structures of the compounds and then locate the functional groups. One compound is an alcohol and one is an ether.

hydroxy group

ethanol

- C–H absorption at ~3000 cm^{-1}
- O–H absorption at 3600–3200 cm^{-1}

no OH group

dimethyl ether

- C–H absorption at ~3000 cm^{-1} **only**

Although both compounds have sp^3 hybridized C–H bonds, ethanol has an OH group that gives a strong absorption at 3600–3200 cm^{-1}, and dimethyl ether does not. This feature distinguishes the two isomers.

Problem B.14 How do the three isomers of molecular formula C_3H_6O (**A**, **B**, and **C**) differ in their IR spectra?

A **B** **C**

More Practice: Try Problems B.24, B.27–B.30.

Problem B.15 What are the major IR absorptions in the functional group region for each compound?

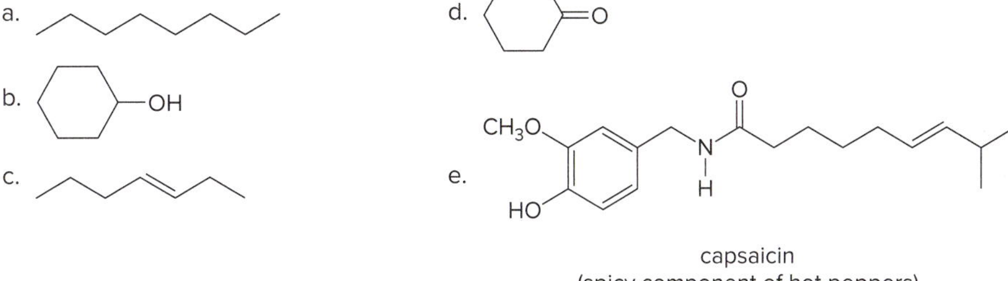

Corn and olive oils are high in unsaturated fatty acids like oleic acid (Problem B.16). *Elite Images/ McGraw Hill*

Problem B.16 What are the major IR absorptions in the functional group region for oleic acid, a common unsaturated fatty acid (Section 10.6A)?

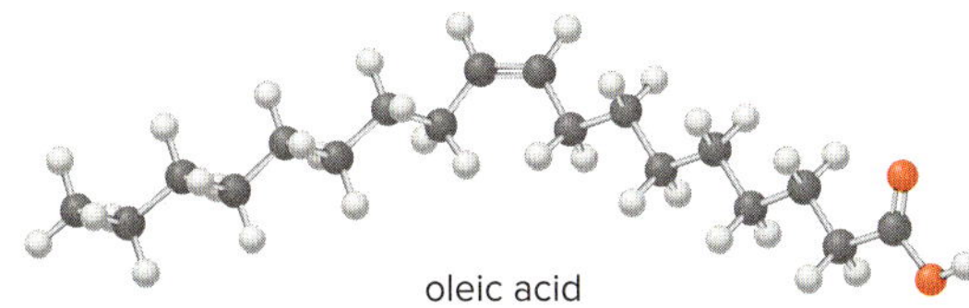

B.5 IR and Structure Determination

Since its introduction, IR spectroscopy has proven to be a valuable tool for determining the functional groups in organic molecules.

In the 1940s, IR spectroscopy played a key role in elucidating the structure of the antibiotic penicillin G, the chapter-opening molecule. **β-Lactams,** four-membered rings that contain an amide, have a carbonyl group that absorbs at ~1760 cm^{-1}, a much higher frequency than that observed for most amides and many other carbonyl groups. Because penicillin G had an IR absorption at this frequency, **A** became the leading candidate for the structure of penicillin rather than **B,** a possibility originally considered more likely. Structure **A** was later confirmed by X-ray analysis.

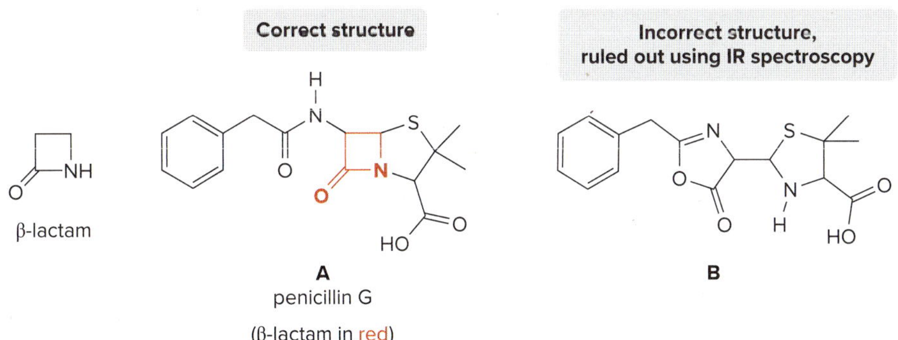

Problem B.17 Imipenem, sold in combination with cilastatin (Problem 10.36c), is a broad-spectrum antibiotic used to treat a number of bacterial infections. List five IR absorptions that imipenem shows in the functional group region.

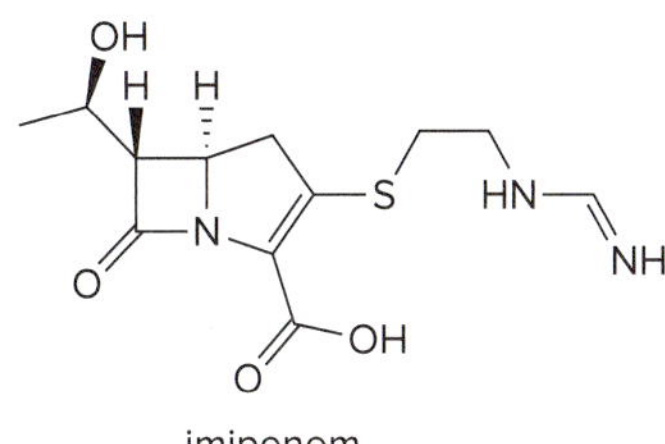

IR spectroscopy is often used to determine the outcome of a chemical reaction. For example, oxidation of the hydroxy group in **C** to form the carbonyl group in periplanone B is accompanied by the disappearance of the OH absorption (3600–3200 cm^{-1}) and the appearance of a carbonyl absorption near 1700 cm^{-1} in the IR spectrum of the product. Periplanone B is the sex pheromone of the female American cockroach.

The absorption at 3600–3200 cm^{-1} disappears.

C —(Cr^{6+} oxidant)→ periplanone B

The absorption at ~1700 cm^{-1} appears.

Problem B.18 How can IR spectroscopy be used to determine when the following reaction is complete?

$\xrightarrow[H_2SO_4]{H_2O}$ (OH)

The combination of IR and mass spectral data provides key information on the structure of an unknown compound. The mass spectrum reveals the molecular weight of the unknown (and the molecular formula if an exact mass is available), and the IR spectrum helps to identify the important functional groups.

How To Use MS and IR for Structure Determination

Example What information is obtained from the mass spectrum and IR spectrum of an unknown compound **X?** Assume **X** contains the elements C, H, and O.

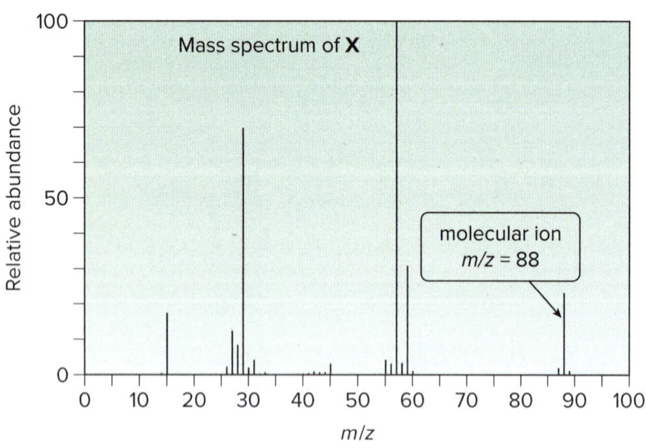

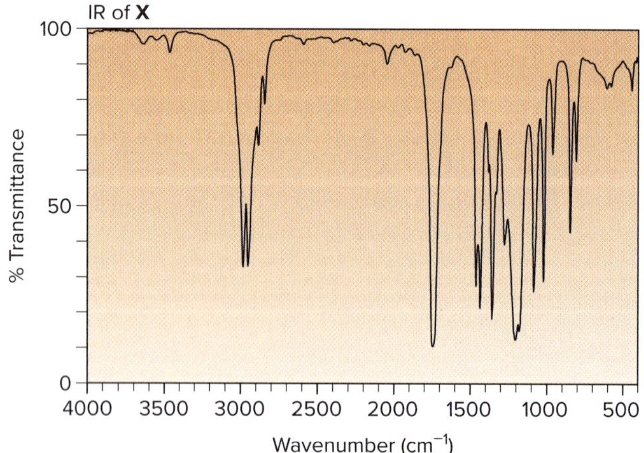

Step [1] **Use the molecular ion to determine possible molecular formulas. Use an exact mass (when available) to determine a molecular formula.**

- Use the procedure outlined in Sample Problem A.2 to calculate possible molecular formulas. For a molecular ion at m/z = 88:

$$\frac{88}{12} = 7\text{ C's (maximum, remainder = 4)} \dashrightarrow C_7H_4 \xrightarrow[+1\,O]{-CH_4} C_6O \xrightarrow[+12\,H's]{-1\,C} C_5H_{12}O \xrightarrow[+1\,O]{-CH_4} C_4H_8O_2 \xrightarrow[+1\,O]{-CH_4} C_3H_4O_3$$

three possible formulas ($C_5H_{12}O$, $C_4H_8O_2$, $C_3H_4O_3$)

—Continued

- Discounting C_7H_4 (a hydrocarbon) and C_6O (because it contains no H's) gives three possible formulas for **X.**
- If high-resolution mass spectral data are available, the molecular formula can be determined directly. If the molecular ion had an exact mass of 88.0580, the molecular formula of **X** is $\mathbf{C_4H_8O_2}$ (exact mass = 88.0524) rather than $C_5H_{12}O$ (exact mass = 88.0888) or $C_3H_4O_3$ (exact mass = 88.0160).

Step [2] **Calculate the number of degrees of unsaturation (Section 10.2).**

- For a compound of molecular formula $C_4H_8O_2$, the maximum number of H's = $2n + 2 = 2(4) + 2 = 10$.
- Because the compound contains only 8 H's, it has $10 - 8 = 2$ H's fewer than the maximum number.
- Because each degree of unsaturation removes 2 H's, **X** has one degree of unsaturation. **X has one ring or one π bond.**

Step [3] **Determine what functional group is present from the IR spectrum.**

- The two major absorptions in the IR spectrum above 1500 cm^{-1} are due to sp^3 hybridized C–H bonds (~3000–2850 cm^{-1}) and a C=O group (1740 cm^{-1}). Thus, the one degree of unsaturation in **X** is due to the presence of the **C=O.**

Mass spectrometry and IR spectroscopy give valuable but limited information on the identity of an unknown. Although the mass spectral and IR data reveal that **X** has a molecular formula of $C_4H_8O_2$ and contains a carbonyl group, more data are needed to determine its complete structure. In Spectroscopy C, we will learn how other spectroscopic data can be used for that purpose.

Problem B.19 Which of the following possible structures for **X** can be excluded on the basis of its IR spectrum?

a. b. HO c. d. OH

Problem B.20 Propose structures consistent with each set of data: (a) a hydrocarbon with a molecular ion at $m/z = 68$ and IR absorptions at 3310, 3000–2850, and 2120 cm^{-1}; (b) a compound containing C, H, and O with a molecular ion at $m/z = 60$ and IR absorptions at 3600–3200 and 3000–2850 cm^{-1}.

Spectroscopy B CHAPTER REVIEW

KEY CONCEPTS

[1] Electromagnetic radiation (B.1)

- The wavelength (λ) and frequency (ν) of electromagnetic radiation are ***inversely*** related (**c** = speed of light):

$$\lambda = c/\nu \quad \text{or} \quad \nu = c/\lambda$$

2 Energy and frequency

- The energy (***E***) of a photon is **proportional** to its frequency (ν); the higher the frequency, the higher the energy [h = Planck's constant (6.63×10^{-34} J · s)]:

$$E = h\nu$$

Try Problems B.1–B.3.

[2] Bond strength and IR absorption (B.3)

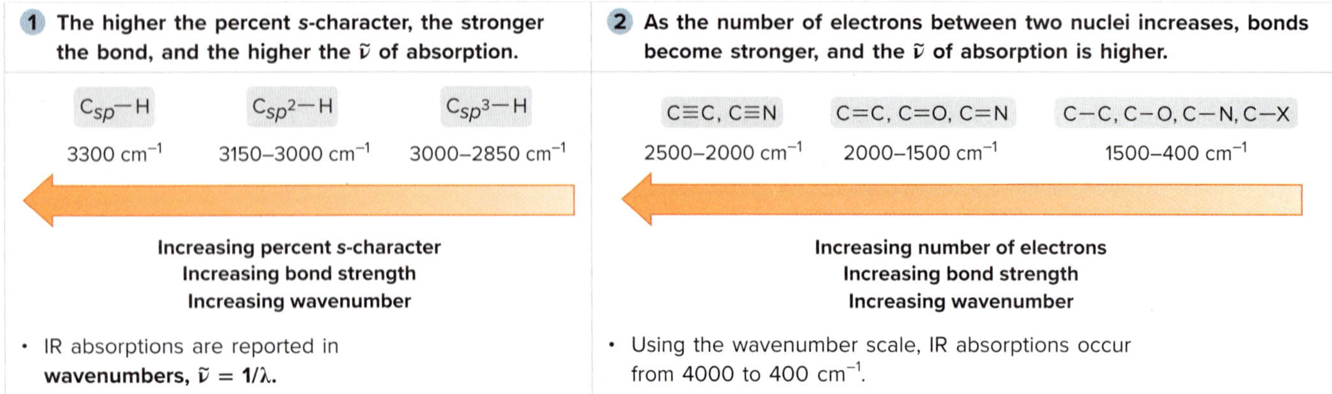

- IR absorptions are reported in **wavenumbers, $\tilde{\nu}$ = 1/λ.**
- Using the wavenumber scale, IR absorptions occur from 4000 to 400 cm^{-1}.

See Table B.1. Try Problems B.4, B.5.

[3] Factors affecting the location of a carbonyl absorption (B.3C, B.4B)

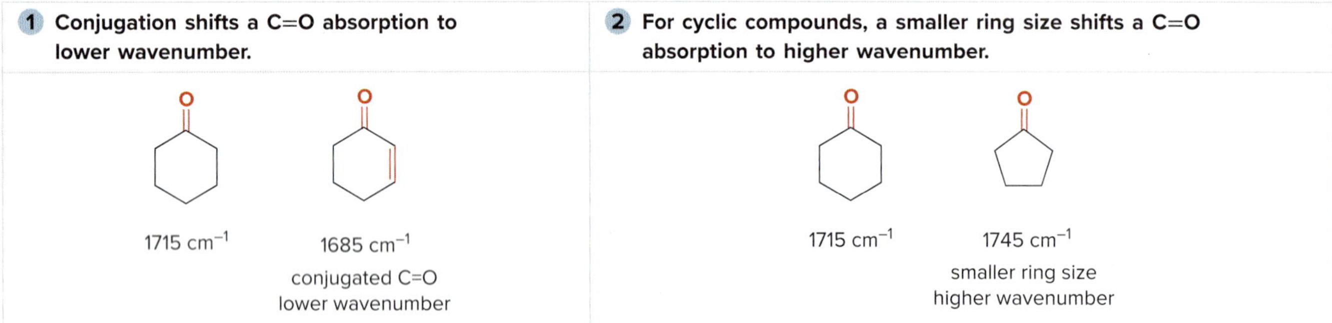

See Sample Problem B.1, Table B.2. Try Problems B.25, B.26.

KEY SKILLS

[1] Using the functional groups to distinguish two compounds by IR spectroscopy (B.4D)

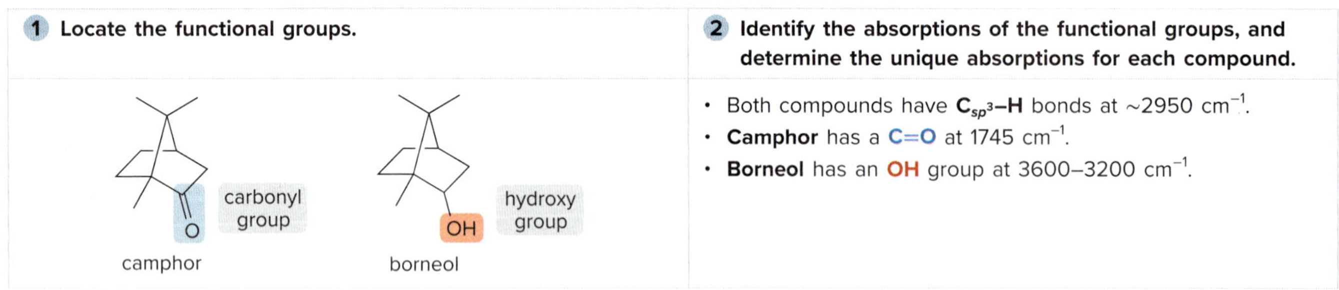

- Both compounds have **C_{sp^3}–H** bonds at ~2950 cm^{-1}.
- **Camphor** has a **C=O** at 1745 cm^{-1}.
- **Borneol** has an **OH** group at 3600–3200 cm^{-1}.

See Sample Problem B.3, Table B.2. Try Problems B.24, B.27–B.29.

[2] Using MS and IR to determine possible structures of a compound that contains C, H, and O (B.5); example: m/z = 86

1 Use the molecular ion to determine the possible molecular formulas.

molecular ion at m/z = 86

[1] $\frac{86}{12}$ = 7 C's maximum (remainder = 2) ⟶ C_7H_2 (hydrocarbon)

[2] $C_7H_2 \xrightarrow[+12\ H\text{'s}]{-1\ C} C_6H_{14}$ (hydrocarbon)

[3] $C_6H_{14} \xrightarrow[+1\ O]{-CH_4} C_5H_{10}O$

[4] $C_5H_{10}O \xrightarrow[+1\ O]{-CH_4} C_4H_6O_2$

2 Calculate the number of degrees of unsaturation (10.2).

$C_5H_{10}O_2$

For n carbons, the maximum number of H's is $2n + 2$; in this example, $2n + 2 = 2(5) + 2 = 12$.

12 H's (maximum) – 10 H's (actual) = 2 H's fewer than the maximum number

$\frac{\text{2 H's fewer than the maximum}}{\text{2 H's per degree of unsaturation}}$

Answer: one degree of unsaturation

$C_4H_6O_2$

For n carbons, the maximum number of H's is $2n + 2$; in this example, $2n + 2 = 2(4) + 2 = 10$.

10 H's (maximum) – 6 H's (actual) = 4 H's fewer than the maximum number

$\frac{\text{4 H's fewer than the maximum}}{\text{2 H's per degree of unsaturation}}$

Answer: two degrees of unsaturation

3 Use an exact mass to determine a molecular formula.

- High-resolution mass spectrometry gives the molecular formula of a compound.
- If the **exact mass** is **86.0775,** the molecular formula of **X** is **$C_5H_{10}O$ (exact mass = 86.0732)** rather than $C_4H_6O_2$ (exact mass = 86.0368).

4 Determine the functional groups by IR.

- **C_{sp^3}–H** bonds at 2973–2877 cm^{-1}
- **C=O** bond (1718 cm^{-1})

pentan-3-one **pentan-2-one** **3-methylbutan-2-one**

Three structures containing a ketone are consistent with the data.

See *How To* (B.5), Table B.2. Try Problems B.31–B.41.

SPECTROSCOPY CHAPTER B MULTIPLE-CHOICE SELF-TEST

The Self-Test consists of multiple-choice questions similar to those found on the American Chemical Society organic chemistry exam. Answers are given at the end of the chapter.

1. In which region of the IR spectrum does $CH_3CH_2CH_2CO_2CH_3$ *not* absorb. (a) 4000–2500 cm^{-1}; (b) 2500 2000 cm^{-1}; (c) 2000–1500 cm^{-1}; (d) < 1500 cm^{-1}?

2. In which compound does the carbonyl group absorb at the highest wavenumber?

a. 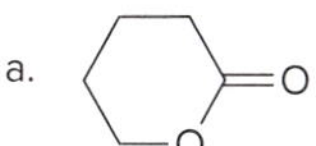

b.

c.

d. 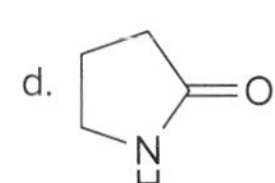

3. Which compound shows IR absorptions at 3300 and 2250 cm^{-1}?

a. 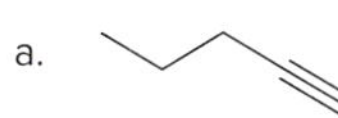

b.

c. 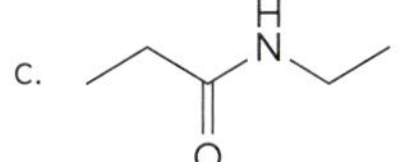

d. 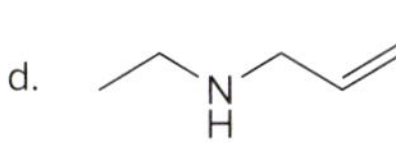

4. Label the C–H bonds that absorb at the highest and lowest wavenumbers.

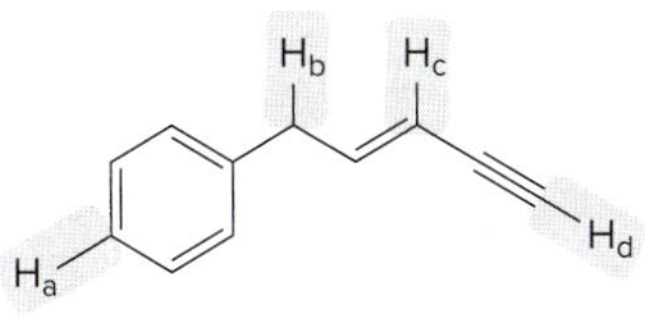

a. C–H_d absorbs at the highest wavenumber and C–H_a absorbs at the lowest wavenumber.

b. C–H_b absorbs at the highest wavenumber and C–H_d absorbs at the lowest wavenumber.

c. C–H_a absorbs at the highest wavenumber and C–H_b absorbs at the lowest wavenumber.

d. C–H_d absorbs at the highest wavenumber and C–H_b absorbs at the lowest wavenumber.

5. What is a possible structure for a compound that has IR absorptions at 3600–3200, 2950, and 1720 cm^{-1}?

a. b. c. d.

6. How would the IR spectrum of a reaction mixture change when **A** is converted to **B?**

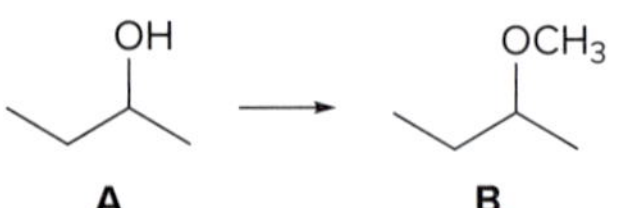

a. An absorption at 3600–3200 cm^{-1} disappears.
b. An absorption at 1720 cm^{-1} disappears.
c. An absorption at 1720 cm^{-1} appears.
d. An absorption at 1650 cm^{-1} appears.

PROBLEMS

Problems that combine mass spectrometry, infrared spectroscopy, and nuclear magnetic resonance spectroscopy are found at the end of Spectroscopy C.

Problem Using Three-Dimensional Models

B.21 What major IR absorptions are present above 1500 cm^{-1} for each compound?

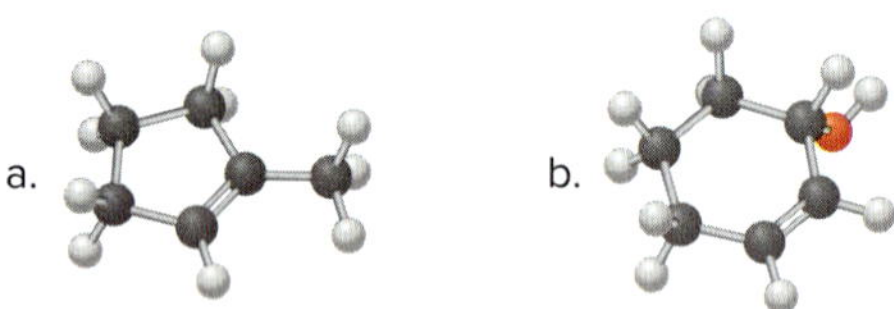

Infrared Spectroscopy

B.22 What major IR absorptions are present above 1500 cm^{-1} for each compound?

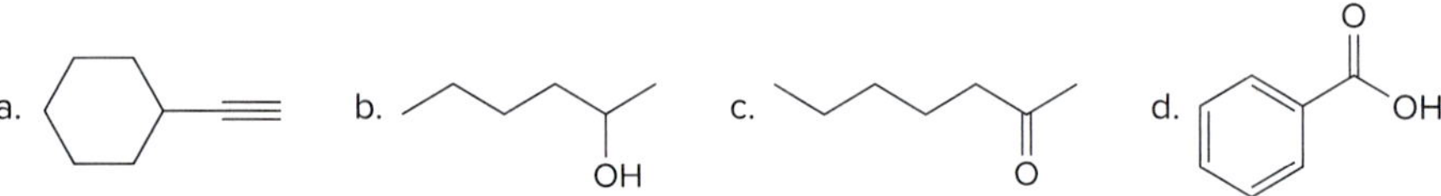

B.23 Estrone is a female sex hormone, and etonogestrel is a synthetic hormone used in contraceptive implants to prevent pregnancy. (a) Identify the prominent IR absorptions resulting from the functional groups in each compound. (b) How do the locations of the carbonyl absorptions in these two compounds compare? Explain your reasoning.

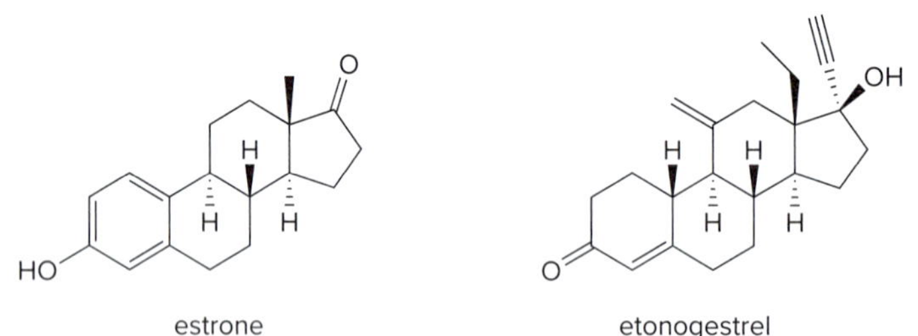

estrone etonogestrel

B.24 Morphine, heroin, and oxycodone are three addicting analgesic narcotics. How could IR spectroscopy be used to distinguish these three compounds from each other?

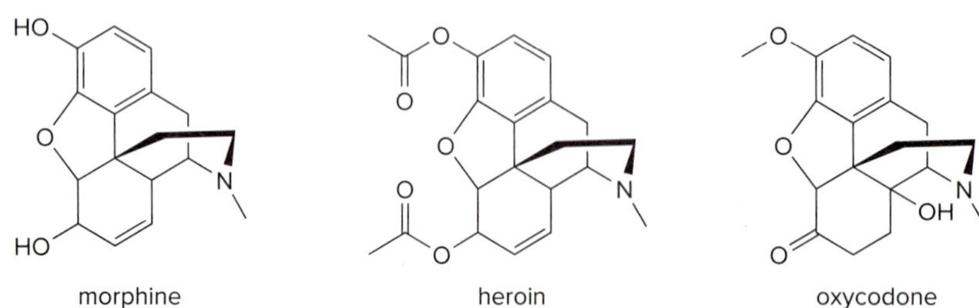

morphine heroin oxycodone

B.25 (a) Which of the following compounds has a C=O that absorbs at the *highest* wavenumber? (b) Which of the following compounds has a C=O that absorbs at the *lowest* wavenumber?

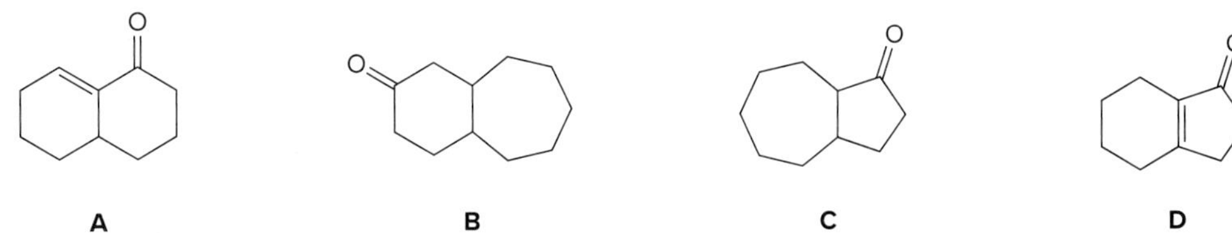

A B C D

B.26 Rank the following compounds in order of increasing wavenumber of the carbonyl absorption in the IR.

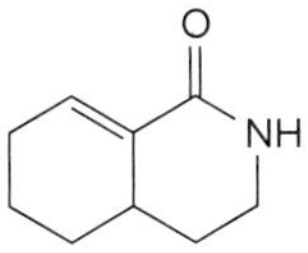

A 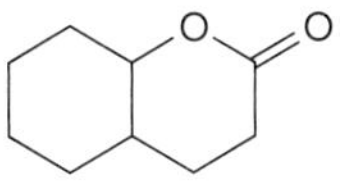 B 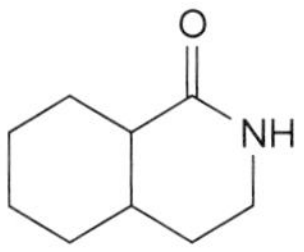 C 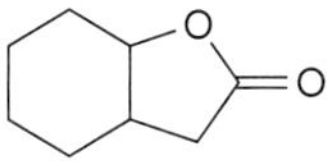 D

B.27 Tell how IR spectroscopy could be used to determine when each reaction is complete.

a. 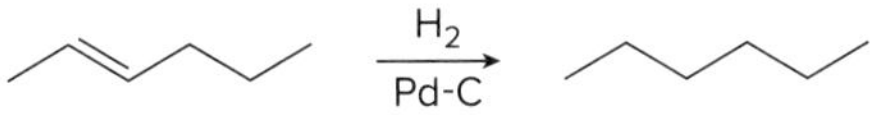 $\xrightarrow[\text{Pd-C}]{H_2}$

b. 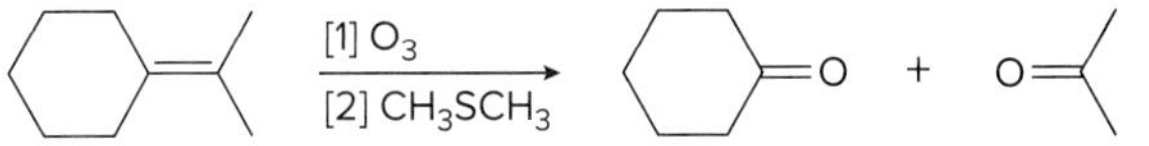 $\xrightarrow[\text{[2] } CH_3SCH_3]{\text{[1] } O_3}$ +

B.28 The compounds involved in the last three reactions in the synthesis of atomoxetine, a drug approved for the treatment of attention deficit hyperactivity syndrome (ADHD), are shown. How would the IR spectra of the reactant and product change in each reaction?

[1] → [2] → [3] →

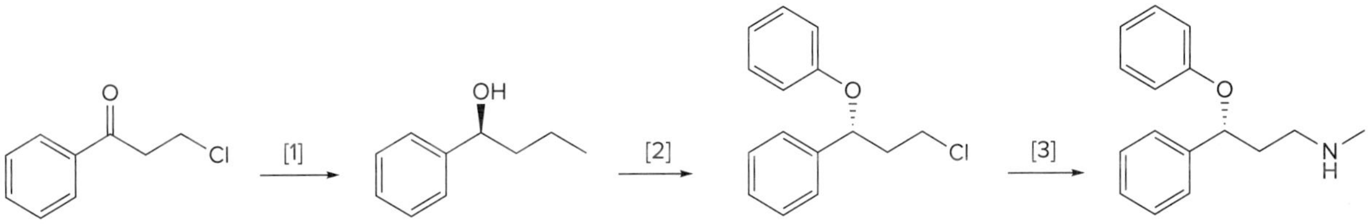

atomoxetine

B.29 The compounds involved in the last three reactions in the synthesis of sarcophytol B, a compound isolated from soft corals found in the Indian and Pacific Oceans, are shown. How would the IR spectra of the reactant and product change in each reaction?

[1] → [2] → [3] →

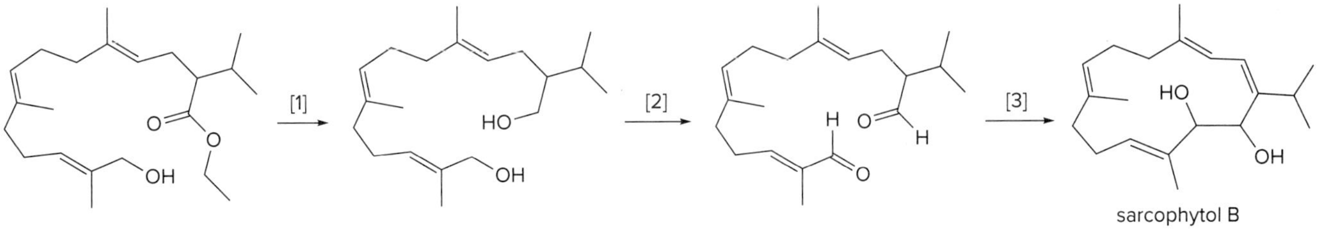

sarcophytol B

B.30 Match each compound to its IR spectrum.

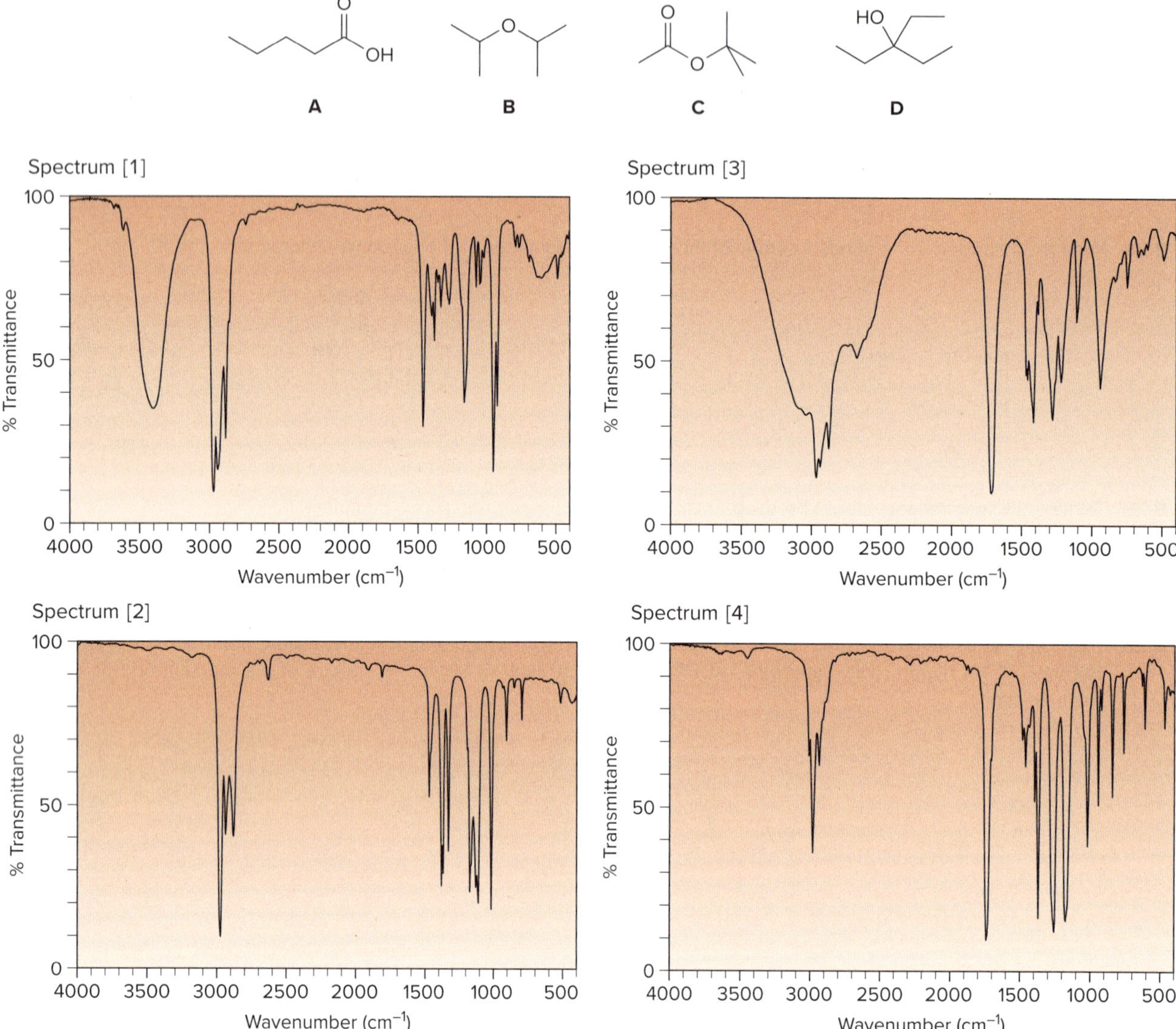

Spectroscopy Problems That Combine Mass Spectrometry and Infrared Spectroscopy

B.31 Propose possible structures consistent with each set of data. Assume each compound has an sp^3 hybridized C—H absorption in its IR spectrum, and that other major IR absorptions above 1500 cm^{-1} are listed.

a. a compound having a molecular ion at 72 and an absorption in its IR spectrum at 1725 cm^{-1}

b. a compound having a molecular ion at 55 and an absorption in its IR spectrum at ~2250 cm^{-1}

c. a compound having a molecular ion at 74 and an absorption in its IR spectrum at 3600–3200 cm^{-1}

B.32 A chiral hydrocarbon **X** exhibits a molecular ion at 82 in its mass spectrum. The IR spectrum of **X** shows peaks at 3300, 3000–2850, and 2250 cm^{-1}. Propose a structure for **X**.

B.33 A chiral compound **Y** has a strong absorption at 2970–2840 cm^{-1} in its IR spectrum and gives the following mass spectrum. Propose a structure for **Y.**

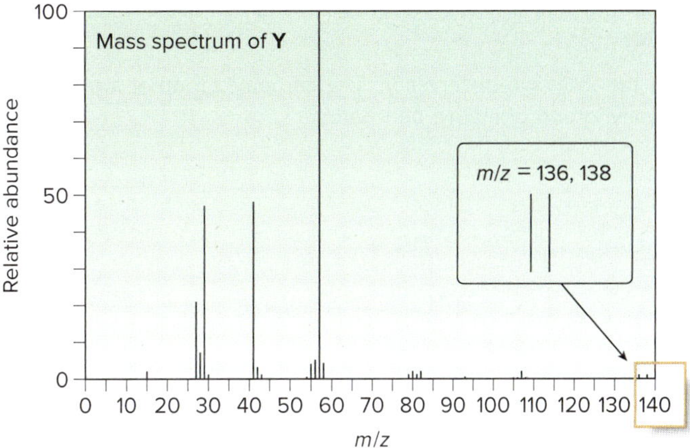

B.34 Treatment of benzoic acid ($C_6H_5CO_2H$) with NaOH followed by 1-iodo-3-methylbutane forms **H. H** has a molecular ion at 192 and IR absorptions at 3064, 3035, 2960–2872, and 1721 cm^{-1}. Propose a structure for **H.**

B.35 Treatment of benzaldehyde (C_6H_5CHO) with Zn(Hg) in aqueous HCl forms a compound **Z** that has a molecular ion at 92 in its mass spectrum. **Z** shows absorptions at 3150–2950, 1605, and 1496 cm^{-1} in its IR spectrum. Give a possible structure for **Z.**

B.36 Reaction of *tert*-butyl pentyl ether [$CH_3CH_2CH_2CH_2CH_2OC(CH_3)_3$] with HBr forms 1-bromopentane ($CH_3CH_2CH_2CH_2CH_2Br$) and compound **B. B** has a molecular ion in its mass spectrum at 56 and gives peaks in its IR spectrum at 3150–3000, 3000–2850, and 1650 cm^{-1}. Propose a structure for **B,** and draw a stepwise mechanism that accounts for its formation.

B.37 Reaction of 2-methylpropanoic acid [$(CH_3)_2CHCO_2H$] with $SOCl_2$ followed by 2-methylpropan-1-ol forms **X. X** has a molecular ion at 144 and IR absorptions at 2965, 2940, and 1739 cm^{-1}. Propose a structure for **X.**

B.38 Reaction of pentanoyl chloride ($CH_3CH_2CH_2CH_2COCl$) with lithium dimethyl cuprate [$LiCu(CH_3)_2$] forms a compound **J** that has a molecular ion in its mass spectrum at 100, as well as fragments at *m/z* = 85, 57, and 43 (base). The IR spectrum of **J** has strong peaks at 2962 and 1718 cm^{-1}. Propose a structure for **J.**

B.39 Benzonitrile (C_6H_5CN) is reduced to two different products depending on the reducing agent used. Treatment with lithium aluminum hydride followed by water forms **K,** which has a molecular ion in its mass spectrum at 107 and the following IR absorptions: 3373, 3290, 3062, 2920, and 1600 cm^{-1}. Treatment with a milder reducing agent forms **L,** which has a molecular ion in its mass spectrum at 106 and the following IR absorptions: 3086, 2820, 2736, 1703, and 1600 cm^{-1}. **L** shows fragments in its mass spectrum at *m/z* = 105 and 77. Propose structures for **K** and **L,** and explain how you arrived at your conclusions.

B.40 Treatment of anisole ($CH_3OC_6H_5$) with Cl_2 and $FeCl_3$ forms **P,** which has peaks in its mass spectrum at *m/z* = 142 (M), 144 (M + 2), 129, and 127. **P** has absorptions in its IR spectrum at 3096–2837 (several peaks), 1582, and 1494 cm^{-1}. Propose possible structures for **P.**

B.41 Reaction of $BrCH_2CH_2CH_2CH_2NH_2$ with NaH forms compound **W,** which gives the IR and mass spectra shown below. Propose a structure for **W** and draw a stepwise mechanism that accounts for its formation.

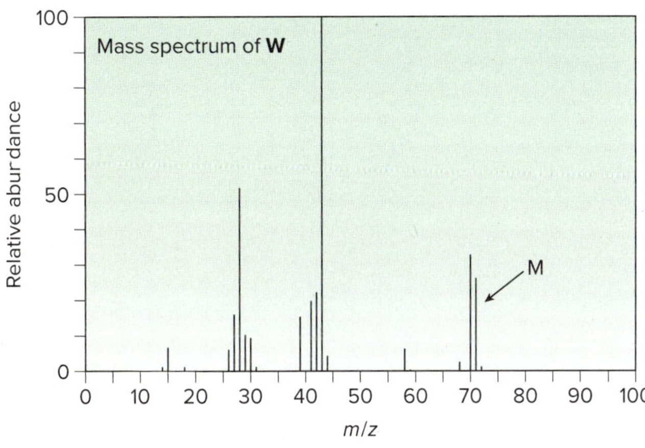

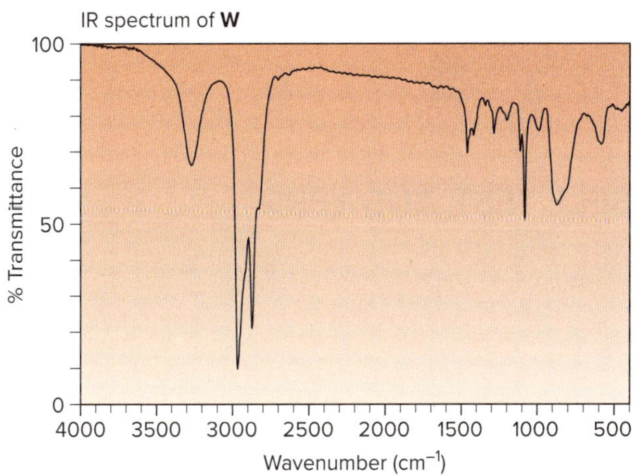

Challenge Problems

B.42 Acid chlorides (RCOCl) constitute another family of compounds that contains a carbonyl group. Would you expect the C=O of an acid chloride to absorb at a higher or lower wavenumber than an ester? Explain your reasoning. We will learn more about acid chlorides in Chapter 20.

B.43 Suggest an explanation for the following observation. The carbonyl group of methyl salicylate absorbs at a significantly lower wavenumber than the carbonyl group of methyl benzoate.

methyl salicylate
$\tilde{\nu}$ = 1680 cm^{-1}

methyl benzoate
$\tilde{\nu}$ = 1728 cm^{-1}

B.44 Explain why the C–O bond of diethyl ether (in red) absorbs at 1050 cm^{-1}, whereas the C–O bond of methyl acetate (in red) absorbs at 1250 cm^{-1}.

diethyl ether

methyl acetate

B.45 Oxidation of citronellol, a constituent of rose and geranium oils, with PCC in the presence of added $NaOCOCH_3$ forms compound **A. A** has a molecular ion in its mass spectrum at 154 and a strong peak in its IR spectrum at 1730 cm^{-1}, in addition to C–H stretching absorptions. Without added $NaOCOCH_3$, oxidation of citronellol with PCC yields isopulegone, which is then converted to **B** with aqueous base. **B** has a molecular ion at 152 and a peak in its IR spectrum at 1680 cm^{-1}, in addition to C–H stretching absorptions.

citronellol

isopulegone

a. Identify the structures of **A** and **B.**
b. Draw a mechanism for the conversion of citronellol to isopulegone.
c. Draw a mechanism for the conversion of isopulegone to **B.**

B.46 The carbonyl absorptions of esters **X** and **Y** differ by 25 cm^{-1}. Which compound absorbs at higher wavenumber and why?

X **Y**

SELF-TEST ANSWERS

1. b 2. b 3. a 4. d 5. b 6. a

Nuclear Magnetic Resonance Spectroscopy

Daniel C. Smith

C.1 An introduction to NMR spectroscopy

C.2 ^{1}H NMR: Number of signals

C.3 ^{1}H NMR: Position of signals

C.4 The chemical shift of protons on sp^2 and sp hybridized carbons

C.5 ^{1}H NMR: Intensity of signals

C.6 ^{1}H NMR: Spin–spin splitting

C.7 More-complex examples of splitting

C.8 Spin–spin splitting in alkenes

C.9 Other facts about ^{1}H NMR spectroscopy

C.10 Using ^{1}H NMR to identify an unknown

C.11 ^{13}C NMR spectroscopy

C.12 Magnetic resonance imaging (MRI)

The nation of Palau, located in the western Pacific Ocean, is composed of over 300 islands surrounded by pristine water that is home to a variety of marine organisms containing intriguing natural products. One such compound is palau'amine, a complex natural product isolated from the sea sponge *Hymeniacidon agminata* (formerly *Stylotella agminata*). The initial structure proposed for palau'amine in 1993 was revised in 2007 using a variety of modern spectroscopic techniques, including nuclear magnetic resonance spectroscopy. The dense array of functional groups in palau'amine and its antitumor and immunosuppressive properties attracted the attention of dozens of organic chemists, leading to its total synthesis in the laboratory in early 2010. In Spectroscopy Part C, we learn how nuclear magnetic resonance spectroscopy plays a key role in structure determination.

Why Study . . .

Nuclear Magnetic Resonance Spectroscopy?

In Spectroscopy C, we continue our study of organic structure determination by learning about **nuclear magnetic resonance (NMR)** spectroscopy. NMR spectroscopy is the most powerful tool for characterizing organic molecules, because it can be used to **identify the carbon–hydrogen framework in a compound.**

C.1 An Introduction to NMR Spectroscopy

Two common types of NMR spectroscopy are used to characterize organic structure:

- **^{1}H NMR (proton NMR)** is used to determine the number and type of hydrogen atoms in a molecule; and
- **^{13}C NMR (carbon NMR)** is used to determine the type of carbon atoms in a molecule.

Before you learn how to use NMR spectroscopy to determine the structure of a compound, it is helpful to understand the physics behind it. Keep in mind, though, that NMR stems from the same basic principle as all other forms of spectroscopy: Energy interacts with a molecule, and absorptions occur only when the incident energy matches the energy difference between two states.

C.1A The Basis of NMR Spectroscopy

The source of energy in NMR is radio waves. Radiation in the radiofrequency region of the electromagnetic spectrum (so-called **RF** radiation) has very long wavelengths, so its corresponding frequency and energy are both low. **When these low-energy radio waves interact with a molecule, they can change the nuclear spins of some elements, including ^{1}H and ^{13}C.**

A spinning proton creates a magnetic field.

When a charged particle such as a proton spins on its axis, it creates a magnetic field. For the purpose of this discussion, therefore, a nucleus is a tiny bar magnet, symbolized by ⍿. Normally these nuclear magnets are randomly oriented in space, but in the presence of an external magnetic field, B_0, they are oriented with or against this applied field. More nuclei are oriented *with* the applied field because this arrangement is lower in energy, but the **energy difference between these two states is very small** (< 0.4 J/mol).

With no external magnetic field...

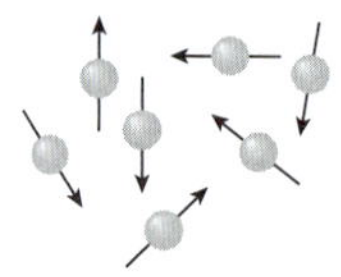

The nuclear magnets are randomly oriented.

In a magnetic field...

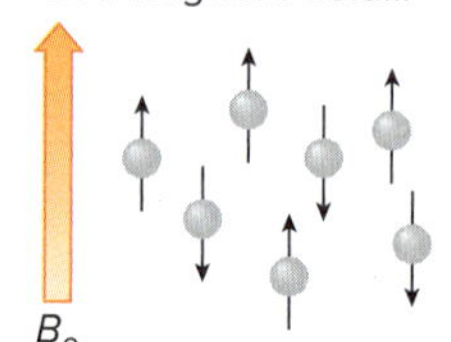

The nuclear magnets are oriented with or against B_0.

In a magnetic field, there are now two different energy states for a proton:

- **In the *lower-energy state* the nucleus is aligned in the *same direction* as B_0.**
- **In the *higher-energy state* the nucleus is aligned *opposed* to B_0.**

When an external energy source ($h\nu$) that matches the energy difference (ΔE) between these two states is applied, energy is absorbed, causing the **nucleus to "spin flip" from one orientation to another.** The energy difference between these two nuclear spin states corresponds to the low-frequency radiation in the RF region of the electromagnetic spectrum.

Absorbing RF radiation causes the nucleus to spin flip.

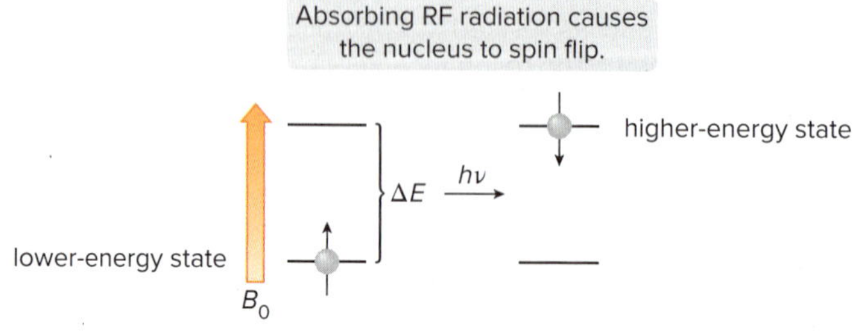

- **A nucleus is in *resonance* when it absorbs RF radiation and "spin flips" to a higher-energy state.**

Thus, two variables characterize NMR:

- **An applied magnetic field, B_0.** Magnetic field strength is measured in tesla (T).
- **The frequency ν of radiation used for resonance,** measured in hertz (Hz) or megahertz (MHz); (1 MHz = 10^6 Hz).

The frequency needed for resonance and the applied magnetic field strength are proportionally related:

$$\nu \propto B_0$$

frequency — applied magnetic field strength

- **The *stronger* the magnetic field, the *larger* the energy difference between the two nuclear spin states, and the *higher* the ν needed for resonance.**

NMR spectrometers are referred to as 300 MHz instruments, 500 MHz instruments, and so forth, depending on the frequency of RF radiation used for resonance.

Early NMR spectrometers used a magnetic field strength of ~1.4 T, which required RF radiation of 60 MHz for resonance. Modern NMR spectrometers use stronger magnets, thus requiring higher frequencies of RF radiation for resonance. For example, a magnetic field strength of 7.05 T requires a frequency of 300 MHz for a proton to be in resonance. These spectrometers use very powerful magnetic fields to create a small, but measurable energy difference between the two possible spin states. A schematic of an NMR spectrometer is shown in Figure C.1.

If all protons absorbed at the same frequency in a given magnetic field, the spectra of all compounds would consist of a single absorption, rendering NMR useless for structure determination. Fortunately, however, this is not the case.

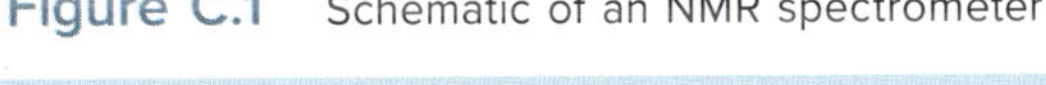

Figure C.1 Schematic of an NMR spectrometer

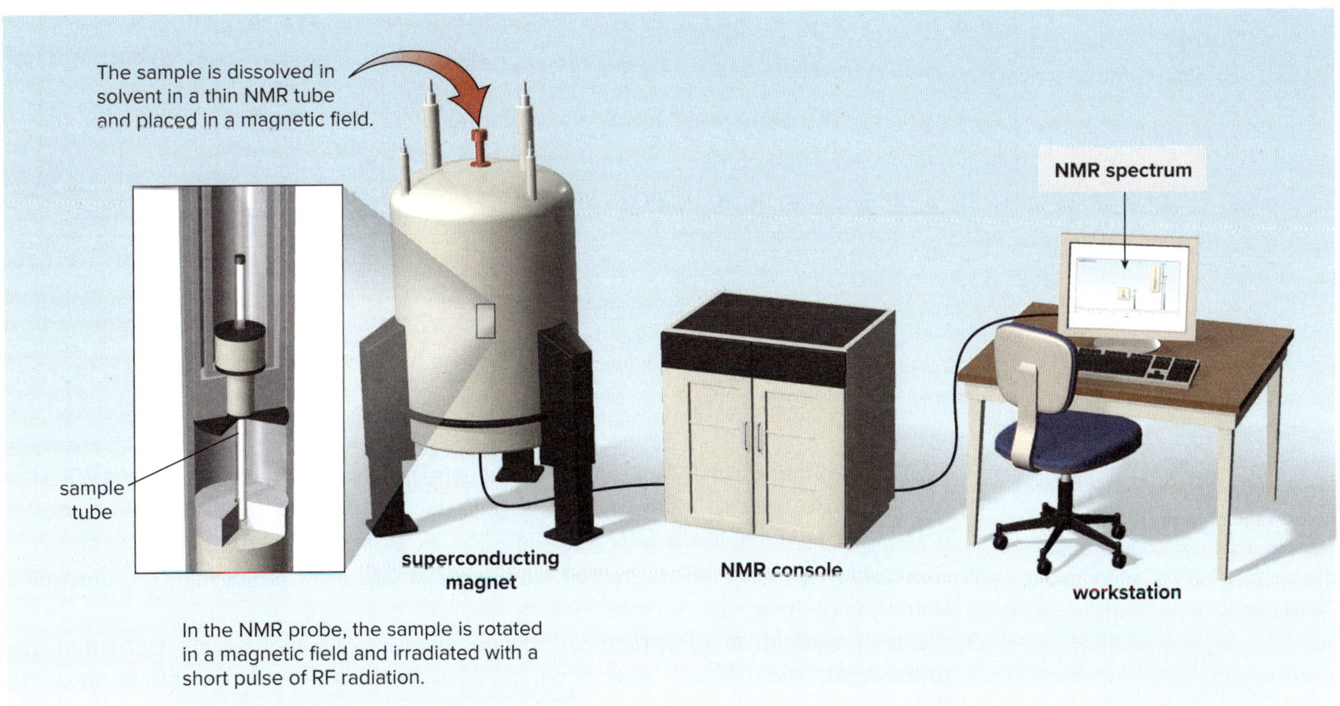

- **An NMR spectrometer.** The sample is dissolved in a solvent, usually $CDCl_3$ (deuterochloroform), and placed in a magnetic field. A radiofrequency generator then irradiates the sample with a short pulse of radiation, causing resonance. When the nuclei fall back to their lower-energy state, the detector measures the energy released, and a spectrum is recorded. The superconducting magnets in modern NMR spectrometers have coils that are cooled in liquid helium and conduct electricity with essentially no resistance.

- **All protons do *not* absorb at the same frequency. Protons in different environments absorb at slightly different frequencies, so they are distinguishable by NMR.**

The frequency at which a particular proton absorbs is determined by its electronic environment, as discussed in Section C.3. Because electrons are moving charged particles, they create a magnetic field opposed to the applied field B_0, and the size of the magnetic field generated by the electrons around a proton determines where it absorbs. Modern NMR spectrometers use a constant magnetic field strength B_0, and then a narrow range of frequencies is applied to achieve the resonance of all protons.

Only nuclei that contain odd mass numbers (such as ^{1}H, ^{13}C, ^{19}F, and ^{31}P) or odd atomic numbers (such as ^{2}H and ^{14}N) give rise to NMR signals. Because both ^{1}H and ^{13}C, the less abundant isotope of carbon, are NMR active, NMR allows us to map the carbon and hydrogen framework of an organic molecule.

C.1B A ^{1}H NMR Spectrum

An NMR spectrum plots the **intensity of a signal** against its **chemical shift** measured in **parts per million (ppm).** The common scale of chemical shifts is called the **δ (delta) scale.** The proton NMR spectrum of *tert*-butyl methyl ether [$CH_3OC(CH_3)_3$] illustrates several important features:

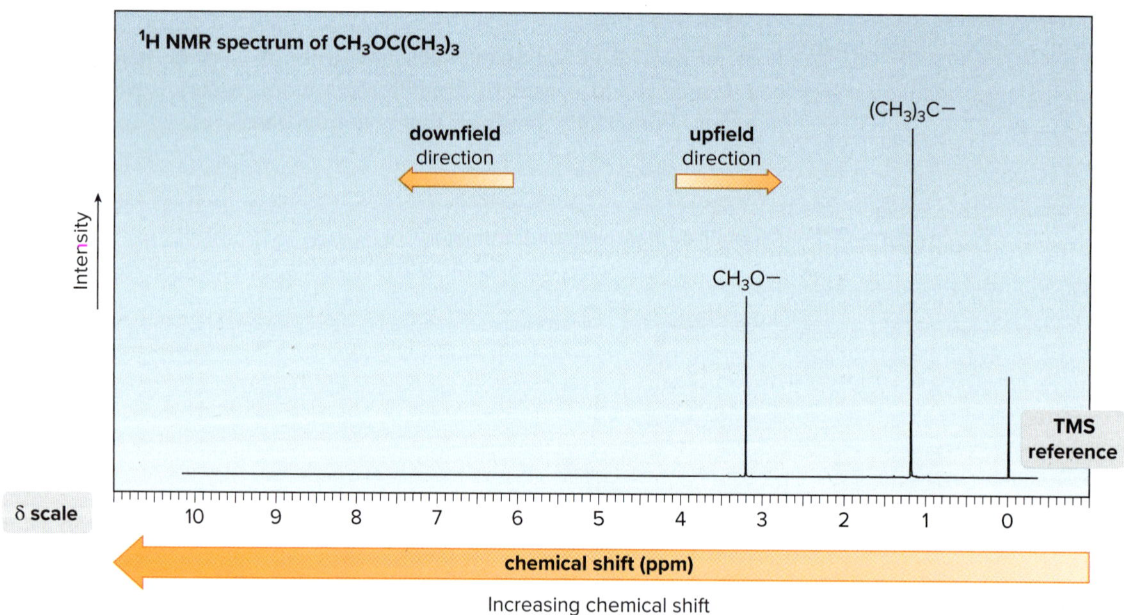

Spectra courtesy of the Chemistry Department of Rutgers University

tert-Butyl methyl ether (MTBE) is the high-octane gasoline additive that has contaminated the water supply in some areas (Section 3.4).

$(CH_3)_4Si$
tetramethylsilane
TMS

- NMR absorptions generally appear as sharp signals. The ^{1}H NMR spectrum of $CH_3OC(CH_3)_3$ consists of two signals: a tall peak at 1.2 ppm due to the $(CH_3)_3C$– group, and a smaller peak at 3.2 ppm due to the CH_3O– group.
- **Increasing chemical shift is plotted from *right to left.*** Most protons absorb somewhere from 0 to 12 ppm.
- The terms **upfield** and **downfield** describe the relative location of signals. **Upfield means to the *right.*** The $(CH_3)_3C$– peak is upfield from the CH_3O– peak. **Downfield means to the *left.*** The CH_3O– peak is downfield from the $(CH_3)_3C$– peak.

NMR absorptions are measured relative to the position of a reference signal at 0 ppm on the δ scale due to **tetramethylsilane (TMS). TMS** is a volatile and inert compound that gives a single peak upfield from other typical NMR absorptions.

Although chemical shifts are measured relative to the TMS signal at 0 ppm, this reference is often not plotted on a spectrum.

The *positive* direction of the δ scale is *downfield* from TMS. A very small number of absorptions occur upfield from the TMS signal, which is defined as the negative direction of the δ scale. (See Problem C.82.)

The **chemical shift** on the *x* axis gives the position of an NMR signal, measured in ppm, according to this equation:

$$\underset{\text{(in ppm on the } \delta \text{ scale)}}{\textbf{chemical shift}} = \frac{\text{observed chemical shift (in Hz) downfield from TMS}}{\nu \text{ of the NMR spectrometer (in MHz)}}$$

Because the frequency of the radiation required for resonance is proportional to the strength of the applied magnetic field, B_0, reporting NMR absorptions in frequency would be meaningless unless the value of B_0 was also reported. By reporting the absorption as a fraction of the NMR operating frequency, though, we get units—ppm—that are independent of the spectrometer.

Sample Problem C.1 Calculating Chemical Shift

Calculate the chemical shift of an absorption that occurs at 1500 Hz downfield from TMS using a 300 MHz NMR spectrometer.

Solution

Use the equation that defines the chemical shift in ppm:

$$\text{chemical shift} = \frac{\text{1500 Hz downfield from TMS}}{\text{300 MHz operating frequency}} = \text{5 ppm}$$

Problem C.1 The ^{1}H NMR spectrum of CH_3OH recorded on a 500 MHz NMR spectrometer consists of two signals, one due to the CH_3 protons at 1715 Hz and one due to the OH proton at 1830 Hz, both measured downfield from TMS. (a) Calculate the chemical shift of each absorption. (b) Do the CH_3 protons absorb upfield or downfield from the OH proton?

More Practice: Try Problems C.42, C.43.

Problem C.2 The ^{1}H NMR spectrum of 1,2-dimethoxyethane ($CH_3OCH_2CH_2OCH_3$) recorded on a 300 MHz NMR spectrometer consists of signals at 1017 Hz and 1065 Hz downfield from TMS. (a) Calculate the chemical shift of each absorption. (b) At what frequency would each absorption occur if the spectrum were recorded on a 500 MHz NMR spectrometer?

Four different features of a ^{1}H NMR spectrum provide information about a compound's structure:

[1] Number of signals (Section C.2)
[2] Position of signals (Sections C.3 and C.4)
[3] Intensity of signals (Section C.5)
[4] Spin–spin splitting of signals (Sections C.6–C.8)

C.2 ^{1}H NMR: Number of Signals

How many ^{1}H NMR signals does a compound exhibit? The number of NMR signals *equals* the number of different types of protons in a compound.

C.2A General Principles

- **Protons in different environments give different NMR signals. Equivalent protons give the same NMR signal.**

Any CH_3 group is different from any CH_2 group, which is different from any CH group in a molecule. Two CH_3 groups may be identical (as in CH_3OCH_3) or different (as in $CH_3OCH_2CH_3$), depending on what each CH_3 group is bonded to.

tert-Butyl methyl ether [$CH_3OC(CH_3)_3$] (Section C.1) exhibits two NMR signals because it contains two different kinds of protons: one CH_3 group is bonded to $–OC(CH_3)_3$, whereas the other three CH_3 groups are each bonded to the same group, $[–C(CH_3)_2]OCH_3$.

In many compounds, deciding whether two protons are in identical or different environments is intuitive.

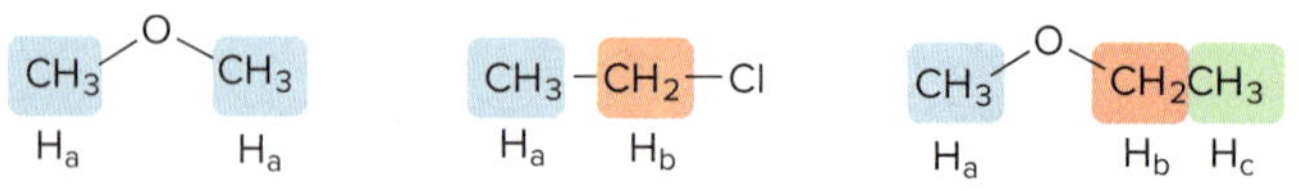

$CH_3–O–CH_3$ (H_a, H_a): **All equivalent H's**, **1** NMR signal

$CH_3–CH_2–Cl$ (H_a, H_b): **2** types of H's, **2** NMR signals

$CH_3–O–CH_2CH_3$ (H_a, H_b, H_c): **3** types of H's, **3** NMR signals

- **CH_3OCH_3:** Each CH_3 group is bonded to the same group ($–OCH_3$), making both CH_3 groups equivalent.
- **CH_3CH_2Cl:** The protons of the CH_3 group are different from those of the CH_2 group.
- **$CH_3OCH_2CH_3$:** The protons of the CH_2 group are different from those in each CH_3 group. The two CH_3 groups are also different from each other; one CH_3 group is bonded to $–OCH_2CH_3$ and the other is bonded to $–CH_2OCH_3$.

In some cases, it is less obvious by inspection if two protons are equivalent or different. To rigorously determine whether two protons are in identical environments (and therefore give rise to one NMR signal), replace each H atom in question by another atom Z (for example, Z = Cl). **If substitution by Z yields the same compound or enantiomers, the two protons are equivalent,** as shown in Sample Problem C.2.

Sample Problem C.2 Determining the Different Types of H's in a Molecule

How many different kinds of H atoms does $CH_3CH_2CH_2CH_2CH_3$ contain?

Solution

In comparing two H atoms, replace each H by Z (for example, Z = Cl), and examine the substitution products that result. The two CH_3 groups are identical because substitution of one H by Cl on each carbon gives the same product, 1-chloropentane.

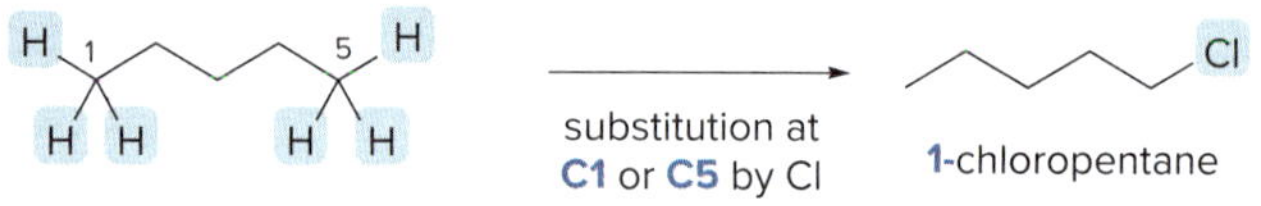

substitution at **C1** or **C5** by Cl → **1**-chloropentane

There are two different types of CH_2 groups. Substitution of Cl for H on C2 or C4 gives the same product, 2-chloropentane, so these H's are identical. Substitution of Cl for H on C3 gives a different product, 3-chloropentane, so this CH_2 group is different from the other two CH_2 groups.

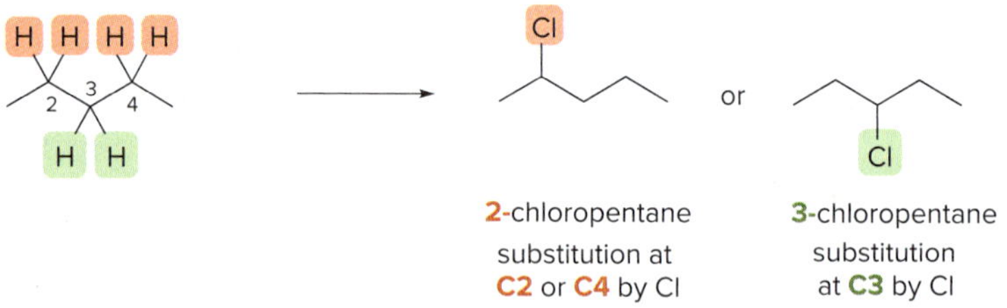

2-chloropentane: substitution at **C2** or **C4** by Cl

or

3-chloropentane: substitution at **C3** by Cl

Thus, $CH_3CH_2CH_2CH_2CH_3$ has three different types of protons and gives three different NMR signals.

$CH_3–CH_2–CH_2–CH_2–CH_3$ (H_a H_b H_c H_b H_a)

Problem C.3 How many ^1H NMR signals does each compound show?

a.  b. 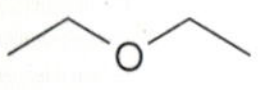 c. 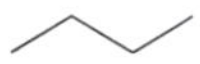 d. 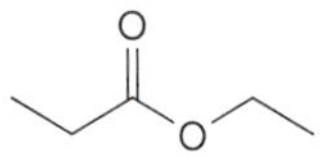 e. 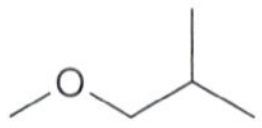 f. 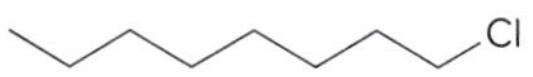 g.  h.

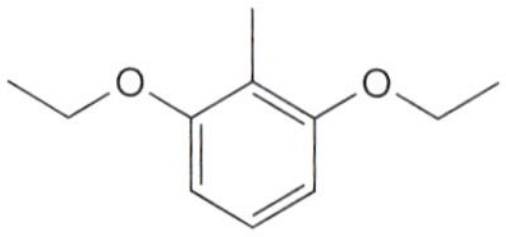

More Practice: Try Problems C.39a, C.40, C.41, C.55d, C.60a, C.61d.

Figure C.2
The number of ^{1}H NMR signals of some representative organic compounds

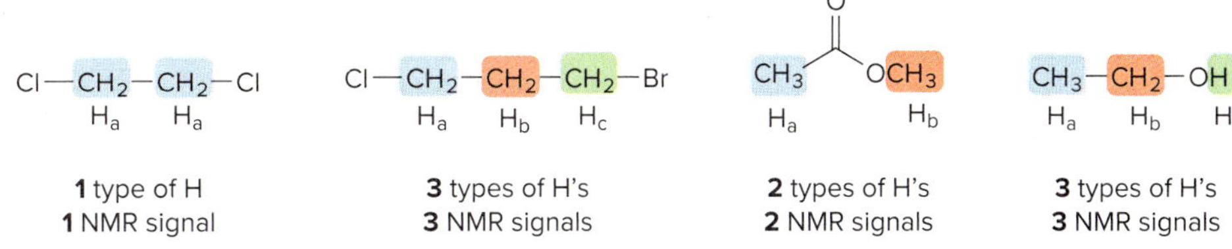

Figure C.2 gives the number of NMR signals exhibited by four additional molecules. All protons—not just protons bonded to carbon atoms—give rise to NMR signals. Ethanol (CH_3CH_2OH), for example, gives three NMR signals, one of which is due to its OH proton.

C.2B Determining Equivalent Protons in Alkenes and Cycloalkanes

To determine equivalent protons in cycloalkanes and alkenes that have restricted bond rotation, always **draw in all bonds to hydrogen.**

Draw [cyclopropane with all H's and Cl drawn] **NOT** [cyclopropane—Cl] **Draw** [Cl, H / H, H on C=C] **NOT** ClCH=CH$_2$

Then, in comparing two H atoms on a ring or double bond, **two protons are equivalent only if they are cis (or trans) to the same groups,** as illustrated with 1,1-dichloroethylene, 1-bromo-1-chloroethylene, and chloroethylene.

1,1-dichloroethylene	1-bromo-1-chloroethylene	chloroethylene
H cis to Cl; H cis to Cl	H_a cis to Cl; H_b cis to Br	H_b cis to Cl; H_c cis to H_a
1 type of H **1** NMR signal	**2** types of H's **2** NMR signals	**3** types of H's **3** NMR signals

- **1,1-Dichloroethylene:** The two H atoms on the C=C are both cis to a Cl atom. Thus, both H atoms are equivalent.
- **1-Bromo-1-chloroethylene:** H_a is cis to a Cl atom and H_b is cis to a Br atom. Thus, H_a and H_b are different, giving rise to two NMR signals.
- **Chloroethylene:** H_a is bonded to the carbon with the Cl atom, making it different from H_b and H_c. Of the remaining two H atoms, H_b is cis to a Cl atom and H_c is cis to a H atom, making them different. All three H atoms in this compound are different.

Proton equivalency in cycloalkanes can be determined similarly.

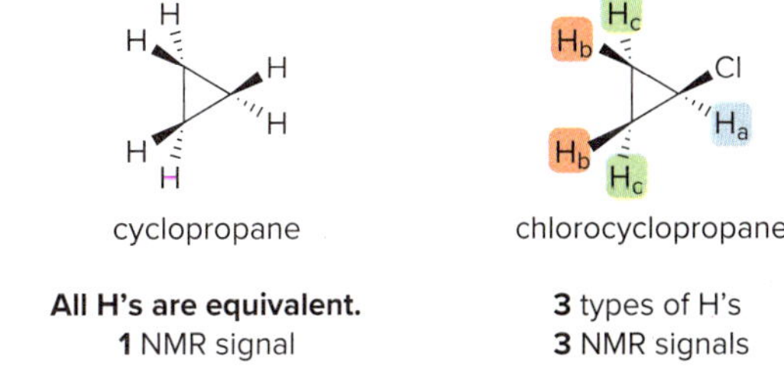

- **Cyclopropane:** All H atoms are equivalent, so there is only one NMR signal.
- **Chlorocyclopropane:** There are now three kinds of H atoms: H_a is bonded to a carbon bonded to a Cl; both H_b protons are cis to the Cl, whereas both H_c protons are cis to another H.

Sample Problem C.3 Determining Proton Equivalency in Cyclic Compounds

How many ^1H NMR signals does **A** exhibit?

OH

A

Solution

Use wedges and dashed wedges to emphasize the relative location of groups on a ring. **Two protons are equivalent only if they are cis (or trans) to the same groups.** Start with the protons that can be assigned most easily. In this example, the OH and the CH on the ring look different from all other protons, so they give two NMR signals. The two CH_3 groups are different from each other because one CH_3 is cis to H_b and one is trans to H_b. Likewise, H_e and H_f are different from each other, because H_e is trans to H_b, whereas H_f is cis to H_b.

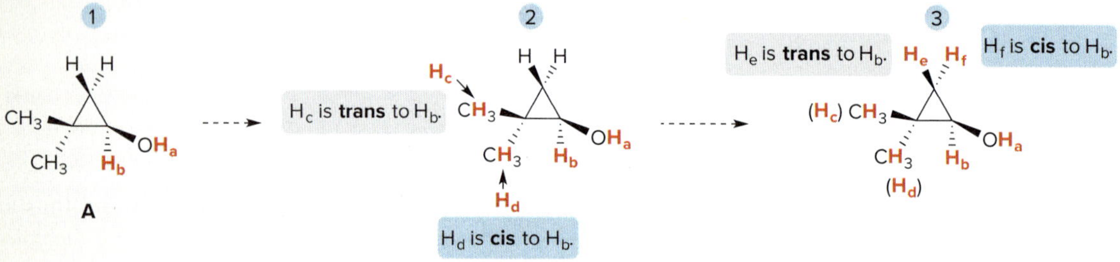

1 H H CH_3 CH_3 H_b OH_a **A**

2 H_c is **trans** to H_b. H_c CH_3 H H OH_a CH_3 H_b H_d H_d is **cis** to H_b.

3 H_e is **trans** to H_b. H_e H_f H_f is **cis** to H_b. (H_c) CH_3 OH_a CH_3 (H_d) H_b

Thus, **A** contains **six** different types of H's and gives **six** ^1H NMR signals.

Problem C.4 How many ^1H NMR signals does each dimethylcyclopropane show?

a.  b.  c.

More Practice: Try Problems C.40h–j, C.55d, C.60a.

Problem C.5 How many ^1H NMR signals does each alkene exhibit?

a. 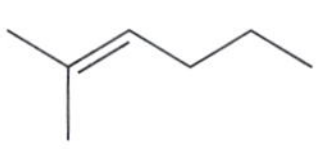 b. 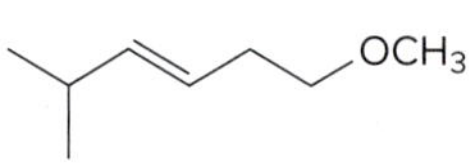 OCH_3 c. 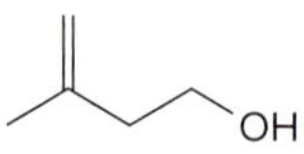 OH

Problem C.6 How many ^1H NMR signals does each compound give?

a. 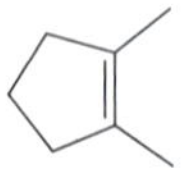 b. 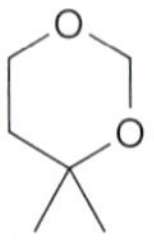 O O c. 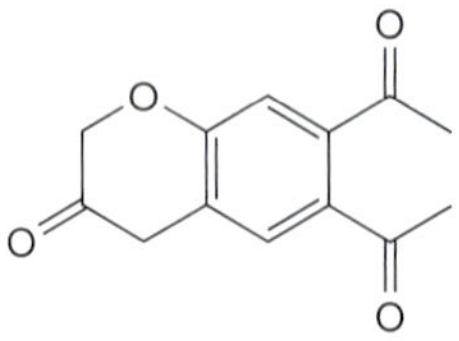 O O O O d. 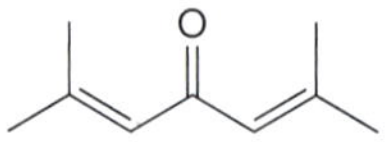 O

C.2C Homotopic, Enantiotopic, and Diastereotopic Protons

Let's look more closely at the protons of a single sp^3 hybridized CH_2 group to determine whether these two protons are always equivalent to *each other*. Three examples illustrate different outcomes.

$CH_3CH_2CH_3$ has two different types of protons—those of the CH_3 groups and those of the CH_2 group—meaning that the two H atoms of the CH_2 group are *equivalent to each other*. Replacement of each H by Z forms the *same* product, so they give *one* NMR signal.

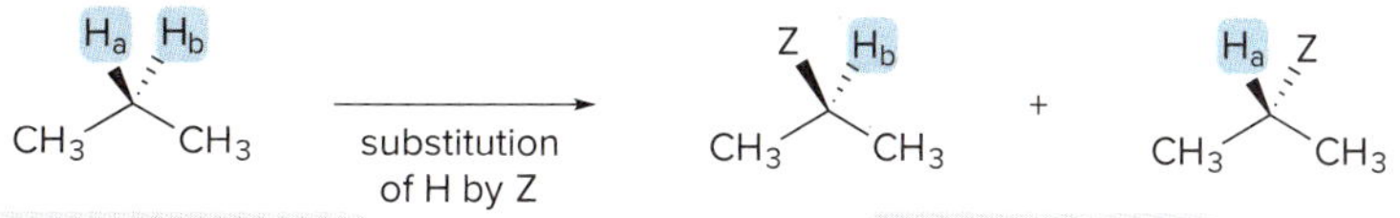

H_a and H_b are **homotopic.** — substitution of H by Z → **identical products**

- **When substitution of two H atoms by Z forms the *same* product, these equivalent hydrogens are called *homotopic* protons.**

CH_3CH_2Cl has two different types of protons—those of the CH_3 group and those of the CH_2 group—meaning that the two H atoms of the CH_2 group are *equivalent to each other.* Replacement of each H of the CH_2 group by an atom Z creates a new stereogenic center, forming two products that are **enantiomers.**

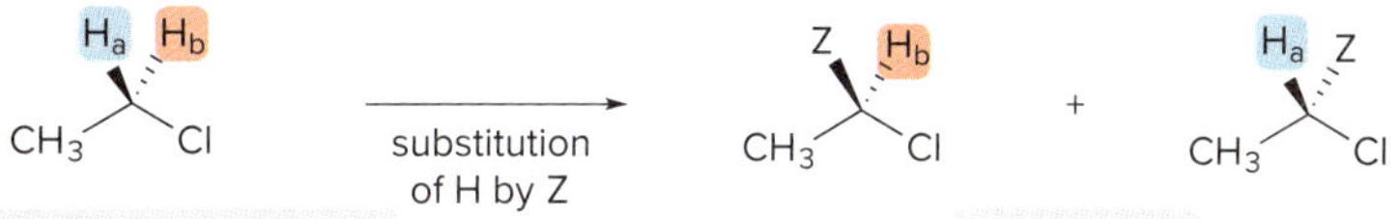

H_a and H_b are **enantiotopic.** — substitution of H by Z → **enantiomers**

- **When substitution of two H atoms by Z forms *enantiomers*, the two H atoms are equivalent and give a single NMR signal. These two H atoms are called *enantiotopic* protons.**

In contrast, the two H atoms of the CH_2 group in (*R*)-2-chlorobutane, which contains one stereogenic center, are *not* equivalent to each other. Substitution of each H by Z forms two **diastereomers,** and thus, these two H atoms give *different* NMR signals.

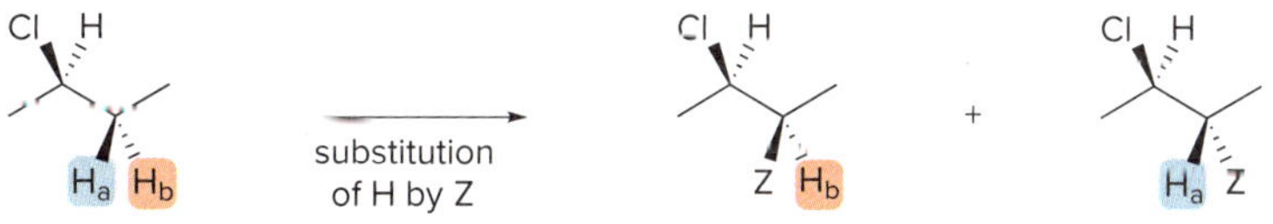

(*R*)-2-chlorobutane

H_a and H_b are **diastereotopic.** — substitution of H by Z → **diastereomers**

- **When substitution of two H atoms by Z forms *diastereomers*, the two H atoms are *not* equivalent, and give two NMR signals. These two H atoms are called *diastereotopic* protons.**

Sample Problem C.4 Classifying Protons as Homotopic, Enantiotopic, or Diastereotopic

Classify the protons in each labeled CH_2 group as homotopic, enantiotopic, or diastereotopic.

a. 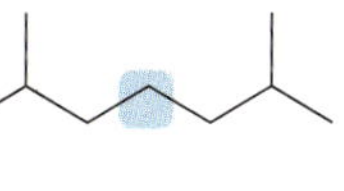 b. 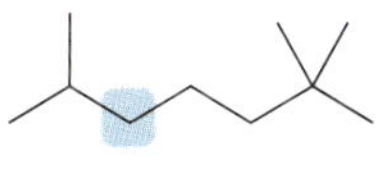 c.

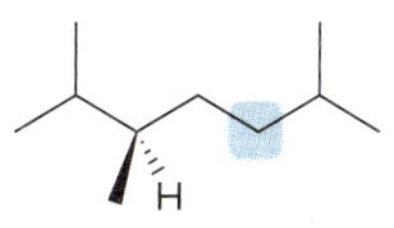

Solution

To determine equivalency in these cases, look for whether the compound has a stereogenic center to begin with and whether a new stereogenic center is formed when H is replaced by Z.

a. The compound is achiral and has no stereogenic center. Replacement of each H on the labeled CH_2 group by Z forms the same product, making them **homotopic.** The H's within the CH_2 group are *equivalent* to each other and give *one* NMR signal.

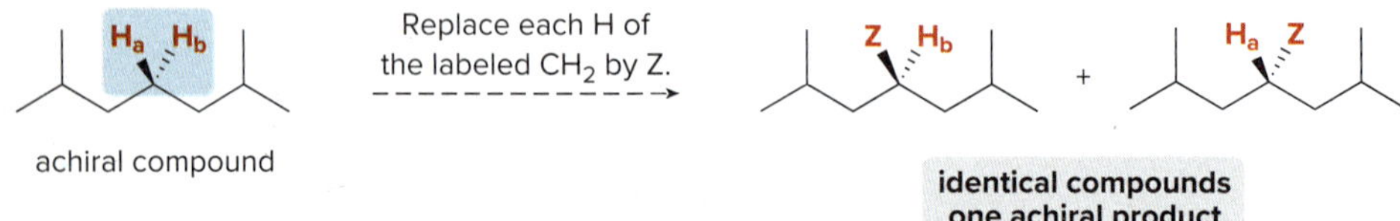

b. The compound is achiral and has no stereogenic center. Because a new stereogenic center is formed on substitution of H by Z, the protons of the CH_2 group are **enantiotopic.** These H's are *equivalent* to each other and give *one* NMR signal.

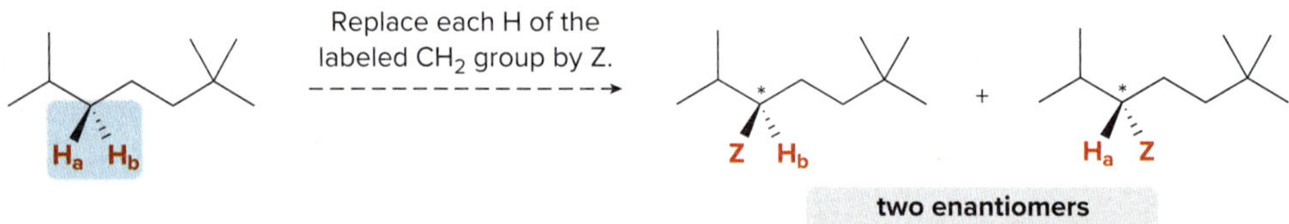

c. The compound has one stereogenic center to begin with. Because a new stereogenic center is formed on substitution of H by Z, the protons are **diastereotopic.** The H's within the CH_2 group are *different* from each other and give *different* NMR signals.

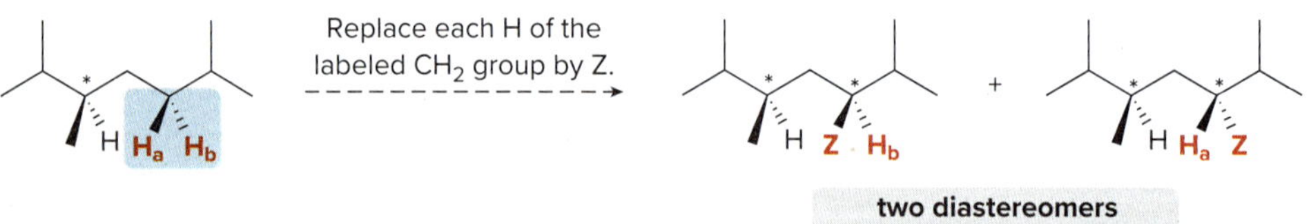

Problem C.7 Label the protons in each highlighted CH_2 group as enantiotopic, diastereotopic, or homotopic.

a. O b. c. d.

Problem C.8 How many ^1H NMR signals would you expect for each compound?

a. Cl b. Cl O c. Br d. Cl Cl

C.3 ^1H NMR: Position of Signals

In the NMR spectrum of *tert*-butyl methyl ether in Section C.1B, why does the $CH_3O–$ group absorb downfield from the $–C(CH_3)_3$ group?

- **Where a particular proton absorbs depends on its electronic environment.**

C.3A Shielding and Deshielding Effects

To understand how the electronic environment around a nucleus affects its chemical shift, recall that in a magnetic field, an electron creates a small magnetic field that opposes the applied magnetic field, B_0. **Electrons are said to *shield* the nucleus from B_0.**

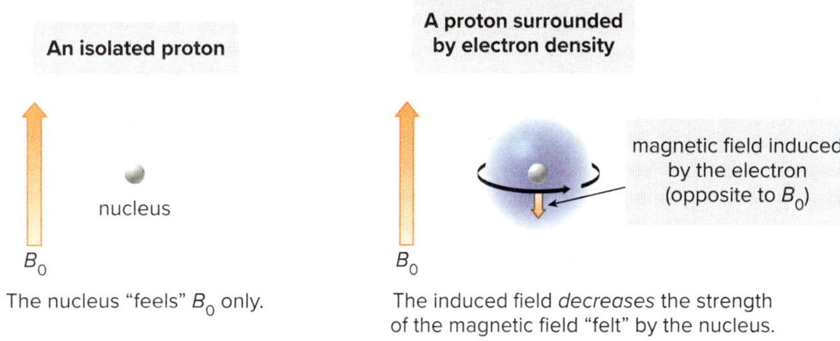

In the vicinity of the nucleus, therefore, the magnetic field generated by the circulating electron *decreases* the external magnetic field that the proton "feels." Because the proton experiences a lower magnetic field strength, it needs a lower frequency to achieve resonance. Lower frequency is to the right in an NMR spectrum, toward lower chemical shift, so **shielding shifts an absorption *upfield,*** as shown in Figure C.3a.

What happens if the electron density around a nucleus is *decreased,* instead? For example, how do the chemical shifts of the protons in CH_4 and CH_3Cl compare?

Figure C.3 How chemical shift is affected by electron density around a nucleus

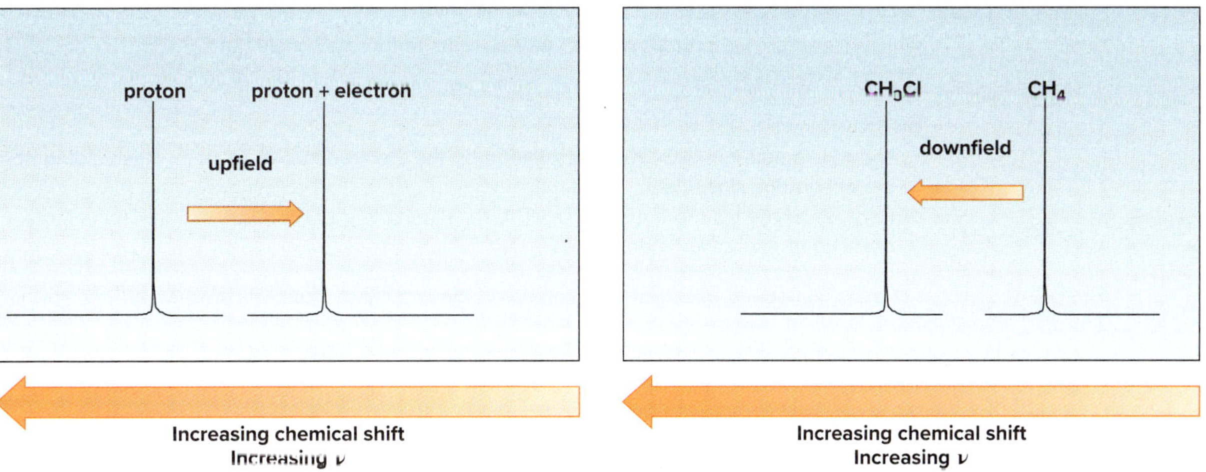

The less shielded the nucleus becomes, the more of the applied magnetic field (B_0) it feels. This *deshielded* nucleus experiences a higher magnetic field strength, so it needs a higher frequency to achieve resonance. Higher frequency is to the *left* in an NMR spectrum, toward higher chemical shift, so **deshielding shifts an absorption downfield,** as shown in Figure C.3b for CH_3Cl versus CH_4. The electronegative Cl atom withdraws electron density from the carbon and hydrogen atoms in CH_3Cl, thus deshielding them relative to those in CH_4.

Remember the trend: ***D**ecreased electron density **d**eshields a nucleus and an absorption moves **d**ownfield.*

- **Protons near electronegative atoms are deshielded, so they absorb downfield.**

Figure C.4 summarizes the effects of shielding and deshielding.

These electron density arguments explain the relative position of NMR signals in many compounds.

CH_3-CH_2-Cl (H_a = CH_3, H_b = CH_2)

- The H_b protons are **deshielded** because they are closer to the electronegative Cl atom, so they absorb **downfield** from H_a.

$Br-CH_2-CH_2-F$ (H_a = CH_2 next to Br, H_b = CH_2 next to F)

- Because F is more electronegative than Br, the H_b protons are more **deshielded** than the H_a protons and absorb farther **downfield.**

$Cl-CH_2-CHCl_2$ (H_a = CH_2, H_b = $CHCl_2$)

- The larger number of electronegative Cl atoms (two vs. one) **deshields** H_b more than H_a, so it absorbs **downfield** from H_a.

Figure C.4 Shielding and deshielding effects

a. A shielded nucleus

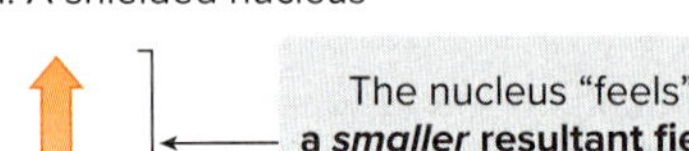
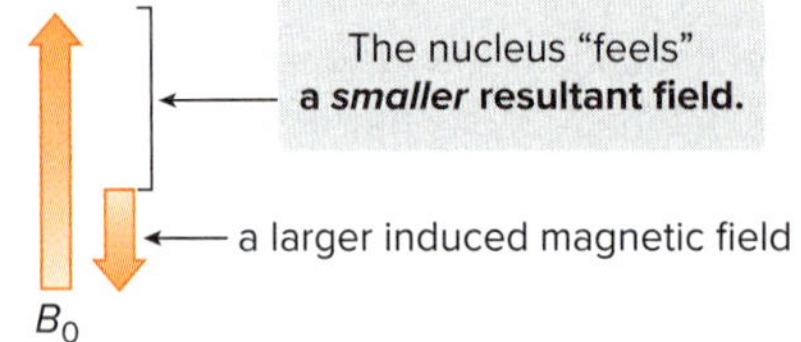

- As the electron density around the nucleus increases, the nucleus feels a *smaller* resultant magnetic field, so a *lower* frequency is needed to achieve resonance.
- **The absorption shifts *upfield.***

b. A deshielded nucleus

The nucleus "feels" a *larger* resultant field.

a smaller induced magnetic field

B_0

- As the electron density around the nucleus decreases, the nucleus feels a *larger* resultant magnetic field, so a *higher* frequency is needed to achieve resonance.
- **The absorption shifts *downfield.***

Sample Problem C.5 Determining Shielding and Deshielding Effects

Which of the labeled protons in each pair absorbs farther downfield: (a) $CH_3CH_2\mathbf{CH_3}$ or $CH_3O\mathbf{CH_3}$; (b) $CH_3O\mathbf{CH_3}$ or $CH_3S\mathbf{CH_3}$?

Solution

a. The CH_3 group in CH_3OCH_3 is deshielded by the electronegative O atom. **Deshielding shifts the absorption downfield.**

b. Because oxygen is more electronegative than sulfur, the CH_3 group in CH_3OCH_3 is more **deshielded** and absorbs **downfield.**

Problem C.9 For each compound, which of the protons on the highlighted carbons absorbs farther downfield?

a. 

b. 

c. 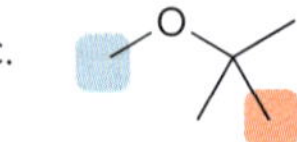

More Practice: Try Problems C.44, C.55b.

C.3B Chemical Shift Values

Not only is the *relative* position of NMR absorptions predictable, but it is also possible to predict the approximate chemical shift value for a given type of proton.

- **Protons in a given environment absorb in a predictable region in an NMR spectrum.**

A more detailed list of characteristic chemical shift values is found in Appendix H.

Table C.1 lists the typical chemical shift values for the most common bonds encountered in organic molecules.

Table C.1 also illustrates that absorptions for a given type of C—H bond occur in a narrow range of chemical shift values, usually 1–2 ppm. For example, all sp^3 hybridized C—H bonds in alkanes and cycloalkanes absorb between 0.9 and 2.0 ppm. By contrast, absorptions due to N—H and O—H protons can occur over a broader range. For example, the OH proton of an alcohol is found anywhere in the 1–5 ppm range. The position of these absorptions is affected by the extent of hydrogen bonding, making it more variable.

Table C.1 Characteristic Chemical Shifts of Common Types of Protons

Type of proton	Chemical shift (ppm)	Type of proton	Chemical shift (ppm)
sp^3 C—H	0.9–2	R(R)C=C—H	4.5–6
• RCH_3	~0.9		
• R_2CH_2	~1.3	aryl C—H (benzene ring—H)	6.5–8
• R_3CH	~1.7		
Z=C—C—H (allylic), Z = C, O, N	1.5–2.5	RC(=O)H	9–10
—C≡C—H	~2.5	RC(=O)OH	10–12
Z—C—H, Z = N, O, X	2.5–4	R—O—H or R—N—H	1–5

The chemical shift of a particular type of C—H bond is also affected by the number of R groups bonded to the carbon atom.

R—CH₃ (~0.9 ppm) → R_2CH_2 (~1.3 ppm) → R_3CH (~1.7 ppm)

Increasing alkyl substitution
Increasing chemical shift

- **The chemical shift of a C—H bond *increases* with *increasing* alkyl substitution.**

Problem C.10 For each compound, first label each different type of proton and then rank the protons in order of increasing chemical shift.

a. Cl—CH₂CH₂CH₂—Br b. CH₃OCH₂OC(CH₃)₃ c. CH₃COCH₂CH₃ d. CH₃OC(=O)OCH₂CH₂CH₃

Problem C.11 Label each statement as True or False.

a. When a nucleus is strongly shielded, the effective field is larger than the applied field and the absorption shifts downfield.
b. When a nucleus is strongly shielded, the effective field is smaller than the applied field and the absorption is shifted upfield.
c. A nucleus that is strongly deshielded requires a lower field strength for resonance.
d. A nucleus that is strongly shielded absorbs at a larger δ value.

C.4 The Chemical Shift of Protons on sp^2 and *sp* Hybridized Carbons

The chemical shift of protons bonded to benzene rings, C–C double bonds, and C–C triple bonds merits additional comment.

Benzene–H	Alkene C=C–H	Alkyne C≡C–H
7.3 ppm	4.5–6 ppm	2.5 ppm

Each of these functional groups contains π bonds with **loosely held π electrons.** When placed in a magnetic field, these π electrons move in a circular path, inducing a new magnetic field. How this induced magnetic field affects the chemical shift of a proton depends on the direction of the induced field *in the vicinity of the absorbing proton.*

Protons on Benzene Rings

In a magnetic field, the six π electrons in **benzene** circulate around the ring, creating a ring current. The magnetic field induced by these moving electrons *reinforces* the applied magnetic field in the vicinity of the protons. The protons thus feel a stronger magnetic field and a higher frequency is needed for resonance, so the **protons are deshielded and the absorption is *downfield.***

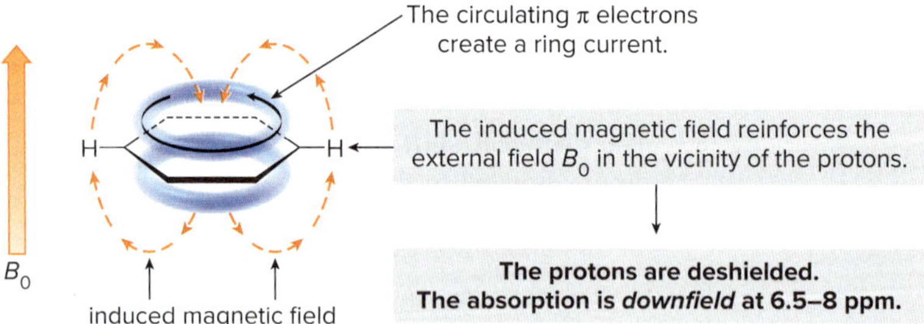

Protons on Carbon–Carbon Double Bonds

A similar phenomenon occurs with protons on carbon–carbon double bonds. In a magnetic field, the loosely held π electrons create a magnetic field that *reinforces* the applied field in the vicinity of the protons. Because the protons now feel a stronger magnetic field, they require a higher frequency for resonance. **The protons are deshielded and the absorption is *downfield.***

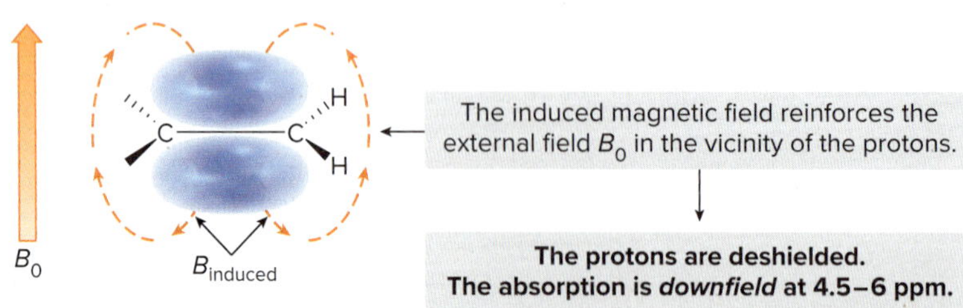

Protons on Carbon–Carbon Triple Bonds

In a magnetic field, the π electrons of a carbon–carbon triple bond are induced to circulate, but in this case the induced magnetic field *opposes* the applied magnetic field (B_0). The proton thus feels a weaker magnetic field, so a lower frequency is needed for resonance. **The nucleus is shielded and the absorption is *upfield*.**

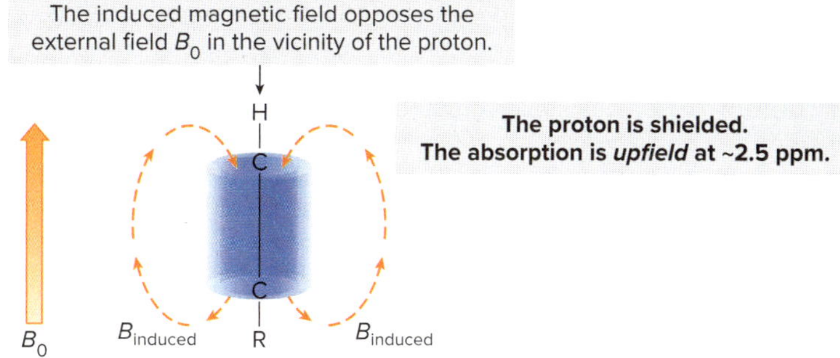

Table C.2 summarizes the shielding and deshielding effects due to circulating π electrons.

Table C.2 Effect of π Electrons on Chemical Shift Values

Proton type	Effect	Chemical shift (ppm)
benzene–H	highly deshielded	6.5–8
C=C–H	deshielded	4.5–6
C≡C–H	shielded	~2.5

To remember the chemical shifts of some common bond types, it is helpful to think of a ^{1}H NMR spectrum as being divided into six different regions (Figure C.5).

Figure C.5 Regions in the ^{1}H NMR spectrum

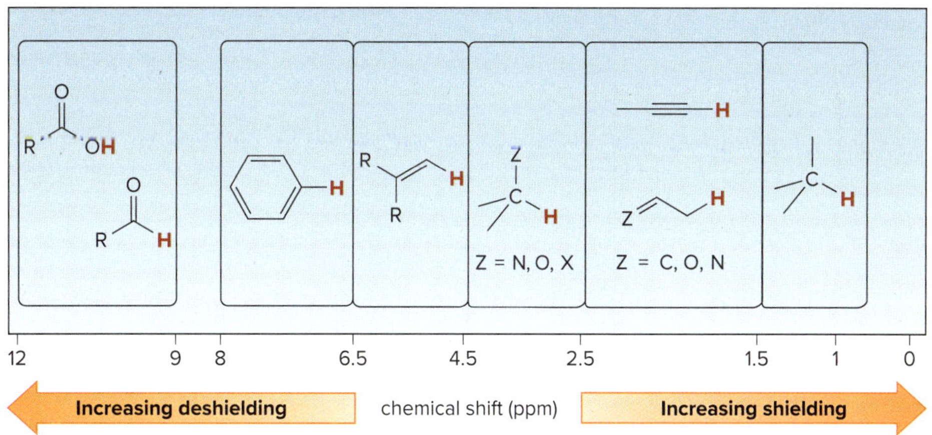

- **Shielded** protons absorb at **lower** chemical shift (to the **right**).
- **Deshielded** protons absorb at **higher** chemical shift (to the **left**).
- Note: The drawn chemical shift scale is not linear.

Sample Problem C.6 Predicting the Relative Chemical Shift of Protons

Rank H_a, H_b, and H_c in order of increasing chemical shift.

H_c
OCH_2CH_3
H_b H_a

Solution

The H_a protons are bonded to an sp^3 hybridized carbon, so they are shielded and absorb upfield compared to H_b and H_c. Because the H_b protons are deshielded by the electronegative oxygen atom on the C to which they are bonded, they absorb downfield from H_a. The H_c proton is deshielded by two factors. The electronegative O atom withdraws electron density from H_c. Moreover, because H_c is bonded directly to a C=C, the magnetic field induced by the π electrons causes further deshielding. Thus, in order of increasing chemical shift, $H_a < H_b < H_c$.

Problem C.12 Rank each group of protons in order of increasing chemical shift.

a. [alkyne with H_a] [alkene with H_b] [alkane with H_c]

b. $CH_3C(=O)OCH_2CH_3$ (H_a = CH_3; H_b = CH_2; H_c = CH_3 of ethyl)

More Practice: Try Problem C.44.

C.5 ¹H NMR: Intensity of Signals

The relative intensity of ¹H NMR signals also provides information about a compound's structure.

- **The area under an NMR signal is proportional to the number of absorbing protons.**

For example, in the ¹H NMR spectrum of $CH_3OC(CH_3)_3$, the ratio of the area under the downfield peak (due to the CH_3O– group) to the upfield peak [due to the $–C(CH_3)_3$ group] is 1:3. An NMR spectrometer automatically integrates the area under the peaks, and prints out a digital display of the *relative* areas of the NMR signals. Older NMR spectrometers print out a stepped curve (an **integral**) on the spectrum. The height of each step is proportional to the area under the peak, which is in turn proportional to the number of absorbing protons.

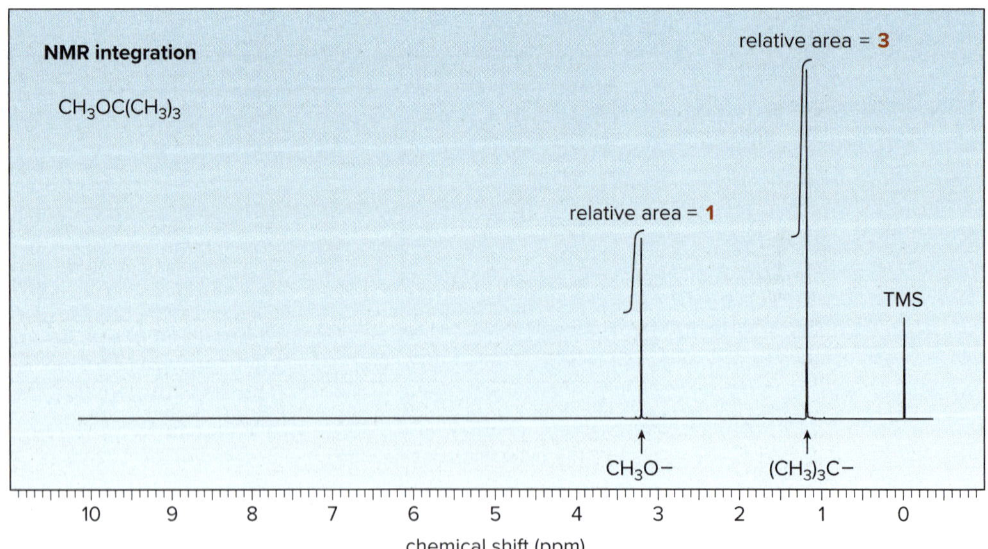

Integrals can be manually measured, but modern NMR spectrometers automatically calculate and plot the value of each integral in arbitrary units. If the heights of two integrals are in a 1:3 ratio, then the ratio of absorbing protons is 1:3, or 2:6, or 3:9, and so forth. This tells the *ratio,* not the absolute number of protons.

Problem C.13 Which compounds give a ¹H NMR spectrum with two signals in a ratio of 2:3?

a. [CH_3CH_2Cl skeletal structure] b. [$CH_3CH_2CH_3$ skeletal structure] c. [$CH_3CH_2OCH_2CH_3$ skeletal structure] d. [$CH_3OCH_2CH_2OCH_3$ skeletal structure]

Problem C.14 Compound **A** exhibits two signals in its ^{1}H NMR spectrum at 2.64 and 3.69 ppm, and the ratio of the absorbing signals is 2:3. Compound **B** exhibits two signals in its ^{1}H NMR spectrum at 2.09 and 4.27 ppm, and the ratio of the absorbing signals is 3:2. Which compound corresponds to dimethyl succinate, and which compound corresponds to ethylene diacetate?

dimethyl succinate

ethylene diacetate

C.6 ^{1}H NMR: Spin–Spin Splitting

The ^{1}H NMR spectra you have seen up to this point have been limited to one or more single absorptions called **singlets.** In the ^{1}H NMR spectrum of $BrCH_2CHBr_2$, however, the two signals for the two different kinds of protons are each split into more than one peak. The splitting patterns, the result of **spin–spin splitting,** can be used to determine how many protons reside on the carbon atoms near the absorbing proton.

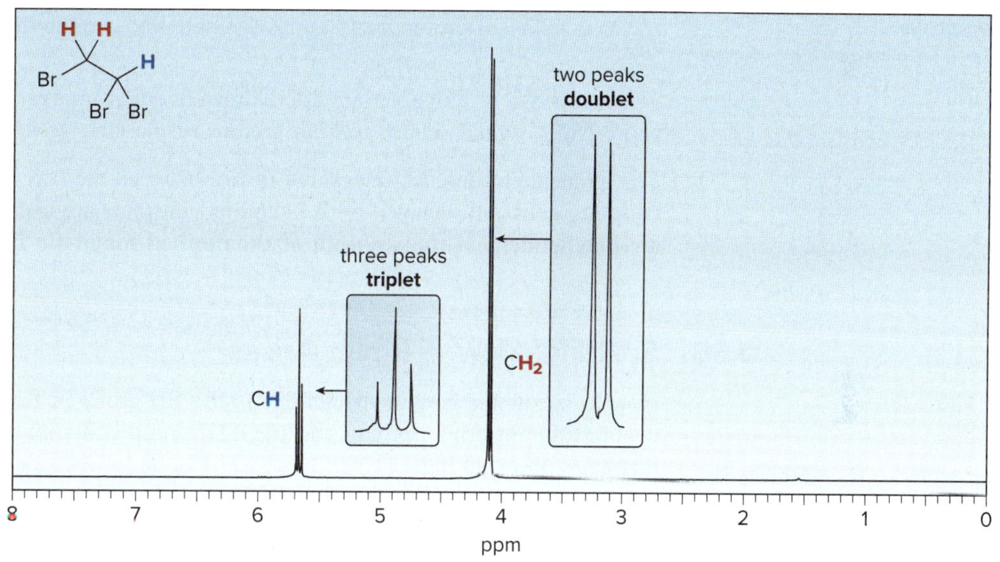

To understand spin–spin splitting, we must distinguish between the **absorbing protons** that give rise to an NMR signal, and the **adjacent protons** that cause the signal to split. **The number of adjacent protons determines the observed splitting pattern.**

- The CH_2 signal appears as **two peaks,** called a ***doublet.*** The relative area under the peaks of a doublet is 1:1.
- The CH signal appears as **three peaks,** called a ***triplet.*** The relative area under the peaks of a triplet is 1:2:1.

Spin–spin splitting occurs between nonequivalent protons on the same carbon or adjacent carbons. To illustrate how spin–spin splitting arises, we'll examine nonequivalent protons on adjacent carbons, the more common example. Spin–spin splitting arises because protons are little magnets that can be aligned with or against an applied magnetic field, and this affects the magnetic field that a nearby proton feels.

C.6A Splitting: How a Doublet Arises

absorbing H's

H H

H adjacent H

Br

Br Br

The adjacent H can be aligned with (↑) or against (↓) B_0.

First, let's examine how the doublet due to the CH_2 group in $BrCH_2CHBr_2$ arises. The CH_2 group contains the absorbing protons and the CH group contains the adjacent proton that causes the splitting.

When placed in an applied magnetic field (B_0), the adjacent proton ($CHBr_2$) can be aligned with (↑) or against (↓) B_0. As a result, the absorbing protons (CH_2Br) feel two slightly different magnetic fields—one slightly larger than B_0 and one slightly smaller than B_0. Because the absorbing protons feel two different magnetic fields, they absorb at two different frequencies in the NMR spectrum, thus splitting a single absorption into a doublet.

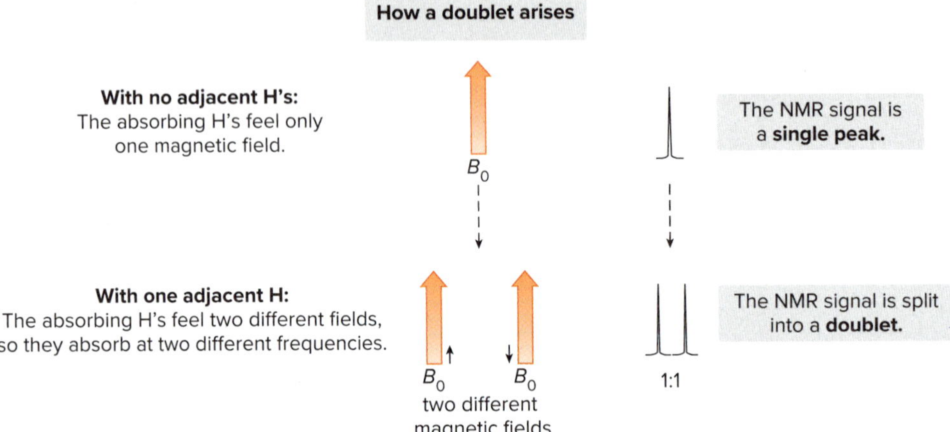

Keep in mind the difference between an **NMR signal** and an **NMR peak.** An NMR signal is the entire absorption due to a particular kind of proton. NMR peaks are contained within a signal. **A doublet constitutes one signal that is split into two peaks.**

- **One adjacent proton splits an NMR signal into a doublet.**

The two peaks of a doublet are approximately equal in area. The area under both peaks—the entire NMR signal—is due to both protons of the CH_2 group of $BrCH_2CHBr_2$.

coupling constant, ***J,*** in Hz

The frequency difference (measured in Hz) between the two peaks of the doublet is called the **coupling constant,** denoted by ***J.*** Coupling constants are usually in the range of 0–18 Hz, and are **independent of the strength of the applied magnetic field, B_0.**

C.6B Splitting: How a Triplet Arises

adjacent H's

H_a H_b H absorbing H

Br Br Br

H_a and H_b can each be aligned with (↑) or against (↓) B_0.

Now let's examine how the triplet due to the CH group in $BrCH_2CHBr_2$ arises. The CH group contains the absorbing proton and the CH_2 group contains the adjacent protons (H_a and H_b) that cause the splitting.

When placed in an applied magnetic field (B_0), the adjacent protons H_a and H_b can each be aligned with (↑) or against (↓) B_0. As a result, the absorbing proton feels three slightly different magnetic fields—one slightly larger than B_0, one slightly smaller than B_0, and one the same strength as B_0.

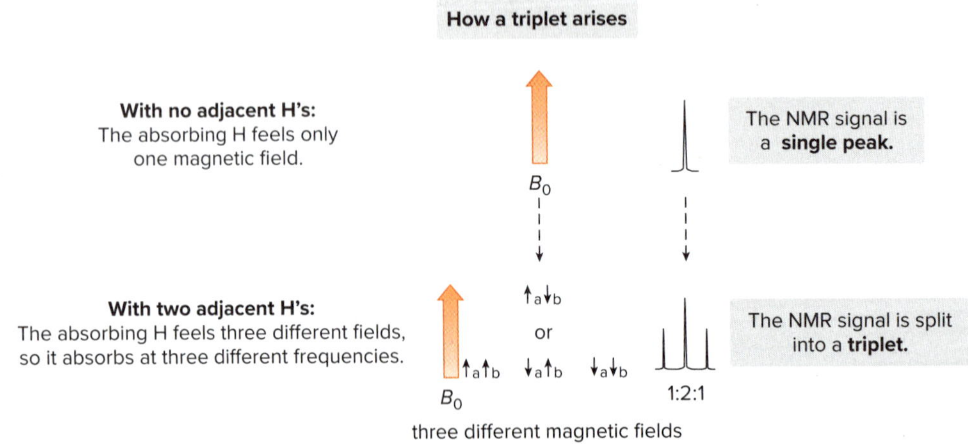

Because the absorbing proton feels three different magnetic fields, it absorbs at three different frequencies in the NMR spectrum, thus splitting a single absorption into a triplet. Because there are two different ways to align one proton with B_0 and one proton against B_0—that is, $\uparrow_a\downarrow_b$ and $\downarrow_a\uparrow_b$—the middle peak of the triplet is twice as intense as the two outer peaks, making the ratio of the areas under the three peaks 1:2:1.

- **Two adjacent protons split an NMR signal into a triplet.**

When two protons split each other's NMR signals, they are said to be ***coupled.*** In $BrCH_2CHBr_2$, the CH proton is coupled to the CH_2 protons. **The spacing between peaks in a split NMR signal, measured by the *J* value, is *equal* for coupled protons.**

C.6C Splitting: The Rules and Examples

Three general rules describe the splitting patterns commonly seen in the ^{1}H NMR spectra of organic compounds.

Rule [1] **Equivalent protons don't split each other's signals.**

Rule [2] **A set of *n* nonequivalent protons splits the signal of a nearby proton into *n* + 1 peaks.**

- In $BrCH_2CHBr_2$, for example, *one* adjacent CH proton splits an NMR signal into *two* peaks (a doublet), and *two* adjacent CH_2 protons split an NMR signal into *three* peaks (a triplet). Names for split NMR signals containing two to seven peaks are given in Table C.3. An NMR signal having more than seven peaks is called a **multiplet.**
- The inside peaks of a split NMR signal are always most intense, with the area under the peaks decreasing from the inner to the outer peaks in a given splitting pattern.

Table C.3 Names for a Given Number of Peaks in an NMR Signal

Number of peaks	Name
1	singlet
2	doublet
3	triplet
4	quartet
5	quintet
6	sextet
7	septet
> 7	multiplet

Rule [3] **Splitting is observed for nonequivalent protons on the same carbon or adjacent carbons.**

If H_a and H_b are not equivalent, splitting is observed in each of the following cases.

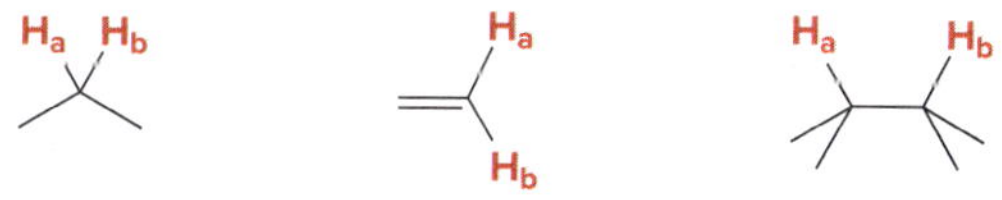

H_a H_b H_a H_b H_a H_b

The splitting of an NMR signal reveals the number of nearby nonequivalent protons. It tells nothing about the absorbing proton itself.

Splitting is not generally observed between protons separated by more than three σ bonds. Although H_a and H_b are not equivalent to each other in butan-2-one and ethyl methyl ether, H_a and H_b are separated by four σ bonds, so they are too far away to split each other's NMR signals.

O σ σ σ σ H_a H_b

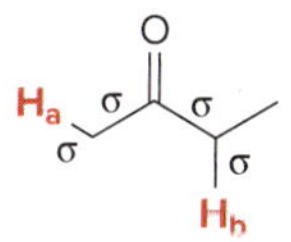

butan-2-one
H_a and H_b are separated by **four** σ bonds.

no splitting between H_a and H_b

σ O σ σ σ H_a H_b

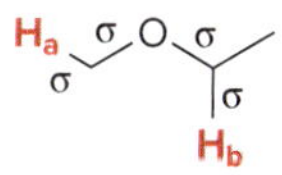

ethyl methyl ether
H_a and H_b are separated by **four** σ bonds.

no splitting between H_a and H_b

Table C.4 illustrates common splitting patterns observed for adjacent nonequivalent protons.

Table C.4 Common Splitting Patterns Observed in ^{1}H NMR

Example	Pattern	Analysis				
[1] H–C–C–H	H H	• **H**: one adjacent **H** proton	--->	two peaks	--->	a **doublet**
		• **H**: one adjacent **H** proton	--->	two peaks	--->	a **doublet**
[2] H–C–CH_2–	H H	• **H**: two adjacent **H** protons	--->	three peaks	--->	a **triplet**
		• **H**: one adjacent **H** proton	--->	two peaks	--->	a **doublet**
[3] –CH_2CH_2–	H H	• **H**: two adjacent **H** protons	--->	three peaks	--->	a **triplet**
		• **H**: two adjacent **H** protons	--->	three peaks	--->	a **triplet**
[4] –CH_2CH_3	H H	• **H**: three adjacent **H** protons	--->	four peaks	--->	a **quartet***
		• **H**: two adjacent **H** protons	--->	three peaks	--->	a **triplet**
[5] H–C–CH_3	H H	• **H**: three adjacent **H** protons	--->	four peaks	--->	a **quartet***
		• **H**: one adjacent **H** proton	--->	two peaks	--->	a **doublet**

*The relative area under the peaks of a quartet is 1:3:3:1.

Predicting splitting is always a two-step process:

- **Determine if two protons are equivalent or different.** Only *nonequivalent* protons split each other.
- **Determine if two nonequivalent protons are close enough to split each other's signals.** Splitting is observed only for nonequivalent protons on the *same* carbon or *adjacent* carbons.

Several examples of spin–spin splitting in specific compounds illustrate the result of this two-step strategy.

H H, Cl, Cl, H H

- All protons are equivalent, so there is no splitting and the NMR signal is one singlet.

H_a H_a, Br, Cl, H_b H_b

- There are two NMR signals. H_a and H_b are nonequivalent protons bonded to adjacent C atoms, so they are close enough to split each other's NMR signals. The H_a signal is split into a triplet by the two H_b protons. The H_b signal is split into a triplet by the two H_a protons.

O, CH_3, CH_2CH_3, H_a, H_b H_c

- There are three NMR signals. H_a has no adjacent nonequivalent protons, so its signal is a singlet. The H_b signal is split into a quartet by the three H_c protons. The H_c signal is split into a triplet by the two H_b protons.

Cl, H_a, Br, H_b

- There are two NMR signals. H_a and H_b are nonequivalent protons on the same carbon, so they are close enough to split each other's NMR signals. The H_a signal is split into a doublet by H_b. The H_b signal is split into a doublet by H_a.

Problem C.15 Into how many peaks will each proton shown in red be split?

a. CH_3CH_2COCl b. CH_3CHBr_2 c. $CH_3COCH_2CH_2Br$ d. (H)(Br)C=C(Cl)(H) e. f. $ClCH_2CH(OCH_3)_2$

Problem C.16 For each compound, give the number of ^1H NMR signals and then determine how many peaks are present for each NMR signal.

a. b. c. d.

Problem C.17 Sketch the NMR spectrum of CH_3CH_2Cl, giving the approximate location of each NMR signal.

C.7 More-Complex Examples of Splitting

Up to now you have studied examples of spin–spin splitting where the absorbing proton has nearby protons on *one* adjacent carbon only. What happens when the absorbing proton has nonequivalent protons on *two* adjacent carbons? Different outcomes are possible, depending on whether the adjacent nonequivalent protons are *equivalent to* or *different from* each other.

For example, 2-bromopropane [$(CH_3)_2CHBr$] has two types of protons—H_a and H_b—so it exhibits two NMR signals, as shown in Figure C.6.

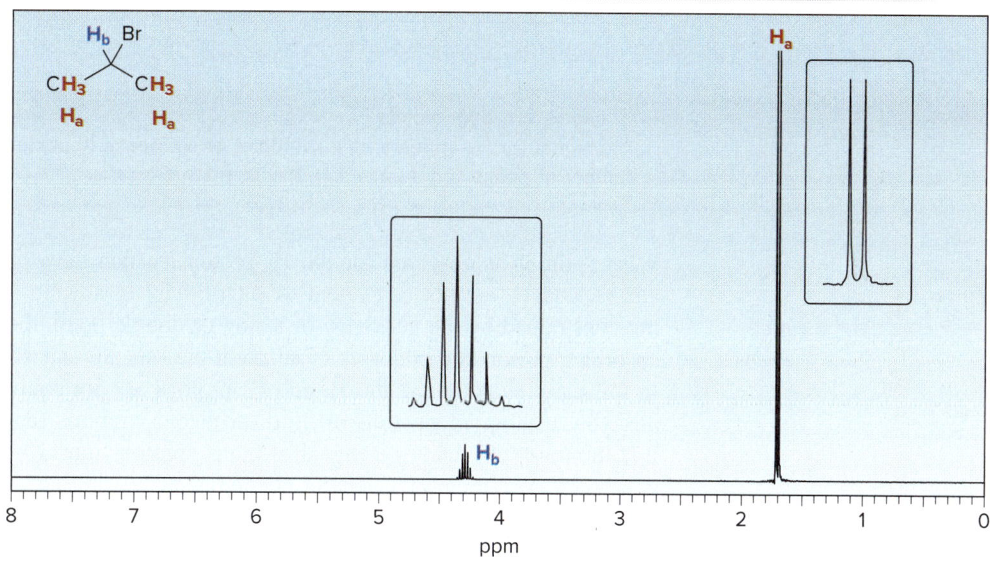

Figure C.6 The ^1H NMR spectrum of 2-bromopropane, $(CH_3)_2CHBr$

- The H_a protons have only one adjacent nonequivalent proton (H_b), so they are split into two peaks, a **doublet.**
- H_b has three H_a protons on each side. Because the six H_a protons are *equivalent to each other,* the $\boldsymbol{n + 1}$ rule can be used to determine splitting: $6 + 1 = 7$ peaks, a **septet.**

This is a specific example of a general rule:

- **Whenever two (or three) sets of adjacent protons are *equivalent to each other*, use the $n + 1$ rule to determine the splitting pattern.**

When an absorbing proton is flanked by two sets of adjacent protons that are *not equivalent to each other,* the outcome depends on the coupling constant (J) between the absorbing proton and its neighboring protons.

Let us begin with the result that occurs in **flexible alkyl chains;** that is, **the absorbing and adjacent protons are *not* bonded to a ring or double bond,** as illustrated with 1-bromopropane, $CH_3CH_2CH_2Br$.

$CH_3CH_2CH_2$—Br
H_a H_b H_c

$CH_3CH_2CH_2Br$ has three different types of protons—H_a, H_b, and H_c— so it exhibits three NMR signals. The H_a and H_c signals are both triplets because they are adjacent to two H_b protons, as shown in Figure C.7.

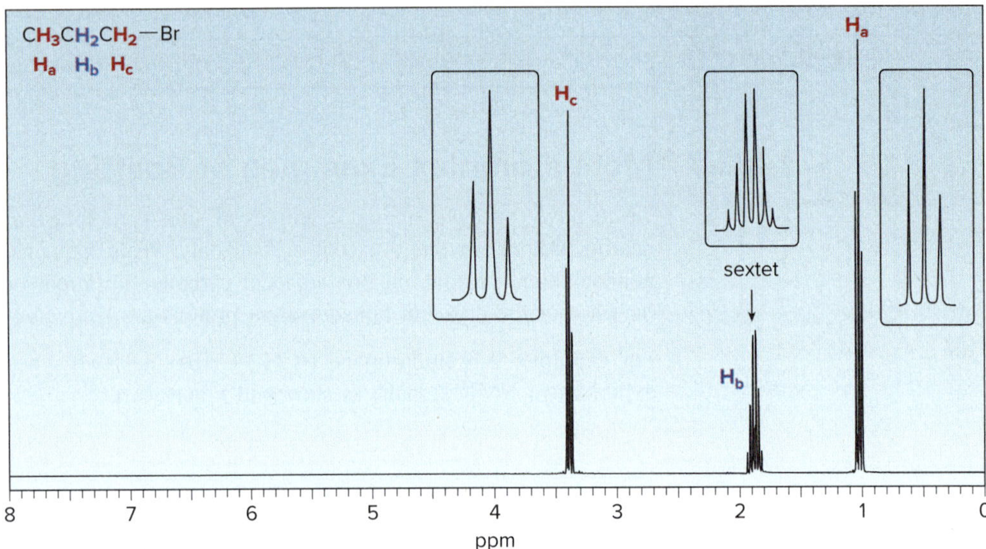

Figure C.7 The ¹H NMR spectrum of 1-bromopropane, $CH_3CH_2CH_2Br$

- H_a and H_c are both triplets.
- **The signal for H_b appears as a multiplet of six peaks (a sextet),** due to peak overlap; the number of peaks = $n + m + 1 = 3 + 2 + 1 = 6$ **peaks.**

What splitting is observed for the H_b protons, which have protons on both adjacent carbons, and H_a and H_c are not equivalent to each other? In acyclic molecules of this sort, which are not constrained by the geometry of a ring or double bond, the coupling constants between the absorbing proton and both sets of adjacent protons are equal (or close to it); that is, $J_{ab} = J_{bc}$. In this case, even though the H_a and H_c protons are not equivalent to each other, **we can just add the number of protons on both adjacent carbons together.** The 3 H_a protons and the 2 H_c protons split the NMR signal of the H_b protons into **3 + 2 + 1 = 6** peaks, a **sextet.** This is a specific example of a general phenomenon:

- **In a flexible alkyl chain, the *n* alkyl protons on one adjacent carbon and the *m* protons on the other adjacent carbon split the observed signal into *n* + *m* + 1 peaks.**

Now let's examine the observed splitting when an absorbing proton (**H_b**) is flanked by two sets of adjacent protons (**H_a** and **H_c**), at least one of which is bonded to a **ring or double bond,** as in (*E*)-1,3-dichloropropene. In this example, $J_{ab} >> J_{bc}$.

Ha Hc Hc
Cl Cl
Hb

(*E*)-1,3-dichloropropene

To determine the splitting of the H_b signal in this case, we must consider the effect of the H_a protons and the H_c protons *separately*. The H_a proton splits the H_b signal into two peaks and the two H_c protons split each of these two peaks into three peaks—that is, the NMR signal due to H_b consists of **2 × 3 = 6 peaks, a doublet of triplets.** Figure C.8 shows a splitting diagram that illustrates how these six peaks arise. This is a specific example of a general phenomenon:

- **When two sets of adjacent protons are *different from each other* (*n* protons on one adjacent carbon and *m* protons on the other), the number of peaks in an NMR signal is $(n + 1)(m + 1)$.**

Figure C.8

A splitting diagram for H_b in (*E*)-1,3-dichloropropene

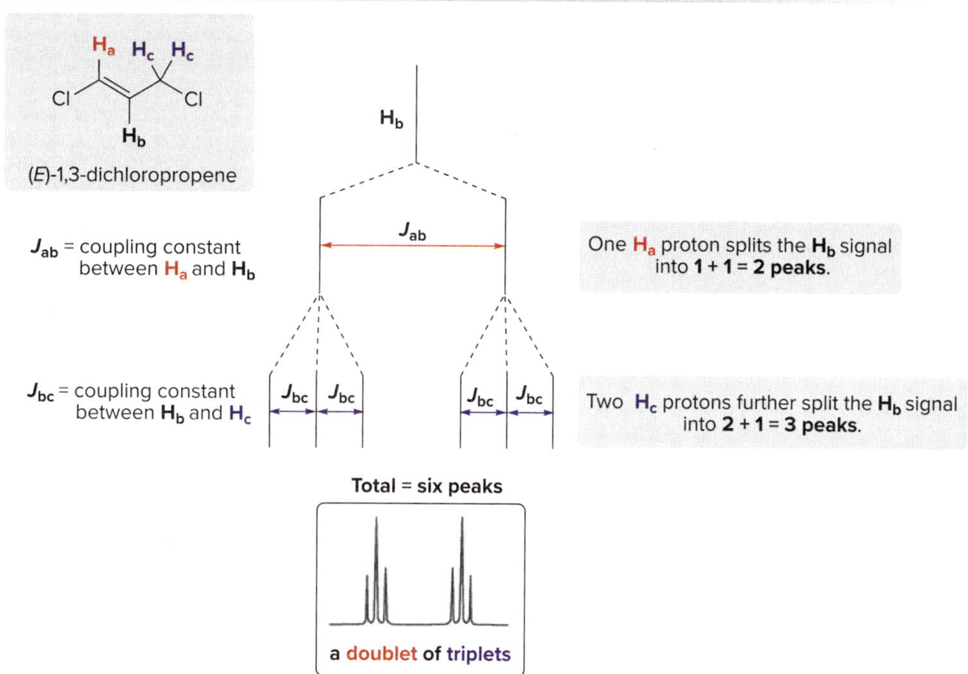

- The H_b signal is split into six peaks, a **doublet of triplets.** The number of peaks actually seen for the signal depends on the relative size of the coupling constants, J_{ab} and J_{bc}. When $J_{ab} >> J_{bc}$, all six lines are observed. When J_{ab} and J_{bc} are more similar in size, peaks overlap and fewer lines are seen.

The three possibilities for determining splitting patterns when an absorbing proton has nonequivalent protons on two adjacent carbons are shown with examples in Sample Problem C.7 and in the Key Skills section of the Chapter Review.

More features of the splitting seen with protons on carbon–carbon double bonds is discussed in Section C.8. Sample Problem C.7 illustrates how to determine splitting in three different compounds.

The $(n + 1)(m + 1)$ rule in splitting always gives the ***maximum*** number of peaks that is possible when an absorbing proton has *n* adjacent protons on one side and *m* protons on the other, and the coupling constants between nearby protons are different. As the difference between *J* values decreases, peaks overlap and fewer than the maximum number of peaks is observed.

Sample Problem C.7 Determining the Number of Peaks in an NMR Signal

How many peaks are present in the NMR signal of the labeled protons of each compound?

a. H H; Cl, Cl

b. H H; Cl, Br

c. H; CH_3, OH, H

Solution

When an absorbing proton is flanked by two sets of adjacent protons, there are three possibilities for determining the splitting pattern, as seen in parts (a), (b), and (c).

a. **5 peaks for H_b**
$(n + 1) = (4\ H_a + 1)$

H_b H_b; Cl, Cl; H_a H_a H_a H_a

4 adjacent H_a protons

- H_b has two H_a protons on each adjacent C. Because the four H_a protons are equivalent to each other, the ***n* + 1** rule can be used to determine splitting: 4 + 1 = **5 peaks,** a quintet.

b.

5 peaks for H_b
$(n + m + 1) = (2\ H_a + 2\ H_c + 1)$

H_b H_b Cl Br H_a H_a H_c H_c

2 H_a and **2 H_c** protons bonded to an alkyl chain

- Even though H_a and H_c are not equivalent to each other, they are bonded to a flexible alkyl chain. In this case, we can add the number of protons on both adjacent carbons together, so the number of peaks for H_b = $\boldsymbol{n + m + 1}$ = 2 + 2 + 1 = **5 peaks.**

c.

8 peaks for H_b
$(n + 1)(m + 1) = (3\ H_a + 1)(1\ H_c + 1)$

H_b CH_3 OH H_a H_c

3 H_a protons and **1 H_c** proton
H_b is bonded to a C=C.

- H_b has three H_a protons on one adjacent C and one H_c proton on the other. Because H_a and H_c are not equivalent to each other, the number of peaks for H_b = $\boldsymbol{(n + 1)(m + 1)}$ = (3 + 1)(1 + 1) = **8 peaks.**

Problem C.18 How many peaks are present in the NMR signal of each labeled proton?

a. 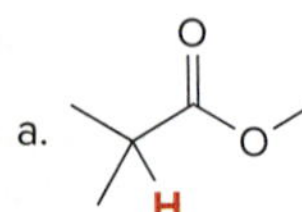

b. 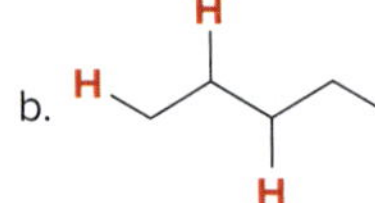

c. 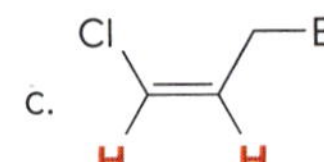

d. 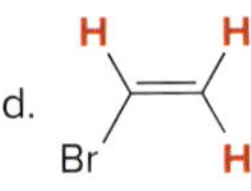

More Practice: Try Problems C.45, C.46, C.60b, C.61e.

Problem C.19 Describe the ^{1}H NMR spectrum of each compound. State how many NMR signals are present, the splitting pattern for each signal, and the approximate chemical shift.

a. b. c. d.

Problem C.20 Which compounds (**A–C**) contain a given splitting pattern in one of their NMR signals? A splitting pattern may be seen in more than one compound.

OH HO OH **A** OCH_3 CH_3O OH **B** O **C**

a. quintet
b. doublet
c. doublet of triplets
d. sextet
e. triplet
f. septet

C.8 Spin–Spin Splitting in Alkenes

In addition to what was presented in Section C.7, protons on cis, trans, and terminal carbon–carbon double bonds often give characteristic splitting patterns. A disubstituted double bond can have two **geminal protons** (on the same carbon atom), two **cis protons,** or two **trans protons.** When these protons are different, each proton splits the NMR signal of the other, so that each proton appears as a doublet. **The magnitude of the coupling constant *J* for these doublets depends on the arrangement of hydrogen atoms.**

geminal H's	<	**cis H's**	<	**trans H's**
$J_{geminal}$		J_{cis}		J_{trans}
0–3 Hz		5–10 Hz		11–18 Hz

Thus, the E and Z isomers of 3-chloropropenoic acid both exhibit two doublets for the two alkenyl protons, but the coupling constant is larger when the protons are trans compared to when the protons are cis, as shown in Figure C.9.

Figure C.9 ^{1}H NMR spectra for the alkenyl protons of (*E*)- and (*Z*)-3-chloropropenoic acid

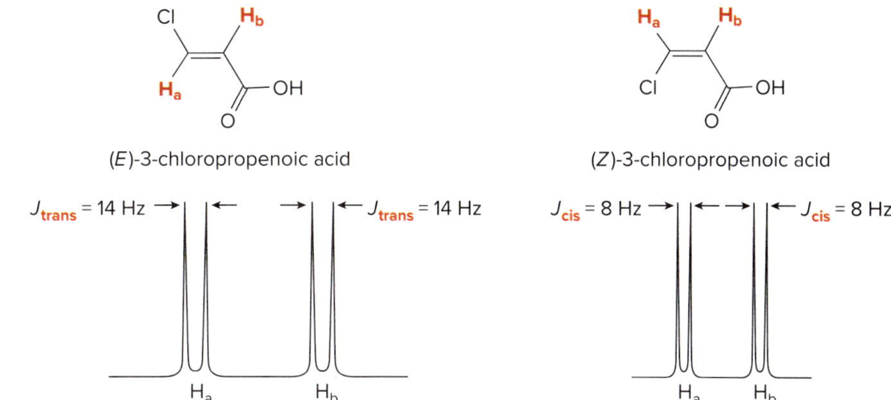

- Although both (*E*)- and (*Z*)-3-chloropropenoic acid show two doublets in their ^{1}H NMR spectra for their alkenyl protons, $J_{trans} > J_{cis}$.

When a double bond is monosubstituted, there are three nonequivalent protons, and the pattern is more complicated because all three protons are *coupled to each other*. For example, vinyl acetate ($CH_2{=}CHOCOCH_3$) has four different types of protons, three of which are bonded to the double bond. Besides the singlet for the CH_3 group, each proton on the double bond is coupled to two other different protons on the double bond, giving the spectrum in Figure C.10. Because the protons are bonded to a double bond, we determine the splitting using the $(n + 1)(m + 1)$ rule.

- H_b has two nearby nonequivalent protons that split its signal, the geminal proton H_c and the trans proton H_d. H_d splits the H_b signal into a doublet, and the H_c proton splits the doublet into two doublets. This pattern of four peaks is called a **doublet of doublets.**
- H_c has two nearby nonequivalent protons that split its signal, the geminal proton H_b and the cis proton H_d. H_d splits the H_c signal into a doublet, and the H_b proton splits the doublet into two doublets, forming another **doublet of doublets.**
- H_d has two nearby nonequivalent protons that split its signal, the trans proton H_b and the cis proton H_c. H_b splits the H_d signal into a doublet, and the H_c proton splits the doublet into two doublets, forming another **doublet of doublets.**

Figure C.10 The ^{1}H NMR spectrum of vinyl acetate ($CH_2{=}CHOCOCH_3$)

Vinyl acetate is polymerized to poly(vinyl acetate) (Problem 13.26), a polymer used in paints, glues, and adhesives.

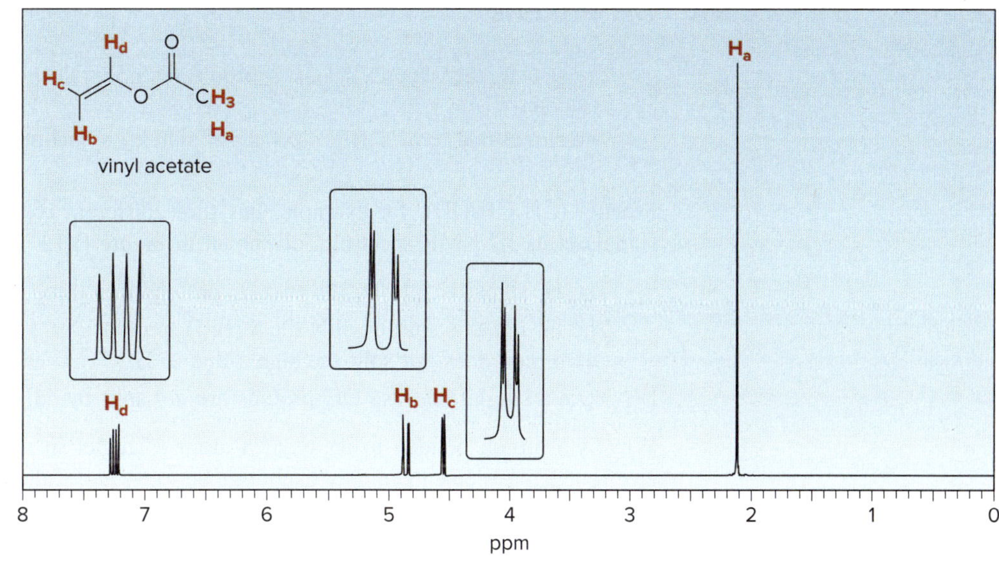

Splitting diagrams for the three alkenyl protons in vinyl acetate are drawn in Figure C.11. Note that each pattern is different in appearance because the magnitude of the coupling constants forming them is different.

Figure C.11 Splitting diagram for the alkenyl protons in vinyl acetate ($CH_2{=}CHOCOCH_3$)

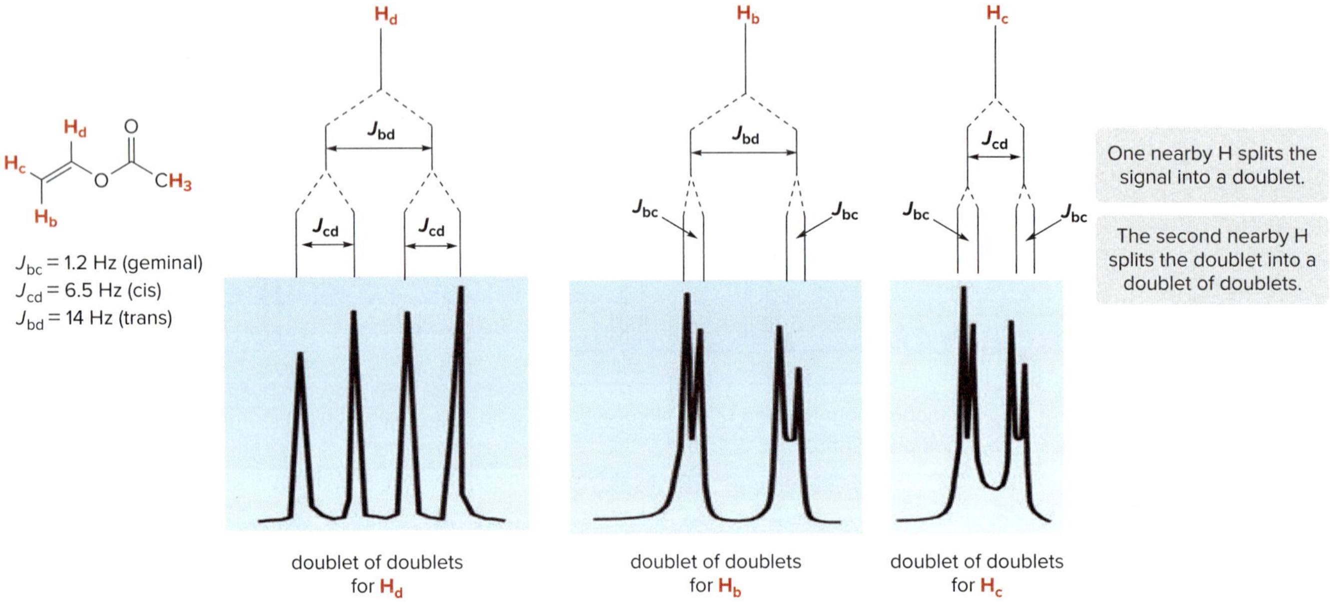

Problem C.21 Identify **A** and **B,** isomers of molecular formula $C_3H_4Cl_2$, from the given ^1H NMR data: Compound **A** exhibits signals at 1.75 (doublet, 3 H, J = 6.9 Hz) and 5.89 (quartet, 1 H, J = 6.9 Hz) ppm. Compound **B** exhibits signals at 4.16 (singlet, 2 H), 5.42 (doublet, 1 H, J = 1.9 Hz), and 5.59 (doublet, 1 H, J = 1.9 Hz) ppm.

Problem C.22 Draw the structure of an alkene of molecular formula C_6H_{12} that fits each description.

a. a compound with three signals for the alkenyl protons, each of which appears as a doublet of doublets

b. a compound that shows one singlet for its alkenyl protons

C.9 Other Facts About ^1H NMR Spectroscopy

C.9A OH Protons

- **Under usual conditions, an OH proton does not split the NMR signal of adjacent protons.**
- **The signal due to an OH proton is not split by adjacent protons.**

Ethanol (CH_3CH_2OH), for example, has three different types of protons, so there are three signals in its ^1H NMR spectrum, as shown in Figure C.12.

- The H_a signal is split by the two H_b protons into three peaks, a **triplet.**
- The H_b signal is split by only the three H_a protons into four peaks, a **quartet.** The adjacent OH proton does *not* split the signal due to H_b.
- H_c is a **singlet** because OH protons are *not* split by adjacent protons.

Why is a proton bonded to an oxygen atom a singlet in a ^1H NMR spectrum? Protons on electronegative elements rapidly **exchange** between molecules in the presence of trace amounts of acid or base. It is as if the CH_2 group in ethanol never "feels" the presence of the OH proton, because the OH proton is rapidly moving from one molecule to another. We therefore

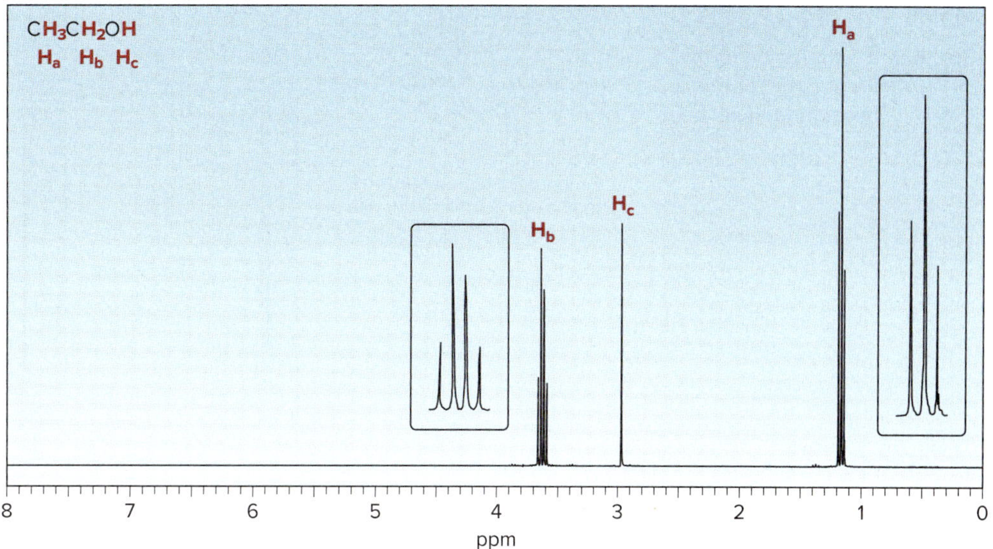

Figure C.12 The ¹H NMR spectrum of ethanol (CH_3CH_2OH)

see a peak due to the OH proton, but it is a single peak with no splitting. This phenomenon usually occurs with NH and OH protons.

Problem C.23 How many signals are present in the ¹H NMR spectrum for each molecule? What splitting is observed in each signal?

a. OH b. OH c. NH_2

C.9B Cyclohexane Conformations

How do the rotation around carbon–carbon σ bonds and the ring flip of cyclohexane rings affect an NMR spectrum? Because these processes are rapid at room temperature, an NMR spectrum records an **average** of all conformations that interconvert.

Thus, even though each cyclohexane carbon has two different types of hydrogens—one axial and one equatorial—the two chair forms of cyclohexane rapidly interconvert them, and an **NMR spectrum shows a single signal for the average environment** that it "sees."

H_a axial, H_b ⇌ H_a equatorial, H_b

Axial and equatorial H's rapidly interconvert. NMR sees an average environment and shows one signal.

Problem C.24 How many NMR signals are present for each compound?

a. b. O c. d. O O

C.9C Protons on Benzene Rings

We will learn more about the spectroscopic absorptions of benzene derivatives in Chapter 15.

Benzene has six equivalent, deshielded protons and exhibits a single peak in its ¹H NMR spectrum at 7.27 ppm. Monosubstituted benzene derivatives—that is, benzene rings with one H atom replaced by another substituent Z—contain five deshielded protons that are no longer all equivalent to each other. The identity of Z determines the appearance of this region of a ¹H NMR spectrum (6.5–8 ppm), as shown in Figure C.13. We will not analyze the splitting patterns observed for the ring protons of monosubstituted benzenes.

Figure C.13

The 6.5–8 ppm region of the ^1H NMR spectrum of three benzene derivatives

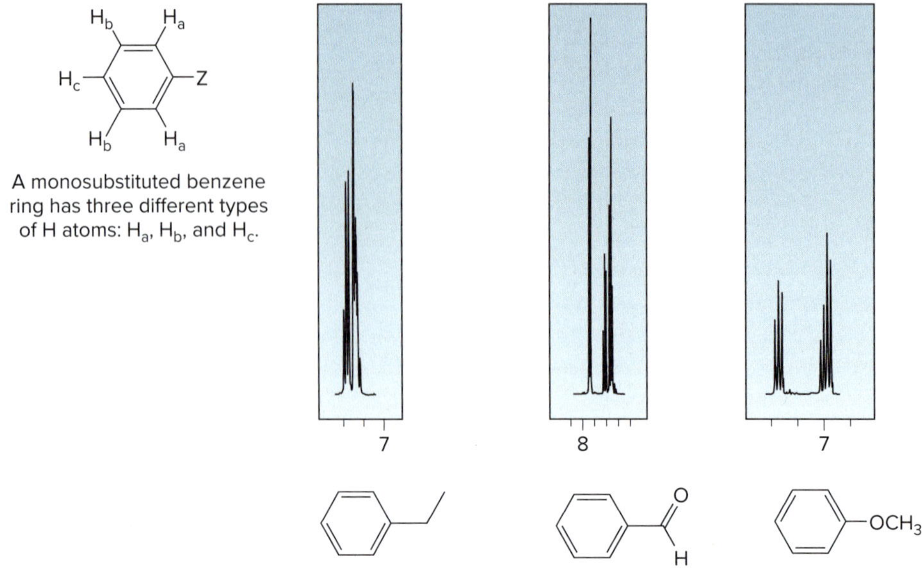

- The appearance of the signals in the 6.5–8 ppm region of the ^1H NMR spectrum depends on the identity of Z in C_6H_5Z.

Problem C.25 What protons in alcohol **A** give rise to each signal in its ^1H NMR spectrum? Explain all splitting patterns observed for absorptions between 0 to 7 ppm.

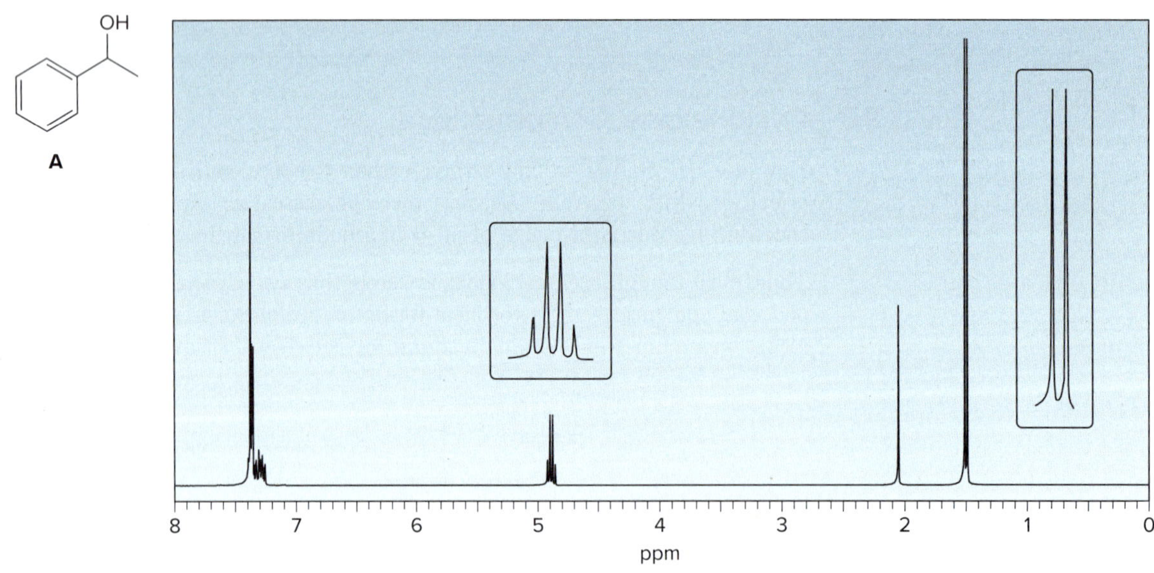

Problem C.26 How many peaks are observed in the ^1H NMR signal for each proton shown in red in palau'amine, the complex chapter-opening molecule?

palau'amine

C.10 Using ^{1}H NMR to Identify an Unknown

^{1}H NMR spectroscopy is a powerful technique that can be used to distinguish between isomers, as shown in Sample Problem C.8.

Sample Problem C.8 Using ^{1}H NMR Spectroscopy to Distinguish Between Compounds

How could ^{1}H NMR spectroscopy be used to distinguish between compounds **X** and **Y?**

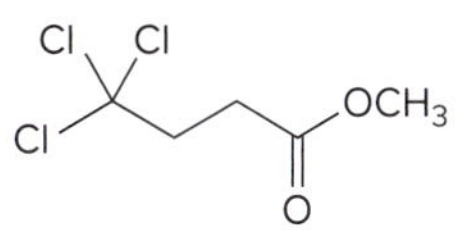

X 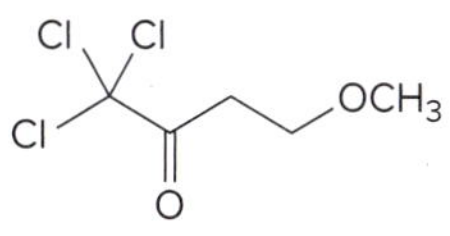 **Y**

Solution

[1] Determine whether the number of signals expected for each compound differs.

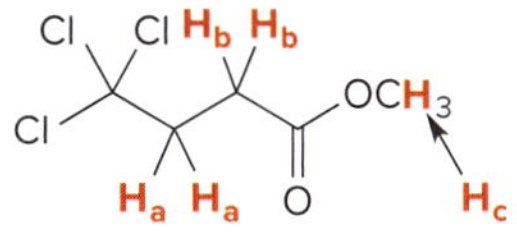

X 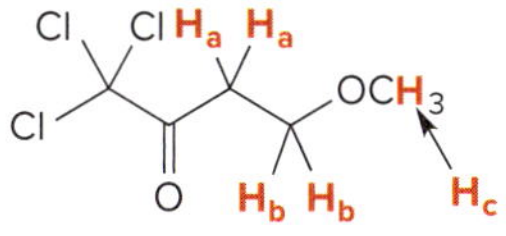 **Y**

In this example, both **X** and **Y** have three types of H's resulting in three signals, so another point of difference must be found.

[2] Determine the splitting pattern for each signal.

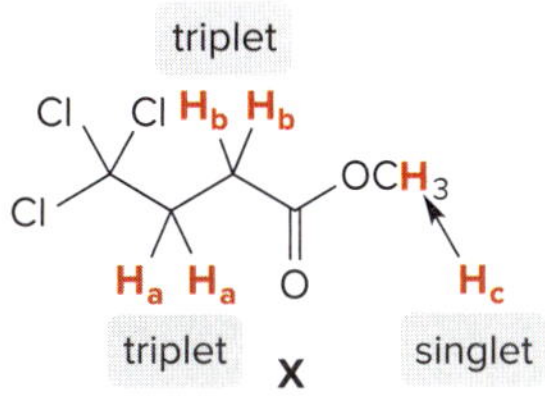

The ^{1}H NMR signals for both compounds consist of a singlet and two triplets.

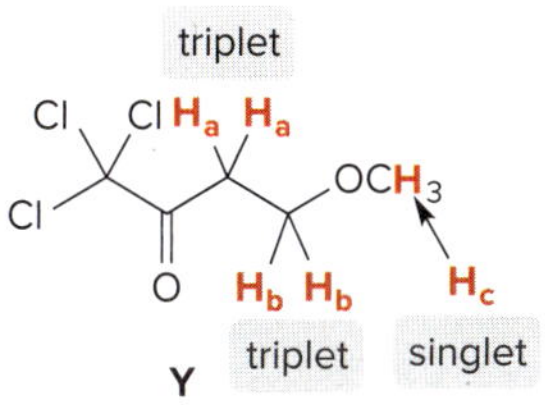

[3] Determine the approximate chemical shift for one or more types of H's.

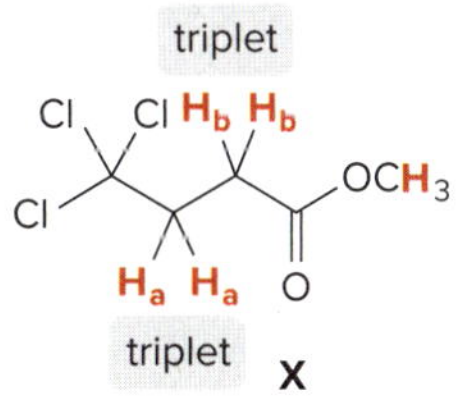

Only one chemical shift difference is needed to distinguish compounds. **Y** has H_b protons on a carbon bonded to an electronegative O atom, so these protons are deshielded and absorb in the 3–4 δ range. Neither of the triplets in **X** is located on a carbon that is also bonded to an electronegative atom, so no triplet in **X** absorbs in this region. Thus approximate chemical shift values can be used to distinguish between these two isomers.

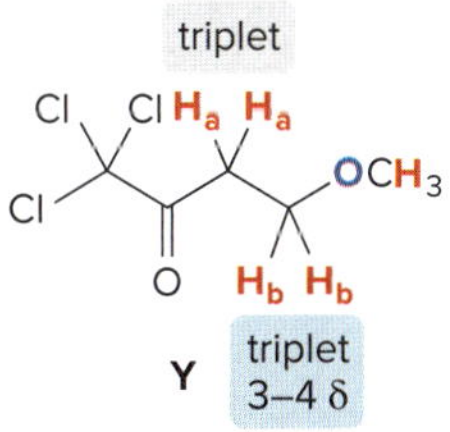

Problem C.27 How could ^{1}H NMR spectroscopy be used to distinguish between each pair of compounds?

a. 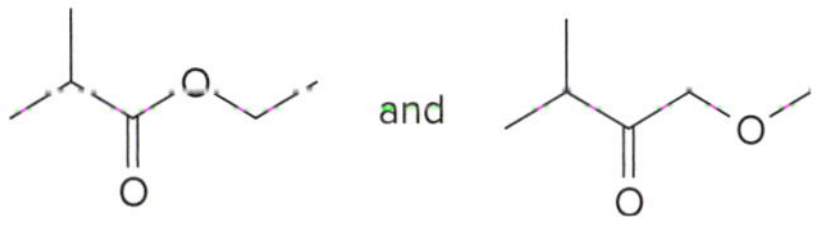 and

c. 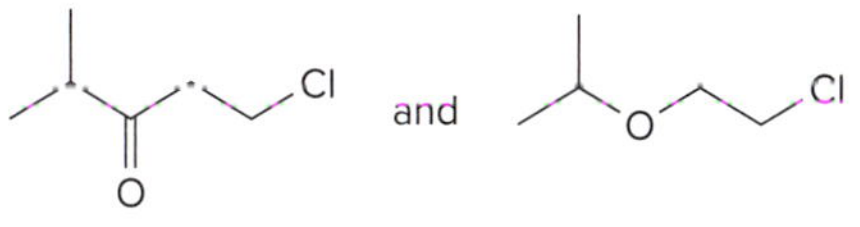 and

b. 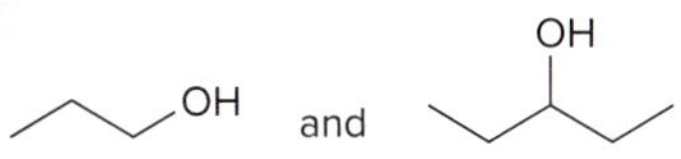 and

More Practice: Try Problems C.56, C.58, C.63, C.68, C.69.

Combined with mass spectrometry (which gives a compound's molecular formula) and infrared spectroscopy (which identifies a compound's functional group), we can then use its ^{1}H NMR spectrum to determine the structure of an unknown. A suggested procedure is illustrated for compound **X,** whose molecular formula ($C_4H_8O_2$) and functional group (C=O) were determined in Section B.5.

How To Use ^{1}H NMR Data to Determine a Structure

Example Using its ^{1}H NMR spectrum, determine the structure of an unknown compound **X** that has molecular formula $C_4H_8O_2$ and contains a C=O absorption in its IR spectrum.

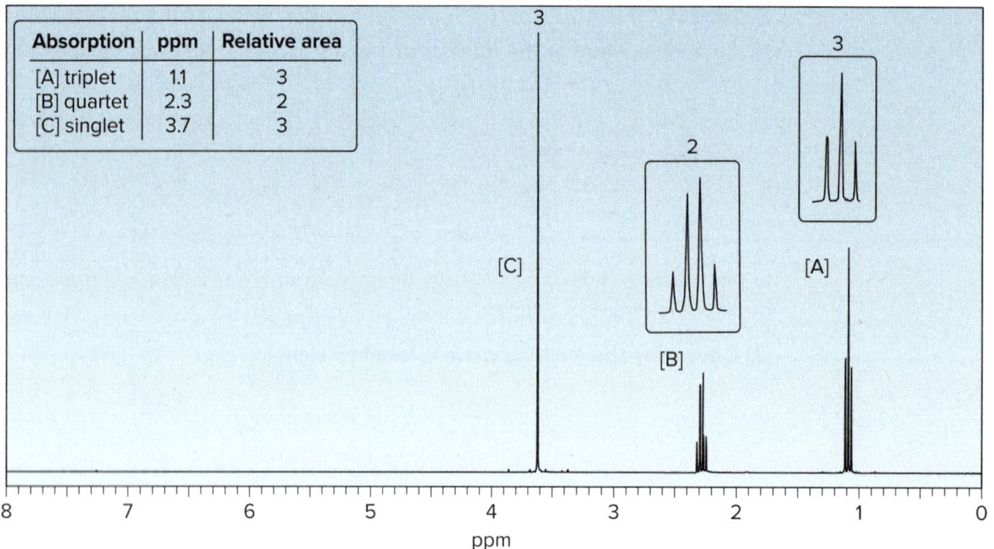

Absorption	ppm	Relative area
[A] triplet	1.1	3
[B] quartet	2.3	2
[C] singlet	3.7	3

Step [1] **Determine the number of different kinds of protons.**

- **The number of NMR signals equals the number of different types of protons.**
- This molecule has three NMR signals ([A], [B], and [C]) and therefore **three** types of protons (H_a, H_b, and H_c).

Step [2] **Use the relative area to determine the number of H atoms giving rise to each signal.**

- The relative area (printed on top of each signal) gives the *ratio* of absorbing protons responsible for each signal. In this case, the ratio is 3:2:3 for the signals from left to right.
- **When the sum of the relative areas *equals* the number of H's in the molecular formula, the relative area gives the number of absorbing H's responsible for the NMR signal.** In this example, the sum of the relative areas is 3 + 2 + 3 = 8, and the unknown has 8 H's, so the signals are due to 3 H's, 2 H's, and 3 H's from left to right in the spectrum.

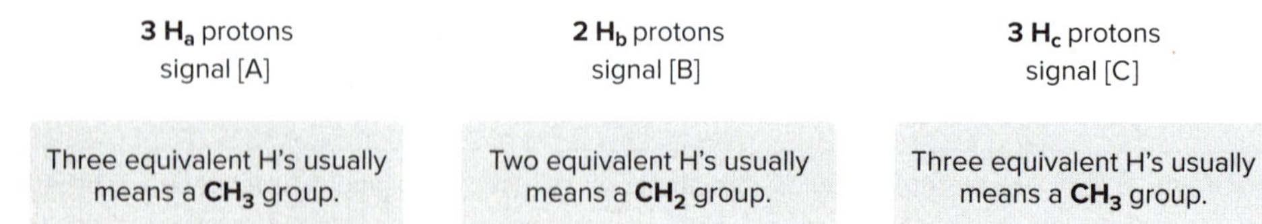

Step [3] **Use individual splitting patterns to determine what carbon atoms are bonded to each other.**

- Start with the singlets. Signal [C] is due to a CH_3 group with no adjacent nonequivalent H atoms. Possible structures include:

CH_3O— or CH_3–C(=O)– or CH_3–C<

—Continued

- Because signal [A] is a **triplet,** there must be **2 H's** (CH_2 group) on the adjacent carbon.
- Because signal [B] is a **quartet,** there must be **3 H's** (CH_3 group) on the adjacent carbon.
- This information suggests that **X** has an **ethyl** group ---→ CH_3CH_2–.

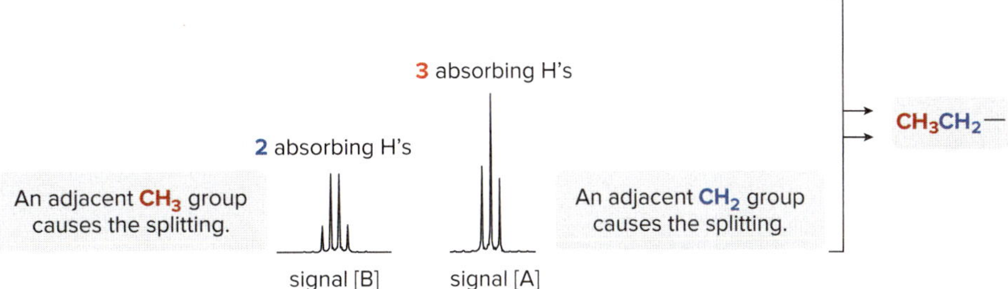

To summarize, **X** contains CH_3–, CH_3CH_2–, and C=O (from the IR). Comparing these atoms with the molecular formula shows that one O atom is missing. Because O atoms do not absorb in a ^{1}H NMR spectrum, their presence can be inferred only by examining the chemical shift of protons near them. O atoms are more electronegative than C, thus deshielding nearby protons and shifting their absorption downfield.

Step [4] **Use chemical shift data to complete the structure.**

- Put the structure together in a manner that preserves the splitting data and is consistent with the reported chemical shifts.
- In this example, two isomeric structures (**A** and **B**) are possible for **X** considering the splitting data only:

Structural pieces

CH_3— (H_c)

CH_3CH_2— (H_a H_b)

Possible structures

A: $CH_3CH_2C(=O)OCH_3$ (H_a H_b ... H_c)

or

B: $CH_3C(=O)OCH_2CH_3$ (H_c ... H_b H_a)

- Chemical shift information distinguishes the two possibilities. **The electronegative O atom deshields adjacent H's, shifting them downfield** between 3 and 4 ppm. If **A** is the correct structure, the singlet due to the CH_3 group (H_c) should occur downfield, whereas if **B** is the correct structure, the quartet due to the CH_2 group (H_b) should occur downfield.
- Because the NMR of **X** has a singlet (not a quartet) at 3.7, **A is the correct structure.**

Problem C.28 Propose a structure for a compound of molecular formula $C_7H_{14}O_2$ with an IR absorption at 1740 cm^{-1} and the following ^{1}H NMR data:

Absorption	ppm	Relative area
singlet	1.2	9
triplet	1.3	3
quartet	4.1	2

Problem C.29 Propose a structure for a compound of molecular formula C_3H_8O with an IR absorption at 3600–3200 cm^{-1} and the following NMR spectrum:

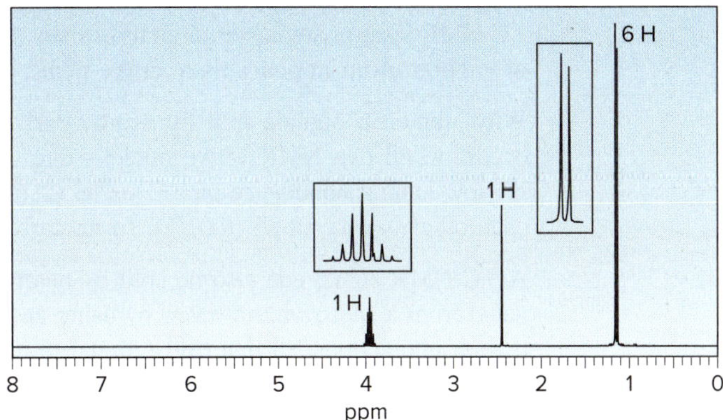

Problem C.30 Identify products **A** and **B** from the given ^{1}H NMR data.

a. Treatment of $CH_2{=}CHCOCH_3$ with one equivalent of HCl forms compound **A. A** exhibits the following absorptions in its ^{1}H NMR spectrum: 2.2 (singlet, 3 H), 3.05 (triplet, 2 H), and 3.6 (triplet, 2 H) ppm. What is the structure of **A?**

b. Treatment of acetone [$(CH_3)_2C{=}O$] with dilute aqueous base forms **B.** Compound **B** exhibits four singlets in its ^{1}H NMR spectrum at 1.3 (6 H), 2.2 (3 H), 2.5 (2 H), and 3.8 (1 H) ppm. What is the structure of **B?**

C.11 ^{13}C NMR Spectroscopy

^{13}C NMR spectroscopy is also an important tool for organic structure analysis. The physical basis for ^{13}C NMR is the same as for ^{1}H NMR. When placed in a magnetic field, B_0, ^{13}C nuclei can align themselves with or against B_0. More nuclei are aligned with B_0 because this arrangement is lower in energy, but these nuclei can be made to spin flip against the applied field by applying RF radiation of the appropriate frequency.

^{13}C NMR spectra, like ^{1}H NMR spectra, plot peak intensity versus chemical shift, using TMS as the reference signal at 0 ppm. ^{13}C occurs in only 1.1% natural abundance, however, so ^{13}C NMR signals are much weaker than ^{1}H NMR signals. To overcome this limitation, modern spectrometers irradiate samples with many pulses of RF radiation and use mathematical tools to increase signal sensitivity and decrease background noise. The spectrum of acetic acid (CH_3COOH) illustrates the general features of a ^{13}C NMR spectrum.

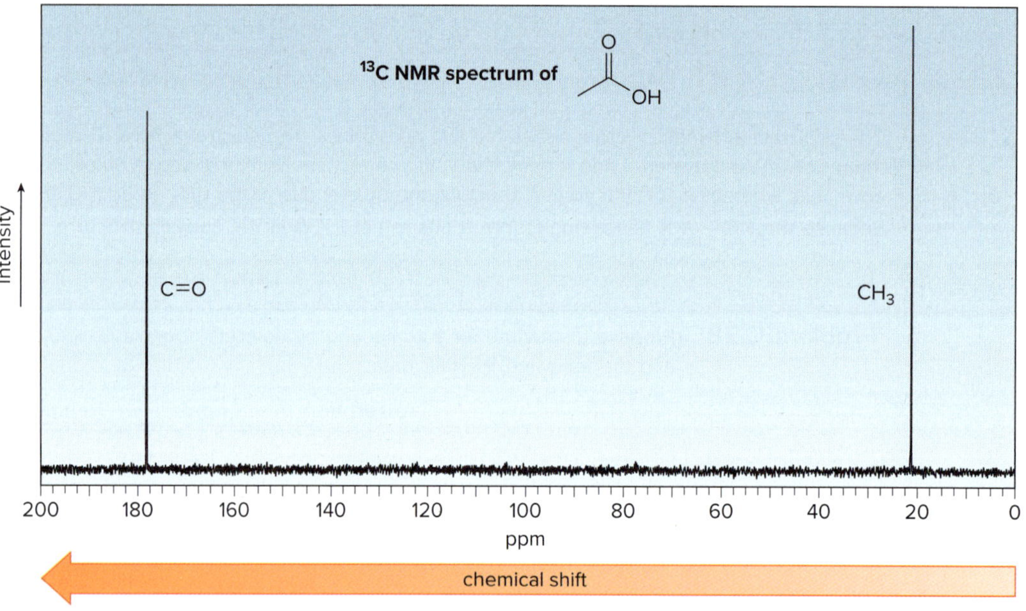

^{13}C NMR spectra are easier to analyze than ^{1}H spectra because signals are not split. **Each type of carbon atom appears as a single peak.**

Why aren't ^{13}C signals split by nearby carbon atoms? Recall from Section C.6 that splitting occurs when two NMR active nuclei—like two protons—are close to each other. Because of the low natural abundance of ^{13}C nuclei (1.1%), the chance of two ^{13}C nuclei being bonded to each other is very small (0.01%), so no carbon–carbon splitting is observed.

A ^{13}C NMR signal can also be split by nearby protons. This ^{1}H–^{13}C splitting is usually eliminated from a spectrum, however, by using an instrumental technique that decouples the proton–carbon interactions, so that every signal in a ^{13}C NMR spectrum is a singlet.

Two features of ^{13}C NMR spectra provide the most structural information: the **number of signals** observed and the **chemical shifts** of those signals.

C.11A ^{13}C NMR: Number of Signals

- **The number of signals in a ^{13}C spectrum gives the number of different types of carbon atoms in a molecule.**

Carbon atoms in the same environment give the same NMR signal, whereas carbons in different environments give different NMR signals. The ^{13}C NMR spectrum of CH_3COOH has two signals because there are two different types of carbon atoms—the C of the CH_3 group and the C of the carbonyl (C=O).

- **Because ^{13}C NMR signals are not split, the number of signals equals the number of lines in the ^{13}C NMR spectrum.**

Thus, the ^{13}C NMR spectra of dimethyl ether, chloroethane, and methyl acetate exhibit one, two, and three lines, respectively, because these compounds contain one, two, and three different types of carbon atoms.

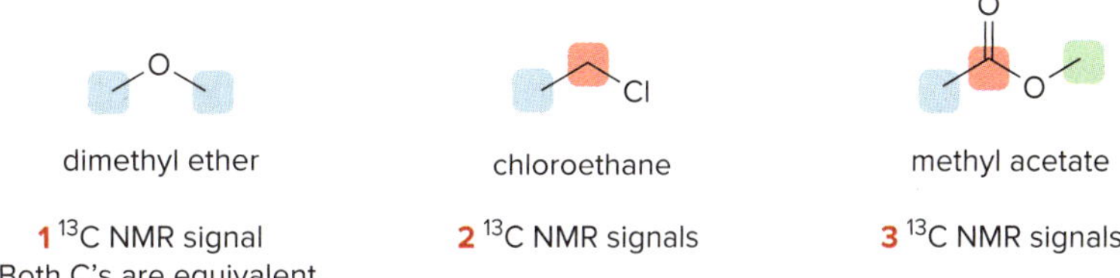

In contrast to what occurs in proton NMR, peak intensity is not proportional to the number of absorbing carbons, so ^{13}C NMR signals are not integrated.

Sample Problem C.9 Determining the Number of Lines in a ^{13}C NMR Spectrum

How many lines are observed in the ^{13}C NMR spectrum of each compound?

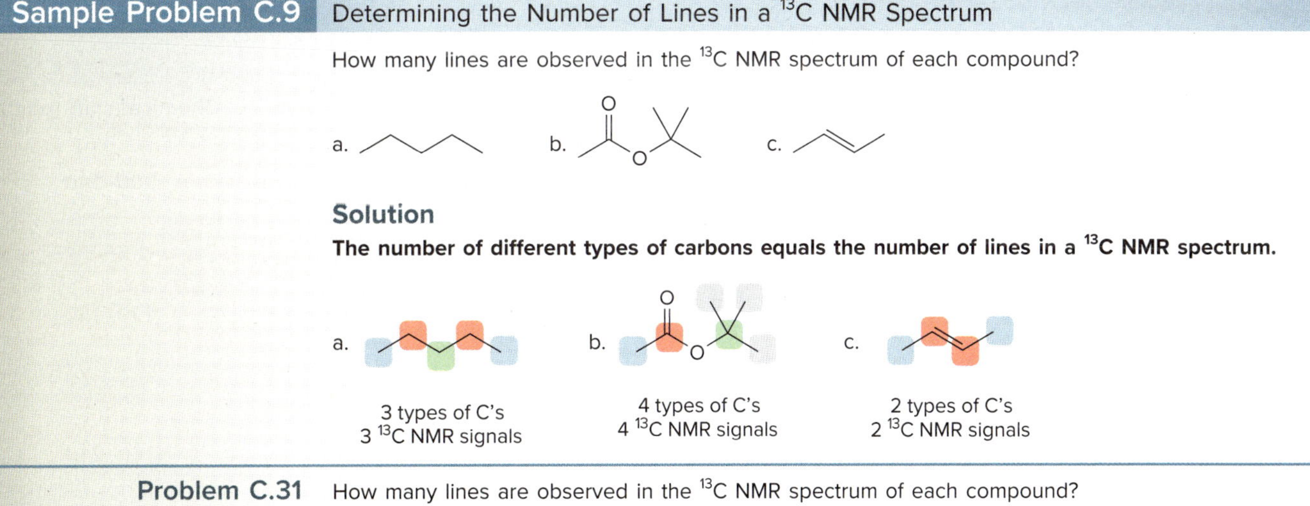

Solution

The number of different types of carbons equals the number of lines in a ^{13}C NMR spectrum.

Problem C.31 How many lines are observed in the ^{13}C NMR spectrum of each compound?

a. b. c. d.

More Practice: Try Problems C.39b, C.50, C.52, C.55c, C.60c, C.61c.

Problem C.32 Draw all constitutional isomers of molecular formula $C_3H_6Cl_2$.

a. How many signals does each isomer exhibit in its ^{1}H NMR spectrum?

b. How many lines does each isomer exhibit in its ^{13}C NMR spectrum?

c. When only the number of signals in both ^{1}H and ^{13}C NMR spectroscopy is considered, is it possible to distinguish all of these constitutional isomers?

Problem C.33 Esters of chrysanthemic acid are naturally occurring insecticides. How many lines are present in the ^{13}C NMR spectrum of chrysanthemic acid?

OH
O
chrysanthemic acid

Esters of chrysanthemic acid (Problem C.33) are obtained from the flowers of *Chrysanthemum cinerariifolium*. Because they are biodegradable and active against numerous insect species, these esters are widely used insecticides (see also Section 24.5).
Gail Whitfield/Alamy Stock Photo

C.11B ^{13}C NMR: Position of Signals

In contrast to the small range of chemical shifts in ^{1}H NMR (0–12 ppm usually), ^{13}C NMR absorptions occur over a much broader range, 0–220 ppm. The chemical shifts of carbon atoms in ^{13}C NMR depend on the same effects as the chemical shifts of protons in ^{1}H NMR:

- **The sp^3 hybridized C atoms of alkyl groups are shielded and absorb upfield.**
- **Electronegative elements like halogen, nitrogen, and oxygen shift absorptions downfield.**
- **The sp^2 hybridized C atoms of alkenes and benzene rings absorb downfield.**
- **Carbonyl carbons are highly deshielded, and absorb farther downfield than other carbon types.**

Table C.5 lists common ^{13}C chemical shift values. The ^{13}C NMR spectra of propan-1-ol ($CH_3CH_2CH_2OH$) and methyl acetate ($CH_3CO_2CH_3$) in Figure C.14 illustrate these principles.

Table C.5 Common ^{13}C Chemical Shift Values

Type of carbon	Chemical shift (ppm)	Type of carbon	Chemical shift (ppm)
C	5–45	C=C	100–140
Z–C, Z = N, O, X	30–80	C—	120–150
—C≡C—	65–100	O=C	160–210

Problem C.34 Which of the highlighted carbon atoms in each molecule absorbs farther downfield?

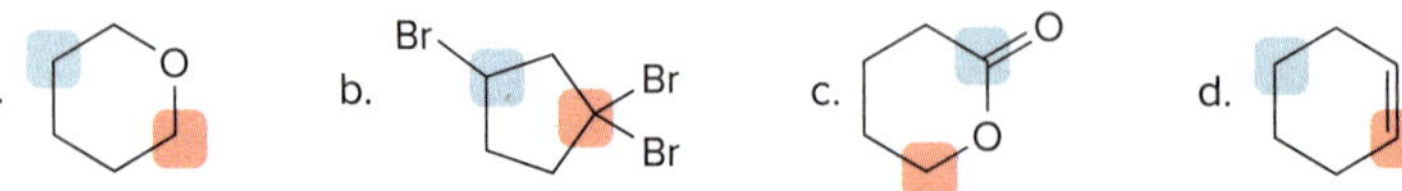

Problem C.35 Rank the highlighted carbons in each molecule in order of increasing chemical shift.

Figure C.14 Representative ^{13}C NMR spectra

a. Propan-1-ol

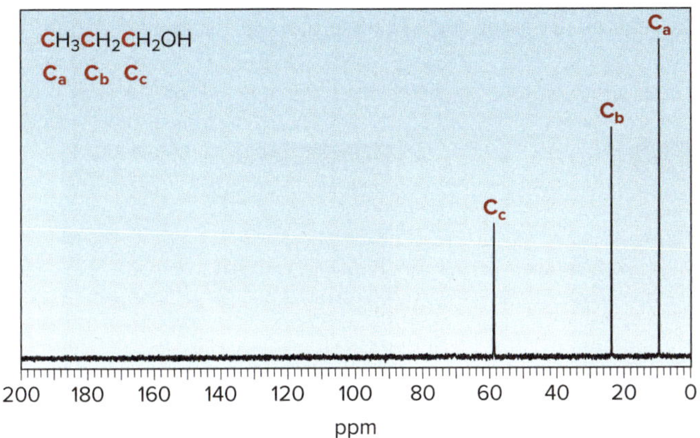

b. Methyl acetate

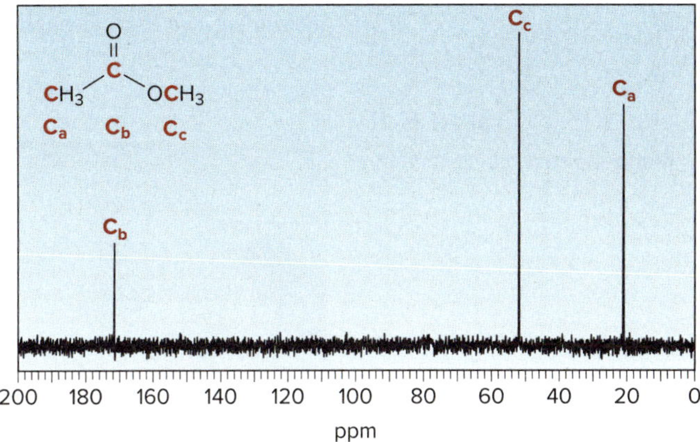

- The three types of C's in propan-1-ol—identified as C_a, C_b, and C_c—give rise to three ^{13}C NMR signals.
- Deshielding increases with increasing proximity to the electronegative O atom, and the absorption shifts downfield; thus, in order of increasing chemical shift: $C_a < C_b < C_c$.

- The three types of C's in methyl acetate—identified as C_a, C_b, and C_c—give rise to three ^{13}C NMR signals.
- **The carbonyl carbon (C_b) is highly deshielded, so it absorbs farthest downfield.**
- C_a, an sp^3 hybridized C that is not bonded to an O atom, is the most shielded, and so it absorbs farthest upfield.
- Thus, in order of increasing chemical shift: $C_a < C_c < C_b$.

Problem C.36 Identify the carbon atoms that give rise to each NMR signal.

a.

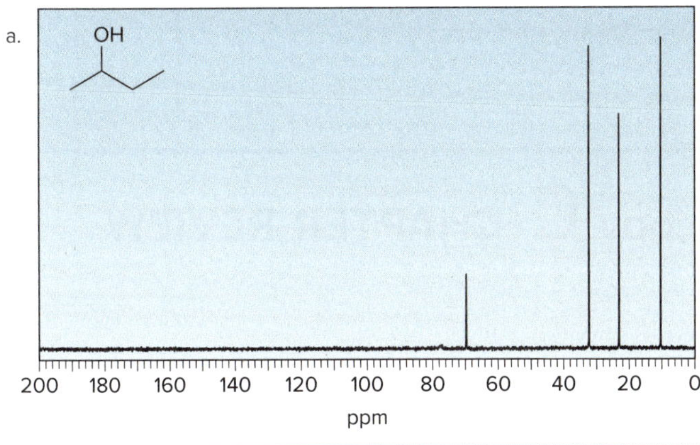

b.

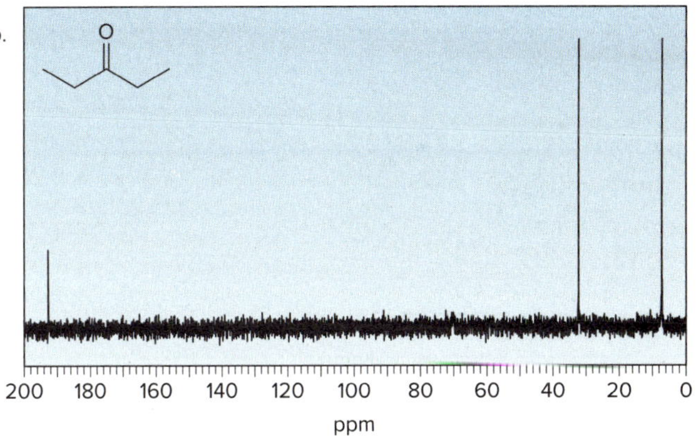

Problem C.37 A compound of molecular formula $C_4H_8O_2$ shows no IR peaks at 3600–3200 or 1700 cm^{-1}. It exhibits one singlet in its ^1H NMR spectrum at 3.69 ppm, and one line in its ^{13}C NMR spectrum at 67 ppm. What is the structure of this unknown?

Problem C.38 Draw the structure of a compound of molecular formula C_4H_8O that has a signal in its ^{13}C NMR spectrum at > 160 ppm. Then draw the structure of an isomer of molecular formula C_4H_8O that has all of its ^{13}C NMR signals at < 160 ppm.

C.12 Magnetic Resonance Imaging (MRI)

Magnetic resonance imaging (MRI)—NMR spectroscopy in medicine—is a powerful diagnostic technique (Figure C.15a). The "sample" is the patient, who is placed in a large cavity in a magnetic field, and then irradiated with RF energy. Because RF energy has very low frequency and low energy, the method is safer than X-rays or computed tomography (CT) scans that employ high-frequency, high-energy radiation that is known to damage living cells.

Living tissue contains protons (especially the H atoms in H_2O) in different concentrations and environments. When irradiated with RF energy, these protons are excited to a higher-energy spin state, and then fall back to the lower-energy spin state. These data are analyzed by a computer that generates a plot that delineates tissues of different proton density (Figure C.15b). MRIs can be recorded in any plane. Moreover, because the calcium present in bones is not NMR active, an MRI instrument can "see through" bones such as the skull and visualize the soft tissue underneath.

Figure C.15 Magnetic resonance imaging

a.

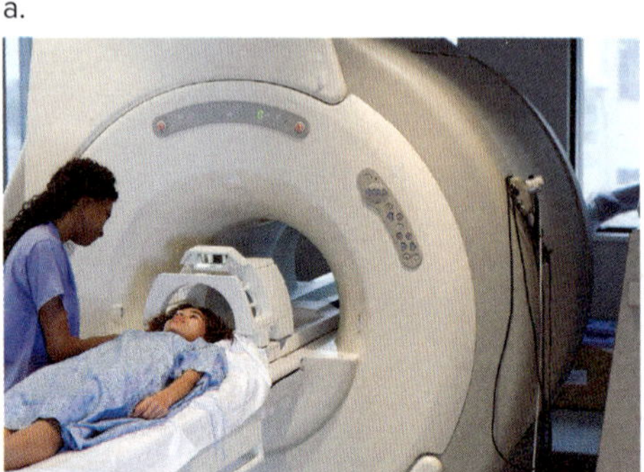

ERproductions Ltd/Blend Images

b.

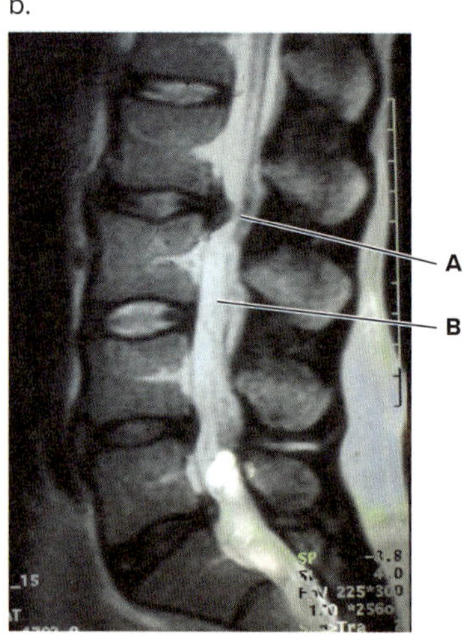

With permission, Daniel C. Smith.

a. An MRI instrument: An MRI instrument is especially useful for visualizing soft tissue. The 2003 Nobel Prize in Physiology or Medicine was awarded to chemist Paul C. Lauterbur and physicist Sir Peter Mansfield for their contributions in developing magnetic resonance imaging.

b. An MRI image of the lower back: **A** labels spinal cord compression from a herniated disc. **B** labels the spinal cord, which would not be visualized with conventional X-rays.

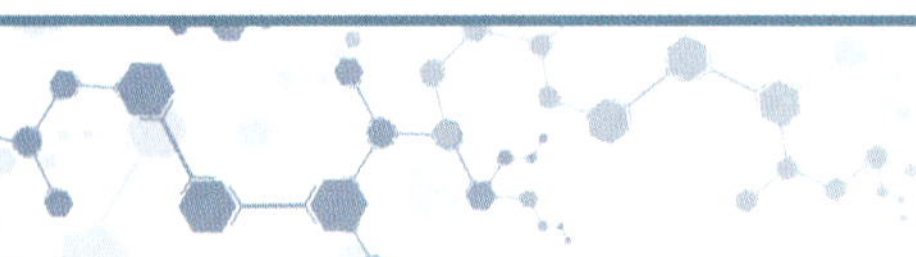

Spectroscopy C CHAPTER REVIEW

KEY CONCEPTS

Homotopic, enantiotopic, and diastereotopic protons (C.2C)

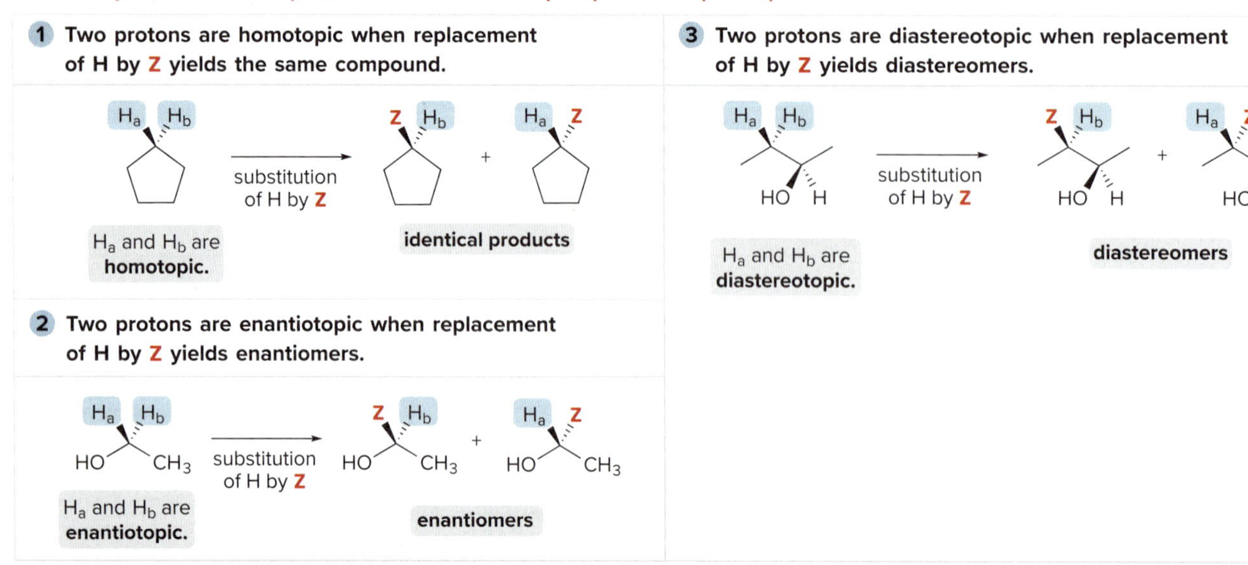

See Sample Problem C.4. Try Problem C.7.

KEY SKILLS

[1] Calculating the chemical shift of an absorption that occurs at 1000 Hz downfield from TMS using a 400 MHz NMR spectrometer (C.1)

1 Use the equation that defines chemical shift in ppm.

$$\text{chemical shift (in ppm on the } \delta \text{ scale)} = \frac{\text{observed chemical shift (in Hz) downfield from TMS}}{\nu \text{ of the NMR spectrometer (in MHz)}}$$

2 Insert the values into the equation.

$$= \frac{\text{1000 Hz downfield from TMS}}{\text{400 MHz operating frequency}} = \textbf{2.5 ppm}$$

See Sample Problem C.1. Try Problems C.42, C.43.

[2] Determining the different types of protons in a compound (C.2A); example: 1,4-dichlorobutane

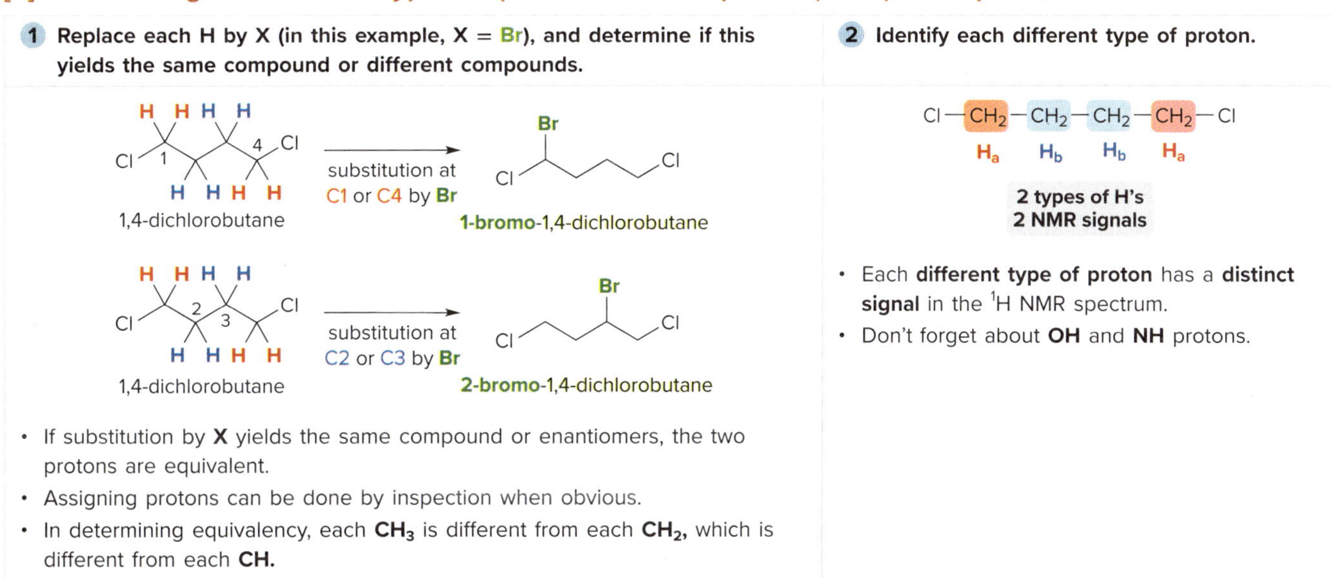

- If substitution by **X** yields the same compound or enantiomers, the two protons are equivalent.
- Assigning protons can be done by inspection when obvious.
- In determining equivalency, each $\mathbf{CH_3}$ is different from each $\mathbf{CH_2}$, which is different from each **CH.**

- Each **different type of proton** has a **distinct signal** in the ^{1}H NMR spectrum.
- Don't forget about **OH** and **NH** protons.

See Sample Problem C.2, Figure C.2. Try Problems C.39a, C.40, C.41, C.55d, C.60a, C.61d.

[3] Determining equivalency in a cycloalkane (C.2B)

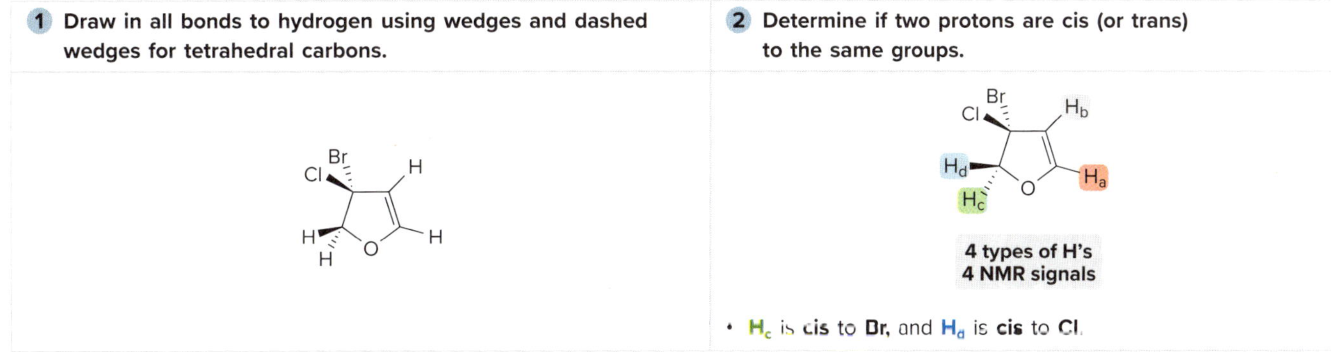

- $\mathbf{H_c}$ is **cis** to **Br**, and $\mathbf{H_d}$ is **cis** to **Cl**.

See Sample Problem C.3. Try Problems C.40h–j, C.60a.

[4] Determining which protons absorb farther downfield; two factors

1 Use the presence of nearby electronegative atoms to determine deshielding effects (C.3A).

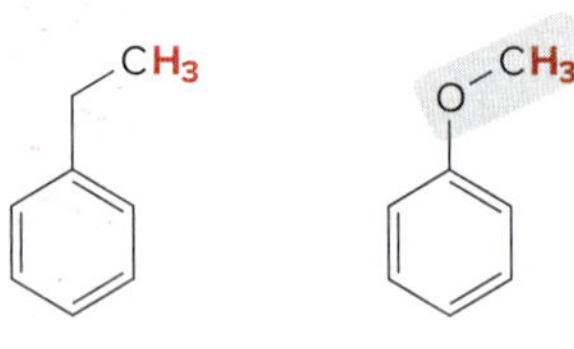

ethylbenzene anisole

- The CH_3 group in anisole is **deshielded** by the **electronegative O atom.**
- Electronegative atoms withdraw electron density, **deshield** a nucleus, and shift an absorption **downfield.**
- **Shielding** shifts an absorption **upfield.**

See Sample Problem C.5.

2 Determine shielding and deshielding effects when protons are bonded to sp^2 and sp hybridized carbons (C.4).

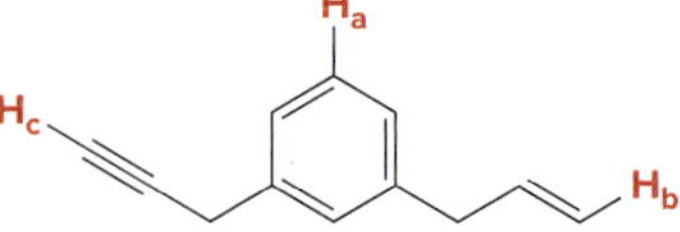

- H_c is **shielded** because it is bonded to an ***sp* hybridized carbon.**
- H_b is **deshielded** because it is bonded to an **sp^2 hybridized carbon.**
- H_a is **highly deshielded** because it is bonded to an **sp^2 hybridized carbon** on a **benzene ring.**

Answer:
In order of increasing chemical shift, $H_c < H_b < H_a$

See Sample Problem C.6.

Try Problems C.44, C.55b.

[5] Determining the ^1H NMR integration ratio for a compound (C.5); example: $CH_3CH_2OCH_3$

1 Identify the nonequivalent protons.

- three types of protons

2 Count the number of protons in each group.

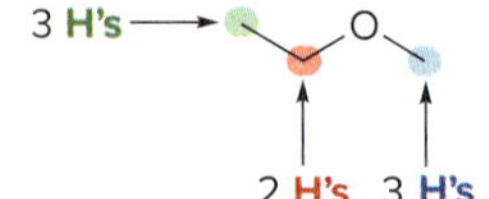

3 Determine the integration ratio.

Answer: 3:2:3

- The area under an NMR signal is proportional to the number of absorbing protons.

Try Problems C.13, C.14.

[6] Determining the splitting pattern for a molecule using the *n* + 1 rule (C.6)

1 Identify the nonequivalent protons.

$HOCH_2CH_2C(=O)OCH_3$

4 types of H's

2 Determine if two sets of nonequivalent protons are close enough to split each other's signals.

no splitting with OH and NH protons

singlet ⟶ $HOCH_2CH_2C(=O)OCH_3$ ← **singlet** (for OCH_3)

nonequivalent protons on adjacent carbons

- Equivalent protons do not split each other's signals.

3 Apply the *n* + 1 rule.

$HOCH_2CH_2C(=O)OCH_3$ (H_a, H_b)

H_a: two adjacent H's
3 peaks
triplet

H_b: two adjacent H's
3 peaks
triplet

- A set of ***n* nonequivalent protons** on the same carbon or adjacent carbons **splits an NMR signal into *n* + 1 peaks.**

See Tables C.3, C.4. Try Problems C.45, C.60b, C.61b.

[7] Determining the number of peaks present in the ^{1}H NMR signal of an alkene using the $(n + 1)(m + 1)$ rule (C.7); example: (*Z*)-1-bromoprop-1-ene

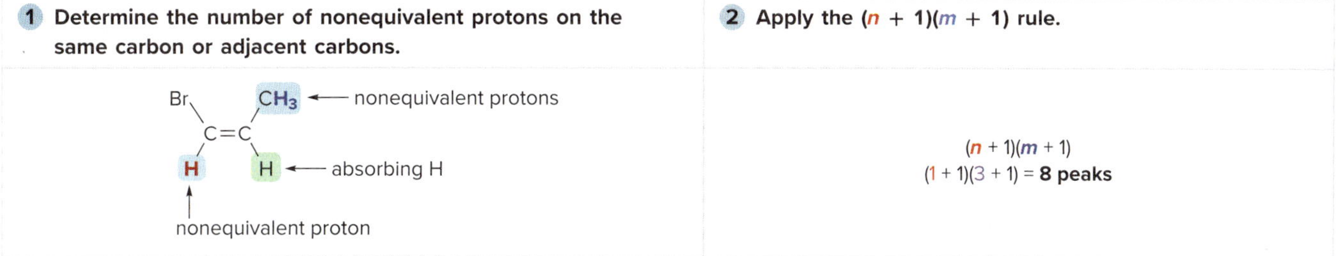

See Figure C.8, Sample Problem C.7. Try Problems C.45i, j; C.46; C.47; C.60b.

[8] Determining splitting patterns when an absorbing proton has nonequivalent protons on two adjacent carbons (C.7–C.8); three possibilities

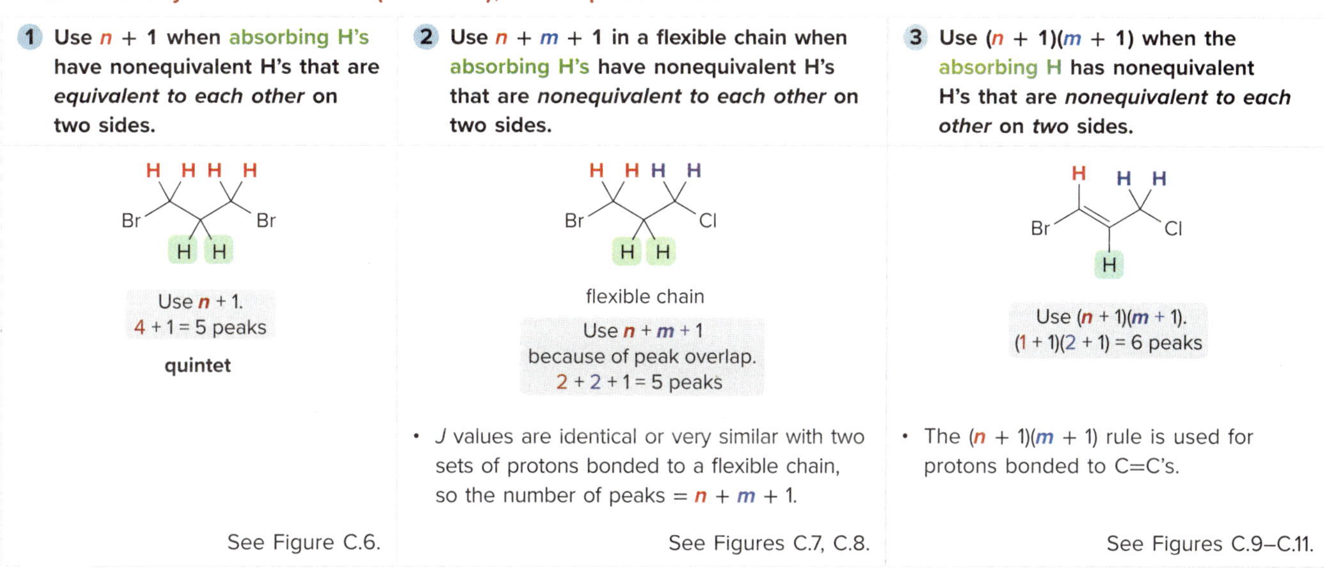

See Sample Problem C.7, Tables C.3, C.4. Try Problems C.45d–j, C.46.

[9] Using a molecular formula and ^{1}H NMR data to determine a structure (C.10); example: $C_4H_{10}O$ with the given ^{1}H NMR data

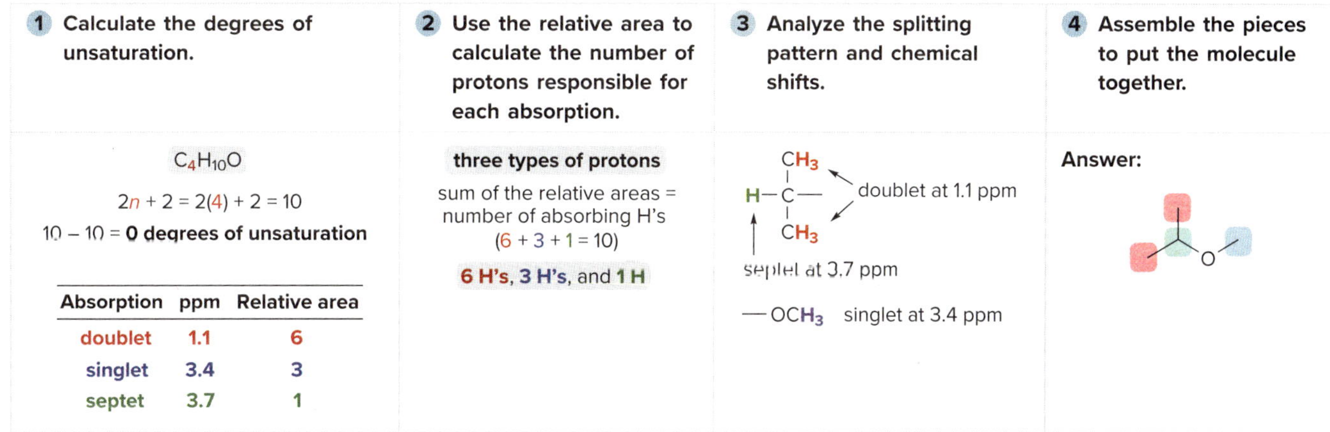

See *How To* (p. 586). Try Problems C.62, C.63, C.68–C.70.

[10] Determining the different types of C atoms in a compound (C.11A)

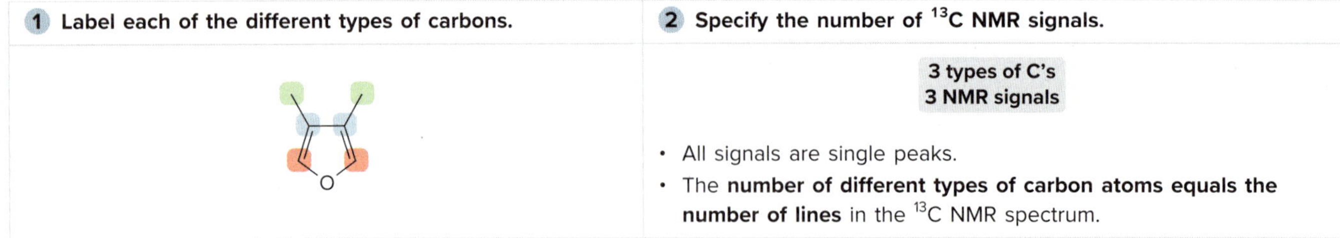

1 Label each of the different types of carbons.	2 Specify the number of ^{13}C NMR signals.
	3 types of C's **3 NMR signals** • All signals are single peaks. • The **number of different types of carbon atoms equals the number of lines** in the ^{13}C NMR spectrum.

See Sample Problem C.9. Try Problems C.50, C.52, C.55c, C.57, C.60c, C.61c.

[11] Determining which C atom absorbs farther downfield (C.11); two factors

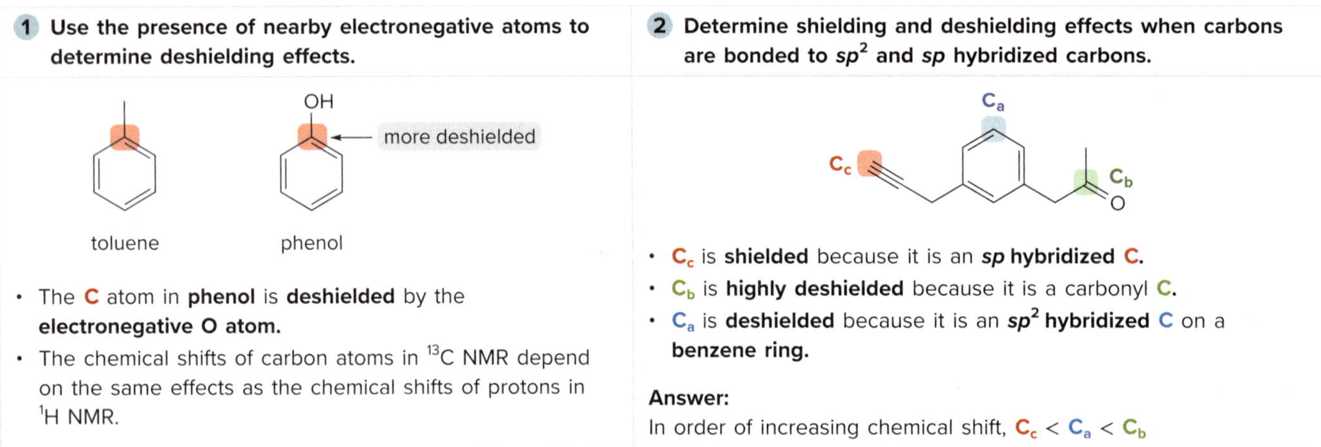

1 Use the presence of nearby electronegative atoms to determine deshielding effects.	2 Determine shielding and deshielding effects when carbons are bonded to sp^2 and sp hybridized carbons.
• The **C** atom in **phenol** is **deshielded** by the **electronegative O atom.** • The chemical shifts of carbon atoms in ^{13}C NMR depend on the same effects as the chemical shifts of protons in ^1H NMR.	• C_c is **shielded** because it is an ***sp* hybridized C.** • C_b is **highly deshielded** because it is a carbonyl **C.** • C_a is **deshielded** because it is an ***sp*^{2} hybridized** C on a **benzene ring.** **Answer:** In order of increasing chemical shift, $C_c < C_a < C_b$

Try Problems C.53, C.54, C.55a.

[12] Using a molecular formula, IR, ^1H NMR, and ^{13}C NMR for structure determination (C.10); example: C_3H_5ClO

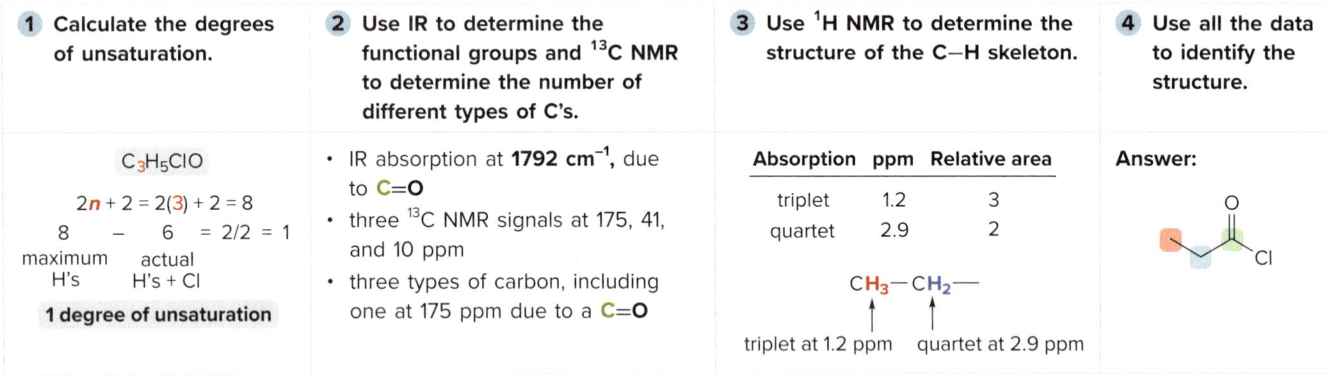

1 Calculate the degrees of unsaturation.	2 Use IR to determine the functional groups and ^{13}C NMR to determine the number of different types of C's.	3 Use ^1H NMR to determine the structure of the C–H skeleton.	4 Use all the data to identify the structure.
C_3H_5ClO $2n + 2 = 2(3) + 2 = 8$ $8 - 6 = 2/2 = 1$ maximum H's; actual H's + Cl **1 degree of unsaturation**	• IR absorption at **1792 cm^{-1}**, due to **C=O** • three ^{13}C NMR signals at 175, 41, and 10 ppm • three types of carbon, including one at 175 ppm due to a **C=O**	**Absorption / ppm / Relative area** triplet 1.2 3 quartet 2.9 2 CH_3-CH_2- triplet at 1.2 ppm; quartet at 2.9 ppm	**Answer:**

Try Problems C.62–C.79.

SPECTROSCOPY CHAPTER C MULTIPLE-CHOICE SELF-TEST

The Self-Test consists of multiple-choice questions similar to those found on the American Chemical Society organic chemistry exam. Answers are given at the end of the chapter.

1. How many different kinds of H atoms does butan-2-ol [$CH_3CH(OH)CH_2CH_3$] contain: (a) three; (b) four; (c) five; (d) six?
2. Which labeled H atom appears as a triplet in an NMR spectrum?

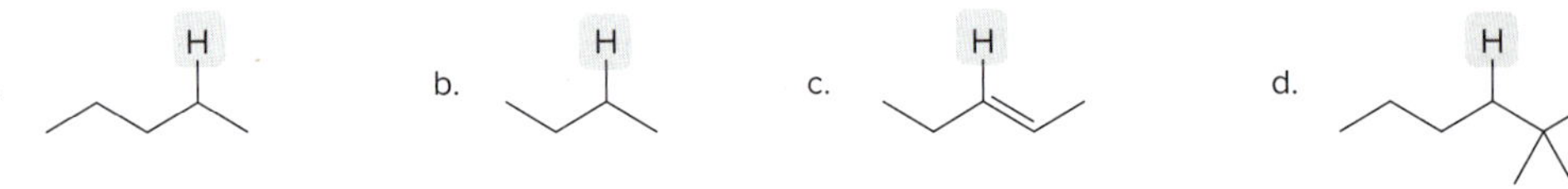

3. Which labeled H atom absorbs farthest downfield?

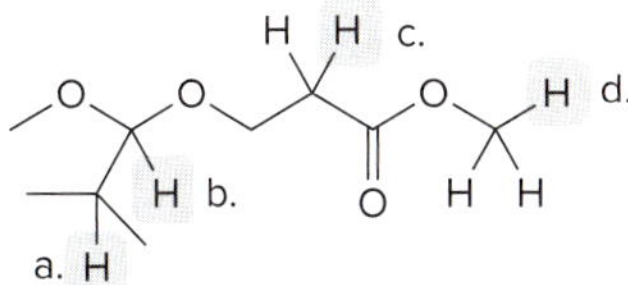

4. Which labeled H atom appears as six peaks in its NMR signal?

a. 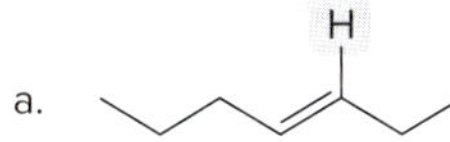

b. 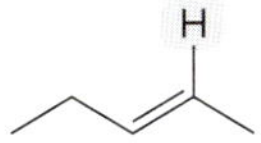 H

c. 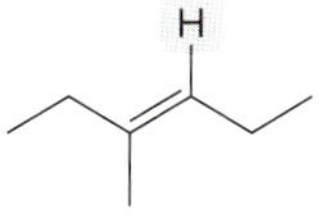 H

d. 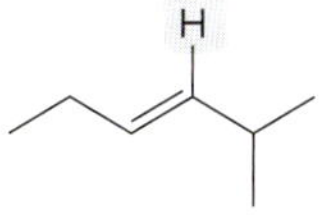 H

5. Which structure is possible for a compound with a peak at ~1720 cm^{-1} in its IR spectrum and two singlets in its NMR spectrum?

a. 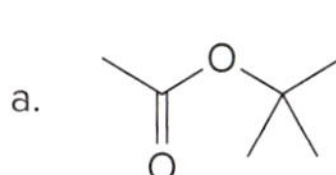

b. 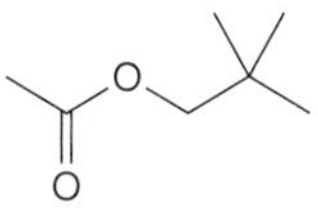 O O

c. 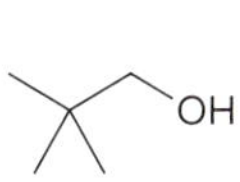 OH

d. 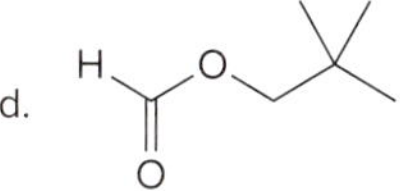

6. How many lines are present in the ^{13}C NMR spectrum of compound **A?**

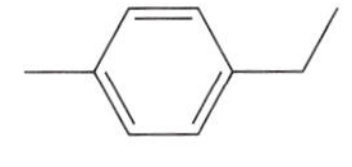

A

a. three b. five c. seven d. nine

7. Which compound has the fewest signals in its NMR spectrum?

a. 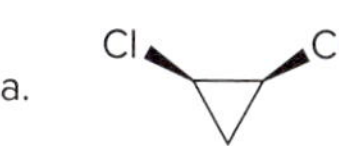

b. 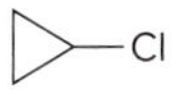 Cl

c. 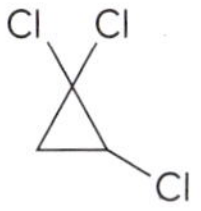 Cl Cl Cl

d.  Cl Cl

8. Which compound has a single signal in both its ^{13}C and ^{1}H NMR spectra?

a. 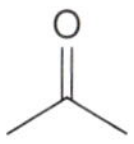 O

b. 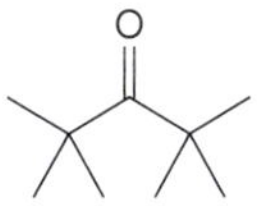 O

c. 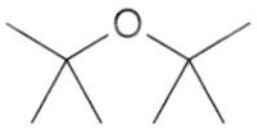 O

d.

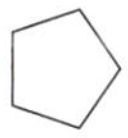

9. Which of the following statements is true?

a. Benzene protons are highly shielded, so they absorb far downfield.

b. A signal at 400 Hz is downfield from a signal at 200 Hz.

c. A quintet is due to a proton that has six adjacent nonequivalent protons.

d. If two signals occur at 1.0 and 1.5 ppm using a 500 MHz NMR spectrometer, the signals are separated by 25 Hz.

10. Which compound has a triplet at ~4.0 ppm in its NMR spectrum?

a. 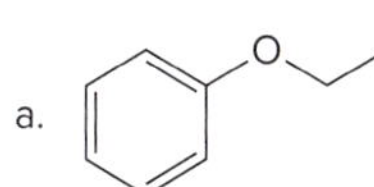

b. 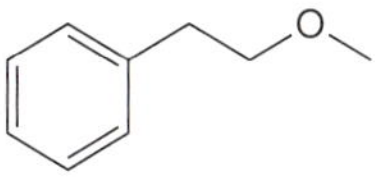 O

c. 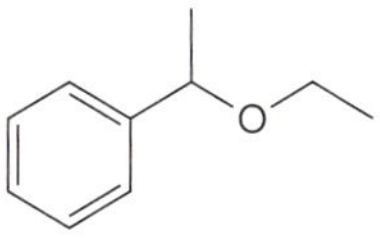 O

d. 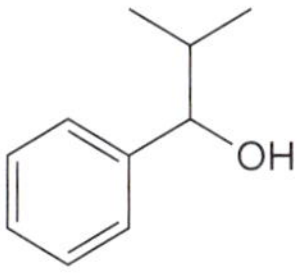 OH

PROBLEMS

Problem Using Three-Dimensional Models

C.39 (a) How many ^{1}H NMR signals does each of the following compounds exhibit? (b) How many ^{13}C NMR signals does each compound exhibit?

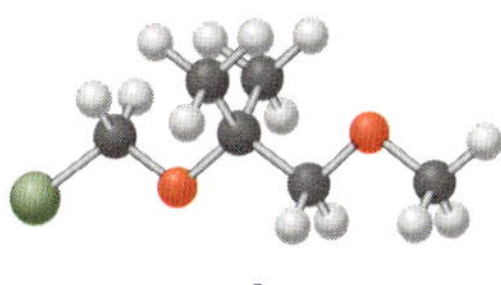

A

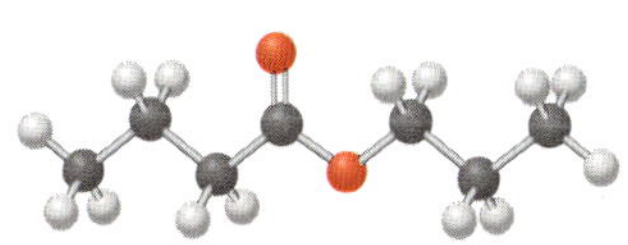

B

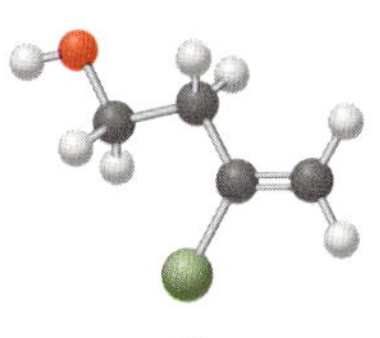

C

^{1}H NMR Spectroscopy—Determining Equivalent Protons

C.40 How many different types of protons are present in each compound?

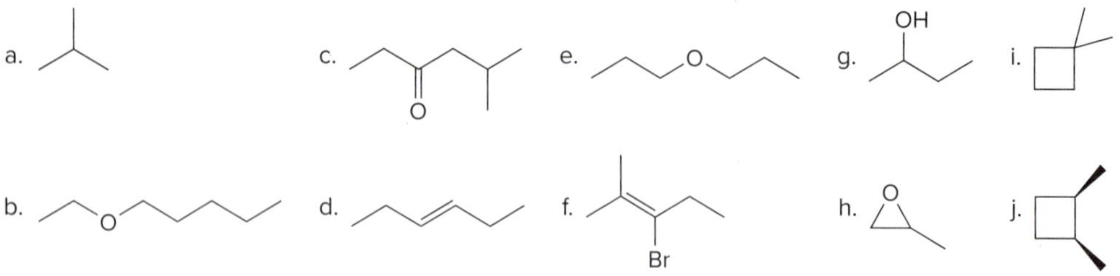

a. b. c. d. e. f. Br g. OH h. O i. j.

C.41 How many ^{1}H NMR signals does each natural product exhibit?

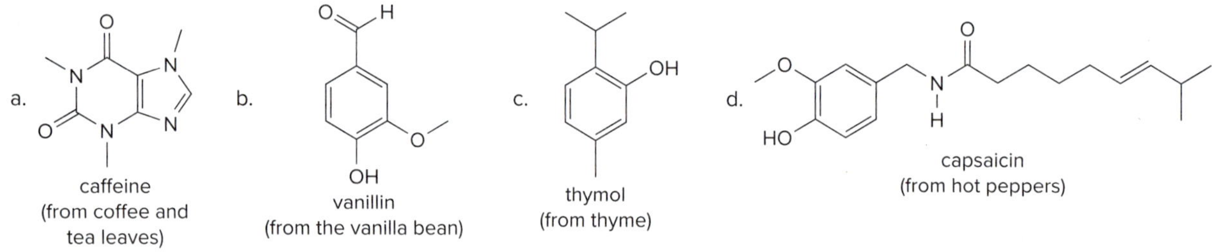

a. caffeine (from coffee and tea leaves)

b. vanillin (from the vanilla bean)

c. thymol (from thyme)

d. capsaicin (from hot peppers)

^{1}H NMR—Chemical Shift

C.42 Using a 300 MHz NMR instrument:

a. How many Hz downfield from TMS is a signal at 2.5 ppm?

b. If a signal comes at 1200 Hz downfield from TMS, at what ppm does it occur?

c. If two signals are separated by 2 ppm, how many Hz does this correspond to?

C.43 What effect does increasing the operating frequency of a ^{1}H NMR spectrum have on each value: (a) the chemical shift in δ; (b) the frequency of an absorption in Hz; (c) the magnitude of a coupling constant *J* in Hz?

C.44 Rank the labeled protons in order of increasing chemical shift.

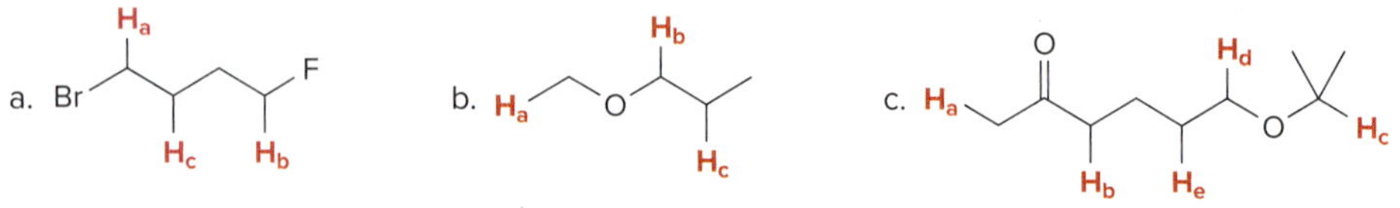

a. Br, F, H_a, H_b, H_c

b. H_a, H_b, H_c

c. H_a, H_b, H_c, H_d, H_e

^{1}H NMR—Splitting

C.45 Into how many peaks will the signal for each of the labeled protons be split?

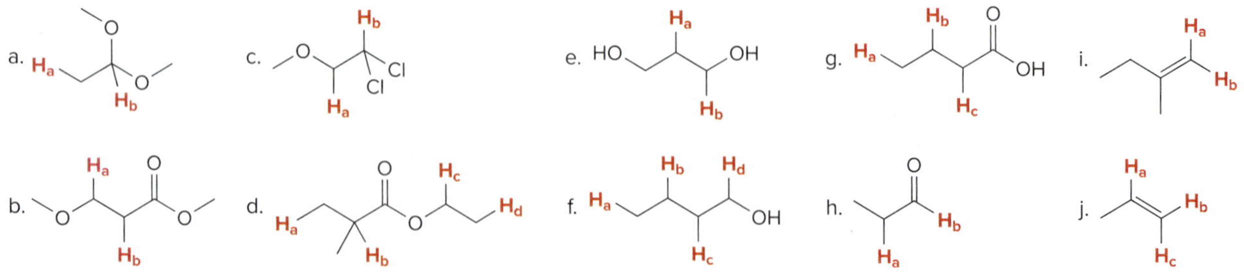

a. H_a, H_b

b. H_a, H_b

c. H_a, H_b

d. H_a, H_b, H_c, H_d

e. HO, OH, H_a, H_b

f. H_a, H_b, H_c, H_d, OH

g. H_a, H_b, H_c, OH

h. H_a, H_b

i. H_a, H_b

j. H_a, H_b, H_c

C.46 What splitting pattern is observed for each proton in the following compounds?

a. H, O, Cl, Br

b. O, Cl

C.47 Label the signals due to H_a, H_b, and H_c in the ^1H NMR spectrum of acrylonitrile ($CH_2{=}CHCN$). Draw a splitting diagram for the absorption due to the H_a proton.

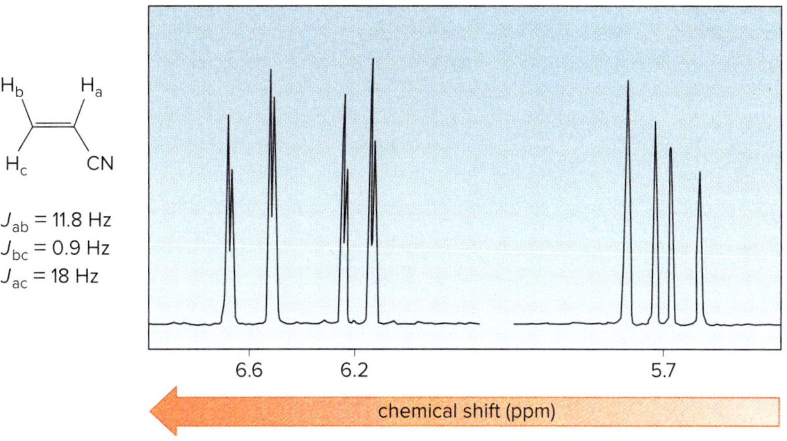

C.48 Draw a splitting diagram for H_b in compound **X** given the following coupling constants: (a) $J_{ab} >> J_{bc}$; (b) $J_{ab} = J_{bc}$. Clearly indicate how many peaks are visible in the H_b signal in each circumstance.

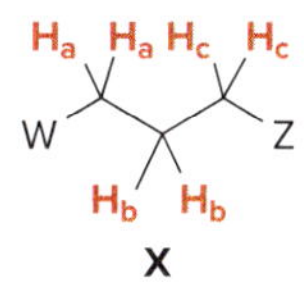

C.49 What is the difference between a sextet and a doublet of triplets?

^{13}C NMR

C.50 Draw the four constitutional isomers having molecular formula C_4H_9Br and indicate how many different kinds of carbon atoms each has.

C.51 Explain why the carbonyl carbon of an aldehyde or ketone absorbs farther downfield than the carbonyl carbon of an ester in a ^{13}C NMR spectrum.

C.52 How many ^{13}C NMR signals does each compound exhibit?

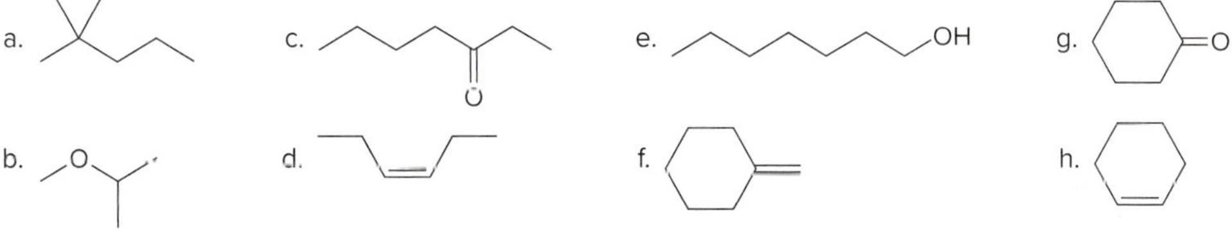

C.53 Rank the highlighted carbon atoms in each compound in order of increasing chemical shift.

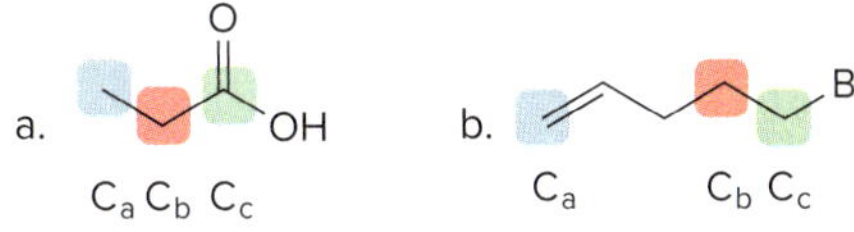

b. Br

C_a C_b C_c

C.54 Identify the carbon atoms that give rise to the signals in the ^{13}C NMR spectrum of each compound.

a. $CH_3CH_2CH_2CH_2OH$; ^{13}C NMR: 14, 19, 35, and 62 ppm

b. $(CH_3)_2CHCHO$; ^{13}C NMR: 16, 41, and 205 ppm

c. $CH_2{=}CHCH(OH)CH_3$; ^{13}C NMR: 23, 69, 113, and 143 ppm

Identifying Isomers Using NMR Spectroscopy

C.55 Answer the following questions about isomers **A, B,** and **C.**

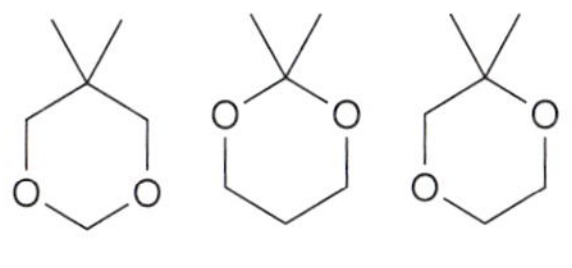

a. Which compound has the carbon that absorbs farthest downfield in its ^{13}C NMR?

b. Which compound has the proton that absorbs farthest downfield in its ^1H NMR?

c. Which compound exhibits the most lines in its ^{13}C NMR spectrum?

d. Which compound has the largest number of signals in its ^1H NMR spectrum?

C.56 How could ^{1}H NMR spectroscopy be used to distinguish among isomers **A, B,** and **C?**

A **B** **C**

C.57 How could ^{13}C NMR spectroscopy be used to distinguish among isomers **X, Y,** and **Z?**

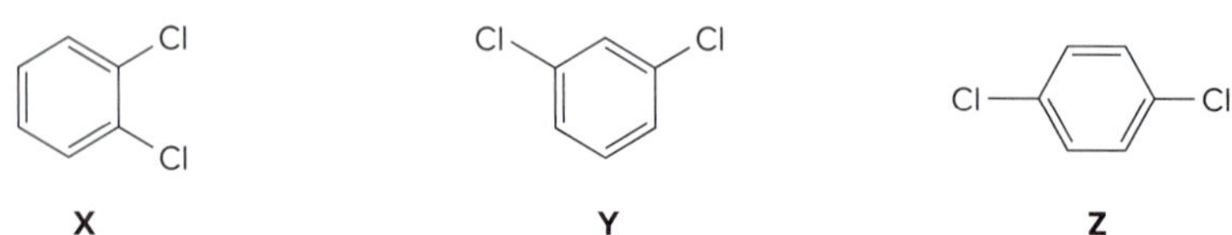

C.58 How could ^{1}H NMR spectroscopy be used to distinguish between each pair of compounds?

a. and

b. and

Combined Spectroscopy Problems

Additional spectroscopy problems are located at the end of Chapters 9–12, 13–21, and 23.

C.59 Which of the following drugs show more ^{13}C NMR signals than ^{1}H NMR signals in their NMR spectra?

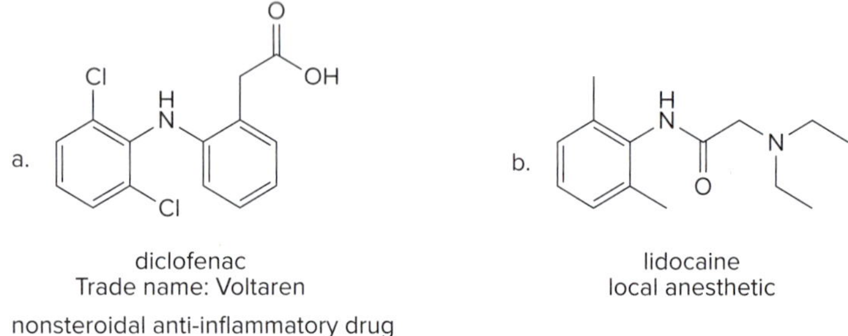

a. diclofenac
Trade name: Voltaren
nonsteroidal anti-inflammatory drug

b. lidocaine
local anesthetic

C.60 Answer the following questions for compounds **L, M,** and **N** drawn below.

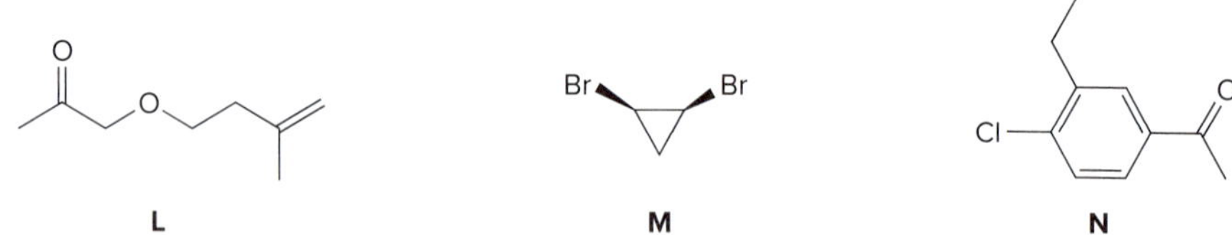

a. How many signals are expected in the ^{1}H NMR spectrum?

b. Into how many peaks is each signal in the ^{1}H NMR spectrum split?

c. How many lines are expected in the ^{13}C NMR spectrum?

C.61 Answer the following questions about each of the hydroxy ketones: 1-hydroxybutan-2-one (**A**) and 4-hydroxybutan-2-one (**B**).

1-hydroxybutan-2-one **A**

4-hydroxybutan-2-one **B**

a. What is the molecular ion in the mass spectrum?

b. What IR absorptions are present in the functional group region?

c. How many lines are observed in the ^{13}C NMR spectrum?

d. How many signals are observed in the ^{1}H NMR spectrum?

e. Give the splitting observed for each type of proton as well as its approximate chemical shift.

C.62 Propose a structure consistent with each set of spectral data:

a. $C_4H_8Br_2$: IR peak at 3000–2850 cm^{-1}; NMR (ppm):
1.87 (singlet, 6 H)
3.86 (singlet, 2 H)

b. $C_3H_6Br_2$: IR peak at 3000–2850 cm^{-1}; NMR (ppm):
2.4 (quintet)
3.5 (triplet)

c. $C_5H_{10}O_2$: IR peak at 1740 cm^{-1}; NMR (ppm):
1.15 (triplet, 3 H) 2.30 (quartet, 2 H)
1.25 (triplet, 3 H) 4.72 (quartet, 2 H)

d. C_3H_6O: IR peak at 1730 cm^{-1}; NMR (ppm):
1.11 (triplet)
2.46 (multiplet)
9.79 (triplet)

C.63 Identify the structures of isomers **A** and **B** (molecular formula $C_9H_{10}O$).

Compound **A:** IR peak at 1742 cm^{-1}; ^1H NMR data (ppm) at 2.15 (singlet, 3 H), 3.70 (singlet, 2 H), and 7.20 (broad singlet, 5 H).

Compound **B:** IR peak at 1688 cm^{-1}; ^1H NMR data (ppm) at 1.22 (triplet, 3 H), 2.98 (quartet, 2 H), and 7.28–7.95 (multiplet, 5 H).

C.64 Reaction of $C_6H_5CH_2CH_2OH$ with CH_3COCl affords compound **W,** which has molecular formula $C_{10}H_{12}O_2$. **W** shows prominent IR absorptions at 3088–2897, 1740, and 1606 cm^{-1}. **W** exhibits the following signals in its ^1H NMR spectrum: 2.02 (singlet), 2.91 (triplet), 4.25 (triplet), and 7.20–7.35 (multiplet) ppm. What is the structure of **W?** We will learn about this reaction in Chapter 20.

C.65 Treatment of 2-methylpropanenitrile [$(CH_3)_2CHCN$] with $CH_3CH_2CH_2MgBr$, followed by aqueous acid, affords compound **V,** which has molecular formula $C_7H_{14}O$. **V** has a strong absorption in its IR spectrum at 1713 cm^{-1}, and gives the following ^1H NMR data: 0.91 (triplet, 3 H), 1.09 (doublet, 6 H), 1.6 (multiplet, 2 H), 2.43 (triplet, 2 H), and 2.60 (septet, 1 H) ppm. What is the structure of **V?** We will learn about this reaction in Chapter 19.

C.66 Compound **C** has a molecular ion in its mass spectrum at 146 and a prominent absorption in its IR spectrum at 1762 cm^{-1}. **C** shows the following ^1H NMR spectral data: 1.47 (doublet, 3 H), 2.07 (singlet, 6 H), and 6.84 (quartet, 1 H) ppm. What is the structure of **C?**

C.67 Treatment of compound **D** with $LiAlH_4$ followed by H_2O forms compound **E. D** shows a molecular ion in its mass spectrum at m/z = 71 and IR absorptions at 3600–3200 and 2263 cm^{-1}. **E** shows a molecular ion in its mass spectrum at m/z = 75 and IR absorptions at 3636 and 3600–3200 cm^{-1}. Propose structures for **D** and **E** from these data and the given ^1H NMR spectra.

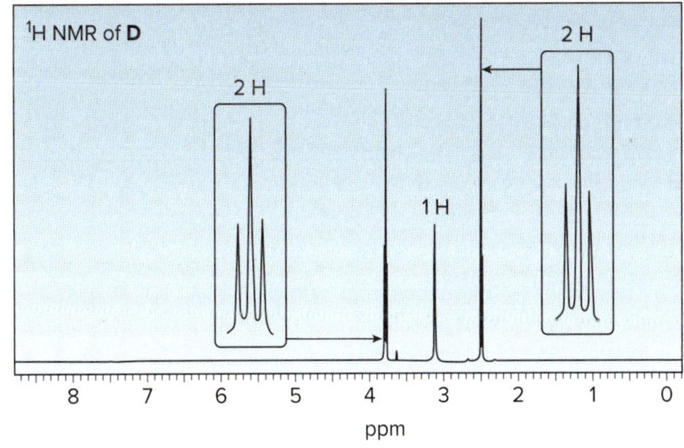

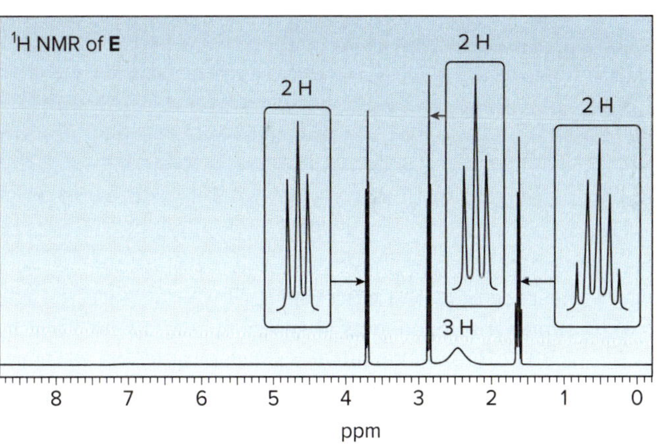

C.68 Identify the structures of isomers **E** and **F** (molecular formula $C_4H_8O_2$). Relative areas are given above each signal.

a. **Compound E:** IR absorption at 1743 cm^{-1}

b. **Compound F:** IR absorption at 1730 cm^{-1}

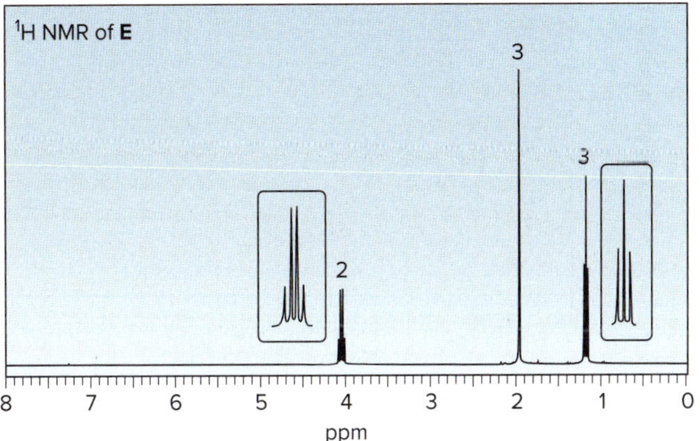

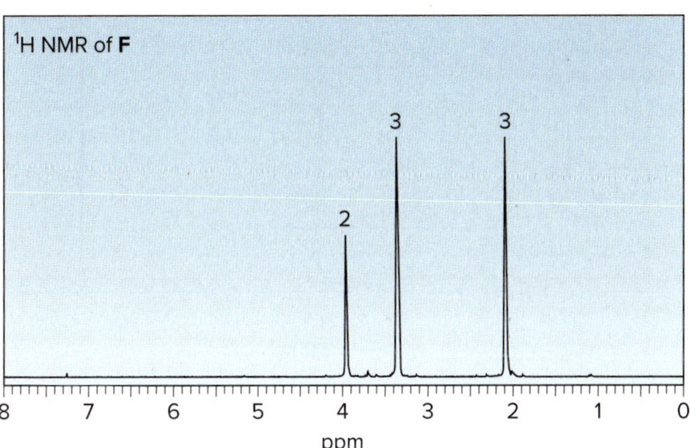

C.69 Identify the structures of isomers **H** and **I** (molecular formula $C_8H_{11}N$).

a. **Compound H:** IR absorptions at 3365, 3284, 3026, 2932, 1603, and 1497 cm^{-1}

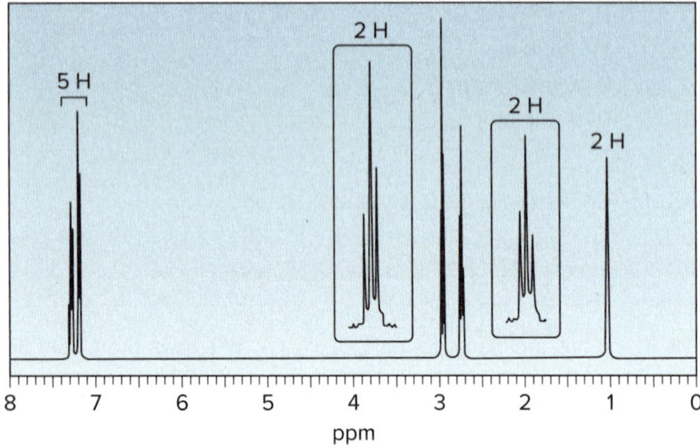

b. **Compound I:** IR absorptions at 3367, 3286, 3027, 2962, 1604, and 1492 cm^{-1}

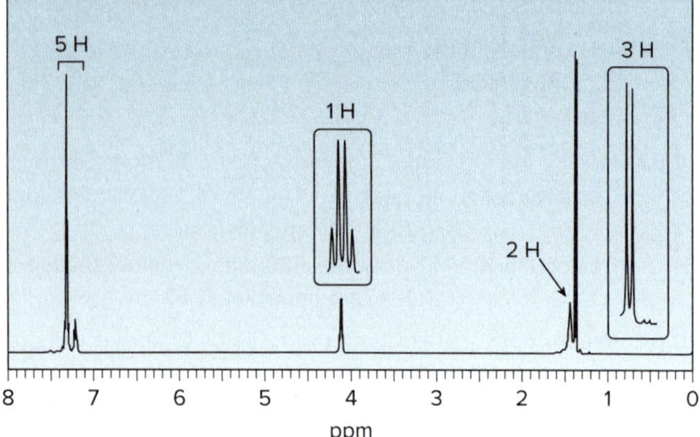

C.70 Propose a structure consistent with each set of data.

a. $C_9H_{10}O_2$: IR absorption at 1718 cm^{-1}

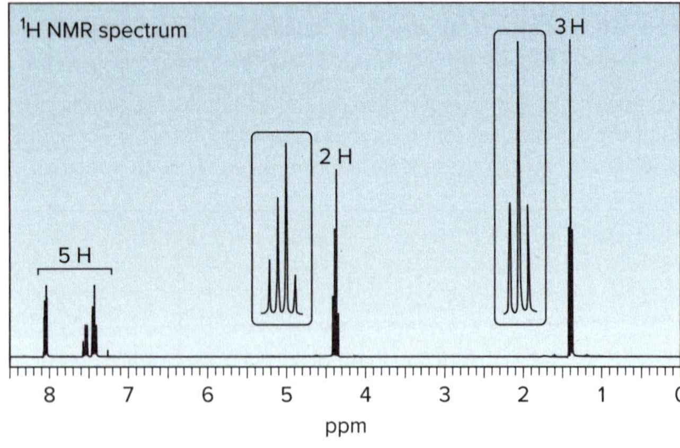

b. C_9H_{12}: IR absorption at 2850–3150 cm^{-1}

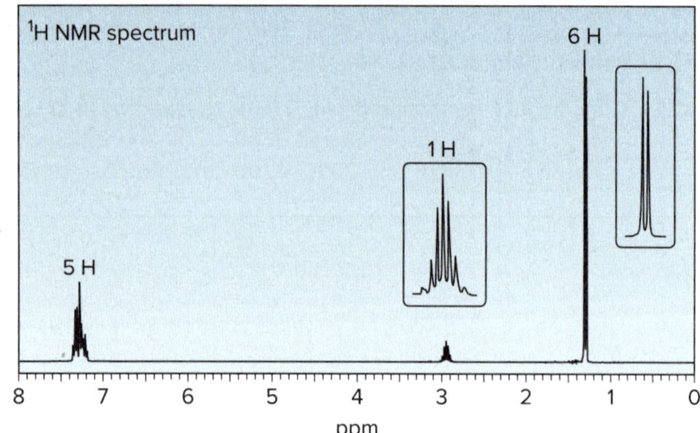

C.71 Reaction of $(CH_3)_3CCHO$ with $(C_6H_5)_3P{=}C(CH_3)OCH_3$, followed by treatment with aqueous acid, affords **R** ($C_7H_{14}O$). **R** has a strong absorption in its IR spectrum at 1717 cm^{-1} and three singlets in its ^1H NMR spectrum at 1.02 (9 H), 2.13 (3 H), and 2.33 (2 H) ppm. What is the structure of **R?** We will learn about this reaction in Chapter 18.

C.72 The treatment of $(CH_3)_2C{=}CHCH_2Br$ with H_2O forms **B** (molecular formula $C_5H_{10}O$) as one of the products. Determine the structure of **B** from its ^1H NMR and IR spectra.

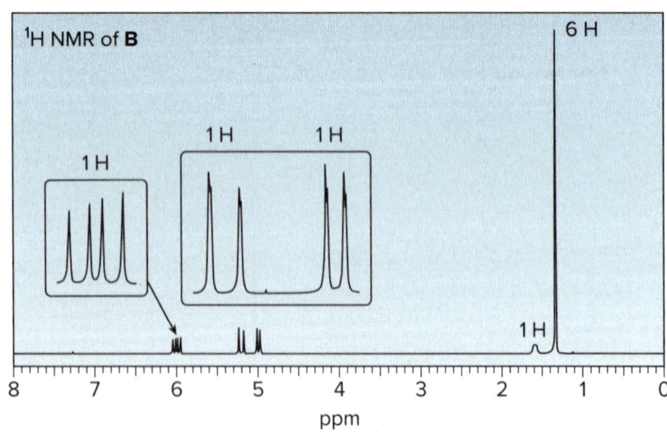

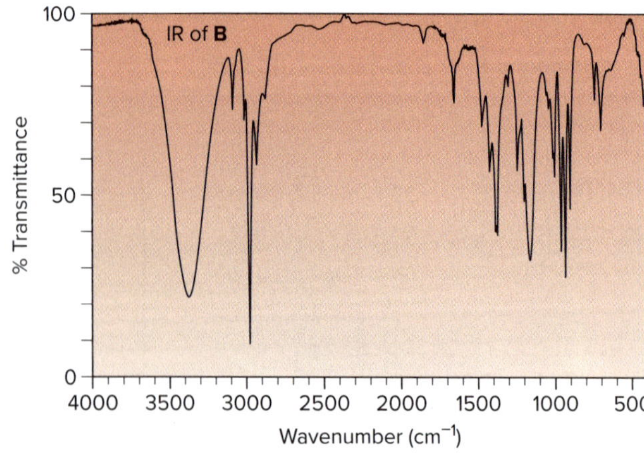

C.73 Propose a structure consistent with each set of data.

a. Compound **J:** molecular ion at 72; IR peak at 1710 cm^{-1}; ^{1}H NMR data (ppm) at 1.0 (triplet, 3 H), 2.1 (singlet, 3 H), and 2.4 (quartet, 2 H)

b. Compound **K:** molecular ion at 88; IR peak at 3600–3200 cm^{-1}; ^{1}H NMR data (ppm) at 0.9 (triplet, 3 H), 1.2 (singlet, 6 H), 1.5 (quartet, 2 H), and 1.6 (singlet, 1 H)

C.74 An unknown compound **D** exhibits a strong absorption in its IR spectrum at 1692 cm^{-1}. The mass spectrum of **D** shows a molecular ion at m/z = 150 and a base peak at 121. The ^{1}H NMR spectrum of **D** is shown below. What is the structure of **D?**

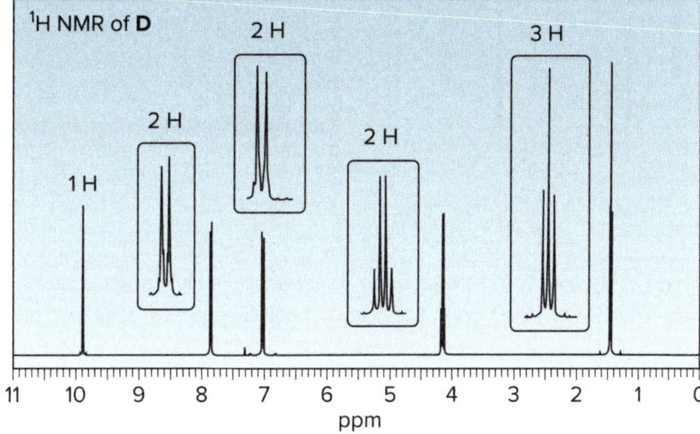

C.75 In the presence of a small amount of acid, a solution of acetaldehyde (CH_3CHO) in methanol (CH_3OH) was allowed to stand and a new compound **L** was formed. **L** has a molecular ion in its mass spectrum at 90 and IR absorptions at 2992 and 2941 cm^{-1}. **L** shows three signals in its ^{13}C NMR at 19, 52, and 101 ppm. The ^{1}H NMR spectrum of **L** is given below. What is the structure of **L?**

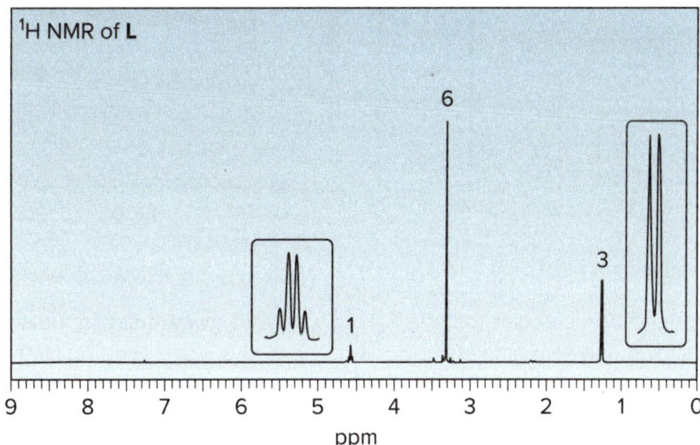

C.76 Compound **O** has molecular formula $C_{10}H_{12}O$ and shows an IR absorption at 1687 cm^{-1}. The ^{1}H NMR spectrum of **O** is given below. What is the structure of **O?**

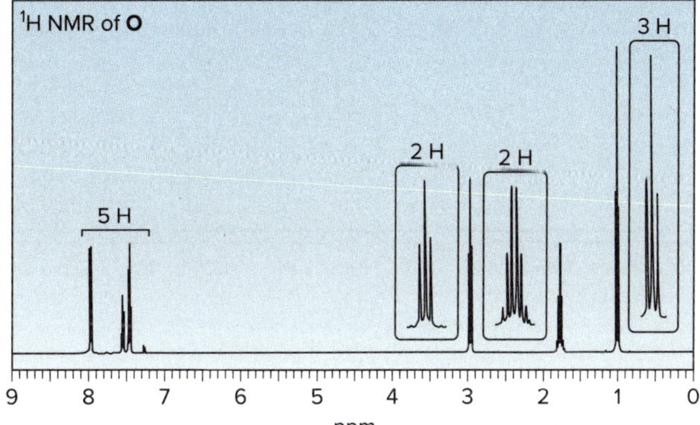

C.77 Compound **P** has molecular formula $C_5H_9ClO_2$. Deduce the structure of **P** from its ^1H and ^{13}C NMR spectra.

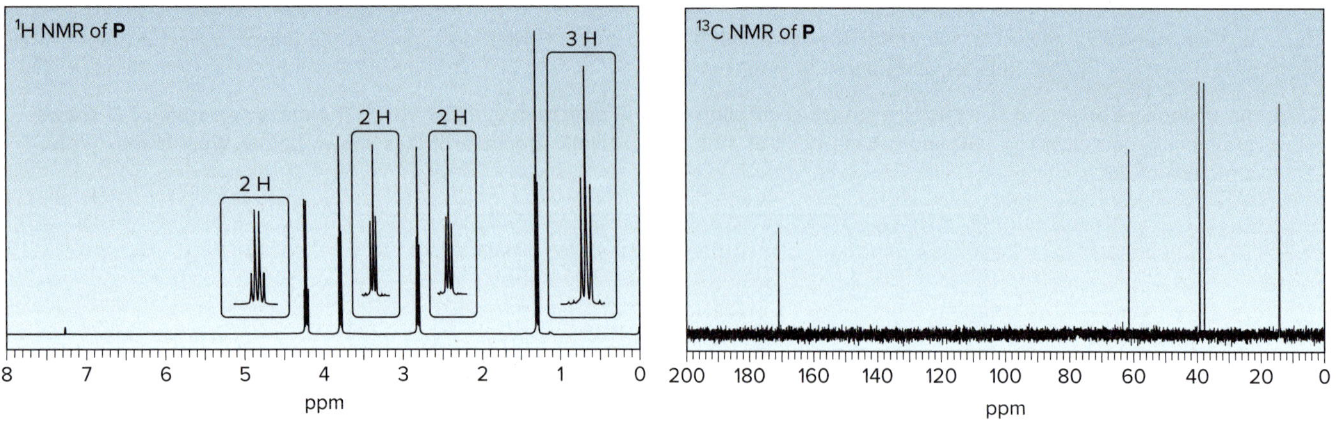

C.78 Treatment of butan-2-one ($CH_3COCH_2CH_3$) with strong base followed by CH_3I forms a compound **Q,** which gives a molecular ion in its mass spectrum at 86. The IR (> 1500 cm^{-1} only) and ^1H NMR spectra of **Q** are given below. What is the structure of **Q?**

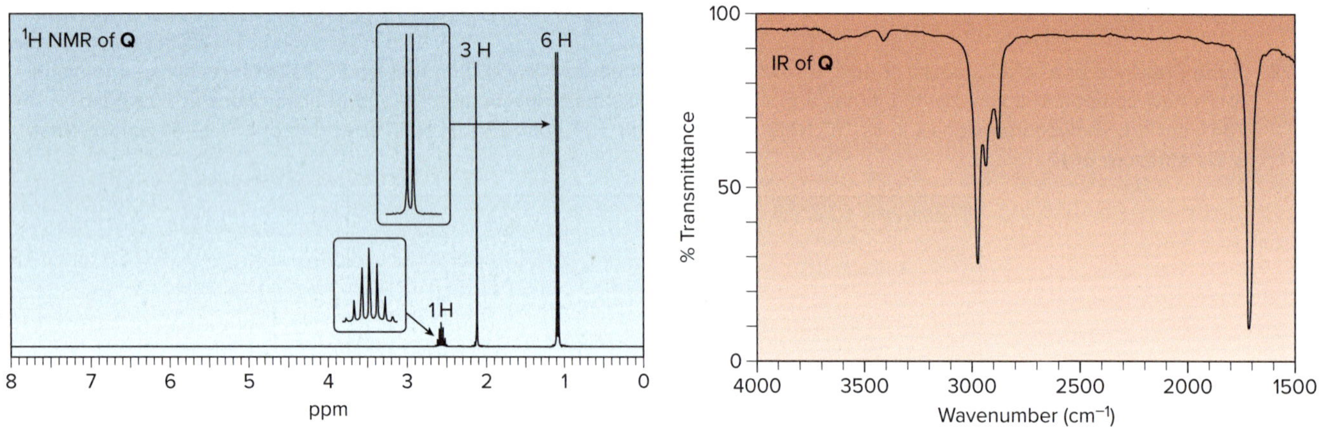

C.79 Propose a structure for compound **X** (molecular formula $C_6H_{12}O_2$), which gives a strong peak in its IR spectrum at 1740 cm^{-1}. The ^1H NMR spectrum of **X** shows only two singlets, including one at 3.5 ppm. The ^{13}C NMR spectrum is given below. Propose a structure for **X.**

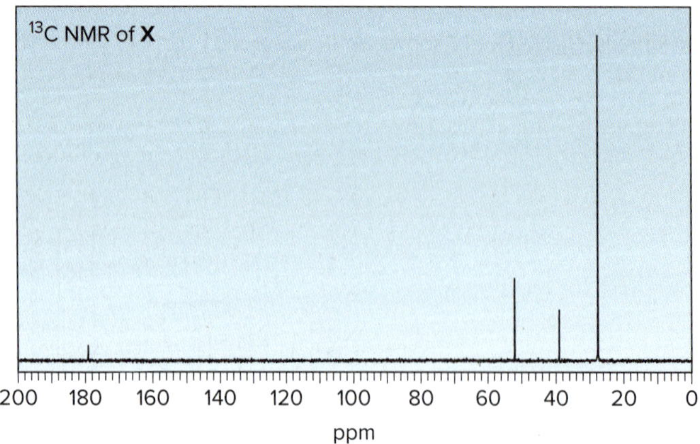

Challenge Problems

C.80 Reaction of unknown **A** with HCl forms chlorohydrin **B** as the major product. **A** shows no absorptions in its IR spectrum at 1700 cm^{-1} or 3600–3200 cm^{-1}, and gives the following ^1H NMR data: 1.4 (doublet, 3 H), 3.0 (quartet of doublets, 1 H), 3.5 (doublet, 1 H), 3.8 (singlet, 3 H), 6.9 (doublet, 2 H), and 7.2 (doublet, 2 H) ppm. (a) Propose a structure for **A,** including stereochemistry. (b) Explain why **B** is the major product in this reaction.

B

C.81 The ^1H NMR spectrum of *N,N*-dimethylformamide shows three singlets at 2.9, 3.0, and 8.0 ppm. Explain why the two CH_3 groups are not equivalent to each other, thus giving rise to two NMR signals.

N,N-dimethylformamide

C.82 18-Annulene shows two signals in its ^1H NMR spectrum, one at 8.9 (12 H) and one at –1.8 (6 H) ppm. Using a similar argument to that offered for the chemical shift of benzene protons, explain why both shielded and deshielded values are observed for 18-annulene.

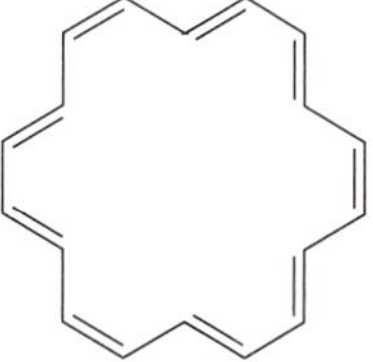

18-annulene

C.83 Explain why the ^{13}C NMR spectrum of 3-methylbutan-2-ol shows five signals.

C.84 Because ^{31}P has an odd mass number, ^{31}P nuclei absorb in the NMR and, in many ways, these nuclei behave similarly to protons in NMR spectroscopy. With this in mind, explain why the ^1H NMR spectrum of methyl dimethylphosphonate, $CH_3PO(OCH_3)_2$, consists of two doublets at 1.5 and 3.7 ppm.

C.85 Cyclohex-2-enone has two protons on its carbon–carbon double bond (labeled H_a and H_b) and two protons on the carbon adjacent to the double bond (labeled H_c). (a) If J_{ab} = 11 Hz and J_{bc} = 4 Hz, sketch the splitting pattern observed for each proton on the sp^2 hybridized carbons. (b) Despite the fact that H_a is located adjacent to an electron-withdrawing C=O, its absorption occurs upfield from the signal due to H_b (6.0 vs. 7.0 ppm). Offer an explanation.

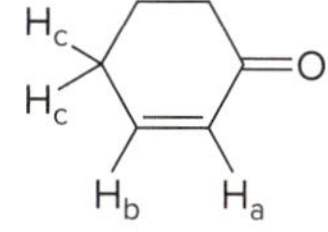

cyclohex-2-enone

SELF-TEST ANSWERS

1. d 2. d 3. b 4. a 5. a 6. c 7. d 8. d 9. b 10. b

13 Radical Reactions

13.1 Introduction
13.2 General features of radical reactions
13.3 Halogenation of alkanes
13.4 The mechanism of halogenation
13.5 Chlorination of other alkanes
13.6 Chlorination versus bromination
13.7 Halogenation as a tool in organic synthesis
13.8 The stereochemistry of halogenation reactions
13.9 Application: The ozone layer and CFCs
13.10 Radical halogenation at an allylic carbon
13.11 Application: Oxidation of unsaturated lipids
13.12 Application: Antioxidants
13.13 Radical addition reactions to double bonds
13.14 Polymers and polymerization

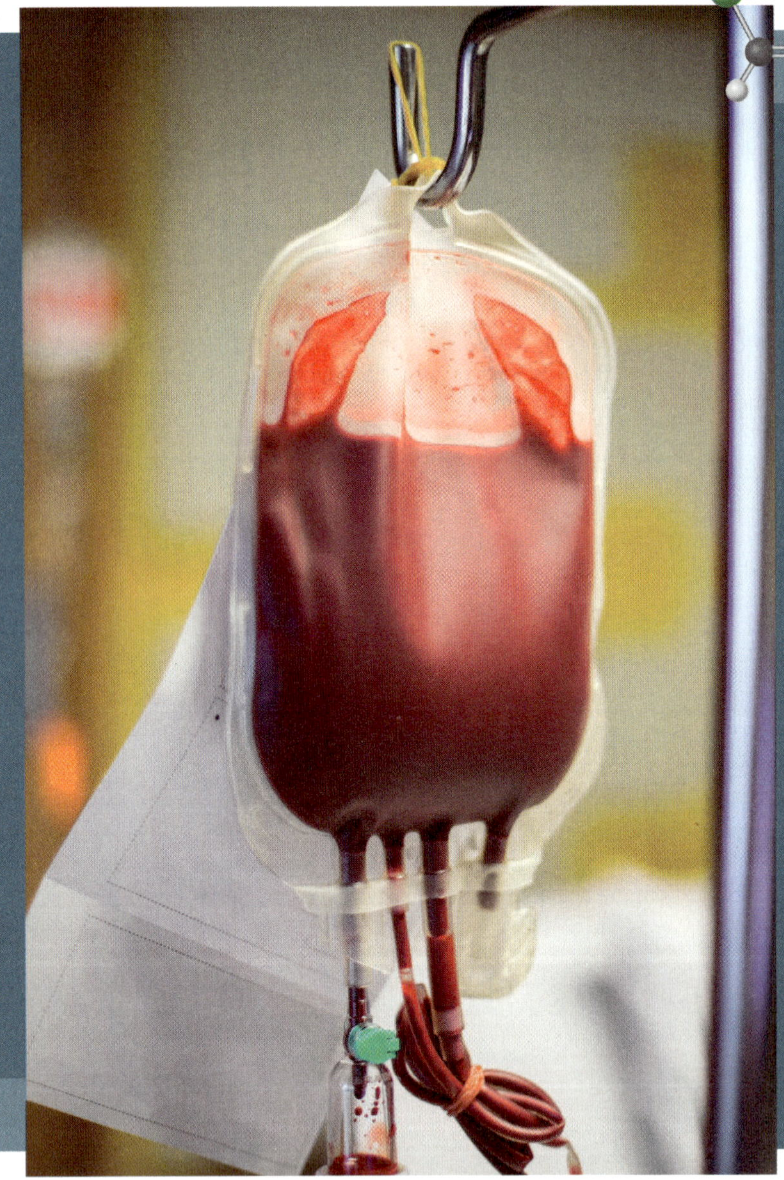

Hin255/iStock/Getty Images

Poly(vinyl chloride) (PVC), a synthetic polymer prepared from the monomer **vinyl chloride,** is used in a wide variety of medical, industrial, and home products. Rigid PVC is found in pipes and bottles, whereas flexible PVC is used in blood bags, tubing, and materials needed for hemodialysis and heart bypass. Because PVC is water insoluble, garden hoses, drainpipes, and rain gear are made of PVC. PVC is lightweight, tear resistant, and easily sterilized, and it can be recycled many times before it is no longer usable. In Chapter 13, we learn how polymers like poly(vinyl chloride) are prepared.

Why Study . . . Radical Reactions?

A small but significant group of reactions involves the homolysis of nonpolar bonds to form highly reactive **radical intermediates.** Although they are unlike other organic reactions, radical transformations are important in many biological and industrial processes. The gases O_2 and NO (nitric oxide) are both radicals. Many oxidation reactions with O_2 involve radical intermediates, and biological processes mediated by NO such as blood clotting and neurotransmission may involve radicals. Many useful industrial products such as Styrofoam and polyethylene are prepared by radical processes.

In Chapter 13 we examine the cleavage of nonpolar bonds by radical reactions.

13.1 Introduction

Radicals were first discussed in Section 6.3.

- **A *radical* is a reactive intermediate with a single unpaired electron, formed by homolysis of a covalent bond.**

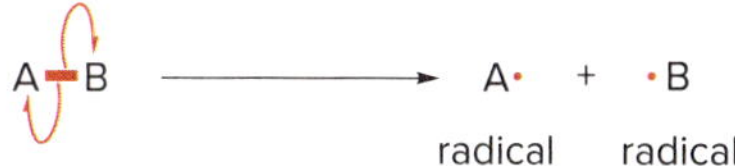

Use half-headed curved arrows in radical reactions.

A radical contains an atom that does not have an octet of electrons, making it reactive and unstable. Radical processes involve single electrons, so half-headed arrows are used to show the movement of electrons. **One half-headed arrow is used for each electron.**

Carbon radicals are classified as **primary (1°), secondary (2°),** or **tertiary (3°)** by the number of R groups bonded to the carbon with the unpaired electron. A carbon radical is ***sp*^{2} hybridized** and **trigonal planar,** like sp^2 hybridized carbocations. The unhybridized p orbital contains the unpaired electron and extends above and below the trigonal planar carbon.

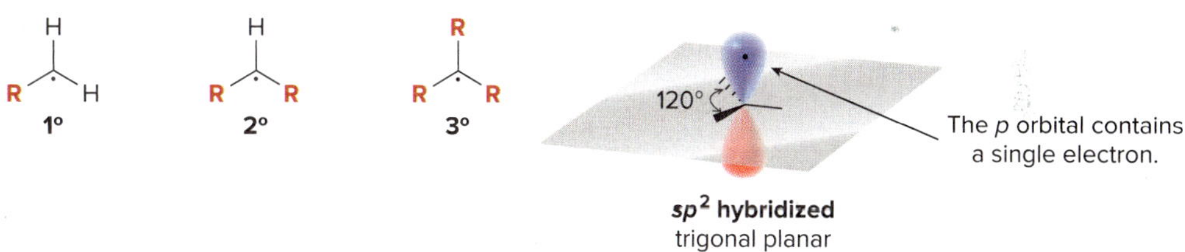

Bond dissociation energies for the cleavage of C–H bonds are used as a measure of radical stability. For example, two different radicals can be formed by cleavage of the C–H bonds in $CH_3CH_2CH_3$.

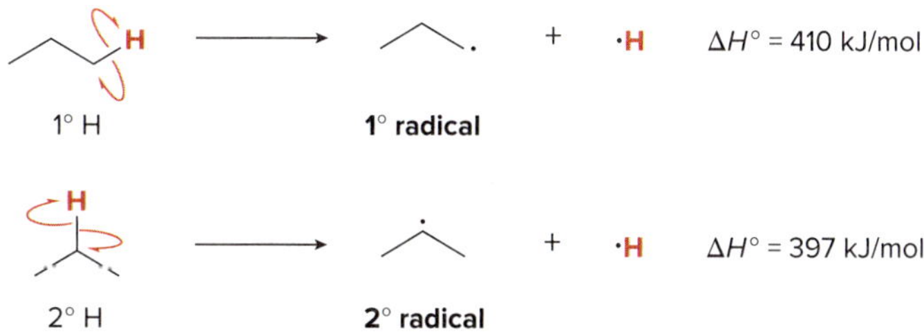

Cleavage of the **stronger 1° C–H** bond to form the 1° radical ($CH_3CH_2CH_2$·) requires *more* energy than cleavage of the **weaker 2° C–H** bond to form the 2° radical [$(CH_3)_2CH$·]—410 versus 397 kJ/mol. This makes the 2° radical more stable, because less energy is required for its formation, as illustrated in Figure 13.1. Thus, **cleavage of the weaker bond forms the more stable radical,** a specific example of a general trend.

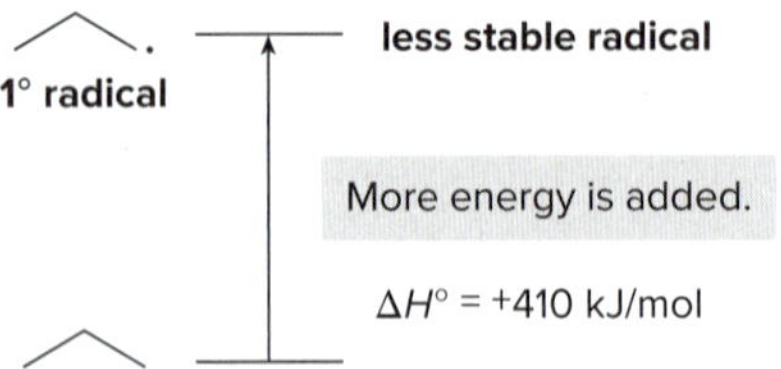

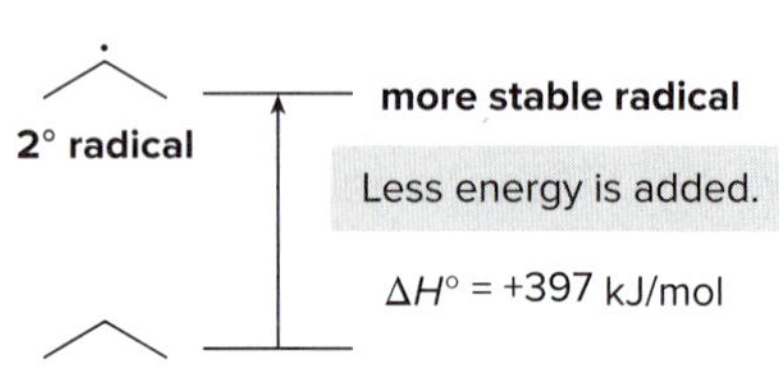

Figure 13.1 The relative stability of 1° and 2° carbon radicals

- **The stability of a radical increases as the number of alkyl groups bonded to the radical carbon increases.**

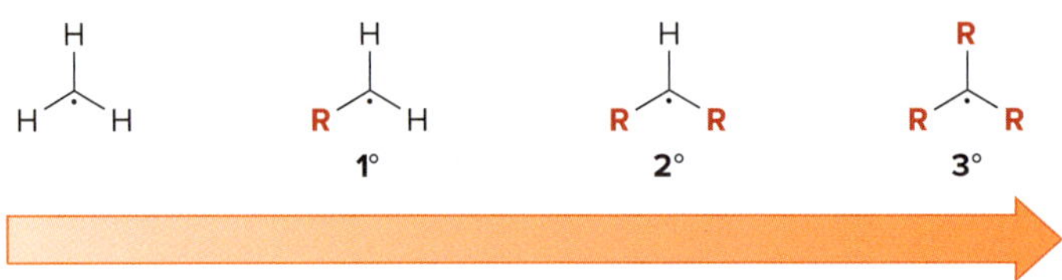

The **lower** the bond dissociation energy for a C–H bond, the **more stable** the resulting carbon radical.

Thus, a 3° radical is more stable than a 2° radical, and a 2° radical is more stable than a 1° radical. Increasing alkyl substitution increases radical stability in the same way it increases carbocation stability. **Alkyl groups are more polarizable than hydrogen atoms,** so they can more easily donate electron density to the electron-deficient carbon radical, thus increasing stability.

Unlike carbocations, however, **less stable radicals generally do *not* rearrange to more stable radicals.** This difference can be used to distinguish between reactions involving radical intermediates and those involving carbocations.

Problem 13.1 Classify each radical as 1°, 2°, or 3°.

a. b. c. d.

Problem 13.2 Draw the most stable radical that can result from cleavage of a C–H bond in each molecule.

a. b. c. d.

13.2 General Features of Radical Reactions

R–Ö· radical **initiator**

Radicals are formed from covalent bonds by adding energy in the form of **heat (Δ)** or **light (*hν*).** Some radical reactions are carried out in the presence of a **radical initiator,** a compound that contains an especially weak bond that serves as a source of radicals. **Peroxides,** compounds with the general structure **RO–OR,** are the most commonly used radical initiators. Heating a peroxide readily causes homolysis of the weak O–O bond, forming two RO· radicals.

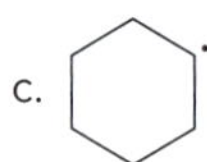

13.2A Two Common Reactions of Radicals

Radicals undergo two main types of reactions: **they react with σ bonds,** and **they add to π bonds,** in both cases achieving an octet of electrons.

[1] Reaction of a Radical X· with a C–H Bond

A radical X· abstracts a hydrogen atom from a C–H σ bond to form H–X and a carbon radical. One electron from the C–H bond is used to form the new H–X bond, and the other electron in the C–H bond remains on carbon. The result is that the original radical X· is now surrounded by an octet of electrons, and a new radical is formed.

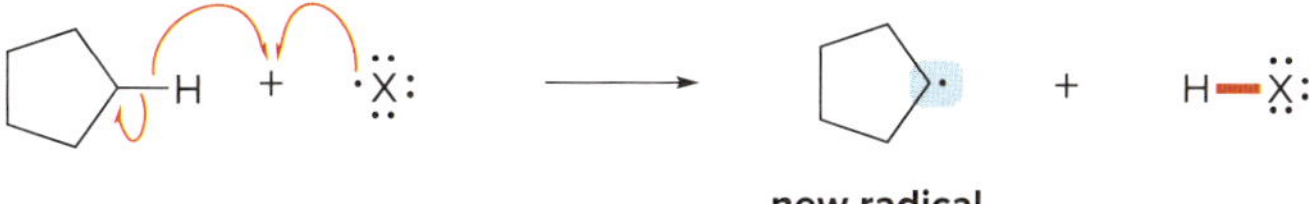

new radical

- One electron in H–X comes from the radical.
- One electron in H–X comes from the C–H bond.

This radical reaction is typically seen with the nonpolar C–H bonds of **alkanes,** which cannot react with polar or ionic electrophiles and nucleophiles.

[2] Reaction of a Radical X· with a C=C

A radical X· also adds to the π bond of a carbon–carbon double bond. One electron from the double bond is used to form a new C–X bond, and the other electron remains on the other carbon originally part of the double bond.

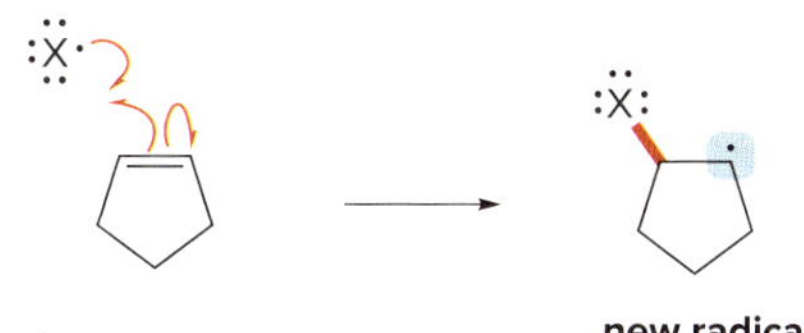

new radical

- One electron in C–X comes from the radical.
- One electron in C–X comes from the π bond.

Whenever a radical reacts with a stable single or double bond, **a new radical is formed** in the products.

Although the electron-rich double bond of an **alkene** reacts with electrophiles by ionic addition mechanisms, it also reacts with radicals because these reactive intermediates are also electron deficient.

13.2B Two Radicals Reacting with Each Other

A radical, once formed, rapidly reacts with whatever is available. Usually that means a stable σ or π bond. Occasionally, however, two radicals come into contact with each other, and they react to form a σ bond.

One electron in X–X comes from each radical.

The reaction of a radical with oxygen, a diradical in its ground-state electronic configuration, is another example of two radicals reacting with each other. In this case, the reaction of O_2 with X· forms a new radical, thus preventing X· from reacting with an organic substrate.

·O–O· + ·X: ⟶ ·O–O–X:

a diradical

Compounds that prevent radical reactions from occurring are called *radical inhibitors* or *radical scavengers*. Besides O_2, vitamin E and related compounds, discussed in Section 13.12, are radical scavengers, too. The fact that these compounds inhibit a reaction often suggests that the reaction occurs via radical intermediates.

Problem 13.3 Draw the products formed when a chlorine atom (Cl·) reacts with each species.

a. (cyclohexane) b. $CH_2{=}CH_2$ c. :Cl· d. O_2

13.3 Halogenation of Alkanes

In the presence of light or heat, alkanes react with halogens to form alkyl halides. Halogenation is a **radical substitution reaction,** because a halogen atom X replaces a hydrogen via a mechanism that involves radical intermediates.

$$\text{R}_3\text{C–H} + X_2 \xrightarrow{h\nu \text{ or } \Delta} \text{R}_3\text{C–X} + \text{H–X}$$

X = Cl or Br — **alkyl halide**

Halogenation of alkanes is useful only with Cl_2 and Br_2. Reaction with F_2 is too violent and reaction with I_2 is too slow to be useful. With an alkane that has more than one type of hydrogen atom, a mixture of alkyl halides may result (Reaction [2]).

When asked to draw the products of halogenation of an alkane, **draw the products of monohalogenation only,** unless specifically directed to do otherwise.

[1] cyclohexane + Br_2 $\xrightarrow{h\nu \text{ or } \Delta}$ bromocyclohexane + H–Br

[2] 2 propane + 2 Cl_2 $\xrightarrow{h\nu \text{ or } \Delta}$ 1-chloropropane + 2-chloropropane + 2 H–Cl

1 : 1

In these examples of halogenation, a halogen has replaced a single hydrogen atom on the alkane. Can the other hydrogen atoms be replaced, too? Figure 13.2 shows that when CH_4 is treated with excess Cl_2, all four hydrogen atoms can be successively replaced by Cl to form CCl_4. **Monohalogenation**—the substitution of a single H by X—can be achieved experimentally by adding halogen X_2 to an excess of alkane.

$$CH_4 \xrightarrow[h\nu]{Cl_2} CH_3Cl + HCl \xrightarrow[h\nu]{Cl_2} CH_2Cl_2 + HCl \xrightarrow[h\nu]{Cl_2} CHCl_3 + HCl \xrightarrow[h\nu]{Cl_2} CCl_4 + HCl$$

Figure 13.2 Complete halogenation of CH_4 using excess Cl_2

Sample Problem 13.1 Drawing the Products of the Chlorination of an Alkane

Draw all the constitutional isomers formed by monohalogenation of $(CH_3)_2CHCH_2CH_3$ with Cl_2 and $h\nu$.

Solution

Substitute Cl for H on every carbon, and then check to see if any products are identical. The starting material has five C's, but replacement of one H atom on two C's gives the same product. Thus, **$(CH_3)_2CHCH_2CH_3$ affords four monochloro substitution products.**

2-methylbutane $\xrightarrow[h\nu]{Cl_2}$

1-chloro-**3**-methylbutane + **2**-chloro-**3**-methylbutane + **2**-chloro-**2**-methylbutane + **1**-chloro-**2**-methylbutane + **1**-chloro-**2**-methylbutane

Same name
Identical compounds

Problem 13.4 Draw all constitutional isomers formed by monochlorination of each alkane.

a. b. c. d.

More Practice: Try Problems 13.28a, 13.35, 13.46a.

Problem 13.5 Compounds **A** and **B** are isomers having molecular formula C_5H_{12}. Heating **A** with Cl_2 gives a single product of monohalogenation, whereas heating **B** under the same conditions forms three constitutional isomers. What are the structures of **A** and **B?**

13.4 The Mechanism of Halogenation

Unlike nucleophilic substitution, which proceeds by two different mechanisms depending on the starting material and reagent, all halogenation reactions of alkanes—regardless of the halogen and alkane used—proceed by the *same* mechanism. Three facts about halogenation suggest that the mechanism involves **radical,** not ionic, intermediates.

Fact	Explanation
[1] Light, heat, or added peroxide is necessary for the reaction.	• Light or heat provides the energy needed for homolytic bond cleavage to form radicals. Breaking the weak O–O bond of peroxides initiates radical reactions as well.
[2] O_2 inhibits the reaction.	• The diradical O_2 removes radicals from a reaction mixture, thus preventing reaction.
[3] No rearrangements are observed.	• **Radicals do *not* rearrange.**

13.4A The Steps of Radical Halogenation

The chlorination of cyclopentane illustrates the **three distinct parts of radical halogenation** (Mechanism 13.1):

$$\text{cyclopentane} + Cl_2 \xrightarrow[h\nu \text{ or } \Delta]{} \text{chlorocyclopentane} + HCl$$

cyclopentane chlorocyclopentane

- **_Initiation:_ Two radicals are formed by homolysis of a σ bond and this begins the reaction.**
- **_Propagation:_ A radical reacts with another reactant to form a new σ bond and another radical.**
- **_Termination:_ Two radicals combine to form a stable bond. Removing radicals from the reaction mixture without generating any new radicals stops the reaction.**

Although initiation generates the Cl· radicals needed to begin the reaction, the **propagation steps ([2] and [3]) form the two reaction products**—chlorocyclopentane and HCl. Once the process has begun, propagation occurs over and over without the need for Step [1] to occur. **A mechanism such as radical halogenation that involves two or more repeating steps is called a *chain mechanism.*** Each propagation step involves a reactive radical abstracting an atom from a stable bond to form a new bond and **another radical that continues the chain.**

Mechanism 13.1 Radical Halogenation of Alkanes

Part [1] Initiation

$$:\ddot{Cl}-\ddot{Cl}: \xrightarrow[1]{h\nu \text{ or } \Delta} :\ddot{Cl}\cdot + \cdot\ddot{Cl}:$$

1 **Bond cleavage forms two radicals.** Homolysis of the weakest bond (Cl–Cl) requires light or heat and forms two chlorine radicals.

Part [2] Propagation

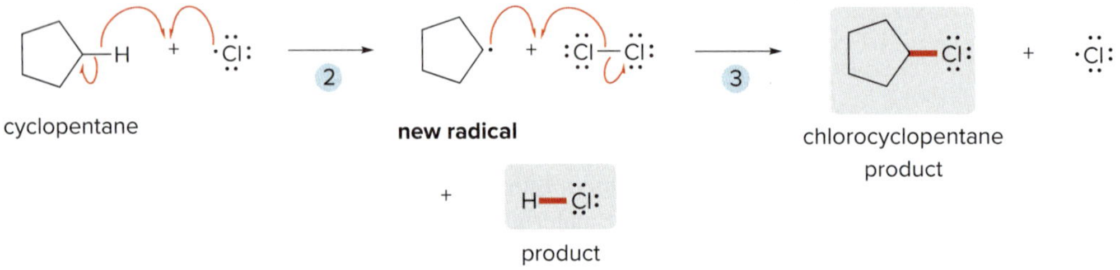

cyclopentane + $\cdot\ddot{Cl}:$ →(2) new radical + H—Cl: (product); new radical + :Cl—Cl: →(3) chlorocyclopentane product + ·Cl:

2 The **Cl· radical abstracts a hydrogen** from cyclopentane to form HCl (a reaction product) and a new carbon radical.

3 The **carbon radical abstracts a chlorine atom** from Cl_2 to form chlorocyclopentane (a reaction product) and Cl·. Because Cl· is a reactant in Step [2], **Steps [2] and [3] can occur repeatedly** without additional initiation (Step [1]).

Part [3] Termination

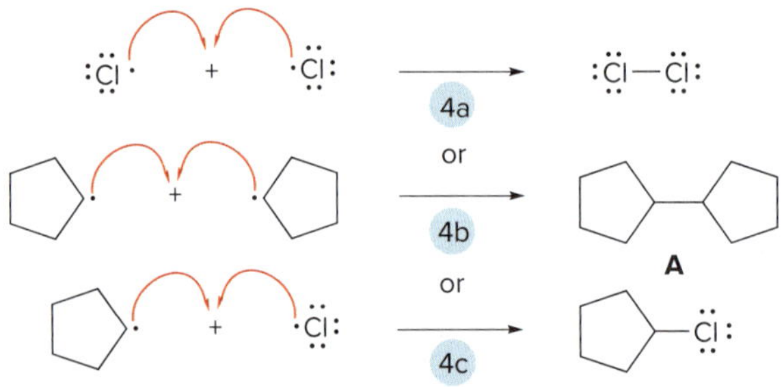

:Cl· + ·Cl: →(4a) :Cl—Cl:

or

cyclopentyl· + ·cyclopentyl →(4b) **A**

or

cyclopentyl· + ·Cl: →(4c) chlorocyclopentane

4 **Termination** of the chain occurs when any two radicals combine to form a bond.

Usually a radical reacts with a stable bond to propagate the chain, but occasionally two radicals combine, and this reaction terminates the chain. Depending on the reaction and the reaction conditions, some radical chain mechanisms can repeat thousands of times before termination occurs.

Termination Step [4a] forms Cl_2, a reactant, whereas Step [4c] forms chlorocyclopentane, one of the reaction products. Termination Step [4b] forms **A,** which is neither a reactant nor a desired product. The formation of a small quantity of **A,** however, is evidence that radicals are formed in the reaction.

The most important steps of radical halogenation are those that lead to product formation—the propagation steps—so subsequent discussion of this reaction concentrates on these steps only.

Problem 13.6 Heating bicyclic alkane **A** with Cl_2 forms **B,** along with other monochlorination products. Write out the two chain-propagating steps for this reaction.

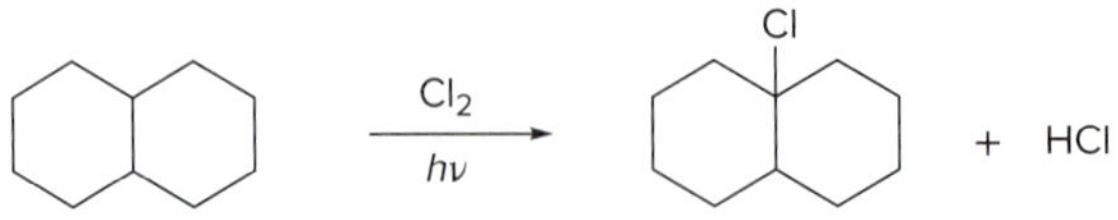

A $\xrightarrow[h\nu]{Cl_2}$ **B** (Cl) + HCl

13.4B Energy Changes During the Chlorination of Ethane

The chlorination of ethane illustrates how bond dissociation energies (Section 6.4) can be used to calculate $\Delta H°$ in chain propagation.

$$CH_3CH_3 + Cl_2 \xrightarrow{h\nu \text{ or } \Delta} CH_3CH_2Cl + HCl$$

Figure 13.3 Energy changes in the propagation steps during the chlorination of ethane

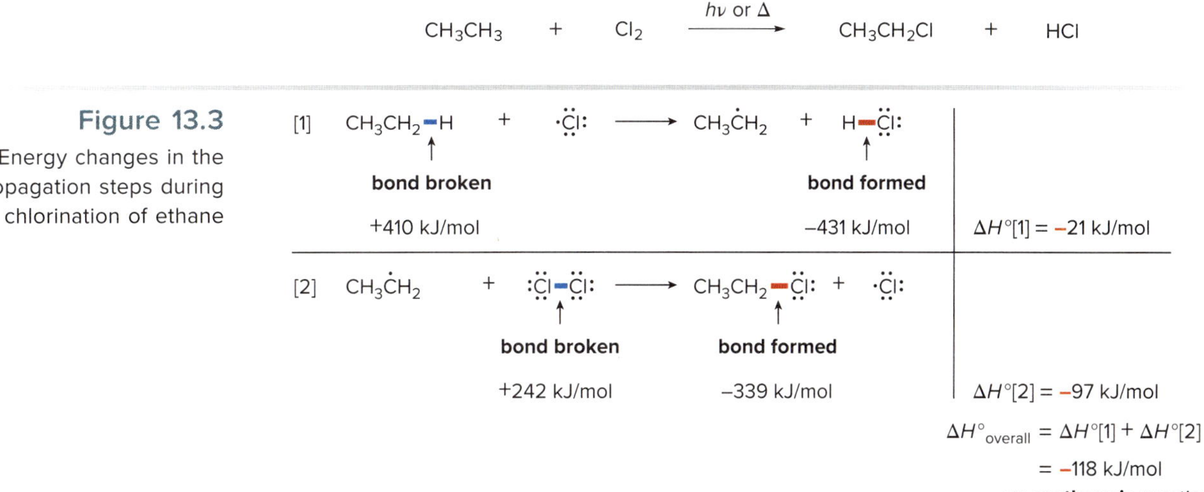

As shown in Figure 13.3, chain propagation consists of the same two steps drawn in Mechanism 13.1: abstraction of a hydrogen atom to form $CH_3CH_2\cdot$ and HCl, followed by abstraction of a chlorine atom by $CH_3CH_2\cdot$ to form CH_3CH_2Cl and a chlorine radical (Cl·). The $\Delta H°$ for each step is negative, making the overall $\Delta H°$ negative and the reaction exothermic. Because the transition state for the first propagation step is higher in energy than the transition state for the second propagation step, the **first step is rate-determining.** Both of these facts are illustrated in the energy diagram in Figure 13.4.

Problem 13.7 Calculate $\Delta H°$ for the rate-determining step of the reaction of CH_4 with I_2. Explain why this result illustrates that this reaction is extremely slow.

Figure 13.4 Energy diagram for the propagation steps in the chlorination of ethane

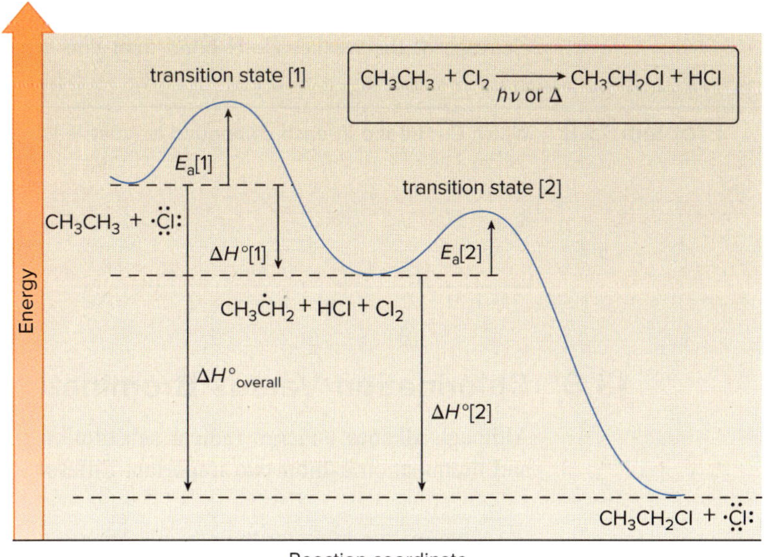

- Because radical halogenation consists of two propagation steps, the energy diagram has two energy barriers.
- The **first step is rate-determining** because its transition state is at higher energy.
- The **reaction is exothermic** because $\Delta H°_{overall}$ is negative.

13.5 Chlorination of Other Alkanes

Recall from Section 13.3 that the chlorination of $CH_3CH_2CH_3$ affords a 1:1 mixture of $CH_3CH_2CH_2Cl$ (formed by removal of a 1° hydrogen) and $(CH_3)_2CHCl$ (formed by removal of a 2° hydrogen).

$CH_3CH_2CH_3$ (1° H's, 2° H's) + Cl_2 $\xrightarrow{h\nu \text{ or } \Delta}$ $CH_3CH_2CH_2Cl$ + $(CH_3)_2CHCl$

	six 1° H's		two 2° H's
expected ratio	3	:	1
observed ratio	1	:	1
	less of this product		**more** of this product

$CH_3CH_2CH_3$ has six 1° hydrogen atoms and only two 2° hydrogens, so the expected product ratio of $CH_3CH_2CH_2Cl$ to $(CH_3)_2CHCl$ (assuming all hydrogens are *equally* reactive) is 3:1. Because the observed ratio is 1:1, however, the 2° C–H bonds must be *more* reactive; that is, **it must be easier to homolytically cleave a 2° C–H bond than a 1° C–H bond.** Recall from Section 13.2 that 2° C–H bonds are *weaker* than 1° C–H bonds. Thus,

- **The *weaker* the C–H bond, the *more readily* the hydrogen atom is removed in radical halogenation.**

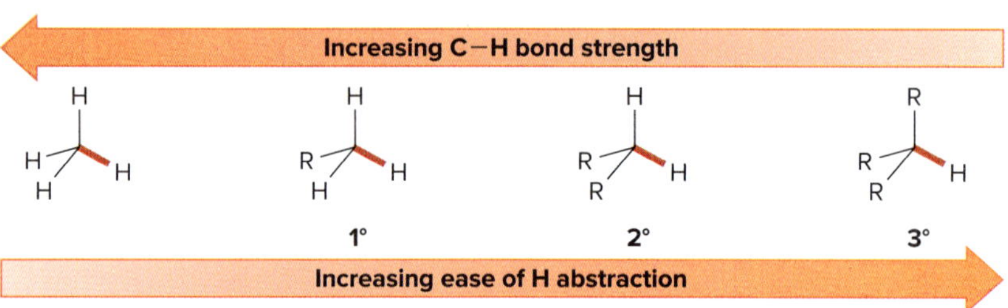

When alkanes react with Cl_2, a mixture of products results, with more product formed by cleavage of the weaker C–H bond than you would expect on statistical grounds.

Problem 13.8 Which C–H bond in each compound is most readily broken during radical halogenation?

13.6 Chlorination Versus Bromination

Although alkanes undergo radical substitution reactions with both Cl_2 and Br_2, chlorination and bromination exhibit two important differences:

- **Chlorination is *faster* than bromination.**
- **Although chlorination is *unselective*, yielding a mixture of products, bromination is often *selective*, yielding one major product.**

For example, propane reacts rapidly with Cl_2 to form a 1:1 mixture of 1° and 2° alkyl chlorides. On the other hand, propane reacts with Br_2 much more slowly and forms 99% $(CH_3)_2CHBr$.

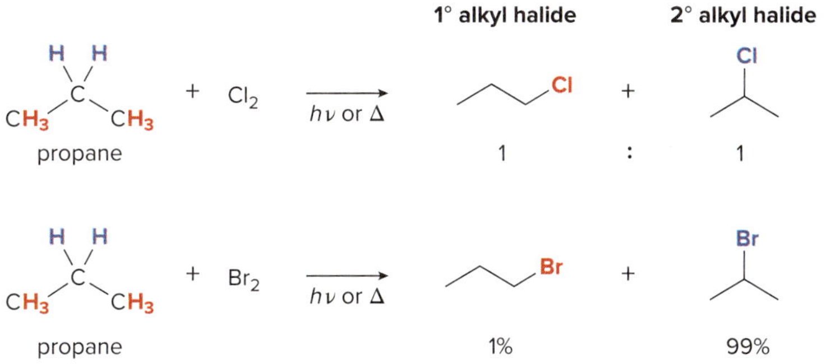

1° alkyl halide | 2° alkyl halide

propane + Cl_2 $\xrightarrow{h\nu \text{ or } \Delta}$ 1 : 1 — **Chlorination is fast and unselective.**

propane + Br_2 $\xrightarrow{h\nu \text{ or } \Delta}$ 1% 99% — **Bromination is slow and selective.**

- **In bromination, the major (and sometimes exclusive) product results from cleavage of the *weakest* C–H bond.**

Sample Problem 13.2 Drawing the Product of Bromination of an Alkane

Draw the major product formed when 3-ethylpentane is heated with Br_2.

Solution

Keep in mind: **the *more substituted* the carbon atom, the *weaker* the C–H bond.** The major bromination product in 3-ethylpentane is formed by cleavage of the sole **3° C–H bond,** its weakest C–H bond.

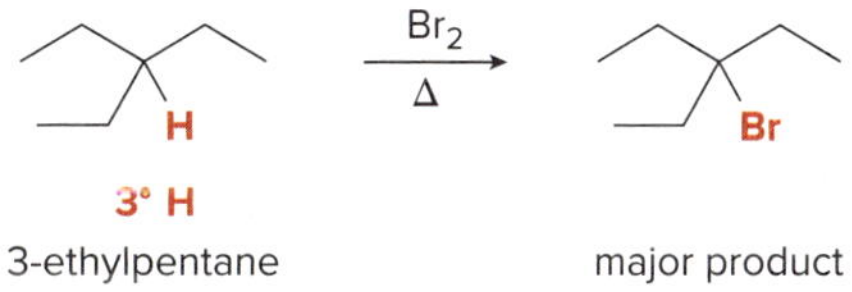

3° H — 3-ethylpentane $\xrightarrow[\Delta]{Br_2}$ major product

Problem 13.9 Draw the major product formed when each cycloalkane is heated with Br_2.

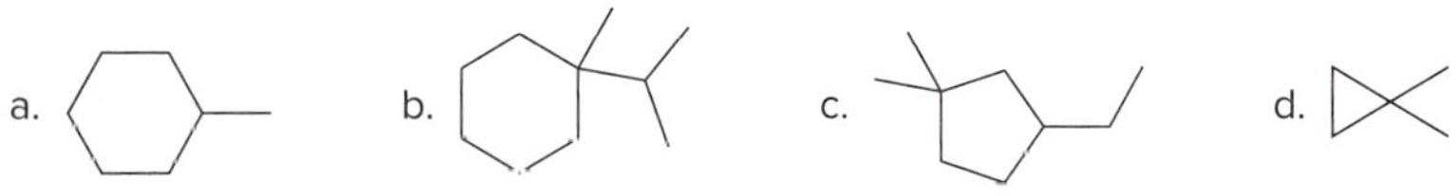

a. b. c. d.

More Practice: Try Problems 13.28b, 13.36, 13.46b

To explain the difference between chlorination and bromination, we return to the Hammond postulate (Section 7.13). The **rate-determining step in halogenation is the abstraction of a hydrogen atom by the halogen radical,** so we must compare these steps for bromination and chlorination. Keep in mind:

- **Transition states in endothermic reactions resemble the *products*. The more stable product is formed faster.**
- **Transition states in exothermic reactions resemble the *starting materials*. The relative stability of the products does not greatly affect the relative energy of the transition states, so a mixture of products often results.**

Bromination: $CH_3CH_2CH_3 + Br_2$

A bromine radical can abstract either a 1° or a 2° hydrogen from propane, generating either a 1° radical or a 2° radical. Calculating $\Delta H°$ using bond dissociation energies reveals that both reactions are ***endothermic***, but **it takes *less energy* to form the *more stable* 2° radical.**

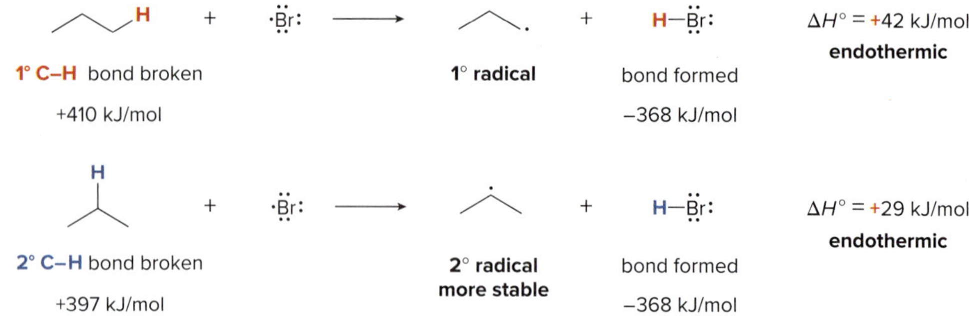

According to the Hammond postulate, the transition state of an endothermic reaction resembles the *products,* so the energy of activation to form the more stable 2° radical is lower and it is formed faster, as shown in the energy diagram in Figure 13.5. Because the 2° radical [$(CH_3)_2CH\cdot$] is converted to 2-bromopropane [$(CH_3)_2CHBr$] in the second propagation step, this **2° alkyl halide is the major product of bromination.**

- **Conclusion: Because the rate-determining step in bromination is *endothermic,* the *more stable* radical is formed faster, and often a single radical halogenation product predominates.**

Figure 13.5
Energy diagram for a selective endothermic reaction

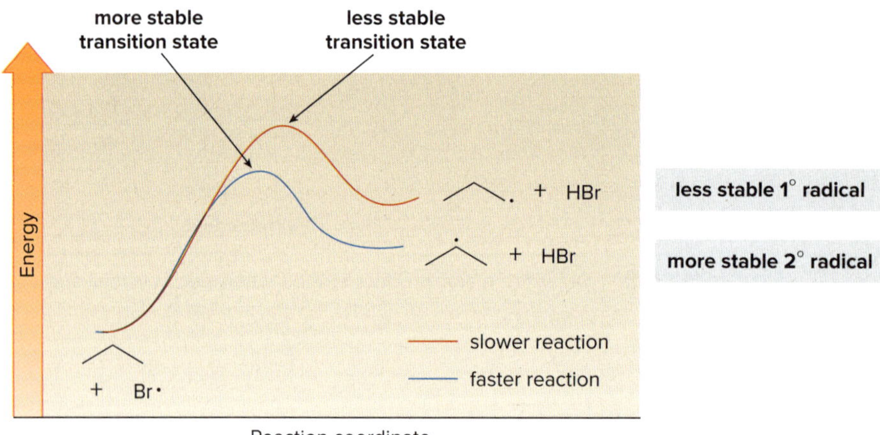

- The transition state to form the less stable 1° radical ($CH_3CH_2CH_2\cdot$) is higher in energy than the transition state to form the more stable 2° radical [$(CH_3)_2CH\cdot$]. Thus, **the 2° radical is formed faster.**

Chlorination: $CH_3CH_2CH_3 + Cl_2$

A chlorine radical can also abstract either a 1° or a 2° hydrogen from propane, generating either a 1° radical or a 2° radical. Calculating $\Delta H°$ using bond dissociation energies reveals that both reactions are **exothermic.**

(propane, 1° H) + ·Cl: ⟶ 1° radical + H–Cl: $\Delta H° = -21$ kJ/mol **exothermic**

1° C–H bond broken +410 kJ/mol; **1° radical**; bond formed −431 kJ/mol

(propane, 2° H) + ·Cl: ⟶ 2° radical + H–Cl: $\Delta H° = -34$ kJ/mol **exothermic**

2° C–H bond broken +397 kJ/mol; **2° radical**; bond formed −431 kJ/mol

Figure 13.6

Energy diagram for a nonselective exothermic reaction

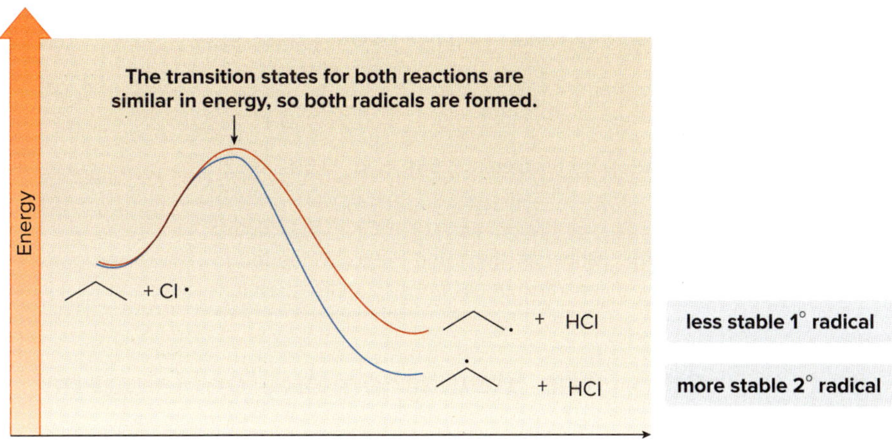

Because chlorination has an *exothermic* rate-determining step, the transition state to form both radicals **resembles the same starting material,** $CH_3CH_2CH_3$. As a result, the relative stability of the two radicals is much less important and **both radicals are formed.** An energy diagram for these processes is drawn in Figure 13.6. Because the 1° and 2° radicals are converted to 1-chloropropane ($CH_3CH_2CH_2Cl$) and 2-chloropropane [$(CH_3)_2CHCl$], respectively, in the second propagation step, **both alkyl halides are formed in chlorination.**

- **Conclusion: Because the rate-determining step in chlorination is *exothermic,* the transition state resembles the starting material, both radicals are formed, and a *mixture* of products results.**

Problem 13.10 Reaction of $(CH_3)_3CH$ with Cl_2 forms two products: $(CH_3)_2CHCH_2Cl$ (63%) and $(CH_3)_3CCl$ (37%). Why is the major product formed by cleavage of the stronger 1° C–H bond?

Problem 13.11 What alkane is needed to make each alkyl halide by radical halogenation?

a. Br b. Br c. Cl

13.7 Halogenation as a Tool in Organic Synthesis

Halogenation is a useful tool because it adds a functional group to a previously unfunctionalized molecule, making an **alkyl halide.** These alkyl halides can then be converted to alkenes by elimination, and to alcohols and ethers by nucleophilic substitution.

Sample Problem 13.3 Using Halogenation in Synthesis

Show how cyclohexane can be converted to cyclohexene by a stepwise sequence.

cyclohexane → cyclohexene

Solution

There is no one-step method to convert an alkane to an alkene. A two-step method is needed:

[1] **Radical halogenation** produces an alkyl halide.

[2] **Elimination of HCl** with a strong base produces cyclohexene.

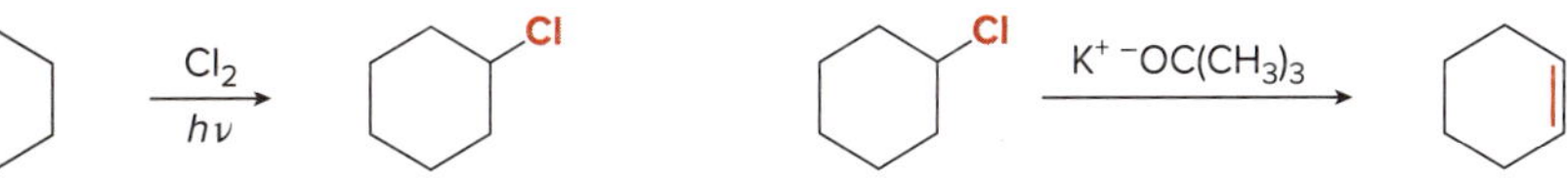

Problem 13.12 Synthesize each compound from $(CH_3)_3CH$.

a. b. OH c. OH

More Practice: Try Problems 13.58a, b, d; 13.60–13.62.

Problem 13.13 Show all steps and reagents needed to convert cyclohexane into each compound: (a) the two enantiomers of *trans*-1,2-dibromocyclohexane; and (b) 1,2-epoxycyclohexane.

13.8 The Stereochemistry of Halogenation Reactions

The stereochemistry of a reaction product depends on whether the reaction occurs at a stereogenic center or at another atom, and whether a new stereogenic center is formed. The rules predicting the stereochemistry of reaction products are summarized in Table 13.1.

Table 13.1 Rules for Predicting the Stereochemistry of Reaction Products

Starting material	Result
Achiral	• An achiral starting material always gives either an achiral or a racemic product.
Chiral	• If a reaction does not occur at a stereogenic center, the configuration at a stereogenic center is *retained* in the product. • If a reaction occurs at a stereogenic center, we must know the *mechanism* to predict the stereochemistry of the product.

13.8A Halogenation of an Achiral Starting Material

Halogenation of the **achiral starting material $CH_3CH_2CH_2CH_3$** forms two constitutional isomers by replacement of either a 1° or 2° hydrogen.

butane + Cl_2 —$h\nu$→ 1-chlorobutane (**achiral product**) + 2-chlorobutane (**new stereogenic center**)

enantiomers

- 1-Chlorobutane ($CH_3CH_2CH_2CH_2Cl$) has no stereogenic center, so it is an **achiral** compound.
- 2-Chlorobutane [$CH_3CH(Cl)CH_2CH_3$] has a new stereogenic center, so an **equal amount of two enantiomers** must form—**a racemic mixture.**

A racemic mixture results when a new stereogenic center is formed because the first propagation step generates a **planar, sp^2 hybridized radical.** Cl_2 then reacts with the planar radical from either the front or back side to form an equal amount of two enantiomers.

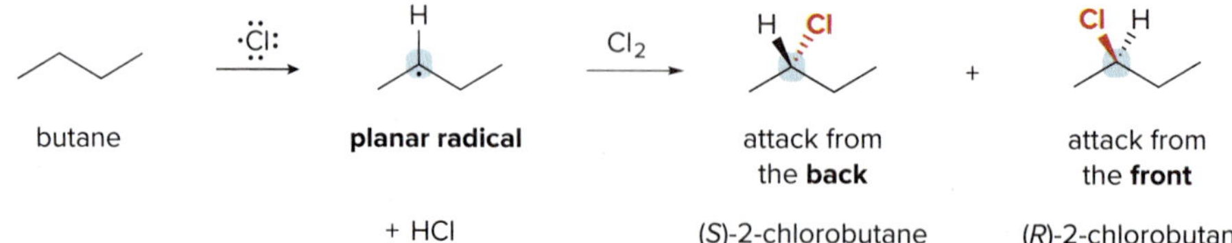

Thus, the achiral starting material butane forms an achiral product (1-chlorobutane) and a racemic mixture of two enantiomers [(*R*)- and (*S*)-2-chlorobutane].

13.8B Halogenation of a Chiral Starting Material

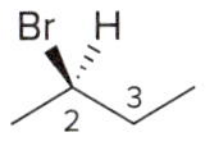

(*R*)-2-bromobutane

Let's now examine chlorination of the chiral starting material (*R*)-2-bromobutane at C2 and C3.

Chlorination at C2 occurs at the stereogenic center. Abstraction of a hydrogen atom at C2 forms a trigonal planar sp^2 hybridized radical that is now *achiral*. This achiral radical then reacts with Cl_2 from either side to form a new stereogenic center, resulting in an **equal amount of two enantiomers—a racemic mixture.**

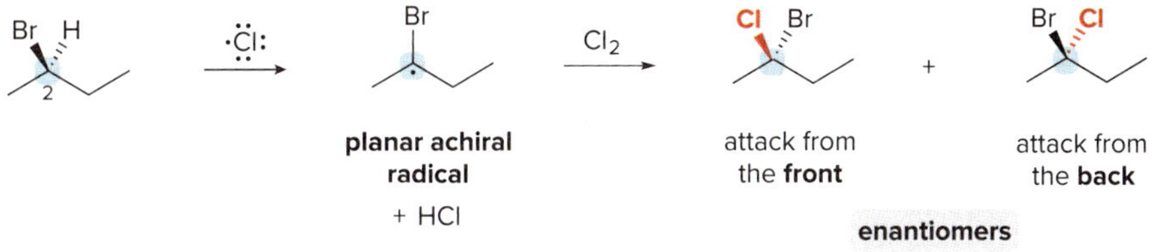

planar achiral radical

+ HCl

attack from the **front**

attack from the **back**

enantiomers

- **Radical halogenation reactions occur with *racemization* at a stereogenic center.**

Chlorination at C3 does *not* occur at the stereogenic center, but it forms a new stereogenic center. Because no bond is broken to the stereogenic center at C2, **its configuration is *retained*** during the reaction. Abstraction of a hydrogen atom at C3 forms a **trigonal planar** sp^2 hybridized radical that still contains this stereogenic center. Reaction of the radical with Cl_2 from either side forms a new stereogenic center, so the products have two stereogenic centers: the configuration at C2 is the *same* in both compounds, but the configuration at C3 is *different,* making them **diastereomers.**

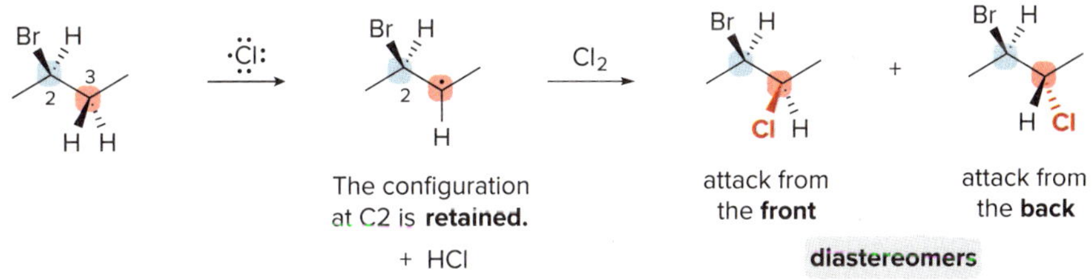

The configuration at C2 is **retained.**

+ HCl

attack from the **front**

attack from the **back**

diastereomers

Thus, four isomers are formed by chlorination of (*R*)-2-bromobutane at C2 and C3. Attack at the stereogenic center (C2) gives a product with one stereogenic center, resulting in a mixture of enantiomers. Attack at C3 forms a new stereogenic center, giving a mixture of diastereomers.

Sample Problem 13.4 Drawing All Stereoisomers Formed by Monochlorination

Draw all stereoisomers formed by monochlorination of **A.**

A

Solution

Look at each C bonded to H's *separately*, and consider whether the reaction occurs *at* a stereogenic center or if it *forms* a stereogenic center. The reactant **A** contains one stereogenic center, making it chiral.

The CH_3 in blue is *not* a stereogenic center, and substitution of a H atom by Cl does *not* form a new stereogenic center. One stereoisomer **B** is formed, and the configuration of the stereogenic center is retained.

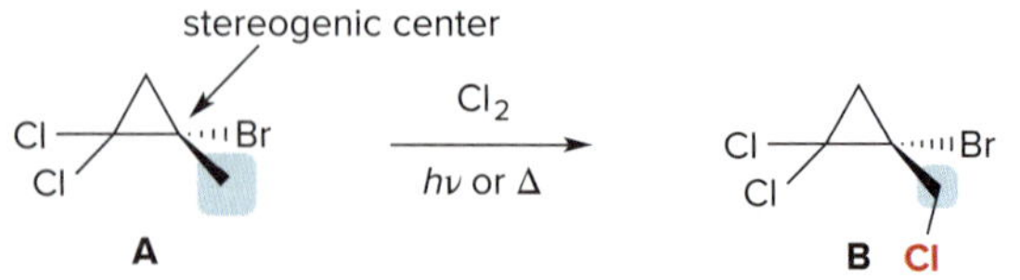

stereogenic center

A $\xrightarrow[h\nu \text{ or } \Delta]{Cl_2}$ **B**

The CH_2 in red is *not* a stereogenic center, but substitution of a H atom by Cl forms a *new* stereogenic center, and the new bond to Cl can form from either above or below the planar radical intermediate. Two products, **C** and **D,** are diastereomers.

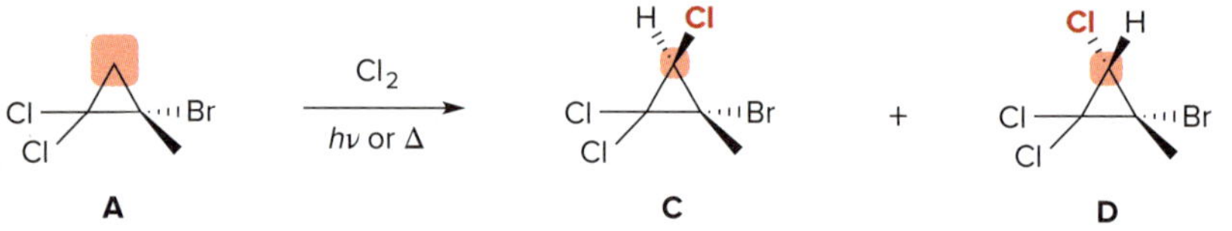

A $\xrightarrow[h\nu \text{ or } \Delta]{Cl_2}$ **C** + **D**

Cl and Br are **trans.** Cl and Br are **cis.**

diastereomers

Thus, chlorination of **A** forms three products, **B, C,** and **D.**

Problem 13.14 What products are formed from monochlorination of (*R*)-2-bromobutane at C1 and C4? Assign *R* and *S* designations to each stereogenic center.

More Practice: Try Problems 13.48a, d; 13.49–13.51.

Problem 13.15 Draw the monochlorination products formed when each compound is heated with Cl_2. Include the stereochemistry at any stereogenic center.

a.  b. 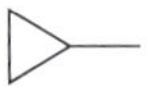 c. 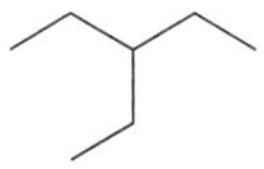 d. 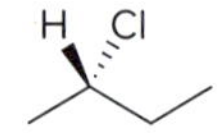 (Consider attack at C2 and C3 only.)

The 1995 Nobel Prize in Chemistry was awarded to Mario Molina, Paul Crutzen, and F. Sherwood Rowland for their work in elucidating the interaction of ozone with CFCs.

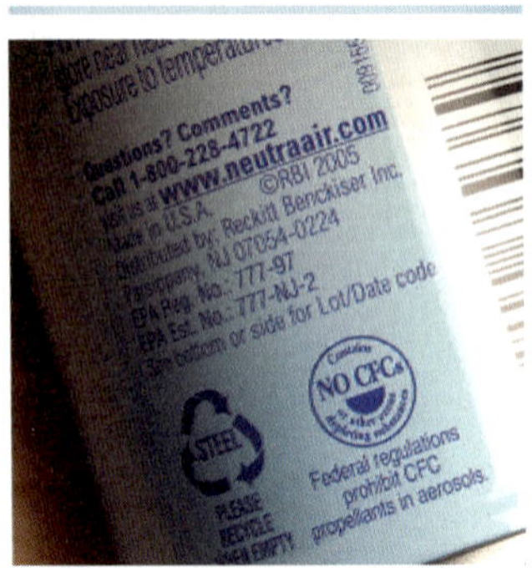

Propane and butane are now used as propellants in spray cans in place of CFCs.
Jill Braaten/McGraw Hill

13.9 Application: The Ozone Layer and CFCs

Ozone is formed in the upper atmosphere by reaction of oxygen molecules with oxygen atoms. Ozone is also decomposed with sunlight back to these same two species. The overall result of these reactions is to convert high-energy ultraviolet light into heat.

Ozone synthesis: $O_2 + \cdot\ddot{O}\cdot \longrightarrow O_3 + \text{heat}$ (ozone)

Ozone decomposition: $O_3 \xrightarrow{h\nu} O_2 + \cdot\ddot{O}\cdot$ (ozone)

Ozone is vital to life; it acts like a shield, protecting the earth's surface from destructive ultraviolet radiation. A decrease in ozone concentration in this protective layer would have some immediate consequences, including an increase in the incidence of skin cancer and eye cataracts. Other long-term effects include a reduced immune response, interference with photosynthesis in plants, and harmful effects on the growth of plankton, the mainstay of the ocean food chain.

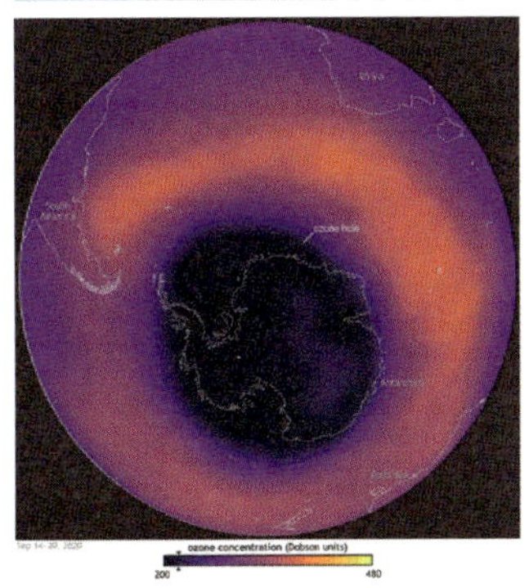

O_3 destruction is most severe in the region of the South Pole, where an ozone hole is visible with satellite imaging. The declining levels of ozone-depleting compounds have had a positive effect on the size of the ozone hole in the southern hemisphere, but much more improvement is needed. *NOAA*

Current research suggests that **chlorofluorocarbons (CFCs)** are responsible for destroying ozone in the upper atmosphere. **CFCs** are simple halogen-containing organic compounds manufactured under the trade name Freons.

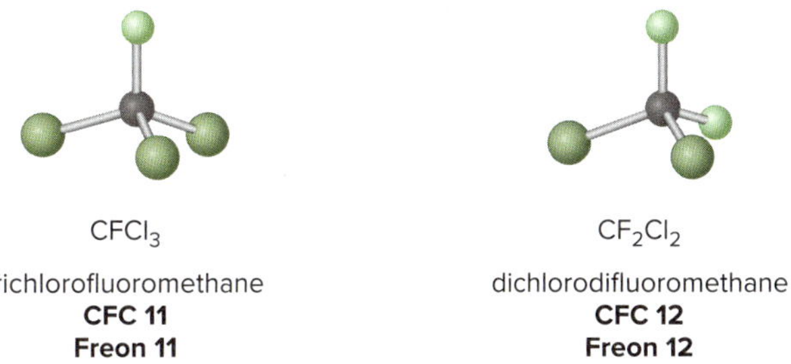

CFCs are inert, odorless, and nontoxic, and they have been used as refrigerants, solvents, and aerosol propellants. Because CFCs are volatile and water insoluble, they readily escape into the upper atmosphere, where they are decomposed by high-energy sunlight to form radicals that destroy ozone by the radical chain mechanism shown in Figure 13.7.

Figure 13.7 CFCs and the destruction of the ozone layer

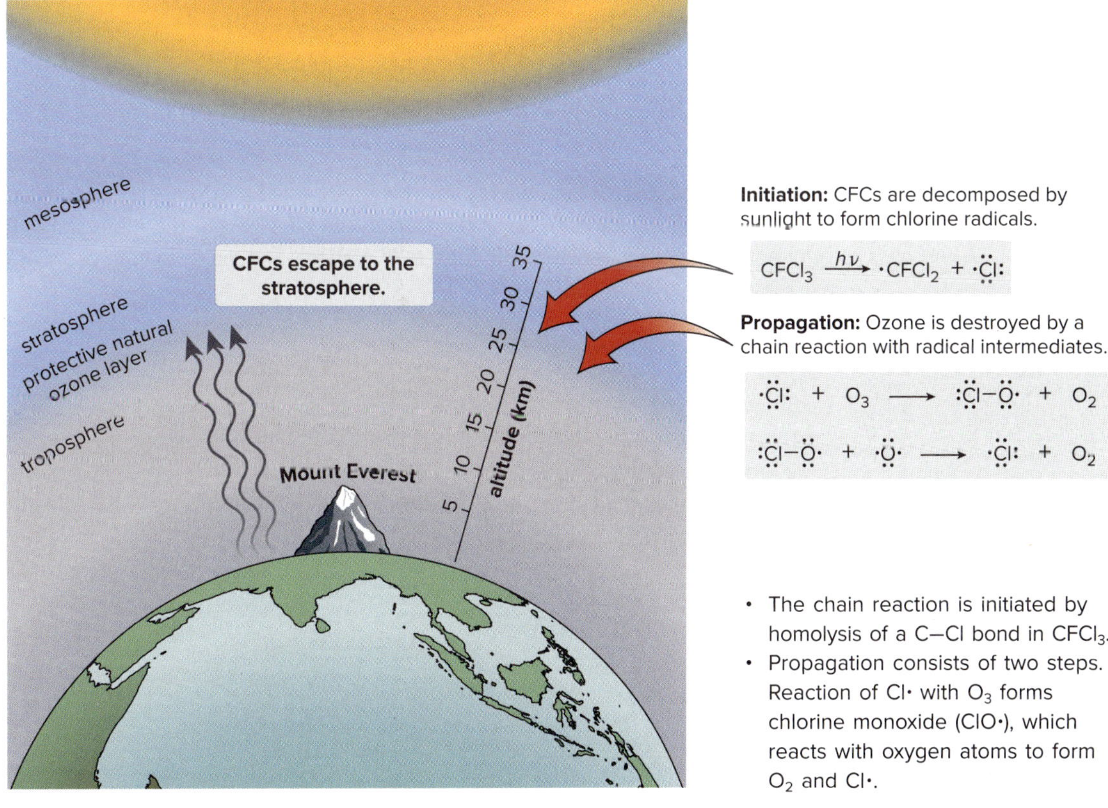

Initiation: CFCs are decomposed by sunlight to form chlorine radicals.

$$CFCl_3 \xrightarrow{h\nu} \cdot CFCl_2 + \cdot\ddot{Cl}:$$

Propagation: Ozone is destroyed by a chain reaction with radical intermediates.

$$\cdot\ddot{Cl}: + O_3 \longrightarrow :\ddot{Cl}-\ddot{O}\cdot + O_2$$

$$:\ddot{Cl}-\ddot{O}\cdot + \cdot\ddot{O}\cdot \longrightarrow \cdot\ddot{Cl}: + O_2$$

- The chain reaction is initiated by homolysis of a C–Cl bond in $CFCl_3$.
- Propagation consists of two steps. Reaction of Cl· with O_3 forms chlorine monoxide (ClO·), which reacts with oxygen atoms to form O_2 and Cl·.

The overall result is that O_3 is consumed as a reactant and O_2 molecules are formed. In this way, a small amount of CFC can destroy a large amount of O_3. These findings led to a ban on the use of CFCs in aerosol propellants in the United States in 1978 and to the phasing out of their use in refrigeration systems.

Newer alternatives to CFCs are **hydrofluorocarbons (HFCs)** such as CH_2FCF_3 and **hydrofluoroolefins (HFOs)** such as $CH_2{=}CFCF_3$. Although HFCs have little impact on the ozone layer, they do have significant global warming potential. As a result, the U.S. Environmental

Protection Agency issued an order in late 2021 phasing out HFCs. Because they are decomposed before reaching the ozone layer and have little global warming potential, HFOs are currently considered the safest CFC alternative.

CH_2FCF_3 **HFC-134a** 2,3,3,3-tetrafluoropropene **HFO-1234yf**

Problem 13.16 Asthma inhalers currently use hydrofluoroalkanes such as 1,1,1,2-tetrafluoroethane (CH_2FCF_3) as propellants. CH_2FCF_3 is decomposed before it reaches the stratosphere by abstraction of a hydrogen atom by the hydroxy radical (·OH). Draw the products of this reaction.

13.10 Radical Halogenation at an Allylic Carbon

Now let's examine radical halogenation at an ***allylic carbon*—the carbon adjacent to a double bond.** Homolysis of the allylic C–H bond of propene generates the **allyl radical,** which has an unpaired electron on the carbon adjacent to the double bond.

propene → allyl radical + ·H $\Delta H° = +364$ kJ/mol

propene allylic C–H (in red) allyl radical

The bond dissociation energy for this process (364 kJ/mol) is even less than that for a 3° C–H bond (381 kJ/mol). Because the weaker the C–H bond, the more stable the resulting radical, an **allyl radical is more stable than a 3° radical,** and the following order of radical stability results:

1° 2° 3° **allyl radical**

Increasing radical stability

The position of the atoms and the σ bonds stays the same in drawing resonance structures. Resonance structures differ in the location of only π bonds and nonbonded electrons.

The allyl radical is more stable than other radicals because two resonance structures can be drawn for it.

two resonance structures for the allyl radical **hybrid**

- **The "true" structure of the allyl radical is a hybrid of the two resonance structures. In the hybrid, the π bond and the unpaired electron are delocalized.**
- **Delocalizing electron density lowers the energy of the hybrid, thus stabilizing the allyl radical.**

Problem 13.17 Draw a second resonance structure for each radical. Then draw the hybrid.

13.10A Selective Bromination at Allylic C–H Bonds

Because allylic C–H bonds are *weaker* than other sp^3 hybridized C–H bonds, the **allylic carbon can be selectively halogenated** by using *N*-bromosuccinimide (**NBS,** Section 10.15) in the presence of light or peroxides. Under these conditions only the allylic C–H bond in cyclohexene reacts to form an allylic halide.

O
N–Br
O

N-bromosuccinimide
NBS

NBS
$h\nu$ or ROOR
allylic C
Br
allylic halide

NBS contains a weak N–Br bond that is homolytically cleaved with light to generate a bromine radical, initiating an allylic halogenation reaction. Propagation then consists of the usual two steps of radical halogenation as shown in Mechanism 13.2.

Mechanism 13.2 Allylic Bromination with NBS

Part [1] Initiation

NBS
$h\nu$
1

1 Homolysis of the weak N–Br bond with light energy forms a **Br· radical** that initiates radical halogenation.

Part [2] Propagation

2
allylic radical
+ H–Br:
(from NBS)
3

2 The Br· radical abstracts an allylic H to afford an **allylic radical.** (Only one resonance structure is drawn.)

3 The allylic radical reacts with Br_2 to form the **allylic halide.** The radical Br· formed in Step [3] can now react in Step [2], so Steps [2] and [3] can repeatedly occur without additional initiation.

Besides acting as a source of Br· to initiate the reaction, **NBS generates a low concentration of Br_2** needed in the second chain propagation step (Step [3] of the mechanism). The HBr formed in Step [2] reacts with NBS to form Br_2, which is then used for halogenation in Step [3] of the mechanism.

NBS + HBr → succinimide + **Br_2**

used in Step [3] of allylic bromination

A **low concentration of Br_2** (from NBS) **favors allylic substitution** (over addition) in part because bromine is needed for only *one* step of the mechanism. When Br_2 adds to a double bond, a low Br_2 concentration would first form a low concentration of bridged bromonium ion (Section 10.13), which must then react with more bromine (in the form of Br^-) in a second step to form a dibromide. **If concentrations of both intermediates—bromonium ion and Br^-—are low, the overall rate of addition is very slow.**

Thus, an alkene with allylic C–H bonds undergoes two different reactions depending on the reaction conditions.

- Treatment of cyclohexene with Br_2 (in an organic solvent like CCl_4) leads to **addition** via **ionic intermediates** (Section 10.13).
- Treatment of cyclohexene with NBS (+ $h\nu$ or ROOR) leads to **allylic substitution,** via **radical intermediates.**

Problem 13.18 Draw the products of each reaction.

13.10B Product Mixtures in Allylic Halogenation

Halogenation at an allylic carbon often results in a mixture of products. For example, bromination of 3-methylbut-1-ene under radical conditions forms a mixture of 3-bromo-3-methylbut-1-ene and 1-bromo-3-methylbut-2-ene.

A mixture is obtained because the reaction proceeds by way of a **resonance-stabilized radical.** Abstraction of an allylic hydrogen from the alkene with a Br· radical (from NBS) forms an allylic radical for which **two different Lewis structures** can be drawn.

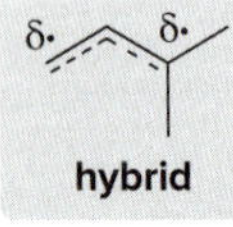

hybrid

As a result, two different C atoms have partial radical character (indicated by δ·), so that Br_2 reacts at two different sites and two allylic halides are formed.

- **Whenever two different resonance structures can be drawn for an allylic radical, two *different* allylic halides are formed by radical substitution.**

Sample Problem 13.5 Drawing the Products of Allylic Halogenation

Draw the products formed when **A** is treated with NBS + $h\nu$.

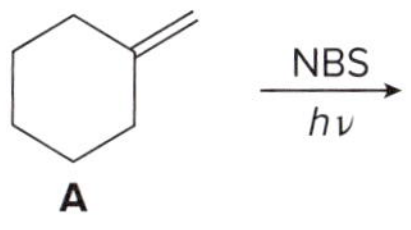

NBS / $h\nu$

A

Solution

Hydrogen abstraction at the allylic C forms a **resonance-stabilized radical** (with two different resonance structures) that reacts with Br_2 to form two constitutional isomers as products.

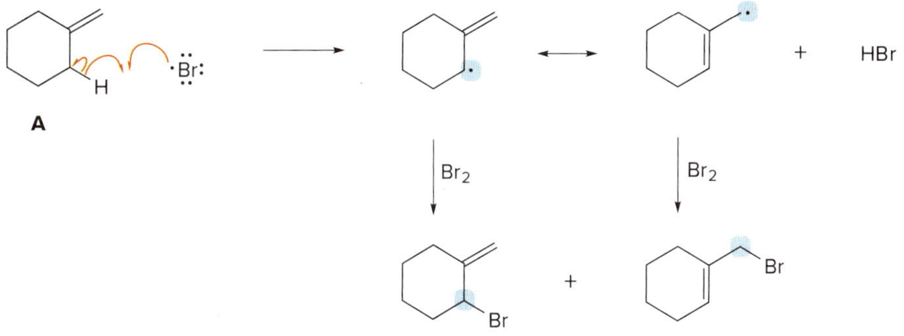

A ·Br: H + HBr Br_2 Br_2 Br + Br

two constitutional isomers

Problem 13.19 Draw all constitutional isomers formed when each alkene is treated with NBS + $h\nu$.

a. b. c.

More Practice: Try Problems 13.43, 13.44, 13.46f.

Problem 13.20 Draw the structure of the four allylic halides formed when 3-methylcyclohexene undergoes allylic halogenation with NBS + $h\nu$.

13.11 Application: Oxidation of Unsaturated Lipids

Oils—triacylglycerols having one or more sites of unsaturation in their long carbon chains—are susceptible to oxidation at their allylic carbon atoms. Oxidation occurs by way of a radical chain mechanism, as shown in Figure 13.8.

- **Step [1]** Oxygen in the air abstracts an allylic hydrogen atom to form an allylic radical because the allylic C–H bond is weaker than the other C–H bonds.
- **Step [2]** The allylic radical reacts with another molecule of O_2 to form a peroxy radical.
- **Step [3]** The peroxy radical abstracts an allylic hydrogen from another lipid molecule to form a hydroperoxide and another allylic radical that continues the chain. Steps [2] and [3] can repeat again and again until some other radical terminates the chain.

The hydroperoxides formed by this process are unstable and decompose to other oxidation products, many of which have a disagreeable odor and taste. **This process turns an oil rancid. Unsaturated lipids are more easily oxidized than saturated ones** because they contain **weak allylic C–H bonds** that are readily cleaved in Step [1] of this reaction, forming resonance-stabilized allylic radicals. Because saturated fats have no double bonds and thus no weak allylic C–H bonds, they are much less susceptible to air oxidation, resulting in increased shelf life of products containing them.

Figure 13.8 The oxidation of unsaturated lipids with O_2

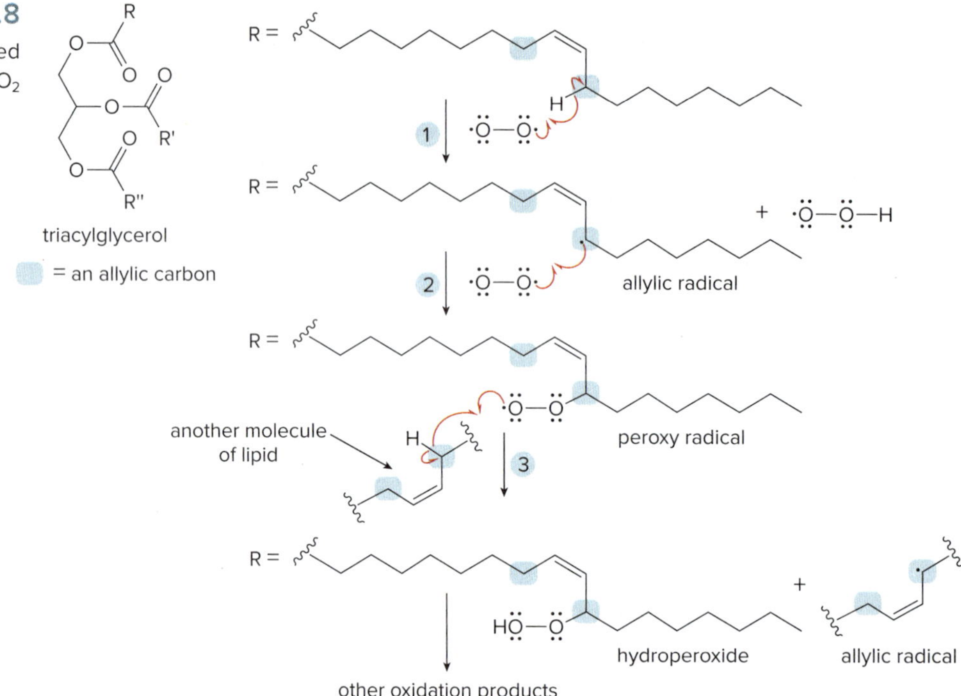

This allylic radical continues the chain. Steps [2] and [3] can be repeated again and again.

- Oxidation is shown at one allylic carbon only. Reaction at the other labeled allylic carbon is also possible.

Problem 13.21 Which C–H bond is most readily cleaved in linoleic acid? Draw all possible resonance structures for the resulting radical. Draw all the hydroperoxides formed by reaction of this resonance-stabilized radical with O_2.

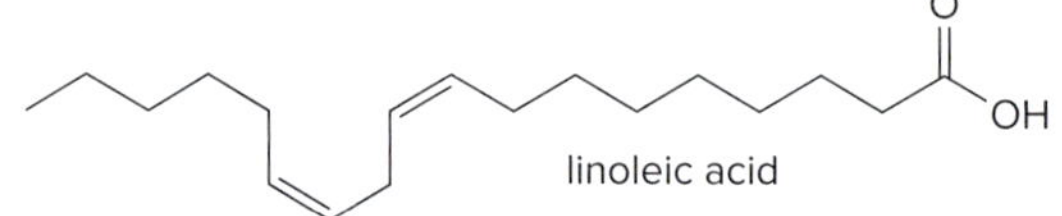

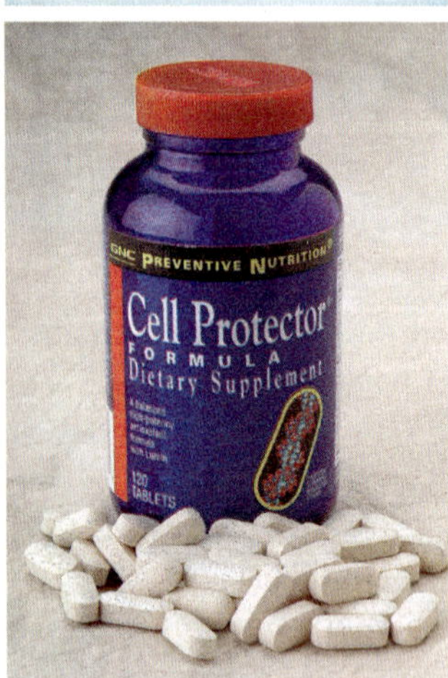

The purported health benefits of antioxidants have made them a popular component in anti-aging formulations.
Elite Images/McGraw Hill

13.12 Application: Antioxidants

An *antioxidant* is a compound that stops an oxidation reaction from occurring.

- Naturally occurring antioxidants such as **vitamin E** prevent radical reactions that can cause cell damage.
- Synthetic antioxidants such as **BHT**—**b**utylated **h**ydroxy **t**oluene—are added to packaged and prepared foods to prevent oxidation and spoilage.

HO — O — **vitamin E**

OH — **BHT** (butylated hydroxy toluene)

Hazelnuts, almonds, and many other types of nuts are an excellent source of the natural antioxidant vitamin E.
Stockbyte/Corbis/Getty Images

Vitamin E and BHT are radical inhibitors, so they terminate radical chain mechanisms by reacting with radicals. How do they trap radicals? Both vitamin E and BHT use a hydroxy group bonded to a benzene ring—a general structure called a **phenol.**

Radicals (R·) abstract a hydrogen atom from the OH group of an antioxidant, forming a new resonance-stabilized radical. **This new radical does not participate in chain propagation,** but rather terminates the chain and halts the oxidation process. All phenols (including vitamin E and BHT) inhibit oxidation by this radical process.

phenol — R· abstracts the H atom from the OH group. → five resonance structures + R—H

R· = a general organic radical

The many nonpolar C–C and C–H bonds of vitamin E make it fat soluble, and thus it dissolves in the nonpolar interior of the cell membrane, where it is thought to inhibit the oxidation of the unsaturated fatty acid residues in the phospholipids. Oxidative damage to lipids in cells via radical mechanisms is thought to play an important role in the aging process. For this reason, many anti-aging formulas with antioxidants like vitamin E are now popular consumer products.

Problem 13.22 Rosmarinic acid is an antioxidant isolated from rosemary. Draw resonance structures for the radical that results from removal of the labeled H atom in rosmarinic acid.

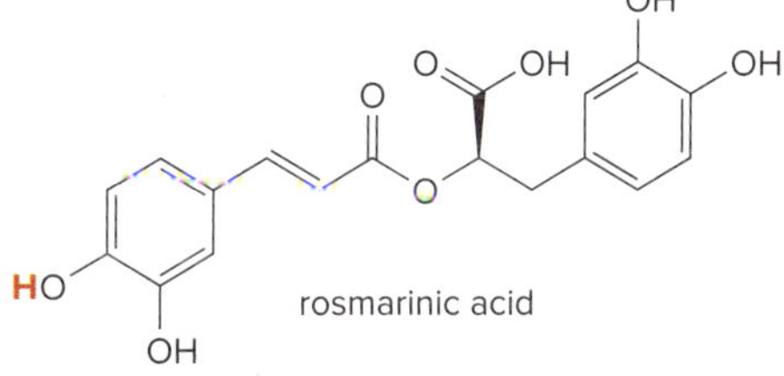

Rosemary extracts contain rosmarinic acid (Problem 13.22), an antioxidant that helps prevent the oxidation of unsaturated vegetable oils.
Daniel C. Smith

13.13 Radical Addition Reactions to Double Bonds

We now turn our attention to the second common reaction of radicals, addition to double bonds. Because an alkene contains an electron-rich, easily broken π bond, it reacts with an electron-deficient radical.

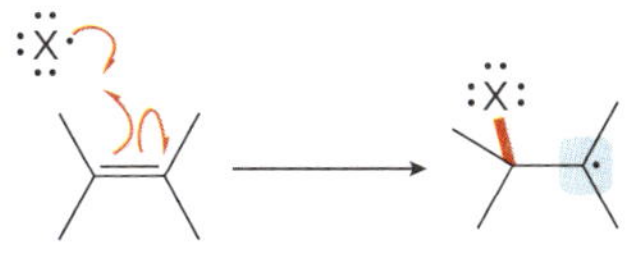

The π bond is broken. **new radical**

Radicals react with alkenes via a radical chain mechanism that consists of initiation, propagation, and termination steps analogous to those discussed previously for radical substitution.

13.13A Addition of HBr

HBr adds to alkenes to form alkyl bromides in the presence of light, heat, or peroxides.

A π bond is broken. + H—Br $\xrightarrow{h\nu,\ \Delta,\text{ or ROOR}}$ alkyl bromide (H, Br)

The regioselectivity of addition to an unsymmetrical alkene is *different* from the addition of HBr without added light, heat, or peroxides.

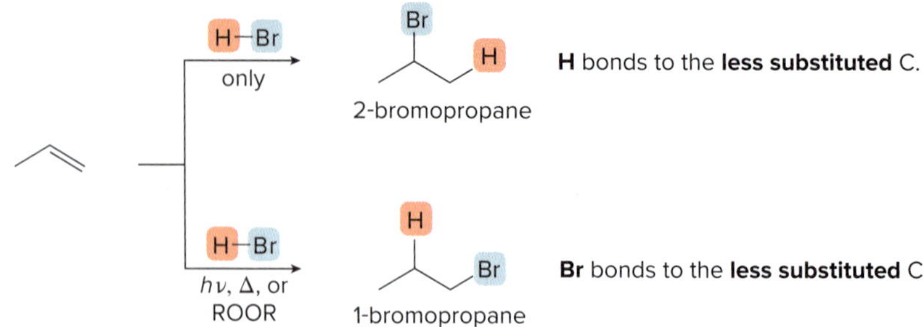

- HBr addition to propene *without* added light, heat, or peroxides gives 2-bromopropane: the **H atom is added to the *less* substituted carbon.** This reaction occurs via **carbocation** intermediates (Section 10.10).
- HBr addition to propene *with* added light, heat, or peroxides gives 1-bromopropane: the **Br atom is added to the *less* substituted carbon.** This reaction occurs via **radical** intermediates.

Problem 13.23 Draw the product(s) formed when each alkene is treated with either [1] HBr alone; or [2] HBr in the presence of peroxides.

a. b.  c.

13.13B The Mechanism of the Radical Addition of HBr to an Alkene

In the presence of added light, heat, or peroxides, HBr addition to an alkene forms radical intermediates and, like other radical reactions, proceeds by a mechanism with three distinct parts: initiation, propagation, and termination. Mechanism 13.3 is written for the reaction of $CH_3CH{=}CH_2$ with HBr and ROOR to form $CH_3CH_2CH_2Br$.

Mechanism 13.3 Radical Addition of HBr to an Alkene

Part [1] Initiation

RO—OR → (1) 2 RO· + H—Br: → (2) ROH + ·Br:

1–2 Initiation with ROOR occurs in two steps—**homolysis of the weak O—O bond** and abstraction of H to form a bromine radical.

Part [2] Propagation

·Br: → (3) **2° radical** (new bond shown in red) + H—Br: → (4) new bond shown in red + ·Br:

3 Addition of Br· to the terminal carbon forms a **2° radical.**

4 Abstraction of H from HBr forms a new C–H bond and a bromine radical, so Steps [3] and [4] can occur repeatedly.

Part [3] Termination

:Br· + ·Br: → (5) :Br—Br:

5 Termination of the chain occurs when any two radicals combine to form a bond.

The first propagation step (Step [3] of the mechanism, the addition of Br• to the double bond) is worthy of note. With propene there are two possible paths for this step, depending on which carbon atom of the double bond forms the new bond to bromine. Path [A] forms a less stable 1° radical, whereas Path [B] forms a more stable 2° radical. **The more stable 2° radical forms faster,** so Path [B] is preferred.

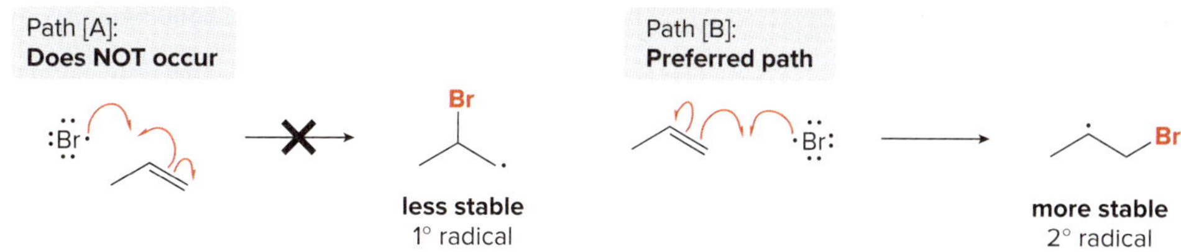

The mechanism also illustrates why the regioselectivity of HBr addition is different depending on the reaction conditions. In both reactions, H and Br add to the double bond, but the *order* of addition depends on the mechanism.

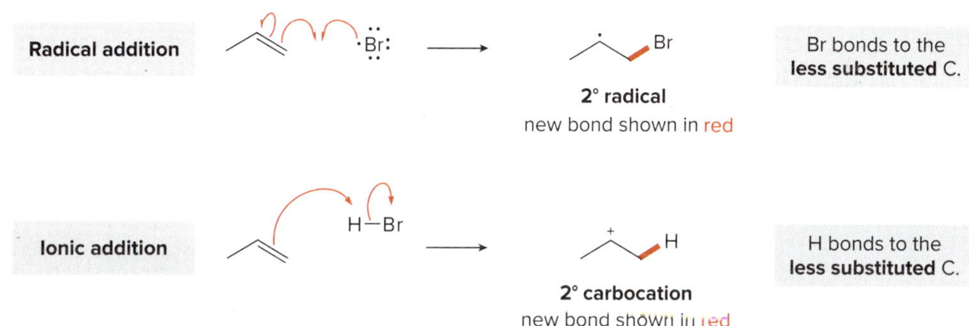

- **In radical addition (HBr with added light, heat, or ROOR),** ***Br• adds first*** **to generate the more stable radical.**
- **In ionic addition (HBr alone),** ***H^+ adds first*** **to generate the more stable carbocation.**

Problem 13.24 When HBr adds to $(CH_3)_2C=CH_2$ under radical conditions, two radicals are possible products in the first step of chain propagation. Draw the structure of both radicals and indicate which one is formed. Then draw the preferred product from HBr addition under radical conditions.

Problem 13.25 What reagents are needed to convert 1-ethylcyclohexene into (a) 1-bromo-2-ethylcyclohexane; (b) 1-bromo-1-ethylcyclohexane; (c) 1,2-dibromo-1-ethylcyclohexane?

13.13C Energy Changes in the Radical Addition of HBr

The energy changes during propagation in the radical addition of HBr to $CH_2=CH_2$ can be calculated from bond dissociation energies, as shown in Figure 13.9.

Both propagation steps for the addition of HBr are exothermic, so propagation is exothermic (energetically favorable) overall. For the addition of HCl or HI, however, one of the chain-propagating steps is quite endothermic and thus too difficult to be part of a repeating chain mechanism. Thus, **HBr adds to alkenes under radical conditions, but HCl and HI do not.**

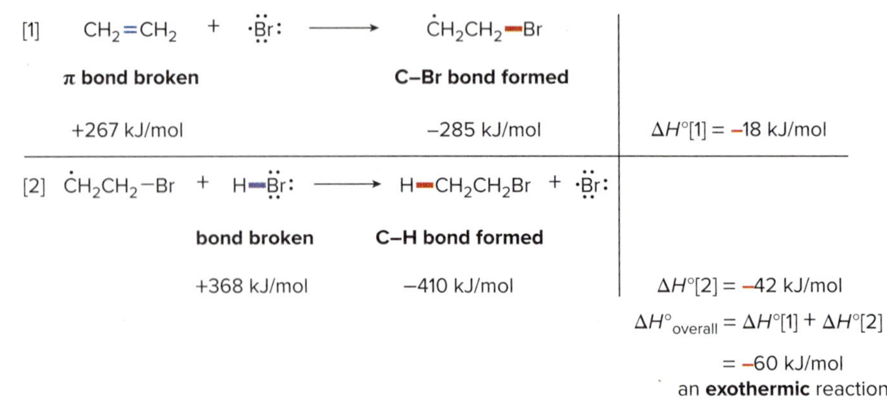

Figure 13.9 Energy changes during the propagation steps: $CH_2{=}CH_2 + HBr \rightarrow CH_3CH_2Br$

13.14 Polymers and Polymerization

HDPE (high-density polyethylene) and **LDPE** (low-density polyethylene) are two common types of polyethylene prepared under different reaction conditions and having different physical properties. HDPE is opaque and rigid, and is used in milk containers and water jugs. LDPE is less opaque and more flexible, and is used in plastic bags and electrical insulation. Products containing HDPE and LDPE (and other plastics) are often labeled with a symbol indicating recycling ease: the lower the number, the easier to recycle.
Jill Braaten/McGraw Hill

Polymers*—large molecules made up of repeating units of smaller molecules called *monomers—include such biologically important compounds as proteins and carbohydrates. They also include such industrially important plastics as polyethylene, poly(vinyl chloride) (PVC, mentioned in the chapter opener), and polystyrene.

13.14A Synthetic Polymers

Many synthetic polymers—that is, those synthesized in the lab—are among the most widely used organic compounds in modern society. Although some synthetic polymers resemble natural substances, many have different and unusual properties that make them more useful than naturally occurring materials. Soft drink bottles, plastic bags, food wrap, compact discs, Teflon, and Styrofoam are all made of synthetic polymers. In this section we examine polymers derived from alkene monomers. Chapter 31 is devoted to a detailed discussion of the synthesis and properties of several different types of synthetic polymers.

- ***Polymerization* is the joining together of monomers to make polymers.**

For example, joining **ethylene monomers** together forms the polymer **polyethylene,** a plastic used in milk containers and sandwich bags.

$CH_2{=}CH_2$ + $CH_2{=}CH_2$ + $CH_2{=}CH_2$ ⟶ (polymerization)

ethylene monomers

polyethylene polymer
new bonds shown in red

Many ethylene derivatives having the general structure **$CH_2{=}CHZ$** are also used as monomers for polymerization. The identity of Z affects the physical properties of the resulting polymer, making some polymers more suitable for one consumer product (e.g., plastic bags or food wrap) than another (e.g., soft drink bottles or compact discs). Polymerization of $CH_2{=}CHZ$ usually affords polymers with the Z groups on every other

Table 13.2 Common Monomers and Polymers Used in Medicine and Dentistry

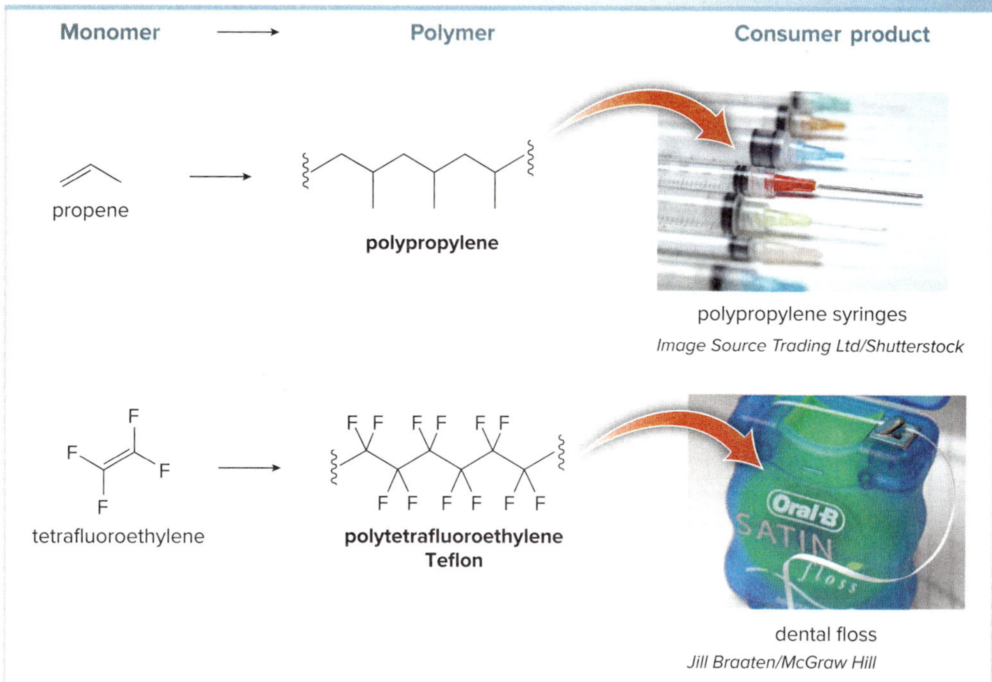

Poly(acrylic acid) (Sample Problem 13.6) is used in disposable diapers because it absorbs 30 times its weight in water. *Daniel C. Smith*

carbon atom in the chain. Table 13.2 lists two common monomers and polymers used in medicine and dentistry.

Z + Z + Z —polymerization→ Z Z Z

new bonds shown in red

Sample Problem 13.6 Drawing the Structure of a Polymer Formed from a Monomer

What polymer is formed when $CH_2{=}CHCO_2H$ (acrylic acid) is polymerized to form poly(acrylic acid)?

Solution

Draw three or more alkene monomers, **break one bond of each double bond, and join the alkenes together with single bonds.** With unsymmetrical alkenes, substituents are bonded to every other carbon.

O OH + O OH + O OH —polymerization→ O OH O OH O OH

Join a C labeled in blue with a C labeled in red.

poly(acrylic acid)

Problem 13.26 (a) Draw the structure of polystyrene, which is formed by polymerizing the monomer styrene, $C_6H_5CH{=}CH_2$. (b) What monomer is used to form polymers **A, B,** and **C?**

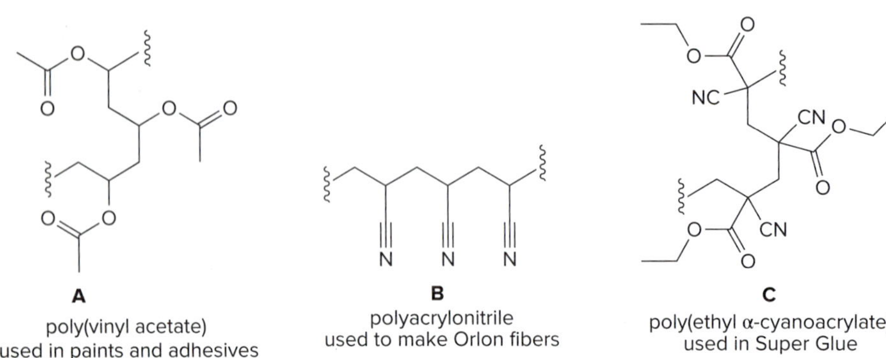

A poly(vinyl acetate) used in paints and adhesives

B polyacrylonitrile used to make Orlon fibers

C poly(ethyl α-cyanoacrylate) used in Super Glue

More Practice: Try Problems 13.68, 13.69, 13.71a.

The polystyrene foam (Problem 13.26a) used in packaging materials and drinking cups for hot beverages is called Styrofoam. Recycled polystyrene can be molded into trays and trash cans. *Jamie Grill/Getty Images*

The alkene monomers used in polymerization are prepared from petroleum.

13.14B Radical Polymerization

The polymers described in Section 13.14A are prepared by polymerization of alkene monomers by **adding a radical to a π bond.** The mechanism resembles the radical addition of HBr to an alkene, except that a **carbon radical rather than a bromine atom is added to the double bond.** Mechanism 13.4 is written with the general monomer $CH_2{=}CHZ$, and again has three parts: initiation, propagation, and termination.

Mechanism 13.4 Radical Polymerization of $CH_2{=}CHZ$

Part [1] Initiation

RÖ–ÖR →(1) 2 RÖ· + $CH_2{=}CHZ$ →(2) RO–CH_2–ĊHZ

(1) – (2) Initiation with ROOR occurs in two steps—homolysis of the **weak O–O bond** and addition of RO· to the alkene to form a carbon radical.

Part [2] Propagation

RO–CH_2–ĊHZ + $CH_2{=}CHZ$ →(3) RO–CH_2CHZ–CH_2–ĊHZ

(3) **Chain propagation consists of a single step.** The carbon radical adds to another alkene to form a new C–C bond and another carbon radical. Addition forms the radical with the unpaired electron on the atom with the Z substituent.

Part [3] Termination

RO–CH_2CHZCH_2ĊHZ + ·CHZCH_2CHZCH_2–OR →(4) RO–CH_2CHZCH_2CHZ–CHZCH_2CHZCH_2–OR

(4) Termination of the chain occurs when any two radicals combine to form a bond.

In radical polymerization, the more substituted radical always adds to the less substituted end of the monomer, a process called **head-to-tail polymerization.**

The **more substituted radical** adds to the **less substituted end** of the double bond.

The new radical is always located on the C bonded to Z.

Problem 13.27 Draw the steps of the mechanism that converts vinyl chloride ($CH_2{=}CHCl$) to poly(vinyl chloride).

Chapter 13 REVIEW

KEY CONCEPTS

Radical stability (13.1)

- A radical is a reactive intermediate with a single unpaired electron.
- The stability of a radical increases as the number of electron-donating groups, such as **alkyl** groups, bonded to the **radical carbon** increases.

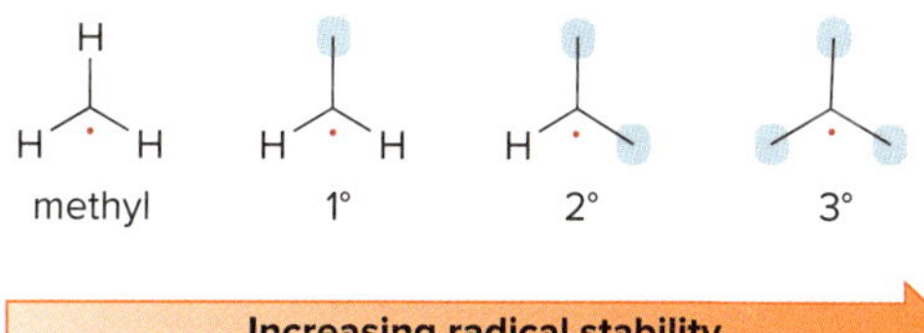

Increasing radical stability

Try Problems 13.30–13.33.

KEY REACTIONS

Radical Reactions

1. Cyclopentane–H → (X_2, $h\nu$ or Δ, X = Cl, Br) → alkyl halide; halogenation (13.4)

2. Cyclopentene (allylic H) → (NBS, $h\nu$ or ROOR) → allylic halide (Br); allylic halogenation (13.10)

3. 1-Methylcyclopentene → (H–Br, $h\nu$, Δ, or ROOR) → alkyl bromide; addition (13.13)

4. Vinyl chloride (CH₂=CHCl) → (ROOR) → polymer (Cl Cl Cl); polymerization (13.14)

Try Problems 13.28, 13.35, 13.36, 13.43, 13.44, 13.46, 13.68, 13.69, 13.71a.

KEY SKILLS

[1] Drawing all the constitutional isomers formed by monochlorination of methylcyclopentane with Cl_2 and heat (13.3)

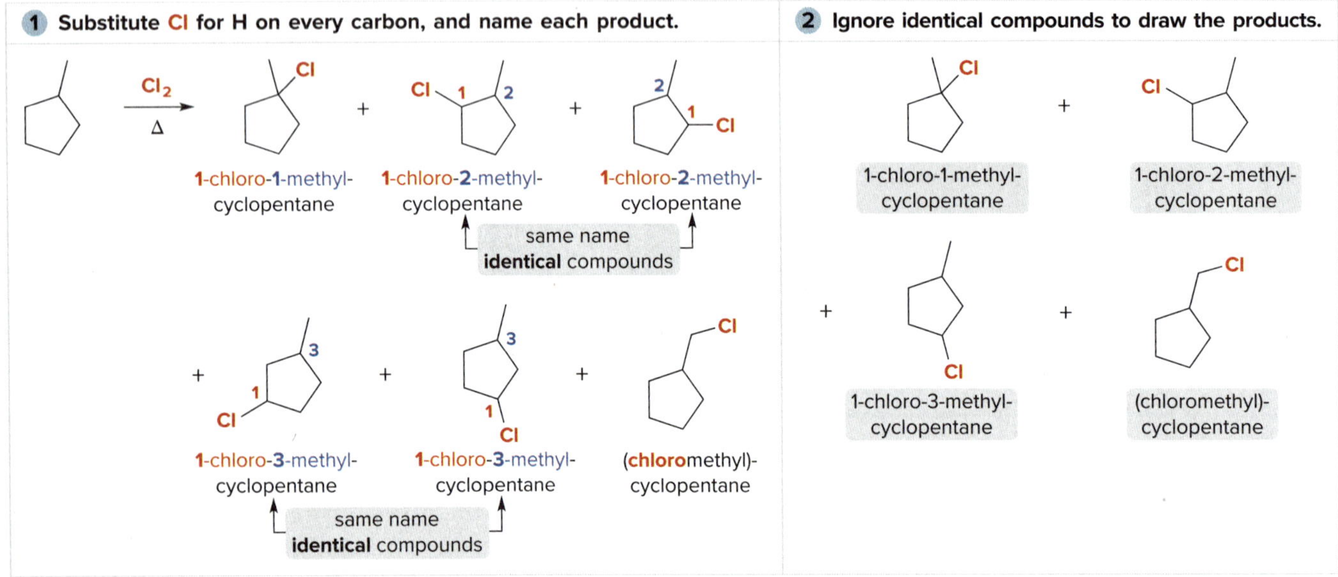

See Sample Problem 13.1. Try Problems 13.28a, 13.35, 13.46a.

[2] Drawing the major product formed by bromination of methylcyclopentane with Br_2 and $h\nu$ (13.6)

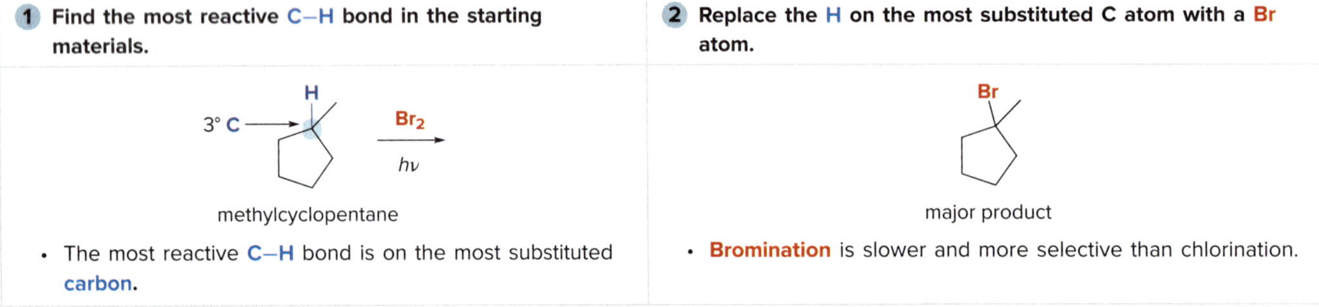

See Sample Problem 13.2. Try Problems 13.28b, 13.36, 13.46b.

[3] Devising a synthesis (13.7); example: $CH_3CH(SCH_3)CH_3$ from $CH_3CH_2CH_3$

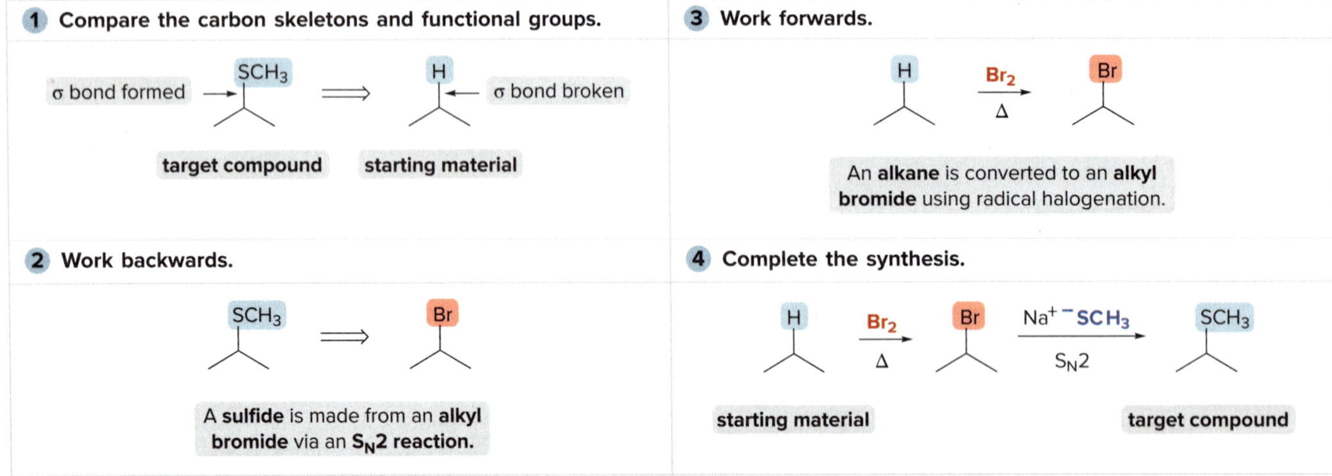

See Sample Problem 13.3. Try Problems 13.58–13.62.

[4] Drawing the stereoisomers formed by the monochlorination of a chiral starting material with Cl_2 and heat (13.8)

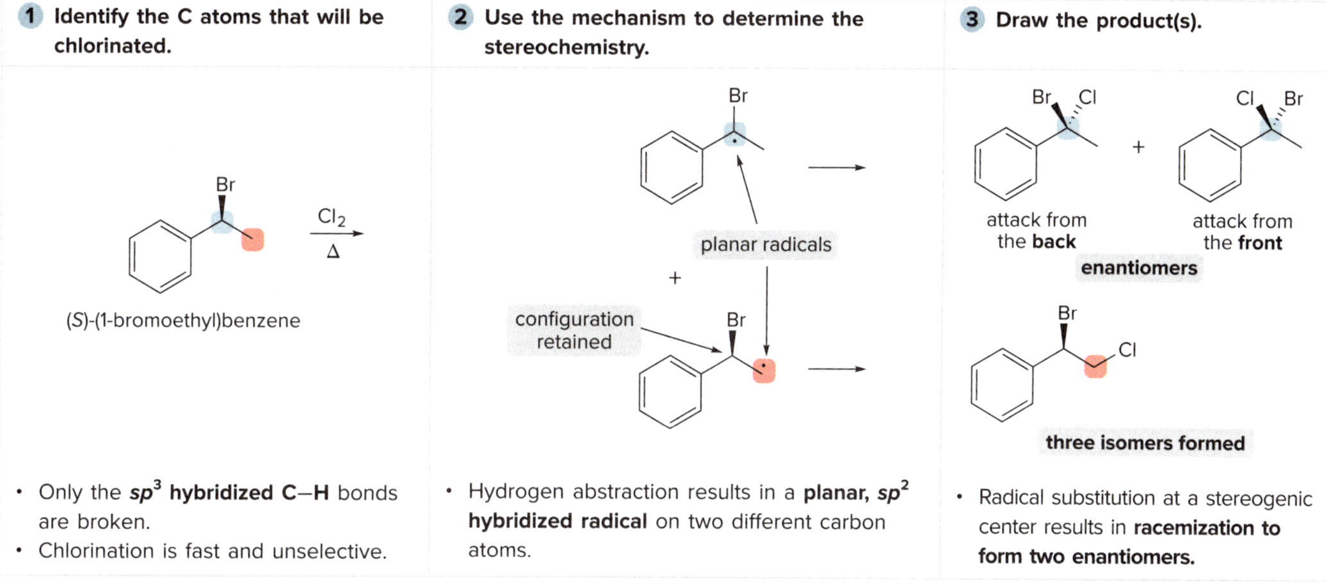

See Sample Problem 13.4. Try Problems 13.48a, d; 13.49–13.51.

[5] Drawing the products formed when methylenecyclopentane is treated with NBS + ROOR (13.10)

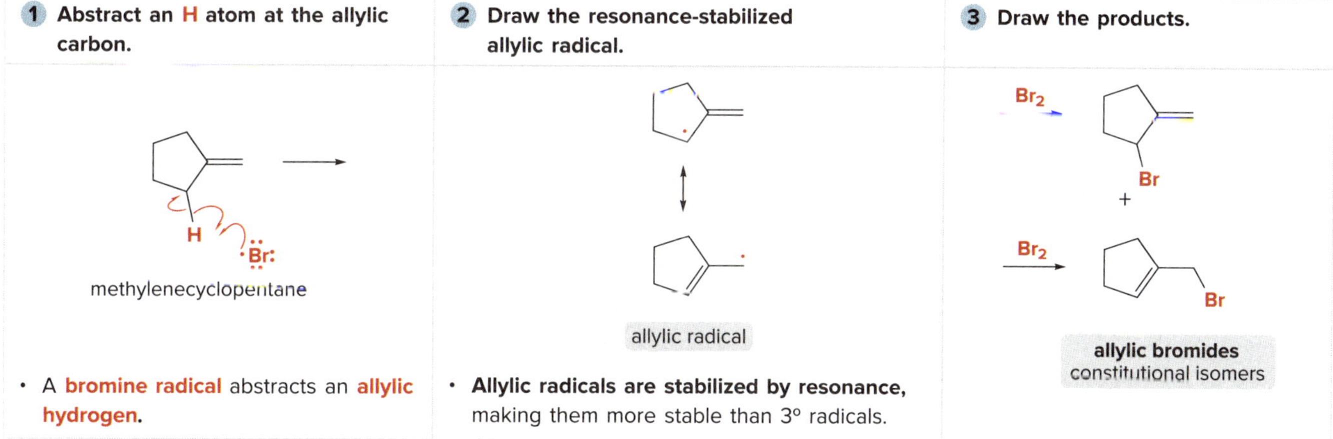

See Sample Problem 13.5. Try Problems 13.43, 13.44, 13.46f.

[6] Drawing the product of a polymerization reaction (13.14); example: polymerization of $CH_2{=}CHCH_3$

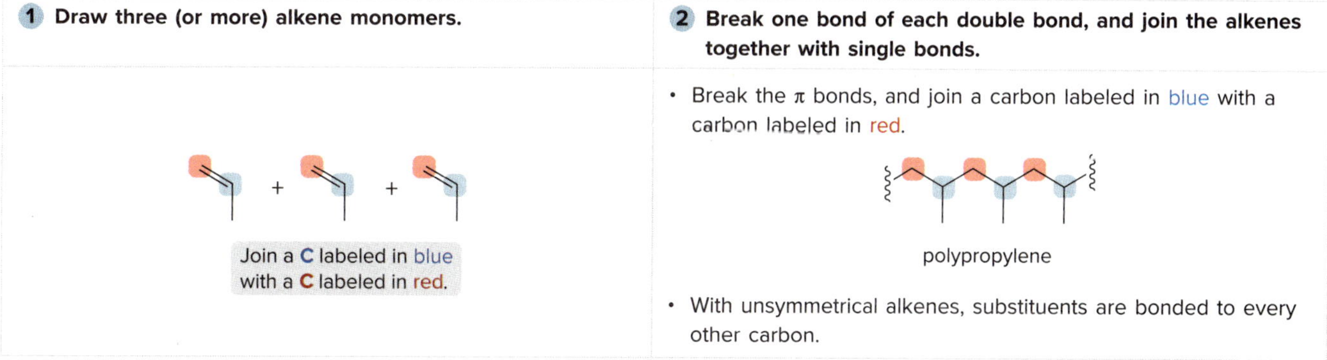

See Sample Problem 13.6. Try Problems 13.68, 13.69, 13.71a.

CHAPTER 13 MULTIPLE-CHOICE SELF-TEST

The Self-Test consists of multiple-choice questions similar to those found on the American Chemical Society organic chemistry exam. Answers are given at the end of the chapter.

1. Label the most stable and least stable radical.

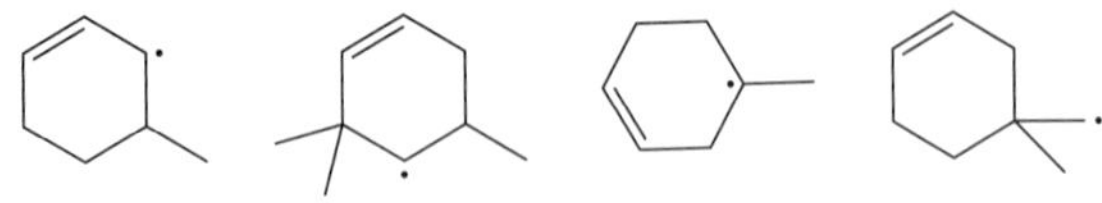

a. **C** is most stable and **D** is least stable.
b. **C** is most stable and **B** is least stable.
c. **B** is most stable and **D** is least stable.
d. **A** is most stable and **D** is least stable.

2. Which H is most easily removed in a radical halogenation reaction?

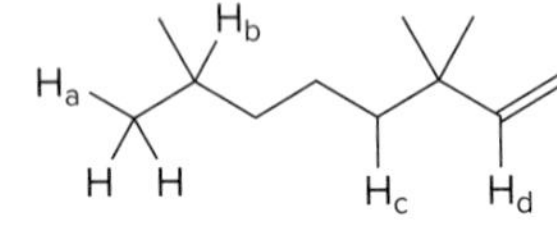

a. H_a b. H_b c. H_c d. H_d

3. What product is *not* formed when **X** is treated with NBS, ROOR?

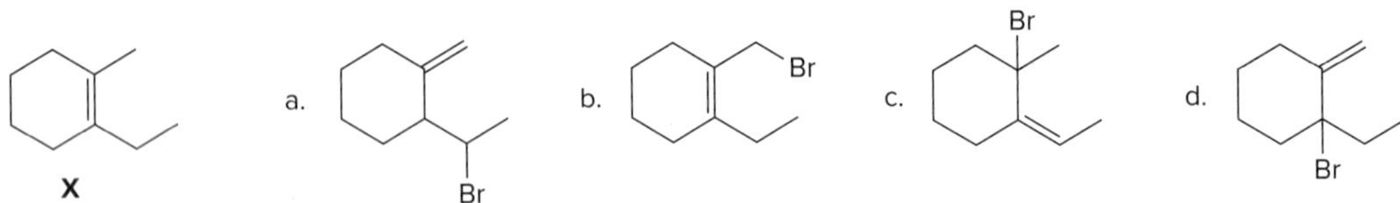

4. Which alkyl halide can be made in good yield by radical halogenation of an alkane?

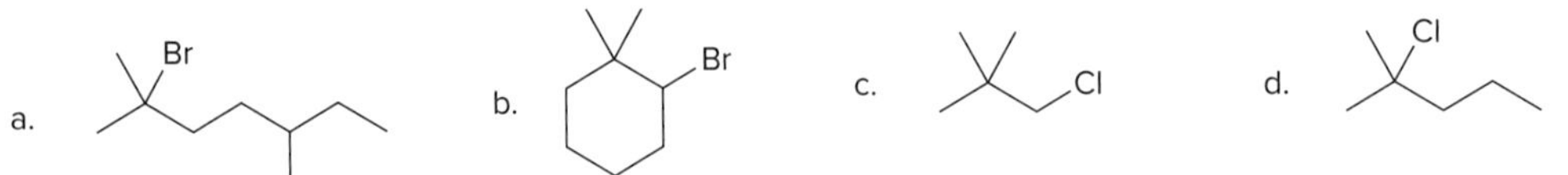

5. How many constitutional isomers of molecular formula $C_6H_{13}Cl$ are formed when $CH_3CH(CH_2CH_3)_2$ is heated with Cl_2: (a) three; (b) four; (c) five; (d) six?

6. In the reaction of 2-methylbut-1-ene with HBr in the presence of ROOR, what intermediate is formed?

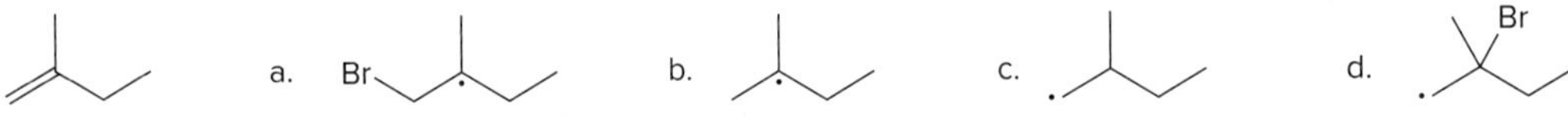

2-methylbut-1-ene

7. Which product is *not* formed when 1,1,1-tribromopropane is heated with Cl_2?

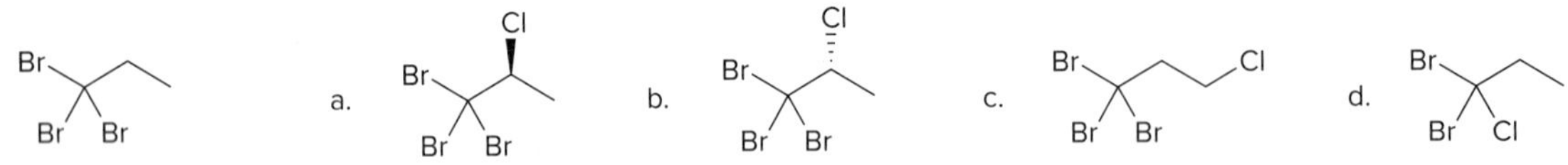

1,1,1-tribromopropane

8. What is the major product formed when **Y** undergoes the following two-step reaction sequence: [1] $KOC(CH_3)_3$; [2] HBr, ROOR?

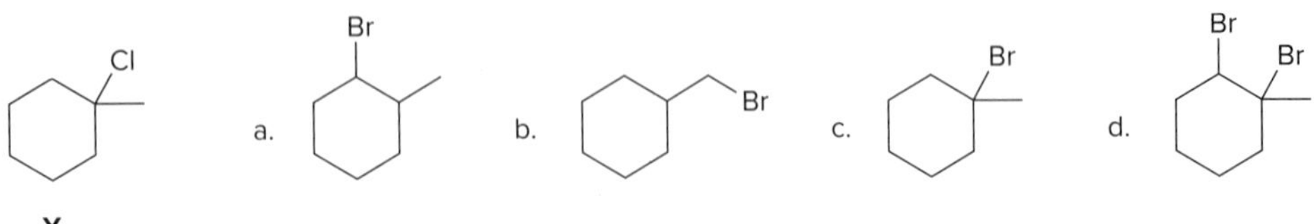

9. What monomer is used to prepare **Z?**

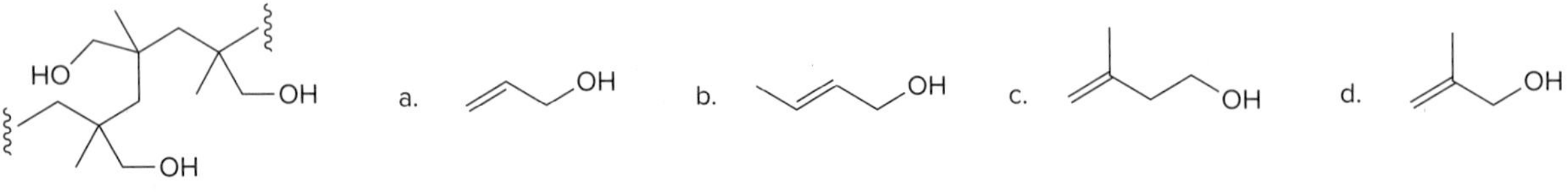

10. What is the major product formed from radical bromination of **A?**

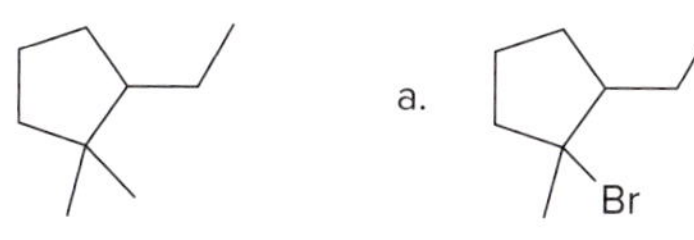

b. 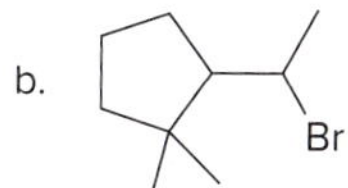

c. 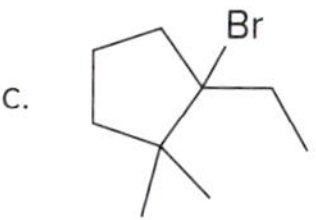

d. 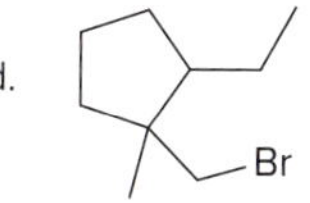

PROBLEMS

Problems Using Three-Dimensional Models

13.28 (a) Draw all constitutional isomers formed by monochlorination of each alkane with Cl_2 and $h\nu$. (b) Draw the major monobromination product formed by heating each alkane with Br_2.

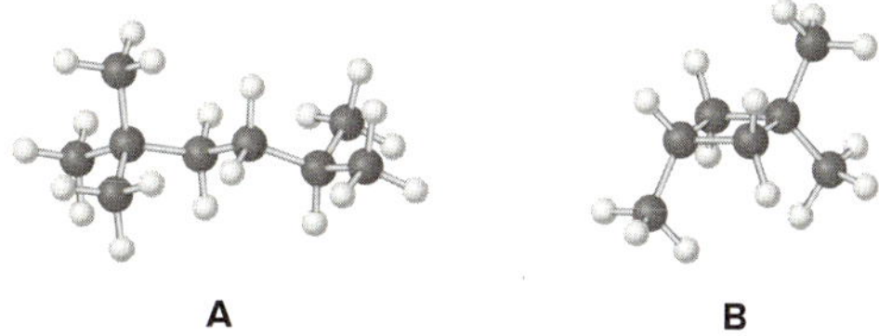

13.29 Draw all resonance structures of the radical that results from abstraction of a hydrogen atom from the antioxidant BHA (**b**utylated **h**ydroxy **a**nisole).

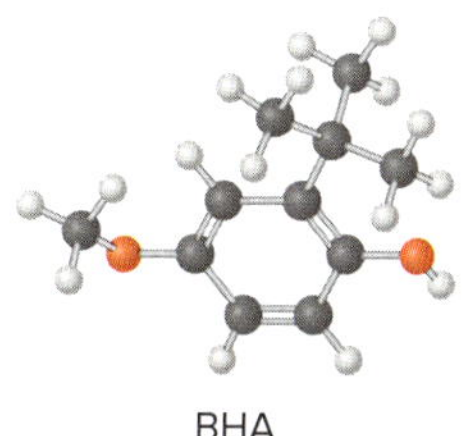

BHA

Radicals and Bond Strength

13.30 With reference to the indicated C–H bonds in the following compound:

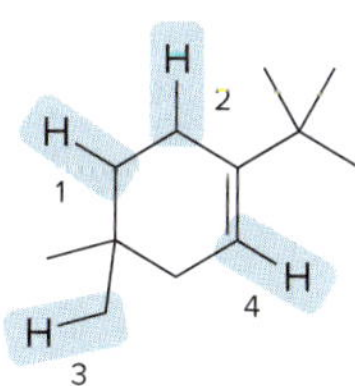

a. Rank the C–H bonds in order of increasing bond strength.
b. Draw the radical resulting from cleavage of each C–H bond, and classify it as 1°, 2°, or 3°.
c. Rank the radicals in order of increasing stability.
d. Rank the C–H bonds in order of increasing ease of H abstraction in a radical halogenation reaction.

13.31 Rank the following radicals in order of increasing stability.

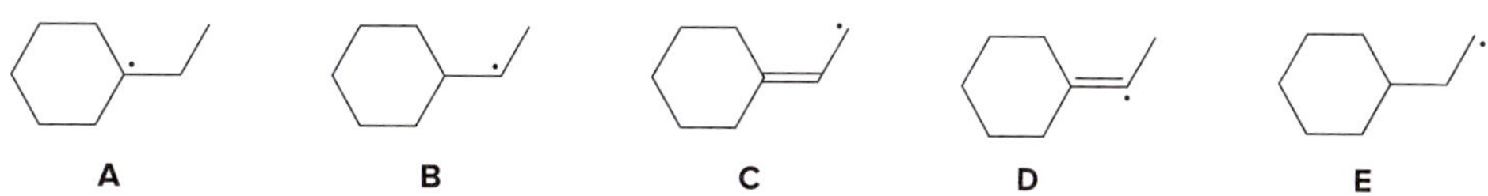

13.32 (a) Use the bond dissociation energies in Appendix E to predict the relative stability of the following radicals. (b) In what type of orbital does the unpaired electron reside in each species?

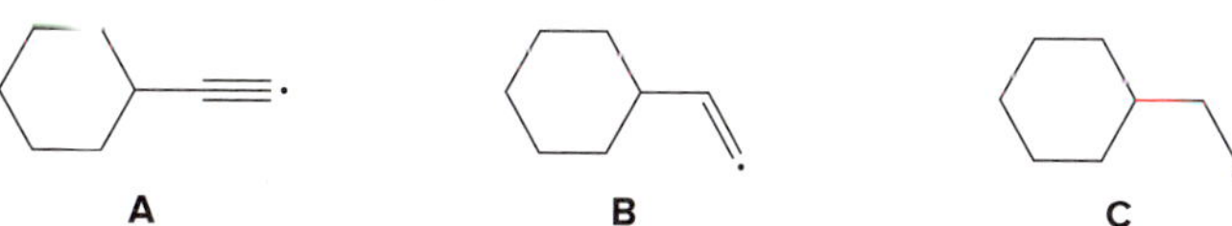

13.33 Explain why radicals **A** and **B** are both more stable than the ethyl radical.

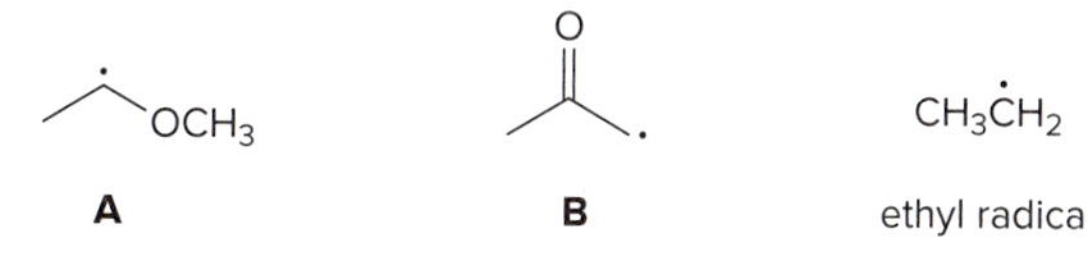

Halogenation of Alkanes

13.34 Rank the indicated hydrogen atoms in order of increasing ease of abstraction in a radical halogenation reaction.

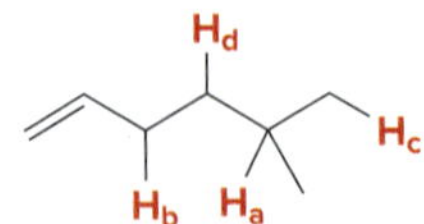

13.35 Draw all constitutional isomers formed by monochlorination of each alkane with Cl_2 and $h\nu$.

a.  b. c. 

13.36 What is the major monobromination product formed by heating each alkane with Br_2?

a. 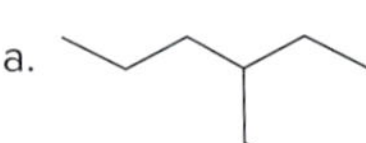 b. c. 

13.37 Five isomeric alkanes (**A–E**) having the molecular formula C_6H_{14} are each treated with $Cl_2 + h\nu$ to give alkyl halides having molecular formula $C_6H_{13}Cl$. **A** yields five constitutional isomers. **B** yields four constitutional isomers. **C** yields two constitutional isomers. **D** yields three constitutional isomers, two of which possess stereogenic centers. **E** yields three constitutional isomers, only one of which possesses a stereogenic center. Identify the structures of **A–E.**

13.38 Which alkyl halides can be prepared in good yield by radical halogenation of an alkane?

a. 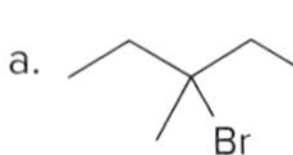

b. Cl
c. Br

13.39 Draw the products of radical chlorination and bromination of each compound. For which compounds is a single constitutional isomer formed for both reactions? What must be true about the structure of a reactant for both reactions to form a single product?

a. 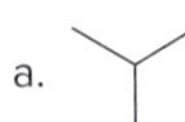 b. c.

13.40 Explain why radical bromination of *p*-xylene forms **C** rather than **D.**

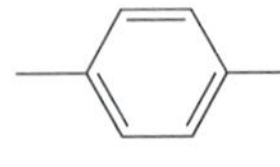
p-xylene

$\xrightarrow[\Delta]{Br_2}$

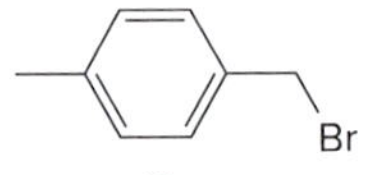

C

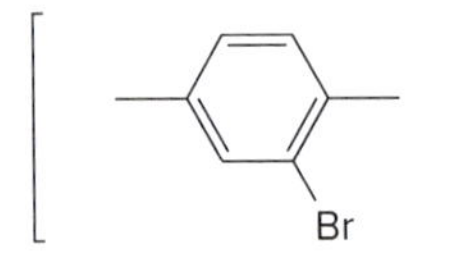

D
NOT formed

13.41 a. What product(s) (excluding stereoisomers) are formed when **Y** is heated with Cl_2?
b. What product(s) (excluding stereoisomers) are formed when **Y** is heated with Br_2?
c. What steps are needed to convert **Y** to the alkene **Z?**

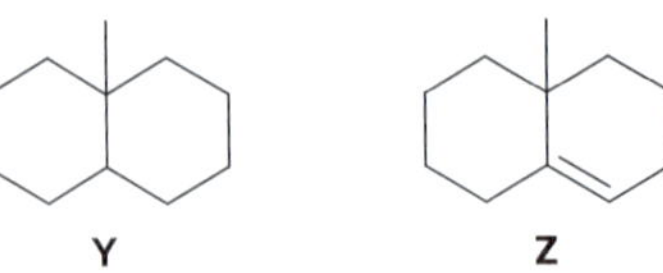

Resonance

13.42 Draw resonance structures for each radical.

a. 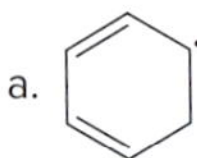 b. 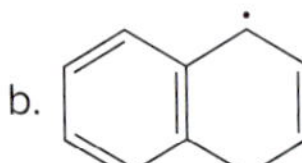 c. d. 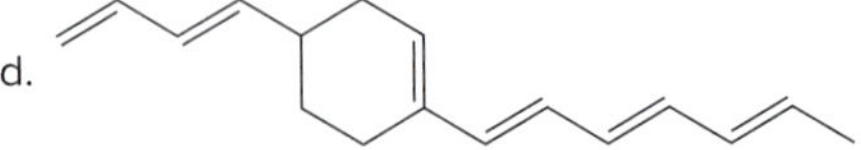

Allylic Halogenation

13.43 Draw the products formed when each alkene is treated with NBS + $h\nu$.

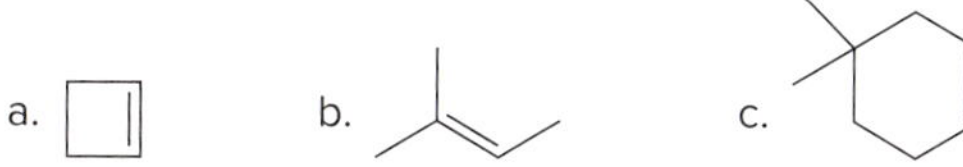

13.44 Draw all constitutional isomers formed when **X** is treated with NBS + $h\nu$.

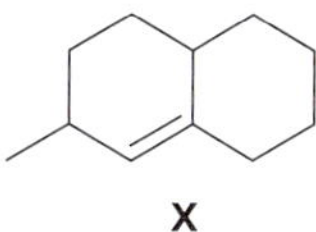

X

13.45 Treatment of propylbenzene with NBS + $h\nu$ affords a single constitutional isomer. Suggest a structure for the product and a reason for its formation.

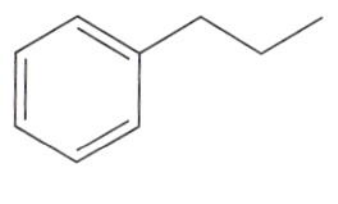

propylbenzene

Reactions

13.46 Draw the organic products formed in each reaction.

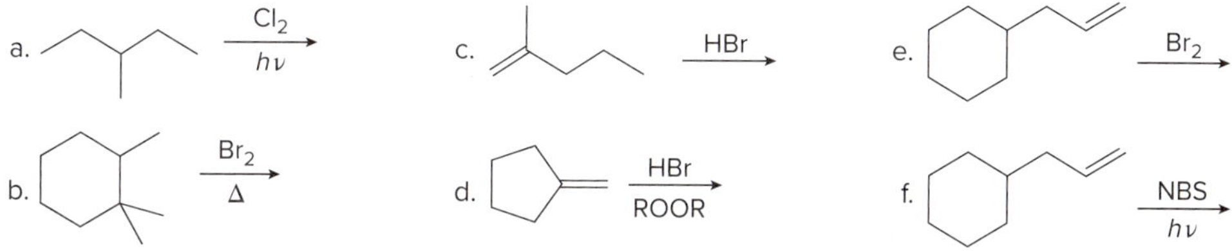

13.47 Treatment of a hydrocarbon **A** (molecular formula C_9H_{18}) with Br_2 in the presence of light forms alkyl halides **B** and **C,** both having molecular formula $C_9H_{17}Br$. Reaction of either **B** or **C** with $KOC(CH_3)_3$ forms compound **D** (C_9H_{16}) as the major product. Ozonolysis of **D** forms cyclohexanone and acetone. Identify the structures of **A–D.**

cyclohexanone acetone

Stereochemistry and Reactions

13.48 Draw the products formed in each reaction and include the stereochemistry around any stereogenic centers.

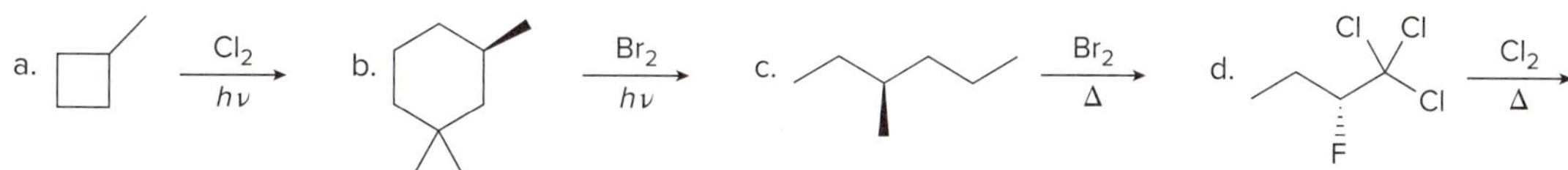

13.49 (a) Draw the products of molecular formula $C_3H_4Cl_2$, including stereoisomers, formed when chlorocyclopropane is heated with Cl_2. (b) Assuming that compounds that have different physical properties are separable, how many fractions would be present if the mixture of products were distilled using an efficient fractional distillation? (c) How many fractions would be optically active?

13.50 (a) Draw all stereoisomers of molecular formula $C_5H_{10}Cl_2$ formed when (*R*)-2-chloropentane is heated with Cl_2. (b) Assuming that products having different physical properties can be separated into fractions by some physical method (such as fractional distillation), how many different fractions would be obtained? (c) Which of these fractions would be optically active?

13.51 (a) Draw all stereoisomers formed by monobromination of the cis and trans isomers of 1,2-dimethylcyclohexane drawn below. (b) How do the products formed from each reactant compare—identical compounds, stereoisomers, or constitutional isomers?

cis-1,2-dimethylcyclohexane *trans*-(1*R*,2*S*)-dimethylcyclohexane

13.52 Draw the six products (including stereoisomers) formed when **A** is treated with NBS + $h\nu$.

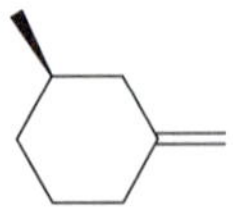

A

13.53 Draw all stereoisomers formed when each compound is treated with HBr in the presence of peroxides.

a. 

b. 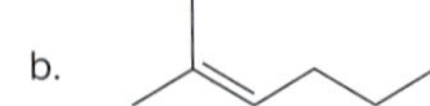

c. 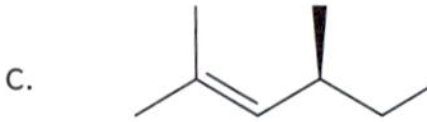

Mechanisms

13.54 Consider the following bromination: $(CH_3)_3CH + Br_2 \xrightarrow{\Delta} (CH_3)_3CBr + HBr$.

a. Calculate $\Delta H°$ for this reaction by using the bond dissociation energies in Table 6.2.

b. Draw out a stepwise mechanism for the reaction, including the initiation, propagation, and termination steps.

c. Calculate $\Delta H°$ for each propagation step.

d. Draw an energy diagram for the propagation steps.

e. Draw the structure of the transition state of each propagation step.

13.55 Draw a stepwise mechanism for the following reaction.

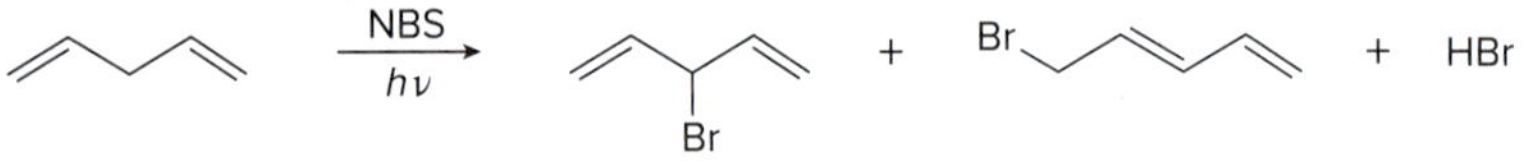

NBS, $h\nu$ → Br + Br + HBr

13.56 Like carbocations, radicals formed from compounds that contain another functional group can undergo intramolecular reactions. Draw a stepwise mechanism for the chain-propagating steps of the following intramolecular reaction.

NBS, $h\nu$ → Br

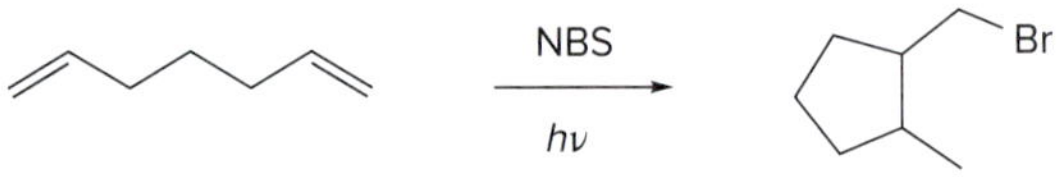

13.57 When 3,3-dimethylbut-1-ene is treated with HBr alone, the major product is 2-bromo-2,3-dimethylbutane. When the same alkene is treated with HBr and peroxide, the sole product is 1-bromo-3,3-dimethylbutane. Explain these results by referring to the mechanisms.

Synthesis

13.58 Devise a synthesis of each compound from cyclopentane and any other required organic or inorganic reagents.

a. 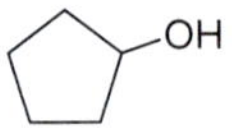 OH b. 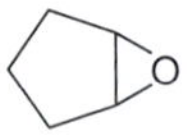 O c. 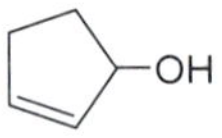 OH d. 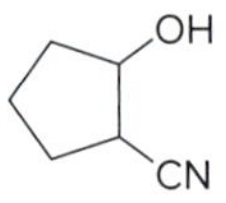 OH, CN

13.59 Devise a synthesis of $CH_3CH_2CH_2CH_2Br$ from HC≡CH. You may use any other required organic compounds or inorganic reagents.

13.60 Devise a synthesis of each compound using CH_3CH_3 as the only source of carbon atoms. You may use any other required organic or inorganic reagents.

a. HC≡CH b.  OH c.  d. 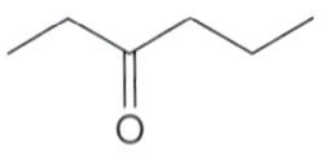 O

13.61 Devise a synthesis of $OHC(CH_2)_4CHO$ from cyclohexane using any required organic or inorganic reagents.

13.62 Devise a synthesis of hexane-2,3-diol from propane as the only source of carbon atoms. You may use any other required organic or inorganic reagents.

Radical Oxidation Reactions

13.63 As described in Section 9.17, the leukotrienes, important components in the asthmatic response, are synthesized from arachidonic acid via the hydroperoxide 5-HPETE. Write a stepwise mechanism for the conversion of arachidonic acid to 5-HPETE with O_2.

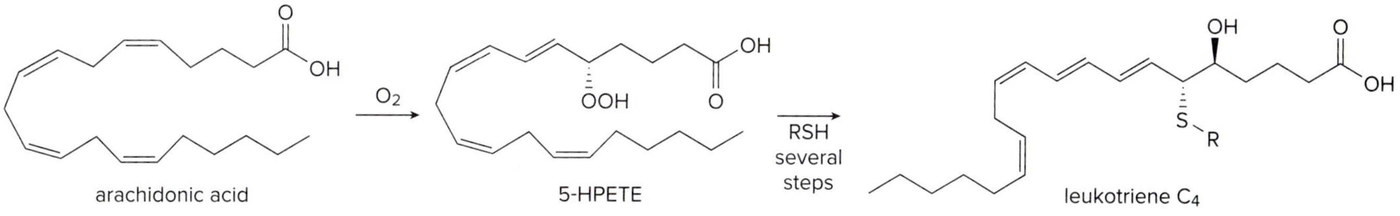

arachidonic acid $\xrightarrow{O_2}$ 5-HPETE $\xrightarrow[\text{several steps}]{RSH}$ leukotriene C_4

13.64 Ethers are oxidized with O_2 to form hydroperoxides that decompose violently when heated. Draw a stepwise mechanism for this reaction.

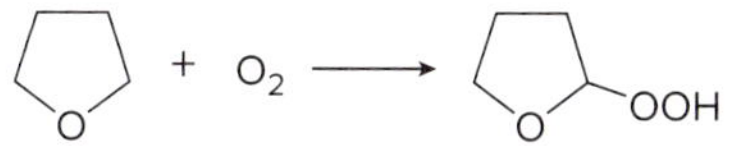

unstable hydroperoxide

Antioxidants

13.65 Resveratrol is an antioxidant found in the skin of red grapes. Its anticancer, anti-inflammatory, and various cardiovascular effects are under active investigation. (a) Draw all resonance structures for the radical that results from homolysis of the OH bond shown in red. (b) Explain why homolysis of this OH bond is preferred to homolysis of either OH bond in the other benzene ring.

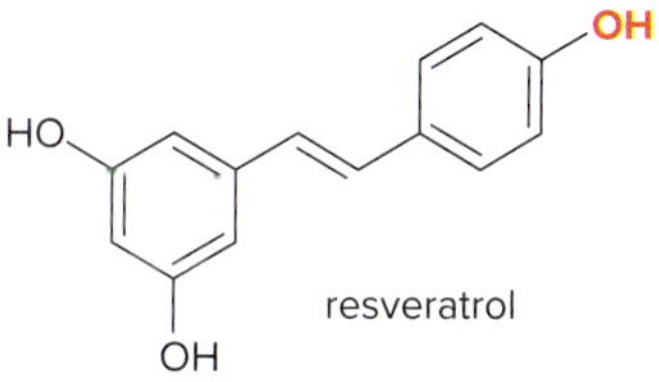

resveratrol

13.66 Vitamin C, which exists primarily as its conjugate base ascorbate in cells, is an antioxidant. When ascorbate reacts with a radical, it loses an electron from its negatively charged oxygen to form a radical. Draw all resonance structures for this radical.

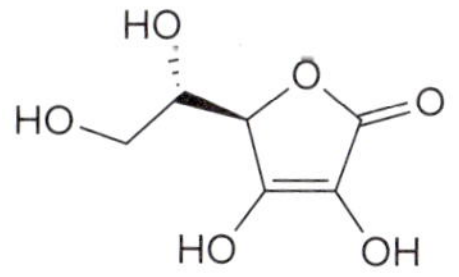

vitamin C
ascorbic acid

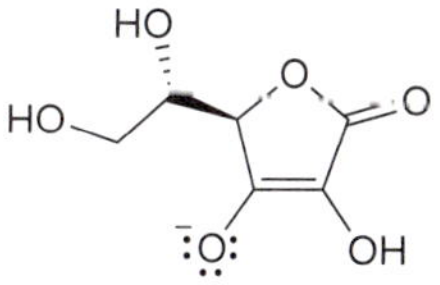

ascorbate

Polymers and Polymerization

13.67 What monomer is needed to form each polymer?

a.  polyisobutylene
(used to make basketballs)

b. 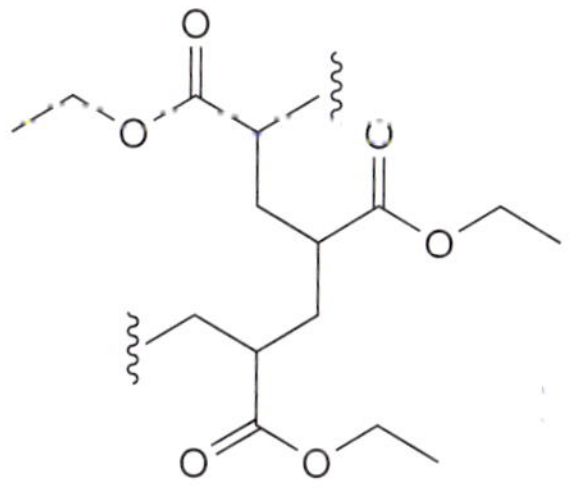 poly(ethyl acrylate)
(used in latex paints)

13.68 What polymer is formed from each monomer? The polymer from butyl α-cyanoacrylate [part (b)] is used in the topical skin adhesive LiquiBand, and vinylpyrrolidone [part (c)] has been used in the manufacture of glue sticks.

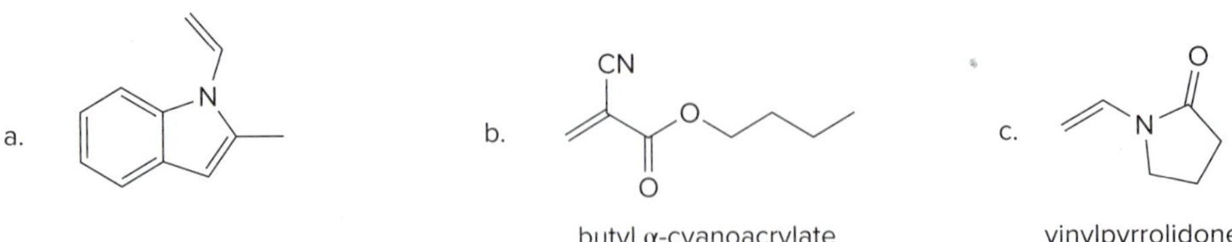

13.69 (a) Hard contact lenses, which first became popular in the 1960s, were made by polymerizing methyl methacrylate [$CH_2{=}C(CH_3)CO_2CH_3$] to form poly(methyl methacrylate) (PMMA). Draw the structure of PMMA. (b) More-comfortable softer contact lenses introduced in the 1970s were made by polymerizing hydroxyethyl methacrylate [$CH_2{=}C(CH_3)CO_2CH_2CH_2OH$] to form poly(hydroxyethyl methacrylate) (poly-HEMA). Draw the structure of poly-HEMA. Because neither polymer allows oxygen from the air to pass through to the retina, newer contact lenses that are both comfortable and oxygen-permeable have now been developed.

13.70 Explain why polystyrene is much more readily oxidized by O_2 in the air than polyethylene is. Which H's in polystyrene are most easily abstracted and why?

polystyrene

polyethylene

13.71 As we will learn in Chapter 31, styrene derivatives such as **A** can be polymerized by way of cationic rather than radical intermediates. Cationic polymerization is an example of electrophilic addition to an alkene involving carbocations.

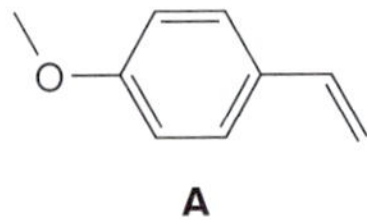

A

a. Draw a short segment of the polymer formed by the polymerization of **A.**

b. Why does **A** react faster than styrene ($C_6H_5CH{=}CH_2$) in a cationic polymerization?

13.72 When two monomers (**X** and **Y**) are polymerized together, a copolymer results. An alternating copolymer is formed when the two monomers **X** and **Y** alternate regularly in the polymer chain. Draw the structure of the alternating copolymer formed when the two monomers, $CH_2{=}CCl_2$ and $CH_2{=}CHC_6H_5$, are polymerized together.

13.73 Polymerization of isoprene under radical conditions forms a mixture of products, two of which (**A** and **B**) are shown. Draw a stepwise mechanism that illustrates how both products are formed. Polyisoprene (**A**) is the main component of commercially available non-latex condoms.

isoprene —radical conditions→ **A** (polyisoprene) + **B**

Spectroscopy

13.74 **A** and **B,** isomers of molecular formula $C_3H_5Cl_3$, are formed by the radical chlorination of a dihalide **C** of molecular formula $C_3H_6Cl_2$.

a. Identify the structures of **A** and **B** from the following ^1H NMR data:

Compound **A:** singlet at 2.23 and singlet at 4.04 ppm

Compound **B:** doublet at 1.69, multiplet at 4.34, and doublet at 5.85 ppm

b. What is the structure of **C?**

13.75 Identify the structure of a minor product formed from the radical chlorination of propane, which has molecular formula $C_3H_6Cl_2$ and exhibits the given ^{1}H NMR spectrum.

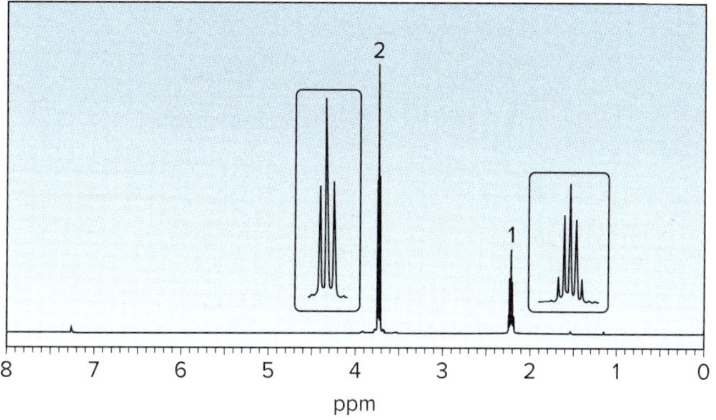

Challenge Problems

13.76 The triphenylmethyl radical is an unusual persistent radical present in solution in equilibrium with its dimer. For 70 years the dimer was thought to be hexaphenylethane, but in 1970, NMR data showed it to be **A**.

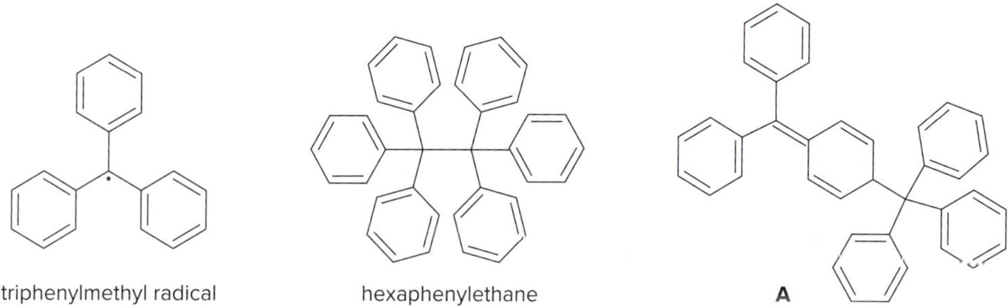

a. Why is the triphenylmethyl radical more stable than most other radicals?
b. Use curved arrow notation to show how two triphenylmethyl radicals dimerize to form **A.**
c. Propose a reason for the formation of **A** rather than hexaphenylethane.
d. How could ^{1}H and ^{13}C NMR spectroscopy be used to distinguish between hexaphenylethane and **A?**

13.77 In the presence of a radical initiator (Z·), tributyltin hydride (R_3SnH, R = $CH_3CH_2CH_2CH_2$) reduces alkyl halides to alkanes: $R'X + R_3SnH \rightarrow R'H + R_3SnX$. The mechanism consists of a radical chain process with an intermediate tin radical:

Initiation: $R_3SnH + Z\cdot \longrightarrow R_3Sn\cdot + HZ$

Propagation:

$R'{-}Br + R_3Sn\cdot \longrightarrow R'\cdot + R_3SnBr$

$R'\cdot + R_3SnH \longrightarrow R'{-}H + R_3Sn\cdot$

This reaction has been employed in many radical cyclization reactions. Draw a stepwise mechanism for the following reaction.

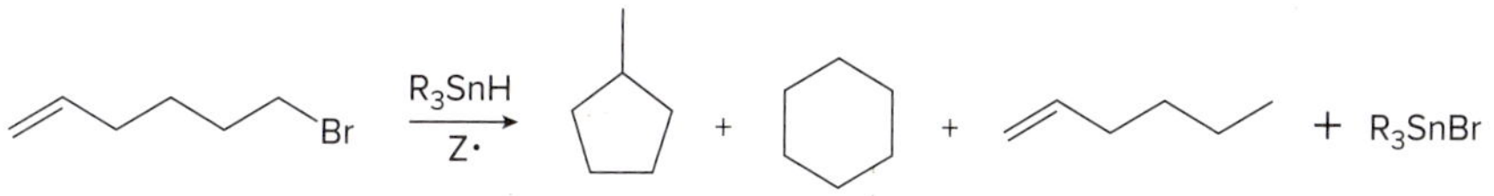

Br $\xrightarrow[Z\cdot]{R_3SnH}$ + + + R_3SnBr

13.78 The naturally occurring lipid hirsutene was synthesized in one step from **A** by a cascade of radical cyclizations based on the tributyltin hydride chemistry described in Problem 13.77. Draw a stepwise mechanism for this process that illustrates how two rings are synthesized in a single reaction.

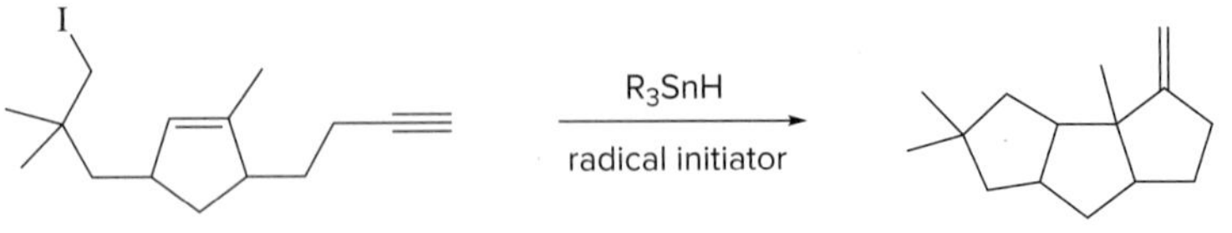

I

R_3SnH

radical initiator

A

hirsutene

13.79 $PGF_{2\alpha}$ (Section 19.5) is synthesized in cells from arachidonic acid ($C_{20}H_{32}O_2$) using a cyclooxygenase enzyme that catalyzes a multistep radical pathway. Part of this process involves the conversion of radical **A** to PGG_2, an unstable intermediate, which is then transformed to $PGF_{2\alpha}$ and other prostaglandins. Draw a stepwise mechanism for the conversion of **A** to PGG_2. (Hint: The mechanism begins with radical addition to a carbon–carbon double bond to form a resonance-stabilized radical.)

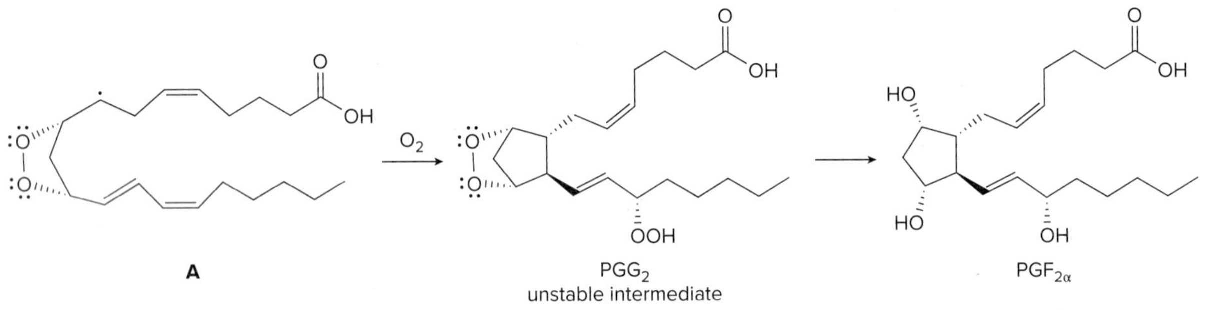

O OH O OH O OH HO HO OOH OH

O_2

A

PGG_2
unstable intermediate

$PGF_{2\alpha}$

SELF-TEST ANSWERS

1. d 2. b 3. a 4. c 5. b 6. a 7. d 8. a 9. d 10. c

Conjugation, Resonance, and Dienes

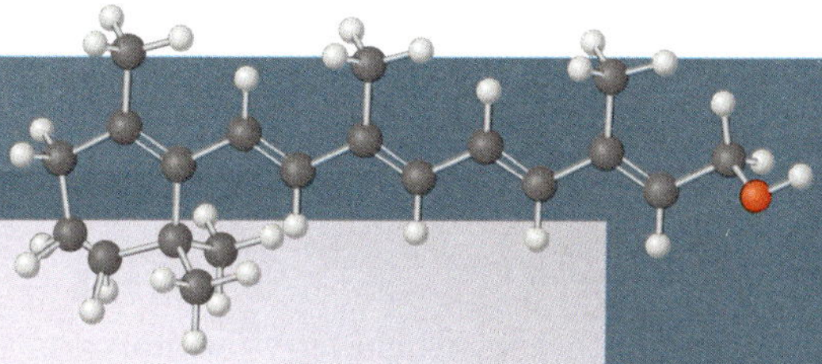

Pixtal/age fotostock

14.1 Conjugation
14.2 Resonance and allylic carbocations
14.3 Common examples of resonance
14.4 The resonance hybrid
14.5 Electron delocalization, hybridization, and geometry
14.6 Conjugated dienes
14.7 Interesting dienes and polyenes
14.8 The carbon–carbon σ bond length in buta-1,3-diene
14.9 Stability of conjugated dienes
14.10 Electrophilic addition: 1,2- versus 1,4-addition
14.11 Kinetic versus thermodynamic products
14.12 The Diels–Alder reaction
14.13 Specific rules governing the Diels–Alder reaction
14.14 Other facts about the Diels–Alder reaction
14.15 Conjugated dienes and ultraviolet light

Vitamin A is a highly conjugated, fat-soluble vitamin obtained from liver, oily fish, and dairy products, and is synthesized from β-carotene, the orange pigment in carrots. In the body, vitamin A is converted to 11-*cis*-retinal, the light-sensitive compound responsible for vision in all vertebrates. A deficiency of vitamin A causes night blindness, as well as dry eyes and skin. In Chapter 14, we learn about the unique chemistry of conjugated dienes and polyenes like vitamin A.

Why Study . . .
Conjugated Systems?

Chapter 14 is the first of three chapters that discuss the chemistry of conjugated molecules—molecules with overlapping *p* orbitals on three or more adjacent atoms. Chapter 14 focuses mainly on acyclic conjugated compounds, whereas Chapters 15 and 16 discuss the chemistry of benzene and related compounds that have a *p* orbital on every atom in a ring.

Much of Chapter 14 is devoted to the properties and reactions of 1,3-dienes, most notably the Diels–Alder reaction, which is widely used in the synthesis of naturally occurring compounds. To understand 1,3-dienes, however, we must first learn about the consequences of having *p* orbitals on three or more adjacent atoms. Because the ability to draw resonance structures is also central to mastering this material, the key aspects of resonance theory are presented in detail.

14.1 Conjugation

***Conjugation* occurs whenever *p* orbitals can overlap on three or more adjacent atoms.** Two common conjugated systems are 1,3-dienes and allylic carbocations.

14.1A 1,3-Dienes

1,3-Dienes such as buta-1,3-diene contain **two carbon–carbon double bonds joined by a single σ bond.** Each carbon atom of a 1,3-diene is bonded to three other atoms and has no nonbonded electron pairs, so each carbon atom is sp^2 hybridized and has one *p* orbital containing an electron. **The four *p* orbitals on adjacent atoms make a 1,3-diene a conjugated system.**

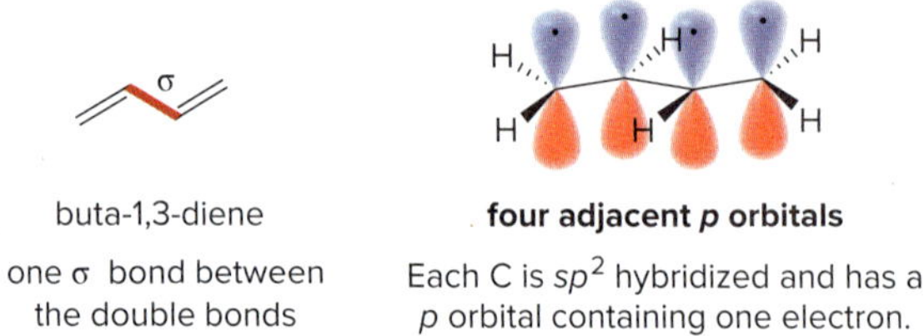

What is special about conjugation? Having three or more *p* orbitals on adjacent atoms allows *p* orbitals to overlap and **electrons to delocalize.**

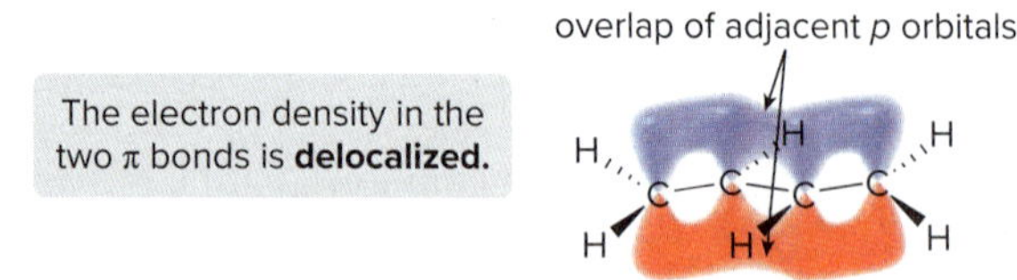

- **When *p* orbitals overlap, the electron density in each of the π bonds is spread out over a larger volume, thus lowering the energy of the molecule and making it more stable.**

Conjugation makes buta-1,3-diene inherently different from penta-1,4-diene, a compound having two double bonds separated by more than one σ bond. The π bonds in penta-1,4-diene are too far apart to be conjugated.

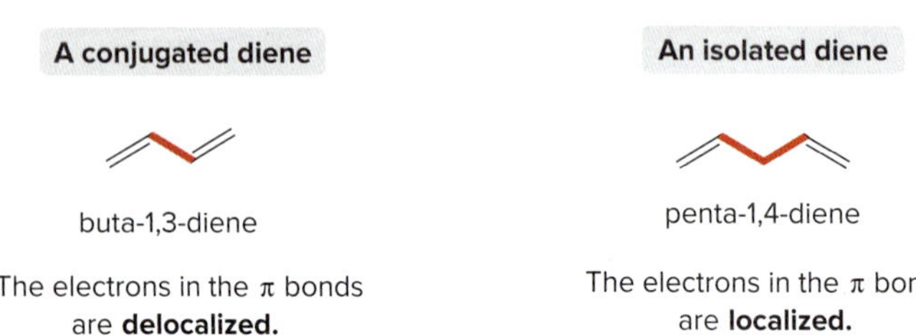

Penta-1,4-diene is an **isolated diene.** The electron density in each π bond of an isolated diene is *localized* between two carbon atoms. In buta-1,3-diene, however, the electron density of both π bonds is *delocalized* over the four atoms of the diene. Electrostatic potential maps in Figure 14.1 clearly indicate the difference between these localized and delocalized π bonds.

Figure 14.1 Electrostatic potential plots for a conjugated and an isolated diene

A conjugated diene

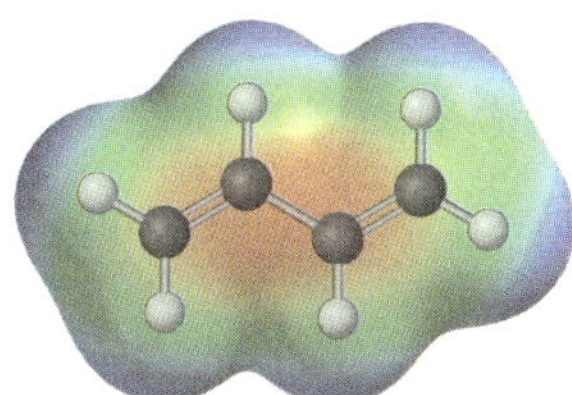

buta-1,3-diene
The red electron-rich region is spread over four adjacent atoms.

An isolated diene

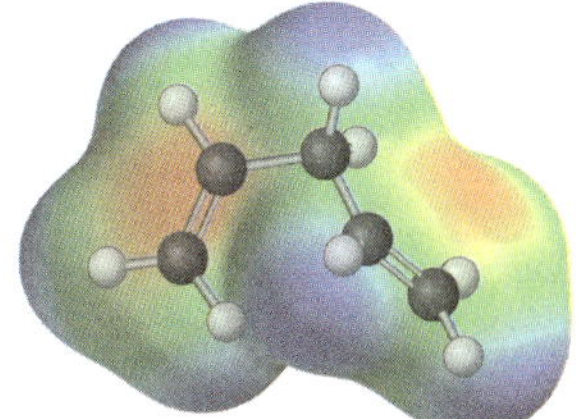

penta-1,4-diene
The red electron-rich regions are **localized** in the π bonds on the two ends of the molecule.

Problem 14.1 Classify each carbon–carbon double bond as isolated or conjugated.

a.

b.

c.

d.

e.

f.

14.1B Allylic Carbocations

The **allyl carbocation** is another example of a conjugated system. The three carbon atoms of the allyl carbocation—the positively charged carbon atom and the two that form the double bond—are sp^2 hybridized and have an unhybridized p orbital. The p orbitals for the double bond carbons each contain an electron, whereas the p orbital for the carbocation is empty.

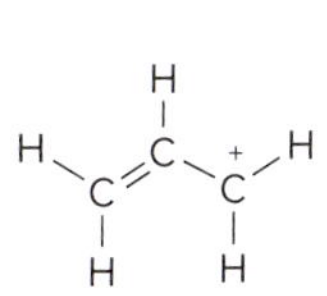

allyl carbocation
Each C is sp^2 hybridized and has a p orbital.

π bond
carbocation

three adjacent *p* orbitals

- **Three *p* orbitals on three adjacent atoms, even if one of the *p* orbitals is empty, make the allyl carbocation conjugated.**

Conjugation stabilizes the allyl carbocation because overlap of three adjacent p orbitals delocalizes the electron density of the π bond over three atoms.

overlap of adjacent p orbitals

Problem 14.2 Which of the following species are conjugated?

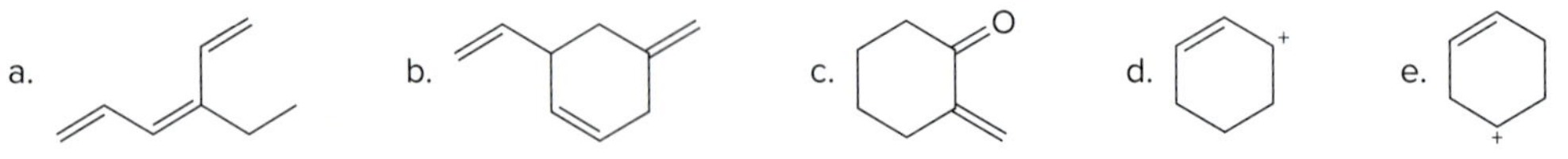

a. b. c. d. e.

14.2 Resonance and Allylic Carbocations

The word *resonance* is used in two different contexts. In NMR spectroscopy, a nucleus is ***in resonance*** when it absorbs energy, promoting it to a higher-energy state. In drawing molecules, there is ***resonance*** when two different Lewis structures can be drawn for the same arrangement of atoms.

Recall from Section 1.6 that resonance structures are two or more different Lewis structures for the same arrangement of atoms. Being able to draw correct resonance structures is crucial to understanding conjugation and the reactions of conjugated dienes.

- **Two resonance structures differ in the placement of π bonds and nonbonded electrons. The placement of atoms and σ bonds stays the same.**

14.2A The Stability of Allylic Carbocations

We have already drawn resonance structures for the acetate anion (Section 2.5C) and the allyl radical (Section 13.10). The **conjugated allyl carbocation** is another example of a species for which two resonance structures can be drawn. Drawing resonance structures for the allyl carbocation is a way to use Lewis structures to illustrate how conjugation delocalizes electrons.

δ+ δ+

resonance structures for the allyl carbocation

hybrid

The true structure of the allyl carbocation is a **hybrid** of the two resonance structures. In the hybrid, the π bond is delocalized over all three atoms. As a result, the positive charge is also delocalized over the two terminal carbons. **Delocalizing electron density lowers the energy of the hybrid,** thus stabilizing the allyl carbocation and making it more stable than a normal 1° carbocation. Experimental data show that its stability is comparable to a more substituted 2° carbocation.

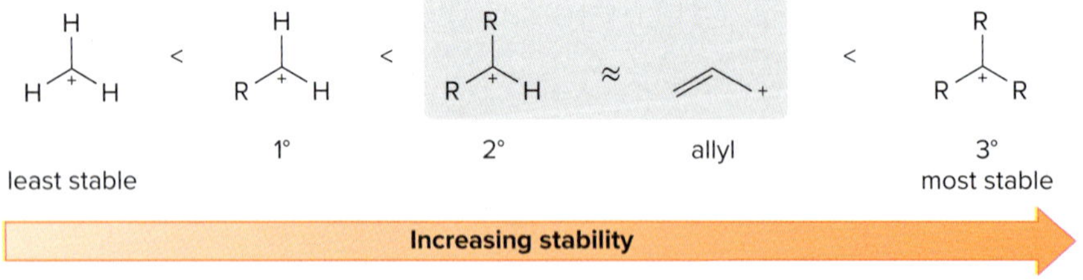

least stable < 1° < 2° ≈ allyl < 3° most stable

Increasing stability

Problem 14.3 Draw a second resonance structure for each carbocation. Then draw the hybrid.

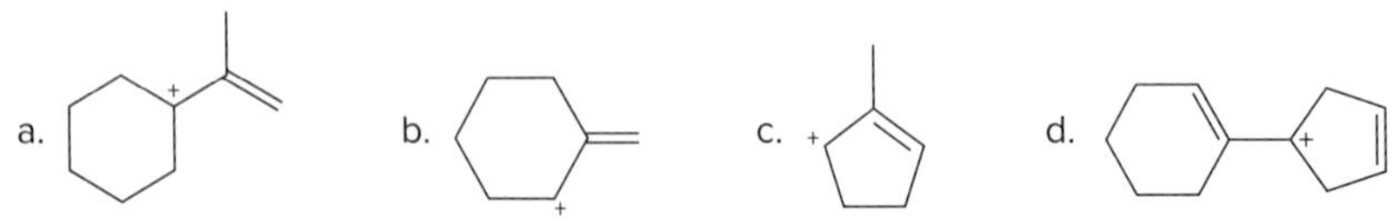

a. b. c. d.

Problem 14.4 Use resonance theory and the Hammond postulate to explain why 3-chloroprop-1-ene ($CH_2{=}CHCH_2Cl$) is more reactive than 1-chloropropane ($CH_3CH_2CH_2Cl$) in S_N1 reactions.

14.2B Allylic Carbocations in Biological Reactions

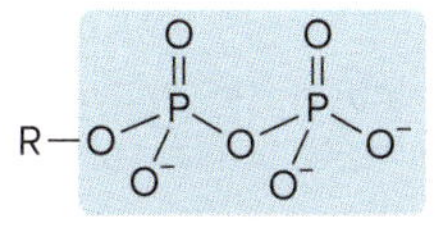

organic diphosphate

R–OPP

diphosphate leaving group
PP_i

Allylic carbocations formed from diphosphates (Section 7.15) are key intermediates in a variety of biological reactions, including the synthesis of geranyl diphosphate from two five-carbon substrates—dimethylallyl diphosphate and isopentenyl diphosphate. Geranyl diphosphate is the precursor of many lipids that occur in plants and animals.

dimethylallyl diphosphate + isopentenyl diphosphate ⟶ geranyl diphosphate

This biological process results in the formation of a new carbon–carbon bond and involves two key steps—loss of a good leaving group (diphosphate, $P_2O_7^{4-}$, abbreviated as PP_i) to form an allylic carbocation, followed by nucleophilic attack with an electron-rich double bond. The steps of the mechanism are shown in Mechanism 14.1.

Mechanism 14.1 Biological Formation of Geranyl Diphosphate

dimethylallyl diphosphate; isopentenyl diphosphate; + PP_i; new bond shown in red; H, :B; geranyl diphosphate + HB^+

1 Loss of the diphosphate leaving group forms an **allylic carbocation.**

2 Nucleophilic attack of isopentenyl diphosphate on the allylic carbocation forms the **new C–C σ bond.**

3 **Loss of a proton** (shown with the general base, B:) forms geranyl diphosphate.

We will learn more about biological reactions involving allylic carbocations derived from diphosphates in Chapter 29.

Problem 14.5 Farnesyl diphosphate is synthesized from isopentenyl diphosphate and **X** by a pathway similar to Mechanism 14.1. Draw the structure of **X.**

isopentenyl diphosphate + **X** ⟶ farnesyl diphosphate

14.3 Common Examples of Resonance

When are resonance structures drawn for a molecule or reactive intermediate? Because resonance involves delocalizing **π bonds** and **nonbonded electrons,** one or both of these structural features must be present to draw additional resonance forms. There are four common bonding patterns for which more than one Lewis structure can be drawn.

Type [1] The Three Atom "Allyl" System, X=Y–Z*

- **For any group of three atoms having a double bond X=Y and an atom Z that contains a *p* orbital with zero, one, or two electrons, two resonance structures are possible:**

$$X{=}Y{-}\overset{*}{Z} \longleftrightarrow \overset{*}{X}{-}Y{=}Z$$

The asterisk [*] corresponds to a charge, a radical, or a lone pair.

* = +, –, ·, or ··

This is called **allyl** type resonance because it can be drawn for allylic carbocations, allylic carbanions, and allylic radicals.

X, Y, and **Z** may all be carbon atoms, as in the case of an allylic carbocation (resonance structures **A** and **B**), or they may be heteroatoms, as in the case of the acetate anion (resonance structures **C** and **D**). The atom **Z** bonded to the multiple bond can be charged (a net positive or negative charge) or neutral (having zero, one, or two nonbonded electrons). **The two resonance structures differ in the location of the double bond, and in the charge, the radical, or the lone pair, generalized by [*].**

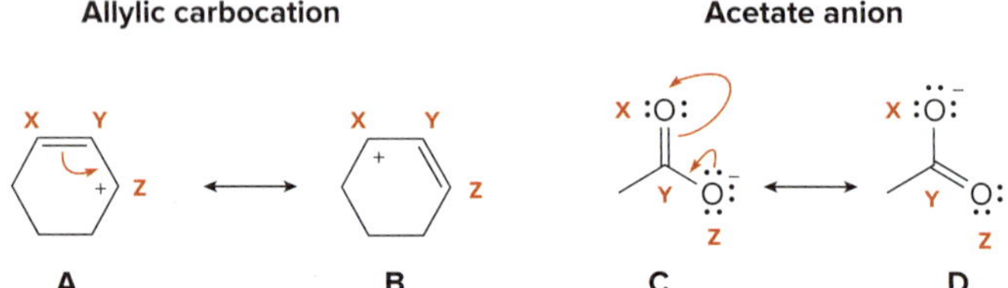

Type [2] Conjugated Double Bonds

Cyclic, completely conjugated rings like benzene have two resonance structures, drawn by moving the electrons in a cyclic manner around the ring. **Three resonance structures can be drawn for conjugated dienes,** two of which involve charge separation.

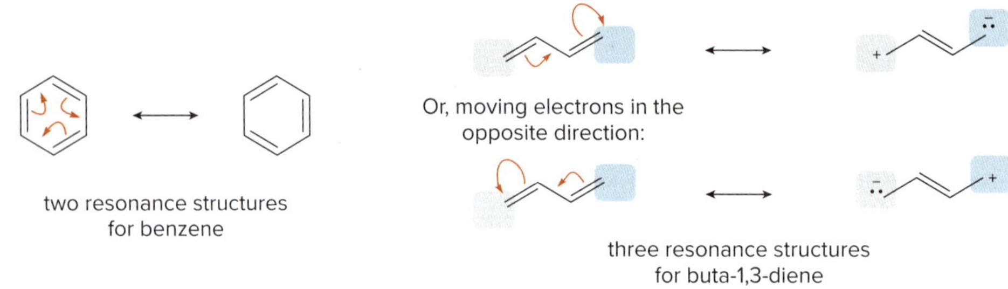

two resonance structures for benzene

three resonance structures for buta-1,3-diene

Type [3] Cations Having a Positive Charge Adjacent to a Lone Pair

- **When a lone pair and a positive charge are located on adjacent atoms, two resonance structures can be drawn.**

$\ddot{X}-\overset{+}{Y} \longleftrightarrow \overset{+}{X}=Y$ $\qquad CH_3\ddot{\ddot{O}}\overset{+}{C}H_2 \longleftrightarrow CH_3\overset{+}{\ddot{O}}=CH_2$

The overall charge is the same in both resonance structures. Based on formal charge, a neutral **X** in one structure must bear a (+) charge in the other.

Type [4] Double Bonds Having One Atom More Electronegative Than the Other

- **For a double bond X=Y in which the electronegativity of Y > X, a second resonance structure can be drawn by moving the π electrons onto Y.**

$X=Y \longleftrightarrow \overset{+}{X}-\overset{-}{\ddot{Y}}:$

Electronegativity of **Y** > **X.**

Sample Problem 14.1 illustrates how to apply these different types of resonance to actual molecules.

Sample Problem 14.1 Drawing Resonance Structures

Draw two more resonance structures for each species.

a. b.

Solution

Mentally breaking a molecule into two- or three-atom units can make it easier to draw additional resonance structures.

a. Think of the top three atoms of the six-membered ring in **A** as an "allyl" unit. Moving the π bond forms a new "allyl" unit in **B,** and moving the π bond in **B** generates a third resonance structure **C.** No new valid resonance structures are generated by moving electrons in **C.**

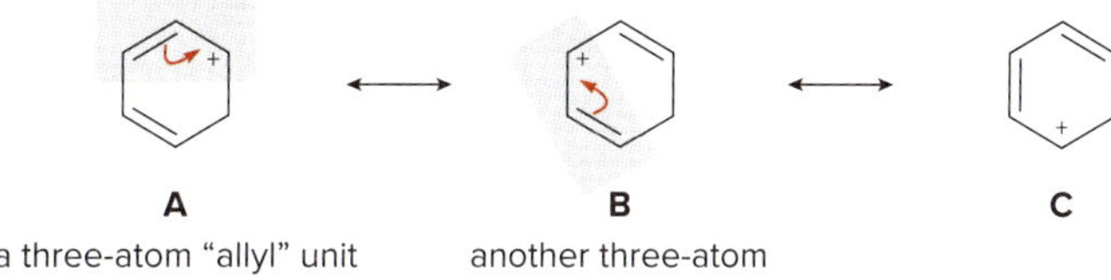

A a three-atom "allyl" unit — **B** another three-atom "allyl" unit — **C**

b. Compound **D** contains a carbonyl group, so moving the electron pair in the double bond to the more electronegative oxygen atom separates the charge and generates structure **E. E** now has a three-atom "allyl" unit, so the remaining π bond can be moved to form structure **F.**

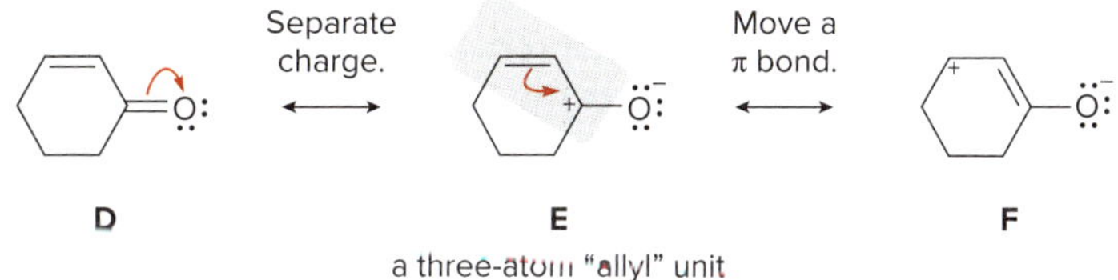

D — **E** a three-atom "allyl" unit — **F**

Problem 14.6 Draw additional resonance structures for each ion.

a. b. c. d.

More Practice: Try Problems 14.35, 14.36.

14.4 The Resonance Hybrid

The lower its energy, the more a resonance structure contributes to the overall structure of the hybrid.

Although the resonance hybrid is some combination of all of its valid resonance structures, the **hybrid more closely resembles the best resonance structure.** Recall from Section 1.6C that the best resonance structure is called the **major contributor** to the hybrid, and other resonance structures are called the **minor contributors.** Two identical resonance structures are equal contributors to the hybrid.

Use three rules to evaluate the relative energies of two or more valid resonance structures.

Rule [1] **Resonance structures with more bonds and fewer charges are better.**

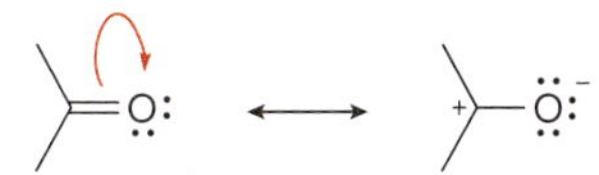

all neutral atoms
one more bond
better resonance structure

charge separation

Rule [2] **Resonance structures in which every atom has an octet are better.**

CH_3–Ö–$\overset{+}{C}H_2$ ⟷ CH_3–$\overset{+}{O}$=CH_2

All second-row elements have an **octet.**
better resonance structure

In this example, the resonance structure in which all atoms have octets is better, even though it places a (+) charge on a more electronegative O atom.

Rule [3] **Resonance structures that place a negative charge on a *more* electronegative atom are better.**

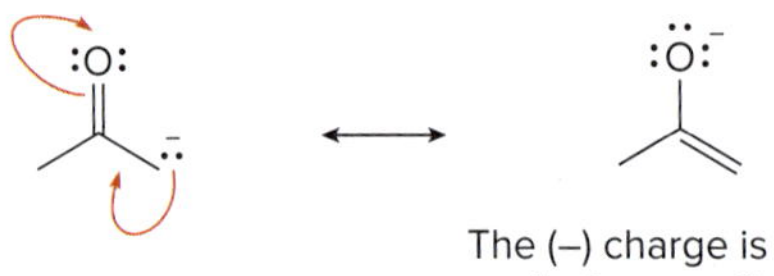

The (–) charge is on the
more electronegative O atom.

better resonance structure

Sample Problem 14.2 illustrates how to determine the relative energy of contributing resonance structures and the hybrid.

Sample Problem 14.2 Determining the Relative Energy of Resonance Structures and the Hybrid

Draw a second resonance structure for carbocation **A,** as well as the hybrid of both resonance structures. Then use Rules [1]–[3] to rank the relative stability of both resonance structures and the hybrid.

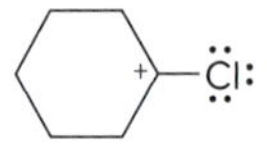

A

Solution

Because **A** contains a positive charge and a lone pair on adjacent atoms, a second resonance structure **B** can be drawn. Because **B** has more bonds and all second-row atoms have octets, **B** is a **better resonance** structure than **A,** making it the **major contributor** to the hybrid **C.** Because the hybrid is more stable than either resonance contributor, the order of stability is:

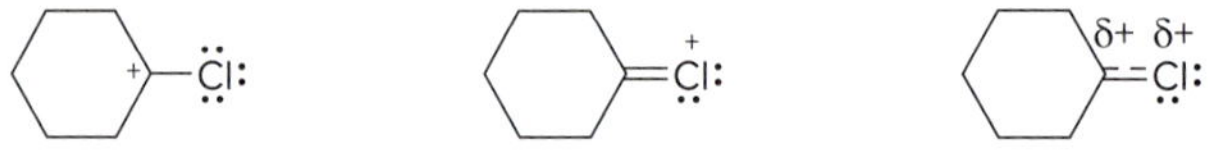

A	**B**	**C**
minor contributor	major contributor	hybrid

δ+ δ+

Increasing stability →

Problem 14.7 Draw a second resonance structure and the hybrid for each species, and then rank the two resonance structures and the hybrid in order of increasing stability.

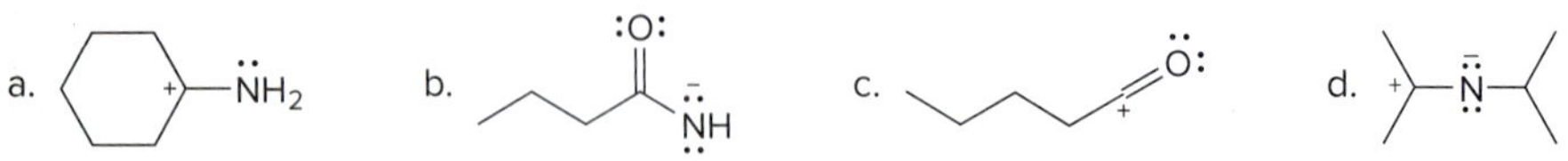

a. $\ddot{N}H_2$ b. :O: $\ddot{N}H$ c. $\ddot{O}$: d. $\ddot{N}$

Problem 14.8 Draw all possible resonance structures for each ion, and indicate which structure makes the largest contribution to the resonance hybrid.

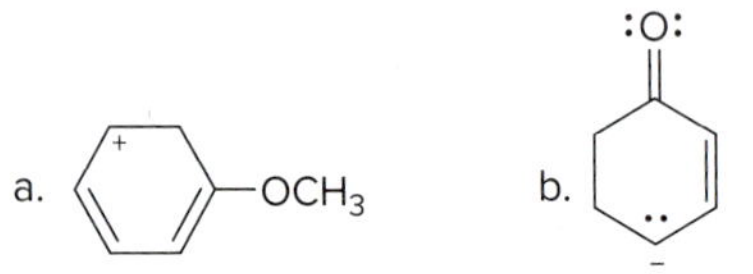

a. OCH_3 b. :O:

14.5 Electron Delocalization, Hybridization, and Geometry

To delocalize nonbonded electrons or electrons in π bonds, there must be p orbitals that can overlap. This may mean that the hybridization of an atom is *different* than would have been predicted using the rules first outlined in Chapter 1.

For example, there are two Lewis structures (**A** and **B**) for the resonance-stabilized anion $(CH_3COCH_2)^-$.

A ⟷ **B**

The labeled C is surrounded by four groups—three atoms and one nonbonded electron pair. **Is it sp^3 hybridized?**

The labeled C is surrounded by three groups—three atoms and no nonbonded electron pairs. **Is it sp^2 hybridized?**

Based on structure **A,** the labeled carbon is sp^3 hybridized, with the lone pair of electrons in an sp^3 hybrid orbital. Based on structure **B,** though, it is sp^2 hybridized with the unhybridized p orbital forming the π portion of the double bond.

Delocalizing electrons stabilizes a molecule. The electron pair on the carbon atom adjacent to the C=O can only be delocalized, though, if it has a p orbital that can overlap with two other p orbitals on two adjacent atoms. Thus, the terminal carbon atom is sp^2 hybridized with trigonal planar geometry. **Three adjacent p orbitals make the anion conjugated.**

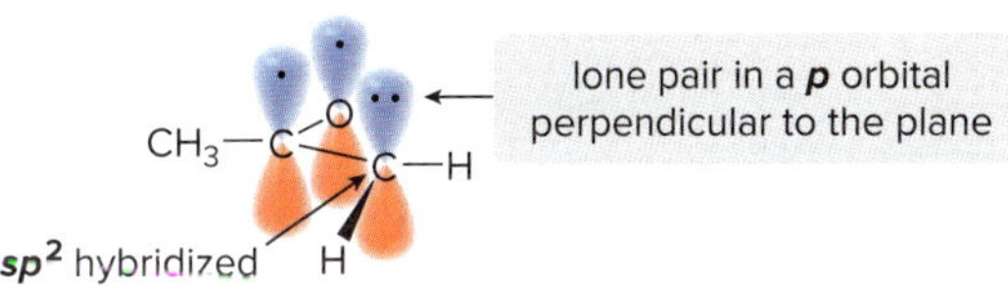

- **In a system X=Y–Z:, Z is generally sp^2 hybridized, and the nonbonded electron pair occupies a p orbital to make the system conjugated.**

Sample Problem 14.3 Determining Hybridization in a Conjugated System

Determine the hybridization around the labeled carbon atom in the following anion.

Solution

Because this is an example of an allyl-type system (X=Y–Z*), a second resonance structure can be drawn that "moves" the lone pair and the π bond. To delocalize the lone pair and make the system conjugated, the **labeled carbon atom must be sp^2 hybridized with the lone pair occupying a p orbital.**

The labeled C atom must be sp^2 hybridized, with the lone pair in a p orbital.

Problem 14.9 Determine the hybridization of the labeled atom in each species.

a. b. c. d.

More Practice: Try Problem 14.37.

14.6 Conjugated Dienes

Compounds with many π bonds are called **polyenes.**

In the remainder of Chapter 14 we examine **conjugated dienes,** compounds having two double bonds joined by one σ bond. Conjugated dienes are also called **1,3-dienes.** Buta-1,3-diene ($CH_2{=}CH{-}CH{=}CH_2$) is the simplest conjugated diene.

Three stereoisomers are possible for 1,3-dienes with alkyl groups bonded to each end carbon of the diene (RCH=CH–CH=CHR).

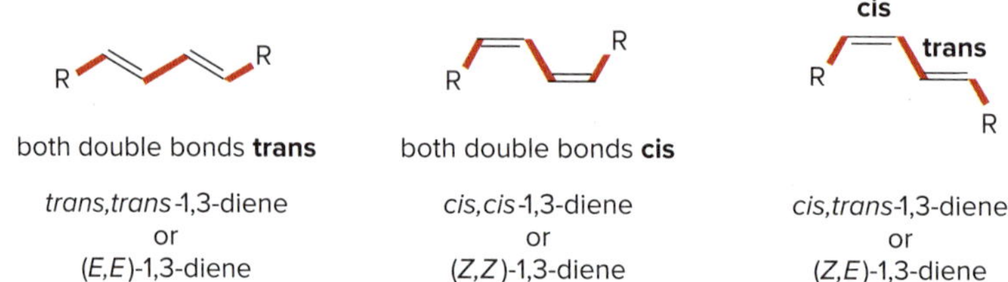

both double bonds **trans** — *trans,trans*-1,3-diene or (*E,E*)-1,3-diene

both double bonds **cis** — *cis,cis*-1,3-diene or (*Z,Z*)-1,3-diene

cis,trans-1,3-diene or (*Z,E*)-1,3-diene

Two possible conformations result from rotation about the C–C bond that joins the two double bonds.

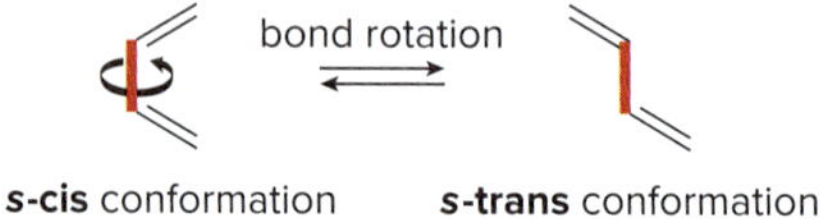

***s*-cis** conformation — ***s*-trans** conformation

- **The *s*-cis conformation has two double bonds on the *same* side of the single bond.**
- **The *s*-trans conformation has two double bonds on *opposite* sides of the single bond.**

Keep in mind that **stereoisomers are discrete molecules,** whereas **conformations interconvert.** Three structures drawn for hexa-2,4-diene illustrate the differences between stereoisomers and conformations in a 1,3-diene:

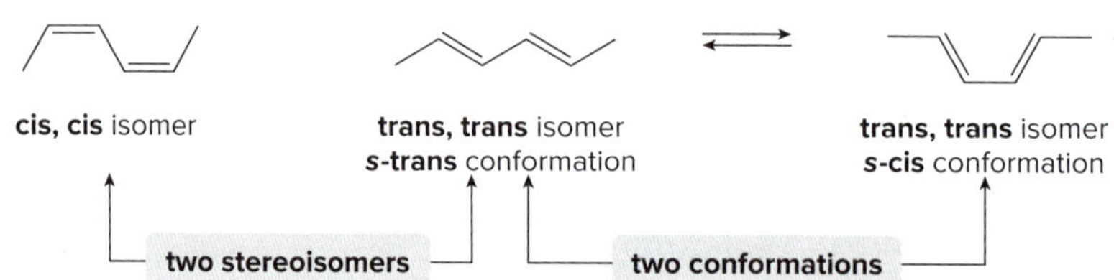

Sample Problem 14.4 Classifying Compounds as Stereoisomers or Different Conformations

Classify each pair of compounds as stereoisomers or conformations: (a) **X** and **Y;** (b) **X** and **Z.**

X Y Z

Solution

- **Stereoisomers are *different* compounds.** Groups on each end of a carbon–carbon double bond are arranged differently.
- **Two conformations are the *same* compound,** which interconvert by bond rotation.

a. **X** and **Y** are **stereoisomers** because the groups around the C=C in blue are arranged differently; in **X** two groups are trans, and in **Y** two groups are cis.

b. Each C=C in **X** and **Z** is bonded to the same groups and has the *E* configuration. **X** has the two double bonds on opposite sides of the C–C in blue, whereas **Z** has two double bonds on the same side of the single bond that joins them together. **X** and **Z** are different **conformations.**

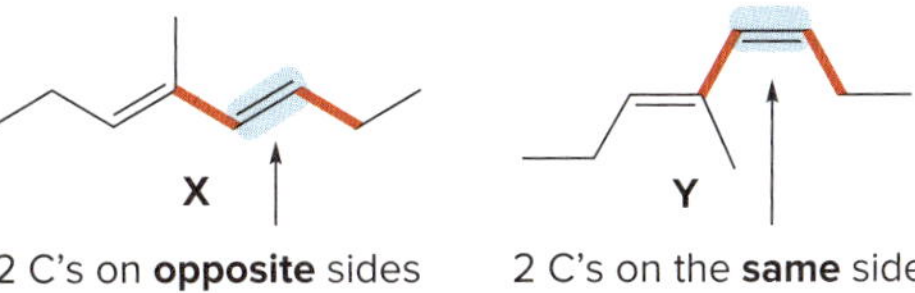

X — 2 C's on **opposite** sides of a C=C

Y — 2 C's on the **same** side of a C=C

stereoisomers

X — 2 C's on **opposite** sides of the C–C *s*-trans

Z — 2 C's on the **same** side of the C–C *s*-cis

conformations

Problem 14.10 Label compounds **B–D** as stereoisomers, conformations, or constitutional isomers of **A.**

A B C D

More Practice: Try Problem 14.41.

Problem 14.11 Draw the structure consistent with each description.

a. (2*E*,4*E*)-octa-2,4-diene in the *s*-trans conformation
b. (3*E*,5*Z*)-nona-3,5-diene in the *s*-cis conformation
c. (3*Z*,5*Z*)-4,5-dimethyldeca-3,5-diene. Draw both the *s*-cis and *s*-trans conformations.

Problem 14.12 Neuroprotectin D1 (NPD1) is synthesized in the body from highly unsaturated essential fatty acids. NPD1 is a potent natural anti-inflammatory agent.

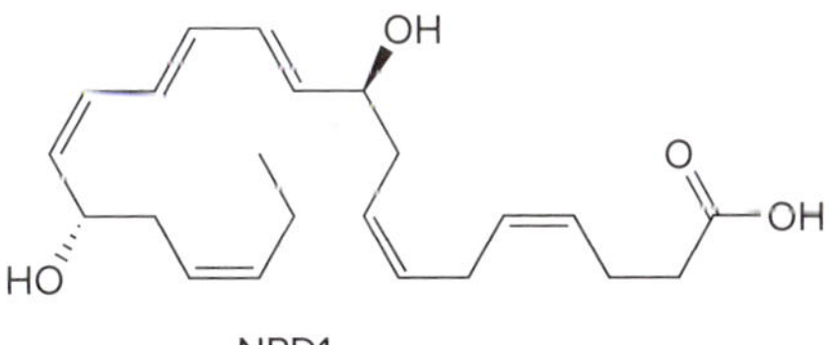

NPD1

a. Label each carbon–carbon double bond as conjugated or isolated.
b. Label each double bond as *E* or *Z*.
c. For each conjugated system, label the given conformation as *s*-cis or *s*-trans.

14.7 Interesting Dienes and Polyenes

Isoprene is a component of the blue haze seen above forested hillsides, such as Virginia's Blue Ridge Mountains.
daveallenphoto/123RF

Isoprene and **lycopene** are two naturally occurring compounds with conjugated double bonds.

isoprene
(2-methylbuta-1,3-diene)

11 conjugated double bonds shown in red

lycopene

Isoprene, the common name for 2-methylbuta-1,3-diene, is given off by plants as the temperature rises, a process thought to increase a plant's tolerance for heat stress.

Lycopene, a naturally occurring molecule responsible for the red color of tomatoes and other fruits, is an antioxidant like vitamin E. The 11 conjugated double bonds of lycopene cause its red color, a phenomenon discussed in Section 14.15A.

Vitamins A and **D** are two fat-soluble vitamins that contain conjugated double bonds. As mentioned in the chapter opener, vitamin A is the starting material for a series of reactions that are responsible for vision in all vertebrates, and vitamin D helps regulate both calcium and phosphorus metabolism. Milk is fortified with vitamin D so that we get enough of this vital nutrient.

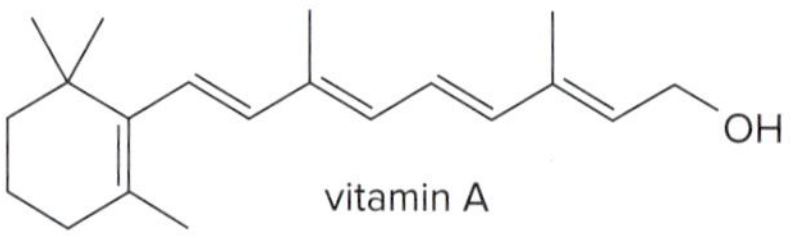

OH

vitamin A

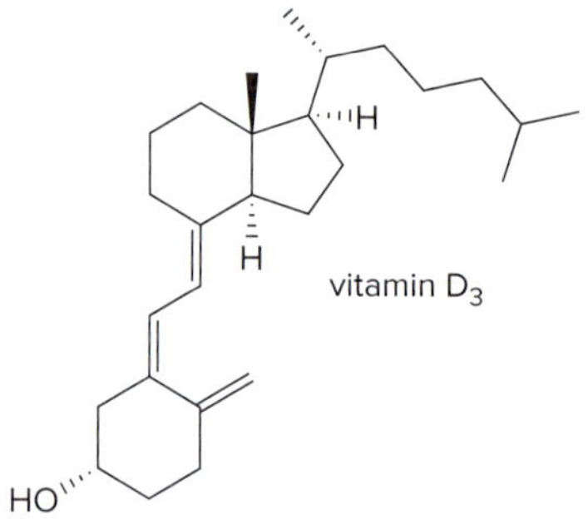

H

H

vitamin D_3

HO

Polyene antibiotics constitute a group of antimicrobial drugs that target fungi. Examples include amphotericin B (Problem 5.65) and nystatin (trade name Mycostatin), which contains six conjugated π bonds. Nystatin, which is used to treat *Candida* infections of the skin, is thought to bind to the steroid ergosterol in the fungal cell membrane, thereby disrupting the normal passage of ions and molecules across the membrane.

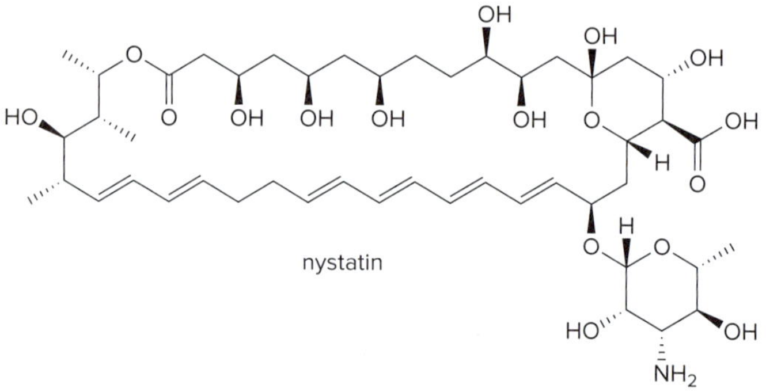

OH OH OH O OH OH OH OH OH O OH H O HO O H O nystatin HO OH NH_2

Problem 14.13 (a) Label the double bonds in vitamin D_3 as *E*, *Z*, or neither. (b) Label the conformation of the C–C bond that joins each pair of double bonds as *s*-cis or *s*-trans.

14.8 The Carbon–Carbon σ Bond Length in Buta-1,3-diene

Four features distinguish conjugated dienes from isolated dienes:

[1] The C–C single bond joining the two double bonds is unusually short.

[2] Conjugated dienes are more stable than similar isolated dienes.

[3] Some reactions of conjugated dienes are different than reactions of isolated double bonds.

[4] Conjugated dienes absorb longer wavelengths of ultraviolet light.

Hybridization can explain why the central carbon–carbon single bond is shorter than the C–C bond in ethane (148 pm vs. 153 pm).

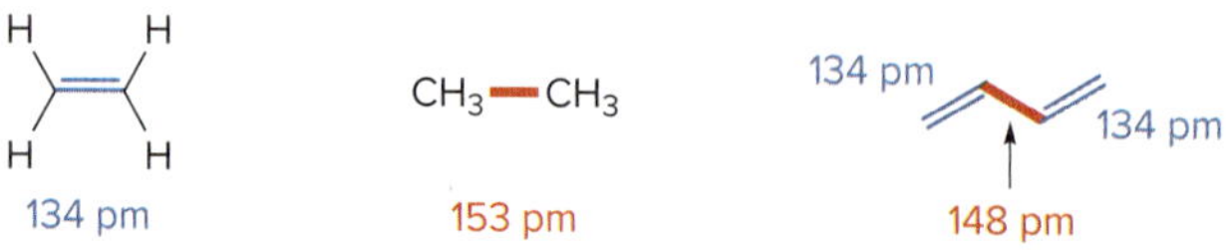

H H H H

134 pm

CH_3—CH_3

153 pm

134 pm 134 pm

148 pm

Each carbon atom in buta-1,3-diene is sp^2 hybridized, so the central C–C single bond is formed by the overlap of **two sp^2 hybridized orbitals,** rather than the sp^3 hybridized orbitals used to form the C–C bond in CH_3CH_3.

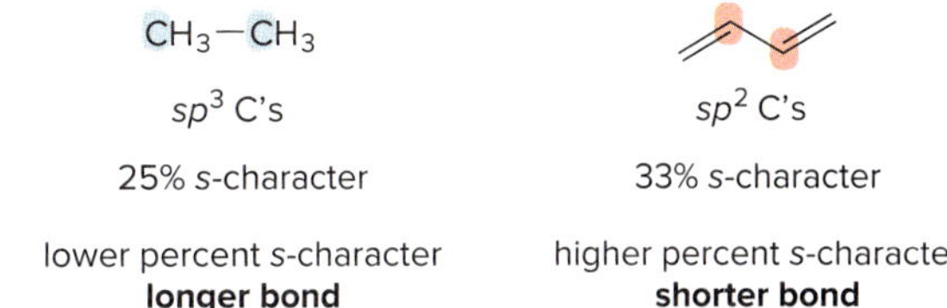

Recall from Section 1.11B that increasing percent *s*-character decreases bond length.

- **Based on hybridization, a C_{sp^2}–C_{sp^2} bond should be shorter than a C_{sp^3}–C_{sp^3} bond because it is formed from orbitals having a *higher* percent *s*-character.**

Problem 14.14 Using hybridization, predict how the bond length of the C–C σ bond in HC≡C–C≡CH should compare with the C–C σ bonds in CH_3CH_3 and CH_2=CH–CH=CH_2.

Problem 14.15 Use resonance theory to explain why the labeled C–O bond lengths are equal in the acetate anion.

:O:

:Ö:⁻

acetate

14.9 Stability of Conjugated Dienes

In Section 12.3, we learned that hydrogen adds to alkenes to form alkanes and that the heat released in this reaction, the **heat of hydrogenation,** can be used as a measure of alkene stability.

+ H_2 —Pd-C→ H H **$\Delta H°$ = heat of hydrogenation**

The relative stability of conjugated and isolated dienes can also be determined by comparing their heats of hydrogenation.

- **When hydrogenation gives the same alkane from two dienes, the more stable diene has the *smaller* heat of hydrogenation.**

For example, both penta-1,4-diene (an isolated diene) and (*E*)-penta-1,3-diene (a conjugated diene) are hydrogenated to pentane with two equivalents of H_2. Because *less* energy is released in converting the conjugated diene to pentane, it must be *lower in energy* (more stable) to begin with. The relative energies of these isomeric pentadienes are illustrated in Figure 14.2.

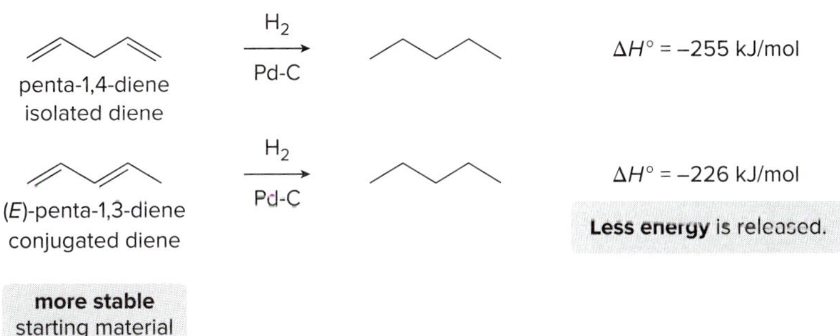

- **A conjugated diene has a smaller heat of hydrogenation and is more stable than a similar isolated diene.**

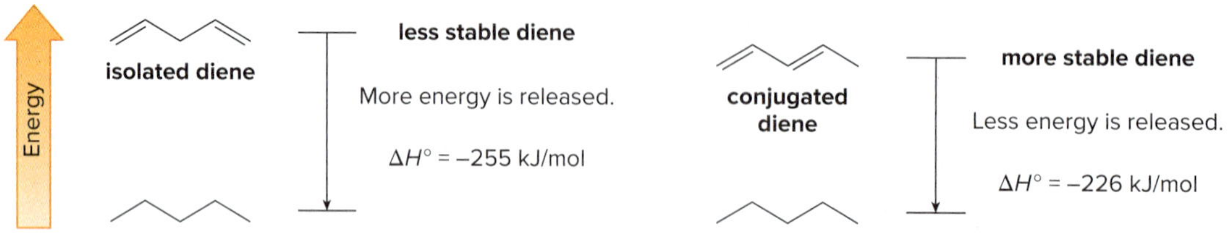

Figure 14.2 Relative energies of an isolated and conjugated diene

In Section 14.1, we learned why a conjugated diene is more stable than an isolated diene. A conjugated diene has overlapping *p* orbitals on four adjacent atoms, so its **π electrons are delocalized over four atoms, thus stabilizing the diene.** This delocalization cannot occur in an isolated diene, so an isolated diene is less stable than a conjugated diene.

Problem 14.16 Which compound in each pair has the larger heat of hydrogenation?

a. 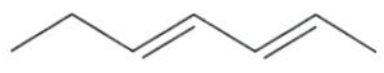 or 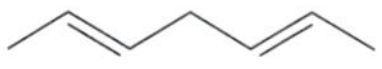 b. 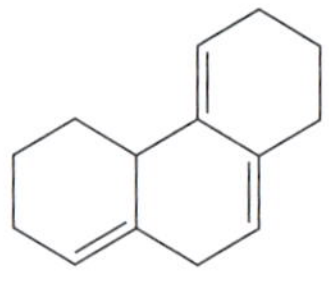 or 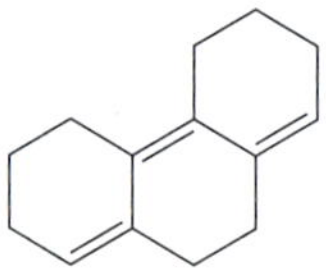

Problem 14.17 Rank the following compounds in order of increasing stability.

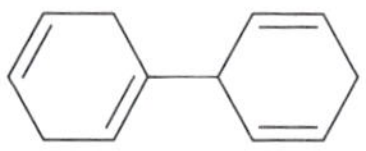
A

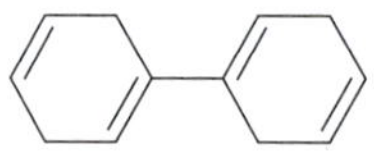
B

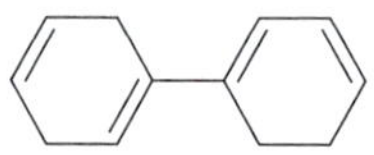
C

14.10 Electrophilic Addition: 1,2- Versus 1,4-Addition

Recall from Chapters 10 and 11 that the characteristic reaction of compounds with π bonds is **addition.** The π bonds in conjugated dienes undergo addition reactions, too, but they differ in two ways from the addition reactions to isolated double bonds.

- **Electrophilic addition in conjugated dienes gives a mixture of products.**
- **Conjugated dienes undergo a unique addition reaction not seen in alkenes or isolated dienes.**

We learned in Chapter 10 that HX adds to the π bond of alkenes to form alkyl halides.

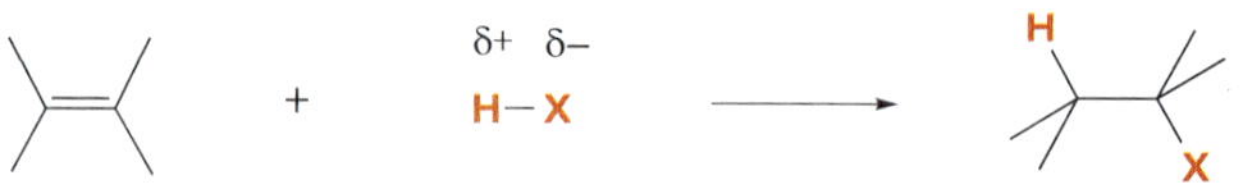

A π bond is broken.

alkyl halide

With an **isolated diene,** electrophilic addition of one equivalent of HBr yields *one* product and Markovnikov's rule is followed. The H atom bonds to the less substituted carbon—that is, the carbon atom of the double bond that had more H atoms to begin with.

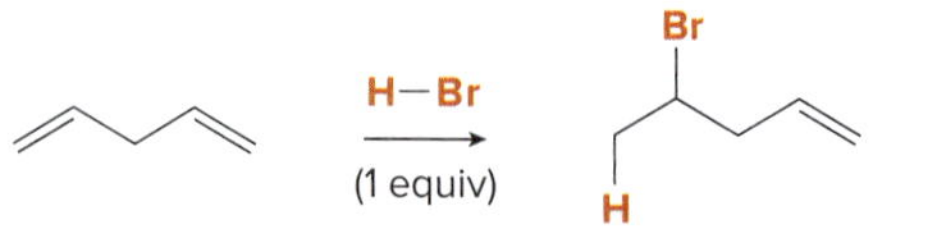

isolated diene

H bonds to the less substituted carbon.

With a conjugated diene, electrophilic addition of one equivalent of HBr affords *two* products.

- The **1,2-addition product** results from Markovnikov addition of HBr across two adjacent carbon atoms (C1 and C2) of the diene.
- The **1,4-addition product** results from addition of HBr to the two end carbons (C1 and C4) of the diene. 1,4-Addition is also called **conjugate addition.**

The ends of the 1,3-diene are called C1 and C4 arbitrarily, without regard to IUPAC numbering.

The mechanism of electrophilic addition of HX involves **two steps:** addition of H^+ (from HX) to form a **resonance-stabilized carbocation,** followed by nucleophilic attack of X^- at either electrophilic end of the carbocation to form two products. Mechanism 14.2 illustrates the reaction of buta-1,3-diene with HBr.

Mechanism 14.2 Electrophilic Addition of HBr to a 1,3-Diene—1,2- and 1,4-Addition

1 H^+ of HBr adds to a terminal carbon of the 1,3-diene to form a **resonance-stabilized allylic carbocation.**

2 **Nucleophilic attack of Br^-** occurs at either site of the resonance-stabilized carbocation that bears a (+) charge, forming the 1,2- and 1,4-addition products.

Like the electrophilic addition of HX to an alkene, the addition of HBr to a conjugated diene forms the more stable carbocation in Step [1], the rate-determining step. In this case, however, the carbocation is both 2° and **allylic,** and thus two Lewis structures can be drawn for it. In the second step, nucleophilic attack of Br^- can then occur at two different electrophilic sites, forming two different products.

- **Addition of HX to a conjugated diene forms 1,2- and 1,4-products because of the resonance-stabilized allylic carbocation intermediate.**

Sample Problem 14.5 Drawing the Products of 1,2- and 1,4-Addition

Draw the products of the following reaction.

Solution

Write the steps of the mechanism to determine the structure of the products. Addition of H^+ forms the more stable 2° allylic carbocation, for which two resonance structures can be drawn. **H^+ always bonds to a terminal carbon of the 1,3-diene,** labeled in blue. Nucleophilic attack of Br^- at either end of the allylic carbocation gives two constitutional isomers, formed by 1,2-addition and 1,4-addition to the diene.

H–Br:

:Br:⁻ H

2° allylic carbocation

1,2-product

1,4-product

Problem 14.18 Draw all products formed when each diene is treated with one equivalent of HCl.

a. b. c. d.

More Practice: Try Problems 14.43; 14.62a, d.

14.11 Kinetic Versus Thermodynamic Products

The amount of 1,2- and 1,4-addition products formed in the electrophilic addition reactions of buta-1,3-diene, a conjugated diene, depends greatly on the reaction conditions.

HBr →

	1,2-product	1,4-product
low temperature (–80 °C)	80%	20%
high temperature (40 °C)	20%	80%

- **At low temperature the major product is formed by 1,2-addition.**
- **At higher temperature the major product is formed by 1,4-addition.**

Moreover, when a mixture containing predominately the 1,2-product is heated, the 1,4-addition product becomes the major product at equilibrium.

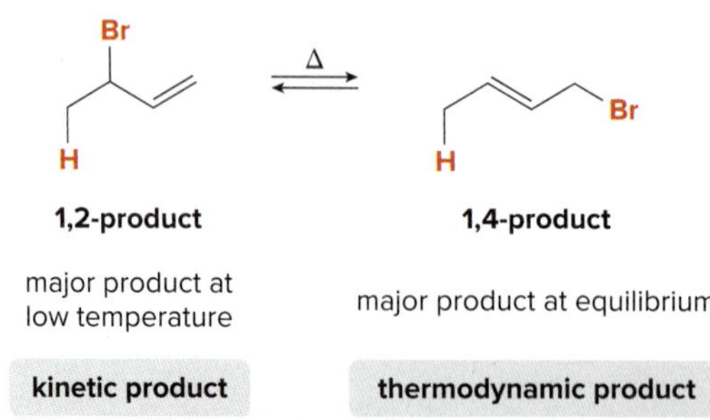

major product at low temperature — **kinetic product**

major product at equilibrium — **thermodynamic product**

- **The 1,2-product must be formed *faster* because it predominates at low temperature. The product that is formed faster is called the *kinetic product.***
- **The 1,4-product must be *more stable* because it predominates at equilibrium. The product that predominates at equilibrium is called the *thermodynamic product.***

In many of the reactions we have learned thus far, the more stable product is formed faster—that is, the kinetic and thermodynamic products are the same. The electrophilic addition of HBr to buta-1,3-diene is different, in that **the more stable product is formed more slowly**—that is, the kinetic and thermodynamic products are *different.* Why is the more stable product formed more slowly?

To answer this question, recall that the **rate of a reaction is determined by its energy of activation (E_a),** whereas the **amount of product present at equilibrium is determined by its stability** (Figure 14.3). When a single starting material **A** forms two different products (**B** and **C**) by two exothermic pathways, the relative height of the energy barriers determines how fast **B** and **C** are formed, whereas the relative energies of **B** and **C** determine the amount of each at equilibrium. In an exothermic reaction, the relative energies of **B** and **C** do not determine the relative energies of activation to form **B** and **C.**

Figure 14.3 How kinetic and thermodynamic products form in a reaction: **A → B + C**

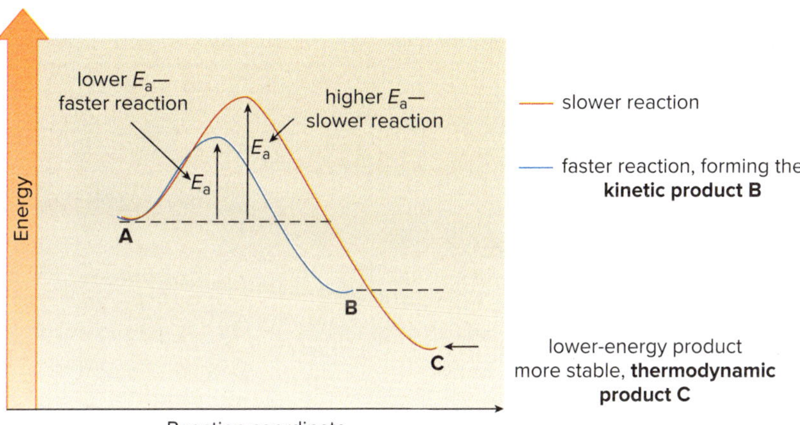

- The conversion of **A → B** is a faster reaction because the energy of activation leading to **B** is *lower.* **B** is the **kinetic product.**
- Because **C** is *lower* in energy, **C** is the **thermodynamic product.**

Why, in the addition of HBr to buta-1,3-diene, is the 1,4-product the more stable thermodynamic product? The 1,4-product (1-bromobut-2-ene) is more stable because it has two alkyl groups bonded to the carbon–carbon double bond, whereas the 1,2-product (3-bromobut-1-ene) has only one.

Br / H	Br / H
3-bromobut-1-ene	1-bromobut-2-ene
1,2-product	**1,4-product**
monosubstituted alkene	disubstituted alkene
less stable	**more stable**
	thermodynamic product

- **The more substituted alkene—1-bromobut-2-ene in this case—is the thermodynamic product.**

A **proximity effect** occurs because one species is close to another.

The 1,2-product is the kinetic product because of a **proximity effect.** When H^+ (from HBr) adds to the double bond, Br^- is *closer* to the adjacent carbon (C2) than it is to C4. Even though the resonance-stabilized carbocation bears a partial positive charge on both C2 and C4, attack at C2 is faster simply because Br^- is closer to this carbon.

1,2-addition product

kinetic product

Br^- is closer to C2 than C4.

- The **kinetic product** is always the **1,2-product.**
- The **thermodynamic product** always has the **more substituted double bond.**

- **The 1,2-product forms faster because of the proximity of Br^- to C2.**

The overall two-step mechanism for addition of HBr to buta-1,3-diene, forming a 1,2-addition product and 1,4-addition product, is illustrated with the energy diagram in Figure 14.4.

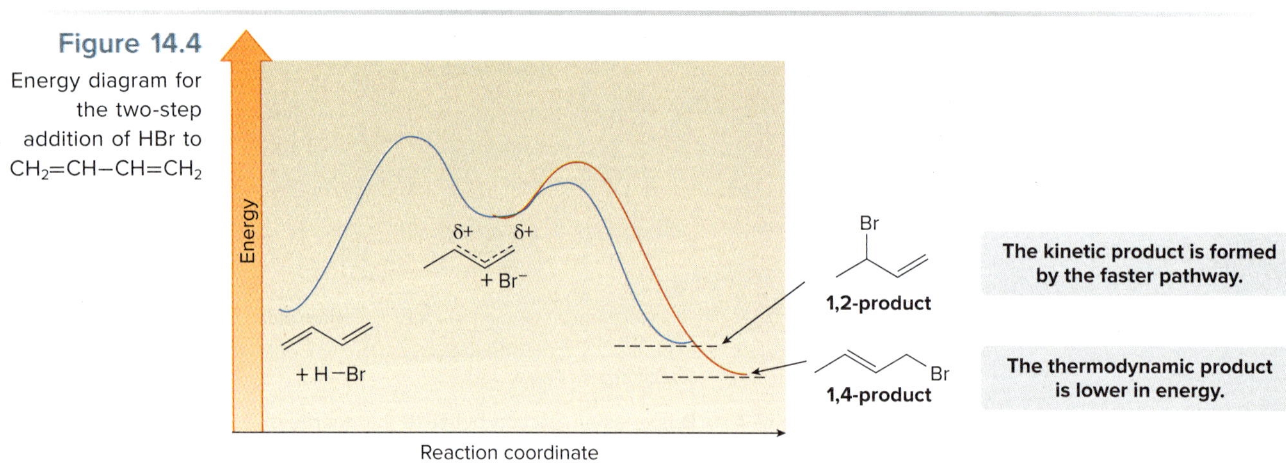

Figure 14.4 Energy diagram for the two-step addition of HBr to $CH_2{=}CH{-}CH{=}CH_2$

Why is the ratio of products temperature dependent?

- **At low temperature, the energy of activation is the more important factor.** Because most molecules do not have enough kinetic energy to overcome the higher energy barrier at lower temperature, they react by the faster pathway, forming the kinetic product.
- **At higher temperature,** most molecules have enough kinetic energy to reach either transition state. The two products are in equilibrium with each other, and the **more stable compound**—which is lower in energy—**becomes the major product.**

Problem 14.19 Label each product in the following reaction as a 1,2-product or a 1,4-product, and decide which is the kinetic product and which is the thermodynamic product.

Problem 14.20 Draw the products formed when each diene is treated with one equivalent of HCl, and designate each compound as a 1,2-product or a 1,4-product. Label the kinetic and thermodynamic products.

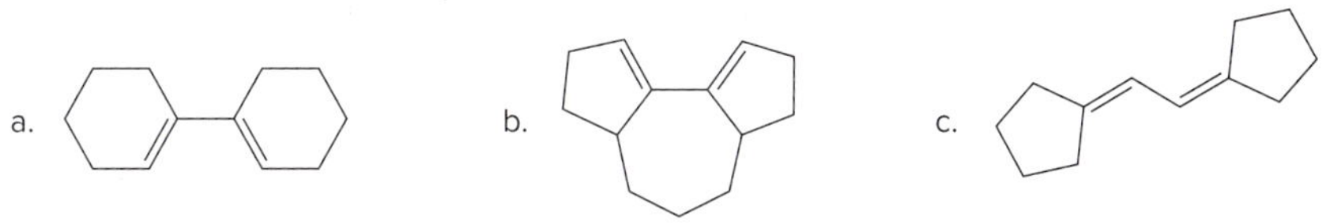

a. b. c.

14.12 The Diels–Alder Reaction

Diels and Alder shared the 1950 Nobel Prize in Chemistry for unraveling the intricate details of this remarkable reaction.

The arrows may be drawn in a clockwise or counterclockwise direction to show the flow of electrons in a Diels–Alder reaction.

The **Diels–Alder reaction,** named for German chemists Otto Diels and Kurt Alder, is an addition reaction between a **1,3-diene** and an alkene called a **dienophile,** to form a new six-membered ring.

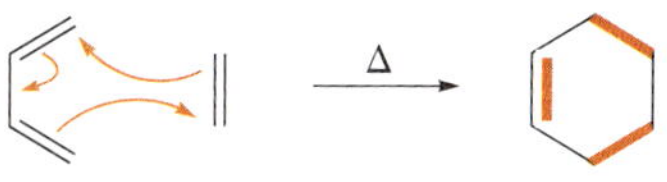

Δ

1,3-diene dienophile

Three curved arrows are needed to show the cyclic movement of electron pairs because three π bonds break and two σ bonds and one π bond form. Because each new σ bond is ~100 kJ/mol stronger than a π bond that is broken, a typical Diels–Alder reaction releases ~200 kJ/mol of energy. The following equations illustrate two examples of the Diels–Alder reaction.

1,3-Diene Dienophile Diels–Alder product

OCH_3 Δ OCH_3

The three new bonds are in red.

Δ

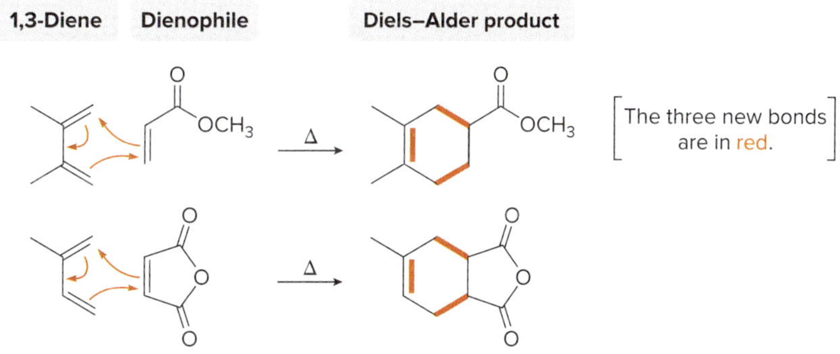

All Diels–Alder reactions have these features in common:

[1] They are initiated by heat; that is, the Diels–Alder reaction is a *thermal* reaction.

[2] They form new six-membered rings.

[3] Three π bonds break, and two new C–C σ bonds and one new C–C π bond form.

[4] They are concerted; that is, all bonds are broken and formed in a single step.

The Diels–Alder reaction forms new carbon–carbon bonds, so it can be used to synthesize larger, more complex molecules from smaller ones. For example, reaction of diene **A** with the dienophile methyl acrylate forms Diels–Alder adduct **B,** which can be converted to shikimic acid in four steps. Shikimic acid is the starting material for the synthesis of oseltamivir (trade name Tamiflu), the antiviral drug used to treat influenza mentioned in Problem 3.10.

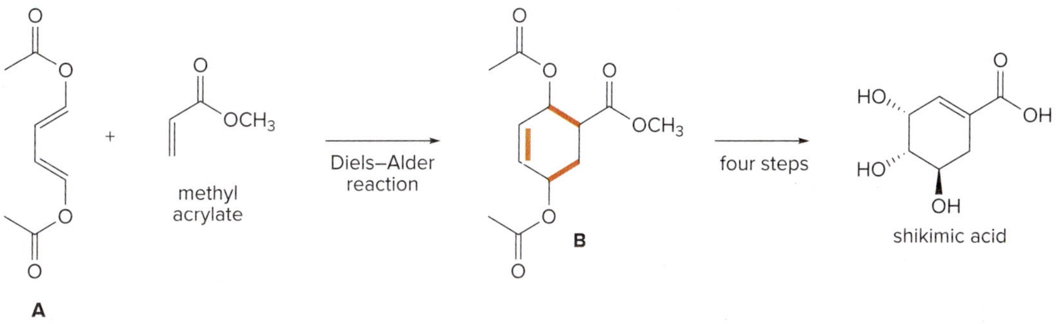

A + methyl acrylate —Diels–Alder reaction→ **B** —four steps→ shikimic acid

Diels–Alder reactions may seem complicated at first, but they are really less complicated than many of the reactions you have already learned, especially those with multistep mechanisms and carbocation intermediates. **The key is to learn how to arrange the starting materials** to more easily visualize the structure of the product.

How To Draw the Product of a Diels–Alder Reaction

Example Draw the product of the following Diels–Alder reaction:

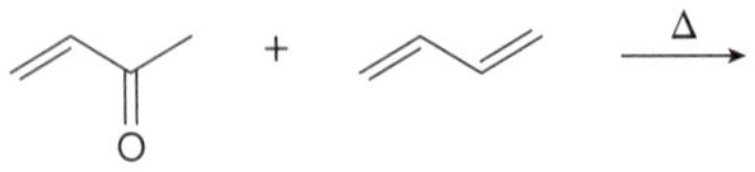

O + Δ

Step [1] **Arrange the 1,3-diene and the dienophile next to each other, with the diene drawn in the *s*-cis conformation.**

- This step is key: **Rotate the diene** so that it is drawn in the **s-cis** conformation, and **place the end C's of the diene close to the double bond of the dienophile.**

rotate

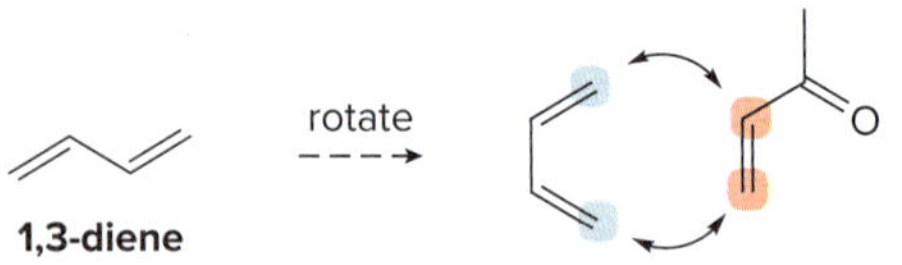

1,3-diene *s*-trans

s-cis **dienophile**

O

Step [2] **Cleave the three π bonds and use arrows to show where the new bonds will be formed.**

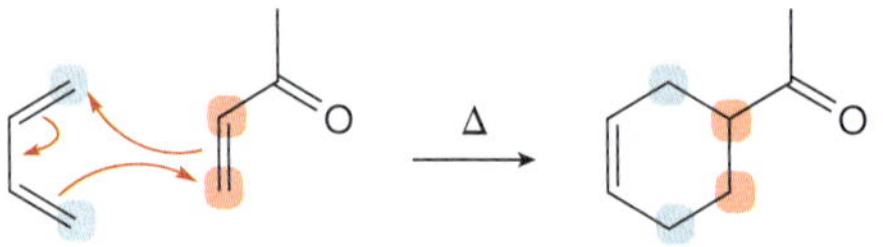

O Δ O

diene **dienophile** **Diels–Alder** product

Problem 14.21 Draw the product formed when each diene and dienophile react in a Diels–Alder reaction.

a. 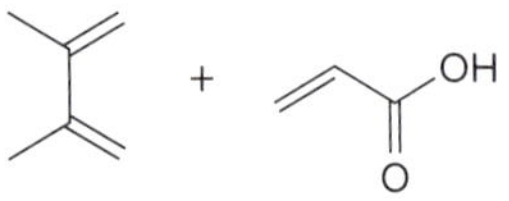 + OH O b. 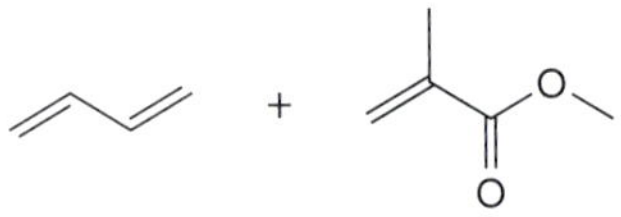 + O O c. 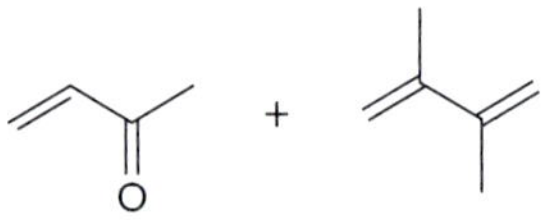 O +

14.13 Specific Rules Governing the Diels–Alder Reaction

Several rules govern the course of the Diels–Alder reaction.

14.13A Diene Reactivity

Rule [1] **The diene can react only when it adopts the *s*-cis conformation.**

Both ends of the conjugated diene must be close to the π bond of the dienophile for reaction to occur. Thus, an acyclic diene in the *s*-trans conformation must rotate about the central C–C σ bond to form the ***s*-cis conformation** before reaction can take place.

s-trans rotate **s-cis** reacting conformation

s-trans rotate **s-cis** reacting conformation

This rotation is prevented in cyclic dienes. As a result:

- **When the two double bonds are constrained in the *s*-cis conformation, the diene is unusually *reactive*.**
- **When the two double bonds are constrained in the *s*-trans conformation, the diene is *unreactive*.**

an **s-cis** 1,3-diene
very reactive

an **s-trans** 1,3-diene
unreactive diene

Zingiberene and β-sesquiphellandrene (Problem 14.23) are trienes obtained from ginger root. Ginger is used as a spice in Indian and Chinese cooking. Ginger candy is sometimes used to treat nausea resulting from seasickness. *Alvis Upitis/ Getty Images*

Problem 14.22 Label each diene as reactive or unreactive in a Diels–Alder reaction.

a. 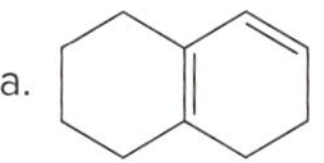 b. 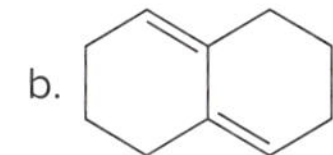 c. 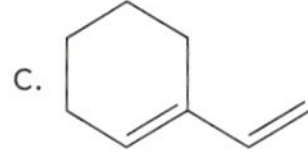 d. 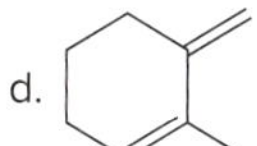 e. 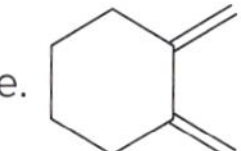

Problem 14.23 Zingiberene and β-sesquiphellandrene, natural products obtained from ginger root, contain conjugated diene units. Which diene reacts faster in the Diels–Alder reaction and why?

zingiberene

β-sesquiphellandrene

14.13B Dienophile Reactivity

Rule [2] **Electron-withdrawing substituents in the dienophile increase the reaction rate.**

In a Diels–Alder reaction, the conjugated diene acts as a nucleophile and the dienophile acts as an electrophile. As a result, **electron-withdrawing groups make the dienophile more electrophilic (and, thus, more reactive)** by withdrawing electron density from the carbon–carbon double bond. If Z is an electron-withdrawing group, then the reactivity of the dienophile increases as follows:

$CH_2{=}CH_2$ Z Z Z

Increasing reactivity

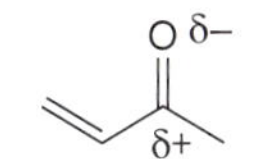

electron-deficient carbonyl carbon

A carbonyl group is an effective electron-withdrawing group because the carbonyl carbon bears a partial positive charge (δ+), which withdraws electron density from the carbon–carbon double bond of the dienophile. Common dienophiles that contain a carbonyl group are shown in Figure 14.5.

Problem 14.24 Rank the following dienophiles in order of increasing reactivity.

OH O **A**

HO OH O O **B**

C

Figure 14.5
Common dienophiles in the Diels–Alder reaction

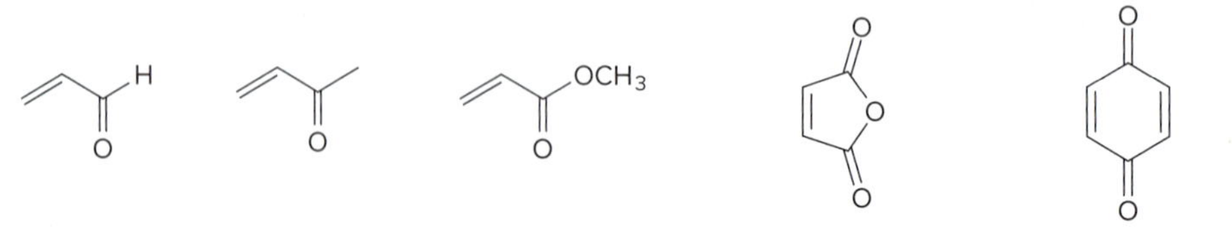

acrolein | methyl vinyl ketone | methyl acrylate | maleic anhydride | benzoquinone

14.13C Stereospecificity

Rule [3] The stereochemistry of the dienophile is retained in the product.

- **A cis dienophile forms a cis-substituted cyclohexene.**
- **A trans dienophile forms a trans-substituted cyclohexene.**

The two **cis** CO_2H groups of maleic acid become two **cis** substituents in a Diels–Alder adduct. The CO_2H groups can be drawn both above or both below the plane to afford a single achiral **meso** compound. The **trans dienophile** fumaric acid yields two enantiomers with **trans** CO_2H groups.

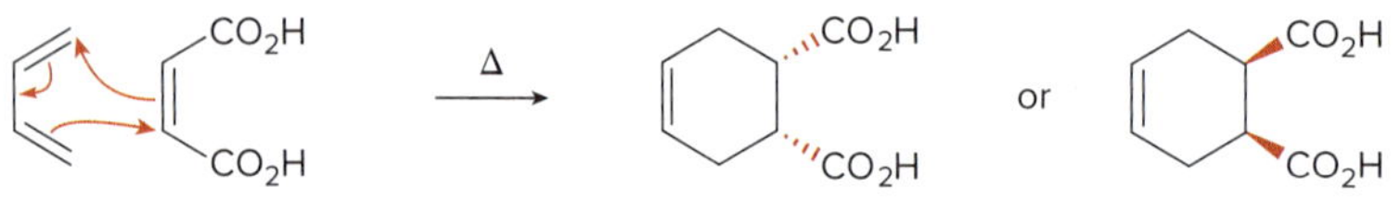

maleic acid
cis dienophile

an achiral meso compound

cis product

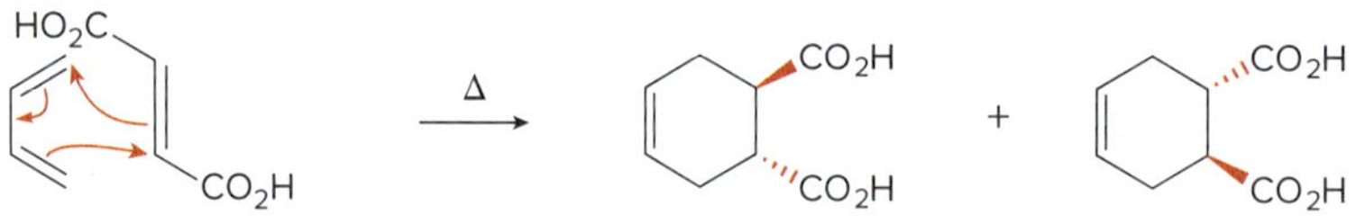

fumaric acid
trans dienophile

enantiomers

trans product

A **cyclic dienophile** forms a **bicyclic product.** A bicyclic system in which the two rings share a common C–C bond is called a **fused ring system.** The two H atoms at the ring fusion must be cis, because they were cis in the starting dienophile. A bicyclic system of this sort is said to be **cis-fused.**

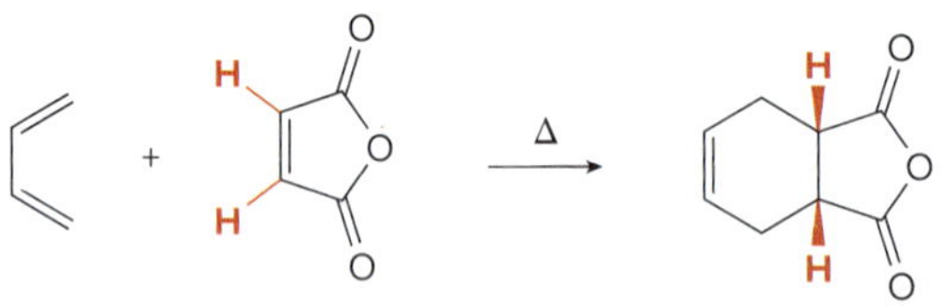

cyclic dienophile
cis H's in the dienophile

bicyclic product
cis H's in the product

Problem 14.25 Draw the Diels–Alder product from each pair of reactants, and indicate the stereochemistry.

a. 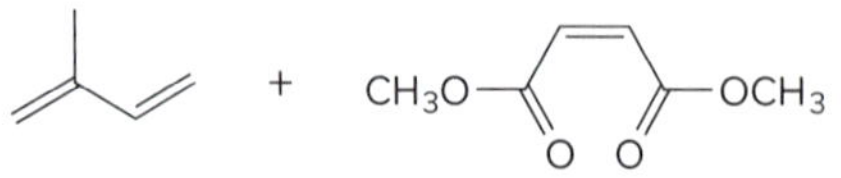

c. 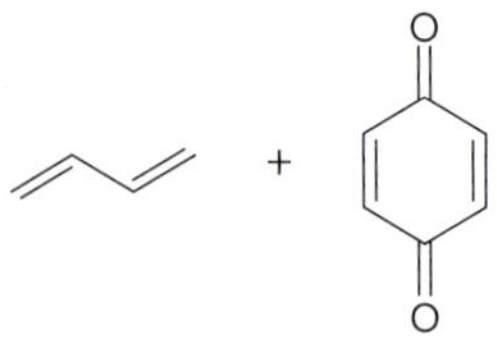

b. 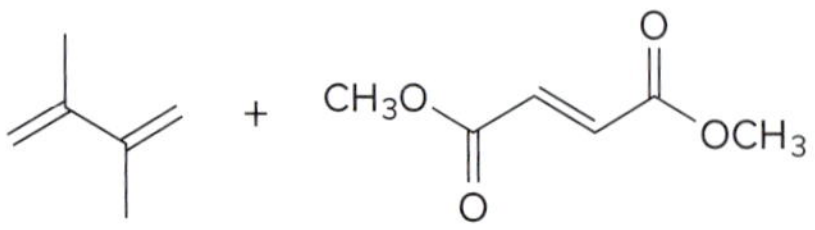

d. 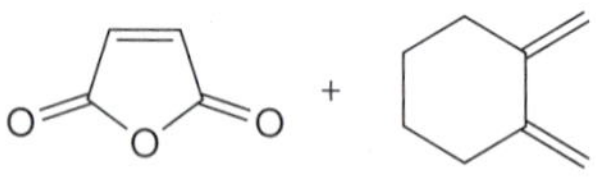

14.13D The Rule of Endo Addition

Rule [4] **When endo and exo products are possible, the endo product is preferred.**

To understand the rule of endo addition, we must first examine Diels–Alder products that result from cyclic 1,3-dienes. When cyclopentadiene reacts with a dienophile such as ethylene, a new six-membered ring forms, and above the ring there is a **one atom "bridge,"** labeled in green. This carbon atom originated as the sp^3 hybridized carbon of the diene that was not involved in the reaction.

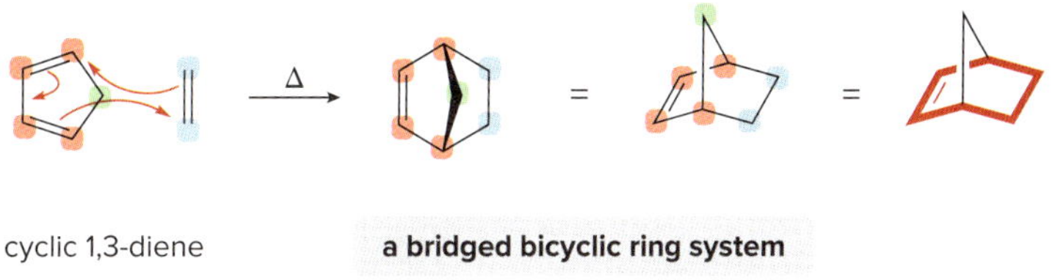

cyclic 1,3-diene

a bridged bicyclic ring system

The product of the Diels–Alder reaction of a cyclic 1,3-diene is bicyclic, but the carbon atoms shared by both rings are *non-adjacent.* Thus, this bicyclic product differs from the fused ring system obtained when the dienophile is cyclic.

- **A bicyclic ring system in which the two rings share non-adjacent carbon atoms is called a *bridged* ring system.**

Fused and bridged bicyclic ring systems are compared in Figure 14.6.

Figure 14.6 Fused and bridged bicyclic ring systems compared

a. A fused bicyclic system

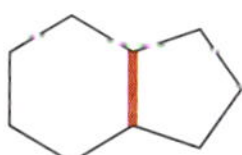

- One bond (in red) is shared by two rings.
- The shared C's are adjacent.

b. A bridged bicyclic system

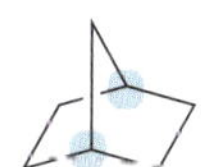

- Two non-adjacent atoms (labeled in blue) are shared by both rings.

When cyclopentadiene reacts with a substituted alkene as the dienophile ($CH_2{=}CHZ$), the substituent Z can be oriented in one of two ways in the product. The terms **endo** and **exo** are used to indicate the position of Z.

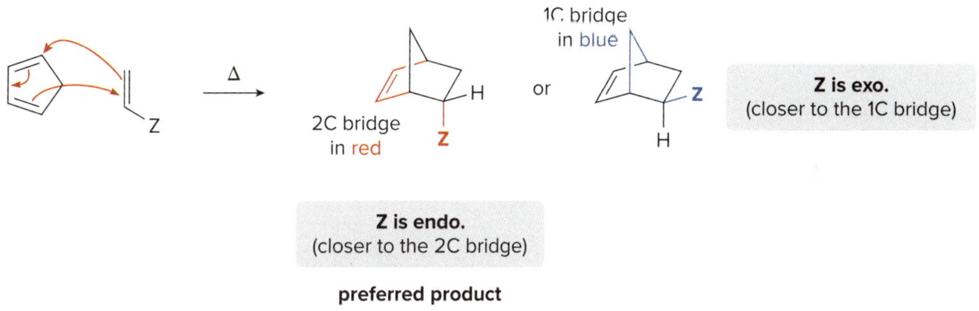

2C bridge in red

1C bridge in blue

Z is endo. (closer to the 2C bridge)

preferred product

Z is exo. (closer to the 1C bridge)

- **A substituent on one bridge is *endo* if it is closer to the *longer* bridge that joins the two carbons common to both rings.**
- **A substituent is *exo* if it is closer to the *shorter* bridge that joins the carbons together.**

To help you distinguish endo and exo, remember that e**n**do is u**n**der the newly formed six-membered ring.

newly formed ring (in red)

H

Z **endo**

In a Diels–Alder reaction, the **endo** product is preferred, as shown in two examples.

Δ

H

O

O

preferred product

Bonds in red are endo.

O

O

O

Δ

H

H

O

O

O

preferred product

More details on the Diels–Alder reaction are given in Section 25.4.

The Diels–Alder reaction is **concerted,** and the reaction occurs with the diene and the dienophile arranged one above the other, as shown in Figure 14.7, not side-by-side. In theory, the substituent Z can be oriented either directly *under* the diene to form the endo product (Pathway [1] in Figure 14.7) or *away* from the diene to form the exo product (Pathway [2] in Figure 14.7). In practice, though, the **endo product is the major product. The transition state leading to the endo product allows more interaction between the electron-rich diene and the electron-withdrawing substituent Z on the dienophile,** an energetically favorable arrangement.

Figure 14.7 How endo and exo products are formed in the Diels–Alder reaction

Pathway [1] With Z oriented under the diene, the endo product is formed.

transition state

+

Z H

Z under the diene

Δ

‡

Z H

flip up

Z H

Z is **below** the two new σ bonds (in red).

=

H

Z

endo product
major product

The electron-withdrawing Z group is closer to the electron-rich diene.

Pathway [2] With Z oriented away from the diene, the exo product is formed.

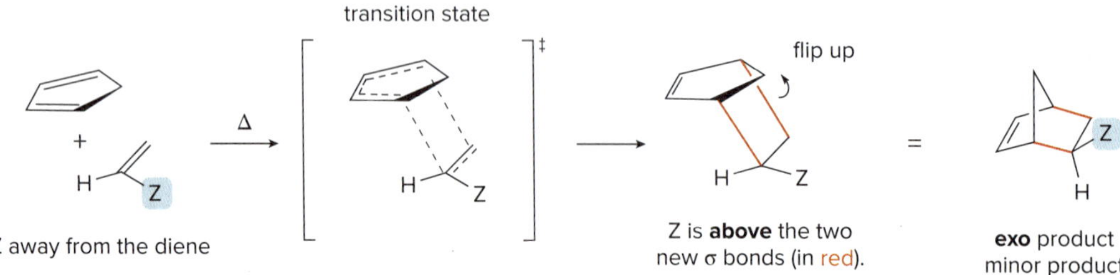

Problem 14.26 Draw the product of each Diels–Alder reaction.

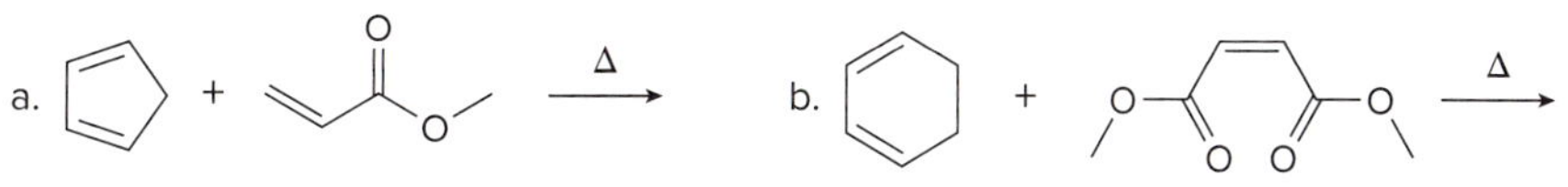

14.14 Other Facts About the Diels–Alder Reaction

14.14A Retrosynthetic Analysis of a Diels–Alder Product

The Diels–Alder reaction is used widely in organic synthesis, so you must be able to look at a compound and determine what conjugated diene and what dienophile were used to make it. To draw the starting materials from a given Diels–Alder adduct:

- **Locate the six-membered ring that contains the C=C.**
- **Draw three arrows around the cyclohexene ring, beginning with the π bond. Each arrow moves two electrons to the adjacent bond, cleaving one π bond and two σ bonds, and forming three π bonds.**
- **Retain the stereochemistry of substituents on the C=C of the dienophile. Cis substituents on the six-membered ring give a cis dienophile.**

This stepwise retrosynthetic analysis gives the 1,3-diene and dienophile needed for any Diels–Alder reaction, as shown in the two examples in Figure 14.8.

Figure 14.8 Finding the diene and dienophile needed for a Diels–Alder reaction

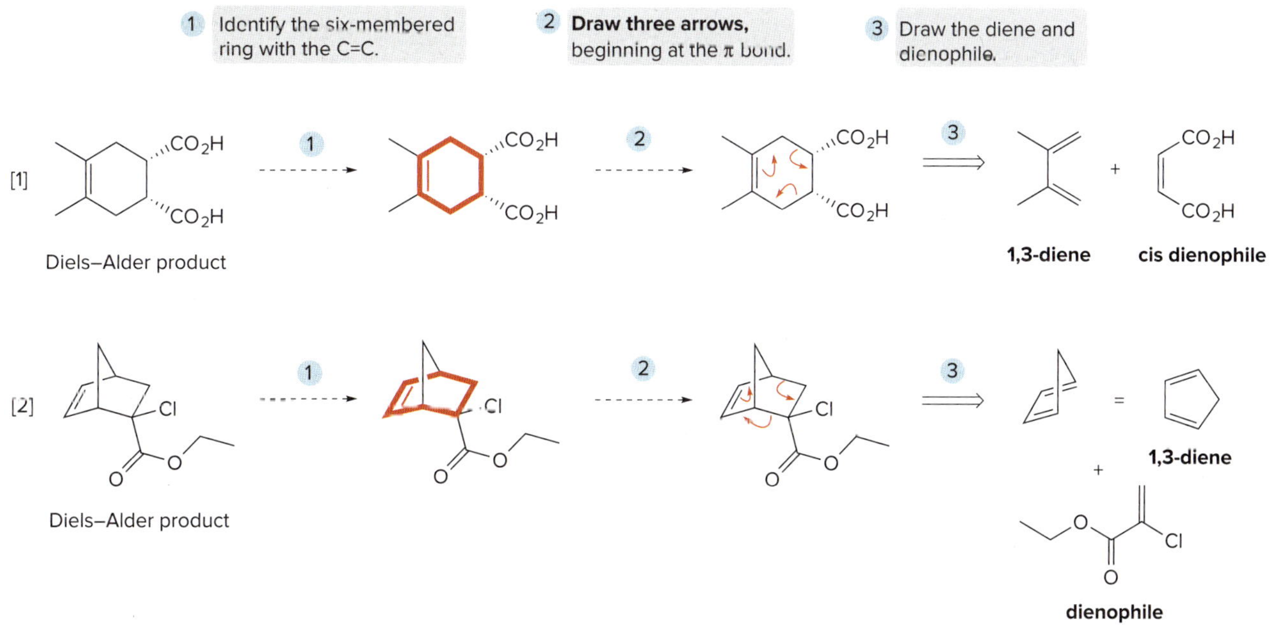

Japanese puffer fish

Stephen Frink/Photodisc/Getty Images

tetrodotoxin

Tetrodotoxin (Problem 14.27d) is a poison isolated from the ovaries and liver of the puffer fish, so named because the fish inflates itself into a ball when alarmed.

Problem 14.27 What diene and dienophile are needed to prepare each product? The compound in part (d) was an intermediate in the synthesis of tetrodotoxin, a complex marine natural product.

a. b. c. d.

14.14B Retro Diels–Alder Reaction

A reactive diene like cyclopenta-1,3-diene readily undergoes a Diels–Alder reaction with *itself;* that is, **cyclopenta-1,3-diene dimerizes because one molecule acts as the diene and another acts as the dienophile.**

diene dienophile → Δ dicyclopentadiene **dimer** = endo product

The formation of dicyclopentadiene is so rapid that it takes only a few hours at room temperature for cyclopentadiene to completely dimerize. How, then, can cyclopentadiene be used in a Diels–Alder reaction if it really exists as a dimer?

When heated, dicyclopentadiene undergoes a **retro Diels–Alder reaction,** and two molecules of cyclopentadiene are re-formed. If cyclopentadiene is immediately treated with a different dienophile, it reacts to form a new Diels–Alder adduct with this dienophile.

dicyclopentadiene → Δ two molecules of cyclopentadiene → Δ

This diene can now be used with a different dienophile.

Problem 14.28 Draw the products of each reaction sequence.

a. [1] Δ [2]

b. [1] Δ [2]

14.14C Application: Diels–Alder Reaction in the Synthesis of Steroids

Recall from Section 4.14 that lipids are water-insoluble biomolecules that have diverse structures.

***Steroids* are tetracyclic lipids containing three six-membered rings and one five-membered ring.** The four rings are designated as **A, B, C,** and **D.**

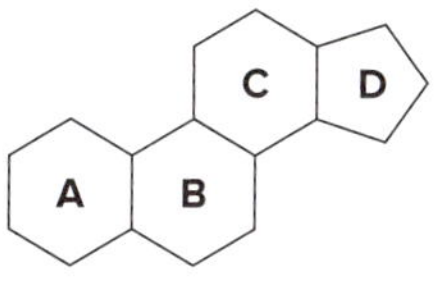

steroid skeleton

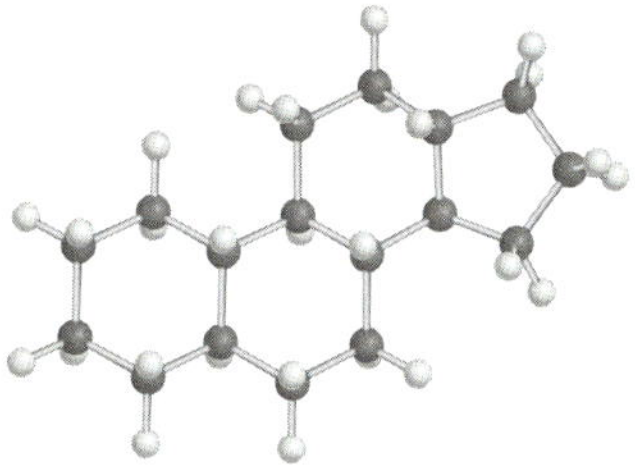

three-dimensional view from above

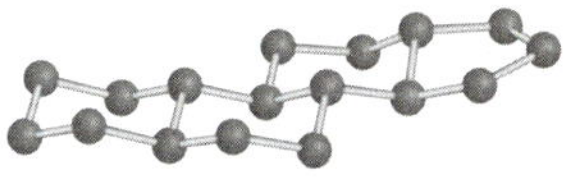

carbon skeleton viewed from the side

Steroids exhibit a wide range of biological properties, depending on the substitution pattern of functional groups on the rings. They include **cholesterol** (a component of cell membranes that is implicated in cardiovascular disease), **estrone** (a female sex hormone responsible for the regulation of the menstrual cycle), and **cortisone** (a hormone responsible for the control of inflammation and the regulation of carbohydrate metabolism).

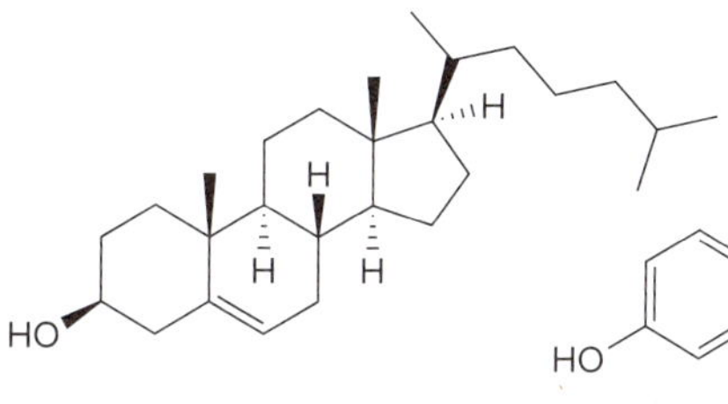

cholesterol

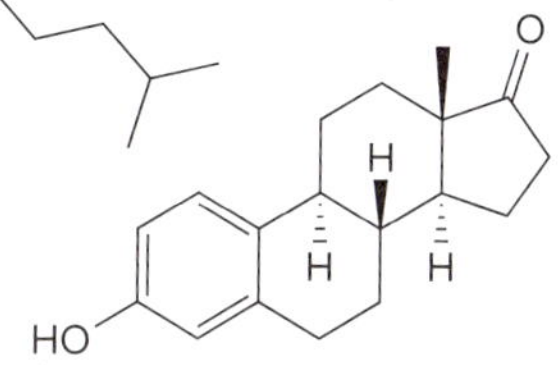

estrone

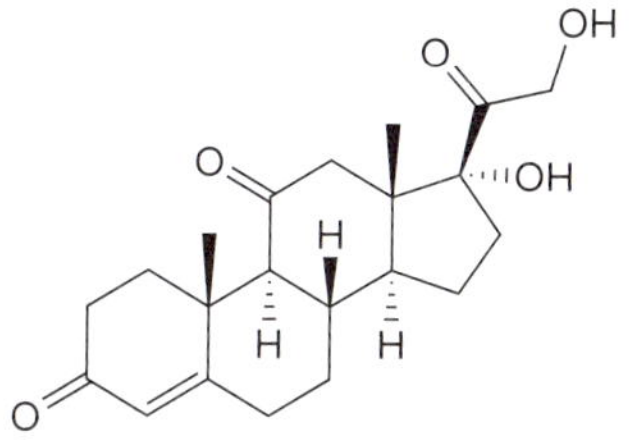

cortisone

Because Diels–Alder reactions form six-membered rings, they have been used widely in the laboratory syntheses of the six-membered rings in steroids. For example, the key Diels–Alder reaction used to prepare the C ring of estrone is drawn.

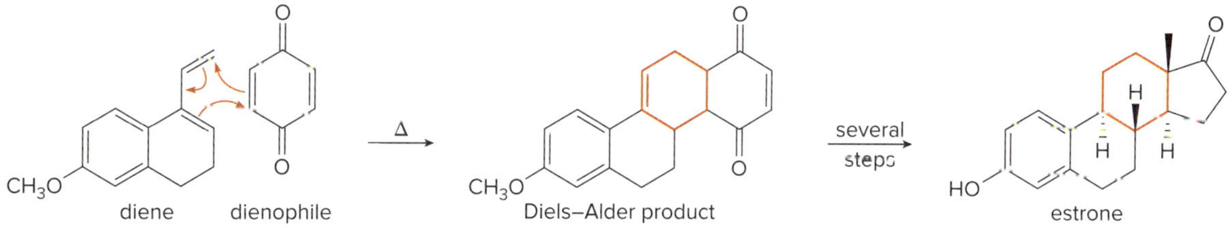

diene + dienophile →(Δ) Diels–Alder product →(several steps) estrone

The analgesic and narcotic effects of opium are largely due to morphine (Problem 14.29) Poppy seed tea, which contains morphine, was used as a folk remedy in parts of England until World War II. *John Foxx/Getty Images*

Problem 14.29 Draw the product (**A**) of the following Diels–Alder reaction. **A** was a key intermediate in the synthesis of the addicting pain reliever morphine.

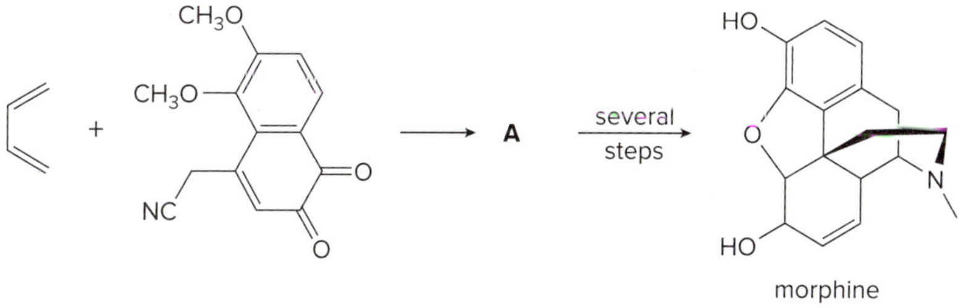

→ **A** →(several steps) morphine

14.15 Conjugated Dienes and Ultraviolet Light

Recall from Spectroscopy Part B that the absorption of infrared light can promote a molecule from a lower vibrational state to a higher one. In a similar fashion, the **absorption of ultraviolet (UV) light can promote an electron from a lower electronic state to a higher one.** Ultraviolet light has a slightly shorter wavelength (and, thus, higher frequency) than visible light. The most useful region of UV light for this purpose is **200–400 nm.**

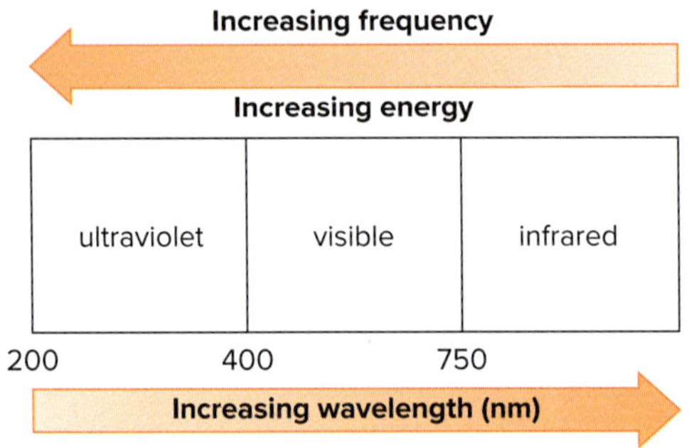

14.15A General Principles

When electrons in a lower-energy state (the **ground state**) absorb light having the appropriate energy, an electron is promoted to a higher electronic state (the **excited state**).

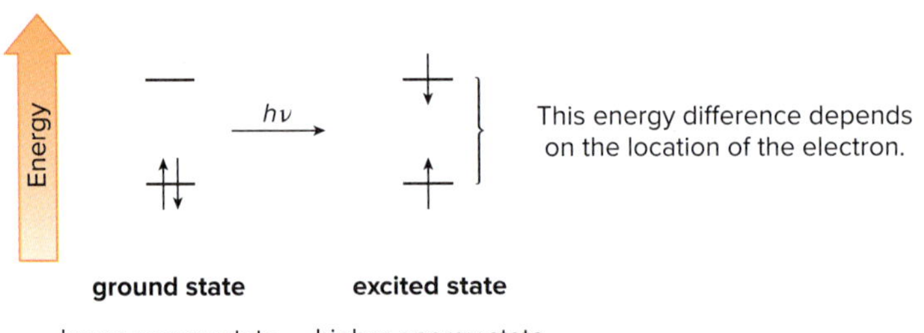

The energy difference between the two states depends on the location of the electron. **The promotion of electrons in σ bonds and unconjugated π bonds requires light having a wavelength of < 200 nm;** that is, it has a shorter wavelength and higher energy than light in the UV region of the electromagnetic spectrum. With conjugated dienes, however, the energy difference between the ground and excited states decreases, so **longer wavelengths of light can be used to promote electrons.** The wavelength of UV light absorbed by a compound is often referred to as its $\boldsymbol{\lambda_{max}}$. Buta-1,3-diene, for example, absorbs UV light at λ_{max} = 217 nm and cyclohexa-1,3-diene has a λ_{max} of 256 nm.

- **Conjugated dienes and polyenes absorb light in the UV region of the electromagnetic spectrum (200–400 nm).**

A UV spectrum is a plot of the absorbance of UV light versus wavelength. A spectrum consists of very broad bands, and the maximum absorbance corresponds to the λ_{max}, as shown in the UV spectrum of isoprene in Figure 14.9.

Figure 14.9
UV spectrum of isoprene

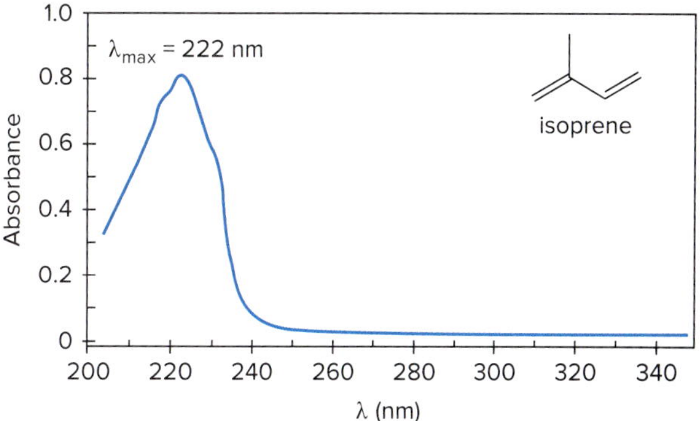

Lycopene is the red pigment found in tomatoes, watermelon, papaya, guava, and pink grapefruit. Lycopene is not destroyed when fruits and vegetables are processed, so tomato juice and ketchup are high in lycopene. *C Squared Studios/Getty Images*

As the number of conjugated π bonds *increases*, the energy difference between the ground and excited state *decreases*, shifting the absorption to *longer* wavelengths.

λ_{max} = 217 nm λ_{max} = 268 nm λ_{max} = 364 nm

Increasing conjugation
Increasing λ_{max}

With molecules having eight or more conjugated π bonds, the absorption shifts from the UV to the visible region and the compound takes on the color of those wavelengths of visible light it does *not* absorb. For example, lycopene absorbs visible light at λ_{max} = 470 nm, in the blue-green region of the visible spectrum. Because it does not absorb light in the red region, lycopene appears bright red (Figure 14.10).

Figure 14.10 Why lycopene appears red

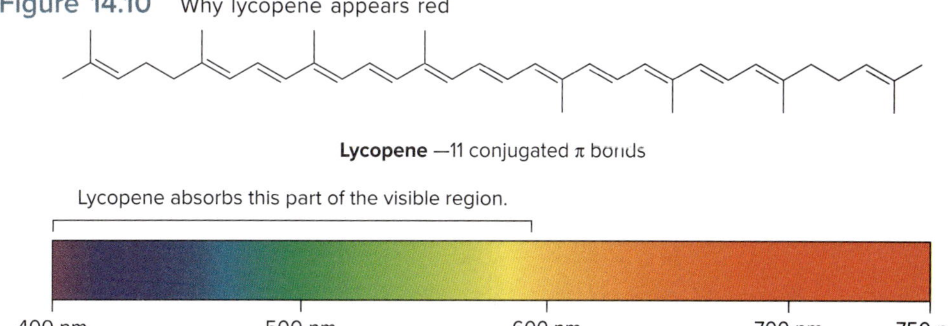

This part of the spectrum is *not* absorbed.

Lycopene appears red.

Problem 14.30 Which compound in each pair absorbs UV light at longer wavelength?

a. or b. or

Problem 14.31 Many compounds used as common food-coloring agents are highly conjugated and absorb light in the visible region of the electromagnetic spectrum. One food color has a λ_{max} at 524 nm and another has a λ_{max} at 630 nm. Which compound appears blue and which appears red?

14.15B Sunscreens

Commercial sunscreens are given an **SPF** rating (sun protection factor), according to the amount of sunscreen present. The higher the number, the greater the protection.
Jill Braaten/McGraw Hill

Ultraviolet radiation from the sun is high enough in energy to cleave bonds, forming radicals that can prematurely age skin and cause skin cancers. The ultraviolet region is often subdivided, based on the wavelength of UV light: UV-A (320–400 nm), UV-B (290–320 nm), and UV-C (< 290 nm). Fortunately, much of the highest-energy UV light (UV-C) is filtered out by the ozone layer, so that only UV light having wavelengths > 290 nm reaches the skin's surface. Much of this UV light is absorbed by **melanin,** the highly conjugated colored pigment in the skin that serves as the body's natural protection against the harmful effects of UV radiation.

Prolonged exposure to the sun can allow more UV radiation to reach your skin than melanin can absorb. Many common commercial sunscreens are conjugated compounds that absorb ultraviolet light and shield your skin (for a time) from harmful UV radiation. Recent research mentioned in the Prologue, however, suggests that some of these sunscreens harm coral reefs, and for this reason, sunscreens containing oxybenzone or octinoxate are now banned in the state of Hawai'i.

oxybenzone

octinoxate

In a search for more environmentally friendly sunscreens from biological sources, scientists have discovered a group of conjugated UV-absorbing compounds in algae, corals, and cyanobacteria. Shinorine is one such compound isolated from the red algae *Porphyra umbilicalis*. It can also be prepared by a faster, higher-yield process from genetically engineered cyanobacteria. Shinorine-containing sunscreens are available in Europe.

shinorine

Problem 14.32 Excluding resonance structures for the carboxy groups, draw all other resonance structures for shinorine involving the conjugated double bonds that are substituted with heteroatoms.

Chapter 14 REVIEW

KEY CONCEPTS

[1] Four common examples of resonance (14.3)

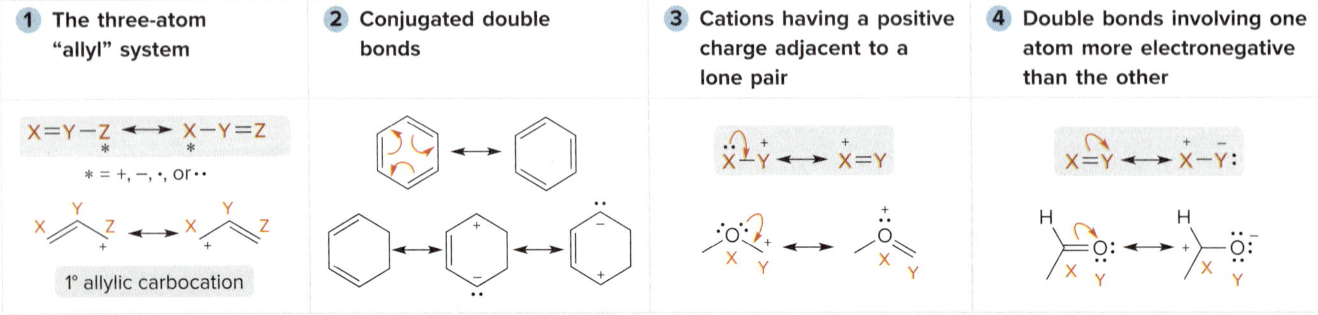

See Sample Problem 14.1. Try Problems 14.35, 14.36.

[2] The unusual properties of conjugated dienes

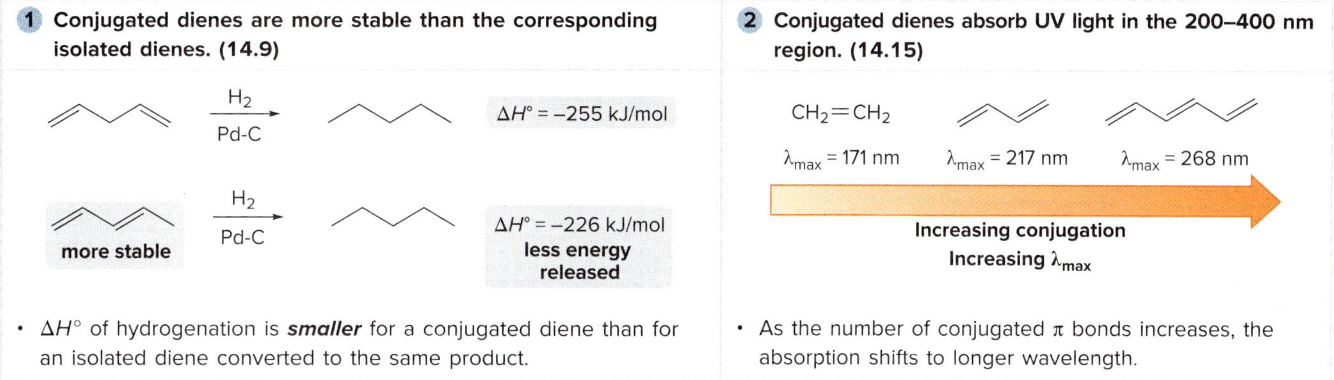

- $\Delta H°$ of hydrogenation is ***smaller*** for a conjugated diene than for an isolated diene converted to the same product.
- As the number of conjugated π bonds increases, the absorption shifts to longer wavelength.

See Figure 14.2. Try Problems 14.42; 14.60a, d; 14.68.

[3] The difference between two conformations and two stereoisomers in 1,3-dienes (14.6)

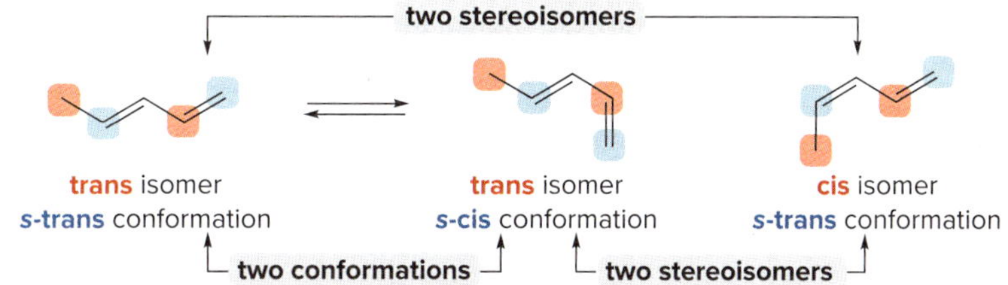

See Sample Problem 14.4. Try Problem 14.41.

[4] Relative reactivity in the Diels–Alder reaction

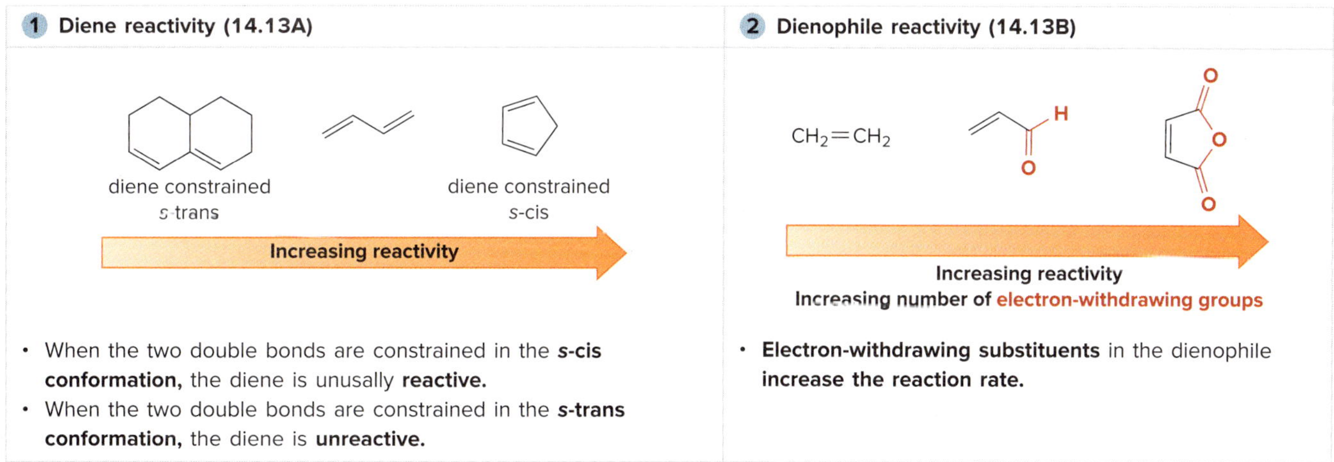

- When the two double bonds are constrained in the ***s*-cis conformation,** the diene is unusally **reactive.**
- When the two double bonds are constrained in the ***s*-trans conformation,** the diene is **unreactive.**
- **Electron-withdrawing substituents** in the dienophile **increase the reaction rate.**

Try Problem 14.60b, c.

KEY REACTIONS

Reactions of conjugated dienes

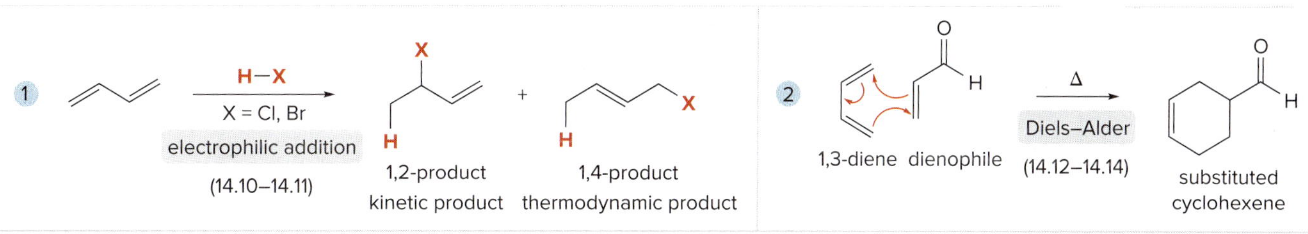

Try Problems 14.43b, c; 14.50; 14.62.

KEY SKILLS

[1] Drawing resonance structures for a conjugated compound (14.3, 14.4)

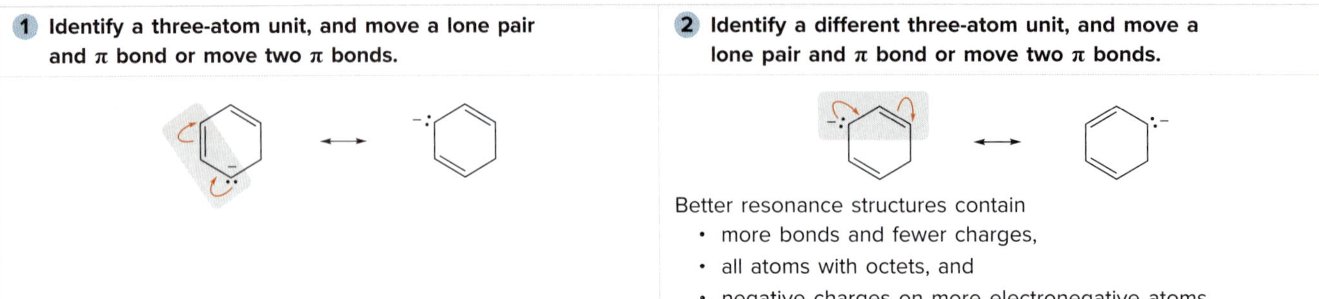

1 **Identify a three-atom unit, and move a lone pair and π bond or move two π bonds.**

2 **Identify a different three-atom unit, and move a lone pair and π bond or move two π bonds.**

Better resonance structures contain
- more bonds and fewer charges,
- all atoms with octets, and
- negative charges on more electronegative atoms.

See Sample Problem 14.1. Try Problems 14.35, 14.36.

[2] Drawing a resonance hybrid from three resonance structures (14.4); example: acrolein

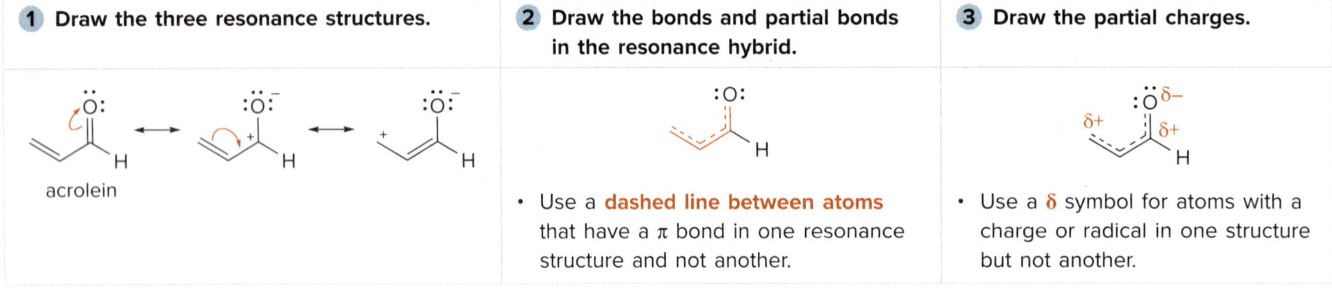

1 **Draw the three resonance structures.**

2 **Draw the bonds and partial bonds in the resonance hybrid.**

- Use a **dashed line between atoms** that have a π bond in one resonance structure and not another.

3 **Draw the partial charges.**

- Use a δ symbol for atoms with a charge or radical in one structure but not another.

See Sample Problem 14.2.

[3] Using two resonance structures to determine the hybridization around a given carbon atom (14.5)

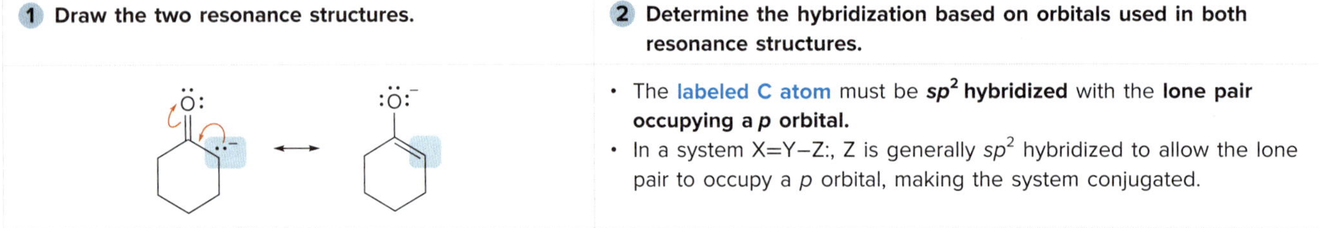

1 **Draw the two resonance structures.**

2 **Determine the hybridization based on orbitals used in both resonance structures.**

- The **labeled C atom** must be sp^2 **hybridized** with the **lone pair occupying a *p* orbital.**
- In a system X=Y–Z:, Z is generally sp^2 hybridized to allow the lone pair to occupy a p orbital, making the system conjugated.

See Sample Problem 14.3. Try Problem 14.37.

[4] Drawing the products from HX addition to a diene (14.10)

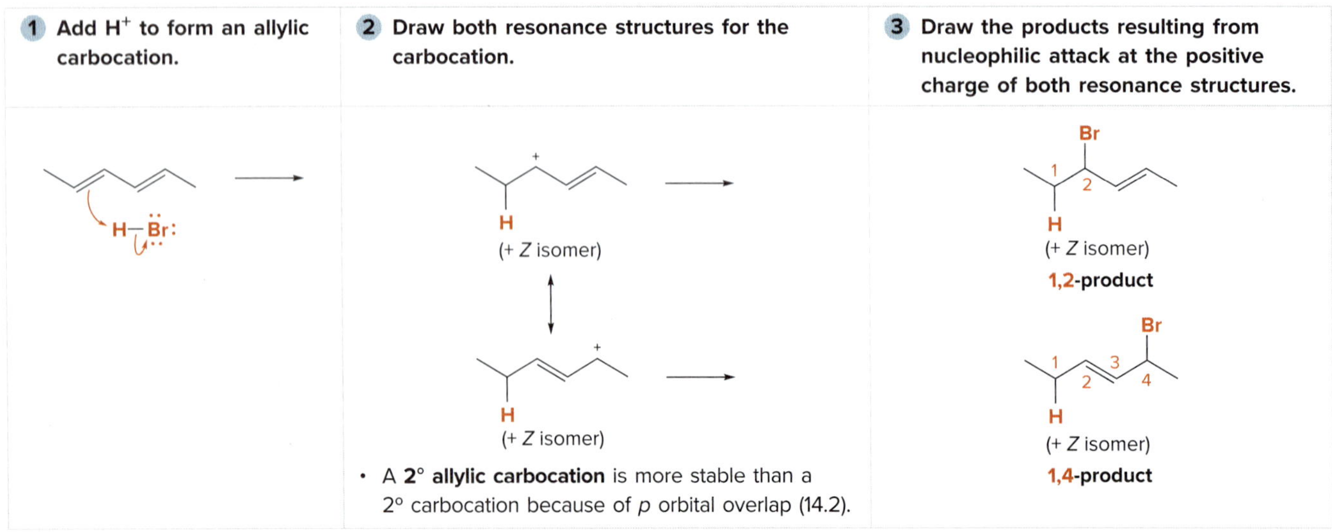

1 **Add H$^+$ to form an allylic carbocation.**

2 **Draw both resonance structures for the carbocation.**

- A 2° **allylic carbocation** is more stable than a 2° carbocation because of p orbital overlap (14.2).

3 **Draw the products resulting from nucleophilic attack at the positive charge of both resonance structures.**

See Sample Problem 14.5. Try Problems 14.43b, c; 14.62a, d.

[5] Determining whether the major product is the kinetic or thermodynamic product (14.11)

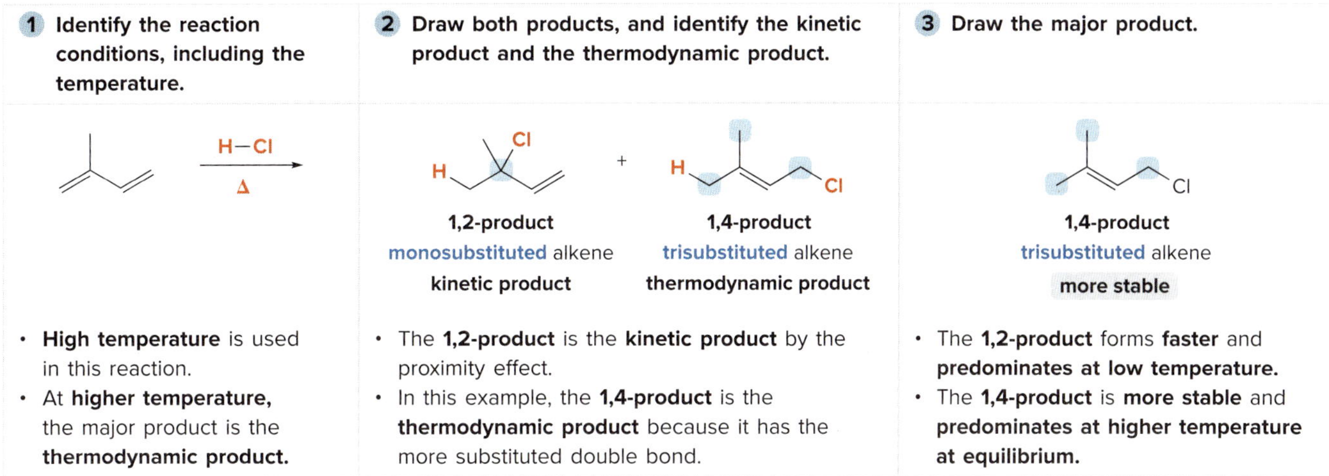

Try Problems 14.47, 14.48.

[6] Drawing the product of a Diels–Alder reaction (14.12)

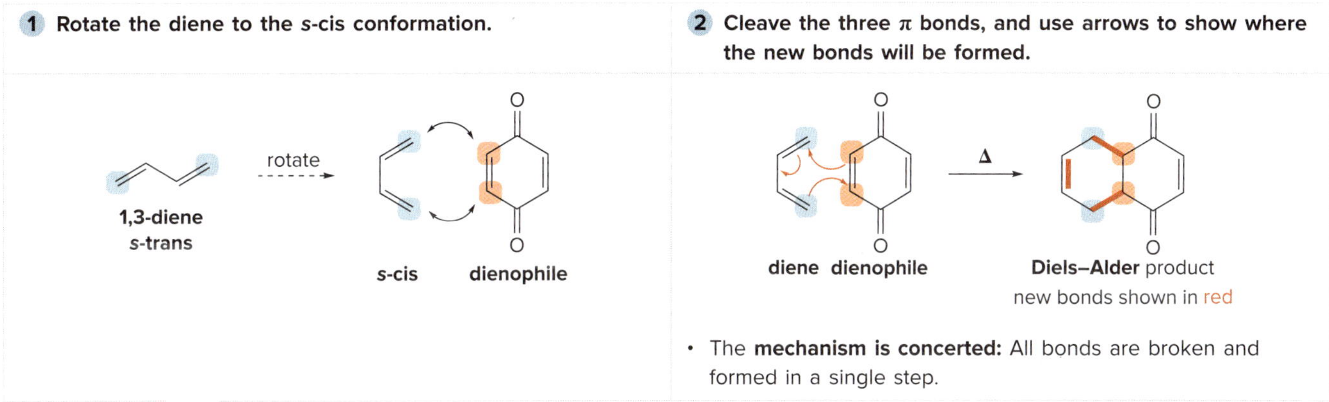

See *How To*, p. 664. Try Problem 14.50.

[7] Finding the diene and dienophile needed for a Diels–Alder reaction (14.14)

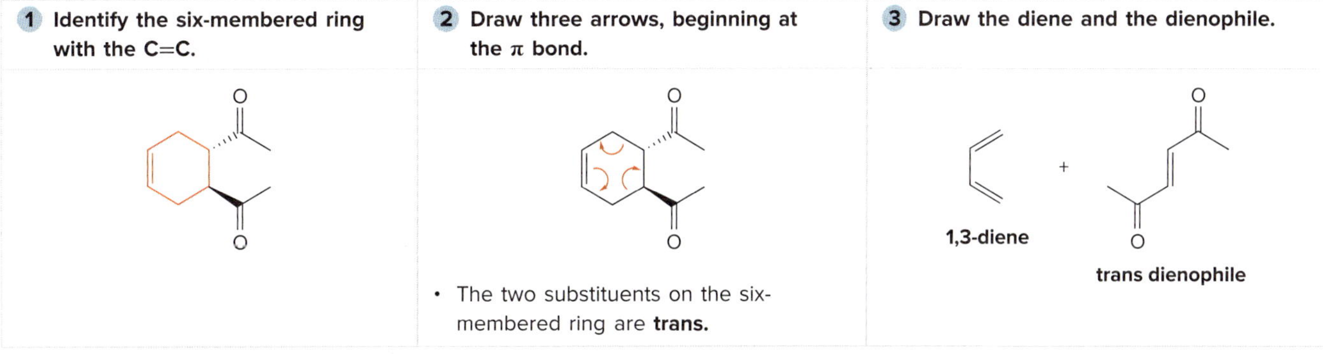

See Figure 14.8. Try Problems 14.51, 14.52, 14.55.

CHAPTER 14 MULTIPLE-CHOICE SELF-TEST

The Self-Test consists of multiple-choice questions similar to those found on the American Chemical Society organic chemistry exam. Answers are given at the end of the chapter.

1. Which species is *not* conjugated?

a. 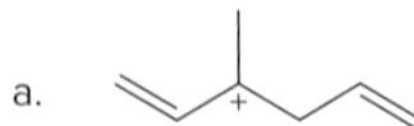
b. 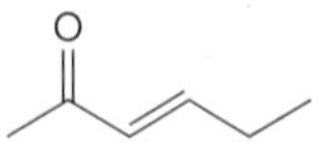
c. 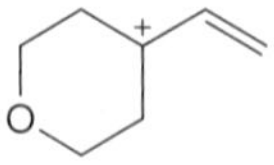
d.

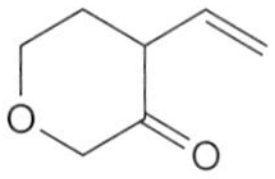

2. How are **A, B,** and **C** related to each other? Choose from constitutional isomers, stereoisomers, or different conformations.

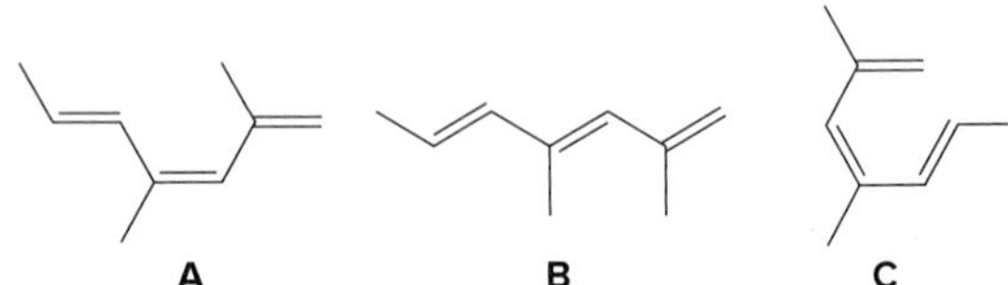

a. **A** and **B** stereoisomers; **B** and **C** different conformations
b. **A** and **B** stereoisomers; **A** and **C** different conformations
c. **A** and **C** different conformations; **B** and **C** constitutional isomers
d. **A, B,** and **C** are all stereoisomers.

3. Which diene is most reactive in the Diels–Alder reaction?

a. 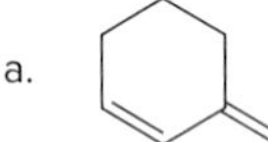
b. 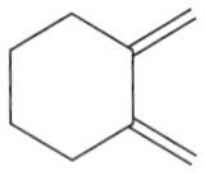
c. 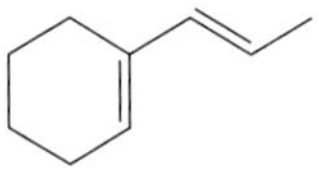
d.

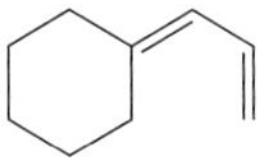

4. Which species is *not* a valid resonance structure for **A?**

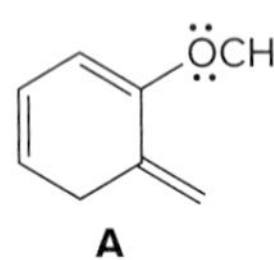

a. 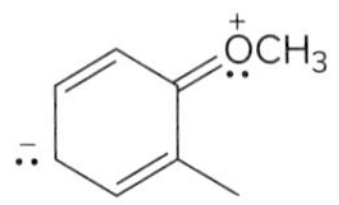
b. 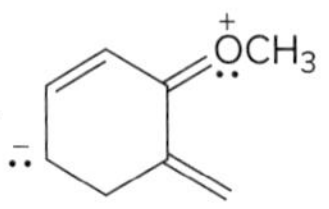
c. 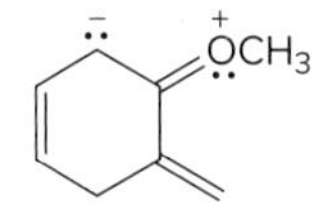
d.

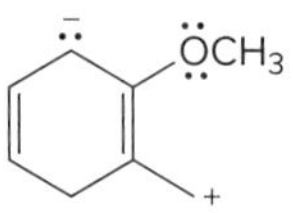

5. Which compound absorbs the longest wavelength of ultraviolet light?

a. 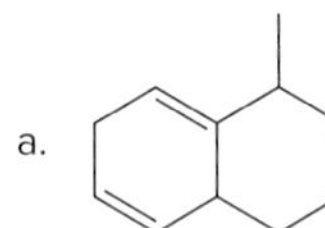
b. 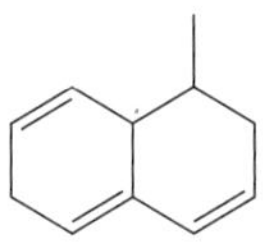
c. 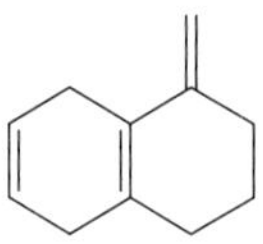
d.

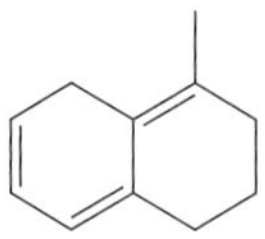

6. Give the IUPAC name of the following diene.

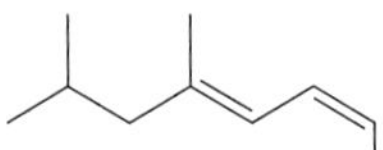

a. (2*Z*,4*E*)-5,7-dimethylocta-2,4-diene
b. (4*E*,6*Z*)-2,4-dimethylocta-4,6-diene
c. (2*Z*,4*Z*)-5,7-dimethylocta-2,4-diene
d. (1*Z*,3*E*)-1,4,6-trimethylhepta-1,3-diene

7. What product is formed in the following Diels–Alder reaction?

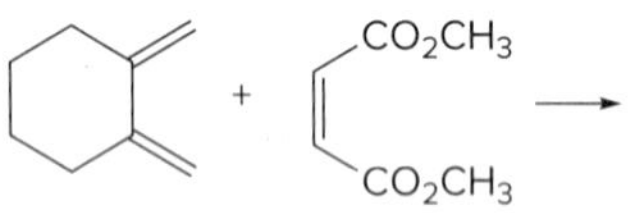

$+$ CO_2CH_3 CO_2CH_3 $\longrightarrow$

a. 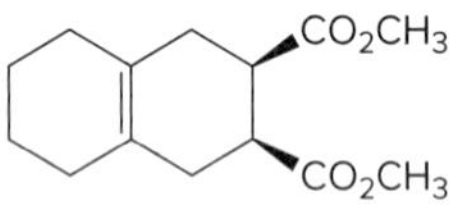 CO_2CH_3 CO_2CH_3
b. 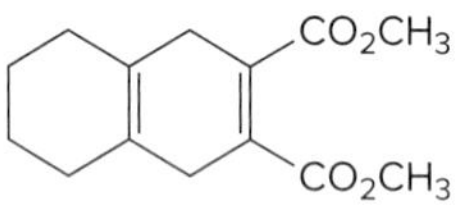 CO_2CH_3 CO_2CH_3
c. 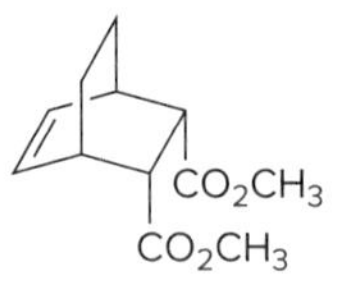 CO_2CH_3 CO_2CH_3
d. 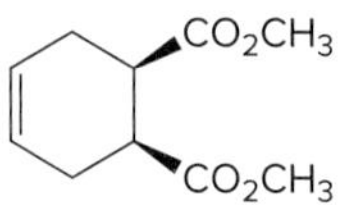 CO_2CH_3 CO_2CH_3

8. Which labeled atoms are sp^2 hybridized?

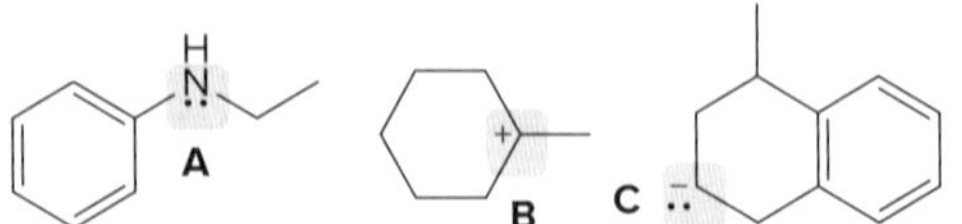

a. **A, B,** and **C**
b. **A** and **B**
c. **B** and **C**
d. **B** only

9. What product is *not* formed when diene **A** reacts with one equivalent of HCl?

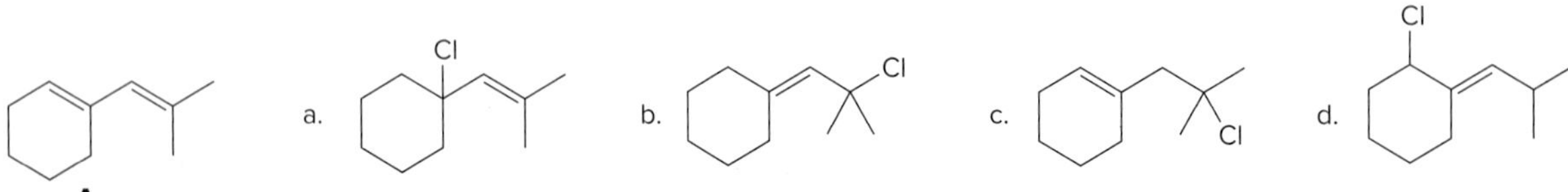

10. What product is formed in the following Diels–Alder reaction?

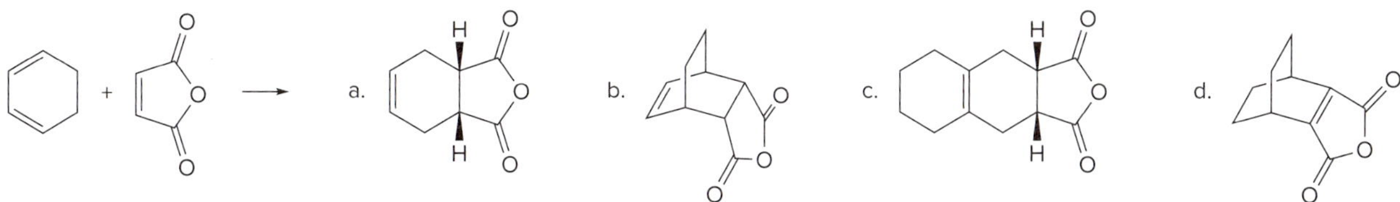

PROBLEMS

Problem Using Three-Dimensional Models

14.33 Name each diene and state whether the ball-and-stick model shows the diene in the *s*-cis or *s*-trans conformation.

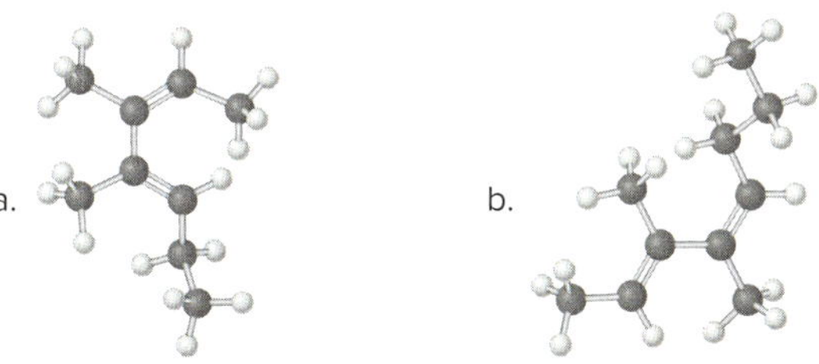

Conjugation

14.34 Which of the following systems are conjugated?

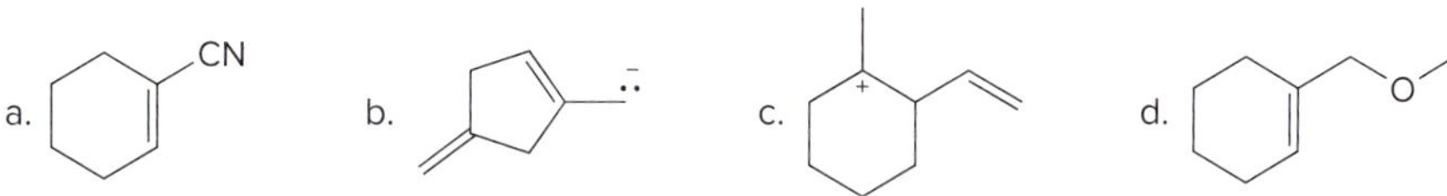

Resonance and Hybridization

14.35 Draw all reasonable resonance structures for each species.

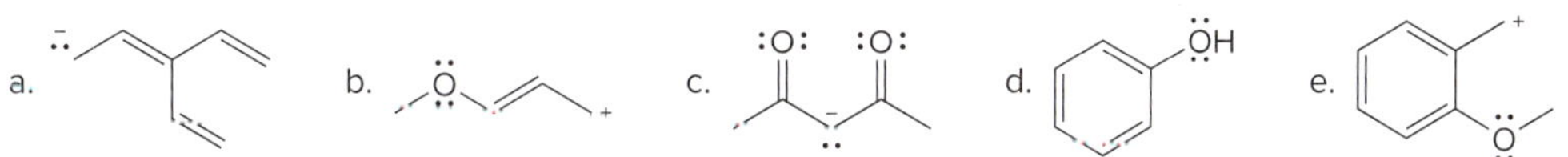

14.36 Draw all reasonable resonance structures for each compound.

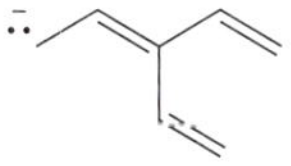

a. 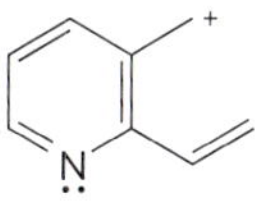 b. 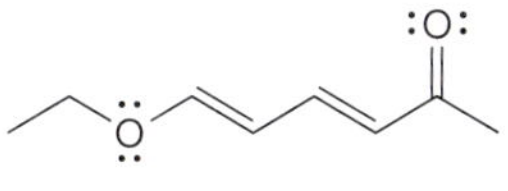 c. 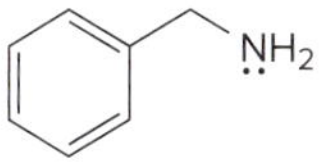 d.

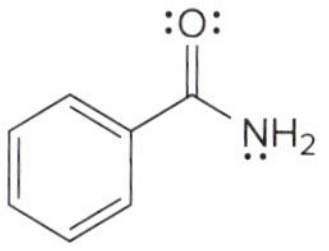

14.37 Determine the hybridization of the labeled atoms in each drug.

a. 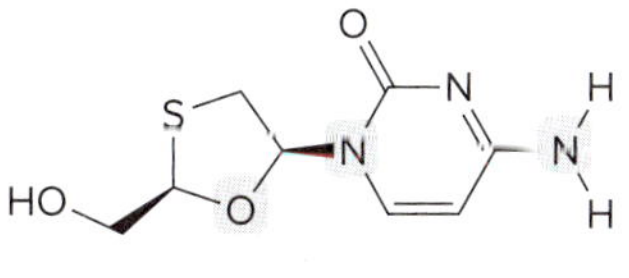 lamivudine
HIV antiviral

b. 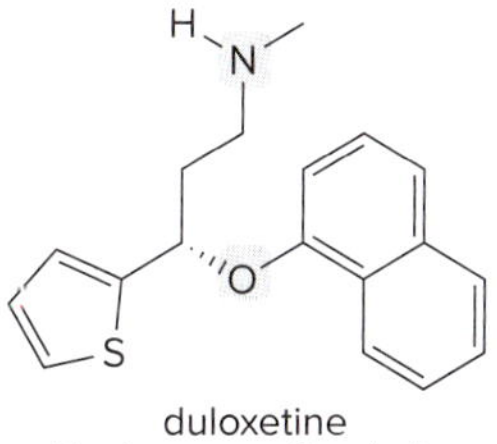 duloxetine
Trade name: Cymbalta
antidepressant

14.38 Explain why the cyclopentadienide anion **A** gives only one signal in its ^{13}C NMR spectrum.

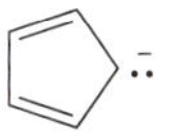

= **A**

Nomenclature and Stereoisomers in Conjugated Dienes

14.39 Draw the structure of each compound.

a. (*Z*)-penta-1,3-diene in the *s*-trans conformation
b. (2*E*,4*Z*)-1-bromo-3-methylhexa-2,4-diene
c. (2*E*,4*E*,6*E*)-octa-2,4,6-triene
d. (2*E*,4*E*)-3-methylhexa-2,4-diene in the *s*-cis conformation

14.40 Name each compound and indicate the conformation around the σ bond that joins the two double bonds.

14.41 Label each pair of compounds as stereoisomers, conformations, or constitutional isomers: (a) **A** and **B;** (b) **A** and **C;** (c) **A** and **D;** (d) **C** and **D.**

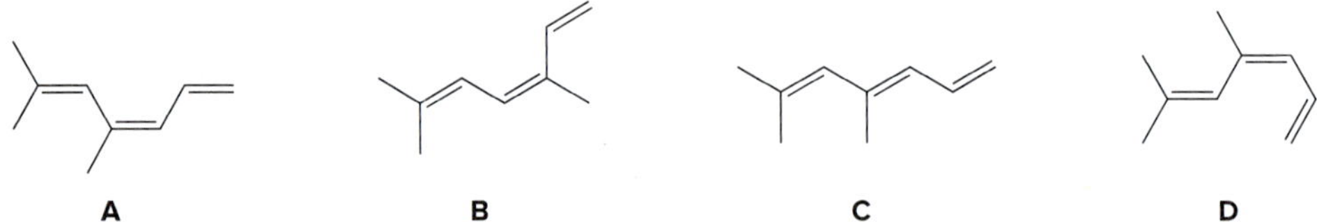

14.42 Rank the following dienes in order of increasing heat of hydrogenation.

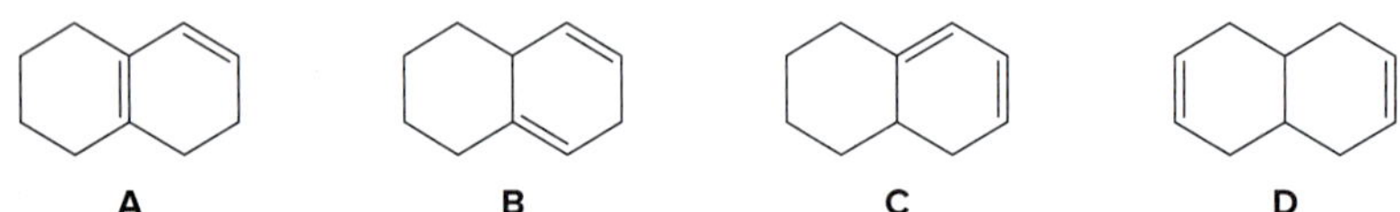

Electrophilic Addition

14.43 Draw the products formed when each compound is treated with one equivalent of HBr.

a. 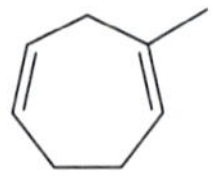 b. 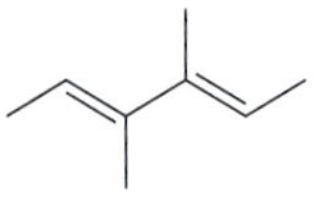 c.

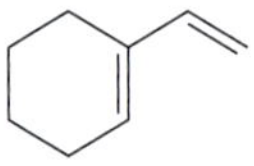

14.44 Ignoring stereoisomers, draw all products that form by addition of HBr to (*E*)-hexa-1,3,5-triene.

14.45 Treatment of alkenes **A** and **B** with HBr gives the same alkyl halide **C.** Draw a mechanism for each reaction, including all reasonable resonance structures for any intermediate.

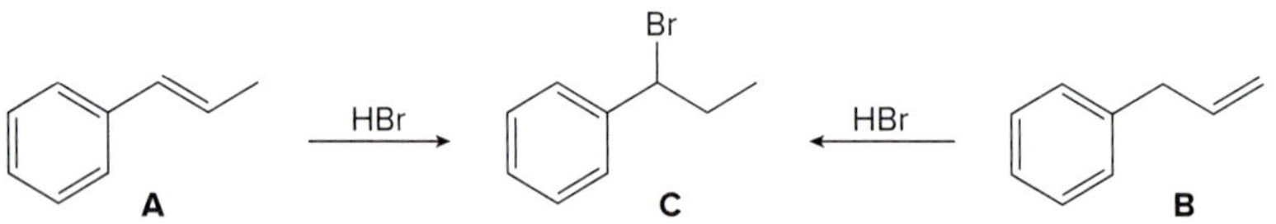

A —HBr→ C (Br) ←HBr— B

14.46 Draw a stepwise mechanism for the following reaction.

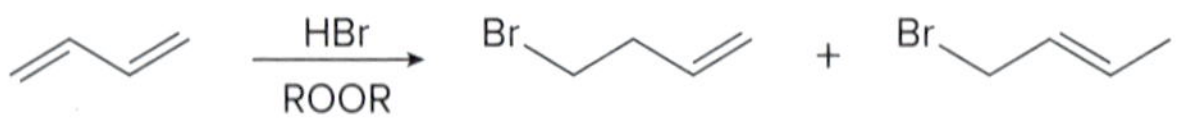

HBr / ROOR → Br + Br

14.47 Addition of HCl to diene **X** forms two alkyl halides **Y** and **Z.**

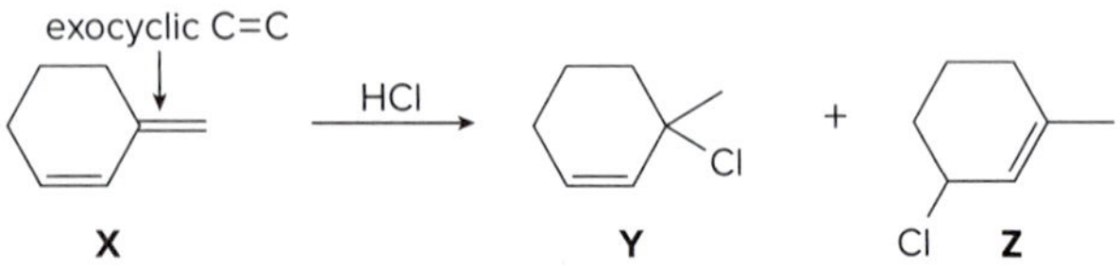

exocyclic C=C

X —HCl→ Y (Cl) + Z (Cl)

a. Label **Y** and **Z** as a 1,2-addition product or a 1,4-addition product.
b. Label **Y** and **Z** as the kinetic or thermodynamic product and explain why.
c. Explain why addition of HCl occurs at the indicated C=C (called an exocyclic double bond), rather than the other C=C (called an endocyclic double bond).

14.48 The major product formed by addition of HBr to $(CH_3)_2C{=}CH{-}CH{=}C(CH_3)_2$ is the same at low and high temperature. Draw the structure of the major product, and explain why the kinetic and thermodynamic products are the same in this reaction.

14.49 From what you have learned about the reaction of conjugated dienes in Section 14.10, predict the products of each of the following electrophilic additions.

a. 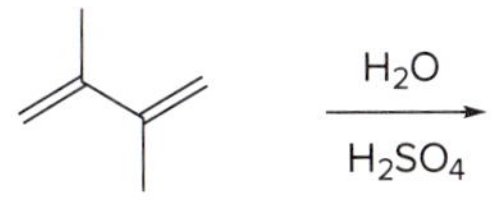 $\xrightarrow[H_2SO_4]{H_2O}$ b. 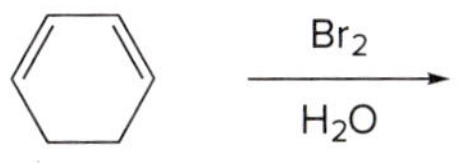 $\xrightarrow[H_2O]{Br_2}$

Diels–Alder Reaction

14.50 Draw the products of the following Diels–Alder reactions. Indicate stereochemistry where appropriate.

a. 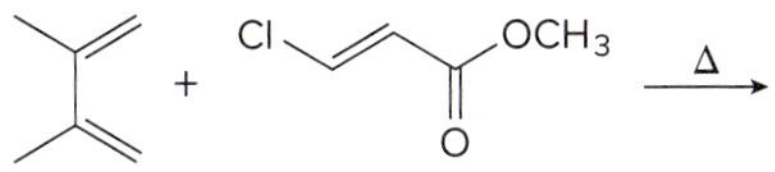 + Cl, OCH$_3$, O $\xrightarrow{\Delta}$

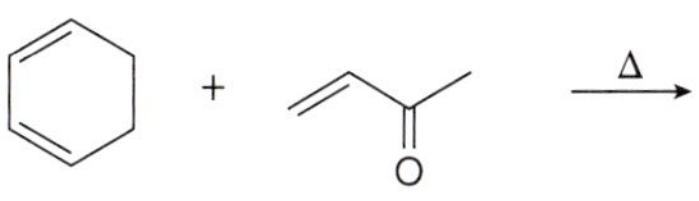

b. 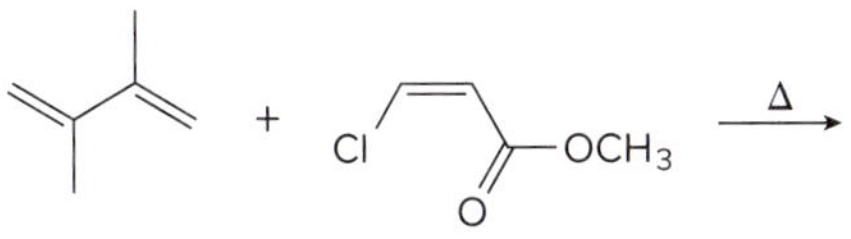 + Cl, OCH$_3$, O $\xrightarrow{\Delta}$

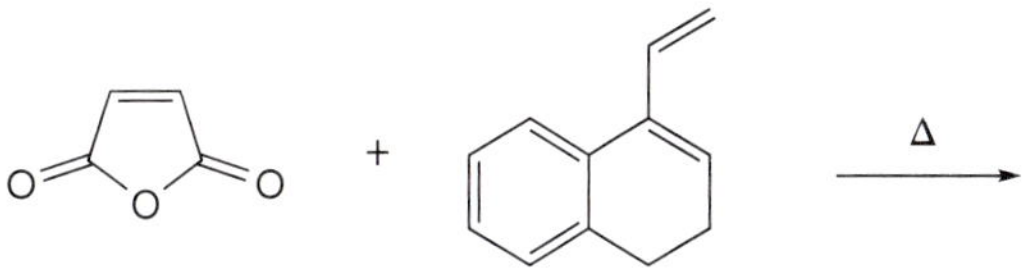

c. 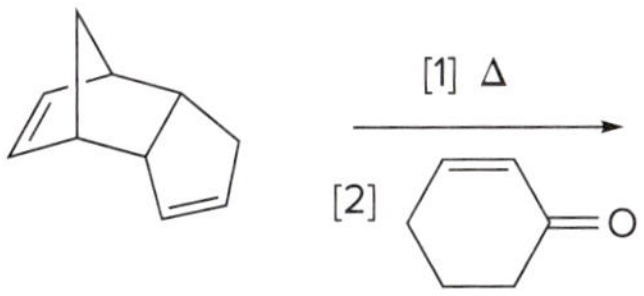 [1] Δ [2] O

d. + O $\xrightarrow{\Delta}$

e. O O O + $\xrightarrow{\Delta}$

f. 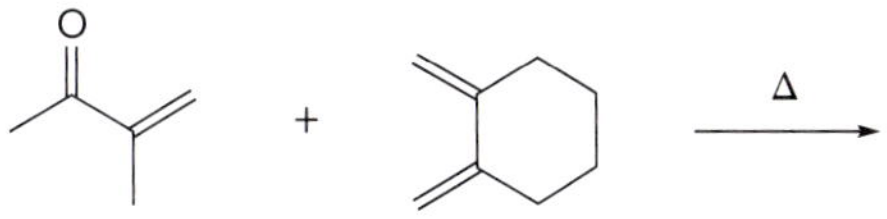 O + $\xrightarrow{\Delta}$

14.51 What diene and dienophile are needed to prepare each Diels–Alder product?

a. 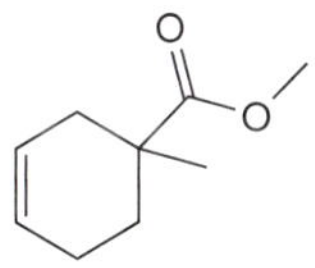 O O

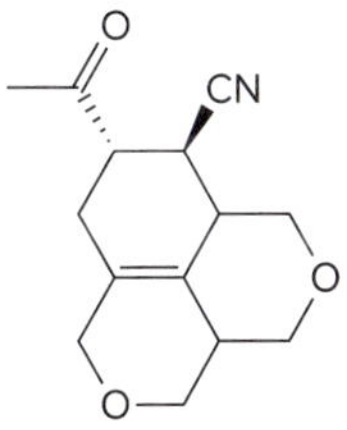

b. 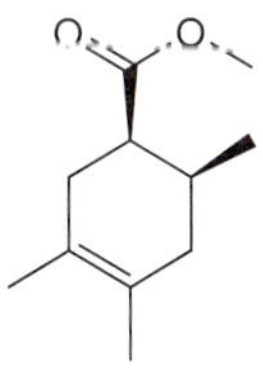 O O

c. Cl O O

d. 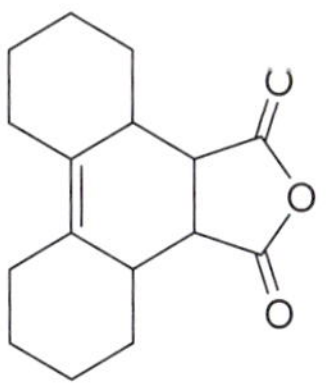 O O O

e. O CN O O

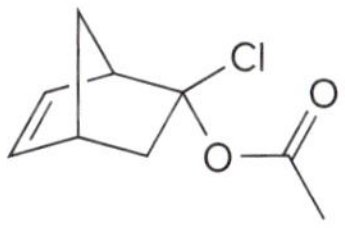

f. 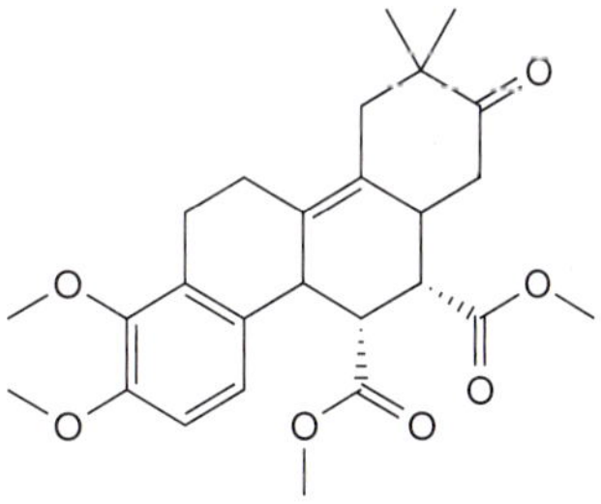 O O O O O O

14.52 Give two different ways to prepare the following compound by the Diels–Alder reaction. Explain which method is preferred.

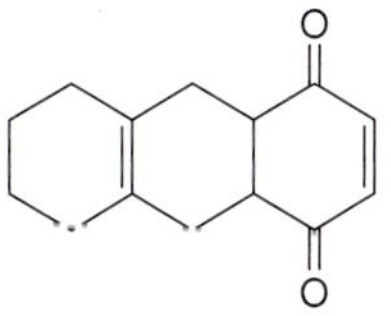

O O

14.53 Compounds containing triple bonds are also Diels–Alder dienophiles. With this in mind, draw the products of each reaction.

a. 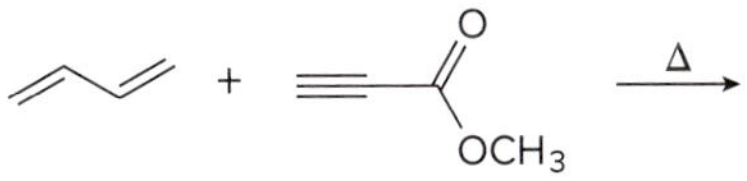 + O OCH$_3$ $\xrightarrow{\Delta}$ b. 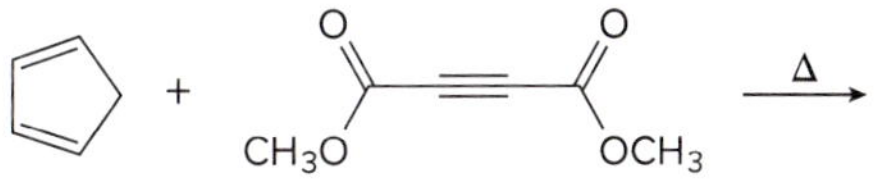 + O O CH$_3$O OCH$_3$ $\xrightarrow{\Delta}$

14.54 Diels–Alder reaction of a monosubstituted diene (such as $CH_2{=}CH{-}CH{=}CHOCH_3$) with a monosubstituted dienophile (such as $CH_2{=}CHCHO$) gives a mixture of products, but the 1,2-disubstituted product often predominates. Draw the resonance hybrid for each reactant, and use the charge distribution of the hybrids to explain why the 1,2-disubstituted product is the major product.

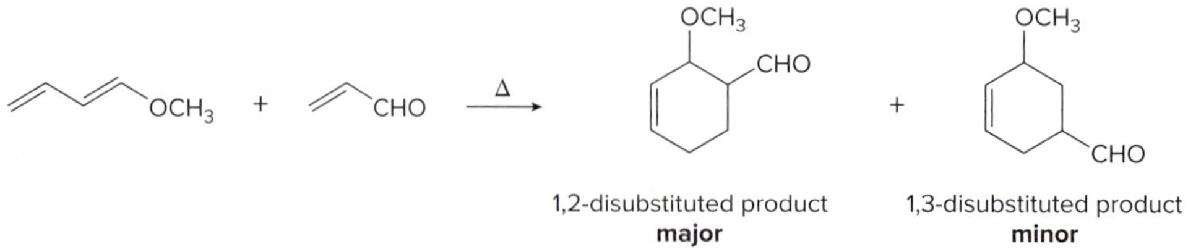

OCH₃ + CHO —Δ→ 1,2-disubstituted product **major** + 1,3-disubstituted product **minor**

14.55 Two steps in the multistep synthesis of the termite soldier pheromone kempene-2 involve Diels–Alder reactions. Give the starting materials needed for each of these reactions.

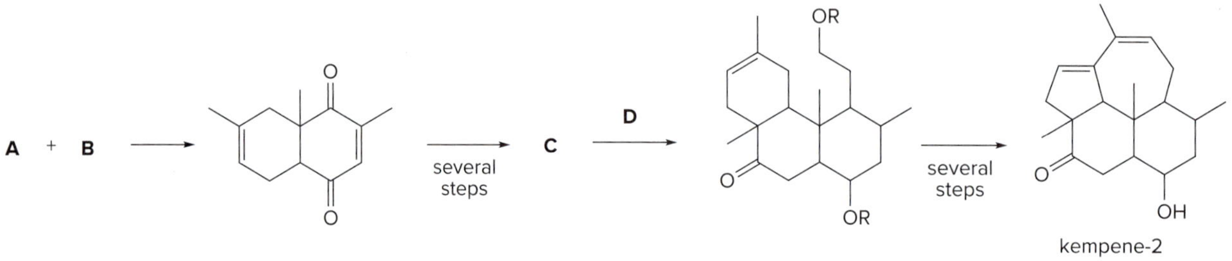

A + **B** → → several steps → **C** —**D**→ → several steps → kempene-2

14.56 Devise a stepwise synthesis of each compound from dicyclopentadiene using a Diels–Alder reaction as one step. You may also use organic compounds having ≤ 4 C's, and any required organic or inorganic reagents.

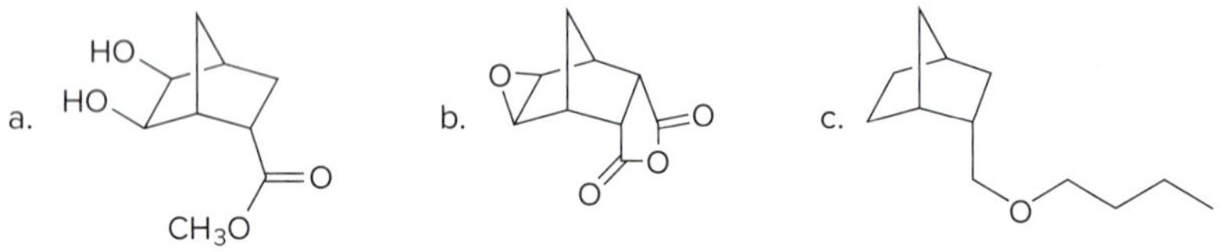

a. b. c.

14.57 Intramolecular Diels–Alder reactions are possible when a substrate contains both a 1,3-diene and a dienophile, as shown in the conversion of **A** to **B.** With this in mind, draw the product when each compound undergoes an intramolecular Diels–Alder reaction.

A —Δ→ **B** (two new rings)

a. b.

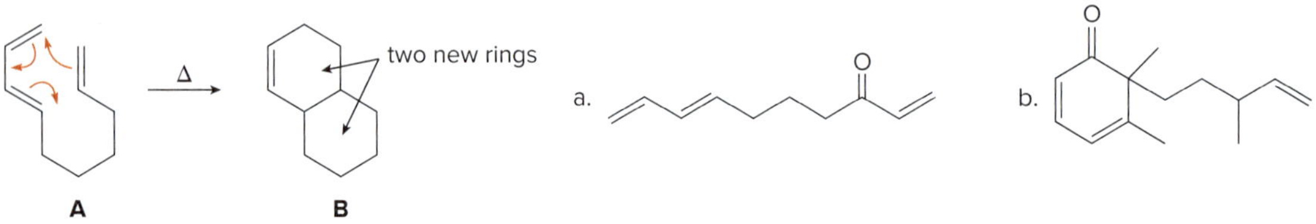

14.58 What starting material was used to make compound **A** by an intramolecular Diels–Alder reaction? **A** is an intermediate in the synthesis of β-erythroidine, an amine isolated from plants of the *Erythrina* species.

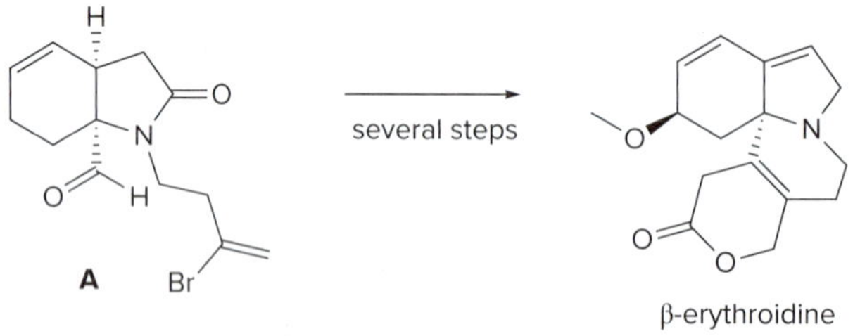

A → several steps → β-erythroidine

14.59 A transannular Diels–Alder reaction is an intramolecular reaction that occurs when the diene and dienophile are contained in one ring, resulting in the formation of a tricyclic ring system. Draw the product formed when the following triene undergoes a transannular Diels–Alder reaction.

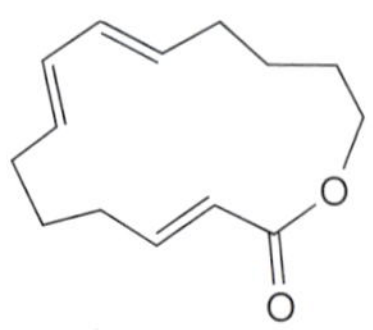

Problems That Combine Concepts

14.60 Consider the four trienes **E–H.**

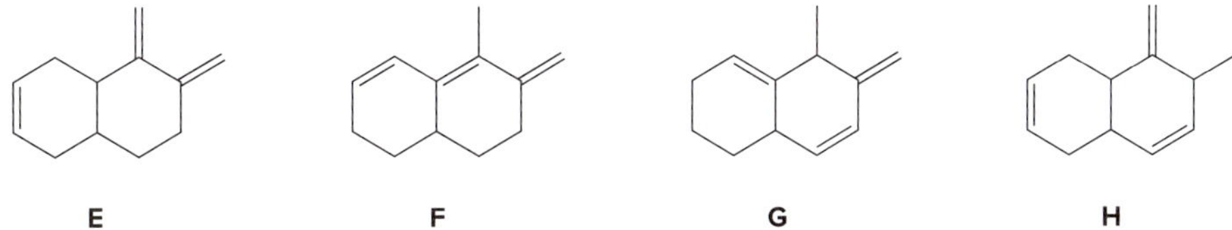

a. Rank compounds **E–H** in order of increasing heat of hydrogenation.
b. Which compound is most reactive in the Diels–Alder reaction?
c. Which compound(s) are unreactive in the Diels–Alder reaction?
d. Which compound absorbs the longest wavelength of ultraviolet light?

14.61 Draw a stepwise mechanism for the following reaction.

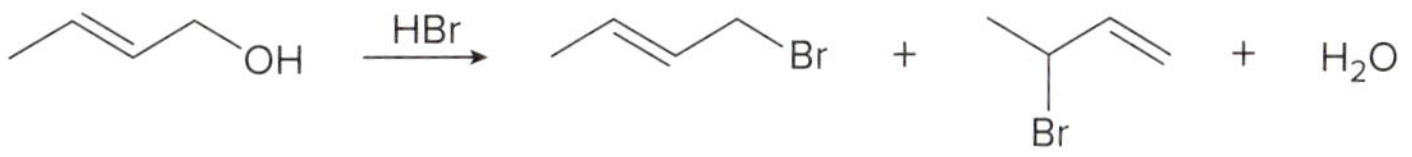

OH $\xrightarrow{\text{HBr}}$ Br + Br + H_2O

14.62 Draw the products of each reaction. Indicate the stereochemistry of Diels–Alder products.

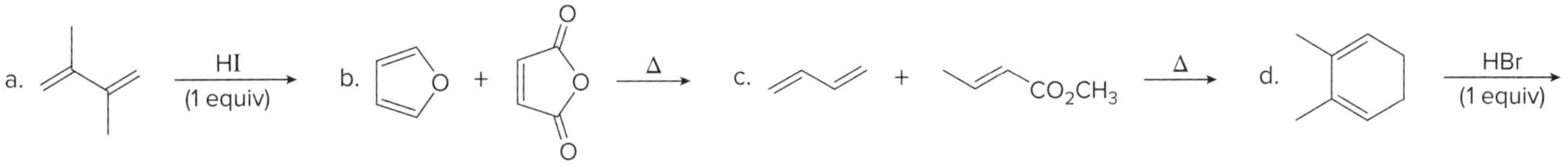

a. $\xrightarrow[\text{(1 equiv)}]{\text{HI}}$ b. O + O, O, O $\xrightarrow{\Delta}$ c. + CO_2CH_3 $\xrightarrow{\Delta}$ d. $\xrightarrow[\text{(1 equiv)}]{\text{HBr}}$

14.63 Draw a stepwise mechanism for the biological conversion of linalyl diphosphate to limonene.

OPP ⟶ + PP_i

linalyl diphosphate limonene

14.64 Which benzylic halide reacts faster in an S_N1 reaction? Explain.

CH_3O Br **A** O Br **B**

14.65 Like alkenes, conjugated dienes can be prepared by elimination reactions. Draw a stepwise mechanism for the acid-catalyzed dehydration of 3-methylbut-2-en-1-ol [$(CH_3)_2C{=}CHCH_2OH$] to isoprene [$CH_2{=}C(CH_3)CH{=}CH_2$].

14.66 (a) Draw the two isomeric dienes formed when $CH_2{=}CHCH_2CH(Cl)CH(CH_3)_2$ is treated with an alkoxide base. (b) Explain why the major product formed in this reaction does not contain the more highly substituted alkene.

Spectroscopy

14.67 The treatment of isoprene [$CH_2{=}C(CH_3)CH{=}CH_2$] with one equivalent of mCPBA forms **A** as the major product. **A** gives a molecular ion at 84 in its mass spectrum, and peaks at 2850–3150 cm^{-1} in its IR spectrum. The ^1H NMR spectrum of **A** is given below. What is the structure of **A?**

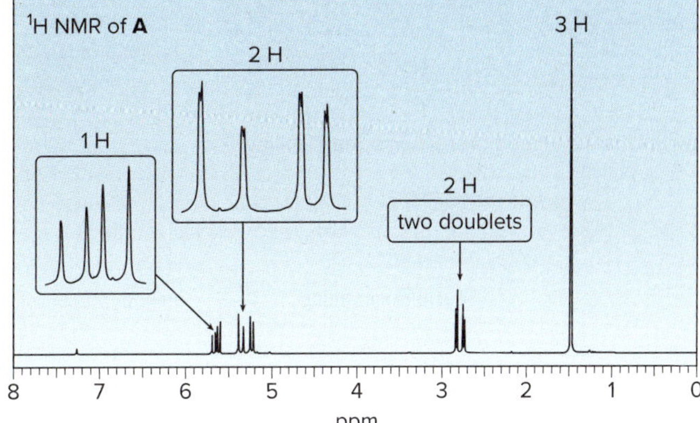

UV Absorption

14.68 Rank the following compounds in order of increasing wavelength of ultraviolet light absorbed.

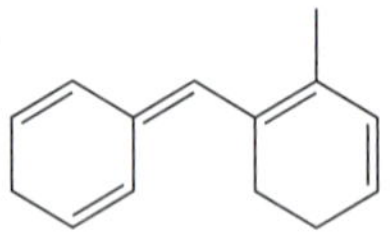

A 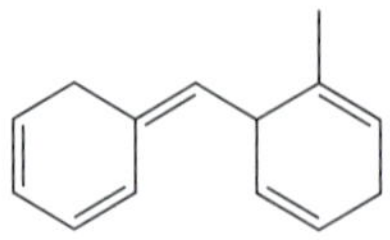 B 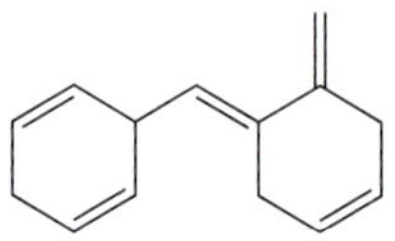 C

14.69 Explain why palythine absorbs light in the UV region of the electromagnetic spectrum, despite the fact that it is not a 1,3-diene. Palythine, which is extracted from red algae, is under investigation as a potential sunscreen component.

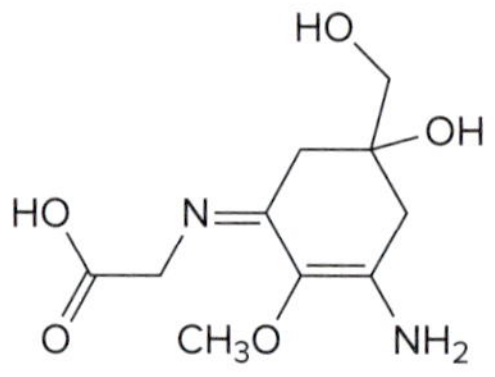

HO OH HO N= O CH_3O NH_2

palythine

14.70 Explain why ferulic acid, a natural product found in rice, oats, and other plants, is both an antioxidant and a sunscreen.

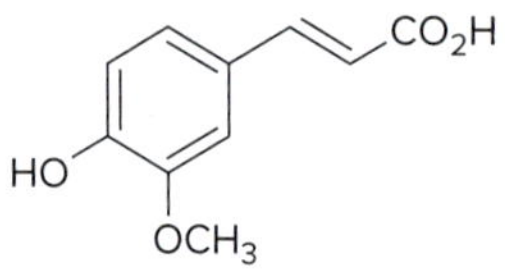

CO_2H HO OCH_3

ferulic acid

Challenge Problems

14.71 In the quest for a viable synthesis of the anticancer agent eleutherobin, which is isolated from marine soft corals of the *Eleutherobia* species, compounds **A** and **B** reacted to form **C** by two successive Diels–Alder reactions. Use curved arrows to illustrate these reactions, which consist of an intermolecular reaction, followed by an intramolecular reaction.

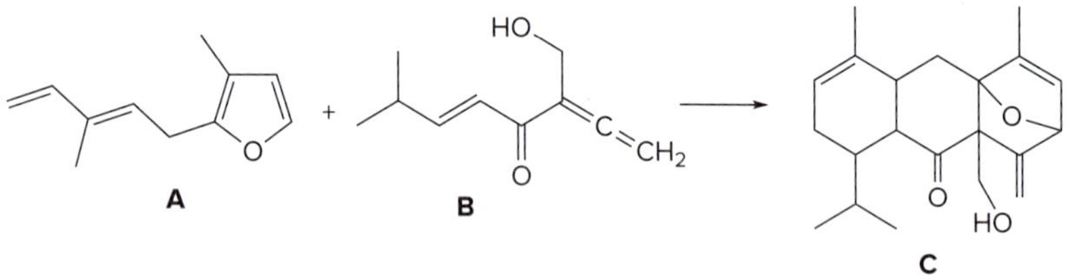

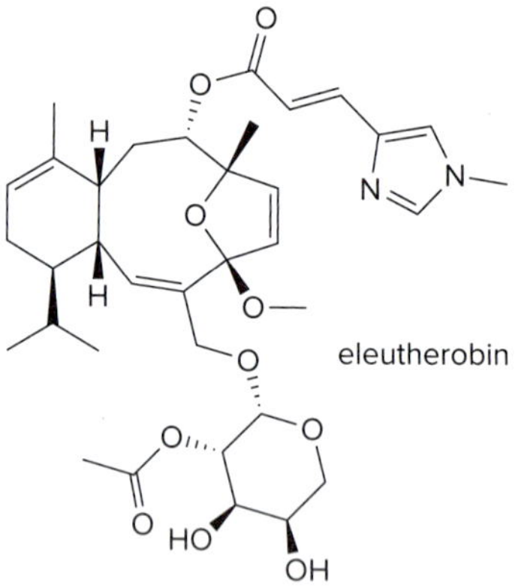

A + B ($C{=}CH_2$) → C

eleutherobin

14.72 Explain why cyclohexanamine is 10^6 times more basic than aniline.

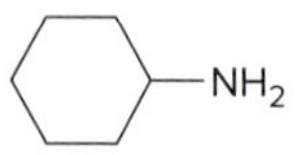

cyclohexanamine 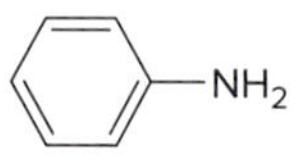 aniline

14.73 Devise a synthesis of **X** from the given starting materials. You may use any organic or inorganic reagents. Account for the stereochemistry observed in **X.**

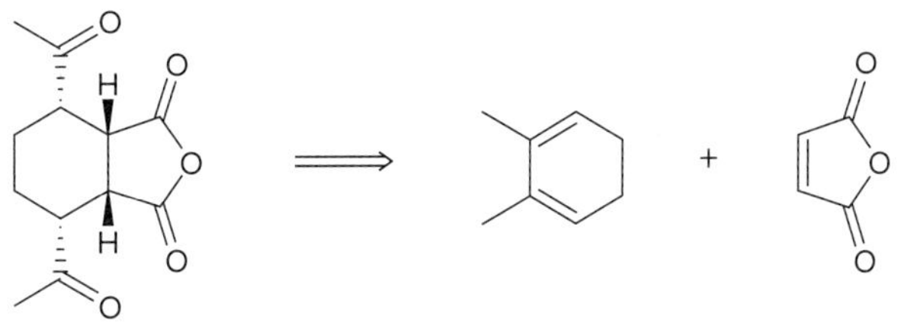

14.74 One step in the synthesis of occidentalol, a natural product isolated from the eastern white cedar tree, involves the following reaction. Identify the structure of **A** and show how **A** is converted to **B.**

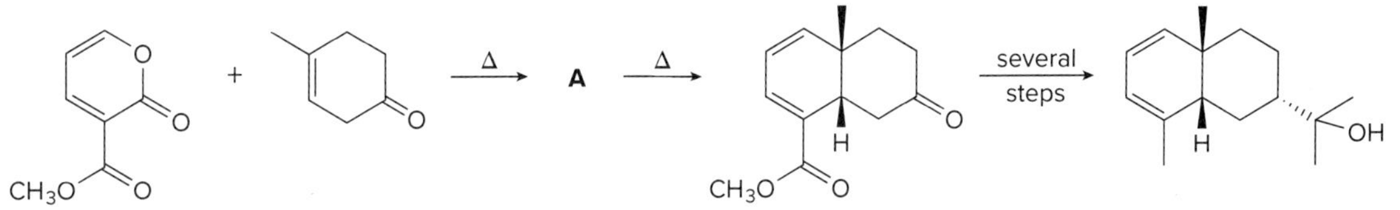

14.75 One step in the synthesis of dodecahedrane (Section 4.10) involves reaction of the tetraene **C** with dimethylacetylene dicarboxylate (**D**) to afford two compounds having molecular formula $C_{16}H_{16}O_4$. This reaction has been called a domino Diels–Alder reaction. Identify the two products formed.

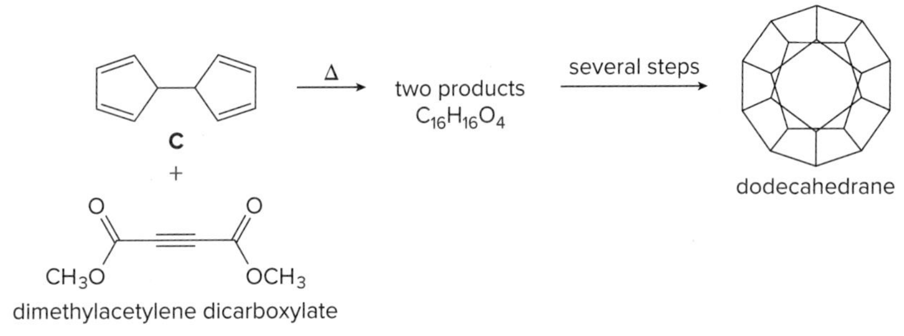

14.76 Devise a stepwise mechanism for the conversion of **M** to **N. N** has been converted in several steps to lysergic acid, a naturally occurring precursor of the hallucinogen LSD (Figure 16.4).

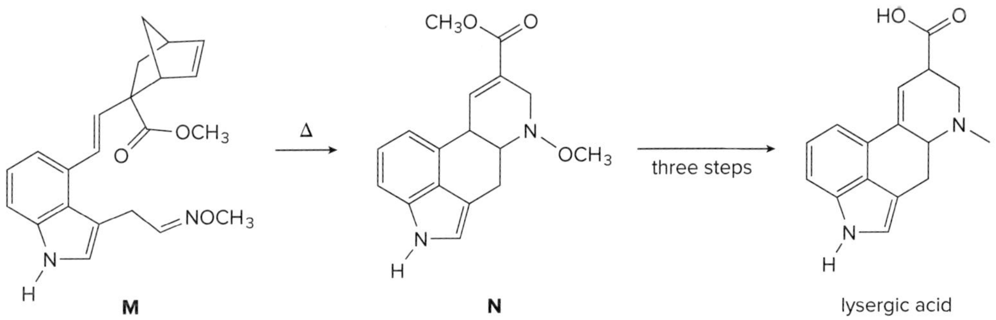

SELF-TEST ANSWERS

1. d 2. b 3. b 4. a 5. d 6. a 7. a 8. b 9. c 10. b

15 Benzene and Aromatic Compounds

Schafer & Hill/Photolibrary/Getty Images

15.1 Background
15.2 The structure of benzene
15.3 Nomenclature of benzene derivatives
15.4 Spectroscopic properties
15.5 Interesting aromatic compounds
15.6 Benzene's unusual stability
15.7 The criteria for aromaticity—Hückel's rule
15.8 Examples of aromatic compounds
15.9 Aromatic heterocycles
15.10 What is the basis of Hückel's rule?
15.11 The inscribed polygon method for predicting aromaticity

The aromatic heterocycle **pilocarpine** is a naturally occurring glaucoma medication and dry-mouth remedy isolated from the leaves of *Pilocarpus microphyllus*, commonly called jaborandi. When the demand for pilocarpine was especially high in the 1970s, jaborandi was overharvested in northern Brazil. More recently, two companies are leading an effort to promote the sustainable harvesting of jaborandi, as well as to establish practices to more fairly compensate the small family farms that produce it. In Chapter 15, we learn about the characteristics of aromatic compounds, including heterocycles like pilocarpine.

Why Study . . .

Aromatic Compounds?

The hydrocarbons we have examined thus far—including the alkanes, alkenes, and alkynes, as well as the conjugated dienes and polyenes of Chapter 14—have been aliphatic hydrocarbons. In Chapter 15, we continue our study of conjugated systems with **aromatic hydrocarbons.**

We begin with **benzene** and then examine other cyclic, planar, and conjugated ring systems to learn the modern definition of what it means to be aromatic. Then, in Chapter 16, we will learn about the reactions of aromatic compounds, highly unsaturated hydrocarbons that do not undergo addition reactions like other unsaturated compounds. An explanation of this behavior relies on an understanding of the structure of aromatic compounds presented in Chapter 15. Many naturally occurring compounds contain aromatic rings, and many useful drugs are aromatic.

15.1 Background

For 6 C's, the maximum number of H's = $2n + 2 = 2(6) + 2 = 14$. Because benzene contains only 6 H's, it has 14 – 6 = 8 H's fewer than the maximum number. This corresponds to 8 H's/2 H's for each degree of unsaturation = **four degrees of unsaturation in benzene.**

Benzene (C_6H_6) is the simplest aromatic hydrocarbon (or arene). Since its isolation by Michael Faraday from the oily residue remaining in the illuminating gas lines in London in 1825, it has been recognized as an unusual compound. Based on the calculation introduced in Section 10.2, **benzene has four degrees of unsaturation, making it a highly unsaturated hydrocarbon.** But, whereas unsaturated hydrocarbons such as alkenes, alkynes, and dienes readily undergo addition reactions, *benzene does not.* For example, bromine adds to ethylene to form a dibromide, but benzene is inert under similar conditions.

$$\text{ethylene } (H_2C{=}CH_2) \xrightarrow{Br_2} BrCH_2CH_2Br \quad \text{addition product}$$

$$\underset{\text{benzene}}{C_6H_6} \xrightarrow{Br_2} \textbf{No reaction}$$

Benzene *does* react with bromine, but only in the presence of $FeBr_3$ (a Lewis acid), and the reaction is a **substitution,** *not* an addition.

$$C_6H_6 \xrightarrow[FeBr_3]{Br_2} C_6H_5Br$$

substitution
Br replaces H.

Thus, any structure proposed for benzene must account for its high degree of unsaturation and its lack of reactivity toward electrophilic addition.

In the last half of the nineteenth century August Kekulé proposed structures that were close to the modern description of benzene. In the Kekulé model, benzene was thought to be a rapidly equilibrating mixture of two compounds, each containing a six-membered ring with three alternating π bonds. These structures are now called **Kekulé structures.** In the Kekulé description, the bond between any two carbon atoms is sometimes a single bond and sometimes a double bond.

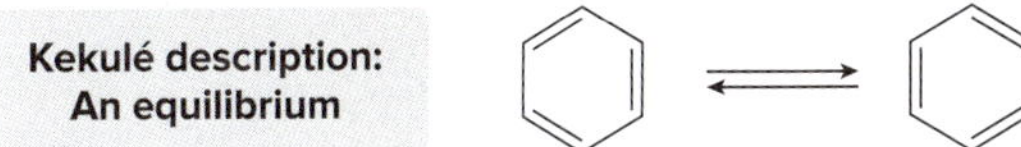

Although benzene is still drawn as a six-membered ring with three alternating π bonds, in reality **there is no equilibrium between two different kinds of benzene molecules.** Instead, current descriptions of benzene are based on resonance and electron delocalization due to orbital overlap, as detailed in Section 15.2

In the nineteenth century, many other compounds having properties similar to those of benzene were isolated from natural sources. Because these compounds possessed strong and characteristic odors, they were called ***aromatic* compounds.** It is their chemical properties, though, not their odor that make these compounds special.

- **Aromatic compounds resemble benzene—they are unsaturated compounds that do not undergo the addition reactions characteristic of alkenes.**

15.2 The Structure of Benzene

Any structure for benzene must account for the following:

- **Benzene contains a six-membered ring and three additional degrees of unsaturation.**
- **Benzene is planar.**
- **All C—C bond lengths are equal.**

Although the Kekulé structures satisfy the first two criteria, they break down with the third, because having three alternating π bonds would mean that benzene should have three short double bonds alternating with three longer single bonds.

This structure implies that the C–C bonds should have **two different lengths.**

- three longer single bonds in red
- three shorter double bonds in black

Resonance

Benzene is conjugated, so we must use resonance and orbitals to describe its structure. The resonance description of benzene consists of two equivalent Lewis structures, each with three double bonds that alternate with three single bonds.

Some texts draw benzene as a hexagon with an inner circle:

The circle represents the **six π electrons,** distributed over the six atoms of the ring.

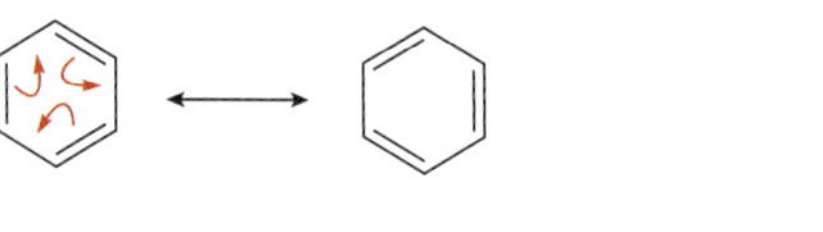

hybrid

The electrons in the π bonds are **delocalized** around the ring.

The resonance description of benzene matches the Kekulé description with one important exception: **The two Kekulé representations are *not* in equilibrium with each other.** Instead, the true structure of benzene is a resonance **hybrid** of the two Lewis structures, with the dashed lines of the hybrid indicating the position of the π bonds.

We will use one of the two Lewis structures and not the hybrid in drawing benzene, because it is easier to keep track of the electron pairs in the π bonds (the π electrons).

- **Because each π bond has two electrons, benzene has six π electrons.**

The resonance hybrid of benzene explains why all C–C bond lengths are the same. Each C–C bond is single in one resonance structure and double in the other, so the actual bond length (139 pm) is *intermediate* between a carbon–carbon single bond (153 pm) and a carbon–carbon double bond (134 pm).

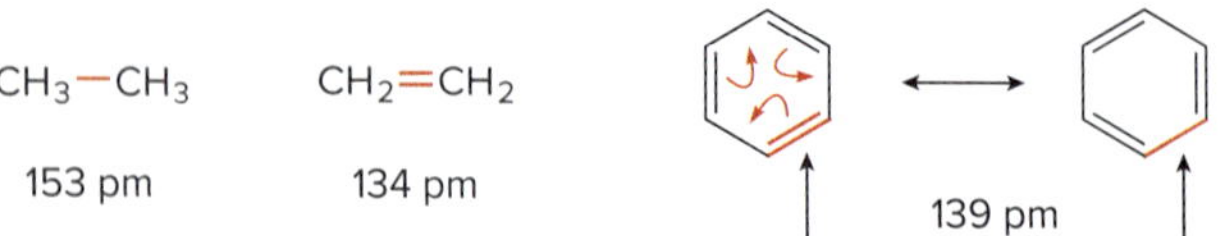

The C–C bonds in benzene are equal and intermediate in length.

Hybridization and Orbitals

Each carbon atom in a benzene ring is surrounded by three atoms and no lone pairs of electrons, making it ***sp*^{2} hybridized and trigonal planar with all bond angles 120°.** Each carbon also has a p orbital with one electron that extends above and below the plane of the molecule.

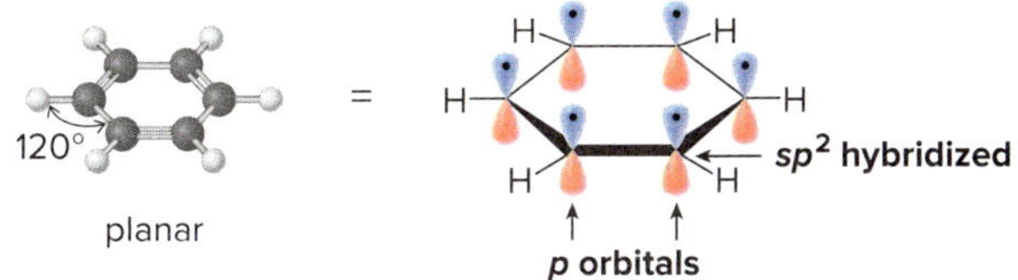

The six adjacent *p* orbitals overlap, delocalizing the six electrons over the six atoms of the ring and making benzene a conjugated molecule. Because each p orbital has two lobes, one above and one below the plane of the benzene ring, the overlap of the p orbitals creates two "doughnuts" of electron density, as shown in Figure 15.1a. The electrostatic potential plot in Figure 15.1b also shows that the electron-rich region is concentrated above and below the plane of the molecule, where the six π electrons are located.

- **Benzene's six π electrons make it electron rich, so it reacts with electrophiles.**

Figure 15.1 Two views of the electron density in a benzene ring

a. View of the p orbital overlap

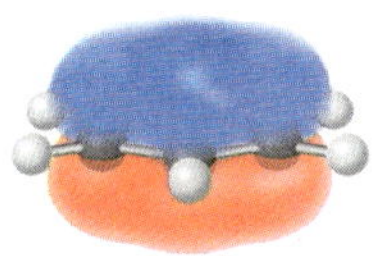

- Overlap of six adjacent p orbitals creates two rings of electron density, one above and one below the plane of the benzene ring.

b. Electrostatic potential plot

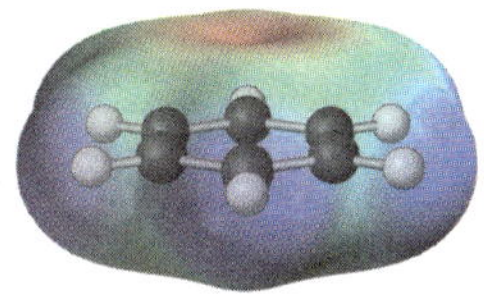

- The electron-rich region (in red) is concentrated above and below the ring carbons, where the six π electrons are located. (The electron-rich region below the plane is hidden from view.)

Problem 15.1 Draw all possible resonance structures for the antihistamine diphenhydramine, the active ingredient in Benadryl.

O
N
diphenhydramine

Problem 15.2 What orbitals are used to form the labeled bonds in the following molecule? Of the labeled C–C bonds, which is the shortest?

H 1
2
3
4
5

15.3 Nomenclature of Benzene Derivatives

Many organic molecules contain a benzene ring with one or more substituents, so we must learn how to name them.

15.3A Monosubstituted Benzenes

To name a benzene ring with one substituent, **name the substituent and add the word *benzene*.** Carbon substituents are named as alkyl groups.

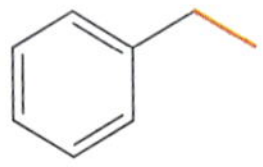

Cl

ethylbenzene 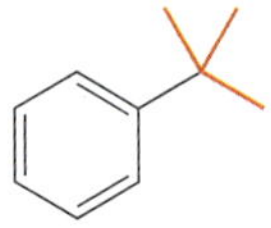 tert-butylbenzene 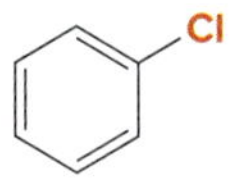 chlorobenzene

Many monosubstituted benzenes, such as those with methyl (CH_3–), hydroxy (–OH), and amino (–NH_2) groups, have common names that you must learn, too.

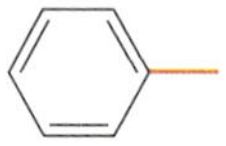

OH NH_2

toluene (methylbenzene) **phenol** (hydroxybenzene) **aniline** (aminobenzene)

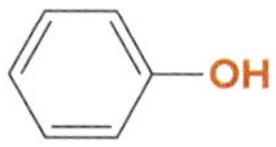

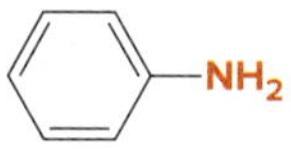

15.3B Disubstituted Benzenes

There are three different ways that two groups can be attached to a benzene ring, so a prefix—**ortho, meta,** or **para**—can be used to designate the relative position of the two substituents. Ortho, meta, and para are also abbreviated as ***o*, *m*,** and ***p*,** respectively.

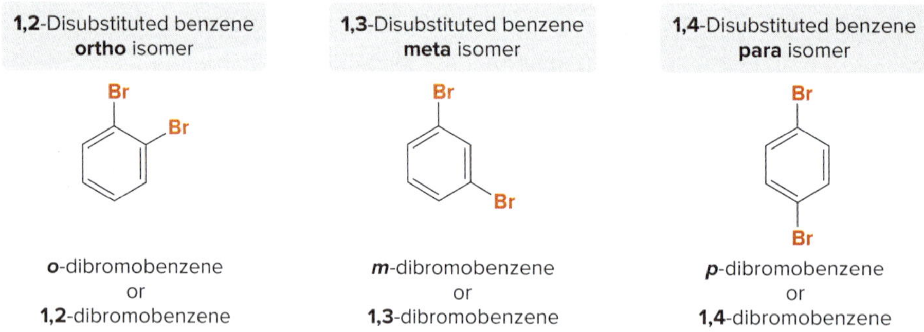

If the two groups on the benzene ring are different, **alphabetize the names of the substituents** preceding the word *benzene*. If one of the substituents is part of a **common root,** name the **molecule as a derivative of that monosubstituted benzene.**

Alphabetize two different substituent names: **Use a common root name:**

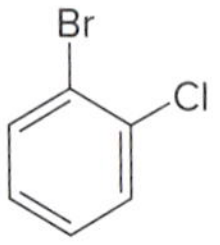

Br Cl NO_2 **nitro** group F toluene Br phenol NO_2 OH

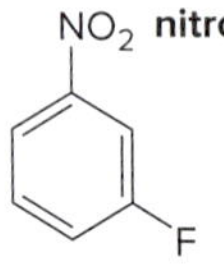

o-**bro**mo**ch**loro-benzene *m*-**f**luoro**n**itro-benzene *p*-bromo**toluene** *o*-nitro**phenol**

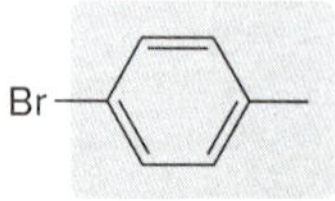

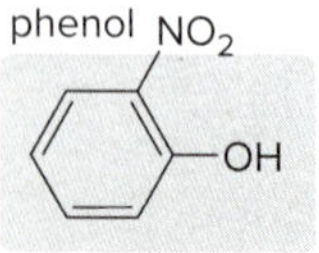

15.3C Polysubstituted Benzenes

For three or more substituents on a benzene ring:

[1] Number to give the lowest possible set of numbers around the ring.

[2] Alphabetize the substituent names.

[3] When substituents are part of common roots, name the molecule as a derivative of that monosubstituted benzene. The substituent that comprises the common root is located at C1.

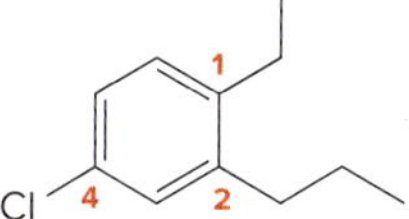

- Assign the lowest possible set of numbers.
- Alphabetize the names of all the substituents.

4-chloro-1-ethyl-2-propylbenzene

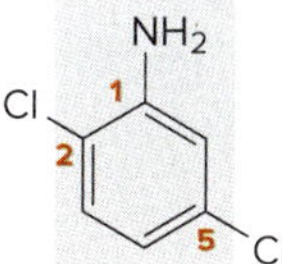

- Name the molecule as a derivative of the common root **aniline.**
- Designate the position of the NH_2 group as "1," and then assign the lowest possible set of numbers to the other substituents.

2,5-dichloroaniline

15.3D Naming Aromatic Rings as Substituents

A benzene substituent (C_6H_5–) is called a **phenyl group,** and it can be abbreviated in a structure as **Ph–.**

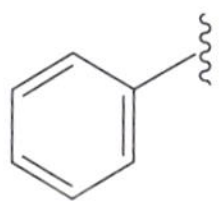

abbreviated as Ph–

phenyl group
C_6H_5–

- **A phenyl group (C_6H_5–) is formed by removing one hydrogen from benzene (C_6H_6).**

The **benzyl** group contains a benzene ring bonded to a CH_2 group. Thus, a benzyl group and a phenyl group differ by the presence of a CH_2 group.

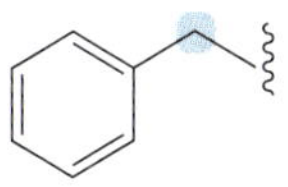

benzyl group
$C_6H_5CH_2$–

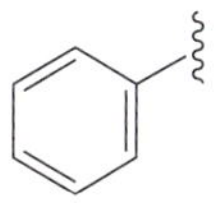

phenyl group
C_6H_5–

Finally, substituents derived from benzene, as well as all other substituted aromatic rings, are collectively called **aryl groups,** abbreviated as Ar–.

Problem 15.3 Give the IUPAC name for each compound.

a.

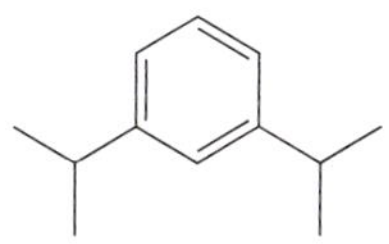

b.

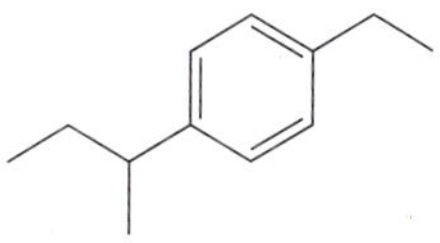

c. 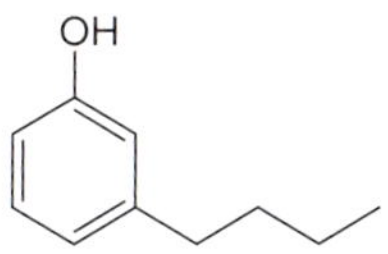 OH

d. 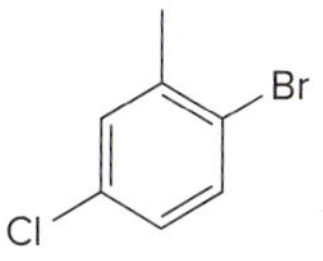 Br Cl

Problem 15.4 Draw the structure corresponding to each name:

a. isobutylbenzene
b. *o*-dichlorobenzene
c. *cis*-1,2-diphenylcyclohexane
d. *m*-bromoaniline
e. 4-chloro-1,2-diethylbenzene
f. 3-*tert*-butyl-2-ethyltoluene

Problem 15.5 What is the structure of propofol, which has the IUPAC name 2,6-diisopropylphenol? Propofol is an intravenous medication used to induce and maintain anesthesia.

15.4 Spectroscopic Properties

The IR spectroscopy of aromatic compounds was discussed in Section B.4A; the NMR spectroscopy of aromatics was presented in Sections C.4 and C.9C. The important IR and NMR absorptions of aromatic compounds are summarized in Table 15.1.

Table 15.1 Characteristic Spectroscopic Absorptions of Benzene Derivatives

Type of spectroscopy	Type of C, H	Absorption
IR absorptions	C_{sp^2}–H C=C (arene)	3150–3000 cm^{-1} 1600, 1500 cm^{-1}
^{1}H NMR absorptions	Ph–H (aryl H)	6.5–8 ppm (highly deshielded protons)
	Ph–CH_2– (benzylic H)	1.5–2.5 ppm (somewhat deshielded C_{sp^3}–H)
^{13}C NMR absorption	C_{sp^2} of arenes	120–150 ppm

^{13}C NMR spectroscopy is used to determine the substitution patterns in disubstituted benzenes, because each line in a spectrum corresponds to a different kind of carbon atom. For example, *o*-, *m*-, and *p*-dibromobenzene each exhibit a different number of lines in its ^{13}C NMR spectrum, as shown in Figure 15.2.

Figure 15.2 ^{13}C NMR absorptions of the three isomeric dibromobenzenes

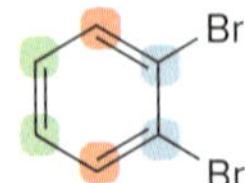

o-dibromobenzene

three types of C's
three ^{13}C NMR signals

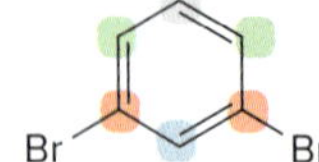

m-dibromobenzene

four types of C's
four ^{13}C NMR signals

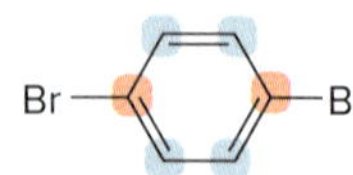

p-dibromobenzene

two types of C's
two ^{13}C NMR signals

- The number of signals (lines) in the ^{13}C NMR spectrum of a disubstituted benzene with two identical groups indicates whether they are ortho, meta, or para to each other.

Problem 15.6 How many ^{13}C NMR signals does each compound exhibit?

a. b. Cl c.

15.5 Interesting Aromatic Compounds

BTX contains **b**enzene, **t**oluene, and **x**ylene (the common name for dimethylbenzene).

Benzene and **toluene,** the simplest aromatic hydrocarbons obtained from petroleum refining, are useful starting materials for synthetic polymers. They are two components of the **BTX** mixture added to gasoline to boost octane ratings.

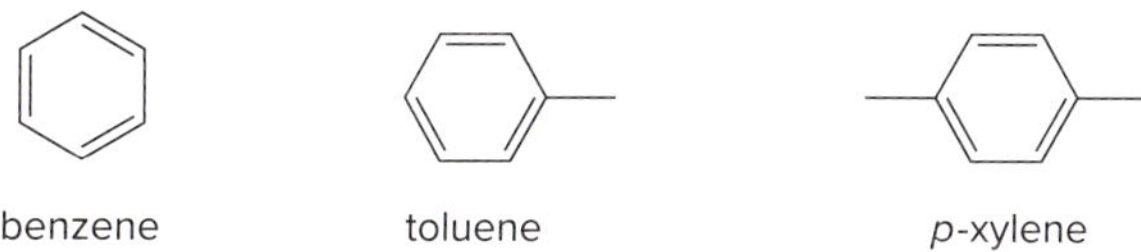

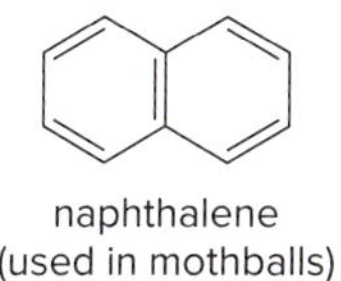

Compounds containing two or more benzene rings that share carbon–carbon bonds are called **polycyclic aromatic hydrocarbons (PAHs).** Naphthalene, the simplest PAH, is present in mothballs.

Benzo[*a*]pyrene, a more complicated PAH shown in Figure 15.3, is formed by the incomplete combustion of organic materials. It is found in cigarette smoke, automobile exhaust, and the fumes from charcoal grills. When ingested or inhaled, benzo[*a*]pyrene and other similar PAHs are oxidized to carcinogenic products, as discussed in Section 9.18.

Figure 15.3 Benzo[*a*]pyrene, a common PAH

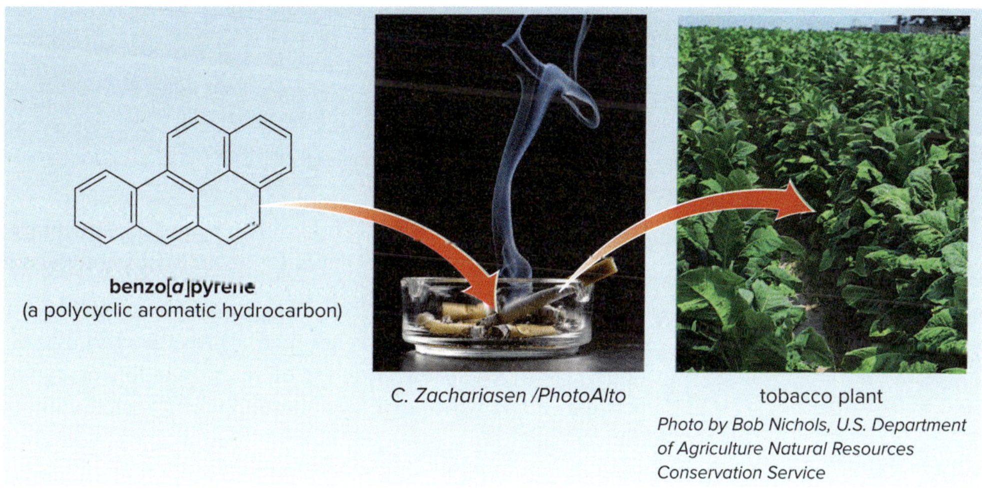

- Benzo[*a*]pyrene, produced by the incomplete oxidation of organic compounds in tobacco, is found in cigarette smoke.

Helicene and **twistoflex** are two synthetic PAHs whose unusual shapes are shown in Figure 15.4. Both helicene and twistoflex are chiral molecules—that is, they are not superimposable on their mirror images, even though neither of them contains a stereogenic center. It's their shape that makes them chiral, not the presence of carbon atoms bonded to four different groups. Each ring system is twisted into a shape that lacks a mirror plane, and each structure is rigid, thus creating the chirality.

Many widely used drugs contain an aromatic ring. Three examples are shown in Figure 15.5.

Figure 15.4 Helicene and twistoflex—Two synthetic polycyclic aromatic hydrocarbons

These two rings are not joined to each other.

helicene = 3-D structure

twistoflex = 3-D structure

- Helicene consists of six benzene rings. Because the rings at both ends are not bonded to each other, all of the rings twist slightly, creating a rigid helical shape that prevents the hydrogen atoms on both ends from crashing into each other. Similarly, to reduce steric hindrance between the hydrogen atoms on nearby benzene rings, twistoflex is also nonplanar.

Figure 15.5 Selected drugs that contain a benzene ring

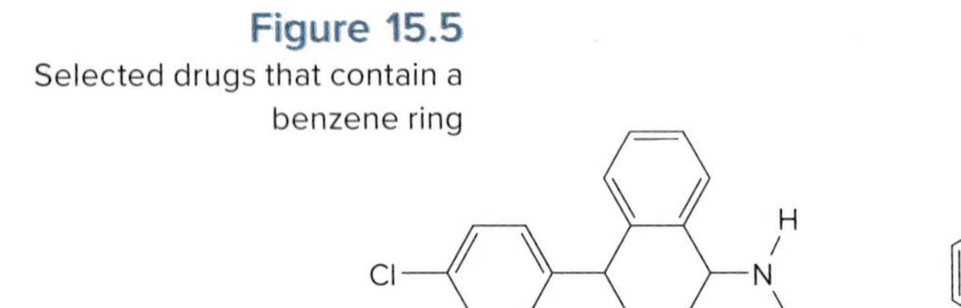

- Trade name: **Zoloft**
- Generic name: **sertraline**
- Use: a psychotherapeutic drug for depression and panic disorders

- Trade name: **Plavix**
- Generic name: **clopidogrel**
- Use: prevention of blood clots

- Trade name: **Nexium**
- Generic name: **esomeprazole**
- Use: anti-ulcer drug

15.6 Benzene's Unusual Stability

Considering benzene as the hybrid of two resonance structures adequately explains its equal C–C bond lengths, but does not account for its unusual stability and lack of reactivity toward addition.

Heats of hydrogenation, which were used in Section 14.9 to show that conjugated dienes are more stable than isolated dienes, can also be used to estimate the stability of benzene. Equations [1]–[3] compare the heats of hydrogenation of cyclohexene, cyclohexa-1,3-diene, and benzene, all of which give cyclohexane when treated with excess hydrogen in the presence of a metal catalyst.

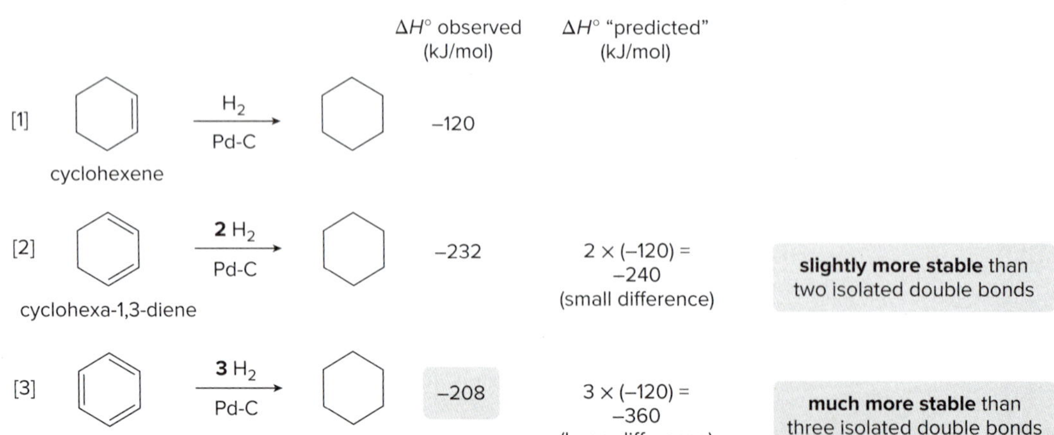

The relative stability of conjugated dienes versus isolated dienes was first discussed in Section 14.9.

The addition of one mole of H_2 to cyclohexene releases –120 kJ/mol of energy (Equation [1]). If each double bond is worth –120 kJ/mol of energy, then the addition of two moles of H_2 to cyclohexa-1,3-diene (Equation [2]) should release 2 × (–120 kJ/mol) = –240 kJ/mol of energy. The observed value, however, is –232 kJ/mol. This is *slightly smaller* than expected because cyclohexa-1,3-diene is a conjugated diene, and **conjugated dienes are more stable than two isolated carbon–carbon double bonds.**

The hydrogenations of cyclohexene and cyclohexa-1,3-diene occur readily at room temperature, but benzene can be hydrogenated only under forcing conditions, and even then the reaction is extremely slow. If each double bond is worth –120 kJ/mol of energy, then the addition of three moles of H_2 to benzene should release 3 × (–120 kJ/mol) = –360 kJ/mol of energy. In fact, the observed heat of hydrogenation is only –208 kJ/mol, which is 152 kJ/mol less than predicted and even *lower* than the observed value for cyclohexa-1,3-diene.

Figure 15.6 compares the hypothetical and observed heats of hydrogenation for benzene.

Figure 15.6 A comparison between the observed and hypothetical heats of hydrogenation for benzene

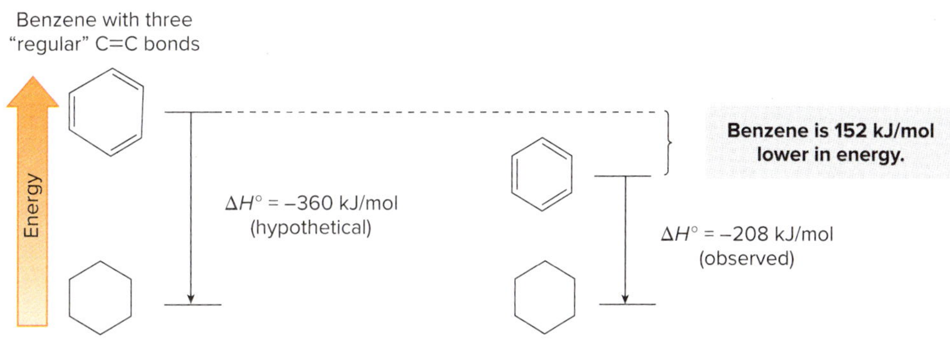

The huge difference between the hypothetical and observed heats of hydrogenation for benzene cannot be explained solely on the basis of resonance and conjugation.

- **The low heat of hydrogenation of benzene means that benzene is *especially stable,* even more so than the conjugated compounds introduced in Chapter 14. This unusual stability is characteristic of aromatic compounds.**

Benzene's unusual behavior in chemical reactions is not limited to hydrogenation. As mentioned in Section 15.1, **benzene does *not* undergo addition reactions typical of other highly unsaturated compounds, including conjugated dienes.** Benzene does not react with Br_2 to yield an addition product. Instead, in the presence of a Lewis acid, bromine *substitutes* for a hydrogen atom, thus yielding a product that retains the benzene ring.

Br_2 (no reaction) — Br, Br

An addition product would no longer contain a benzene ring.

H — Br_2, $FeBr_3$ — Br

A substitution product still contains a benzene ring.

This behavior is characteristic of aromatic compounds. The structural features that distinguish aromatic compounds from the rest are discussed in Section 15.7.

Problem 15.7 Compounds **A** and **B** are both hydrogenated to methylcyclohexane. Which compound has the larger heat of hydrogenation? Which compound is more stable?

15.7 The Criteria for Aromaticity—Hückel's Rule

Four structural criteria must be satisfied for a compound to be aromatic:

- **A molecule must be cyclic, planar, completely conjugated, and contain a particular number of π electrons.**

[1] A molecule must be cyclic.

- **To be aromatic, each *p* orbital must overlap with *p* orbitals on two adjacent atoms.**

The *p* orbitals on all six carbons of benzene continuously overlap, so benzene is aromatic. Hexa-1,3,5-triene has six *p* orbitals, too, but the two on the terminal carbons cannot overlap with each other, so **hexa-1,3,5-triene is *not* aromatic.**

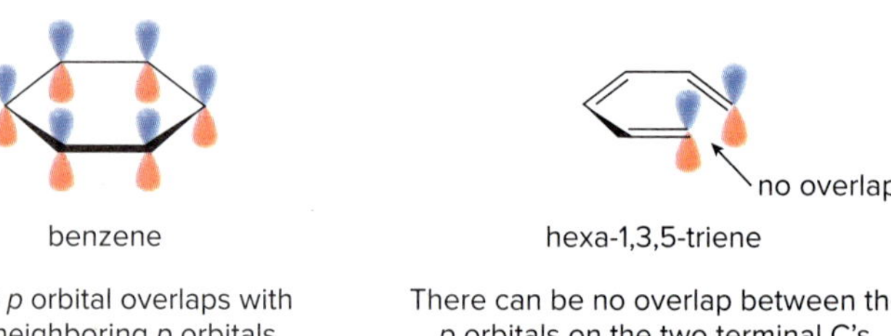

benzene

Every *p* orbital overlaps with two neighboring *p* orbitals.

aromatic

hexa-1,3,5-triene

There can be no overlap between the *p* orbitals on the two terminal C's.

not aromatic

[2] A molecule must be planar.

- **All adjacent *p* orbitals must be aligned so that the π electron density can be delocalized.**

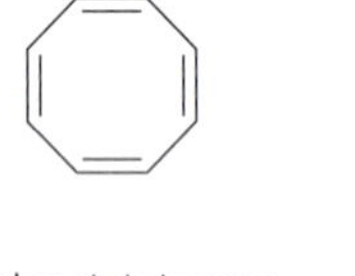

cyclooctatetraene
not aromatic

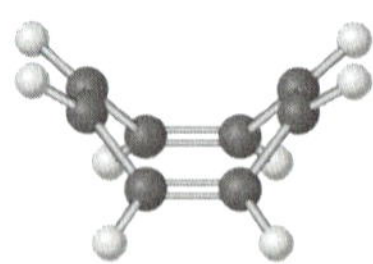

a tub-shaped, eight-membered ring

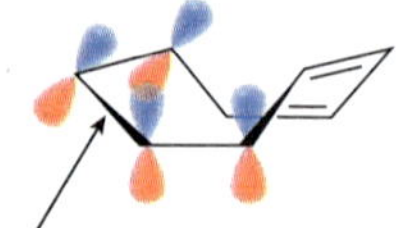

Adjacent *p* orbitals cannot overlap. Electrons cannot delocalize.

Cyclooctatetraene resembles benzene in that it is a cyclic molecule with alternating double and single bonds. Cyclooctatetraene is tub shaped, however, **not planar,** so overlap between adjacent π bonds is impossible. **Cyclooctatetraene, therefore, is *not* aromatic,** so it undergoes addition reactions like those of other alkenes.

Br_2 → Br, Br

cyclooctatetraene → **addition** product

[3] A molecule must be completely conjugated.

- **Aromatic compounds must have a *p* orbital on every atom in the ring.**

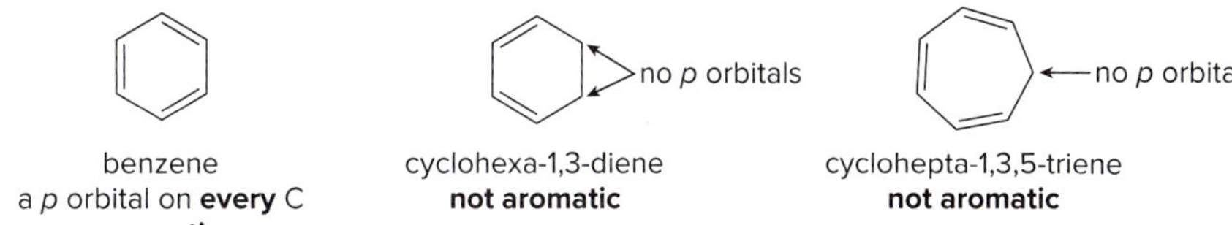

benzene
a *p* orbital on **every** C
aromatic

cyclohexa-1,3-diene
not aromatic

cyclohepta-1,3,5-triene
not aromatic

Both cyclohexa-1,3-diene and cyclohepta-1,3,5-triene contain at least one carbon atom that does not have a *p* orbital, so they are not completely conjugated and therefore ***not* aromatic.**

[4] A molecule must satisfy Hückel's rule, and contain a particular number of π electrons.

Some compounds satisfy the first three criteria for aromaticity, but still they show none of the stability typical of aromatic compounds. For example, **cyclobutadiene** is so highly reactive that it can be prepared only at extremely low temperatures.

cyclobutadiene

a planar, cyclic, completely conjugated molecule that is *not* aromatic

Hückel's rule refers to the number of π electrons, *not* the number of atoms in a particular ring.

It turns out that in addition to being cyclic, planar, and completely conjugated, a compound needs a particular number of π electrons to be aromatic. Erich Hückel first recognized in 1931 that the following criterion, expressed in two parts and now known as **Hückel's rule,** had to be satisfied, as well:

- **An aromatic compound must contain $4n + 2$ π electrons ($n = 0, 1, 2$, and so forth).**
- **Cyclic, planar, and completely conjugated compounds that contain $4n$ π electrons are especially unstable, and are said to be *antiaromatic.***

Table 15.2
The Number of π Electrons That Satisfy Hückel's Rule

n	$4n + 2$
0	2
1	6
2	10
3	14
4, etc.	18

Thus, compounds that contain 2, 6, 10, 14, 18, and so forth π electrons are aromatic, as shown in Table 15.2. **Benzene is aromatic and especially stable because it contains 6 π electrons. Cyclobutadiene is antiaromatic and especially unstable because it contains 4 π electrons.**

Benzene
An aromatic compound

Cyclobutadiene
An antiaromatic compound

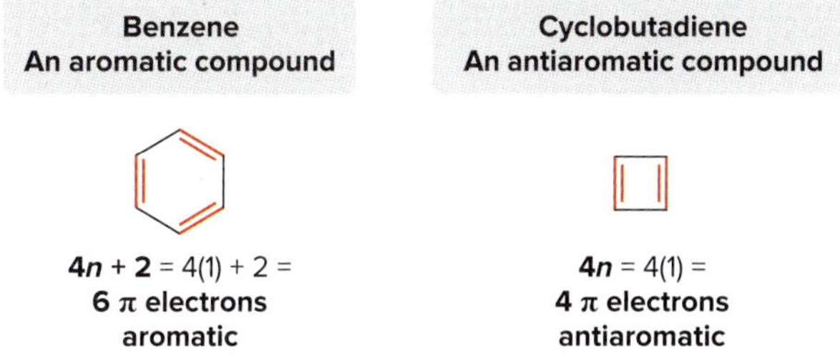

$\mathbf{4n + 2} = 4(1) + 2 =$
6 π electrons
aromatic

$\mathbf{4n} = 4(1) =$
4 π electrons
antiaromatic

Considering aromaticity, all compounds can be classified in one of three ways:

[1]	Aromatic	• A cyclic, planar, completely conjugated compound with $4n + 2$ π electrons
[2]	Antiaromatic	• A cyclic, planar, completely conjugated compound with $4n$ π electrons
[3]	Not aromatic or nonaromatic	• A compound that lacks one (or more) of the four requirements to be aromatic or antiaromatic

Many compounds in addition to benzene are aromatic. Several examples are presented in Sections 15.8 and 15.9.

15.8 Examples of Aromatic Compounds

In Section 15.8, we look at many different types of aromatic compounds.

15.8A Aromatic Compounds with a Single Ring

Benzene is the most common aromatic compound having a single ring. **Completely conjugated rings larger than benzene are also aromatic if they are planar and have $4n + 2$ π electrons.**

- Hydrocarbons containing a single ring with alternating double and single bonds are called *annulenes.*

To name an annulene, indicate the number of atoms in the ring in brackets and add the word *annulene.* Thus, benzene is [6]-annulene. Both **[14]-annulene** and **[18]-annulene** are cyclic, planar, completely conjugated molecules that follow Hückel's rule, so they are aromatic.

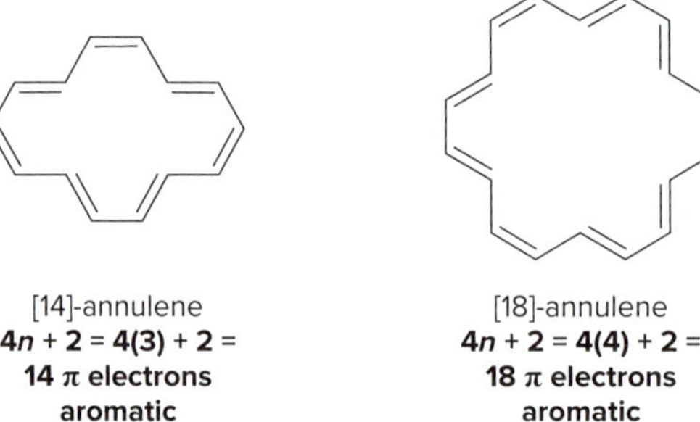

[10]-Annulene has 10 π electrons, which satisfies Hückel's rule, but a planar molecule would place the two H atoms inside the ring too close to each other, so the ring puckers to relieve this strain. Because **[10]-annulene is not planar,** the 10 π electrons can't delocalize over the entire ring and it is **not aromatic.**

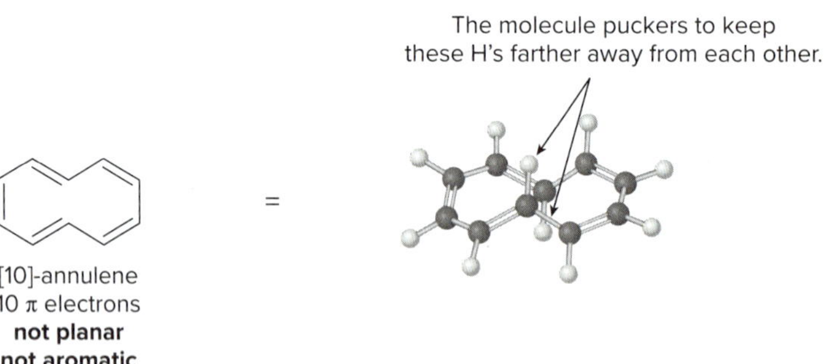

Problem 15.8 Would [16]-, [20]-, or [22]-annulene be aromatic if each ring is planar?

Problem 15.9 Compound **A** exhibits a peak in its ^1H NMR spectrum at 7.6 ppm, indicating that it is aromatic. (a) How are the carbon atoms of the triple bonds hybridized? (b) In what type of orbitals are the π electrons of the triple bonds contained? (c) How many π electrons are delocalized around the ring in **A?**

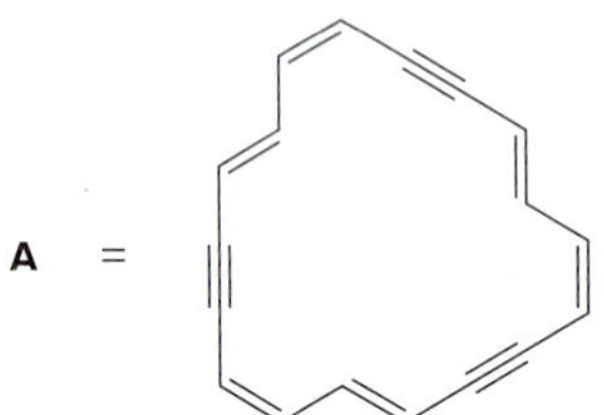

15.8B Aromatic Compounds with More Than One Ring

Hückel's rule for determining aromaticity can be applied only to monocyclic systems, but many aromatic compounds containing several benzene rings joined together are also known. Two or more six-membered rings with alternating double and single bonds can be fused together to form **polycyclic aromatic hydrocarbons (PAHs).** Joining two benzene rings together forms **naphthalene.** There are two different ways to join three rings together, forming **anthracene** and **phenanthrene,** and many more complex hydrocarbons are known.

naphthalene **10** π electrons | anthracene **14** π electrons | phenanthrene **14** π electrons

As the number of fused benzene rings increases, the number of resonance structures increases as well. Although two resonance structures can be drawn for benzene, naphthalene is a hybrid of three resonance structures.

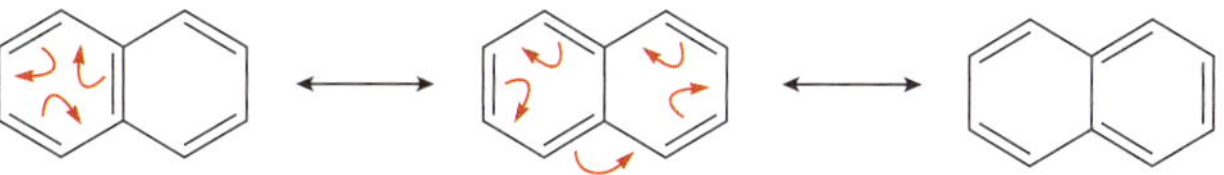

Problem 15.10 Draw the four resonance structures for anthracene.

15.8C Charged Aromatic Compounds

Both negatively and positively charged ions can also be aromatic if they satisfy all the necessary criteria.

Cyclopentadienyl Anion

The **cyclopentadienyl anion** is a cyclic and planar anion with two double bonds and a nonbonded electron pair. The two π bonds contribute four electrons and the lone pair contributes two more, for a total of six. By Hückel's rule, having **six π electrons confers aromaticity.** The **negatively charged carbon atom must be sp^2 hybridized,** and the **nonbonded electron pair must occupy a p orbital** for the ring to be completely conjugated.

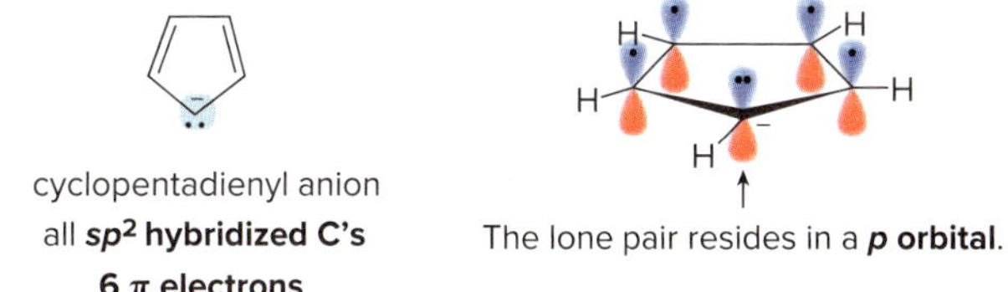

cyclopentadienyl anion
all ***sp*^{2} hybridized C's**
6 π electrons

The lone pair resides in a ***p* orbital**.

- **The cyclopentadienyl anion is aromatic because it is cyclic, planar, completely conjugated, and has six π electrons.**

We can draw **five equivalent resonance structures for the cyclopentadienyl anion,** delocalizing the negative charge over every carbon atom of the ring.

Although five resonance structures can also be drawn for both the **cyclopentadienyl cation** and **radical,** only the cyclopentadienyl anion has six π electrons, a number that satisfies Hückel's rule. The cyclopentadienyl cation has four π electrons, making it antiaromatic and especially unstable. The cyclopentadienyl radical has five π electrons, so it is neither aromatic nor antiaromatic. **Having the "right" number of electrons is necessary for a species to be unusually stable by virtue of aromaticity.**

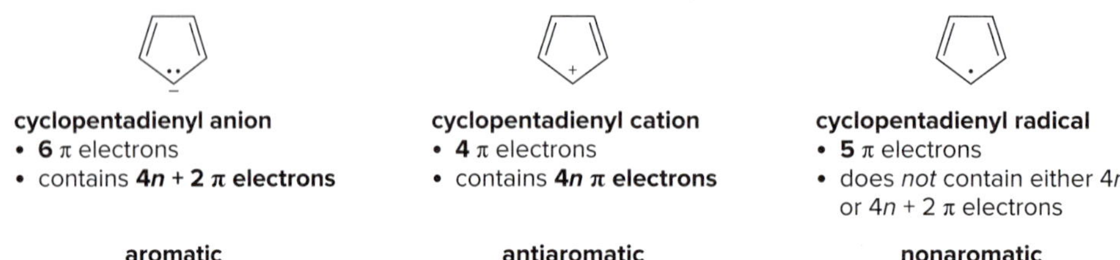

The cyclopentadienyl anion is readily formed from cyclopentadiene by a Brønsted–Lowry acid–base reaction.

H H + :B → H + H–B⁺

cyclopentadiene
not aromatic
$pK_a = 15$

cyclopentadienyl anion
aromatic
a stabilized conjugate base

Cyclopentadiene itself is not aromatic because it is not fully conjugated. **The cyclopentadienyl anion, however, is aromatic, so it is a very stable base.** As such, it makes cyclopentadiene more acidic than other hydrocarbons. In fact, the pK_a of cyclopentadiene is 15, much *lower* (more acidic) than the pK_a of any C–H bond discussed thus far.

- **Cyclopentadiene is more acidic than many hydrocarbons because its conjugate base is aromatic.**

Problem 15.11 Draw the product formed when cyclohepta-1,3,5-triene (pK_a = 39) is treated with a strong base. Why is its pK_a so much higher than the pK_a of cyclopentadiene?

H H :B →

cyclohepta-1,3,5-triene
$pK_a = 39$

Problem 15.12 Rank the following compounds in order of increasing acidity.

A **B** **C**

Tropylium Cation

The **tropylium cation** is a planar carbocation with three double bonds and a positive charge contained in a seven-membered ring. This carbocation is completely conjugated, because the

The cyclopentadienyl anion and the tropylium cation both illustrate an important principle: The **number of π electrons determines aromaticity,** not the number of atoms in a ring or the number of *p* orbitals that overlap. The cyclopentadienyl anion and tropylium cation are aromatic because they each have six π electrons.

positively charged carbon is sp^2 hybridized and has a vacant *p* orbital that overlaps with the six *p* orbitals from the carbons of the three double bonds. **Because the tropylium cation has three π bonds and no other nonbonded electron pairs, it contains six π electrons,** thereby satisfying Hückel's rule.

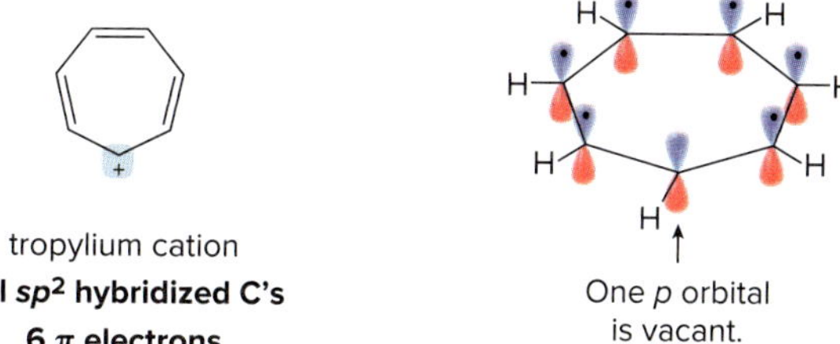

tropylium cation
all sp^2 hybridized C's
6 π electrons

- **The tropylium cation is aromatic because it is cyclic, planar, completely conjugated, and has six π electrons delocalized over the seven atoms of the ring.**

Problem 15.13 Draw the seven resonance structures for the tropylium cation.

Problem 15.14 Assuming the rings are planar, label each ion as aromatic, antiaromatic, or not aromatic.

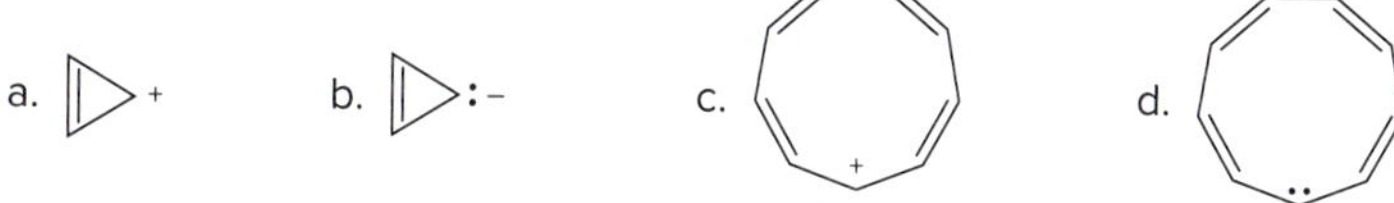

15.9 Aromatic Heterocycles

Recall from Section 9.3 that a **heterocycle** is a ring that contains at least one heteroatom.

Heterocycles containing oxygen, nitrogen, or sulfur—atoms that also have at least one lone pair of electrons—can also be aromatic. With heteroatoms, we must always **determine whether the lone pair is localized on the heteroatom or part of the delocalized π system.**

15.9A Pyridine, Pyrrole, and Histamine

Pyridine

Pyridine is a heterocycle containing a six-membered ring with three π bonds and one nitrogen atom. Similar to benzene, two resonance structures (with all neutral atoms) can be drawn.

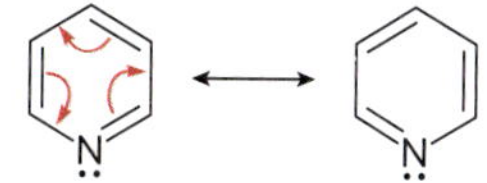

two resonance structures for pyridine
6 π electrons

Pyridine is cyclic, planar, and completely conjugated, because the three single and three double bonds alternate around the ring. **Pyridine has six π electrons, two from each π bond, thus satisfying Hückel's rule and making pyridine aromatic.** The nitrogen atom of pyridine also has a nonbonded electron pair, which is *localized* on the N atom, so it is *not* part of the delocalized π electron system of the aromatic ring.

How is the nitrogen atom of the pyridine ring hybridized? The N atom is surrounded by three groups (two atoms and a lone electron pair), making it sp^2 **hybridized,** and leaving one unhybridized *p* orbital with one electron that overlaps with adjacent *p* orbitals. The lone pair on N resides in an sp^2 hybrid orbital that is perpendicular to the delocalized π electrons.

sp^2 hybridized N

The lone pair occupies an $\mathbf{sp^2}$ hybrid orbital, perpendicular to the direction of the six *p* orbitals.

A *p* orbital on N overlaps with adjacent *p* orbitals, making the ring **completely conjugated.**

Pyrrole

Pyrrole contains a five-membered ring with two π bonds and one nitrogen atom. The N atom also has a lone pair of electrons.

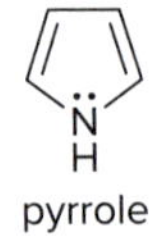

pyrrole

Pyrrole is cyclic and planar, with a total of four π electrons from the two π bonds. Is the nonbonded electron pair localized on N or part of a delocalized π electron system? The lone pair on N is *adjacent* to a double bond. Recall the following general rule from Section 14.5:

- **In a system X=Y–Z:, Z is generally sp^2 hybridized and the lone pair occupies a *p* orbital to make the system conjugated.**

If the lone pair on the N atom occupies a *p* orbital:

- **Pyrrole has a *p* orbital on every adjacent atom, so it is completely conjugated.**
- **Pyrrole has six π electrons—four from the π bonds and two from the lone pair.**

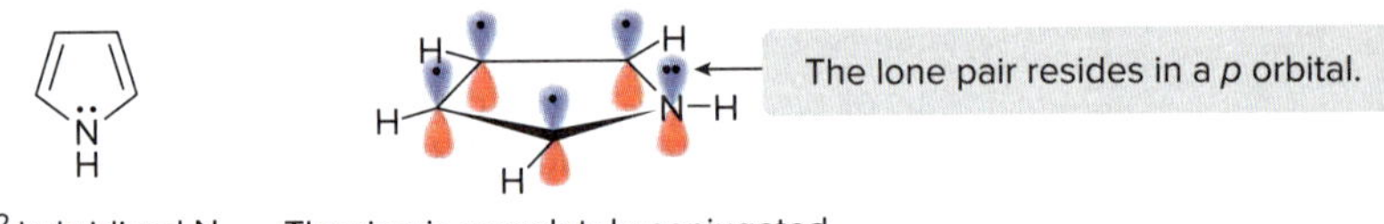

sp^2 hybridized N The ring is completely conjugated with **6 π electrons.**

Because pyrrole is cyclic, planar, completely conjugated, and has $4n + 2$ π electrons, **pyrrole is aromatic. The number of electrons—not the size of the ring—determines whether a compound is aromatic.**

A fundamental difference exists between the N atoms in pyridine and pyrrole.

- **When a heteroatom is already part of a double bond (as in the N of pyridine), its lone pair *cannot* occupy a *p* orbital, so it *cannot* be delocalized over the ring.**
- **When a heteroatom is *not* part of a double bond (as in the N of pyrrole), its lone pair can be located in a *p* orbital and *delocalized* over a ring to make it aromatic.**

Scombroid fish poisoning, which is caused by inadequately refrigerated fish, results when bacteria convert the amino acid histidine in the tissues of the fish into the neurotransmitter histamine.

Daniel C. Smith

Histamine

Histamine has an aromatic heterocycle with two N atoms, one of which is similar to the N atom of pyridine and one of which is similar to the N atom of pyrrole.

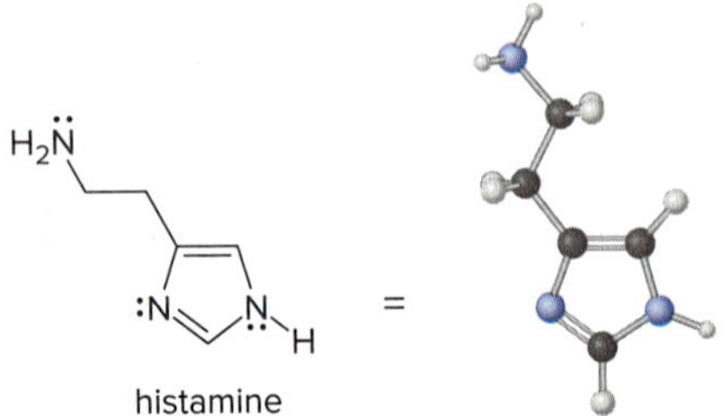

histamine

Histamine has a five-membered ring with two π bonds and two nitrogen atoms, each of which contains a lone pair of electrons. The heterocycle has four π electrons from the two double bonds. **The lone pair on the N in red also occupies a *p* orbital,** making the heterocycle completely conjugated and giving it a total of six π electrons. The lone pair on this N atom is thus delocalized over the five-membered ring and the heterocycle is aromatic. **The lone pair on the N in blue occupies an sp^2 hybrid orbital** perpendicular to the delocalized π electrons.

N: The lone pair resides in a ***p* orbital**.

N: The lone pair resides in an ***sp^2* hybrid orbital**.

- N (in red) resembles the N atom of pyrrole.
- N (in blue) resembles the N atom of pyridine.

Antihistamines that block the action of histamine on the H1 histamine receptor are used to treat the runny nose and watery eyes of an allergic reaction.
Bob London/Alamy Stock Photo

Histamine produces a wide range of physiological effects in the body. Excess histamine is responsible for the runny nose and watery eyes symptomatic of hay fever. It also stimulates the overproduction of stomach acid and contributes to the formation of hives. These effects result from the interaction of histamine with two different cellular receptors. We will learn more about antihistamines and antiulcer drugs, compounds that block the effects of histamine, in Section 23.5.

Sample Problem 15.1 Determining the Hybridization of a Heteroatom in an Aromatic Heterocycle

How is each N atom in pilocarpine, the chapter-opening molecule, hybridized, and in what type of orbital does the lone pair on each N reside?

pilocarpine

Solution

***sp^2* hybridized N**
lone pair in an ***sp^2* hybrid orbital**

This lone pair is **localized** on the N.

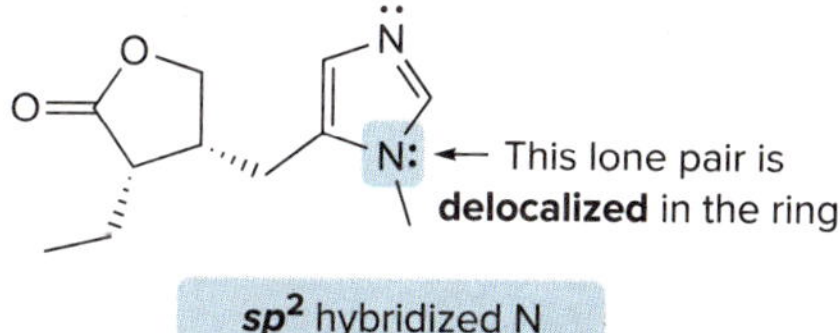

***sp^2* hybridized N**
lone pair in a ***p* orbital**

The N atom labeled in red is already part of a double bond, so it is ***sp^2* hybridized** and its unhybridized *p* orbital is used to form the π bond. As a result, the **lone pair on N occupies one of the *sp^2* hybrid orbitals.**

The N atom labeled in blue is ***sp^2* hybridized,** so its **lone pair can occupy a *p* orbital** and delocalize in the five-membered ring. Delocalization gives the ring six π electrons and makes the ring aromatic.

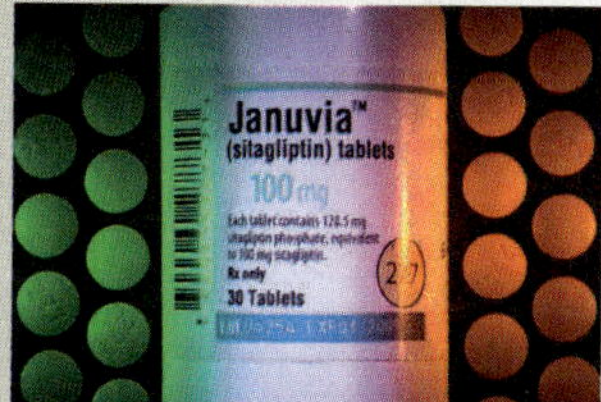

Januvia (Problem 15.15) increases the body's ability to lower blood sugar levels, so it is used alone or in combination with other drugs to treat type 2 diabetes.
Jb Reed/Bloomberg/Getty Images

Problem 15.15 Januvia, the trade name for sitagliptin, was introduced in 2006 for the treatment of type 2 diabetes. (a) Explain why the five-membered ring in sitagliptin is aromatic. (b) Determine the hybridization of each N atom. (c) In what type of orbital does the lone pair on each N atom reside?

sitagliptin

More Practice: Try Problems 15.34; 15.35; 15.56b, c; 15.57a; 15.59b.

Problem 15.16 Which heterocycles are aromatic?

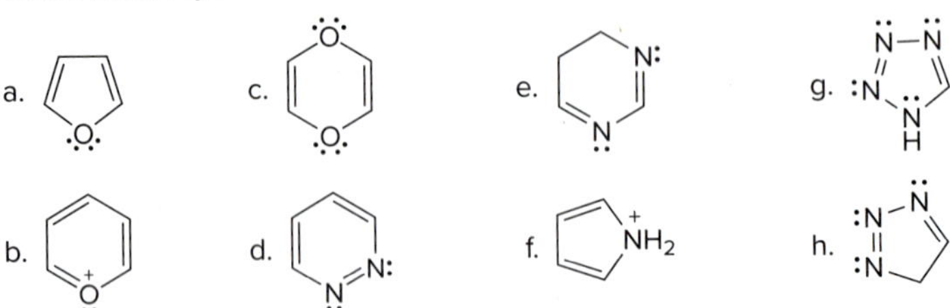

Sample Problem 15.2 Characterizing a Compound as Aromatic, Antiaromatic, or Not Aromatic

Label each compound as aromatic, antiaromatic, or not aromatic. Assume all completely conjugated rings are planar.

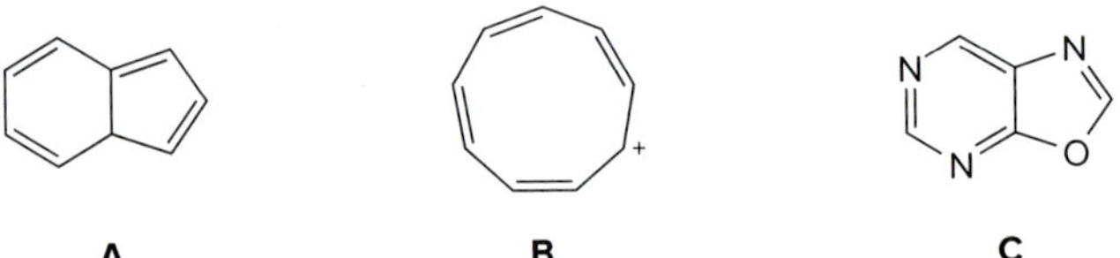

Solution

Each compound is cyclic, and from the problem statement, we assume that completely conjugated rings are planar, so we must answer just two questions to decide on aromaticity:

- **Is the compound completely conjugated?** If a compound is not completely conjugated, it is *not* aromatic.
- **How many π electrons does the compound contain?** A compound with **$4n + 2$ π** electrons is *aromatic;* a compound with **$4n$ π** electrons is *antiaromatic.* A compound with an odd number of π electrons satisfies neither equation and is *not* aromatic.

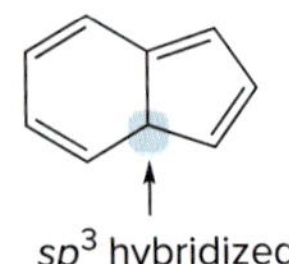

sp^3 hybridized

A

- The C labeled in blue is sp^3 hybridized, so there is no *p* orbital for overlap and the system is **not completely conjugated. A is not aromatic.**

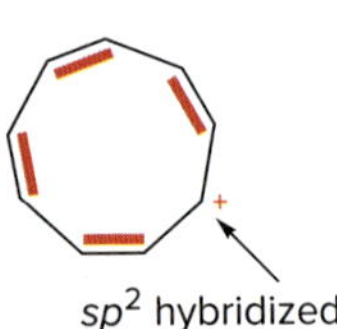

sp^2 hybridized

B

- The ring is completely conjugated because each C of the C=C's and the positively charged C are sp^2 hybridized, so each C has an unhybridized *p* orbital. The ring has **8 π electrons** from the four C=C's, so it has **$4n$ π** electrons. **B is antiaromatic.**

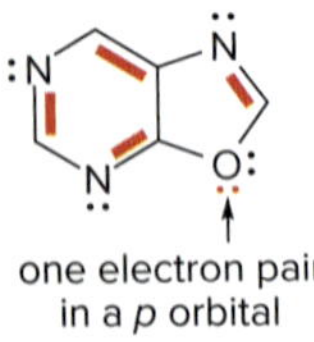

one electron pair in a *p* orbital

C

- For the ring to be completely conjugated, the O atom must be sp^2 hybridized and one electron pair on O must occupy a *p* orbital. The ring has **10 π electrons** from the four C=C's and the O atom, so it has **$4n + 2$ π** electrons. **C is aromatic.**

Problem 15.17 Label each compound as aromatic, antiaromatic, or not aromatic. Assume all completely conjugated rings are planar.

More Practice: Try Problems 15.23, 15.29–15.31, 15.35a, 15.56a.

15.9B Deoxyribonucleic Acid—DNA

A detailed discussion of the structure and properties of DNA is found in Sections 28.2 and 28.3.

As mentioned in Section 3.9, deoxyribonucleic acid (DNA) is the high-molecular-weight polymer that stores the genetic information of an organism. A key element of the structure of DNA is the presence of **aromatic nitrogen heterocycles called bases,** which form a network of hydrogen bonds that hold two chains of DNA together in a double helix.

DNA contains four different bases, derived from the heterocycles **pyrimidine** and **purine.** The π electrons highlighted in red on pyrimidine and purine are delocalized and part of the aromatic system.

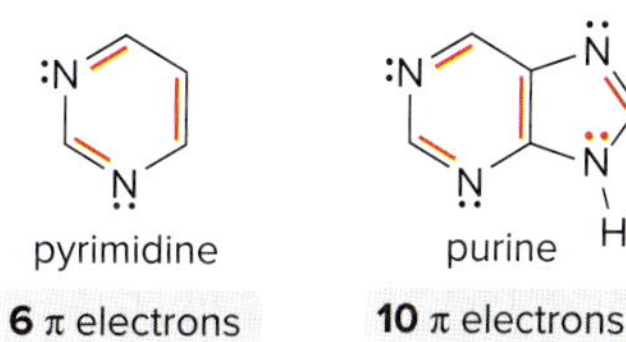

- **The lone pair electrons on both N atoms in pyrimidine, like pyridine, reside in sp^2 hybridized orbitals perpendicular to the aromatic π system.**
- **Purine, a bicyclic compound that contains four sp^2 hybridized N atoms, has a lone pair (highlighted in red) that resides in a p orbital and is part of the aromatic system.**

Two DNA bases—**cytosine** and **thymine**—are derived from pyrimidine and contain six π electrons, and two bases—**adenine** and **guanine**—are derived from purine and contain 10 π electrons.

cytosine
6 π electrons

thymine
6 π electrons

pyrimidines

adenine
10 π electrons

guanine
10 π electrons

purines

Three of the bases are typically drawn in a form that contains a C=O, so it may not be obvious that each is aromatic. Because these bases contain a lone pair on N that can be delocalized on the C=O, however, a resonance structure can be drawn that clearly shows that six π electrons are delocalized in the ring. Two resonance structures are shown for cytosine.

cytosine

6 π electrons

Problem 15.18 Answer each question for the four bases that comprise DNA. (a) How is each N atom hybridized? (b) In what type of orbital does each lone pair on a N reside?

Problem 15.19 Draw resonance structures for the six-membered thymine and guanine rings that clearly show the π electrons delocalized within the aromatic rings.

15.10 What Is the Basis of Hückel's Rule?

Why does the number of π electrons determine whether a compound is aromatic? Cyclobutadiene is cyclic, planar, and completely conjugated, just like benzene, but why is benzene aromatic and cyclobutadiene antiaromatic?

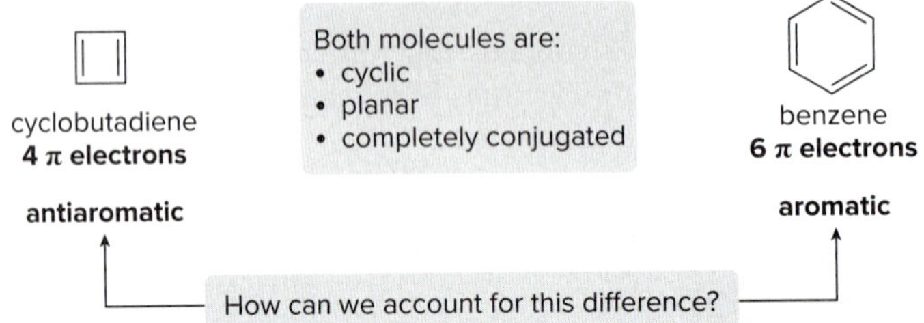

A complete explanation is beyond the scope of an introductory organic chemistry text, but nevertheless, you can better understand the basis of aromaticity by learning more about orbitals and bonding.

15.10A Bonding and Antibonding Orbitals

So far we have used these basic concepts to describe how bonds are formed:

- **Hydrogen uses its 1s orbital to form σ bonds with other elements.**
- **Second-row elements use hybrid orbitals (*sp*, sp^2, or sp^3) to form σ bonds.**
- **Second-row elements use *p* orbitals to form π bonds.**

This description of bonding is called **valence bond theory.** In valence bond theory, a covalent bond is formed by the overlap of two atomic orbitals, and the electron pair in the resulting bond is shared by both atoms. Thus, a carbon–carbon double bond consists of a σ bond, formed by overlap of two sp^2 hybrid orbitals, each containing one electron, and a π bond, formed by overlap of two *p* orbitals, each containing one electron.

This description of bonding works well for most of the organic molecules we have encountered thus far. Unfortunately, it is inadequate for describing systems with many adjacent *p* orbitals that overlap, as there are in aromatic compounds. To more fully explain the bonding in these systems, we must utilize **molecular orbital (MO) theory.**

MO theory describes bonds as the mathematical combination of atomic orbitals that form a new set of orbitals called **molecular orbitals (MOs).** A molecular orbital occupies a region of space *in a molecule* where electrons are likely to be found. When forming molecular orbitals from atomic orbitals, keep in mind:

- **A set of *n* atomic orbitals forms *n* molecular orbitals.**

If *two* atomic orbitals combine, *two* molecular orbitals are formed. This is fundamentally different than valence bond theory. Because aromaticity is based on *p* orbital overlap, what does MO theory predict will happen when two *p* (atomic) orbitals combine?

The two lobes of each *p* orbital are opposite in phase, with a node of electron density at the nucleus. When two *p* orbitals combine, two molecular orbitals should form. The two *p* orbitals can add together constructively—that is, with like phases interacting—or destructively—that is, with opposite phases interacting.

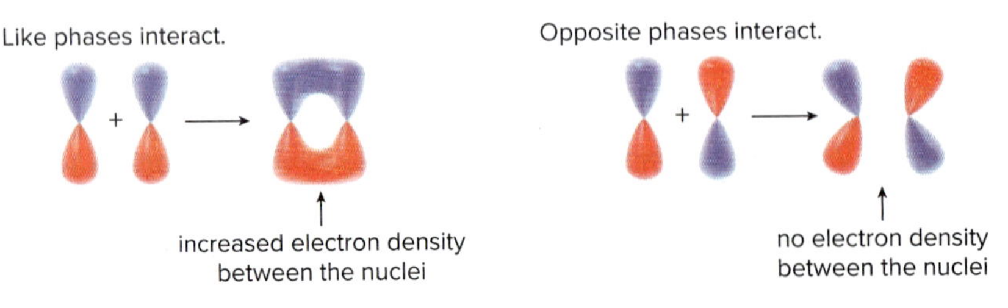

- **When two *p* orbitals of similar phase overlap side-by-side, a π bonding molecular orbital results.**
- **When two *p* orbitals of opposite phase overlap side-by-side, a π* antibonding molecular orbital results.**

A π bonding MO is lower in energy than the two atomic *p* orbitals from which it is formed because a stable bonding interaction results when orbitals of similar phase combine. A bonding interaction holds nuclei together. Similarly, a π* antibonding MO is higher in energy because a destabilizing node results when orbitals of opposite phase combine. A destabilizing interaction pushes nuclei apart.

If two atomic *p* orbitals each have one electron and then combine to form MOs, the two electrons will occupy the lower-energy π bonding MO, as shown in Figure 15.7.

Figure 15.7
Combination of two *p* orbitals to form π and π* molecular orbitals

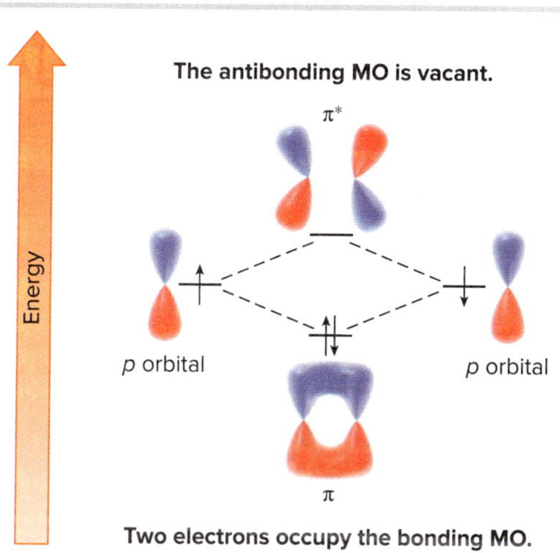

- Two atomic *p* orbitals combine to form two molecular orbitals. The bonding π MO is lower in energy than the two *p* orbitals from which it was formed, and the antibonding π* MO is higher in energy than the two *p* orbitals from which it was formed.
- Two electrons fill the lower-energy bonding MO first.

15.10B Molecular Orbitals Formed When More Than Two *p* Orbitals Combine

The molecular orbital description of benzene is much more complex than the two MOs formed in Figure 15.7. Because each of the six carbon atoms of benzene has a *p* orbital, **six atomic *p* orbitals combine to form six π molecular orbitals,** as shown in Figure 15.8. A description of the exact appearance and energies of these six MOs requires more sophisticated mathematics and understanding of MO theory than is presented in this text. Nevertheless, note that the six MOs are labeled ψ_1–ψ_6, with ψ_1 being the lowest in energy and ψ_6 the highest.

The most important features of the six benzene MOs are as follows:

- **The larger the number of bonding interactions, the lower in energy the MO.** The lowest-energy molecular orbital (ψ_1) has all bonding interactions between the *p* orbitals.
- **The larger the number of nodes, the higher in energy the MO.** The highest-energy MO (ψ_6*) has all nodes between the *p* orbitals.
- Three MOs are lower in energy than the starting *p* orbitals, making them bonding MOs (ψ_1, ψ_2, and ψ_3), whereas three MOs are higher in energy than the starting *p* orbitals, making them antibonding MOs (ψ_4*, ψ_5*, and ψ_6*).

Figure 15.8

How the six *p* orbitals of benzene overlap to form six molecular orbitals

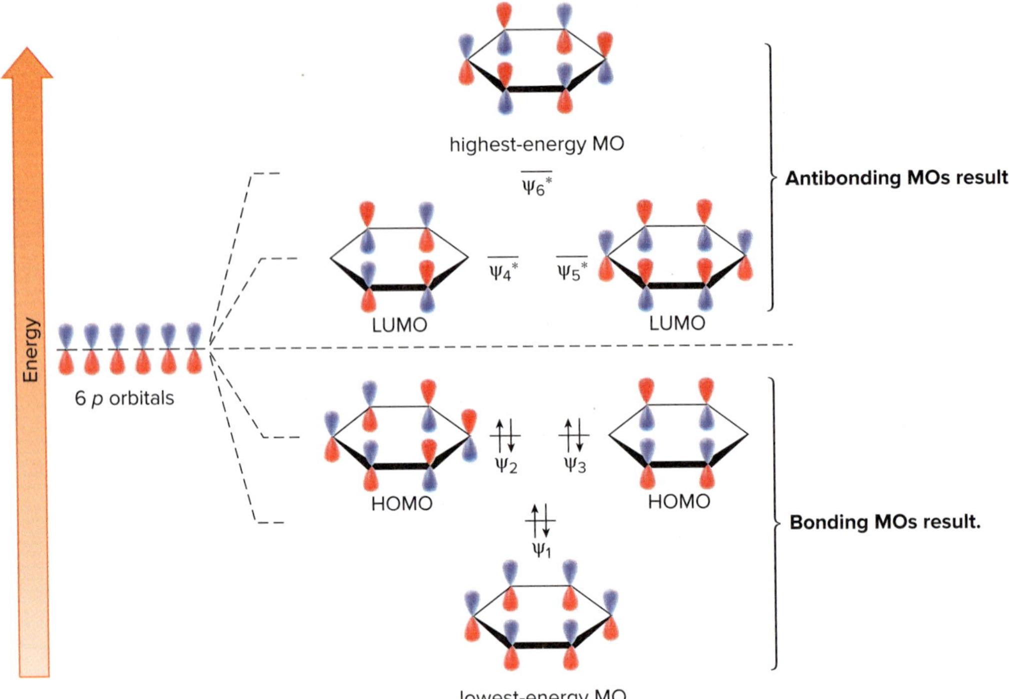

- Depicted in this diagram are the interactions of the six atomic *p* orbitals of benzene, which form six molecular orbitals. When orbitals of like phase combine, a bonding interaction results. When orbitals of opposite phase combine, a destabilizing node results.

- The two pairs of MOs (ψ_2 and ψ_3; ψ_4* and ψ_5*) with the same energy are called **degenerate orbitals.**
- **The highest-energy orbital that contains electrons is called the *highest occupied molecular orbital* (HOMO).** For benzene, the degenerate orbitals ψ_2 and ψ_3 are the HOMOs.
- **The lowest-energy orbital that does *not* contain electrons is called the *lowest unoccupied molecular orbital* (LUMO).** For benzene, the degenerate orbitals ψ_4* and ψ_5* are the LUMOs.

To fill the MOs, the six electrons are added, two to an orbital, beginning with the lowest-energy orbital. As a result, **the six electrons completely fill the bonding MOs, leaving the antibonding MOs empty.** This is what gives benzene and other aromatic compounds their special stability, and this is why six π electrons satisfies Hückel's $4n + 2$ rule.

- **All bonding MOs (and HOMOs) are completely filled in aromatic compounds. No π electrons occupy antibonding MOs.**

15.11 The Inscribed Polygon Method for Predicting Aromaticity

An inscribed polygon is also called a **Frost circle.**

To predict whether a compound has π electrons completely filling bonding MOs, we must know how many bonding molecular orbitals and how many π electrons it has. It is possible to predict the relative energies of cyclic, completely conjugated compounds, without sophisticated math (or knowing what the resulting MOs look like) by using the **inscribed polygon method.**

How To Use the Inscribed Polygon Method to Determine the Relative Energies of MOs for Cyclic, Completely Conjugated Compounds

Example Plot the relative energies of the MOs of benzene.

Step [1] **Draw the polygon in question inside a circle with its vertices touching the circle and one of the vertices pointing down. Mark the points at which the polygon intersects the circle.**

- Inscribe a hexagon inside a circle for benzene. The six vertices of the hexagon form six points of intersection, corresponding to the six MOs of benzene. The pattern—a single MO having the lowest energy, two degenerate pairs of MOs, and a single highest-energy MO—matches that found in Figure 15.8.

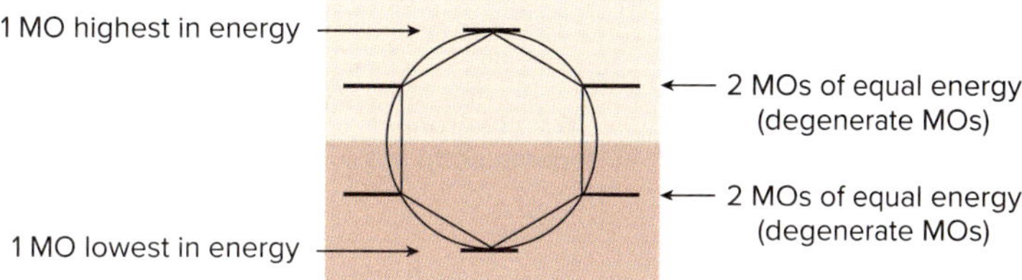

Step [2] **Draw a line horizontally through the center of the circle and label MOs as bonding, nonbonding, or antibonding.**

- **MOs below this line are bonding,** and lower in energy than the *p* orbitals from which they were formed. Benzene has three bonding MOs.
- **MOs at this line are nonbonding,** and equal in energy to the *p* orbitals from which they were formed. Benzene has no nonbonding MOs.
- **MOs above this line are antibonding,** and higher in energy than the *p* orbitals from which they were formed. Benzene has three antibonding MOs.

Step [3] **Add the electrons, beginning with the lowest-energy MO.**

- **All the bonding MOs (and the HOMOs) are completely filled in aromatic compounds. No π electrons occupy antibonding MOs.**
- Benzene is aromatic because it has six π electrons that completely fill the bonding MOs.

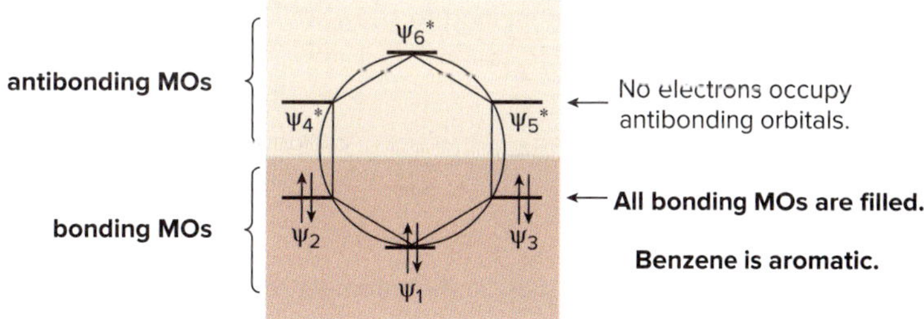

This method works for all monocyclic, completely conjugated hydrocarbons regardless of ring size. Figure 15.9 illustrates MOs for completely conjugated five- and seven-membered rings using this method. The total number of MOs always equals the number of vertices of the polygon. **Because both systems have three bonding MOs, each needs six π electrons to fully occupy them,** making the cyclopentadienyl anion and the tropylium cation aromatic, as we learned in Section 15.8C.

The inscribed polygon method is consistent with **Hückel's $4n + 2$ rule**; that is, **there is always one lowest-energy bonding MO** that can hold two π electrons and the **other bonding MOs come in degenerate pairs** that can hold a total of four π electrons. For the compound to be aromatic, these MOs must be completely filled with electrons, so the "magic numbers" for aromaticity fit Hückel's $4n + 2$ rule (Figure 15.10).

Figure 15.9
Using the inscribed polygon method for five- and seven-membered rings

Five-membered ring **Seven-membered ring**

Always draw the polygon with a vertex pointing **down:**

three bonding MOs

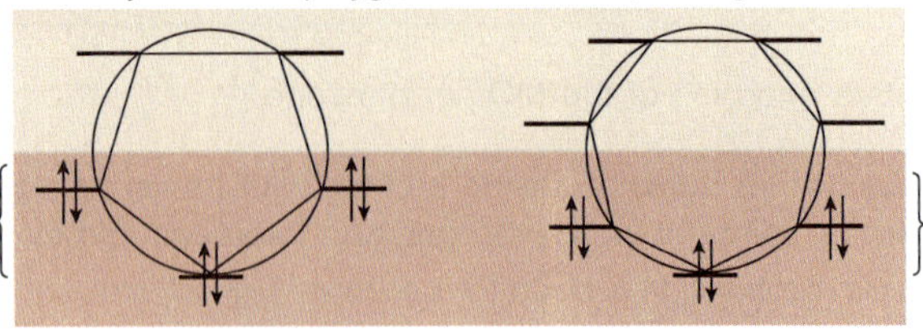

three bonding MOs

- Both systems have **three** bonding MOs.
- Both systems need **six** π electrons to be aromatic.

6 π electrons
cyclopentadienyl anion

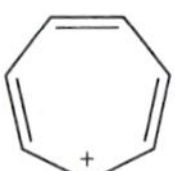

6 π electrons
tropylium cation

Figure 15.10
MO patterns for cyclic, completely conjugated systems

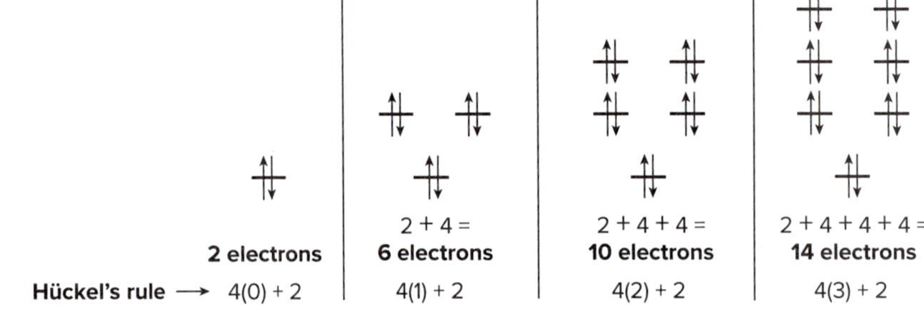

Sample Problem 15.3 Using the Inscribed Polygon Method in Determining Aromaticity

Use the inscribed polygon method to show why cyclobutadiene is not aromatic.

cyclobutadiene
4 π electrons

Solution

Cyclobutadiene has four MOs (formed from its four *p* orbitals), to which its four π electrons must be added.

Step [1] Inscribe a square with a vertex down and mark its four points of intersection with the circle.

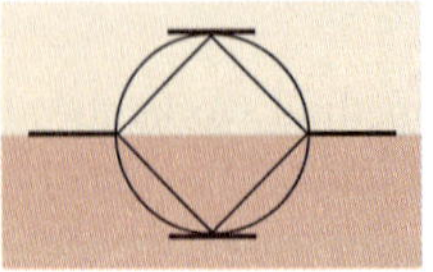

- The four points of intersection correspond to the four MOs of cyclobutadiene.

Steps [2] and [3] Draw a line through the center of the circle, label the MOs, and add the electrons.

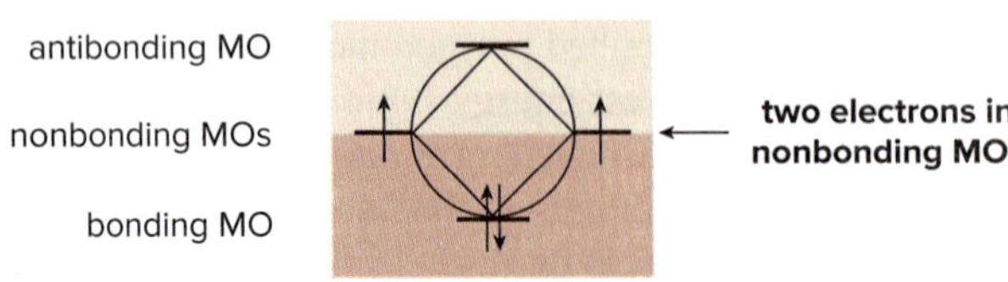

- Cyclobutadiene has four MOs—one bonding, two nonbonding, and one antibonding.
- Adding cyclobutadiene's **four π electrons to these orbitals places two in the lowest-energy bonding MO and one each in the two nonbonding MOs.**
- Separating electrons in two degenerate MOs keeps **like charges farther away from each other.**

Conclusion: Cyclobutadiene is not aromatic because its HOMOs, **two degenerate nonbonding MOs, are not completely filled.**

Problem 15.20 Use the inscribed polygon method to show why the following cation is aromatic.

More Practice: Try Problems 15.48, 15.49.

Problem 15.21 Use the inscribed polygon method to show why the cyclopentadienyl cation and radical are not aromatic.

The procedure followed in Sample Problem 15.3 also illustrates why cyclobutadiene is antiaromatic. Having the two unpaired electrons in nonbonding MOs suggests that cyclobutadiene should be a highly unstable diradical. In fact, antiaromatic compounds resemble cyclobutadiene because their HOMOs contain two unpaired electrons, making them especially unstable.

Chapter 15 REVIEW

KEY CONCEPTS

[1] Nomenclature of disubstituted and trisubstituted benzenes (15.3)

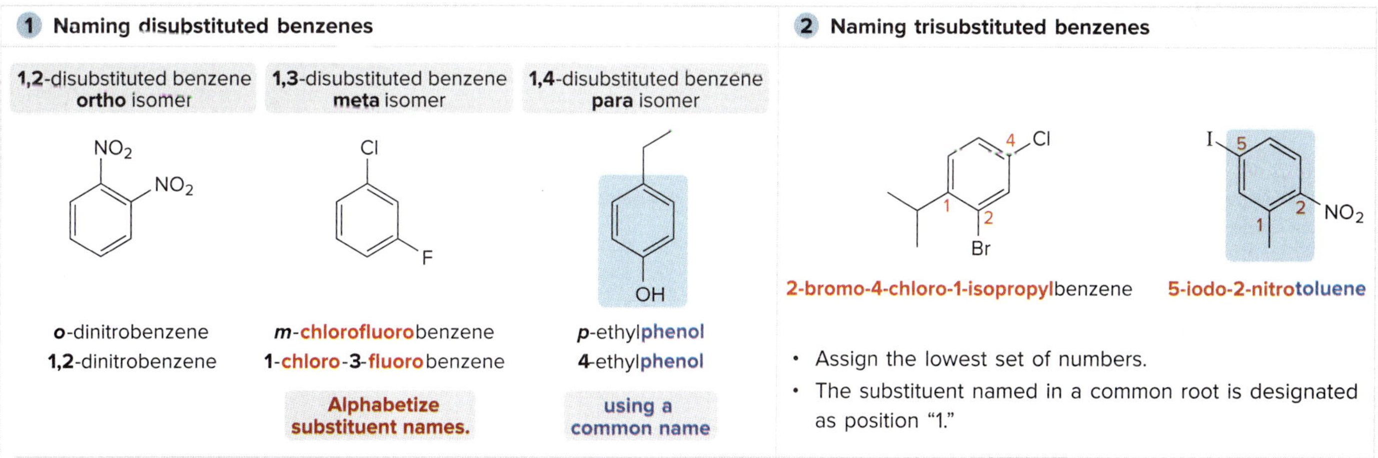

Try Problems 15.22, 15.25, 15.26, 15.27b.

[2] Examples of aromatic, nonaromatic, and antiaromatic compounds (15.7–15.9)

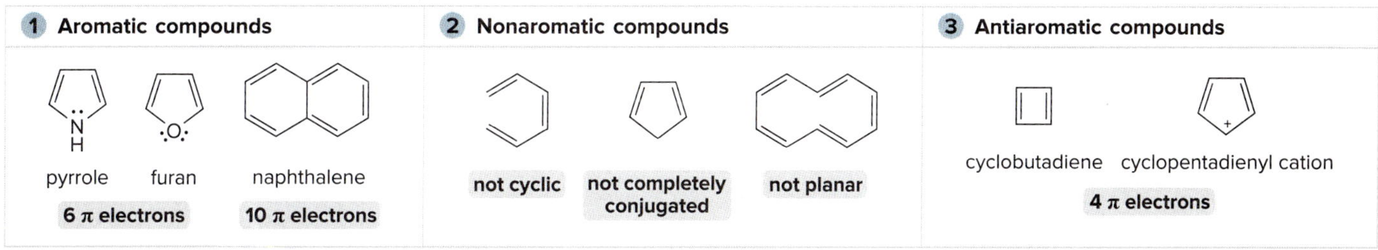

KEY SKILLS

[1] Determining if a cyclic, planar compound is aromatic, antiaromatic, or not aromatic (15.7, 15.8); example: cyclopentadienyl cation

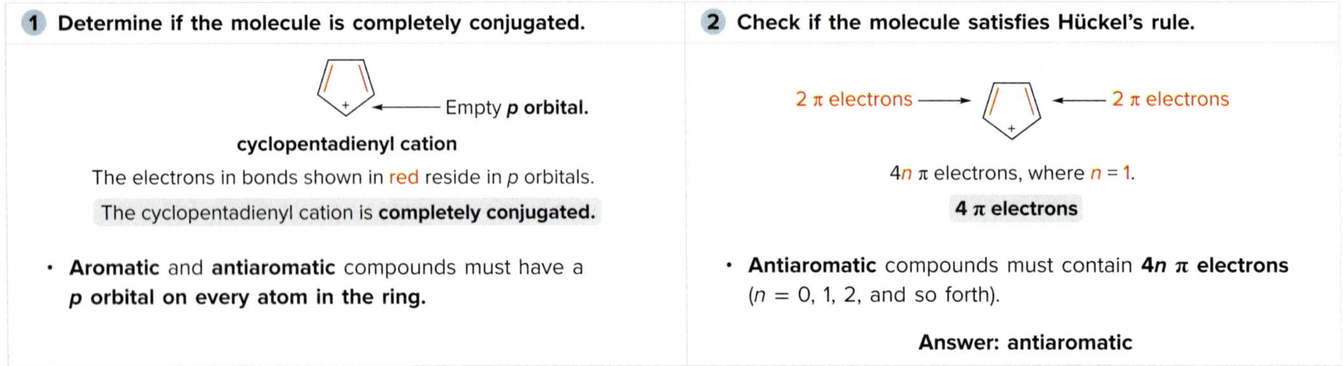

See Sample Problem 15.2. Try Problems 15.23, 15.29–15.31.

[2] Determining if a planar heterocyclic compound is aromatic, antiaromatic, or not aromatic (15.7, 15.9); example: furan

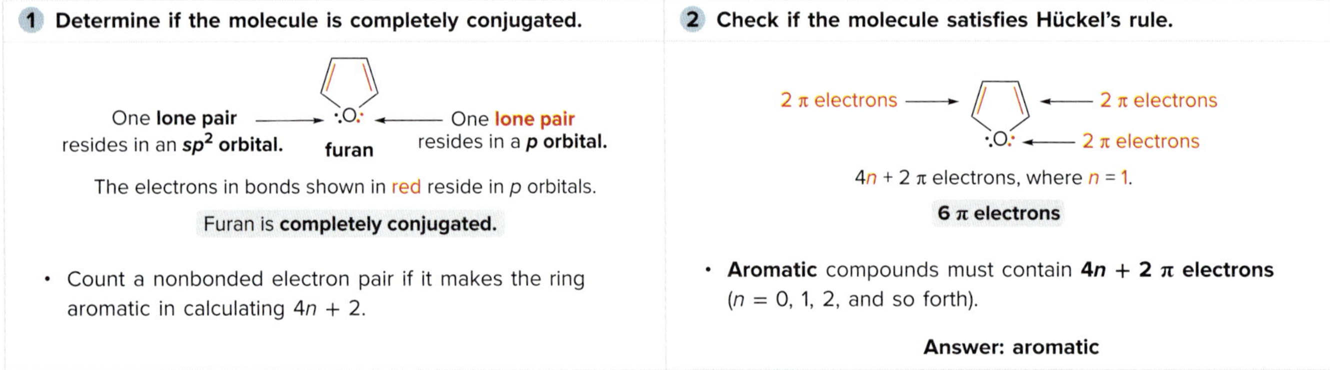

See Sample Problems 15.1, 15.2. Try Problems 15.23b, c; 15.30; 15.31e, f; 15.35a; 15.56a; 15.57b; 15.59a.

[3] Using the inscribed polygon method to determine if a compound is aromatic (15.11); example: the tropylium radical

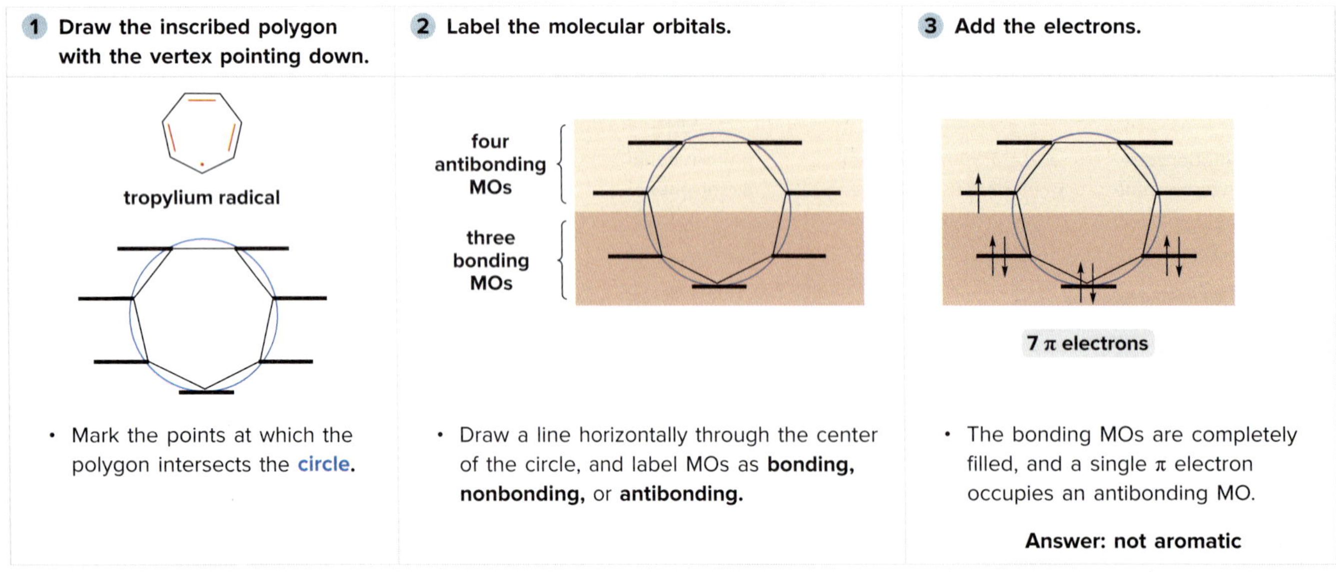

See *How To* p. 709; Figures 15.9, 15.10; Sample Problem 15.3. Try Problems 15.48, 15.49.

CHAPTER 15 MULTIPLE-CHOICE SELF-TEST

The Self-Test consists of multiple-choice questions similar to those found on the American Chemical Society organic chemistry exam. Answers are given at the end of the chapter.

1. Which compound has two groups para to each other on the benzene ring?

a. 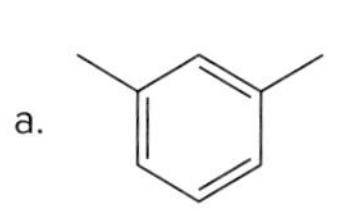

b.

c. 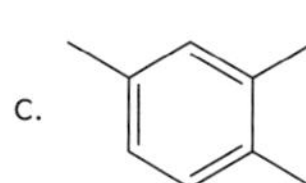

d. 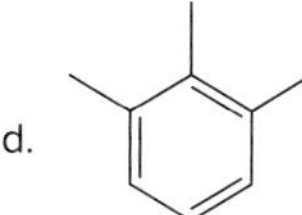

2. Which species is aromatic if the ring is planar?

a. Cl

b. HN +

c. 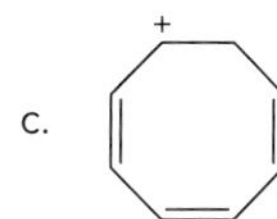

d. Cl Cl

3. Which labeled H atom is most acidic?

H_c H_d H_a H_b

a. H_a b. H_b c. H_c d. H_d

4. Give the IUPAC name for the following compound.

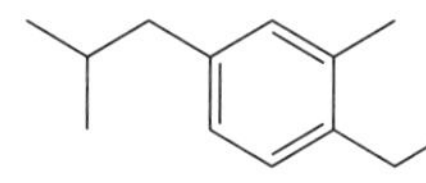

a. 1-ethyl-4-isobutyl-2-methylbenzene
b. 5-*sec*-butyl-2-ethyl-1-methylbenzene
c. 5-isobutyl-2-ethyltoluene
d. 2-ethyl-5-isobutyltoluene

5. What is the hybridization of each labeled N atom and in what type of orbital does its lone pair reside?

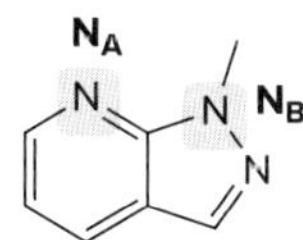

a. N_A: sp^2 hybridized with the lone pair in an sp^2 hybrid orbital; N_B: sp^3 hybridized with the lone pair in an sp^3 hybrid orbital
b. N_A: sp^2 hybridized with the lone pair in an sp^2 hybrid orbital; N_B: sp^2 hybridized with the lone pair in a *p* orbital
c. N_A: sp^2 hybridized with the lone pair in a *p* orbital; N_B: sp^2 hybridized with the lone pair in a *p* orbital
d. N_A: sp^2 hybridized with the lone pair in an sp^2 hybrid orbital; N_B: sp^2 hybridized with the lone pair in an sp^2 hybrid orbital

6. Which statement is *false* about the molecular orbitals in cyclooctatetraene?
a. There are three bonding MOs all filled with electrons.
b. There are four antibonding MOs that contain a total of two electrons.
c. There are two nonbonding MOs that contain a total of two electrons.
d. There are three antibonding MOs that contain no electrons.

7. Which compound is aromatic if the ring is planar: (a) [12]-annulene; (b) [14]-annulene; (c) [16]-annulene; (d) [24]-annulene?

8. Rank the labeled bonds in order of increasing bond length.

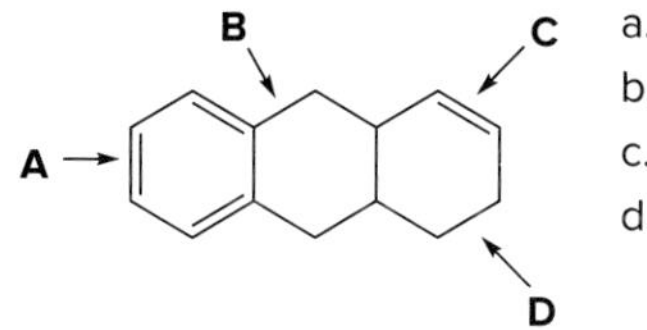

a. **A** < **C** < **B** < **D**
b. **C** < **A** < **D** < **B**
c. **A** < **C** < **D** < **B**
d. **C** < **A** < **B** < **D**

PROBLEMS

Problems Using Three-Dimensional Models

15.22 Name each compound and state how many lines are observed in its ^{13}C NMR spectrum.

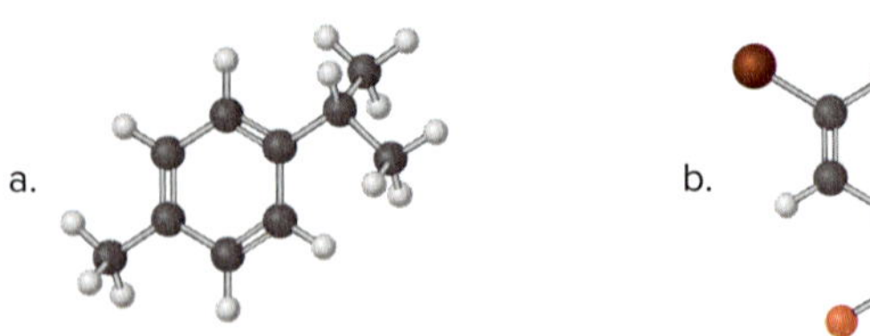

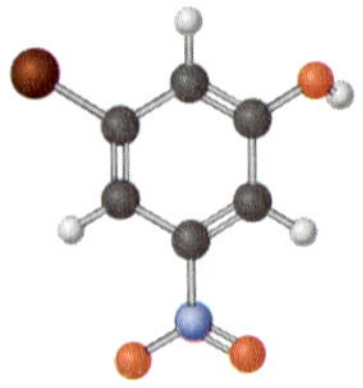

15.23 Classify each compound as aromatic, antiaromatic, or not aromatic.

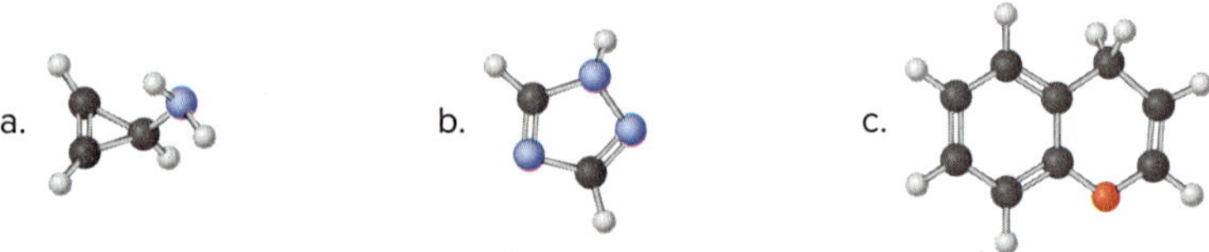

Benzene Structure and Nomenclature

15.24 Draw all aromatic hydrocarbons that have molecular formula C_8H_{10}. For each compound, determine how many isomers of molecular formula C_8H_9Br would be formed if one H atom on the benzene ring were replaced by a Br atom.

15.25 Give the IUPAC name for each compound.

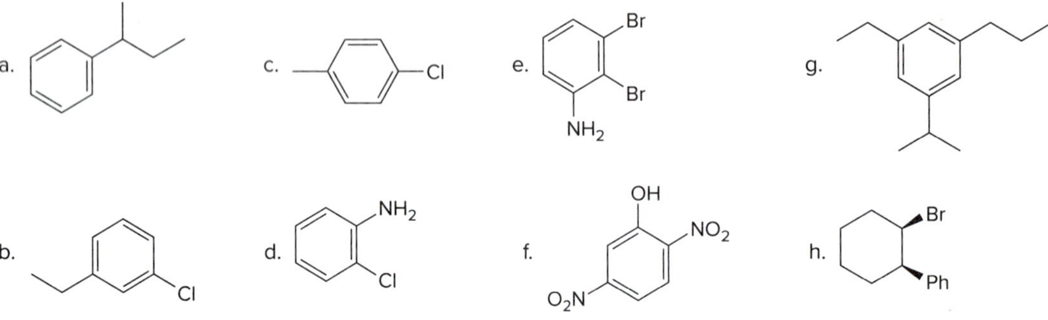

15.26 Draw a structure corresponding to each name.

a. *p*-dichlorobenzene
b. *p*-iodoaniline
c. *o*-bromonitrobenzene
d. 2,6-dimethoxytoluene
e. 2-phenylprop-2-en-1-ol
f. *trans*-1-benzyl-3-phenylcyclopentane

15.27 a. Draw the 14 constitutional isomers of molecular formula C_8H_9Cl that contain a benzene ring.
b. Name all compounds that contain a trisubstituted benzene ring.
c. For which compound(s) are stereoisomers possible? Draw all possible stereoisomers.

Aromaticity

15.28 How many π electrons are contained in each molecule?

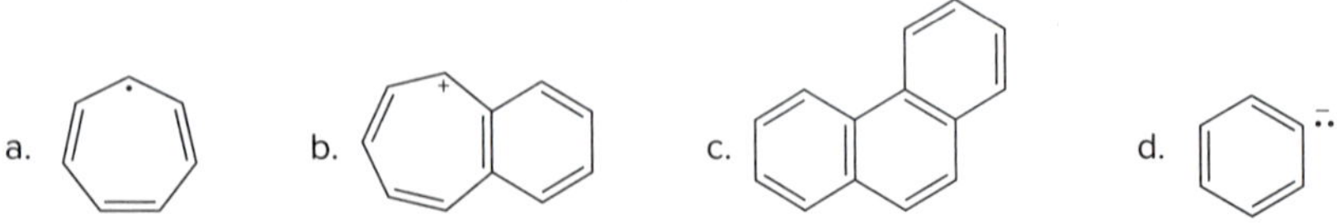

15.29 Which compounds are aromatic? For any compound that is not aromatic, state why this is so.

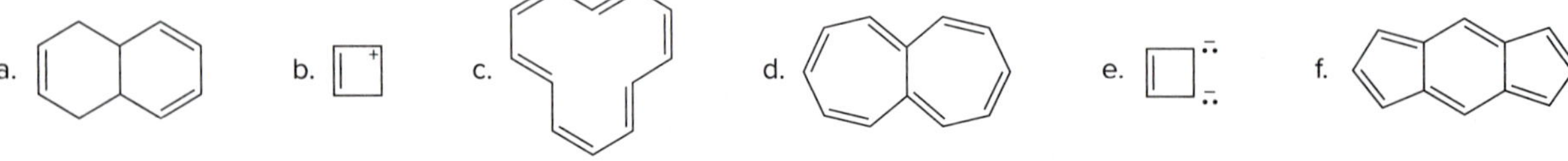

15.30 Label each heterocycle as aromatic, antiaromatic, or not aromatic.

a. 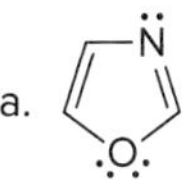
b. 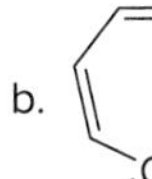
c. 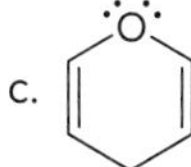
d. 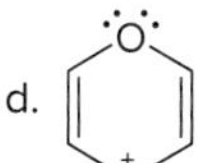
e. 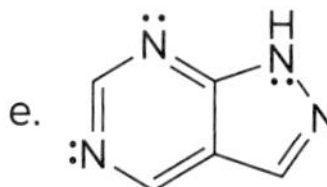
f. 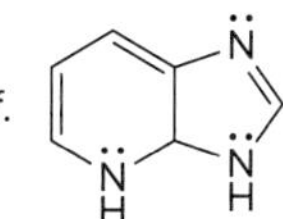

15.31 Label each compound as aromatic, antiaromatic, or not aromatic. Assume all completely conjugated rings are planar.

a. 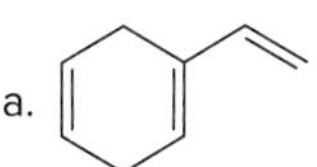
b. 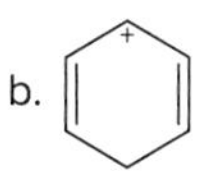
c. 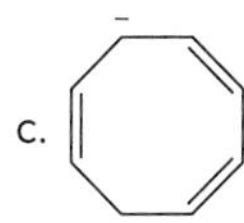
d. 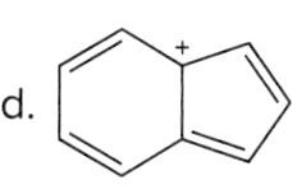
e. 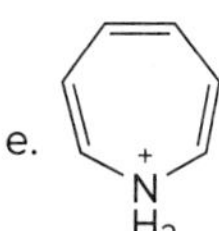
f. 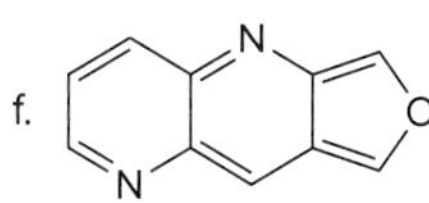

15.32 Hydrocarbon **A** possesses a significant dipole, even though it is composed of only C–C and C–H bonds. Explain why the dipole arises and use resonance structures to illustrate the direction of the dipole. Which ring is more electron rich?

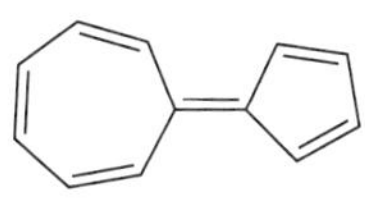

A

15.33 Pentalene, azulene, and heptalene are conjugated hydrocarbons that do not contain a benzene ring. Assuming the rings are flat, which hydrocarbons are especially stable or unstable based on the number of π electrons they contain? Explain your choices.

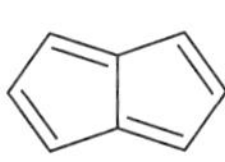
pentalene

azulene

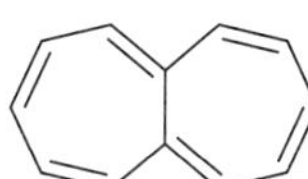
heptalene

15.34 Tofacitinib is a drug used to treat rheumatoid arthritis.

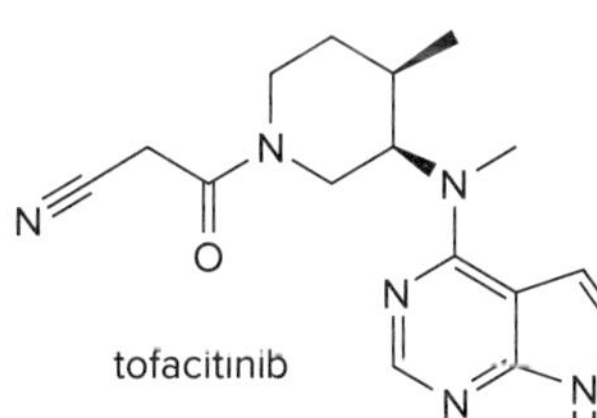
tofacitinib

a. Determine the hybridization of each N atom in tofacitinib.
b. In what type of orbital does the lone pair on each N atom reside?

15.35 Eszopiclone (trade name Lunesta) is an oral medication used to treat insomnia.

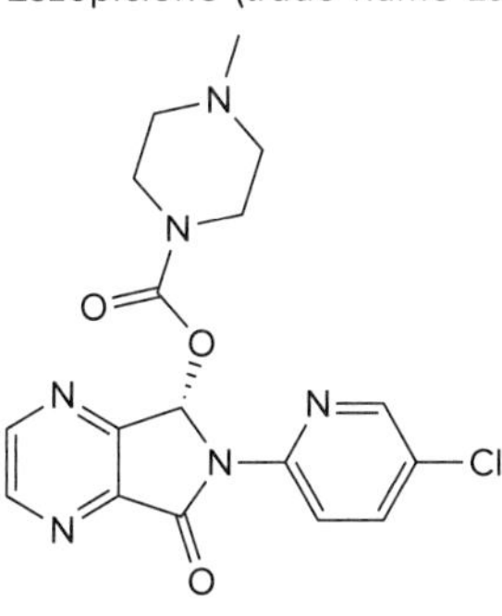
eszopiclone

a. How many aromatic rings does eszopiclone contain?
b. What is the hybridization of each N atom?
c. In what type of orbital does the lone pair on each N atom reside?
d. How many sp^2 hybridized C atoms does eszopiclone contain?

15.36

C

a. How many π electrons does **C** contain?
b. How many π electrons are delocalized in the ring?
c. Explain why **C** is aromatic.

15.37 Why is caffeine considered to be an aromatic compound?

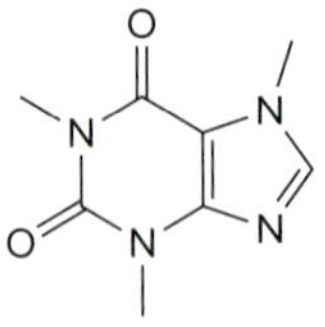

caffeine

15.38 Rank alkyl chlorides **A, B,** and **C** in order of increasing reactivity in an S_N1 reaction, and explain your choice.

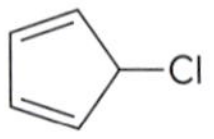

A 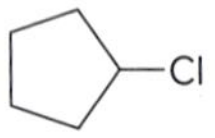 **B** 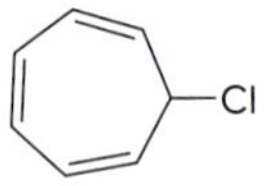 **C**

15.39 Draw a stepwise mechanism for the following reaction.

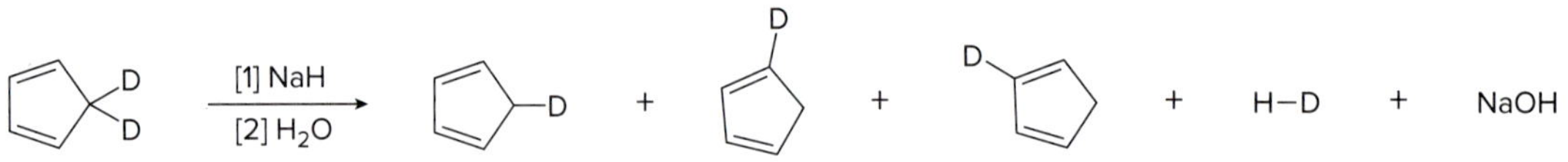

[1] NaH, [2] H_2O → + + + H–D + NaOH

15.40 Explain why α-pyrone reacts with Br_2 to yield a substitution product (like benzene does), rather than an addition product to one of its C=C bonds.

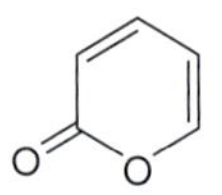

α-pyrone

Resonance

15.41 Draw additional resonance structures for each species.

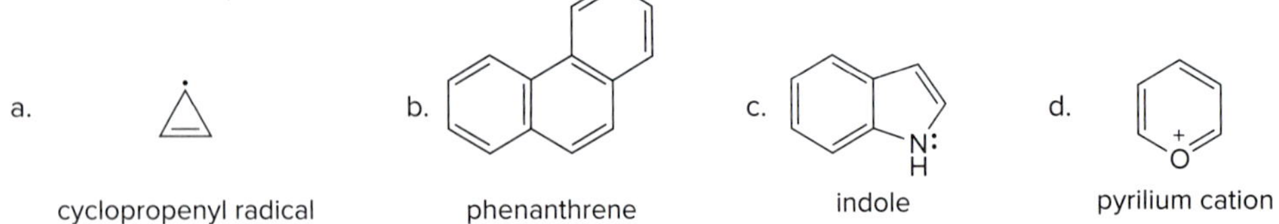

cyclopropenyl radical phenanthrene indole pyrilium cation

15.42 The carbon–carbon bond lengths in naphthalene are not equal. Use a resonance argument to explain why bond (a) is shorter than bond (b).

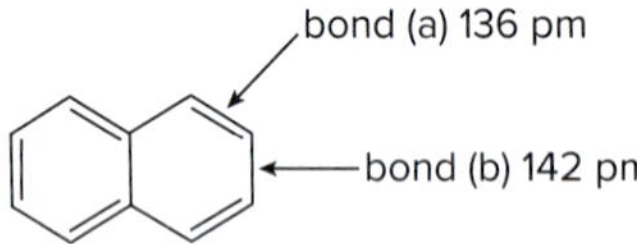

15.43 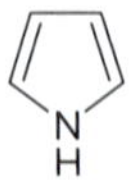 pyrrole  furan

a. Draw all reasonable resonance structures for pyrrole, and explain why pyrrole is less resonance stabilized than benzene.

b. Draw all reasonable resonance structures for furan, and explain why furan is less resonance stabilized than pyrrole.

Acidity

15.44 Rank the following compounds in order of increasing acidity.

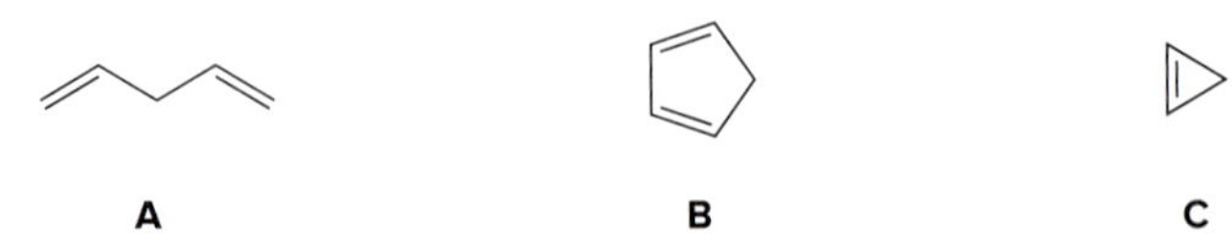

15.45 Treatment of indene with $NaNH_2$ forms its conjugate base in a Brønsted–Lowry acid–base reaction. Draw all reasonable resonance structures for indene's conjugate base, and explain why the pK_a of indene is lower than the pK_a of most hydrocarbons.

indene + $NaNH_2$ ⟶ (conjugate base) Na^+ + NH_3

indene
pK_a = 20

15.46 Draw the conjugate bases of pyrrole and cyclopentadiene. Explain why the sp^3 hybridized C–H bond of cyclopentadiene is more acidic than the N–H bond of pyrrole.

15.47 a. Explain why protonation of pyrrole occurs at C2 to form **A,** rather than on the N atom to form **B.**

b. Explain why **A** is more acidic than **C,** the conjugate acid of pyridine.

pyrrole	pK_a = 0.4		pK_a = 5.3
	A (C2)	**B**	**C**

Inscribed Polygon Method

15.48 Use the inscribed polygon method to show the pattern of molecular orbitals in cyclooctatetraene.

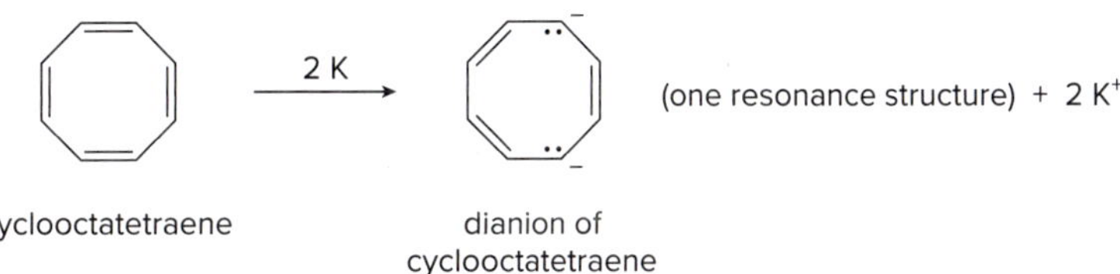

a. Label the MOs as bonding, antibonding, or nonbonding.

b. Indicate the arrangement of electrons in these orbitals for cyclooctatetraene, and explain why cyclooctatetraene is not aromatic.

c. Treatment of cyclooctatetraene with potassium forms a dianion. How many π electrons does this dianion contain?

d. How are the π electrons in this dianion arranged in the molecular orbitals?

e. Classify the dianion of cyclooctatetraene as aromatic, antiaromatic, or not aromatic, and explain why this is so.

15.49 Use the inscribed polygon method to show the pattern of molecular orbitals in cyclonona-1,3,5,7-tetraene, and use it to label its cation, radical, and anion as aromatic, antiaromatic, or not aromatic.

Spectroscopy

15.50 How many ^{13}C NMR signals does each compound exhibit?

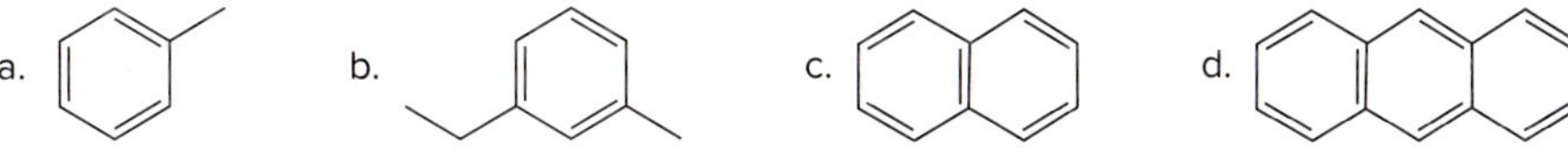

15.51 Which of the diethylbenzene isomers (ortho, meta, or para) corresponds to each set of ^{13}C NMR spectral data?

[A] ^{13}C NMR signals: 16, 29, 125, 127.5, 128.4, and 144 ppm

[B] ^{13}C NMR signals: 15, 26, 126, 128, and 142 ppm

[C] ^{13}C NMR signals: 16, 29, 128, and 141 ppm

15.52 Propose a structure consistent with each set of data.

a. $C_{10}H_{14}$: IR absorptions at 3150–2850, 1600, and 1500 cm^{-1}

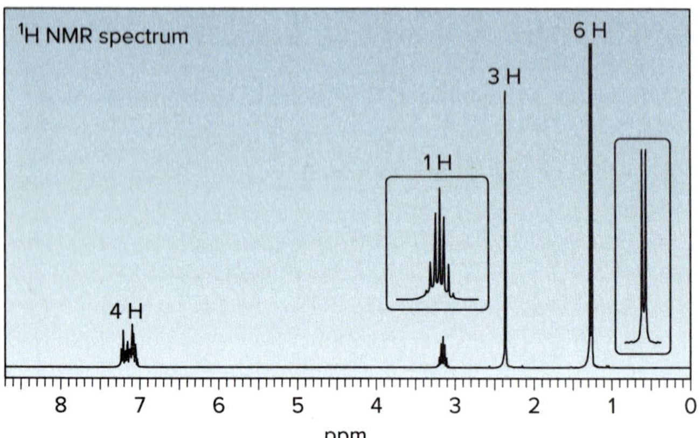

b. C_9H_{12}: ^{13}C NMR signals at 21, 127, and 138 ppm

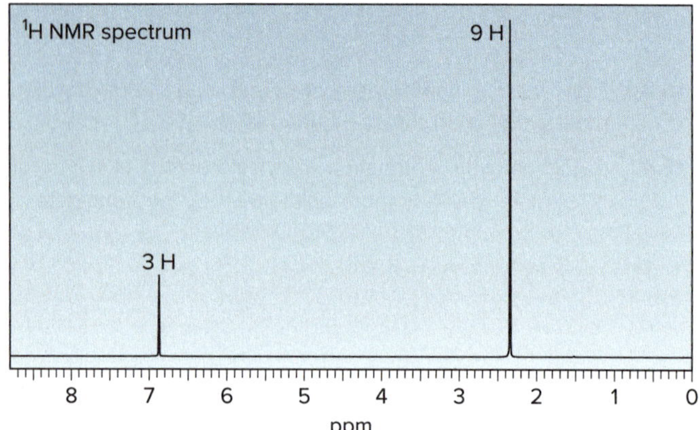

c. C_8H_{10}: IR absorptions at 3108–2875, 1606, and 1496 cm^{-1}

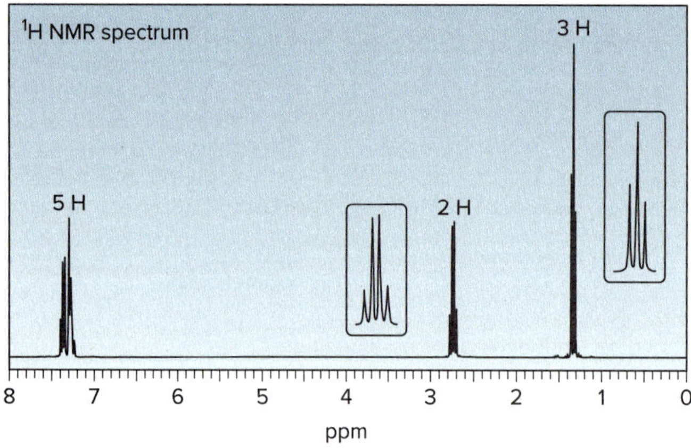

15.53 Thymol (molecular formula $C_{10}H_{14}O$) is the major component of the oil of thyme. Thymol shows IR absorptions at 3500–3200, 3150–2850, 1621, and 1585 cm^{-1}. The ^1H NMR spectrum of thymol is given below. Propose a possible structure for thymol.

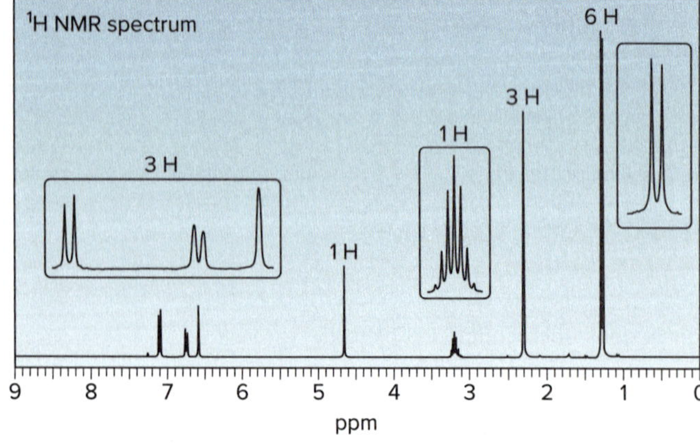

15.54 You have a sample of a compound of molecular formula $C_{11}H_{15}NO_2$, which has a benzene ring substituted by two groups, $(CH_3)_2N-$ and $-CO_2CH_2CH_3$, and exhibits the given ^{13}C NMR. What disubstituted benzene isomer corresponds to these ^{13}C data?

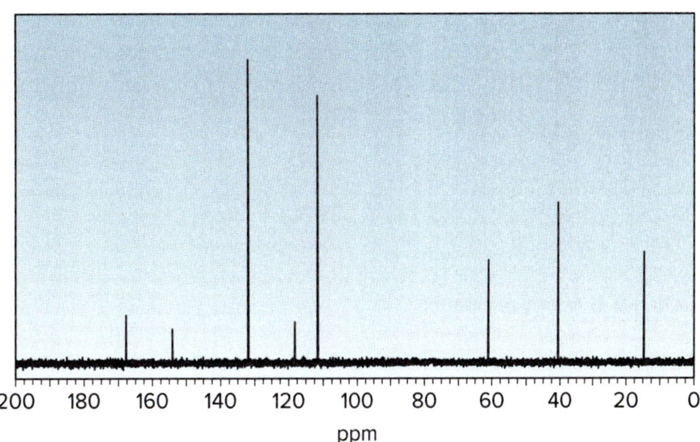

Problems That Combine Concepts

15.55 Explain why tetrahydrofuran has a higher boiling point and is much more water soluble than furan, even though both compounds are cyclic ethers containing four carbons.

15.56 Rizatriptan (trade name Maxalt) is a prescription drug used for the treatment of migraines.

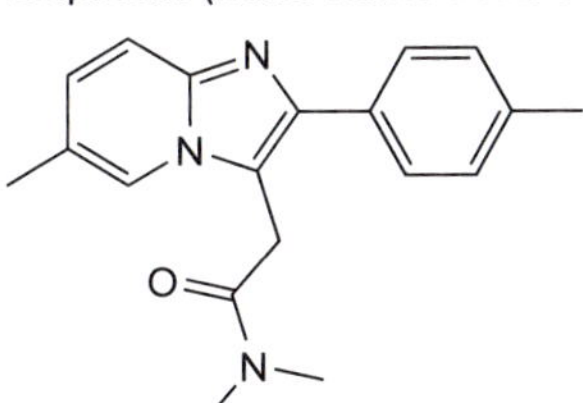

rizatriptan

a. How many aromatic rings does rizatriptan contain?
b. Determine the hybridization of each N atom.
c. In what type of orbital does the lone pair on each N reside?
d. Draw all the resonance structures for rizatriptan that contain only neutral atoms.
e. Draw all reasonable resonance structures for the five-membered ring that contains three N atoms.

15.57 Zolpidem (trade name Ambien) promotes the rapid onset of sleep, making it a widely prescribed drug for treating insomnia.

N
N
O
N
zolpidem

a. In what type of orbital does the lone pair on each N atom in the heterocycle reside?
b. Explain why the bicyclic ring system that contains both N atoms is aromatic.
c. Draw all reasonable resonance structures for the bicyclic ring system.

15.58 Answer the following questions about curcumin, a yellow pigment isolated from turmeric, a tropical perennial in the ginger family and a principal ingredient in curry powder.

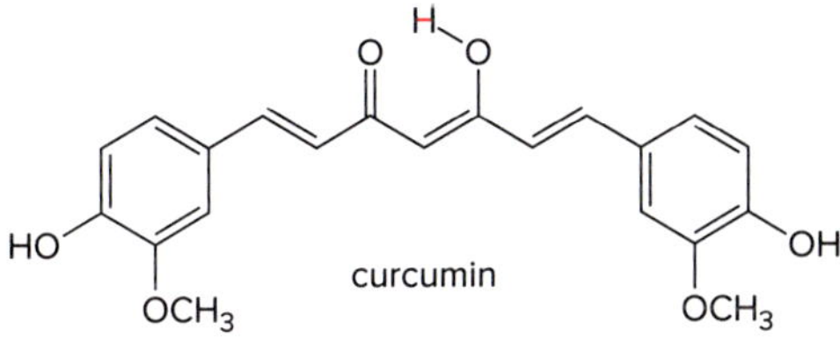

curcumin

a. In Chapter 11, we learned that most enols, compounds that contain a hydroxy group bonded to a C=C, are unstable and tautomerize to carbonyl groups. Draw the keto form of the enol of curcumin, and explain why the enol is more stable than many other enols.
b. Explain why the enol O–H proton is more acidic than an alcohol O–H proton.
c. Why is curcumin colored?
d. Explain why curcumin is an antioxidant.

15.59 Stanozolol is an anabolic steroid that promotes muscle growth. Although stanozolol has been used by athletes and body builders, many physical and psychological problems result from prolonged use and it is banned in competitive sports.

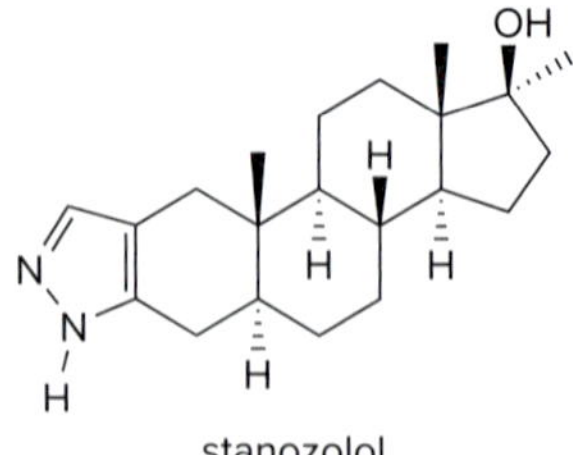

stanozolol

a. Explain why the nitrogen heterocycle—a pyrazole ring—is aromatic.
b. In what type of orbital is the lone pair on each N atom contained?
c. Draw all reasonable resonance structures for stanozolol.
d. Explain why the pK_a of the N–H bond in the pyrazole ring is comparable to the pK_a of the O–H bond, making it considerably more acidic than amines such as CH_3NH_2 (pK_a = 40).

Challenge Problems

15.60 Explain why **A** is aromatic but **B** is not aromatic.

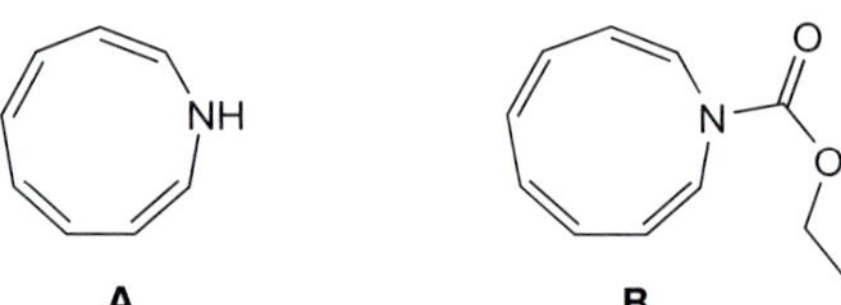

15.61 Use the observed ^{1}H NMR data to decide whether **C** and its dianion are aromatic, antiaromatic, or not aromatic. **C** shows NMR signals at –4.25 (6 H) and 8.14–8.67 (10 H) ppm. The dianion of **C** shows NMR signals at –3 (10 H) and 21 (6 H) ppm. Why are the signals shifted upfield (or downfield) to such a large extent?

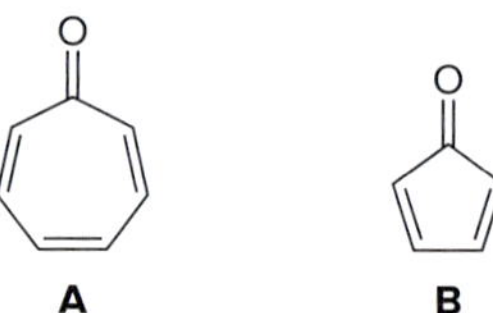
C

15.62 Explain why compound **A** is much more stable than compound **B.**

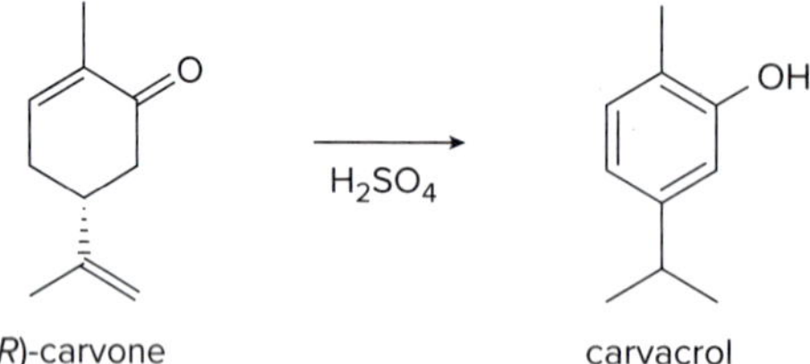

15.63 (*R*)-Carvone, the major component of the oil of spearmint, undergoes acid-catalyzed isomerization to carvacrol, a major component of the oil of thyme. Draw a stepwise mechanism and explain why this isomerization occurs.

15.64 Explain why triphenylene resembles benzene in that it does not undergo addition reactions with Br_2, but phenanthrene reacts with Br_2 to yield the addition product drawn. (Hint: Draw resonance structures for both triphenylene and phenanthrene, and use them to determine how delocalized each π bond is.)

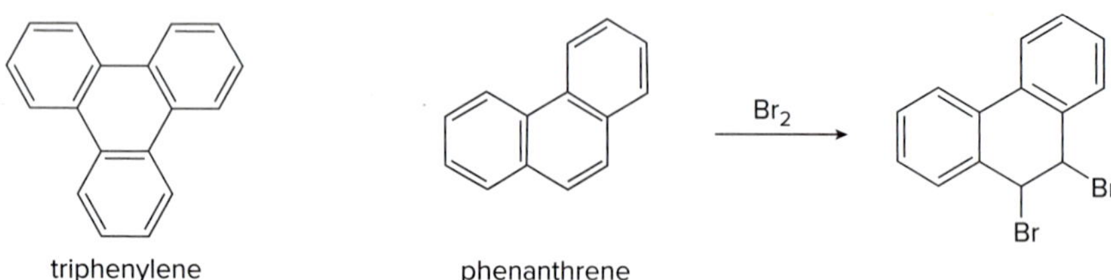

15.65 Although benzene itself absorbs at 128 ppm in its ^{13}C NMR spectrum, the carbons of substituted benzenes absorb either upfield or downfield from this value depending on the substituent. Explain the observed values for the carbon ortho to the given substituent in the monosubstituted benzene derivatives **X** and **Y.**

X: 113 ppm   Y: 130 ppm

SELF-TEST ANSWERS

1. c 2. b 3. a 4. d 5. b 6. b 7. b 8. d

16 Reactions of Aromatic Compounds

16.1 Electrophilic aromatic substitution
16.2 The general mechanism
16.3 Halogenation
16.4 Nitration and sulfonation
16.5 Friedel–Crafts alkylation and Friedel–Crafts acylation
16.6 Substituted benzenes
16.7 Electrophilic aromatic substitution of substituted benzenes
16.8 Why substituents activate or deactivate a benzene ring
16.9 Orientation effects in substituted benzenes
16.10 Limitations on electrophilic substitution reactions with substituted benzenes
16.11 Disubstituted benzenes
16.12 Synthesis of benzene derivatives
16.13 Nucleophilic aromatic substitution
16.14 Reactions of substituted benzenes
16.15 Multistep synthesis

Comstock/Getty Images

Vitamin K_1, phylloquinone, is a fat-soluble vitamin that regulates the synthesis of proteins needed for blood to clot. Dietary sources of vitamin K_1 include cauliflower, broccoli, soybeans, leafy greens, and green tea. A severe deficiency of vitamin K_1 leads to excessive and sometimes fatal bleeding because of inadequate blood clotting. Vitamin K_1 is synthesized by a biological Friedel–Crafts reaction, one of the many examples of electrophilic aromatic substitution, a key reaction of aromatic compounds presented in Chapter 16.

Why Study . . .

Reactions of Aromatic Compounds?

Chapter 16 discusses the chemical reactions of benzene and other aromatic compounds. Although aromatic rings are unusually stable, making benzene unreactive in most of the reactions discussed so far, benzene acts as a nucleophile with certain electrophiles, yielding substitution products with an intact aromatic ring.

We begin with the basic features and mechanism of electrophilic aromatic substitution (Sections 16.1–16.5), the most prevalent reaction of benzene. Next, we discuss the electrophilic aromatic substitution of substituted benzenes (Sections 16.6–16.12), and conclude with nucleophilic aromatic substitution and other useful reactions of benzene derivatives (Sections 16.13 and 16.14). These reactions have been used to prepare antidepressants, antipsychotics, and drugs to treat diabetes.

16.1 Electrophilic Aromatic Substitution

Based on its structure and properties, what kinds of reactions should benzene undergo? Are any of its bonds particularly weak? Does it have electron-rich or electron-deficient atoms?

- Benzene has six π electrons delocalized in six *p* orbitals that overlap above and below the plane of the ring. These loosely held π electrons make the benzene ring electron rich, so it reacts with *electrophiles.*
- Because benzene's six π electrons satisfy Hückel's rule, benzene is especially stable. Reactions that keep the aromatic ring *intact* are therefore favored.

As a result, **the characteristic reaction of benzene is *electrophilic aromatic substitution*—a hydrogen atom is replaced by an electrophile.**

Electrophilic aromatic substitution

C_6H_5–H + E^+ ⟶ C_6H_5–E + H^+

aromatic product

As we learned in Section 15.6, benzene does *not* undergo addition reactions like other unsaturated hydrocarbons, because addition would yield a product that is not aromatic. Substitution of a hydrogen, on the other hand, keeps the aromatic ring intact.

Five specific examples of electrophilic aromatic substitution are shown in Figure 16.1. The basic mechanism, discussed in Section 16.2, is the same in all five cases. The reactions differ only in the identity of the electrophile, E^+.

Problem 16.1 Why is benzene less reactive toward electrophiles than an alkene, even though it has more π electrons than an alkene (six versus two)?

Figure 16.1 Five examples of electrophilic aromatic substitution

Reaction	Electrophile
[1] Halogenation—Replacement of H by X (Cl or Br)	
benzene + X_2 / FeX_3 (X = Cl, X = Br) → aryl halide	$E^+ = Cl^+$ or Br^+
[2] Nitration—Replacement of H by NO_2	
benzene + HNO_3 / H_2SO_4 → nitrobenzene	$E^+ = \overset{+}{N}O_2$
[3] Sulfonation—Replacement of H by SO_3H	
benzene + SO_3 / H_2SO_4 → benzenesulfonic acid	$E^+ = \overset{+}{S}O_3H$
[4] Friedel–Crafts alkylation—Replacement of H by R	
benzene + RCl / $AlCl_3$ → alkyl benzene (arene)	$E^+ = R^+$
[5] Friedel–Crafts acylation—Replacement of H by RCO	
benzene + RCOCl / $AlCl_3$ → ketone	$E^+ = R-\overset{+}{C}=\ddot{O}:$

Friedel–Crafts alkylation and acylation, named for Charles Friedel and James Crafts, who discovered the reactions in the nineteenth century, form new carbon–carbon bonds.

16.2 The General Mechanism

No matter what electrophile is used, all electrophilic aromatic substitution reactions occur via a **two-step mechanism:** addition of the electrophile E^+ to form a resonance-stabilized carbocation, followed by deprotonation with base, as shown in Mechanism 16.1.

Mechanism 16.1 General Mechanism—Electrophilic Aromatic Substitution

1 Addition of the electrophile E^+ forms a new C–E bond and a **resonance-stabilized carbocation.** This step is rate-determining because the aromaticity of the benzene ring is lost.

2 A base removes the proton ***on the carbon bonded to the electrophile,*** re-forming the aromatic ring. Any resonance structure can be used to draw the product.

The first step in electrophilic aromatic substitution forms a carbocation, for which three resonance structures can be drawn. To help keep track of the location of the positive charge:

- **Always draw in the H atom on the carbon bonded to E. This serves as a reminder that it is the only sp^3 hybridized carbon in the carbocation intermediate.**
- **Notice that the positive charge in a given resonance structure is always located ortho or para to the new C–E bond. In the hybrid, therefore, the charge is delocalized over three atoms of the ring.**

(+) ortho to E ⟷ **(+) para to E** ⟷ **(+) ortho to E** **hybrid**

This two-step mechanism for electrophilic aromatic substitution applies to all of the electrophiles in Figure 16.1. **The net result of addition of an electrophile (E^+) followed by elimination of a proton (H^+) is substitution of E for H.**

The energy changes in electrophilic aromatic substitution are shown in Figure 16.2. Because the transition state of the first step is higher in energy, it is rate-determining.

Problem 16.2 In Step [2] of Mechanism 16.1, loss of a proton to form the substitution product was drawn using only one resonance structure. Use curved arrows to show how the other two resonance structures can be converted to the substitution product (PhE) by removal of a proton with :B.

Figure 16.2 Energy diagram for electrophilic aromatic substitution: $PhH + E^+ \rightarrow PhE + H^+$

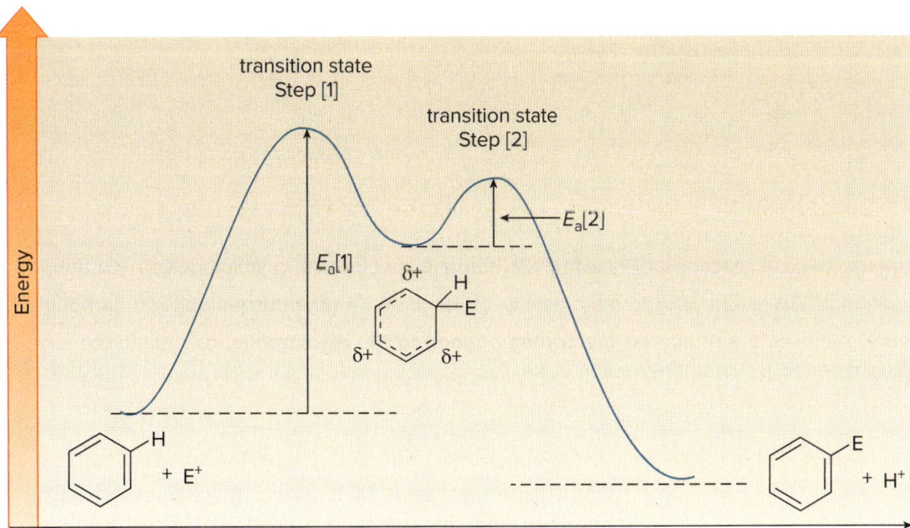

- The mechanism has two steps, so there are two energy barriers.
- **Step [1] is rate-determining;** its transition state is at higher energy.

16.3 Halogenation

The general mechanism outlined in Mechanism 16.1 can now be applied to each of the five specific examples of electrophilic aromatic substitution shown in Figure 16.1. For each mechanism we must learn how to generate a specific electrophile. This step is *different* with each electrophile. Then, the electrophile reacts with benzene by the two-step process of Mechanism 16.1. These two steps are the *same* for all five reactions.

In **halogenation,** benzene reacts with Cl_2 or Br_2 in the presence of a Lewis acid catalyst, such as $FeCl_3$ or $FeBr_3$, to give the **aryl halides** chlorobenzene or bromobenzene, respectively. Analogous reactions with I_2 and F_2 are not synthetically useful because I_2 is too unreactive and F_2 reacts too violently.

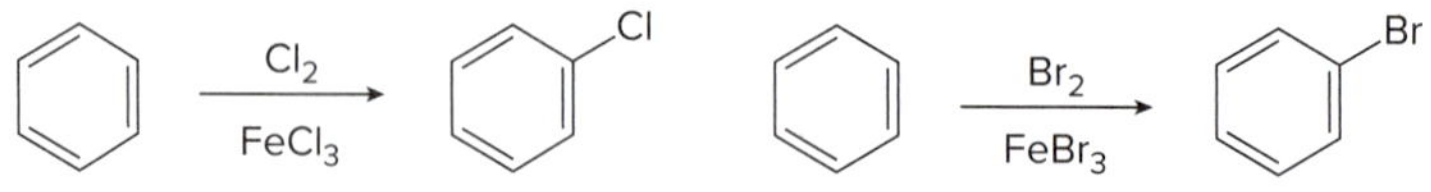

Cl2 / FeCl3 → chlorobenzene

Br2 / FeBr3 → bromobenzene

In bromination (Mechanism 16.2), the Lewis acid $FeBr_3$ reacts with Br_2 to form a **Lewis acid–base complex** that weakens and polarizes the Br–Br bond, making it more electrophilic. This reaction is Step [1] of the mechanism for the bromination of benzene. The remaining two steps follow directly from the general mechanism for electrophilic aromatic substitution: addition of the electrophile (Br^+ in this case) forms a resonance-stabilized carbocation, and loss of a proton regenerates the aromatic ring.

Mechanism 16.2 Bromination of Benzene

:Br–Br: + FeBr3

1

:Br–Br–FeBr3

2

resonance-stabilized carbocation

+ FeBr4⁻

:Br–FeBr3

3

+ HBr

+ FeBr3

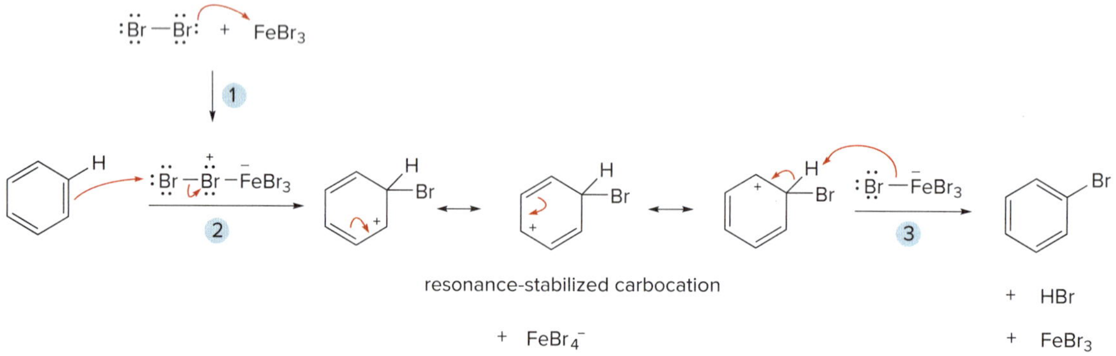

1 Lewis acid–base reaction of Br_2 with $FeBr_3$ forms a species with a weakened Br–Br bond that serves as source of Br^+.

2 Addition of the electrophile forms a new C–Br bond and a **resonance-stabilized carbocation.**

3 $FeBr_4^-$ removes the proton ***on the carbon bonded to the electrophile,*** re-forming the aromatic ring. The Lewis acid catalyst $FeBr_3$ is regenerated for another reaction cycle.

Chlorination proceeds by a similar mechanism. Reactions that introduce a halogen substituent on a benzene ring are widely used, and many halogenated aromatic compounds with a range of biological activity have been synthesized, as shown in Figure 16.3.

Problem 16.3 Which species is *not* a valid Lewis structure for the cation that results when benzene reacts with Cl_2, $FeCl_3$?

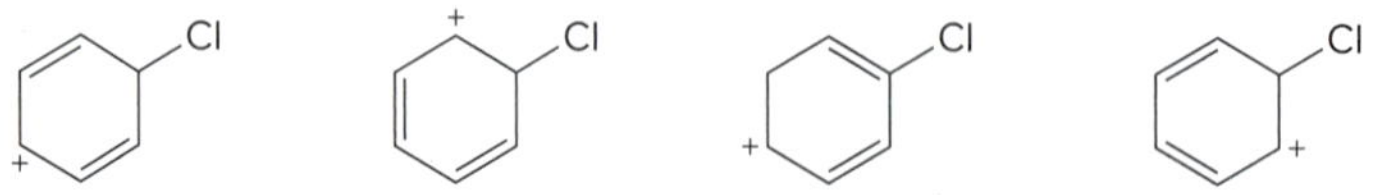

A B C D

Figure 16.3
Examples of biologically active aryl chlorides

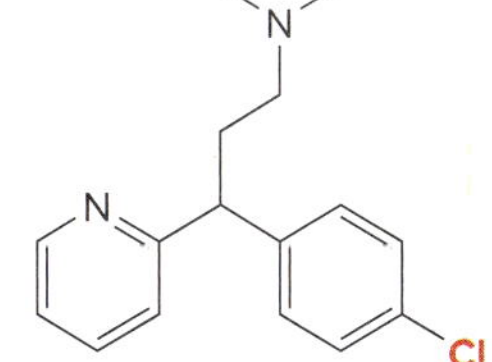

Generic name: **bupropion**
Trade names: **Wellbutrin, Zyban**
antidepressant,
also used to reduce nicotine cravings

chlorpheniramine
antihistamine

Herbicides were used extensively during the Vietnam War to defoliate dense jungle areas. The concentration of certain herbicide by-products in the soil remains high today.

Source: National Archives and Records Administration [NWDNS-111-C-CC59950]

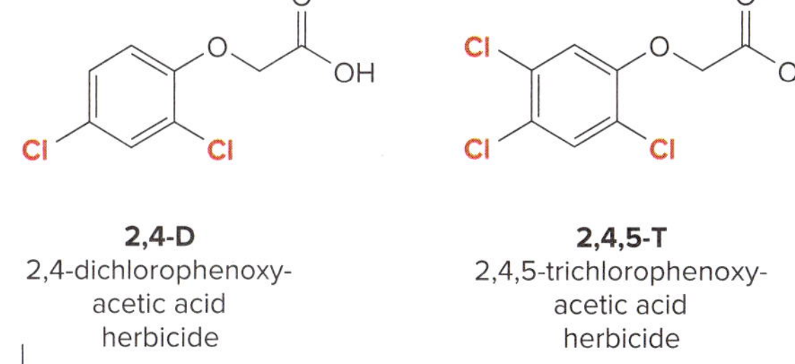

2,4-D
2,4-dichlorophenoxy-acetic acid
herbicide

2,4,5-T
2,4,5-trichlorophenoxy-acetic acid
herbicide

the active components in **Agent Orange,** a defoliant used in the Vietnam War

16.4 Nitration and Sulfonation

Nitration and **sulfonation** of benzene introduce two different functional groups on an aromatic ring. Nitration is an especially useful reaction because a nitro group can then be reduced to an NH_2 group, a common benzene substituent, in a reaction discussed in Section 16.14.

HNO_3, H_2SO_4 → nitrobenzene → [Section 16.14D] → aniline

SO_3, H_2SO_4 → benzenesulfonic acid

Generation of the electrophile in both nitration and sulfonation requires strong acid. In **nitration,** the electrophile is $^+NO_2$ (the **nitronium ion**), formed by protonation of HNO_3 followed by loss of water (Mechanism 16.3).

Mechanism 16.3 Formation of the Nitronium Ion ($^+NO_2$) for Nitration

$H-\ddot{O}-NO_2$ + $H-OSO_3H$ →[1] $H-\overset{+}{O}(H)-NO_2$ + HSO_4^- →[2] $H_2\ddot{O}$: + $\overset{+}{N}O_2$ = $:\ddot{O}=\overset{+}{N}=\ddot{O}:$

electrophile

In **sulfonation,** protonation of sulfur trioxide, SO_3, forms a positively charged sulfur species ($^+SO_3H$) that acts as an electrophile (Mechanism 16.4).

Mechanism 16.4 Formation of the Electrophile $^+SO_3H$ for Sulfonation

SO_3 + H–OSO_3H ⟶ $^+SO_2$–OH = $^+SO_3H$ (electrophile) + HSO_4^-

These steps illustrate how to generate the electrophile E^+ for nitration and sulfonation, the process that begins any mechanism for electrophilic aromatic substitution. To complete either of these mechanisms, you must replace the electrophile E^+ by either $^+NO_2$ or $^+SO_3H$ in the general mechanism (Mechanism 16.1). Thus, **the two-step sequence that replaces H by E is the same regardless of E^+.** This is shown in Sample Problem 16.1 using the reaction of benzene with the nitronium ion.

Sample Problem 16.1 Drawing the Mechanism for Nitration of Benzene

Draw a stepwise mechanism for the nitration of a benzene ring.

benzene $\xrightarrow[H_2SO_4]{HNO_3}$ nitrobenzene ($C_6H_5NO_2$)

Solution

We must first generate the electrophile and then write the two-step mechanism for electrophilic aromatic substitution using it.

Part [1] Generation of the electrophile $^+NO_2$

HO–NO_2 + H–OSO_3H $\xrightarrow{[1]}$ H_2O^+–NO_2 + HSO_4^- $\xrightarrow{[2]}$ H_2O + $^+NO_2$

Part [2] Two-step mechanism for electrophilic aromatic substitution

benzene + $^+NO_2$ $\xrightarrow{[3]}$ carbocation (H, NO_2) [+ two resonance structures] + HSO_4^- $\xrightarrow{[4]}$ nitrobenzene + H_2SO_4

Any species with a lone pair of electrons can be used to remove the proton in the last step. In this case, the mechanism is drawn with HSO_4^-, formed when $^+NO_2$ is generated as the electrophile.

Problem 16.4 Draw a stepwise mechanism for the sulfonation of an alkyl benzene such as **A** to form a substituted benzenesulfonic acid **B.** Treatment of **B** with base forms a sodium salt **C** that can be used as a synthetic detergent to clean away dirt.

A $\xrightarrow[H_2SO_4]{SO_3}$ **B** (–SO_3H) $\xrightarrow{NaOH}$ synthetic detergent **C** (–SO_3^- Na^+)

16.5 Friedel–Crafts Alkylation and Friedel–Crafts Acylation

Friedel–Crafts alkylation and **Friedel–Crafts acylation** form new carbon–carbon bonds.

16.5A General Features

In **Friedel–Crafts alkylation,** treatment of benzene with an alkyl halide and a Lewis acid ($AlCl_3$) forms an alkyl benzene. This reaction is an **alkylation** because it results in transfer of an alkyl group from one atom to another (from Cl to benzene).

Friedel–Crafts alkylation

RCl, $AlCl_3$ → alkyl benzene + HCl

$AlCl_3$ + HCl

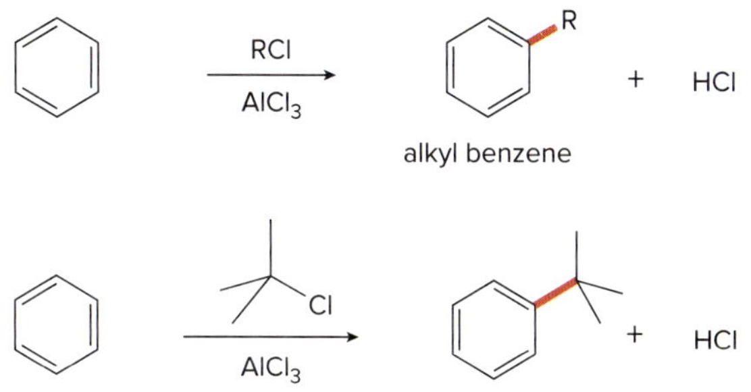

In **Friedel–Crafts acylation,** a benzene ring is treated with an **acid chloride** (RCOCl) and $AlCl_3$ to form a ketone. Because the new group bonded to the benzene ring is called an **acyl group,** the transfer of an acyl group from one atom to another is an **acylation.**

Friedel–Crafts acylation

+ acid chloride → $AlCl_3$ → ketone (acyl group) + HCl

+ → $AlCl_3$ → + HCl

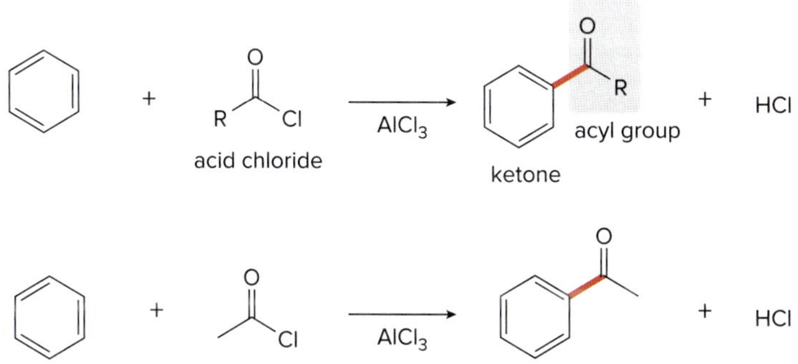

Acid chlorides are also called **acyl chlorides.**

Problem 16.5 What product is formed when benzene is treated with each organic halide in the presence of $AlCl_3$?

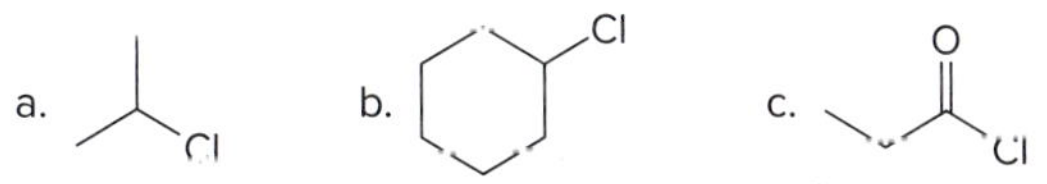

a. b. c.

Problem 16.6 What acid chloride would be needed to prepare each of the following ketones from benzene using a Friedel–Crafts acylation?

a. b. c.

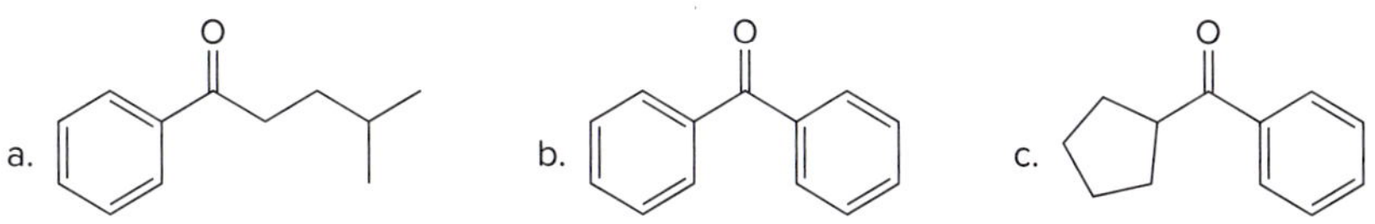

16.5B Mechanism

The mechanisms of alkylation and acylation proceed in a manner analogous to those for halogenation, nitration, and sulfonation. The unique feature in each reaction is how the electrophile is generated.

In **Friedel–Crafts alkylation,** the Lewis acid $AlCl_3$ reacts with the alkyl chloride to form a **Lewis acid–base complex,** illustrated with CH_3CH_2Cl and $(CH_3)_3CCl$ as alkyl chlorides.

The identity of the alkyl chloride determines the exact course of the reaction as shown in Mechanism 16.5.

Mechanism 16.5 Formation of the Electrophile in Friedel–Crafts Alkylation—Two Possibilities

Possibility [1] For **CH_3Cl** and **1° RCl**

electrophile

Lewis acid–base complex

Possibility [2] For **2°** and **3° RCl**

Lewis acid–base complex

electrophile

- **For CH_3Cl and 1° RCl, the Lewis acid–base complex itself serves as the electrophile for electrophilic aromatic substitution.**
- **With 2° and 3° RCl, the Lewis acid–base complex reacts further to give a 2° or 3° carbocation, which serves as the electrophile. Carbocation formation occurs only with 2° and 3° alkyl chlorides, because they afford more stable carbocations.**

In either case, the electrophile goes on to react with benzene in the two-step mechanism characteristic of electrophilic aromatic substitution, illustrated in Mechanism 16.6 using the 3° carbocation, $(CH_3)_3C^+$.

Mechanism 16.6 Friedel–Crafts Alkylation Using a 3° Carbocation

carbocation

[+ two resonance structures]

+ HCl + $AlCl_3$

1 Addition of the carbocation electrophile forms a **new carbon–carbon bond.**

2 $AlCl_4^-$ removes a proton on the carbon bearing the new substituent to re-form the aromatic ring.

In **Friedel–Crafts acylation,** the Lewis acid $AlCl_3$ ionizes the carbon–halogen bond of the acid chloride, thus forming a positively charged carbon electrophile called an **acylium ion,** which is resonance stabilized (Mechanism 16.7). The positively charged carbon atom of the acylium ion then goes on to react with benzene in the two-step mechanism of electrophilic aromatic substitution.

Mechanism 16.7 Formation of the Electrophile in Friedel–Crafts Acylation

Lewis acid–base complex

resonance-stabilized acylium ion

To complete the mechanism for acylation, insert the electrophile into the general mechanism and draw the last two steps, as illustrated in Sample Problem 16.2.

Sample Problem 16.2 Drawing a Mechanism for a Friedel–Crafts Reaction

Draw a stepwise mechanism for the following Friedel–Crafts acylation.

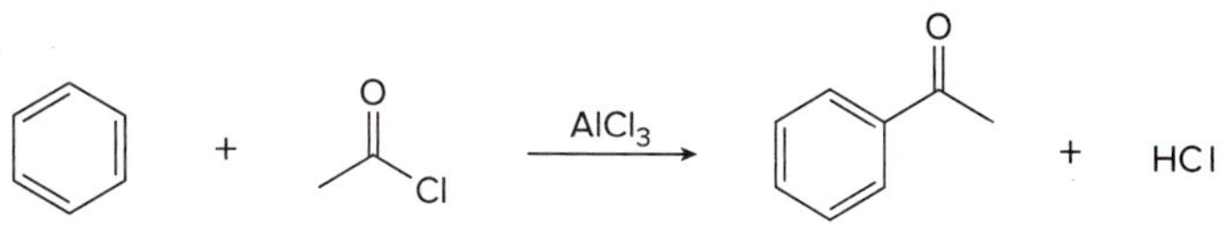
O
+
O
Cl
$AlCl_3$
+
HCl

Solution

First generate the **acylium ion,** and then write the two-step mechanism for electrophilic aromatic substitution using it for the electrophile.

Part [1] Generation of the electrophile $(CH_3CO)^+$

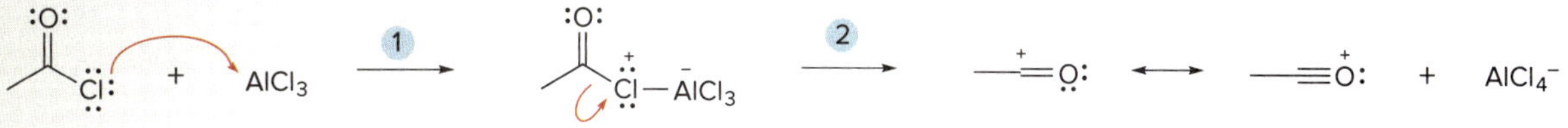
:O:
Cl
+
$AlCl_3$
1
:O:
Cl
$AlCl_3$
2
O:
O:
+
$AlCl_4^-$

Lewis acid–base complex

resonance-stabilized acylium ion

Part [2] Two-step mechanism for electrophilic aromatic substitution

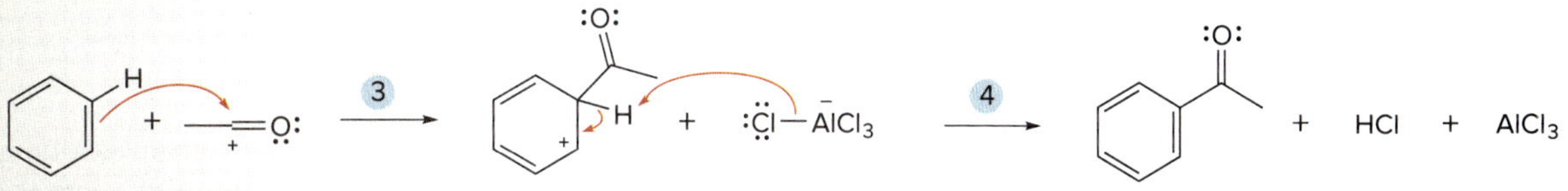
H
+
O:
3
:O:
H
+
:Cl
$AlCl_3$
4
:O:
+
HCl
+
$AlCl_3$

[+ two resonance structures]

Problem 16.7 What acylium ion is formed from each acid chloride?

a. 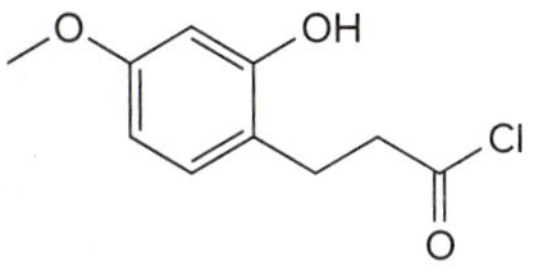
O
OH
Cl
O

b. 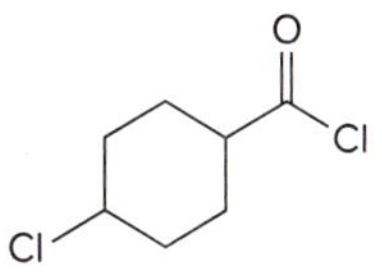
O
Cl
Cl

c. 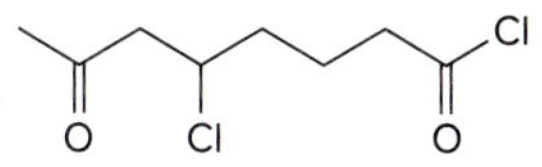
Cl
O
Cl
O

More Practice: Try Problems 16.11, 16.61.

16.5C Other Facts About Friedel–Crafts Alkylation

Three additional facts about Friedel–Crafts alkylations must be kept in mind.

[1] Vinyl halides and aryl halides do *not* react in Friedel–Crafts alkylation.

Most Friedel–Crafts reactions involve carbocation electrophiles. Because the carbocations derived from vinyl halides and aryl halides are highly unstable and do not readily form, these organic halides do *not* undergo Friedel–Crafts alkylation.

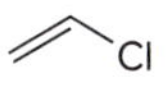
Cl

vinyl halide
unreactive

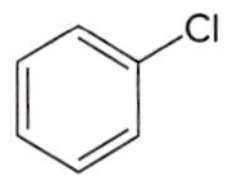
Cl

aryl halide
unreactive

Problem 16.8 Which halides are unreactive in a Friedel–Crafts alkylation reaction?

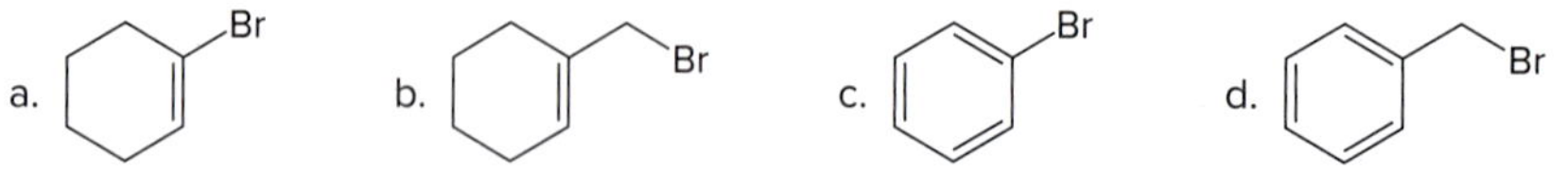

a. b. c. d.

Br Br Br Br

[2] Rearrangements can occur.

The Friedel–Crafts reaction can yield products having rearranged carbon skeletons when 1° and 2° alkyl halides are used as starting materials, as shown in Equations [1] and [2]. In both reactions, the carbon atom bonded to the halogen in the starting material (labeled in blue) is not bonded to the benzene ring in the product, thus indicating that a rearrangement has occurred.

Recall from Section 9.9 that a 1,2-shift converts a less stable carbocation to a more stable carbocation by shift of a hydrogen atom or an alkyl group.

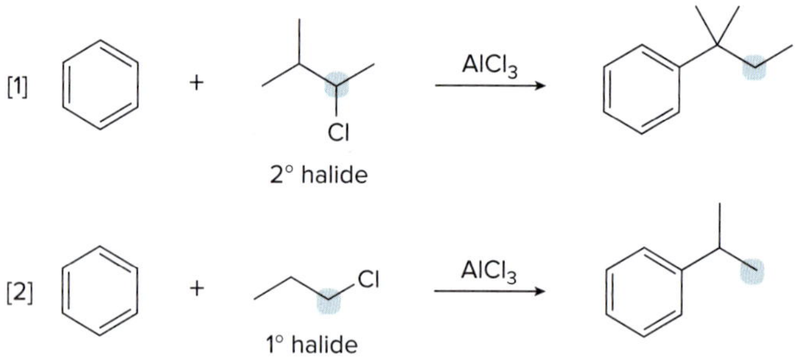

[1] + Cl $\xrightarrow{AlCl_3}$

2° halide

[2] + Cl $\xrightarrow{AlCl_3}$

1° halide

The result in Equation [1] is explained by a carbocation rearrangement involving a 1,2-hydride shift: **the less stable 2° carbocation (formed from the 2° halide) rearranges to a more stable 3° carbocation,** as illustrated in Mechanism 16.8.

Mechanism 16.8 Friedel–Crafts Alkylation Involving Carbocation Rearrangement

Part [1] Formation of a 2° carbocation and rearrangement

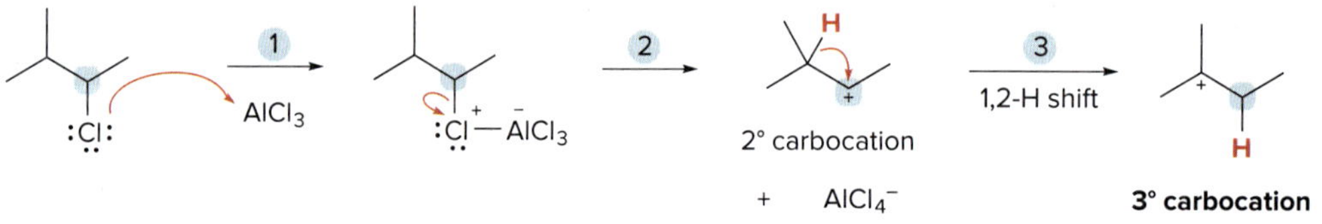

:Cl: 1 $AlCl_3$:Cl—$AlCl_3$ 2 H 2° carbocation + $AlCl_4^-$ 3 1,2-H shift H 3° carbocation

1 – 2 Lewis acid–base reaction of the alkyl chloride with $AlCl_3$ and cleavage of the C–Cl bond form a 2° carbocation.

3 **1,2-Hydride shift** converts a 2° carbocation to a **more stable 3° carbocation.**

Part [2] Two-step mechanism for electrophilic aromatic substitution

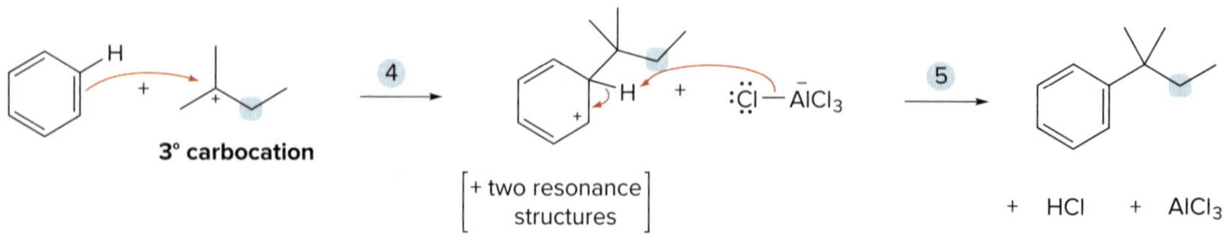

H 3° carbocation 4 H + :Cl—$AlCl_3$ [+ two resonance structures] 5 + HCl + $AlCl_3$

4 Addition of the 3° carbocation forms a new carbon–carbon bond and a **resonance-stabilized carbocation.**

5 $AlCl_4^-$ removes a proton on the carbon bearing the new substituent to re-form the aromatic ring.

Rearrangements can occur even when no free carbocation is formed initially. For example, the 1° alkyl chloride in Equation [2] forms a complex with $AlCl_3$, which does *not* decompose to an unstable 1° carbocation, as shown in Mechanism 16.9. Instead, a **1,2-hydride shift** forms a 2° carbocation, which then serves as the electrophile in the two-step mechanism for electrophilic aromatic substitution.

Mechanism 16.9 A Rearrangement Reaction Beginning with a 1° Alkyl Chloride

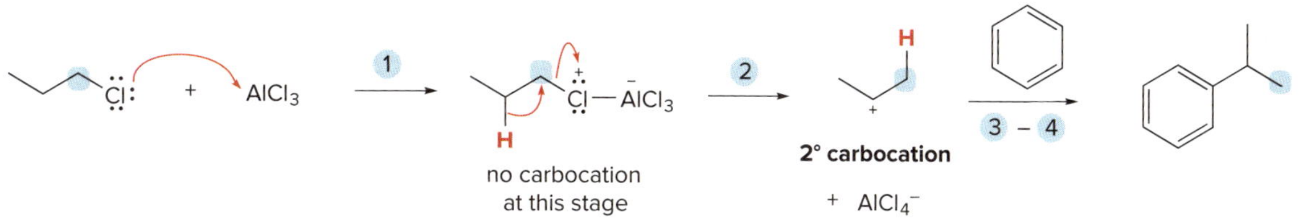

Problem 16.9

Draw a stepwise mechanism for the following reaction.

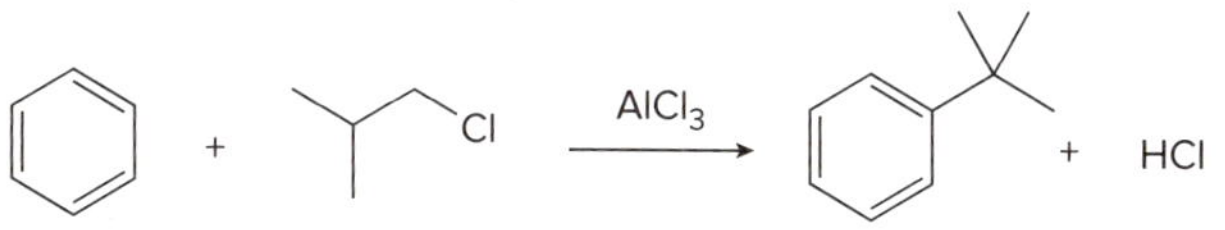

[3] Other functional groups that form carbocations can also be used as starting materials.

Although Friedel–Crafts alkylation works well with alkyl halides, any compound that readily forms a carbocation can be used instead. The two most common alternatives are alkenes and alcohols, both of which afford carbocations in the presence of strong acid.

- **Protonation of an alkene forms a carbocation, which can then serve as an electrophile in a Friedel–Crafts alkylation.**
- **Protonation of an alcohol, followed by loss of water, likewise forms a carbocation.**

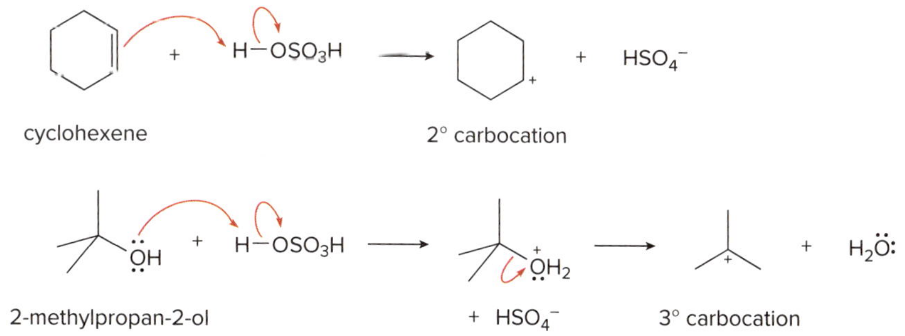

Each carbocation can then go on to react with benzene to form a product of electrophilic aromatic substitution. For example:

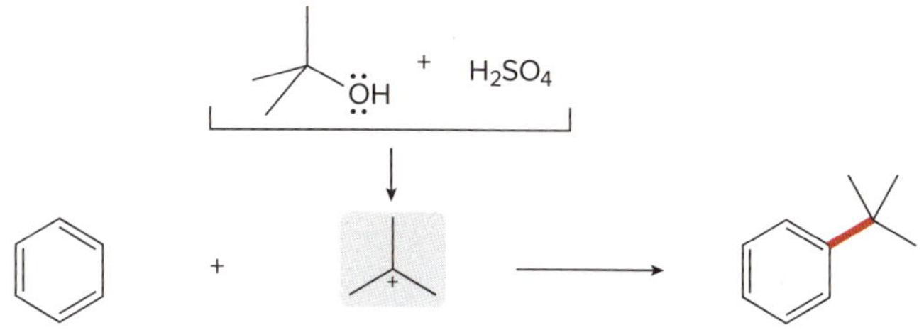

Problem 16.10 Draw the product of each reaction.

a. benzene + cyclohexene $\xrightarrow{H_2SO_4}$

b. benzene + 2-methylpropene $\xrightarrow{H_2SO_4}$

c. benzene + (OH alcohol) $\xrightarrow{H_2SO_4}$

d. benzene + (cyclohexyl, OH) $\xrightarrow{H_2SO_4}$

16.5D Intramolecular Friedel–Crafts Reactions

All of the Friedel–Crafts reactions discussed thus far have resulted from intermolecular reaction of a benzene ring with an electrophile. Starting materials that contain both units are capable of **intramolecular reaction,** and this forms a new ring. Treatment of compound **A,** which contains both a benzene ring and an acid chloride, with $AlCl_3$, forms α-tetralone by an intramolecular Friedel–Crafts acylation reaction.

A $\xrightarrow{AlCl_3}$ α-tetralone + HCl

new C–C bond in red

Such an intramolecular Friedel–Crafts acylation was a key step in the synthesis of the hallucinogen LSD, as shown in Figure 16.4.

Figure 16.4 Intramolecular Friedel–Crafts acylation in the synthesis of LSD

Ergot-infected grain, the source of lysergic acid. *Carmen Rieb/ Shutterstock*

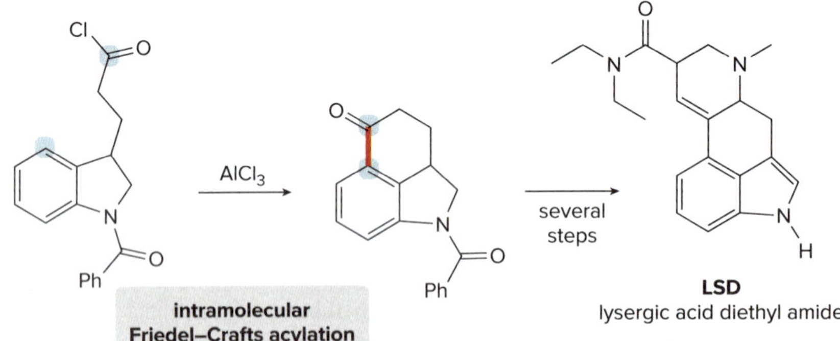

- **Intramolecular Friedel–Crafts acylation** at the labeled carbons formed a product containing a new six-membered ring, which was converted to LSD in several steps.
- LSD was first prepared by Swiss chemist Albert Hofmann in 1938 from a related organic compound isolated from the ergot fungus that attacks rye and other grains. Ergot has a long history as a dreaded poison, affecting individuals who become ill from eating ergot-contaminated bread. The hallucinogenic effects of LSD were first discovered when Hofmann accidentally absorbed a small amount of the drug through his fingertips.

Problem 16.11 Draw a stepwise mechanism for the intramolecular Friedel–Crafts acylation of compound **A** to form **B. B** can be converted in one step to the antidepressant sertraline.

Cl, Cl, Cl, =O — $AlCl_3$ → Cl, Cl, =O — one step → Cl, Cl, H, N

A **B** sertraline (Zoloft)

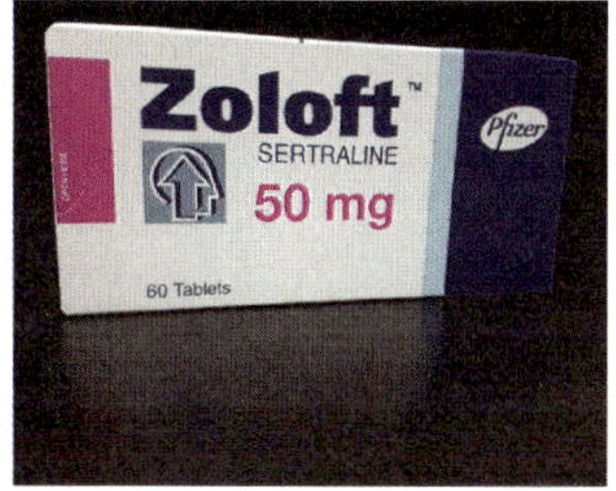

Sertraline (trade name Zoloft, Problem 16.11) is an effective antidepressant because it increases the concentration of the neurotransmitter serotonin in the brain. *Omeletzz/Shutterstock*

Problem 16.12 Intramolecular reactions are also observed in Friedel–Crafts alkylation. Draw the intramolecular alkylation product formed from each of the following reactants. (Watch out for rearrangements!)

a. b. OH c. Cl

16.5E Biological Friedel–Crafts Reactions

Biological Friedel–Crafts reactions occur as well. As we learned in Section 14.2, allylic diphosphates contain a good leaving group, so they can serve as a source of allylic carbocations. A key step in the biological synthesis of vitamin K_1, the chapter-opening molecule, involves Friedel–Crafts reaction of 1,4-dihydroxynaphthoic acid with phytyl diphosphate to form **X,** which is converted to vitamin K_1 in several steps, as shown in Figure 16.5.

Figure 16.5 Friedel–Crafts reaction in the synthesis of vitamin K_1

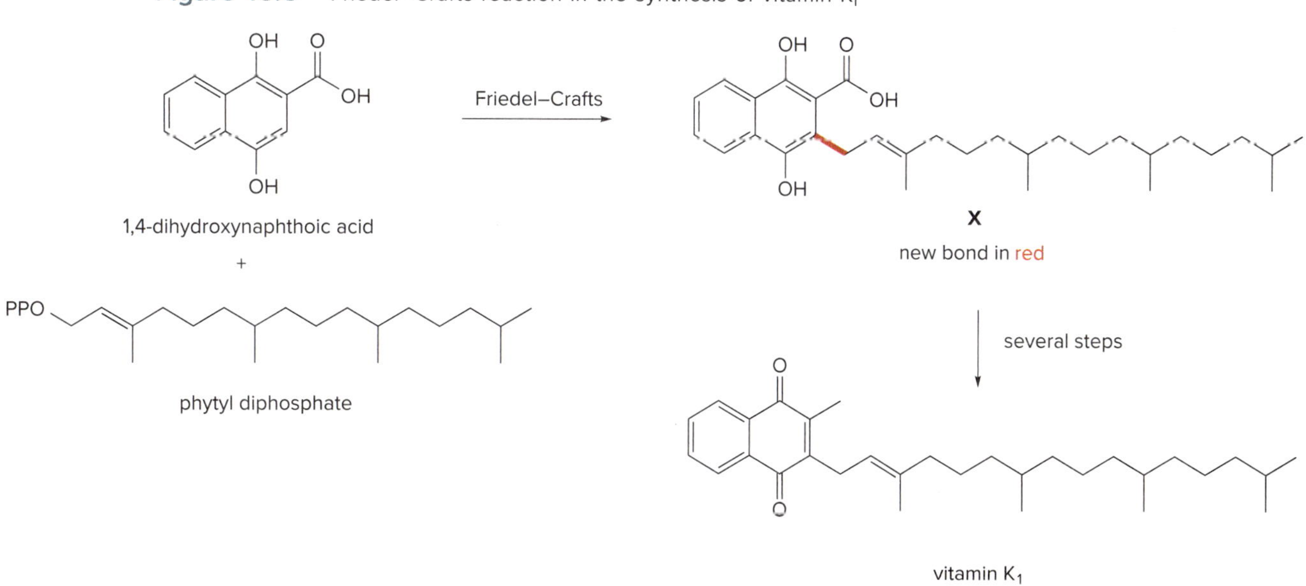

Problem 16.13 (a) Draw resonance structures for the carbocation formed after loss of a leaving group from phytyl diphosphate. (b) Draw the two-step mechanism for Friedel–Crafts alkylation of 1,2-dihydroxynaphthoic acid with this carbocation to form **X.**

16.6 Substituted Benzenes

Many substituted benzene rings undergo electrophilic aromatic substitution. Common substituents include halogens, OH, NH_2, alkyl, and many functional groups that contain a carbonyl. Each substituent either increases or decreases the electron density in the benzene ring, and this affects the course of electrophilic aromatic substitution, as we will learn in Section 16.7.

What makes a substituent on a benzene ring electron donating or electron withdrawing? The answer is **inductive effects** and **resonance effects,** both of which can add or remove electron density.

Inductive Effects

Inductive effects stem from the **electronegativity** of the atoms in the substituent and the **polarizability** of the substituent group.

Inductive and resonance effects were first discussed in Sections 2.5B and 2.5C, respectively.

- **Atoms more electronegative than carbon—including N, O, and X—pull electron density away from carbon and thus exhibit an electron-*withdrawing* inductive effect.**
- **Polarizable alkyl groups donate electron density, and thus exhibit an electron-*donating* inductive effect.**

Considering inductive effects *only,* an NH_2 group withdraws electron density and CH_3 donates electron density.

Electron-withdrawing inductive effect

NH_2

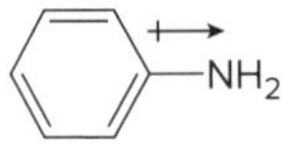

- N is ***more* electronegative** than C.
- N inductively **withdraws** electron density.

Electron-donating inductive effect

CH_3

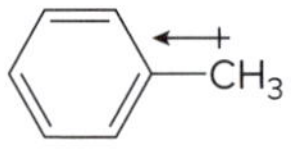

- Alkyl groups are **polarizable,** making them electron-**donating** groups.

Problem 16.14 Which substituents have an electron-withdrawing and which have an electron-donating inductive effect: (a) $CH_3CH_2CH_2CH_2$–; (b) Br–; (c) CH_3CH_2O–?

Resonance Effects

Resonance effects can either donate or withdraw electron density, depending on whether they place a positive or negative charge on the benzene ring.

- **A resonance effect is electron *donating* when resonance structures place a *negative* charge on carbons of the benzene ring.**
- **A resonance effect is electron *withdrawing* when resonance structures place a *positive* charge on carbons of the benzene ring.**

An electron-donating resonance effect is observed whenever an atom Z having a lone pair of electrons is bonded directly to a benzene ring (general structure—$\mathbf{C_6H_5–Z:}$). Common examples of Z include N, O, and halogen. For example, five resonance structures can be drawn for aniline ($C_6H_5NH_2$). Because three of them place a *negative* charge on a carbon atom of the benzene ring, an **NH_2 group *donates* electron density to a benzene ring by a resonance effect.**

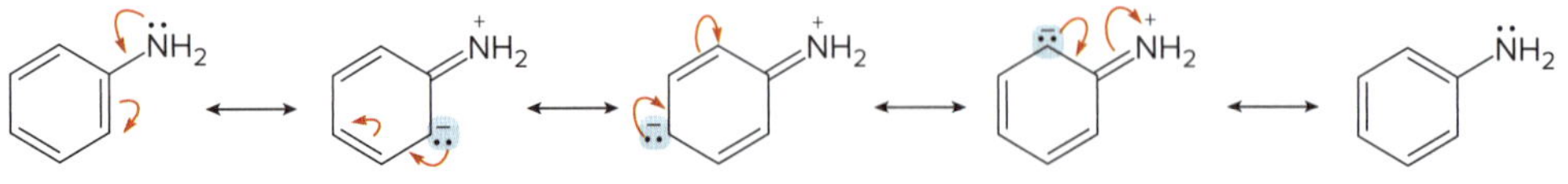

$\ddot{N}H_2$ ⟷ $\overset{+}{N}H_2$ ⟷ $\overset{+}{N}H_2$ ⟷ $\overset{+}{N}H_2$ ⟷ $\ddot{N}H_2$

aniline

Three resonance structures place a (–) charge on atoms in the ring.

In contrast, **an electron-withdrawing resonance effect is observed in substituted benzenes having the general structure C_6H_5—Y=Z,** where Z is more electronegative than Y. For example, seven resonance structures can be drawn for benzaldehyde (C_6H_5CHO). Because three of them place a *positive* charge on a carbon atom of the benzene ring, a CHO group *withdraws* electron density from a benzene ring by a resonance effect.

benzaldehyde

Three resonance structures place a (+) charge on atoms in the ring.

Problem 16.15 (a) For which compounds **A–E** can resonance structures that place a charge on the carbons of the benzene ring be drawn? (b) For those compounds for which resonance structures can be drawn, draw all resonance structures, and determine whether the substituent has an electron-donating or electron-withdrawing resonance effect.

A **B** **C** **D** **E**

Considering Both Inductive and Resonance Effects

To predict whether a substituted benzene is more or less electron rich than benzene itself, we must consider the **net balance of *both* the inductive and the resonance effects.** Alkyl groups, for instance, donate electrons by an inductive effect, but they have no resonance effect because they lack nonbonded electron pairs or π bonds. As a result,

- **An alkyl group is an electron-*donating* group and an alkyl benzene is more electron rich than benzene.**

When electronegative atoms, such as N, O, or halogen, are bonded to the benzene ring, they inductively *withdraw* electron density from the ring. All of these groups also have a nonbonded pair of electrons, so they *donate* electron density to the ring by resonance. **The *identity of the element* determines the net balance of these opposing effects.**

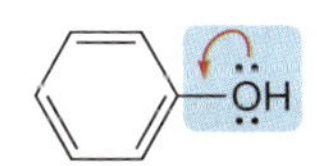

OH is electron donating. The resonance effect predominates.

Cl is electron withdrawing. The inductive effect predominates.

Z = N, O, X

Induction and resonance have opposite effects.
- Z inductively *withdraws* electron density.
- Z *donates* electron density by resonance.

- **When a neutral O or N atom is bonded directly to a benzene ring, the resonance effect dominates and the net effect is *electron donation*.**
- **When a halogen X is bonded to a benzene ring, the inductive effect dominates and the net effect is *electron withdrawal*.**

Thus, **NH_2 and OH are electron-*donating* groups** because the resonance effect predominates, whereas **Cl and Br are electron-*withdrawing* groups** because the inductive effect predominates.

C=O is electron withdrawing.

Finally, the inductive and resonance effects in compounds having the general structure **C_6H_5—Y=Z** (with Z more electronegative than Y) are **both electron withdrawing;** in other words, the two effects *reinforce* each other. This is true for benzaldehyde (C_6H_5CHO) and all other compounds that contain a carbonyl group bonded directly to the benzene ring.

Figure 16.6
The effect of substituents on the electron density in substituted benzenes

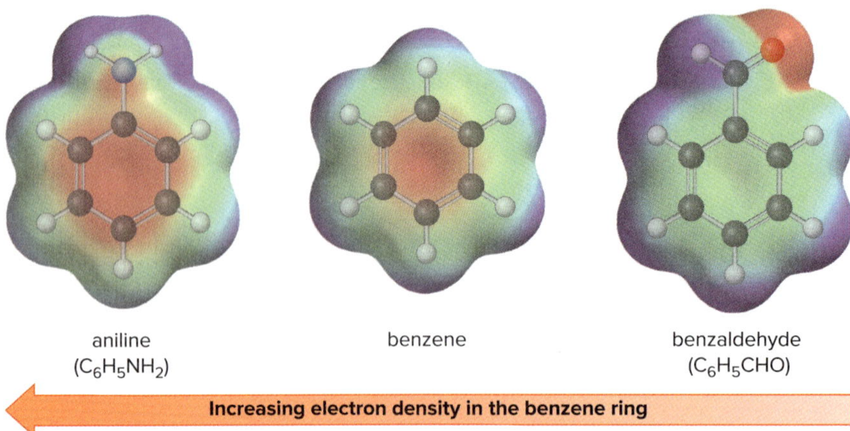

- The NH_2 group donates electron density, making the benzene ring more electron rich (redder), whereas the CHO group withdraws electron density, making the benzene ring less electron rich (greener).

Thus, on balance, an **NH_2 group is electron *donating,*** so the benzene ring of aniline ($C_6H_5NH_2$) has more electron density than benzene. An **aldehyde group (CHO), on the other hand, is electron *withdrawing,*** so the benzene ring of benzaldehyde (C_6H_5CHO) has less electron density than benzene. These effects are illustrated in the electrostatic potential maps in Figure 16.6. These compounds represent examples of the general structural features in electron-donating and electron-withdrawing substituents:

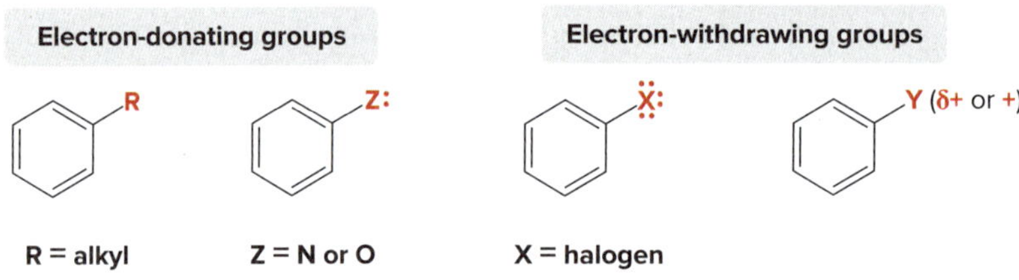

- **Common electron-donating groups are alkyl groups or groups with an N or O atom (with a lone pair) bonded to the benzene ring.**
- **Common electron-withdrawing groups are halogens or groups with an atom Y bearing a full or partial positive charge (+ or δ+) bonded to the benzene ring.**

The net effect of electron donation and withdrawal on the reactions of substituted aromatics is discussed in Sections 16.7–16.9.

Sample Problem 16.3 Classifying a Substituent as Electron Donating or Electron Withdrawing

Classify each substituent as electron donating or electron withdrawing.

a. [benzene ring bonded to O–C(=O)–CH₃] b. [benzene ring bonded to CN]

Solution

If necessary, draw out the atoms and bonds of the substituent to clearly see lone pairs and multiple bonds. **Always look at the atom bonded directly to the benzene ring** to determine electron-donating or electron-withdrawing effects.

- **An O or N atom with a lone pair of electrons makes a substituent electron donating.**
- **A halogen or an atom with a partial positive charge makes a substituent electron withdrawing.**

a.

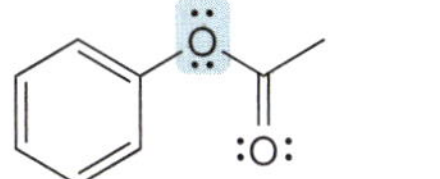

- An O atom with a lone pair is bonded directly to the benzene ring.

an electron-donating group

b.

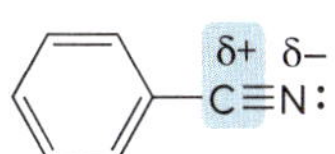

- An atom with a partial (+) charge is bonded directly to the benzene ring.

an electron-withdrawing group

Problem 16.16 Classify each substituent as electron donating or electron withdrawing.

a. OCH_3 b. I c. 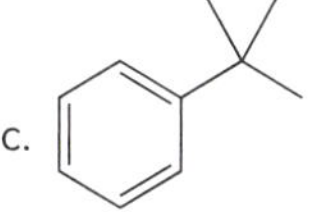 d. N H

More Practice: Try Problem 16.52.

16.7 Electrophilic Aromatic Substitution of Substituted Benzenes

Electrophilic aromatic substitution is a general reaction of *all* aromatic compounds, including polycyclic aromatic hydrocarbons, heterocycles, and substituted benzene derivatives. A substituent affects two aspects of electrophilic aromatic substitution:

- **The rate of reaction:** A substituted benzene reacts faster or slower than benzene itself.
- **The orientation:** The new group is located either ortho, meta, or para to the existing substituent. The identity of the first substituent determines the position of the second substituent.

Toluene ($C_6H_5CH_3$) and nitrobenzene ($C_6H_5NO_2$) illustrate two possible outcomes.

[1] Toluene

Toluene reacts **faster** than benzene in all substitution reactions. Thus, its **electron-donating CH_3 group *activates* the benzene ring** to electrophilic attack. Although three products are possible, compounds with the new group ortho or para to the CH_3 group predominate. The CH_3 group is therefore called an **ortho, para director.**

Br_2 / $FeBr_3$ → Br + Br + Br

ortho 40% | meta trace | **para** 60%

[2] Nitrobenzene

Nitrobenzene reacts **more slowly** than benzene in all substitution reactions. Thus, its **electron-withdrawing NO_2 group *deactivates* the benzene ring** to electrophilic attack. Although three products are possible, the compound with the new group meta to the NO_2 group predominates. The NO_2 group is called a **meta director.**

NO_2 $\xrightarrow[H_2SO_4]{HNO_3}$ ortho (NO_2, NO_2) + meta (NO_2, NO_2) + para (O_2N, NO_2)

ortho 7% · **meta** 93% · para trace

Substituents either activate or deactivate a benzene ring toward electrophiles, and direct selective substitution at specific sites on the ring. **All substituents can be divided into three general types.**

[1] Ortho, para directors and activators

- **Substituents that *activate* a benzene ring and direct substitution ortho and para**

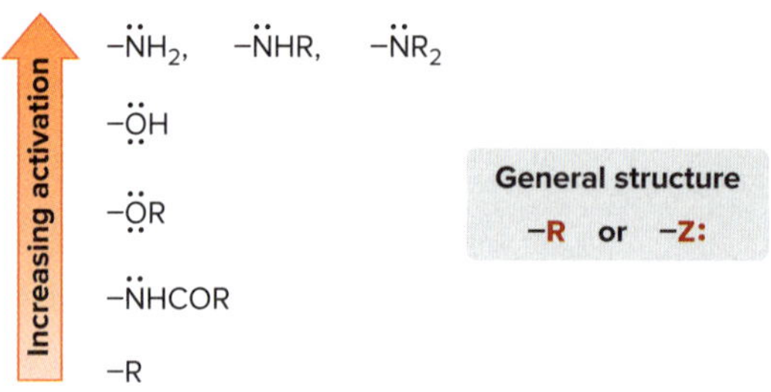

[2] Ortho, para deactivators

- **Substituents that *deactivate* a benzene ring and direct substitution ortho and para**

–F: –Cl: –Br: –I:

[3] Meta directors

- **Substituents that direct substitution meta**
- **All meta directors *deactivate* the ring.**

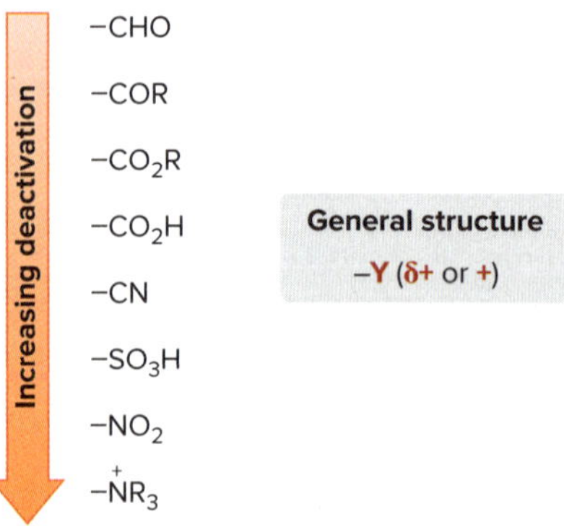

To learn these lists: **Keep in mind that the halogens are in a class by themselves.** Then learn the general structures for each type of substituent.

- **All ortho, para directors are R groups or have a nonbonded electron pair on the atom bonded to the benzene ring.**

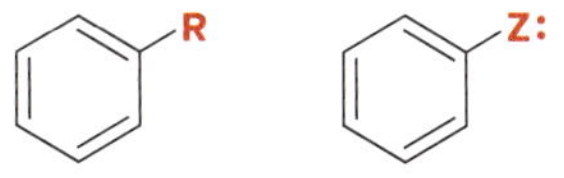

R — **ortho, para director**

Z: — **ortho, para director**

Z = N or O → The ring is **activated.**
Z = halogen → The ring is **deactivated.**

- **All meta directors have a full or partial positive charge on the atom bonded to the benzene ring.**

Y (δ+ or +)

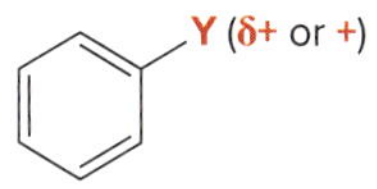

meta director

Sample Problem 16.4 shows how this information can be used to predict the products of electrophilic aromatic substitution reactions.

Sample Problem 16.4 Drawing the Electrophilic Substitution Products of a Substituted Benzene

Draw the products of each reaction, and state whether the reaction is faster or slower than a similar reaction with benzene.

a. 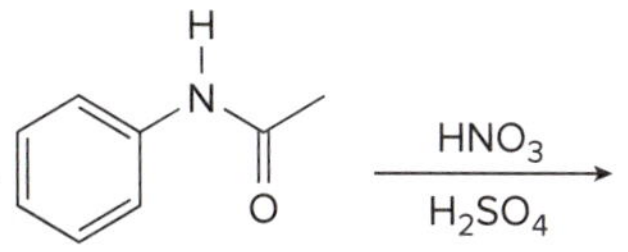 $\xrightarrow[H_2SO_4]{HNO_3}$

b. 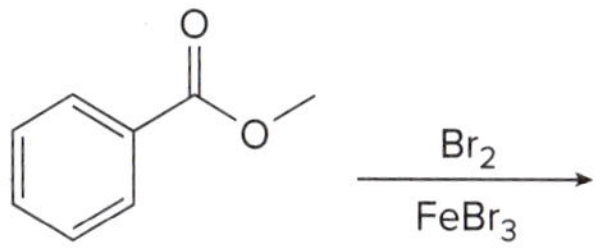 $\xrightarrow[FeBr_3]{Br_2}$

Solution

To draw the products:

- Draw the Lewis structure for the substituent to see if it has a **lone pair** or **partial positive charge** on the atom bonded to the benzene ring.
- **Classify the substituent**—ortho, para activating; ortho, para deactivating; or meta deactivating—and draw the products.

a. 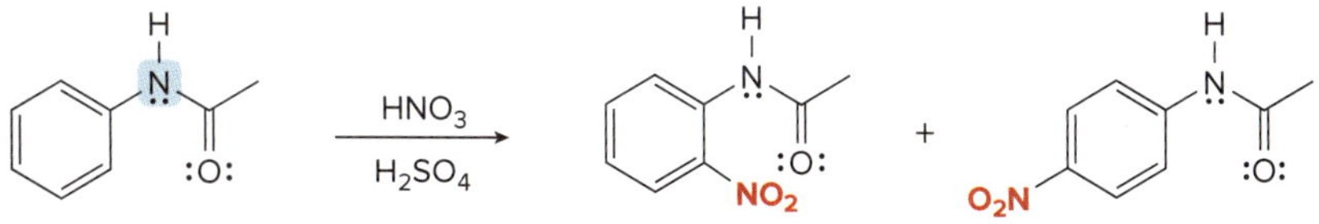 $\xrightarrow[H_2SO_4]{HNO_3}$ NO_2 **ortho** + O_2N **para**

The lone pair on N makes this group an **ortho, para activator. This compound reacts *faster* than benzene.**

b. 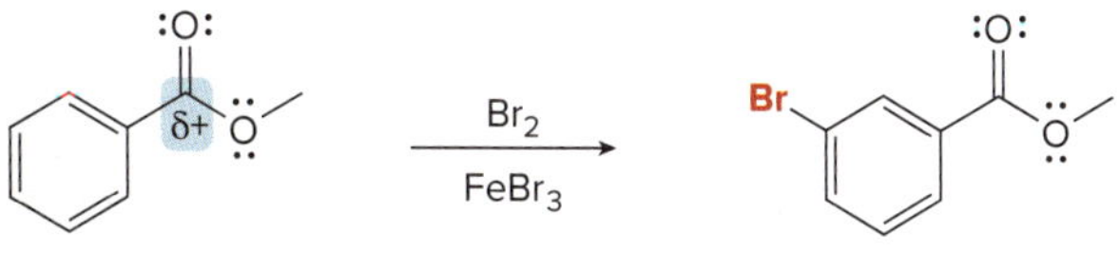 δ+ $\xrightarrow[FeBr_3]{Br_2}$ Br meta

The δ+ on the C bonded to the benzene ring makes the group a **meta deactivator. This compound reacts *more slowly* than benzene.**

Problem 16.17 Draw the products formed when each compound is treated with HNO_3 and H_2SO_4. State whether the reaction occurs faster or slower than a similar reaction with benzene.

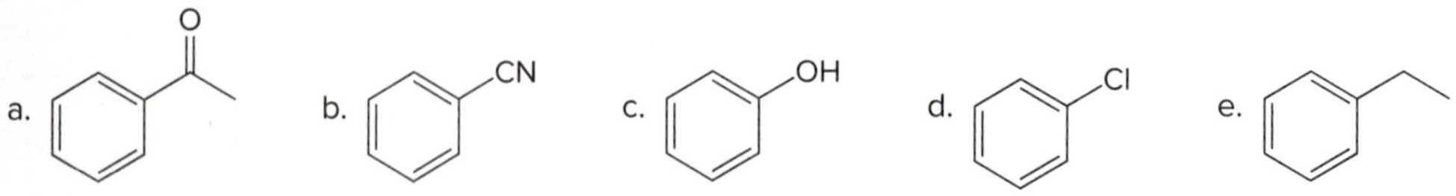

a. b. CN c. OH d. Cl e.

More Practice: Try Problems 16.41[1], 16.55.

Sample Problem 16.5 Determining Which Ring in a Polycyclic Compound Is More Reactive in Electrophilic Aromatic Substitution

Which ring in each compound is more reactive toward electrophiles?

a. b.

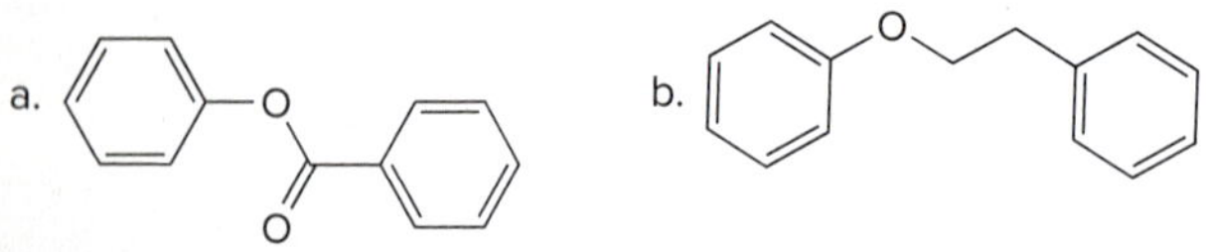

Solution

Look at the atom bonded *directly* to the aromatic ring to decide on reactivity in electrophilic aromatic substitution.

- **An N or O atom with a lone pair makes a ring *more* reactive.**
- **An alkyl group makes a ring *somewhat more* reactive.**
- **An atom with a full or partial positive charge makes a ring *less* reactive.**

a. Ring **A** is bonded to an O atom with a lone pair, whereas ring **B** is bonded to a C that bears a δ+. Ring **A** is more electron rich and more reactive.

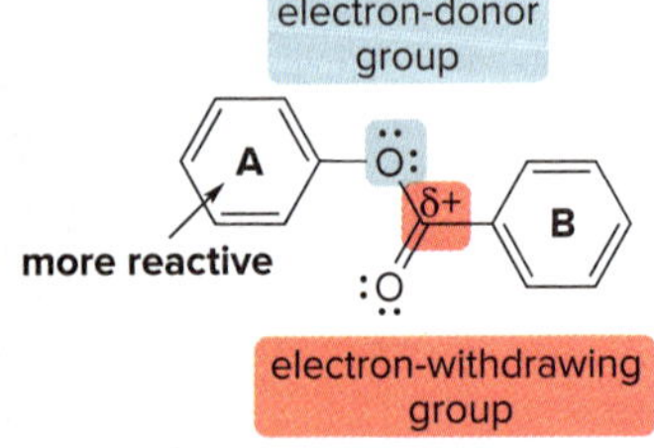

b. Ring **C** is bonded to an O atom with a lone pair, whereas ring **D** is bonded to an alkyl carbon. Ring **C** is more electron rich and more reactive.

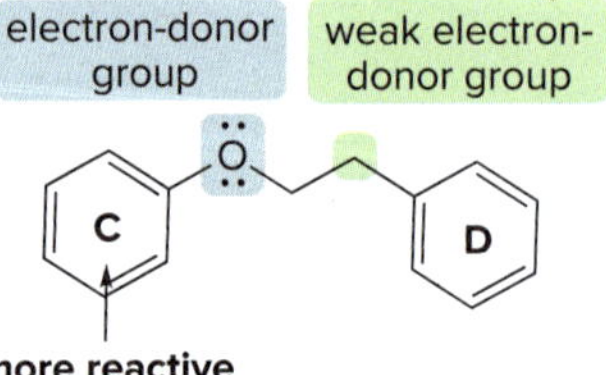

Problem 16.18 Determine which ring in each compound is more reactive toward electrophiles, and explain your choice.

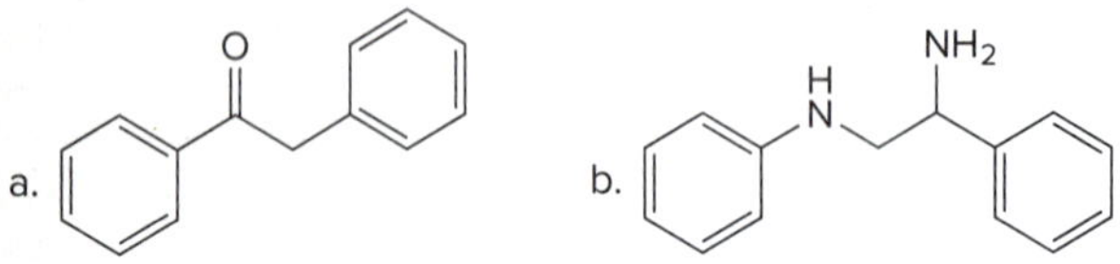

a. b. NH2

More Practice: Try Problems 16.53–16.55.

Problem 16.19 Consider the tetracyclic compound with rings labeled **A–D.** (a) Which ring is the *most* reactive in electrophilic aromatic substitution? (b) Which ring is the *least* reactive in electrophilic aromatic substitution?

O
O
A B C D
O
OCH_3

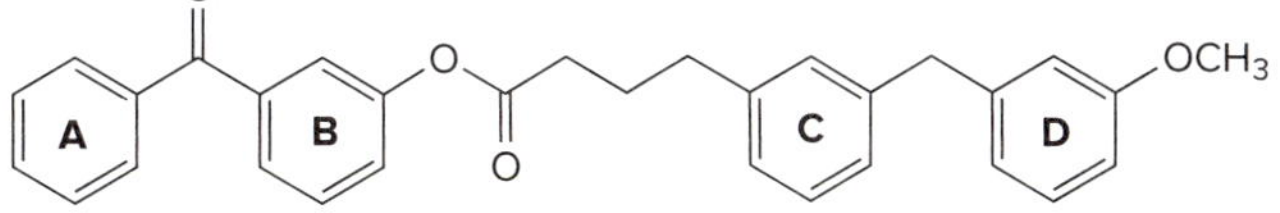

16.8 Why Substituents Activate or Deactivate a Benzene Ring

- **Why do substituents activate or deactivate a benzene ring?**
- **Why are particular orientation effects observed?** Why are some groups ortho, para directors and some groups meta directors?

To understand why some substituents make a benzene ring react *faster* than benzene itself (activators), whereas others make it react *slower* (deactivators), we must evaluate the rate-determining step (the first step) of the mechanism. Recall from Section 16.2 that the first step in electrophilic aromatic substitution is the addition of an electrophile (E^+) to form a resonance-stabilized carbocation. The Hammond postulate (Section 7.13) makes it possible to predict the relative rate of the reaction by looking at the stability of the carbocation intermediate.

- **The more stable the carbocation, the lower in energy the transition state that forms it, and the faster the reaction.**

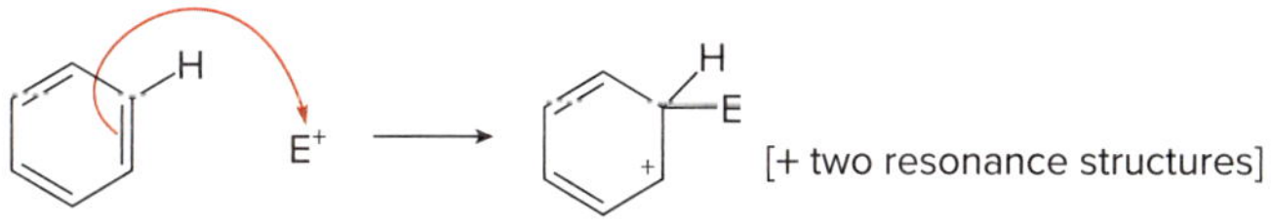

H
E^+
H
E
+
[+ two resonance structures]

Stabilizing the carbocation makes the reaction faster.

The principles of inductive effects and resonance effects, first introduced in Section 16.6, can now be used to predict carbocation stability.

- **Electron-donating groups stabilize the carbocation and *activate* a benzene ring toward electrophilic attack. All activators are R groups, or they have an N or O atom with a lone pair bonded directly to the benzene ring.**
- **Electron-withdrawing groups destabilize the carbocation and *deactivate* a benzene ring toward electrophilic attack. All deactivators are halogens, or they have an atom with a full or partial positive charge bonded directly to the benzene ring.**

The energy diagrams in Figure 16.7 illustrate the effect of electron-donating and electron-withdrawing groups on the energy of the transition state of the rate-determining step in electrophilic aromatic substitution.

Problem 16.20 Label each compound as more or less reactive than benzene in electrophilic aromatic substitution.

a. b. c. d.

Figure 16.7 Energy diagrams comparing the rate of electrophilic aromatic substitution of substituted benzenes

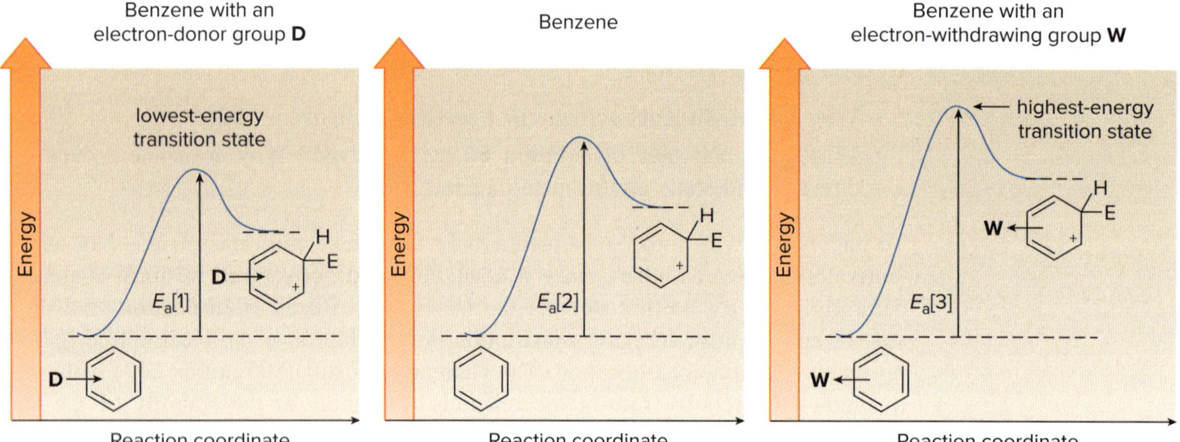

- Electron-donor groups **D** *stabilize* the carbocation intermediate, lower the energy of the transition state, and *increase* the rate of reaction.
- Electron-withdrawing groups **W** *destabilize* the carbocation intermediate, raise the energy of the transition state, and *decrease* the rate of reaction.

Problem 16.21 Rank compounds **A, B,** and **C** in order of increasing reactivity in electrophilic aromatic substitution.

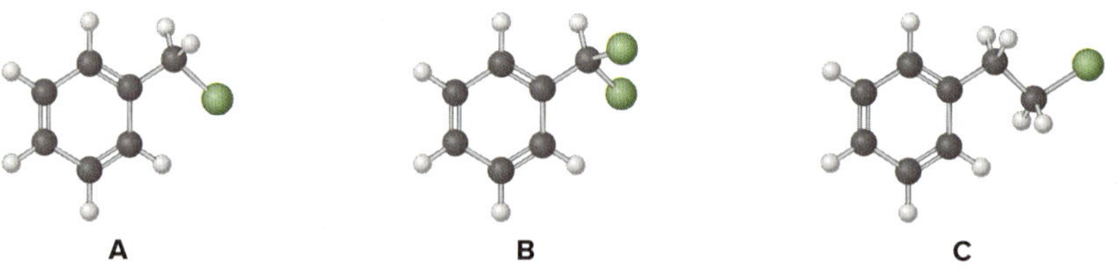

Problem 16.22 Rank the following compounds in order of increasing reactivity in electrophilic aromatic substitution: (a) $C_6H_5OCH_3$; (b) $C_6H_5CO_2CH_3$; (c) $C_6H_5CH_2OCH_3$; (d) $C_6H_5CH_2CH_2OCH_3$.

16.9 Orientation Effects in Substituted Benzenes

To understand why particular orientation effects arise, you must keep in mind the general structures for ortho, para directors and for meta directors already given in Section 16.7.

- **All ortho, para directors are R groups or have a nonbonded electron pair on the atom bonded to the benzene ring.**
- **All meta directors have a full or partial positive charge on the atom bonded to the benzene ring.**

To evaluate the directing effects of a given substituent, we can follow a stepwise procedure.

How To Determine the Directing Effects of a Particular Substituent

Step [1] **Draw all resonance structures for the carbocation formed from attack of an electrophile E^+ at the ortho, meta, and para positions of a substituted benzene (C_6H_5–A).**

A
ortho
meta
para

- There are at least three resonance structures for each site of reaction.
- Each resonance structure places a positive charge **ortho** or **para** to the new C–E bond.

Step [2] **Evaluate the stability of the intermediate resonance structures. The electrophile attacks at those positions that give the *most stable* carbocation.**

Sections 16.9A–C show how this two-step procedure can be used to evaluate the directing effects of the CH_3 group in toluene, the NH_2 group in aniline, and the NO_2 group in nitrobenzene, respectively.

16.9A The CH_3 Group—An ortho, para Director

To understand why a **CH_3 group directs electrophilic aromatic substitution to the ortho and para positions,** first draw all resonance structures that result from electrophilic attack at the ortho, meta, and para positions to the CH_3 group.

Always draw in the H atom at the site of electrophilic attack. This will help you keep track of where the charges go.

ortho attack
E^+
H
E
CH$_3$ stabilizes the (+) charge.
preferred product

meta attack
E^+
H
E

para attack
E^+
H
E
CH$_3$ stabilizes the (+) charge.
preferred product

The positive charge in all resonance structures is always **ortho or para to the new C–E bond.** It is *not* necessarily ortho or para to the CH_3 group.

To evaluate the stability of the resonance structures, determine whether any are especially stable or unstable. In this example, **attack ortho or para to CH_3 generates a resonance structure that places a positive charge on a carbon atom with the CH_3 group.** The electron-donating CH_3 group *stabilizes* the adjacent positive charge. In contrast, attack meta to the CH_3 group does *not* generate any resonance structure stabilized by electron donation. Other alkyl groups are ortho, para directors for the same reason.

- **The CH_3 group directs electrophilic attack ortho and para to itself because an electron-donating inductive effect stabilizes the carbocation intermediate.**

Problem 16.23 Draw all resonance structures for the carbocation formed by attack of an electrophile E^+ at the indicated position of *m*-diethylbenzene. Label any resonance structure that is especially stable.

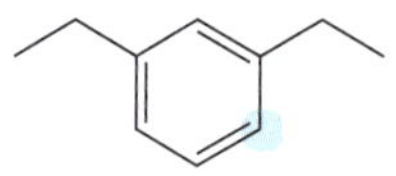

m-diethylbenzene

16.9B The NH_2 Group—An ortho, para Director

To understand why an **amino group (NH_2) directs electrophilic aromatic substitution to the ortho and para positions,** follow the same procedure.

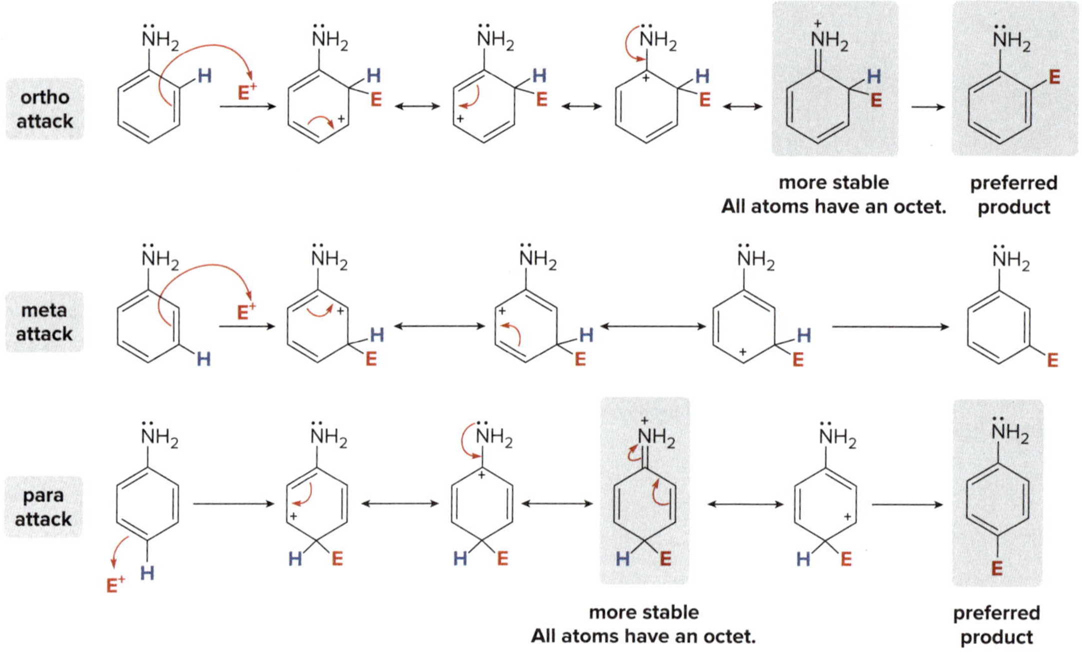

ortho attack

more stable
All atoms have an octet.

preferred product

meta attack

para attack

more stable
All atoms have an octet.

preferred product

Attack at the meta position generates the usual three resonance structures. Because of the lone pair on the N atom, attack at the ortho and para positions generates a fourth resonance structure, which is stabilized because **every atom has an octet of electrons. This additional resonance structure can be drawn for all substituents that have an N, O, or halogen atom bonded directly to the benzene ring.**

- **The NH_2 group directs electrophilic attack ortho and para to itself because the carbocation intermediate has additional resonance stabilization.**

Problem 16.24 Which species are valid resonance structures for the cation that results when *o*-dimethoxybenzene reacts with Br_2, $FeBr_3$ at the indicated position?

o-dimethoxybenzene

A B C D

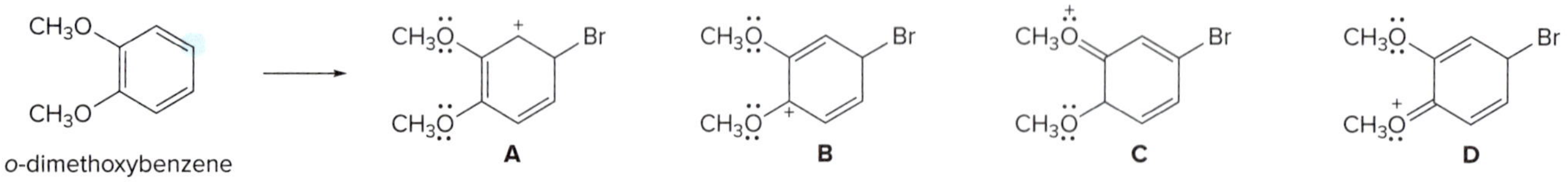

16.9C The NO_2 Group—A meta Director

To understand why a **nitro group (NO_2) directs electrophilic aromatic substitution to the meta position,** follow the same procedure.

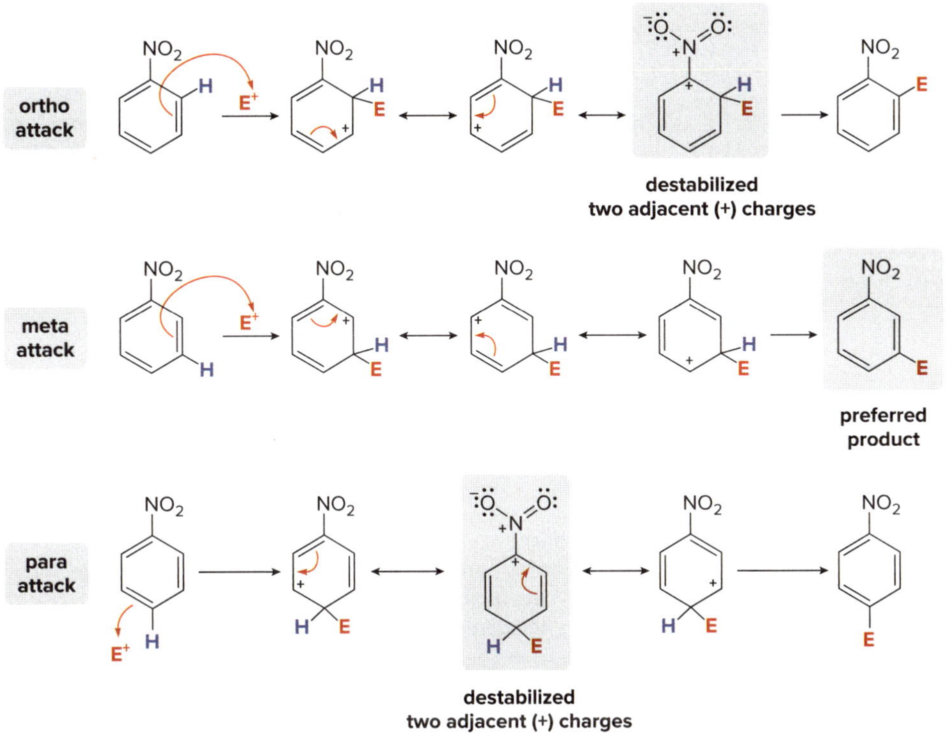

ortho attack

destabilized
two adjacent (+) charges

meta attack

preferred product

para attack

destabilized
two adjacent (+) charges

Attack at each position generates three resonance structures. One resonance structure resulting from attack at the ortho and para positions is especially *destabilized,* because it contains a positive charge on two adjacent atoms. Attack at the meta position does not generate any particularly unstable resonance structures.

- **With the NO_2 group (and all meta directors), meta attack occurs because attack at the ortho or para position gives a destabilized carbocation intermediate.**

Problem 16.25 Draw all resonance structures for the carbocation formed by ortho attack of the electrophile $^+NO_2$ on each starting material. Label any resonance structures that are especially stable or unstable.

a. b. OH c. O H

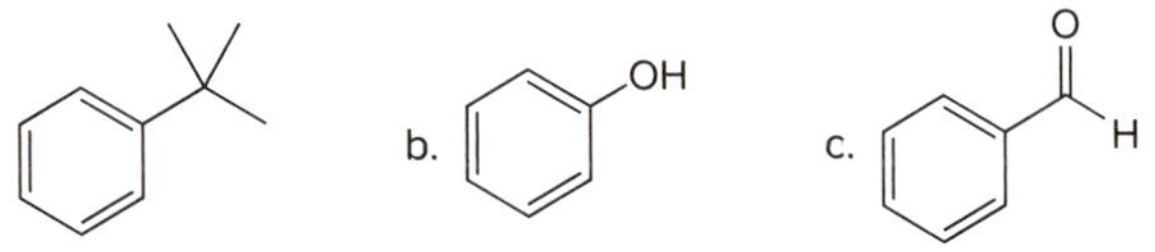

Figure 16.8 summarizes the reactivity and directing effects of the common substituents on benzene rings.

Figure 16.8

The reactivity and directing effects of common substituted benzenes

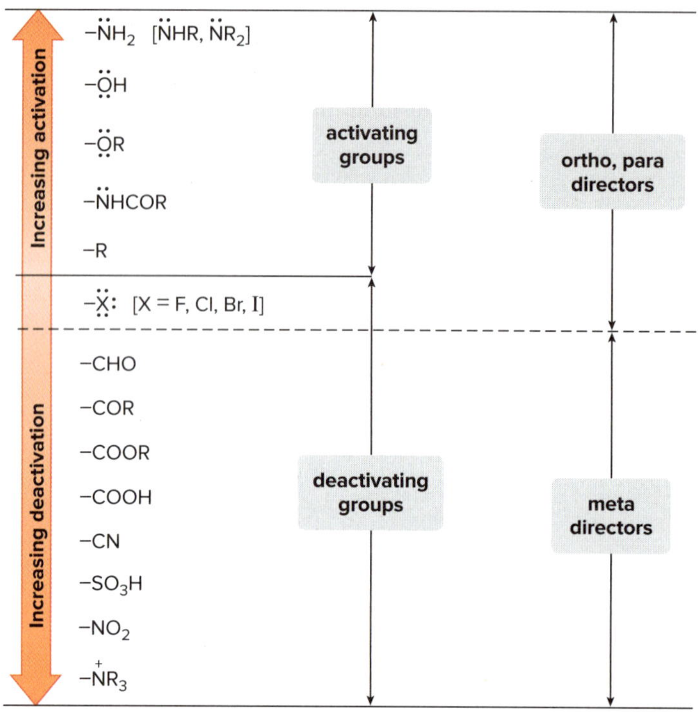

In summary:

[1] All ortho, para directors except the halogens activate the benzene ring.

[2] All meta directors deactivate the benzene ring.

[3] The halogens deactivate the benzene ring.

16.10 Limitations on Electrophilic Substitution Reactions with Substituted Benzenes

Although electrophilic aromatic substitution works well with most substituted benzenes, halogenation and the Friedel–Crafts reactions have some additional limitations that must be kept in mind.

16.10A Halogenation of Activated Benzenes

Considering all electrophilic aromatic substitution reactions, halogenation occurs the most readily. As a result, benzene rings activated by strong electron-donating groups—OH, NH_2, and their alkyl derivatives (OR, NHR, and NR_2)—undergo **polyhalogenation** when treated with X_2 and FeX_3. Aniline ($C_6H_5NH_2$) and phenol (C_6H_5OH) both give a tribromo derivative when treated with Br_2 and $FeBr_3$. **Substitution occurs at all hydrogen atoms ortho and para to the NH_2 and OH groups.**

phenol $\xrightarrow[FeBr_3]{Br_2}$ 2,4,6-tribromophenol

Monosubstitution of H by Br occurs with Br_2 *alone* without added catalyst to form a mixture of ortho and para products.

phenol $\xrightarrow{Br_2}$ o-bromophenol + p-bromophenol

Problem 16.26 Draw the products of each reaction.

a. phenol $\xrightarrow[FeCl_3]{Cl_2}$ b. phenol $\xrightarrow{Cl_2}$ c. toluene $\xrightarrow[FeCl_3]{Cl_2}$

16.10B Limitations in Friedel–Crafts Reactions

Friedel–Crafts reactions are the most difficult electrophilic aromatic substitution reactions to carry out in the laboratory. **They do not occur when the benzene ring is substituted with NO_2 (or any meta deactivator) or with NH_2, NHR, or NR_2 (strong activators).**

A benzene ring deactivated by a strong electron-withdrawing group—that is, any of the **meta directors**—is not electron rich enough to undergo Friedel–Crafts reactions.

$C_6H_5NO_2$ (strong deactivator) $\xrightarrow[AlCl_3]{RCl}$ **No reaction**

Friedel–Crafts reactions also do not occur with NH_2 groups, which are strong activating groups. **NH_2 groups are strong Lewis bases** (due to the nonbonded electron pair on N), so they react with $AlCl_3$, the Lewis acid needed for alkylation or acylation. The resulting product contains a positive charge adjacent to the benzene ring, so the **ring is now strongly deactivated** and therefore unreactive in Friedel–Crafts reactions.

$C_6H_5\ddot{N}H_2$ (Lewis base) + $AlCl_3$ (Lewis acid) $\longrightarrow$ $C_6H_5N^+H_2-AlCl_3^-$ (strong deactivator) $\xrightarrow[AlCl_3]{RCl}$ **No reaction**

Problem 16.27 Which of the following compounds undergo Friedel–Crafts alkylation with CH_3Cl and $AlCl_3$? Draw the products formed when a reaction occurs.

a. $C_6H_5-SO_3H$ b. C_6H_5-Cl c. $C_6H_5-N(CH_3)_2$ d. $C_6H_5-NH-C(=O)CH_3$

Another limitation of the Friedel–Crafts alkylation arises because of **polyalkylation.** Treatment of benzene with an alkyl halide and $AlCl_3$ places an electron-donor R group on the ring. Because R groups *activate* a ring, the alkylated product (C_6H_5R) is now *more reactive* than benzene itself toward further substitution, and it reacts again with RCl to give products of polyalkylation.

To minimize polyalkylation, a large excess of benzene is used relative to the amount of alkyl halide.

benzene $\xrightarrow[AlCl_3]{RCl}$ C_6H_5-R (electron donor) $\xrightarrow[AlCl_3]{RCl}$ o-$C_6H_4R_2$ + p-$C_6H_4R_2$ (polyalkylation products)

Polysubstitution does not occur with Friedel–Crafts acylation, because the product now has an electron-withdrawing group that deactivates the ring toward another electrophilic substitution.

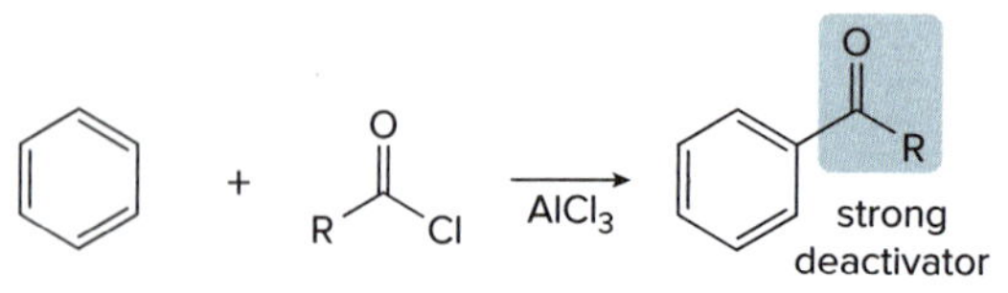

$AlCl_3$

strong deactivator

16.11 Disubstituted Benzenes

What happens in electrophilic aromatic substitution when a disubstituted benzene ring is used as starting material? **To predict the products, look at the directing effects of *both* substituents and then determine the net result,** using three guidelines.

Rule [1] **When the directing effects of two groups *reinforce*, the new substituent is located on the position directed by both groups.**

The CH_3 group in *p*-nitrotoluene is an ortho, para director and the NO_2 group is a meta director. These two effects reinforce each other so that one product is formed on treatment with Br_2 and $FeBr_3$. The position para to the CH_3 group is "blocked" by a nitro group, so no substitution can occur on that carbon.

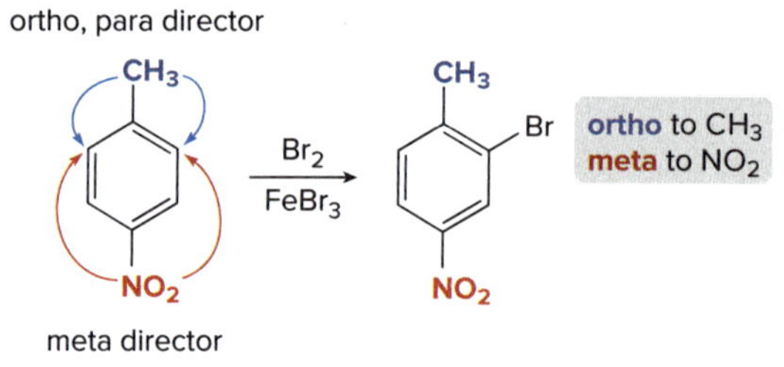

ortho, para director

Br_2 / $FeBr_3$

ortho to CH_3
meta to NO_2

meta director

p-nitrotoluene

Rule [2] **If the directing effects of two groups *oppose* each other, the more powerful activator "wins out."**

In compound **A,** the $NHCOCH_3$ group activates its two ortho positions, and the CH_3 group activates its two ortho positions to reaction with electrophiles. Because the $NHCOCH_3$ is a stronger activator, substitution occurs ortho to it.

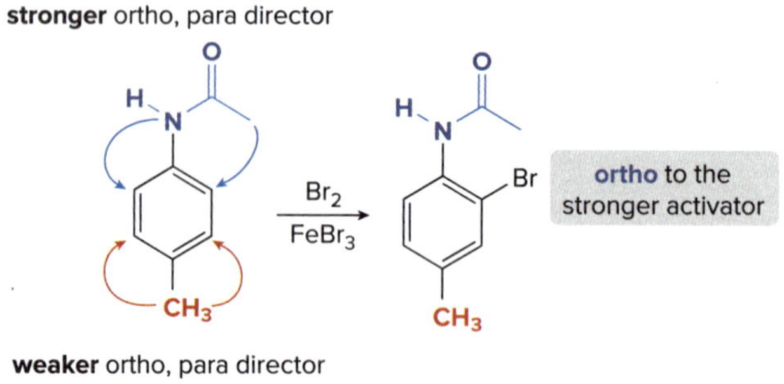

stronger ortho, para director

Br_2 / $FeBr_3$

ortho to the stronger activator

weaker ortho, para director

A

Rule [3] **No substitution occurs between two meta substituents because of crowding.**

For example, no substitution occurs at the carbon atom between the two CH_3 groups in *m*-xylene, even though two CH_3 groups activate that position.

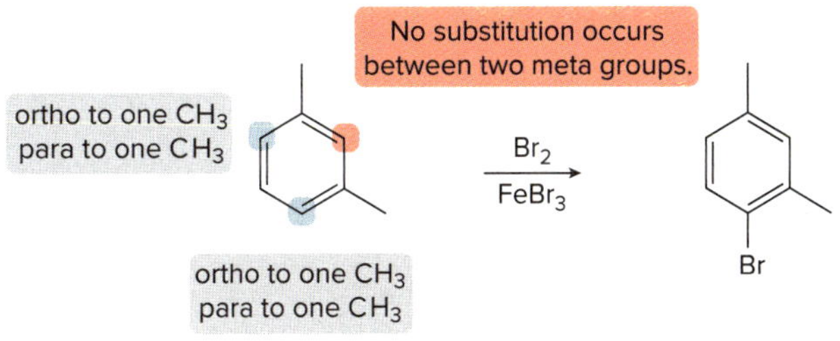

ortho to one CH_3
para to one CH_3

No substitution occurs between two meta groups.

$\xrightarrow[FeBr_3]{Br_2}$

Br

ortho to one CH_3
para to one CH_3

m-xylene
(1,3-dimethylbenzene)

Sample Problem 16.6 Drawing the Substitution Products from a Disubstituted Benzene

Draw the products formed from nitration of each compound.

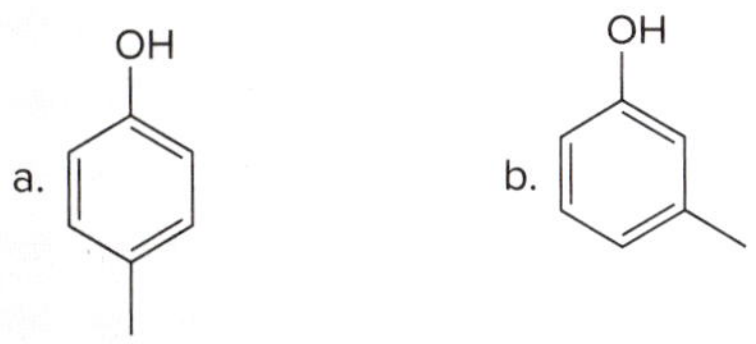

a. OH (p-cresol)   b. OH (m-cresol)

Solution

a. Both the OH and CH_3 groups are ortho, para directors. Because the **OH group is a stronger activator,** substitution occurs ortho to it.

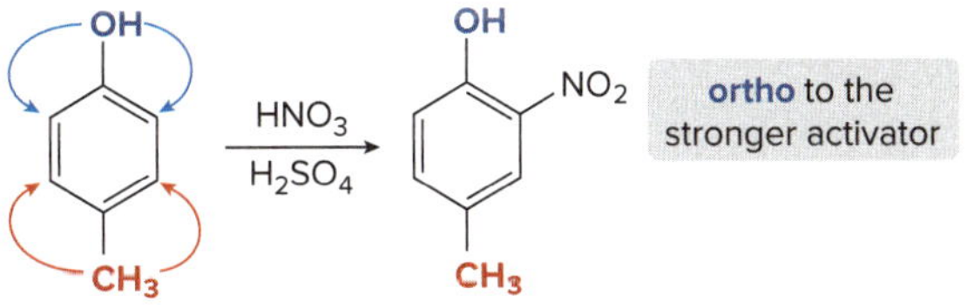

OH, CH_3 $\xrightarrow[H_2SO_4]{HNO_3}$ OH, NO_2, CH_3 — **ortho** to the stronger activator

b. Both the OH and CH_3 groups are ortho, para directors whose directing effects reinforce each other in this case. **No substitution occurs between the two meta substituents,** however, so two products result.

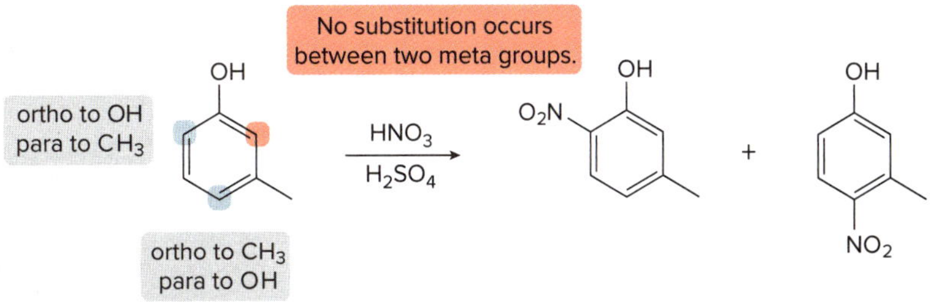

ortho to OH
para to CH_3

No substitution occurs between two meta groups.

OH $\xrightarrow[H_2SO_4]{HNO_3}$ OH, O_2N + OH, NO_2

ortho to CH_3
para to OH

Problem 16.28 Draw the products formed when each compound is treated with HNO_3 and H_2SO_4.

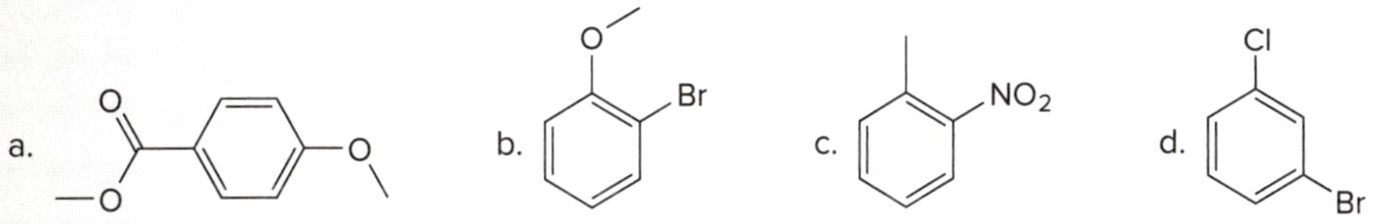

a. O, O, O   b. O, Br   c. NO_2 d. Cl, Br

More Practice: Try Problems 16.40, 16.43a–e, 16.45a–c.

16.12 Synthesis of Benzene Derivatives

To synthesize benzene derivatives with more than one substituent, we must always take into account the directing effects of each substituent. In a disubstituted benzene, **the directing effects indicate which substituent must be added to the ring first.**

For example, the Br group in *p*-bromonitrobenzene is an ortho, para director and the NO_2 group is a meta director. Because the two substituents are para to each other, the ortho, para director must be introduced *first* when synthesizing this compound from benzene.

Br ortho, para director

NO_2 meta director

p-bromonitrobenzene

With two para substituents, add the ortho, para director first.

Thus, Pathway [1], in which bromination precedes nitration, yields the **desired para product,** whereas Pathway [2], in which nitration precedes bromination, yields the **undesired meta isomer.**

Pathway [1] Bromination before nitration: The **desired para product** is formed.

ortho, para director

Br

Br_2 / $FeBr_3$

HNO_3 / H_2SO_4

Br, NO_2 — desired product

\+

Br, NO_2 — The ortho isomer can be separated from the mixture.

Pathway [2] Nitration before bromination: The **undesired meta isomer** is formed.

HNO_3 / H_2SO_4

NO_2 — meta director

Br_2 / $FeBr_3$

Br, NO_2 — meta isomer

Pathway [1] yields both the desired para product and the undesired ortho isomer. Because these compounds are constitutional isomers, they are separable. Obtaining such a mixture of ortho and para isomers is often unavoidable.

Sample Problem 16.7 Synthesizing a Disubstituted Benzene

Devise a synthesis of *o*-nitrotoluene from benzene.

NO_2

o-nitrotoluene ⟹ benzene

Solution

The CH_3 group in *o*-nitrotoluene is an ortho, para director and the NO_2 group is a meta director. Because the two substituents are ortho to each other, the **ortho, para director must be introduced first.** The synthesis thus involves two steps: Friedel–Crafts alkylation followed by nitration.

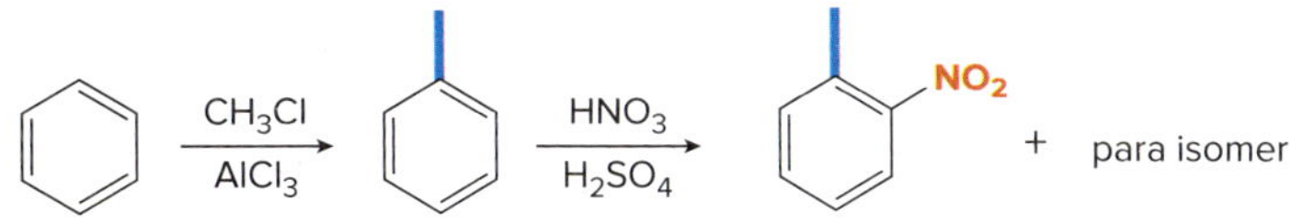

CH_3Cl / $AlCl_3$ → HNO_3 / H_2SO_4 → NO_2 + para isomer

o-nitrotoluene

Problem 16.29 Devise a synthesis of each compound from the indicated starting material.

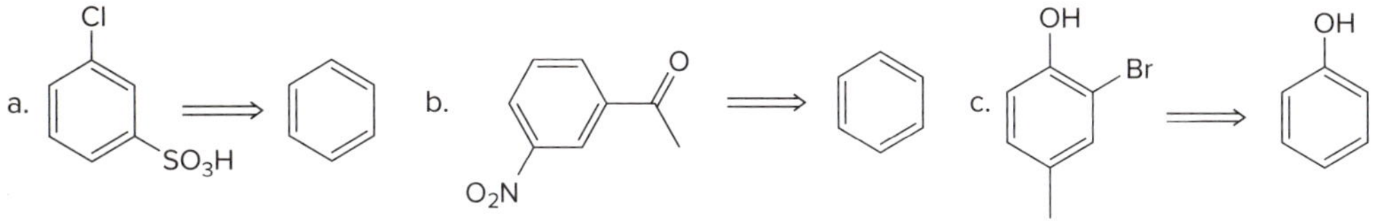

a. Cl, SO_3H ⟹ benzene b. O, O_2N ⟹ benzene c. OH, Br ⟹ OH

More Practice: Try Problems 16.41[1], 16.45a.

16.13 Nucleophilic Aromatic Substitution

Although most reactions of aromatic compounds occur by way of electrophilic aromatic substitution, **aryl halides undergo a limited number of substitution reactions with strong nucleophiles.**

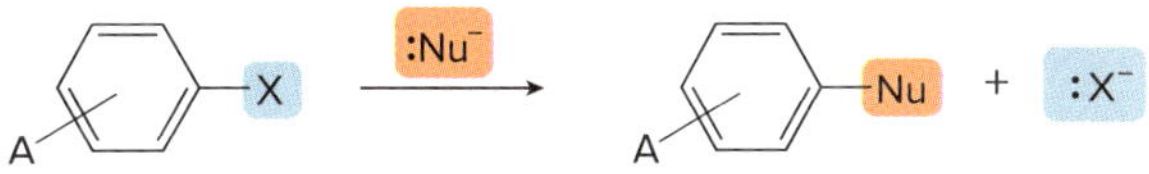

A, X —:Nu^-→ A, Nu + :X^-

X = F, Cl, Br, I
A = H or electron-withdrawing group

- **Nucleophilic aromatic substitution results in the substitution of a halogen X on a benzene ring by a nucleophile (:Nu^-).**

As we learned in Section 7.16, these reactions *cannot* occur by an S_N1 or S_N2 mechanism, which take place only at sp^3 hybridized carbons. Instead, two different mechanisms are proposed to explain the results: **addition–elimination** (Section 16.13A) and **elimination–addition** (Section 16.13B).

16.13A Nucleophilic Aromatic Substitution by Addition–Elimination

Aryl halides with strong electron-withdrawing groups (such as NO_2) on the ortho or para positions react with nucleophiles to afford substitution products. Treatment of *p*-chloronitrobenzene with hydroxide (^-OH) affords *p*-nitrophenol by replacement of Cl by OH.

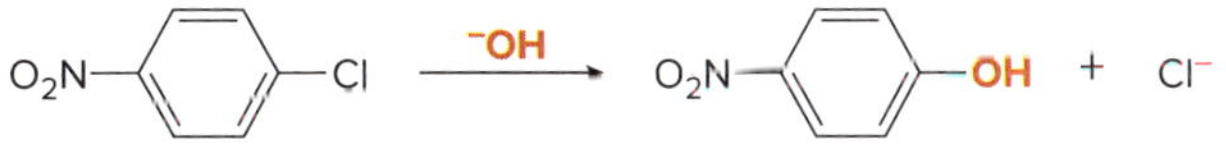

O_2N–C₆H₄–Cl —^-OH→ O_2N–C₆H₄–**OH** + Cl^-

p-chloronitrobenzene *p*-nitrophenol

Nucleophilic aromatic substitution occurs with a variety of strong nucleophiles, including ^-OH, ^-OR, $^-NH_2$, ^-SR, and in some cases, neutral nucleophiles such as NH_3 and RNH_2. The mechanism of these reactions has two steps: **addition of the nucleophile** to form a resonance-stabilized carbanion, followed by **elimination of the halogen leaving group.** Mechanism 16.10 is drawn with an aryl chloride containing a general electron-withdrawing group W.

Mechanism 16.10 Nucleophilic Aromatic Substitution by Addition–Elimination

1 Addition of the nucleophile forms a **resonance-stabilized carbanion** and a new C–Nu bond in the rate-determining step.

2 Loss of the leaving group re-forms the aromatic ring.

In nucleophilic aromatic substitution, the following trends in reactivity are observed.

- **Increasing the number of electron-withdrawing groups *increases* the reactivity of the aryl halide. Electron-withdrawing groups stabilize the intermediate carbanion and, by the Hammond postulate, lower the energy of the transition state that forms it.**
- **Increasing the electronegativity of the halogen *increases* the reactivity of the aryl halide. A more electronegative halogen stabilizes the intermediate carbanion by an inductive effect, making aryl fluorides (ArF) much *more* reactive than other aryl halides, which contain less electronegative halogens.**

Thus, aryl chloride **B** is more reactive than *o*-chloronitrobenzene (**A**) because it contains *two* electron-withdrawing NO_2 groups. Aryl fluoride **C** is more reactive than **B** because **C** contains the *more electronegative* halogen, fluorine.

The location of the electron-withdrawing group greatly affects the rate of nucleophilic aromatic substitution. When a nitro group is located ortho or para to the halogen, the negative charge of the intermediate carbanion can be delocalized onto the NO_2 group, thus stabilizing it. With a meta NO_2 group, no such additional delocalization onto the NO_2 group occurs.

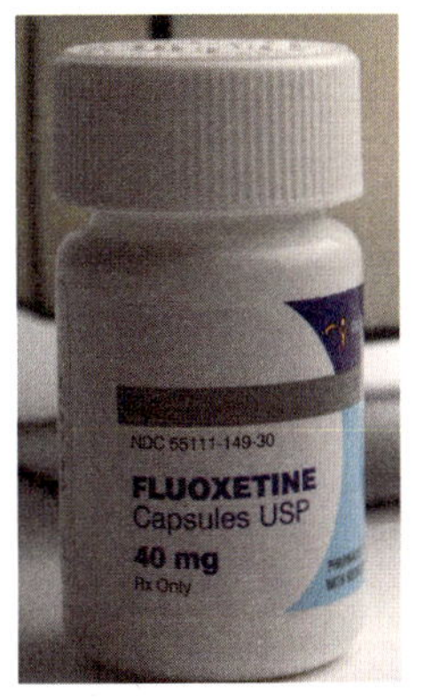

One method to synthesize the antidepressant fluoxetine involves nucleophilic aromatic substitution (Problem 16.31).
Jill Braaten/McGraw Hill

Thus, **nucleophilic aromatic substitution by an addition–elimination mechanism occurs only with aryl halides that contain electron-withdrawing substituents at the ortho or para position.**

Problem 16.30 Draw the products of each reaction.

a. 2,4-dinitrochlorobenzene (O_2N, NO_2, Cl) $\xrightarrow{NaOCH_3}$

b. 4-fluoroacetophenone (F, O) $\xrightarrow{^{-}OH}$

Problem 16.31 Draw a stepwise mechanism for the following reaction that forms ether **D**. **D** can be converted to the antidepressant fluoxetine (trade name Prozac) in a single step.

F_3C-C$_6$H$_4$-Cl + HO-CH(Ph)CH$_2$CH$_2$N(CH$_3$)$_2$ $\xrightarrow{KOC(CH_3)_3}$ **D** [F_3C-C$_6$H$_4$-O-CH(Ph)CH$_2$CH$_2$N(CH$_3$)$_2$] $\xrightarrow{\text{one step}}$ fluoxetine [F_3C-C$_6$H$_4$-O-CH(Ph)CH$_2$CH$_2$NHCH$_3$]

16.13B Nucleophilic Aromatic Substitution by Elimination–Addition: Benzyne

Aryl halides that do not contain an electron-withdrawing group generally do *not* react with nucleophiles. **Under extreme reaction conditions, however, nucleophilic aromatic substitution can occur with aryl halides.** For example, heating chlorobenzene with NaOH above 300 °C and 170 atmospheres of pressure affords phenol.

chlorobenzene (C$_6$H$_5$–Cl) $\xrightarrow[\text{[2] } H_3O^+]{\text{[1] NaOH, 300 °C, 170 atm}}$ phenol (C$_6$H$_5$–**OH**) + NaCl

The mechanism proposed to explain this result involves formation of a **benzyne** intermediate (C_6H_4) by **elimination–addition.** As shown in Mechanism 16.11, **benzyne is a highly reactive, unstable intermediate formed by elimination of HX from an aryl halide.**

Mechanism 16.11 Nucleophilic Aromatic Substitution by Elimination–Addition: Benzyne

Chlorobenzene + :ÖH$^-$ —[1]→ aryl carbanion (Cl) + $H_2\ddot{O}$: —[2]→ **benzyne** + :Cl:$^-$; benzyne + :Nu$^-$ —[3]→ aryl carbanion (Nu) —[4] H–ÖH→ substituted benzene (Nu, H) + :ÖH$^-$

1 – 2 Elimination of H and X from two adjacent atoms forms a reactive **benzyne** intermediate.

3 – 4 Nucleophilic attack and protonation form the substitution product.

Formation of a benzyne intermediate explains why substituted aryl halides form **mixtures** of products. **Nucleophilic aromatic substitution by an elimination–addition mechanism affords substitution on the carbon bonded directly to the leaving group and the carbon**

adjacent to it. As an example, treatment of *p*-chlorotoluene with $NaNH_2$ forms para- and meta-substitution products.

$NaNH_2$ / NH_3

p-chlorotoluene → *p*-methylaniline (*p*-toluidine) + *m*-methylaniline (*m*-toluidine)

This result is explained by the fact that nucleophilic attack on the benzyne intermediate may occur at either C3 to form *m*-methylaniline, or C4 to form *p*-methylaniline.

p-chlorotoluene → ($NaNH_2$, two steps) → benzyne + $:\ddot{N}H_2^-$ → (nucleophilic attack at C3) → (NH_3) → *m*-methylaniline

or

benzyne + $:\ddot{N}H_2^-$ → (nucleophilic attack at C4) → (NH_3) → *p*-methylaniline

As you might expect, the triple bond in benzyne is unusual. Each carbon of the six-membered ring is sp^2 hybridized, and as a result, the σ bond and two π bonds of the triple bond are formed with the following orbitals:

- **The σ bond is formed by overlap of two sp^2 hybrid orbitals.**
- **One π bond is formed by overlap of two *p* orbitals perpendicular to the plane of the molecule.**
- **The second π bond is formed by overlap of two sp^2 hybrid orbitals.**

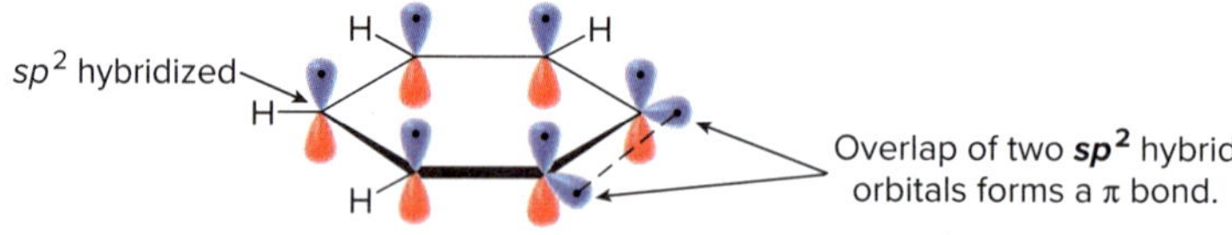

Thus, the second π bond of benzyne differs from all other π bonds seen thus far, because **it is formed by the side-by-side overlap of sp^2 *hybrid orbitals,* not *p* orbitals.** This π bond, located in the plane of the molecule, is extremely weak.

Problem 16.32 Draw the products of each reaction.

a. chlorobenzene $\xrightarrow[NH_3]{NaNH_2}$

b. CH_3O–C_6H_4–Cl $\xrightarrow[H_2O,\ \Delta]{NaOH}$

c. 2-ethylchlorobenzene $\xrightarrow[NH_3]{KNH_2}$

Problem 16.33 Draw all products formed when *m*-chlorotoluene is treated with KNH_2 in NH_3.

16.14 Reactions of Substituted Benzenes

We finish Chapter 16 by learning some additional reactions of substituted benzenes that greatly expand the ability to synthesize benzene derivatives. In Section 16.14A we return to radical halogenation, and in Sections 16.14B–D, we examine useful oxidation and reduction reactions. Only reagents and reactions are presented, without reference to detailed mechanisms.

16.14A Halogenation of Alkyl Benzenes

With proper choice of reaction conditions, an **alkyl benzene undergoes selective bromination at the benzylic C–H bond** under radical conditions to form a **benzylic halide.** For example, radical bromination of ethylbenzene using either Br_2 (in the presence of light or heat) or *N*-bromosuccinimide (NBS, in the presence of light or peroxides) forms a benzylic bromide as the sole product. The mechanism for halogenation at the benzylic position resembles other radical halogenation reactions discussed in Chapter 13.

ethylbenzene —[Br_2, $h\nu$ or Δ; or NBS, $h\nu$ or ROOR]→ a benzylic bromide + HBr

radical conditions

Thus, an alkyl benzene undergoes two different reactions with Br_2, depending on the reaction conditions.

Br_2, $FeBr_3$ → ortho isomer + para isomer — **Ionic conditions**

Br_2, $h\nu$ or Δ → (benzylic Br product) — **Radical conditions**

- **With Br_2 and $FeBr_3$ (ionic conditions), electrophilic aromatic substitution occurs,** resulting in replacement of H by Br on the aromatic ring to form ortho and para isomers.
- **With Br_2 and light or heat (radical conditions), substitution of H by Br occurs** at the *benzylic* carbon of the alkyl group.

The radical bromination of alkyl benzenes is a useful reaction because the resulting benzylic halide can serve as starting material for a variety of substitution and elimination reactions, thus making it possible to form many new substituted benzenes. Sample Problem 16.8 illustrates one possibility.

Sample Problem 16.8 Using Benzylic Bromination to Introduce a Double Bond

Design a synthesis of styrene from ethylbenzene.

styrene ⟹ ethylbenzene

Solution

The double bond can be introduced by a two-step reaction sequence: **bromination** at the benzylic position under radical conditions, followed by **elimination of HBr** with strong base to form the π bond.

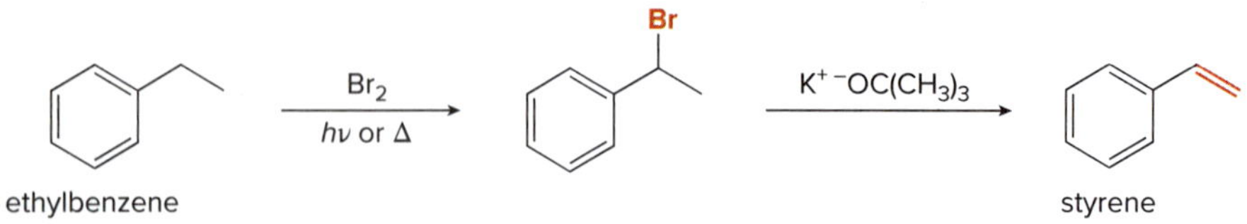

ethylbenzene —[Br_2, $h\nu$ or Δ]→ (Br) —[$K^+ \ ^-OC(CH_3)_3$]→ styrene

Problem 16.34 How could you use ethylbenzene to prepare each compound? More than one step is required.

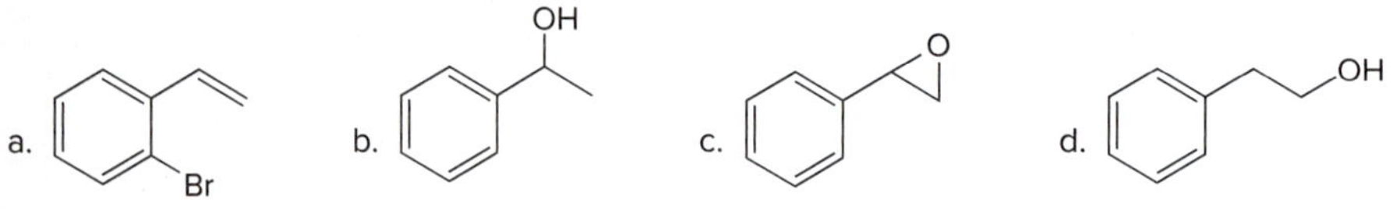

a. (Br) b. (OH) c. (O) d. (OH)

More Practice: Try Problems 16.68a, e, f; 16.69a, c; 16.70.

16.14B Oxidation of Alkyl Benzenes

Arenes containing at least one benzylic C–H bond are oxidized with $KMnO_4$ to benzoic acid, a carboxylic acid with the carboxy group (COOH) bonded directly to the benzene ring. With some alkyl benzenes, this also results in the cleavage of carbon–carbon bonds, so the product has fewer carbon atoms than the starting material.

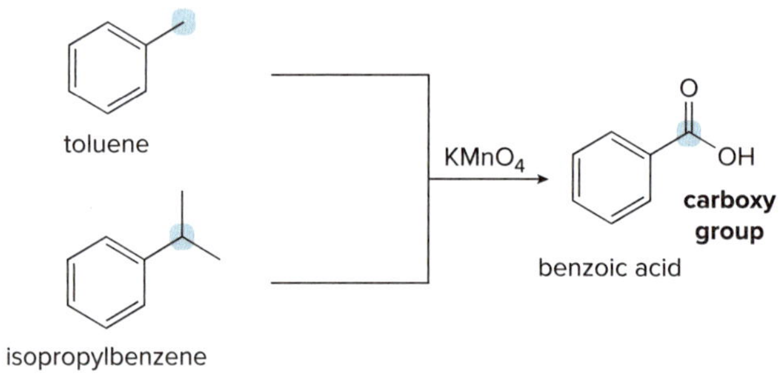

toluene, isopropylbenzene —[$KMnO_4$]→ benzoic acid (O, OH, **carboxy group**)

Substrates with more than one alkyl group are oxidized to dicarboxylic acids. **Compounds without a benzylic C–H bond are inert to oxidation.**

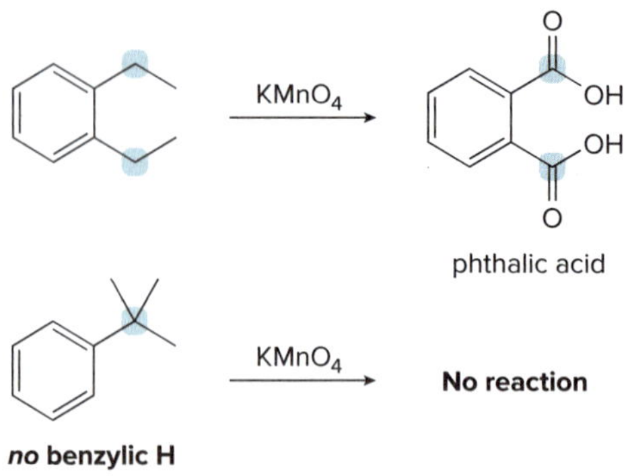

—[$KMnO_4$]→ phthalic acid (O, OH, OH, O)

***no* benzylic H** —[$KMnO_4$]→ **No reaction**

16.14C Reduction of Aryl Ketones to Alkyl Benzenes

Ketones formed as products in Friedel–Crafts acylation can be reduced to alkyl benzenes by two different methods.

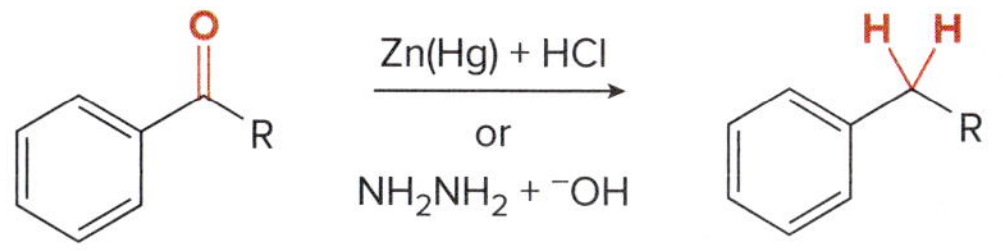

Zn(Hg) + HCl
or
NH_2NH_2 + ^{-}OH

- The **Clemmensen reduction** uses zinc and mercury in the presence of strong acid.
- The **Wolff–Kishner reduction** uses hydrazine (NH_2NH_2) and strong base (KOH).

Because both C–O bonds in the starting material are converted to C–H bonds in the product, the reduction is difficult and the reaction conditions must be harsh.

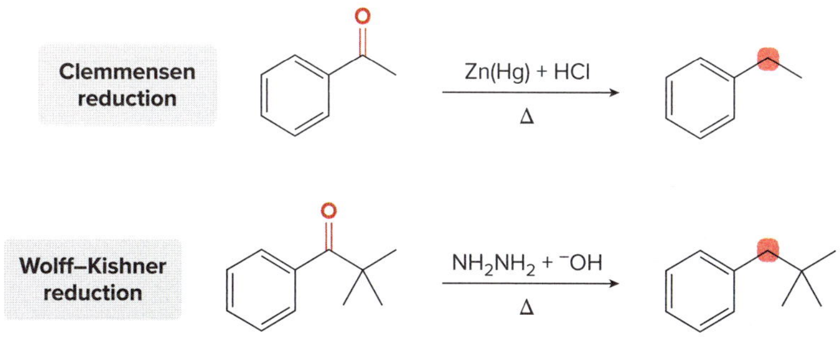

Clemmensen reduction — Zn(Hg) + HCl, Δ

Wolff–Kishner reduction — NH_2NH_2 + ^{-}OH, Δ

We now know two different ways to introduce an alkyl group on a benzene ring (Figure 16.9):

- **A one-step method using Friedel–Crafts alkylation**
- **A two-step method using Friedel–Crafts acylation to form a ketone, followed by reduction**

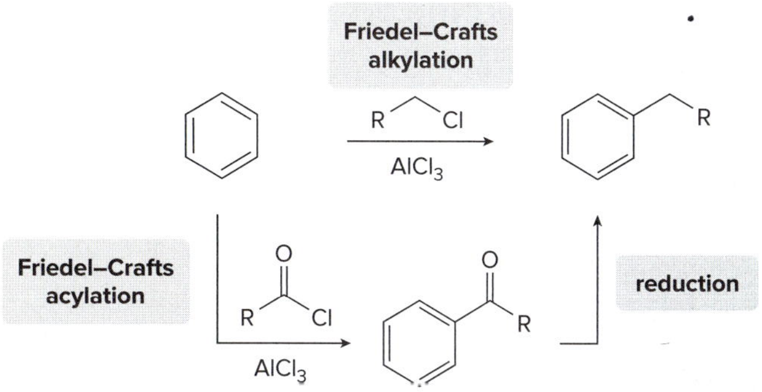

Figure 16.9 Two methods to prepare an alkyl benzene

Friedel–Crafts alkylation: RCH₂Cl, $AlCl_3$

Friedel–Crafts acylation: RCOCl, $AlCl_3$

reduction

Although the two-step method seems more roundabout, it must be used to synthesize certain alkyl benzenes that cannot be prepared by the one-step Friedel–Crafts alkylation because of rearrangements.

Recall from Section 16.5C that propylbenzene cannot be prepared by a Friedel–Crafts alkylation. Instead, when benzene is treated with 1-chloropropane and $AlCl_3$, isopropylbenzene is formed by a rearrangement reaction. Propylbenzene can be made, however, by a two-step procedure using Friedel–Crafts acylation followed by reduction.

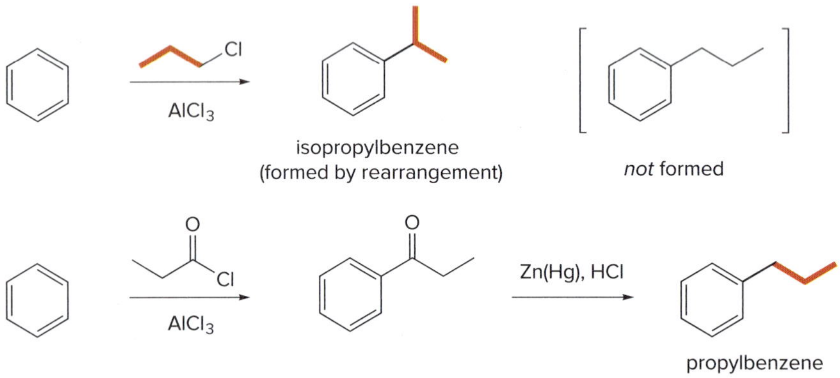

Cl
$AlCl_3$
isopropylbenzene
(formed by rearrangement)
not formed

O
Cl
$AlCl_3$
O
Zn(Hg), HCl
propylbenzene

Problem 16.35 Write out the two-step sequence that converts benzene to each compound.

a. 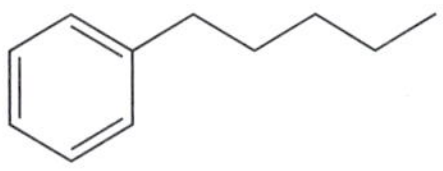 b.

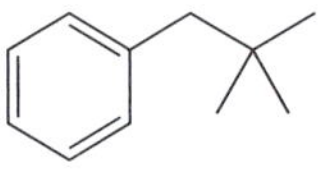

Problem 16.36 What steps are needed to convert benzene to *p*-isobutylacetophenone, a synthetic intermediate used in the synthesis of the anti-inflammatory agent ibuprofen.

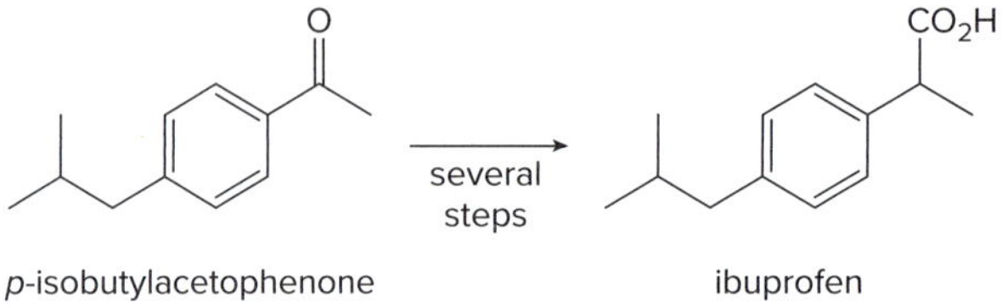

O
several steps
CO_2H
p-isobutylacetophenone
ibuprofen

16.14D Reduction of Nitro Groups

A nitro group (NO_2) is easily introduced on a benzene ring by nitration with strong acid (Section 16.4). This process is useful because the **nitro group is readily reduced to an amino group (NH_2)** under a variety of conditions. The most common methods use H_2 and a catalyst, or a metal (such as Fe or Sn) and a strong acid like HCl.

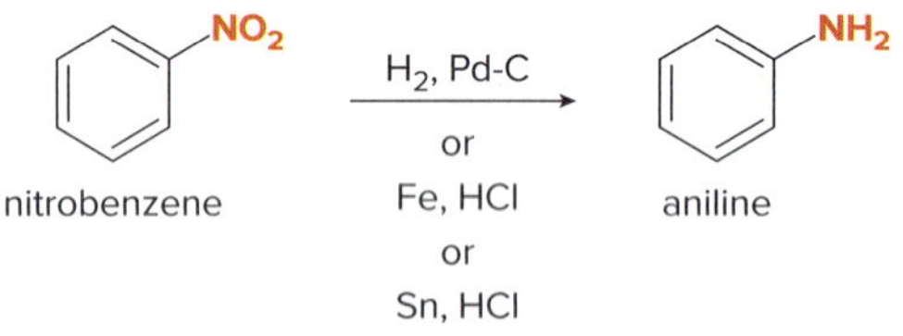

NO_2
nitrobenzene
H_2, Pd-C
or
Fe, HCl
or
Sn, HCl
NH_2
aniline

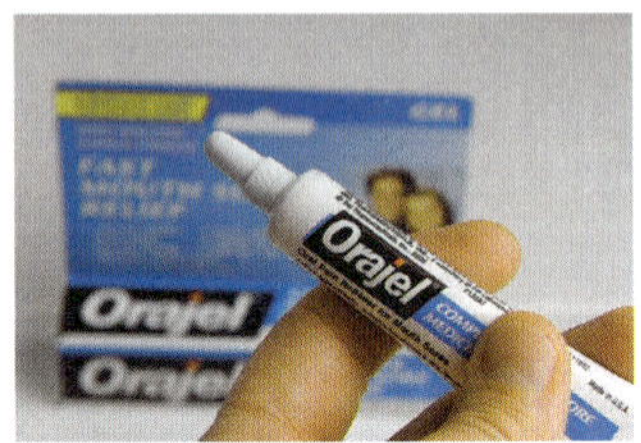

Benzocaine is the active ingredient in the over-the-counter topical anesthetic Orajel. *Jill Braaten/McGraw Hill*

For example, reduction of ethyl *p*-nitrobenzoate with H_2 and a palladium catalyst forms ethyl *p*-aminobenzoate, a local anesthetic commonly called benzocaine.

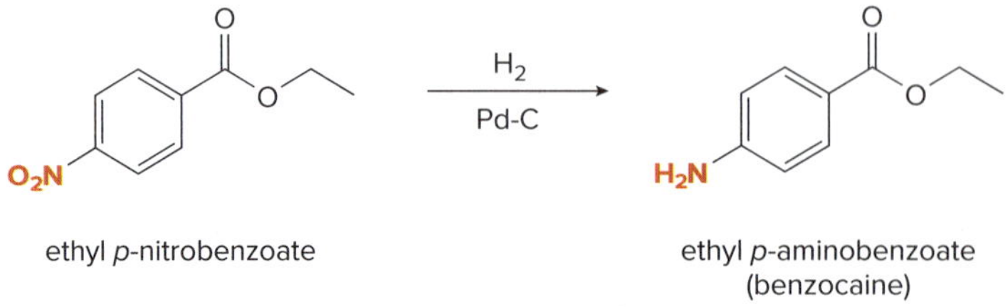

O
O
O_2N
ethyl *p*-nitrobenzoate
H_2
Pd-C
O
O
H_2N
ethyl *p*-aminobenzoate
(benzocaine)

Sample Problem 16.9 illustrates the utility of this process in a short synthesis.

Sample Problem 16.9 Introducing an NH_2 Group in a Synthesis

Design a synthesis of *m*-bromoaniline from benzene.

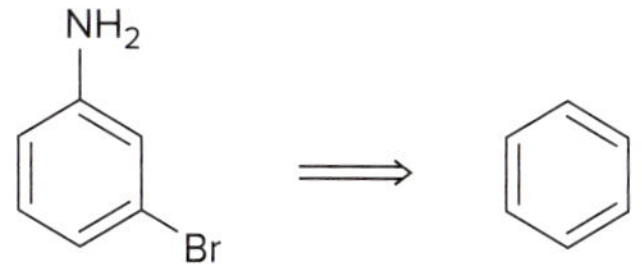

m-bromoaniline

Solution

To devise a retrosynthetic plan, keep in mind:

- The NH_2 group cannot be introduced directly on the ring by electrophilic aromatic substitution. It must be added by a two-step process: **nitration followed by reduction.**
- Both the Br and NH_2 groups are ortho, para directors, but they are located meta to each other on the ring. However, an **NO_2 group (from which an NH_2 group is made) *is* a meta director,** and we can use this fact to our advantage.

Retrosynthetic Analysis

Working backwards gives the following **three-step retrosynthetic analysis:**

reduction [1] nitration [2] [3] bromination

m-bromoaniline

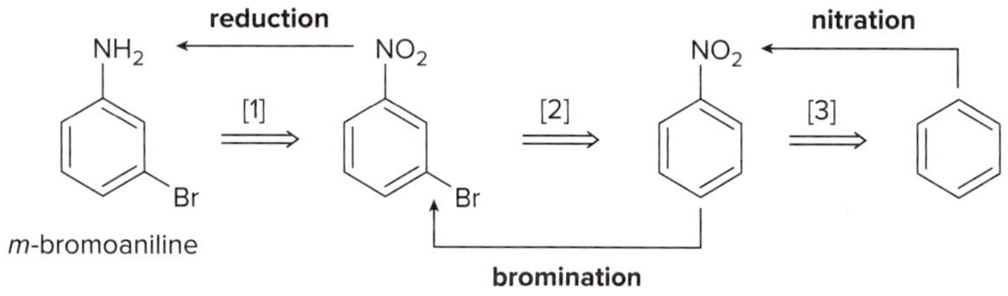

- [1] Form the NH_2 group by reduction of NO_2.
- [2] Introduce the Br group meta to the NO_2 group by halogenation.
- [3] Add the NO_2 group by nitration.

Synthesis

The synthesis involves three steps, and the order is crucial for success. Halogenation (Step [2] of the synthesis) must occur *before* reduction (Step [3]) in order to form the meta-substitution product.

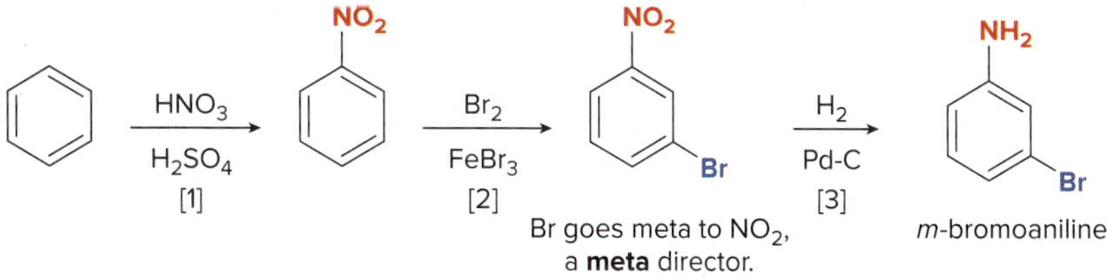

HNO_3, H_2SO_4 [1]; Br_2, $FeBr_3$ [2]; H_2, Pd-C [3]

Br goes meta to NO_2, a **meta** director.

m-bromoaniline

Problem 16.37 Synthesize each compound from benzene.

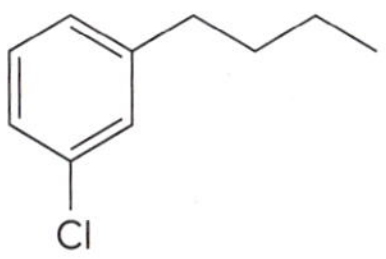

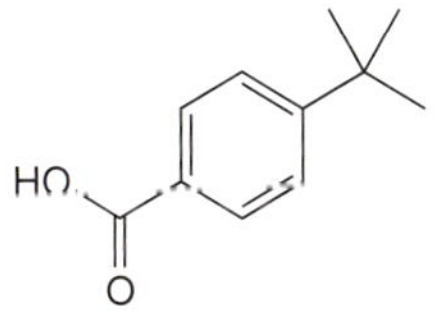

a. 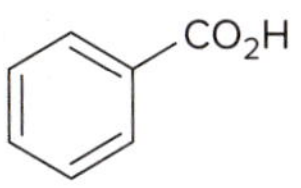 b. 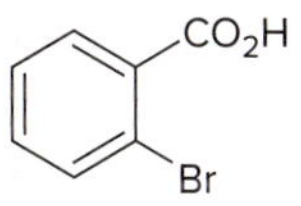 c. d. 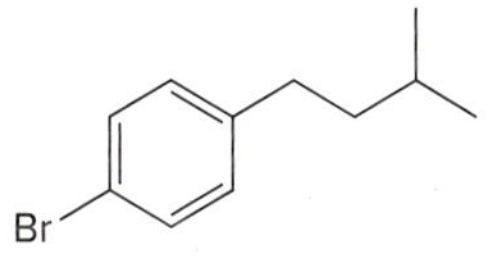 e. f.

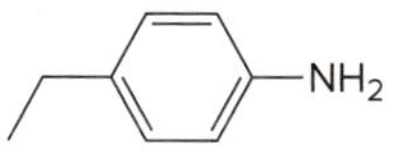

More Practice: Try Problems 16.68b–d, 16.69b.

16.15 Multistep Synthesis

The reactions learned in Chapter 16 make it possible to synthesize a wide variety of substituted benzenes, as shown in Sample Problems 16.10 and 16.11.

Sample Problem 16.10 Designing a Multistep Synthesis

Synthesize *p*-nitrobenzoic acid from benzene.

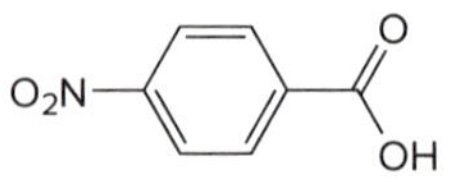

p-nitrobenzoic acid

Solution

Both groups on the ring (NO_2 and COOH) are meta directors. To place these two groups para to each other, remember that the **COOH group is prepared by oxidizing an alkyl group, which is an ortho, para director.**

Retrosynthetic Analysis

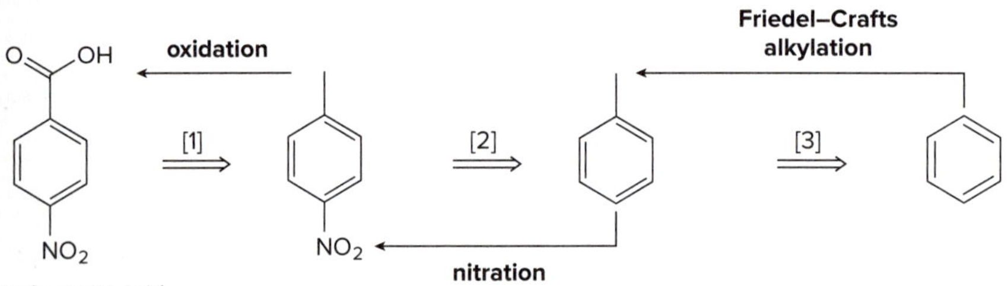

oxidation [1] ⟹ [2] ⟹ nitration [3] ⟹ Friedel–Crafts alkylation

p-nitrobenzoic acid

Working backwards:

- [1] Form the COOH group by oxidation of an alkyl group.
- [2] Introduce the NO_2 group para to the CH_3 group (an ortho, para director) by nitration.
- [3] Add the CH_3 group by Friedel–Crafts alkylation.

Synthesis

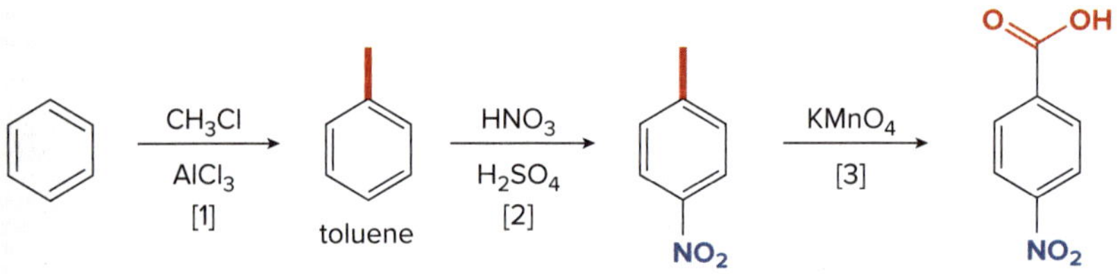

CH_3Cl, $AlCl_3$ [1] → toluene → HNO_3, H_2SO_4 [2] → NO_2 [+ ortho isomer] → $KMnO_4$ [3] → NO_2 *p*-nitrobenzoic acid

- **Friedel–Crafts alkylation** with CH_3Cl and $AlCl_3$ forms toluene in Step [1]. Because CH_3 is an ortho, para director, nitration yields the desired para product, which can be separated from its ortho isomer (Step [2]).
- **Oxidation with $KMnO_4$** converts the CH_3 group to a COOH group, giving the desired product in Step [3].

Problem 16.38 Synthesize each compound from benzene.

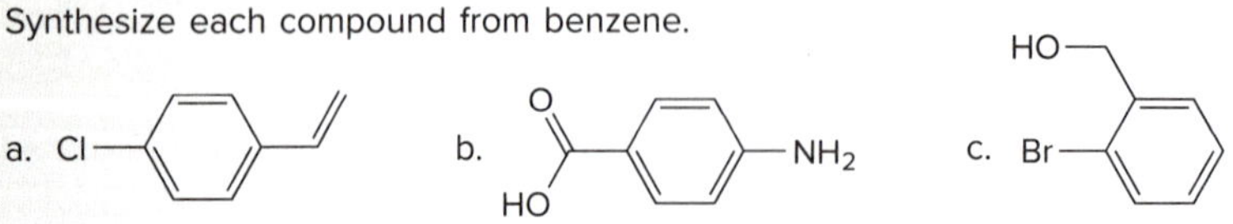

a. Cl–(para-substituted styrene)

b. HO–C(=O)–C$_6$H$_4$–NH_2
PABA
sunscreen component

c. Br, HO (ortho-substituted benzyl alcohol)

More Practice: Try Problem 16.71.

Sample Problem 16.11 Synthesizing a Trisubstituted Benzene

Synthesize the trisubstituted benzene **A** from benzene.

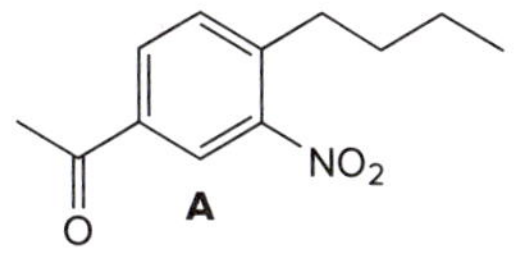

NO_2
O **A**

Solution

Two groups (CH_3CO and NO_2) in **A** are meta directors located meta to each other, and the third substituent, an alkyl group, is an ortho, para director.

Retrosynthetic Analysis

With three groups on the benzene ring, **begin by determining the possible disubstituted benzenes that are immediate precursors of the target compound,** and then eliminate any that cannot be converted to the desired product. For example, three different disubstituted benzenes (**B–D**) can theoretically be precursors to **A**. However, conversion of compounds **B** or **D** to **A** would require a Friedel–Crafts reaction on a deactivated benzene ring, a reaction that does not occur. Thus, only **C** is a feasible precursor of **A**.

Target compound

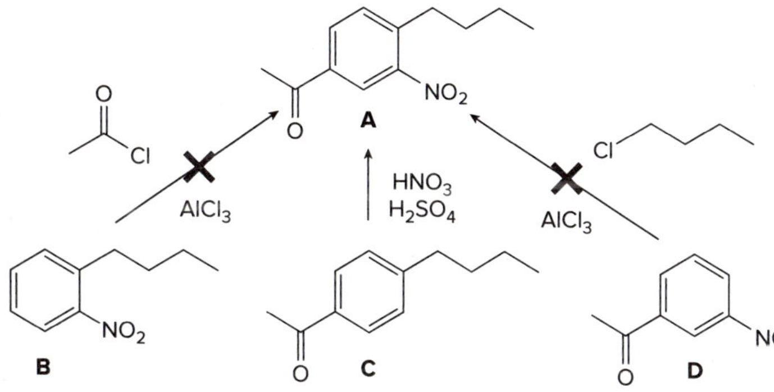

O Cl $AlCl_3$ | NO_2 O **A** | HNO_3 H_2SO_4 | Cl $AlCl_3$

NO_2 **B** | O **C** | O **D** NO_2

no Friedel–Crafts on a strongly deactivated benzene ring

Only this pathway works.

no Friedel–Crafts on a strongly deactivated benzene ring

To complete the retrosynthetic analysis, prepare **C** from benzene:

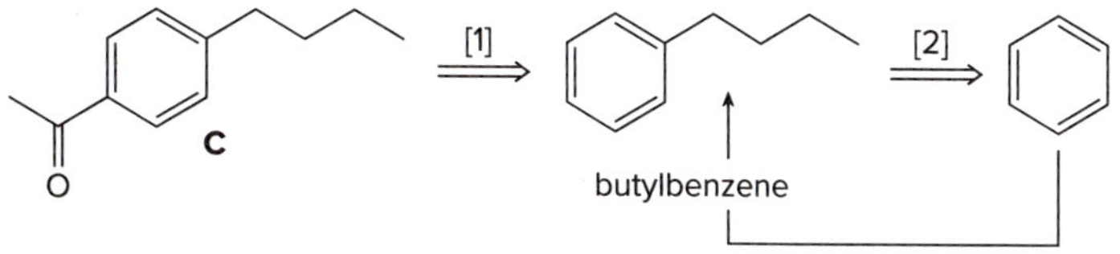

O **C** [1] ⟹ butylbenzene [2] ⟹

two steps

- [1] Add the ketone by Friedel–Crafts acylation.
- [2] Add the alkyl group by the two-step process—Friedel–Crafts acylation followed by reduction. It is not possible to prepare butylbenzene by a one-step Friedel–Crafts alkylation because of a rearrangement reaction (Section 16.14C).

Synthesis

benzene + Cl(C=O)CH₂CH₂CH₃ → $AlCl_3$ [1] → Zn(Hg), HCl [2] → butylbenzene → [3] CH_3COCl + $AlCl_3$ → **C** [+ ortho isomer] → HNO_3, H_2SO_4 [4] → **A** (NO_2)

- Friedel–Crafts acylation followed by reduction with Zn(Hg), HCl yields butylbenzene (Steps [1]–[2]).
- Friedel–Crafts acylation gives the para product **C,** which can be separated from its ortho isomer (Step [3]).
- Nitration in Step [4] introduces the NO_2 group ortho to the alkyl group (an ortho, para director) and meta to the CH_3CO group (a meta director).

Problem 16.39 Synthesize each compound from benzene.

a. (structure with SO_3H) b. (structure with Br) c. (structure with Cl and H)

More Practice: Try Problems 16.68–16.71.

Chapter 16 REVIEW

KEY CONCEPTS

[1] Three rules describing the reactivity and directing effects of common substituents (16.7–16.9)

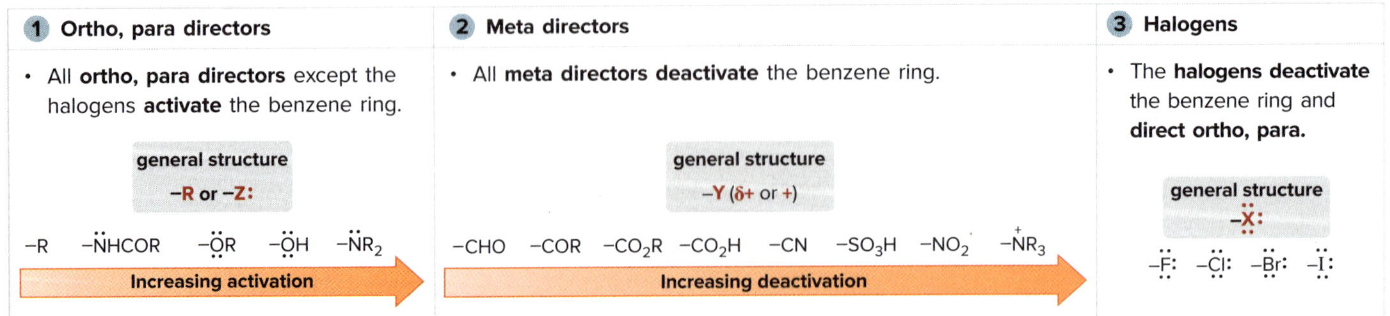

Try Problem 16.52.

[2] Summary of substituent effects in electrophilic aromatic substitution (16.6–16.9)

1 Substituent	2 Inductive effect	3 Resonance effect	4 Reactivity	5 Directing effect
R R = alkyl	donating	none	activating	ortho, para
Z: Z = N or O	withdrawing	donating	activating	ortho, para
X: X = halogen	withdrawing	donating	deactivating	ortho, para
Y (δ+ or +)	withdrawing	withdrawing	deactivating	meta

Try Problems 16.51, 16.52.

KEY REACTIONS

[1] Electrophilic aromatic substitution

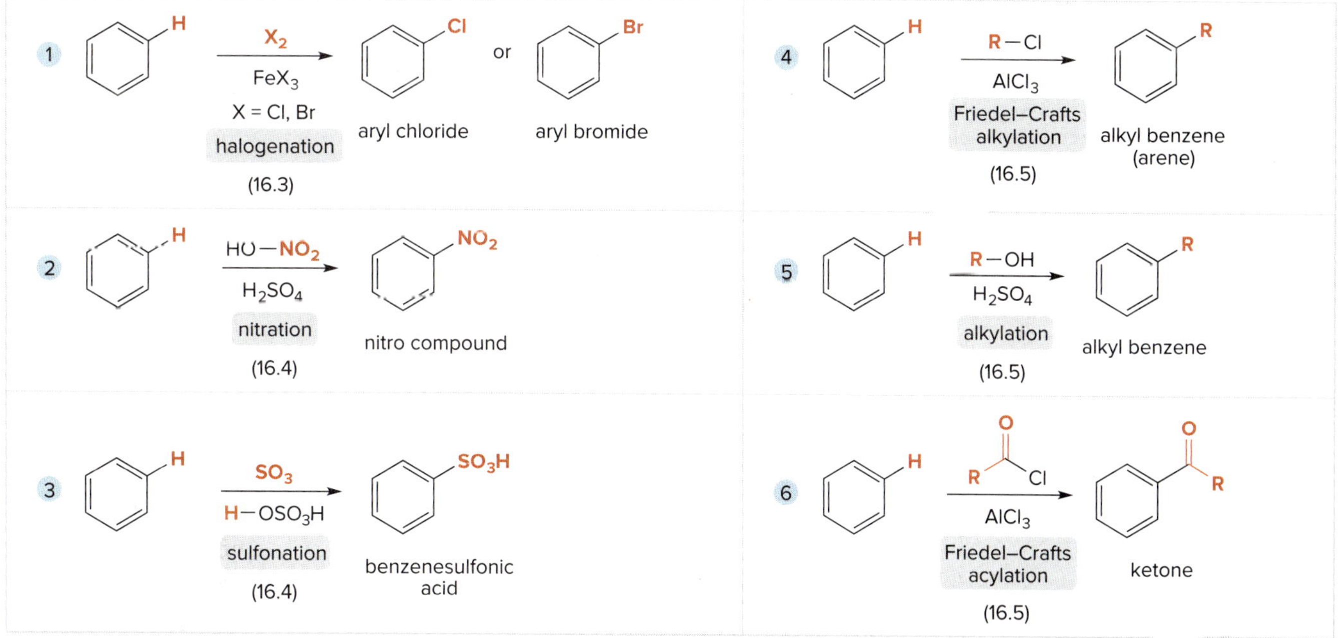

Try Problems 16.40, 16.43a–e.

[2] Nucleophilic aromatic substitution

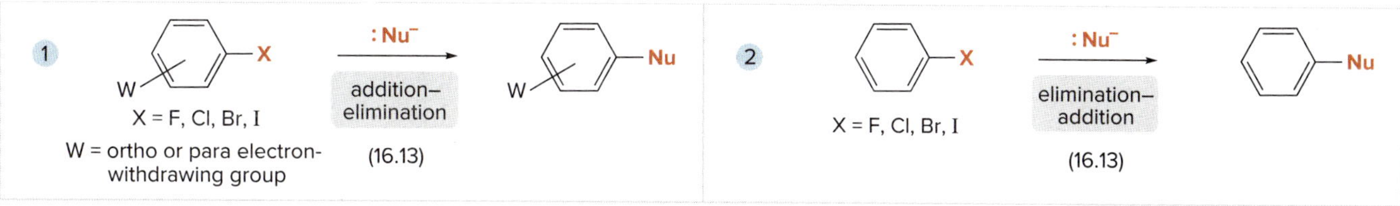

Try Problems 16.43f, 16.45d, 16.47.

[3] Other reactions of benzene derivatives

Reaction	Starting material	Reagents	Product
1	$C_6H_5CH(H)R$	Br_2, $h\nu$ or Δ or NBS, $h\nu$ or ROOR — **benzylic bromination** (16.14A)	$C_6H_5CH(Br)R$ — benzylic bromide
2	$C_6H_5CH_2R$	$KMnO_4$ — **oxidation** (16.14B)	C_6H_5COOH — benzoic acid
3	C_6H_5COR	Zn(Hg) + HCl or NH_2NH_2 + ^{-}OH — **reduction** (16.14C)	$C_6H_5CH_2R$ — alkyl benzene
4	$C_6H_5NO_2$	H_2, Pd-C or Fe, HCl or Sn, HCl — **reduction** (16.14D)	$C_6H_5NH_2$ — aniline

Try Problems 16.41, 16.45a–c.

KEY SKILLS

[1] Classifying substituents as electron donating or electron withdrawing (16.6); two considerations

Draw out the atoms, bonds, and electrons of the substituent, and look at the atom bonded directly to the benzene ring.

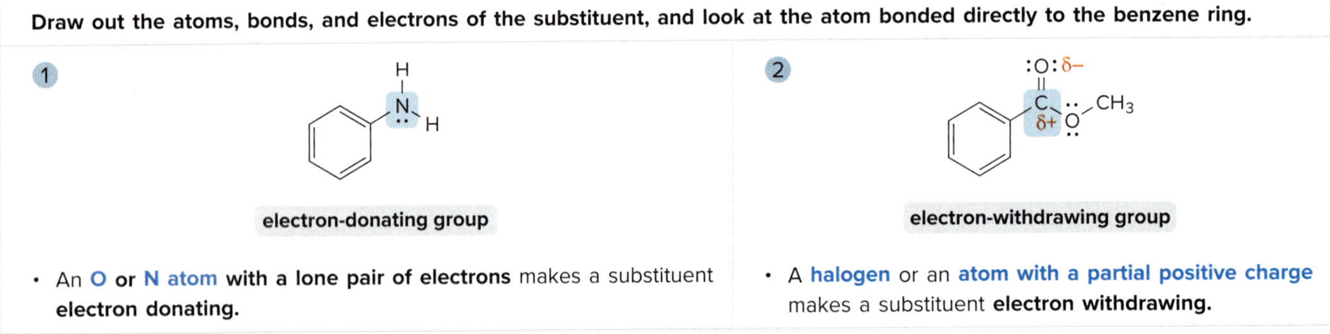

- An **O** or **N atom with a lone pair of electrons** makes a substituent **electron donating.**
- A **halogen** or an **atom with a partial positive charge** makes a substituent **electron withdrawing.**

See Sample Problem 16.3. Try Problems 16.52–16.55.

[2] Drawing the product(s) from reaction of a monosubstituted benzene with an electrophile (16.7)

1 Evaluate the directing effect of the substituent.

2 Classify the substituent, draw the products, and identify whether the reaction is faster or slower than a reaction with benzene.

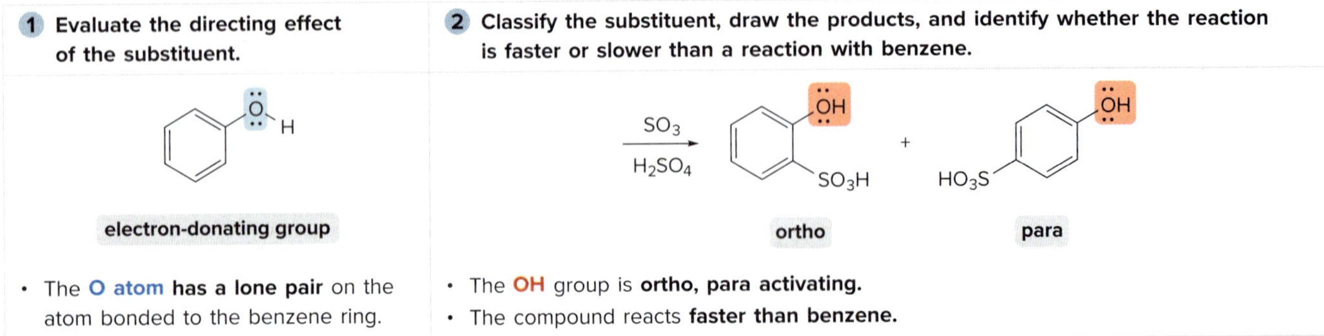

- The **O atom has a lone pair** on the atom bonded to the benzene ring.
- The **OH** group is **ortho, para activating.**
- The compound reacts **faster than benzene.**

See Sample Problem 16.4. Try Problems 16.41, 16.42, 16.45a.

[3] Drawing the product(s) from reaction of a disubstituted benzene with an electrophile (16.11)

1 Evaluate the directing effects, and classify the substituents.

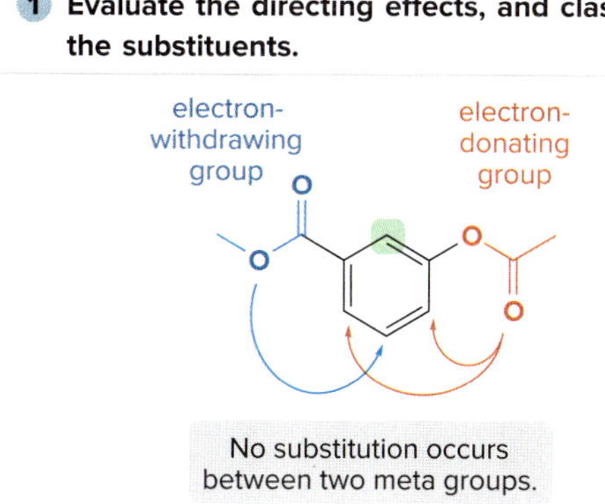

No substitution occurs between two meta groups.

- The $OCOCH_3$ group is an **ortho, para director.**
- The CO_2CH_3 group is a **meta director.**

2 Determine the net result.

Cl_2 / $FeCl_3$ → ortho to the activator + para to the activator

- If the directing effects of two groups oppose each other, **the more powerful activator "wins."**
- **No substitution occurs between two meta substituents** because of crowding.

See Sample Problem 16.6. Try Problems 16.40, 16.43a–e, 16.45c.

KEY MECHANISM CONCEPTS

1 Electrophilic aromatic substitution

(+) ortho to E ⟷ (+) para to E ⟷ (+) ortho to E

The intermediate **carbocation** is **stabilized by resonance.**

- See Mechanisms 16.1, 16.2, and 16.6.
- The mechanism has **two steps.**
- The **first step** is **rate-determining.**

3 Nucleophilic aromatic substitution by addition–elimination

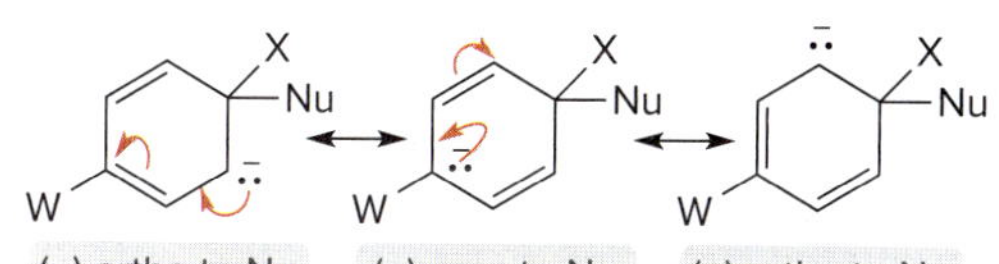

(–) ortho to Nu ⟷ (–) para to Nu ⟷ (–) ortho to Nu

The intermediate **carbanion** is **stabilized by resonance.**

- See Mechanism 16.10.
- The mechanism has **two steps.**
- **Strong electron-withdrawing groups (W)** at the **ortho** and **para** positions are required.
- The **rate is increased** by increasing the number of **electron-withdrawing groups** and increasing the **electronegativity of the halogen (X).**

2 Friedel–Crafts alkylation involving a rearrangement reaction

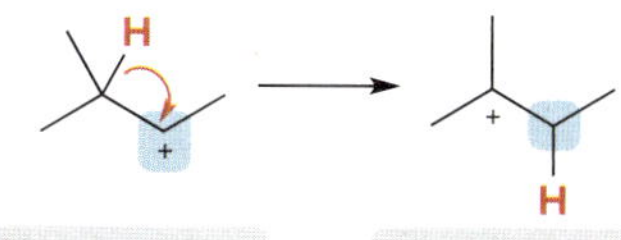

2° carbocation → 3° carbocation

- See Mechanism 16.8.

Cl–$AlCl_3$ →

No 1° carbocation is formed. → 2° carbocation

- See Mechanism 16.9.

4 Nucleophilic aromatic substitution by elimination–addition

benzyne + :Nu^- →

- See Mechanism 16.11.
- Reaction conditions are harsh.
- Product mixtures may result.

Try Problems 16.60–16.67.

CHAPTER 16 MULTIPLE-CHOICE SELF-TEST

The Self-Test consists of multiple-choice questions similar to those found on the American Chemical Society organic chemistry exam. Answers are given at the end of the chapter.

1. What product is formed when isopropylbenzene reacts with $KMnO_4$?

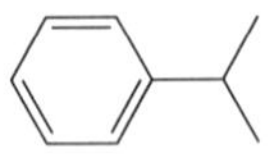

isopropylbenzene a. 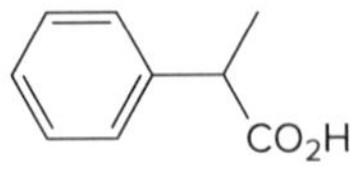 CO_2H b. 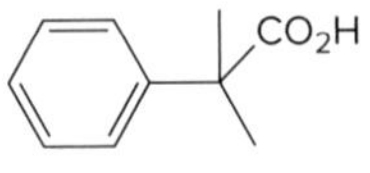 CO_2H c. 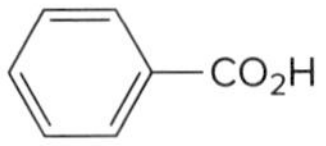 CO_2H d. 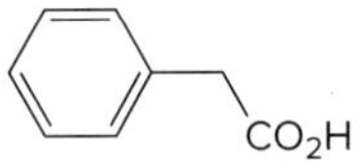 CO_2H

2. Which ring in the given compound is *most* reactive in electrophilic aromatic substitution?

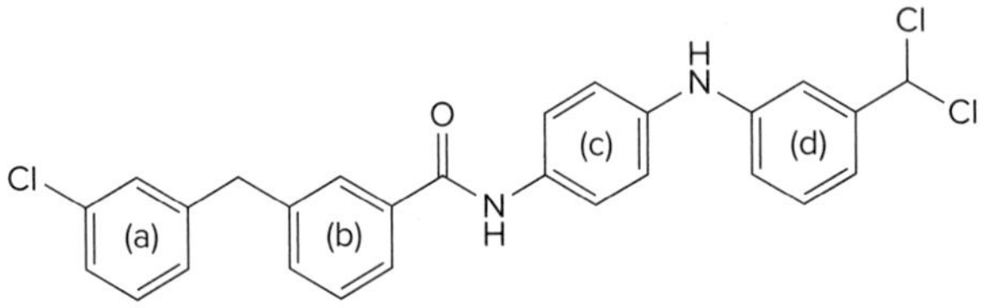

Cl (a) (b) O N H (c) H N (d) Cl Cl

3. Which compound is least reactive when treated with $NaOCH_3$?

a. 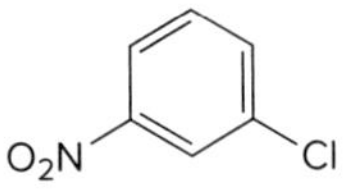 O_2N Cl b. 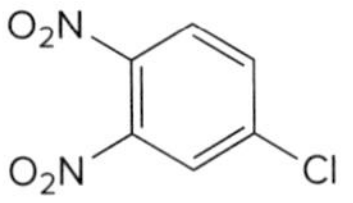 O_2N O_2N Cl c. 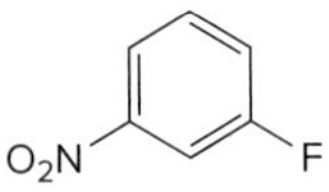 O_2N F d. 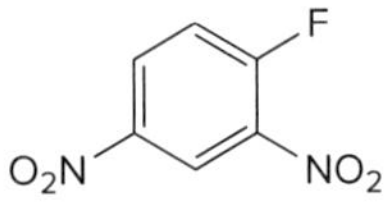 F O_2N NO_2

4. What reactants are needed to convert benzene to styrene?

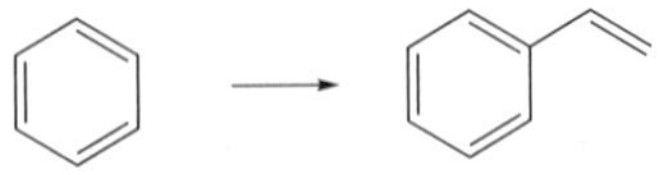

benzene → styrene

a. $CH_2{=}CHCl + AlCl_3$
b. [1] CH_3CH_2Cl, $AlCl_3$; [2] $KOC(CH_3)_3$
c. $CH_2{=}CHBr + AlCl_3$
d. [1] CH_3CH_2Cl, $AlCl_3$; [2] Br_2, *hν*; [3] $KOC(CH_3)_3$

5. Which of the following statements is *true* about an $-N(CH_3)_2$ group on a benzene ring?

a. $N(CH_3)_2$ increases the rates of both electrophilic substitution and nucleophilic substitution.
b. $N(CH_3)_2$ decreases the rates of both electrophilic substitution and nucleophilic substitution.
c. $N(CH_3)_2$ increases the rate of electrophilic substitution and decreases the rate of nucleophilic substitution.
d. $N(CH_3)_2$ decreases the rate of electrophilic substitution and increases the rate of nucleophilic substitution.

6. Which species is *not* a valid resonance structure for the intermediate that results when anisole reacts with Br_2, $FeBr_3$ at the para position?

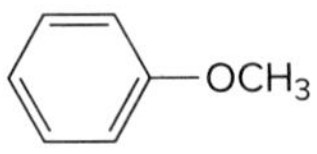

OCH_3 anisole a. 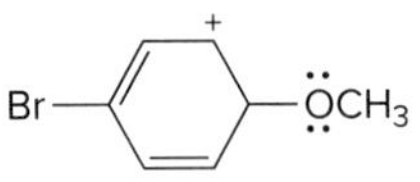 Br + $\ddot{O}CH_3$ b. 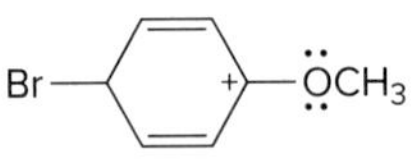 Br + $\ddot{O}CH_3$ c. 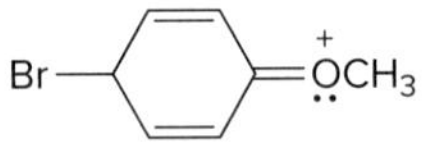 Br $=\overset{+}{O}CH_3$ d. 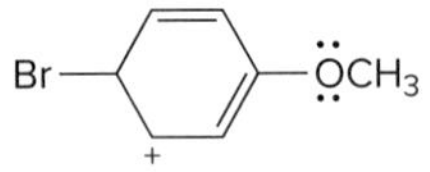 Br + $\ddot{O}CH_3$

7. What product is formed when toluene ($C_6H_5CH_3$) reacts with $CH_3CH_2CH_2Cl$ in the presence of $AlCl_3$?

a. 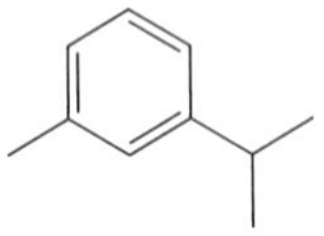 b. 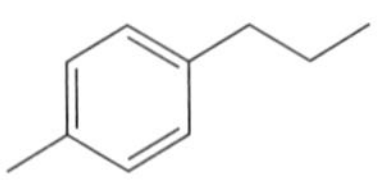 c. 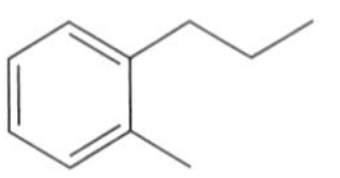 d.

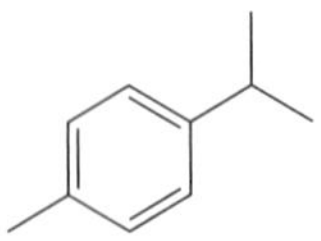

8. Rank **A–D** in order of increasing reactivity in a Friedel–Crafts acylation reaction?

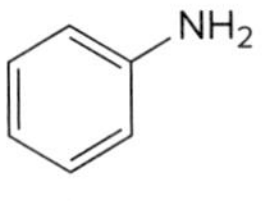

NH_2 **A** Cl **B** **C** O **D**

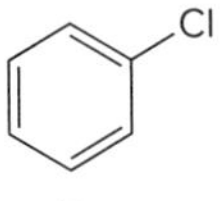

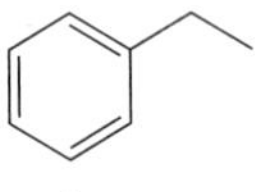

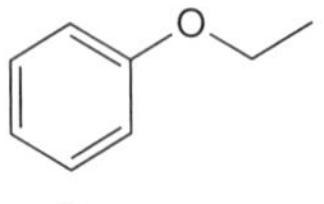

a. **B** < **C** < **D** < **A**
b. **A** < **B** < **C** < **D**
c. **B** < **C** < **A** < **D**
d. **C** < **B** < **D** < **A**

9. Which of the following compounds is formed when **A** is treated with Br_2, $FeBr_3$?

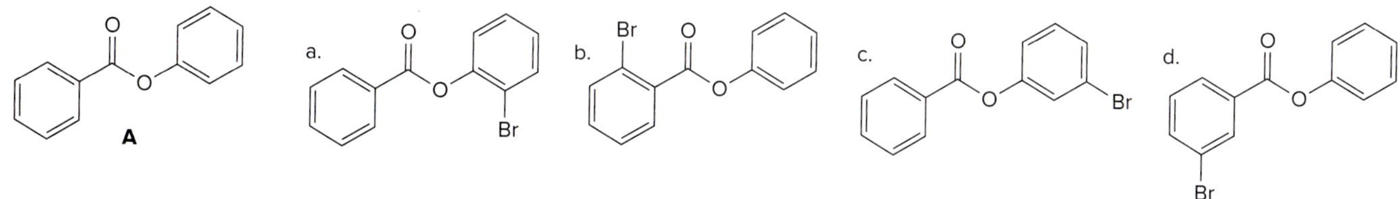

10. Which statement is *false* about an $–OCH_2CH(CH_3)_2$ substituent on a benzene ring?

a. $OCH_2CH(CH_3)_2$ is an ortho, para director.
b. $OCH_2CH(CH_3)_2$ has an electron-donating resonance effect.
c. $OCH_2CH(CH_3)_2$ makes a benzene ring more electron rich than benzene itself.
d. $OCH_2CH(CH_3)_2$ has an electron-donating inductive effect.

PROBLEMS

Problem Using Three-Dimensional Models

16.40 Draw the products formed when **A** and **B** are treated with each of the following reagents: (a) Br_2, $FeBr_3$; (b) HNO_3, H_2SO_4; (c) CH_3CH_2COCl, $AlCl_3$.

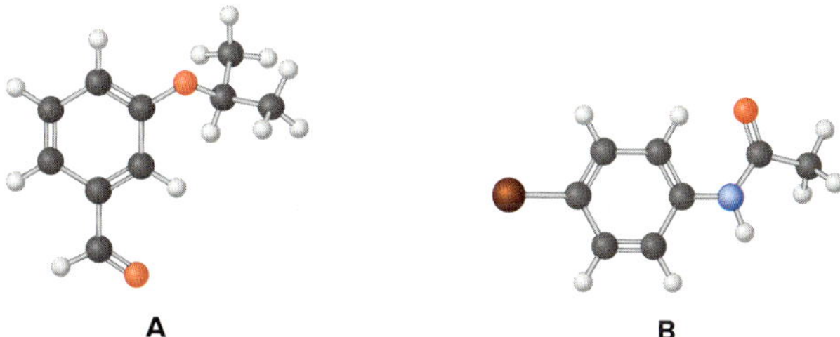

Reactions

16.41 Draw the products formed when phenol (C_6H_5OH) is treated with each set of reagents.

a. [1] HNO_3, H_2SO_4; [2] Sn, HCl
b. [1] $(CH_3CH_2)_2CHCOCl$, $AlCl_3$; [2] Zn(Hg), HCl
c. [1] CH_3CH_2Cl, $AlCl_3$; [2] Br_2, $h\nu$
d. [1] $(CH_3)_2CHCl$, $AlCl_3$; [2] $KMnO_4$

16.42 Draw the products formed when each compound is treated with CH_3CH_2COCl, $AlCl_3$.

a. 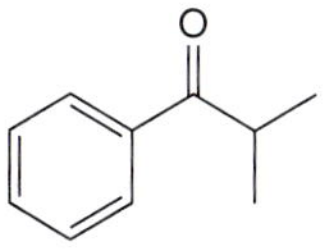 O
b. 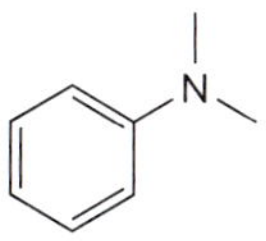 N
c. 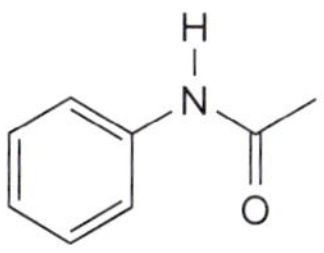 H N O

16.43 Draw the products of each reaction.

a. 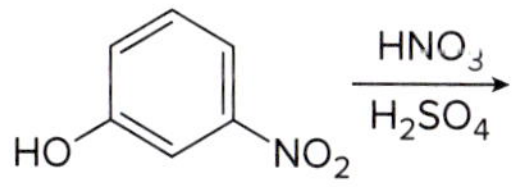 HO, NO_2 — HNO_3, H_2SO_4 →

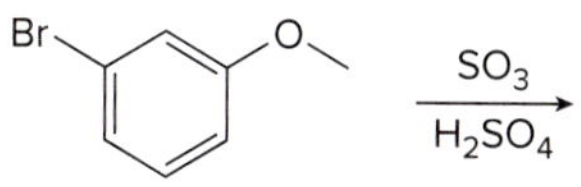

b. 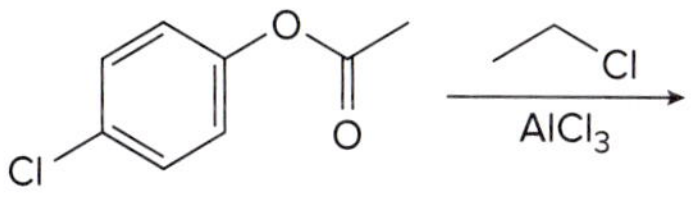 Cl, O, O — Cl, $AlCl_3$ →

c. O, H — Br_2, $FeBr_3$ →

d. 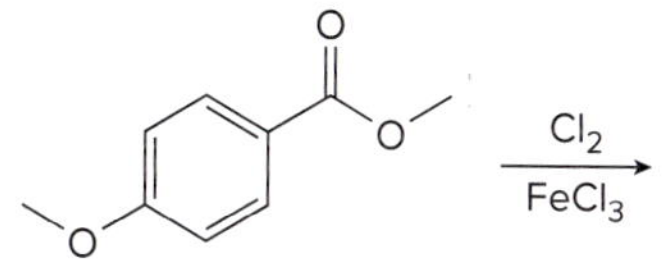 O, O, O — Cl_2, $FeCl_3$ →

e. Br, O — SO_3, H_2SO_4 →

f. 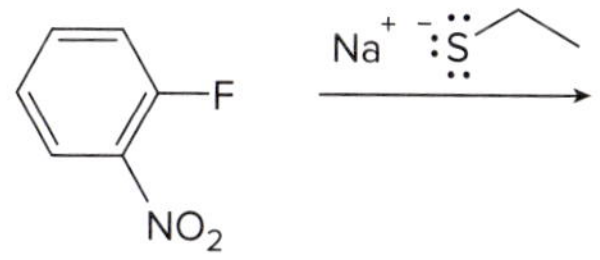 F, NO_2 — Na^+ $^-$:S: →

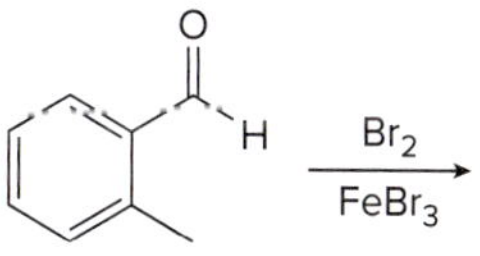

16.44 What products are formed when benzene is treated with each alkyl chloride and $AlCl_3$?

a. 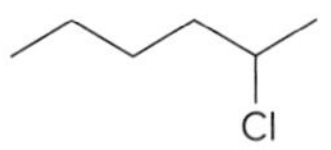 b. 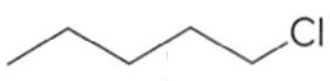 Cl c. 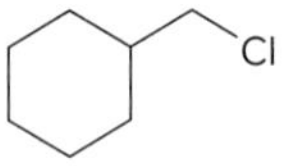 Cl
Cl

16.45 Draw the products of each reaction.

a. 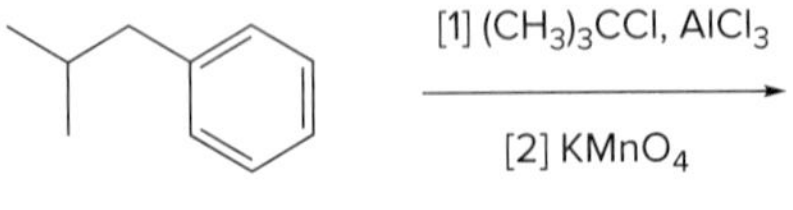 [1] $(CH_3)_3CCl$, $AlCl_3$ [2] $KMnO_4$

c. 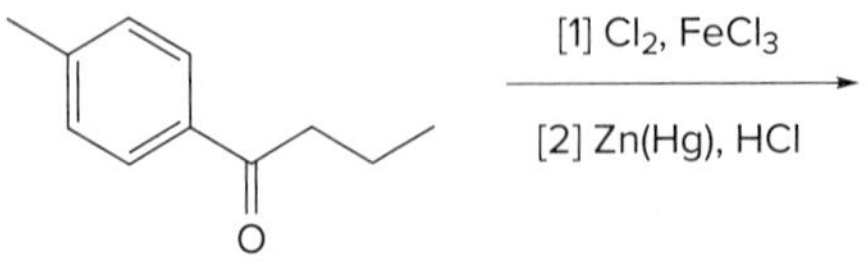 [1] Cl_2, $FeCl_3$ [2] Zn(Hg), HCl
O

b. 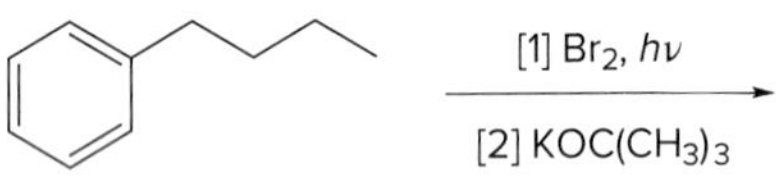 [1] Br_2, $h\nu$ [2] $KOC(CH_3)_3$

d. 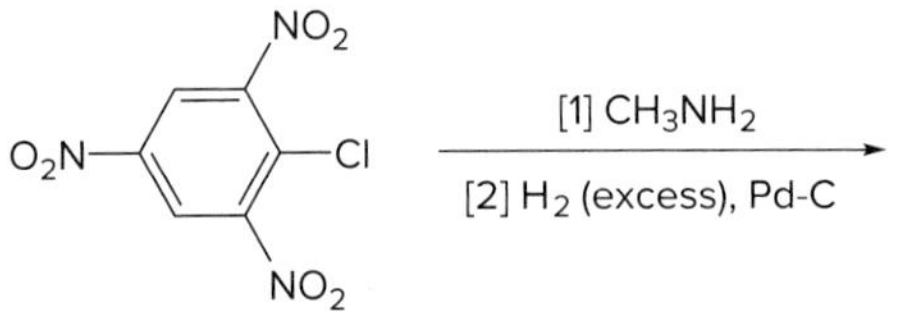 O_2N NO_2 Cl NO_2 [1] CH_3NH_2 [2] H_2 (excess), Pd-C

16.46 You have learned two ways to make an alkyl benzene: Friedel–Crafts alkylation, and Friedel–Crafts acylation followed by reduction. Although some alkyl benzenes can be prepared by both methods, it is often true that only one method can be used to prepare a given alkyl benzene. Which method(s) can be used to prepare each of the following compounds from benzene? Show the steps that would be used.

a. 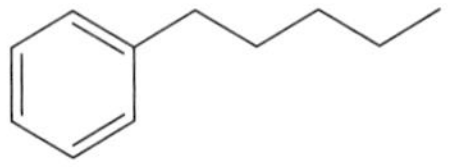 b. 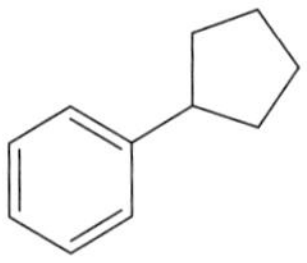 c. 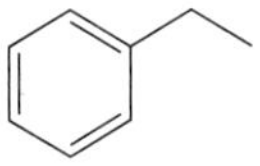 d. 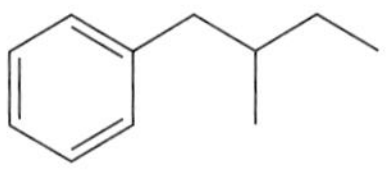 e.

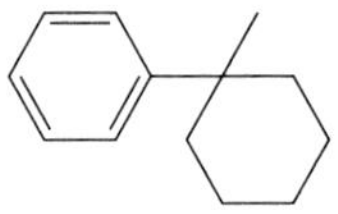

16.47 Identify **X** and **Y,** the products of key steps in two syntheses of pioglitazone, a drug used to treat diabetes.

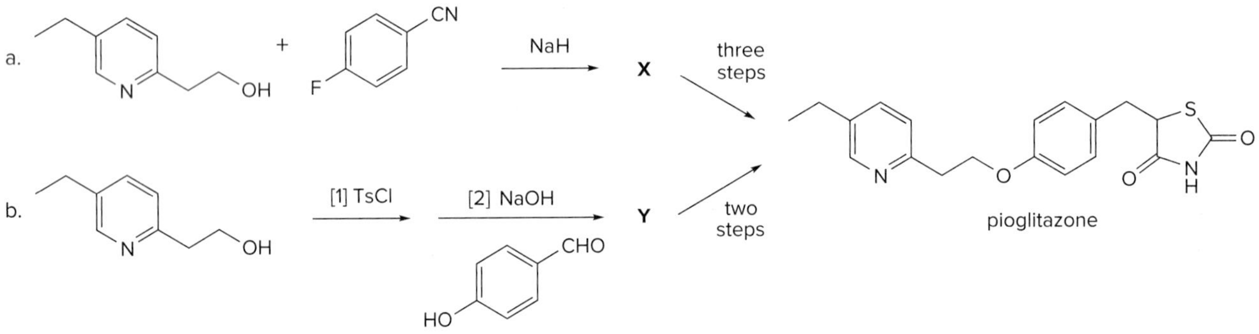

a. N OH + F CN NaH **X** three steps

b. N OH [1] TsCl [2] NaOH HO CHO **Y** two steps

N O S O O N H
pioglitazone

16.48 Identify **M** and **N** in the following reaction sequence, two steps in the original synthesis of the non-sedating antihistamine fexofenadine (Section 23.5B).

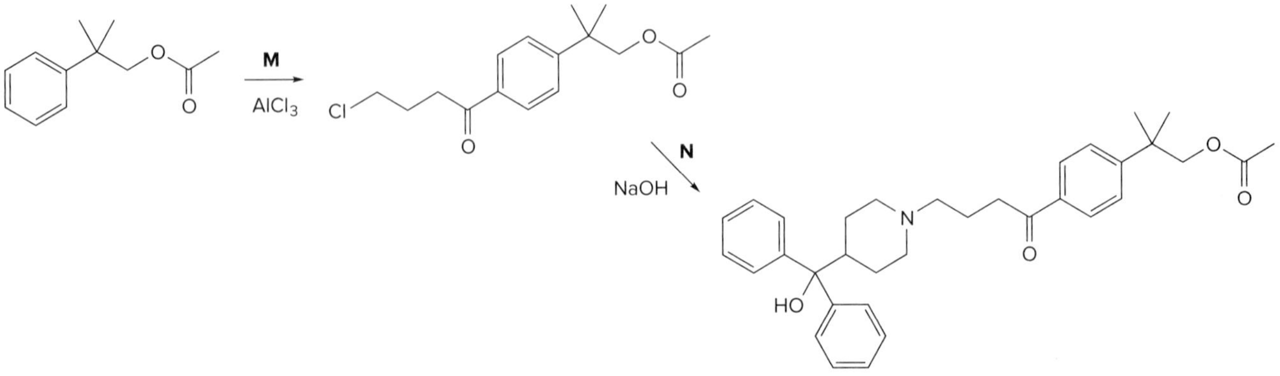

O O **M** $AlCl_3$ Cl O O O

N NaOH

HO N O O O

16.49 Draw the structure of **A,** an intermediate in the synthesis of the antipsychotic drug risperidone. Explain why three rings in risperidone are considered aromatic.

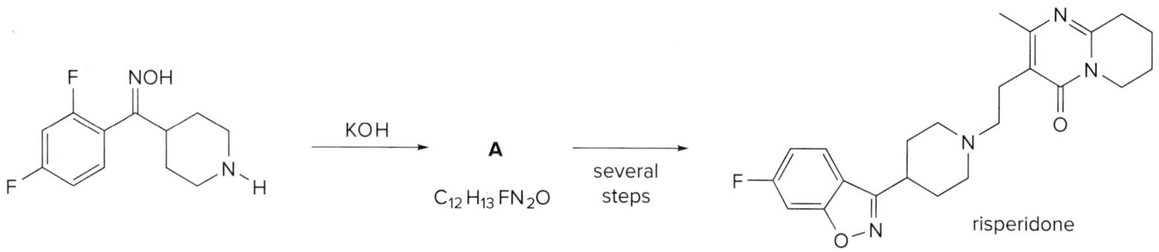

KOH

A

$C_{12}H_{13}FN_2O$

several steps

risperidone

16.50 **D** is an intermediate in the synthesis of rosiglitazone (trade name Avandia), a drug used to treat type 2 diabetes. Suggest two different methods to prepare the ether in **D** by substitution reactions.

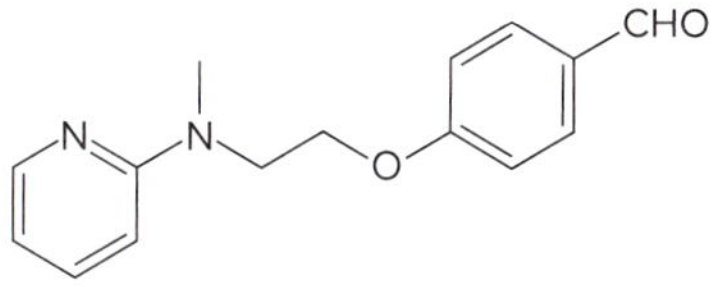

D

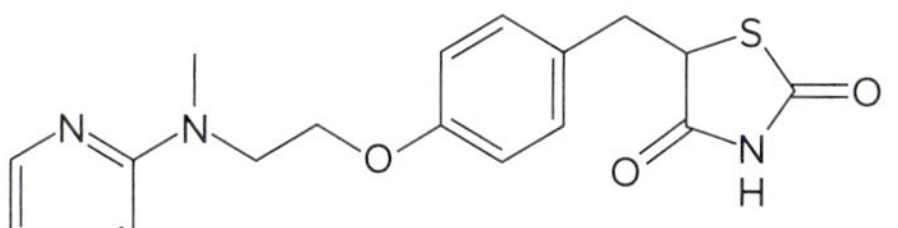

rosiglitazone

Substituent Effects

16.51 Rank the compounds in each group in order of increasing reactivity in electrophilic aromatic substitution: (a) C_6H_6, C_6H_5Cl, C_6H_5CHO, $C_6H_5OCH_3$; (b) $C_6H_5CH_3$, $C_6H_5NH_2$, $C_6H_5CH_2NH_2$, $C_6H_5CONH_2$.

16.52 For each of the following substituted benzenes: [1] C_6H_5Br; [2] C_6H_5CN; [3] $C_6H_5OCOCH_3$:

a. Does the substituent donate or withdraw electron density by an inductive effect?

b. Does the substituent donate or withdraw electron density by a resonance effect?

c. On balance, does the substituent make a benzene ring more or less electron rich than benzene itself?

d. Does the substituent activate or deactivate the benzene ring in electrophilic aromatic substitution?

16.53 Determine which ring in each compound is more reactive in electrophilic aromatic substitution, and draw the product(s) formed when each compound is treated with the general electrophile E^+.

a.

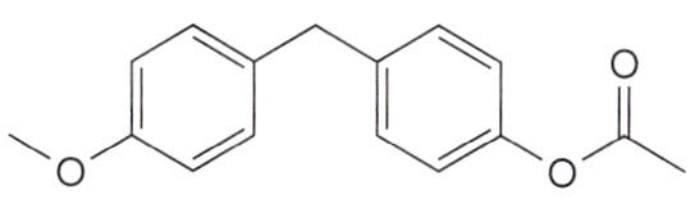

b.

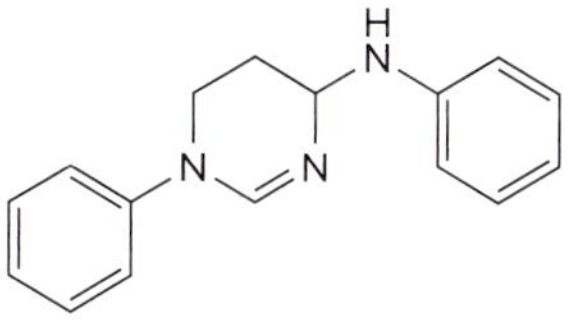

16.54 Consider the tetracyclic aromatic compound drawn below, with rings labeled as **A, B, C,** and **D.** (a) Which of the four rings is *most* reactive in electrophilic aromatic substitution? (b) Which of the four rings is *least* reactive in electrophilic aromatic substitution? (c) What are the major product(s) formed when this compound is treated with one equivalent of Br_2?

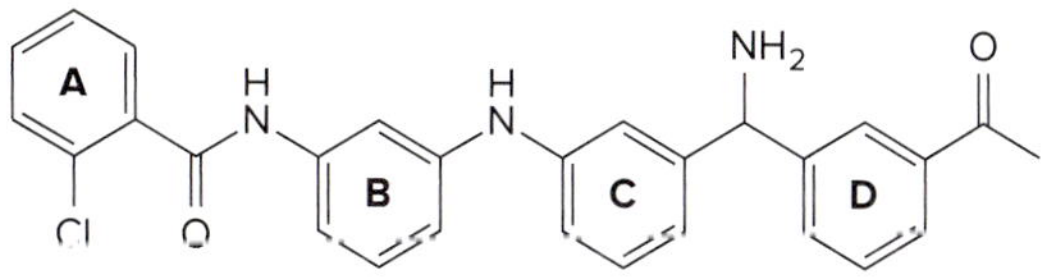

16.55 For each N-substituted benzene, predict whether the compound reacts faster than, slower than, or at a similar rate to benzene in electrophilic aromatic substitution. Then draw the major product(s) formed when each compound reacts with a general electrophile E^+.

a.

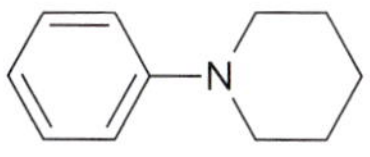

b. 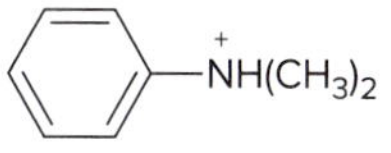 $-\overset{+}{N}H(CH_3)_2$

c. 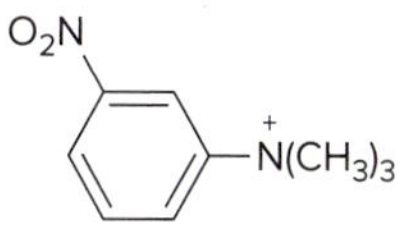 $-\overset{+}{N}(CH_3)_3$

d.

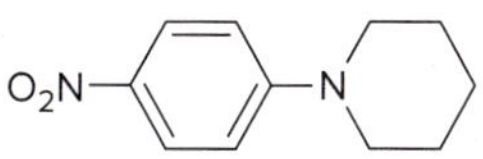

16.56 What is the major product of electrophilic addition of HBr to the following alkene? Explain your choice.

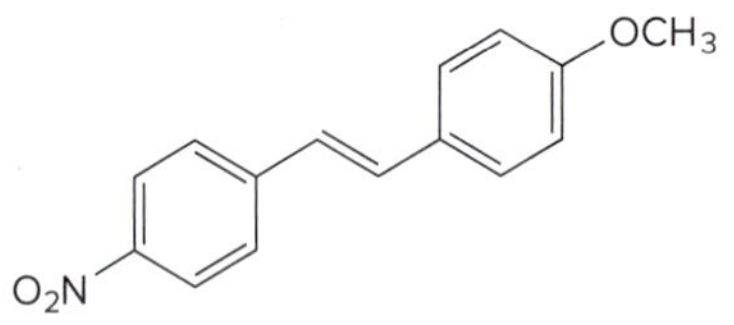

OCH_3

O_2N

16.57 Using resonance structures, explain why a nitroso group (–NO) is an ortho, para director that deactivates a benzene ring toward electrophilic attack.

16.58 Explain this observation: Ethyl 3-phenylpropanoate ($C_6H_5CH_2CH_2CO_2CH_2CH_3$) reacts with electrophiles to afford ortho- and para-disubstituted arenes, but ethyl 3-phenylprop-2-enoate ($C_6H_5CH{=}CHCO_2CH_2CH_3$) reacts with electrophiles to afford meta-disubstituted arenes.

16.59 Rank the aryl halides in each group in order of increasing reactivity in nucleophilic aromatic substitution by an addition–elimination mechanism.

a. chlorobenzene, *p*-fluoronitrobenzene, *m*-fluoronitrobenzene
b. 1-fluoro-2,4-dinitrobenzene, 1-fluoro-3,5-dinitrobenzene, 1-fluoro-3,4-dinitrobenzene
c. 1-fluoro-2,4-dinitrobenzene, 4-chloro-3-nitrotoluene, 4-fluoro-3-nitrotoluene

Mechanisms

16.60 Draw a stepwise, detailed mechanism for the following reaction.

CH_3O ... OH $\xrightarrow{H_2SO_4}$ CH_3O ... + H_2O

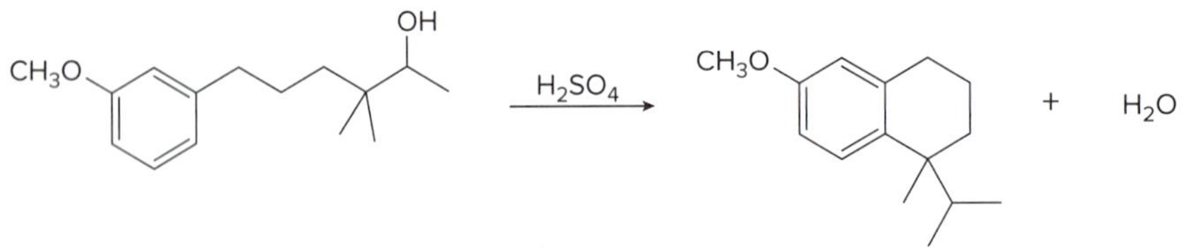

16.61 Draw a stepwise mechanism for the following reaction, which involves two Friedel–Crafts reactions. **B** was an intermediate in the synthesis of the antidepressant sertraline (Problem 16.11).

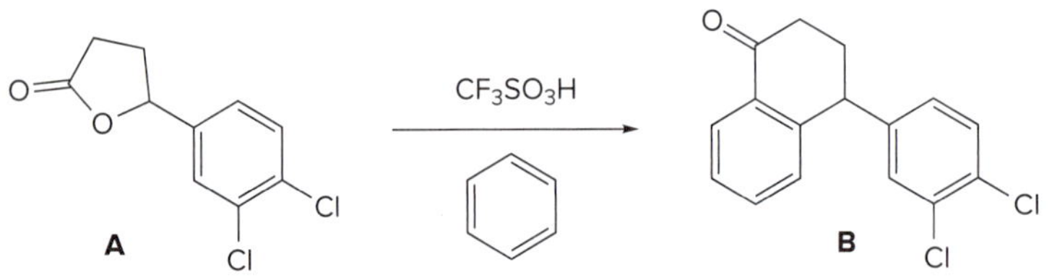

A (O, Cl, Cl) $\xrightarrow{CF_3SO_3H}$ (benzene) **B** (O, Cl, Cl)

16.62 Friedel–Crafts alkylation of benzene with (*R*)-2-chlorobutane and $AlCl_3$ affords *sec*-butylbenzene.

a. How many stereogenic centers are present in the product?
b. Would you expect the product to exhibit optical activity? Explain, with reference to the mechanism.

16.63 Draw a stepwise mechanism for the following reaction, one step in the synthesis of the oral contraceptive desogestrel (Section 11.4).

CH_3O, O, OR, OCH_3 $\xrightarrow{CF_3CO_2H}$ CH_3O, O, OR, CH_3O $\xrightarrow{\text{several steps}}$ desogestrel (Section 11.4)

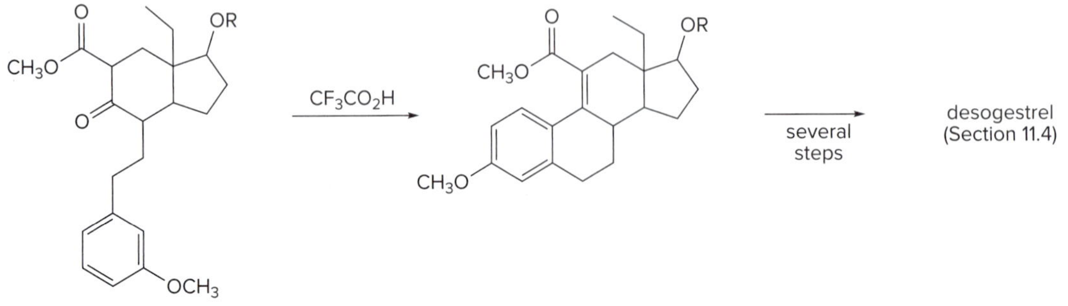

16.64 Draw a stepwise mechanism for the following reaction.

H
N
strong
base
N
Cl

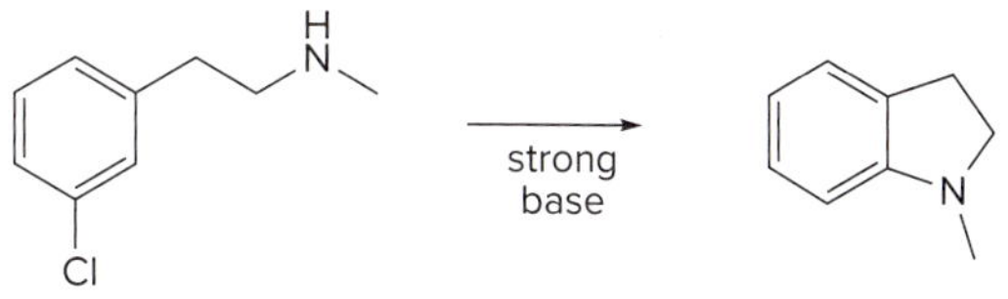

16.65 Although two products (**A** and **B**) are possible when naphthalene undergoes electrophilic aromatic substitution, only **A** is formed. Draw resonance structures for the intermediate carbocation to explain why this is observed.

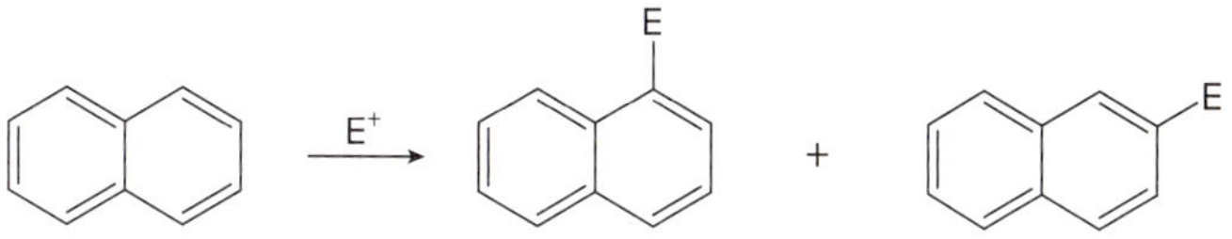

E
E^+
+
E
naphthalene
A
This product is formed.
B
This product is *not* formed.

16.66 Draw a stepwise mechanism for the following reaction, which results in the synthesis of bisphenol F (R = H), an additive used in a variety of packaging materials. Bisphenol F is related to BPA (bisphenol A, R = CH_3), a reagent used to harden some plastics, now removed from certain baby products because of its estrogen-like activity that can disrupt endocrine pathways.

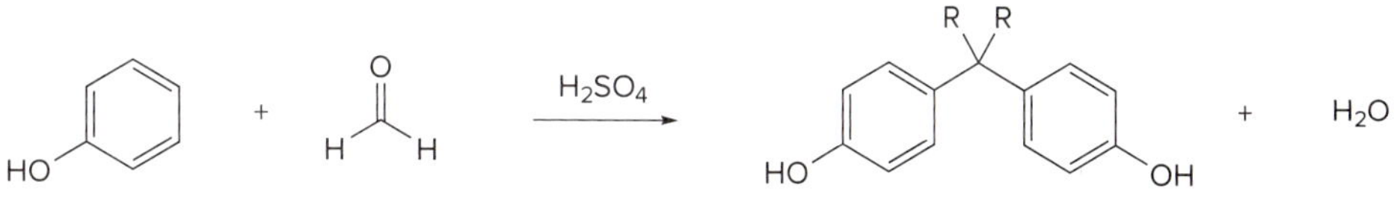

R R
O
+
H_2SO_4
+ H_2O
HO
H H
HO
OH
bisphenol F
R = H

16.67 Benzyl bromide ($C_6H_5CH_2Br$) reacts rapidly with CH_3OH to afford benzyl methyl ether ($C_6H_5CH_2OCH_3$). Draw a stepwise mechanism for the reaction, and explain why this 1° alkyl halide reacts rapidly with a weak nucleophile under conditions that favor an S_N1 mechanism. Would you expect the para-substituted benzylic halides $CH_3OC_6H_4CH_2Br$ and $O_2NC_6H_4CH_2Br$ to each be more or less reactive than benzyl bromide in this reaction? Explain your reasoning.

Synthesis

16.68 Synthesize each compound from benzene, organic halides with < 5 C's, and any other organic or inorganic reagents.

a. 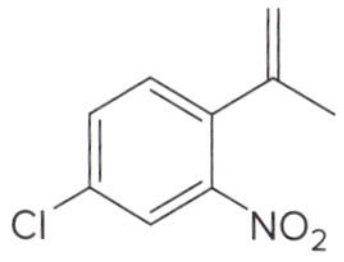 Cl, NO_2

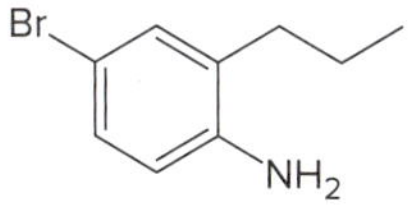

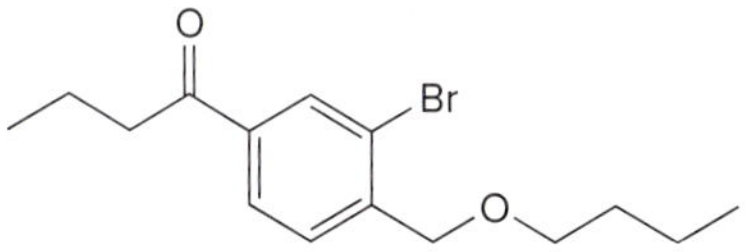

b. 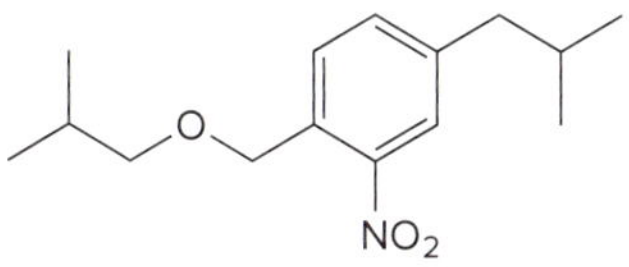 O, NO_2

c. Br, NH_2

d. 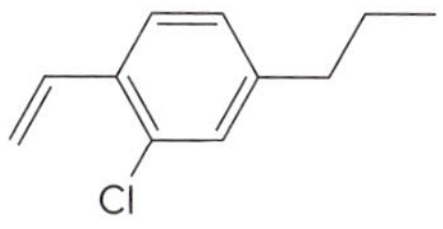 Cl

e. O, Br, O

f. 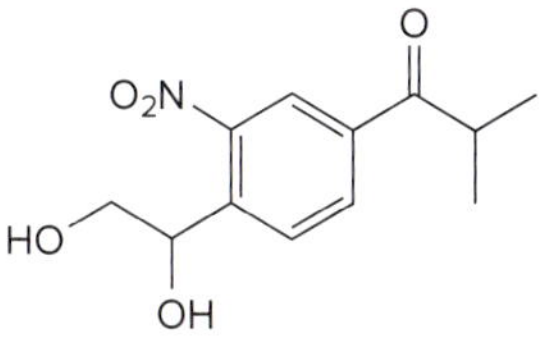 O_2N, O, HO, OH

16.69 Synthesize each compound from toluene ($C_6H_5CH_3$) and any other organic or inorganic reagents.

a. 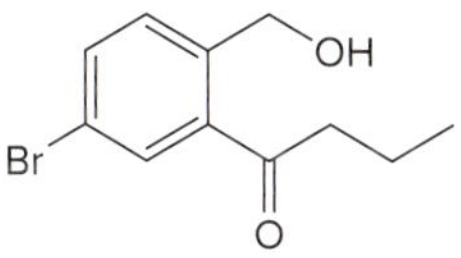 OH, Br, O

b. 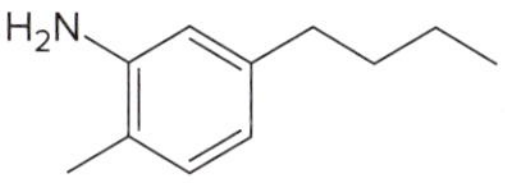 H_2N

c. 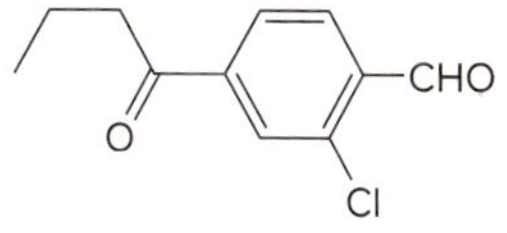 O, CHO, Cl

16.70 Use the reactions in this chapter along with those learned in Chapters 11 and 12 to synthesize each compound. You may use benzene, acetylene (HC≡CH), ethanol, ethylene oxide, and any inorganic reagents.

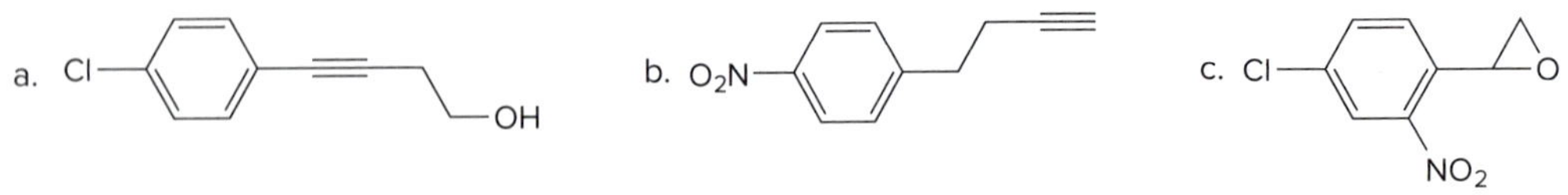

a. Cl … OH b. O_2N … c. Cl … O, NO_2

16.71 Carboxylic acid **X** is an intermediate in the multistep synthesis of proparacaine, a local anesthetic. Devise a synthesis of **X** from phenol and any needed organic or inorganic reagents.

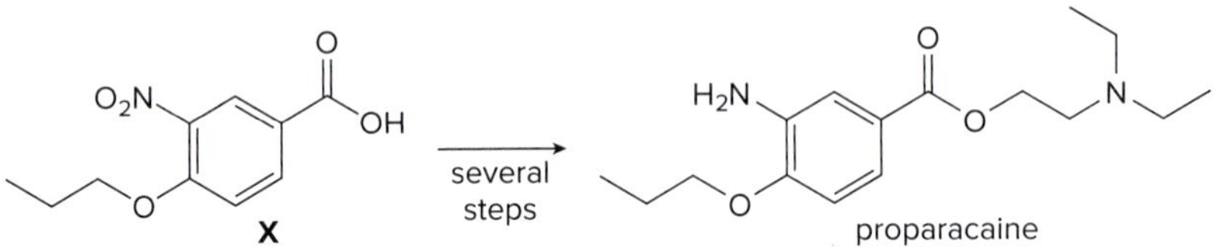

X → several steps → proparacaine

Spectroscopy

16.72 The location of the C=O absorption in the IR of benzene derivatives that contain a conjugated carbonyl group depends on the electron-donating or electron-withdrawing effects of other substituents on the benzene ring. Rank compounds **A–C** in order of increasing wavenumber of their C=O absorption and suggest a reason for your choice.

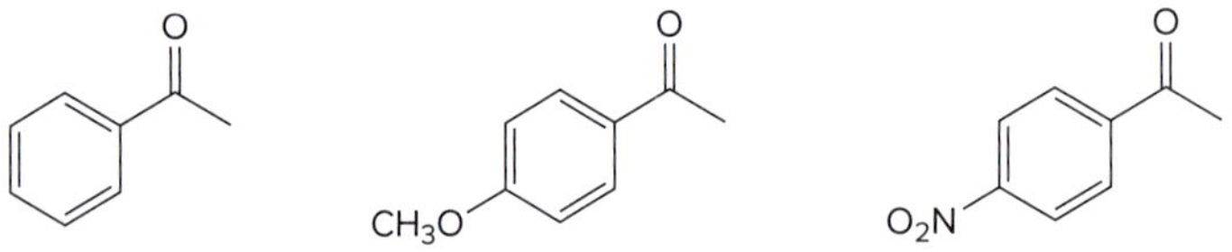

A **B** (CH_3O) **C** (O_2N)

16.73 Identify the structures of isomers **A** and **B** (molecular formula C_8H_9Br).

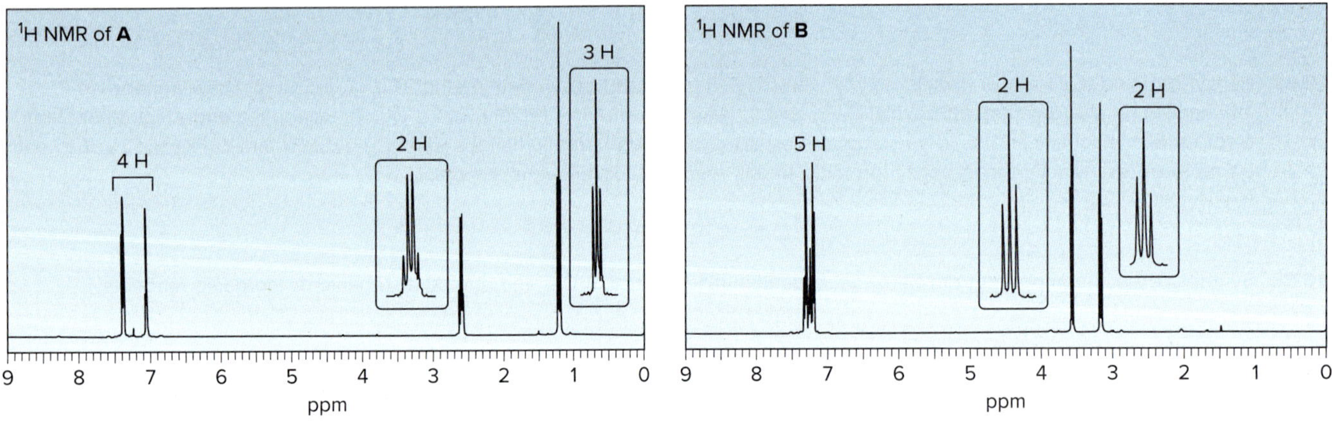

16.74 Propose a structure of compound **C** (molecular formula $C_{10}H_{12}O$) consistent with the following data. **C** is partly responsible for the odor and flavor of raspberries.

Compound **C:** IR absorption at 1717 cm^{-1}

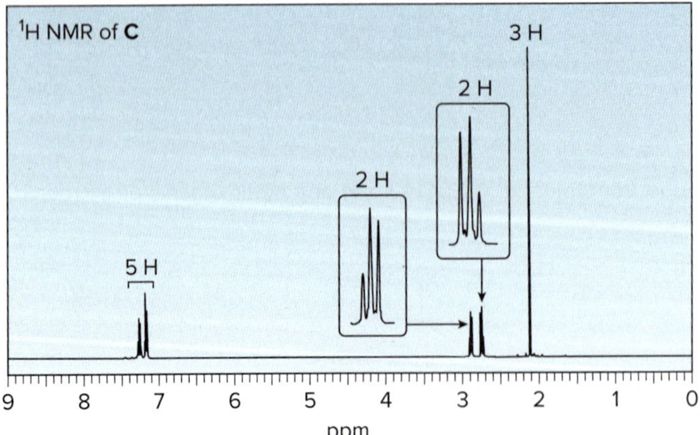

16.75 Compound **X** (molecular formula $C_{10}H_{12}O$) was treated with NH_2NH_2, ^-OH to yield compound **Y** (molecular formula $C_{10}H_{14}$). Based on the ^1H NMR spectra of **X** and **Y** given below, what are the structures of **X** and **Y?**

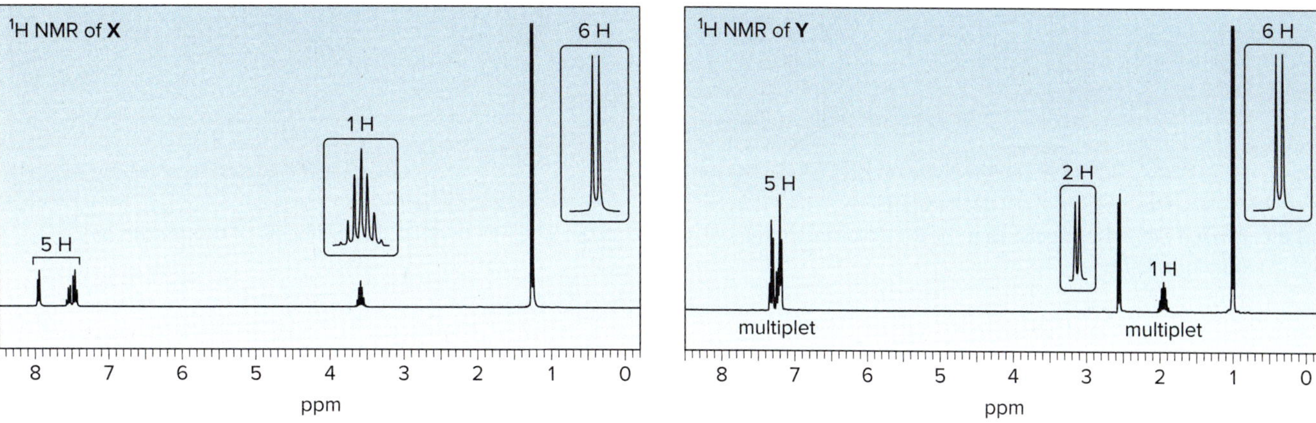

16.76 Reaction of *p*-cresol with two equivalents of 2-methylprop-1-ene affords BHT, a preservative with molecular formula $C_{15}H_{24}O$. BHT gives the following ^1H NMR spectral data: 1.4 (singlet, 18 H), 2.27 (singlet, 3 H), 5.0 (singlet, 1 H), and 7.0 (singlet, 2 H) ppm. What is the structure of BHT? Draw a stepwise mechanism illustrating how it is formed.

$$\text{p-cresol} + \text{2-methylprop-1-ene (2 equiv)} \xrightarrow{H_2SO_4} \text{BHT } (C_{15}H_{24}O)$$

Challenge Problems

16.77 Devise a synthesis of optically active (*S*)-fluoxetine (trade name Prozac) from the given starting materials and any other needed reagents.

(*S*)-fluoxetine ⟹ (cinnamyl alcohol, OH) + (4-chloro-CF$_3$-benzene) + CH_3NH_2

16.78 Clenbuterol is a banned performance-enhancing drug available in some countries as a prescription medication to treat asthma. Clenbuterol has been used to promote livestock growth, but the U.S. Food and Drug Administration has banned the use of clenbuterol in any animal raised for human consumption. Devise a synthesis of clenbuterol from the given starting materials and any other needed reagents.

clenbuterol ⟹ benzene + H_2N-C(CH$_3$)$_3$ + CH$_3$CH$_2$Cl

16.79 The ^1H NMR spectrum of phenol (C_6H_5OH) shows three absorptions in the aromatic region: 6.70 (2 ortho H's), 7.14 (2 meta H's), and 6.80 (1 para H) ppm. Explain why the ortho and para absorptions occur at lower chemical shift than the meta absorption.

16.80 Explain the reactivity and orientation effects observed in each heterocycle.

pyridine $\xrightarrow{E^+}$ 3-substituted product (C3); pyrrole $\xrightarrow{E^+}$ 2-substituted product (C2)

a. Pyridine is less reactive than benzene in electrophilic aromatic substitution and yields 3-substituted products.
b. Pyrrole is more reactive than benzene in electrophilic aromatic substitution and yields 2-substituted products.

16.81 Draw a stepwise mechanism for the dienone–phenol rearrangement, a reaction that forms alkyl-substituted phenols from cyclohexadienones.

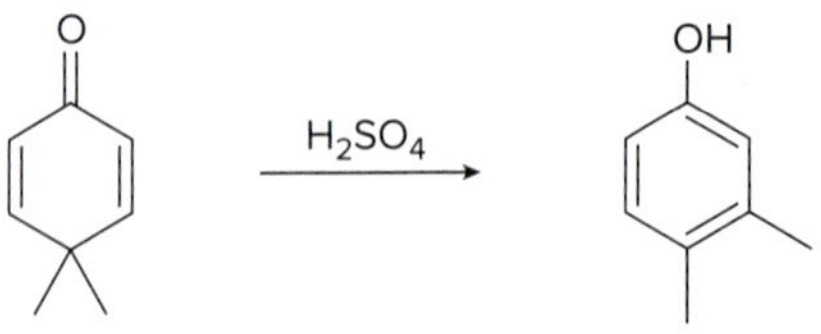
O
H_2SO_4
OH

16.82 Draw a stepwise mechanism for the following intramolecular reaction, which is used in the synthesis of the female sex hormone estrone.

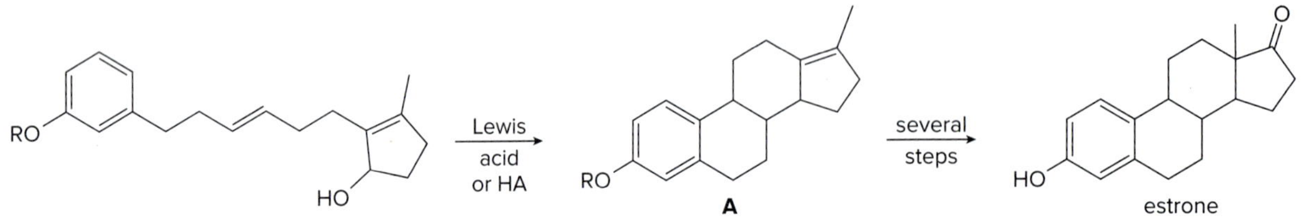
RO
HO
Lewis acid or HA
RO
A
several steps
HO
O
estrone

16.83 The bicyclic heterocycle quinoline undergoes electrophilic aromatic substitution to give the products shown. (a) Explain why electrophilic substitution occurs on the ring without the N atom. (b) Explain why electrophilic substitution occurs more readily at C8 than C7.

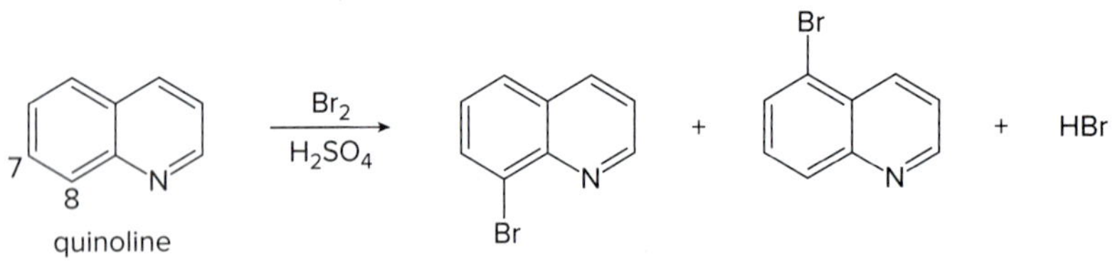
7
8
N
quinoline
Br_2
H_2SO_4
N
Br
+
Br
N
+
HBr

16.84 Draw a stepwise mechanism for the conversion of **A** to **B,** a reaction that involves both double bond isomerization and ring closure. This process was a key step in the synthesis of pygmaeocin C, a natural product isolated from the roots of *Pygmaeopremia herbacea,* a shrub that grows in northern India and Tibet.

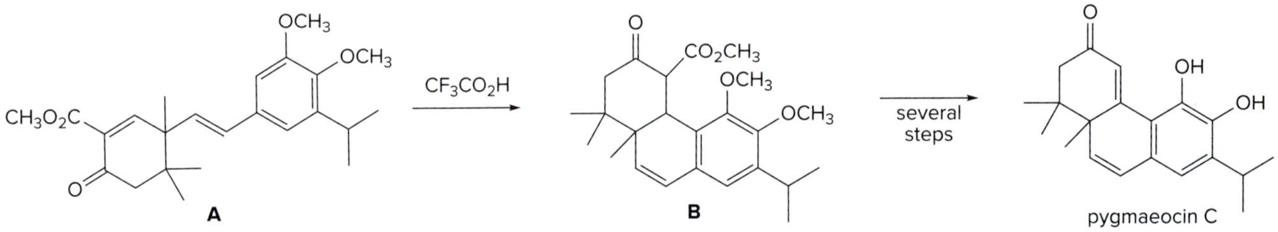
OCH_3
OCH_3
CH_3O_2C
O
A
CF_3CO_2H
O
CO_2CH_3
OCH_3
OCH_3
B
several steps
O
OH
OH
pygmaeocin C

SELF-TEST ANSWERS

1. c 2. c 3. a 4. d 5. c 6. a 7. d 8. b 9. a 10. d

Introduction to Carbonyl Chemistry; Organometallic Reagents; Oxidation and Reduction

17

Folio Images/Alamy Stock Photo

17.1 Introduction
17.2 General reactions of carbonyl compounds
17.3 A preview of oxidation and reduction
17.4 Reduction of aldehydes and ketones
17.5 The stereochemistry of carbonyl reduction
17.6 Enantioselective carbonyl reductions
17.7 Reduction of carboxylic acids and their derivatives
17.8 Oxidation of aldehydes
17.9 Organometallic reagents
17.10 Reaction of organometallic reagents with aldehydes and ketones
17.11 Retrosynthetic analysis of Grignard products
17.12 Protecting groups
17.13 Reaction of organometallic reagents with carboxylic acid derivatives
17.14 Reaction of organometallic reagents with other compounds
17.15 α,β-Unsaturated carbonyl compounds
17.16 Summary—The reactions of organometallic reagents
17.17 Synthesis

Ginkgolide B ($C_{20}H_{24}O_{10}$) is a complex organic compound obtained from the ginkgo tree *Ginkgo biloba*, the oldest seed-producing plant that currently lives on earth. Ginkgolide B was one of four components isolated from ginkgo extracts in 1932, and its structure was determined in 1967. Extracts from the roots, bark, leaves, and seeds of the ginkgo tree have been used in traditional Chinese medicine to treat asthma and improve blood circulation. Ginkgolide B was synthesized in the laboratory of Nobel Laureate E. J. Corey at Harvard University in 1988 using an asymmetric reduction, one of the widely used reactions discussed in Chapter 17.

Why Study . . .

Carbonyl Compounds and Their Reactions?

Chapters 17 through 22 of this text discuss carbonyl compounds—aldehydes, ketones, acid halides, esters, amides, and carboxylic acids. **The carbonyl group is perhaps the most important functional group in organic chemistry,** because its electron-deficient carbon and easily broken π bond make it susceptible to a wide variety of useful reactions.

We begin by examining the similarities and differences between two broad classes of carbonyl compounds. We will then spend the remainder of Chapter 17 on reactions that are especially important in organic synthesis. Chapters 18 and 20 present specific reactions that occur at the carbonyl carbon, and Chapters 21 and 22 concentrate on reactions occurring at the α carbon to the carbonyl group. Chapter 19 covers carboxylic acids, which can react at both their OH and C=O groups, and nitriles (RCN), which undergo reactions similar to those of carbonyl compounds.

Although Chapter 17 is "jam-packed" with reactions, most of them follow one of two general pathways, so they can be classified in a well-organized fashion, provided you remember a few basic principles. Keep in mind these fundamental themes about reactions:

- **Nucleophiles attack electrophiles.**
- **π Bonds are easily broken.**
- **Bonds to good leaving groups are easily cleaved.**

17.1 Introduction

Two broad classes of compounds contain a ***carbonyl group:***

[1] Compounds that have only carbon and hydrogen atoms bonded to the carbonyl group

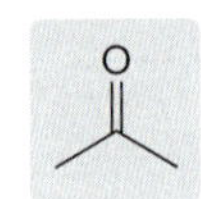

carbonyl group

- **An aldehyde** has at least one H atom bonded to the carbonyl group.
- **A ketone** has two alkyl or aryl groups bonded to the carbonyl group.

[2] Compounds that contain an electronegative atom bonded to the carbonyl group

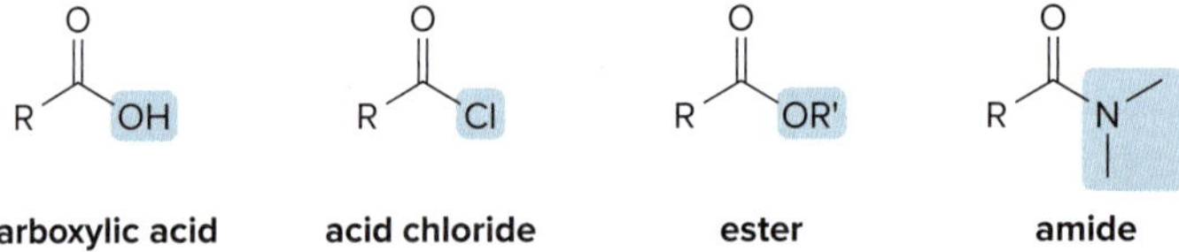

These include **carboxylic acids, acid chlorides, esters,** and **amides,** as well as other similar compounds discussed in Chapter 20. Each of these compounds contains an atom (Cl, O, or N) more electronegative than carbon, capable of acting as a **leaving group.** Acid chlorides, esters, and amides are often called **carboxylic acid derivatives,** because they can be synthesized from carboxylic acids (Chapter 20). Each compound contains an acyl group (RCO–), so they are also called **acyl derivatives.**

- The presence or absence of a leaving group on the carbonyl carbon determines the type of reactions these compounds undergo (Section 17.2).

The carbonyl carbon atom is ***sp*^{2} hybridized** and **trigonal** planar, and all bond angles are ~120°. The double bond of a carbonyl group consists of one σ bond and one π bond. The π bond is formed by the overlap of two *p* orbitals, and extends above and below the plane. In

The aldehyde α-sinensal (Problem 17.1) is the major compound responsible for the orange-like odor of mandarin oil, obtained from the mandarin tree in southern China. *Carr Botanical Consultation*

these features the carbonyl group resembles the trigonal planar, sp^2 hybridized carbons of a C–C double bond.

π bond

120°

trigonal planar

σ bond

In one important way, though, a C=O and a C=C are very different. **The electronegative oxygen atom in the carbonyl group means that the bond is polarized, making the carbonyl carbon electron deficient.** Using a resonance description, the carbonyl group is represented by two resonance structures, with a charge-separated resonance structure a minor contributor to the hybrid.

the major contributor to the hybrid ⟷ a minor contributor to the hybrid

hybrid polarized carbonyl (δ+ δ–)

Problem 17.1

α-sinensal ([1], [2], [3])

a. What orbitals are used to form the indicated bonds in α-sinensal?

b. In what type of orbitals do the lone pairs on O reside?

17.2 General Reactions of Carbonyl Compounds

With what types of reagents should a carbonyl group react? The electronegative oxygen makes the carbonyl carbon **electrophilic,** and because it is trigonal planar, a carbonyl carbon is **uncrowded.** Moreover, a carbonyl group has an **easily broken π bond.**

δ– π bond → **electrophilic carbon** δ+

uncrowded
sp^2 hybridized carbon

As a result, **carbonyl compounds react with nucleophiles.** The outcome of nucleophilic attack, however, depends on the identity of the carbonyl starting material.

- **Aldehydes and ketones undergo nucleophilic *addition*.**

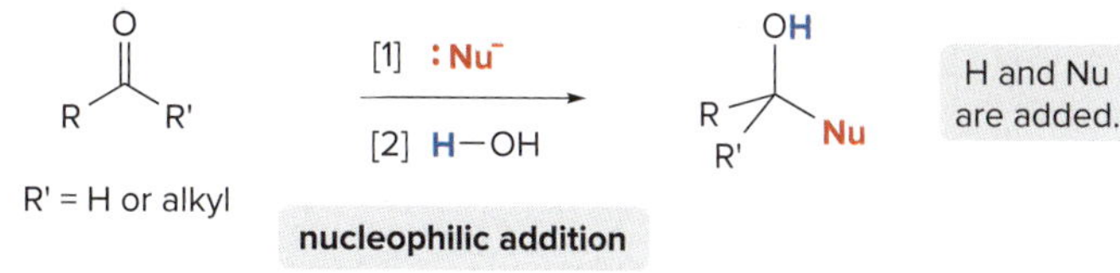

- **Carbonyl compounds that contain leaving groups undergo nucleophilic *substitution*.**

RCOZ —(:Nu⁻)→ RCONu

nucleophilic substitution

Nu replaces Z.

Z = OH, Cl, OR, NH_2

Let's examine each of these general reactions individually.

17.2A Nucleophilic Addition to Aldehydes and Ketones

Aldehydes and ketones react with nucleophiles to form addition products by the two-step process shown in Mechanism 17.1: **nucleophilic attack** followed by **protonation.**

Mechanism 17.1 Nucleophilic Addition—A Two-Step Process

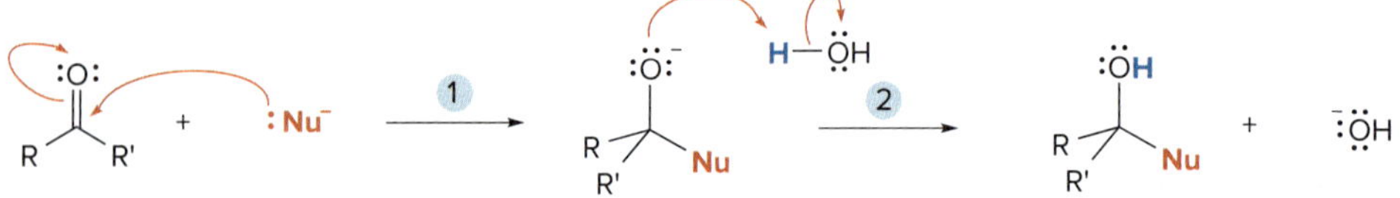

1 **The nucleophile attacks the electrophilic carbonyl**. The π bond is broken, moving an electron pair out on oxygen and forming an sp^3 hybridized carbon.

2 Protonation of the negatively charged oxygen by H_2O forms the **addition product.**

More examples of nucleophilic addition to aldehydes and ketones are discussed in Chapter 18.

The net result is that the π bond is broken, two new σ bonds are formed, and the elements of H and Nu are *added* across the π bond. Nucleophilic addition with two different nucleophiles—**hydride (H:⁻)** and **carbanions (R:⁻)**—is discussed in Chapter 17.

Aldehydes are more reactive than ketones toward nucleophilic attack for both steric and electronic reasons.

aldehyde (δ– :O:, δ+, R, H)	**ketone** (δ– :O:, δ+, R, R')
• less crowded	• more crowded
• less stable	• more stable
• more reactive	• less reactive

- **The two R groups bonded to the ketone carbonyl group make it *more crowded,* so nucleophilic attack is more difficult.**
- **The two electron-donor R groups stabilize the partial charge on the carbonyl carbon of a ketone, making it *more stable* and less reactive.**

17.2B Nucleophilic Substitution of RCOZ (Z = Leaving Group)

Carbonyl compounds with leaving groups react with nucleophiles to form substitution products by the two-step process shown in Mechanism 17.2: **nucleophilic attack,** followed by **loss of the leaving group.**

Mechanism 17.2 Nucleophilic Substitution—A Two-Step Process

$Z = OH, Cl, OR', NH_2$

1 **The nucleophile attacks the electrophilic carbonyl.** The π bond is broken, moving an electron pair out on oxygen and forming an sp^3 hybridized carbon.

2 An electron pair on oxygen re-forms the π bond and **Z comes off as a leaving group** with the electron pair in the C–Z bond.

The net result is that Nu replaces Z—a nucleophilic substitution reaction. This reaction is often called **nucleophilic *acyl* substitution** to distinguish it from the nucleophilic substitution reactions at sp^3 hybridized carbons discussed in Chapter 7. Nucleophilic substitution with two different nucleophiles—**hydride (H:⁻)** and **carbanions (R:⁻)**—is discussed in Chapter 17. Other nucleophiles are examined in Chapter 20.

Carboxylic acid derivatives differ greatly in their reactivity toward nucleophiles. The order in which they react parallels the leaving group ability of the group Z bonded to the carbonyl carbon.

- **The *better* the leaving group Z, the *more reactive* RCOZ is in nucleophilic acyl substitution.**

Recall from Section 7.7 that the *weaker* the base, the *better* the leaving group.

Thus, the following trends result:

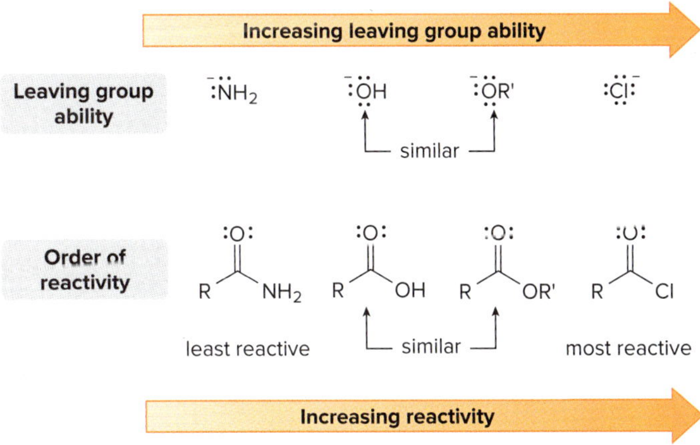

- **Acid chlorides (RCOCl), which have the best leaving group (Cl^-), are the most reactive carboxylic acid derivatives, and amides ($RCONH_2$), which have the worst leaving group ($^-NH_2$), are the least reactive.**
- **Carboxylic acids (RCOOH) and esters (RCOOR'), which have leaving groups of similar basicity (^-OH and $^-OR'$), fall in the middle.**

Nucleophilic addition and nucleophilic acyl substitution involve the *same* first step—**nucleophilic attack on the electrophilic carbonyl group** to form a tetrahedral intermediate. The difference between them is what then happens to this intermediate. **Aldehydes and ketones cannot undergo substitution because they have no leaving group** bonded to the newly formed sp^3 hybridized carbon. Nucleophilic substitution with an aldehyde, for example, would form H:$^-$, an extremely strong base and therefore a very poor (and highly unlikely) leaving group.

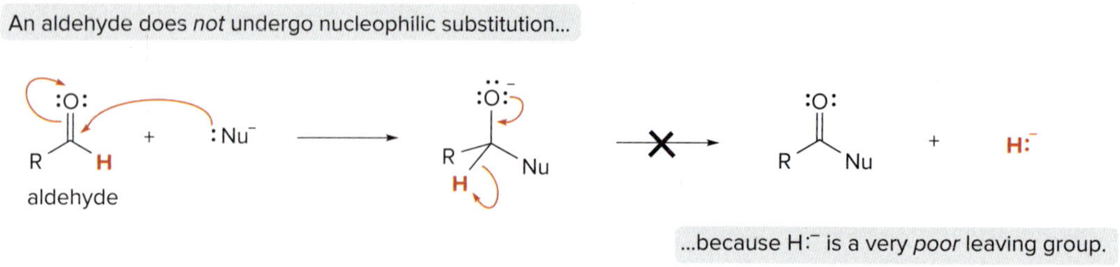

An aldehyde does *not* undergo nucleophilic substitution...

:O: R H aldehyde + :Nu$^-$ → :O:$^-$ R H Nu —✕→ :O: R Nu + H:$^-$

...because H:$^-$ is a very *poor* leaving group.

Problem 17.2 Which carbonyl groups in the anticancer drug Taxol (Section 5.5) will undergo nucleophilic addition, and which will undergo nucleophilic substitution?

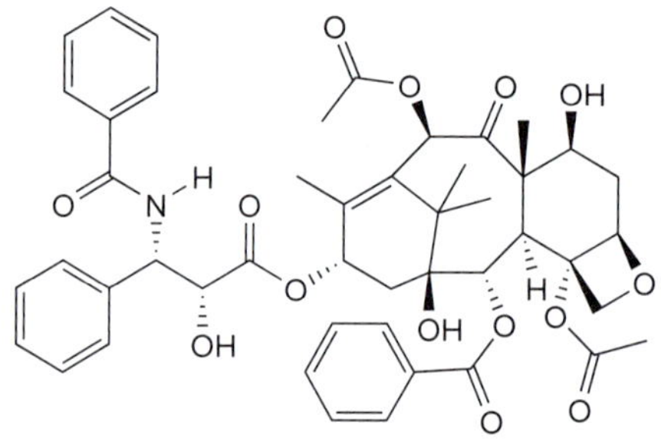

Taxol

Problem 17.3 Rank the compounds in each group in order of increasing reactivity toward nucleophilic attack.

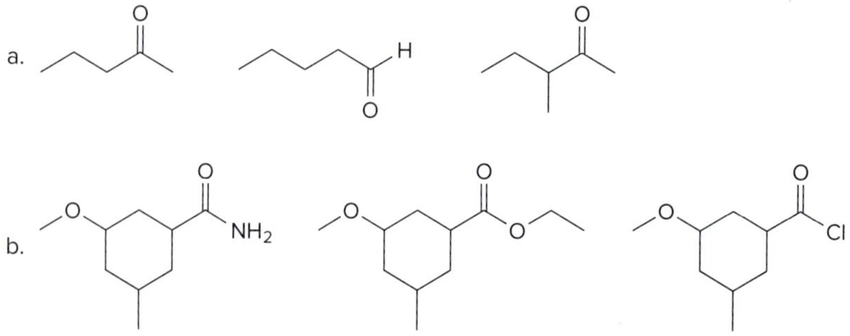

a.

b.

To show how these general principles of nucleophilic substitution and addition apply to carbonyl compounds, we are going to discuss oxidation and reduction reactions, and reactions with organometallic reagents—compounds that contain carbon–metal bonds. We begin with reduction to build on what you learned previously in Chapter 12.

17.3 A Preview of Oxidation and Reduction

Recall the definitions of oxidation and reduction presented in Section 12.1:

- **Oxidation results in an *increase* in the number of C–Z bonds (usually C–O bonds) or a *decrease* in the number of C–H bonds.**
- **Reduction results in a *decrease* in the number of C–Z bonds (usually C–O bonds) or an *increase* in the number of C–H bonds.**

Carbonyl compounds are either reactants or products in many of these reactions, as illustrated in the accompanying diagram. For example, because aldehydes fall in the middle of this scheme, they can be both oxidized and reduced. Carboxylic acids and their derivatives (RCOZ), on the other hand, are already highly oxidized, so their only useful reaction is reduction.

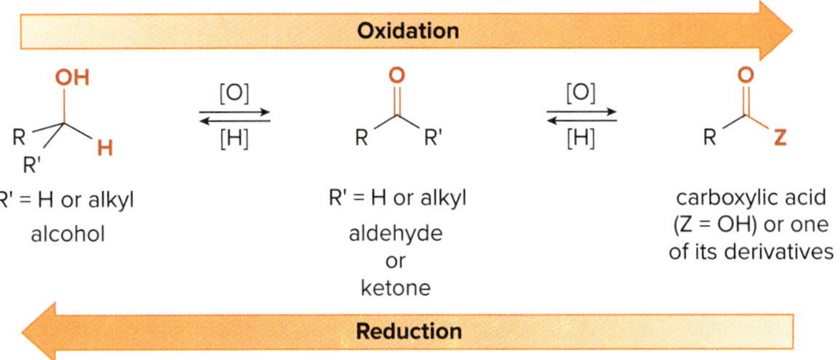

The three most useful oxidation and reduction reactions of carbonyl starting materials can be summarized as follows:

[1] Reduction of aldehydes and ketones to alcohols (Sections 17.4–17.6)

R' = H or alkyl
aldehyde or ketone —[H]→ **1° or 2° alcohol**

Aldehydes and ketones are reduced to 1° and 2° alcohols, respectively.

[2] Reduction of carboxylic acids and their derivatives (Section 17.7)

RCOZ —[H]→ **aldehyde** or **1° alcohol**

The reduction of carboxylic acids and their derivatives gives a variety of products, depending on the identity of Z and the nature of the reducing agent. The usual products are aldehydes or 1° alcohols.

[3] Oxidation of aldehydes to carboxylic acids (Section 17.8)

aldehyde —[O]→ **carboxylic acid**

The most useful oxidation reaction of carbonyl compounds is the oxidation of aldehydes to carboxylic acids.

We begin with reduction, because the mechanisms of reduction reactions follow directly from the general mechanisms for nucleophilic addition and substitution.

17.4 Reduction of Aldehydes and Ketones

$LiAlH_4$ and $NaBH_4$ serve as a source of H:⁻, but there are no free H:⁻ ions present in reactions with these reagents.

The most useful reagents for reducing aldehydes and ketones are the metal hydride reagents (Section 12.2). The two most common metal hydride reagents are **sodium borohydride ($NaBH_4$)** and **lithium aluminum hydride ($LiAlH_4$).** These reagents contain a polar metal–hydrogen bond that serves as a source of the nucleophile hydride, **H:⁻. $LiAlH_4$ is a stronger reducing agent than $NaBH_4$,** because the Al–H bond is more polar than the B–H bond.

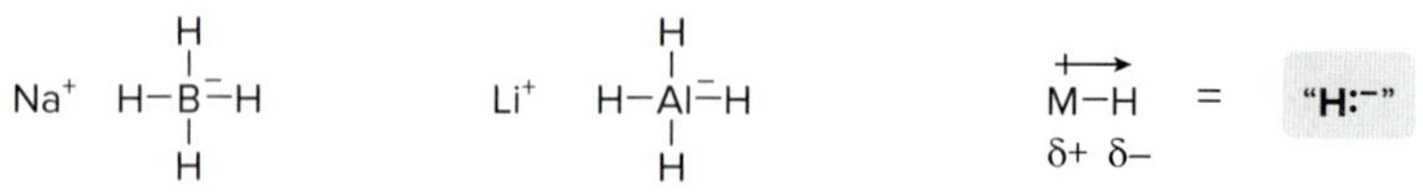

sodium borohydride **lithium aluminum hydride** a polar metal–hydrogen bond

17.4A Reduction with Metal Hydride Reagents

Treating an aldehyde or a ketone with $NaBH_4$ or $LiAlH_4$, followed by water or some other proton source, affords an **alcohol.** This is an addition reaction because **the elements of H_2 are added across the π bond**, but it is also a **reduction** because the product alcohol has fewer C–O bonds than the starting carbonyl compound.

R' = H or alkyl

aldehyde or ketone

$NaBH_4$ or $LiAlH_4$; H_2O

1° or 2° alcohol

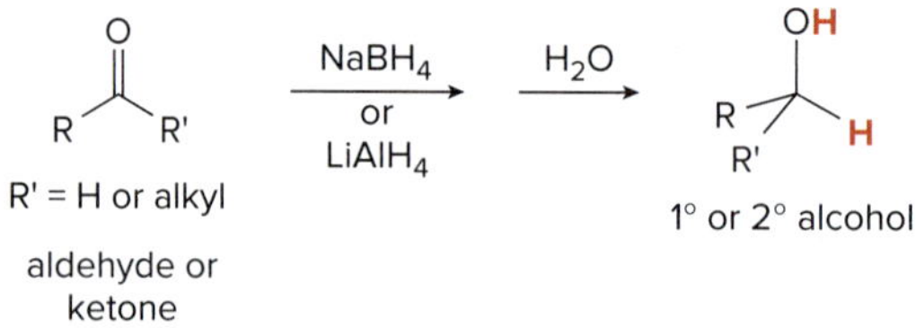

$LiAlH_4$ reductions must be carried out under anhydrous conditions, because water reacts violently with the reagent. Water is added to the reaction mixture (to serve as a proton source) *after* the reduction with $LiAlH_4$ is complete.

The product of this reduction reaction is a **1° alcohol** when the starting carbonyl compound is an aldehyde, and a **2° alcohol** when it is a ketone.

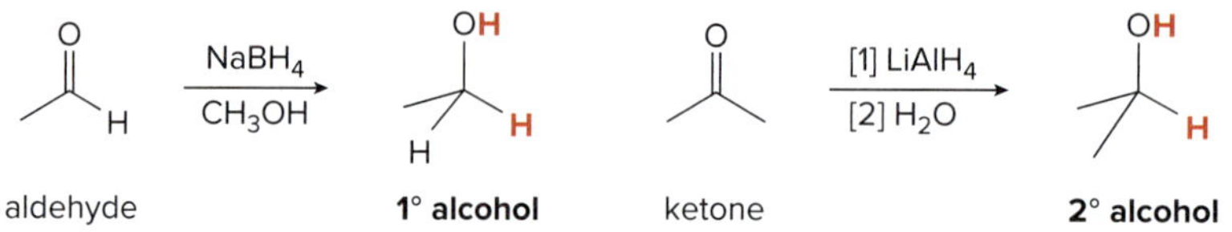

aldehyde → ($NaBH_4$, CH_3OH) → **1° alcohol**

ketone → ([1] $LiAlH_4$ [2] H_2O) → **2° alcohol**

$NaBH_4$ selectively reduces aldehydes and ketones in the presence of most other functional groups. Reductions with $NaBH_4$ are typically carried out in CH_3OH as solvent. $LiAlH_4$ reduces aldehydes and ketones and many other functional groups as well (Sections 12.6 and 17.7).

Problem 17.4 What alcohol is formed when each compound is treated with $NaBH_4$ in CH_3OH?

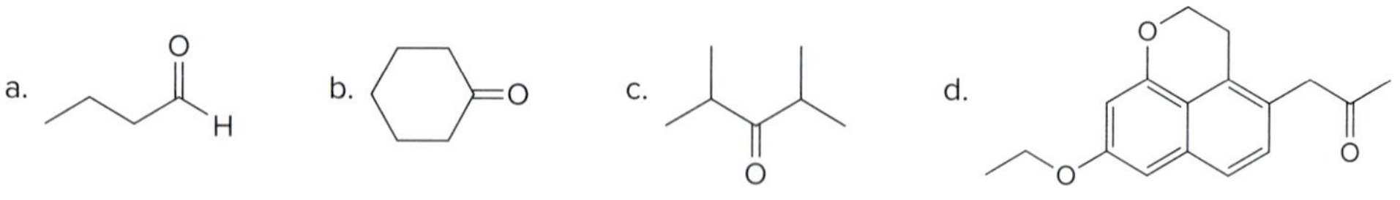

a. b. c. d.

Problem 17.5 What aldehyde or ketone is needed to prepare each alcohol by metal hydride reduction?

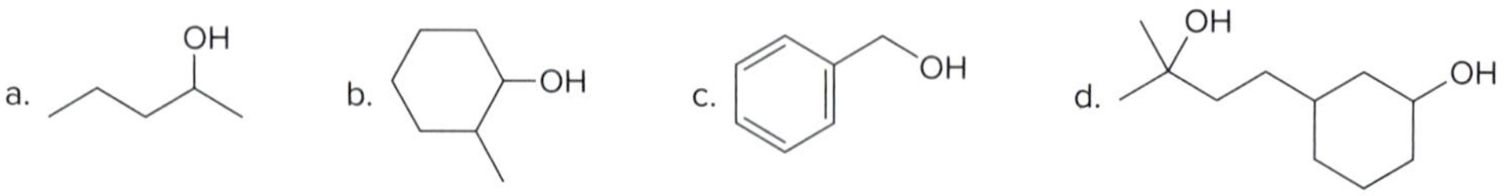

a. b. c. d.

17.4B The Mechanism of Hydride Reduction

Hydride reduction of aldehydes and ketones occurs via the general mechanism of nucleophilic addition—that is, **nucleophilic attack** followed by **protonation.** Mechanism 17.3 is shown using $LiAlH_4$, but an analogous mechanism can be written for $NaBH_4$.

Mechanism 17.3 $LiAlH_4$ Reduction of RCHO and $R_2C=O$

1 **The nucleophile (AlH_4^-) donates $H:^-$ to the carbonyl group,** breaking the π bond and moving an electron pair out on oxygen. This forms a new C–H bond.

2 Protonation of the negatively charged oxygen by H_2O (or CH_3OH) forms the **reduction product** with a new O–H bond.

- **The net result of adding $H:^-$ (from $NaBH_4$ or $LiAlH_4$) and H^+ (from H_2O) is the addition of the elements of H_2 to the carbonyl π bond.**

17.4C Catalytic Hydrogenation of Aldehydes and Ketones

Catalytic hydrogenation also reduces aldehydes and ketones to 1° and 2° alcohols, respectively, using H_2 and Pd-C (or another metal catalyst). H_2 adds to the C=O in much the same way that it adds to the C=C of an alkene (Section 12.3). The metal catalyst (Pd-C) provides a surface that binds the carbonyl starting material and H_2, and two H atoms are sequentially transferred with cleavage of the π bond.

aldehyde → (H_2, Pd-C) → 1° alcohol   ketone → (H_2, Pd-C) → 2° alcohol

When a compound contains both a carbonyl group and a carbon–carbon double bond, selective reduction of one functional group can be achieved by proper choice of reagent.

- **A C=C is reduced faster than a C=O with H_2 (Pd-C).**
- **A C=O is readily reduced with $NaBH_4$ and $LiAlH_4$, but a C=C is inert.**

Thus, cyclohex-2-enone, a compound that contains both a carbon–carbon double bond and a carbonyl group, can be reduced to three different compounds—an allylic alcohol, a carbonyl compound, or an alcohol—depending on the reagent.

cyclohex-2-enone → ($NaBH_4$, CH_3OH) → allylic alcohol
- **$NaBH_4$ reduces the C=O** selectively to form an allylic alcohol.

cyclohex-2-enone → (H_2 (1 equiv), Pd-C) → ketone
- One equivalent of **H_2 reduces the C=C** selectively to form a ketone.

cyclohex-2-enone → (H_2 (excess), Pd-C) → alcohol
- **Excess H_2 reduces both π bonds** to form an alcohol.

Problem 17.6 Draw the products formed when $CH_3COCH_2CH_2CH=CH_2$ is treated with each reagent: (a) $LiAlH_4$, then H_2O; (b) $NaBH_4$ in CH_3OH; (c) H_2 (1 equiv), Pd-C; (d) H_2 (excess), Pd-C; (e) $NaBH_4$ (excess) in CH_3OH; (f) $NaBD_4$ in CH_3OH.

Figure 17.1 A $NaBH_4$ reduction used in organic synthesis

O → $\xrightarrow[CH_3OH]{NaBH_4}$ OH → four steps → muscone

muscone

odor of musk (perfume component)

The male musk deer, a small antlerless deer found in the mountain regions of China and Tibet, has long been hunted for its musk, a strongly scented liquid used in early medicine and later in perfumery. *Aleksey Suvorov/ Alamy Stock Photo*

- **Muscone** is the major compound in musk, one of the oldest known ingredients in perfumes. Musk was originally isolated from the male musk deer, but it can now be prepared synthetically in the laboratory in a variety of ways.

The reduction of aldehydes and ketones is a common reaction used in the synthesis of many natural products, including the fragrant compound muscone shown in Figure 17.1.

17.5 The Stereochemistry of Carbonyl Reduction

Recall from Section 9.16 that an achiral starting material gives a racemic mixture when a new stereogenic center is formed.

The stereochemistry of carbonyl reduction follows the same principles we have previously learned. Reduction converts a **planar sp^2 hybridized carbonyl carbon to a tetrahedral sp^3 hybridized carbon.** What happens when a new stereogenic center is formed in this process? With an achiral reagent like $NaBH_4$ or $LiAlH_4$, **a racemic product is obtained.** For example, $NaBH_4$ in CH_3OH solution reduces butan-2-one, an achiral ketone, to butan-2-ol, an alcohol that contains a new stereogenic center. Both enantiomers of butan-2-ol are formed in equal amounts.

butan-2-one $\xrightarrow[CH_3OH]{NaBH_4}$ butan-2-ol = (S)-butan-2-ol + (R)-butan-2-ol

butan-2-one — achiral starting material

butan-2-ol — new stereogenic center

(S)-butan-2-ol (R)-butan-2-ol — Two enantiomers are formed.

Why is a racemic mixture formed? Because the carbonyl carbon is sp^2 hybridized and planar, hydride can approach the double bond with equal probability from both sides of the plane, forming two alkoxides, which are **enantiomers** of each other. Protonation of the alkoxides gives an equal amount of two alcohols, which are also **enantiomers.**

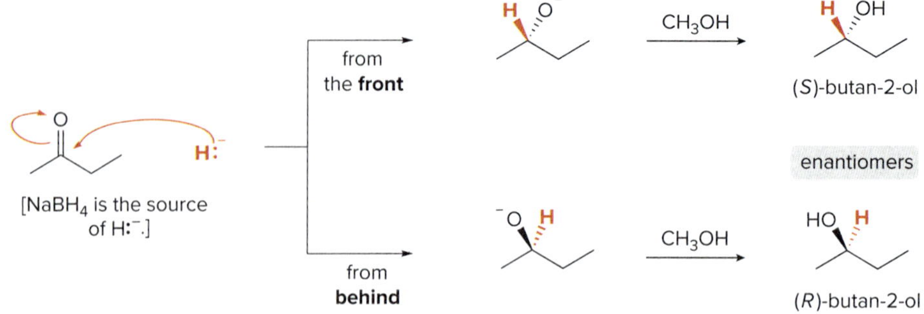

- **Conclusion: Hydride reduction of an achiral ketone with $LiAlH_4$ or $NaBH_4$ gives a racemic mixture of two alcohols when a new stereogenic center is formed.**

Problem 17.7 Draw the products formed (including stereoisomers) when each compound is reduced with $NaBH_4$ in CH_3OH.

a. 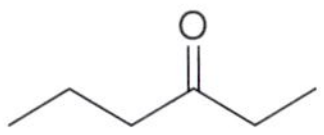 b. 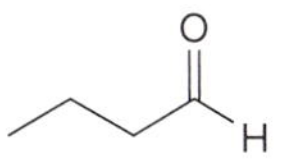 c.

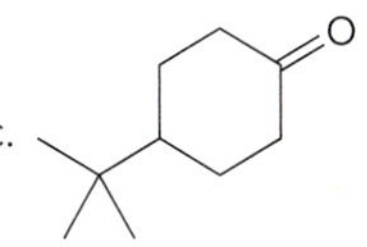

17.6 Enantioselective Carbonyl Reductions

17.6A CBS Reagents

One enantiomer can be formed selectively from the reduction of a carbonyl group, provided a **chiral reducing agent** is used. This strategy is identical to that employed in the Sharpless asymmetric epoxidation reaction (Section 12.15). A reduction that forms one enantiomer predominantly or exclusively is an **enantioselective** or **asymmetric reduction.**

Many different chiral reducing agents have now been prepared for this purpose. One such reagent, formed by reacting borane (**BH_3**) with a heterocycle called an **oxazaborolidine,** has one stereogenic center (and thus two enantiomers).

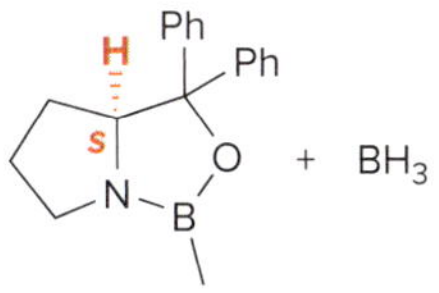

(*S*)-2-methyl-**CBS**-oxazaborolidine + BH_3
(*S*)-CBS reagent

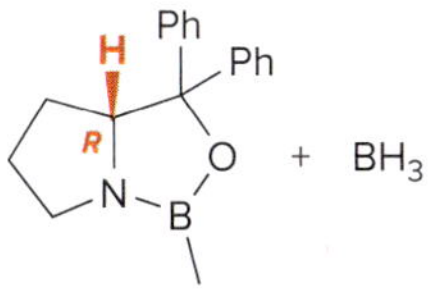

(*R*)-2-methyl-**CBS**-oxazaborolidine + BH_3
(*R*)-CBS reagent

These reagents are called the **(*S*)-CBS reagent** and the **(*R*)-CBS reagent,** named for *C*orey, *B*akshi, and *S*hibata, the chemists who developed these versatile reagents. One B–H bond of BH_3 serves as the source of hydride in this reduction. The stereochemistry of the new stereogenic center in the product is often predictable. For ketones having the general structure C_6H_5COR, draw the starting material with the aryl group on the left side of the carbonyl, as shown with acetophenone. Then, to draw the product, keep in mind:

- **The (*S*)-CBS reagent delivers hydride (H:⁻) from the *front* side of the C=O. This generally affords the *R* alcohol as the major product.**
- **The (*R*)-CBS reagent delivers hydride (H:⁻) from the *back* side of the C=O. This generally affords the *S* alcohol as the major product.**

acetophenone

[1] (*S*)-CBS reagent
[2] H_2O
H:⁻ from the **front**
major product
***R* isomer**

[1] (*R*)-CBS reagent
[2] H_2O
H:⁻ from **behind**
major product
***S* isomer**

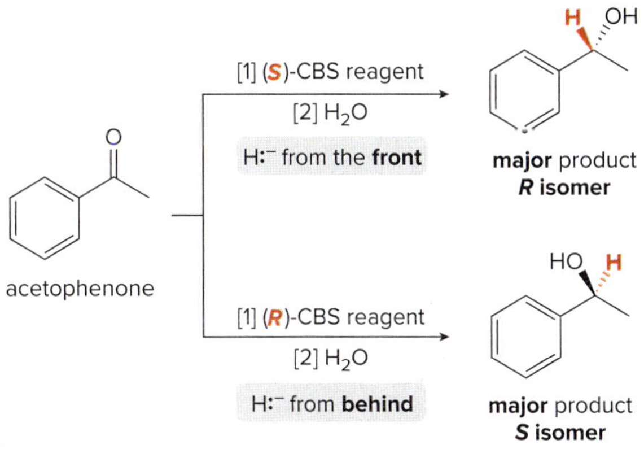

These reagents are highly enantioselective. Treatment of propiophenone with the (*S*)-CBS reagent forms the ***R*** alcohol in 97% enantiomeric excess (*ee*). Enantioselective reductions are key steps in the synthesis of several natural products, including ginkgolide B, the chapter-opening molecule present in extracts from the ginkgo tree shown in Figure 17.2. This new technology provides access to single enantiomers of biologically active compounds, often previously available only as a racemic mixture.

propiophenone → [1] (*S*)-CBS reagent; [2] H_2O; 97% *ee* → ***R*** isomer 98.5% + ***S*** isomer 1.5%

Figure 17.2 Enantioselective reduction in the synthesis of ginkgolide B

A → [1] (*S*)-CBS reagent; [2] HCl, CH_3OH → **B** (93% *ee*) → several steps → ginkgolide B

Jill Braaten

- Reduction of ketone **A** with the (*S*)-CBS reagent yielded allylic alcohol **B** in 93% *ee*. The five-membered ring of **B** forms the ring highlighted in red of ginkgolide B in an enantioselective synthesis.

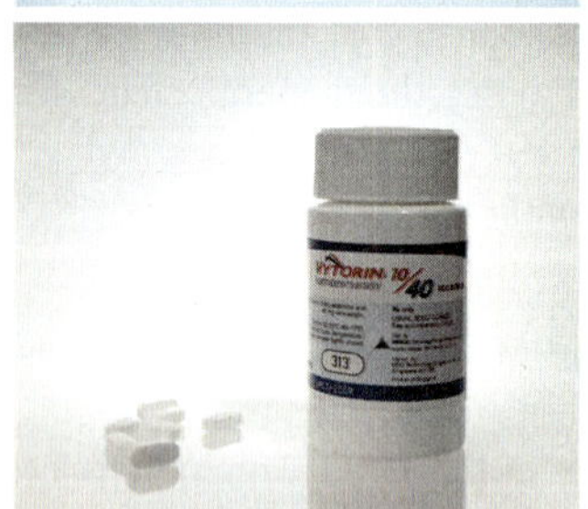

Ezetimibe (Problem 17.9) is sold as a single medication under the trade name of Zetia, or in combination with simvastatin, another cholesterol-lowering medication, and marketed as Vytorin. These drugs are prescribed for individuals who cannot tolerate or derive no benefit from other cholesterol-lowering medications.

K. L. Howard/Alamy Stock Photo

Problem 17.8 What product is formed when ketone **W** is treated with the (*R*)-CBS reagent? This reaction was a key step in the enantioselective synthesis of the long-acting bronchodilator salmeterol (Problem 17.74).

W → [1] (*R*)-CBS reagent; [2] H_2O

Problem 17.9 What carbonyl compound and CBS reagent are needed to prepare **X,** an intermediate in the synthesis of ezetimibe (trade name Zetia), a drug that lowers cholesterol levels by inhibiting its absorption in the intestines?

X → one step → ezetimibe

17.6B Enantioselective Biological Reduction

Although laboratory reduction reactions often do not proceed with 100% enantioselectivity, biological reductions that occur in cells *always* proceed with complete selectivity, forming a single enantiomer. In cells, the reducing agent is **NADH,** nicotinamide adenine dinucleotide (reduced form), the coenzyme introduced in Section 12.14.

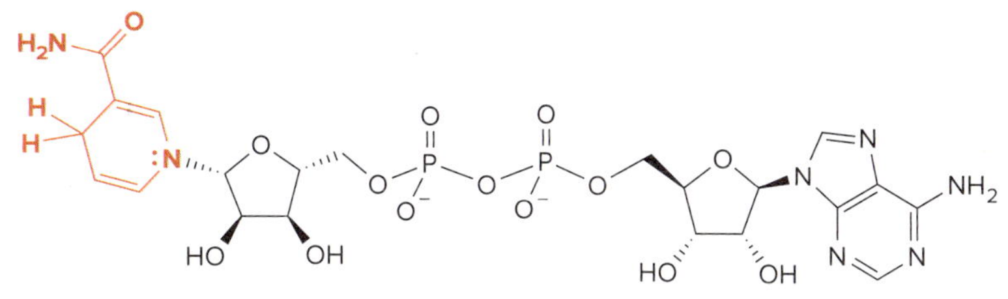

nicotinamide adenine dinucleotide
(reduced form)
NADH

In biological reduction, **NADH donates H:$^-$,** in much the same way as a metal hydride reagent. Nucleophilic attack of hydride and protonation thus form an alcohol from a carbonyl group, and **NADH is converted to NAD$^+$.**

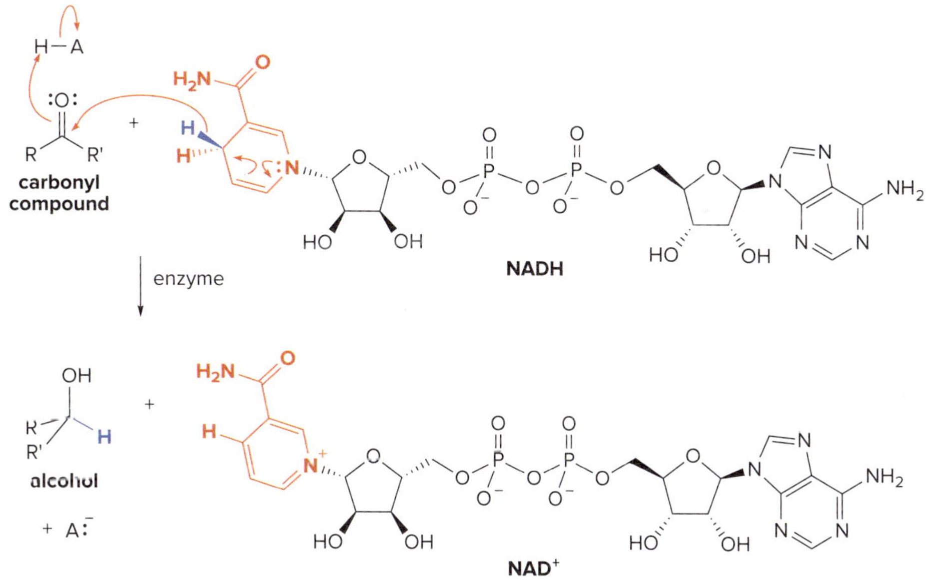

carbonyl compound + NADH —enzyme→ alcohol + A:$^-$ + NAD$^+$

This reaction is completely enantioselective. Reduction of pyruvic acid with NADH catalyzed by lactate dehydrogenase affords a single enantiomer of lactic acid with the *S* configuration. NADH reduces a variety of different carbonyl compounds in biological systems. The configuration of the product (*R* or *S*) depends on the enzyme used to catalyze the process.

Pyruvic acid is formed during the metabolism of glucose. During periods of strenuous exercise, when there is insufficient oxygen to metabolize pyruvic acid to CO_2, pyruvic acid is reduced to lactic acid. The tired feeling of sore muscles is a result of lactic acid accumulation.

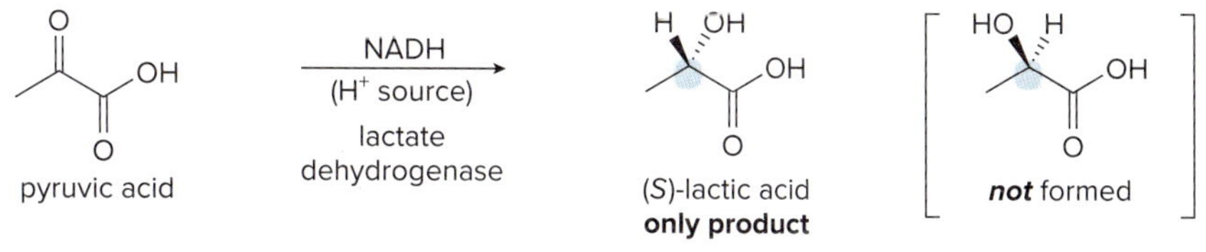

pyruvic acid —NADH (H^+ source), lactate dehydrogenase→ (*S*)-lactic acid **only product**; [**not** formed]

As we learned in Section 12.14, **NAD$^+$, the oxidized form of NADH, is a biological oxidizing agent** capable of oxidizing alcohols to carbonyl compounds, forming NADH in the process.

Niacin can be obtained from foods such as soybeans, which contain it naturally, and from breakfast cereals, which are fortified with it to help people consume their recommended daily allowance of this B vitamin. *C Squared Studios/Getty Images*

NAD^+ is synthesized from the vitamin niacin, which can be obtained from soybeans among other dietary sources.

niacin
vitamin B_3

Problem 17.10 Draw a stepwise mechanism for the following reaction, the last step in the biosynthesis of the amino acid proline.

NADH + H^+ → NAD^+

proline

17.7 Reduction of Carboxylic Acids and Their Derivatives

The reduction of carboxylic acids and their derivatives (**RCOZ**) is complicated because the products obtained depend on the identity of both the leaving group (Z) and the reducing agent. Metal hydride reagents are the most useful reducing reagents. **Lithium aluminum hydride is a strong reducing agent that reacts with *all* carboxylic acid derivatives.** Two other related but more selective reducing agents are also used:

[1] **Diisobutylaluminum hydride, $[(CH_3)_2CHCH_2]_2AlH$,** abbreviated as **DIBAL-H,** has two bulky isobutyl groups, which make this reagent less reactive than $LiAlH_4$.

[2] **Lithium tri-*tert*-butoxyaluminum hydride, $LiAlH[OC(CH_3)_3]_3$,** has three electronegative oxygen atoms bonded to aluminum, which make this reagent less nucleophilic than $LiAlH_4$.

$LiAlH_4$ is a strong, nonselective reducing agent. DIBAL-H and $LiAlH[OC(CH_3)_3]_3$ are milder, more selective reducing agents.

Al–H = $[(CH_3)_2CHCH_2]_2AlH$
diisobutylaluminum hydride
DIBAL-H

Li^+ H–Al–O = $LiAlH[OC(CH_3)_3]_3$
lithium tri-*tert*-butoxyaluminum hydride

In both reagents, the **single H atom bonded to Al is donated as H:$^-$** in hydride reductions.

17.7A Reduction of Acid Chlorides and Esters

Acid chlorides and esters can be reduced to either aldehydes or alcohols, depending on the reagent.

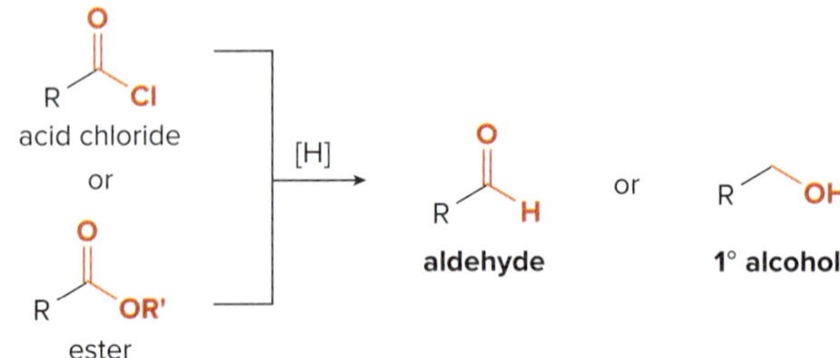

- $LiAlH_4$ converts RCOCl and RCOOR' to alcohols.
- A milder reducing agent (DIBAL-H or $LiAlH[OC(CH_3)_3]_3$) converts RCOCl or RCOOR' to RCHO at low temperatures.

In the reduction of an acid chloride, Cl^- comes off as the leaving group.

In the reduction of the ester, CH_3O^- comes off as the leaving group, which is then protonated by H_2O to form CH_3OH.

Mechanism 17.4 illustrates why two different products are possible. It can be conceptually divided into two parts: **nucleophilic substitution** to form an aldehyde (Steps [1] and [2]), followed by **nucleophilic addition** to the aldehyde to form an alcohol (Steps [3] and [4]). A general mechanism is drawn using $LiAlH_4$ as reducing agent.

Mechanism 17.4 Reduction of RCOCl and RCOOR' with a Metal Hydride Reagent

1. **Nucleophilic attack of H:⁻** forms a tetrahedral intermediate with a leaving group Z.
2. The π bond is re-formed and **the leaving group Z departs.** The overall result of addition of H:⁻ and elimination of Z:⁻ is **substitution of H for Z.**
3. **Nucleophilic attack of H:⁻** forms an alkoxide with no leaving group.
4. Protonation of the alkoxide by H_2O forms the **alcohol** reduction product. The overall result of Steps [3] and [4] is **addition of H_2.**

With less nucleophilic reducing agents such as DIBAL-H and $LiAlH[OC(CH_3)_3]_3$, the process stops after reaction with one equivalent of $H:^-$ and the aldehyde is formed as product (Steps [1] and [2] of Mechanism 17.4). With a stronger reducing agent like $LiAlH_4$, two equivalents of $H:^-$ are added and an alcohol is formed.

Problem 17.11 Draw a stepwise mechanism for the following reaction.

[1] $LiAlH_4$
[2] H_2O

Palau'amine was isolated from the sea sponge *Hymeniacidon agminata* collected in the Pacific Ocean near the island of Palau. *Daniel C. Smith*

Problem 17.12 Draw the structure of both an acid chloride and an ester that can be used to prepare each compound by reduction.

a. b. c.

Selective reductions are routinely used in the synthesis of highly complex natural products such as palau'amine, the complex marine natural product described in the Spectroscopy Part C opener. One reaction in a synthesis of palau'amine involved the reduction of a diester to a diol using $LiAlH_4$, as shown in Figure 17.3.

Figure 17.3 $LiAlH_4$ reduction in the synthesis of palau'amine

A → [1] $LiAlH_4$, [2] H_2O → B → several steps → palau'amine

- One step in a lengthy synthesis of the marine natural product palau'amine involved reduction of a diester to a diol.

Problem 17.13 In a synthesis of trabectedin (ecteinascidin 743), compound **A** was converted to compound **B.** What reagent would you choose to carry out this transformation? Trabectedin is used to treat soft tissue cancers that cannot be removed surgically.

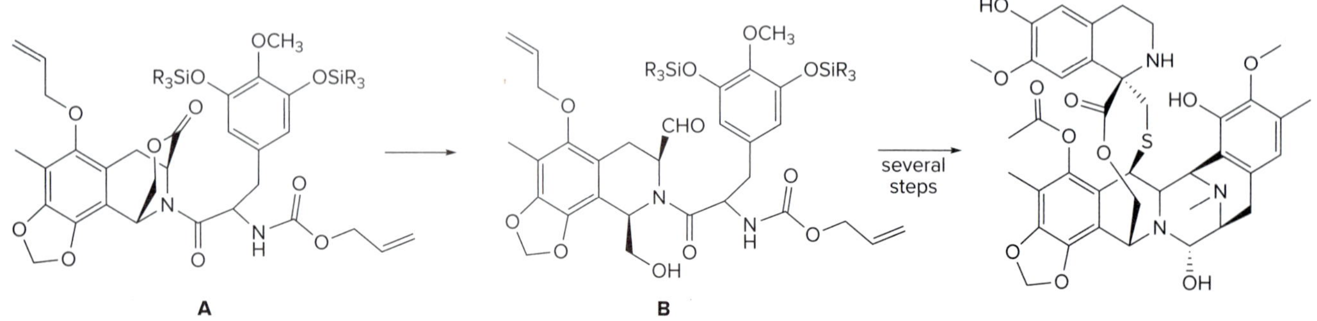

ecteinascidin 743
trabectedin
Trade name: Yondelis

17.7B Reduction of Carboxylic Acids and Amides

The sea squirt *Ecteinascidia turbinata* is the source of the anticancer drug trabectedin (Problem 17.13). *Florent Charpin 2004–2016. All Rights Reserved*

Carboxylic acids are reduced to alcohols with $LiAlH_4$. $LiAlH_4$ is too strong a reducing agent to stop the reaction at the aldehyde stage, but milder reagents are not strong enough to initiate the reaction in the first place, so this is the only useful reduction reaction of carboxylic acids.

carboxylic acid → [1] $LiAlH_4$ [2] H_2O → 1° alcohol

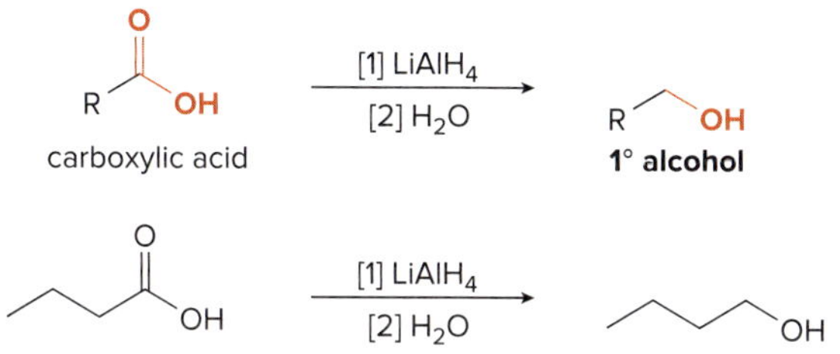

Unlike the $LiAlH_4$ reduction of all other carboxylic acid derivatives, which affords alcohols, the **$LiAlH_4$ reduction of amides forms amines.**

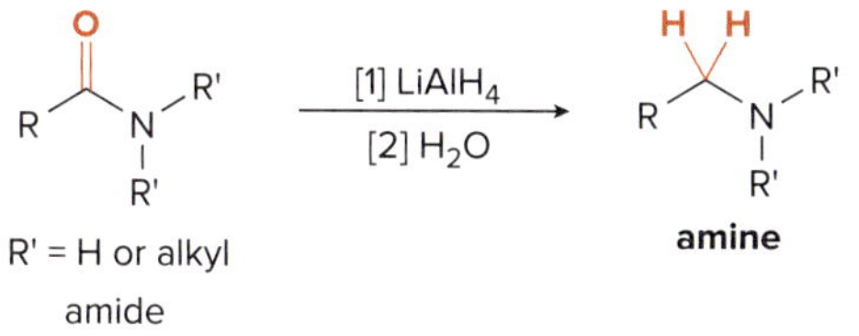

R' = H or alkyl

amide → [1] $LiAlH_4$ [2] H_2O → amine

Both C–O bonds are reduced to C–H bonds by $LiAlH_4$, and any H atom or R group bonded to the amide nitrogen atom remains bonded to it in the product. Because $^-NH_2$ (or ^-NHR or $^-NR_2$) is a *poorer* leaving group than Cl^- or ^-OR, **$^-NH_2$ is never lost during reduction,** and therefore it forms an amine in the final product.

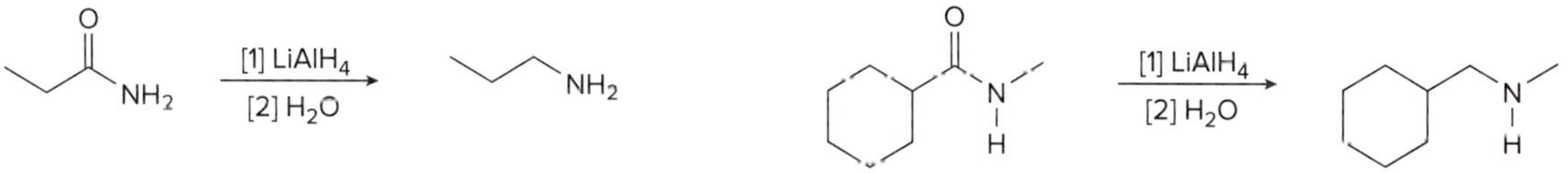

[1] $LiAlH_4$ [2] H_2O

Imines and related compounds are discussed in Chapter 18.

The mechanism, illustrated in Mechanism 17.5 with $RCONH_2$ as starting material, is somewhat different than the previous reductions of carboxylic acid derivatives. Amide reduction proceeds with formation of an intermediate ***imine*****, a compound containing a C–N double bond,** which is then further reduced to an amine.

Problem 17.14 Draw the products formed from $LiAlH_4$ reduction of each compound.

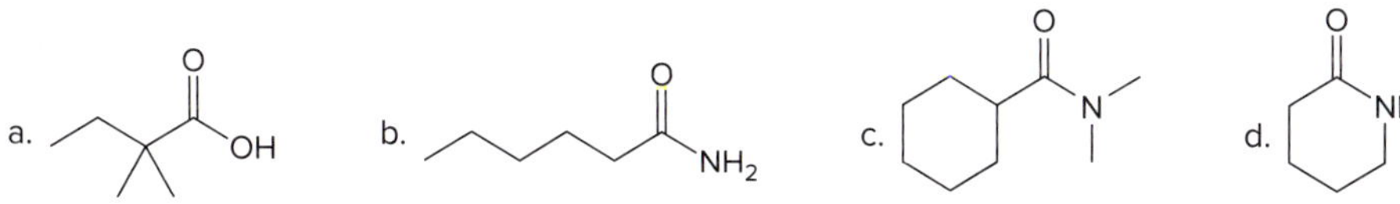

a. b. c. d.

Mechanism 17.5 Reduction of an Amide to an Amine with $LiAlH_4$

Part [1] Reduction of an amide to an imine

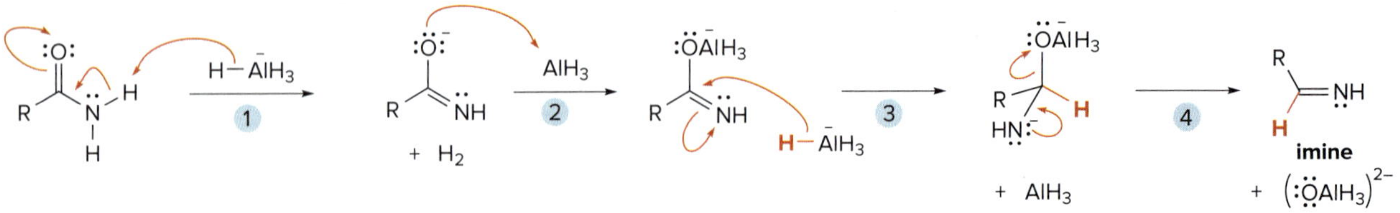

1 – 2 AlH_4^- removes a proton from the amide to form a Lewis base that complexes with AlH_3 in Step [2].

3 – 4 **Nucleophilic attack of H:⁻ and loss of a leaving group, $(OAlH_3)^{2-}$, form an imine.**

Part [2] Reduction of an imine to an amine

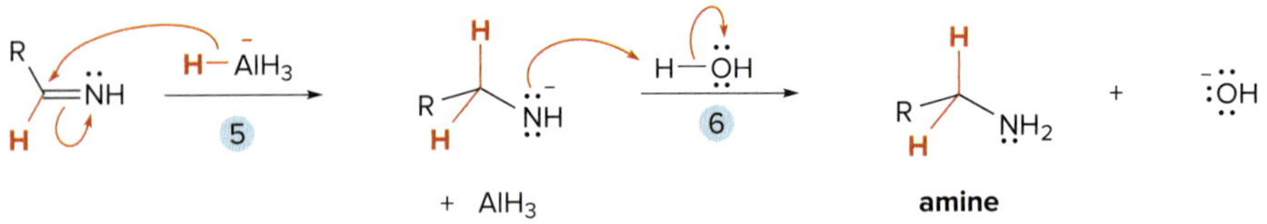

5 – 6 **Nucleophilic addition of H:⁻** and protonation form the **amine.**

Sample Problem 17.1 Determining the Amide That Forms an Amine by Reduction

What amide(s) form amine **X** on treatment with $LiAlH_4$?

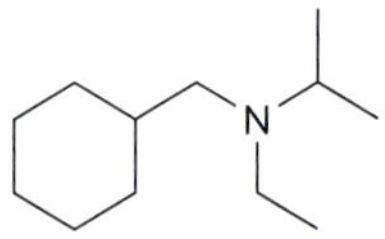

X

Solution

$LiAlH_4$ reduction of an amide converts a **C=O** to a **CH_2** group, so we must examine the alkyl groups bonded to the amine nitrogen. Any alkyl group with a CH_2 group bonded *directly* to the N can be formed by reduction of a C=O. An alkyl group with zero or one H atom *cannot* be formed by reduction of a C=O. For amine **X,** the CH_2 groups in blue can be formed from C=O's, but the CH group labeled in red cannot be formed from a C=O.

A **B**

only 1 H on C

The CH bond in red *cannot* be formed from a C=O.

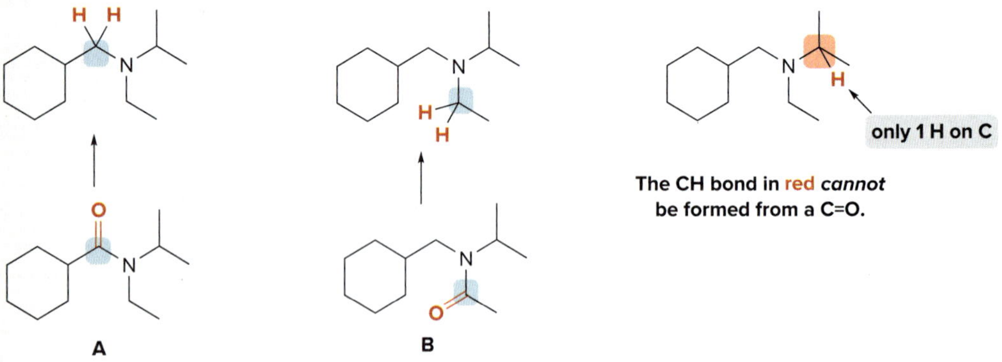

Thus, amides **A** and **B** can be converted to amine **X** by reduction of a C=O with $LiAlH_4$.

Problem 17.15 What amide(s) will form each of the following amines on treatment with $LiAlH_4$?

a. 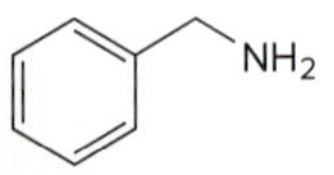 b. 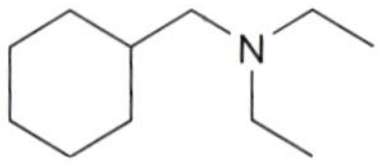 c.

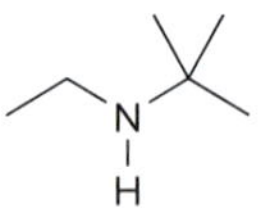

More Practice: Try Problem 17.56.

17.7C A Summary of the Reagents for Reduction

The many available metal hydride reagents reduce a wide variety of functional groups. Keep in mind that **$LiAlH_4$ is such a strong reducing agent that it *nonselectively* reduces most polar functional groups**. All other metal hydride reagents are more selective, and each has its particular reactions that best utilize its reduced reactivity. The reagents and their uses are summarized in Table 17.1.

Problem 17.16 What product is formed when each compound is treated with either $LiAlH_4$ (followed by H_2O), or $NaBH_4$ in CH_3OH?

a. [structure] b. [structure with OH] c. [structure with O]

Table 17.1 A Summary of Metal Hydride Reducing Agents

	Reagent	Starting material	→	Product
Strong reagent	$LiAlH_4$	RCHO	→	RCH_2OH
		R_2CO	→	R_2CHOH
		RCOOH	→	RCH_2OH
		RCOOR'	→	RCH_2OH
		RCOCl	→	RCH_2OH
		$RCONH_2$	→	RCH_2NH_2
Milder reagents	$NaBH_4$	RCHO	→	RCH_2OH
		R_2CO	→	R_2CHOH
	$LiAlH[OC(CH_3)_3]_3$	RCOCl	→	RCHO
	DIBAL-H	RCOOR'	→	RCHO

Aldehydes give a positive Tollens test; that is, they react with Ag^+ to form RCOOH and Ag. When the reaction is carried out in a glass flask, a silver mirror is formed on its walls. Other functional groups give a negative Tollens test, because no silver mirror forms.
Charles D. Winters/McGraw Hill

17.8 Oxidation of Aldehydes

The most common oxidation reaction of carbonyl compounds is the oxidation of **aldehydes to carboxylic acids.** A variety of oxidizing agents can be used, including CrO_3, $Na_2Cr_2O_7$, $K_2Cr_2O_7$, and $KMnO_4$. Cr^{6+} reagents are also used to oxidize 1° and 2° alcohols, as discussed in Section 12.12. Because ketones have no H on the carbonyl carbon, they do *not* undergo this oxidation reaction.

Aldehydes are oxidized selectively in the presence of other functional groups using **silver(I) oxide in aqueous ammonium hydroxide (Ag_2O in NH_4OH).** This is called **Tollens reagent.** Oxidation with Tollens reagent provides a distinct color change, because the Ag^+ reagent is reduced to silver metal (Ag), which precipitates out of solution.

[aldehyde, H] $\xrightarrow[H_2SO_4,\ H_2O]{CrO_3}$ [carboxylic acid, OH] + Cr^{3+}

[HO-cyclohexyl aldehyde, H] $\xrightarrow{Ag_2O,\ NH_4OH}$ [HO-cyclohexyl carboxylic acid, OH] + Ag (silver mirror)

Only the aldehyde is oxidized.

Problem 17.17 What product is formed when each compound is treated with either Ag_2O, NH_4OH or $Na_2Cr_2O_7$, H_2SO_4, H_2O?

a. OH b. OH O H

Problem 17.18 Review the oxidation reactions using Cr^{6+} reagents in Section 12.12. Then draw the product formed when compound **B** is treated with each reagent.

OH CHO HO **B**

a. $NaBH_4$, CH_3OH
b. [1] $LiAlH_4$; [2] H_2O
c. PCC
d. Ag_2O, NH_4OH
e. CrO_3, H_2SO_4, H_2O

17.9 Organometallic Reagents

We will now discuss the reactions of carbonyl compounds with organometallic reagents, another class of nucleophiles.

- ***Organometallic reagents*** **contain a carbon atom bonded to a metal.**

C δ− M δ+ = R—M Most common metals: M = Li, Mg, Cu

organometallic reagent M = metal

Lithium, magnesium, and copper are the most commonly used metals in organometallic reagents, but others (such as Sn, Si, Tl, Al, Ti, and Hg) are known. General structures of the three common organometallic reagents are shown. R can be alkyl, aryl, allyl, benzyl, *sp*^{2} hybridized, and with M = Li or Mg, *sp* hybridized. Because metals are *more electropositive* (less electronegative) than carbon, they donate electron density toward carbon, so that **carbon bears a partial negative charge.**

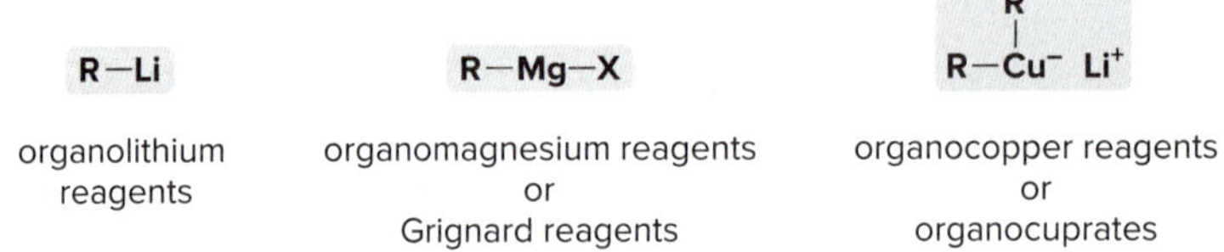

organolithium reagents | organomagnesium reagents or Grignard reagents | organocopper reagents or organocuprates

- **The *more polar* the carbon–metal bond, the *more reactive* the organometallic reagent.**

Because both Li and Mg are very electropositive metals, **organolithium (RLi)** and **organomagnesium reagents (RMgX) contain very polar carbon–metal bonds and are therefore *very reactive* reagents.** Organomagnesium reagents are called **Grignard reagents,** after Victor Grignard, who received the Nobel Prize in Chemistry in 1912 for his work with them.

Electronegativity values for carbon and the common metals in R–M reagents are C (2.5), Li (1.0), Mg (1.3), and Cu (1.8).

Organocopper reagents (R_2CuLi), also called organocuprates, have a less polar carbon–metal bond and are therefore *less reactive.* Although organocuprates contain two alkyl groups bonded to copper, only one R group is utilized in a reaction.

Regardless of the metal, organometallic reagents are useful synthetically because they react as if they were free carbanions; that is, carbon bears a partial *negative* charge, so the **reagents react as bases and nucleophiles.**

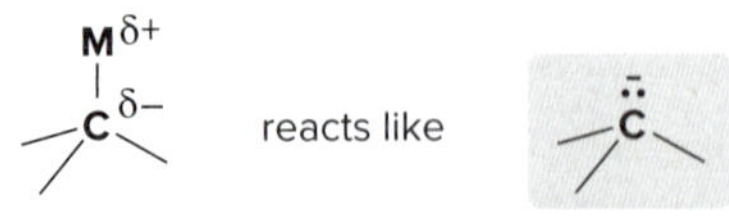

carbanion

a base and a nucleophile

17.9A Preparation of Organometallic Reagents

Organolithium and Grignard reagents are typically prepared by reaction of an organic halide with the corresponding metal, as shown in the accompanying equations.

$$R{-}X + 2\,Li \longrightarrow R{-}Li + LiX$$

organolithium reagent

$$CH_3{-}Br + 2\,Li \longrightarrow CH_3{-}Li + LiBr$$

methyllithium

$$R{-}X + Mg \xrightarrow[\text{diethyl ether}]{} R{-}Mg{-}X$$

Grignard reagent

$$CH_3{-}Br + Mg \xrightarrow[\text{diethyl ether}]{} CH_3{-}Mg{-}Br$$

methylmagnesium bromide

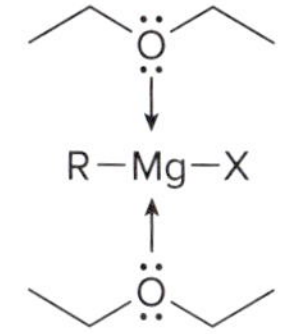

Two molecules of diethyl ether complex with the Mg atom of the Grignard reagent.

With lithium, the halogen and metal exchange to form the organolithium reagent. With magnesium, the metal inserts in the carbon–halogen bond, forming the Grignard reagent. Grignard reagents are usually prepared in diethyl ether ($CH_3CH_2OCH_2CH_3$) as solvent. It is thought that two ether oxygen atoms complex with the magnesium atom, stabilizing the reagent.

Organocuprates are prepared from organolithium reagents by reaction with a Cu^+ salt, often CuI.

$$2\,R{-}Li + CuI \longrightarrow R_2Cu^-\,Li^+ + LiI$$

organocopper reagent

$$2\,CH_3{-}Li + CuI \longrightarrow (CH_3)_2Cu^-\,Li^+ + LiI$$

lithium dimethyl cuprate

Problem 17.19 Write the step(s) needed to convert CH_3CH_2Br to each reagent: (a) CH_3CH_2Li; (b) CH_3CH_2MgBr; (c) $(CH_3CH_2)_2CuLi$.

17.9B Acetylide Anions

The **acetylide anions** discussed in Chapter 11 are another example of organometallic compounds. These reagents are prepared by an acid–base reaction of an alkyne with a base such as $NaNH_2$ or NaH. We can think of these compounds as **organosodium** reagents. Because sodium is even more electropositive (less electronegative) than lithium, the C–Na bond of these organosodium compounds is best described as **ionic,** rather than polar covalent.

$$R{-}C{\equiv}C{-}H + Na^+\ {:}\ddot{N}H_2^- \rightleftharpoons R{-}C{\equiv}C{:}^-\ Na^+ + {:}NH_3$$

acetylide anion
an organosodium compound

An acid–base reaction can also be used to prepare *sp* hybridized organolithium compounds. Treatment of a terminal alkyne with CH_3Li affords a lithium acetylide. Equilibrium favors the products because the *sp* hybridized C–H bond of the terminal alkyne is more acidic than the sp^3 hybridized conjugate acid, CH_4, that is formed.

$$R{-}C{\equiv}C{-}H + CH_3{-}Li \rightleftharpoons R{-}C{\equiv}C{-}Li + CH_3{-}H$$

$R{-}C{\equiv}C{-}H$	$CH_3{-}Li$	$R{-}C{\equiv}C{-}Li$	$CH_3{-}H$
$pK_a \approx 25$	base	**a lithium acetylide**	$pK_a = 50$
stronger acid			**weaker acid**

Problem 17.20 Which of the following species represent organometallic compounds: (a) $BrMgC{\equiv}CCH_2CH_3$; (b) $NaOCH_2CH_3$; (c) $KOC(CH_3)_3$; (d) PhLi?

17.9C Reaction as a Base

- **Organometallic reagents are strong bases that readily abstract a proton from water to form hydrocarbons.**

The electron pair in the carbon–metal bond is used to form a new bond to the proton. Equilibrium favors the products of this acid–base reaction because H_2O is a much stronger acid than the alkane product.

CH_3Li (base) + H–OH (acid, $pK_a = 14$, stronger acid) ⇌ CH_4 ($pK_a = 50$, a very weak acid) + Li^+ ^-OH

Similar reactions occur for the same reason with the O–H proton in alcohols and carboxylic acids, and the N–H protons of amines.

C_6H_5MgBr (strong base) + H–OCH_3 (acid) ⟶ C_6H_5–H + CH_3O^- $(MgBr)^+$

$CH_3CH_2CH_2Li$ (strong base) + H–$OCOCH_3$ (acid) ⟶ $CH_3CH_2CH_2$–H + Li^+ $^-OCOCH_3$

Because organolithium and Grignard reagents are themselves prepared from alkyl halides, a two-step method converts an alkyl halide to an alkane (or another hydrocarbon).

$$\underset{\text{alkyl halide}}{R{-}X} \xrightarrow{M} R{-}M \xrightarrow{H_2O} \underset{\textbf{alkane}}{R{-}H}$$

Problem 17.21 Draw the product formed when each organometallic reagent is treated with H_2O.

a. cyclohexyl–Li b. $(CH_3)_3C$–MgBr c. $C_6H_5CH_2$–MgBr d. $CH_3CH_2C{\equiv}C$–Li

17.9D Reaction as a Nucleophile

Organometallic reagents are also strong nucleophiles that react with electrophilic carbon atoms to form new carbon–carbon bonds. These reactions are very valuable in forming the carbon skeletons of complex organic molecules. The following reactions of organometallic reagents are examined in Sections 17.10, 17.13, and 17.14:

[1] Reaction of R–M with aldehydes and ketones to afford alcohols (Section 17.10)

$$\underset{\substack{R' = H \text{ or alkyl} \\ \text{aldehyde or ketone}}}{RCOR'} \xrightarrow[\text{[2] } H_2O]{\text{[1] } R''{-}M} \underset{\textbf{1°, 2°, or 3° alcohol}}{RR'C(OH)R''}$$

Aldehydes and ketones are converted to 1°, 2°, or 3° alcohols with R"Li or R"MgX.

[2] Reaction of R–M with carboxylic acid derivatives (Section 17.13)

RCOZ → [1] R"–M, [2] H_2O → RCOR" (**ketone**) or RC(OH)R"R" (**3° alcohol**)

Z = Cl or OR'

Acid chlorides and esters can be converted to ketones or 3° alcohols with organometallic reagents. The identity of the product depends on the identity of R"–M and the leaving group Z.

[3] Reaction of R–M with other electrophilic functional groups (Section 17.14)

R–M → [1] CO_2, [2] H_3O^+ → RCOOH (carboxylic acid)

R–M → [1] ethylene oxide, [2] H_2O → RCH_2CH_2OH (alcohol)

Organometallic reagents also react with CO_2 to form carboxylic acids and with epoxides to form alcohols.

17.10 Reaction of Organometallic Reagents with Aldehydes and Ketones

Treatment of an aldehyde or ketone with either an organolithium or Grignard reagent followed by water forms an alcohol with a new carbon–carbon bond. This reaction is an **addition reaction** because the elements of R" and H are added across the π bond.

RCOR' (R' = H or alkyl; aldehyde or ketone) → R"MgX or R"Li → H–OH → RR'C(OH)R" (**1°, 2°, or 3° alcohol**; new C–C bond)

17.10A General Features

This reaction follows the general mechanism for nucleophilic addition (Section 17.2A)—that is, **nucleophilic attack** by a carbanion followed by **protonation.** Mechanism 17.6 is shown using R"MgX, but the same steps occur with organolithium reagents and acetylide anions.

Mechanism 17.6 Nucleophilic Addition of R"MgX to RCHO and $R_2C=O$

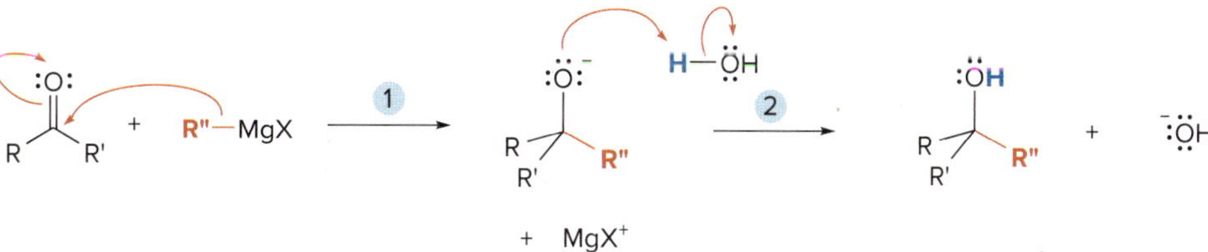

1 **The nucleophile (R")⁻ attacks the carbonyl group,** breaking the π bond and yielding an alkoxide. This forms a new carbon–carbon bond.

2 Protonation of the alkoxide by H_2O forms the **addition product** with a new O–H bond. The overall result is addition of R" and H to the carbonyl group.

This reaction is used to prepare 1°, 2°, and 3° alcohols, depending on the number of alkyl groups bonded to the carbonyl carbon of the aldehyde or ketone.

[1] formaldehyde + R"–MgX → alkoxide → (H_2O) → **1° alcohol**

[2] **aldehyde** (R ≠ H) + R"–MgX → alkoxide → (H_2O) → **2° alcohol**

[3] **ketone** + R"–MgX → alkoxide → (H_2O) → **3° alcohol**

- [1] Addition of R"MgX to formaldehyde ($CH_2{=}O$) forms a 1° alcohol.
- [2] Addition of R"MgX to all other aldehydes forms a 2° alcohol.
- [3] Addition of R"MgX to ketones forms a 3° alcohol.

Each reaction results in addition of one new alkyl group to the carbonyl carbon, and forms one new carbon–carbon bond. The reaction is general for all organolithium and Grignard reagents, and works for acetylide anions as well, as illustrated in Equations [1]–[3].

[1] formaldehyde —[1] CH_3–MgX, [2] H_2O→ **1° alcohol**

[2] benzaldehyde —[1] CH₃CH₂Li, [2] H_2O→ **2° alcohol**

[3] cyclohexanone —[1] HC≡C–Li, [2] H_2O→ **3° alcohol**

Because organometallic reagents are strong bases that rapidly react with H_2O (Section 17.9C), the addition of the new alkyl group must be carried out under anhydrous conditions to prevent traces of water from reacting with the reagent, thus reducing the yield of the desired alcohol. Water is added *after* the addition to protonate the alkoxide.

Problem 17.22 Draw the product of each reaction.

a. 3-pentanone —[1] CH₃CH₂CH₂Li, [2] H_2O→

b. formaldehyde —[1] cyclohexyl–Li, [2] H_2O→

c. cyclopentanone —[1] C_6H_5Li, [2] H_2O→

d. alkynyllithium (Li) —[1] $CH_2{=}O$, [2] H_2O→

17.10B Stereochemistry

Like reduction, addition of organometallic reagents converts an sp^2 hybridized carbonyl carbon to a tetrahedral sp^3 hybridized carbon. Addition of R–M always occurs from both sides of the trigonal planar carbonyl group. **When a new stereogenic center is formed from an achiral starting material, an equal mixture of enantiomers results,** as shown in Sample Problem 17.2.

Sample Problem 17.2 Drawing the Stereoisomers Formed During Grignard Addition

Draw all stereoisomers formed in the following reaction.

[1] ethyl MgBr
[2] H_2O

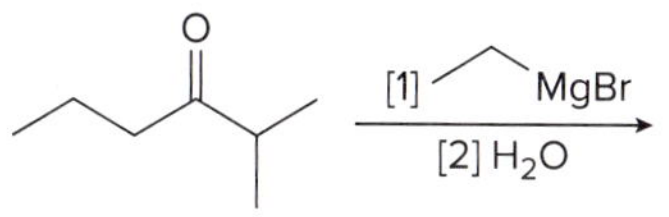

Solution

The Grignard reagent adds from both sides of the trigonal planar carbonyl group, forming two alkoxides, each containing a new stereogenic center labeled in blue. Protonation with water yields **an equal amount of two enantiomers—a racemic mixture.**

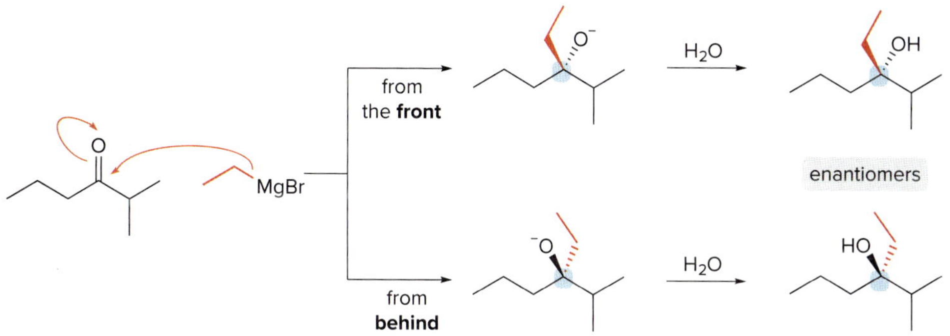

MgBr
from the **front**
from **behind**
H_2O
H_2O
OH
HO
enantiomers

Problem 17.23 Draw the products (including stereochemistry) of the following reactions.

a. 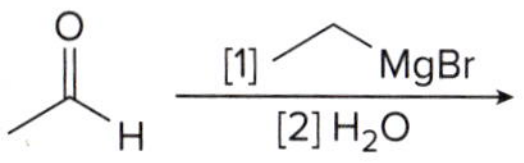 [1] ethyl MgBr, [2] H_2O

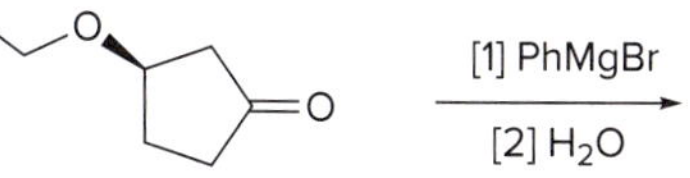

b. 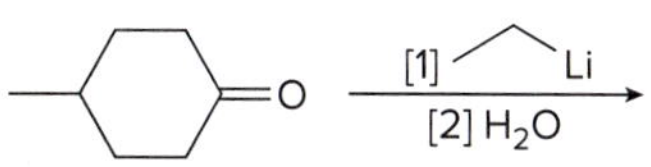 [1] ethyl Li, [2] H_2O

c. [1] PhMgBr, [2] H_2O

d.  + MgBr → H_2O

More Practice: Try Problem 17.46a.

17.10C Applications in Synthesis

Many syntheses of useful compounds utilize the nucleophilic addition of a Grignard or organolithium reagent to form carbon–carbon bonds. For example, the first step in the synthesis of the non-sedating antihistamine cetirizine (trade name Zyrtec) is the Grignard reaction shown in Figure 17.4.

Figure 17.4 A Grignard reaction in the synthesis of the antihistamine cetirizine

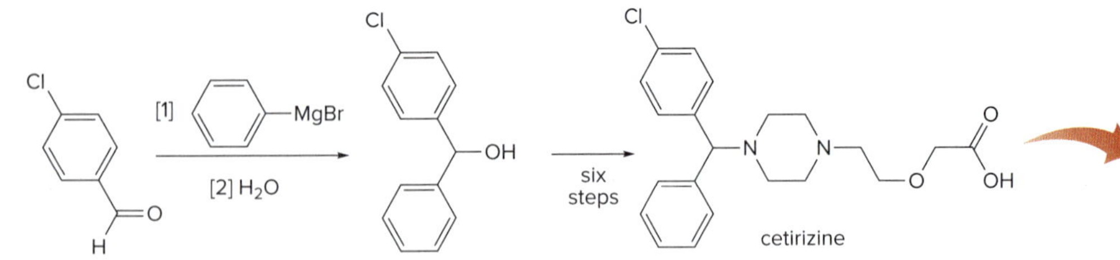

[1] C_6H_5MgBr; [2] H_2O — *p*-chlorobenzaldehyde → 2° alcohol → (six steps) → cetirizine

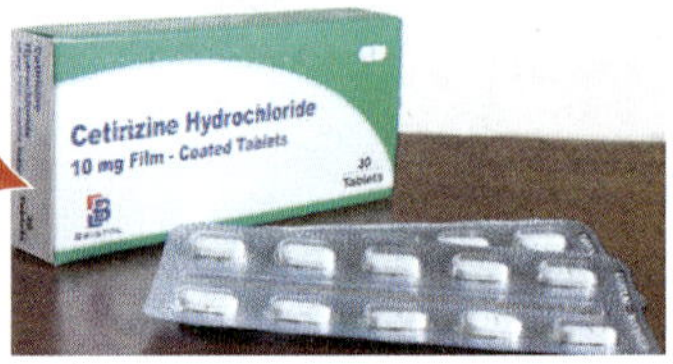

Elizabeth Leyden/Alamy Stock Photo

- Grignard reaction of *p*-chlorobenzaldehyde with phenylmagnesium bromide formed a 2° alcohol, which was converted to the non-sedating antihistamine cetirizine. The last three steps of the synthesis were illustrated in Problem 7.57.

The biotoxin azaspiracid-1 (Figure 17.5) was first isolated from contaminated mussels of the species *Mytilus edulis*. *Matt Bain/NHPA/Photoshot*

The synthesis of **azaspiracid-1,** a biotoxin that causes a common type of shellfish poisoning, illustrates that this nucleophilic addition reaction can occur with a functionalized organolithium reagent and a complex aldehyde. Four of the nine rings of azaspiracid-1 are identified with the letters **A–D,** as shown.

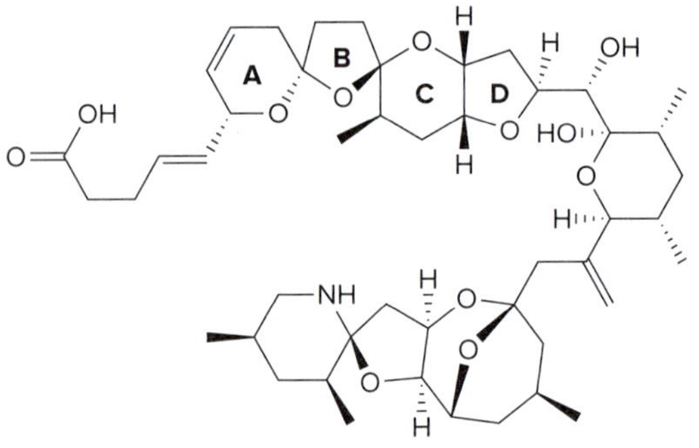

azaspiracid-1

Azaspiracid-1 is a polycyclic marine toxin produced by algae that can accumulate in shellfish and cause food poisoning in humans who consume contaminated shellfish. Azaspiracid-1 was first identified when incidents of shellfish poisoning were linked to the consumption of contaminated mussels of the species *Mytilus edulis* in the 1990s. To obtain sufficient quantities of azaspiracid-1 for biological testing and to gain an understanding of the mechanism of action of this biotoxin, azaspiracid-1 was synthesized in the laboratory. A key step in an early synthesis was the addition of an organolithium reagent to an aldehyde, as shown in Figure 17.5.

Figure 17.5 A key step in the synthesis of azaspiracid-1

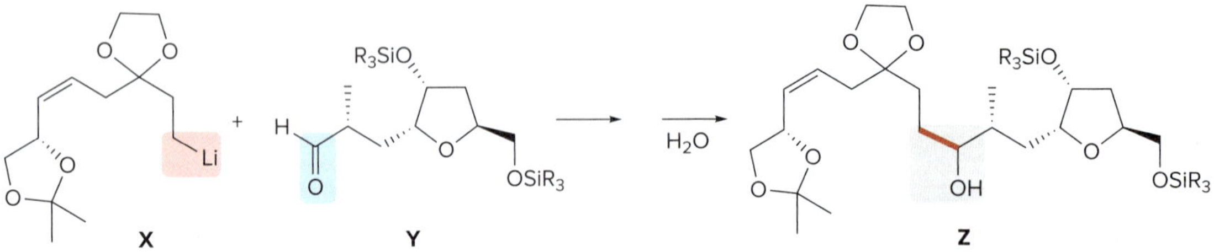

X + **Y** → (H_2O) → **Z**

- Reaction of organolithium reagent **X** with aldehyde **Y** forms alcohol **Z,** with a new carbon–carbon bond shown in red.
- The carbon atoms of **Z** form rings **A–D** in azaspiracid-1.

17.11 Retrosynthetic Analysis of Grignard Products

To use the Grignard addition in synthesis, you must be able to determine what carbonyl and Grignard components are needed to prepare a given compound—that is, **you must work backwards, in the retrosynthetic direction.** This involves a two-step process:

Step [1] Find the carbon bonded to the OH group in the product.

Step [2] Break the molecule into two components: One alkyl group bonded to the carbon with the OH group comes from the organometallic reagent. The rest of the molecule comes from the carbonyl component.

HO R O + R−MgX

Grignard product

two reactants

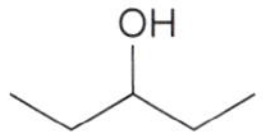

OH

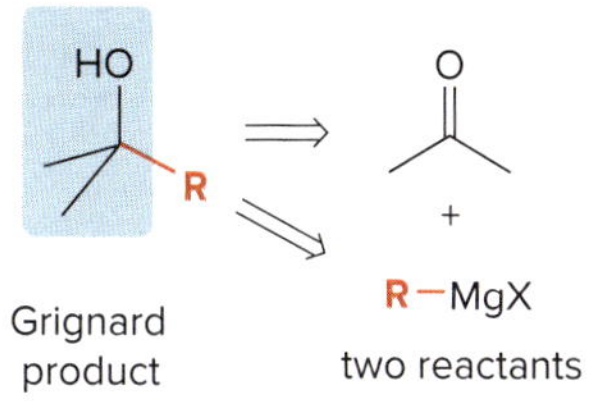

pentan-3-ol

To synthesize pentan-3-ol [$(CH_3CH_2)_2CHOH$] by a Grignard reaction, locate the carbon bonded to the OH group, and then break the molecule into two components at this carbon. Thus, retrosynthetic analysis shows that one of the ethyl groups on this carbon comes from a Grignard reagent (CH_3CH_2MgX), and the rest of the molecule comes from the carbonyl component, a three-carbon aldehyde.

Retrosynthetic analysis

O H HO Form this bond by Grignard addition. MgBr

three-carbon aldehyde

pentan-3-ol

two-carbon Grignard reagent

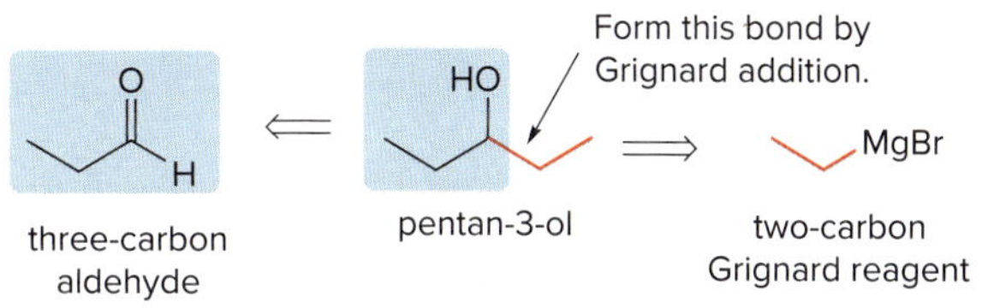

Then, writing the reaction in the synthetic direction—that is, from starting material to product—shows whether the analysis is correct. In this example, a three-carbon aldehyde reacts with CH_3CH_2MgBr to form an alkoxide, which can then be protonated by H_2O to form pentan-3-ol, the desired alcohol.

In the synthetic direction:

O H + MgBr ⟶ O^- $\xrightarrow{H_2O}$ OH

pentan-3-ol

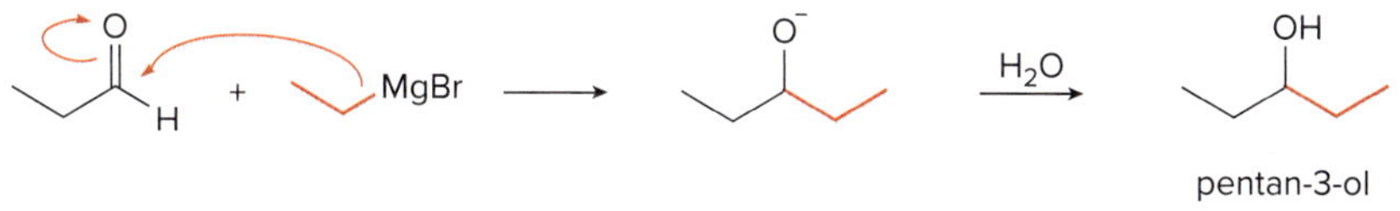

There is often more than one way to synthesize a 2° alcohol by Grignard addition, as shown in Sample Problem 17.3.

Sample Problem 17.3 Determining the Starting Materials in a Grignard Synthesis

Show two different methods to synthesize alcohol **A** using a Grignard reaction.

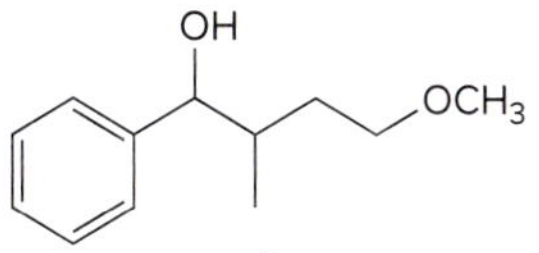

OH OCH_3

A

Solution

Because **A** has two different R groups bonded to the carbon bearing the OH group, there are two different ways to form a new carbon–carbon bond by Grignard addition.

Possibility [1] Use C_6H_5MgBr and aldehyde **B.**

Possibility [2] Use Grignard reagent **C** and benzaldehyde.

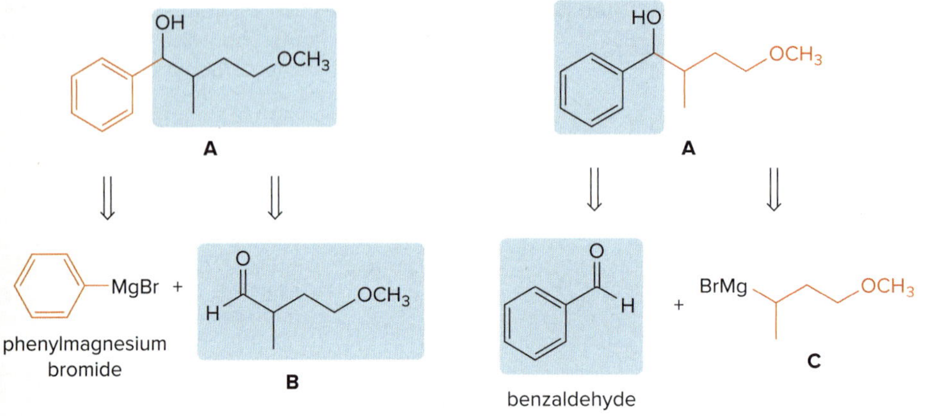

OH
OCH_3
A

HO
OCH_3
A

MgBr + H O OCH_3
phenylmagnesium bromide
B

O H + BrMg OCH_3
benzaldehyde
C

Both methods give the desired product **A,** as can be seen by writing the reactions from starting material to product.

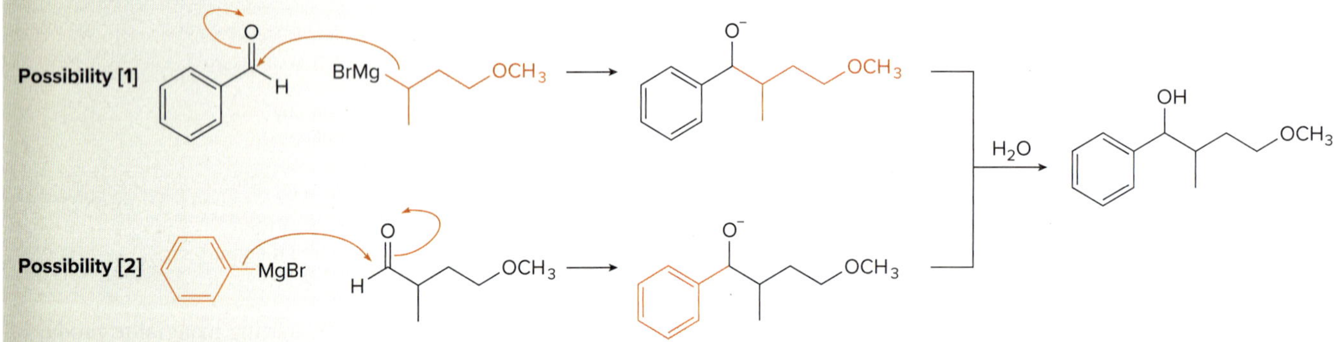

Possibility [1] O H BrMg OCH_3 ⟶ O^- OCH_3

H_2O ⟶ OH OCH_3

Possibility [2] MgBr H O OCH_3 ⟶ O^- OCH_3

Problem 17.24 What Grignard reagent and carbonyl compound are needed to prepare each alcohol? As shown in part (d), 3° alcohols with three different R groups on the carbon bonded to the OH group can be prepared by three different Grignard reactions.

a. 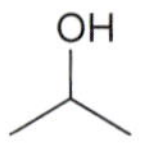 OH

b. 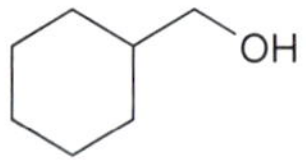 OH

c. 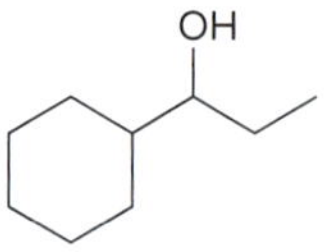 OH (two methods)

d. 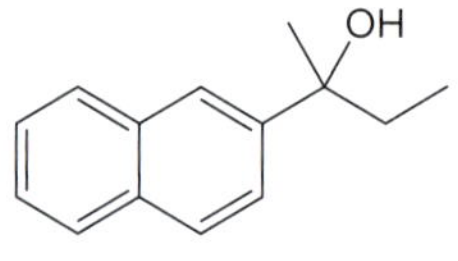 OH (three methods)

More Practice: Try Problems 17.57, 17.58, 17.60.

Problem 17.25 Linalool and lavandulol are two of the major components of lavender oil. (a) What organolithium reagent and carbonyl compound can be used to make each alcohol? (b) How might lavandulol be formed by reduction of a carbonyl compound? (c) Why can't linalool be prepared by a similar pathway?

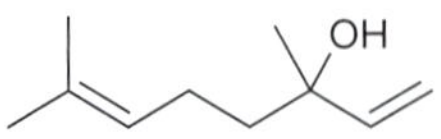

OH
linalool
(three methods)

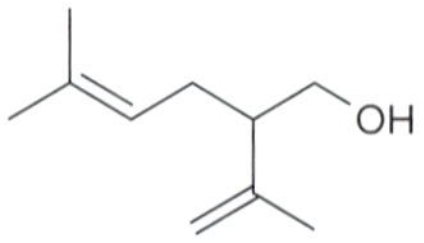

OH
lavandulol

Problem 17.26 What Grignard reagent and carbonyl compound can be used to prepare the antidepressant venlafaxine (trade name Effexor)? (See also Problem 15.51.)

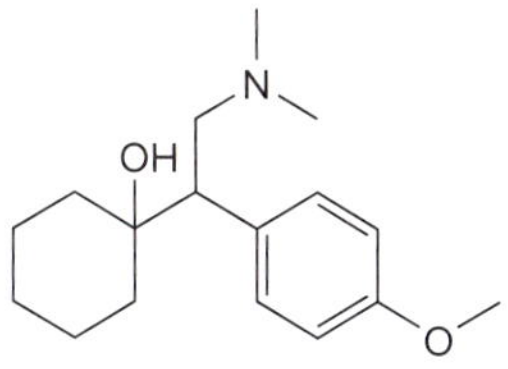

venlafaxine

17.12 Protecting Groups

Rapid acid–base reactions occur between organometallic reagents and all of the following functional groups: ROH, RCOOH, RNH_2, R_2NH, $RCONH_2$, RCONHR, and RSH.

Although the addition of organometallic reagents to carbonyls is a very versatile reaction, it cannot be used with molecules that contain both a carbonyl group and N–H or O–H bonds.

- **Carbonyl compounds that also contain N–H or O–H bonds undergo an acid–base reaction with organometallic reagents, *not* nucleophilic addition.**

Suppose, for example, that you wanted to add methylmagnesium chloride (CH_3MgCl) to the carbonyl group of 5-hydroxypentan-2-one to form a diol. Nucleophilic addition will *not* occur with this substrate. Instead, **because Grignard reagents are strong bases and proton transfer reactions are fast, CH_3MgCl removes the O–H proton before nucleophilic addition takes place.** The stronger acid and base react to form the weaker conjugate acid and conjugate base, as we learned in Section 17.9C.

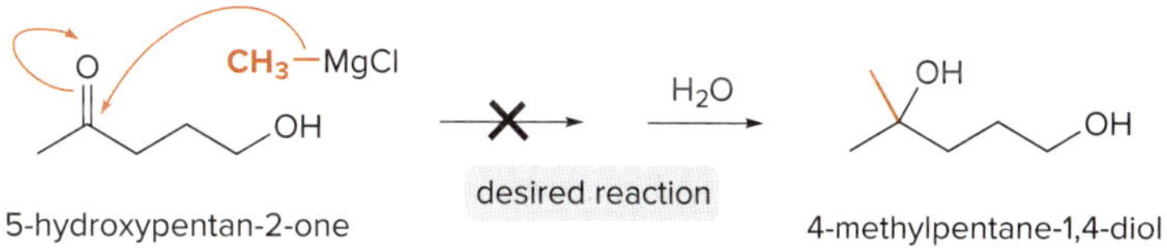

CH_3–MgCl; H_2O
5-hydroxypentan-2-one — desired reaction — 4-methylpentane-1,4-diol

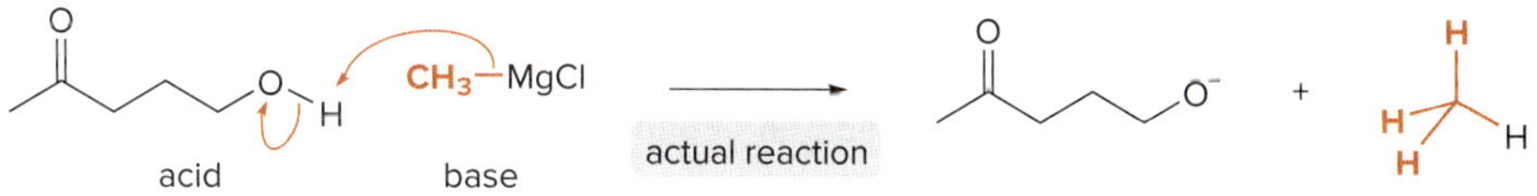

acid; base — actual reaction — products of proton transfer

Solving this problem requires a three-step strategy:

Step [1] Convert the OH group to another functional group that does not interfere with the desired reaction. This new blocking group is called a **protecting group,** and the reaction that creates it is called ***protection.***

Step [2] Carry out the desired reaction.

Step [3] Remove the protecting group. This reaction is called ***deprotection.***

Application of the general strategy to the Grignard addition of CH_3MgCl to 5-hydroxypentan-2-one is illustrated in Figure 17.6.

Figure 17.6 General strategy for using a protecting group

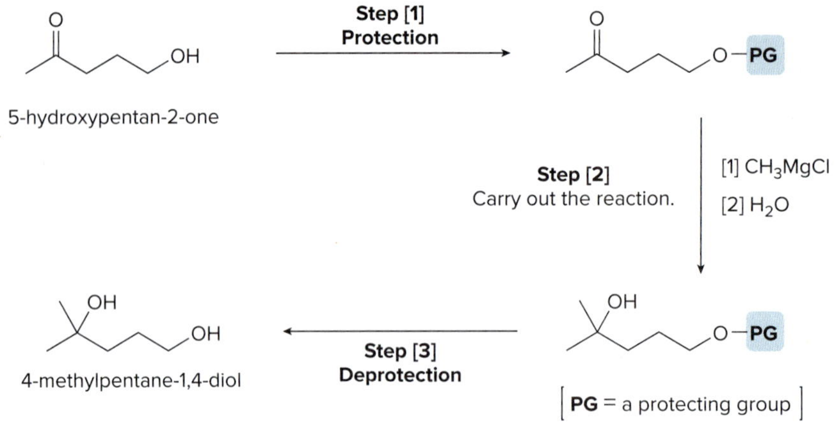

5-hydroxypentan-2-one → Step [1] Protection → O–PG

Step [2] Carry out the reaction. [1] CH_3MgCl [2] H_2O

4-methylpentane-1,4-diol ← Step [3] Deprotection

[**PG** = a protecting group]

- In Step [1], the OH proton in 5-hydroxypentan-2-one is replaced with a protecting group, written as **PG.** Because the product of Step [1] no longer has an OH proton, it can now undergo nucleophilic addition.
- In Step [2], CH_3MgCl adds to the carbonyl group to yield a 3° alcohol after protonation with water.
- Removal of the protecting group in Step [3] forms the desired product, 4-methylpentane-1,4-diol.

A common OH protecting group is a **silyl ether.** A silyl ether has a new O–Si bond in place of the O–H bond of the alcohol. The most widely used silyl ether protecting group is the ***tert*-butyldimethylsilyl ether,** abbreviated as **TBS.**

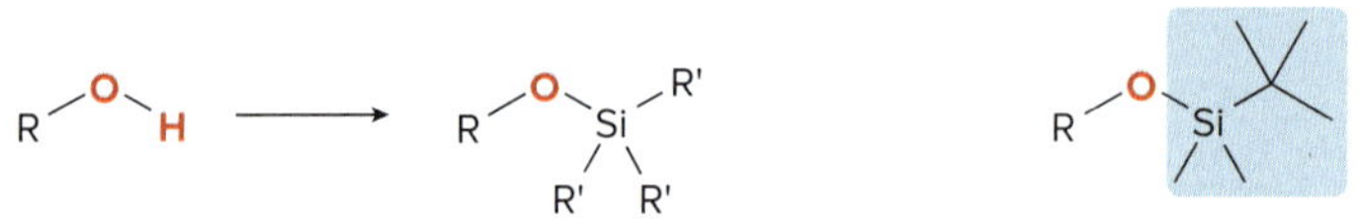

silyl ether

tert-butyldimethylsilyl ether

RO–**TBS**

tert-Butyldimethylsilyl ethers are prepared from alcohols by reaction with *tert*-butyldimethylsilyl chloride and an amine base, usually imidazole.

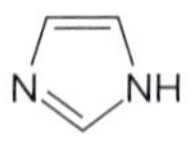

imidazole

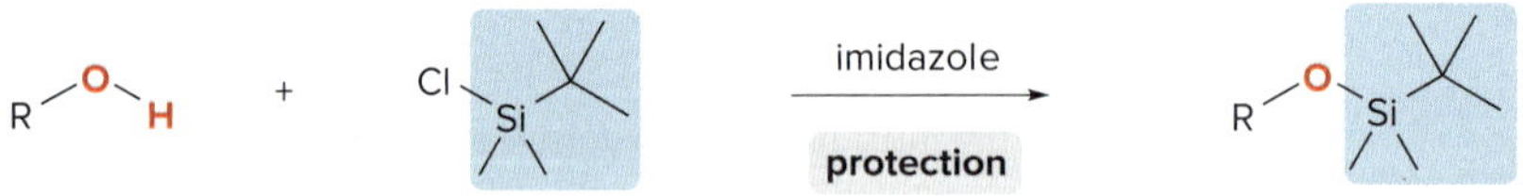

imidazole

protection

tert-butyldimethylsilyl chloride

TBS–Cl

tert-butyldimethylsilyl ether

RO–**TBS**

The silyl ether is typically removed with a fluoride salt, usually **tetrabutylammonium fluoride** $(CH_3CH_2CH_2CH_2)_4N^+F^-$, drawn as $Bu_4N^+F^-$ (Bu = butyl).

$Bu_4N^+F^-$

deprotection

tert-butyldimethylsilyl ether

The alcohol is regenerated.

The use of a *tert*-butyldimethylsilyl ether as a protecting group makes possible the synthesis of 4-methylpentane-1,4-diol by a three-step sequence.

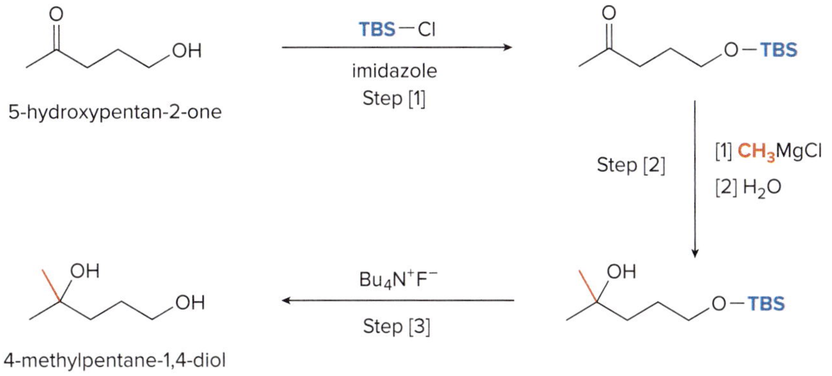

5-hydroxypentan-2-one

TBS—Cl
imidazole
Step [1]

Step [2] [1] CH_3MgCl [2] H_2O

$Bu_4N^+F^-$
Step [3]

4-methylpentane-1,4-diol

- **Step [1] Protect the OH group** as a *tert*-butyldimethylsilyl ether by reaction with *tert*-butyldimethylsilyl chloride and imidazole.
- **Step [2] Carry out nucleophilic addition** by using CH_3MgCl, followed by protonation.
- **Step [3] Remove the protecting group** with tetrabutylammonium fluoride to form the desired addition product.

Protecting groups block interfering functional groups, and in this way, a wider variety of reactions can take place with a particular substrate. For more on protecting groups, see the discussion of acetals in Section 18.14.

Problem 17.27 Using protecting groups, show how estrone can be converted to ethynylestradiol, a widely used oral contraceptive.

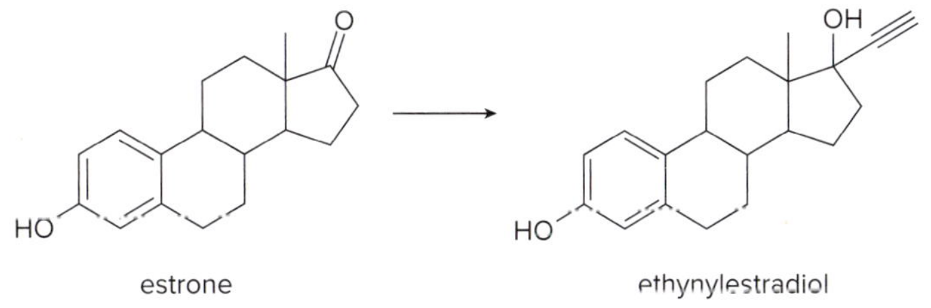

estrone → ethynylestradiol

17.13 Reaction of Organometallic Reagents with Carboxylic Acid Derivatives

Organometallic reagents react with carboxylic acid derivatives (RCOZ) to form two different products, depending on the identity of both the leaving group Z and the reagent R—M. The most useful reactions are carried out with esters and acid chlorides, forming either **ketones** or **3° alcohols.**

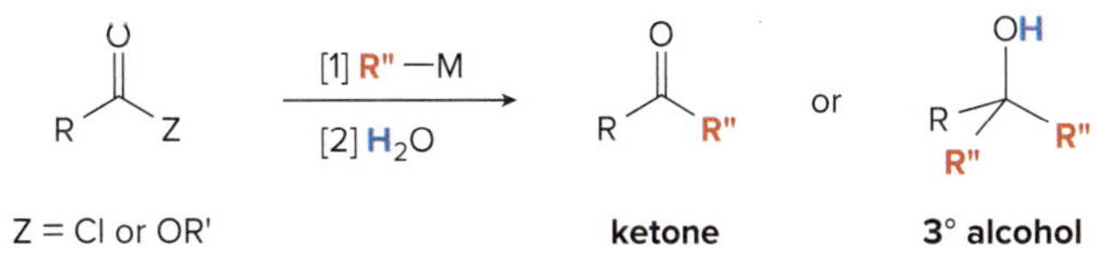

Z = Cl or OR' —[1] R''—M, [2] H_2O→ **ketone** or **3° alcohol**

- **Keep in mind that RLi and RMgX are very reactive reagents, whereas R_2CuLi is much less reactive. This reactivity difference makes selective reactions possible.**

17.13A Reaction of RLi and RMgX with Esters and Acid Chlorides

Both esters and acid chlorides form 3° alcohols when treated with two equivalents of either Grignard or organolithium reagents. Two new carbon–carbon bonds are formed in the product.

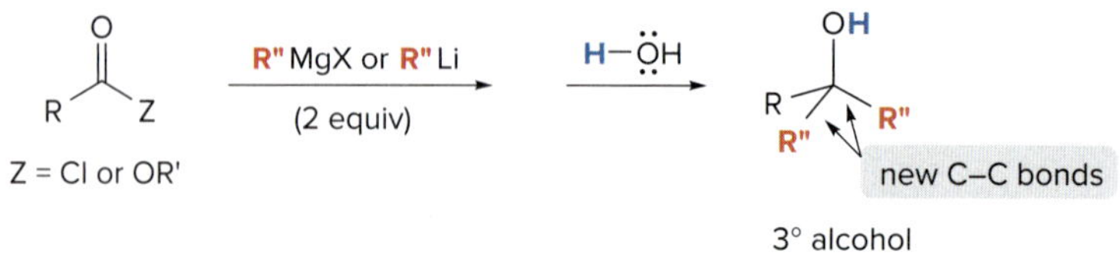

Z = Cl or OR'; R"MgX or R"Li (2 equiv); H–OH; 3° alcohol; new C–C bonds

Two examples using Grignard reagents are shown.

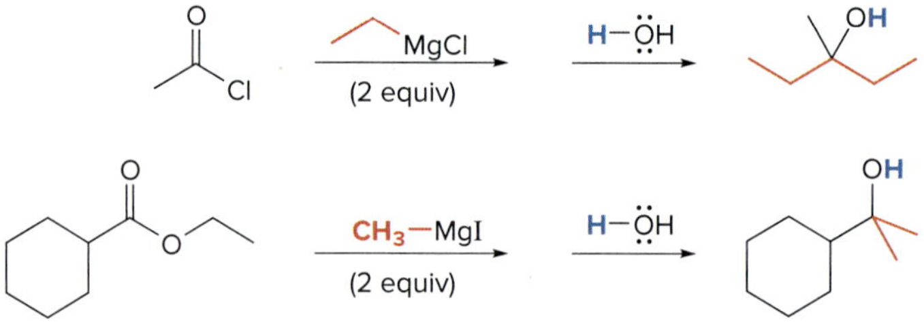

(2 equiv) MgCl; H–OH

CH_3–MgI (2 equiv); H–OH

Problem 17.28 Draw the product formed when each compound is treated with two equivalents of $CH_3CH_2CH_2CH_2MgBr$ followed by H_2O.

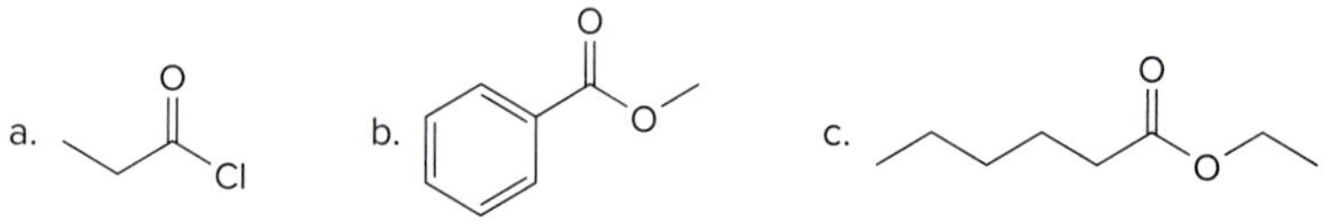

a. b. c.

The mechanism for this addition reaction resembles the mechanism for the metal hydride reduction of acid chlorides and esters discussed in Section 17.7A. The mechanism is conceptually divided into two parts: **nucleophilic substitution** to form a ketone (Steps [1] and [2]), followed by **nucleophilic addition** to form a 3° alcohol (Steps [3] and [4]), as shown in Mechanism 17.7.

Mechanism 17.7 Reaction of R"MgX or R"Li with RCOCl and RCOOR'

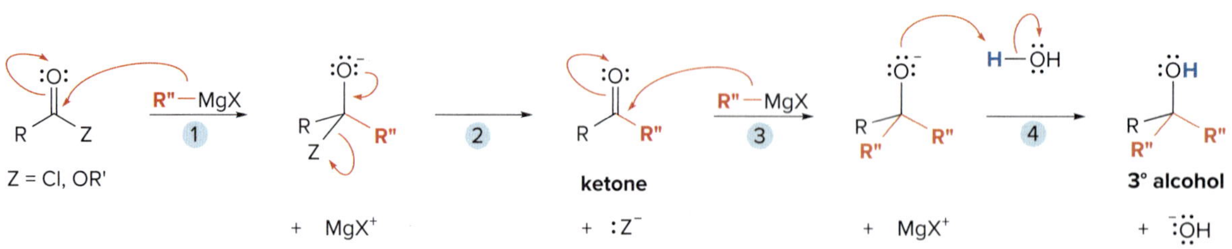

Z = Cl, OR'; R"–MgX [1]; + MgX^+; [2]; ketone; + $:Z^-$; R"–MgX [3]; + MgX^+; H–OH [4]; 3° alcohol; + $:OH^-$

1. **Nucleophilic attack of (R")⁻** forms a tetrahedral intermediate with a leaving group Z.
2. The π bond is re-formed and the **leaving group Z departs** to form a ketone. The overall result of addition of (R")⁻ and elimination of Z:⁻ is **substitution of R" for Z.**
3. **Nucleophilic attack of (R")⁻** forms an alkoxide with no leaving group.
4. Protonation of the alkoxide by H_2O forms a **3° alcohol.**

Organolithium and Grignard reagents afford 3° alcohols when they react with esters and acid chlorides. As soon as the ketone forms by addition of one equivalent of reagent to RCOZ (Steps [1] and [2] of the mechanism), it reacts with a second equivalent of reagent to form the 3° alcohol.

This reaction is more limited than the Grignard addition to aldehydes and ketones, because only 3° alcohols having **two identical alkyl groups** can be prepared. Nonetheless, it is still a valuable reaction because it forms two new carbon–carbon bonds.

Sample Problem 17.4 Identifying the Ester and Grignard Reagent Needed to Prepare an Alcohol

What ester and Grignard reagent are needed to prepare the following alcohol?

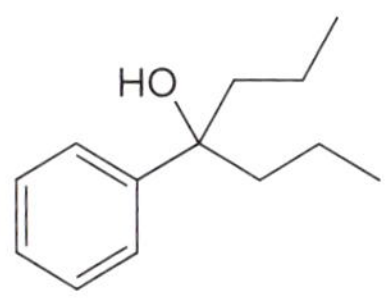

HO

Solution

A 3° alcohol formed from an ester and Grignard reagent must have **two identical R groups,** and these **R groups come from RMgX. The remainder of the molecule comes from the ester.** The carbon (labeled in blue) bonded to the OH group comes from the carbonyl carbon.

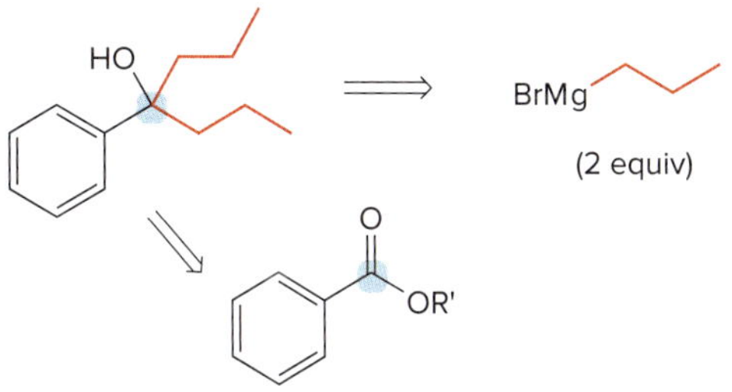

HO ⟹ BrMg (2 equiv)

O, OR'

R' = any alkyl group

Checking in the synthetic direction:

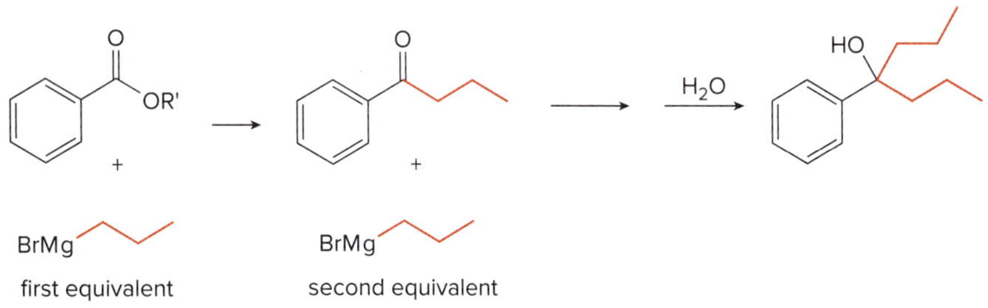

O, OR' + BrMg (first equivalent) ⟶ O + BrMg (second equivalent) ⟶ $\xrightarrow{H_2O}$ HO

Problem 17.29 What ester and Grignard reagent are needed to prepare each alcohol?

a. 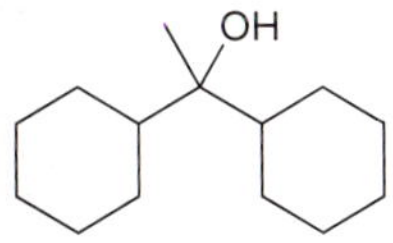 OH b. 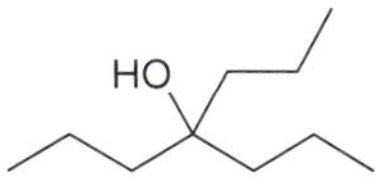 HO c. 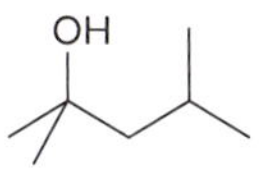 OH

More Practice: Try Problem 17.59.

17.13B Reaction of R_2CuLi with Acid Chlorides

To form a ketone from a carboxylic acid derivative, a less reactive organometallic reagent—namely, an **organocuprate**—is needed. **Acid chlorides, which have the best leaving group (Cl^-) of the carboxylic acid derivatives, react with R'_2CuLi, to give a ketone as product.** Esters, which contain a poorer leaving group (^-OR), do *not* react with R'_2CuLi.

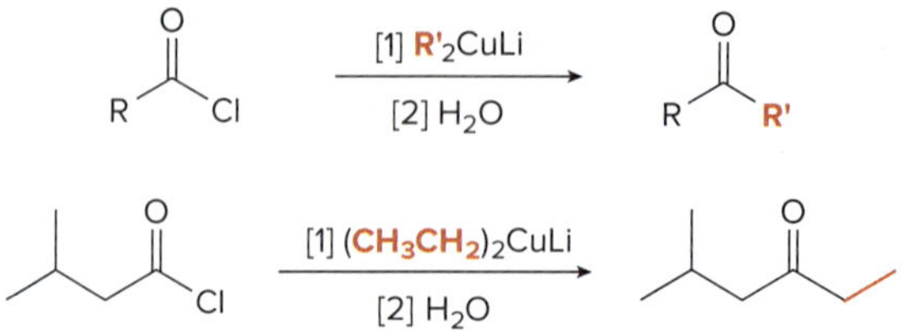

[1] R'_2CuLi
[2] H_2O

[1] $(CH_3CH_2)_2CuLi$
[2] H_2O

This reaction results in nucleophilic substitution of an alkyl group R' for the leaving group Cl, forming one new carbon–carbon bond.

Problem 17.30 What organocuprate reagent is needed to convert CH_3CH_2COCl to each ketone?

a. 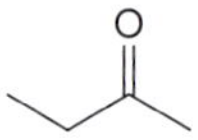 b. 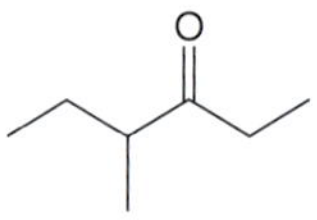 c. 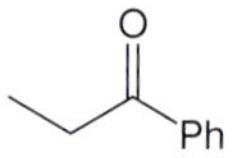 Ph

Problem 17.31 What reagent is needed to convert $(CH_3)_2CHCH_2COCl$ to each compound?

a. 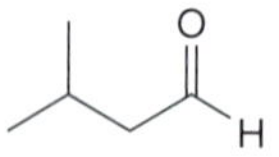 H b. 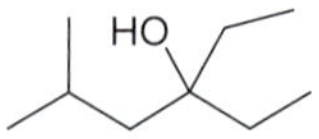 HO c. 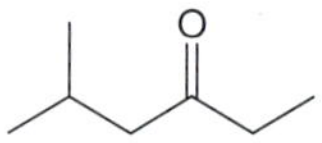 d. 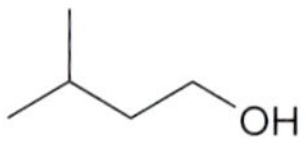 OH

A ketone with two different R groups bonded to the carbonyl carbon can be made by two different methods, as illustrated in Sample Problem 17.5.

Sample Problem 17.5 Determining the Acid Chloride and Organocuprate Needed to Prepare a Ketone

Show two different ways to prepare pentan-2-one from an acid chloride and an organocuprate reagent.

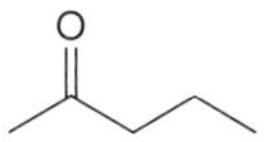

pentan-2-one

Solution

In each case, one alkyl group comes from the organocuprate and one comes from the acid chloride.

Possibility [1] Use $(CH_3)_2CuLi$ and a four-carbon acid chloride.

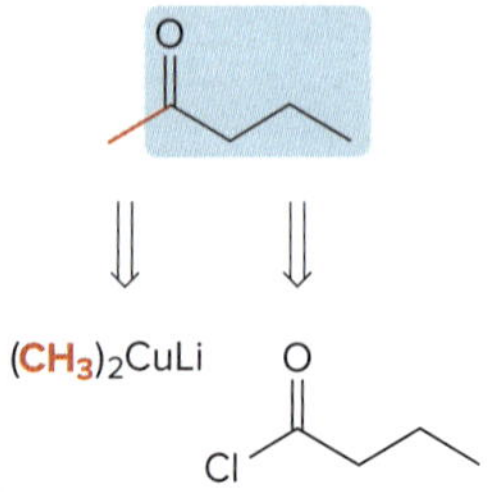

$(CH_3)_2CuLi$

Possibility [2] Use $(CH_3CH_2CH_2)_2CuLi$ and a two-carbon acid chloride.

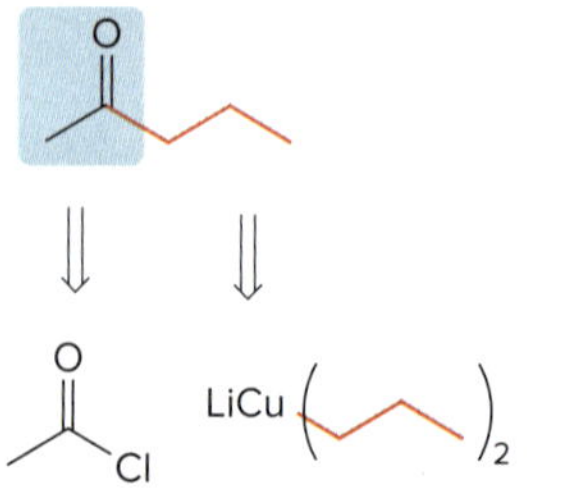

LiCu( )$_2$

Problem 17.32 Draw two different ways to prepare each ketone from an acid chloride and an organocuprate reagent.

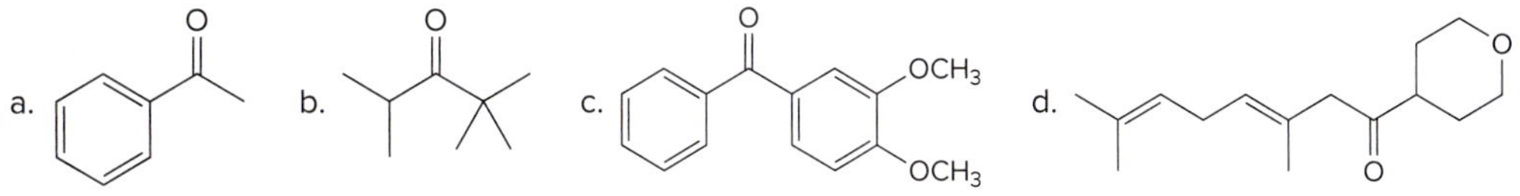

a. b. c. d.

More Practice: Try Problem 17.40a.

17.14 Reaction of Organometallic Reagents with Other Compounds

Because organometallic reagents are strong nucleophiles, they react with many other electrophiles in addition to carbonyl groups. Because these reactions always lead to the formation of new carbon–carbon bonds, they are also valuable in organic synthesis. In Section 17.14, we examine the reactions of organometallic reagents with **carbon dioxide** and **epoxides.**

17.14A Reaction of Grignard Reagents with Carbon Dioxide

Grignard reagents react with CO_2 to give carboxylic acids after protonation with aqueous acid. This reaction, called **carboxylation,** forms a carboxylic acid with one more carbon atom than the Grignard reagent from which it is prepared.

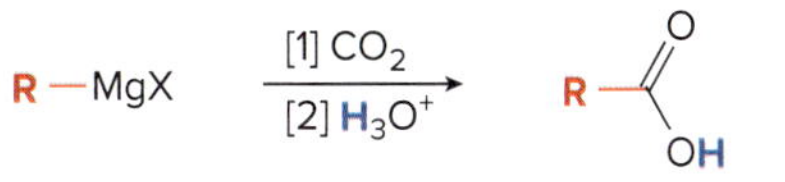

R—MgX → [1] CO_2 [2] H_3O^+ → R–COOH

carboxylic acid

Because Grignard reagents are made from alkyl or aryl halides, an organic halide can be converted to a **carboxylic acid having one more carbon atom** by a two-step reaction sequence: **formation of a Grignard reagent,** followed by **reaction with CO_2.**

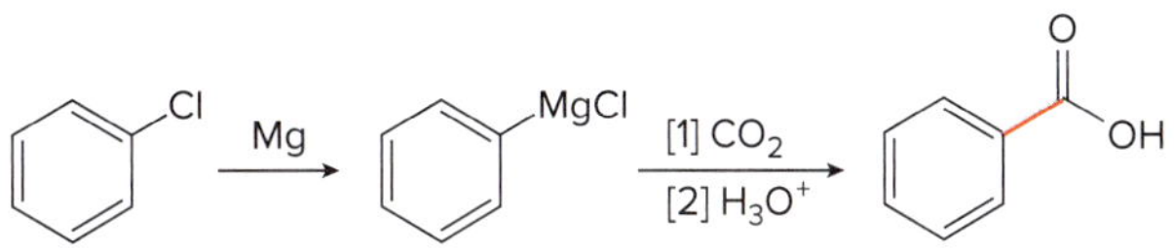

Cl → Mg → MgCl → [1] CO_2 [2] H_3O^+ → OH

new C–C bond in red

The mechanism resembles earlier reactions of nucleophilic Grignard reagents with carbonyl groups, as shown in Mechanism 17.8.

Mechanism 17.8 Carboxylation—Reaction of RMgX with CO_2

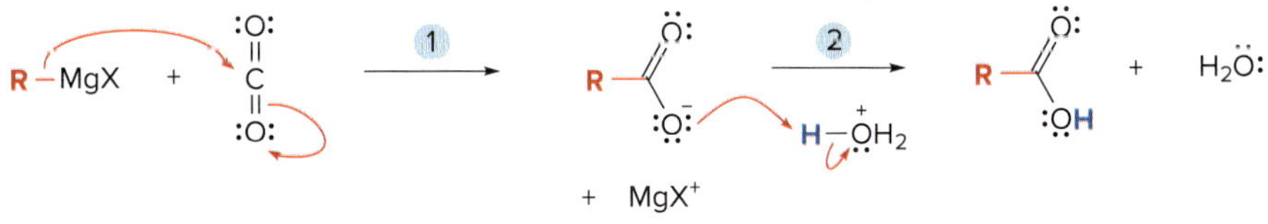

R—MgX + :O=C=O: → (1) → R–C(=O)–O:⁻ + MgX^+ → (2) $H–OH_2^+$ → R–C(=O)–OH + H_2O:

1 The nucleophilic Grignard reagent attacks the electrophilic carbon of CO_2, cleaving the π bond and forming a new carbon–carbon bond.

2 Protonation of the carboxylate anion with aqueous acid forms the carboxylic acid.

Problem 17.33 What carboxylic acid is formed from each alkyl halide on treatment with [1] Mg; [2] CO_2; [3] H_3O^+?

a. 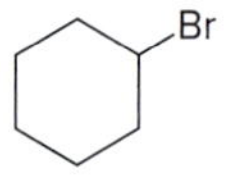 b.  c.

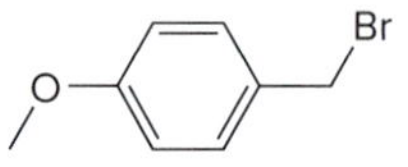

17.14B Reaction of Organometallic Reagents with Epoxides

Like other strong nucleophiles, **organometallic reagents**—RLi, RMgX, and R_2CuLi—**open epoxide rings to form alcohols.**

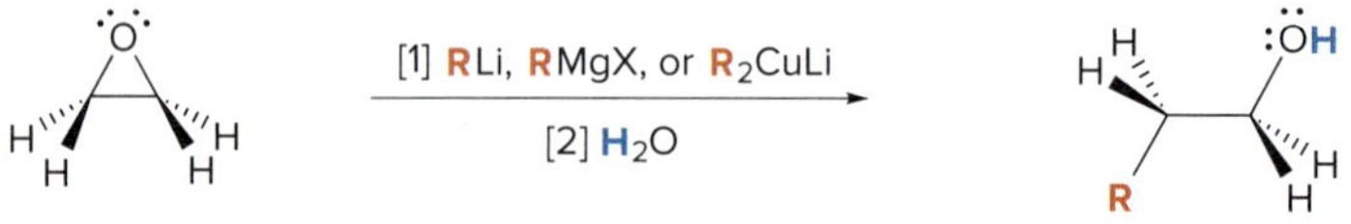

[1] RLi, RMgX, or R_2CuLi
[2] H_2O

alcohol

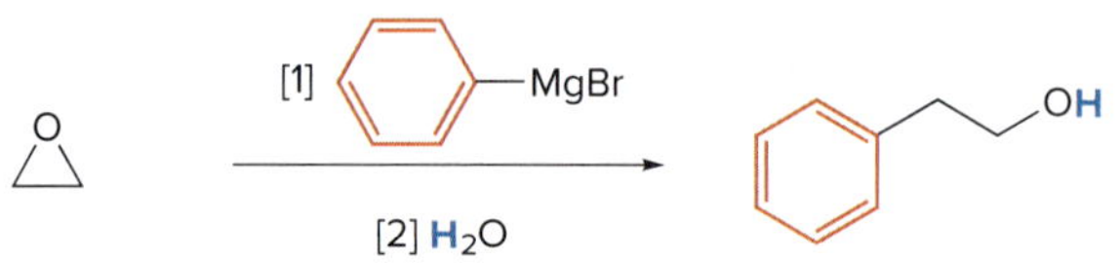

[1] ⟨benzene⟩—MgBr
[2] H_2O

The opening of epoxide rings with negatively charged nucleophiles was discussed in Section 9.16A.

The reaction follows the same two-step process as the opening of epoxide rings with other negatively charged nucleophiles—that is, **nucleophilic attack from the back side of the epoxide ring, followed by protonation of the resulting alkoxide.** In unsymmetrical epoxides, nucleophilic attack occurs at the *less* substituted carbon atom.

[1]
backside attack at the *less* substituted C

H_2O
[2]

Problem 17.34 What epoxide is needed to convert CH_3CH_2MgBr to each of the following alcohols, after quenching with water?

a. 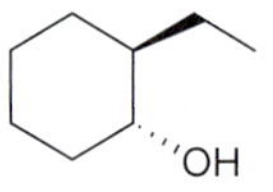 (+ enantiomer) b. 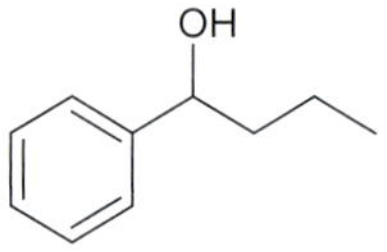 c. 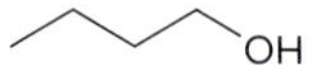 d.

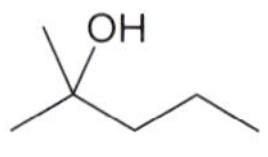

17.15 α,β-Unsaturated Carbonyl Compounds

α,β-Unsaturated carbonyl compounds are conjugated molecules containing a carbonyl group and a carbon–carbon double bond, separated by a single σ bond.

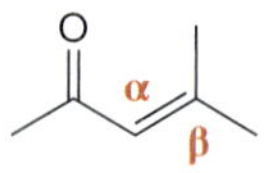

α,β-unsaturated carbonyl compound

Both functional groups of α,β-unsaturated carbonyl compounds have π bonds, but individually, they react with very different kinds of reagents. Carbon–carbon double bonds react with electrophiles (Chapter 10) and carbonyl groups react with nucleophiles (Section 17.2). What happens, then, when these two functional groups having opposite reactivity are in close proximity?

Because the two π bonds are conjugated, the electron density in an α,β-unsaturated carbonyl compound is *delocalized over four atoms*. Three resonance structures show that the carbonyl carbon and the β carbon bear a partial positive charge. This means that **α,β-unsaturated carbonyl compounds can react with nucleophiles at two different sites.**

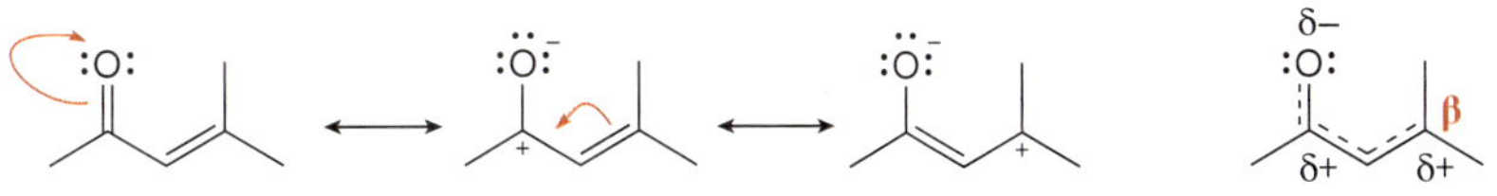

three resonance structures for an α,β-unsaturated carbonyl compound

hybrid
two electrophilic sites

- **Addition of a nucleophile to the carbonyl carbon, called 1,2-addition, adds the elements of H and Nu across the C=O, forming an allylic alcohol.**

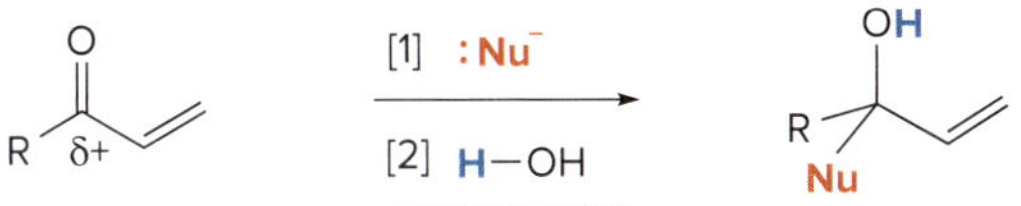

[1] :Nu⁻
[2] H–OH
1,2-addition

allylic alcohol

- **Addition of a nucleophile to the β carbon, called 1,4-addition or conjugate addition, forms a carbonyl compound.**

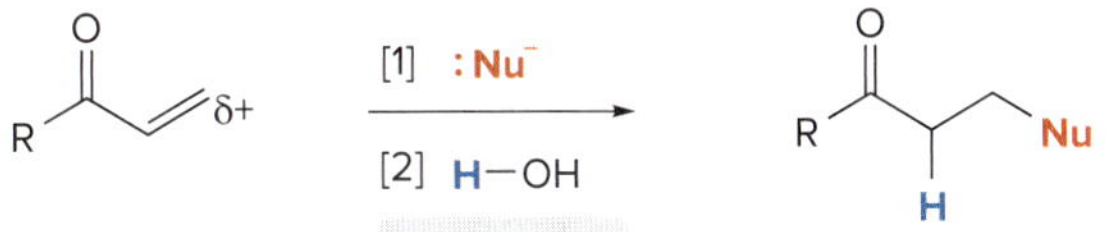

[1] :Nu⁻
[2] H–OH
1,4-addition

a carbonyl compound with a new substituent on the **β carbon**

Both 1,2- and 1,4-addition result in nucleophilic **addition of the elements of H and Nu.**

17.15A The Mechanisms for 1,2-Addition and 1,4-Addition

The steps for the mechanism of 1,2-addition are exactly the same as those for the nucleophilic addition to an aldehyde or ketone—that is, **nucleophilic attack,** followed by **protonation** (Mechanism 17.6).

The mechanism for 1,4-addition also begins with nucleophilic attack, and then protonation and tautomerization add the elements of H and Nu to the α and β carbons of the carbonyl compound, as shown in Mechanism 17.9.

Mechanism 17.9 1,4-Addition to an α,β-Unsaturated Carbonyl Compound

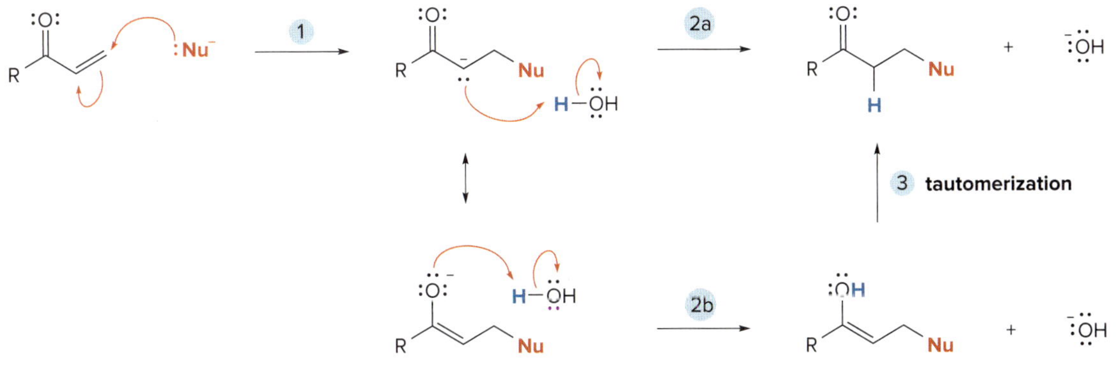

1

2a

3 **tautomerization**

2b

resonance-stabilized enolate

enol

1 Nucleophilic attack at the electrophilic β carbon forms a **resonance-stabilized enolate anion,** which can react on either carbon or oxygen in the second step.

2a Protonation of the carbon end of the enolate forms the 1,4-addition product directly.

2b – 3 Protonation of the oxygen end of the enolate forms an **enol,** which undergoes **tautomerization** by the two-step process described in Section 11.9. This forms the same 1,4-addition product that results from protonation on carbon.

17.15B Reaction of α,β-Unsaturated Carbonyl Compounds with Organometallic Reagents

The **identity of the metal** in an organometallic reagent determines whether it reacts with an α,β-unsaturated aldehyde or ketone by 1,2-addition or 1,4-addition.

- **Organolithium and Grignard reagents form 1,2-addition products.**

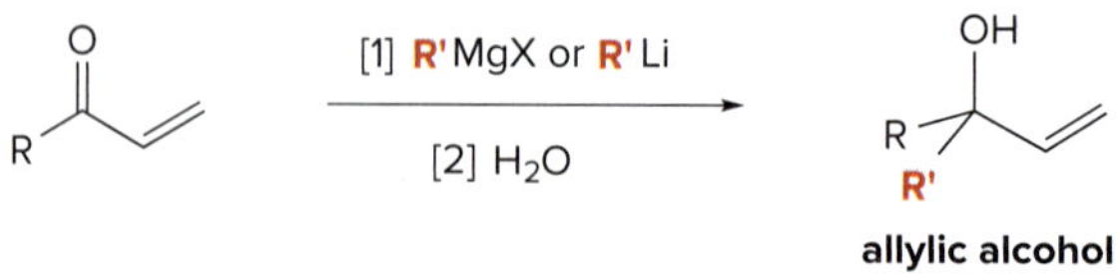
[1] R'MgX or R'Li
[2] H_2O
OH
R
R'
allylic alcohol

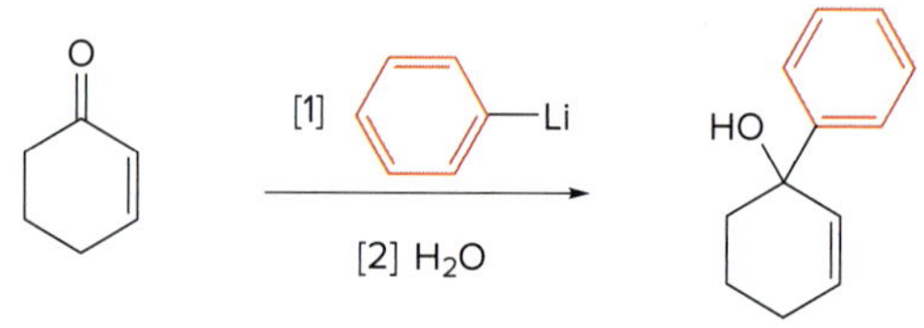
[1]
Li
[2] H_2O
HO

Why is conjugate addition also called 1,4-addition? If the atoms of the enol are numbered beginning with the O atom, then the elements of H and Nu are bonded to atoms "1" and "4," respectively.

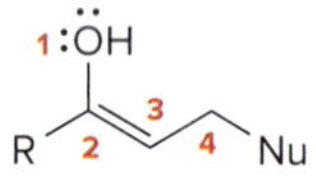
1 :ÖH
3
R 2 4 Nu

- **Organocuprate reagents form 1,4-addition products.**

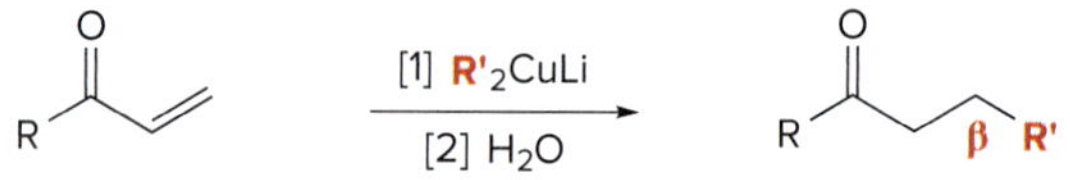
[1] R'_2CuLi
[2] H_2O
R
β R'

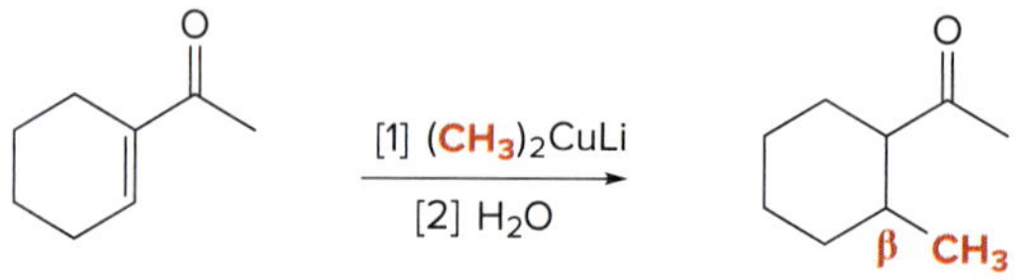
[1] $(CH_3)_2CuLi$
[2] H_2O
β CH_3

Sample Problem 17.6 Drawing the Products of 1,2- and 1,4-Addition

Draw the products of each reaction.

a. 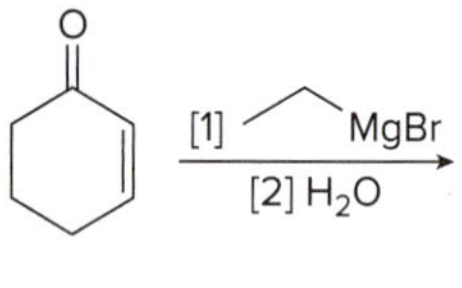
[1] MgBr
[2] H_2O

b. 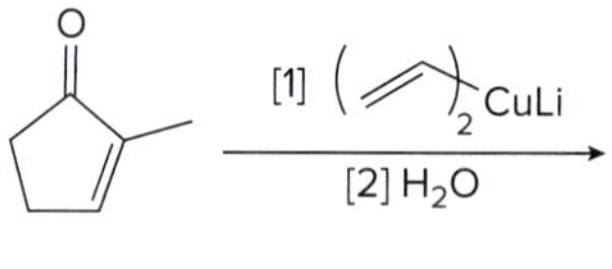
[1] CuLi
2
[2] H_2O

Solution

The characteristic reaction of α,β-unsaturated carbonyl compounds is nucleophilic addition. The reagent determines the mode of addition (1,2- or 1,4-).

a. **Grignard reagents undergo 1,2-addition.** CH_3CH_2MgBr adds a new CH_3CH_2 group at the carbonyl carbon.

b. **Organocuprate reagents undergo 1,4-addition.** The cuprate reagent adds a new vinyl group ($CH_2{=}CH$) at the β carbon.

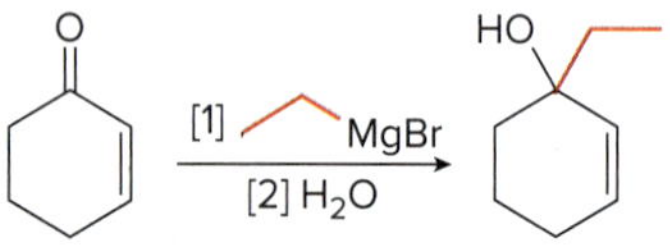
[1] MgBr
[2] H_2O
HO

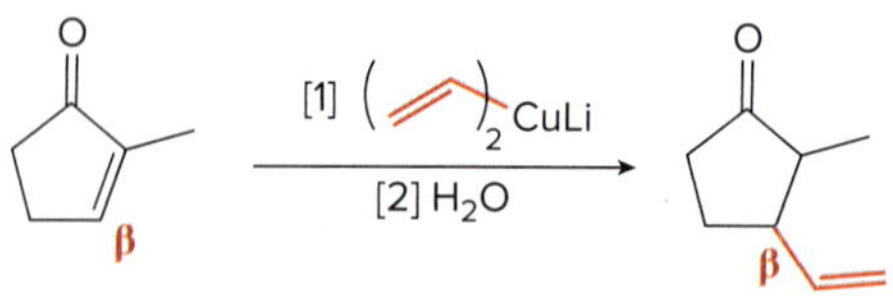
β
[1] CuLi
2
[2] H_2O
β

Problem 17.35 Draw the product when each compound is treated with either $(CH_3)_2CuLi$, followed by H_2O, or $HC{\equiv}CLi$, followed by H_2O.

a. 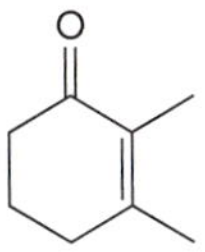 b. 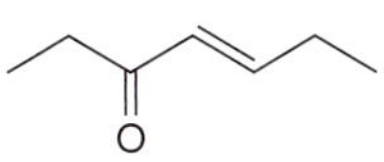 c.

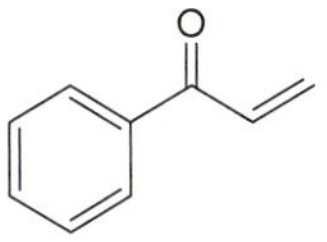

More Practice: Try Problems 17.40c; 17.42d–f; 17.45b, e.

17.16 Summary—The Reactions of Organometallic Reagents

We have now seen many different reactions of organometallic reagents with a variety of functional groups, and you may have some difficulty keeping them all straight. Rather than memorizing them all, keep in mind the following three concepts:

[1] Organometallic reagents (R–M) attack electrophilic carbon atoms, especially the carbonyl carbon.

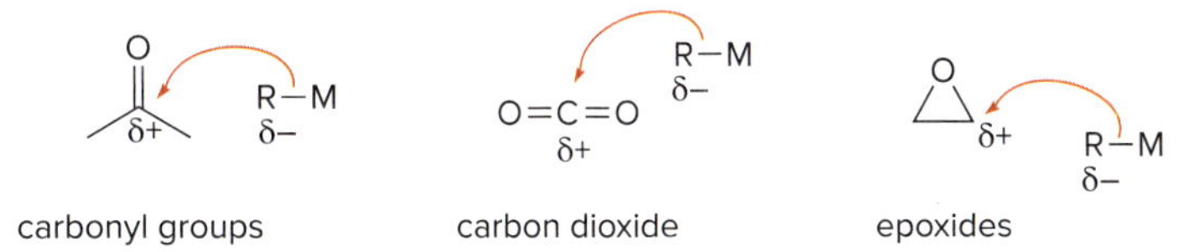

carbonyl groups   carbon dioxide   epoxides

[2] After an organometallic reagent adds to a carbonyl group, the fate of the intermediate depends on the presence or absence of a leaving group.

- Without a leaving group, the characteristic reaction is ***nucleophilic addition.***
- With a leaving group, the reaction is ***nucleophilic substitution.***

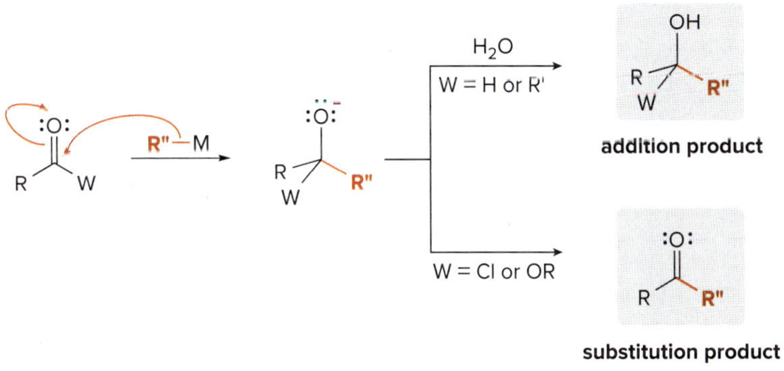

H_2O, W = H or R' → **addition product**

W = Cl or OR → **substitution product**

[3] The polarity of the R–M bond determines the reactivity of the reagents.

- RLi and RMgX are very reactive reagents.
- R_2CuLi is much less reactive.

17.17 Synthesis

The reactions learned in Chapter 17 have proven extremely useful in organic synthesis. Oxidation and reduction reactions interconvert two functional groups that differ in oxidation state. Organometallic reagents form new carbon–carbon bonds.

Synthesis is perhaps the most difficult aspect of organic chemistry. It requires you to remember both the new reactions you've just learned and the ones you've encountered in previous chapters. In a successful synthesis, you must also put these reactions in a logical order. Don't be discouraged. Learn the basic reactions and then practice them over and over again with synthesis problems.

In Sample Problems 17.7 and 17.8 that follow, keep in mind that the products formed by the reactions of Chapter 17 can themselves be transformed into many other functional groups. For example, hexan-2-ol, the product of Grignard addition of butylmagnesium chloride to acetaldehyde, can be transformed into a variety of other compounds, as shown in Figure 17.7.

Cl —Mg→ MgCl —[1] CH_3CHO, [2] H_2O→ hexan-2-ol

Figure 17.7 Conversion of hexan-2-ol to other compounds

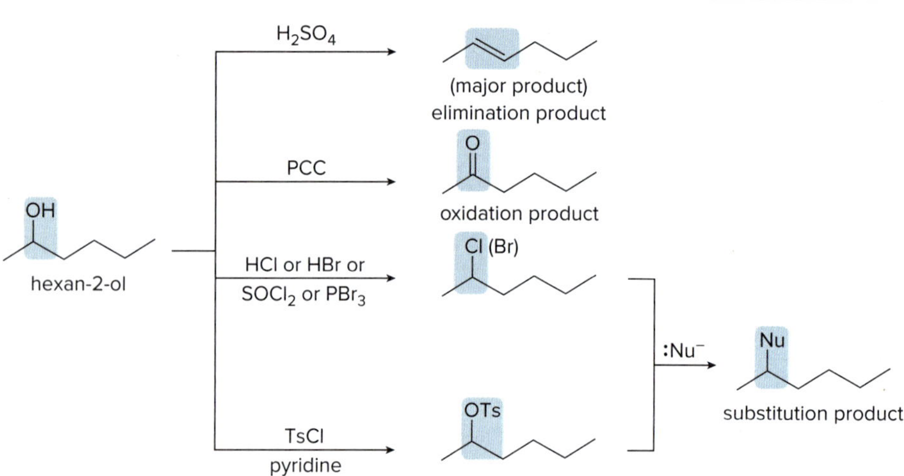

Before proceeding with Sample Problems 17.7 and 17.8, you should review the stepwise strategy for designing a synthesis found in Section 11.12.

Sample Problem 17.7 Devising a Synthesis with a Carbon–Carbon Bond-Forming Reaction

Synthesize 2,4-dimethylhexan-3-one from four-carbon alcohols.

2,4-dimethylhexan-3-one ⟹ alcohols containing 4 C's

Retrosynthetic Analysis

ketone ⟹[1] 2° alcohol ⟹[2] MgX + H(C=O) aldehyde

Synthesize each of these components.

Thinking backwards:

- [1] Form the ketone by oxidation of a 2° alcohol.
- [2] Make the 2° alcohol by Grignard addition to an aldehyde. Both of these compounds have 4 C's, and each must be synthesized from an alcohol.

Synthesis

First, make both components needed for the Grignard reaction.

HCl or $SOCl_2$; Mg; OH; PCC; H; O; OH; Cl; MgCl

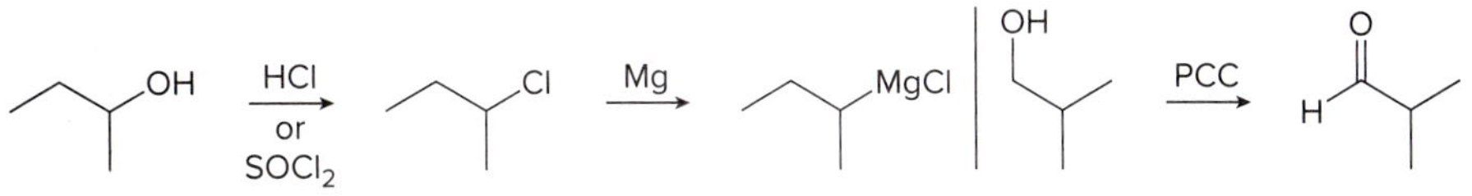

Then complete the synthesis with Grignard addition, followed by oxidation of the alcohol to the ketone.

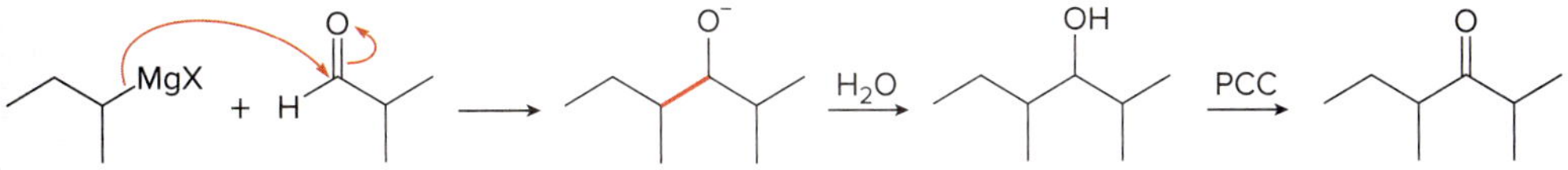

MgX + H; O; O^-; H_2O; OH; PCC; O

new C–C bond in red

Problem 17.36 Convert propan-2-ol [$(CH_3)_2CHOH$] to each compound. You may use any other organic or inorganic compounds.

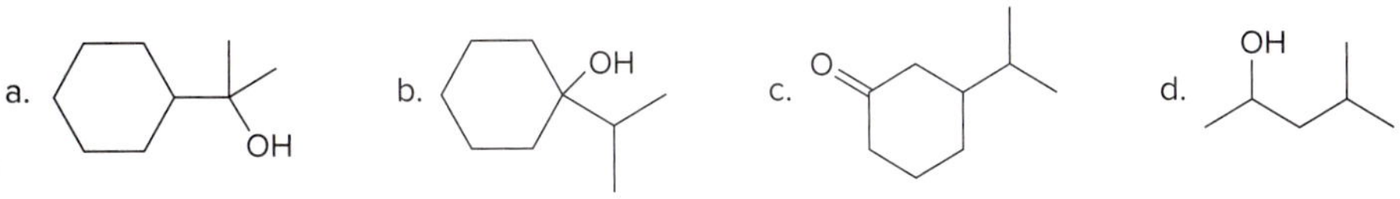

a. OH b. OH c. O d. OH

More Practice: Try Problems 17.38; 17.63a, d; 17.64a, c.

Sample Problem 17.8 Devising a Synthesis with a Grignard Addition

Synthesize isopropylcyclopentane from alcohols having ≤ 5 C's.

isopropylcyclopentane ⟹ alcohols having ≤ 5 C's

Retrosynthetic Analysis

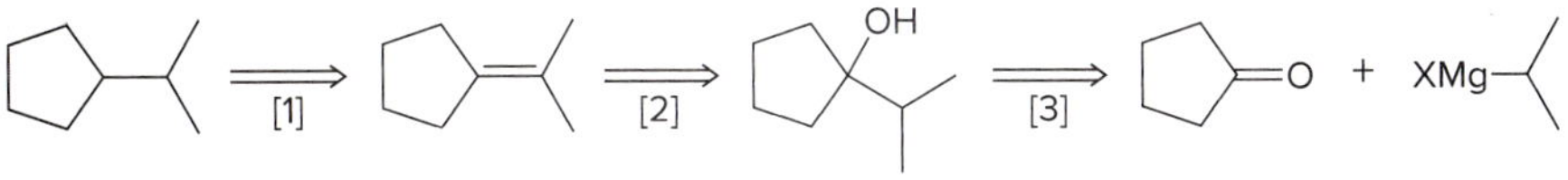

[1] [2] OH [3] O + XMg

Thinking backwards:

- [1] Form the alkane by hydrogenation of an alkene.
- [2] Introduce the double bond by dehydration of an alcohol.
- [3] Form the 3° alcohol by Grignard addition to a ketone. Both components of the Grignard reaction must then be synthesized.

Synthesis

First, make both components needed for the Grignard reaction.

OH PCC O HO PBr_3 Br Mg BrMg

Complete the synthesis with Grignard addition, dehydration, and hydrogenation.

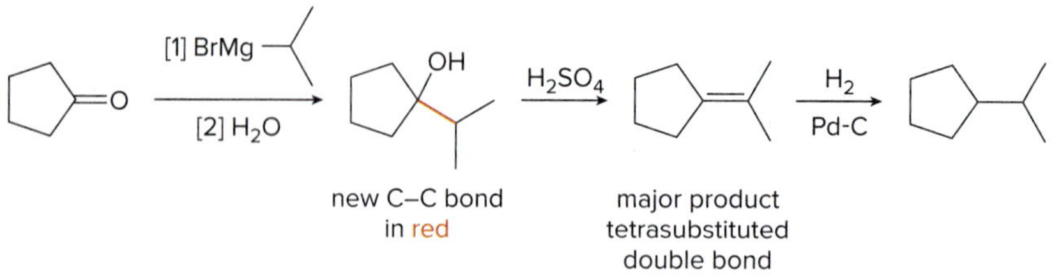

[1] BrMg– ; [2] H_2O → (new C–C bond in red) → H_2SO_4 → (major product tetrasubstituted double bond) → H_2, Pd-C →

Problem 17.37 Synthesize each compound from cyclohexanol, ethanol, and any other needed reagents.

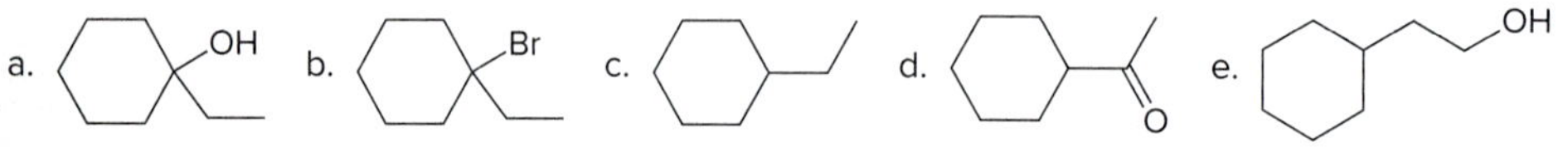

a. b. c. d. e.

More Practice: Try Problems 17.62; 17.63b, c; 17.64b, c; 17.65–17.68.

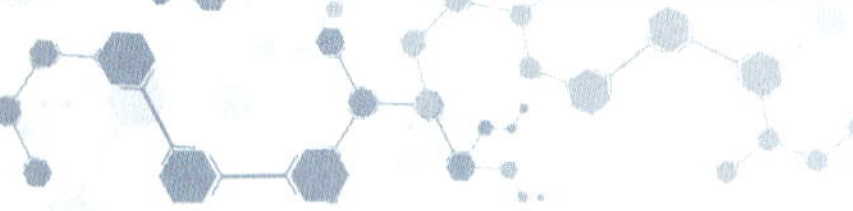

Chapter 17 REVIEW

KEY REACTIONS

Reduction Reactions

[1] Reduction of aldehydes and ketones

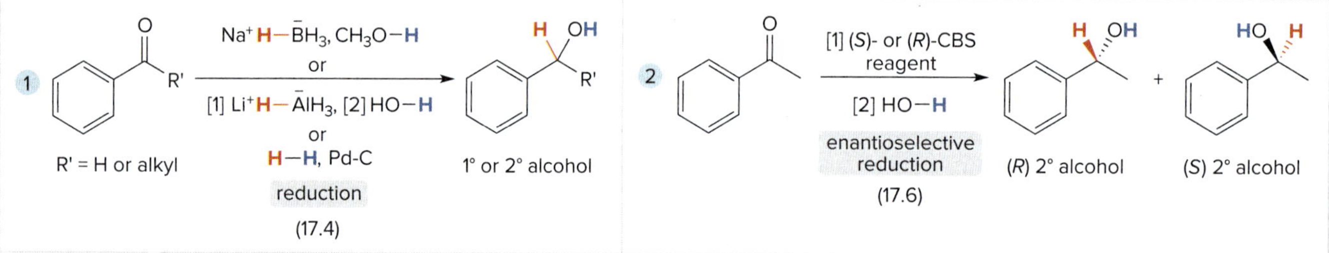

1. R' = H or alkyl → $Na^+ H–\bar{B}H_3$, $CH_3O–H$ or [1] $Li^+ H–\bar{A}lH_3$, [2] HO–H or H–H, Pd-C → 1° or 2° alcohol. reduction (17.4)

2. [1] (*S*)- or (*R*)-CBS reagent, [2] HO–H → (*R*) 2° alcohol + (*S*) 2° alcohol. enantioselective reduction (17.6)

Try Problems 17.37a, b, c; 17.41; 17.46c.

[2] Reduction of α,β-unsaturated aldehydes and ketones

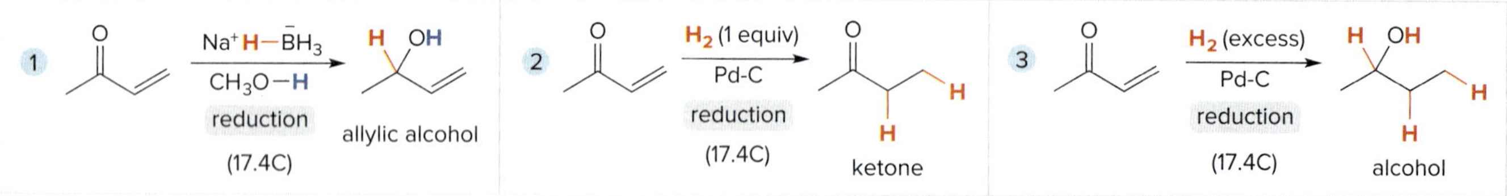

1. $Na^+ H–\bar{B}H_3$, $CH_3O–H$ → allylic alcohol. reduction (17.4C)

2. H_2 (1 equiv), Pd-C → ketone. reduction (17.4C)

3. H_2 (excess), Pd-C → alcohol. reduction (17.4C)

Try Problem 17.42a, b, c.

[3] Reduction of acid chlorides

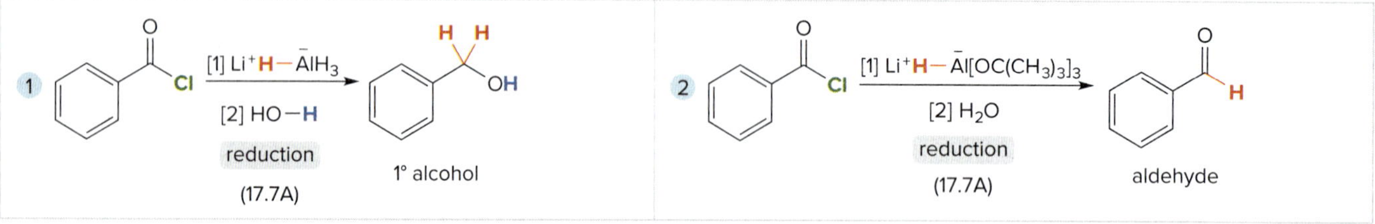

1. [1] $Li^+ H–\bar{A}lH_3$, [2] HO–H → 1° alcohol. reduction (17.7A)

2. [1] $Li^+ H–\bar{A}l[OC(CH_3)_3]_3$, [2] H_2O → aldehyde. reduction (17.7A)

Try Problem 17.43d.

[4] Reduction of esters

1. Methyl benzoate ($C_6H_5CO-OCH_3$) → [1] $Li^+H-\bar{A}lH_3$; [2] HO−H; reduction (17.7A) → $C_6H_5CH_2OH$ (1° alcohol)

2. Methyl benzoate ($C_6H_5CO-OCH_3$) → [1] DIBAL-H; [2] H_2O; reduction (17.7A) → C_6H_5CHO (aldehyde)

Try Problem 17.43a, b.

[5] Reduction of carboxylic acids and amides

1. C_6H_5COOH → [1] $Li^+H-\bar{A}lH_3$; [2] HO−H; reduction (17.7B) → $C_6H_5CH_2OH$ (1° alcohol)

2. $C_6H_5CONR'_2$ (R' = H or alkyl) → [1] $Li^+H-\bar{A}lH_3$; [2] H_2O; reduction (17.7B) → $C_6H_5CH_2NR'_2$ (amine)

Try Problems 17.43b, c; 17.56.

Oxidation of Aldehydes to Carboxylic Acids

C_6H_5CHO (aldehyde) → CrO_3, $Na_2Cr_2O_7$, $K_2Cr_2O_7$, $KMnO_4$, or Ag_2O, NH_4OH (17.8) → C_6H_5COOH (carboxylic acid)

Try Problems 17.39d–f, 17.44c–e.

Preparation of Organometallic Reagents

1. C_6H_5-X + 2 Li → (17.9A) → C_6H_5-Li + LiX (organolithium reagent)

2. C_6H_5-X + Mg → $(CH_3CH_2)_2O$ (17.9A) → C_6H_5-Mg-X (Grignard reagent)

3. CH_3CH_2-X + 2 Li → CH_3CH_2-Li + LiX

2 CH_3CH_2-Li + CuI → (17.9A) → $(CH_3CH_2)_2Cu^-Li^+$ + LiI (organocuprate reagent)

4. $C_6H_5-C\equiv C-H$ → $Na^+\ ^-NH_2$ (17.9B) → $C_6H_5-C\equiv C^-Na^+$ + $H-NH_2$ (a sodium acetylide)

$C_6H_5-C\equiv C-H$ → CH_3-Li (17.9B) → $C_6H_5-C\equiv C-Li$ + CH_3-H (a lithium acetylide)

Reactions with Organometallic Reagents

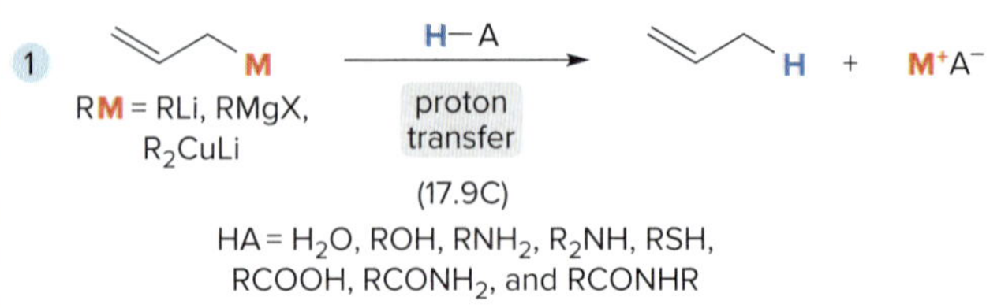

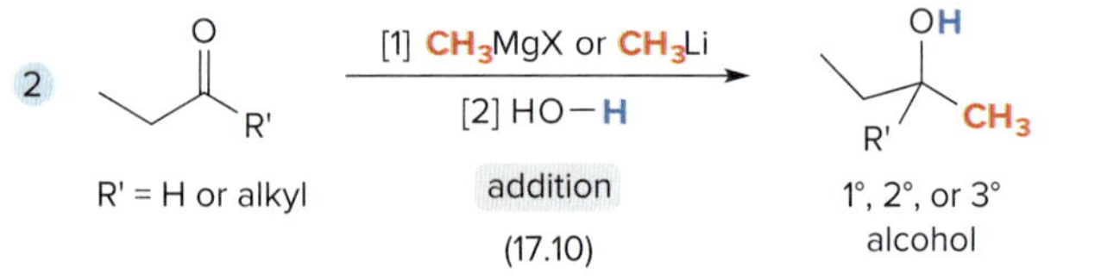

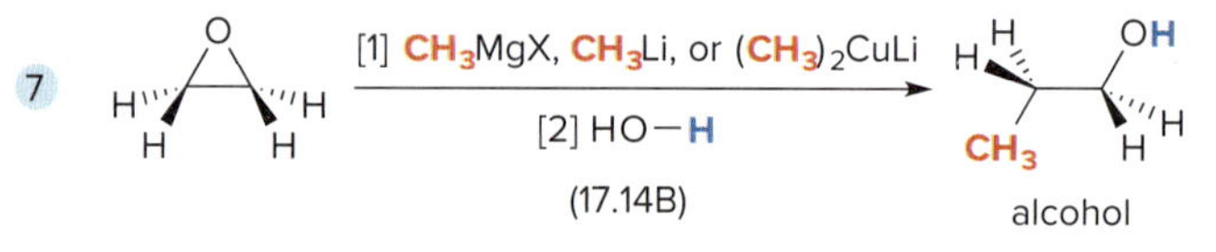

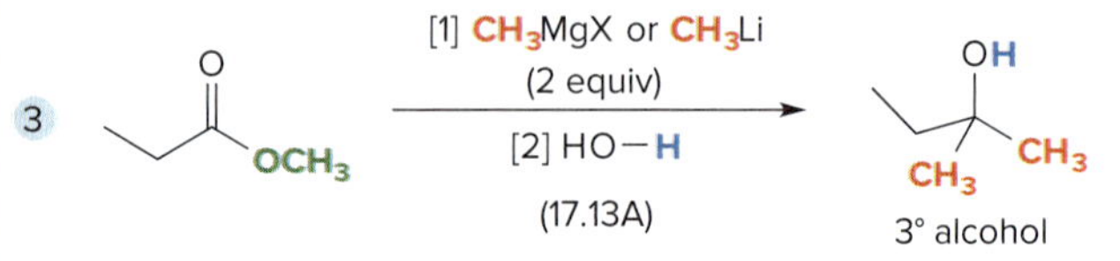

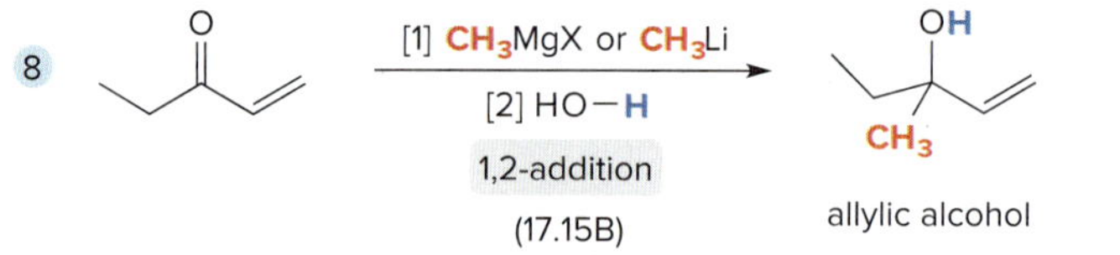

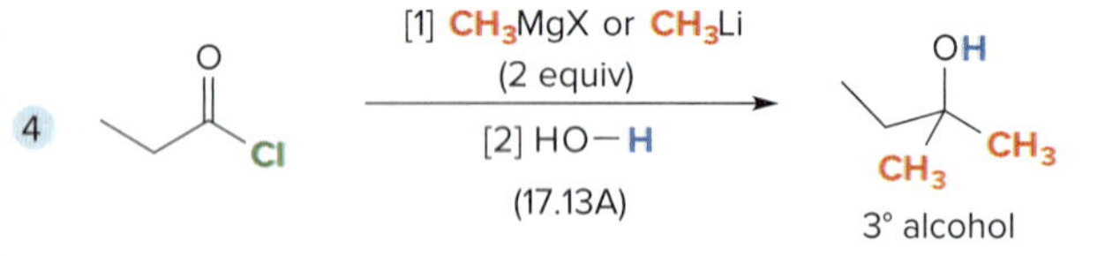

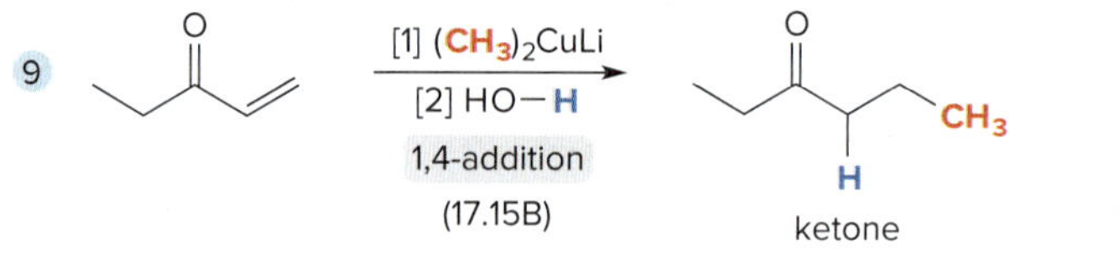

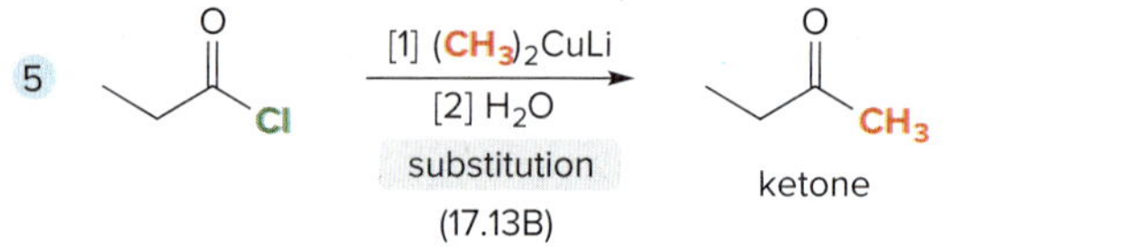

1 RM = RLi, RMgX, R_2CuLi — H–A — proton transfer (17.9C) — H + M^+A^-

HA = H_2O, ROH, RNH_2, R_2NH, RSH, RCOOH, $RCONH_2$, and RCONHR

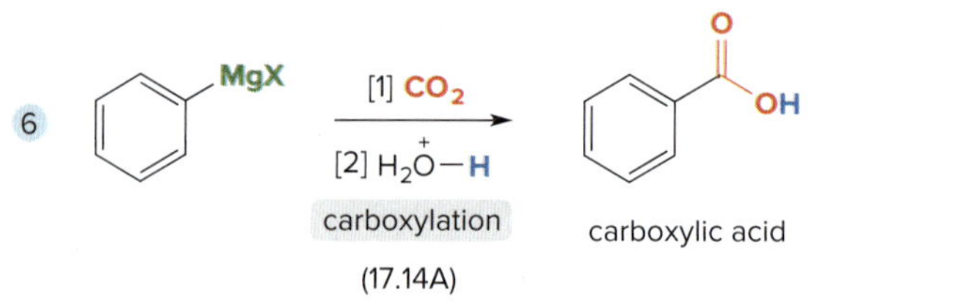

2 R' = H or alkyl — [1] CH_3MgX or CH_3Li [2] HO–H — addition (17.10) — 1°, 2°, or 3° alcohol

3 OCH$_3$ — [1] CH_3MgX or CH_3Li (2 equiv) [2] HO–H (17.13A) — 3° alcohol

4 Cl — [1] CH_3MgX or CH_3Li (2 equiv) [2] HO–H (17.13A) — 3° alcohol

5 Cl — [1] $(CH_3)_2CuLi$ [2] H_2O — substitution (17.13B) — ketone

6 MgX — [1] CO_2 [2] $H_2\overset{+}{O}$–H — carboxylation (17.14A) — carboxylic acid

7 [1] CH_3MgX, CH_3Li, or $(CH_3)_2CuLi$ [2] HO–H (17.14B) — alcohol

8 [1] CH_3MgX or CH_3Li [2] HO–H — 1,2-addition (17.15B) — allylic alcohol

9 [1] $(CH_3)_2CuLi$ [2] HO–H — 1,4-addition (17.15B) — ketone

Try Problems 17.39g–k; 17.40; 17.42d–f; 17.45; 17.46a, b, d.

Protecting Groups

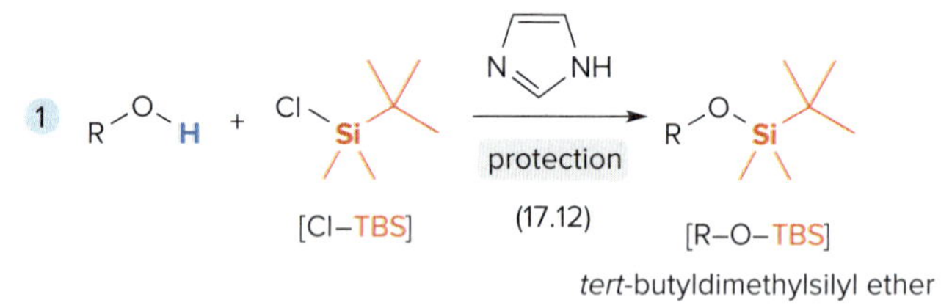

1 R–O–H + [Cl–TBS] — imidazole (N, NH) — protection (17.12) — [R–O–TBS] *tert*-butyldimethylsilyl ether

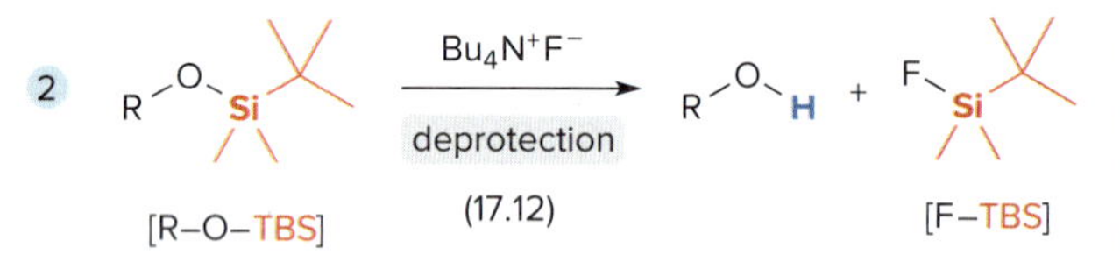

2 [R–O–TBS] — $Bu_4N^+F^-$ — deprotection (17.12) — R–O–H + [F–TBS]

Try Problems 17.39l, 17.47.

KEY SKILLS

[1] Drawing all stereoisomers that form in a Grignard reaction (17.10B)

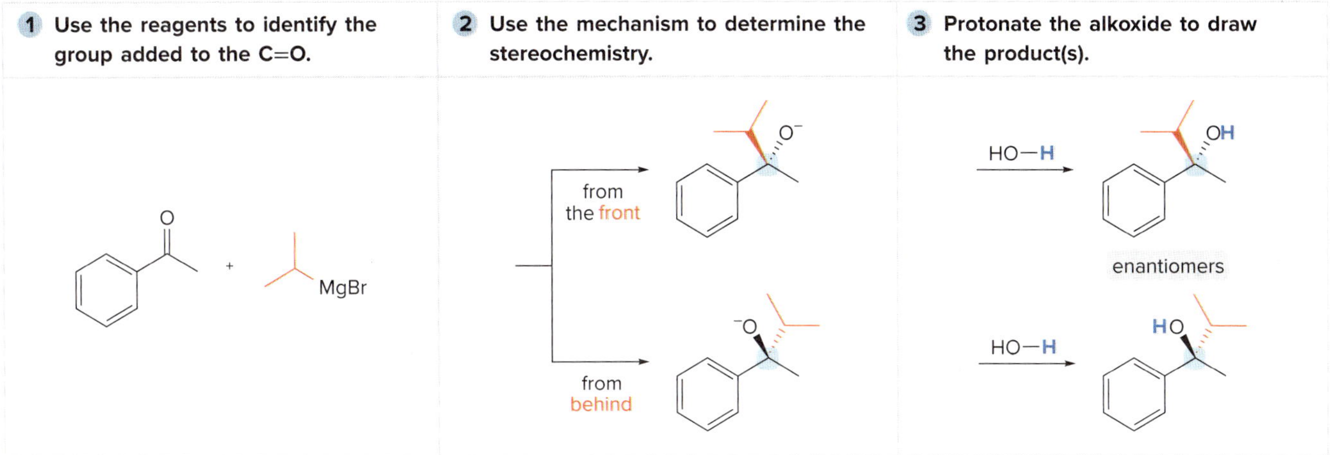

See Sample Problem 17.2. Try Problem 17.46a.

[2] Determining the starting materials for the preparation of an alcohol from an organolithium reagent and an ester (17.13); example: 3-ethyl-2-methylpentan-3-ol

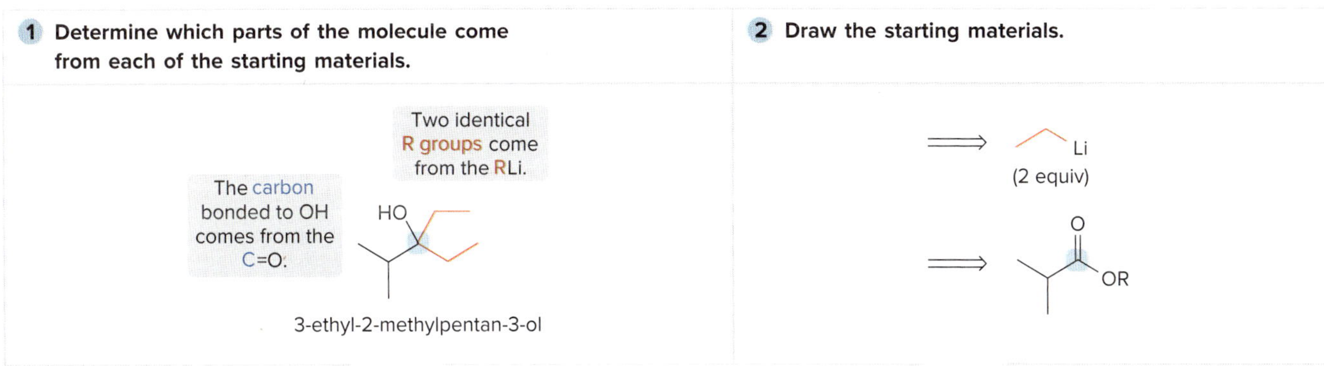

See Sample Problem 17.4. Try Problem 17.59.

[3] Using a protecting group (17.12)

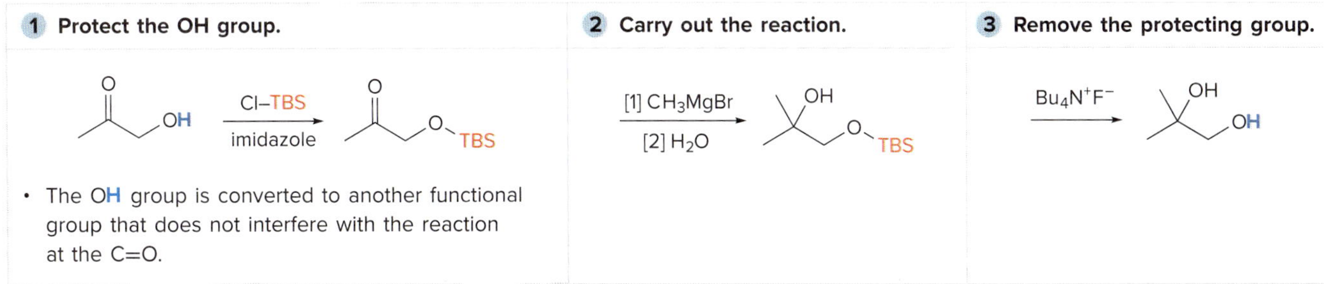

- The OH group is converted to another functional group that does not interfere with the reaction at the C=O.

See Figure 17.6. Try Problem 17.47.

[4] Drawing the product that forms in the reaction of an α,β-unsaturated carbonyl compound with an organometallic reagent (17.10B)

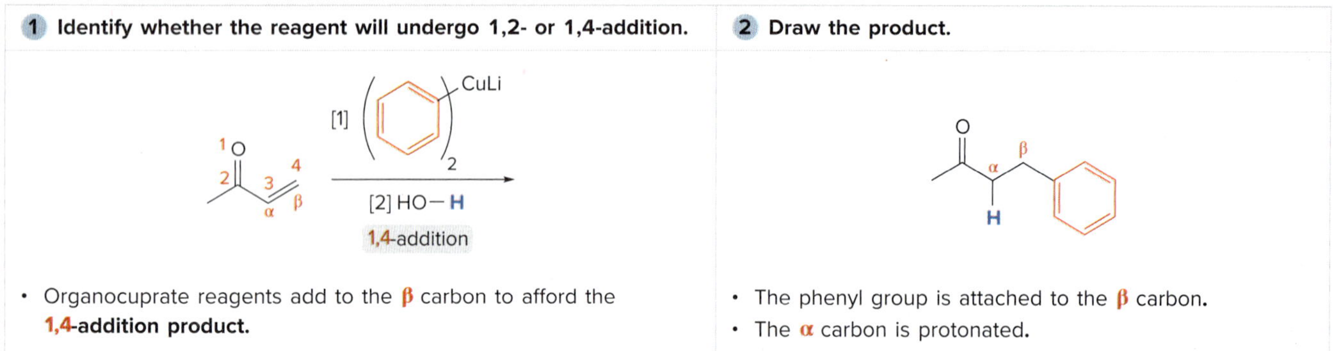

- Organocuprate reagents add to the β carbon to afford the **1,4-addition product.**
- The phenyl group is attached to the β carbon.
- The α carbon is protonated.

See Sample Problem 17.6. Try Problems 17.40c; 17.42d–f; 17.45b, e.

KEY MECHANISM CONCEPTS

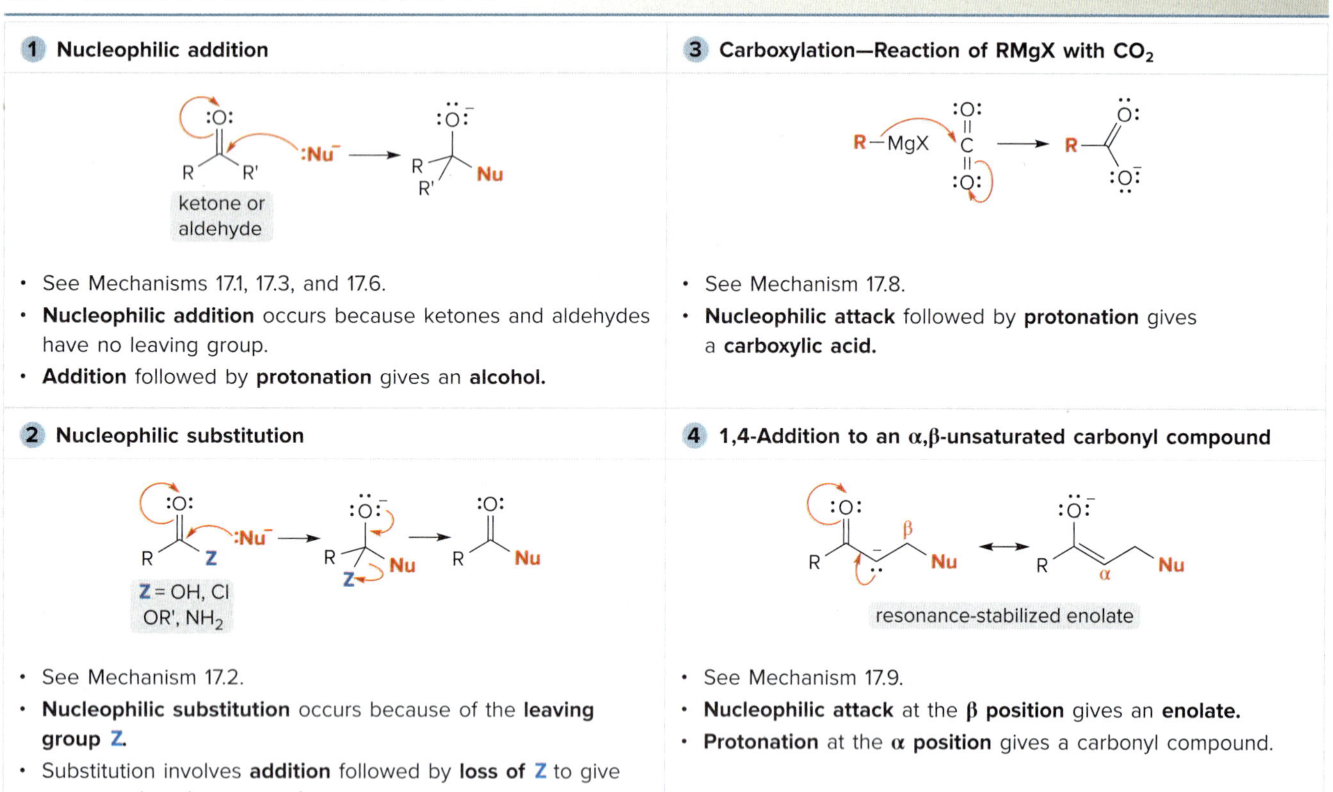

1 Nucleophilic addition

- See Mechanisms 17.1, 17.3, and 17.6.
- **Nucleophilic addition** occurs because ketones and aldehydes have no leaving group.
- **Addition** followed by **protonation** gives an **alcohol.**

3 Carboxylation—Reaction of RMgX with CO_2

- See Mechanism 17.8.
- **Nucleophilic attack** followed by **protonation** gives a **carboxylic acid.**

2 Nucleophilic substitution

- See Mechanism 17.2.
- **Nucleophilic substitution** occurs because of the **leaving group Z.**
- Substitution involves **addition** followed by **loss of Z** to give a new carbonyl compound.

4 1,4-Addition to an α,β-unsaturated carbonyl compound

- See Mechanism 17.9.
- **Nucleophilic attack** at the **β position** gives an **enolate.**
- **Protonation** at the **α position** gives a carbonyl compound.

Try Problems 17.51–17.55.

CHAPTER 17 MULTIPLE-CHOICE SELF-TEST

The Self-Test consists of multiple-choice questions similar to those found on the American Chemical Society organic chemistry exam. Answers are given at the end of the chapter.

1. Which carbonyl groups undergo nucleophilic addition and which undergo nucleophilic substitution in the given compound?

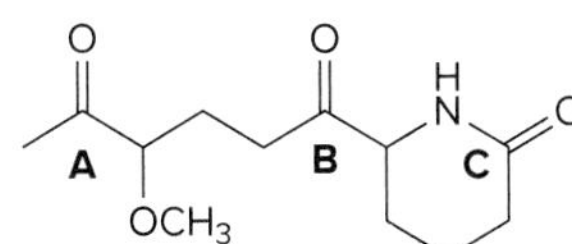

a. **A** addition; **B** and **C** substitution
b. **A** and **B** addition; **C** substitution
c. **A** and **C** addition; **B** substitution
d. **A, B,** and **C** substitution

2. Which compounds can be converted to **X** by a reduction with $LiAlH_4$?

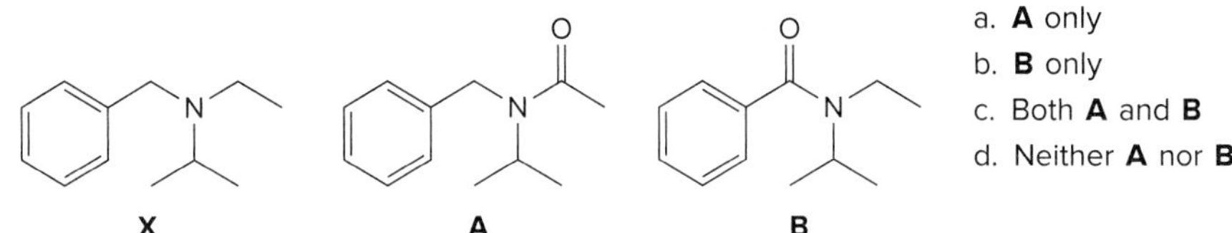

a. **A** only
b. **B** only
c. Both **A** and **B**
d. Neither **A** nor **B**

3. What product is formed when **Y** reacts with DIBAL-H?

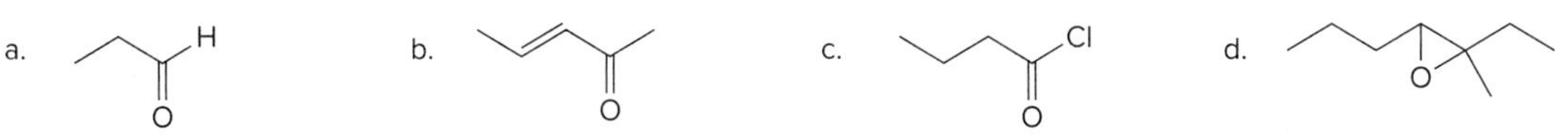

4. Which compound does not react with $(CH_3)_2CuLi$?

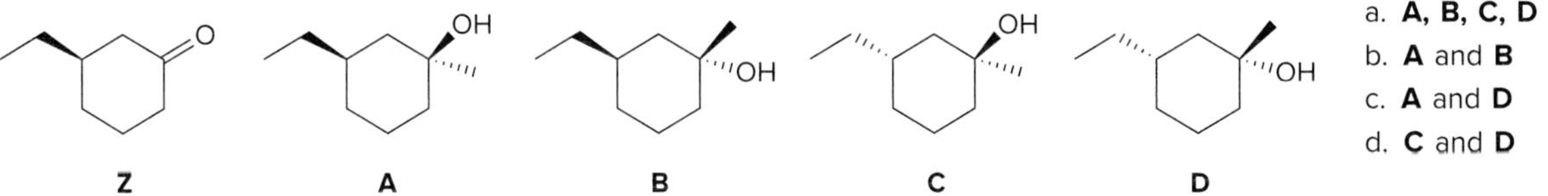

5. What products are formed when **Z** reacts with CH_3MgBr, followed by protonation with H_2O?

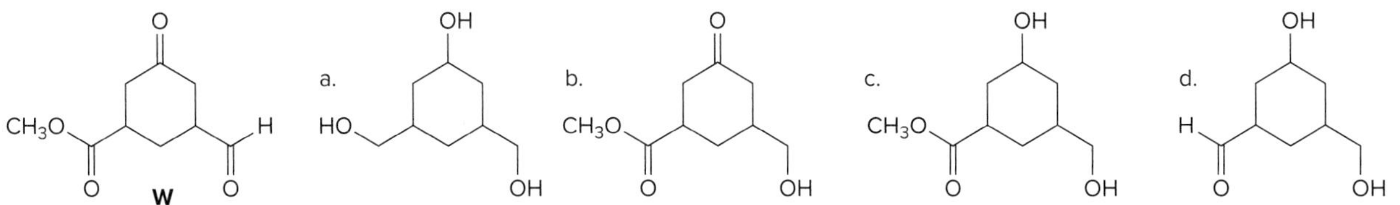

a. **A, B, C, D**
b. **A** and **B**
c. **A** and **D**
d. **C** and **D**

6. What product is formed when **W** reacts with $NaBH_4$ in CH_3OH?

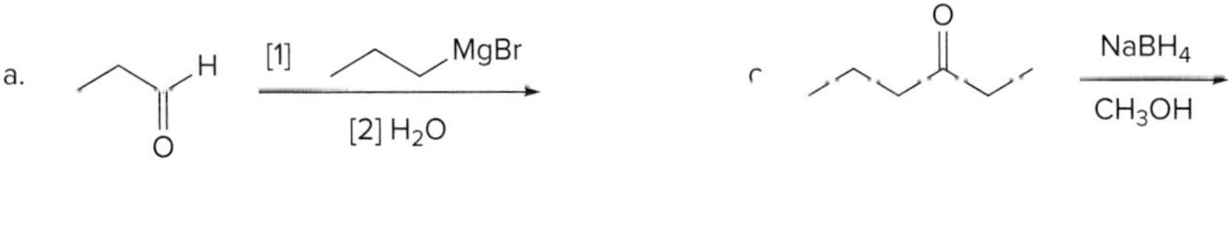

7. Which method cannot be used to prepare hexan-3-ol $[CH_3CH_2CH(OH)CH_2CH_2CH_3]$?

a. (propanal) [1] $CH_3CH_2CH_2MgBr$; [2] H_2O

c. (hexan-3-one) $NaBH_4$, CH_3OH

b. (butanal) [1] CH_3CH_2MgBr; [2] H_2O

d. (methyloxirane) [1] $CH_3CH_2CH_2MgBr$; [2] H_2O

8. Which method cannot be used to prepare alcohol **V?**

OH
V

a. OCH_3 — [1] MgBr (2 equiv) [2] H_2O

b. O — [1] MgBr [2] H_2O

c. OH — [1] MgBr (2 equiv) [2] H_2O

d. O — [1] MgBr [2] H_2O

9. What reagent is best to carry out the following transformation?

HO ... H (O) ⟶ HO ... OH (O)

a. Ag_2O, NH_4OH
b. CrO_3, H_2SO_4
c. PCC
d. $KMnO_4$

10. Label the most reactive and least reactive compound in a nucleophilic acyl substitution reaction.

A (Cl) **B** (OH) **C** (NH_2) **D** (OCH_3)

a. **A** most reactive; **B** least reactive
b. **D** most reactive; **C** least reactive
c. **D** most reactive; **B** least reactive
d. **A** most reactive; **C** least reactive

PROBLEMS

Problem Using Three-Dimensional Models

17.38 Devise a synthesis of each alcohol from organic alcohols having one or two carbons and any required reagents.

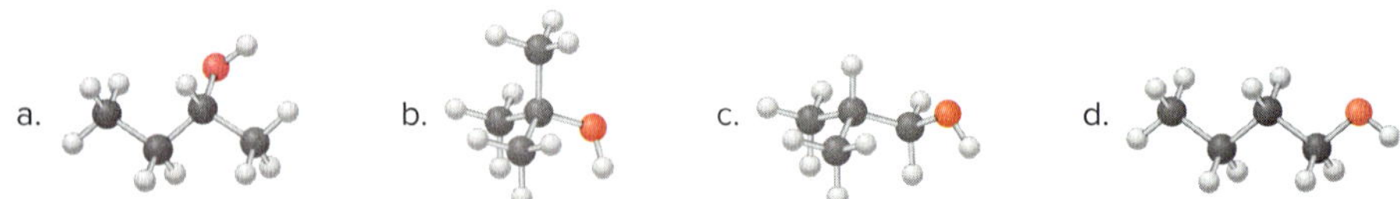

Reactions and Reagents

17.39 Draw the product formed when pentanal ($CH_3CH_2CH_2CH_2CHO$) is treated with each reagent. With some reagents, no reaction occurs.

a. $NaBH_4$, CH_3OH
b. [1] $LiAlH_4$; [2] H_2O
c. H_2, Pd-C
d. PCC
e. $Na_2Cr_2O_7$, H_2SO_4, H_2O
f. Ag_2O, NH_4OH
g. [1] CH_3MgBr; [2] H_2O
h. [1] C_6H_5Li; [2] H_2O
i. [1] $(CH_3)_2CuLi$; [2] H_2O
j. [1] HC≡CNa; [2] H_2O
k. [1] $CH_3C{\equiv}CLi$; [2] H_2O
l. The product in (a), then TBS–Cl, imidazole

17.40 Draw the product formed when $(CH_3CH_2CH_2CH_2)_2CuLi$ is treated with each compound. In some cases, no reaction occurs.

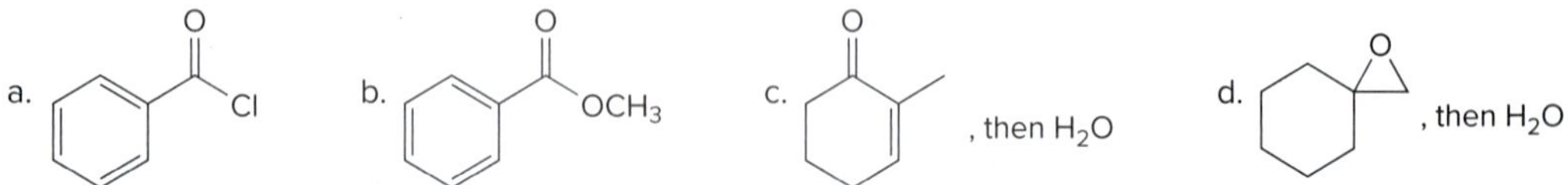

17.41 The stereochemistry of the products of reduction depends on the reagent used, as you learned in Sections 17.5 and 17.6. With this in mind, how would you convert 3,3-dimethylbutan-2-one [$CH_3COC(CH_3)_3$] to: (a) racemic 3,3-dimethylbutan-2-ol [$CH_3CH(OH)C(CH_3)_3$]; (b) only (R)-3,3-dimethylbutan-2-ol; (c) only (S)-3,3-dimethylbutan-2-ol?

17.42 Draw the product formed when the α,β-unsaturated ketone **A** is treated with each reagent.

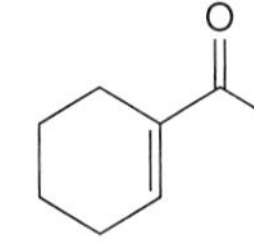

A

a. $NaBH_4$, CH_3OH
b. H_2 (1 equiv), Pd-C
c. H_2 (excess), Pd-C
d. [1] CH_3Li; [2] H_2O
e. [1] CH_3CH_2MgBr; [2] H_2O
f. [1] $(CH_2{=}CH)_2CuLi$; [2] H_2O

17.43 Draw the products of each reduction reaction.

a. 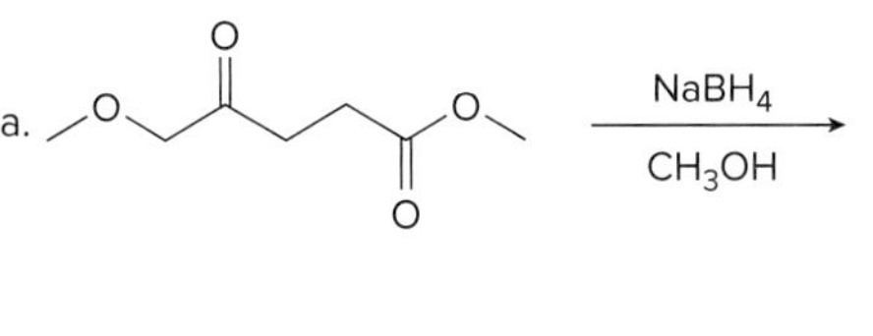

b. 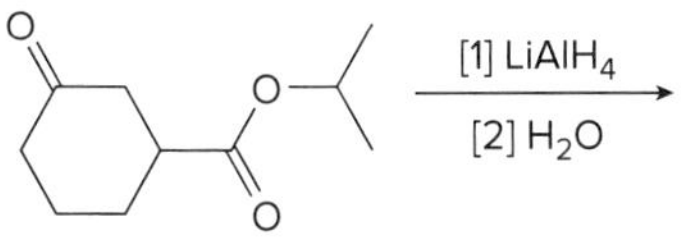 [1] $LiAlH_4$; [2] H_2O

c. 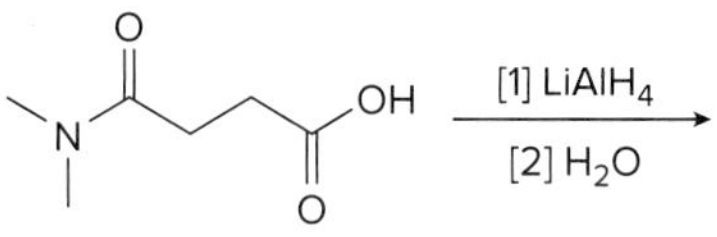 [1] $LiAlH_4$; [2] H_2O

d. 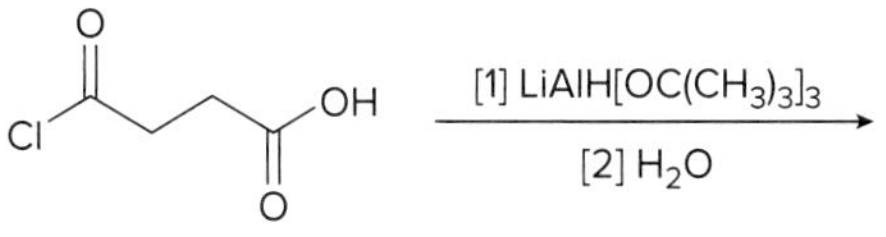 [1] $LiAlH[OC(CH_3)_3]_3$; [2] H_2O

17.44 Draw the product(s) formed when **A** is treated with each reagent.

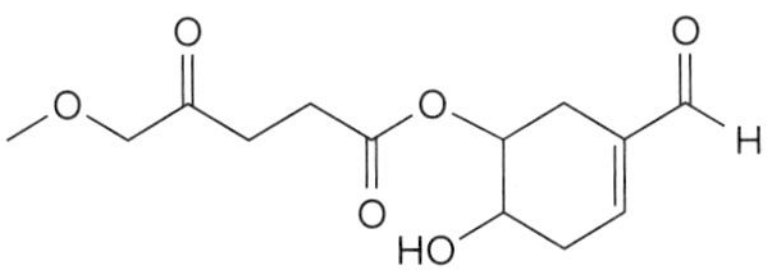

A

a. $NaBH_4$, CH_3OH
b. $LiAlH_4$, then H_2O
c. Ag_2O, NH_4OH
d. CrO_3, H_2SO_4, H_2O
e. PCC

17.45 Draw the products of the following reactions with organometallic reagents.

a. 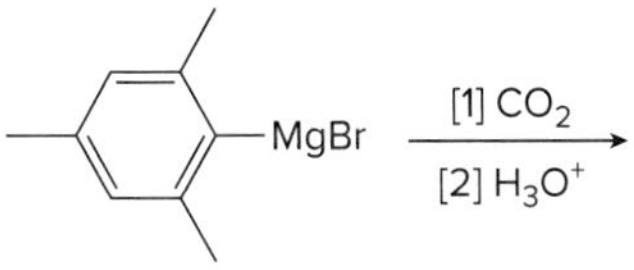 MgBr — [1] CO_2; [2] H_3O^+

b. 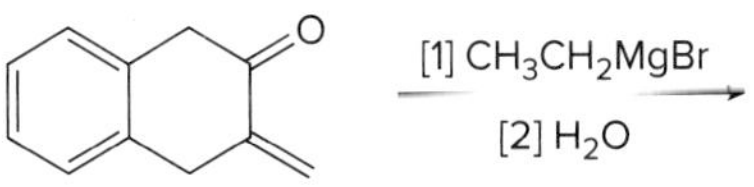 [1] CH_3CH_2MgBr; [2] H_2O

c. 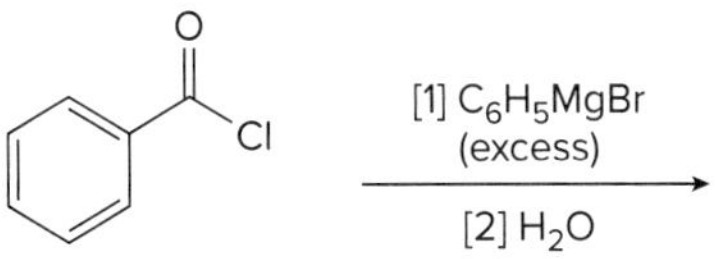 Cl — [1] C_6H_5MgBr (excess); [2] H_2O

d. 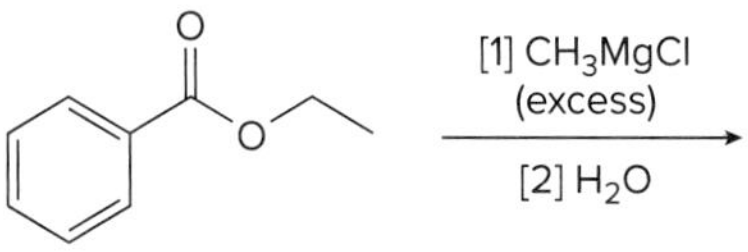 [1] CH_3MgCl (excess); [2] H_2O

e. 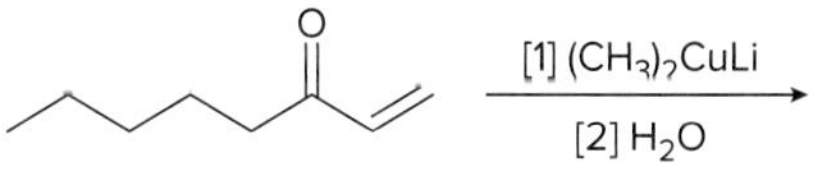 [1] $(CH_3)_2CuLi$; [2] H_2O

f. 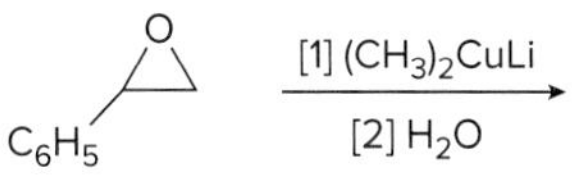 C_6H_5 — [1] $(CH_3)_2CuLi$; [2] H_2O

17.46 Draw all stereoisomers formed in each reaction.

a. 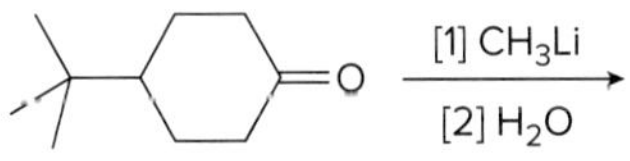 [1] CH_3Li; [2] H_2O

b. 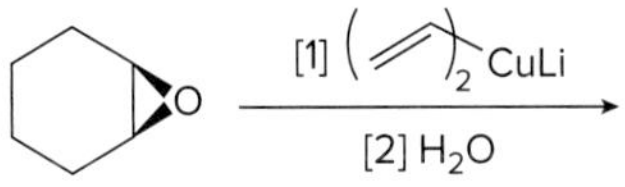 [1] $(CH_2{=}CH)_2CuLi$; [2] H_2O

c. 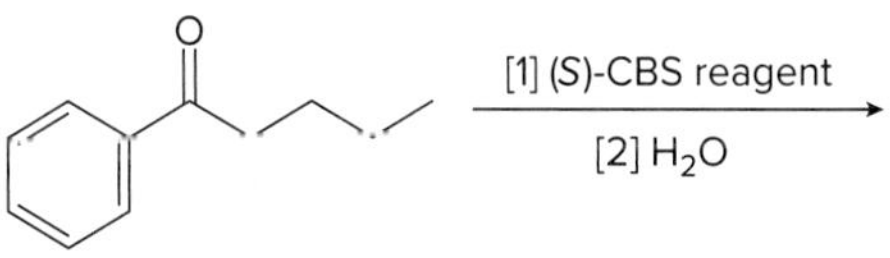 [1] (*S*)-CBS reagent; [2] H_2O

d. 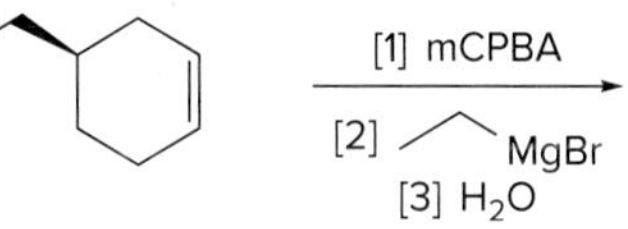 [1] mCPBA; [2] CH_3CH_2MgBr; [3] H_2O

17.47 A student tried to carry out the following reaction sequence, but none of diol **A** was formed. Explain what was wrong with this plan, and design a successful stepwise synthesis of **A.**

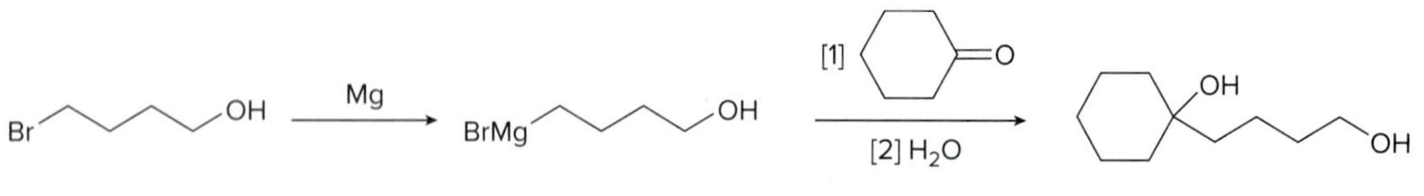

Br–(CH₂)₄–OH → (Mg) → BrMg–(CH₂)₄–OH → ([1] cyclohexanone; [2] H_2O) → **A**

17.48 Identify the lettered compounds in the following reaction scheme. Compounds **F, G,** and **K** are isomers of molecular formula $C_{13}H_{18}O$. How could ^1H NMR spectroscopy distinguish these three compounds from each other?

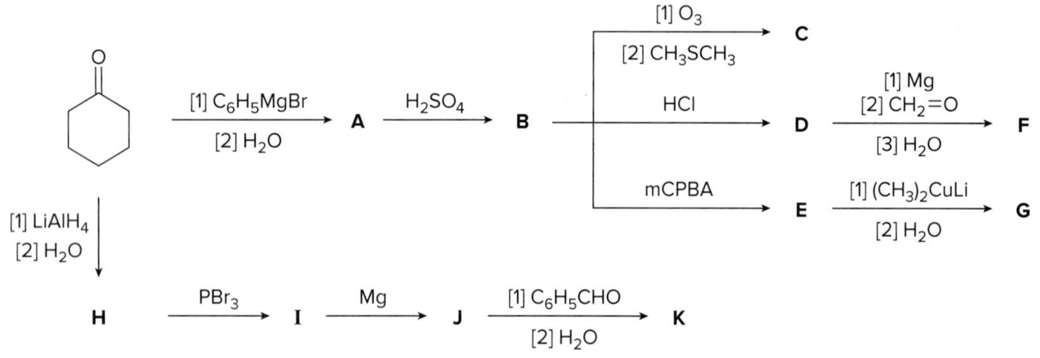

cyclohexanone → ([1] C_6H_5MgBr; [2] H_2O) → **A** → (H_2SO_4) → **B**

B → ([1] O_3; [2] CH_3SCH_3) → **C**

B → (HCl) → **D** → ([1] Mg; [2] $CH_2{=}O$; [3] H_2O) → **F**

B → (mCPBA) → **E** → ([1] $(CH_3)_2CuLi$; [2] H_2O) → **G**

cyclohexanone → ([1] $LiAlH_4$; [2] H_2O) → **H** → (PBr_3) → **I** → (Mg) → **J** → ([1] C_6H_5CHO; [2] H_2O) → **K**

17.49 Several steps in the synthesis of optically active duloxetine, an antidepressant sold under the trade name Cymbalta, are shown. Identify the structure of intermediates **A–C** and the final product duloxetine, including stereochemistry, in this reaction sequence.

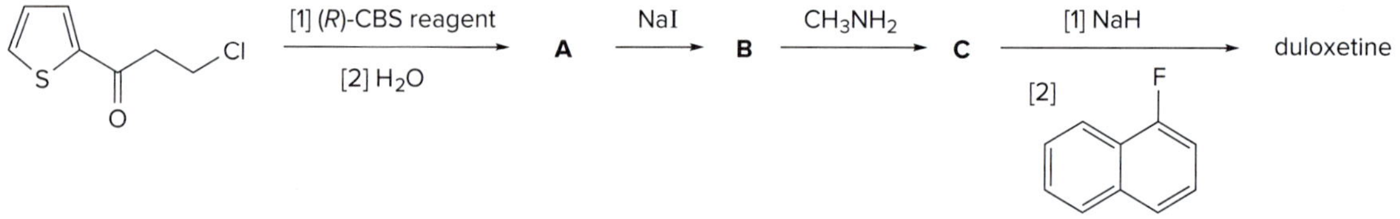

2-thienyl–CO–CH₂CH₂–Cl → ([1] (*R*)-CBS reagent; [2] H_2O) → **A** → (NaI) → **B** → (CH_3NH_2) → **C** → ([1] NaH; [2] 1-fluoronaphthalene) → duloxetine

17.50 In Figure 17.3 you learned about the conversion of **A** to **B,** one step in a multistep synthesis of palau'amine. Answer the following questions about other steps in the synthesis. (a) What diene and dienophile are needed to synthesize **A** by a Diels–Alder reaction? (b) **B** was converted to **C** by a series of reactions. Draw the structure of **C,** and identify the intermediate formed after each step.

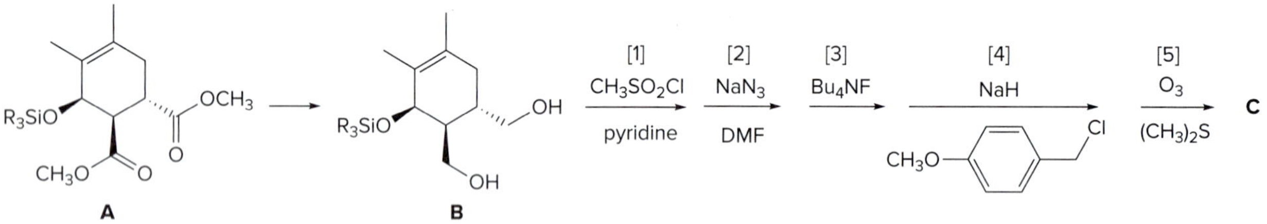

A (R_3SiO, OCH_3, CH_3O) → **B** (R_3SiO, OH, OH) → ([1] CH_3SO_2Cl, pyridine) → ([2] NaN_3, DMF) → ([3] Bu_4NF) → ([4] NaH, CH_3O–C₆H₄–CH₂Cl) → ([5] O_3, $(CH_3)_2S$) → **C**

Mechanism

17.51 Draw a stepwise mechanism for the following reaction, which synthesizes the antidepressant venlafaxine.

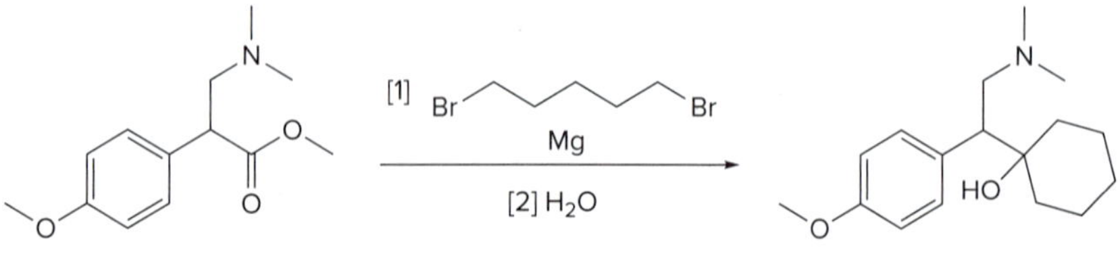

→ ([1] Br–(CH₂)₅–Br, Mg; [2] H_2O) → venlafaxine

17.52 Draw a stepwise mechanism for the following reaction, a key step in the synthesis of the female beetle pheromone lineatin. Lineatin has potential as a pest control agent because it can be used as a lure to mass-trap insects.

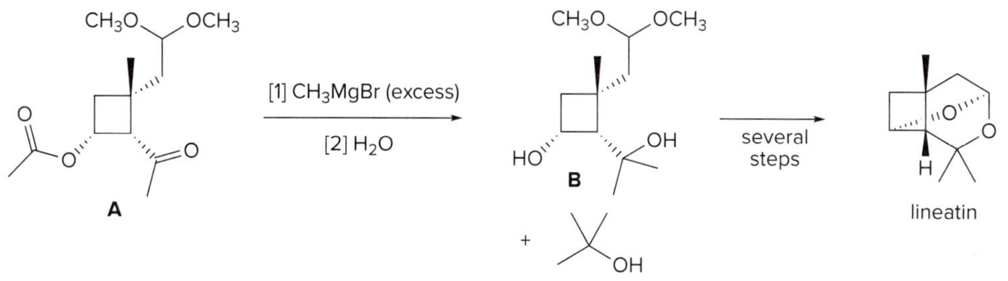

CH_3O OCH_3 A [1] CH_3MgBr (excess) [2] H_2O CH_3O OCH_3 HO OH B + OH several steps H O O lineatin

17.53 Draw a stepwise mechanism for the following reaction.

O OCH_3 Cl MgBr R H O OCH_3 O R + MgBrCl

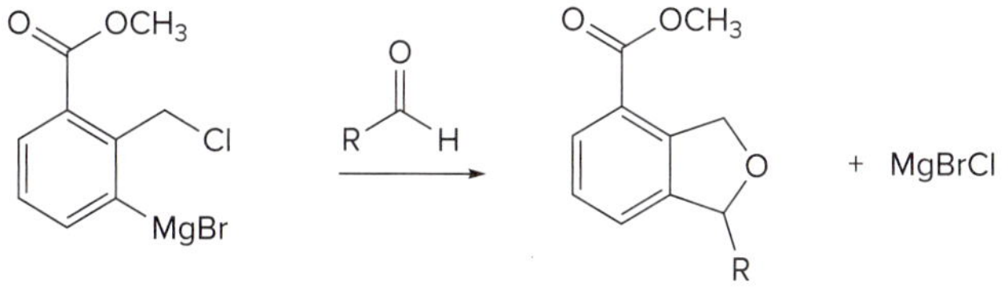

17.54 Slow addition of organolithium reagent **A** to **B** afforded **C,** an intermediate in the synthesis of resiniferatoxin, a compound isolated from a flowering cactus that is more potent than the capsaicin of chili peppers in producing a burning and numbing sensation in the mouth. Draw a stepwise mechanism for this process.

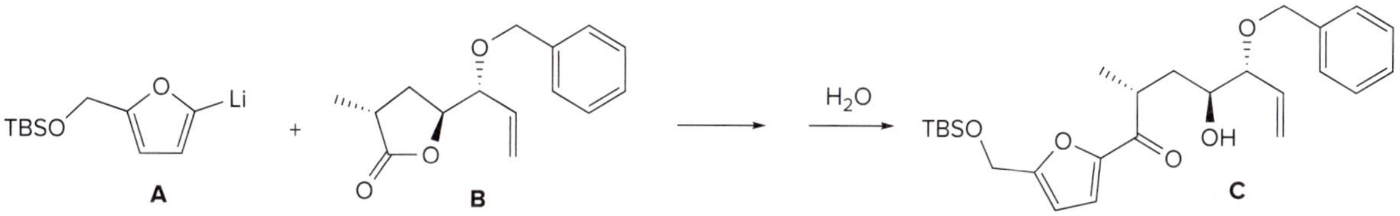

TBSO O Li A + O O O B H_2O TBSO O O O OH C

17.55 Draw a stepwise mechanism for the following reaction.

O O O O [1] $LiAlH_4$ [2] H_2O HO OH OH OH

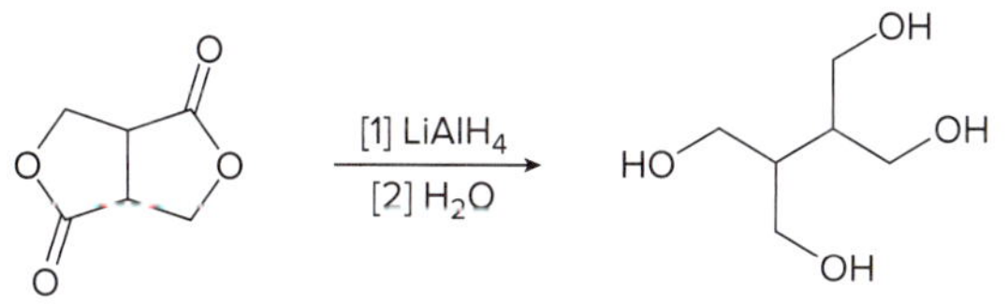

Synthesis

17.56 What amides will form each amine on treatment with $LiAlH_4$?

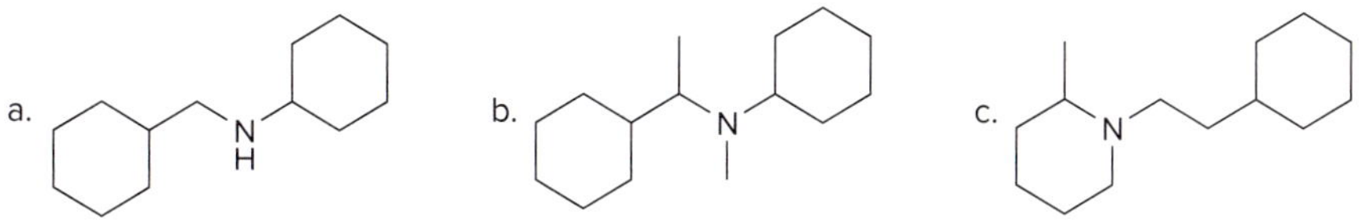

a. N H b. N c. N

17.57 What Grignard reagent and aldehyde (or ketone) are needed to prepare each alcohol? Show all possible routes.

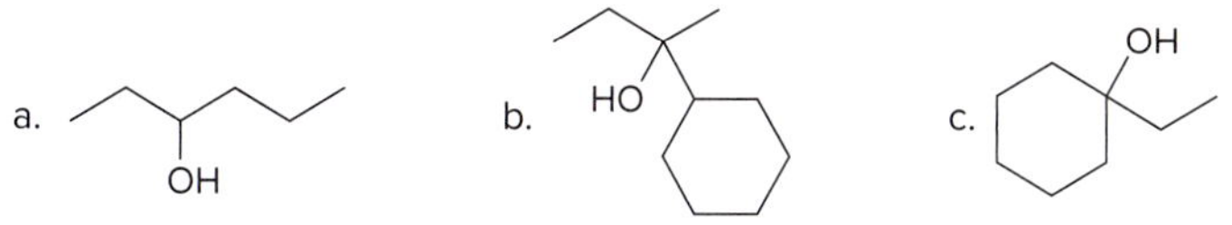

a. OH b. HO c. OH

17.58 Procyclidine is a drug that has been used to treat the uncontrolled body movements associated with Parkinson's disease. Draw three different methods to prepare procyclidine using a Grignard reagent.

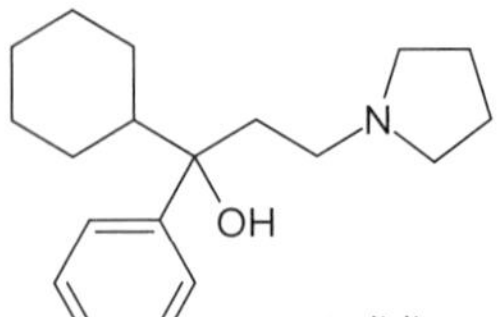

N OH

procyclidine

17.59 What ester and Grignard reagent are needed to synthesize each alcohol?

a. 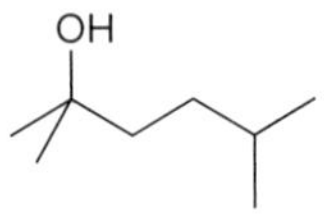 OH b. 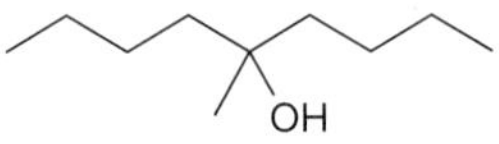 OH

17.60 What organolithium reagent and carbonyl compound can be used to prepare each of the following compounds? You may use aldehydes, ketones, or esters as carbonyl starting materials.

a. 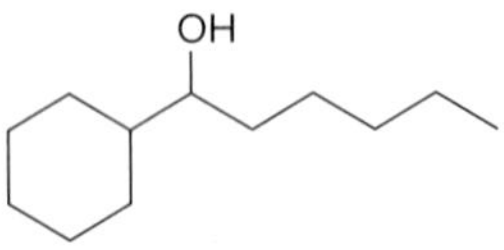 OH (two ways)

c. 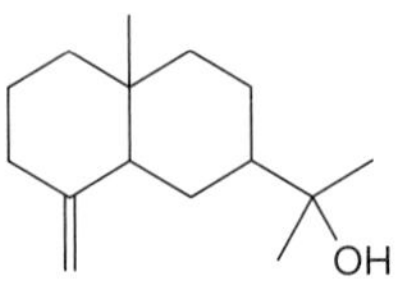 OH β-eudesmol (three ways)

b. 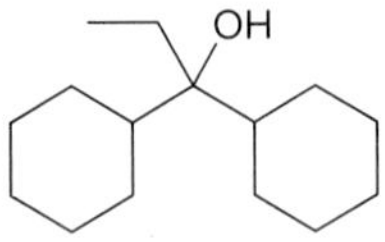 OH (three ways)

d. 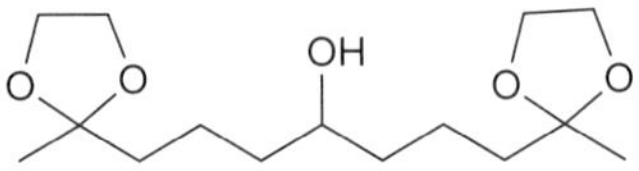 O O OH O O (two ways)

17.61 What epoxide and organometallic reagent are needed to synthesize each alcohol?

a. 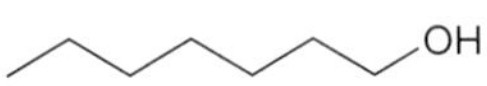 OH b. 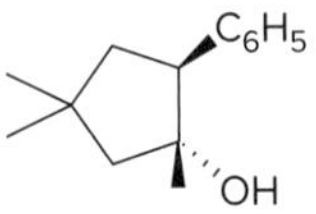 C_6H_5 OH c. HO H H

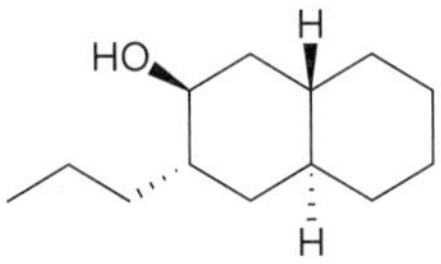

17.62 Propose at least three methods to convert $C_6H_5CH_2CH_2Br$ to $C_6H_5CH_2CH_3$.

17.63 Synthesize each compound from cyclohexanol using any other organic or inorganic compounds.

a. 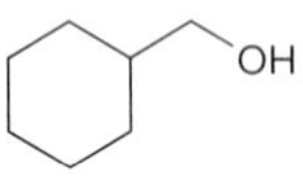 OH b. 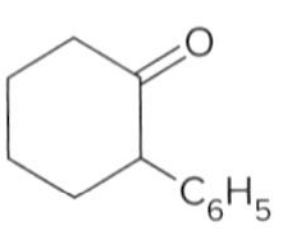 O C_6H_5 c. 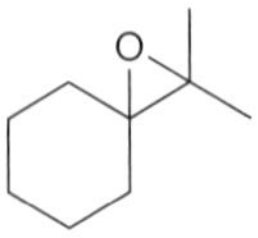 O d. 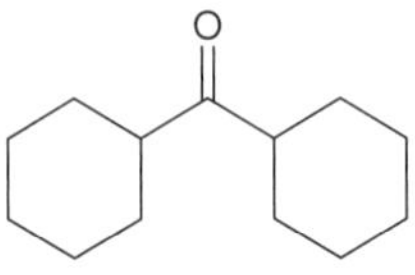 O

(Each cyclohexane ring must come from cyclohexanol.)

17.64 Convert benzene into each compound. You may also use any inorganic reagents and organic alcohols having four or fewer carbons. One step of the synthesis must use a Grignard reagent.

a. 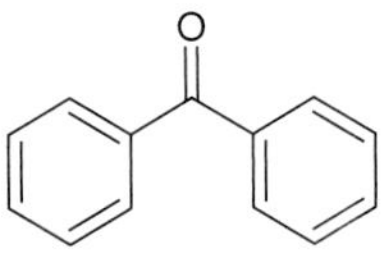 O b. 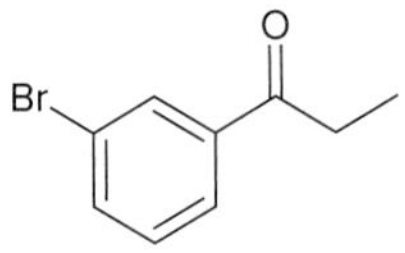 Br O c. 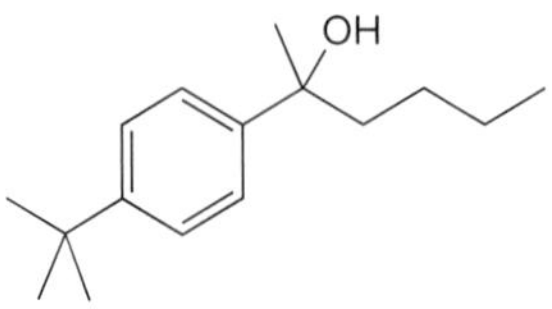 OH

17.65 Design a synthesis of each compound from alcohols having four or fewer carbons as the only organic starting materials. You may use any other inorganic reagents you choose.

a. b. CO_2H c.

17.66 Devise a synthesis of each alkyne. You may use acetylene, benzene, organic halides, ethylene oxide, and any other required inorganic reagents.

a. HO OH b. O

17.67 Devise a synthesis of each compound from cyclohex-2-enone and organic halides having one or two carbons. You may use any other required inorganic reagents.

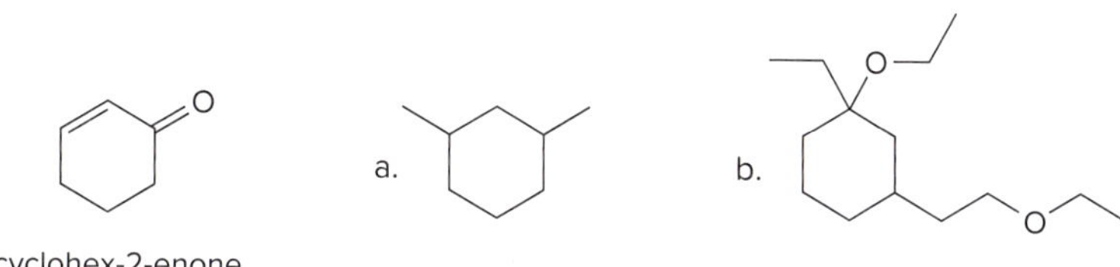

cyclohex-2-enone

17.68 Devise a synthesis of (*E*)-tetradec-11-enal, a sex pheromone of the spruce budworm, a pest that destroys fir and spruce forests, from acetylene, $Br(CH_2)_{10}OH$, and any needed organic compounds or inorganic reagents.

O H

(*E*)-tetradec-11-enal

Spectroscopy

17.69 An unknown compound **A** (molecular formula $C_7H_{14}O$) was treated with $NaBH_4$ in CH_3OH to form compound **B** (molecular formula $C_7H_{16}O$). Compound **A** has a strong absorption in its IR spectrum at 1716 cm^{-1}. Compound **B** has a strong absorption in its IR spectrum at 3600–3200 cm^{-1}. The ^1H NMR spectra of **A** and **B** are given. What are the structures of **A** and **B?**

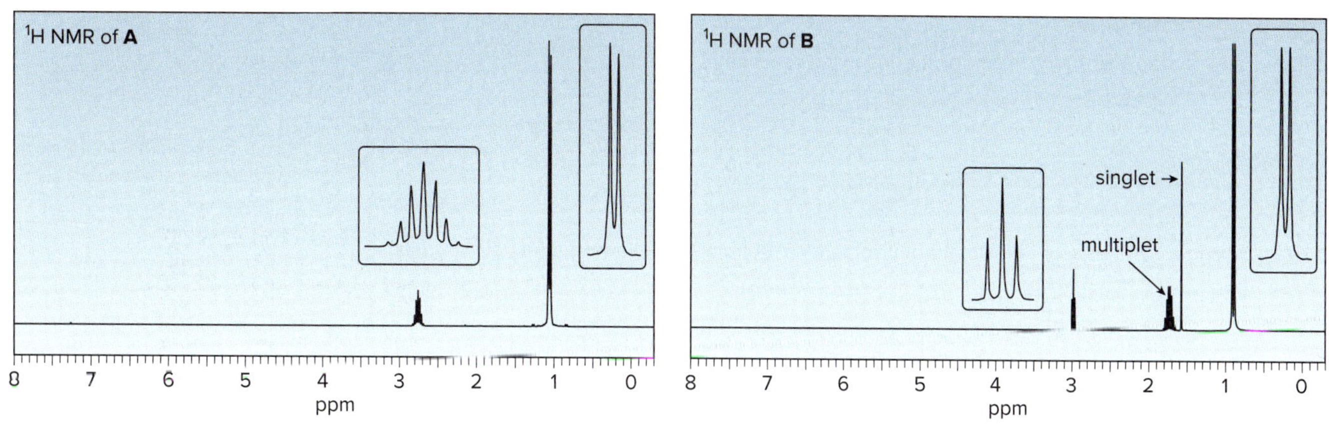

17.70 Treatment of compound **E** (molecular formula $C_4H_8O_2$) with excess CH_3CH_2MgBr yields compound **F** (molecular formula $C_6H_{14}O$) after protonation with H_2O. **E** shows a strong absorption in its IR spectrum at 1743 cm^{-1}. **F** shows a strong IR absorption at 3600–3200 cm^{-1}. The ^1H NMR spectral data of **E** and **F** are given. What are the structures of **E** and **F?**

Compound **E** signals at 1.2 (triplet, 3 H), 2.0 (singlet, 3 H), and 4.1 (quartet, 2 H) ppm

Compound **F** signals at 0.9 (triplet, 6 H), 1.1 (singlet, 3 H), 1.5 (quartet, 4 H), and 1.55 (singlet, 1 H) ppm

17.71 Reaction of butanenitrile ($CH_3CH_2CH_2CN$) with methylmagnesium bromide (CH_3MgBr), followed by treatment with aqueous acid, forms compound **G. G** has a molecular ion in its mass spectrum at $m/z = 86$ and a base peak at $m/z = 43$. **G** exhibits a strong absorption in its IR spectrum at 1721 cm^{-1} and has the ^1H NMR spectrum given below. What is the structure of **G?** We will learn about the details of this reaction in Chapter 19.

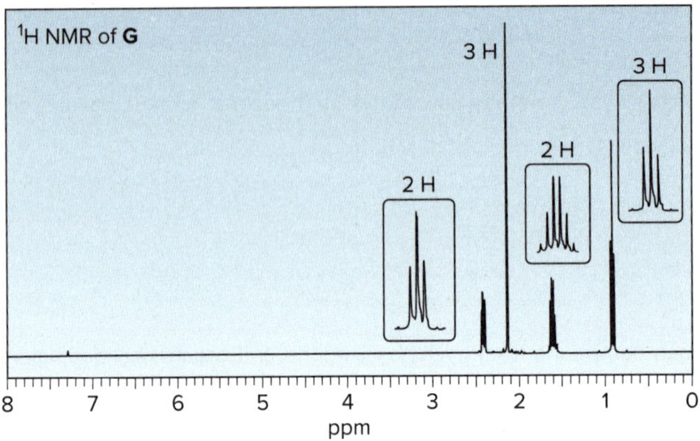

17.72 Treatment of isobutene [$(CH_3)_2C{=}CH_2$] with $(CH_3)_3CLi$ forms a carbanion that reacts with $CH_2{=}O$ to form **H** after water is added to the reaction mixture. **H** has a molecular ion in its mass spectrum at $m/z = 86$, and shows fragments at 71 and 68. **H** exhibits absorptions in its IR spectrum at 3600–3200 and 1651 cm^{-1}, and has the ^1H NMR spectrum given below. What is the structure of **H?**

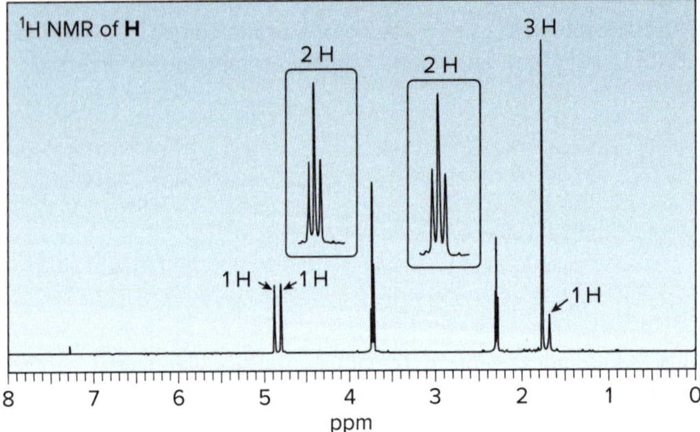

Challenge Problems

17.73 Draw a stepwise mechanism for the following reaction.

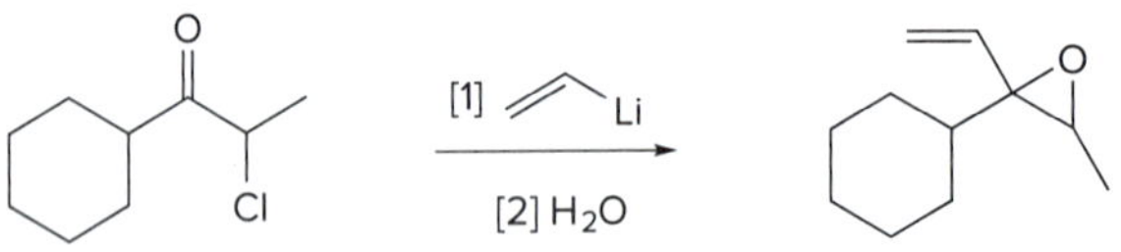

[1] vinyllithium (Li)
[2] H_2O

17.74 Design a synthesis of (*R*)-salmeterol (Problem 17.8) from the following starting materials.

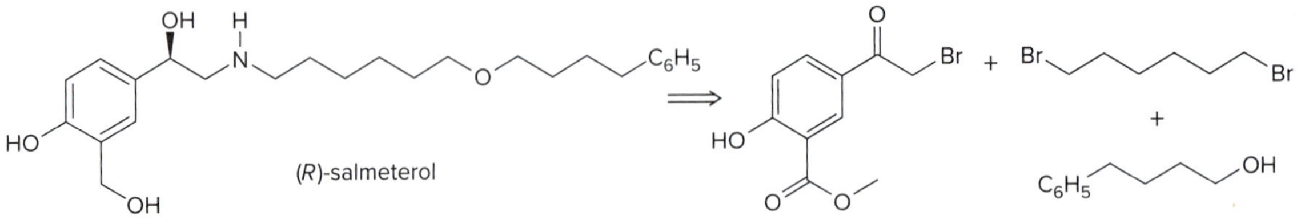

(*R*)-salmeterol ⟹ + Br~~~~~~Br + C_6H_5~~~~OH

17.75 Explain why the β carbon of an α,β-unsaturated carbonyl compound absorbs farther downfield in the ^{13}C NMR spectrum than the α carbon, even though the α carbon is closer to the electron-withdrawing carbonyl group. For example, the β carbon of mesityl oxide absorbs at 150.5 ppm, whereas the α carbon absorbs at 122.5 ppm.

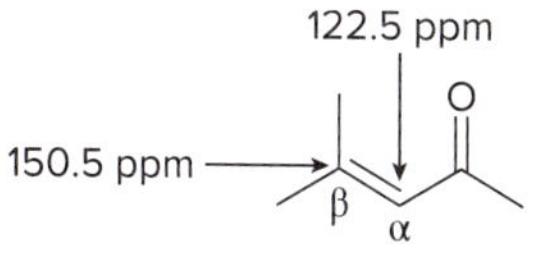

122.5 ppm

150.5 ppm

β α

mesityl oxide

17.76 Reaction of benzylmagnesium chloride with formaldehyde yields alcohols **N** and **P** after protonation. Draw a stepwise mechanism that shows how both products are formed.

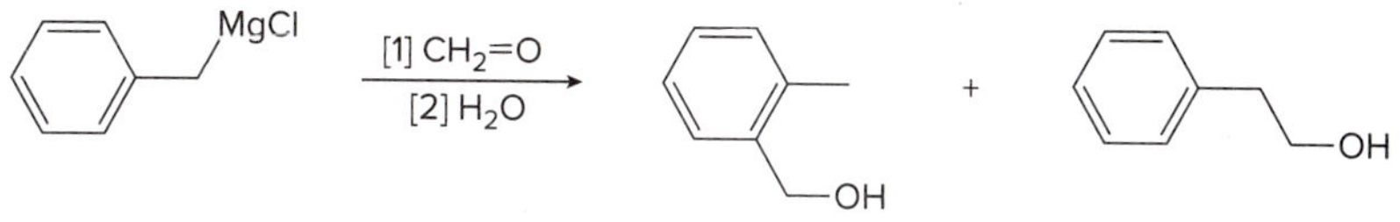

MgCl

[1] $CH_2{=}O$
[2] H_2O

OH + OH

N major product

P minor product

17.77 Draw a stepwise mechanism for the following reaction. (Hint: Conjugate addition can occur with heteroatoms as well as carbon nucleophiles.)

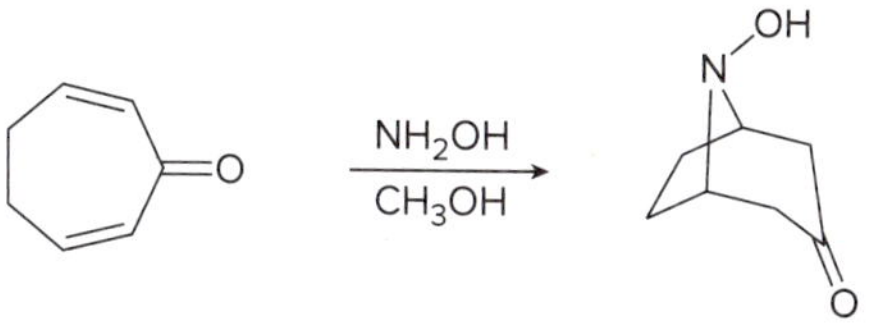

O

NH_2OH
CH_3OH

N OH

O

17.78 Draw a stepwise mechanism for the following reaction of a Grignard reagent with a cyclic amide.

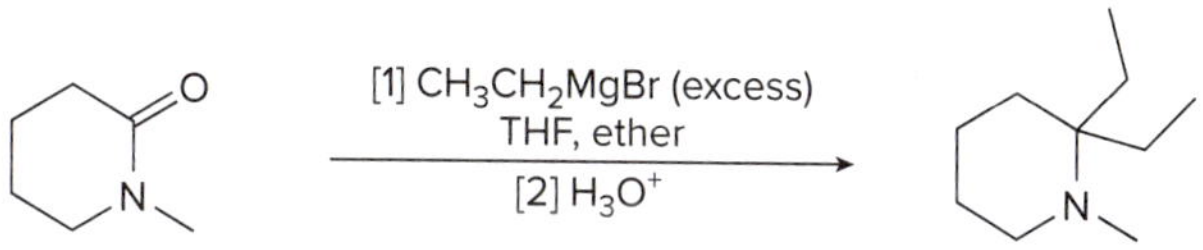

O N

[1] CH_3CH_2MgBr (excess) THF, ether
[2] H_3O^+

N

SELF-TEST ANSWERS

1. b 2. c 3. a 4. a 5. b 6. c 7. d 8. c 9. a 10. d

18 Aldehydes and Ketones—Nucleophilic Addition

18.1 Introduction
18.2 Nomenclature
18.3 Properties of aldehydes and ketones
18.4 Interesting aldehydes and ketones
18.5 Preparation of aldehydes and ketones
18.6 Reactions of aldehydes and ketones—General considerations
18.7 Nucleophilic addition of H^- and R^-—A review
18.8 Nucleophilic addition of ^-CN
18.9 The Wittig reaction
18.10 Addition of 1° amines
18.11 Addition of 2° amines
18.12 Addition of H_2O—Hydration
18.13 Addition of alcohols—Acetal formation
18.14 Acetals as protecting groups
18.15 Cyclic hemiacetals
18.16 An introduction to carbohydrates

Brett Hondow/Alamy Stock Photo

Rebaudioside A, one of the sweet compounds derived from the alcohol **steviol,** whose structure is shown in the ball-and-stick model, is obtained from the leaves of the stevia plant, a shrub native to Central and South America. The stevia plant has been used for centuries in Paraguay to sweeten food, and today, rebaudioside A, marketed under the name Truvia, is a commonly used commercial sweetener 400 times sweeter than sucrose. The structure of rebaudioside A contains four acetal units, a functional group derived from an aldehyde or ketone by nucleophilic addition, the principal reaction discussed in Chapter 18.

Why Study . . . Aldehydes and Ketones?

In Chapter 18, we continue the study of carbonyl compounds with a detailed look at **aldehydes** and **ketones.** We will first learn about the nomenclature, physical properties, and spectroscopic absorptions that characterize aldehydes and ketones. The remainder of Chapter 18 is devoted to **nucleophilic addition** reactions. Although we have already learned two examples of this reaction in Chapter 17, nucleophilic addition to aldehydes and ketones is a general reaction that occurs with many nucleophiles, forming a wide variety of products, including carbohydrates and molecules central to the process of vision.

Every new reaction in Chapter 18 involves nucleophilic addition, so the challenge lies in learning the specific reagents and mechanisms that characterize each reaction.

18.1 Introduction

An aldehyde is often written as **RCHO.** Remember that the **H atom is bonded to the carbon atom,** *not* the oxygen. Likewise, a ketone is written as **RCOR** or, if both alkyl groups are the same, **R_2CO.** Each structure must contain a C=O for every atom to have an octet.

As we learned in Chapter 17, **aldehydes and ketones contain a carbonyl group.** An aldehyde contains at least one H atom bonded to the carbonyl carbon, whereas a ketone has two alkyl or aryl groups bonded to it.

carbonyl group | aldehyde | ketone

Two structural features determine the chemistry and properties of aldehydes and ketones.

sp^2 hybridized
~120°
δ+ δ−
trigonal planar | electrophilic carbon

- The carbonyl group is sp^2 hybridized and trigonal planar, making it relatively *uncrowded.*
- The electronegative oxygen atom polarizes the carbonyl group, making the carbonyl carbon *electrophilic.*

As a result, **aldehydes and ketones react with nucleophiles.** The relative reactivity of the carbonyl group is determined by the number of R groups bonded to it. **As the number of R groups around the carbonyl carbon *increases,* the reactivity of the carbonyl compound *decreases,*** resulting in the following order of reactivity:

Increasing the number of alkyl groups on the carbonyl carbon decreases reactivity for both steric and electronic reasons, as discussed in Section 17.2B.

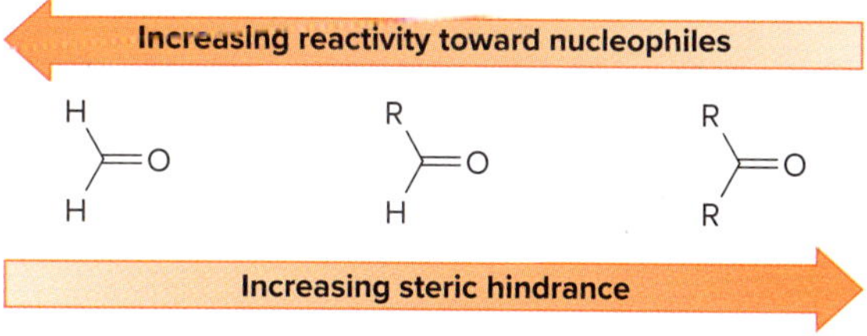

Problem 18.1 Rank the following compounds in order of increasing reactivity toward nucleophilic attack.

Problem 18.2 Explain why benzaldehyde is less reactive than cyclohexanecarbaldehyde toward nucleophilic attack.

benzaldehyde | cyclohexanecarbaldehyde

18.2 Nomenclature

Both IUPAC and common names are used for aldehydes and ketones.

18.2A Naming Aldehydes in the IUPAC System

In IUPAC nomenclature, aldehydes are identified by a suffix added to the parent name of the longest chain. Two different suffixes are used, depending on whether the CHO group is bonded to a chain or a ring.

To name an aldehyde using the IUPAC system:

[1] If the CHO is bonded to a chain of carbons, find the longest chain containing the CHO group, and change the ***-e*** ending of the parent alkane to the suffix ***-al.*** If the CHO group is bonded to a ring, name the ring and add the suffix ***-carbaldehyde.***

[2] Number the chain or ring to put the CHO group at C1, but omit this number from the name. Apply all of the other usual rules of nomenclature.

Sample Problem 18.1 Naming an Aldehyde Using the IUPAC System

Give the IUPAC name for each compound.

a. 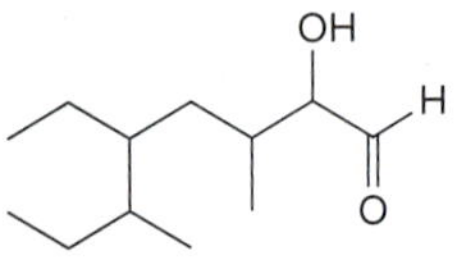 OH H O

b. 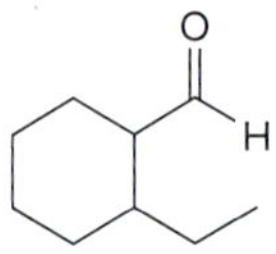 O H

Solution

a. [1] Find and name the longest chain containing the CHO:

OH H O

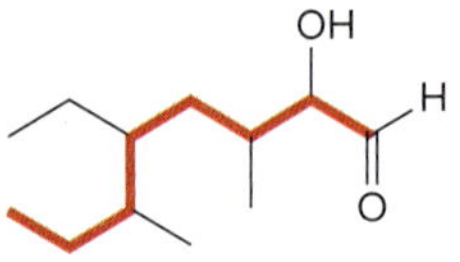

octane (8 C's) ⟶ octan*al*

[2] Number and name substituents:

OH 3 2 1 H 5 6 O

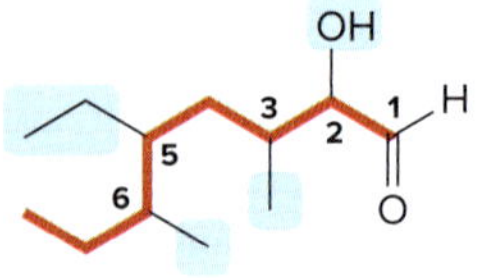

Answer:
5-ethyl-2-hydroxy-3,6-dimethyloctanal

b. [1] Find and name the ring bonded to the CHO group:

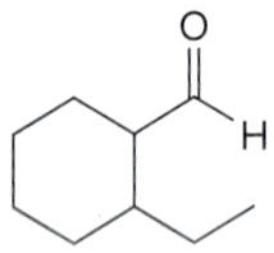

O H

cyclohexane + carbaldehyde (6 C's)

[2] Number and name substituents:

O 1 H 2

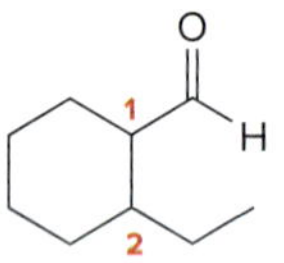

Answer:
2-ethylcyclohexanecarbaldehyde

Problem 18.3 Give the IUPAC name for each aldehyde.

a. 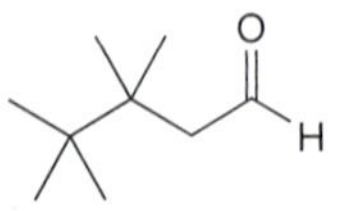 O H

b. 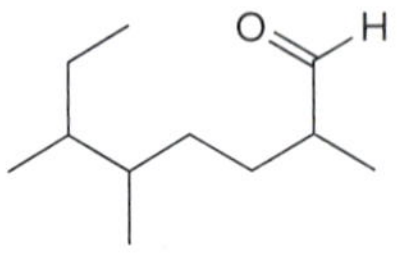 O H

c. 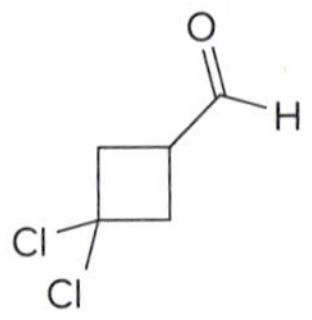 O H Cl Cl

More Practice: Try Problems 18.39a (**A**); 18.41b, d; 18.42a, b, f.

Problem 18.4 Give the structure corresponding to each IUPAC name.

a. 2-isobutyl-3-isopropylhexanal

b. *trans*-3-methylcyclopentanecarbaldehyde

c. 1-methylcyclopropanecarbaldehyde

d. 3,6-diethylnonanal

18.2B Common Names for Aldehydes

Many simple aldehydes have common names that are widely used.

- **A common name for an aldehyde is formed by taking the common parent name and adding the suffix *-aldehyde.***

Table 18.1 lists common parent names for some simple aldehydes. These parent names are used in the nomenclature of many other carbonyl compounds (Chapters 19 and 20). The common names **formaldehyde, acetaldehyde,** and **benzaldehyde** are generally used instead of their IUPAC names.

Table 18.1 Common Names for Some Simple Aldehydes

Number of C atoms	Structure	Parent name	Common name
1	HCHO	**form-**	formaldehyde
2	CH_3CHO	**acet-**	acetaldehyde
3	CH_3CH_2CHO	**propion-**	propionaldehyde
4	$CH_3(CH_2)_2CHO$	**butyr-**	butyraldehyde
5	$CH_3(CH_2)_3CHO$	**valer-**	valeraldehyde
6	$CH_3(CH_2)_4CHO$	**capro-**	caproaldehyde
	C_6H_5CHO	**benz-**	benzaldehyde

Greek letters are used to designate the location of substituents in common names.

- **The carbon adjacent to the CHO is called the α carbon.**
- **The carbon bonded to the α carbon is the β carbon, followed by the γ (gamma) carbon, the δ (delta) carbon, and so forth down the chain. The last carbon in the chain is sometimes called the Ω (omega) carbon.**

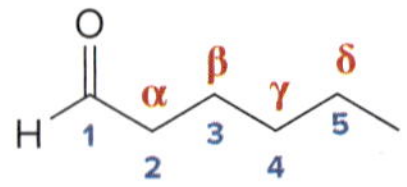

IUPAC numbering begins at the C=O.
Greek lettering begins at the C bonded to the C=O.

Figure 18.1 gives the common and IUPAC names for three aldehydes.

Figure 18.1 Three examples of aldehyde nomenclature

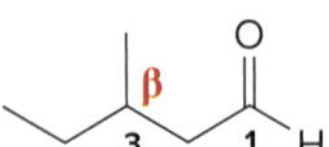

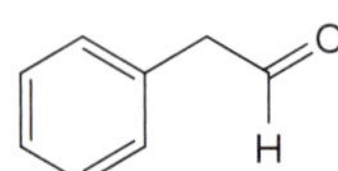

2-chloropropanal (α-chloropropionaldehyde)

3-methylpentanal (β-methylvaleraldehyde)

phenylethanal (phenylacetaldehyde)

(Common names are in parentheses.)

18.2C Naming Ketones in the IUPAC System

- **In the IUPAC system, all ketones are identified by the suffix *-one*.**

To name an acyclic ketone using IUPAC rules:

[1] Find the longest chain containing the carbonyl group, and change the ***-e*** ending of the parent alkane to the suffix ***-one.***

[2] Number the carbon chain to give the carbonyl carbon the lower number. Apply all of the other usual rules of nomenclature.

With cyclic ketones, numbering always begins at the carbonyl carbon, but the "1" is usually omitted from the name. The ring is then numbered clockwise or counterclockwise to give the *first* substituent the lower number.

Sample Problem 18.2 Naming a Ketone Using the IUPAC System

Give the IUPAC name for each ketone.

a. 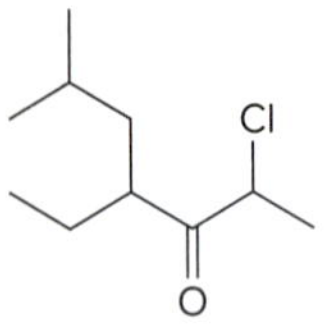 b.

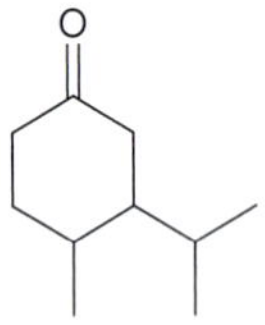

Solution

a. [1] Find and name the longest chain containing the carbonyl group:

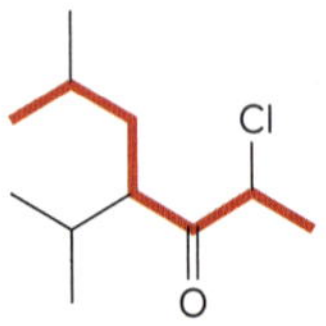

heptane (7 C's) ⟶ hept*anone*

[2] Number and name substituents:

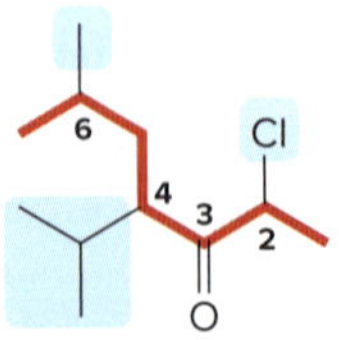

Answer: 2-chloro-4-isopropyl-6-methylheptan-3-one

b. [1] Name the ring:

cyclohexane ⟶ cyclohexan*one*
(6 C's)

[2] Number and name substituents:

Answer:
3-isopropyl-4-methylcyclohexanone

Problem 18.5 Give the IUPAC name for each ketone.

a. b. c.

More Practice: Try Problems 18.39a (**B**); 18.41a, c; 18.42e.

18.2D Common Names for Ketones

Most common names for ketones are formed by **naming both alkyl groups** on the carbonyl carbon, **arranging them alphabetically,** and adding the word ***ketone.*** Using this method, the common name for butan-2-one becomes ethyl methyl ketone.

IUPAC name: **butan-2-one**

CH_3 (methyl), CH_2CH_3 (ethyl)

Common name: **ethyl methyl ketone**

Three widely used common names for some simple ketones do not follow this convention:

acetone acetophenone benzophenone

Figure 18.2 gives acceptable names for two ketones.

Figure 18.2 Two examples of ketone nomenclature

IUPAC name: 2-methylpentan-3-one
Common name: ethyl isopropyl ketone

m-bromoacetophenone
or
3-bromoacetophenone

18.2E Additional Nomenclature Facts

Do not confuse a **benzyl** group with a **benzoyl** group.

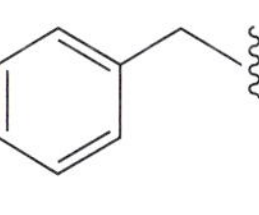
benz*yl* group

Sometimes **acyl groups (RCO–)** must be named as substituents. To name an acyl group, take either the IUPAC or common parent name and add the suffix ***-yl*** or ***-oyl.*** The three most common acyl groups are drawn below.

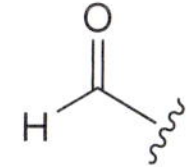
formyl group

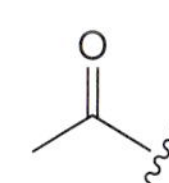
acetyl group

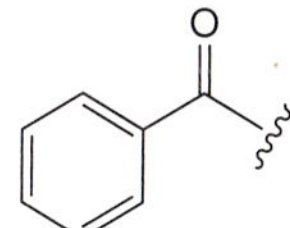
benzoyl group

Compounds containing both a C–C double bond and an aldehyde are named as **enals,** and compounds that contain both a C–C double bond and a ketone are named as **enones.** The chain is numbered to **give the carbonyl group the *lower* number.**

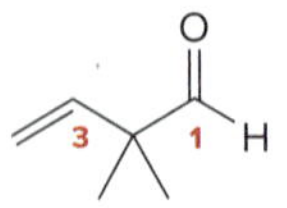

2,2-dimethylbut-3-enal

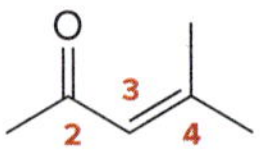

4-methylpent-3-en-2-one

Problem 18.6 Give the structure corresponding to each name: (a) *sec*-butyl ethyl ketone; (b) methyl vinyl ketone; (c) *p*-ethylacetophenone; (d) 3-benzoyl-2-benzylcyclopentanone; (e) 6,6-dimethylcyclohex-2-enone; (f) 3-ethylhex-5-enal.

Problem 18.7 Give the IUPAC name (including any *E,Z* designation) for each unsaturated aldehyde.

a. 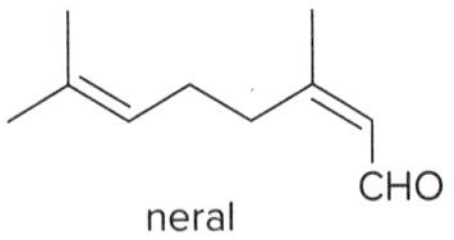 CHO
neral
found in lemongrass

b. 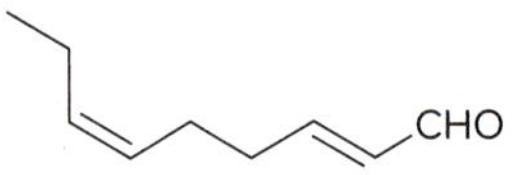 CHO
cucumber aldehyde

c. 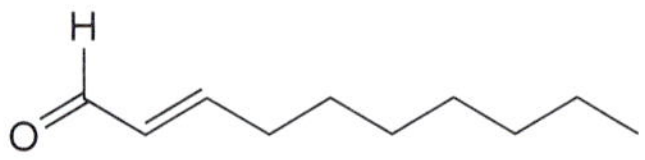 H O
found in stinkbugs and cilantro

18.3 Properties of Aldehydes and Ketones

18.3A Physical Properties

Aldehydes and ketones exhibit dipole–dipole interactions as a result of their polar carbonyl group. Because their oxygen atom can hydrogen bond to water, aldehydes and ketones with $\leq$ 5 C's are water soluble.

Aldehydes and ketones have no O–H bond, so two molecules of RCHO or RCOR are incapable of intermolecular hydrogen bonding, making them *less polar* and lower boiling than alcohols of comparable size.

MW = 72
bp 76 °C

OH
MW = 74
bp 118 °C

Increasing strength of intermolecular forces
Increasing boiling point

18.3B Spectroscopic Properties

Many details of the spectroscopy of aldehydes and ketones have been presented in Spectroscopy Parts A, B, and C:

- Fragmentation patterns in mass spectra: Section A.4A and Sample Problem A.7
- The carbonyl absorption in infrared spectra: Sections B.3C and B.4B
- ^{1}H and ^{13}C NMR absorptions: Section C.11B and Tables C.1 and C.5

Key NMR and IR absorptions for aldehydes and ketones are summarized in Table 18.2, and Figure 18.3 illustrates ^{1}H and ^{13}C NMR spectra for a simple aldehyde.

Table 18.2 Characteristic Spectroscopic Absorptions of Aldehydes and Ketones

Type of spectroscopy	Type of C, H	Absorption
IR absorptions	R–C(=O)–**H**	2700–2830 cm^{-1} (one or two peaks)
	R–C(=**O**)–R(H)	~1700 cm^{-1} (increasing ν with decreasing ring size)
	R–C(=**O**)–CH=CH$_2$ conjugated	1680 cm^{-1}
^{1}H NMR absorptions	R–C(=O)–**H**	9–10 ppm
	R–C(=O)–CH$_2$–**H**	2–2.5 ppm
^{13}C NMR absorption	R–C(=**O**)–R(H)	190–215 ppm

Figure 18.3 The ^{1}H and ^{13}C NMR spectra of propanal, CH_3CH_2CHO

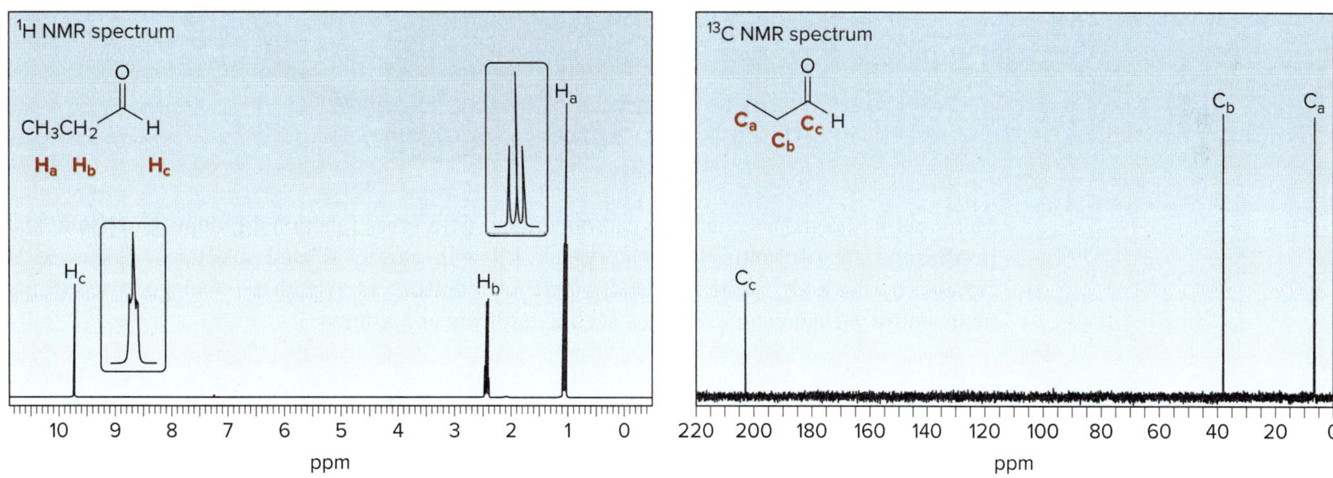

- **^{1}H NMR:** There are three signals due to the three different kinds of hydrogens, labeled H_a, H_b, and H_c. The **deshielded CHO proton** occurs downfield at 9.8 ppm. The H_c signal is split into a triplet by the adjacent CH_2 group, but the coupling constant is small.
- **^{13}C NMR:** There are three signals due to the three different kinds of carbons, labeled C_a, C_b, and C_c. The **deshielded carbonyl carbon** absorbs downfield at 203 ppm.

Problem 18.8 Rank the following compounds in order of increasing frequency of their carbonyl absorption in the infrared.

A B C

18.4 Interesting Aldehydes and Ketones

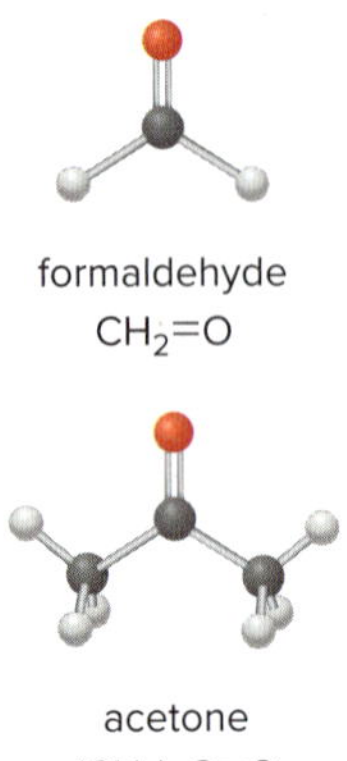

Because it is a starting material for the synthesis of many resins and plastics, billions of pounds of **formaldehyde** are produced annually in the United States by the oxidation of methanol (CH_3OH). Formaldehyde is also sold as a 37% aqueous solution called **formalin,** which has been used as a disinfectant, antiseptic, and preservative for biological specimens. Formaldehyde, a product of the incomplete combustion of coal and other fossil fuels, is partly responsible for the irritation caused by smoggy air.

Acetone is an industrial solvent and a starting material in the synthesis of some organic polymers. Acetone is produced in vivo during the breakdown of fatty acids. In diabetes, a common endocrine disease in which normal metabolic processes are altered because of the inadequate secretion of insulin, individuals often have unusually high levels of acetone in their bloodstreams. The characteristic odor of acetone can be detected on the breath of diabetic patients when their disease is poorly controlled.

Many aldehydes with characteristic odors occur in nature, including vanillin from vanilla beans and cinnamaldehyde from cinnamon.

Many steroid hormones contain a carbonyl along with other functional groups. **Cortisone** and **prednisone** are two anti-inflammatory steroids with closely related structures. Cortisone is secreted by the body's adrenal gland, whereas prednisone is a synthetic analogue used in the treatment of inflammatory diseases such as arthritis and asthma.

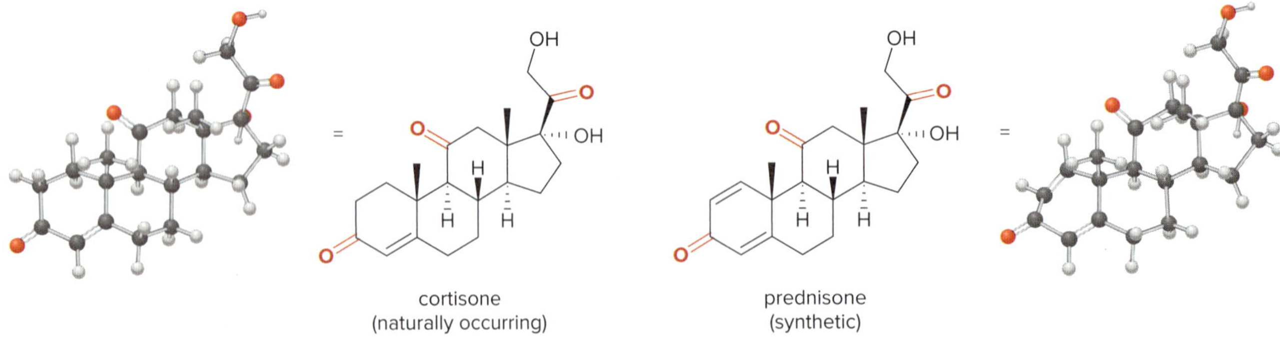

18.5 Preparation of Aldehydes and Ketones

Aldehydes and ketones can be prepared by a variety of methods. Because these reactions are needed for many multistep syntheses, Section 18.5 briefly summarizes earlier reactions that synthesize an aldehyde or ketone.

Aldehydes are prepared from 1° alcohols, esters, acid chlorides, and alkynes (Table 18.3).

Table 18.3 Common Methods to Synthesize Aldehydes

Method	Reaction
[1] Oxidation of 1° alcohols with PCC (Section 12.12B)	RCH_2OH (1° alcohol) —PCC→ RCHO
[2] Reduction of esters (Section 17.7A)	$RCOOR'$ (ester) —[1] DIBAL-H; [2] H_2O→ RCHO
[3] Reduction of acid chlorides (Section 17.7A)	RCOCl (acid chloride) —[1] $LiAlH[OC(CH_3)_3]_3$; [2] H_2O→ RCHO
[4] Hydroboration–oxidation of an alkyne (Section 11.10)	R—C≡C—H (alkyne) —[1] R_2BH; [2] H_2O_2, ^-OH→ RCH_2CHO

Ketones are prepared from 2° alcohols, acid chlorides, and alkynes (Table 18.4).

Table 18.4 Common Methods to Synthesize Ketones

Method	Reaction
[1] Oxidation of 2° alcohols with Cr^{6+} reagents (Section 12.12A)	RR'CH(OH) (2° alcohol) —CrO_3 or $Na_2Cr_2O_7$ or $K_2Cr_2O_7$ or PCC→ RCOR'
[2] Reaction of acid chlorides with organocuprates (Section 17.13)	RCOCl (acid chloride) —[1] R'_2CuLi; [2] H_2O→ RCOR'
[3] Friedel–Crafts acylation (Section 16.5)	benzene + RCOCl (acid chloride) —$AlCl_3$→ C_6H_5COR
[4] Hydration of an alkyne (Section 11.9)	R—C≡C—H (alkyne) —H_2O, H_2SO_4, $HgSO_4$→ $RCOCH_3$

Aldehydes and ketones are also both obtained as products of the oxidative cleavage of alkenes (Section 12.10).

$$R_2C{=}CHR' \text{ (alkene)} \xrightarrow{O_3} \xrightarrow[\text{or } CH_3SCH_3]{Zn,\ H_2O} R_2C{=}O \text{ (ketone)} + O{=}CHR' \text{ (aldehyde)}$$

Problem 18.9 What reagents are needed to convert each compound to butanal ($CH_3CH_2CH_2CHO$)?

a. 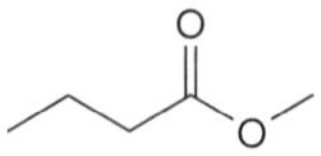 b. 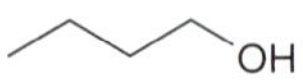 OH c. 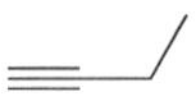 d.

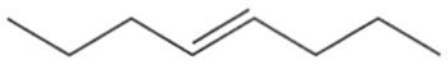

Problem 18.10 What reagents are needed to convert each compound to acetophenone ($C_6H_5COCH_3$)?

a.  b. 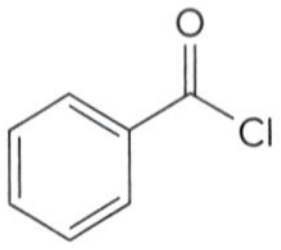 Cl c.

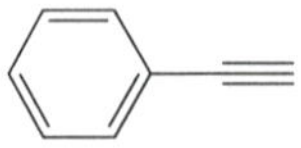

18.6 Reactions of Aldehydes and Ketones—General Considerations

Let's begin our discussion of carbonyl reactions by looking at the two general kinds of reactions that aldehydes and ketones undergo.

[1] Reaction at the carbonyl carbon

Recall from Chapter 17 that the uncrowded, electrophilic carbonyl carbon makes aldehydes and ketones susceptible to **nucleophilic addition** reactions.

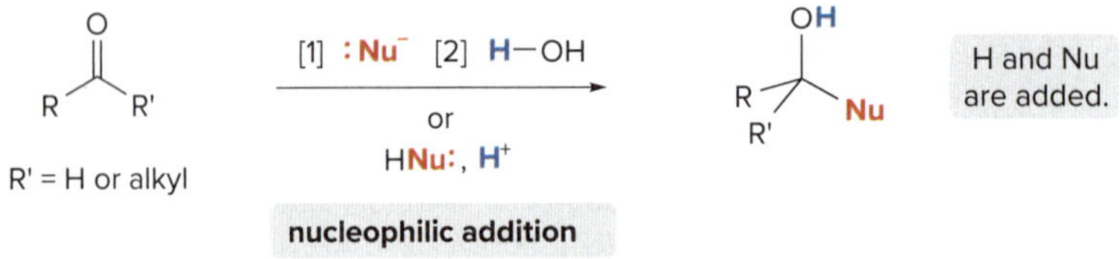

R' = H or alkyl

[1] :Nu⁻ [2] H—OH or HNu:, H^+

nucleophilic addition

H and Nu are added.

The elements of H and Nu are added to the carbonyl group. In Chapter 17, you learned about this reaction with hydride ($H:^-$) and carbanions ($R:^-$) as nucleophiles. In Chapter 18, we will discuss similar reactions with other nucleophiles.

[2] Reaction at the α carbon

A second general reaction of aldehydes and ketones involves reaction at the **α carbon.** A C—H bond on the α carbon to a carbonyl group is more acidic than many other C—H bonds, because reaction with base forms a resonance-stabilized enolate anion.

- **Enolates are nucleophiles, so they react with electrophiles (E^+) to form new bonds on the α carbon.**

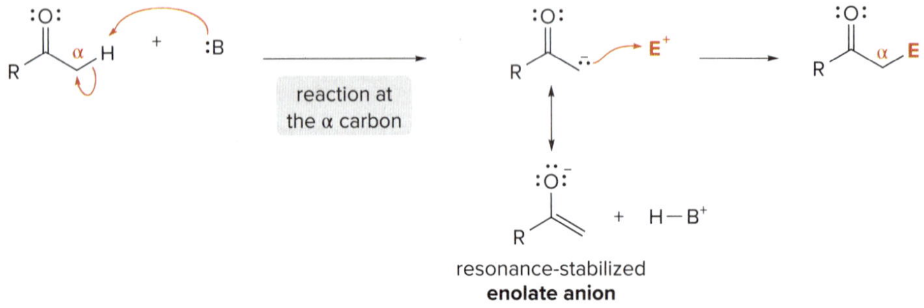

reaction at the α carbon

+ H—B^+

resonance-stabilized
enolate anion

Chapters 21 and 22 are devoted to reactions at the α carbon to a carbonyl group.

- **Aldehydes and ketones react with nucleophiles at the carbonyl carbon.**
- **Aldehydes and ketones form enolates that react with electrophiles at the α carbon.**

18.6A The General Mechanism of Nucleophilic Addition

Two general mechanisms are usually drawn for nucleophilic addition, depending on the nucleophile (negatively charged versus neutral) and the presence or absence of an acid catalyst. With negatively charged nucleophiles, nucleophilic addition follows the two-step process first discussed in Chapter 17—**nucleophilic attack** followed by **protonation,** as shown in Mechanism 18.1.

Mechanism 18.1 General Mechanism—Nucleophilic Addition

R' = H or alkyl

1 The **nucleophile attacks** the electrophilic carbonyl. The π bond is broken, moving an electron pair out on oxygen and forming an sp^3 hybridized carbon.

2 Protonation of the negatively charged oxygen by H_2O forms the **addition product.**

In this mechanism, **nucleophilic attack** ***precedes*** **protonation.** This process occurs with strong neutral or negatively charged nucleophiles.

With some neutral nucleophiles, however, nucleophilic addition does not occur unless an **acid catalyst** is added. The general mechanism for this reaction consists of three steps (not two), but the same product results because H and Nu add across the carbonyl π bond. In this mechanism, **protonation** ***precedes*** **nucleophilic attack.** Mechanism 18.2 is shown with the neutral nucleophile H—Nu: and a general acid H—A.

Mechanism 18.2 General Mechanism—Acid-Catalyzed Nucleophilic Addition

R' = H or alkyl

resonance-stabilized cation

1 Protonation of the carbonyl oxygen forms a **resonance-stabilized cation.**

2–3 Nucleophilic attack and deprotonation form the neutral addition product. The overall result is **addition of H and Nu** to the carbonyl group.

The effect of protonation is to convert a neutral carbonyl group to one having a net positive charge. **This protonated carbonyl group is much more electrophilic,** and much more susceptible to attack by a nucleophile. This step is unnecessary with strong nucleophiles like hydride (H:$^-$) that were used in Chapter 17. With weaker nucleophiles, however, nucleophilic attack does not occur unless the carbonyl group is first protonated.

no net charge, **less electrophilic**

net (+) charge, **more electrophilic**

This step is a specific example of a general phenomenon:

- **Any reaction involving a carbonyl group and a strong acid begins with the same first step—protonation of the carbonyl oxygen.**

18.6B The Nucleophile

What nucleophiles add to carbonyl groups? This cannot be predicted solely on the trends in nucleophilicity learned in Chapter 7. Only *some* of the nucleophiles that react well in nucleophilic substitution at sp^3 hybridized carbons give reasonable yields of nucleophilic addition products.

Cl$^-$, Br$^-$, and I$^-$ are good nucleophiles in substitution reactions at sp^3 hybridized carbons, but they are *ineffective* nucleophiles in addition. Addition of Cl$^-$ to a carbonyl group, for example, would cleave the C–O π bond, forming an alkoxide. Because Cl$^-$ is a much *weaker* base than the alkoxide formed, equilibrium favors the starting materials (the weaker base, Cl$^-$), *not* the addition product.

weaker base

stronger base

The situation is further complicated because some of the initial nucleophilic addition adducts are unstable and undergo elimination to form a stable product. For example, amines (RNH_2) add to carbonyl groups in the presence of mild acid to form unstable **carbinolamines,** which readily lose water to form **imines. This addition–elimination sequence replaces a C=O by a C=N.** The details of this process are discussed in Section 18.10A.

RNH_2 addition → carbinolamine → $-H_2O$ elimination → imine

Figure 18.4 lists nucleophiles that add to a carbonyl group, as well as the products obtained from nucleophilic addition using cyclohexanone as a representative ketone. These reactions are discussed in the remaining sections of Chapter 18. In cases in which the initial addition adduct is unstable, it is enclosed within brackets, followed by the final product.

Figure 18.4

Specific examples of nucleophilic addition

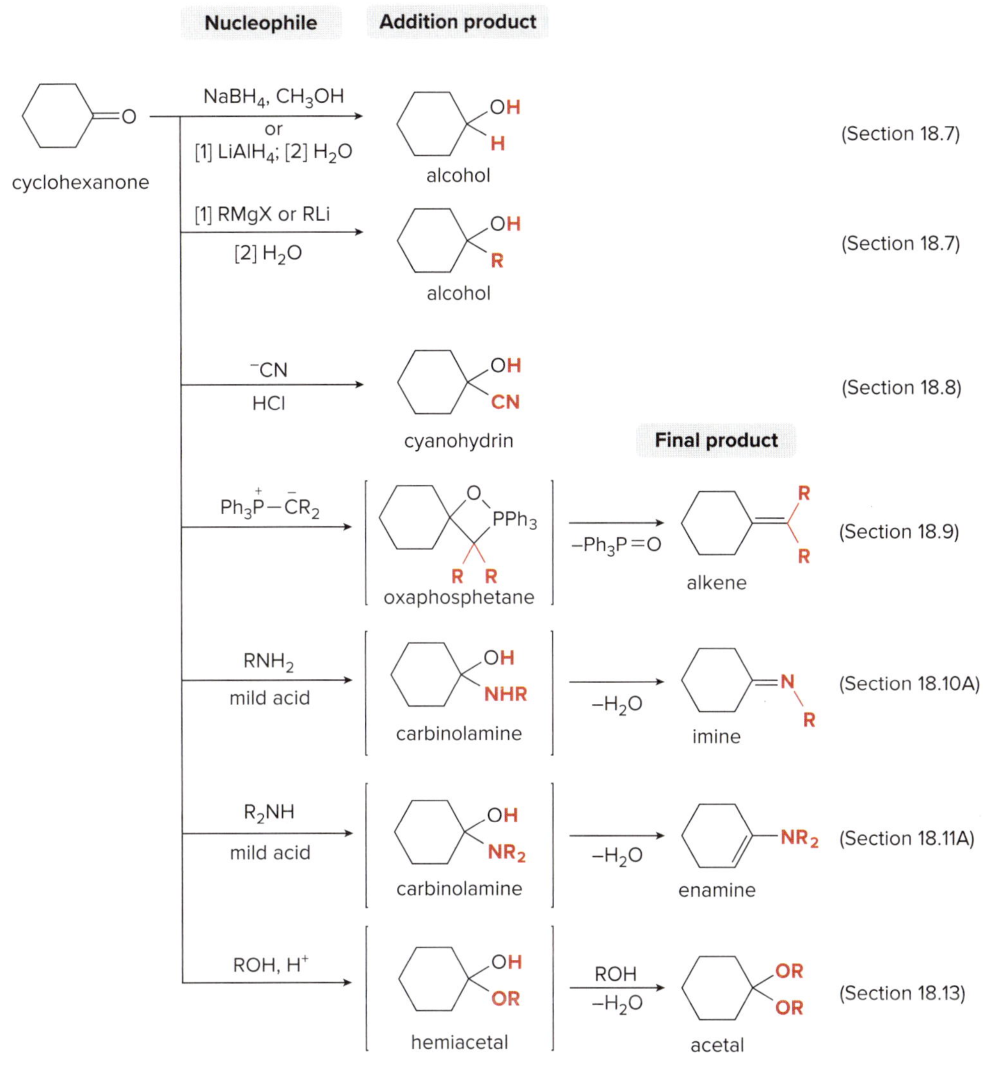

18.7 Nucleophilic Addition of H⁻ and R⁻—A Review

We begin our study of nucleophilic additions to aldehydes and ketones by briefly reviewing nucleophilic addition of hydride and carbanions, two reactions examined in Sections 17.4 and 17.10, respectively.

Treatment of an aldehyde or ketone with either $NaBH_4$ or $LiAlH_4$ followed by protonation forms a 1° or 2° alcohol. $NaBH_4$ and $LiAlH_4$ serve as a source of **hydride, H:⁻—the nucleophile**—and the reaction results in addition of the elements of H_2 across the C–O π bond. Addition of H_2 reduces the carbonyl group to an alcohol.

O
R R'
R' = H or alkyl

$NaBH_4$ or $LiAlH_4$ (source of H:⁻) → H–OH →

OH
R H
R'
1° or 2° alcohol

Hydride reduction of aldehydes and ketones occurs via the two-step mechanism of nucleophilic addition—that is, **nucleophilic attack of H:⁻ followed by protonation**—shown in Section 17.4B.

:O: H H–$\bar{A}$lH$_3$ Li$^+$ [1] :O:⁻ H H H–ÖH + AlH$_3$ [2] :ÖH H H + Li$^+$:ÖH⁻

Treatment of an aldehyde or ketone with either an organolithium (R''Li) or Grignard reagent (R''MgX) followed by water forms a 1°, 2°, or 3° alcohol containing a new carbon–carbon bond. R''Li and R''MgX serve as a source of a **carbanion (R'')⁻—the nucleophile**—and the reaction results in addition of the elements of R'' and H across the C–O π bond.

O R R' R' = H or alkyl; R''MgX or R''Li; H–OH; OH R R' R'' new C–C bond; **1°, 2°, or 3° alcohol**

The stereochemistry of hydride reduction and Grignard addition was discussed in Sections 17.5 and 17.10B, respectively.

The nucleophilic addition of carbanions to aldehydes and ketones occurs via the two-step mechanism of nucleophilic addition—that is, **nucleophilic attack of (R'')⁻ followed by protonation**—shown in Section 17.10A.

:O: H Li [1] :O:⁻ H H–ÖH [2] :ÖH H + Li$^+$:ÖH⁻

2° alcohol

In both reactions, the nucleophile—either hydride or a carbanion—attacks the trigonal planar sp^2 hybridized carbonyl from both sides, so that when a new stereogenic center is formed, a mixture of stereoisomers results, as was illustrated in Sample Problem 17.2.

Problem 18.11 Draw the products of each reaction. Include all stereoisomers formed.

a. O NaBH$_4$ CH$_3$OH

b. O [1] MgBr [2] H$_2$O

18.8 Nucleophilic Addition of ⁻CN

Treatment of an aldehyde or ketone with NaCN and a strong acid such as HCl adds the elements of HCN across the carbon–oxygen π bond, forming a **cyanohydrin.**

O R R' R' = H or alkyl; NaCN HCl **"HCN"**; OH R R' CN cyanohydrin

This reaction adds one carbon to the aldehyde or ketone, forming a **new carbon–carbon bond.**

NaCN / HCl

acetaldehyde cyanohydrin
new C–C bond in red

18.8A The Mechanism

The mechanism of cyanohydrin formation involves the usual two steps of nucleophilic addition: **nucleophilic attack followed by protonation** as shown in Mechanism 18.3.

Mechanism 18.3 Nucleophilic Addition of ⁻CN—Cyanohydrin Formation

R' = H or alkyl

new C–C bond in red

1. Nucleophilic attack of ⁻CN forms a **new carbon–carbon bond** with cleavage of the C–O π bond.
2. Protonation of the negatively charged oxygen by HCN forms the **addition product.** The HCN used in this step is formed by the acid–base reaction of ⁻CN with the strong acid, HCl.

This reaction does not occur with HCN alone. The **cyanide anion** makes addition possible because it is a **strong nucleophile** that attacks the carbonyl group.

Cyanohydrins can be reconverted to carbonyl compounds by treatment with base. This process is just the reverse of the addition of HCN: **deprotonation followed by elimination of ⁻CN.**

$+ H_2\ddot{O}:$

Note the difference between two similar terms. **Hydration** results in *adding* water to a compound. **Hydrolysis** results in *cleaving bonds* with water.

The cyano group (CN) of a cyanohydrin is readily hydrolyzed to a carboxy group (COOH) by heating with aqueous acid or base. **Hydrolysis replaces the three C–N bonds by three C–O bonds.**

H_2O (H^+ or ^-OH) Δ

Problem 18.12 Draw the products of each reaction. Include all stereoisomers.

a. CHO — NaCN / HCl

b. OH, CN — H_3O^+, Δ

c. NaCN / HCl

18.8B Application: Naturally Occurring Cyanohydrin Derivatives

Peach and apricot pits are a natural source of the cyanohydrin derivative amygdalin. *Jill Braaten/McGraw Hill*

Although the cyanohydrin is an uncommon functional group, **linamarin** and **amygdalin** are two naturally occurring cyanohydrin derivatives. Both contain a carbon atom bonded to both an oxygen atom and a cyano group, analogous to a cyanohydrin.

linamarin amygdalin laetrile

Cassava is a widely grown root crop, first introduced to Africa by Portuguese traders from Brazil in the sixteenth century. The peeled root is eaten after boiling or roasting. If the root is eaten without processing, illness and even death can result from high levels of HCN. *Daniel C. Smith*

Linamarin is isolated from cassava, a woody shrub grown as a root crop in the humid tropical regions of South America and Africa. **Amygdalin** is present in the seeds and pits of apricots, peaches, and wild cherries. Amygdalin and the related synthetic compound **laetrile** were once touted as anticancer drugs, although their effectiveness is unproven.

Linamarin, amygdalin, and laetrile are toxic compounds because they are metabolized to cyanohydrins, which are hydrolyzed to carbonyl compounds and **toxic HCN gas,** a cellular poison with a characteristic almond odor. This second step is merely the reconversion of a cyanohydrin to a carbonyl compound, a process that occurs with base in reactions run in the laboratory (Section 18.8A). If cassava root is processed with care, linamarin is enzymatically metabolized by this reaction sequence and the toxic HCN is released before the root is ingested, making it safe to eat.

linamarin (cyanohydrin derivative) $\xrightarrow{\text{enzyme}}$ acetone cyanohydrin $\xrightarrow{\text{enzyme}}$ acetone + HCN (toxic by-product)

Problem 18.13 Taxiphyllin is a cyanohydrin derivative found in the young shoots of some species of bamboo. What cyanohydrin and carbonyl compound are formed when taxiphyllin is metabolized in a similar manner to linamarin?

taxiphyllin

18.9 The Wittig Reaction

The additions of H^-, R^-, and ^-CN all involve the same two steps—**nucleophilic attack followed by protonation.** Other examples of nucleophilic addition in Chapter 18 are somewhat different. Although they still involve attack of a nucleophile, the initial addition adduct is converted to another product by one or more reactions.

The first reaction in this category is the **Wittig reaction,** named for German chemist Georg Wittig, who was awarded the Nobel Prize in Chemistry in 1979 for its discovery. The Wittig reaction uses a carbon nucleophile, the **Wittig reagent,** to form **alkenes.** When a carbonyl compound is treated with a Wittig reagent, the carbonyl oxygen atom is replaced by the negatively charged alkyl group bonded to the phosphorus—that is, **the C=O is converted to a C=C.**

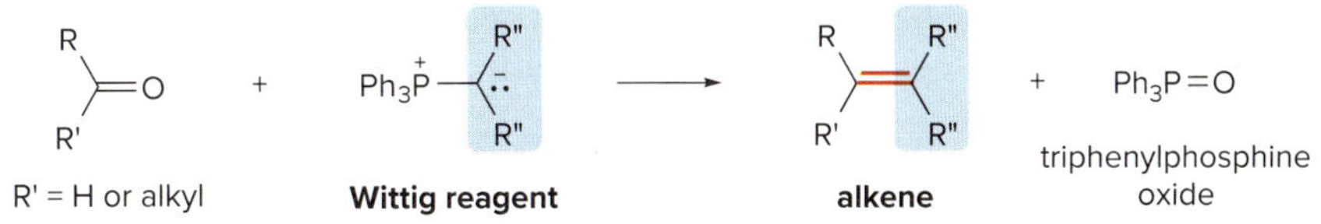

R' = H or alkyl — **Wittig reagent** — **alkene** — triphenylphosphine oxide

- **A Wittig reaction forms two new carbon–carbon bonds—one new σ bond and one new π bond—as well as a phosphorus by-product, $Ph_3P{=}O$ (triphenylphosphine oxide).**

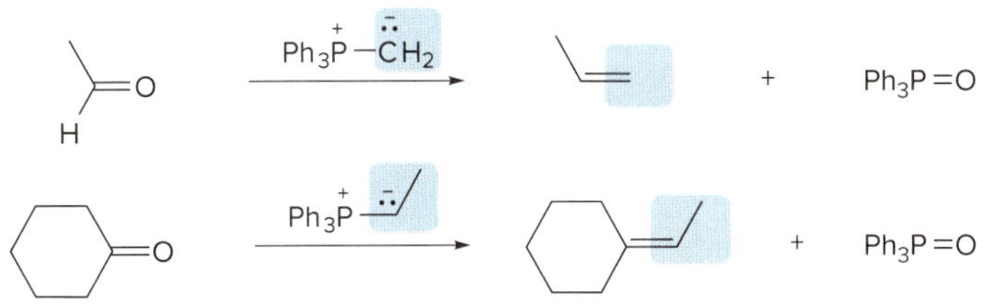

18.9A The Wittig Reagent

A **Wittig reagent** is an **organophosphorus reagent**—a reagent that contains a carbon–phosphorus bond. A typical Wittig reagent has a phosphorus atom bonded to three phenyl groups, plus another alkyl group that bears a negative charge.

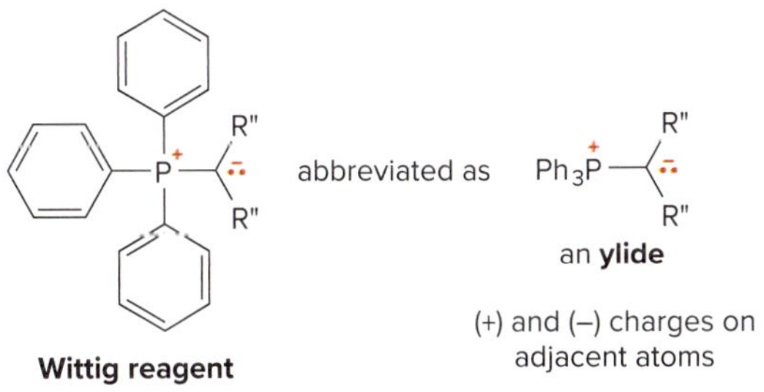

Wittig reagent abbreviated as **an ylide**

(+) and (–) charges on adjacent atoms

Phosphorus ylides are also called **phosphoranes.**

A Wittig reagent is an ***ylide,*** **a species that contains two oppositely charged atoms bonded to each other, and both atoms have octets.** In a Wittig reagent, a negatively charged carbon atom is bonded to a positively charged phosphorus atom.

Because phosphorus is a third-row element, it can be surrounded by more than eight electrons. As a result, a second resonance structure can be drawn that places a double bond between carbon and phosphorus. Regardless of which resonance structure is drawn, a **Wittig reagent has no net charge.** In one resonance structure, though, the **carbon atom bonded to phosphorus (labeled in blue) bears a net negative charge, so it is *nucleophilic.***

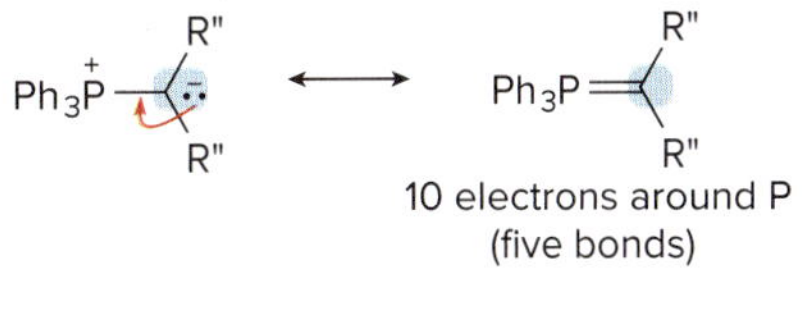

10 electrons around P (five bonds)

Wittig reagents are synthesized by a two-step procedure.

Step [1] **S_N2 reaction of triphenylphosphine with an alkyl halide forms a phosphonium salt.**

$Ph_3P:$ + RCH_2X (triphenylphosphine, **nucleophile**) $\xrightarrow{S_N2}$ $Ph_3\overset{+}{P}CH_2R$ $:X:^-$ (phosphonium salt)

Because phosphorus is located below nitrogen in the periodic table, a neutral phosphorus atom with three bonds also has a lone pair of electrons.

Triphenylphosphine (Ph_3P:), which contains a lone pair of electrons on P, is the nucleophile. Because the reaction follows an S_N2 mechanism, it works best with **unhindered CH_3X and 1° alkyl halides (RCH_2X).** Secondary alkyl halides (R_2CHX) can also be used, although yields are often lower.

Step [2] **Deprotonation of the phosphonium salt with a strong base (:B) forms the ylide.**

$Ph_3\overset{+}{P}CH(H)R$ $:X:^-$ (phosphonium salt) + :B (strong base) $\longrightarrow$ $Ph_3\overset{+}{P}\overset{-}{C}HR$ (**ylide**) + $H{-}B^+$

Bu—Li

strong base

Because removal of a proton from a carbon bonded to phosphorus generates a resonance-stabilized carbanion (the ylide), this proton is somewhat more acidic than other protons on an alkyl group in the phosphonium salt. Very strong bases are still needed, though, to favor the products of this acid–base reaction. Common bases used for this reaction are the organolithium reagents such as **butyllithium, $CH_3CH_2CH_2CH_2Li$,** abbreviated as **BuLi.**

Section 17.9C discussed the reaction of organometallic reagents as strong bases.

To synthesize the Wittig reagent, $Ph_3P{=}CH_2$, use these two steps:

$Ph_3P:$ + $CH_3{-}Br:$ $\xrightarrow[{[1]}]{S_N2}$ $Ph_3\overset{+}{P}{-}CH_3$ $:Br:^-$ (methyltriphenyl-phosphonium bromide) + Bu—Li $\xrightarrow{[2]}$ $Ph_3\overset{+}{P}{-}\overset{-}{C}H_2$ $\longleftrightarrow$ $Ph_3P{=}CH_2$ (**two resonance structures for the ylide**) + Bu—H (butane) + LiBr

- **Step [1]** Form the **phosphonium salt** by S_N2 reaction of Ph_3P**:** and CH_3Br.
- **Step [2]** Form the **ylide** by removal of a proton using BuLi as a strong base.

Problem 18.14 Draw the products of the following Wittig reactions.

a. $(CH_3)_2C{=}O$ + $Ph_3P{=}CH_2$ $\longrightarrow$

b. cyclopentanone + $Ph_3P{=}CHCH_2CH_2CH_3$ $\longrightarrow$

Problem 18.15 Outline a synthesis of each Wittig reagent from Ph_3P and an alkyl halide.

a. $Ph_3P{=}CHCH_3$ b. $Ph_3P{=}C(CH_3)_2$ c. $Ph_3P{=}CHC_6H_5$

18.9B Mechanism of the Wittig Reaction

The currently accepted mechanism of the Wittig reaction involves two steps. Like other nucleophiles, the Wittig reagent attacks an electrophilic carbonyl carbon, but then the initial addition adduct undergoes elimination to form an alkene. Mechanism 18.4 is drawn using $Ph_3P{=}CH_2$.

Mechanism 18.4 The Wittig Reaction

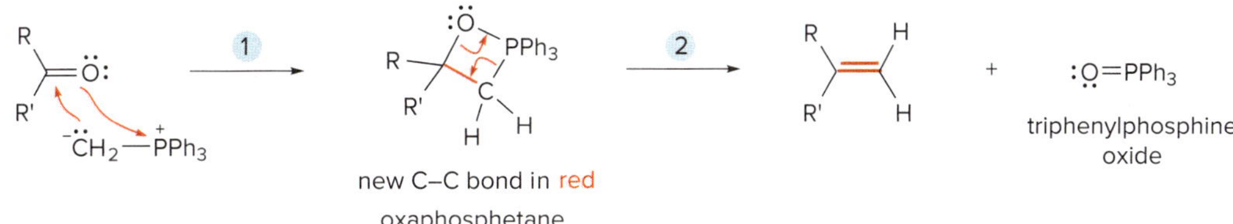

1 The negatively charged carbon of the ylide attacks the carbonyl carbon as the carbonyl oxygen attacks the positively charged P atom. This step forms **two bonds** and generates a **four-membered ring** called an **oxaphosphetane.**

2 **Elimination of triphenylphosphine oxide forms two new π bonds.** The formation of the strong P=O provides the driving force for the Wittig reaction.

One limitation of the Wittig reaction is that a mixture of alkene stereoisomers sometimes forms. For example, reaction of propanal (CH_3CH_2CHO) with a Wittig reagent forms the mixture of *E* and *Z* isomers shown.

E isomer 59%

Z isomer 41%

Because the Wittig reaction forms two carbon–carbon bonds in a single reaction, it has been used to synthesize many natural products, including vitamin A, shown in Figure 18.5.

Figure 18.5 A Wittig reaction used in the industrial synthesis of vitamin A

NaOCH3 / CH3OH

vitamin A

- Often when a Wittig reaction forms a conjugated system, as is the case here, the more stable *E* alkene is the major product.

Problem 18.16 Draw the products (including stereoisomers) formed when benzaldehyde (C_6H_5CHO) is treated with each Wittig reagent.

a. Ph_3P b. Ph_3P c. Ph_3P

18.9C Retrosynthetic Analysis

To use the Wittig reaction in synthesis, you must be able to determine what carbonyl compound and Wittig reagent are needed to prepare a given compound—that is, **you must work backwards, in the retrosynthetic direction.** There can be two different Wittig routes to a given alkene, but one is often preferred on steric grounds.

How To Determine the Starting Materials for a Wittig Reaction Using Retrosynthetic Analysis

Example What starting materials are needed to synthesize alkene **X** by a Wittig reaction?

X

Step [1] **Cleave the carbon–carbon double bond into two components.**

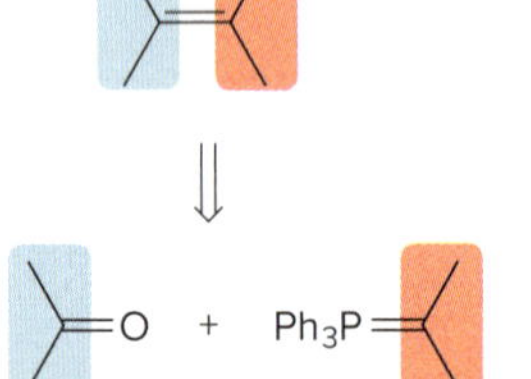

- Part of the molecule becomes the carbonyl component, and the other part becomes the Wittig reagent.

There are usually two routes to a given alkene using a Wittig reaction:

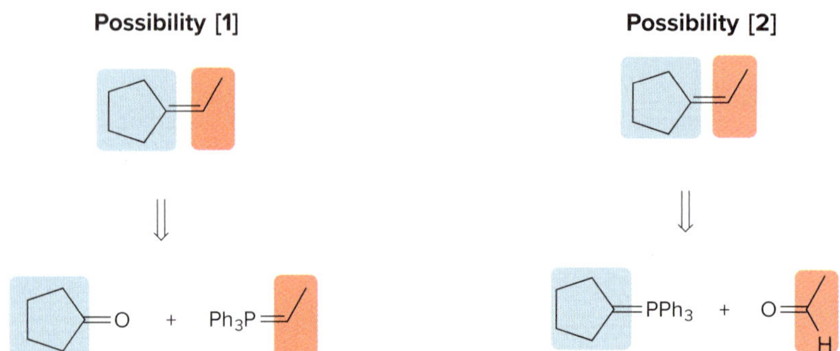

Step [2] **Compare the Wittig reagents. The preferred pathway uses a Wittig reagent derived from an unhindered alkyl halide—CH_3X or RCH_2X.**

Determine what alkyl halide is needed to prepare each Wittig reagent:

Possibility [1] Ph_3P= ⟹ $Ph_3\overset{+}{P}$— X^- ⟹ Ph_3P: + X— **1° halide preferred pathway**

Possibility [2] =PPh_3 ⟹ —$\overset{+}{P}Ph_3$ X^- ⟹ —X + :PPh_3 2° halide

Because the synthesis of the Wittig reagent begins with an **S_N2** reaction, **the preferred pathway begins with an unhindered methyl halide or 1° alkyl halide.** In this example, retrosynthetic analysis of both Wittig reagents indicates that only one of them ($Ph_3P{=}CHCH_3$) can be synthesized from a **1° alkyl halide,** making Possibility [1] the preferred pathway.

Problem 18.17 What starting materials are needed to prepare each alkene by a Wittig reaction? When there are two possible routes, indicate which route, if any, is preferred.

a. 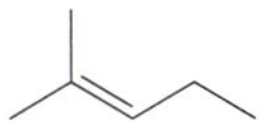 c.

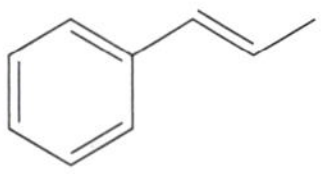

b. 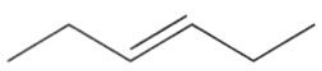 d.

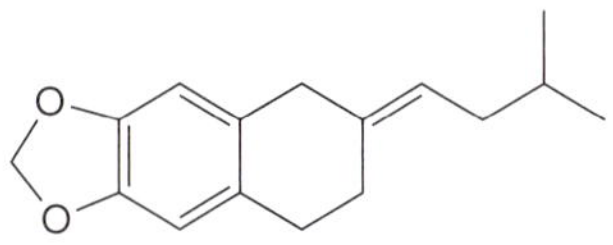

18.9D Comparing Methods of Alkene Synthesis

An advantage in using the Wittig reaction over other elimination methods to synthesize alkenes is that **you always know the location of the double bond.** Whereas other methods of alkene synthesis often give a mixture of constitutional isomers, the **Wittig reaction always gives a *single* constitutional isomer.**

For example, two methods can be used to convert cyclohexanone into alkene **B** (methylenecyclohexane): **a two-step method consisting of Grignard addition followed by dehydration, or a one-step Wittig reaction.**

cyclohexanone → **B**

Recall from Section 9.8 that the major product formed in acid-catalyzed dehydration of an alcohol is the more substituted alkene.

In a two-step method, treatment of cyclohexanone with CH_3MgBr forms a 3° alcohol after protonation. Dehydration of the alcohol with H_2SO_4 forms a mixture of alkenes, in which the desired disubstituted alkene is the minor product.

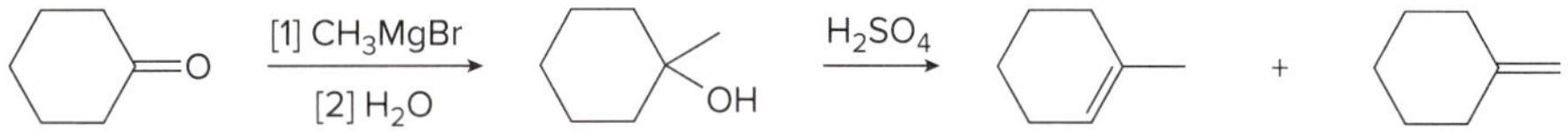

cyclohexanone —[1] CH_3MgBr, [2] H_2O→ 3° alcohol —H_2SO_4→ trisubstituted C=C **major product** + **B** disubstituted C=C **minor product**

By contrast, reaction of cyclohexanone with $Ph_3P{=}CH_2$ affords the desired alkene as the only product. The newly formed double bond always joins the carbonyl carbon with the negatively charged carbon of the Wittig reagent. In other words, **the position of the double bond is always unambiguous in the Wittig reaction.** This makes the Wittig reaction an especially attractive method for preparing many alkenes.

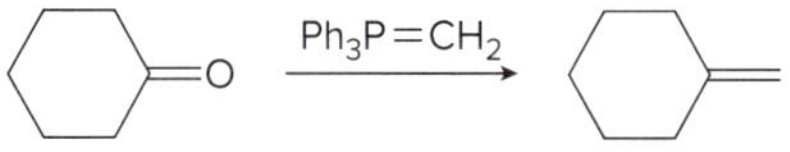

cyclohexanone —$Ph_3P{=}CH_2$→ **B** **only product**

Problem 18.18 Show two methods to synthesize each alkene: a one-step method using a Wittig reagent, and a two-step method that forms a carbon–carbon bond with an organometallic reagent in one of the steps.

a. 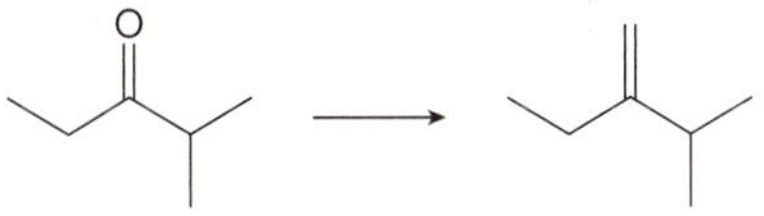 → b. 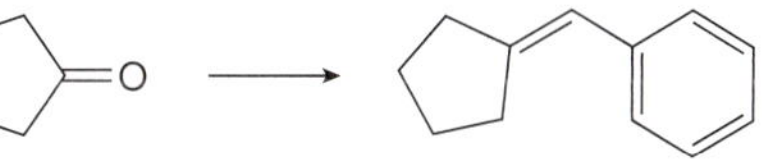 →

18.10 Addition of 1° Amines

We now move on to the reaction of aldehydes and ketones with nitrogen and oxygen heteroatoms. **Amines are organic nitrogen compounds that contain a nonbonded electron pair on the N atom.** As we learned in Section 3.2, amines are classified as 1°, 2°, or 3° by the number of alkyl groups bonded to the *nitrogen* atom.

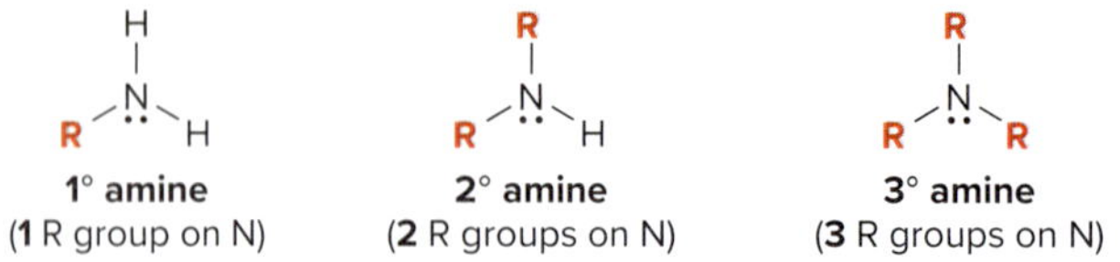

1° amine (**1** R group on N) | **2° amine** (**2** R groups on N) | **3° amine** (**3** R groups on N)

Both 1° and 2° amines react with aldehydes and ketones. We begin by examining the reaction of aldehydes and ketones with 1° amines.

18.10A Formation of Imines

Treatment of an aldehyde or ketone with a 1° amine affords an **imine** (also called a **Schiff base**). Nucleophilic attack of the 1° amine on the carbonyl group forms an unstable **carbinolamine,** which loses water to form an imine. The overall reaction results in **replacement of C=O by C=NR.**

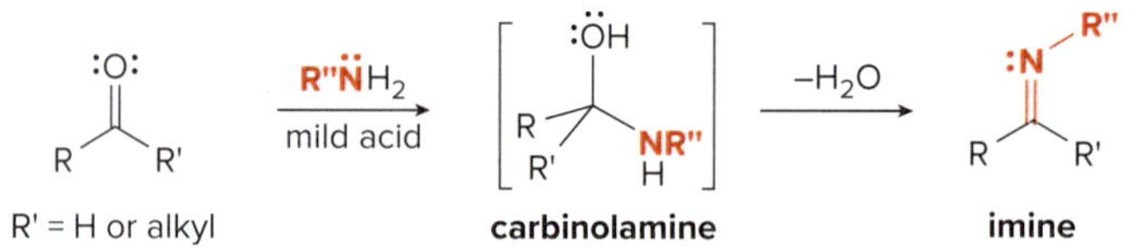

R' = H or alkyl — R"NH$_2$, mild acid → [**carbinolamine**] —$-H_2O$→ **imine**

Because the N atom of an imine is surrounded by three groups (two atoms and a lone pair), it is sp^2 hybridized, making the C–N–R" bond angle ~120° (*not* 180°). Imine formation is fastest when the reaction medium is weakly acidic.

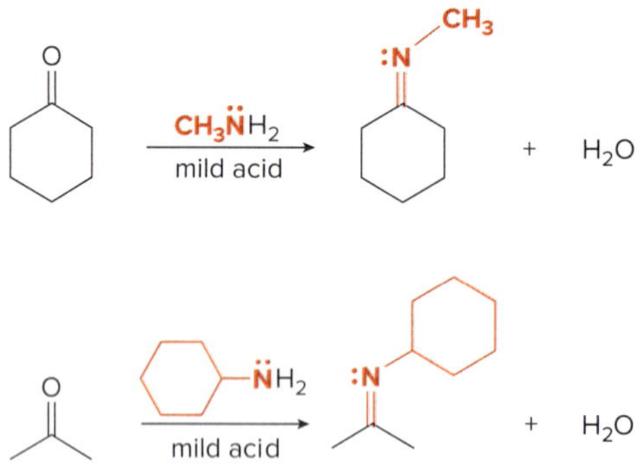

CH$_3$NH$_2$, mild acid → + H$_2$O

mild acid → + H$_2$O

The mechanism of imine formation (Mechanism 18.5) can be divided into two distinct parts: **nucleophilic addition of the 1° amine (Steps [1] and [2]), followed by elimination of H_2O (Steps [3]–[5]).** Each step involves a reversible equilibrium, so that the reaction is driven to completion by removing H_2O.

Imine formation is most rapid at pH 4–5. Mild acid is needed for protonation of the hydroxy group in Step [3] to form a **good leaving group.** Under strongly acidic conditions, the reaction rate decreases because the amine nucleophile is protonated. With no free electron pair, it is no longer a nucleophile, and so nucleophilic addition cannot occur.

Mechanism 18.5 Imine Formation from an Aldehyde or a Ketone

1 – 2 **Nucleophilic attack of the amine** followed by proton transfer forms the **carbinolamine.**

3 Protonation of the OH group forms a **good leaving group.**

4 Loss of H_2O forms a **resonance-stabilized iminium ion.**

5 Loss of a proton forms the **imine.**

Problem 18.19 Draw the product formed when $CH_3CH_2CH_2CH_2NH_2$ reacts with each carbonyl compound in the presence of mild acid.

a. (benzene ring)–CHO b. (acetone) c. (cyclopentanone)

Problem 18.20 What 1° amine and carbonyl compound are needed to prepare each imine?

a. b. c. d.

18.10B Application: Retinal, Rhodopsin, and the Chemistry of Vision

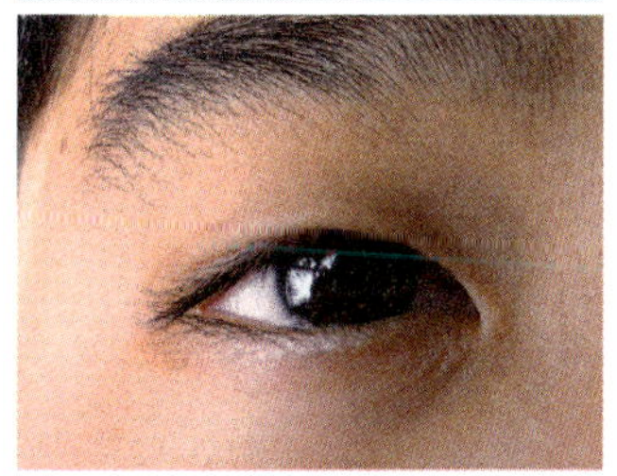

11-*cis*-Retinal is the light-sensitive aldehyde that plays a key role in the chemistry of vision for all vertebrates, arthropods, and mollusks. *Daniel C. Smith*

Many imines play vital roles in biological systems. A key molecule in the chemistry of vision is the highly conjugated imine **rhodopsin,** which is synthesized in the rod cells of the eye from **11-*cis*-retinal** and a 1° amine in the protein **opsin.**

The central role of rhodopsin in the visual process was delineated by Nobel Laureate George Wald of Harvard University.

The complex process of vision centers around this imine derived from retinal (Figure 18.6). The 11-cis double bond in rhodopsin creates crowding in the rather rigid side chain. When light strikes the rod cells of the retina, it is absorbed by the conjugated double bonds of rhodopsin, and the **11-cis double bond is isomerized to the 11-trans arrangement.** This isomerization is accompanied by a drastic change in shape in the protein, altering the concentration of Ca^{2+} ions moving across the cell membrane, and sending a nerve impulse to the brain, which is then processed into a visual image.

Figure 18.6 The key reaction in the chemistry of vision

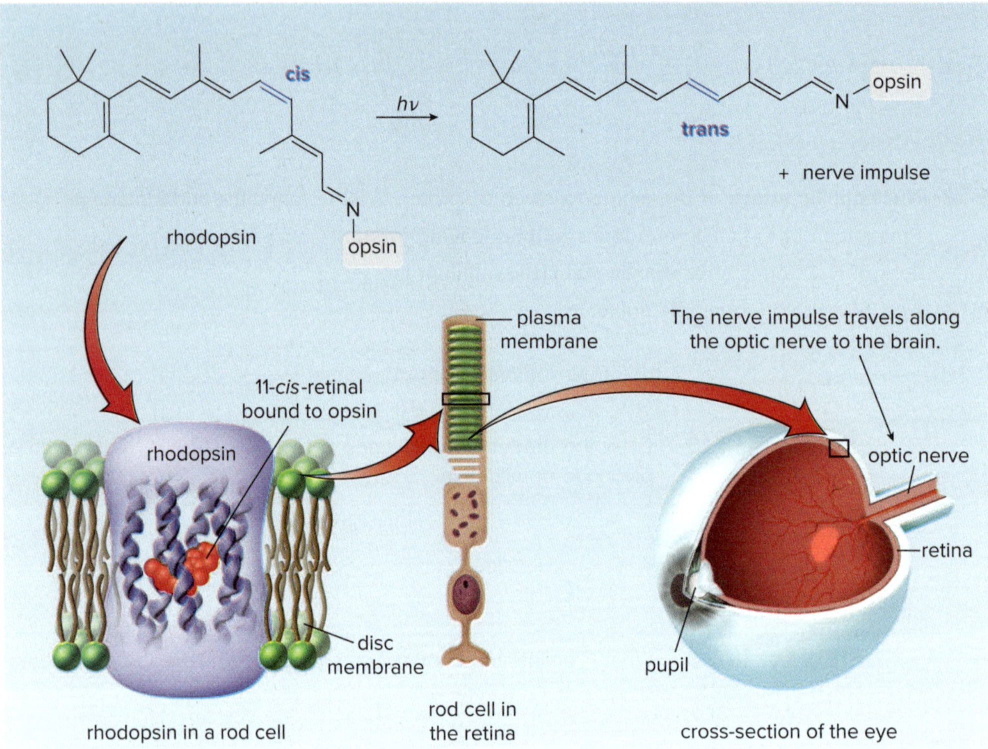

- Rhodopsin is a light-sensitive compound located in the membrane of the rod cells in the retina of the eye. Rhodopsin contains the protein opsin bonded to 11-*cis*-retinal via an imine linkage. When light strikes this molecule, the **crowded 11-cis double bond isomerizes to the 11-trans isomer,** and a nerve impulse is transmitted to the brain by the optic nerve.

18.11 Addition of 2° Amines

18.11A Formation of Enamines

A 2° amine reacts with an aldehyde or a ketone to give an **enamine.** ***Enamines* have a nitrogen atom bonded to a double bond** (alk*ene* + *amine* = *enamine*).

:O: R' (R' = H or alkyl) —$R_2\ddot{N}H$→ [:ÖH R' NR₂] **carbinolamine** —$-H_2O$→ :NR₂ R' **enamine**

Like imines, enamines are also formed by the addition of a nitrogen nucleophile to a carbonyl group followed by elimination of water. In this case, however, **elimination occurs across two adjacent *carbon* atoms** to form a new carbon–carbon π bond.

mild acid + H_2O

mild acid + H_2O

The mechanism for enamine formation (Mechanism 18.6) is identical to the mechanism for imine formation except for the *last step,* involving formation of the π bond. The mechanism can be divided into two distinct parts: **nucleophilic addition of the 2° amine (Steps [1] and [2]), followed by elimination of H_2O (Steps [3]–[5]).** Each step involves a reversible equilibrium once again, so that the reaction is driven to completion by removing H_2O.

Mechanism 18.6 Enamine Formation from an Aldehyde or a Ketone

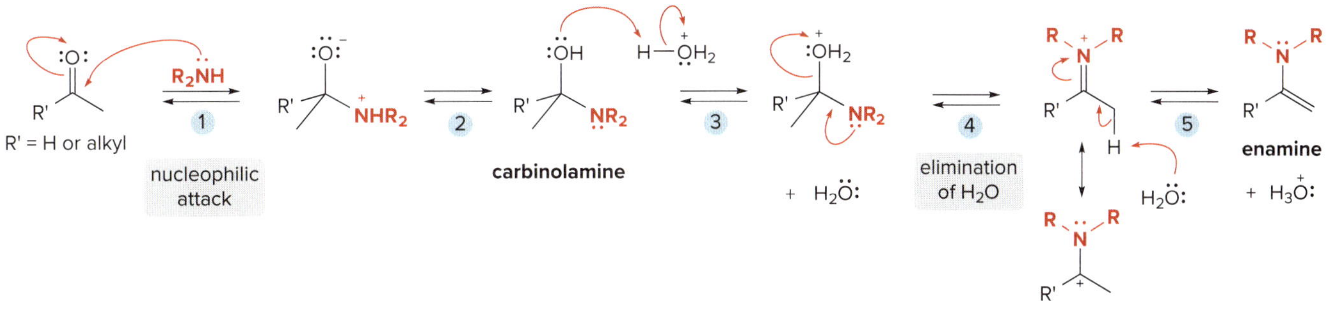

1–2 **Nucleophilic attack of the amine** followed by proton transfer forms the **carbinolamine.**

3 Protonation of the OH group forms a **good leaving group.**

4 Loss of H_2O forms a **resonance-stabilized iminium ion.**

5 Loss of a proton from the adjacent C–H bond forms the **enamine.**

The mechanisms illustrate why **the reaction of 1° amines with carbonyl compounds forms *imines,* but the reaction with 2° amines forms *enamines.*** In Figure 18.7, the last step of both mechanisms is compared using cyclohexanone as starting material. The position of the double bond depends on which proton is removed in the last step. Removal of an N–H proton forms a C=N, whereas removal of a C–H proton forms a C=C.

Problem 18.21 What 2° amine and carbonyl compound are needed to prepare each enamine?

a. b. c.

Problem 18.22 What two enamines are formed when 2-methylcyclohexanone is treated with $(CH_3)_2NH$?

Figure 18.7

The formation of imines and enamines compared

$R\ddot{N}H_2$ 1° amine — The N–H proton is removed. — imine

$R_2\ddot{N}H$ 2° amine — The C–H proton is removed. — enamine

- With a **1° amine,** the intermediate iminium ion still has a proton on the N atom that may be removed to form a C=N.
- With a **2° amine,** the intermediate iminium ion has *no* proton on the N atom. A proton must be removed from an adjacent C–H bond, and this forms a C=C.

18.11B Imine and Enamine Hydrolysis

Because imines and enamines are formed by a set of reversible reactions, **both can be converted back to carbonyl compounds by hydrolysis with mild acid.**

- **Hydrolysis of imines and enamines forms aldehydes and ketones.**

imine $\xrightarrow{H_3O^+}$ ketone + H_2N–R

enamine $\xrightarrow{H_3O^+}$ ketone + amine

The mechanism of these reactions is exactly the *reverse* of the mechanism written for the formation of imines and enamines. In the hydrolysis of enamines shown in Mechanism 18.7, the carbonyl carbon in the product comes from the sp^2 hybridized carbon bonded to the N atom in the starting material.

Mechanism 18.7 Hydrolysis of an Enamine

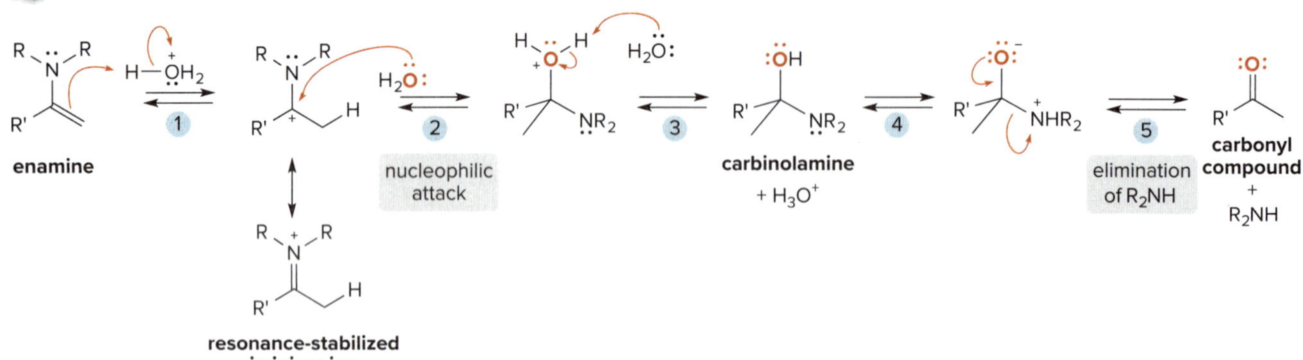

1 Protonation of the enamine forms a **resonance-stabilized iminium ion.**

2–3 **Nucleophilic attack of H_2O** and deprotonation form a **carbinolamine.**

4–5 Proton transfer and loss of R_2NH form the **carbonyl group.**

Sample Problem 18.3 Drawing the Products of Imine and Enamine Hydrolysis

Draw the products formed by the hydrolysis of each compound.

Solution

- An imine contains a C=N, which is converted to a **C=O and a 1° amine** during hydrolysis.
- An enamine, which contains a N atom bonded to a C=C, is hydrolyzed to a **2° amine and a carbonyl compound.**

The carbon in **A, B,** and **C** labeled in blue is converted to the carbonyl carbon.

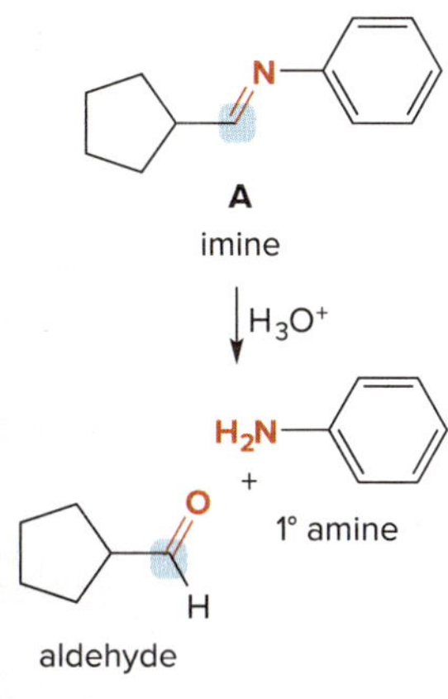

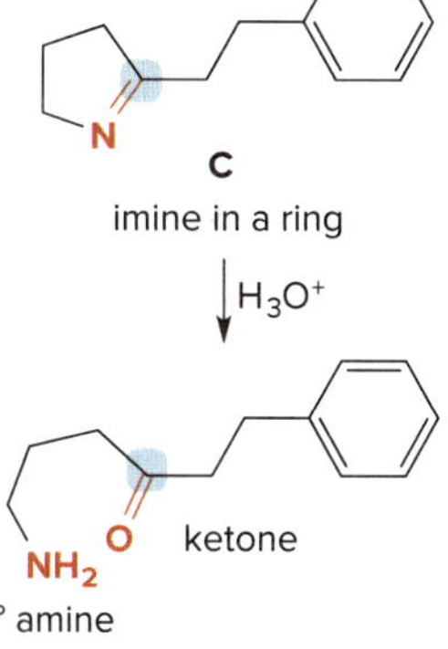

- The imine **A** is converted to a 1° amine and an aldehyde.
- The enamine **B** is converted to an aldehyde and a 2° amine. The alkenyl carbon bonded to N is converted to the carbon of the C=O.
- The imine **C** is converted to a 1° amine and a ketone. Because **C** is cyclic, both functional groups end up in the *same* compound.

Problem 18.23 What carbonyl compound and amine are formed by the hydrolysis of each compound?

a. b. c. d.

More Practice: Try Problems 18.44c, g; 18.51; 18.53.

Problem 18.24 Draw a stepwise mechanism for the following imine hydrolysis.

18.12 Addition of H_2O—Hydration

Treatment of a carbonyl compound with H_2O in the presence of an acid or base catalyst **adds the elements of H and OH across the carbon–oxygen π bond,** forming a ***gem*-diol** or **hydrate.**

$$\mathrm{RC(=O)R'} \xrightarrow[\mathrm{H^+\ or\ ^-OH}]{\mathrm{H_2O}} \mathrm{RR'C(OH)_2}$$

R' = H or alkyl

***gem*-diol (hydrate)**

Hydration of a carbonyl group gives a good yield of *gem*-diol only with an **unhindered aldehyde** like formaldehyde, and with aldehydes containing nearby **electron-withdrawing groups.**

$$\mathrm{HCHO} \xrightarrow{\mathrm{H_2O}} \mathrm{HOCH_2OH}$$

formaldehyde → formaldehyde hydrate

$$\mathrm{Cl_3CCHO} \xrightarrow{\mathrm{H_2O}} \mathrm{Cl_3CCH(OH)_2}$$

chloral → chloral hydrate

18.12A The Thermodynamics of Hydrate Formation

Whether addition of H_2O to a carbonyl group affords a good yield of the *gem*-diol depends on the relative energies of the starting material and the product. With *less stable* carbonyl starting materials, equilibrium favors the *hydrate* product, whereas with *more stable* carbonyl starting materials, equilibrium favors the *carbonyl starting material.* Because **alkyl groups stabilize a carbonyl group** (Section 17.2B):

- *Increasing* the number of alkyl groups on the carbonyl carbon *decreases* the amount of hydrate at equilibrium.

This can be illustrated by comparing the amount of hydrate formed from formaldehyde, acetaldehyde, and acetone.

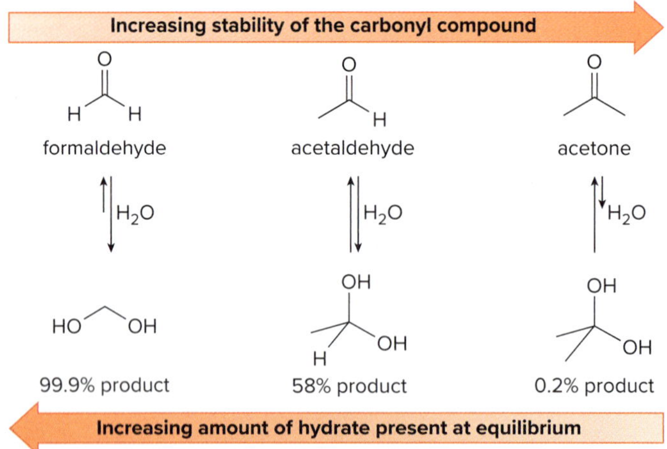

Formaldehyde, the least stable carbonyl compound, forms the largest percentage of hydrate. On the other hand, acetone and other ketones, which have two electron-donor R groups, form $< 1\%$ of the hydrate at equilibrium. Other electronic factors come into play as well:

- Electron-*donating* groups near the carbonyl carbon stabilize the carbonyl group, *decreasing* the amount of the hydrate at equilibrium.
- Electron-*withdrawing* groups near the carbonyl carbon destabilize the carbonyl group, *increasing* the amount of hydrate at equilibrium.

Chloral hydrate, a sedative sometimes administered to calm a patient prior to a surgical procedure, has also been used for less reputable purposes. Adding it to an alcoholic beverage makes a so-called knock-out drink, causing an individual who drinks it to pass out. Because it is addictive and care must be taken in its administration, chloral hydrate is a controlled substance.

This explains why chloral (trichloroacetaldehyde) forms a large amount of hydrate at equilibrium. Three electron-withdrawing Cl atoms place a partial positive charge on the α carbon to the carbonyl, destabilizing the carbonyl group, and therefore increasing the amount of hydrate at equilibrium.

chloral

Adjacent like charges (δ+) *destabilize* the carbonyl and *increase* the amount of hydrate.

Problem 18.25 Rank the following carbonyl compounds in order of increasing percentage of hydrate present at equilibrium.

A **B** **C**

18.12B The Kinetics of Hydrate Formation

Although H_2O itself adds slowly to a carbonyl group, both acid and base catalyze the addition. In base, the nucleophile is ^-OH, and the mechanism follows the usual two steps for nucleophilic addition: **nucleophilic attack followed by protonation,** as shown in Mechanism 18.8.

Mechanism 18.8 Base-Catalyzed Addition of H_2O to a Carbonyl Group

R' = H or alkyl

gem-diol

1 **The nucleophile ($^-$OH) attacks the carbonyl,** breaking the π bond and moving an electron pair out on oxygen.

2 Protonation of the negatively charged oxygen by H_2O forms the **hydration product.**

The acid-catalyzed addition follows the general mechanism presented in Section 18.6A. For a poorer nucleophile like H_2O to attack a carbonyl group, the **carbonyl must be protonated by acid first; thus, protonation *precedes* nucleophilic attack.** The overall mechanism has three steps, as shown in Mechanism 18.9.

Mechanism 18.9 Acid-Catalyzed Addition of H_2O to a Carbonyl Group

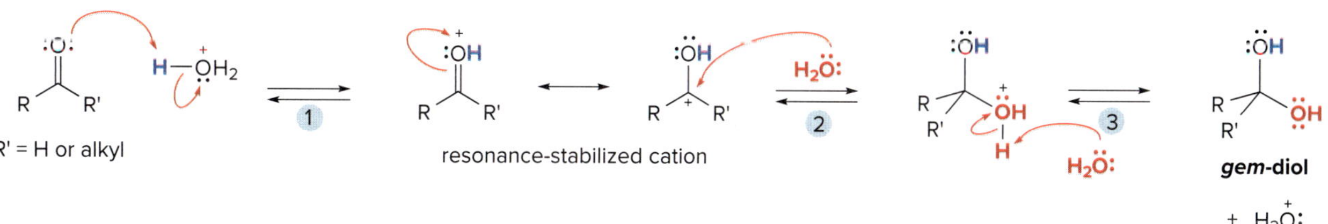

1 Protonation of the carbonyl oxygen forms a **resonance-stabilized cation.**

2 – 3 Nucleophilic attack and deprotonation form the ***gem*-diol.** The overall result is addition of H and OH to the carbonyl group.

Acid and base increase the rate of reaction for different reasons:

- **Base converts H_2O to ^-OH, a *stronger nucleophile.***
- **Acid protonates the carbonyl group, making it *more electrophilic* toward nucleophilic attack.**

These catalysts increase the rate of the reaction, but they do not affect the equilibrium constant. Starting materials that give a low yield of *gem*-diol do so whether or not a catalyst is present. Because these reactions are reversible, the conversion of *gem*-diols to aldehydes and ketones is also catalyzed by acid and base, and the steps of the mechanism are reversed.

Problem 18.26 Draw a stepwise mechanism for the following reaction.

OH, OH $\xrightarrow{H_3O^+}$ =O + H_2O

18.13 Addition of Alcohols—Acetal Formation

Aldehydes and ketones react with *two* equivalents of alcohol to form acetals. In an acetal, the carbonyl carbon from the aldehyde or ketone is now singly bonded to **two OR" (alkoxy) groups.**

O, R, R' + R"OH (2 equiv) $\overset{H^+}{\rightleftharpoons}$ OR", R, R', OR" + H_2O

R' = H or alkyl — **acetal**

The term *acetal* refers to any compound derived from an aldehyde or ketone, having two OR groups bonded to a single carbon. The term *ketal* is sometimes used when the starting carbonyl compound is a ketone; that is, the carbon bonded to the alkoxy groups is *not* bonded to a H atom and the general structure is $R_2C(OR')_2$. Because ketals are considered a subclass of acetals in the IUPAC system, we will use the single general term *acetal* for any compound having two OR groups on a carbon atom.

This reaction differs from other additions we have seen thus far, because **two equivalents of alcohol are added to the carbonyl group,** and two new C–O σ bonds are formed. Acetal formation is catalyzed by acids, commonly *p*-toluenesulfonic acid (TsOH).

O, H + CH_3OH (2 equiv) $\overset{TsOH}{\rightleftharpoons}$ OCH_3, H, OCH_3 + H_2O

acetal

When a diol such as ethylene glycol is used in place of two equivalents of ROH, a cyclic acetal is formed. Both oxygen atoms in the cyclic acetal come from the diol.

O + HO–OH (ethylene glycol) $\overset{TsOH}{\rightleftharpoons}$ O, O + H_2O

a cyclic acetal

Acetals are *not* ethers, even though both functional groups contain a C–O σ bond. Having two C–O σ bonds on the *same* carbon atom makes an acetal very different from an ether.

R–C(OR)(OR)–R (acetal) ≠ R–O–R (ether)

Like *gem*-diol formation, the synthesis of acetals is reversible, and often the equilibrium favors reactants, not products. In acetal synthesis, however, water is formed as a by-product, so the equilibrium can be driven to the right by **removing the water as it is formed.** This can be done in a variety of ways in the laboratory. A drying agent can be added that reacts with the water, or more commonly, the water can be distilled from the reaction mixture as it is formed. Driving an equilibrium to the right by removing one of the products is an application of Le Châtelier's principle (see Section 9.8).

Sample Problem 18.4 Locating Ethers and Acetals in a Complex Compound

Label the ethers and acetals in spirastrellolide A, a natural product isolated from the Caribbean marine sponge *Spirastrella coccinea*.

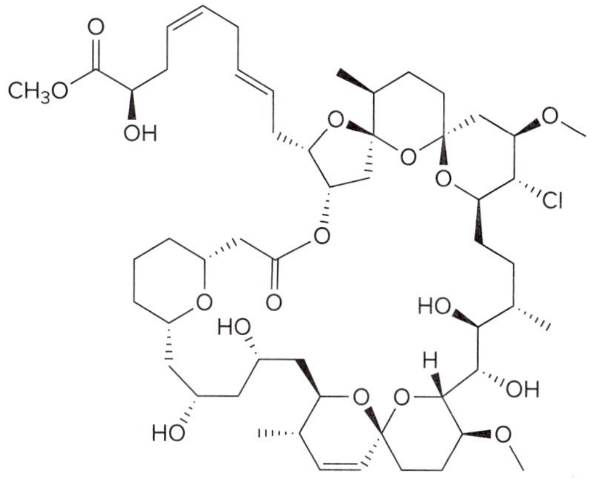

spirastrellolide A

Solution

In a complex compound, examine each oxygen-containing functional group *individually*. Ethers and acetals can both occur in chains or be part of rings.

- **An acetal contains two C–O σ bonds on the *same* carbon atom.**
- **An ether contains two R groups bonded to a *single* O atom.**

Spirastrellolide A contains three acetals (O atoms labeled in red and acetal carbons labeled with a gray box) and three ethers (O atoms labeled in blue).

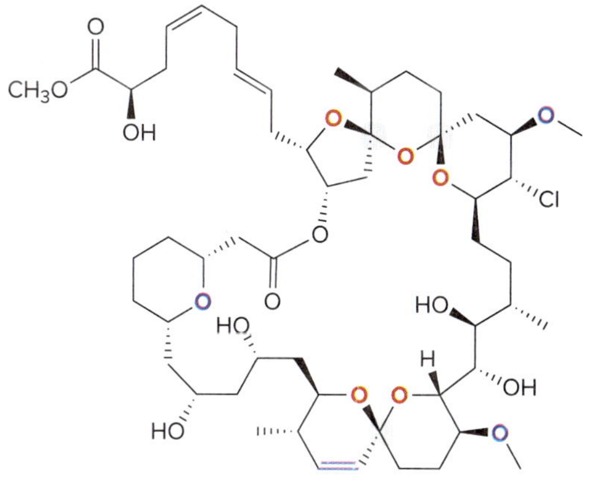

spirastrellolide A

Problem 18.27 Label the ethers and acetals in nigericine, an antibiotic active against gram-positive bacteria.

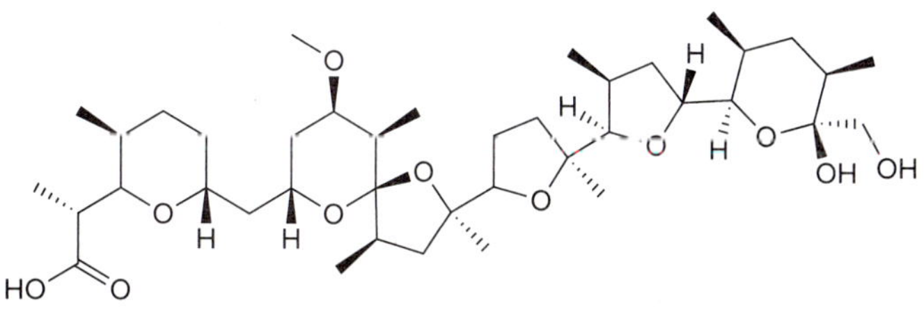

nigericine

More Practice: Try Problems 18.52a, 18.54a.

Problem 18.28 Draw the products of each reaction.

a. (cyclopentanone) $\xrightarrow[\text{TsOH}]{2\ CH_3OH}$

b. (2-butanone) + HOCH$_2$CH$_2$OH $\xrightarrow{\text{TsOH}}$

18.13A The Mechanism

The mechanism for acetal formation can be divided into two parts: **the addition of one equivalent of alcohol** to form a **hemiacetal,** followed by the **conversion of the hemiacetal** to the **acetal.** A **hemiacetal** has a carbon atom bonded to one OH group and one OR group.

R(C=O)R $\underset{\text{Part [1]}}{\overset{\text{R'OH, H}^+}{\rightleftharpoons}}$ R$_2$C(OH)(OR') **hemiacetal** $\underset{\text{Part [2]}}{\overset{\text{R'OH, H}^+}{\rightleftharpoons}}$ R$_2$C(OR')(OR') **acetal** + H_2O

Like *gem*-diols, hemiacetals are often higher in energy than their carbonyl starting materials, making the direction of equilibrium unfavorable for hemiacetal formation. The elimination of H_2O, which can be removed from the reaction mixture to drive the equilibrium to favor product, occurs during the conversion of the hemiacetal to the acetal. This explains why two equivalents of ROH react with a carbonyl compound, forming the acetal as product.

Mechanism 18.10 is written in two parts with a general acid HA.

Mechanism 18.10 Acetal Formation

Part [1] Formation of a hemiacetal

:O: + H—A ⇌ [1] :OH$^+$ ↔ :OH (C$^+$) **resonance-stabilized cation** + :A$^-$ ⇌ [2] R'ÖH, nucleophilic attack ⇌ R$_2$C(OH)(O$^+$(H)R') + :A$^-$ ⇌ [3] R$_2$C(OH)(OR') **hemiacetal** + H—A

1 Protonation of the carbonyl oxygen forms a **resonance-stabilized cation.**

2 – 3 Nucleophilic attack by R'OH and deprotonation form the **hemiacetal.** The overall result is addition of H and OR' to the carbonyl group.

Part [2] Formation of an acetal

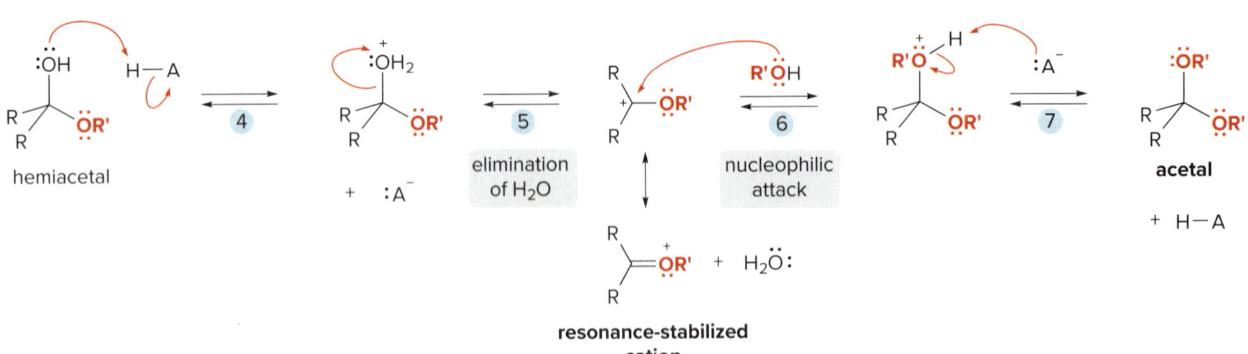

4 Protonation of the OH group of the hemiacetal forms a **good leaving group.**

5 Loss of H_2O forms a **resonance-stabilized cation.**

6 – 7 Nucleophilic attack by R'OH followed by loss of a proton forms the **acetal.** The overall result of Part [2] is the addition of a second OR' group to the carbonyl.

Although this mechanism is lengthy—there are seven steps—there are only three different kinds of reactions: **addition of a nucleophile, elimination of a leaving group,** and **proton transfer.** Steps [2] and [6] involve nucleophilic attack, and Step [5] eliminates H_2O. The other four steps in the mechanism shuffle protons from one oxygen atom to another, to make a better leaving group or a more electrophilic carbonyl group.

Problem 18.29 Draw a stepwise mechanism for the following reaction.

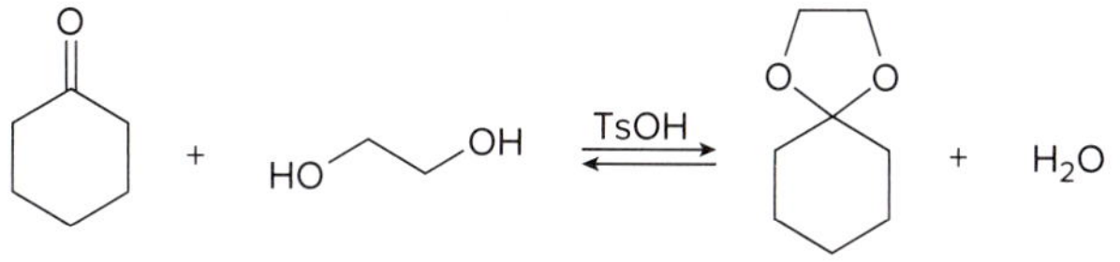

cyclohexanone + HOCH₂CH₂OH ⇌ (TsOH) cyclic acetal + H_2O

18.13B Hydrolysis of Acetals

Conversion of an aldehyde or ketone to an acetal is a **reversible reaction,** so **an acetal can be hydrolyzed to an aldehyde or ketone by treatment with aqueous acid.** Because this reaction is also an equilibrium process, it is driven to the right by using a large excess of water for hydrolysis.

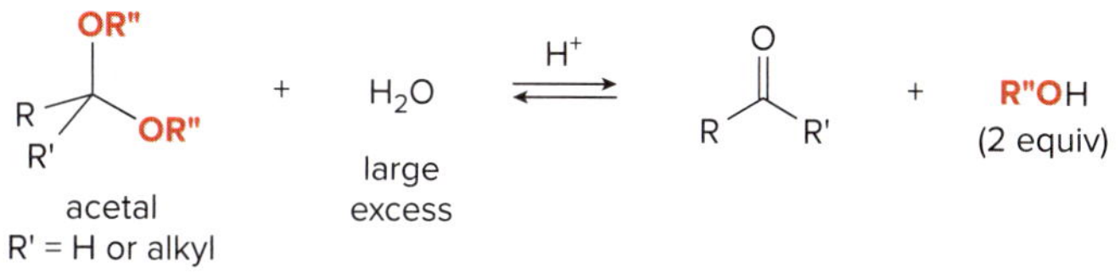

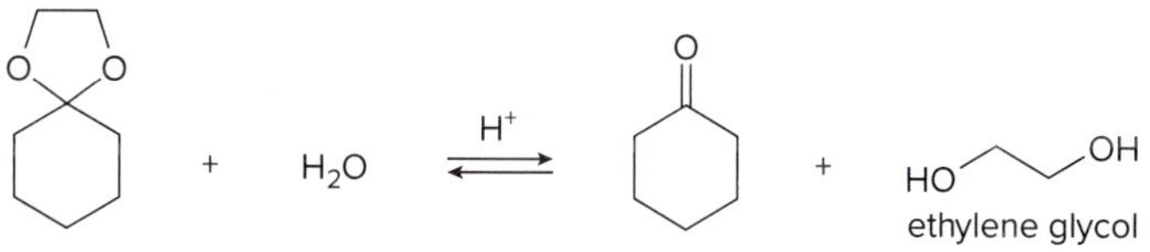

acetal (R' = H or alkyl) + H_2O (large excess) ⇌ (H^+) RCOR' + R"OH (2 equiv)

cyclic acetal + H_2O ⇌ (H^+) cyclohexanone + HOCH₂CH₂OH (ethylene glycol)

The mechanism for this reaction is the reverse of acetal synthesis, as shown in Mechanism 18.11.

Acetal hydrolysis requires a strong acid to make a good leaving group (ROH). **Acetal hydrolysis does *not* occur in base.**

Sample Problem 18.5 Drawing the Products of Acetal Hydrolysis

Identify the acetal carbons in **A**, and draw the products formed by hydrolysis of **A** with aqueous acid.

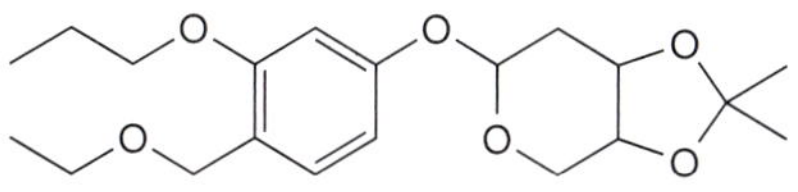

A

Solution

Determine the identity of the functional group that contains each O atom.

- **An acetal contains *two* oxygen atoms bonded to the *same* carbon.**
- **An ether contains *one* oxygen atom bonded to *two* carbons.**

A contains two ethers (in gray) and two acetals (with O atoms in red and acetal carbons labeled in blue).

- The **acetal C bonded to two O's is converted to a carbonyl C** during hydrolysis.
- The **acetal O's are converted to OH groups.**

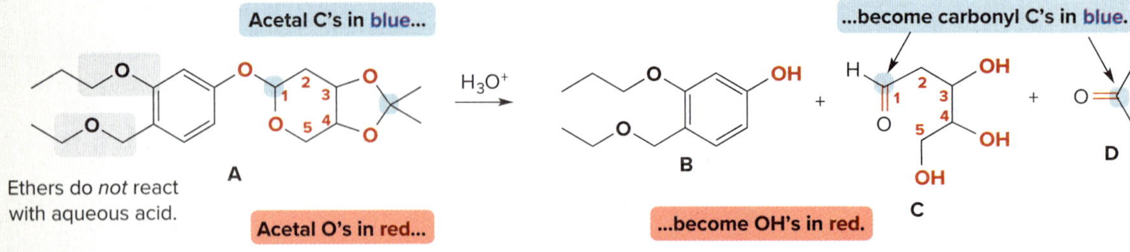

All C–O bonds of the acetals are broken and three products are formed.

Problem 18.30

Draw the products formed when each acetal is treated with aqueous acid.

a. b. c. d. e. f.

More Practice: Try Problems 18.46, 18.50, 18.52b, 18.54.

Mechanism 18.11 Acetal Hydrolysis

Part [1] Conversion of an acetal to a hemiacetal

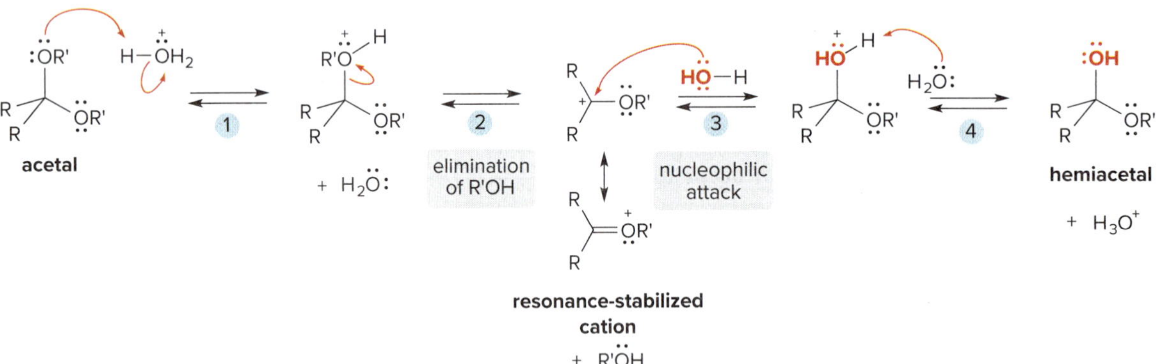

1 Protonation of one OR' group of the acetal forms a **good leaving group.**

2 Loss of R'OH forms a **resonance-stabilized cation.**

3 – 4 Nucleophilic attack by H_2O followed by loss of a proton forms the **hemiacetal.** The overall result of Part [1] is the **substitution** of one OR' group by OH.

Part [2] Conversion of a hemiacetal to a carbonyl group

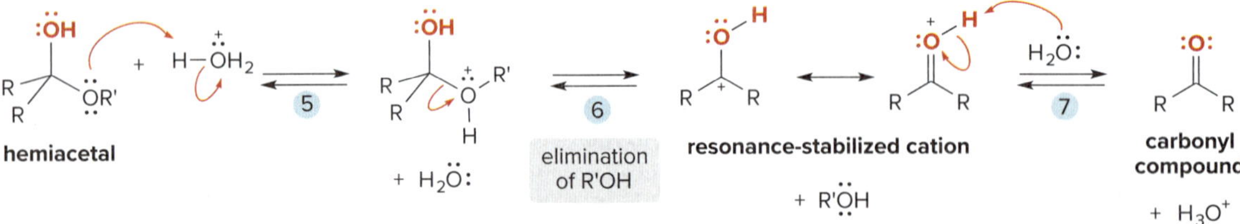

5 Protonation of the OR' group of the hemiacetal forms a **good leaving group.**

6 – 7 **Loss of R'OH** and deprotonation form a **carbonyl compound.**

Problem 18.31 Safrole is a naturally occurring acetal isolated from sassafras plants. Once used as a common food additive in root beer and other beverages, it is now banned because it is carcinogenic. What compounds are formed when safrole is hydrolyzed with aqueous acid?

Sassafras, source of safrole (Problem 18.31) *kj2011/iStock/Getty Images*

safrole

Problem 18.32 Identify the acetal in oleandrin, one of the toxins present in the sap of the oleander plant, and draw the products formed by acid-catalyzed hydrolysis of the acetal.

Oleandrin (Problem 18.32) is obtained from the oleander plant *Nerium oleander*. *Alessandro0770/Getty Images*

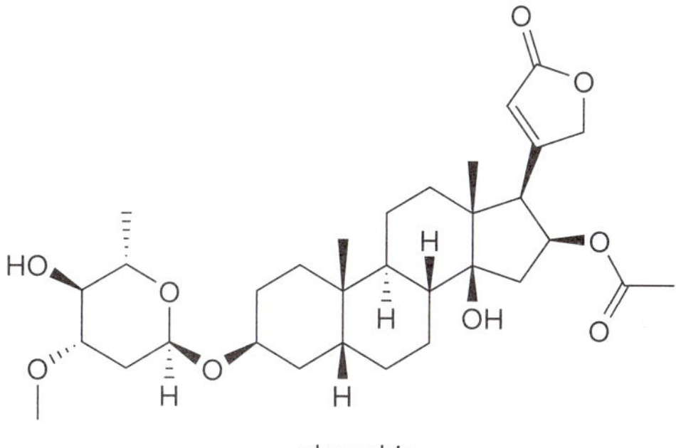

oleandrin

18.14 Acetals as Protecting Groups

Just as the *tert*-butyldimethylsilyl ethers are used as protecting groups for alcohols (Section 17.12), **acetals are valuable protecting groups for aldehydes and ketones.**

Suppose a starting material **A** contains both a ketone and an ester, and it is necessary to selectively reduce the ester to an alcohol (6-hydroxyhexan-2-one), leaving the ketone untouched. Such a selective reduction is *not* possible in one step. Because ketones are more readily reduced, methyl 5-hydroxyhexanoate is formed instead.

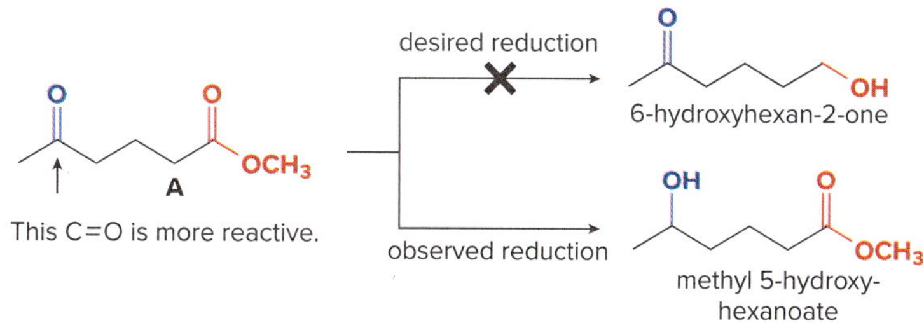

To solve this problem, we can use a protecting group to block the more reactive ketone carbonyl group. The overall process requires three steps.

[1] Protect the interfering functional group—the ketone carbonyl.

[2] Carry out the desired reaction—reduction.

[3] Remove the protecting group.

The following three-step sequence using a cyclic acetal leads to the desired product.

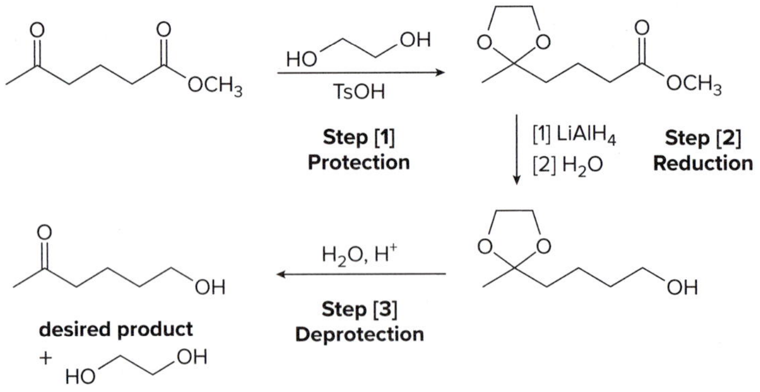

- **Step [1]** The ketone carbonyl is protected as a cyclic acetal by reaction of the starting material with $HOCH_2CH_2OH$ and TsOH.
- **Step [2]** Reduction of the ester is then carried out with $LiAlH_4$, followed by treatment with H_2O.
- **Step [3]** The acetal is then converted back to a ketone carbonyl group with aqueous acid.

Acetals are widely used protecting groups for aldehydes and ketones because they are easy to add and easy to remove, and they are stable to a wide variety of reaction conditions. Acetals do *not* react with base, oxidizing agents, reducing agents, or nucleophiles. Good protecting groups must survive a variety of reaction conditions that take place at other sites in a molecule, but they must also be selectively removed under mild conditions when needed.

Problem 18.33 How would you use a protecting group to carry out the following transformation?

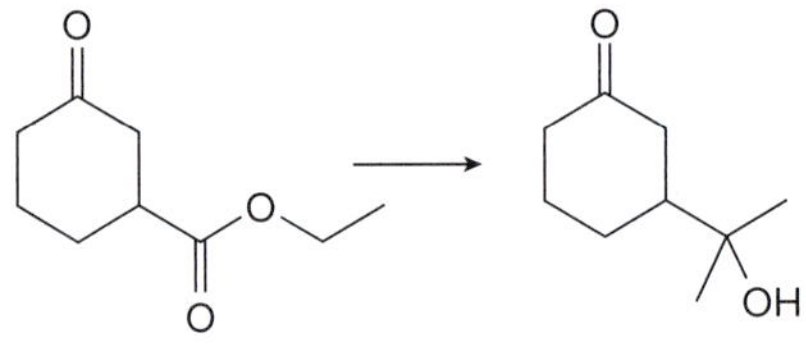

18.15 Cyclic Hemiacetals

Cyclic hemiacetals are also called **lactols.**

Although acyclic hemiacetals are generally unstable and therefore not present in appreciable amounts at equilibrium, **cyclic hemiacetals containing five- and six-membered rings are stable compounds** that are readily isolated.

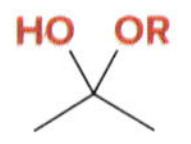

hemiacetal

One C is bonded to:
- an **OH** group
- an **OR** group

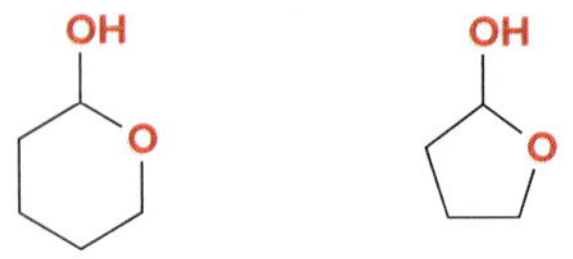

cyclic hemiacetals

Each C is bonded to:
- an **OH** group
- an **OR** group that is part of a ring

18.15A Forming Cyclic Hemiacetals

All hemiacetals are formed by nucleophilic addition of a hydroxy group to a carbonyl group. In the same way, cyclic hemiacetals are formed by **intramolecular cyclization of hydroxy aldehydes.**

5-hydroxypentanal ⇌ (6%) (94%)

4-hydroxybutanal ⇌ (11%) (89%)

[Equilibrium proportions of each compound are given.]

Such intramolecular reactions to form five- and six-membered rings are faster than the corresponding intermolecular reactions. The two reacting functional groups, in this case OH and C=O, are held in close proximity, increasing the probability of reaction.

Problem 18.34 What lactol (cyclic hemiacetal) is formed from intramolecular cyclization of each hydroxy aldehyde?

a. b.

Hemiacetal formation is catalyzed by both acid and base. The acid-catalyzed mechanism is identical to Part [1] of Mechanism 18.10, except that the reaction occurs in an **intramolecular** fashion, as shown for the acid-catalyzed cyclization of 5-hydroxypentanal to form a six-membered cyclic hemiacetal in Mechanism 18.12.

Mechanism 18.12 Acid-Catalyzed Cyclic Hemiacetal Formation

[1] → [2] nucleophilic attack → [3] → hemiacetal + H—A

1 – 2 Protonation of the carbonyl oxygen followed by **intramolecular nucleophilic attack** forms the six-membered ring.

3 Deprotonation forms the neutral **cyclic hemiacetal.**

Intramolecular cyclization of a hydroxy aldehyde forms a **hemiacetal with a new stereogenic center, so that an equal amount of two enantiomers** results.

new stereogenic center

enantiomers

18.15B The Conversion of Hemiacetals to Acetals

Cyclic hemiacetals can be converted to acetals by treatment with an alcohol and acid. This reaction converts the OH group that is part of the hemiacetal to an OR group.

OH, O (hemiacetal) $\xrightleftharpoons{CH_3OH,\ H^+}$ OCH_3, O (acetal) + H_2O

hemiacetal acetal

Mechanism 18.13, which is similar to Part [2] of Mechanism 18.10, illustrates the conversion of a cyclic hemiacetal to an acetal.

Mechanism 18.13 A Cyclic Acetal from a Cyclic Hemiacetal

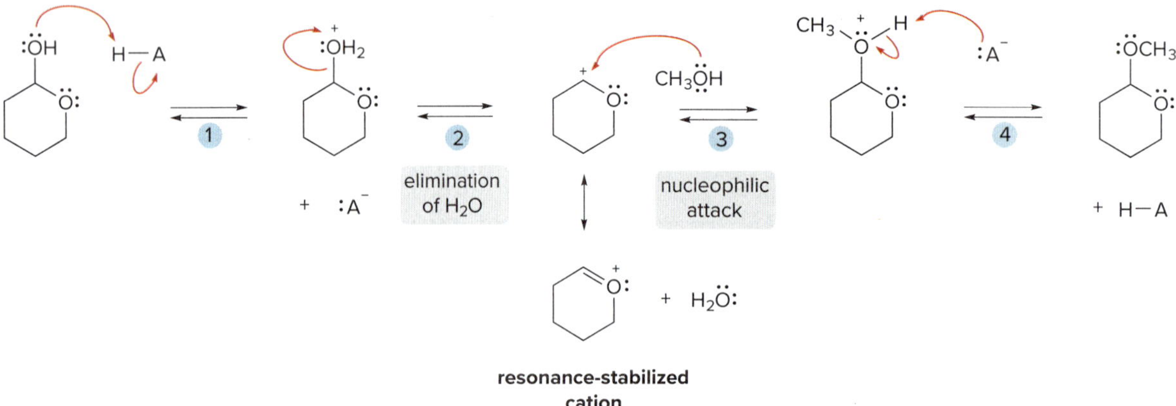

1 Protonation of the OH group of the hemiacetal forms a **good leaving group.**

2 Loss of H_2O forms a **resonance-stabilized cation.**

3 – 4 Nucleophilic attack by CH_3OH followed by loss of a proton forms the **acetal.**

The overall result of this reaction is the **replacement of the hemiacetal OH group by an OCH_3 group.** This substitution reaction readily occurs because the carbocation formed in Step [2] is stabilized by resonance. This fact makes the OH group of a hemiacetal different from the hydroxy group in other alcohols.

Thus, when a compound that contains both an alcohol OH group and a hemiacetal OH group is treated with an alcohol and acid, **only the hemiacetal OH group reacts** to form an acetal. The alcohol OH group does *not* react.

The conversion of cyclic hemiacetals to acetals is an important reaction in carbohydrate chemistry, as discussed in Chapter 26.

OH, O, OH $\xrightleftharpoons{CH_3OH,\ H^+}$ OCH_3, O, OH + H_2O

Only the hemiacetal OH reacts.

Problem 18.35 Monensin, a polyether antibiotic produced by *Streptomyces cinnamonensis,* is used as an additive in cattle feed. Label each acetal, hemiacetal, and ether in monensin.

monensin

Problem 18.36 Draw the products of each reaction.

a. b.

Problem 18.37 Identify the acetals in rebaudioside A, the molecule that opened this chapter, and draw the products formed when all the acetals are hydrolyzed with aqueous acid.

The stevia plant is the source of rebaudioside A (Problem 18.37), sold as the sweetener Truvia. *Linda Hall/Shutterstock*

rebaudioside A
Trade name: Truvia

18.16 An Introduction to Carbohydrates

Carbohydrates, commonly referred to as sugars and starches, are polyhydroxy aldehydes and ketones, or compounds that can be hydrolyzed to them. Along with proteins, fatty acids, and nucleotides, they form one of the four main groups of biomolecules responsible for the structure and function of all living cells.

Glucose is the carbohydrate that is transported in the blood to individual cells. The hormone insulin regulates the level of glucose in the blood. Diabetes is a common disease that results from a deficiency of insulin, resulting in increased glucose levels in the blood and other metabolic abnormalities. Insulin injections control glucose levels.

Many carbohydrates contain cyclic acetals or hemiacetals. Examples include **glucose,** the most common simple sugar, and **lactose,** the principal carbohydrate in milk. Hemiacetal carbons are labeled in blue, whereas the acetal carbon is labeled in red.

3-D structure

β-D-glucose
(one form of glucose)

lactose

3-D structure

Hemiacetals in sugars are formed in the same way that other hemiacetals are formed—that is, by **cyclization of hydroxy aldehydes.** Thus, the hemiacetal of glucose is formed by cyclization of an acyclic *poly*hydroxy aldehyde **A,** as shown in the accompanying equation. This process illustrates two important features.

A ⇌ β-D-glucose 63% (equatorial **OH**) + α-D-glucose 37% (axial **OH**)

- When the OH group on C5 is the nucleophile, **cyclization yields a six-membered ring,** and this ring size is preferred.
- **Cyclization forms a new stereogenic center** (labeled in blue), exactly analogous to the cyclization of the simpler hydroxy aldehyde (5-hydroxypentanal) in Section 18.15A. **The new OH group of the hemiacetal can occupy either the equatorial or axial position.**

For glucose, this results in two cyclic forms, called **β-D-glucose** (having an equatorial OH group) and **α-D-glucose** (having an axial OH group). Because β-D-glucose has the new OH group in the more roomy equatorial position, this cyclic form of glucose is the major product. At equilibrium, only a trace of the acyclic hydroxy aldehyde **A** is present.

Many more details on this process and other aspects of carbohydrate chemistry are presented in Chapter 26.

Problem 18.38

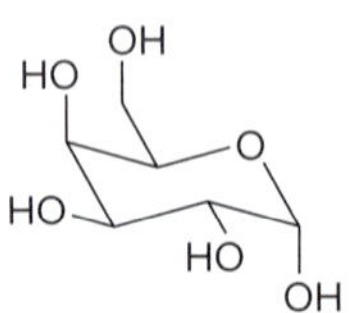

α-D-galactose

a. How many stereogenic centers are present in α-D-galactose?
b. Label the hemiacetal carbon in α-D-galactose.
c. Draw the structure of β-D-galactose.
d. Draw the structure of the polyhydroxy aldehyde that cyclizes to α- and β-D-galactose.
e. From what you learned in Section 18.15B, what product(s) is (are) formed when α-D-galactose is treated with CH_3OH and an acid catalyst?

Chapter 18 REVIEW

KEY CONCEPTS

The relationship between the stability of a carbonyl compound and hydrate formation (18.12)

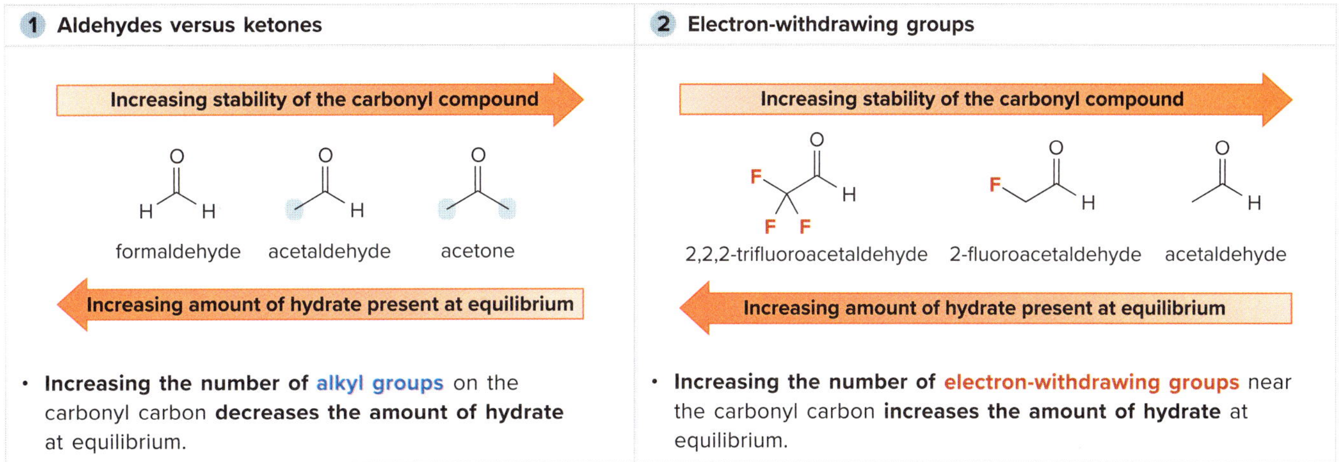

- **Increasing the number of alkyl groups** on the carbonyl carbon **decreases the amount of hydrate** at equilibrium.
- **Increasing the number of electron-withdrawing groups** near the carbonyl carbon **increases the amount of hydrate** at equilibrium.

Try Problem 18.55a, b.

KEY REACTIONS

Nucleophilic Addition Reactions

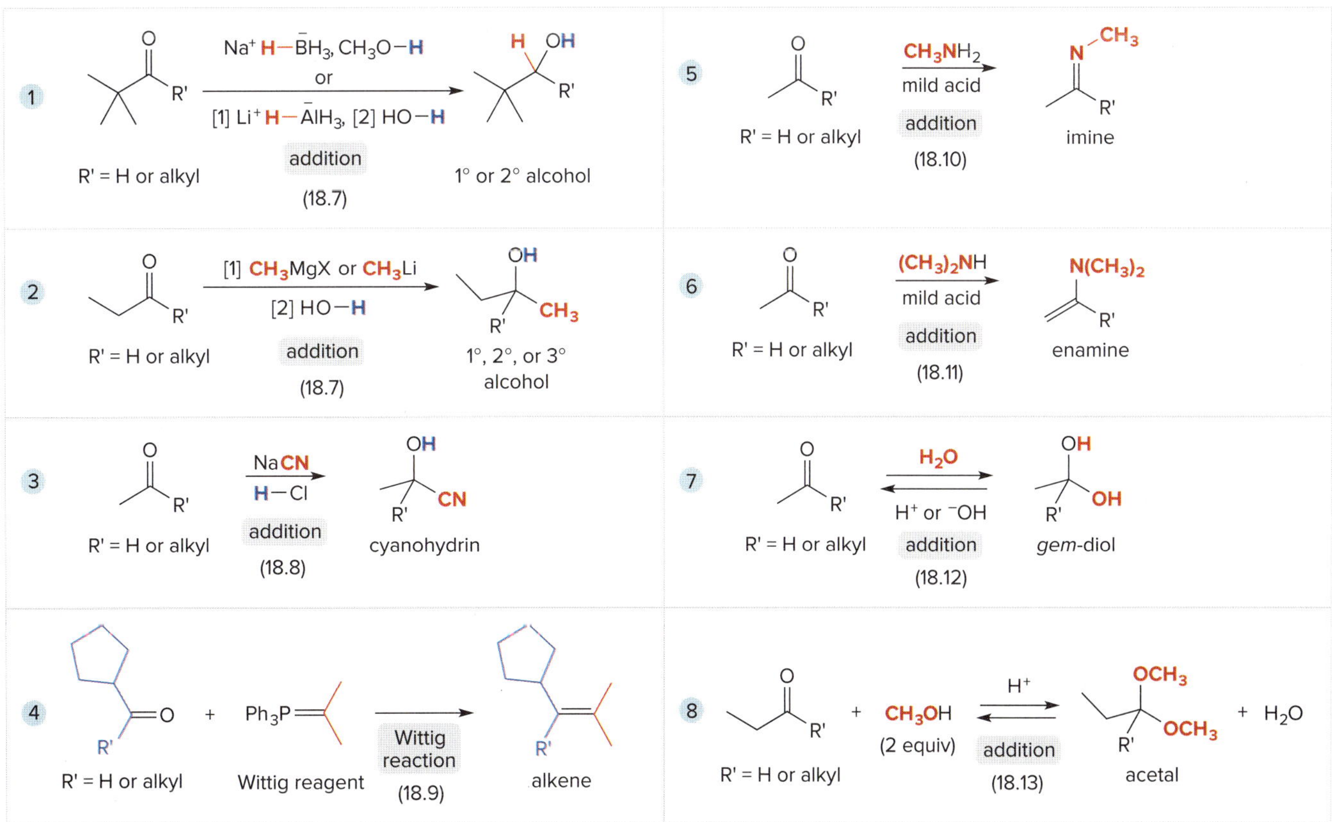

Try Problems 18.39b; 18.43; 18.44a, b, d, f, h; 18.47; 18.48.

Other Reactions

1 X → [1] Ph_3P: [2] Bu–Li (18.9A) → PPh_3 Wittig reagent ylide	4 N imine → H_2O, H^+ hydrolysis (18.11) → O ketone + $-NH_2$
2 OH CN R' cyanohydrin R' = H or alkyl → ^-OH elimination (18.8) → O R' aldehyde or ketone + HO–H + ^-CN	5 N enamine → H_2O, H^+ hydrolysis (18.11) → O ketone + H N
3 HO C≡N R' cyanohydrin R' = H or alkyl → H_2O H^+ or ^-OH Δ hydrolysis (18.8) → HO OH O R' α-hydroxy carboxylic acid (with acid) or HO O^- O R' α-hydroxy carboxylate (with base)	6 OCH_3 OCH_3 R' R' = H or alkyl acetal + H_2O ⇌ H^+ hydrolysis (18.13) → O R' aldehyde or ketone + CH_3OH (2 equiv)

See Sample Problem 18.3. Try Problems 18.44c, e, g; 18.46; 18.50; 18.51; 18.52; 18.53; 18.54b.

KEY SKILLS

[1] Synthesizing Wittig reagents by a two-step procedure (18.9A)

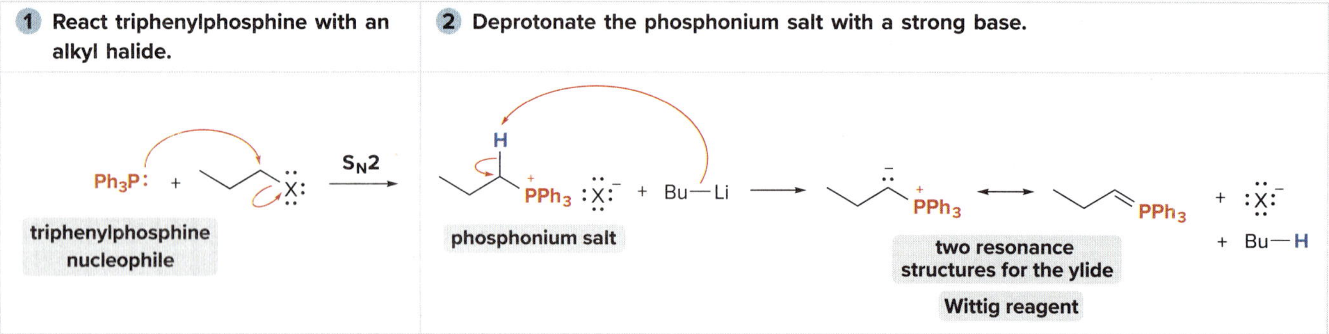

Try Problem 18.53.

[2] Determining the starting materials for a Wittig reaction using retrosynthetic analysis (18.9C)

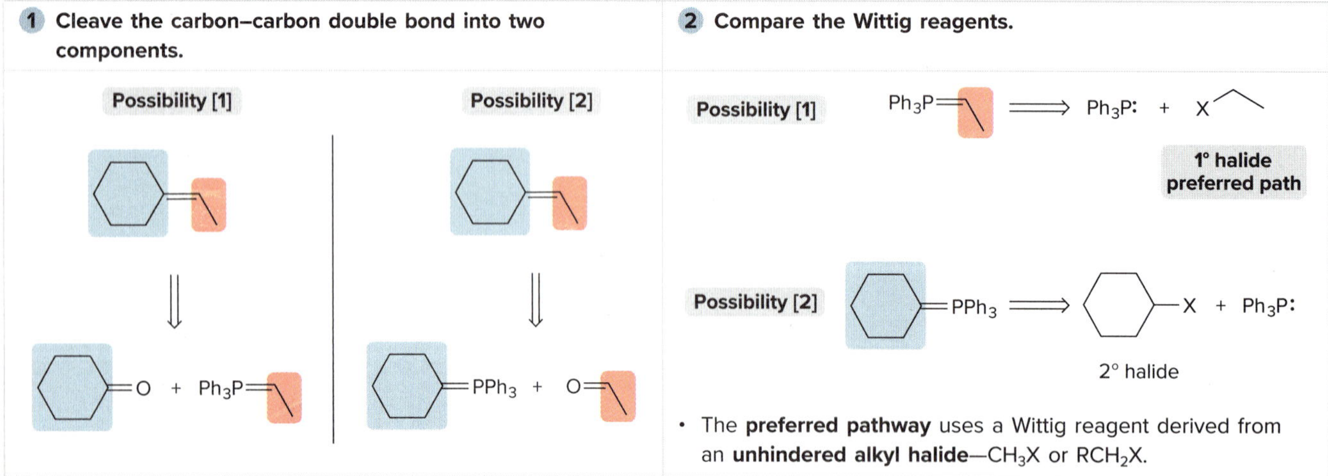

- The **preferred pathway** uses a Wittig reagent derived from an **unhindered alkyl halide**—CH_3X or RCH_2X.

See *How To*, p. 852. Try Problems 18.56, 18.59.

[3] Using an acetal as a protecting group (18.14)

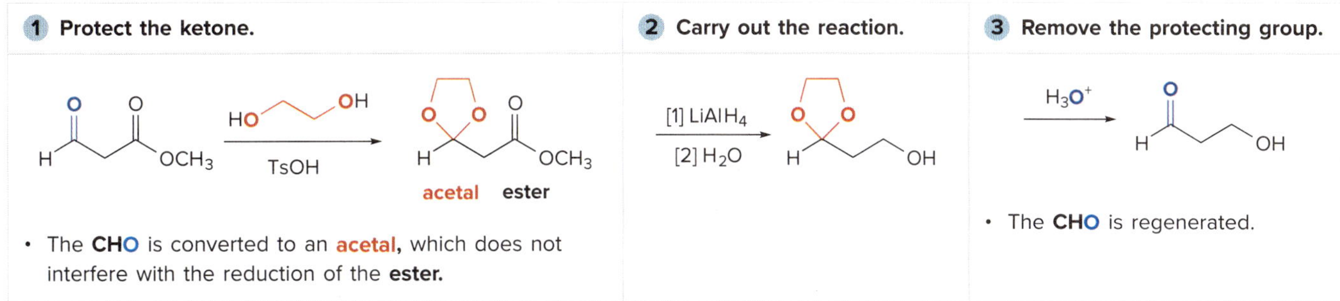

- The **CHO** is converted to an **acetal,** which does not interfere with the reduction of the **ester.**
- The **CHO** is regenerated.

Try Problems 18.66, 18.67.

[4] Drawing the stereoisomers that form in the intramolecular cyclization of a hydroxy aldehyde (18.15A)

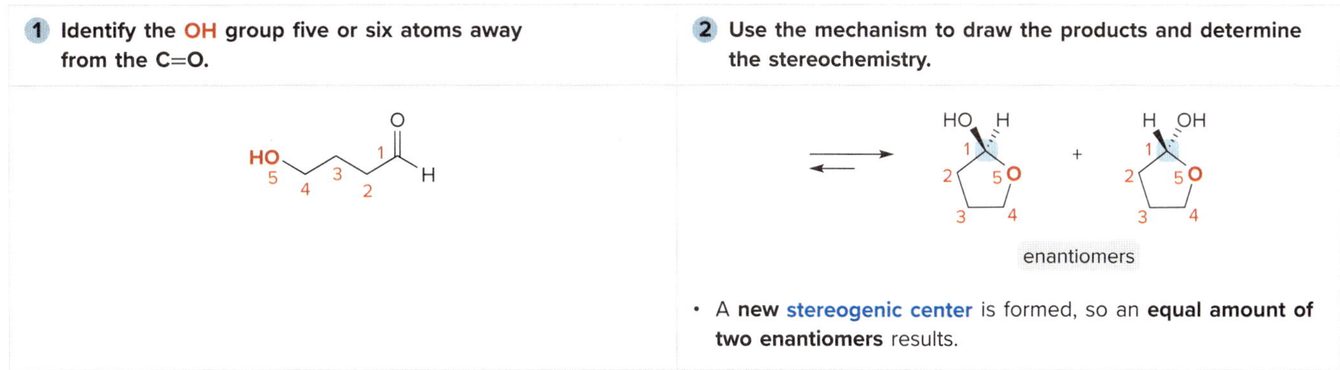

- A **new stereogenic center** is formed, so an **equal amount of two enantiomers** results.

Try Problems 18.49, 18.82.

[5] Determining the reactive OH group that forms an acetal when treated with an alcohol and acid (18.15B)

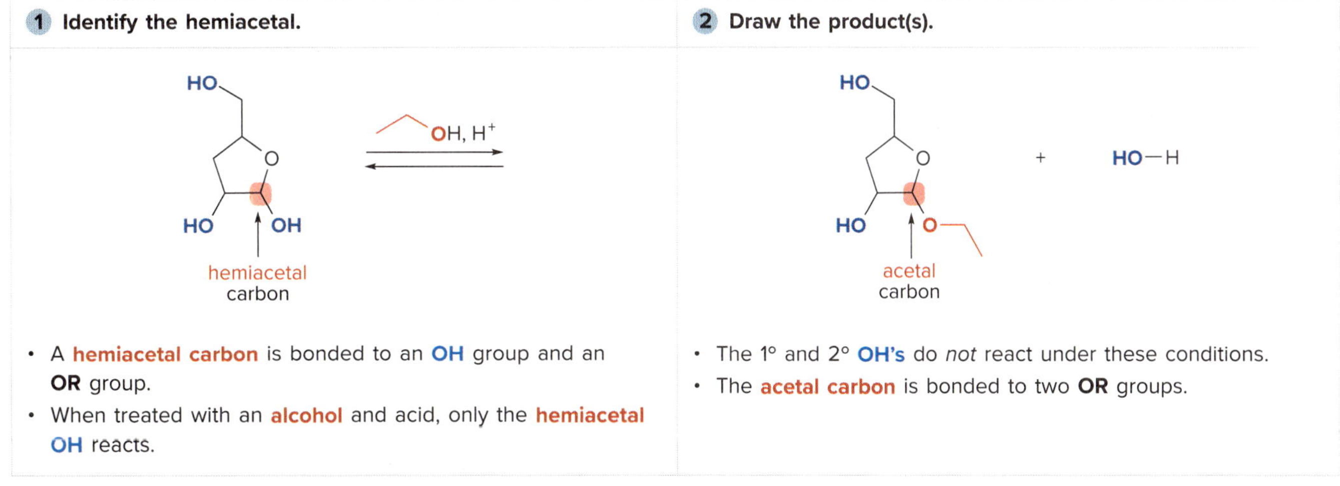

- A **hemiacetal carbon** is bonded to an **OH** group and an **OR** group.
- When treated with an **alcohol** and acid, only the **hemiacetal OH** reacts.
- The 1° and 2° **OH's** do *not* react under these conditions.
- The **acetal carbon** is bonded to two **OR** groups.

Try Problem 18.47d.

CHAPTER 18 MULTIPLE-CHOICE SELF-TEST

The Self-Test consists of multiple-choice questions similar to those found on the American Chemical Society organic chemistry exam. Answers are given at the end of the chapter.

1. How many acetals, hemiacetals, and ethers does the given compound contain?

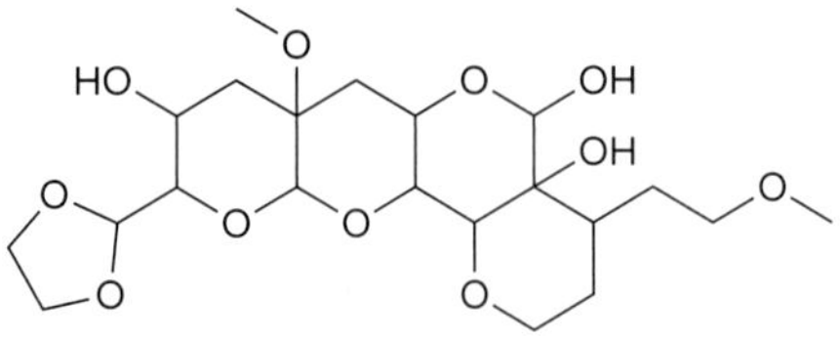

a. one acetal, two hemiacetals, two ethers
b. two acetals, two hemiacetals, two ethers
c. two acetals, one hemiacetal, two ethers
d. two acetals, one hemiacetal, three ethers

2. Give the IUPAC name for the following compound.

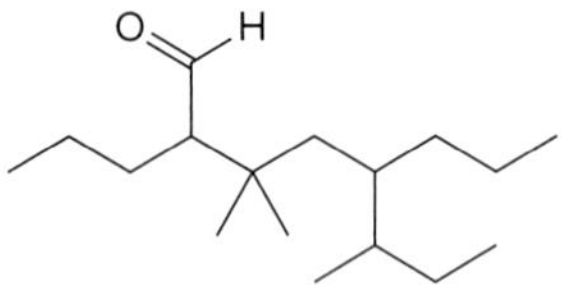

a. 5-*sec*-butyl-3,3-dimethyloctanal
b. 2,5-dipropyl-3,3,6-trimethyloctan-1-al
c. 4-*sec*-butyl-7-formyl-6,6-dimethyldecanal
d. 3,3,6-trimethyl-2,5-dipropyloctanal

3. What product(s) are formed when **A** and **B** react in the presence of acid?

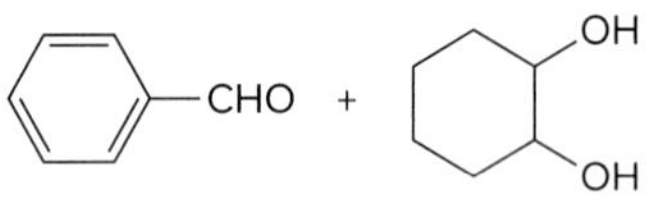

A: CHO + B: OH, OH

a. 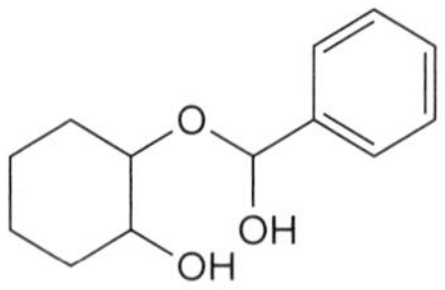 b. 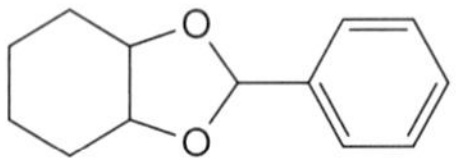 c. 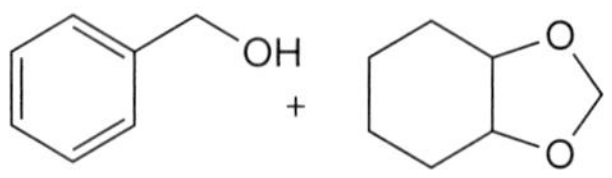 d.

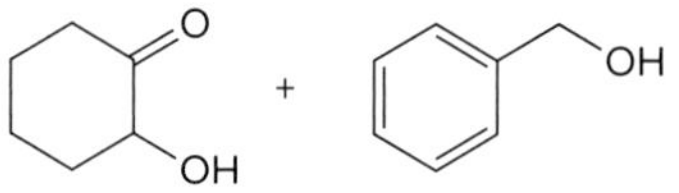

4. What product is formed in the following reaction?

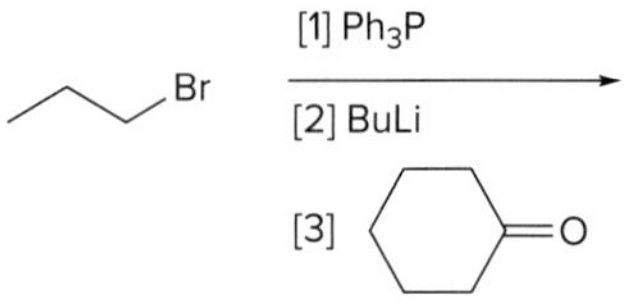

Br [1] Ph_3P [2] BuLi [3] (cyclohexanone)

a. 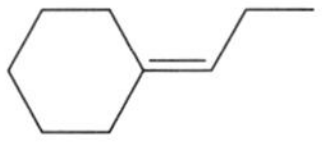 b. 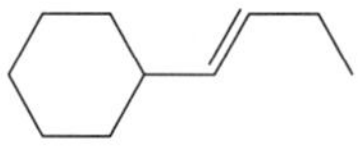 c. 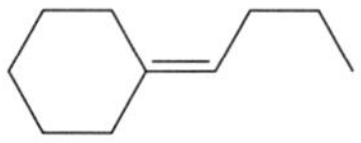 d.

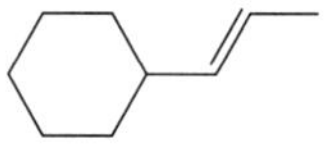

5. What products are formed when the given enamine is hydrolyzed in aqueous acid?

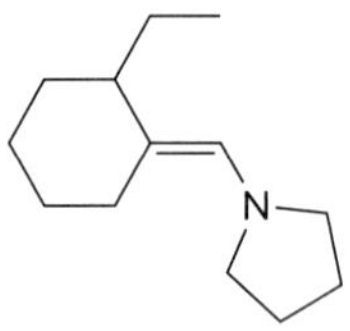

a. 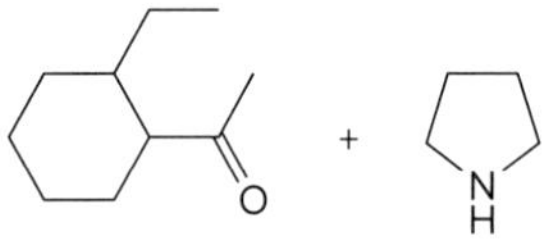 b. 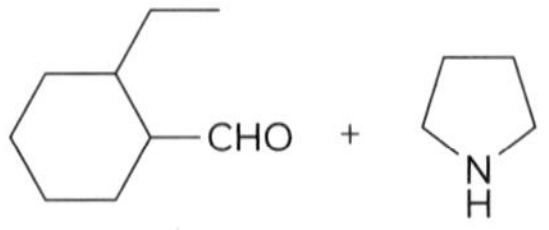 c. d.

6. Which carbonyl compound forms the largest amount of hydrate?

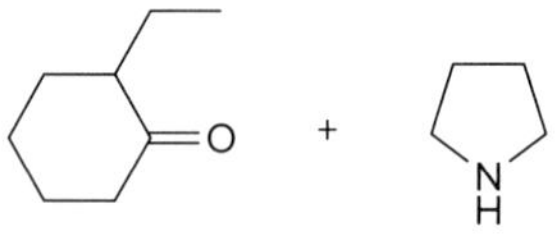

a. 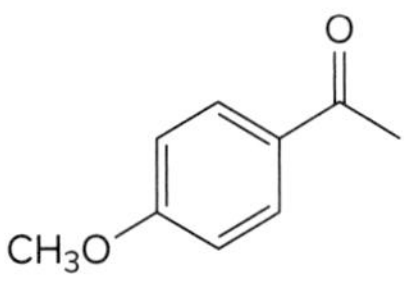 (CH_3O) b. 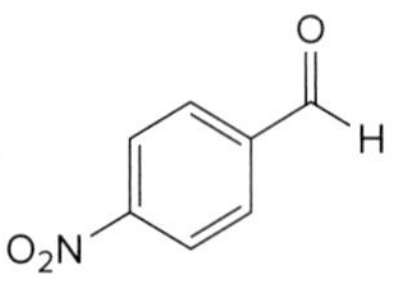 (O_2N) c. 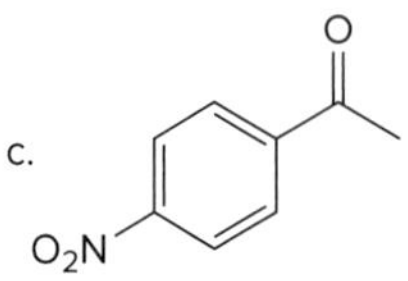 (O_2N) d. 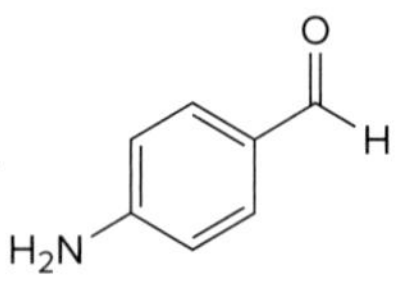 (H_2N)

7. What product is formed in the given two-step reaction sequence?

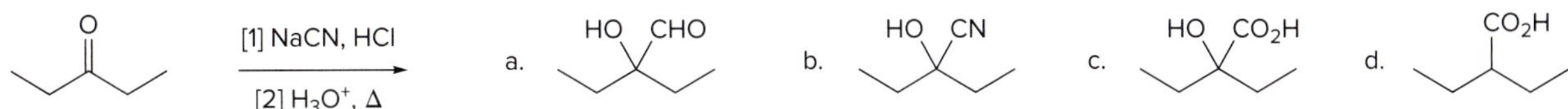

8. What product is formed in the following reaction?

O OH OH CH_3OH H^+ a. O OCH_3 OH b. O OCH_3 OCH_3 c. CH_3O CHO OH d. HO CHO OCH_3

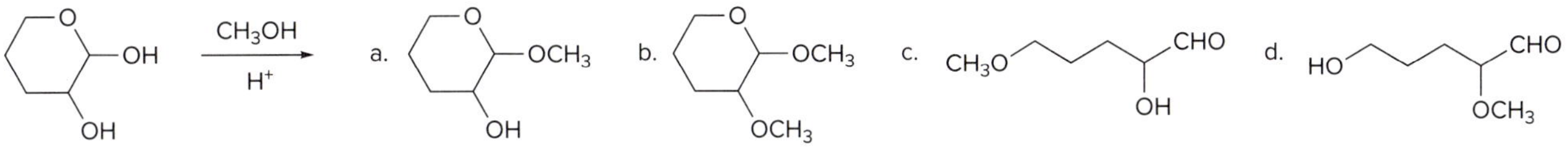

9. What compounds are formed when **A** undergoes hydrolysis in aqueous acid?

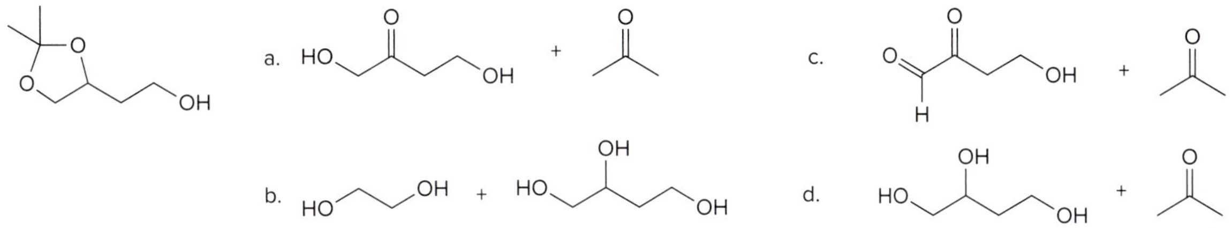

O O OH a. HO O OH + O c. O O H OH + O

b. HO OH + HO OH OH d. HO OH OH + O

10. What product is formed in the following reaction?

NH_2 + O mild H^+ a. N b. NH c. NH d. HO NH

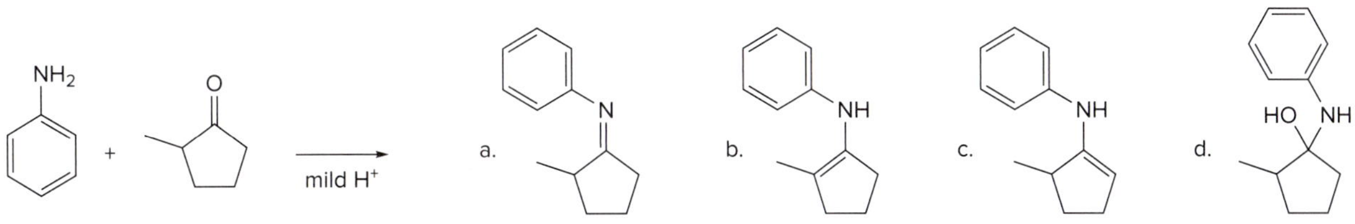

PROBLEMS

Problems Using Three-Dimensional Models

18.39 (a) Give the IUPAC name for **A** and **B.** (b) Draw the product formed when **A** or **B** is treated with each reagent: [1] $NaBH_4$, CH_3OH; [2] CH_3MgBr, then H_2O; [3] $Ph_3P{=}CHOCH_3$; [4] $CH_3CH_2CH_2NH_2$, mild acid; [5] $HOCH_2CH_2CH_2OH$, H^+.

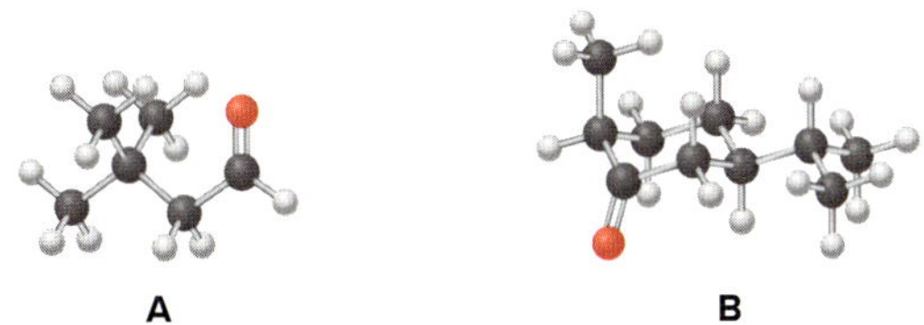

18.40 What carbonyl compound and diol are needed to prepare each compound?

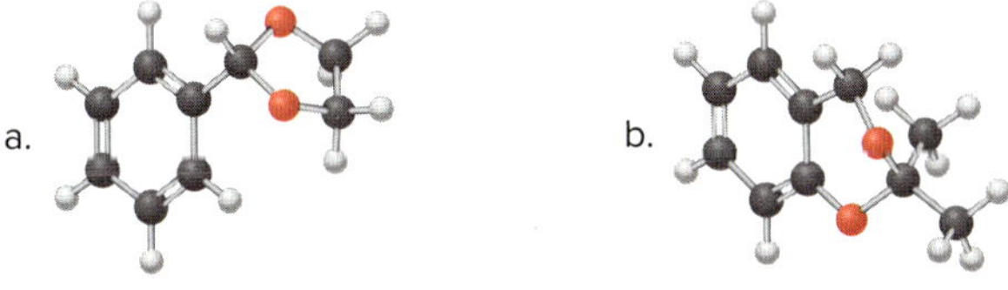

Nomenclature

18.41 Give the IUPAC name for each compound.

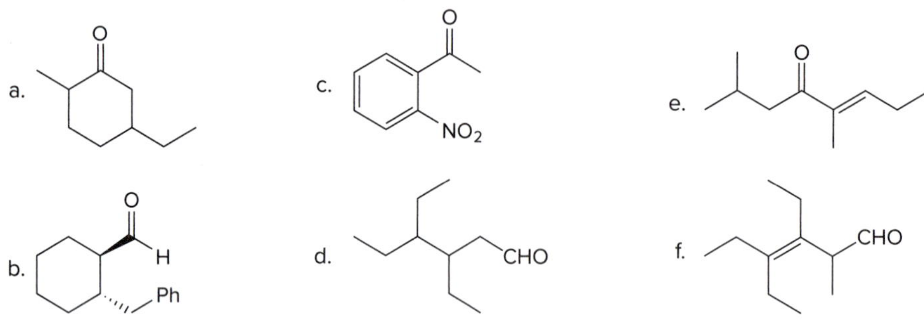

18.42 Give the structure corresponding to each name.

a. 2-methyl-3-phenylbutanal
b. 3,3-dimethylcyclohexanecarbaldehyde
c. 3-benzoylcyclopentanone
d. 2-formylcyclopentanone
e. (*R*)-3-methylheptan-2-one
f. *m*-acetylbenzaldehyde
g. 2-*sec*-butylcyclopent-3-enone
h. 5,6-dimethylcyclohex-1-enecarbaldehyde

Reactions

18.43 Draw the products formed in each reaction sequence.

a. Br; [1] Ph_3P; [2] BuLi; [3] (aldehyde, H)

b. Cl; [1] Ph_3P; [2] BuLi; [3] (aldehyde, H)

18.44 Draw the products of each reaction.

a. (aldehyde, H) + H_2N–(cyclohexyl) $\xrightarrow{\text{mild acid}}$

b. (ketone) $\xrightarrow[\text{H}^+]{\text{HO}\frown\text{OH}}$

c. (imine, N) $\xrightarrow{H_3O^+}$

d. C_6H_5 (ketone) + (pyrrolidine, NH) $\xrightarrow{\text{mild acid}}$

e. HO CN (diphenyl cyanohydrin) $\xrightarrow{H_3O^+, \Delta}$

f. (cyclic hemiacetal, O, OH) $\xrightarrow[\text{H}^+]{CH_3CH_2OH}$

g. (enamine, N) $\xrightarrow{H_3O^+}$

h. (cyclopentanone, O) $\xrightarrow{Ph_3P{=}\text{CH}\ldots\text{CO}OCH_3}$

18.45 In a synthesis of the non-sedating antihistamine fexofenadine (Section 23.5), compounds **A** and **B** were converted to acetal **C,** which was treated with aqueous acid to form **D.** Draw the structures of **B** and **D.**

A (NH, OH) $\xrightarrow[\text{NaHCO}_3]{\textbf{B}}$ **C** (N, O, O, OH) $\xrightarrow{H_3O^+}$ **D**

18.46 What carbonyl compound and alcohol are formed by hydrolysis of each acetal?

a. 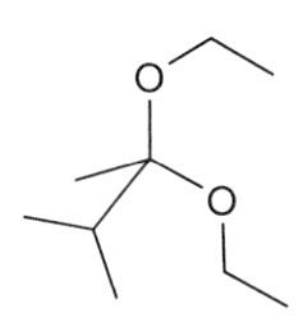 b. 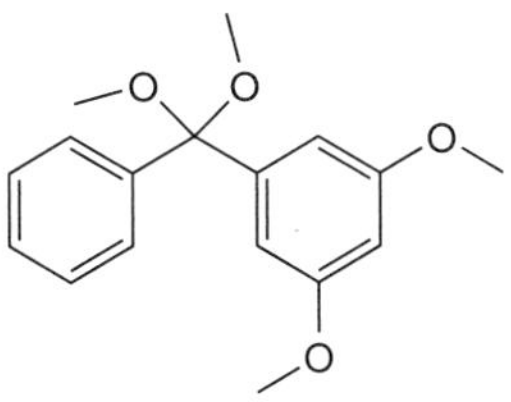 c.

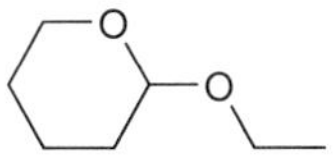

18.47 Draw all stereoisomers formed in each reaction.

a. 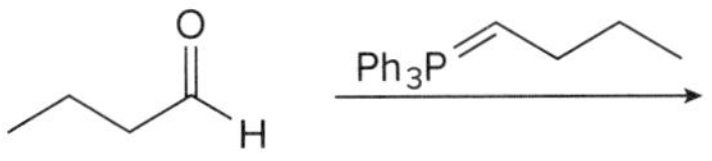 Ph_3P →

c. 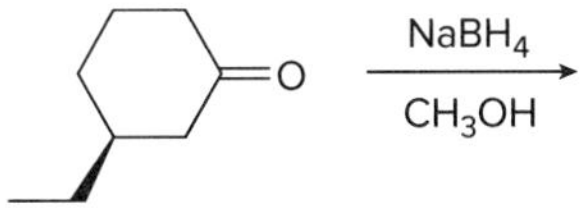 $NaBH_4$ / CH_3OH →

b. 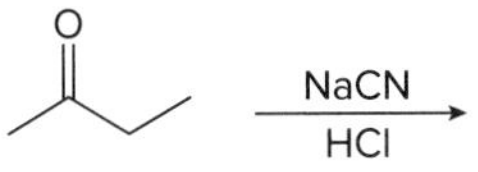 NaCN / HCl →

d. 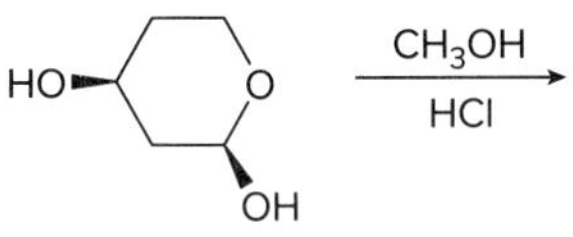 HO ... O ... OH; CH_3OH / HCl →

18.48 What product is formed when each compound undergoes an intramolecular reaction in the presence of acid?

a. 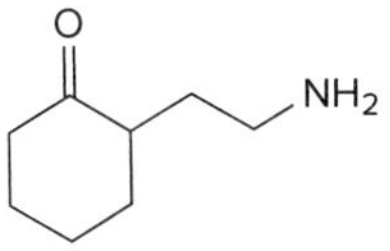 NH_2 b. 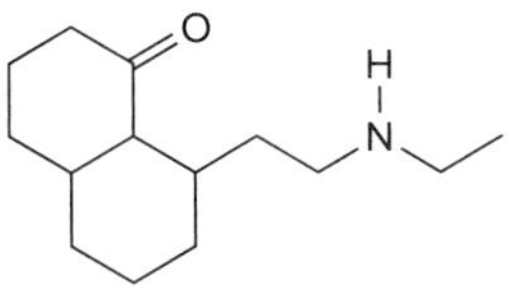 H N c. H N

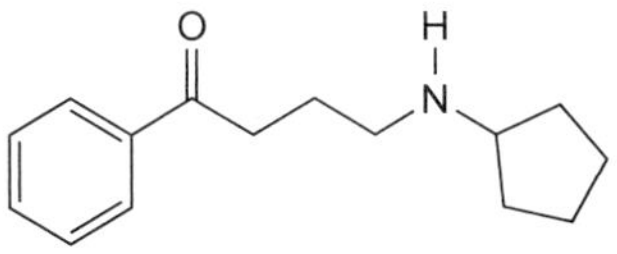

18.49 Hydroxy aldehydes **A** and **B** readily cyclize to form hemiacetals. Draw the stereoisomers formed in this reaction from both **A** and **B.** Explain why this process gives an optically inactive product mixture from **A** and an optically active product mixture from **B.**

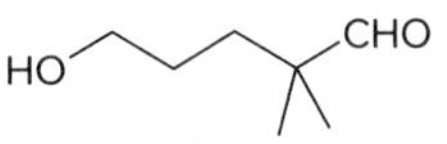

HO CHO **A** HO H CHO **B**

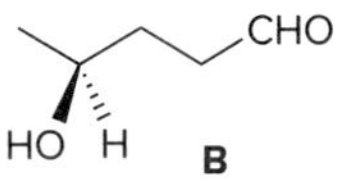

18.50 What products are formed when each acetal is hydrolyzed with aqueous acid?

a.  b. 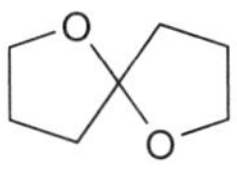 c.

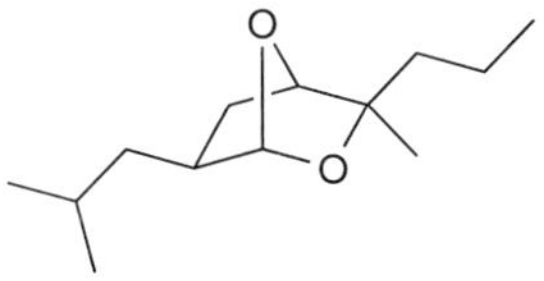

18.51 What hydrolysis products are formed when the following compound is treated with aqueous acid?

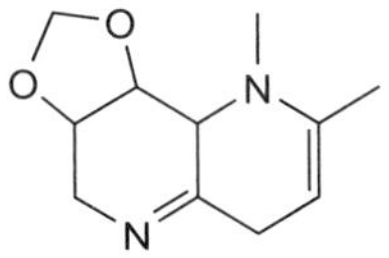

18.52 Attenol A and pinnatoxin A are natural products isolated from marine sources. (a) Locate the acetals, hemiacetals, imines, and enamines in both compounds. (b) Draw the hydrolysis product formed when attenol A is treated with aqueous acid. Include stereochemistry at all stereogenic centers.

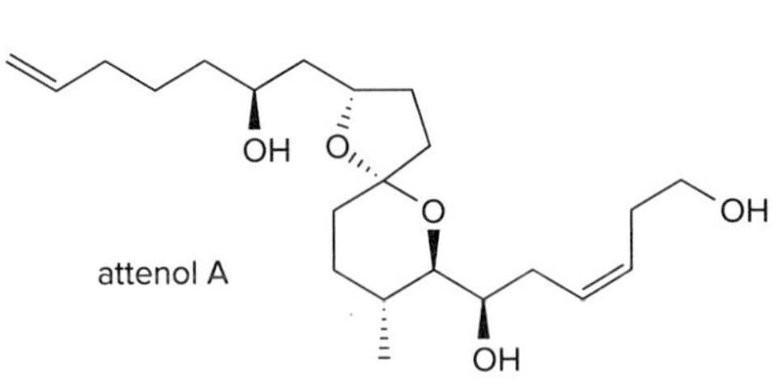

attenol A

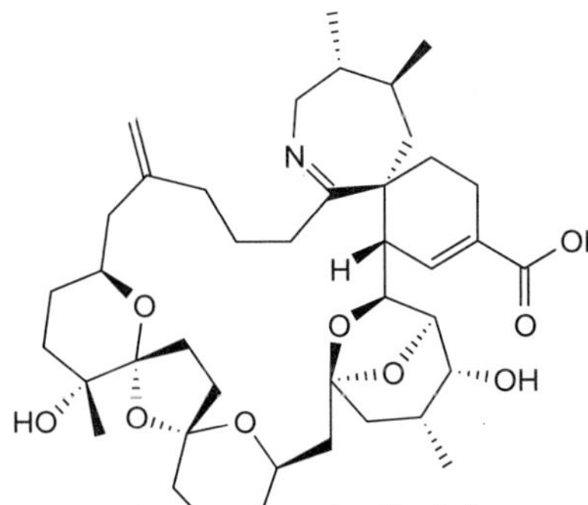

pinnatoxin A

18.53 What products are formed by hydrolysis of each imine or enamine?

a. 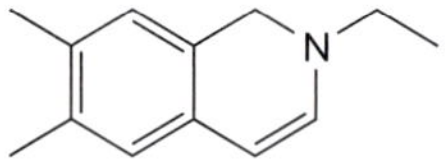 b. 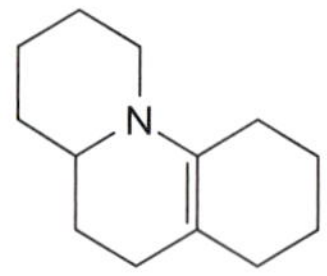 c.

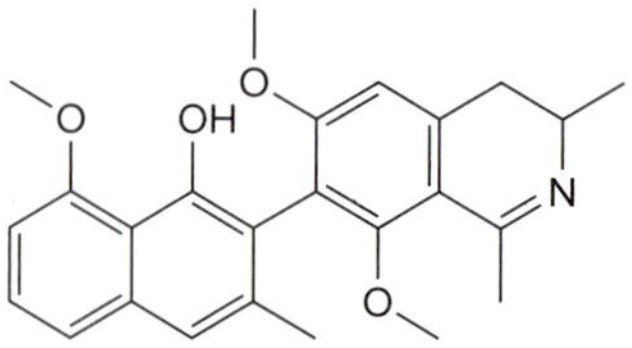

ancistrociadidine
antimalarial natural product

18.54 Two commercial drugs that contain acetal moieties are etoposide and ertugliflozin. Etoposide, sold as a phosphate derivative under the trade name Etopophos, is used for the treatment of lung cancer, testicular cancer, and lymphomas. Ertugliflozin, sold under the trade name Stegaltro, is used for the treatment of type 2 diabetes. (a) Locate the acetals in each drug. (b) What products are formed when all the acetals are hydrolyzed with aqueous acid?

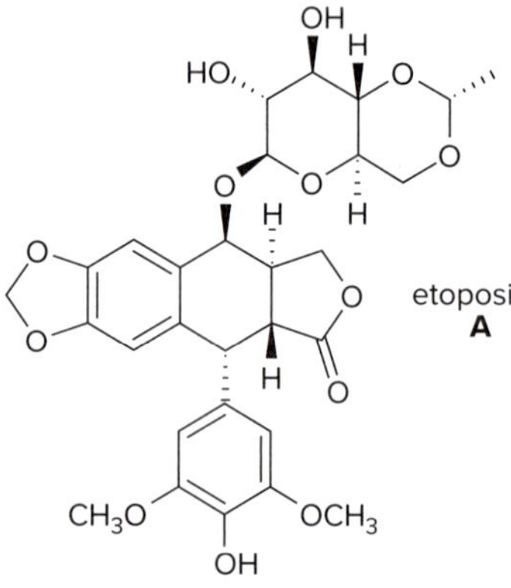

etoposide
A

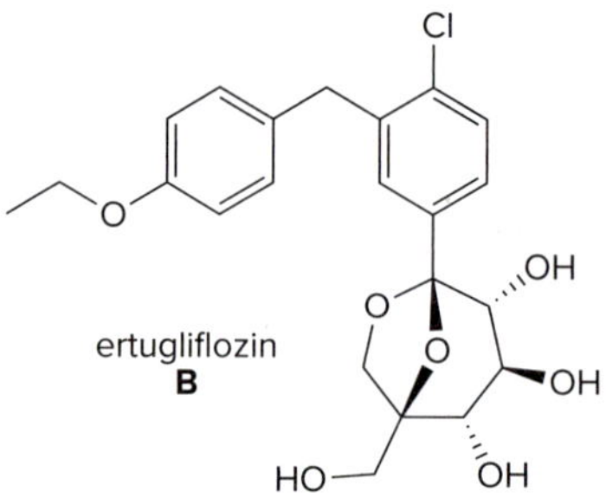

ertugliflozin
B

Properties of Aldehydes and Ketones

18.55 Consider carbonyl compounds **A–E** drawn below. (a) Rank **A–E** in order of increasing stability. (b) Rank **A–E** in order of increasing amount of hydrate formed when treated with aqueous acid. (c) Which compound is most reactive in nucleophilic addition? (d) From what you learned about the position of the carbonyl absorption in the IR in Sections B.3C and B.4B, which compound has a carbonyl absorption at lowest frequency?

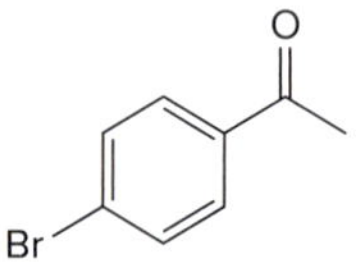

A 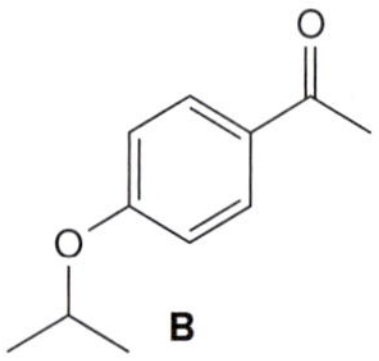 **B** 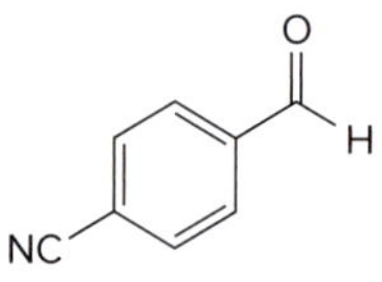 **C** 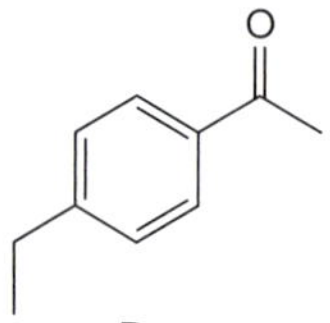 **D** 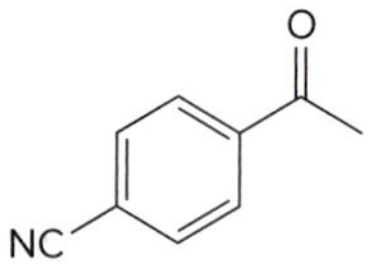 **E**

Synthesis

18.56 What Wittig reagent and carbonyl compound are needed to prepare each alkene? When two routes are possible, indicate which route, if any, is preferred.

a. 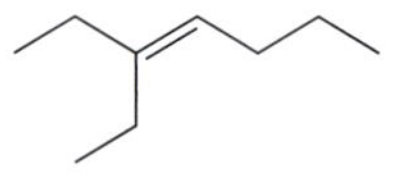 b. 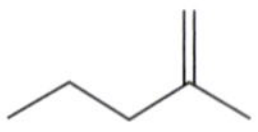 c.

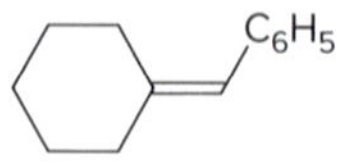

18.57 Itraconazole, sold under the trade names Sporanox, Sporaz, and others, is an antifungal drug first available in the United States in 1992 to treat a variety of fungal infections. What carbonyl compound and alcohol are needed to synthesize **A,** which was converted to itraconazole in several steps?

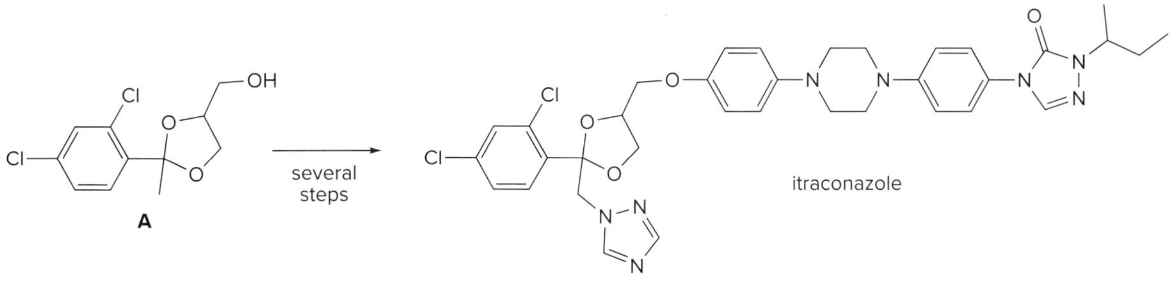

A → several steps → itraconazole

18.58 What carbonyl compound and amine or alcohol are needed to prepare each product?

a. 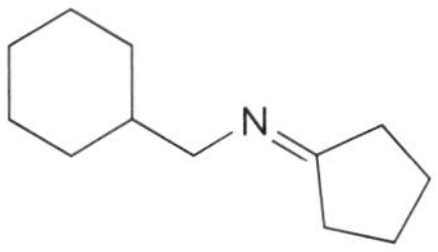 b. 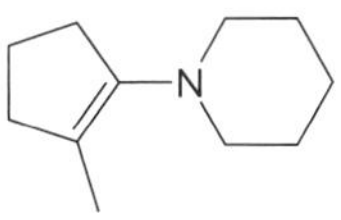 c. 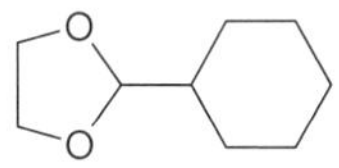 d.

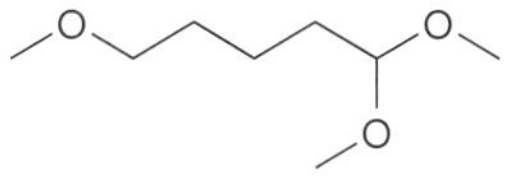

18.59 Show two different methods to carry out the following transformation: a one-step method using a Wittig reagent, and a two-step method using a Grignard reagent. Which route, if any, is preferred?

CHO →

18.60 Devise a synthesis of each alkene using a Wittig reaction to form the double bond. You may use benzene and organic alcohols having four or fewer carbons as starting materials and any required reagents.

a. 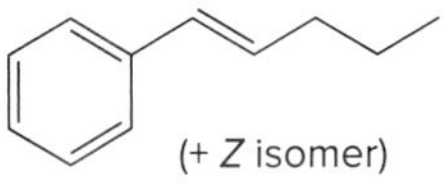 (+ *Z* isomer) b.

18.61 Devise a synthesis of **X,** an imine with antibacterial activity, from phenol and any needed organic compounds and reagents.

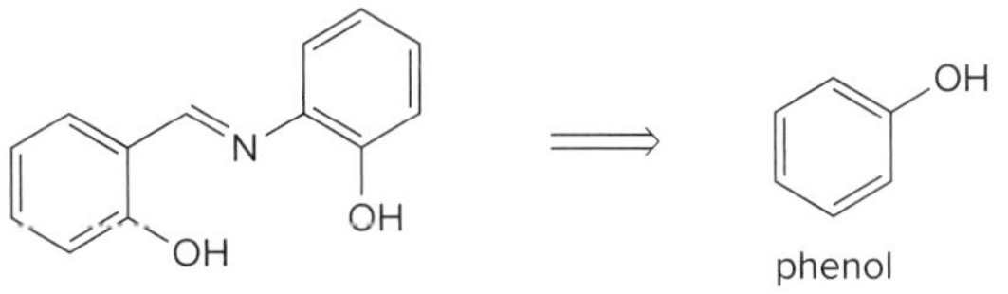

N-(salicylidene)-2-hydroxyaniline
X

⟹ phenol

18.62 Devise a synthesis of each acetal from 1-bromo-2-methylhexane, alcohols (and diols) containing one or two carbons, and any needed inorganic reagents.

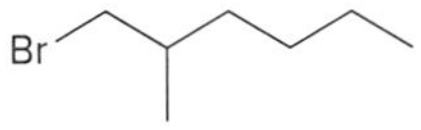

1-bromo-2-methylhexane

a. 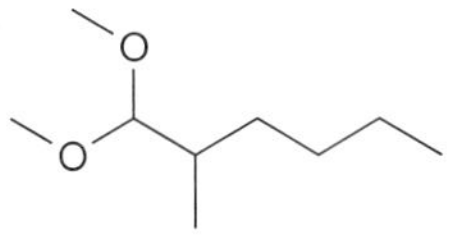 b. 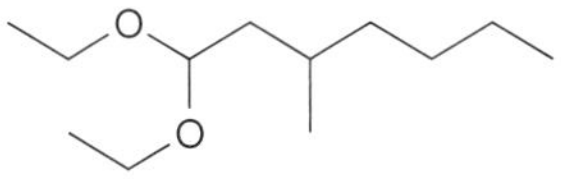 c.

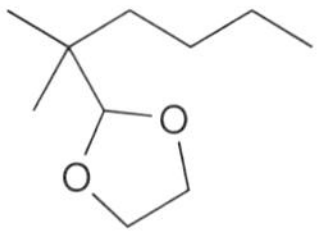

18.63 Devise a synthesis of each compound from cyclohexene and organic alcohols. You may use any other required organic or inorganic reagents.

a. 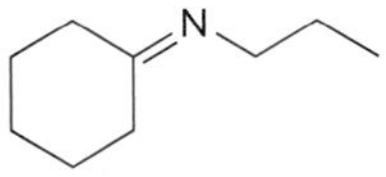 b.

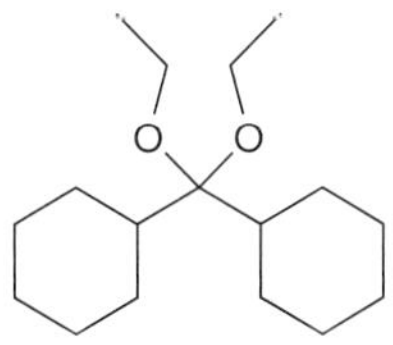

18.64 Devise a synthesis of each compound from the given starting materials. You may also use organic alcohols having four or fewer carbons, and any organic or inorganic reagents.

a. 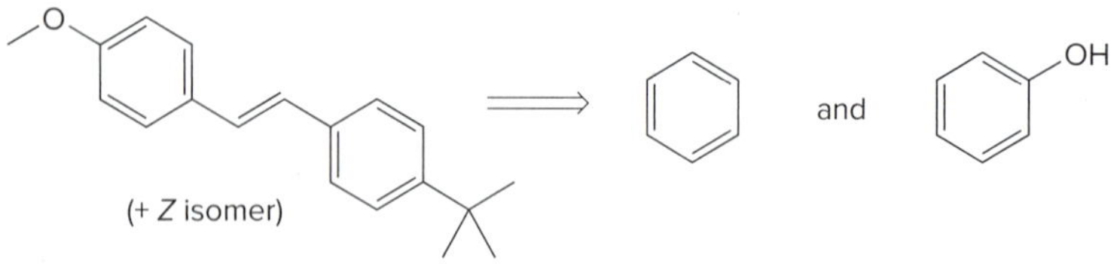 (+ *Z* isomer) ⟹ and OH

b. 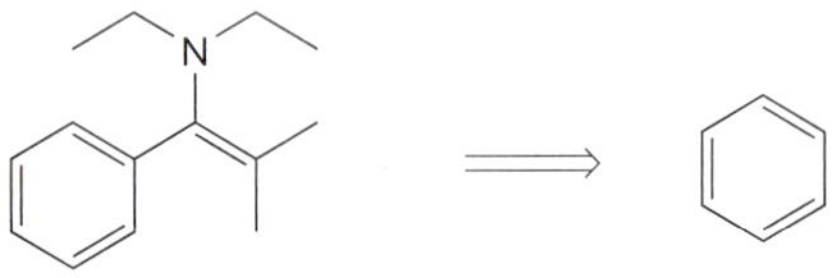 N ⟹

18.65 Devise a synthesis of each compound from ethanol (CH_3CH_2OH) as the only source of carbon atoms. You may use any other organic or inorganic reagents you choose.

a. 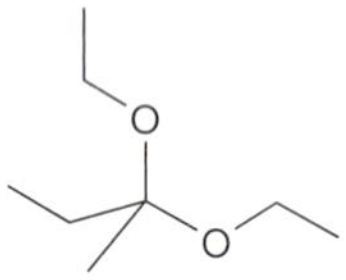 O O b. 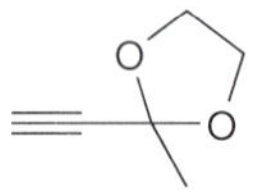 O O

Protecting Groups

18.66 Design a stepwise synthesis to convert cyclopentanone and 4-bromobutanal to hydroxy aldehyde **A.**

=O + Br CHO ⟶ OH CHO

cyclopentanone 4-bromobutanal **A**

18.67 Besides the *tert*-butyldimethylsilyl ethers introduced in Chapter 17, there are many other widely used alcohol protecting groups. For example, an alcohol such as cyclohexanol can be converted to a **meth**o**x**y **m**ethyl ether (a MOM protecting group) by treatment with base and chloromethyl methyl ether, $ClCH_2OCH_3$. The protecting group can be removed by treatment with aqueous acid.

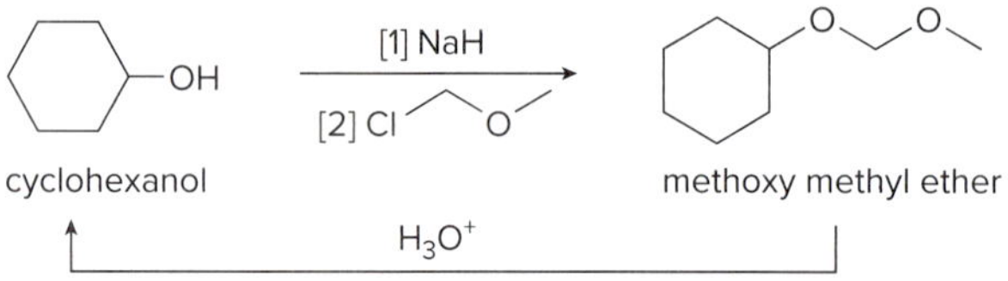

OH [1] NaH [2] Cl O ⟶ O O

cyclohexanol methoxy methyl ether

H_3O^+

a. Write a stepwise mechanism for the formation of a MOM ether from cyclohexanol.
b. What functional group comprises a MOM ether?
c. Besides cyclohexanol, what other products are formed by aqueous hydrolysis of the MOM ether? Draw a stepwise mechanism that accounts for formation of each product.

Mechanism

18.68 Draw a stepwise mechanism for the following reaction.

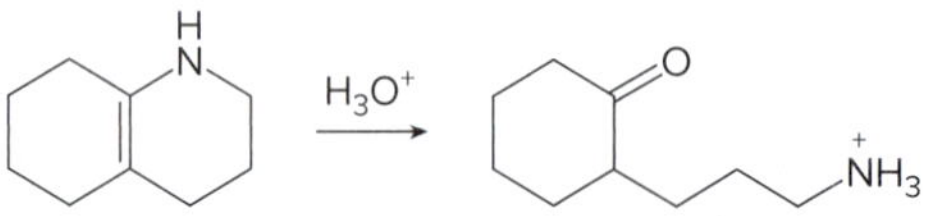

H N $\xrightarrow{H_3O^+}$ O $\overset{+}{N}H_3$

18.69 One acetal can be converted to a different acetal by reaction with a diol in the presence of acid, a process called transacetalization. Draw a stepwise mechanism for the following transacetalization.

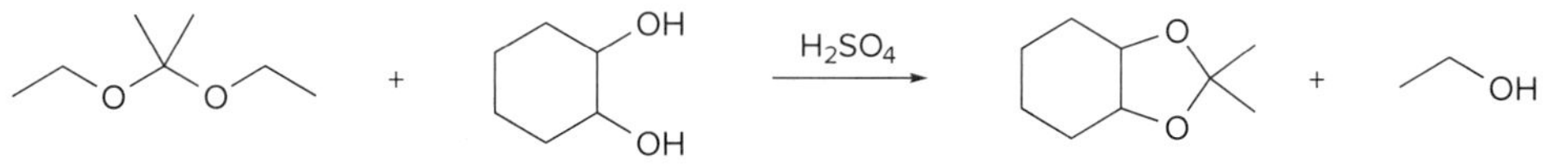

18.70 Draw a stepwise mechanism for the following reaction, a key step in the synthesis of the anti-inflammatory drug celecoxib (trade name Celebrex).

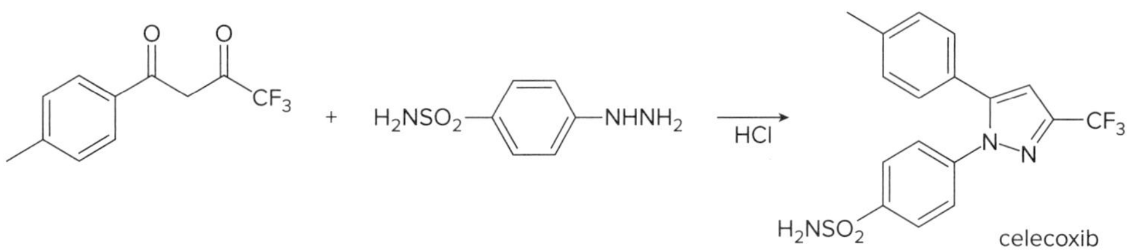

18.71 Treatment of $(HOCH_2CH_2CH_2CH_2)_2CO$ with acid forms a product of molecular formula $C_9H_{16}O_2$ and a molecule of water. Draw the structure of the product and explain how it is formed.

18.72 Draw a stepwise mechanism for the following reaction.

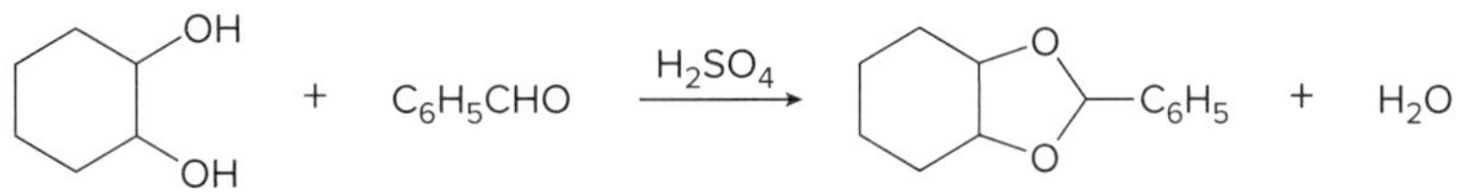

18.73 Draw a stepwise mechanism for the following reaction, a key step in the synthesis of ticlopidine, a drug that inhibits platelet aggregation. Ticlopidine has been used to reduce the risk of stroke in patients who cannot tolerate aspirin.

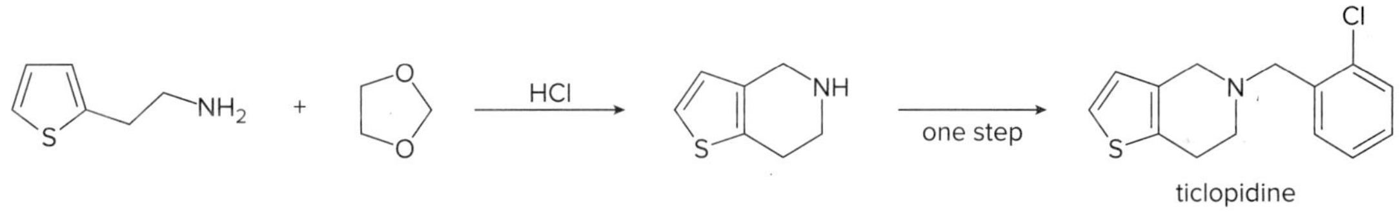

18.74 Explain how the following reaction occurs, even though the starting material is drawn with no carbonyl group.

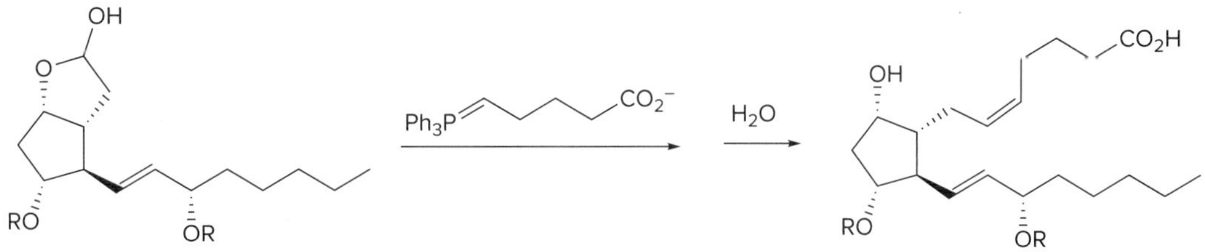

18.75 Salsolinol is a naturally occurring compound found in bananas, chocolate, and several foods derived from plant sources. Salsolinol is also formed in the body when acetaldehyde, an oxidation product of the ethanol ingested in an alcoholic beverage, reacts with dopamine, a neurotransmitter. Draw a stepwise mechanism for the formation of salsolinol in the following reaction.

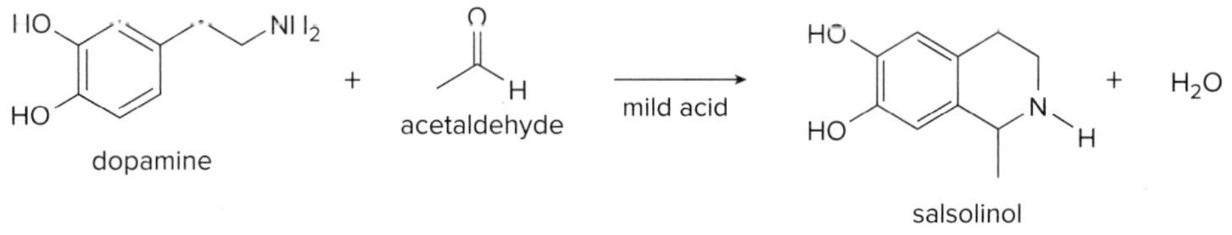

18.76 Reaction of 5,5-dimethoxypentan-2-one with methylmagnesium iodide followed by treatment with aqueous acid forms cyclic hemiacetal **Y.** Draw a stepwise mechanism that illustrates how **Y** is formed.

OCH_3 OCH_3 [1] CH_3MgI [2] H_3O^+ OH O

5,5-dimethoxypentan-2-one **Y**

Spectroscopy

18.77 Although the carbonyl absorption of cyclic ketones generally shifts to higher wavenumber with decreasing ring size (Section B.4B), the C=O of cyclopropenone absorbs at lower wavenumber in its IR spectrum than the C=O of cyclohex-2-enone. Explain this observation by using the principles of aromaticity learned in Chapter 15.

cyclopropenone ($1640\ cm^{-1}$) cyclohex-2-enone ($1685\ cm^{-1}$)

18.78 Use the ^1H NMR and IR data to determine the structure of each compound.

Compound **A**	Molecular formula:	$C_{10}H_{12}O$
	IR absorption at	$1686\ cm^{-1}$
	^1H NMR data:	1.21 (triplet, 3 H), 2.39 (singlet, 3 H), 2.95 (quartet, 2 H), 7.24 (doublet, 2 H), and 7.85 (doublet, 2 H) ppm
Compound **B**	Molecular formula:	$C_{10}H_{12}O$
	IR absorption at	$1719\ cm^{-1}$
	^1H NMR data:	1.02 (triplet, 3 H), 2.45 (quartet, 2 H), 3.67 (singlet, 2 H), and 7.06–7.48 (multiplet, 5 H) ppm

18.79 A solution of acetone [$(CH_3)_2C=O$] in ethanol (CH_3CH_2OH) in the presence of a trace of acid was allowed to stand for several days, and a new compound of molecular formula $C_7H_{16}O_2$ was formed. The IR spectrum showed only one major peak in the functional group region around $3000\ cm^{-1}$, and the ^1H NMR spectrum is given here. What is the structure of the product?

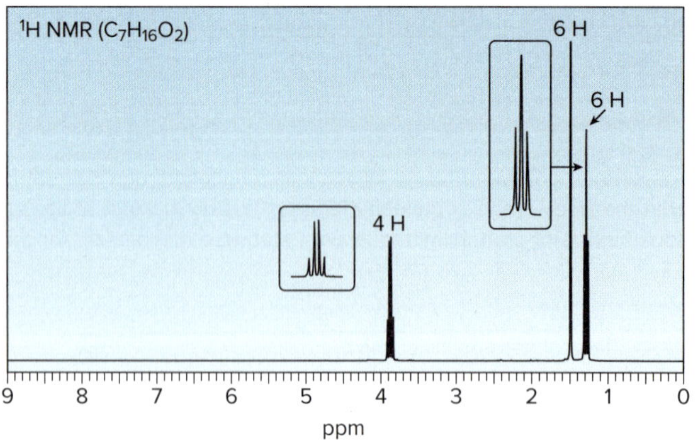

18.80 Identify the structure of compound **A** (molecular formula $C_9H_{10}O$) from the ^1H NMR and IR spectra given.

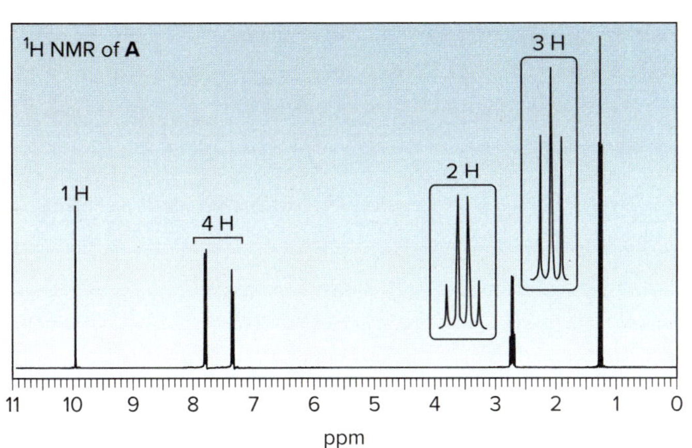

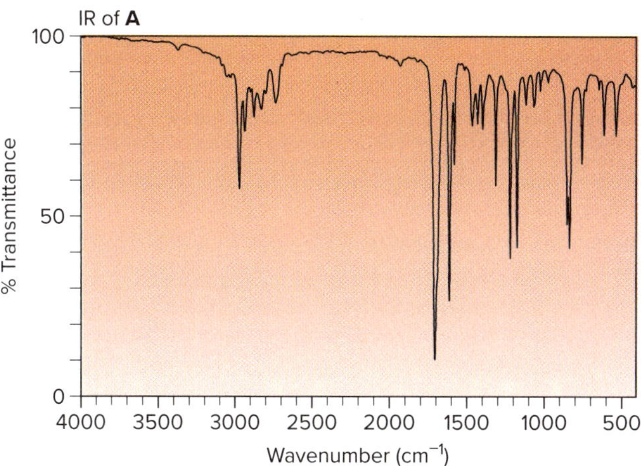

18.81 An unknown compound **C** of molecular formula $C_6H_{12}O_3$ exhibits a strong absorption in its IR spectrum at 1718 cm^{-1} and the given ^1H NMR spectrum. What is the structure of **C?**

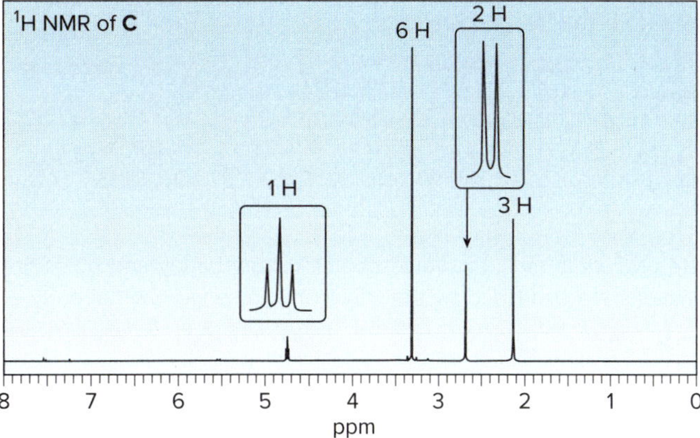

Carbohydrates

18.82 Draw the structure of the acyclic polyhydroxy aldehyde that cyclizes to each hemiacetal.

a. HO, HO, OH, OH, OH

b. HO, HO, HO, OH, OH

18.83 β-D-Glucose, a hemiacetal, can be converted to a mixture of acetals on treatment with CH_3OH in the presence of acid. Draw a stepwise mechanism for this reaction. Explain why two acetals are formed from a single starting material.

β-D-glucose $\xrightarrow{CH_3OH,\ HCl}$ (β acetal, OCH_3) + (α acetal, OCH_3) + H_2O

Challenge Problems

18.84 Draw a stepwise mechanism for the following isomerization.

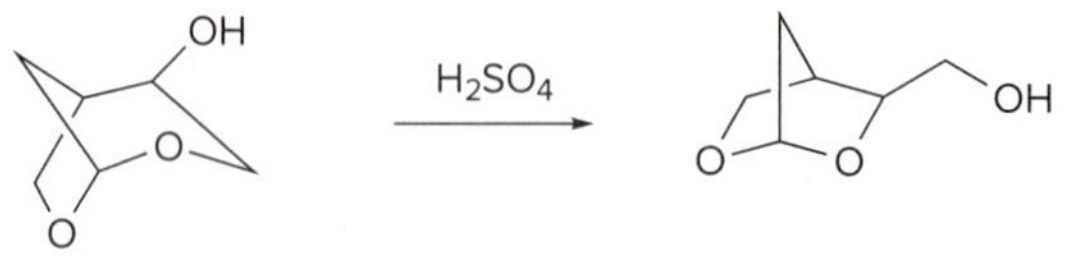

H_2SO_4

18.85 Devise a stepwise mechanism for the following reaction.

[1] BuLi
[2]
[3] *sec*-BuLi
[4] $CH_2{=}O$
[5] H_2O

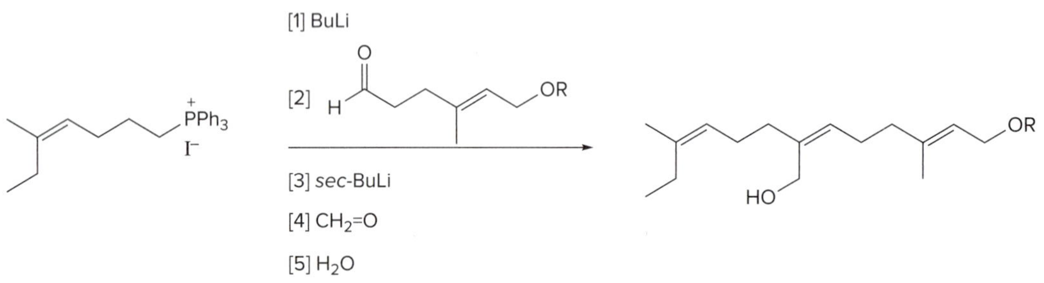

18.86 Draw a stepwise mechanism for the following reaction.

H^+

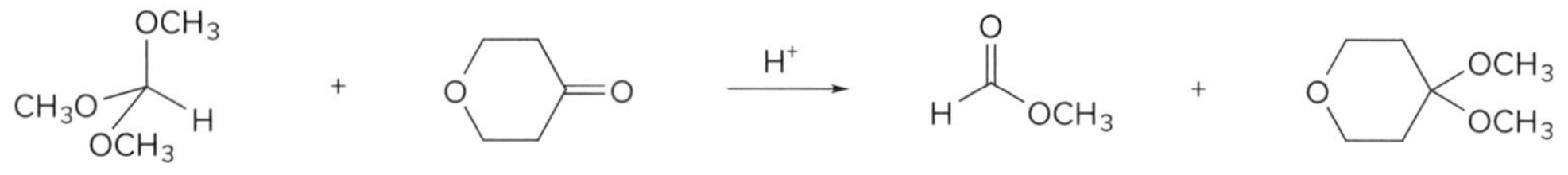

18.87 Maltose is a carbohydrate present in malt, the liquid obtained from barley and other grains. Although maltose has numerous functional groups, its reactions are explained by the same principles we have already encountered.

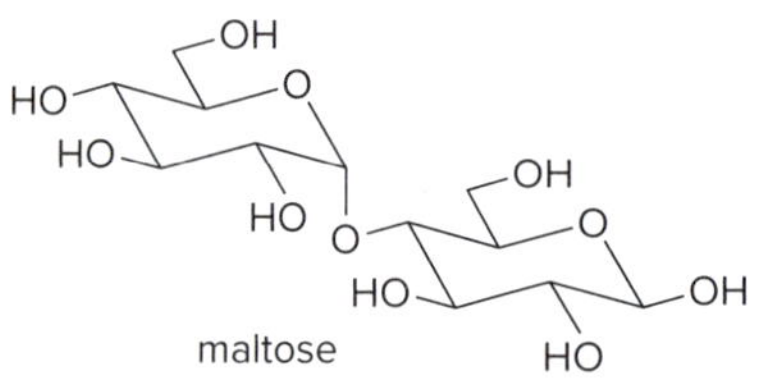

maltose

a. Label the acetal and hemiacetal carbons.
b. What products are formed when maltose is treated with each of these reagents: [1] H_3O^+; [2] CH_3OH and HCl; [3] excess NaH, then excess CH_3I?
c. Draw the products formed when the compound formed in Reaction [3] of part (b) is treated with aqueous acid.

The reactions in parts (b) and (c) are used to determine structural features of carbohydrates like maltose. We will learn much more about maltose and similar carbohydrates in Chapter 26.

18.88 Identify **R** and **S** in the following reaction sequence, and draw a mechanism for the conversion of **R** to **S** (molecular formula $C_6H_{10}O_3$). **S** was used in the synthesis of darunavir (trade name Prezista), used to treat HIV.

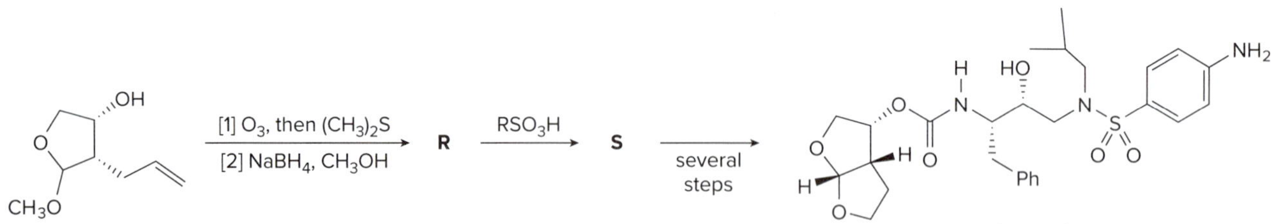

[1] O_3, then $(CH_3)_2S$
[2] $NaBH_4$, CH_3OH
→ **R** —RSO_3H→ **S** —several steps→ darunavir

18.89 Draw a stepwise mechanism for the following reaction, a key step in the synthesis of conivaptan (trade name Vaprisol), a drug used in the treatment of low sodium levels.

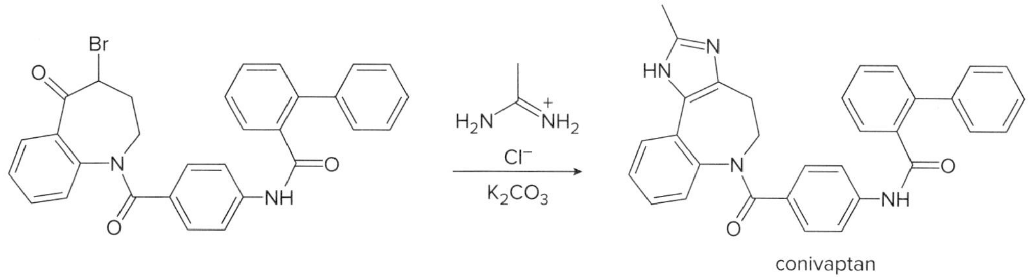

Br O N O NH O $\xrightarrow[\text{K}_2\text{CO}_3]{\text{H}_2\text{N}-\text{C}(\text{CH}_3)=\overset{+}{\text{N}}\text{H}_2\ \ \text{Cl}^-}$ HN N N O NH O

conivaptan

SELF-TEST ANSWERS

1. d 2. d 3. b 4. a 5. b 6. b 7. c 8. a 9. d 10. a

19 Carboxylic Acids and Nitriles

19.1 Structure and bonding
19.2 Nomenclature
19.3 Physical and spectroscopic properties
19.4 Interesting carboxylic acids and nitriles
19.5 Aspirin, arachidonic acid, and prostaglandins
19.6 Preparation of carboxylic acids
19.7 Carboxylic acids—Strong organic Brønsted–Lowry acids
19.8 Inductive effects in aliphatic carboxylic acids
19.9 Substituted benzoic acids
19.10 Extraction
19.11 Amino acids
19.12 Nitriles

Daniel C. Smith

Aspirin, **acetylsalicylic acid,** is one of the most widely used over-the-counter drugs. Although not naturally occurring, it is structurally similar to salicin in willow bark and salicylic acid in meadowsweet. Aspirin is an analgesic (pain-relieving), antipyretic (fever-reducing), and anti-inflammatory drug that is now also used as an antiplatelet agent to treat and prevent heart attacks and strokes. In Chapter 19, we will learn about carboxylic acids like aspirin, as well as the mechanism that explains how aspirin relieves pain and reduces inflammation.

Why Study . . .

Carboxylic Acids and Nitriles?

Chapter 19 concentrates on two classes of compounds, **carboxylic acids (RCO_2H)** and **nitriles (RCN).** With a polarized C=O and an acidic O–H bond, carboxylic acids undergo a variety of reactions. In this chapter we concentrate on one feature only—the **acidity of carboxylic acids.** Aspirin and naturally occurring fatty acids and prostaglandins are all carboxylic acids.

Nitriles are less common, but this useful functional group can be transformed into many other common functional groups. Moreover, several drugs that contain one or more cyano groups (C≡N) are used in the treatment of breast cancer and depression.

19.1 Structure and Bonding

The word ***carboxy*** (for a COOH group) is derived from ***carb***onyl (C=O) + hydr***oxy*** (OH).

***Carboxylic acids* are organic compounds containing a carboxy group (COOH).** Although the structure of a carboxylic acid is often abbreviated as **RCOOH** or **RCO_2H,** keep in mind that the central carbon atom of the functional group is doubly bonded to one oxygen atom and singly bonded to another.

carboxylic acid   carboxy group

The carbon atom of a carboxy group is surrounded by three groups, making it ***sp^2* hybridized** and **trigonal planar,** with bond angles of approximately 120°. The C=O of a carboxylic acid is *shorter* than its C–O. Because oxygen is more electronegative than either carbon or hydrogen, **the C–O and O–H bonds are polar.**

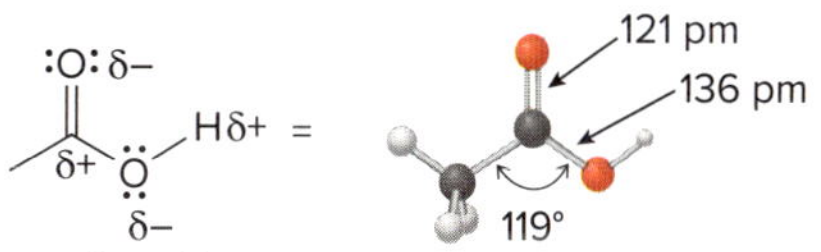

acetic acid

121 pm
136 pm
119°

***Nitriles* are compounds that contain a cyano group, C≡N, bonded to an alkyl group.** Nitriles have no carbonyl group, so they are structurally distinct from carboxylic acids. The carbon atom of the cyano group, however, has the same oxidation state as the carbonyl carbon of a carboxylic acid, so there are certain parallels in their chemistry.

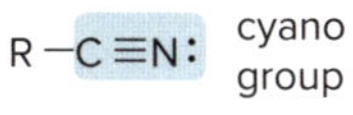

R–C≡N: cyano group

nitrile

Each labeled C has three bonds to a more electronegative element.

The structure and bonding in nitriles is very different from that in carboxylic acids, and it resembles the carbon–carbon triple bond of alkynes. Unlike alkynes, however, **nitriles contain an electrophilic carbon atom,** making them susceptible to nucleophilic attack.

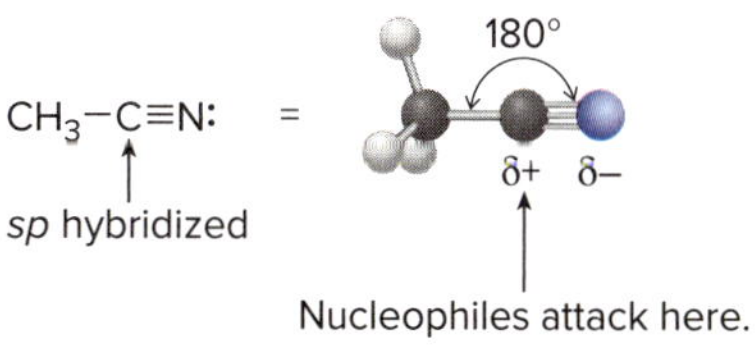

CH_3–C≡N:

sp hybridized

180°

δ+ δ–

Nucleophiles attack here.

- **The carbon atom of the C≡N group is *sp* hybridized, making it linear with a bond angle of 180°.**
- **The triple bond consists of one σ and two π bonds.**

19.2 Nomenclature

Both IUPAC and common names are used for carboxylic acids and nitriles.

19.2A Naming Carboxylic Acids

In IUPAC nomenclature, carboxylic acids are identified by a suffix added to the parent name of the longest chain, and two different endings are used depending on whether the carboxy group is bonded to a chain or a ring.

To name a carboxylic acid using the IUPAC system:

[1] If the COOH is bonded to a *chain* of carbons, find the longest chain containing the COOH group, and change the ***-e*** ending of the parent alkane to the suffix ***-oic acid.*** If the COOH group is bonded to a *ring,* name the ring and add the words ***carboxylic acid.***

[2] Number the carbon chain or ring to put the **COOH group at C1,** but omit this number from the name. Apply all the other usual rules of nomenclature.

Compounds that contain both a C=C and a CO_2H are named as **enoic acids,** and the chain is numbered to put the CO_2H group at C1.

Sample Problem 19.1 Naming a Carboxylic Acid Using the IUPAC System

Give the IUPAC name of each compound.

a. 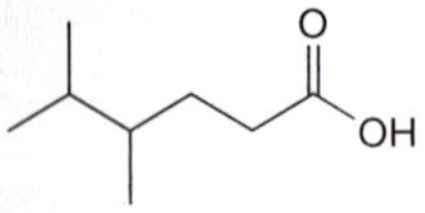 O, OH

b. 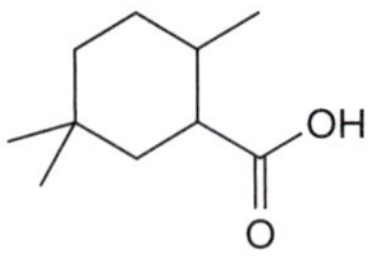 OH, O

Solution

a. [1] Find and name the longest chain containing COOH:

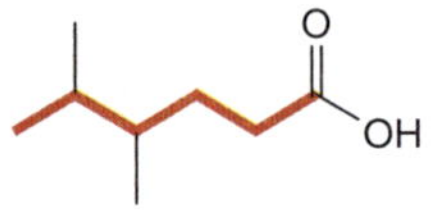

O, OH

hexane ⟶ **hexan*oic acid***
(6 C's)

The COOH contributes one C to the longest chain.

[2] Number and name the substituents:

O, 5, 4, 1, OH

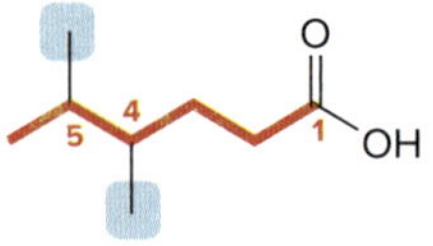

two methyl substituents on C4 and C5

Answer: 4,5-dimethylhexanoic acid

b. [1] Find and name the ring bonded to COOH.

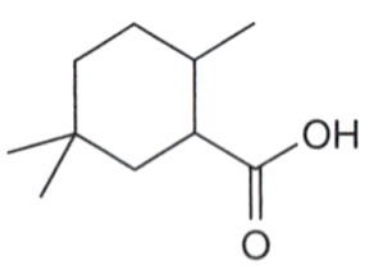

OH, O

cyclohexane + carboxylic acid
(6 C's)

[2] Number and name the substituents:

2, 5, 1, OH, O

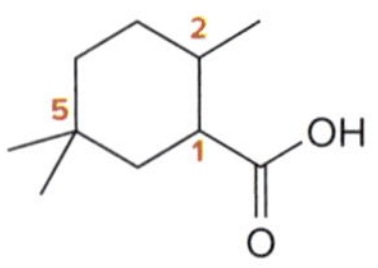

Number to put COOH at C1 and give the second substituent (CH_3) the lower number (C2).

Answer: 2,5,5-trimethylcyclohexanecarboxylic acid

Problem 19.1 Give the IUPAC name for each compound.

a. 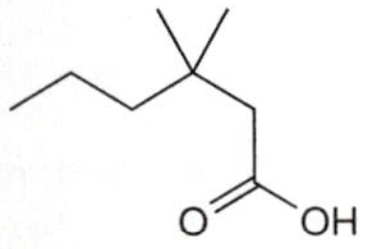 O, OH

b. 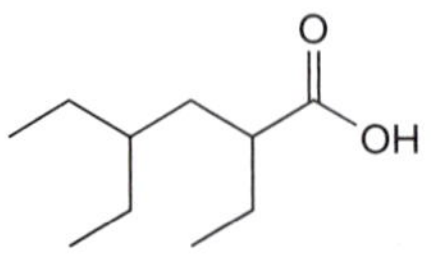 O, OH

c. 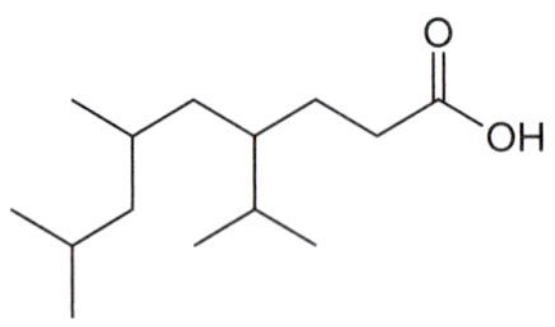 O, OH

More Practice: Try Problems 19.33a; 19.35a, c, d, f, i; 19.36a–d.

Problem 19.2 Name each carboxylic acid, two volatile compounds that contribute to underarm odor.

a. b.

Problem 19.3 Give the structure corresponding to each IUPAC name.

a. 2-bromobutanoic acid
b. 2,3-dimethylpentanoic acid
c. 3,3,4-trimethylheptanoic acid
d. 2-*sec*-butyl-4,4-diethylnonanoic acid
e. 3,4-diethylcyclohexanecarboxylic acid
f. 1-isopropylcyclobutanecarboxylic acid

Most simple carboxylic acids have common names that are more widely used than their IUPAC names.

- **A common name is formed by using a common parent name followed by the suffix *-ic acid.***

The common parent names for simple carboxylic acids are similar to those used for aldehydes (Table 18.1). The common names formic acid, acetic acid, and benzoic acid are virtually always used instead of their IUPAC names.

formic acid (methanoic acid)

acetic acid (ethanoic acid)

benzoic acid (benzenecarboxylic acid)

Problem 19.4 Draw the structure corresponding to each common name:

a. α-methoxyvaleric acid
b. β-phenylpropionic acid
c. α,β-dimethylcaproic acid
d. α-chloro-β-methylbutyric acid

19.2B Naming Dicarboxylic Acids and Carboxylates

Many compounds containing two carboxy groups are also known. In the IUPAC system, **diacids** are named by adding the suffix ***-dioic acid*** to the name of the parent alkane. The three simplest diacids are most often identified by their common names, as shown.

oxalic acid (ethanedioic acid)

malonic acid (propanedioic acid)

succinic acid (butanedioic acid)

Metal salts of carboxylate anions are formed from carboxylic acids in many reactions in Chapter 19. To name the **metal salt of a carboxylate anion,** change the *-ic acid* ending of the carboxylic acid to the suffix ***-ate*** and put three parts together:

name of the metal cation + parent + suffix

parent: common or IUPAC; suffix: ***-ate***

Two examples are shown in Figure 19.1.

Figure 19.1 Naming the metal salts of carboxylate anions

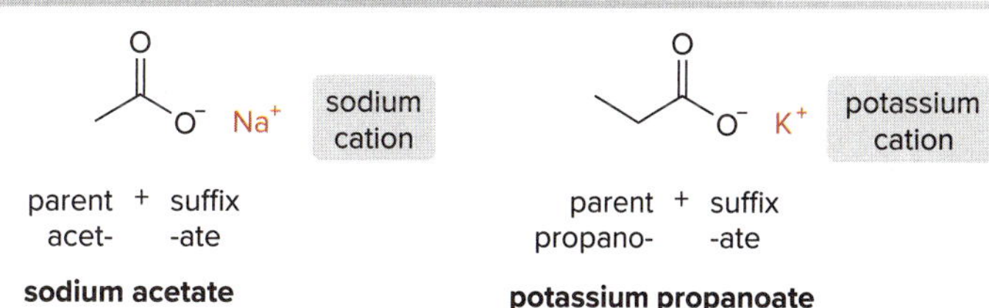

Problem 19.5 Give the IUPAC name for each compound.

a. O, O^- Li^+

b. CO_2H, CO_2H

c. O, O^- K^+

d. O, O^- Na^+, Br

Problem 19.6 Depakote, a drug used to treat seizures and bipolar disorder, consists of a mixture of valproic acid $[(CH_3CH_2CH_2)_2CHCO_2H]$ and its sodium salt. Give IUPAC names for each of these compounds.

19.2C Naming Nitriles

In contrast to the carboxylic acids, **nitriles are named as alkane derivatives.** To name a nitrile using IUPAC rules:

In naming a nitrile, the CN carbon is one carbon atom of the longest chain. CH_3CH_2CN is propanenitrile, *not* ethanenitrile.

- **Find the longest chain that contains the CN and add the word *nitrile* to the name of the parent alkane. Number the chain to put CN at C1, but omit this number from the name.**

Common names for nitriles are derived from the names of the carboxylic acid having the same number of carbon atoms by replacing the ***-ic acid*** ending of the carboxylic acid by the suffix ***-onitrile.***

When CN is named as a substituent, it is called a ***cyano*** group. Figure 19.2 illustrates features of nitrile nomenclature.

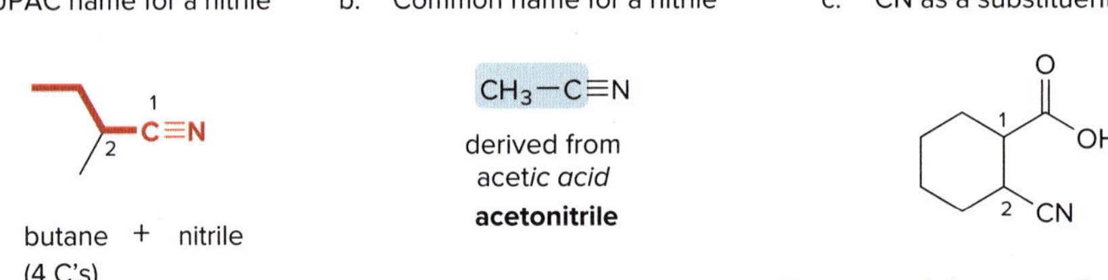

Figure 19.2 Summary of nitrile nomenclature

Sample Problem 19.2 Naming a Nitrile Using the IUPAC System

Give the IUPAC name for the following nitrile.

CN

Solution

[1] Find and name the longest chain containing the CN.

CN

hexane + nitrile
(6 C's)

The CN contributes one C to the longest chain.

[2] Number and name the substituents.

1 CN, 2, 4, 5

methyls at C4 and C5
propyl at C2

Answer: 4,5-dimethyl-2-propylhexanenitrile

Problem 19.7 Give the IUPAC name for each nitrile.

a. b. c.

More Practice: Try Problems 19.34a; 19.35g, h; 19.36i, j.

19.3 Physical and Spectroscopic Properties

19.3A Physical Properties

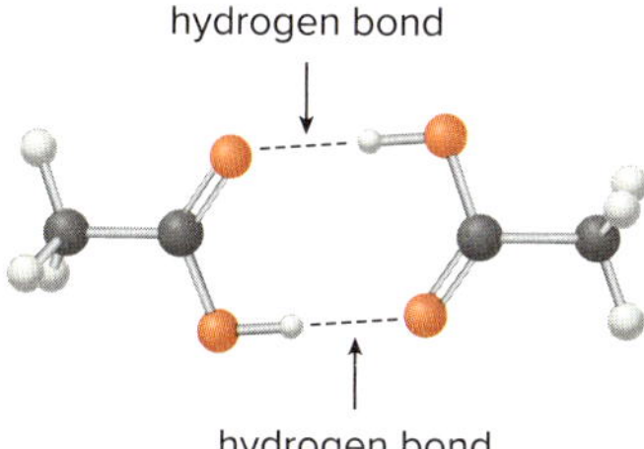

Carboxylic acids and nitriles exhibit **dipole–dipole** interactions because they have polar C–O, C–N, and O–H bonds. Carboxylic acids also exhibit intermolecular **hydrogen bonding** because they possess a hydrogen atom bonded to an electronegative oxygen atom. Carboxylic acids often exist as **dimers,** held together by *two* intermolecular hydrogen bonds between the carbonyl oxygen atom of one molecule and the OH hydrogen atom of another molecule. Carboxylic acids are the **most polar** organic compounds we have studied so far.

As a result, carboxylic acids have higher boiling points and melting points than other neutral organic compounds of comparable molecular weight.

Fatty acids (Section 10.6) are water-insoluble carboxylic acids with > 5 carbons.

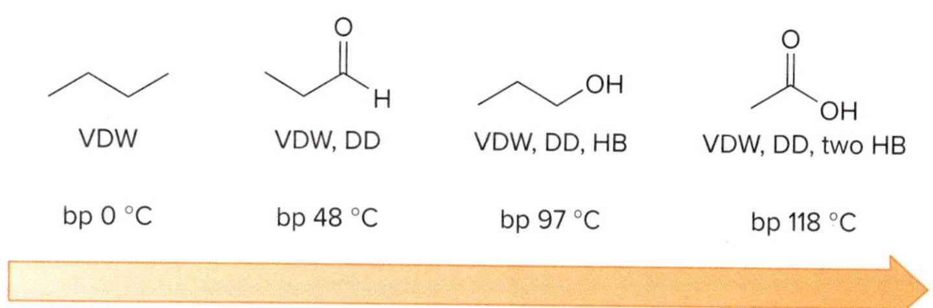

19.3B Spectroscopic Properties

Many details of the spectroscopy of carboxylic acids and nitriles have been presented in Spectroscopy Parts B and C:

- The infrared absorptions of carboxylic acids: Section B.4B, Figure B.9, and Table B.2
- The infrared absorption of nitriles: Section B.4C, Figure B.10, and Table B.2
- ^{1}H and ^{13}C NMR absorptions: Tables C.1 and C.5

Key NMR and IR absorptions for carboxylic acids and nitriles are summarized in Table 19.1, and Figure 19.3 illustrates ^{1}H and ^{13}C NMR spectra for a simple carboxylic acid.

Problem 19.8 Explain how you could use IR spectroscopy to distinguish among the following three compounds.

A **B** **C**

Table 19.1 Characteristic Spectroscopic Absorptions of Carboxylic Acids and Nitriles

Compound	Type of spectroscopy	Type of C, H	Absorption
Carboxylic acid	**IR absorptions**	RCO**OH**	2500–3500 cm^{-1} (very broad, strong)
		RC(=**O**)OH	1710 cm^{-1} (strong)
	^1H NMR absorptions	RCO**OH**	10–12 ppm
		HOOC–C**H**R	2–2.5 ppm
	^{13}C NMR absorption	RC(=**O**)OH	170–210 ppm
Nitrile	**IR absorption**	—**C≡N**	2250 cm^{-1}
	^{13}C NMR absorption	—**C**≡N	115–120 ppm

Figure 19.3 The ^1H and ^{13}C NMR spectra of propanoic acid

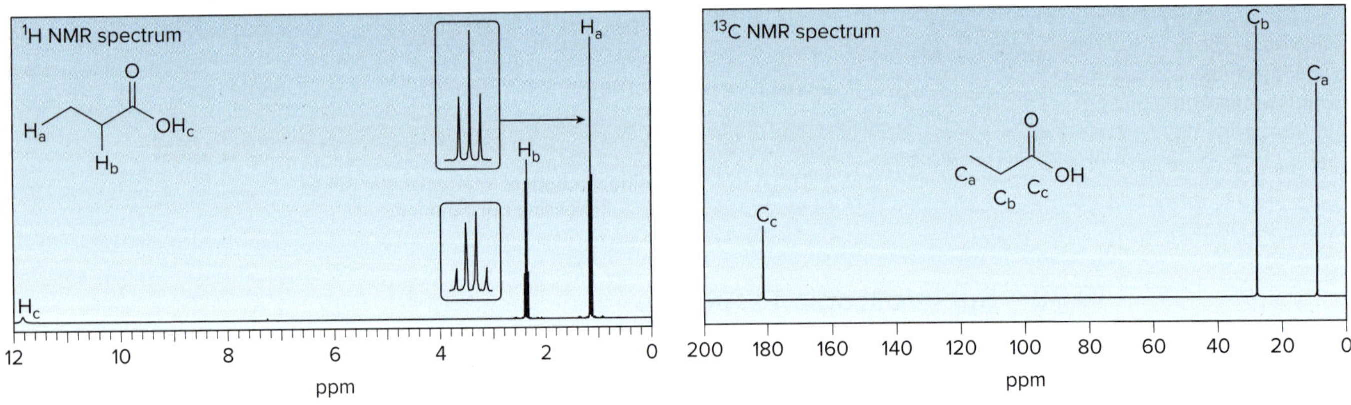

- **^1H NMR spectrum:** There are three signals due to three different kinds of H atoms. The H_a and H_b signals are split into a triplet and quartet, respectively. The H_c signal, a singlet, is due to the highly deshielded OH proton.
- **^{13}C NMR spectrum:** There are three signals due to three different kinds of carbon atoms. The carbonyl carbon is highly deshielded.

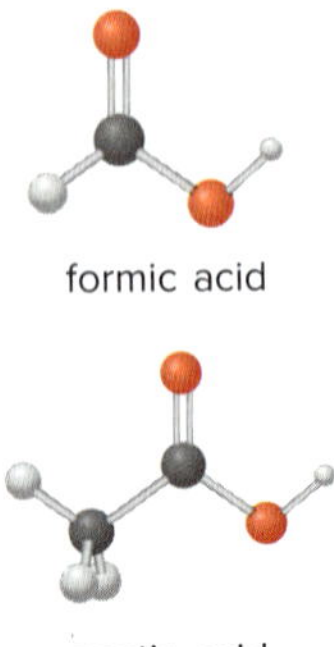

19.4 Interesting Carboxylic Acids and Nitriles

Several simple carboxylic acids have characteristic odors and flavors.

Formic acid (HCOOH), a carboxylic acid with an acrid odor and a biting taste, is responsible for the sting of some types of ants. The name is derived from the Latin word *formica,* meaning "ant."

Acetic acid (CH_3COOH) is the sour-tasting component of vinegar. The name comes from the Latin word *acetum,* meaning "vinegar." The air oxidation of ethanol to acetic acid is the process that makes "bad" wine taste sour. Acetic acid is an industrial starting material for polymers used in paints and adhesives. Pure acetic acid is often called *glacial* acetic acid because it freezes just below room temperature (mp = 17 °C), forming white crystals reminiscent of the ice in a glacier.

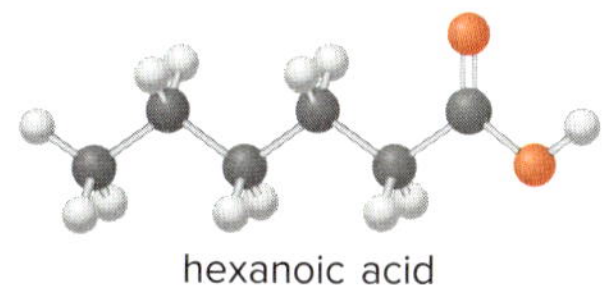
hexanoic acid

Hexanoic acid [$CH_3(CH_2)_4COOH$] is a low-molecular-weight carboxylic acid with the foul odor of dirty socks and locker rooms. Its common name, caproic acid, is derived from the Latin word *caper,* meaning "goat." The fleshy coat of seeds that are produced by female ginkgo trees contains hexanoic acid, giving the seeds an unpleasant and even repulsive odor.

Female ginkgo trees produce seeds with an unpleasant odor due to the presence of hexanoic acid. *Dr. Steven J. Wolf*

Oxalic acid and **lactic acid** are simple carboxylic acids quite prevalent in nature. Oxalic acid occurs naturally in spinach and rhubarb. Lactic acid gives sour milk its distinctive taste.

oxalic acid

lactic acid

Although oxalic acid is toxic, you would have to eat about 9 pounds of spinach at one time to ingest a fatal dose. *Katarzyna Bialasiewicz/123RF*

Salts of carboxylic acids are commonly used as preservatives. Sodium benzoate, a fungal growth inhibitor, is a preservative used in soft drinks, and potassium sorbate is an additive that prolongs the shelf life of baked goods and other foods.

Soaps, the sodium salts of fatty acids, were discussed in Section 3.6.

sodium benzoate

potassium sorbate

Although nitriles are much less common than carboxylic acids, the naturally occurring cyanohydrin derivatives discussed in Section 18.8 constitute one group of compounds that contain a nitrile. In addition, several widely used drugs contain one or more cyano groups, including anastrozole, used to reduce the recurrence of breast cancer in women whose tumors are estrogen positive, and escitalopram, used to treat depression and anxiety. PF-07321332 (generic name nirmatrelvir, Problem 3.60) is an experimental oral antiviral drug developed by the pharmaceutical company Pfizer that has been shown to reduce the risk of hospitalization and death from COVID-19 in clinical trials.

Generic name: anastrozole
Trade name: Arimidex

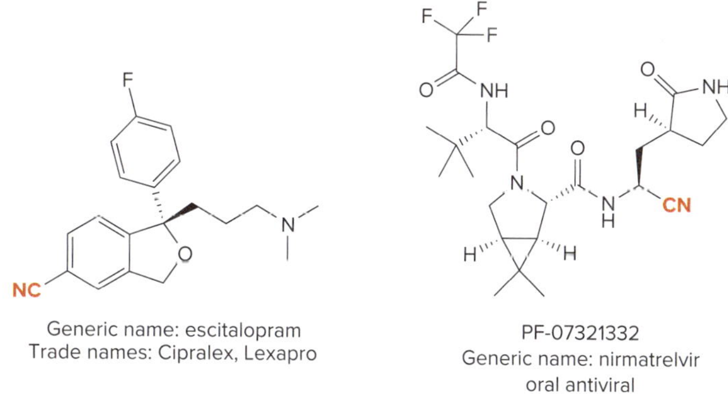

Generic name: escitalopram
Trade names: Cipralex, Lexapro

PF-07321332
Generic name: nirmatrelvir
oral antiviral

Anastrozole is called an **aromatase inhibitor** because it blocks the activity of the aromatase enzyme, which is responsible for estrogen synthesis. This inhibits tumor growth in those forms of breast cancer that are stimulated by estrogen.

19.5 Aspirin, Arachidonic Acid, and Prostaglandins

As mentioned in the chapter opener, **aspirin (acetylsalicylic acid)** is a synthetic carboxylic acid, similar in structure to **salicin,** a naturally occurring compound isolated from willow bark, and **salicylic acid,** found in meadowsweet.

aspirin
(acetylsalicylic acid)

salicin
(isolated from
willow bark)

salicylic acid
(isolated from
meadowsweet)

sodium salicylate
(sweet carboxylate salt)

Both salicylic acid and sodium salicylate (its sodium salt) were widely used analgesics in the nineteenth century, but both had undesirable side effects. Salicylic acid irritated the mucous membranes of the mouth and stomach, and sodium salicylate was too sweet for most patients. Aspirin, a synthetic compound, was first sold in 1899 after Felix Hoffmann, a German chemist at Bayer Company, developed a feasible commercial synthesis. Hoffmann's work was motivated by personal reasons: his father suffered from rheumatoid arthritis and was unable to tolerate the sweet taste of sodium salicylate.

The word *aspirin* is derived from the prefix ***a***- for *acetyl* + ***spir*** from the Latin name *spirea* for the meadowsweet plant.

How does aspirin relieve pain and reduce inflammation? Aspirin blocks the synthesis of **prostaglandins,** 20-carbon fatty acids with a five-membered ring that are responsible for pain, inflammation, and a wide variety of other biological functions. **$PGF_{2\alpha}$** contains the typical carbon skeleton of a prostaglandin.

Aspirin is the most widely used pain reliever and anti-inflammatory agent in the world, yet its mechanism of action remained unknown until the 1970s. John Vane, Bengt Samuelsson, and Sune Bergstrom shared the 1982 Nobel Prize in Physiology or Medicine for unraveling the details of its mechanism.

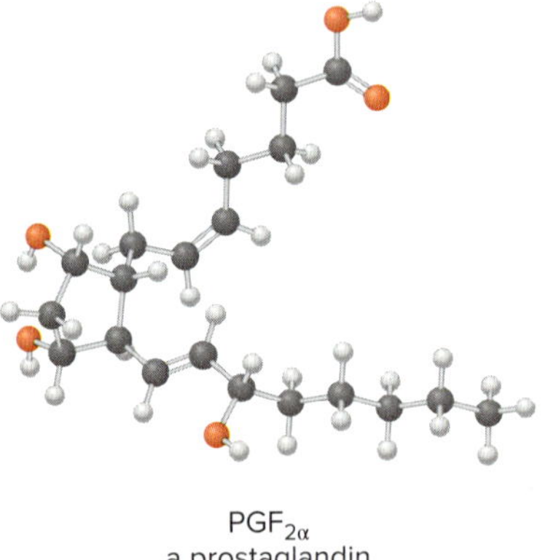

$PGF_{2\alpha}$
a prostaglandin

Prostaglandins are not stored in cells. Rather, they are synthesized from arachidonic acid, a polyunsaturated fatty acid having four cis double bonds. Unlike hormones, which are transported in the bloodstream to their sites of action, prostaglandins act where they are synthesized. Aspirin acts by blocking the synthesis of prostaglandins from arachidonic acid. Aspirin inactivates cyclooxygenase, an enzyme that converts arachidonic acid to PGG_2, an unstable precursor of $PGF_{2\alpha}$ and other prostaglandins. **Aspirin lessens pain and decreases inflammation because it prevents the synthesis of prostaglandins, the compounds responsible for both of these physiological responses.**

arachidonic acid → (cyclo-oxygenase) → [PGG_2 unstable intermediate] → $PGF_{2\alpha}$ and other prostaglandins

Although prostaglandins have a wide range of biological activity, their inherent instability often limits their usefulness as drugs. Consequently, more-stable analogues with useful medicinal properties have been synthesized. For example, latanoprost (trade name Xalatan) and bimatoprost (trade name Lumigan) are prostaglandin analogues used to reduce eye pressure in individuals with glaucoma.

latanoprost

bimatoprost

Problem 19.9 How many tetrahedral stereogenic centers does $PGF_{2\alpha}$ contain? Draw its enantiomer. How many of its double bonds can exhibit cis-trans isomerism? Considering both its double bonds and its tetrahedral stereogenic centers, how many stereoisomers are possible for $PGF_{2\alpha}$?

19.6 Preparation of Carboxylic Acids

We begin our study of the reactions involving carboxylic acids and nitriles by summarizing methods that introduce a carboxy group presented in earlier chapters. In Sections 19.7–19.11, we then concentrate on the acidity of carboxylic acids, and in Section 19.12, we examine the preparation and reactions of nitriles.

Where have we encountered carboxylic acids as reaction products before? The carbonyl carbon is highly oxidized because it has three C–O bonds, so **carboxylic acids are often prepared by oxidation reactions.** Four oxidation methods and one carbon–carbon bond-forming reaction are listed in Table 19.2.

Table 19.2 Methods That Synthesize Carboxylic Acids

Method	Reaction
[1] **Oxidation of 1° alcohols** (Section 12.12B)	RCH_2OH (1° alcohol) $\xrightarrow[H_2SO_4,\ H_2O]{CrO_3}$ RCOOH
[2] **Oxidation of aldehydes** (Section 17.8)	RCHO (aldehyde) $\xrightarrow{CrO_3,\ H_2SO_4,\ H_2O \text{ or } Ag_2O,\ NH_4OH}$ RCOOH
[3] **Carboxylation of Grignard reagents** (Section 17.14)	R–MgX (Grignard reagent) $\xrightarrow[[2]\ H_3O^+]{[1]\ CO_2}$ RCOOH
[4] **Oxidation of alkyl benzenes** (Section 16.14B)	$C_6H_5CHR'_2$ (alkyl benzene, R' = H or alkyl) $\xrightarrow{KMnO_4}$ C_6H_5COOH
[5] **Oxidative cleavage of alkynes** (Section 12.11)	R–C≡C–R' (internal alkyne) $\xrightarrow[[2]\ H_2O]{[1]\ O_3}$ RCOOH + R'COOH R–C≡C–H (terminal alkyne) $\xrightarrow[[2]\ H_2O]{[1]\ O_3}$ RCOOH + CO_2

Problem 19.10 What alcohol can be oxidized to each carboxylic acid?

a. $CH_3CH_2CH_2CH_2CH_2COOH$ b. $(CH_3)_2CHCH_2COOH$ c. $C_6H_{11}CH_2COOH$ (cyclohexylacetic acid)

Problem 19.11 Identify **A–F** in the following reactions.

a. **A** —(CrO_3 / H_2SO_4, H_2O)→ benzoic acid (OH)

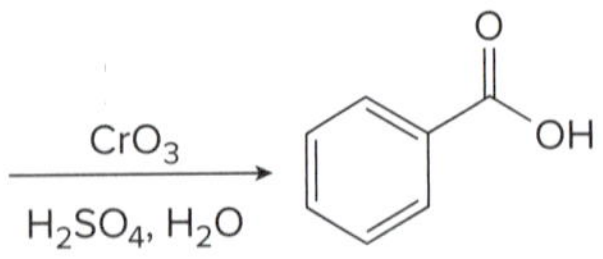

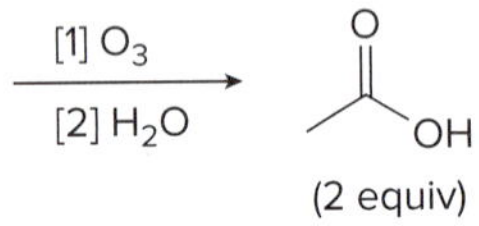

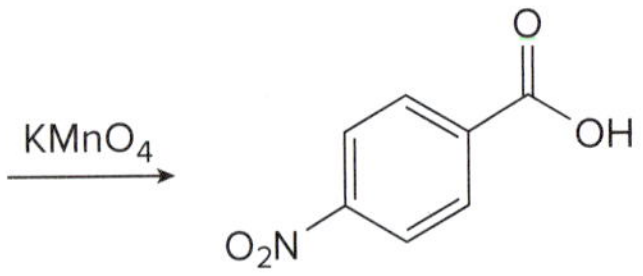

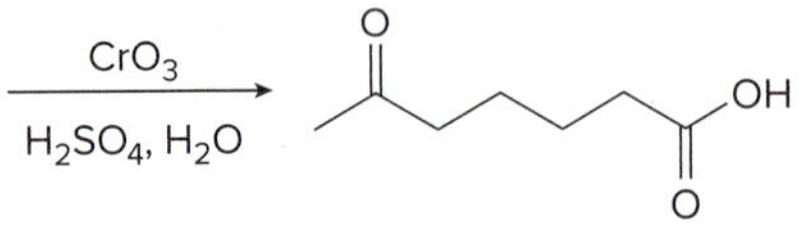

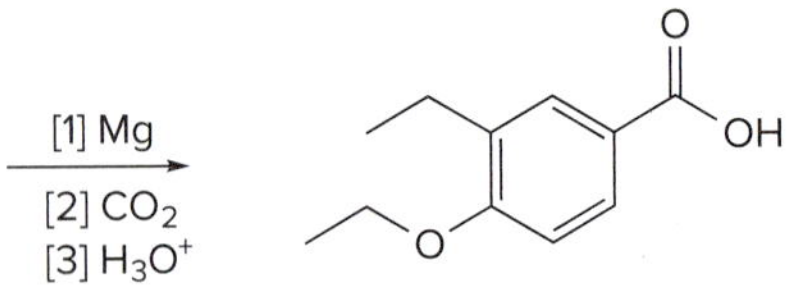

b. **B** —([1] O_3 / [2] H_2O)→ acetic acid (OH) (2 equiv)

c. **C** —($KMnO_4$)→ O_2N-substituted benzoic acid (OH)

d. **D** —(CrO_3 / H_2SO_4, H_2O)→ keto acid (O, OH, O)

e. **E** —([1] Mg / [2] CO_2 / [3] H_3O^+)→ substituted benzoic acid (O, OH, O)

f. **F** —(Ag_2O, NH_4OH)→ product (O, OH, O, OH)

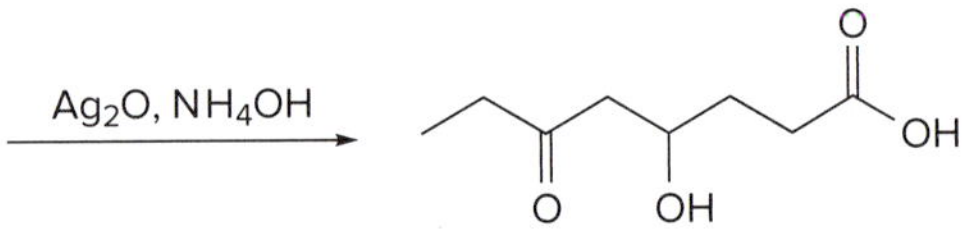

19.7 Carboxylic Acids—Strong Organic Brønsted–Lowry Acids

The polar C–O and O–H bonds, nonbonded electron pairs on oxygen, and the π bond give a carboxylic acid many reactive sites, complicating its chemistry somewhat. By far, **the most important reactive feature of a carboxylic acid is its polar O–H bond, which is readily cleaved with base.**

- **Carboxylic acids are strong organic acids,** and as such, readily react with Brønsted–Lowry bases to form carboxylate anions.

Recall from Section 2.3 that **the lower the pK_a, the stronger the acid.**

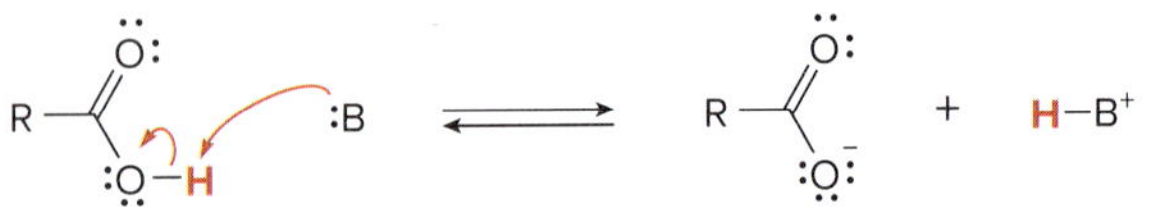

RCOOH + :B ⇌ RCOO⁻ (**carboxylate anion**) + $H{-}B^+$

What bases are used to deprotonate a carboxylic acid? As we learned in Section 2.3, **equilibrium favors the products of an acid–base reaction when the weaker base and acid are formed.** Because a weaker acid has a higher pK_a, this general rule results:

- **An acid can be deprotonated by a base that has a conjugate acid with a *higher* pK_a.**

Because the pK_a values of many carboxylic acids are ~5, bases that have conjugate acids with pK_a values *higher* than 5 are strong enough to deprotonate them. Thus, acetic acid (pK_a = 4.8) and benzoic acid (pK_a = 4.2) can be deprotonated with NaOH and $NaHCO_3$, as shown in the following equations.

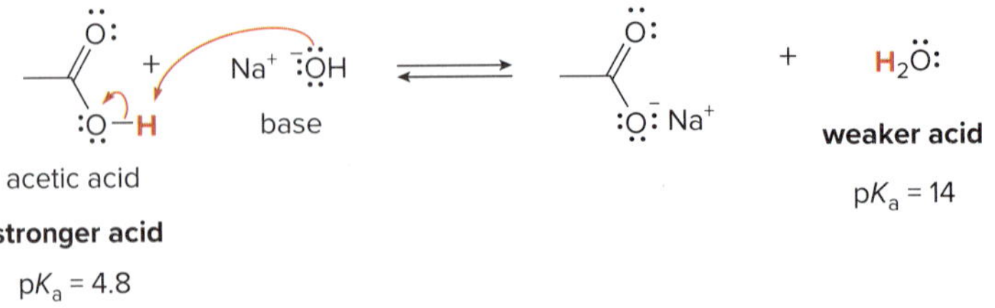

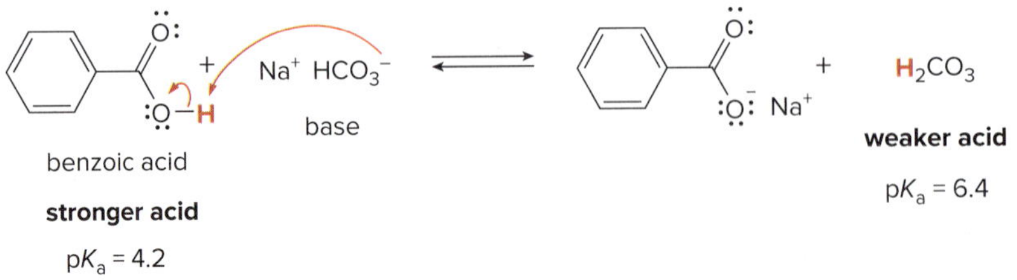

acetic acid (**stronger acid**, pK_a = 4.8) + Na^+ $^-$:OH (base) ⇌ acetate Na^+ + H_2O (**weaker acid**, pK_a = 14)

benzoic acid (**stronger acid**, pK_a = 4.2) + Na^+ HCO_3^- (base) ⇌ benzoate Na^+ + H_2CO_3 (**weaker acid**, pK_a = 6.4)

Table 19.3 lists common bases that can be used to deprotonate carboxylic acids. It is noteworthy that even a weak base like $NaHCO_3$ is strong enough to remove a proton from RCOOH.

Table 19.3 Common Bases Used to Deprotonate Carboxylic Acids

	Base	Conjugate acid (pK_a)
Increasing basicity ↓	Na^+ HCO_3^-	H_2CO_3 (6.4)
	NH_3	NH_4^+ (9.3)
	Na_2CO_3	HCO_3^- (10.2)
	Na^+ ^-OH	H_2O (14)
	Na^+ $^-OCH_3$	CH_3OH (15.5)
	Na^+ $^-OCH_2CH_3$	CH_3CH_2OH (16)
	Na^+ H^-	H_2 (35)

Why are carboxylic acids such strong organic acids? Remember that a strong acid has a weak, stabilized conjugate base. **Deprotonation of a carboxylic acid forms a resonance-stabilized conjugate base—a carboxylate anion.** Two equivalent resonance structures can be drawn for acetate (the conjugate base of acetic acid), both of which place a negative charge on an electronegative O atom. In the resonance hybrid, therefore, the negative charge is delocalized over two oxygen atoms.

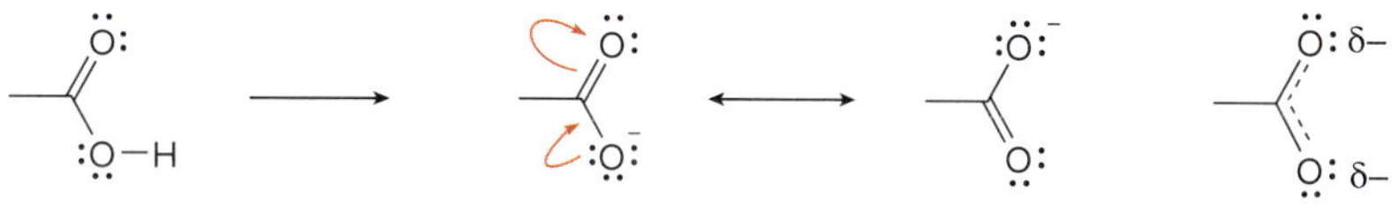

acetic acid

two resonance structures for acetate, the conjugate base

hybrid

How resonance affects acidity was first discussed in Section 2.5C.

Experimental data support this resonance description of acetate. **The acetate anion has two C–O bonds of equal length** (127 pm) and intermediate between the length of a C–O single bond (136 pm) and C=O (121 pm).

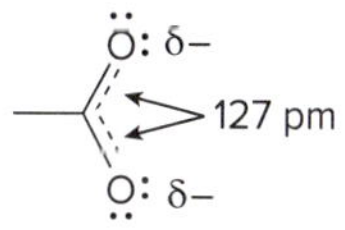

δ–
127 pm
δ–

acetate hybrid

Resonance stabilization accounts for why carboxylic acids are more acidic than other compounds with O–H bonds—namely, alcohols and phenols. For example, the pK_a values of ethanol (CH_3CH_2OH) and phenol (C_6H_5OH) are 16 and 10, respectively, both higher than the pK_a of acetic acid (4.8).

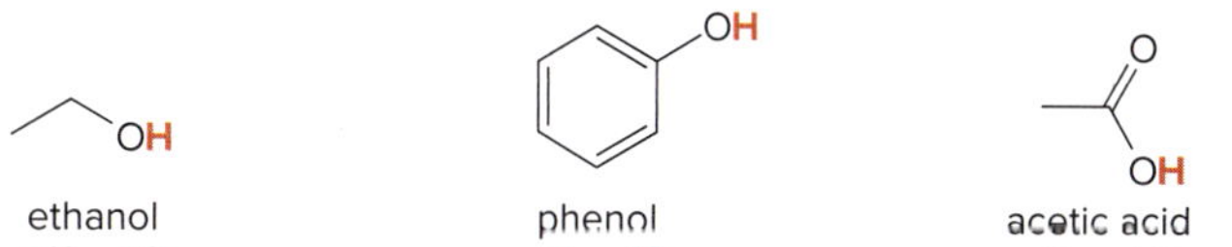

ethanol
$pK_a = 16$

phenol
$pK_a = 10$

acetic acid
$pK_a = 4.8$

Increasing acidity →

To understand the relative acidity of ethanol, phenol, and acetic acid, we must compare the stability of their conjugate bases and use this rule:

- **Anything that stabilizes a conjugate base A:$^-$ makes the starting acid H–A more acidic.**

Ethoxide, the conjugate base of ethanol, bears a negative charge on an oxygen atom, but there are no additional factors to further stabilize the anion. Because ethoxide is less stable than acetate, **ethanol is a weaker acid than acetic acid.**

ethanol → ethoxide

The resonance hybrid of phenoxide illustrates that its negative charge is dispersed over four atoms—three C atoms and one O atom.

Like acetate, **phenoxide** ($C_6H_5O^-$, the conjugate base of phenol) is also resonance stabilized. In the case of phenoxide, however, there are *five* resonance structures that disperse the negative charge over a total of *four* different atoms (three different carbons and the oxygen).

hybrid — phenol → 1 ↔ 2 ↔ 3 ↔ 4 ↔ 5

Phenoxide is more stable than ethoxide, but less stable than acetate, because acetate has two electronegative oxygen atoms upon which to delocalize the negative charge, whereas phenoxide has only one. Additionally, phenoxide resonance structures **2–4** have the negative charge on a carbon, a less electronegative element than oxygen. As a result, structures **2–4** are less stable than structures **1** and **5,** which have the negative charge on oxygen.

Moreover, resonance structures **1** and **5** have intact aromatic rings, whereas structures **2–4** do not. This, too, makes structures **2–4** less stable than **1** and **5.** Figure 19.4 summarizes this information about phenoxide by displaying the approximate relative energies of its five resonance structures and its hybrid.

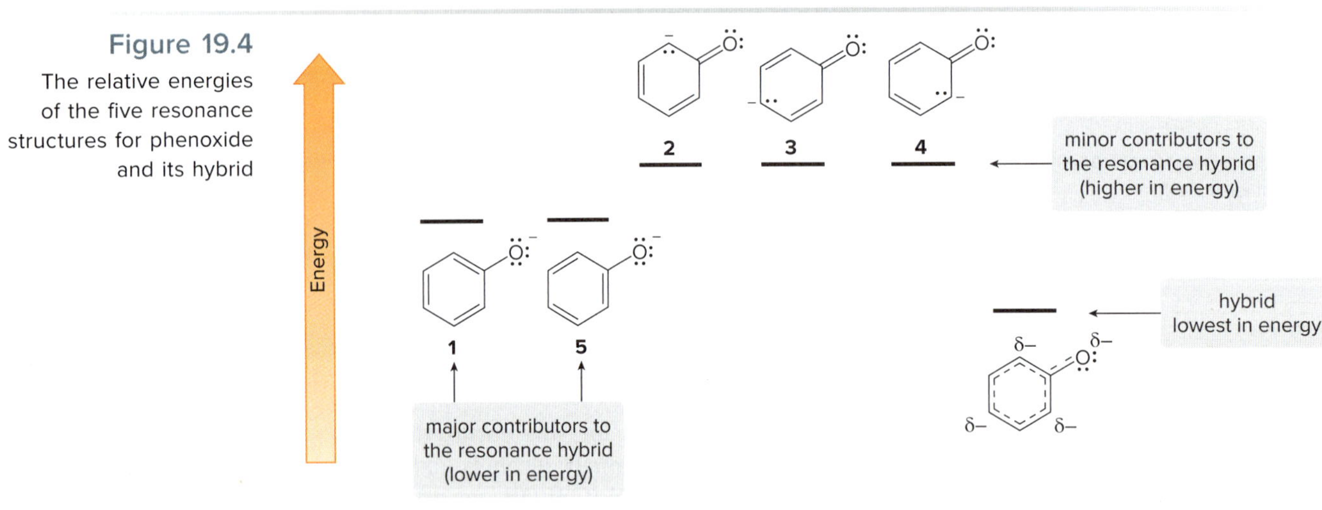

Figure 19.4 The relative energies of the five resonance structures for phenoxide and its hybrid

As a result, resonance stabilization of the conjugate base is important in determining acidity, but **the absolute number of resonance structures alone is not what's important.** We must evaluate their relative contributions to predict the relative stability of the conjugate bases.

- **Because of their O–H bond, RCOOH, ROH, and C_6H_5OH are *more acidic* than most organic hydrocarbons.**
- **A carboxylic acid is a *stronger* acid than an alcohol or a phenol because its conjugate base is more effectively resonance stabilized.**

Keep in mind that although carboxylic acids are strong organic acids, they are still *much weaker* than strong inorganic acids like HCl and H_2SO_4, which have pK_a values < 0.

Because alcohols and phenols are weaker acids than carboxylic acids, stronger bases are needed to deprotonate them. To deprotonate C_6H_5OH ($pK_a = 10$), a base whose conjugate acid has a $pK_a > 10$ is needed. Thus, of the bases listed in Table 19.4, $NaOCH_3$, NaOH, $NaOCH_2CH_3$, and NaH are strong enough. To deprotonate CH_3CH_2OH ($pK_a = 16$), only NaH is strong enough.

Problem 19.12 Draw the products of each acid–base reaction.

a. 4-methylphenol $\xrightarrow{NaOCH_3}$

b. *tert*-butyl alcohol $\xrightarrow{NaH}$

c. benzoic acid $\xrightarrow{NaHCO_3}$

d. albuterol (bronchodilator) $\xrightarrow{NaOH}$

albuterol
bronchodilator

Problem 19.13 Given the pK_a values in Appendix C, which of the following bases are strong enough to deprotonate CH_3COOH: (a) F^-; (b) $(CH_3)_3CO^-$; (c) CH_3^-; (d) $^-NH_2$; (e) Cl^-?

Problem 19.14 (a) Rank the labeled protons (H_a–H_c) in glucocaffeic acid, a naturally occurring compound found in blueberries and blackcurrants, in order of increasing acidity. (b) If glucocaffeic acid is treated with three equivalents of NaOH, what product is formed?

glucocaffeic acid

19.8 Inductive Effects in Aliphatic Carboxylic Acids

The pK_a of a carboxylic acid is affected by nearby groups that inductively donate or withdraw electron density.

- **Electron-withdrawing groups *stabilize* a conjugate base, making a carboxylic acid *more* acidic.**
- **Electron-donating groups *destabilize* the conjugate base, making a carboxylic acid *less* acidic.**

We first learned about inductive effects and acidity in Section 2.5B.

The relative acidity of CH_3COOH, $ClCH_2COOH$, and $(CH_3)_3CCOOH$ illustrates these principles in the following equations.

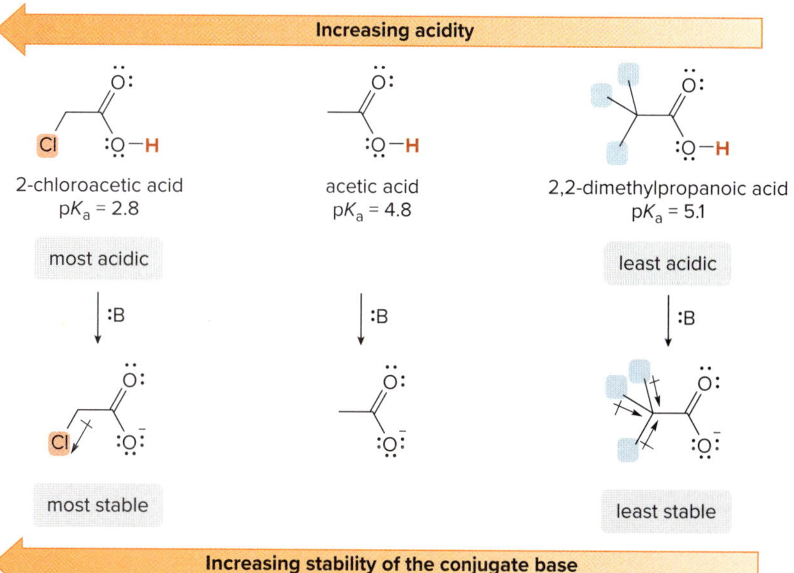

- $ClCH_2COOH$ is *more* acidic (pK_a = 2.8) than CH_3COOH (pK_a = 4.8) because its conjugate base is stabilized by the **electron-withdrawing inductive effect of the electronegative Cl.**
- $(CH_3)_3CCOOH$ is *less* acidic (pK_a = 5.1) than CH_3COOH because the **three polarizable CH_3 groups donate electron density and destabilize the conjugate base.**

The number, electronegativity, and location of substituents also affect acidity.

- **The larger the number of electronegative substituents, the stronger the acid.**

Cl
OH
O

pK_a = 2.8

Cl
Cl
OH
O

pK_a = 1.3

Cl
Cl
Cl
OH
O

pK_a = 0.9

Increasing acidity
Increasing number of electronegative Cl atoms

- **The more electronegative the substituent, the stronger the acid.**

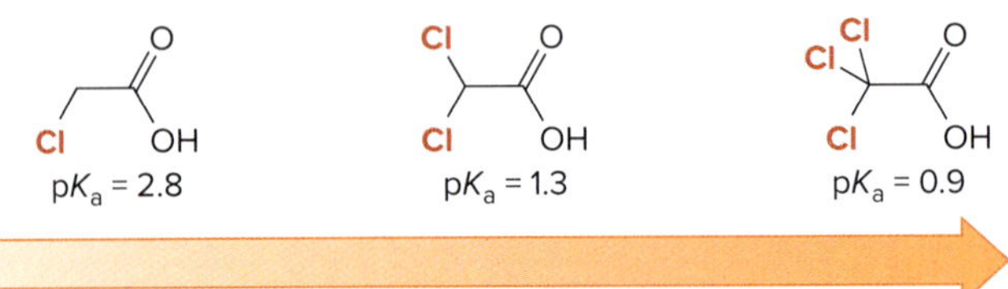

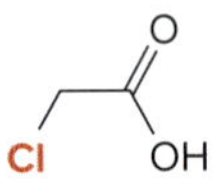

Cl
OH
O

pK_a = 2.8

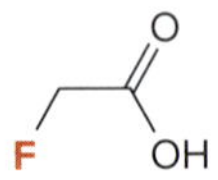

F
OH
O

pK_a = 2.6

F is more electronegative than Cl.
stronger acid

- **The closer the electron-withdrawing group to the COOH, the stronger the acid.**

Cl
O
OH

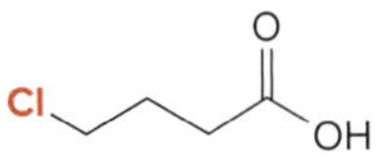

4-chlorobutanoic acid
pK_a = 4.5

Cl
O
OH

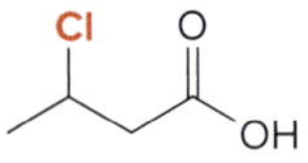

3-chlorobutanoic acid
pK_a = 4.1

O
OH
Cl

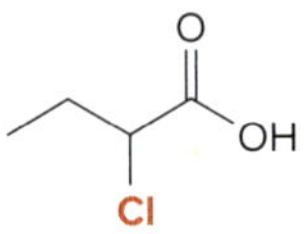

2-chlorobutanoic acid
pK_a = 2.9

Increasing acidity
Increasing proximity of Cl to COOH

Problem 19.15 Match each of the following pK_a values (3.2, 4.9, and 0.5) to the appropriate carboxylic acid: (a) CH_3CH_2COOH; (b) CF_3COOH; (c) ICH_2COOH.

Problem 19.16 Rank **A, B,** and **C** in order of increasing acidity.

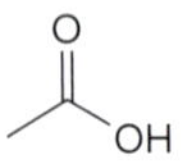

O
OH

A

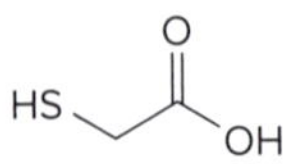

HS
O
OH

B

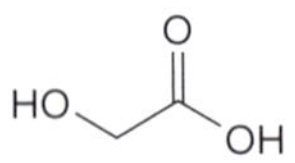

HO
O
OH

C

19.9 Substituted Benzoic Acids

Recall from Section 16.6 that substituents on a benzene ring either donate or withdraw electron density, depending on the balance of their inductive and resonance effects. These same effects also determine the acidity of substituted benzoic acids. There are two rules to keep in mind.

Rule [1] Electron-donor groups *destabilize* a conjugate base, making an acid *less* acidic.

An electron-donor group destabilizes a conjugate base by donating electron density onto a negatively charged carboxylate anion. A benzoic acid substituted by an electron-donor group has a *higher* pK_a than benzoic acid (pK_a = 4.2).

D = Electron-donor group

This acid is *less acidic* than benzoic acid.

pK_a > 4.2

D *destabilizes* the carboxylate anion.

The electron-donor groups that activate benzene to electrophilic attack make a benzoic acid *less* acidic.

Rule [2] Electron-withdrawing groups *stabilize* a conjugate base, making an acid *more* acidic.

An electron-withdrawing group stabilizes a conjugate base by removing electron density from the negatively charged carboxylate anion. A benzoic acid substituted by an electron-withdrawing group has a *lower* pK_a than benzoic acid (pK_a = 4.2).

W = Electron-withdrawing group

This acid is *more acidic* than benzoic acid.

pK_a < 4.2

W *stabilizes* the carboxylate anion.

The electron-withdrawing groups that deactivate benzene to electrophilic attack make a benzoic acid *more* acidic.

Figure 19.5 illustrates how common electron-donating and electron-withdrawing groups affect both the rate of reaction of a benzene ring toward electrophiles and the acidity of substituted benzoic acids.

Figure 19.5

How common substituents affect the reactivity of a benzene ring toward electrophiles and the acidity of substituted benzoic acids

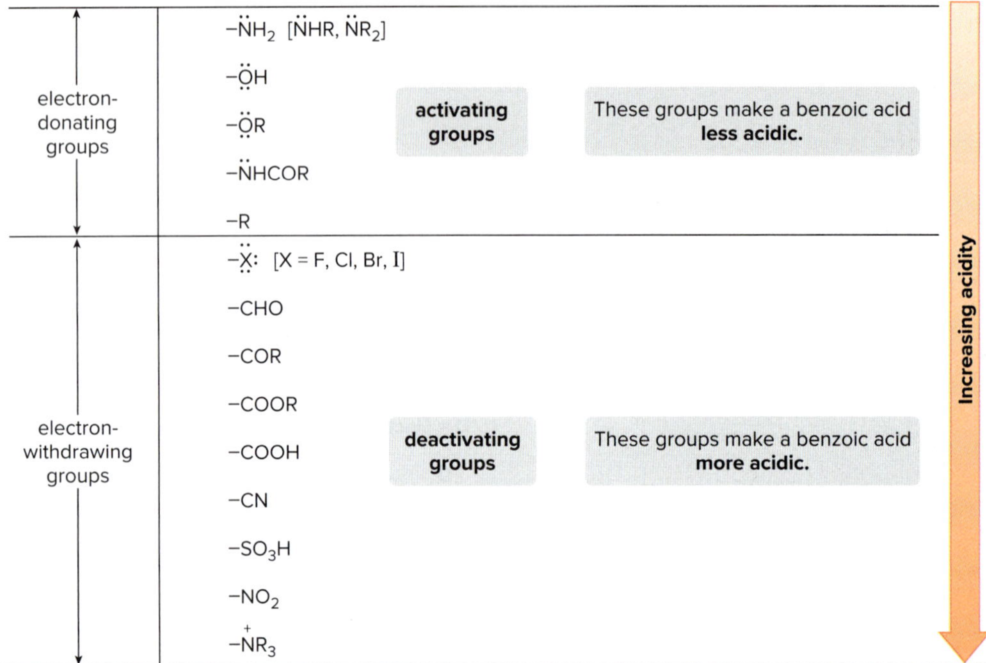

- **Groups that *donate* electron density *activate* a benzene ring toward electrophilic attack and make a benzoic acid *less* acidic.** Common electron-donating groups are R groups, or groups that have a N or O atom (with a lone pair) bonded to the benzene ring.
- **Groups that *withdraw* electron density *deactivate* a benzene ring toward electrophilic attack, and make a benzoic acid *more* acidic.** Common electron-withdrawing groups are the halogens, or groups with an atom Y (with a full or partial positive charge) bonded to the benzene ring.

Sample Problem 19.3 Determining the Relative Acidity of Substituted Benzoic Acids

Rank the following three carboxylic acids in order of increasing acidity.

A benzoic acid | **B** *p*-methoxybenzoic acid | **C** *p*-nitrobenzoic acid

Solution

***p*-Methoxybenzoic acid (B):** The CH_3O group is an electron-donor group because its electron-donating resonance effect is stronger than its electron-withdrawing inductive effect (Section 16.6). This *destabilizes* the conjugate base by donating electron density to the negatively charged carboxylate anion, making **B** *less acidic* than benzoic acid **A.** Two of the possible resonance structures for **B**'s conjugate base are drawn.

B *p*-methoxybenzoic acid → **electron-donor group** ↔

Having two (–) charges on nearby atoms *destabilizes* the conjugate base.

***p*-Nitrobenzoic acid (C):** The NO_2 group is an electron-withdrawing group because of both inductive effects and resonance (Section 16.6). This *stabilizes* the conjugate base by removing electron density from the negatively charged carboxylate anion, making **C** *more acidic* than benzoic acid **A.** Two of the possible resonance structures for **C**'s conjugate base are drawn.

C
p-nitrobenzoic acid

electron-withdrawing group

Having unlike charges on nearby atoms *stabilizes* the conjugate base.

By this analysis, the order of acidity is **B** < **A** < **C.**

Problem 19.17 Rank the compounds in each group in order of increasing acidity.

a.

b.

More Practice: Try Problems 19.39, 19.44.

Problem 19.18 Substituted phenols show substituent effects similar to substituted benzoic acids. Should the pK_a of phenol **A,** one of the naturally occurring phenols called urushiols isolated from poison ivy, be higher or lower than the pK_a of phenol (C_6H_5OH, pK_a = 10)? Explain.

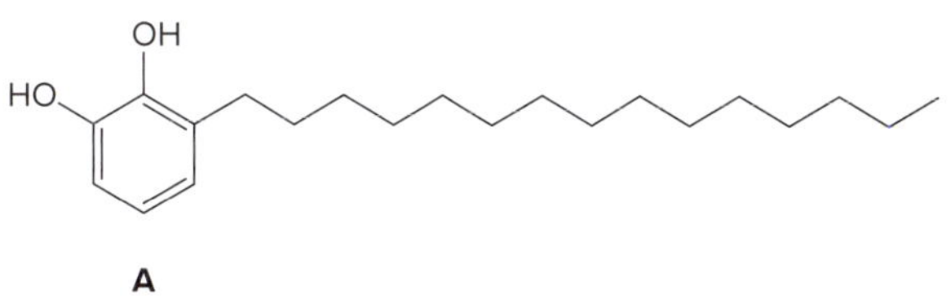

Poison ivy contains the irritant urushiol (Problem 19.18).
Ken Samuelsen/Photodisc/Getty Images

Extraction has long been and remains the first step in isolating a natural product from its source.

19.10 Extraction

An organic chemist in the laboratory must separate and purify mixtures of compounds. One particularly useful technique is **extraction,** which uses solubility differences and acid–base principles to separate and purify compounds.

Two solvents are used in extraction: water or an aqueous solution such as 10% $NaHCO_3$ or 10% NaOH; and an organic solvent such as dichloromethane (CH_2Cl_2), diethyl ether, or hexane. **Compounds are separated by their solubility differences in an aqueous and organic solvent.**

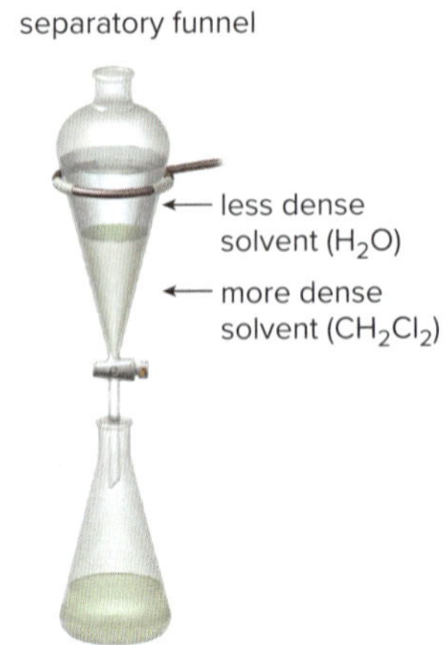

When two insoluble liquids are added to a separatory funnel, two layers are visible. To separate the layers, the lower layer can be drained from the bottom of the separatory funnel by opening the stopcock. The top layer can then be poured out the top neck of the funnel.

An item of glassware called a **separatory funnel** is used for the extraction. When two insoluble liquids are added to the separatory funnel, two layers form, with the less dense liquid on top and the more dense liquid on the bottom.

Suppose a mixture of benzoic acid (C_6H_5COOH) and NaCl is added to a separatory funnel containing H_2O and CH_2Cl_2. The benzoic acid would dissolve in the organic layer, and the NaCl would dissolve in the water layer. Separating the organic and aqueous layers and placing them in different flasks separates the benzoic acid and NaCl from each other.

How could we separate a mixture of benzoic acid and cyclohexanol? **Both compounds are organic, and as a result, both are soluble in an organic solvent such as CH_2Cl_2 and insoluble in water.** If a mixture of benzoic acid and cyclohexanol were added to a separatory funnel with CH_2Cl_2 and water, both would dissolve in the CH_2Cl_2 layer, and the two compounds would *not* be separated from each other. Is it possible to use extraction to separate two compounds of this sort that have similar solubility properties?

benzoic acid
- insoluble in water
- soluble in CH_2Cl_2

similar solubility properties

cyclohexanol
- insoluble in water
- soluble in CH_2Cl_2

If a carboxylic acid is one of the compounds, the answer is *yes,* because we can use acid–base chemistry to change its solubility properties.

When benzoic acid (a strong organic acid) is treated with aqueous NaOH, benzoic acid is deprotonated, forming sodium benzoate. **Because sodium benzoate is ionic, it is *soluble* in water, but *insoluble* in organic solvents.**

benzoic acid ($pK_a = 4.2$) + $Na^+ \ ^-OH$ (base) ⇌ sodium benzoate + H_2O ($pK_a = 14$)

benzoic acid:
- insoluble in water
- soluble in CH_2Cl_2

different solubility properties

sodium benzoate:
- soluble in water
- insoluble in CH_2Cl_2

A similar acid–base reaction does *not* occur when cyclohexanol is treated with NaOH because organic alcohols are much weaker organic acids, so they can be deprotonated only by a *very strong base* such as NaH. **NaOH is not strong enough to form significant amounts of the sodium alkoxide.**

cyclohexanol ($pK_a \sim 17$) + $Na^+ \ ^-OH$ (base) ⇌ sodium alkoxide + H_2O ($pK_a = 14$)

Because equilibrium favors the starting materials, little alkoxide is formed.

This difference in acid–base chemistry can be used to separate benzoic acid and cyclohexanol by the stepwise extraction procedure illustrated in Figure 19.6. This extraction scheme relies on two principles:

- **Extraction can separate only compounds having different solubility properties. One compound must dissolve in the aqueous layer and one must dissolve in the organic layer.**
- **A carboxylic acid can be separated from other organic compounds by converting it to a water-soluble carboxylate anion by an acid–base reaction.**

Thus, the water-soluble salt, $C_6H_5CO_2^-Na^+$ (derived from $C_6H_5CO_2H$ by an acid–base reaction), can be separated from water-insoluble cyclohexanol by an extraction procedure.

Figure 19.6 Separation of benzoic acid and cyclohexanol by an extraction procedure

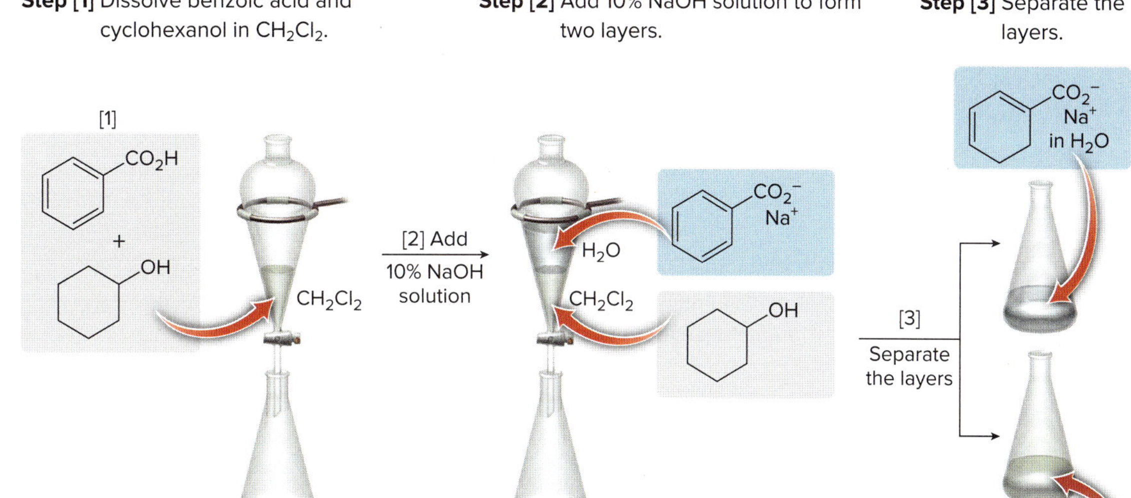

- Both compounds dissolve in the organic solvent CH_2Cl_2.

- Adding 10% aqueous NaOH solution forms two layers. When the two layers are mixed, the **NaOH deprotonates $C_6H_5CO_2H$ to form $C_6H_5CO_2^-Na^+$, which dissolves in the aqueous layer.**
- The cyclohexanol remains in the CH_2Cl_2 layer.

- Draining the lower layer out the bottom stopcock separates the two layers, and the separation process is complete.
- Cyclohexanol (dissolved in CH_2Cl_2) is in one flask. The sodium salt of benzoic acid, $C_6H_5CO_2^-Na^+$ (dissolved in water) is in another flask.

Sample Problem 19.4 Separating Compounds by Extraction

A mixture of **A**, **B**, and **C** was added to a separatory funnel containing CH_2Cl_2 and 10% aqueous NaOH solution. Which compound(s) are present in the aqueous layer, and which compound(s) are present in the organic layer?

A **B** **C**

Solution

Recall the principles of solubility:

- **Organic compounds are soluble in organic solvents.**
- **Organic compounds that can hydrogen bond to H_2O are water soluble if they have $\leq$ 5 C's.**
- **Uncharged organic compounds with $>$ 5 C's are not water soluble.**
- **Ionic compounds are water soluble.**

A, B, and **C** are uncharged organic compounds, so they are *soluble* in CH_2Cl_2, and because they each have $>$ 5 C's, they are *insoluble* in H_2O. **C,** however, has a CO_2H group with an acidic H atom that can be removed with NaOH. Deprotonation forms a **carboxylate anion** that is now *water soluble.*

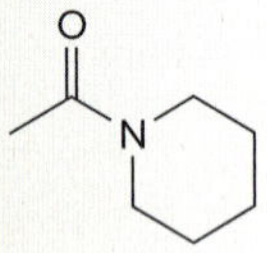

A
• insoluble in water
• soluble in CH_2Cl_2

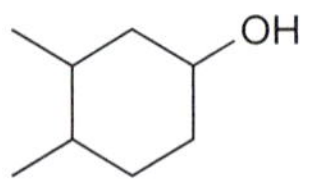

B
• insoluble in water
• soluble in CH_2Cl_2

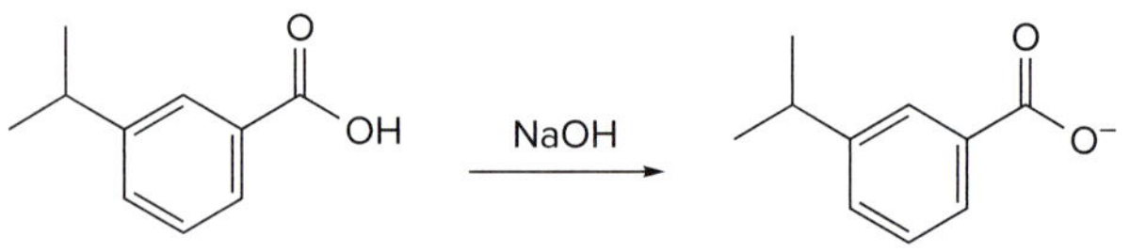

C
• insoluble in water
• soluble in CH_2Cl_2

NaOH →

carboxylate anion
• soluble in water
• insoluble in CH_2Cl_2

As a result, in a separatory funnel with CH_2Cl_2 and 10% aqueous NaOH solution, **A** and **B** are soluble in the CH_2Cl_2 layer, and **C** is deprotonated to form a carboxylate anion that is now soluble in the aqueous layer.

Problem 19.19 Considering compounds **A–D,** which of the following pairs of compounds are separable by an aqueous extraction procedure?

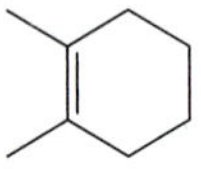

A 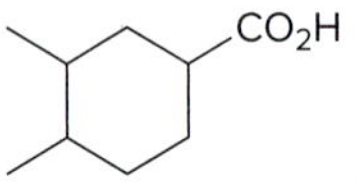 B 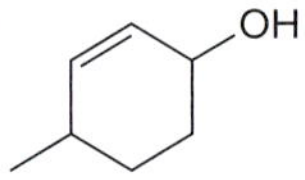 C 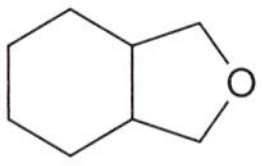 D

a. **A** and **B**
b. **A** and **C**
c. **C** and **D**
d. **B** and **D**

More Practice: Try Problems 19.55–19.57.

19.11 Amino Acids

Chapter 27 discusses the conversion of amino acids to proteins.

Amino acids, one of four kinds of small biomolecules that have important biological functions in the cell (Section 3.9), also undergo proton transfer reactions.

19.11A Introduction

Amino acids contain two functional groups—an amino group (NH_2) and a carboxy group (COOH). In most naturally occurring amino acids, the amino group is bonded to the α carbon, so they are called **α-amino acids.** Amino acids are the building blocks of proteins, biomolecules that comprise muscle, hair, fingernails, and many other biological tissues.

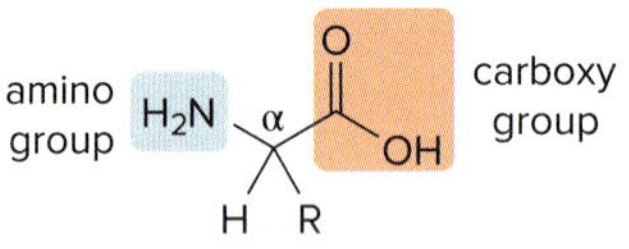

amino group H_2N α carboxy group O OH H R

α-amino acid

The 20 amino acids that occur naturally in proteins differ in the identity of the R group bonded to the α carbon. **The simplest amino acid, called glycine, has R = H.** When the R group is any other substituent, **the α carbon is a stereogenic center,** and there are two possible enantiomers.

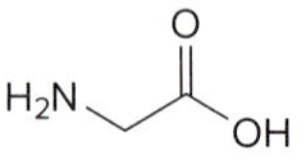

glycine
no stereogenic centers

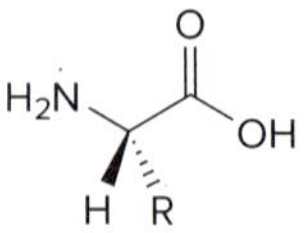

L amino acid
Only this isomer occurs in proteins.

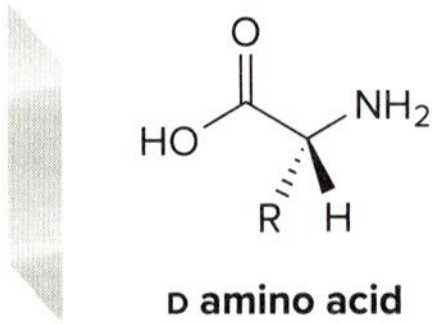

D amino acid

Humans can synthesize only 10 of the 20 amino acids needed for protein synthesis. The remaining 10, called **essential amino acids,** must be obtained from the diet and consumed on a regular, almost daily basis. Vegetarian diets must be carefully balanced to obtain all the essential amino acids. Grains—wheat, rice, and corn—are low in lysine, and legumes—beans, peas, and peanuts—are low in methionine, but a combination of these foods provides all the needed amino acids. Thus, a diet of corn tortillas and beans, or rice and tofu, provides all essential amino acids. A peanut butter sandwich on wheat bread does, too. *Brent Hofacker/ Shutterstock*

Amino acids exist in nature as only one of these enantiomers. Except when the R group is CH_2SH, the stereogenic center on the α carbon has the *S* configuration. An older system of nomenclature names the **naturally occurring enantiomer of an amino acid as the L isomer, and its unnatural enantiomer the D isomer.**

The R group of an amino acid can be H, alkyl, aryl, or an alkyl chain containing a N, O, or S atom. Representative examples are listed in Table 19.4. All amino acids have common names, which are abbreviated by a three-letter or one-letter designation. For example, glycine is often written as the three-letter abbreviation **Gly** or the one-letter abbreviation **G.** These abbreviations are also given in Table 19.4. A complete list of the 20 naturally occurring amino acids is found in Figure 27.2.

Table 19.4 Representative Amino Acids

General structure: H_2N–CH(R)–C(=O)OH

R group	Name	Three-letter abbreviation	One-letter abbreviation
H	glycine	Gly	G
CH_3	alanine	Ala	A
$CH_2C_6H_5$	phenylalanine	Phe	F
CH_2OH	serine	Ser	S
CH_2SH	cysteine	Cys	C
$CH_2CH_2SCH_3$	methionine	Met	M
CH_2CH_2COOH	glutamic acid	Glu	E
$(CH_2)_4NH_2$	lysine	Lys	K

Problem 19.20 Draw both enantiomers of each amino acid and label them as *R* or *S:* (a) phenylalanine; (b) methionine.

19.11B Acid–Base Properties

An amino acid is both an acid and a base.

- **The NH_2 group has a nonbonded electron pair, making it a base.**
- **The COOH group has an acidic proton, making it an acid.**

Amino acids are never uncharged neutral compounds. They exist as salts, so they have very high melting points and are very soluble in water.

- **Proton transfer from the acidic carboxy group to the basic amino group forms a salt called a *zwitterion,* which contains both a positive and a negative charge.**

basic site $H_2\ddot{N}$–CH(R)–C(=O)–$\ddot{O}$H acidic proton —proton transfer→ $H_3\overset{+}{N}$–CH(R)–C(=O)–$\ddot{O}:^-$

This neutral form of an amino acid does *not* exist.

zwitterion

This salt is the neutral form of an amino acid.

In actuality, an amino acid can exist in three different forms, depending on the pH of the aqueous solution in which it is dissolved.

When the pH of a solution is ~6, alanine (R = CH_3) exists in its zwitterionic form (**A**), having no net charge. In this form the carboxy group bears a negative charge—it is a **carboxylate anion**—and the amino group bears a net positive charge (an **ammonium cation**).

ammonium cation
carboxylate anion

alanine
A
a neutral zwitterion
This form exists at pH ≈ 6.

When strong acid is added to lower the pH (≤ 2), the carboxylate anion is protonated and the **amino acid has a net positive charge** (form **B**).

A

B
overall (**+1**) charge
This form exists at pH ≤ 2.

When strong base is added to **A** to raise the pH (≥ 10), the ammonium cation is deprotonated and the **amino acid has a net negative charge** (form **C**).

A

C
overall (**–1**) charge
This form exists at pH ≥ 10.

Thus, **alanine exists in one of three different forms depending on the pH of the solution in which it is dissolved.** If the pH of a solution is gradually increased from 2 to 10, the following process occurs.

- **At low pH, alanine has a net (+) charge (form B).**
- **As the pH is increased to ~6, the carboxy group is deprotonated, and the amino acid exists as a zwitterion with no overall charge (form A).**
- **At high pH, the ammonium cation is deprotonated, and the amino acid has a net (–) charge (form C).**

These reactions are summarized in Figure 19.7.

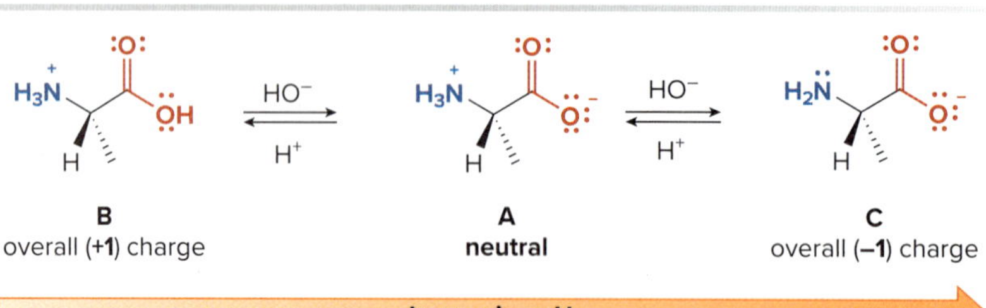

Figure 19.7 Summary of the acid–base reactions of alanine

Problem 19.21 Explain why amino acids, unlike most other organic compounds, are insoluble in organic solvents like diethyl ether.

Problem 19.22 Draw the positively charged, neutral, and negatively charged forms for the amino acid glycine. Which species predominates at pH 11? Which species predominates at pH 1?

19.11C Isoelectric Point

Because a protonated amino acid has at least two different protons that can be removed, a pK_a value is reported for each of these protons. For example, the pK_a of the carboxy proton of alanine is 2.35 and the pK_a of the ammonium proton is 9.87. Table 27.1 lists these values for all 20 amino acids.

- **The pH at which the amino acid exists primarily in its neutral form is called its *isoelectric point,* abbreviated as p*I*.**

More information on the isoelectric point can be found in Section 27.1.

For amino acids without other acidic or basic functional groups, the isoelectric point is the average of both pK_a values of an amino acid:

$$\textbf{Isoelectric point} = \mathbf{p}\boldsymbol{I} = \frac{pK_a\,(\text{COOH}) + pK_a\,(\text{NH}_3^{+})}{2}$$

$$\text{For alanine: } \mathbf{p}\boldsymbol{I} = \frac{2.35 + 9.87}{2} = \underset{pI\ (\text{alanine})}{6.12}$$

Problem 19.23 The pK_a values for the carboxy and ammonium protons of phenylalanine are 2.58 and 9.24, respectively. What is the isoelectric point of phenylalanine? Draw the structure of phenylalanine at its isoelectric point.

Problem 19.24 Explain why the pK_a of the COOH group of glycine is much lower than the pK_a of the COOH of acetic acid (2.35 compared to 4.8).

19.12 Nitriles

We end Chapter 19 with the chemistry of nitriles. Nitriles are readily prepared by $\mathbf{S_N2}$ substitution reactions of unhindered methyl and 1° alkyl halides with $^{-}$CN. This reaction adds one carbon to the alkyl halide and **forms a new carbon–carbon bond.**

Because a nitrile contains an electrophilic carbon atom that is part of a multiple bond but no leaving group, a nitrile reacts with nucleophiles by a **nucleophilic addition reaction.** The nature of the nucleophile determines the structure of the product.

The reactions of nitriles with water, hydride, and organometallic reagents as nucleophiles are as follows:

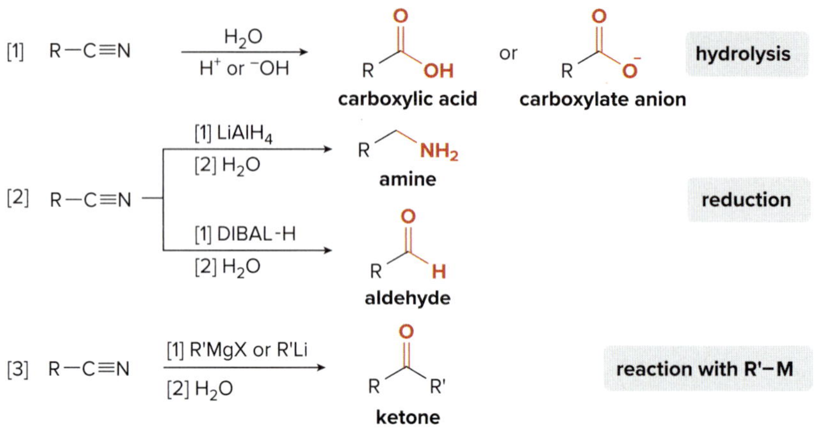

[1] R—C≡N —(H_2O; H^+ or ^-OH)→ carboxylic acid or carboxylate anion — **hydrolysis**

[2] R—C≡N —([1] $LiAlH_4$ [2] H_2O)→ amine; —([1] DIBAL-H [2] H_2O)→ aldehyde — **reduction**

[3] R—C≡N —([1] R'MgX or R'Li [2] H_2O)→ ketone — **reaction with R'–M**

19.12A Hydrolysis of Nitriles

Nitriles are hydrolyzed with water in the presence of acid or base to yield **carboxylic acids** or **carboxylate anions.** In this reaction, the three C–N bonds are replaced by three C–O bonds.

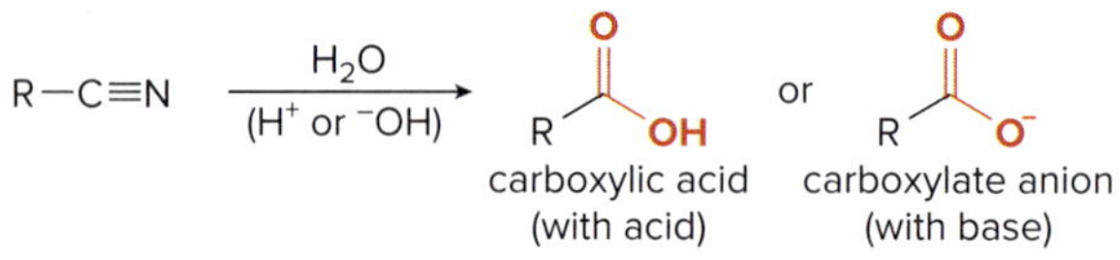

R—C≡N —(H_2O (H^+ or ^-OH))→ carboxylic acid (with acid) or carboxylate anion (with base)

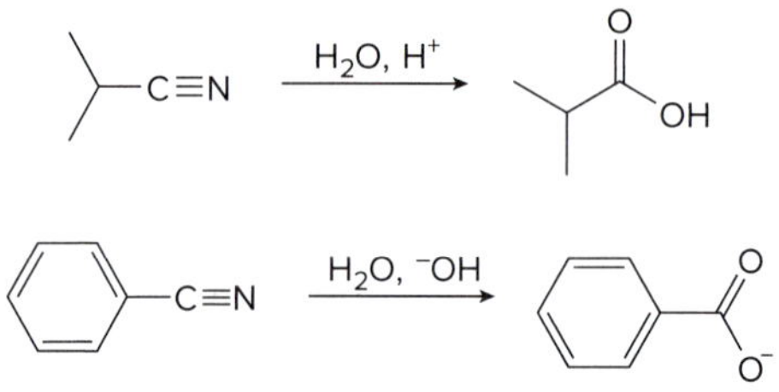

(CH$_3$)$_2$CH—C≡N —(H_2O, H^+)→ (CH$_3$)$_2$CHCOOH

C$_6$H$_5$—C≡N —(H_2O, ^-OH)→ C$_6$H$_5$COO$^-$

The mechanism of this reaction involves the formation of an **amide tautomer.** Two tautomers can be drawn for any carbonyl compound, and those for a 1° amide are as follows:

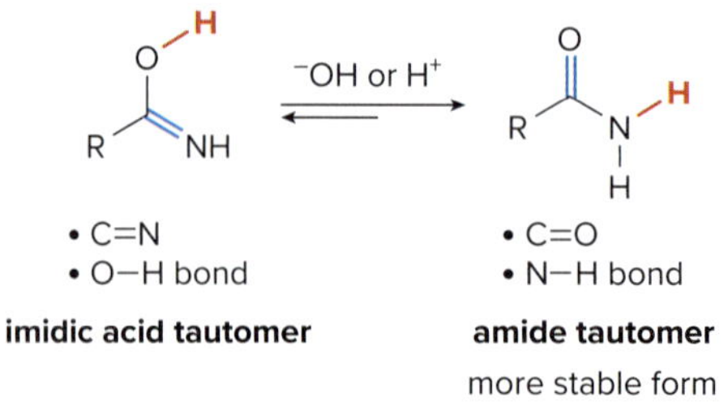

imidic acid tautomer (• C=N • O–H bond) ⇌ (^-OH or H^+) amide tautomer (• C=O • N–H bond), more stable form

Recall from Chapter 11 that tautomers are constitutional isomers that differ in the location of a double bond and a proton.

- **The amide form is the more stable tautomer, having a C=O and an N–H bond.**
- **The imidic acid tautomer is the less stable form, having a C=N and an O–H bond.**

The imidic acid and amide tautomers are interconverted by treating with acid or base, analogous to the keto–enol tautomers of other carbonyl compounds. In fact, the two amide tautomers are exactly the same as keto–enol tautomers except that a nitrogen atom replaces a carbon atom bonded to the carbonyl group.

Problem 19.25 Draw a tautomer of each compound.

a. [structure: H_2N–C(=O)– attached to cyclohexene ring]
b. [structure: OH, O, N–H]
c. [structure: NH, OH]

The mechanism of nitrile hydrolysis in both acid and base consists of two parts: [1] **nucleophilic addition** to form the imidic acid tautomer followed by **tautomerization** to form the amide, and [2] **hydrolysis of the amide** to form RCO_2H or RCO_2^-. The mechanism is shown for the basic hydrolysis of RCN to RCO_2^- (Mechanism 19.1).

Mechanism 19.1 Hydrolysis of a Nitrile in Base

Part [1] Conversion of a nitrile to a 1° amide

R–C≡N: + $^-$:OH →(1) → H–OH →(2) imidic acid →(3) →(4) amide + $^-$OH

imidic acid

amide

1 – 2 Nucleophilic attack of $^-$OH followed by protonation forms an **imidic acid.**

3 – 4 **Tautomerization** occurs by a two-step sequence—deprotonation followed by protonation.

Part [2] Hydrolysis of the 1° amide to a carboxylate anion

amide →($^-$OH, H_2O; three steps (Mechanism 20.9)) **carboxylate anion** + :NH_3

Conversion of the amide to the carboxylate occurs by a three-step sequence that will be discussed in Chapter 20 (Mechanism 20.9).

Problem 19.26 Draw the products of each reaction.

a. [propyl bromide] →(NaCN)

b. [1,2-dicyanobenzene] →(H_2O, H^+)

c. [2-cyanohexane] →(H_2O, $^-$OH)

d. [cyclohexylmethyl chloride] →([1] NaCN; [2] H_2O, H^+)

Problem 19.27 (a) What product is formed when 3,3-dimethylhexan-1-ol is treated with the following set of reagents: [1] PBr_3; [2] NaCN; [3] H_3O^+? (b) How does this compound differ from the product that results when 3,3-dimethylhexan-1-ol is treated with CrO_3 and H_2SO_4?

19.12B Reduction of Nitriles

Nitriles are reduced with metal hydride reagents to form either 1° amines or aldehydes, depending on the reducing agent.

- **Treatment of a nitrile with $LiAlH_4$ followed by H_2O adds two equivalents of H_2 across the triple bond, forming a 1° amine.**

—C≡N → [1] $LiAlH_4$ / [2] H_2O → H H, NH_2

- **Treatment of a nitrile with a milder reducing agent such as DIBAL-H followed by H_2O forms an aldehyde.**

—C≡N → [1] DIBAL-H / [2] H_2O → O, H

The mechanism of both reactions involves **nucleophilic addition of hydride (H^-) to the polarized C–N triple bond.** Mechanism 19.2 illustrates that reduction of a nitrile to an amine requires addition of two equivalents of $H:^-$ from $LiAlH_4$. It is likely that intermediate nitrogen anions complex with AlH_3 (formed in situ) to facilitate the addition. Protonation of the dianion in Step [4] forms the amine.

Mechanism 19.2 Reduction of a Nitrile with $LiAlH_4$

R–C≡N: → ($H-\bar{A}lH_3$, 1) → H, R, $\bar{N}$:, AlH_3 → (2) → H, R, N, $\bar{A}lH_3$, $H-\bar{A}lH_3$, AlH_3 → (3) → H H, R, N, $\bar{A}lH_3$, $^-AlH_3$, + AlH_3 → (2 H_2O, 4) → H H, R, NH_2 + 2 $H_3\bar{A}lOH$

1 – 2 Addition of one equivalent of $H:^-$ from $LiAlH_4$ forms an intermediate with **one new C–H bond,** which complexes with AlH_3.

3 – 4 Nucleophilic attack of a second equivalent of $H:^-$ and complexation with AlH_3 form a dianion, which reacts with water to form **two new N–H bonds,** giving the **1° amine.**

With **DIBAL-H,** nucleophilic addition of one equivalent of hydride forms an anion (Step [1]), which is protonated with water to generate an **imine,** as shown in Mechanism 19.3. As described in Section 18.11, imines are hydrolyzed in water to form aldehydes. Mechanism 19.3 is written without complexation of aluminum with the anion formed in Step [1], to emphasize the identity of intermediates formed during reduction.

Mechanism 19.3 Reduction of a Nitrile with DIBAL-H

R–C≡N: → ($H-AlR_2$, 1) → H, R, =$\bar{N}$: → (H–ÖH, 2) → H, R, =N, H **imine** + :ÖH⁻ → ($H_2\ddot{O}$:, (Mechanism 18.5)) → H, R, =Ö:

1 Addition of $H:^-$ from DIBAL-H (drawn as R_2AlH) forms the new C–H bond.

2 Protonation forms an **imine,** which is hydrolyzed to an aldehyde by a stepwise sequence that is the reverse of Mechanism 18.5.

Problem 19.28 Draw the product of each reaction.

a. (3-methoxyphenyl)ethyl bromide → [1] NaCN [2] $LiAlH_4$ [3] H_2O

b. cyclohexyl–C≡N → [1] DIBAL-H [2] H_2O

19.12C Addition of Grignard and Organolithium Reagents to Nitriles

Both Grignard and organolithium reagents react with nitriles to form ketones with a new carbon–carbon bond.

C₆H₅–C≡N → [1] CH₃CH₂MgBr [2] H_2O → C₆H₅COCH₂CH₃

The reaction occurs by nucleophilic addition of the organometallic reagent to the polarized C–N triple bond to form an anion (Step [1]), which is protonated with water to form an **imine.** Water then hydrolyzes the imine, replacing the C=N by C=O as described in Section 18.11. The final product is a ketone with a new carbon–carbon bond (Mechanism 19.4).

Mechanism 19.4 Addition of Grignard and Organolithium Reagents (R–M) to Nitriles

R–C≡N: + R'–M (step 1) → R'(R)C=N:⁻ → H–ÖH (step 2) → R'(R)C=N–H (imine) + :ÖH⁻ → H_2O (Mechanism 18.5) → R'(R)C=O

1 Addition of R:⁻ from R'M (M = MgX or Li) forms a **new C–C bond.**

2 Protonation forms an **imine,** which is hydrolyzed to a ketone by a stepwise sequence that is the reverse of Mechanism 18.5.

Problem 19.29 Draw the products of each reaction.

a. 2-methoxybenzonitrile (OCH_3, CN) → [1] CH₃CH₂MgCl [2] H_2O

b. 3-cyanopentane derivative (CN) → [1] C₆H₅Li [2] H_2O

Problem 19.30 What reagents are needed to convert phenylacetonitrile ($C_6H_5CH_2CN$) to each compound: (a) $C_6H_5CH_2COCH_3$; (b) $C_6H_5CH_2COC(CH_3)_3$; (c) $C_6H_5CH_2CHO$; (d) $C_6H_5CH_2COOH$?

Problem 19.31 Outline two different ways that butan-2-one can be prepared from a nitrile and a Grignard reagent.

Problem 19.32 Fadrozole, trade name Afema, is used in Japan for the treatment of breast cancer. What product is formed when fadrozole undergoes each reaction?

a. [1] $LiAlH_4$ [2] H_2O
b. H_3O^+
c. [1] CH_3Li [2] H_2O
d. [1] DIBAL-H [2] H_2O
e. ^-OH, H_2O
f. [1] PhMgBr [2] H_2O

fadrozole

Chapter 19 REVIEW

KEY CONCEPTS

[1] Acidity and resonance effects (19.7)

- A **carboxylic acid** is **more acidic than phenol** and **ethanol** because its **conjugate base** is **more stabilized by resonance.**

ethanol $pK_a = 16$ | phenol $pK_a = 10$ | **acetic acid** $pK_a = 4.8$

Increasing acidity →

Try Problems 19.37, 19.40.

[2] Other factors that affect acidity

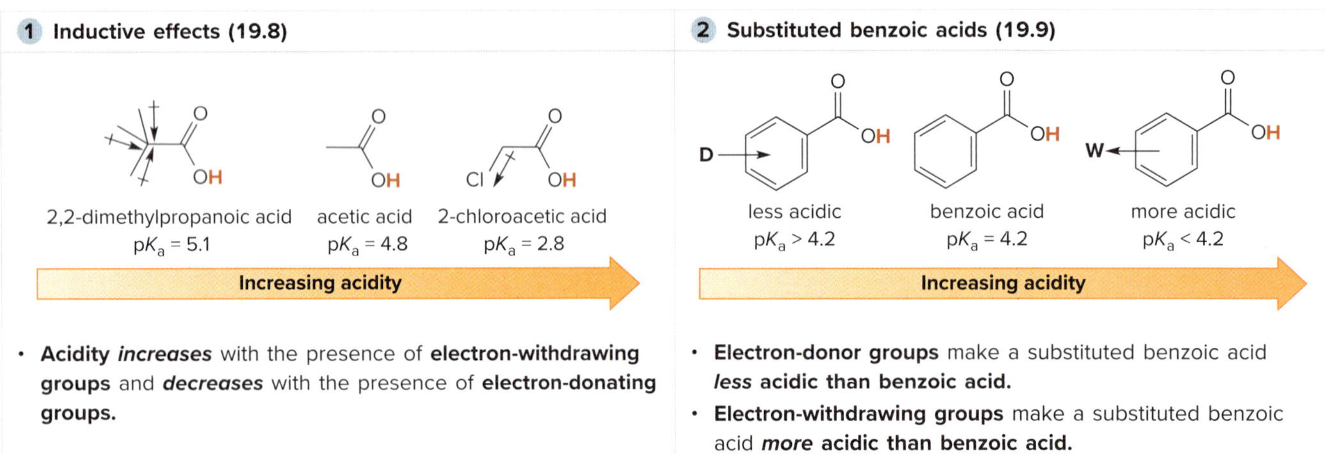

1 Inductive effects (19.8)

- **Acidity *increases*** with the presence of **electron-withdrawing groups** and ***decreases*** with the presence of **electron-donating groups.**

2 Substituted benzoic acids (19.9)

- **Electron-donor groups** make a substituted benzoic acid ***less* acidic than benzoic acid.**
- **Electron-withdrawing groups** make a substituted benzoic acid ***more* acidic than benzoic acid.**

Try Problem 19.44.

[3] Positively charged, neutral, and negatively charged forms of an amino acid (19.11); example: phenylalanine

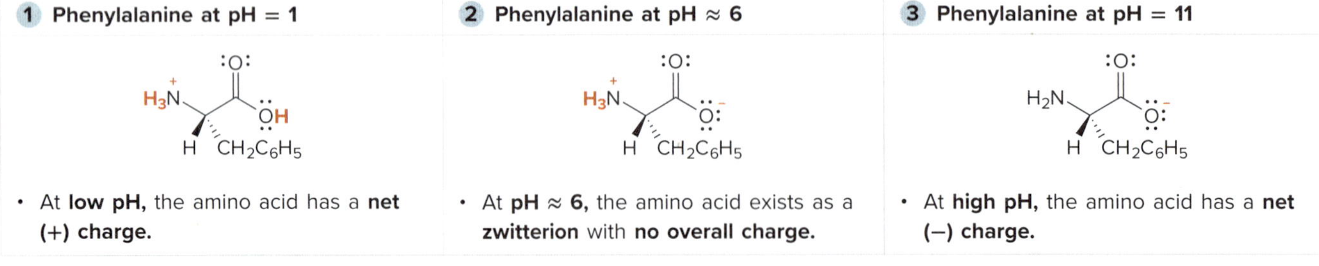

1 Phenylalanine at pH = 1

- At **low pH,** the amino acid has a **net (+) charge.**

2 Phenylalanine at pH ≈ 6

- At **pH ≈ 6,** the amino acid exists as a **zwitterion** with **no overall charge.**

3 Phenylalanine at pH = 11

- At **high pH,** the amino acid has a **net (–) charge.**

See Figure 19.7. Try Problems 19.58c, 19.60.

KEY REACTIONS

[1] Nitrile Synthesis

R–X (X = Cl, Br, I) $\xrightarrow[\text{S}_N2\ (19.12)]{\text{NaCN}}$ R–CN (nitrile) + X^-

Try Problems 19.51f, 19.55a.

[2] Reactions of Nitriles

1. nitrile (R–C≡N) $\xrightarrow[\Delta]{H_2O,\ H^+ \text{ or } ^-OH}$ carboxylic acid (with acid) or carboxylate (with base) — hydrolysis (19.12A)

2. nitrile (R–C≡N) $\xrightarrow[\text{[2] } 2\ HO–H]{\text{[1] } Li^+\ H–\bar{Al}H_3}$ 1° amine (R–CH$_2$NH$_2$) — reduction (19.12B)

3. nitrile (R–C≡N) $\xrightarrow[\text{[2] } H_2O]{\text{[1] DIBAL-H}}$ aldehyde — reduction (19.12B)

4. nitrile (R–C≡N) $\xrightarrow[\text{[2] } H_2O]{\text{[1] } CH_3MgX \text{ or } CH_3Li}$ ketone (R–CO–CH$_3$) (19.12C)

Try Problems 19.34b; 19.51e, f; 19.52d; 19.54a, c, d.

KEY SKILLS

[1] Drawing the products of an acid–base reaction involving a carboxylic acid (19.7)

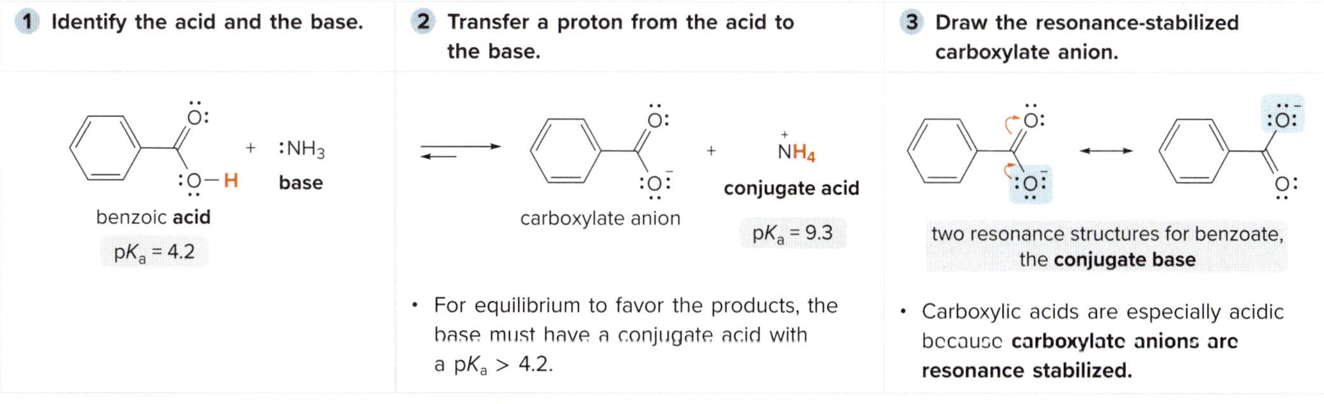

See Table 19.3. Try Problems 19.33b, 19.38, 19.53.

[2] Ranking benzoic acids in order of increasing acidity (19.9)

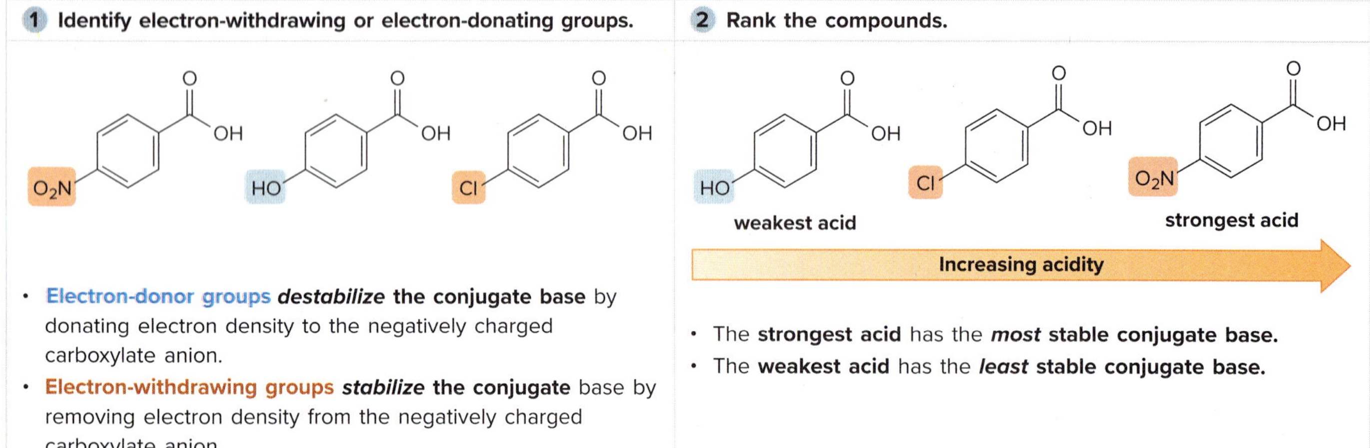

1 Identify electron-withdrawing or electron-donating groups.

- Electron-donor groups ***destabilize*** **the conjugate base** by donating electron density to the negatively charged carboxylate anion.
- Electron-withdrawing groups ***stabilize*** **the conjugate** base by removing electron density from the negatively charged carboxylate anion.

2 Rank the compounds.

- The **strongest acid** has the ***most*** **stable conjugate base.**
- The **weakest acid** has the ***least*** **stable conjugate base.**

See Sample Problem 19.3. Try Problem 19.39.

[3] Separating a carboxylic acid from an alcohol by extraction (19.10)

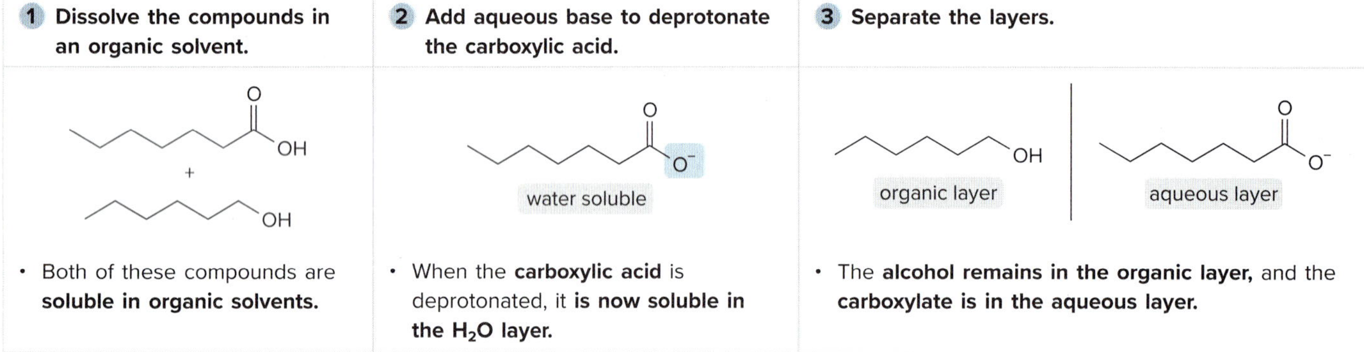

1 Dissolve the compounds in an organic solvent.

- Both of these compounds are **soluble in organic solvents.**

2 Add aqueous base to deprotonate the carboxylic acid.

- When the **carboxylic acid** is deprotonated, it **is now soluble in the H_2O layer.**

3 Separate the layers.

- The **alcohol remains in the organic layer,** and the **carboxylate is in the aqueous layer.**

See Sample Problem 19.4, Figure 19.6. Try Problems 19.55–19.57.

CHAPTER 19 MULTIPLE-CHOICE SELF-TEST

The Self-Test consists of multiple-choice questions similar to those found on the American Chemical Society organic chemistry exam. Answers are given at the end of the chapter.

1. Give the IUPAC name for the following compound.

O
OH

a. 2-ethyl-5,5-dimethylcyclohexane-1-carboxylic acid
b. 5,5-dimethyl-2-ethylcyclohexanecarboxylic acid
c. 6-ethyl-3,3-dimethylcyclohexanecarboxylic acid
d. 2-ethyl-5,5-dimethylcyclohexanecarboxylic acid

2. Which compound has the lowest pK_a?

a. CO_2H N H
b. CO_2H O
c. CO_2H
d. CO_2H O

3. Which anion is the strongest base?

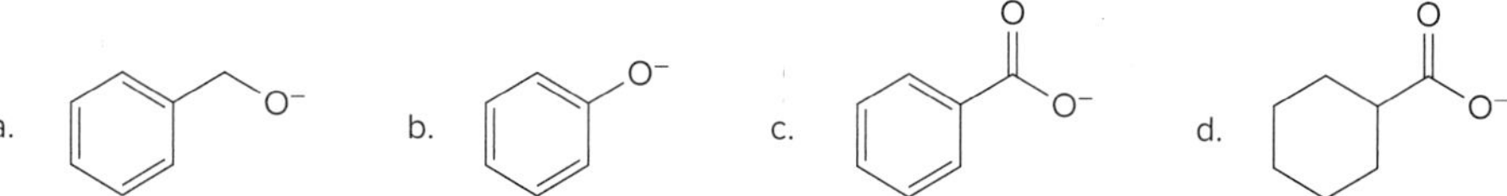

4. What product is formed when **X** is treated with two equivalents of NaOH?

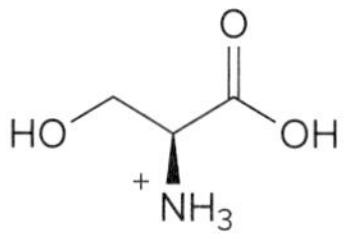

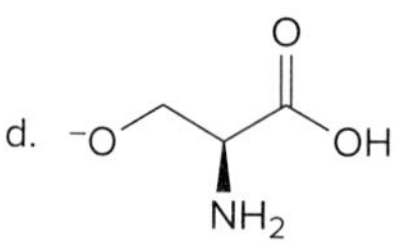

5. What reagent is needed to carry out the following transformation?

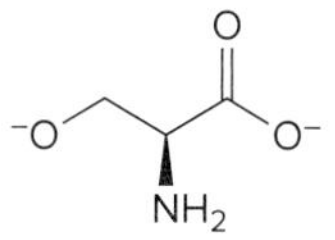

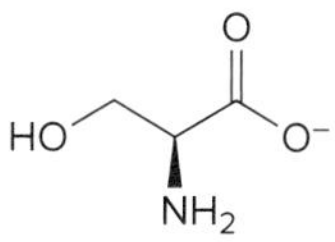

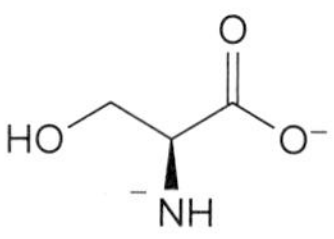

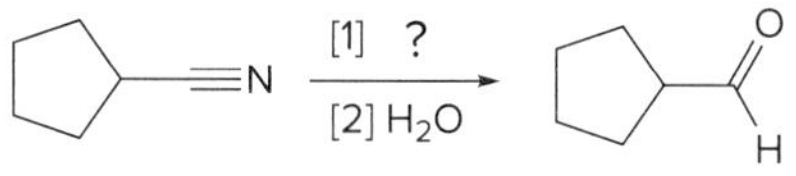

[1] ?
[2] H_2O

a. $LiAlH_4$ b. DIBAL-H c. CH_3MgBr d. $NaBH_4$

6. When compounds **A, B,** and **C** are added to a separatory funnel containing CH_2Cl_2 and 10% $NaHCO_3$ solution, what compounds are dissolved in the organic layer?

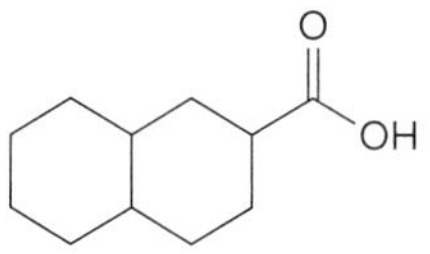

a. **A, B,** and **C**
b. **A** and **B**
c. **B** and **C**
d. **B** only

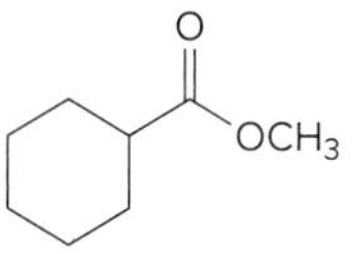

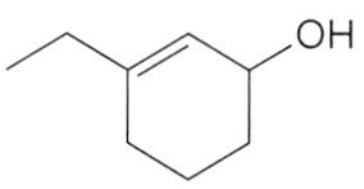

7. Give the IUPAC name for the following compound.

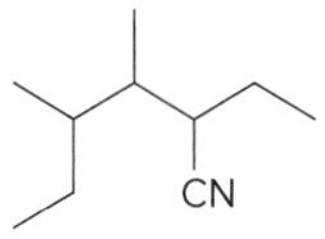

a. 2-ethyl-3,4-dimethylhexanenitrile
b. 4,5-dimethylheptane-3-nitrile
c. 3-cyano-4,5-dimethylheptane
d. 2,3-dimethyl-1-ethylhexanenitrile

8. What product is formed when **A** is treated with excess Na_2CO_3?

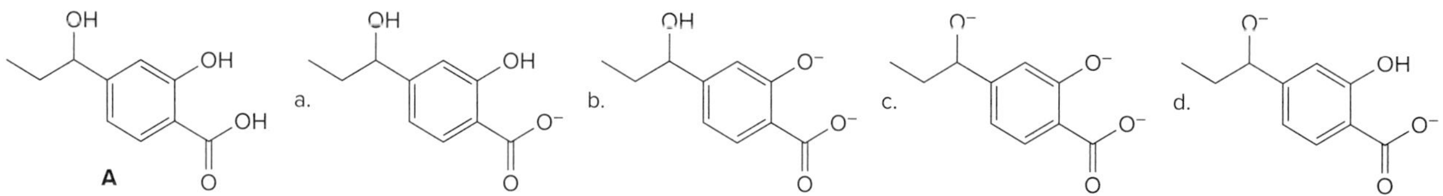

9. Which statement about phenol (PhOH) is *false*?
a. PhOH is more acidic than $PhCH_2OH$.
b. The conjugate base of PhOH is less effectively resonance stabilized than the conjugate base of $PhCO_2H$.
c. $NaOCH_3$ is *not* a strong enough base to remove a proton from PhOH.
d. $NaNH_2$ is a strong enough base to remove a proton from PhOH.

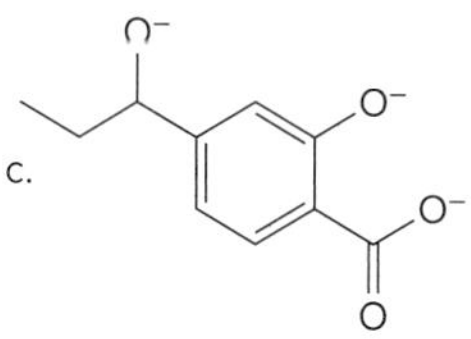

10. What product is formed in the following reaction sequence?

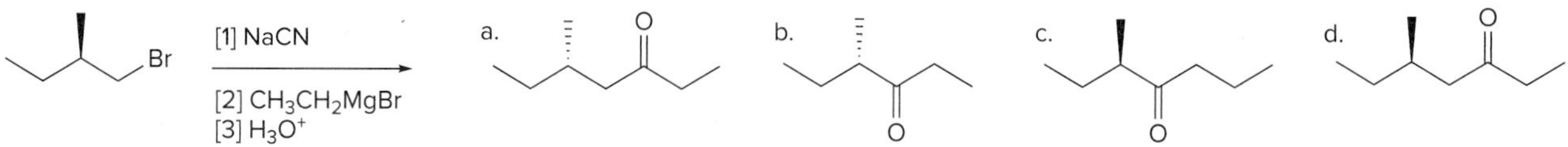

PROBLEMS

Problems Using Three-Dimensional Models

19.33 Answer each question for **A** and **B** depicted in the ball-and-stick models.

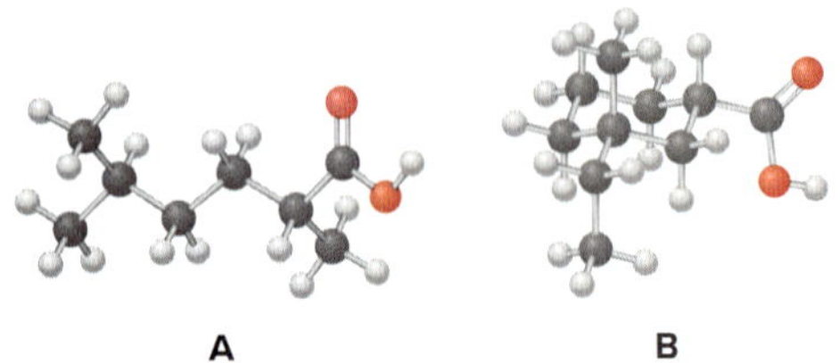

a. What is the IUPAC name for each compound?
b. What product is formed when each compound is treated with NaOH?
c. Name the products formed in part (b).
d. Draw the structure of an isomer that is at least 10^5 times less acidic than each compound.

19.34 (a) Give an acceptable name for compound **C.** (b) Draw the organic products formed when **C** is treated with each reagent: [1] H_3O^+; [2] ^-OH, H_2O; [3] $CH_3CH_2CH_2MgBr$ (excess), then H_2O; [4] $LiAlH_4$, then H_2O.

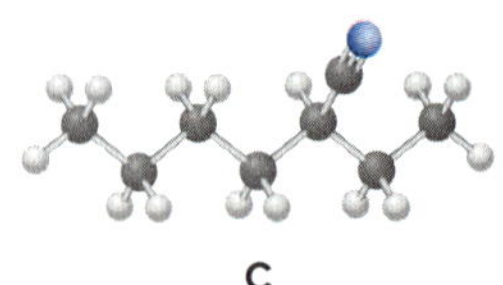

Nomenclature

19.35 Give the IUPAC name for each compound.

a.  O, OH
d. 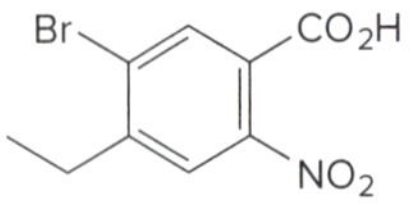 Br, CO_2H, NO_2
g. 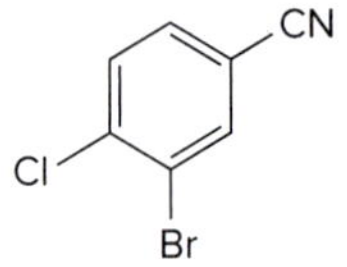 CN, Cl, Br
b. 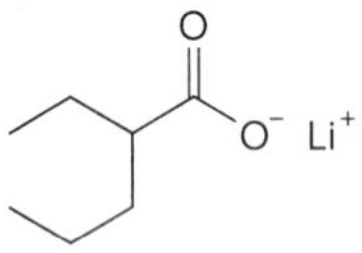 O, O^- Li^+
e. 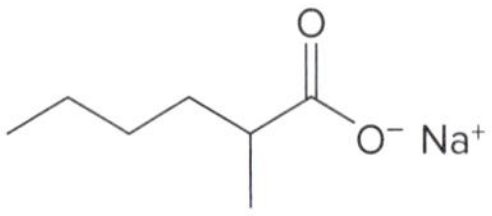 O, O^- Na^+
h. 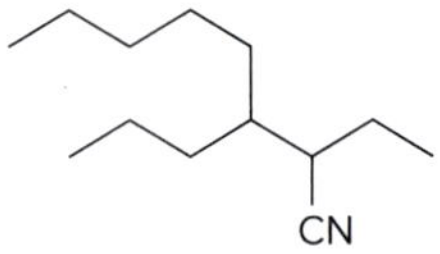 CN
c. 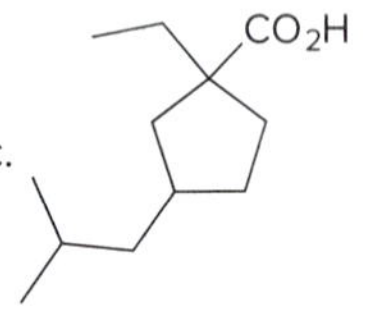 CO_2H
f. 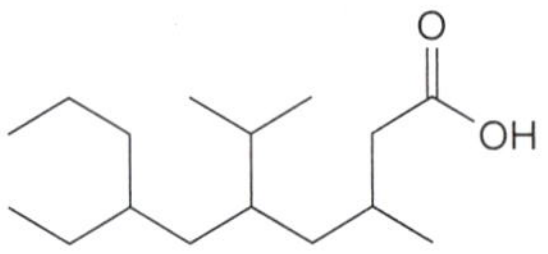 O, OH
i. 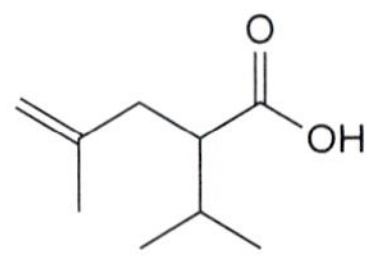 O, OH

19.36 Draw the structure corresponding to each name.

a. 3,3-dimethylpentanoic acid
b. 4-chloro-3-phenylheptanoic acid
c. (*R*)-2-chloropropanoic acid
d. *m*-hydroxybenzoic acid
e. potassium acetate
f. sodium α-bromobutyrate
g. 2,2-dichloropentanedioic acid
h. 4-isopropyl-2-methyloctanedioic acid
i. 3,3-dimethylpentanenitrile
j. 4,5-diethyl-2-isopropylnonanenitrile

Acid–Base Reactions; General Questions on Acidity

19.37 Using the pK_a table in Appendix C, determine whether each of the following bases is strong enough to deprotonate the three compounds listed below. Bases: [1] ^-OH; [2] $CH_3CH_2^-$; [3] $^-NH_2$; [4] NH_3; [5] $HC{\equiv}C^-$.

a. 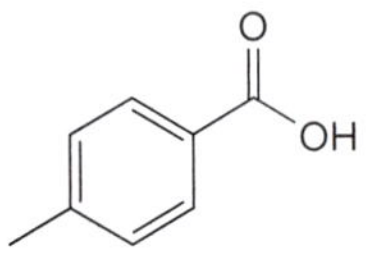 O, OH — $pK_a = 4.3$
b.  Cl, OH — $pK_a = 9.4$
c. 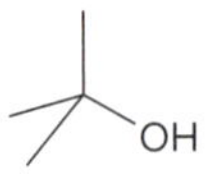 OH — $pK_a = 18$

19.38 Draw the products of each acid–base reaction, and using the pK_a table in Appendix C, determine if equilibrium favors the reactants or products.

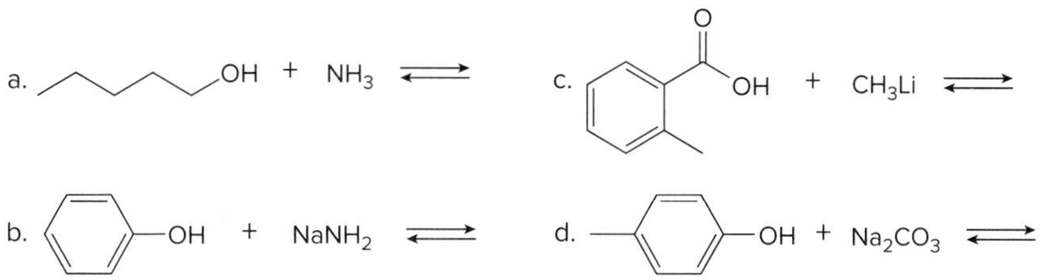

a. OH + NH_3 ⇌

b. OH + $NaNH_2$ ⇌

c. OH + CH_3Li ⇌

d. OH + Na_2CO_3 ⇌

19.39 (a) Rank the following compounds in order of increasing acidity. (b) Which compound forms the strongest conjugate base?

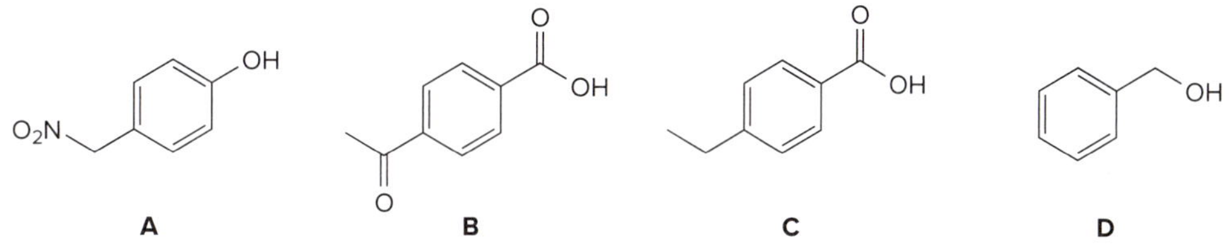

O_2N OH OH OH OH

A **B** **C** **D**

19.40 Caftaric acid is found in grapes, wine, and raisins. Rank the labeled protons in caftaric acid in order of increasing acidity.

HO O H_b O O OH_d HO OH_c OH_a

caftaric acid

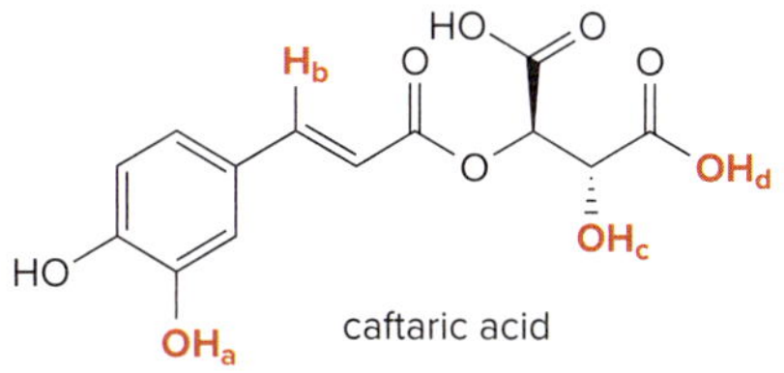

19.41 Rank compounds **A–D** in order of increasing basicity.

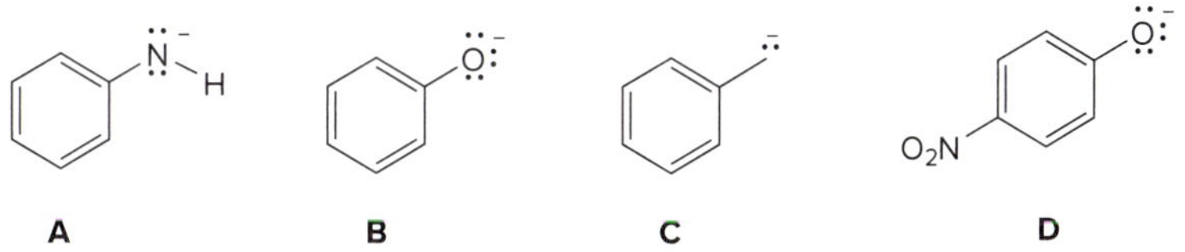

H O_2N

A **B** **C** **D**

19.42 Although codeine occurs in low concentration in the opium poppy, most of the codeine used in medicine is prepared from morphine (the principal component of opium) by the following reaction. Explain why selective methylation occurs at only one OH in morphine to give codeine. Codeine is a less potent and less addictive analgesic than morphine.

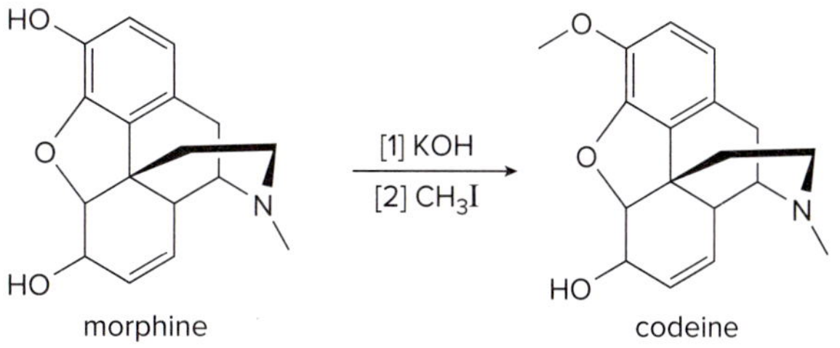

HO O N HO [1] KOH [2] CH_3I O O N HO

morphine codeine

19.43 Explain each statement.

a. The pK_a of *p*-nitrophenol is lower than the pK_a of phenol (7.2 vs. 10).

b. The pK_a of *p*-nitrophenol is lower than the pK_a of *m*-nitrophenol (7.2 vs. 8.3).

19.44 Explain this statement: Although 2-methoxyacetic acid (CH_3OCH_2COOH) is a stronger acid than acetic acid (CH_3COOH), *p*-methoxybenzoic acid ($CH_3OC_6H_4COOH$) is a weaker acid than benzoic acid (C_6H_5COOH).

19.45 The pK_a of *p*-methylthiophenol ($CH_3SC_6H_4OH$) is 9.53. Is *p*-methylthiophenol more or less reactive in electrophilic aromatic substitution than phenol?

19.46 Explain why the pK_a of compound **A** is lower than the pK_a's of both compounds **B** and **C.**

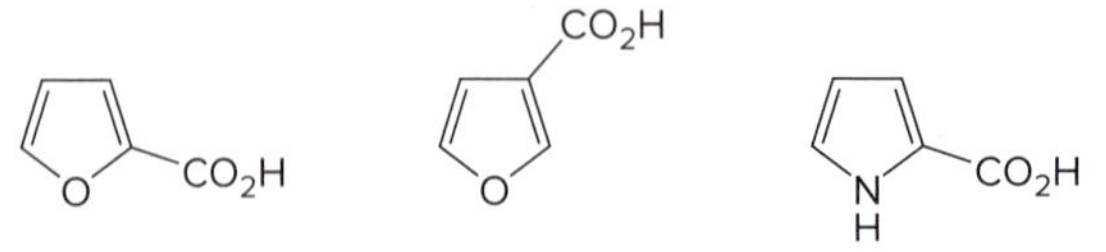

CO_2H CO_2H CO_2H

N H

A pK_a = 3.2 **B** pK_a = 3.9 **C** pK_a = 4.4

19.47 Phthalic acid and isophthalic acid have protons on two carboxy groups that can be removed with base. (a) Explain why the pK_a for loss of the first proton (pK_{a1}) is lower for phthalic acid than isophthalic acid. (b) Explain why the pK_a for loss of the second proton (pK_{a2}) is higher for phthalic acid than isophthalic acid.

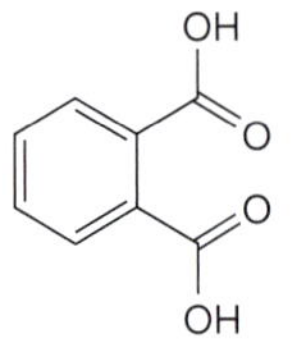

phthalic acid
pK_{a1} = 2.9
pK_{a2} = 5.4

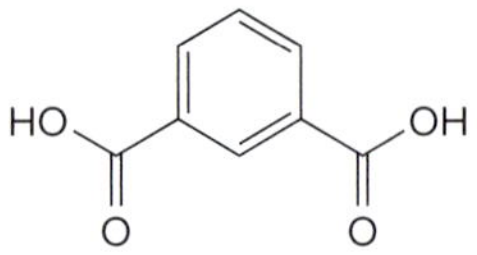

isophthalic acid
pK_{a1} = 3.7
pK_{a2} = 4.6

19.48 Explain this result: Acetic acid (CH_3COOH), labeled at its OH oxygen with the uncommon ^{18}O isotope (shown in red), was treated with aqueous base, and then the solution was acidified. Two products having the ^{18}O label at different locations were formed.

[1] NaOH
[2] H_3O^+

19.49 Draw all resonance structures of the conjugate bases formed by removal of the labeled protons (H_a, H_b, and H_c) in cyclohexane-1,3-dione and acetanilide. For each compound, rank these protons in order of increasing acidity and explain the order you chose.

a. 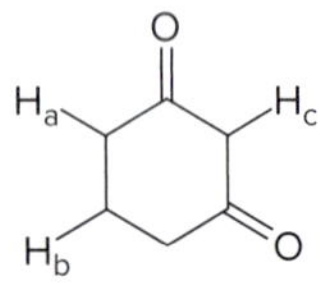 cyclohexane-1,3-dione

b. 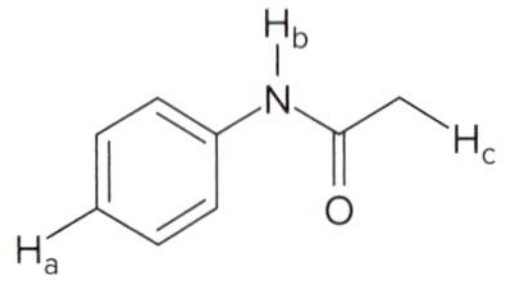 acetanilide

19.50 The pK_a of acetamide (CH_3CONH_2) is 16. Draw the structure for its conjugate base, and explain why acetamide is less acidic than CH_3COOH.

General Reactions

19.51 Draw the organic products formed in each reaction.

a. 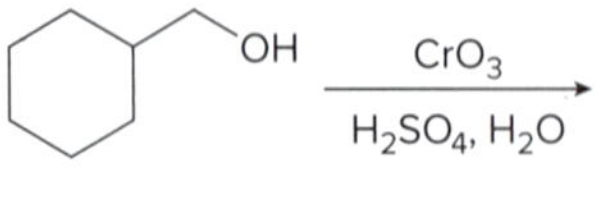 $\xrightarrow[H_2SO_4,\ H_2O]{CrO_3}$

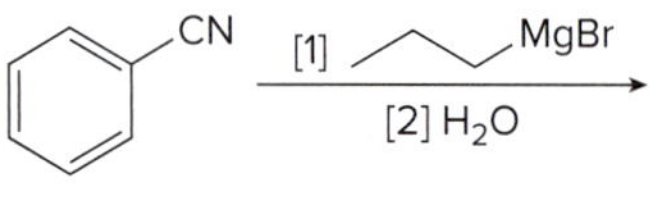

b. 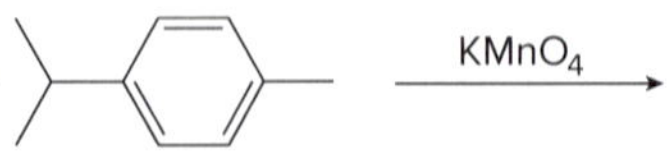 $\xrightarrow{KMnO_4}$

c. $\xrightarrow[[2]\ H_2O]{[1]\ O_3}$

d. 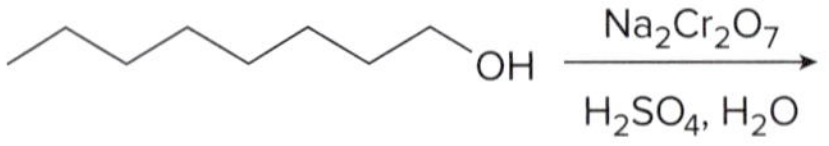 $\xrightarrow[H_2SO_4,\ H_2O]{Na_2Cr_2O_7}$

e. $\xrightarrow[[2]\ H_2O]{[1]\ \text{CH}_3\text{CH}_2\text{CH}_2\text{MgBr}}$

f. 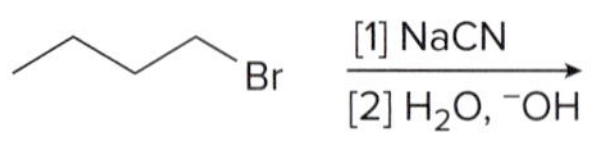 $\xrightarrow[[2]\ H_2O,\ ^-OH]{[1]\ NaCN}$

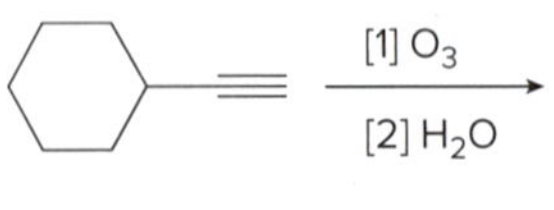

19.52 Identify the lettered compounds in each reaction sequence.

a. 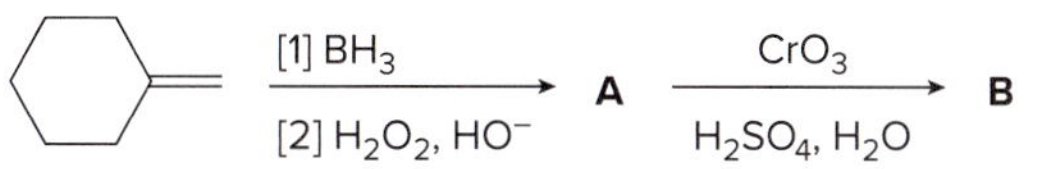 [1] BH_3 [2] H_2O_2, HO^- → **A**; CrO_3, H_2SO_4, H_2O → **B**

b. $HC{\equiv}CH$ [1] $NaNH_2$ [2] CH_3I → **C**; [1] $NaNH_2$ [2] CH_3CH_2I → **D**; [1] O_3 [2] H_2O → **E** + **F**

c. 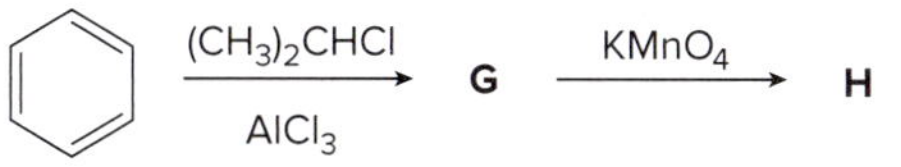 $(CH_3)_2CHCl$, $AlCl_3$ → **G**; $KMnO_4$ → **H**

d. 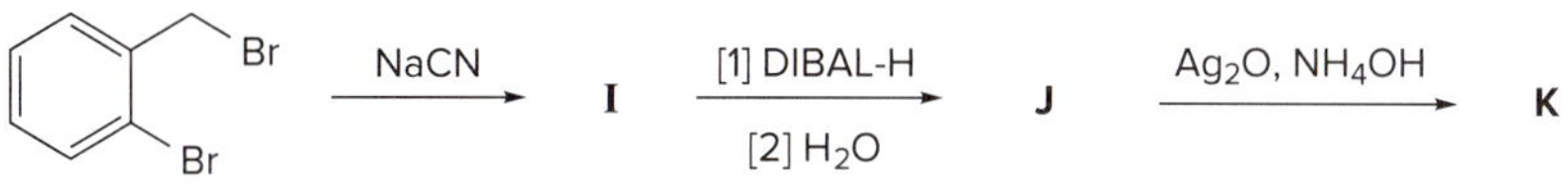 Br, Br; NaCN → **I**; [1] DIBAL-H [2] H_2O → **J**; Ag_2O, NH_4OH → **K**

19.53 Identify **A** and **B** in the following reaction sequence.

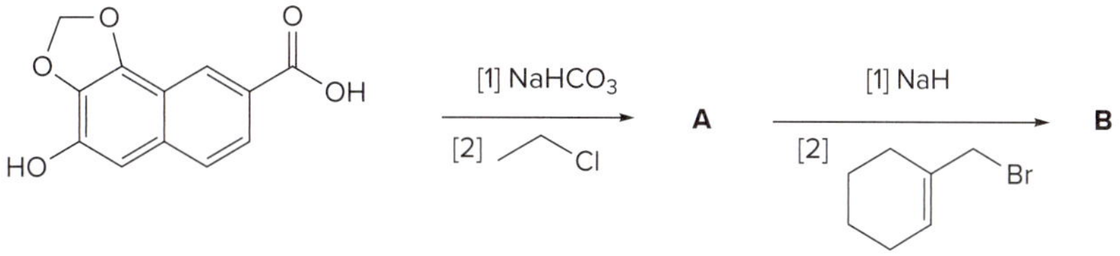

O, O, O, OH, HO; [1] $NaHCO_3$ [2] Cl → **A**; [1] NaH [2] Br → **B**

19.54 Draw the products of each reaction, and indicate the stereochemistry at all stereogenic centers.

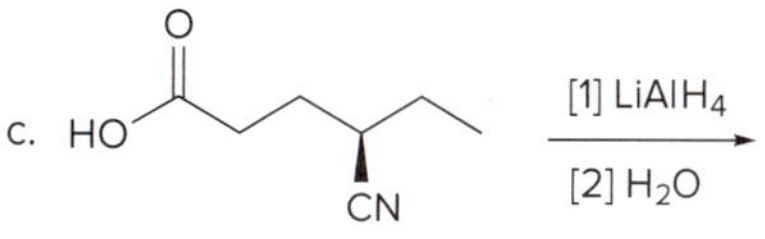

a. 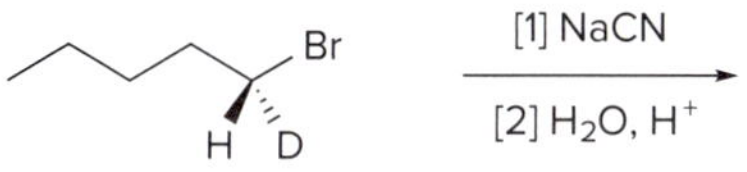 Br, H, D; [1] NaCN [2] H_2O, H^+ →

b. 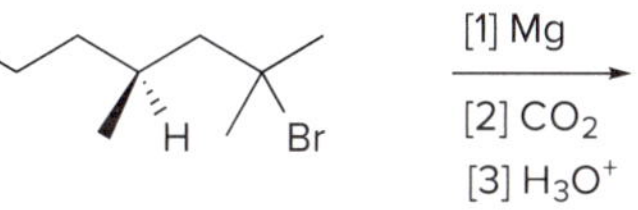 H, Br; [1] Mg [2] CO_2 [3] H_3O^+ →

c. HO, O, CN; [1] $LiAlH_4$ [2] H_2O →

d. CN, OCH_3; [1] DIBAL-H [2] H_2O [3] NH_2, mild H^+ →

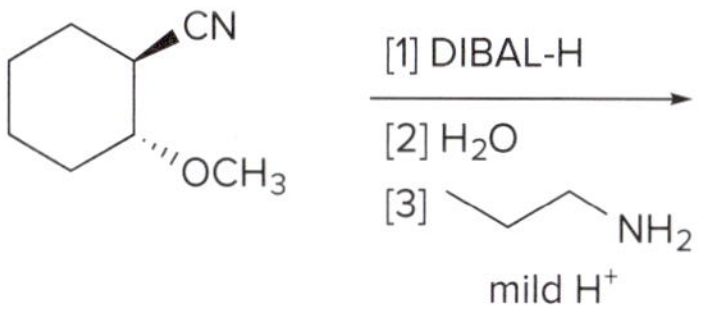

Extraction

19.55 Write out the steps needed to separate hydrocarbon **A** and carboxylic acid **B** by using an extraction procedure.

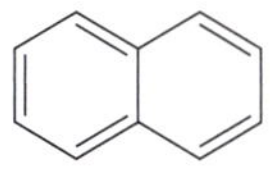

COOH

A 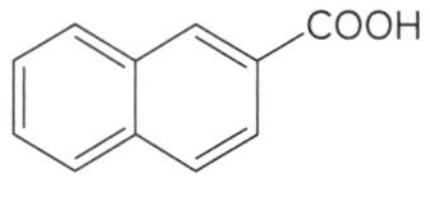 **B**

19.56 Because phenol (C_6H_5OH) is less acidic than a carboxylic acid, it can be deprotonated by NaOH but not by the weaker base $NaHCO_3$. Using this information, write out an extraction sequence that can be used to separate C_6H_5OH, benzoic acid, and cyclohexanol. Show what compound is present in each layer at each stage of the process, and if it is present in its neutral or ionic form.

19.57 A mixture of **A, B,** and **C** was added to a separatory funnel containing CH_2Cl_2, and an aqueous layer was added. In which layer is each compound dissolved when the aqueous layer consists of (a) pure water; (b) 10% NaOH solution; (c) 10% $NaHCO_3$ solution?

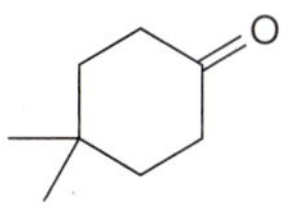

O, O, OH, OH

A 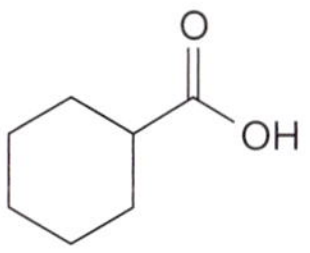 **B** 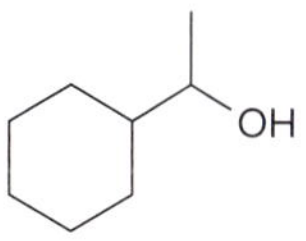 **C**

Amino Acids

19.58 Threonine is a naturally occurring amino acid that has two stereogenic centers.

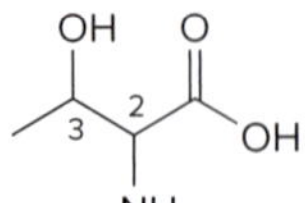

threonine

a. Draw the four possible stereoisomers using wedges and dashed wedges.
b. The naturally occurring amino acid has the 2*S*,3*R* configuration at its two stereogenic centers. Which structure does this correspond to?
c. Draw the naturally occurring isomer in its zwitterionic form.

19.59 Kainic acid, an amino acid isolated from red seaweed, is used by laboratory scientists to study various neurological conditions. Kainic acid is drawn in the form in which it exists at pH = 1. Rank the labeled protons in kainic acid in order of increasing acidity and explain your choice.

OH_a O O H_c H OH_b

kainic acid

19.60 Hypoglycin A, an amino acid derivative found in unripened lychee, is an acutely toxic compound that produces seizures, coma, and sometimes death in undernourished children when ingested on an empty stomach (Problem 5.25). (a) Draw the neutral, positively charged, and negatively charged forms of hypoglycin A. (b) Which form predominates at pH = 1, 6, and 11? (c) What is the structure of hypoclycin A at its isoelectric point?

hypoglycin A

19.61 Calculate the isoelectric point for each amino acid.
a. asparagine: pK_a (COOH) = 2.02; pK_a (α-NH_3^+) = 8.80
b. methionine: pK_a (COOH) = 2.28; pK_a (α-NH_3^+) = 9.21

19.62 Lysine and tryptophan are two amino acids that contain an additional N atom in the R group bonded to the α carbon. While lysine is classified as a basic amino acid because it contains an additional basic N atom, tryptophan is classified as a neutral amino acid. Explain why this difference in classification occurs.

lysine

tryptophan

19.63 Glutamic acid is a naturally occurring α-amino acid that contains a carboxy group in its R group side chain (Table 19.4). (Glutamic acid is drawn in its neutral form with no charged atoms, a form that does not actually exist at any pH.)

glutamic acid

a. What form of glutamic acid exists at pH = 1?
b. If the pH is gradually increased, what form of glutamic acid exists after one equivalent of base is added? After two equivalents? After three equivalents?
c. Propose a structure of monosodium glutamate, the common flavor enhancer known as MSG.

Synthesis

19.64 Two methods convert an alkyl halide to a carboxylic acid having one more carbon atom.

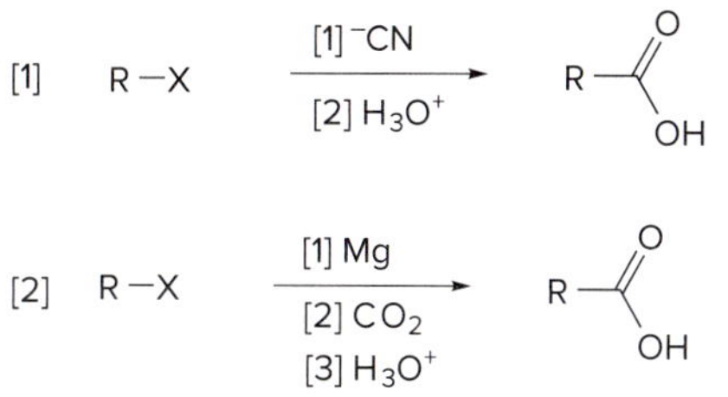

[1] R−X —[1] ^{-}CN; [2] H_3O^+→ RCOOH (Section 19.12)

[2] R−X —[1] Mg; [2] CO_2; [3] H_3O^+→ RCOOH (Section 17.14A)

Depending on the structure of the alkyl halide, one or both of these methods may be employed. For each alkyl halide, write out a stepwise sequence that converts it to a carboxylic acid with one more carbon atom. If both methods work, draw both routes. If one method cannot be used, state why it can't.

a. CH_3Cl b. 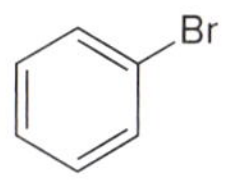 [bromobenzene] c. 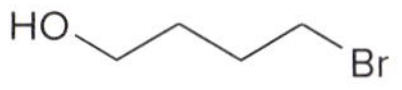 HO–(CH$_2$)$_5$–Br

19.65 Synthesize each compound from benzonitrile (C_6H_5CN) as the only organic starting material; that is, every carbon in the product must originate in benzonitrile.

a. 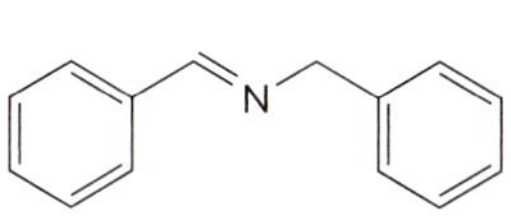 [structure: C$_6$H$_5$CH=NCH$_2$C$_6$H$_5$] b. 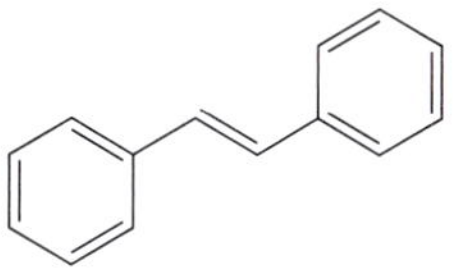 [structure: C$_6$H$_5$CH=CHC$_6$H$_5$] c. [structure: C$_6$H$_5$C(OH)(CH$_2$C$_6$H$_5$)$_2$]

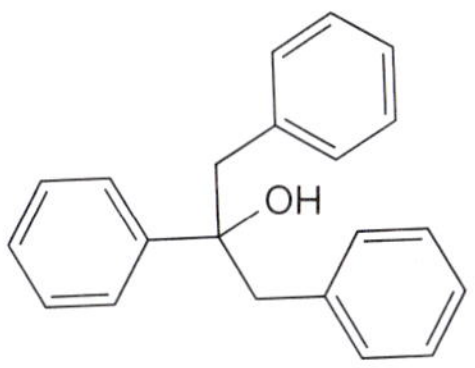

19.66 Devise a synthesis of each compound from the indicated starting material. You may also use any organic compounds with one or two carbons and any needed inorganic reagents.

a. 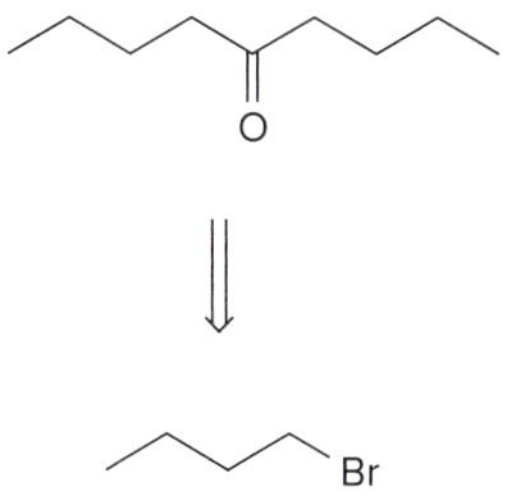 [structure: nonan-5-one] ⟸ [1-bromobutane]

b. 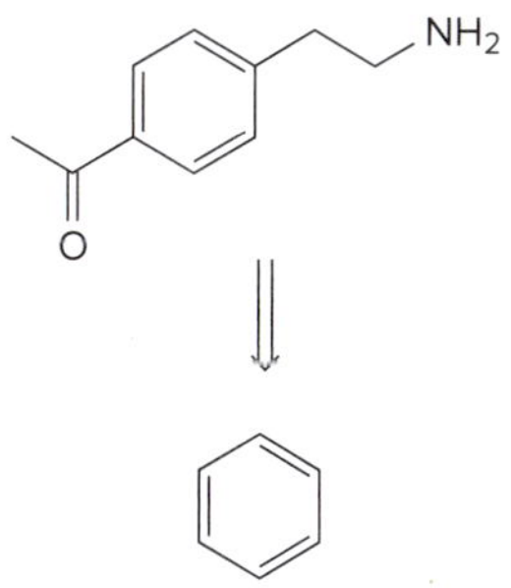 [structure: 4-(2-aminoethyl)acetophenone, with NH$_2$] ⟸ [benzene]

c. 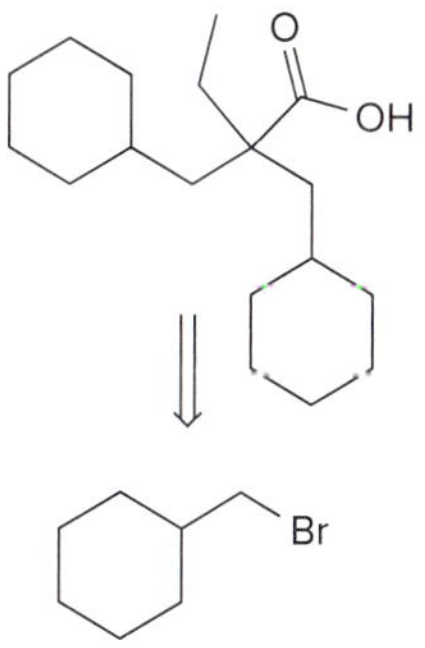 [structure: 2-(cyclohexylmethyl)-2-ethyl-3-cyclohexylpropanoic acid, with OH] ⟸ [(bromomethyl)cyclohexane, with Br]

Spectroscopy

19.67 Identify each compound from its spectral data.

a. Molecular formula: $C_8H_8O_3$
IR: 3500–2500 cm^{-1}, 1710 cm^{-1}
^{1}H NMR data: 4.7 (singlet, 2 H), 6.9–7.3 (multiplet, 5 H), and 11.3 (singlet, 1 H) ppm

b. Molecular formula: C_4H_7N
IR: 2250 cm^{-1}
^{1}H NMR data: 1.08 (triplet, 3 H), 1.70 (multiplet, 2 H), and 2.34 (triplet, 2 H) ppm

19.68 Use the ^1H NMR and IR spectra given below to identify the structure of compound **B** (molecular formula $C_4H_8O_2$).

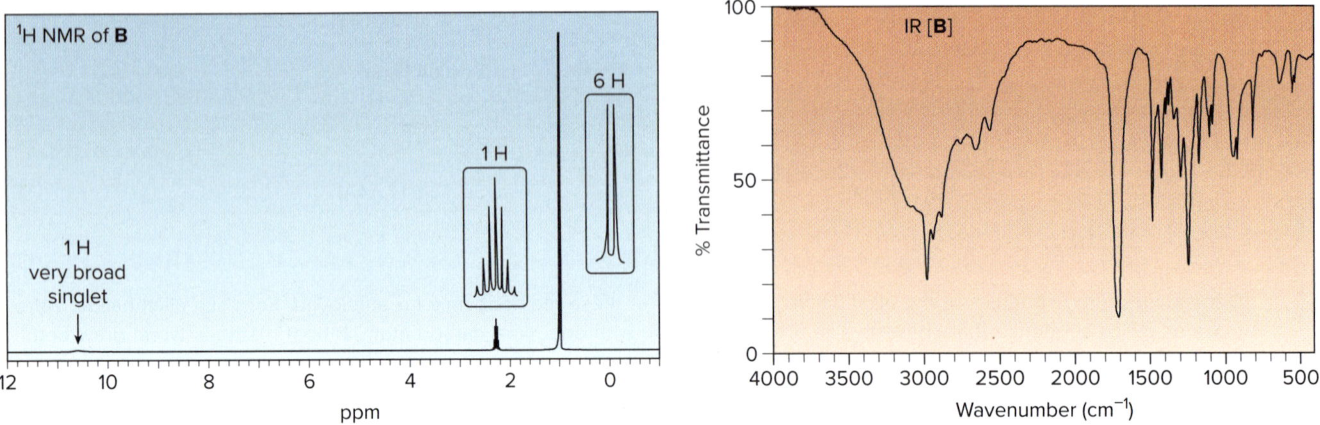

19.69 An unknown compound **C** (molecular formula $C_4H_8O_3$) exhibits IR absorptions at 3600–2500 and 1734 cm^{-1}, as well as the following ^1H NMR spectrum. What is the structure of **C?**

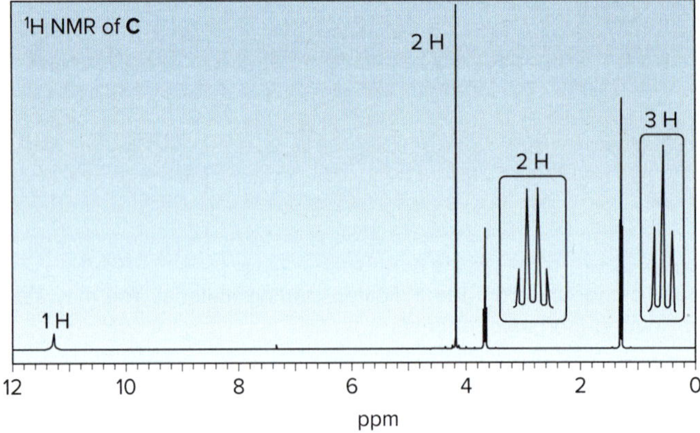

19.70 Propose a structure for **D** (molecular formula $C_9H_9ClO_2$) consistent with the given spectroscopic data.

^{13}C NMR signals at 30, 36, 128, 130, 133, 139, and 179 ppm

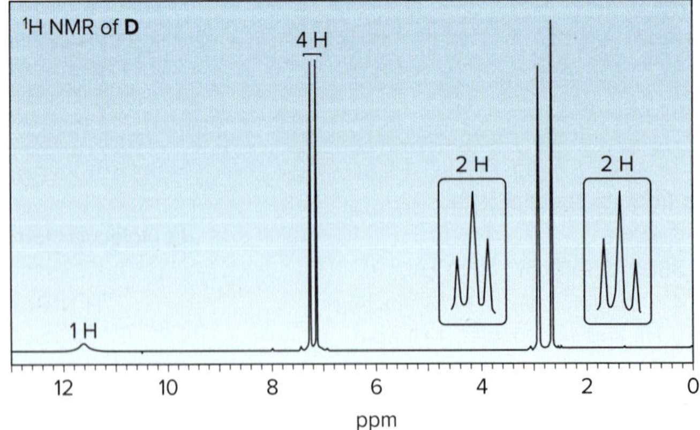

Challenge Problems

19.71 Explain why using one or two equivalents of NaH results in different products in the following reactions.

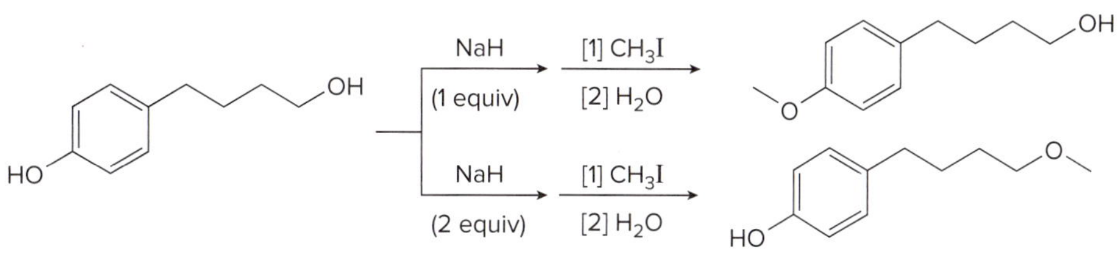

NaH (1 equiv) → [1] CH_3I [2] H_2O

NaH (2 equiv) → [1] CH_3I [2] H_2O

19.72 Although *p*-hydroxybenzoic acid is less acidic than benzoic acid, *o*-hydroxybenzoic acid is slightly more acidic than benzoic acid. Explain this result.

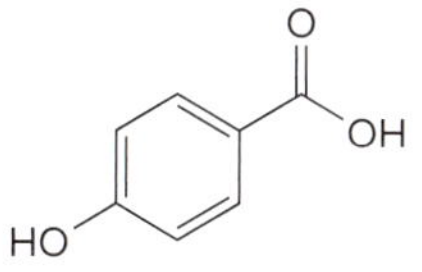

p-hydroxybenzoic acid

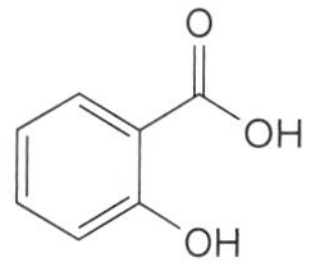

o-hydroxybenzoic acid

19.73 2-Hydroxybutanedioic acid occurs naturally in apples and other fruits. Rank the labeled protons (H_a–H_e) in order of increasing acidity, and explain in detail the order you chose.

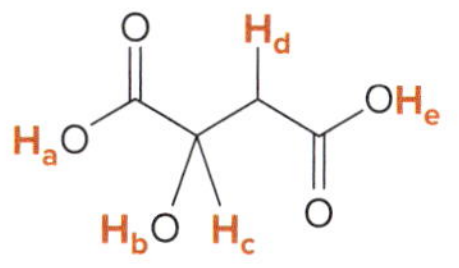

2-hydroxybutanedioic acid

19.74 Explain why tropolone (pK_a = 6.7) is more acidic than ethanol and phenol.

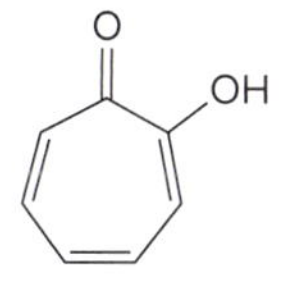

tropolone

SELF-TEST ANSWERS

1. d 2. d 3. a 4. b 5. b 6. c 7. a 8. a 9. c 10. d

20 Carboxylic Acids and Their Derivatives—Nucleophilic Acyl Substitution

Likit Supasai/Shutterstock

20.1 Introduction
20.2 Structure and bonding
20.3 Nomenclature
20.4 Physical and spectroscopic properties
20.5 Interesting esters and amides
20.6 Introduction to nucleophilic acyl substitution
20.7 Reactions of acid chlorides
20.8 Reactions of anhydrides
20.9 Reactions of carboxylic acids
20.10 Reactions of esters
20.11 Application: Lipid hydrolysis
20.12 Reactions of amides
20.13 Application: The mechanism of action of β-lactam antibiotics
20.14 Summary of nucleophilic acyl substitution reactions
20.15 Natural and synthetic fibers
20.16 Biological acylation reactions

Cocaine is an addictive stimulant obtained from the leaves of the coca plant, *Erythroxylon coca*. Chewing coca leaves for pleasure has been practiced by the indigenous peoples of South America for over a thousand years, and coca leaves were a very minor ingredient in Coca-Cola for the first 20 years of its production. Cocaine is a widely abused recreational drug, and the possession and use of cocaine is currently illegal in most countries. Cocaine contains two esters, carboxylic acid derivatives discussed in Chapter 20.

Why Study . . . Carboxylic Acid Derivatives?

Chapter 20 continues the study of carbonyl compounds with a detailed look at **nucleophilic acyl substitution,** a key reaction of carboxylic acids and their derivatives. Substitution at sp^2 hybridized carbon atoms was introduced in Chapter 17 with reactions involving carbon and hydrogen nucleophiles. In Chapter 20, we learn that nucleophilic acyl substitution is a general reaction that occurs with a variety of heteroatomic nucleophiles. ***Every* reaction in Chapter 20 that begins with a carbonyl compound involves nucleophilic substitution.** Nucleophilic acyl substitutions are useful reactions in both the laboratory and biological systems. Penicillin is an effective antibiotic because it kills bacteria by a nucleophilic substitution mechanism.

20.1 Introduction

Chapter 20 focuses on carbonyl compounds that contain an **acyl group bonded to an electronegative atom.** These include the **carboxylic acids,** as well as carboxylic acid derivatives that can be prepared from them: **acid chlorides, anhydrides, esters,** and **amides.**

:O:
R Z:
acyl group

Z = an atom more electronegative than carbon

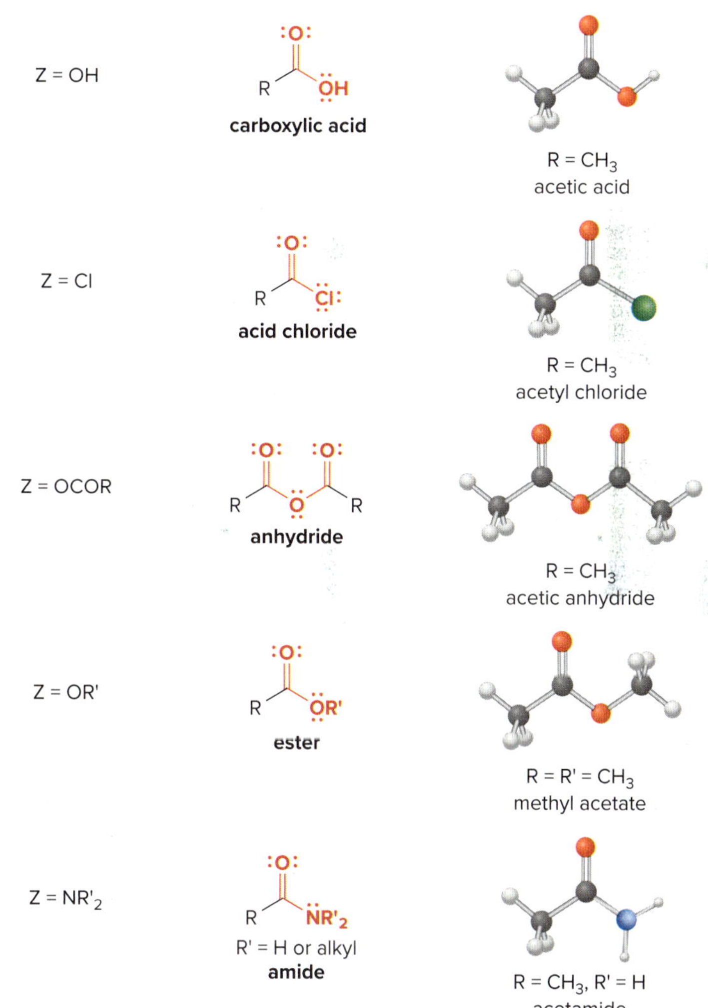

Anhydrides contain two carbonyl groups joined by a single oxygen atom. **Symmetrical anhydrides** have two identical alkyl groups bonded to the carbonyl carbons, and **mixed anhydrides** have two different alkyl groups. **Cyclic anhydrides** are also known.

symmetrical anhydride mixed anhydride cyclic anhydride

As we learned in Section 3.2, **amides** are classified as **1°, 2°,** or **3°** depending on the number of carbon atoms bonded directly to the *nitrogen* atom.

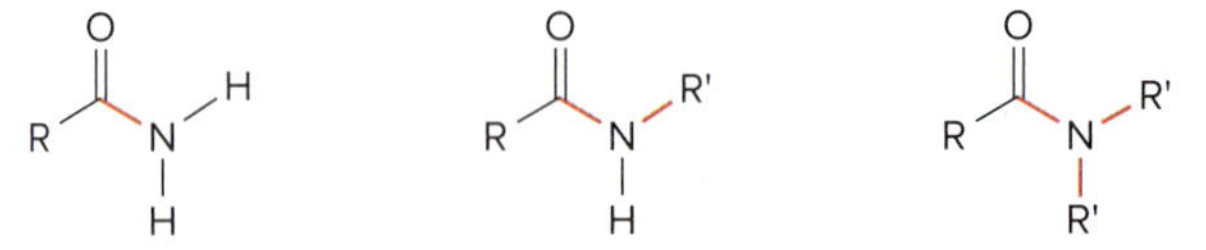

1° amide
1 C—N bond

2° amide
2 C—N bonds

3° amide
3 C—N bonds

Cyclic esters and amides are called **lactones** and **lactams,** respectively. The ring size of the heterocycle is indicated by a Greek letter. An amide in a four-membered ring is called a **β-lactam,** because the β carbon to the carbonyl is bonded to the heteroatom. An ester in a five-membered ring is called a **γ-lactone.**

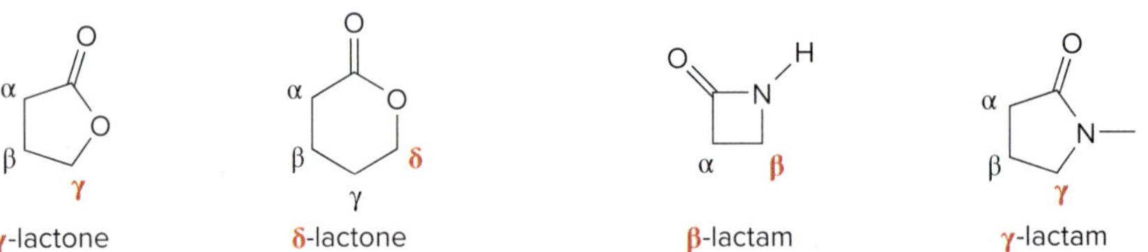

All of these compounds contain an acyl group bonded to an electronegative atom Z that can serve as a **leaving group.** As a result, these compounds undergo **nucleophilic acyl substitution.** Recall from Chapters 17 and 18 that aldehydes and ketones do *not* undergo nucleophilic substitution because they have no leaving group on the carbonyl carbon.

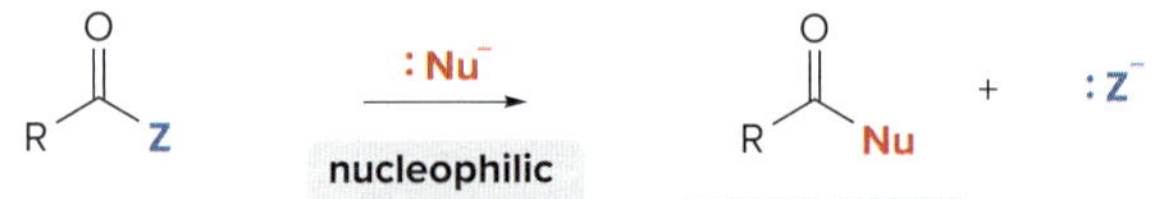

Nucleophilic acyl substitution was first discussed in Chapter 17 with R^- and H^- as the nucleophiles. This substitution reaction is general for a variety of nucleophiles, making it possible to form many different substitution products, as discussed in Sections 20.7–20.12.

Z = OH, Cl, OCOR, OR', NR'_2

$:Nu^-$

nucleophilic substitution

Nu replaces **Z.**

Problem 20.1 Oxytocin, sold under the trade name Pitocin, is a naturally occurring hormone used to stimulate uterine contractions and induce labor. Classify each amide in oxytocin as 1°, 2°, or 3°.

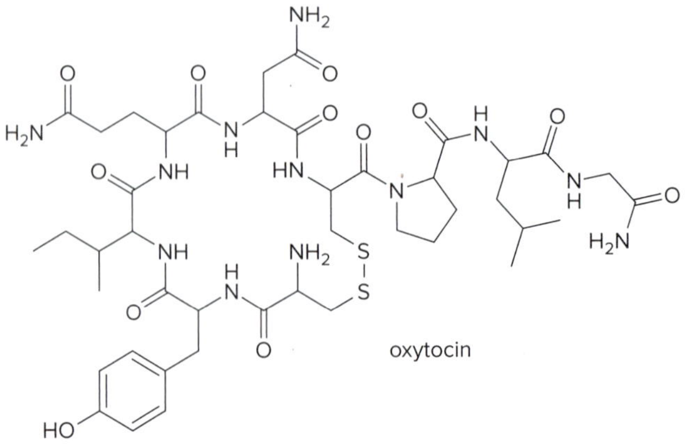

oxytocin

20.2 Structure and Bonding

The two most important features of any carbonyl group, regardless of the other groups bonded to it, are the following:

sp^2 hybridized

~120°

trigonal planar

$\delta+$ $\delta-$

electrophilic carbon

- The carbonyl carbon is sp^2 hybridized and trigonal planar, making it relatively *uncrowded*.
- The electronegative oxygen atom polarizes the carbonyl group, making the carbonyl carbon *electrophilic*.

As we learned in Section B.3C, three resonance structures can be drawn for RCOZ, compared to just two for aldehydes and ketones (Section 17.1). These three resonance structures stabilize RCOZ by delocalizing electron density. In fact, **the more resonance structures 2 and 3 contribute to the resonance hybrid, the *more* stable RCOZ is.**

1 ⟷ **2** ⟷ **3** **hybrid**

- The *more* basic Z is, the *more* it donates its electron pair, and the *more* resonance structure **3** contributes to the hybrid.

To determine the relative basicity of the leaving group Z, we compare the pK_a values of the conjugate acids HZ, given in Table 20.1. The following order of basicity results:

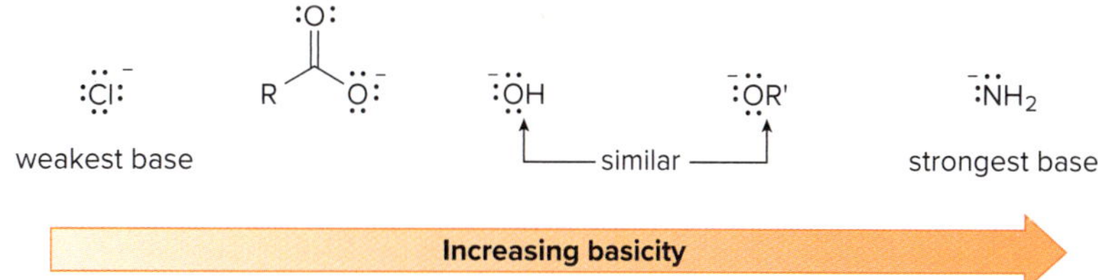

Table 20.1 pK_a Values of the Conjugate Acids (HZ) for Common Z Groups of Acyl Compounds (RCOZ)

Structure	Leaving group (Z^-)	Conjugate acid (HZ)	pK_a
RCOCl acid chloride	Cl^-	HCl	–7
$(RCO)_2O$ anhydride	RCO_2^-	RCO_2H	3–5
RCO_2H carboxylic acid	^-OH	H_2O	14
RCO_2R' ester	$^-OR'$	R'OH	15.5–18
$RCONR'_2$ amide	$^-NR'_2$	R'_2NH	38–40

Increasing basicity of Z (downward); Increasing acidity of HZ (upward)

Because the basicity of Z determines the relative stability of the carboxylic acid derivatives, the following **order of stability** results:

acid chlorides | anhydrides | carboxylic acids ~ esters | amides

(carboxylic acids and esters: similar)

Increasing stability →

Thus, **an acid chloride is the *least* stable carboxylic acid derivative** because Cl^- is the weakest base. **An amide is the *most* stable carboxylic acid derivative** because $^-NR'_2$ is the strongest base.

- **In summary: As the basicity of Z *increases,* the stability of RCOZ *increases* because of added resonance stabilization.**

Problem 20.2 Draw the three possible resonance structures for an acid bromide, RCOBr. Then, using the pK_a values in Appendix C, decide if RCOBr is more or less stabilized by resonance than a carboxylic acid (RCOOH).

Problem 20.3 How do the following experimental results support the resonance description of the relative stability of acid chlorides compared to amides? The C–Cl bond lengths in CH_3Cl and CH_3COCl are identical (178 pm), but the C–N bond in $HCONH_2$ is shorter than the C–N bond in CH_3NH_2 (135 pm versus 147 pm).

20.3 Nomenclature

The names of carboxylic acid derivatives are formed from the names of the parent carboxylic acids discussed in Section 19.2. Keep in mind that the common names **formic acid, acetic acid,** and **benzoic acid** are generally used for the parent acid, so these common parent names are used for their derivatives as well.

20.3A Naming an Acid Chloride—RCOCl

Acid chlorides are named by naming the acyl group and adding the word ***chloride.*** Two different methods are used.

[1] For acyclic acid chlorides: Change the suffix *-ic acid* of the parent carboxylic acid to the suffix *-yl chloride;* or

[2] When the –COCl group is bonded to a ring: Change the suffix *-carboxylic acid* to *-carbonyl chloride.*

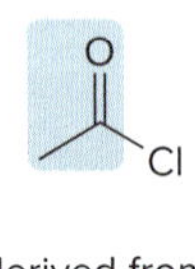

derived from acet*ic acid*

acetyl chloride

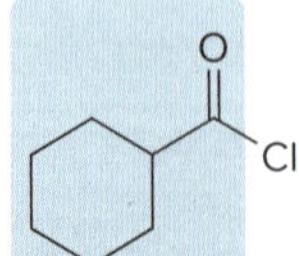

derived from cyclohexane*carboxylic acid*

cyclohexanecarbonyl chloride

derived from 2-methylbutano*ic acid*

2-methylbutanoyl chloride

20.3B Naming an Anhydride

Symmetrical anhydrides are named by changing the ***acid*** ending of the parent carboxylic acid to the word ***anhydride.*** **Mixed anhydrides,** which are derived from two different carboxylic acids, are named by alphabetizing the names for both acids and replacing the word ***acid*** by the word ***anhydride.***

The word *anhydride* means "without water." Removing one molecule of water from two molecules of carboxylic acid forms an anhydride.

R–COOH + HO–CO–R → ($-H_2O$) → RCO–O–COR

anhydride

derived from acetic acid

acetic anhydride

derived from acetic acid and benzoic acid

acetic benzoic anhydride

20.3C Naming an Ester—RCOOR'

Esters are often written as RCOOR', where the alkyl group (R') is written *last.* When an ester is named, however, the R' group appears *first* in the name.

An ester has two parts to its structure, each of which must be named: an **acyl group (RCO–)** and an **alkyl group** (designated as **R'**) bonded to an oxygen atom.

- **In the IUPAC system, esters are identified by the suffix *-ate.***

How To Name an Ester (RCO_2R') Using the IUPAC System

Example Give a systematic name for each ester:

a. b.

Step [1] **Name the R' group bonded to the oxygen atom as an alkyl group.**

- The name of the alkyl group, ending in the suffix *-yl,* becomes the ***first*** part of the ester name.

ethyl group

***tert*-butyl** group

Step [2] **Name the acyl group (RCO–) by changing the *-ic acid* ending of the parent carboxylic acid to the suffix *-ate.***

- The name of the acyl group becomes the **second** part of the name.

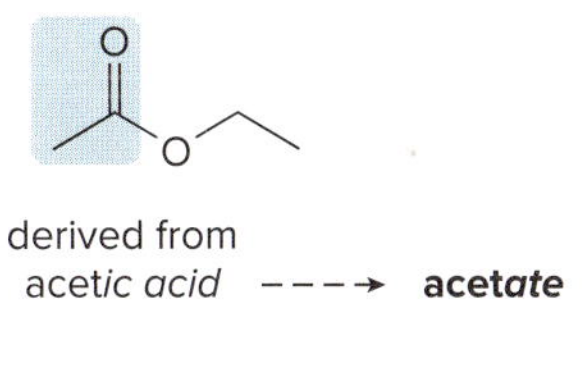

derived from acet*ic acid* ---→ **acet*ate***

Answer: ethyl acetate

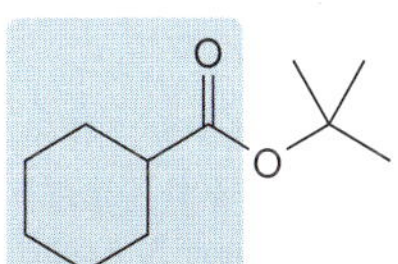

derived from cyclohexanecarboxyl*ic acid* ---→ **cyclohexanecarboxyl*ate***

Answer: *tert*-butyl cyclohexanecarboxylate

20.3D Naming an Amide

All 1° amides are named by replacing the ***-ic acid, -oic acid,*** or ***-ylic acid*** ending of the parent carboxylic acid with the suffix ***amide.***

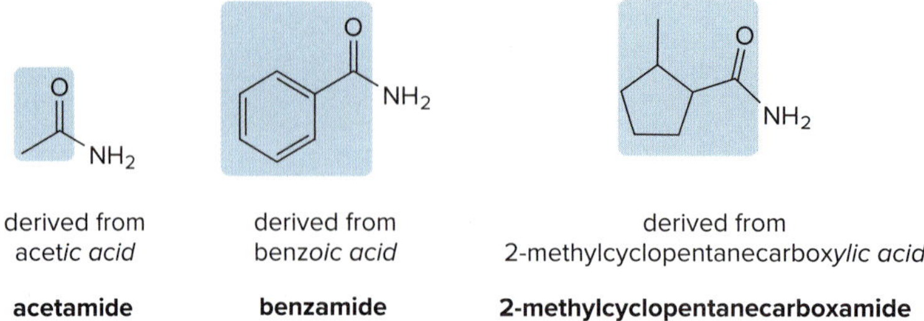

A 2° or 3° amide has two parts to its structure: an **acyl group** that contains the carbonyl group **(RCO–)** and one or two **alkyl groups** bonded to the nitrogen atom.

How To Name a 2° or 3° Amide

Example Give a systematic name for each amide:

a. H–C(=O)–NH–CH$_2$CH$_3$ b. C$_6$H$_5$–C(=O)–N(CH$_3$)$_2$

Step [1] **Name the alkyl group (or groups) bonded to the N atom of the amide. Use the prefix "*N*-" preceding the name of each alkyl group.**

- The names of the alkyl groups form the ***first*** part of each amide name.
- For 3° amides, use the prefix ***di-*** if the two alkyl groups on N are the same. If the two alkyl groups are different, **alphabetize** their names. One "***N*-**" is needed for each alkyl group, even if both R groups are identical.

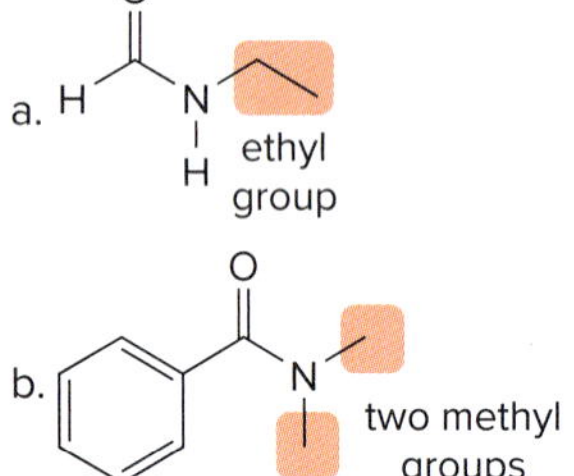

- The compound is a 2° amide with one ethyl group → ***N*-ethyl.**

- The compound is a 3° amide with two methyl groups.
- Use the prefix *di-* and two "*N*-" to begin the name → ***N,N*-dimethyl.**

Step [2] **Name the acyl group (RCO–) with the suffix *-amide.***

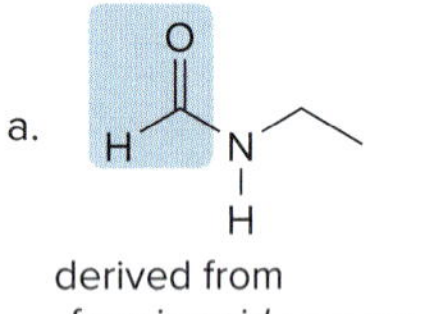

derived from
formic acid - - - - -> form***amide***

- Change the *-ic acid* or *-oic acid* suffix of the parent carboxylic acid to the suffix ***-amide.***
- Put the two parts of the name together.
- **Answer: *N*-ethylformamide**

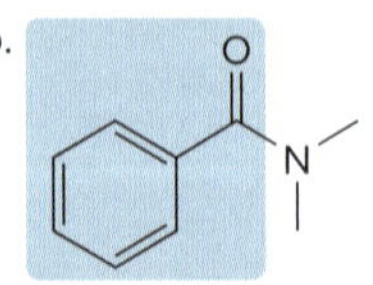

derived from
benzoic acid - - -> benz***amide***

- Change benz*oic acid* to **benz*amide*.**
- Put the two parts of the name together.
- **Answer: *N,N*-dimethylbenzamide**

Table 20.2 summarizes the most important points about the nomenclature of carboxylic acid derivatives.

Table 20.2 Summary: Nomenclature of Carboxylic Acid Derivatives

Compound	Name ending	Example	Name
acid chloride	**-yl chloride** or **-carbonyl chloride**	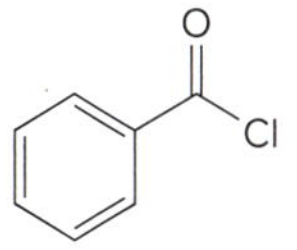	benzoyl chloride
anhydride	**anhydride**	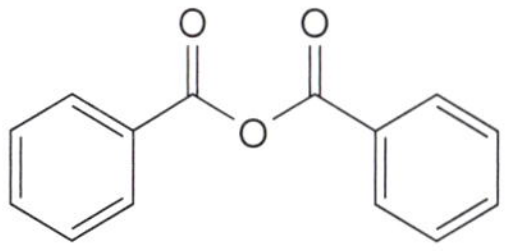	benzoic anhydride
ester	**-ate**	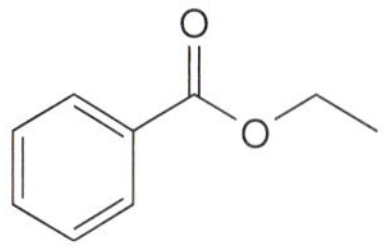	ethyl benzoate
amide	**-amide**	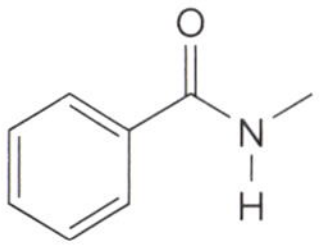	*N*-methylbenzamide

Sample Problem 20.1 Naming Carboxylic Acid Derivatives

Give the IUPAC name for each compound.

a. 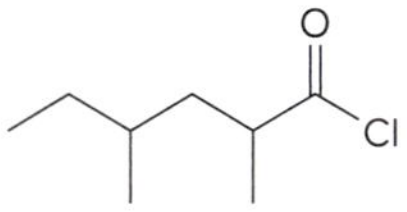 b.

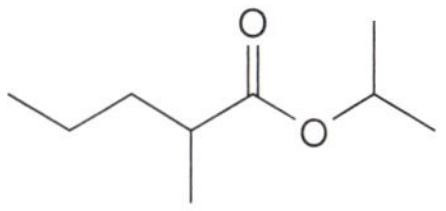

Solution

a. The functional group is an acid chloride bonded to a chain of atoms, so the name ends in ***-yl chloride.***

[1] Find and name the longest chain containing the COCl:

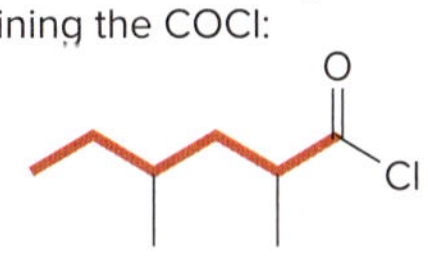

hexano*ic acid* ⟶ **hexano*yl chloride***
(6 C's)

[2] Number and name the substituents:

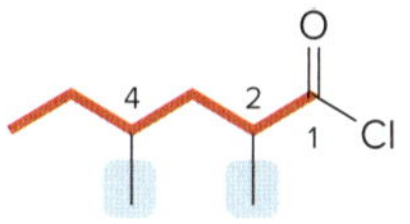

Answer:
2,4-dimethylhexanoyl chloride

b. The functional group is an ester, so the name ends in ***-ate.***

[1] Find and name the longest chain containing the carbonyl group:

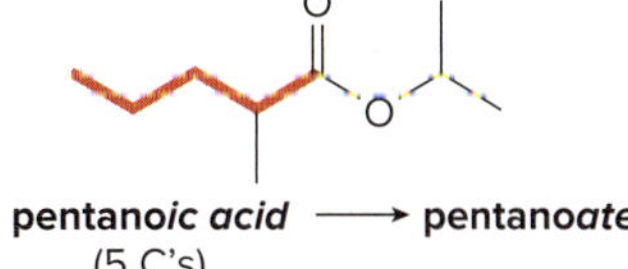

pentano*ic acid* ⟶ **pentano*ate***
(5 C's)

[2] Number and name the substituents:

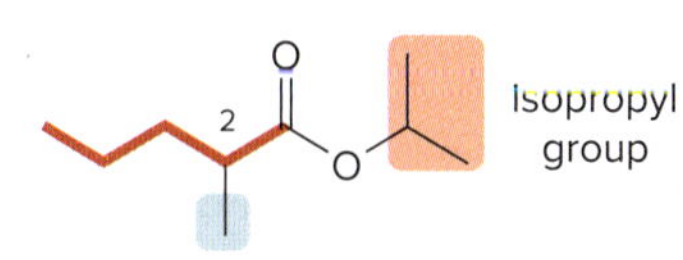

Answer: isopropyl 2-methylpentanoate

The name of the alkyl group on the O atom goes ***first*** in the name.

Problem 20.4 Give an IUPAC or common name for each compound.

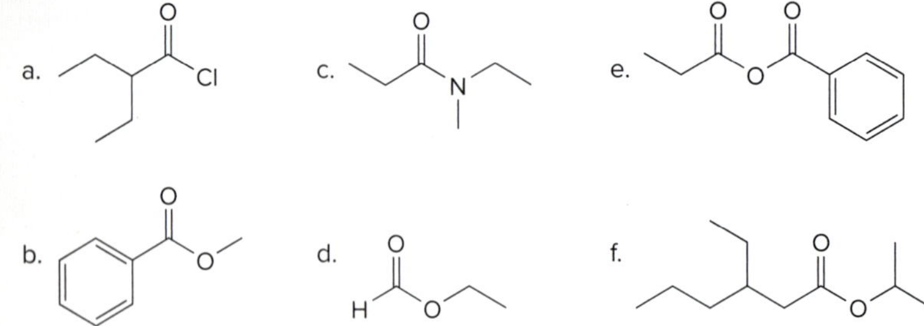

More Practice: Try Problems 20.34a, 20.36, 20.37.

Problem 20.5 Draw the structure corresponding to each name.

a. 5-methylheptanoyl chloride
b. isopropyl propanoate
c. acetic formic anhydride
d. *N*-isobutyl-*N*-methylbutanamide
e. *sec*-butyl 2-methylhexanoate
f. *N*-ethylhexanamide

20.4 Physical and Spectroscopic Properties

20.4A Physical Properties

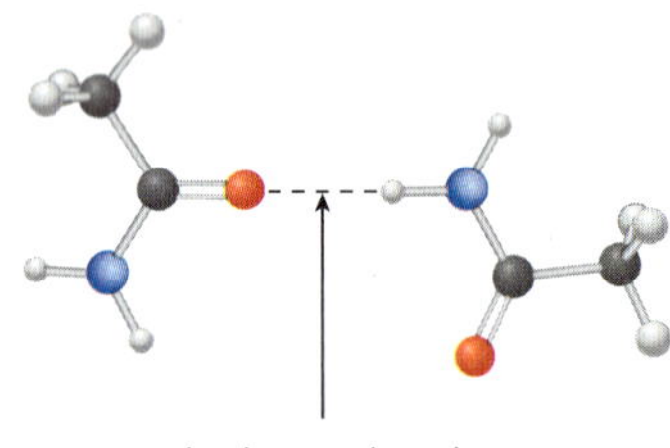

hydrogen bond

Because all carbonyl compounds have a polar carbonyl group, they exhibit **dipole–dipole interactions.** Primary (1°) and 2° amides are capable of intermolecular hydrogen bonding because they contain one or two N–H bonds. The N–H bond of one amide intermolecularly hydrogen bonds to the C=O of another amide, as shown using two acetamide molecules (CH_3CONH_2). As a result, **1° and 2° amides have *higher* boiling points and melting points** than compounds of comparable molecular weight.

MW = 78.5 bp 52 °C	~	MW = 74 bp 58 °C	~	MW = 72 bp 80 °C	<	MW = 73 bp 213 °C
		similar boiling points				**higher boiling point** **1° amide**

Problem 20.6 Explain why the boiling point of CH_3CONH_2 (221 °C) is significantly higher than the boiling point of CH_3CO_2H (118 °C).

20.4B Spectroscopic Properties

Many details of the spectroscopy of carboxylic acid derivatives have been presented in Spectroscopy Parts B and C.

- The infrared absorption of the carbonyl group of carboxylic acid derivatives: Sections B.3C and B.4B, Sample Problem B.1, and Table B.2
- ^{1}H and ^{13}C NMR absorptions: Tables C.1 and C.5

Key NMR and IR absorptions for carboxylic acid derivatives are summarized in Table 20.3. Recall from Section B.3C that the location of the carbonyl absorption depends on the identity of Z in RCOZ.

- **As the basicity of Z *increases*, resonance stabilization of RCOZ *increases*, and the C=O absorption shifts to *lower* frequency.**

Table 20.3 Characteristic Spectroscopic Absorptions of Carboxylic Acid Derivatives

Type of spectroscopy	Compound	Type of C, H	Absorption
IR absorptions	**Acid chloride**	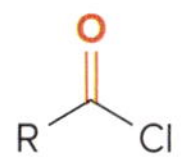 RCOCl	$1800\ cm^{-1}$
	Anhydride	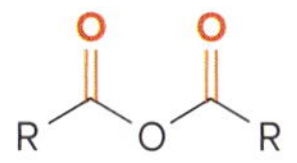 RCOOCOR	1820 and $1760\ cm^{-1}$ (two peaks)
	Ester	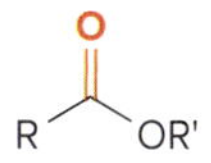 RCOOR'	$1735–1745\ cm^{-1}$
	Amide	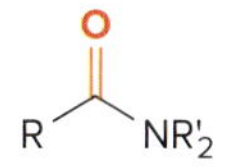 $RCONR'_2$ R' = H or alkyl	$1630–1680\ cm^{-1}$
		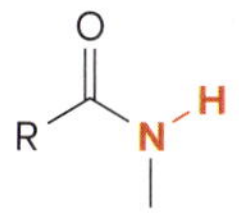 RCON–H	$3200–3400\ cm^{-1}$ (one or two N–H stretching peaks) $1640\ cm^{-1}$ (N–H bending)
^{1}H NMR absorptions	**All acyl derivatives**	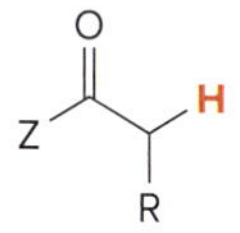 ZCOCH(R)–**H**	2–2.5 ppm
	Amide (1° and 2°)	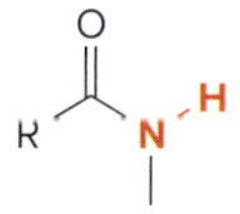 RCON–**H**	7.5–8.5 ppm
^{13}C NMR absorption	**All acyl derivatives**	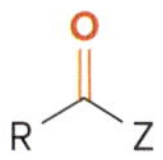 RCOZ	160–180 ppm

Problem 20.7 Rank compounds **A–D** in order of increasing frequency of the C=O absorption in their IR spectra.

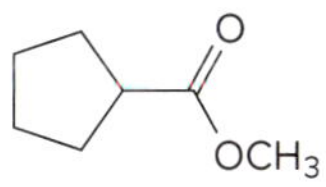

A (methyl cyclopentanecarboxylate, OCH_3) B (lactone with isopropenyl group) C (benzoyl piperidine, N) D (methyl benzoate, OCH_3)

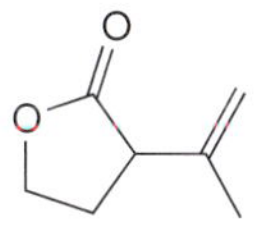

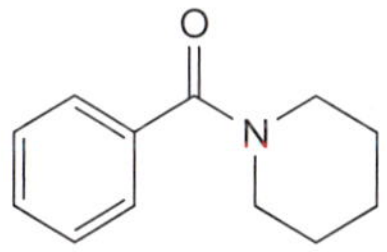

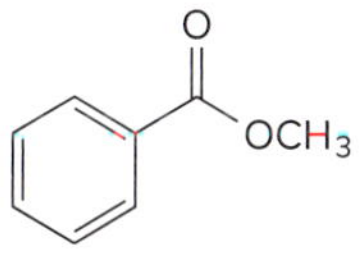

20.5 Interesting Esters and Amides

20.5A Esters

The characteristic odor of many fruits is due to low-molecular-weight esters. *Jill Braaten*

Many low-molecular-weight esters have pleasant and very characteristic odors.

isoamyl acetate
odor of banana

ethyl butanoate
odor of mango

methyl 2-methylbutanoate
odor of pineapple

Several esters, including vitamin C and molnupiravir, have important biological activities.

vitamin C

molnupiravir

Vitamin C (or **ascorbic acid**) is a water-soluble vitamin containing a five-membered lactone that we first discussed in Section 3.5B. Although vitamin C is synthesized in plants, humans do not have the necessary enzymes to make it, so they must obtain it from their diet.

Molnupiravir, sold under the trade name Lagevrio, is an oral antiviral drug originally developed to treat influenza, and as of December 2021, given an emergency use authorization to treat individuals with mild to moderate COVID-19 symptoms. Molnupiravir inhibits the replication of SARS-CoV-2, the virus that causes COVID-19, by triggering a series of lethal mutations in the virus.

20.5B Amides

An important group of naturally occurring amides consists of ***proteins,*** **polymers of amino acids joined together by amide linkages.** Proteins differ in the length of the polymer chain, as well as in the identity of the R groups bonded to it. The word *protein* is usually reserved for high-molecular-weight polymers composed of 40 or more amino acid units, whereas the designation *peptide* is given to polymers of lower molecular weight.

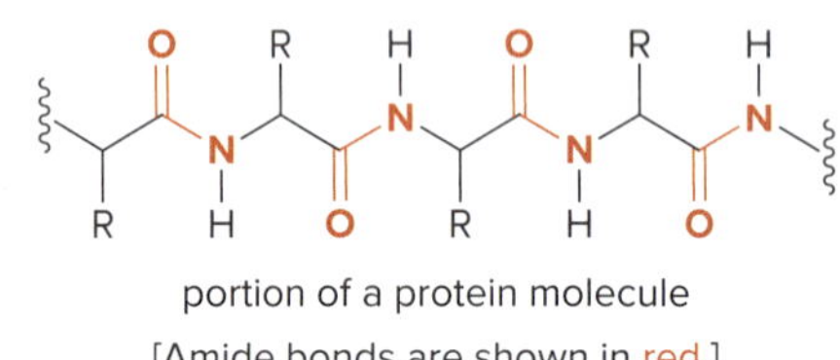

portion of a protein molecule
[Amide bonds are shown in red.]

Peptides and proteins are discussed in detail in Chapter 27.

Proteins and peptides have diverse functions in the cell. They form the structural components of muscle, connective tissue, hair, and nails. They catalyze reactions and transport ions and molecules across cell membranes. **Met-enkephalin,** for example, a peptide with four amide bonds found predominately in nerve tissue cells, relieves pain and acts as an opiate by producing morphine-like effects.

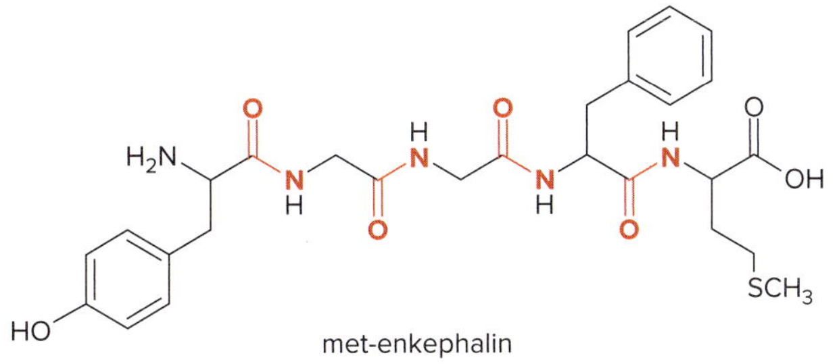

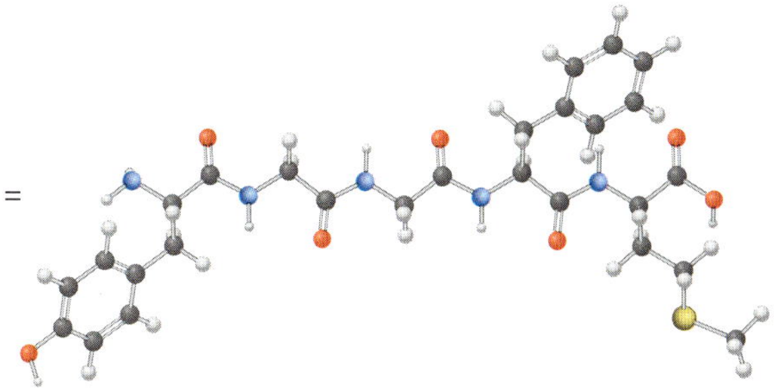

met-enkephalin
[The four amide bonds are shown in red.]

Penicillins are a group of structurally related antibiotics, known since the pioneering work of Sir Alexander Fleming led to the discovery of penicillin G in the 1920s. All penicillins contain a strained β-lactam fused to a five-membered ring, as well as a second amide located α to the β-lactam carbonyl group. Particular penicillins differ in the identity of the R group in the amide side chain.

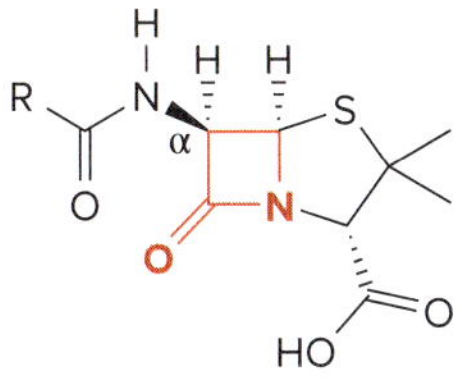

penicillin
β-lactam shown in red

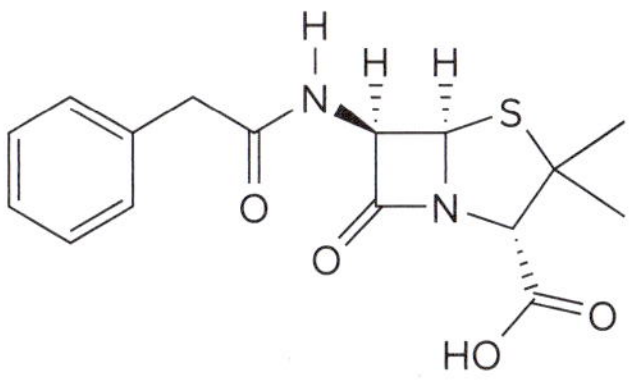

penicillin G

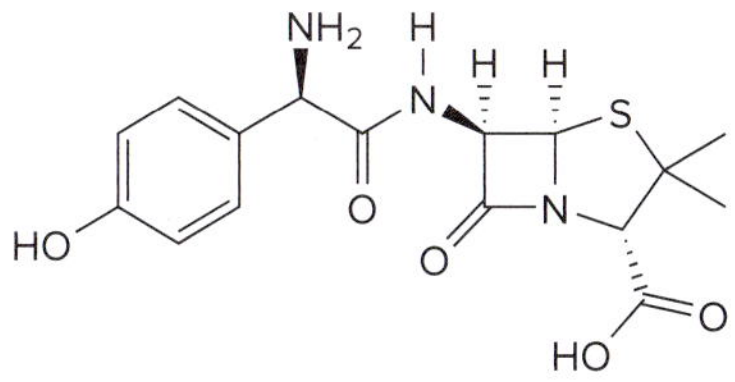

amoxicillin

Cephalosporins represent a second group of β-lactam antibiotics that contain a four-membered ring fused to a six-membered ring. Cephalosporins are generally active against a broader range of bacteria than penicillins.

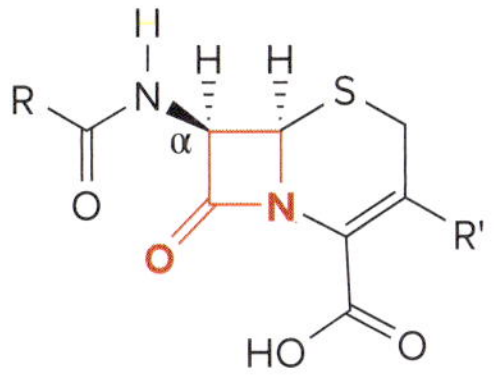

cephalosporin
β-lactam shown in red

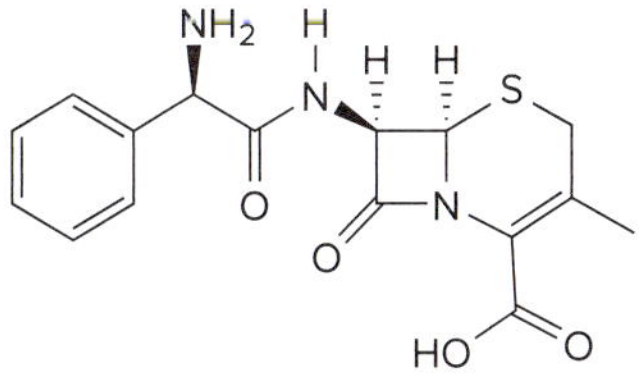

cephalexin
Trade name: Keflex

20.6 Introduction to Nucleophilic Acyl Substitution

The characteristic reaction of carboxylic acid derivatives is *nucleophilic acyl substitution.* This is a general reaction that occurs with both negatively charged nucleophiles ($Nu:^-$) and neutral nucleophiles (HNu:).

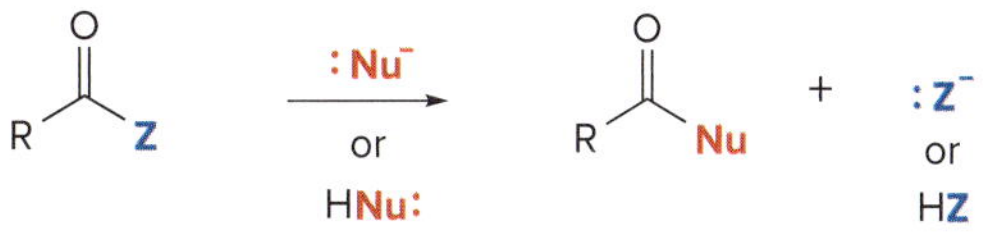

$$RCOZ \xrightarrow[\text{or HNu:}]{:Nu^-} RCONu + :Z^- \text{ or } HZ$$

Nu replaces Z.

- **Carboxylic acid derivatives (RCOZ) react with nucleophiles because they contain an electrophilic, unhindered carbonyl carbon.**
- **Substitution, *not* addition, occurs because carboxylic acid derivatives (RCOZ) have a leaving group Z on the carbonyl carbon.**

The mechanism for nucleophilic acyl substitution was first presented in Section 17.2.

20.6A The Mechanism

The general mechanism for nucleophilic acyl substitution is a two-step process: **nucleophilic attack** followed by **loss of the leaving group,** as shown in Mechanism 20.1.

Mechanism 20.1 General Mechanism—Nucleophilic Acyl Substitution

$$RCOZ + :Nu^- \xrightarrow{1} R(Z)C(O^-)Nu \xrightarrow{2} RCONu + :Z^-$$

Z = OH, Cl, OCOR, OR', NH_2

1 **The nucleophile attacks the electrophilic carbonyl group.** The π bond is broken, moving an electron pair out on oxygen and forming an sp^3 hybridized carbon.

2 An electron pair on oxygen re-forms the π bond and **Z comes off as a leaving group** with the electron pair in the C–Z bond.

The overall result of addition of a nucleophile and elimination of a leaving group is *substitution* of the nucleophile for the leaving group. Recall from Chapter 17 that nucleophilic substitution occurs with carbanions (R^-) and hydride (H^-) as nucleophiles. A variety of oxygen and nitrogen nucleophiles also participate in this reaction.

Oxygen nucleophiles			
$^-$:ÖH	$H_2\ddot{O}$:	RÖH	RCO_2^-

Nitrogen nucleophiles		
$\ddot{N}H_3$	$R\ddot{N}H_2$	$R_2\ddot{N}H$

Nucleophilic acyl substitution using heteroatomic nucleophiles results in the conversion of one carboxylic acid derivative another, as shown in two examples.

$$CH_3COCl \xrightarrow{H-NH_2} CH_3CONH_2 + H-Cl$$

1° amide

$$C_6H_5COOH \xrightarrow{H-OCH_3,\ H^+} C_6H_5COOCH_3 + H-OH$$

ester

Each reaction results in the replacement of the leaving group by the nucleophile, regardless of the identity of or charge on the nucleophile. To draw any nucleophilic acyl substitution product:

- **Find the sp^2 hybridized carbon with the leaving group.**
- **Identify the nucleophile.**
- **Substitute the nucleophile for the leaving group. With a neutral nucleophile, a proton must be lost to obtain a neutral substitution product.**

20.6B Relative Reactivity of Carboxylic Acids and Their Derivatives

As discussed in Section 17.2B, carboxylic acids and their derivatives differ greatly in reactivity toward nucleophiles. The order of reactivity parallels the leaving group ability of the group Z.

- **The better the leaving group, the more reactive RCOZ is in nucleophilic acyl substitution.**

Recall that the **best leaving group is the weakest base.** The relative basicity of the common leaving groups, Z, is given in Table 20.1.

Thus, the following trends result:

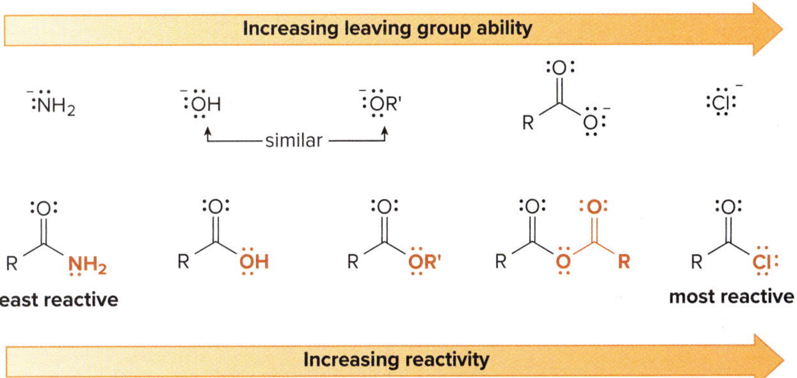

Based on this order of reactivity, ***more reactive*** **acyl compounds (acid chlorides and anhydrides) can be converted to *less reactive* ones (carboxylic acids, esters, and amides). The reverse is not usually true.**

To see why this is so, recall that nucleophilic addition to a carbonyl group forms a tetrahedral intermediate with two possible leaving groups, Z^- or $:Nu^-$. The group that is subsequently eliminated is the *better* of the two leaving groups. For a reaction to form a substitution product, therefore, Z^- must be the *better leaving group*, making the starting material RCOZ a *more reactive* acyl compound.

two possible leaving groups

For a reaction to occur, Z must be a better leaving group than Nu.

To evaluate whether a nucleophilic substitution reaction will occur, **compare the leaving group ability of the incoming nucleophile and the departing leaving group,** as shown in Sample Problem 20.2.

Sample Problem 20.2 Using Basicity to Determine Whether a Nucleophilic Acyl Substitution Might Occur

Determine whether each nucleophilic acyl substitution is likely to occur.

a. $CH_3COCl \xrightarrow{?} CH_3COOCH_2CH_3$ b. $C_6H_5CONH_2 \xrightarrow{?} (C_6H_5CO)_2O$

Solution

a. Conversion of CH_3COCl to $CH_3COOCH_2CH_3$ requires the **substitution of Cl^- by $^-OCH_2CH_3$.** Because **Cl^- is a weaker base** and therefore a better leaving group than $^-OCH_2CH_3$, **this reaction occurs.**

b. Conversion of $C_6H_5CONH_2$ to $(C_6H_5CO)_2O$ requires the **substitution of $^-NH_2$ by $^-OCOC_6H_5$.** Because **$^-NH_2$ is a stronger base** and therefore a poorer leaving group than $^-OCOC_6H_5$, **this reaction does *not* occur.**

Problem 20.8 Without reading ahead in Chapter 20, state whether it should be possible to carry out each of the following nucleophilic substitution reactions.

a. $CH_3COCl \rightarrow CH_3COOH$

b. $CH_3CONHCH_3 \rightarrow CH_3COOCH_3$

c. $CH_3COOCH_2CH_3 \rightarrow CH_3COCl$

d. $(CH_3CO)_2O \rightarrow CH_3CONH_2$

More Practice: Try Problems 20.35, 20.39.

Learn the order of reactivity of carboxylic acid derivatives. Keeping this in mind allows you to organize a very large number of reactions.

To summarize:

- **Nucleophilic substitution occurs when the leaving group Z^- is a *weaker* base and therefore *better* leaving group than the attacking nucleophile $:Nu^-$.**
- ***More* reactive acyl compounds can be converted to *less* reactive acyl compounds by nucleophilic substitution.**

Problem 20.9 Rank the compounds in each group in order of increasing reactivity in nucleophilic acyl substitution.

a. $C_6H_5CO_2CH_3$, C_6H_5COCl, $C_6H_5CONH_2$

b. $CH_3CH_2CO_2H$, $(CH_3CH_2CO)_2O$, $CH_3CH_2CONHCH_3$

Problem 20.10 Explain why trichloroacetic anhydride [$(Cl_3CCO)_2O$] is more reactive than acetic anhydride [$(CH_3CO)_2O$] in nucleophilic acyl substitution reactions.

20.6C A Preview of Specific Reactions

Sections 20.7–20.13 are devoted to specific examples of nucleophilic acyl substitution using heteroatoms as nucleophiles. There are a great many reactions, and it is easy to confuse them unless you learn the general order of reactivity of carboxylic acid derivatives. **Keep in mind that every reaction that begins with an acyl starting material involves nucleophilic substitution.**

We begin with the reactions of acid chlorides, the most reactive acyl compounds, then proceed to less and less reactive carboxylic acid derivatives, ending with amides. Acid chlorides undergo many reactions, because they have the best leaving group of all acyl compounds, whereas amides undergo only one reaction, which must be carried out under harsh reaction conditions, because amides have a poor leaving group.

In general, we will examine nucleophilic acyl substitution with four different nucleophiles, as shown in the following equations.

RCOZ reacts with:
- $^{-}$:O–C(O)R → anhydride (RCO–O–COR)
- $^{-}$:OH or H–OH → carboxylic acid (RCOOH)
- H–OR' → ester (RCOOR')
- H–NH$_2$ or H–NHR' or H–NR'$_2$ → RCONH$_2$ (1° amide) or RCONHR' (2° amide) or RCONR'$_2$ (3° amide)

These reactions are used to make anhydrides, carboxylic acids, esters, and amides, but not acid chlorides, from other acyl compounds. Acid chlorides are the most reactive acyl compounds (they have the best leaving group), so they are not easily formed as a product of nucleophilic substitution reactions. **Acid chlorides can only be prepared from carboxylic acids using special reagents,** as discussed in Section 20.9A.

20.7 Reactions of Acid Chlorides

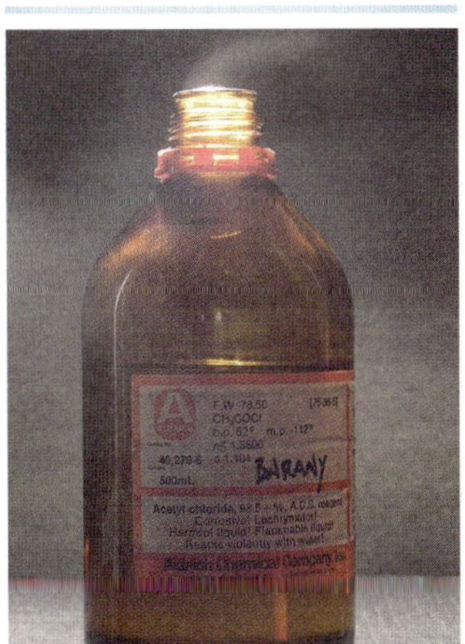

The reaction of acid chlorides with water is rapid. Exposure of an acid chloride to moist air on a humid day leads to some hydrolysis, giving the acid chloride a very acrid odor, due to the HCl formed as a by-product. *Joe Franek/McGraw Hill*

Acid chlorides readily react with nucleophiles to form nucleophilic substitution products, with HCl usually formed as a reaction by-product. A weak base like pyridine is added to the reaction mixture to remove this strong acid, forming an ammonium salt.

RCOCl + H–Nu: → RCONu + H–Cl (by-product)

H–Cl + pyridine → pyridinium (N–H^{+}) Cl^{-}

Acid chlorides react with oxygen nucleophiles to form anhydrides, carboxylic acids, and esters.

[1] RCOCl + $^{-}$OCOR → RCO–O–COR (**anhydride**) + Cl^{-}

[2] RCOCl + H–OR' (R' = H or alkyl) —pyridine→ RCOOR' (**carboxylic acid,** R' = H; **ester,** R' = alkyl) + pyridinium H Cl^{-}

Insect repellents containing DEET have become particularly popular because of the recent spread of many insect-borne diseases such as West Nile virus and Lyme disease. DEET does not kill insects—it repels them. It is thought that DEET somehow confuses insects so that they can no longer sense the warm moist air that surrounds a human body.
Scott Bauer/USDA-ARS

Acid chlorides also react with ammonia and 1° and 2° amines to form 1°, 2°, and 3° amides, respectively. Two equivalents of NH_3 or amine are used. One equivalent acts as a nucleophile to replace Cl and form the substitution product, while the second equivalent reacts as a base with the HCl by-product to form an ammonium salt.

$$RCOCl + HNR'R'' \longrightarrow RCONR'R'' + HCl$$

R', R'' = H or alkyl (2 equiv)

1° amide, R', R'' = H
2° amide, R' = H, R'' = alkyl
3° amide, R', R'' = alkyl

HCl + HNR'R'' → $H_2N^+R'R''$ Cl^-

As an example, reaction of an acid chloride with diethylamine forms the 3° amide *N,N*-diethyl-*m*-toluamide, popularly known as **DEET.** DEET, the active ingredient in the most widely used insect repellents, is effective against mosquitoes, fleas, and ticks.

m-toluoyl chloride + diethylamine (excess) → *N,N*-diethyl-*m*-toluamide (**DEET**) + diethylammonium Cl^-

Problem 20.11 Draw the products formed when benzoyl chloride (C_6H_5COCl) is treated with each nucleophile: (a) H_2O, pyridine; (b) CH_3COO^-; (c) NH_3 (excess); (d) $(CH_3)_2NH$ (excess).

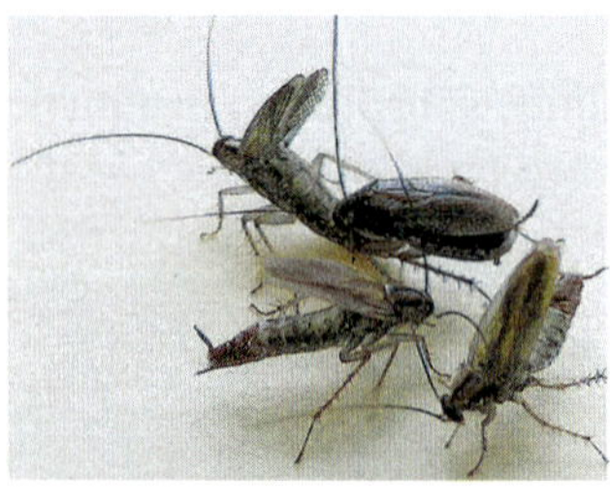

A short laboratory synthesis of blattellaquinone (Problem 20.12), the sex pheromone of the female German cockroach, opens new possibilities for cockroach population control using pheromone-baited traps.
Coby Schal

With a carboxylate nucleophile the mechanism follows the general, two-step mechanism discussed in Section 20.6A: **nucleophilic attack followed by loss of the leaving group,** as shown in Mechanism 20.2.

Mechanism 20.2 Conversion of Acid Chlorides to Anhydrides

$$RCOCl + RCOO^- \xrightarrow{1} \text{tetrahedral intermediate} \xrightarrow{2} RCOOCOR \text{ (anhydride)} + Cl^-$$

1. **The nucleophilic carboxylate anion attacks the carbonyl group,** forming an sp^3 hybridized carbon.
2. Elimination of the leaving group (Cl^-) forms the **substitution product,** an anhydride.

Problem 20.12 Draw a stepwise mechanism for the formation of **A** from an alcohol and acid chloride. **A** was converted in one step to blattellaquinone, the sex pheromone of the female German cockroach, *Blattella germanica.*

2,5-dimethoxybenzyl alcohol + 3-methylbutanoyl chloride $\xrightarrow{\text{pyridine}}$ **A** → blattellaquinone

20.8 Reactions of Anhydrides

Nucleophilic substitution occurs only when the leaving group is a weaker base and therefore a better leaving group than the attacking nucleophile.

Although somewhat less reactive than acid chlorides, anhydrides nonetheless readily react with most nucleophiles to form substitution products. Nucleophilic substitution reactions of anhydrides are no different than the reactions of other carboxylic acid derivatives, even though anhydrides contain two carbonyl groups. **Nucleophilic attack occurs at one carbonyl group, while the second carbonyl becomes part of the leaving group.**

$$(RCO)_2O + H{-}Nu: \longrightarrow RCONu + HOCOR \text{ (by-product)}$$

Anhydrides can't be used to make acid chlorides, because $RCOO^-$ is a stronger base and therefore a poorer leaving group than Cl^-. Anhydrides can be used to make all other acyl derivatives, however. Reaction with water and alcohols yields **carboxylic acids** and **esters,** respectively. Reaction with two equivalents of NH_3 or amines forms **1°, 2°, and 3° amides.** A molecule of carboxylic acid (or a carboxylate salt) is always formed as a by-product.

$$(RCO)_2O + H{-}OR' \longrightarrow RCOOR' + HOCOR$$

R' = H or alkyl

carboxylic acid, R' = H; **ester,** R' = alkyl; by-product

$$(RCO)_2O + H{-}NH_2 \text{ (2 equiv)} \longrightarrow RCONH_2 + NH_4^+\ {}^-OCOR$$

1° amide; by-product

Problem 20.13 Draw the products formed when benzoic anhydride $[(C_6H_5CO)_2O]$ is treated with each nucleophile: (a) H_2O; (b) CH_3OH; (c) NH_3 (excess); (d) $(CH_3)_2NH$ (excess).

The conversion of an anhydride to an amide illustrates the mechanism of nucleophilic acyl substitution with an anhydride as starting material (Mechanism 20.3). Besides the usual steps of **nucleophilic addition** and **elimination of the leaving group,** an additional proton transfer is needed.

Mechanism 20.3 Conversion of an Anhydride to an Amide

:NH3 —1→ —2→ + NH_4^+ —3→ R–CO–NH_2 (**1° amide**) + $^-$:O–CO–R

1 **The nucleophile (NH_3) attacks the carbonyl,** forming an sp^3 hybridized carbon.

2–3 Loss of a proton and elimination of the leaving group (RCO_2^-) form the **substitution product,** a 1°amide.

Acetaminophen reduces pain and fever, but it is not anti-inflammatory, so it is ineffective in treating conditions like arthritis, which have a significant inflammatory component. In large doses, acetaminophen causes liver damage, so dosage recommendations must be carefully followed.

Anhydrides react with alcohols and amines with ease, so they are often used in the laboratory to prepare esters and amides. For example, acetic anhydride is used to prepare two analgesics, **acetylsalicylic acid** (aspirin) and **acetaminophen** (the active ingredient in Tylenol).

acetylsalicylic acid
(aspirin)

acetaminophen
(active ingredient in Tylenol)

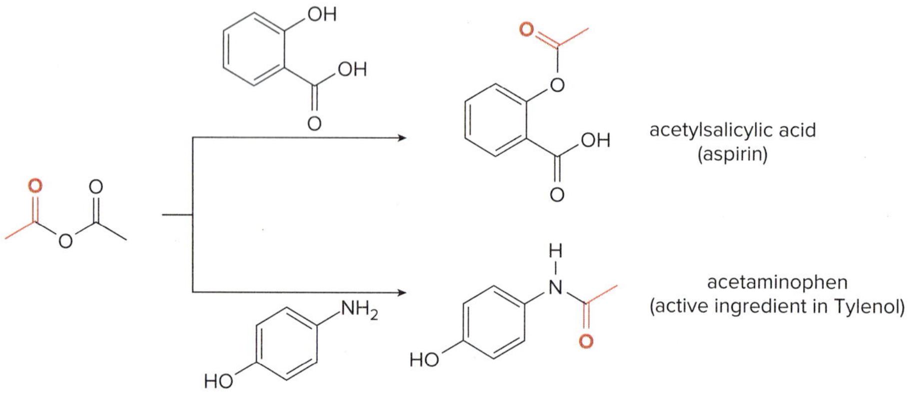

These are called **acetylation** reactions because they result in the transfer of an acetyl group, CH_3CO–, from one heteroatom to another.

Heroin is prepared by the acetylation of morphine, an analgesic compound isolated from the opium poppy. Both OH groups of morphine are readily acetylated with acetic anhydride to form the diester present in heroin.

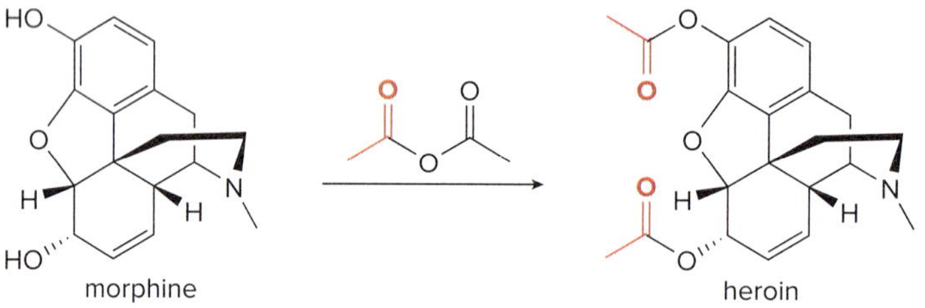

morphine

heroin

Problem 20.14 If anhydrides react like acid chlorides with the nucleophiles described in Chapter 17, draw the products formed when each of the following nucleophiles reacts with benzoic anhydride [$(C_6H_5CO)_2O$]: (a) CH_3MgBr (2 equiv), then H_2O; (b) $LiAlH_4$, then H_2O; (c) $LiAlH[OC(CH_3)_3]_3$, then H_2O.

20.9 Reactions of Carboxylic Acids

Carboxylic acids are strong organic acids. Because acid–base reactions proceed rapidly, any nucleophile that is also a strong base will react with a carboxylic acid by removing a proton ***first,*** before any nucleophilic substitution reaction can take place.

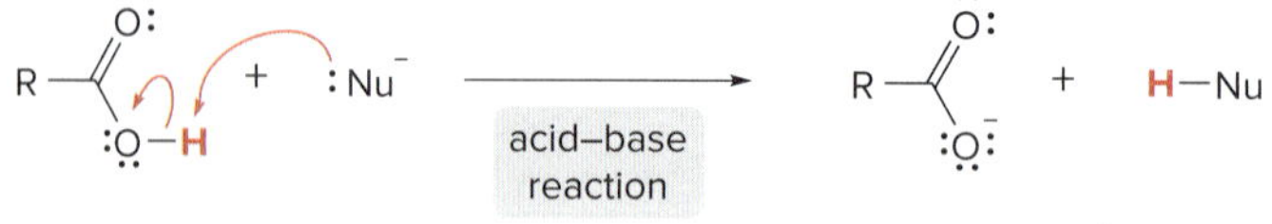

acid–base
reaction

An acid–base reaction occurs with $^-$OH, NH_3, and amines, all common nucleophiles used in nucleophilic acyl substitution reactions. Nonetheless, carboxylic acids do undergo nucleophilic acyl substitution and can be converted to a variety of other acyl derivatives using special reagents, with acid catalysis or, sometimes, by using rather forcing reaction conditions. These reactions are summarized in Figure 20.1 and detailed in Sections 20.9A–20.9D.

Figure 20.1
Nucleophilic acyl substitution reactions of carboxylic acids

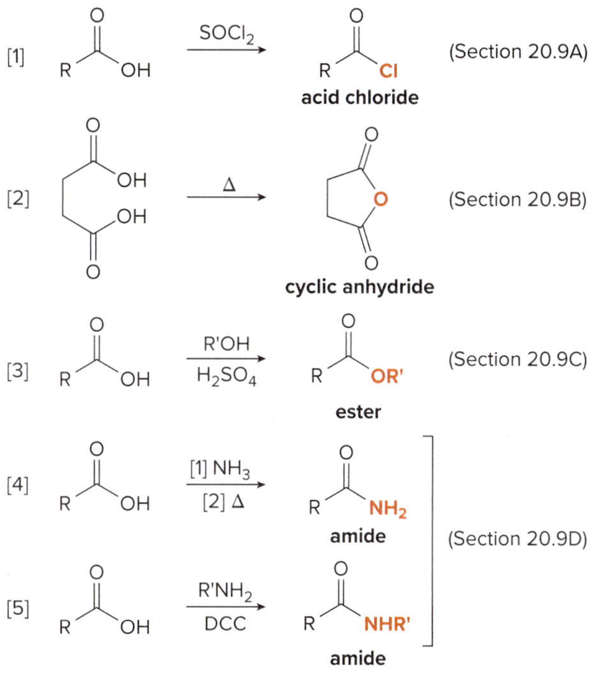

20.9A Conversion of RCOOH to RCOCl

Carboxylic acids can't be converted to acid chlorides by using Cl^- as a nucleophile, because the attacking nucleophile Cl^- is a weaker base than the departing leaving group, ^-OH. But carboxylic acids *can* be converted to acid chlorides using thionyl chloride, **$SOCl_2$,** a reagent that was introduced in Section 9.12 to convert alcohols to alkyl chlorides.

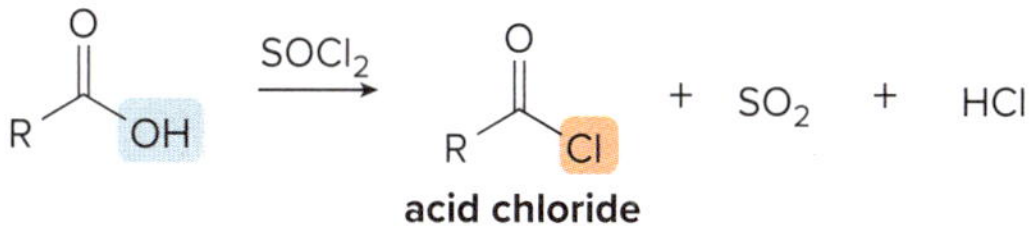

Treatment of benzoic acid with $SOCl_2$ forms benzoyl chloride. This reaction converts a less reactive acyl derivative (a carboxylic acid) to a more reactive one (an acid chloride). This is possible because **thionyl chloride converts the OH group of the acid to a better leaving group, and because it provides the nucleophile (Cl^-) to displace the leaving group.** The steps in the process are illustrated in Mechanism 20.4.

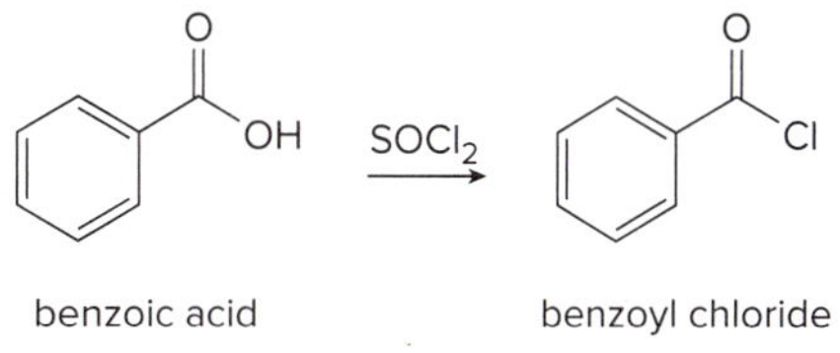

Mechanism 20.4 Conversion of Carboxylic Acids to Acid Chlorides

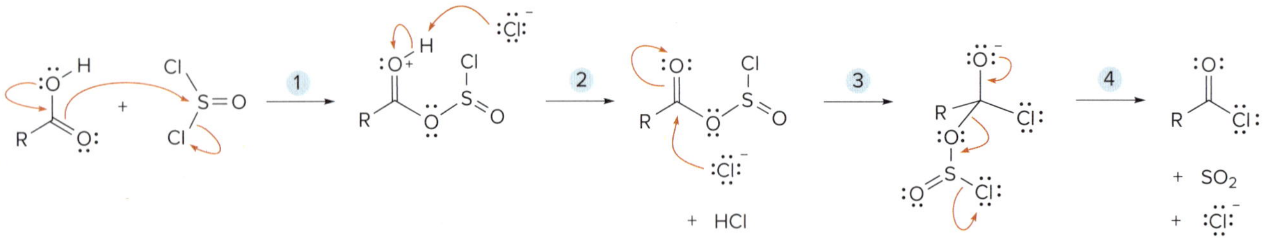

+ HCl

+ SO_2

+ $:\ddot{Cl}:^-$

1 – 2 Reaction of the carboxylic acid with $SOCl_2$ and loss of a proton convert the OH group to OSOCl, a **good leaving group.**

3 – 4 Nucleophilic attack of chloride generates a tetrahedral intermediate, and loss of the leaving group (SO_2 and Cl^-) forms the **acid chloride.**

Problem 20.15 Draw the products of each reaction.

a. 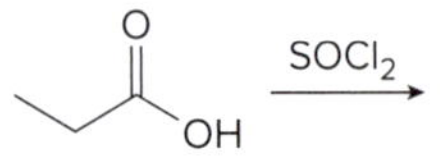 $\xrightarrow{SOCl_2}$

b. 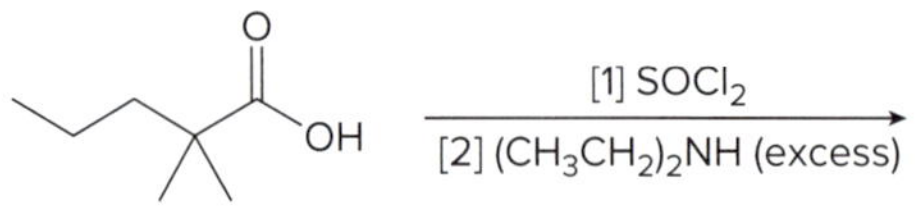 $\xrightarrow[\text{[2] } (CH_3CH_2)_2NH \text{ (excess)}]{\text{[1] } SOCl_2}$

20.9B Conversion of RCOOH to $(RCO)_2O$

Carboxylic acids cannot be readily converted to anhydrides, but dicarboxylic acids can be converted to cyclic anhydrides by heating to high temperatures. This is a **dehydration** reaction because a water molecule is lost from the diacid.

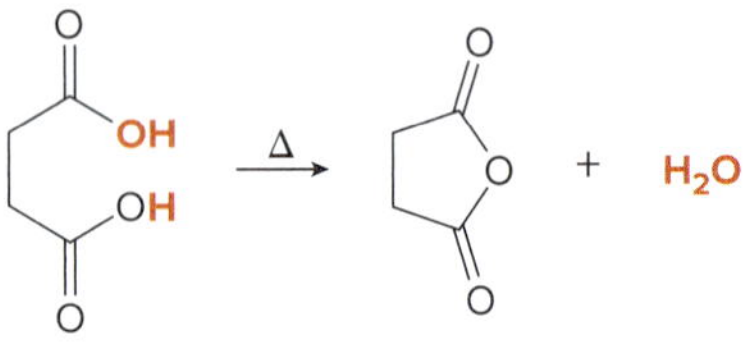

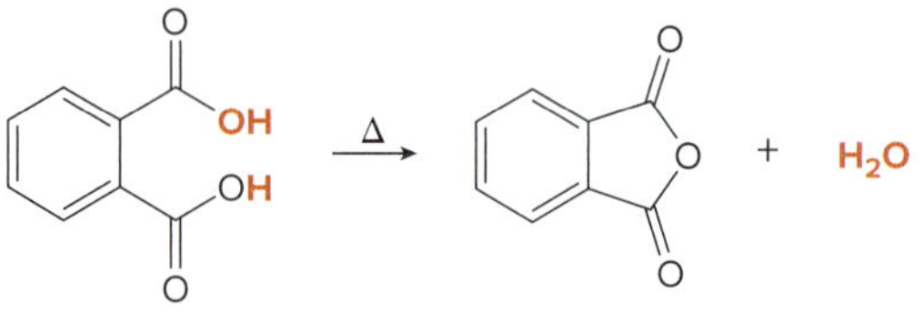

20.9C Conversion of RCOOH to RCOOR'

Treatment of a carboxylic acid with an alcohol in the presence of an acid catalyst forms an ester. This reaction is called a **Fischer esterification.**

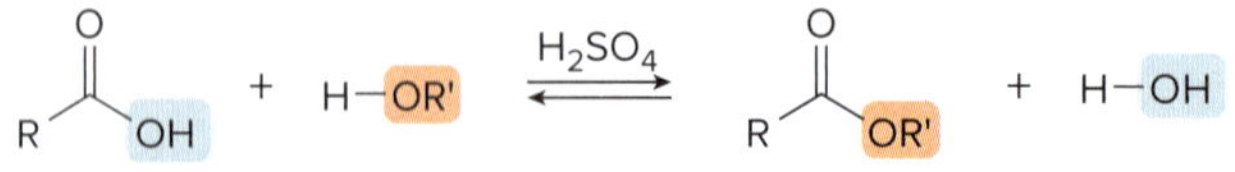

$$RCOOH + H{-}OR' \overset{H_2SO_4}{\rightleftharpoons} RCOOR' + H{-}OH$$

ester

This reaction is an equilibrium. According to Le Châtelier's principle (Section 9.8D), it is driven to the right by using excess alcohol or by removing the water as it is formed.

Ethyl acetate is a common organic solvent with a characteristic odor. It is used in nail polish remover and model airplane glue.

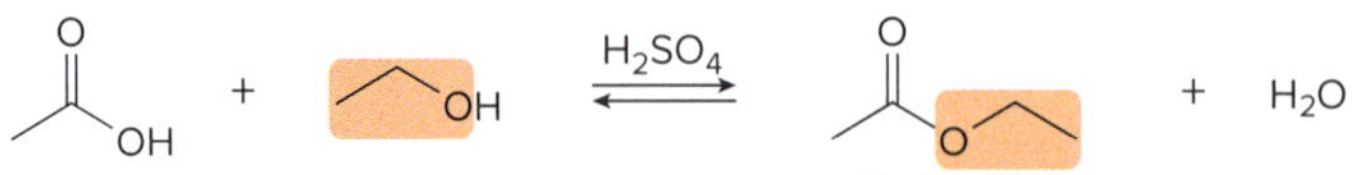

$$\text{acetic acid} + \text{ethanol} \overset{H_2SO_4}{\rightleftharpoons} \text{ethyl acetate} + H_2O$$

ethyl acetate

The mechanism for the Fischer esterification involves the usual two steps of nucleophilic acyl substitution—that is, **addition of a nucleophile followed by elimination of a leaving group.** Because the reaction is acid catalyzed, however, there are additional protonation and deprotonation steps. As always, though, the first step of any mechanism with an oxygen-containing starting material and an acid is to **protonate an oxygen atom** as shown with a general acid HA in Mechanism 20.5.

Mechanism 20.5 Fischer Esterification—Acid-Catalyzed Conversion of Carboxylic Acids to Esters

Part [1] Addition of the nucleophile R'OH

1 Protonation of the carbonyl oxygen makes the carbonyl more electrophilic.

2 – 3 **Nucleophilic attack** by R'OH forms a tetrahedral intermediate, and deprotonation gives the addition product.

Part [2] Elimination of the leaving group H_2O

4 Protonation of the OH group forms a **good leaving group.**

5 – 6 Loss of H_2O and deprotonation give the **ester.**

Esterification of a carboxylic acid occurs in the presence of acid but not in the presence of base. Base removes a proton from the carboxylic acid, forming an electron-rich carboxylate anion, which does not react with an electron-rich nucleophile.

Intramolecular esterification of γ- and δ-hydroxy carboxylic acids forms five- and six-membered lactones.

Problem 20.16 Draw the products of each reaction.

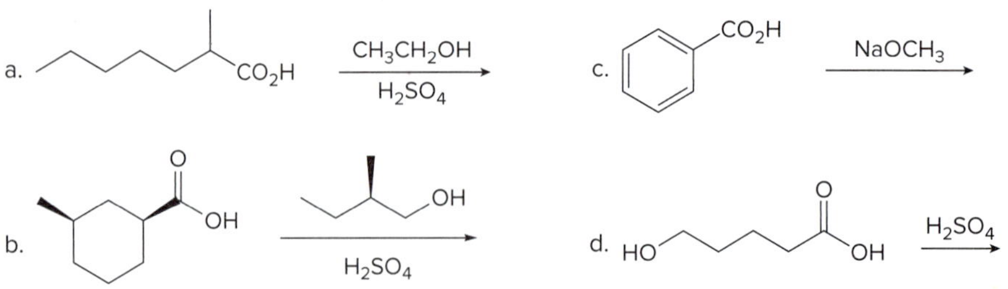

a. CO_2H — CH_3CH_2OH, H_2SO_4 →

b. OH — OH, H_2SO_4 →

c. CO_2H — $NaOCH_3$ →

d. HO — OH — H_2SO_4 →

Problem 20.17 Draw the products formed when benzoic acid ($C_6H_5CO_2H$) is treated with CH_3OH having its O atom labeled with ^{18}O ($CH_3{}^{18}OH$). Indicate where the labeled oxygen atom resides in the products.

Problem 20.18 Draw a stepwise mechanism for the following reaction.

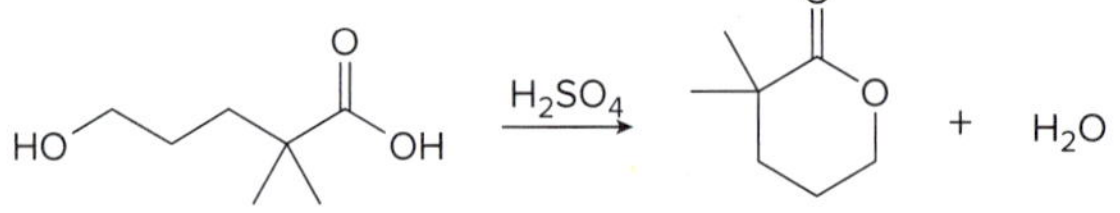

HO — OH — H_2SO_4 → O + H_2O

20.9D Conversion of RCOOH to RCONR'$_2$

The direct conversion of a carboxylic acid to an amide with NH_3 or an amine is very difficult, even though a more reactive acyl compound is being transformed into a less reactive one. The problem is that carboxylic acids are strong organic acids and NH_3 and amines are bases, so they undergo an **acid–base reaction to form an ammonium salt** before any nucleophilic substitution occurs.

Amides are much more easily prepared from acid chlorides and anhydrides, as discussed in Sections 20.7 and 20.8.

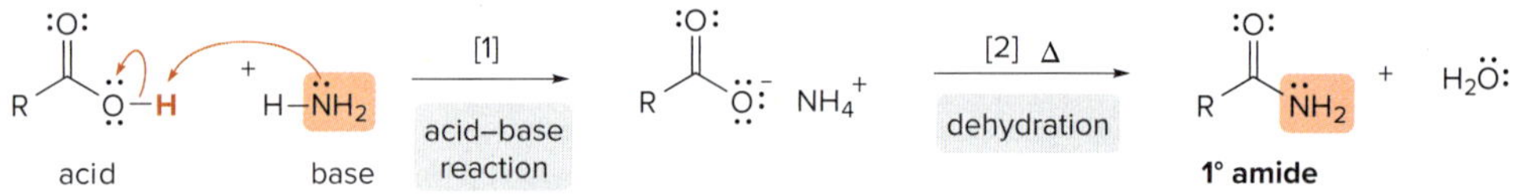

R–COOH (acid) + H–NH_2 (base) —[1] acid–base reaction→ $RCOO^-$ NH_4^+ —[2] Δ dehydration→ $RCONH_2$ (1° amide) + H_2O

Heating at high temperature (>100 °C) dehydrates the resulting ammonium salt of the carboxylate anion to form an amide, though the yield can be low.

Therefore, the overall conversion of RCOOH to $RCONH_2$ requires two steps:

[1] **Acid–base reaction of RCOOH with NH_3 to form an ammonium salt**

[2] **Dehydration at high temperature (>100 °C)**

C_6H_5COOH —[1] NH_3; [2] Δ→ $C_6H_5CONH_2$ + H_2O

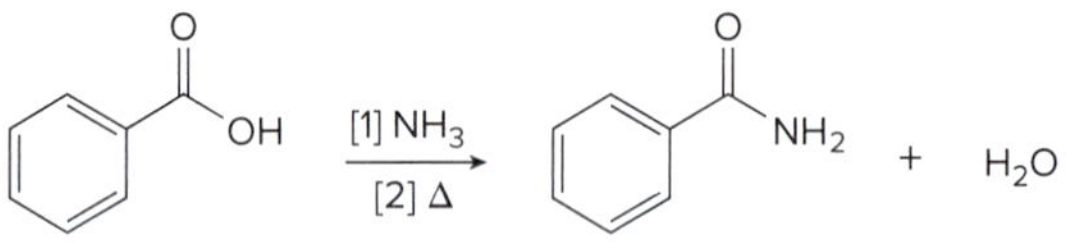

A carboxylic acid and an amine readily react to form an amide in the presence of an additional reagent, **dicyclohexylcarbodiimide (DCC),** which is converted to the by-product dicyclohexylurea in the course of the reaction.

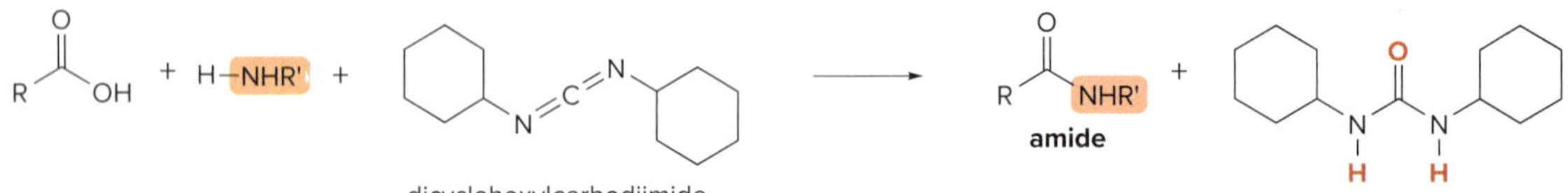

RCOOH + H–NHR' + dicyclohexylcarbodiimide (**DCC**, a dehydrating agent) → RCONHR' (**amide**) + dicyclohexylurea

H_2O has been added.

DCC is a dehydrating agent. The dicyclohexylurea by-product is formed by adding the elements of H_2O to DCC. DCC promotes amide formation by converting the carboxy OH group to a better leaving group.

CH_3NH_2 DCC

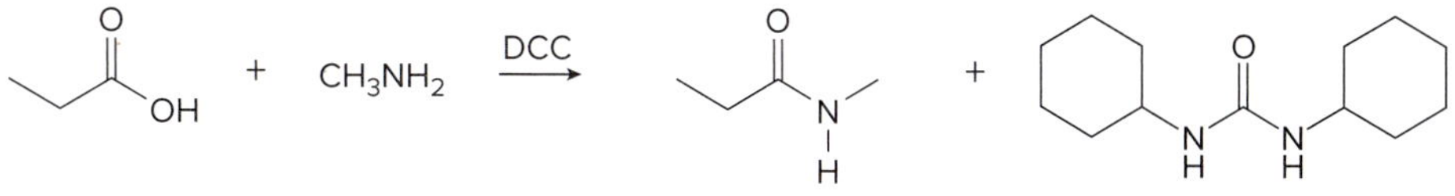

The mechanism consists of two parts: [1] conversion of the OH group to a better leaving group, followed by [2] **addition of the nucleophile and loss of the leaving group** to form the product of nucleophilic acyl substitution (Mechanism 20.6).

Mechanism 20.6 Conversion of Carboxylic Acids to Amides with DCC

Part [1] Conversion of OH to a better leaving group

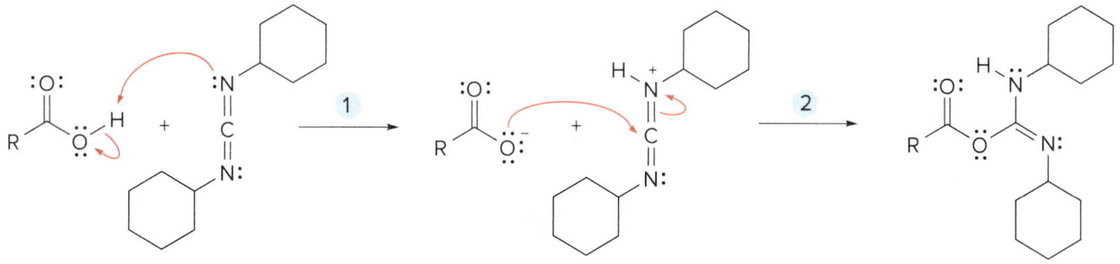

1. Acid–base reaction results in transfer of a proton from the carboxylic acid to DCC.
2. Nucleophilic attack of RCO_2^- on the conjugate acid of DCC forms an addition product. The overall result of Steps [1] and [2] is conversion of OH to a **better leaving group.**

Part [2] Addition of the nucleophile and loss of the leaving group

3 nucleophilic attack

4

5 loss of the leaving group

amide

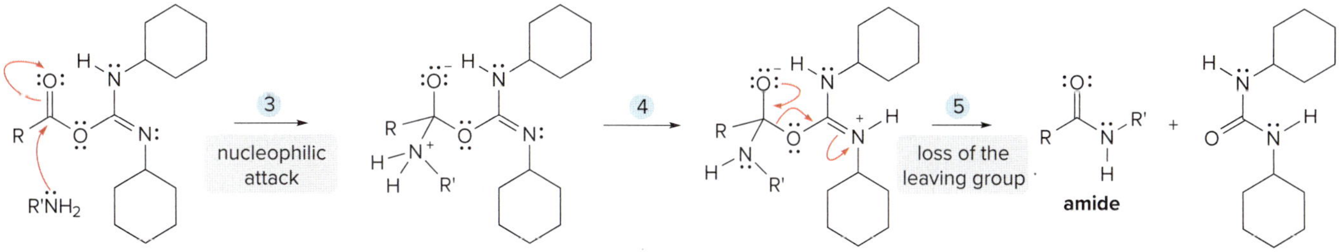

3 Nucleophilic attack of the amine on the activated carboxy group forms a tetrahedral intermediate.

4 – 5 Proton transfer and elimination of dicyclohexylurea as the leaving group form the **amide.**

The reaction of an acid and an amine with DCC is often used in the laboratory to form the amide bond in peptides, as is discussed in Chapter 27.

Problem 20.19 What product is formed when acetic acid is treated with each reagent: (a) CH_3NH_2; (b) CH_3NH_2, then heat; (c) CH_3NH_2 + DCC?

Problem 20.20 Draw the product of the following reaction, which demonstrates a method commonly used in the synthesis of the peptides discussed further in Chapter 27.

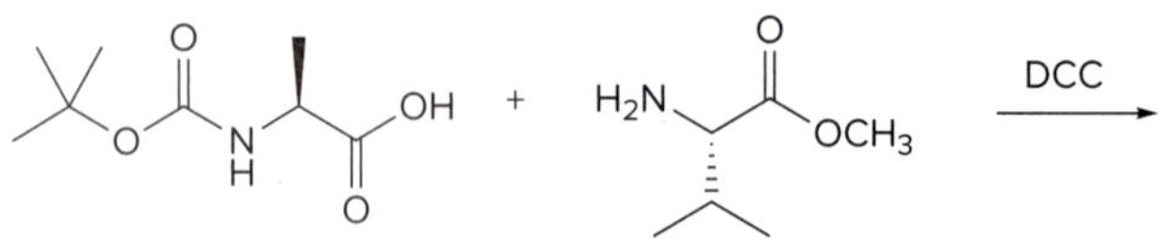

OH + H_2N OCH_3 DCC →

20.10 Reactions of Esters

Esters can be converted to carboxylic acids and amides.

- **Esters are hydrolyzed with water in the presence of either acid or base to form carboxylic acids or carboxylate anions.**

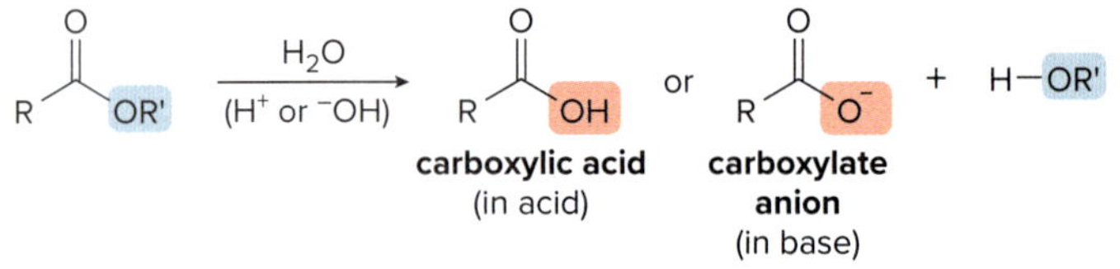

R OR' —H_2O (H^+ or ^-OH)→ R OH **carboxylic acid** (in acid) or R O^- **carboxylate anion** (in base) + H—OR'

- **Esters react with NH_3 and amines to form 1°, 2°, or 3° amides.**

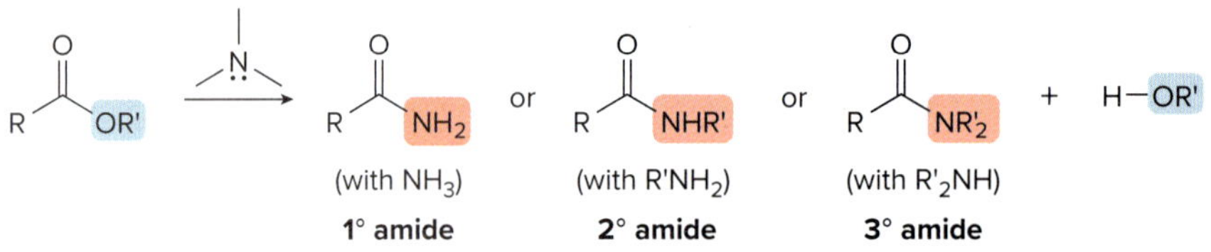

R OR' → R NH_2 (with NH_3) **1° amide** or R NHR' (with $R'NH_2$) **2° amide** or R NR'_2 (with R'_2NH) **3° amide** + H—OR'

20.10A Ester Hydrolysis in Aqueous Acid

The first step in acid-catalyzed ester hydrolysis is **protonation on oxygen,** the same first step of any mechanism involving an oxygen-containing starting material and an acid.

The hydrolysis of esters in aqueous acid is a reversible equilibrium reaction that is driven to the right by using a large excess of water.

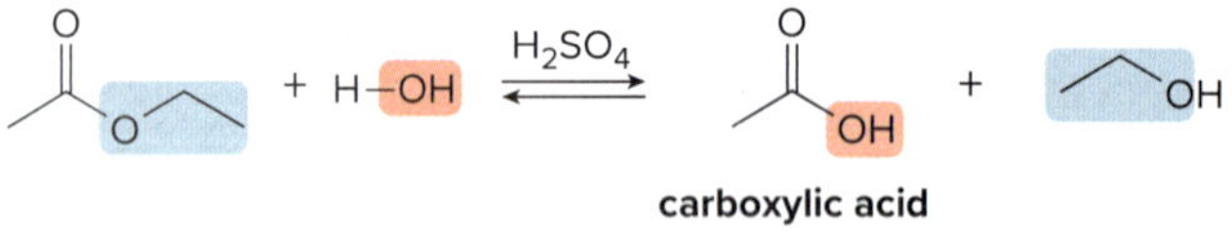

+ H—OH ⇌(H_2SO_4) OH **carboxylic acid** + OH

The mechanism of ester hydrolysis in acid (shown in Mechanism 20.7) is the reverse of the mechanism of ester synthesis from carboxylic acids (Mechanism 20.5). Thus, the mechanism consists of the **addition of the nucleophile and the elimination of the leaving group,** the two steps common to all nucleophilic acyl substitutions, as well as several proton transfers, because the reaction is acid-catalyzed.

Problem 20.21 Draw all additional resonance structures for any cation in Mechanism 20.7 that is resonance stabilized.

Mechanism 20.7 Acid-Catalyzed Hydrolysis of an Ester to a Carboxylic Acid

Part [1] Addition of the nucleophile H_2O

1 Protonation of the carbonyl oxygen makes the carbonyl more electrophilic.

2 – 3 **Nucleophilic attack by H_2O** forms a tetrahedral intermediate, and deprotonation gives the addition product.

Part [2] Elimination of the leaving group R'OH

4 Protonation of the OR' group forms a **good leaving group.**

5 – 6 Loss of R'OH and deprotonation give the **carboxylic acid.**

20.10B Ester Hydrolysis in Aqueous Base

The word ***saponification*** comes from the Latin *sapo,* meaning "soap." Soap is prepared by hydrolyzing esters in fats with aqueous base, as explained in Section 20.11B.

Esters are hydrolyzed in aqueous base to form carboxylate anions. Basic hydrolysis of an ester is called **saponification.**

carboxylate anion

The mechanism for this reaction has the usual two steps of the general mechanism for nucleophilic acyl substitution presented in Section 20.6A—**addition of the nucleophile** followed by **loss of a leaving group**—plus an additional step involving proton transfer (Mechanism 20.8).

Mechanism 20.8 Base-Promoted Hydrolysis of an Ester to a Carboxylate Anion

1 – 2 **Addition of the nucleophile ($^-$OH)** followed by **elimination of the leaving group ($^-$OR')** form a carboxylic acid. These two steps are reversible.

3 Because the carboxylic acid is a strong organic acid and the leaving group ($^-$OR') is a strong base, an acid–base reaction forms the **carboxylate anion.**

The carboxylate anion is resonance stabilized, and this drives the equilibrium in its favor. Once the reaction is complete and the carboxylate anion is formed, it can be protonated with strong acid to form the neutral carboxylic acid.

Hydrolysis is base promoted, *not* base catalyzed, because the base (^-OH) is the nucleophile that adds to the ester and forms part of the product. It participates in the reaction and is not regenerated later.

Where do the oxygen atoms in the product come from? **The C–OR' bond in the ester is cleaved,** so the OR' group becomes the alcohol by-product (R'OH) and **one of the oxygens in the carboxylate anion product comes from ^-OH** (the nucleophile).

Problem 20.22 Draw the products of the following reactions.

Problem 20.23 The antiviral agent molnupiravir (Section 20.5A) is a prodrug, a compound with little pharmacological activity, which is metabolized to an active drug. The first step in the metabolism of molnupiravir is the hydrolysis of the ester. What products are formed in this reaction?

20.11 Application: Lipid Hydrolysis

20.11A Triacylglycerol Hydrolysis with Enzymes or Acid

The most prevalent naturally occurring esters are the **triacylglycerols,** which were first discussed in Section 10.6. **Triacylglycerols are the lipids that comprise animal fats and vegetable oils.**

- **Each triacylglycerol is a triester, containing three long hydrocarbon side chains.**
- **Unsaturated triacylglycerols have one or more double bonds in their long hydrocarbon chains, whereas saturated triacylglycerols have none.**

Figure 20.2 contains a ball-and-stick model of a saturated fat.

Figure 20.2 The three-dimensional structure of a saturated triacylglycerol

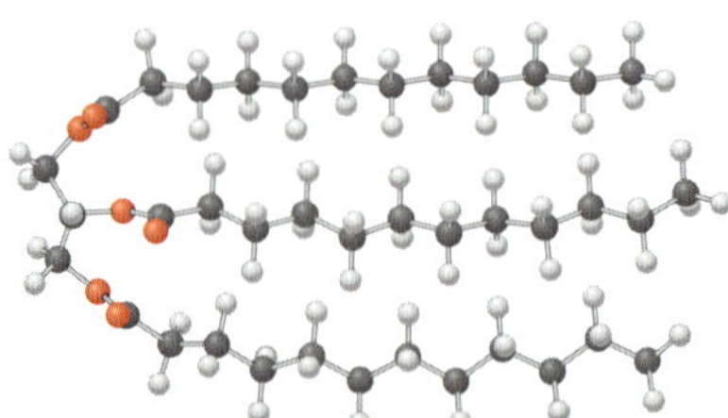

- This triacylglycerol has no double bonds in the three R groups (each with 11 C's) bonded to the ester carbonyls, making it a saturated fat.

Animals store energy in the form of triacylglycerols, kept in a layer of fat cells below the surface of the skin. This fat serves to insulate the organism, as well as provide energy for its metabolic needs for long periods. The first step in the metabolism of a triacylglycerol is **hydrolysis of the ester bonds to form glycerol and three fatty acids.** In cells, this reaction is carried out with enzymes called **lipases.**

triacylglycerol —(3 H—OH, lipase)→ glycerol + HO(C=O)R + HO(C=O)R' + HO(C=O)R''

Three carboxylic acids containing 12–20 C's are formed as products.

[The three bonds drawn in red are cleaved in hydrolysis.]

The fatty acids produced on hydrolysis are then oxidized in a stepwise fashion, ultimately yielding CO_2 and H_2O, as well as a great deal of energy. Oxidation of fats yields twice as much energy per gram as oxidation of an equivalent weight of carbohydrate.

Problem 20.24 Draw the products formed when the given triacylglycerol is hydrolyzed in aqueous acid.

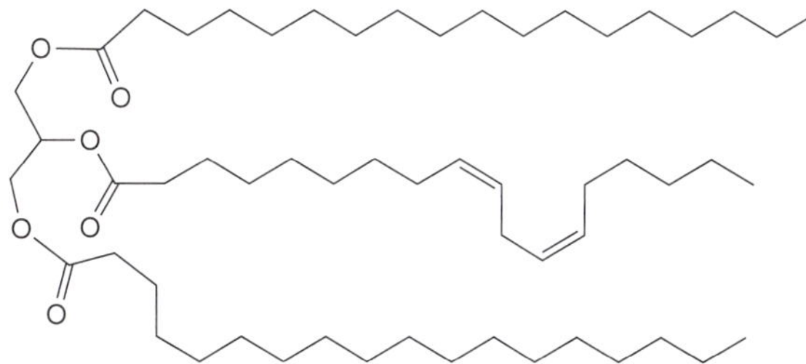

Soap was discussed in Section 3.6.

All soaps are salts of fatty acids. The main difference between soaps is the addition of other ingredients that do not alter their cleaning properties: dyes for color, scents for a pleasing odor, and oils for lubrication. Soaps that float are aerated, so that they are less dense than water. *Jill Braaten/ McGraw Hill*

20.11B The Synthesis of Soap

Soap is prepared by the basic hydrolysis or saponification of a triacylglycerol. Heating an animal fat or vegetable oil with aqueous base hydrolyzes the three esters to form glycerol and sodium salts of three fatty acids. These carboxylate salts are **soaps,** which clean away dirt because of their two structurally different regions. The nonpolar tail dissolves grease and oil and the polar head makes it soluble in water (Figure 3.3).

triacylglycerol —(NaOH, H_2O)→ glycerol + Na^+ ^-O(C=O)R + Na^+ ^-O(C=O)R' + Na^+ ^-O(C=O)R''

Soaps are carboxylate salts derived from fatty acids.

For example:

Na^+ ^-O — polar head; nonpolar tail

3-D structure

Soaps are typically made from lard (from hogs), tallow (from cattle or sheep), coconut oil, or palm oil. Most triacylglycerols have two or three different R groups in their hydrocarbon chains, so soaps are usually mixtures of two or three different carboxylate salts.

Problem 20.25 What is the composition of the soap prepared by hydrolysis of the following triacylglycerol?

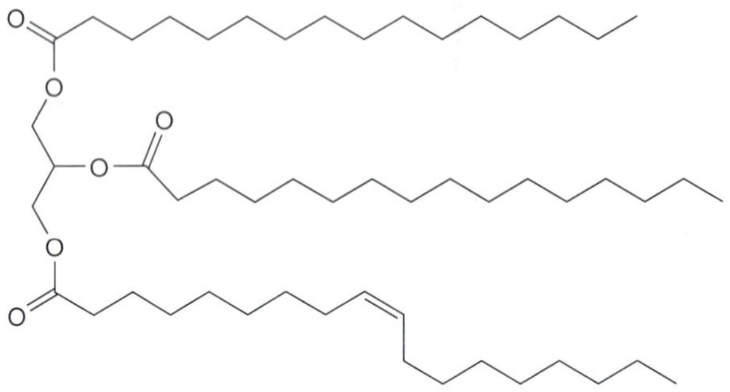

20.12 Reactions of Amides

Because amides have the poorest leaving group of all the carboxylic acid derivatives, they are the least reactive. Under strenuous reaction conditions, **amides are hydrolyzed in acid or base to form carboxylic acids or carboxylate anions.**

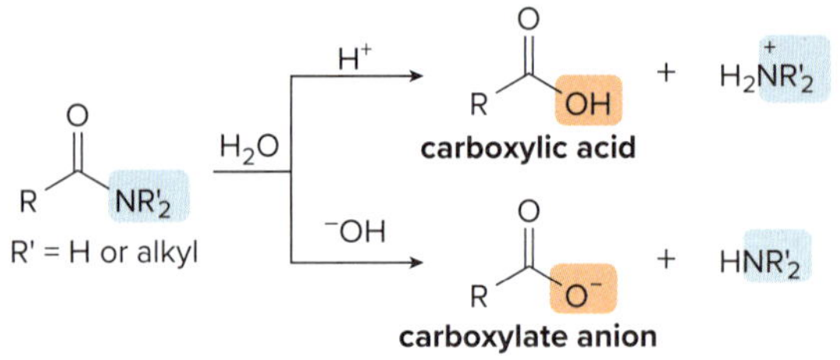

R' = H or alkyl

carboxylic acid

carboxylate anion

In acid, the amine by-product is protonated as an ammonium ion, whereas in base, a neutral amine is formed.

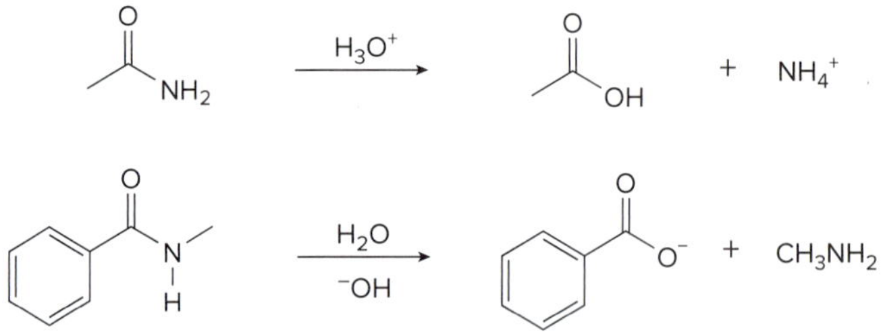

The relative lack of reactivity of the amide bond is notable in proteins, which are polymers of amino acids connected by amide linkages (Section 20.5B). Proteins are stable in aqueous solution in the absence of acid or base, so they can perform their various functions in the aqueous cellular environment without breaking down. The hydrolysis of the amide bonds in proteins requires a variety of specific enzymes.

The mechanism of amide hydrolysis in acid is exactly the same as the mechanism of ester hydrolysis in aqueous acid (Section 20.10A) except that the leaving group is different.

The mechanism of amide hydrolysis in base has the usual two steps of the general mechanism for nucleophilic acyl substitution—**addition of the nucleophile** followed by **loss of a leaving group**—plus an additional proton transfer. The initially formed carboxylic acid reacts further under basic conditions to form the resonance-stabilized carboxylate anion, and this drives the reaction to completion. Mechanism 20.9 is written for a 1° amide.

Mechanism 20.9 Amide Hydrolysis in Base

1 – 2 **Addition of the nucleophile ($^-$OH)** followed by **elimination of the leaving group ($^-NH_2$)** forms a carboxylic acid. These two steps are reversible.

3 Because the carboxylic acid is a strong organic acid and the leaving group ($^-NH_2$) is a strong base, an acid–base reaction forms the **carboxylate anion.**

Step [2] of Mechanism 20.9 deserves additional comment. For amide hydrolysis to occur, the tetrahedral intermediate must lose $^-NH_2$, a *stronger* base and therefore *poorer* leaving group than $^-$OH. This means that loss of $^-NH_2$ does not often happen. Instead, $^-$OH is lost as the leaving group most of the time, and the starting material is regenerated. But, when $^-NH_2$ is occasionally eliminated, the carboxylic acid product is converted to a lower-energy carboxylate anion in Step [3], and this drives the equilibrium to favor its formation.

Problem 20.26 Draw the products formed in each reaction.

Problem 20.27 With reference to the structures of acetylsalicylic acid (aspirin) and acetaminophen (the active ingredient in Tylenol), explain why acetaminophen tablets can be stored in the medicine cabinet for years, but aspirin tablets slowly decompose over time.

20.13 Application: The Mechanism of Action of β-Lactam Antibiotics

Penicillin and related β-lactams kill bacteria by a nucleophilic acyl substitution reaction. All penicillins have an unreactive amide side chain and a very reactive amide that is part of a β-lactam. The β-lactam is more reactive than other amides because it is part of a strained, four-membered ring that is readily opened with nucleophiles.

Unlike mammalian cells, bacterial cells are surrounded by a fairly rigid cell wall, which allows the bacterium to live in many different environments. This protective cell wall is composed of carbohydrates linked together by peptide chains containing amide linkages, formed using the enzyme **glycopeptide transpeptidase.**

Penicillin interferes with the synthesis of the bacterial cell wall. A nucleophilic OH group of the glycopeptide transpeptidase enzyme cleaves the β-lactam ring of penicillin by a **nucleophilic acyl substitution reaction.** The opened ring of the penicillin molecule remains covalently bonded to the enzyme, thus deactivating the enzyme, halting cell wall construction, and killing the bacterium. Penicillin has no effect on mammalian cells because they are surrounded by a flexible membrane composed of a lipid bilayer (Chapter 3) and not a cell wall.

nucleophilic attack [1]

opening of the β-lactam ring [2] proton transfer

active enzyme — enzyme — inactive enzyme

The enzyme is now inactive.
Cell wall construction stops.

Thus, penicillin and other β-lactam antibiotics are biologically active precisely because they undergo a nucleophilic acyl substitution reaction with an important bacterial enzyme.

Problem 20.28 Some penicillins cannot be administered orally because their β-lactam is rapidly hydrolyzed by the acidic environment of the stomach. What product is formed in the following hydrolysis reaction?

$\xrightarrow{H_3O^+}$

20.14 Summary of Nucleophilic Acyl Substitution Reactions

To help you organize and remember all of the nucleophilic acyl substitution reactions that can occur at a carbonyl carbon, keep in mind these two principles:

- **The *better* the leaving group, the *more reactive* the carboxylic acid derivative.**
- **More reactive acyl compounds can always be converted to less reactive ones. The reverse is not usually true.**

This results in the following order of reactivity:

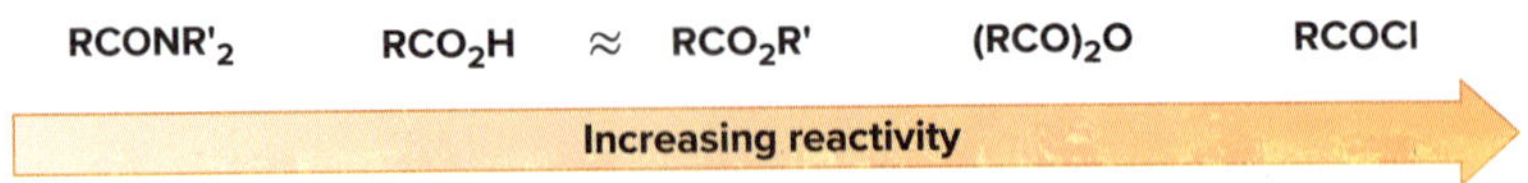

Table 20.4 summarizes the specific nucleophilic acyl substitution reactions. Use it as a quick reference to remind you which products can be formed from a given starting material.

Table 20.4 Summary of the Nucleophilic Substitution Reactions of Carboxylic Acids and Their Derivatives

Starting material		Product: RCOCl	$(RCO)_2O$	RCO_2H	RCO_2R'	$RCONR'_2$
[1] RCOCl	→	–	✓	✓	✓	✓
[2] $(RCO)_2O$	→	✗	–	✓	✓	✓
[3] RCO_2H	→	✓	✓	–	✓	✓
[4] RCO_2R'	→	✗	✗	✓	–	✓
[5] $RCONR'_2$	→	✗	✗	✓	✗	–

Table key: ✓ = A reaction occurs.
✗ = No reaction occurs.

Problem 20.29 Draw the organic products formed in each reaction.

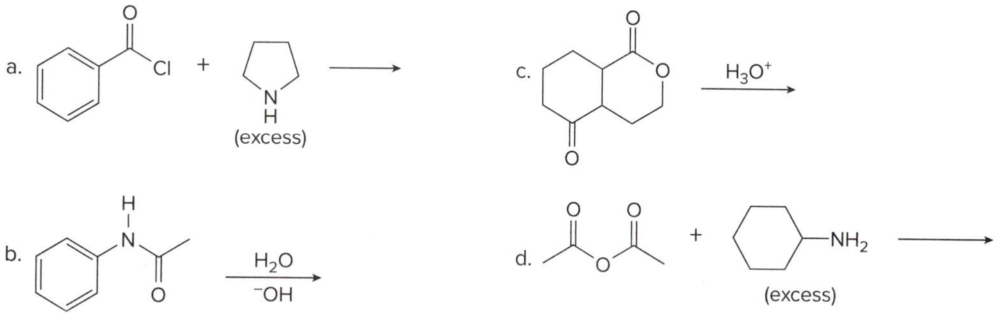

a. [benzoyl chloride] + [pyrrolidine] (excess) →

b. [acetanilide] $\xrightarrow[^{-}OH]{H_2O}$

c. [bicyclic keto lactone] $\xrightarrow{H_3O^+}$

d. [acetic anhydride] + [cyclohexylamine, NH_2] (excess) →

20.15 Natural and Synthetic Fibers

All natural and synthetic fibers are high-molecular-weight polymers. Natural fibers are obtained from either plant or animal sources, and this determines the fundamental nature of their chemical structure. Fibers like **wool and silk obtained from animals are proteins,** so they are formed from amino acids joined together by many amide linkages. **Cotton and linen,** on the other hand, are derived from plants, so they are **carbohydrates having the general structure of cellulose,** formed from glucose monomers. General structures for these polymers are shown in Figure 20.3.

Figure 20.3 The general structure of the common natural fibers

a. Wool and silk—Proteins with many amide bonds

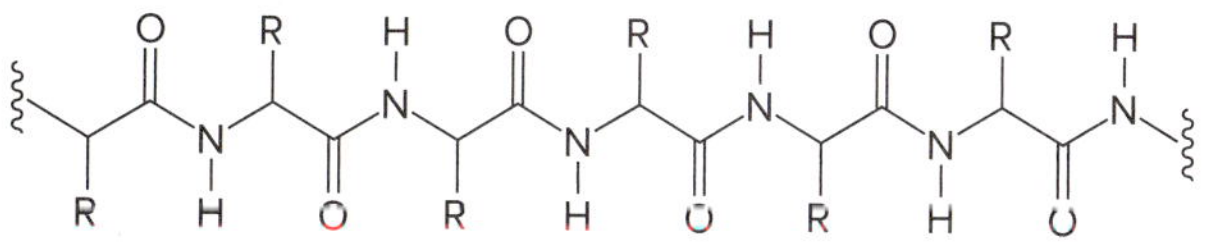

b. Cotton and linen—Carbohydrates like cellulose

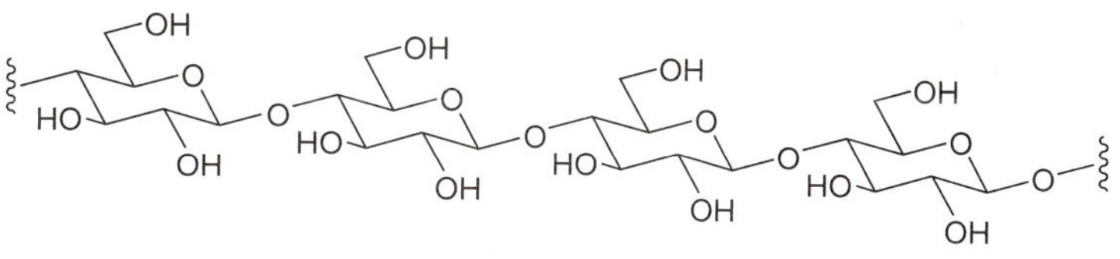

An important practical application of organic chemistry has been the preparation of synthetic fibers, many of which have properties that are different from and sometimes superior to their naturally occurring counterparts. The two most common classes of synthetic polymers are based on **polyamides** and **polyesters.**

DuPont built the first commercial nylon plant in 1938. Although it was initially used by the military to make parachutes, nylon quickly replaced silk in many common products after World War II. *Jeff Morgan 14/Alamy Stock Photo*

20.15A Nylon—A Polyamide

The search for a synthetic fiber led to the discovery of **nylon,** a **polyamide** that is strong and durable and resembles the silk produced by silkworms. There are several different kinds of nylon, but the most well known is called nylon 6,6.

[The amide bonds are labeled in red.]

nylon 6,6

Nylon 6,6 can be synthesized from two six-carbon monomers (hence its name)—adipoyl chloride **($ClOCCH_2CH_2CH_2CH_2COCl$)** and hexamethylenediamine **($H_2NCH_2CH_2CH_2CH_2CH_2CH_2NH_2$).** This diacid chloride and diamine react together to form new amide bonds, yielding the polymer. Nylon is called a **condensation polymer** because a small molecule, in this case HCl, is eliminated during its synthesis.

H_2N–(CH₂)₆–NH_2 + Cl–CO–(CH₂)₄–CO–Cl + H_2N–(CH₂)₆–NH_2 + Cl–CO–(CH₂)₄–CO–Cl

↓

nylon 6,6 + 3 HCl

Problem 20.30 What two monomers are needed to prepare nylon 6,10?

nylon 6,10

20.15B Polyesters

Polyesters constitute a second major class of condensation polymers. The most common polyester is polyethylene terephthalate **(PET),** which is sold under a variety of trade names (Dacron, Terylene, and Mylar) depending on its use.

polyethylene terephthalate
PET
(Dacron, Terylene, and Mylar)

Ester bonds (in red) join the carbon skeleton together.

As we will learn in Section 31.9, PET is more easily recycled than other common polymers. For example, recycled PET is used to make reusable shopping bags. *Jill Braaten*

One method of synthesizing a polyester is by acid-catalyzed esterification of a diacid with a diol (Fischer esterification).

terephthalic acid + ethylene glycol → (acid catalyst) polyester + $3\ H_2O$

Because these polymers are easily and cheaply prepared and form strong and chemically stable materials, they have been used in clothing, films, tires, and many other products.

Problem 20.31 Draw the structure of Kodel, a polyester formed from 1,4-dihydroxymethylcyclohexane and terephthalic acid. Explain why fabrics made from Kodel are stiff and crease resistant.

1,4-dihydroxymethylcyclohexane + terephthalic acid

Problem 20.32 Poly(lactic acid) (PLA) has received much recent attention because the lactic acid monomer [$CH_3CH(OH)COOH$] from which it is made can be obtained from carbohydrates rather than petroleum. This makes PLA a more "environmentally friendly" polyester. (A further discussion of green polymer synthesis is presented in Chapter 31.) Draw the structure of PLA.

20.16 Biological Acylation Reactions

Nucleophilic acyl substitution is a common reaction in biological systems. These acylation reactions are called **acyl transfer reactions** because they result in the transfer of an acyl group from one atom to another (from Z to Nu in this case).

RCOZ + $:Nu^-$ → (acyl transfer reaction) RCONu + :Z

thioester (RCOSR')

In cells, such acylations occur with the sulfur analogue of an ester, called a **thioester,** having the general structure **RCOSR'.** The most common thioester is called **acetyl coenzyme A,** often referred to as **acetyl CoA.**

acetyl group — acetyl coenzyme A = acetyl CoA ($CH_3COSCoA$)

- A thioester (RCOSR') has a good leaving group ($^-SR'$), so, like other acyl compounds, it undergoes substitution reactions with other nucleophiles. With acetyl CoA, an acetyl group is transferred from SCoA to a nucleophile, $:Nu^-$.

$CH_3C(=O)SCoA$ + H–Nu $\xrightarrow{\text{enzyme}}$ $CH_3C(=O)Nu$ + H–SCoA

For example, acetyl CoA undergoes enzyme-catalyzed nucleophilic acyl substitution with choline, forming acetylcholine, a charged compound that transmits nerve impulses between nerve cells.

$CH_3C(=O)SCoA$ (acetyl CoA) + $HOCH_2CH_2N^+(CH_3)_3$ (choline) $\xrightarrow{\text{enzyme}}$ $CH_3C(=O)OCH_2CH_2N^+(CH_3)_3$ (acetylcholine (neurotransmitter)) + H–SCoA

Many other acyl transfer reactions are important cellular processes. Thioesters of fatty acids react with cholesterol, forming **cholesteryl esters** in an enzyme-catalyzed reaction (Figure 20.4). These esters are the principal form in which cholesterol is stored and transported in the body. Because cholesterol is a lipid, insoluble in the aqueous environment of the blood, it travels through the bloodstream in particles that also contain proteins and phospholipids. These particles are classified by their density.

- **LDL particles** (low-density lipoproteins) transport cholesterol from the liver to the tissues.
- **HDL particles** (high-density lipoproteins) transport cholesterol from the tissues back to the liver, where it is metabolized or converted to other steroids.

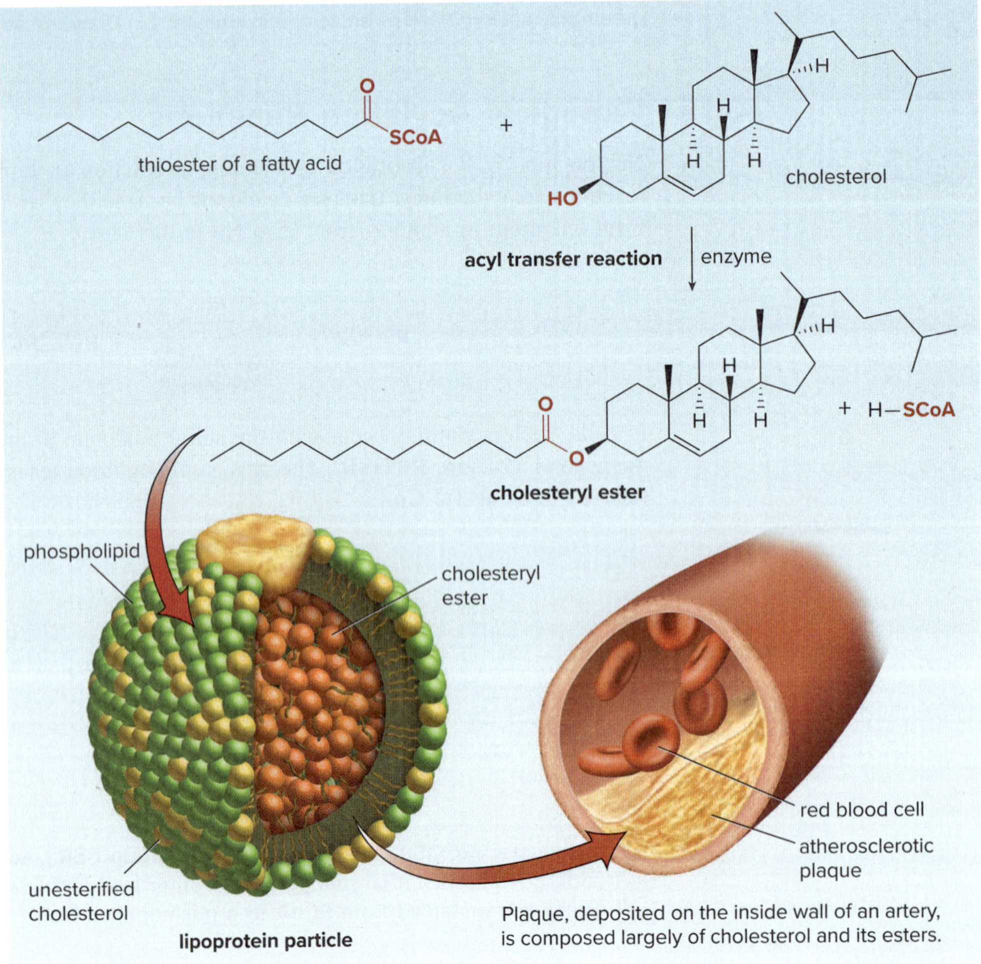

Figure 20.4 Cholesteryl esters and lipoprotein particles

Atherosclerosis is a disease that results from the buildup of fatty deposits on the walls of arteries, forming deposits called **plaque.** Plaque is composed largely of the cholesterol (esterified as an ester) of LDL particles. LDL is often referred to as "bad cholesterol" for this reason. In contrast, HDL particles are called "good cholesterol" because they reduce the amount of cholesterol in the bloodstream by transporting it back to the liver.

Problem 20.33 Glucosamine is a dietary supplement available in many over-the-counter treatments for osteoarthritis. Reaction of acetyl CoA with glucosamine forms NAG, *N*-acetylglucosamine, the monomer used to form chitin, the carbohydrate that forms the rigid shells of lobsters and crabs. What is the structure of NAG?

glucosamine + $CH_3COSCoA$ ⟶

Chapter 20 REVIEW

KEY CONCEPTS

The properties of RCOZ

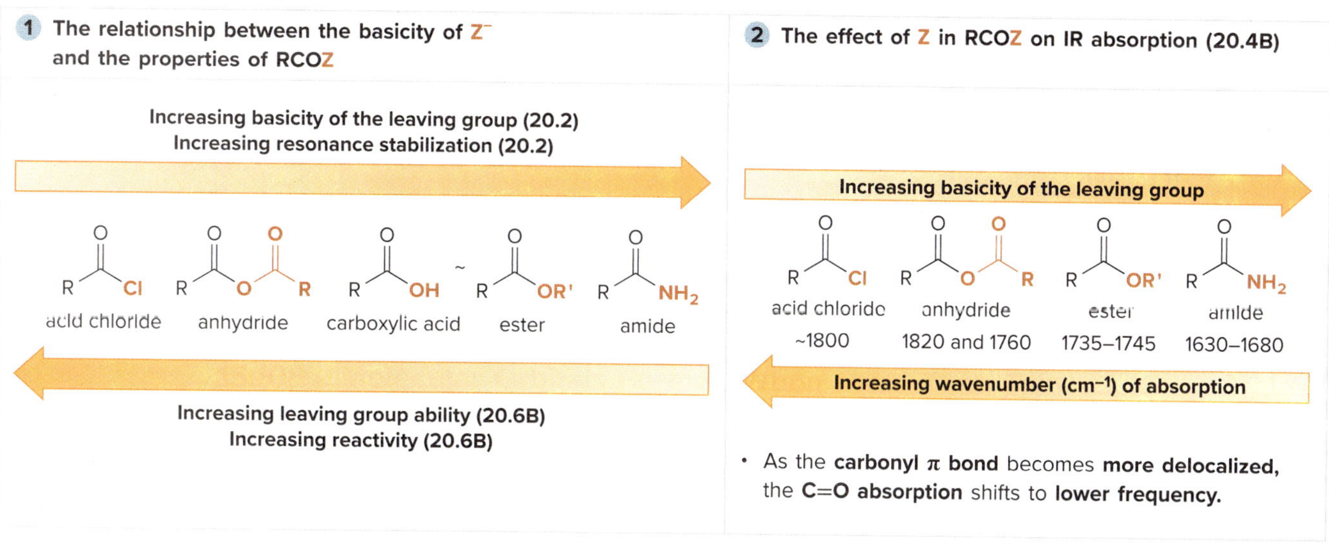

Try Problem 20.39.

KEY REACTIONS

Nucleophilic Acyl Substitution Reactions

[1] Reactions that produce acid chlorides (RCOCl)

carboxylic acid $\xrightarrow[(20.9A)]{SOCl_2}$ acid chloride + SO_2 + HCl

Try Problem 20.40c.

[2] Reactions that produce anhydrides $[(RCO)_2O]$

1. acid chloride + ^-O(C=O)R → (20.7) → anhydride + Cl^-

2. dicarboxylic acid → Δ (20.9B) → anhydride + H_2O

Try Problems 20.40i, 20.41d.

[3] Reactions that produce carboxylic acids (RCOOH) and carboxylates ($RCOO^-$)

1. acid chloride + H_2O → pyridine (20.7) → carboxylic acid + pyridinium ($^+$N–H) Cl^-

2. anhydride + H_2O → (20.8) → 2 carboxylic acid

3. ester (OCH_3) → H_2O, H^+ (20.10) → carboxylic acid + CH_3OH

4. ester (OCH_3) → ^-OH, H_2O (20.10) → carboxylate + CH_3OH

5. amide (NR'_2), R' = H or alkyl → H_2O, H^+ (20.12) → carboxylic acid + $R'_2\overset{+}{N}H_2$

6. amide (NR'_2), R' = H or alkyl → ^-OH, H_2O (20.12) → carboxylate + R'_2NH

Try Problems 20.34b [1], [2]; 20.41a; 20.44; 20.45; 20.47b; 20.48.

[4] Reactions that produce esters (RCOOR')

1. acid chloride → CH_3OH, pyridine (20.7) → ester (OCH_3) + pyridinium ($^+$N–H) Cl^-

2. anhydride → ethanol (OH) (20.8) → ester + carboxylic acid

3. carboxylic acid → CH_3OH, H_2SO_4, Fisher esterification (20.9C) → ester (OCH_3) + H_2O

Try Problems 20.40g, l; 20.47a, c.

[5] Reactions that produce amides ($RCONR'_2$)

1. acid chloride → NH_3 (2 equiv) (20.7) → amide + $NH_4^+Cl^-$

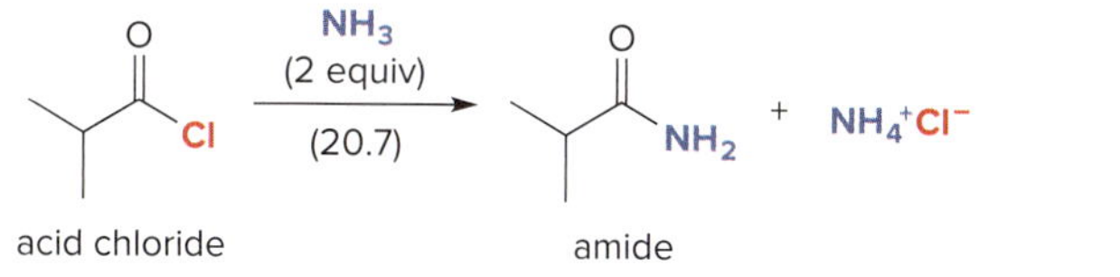

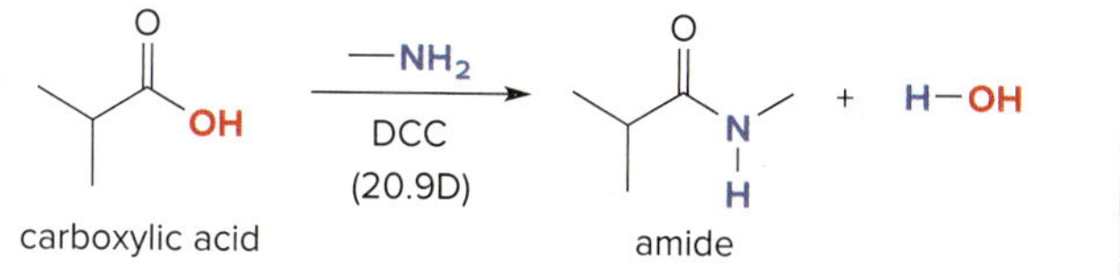

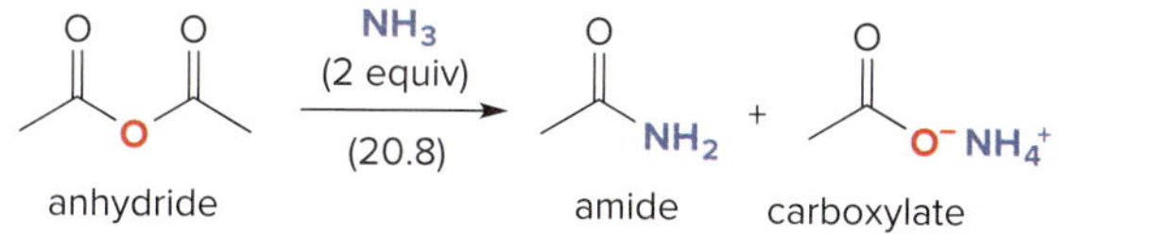

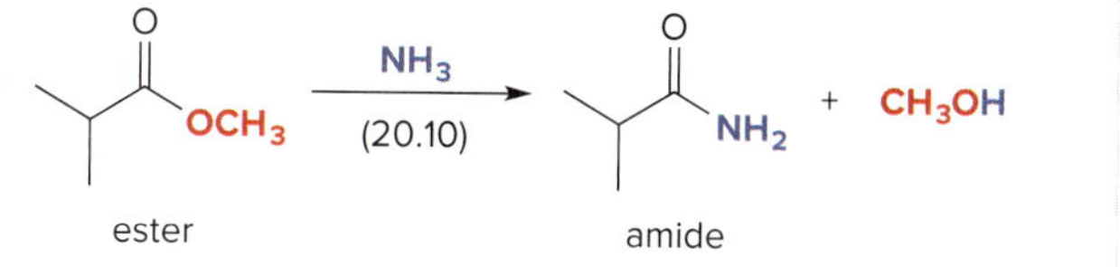

2. anhydride → NH_3 (2 equiv) (20.8) → amide + carboxylate ($O^- NH_4^+$)

3. carboxylic acid → [1] NH_3 [2] Δ (20.9D) → amide + H–OH

4. carboxylic acid → $-NH_2$, DCC (20.9D) → amide + H–OH

5. ester (OCH_3) → NH_3 (20.10) → amide + CH_3OH

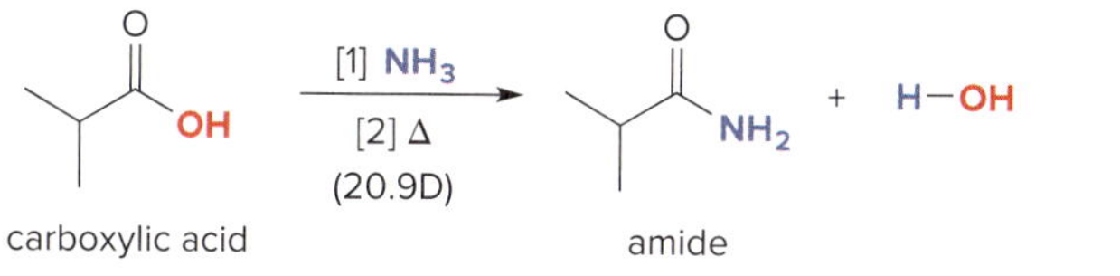

Try Problems 20.40f, j, k; 20.43; 20.47d.

KEY SKILLS

[1] Determining whether a nucleophilic acyl substitution will occur (20.6B)

1 **Identify the different groups attached to the C=O.**

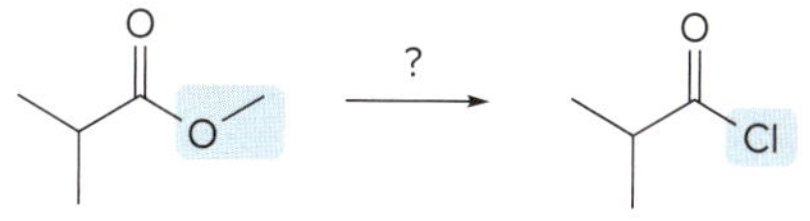

- This conversion requires the substitution of $^-OCH_3$ by Cl^-.

2 **Compare basicity of the leaving group and nucleophile to determine if the reaction will occur.**

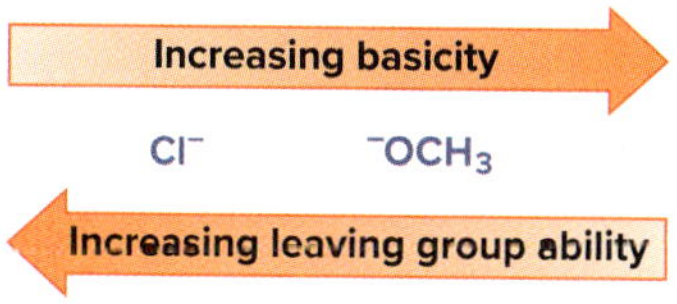

Increasing basicity →

Cl^- $^-OCH_3$

← Increasing leaving group ability

- Because $^-OCH_3$ is a **stronger base** and therefore a **poorer leaving group** than Cl^-, **this reaction does *not* occur.**

See Sample Problem 20.2. Try Problem 20.40d.

[2] Determining the carboxylic acid and alcohol needed for a Fischer esterification (20.9C)

1 **Cleave the carbon–oxygen single bond attached to the carbonyl.**

2 **Draw the RCOOH, which becomes the RC=O.**

3 **Draw the HOR, which becomes the OR group.**

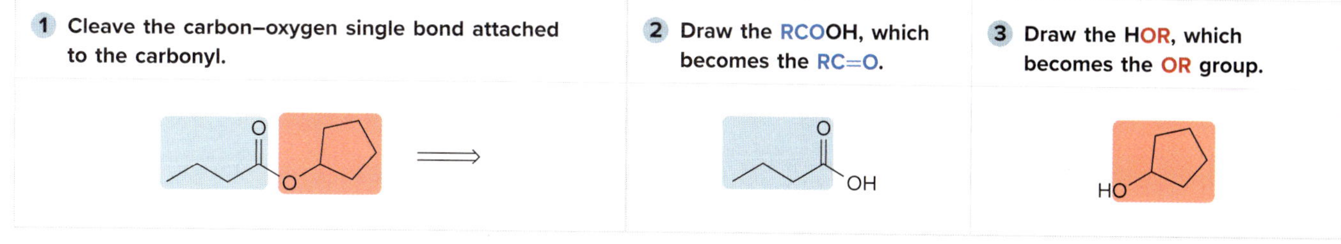

Try Problem 20.57.

CHAPTER 20 MULTIPLE-CHOICE SELF-TEST

The Self-Test consists of multiple-choice questions similar to those found on the American Chemical Society organic chemistry exam. Answers are given at the end of the chapter.

1. Which functional group is most reactive in nucleophilic acyl substitution?

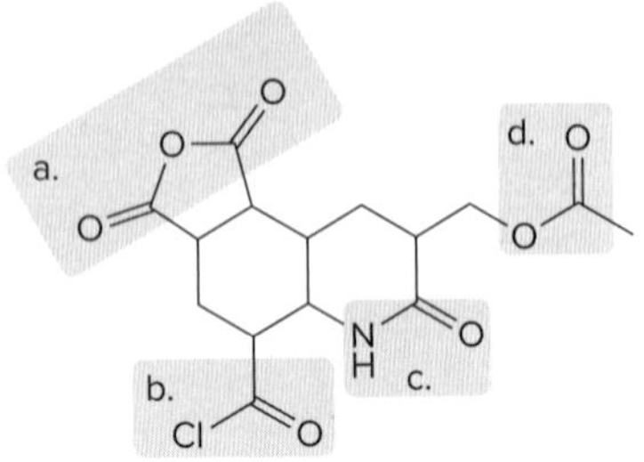

a. b. c. d.

2. What compound is *not* formed when **A** is treated with aqueous acid?

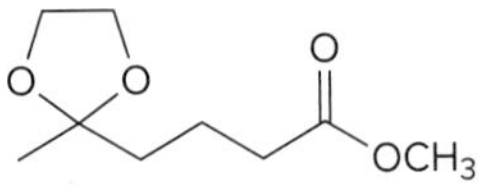

A

a. 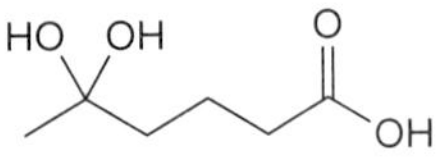 b. 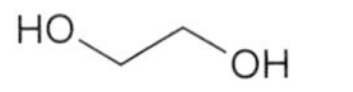 c. CH_3OH d.

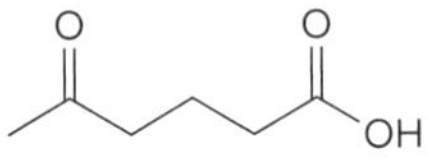

3. Which method can*not* be used to form **B**?

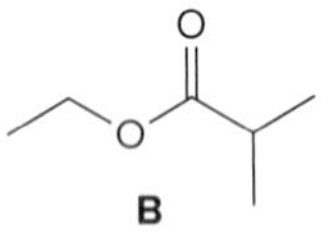

B

a. 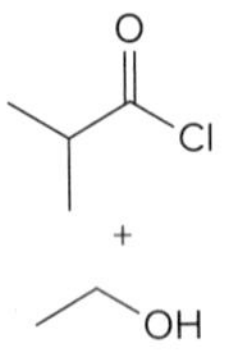 + b. 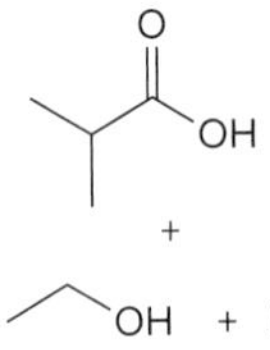 + + NaOH c. 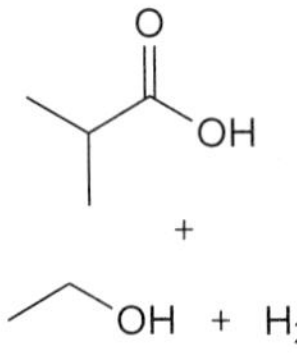 + + H_2SO_4 d. 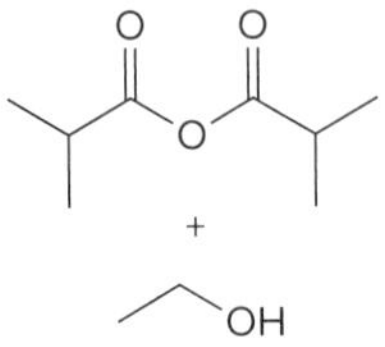 +

4. What is the IUPAC name for the following compound?

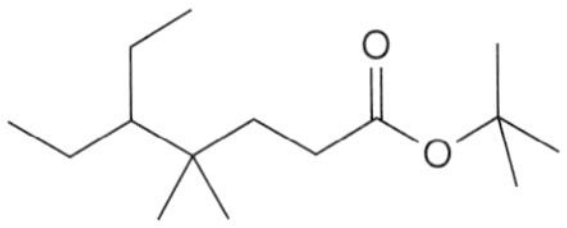

a. *tert*-butoxy 5-ethyl-4,4-dimethylheptanoate
b. *tert*-butyl 5-ethyl-4,4-dimethylheptanecarboxylate
c. *tert*-butyl 3-ethyl-4,4-dimethylheptanoate
d. *tert*-butyl 5-ethyl-4,4-dimethylheptanoate

5. Which compound has a carbonyl absorption at the lowest wavenumber?

a. 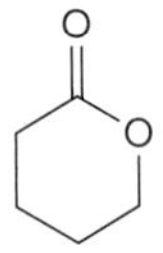 b. 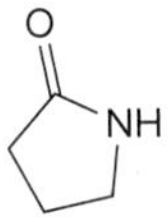 c. 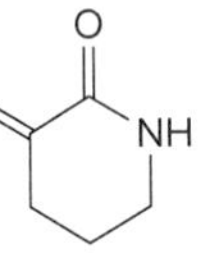 d.

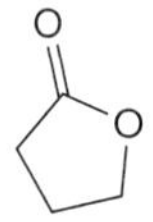

6. Which reactants can*not* be used to prepare **C**?

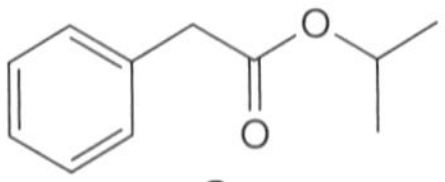

C

a. 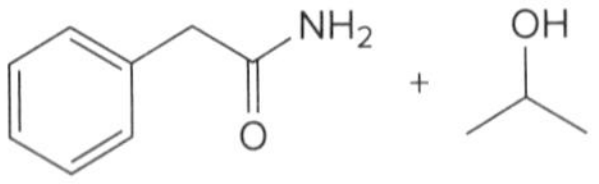 + b. 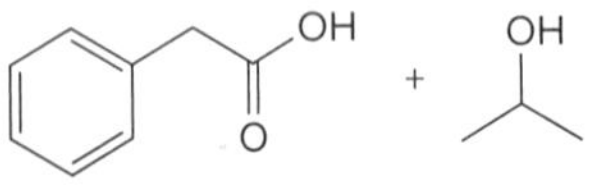 + c. + d. 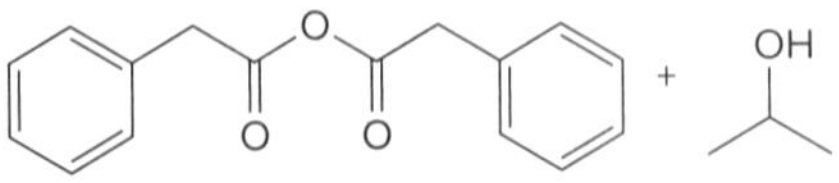 +

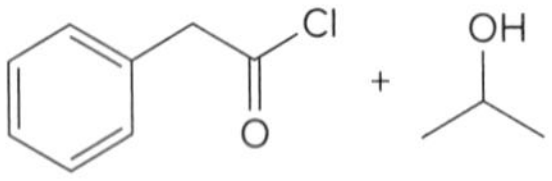

7. Which method can*not* be used to convert $CH_3CH_2CH_2CO_2H$ to $CH_3CH_2CH_2CH_2OH$?

a. [1] $LiAlH_4$; [2] H_2O
b. [1] CH_3OH, H^+; [2] $LiAlH_4$; [3] H_2O
c. [1] $SOCl_2$; [2] $LiAlH[OC(CH_3)_3]_3$; [3] H_2O
d. [1] $SOCl_2$; [2] CH_3OH, H^+; [3] $LiAlH_4$; [4] H_2O

8. What reactants can be used to make a polyester?

a. 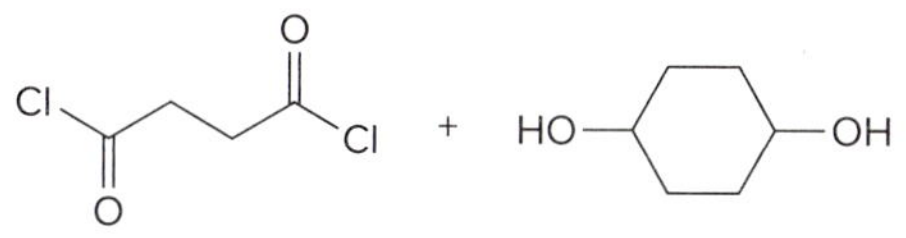 $ClCO(CH_2)_2COCl$ + HO–(cyclohexane)–OH

c. 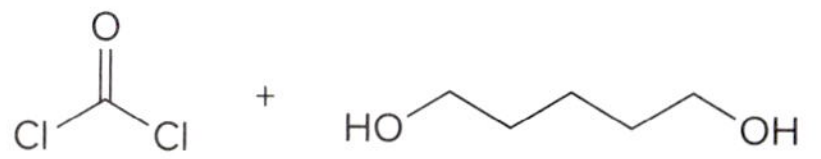 $ClCOCl$ + $HO(CH_2)_6OH$

b. 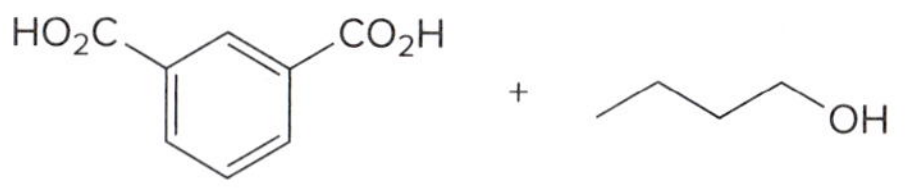 HO_2C–(benzene)–CO_2H + $CH_3CH_2CH_2CH_2OH$

d.  (furan)–CO_2H + $HOCH_2CH_2OH$

9. Which statement about $CH_3(CH_2)_3CON(CH_3)CH_2CH_3$ is *false?*

a. The amide can be prepared from $CH_3(CH_2)_3CO_2H$ + $CH_3NHCH_2CH_3$ + DCC.

b. The amide can be prepared from $CH_3(CH_2)_3CO_2H$ + $CH_3NHCH_2CH_3$.

c. The amide is hydrolyzed in base to $CH_3(CH_2)_3CO_2^-$ + $CH_3NHCH_2CH_3$.

d. The IUPAC name is *N*-ethyl-*N*-methylpentanamide.

10. In the hydrolysis of **L** to **M,** which intermediate is *not* formed?

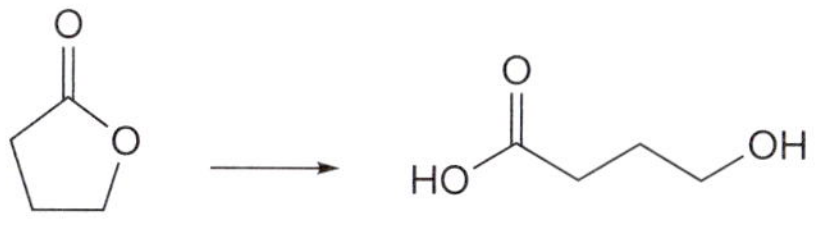

L ⟶ **M**

a. 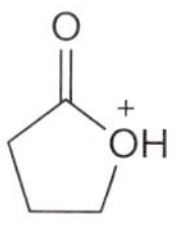 b. 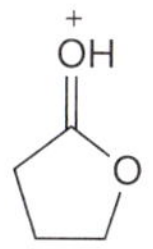 c. 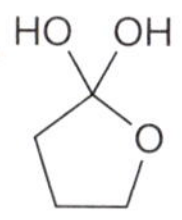 d.

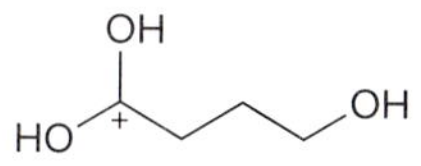

PROBLEMS

Problems Using Three-Dimensional Models

20.34 (a) Give an acceptable name for compound **A.** (b) Draw the organic products formed when **A** is treated with each reagent: [1] H_3O^+; [2] ^-OH, H_2O; [3] $CH_3CH_2CH_2MgBr$ (excess), then H_2O; [4] $LiAlH_4$, then H_2O.

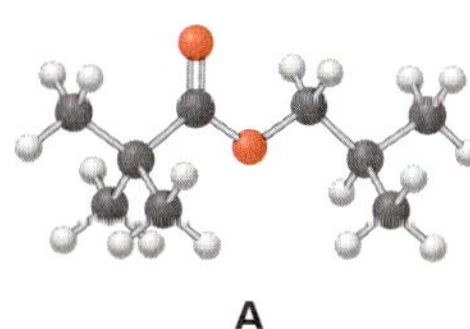

A

20.35 Which ester, **C** or **D,** is more reactive in nucleophilic acyl substitution? Explain your reasoning.

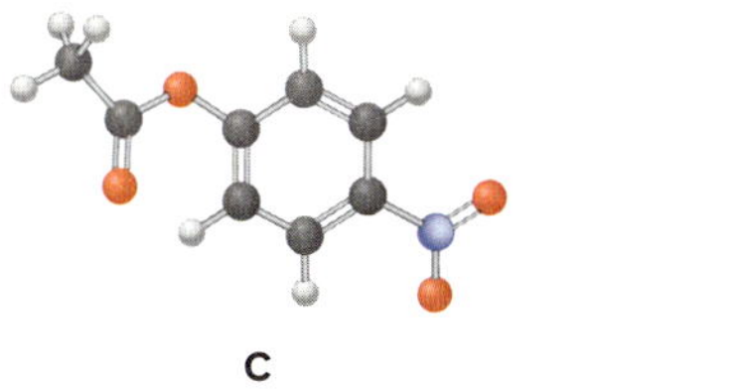

C

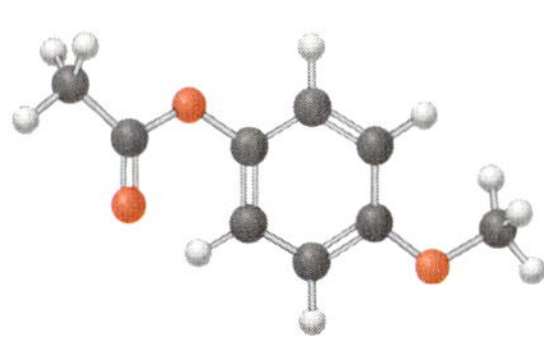

D

Nomenclature

20.36 Give the IUPAC or common name for each compound.

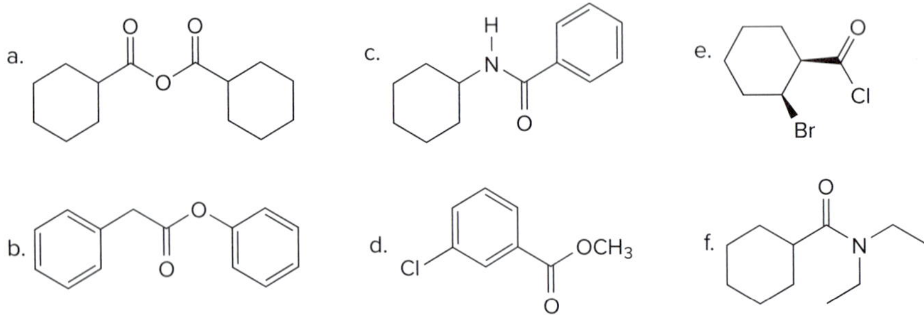

20.37 Give the structure corresponding to each name.

a. cyclohexyl propanoate
b. cyclohexanecarboxamide
c. benzoic propanoic anhydride
d. 3-methylhexanoyl chloride
e. octyl butanoate
f. *N,N*-dibenzylformamide

Properties of Carboxylic Acid Derivatives

20.38 Explain why imidazolides are much more reactive than other amides in nucleophilic acyl substitution.

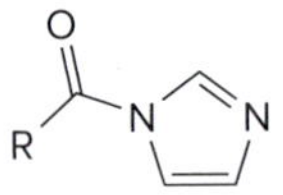

imidazolide

20.39 (a) Propose an explanation for the difference in the frequency of the carbonyl absorptions of phenyl acetate (1765 cm^{-1}) and cyclohexyl acetate (1738 cm^{-1}). (b) Which carbonyl group is more effectively stabilized by resonance? (c) Which ester reacts faster when treated with aqueous base?

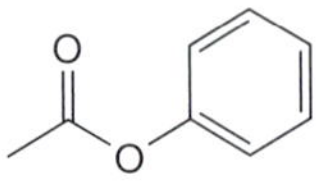

phenyl acetate 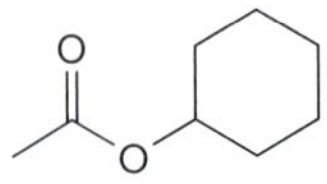 cyclohexyl acetate

Reactions

20.40 Draw the product formed when phenylacetic acid ($C_6H_5CH_2COOH$) is treated with each reagent. With some reagents, no reaction occurs.

a. $NaHCO_3$
b. $NaOH$
c. $SOCl_2$
d. $NaCl$
e. NH_3 (1 equiv)
f. NH_3, Δ
g. CH_3OH, H_2SO_4
h. CH_3OH, ^-OH
i. [1] $NaOH$; [2] CH_3COCl
j. CH_3NH_2, DCC
k. [1] $SOCl_2$; [2] $CH_3CH_2CH_2NH_2$ (excess)
l. [1] $SOCl_2$; [2] $(CH_3)_2CHOH$

20.41 Draw the organic products formed in each reaction.

a. 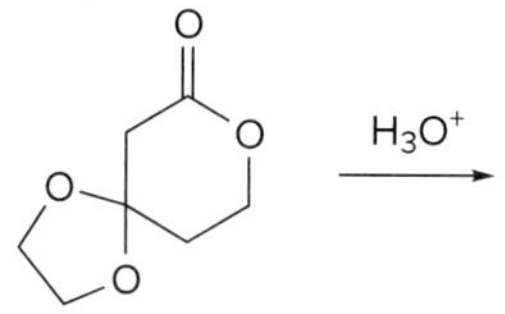
O
O
O
O
H_3O^+

c. 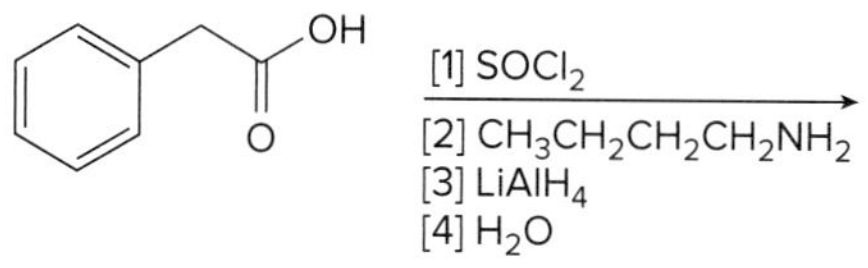
OH
O
[1] $SOCl_2$
[2] $CH_3CH_2CH_2CH_2NH_2$
[3] $LiAlH_4$
[4] H_2O

b. 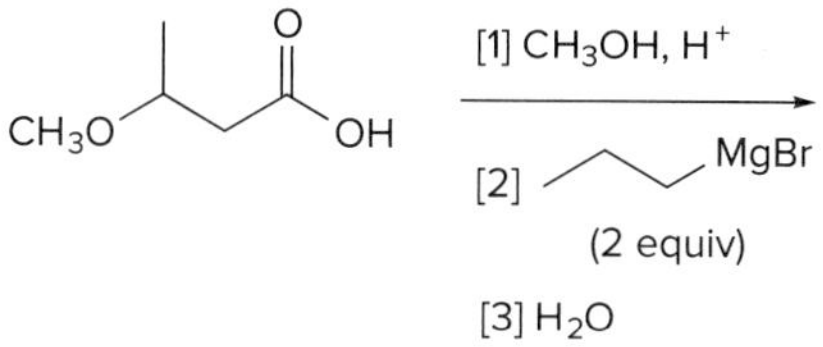
CH_3O
O
OH
[1] CH_3OH, H^+
[2] MgBr
(2 equiv)
[3] H_2O

d. 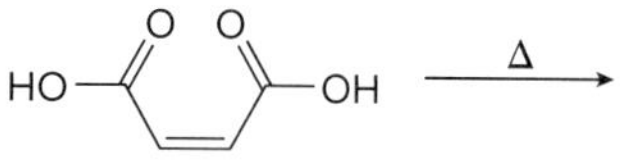
HO
O
O
OH
Δ

20.42 Cinnamoylcocaine, a natural product that occurs in coca leaves, can be converted to cocaine, the chapter-opening molecule, by the following reaction sequence. Identify the structure of cinnamoylcocaine, as well as intermediates **X** and **Y.**

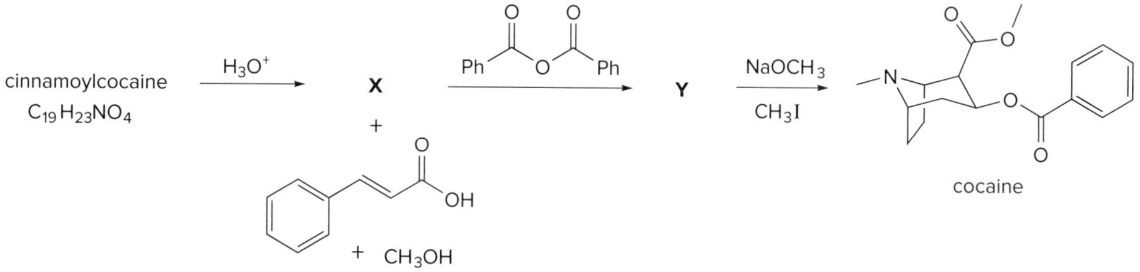
cinnamoylcocaine
$C_{19}H_{23}NO_4$
H_3O^+
X
+
OH
+ CH_3OH
Ph
O
Ph
Y
$NaOCH_3$
CH_3I
N
cocaine

20.43 Identify the product **X** in the following reaction, one step in the synthesis of the blood-pressure-lowering drug captopril (Problem 5.61).

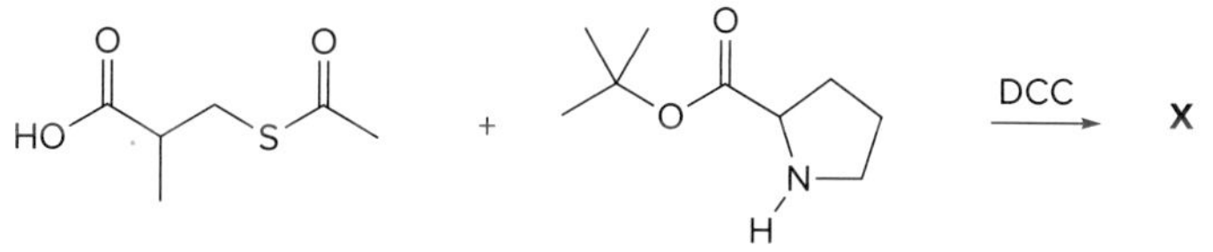
HO
S
+
O
N
H
DCC
X

20.44 Draw the products formed when the peptide angiotensin II is hydrolyzed. As you learned in the Chapter 1 opener, angiotensin increases blood pressure by narrowing blood vessels, so drugs that decrease the amount of angiotensin lower blood pressure.

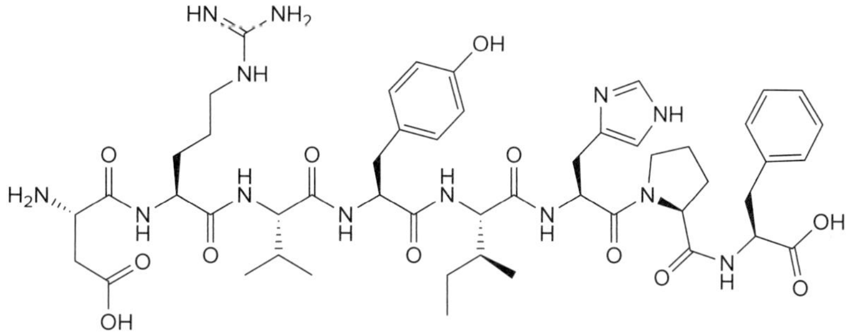
HN
NH_2
NH
OH
N
NH
H_2N
OH
OH

angiotensin II

20.45 What products are formed by hydrolysis of each lactone or lactam with acid?

a. 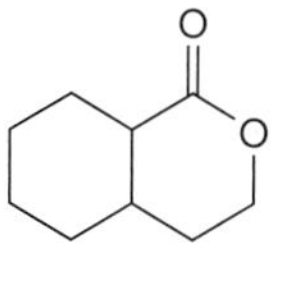

b. 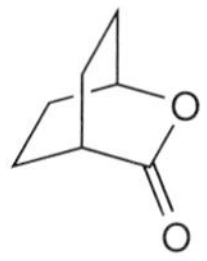

c. 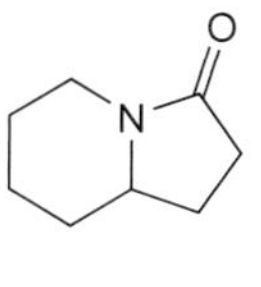

d. 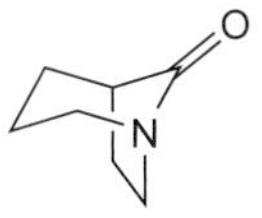

20.46 Identify compounds **A–M** in the following reaction sequence.

Br; NaCN; A; H_3O^+; B; $SOCl_2$; D; [1] $(CH_3)_2CuLi$ [2] H_2O; E; [1] $LiAlH_4$ [2] H_2O; C; CH_3OH, H^+; F; [1] DIBAL-H [2] H_2O; G; [1] CH_3Li [2] H_2O; H; PCC; $(CH_3CO)_2O$; I; [1] $LiAlH_4$ [2] H_2O; J; TsCl, pyridine; K; NaCN; L; [1] CH_3MgBr [2] H_2O; M

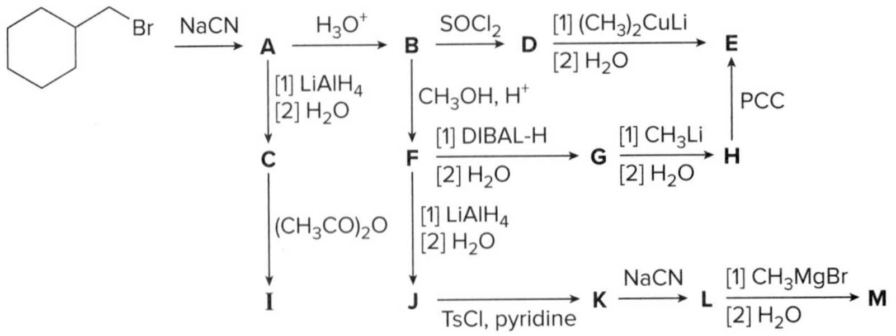

20.47 Draw the products of each reaction and indicate the stereochemistry at any stereogenic centers.

a. 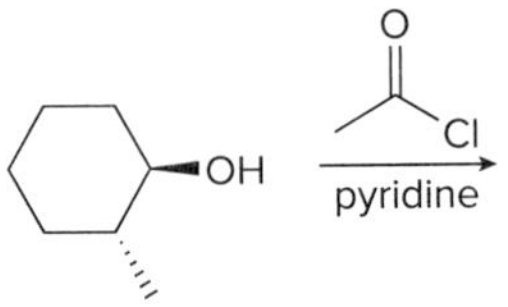 OH; Cl; O; pyridine

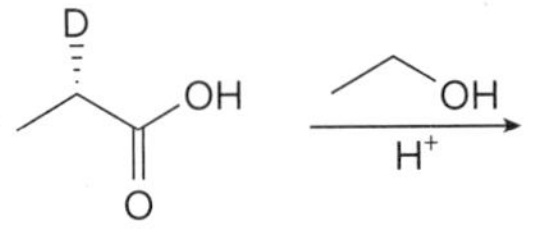

b. 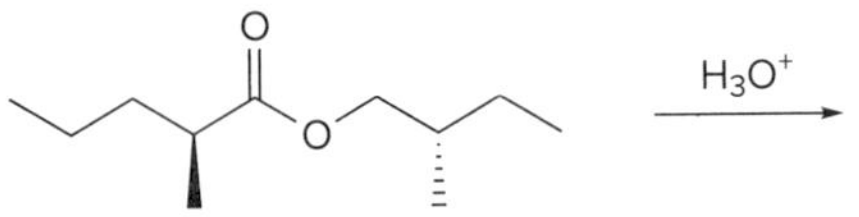 O; O; H_3O^+

c. D; OH; O; OH; H^+

d. O; Cl; +; H; NH_2; (2 equiv)

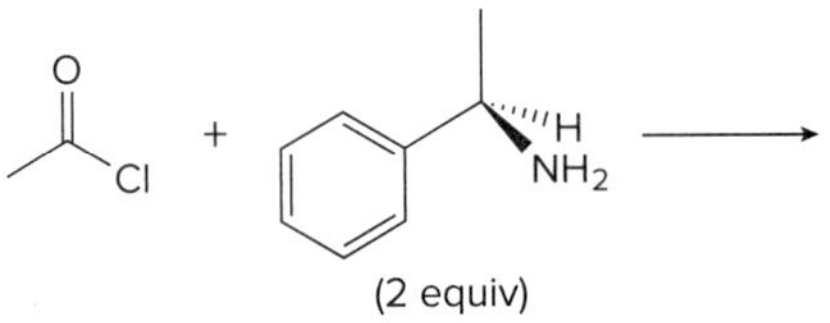

20.48 What products are formed when all of the amide and ester bonds are hydrolyzed in each of the following compounds? **Tamiflu** [part (a)] is the trade name of the antiviral agent oseltamivir, thought to be the most effective agent in treating influenza. **Dutasteride** [part (b), trade name Avodart] was approved in 2001 for the treatment of an enlarged prostate. **Aspartame** [part (c)] is the artificial sweetener used in Equal and many diet beverages. One of the products of this hydrolysis reaction is the amino acid phenylalanine. Infants afflicted with phenylketonuria cannot metabolize this amino acid, so it accumulates, causing mental retardation. When the affliction is identified early, a diet limiting the consumption of phenylalanine (and compounds like aspartame that are converted to it) can make a normal life possible.

a. 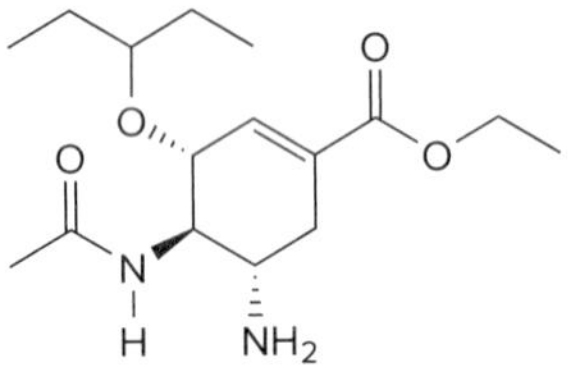 O; O; O; N; H; NH_2

oseltamivir

b. 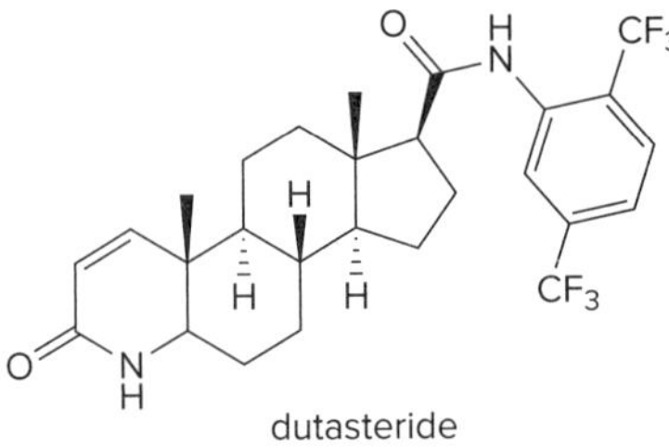 O; H; N; CF_3; H; H; H; CF_3; O; N; H

dutasteride

c. 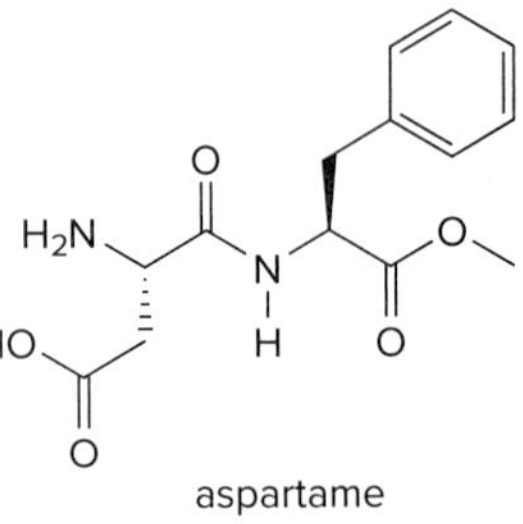 H_2N; O; N; H; O; O; HO; O

aspartame

20.49 Identify **F** in the following reaction sequence. **F** was converted in several steps to the antidepressant paroxetine (trade name Paxil; see also Problem 9.8).

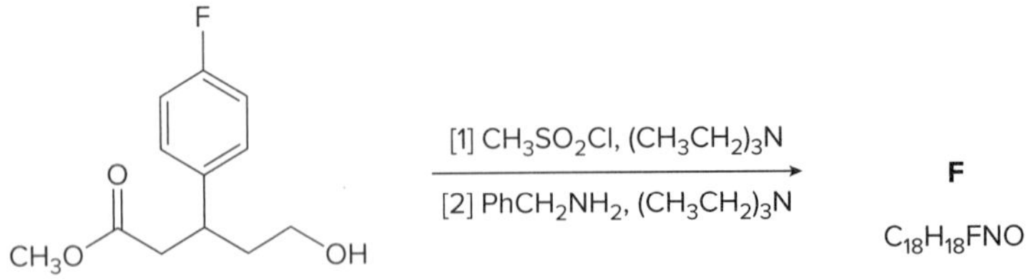

F; O; CH_3O; OH; [1] CH_3SO_2Cl, $(CH_3CH_2)_3N$ [2] $PhCH_2NH_2$, $(CH_3CH_2)_3N$; **F** $C_{18}H_{18}FNO$

Mechanism

20.50 Aspirin is an anti-inflammatory agent because it inhibits the conversion of arachidonic acid to prostaglandins by the transfer of its acetyl group ($CH_3CO–$) to an OH group at the active site of an enzyme (Section 19.5). This reaction, called transesterification, results in the conversion of one ester to another by a nucleophilic acyl substitution reaction. Draw a stepwise mechanism for the given transesterification.

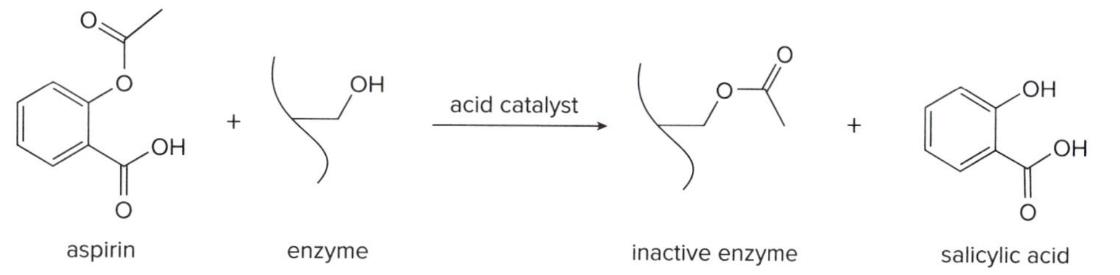

acid catalyst

aspirin + enzyme → inactive enzyme + salicylic acid

20.51 Draw a stepwise mechanism for the following reaction, one step in the synthesis of the cholesterol-lowering drug ezetimibe (Section 17.6).

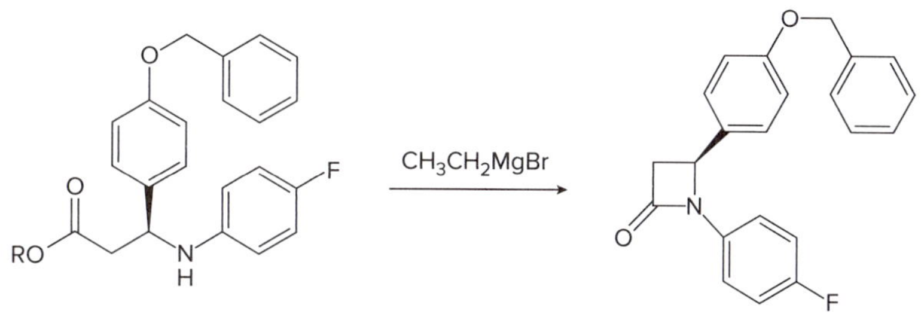

CH_3CH_2MgBr

20.52 Draw a stepwise mechanism for the following reaction, which involves both a Diels–Alder reaction and a nucleophilic acyl substitution.

Δ

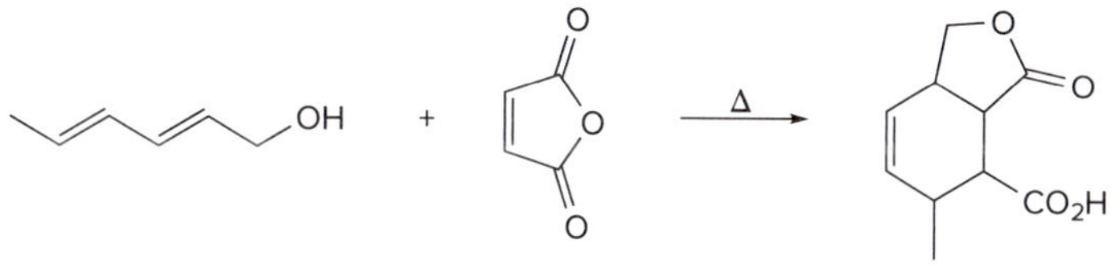

20.53 Draw a stepwise mechanism for the conversion of lactone **C** to carboxylic acid **D. C** is a key intermediate in the synthesis of prostaglandins (Section 19.5) by Nobel Laureate E. J. Corey and co-workers at Harvard University.

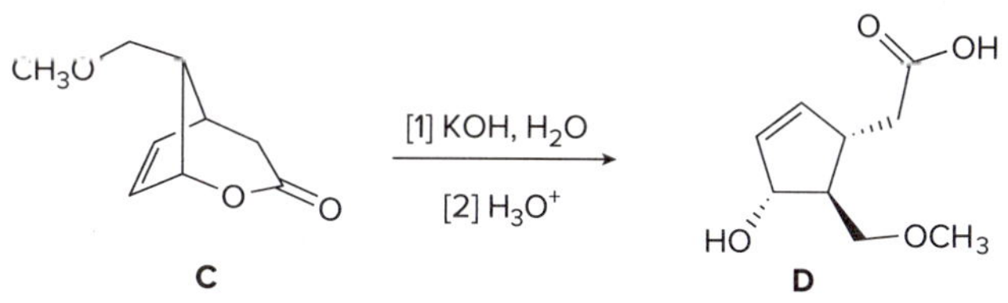

[1] KOH, H_2O
[2] H_3O^+

C → **D**

20.54 Two steps in the synthesis of tadalafil, a drug sold under the trade name Cialis for the treatment of erectile dysfunction, are shown. Identify intermediate **A,** and draw a mechanism for the conversion of **A** to tadalafil.

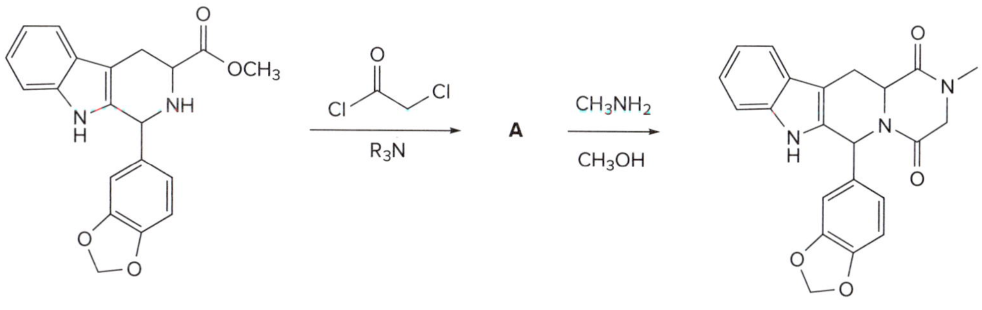

$ClCOCH_2Cl$, R_3N → **A** → CH_3NH_2, CH_3OH → tadalafil

20.55 Draw a stepwise mechanism for the following reaction.

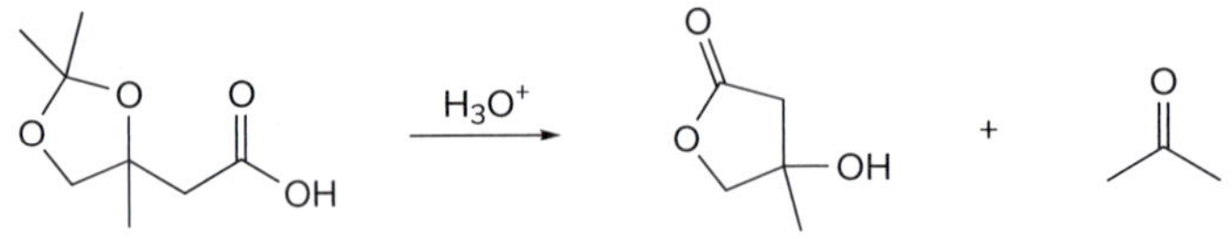

H_3O^+

20.56 Three steps in the synthesis of the anticancer drug Taxol (paclitaxel, Chapter 5 opening molecule) involve the conversion of **A** to **B.** Draw stepwise mechanisms for Steps [2] and [3] in this reaction scheme.

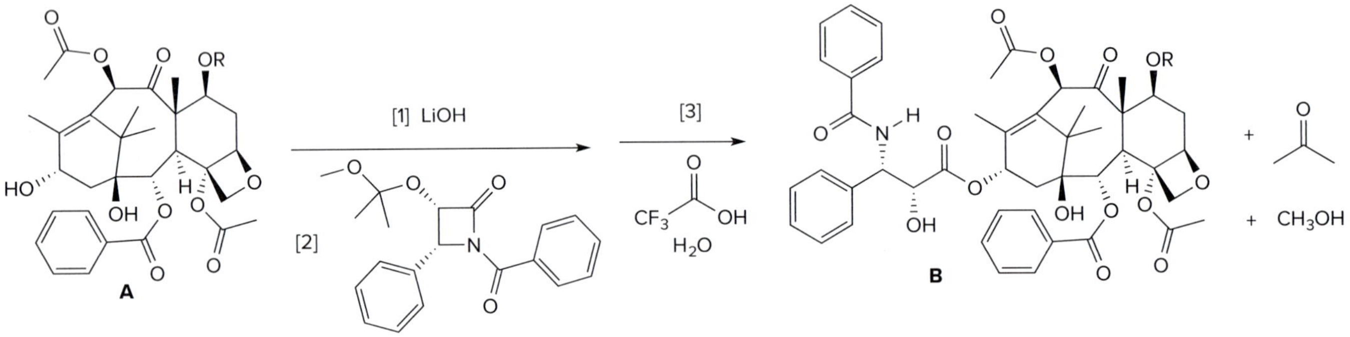

[1] LiOH

[2]

[3] CF_3CO_2H, H_2O

A

B

+ CH_3OH

Synthesis

20.57 What carboxylic acid and alcohol are needed to prepare each ester by Fischer esterification?

a.

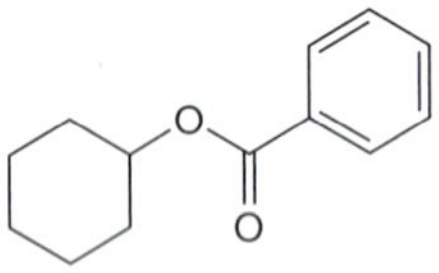

b.

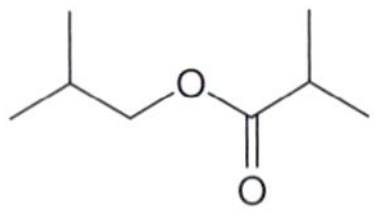

c. 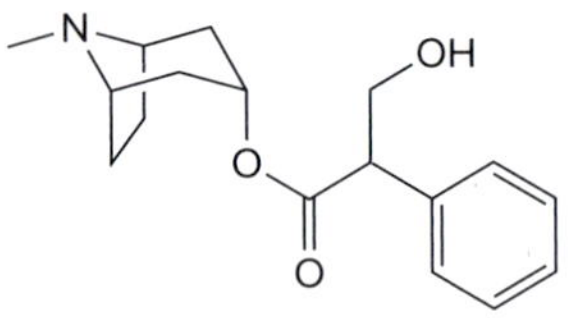 atropine
biologically active amine
from *Atropa belladonna*

d. 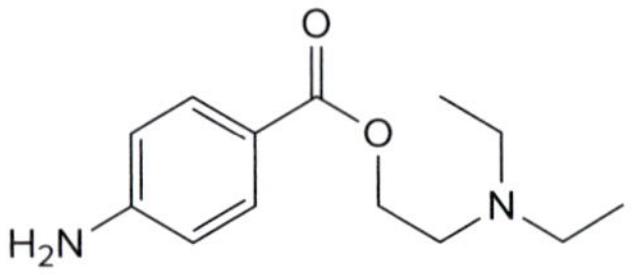 procaine
local anesthetic

20.58 Devise a synthesis of each compound using 1-bromobutane ($CH_3CH_2CH_2CH_2Br$) as the only organic starting material. You may use any other inorganic reagents.

a.

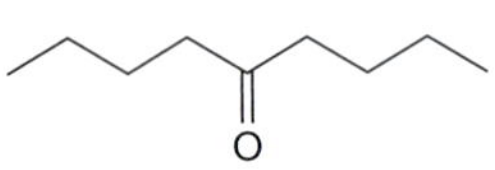

b.

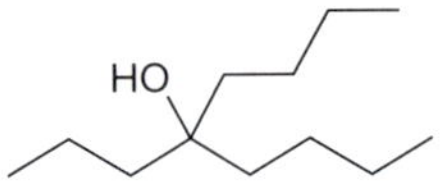

c.

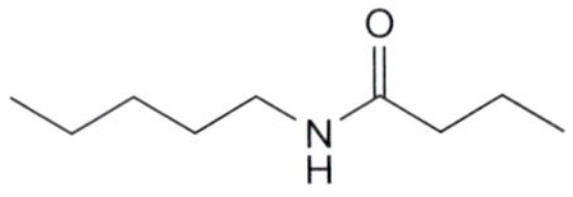

20.59 Devise a synthesis of benzocaine, ethyl *p*-aminobenzoate ($H_2NC_6H_4CO_2CH_2CH_3$), from benzene, organic alcohols, and any needed organic or inorganic reagents. Benzocaine is the active ingredient in the topical anesthetic Orajel (Section 16.14D).

20.60 Devise a synthesis of each compound from benzene and organic alcohols containing four or fewer carbons. You may also use any required organic or inorganic reagents.

a.

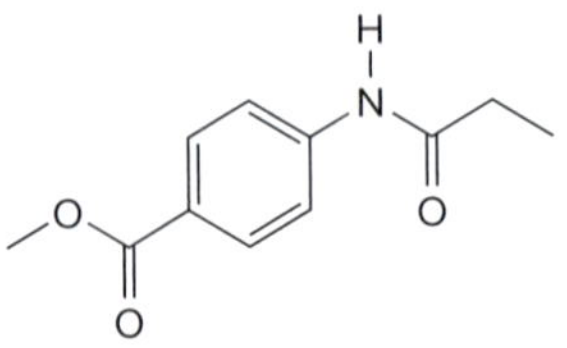

b.

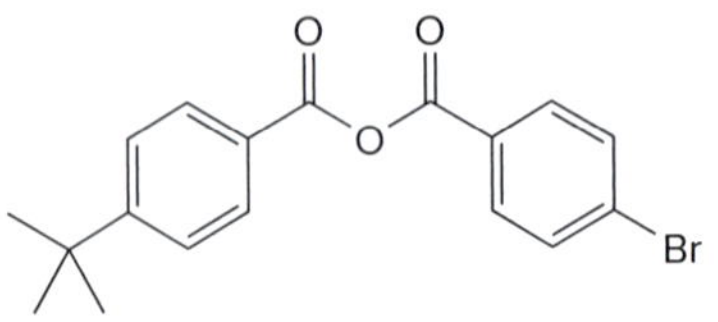

c.

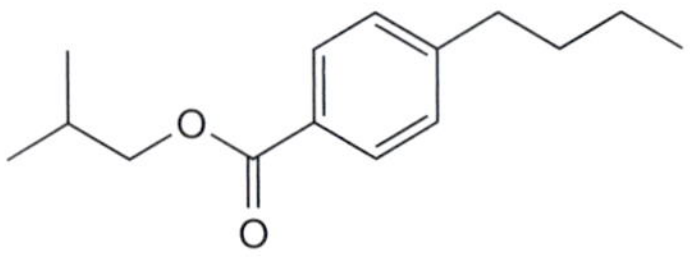

20.61 How would you convert benzoic acid ($C_6H_5CO_2H$) to each compound?

a.

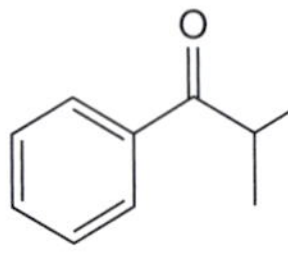

b.

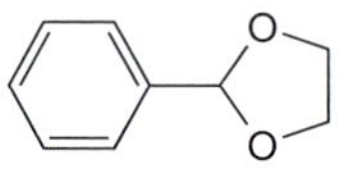

c.

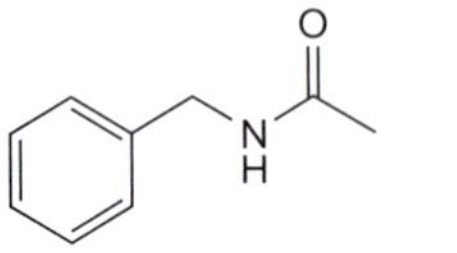

d.

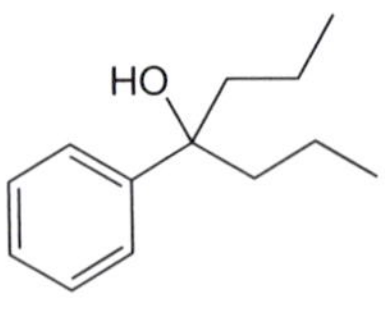

Polymers

20.62 What polyester or polyamide can be prepared from each pair of monomers?

a. HO–(cyclohexane)–OH and HO(C=O)CH₂CH₂(C=O)OH b. ClC(=O)–(benzene)–C(=O)Cl and H_2N(CH₂)₄NH_2

20.63 What two monomers are needed to prepare each polymer?

a. b.

Spectroscopy

20.64 Identify the structure of each compound from the given data.

a. Molecular formula	$C_6H_{12}O_2$
IR absorption:	1738 cm^{-1}
^1H NMR:	1.12 (triplet, 3 H), 1.23 (doublet, 6 H), 2.28 (quartet, 2 H), and 5.00 (septet, 1 H) ppm
b. Molecular formula	C_8H_9NO
IR absorptions:	3328 and 1639 cm^{-1}
^1H NMR:	2.95 (singlet, 3 H), 6.95 (singlet, 1 H), and 7.3–7.7 (multiplet, 5 H) ppm

20.65 Identify the structures of **A** and **B,** isomers of molecular formula $C_{10}H_{12}O_2$, from their IR data and ^1H NMR spectra.

a. IR absorption for **A** at 1718 cm^{-1}

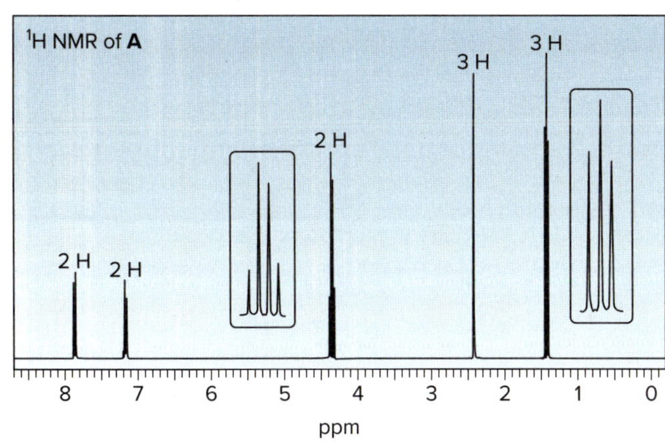

b. IR absorption for **B** at 1740 cm^{-1}

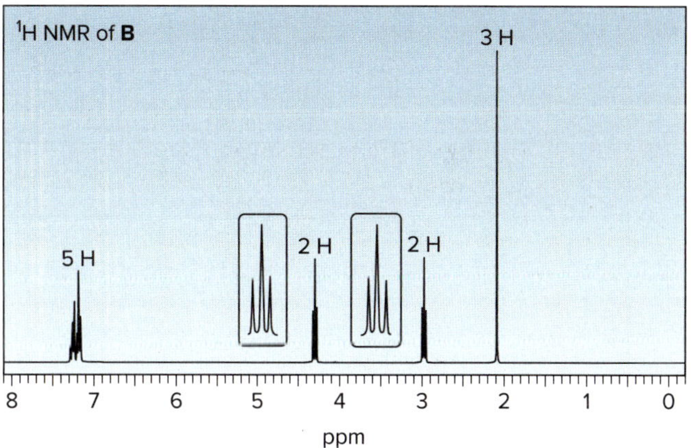

20.66 Phenacetin is an analgesic compound having molecular formula $C_{10}H_{13}NO_2$. Once a common component in over-the-counter pain relievers such as APC (**a**spirin, **p**henacetin, **c**affeine), phenacetin is no longer used because of its liver toxicity. Deduce the structure of phenacetin from its ^1H NMR and IR spectra.

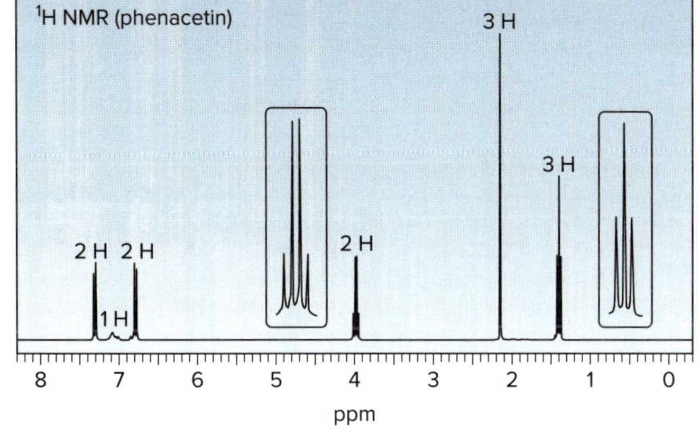

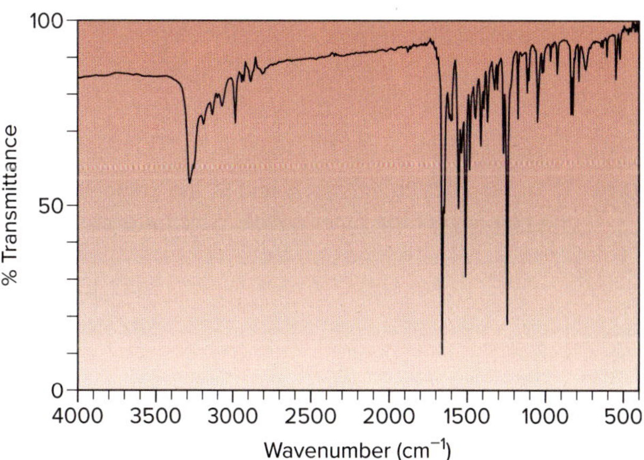

20.67 Identify the structure of compound **C** (molecular formula $C_{11}H_{15}NO_2$), which has an IR absorption at 1699 cm^{-1} and the ^1H NMR spectrum shown below.

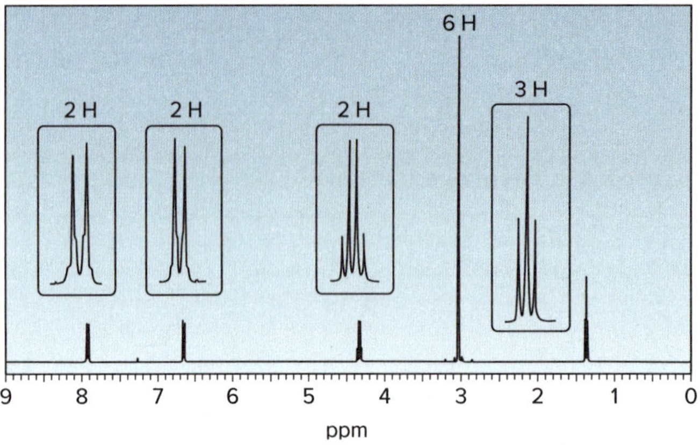

20.68 Identify the structures of **D** and **E,** isomers of molecular formula $C_6H_{12}O_2$, from their IR and ^1H NMR data. Signals at 1.35 and 1.60 ppm in the ^1H NMR spectrum of **D** and 1.90 ppm in the ^1H NMR spectrum of **E** are multiplets.

a. IR absorption for **D** at 1743 cm^{-1}

b. IR absorption for **E** at 1746 cm^{-1}

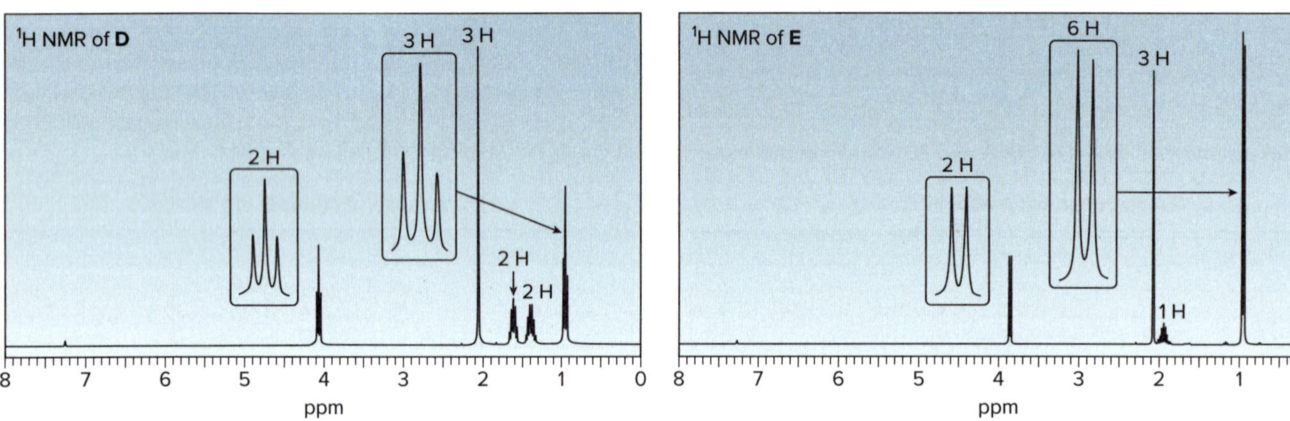

Challenge Problems

20.69 One step in the synthesis of aliskiren, a drug used to treat hypertension (Problems 5.6 and 12.63), involves the conversion of **A** to **B.** Draw a stepwise mechanism for this process that explains the observed stereochemistry.

[1] LiOH
[2] HCl

A **B**

20.70 With reference to amides **A** and **B,** the carbonyl of one amide absorbs at a much higher wavenumber in its IR spectrum than the carbonyl of the other amide. Which absorbs at a higher wavenumber and why?

A **B**

20.71 The ^1H NMR spectrum of 2-chloroacetamide ($ClCH_2CONH_2$) shows three signals at 4.02, 7.35, and 7.60 ppm. What protons give rise to each signal? Explain why three signals are observed.

20.72 Benzhydrocodone is an opiate prodrug (Problem 20.23) approved for the short-term treatment of acute pain. The first step in the metabolism of benzhydrocodone is hydrolysis to form hydrocodone, the active drug. Draw a stepwise mechanism for this reaction, assuming that it occurs in the presence of acid.

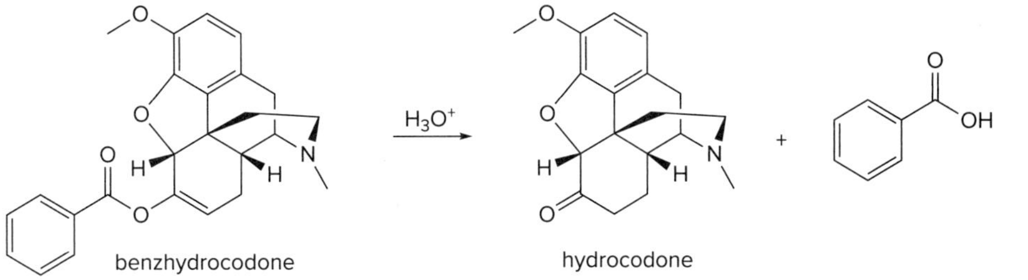

H_3O^+

benzhydrocodone

hydrocodone

20.73 Draw a stepwise mechanism for the following reaction, the last step in a five-step industrial synthesis of vitamin C that begins with the simple carbohydrate glucose.

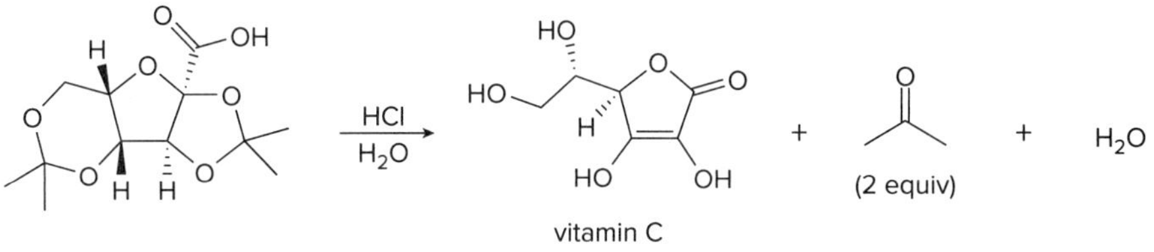

HCl, H_2O

vitamin C

(2 equiv)

H_2O

20.74 Draw a stepwise mechanism for the following reaction, a key step in the synthesis of linezolid, an antibacterial agent.

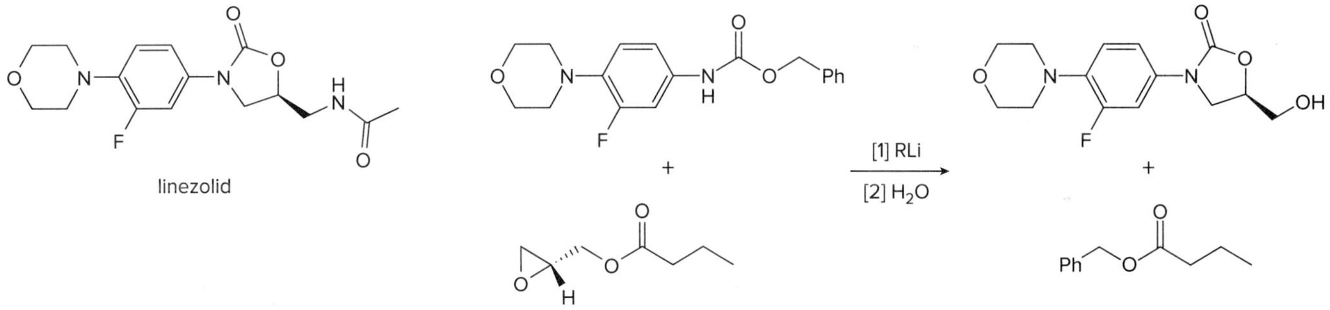

linezolid

[1] RLi
[2] H_2O

SELF-TEST ANSWERS

1. b 2. a 3. b 4. d 5. c 6. a 7. c 8. a 9. b 10. a

21 Substitution Reactions of Carbonyl Compounds at the α Carbon

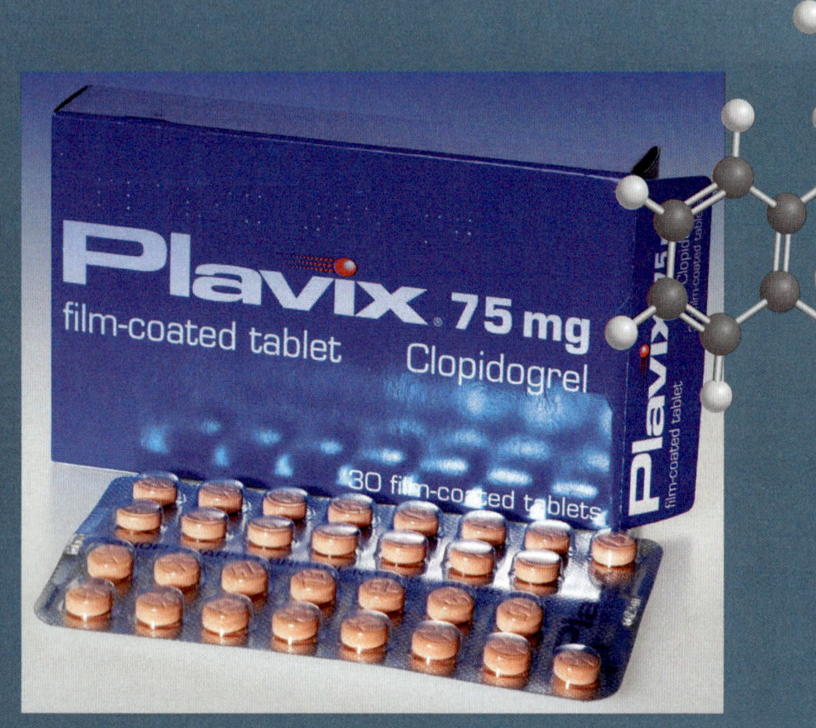

Liquid Light/Alamy Stock Photo

21.1 Introduction
21.2 Enols
21.3 Enolates
21.4 Enolates of unsymmetrical carbonyl compounds
21.5 Racemization at the α carbon
21.6 A preview of reactions at the α carbon
21.7 Halogenation at the α carbon
21.8 Direct enolate alkylation
21.9 Malonic ester synthesis
21.10 Acetoacetic ester synthesis

Clopidogrel, sold under the trade name Plavix, is an oral drug prescribed to prevent blood clots in patients who are at risk of heart attack and stroke. Approved for medical use in the United States in 1997, clopidogrel is now available in over 100 countries. In combination with aspirin, it is used as an antiplatelet agent for patients who have had surgery to insert a coronary stent. Clopidogrel has been synthesized by reaction of an α-halo carbonyl compound with a nucleophile, one of the many useful reactions described in Chapter 21.

Why Study . . .

Reactions at the α Carbon of a Carbonyl Group?

Chapters 21 and 22 focus on reactions that occur at the α carbon to a carbonyl group. These reactions are different from the reactions of Chapters 17, 18, and 20, all of which involved nucleophilic attack at the electrophilic carbonyl carbon. In reactions at the α carbon, the carbonyl compound serves as a *nucleophile* that reacts with a carbon or halogen electrophile to form a new bond to the α carbon.

Chapter 21 concentrates on **substitution reactions at the α carbon,** whereas Chapter 22 concentrates on reactions between two carbonyl compounds, one of which serves as the nucleophile and one of which is the electrophile. Many of the reactions in Chapter 21 form new carbon–carbon bonds, thus adding to your repertoire of reactions that can be used to synthesize more-complex organic molecules from simple precursors. As you will see, the reactions introduced in Chapter 21 have been used to prepare a wide variety of interesting and useful compounds.

21.1 Introduction

Up to now, the discussion of carbonyl compounds has centered on their reactions with nucleophiles at the electrophilic carbonyl carbon. **Two general reactions are observed,** depending on the structure of the carbonyl starting material.

- ***Nucleophilic addition*** **occurs when there is no electronegative atom Z on the carbonyl carbon (as with aldehydes and ketones).**

aldehyde or ketone —[1] :Nu⁻ [2] H—OH→ (nucleophilic addition)

With no leaving group, H and Nu are added.

- ***Nucleophilic acyl substitution*** **occurs when there is an electronegative atom Z on the carbonyl carbon (as with carboxylic acids and their derivatives).**

:Nu⁻ → nucleophilic substitution; + :Z⁻

Z = electronegative element

With a leaving group, Nu replaces Z.

Reactions can also occur at the α carbon to the carbonyl group. These reactions proceed by way of **enols** or **enolates,** two electron-rich intermediates that react with electrophiles, forming a new bond on the α carbon. This reaction results in the **substitution of the electrophile E for hydrogen.**

Hydrogen atoms on the α carbon are called **α hydrogens.**

reaction at the α carbon → enol / enolate → E^+ → E replaces H on the α carbon.

21.2 Enols

Recall from Chapter 11 that **enol and keto forms are tautomers of the carbonyl group that differ in the position of a double bond and a proton.** These constitutional isomers are in equilibrium with each other.

- A keto tautomer has a C=O and an additional C–H bond.
- An enol tautomer has an O–H group bonded to a C=C.

Equilibrium favors the keto form for most carbonyl compounds largely because a C=O is much stronger than a C=C. For simple carbonyl compounds, < 1% of the enol is present at equilibrium. With unsymmetrical ketones, moreover, two different enols are possible, yet they still total < 1%.

With compounds containing two carbonyl groups separated by a single carbon (called β-dicarbonyl compounds or 1,3-dicarbonyl compounds), however, the concentration of the enol form sometimes exceeds the concentration of the keto form.

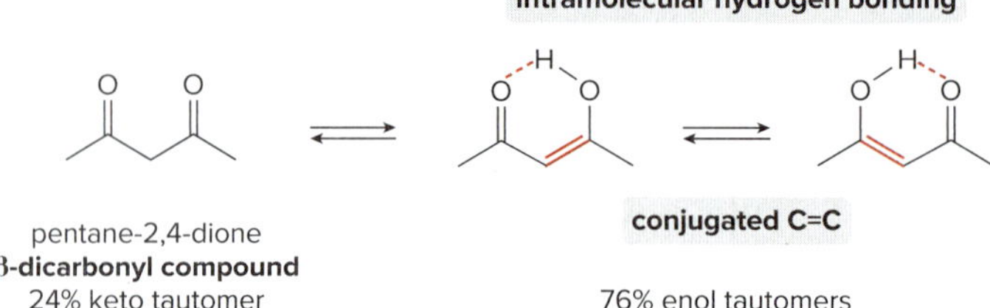

Two factors stabilize the enol of β-dicarbonyl compounds: **conjugation** and **intramolecular hydrogen bonding.** The C=C of the enol is conjugated with the carbonyl group, allowing delocalization of the electron density in the π bonds. Moreover, the OH of the enol can hydrogen bond to the oxygen of the nearby carbonyl group. Such intramolecular hydrogen bonds are especially stabilizing when they form a six-membered ring, as in this case.

Sample Problem 21.1 Interconverting Keto and Enol Tautomers

Convert each compound to its enol or keto tautomer.

Solution

a. To convert a carbonyl compound to its enol tautomer, **draw a double bond between the carbonyl carbon and the α carbon, and change the C=O to C–OH.** In this case, both α carbons are identical, so only one enol is possible.

b. To convert an enol to its keto tautomer, **change the C–OH to C=O and add a proton to the other end of the C=C.**

Problem 21.1 Draw the enol or keto tautomer(s) of each compound.

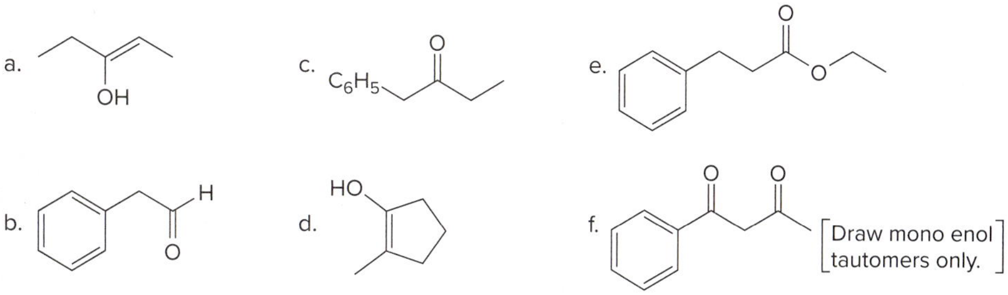

a. OH

b. H O

c. C_6H_5 O

d. HO

e. O O

f. O O [Draw mono enol tautomers only.]

More Practice: Try Problems 21.33, 21.40a, 21.42.

Problem 21.2 Which carbonyl compound in each pair exhibits the higher percentage of the enol tautomer?

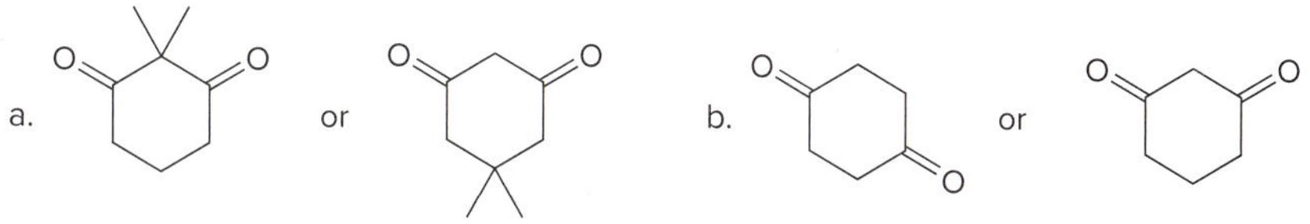

a. O O or O O

b. O O or O O

Problem 21.3 Leptospermone is a herbicide produced by the bottlebrush plant. Draw all possible mono enol tautomers of leptospermone, ignoring stereoisomers. Determine if all the tautomers are similar in stability, or if one tautomer is more or less stable than the others.

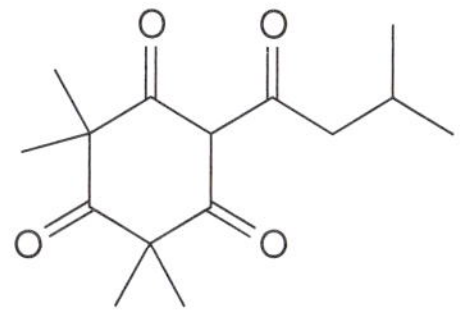

leptospermone

Callistimon citrinus, commonly called bottlebrush, is a plant native to Australia and the source of leptospermone (Problem 21.3). *Rafael Santos Rodriguez/Shutterstock*

21.2A The Mechanism of Tautomerization

Tautomerization, the process of converting one tautomer to another, is catalyzed by both acid and base. Tautomerization always requires two steps (**protonation** and **deprotonation**), but the order of these steps depends on whether the reaction takes place in acid or base. In Mechanisms 21.1 and 21.2 for tautomerization, the keto form is converted to the enol form. All of the steps are reversible, though, so they equally apply to the conversion of the enol form to the keto form.

Mechanism 21.1 Tautomerization in Acid

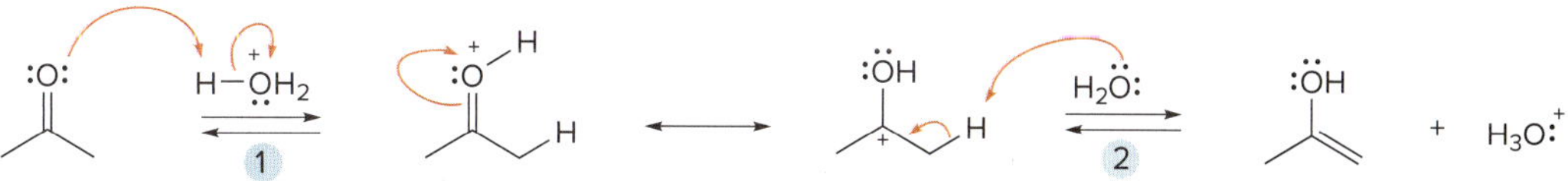

resonance-stabilized carbocation

1. With acid, **protonation *precedes* deprotonation.** Protonation of the carbonyl forms a resonance-stabilized carbocation.
2. Removal of a proton forms the enol.

Mechanism 21.2 Tautomerization in Base

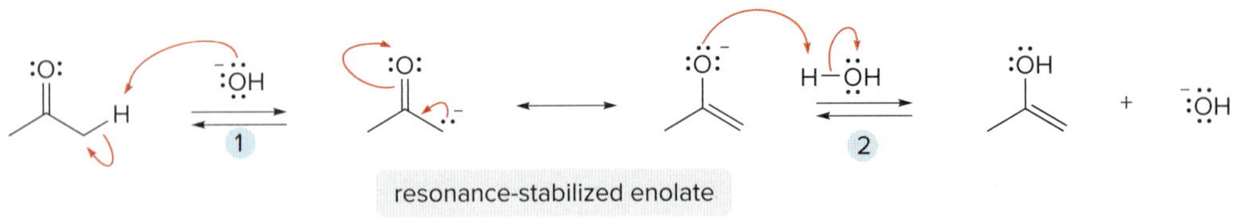

resonance-stabilized enolate

1 With base, **deprotonation *precedes* protonation.** Removal of a proton on the α carbon forms a resonance-stabilized enolate.

2 Protonation of the enolate forms the enol.

21.2B Enols in Biological Systems

Key reactions in carbohydrate metabolism involve tautomerizations and result in the interconversion of α-hydroxy ketones and α-hydroxy aldehydes. In this case, tautomerization generates an **enediol,** because two OH groups are bonded to the C=C.

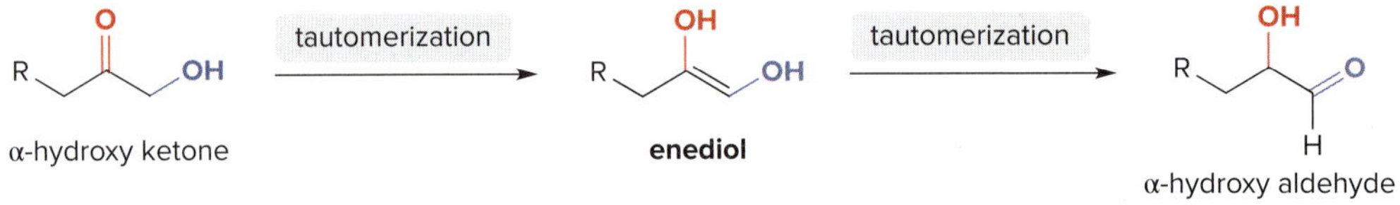

α-hydroxy ketone → (tautomerization) → **enediol** → (tautomerization) → α-hydroxy aldehyde

For example, in the metabolic breakdown of glucose, dihydroxyacetone phosphate is converted to glyceraldehyde 3-phosphate by **two keto–enol tautomerizations.** Although each reaction involves both protonation and deprotonation, both processes are written as a single step in a biological tautomerization.

dihydroxyacetone phosphate ⇌ (keto–enol tautomerization) ⇌ **enediol** ⇌ (enol–keto tautomerization) ⇌ glyceraldehyde 3-phosphate

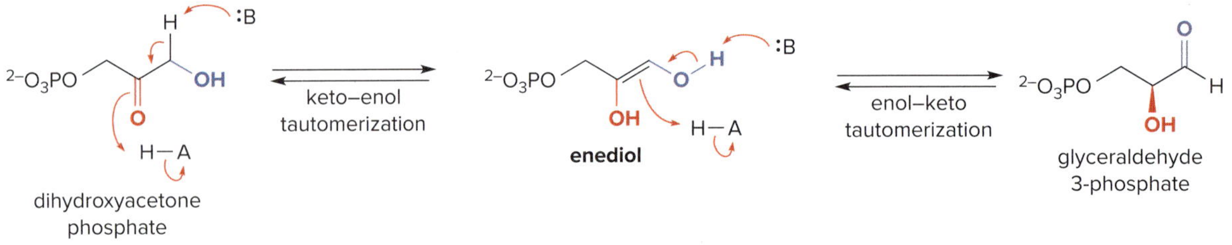

Problem 21.4 One step in the metabolism of glucose involves the isomerization of glucose 6-phosphate to fructose 6-phosphate. (a) Draw a stepwise mechanism for this process if it is carried out in the presence of acid, as illustrated in Mechanism 21.1. (b) Use curved arrows to write the reaction as two successive biological tautomerizations using HA as an acid and B: as a base, as shown above.

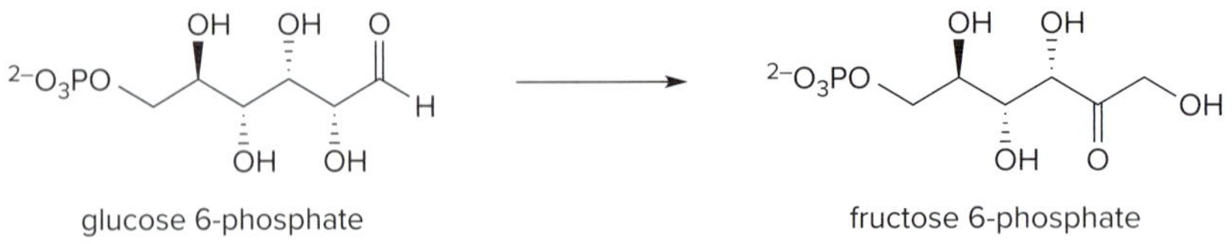

glucose 6-phosphate → fructose 6-phosphate

21.2C How Enols React

Like other compounds with carbon–carbon double bonds, **enols are electron rich, so they react as nucleophiles. Enols are even more electron rich than alkenes, though, because the OH group has a powerful electron-donating resonance effect.** A second resonance structure can be drawn for the enol that places a negative charge on one of the carbon atoms. As a result, this carbon atom is especially nucleophilic, and it can react with an electrophile E^+ to form a new bond to carbon. Loss of a proton then forms a neutral product.

two resonance structures for an enol

loss of H^+

- **Reaction of an enol with an electrophile E^+ forms a new C–E bond on the α carbon. The net result is substitution of H by E on the α carbon.**

tautomerization

reaction with E^+

Problem 21.5 When phenylacetaldehyde ($C_6H_5CH_2CHO$) is dissolved in D_2O with added DCl, the hydrogen atoms α to the carbonyl are gradually replaced by deuterium atoms. Write a mechanism for this process that involves enols as intermediates.

21.3 Enolates

Enolates are formed when a base removes a proton on the α carbon to a carbonyl group. A C–H bond on the α carbon is more acidic than many other sp^3 hybridized C–H bonds, because **the resulting enolate is resonance stabilized.** Moreover, one of the resonance structures is especially stable because it places a negative charge on an electronegative oxygen atom.

Forming enolates from carbonyl compounds was first discussed in Section 18.6.

resonance-stabilized enolate anion

+ HB^+

Enolates are always formed by removal of a proton on the **α carbon.**

cyclohexanone

+ HB^+

The pK_a of the α hydrogen in an aldehyde or ketone is ~20. As shown in Table 21.1, this makes it considerably more acidic than the C–H bonds in CH_3CH_3 and $CH_3CH{=}CH_2$. Although C–H bonds α to a carbonyl are *more acidic* than many other C–H bonds, they are still *less acidic* than O–H bonds that always place the negative charge of the conjugate base on an electronegative oxygen atom (c.f. CH_3CH_2OH and CH_3COOH in Table 21.1).

Table 21.1 A Comparison of pK_a Values

Increasing acidity / Increasing stability of the conjugate base (top to bottom)

Compound	pK_a	Conjugate base	Structural features of the conjugate base
CH_3CH_3	50	$CH_3\ddot{C}H_2^-$	• The conjugate base has a (–) charge on C, but is not resonance stabilized.
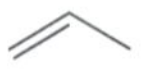 propene	43	 allyl anion ⟷ allyl anion	• The conjugate base has a (–) charge on C, and is resonance stabilized.
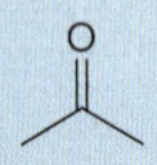 acetone	19.2	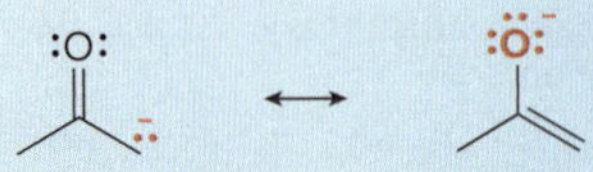 enolate (two resonance structures)	• **The conjugate base has two resonance structures, one of which has a (–) charge on O.**
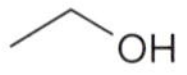 ethanol (OH)	16	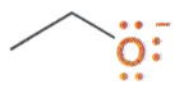 ethoxide	• The conjugate base has a (–) charge on O, but is not resonance stabilized.
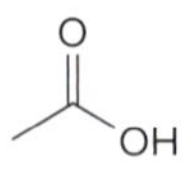 acetic acid (OH)	4.8	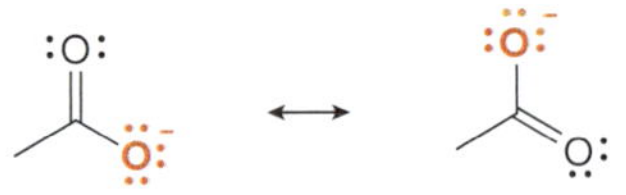 acetate (two resonance structures)	• The conjugate base has two resonance structures, both of which have a (–) charge on O.

- **Resonance stabilization of the conjugate base *increases* acidity.**
  - $CH_2{=}CHCH_3$ is more acidic than CH_3CH_3.
  - CH_3COOH is more acidic than CH_3CH_2OH.
- **Placing a negative charge on O in the conjugate base *increases* acidity.**
  - CH_3CH_2OH is more acidic than CH_3CH_3.
  - CH_3COCH_3 is more acidic than $CH_2{=}CHCH_3$.
  - CH_3COOH (with two O atoms) is more acidic than CH_3COCH_3.

21.3A Examples of Enolates and Related Anions

In addition to enolates from aldehydes and ketones, **enolates from esters and 3° amides can be formed,** although the α hydrogen is somewhat less acidic. **Nitriles** also have acidic protons on the carbon atom adjacent to the cyano group, because the negative charge of the conjugate base is stabilized by delocalization onto an electronegative nitrogen atom.

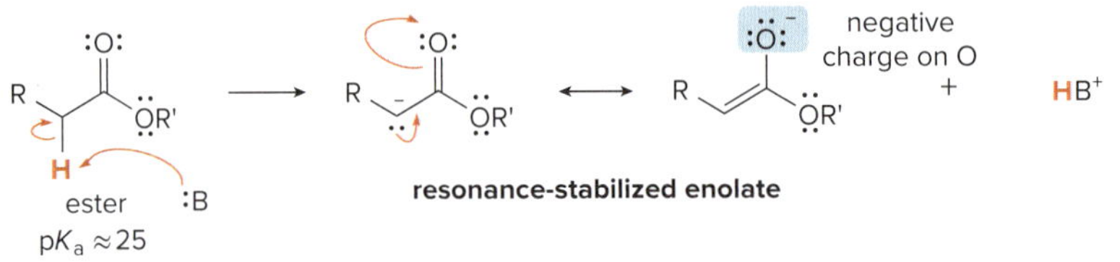

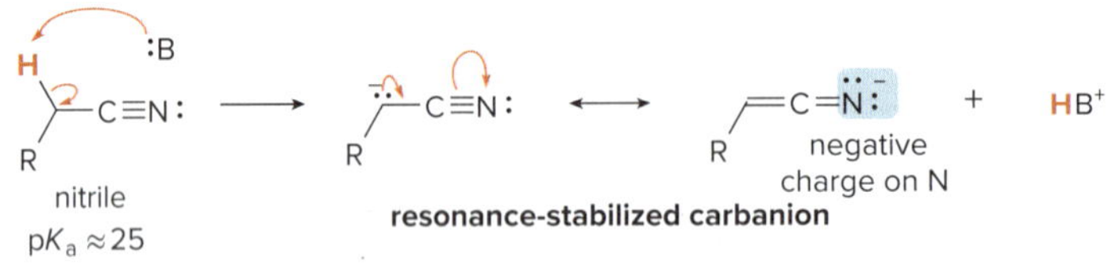

The protons on the carbon between the two carbonyl groups of a β-dicarbonyl compound are especially acidic because resonance delocalizes the negative charge on two different oxygen atoms. Table 21.2 lists pK_a values for β-dicarbonyl compounds as well as other carbonyl compounds and nitriles.

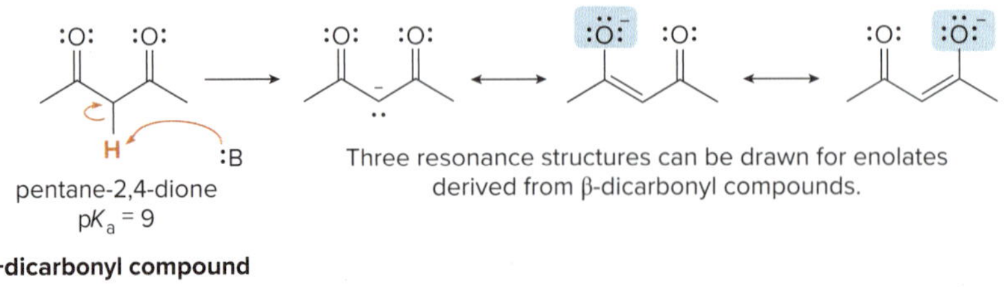

resonance structure can be drawn for the enol that places a negative charge on one of the carbon atoms. As a result, this carbon atom is especially nucleophilic, and it can react with an electrophile E^+ to form a new bond to carbon. Loss of a proton then forms a neutral product.

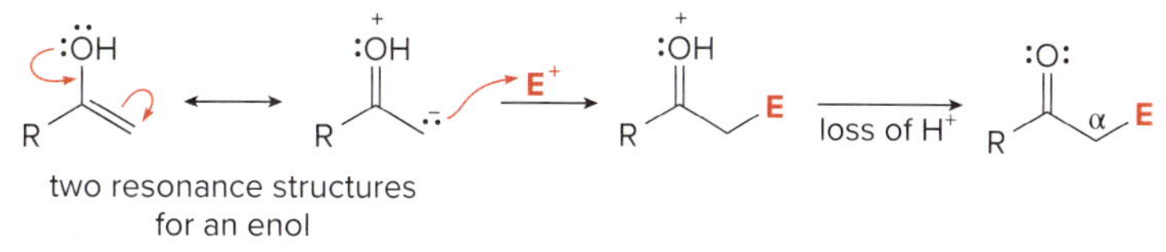

two resonance structures for an enol

loss of H^+

- **Reaction of an enol with an electrophile E^+ forms a new C–E bond on the α carbon. The net result is substitution of H by E on the α carbon.**

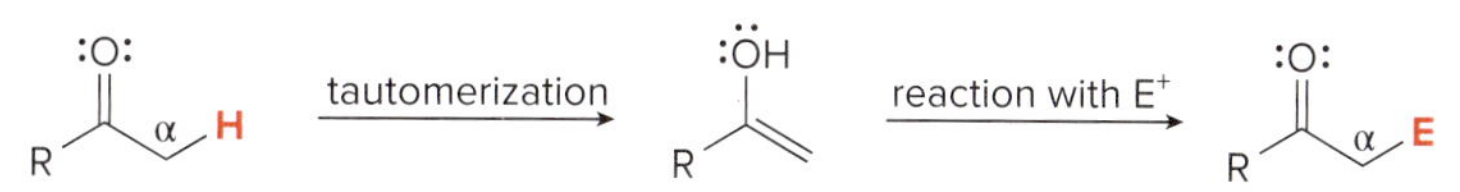

tautomerization

reaction with E^+

Problem 21.5 When phenylacetaldehyde ($C_6H_5CH_2CHO$) is dissolved in D_2O with added DCl, the hydrogen atoms α to the carbonyl are gradually replaced by deuterium atoms. Write a mechanism for this process that involves enols as intermediates.

21.3 Enolates

Enolates are formed when a base removes a proton on the α carbon to a carbonyl group. A C–H bond on the α carbon is more acidic than many other sp^3 hybridized C–H bonds, because **the resulting enolate is resonance stabilized.** Moreover, one of the resonance structures is especially stable because it places a negative charge on an electronegative oxygen atom.

Forming enolates from carbonyl compounds was first discussed in Section 18.6.

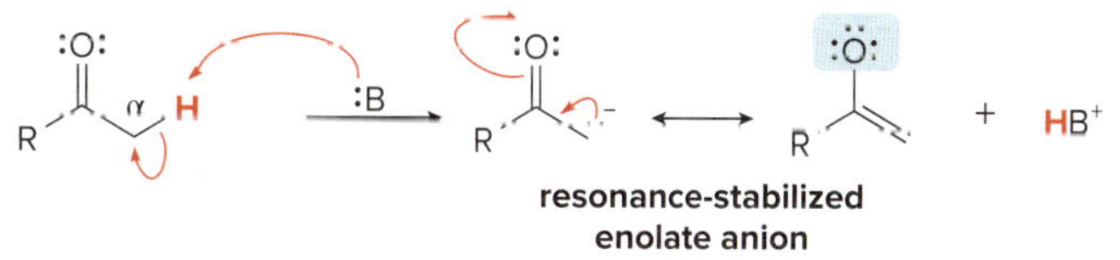

resonance-stabilized enolate anion

Enolates are always formed by removal of a proton on the **α carbon.**

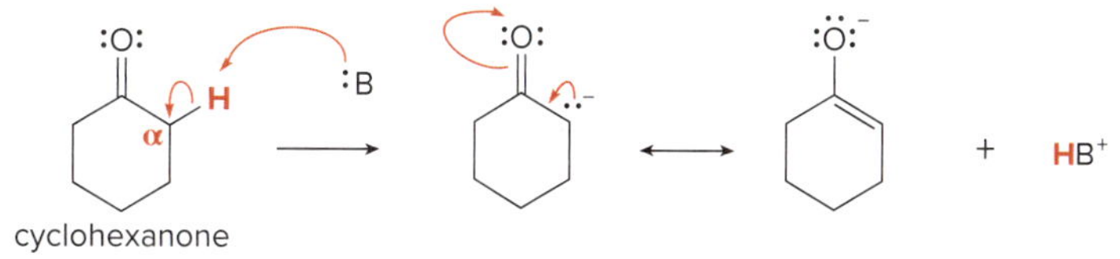

cyclohexanone

The pK_a of the α hydrogen in an aldehyde or ketone is ~20. As shown in Table 21.1, this makes it considerably more acidic than the C–H bonds in CH_3CH_3 and $CH_3CH{=}CH_2$. Although C–H bonds α to a carbonyl are *more acidic* than many other C–H bonds, they are still *less acidic* than O–H bonds that always place the negative charge of the conjugate base on an electronegative oxygen atom (c.f. CH_3CH_2OH and CH_3COOH in Table 21.1).

Table 21.1 A Comparison of pK_a Values

Compound	pK_a	Conjugate base	Structural features of the conjugate base
CH_3CH_3	50	$CH_3\ddot{C}H_2^-$	• The conjugate base has a (–) charge on C, but is not resonance stabilized.
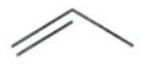 $CH_2{=}CHCH_3$	43	 $CH_2{=}CH\ddot{C}H_2^-$ ⟷ $^-\ddot{C}H_2CH{=}CH_2$	• The conjugate base has a (–) charge on C, and is resonance stabilized.
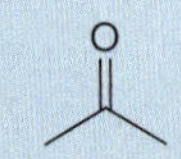 CH_3COCH_3	19.2	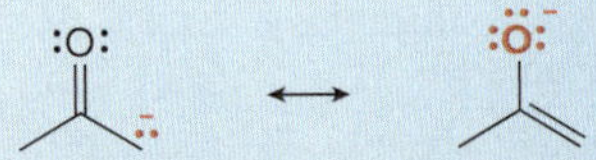 $CH_3COCH_2^-$ ⟷ $CH_3C(O^-){=}CH_2$	• **The conjugate base has two resonance structures, one of which has a (–) charge on O.**
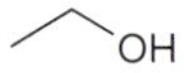 CH_3CH_2OH	16	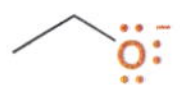 $CH_3CH_2O^-$	• The conjugate base has a (–) charge on O, but is not resonance stabilized.
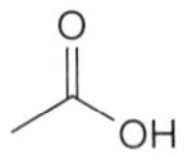 CH_3COOH	4.8	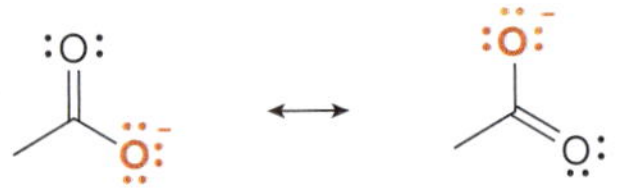 $CH_3C(=O)O^-$ ⟷ $CH_3C(O^-){=}O$	• The conjugate base has two resonance structures, both of which have a (–) charge on O.

Increasing acidity / Increasing stability of the conjugate base (arrow pointing down the table)

- **Resonance stabilization of the conjugate base *increases* acidity.**
  - $CH_2{=}CHCH_3$ is more acidic than CH_3CH_3.
  - CH_3COOH is more acidic than CH_3CH_2OH.
- **Placing a negative charge on O in the conjugate base *increases* acidity.**
  - CH_3CH_2OH is more acidic than CH_3CH_3.
  - CH_3COCH_3 is more acidic than $CH_2{=}CHCH_3$.
  - CH_3COOH (with two O atoms) is more acidic than CH_3COCH_3.

21.3A Examples of Enolates and Related Anions

In addition to enolates from aldehydes and ketones, **enolates from esters and 3° amides can be formed,** although the α hydrogen is somewhat less acidic. **Nitriles** also have acidic protons on the carbon atom adjacent to the cyano group, because the negative charge of the conjugate base is stabilized by delocalization onto an electronegative nitrogen atom.

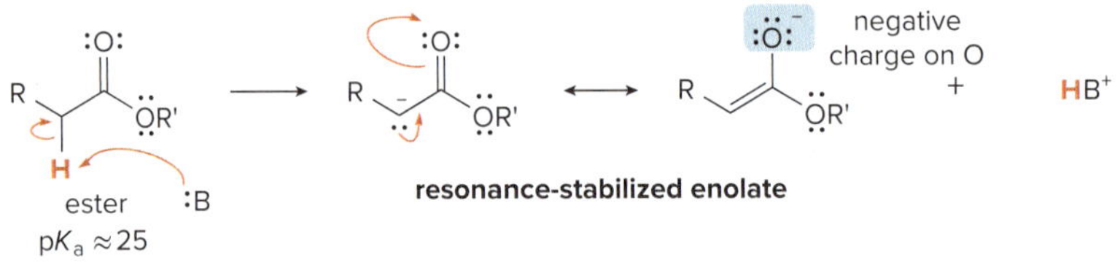

ester
$pK_a \approx 25$

negative charge on O + HB^+

resonance-stabilized enolate

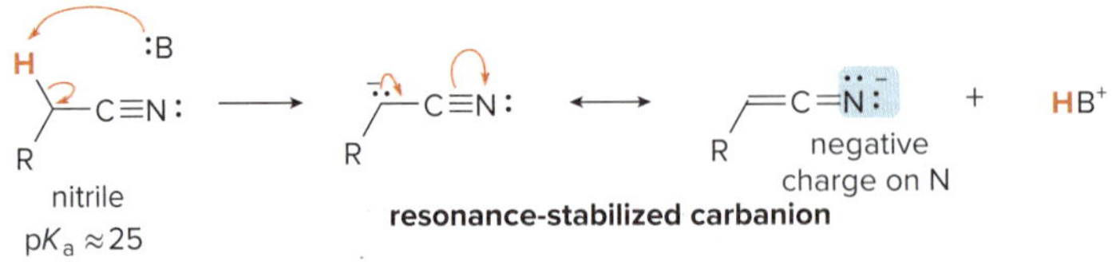

nitrile
$pK_a \approx 25$

negative charge on N + HB^+

resonance-stabilized carbanion

The protons on the carbon between the two carbonyl groups of a β-dicarbonyl compound are especially acidic because resonance delocalizes the negative charge on two different oxygen atoms. Table 21.2 lists pK_a values for β-dicarbonyl compounds as well as other carbonyl compounds and nitriles.

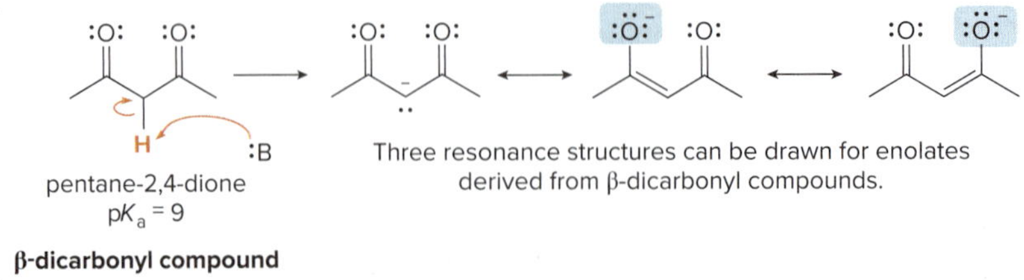

pentane-2,4-dione
$pK_a = 9$

β-dicarbonyl compound

Three resonance structures can be drawn for enolates derived from β-dicarbonyl compounds.

Table 21.2 pK_a Values for Some Carbonyl Compounds and Nitriles

Compound type	Example	pK_a	Compound type	Example	pK_a
[1] Amide	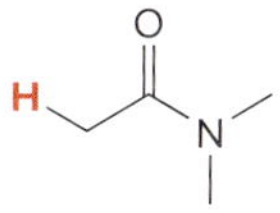	30	[6] 1,3-Diester	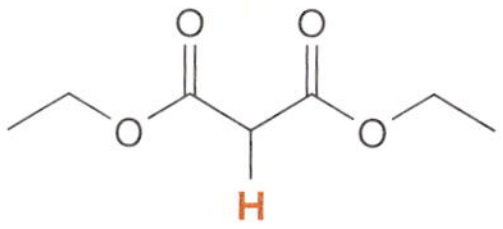	13.3
[2] Nitrile	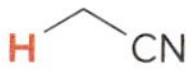	25	[7] 1,3-Dinitrile	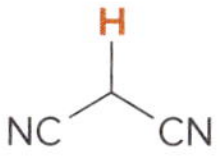	11
[3] Ester	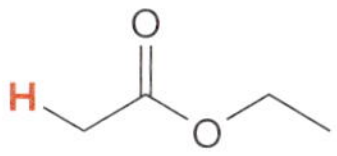	25	[8] β-Keto ester	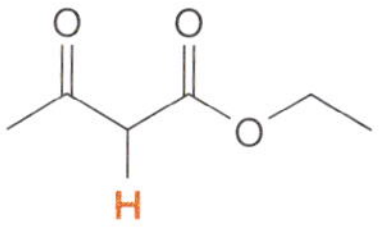	10.7
[4] Ketone	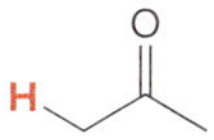	19.2	[9] β-Diketone	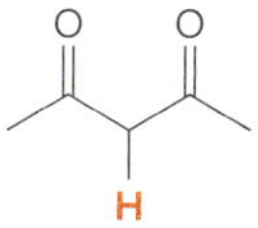	9
[5] Aldehyde	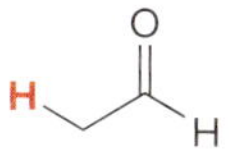	17			

Problem 21.6 Draw additional resonance structures for each anion.

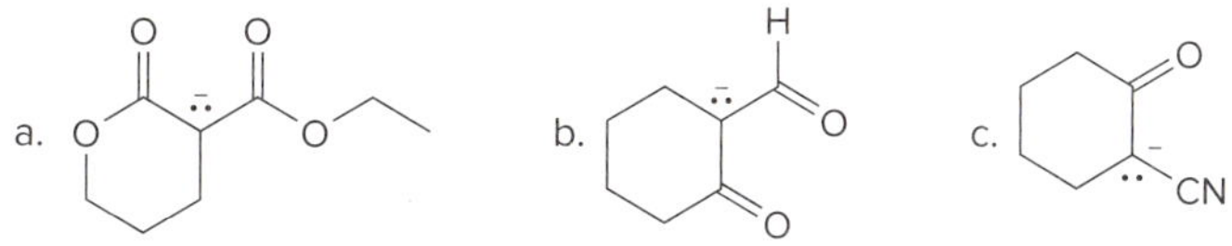

a. b. c.

Problem 21.7 Which C–H bonds in the following molecules are acidic because the resulting conjugate base is resonance stabilized?

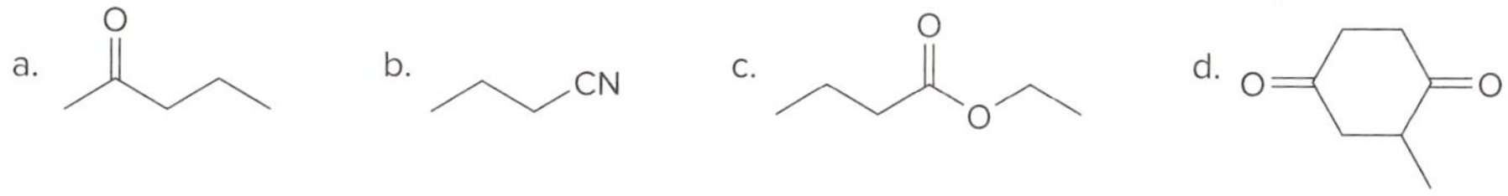

a. b. c. d.

Problem 21.8 Rank the labeled protons in the following compound in order of increasing acidity.

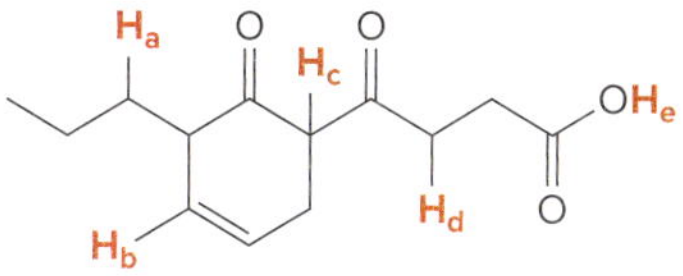

21.3B The Base

The formation of an enolate is an acid–base equilibrium, so the ***stronger*** **the base, the *more* enolate that forms.**

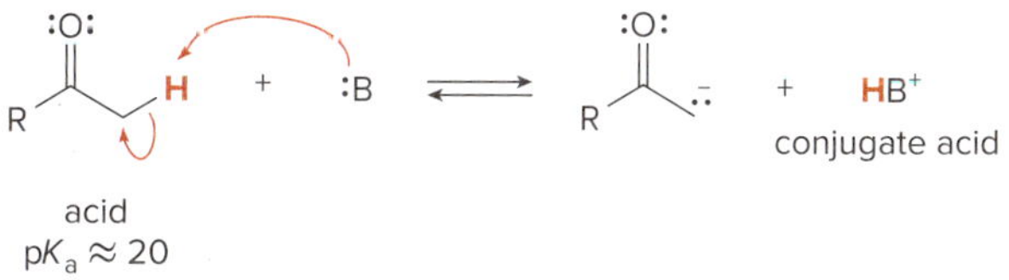

acid
p$K_a \approx 20$

conjugate acid

Stronger **bases drive the equilibrium to the *right*.**

We can predict the extent of an acid–base reaction by comparing the pK_a of the starting acid (the carbonyl compound in this case) with the pK_a of the conjugate acid formed. **The equilibrium favors the side with the *weaker* acid (the acid with the *higher* pK_a value).** The pK_a of many carbonyl compounds is ~20, so a significant amount of enolate will form only if the pK_a of the conjugate acid is > 20.

We have now used the term ***amide*** in two different ways—first as a functional group ($RCONH_2$) and now as a base (e.g., $^-NH_2$, which can be purchased as a sodium or lithium salt, $NaNH_2$ or $LiNH_2$, respectively). In Chapter 21 we will use dialkylamides, $^-NR_2$, in which the two H atoms of $^-NH_2$ have been replaced by R groups.

The common bases used to form enolates are hydroxide (^-OH), various alkoxides (^-OR), hydride (H^-), and dialkylamides ($^-NR_2$). How much enolate is formed using each of these bases is indicated in Table 21.3.

Table 21.3 Enolate Formation with Various Bases: $RCOCH_3$ (pK_a ≈ 20) + B: → $RCOCH_2^-$ + HB^+

	Base (B:)	Conjugate acid (HB^+)	pK_a of HB^+	% Enolate
[1]	$Na^+ \, ^-OH$	H_2O	14	< 1%
[2]	$Na^+ \, ^-OCH_2CH_3$	CH_3CH_2OH	16	< 1%
[3]	$K^+ \, ^-OC(CH_3)_3$	$(CH_3)_3COH$	18	1–10%
[4]	Na^+H^-	H_2	35	100%
[5]	$Li^+ \, ^-N[CH(CH_3)_2]_2$	$HN[CH(CH_3)_2]_2$	40	100%

Enolate formation with LDA is typically carried out at −78 °C, a convenient temperature to maintain in the laboratory because it is the temperature at which dry ice (solid CO_2) sublimes. Immersing a reaction flask in a cooling bath containing dry ice and acetone keeps its contents at a constant low temperature. *Joe Franek/McGraw Hill Education*

When the pK_a of the conjugate acid is < 20, as it is for ^-OH and all ^-OR (entries 1–3), only a small amount of enolate is formed at equilibrium. These bases are more useful in forming enolates when more acidic 1,3-dicarbonyl compounds are used as starting materials. They are also used when both the enolate and the carbonyl starting material are involved in the reaction, as is the case for reactions described in Chapter 22.

To form an enolate in essentially 100% yield, a much stronger base such as lithium diisopropylamide, **$Li^+ \, ^-N[CH(CH_3)_2]_2$**, abbreviated as **LDA,** is used (entry 5). **LDA is a strong nonnucleophilic base.** Like the other nonnucleophilic bases (Sections 7.8B and 8.1), its bulky isopropyl groups make the nitrogen atom too hindered to serve as a nucleophile. It is still able, though, to remove a proton in an acid–base reaction.

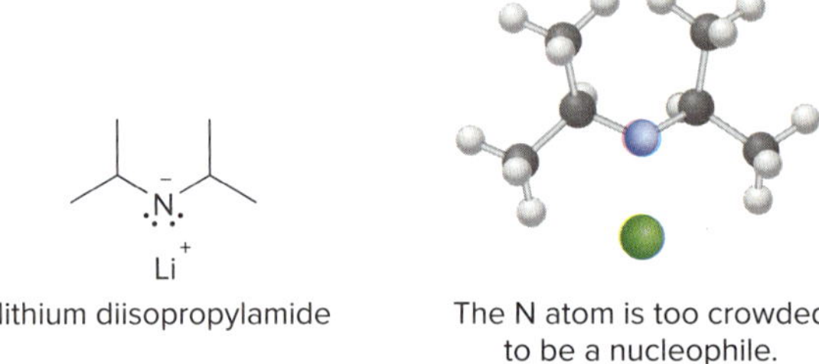

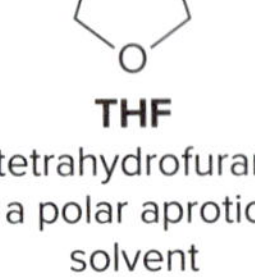

THF
tetrahydrofuran
a polar aprotic solvent

LDA quickly deprotonates essentially all of the carbonyl starting material, even at −78 °C, to form the enolate product. THF is the typical solvent for these reactions.

pK_a = 20 + LDA ⇌ (THF, −78 °C) enolate + diisopropylamine pK_a = 40

Equilibrium greatly favors the products.
Essentially all of the ketone is converted to enolate.

LDA can be prepared by deprotonating diisopropylamine with an organolithium reagent such as butyllithium, and then used immediately in a reaction.

butyllithium (Li) + H–N: diisopropylamine → butane + Li^+ :N: **LDA**

Problem 21.9 Draw the product formed when each starting material is treated with LDA in THF solution at −78 °C.

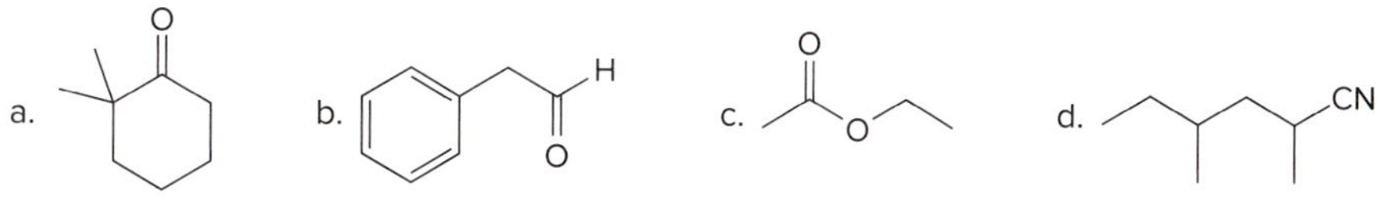

a. b. c. d.

Problem 21.10 As we learned in Chapter 17, organolithium reagents (RLi) are strong bases that readily react with acidic protons. Why aren't organolithium reagents used to generate enolates?

21.3C General Reactions of Enolates

Enolates are nucleophiles, and as such they react with many electrophiles. Because an enolate is resonance stabilized, however, it has two reactive sites—the carbon and oxygen atoms that bear the negative charge. **A nucleophile with two reactive sites is called an *ambident nucleophile.*** In theory, each of these atoms could react with an electrophile to form two different products, one with a new bond to carbon and one with a new bond to oxygen.

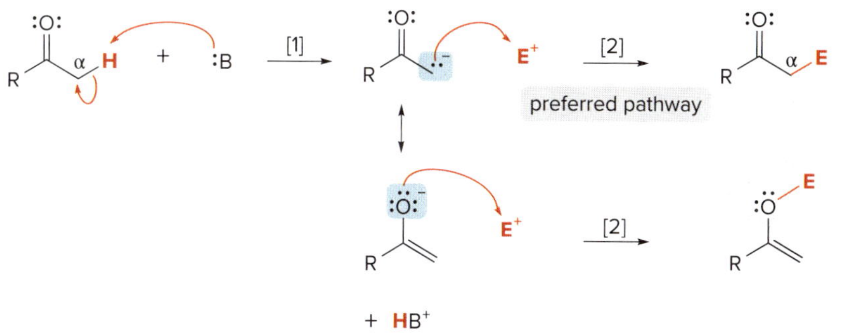

[1] [2] [2] preferred pathway + HB^+

Because enolates usually react at carbon instead of oxygen, the resonance structure that places the negative charge on oxygen will often be omitted in multistep mechanisms.

An enolate usually reacts at the carbon end, however, because this site is more nucleophilic. Thus, **enolates generally react with electrophiles on the α carbon,** so that many reactions in Chapter 21 follow a two-step path:

- [1] Reaction of a carbonyl compound with base forms an enolate.
- [2] Reaction of the enolate with an electrophile forms a new bond on the α carbon.

21.4 Enolates of Unsymmetrical Carbonyl Compounds

What happens when an unsymmetrical carbonyl compound like 2-methylcyclohexanone is treated with base? **Two enolates are possible,** one formed by removal of a 2° hydrogen, and one formed by removal of a 3° hydrogen.

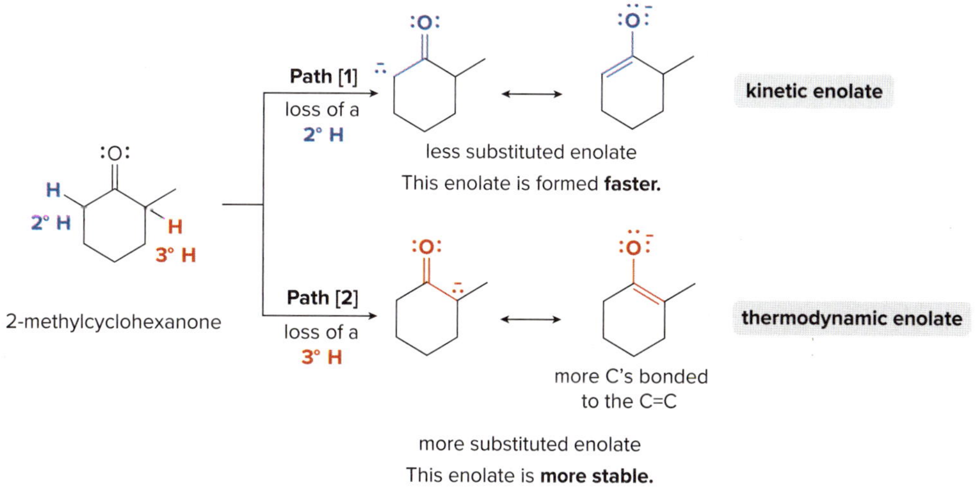

2° H 3° H
2-methylcyclohexanone

Path [1] loss of a 2° H

less substituted enolate
This enolate is formed **faster.**

kinetic enolate

Path [2] loss of a 3° H

more C's bonded to the C=C

more substituted enolate
This enolate is **more stable.**

thermodynamic enolate

Path [1] occurs *faster* than Path [2] because it results in removal of the less hindered 2° hydrogen, forming an enolate on the less substituted α carbon. Path [2] results in removal of a 3° hydrogen, forming the *more stable* enolate with the more substituted double bond. This enolate predominates at equilibrium.

- **The kinetic enolate is formed faster because it results from removal of the more accessible hydrogen. The kinetic enolate is the *less substituted* enolate.**
- **The thermodynamic enolate is lower in energy because it is the *more substituted* enolate.**

It is possible to regioselectively form one or the other enolate by the proper use of reaction conditions, because the base, solvent, and reaction temperature all affect the identity of the enolate formed.

Kinetic Enolates

The kinetic enolate forms faster, so mild reaction conditions favor it over slower processes with higher energies of activation. It is the less stable enolate, so it must not be allowed to equilibrate to the more stable thermodynamic enolate. **The kinetic enolate is favored by**

[1] **A strong nonnucleophilic base.** A strong base ensures that the enolate is formed rapidly. A **bulky base like LDA removes the more accessible proton on the less substituted carbon** much faster than a more hindered proton.

[2] **Polar aprotic solvent.** The solvent must be polar to dissolve the polar starting materials and intermediates. It must be aprotic so that it does not protonate any enolate that is formed. **THF** is both polar and aprotic.

[3] **Low temperature.** The temperature must be low **(–78 °C)** to prevent the kinetic enolate from equilibrating to the thermodynamic enolate.

:O: H LDA THF –78 °C :O:

major product

kinetic enolate

- **A kinetic enolate is formed with a strong, nonnucleophilic base (LDA) in a polar aprotic solvent (THF) at low temperature (–78 °C).**

Thermodynamic Enolates

A thermodynamic enolate is favored by equilibrating conditions. This is often achieved using a **strong base in a protic solvent.** A strong base yields both enolates, but in a protic solvent, enolates can also be protonated to re-form the carbonyl starting material. At equilibrium, the lower-energy intermediate always wins out, so that **the more stable, more substituted enolate is present in higher concentration.** Thus, the **thermodynamic enolate is favored by**

[1] **A strong base.** $Na^+\ {}^-OCH_2CH_3$, $K^+\ {}^-OC(CH_3)_3$, or other alkoxides are common.

[2] **Protic solvent.** CH_3CH_2OH or other alcohols.

[3] **Room temperature (25 °C).**

To simplify structures, we use abbreviations:
Me = CH_3, so $NaOCH_3$ = NaOMe
Et = CH_2CH_3, so $NaOCH_2CH_3$ = NaOEt
*t*Bu = $C(CH_3)_3$, so $KOC(CH_3)_3$ = KO*t*Bu

:O: H NaOEt EtOH 25 °C :O:

major product

thermodynamic enolate

- **A thermodynamic enolate is formed with a strong base (RO^-) in a polar protic solvent (ROH) at room temperature.**

Sample Problem 21.2 Determining the Enolate Formed from an Unsymmetrical Ketone

What is the major enolate formed in each reaction?

a. $\xrightarrow[\text{THF, }-78\ ^\circ\text{C}]{\text{LDA}}$ b. $\xrightarrow[\text{EtOH, }25\ ^\circ\text{C}]{\text{NaOEt}}$

Solution

a. **LDA is a strong, nonnucleophilic base** that removes a proton on the less substituted α carbon to form the **kinetic enolate.**

less substituted C

LDA, THF
−78 °C

kinetic enolate

b. **$NaOCH_2CH_3$ (a strong base) and CH_3CH_2OH (a protic solvent)** favor removal of a proton from the more substituted α carbon to form the **thermodynamic enolate.**

more substituted C

NaOEt
EtOH, 25 °C

thermodynamic enolate

Problem 21.11 What enolate is formed when each ketone is treated with LDA in THF solution? What enolate is formed when these same ketones are treated with $NaOCH_3$ in CH_3OH solution?

a. b. c.

More Practice: Try Problem 21.36.

21.5 Racemization at the α Carbon

Recall from Section 14.5 that an enolate can be stabilized by the delocalization of electron density only if it possesses the proper geometry and hybridization.

- **The electron pair on the carbon adjacent to the C=O must occupy a *p* orbital that overlaps with the two other *p* orbitals of the C=O, making an enolate conjugated.**

- **Thus, all three atoms of the enolate are *sp*^{2} hybridized and trigonal planar.**

These bonding features are shown in the acetone enolate in Figure 21.1.

Figure 21.1
The hybridization and geometry of the acetone enolate $(CH_3COCH_2)^-$

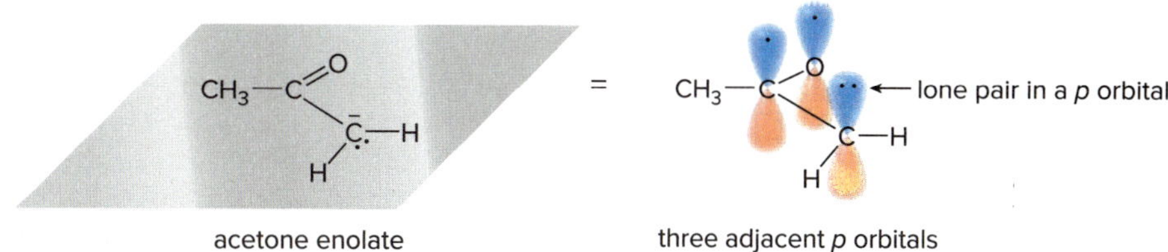

acetone enolate three adjacent *p* orbitals

- The O atom and both C's of the enolate are ***sp*^{2} hybridized** and lie in a plane.
- Each atom has a *p* orbital extending above and below the plane; these orbitals overlap to delocalize electron density.

When the α carbon to the carbonyl is a stereogenic center, treatment with aqueous base leads to **racemization** by a two-step process: **deprotonation to form an enolate and protonation to re-form the carbonyl compound.** For example, chiral ketone **A** reacts with aqueous ^{-}OH to form an achiral enolate having an sp^2 hybridized α carbon. Because the enolate is planar, it can be protonated with H_2O with equal probability from both directions, yielding a racemic mixture of two ketones.

H_2O front
enantiomers
^{-}OH
H_2O
$+ H_2O$
A
chiral starting material
achiral enolate
H_2O behind
racemic mixture

Problem 21.12 Explain each observation: (a) When (*R*)-2-methylcyclohexanone is treated with NaOH in H_2O, the optically active solution gradually loses optical activity. (b) When (*R*)-3-methylcyclohexanone is treated with NaOH in H_2O, the solution remains optically active.

21.6 A Preview of Reactions at the α Carbon

Having learned about the synthesis and properties of enolates, we can now turn our attention to their reactions. Like enols, **enolates are nucleophiles,** but because they are negatively charged, **enolates are much more nucleophilic than neutral enols.** Consequently, they undergo a wider variety of reactions.

Two general types of reactions of enolates—**substitutions** and **reactions with other carbonyl compounds**—will be discussed in the remainder of Chapter 21 and in Chapter 22. Both reactions form new bonds to the carbon α to the carbonyl.

- **Enolates react with electrophiles to afford substitution products.**

enolate
X–X → + X^- **halogenation**
R–X → + X^- **alkylation**
new bond in red

Two different kinds of substitution reactions are examined: **halogenation** with X_2 and **alkylation** with alkyl halides RX. These reactions are detailed in Sections 21.7–21.10.

- **Enolates react with other carbonyl groups at the electrophilic carbonyl carbon.**

enolate
δ−
δ+
new C–C bond in red

These reactions are more complicated because the initial addition adduct goes on to form different products depending on the structure of the carbonyl group. These reactions form the subject of Chapter 22.

21.7 Halogenation at the α Carbon

The first substitution reaction we examine is **halogenation.** Treatment of a ketone or aldehyde with halogen and either acid or base results in **substitution of X for H on the α carbon,** forming an **α-halo aldehyde or ketone.** Halogenation readily occurs with Cl_2, Br_2, and I_2.

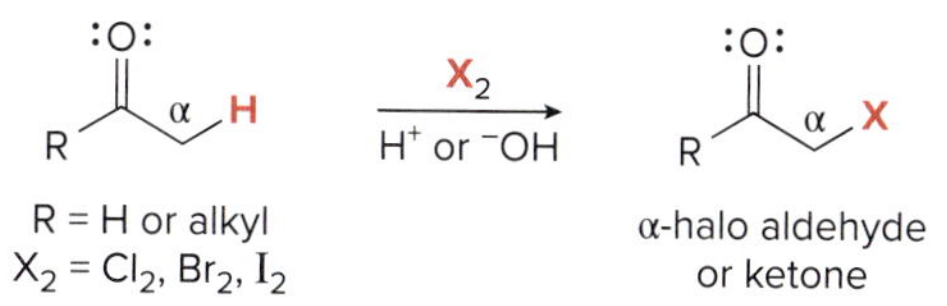

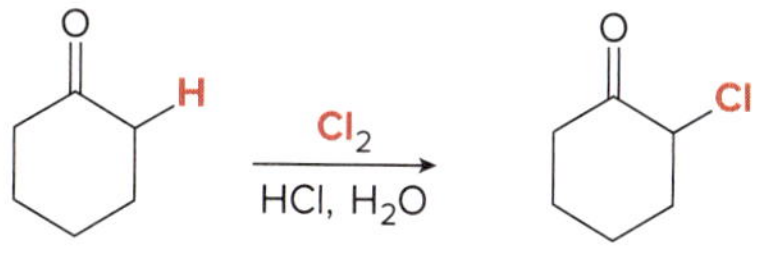

R = H or alkyl
$X_2 = Cl_2, Br_2, I_2$

X_2
H^+ or ^-OH

α-halo aldehyde or ketone

Cl_2
HCl, H_2O

The mechanisms of halogenation in acid and base are somewhat different.

- Reactions done in acid generally involve *enol* intermediates.
- Reactions done in base generally involve *enolate* intermediates.

21.7A Halogenation in Acid

Halogenation is often carried out by treating a carbonyl compound with a halogen in acetic acid. In this way, acetic acid is both the solvent and the acid catalyst for the reaction.

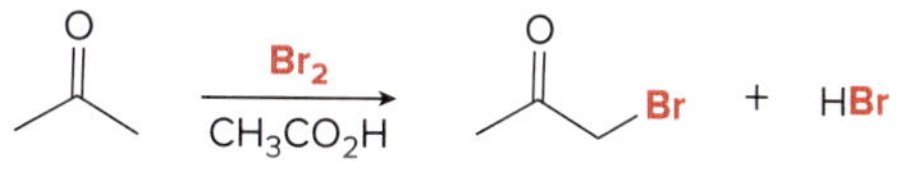

Br_2
CH_3CO_2H
\+ HBr

The mechanism of acid-catalyzed halogenation consists of two parts: **tautomerization** of the carbonyl compound to the enol form, and **reaction of the enol with halogen.** Mechanism 21.3 illustrates the reaction of $(CH_3)_2C{=}O$ with Br_2 in CH_3CO_2H.

Mechanism 21.3 Acid Catalyzed Halogenation at the α Carbon

$H{-}O_2CCH_3$ (1) ; $^-O_2CCH_3$ (2) ; **enol** + CH_3CO_2H ; $:\ddot{B}r{-}\ddot{B}r:$ (3) ; new bond in red ; (4) ; + $H{-}\ddot{B}r:$

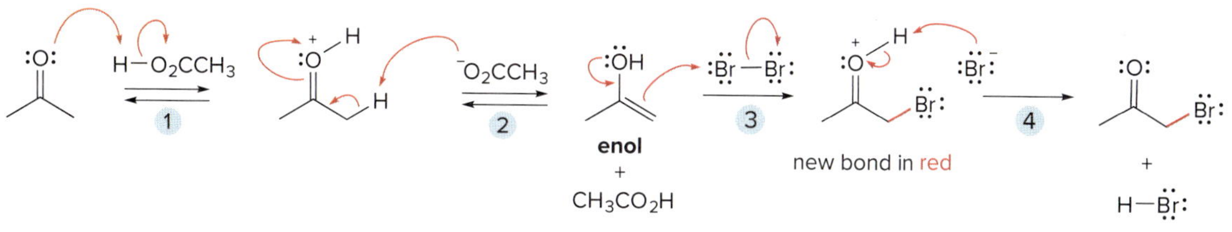

1 – 2 The ketone is converted to its **enol tautomer** by the two-step process of protonation followed by deprotonation.

3 – 4 Addition of the halogen to the enol forms a new bond to Br on the α carbon, and deprotonation yields the substitution product.

Problem 21.13 Draw the products of each reaction.

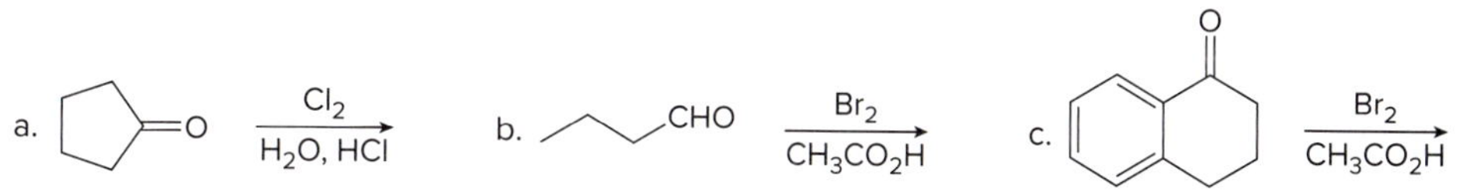

a. Cl_2 / H_2O, HCl

b. CHO Br_2 / CH_3CO_2H

c. Br_2 / CH_3CO_2H

21.7B Halogenation in Base

Halogenation in base is much less useful, because it is often difficult to stop the reaction after addition of just one halogen atom to the α carbon. For example, treatment of propiophenone with Br_2 and aqueous ^{-}OH yields a dibromo ketone.

Reactions of carbonyl compounds with base invariably involve enolates because the α hydrogens of the carbonyl compound are easily removed.

propiophenone $\xrightarrow[^{-}OH]{Br_2}$ (dibromo ketone: Br Br)

The mechanism for introduction of each Br atom involves the same two steps: **deprotonation with base followed by reaction with Br_2** to form a new C–Br bond, as shown in Mechanism 21.4.

Mechanism 21.4 Halogenation at the α Carbon in Base

[Step 1: ketone + :ÖH ⇌ enolate + H_2O; Step 2: enolate + Br–Br → α-bromo ketone + $:Br:^{-}$]

new bond in red

1 Treatment of the ketone with base forms a **nucleophilic enolate.**

2 Reaction of the enolate with Br_2 forms the substitution product in which one H is replaced by Br on the α carbon.

Only a small amount of the enolate forms at equilibrium using ^{-}OH as base, but the enolate is such a strong nucleophile that it readily reacts with Br_2, thus driving the equilibrium to the right. Then, the same two steps introduce the second Br atom on the α carbon: **deprotonation** followed by **nucleophilic attack.**

α-bromopropiophenone ⇌ (1) enolate + H_2O → (2) disubstitution product + $:Br:^{-}$

The electronegative Br stabilizes the negative charge.

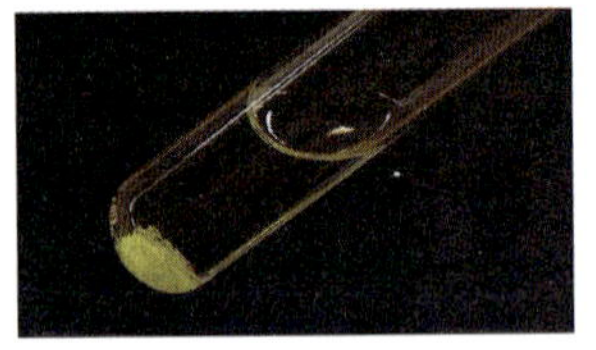

Although all ketones with α hydrogens react with base and I_2, only **methyl** ketones form CHI_3 (iodoform), a pale yellow solid that precipitates from the reaction mixture. This reaction is the basis of the **iodoform test,** once a common chemical method to detect methyl ketones. Methyl ketones give a positive iodoform test (appearance of a yellow solid), whereas other ketones give a negative iodoform test (no change in the reaction mixture). *Joe Franek/McGraw Hill*

It is difficult to stop this reaction after the addition of one Br atom because the electron-withdrawing inductive effect of Br stabilizes the second enolate. As a result, the α H of α-bromopropiophenone is more acidic than the α H atoms of propiophenone, making it easier to remove with base.

Halogenation of a **methyl ketone** with excess halogen, called the **haloform reaction,** results in cleavage of a carbon–carbon σ bond and formation of two products, a carboxylate anion and CHX_3 (commonly called **haloform**).

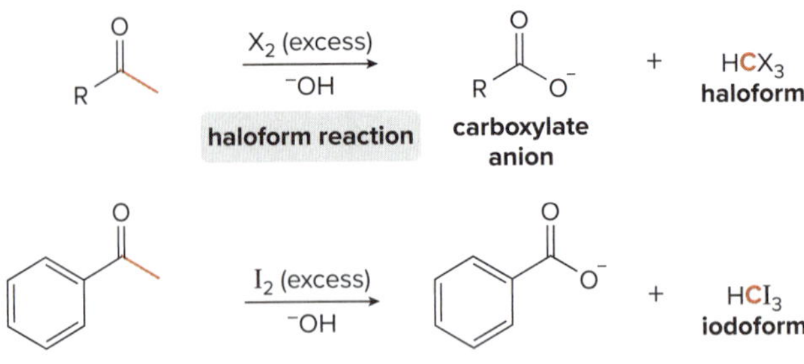

In the haloform reaction, the three H atoms of the CH_3 group are successively replaced by X to form an intermediate that is oxidatively cleaved with base. Mechanism 21.5 is written with I_2 as halogen, forming CHI_3 (iodoform) as product.

Mechanism 21.5 The Haloform Reaction

Part [1] Halogenation at the α carbon

1 – 2 Removal of a proton forms an enolate that reacts with I_2 to yield an α-iodo ketone. Steps [1] and [2] are repeated twice more to form the triiodo substitution product.

Part [2] Oxidative cleavage with ^-OH

3 Nucleophilic addition of ^-OH forms a tetrahedral intermediate.

4 **Elimination of $^-CI_3$ cleaves a carbon–carbon bond,** forming the substitution product.

5 Proton transfer generates the carboxylate anion and HCI_3, iodoform.

Steps [3] and [4] result in a **nucleophilic *substitution*** reaction of a *ketone*. Because ketones normally undergo **nucleophilic *addition*,** this two-step sequence makes the haloform reaction unique. Substitution occurs because the three electronegative halogen atoms make CX_3 (CI_3 in the example) a good leaving group.

The three possible outcomes of halogenation at the α carbon are summarized in Key Reactions in the end-of-chapter review.

Problem 21.14 Draw the products of each reaction. Assume excess halogen is present.

a. $\xrightarrow{Br_2,\ ^-OH}$ b. $\xrightarrow{I_2,\ ^-OH}$ c. $\xrightarrow{I_2,\ ^-OH}$

21.7C Reactions of α-Halo Carbonyl Compounds

α-Halo carbonyl compounds undergo two useful reactions—**elimination** with base and **substitution** with nucleophiles.

For example, treatment of 2-bromocyclohexanone with the base Li_2CO_3 in the presence of LiBr in the polar aprotic solvent DMF [$HCON(CH_3)_2$] affords cyclohex-2-enone by **elimination of the elements of Br and H from the α and β carbons,** respectively. Thus, a two-step method can convert a carbonyl compound such as cyclohexanone to an **α,β-unsaturated carbonyl compound** such as cyclohex-2-enone.

cyclohexanone → (Br_2, CH_3CO_2H; halogenation) → 2-bromocyclohexanone → (Li_2CO_3, LiBr, DMF; elimination) → cyclohex-2-enone

A new π bond is formed in two steps.

α,β-Unsaturated carbonyl compounds undergo a variety of 1,2- and 1,4-addition reactions as discussed in Section 17.15.

[1] Bromination at the α carbon is accomplished with Br_2 in CH_3CO_2H.

[2] Elimination of Br and H occurs with Li_2CO_3 and LiBr in DMF.

α-Halo carbonyl compounds also react with nucleophiles by S_N2 reactions. For example, reaction of 2-bromocyclohexanone with CH_3CH_2SH affords the substitution product **A.**

2-bromocyclohexanone → (SH, S_N2) → **A**

Sample Problem 21.3 Using α-Halo Carbonyl Compounds in Synthesis

What steps are needed to convert ketone **A** to **B** and **C?**

A **B** **C**

Solution

To introduce the N atom on the α carbon to form **B** or the double bond between the α and β carbons to form **C,** we must first convert **A** to an **α-halo ketone, D.** The halogen can then act as a leaving group in a **substitution** reaction to form **B,** or an **elimination** reaction to form **C.**

A → (Br_2, CH_3CO_2H) → **D** → (nucleophile H_2N) → **B**

D → (base: Li_2CO_3, LiBr, DMF) → **C**

- Reaction of **D** with an amine nucleophile forms the substitution product **B** by an S_N2 mechanism.
- Reaction of **D** with a base forms the elimination product **C.** The elements of Br and H are removed from the α and β carbons to form a π bond.

Problem 21.15 Draw the organic products formed when 2-bromopentan-3-one ($CH_3CH_2COCHBrCH_3$) is treated with each reagent: (a) Li_2CO_3, LiBr, DMF; (b) $CH_3CH_2NH_2$; (c) CH_3SH.

More Practice: Try Problems 21.51b, c; 21.53; 21.55.

Problem 21.16 Identify **A** in the following reaction, one step in the synthesis of bosentan, a drug used to treat a chronic connective tissue disorder that can cause pulmonary hypertension and open wounds on the fingertips (digital ulcers). Identify the atoms in bosentan that originate in **A.**

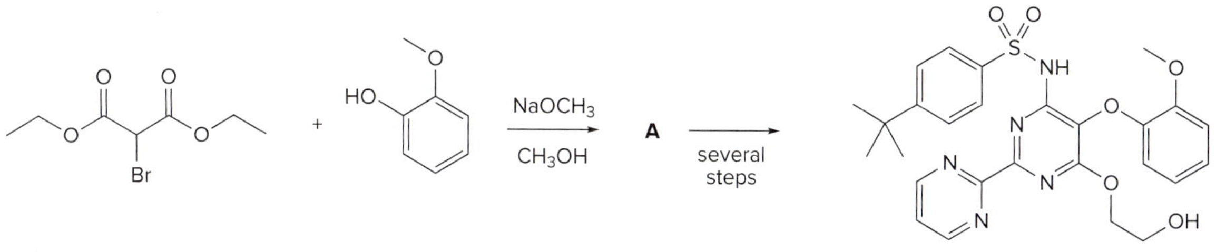

NaOCH$_3$ / CH$_3$OH → A → several steps → bosentan

21.8 Direct Enolate Alkylation

Treatment of an aldehyde or ketone with base and an alkyl halide (RX) results in ***alkylation*—the substitution of R for H on the α carbon atom.** Alkylation forms a new carbon–carbon bond on the α carbon.

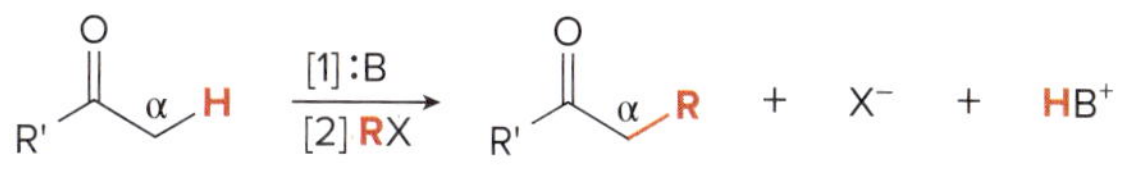

[1] :B, [2] RX; + X$^-$ + HB$^+$

new bond in red

21.8A General Features

We will begin with the most direct method of alkylation, and then (in Sections 21.9 and 21.10) examine two older, multistep methods that are still used today. Direct alkylation is carried out by a two-step process:

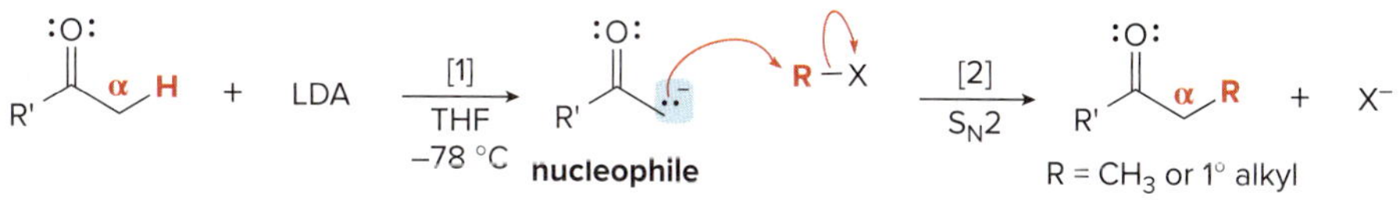

+ LDA —[1] THF, −78 °C→ nucleophile + R–X —[2] S_N2→ + X$^-$

R = CH$_3$ or 1° alkyl

[1] **Deprotonation:** Base removes a proton from the α carbon to generate an enolate. The reaction works best with a strong nonnucleophilic base like LDA in THF solution at low temperature (–78 °C).

[2] **Nucleophilic attack:** The nucleophilic enolate attacks the alkyl halide, displacing the halide (a good leaving group) and forming the alkylation product by an S_N2 reaction.

Because Step [2] is an **S_N2 reaction,** it works best with **unhindered methyl and 1° alkyl halides.** Hindered alkyl halides and those with halogens bonded to *sp*^{2} hybridized carbons do *not* undergo substitution.

R_3CX, $CH_2{=}CHX$, and C_6H_5X do *not* undergo alkylation reactions with enolates, because they are unreactive in S_N2 reactions.

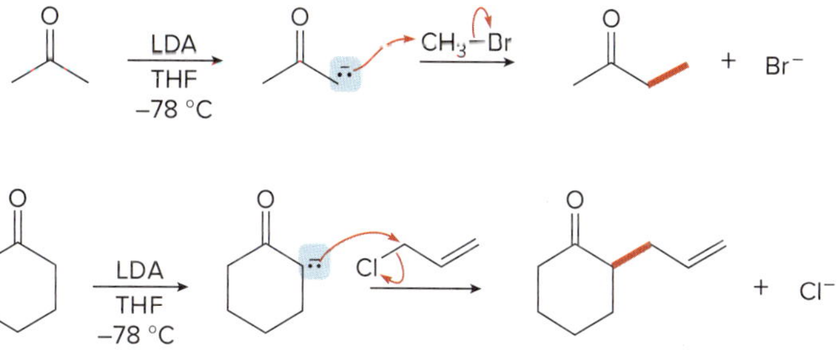

LDA, THF, −78 °C; CH_3–Br → + Br$^-$

LDA, THF, −78 °C; Cl → + Cl$^-$

new C–C bond in red

Ester enolates and carbanions derived from nitriles are also alkylated under these conditions.

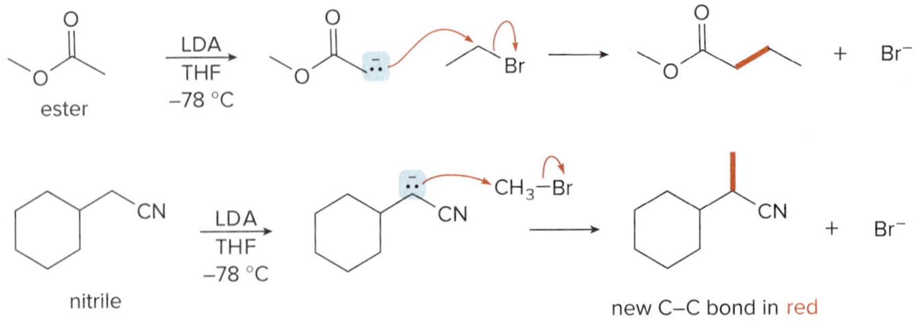

ester — LDA, THF, −78 °C → (enolate) + Br → product + Br^-

nitrile — LDA, THF, −78 °C → (carbanion) + CH_3–Br → product + Br^-

new C–C bond in red

Problem 21.17 What product is formed when each compound is treated first with LDA in THF solution at low temperature, followed by CH_3CH_2I?

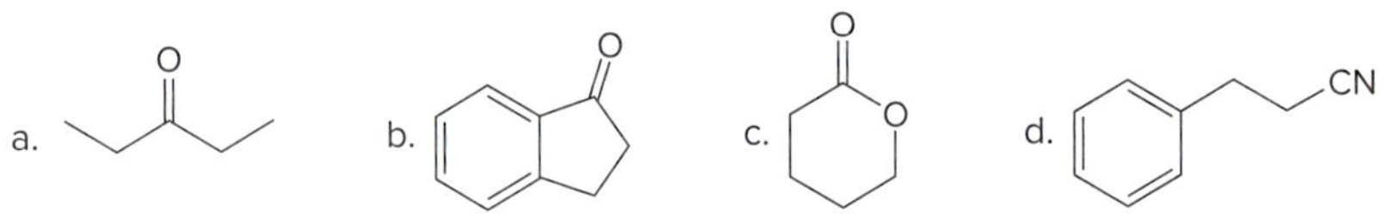

a. b. c. d.

The stereochemistry of enolate alkylation follows the general rule governing the stereochemistry of reactions: **an achiral starting material yields an achiral or racemic product.** For example, when cyclohexanone (an achiral starting material) is converted to 2-ethylcyclohexanone by treatment with base and CH_3CH_2I, a new stereogenic center (labeled in blue) is introduced, and both enantiomers of the product are formed in equal amounts—that is, a **racemic mixture.**

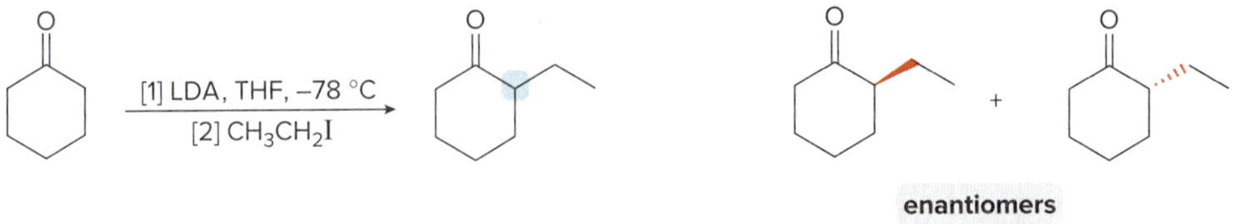

[1] LDA, THF, −78 °C
[2] CH_3CH_2I

enantiomers

Problem 21.18 Draw the products obtained (including stereochemistry) when each compound is treated with LDA, followed by CH_3I.

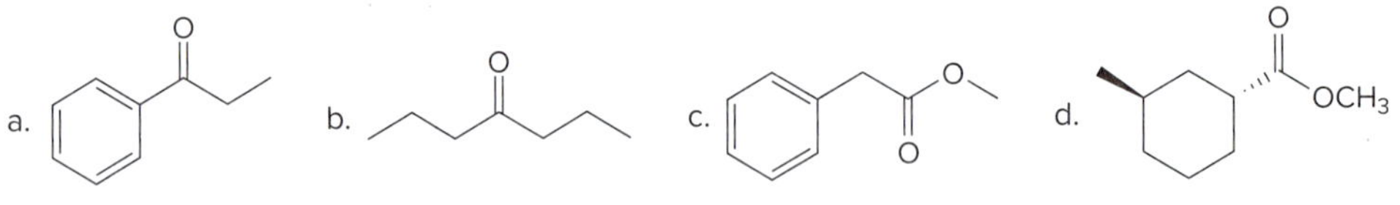

a. b. c. d. OCH_3

Problem 21.19 The analgesic naproxen can be prepared by a stepwise reaction sequence from ester **A.** Using enolate alkylation in one step, what reagents are needed to convert **A** to naproxen? Draw the structure of each intermediate. Explain why a racemic product is formed.

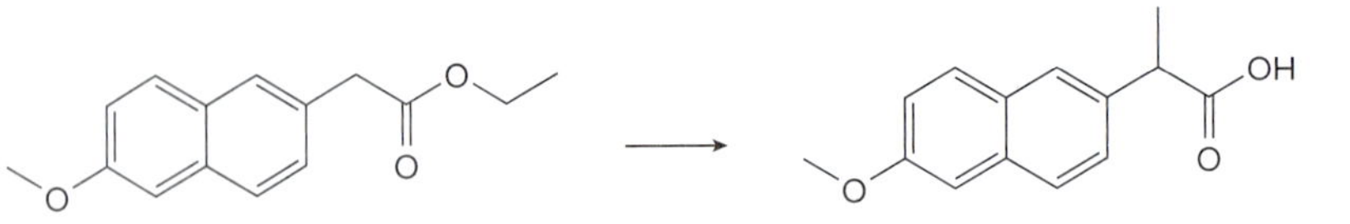

A → naproxen

21.8B Alkylation of Unsymmetrical Ketones

An unsymmetrical ketone can be regioselectively alkylated to yield one major product. The strategy depends on the use of the appropriate base, solvent, and temperature to form the kinetic or thermodynamic enolate (Section 21.4), which is then treated with an alkyl halide to form the alkylation product.

An example of this strategy is seen in the intramolecular alkylation of bromo ketone **A** to form either **B** or **C,** depending on the reaction conditions.

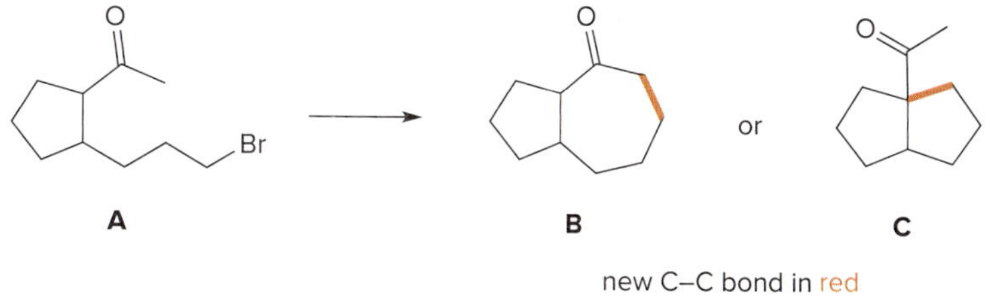

A → B or C

new C–C bond in red

- **Treatment of A with LDA in THF at −78° gives the less substituted enolate, which undergoes an intramolecular S_N2 reaction to form the seven-membered ring in B.**

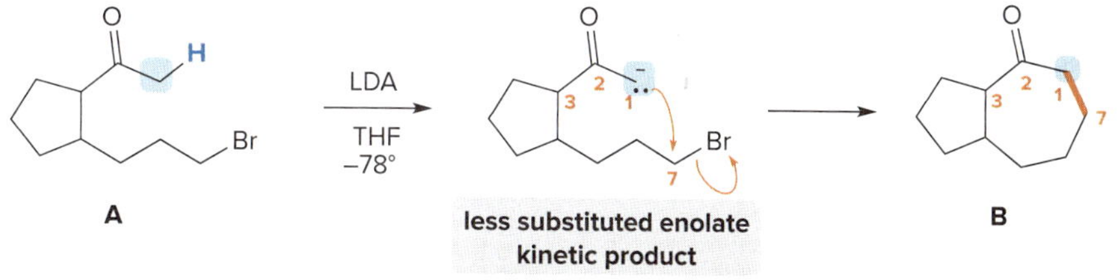

A —LDA, THF, −78°→ less substituted enolate (kinetic product) → B

- **Treatment of A with $KOC(CH_3)_3$ in $(CH_3)_3COH$ at room temperature gives the *more* substituted enolate, which undergoes an intramolecular S_N2 reaction to form the five-membered ring in C.**

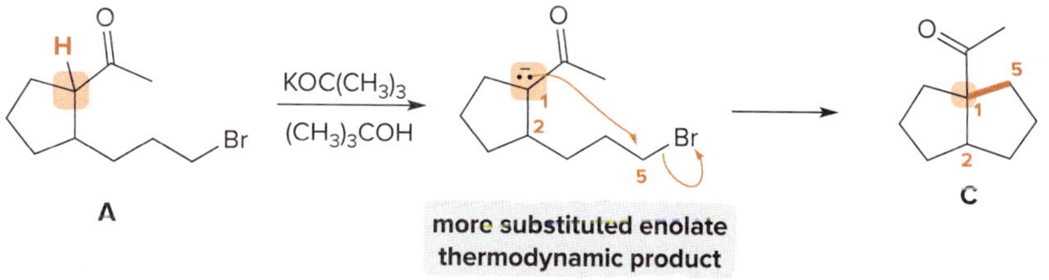

A —$KOC(CH_3)_3$, $(CH_3)_3COH$→ more substituted enolate (thermodynamic product) → C

Finally, while enolate alkylation at the less substituted α carbon using LDA is a reliable regioselective reaction, enolate alkylation at the more substituted α carbon with $KOC(CH_3)_3$ may lead to mixtures of products. Regioselectivity depends on the identity of the substrate and the experimental parameters, which sometimes must be carefully monitored to maximize the yield of the desired alkylation product.

Problem 21.20 How can pentan-2-one be converted to each compound?

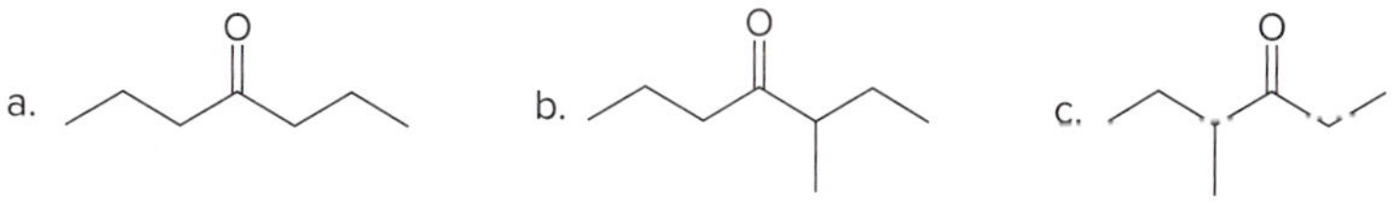

a.

b.

c.

21.8C Application of Enolate Alkylation: Tamoxifen Synthesis

Tamoxifen is a potent anticancer drug that has been used to treat certain forms of breast cancer for many years. One step in the synthesis of tamoxifen involves the treatment of ketone **A** with NaH as base to form an enolate. Alkylation of this enolate with CH_3CH_2I forms **B** in

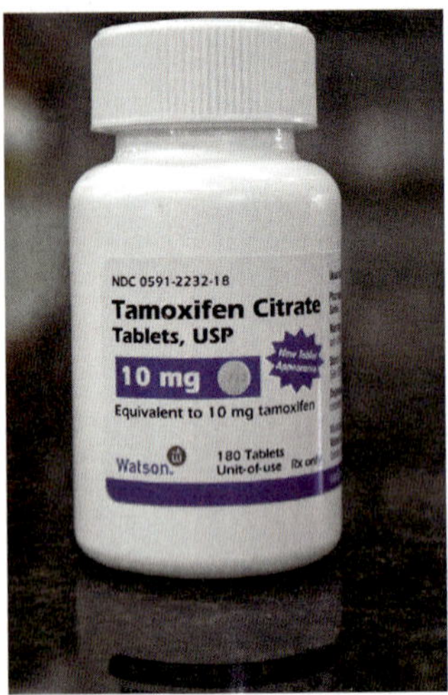

Tamoxifen has been commercially available since the 1970s, sold under the brand name of Nolvadex.
Mary Reeg/McGraw Hill

high yield. **B** is converted to tamoxifen in several steps, some of which are reactions you have already learned.

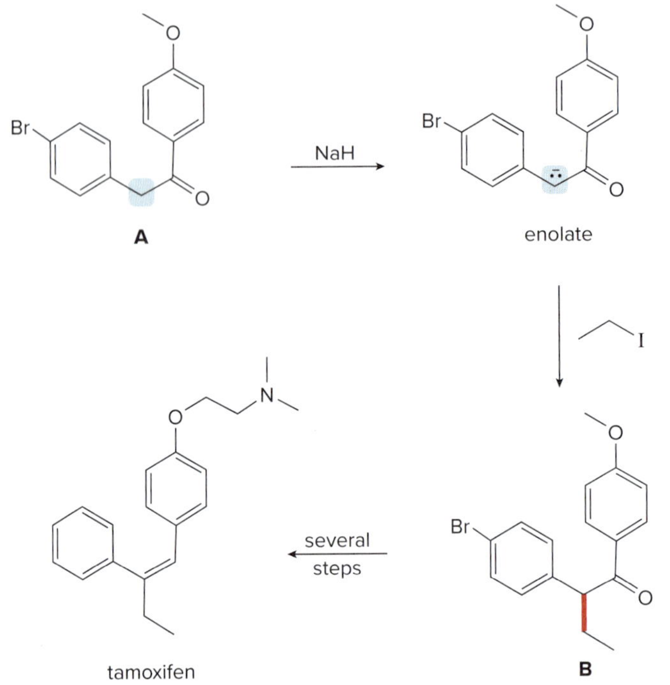

A —NaH→ enolate —(CH₃CH₂I)→ B —several steps→ tamoxifen

A

enolate

B

new C–C bond in red

tamoxifen

Only the *Z* isomer of the C=C provides beneficial effects.

Problem 21.21 Identify **A, B,** and **C,** intermediates in the synthesis of the five-membered ring called an α-methylene-γ-butyrolactone. This heterocyclic ring system is present in some antitumor agents.

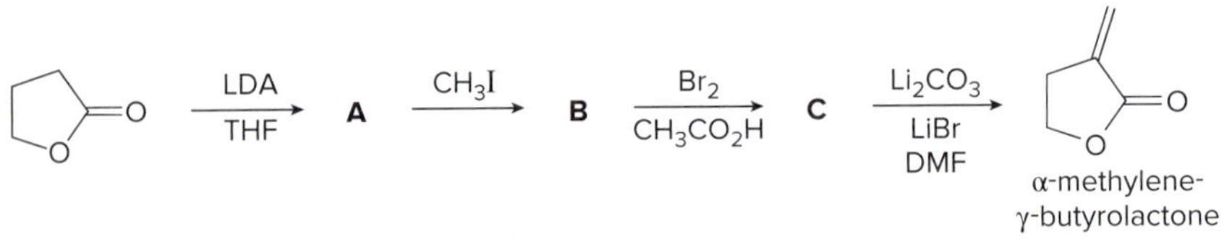

γ-butyrolactone —LDA, THF→ **A** —CH_3I→ **B** —Br_2, CH_3CO_2H→ **C** —Li_2CO_3, LiBr, DMF→ α-methylene-γ-butyrolactone

Problem 21.22 A key step in the synthesis of laurencin, a marine natural product isolated from red algae of the genus *Laurencia* (Problem 7.5), involves an intramolecular enolate alkylation. Identify the product **A** in the following reaction.

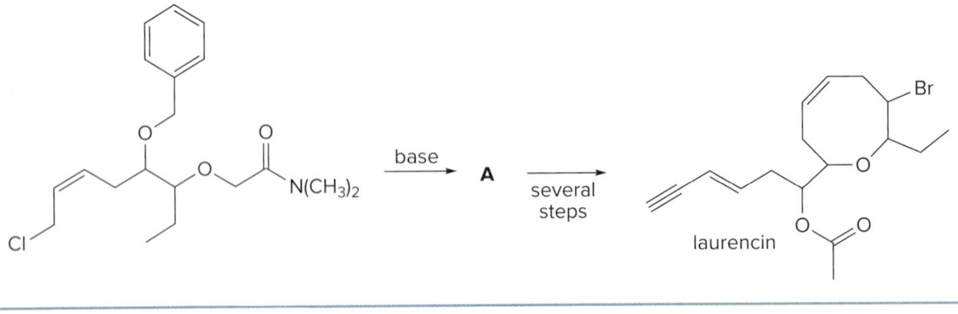

Cl ... N(CH₃)₂ —base→ **A** —several steps→ laurencin

21.9 Malonic Ester Synthesis

Besides the direct method of enolate alkylation discussed in Section 21.8, a new alkyl group can also be introduced on the α carbon using the malonic ester synthesis and the acetoacetic ester synthesis.

- **The malonic ester synthesis prepares carboxylic acids** having two general structures:

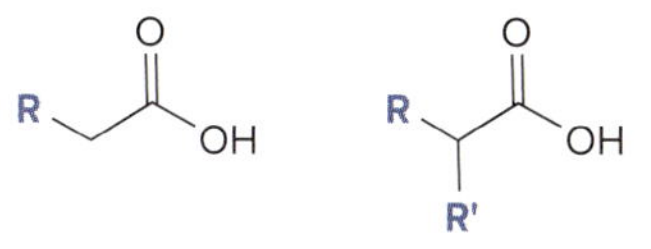

- **The acetoacetic ester synthesis prepares methyl ketones** having two general structures:

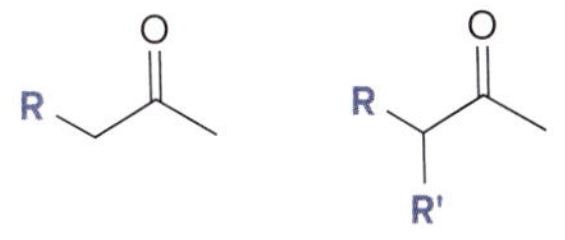

21.9A Background for the Malonic Ester Synthesis

- The malonic ester synthesis is a stepwise method for converting diethyl malonate to a carboxylic acid having one or two alkyl groups on the α carbon.

To simplify the structures, the CH_3CH_2 groups of the esters are abbreviated as Et.

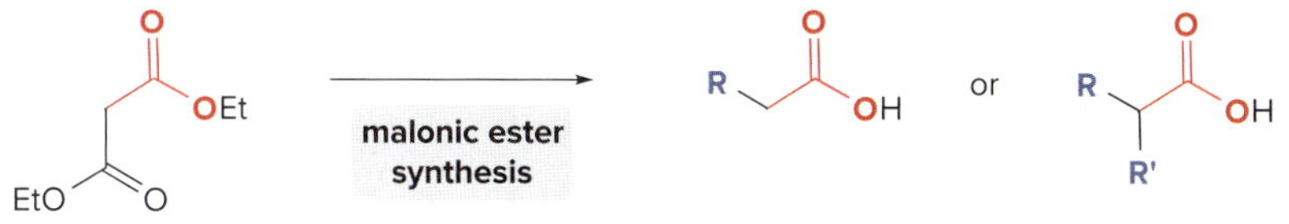

malonic ester synthesis

diethyl malonate
Et = CH_3CH_2

Before writing out the steps in the malonic ester synthesis, recall from Section 20.10 that esters are hydrolyzed by aqueous acid. Thus, heating diethyl malonate with acid and water hydrolyzes both esters to carboxy groups, forming a β-diacid (1,3-diacid).

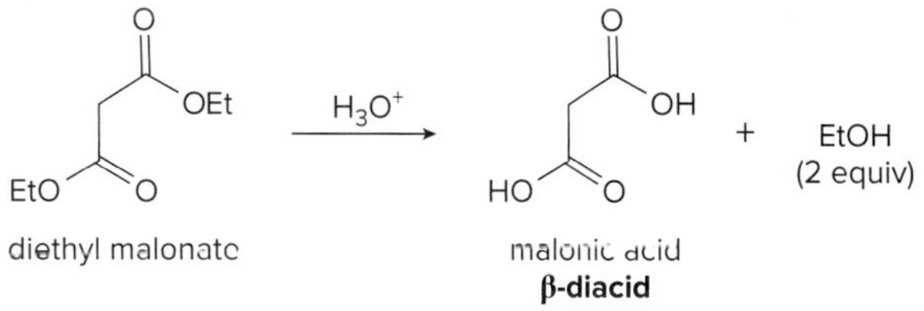

H_3O^+ — diethyl malonate → malonic acid (β-diacid) + EtOH (2 equiv)

The resulting β-diacids are unstable to heat. They decarboxylate (lose CO_2), resulting in cleavage of a carbon–carbon bond and formation of a carboxylic acid. Decarboxylation is *not* a general reaction of all carboxylic acids. It occurs with β-diacids, however, because CO_2 can be eliminated through a cyclic, six-atom transition state. This forms an enol of a carboxylic acid, which in turn tautomerizes to the more stable keto form.

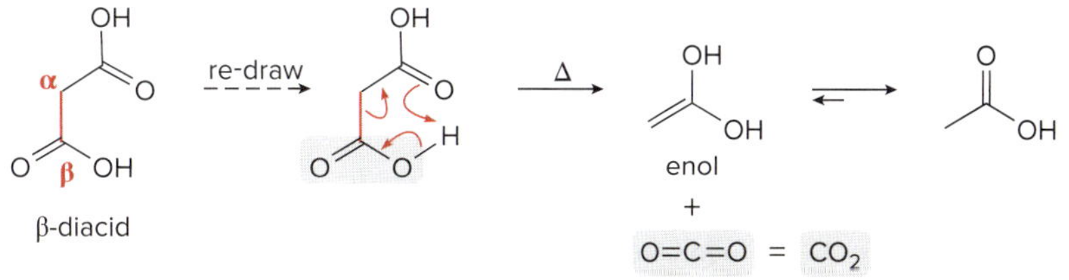

β-diacid — re-draw — Δ → enol + O=C=O = CO_2

The net result of decarboxylation is cleavage of a carbon–carbon bond on the α carbon, with loss of CO_2.

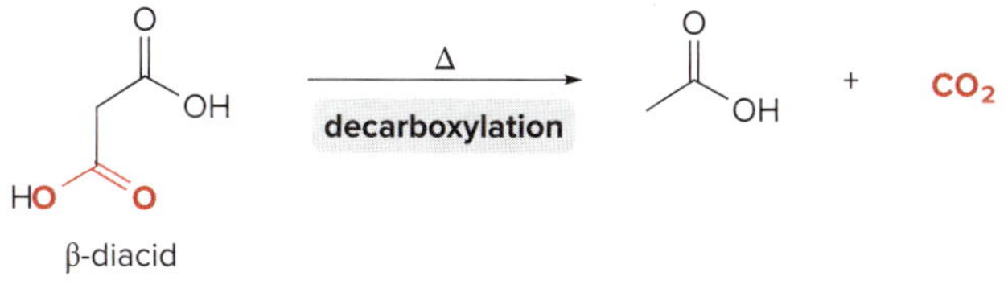

β-diacid — Δ, decarboxylation → + CO_2

Decarboxylation occurs readily whenever a carboxy group (COOH) is bonded to the α carbon of another carbonyl group. For example, β-keto acids also readily lose CO_2 on heating to form ketones.

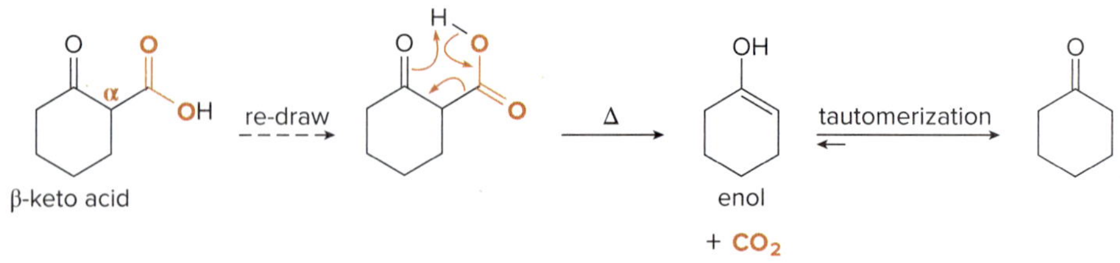

β-keto acid —re-draw→ —Δ→ enol + CO_2 —tautomerization⇄

Sample Problem 21.4 Determining Which Carbonyl Compounds Undergo Decarboxylation

Which of the following compounds will readily lose CO_2 when heated?

a. b.

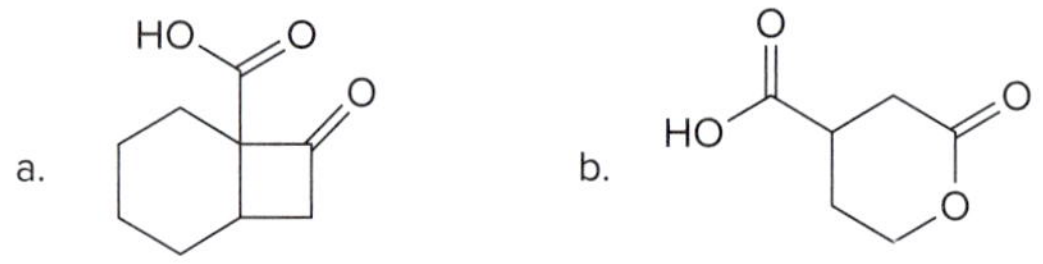

Solution

Decarboxylation occurs readily only when a CO_2H group is bonded to the α carbon of another carbonyl group.

a. —Δ→ + CO_2 b. —Δ→ No reaction

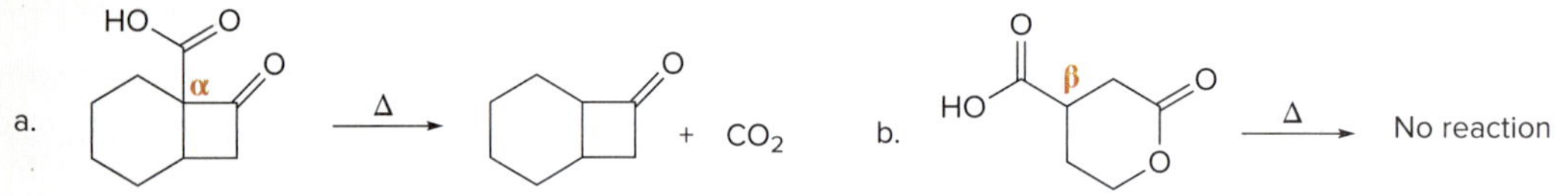

In part (a) the CO_2H group is located α to a carbonyl, so CO_2 is readily lost to form a ketone. In part (b) the CO_2H group is located β to a carbonyl, so CO_2 is not readily lost.

Problem 21.23 Which of the following compounds will readily lose CO_2 when heated?

a. b. c. d.

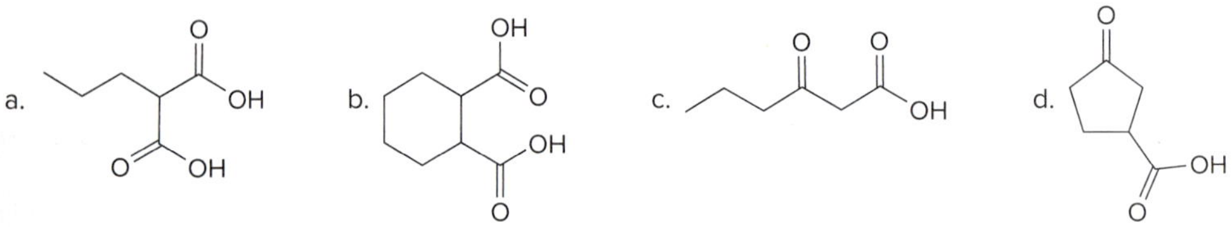

More Practice: Try Problems 21.51a, 21.54.

21.9B Steps in the Malonic Ester Synthesis

The malonic ester synthesis converts diethyl malonate to a carboxylic acid in three steps.

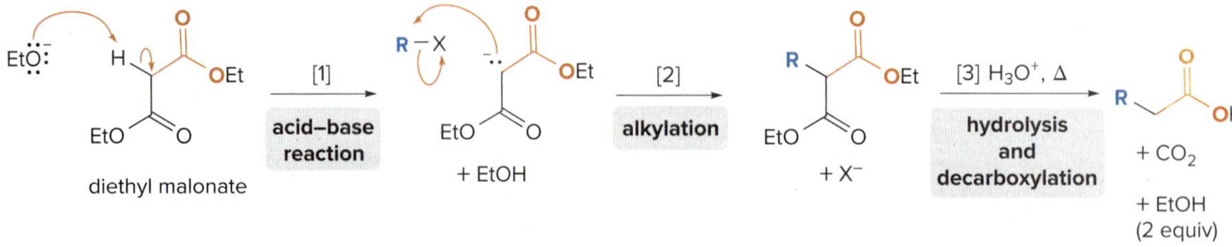

diethyl malonate —[1] acid–base reaction→ + EtOH —[2] alkylation→ + X^- —[3] H_3O^+, Δ hydrolysis and decarboxylation→ + CO_2 + EtOH (2 equiv)

[1] **Deprotonation.** Treatment of diethyl malonate with $^{-}$OEt removes the acidic α proton between the two carbonyl groups. Recall from Section 21.3A that these protons are more acidic than other α protons because **three resonance structures can be drawn for the enolate,** instead of the usual two. Thus, $^{-}$OEt, rather than the stronger base LDA, can be used for this reaction.

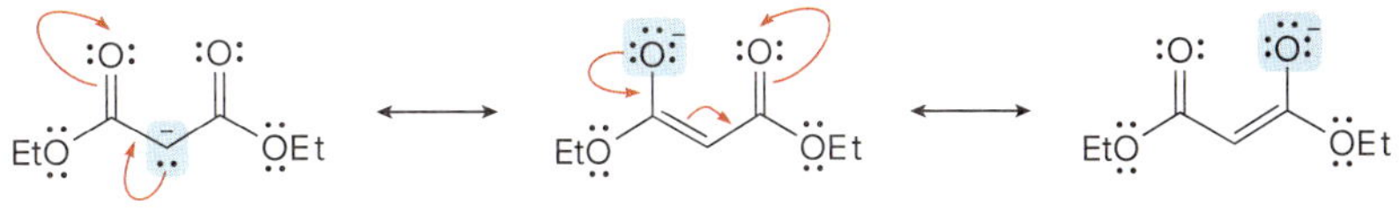

three resonance structures for the conjugate base

[2] **Alkylation.** The nucleophilic enolate reacts with an alkyl halide in an S_N2 reaction to form a substitution product. Because the mechanism is S_N2, the yields are higher when R is CH_3 or a 1° alkyl group.

[3] **Hydrolysis and decarboxylation.** Heating the diester with aqueous acid hydrolyzes the diester to a β-diacid, which loses CO_2 to form a carboxylic acid.

The synthesis of butanoic acid ($CH_3CH_2CH_2COOH$) from diethyl malonate illustrates the basic process:

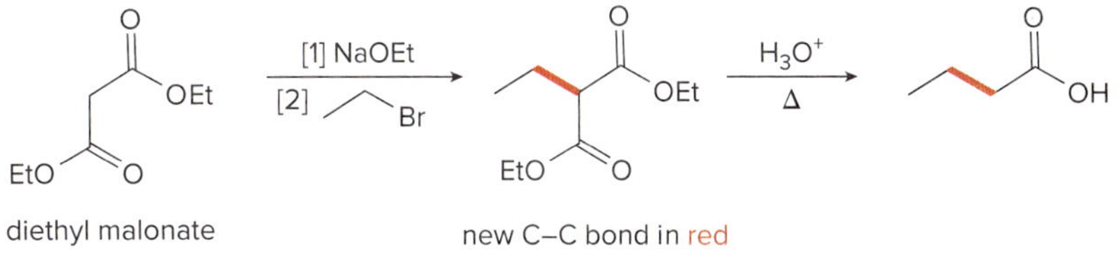

If the first two steps of the reaction sequence are repeated *prior* to hydrolysis and decarboxylation, then a carboxylic acid having *two new alkyl groups* on the α carbon can be synthesized. This is illustrated in the synthesis of 2-benzylbutanoic acid [$CH_3CH_2CH(CH_2C_6H_5)COOH$] from diethyl malonate:

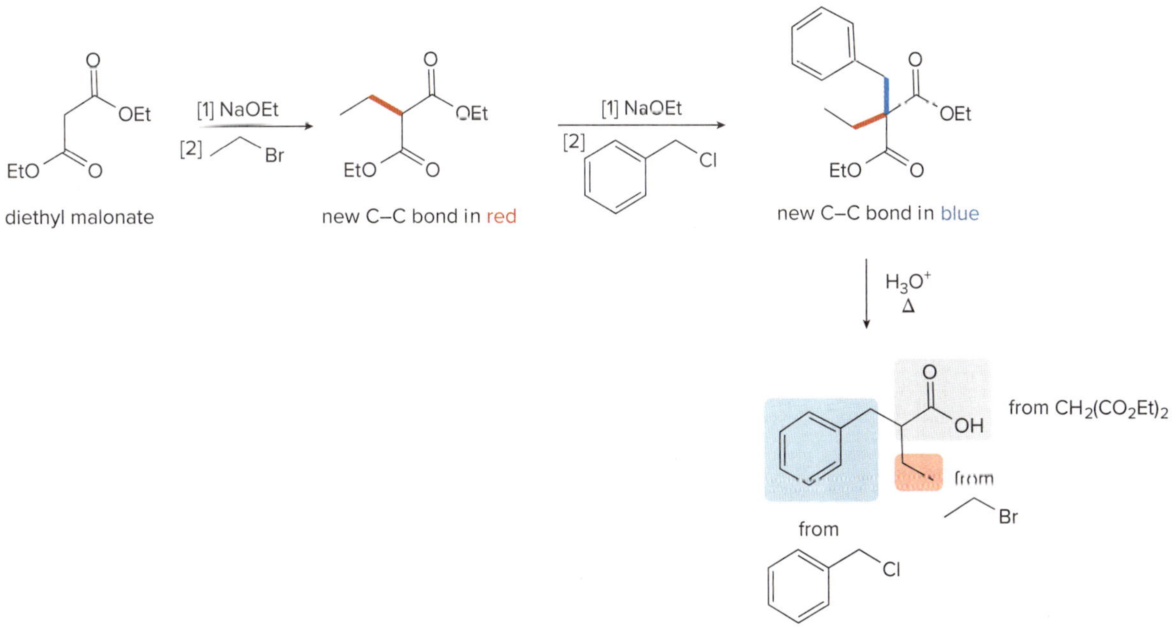

An intramolecular malonic ester synthesis can be used to form rings having three to six atoms, provided the appropriate dihalide is used as starting material. For example, cyclopentanecarboxylic acid can be prepared from diethyl malonate and 1,4-dibromobutane ($BrCH_2CH_2CH_2CH_2Br$) by this sequence of reactions:

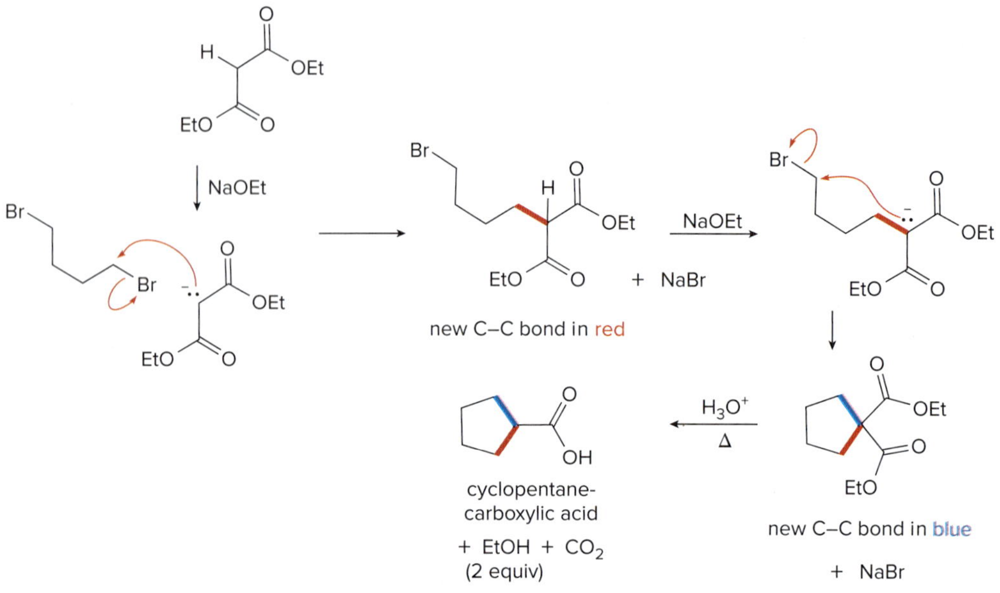

NaOEt

new C–C bond in red

NaOEt

+ NaBr

H_3O^+, Δ

cyclopentanecarboxylic acid + EtOH + CO_2 (2 equiv)

new C–C bond in blue

+ NaBr

Problem 21.24 Draw the products of each reaction.

a. $CH_2(CO_2Et)_2$ → [1] NaOEt; [2] (bromomethyl)cyclohexane → H_3O^+, Δ

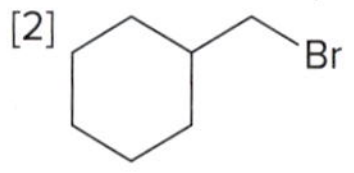

b. $CH_2(CO_2Et)_2$ → [1] NaOEt; [2] CH_3Br → [1] NaOEt; [2] CH_3Br → H_3O^+, Δ

Problem 21.25 What cyclic product is formed from each dihalide using the malonic ester synthesis?

a.  $ClCH_2CH_2CH_2CH_2Cl$ b. 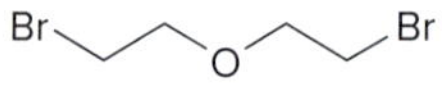 $BrCH_2CH_2OCH_2CH_2Br$

21.9C Retrosynthetic Analysis

To use the malonic ester synthesis, you must be able to determine what starting materials are needed to prepare a given compound—that is, you must **work backwards in the retrosynthetic direction.** This involves a two-step process:

[1] Locate the α carbon to the COOH group, and identify all alkyl groups bonded to the α carbon.

[2] Break the molecule into two (or three) components: Each alkyl group bonded to the α carbon comes from an alkyl halide. The remainder of the molecule comes from $CH_2(COOEt)_2$.

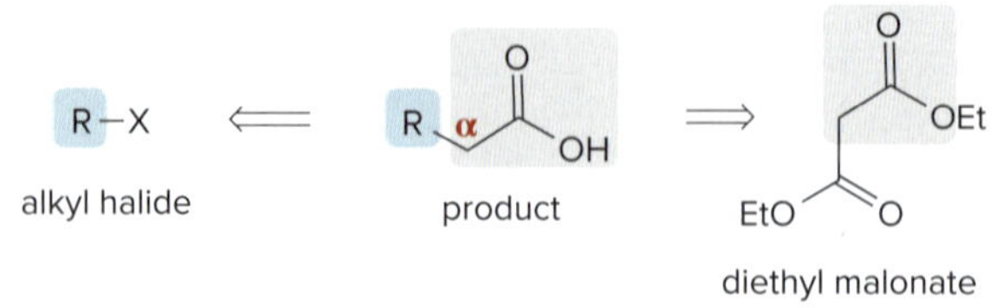

R–X ⟸ RCH_2COOH ⟹ $CH_2(COOEt)_2$

alkyl halide — product — diethyl malonate

Sample Problem 21.5 Determining the Starting Materials in a Malonic Ester Synthesis

What starting materials are needed to prepare 2-methylhexanoic acid [$CH_3CH_2CH_2CH_2CH(CH_3)COOH$] using a malonic ester synthesis?

Solution

The target molecule has two different alkyl groups bonded to the α carbon, so three components are needed for the synthesis:

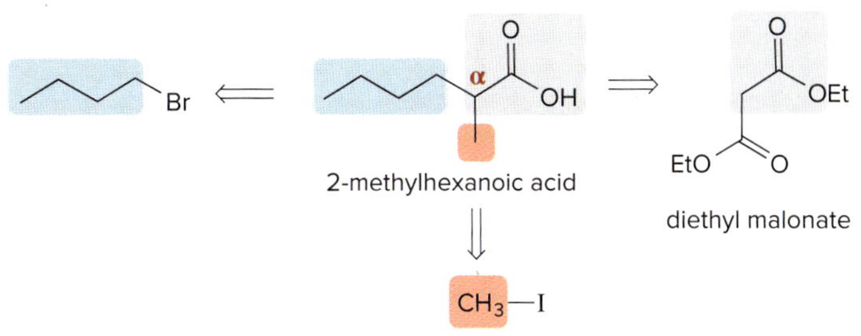

2-methylhexanoic acid

diethyl malonate

CH_3—I

Writing the synthesis in the synthetic direction:

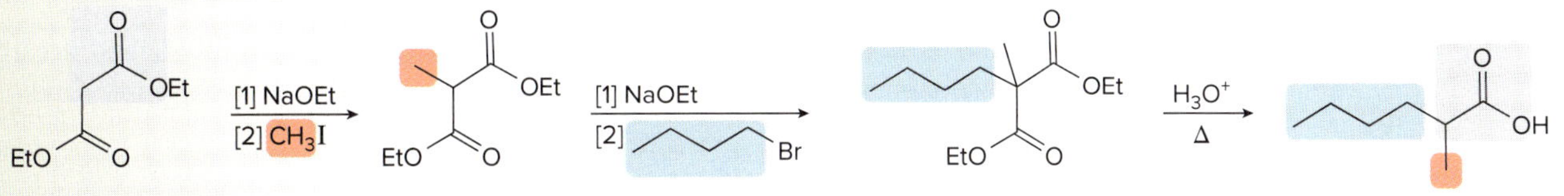

[1] NaOEt [2] CH_3I

[1] NaOEt [2] Br

H_3O^+ Δ

Problem 21.26 What alkyl halides are needed to prepare each carboxylic acid by the malonic ester synthesis?

a. b. c.

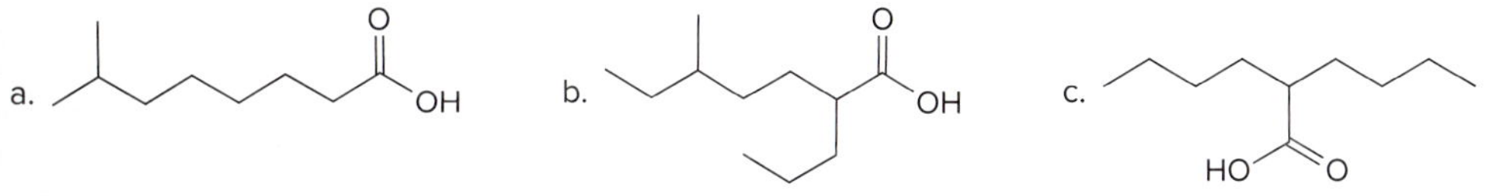

More Practice: Try Problems 21.44–21.46.

Problem 21.27 Explain why each of the following carboxylic acids cannot be prepared by a malonic ester synthesis.

a. b. c.

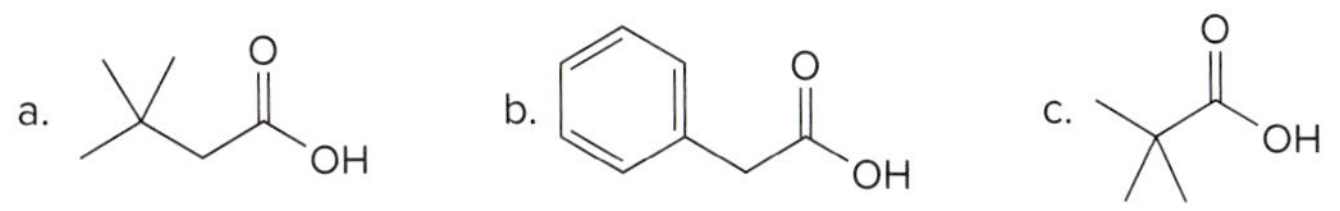

21.10 Acetoacetic Ester Synthesis

- **The acetoacetic ester synthesis is a stepwise method for converting ethyl acetoacetate to a ketone having one or two alkyl groups on the α carbon.**

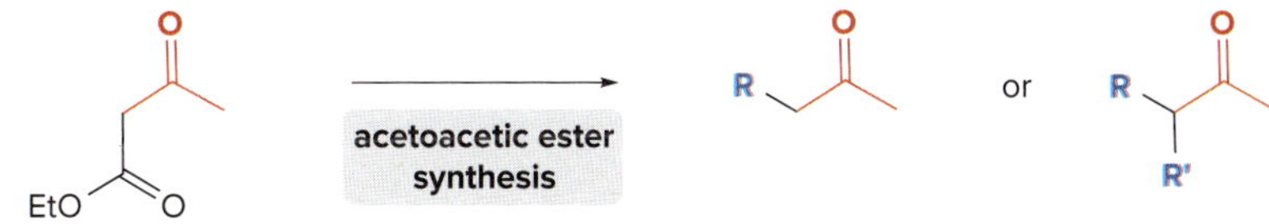

ethyl acetoacetate

21.10A Steps in the Acetoacetic Ester Synthesis

β-keto ester

The steps in the acetoacetic ester synthesis are exactly the same as those in the malonic ester synthesis. Because the starting material, CH_3COCH_2COOEt, is a β-keto ester, the final product is a **ketone,** not a carboxylic acid.

ethyl acetoacetate —[1] acid–base reaction→ (+ EtOH) —[2] alkylation→ (+ X^-) —[3] H_3O^+, Δ hydrolysis and decarboxylation→ + CO_2 + EtOH

[1] **Deprotonation.** Treatment of ethyl acetoacetate with $^-$OEt removes the acidic proton between the two carbonyl groups.

[2] **Alkylation.** The nucleophilic enolate reacts with an alkyl halide (RX) in an S_N2 reaction to form a substitution product. Because the mechanism is S_N2, the yields are higher when R is CH_3 or a 1° alkyl group.

[3] **Hydrolysis and decarboxylation.** Heating the β-keto ester with aqueous acid hydrolyzes the ester to a β-keto acid, which loses CO_2 to form a ketone.

If the first two steps of the reaction sequence are repeated *prior* to hydrolysis and decarboxylation, then a ketone having *two new alkyl groups* on the α carbon can be synthesized.

ethyl acetoacetate —[1] NaOEt, [2] RX→ new C–C bond in red —[1] NaOEt, [2] R'X→ new C–C bond in blue —H_3O^+, Δ→

Problem 21.28 What ketones are prepared by the following reactions?

a. [1] NaOEt; [2] CH_3I; [3] H_3O^+, Δ

b. [1] NaOEt; [2] $CH_3CH_2CH_2Br$; [3] NaOEt; [4] $C_6H_5CH_2I$; [5] H_3O^+, Δ

21.10B Retrosynthetic Analysis

To determine what starting materials are needed to prepare a given ketone using the acetoacetic ester synthesis, you must again work in the **retrosynthetic** direction. This involves a two-step process:

[1] Identify the alkyl groups bonded to the α carbon to the carbonyl group.

[2] Break the molecule into two (or three) components: Each alkyl group bonded to the α carbon comes from an alkyl halide. The remainder of the molecule comes from CH_3COCH_2COOEt.

R–X alkyl halide ⟸ product ⟹ ethyl acetoacetate

⇓

R'–X alkyl halide

For a ketone with two R groups on the α carbon, three components are needed.

Sample Problem 21.6 Determining the Starting Materials in an Acetoacetic Ester Synthesis

What starting materials are needed to synthesize heptan-2-one using the acetoacetic ester synthesis?

heptan-2-one

Solution

Heptan-2-one has only one alkyl group bonded to the α carbon, so only one alkyl halide is needed in the acetoacetic ester synthesis.

Br ⟸ heptan-2-one (α) ⟹ ethyl acetoacetate

Writing the acetoacetic ester synthesis in the synthetic direction:

ethyl acetoacetate → [1] NaOEt [2] (1-bromobutane) → (alkylated β-keto ester) → H_3O^+, Δ → heptan-2-one

Problem 21.29 What alkyl halides are needed to prepare each ketone using the acetoacetic ester synthesis?

a. b. c. d.

More Practice: Try Problems 21.49, 21.50.

Problem 21.30 Treatment of ethyl acetoacetate with NaOEt (2 equiv) and $BrCH_2CH_2Br$ forms compound **X.** This reaction is the first step in the synthesis of illudin-S, an antitumor substance isolated from the jack-o'-lantern, a poisonous, saffron-colored mushroom. What is the structure of **X?**

The jack-o'-lantern, source of the antitumor agent illudin-S (Problem 21.30) *AwakenedEye/ Getty Images*

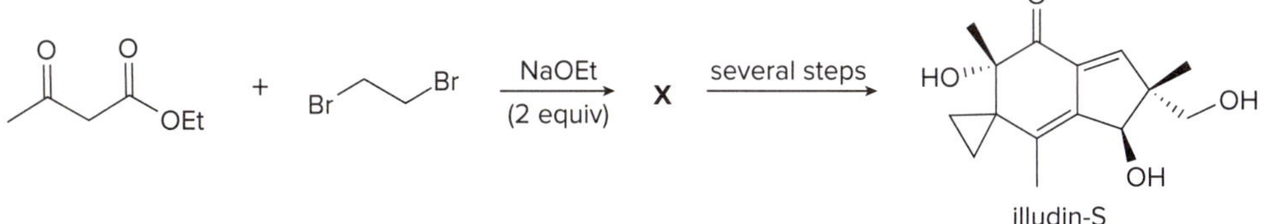

The acetoacetic ester synthesis and direct enolate alkylation are two different methods that prepare similar ketones. Butan-2-one, for example, can be synthesized from acetone by direct enolate alkylation with CH_3I (Method [1]), or by alkylation of ethyl acetoacetate followed by hydrolysis and decarboxylation (Method [2]).

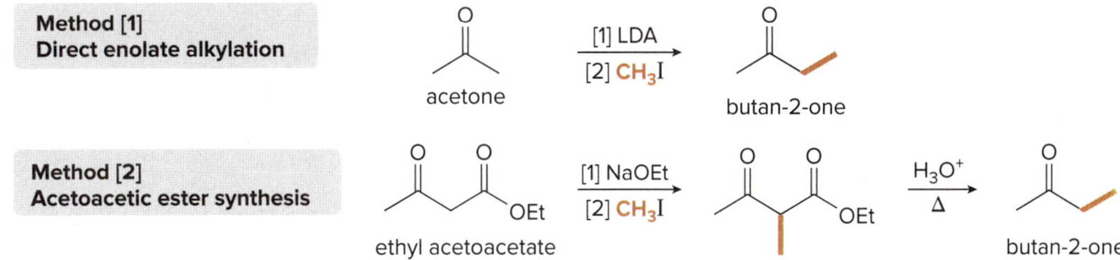

Why would you ever make butan-2-one from ethyl acetoacetate when you could make it in fewer steps from acetone? There are many factors to consider. First of all, synthetic organic chemists like to have a variety of methods to accomplish a single kind of reaction. Sometimes subtle changes in the structure of a starting material make one reaction work better than another.

In the chemical industry, moreover, cost is an important issue. Any reaction needed to make a large quantity of a useful drug or other consumer product must use cheap starting materials. Direct enolate alkylation usually requires a very strong base like LDA to be successful, whereas the acetoacetic ester synthesis utilizes NaOEt. NaOEt can be prepared from cheaper starting materials, and this makes the acetoacetic ester synthesis an attractive method, even though it involves more steps.

Thus, each method has its own advantages and disadvantages, depending on the starting material, the availability of reagents, the cost, and the occurrence of side reactions.

Problem 21.31 Nabumetone is a pain reliever and anti-inflammatory agent sold under the brand name of Relafen.

CH_3O
nabumetone

a. Write out a synthesis of nabumetone from ethyl acetoacetate.
b. What ketone and alkyl halide are needed to synthesize nabumetone by direct enolate alkylation?

Chapter 21 REVIEW

KEY REACTIONS

[1] Halogenation at the α carbon

1. R–CO–CH(α)–H → (X_2, CH_3COOH; $X_2 = Cl_2, Br_2, I_2$) → R–CO–CH(α)–X
R = H or alkyl; halogenation in acid (21.7A); α-halo aldehyde or ketone

2. PhCO–CH(α)H₂ → (X_2 (excess), ^-OH; $X_2 = Cl_2, Br_2, I_2$) → PhCO–C(α)X₂
halogenation in base (21.7B)

3. methyl ketone → (X_2 (excess), ^-OH; $X_2 = Cl_2, Br_2, I_2$) → PhCOO⁻ + HCX_3 (haloform)
halogenation (21.7B)

Try Problems 21.43; 21.51c, d, f.

[2] Reactions of α-halo carbonyl compounds

1. PhCO–CH(α)Br–CH₃ → (Li_2CO_3, LiBr; DMF) → PhCO–CH(α)=CH(β)
elimination (21.7C)

2. PhCO–CH₂(α)–Br → (CH_3SH; S_N2) → PhCO–CH₂(α)–SCH_3
(21.7C)

See Sample Problem 21.3. Try Problems 21.51b, c; 21.53; 21.55.

[3] Alkylation reactions at the α carbon

1 [1] LDA or NaOR; [2] $CH_3—X$; X = Cl, Br, I → product with α CH_3 + X^-

alkylation

(21.8)

2 EtO, α, OEt; [1] NaOEt; [2] X; [3] H_3O^+, Δ; X = Cl, Br, I → α, OH

malonic ester synthesis

(21.9)

3 EtO, α; [1] NaOEt; [2] X; [3] H_3O^+, Δ; X = Cl, Br, I → α

acetoacetic ester synthesis

(21.10)

Try Problems 21.51e, 21.52, 21.56, 21.58.

KEY SKILLS

[1] Converting an enol to a keto tautomer in acid (21.2)

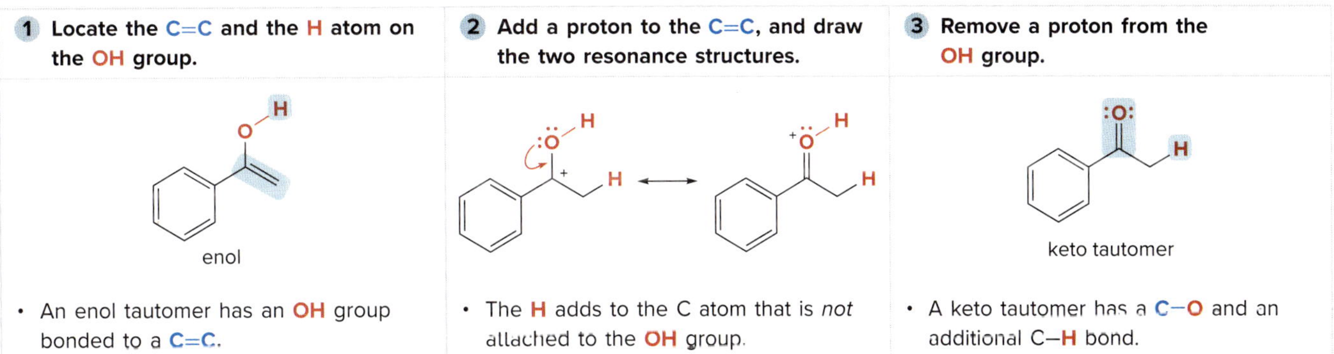

See Sample Problem 21.1. Try Problems 21.33, 21.40a, 21.42.

[2] Determining the major enolate formed in a reaction (21.4)

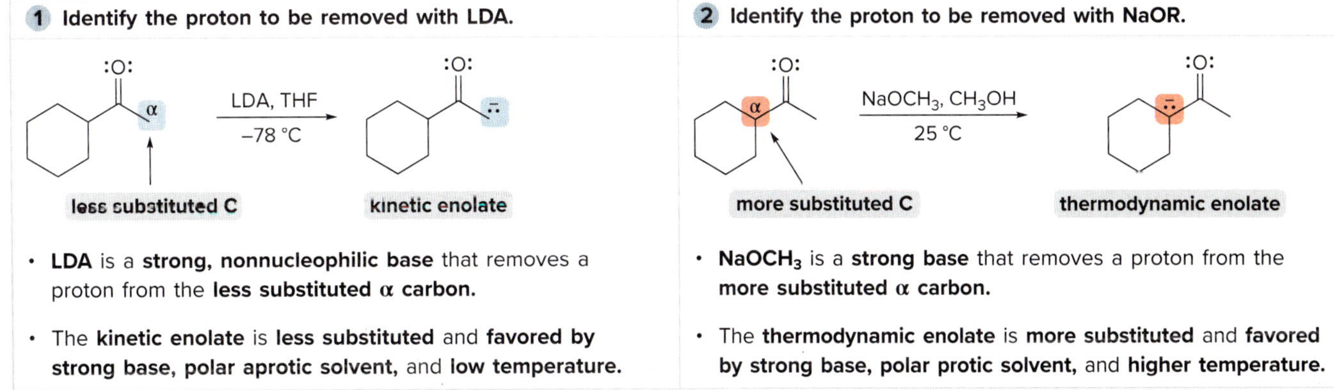

See Sample Problem 21.2. Try Problem 21.36.

[3] Preparing a carboxylic acid using a malonic ester synthesis (21.9B); example: 2,4-dimethylpentanoic acid

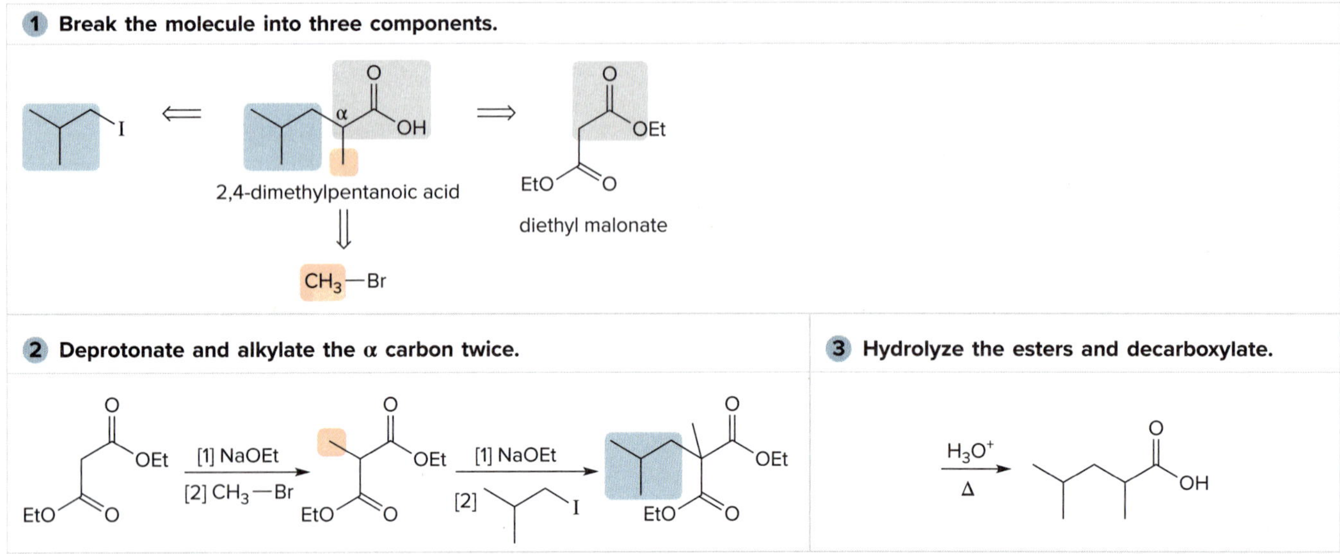

See Sample Problem 21.5. Try Problems 21.44–21.46.

[4] Preparing a ketone using the acetoacetic ester synthesis (21.10B); example: hexan-2-one

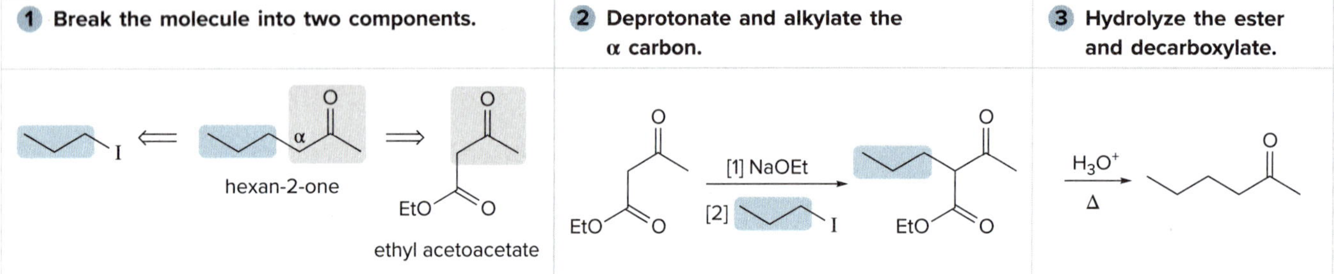

See Sample Problem 21.6. Try Problems 21.49, 21.50.

CHAPTER 21 MULTIPLE-CHOICE SELF-TEST

The Self-Test consists of multiple-choice questions similar to those found on the American Chemical Society organic chemistry exam. Answers are given at the end of the chapter.

1. Which compound will readily lose CO_2 when heated?

a. b. c. d.

2. What product is formed in the following reaction sequence?

[1] Br_2, CH_3CO_2H

[2] Li_2CO_3, LiBr, DMF

a. b. c. d.

3. What anion is formed when **X** is treated with LDA?

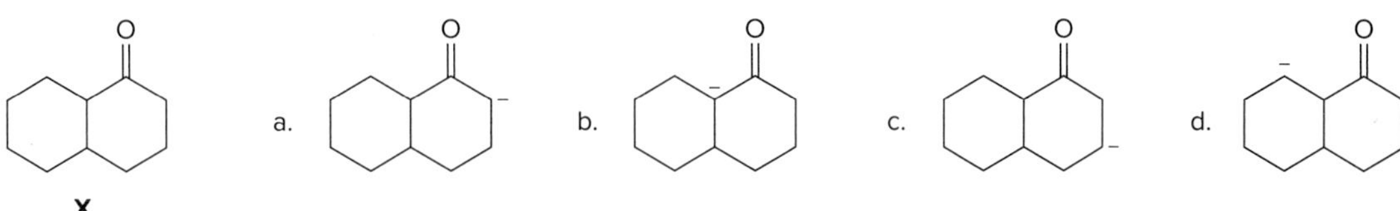

4. Which labeled proton is most acidic?

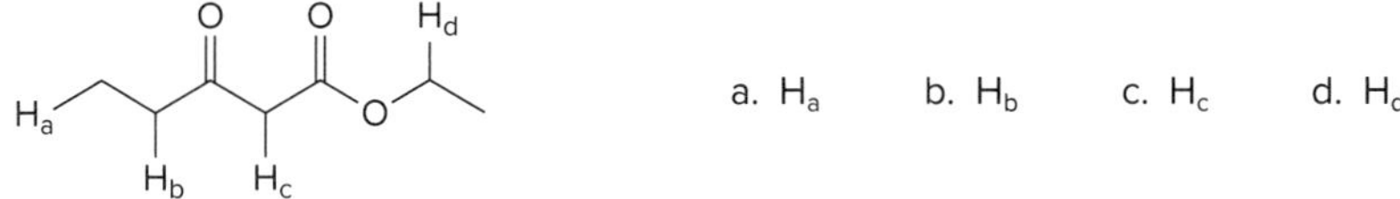

a. H_a b. H_b c. H_c d. H_d

5. What product is formed in the following reaction sequence?

[1] LDA, [2] CH_3I; [1] LDA, [2] CH_3I

a. b. c. d.

6. Which reaction conditions do *not* favor formation of a thermodynamic enolate?

a. 25 °C b. CH_3CH_2OH solvent c. $NaOCH_2CH_3$ d. LDA

7. Which compound is *not* an enol of **Y?**

Y a. b. c. d.

8. Besides $CH_2(CO_2CH_2CH_3)_2$, what starting materials are needed to make **Z?**

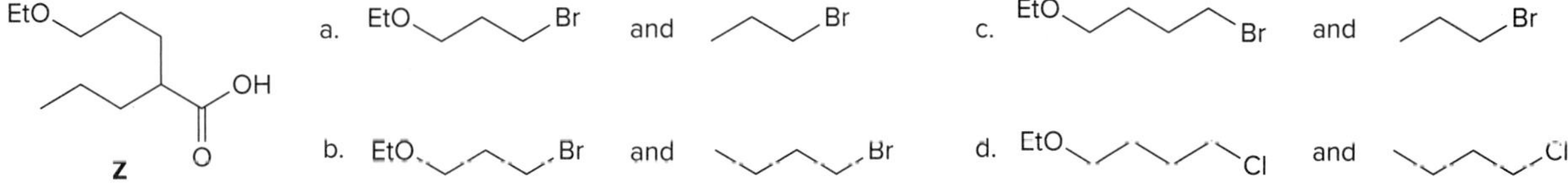

9. What product is formed in the following reaction sequence?

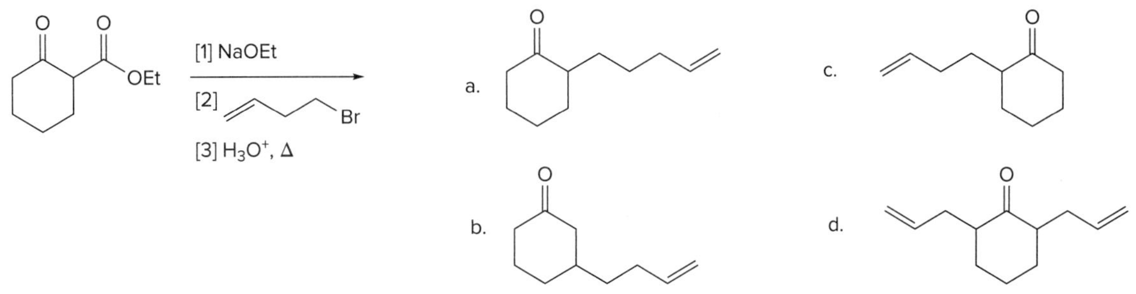

10. Which ketone can be synthesized from $CH_3COCH_2CO_2Et$ using the acetoacetic ester synthesis?

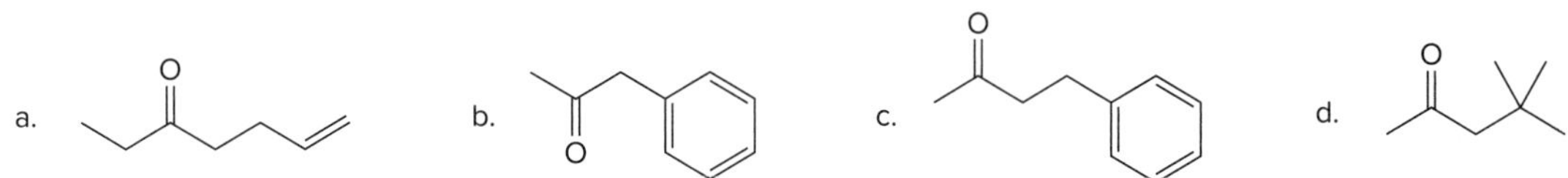

PROBLEMS

Problem Using Three-Dimensional Models

21.32 The cis ketone **A** is isomerized to a trans ketone **B** with aqueous NaOH. A similar isomerization does not occur with ketone **C.** (a) Draw the structure of **B** using a chair cyclohexane. (b) Label the substituents in **C** as cis or trans, and explain the difference in reactivity.

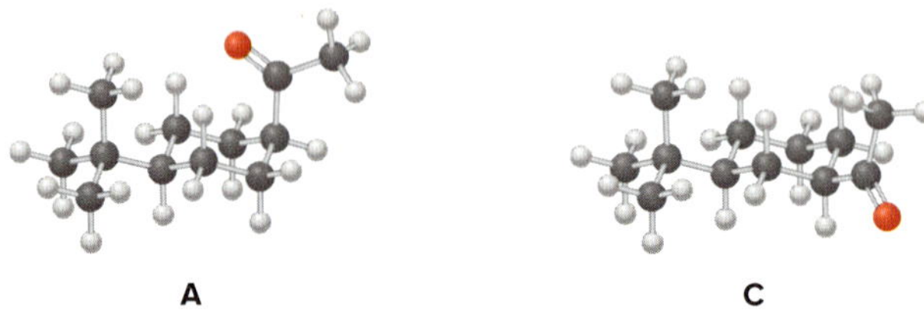

Enols, Enolates, and Acidic Protons

21.33 Draw enol tautomer(s) for each compound.

a. 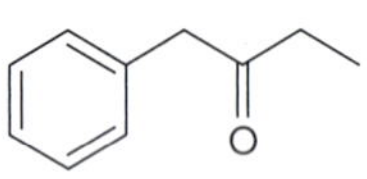 b. 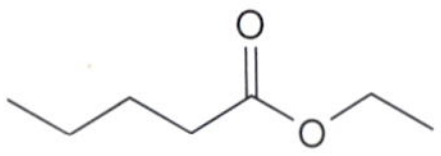 c. 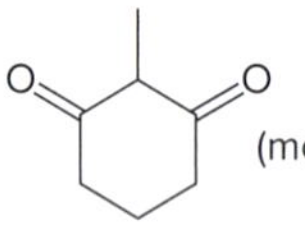 (mono enol form)

21.34 What hydrogen atoms in each compound have a $pK_a \leq 25$?

a. 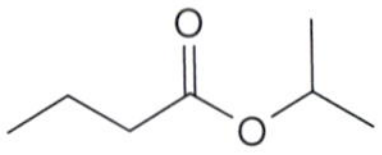 b. 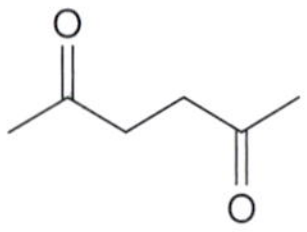 c.

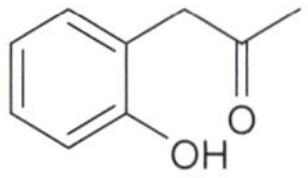

21.35 Rank the labeled protons in each compound in order of increasing acidity.

a. 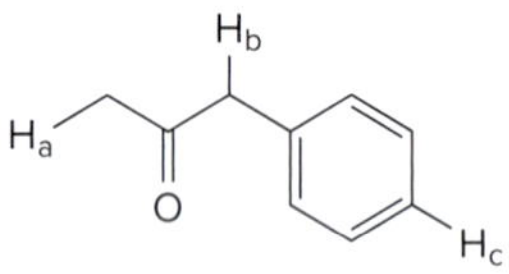 H_a, H_b, H_c b. H_a, H_b, H_c

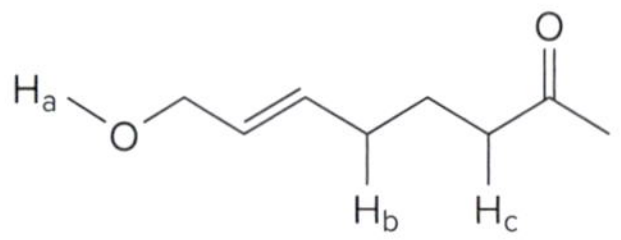

21.36 What is the major enolate (or carbanion) formed when each compound is treated with LDA?

a. 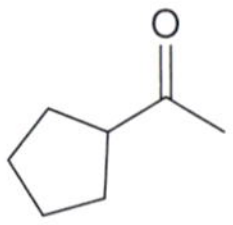 b. 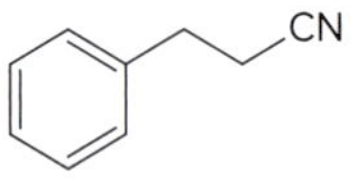 c. 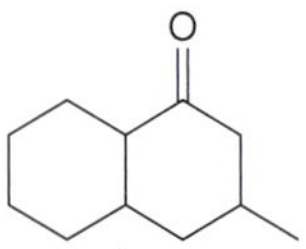 d.

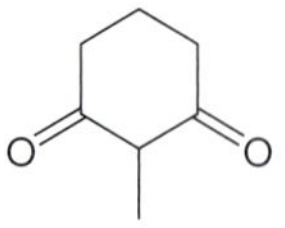

21.37 The following reaction was used in the synthesis of the naturally occurring antimalarial compound artemisinin (Figure 10.12). What intermediate is formed after addition of (a) one equivalent of LDA? (b) two equivalents of LDA? (c) Explain why alkylation occurs at the labeled α carbon.

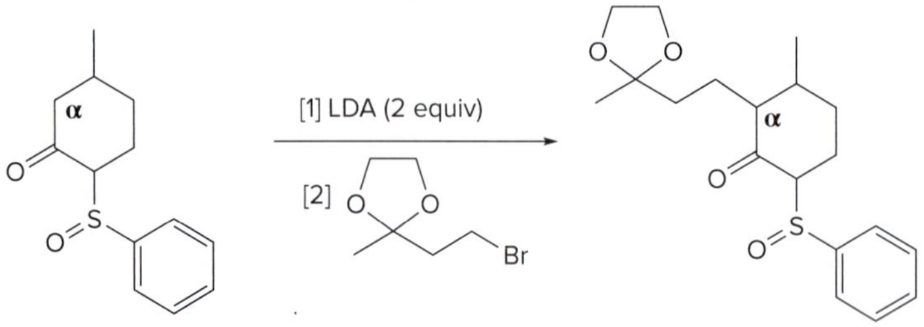

[1] LDA (2 equiv)
[2] (bromide reagent) Br

21.38 Explain why the α protons of an ester are significantly more acidic than the α protons of an amide [$CH_3CO_2CH_2CH_3$, pK_a = 24.5 versus $CH_3CON(CH_3)_2$, pK_a = 30].

21.39 Why is the pK_a of the H_a protons in 1-acetylcyclohexene higher than the pK_a of the H_b protons?

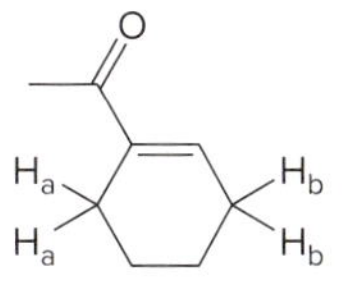

1-acetylcyclohexene

21.40 Acyclovir is an effective antiviral agent used to treat the herpes simplex virus. (a) Draw the enol form of acyclovir, and explain why it is aromatic. (b) Why is acyclovir typically drawn in its keto form, despite the fact that its enol is aromatic?

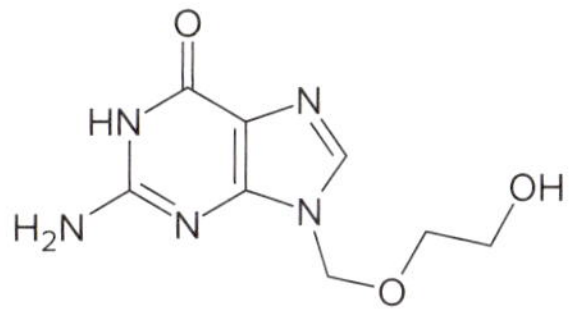

acyclovir

21.41 Explain why pentane-2,4-dione forms two different alkylation products (**A** or **B**) when the number of equivalents of base is increased from one to two.

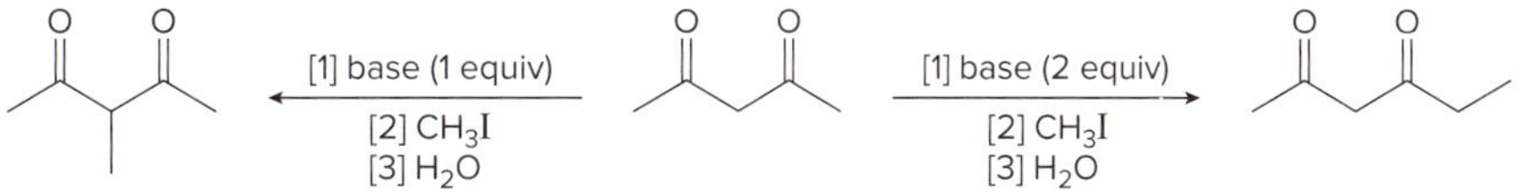

[1] base (1 equiv)
[2] CH_3I
[3] H_2O

[1] base (2 equiv)
[2] CH_3I
[3] H_2O

A pentane-2,4-dione **B**

21.42 Vitamin C is a stable enediol. Draw the structure of the two keto tautomers in equilibrium with the enediol, and explain why the enediol is more stable than the other tautomers.

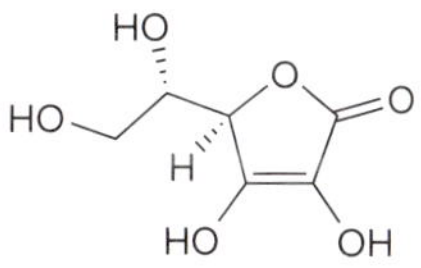

vitamin C
an enediol

Halogenation

21.43 Draw a stepwise mechanism for the following reaction.

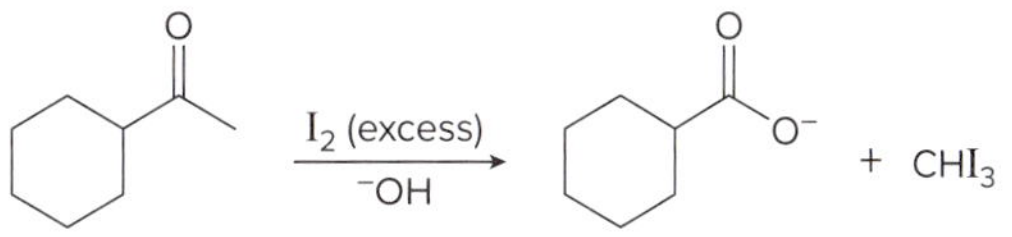

I_2 (excess)
^-OH
+ CHI_3

Malonic Ester Synthesis

21.44 What alkyl halides are needed to prepare each carboxylic acid using the malonic ester synthesis?

a. 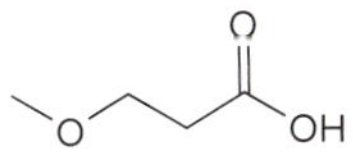 b. 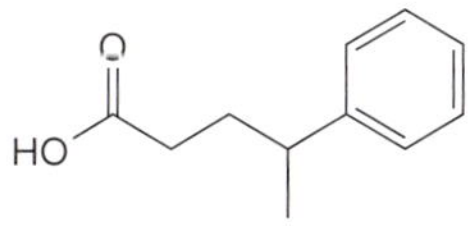 c.

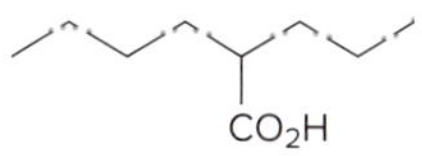

21.45 Use the malonic ester synthesis to prepare each carboxylic acid.

a. 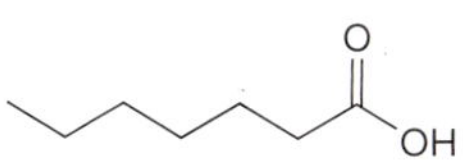 b. 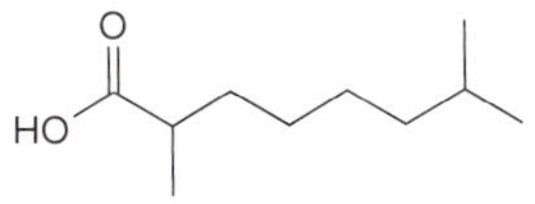 c.

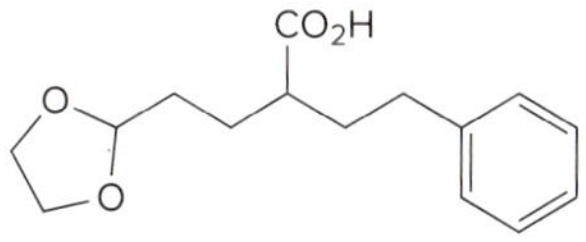

21.46 Devise a synthesis of valproic acid [$(CH_3CH_2CH_2)_2CHCO_2H$], a medicine used to treat epileptic seizures, using the malonic ester synthesis.

21.47 Synthesize **A** from diethyl malonate and any needed organic compounds and inorganic reagents.

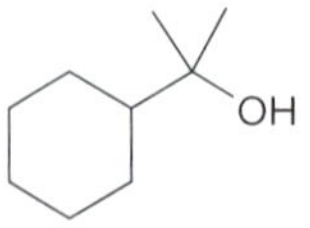

OH

A

21.48 The enolate derived from diethyl malonate reacts with a variety of electrophiles (not just alkyl halides) to form new carbon–carbon bonds. With this in mind, draw the products formed when $Na^+ \ ^-CH(CO_2Et)_2$ reacts with each electrophile, followed by treatment with H_2O.

a.  b. 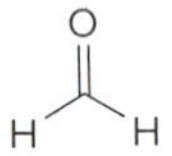 c. 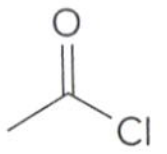 d.

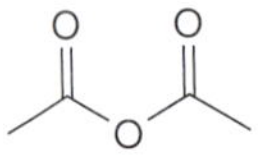

Acetoacetic Ester Synthesis

21.49 What alkyl halides are needed to prepare each ketone using the acetoacetic ester synthesis?

a. 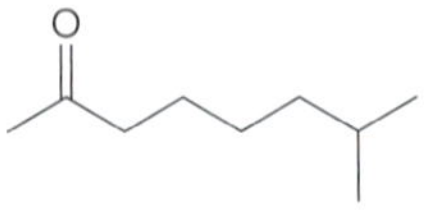 b. 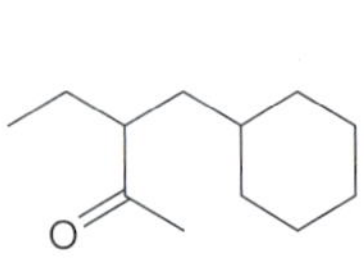 c.

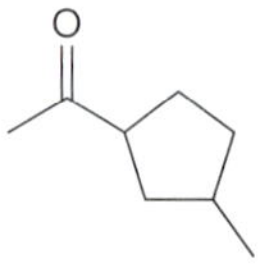

21.50 Synthesize each compound from ethyl acetoacetate. You may use any other organic compounds or inorganic reagents.

a. 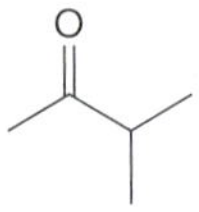 b. 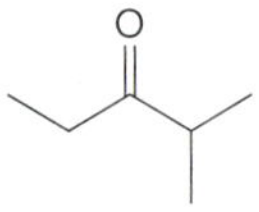 c.

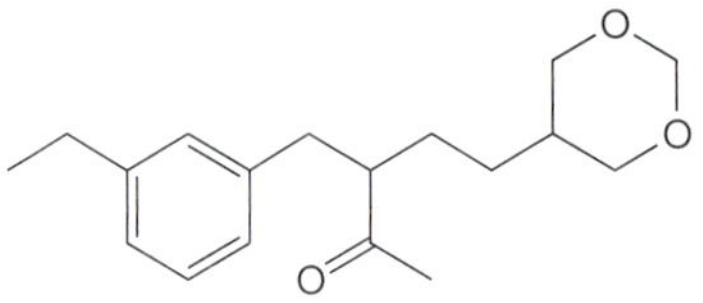

Reactions

21.51 Draw the organic products formed in each reaction.

a. 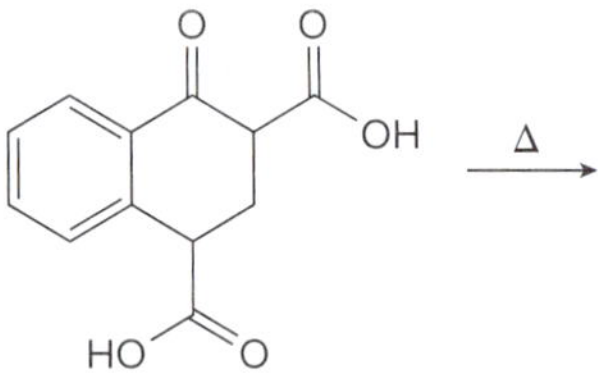 $\xrightarrow{\Delta}$

c. 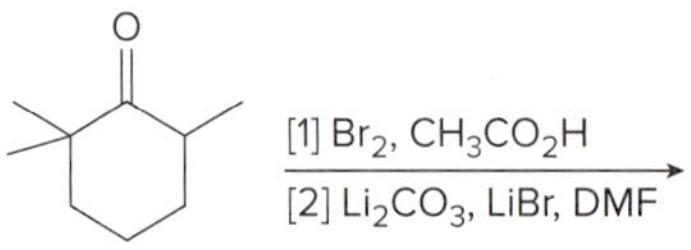 [1] Br_2, CH_3CO_2H [2] Li_2CO_3, LiBr, DMF

e.  Cl⁀⁀⁀CN $\xrightarrow{NaH}$ C_6H_9N

b. 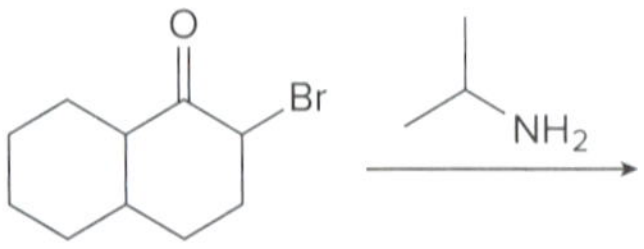 $\xrightarrow{(CH_3)_2CHNH_2}$

d. 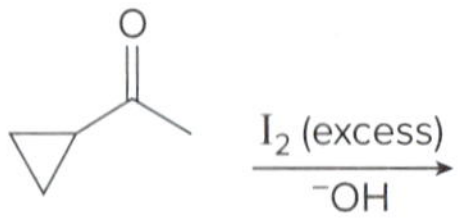 $\xrightarrow[^-OH]{I_2\text{ (excess)}}$

f. 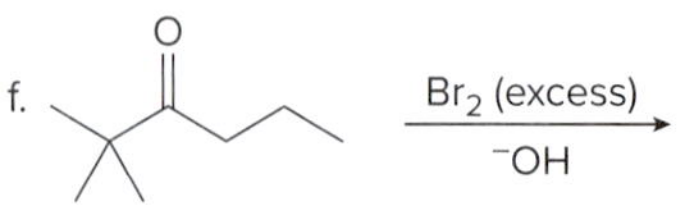 $\xrightarrow[^-OH]{Br_2\text{ (excess)}}$

21.52 Draw the products formed (including stereoisomers) in each reaction.

a. 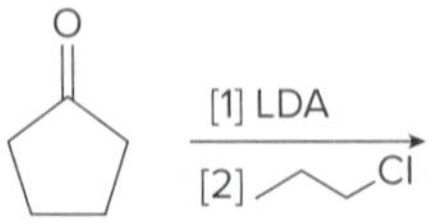 [1] LDA [2] ⁀⁀Cl

b. 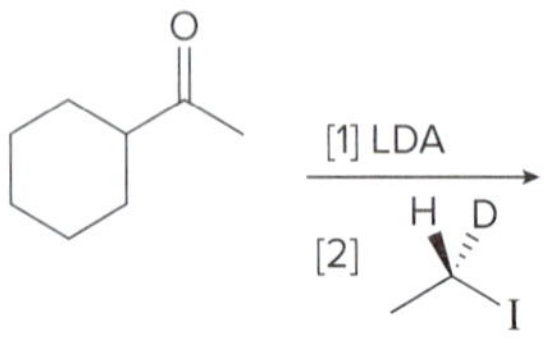 [1] LDA [2] (H, D, I stereocenter)

c. 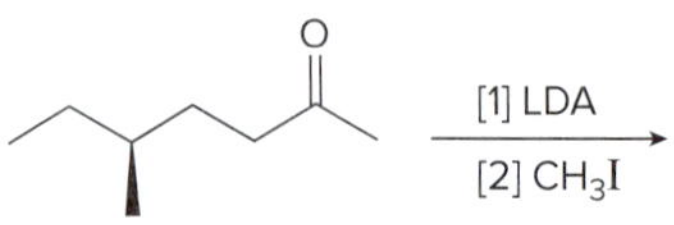 [1] LDA [2] CH_3I

21.53 Identify **C** and **D** in the following reaction scheme, two steps in the synthesis of the cholesterol-lowering drug atorvastatin (trade name Lipitor, Section 29.8).

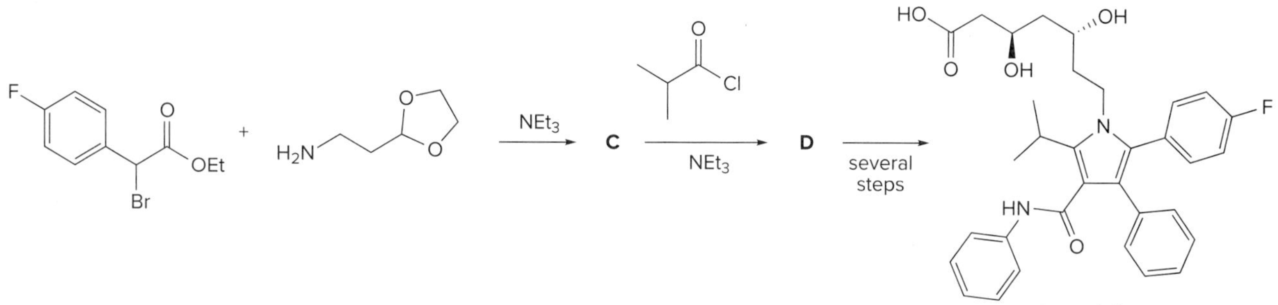

+ H_2N NEt_3 **C** NEt_3 **D** several steps

atorvastatin

21.54 Identify the product in each reaction, and explain why starting materials with identical functional groups give different products.

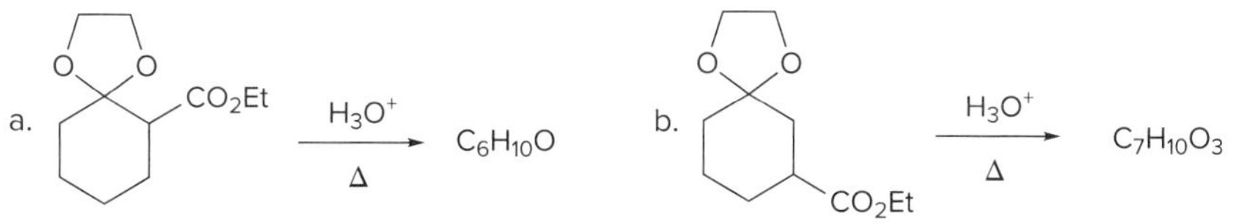

a. CO_2Et H_3O^+, Δ → $C_6H_{10}O$

b. CO_2Et H_3O^+, Δ → $C_7H_{10}O_3$

21.55 Identify the structures of **A, B,** and **C,** intermediates in the synthesis of the chapter-opening molecule clopidogrel.

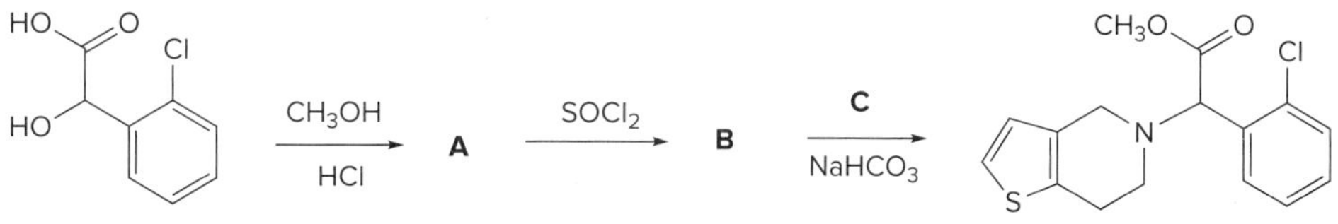

CH_3OH, HCl → **A** $SOCl_2$ → **B** **C**, $NaHCO_3$ →

clopidogrel

21.56 Draw the product formed in each intramolecular reaction. The reaction in part (a) was used in the synthesis of β-pinene, a lipid with a pine-like odor used in fragrances. The reaction in part (b) was used to prepare grandisol, the sex attractant of the cotton boll weevil.

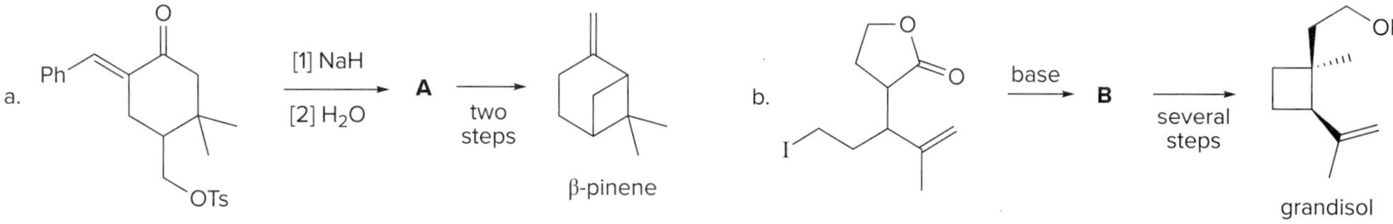

a. [1] NaH [2] H_2O → **A** two steps → β-pinene

b. base → **B** several steps → grandisol

21.57 Direct alkylation of **D** by treatment with one equivalent of LDA and CH_3I does not form ibuprofen. Identify the product of this reaction and explain how it is formed.

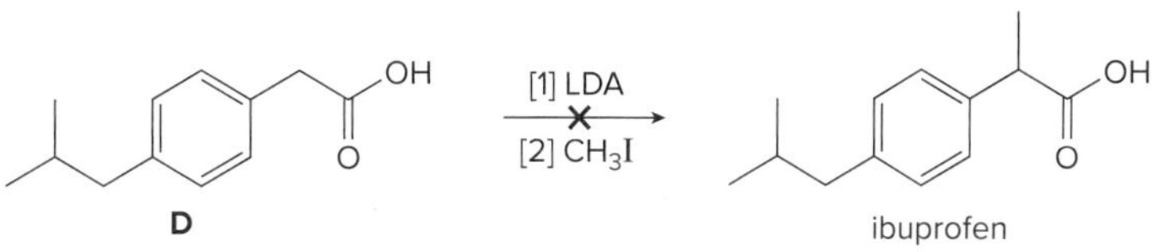

D [1] LDA [2] CH_3I ✗→ ibuprofen

21.58 A key step in the synthesis of the narcotic analgesic meperidine (trade name Demerol) is the conversion of phenylacetonitrile to **X.** (a) What is the structure of **X?** (b) What reactions convert **X** to meperidine?

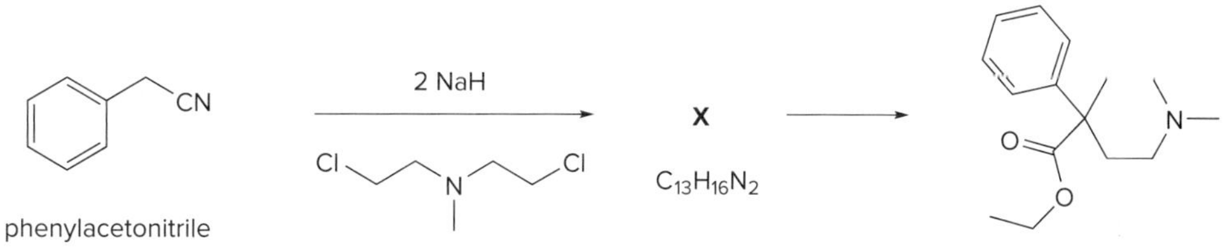

phenylacetonitrile 2 NaH → **X** ($C_{13}H_{16}N_2$) → meperidine

Mechanism

21.59 Although ibuprofen is sold as a racemic mixture, only the *S* enantiomer acts as an analgesic. In the body, however, some of the *R* enantiomer is converted to the *S* isomer by tautomerization to an enol and then protonation to regenerate the carbonyl compound. Write a stepwise mechanism for this isomerization.

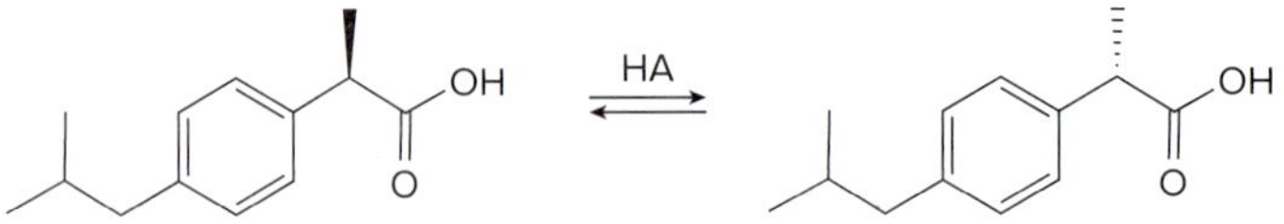

OH HA OH

R isomer
inactive enantiomer

S isomer
active enantiomer

21.60 Draw a stepwise mechanism showing how two alkylation products are formed in the following reaction.

[1] LDA
[2] CH_3I
+ + I^-

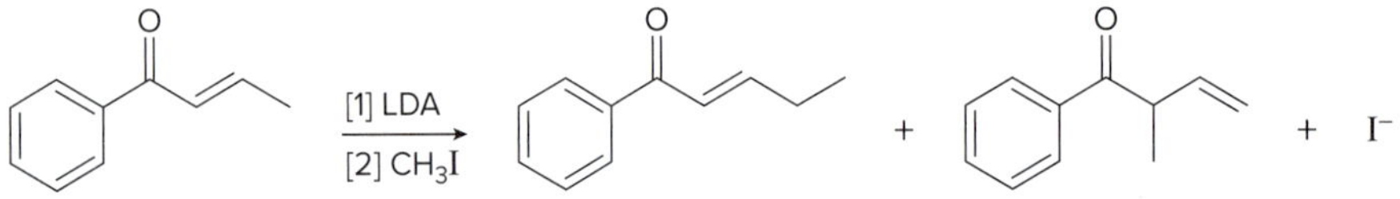

21.61 Draw a stepwise mechanism for the following reaction.

EtO OEt
[1] NaOEt
[2]
[3] H_3O^+, Δ

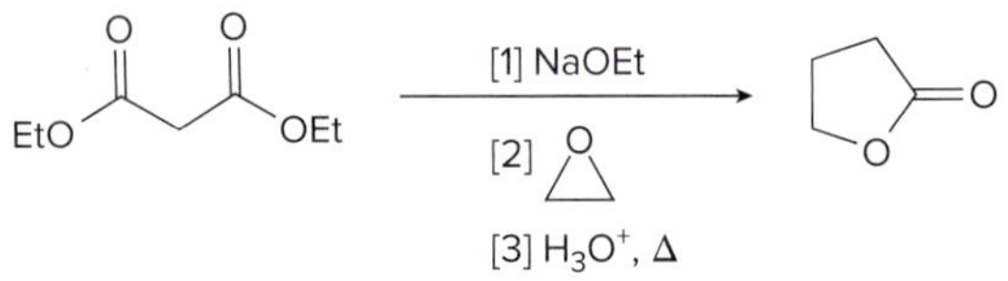

21.62 Draw stepwise mechanisms illustrating how each product is formed.

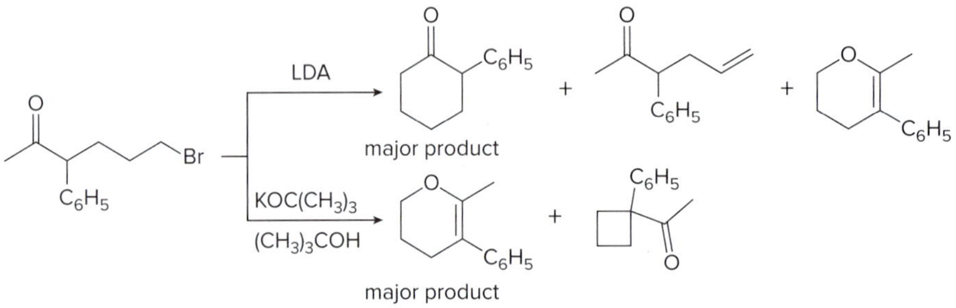

Br
C_6H_5
LDA → C_6H_5 + C_6H_5 + C_6H_5
major product
$KOC(CH_3)_3$, $(CH_3)_3COH$ → C_6H_5 + C_6H_5
major product

Synthesis

21.63 (a) Draw two different halo ketones that can form **A** by an intramolecular alkylation reaction. (b) How can **A** be synthesized by an acetoacetic ester synthesis?

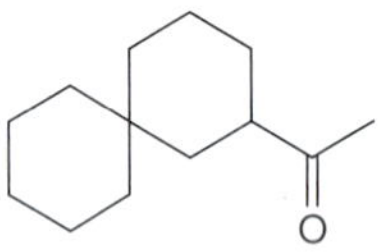

A

21.64 Synthesize each compound from cyclohexanone and organic halides having ≤ 4 C's. You may use any other inorganic reagents.

a. 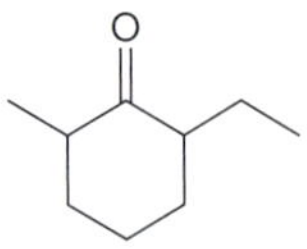 b. 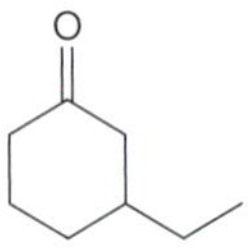 c.

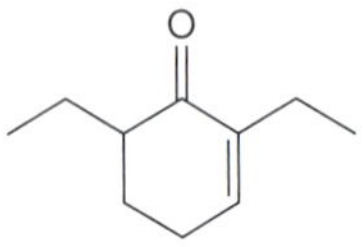

21.65 Bupropion, sold under the trade name of Zyban, is an antidepressant that was approved to aid smoking cessation in 1997. Devise a synthesis of bupropion from benzene, organic compounds that have fewer than five carbons, and any required inorganic reagents.

+ organic compounds with < 5 C's

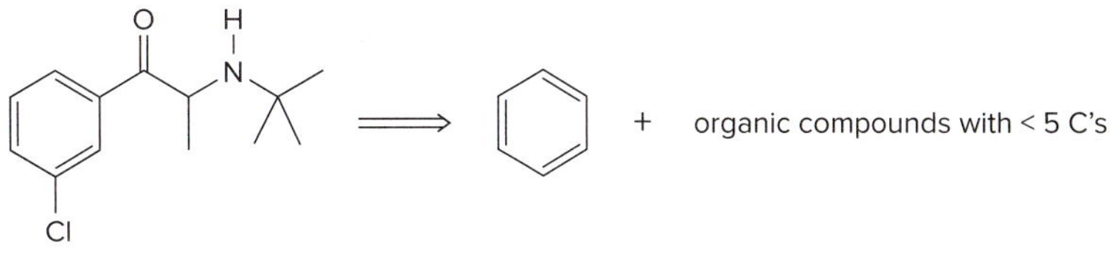

bupropion

21.66 Devise a synthesis of anastrozole, a drug used to reduce the recurrence of breast cancer (Section 19.4), from the given compounds. You may use any other needed organic compounds or inorganic reagents.

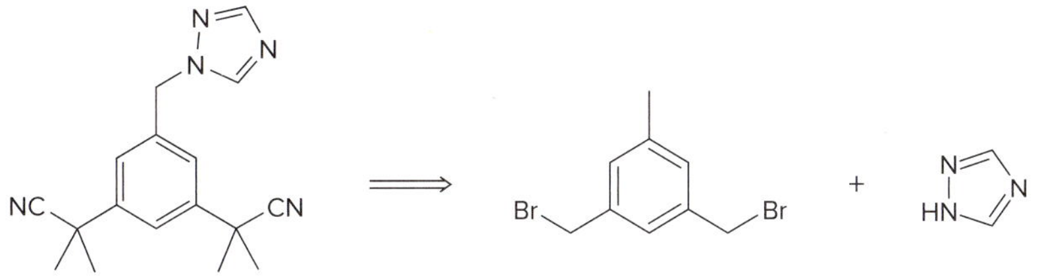

anastrozole

21.67 Synthesize (*Z*)-hept-5-en-2-one from ethyl acetoacetate ($CH_3COCH_2CO_2Et$) and the given starting material. You may also use any other organic compounds or required inorganic reagents.

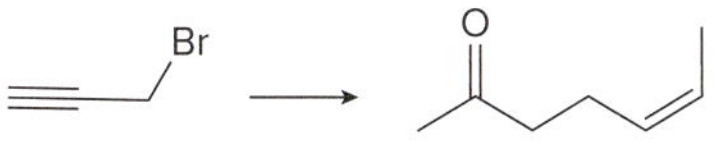

(*Z*)-hept-5-en-2-one

21.68 Oleocanthal is a naturally occurring antioxidant isolated from olive oil. A published synthesis involved the conversion of ketone **X** to ester **Y,** which was converted in two steps to oleocanthal. Devise a synthesis of **Y** from **X,** phenol (C_6H_5OH), organic compounds with four or fewer carbons, and any needed organic and inorganic reagents.

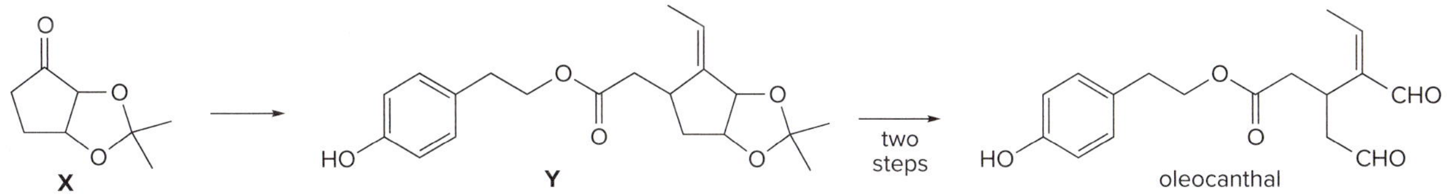

X → Y → two steps → oleocanthal

Spectroscopy

21.69 Treatment of **W** with CH_3Li, followed by CH_3I, affords compound **Y** ($C_7H_{14}O$) as the major product. **Y** shows a strong absorption in its IR spectrum at 1713 cm^{-1}, and its ^1H NMR spectrum is given below. (a) Propose a structure for **Y.** (b) Draw a stepwise mechanism for the conversion of **W** to **Y.**

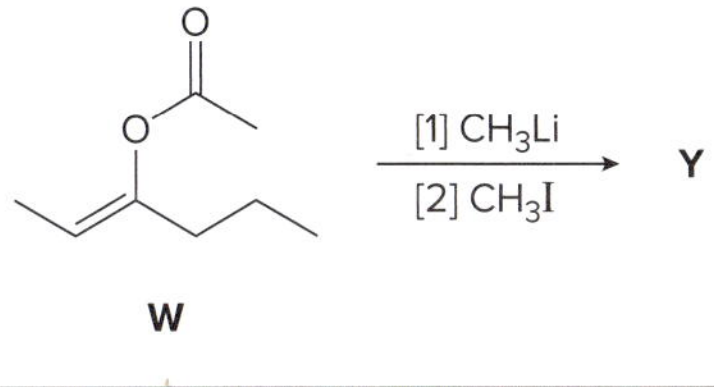

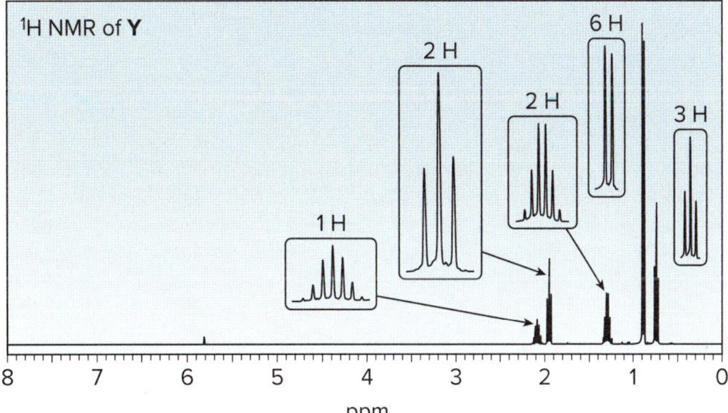

Challenge Problems

21.70 Explain why H_a is much less acidic than H_b. Then draw a mechanism for the following reaction.

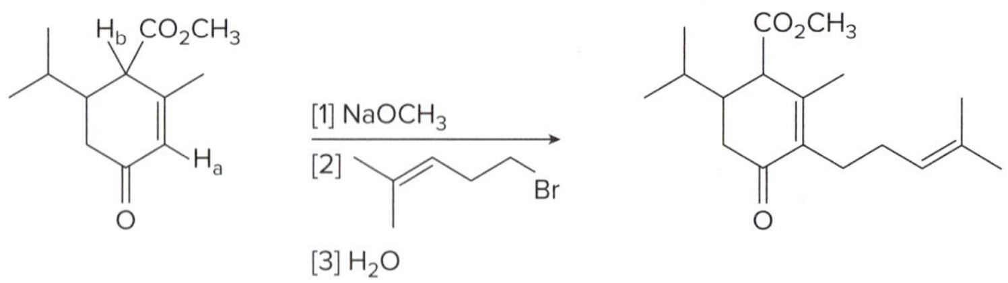
H$_b$ CO$_2$CH$_3$
H$_a$
O
[1] NaOCH$_3$
[2]
Br
[3] H$_2$O
CO$_2$CH$_3$
O

21.71 Devise a stepwise mechanism for the following reaction.

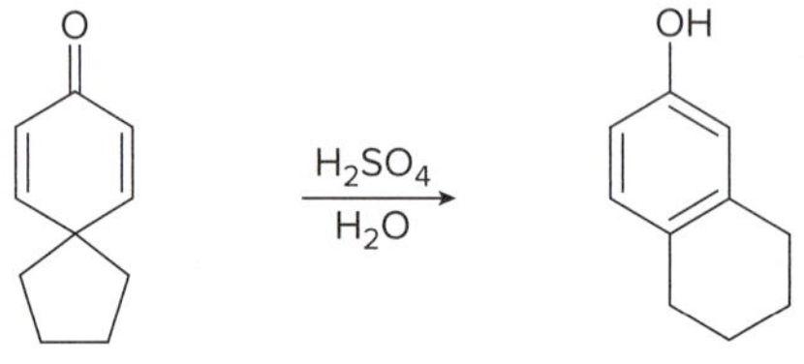
O
H$_2$SO$_4$
H$_2$O
OH

21.72 The last step in the synthesis of β-vetivone, a major constituent of vetiver, a perennial grass found in tropical and subtropical regions of the world, involves treatment of **C** with CH_3Li to form an intermediate **X,** which forms β-vetivone with aqueous acid. Identify the structure of **X** and draw a mechanism for converting **X** to β-vetivone.

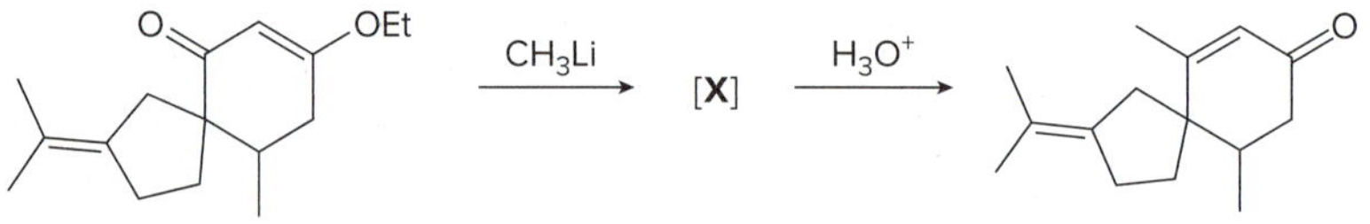
O
OEt
CH$_3$Li
[X]
H$_3$O$^+$
O

C β-vetivone

21.73 Keeping in mind the mechanism for the dissolving metal reduction of alkynes to trans alkenes in Chapter 12, write a stepwise mechanism for the following reaction, which involves the conversion of an α,β-unsaturated carbonyl compound to a carbonyl compound with a new alkyl group on the α carbon.

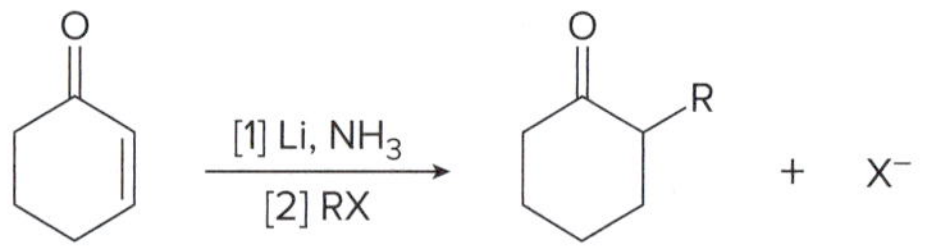
O
[1] Li, NH$_3$
[2] RX
O
R
+ X$^-$

21.74 Compounds **A** and **B** undergo an intramolecular reaction with base to form isomeric products **C** and **D,** respectively. Both **C** and **D** exhibit a strong IR absorption at < 1700 cm^{-1}. The ^1H NMR spectrum of **C** shows two nonequivalent alkenyl protons, and the ^1H NMR spectrum of **D** shows three nonequivalent alkenyl protons. Draw the structures of **C** and **D,** and use curved arrows to illustrate how **D** is formed from **B.**

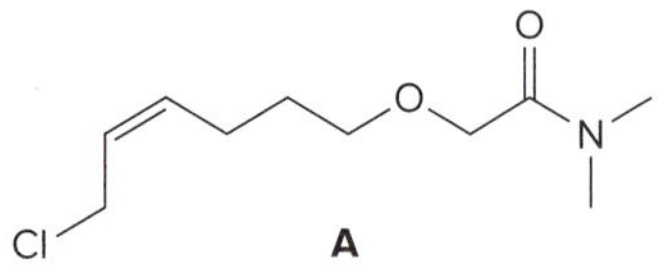
O
O
N
Cl

A

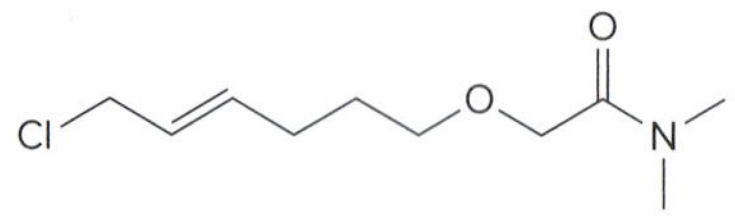
O
Cl
O
N

B

SELF-TEST ANSWERS

1. c 2. b 3. a 4. c 5. d 6. d 7. c 8. a 9. c 10. c

Carbonyl Condensation Reactions

22

Corbis/SuperStock

22.1 The aldol reaction
22.2 Crossed aldol reactions
22.3 Directed aldol reactions
22.4 Intramolecular aldol reactions
22.5 The Claisen reaction
22.6 The crossed Claisen and related reactions
22.7 The Dieckmann reaction
22.8 The Michael reaction
22.9 The Robinson annulation

ar-Turmerone is isolated from turmeric, a tropical flowering perennial in the ginger family grown primarily in Southeast Asia and India. The dried and ground root of the turmeric plant is an essential ingredient in curry. *ar*-Turmerone is an α,β-unsaturated carbonyl compound that can be prepared by a directed aldol reaction between two different carbonyl compounds. In Chapter 22, we learn about carbon–carbon bond-forming reactions between the α carbon of one carbonyl compound and the carbonyl group of another.

Why Study . . .

Carbonyl Condensation Reactions?

In Chapter 22, we examine **carbonyl condensations**—that is, reactions between two carbonyl compounds—a second type of reaction that occurs at the α carbon of a carbonyl group. Much of what is presented in Chapter 22 applies principles you have already learned. Many of the reactions may look more complicated than those in previous chapters, but they are fundamentally the same. Nucleophiles attack electrophilic carbonyl groups to form the products of nucleophilic addition or substitution, depending on the structure of the carbonyl starting material.

The reactions in Chapter 22 form new carbon–carbon bonds at the α carbon to a carbonyl group, so these reactions are extremely useful in the synthesis of complex natural products.

22.1 The Aldol Reaction

Chapter 22 concentrates on the second general reaction of enolates—**reaction with other carbonyl compounds.** In these reactions, one carbonyl component serves as the nucleophile and one serves as the electrophile, and a new carbon–carbon bond is formed.

enolate | second carbonyl component | new C–C bond in red

nucleophile | **electrophile**

The presence or absence of a leaving group on the electrophilic carbonyl carbon determines the structure of the product. Even though they appear somewhat more complicated, these reactions are often reminiscent of the nucleophilic addition and nucleophilic acyl substitution reactions of Chapters 18 and 20. Four types of reactions are examined:

- **Aldol reaction** (Sections 22.1–22.4)
- **Claisen reaction** (Sections 22.5–22.7)
- **Michael reaction** (Section 22.8)
- **Robinson annulation** (Section 22.9)

22.1A General Features of the Aldol Reaction

In the **aldol reaction,** two molecules of an aldehyde or ketone react with each other in the presence of base to form a **β-hydroxy carbonyl compound.** For example, treatment of acetaldehyde with aqueous ^{-}OH forms 3-hydroxybutanal, a **β-hydroxy aldehyde.**

Many aldol products contain an ***ald***ehyde and an alcoh***ol***—hence the name ***aldol.***

2 acetaldehyde $\xrightarrow{^{-}OH,\ H_2O}$ (**aldol reaction**) 3-hydroxybutanal — **β-hydroxy carbonyl compound**

new C–C bond in red

The mechanism of the aldol reaction has **three steps,** as shown in Mechanism 22.1. Carbon–carbon bond formation occurs in Step [2], when the nucleophilic enolate reacts with the electrophilic carbonyl carbon.

The aldol reaction is a reversible equilibrium, so the position of the equilibrium depends on the base and the carbonyl compound. **$^{-}$OH is the base** typically used in an aldol reaction. Recall from Section 21.3B that only a small amount of enolate forms with $^{-}$OH. In this case, that's appropriate because the starting aldehyde is needed to react with the enolate in the second step of the mechanism.

Mechanism 22.1 The Aldol Reaction

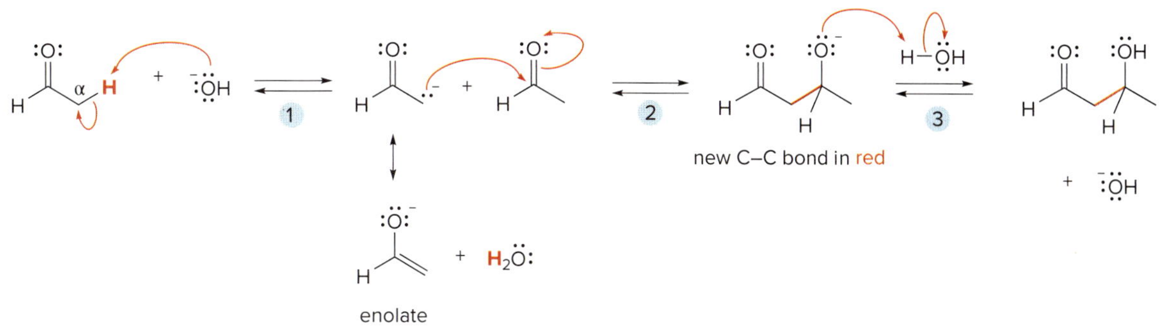

1 The base removes a proton on the α carbon to form a **resonance-stabilized enolate.**

2 **Nucleophilic attack** of the enolate on an electrophilic carbonyl in another molecule of aldehyde forms a new C–C bond.

3 Protonation of the alkoxide forms the **β-hydroxy aldehyde.**

Aldol reactions can be carried out with either aldehydes or ketones. With aldehydes, the equilibrium usually favors the products, but with ketones the equilibrium favors the starting materials. There are ways of driving this equilibrium to the right, however, so we will write aldol products whether the substrate is an aldehyde or a ketone.

- **The characteristic reaction of aldehydes and ketones is *nucleophilic addition* (Section 18.7). An aldol reaction is a nucleophilic addition in which an enolate is the nucleophile. See the comparison in Figure 22.1.**

Figure 22.1 The aldol reaction—An example of nucleophilic addition

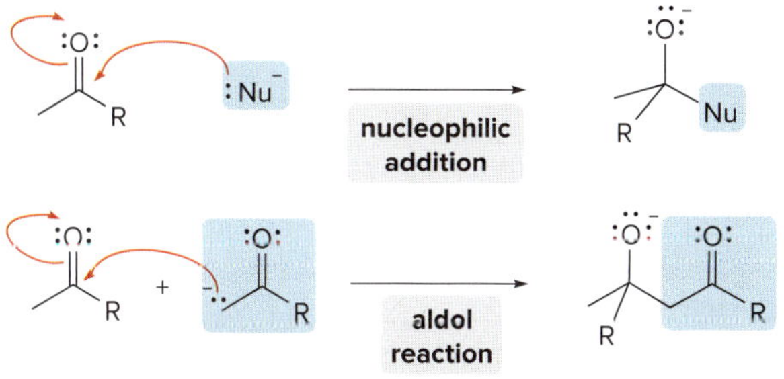

- Aldehydes and ketones react by nucleophilic addition. In an aldol reaction, **an enolate is the nucleophile** that adds to the carbonyl group.

A **second example of an aldol** reaction is shown with propanal as starting material. The two molecules of the aldehyde that participate in the aldol reaction react in opposite ways:

- **One molecule of propanal becomes an enolate—an electron-rich *nucleophile.***
- **One molecule of propanal serves as the *electrophile* because its carbonyl carbon is electron deficient.**

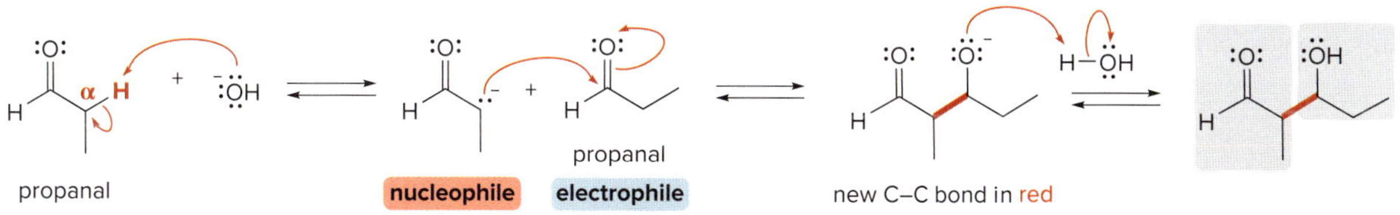

These two examples illustrate the general features of the aldol reaction. **The α carbon of one carbonyl component becomes bonded to the carbonyl carbon of the other component.**

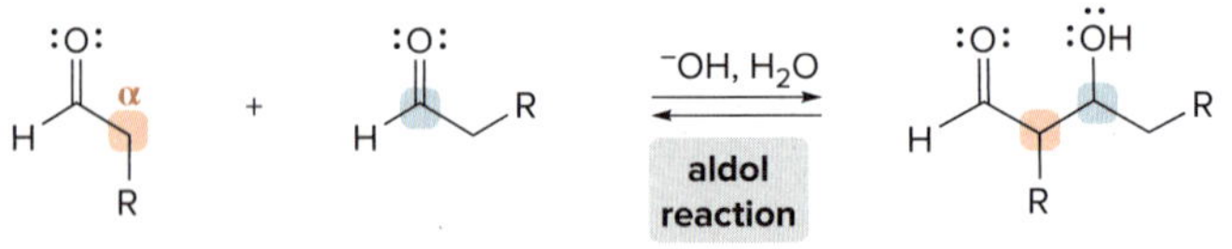

:O: α H R + :O: H R ⇌ (^-OH, H_2O; **aldol reaction**) :O: :OH H R R

Problem 22.1 Draw the aldol product formed from each compound.

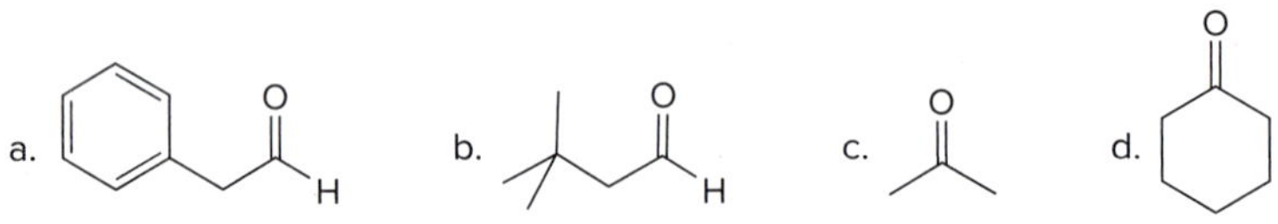

a. b. c. d.

Problem 22.2 Which carbonyl compounds do *not* undergo an aldol reaction when treated with ^-OH in H_2O?

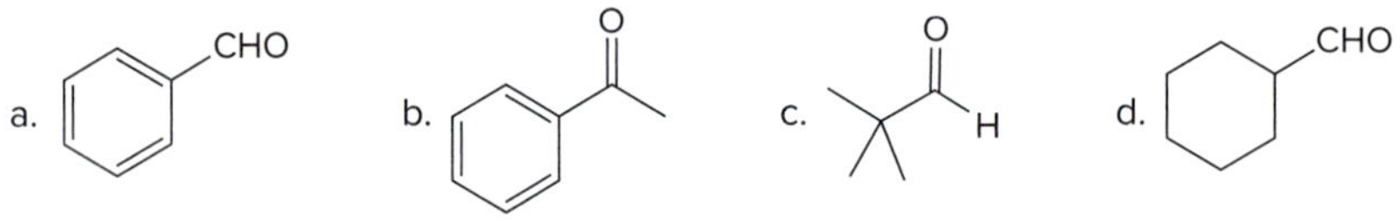

a. CHO b. c. H d. CHO

22.1B Retro-Aldol Reaction

Because an aldol reaction is a reversible equilibrium, the β-hydroxy carbonyl products can be re-converted to carbonyl starting materials with heat in the presence of base by a **retro-aldol reaction.** The conversion of 3-hydroxybutanal to acetaldehyde is a retro-aldol reaction, which results in cleavage of the carbon–carbon bond between the α and β carbons.

3-hydroxybutanal ⇌ (^-OH, H_2O; retro-aldol reaction) 2 acetaldehyde

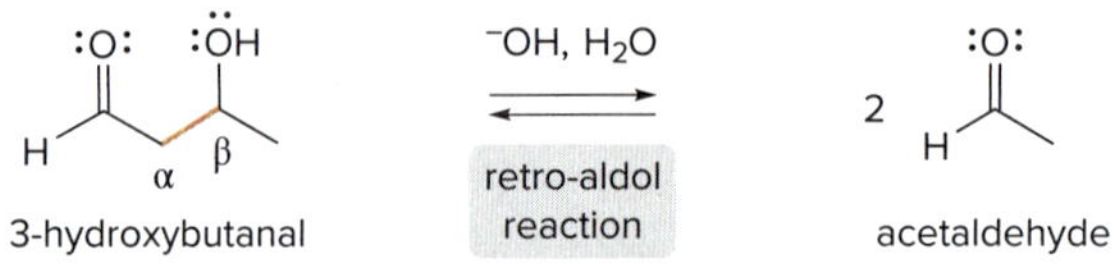

The three-step mechanism of a retro-aldol reaction is just the reverse of an aldol reaction, as shown in Mechanism 22.2.

Mechanism 22.2 The Retro-Aldol Reaction

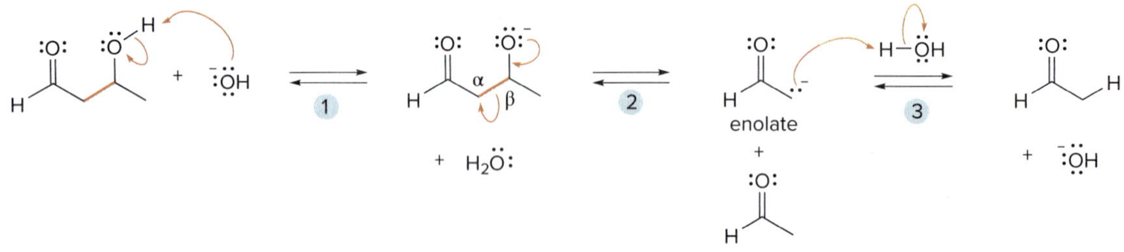

[1] → + $H_2\ddot{O}$: → [2] → enolate + aldehyde → [3] ($H-\ddot{O}H$) → + $^-:\ddot{O}H$

1 The base removes the OH proton to form an alkoxide.

2 An electron pair of the alkoxide is used to form a C=O and the **carbon–carbon bond between the α and β carbons is cleaved.** This process forms an **enolate** and a **molecule of aldehyde.**

3 Protonation of the enolate forms another molecule of aldehyde.

Problem 22.3 What ketone or aldehyde is obtained when each compound is heated in the presence of aqueous base?

a. b.

Problem 22.4 A key step in the metabolism of glucose (Section 30.4) is the retro-aldol reaction of fructose 1,6-bisphosphate, which is catalyzed by an aldolase enzyme. Draw the products of this reaction, including the stereochemistry at any stereogenic centers.

fructose 1,6-bisphosphate

22.1C Dehydration of the Aldol Product

The β-hydroxy carbonyl compounds formed in the aldol reaction dehydrate more readily than other alcohols. In fact, under the basic reaction conditions, the initial aldol product is often not isolated. Instead, **it loses the elements of H_2O from the α and β carbons to form an α,β-unsaturated carbonyl compound, a conjugated product.**

All alcohols—including β-hydroxy carbonyl compounds—dehydrate in the presence of *acid.* **Only β-hydroxy carbonyl compounds dehydrate in the presence of base.**

An aldol reaction is often called an **aldol condensation,** because the β-hydroxy carbonyl compound that is initially formed loses H_2O by dehydration. **A *condensation reaction* is one in which a small molecule, in this case H_2O, is eliminated during a reaction.**

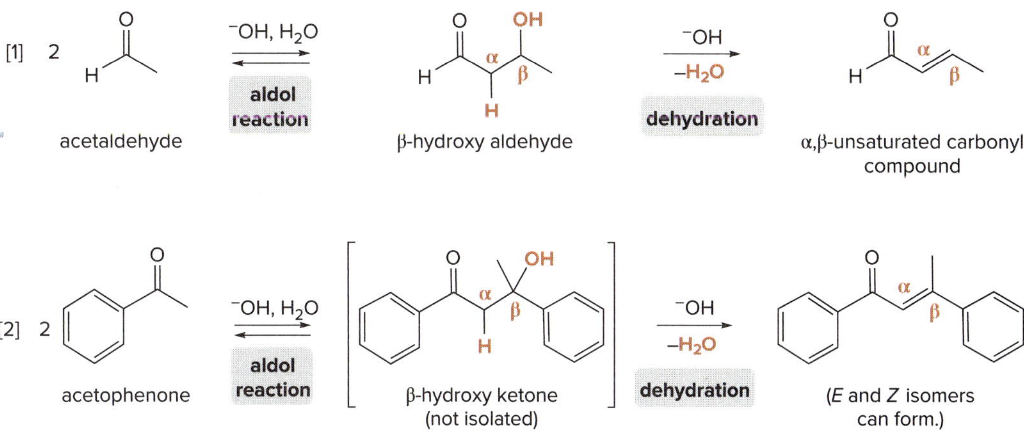

It may or may not be possible to isolate the β-hydroxy carbonyl compound under the conditions of the aldol reaction.

- **When the α,β-unsaturated carbonyl compound is *also conjugated* with a carbon–carbon double bond or a benzene ring, as in the case of Reaction [2], elimination of H_2O is spontaneous and the β-hydroxy carbonyl compound cannot be isolated.**

The mechanism of dehydration consists of two steps: **deprotonation followed by loss of $^{-}$OH,** as shown in Mechanism 22.3.

Mechanism 22.3 Dehydration of β-Hydroxy Carbonyl Compounds with Base

resonance-stabilized enolate

1 The base removes a proton on the α carbon to form a **resonance-stabilized enolate.**

2 $^{-}$OH is eliminated as the electron pair of the enolate forms the **new π bond.**

Like E1 elimination, E1cB requires **two steps.** Unlike E1, though, the intermediate in E1cB is a *carbanion,* not a carbocation. E1cB stands for **Elimination, unimolecular, conjugate base.**

This elimination mechanism, called the **E1cB mechanism,** differs from the two more general mechanisms of elimination, E1 and E2, which were discussed in Chapter 8. The E1cB mechanism involves two steps and proceeds by way of an **anionic** intermediate.

Regular alcohols dehydrate only in the presence of acid but not base, because hydroxide is a poor leaving group. When the hydroxy group is β to a carbonyl group, however, loss of H and OH from the α and β carbons forms a **conjugated double bond,** and the stability of the conjugated system makes up for having such a poor leaving group.

Dehydration of the initial β-hydroxy carbonyl compound drives the equilibrium of an aldol reaction to the right, thus favoring product formation. Once the conjugated α,β-unsaturated carbonyl compound forms, it is *not* re-converted to the β-hydroxy carbonyl compound.

The E1cB mechanism is especially common in biological pathways. For example, the dehydration of β-hydroxy thioesters to α,β-unsaturated thioesters, a process that occurs during the biosynthesis of fatty acids, follows an E1cB mechanism.

β-hydroxy thioester —E1cB mechanism→ α,β-unsaturated thioester

Problem 22.5 What unsaturated carbonyl compound is formed by dehydration of each β-hydroxy carbonyl compound?

a. b. c.

22.1D Retrosynthetic Analysis

To utilize the aldol reaction in synthesis, you must be able to determine which aldehyde or ketone is needed to prepare a particular β-hydroxy carbonyl compound or α,β-unsaturated carbonyl compound—that is, you must be able to **work backwards, in the retrosynthetic direction.**

How To Synthesize a Compound Using the Aldol Reaction

Example What starting material is needed to prepare each compound by an aldol reaction?

a. b.

Step [1] **Locate the α and β carbons of the carbonyl group.**

- When a carbonyl group has two different α carbons, **choose the side that contains the OH group** (in a β-hydroxy carbonyl compound) **or is part of the C=C** (in an α,β-unsaturated carbonyl compound).

Step [2] **Break the molecule into two components between the α and β carbons.**

- The α carbon and all remaining atoms bonded to it belong to one carbonyl component. The β carbon and all remaining atoms bonded to it belong to the other carbonyl component. Both components are identical in all aldols we have thus far examined.

a. Break the molecule into two halves at the labeled bond.

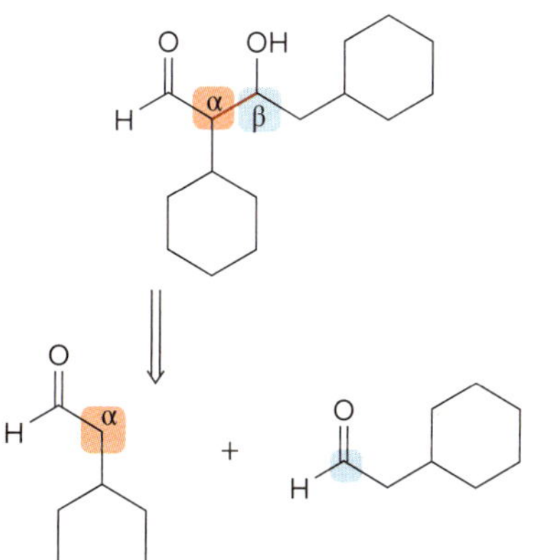

two molecules of the same aldehyde

b. Break the molecule into two halves at the labeled bond.

two molecules of cyclohexanone

Problem 22.6 What aldehyde or ketone is needed to prepare each compound by an aldol reaction?

a. b. c.

22.2 Crossed Aldol Reactions

In all of the aldol reactions discussed so far, the electrophilic carbonyl and the nucleophilic enolate have originated from the *same* aldehyde or ketone. Sometimes, though, it is possible to carry out an aldol reaction between two *different* carbonyl compounds.

- **An aldol reaction between two different carbonyl compounds is called a *crossed aldol* or *mixed aldol reaction.***

22.2A A Crossed Aldol Reaction with Two Different Aldehydes, Both Having α H Atoms

When two different aldehydes, both having α H atoms, are combined in an aldol reaction, *four* different β-hydroxy carbonyl compounds are formed. Four products form, not one, because *both* aldehydes can lose an acidic α hydrogen atom and form an enolate in the presence of base. *Both* enolates can then react with *both* carbonyl compounds, as shown for acetaldehyde and propanal in the following reaction scheme.

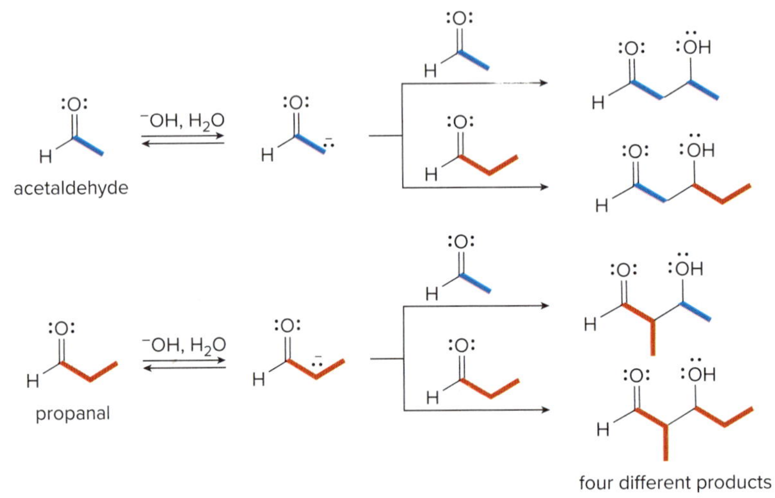

acetaldehyde —^{-}OH, H_2O→ enolate

propanal —^{-}OH, H_2O→ enolate

four different products

- Conclusion: When two different aldehydes have α hydrogens, a crossed aldol reaction is *not* synthetically useful.

22.2B Synthetically Useful Crossed Aldol Reactions

Crossed aldols are synthetically useful in two different situations.

- A crossed aldol occurs when only *one* carbonyl component has α H atoms.

When one carbonyl compound has no α hydrogens, a crossed aldol reaction often leads to one product. Two common carbonyl compounds with no α hydrogens used for this purpose are **formaldehyde ($CH_2{=}O$)** and **benzaldehyde (C_6H_5CHO).**

For example, reaction of C_6H_5CHO (as the electrophile) with acetaldehyde (CH_3CHO) in the presence of base forms a single α,β-unsaturated carbonyl compound after dehydration.

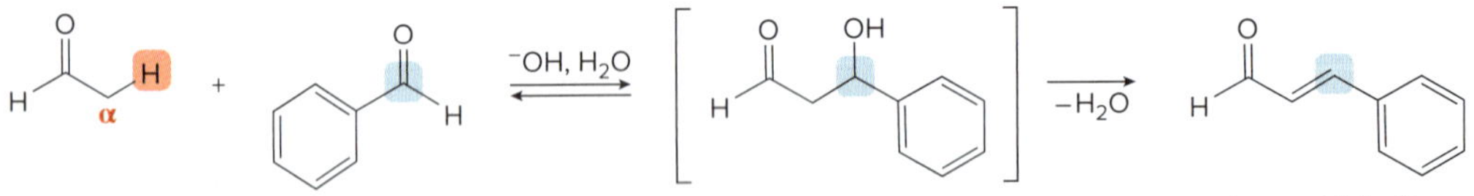

^{-}OH, H_2O ; $-H_2O$

cinnamaldehyde
(component of cinnamon)
(+ *Z* isomer)

The yield of a single crossed aldol product is increased further if the electrophilic carbonyl component is relatively unhindered (as is the case with most aldehydes) and if it is used in excess.

Problem 22.7 2-Pentylcinnamaldehyde, commonly called flosal, is a perfume ingredient with a jasmine-like odor. Flosal is an α,β-unsaturated aldehyde made by a crossed aldol reaction between benzaldehyde (C_6H_5CHO) and heptanal ($CH_3CH_2CH_2CH_2CH_2CH_2CHO$), followed by dehydration. Draw a stepwise mechanism for the following reaction that prepares flosal.

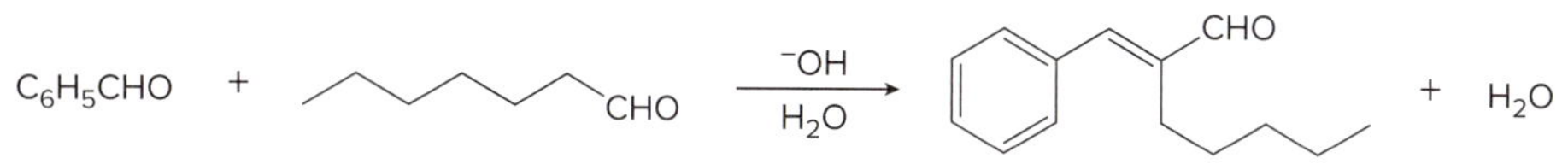

C_6H_5CHO + [heptanal] CHO $\xrightarrow[H_2O]{^-OH}$ [flosal] CHO + H_2O

flosal
(perfume component)

Problem 22.8 Draw the products formed in each crossed aldol reaction.

a. 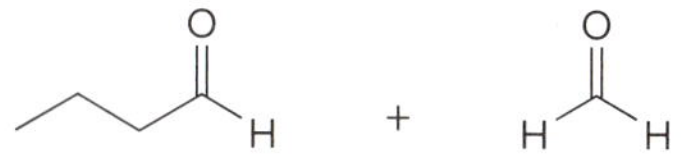 [butanal] + [formaldehyde]

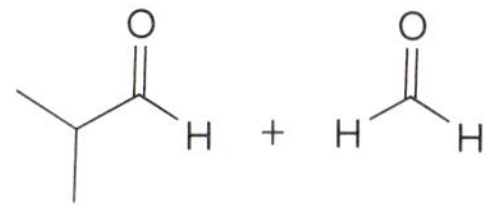

b. 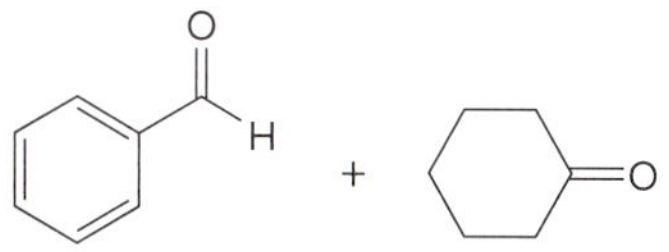 [benzaldehyde] + [cyclohexanone]

c. [2-methylpropanal] + [formaldehyde]

d. 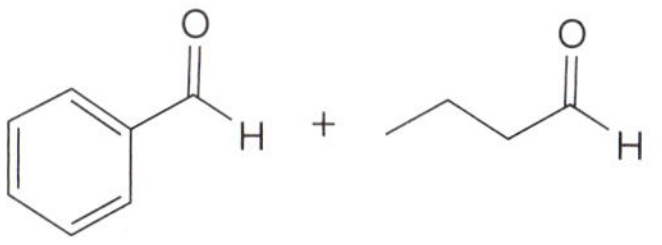 [benzaldehyde] + [butanal]

- **A crossed aldol occurs when one carbonyl component has especially acidic α H atoms.**

A useful crossed aldol reaction takes place between an aldehyde or ketone and a β-dicarbonyl (or similar) compound.

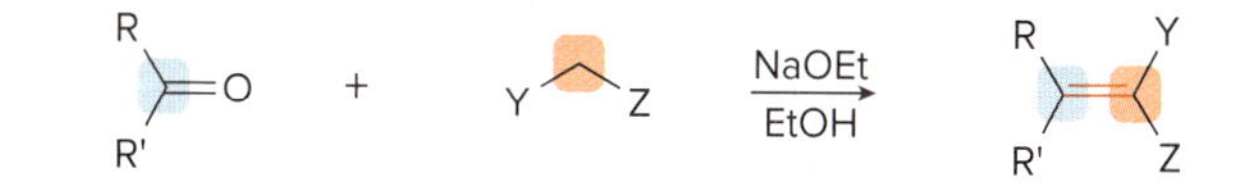

R(R')C=O + Y–CH₂–Z $\xrightarrow[EtOH]{NaOEt}$ R(R')C=C(Y)(Z)

R' = H or alkyl

Y, Z = COOEt, CHO, COR, CN
β-dicarbonyl compound
(and related compounds)

new C–C σ and π bonds in red

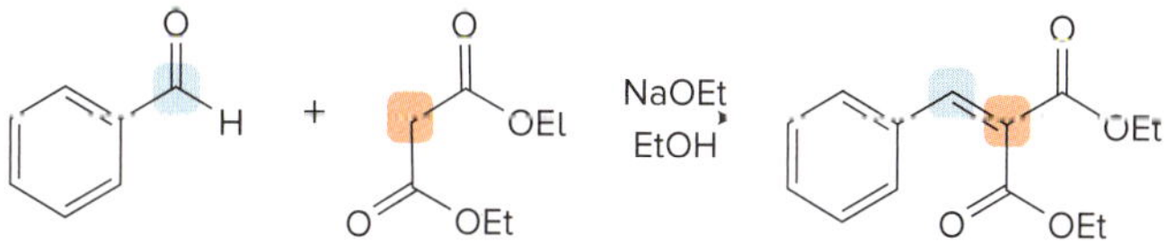

benzaldehyde + diethyl malonate $\xrightarrow[EtOH]{NaOEt}$ [product]

As we learned in Section 21.3, the α hydrogens between two carbonyl groups are especially acidic, so they are more readily removed than other α H atoms. As a result, **the β-dicarbonyl compound always becomes the enolate component of the aldol reaction.** Figure 22.2 shows the steps for the crossed aldol reaction between diethyl malonate and benzaldehyde. In this type of crossed aldol reaction, the initial β-hydroxy carbonyl compound *always* loses water to form the highly conjugated product.

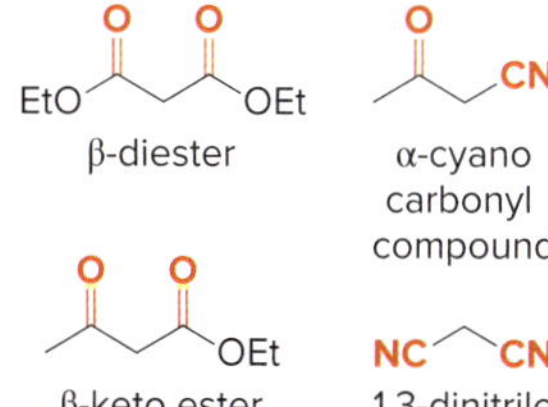

β-Dicarbonyl compounds are sometimes called **active methylene compounds** because they are more reactive toward base than other carbonyl compounds. **1,3-Dinitriles** and **α-cyano carbonyl compounds** are also active methylene compounds.

Figure 22.2 Crossed aldol reaction between benzaldehyde and $CH_2(COOEt)_2$

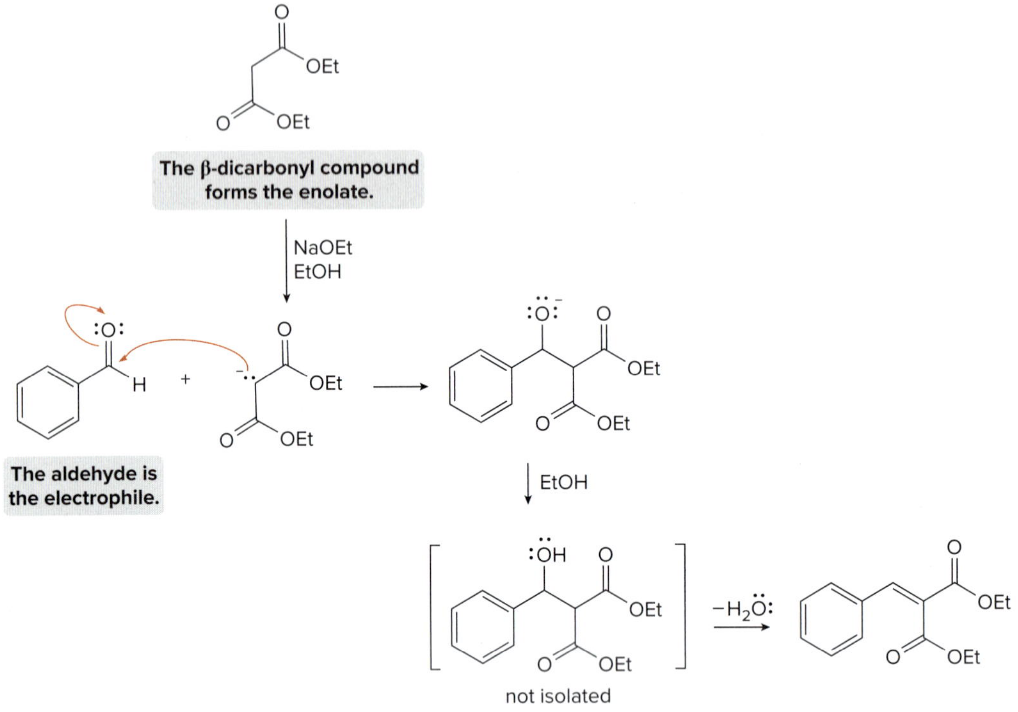

Problem 22.9 Draw the products formed in the crossed aldol reaction of phenylacetaldehyde ($C_6H_5CH_2CHO$) with each compound: (a) $CH_2(COOEt)_2$; (b) $CH_2(COCH_3)_2$; (c) CH_3COCH_2CN.

Problem 22.10 Draw the product, a potent irritant (referred to as CS) present in pepper balls, formed in the following reaction. The use of CS as a riot-control agent has come under scrutiny, in part because the high temperatures required in its dispersal result in decomposition to a variety of unknown, and potentially highly toxic, products.

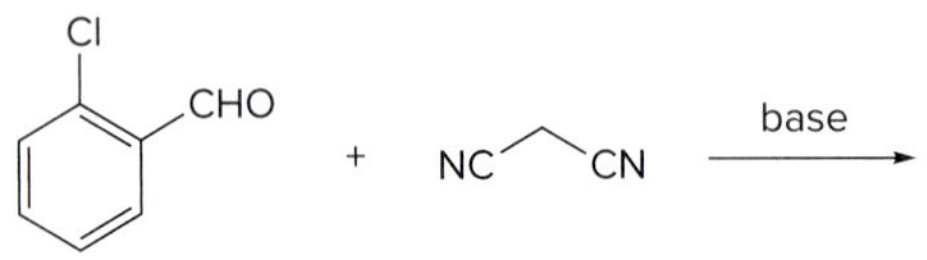

22.2C Enantioselective Aldol Reactions

David MacMillan and Benjamin List were awarded the 2021 Nobel Prize in Chemistry for their pioneering work using chiral organic molecules to catalyze asymmetric organic reactions.

Like the Sharpless epoxidation (Section 12.15) and the enantioselective ketone reduction with CBS reagents (Section 17.6A), so, too, enantioselective aldol reactions can be carried out. In this case, a small chiral molecule is used as an **organocatalyst** to selectively form one enantiomer of a β-hydroxy carbonyl product. For example, reaction of acetone with 2-methylpropanal in the presence of the amino acid proline forms the aldol product in 96% enantiomeric excess.

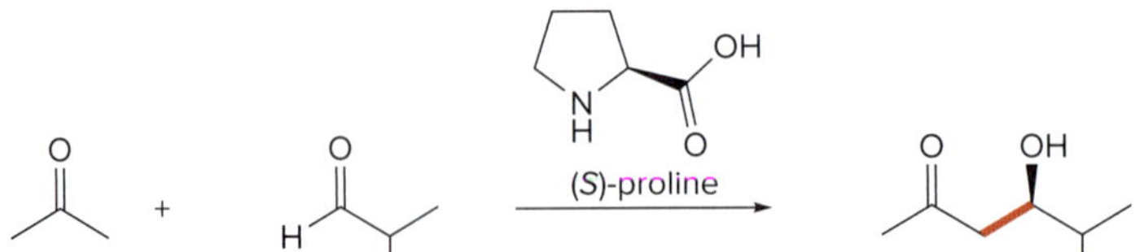

acetone 2-methylpropanal 97% yield 96% *ee*

This type of reaction, an example of **asymmetric organocatalysis,** is considered more environmentally friendly than methods that employ toxic metal catalysts, so it is widely used in the synthesis of pharmaceuticals.

Problem 22.11 Draw the aldol product (including stereochemistry) formed when acetone reacts with each aldehyde in the presence of proline. The *R* isomer is the major enantiomer in each reaction.

a. 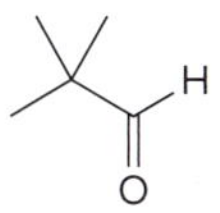 b. 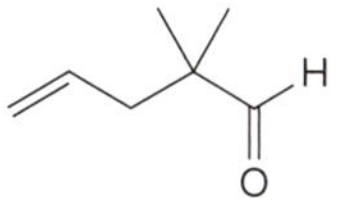 c.

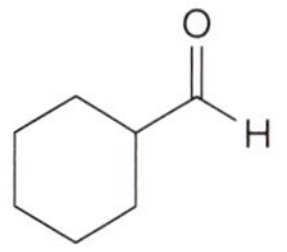

22.3 Directed Aldol Reactions

A **directed aldol reaction** is a variation of the crossed aldol reaction that clearly defines which carbonyl compound becomes the nucleophilic enolate and which reacts at the electrophilic carbonyl carbon. The strategy of a directed aldol reaction is as follows:

[1] Prepare the enolate of one carbonyl component with LDA.

[2] Add the second carbonyl compound (the electrophile) to this enolate.

Because the steps are done sequentially and a strong nonnucleophilic base is used to form the enolate of only one carbonyl component, a variety of carbonyl substrates can be used in the reaction. Both carbonyl components can have α hydrogens because only one enolate is prepared with LDA. Also, when an unsymmetrical ketone is used, LDA selectively forms the **less substituted, kinetic enolate.**

Sample Problem 22.1 illustrates the steps of a directed aldol reaction between a ketone and an aldehyde, both of which have α hydrogens.

Sample Problem 22.1 Determining the Product of a Directed Aldol Reaction

Draw the product of the following directed aldol reaction.

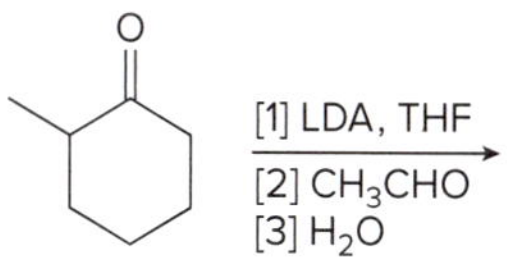

2-methylcyclohexanone → [1] LDA, THF; [2] CH_3CHO; [3] H_2O

Solution

2-Methylcyclohexanone forms an enolate on the less substituted carbon, which then reacts with the electrophile, CH_3CHO.

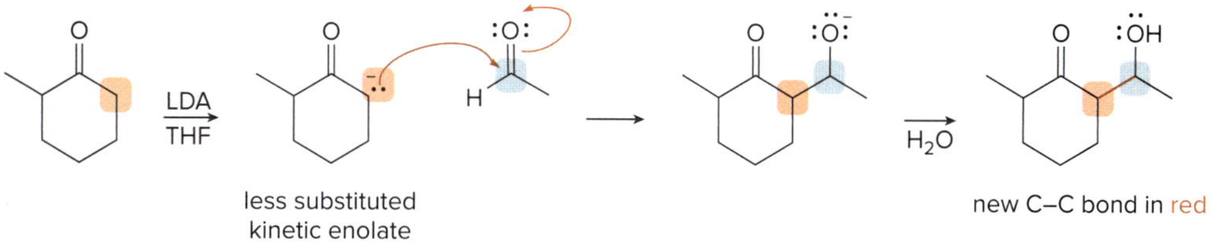

LDA, THF → less substituted kinetic enolate → → H_2O → new C–C bond in red

Problem 22.12 Draw the product of each directed aldol reaction.

a. 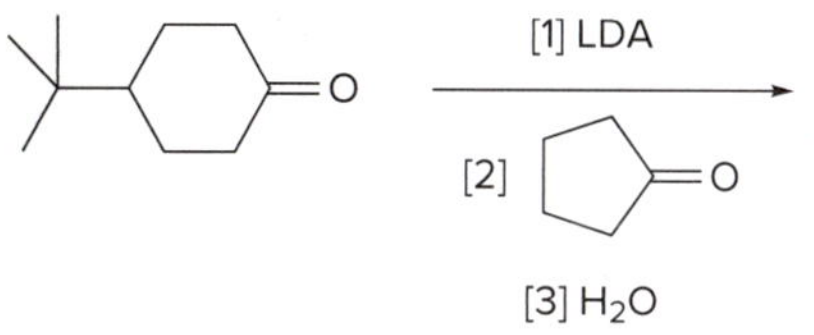 [1] LDA; [2] cyclopentanone; [3] H_2O

b. 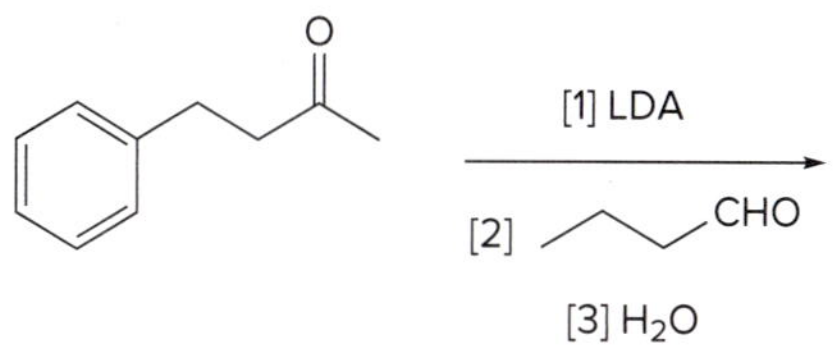 [1] LDA; [2] $CH_3CH_2CH_2CHO$ (CHO); [3] H_2O

More Practice: Try Problems 22.31, 22.45c.

Problem 22.13 Identify the product **A** formed in the following directed aldol reaction, one step in the synthesis of periplanone B, the sex pheromone of the female American cockroach.

[1] LDA
[2] (E)-but-2-enal
[3] H_2O
→ **A** → several steps → periplanone B

Periplanone B (Problem 22.13) is a pheromone secreted by the female American cockroach, *Periplaneta americana*. *George Grall/National Geographic Creative/Alamy Stock Photo*

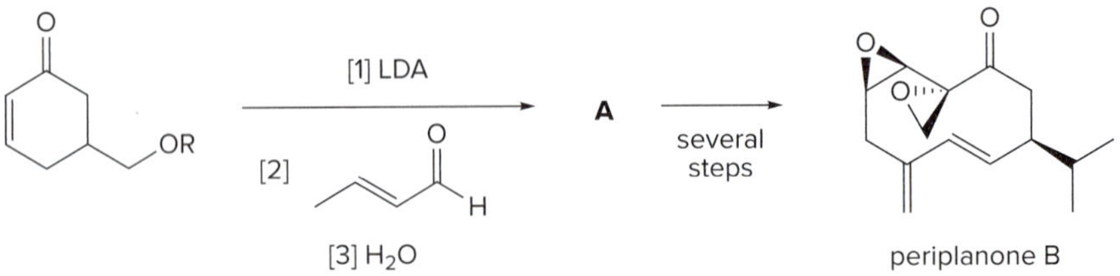

To determine the needed carbonyl components for a directed aldol, follow the same strategy used for a regular aldol reaction in Section 22.1D, as shown in Sample Problem 22.2.

Sample Problem 22.2 Determining the Starting Materials of a Directed Aldol Reaction

What starting materials are needed to prepare *ar*-turmerone, the chapter-opening molecule, using a directed aldol reaction?

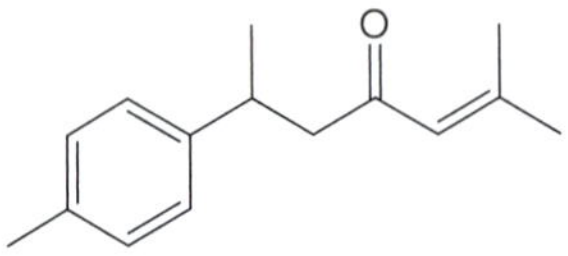

ar-turmerone

Solution

When the desired product is an α,β-unsaturated carbonyl compound, identify the α and β carbons that are part of the C=C, and break the molecule into two components between these carbons.

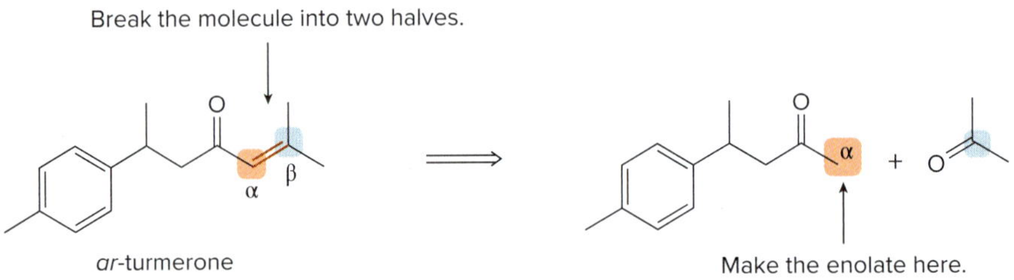

Break the molecule into two halves.

ar-turmerone ⟹ +

Make the enolate here.

Problem 22.14 What carbonyl starting materials are needed to prepare each compound using a directed aldol reaction?

a. b. c.

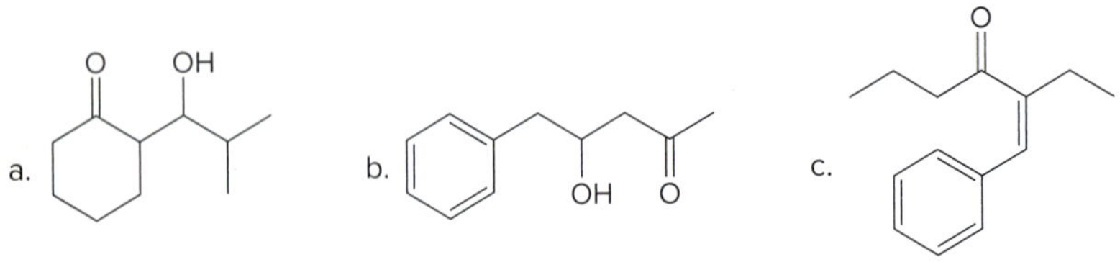

More Practice: Try Problem 22.33.

Problem 22.15 A key step in the synthesis of donepezil is a directed aldol reaction that forms α,β-unsaturated carbonyl compound **X.** What carbonyl starting materials are needed to prepare **X** using a directed aldol reaction? What reagents are needed to convert **X** to donepezil?

CH_3O CH_3O **X**

CH_3O CH_3O donepezil

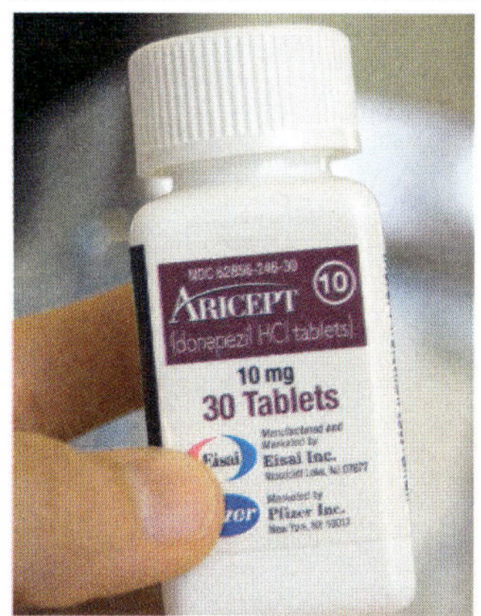

Donepezil (trade name Aricept, Problem 22.15) is a drug used to improve cognitive function in patients suffering from Alzheimer's disease and other types of dementia. *Jill Braaten*

22.4 Intramolecular Aldol Reactions

Aldol reactions with dicarbonyl compounds can be used to make five- and six-membered rings. The enolate formed from one carbonyl group is the nucleophile, and the carbonyl carbon of the other carbonyl group is the electrophile. For example, treatment of hexane-2,5-dione with base forms a five-membered ring.

Hexane-2,5-dione is called a **1,4-dicarbonyl compound** to emphasize the relative position of its carbonyl groups. 1,4-Dicarbonyl compounds are starting materials for synthesizing **five-membered rings.**

hexane-2,5-dione — re-draw → NaOEt / EtOH → new C–C σ and π bonds

The steps in this process, shown in Mechanism 22.4, are no different from the general mechanisms of the aldol reaction and dehydration described in Section 22.1.

Mechanism 22.4 The Intramolecular Aldol Reaction

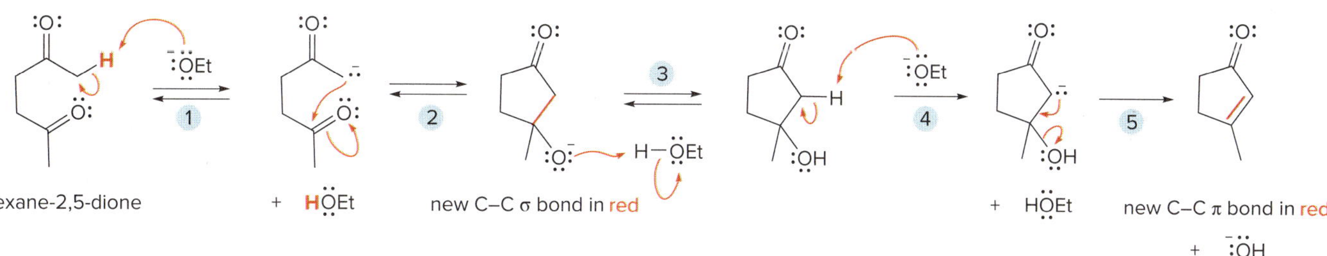

1 The base removes a proton on the α carbon to form a **resonance-stabilized enolate.**

2 **Nucleophilic attack** of the enolate on the electrophilic carbonyl *in the same molecule* forms a new C–C σ bond, generating the five-membered ring.

3 Protonation of the alkoxide forms the **β-hydroxy carbonyl compound.**

4 – 5 **Dehydration occurs by the two-step E1cB mechanism**—loss of a proton to form an enolate and elimination of ^{-}OH to form a π bond.

When hexane-2,5-dione is treated with base in Step [1], two different enolates are possible—enolates **A** and **B,** formed by removal of H_a and H_b, respectively. Although enolate **A** goes on to form the five-membered ring, intramolecular cyclization using enolate **B** would lead to a strained three-membered ring.

hexane-2,5-dione; NaOEt, EtOH; loss of H_a → A → Mechanism 22.4 → The more stable five-membered ring forms.

loss of H_b → B ⇌ The strained three-membered ring does *not* form.

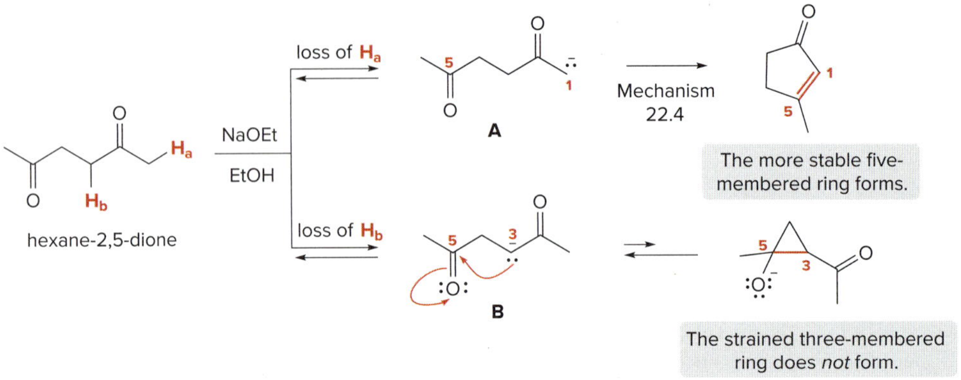

Because the three-membered ring is much higher in energy than the enolate starting material, equilibrium greatly favors the starting materials and the **three-membered ring does not form.** Under the reaction conditions, enolate **B** is re-protonated to form hexane-2,5-dione, because all steps except dehydration are equilibria. **Thus, equilibrium favors formation of the more stable five-membered ring over the much less stable three-membered ring.**

In a similar fashion, six-membered rings can be formed from the intramolecular aldol reaction of **1,5-dicarbonyl compounds.**

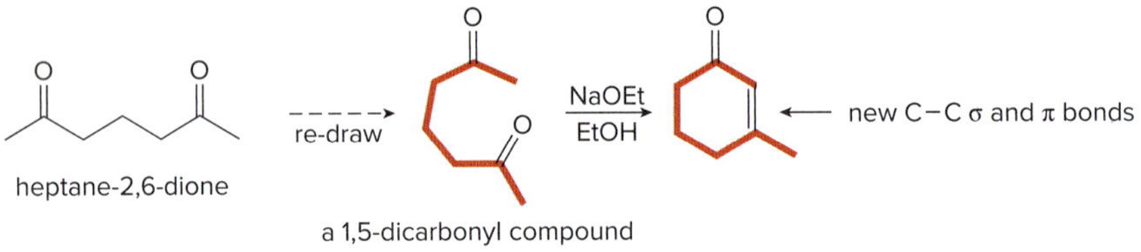

heptane-2,6-dione --re-draw--> a 1,5-dicarbonyl compound --NaOEt, EtOH--> ← new C–C σ and π bonds

The synthesis of the female sex hormone **progesterone** by W. S. Johnson and co-workers at Stanford University is considered one of the classics in total synthesis. The last six-membered ring needed in the steroid skeleton was prepared by a two-step sequence using an intramolecular aldol reaction, as shown in Figure 22.3.

Figure 22.3 The synthesis of progesterone using an intramolecular aldol reaction

[1] O_3; [2] Zn, H_2O → 1,5-dicarbonyl compound → ^-OH, H_2O → progesterone

Ozone oxidatively cleaves the C=C.

Intramolecular aldol reaction forms the six-membered ring.

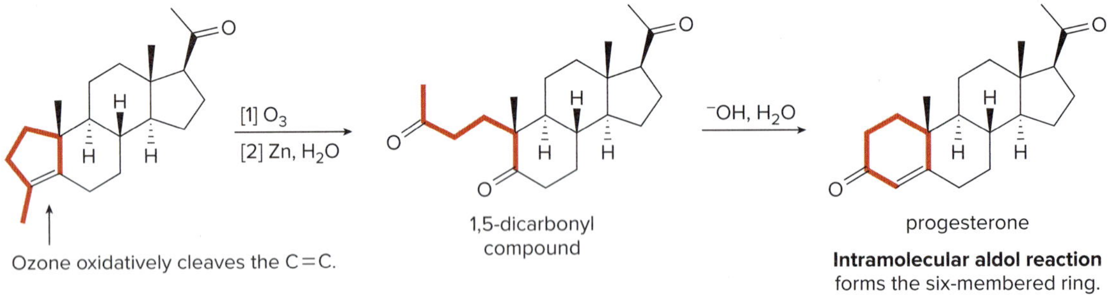

- Oxidative cleavage of the alkene with O_3, followed by Zn, H_2O (Section 12.10), gives the 1,5-dicarbonyl compound.
- Intramolecular aldol reaction of the 1,5-dicarbonyl compound with dilute ^-OH in H_2O solution forms progesterone.
- **This two-step reaction sequence converts a five-membered ring to a six-membered ring.** Reactions that synthesize larger rings from smaller ones are called **ring expansion reactions.**

Problem 22.16 Draw a stepwise mechanism for the conversion of heptane-2,6-dione to 3-methylcyclohex-2-enone with NaOEt, EtOH.

Sample Problem 22.3 Drawing the Major Product of an Intramolecular Aldol Reaction

What is the major intramolecular aldol product formed when dicarbonyl compound **A** is treated with base?

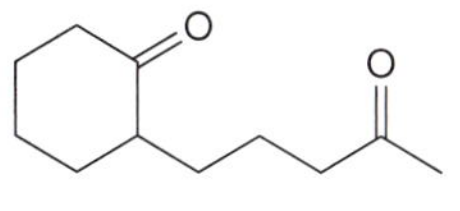

A

Solution

To draw the products of an intramolecular aldol reaction:

- **Identify all the α carbons** that are bonded to H's.
- Determine how far each α carbon is from the other carbonyl carbon. Reactions that yield **five- or six-membered rings** are favored.
- **α,β-Unsaturated carbonyl systems are favored** over β-hydroxy carbonyl compounds that cannot dehydrate to a conjugated product.

A contains four α carbons that can form enolates. First, consider enolates from the cyclohexanone (at C5 and C7 below) reacting with the acyclic carbonyl (C1). Only the enolate at C5 forms a five-membered ring by intramolecular aldol, but the β-hydroxy ketone **B** can*not* dehydrate to a conjugated system because the α carbon has no H for dehydration. Thus, this path is *not* favored.

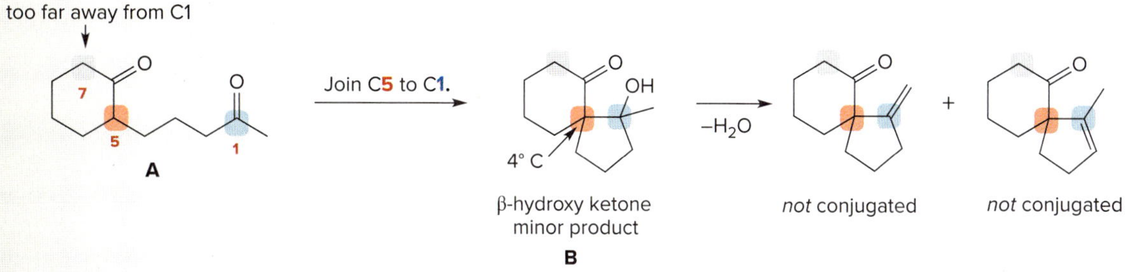

too far away from C1

7 5 1 A

Join C**5** to C**1**.

OH 4° C

β-hydroxy ketone
minor product
B

$-H_2O$

+

not conjugated

not conjugated

Then, consider enolates from the acyclic ketone (at C5 and C7 below) reacting with the cyclohexanone carbonyl (C1). Only the enolate at C5 forms a five-membered ring by intramolecular aldol, and dehydration forms an α,β-unsaturated carbonyl compound **C**, so **C is the major product.**

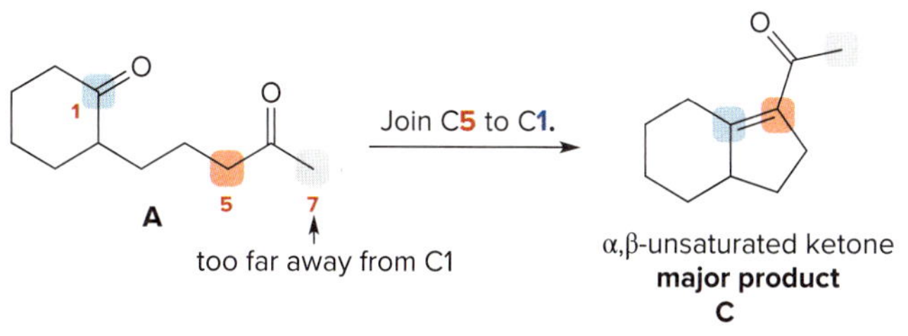

1 A 5 7

too far away from C1

Join C**5** to C**1**.

α,β-unsaturated ketone
major product
C

Problem 22.17 What cyclic product is formed when each dicarbonyl compound is treated with aqueous ^{-}OH?

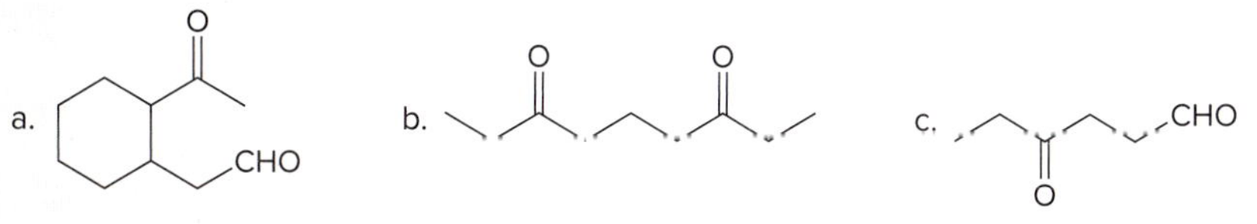

a. CHO

b.

c. CHO

More Practice: Try Problem 22.32, 22.35.

Problem 22.18 Following the two-step reaction sequence depicted in Figure 22.3, write out the steps needed to convert **A** to **B.**

A → B

22.5 The Claisen Reaction

The **Claisen reaction** is the second general reaction of enolates with other carbonyl compounds. In the Claisen reaction, two molecules of an ester react with each other in the presence of an alkoxide base to form a **β-keto ester.** For example, treatment of ethyl acetate with NaOEt forms ethyl acetoacetate after protonation with aqueous acid.

Unlike the aldol reaction, which is base-catalyzed, a full equivalent of base is needed to deprotonate the β-keto ester formed in Step [3] of the Claisen reaction.

2 ethyl acetate —[1] NaOEt; [2] H_3O^+ (**Claisen reaction**)→ ethyl acetoacetate (**β-keto ester**)

new C–C bond in red

The mechanism for the Claisen reaction (Mechanism 22.5) resembles the mechanism of an aldol reaction in that it involves nucleophilic addition of an enolate to an electrophilic carbonyl group. Because esters have a leaving group on the carbonyl carbon, however, **loss of a leaving group occurs to form the product of substitution, *not* addition.**

Mechanism 22.5 The Claisen Reaction

Part [1] Formation of the β-keto ester

EtO–C(=O)–CH₃ (α H) + ⁻:OEt —1→ enolate + HOEt —2 (**nucleophilic attack**)→ tetrahedral intermediate (new C–C bond in red) —3 (**elimination of ⁻OEt**)→ β-keto ester + ⁻:OEt

enolate

1 The base removes a proton on the α carbon to form a **resonance-stabilized enolate.**

2 **Nucleophilic attack** of the enolate on an electrophilic carbonyl in another molecule of ester forms a new C–C bond.

3 Loss of the leaving group (⁻OEt) forms a **β-keto ester.**

Part [2] Deprotonation and protonation

β-keto ester + ⁻:OEt —4→ resonance-stabilized enolate + HOEt —5 ($H–OH_2^+$)→ **β-keto ester** + H_2O:

4 Because the β-keto ester formed in Step [3] has especially acidic protons between its two carbonyl groups, the base removes a proton to form a **resonance-stabilized enolate.**

5 Protonation of the enolate with strong acid re-forms the **β-keto ester.**

Because the generation of a resonance-stabilized enolate from the product β-keto ester drives the Claisen reaction (Step [4] of Mechanism 22.5), **only esters with two or three hydrogens on the α carbon undergo this reaction;** that is, esters must have the general structure CH_3CO_2R' or RCH_2CO_2R'.

- **Keep in mind: The characteristic reaction of esters is nucleophilic substitution. A Claisen reaction is a nucleophilic substitution in which an enolate is the nucleophile.**

Figure 22.4 compares the general reaction for nucleophilic substitution of an ester with the Claisen reaction. Sample Problem 22.4 reinforces the basic features of the Claisen reaction.

Figure 22.4 The Claisen reaction—An example of nucleophilic substitution

nucleophilic attack

loss of the leaving group

Claisen reaction

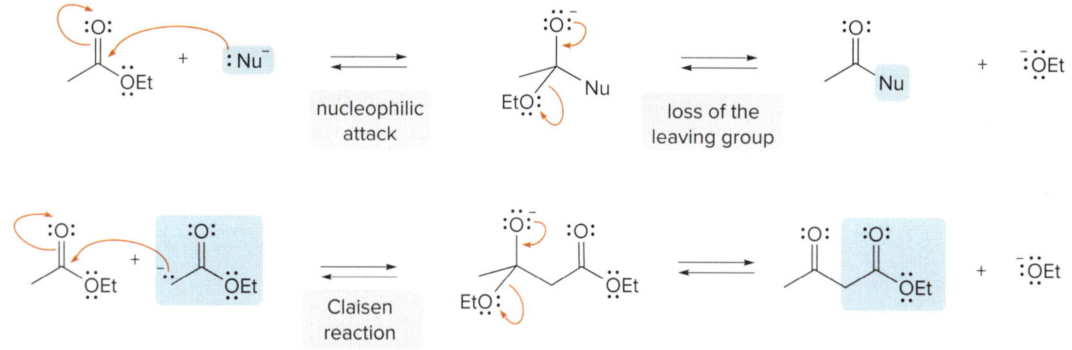

- Esters react by **nucleophilic substitution.** In a Claisen reaction, an **enolate is the nucleophile** that adds to the carbonyl group.

Sample Problem 22.4 Determining the Product of a Claisen Reaction

Draw the product of the following Claisen reaction.

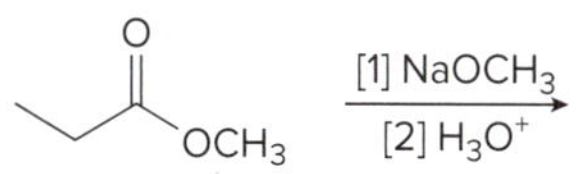

[1] $NaOCH_3$
[2] H_3O^+

Solution

To draw the product of any Claisen reaction, form a new carbon–carbon bond between the α carbon of one ester and the carbonyl carbon of another ester, with elimination of the leaving group ($^-OCH_3$ in this case).

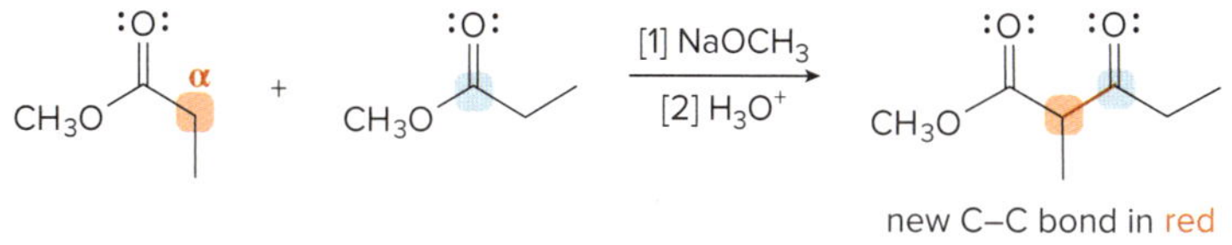

[1] $NaOCH_3$
[2] H_3O^+

new C–C bond in red

Next, write out the steps of the reaction to verify this product.

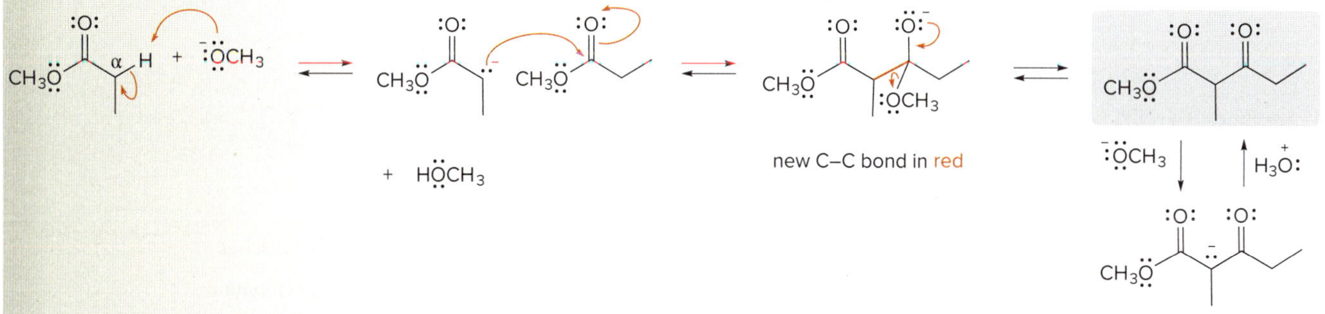

new C–C bond in red

Problem 22.19 What β-keto ester is formed when each ester is used in a Claisen reaction?

a. 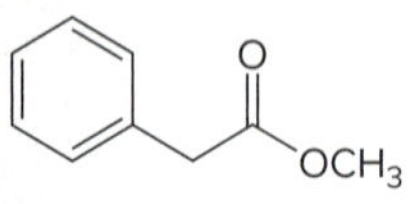 (structure: OCH$_3$)
c. 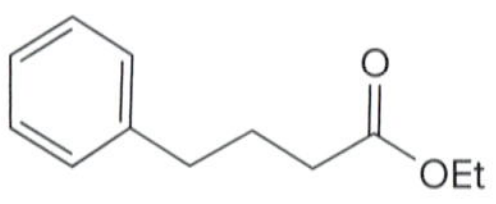 (structure: OEt)
b. 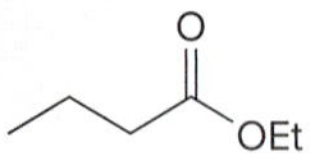 (structure: OEt)
d. 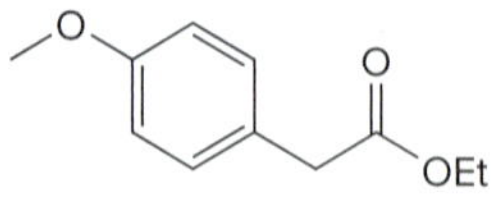 (structure: OEt)

22.6 The Crossed Claisen and Related Reactions

Like the aldol reaction, it is sometimes possible to carry out a Claisen reaction with two different carbonyl components as starting materials.

- **A Claisen reaction between two different carbonyl compounds is called a *crossed Claisen reaction.***

22.6A Two Useful Crossed Claisen Reactions

A crossed Claisen reaction is synthetically useful in two different instances.

- **A crossed Claisen occurs between two different esters when only one has α hydrogens.**

When one ester has no α hydrogens, a crossed Claisen reaction often leads to one product. Common esters with no α H atoms include ethyl formate (**HCO_2Et**) and ethyl benzoate (**$C_6H_5CO_2Et$**). For example, the reaction of ethyl benzoate (as the electrophile) with ethyl acetate (which forms the enolate) in the presence of base forms predominately one β-keto ester.

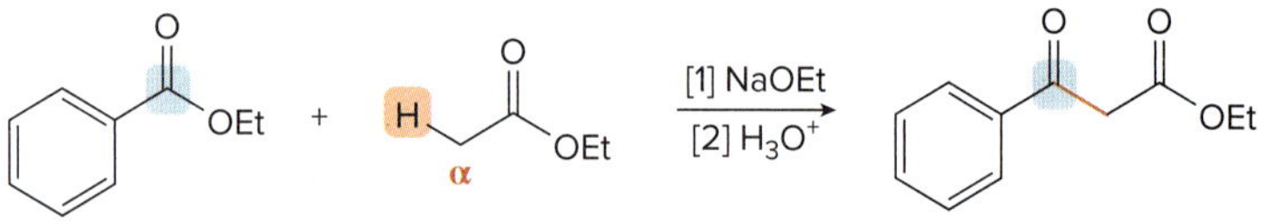

ethyl benzoate + ethyl acetate (α) →[1] NaOEt, [2] H_3O^+ → β-keto ester

OEt
H
α
ethyl benzoate
ethyl acetate
Only this ester can form an enolate.
[1] NaOEt
[2] H_3O^+
new C–C bond in red
β-keto ester

- **A crossed Claisen occurs between a ketone and an ester.**

The reaction of a ketone and an ester in the presence of base also forms the product of a crossed Claisen reaction. The enolate is generally formed from the ketone component, and the reaction works best when the ester has no α hydrogens. The product of this crossed Claisen reaction is a **β-dicarbonyl compound,** but ***not*** a β-keto ester.

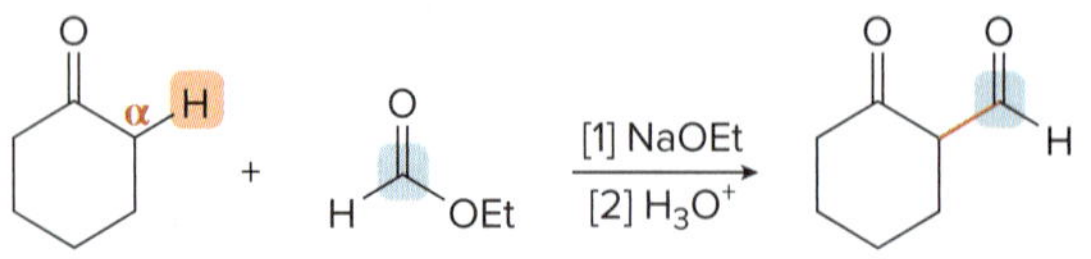

α H
H OEt
[1] NaOEt
[2] H_3O^+
new C–C bond in red
β-dicarbonyl compound

Problem 22.20 What crossed Claisen product is formed from each pair of compounds?

a. OEt and H OEt b. and OEt

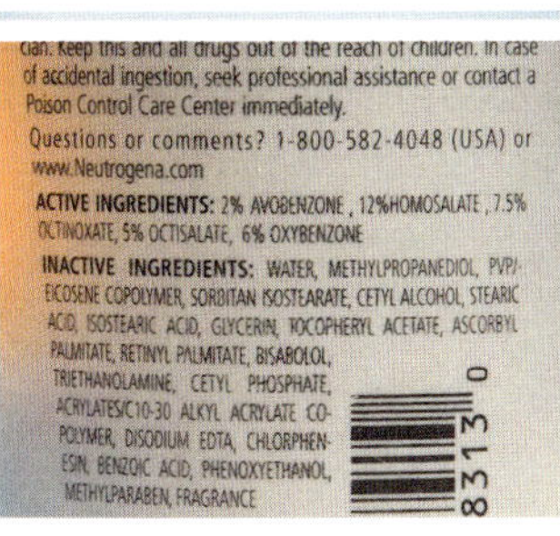

Sunscreen ingredients (Problem 22.21) *Jill Braaten/ McGraw Hill*

Problem 22.21 Avobenzone is a conjugated compound that absorbs ultraviolet light with wavelengths in the 320–400 nm region, so it is a common ingredient in commercial sunscreens. Write out two different crossed Claisen reactions that form avobenzone.

avobenzone

22.6B Other Useful Variations of the Crossed Claisen Reaction

β-Dicarbonyl compounds are also prepared by reacting an enolate with **ethyl chloroformate** and **diethyl carbonate.**

ethyl chloroformate **diethyl carbonate**

These reactions resemble a Claisen reaction because they involve the same three steps:

[1] **Formation of an enolate**

[2] **Nucleophilic addition to a carbonyl group**

[3] **Elimination of a leaving group**

For example, reaction of an ester enolate with diethyl carbonate yields a **β-diester** (Reaction [1]), whereas reaction of a ketone enolate with ethyl chloroformate forms a **β-keto ester** (Reaction [2]). New carbon–carbon bonds are shown in red.

[1] OEt —[1] NaOEt; [2] EtO(C=O)OEt→ EtO **β-diester** OEt + $^{-}$OEt

[2] —[1] NaOEt; [2] Cl(C=O)OEt→ **β-keto ester** OEt + Cl^-

Reaction [2] is noteworthy because it provides easy access to **β-keto esters,** which are useful starting materials in the acetoacetic ester synthesis (Section 21.10). In this reaction, Cl^- is eliminated rather than ^{-}OEt in Step [3], because Cl^- is a better leaving group, as shown in the following steps.

[1] + EtOH [2] [3] **β-keto ester** + Cl^-

Problem 22.22 Draw the products of each reaction.

a. [cyclohexanone] $\xrightarrow[\text{[2] }(EtO)_2C{=}O]{\text{[1] NaOEt}}$

b. [ethyl phenylacetate] $\xrightarrow[\text{[2] }ClCO_2Et]{\text{[1] NaOEt}}$

Problem 22.23 Two steps in a synthesis of the analgesic ibuprofen include a carbonyl condensation reaction, followed by an alkylation reaction. Identify intermediates **A** and **B** in the synthesis of ibuprofen.

[ester] $\xrightarrow[\text{[2] }(EtO)_2C{=}O]{\text{[1] NaOEt}}$ **A** $\xrightarrow[\text{[2] }CH_3I]{\text{[1] NaOEt}}$ **B** $\xrightarrow[\Delta]{H_3O^+}$ ibuprofen

22.7 The Dieckmann Reaction

Intramolecular Claisen reactions of diesters form five- and six-membered rings. The enolate of one ester is the nucleophile, and the carbonyl carbon of the other is the electrophile. An intramolecular Claisen reaction is called a **Dieckmann reaction.** Two types of diesters give good yields of cyclic products.

Like aspirin (Section 19.5), ibuprofen (Problem 22.23) acts as an anti-inflammatory agent by blocking the conversion of arachidonic acid to prostaglandins. *David A. Tietz/McGraw Hill*

- **1,6-Diesters yield five-membered rings by the Dieckmann reaction.**

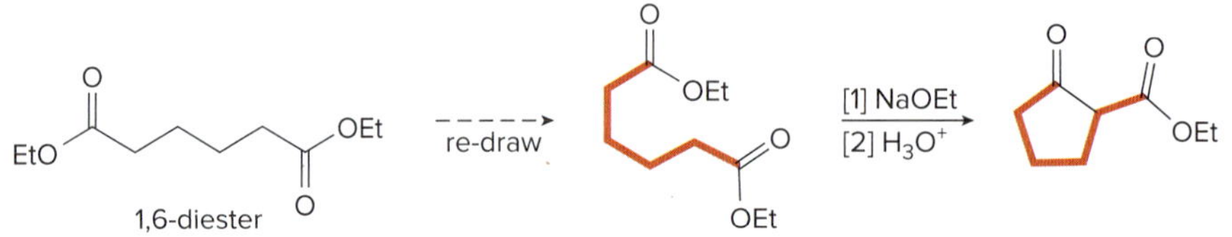

- **1,7-Diesters yield six-membered rings by the Dieckmann reaction.**

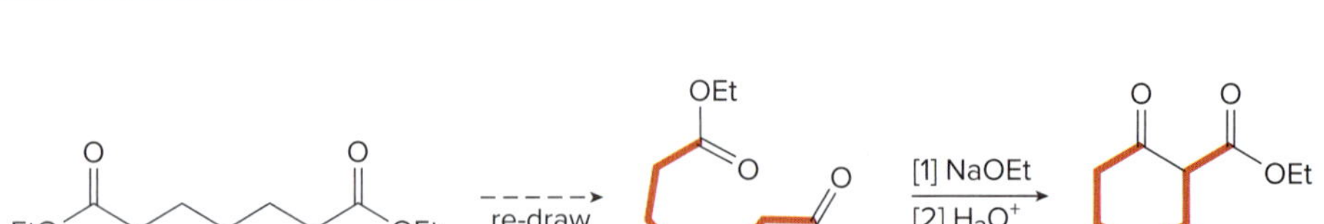

The mechanism of the Dieckmann reaction is exactly the same as the mechanism of an intermolecular Claisen reaction. It is illustrated in Mechanism 22.6 for the formation of a six-membered ring.

Problem 22.24 What two β-keto esters are formed in the Dieckmann reaction of the following diester?

[diester: EtO … OEt]

Mechanism 22.6 The Dieckmann Reaction

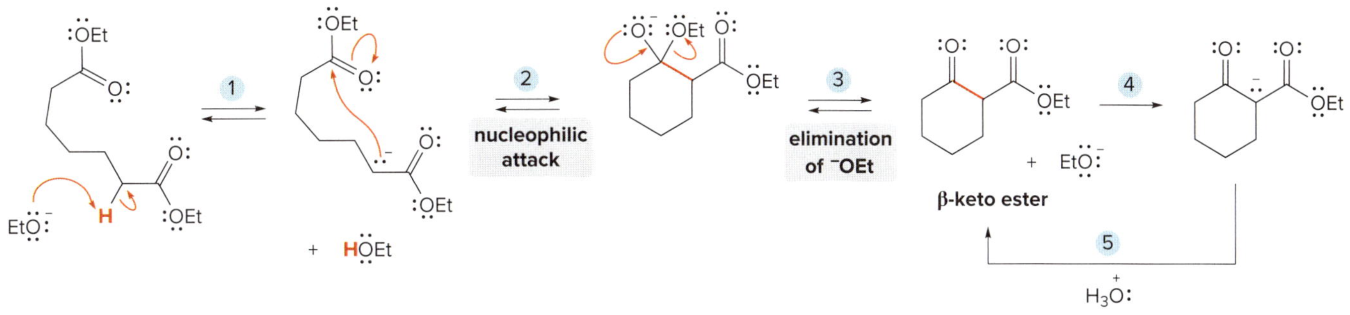

1 – 2 The base removes a proton on the α carbon to form an **enolate,** which attacks the electrophilic carbonyl of the other ester to form a new C–C bond.

3 **Elimination of $^{-}$OEt** forms the β-keto ester.

4 – 5 Under the basic reaction conditions, the proton between the two carbonyls is removed with base to form an enolate, which is protonated with acid to re-form the β-keto ester.

22.8 The Michael Reaction

Like the aldol and Claisen reactions, the **Michael reaction involves two carbonyl components—the enolate of one carbonyl compound and an α,β-unsaturated carbonyl compound.**

Two components of a Michael reaction

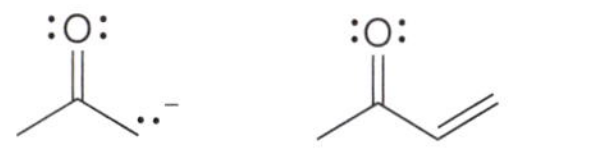

enolate

α,β-unsaturated carbonyl compound

Recall from Section 17.15 that α,β-unsaturated carbonyl compounds are resonance stabilized and have **two electrophilic sites—the carbonyl carbon and the β carbon.**

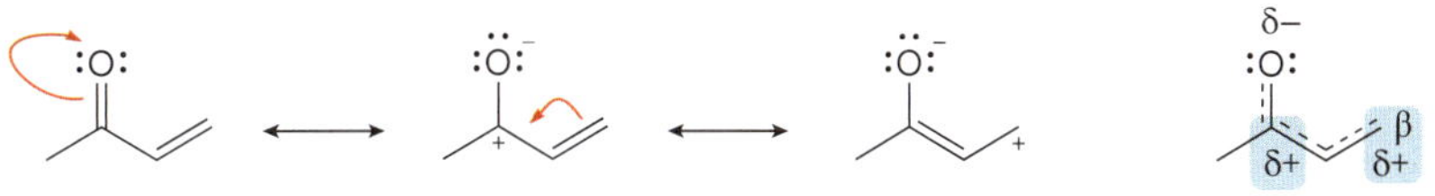

three resonance structures for an α,β-unsaturated carbonyl compound

hybrid

two electrophilic sites

- **The Michael reaction involves the conjugate addition (1,4-addition) of a resonance-stabilized enolate to the β carbon of an α,β-unsaturated carbonyl system.**

All conjugate additions add the **elements of H and Nu across the α and β carbons.**

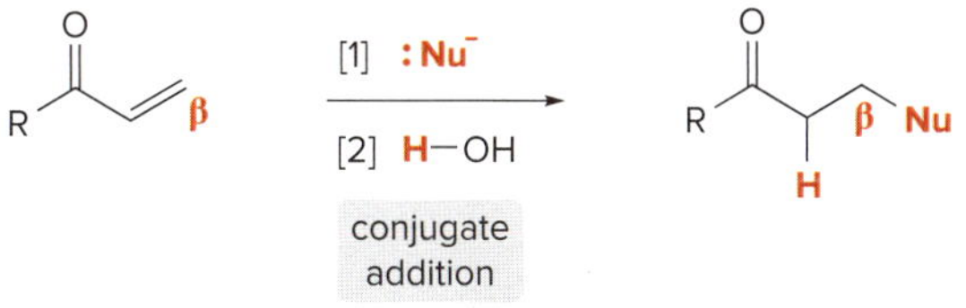

conjugate addition

In the Michael reaction, the **nucleophile is an enolate.** Enolates of active methylene compounds are particularly common. The α,β-unsaturated carbonyl component is often called a **Michael acceptor.**

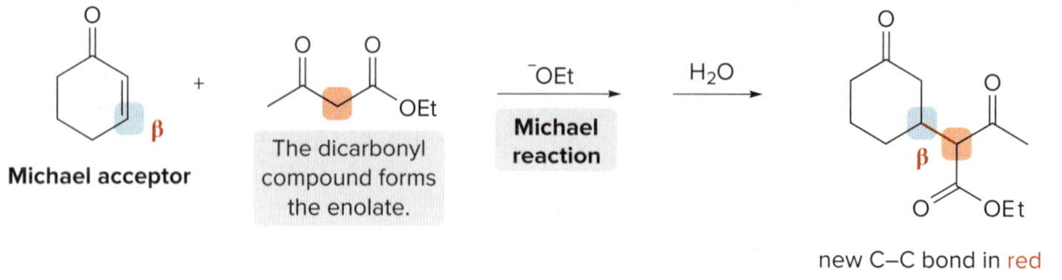

Michael acceptor

The dicarbonyl compound forms the enolate.

$^-$OEt

Michael reaction

H_2O

new C–C bond in red

Problem 22.25 Which of the following compounds can serve as Michael acceptors?

a. b. c. d. CH_3O

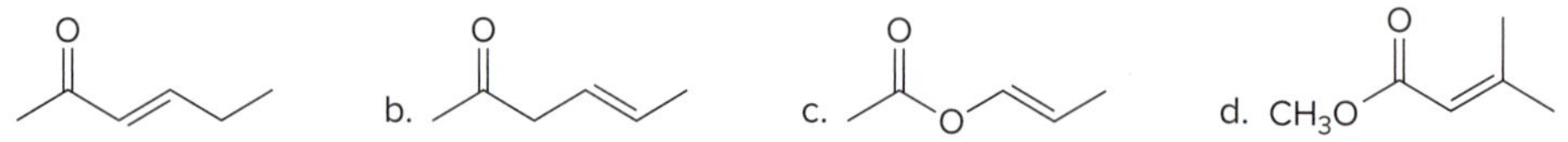

The Michael reaction always forms a new carbon–carbon bond on the β carbon of the Michael acceptor. The mechanism of the Michael reaction is shown in Mechanism 22.7. **The key step is nucleophilic addition of the enolate to the β carbon of the Michael acceptor in Step [2].**

Mechanism 22.7 The Michael Reaction

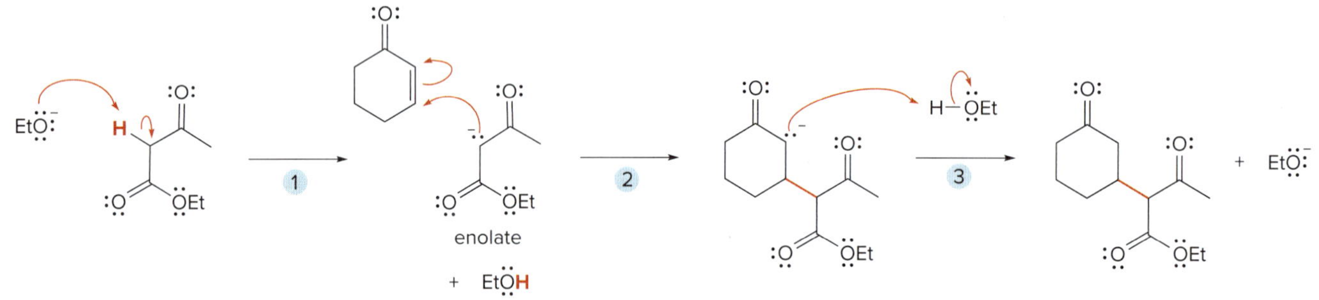

enolate

+ EtOH

1 The base removes a proton on the carbon between the two carbonyl groups to form an **enolate.**

2 **Nucleophilic addition of the enolate to the β carbon** of the α,β-unsaturated carbonyl compound forms a new carbon–carbon bond and another enolate.

3 Protonation of the enolate forms the **1,4-addition product.**

When the product of a Michael reaction is also a β-keto ester, it can be hydrolyzed and decarboxylated by heating in aqueous acid, as discussed in Section 21.9. This forms a **1,5-dicarbonyl compound.** Figure 22.5 shows a Michael reaction that was a key step in the synthesis of **estrone,** a female sex hormone.

1,5-Dicarbonyl compounds are starting materials for intramolecular aldol reactions, as described in Section 22.4.

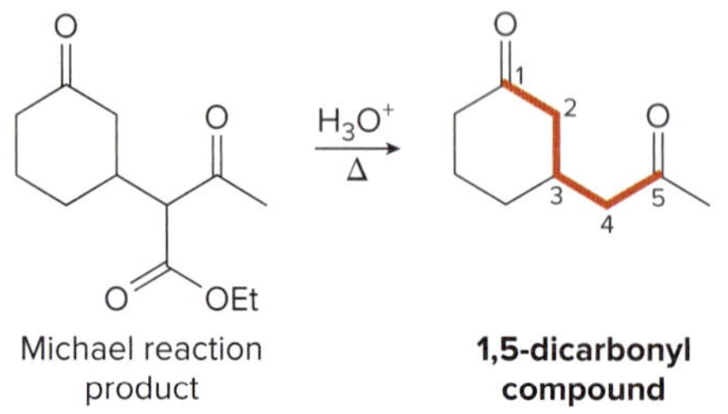

H_3O^+, Δ

Michael reaction product

1,5-dicarbonyl compound

Figure 22.5 Using a Michael reaction in the synthesis of the steroid estrone

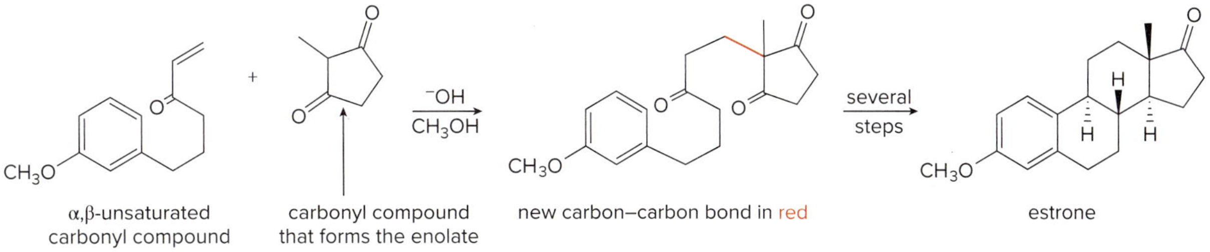

Problem 22.26 What product is formed when each pair of compounds is treated with NaOEt in ethanol?

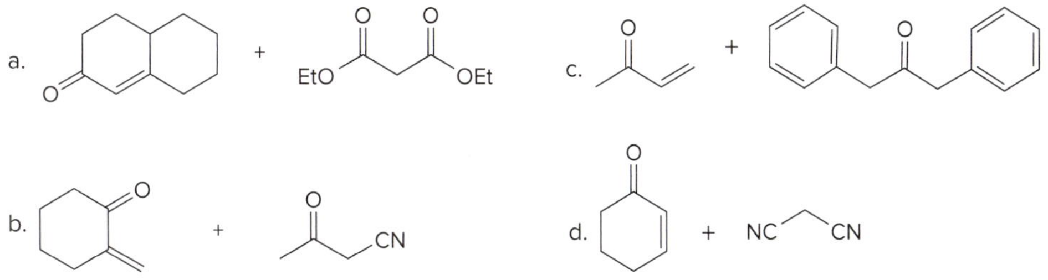

Problem 22.27 What starting materials are needed to prepare each compound by the Michael reaction?

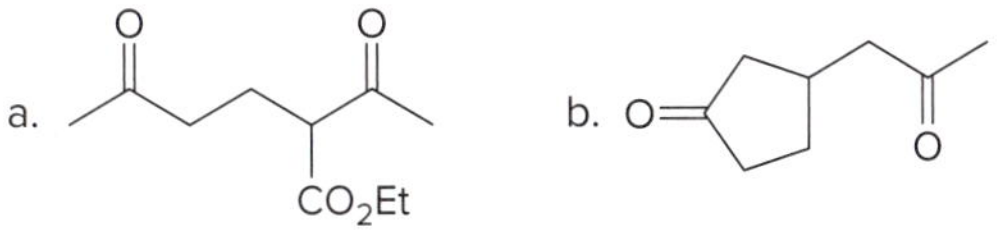

22.9 The Robinson Annulation

The word ***annulation*** comes from the Greek word *annulus* for "ring." The Robinson annulation is named for English chemist Sir Robert Robinson, who was awarded the 1947 Nobel Prize in Chemistry.

The Robinson annulation is a ring-forming reaction that combines a Michael reaction with an intramolecular aldol reaction. Like the other reactions in Chapter 22, it involves enolates and it forms carbon carbon bonds. The two starting materials for a Robinson annulation are an α,β-unsaturated carbonyl compound and an enolate.

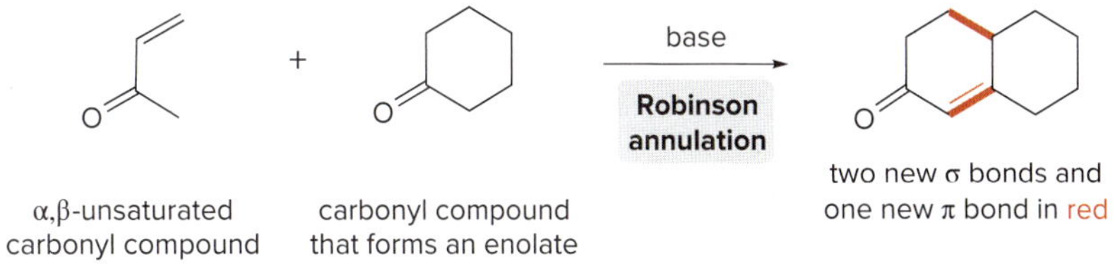

The Robinson annulation forms a six-membered ring and three new carbon–carbon bonds—two σ bonds and one π bond. The product contains an α,β-unsaturated ketone in a cyclohexane ring—that is, a **cyclohex-2-enone** ring. To generate the enolate component of the Robinson annulation, $^{-}$OH in H_2O and $^{-}$OEt in EtOH are typically used.

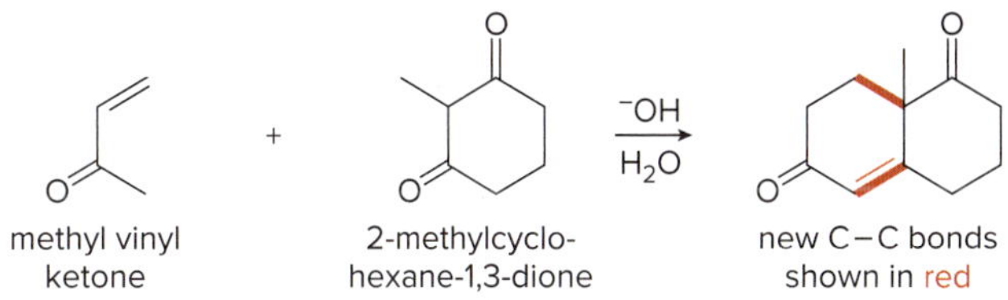

The mechanism of the Robinson annulation consists of a **Michael addition** to the α,β-unsaturated carbonyl compound to form a 1,5-dicarbonyl compound, followed by an **intramolecular aldol reaction** to form the six-membered ring. The mechanism is written out in three parts in Mechanism 22.8 for the reaction between methyl vinyl ketone and 2-methylcyclohexane-1,3-dione.

Mechanism 22.8 The Robinson Annulation

Part [1] Michael addition

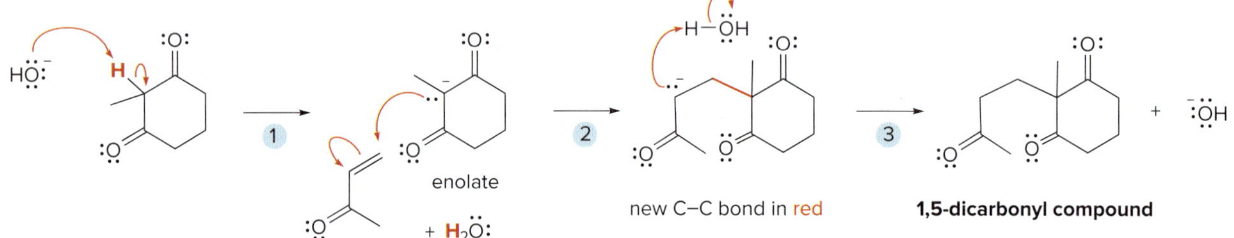

1 – 2 Base removes the most acidic proton—the proton between the two carbonyl groups—to form an **enolate. Conjugate addition** of the enolate to the α,β-unsaturated carbonyl compound forms a new carbon–carbon bond, generating an enolate.

3 Protonation of the enolate forms a **1,5-dicarbonyl compound.**

Part [2] Intramolecular aldol reaction

new C–C bond in red

β-hydroxy carbonyl compound

4 – 5 The base removes a proton to form an **enolate,** which attacks a carbonyl group to form a new C–C σ bond, generating the six-membered ring.

6 Protonation of the alkoxide forms the **β-hydroxy carbonyl compound.**

Part [3] Dehydration of the β-hydroxy carbonyl compound

new C–C π bond in red

7 – 8 **Dehydration occurs by the two-step E1cB mechanism**—loss of a proton to form an enolate and elimination of ^{-}OH to form a π bond.

The mechanism begins with the three-step **Michael addition** that forms the first carbon–carbon σ bond, generating the 1,5-dicarbonyl compound (Part [1]). An **intramolecular aldol reaction** (Part [2]) forms the second carbon–carbon σ bond, and **dehydration** of the β-hydroxy ketone (Part [3]) forms the π bond. All of the parts of this mechanism have been discussed in previous sections of Chapter 22. However, the end result of the Robinson annulation—the formation of a cyclohex-2-enone ring—is new.

To draw the product of Robinson annulation without writing out the mechanism each time, **place the α carbon of the compound that becomes the enolate next to the β carbon of the α,β-unsaturated carbonyl compound.** Then, join the appropriate carbons together as shown. If you follow this method of drawing the starting materials, the double bond in the product always ends up in the same position in the six-membered ring.

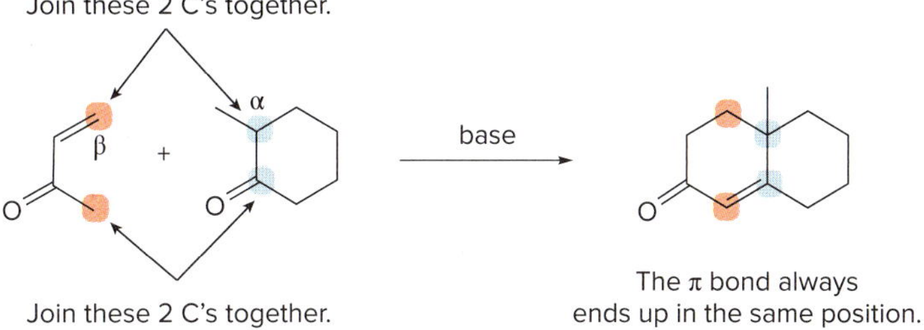

Sample Problem 22.5 Drawing the Product of a Robinson Annulation

Draw the Robinson annulation product formed from the following starting materials.

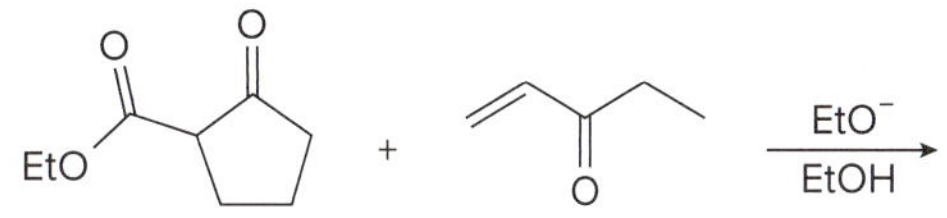

EtO O O + O $\xrightarrow[\text{EtOH}]{\text{EtO}^-}$

Solution

Arrange the starting materials to put the reactive atoms next to each other. For example:

- Place the α,β-unsaturated carbonyl compound *to the left* of the carbonyl compound.
- Determine which α carbon will become the enolate. **The most acidic H is always removed with base first,** which in this case is the H (in red) on the α carbon between the two carbonyl groups. **This α carbon is drawn adjacent to the β carbon of the α,β-unsaturated carbonyl compound.**

Then draw the bonds to form the new six-membered ring.

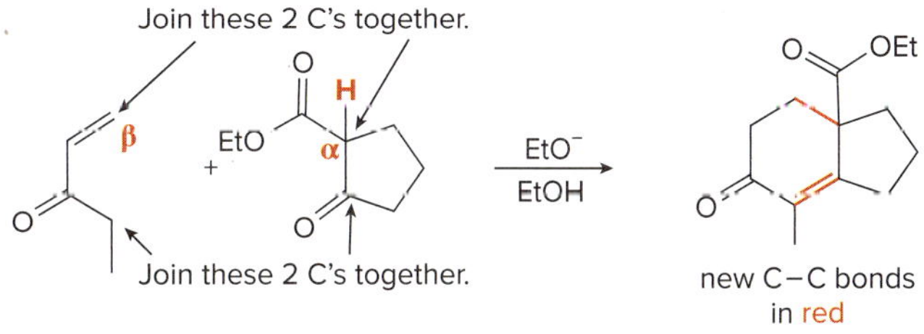

Join these 2 C's together.

β + EtO H α $\xrightarrow[\text{EtOH}]{\text{EtO}^-}$ O OEt

Join these 2 C's together.

new C–C bonds in red

Problem 22.28 Draw the products when each pair of compounds is treated with $CH_3CH_2O^-$, CH_3CH_2OH in a Robinson annulation reaction.

a. 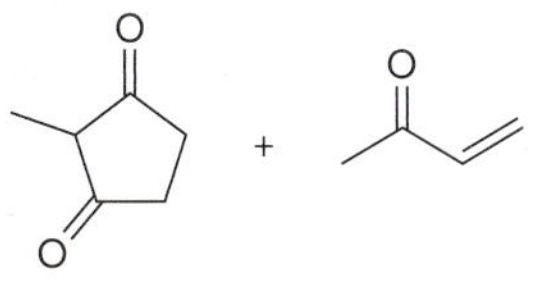 + c. 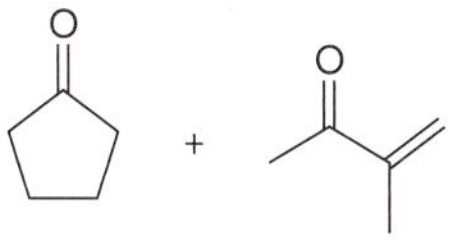 +

b.  OEt + d. 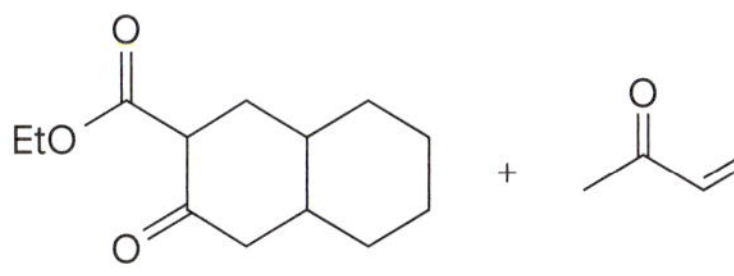 EtO +

More Practice: Try Problems 22.43, 22.62d.

To use the Robinson annulation in synthesis, you must be able to determine what starting materials are needed to prepare a given compound, by working in the retrosynthetic direction.

How To Synthesize a Compound Using the Robinson Annulation

Example What starting materials are needed to synthesize the following compound using a Robinson annulation?

CH_3O

Step [1] **Locate the cyclohex-2-enone ring and re-draw the target molecule if necessary.**

- To most easily determine the starting materials, always arrange the α,β-unsaturated carbonyl system in the same location. The target compound may have to be flipped or rotated, and you must be careful not to move any bonds to the wrong location during this process.

CH_3O flip OCH_3

Synthesize this ring.

Arrange the C=O and C=C in the same positions as in previous examples of the Robinson annulation.

Step [2] **Break the cyclohex-2-enone ring into two components.**

- Break the C=C. One half becomes the carbonyl group of the enolate component.
- Break the bond between the β carbon and the carbon to which it is bonded.

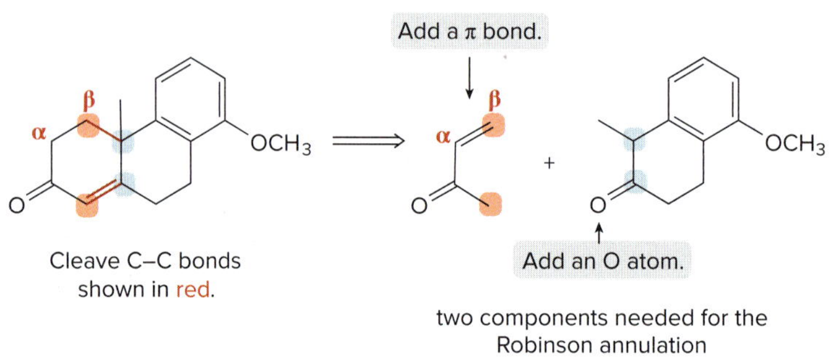

Problem 22.29 Which of the following bicyclic ring systems can be prepared by an intermolecular Robinson annulation?

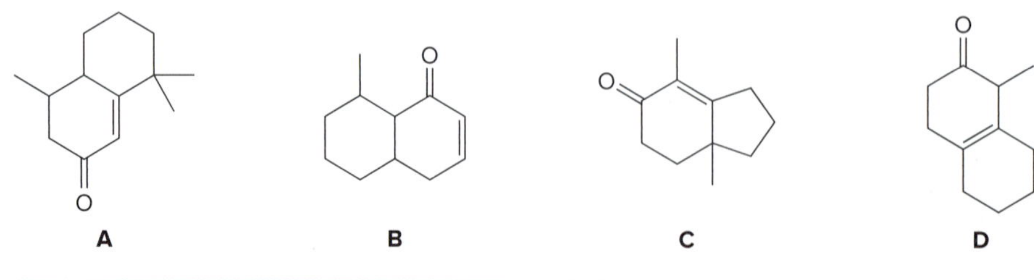

Problem 22.30 What starting materials are needed to synthesize each compound by a Robinson annulation?

a. b. c.

Chapter 22 REVIEW

KEY REACTIONS

[1] The four major carbonyl condensation reactions

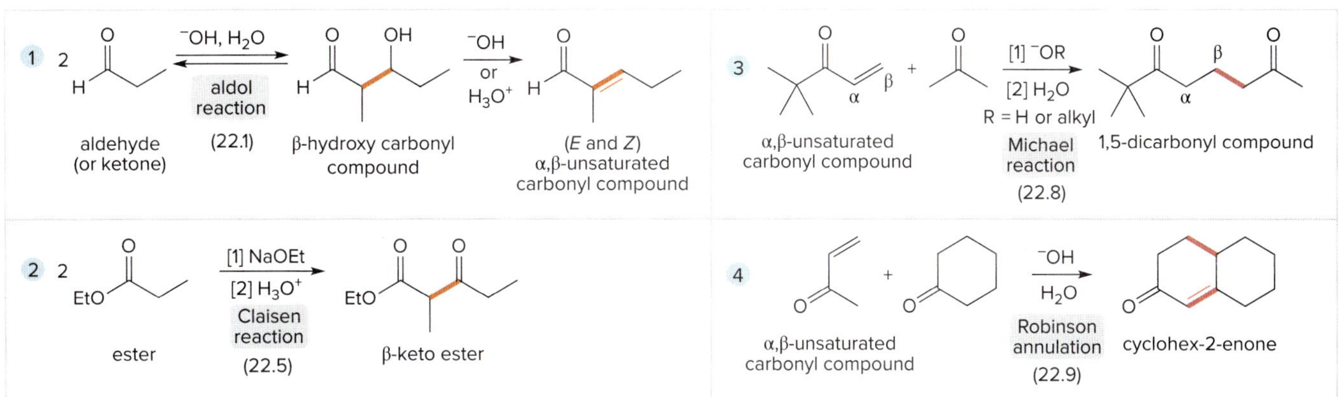

Try Problems 22.34; 22.38; 22.43; 22.45a, b, d, e; 22.47.

[2] Useful variations

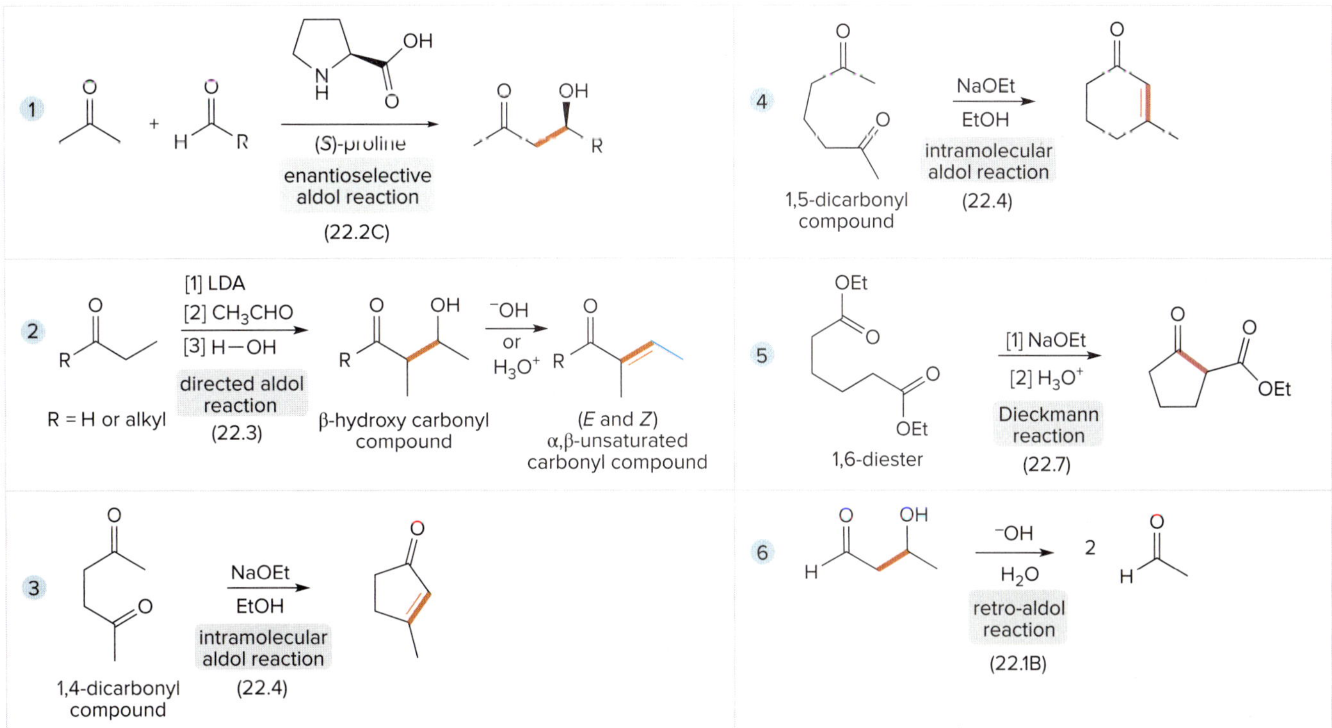

Try Problems 22.31; 22.32; 22.45c, f; 22.46; 22.48.

KEY SKILLS

[1] Drawing the product of a directed aldol reaction (22.3)

1. Prepare the enolate of one carbonyl component with LDA.
2. Add the second carbonyl compound to this enolate.
3. Add H_2O, and draw the product.

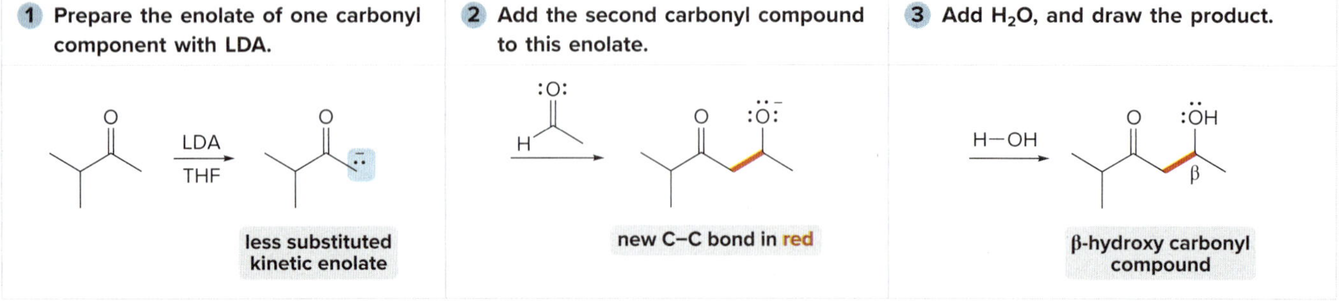

See Sample Problem 22.1. Try Problems 22.31, 22.45c.

[2] Identifying the starting materials to synthesize an α,β-unsaturated carbonyl compound using a directed aldol reaction (22.3)

1. Identify the α and β carbons that are part of the C=C.
2. Break the molecule into two components between these carbons, and add a double bond to oxygen at the β carbon.

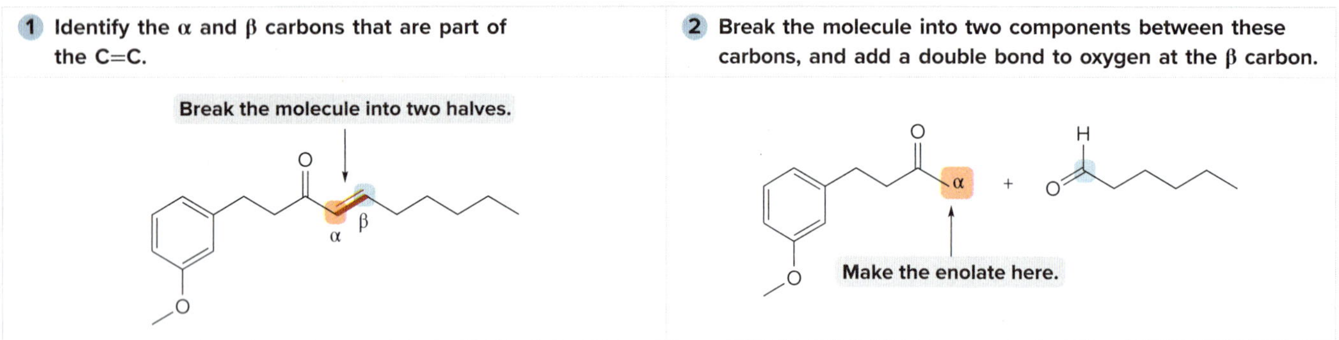

See Sample Problem 22.2. Try Problem 22.33.

[3] Converting a six-membered ring to a five-membered ring using an intramolecular aldol reaction (22.4)

1. Treat the alkene with O_3, followed by Zn and H_2O.
2. React the 1,6-dicarbonyl compound with base.
3. Draw the product.

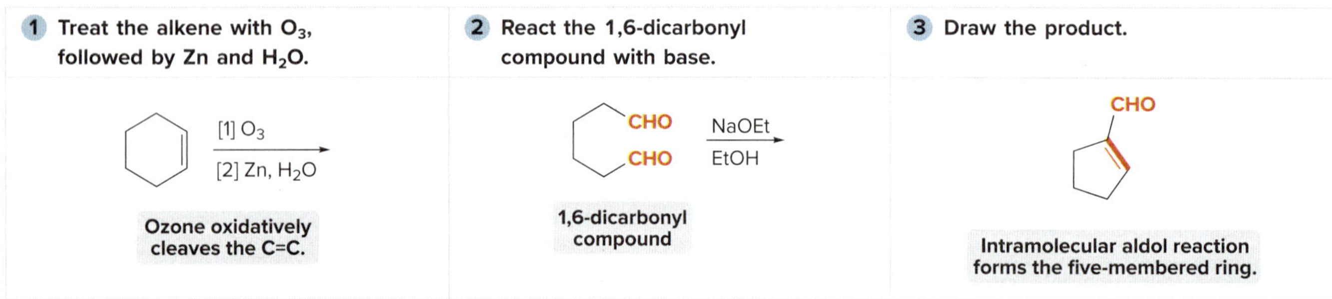

See Figure 22.3. Try Problems 22.36, 22.49.

[4] Drawing the product of a Claisen reaction (22.5)

1. Identify the α carbon of one ester and the carbonyl carbon of the other.
2. Draw the product.

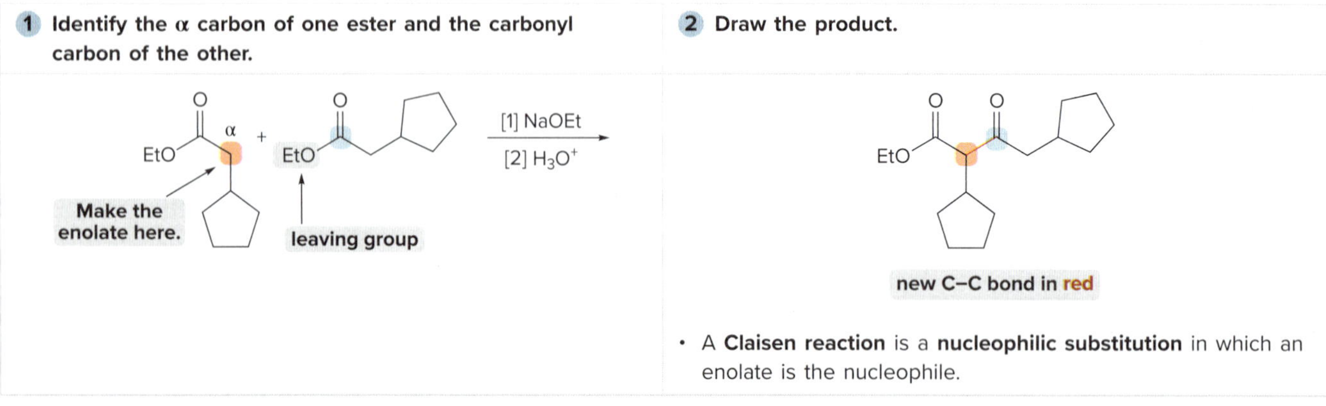

- A **Claisen reaction** is a **nucleophilic substitution** in which an enolate is the nucleophile.

See Sample Problem 22.4. Try Problem 22.38.

[5] Drawing the product of a Robinson annulation (22.9)

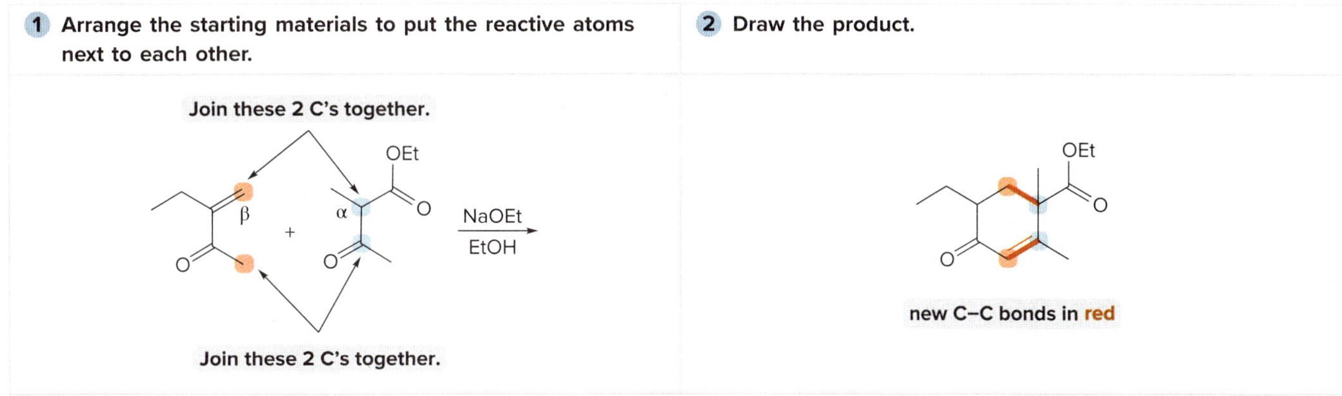

See Sample Problem 22.5. Try Problems 22.43, 22.62d.

[6] Identifying the starting materials to synthesize a compound using the Robinson annulation (22.9)

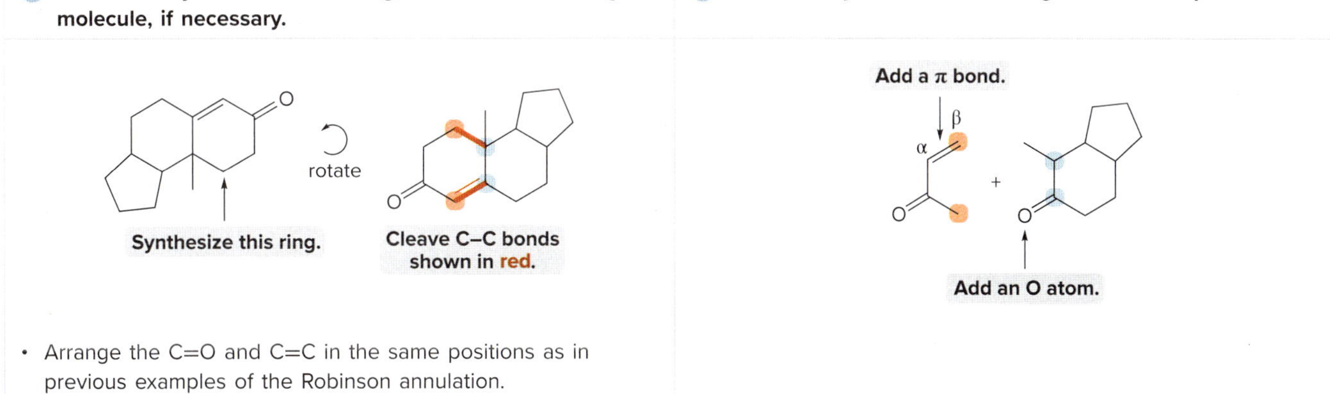

See *How To*, p. 1040. Try Problem 22.44.

CHAPTER 22 MULTIPLE-CHOICE SELF-TEST

The Self-Test consists of multiple-choice questions similar to those found on the American Chemical Society organic chemistry exam. Answers are given at the end of the chapter.

1. Which compound undergoes dehydration in both acid and base?

a. O, OH b. O, OH c. O, HO d. HO, O

2. What is the major product formed by an intramolecular aldol reaction of **A**?

O, CHO
A

a. O b. CHO c. O d. HO, CHO

3. Which two compounds can undergo a crossed aldol reaction to form one major product?

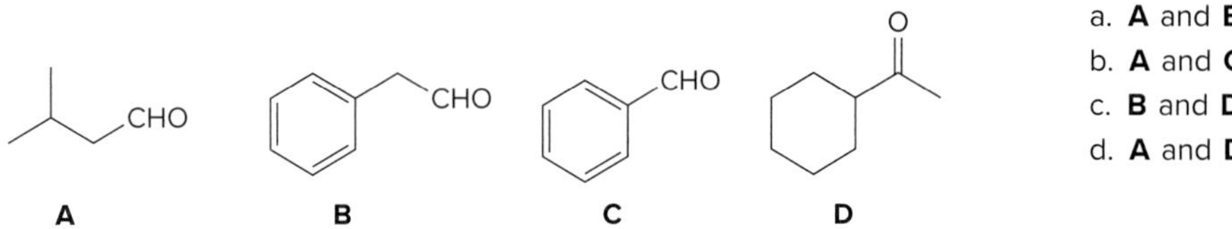

a. **A** and **B**
b. **A** and **C**
c. **B** and **D**
d. **A** and **D**

4. What starting materials are needed to synthesize **B** by a Robinson annulation?

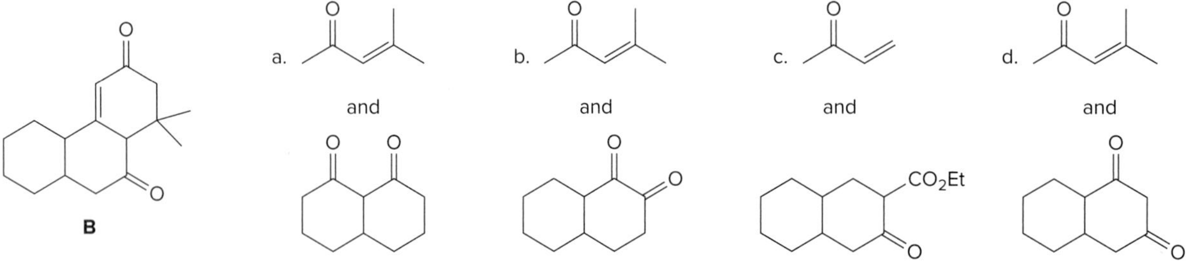

5. Which product is formed in the Dieckmann reaction of diester **E?**

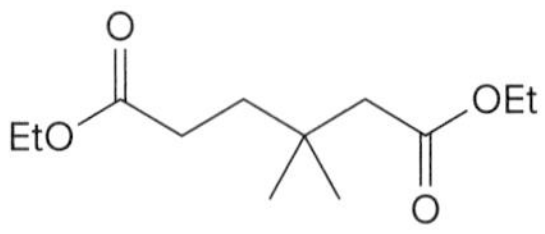

EtO OEt **E**

a. 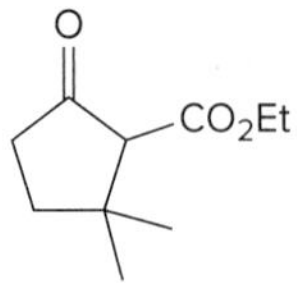 CO_2Et b. 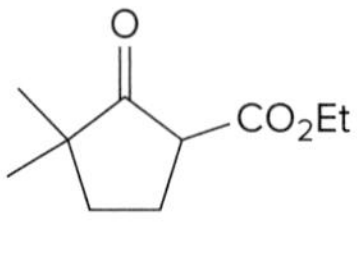 CO_2Et c. 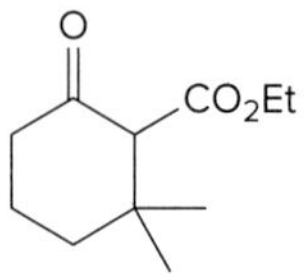 CO_2Et d. 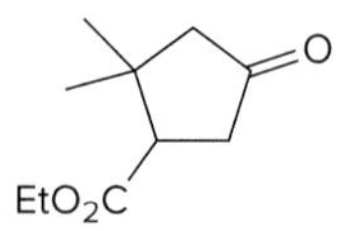 EtO_2C

6. Which compound is *not* a Michael acceptor?

a. 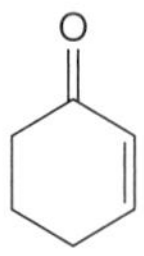 b. 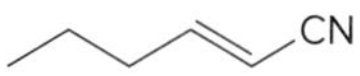 CN c. 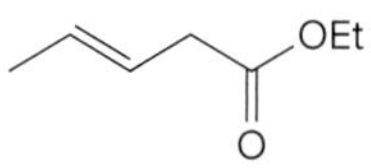 OEt d.

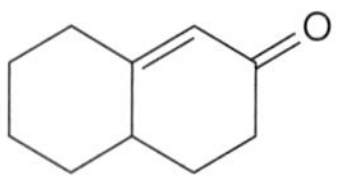

7. What product is formed in the following reaction?

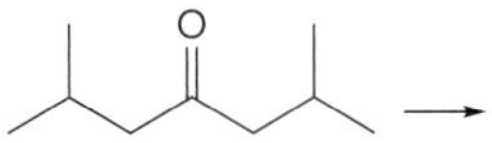

+ HCO_2Et + NaOEt →

a. 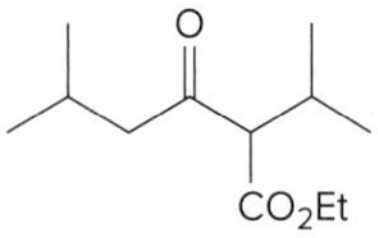 CO_2Et b. 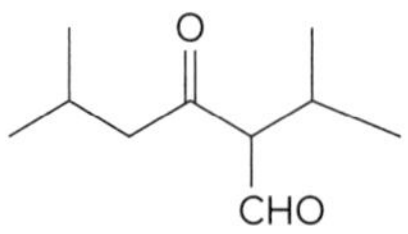 CHO c. 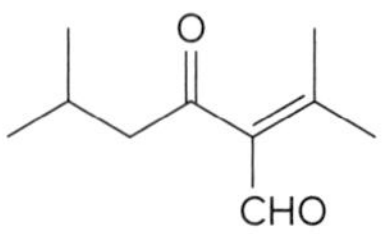 CHO d. 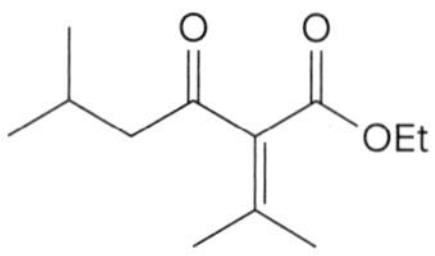 OEt

8. Which compound can*not* be formed by an aldol reaction?

a. 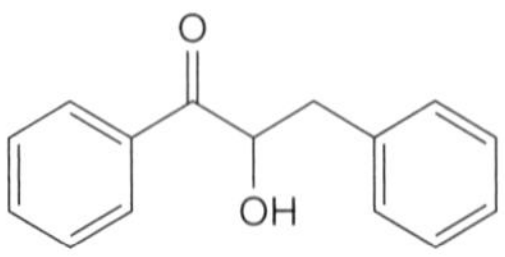 OH b. 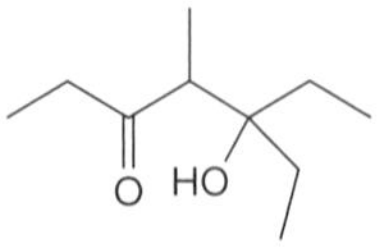 O HO c. 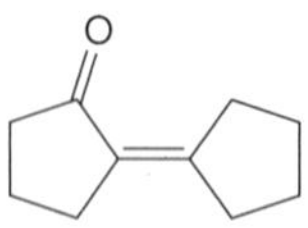 d.

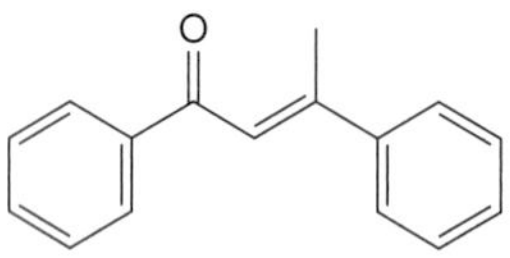

9. Which compounds can*not* be used to form **F?**

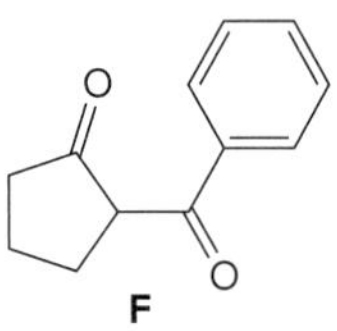

F

a.  + COCl

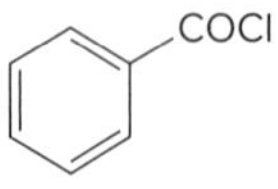

b. 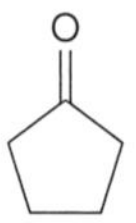 + CO_2Et

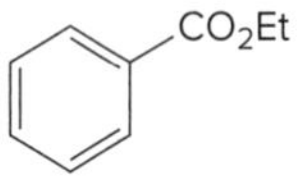

c. 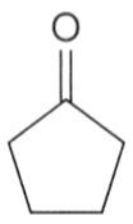 + CHO

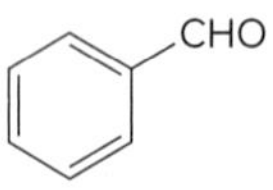

d. 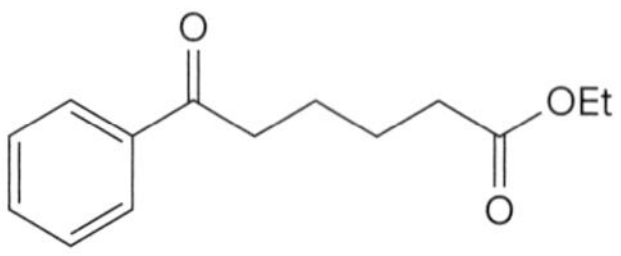 OEt

10. Which compound undergoes an aldol reaction in the presence of aqueous base?

a. 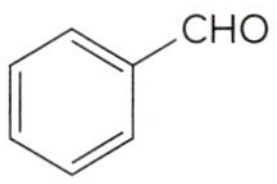 CHO b. 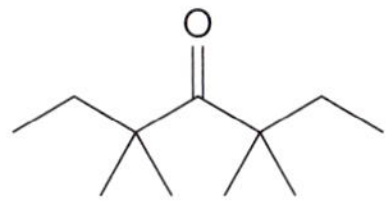 O c. 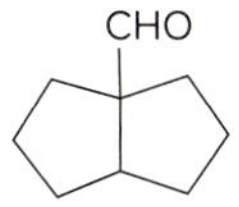 CHO d. 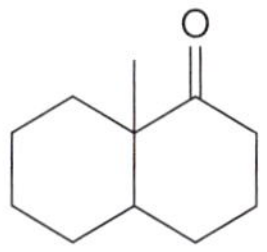 O

PROBLEMS

The Aldol Reaction

22.31 Draw the product formed in each directed aldol reaction.

a. 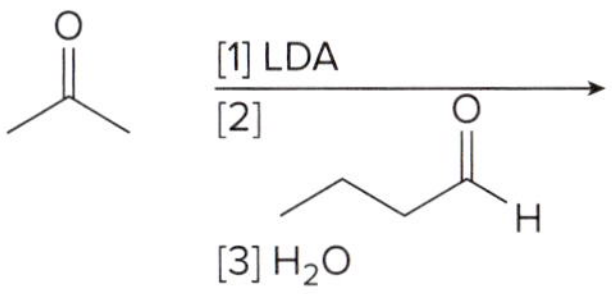 O [1] LDA [2] O H [3] H_2O

b. 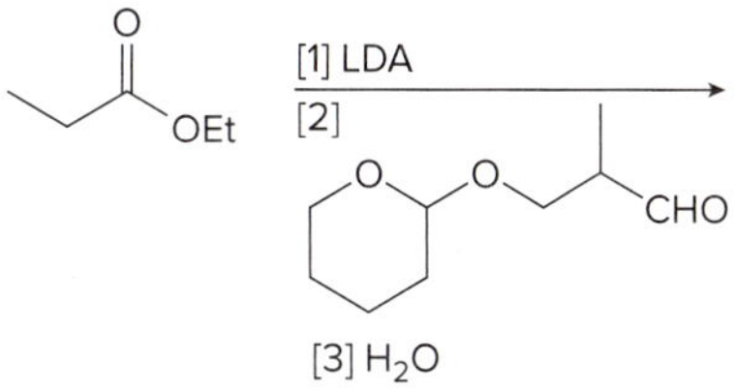 O OEt [1] LDA [2] O O CHO [3] H_2O

22.32 Draw the product formed when each dicarbonyl compound undergoes an intramolecular aldol reaction followed by dehydration, when possible.

a.  OHC CHO b. 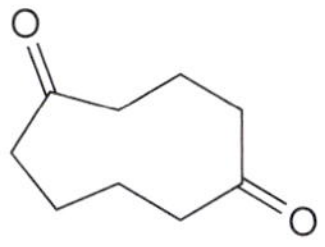 O O c. 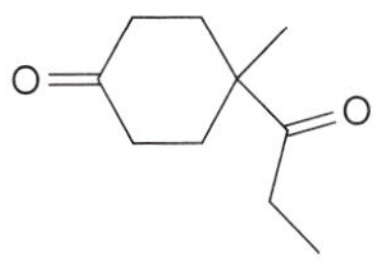 O O

22.33 What starting materials are needed to synthesize each compound using an aldol or similar reaction?

a. 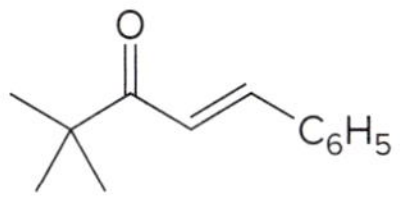 O C_6H_5 b. O c. O C_6H_5 C_6H_5 d. 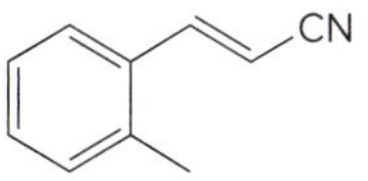 CN

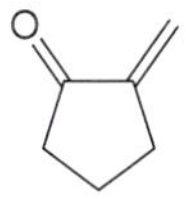

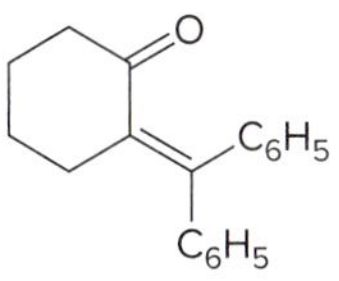

22.34 What product is formed when a solution of **A** and **B** is treated with mild base? This reaction is the first step in the synthesis of rosuvastatin (sold as a calcium salt under the trade name Crestor), a drug used to treat patients with high cholesterol.

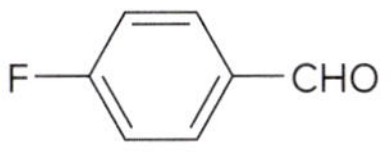

F CHO + O O OEt

A 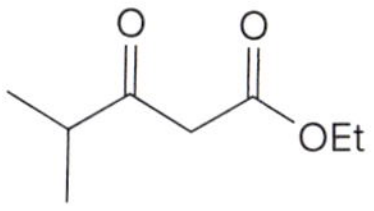 **B**

OH OH CO_2H N CH_3SO_2 N N F

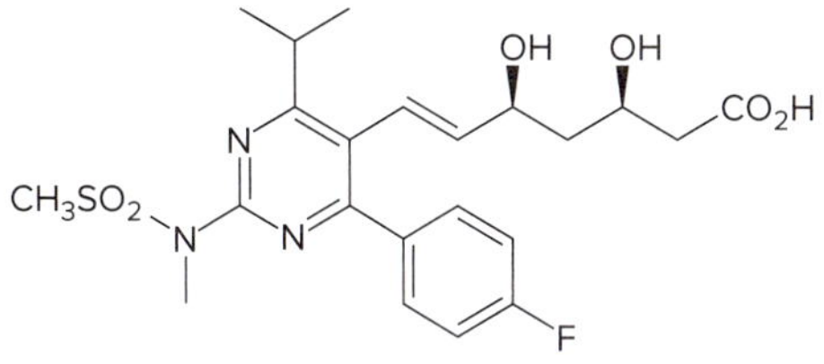

rosuvastatin

22.35 What dicarbonyl compound is needed to prepare each compound by an intramolecular aldol reaction?

a. 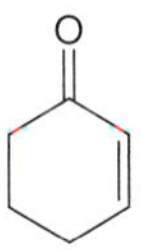 O b. 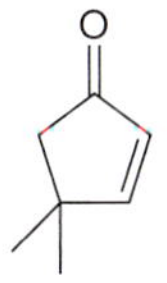 O c. 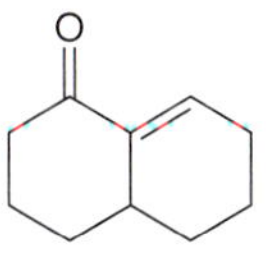 O d. 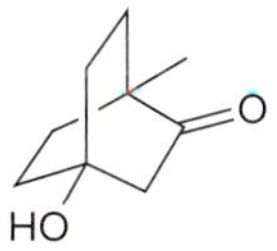 O HO

22.36 Identify the structures of **C** and **D** in the following reaction sequence.

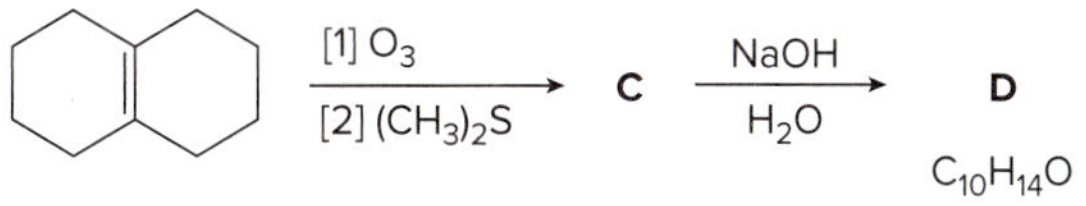

[1] O_3 [2] $(CH_3)_2S$ → **C** NaOH, H_2O → **D** $C_{10}H_{14}O$

22.37 Explain why ketone **K** undergoes aldol reactions but ketone **J** does not.

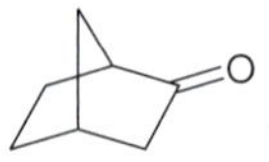

K 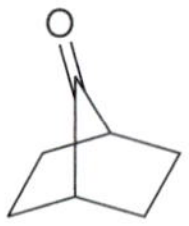 **J**

The Claisen and Dieckmann Reactions

22.38 Draw the product formed from a Claisen reaction with the given starting materials using ^{-}OEt, EtOH.

a. 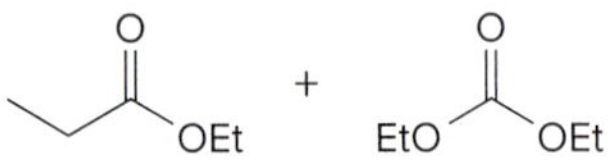 OEt + EtO OEt b. 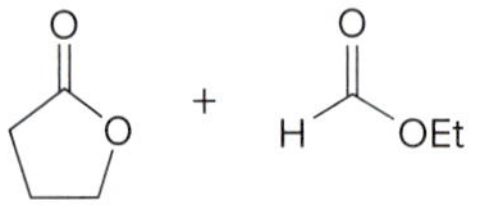 O + H OEt c. 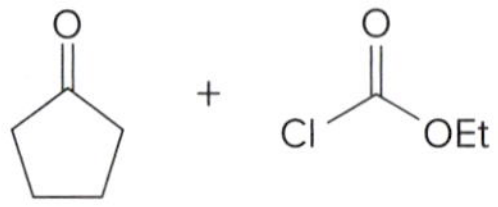 + Cl OEt

22.39 What starting materials are needed to synthesize each compound by a crossed Claisen reaction?

a. 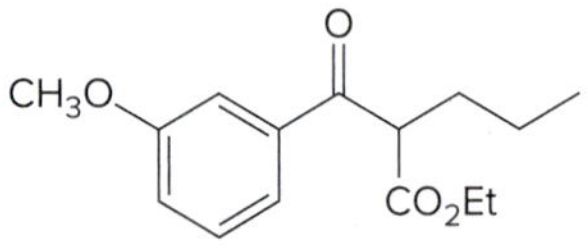 CH_3O CO_2Et b. C_6H_5 c. CHO

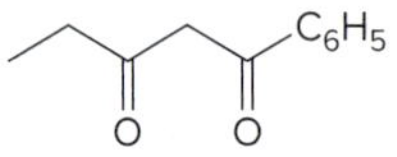

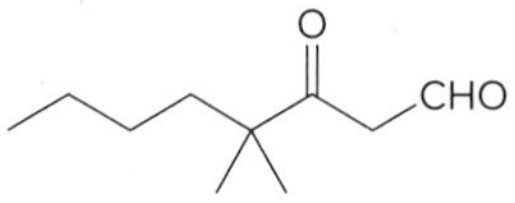

22.40 Even though **B** contains three ester groups, a single Dieckmann product results when **B** is treated with $NaOCH_3$ in CH_3OH, followed by H_3O^+. Draw the structure and explain why it is the only product formed.

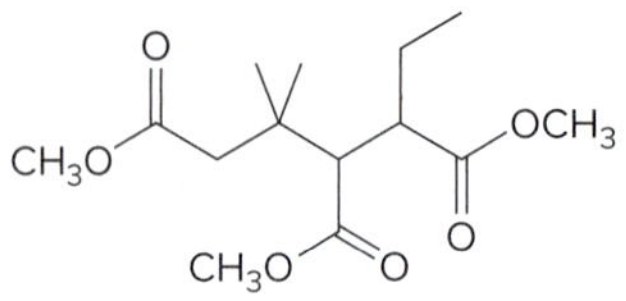

CH_3O OCH_3 CH_3O

B

Michael Reaction

22.41 What starting materials are needed to prepare each compound using a Michael reaction?

a. 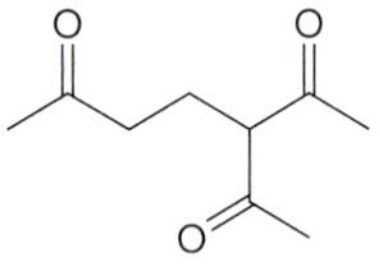 b. 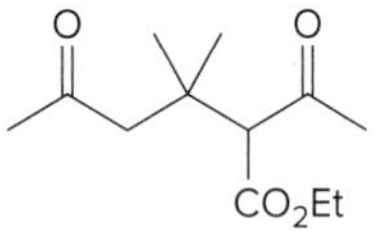 CO_2Et c. 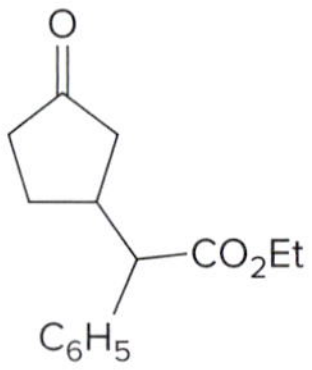 CO_2Et C_6H_5

22.42 β-Vetivone is isolated from vetiver, a perennial grass that yields a variety of compounds used in traditional eastern medicine, pest control, and fragrance. In one synthesis, ketone **A** is converted to β-vetivone by a two-step process: Michael reaction, followed by intramolecular aldol reaction. (a) What Michael acceptor is needed for the conjugate addition? (b) Draw a stepwise mechanism for the aldol reaction, which forms the six-membered ring.

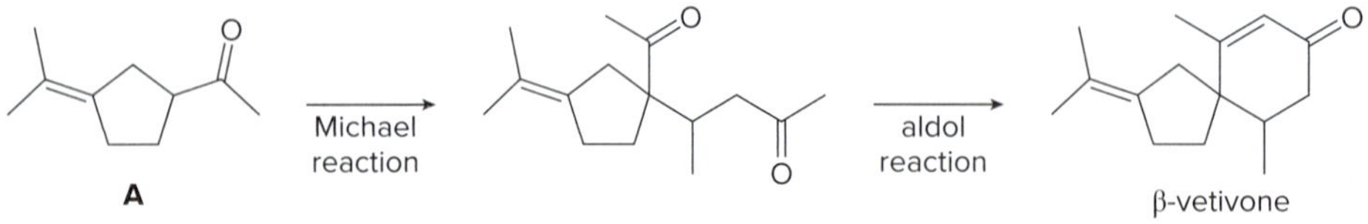

A → Michael reaction → aldol reaction → β-vetivone

Robinson Annulation

22.43 Draw the product of each Robinson annulation from the given starting materials using ^{-}OH in H_2O solution.

a. 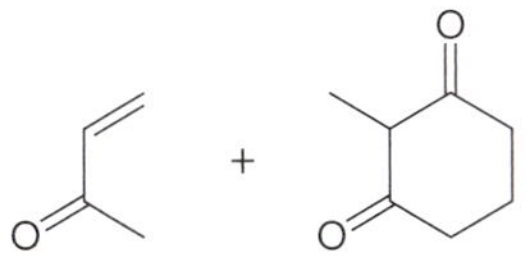 +

b. 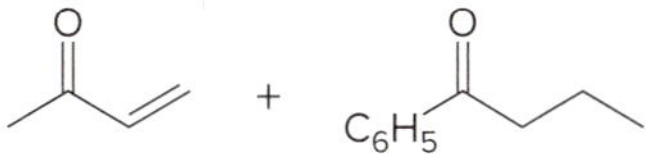 + C_6H_5

c. 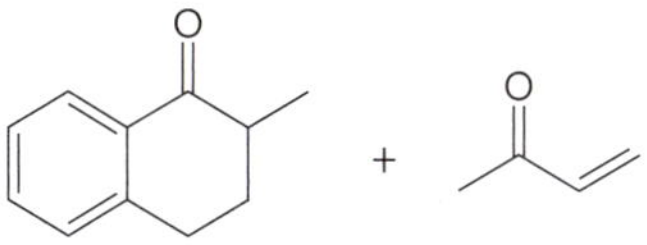 +

22.44 What starting materials are needed to synthesize each compound using a Robinson annulation?

a.

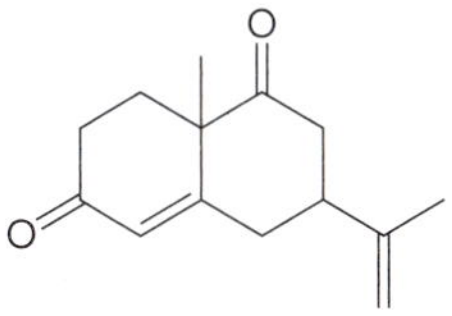

b.

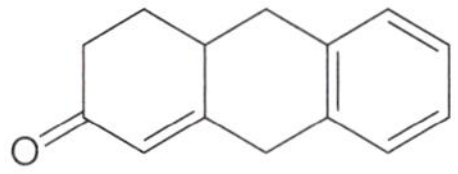

c.

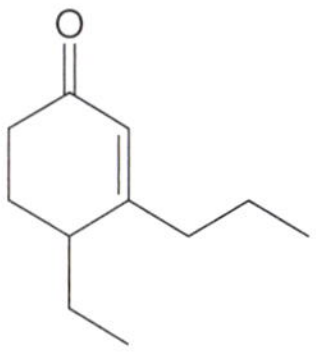

Reactions

22.45 Draw the organic products formed in each reaction.

a. 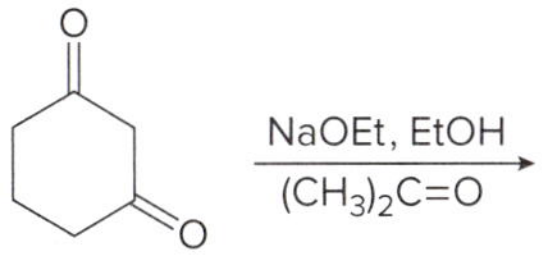 NaOEt, EtOH; $(CH_3)_2C{=}O$

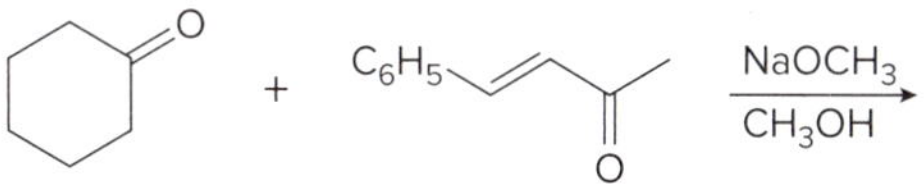

b. 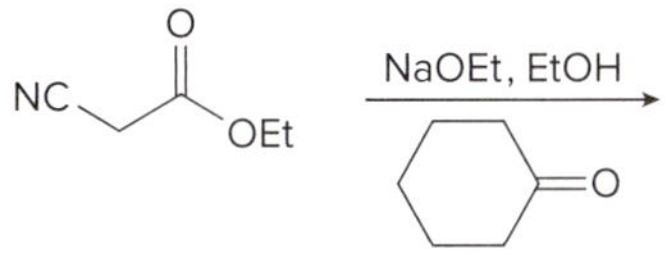 NC–CH_2–COOEt; NaOEt, EtOH; cyclohexanone

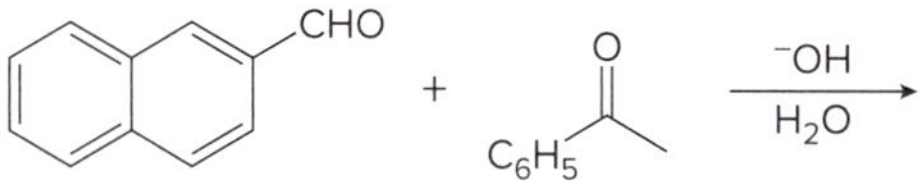

c. 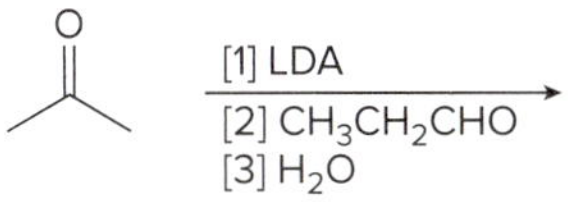 [1] LDA [2] CH_3CH_2CHO [3] H_2O

d. + C_6H_5; $NaOCH_3$, CH_3OH

e. CHO + C_6H_5; ^{-}OH, H_2O

f. 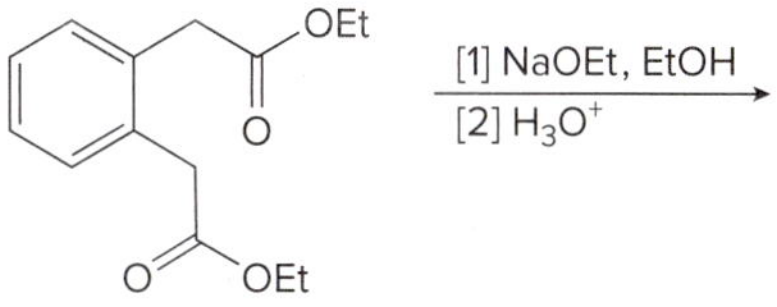 OEt, OEt; [1] NaOEt, EtOH [2] H_3O^+

22.46 What products would be formed by a retro-aldol reaction of **X?**

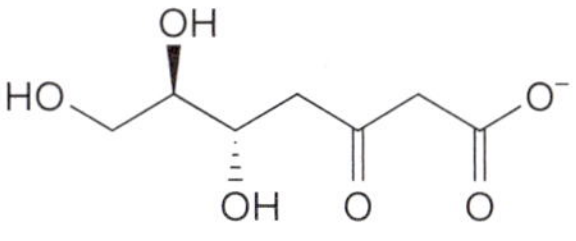

HO, OH, OH, O, O, O^-

X

22.47 As we show in Chapter 30, crossed aldol reactions are common in biological systems. For example, the first step in the citric acid cycle (Section 30.6) is the reaction of the α carbon of acetyl CoA (Section 20.16) with the ketone carbonyl of oxaloacetate. Draw the product of this reaction, a single enantiomer with the *S* configuration at the new stereogenic center.

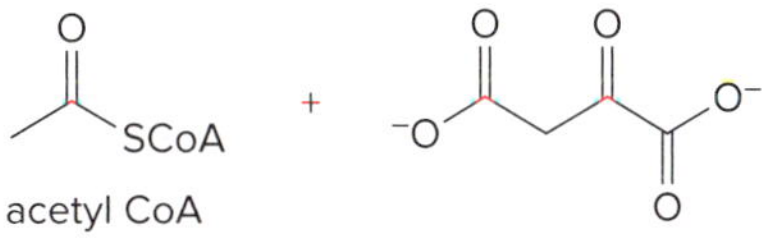

SCoA + ^{-}O … O^-

acetyl CoA

oxaloacetate

22.48 What product (including stereochemistry) is formed in each of the following intramolecular reactions?

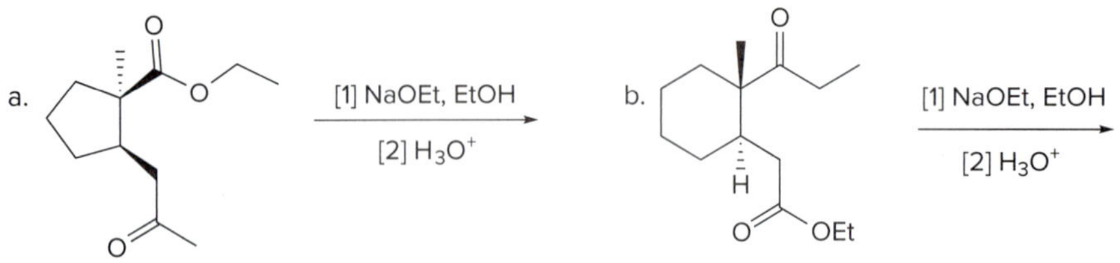

a. [1] NaOEt, EtOH; [2] H_3O^+

b. [1] NaOEt, EtOH; [2] H_3O^+

H

OEt

22.49 Identify compounds **A** and **B,** two synthetic intermediates in the 1979 synthesis of the plant growth hormone gibberellic acid by Corey and Smith. Gibberellic acid induces cell division and elongation, thus making plants tall and leaves large.

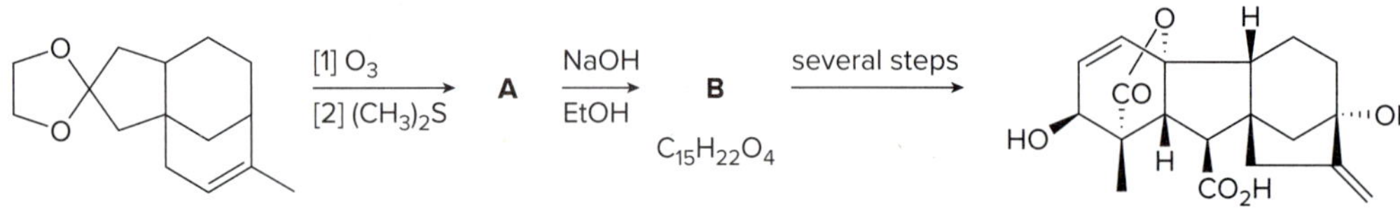

[1] O_3 [2] $(CH_3)_2S$ → **A** → NaOH, EtOH → **B** $C_{15}H_{22}O_4$ → several steps →

gibberellic acid

Mechanisms

22.50 In theory, the intramolecular aldol reaction of 6-oxoheptanal could yield the three compounds shown. It turns out, though, that 1-acetylcyclopentene is by far the major product. Why are the other two compounds formed in only minor amounts? Draw a stepwise mechanism to show how all three products are formed.

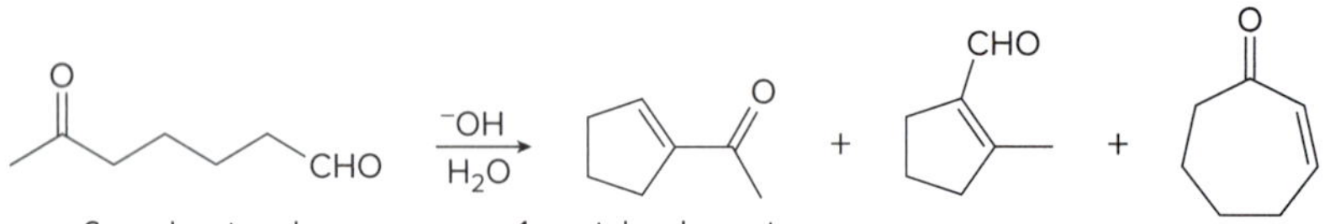

6-oxoheptanal → ^-OH, H_2O → 1-acetylcyclopentene major product + CHO +

22.51 Draw a stepwise mechanism that illustrates how both products are formed in the following reaction.

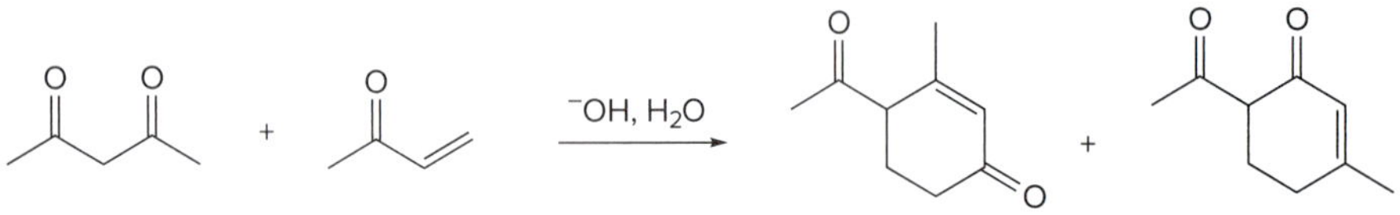

^-OH, H_2O

22.52 Biyouyanagin A is an anti-HIV agent isolated from the leaves of a plant of the genus *Hypericum* that is used in traditional Japanese medicine. The six-membered ring in biyouyanagin A was formed in the given reaction. Draw a stepwise mechanism for this process.

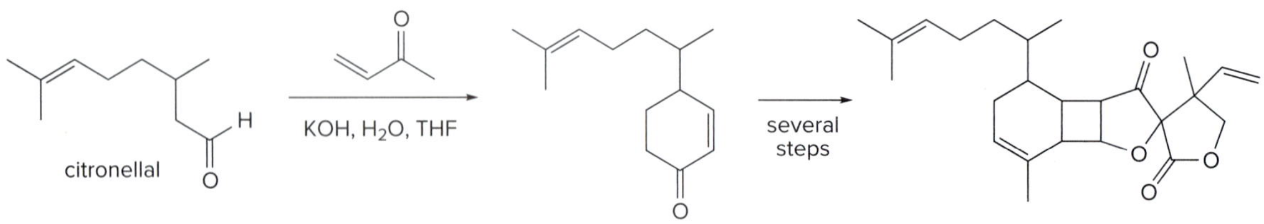

citronellal → KOH, H_2O, THF → → several steps →

biyouyanagin A

22.53 Jiadifenin is a natural product isolated from the fruit of the Chinese plant *Illicium jiadifengpi,* which has potential for use in treating neurodegenerative disease. The lactone in jiadifenin is formed in the following two-step reaction. Write a stepwise mechanism for the conversion of **M** to **N.**

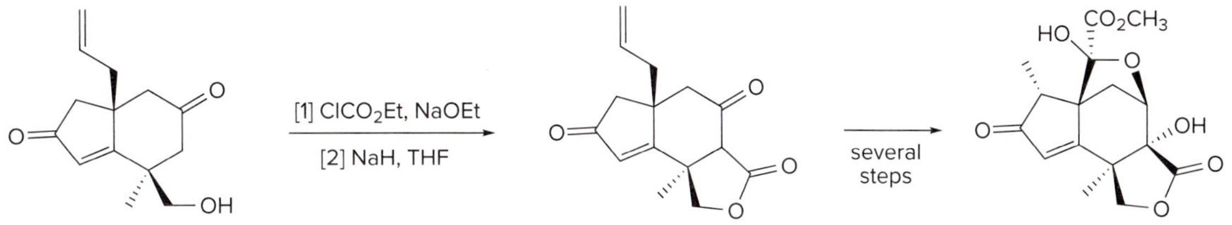

[1] $ClCO_2Et$, NaOEt
[2] NaH, THF
several steps
OH
CO_2CH_3
HO

M **N** jiadifenin

22.54 Reaction of **X** and phenylacetic acid forms an intermediate **Y,** which undergoes an intramolecular reaction to yield rofecoxib. Rofecoxib is a nonsteroidal anti-inflammatory agent once marketed under the trade name Vioxx, now withdrawn from the market because of increased risk of heart attacks from long-term use in some patients. Identify **Y** and draw a stepwise mechanism for its conversion to rofecoxib.

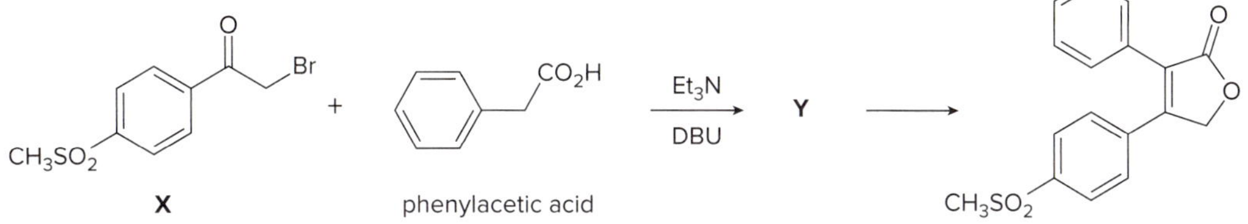

CH_3SO_2 Br + CO_2H Et_3N DBU **Y** CH_3SO_2

X phenylacetic acid rofecoxib

22.55 Coumarin, a naturally occurring compound isolated from lavender, sweet clover, and tonka bean, is made in the laboratory from *o*-hydroxybenzaldehyde by the reaction depicted below. Draw a stepwise mechanism for this reaction. Coumarin derivatives are useful synthetic anticoagulants.

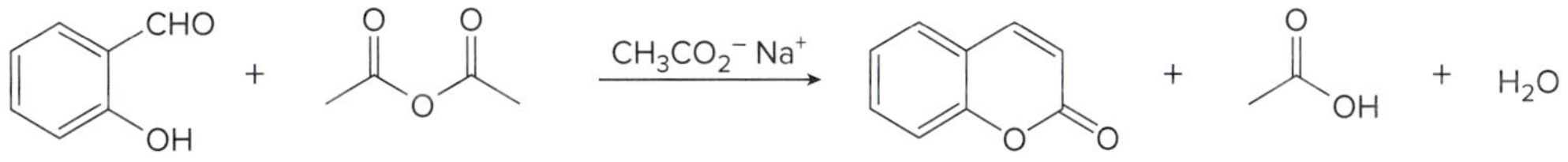

CHO OH + $CH_3CO_2^-$ Na^+ + OH + H_2O

o-hydroxybenzaldehyde coumarin

22.56 When **A** is treated with aqueous ^-OH, the major product is compound **B,** which undergoes ester hydrolysis and decarboxylation to form **C.** Draw a stepwise mechanism for the conversion of **A** to **B.**

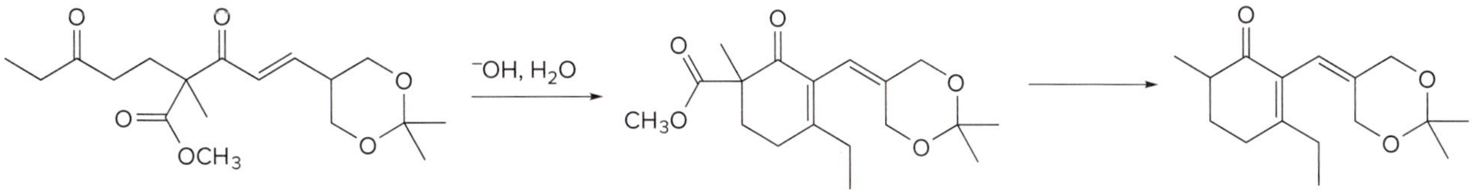

OCH_3 ^-OH, H_2O CH_3O

A **B** **C**

Synthesis

22.57 Devise a synthesis of each compound from cyclopentanone, benzene, and organic alcohols having ≤ 3 C's. You may also use any required organic or inorganic reagents.

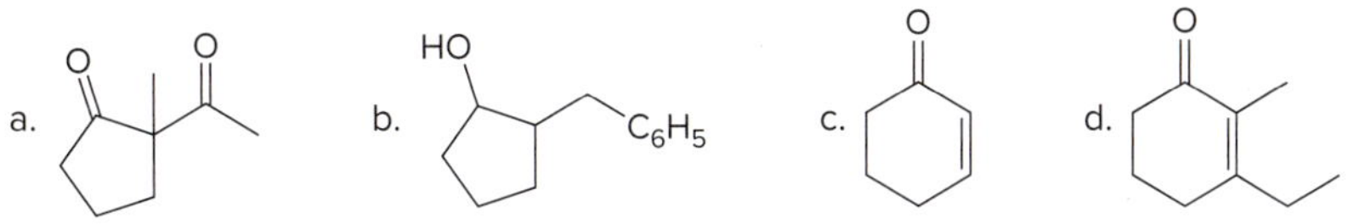

a. b. HO C_6H_5 c. d.

22.58 How would you convert cyclohexanone to each of the following compounds?

a. 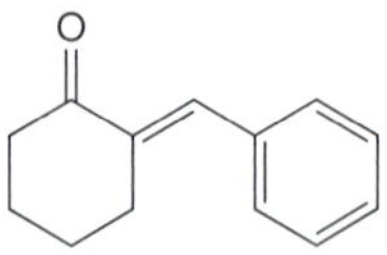 b. 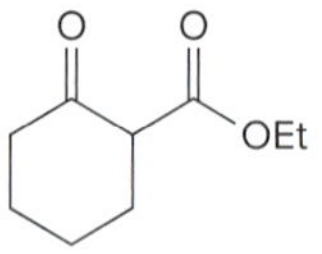 OEt c. 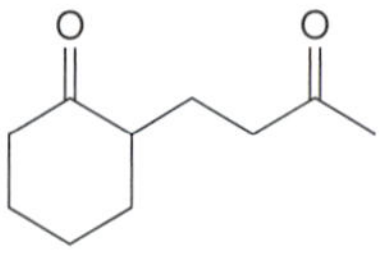 d. 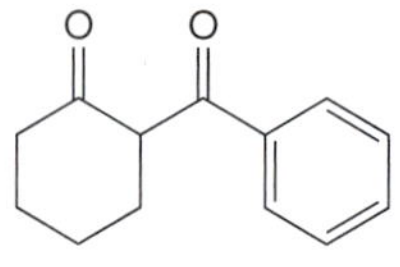 e.

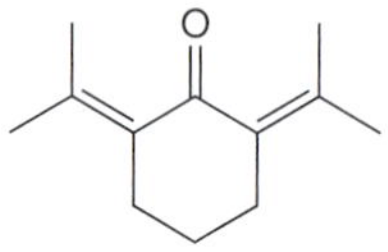

22.59 Devise a synthesis of each compound from $CH_3CH_2CH_2CO_2Et$, benzene, and alcohols having ≤ 2 C's. You may also use any required organic or inorganic reagents.

a. 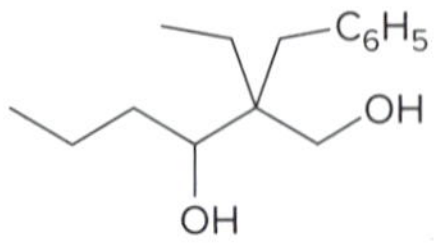 C_6H_5 OH OH b. 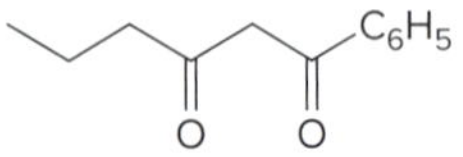 C_6H_5

22.60 Devise a synthesis of each compound using acetone [$(CH_3)_2C{=}O$] as the only source of carbon atoms. You may use any needed organic or inorganic reagents.

a. 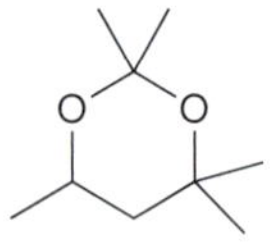 b. 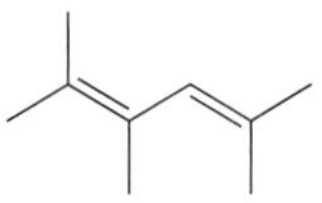 c.

22.61 Octinoxate is an unsaturated ester used as an active ingredient in sunscreens. (a) What carbonyl compounds are needed to synthesize this compound using a condensation reaction? (b) Devise a synthesis of octinoxate from the given organic starting materials and any other needed reagents.

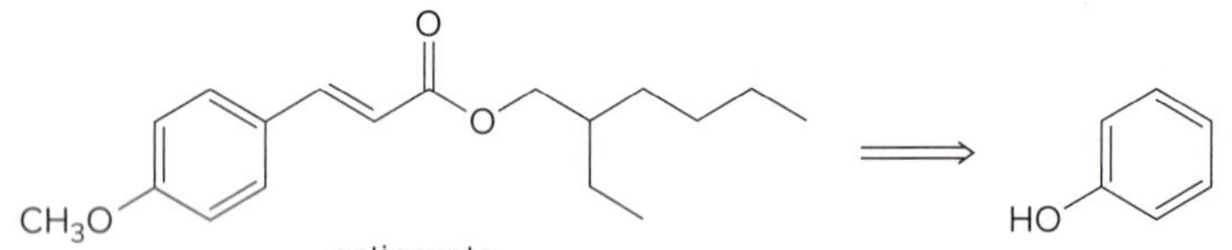

CH_3O octinoxate ⟹ HO + alcohols with < 5 C's

Problem That Combines Concepts

22.62 Answer the following questions about 2-acetylcyclopentanone.

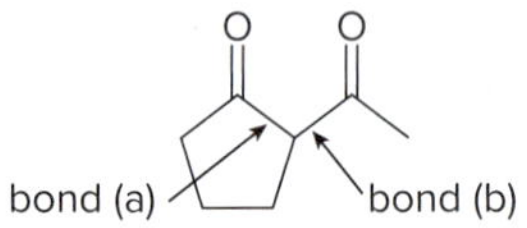

bond (a) bond (b)

2-acetylcyclopentanone

a. What starting materials are needed to form 2-acetylcyclopentanone by a Claisen reaction that forms bond (a)?
b. What starting materials are needed to form 2-acetylcyclopentanone by a Claisen reaction that forms bond (b)?
c. What product is formed when 2-acetylcyclopentanone is treated with $NaOCH_2CH_3$, followed by CH_3I?
d. Draw the Robinson annulation product(s) formed by reaction of 2-acetylcyclopentanone with methyl vinyl ketone ($CH_2{=}CHCOCH_3$).
e. Draw the structure of the most stable enol tautomer(s).

Challenge Problems

22.63 Draw a stepwise mechanism for the following reaction, which was used in the synthesis of ezetimibe (Section 17.6), a drug used to treat patients with high cholesterol.

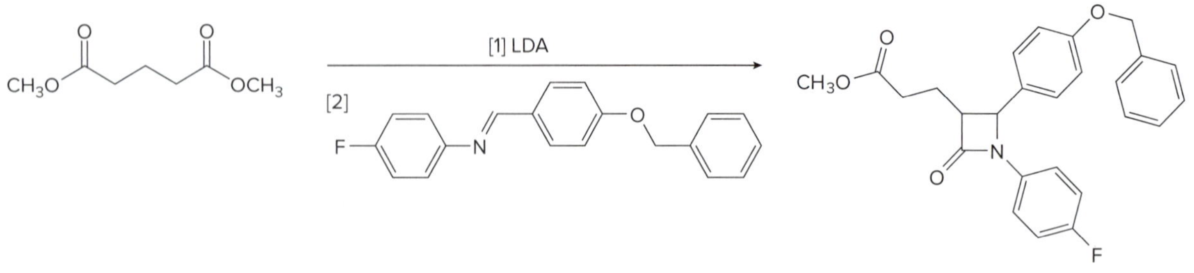

CH_3O ... OCH_3 → [1] LDA; [2] F–C₆H₄–N=CH–C₆H₄–O–CH₂–C₆H₅ → CH_3O ... N ... F

22.64 A key step in a reported synthesis of morphine (Section 2.2), the addictive opiate used to treat severe pain, involves the conversion of **A** to **B.** Draw a stepwise mechanism for this process, which involves both an intramolecular alkylation and an intramolecular aldol reaction.

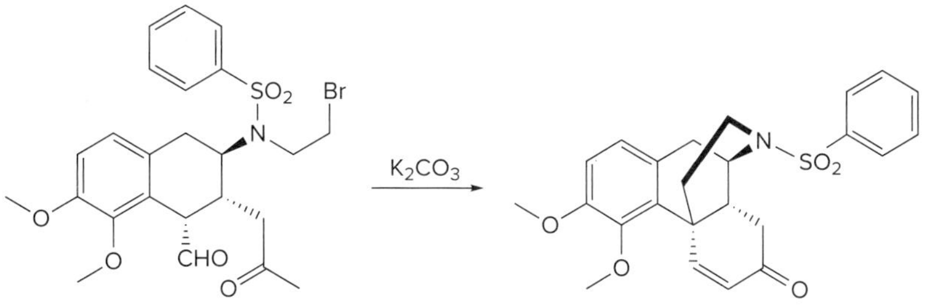

SO_2 Br N CHO O O O A K_2CO_3 N SO_2 O O O B

22.65 Isophorone is formed from three molecules of acetone [$(CH_3)_2C{=}O$] in the presence of base. Draw a mechanism for this process.

O

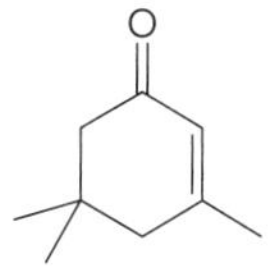

isophorone

22.66 Devise a stepwise mechanism for the following reaction. (Hint: The mechanism begins with the conjugate addition of ^-OH.)

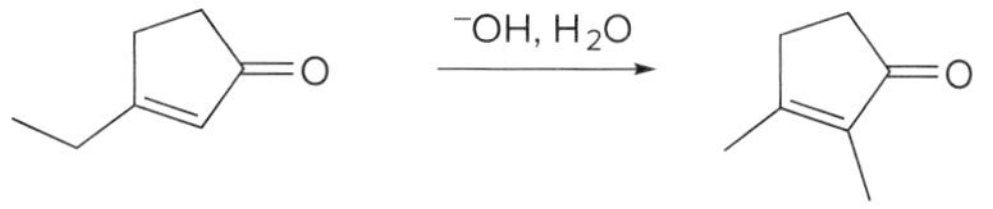

O ^-OH, H_2O O

22.67 Draw a stepwise mechanism for the following reaction. (Hint: Two Michael reactions are needed.)

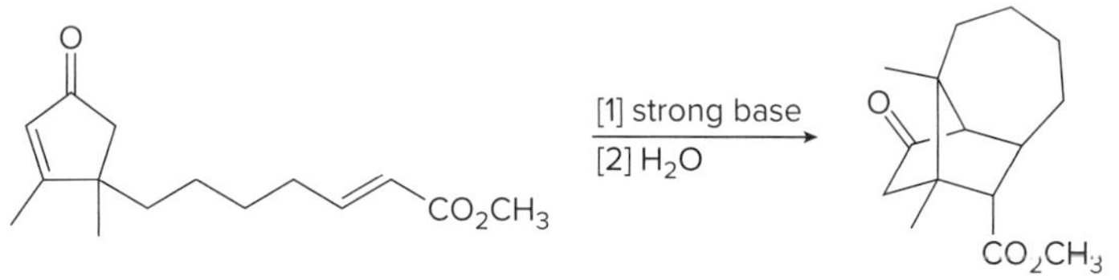

O CO_2CH_3 [1] strong base [2] H_2O O CO_2CH_3

22.68 4-Methylpyridine reacts with benzaldehyde (C_6H_5CHO) in the presence of base to form **A.** (a) Draw a stepwise mechanism for this reaction. (b) Would you expect 2-methylpyridine or 3-methylpyridine to undergo a similar type of condensation reaction? Explain why or why not.

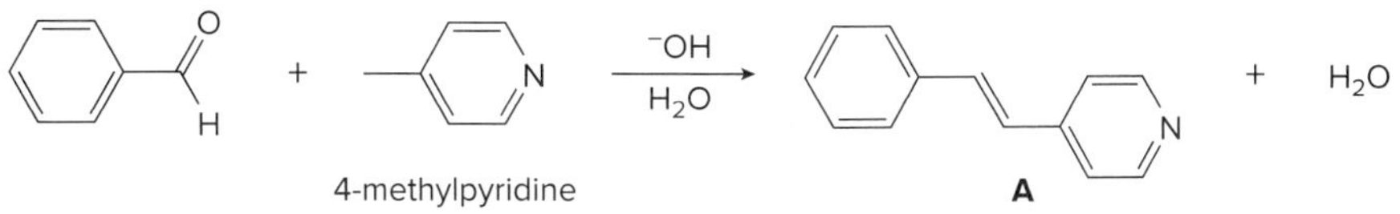

O H + N 4-methylpyridine ^-OH H_2O N A + H_2O

22.69 Draw a stepwise mechanism for the following reaction, one step in the synthesis of the cholesterol-lowering drug pitavastatin, marketed in Japan as a calcium salt under the name Livalo.

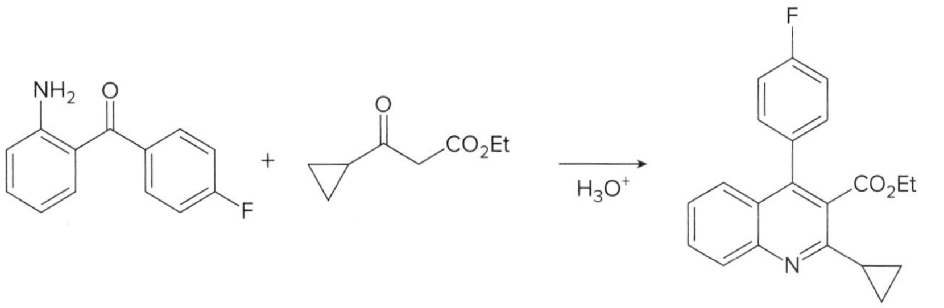

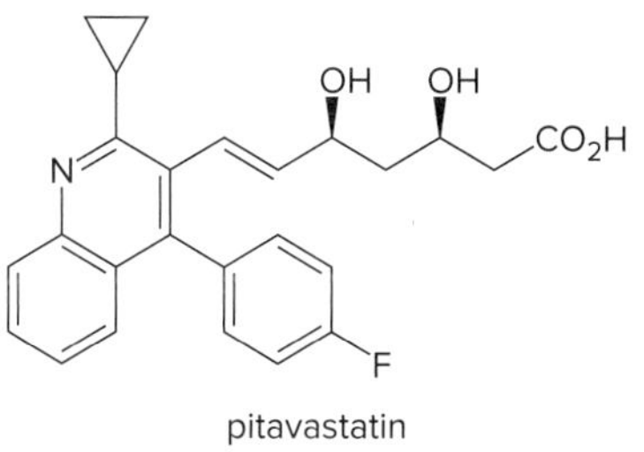

NH_2 O F + O CO_2Et H_3O^+ F CO_2Et N

N OH OH CO_2H F

pitavastatin

22.70 Devise a stepwise mechanism for the following reaction, a key step in the synthesis of the antibiotic abyssomicin C. Abyssomicin C was isolated from sediment collected from almost 1000 ft below the surface in the Sea of Japan. (Hint: The mechanism begins with a Dieckmann reaction.)

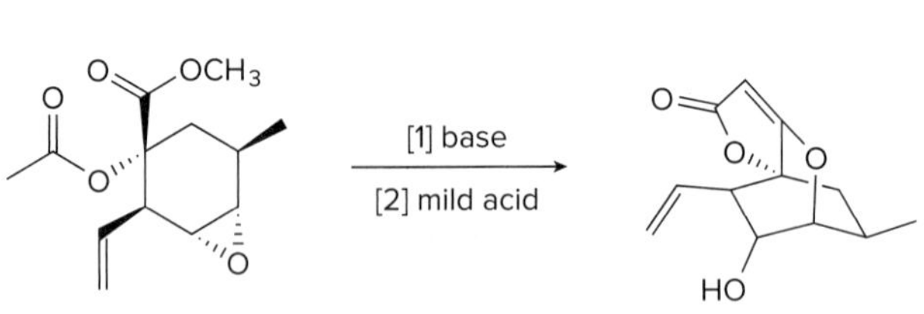

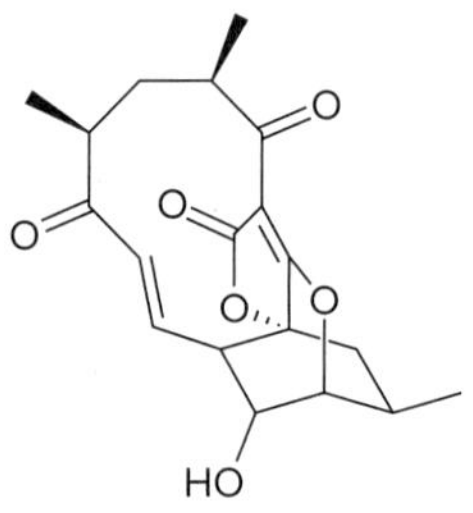

OCH_3 O O O O

[1] base

[2] mild acid

O O O HO

O O O O O HO

abyssomicin C

SELF-TEST ANSWERS

1. c 2. a 3. b 4. d 5. a 6. c 7. b 8. a 9. c 10. d

Amines

23

kai4107/Shutterstock

23.1 Introduction
23.2 Structure and bonding
23.3 Nomenclature
23.4 Physical and spectroscopic properties
23.5 Interesting and useful amines
23.6 Preparation of amines
23.7 Reactions of amines—General features
23.8 Amines as bases
23.9 Relative basicity of amines and other compounds
23.10 Amines as nucleophiles
23.11 Hofmann elimination
23.12 Reaction of amines with nitrous acid
23.13 Substitution reactions of aryl diazonium salts
23.14 Coupling reactions of aryl diazonium salts
23.15 Application: Synthetic dyes and sulfa drugs

Nicotine is an addictive and highly toxic amine isolated from tobacco. In small doses it acts as a stimulant, but in large doses it causes depression, nausea, and even death. Nicotine is synthesized in plants as a defense against insect predators and is used commercially as an insecticide. In Chapter 23, we learn about the properties and reactions of amines like nicotine.

Why Study . . . Amines?

We now leave the chemistry of carbonyl compounds to concentrate on **amines,** organic derivatives of ammonia (NH_3), formed by replacing one or more hydrogen atoms by alkyl or aryl groups. **Amines are stronger bases and better nucleophiles than other neutral organic compounds,** so much of Chapter 23 focuses on these properties.

Like that of alcohols, the chemistry of amines does not fit neatly into one reaction class, and this can make learning the reactions of amines challenging. Many interesting natural products and widely used drugs are amines, so you also need to know how to introduce this functional group into organic molecules.

23.1 Introduction

***Amines* are organic nitrogen compounds,** formed by replacing one or more hydrogen atoms of ammonia (NH_3) with alkyl groups. As discussed in Section 3.2, amines are classified as 1°, 2°, or 3° by the number of alkyl groups bonded to the *nitrogen* atom.

1° amine	**2° amine**	**3° amine**
(**1 R** group on N)	(**2 R** groups on N)	(**3 R** groups on N)

Like ammonia, **the amine nitrogen atom has a nonbonded electron pair,** making it both a base and a nucleophile. As a result, amines react with electrophiles to form **ammonium salts**—compounds with a positively charged ammonium ion and an anionic counterion.

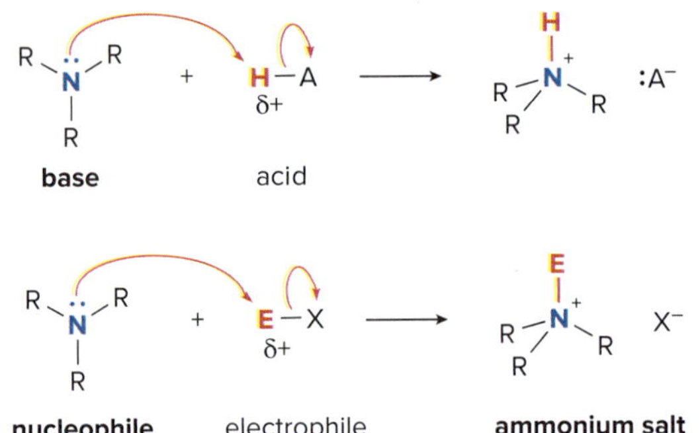

- **The chemistry of amines is dominated by the nonbonded electron pair on the nitrogen atom.**

23.2 Structure and Bonding

An amine nitrogen atom is surrounded by three atoms and one nonbonded electron pair, making the N atom ***sp*³ hybridized** and **trigonal pyramidal,** with bond angles of approximately 109.5°.

147 pm

108°

CH_3NH_2

The electrostatic potential map of CH_3NH_2 clearly shows that the electron-rich region (in red) is on the N atom.

Because nitrogen is much more electronegative than carbon or hydrogen, **the C–N and N–H bonds are all polar,** with the N atom electron rich and the C and H atoms electron poor, as seen in the electrostatic potential map of CH_3NH_2 (methylamine).

An amine nitrogen atom bonded to an electron pair and three different groups is technically a stereogenic center, so two nonsuperimposable trigonal pyramids can be drawn.

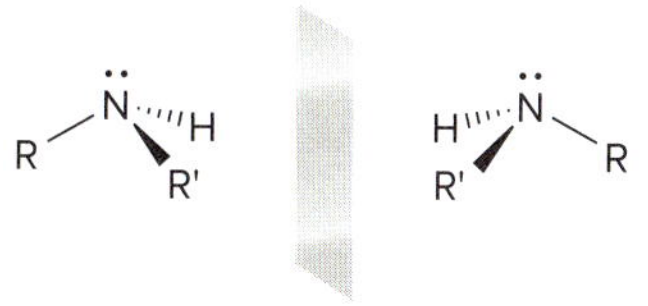

nonsuperimposable mirror images

This does not mean, however, that such an amine exists as two different enantiomers, because one is rapidly converted to the other at room temperature. The amine flips inside out, passing through a trigonal planar (achiral) transition state. **Because the two enantiomers interconvert, we can ignore the chirality of the amine nitrogen.**

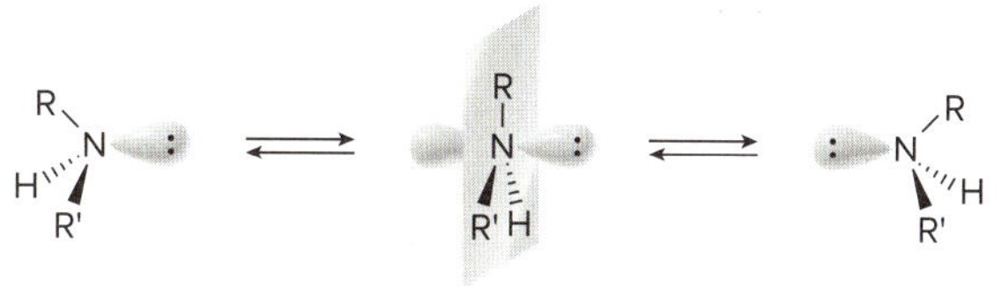

planar transition state

In contrast, **the chirality of an ammonium ion with four different groups on N *cannot* be ignored.** Because there is no nonbonded electron pair on the nitrogen atom, **interconversion cannot occur,** and the N atom is just like a carbon atom with four different groups around it.

chiral ammonium ion

- **The N atom of an ammonium ion is a stereogenic center when N is surrounded by four different groups.**

Problem 23.1 Label the stereogenic centers in each compound.

a.

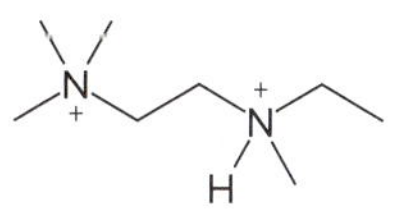

b. 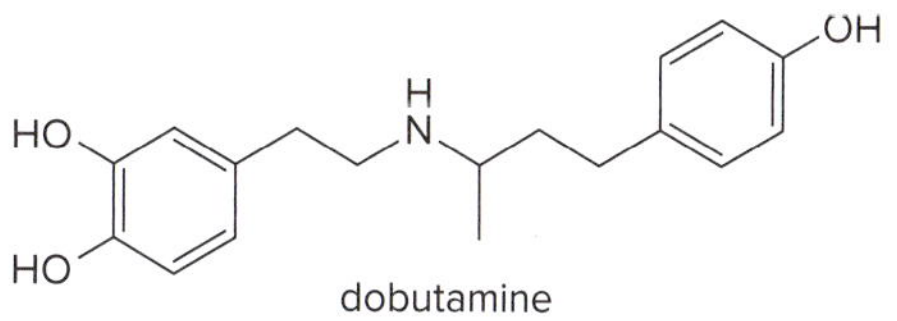 dobutamine
(heart stimulant used in stress tests to measure cardiac fitness)

23.3 Nomenclature

23.3A Primary Amines

Primary amines are named using either systematic or common names.

- **To assign the systematic name, find the longest continuous carbon chain bonded to the amine nitrogen, and change the *-e* ending of the parent alkane to the suffix *-amine.* Then use the usual rules of nomenclature to number the chain and name the substituents.**
- **To assign a common name, name the alkyl group bonded to the nitrogen atom and add the suffix *-amine,* forming a single word.**

CH_3NH_2
Systematic name: **methanamine**
Common name: **methylamine**

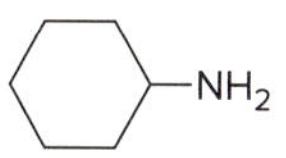

Systematic name: **cyclohexanamine**
Common name: **cyclohexylamine**

23.3B Secondary and Tertiary Amines

Secondary and tertiary amines having identical alkyl groups are named by using the prefix ***di-*** or ***tri-*** with the name of the primary amine.

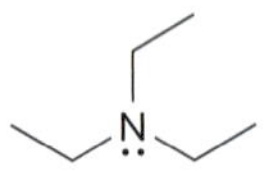

triethylamine

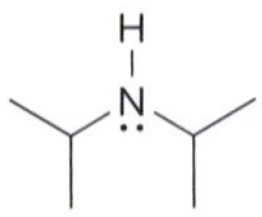

diisopropylamine

Secondary and tertiary amines having more than one kind of alkyl group are named as ***N*-substituted primary amines,** using the following procedure.

How To Name 2° and 3° Amines with Different Alkyl Groups

Example Name the following 2° amine: $(CH_3)_2CHNHCH_3$.

Step [1] **Designate the longest alkyl chain (or largest ring) bonded to the N atom as the parent amine and assign a systematic name.**

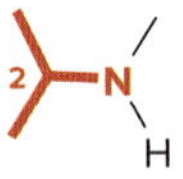

3 C's in the longest chain ---→ propan-2-amine

Step [2] **Name the other groups on the N atom as alkyl groups, alphabetize the names, and put the prefix *N*- before the name.**

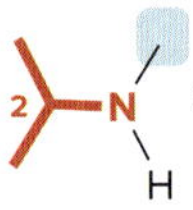

methyl substituent **Answer: *N*-methylpropan-2-amine**

Sample Problem 23.1 Naming an Amine

Name each amine.

a. 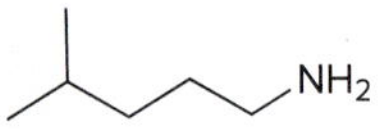 b.

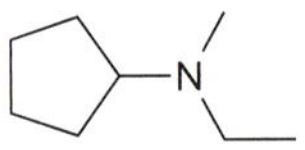

Solution

a. [1] A 1° amine: Find and name the longest chain containing the amine nitrogen.

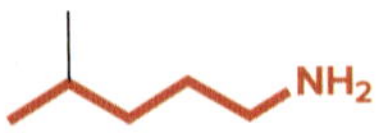

pentane (5 C's) ---→ **pentan*amine***

[2] Number and name the substituents.

You must use a number to show the location of the NH_2 group.

Answer: 4-methylpentan-1-amine

b. For a 3° amine, one alkyl group on N is the principal R group and the others are substituents.

[1] Name the ring bonded to the N.

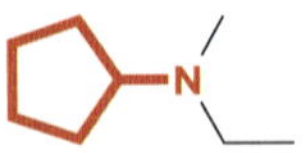

cyclopentanamine

[2] Name the substituents.

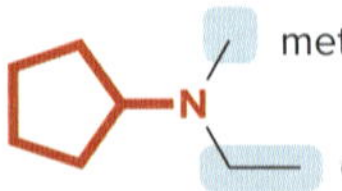

methyl

ethyl

Two N's are needed, one for each alkyl group.

Answer: *N*-ethyl-*N*-methylcyclopentanamine

Problem 23.2 Name each amine.

a. NH_2

b. N–H

c. N

d. NH_2

e. N–H

f. N–H

More Practice: Try Problems 23.37, 23.38.

23.3C Aromatic Amines

Aromatic amines are named as derivatives of aniline.

aniline **N-ethylaniline** **o-bromoaniline**

23.3D Miscellaneous Nomenclature Facts

An NH_2 group named as a substituent is called an **amino group.**

There are many different **nitrogen heterocycles,** and each ring type is named differently depending on the number of N atoms in the ring, the ring size, and whether it is aromatic or not. The structures and names of common nitrogen heterocycles are shown in Figure 23.1.

Figure 23.1 Common nitrogen heterocycles

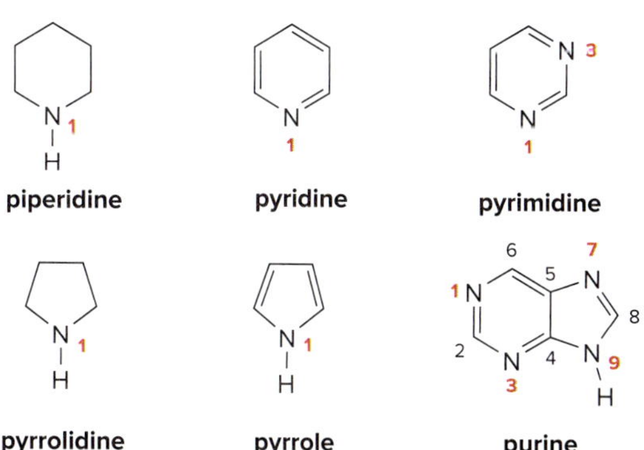

- Heterocycles with one N atom are numbered to place the N atom at the "1" position.
- Heterocycles with two N atoms are numbered to place one N atom at the "1" position and give the second N atom the lower number.

Problem 23.3 Draw a structure corresponding to each name.

a. 2,4-dimethylhexan-3-amine
b. *N*-methylpentan-1-amine
c. *N*-isopropyl-*p*-nitroaniline
d. *N*-methylpiperidine
e. *N,N*-dimethylethanamine
f. 2-aminocyclohexanone
g. *N*-methylaniline
h. *m*-ethylaniline

23.4 Physical and Spectroscopic Properties

23.4A Physical Properties

Intermolecular hydrogen bonding in a 1° amine

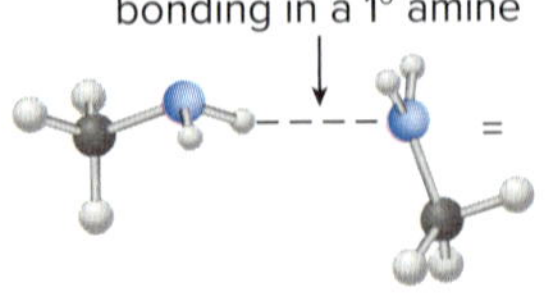

Amines exhibit dipole–dipole interactions because of the polar C–N and N–H bonds. **Primary and secondary amines are also capable of intermolecular hydrogen bonding,** because they contain N–H bonds. Because nitrogen is less electronegative than oxygen, however, intermolecular hydrogen bonds between N and H are *weaker* than those between O and H.

Primary (1°) and 2° amines have higher boiling points than similar compounds (like ethers) incapable of hydrogen bonding, but lower boiling points than alcohols that have stronger intermolecular hydrogen bonds. Tertiary (3°) amines have lower boiling points than 1° and 2° amines of comparable molecular weight, because they have no N–H bonds and are incapable of hydrogen bonding.

bp 38 °C **no N–H bond** | bp 78 °C **N–H bond** | bp 118 °C **O–H bond**

Increasing intermolecular forces
Increasing boiling point

Problem 23.4 Arrange the compounds in order of increasing boiling point.

A (cyclohexyl–NH_2) B C D (cyclohexyl–OH)

23.4B Spectroscopic Properties

The spectroscopic properties of amines have been detailed in Spectroscopy Parts A, B, and C.

- Mass spectra: The odd molecular ion in Section A.1B and fragmentation patterns in Section A.4C
- Infrared absorptions: Section B.4C, Figure B.10, and Table B.2
- ^1H and ^{13}C NMR absorptions: Section C.9A and Tables C.1 and C.5

The general molecular formula for an amine with one N atom is $C_nH_{2n+3}N$.

Key NMR and IR absorptions for amines are summarized in Table 23.1. Figure 23.2 illustrates that the number of N–H peaks in an IR spectrum can be used to distinguish 1°, 2°, and 3° amines.

- **1° Amines show *two* N–H absorptions at 3300–3500 cm^{-1}.**
- **2° Amines show *one* N–H absorption at 3300–3500 cm^{-1}.**
- **3° Amines do *not* absorb at 3300–3500 cm^{-1} because 3° amines have no N–H bonds.**

Table 23.1 Characteristic Spectroscopic Absorptions of Amines

Type of spectroscopy	Type of C, H	Absorption
IR absorption	N–H	3300–3500 cm^{-1} (one or two peaks)
^1H NMR absorptions	N–H	0.5–5.0 ppm
	N–C–H	2.3–3.0 ppm
^{13}C NMR absorption	C–N	30–50 ppm

Figure 23.2 The single bond region of the IR spectra for a 1°, 2°, and 3° amine

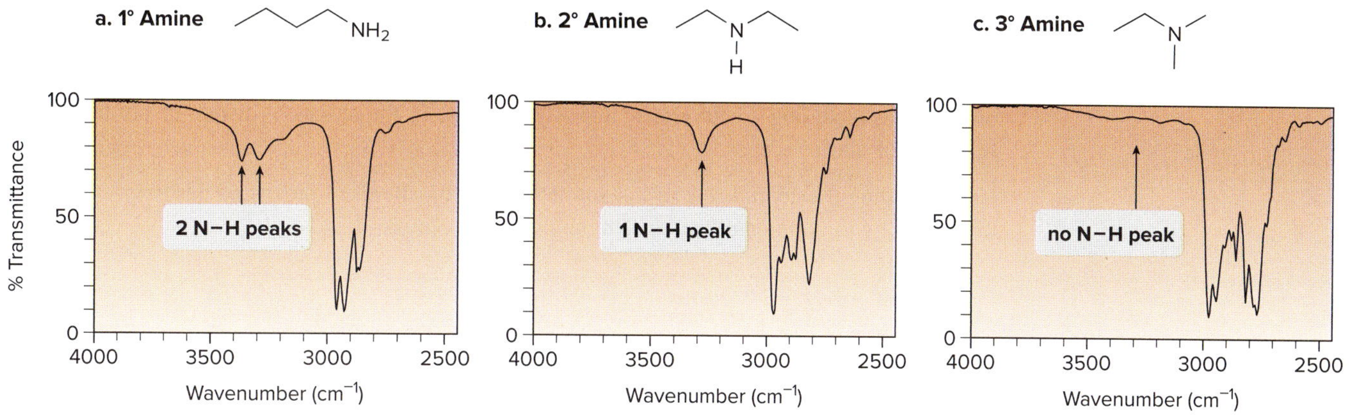

a. 1° Amine
NH2
b. 2° Amine
N
H
c. 3° Amine
N
% Transmittance
100
50
0
2 N–H peaks
1 N–H peak
no N–H peak
4000
3500
3000
2500
Wavenumber (cm−1)

23.5 Interesting and Useful Amines

A great many simple and complex amines occur in nature, and others with biological activity have been synthesized in the lab.

23.5A Simple Amines and Alkaloids

Many low-molecular-weight amines have *very* foul odors. **Trimethylamine** [$(CH_3)_3N$], formed when enzymes break down certain fish proteins, has the characteristic odor of rotting fish. **Putrescine** ($NH_2CH_2CH_2CH_2CH_2NH_2$) and **cadaverine** ($NH_2CH_2CH_2CH_2CH_2CH_2NH_2$) are both poisonous diamines with putrid odors. They, too, are present in rotting fish and are partly responsible for the odors of semen, urine, and bad breath.

The word ***alkaloid*** is derived from the word *alkali,* because aqueous solutions of alkaloids are slightly basic.

Naturally occurring amines derived from plant sources are called **alkaloids.** Alkaloids previously encountered in the text include **pilocarpine** (Chapter 15 opener), **morphine** (Section 20.8), and **cocaine** (Chapter 20 opener). Two other common alkaloids are **atropine** and **scopolamine,** illustrated in Figure 23.3.

Figure 23.3 Two common alkaloids—Atropine and scopolamine

N
OH
O
O

atropine

nightshade

- **Atropine** is an alkaloid isolated from *Atropa belladonna*, the deadly nightshade plant. In the Renaissance, women used the juice of the berries of the nightshade to enlarge the pupils of their eyes for cosmetic reasons. Atropine causes an increase in heart rate, relaxes smooth muscles, and interferes with nerve impulses transmitted by acetylcholine. In higher doses atropine is poisonous, leading to convulsions, coma, and death.

N
OH
O
O
O

scopolamine

angel's trumpet

- **Scopolamine** is a complex amine that occurs in angel's trumpets, ornamental plants with large trumpet-shaped flowers native to South America. Transdermal patches that release a minute dose of scopolamine are used to treat the nausea and vomiting associated with motion sickness.

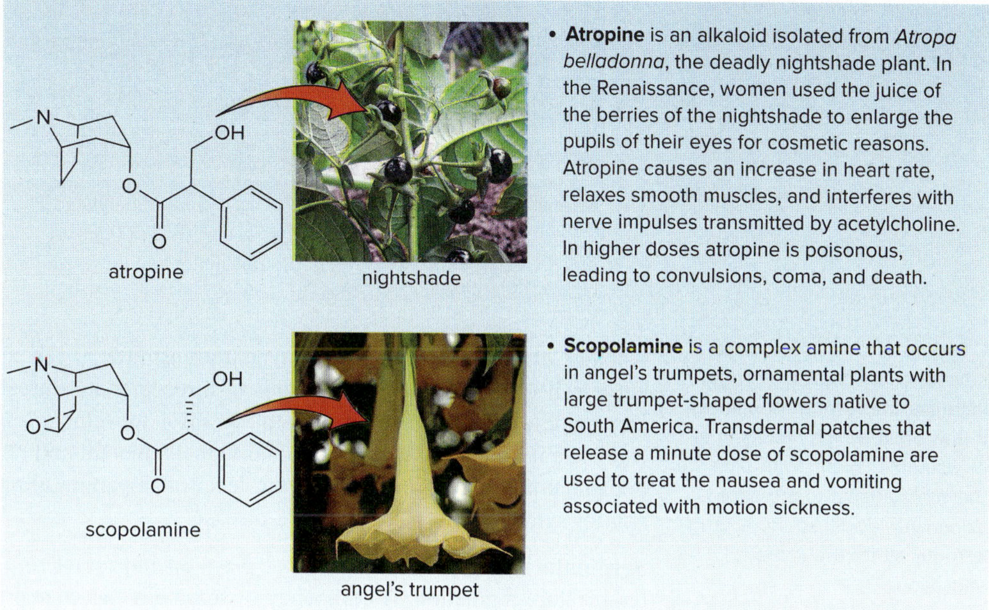

Werner Arnold; James Forte/Getty Images

23.5B Histamine and Antihistamines

histamine

Histamine, a simple triamine first discussed in Section 15.9, is responsible for a wide variety of physiological effects. Histamine is a vasodilator (it dilates capillaries), so it is released at the site of an injury or infection to increase blood flow. It is also responsible for the symptoms of allergies, including a runny nose and watery eyes. In the stomach, histamine stimulates the secretion of acid.

Understanding the central role of histamine in these biochemical processes has helped chemists design drugs to counteract some of its undesirable effects.

fexofenadine
antihistamine

cimetidine
(Tagamet)
antiulcer drug

Antihistamines bind to the same active site of the enzyme that binds histamine in the cell, but they evoke a different response. An antihistamine like **fexofenadine** (trade name Allegra), for example, inhibits vasodilation, so it is used to treat the symptoms of the common cold and allergies. Unlike many antihistamines, fexofenadine does not cause drowsiness because it binds to histamine receptors but does not cross the blood–brain barrier, so it does not affect the central nervous system. **Cimetidine** (trade name Tagamet) is a histamine mimic that blocks the secretion of hydrochloric acid in the stomach, so it is used to treat individuals with ulcers.

23.5C Derivatives of 2-Phenylethanamine

A large number of physiologically active compounds are derived from **2-phenylethanamine, $C_6H_5CH_2CH_2NH_2$.** Some of these compounds are synthesized in cells and needed to maintain healthy mental function. Others are isolated from plant sources or are synthesized in the laboratory and have a profound effect on the brain because they interfere with normal neurochemistry. These compounds include **adrenaline, noradrenaline,** and **methamphetamine.** Each contains a benzene ring bonded to a two-carbon unit with a nitrogen atom (shown in red).

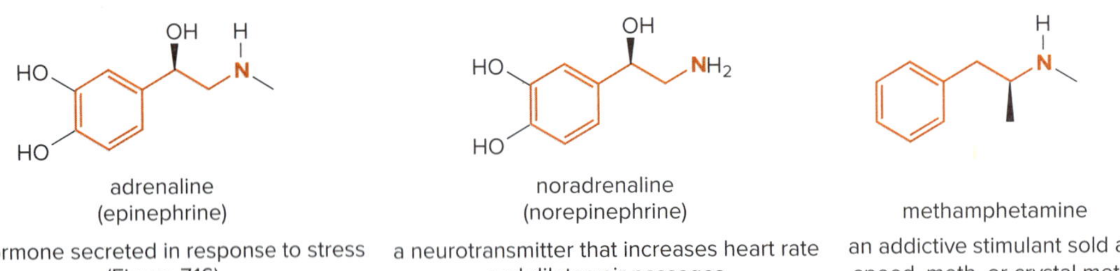

adrenaline (epinephrine): a hormone secreted in response to stress (Figure 7.16)

noradrenaline (norepinephrine): a neurotransmitter that increases heart rate and dilates air passages

methamphetamine: an addictive stimulant sold as speed, meth, or crystal meth

Cocaine, amphetamines, and several other addictive drugs increase the level of dopamine in the brain, which results in a pleasurable "high." With time, the brain adapts to increased dopamine levels, so more drug is required for the same sensation.

Another example, **dopamine,** is a neurotransmitter, a chemical messenger released by one nerve cell (neuron), which then binds to a receptor in a neighboring target cell (Figure 23.4). Dopamine affects brain processes that control movement and emotions, so proper dopamine levels are necessary to maintain an individual's mental and physical health. For example, when dopamine-producing neurons die, the level of dopamine drops, resulting in the loss of motor control symptomatic of Parkinson's disease.

Serotonin is a neurotransmitter that plays an important role in mood, sleep, perception, and temperature regulation. A deficiency of serotonin causes depression. Understanding the central

Figure 23.4 Dopamine—A neurotransmitter

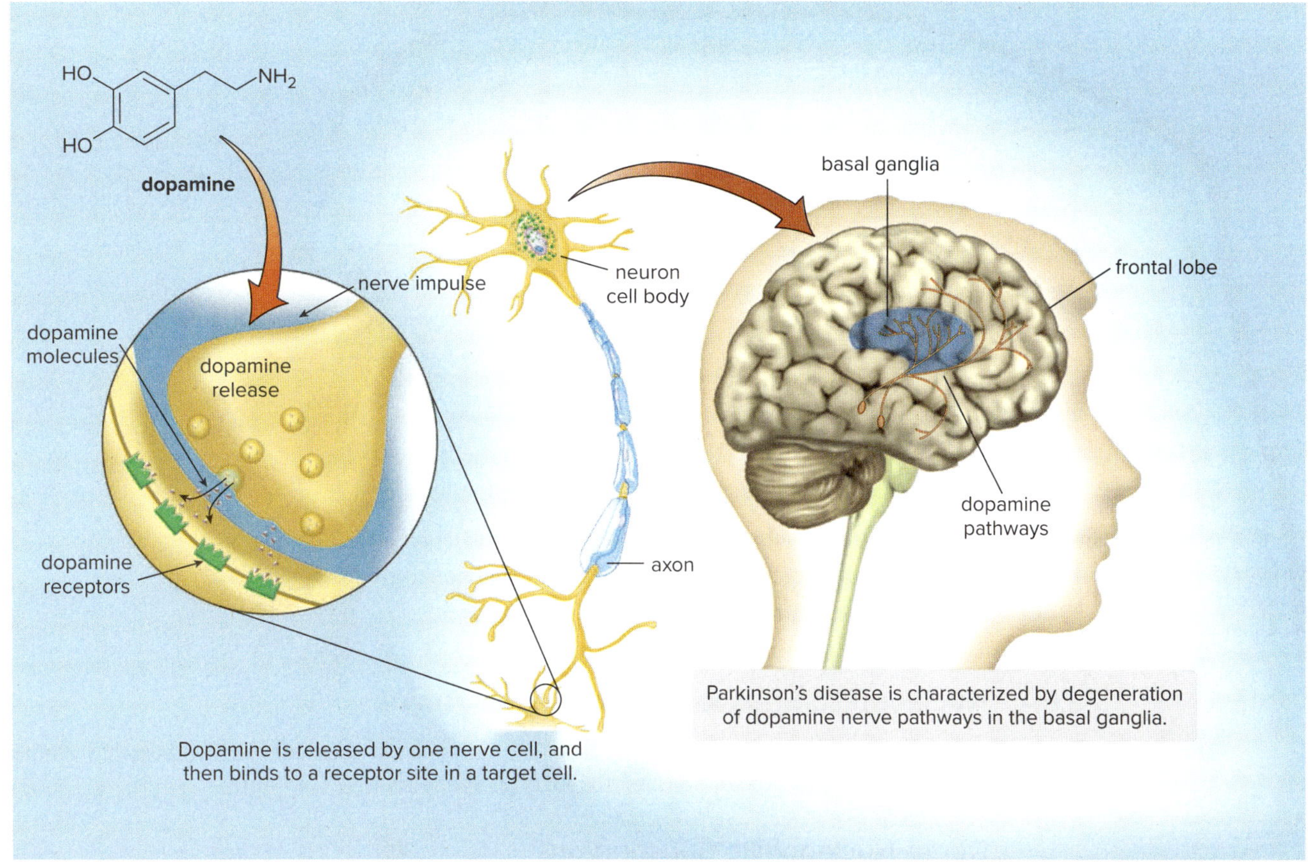

role of serotonin in determining one's mood has led to the development of a variety of drugs for the treatment of depression. The most widely used antidepressants today are selective serotonin reuptake inhibitors (SSRIs). These drugs act by inhibiting the reuptake of serotonin by the neurons that produce it, thus effectively increasing its concentration. Fluoxetine (trade name Prozac) is a common antidepressant that acts in this way.

serotonin   fluoxetine

Bufo toads from the Amazon jungle are the source of the hallucinogen bufotenin.
Daniel C. Smith

Drugs that interfere with the metabolism of serotonin have a profound effect on mental state. For example, bufotenin, isolated from *Bufo* toads from the Amazon jungle, and psilocin, isolated from *Psilocybe* mushrooms, are very similar in structure to serotonin and both cause intense hallucinations.

bufotenin   psilocin

Problem 23.5 Identify the atoms of 2-phenylethanamine in each of the following compounds. Mescaline is a hallucinogen from the peyote cactus, LSD is a hallucinogen, and codeine is a narcotic.

a. 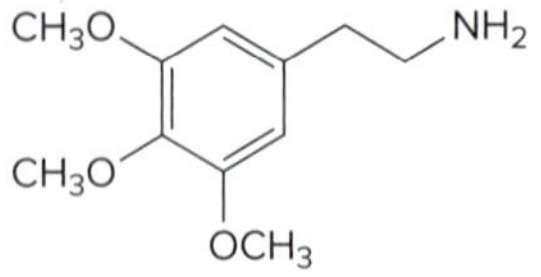 mescaline

b.  **LSD** lysergic acid diethylamide

c. 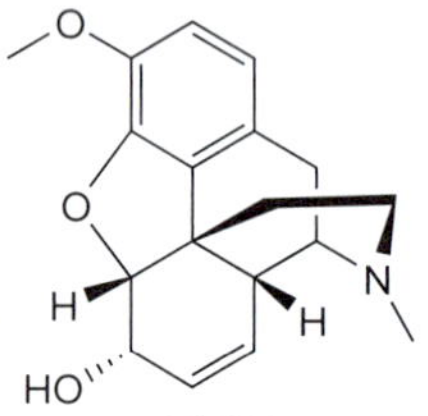 codeine

23.6 Preparation of Amines

In the preparations of a given functional group, many different starting materials form a common product (amines, in this case).

Three types of reactions are used to prepare an amine:

[1] **Nucleophilic substitution** using nitrogen nucleophiles

[2] **Reduction** of other nitrogen-containing functional groups

[3] **Reductive amination** of aldehydes and ketones

23.6A Nucleophilic Substitution Routes to Amines

Nucleophilic substitution is the key step in two different methods for synthesizing amines: direct nucleophilic substitution and the Gabriel synthesis of 1° amines.

Direct Nucleophilic Substitution

Conceptually, the simplest method to synthesize an amine is by **S_N2 reaction of an alkyl halide with NH_3 or an amine.** The method requires two steps (Equation [1]):

[1] **Nucleophilic attack** of the nitrogen nucleophile forms an ammonium salt.

[2] **Removal of a proton** on N forms the amine.

[1] $R{-}X + NH_3 \xrightarrow{S_N2} RNH_3^+ \; X^- \xrightarrow{NH_3} RNH_2 + NH_4^+$ (**1° amine**)

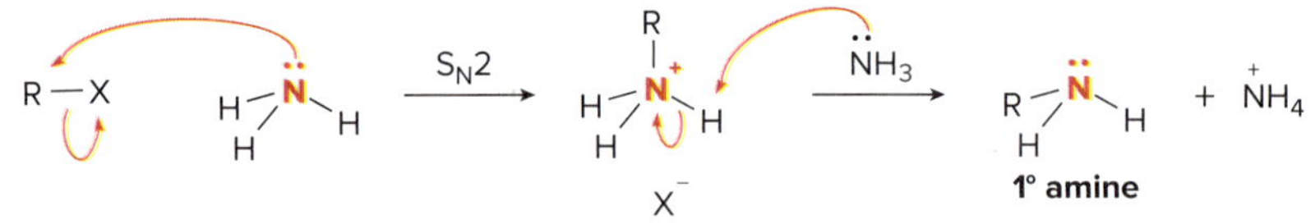

[2] RNH_2 (**1° amine**) $\xrightarrow[\text{[2] } RNH_2]{\text{[1] RX}}$ R_2NH (**2° amine**) $\xrightarrow[\text{[2] } RNH_2]{\text{[1] RX}}$ R_3N (**3° amine**) $\xrightarrow[\text{[2] } RNH_2]{\text{[1] RX}}$ $R_4N^+ \; X^-$ (**quaternary ammonium salt**)

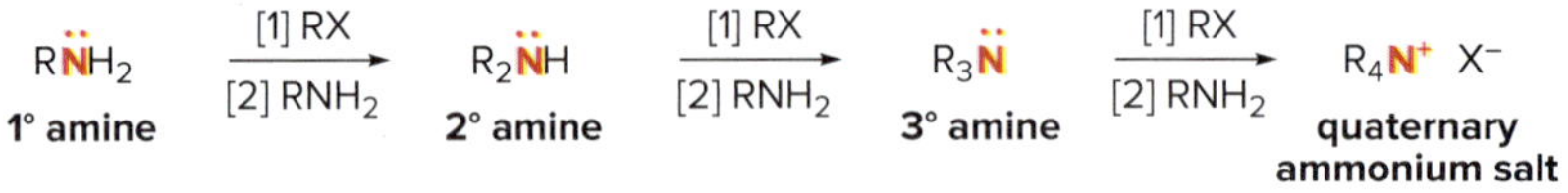

The identity of the nitrogen nucleophile determines the type of amine or ammonium salt formed as product. **One new carbon–nitrogen bond is formed in each reaction.** Because the reaction follows an S_N2 mechanism, the alkyl halide must be unhindered—that is, CH_3X or RCH_2X.

Although this process seems straightforward, polyalkylation of the nitrogen nucleophile limits its usefulness (Equation [2]). **Any amine formed by nucleophilic substitution still has a nonbonded electron pair, making it a nucleophile as well.** It will react with remaining alkyl halide to form a more substituted amine. Because of this, a mixture of 1°, 2°, and 3° amines often results. Only the final product—called a **quaternary ammonium salt** because it has four alkyl groups on N—cannot react further, and so the reaction stops.

As a result, this reaction is most useful for preparing 1° amines by using a very large excess of NH_3 (a relatively inexpensive starting material) and for preparing quaternary ammonium salts by alkylating any nitrogen nucleophile with one or more equivalents of alkyl halide.

Br + NH_3 (excess) ⟶ NH_2 (1° amine) + $NH_4^+ Br^-$

CH_3–Br (excess) + H_2N– ⟶ N^+, Br^- (quaternary ammonium salt)

Problem 23.6 Draw the product of each reaction.

a. Cl + NH_3 (excess) ⟶

b. H–N + Br (excess) ⟶

The Gabriel Synthesis of 1° Amines

To avoid polyalkylation, a nitrogen nucleophile can be used that reacts in a single nucleophilic substitution reaction—that is, the reaction forms a product that does *not* contain a nucleophilic nitrogen atom capable of reacting further.

The **Gabriel synthesis** consists of two steps and uses a resonance-stabilized nitrogen nucleophile to synthesize 1° amines via nucleophilic substitution. The Gabriel synthesis begins with **phthalimide,** one of a group of compounds called **imides.** The **N–H bond of an imide is especially acidic** because the resulting anion is resonance stabilized by the two flanking carbonyl groups.

phthalimide
$pK_a = 10$

resonance-stabilized anion

In the Gabriel synthesis, treatment of phthalimide with ^-OH forms a nucleophilic anion that can react with an unhindered alkyl halide—that is, CH_3X or RCH_2X—in an **S_N2 reaction** to form a substitution product. This alkylated imide is then hydrolyzed with aqueous base to give a 1° amine and a dicarboxylate. This reaction is similar to the hydrolysis of amides to afford carboxylate anions and amines, as discussed in Section 20.12. The overall result of this two-step sequence is **nucleophilic substitution of X by NH_2,** so the Gabriel synthesis can be used to prepare 1° amines only.

Steps in the Gabriel synthesis

R–X (R = CH_3 or 1° alkyl) + nucleophile —S_N2 [1] (nucleophilic substitution)⟶ alkylated imide + X^- —^-OH, H_2O [2] (hydrolysis)⟶ R–NH_2 (1° amine) + dicarboxylate by-product

- **The Gabriel synthesis converts an alkyl halide to a 1° amine by a two-step process: nucleophilic substitution followed by hydrolysis.**

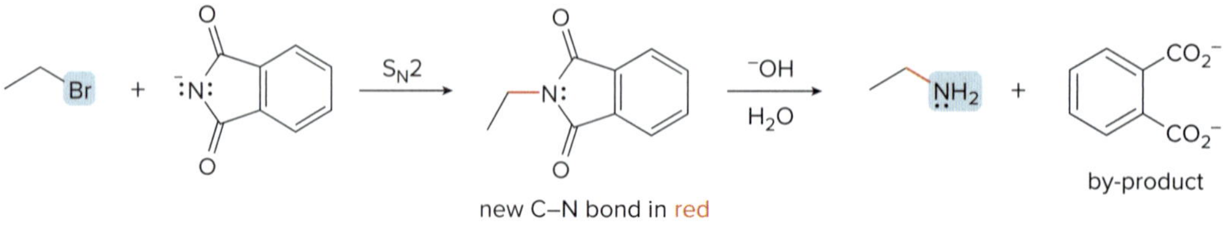

CH_3CH_2Br + phthalimide anion (:N:) $\xrightarrow{S_N2}$ N-ethylphthalimide $\xrightarrow[H_2O]{^-OH}$ $CH_3CH_2\ddot{N}H_2$ + phthalate (CO_2^-, CO_2^-)

new C–N bond in red

by-product

Problem 23.7 Draw the products of the following reactions.

a. 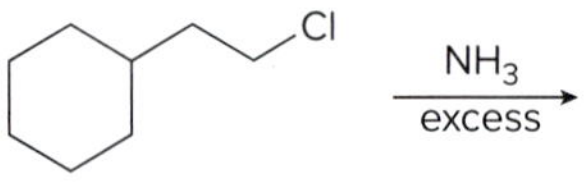 2-cyclohexylethyl chloride (Cl) $\xrightarrow[\text{excess}]{NH_3}$

b. 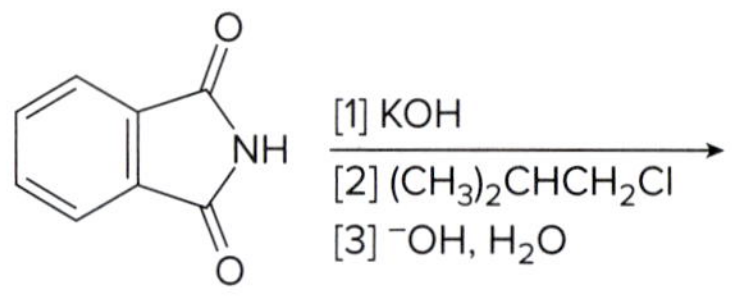 phthalimide (NH) $\xrightarrow[\text{[2] }(CH_3)_2CHCH_2Cl\text{ [3] }^-OH,\ H_2O]{\text{[1] KOH}}$

Problem 23.8 Which amines cannot be prepared by the Gabriel synthesis? Explain your choices.

a. 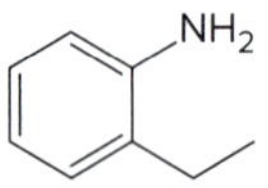 2-ethylaniline (NH_2) b. 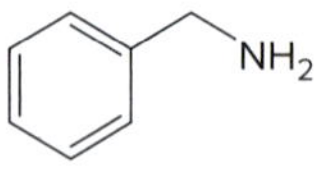 benzylamine (NH_2) c. 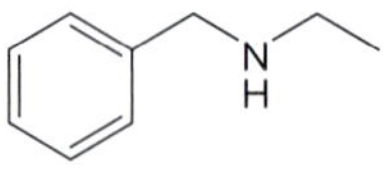 N-ethylbenzylamine (N–H) d. 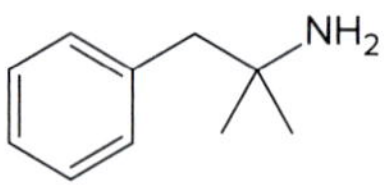 1-phenyl-2-methyl-2-propanamine (NH_2)

23.6B Reduction of Other Functional Groups That Contain Nitrogen

Amines can be prepared by reduction of nitro compounds, nitriles, and amides. Because the details of these reactions have been discussed previously, they are presented here in summary form only.

[1] From nitro compounds (Section 16.14D)

Nitro groups are reduced to 1° amines using a variety of reducing agents.

$$R-NO_2 \xrightarrow[\text{Fe, HCl or Sn, HCl}]{H_2,\text{ Pd-C or}} R-NH_2$$

1° amine

[2] From nitriles (Section 19.12B)

Nitriles are reduced to 1° amines with $LiAlH_4$.

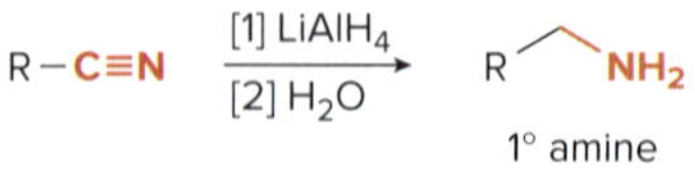

$$R-C\equiv N \xrightarrow[\text{[2] }H_2O]{\text{[1] }LiAlH_4} RCH_2NH_2$$

1° amine

Because a cyano group is readily introduced by S_N2 substitution of alkyl halides with ^-CN, this provides **a two-step method to convert an alkyl halide to a 1° amine with one more carbon atom.** The conversion of $(CH_3)_2CHCH_2CH_2Br$ to $(CH_3)_2CHCH_2CH_2CH_2NH_2$ illustrates this two-step sequence.

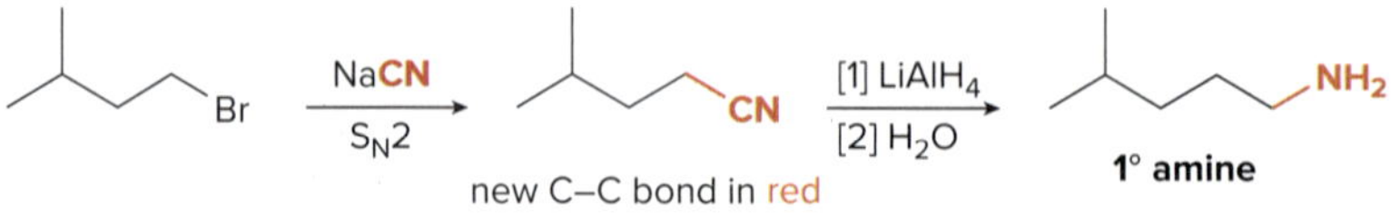

$$(CH_3)_2CHCH_2CH_2Br \xrightarrow[S_N2]{Na\mathbf{CN}} (CH_3)_2CHCH_2CH_2CN \xrightarrow[\text{[2] }H_2O]{\text{[1] }LiAlH_4} (CH_3)_2CHCH_2CH_2CH_2NH_2$$

new C–C bond in red

1° amine

[3] From amides (Section 17.7B)

Primary (1°), 2°, and 3° amides are reduced to 1°, 2°, and 3° amines, respectively, by using $LiAlH_4$.

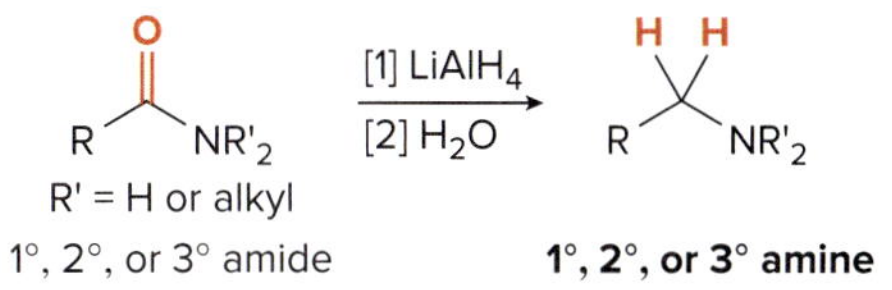

R–C(=O)–NR'$_2$ (R' = H or alkyl; 1°, 2°, or 3° amide) —[1] $LiAlH_4$ [2] H_2O→ R–CH$_2$–NR'$_2$ (1°, 2°, or 3° amine)

Problem 23.9 What nitro compound, nitrile, and amide are reduced to each compound?

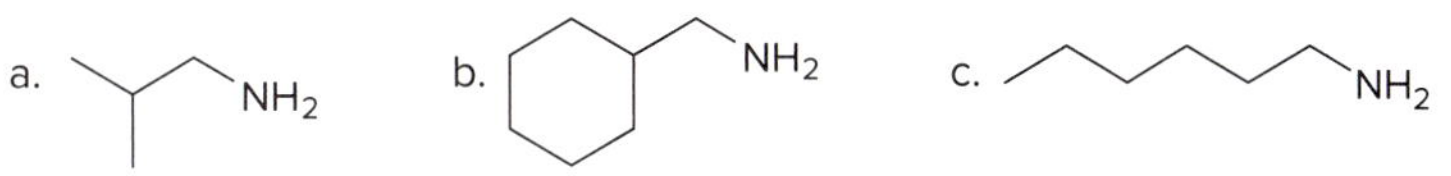

a. b. c.

Problem 23.10 What amine is formed by reduction of each amide?

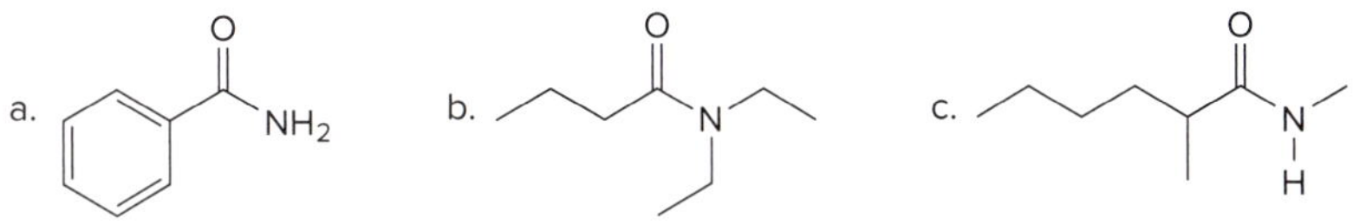

a. b. c.

Problem 23.11 Which amines cannot be prepared by reduction of an amide?

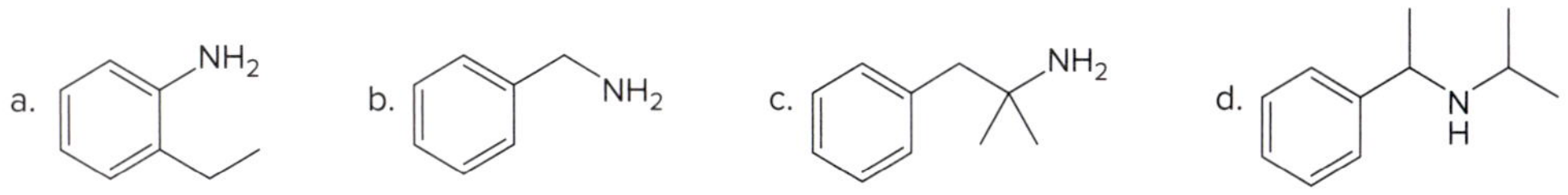

a. b. c. d.

23.6C Reductive Amination of Aldehydes and Ketones

Reductive amination is a two-step method that converts aldehydes and ketones to 1°, 2°, and 3° amines. Let's first examine this method using NH_3 to prepare 1° amines. There are two distinct parts in reductive amination:

[1] **Nucleophilic attack of NH_3 on the carbonyl group forms an imine** (Section 18.10A), which is not isolated; then,

[2] **Reduction of the imine forms an amine** (Section 17.7B).

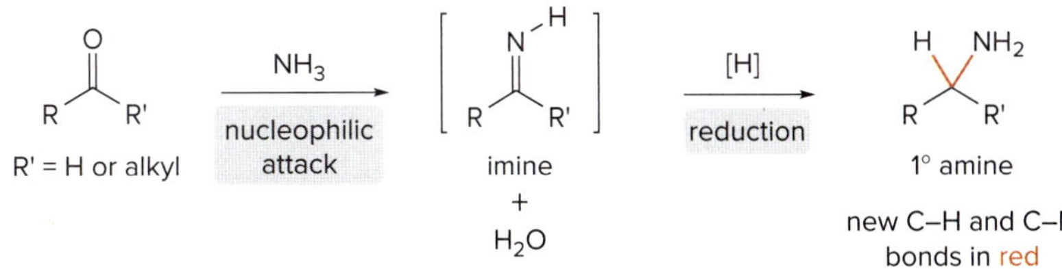

R–C(=O)–R' (R' = H or alkyl) —NH_3 (nucleophilic attack)→ [R–C(=NH)–R'] imine + H_2O —[H] (reduction)→ R–CH(NH$_2$)–R' 1° amine

new C–H and C–N bonds in red

• **Reductive amination replaces a C=O by a C–H and C–N bond.**

$NaBH_3CN$
sodium cyanoborohydride

The most effective reducing agent for this reaction is sodium cyanoborohydride ($NaBH_3CN$). This hydride reagent is a derivative of sodium borohydride ($NaBH_4$), formed by replacing one H atom by CN.

Reductive amination combines two reactions we have already learned in a different way. Two examples are shown. The second reaction is noteworthy because the product is **amphetamine,** a potent central nervous system stimulant.

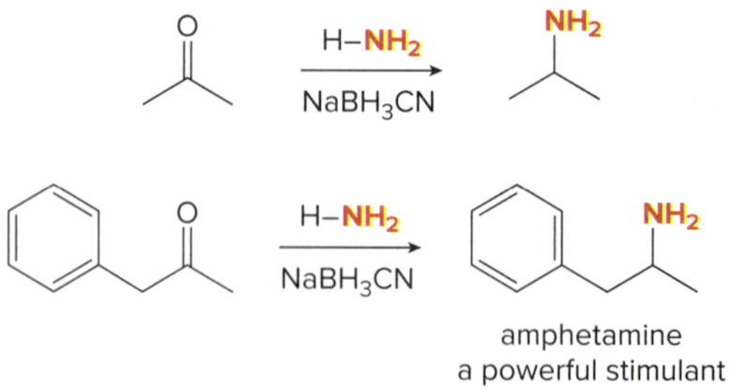
H–NH2
NaBH3CN
NH2
H–NH2
NaBH3CN
NH2
amphetamine
a powerful stimulant

With a 1° or 2° amine as starting material, reductive amination is used to prepare 2° and 3° amines, respectively. Note the result: **Reductive amination uses an aldehyde or ketone to replace one H atom on a nitrogen atom by an alkyl group,** making a more substituted amine.

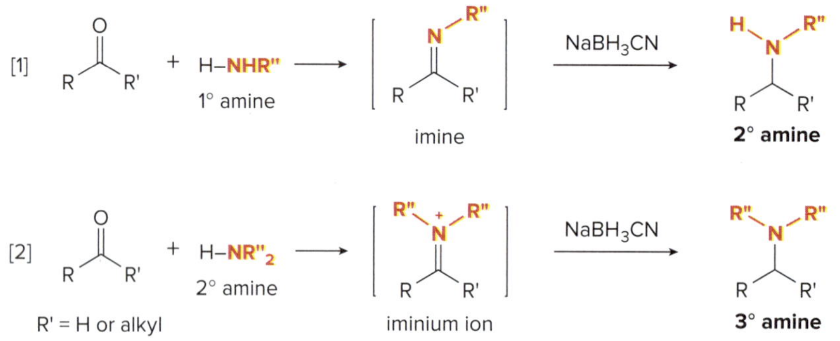
[1]
O
R
R'
+ H–NHR"
1° amine
N
R"
R
R'
imine
NaBH3CN
H
R"
N
R
R'
2° amine
[2]
O
R
R'
+ H–NR"2
2° amine
R' = H or alkyl
R"
+
N
R"
R
R'
iminium ion
NaBH3CN
R"
R"
N
R
R'
3° amine

The synthesis of methamphetamine (Section 23.5C) by reductive amination is illustrated in Figure 23.5.

Figure 23.5 Synthesis of methamphetamine by reductive amination

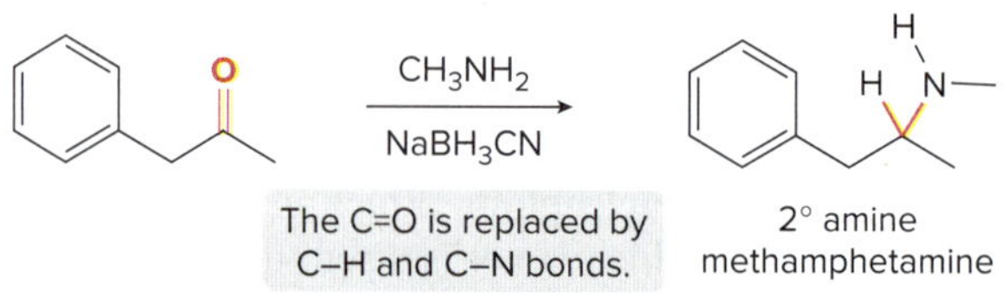
O
CH3NH2
NaBH3CN
H
H
N
The C=O is replaced by
C–H and C–N bonds.
2° amine
methamphetamine

- In reductive amination, one of the H atoms bonded to N is replaced by an alkyl group. As a result, a 1° amine is converted to a 2° amine and a 2° amine is converted to a 3° amine. In this reaction, CH_3NH_2 (a 1° amine) is converted to methamphetamine (a 2° amine).

Problem 23.12 Draw the product of each reaction.

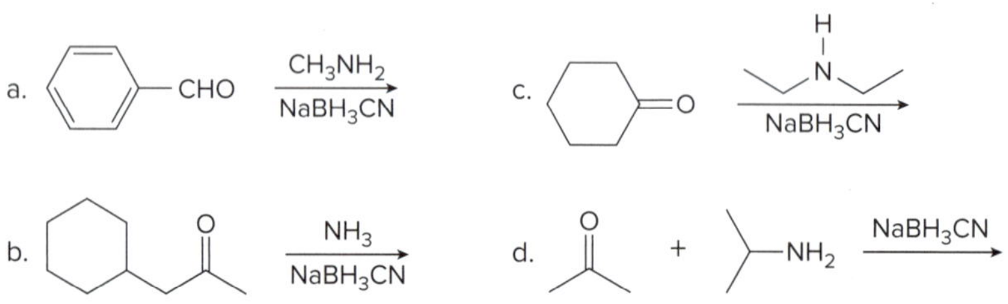
a.
CHO
CH3NH2
NaBH3CN
c.
O
H
N
NaBH3CN
b.
O
NH3
NaBH3CN
d.
O
+
NH2
NaBH3CN

Problem 23.13 Identify the product of the reductive amination of ketone **A** with amine **B.** This reaction is an early step in the synthesis of matrine, an alkaloid isolated from the flowering plant lupin.

Lupin (Problem 23.13) is a flowering plant in the bean family that is abundant in the Andes of Peru and along the roadsides of Alaska.
Daniel C. Smith

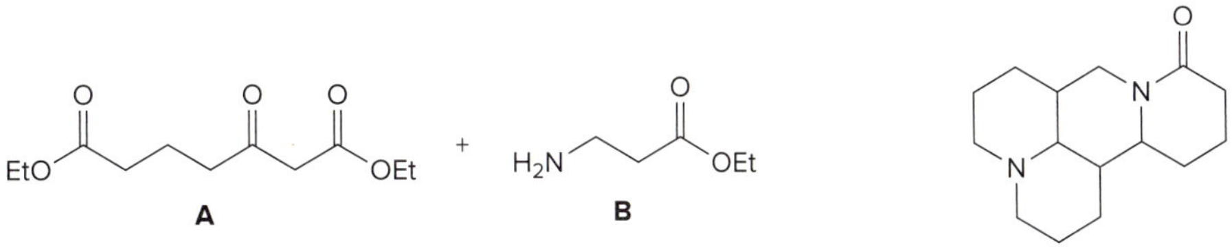

EtO O O O OEt + H_2N O OEt

A **B**

matrine

To use reductive amination in synthesis, you must be able to determine what aldehyde or ketone and nitrogen compound are needed to prepare a given amine—that is, you must work backwards in the retrosynthetic direction. Keep in mind these two points:

- **One alkyl group on N comes from the carbonyl compound.**
- **The remainder of the molecule comes from NH_3 or an amine.**

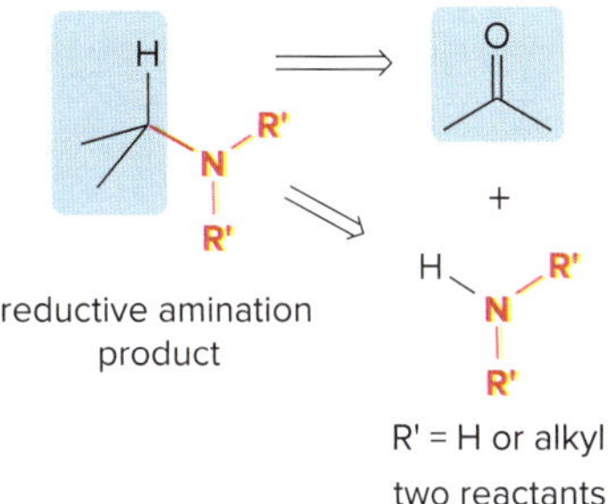

For example, 2-phenylethanamine is a 1° amine, so it has only one alkyl group bonded to N. This alkyl group must come from the carbonyl compound, and the rest of the molecule then comes from the nitrogen component. **For a 1° amine, the nitrogen component must be NH_3.**

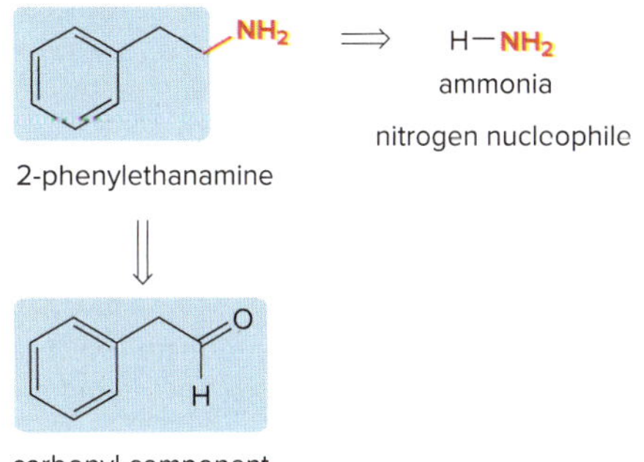

There is usually more than one way to use reductive amination to synthesize 2° and 3° amines, as shown in Sample Problem 23.2 for a 2° amine.

Sample Problem 23.2 Determining the Starting Materials in a Reductive Amination

What aldehyde or ketone and nitrogen component are needed to synthesize *N*-ethylcyclohexanamine by a reductive amination reaction?

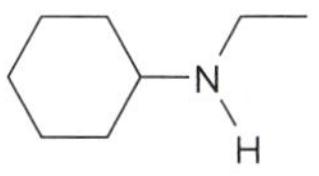

N
H

N-ethylcyclohexanamine

Solution

Because *N*-ethylcyclohexanamine has two different alkyl groups bonded to the N atom, either R group can come from the carbonyl component and there are two different ways to form a C–N bond by reductive amination.

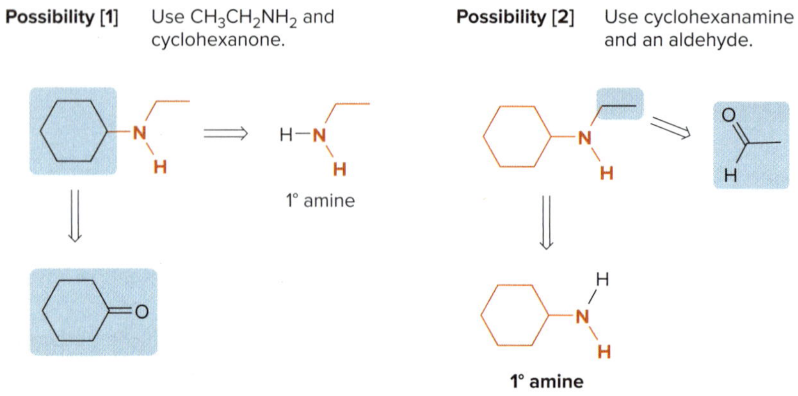

Possibility [1] Use $CH_3CH_2NH_2$ and cyclohexanone.

1° amine

Possibility [2] Use cyclohexanamine and an aldehyde.

1° amine

Because **reductive amination adds one R group to a nitrogen atom,** both routes to form the 2° amine begin with a 1° amine.

Problem 23.14 What starting materials are needed to prepare each drug using reductive amination? Give all possible pairs of compounds when more than one route is possible. In part (c), consider only the acyclic amine.

a.

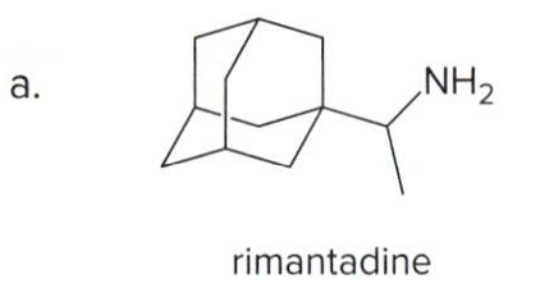

rimantadine
antiviral used to treat influenza

b.

pseudoephedrine
nasal decongestant

c.

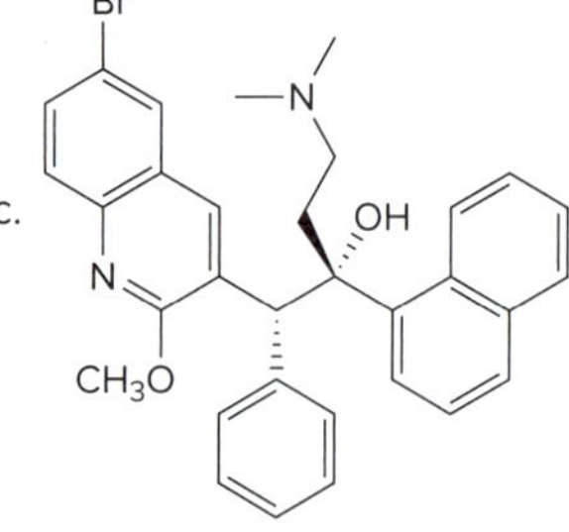

bedaquiline
drug to treat tuberculosis

More Practice: Try Problems 23.47, 23.54b, 23.55b.

Problem 23.15 (a) Explain why phentermine [$PhCH_2C(CH_3)_2NH_2$] can't be made by a reductive amination reaction. (b) Give a systematic name for phentermine, one of the components of the banned diet drug fen–phen.

23.7 Reactions of Amines—General Features

- **The chemistry of amines is dominated by the lone pair of electrons on nitrogen.**

Only three elements in the second row of the periodic table have nonbonded electron pairs in neutral organic compounds: nitrogen, oxygen, and fluorine. Because basicity and nucleophilicity decrease across the row, **nitrogen is the most basic and most nucleophilic** of these elements.

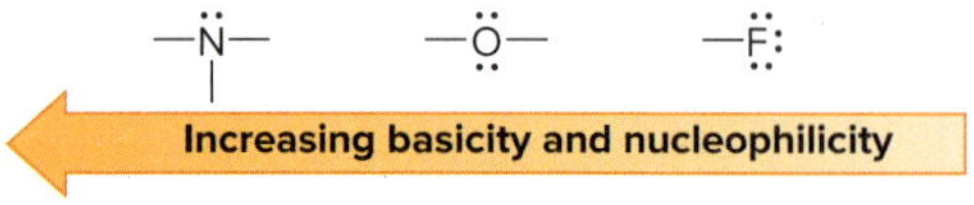

Increasing basicity and nucleophilicity

- **Amines are stronger bases and nucleophiles than other neutral organic compounds.**

$R_3N{:}$ + H–A (δ+) ⟶ $R_3\overset{+}{N}$–H + :A⁻

base acid

$R_3N{:}$ + E–X (δ+) ⟶ $R_3\overset{+}{N}$–E + X⁻

nucleophile electrophile

- Amines react as *bases* with compounds that contain acidic protons.
- Amines react as *nucleophiles* with compounds that contain electrophilic carbons.

23.8 Amines as Bases

Amines react as bases with a variety of organic and inorganic acids.

R–$\ddot{N}H_2$ (base) + H–A (acid) ⇌ R–$\overset{+}{N}H_3$ (conjugate acid, $pK_a \approx 10–11$) + :A⁻

To favor the products, the **pK_a of HA must be < 10.**

What acids can be used to protonate an amine? Equilibrium favors the products of an acid–base reaction when the weaker acid and base are formed. Because the pK_a of many protonated amines is 10–11, the **pK_a of the starting acid must be less than 10** for equilibrium to favor the products. Amines are thus readily protonated by strong inorganic acids like HCl and H_2SO_4, and by carboxylic acids as well.

$CH_3CH_2NH_2$ + H–Cl ($pK_a = -7$) ⟶ $CH_3CH_2\overset{+}{N}H_3$ ($pK_a = 10.8$) + :Cl:⁻

$(CH_3CH_2)_3N$ + CH_3COOH ($pK_a = 4.8$) ⟶ $(CH_3CH_2)_3\overset{+}{N}H$ ($pK_a = 11.0$) + CH_3COO^-

Equilibrium favors the products.

The principles used in an extraction procedure were detailed in Section 19.10.

Because amines are protonated by aqueous acid, they can be separated from other organic compounds by extraction using a separatory funnel. **Extraction separates compounds based on solubility differences.** When an amine is protonated by aqueous acid, its solubility properties change.

For example, when cyclohexanamine is treated with aqueous HCl, it is protonated, forming an ammonium salt. **Because the ammonium salt is ionic, it is soluble in water,** but insoluble in organic solvents. A similar acid–base reaction does not occur with other organic compounds like alcohols, which are much less basic.

C_6H_{11}–$\ddot{N}H_2$ + H–Cl ⟶ C_6H_{11}–$\overset{+}{N}H_3$ + Cl⁻

cyclohexanamine
- insoluble in H_2O
- soluble in CH_2Cl_2

cyclohexanammonium chloride
- soluble in H_2O
- insoluble in CH_2Cl_2

This difference in acid–base chemistry can be used to separate cyclohexanamine and cyclohexanol by the stepwise extraction procedure illustrated in Figure 23.6.

Figure 23.6 Separation of cyclohexanamine and cyclohexanol by an extraction procedure

Step [1] Dissolve cyclohexanamine and cyclohexanol in CH_2Cl_2.

Step [2] Add 10% HCl solution to form two layers.

Step [3] Separate the layers.

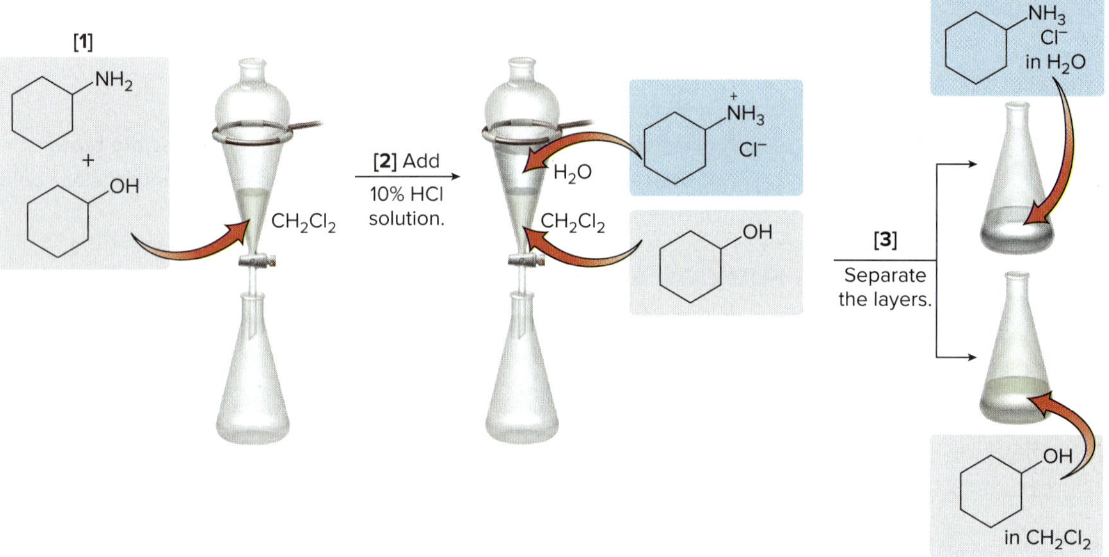

- Both compounds dissolve in the organic solvent CH_2Cl_2.

- Adding 10% aqueous HCl solution forms two layers. When the two layers are mixed, the **HCl protonates the amine (RNH_2) to form $RNH_3^+Cl^-$**, which dissolves in the aqueous layer.
- The cyclohexanol remains in the CH_2Cl_2 layer.

- Draining the lower layer out the bottom stopcock separates the two layers.
- Cyclohexanol (dissolved in CH_2Cl_2) is in one flask. The ammonium salt, $RNH_3^+Cl^-$ (dissolved in water), is in another flask.

- **An amine can be separated from other organic compounds by converting it to a water-soluble ammonium salt by an acid–base reaction.**

Thus, the water-soluble salt $C_6H_{11}NH_3^+Cl^-$ (obtained by protonation of $C_6H_{11}NH_2$) can be separated from water-insoluble cyclohexanol by an aqueous extraction procedure.

Problem 23.16 Draw the products of each acid–base reaction. Indicate whether equilibrium favors the reactants or products.

a. NH_2 + HCl ⇌

b. OH + N–H ⇌

c. N–H + H_2O ⇌

d. MDMA "Ecstasy" illegal stimulant + CH_3CO_2H ⇌

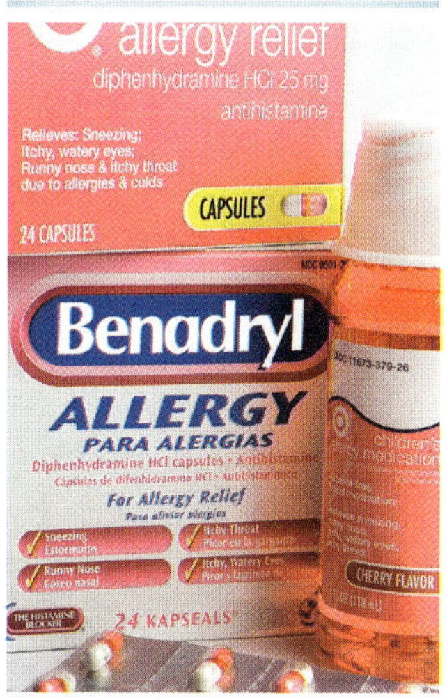

Many antihistamines and decongestants are sold as their ammonium salts.

Jill Braaten/McGraw Hill

Many water-insoluble amines with useful medicinal properties are sold as their water-soluble ammonium salts, which are more easily transported through the body in the aqueous medium of the blood. Benadryl, formed by treating diphenhydramine with HCl, is an over-the-counter antihistamine that is used to relieve the itch and irritation of skin rashes and hives.

diphenhydramine + H–Cl: → ammonium salt + :Cl:⁻

diphenhydramine
water insoluble

ammonium salt
Benadryl
(diphenhydramine hydrochloride)
water soluble

Problem 23.17 Write out steps to show how each of the following pairs of compounds can be separated by an extraction procedure.

a. cyclohexyl–NH_2 and methylcyclohexane b. tributylamine (N) and dibutyl ether (O)

23.9 Relative Basicity of Amines and Other Compounds

The relative acidity of different compounds can be compared using their pK_a values. **The relative *basicity* of different compounds (such as amines) can be compared using the pK_a values of their *conjugate acids.***

- The *weaker* the conjugate acid, the *higher* its pK_a and the *stronger* the base.

$$R{-}\ddot{N}H_2 + H{-}A \rightleftharpoons R{-}\overset{+}{N}H_3 + :A^-$$

base — conjugate acid

***stronger* base** ← **The *weaker* the acid, the *higher* the pK_a.**

To compare the basicity of two compounds, keep in mind the following:

- Any factor that *increases* the electron density on the N atom *increases* an amine's basicity.
- Any factor that *decreases* the electron density on N *decreases* an amine's basicity.

23.9A Comparing an Amine and NH_3

Because **alkyl groups are electron donating,** they increase the electron density on nitrogen, which makes an amine like $CH_3CH_2NH_2$ more basic than NH_3. In fact, the pK_a of $CH_3CH_2NH_3^+$ is *higher* than the pK_a of NH_4^+, so **$CH_3CH_2NH_2$ is a *stronger* base than NH_3.**

pK_a = 9.3	$H{-}\overset{+}{N}H_3$ lower pK_a stronger acid	$\ddot{N}H_3$ **weaker base**
pK_a = 10.8	$CH_3CH_2\overset{+}{N}H_3$ higher pK_a weaker acid	$CH_3CH_2\ddot{N}H_2$ **stronger base**

One electron-donor group makes the amine more basic.

The relative basicity of 1°, 2°, and 3° amines depends on additional factors, and will not be considered in this text.

- Primary (1°), 2°, and 3° alkylamines are *more basic* than NH_3 because of the electron-donating inductive effect of the R groups.

Problem 23.18 Which compound in each pair is more basic: (a) $(CH_3)_2NH$ and NH_3; (b) $CH_3CH_2NH_2$ and $ClCH_2CH_2NH_2$?

23.9B Comparing an Alkylamine and an Arylamine

To compare an alkylamine ($CH_3CH_2NH_2$) and an arylamine ($C_6H_5NH_2$, aniline), we must look at the **availability of the nonbonded electron pair on N.** With $CH_3CH_2NH_2$, the electron pair is localized on the N atom. With an arylamine, however, the electron pair is now delocalized on the benzene ring. This *decreases* the electron density on N and makes $C_6H_5NH_2$ less basic than $CH_3CH_2NH_2$.

$\ddot{N}H_2$

The electron pair is **localized** on N.

stronger base

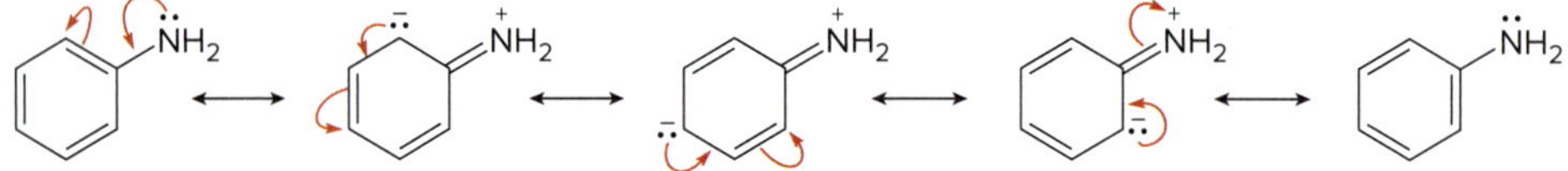

The electron pair is **delocalized** on the benzene ring.

weaker base

The pK_a values support this reasoning. Because the pK_a of $CH_3CH_2NH_3^+$ is *higher* than the pK_a of $C_6H_5NH_3^+$ (10.8 vs. 4.9), **$CH_3CH_2NH_2$ is a *stronger* base than $C_6H_5NH_2$.**

- **Arylamines are *less basic* than alkylamines because the electron pair on N is delocalized.**

Substituted anilines are more or less basic than aniline depending on the nature of the substituent.

- **Electron-donor groups *add* electron density to the benzene ring, making the arylamine *more basic* than aniline.**

D = electron-donor group

D → $\ddot{N}H_2$

D makes the amine *more basic* than aniline.

D
–NH2
–OH
–OR
–NHCOR
–R

- **Electron-withdrawing groups *remove* electron density from the benzene ring, making the arylamine *less basic* than aniline.**

W = electron-withdrawing group

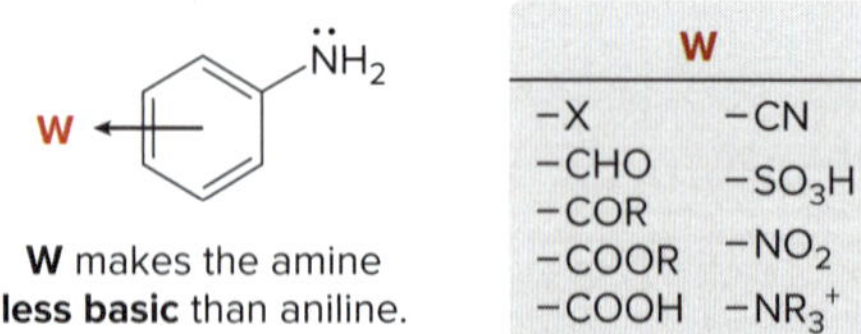

W makes the amine **less basic** than aniline.

W	
–X	–CN
–CHO	–SO3H
–COR	–NO2
–COOR	–NR3+
–COOH	

The effect of electron-donating and electron-withdrawing groups on the acidity of substituted benzoic acids was discussed in Section 19.9.

Whether a substituent donates or withdraws electron density depends on the balance of its inductive and resonance effects (Section 16.6 and Figure 16.8).

Sample Problem 23.3 Determining the Relative Basicity of Anilines

Rank the following compounds in order of increasing basicity.

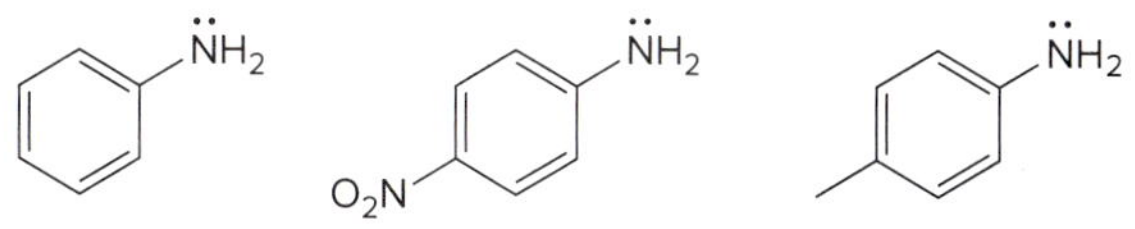

aniline — p-nitroaniline — p-methylaniline (p-toluidine)

Solution

p-Nitroaniline: NO_2 is an electron-withdrawing group, making the amine *less basic* than aniline.

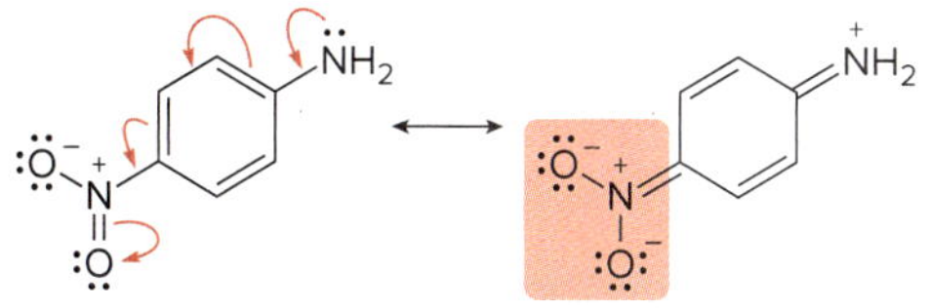

p-Methylaniline: CH_3 has an electron-donating inductive effect, making the amine *more basic* than aniline.

The lone pair on N is **delocalized** on the O atom, *decreasing* the basicity of the amine.

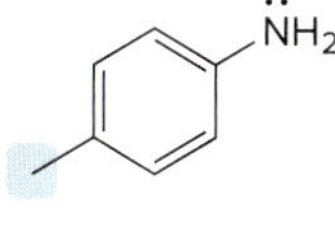

CH_3 inductively **donates** electron density, *increasing* the basicity of the amine.

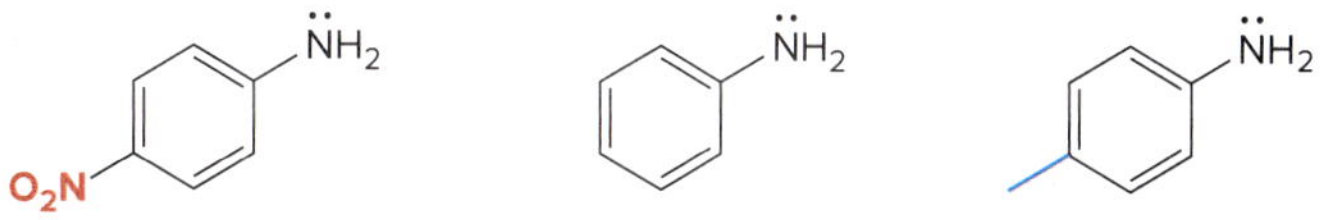

p-nitroaniline — aniline — p-methylaniline (p-toluidine)

Increasing basicity

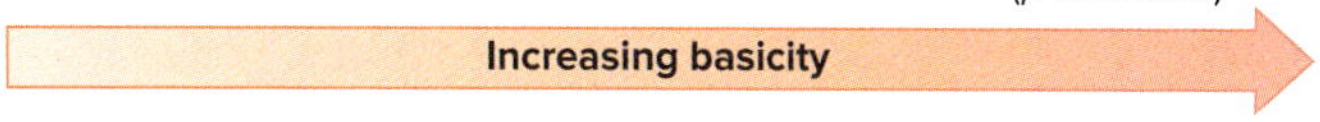

Problem 23.19 Rank compounds **A–D** in order of increasing basicity.

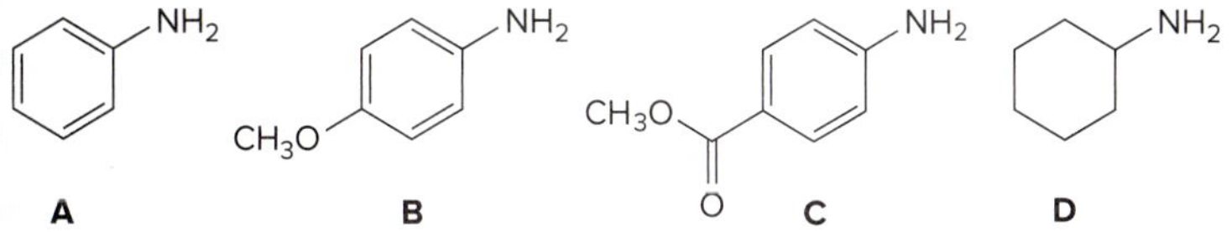

A **B** **C** **D**

More Practice: Try Problems 23.39, 23.44.

23.9C Comparing an Alkylamine and an Amide

To compare the basicity of an alkylamine (RNH_2) and an amide ($RCONH_2$), we must once again compare the availability of the nonbonded electron pair on nitrogen. With RNH_2, the electron pair is localized on the N atom. With an amide, however, the electron pair is *delocalized* on the carbonyl oxygen by resonance. This *decreases* the electron density on N, making **an amide much *less basic* than an alkylamine.**

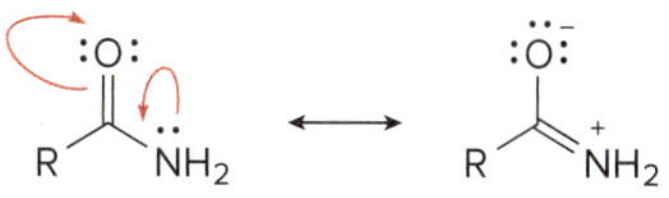

The electron pair on N is **delocalized** on O by resonance.

- **Amides are much less basic than amines because the electron pair on N is delocalized.**

Amides are not much more basic than any carbonyl compound. When an amide is treated with acid, **protonation occurs at the carbonyl oxygen, *not* the nitrogen,** because the resulting cation is resonance stabilized.

three resonance structures for the conjugate acid

Problem 23.20 Rank the following compounds in order of increasing basicity.

23.9D Heterocyclic Aromatic Amines

To determine the relative basicity of nitrogen heterocycles that are also aromatic, you must know **whether the nitrogen lone pair is part of the aromatic π system.**

For example, pyridine and pyrrole are both aromatic, but the nonbonded electron pair on the N atom in these compounds is located in different orbitals. Recall from Section 15.9 that the **lone pair of electrons in pyridine occupies an sp^2 hybridized orbital,** perpendicular to the π bonds of the molecule, so it is *not* part of the aromatic system, whereas that of pyrrole resides in a *p* orbital, making it part of the aromatic system. **The lone pair on pyrrole, therefore, is delocalized on all of the atoms of the five-membered ring,** making pyrrole a much *weaker base* than pyridine.

The lone pair resides in an sp^2 hybrid orbital.

The lone pair resides in a *p* orbital and is *delocalized* in the ring.

Protonation of pyrrole occurs at a ring *carbon,* not the N atom, as noted in Problem 15.47.

As a result, the pK_a of the conjugate acid of pyrrole is much less than that of the conjugate acid of pyridine.

- Pyrrole is much *less basic* than pyridine because its lone pair of electrons is part of the aromatic π system.

23.9E Hybridization Effects

The effect of hybridization on the acidity of an H–A bond was first discussed in Section 2.5D.

The hybridization of the orbital that contains an amine's lone pair also affects its basicity. This is illustrated by comparing the basicity of **piperidine** and **pyridine,** two nitrogen heterocycles. The lone pair in piperidine resides in an sp^3 hybrid orbital that has 25% *s*-character. The lone pair in pyridine resides in an sp^2 hybrid orbital that has 33% *s*-character.

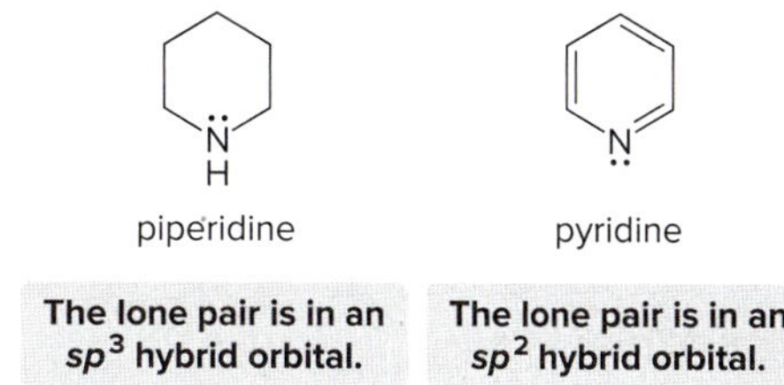

- The *higher* the percent *s*-character of the orbital containing the lone pair, the more tightly the lone pair is held and the *weaker* the base.

Pyridine is a weaker base than piperidine because its nonbonded pair of electrons resides in an sp^2 hybrid orbital. Although pyridine is an aromatic amine, its lone pair is *not* part of the delocalized π system, so its **basicity is determined by the hybridization of its N atom.** As a result, the pK_a value of the conjugate acid of pyridine is much *lower* than that of the conjugate acid of piperidine, making pyridine the *weaker* base.

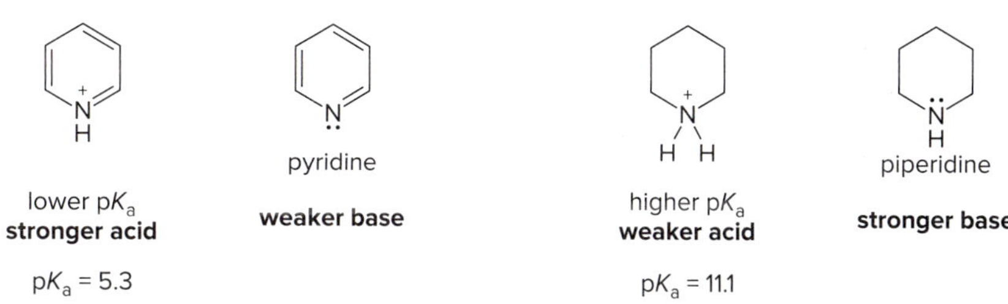

Problem 23.21 Rank the following ammonium ions in order of increasing pK_a.

$\overset{+}{N}H_3$ O_2N $\overset{+}{N}H_3$ $\overset{+}{N}H_3$ Cl $\overset{+}{N}H_3$

A **B** **C** **D**

23.9F Summary of the Factors That Determine Amine Basicity

Acid–base chemistry is central to many processes in organic chemistry, so it has been a constant theme throughout this text. Tables 23.2 and 23.3 organize and summarize the acid–base principles discussed in Section 23.9. The principles in these tables can be used to determine the most basic site in a molecule that has more than one nitrogen atom, as shown in Sample Problem 23.4.

Table 23.2 Factors That Determine Amine Basicity

Factor	Example
[1] **Inductive effects:** Electron-donating groups bonded to N *increase* basicity.	• RNH_2, R_2NH, and R_3N are more basic than NH_3.
[2] **Resonance effects:** Delocalizing the lone pair on N *decreases* basicity.	• Arylamines ($C_6H_5NH_2$) are less basic than alkylamines (RNH_2). • Amides ($RCONH_2$) are much less basic than amines (RNH_2).
[3] **Aromaticity:** Having the lone pair on N as part of the aromatic π system *decreases* basicity.	• Pyrrole is less basic than pyridine. [pyrrole] less basic; [pyridine] more basic
[4] **Hybridization effects:** Increasing the percent *s*-character in the orbital with the lone pair *decreases* basicity.	• Pyridine is less basic than piperidine. [pyridine] less basic; [piperidine] more basic

Table 23.3 Table of pK_a Values of Some Representative Organic Nitrogen Compounds

	Compound	pK_a of the conjugate acid
Ammonia	NH_3	9.3
Alkylamines[a]	piperidine (NH)	11.1
	$(CH_3CH_2)_2NH$	11.1
	$(CH_3CH_2)_3N$	11.0
	$CH_3CH_2NH_2$	10.8
Arylamines[b]	*p*-$CH_3OC_6H_4NH_2$	5.3
	p-$CH_3C_6H_4NH_2$	5.1
	$C_6H_5NH_2$	4.9
	p-$NO_2C_6H_4NH_2$	1.0
Heterocyclic aromatic amines[c]	pyridine (N)	5.3
	pyrrole (NH)	0.4
Amides	$RCONH_2$	−1

[a] Alkylamines have pK_a values of ~10–11.
[b] The pK_a *decreases* as the electron density of the benzene ring *decreases*.
[c] The pK_a depends on whether the lone pair of N is *localized* or *delocalized*.

Sample Problem 23.4 Determining Which Nitrogen Atom Is the Strongest Base

Hydroxychloroquinine is used to prevent and treat malaria, a disease caused by a protozoan parasite spread by the *Anopheles* mosquito. Although there was speculation in 2020 that hydroxychloroquinine could be used to treat COVID-19, extensive human trials have shown that hydroxychloroquinine is ineffective for this purpose.
CDC/James Gathany/

Which N atom in hydroxychloroquine is the strongest base?

hydroxychloroquine

Solution

Examine the nitrogen atoms in hydroxychloroquine, labeled in red, blue, and green, and recall that decreasing the electron density on N decreases basicity.

strongest base

- N is bonded to an aromatic ring, so its lone pair is delocalized in the ring like aniline, *decreasing* basicity.
- The lone pair is localized on N, but N is sp^2 hybridized. *Increasing* percent *s*-character *decreases* basicity.
- N has a localized lone pair and is $\boldsymbol{sp^3}$ hybridized, making it the *most basic* site in the molecule.

Problem 23.22 Which N atom in each compound is more basic? What product is formed when each compound is treated with HCl? Matrine is an alkaloid isolated from lupin, quinine is an antimalarial drug obtained from the bark of the cinchona tree, and vismodegib (trade name Erivedge) is a drug used to treat skin cancer.

a. matrine
alkaloid in lupin

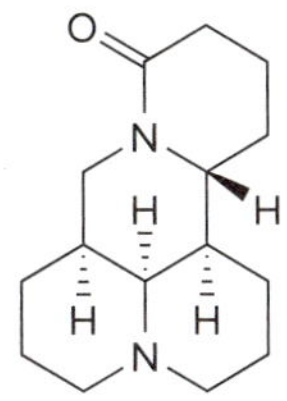

b. quinine
antimalarial drug

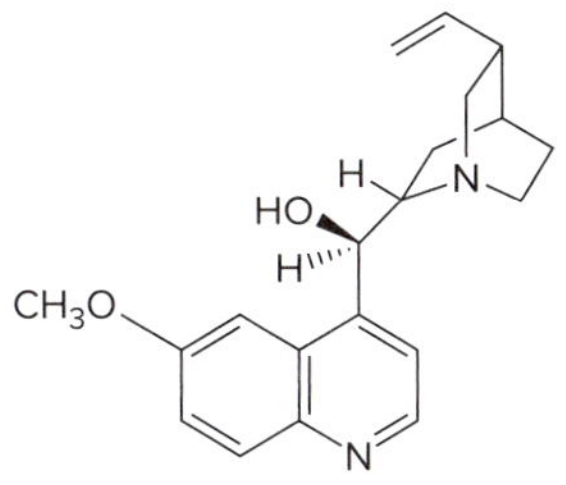

c. vismodegib
Trade name: Erivedge
drug to treat skin cancer

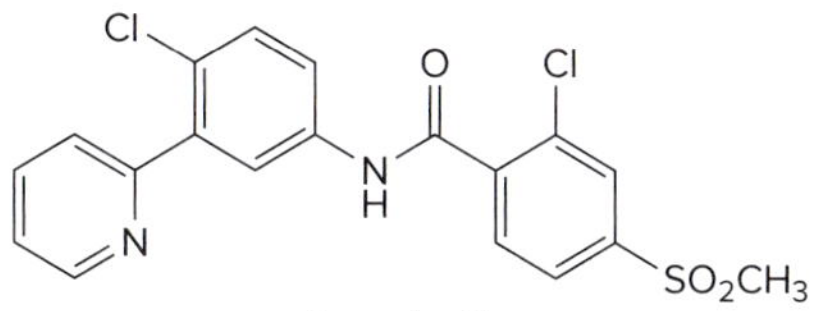

More Practice: Try Problems 23.36a, 23.40–23.42.

23.10 Amines as Nucleophiles

Amines react as nucleophiles with electrophilic carbon atoms. The details of these reactions have been described in Chapters 18 and 20, so they are summarized here only to emphasize the similar role that the amine nitrogen plays.

- **Amines attack carbonyl groups to form products of nucleophilic addition or substitution.**

The nature of the product depends on the carbonyl electrophile. These reactions are limited to 1° and 2° amines, because only these compounds yield neutral organic products.

[1] Reaction of 1° and 2° amines with aldehydes and ketones (Sections 18.10–18.11)

Aldehydes and ketones react with 1° amines to form imines and with 2° amines to form enamines. Both reactions involve **nucleophilic addition** of the amine to the carbonyl group to form a carbinolamine, which then loses water to form the final product.

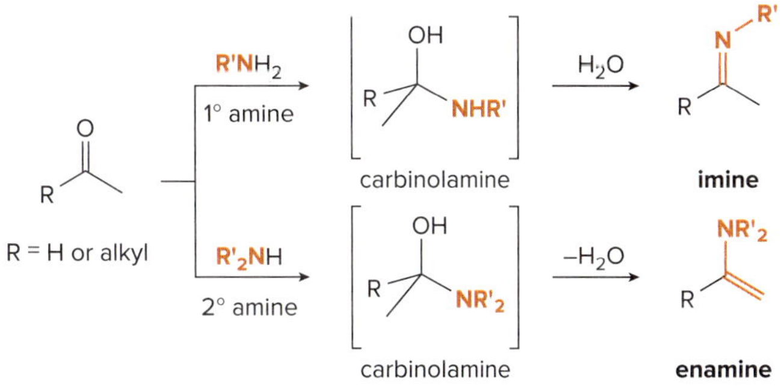

R = H or alkyl

$R'NH_2$ (1° amine) → carbinolamine → (H_2O) → **imine**

R'_2NH (2° amine) → carbinolamine → ($-H_2O$) → **enamine**

[2] Reaction of NH_3 and 1° and 2° amines with acid chlorides and anhydrides (Sections 20.7–20.8)

Acid chlorides and anhydrides react with NH_3, 1° amines, and 2° amines to form 1°, 2°, and 3° amides, respectively. These reactions involve attack of the nitrogen nucleophile on the carbonyl group followed by elimination of a leaving group (Cl^- or RCO_2^-). The overall result of this reaction is **substitution** of the leaving group by the nitrogen nucleophile.

$RCOZ$ + $H{-}NR'_2$ (2 equiv) ⟶ $RCONR'_2$ (**amide**) + $\overset{+}{N}H_4\ Z^-$

Z = Cl or OCOR   R' = H or alkyl

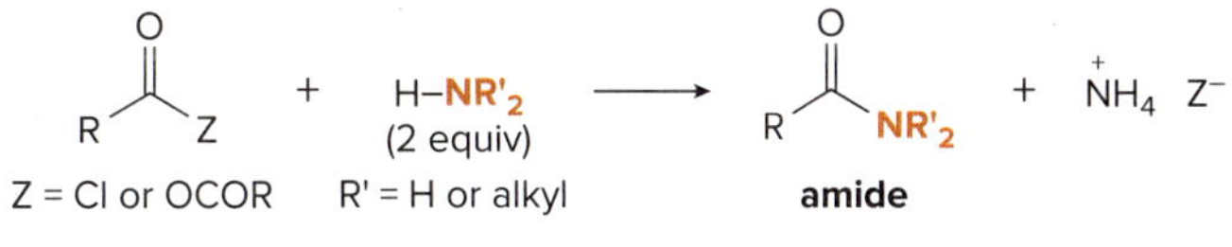

Problem 23.23 Draw the products formed when each carbonyl compound reacts with the following amines: [1] $CH_3CH_2CH_2NH_2$; [2] $(CH_3CH_2)_2NH$.

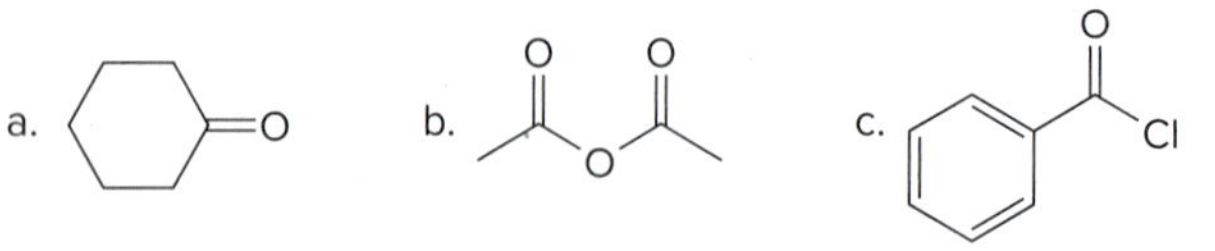

a. b. c.

The conversion of amines to amides is useful in the synthesis of substituted anilines. For example, aniline itself does not undergo Friedel–Crafts reactions (Section 16.10B). Instead, its basic lone pair on N reacts with the Lewis acid ($AlCl_3$) to form a deactivated complex that does not undergo further reaction.

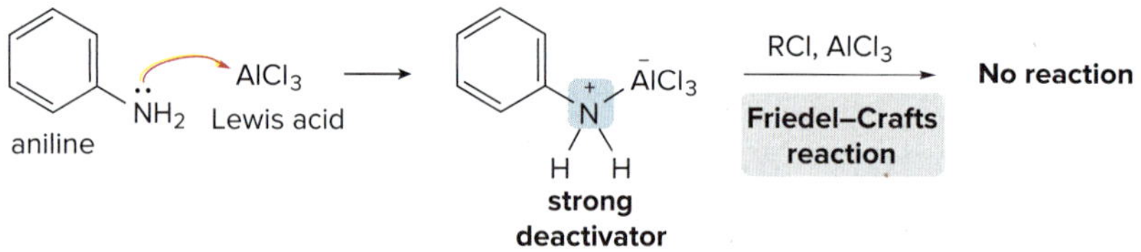

The N atom of an amide, however, is much less basic than the N atom of an amine, so it does not undergo a similar Lewis acid–base reaction with $AlCl_3$. A three-step reaction sequence involving an intermediate amide can thus be used to form the products of the Friedel–Crafts reaction.

[1] **Convert the amine (aniline) into an amide (acetanilide).**

[2] **Carry out the Friedel–Crafts reaction.**

[3] **Hydrolyze the amide** to generate the free amino group.

This three-step procedure is illustrated in Figure 23.7. In this way, **the amide serves as a protecting group for the NH_2 group,** in much the same way that *tert*-butyldimethylsilyl ethers and acetals are used to protect alcohols and carbonyls, respectively (Sections 17.12 and 18.14).

Figure 23.7 An amide as a protecting group for an amine

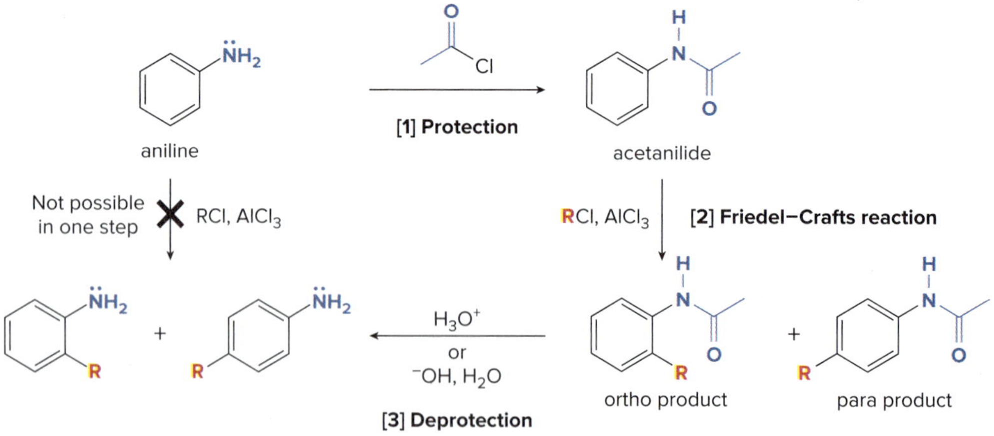

A three-step sequence uses an amide as a protecting group.

[1] Treatment of aniline with acetyl chloride (CH_3COCl) forms an **amide** (acetanilide).

[2] Acetanilide, having a much less basic N atom compared to aniline, undergoes **electrophilic aromatic substitution** under Friedel–Crafts conditions, forming a mixture of ortho and para products.

[3] **Hydrolysis of the amide** forms the Friedel–Crafts substitution products.

Problem 23.24 Devise a synthesis of each compound from aniline ($C_6H_5NH_2$).

a. b.

23.11 Hofmann Elimination

Amines, like alcohols, contain a poor leaving group. To undergo a β elimination reaction, for example, a 1° amine would need to lose the elements of NH_3 across two adjacent atoms. The leaving group, $^-NH_2$, is such a strong base, however, that this reaction does *not* occur.

The only way around this obstacle is to **convert $^-NH_2$ to a better leaving group.** The most common method to accomplish this is called a **Hofmann elimination,** which converts an amine to a quaternary ammonium salt prior to β elimination.

23.11A Details of the Hofmann Elimination

The **Hofmann elimination** converts an amine to an alkene.

[1] CH_3I (excess)
[2] Ag_2O
[3] Δ

Hofmann elimination

+ H_2O + $N(CH_3)_3$ + AgI

by-products

The Hofmann elimination consists of three steps, as shown for the conversion of propan-1-amine to propene.

propan-1-amine → CH_3I (excess) [1] → $\overset{+}{N}(CH_3)_3$ I^- **quaternary ammonium salt** → Ag_2O [2] → $\overset{+}{N}(CH_3)_3$ + ^-OH + AgI → Δ [3] → propene + H_2O + $N(CH_3)_3$ leaving group

- In Step [1], the amine reacts as a nucleophile in an S_N2 reaction with excess CH_3I to form a quaternary ammonium salt. **The $N(CH_3)_3$ group thus formed is a much better leaving group than $^-NH_2$.**
- Step [2] converts one ammonium salt to another one with a different anion. The silver(I) oxide, Ag_2O, replaces the I^- anion with ^-OH, a strong base.
- When the ammonium salt is heated in Step [3], **^-OH removes a proton from the β carbon atom,** forming the new π bond of the alkene. The mechanism of elimination is **E2,** so

- All bonds are broken and formed in a single step.
- Elimination occurs through an anti periplanar geometry—that is, H and $N(CH_3)_3$ are oriented on opposite sides of the molecule.

The general E2 mechanism for the Hofmann elimination is shown in Mechanism 23.1.

Mechanism 23.1 The E2 Mechanism for the Hofmann Elimination

Elimination occurs with an anti periplanar arrangement of H and $N(CH_3)_3$. **Base removes a proton on the β carbon,** the electron pair in the C–H bond forms the π bond, and **$N(CH_3)_3$ comes off as the leaving group.**

All Hofmann elimination reactions result in the formation of a new π bond between the α and β carbon atoms.

To help remember the reagents needed for the steps of the Hofmann elimination, keep in mind what happens in each step.

- **Step [1]** makes a **good leaving group** by forming a quaternary ammonium salt.
- **Step [2]** provides the **strong base,** ^-OH, needed for elimination.
- **Step [3]** is the **E2 elimination** that forms the new π bond.

Problem 23.25 Draw the product formed by treating each compound with excess CH_3I, followed by Ag_2O, and then heat.

23.11B Regioselectivity of the Hofmann Elimination

There is one major difference between a Hofmann elimination and other E2 eliminations.

- **When constitutional isomers are possible, the major alkene has the *less* substituted double bond in a Hofmann elimination.**

For example, Hofmann elimination of the elements of H and $N(CH_3)_3$ from 2-methylcyclopentanamine, which has two different β carbons (labeled $β_1$ and $β_2$), yields two constitutional isomers: the disubstituted alkene **A** (the major product) and the trisubstituted alkene **B** (the minor product).

This regioselectivity distinguishes a Hofmann elimination from other E2 eliminations, which form the *more* substituted double bond by the Zaitsev rule (Section 8.5). This result is sometimes explained by the size of the leaving group, $N(CH_3)_3$. **In a Hofmann elimination, the base removes a proton from the *less* substituted, more accessible β carbon atom, because of the bulky leaving group on the nearby α carbon.**

Sample Problem 23.5 Drawing the Major Product of a Hofmann Elimination

Draw the major product formed from Hofmann elimination of the following amine.

NH_2 — [1] CH_3I (excess) [2] Ag_2O [3] Δ →

Solution

The amine has three β carbons but two of them are identical, so two alkenes are possible. **Draw elimination products by forming alkenes having a C=C between the α and β carbons.** The major product has the **less substituted double bond**—that is, the alkene with the C=C between the α and β_1 carbons in this example.

β_2, β_1, α, NH_2, β_2 — [1] CH_3I (excess) [2] Ag_2O [3] Δ → α, β_1 + α, β_2

major product disubstituted alkene

minor product trisubstituted alkene

Problem 23.26 Draw the major product formed by treating each amine with excess CH_3I, followed by Ag_2O, and then heat.

a. NH_2 b. H_2N c. N H

More Practice: Try Problems 23.53, 23.54c, 23.55c, 23.56f, 23.57.

Hygrine (Problem 23.27), cocaine, and several other biologically active alkaloids are obtained from coca leaves (*Erythroxylon coca*).

Lew Robertson/Getty Images

Problem 23.27 Draw the products formed from a Hofmann elimination of hygrine, an alkaloid isolated from coca leaves. Label the major product.

N O hygrine

Figure 23.8 contrasts the products formed by E2 elimination reactions using an alkyl halide and an amine as starting materials. Treatment of the alkyl halide (2-bromopentane) with base forms the *more* substituted alkene as the major product, following the **Zaitsev rule.** In contrast, the three-step Hofmann sequence of an amine (pentan-2-amine) forms the *less* substituted alkene as major product.

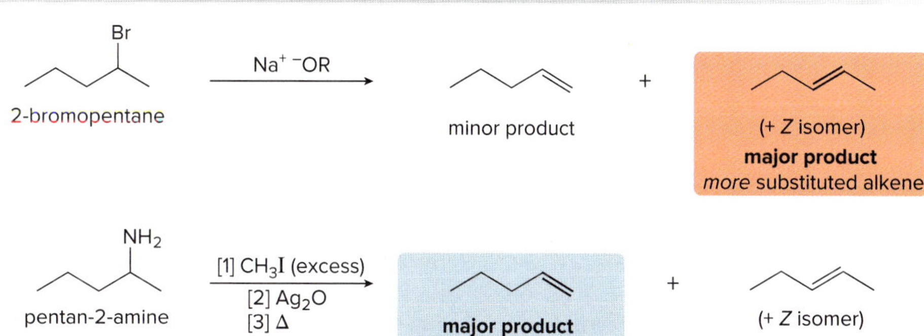

Figure 23.8 A comparison of E2 elimination reactions using alkyl halides and amines

Problem 23.28 Draw the major product formed in each reaction.

a. (2-bromo-3-methylhexane skeleton, Br) $\xrightarrow{Na^+ \ ^-OR}$

b. (NH$_2$ analog) $\xrightarrow{[1]\ CH_3I\ (excess);\ [2]\ Ag_2O;\ [3]\ \Delta}$

c. (cyclopentane with Cl and propyl) $\xrightarrow{Na^+ \ ^-OR}$

d. (cyclopentane with NH$_2$ and propyl) $\xrightarrow{[1]\ CH_3I\ (excess);\ [2]\ Ag_2O;\ [3]\ \Delta}$

23.12 Reaction of Amines with Nitrous Acid

Nitrous acid, HNO_2, is a weak, unstable acid formed from $NaNO_2$ and a strong acid like HCl.

$$H{-}Cl: \ + \ Na^+ \ ^-:O{-}N{=}O: \longrightarrow HO{-}N{=}O: \ + \ Na^+ \ :Cl:^-$$

nitrous acid

In the presence of acid, nitrous acid decomposes to **^+NO, the nitrosonium ion.** This electrophile then goes on to react with the nucleophilic nitrogen atom of amines to form **diazonium salts ($RN_2^+Cl^-$)** from 1° amines and ***N*-nitrosamines ($R_2NN{=}O$)** from 2° amines.

$$H{-}Cl: \ + \ HO{-}N{=}O: \longrightarrow H_2\overset{+}{O}{-}N{=}O: \ + \ :Cl:^- \longrightarrow H_2O: \ + \ ^+N{=}O:$$

nitrous acid — nitrosonium ion — electrophile

23.12A Reaction of ^+NO with 1° Amines

Nitrous acid reacts with 1° alkylamines and arylamines to form diazonium salts. This reaction is called **diazotization.**

$$R{-}NH_2 \xrightarrow[HCl]{NaNO_2} R{-}\overset{+}{N}{\equiv}N: \ Cl^- \qquad C_6H_5{-}NH_2 \xrightarrow[HCl]{NaNO_2} C_6H_5{-}\overset{+}{N}{\equiv}N: \ Cl^-$$

alkyl diazonium salt — **aryl diazonium salt**

The mechanism for this reaction begins with nucleophilic attack of the amine on the nitrosonium ion, and it can conceptually be divided into two parts: formation of an ***N*-nitrosamine,** followed by loss of H_2O, as shown in Mechanism 23.2.

Alkyl diazonium salts are generally not useful compounds. They readily decompose below room temperature to form carbocations with loss of N_2, a very good leaving group. These carbocations usually form a complex mixture of substitution, elimination, and rearrangement products.

$$(CH_3)_3C{-}NH_2 \xrightarrow[HCl]{NaNO_2} \left[(CH_3)_3C{-}\overset{+}{N}{\equiv}N: \ Cl^-\right] \longrightarrow (CH_3)_3C^+ \longrightarrow$$

products of substitution, elimination, and rearrangement (with some reactants)

1° alkylamine — unstable diazonium salt — **carbocation** + N_2 good leaving group

Mechanism 23.2 Formation of a Diazonium Salt from a 1° Amine

Part [1] Formation of an *N*-nitrosamine

1 – 2 Nucleophilic attack of the amine on the nitrosonium ion (^{+}NO), followed by loss of a proton, forms an ***N*-nitrosamine.**

Part [2] Formation of a diazonium salt

3 – 5 Three proton transfers form an intermediate with a **good leaving group (H_2O).**

6 Loss of water forms a **diazonium ion (RN_2^+).** The diazonium salt formed in this reaction consists of the diazonium ion (RN_2^+) and a chloride anion (Cl^-).

Care must be exercised in handling diazonium salts, because they can explode if allowed to dry.

On the other hand, **aryl diazonium salts are very useful synthetic intermediates.** Although they are rarely isolated and are generally unstable above 0 °C, they are useful starting materials in two general kinds of reactions described in Section 23.13.

23.12B Reaction of ^{+}NO with 2° Amines

Secondary alkylamines and arylamines react with nitrous acid to form *N*-nitrosamines.

N-nitrosamine in tobacco smoke

Many *N*-nitrosamines are potent carcinogens found in some food and tobacco smoke. Nitrosamines in food can be formed in the same way they are formed in the laboratory: **reaction of a 2° amine with the nitrosonium ion,** formed from nitrous acid (HNO_2). The mechanism for this reaction follows the two steps of Part [1] of Mechanism 23.2.

Problem 23.29 Draw the product formed when each compound is treated with $NaNO_2$ and HCl.

23.13 Substitution Reactions of Aryl Diazonium Salts

Aryl diazonium salts undergo two general reactions.

- **Substitution of N_2 by an atom or a group of atoms Z forms a variety of substituted benzenes.**

- **Coupling of a diazonium salt with another benzene derivative forms an azo compound, a compound containing a nitrogen–nitrogen double bond.**

$$C_6H_5N_2^+ Cl^- + C_6H_5Y \longrightarrow C_6H_5N{=}NC_6H_4Y + HCl$$

azo compound

Y = NH_2, NHR, NR_2, OH (a strong electron-donor group)

23.13A Specific Substitution Reactions

Aryl diazonium salts react with a variety of reagents to form products in which **Z (an atom or group of atoms) replaces N_2, a very good leaving group.** The mechanism of these reactions varies with the identity of Z, so we will concentrate on the products of the reactions, not the mechanisms.

$$C_6H_5N_2^+ Cl^- \xrightarrow{Z} C_6H_5Z + N_2 + Cl^-$$

N_2: good leaving group

[1] Substitution by OH—Synthesis of phenols

$$C_6H_5N_2^+ Cl^- \xrightarrow{H_2O} C_6H_5OH$$

phenol

A diazonium salt reacts with H_2O to form a **phenol.**

[2] Substitution by Cl or Br—Synthesis of aryl chlorides and bromides

$$C_6H_5N_2^+ Cl^- \xrightarrow{CuCl} C_6H_5Cl \qquad C_6H_5N_2^+ Cl^- \xrightarrow{CuBr} C_6H_5Br$$

aryl chloride **aryl bromide**

A diazonium salt reacts with copper(I) chloride or copper(I) bromide to form an **aryl chloride** or **aryl bromide,** respectively. This is called the **Sandmeyer reaction.** It provides an alternative to direct chlorination and bromination of an aromatic ring using Cl_2 or Br_2 and a Lewis acid catalyst.

[3] Substitution by F—Synthesis of aryl fluorides

$$C_6H_5N_2^+ Cl^- \xrightarrow{HBF_4} C_6H_5F$$

aryl fluoride

A diazonium salt reacts with fluoroboric acid (HBF_4) to form an **aryl fluoride.** This is a useful reaction because aryl fluorides cannot be produced by direct fluorination with F_2 and a Lewis acid catalyst, because F_2 reacts too violently (Section 16.3).

[4] Substitution by I—Synthesis of aryl iodides

N_2^+ Cl^- → NaI or KI → I

aryl iodide

A diazonium salt reacts with sodium or potassium iodide to form an **aryl iodide.** This, too, is a useful reaction because aryl iodides cannot be produced by direct iodination with I_2 and a Lewis acid catalyst, because I_2 reacts too slowly (Section 16.3).

[5] Substitution by CN—Synthesis of benzonitriles

N_2^+ Cl^- → CuCN → CN

benzonitrile

A diazonium salt reacts with copper(I) cyanide to form a **benzonitrile.** Because a cyano group can be hydrolyzed to a carboxylic acid, reduced to an amine or aldehyde, or converted to a ketone with organometallic reagents, this reaction provides easy access to a wide variety of benzene derivatives using chemistry described in Section 19.12.

[6] Substitution by H—Synthesis of benzene

N_2^+ Cl^- → H_3PO_2 → H

benzene

A diazonium salt reacts with hypophosphorus acid (H_3PO_2) to form **benzene.** This reaction has limited utility because it reduces the functionality of the benzene ring by replacing N_2 with a hydrogen atom. Nonetheless, this reaction *is* useful in synthesizing compounds that have substitution patterns that are not available by other means.

For example, it is not possible to synthesize 1,3,5-tribromobenzene from benzene by direct bromination. Because Br is an ortho, para director, bromination with Br_2 and $FeBr_3$ will not add Br substituents meta to each other on the ring.

Br Br Br

1,3,5-tribromobenzene

It is possible, however, to add three Br atoms meta to each other when aniline is the starting material. Because an NH_2 group is a very powerful ortho, para director, three Br atoms are introduced in a single step on halogenation (Section 16.10A). Then, the NH_2 group can be removed by diazotization and reaction with H_3PO_2.

NH_2

Add three Br's ortho and para to the NH_2 group...

NH_2 Br Br Br

...and then remove the NH_2 in two steps.

H Br Br Br

The complete synthesis of 1,3,5-tribromobenzene from benzene is outlined in Figure 23.9.

Figure 23.9 The synthesis of 1,3,5-tribromobenzene from benzene

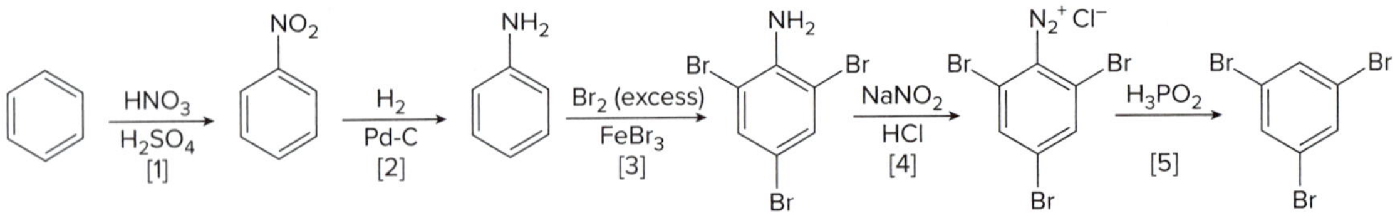

- Nitration followed by reduction forms aniline ($C_6H_5NH_2$) from benzene (Steps [1] and [2]).
- Bromination of aniline yields the tribromo derivative in Step [3].
- The NH_2 group is removed by a two-step process: diazotization with $NaNO_2$ and HCl (Step [4]), followed by substitution of the diazonium ion by H with H_3PO_2.

Problem 23.30 Draw the product formed in each reaction.

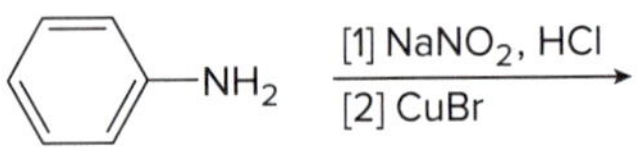

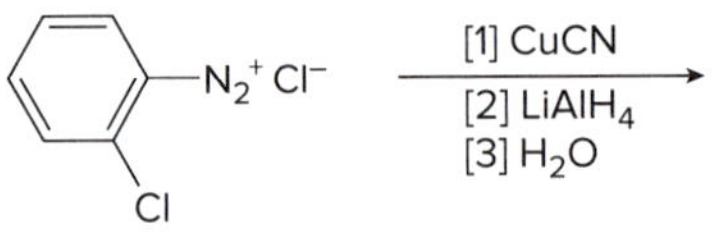

23.13B Using Diazonium Salts in Synthesis

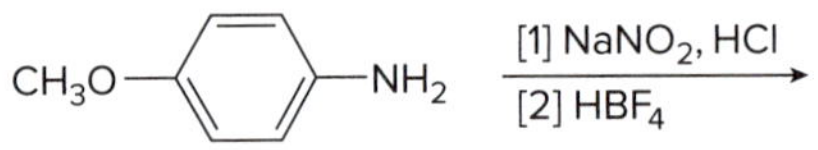

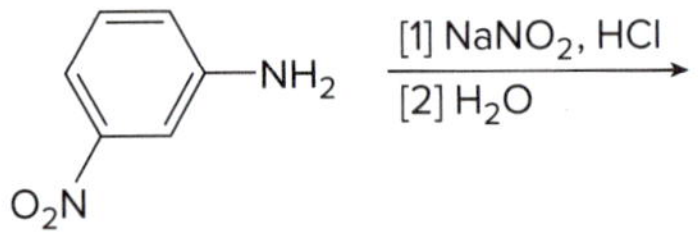

Diazonium salts provide easy access to many different benzene derivatives. Keep in mind the following four-step sequence, because it will be used to synthesize many substituted benzenes.

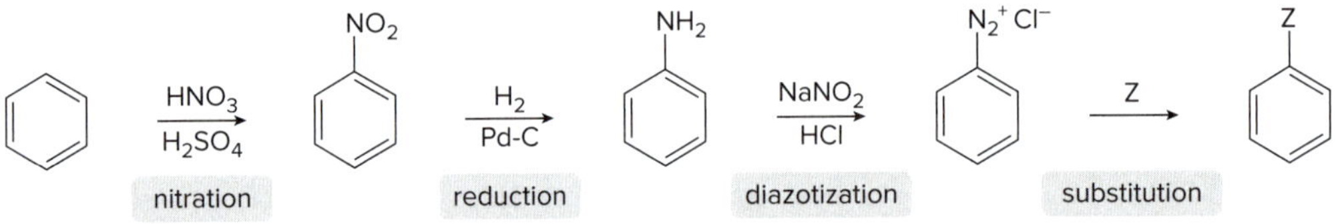

Sample Problems 23.6 and 23.7 apply these principles to two different multistep syntheses.

Sample Problem 23.6 Using Diazonium Salts in Synthesis

Synthesize *m*-chlorophenol from benzene.

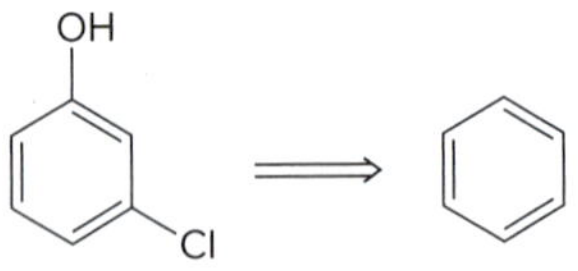

Solution

Both OH and Cl are ortho, para directors, but they are located *meta* to each other. The OH group must be formed from a diazonium salt, which can be made from an NO_2 group (a meta director) by a stepwise method.

Retrosynthetic Analysis

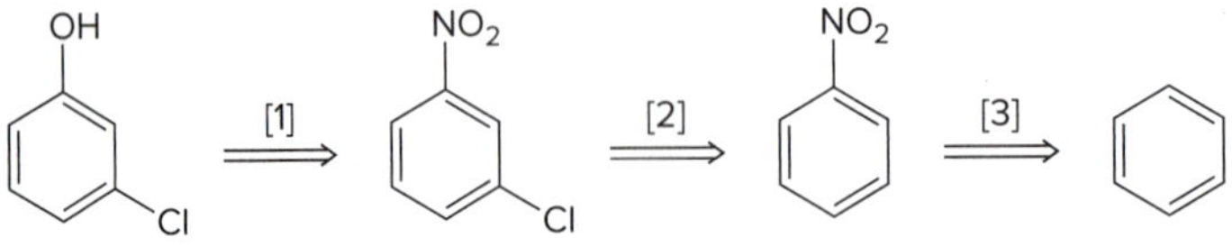

Working backwards:

- [1] Form the OH group from NO_2 by a three-step procedure using a diazonium salt.
- [2] Introduce Cl meta to NO_2 by halogenation.
- [3] Add the NO_2 group by nitration.

Synthesis

benzene —(HNO_3, H_2SO_4 [1])→ nitrobenzene —(Cl_2, $FeCl_3$ [2])→ m-chloronitrobenzene —(H_2, Pd-C [3])→ m-chloroaniline (NH_2) —($NaNO_2$, HCl [4])→ m-chlorobenzenediazonium chloride ($N_2^+ Cl^-$) —(H_2O [5])→ m-chlorophenol (OH, Cl)

- Nitration followed by chlorination meta to the NO_2 group forms the **meta disubstituted benzene** (Steps [1]–[2]).
- **Reduction of the nitro group** (Step [3]) followed by **diazotization** forms the **diazonium salt** in Step [4], which is then converted to the desired phenol by treatment with H_2O (Step [5]).

Problem 23.31 Devise a synthesis of each compound from benzene.

a. fluorobenzene (F) b. HO, OH (m-dihydroxybenzene) c. o-iodotoluene (I) d. 1,3,5-trichlorobenzene (Cl, Cl, Cl)

More Practice: Try Problems 23.65a, b; 23.66a, b; 23.67, 23.68.

Sample Problem 23.7 Devising a Synthesis with Diazonium Salts

Synthesize *p*-bromobenzaldehyde from benzene.

p-bromobenzaldehyde (CHO, Br) ⟹ benzene

Solution

Because the two groups are located para to each other and Br is an ortho, para director, Br should be added to the ring *first*. **To add the CHO group, recall that it can be formed from CN by reduction.**

Retrosynthetic Analysis

p-BrC$_6$H$_4$CHO (CHO, Br) ⟹[1] p-BrC$_6$H$_4$CN (CN, Br) ⟹[2] p-BrC$_6$H$_4$NO$_2$ (NO_2, Br) ⟹[3] bromobenzene (Br) ⟹[4] benzene

Working backwards:

- [1] Form the CHO group by reduction of CN.
- [2] Prepare the CN group from an NO_2 group by a three-step sequence using a diazonium salt.
- [3] Introduce the NO_2 group by nitration, para to the Br atom.
- [4] Introduce Br by bromination with Br_2 and $FeBr_3$.

Synthesis

Benzene → (Br_2, $FeBr_3$ [1]) → bromobenzene → (HNO_3, H_2SO_4 [2]) → p-bromonitrobenzene (NO_2, Br) → (H_2, Pd-C [3]) → p-bromoaniline (NH_2, Br) → ($NaNO_2$, HCl [4]) → $N_2^+ Cl^-$, Br → (CuCN [5]) → CN, Br → ([1] DIBAL-H, [2] H_2O [6]) → CHO, Br

- **Bromination followed by nitration forms a disubstituted benzene** with two para substituents (Steps [1]–[2]), which can be separated from its undesired ortho isomer.
- **Reduction of the NO_2 group** (Step [3]) followed by **diazotization** forms the diazonium salt in Step [4], which is converted to a nitrile by reaction with CuCN (Step [5]).
- **Reduction of the CN group with DIBAL-H** (a mild reducing agent) **forms the CHO group** and completes the synthesis.

Problem 23.32 Devise a synthesis of each compound from benzene. You may use any other organic or inorganic reagents.

a. I–C₆H₄–CH₃ (para) b. NC–C₆H₄–Br (meta) c. HO–C₆H₄–COOH (para)

More Practice: Try Problems 23.69a, b; 23.70; 23.71.

23.14 Coupling Reactions of Aryl Diazonium Salts

The second general reaction of diazonium salts is **coupling.** When a diazonium salt is treated with an aromatic compound that contains a strong electron-donor group, the two rings join together to form an **azo compound,** a compound with a nitrogen–nitrogen double bond.

$C_6H_5N_2^+ Cl^-$ + C_6H_5–Y → (azo coupling) C_6H_5–N=N–C_6H_4–Y + HCl

Y = NH_2, NHR, NR_2, OH (a strong electron-donor group)

azo compound

Synthetic dyes are described in more detail in Section 23.15A.

Azo compounds are highly conjugated, rendering them colored (Section 14.15). Many of these compounds, such as the azo compound "butter yellow," are synthetic dyes. Butter yellow was once used to color margarine.

$C_6H_5N_2^+ Cl^-$ + $C_6H_5N(CH_3)_2$ → C_6H_5–N=N–C_6H_4–$N(CH_3)_2$

a yellow azo dye "butter yellow"

This reaction is another example of **electrophilic aromatic substitution,** with the **diazonium salt acting as the electrophile.** Like all electrophilic substitutions (Section 16.2), the mechanism has two steps: **addition of the electrophile** (the diazonium ion) to form a **resonance-stabilized carbocation,** followed by deprotonation, as shown in Mechanism 23.3.

Mechanism 23.3 Azo Coupling

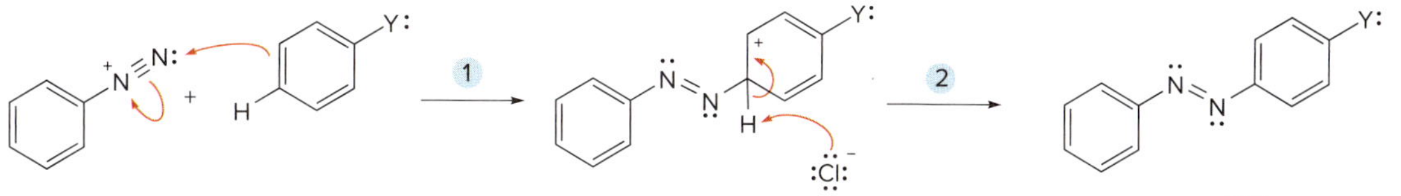

(+ three additional resonance structures)
resonance-stabilized carbocation

+ HCl

1 The diazonium ion reacts with the benzene ring to form a **resonance-stabilized carbocation.**

2 Loss of a proton regenerates the aromatic ring.

Because a diazonium salt is weakly electrophilic, the reaction occurs only when the benzene ring has **a strong electron-donor group Y, where Y = NH_2, NHR, NR_2, or OH.** Although these groups activate both the ortho and para positions, para substitution occurs unless the para position already has another substituent present.

Problem 23.33 Draw the product formed when $C_6H_5N_2^+Cl^-$ reacts with each compound.

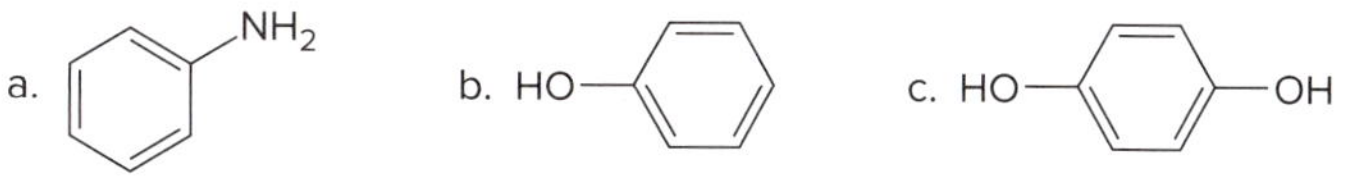

a. b. HO– c. HO– –OH

To determine what starting materials are needed to synthesize a particular azo compound, always divide the molecule into two components: **one has a benzene ring with a diazonium ion, and one has a benzene ring with a very strong electron-donor group.**

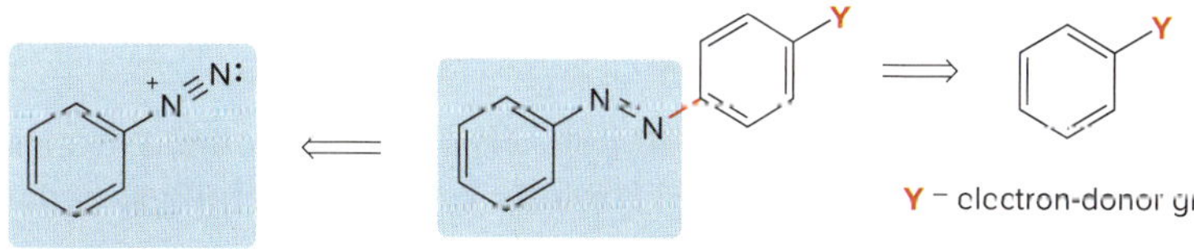

Y = electron-donor group

Sample Problem 23.8 Synthesizing an Azo Compound

What starting materials are needed to synthesize the following azo compound?

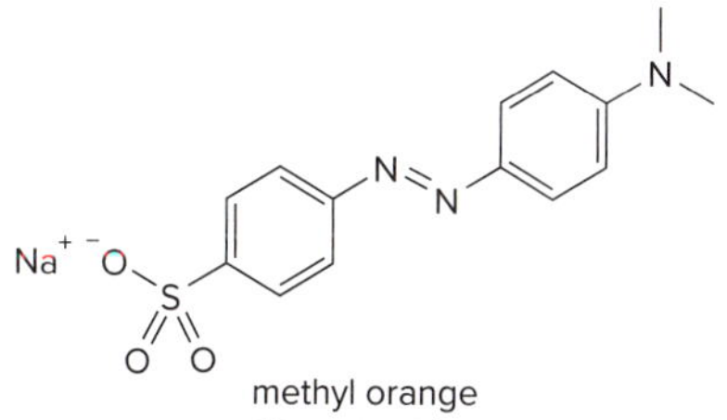

methyl orange
an orange dye

Solution

Both benzene rings in methyl orange have a substituent, but only one group, $N(CH_3)_2$, is a strong electron donor. In determining the two starting materials, the **diazonium ion must be bonded to the ring that is *not* bonded to $N(CH_3)_2$.**

The diazonium ion is in one compound.

Break the molecule into two components at this C–N bond.

The electron-donor group is in the other compound.

Problem 23.34 What starting materials are needed to synthesize each azo compound?

a. b.

More Practice: Try Problems 23.65c, 23.66c, 23.69c.

23.15 Application: Synthetic Dyes and Sulfa Drugs

Azo compounds have two important applications: as dyes and as sulfa drugs, the first synthetic antibiotics.

23.15A Natural and Synthetic Dyes

Until 1856, all dyes were natural in origin, obtained from plants, animals, or minerals. Three natural dyes known for centuries are **indigo, tyrian purple,** and **alizarin.**

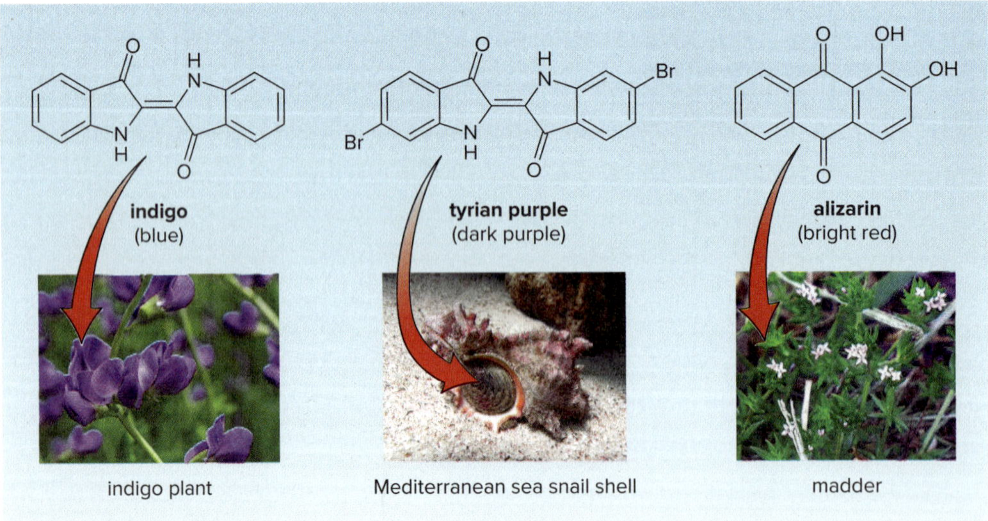

Michael G McKinne/Shutterstock; Kristina Vackova/Shutterstock; Bob Gibbons/Alamy Stock Photo

The blue dye **indigo,** derived from the plant *Indigofera tinctoria,* has been used in India for thousands of years. Traders introduced it to the Mediterranean area and then to Europe. **Tyrian purple,** a natural dark purple dye obtained from the mucous gland of a Mediterranean snail of the genus *Murex,* was a symbol of royalty before the collapse of the Roman Empire. **Alizarin,** a bright red dye obtained from madder root (*Rubia tinctorum*), a plant native to India and northeastern Asia, has been found in cloth entombed with Egyptian mummies.

major component of Perkin's mauveine

Because all three of these dyes were derived from natural sources, they were difficult to obtain, making them expensive and available only to the privileged. This all changed in 1856 when William Henry Perkin, an 18-year-old student with a makeshift home laboratory, serendipitously prepared a purple dye, which would later be called mauveine, during his failed attempt to synthesize the antimalarial drug quinine. Mauveine is a mixture of four compounds that differ in the number and location of methyl groups on the aromatic rings.

A purple shawl dyed with Perkin's mauveine *Science & Society Picture Library/Getty Images*

Perkin's discovery marked the beginning of the chemical industry. He patented the dye and went on to build a factory to commercially produce it on a large scale. This event began the surge of research in organic chemistry, not just in the synthesis of dyes, but in the production of perfumes, anesthetics, inks, and drugs as well. Perkin was a wealthy man when he retired at the age of 36 to devote the rest of his life to basic chemical research. The most prestigious award given by the American Chemical Society is named the Perkin Medal in his honor.

Many common synthetic dyes, such as para red and Congo red, are **azo compounds,** prepared by the diazonium coupling reaction described in Section 23.14.

para red

Congo red

Although natural and synthetic dyes are quite varied in structure, **all of them are colored because they are highly conjugated.** A molecule with eight or more π bonds in conjugation absorbs light in the visible region of the electromagnetic spectrum (Section 14.15A), taking on the color from the visible spectrum that it does *not* absorb.

Problem 23.35 What two components are needed to prepare para red by azo coupling?

23.15B Sulfa Drugs

Although they may seem quite unrelated, the synthesis of colored dyes led to the development of the first synthetic antibiotics. Much of the early effort in this field was done by the German chemist Paul Ehrlich, who worked with synthetic dyes and used them to stain tissues. This led him on a search for dyes that were lethal to bacteria without affecting other tissue cells, hoping that these dyes could treat bacterial infections. For many years this effort was unsuccessful.

Then, in 1935, Gerhard Domagk, a German physician working for a dye manufacturer, first used a synthetic dye as a drug to kill bacteria. His daughter had contracted a streptococcal infection, and as she neared death, he gave her **prontosil,** an azo dye that inhibited the growth of certain bacteria in mice. His daughter recovered, and the modern era of synthetic antibiotics was initiated. For his pioneering work, Domagk was awarded the Nobel Prize in Physiology or Medicine in 1939.

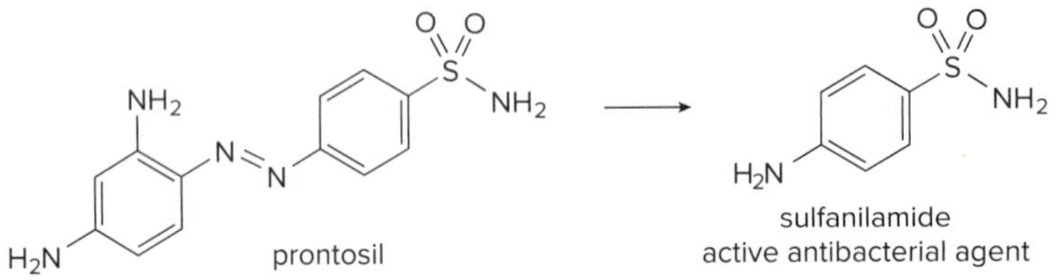

prontosil → sulfanilamide
active antibacterial agent

Prontosil and other sulfur-containing antibiotics are collectively called **sulfa drugs.** Prontosil is not the active agent itself. In cells, it is metabolized to **sulfanilamide,** the active drug. To understand how sulfanilamide functions as an antibacterial agent we must examine **folic acid,** which microorganisms synthesize from *p*-aminobenzoic acid.

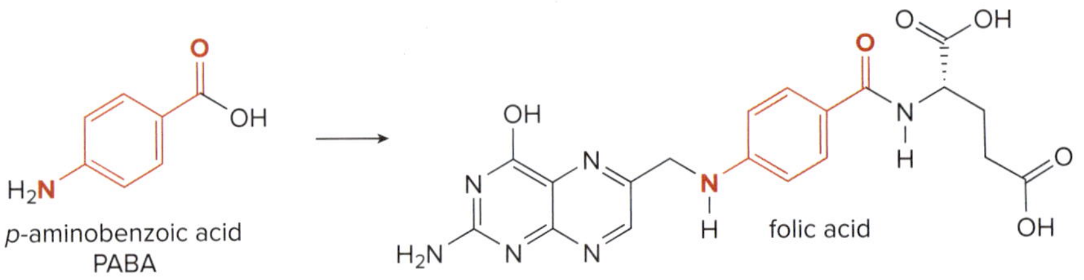

p-aminobenzoic acid
PABA → folic acid

Sulfanilamide and *p*-aminobenzoic acid are similar in size and shape and have related functional groups. Thus, when sulfanilamide is administered, bacteria attempt to use it in place of *p*-aminobenzoic acid to prepare folic acid, and this derails folic acid synthesis, so bacteria cannot grow and reproduce. Sulfanilamide affects only bacterial cells, though, because humans do not synthesize folic acid and must obtain it from their diets.

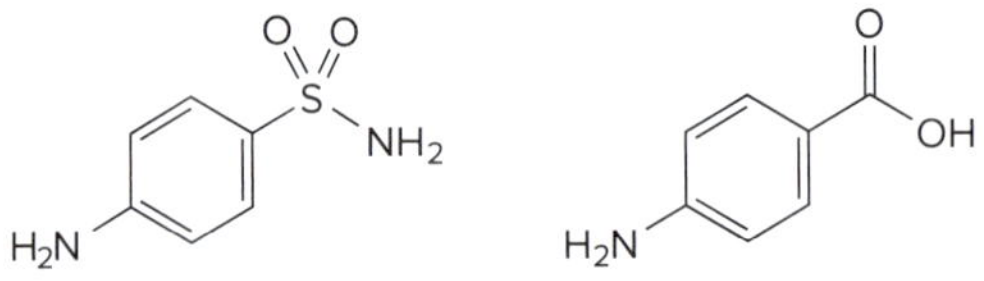

sulfanilamide **p-aminobenzoic acid**

Many other compounds of similar structure have been prepared and are still widely used as antibiotics. The structures of two other sulfa drugs are shown in Figure 23.10.

Figure 23.10
Two common sulfa drugs

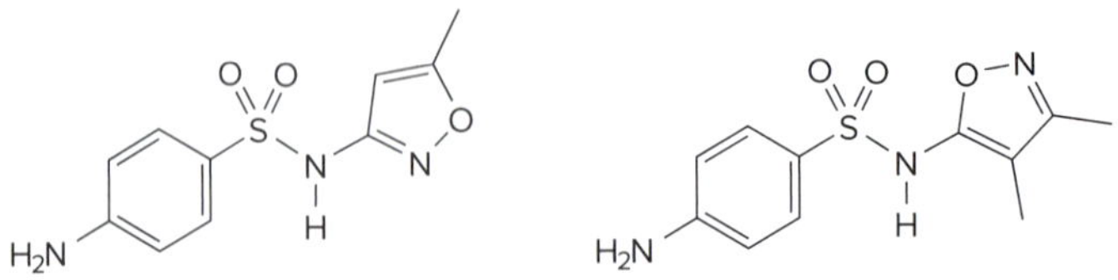

sulfamethoxazole sulfisoxazole

- Sulfamethoxazole is the sulfa drug in Bactrim, and sulfisoxazole is sold as Gantrisin. Both drugs are commonly used in the treatment of ear and urinary tract infections.

Chapter 23 REVIEW

KEY CONCEPTS

The basicity of amines (23.10)

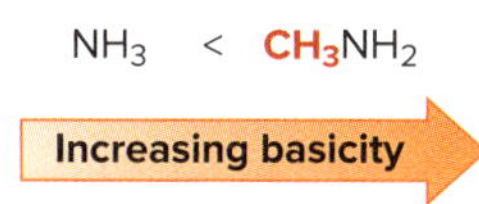

- **Alkylamines are more basic than NH_3** because of the **electron-donating R groups.**

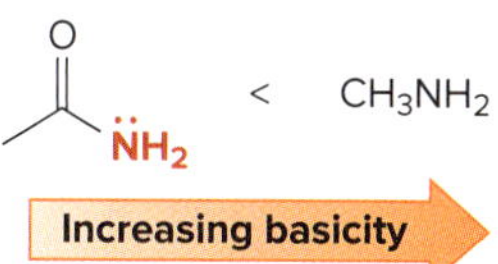

- **Alkylamines are more basic than amides,** which have a **delocalized lone pair** from the N atom.

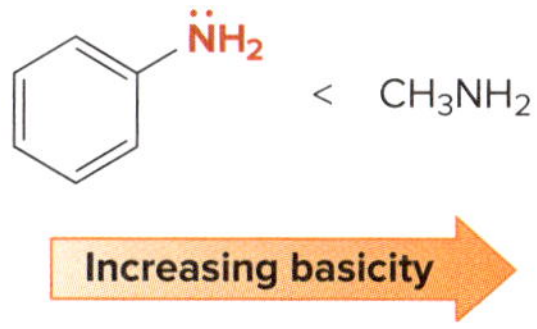

$C_6H_5NH_2$ < CH_3NH_2

Increasing basicity

- **Alkylamines are more basic than arylamines,** which have a **delocalized lone pair** from the N atom.

pyridine (N) < piperidine (NH)

Increasing basicity

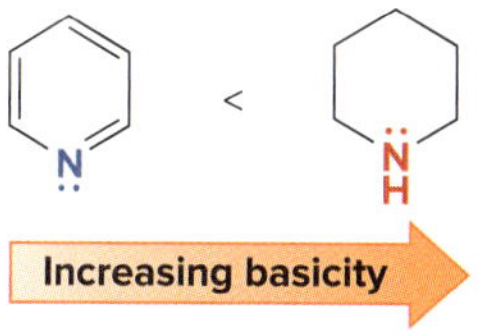

- **Alkylamines** with a **lone pair in an sp^3 hybrid orbital are more basic** than those with a **lone pair in an sp^2 hybrid orbital.**

Try Problems 23.36a, 23.39–23.42.

KEY REACTIONS

Preparation of Amines

[1] Nucleophilic substitution

1. $CH_3—X + NH_3 \xrightarrow[(23.6A)]{S_N2} CH_3—NH_2 + NH_4^+X^-$
   X = Cl, Br, I; NH_3 excess; product: 1° amine

2. 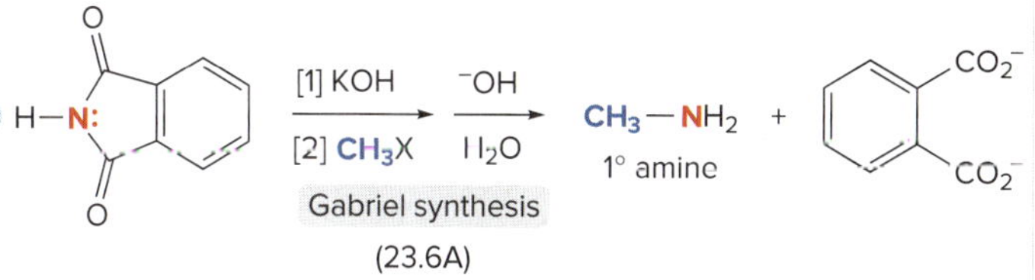 H—N: (phthalimide) $\xrightarrow{[1]\ KOH,\ [2]\ CH_3X}$ $\xrightarrow{^-OH,\ H_2O}$ $CH_3—NH_2$ (1° amine) + $C_6H_4(CO_2^-)_2$
   Gabriel synthesis (23.6A)

Try Problem 23.52d.

[2] Methods that involve reduction

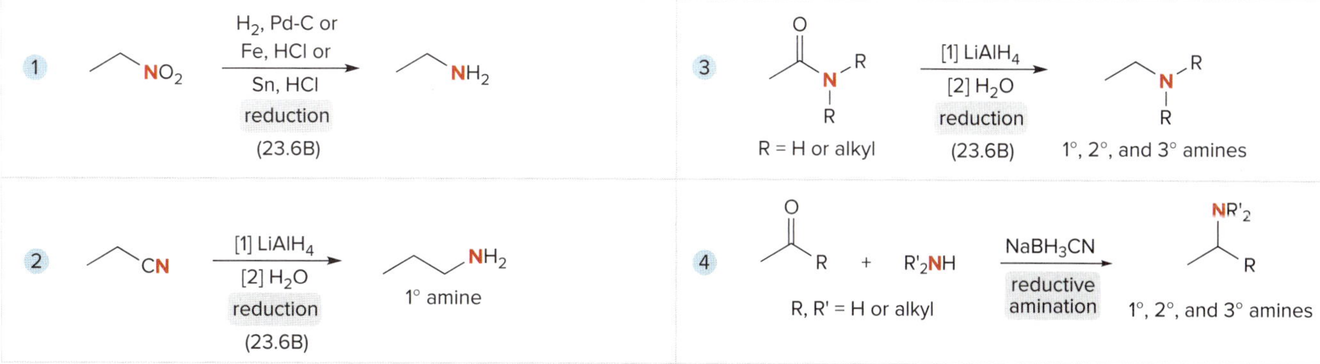

Try Problems 23.48–23.50; 23.52j; 23.56a, b, d.

Reactions of Amines

[1] Reaction as a base

$CH_3CH_2\ddot{N}H_2$ + H–A ⇌ $CH_3CH_2\overset{+}{N}H_3$ + :A^-

base — acid — (23.8) — conjugate acid

Try Problems 23.36b; 23.40; 23.42; 23.52a, g.

[2] Nucleophilic addition to aldehydes and ketones

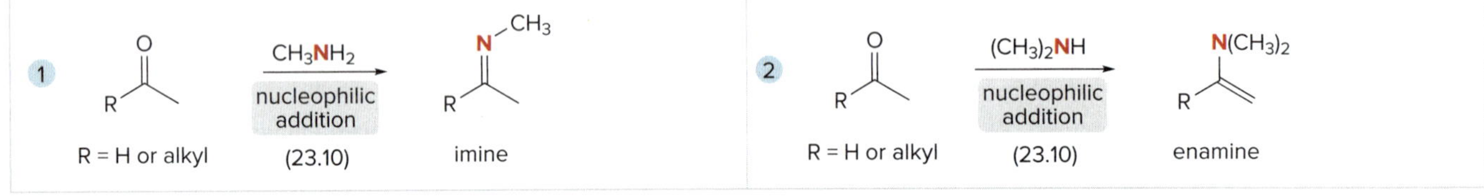

Try Problems 23.52e, 23.56e.

[3] Nucleophilic substitution with acid chlorides and anhydrides

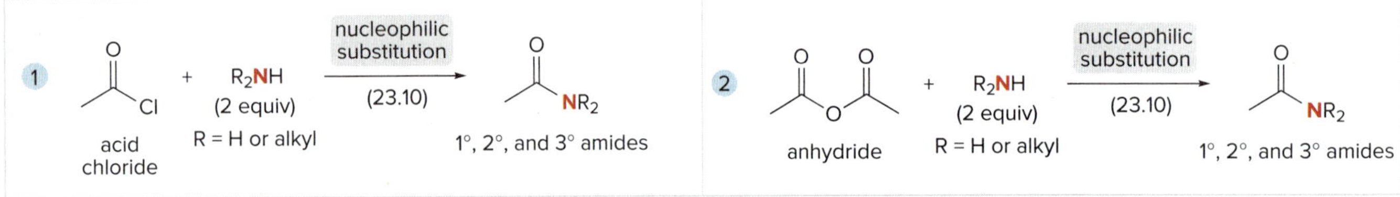

Try Problem 23.52b, c.

[4] Hofmann elimination

[1] CH_3I (excess); [2] Ag_2O; [3] Δ — Hofmann elimination (23.11)

NH_2 → major product + $N(CH_3)_3$ + H_2O + AgI

Try Problems 23.53, 23.54c, 23.55c, 23.56f, 23.57.

[5] Reaction with nitrous acid

Try Problems 23.52h, 23.56c.

Reactions of Diazonium Salts

[1] Substitution reactions

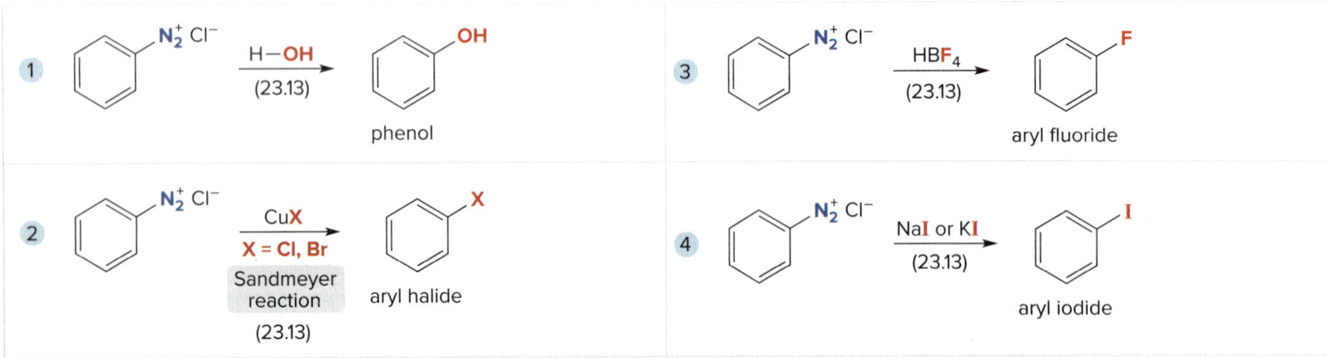

5 $C_6H_5N_2^+ Cl^-$ —CuCN (23.13)→ benzonitrile

6 $C_6H_5N_2^+ Cl^-$ —H_3PO_2 (23.13)→ benzene

Try Problem 23.60.

[2] Coupling to form azo compounds

$C_6H_5N_2^+ Cl^-$ + C_6H_5–Y: —azo coupling (23.14)→ azo compound + HCl

Y = NH_2, NHR, NR_2, OH (a strong electron-donor group)

Try Problem 23.60c.

KEY SKILLS

[1] Using retrosynthetic analysis in a reductive amination (23.6C); two possibilities

1 Break the molecule into two components, using one alkyl group on N to form the carbonyl component.

2 Break the molecule into two components, using the other alkyl group on N to form the carbonyl component.

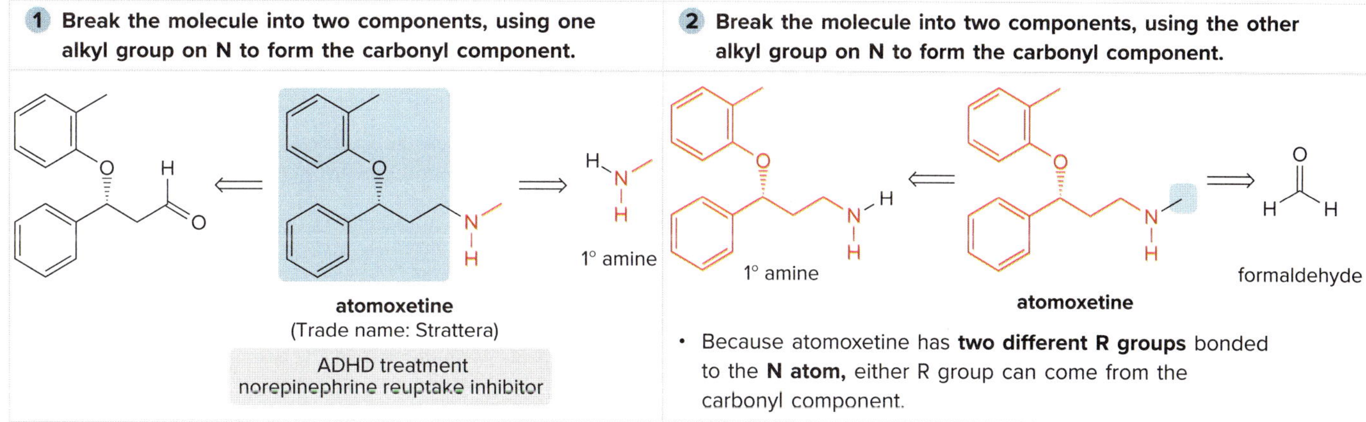

- Because atomoxetine has **two different R groups** bonded to the **N atom,** either R group can come from the carbonyl component.

See Sample Problem 23.2. Try Problems 23.47, 23.54b, 23.55b.

[2] Ranking arylamines in order of increasing basicity (23.9)

1 Identify whether the substituents are electron donating or electron withdrawing.

2 Rank the compounds.

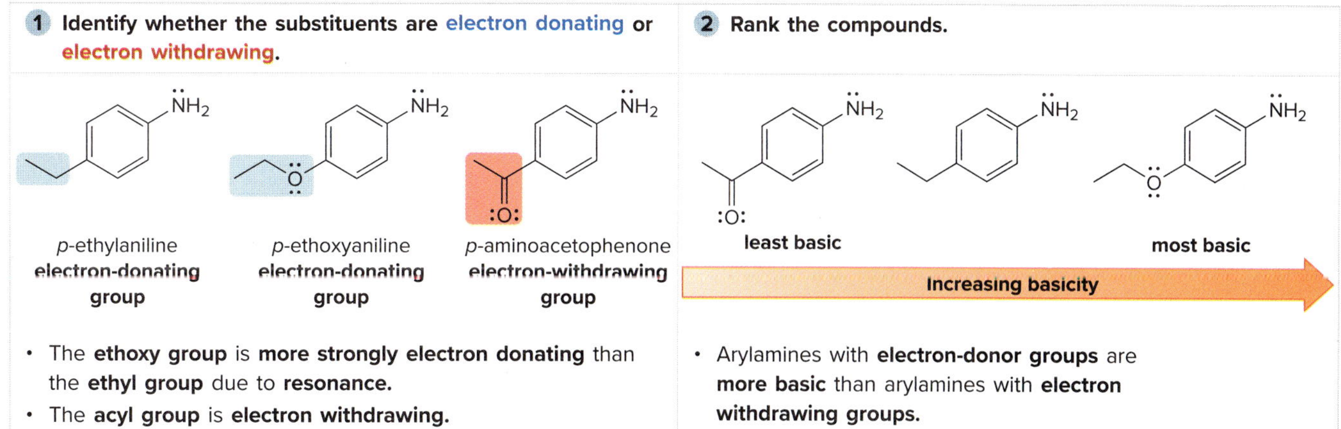

- The **ethoxy group** is **more strongly electron donating** than the **ethyl group** due to **resonance.**
- The **acyl group** is **electron withdrawing.**

- Arylamines with **electron-donor groups** are **more basic** than arylamines with **electron withdrawing groups.**

See Sample Problem 23.3. Try Problem 23.39.

[3] Ranking N atoms in order of increasing basicity; example: manzamine C (23.9)

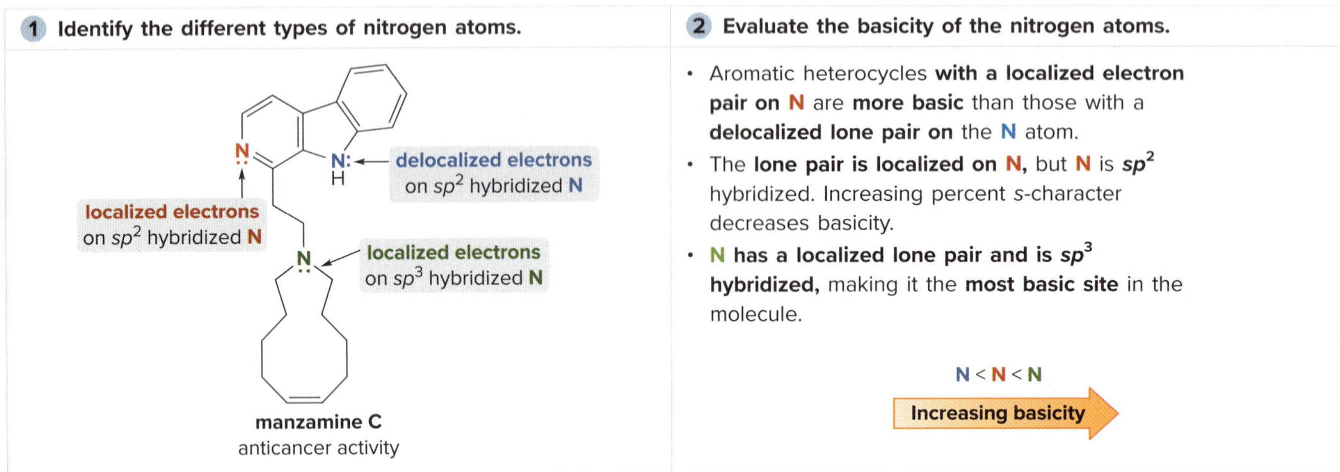

See Sample Problem 23.4. Try Problems 23.36a, 23.40–23.42.

[4] Drawing the major and minor product formed from Hofmann elimination (23.11)

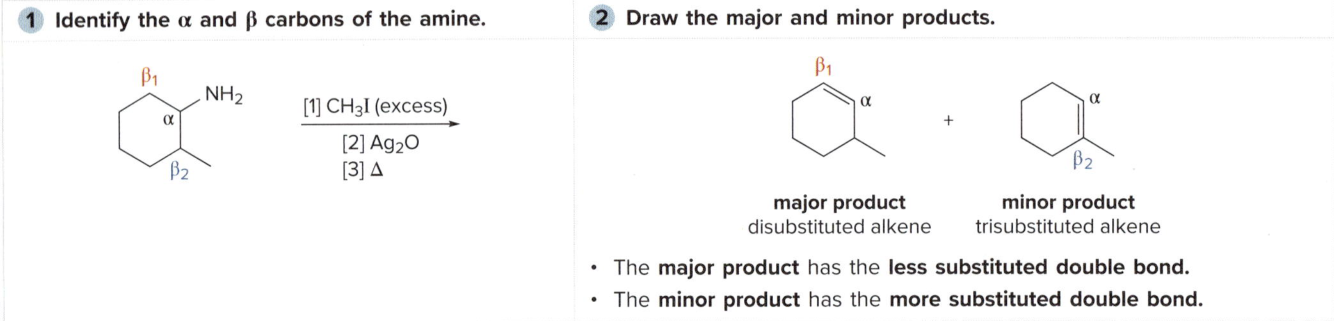

See Sample Problem 23.5. Try Problems 23.53, 23.54c, 23.55c, 23.56f, 23.57.

[5] Drawing the starting materials needed to synthesize an azo compound (23.14)

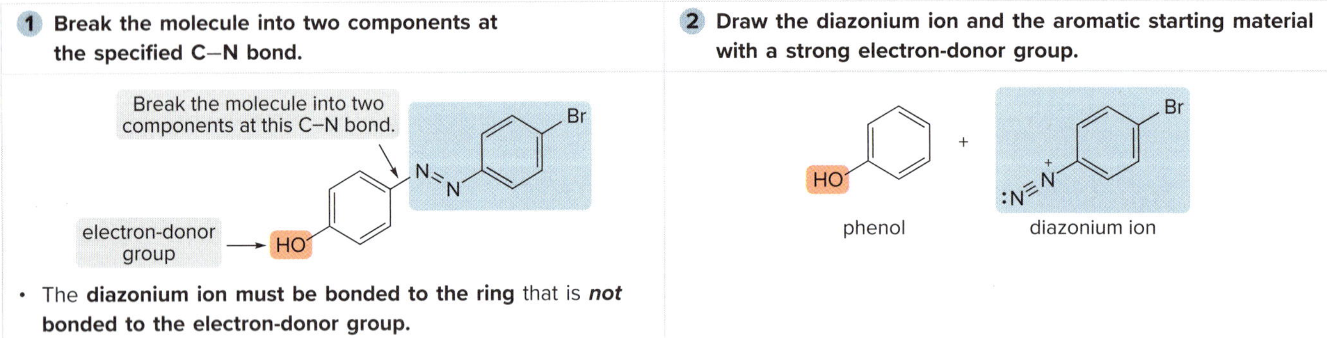

See Sample Problem 23.8. Try Problems 23.65c, 23.66c, 23.69c.

CHAPTER 23 MULTIPLE-CHOICE SELF-TEST

The Self-Test consists of multiple-choice questions similar to those found on the American Chemical Society organic chemistry exam. Answers are given at the end of the chapter.

1. Give a systematic name for the following compound.

a. 2,5-dimethyl-*N*-ethylhexan-3-amine
b. 3-*N*-ethylamino-2,5-dimethylhexane
c. *N*-ethyl-2,5-dimethylhexan-3-amine
d. *N*-ethyl-1-isopropyl-3-methylbutan-1-amine

2. What reaction sequence converts **A** to **B?**

a. [1] CH_3Cl, $AlCl_3$; [2] $NaNO_2$, HCl; [3] CuCl
b. [1] $NaNO_2$, HCl; [2] CuCl; [3] CH_3Cl, $AlCl_3$
c. [1] $NaNO_2$, HCl; [2] Cl_2, $FeCl_3$; [3] CH_3Cl, $AlCl_3$
d. [1] $NaNO_2$, HCl; [2] Cl_2, $FeCl_3$; [3] Cl_2, $FeCl_3$

3. Which labeled N atom is most basic?

4. What starting materials are needed to synthesize **C** by reductive amination?

5. What method can be used to prepare **D** by a reduction reaction?

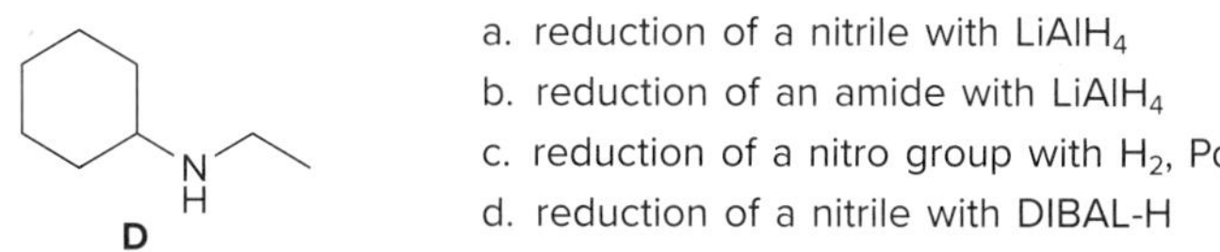

a. reduction of a nitrile with $LiAlH_4$
b. reduction of an amide with $LiAlH_4$
c. reduction of a nitro group with H_2, Pd
d. reduction of a nitrile with DIBAL-H

6. What is the major product formed by the Hofmann elimination of amine **E?**

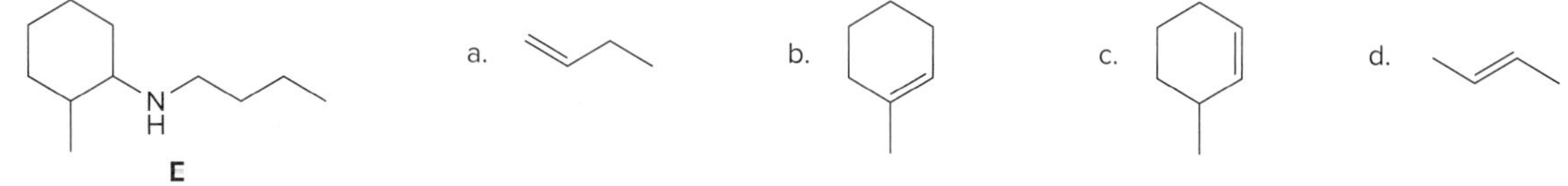

7. In a separatory funnel containing diethyl ether and 10% aqueous HCl solution, which compounds are present in the organic layer?

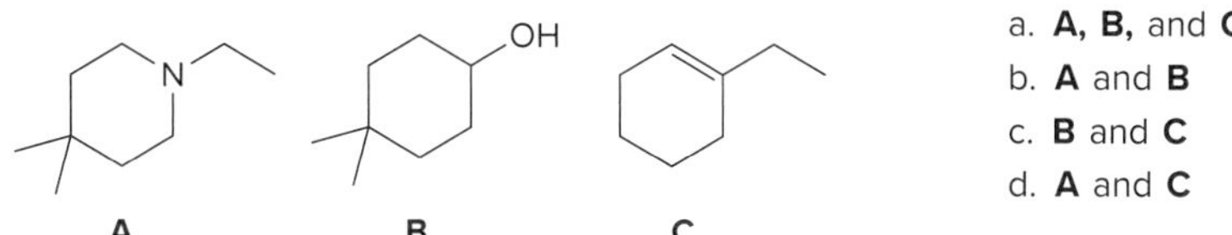

a. **A, B,** and **C**
b. **A** and **B**
c. **B** and **C**
d. **A** and **C**

8. Rank **A–D** in order of decreasing boiling point.

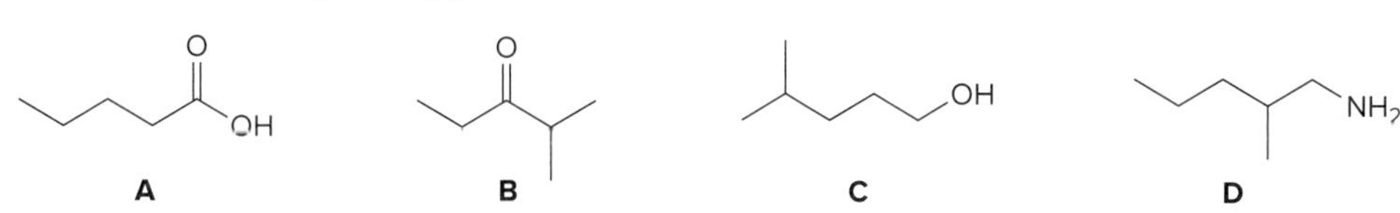

a. **B** > **D** > **C** > **A** b. **A** > **C** > **D** > **B** c. **A** > **D** > **C** > **B** d. **B** > **C** > **D** > **A**

9. What product is formed when the given ketone and amine are treated with $NaBH_3CN$?

NH =O a. N c. $\overset{+}{N}$

b. N d. N

10. What reaction sequence converts benzene into *m*-bromochlorobenzene?

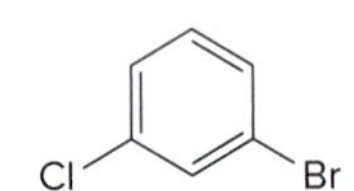

m-bromochlorobenzene

a. [1] HNO_3, H_2SO_4; [2] Cl_2, $FeCl_3$; [3] H_2, Pd-C; [4] $NaNO_2$, HCl; [5] CuBr
b. [1] Cl_2, $FeCl_3$; [2] HNO_3, H_2SO_4; [3] H_2, Pd-C; [4] $NaNO_2$, HCl; [5] CuBr
c. [1] HNO_3, H_2SO_4; [2] H_2, Pd-C; [3] Cl_2, $FeCl_3$; [4] $NaNO_2$, HCl; [5] CuBr
d. [1] HNO_3, H_2SO_4; [2] CuBr; [3] Cl_2, $FeCl_3$

PROBLEMS

Problem Using Three-Dimensional Models

23.36 Varenicline (trade name Chantix) is a drug used to help smokers quit their habit.

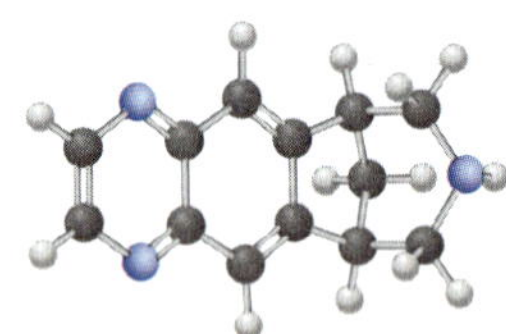
varenicline

a. Which N atom in varenicline is most basic? Explain your choice.
b. What product is formed when varenicline is treated with HCl?

Nomenclature

23.37 Give a systematic or common name for each compound.

a. H–N

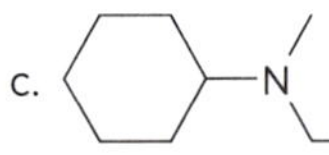

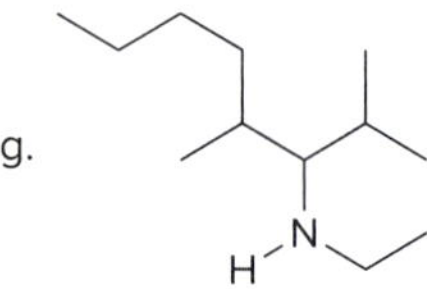

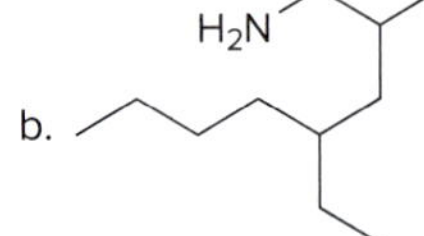

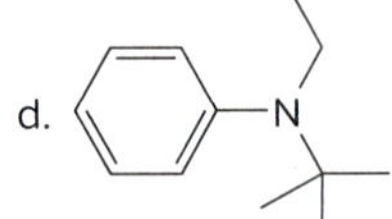

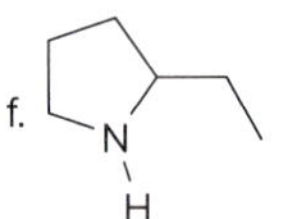

h. N H

23.38 Draw the structure that corresponds to each name.

a. *N*-isobutylcyclopentanamine
b. tri-*tert*-butylamine
c. *N*,*N*-diisopropylaniline
d. *N*-methylpyrrole
e. *N*-methylcyclopentanamine
f. 3-methylhexan-2-amine
g. 2-*sec*-butylpiperidine
h. (*S*)-heptan-2-amine

Basicity

23.39 Rank compounds **A–E** in order of increasing basicity.

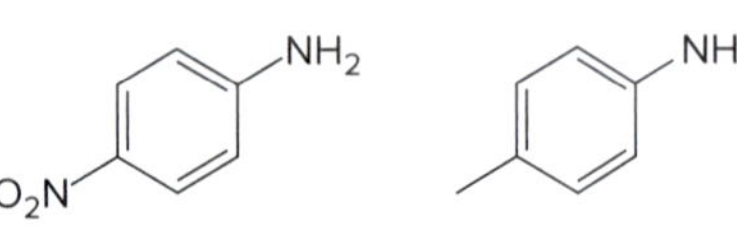

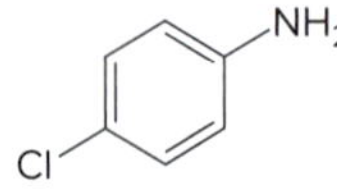

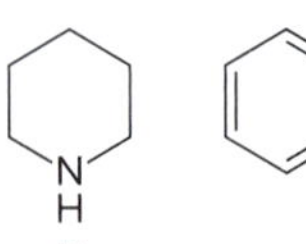

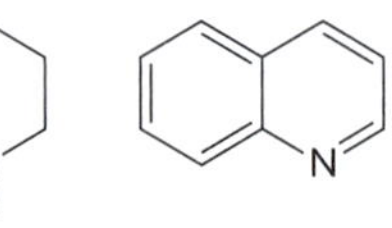

23.40 Decide which N atom in each molecule is most basic, and draw the product formed when each compound is treated with CH_3CO_2H. Zolpidem (trade name Ambien) is used to treat insomnia; aripiprazole (trade name Abilify) is used to treat depression, schizophrenia, and bipolar disorders; and apixaban (trade name Eliquis) is an anticoagulant.

a. 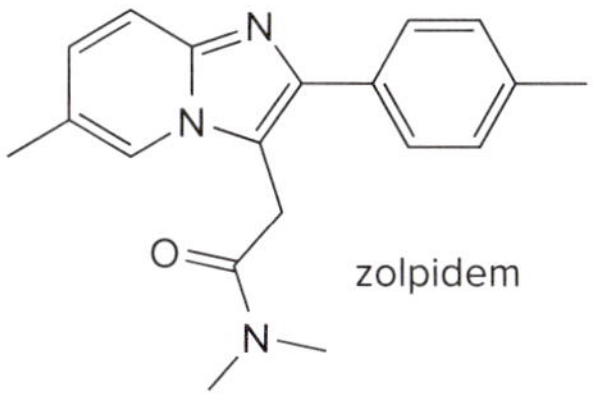 zolpidem b. 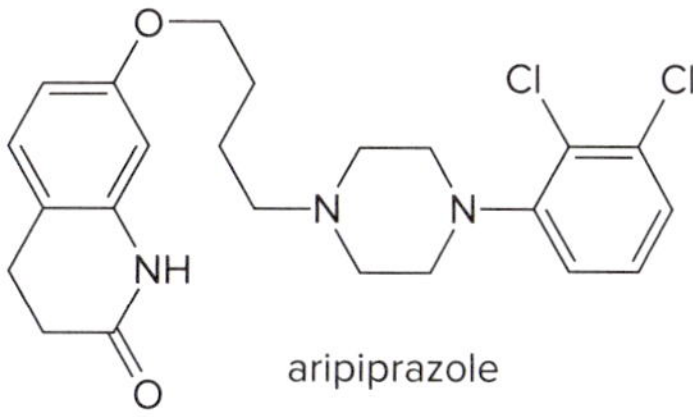 aripiprazole c. 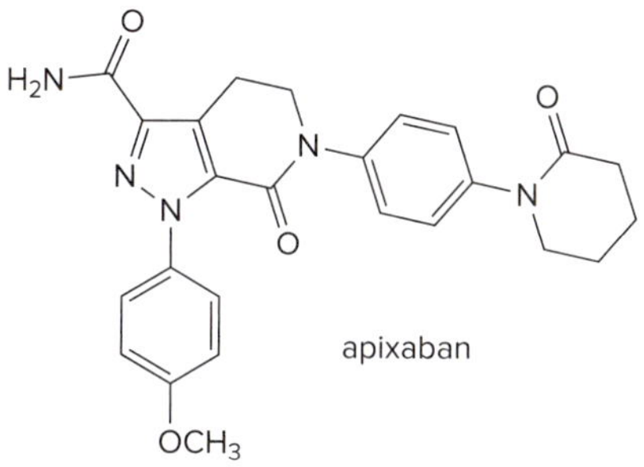 apixaban

23.41 Rank the labeled N atoms in each compound in order of increasing basicity. Imatinib, sold under the trade name Gleevec, is used for the treatment of chronic myeloid leukemia as well as certain gastrointestinal tumors; histamine (Section 23.5B) causes the runny nose and watery eyes associated with allergies; and trazodone is a drug used as a sedative and antidepressant.

a. 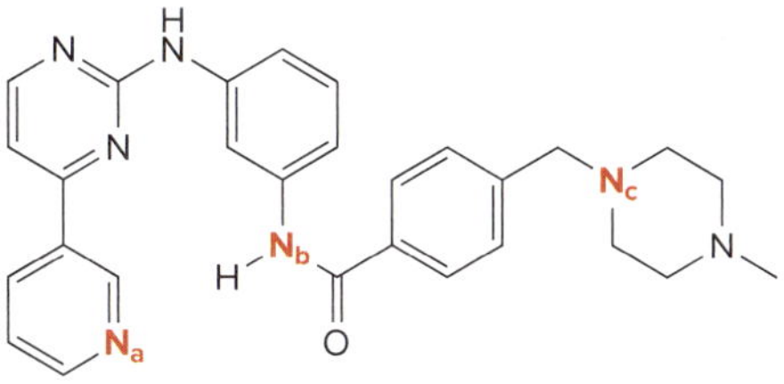 imatinib b. 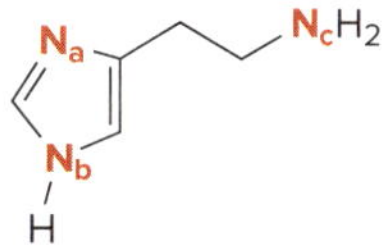 histamine c. 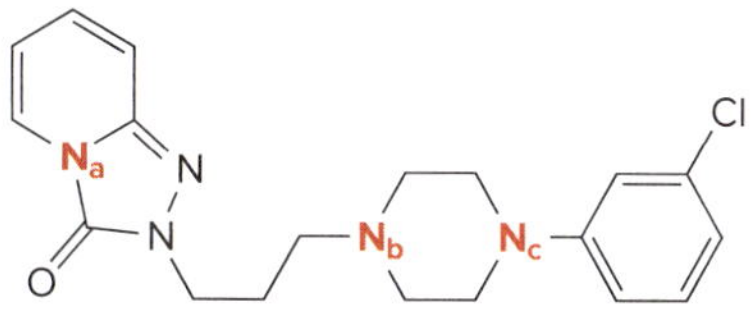 trazodone

23.42 Use the principles in Section 23.9 to determine which N atom in each of the following drugs is most basic. Then draw the product that results when each compound is treated with HCl. Theophylline is used to treat asthma and COPD. Niraparib (trade name Zejula) was approved in the United States in 2017 to treat ovarian cancer.

a. 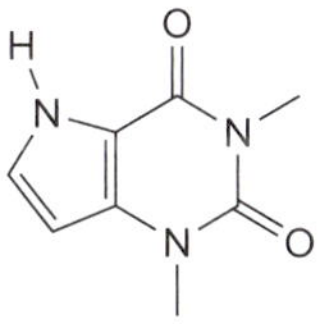 theophylline b. 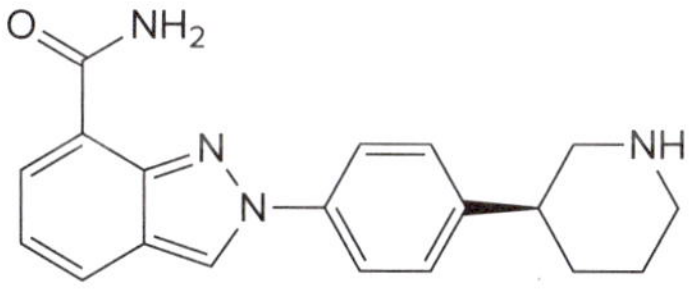 niraparib

23.43 Explain why pyrimidine is less basic than pyridine.

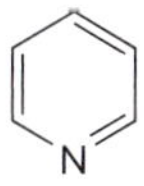

pyridine 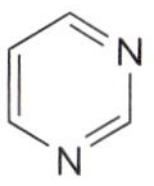 pyrimidine

23.44 Explain why *m*-nitroaniline is a stronger base than *p*-nitroaniline.

23.45 Why is pyrrole more acidic than pyrrolidine?

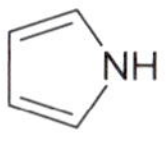

pyrrole $pK_a = 23$

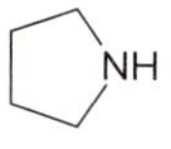

pyrrolidine $pK_a = 44$

Preparation of Amines

23.46 What amide(s) can be used to prepare each amine by reduction?

a.  NH_2 b. 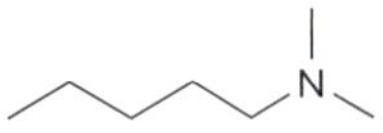 N c. 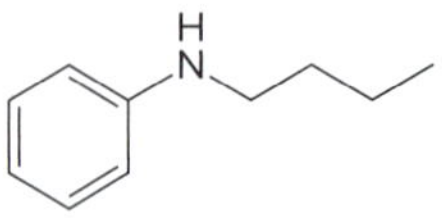 H N

23.47 What carbonyl and nitrogen compounds are needed to make each compound by reductive amination? When more than one set of starting materials is possible, give all possible methods.

a. 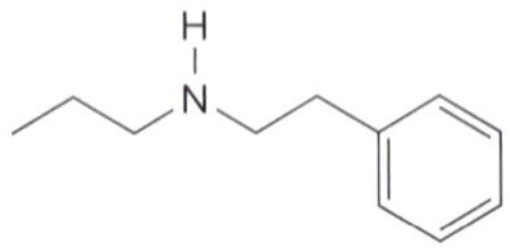 H N b. 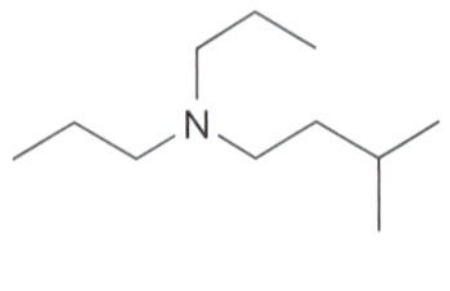 N c. 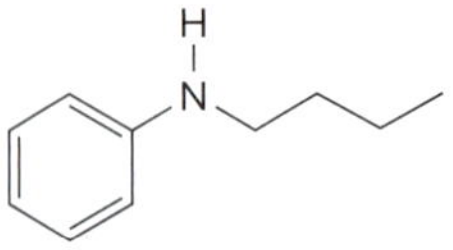 H N

23.48 Draw the product of each reductive amination reaction.

a. 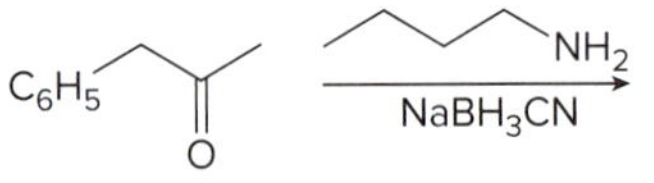 C_6H_5 O $\xrightarrow[NaBH_3CN]{NH_2}$ c. C_6H_5 CHO $\xrightarrow[NaBH_3CN]{NH_3}$

b. 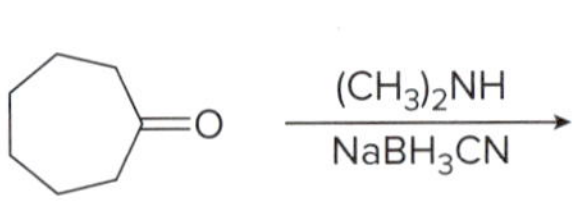 =O $\xrightarrow[NaBH_3CN]{(CH_3)_2NH}$ d. 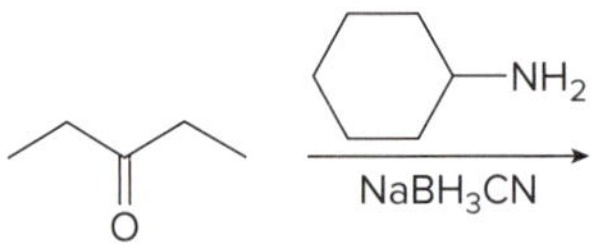 O $\xrightarrow[NaBH_3CN]{NH_2}$

23.49 One step in the synthesis of lisinopril, a drug used to treat high blood pressure, involves the reaction of **A** with **B** in the presence of a reducing agent to form **C.** What is the structure of **C?**

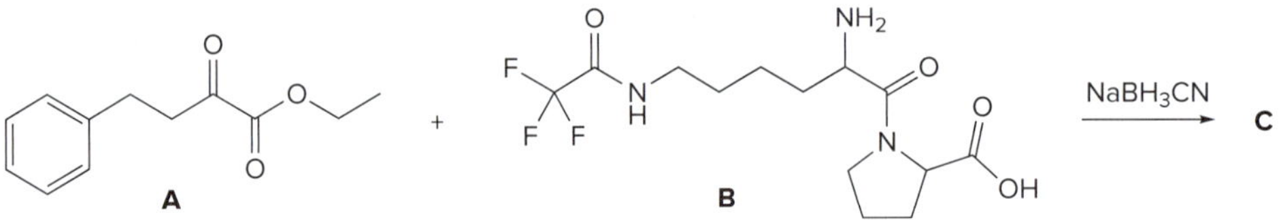

O O O **A** + O F F F N H NH_2 O N O OH **B** $\xrightarrow{NaBH_3CN}$ **C**

23.50 Intramolecular reductive amination of **A** forms pumiliotoxin C, a toxin present in the skin of poison dart frogs. What is the structure of pumiliotoxin C?

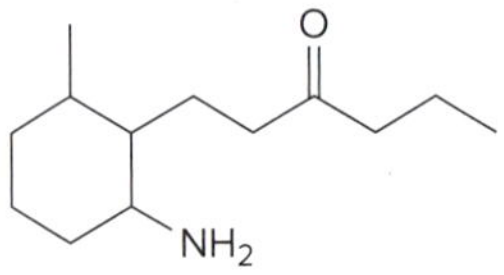

O NH_2 **A**

Extraction

23.51 How would you separate toluene ($C_6H_5CH_3$), benzoic acid ($C_6H_5CO_2H$), and aniline ($C_6H_5NH_2$) by an extraction procedure?

Reactions

23.52 Draw the products formed when *p*-methylaniline (*p*-$CH_3C_6H_4NH_2$) is treated with each reagent.

a. HCl
b. CH_3COCl
c. $(CH_3CO)_2O$
d. excess CH_3I
e. $(CH_3)_2C{=}O$
f. CH_3COCl, $AlCl_3$
g. CH_3CO_2H
h. $NaNO_2$, HCl
i. Part (b), then CH_3COCl, $AlCl_3$
j. CH_3CHO, $NaBH_3CN$

23.53 Draw the products formed when each amine is treated with [1] CH_3I (excess); [2] Ag_2O; [3] Δ. Indicate the major product when a mixture results.

a. 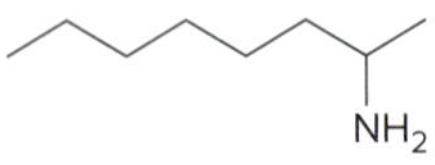 NH_2 b. 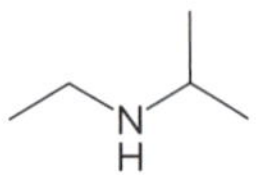 N H c. 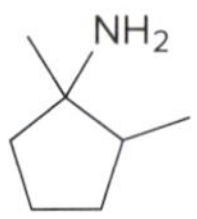 NH_2 d. 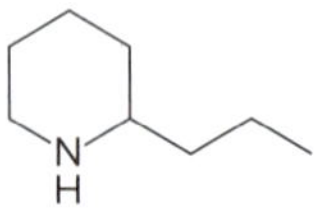 N H

23.54 Answer the following questions about amine **X,** an intermediate in the synthesis of galantamine, a drug used to treat mild to moderate dementia.

HO, O, N, H, Br, OH — **X**

OH, H, O, N, O — galantamine

a. What amides can be reduced to form **X?**
b. What starting materials can be used to form **X** by reductive amination? Draw all possible methods.
c. What products are formed by Hofmann elimination from **X?**

23.55 Answer the following questions about atomoxetine, a drug used to treat attention deficit hyperactivity disorder (ADHD).

atomoxetine

a. What amides can be reduced to form atomoxetine?
b. What starting materials can be used to form atomoxetine by reductive amination? Draw all possible methods.
c. What products are formed by Hofmann elimination of atomoxetine?

23.56 Draw the organic products formed in each reaction.

a. Br–C_6H_4–NO_2 $\xrightarrow[HCl]{Sn}$

b. (CN compound) $\xrightarrow[[2]\ H_2O]{[1]\ LiAlH_4}$

c. (pyrrolidine, NH) $\xrightarrow[HCl]{NaNO_2}$

d. (pyrrolidine, NH) + (benzaldehyde, O, H) $\xrightarrow{NaBH_3CN}$

e. (piperidine, N–H) + (cyclohexanone, =O) $\longrightarrow$

f. (H–N amine) $\xrightarrow{[1]\ CH_3I\ (excess)}$ [2] Ag_2O [3] Δ

23.57 What is the major Hofmann elimination product formed from each amine?

a. CH_3, C_6H_5, NH_2, H, CH_2CH_3, $C(CH_3)_3$

b. NH_2, H, CH_3, C_6H_5, $C(CH_3)_3$, CH_2CH_3

c. C_6H_5, CH_3, H, $(CH_3)_3C$, NH_2, CH_2CH_3

23.58 Identify **A, B,** and **C,** three intermediates in the synthesis of the pain reliever and anesthetic fentanyl.

(Br compound) + (H–N piperidinone, =O) $\longrightarrow$ **A** $\xrightarrow[\text{mild acid}]{C_6H_5NH_2}$ **B** $\xrightarrow{NaBH_3CN}$ **C** $\xrightarrow[\text{pyridine}]{CH_3CH_2COCl}$ fentanyl

23.59 Aprepitant (trade name Emend) is used to prevent the acute nausea and vomiting caused by chemotherapy. Identify **E** and **F**, intermediates in the synthesis of aprepitant.

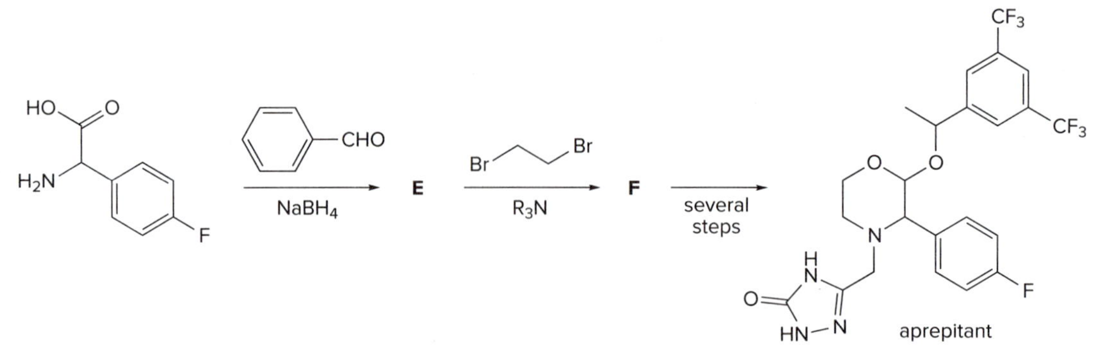

HO, O, H_2N, F — C_6H_5CHO / $NaBH_4$ → **E** — $BrCH_2CH_2Br$ / R_3N → **F** — several steps → aprepitant (CF_3, CF_3, O, O, N, F, HN, N, O, HN—N)

23.60 Draw the product formed when **A** is treated with each series of reagents.

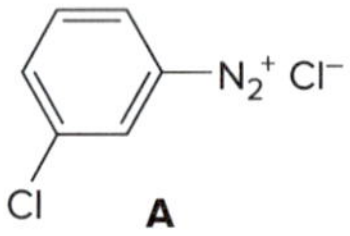

A: Cl, N_2^+ Cl^-

a. [1] H_2O; [2] NaH; [3] CH_3Br
b. [1] CuCN; [2] DIBAL-H; [3] H_2O
c. [1] $C_6H_5NH_2$; [2] CH_3COCl

23.61 A chiral amine **A** having the *R* configuration undergoes Hofmann elimination to form an alkene **B** as the major product. **B** is oxidatively cleaved with ozone, followed by CH_3SCH_3, to form $CH_2{=}O$ and $CH_3CH_2CH_2CHO$. What are the structures of **A** and **B?**

Mechanism

23.62 Draw a stepwise mechanism for each reaction.

a. 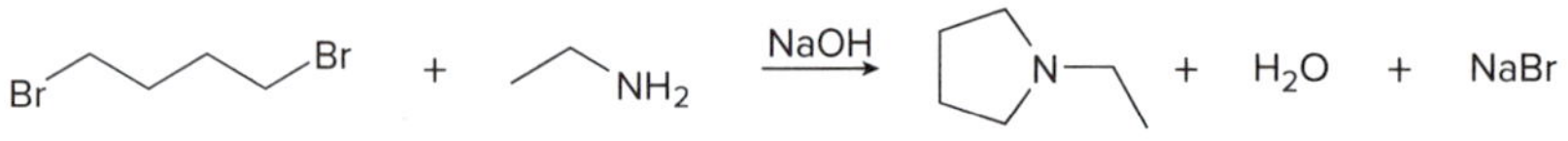 Br(CH₂)₄Br + $CH_3CH_2NH_2$ —NaOH→ N-ethylpyrrolidine + H_2O + NaBr

b. 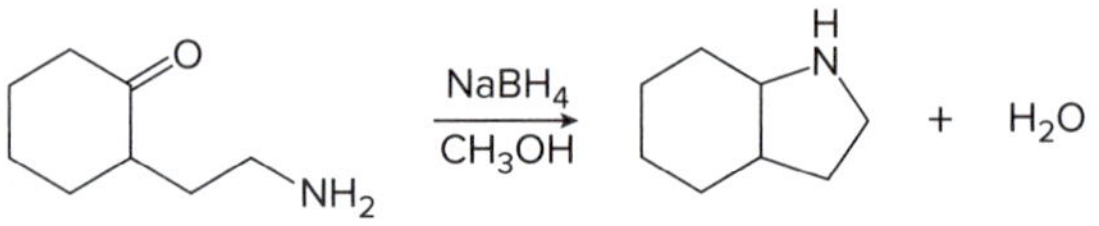 (cyclohexanone with $CH_2CH_2NH_2$ side chain) —$NaBH_4$ / CH_3OH→ (H–N bicyclic amine) + H_2O

23.63 Draw a stepwise mechanism for the following reaction.

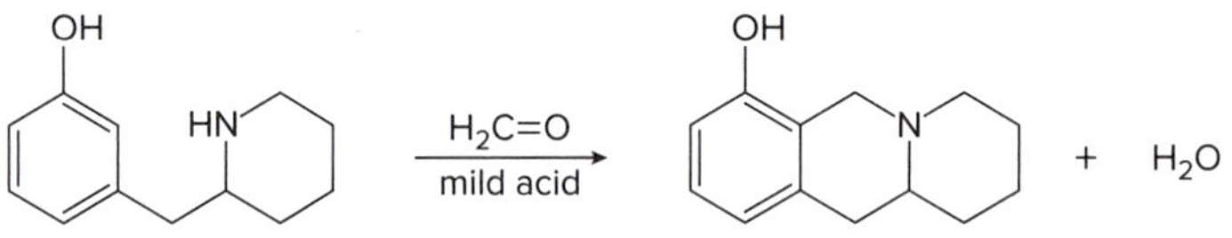

(OH, HN) —$H_2C{=}O$ / mild acid→ (OH, N) + H_2O

23.64 One synthesis of the cholesterol-lowering drug atorvastatin (trade name Lipitor, Section 29.8, Problem 21.53) involves the construction of the pyrrole by reaction of diketone **X** with amine **Y.** Draw a stepwise mechanism for this reaction.

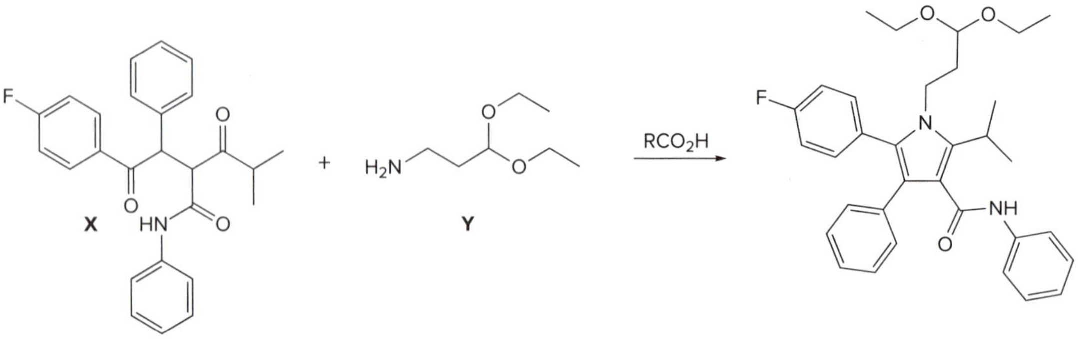

X (F, O, O, HN, O) + **Y** (H_2N, O, O) —RCO_2H→ (F, N, O, O, NH, O)

Synthesis

23.65 Devise a synthesis of each compound from benzene. You may use alcohols with one or two carbons and any inorganic reagents.

23.66 Devise a synthesis of each compound from aniline ($C_6H_5NH_2$) as starting material.

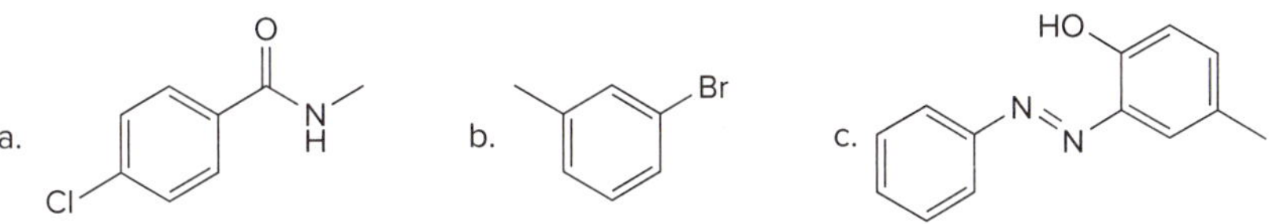

23.67 Devise at least three different methods to prepare *N*-methylbenzylamine ($PhCH_2NHCH_3$) from benzene, any one-carbon organic compounds, and any required reagents.

23.68 Safrole, which is isolated from sassafras (Problem 18.31), can be converted to the illegal stimulant MDMA (3,4-methylenedioxymethamphetamine, "Ecstasy") by a variety of methods. (a) Devise a synthesis that begins with safrole and uses a nucleophilic substitution reaction to introduce the amine. (b) Devise a synthesis that begins with safrole and uses reductive amination to introduce the amine.

MDMA ⟹ safrole

23.69 Synthesize each compound from benzene. Use a diazonium salt as one of the synthetic intermediates.

a. b. c.

23.70 Devise a synthesis of each biologically active compound from benzene.

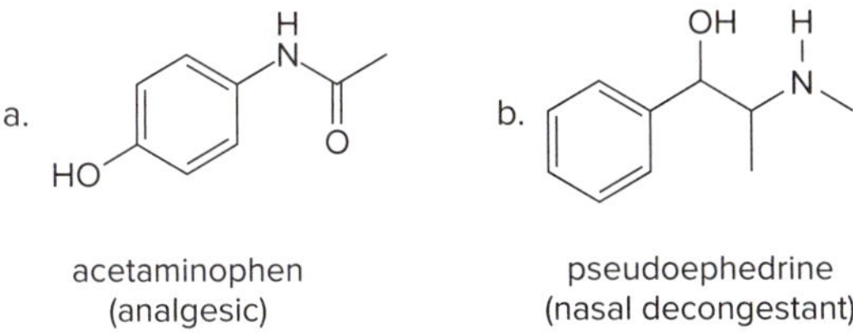

acetaminophen (analgesic)

pseudoephedrine (nasal decongestant)

23.71 Devise a synthesis of each compound from benzene, any organic alcohols having four or fewer carbons, and any required reagents.

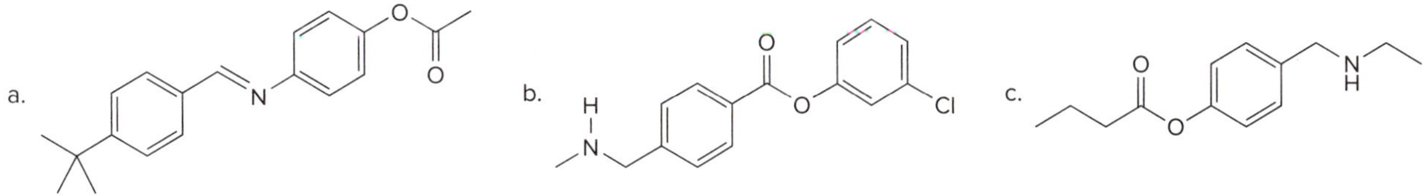

Spectroscopy

23.72 Identify the parent and propose a structure for the base peak in the mass spectrum of butan-1-amine.

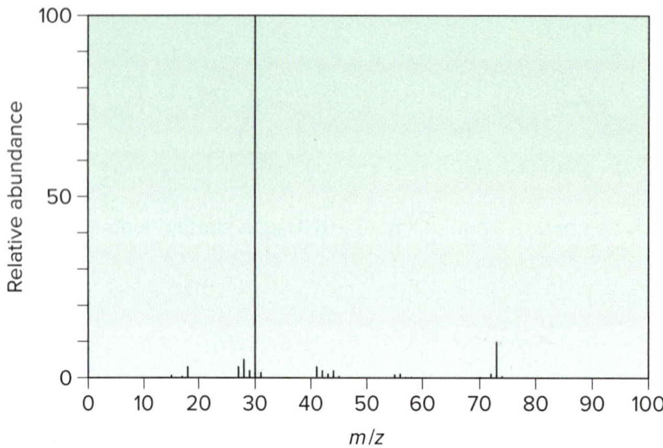

23.73 Three isomeric compounds, **A, B,** and **C,** all have molecular formula $C_8H_{11}N$. The ^1H NMR and IR spectral data of **A, B,** and **C** are given below. What are their structures?

Compound **A:** IR peak at 3400 cm^{-1}

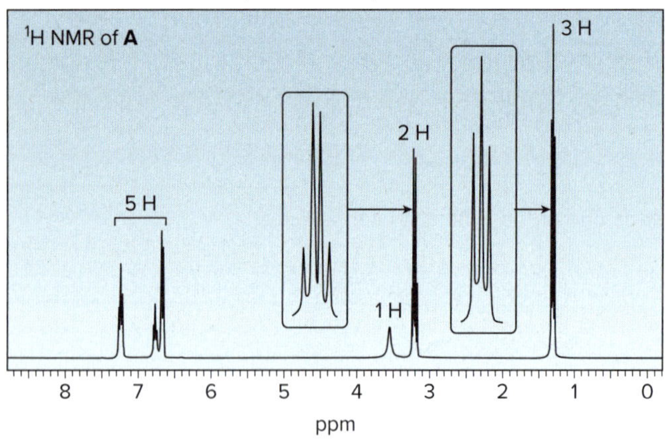

Compound **B:** IR peak at 3310 cm^{-1}

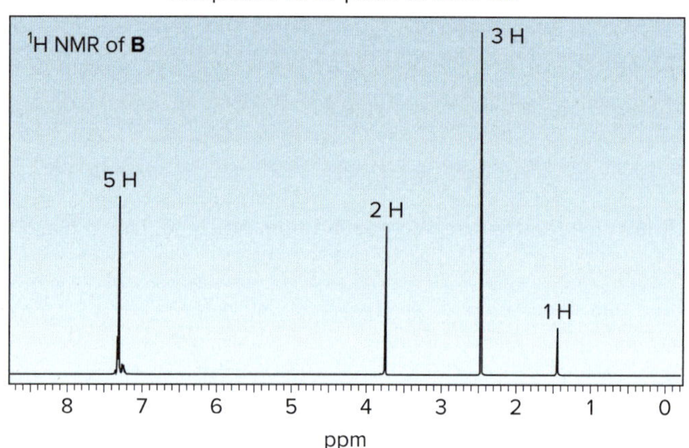

Compound **C:** IR peaks at 3430 and 3350 cm^{-1}

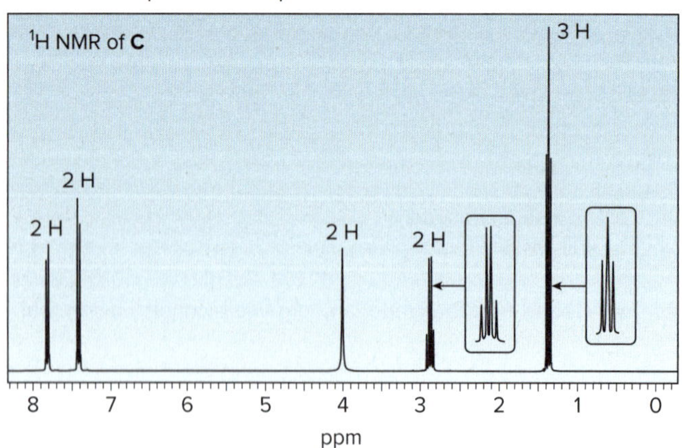

Challenge Problems

23.74 The pK_a of the conjugate acid of guanidine is 13.6, making it one of the strongest neutral organic bases. Offer an explanation.

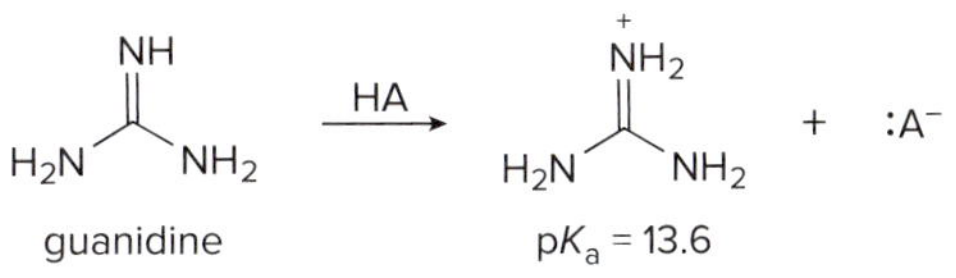

NH / H_2N NH_2 (guanidine) —HA→ $\overset{+}{N}H_2$ / H_2N NH_2 (pK_a = 13.6) + :A^-

23.75 Rank the following compounds in order of increasing basicity and explain the order you chose.

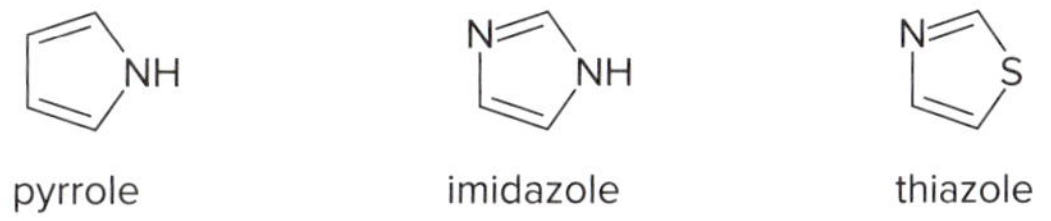

pyrrole imidazole thiazole

23.76 Draw the product **Y** of the following reaction sequence. **Y** was an intermediate in the remarkable synthesis of cyclooctatetraene by Richard Willstatter in 1911.

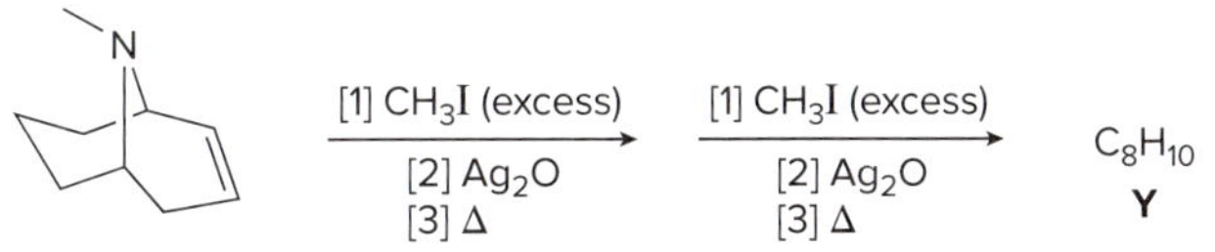

[1] CH_3I (excess) [2] Ag_2O [3] Δ → [1] CH_3I (excess) [2] Ag_2O [3] Δ → C_8H_{10} **Y**

23.77 Devise a synthesis of each compound from the given starting material(s). Albuterol is a bronchodilator and proparacaine is a local anesthetic.

a. albuterol ⟹

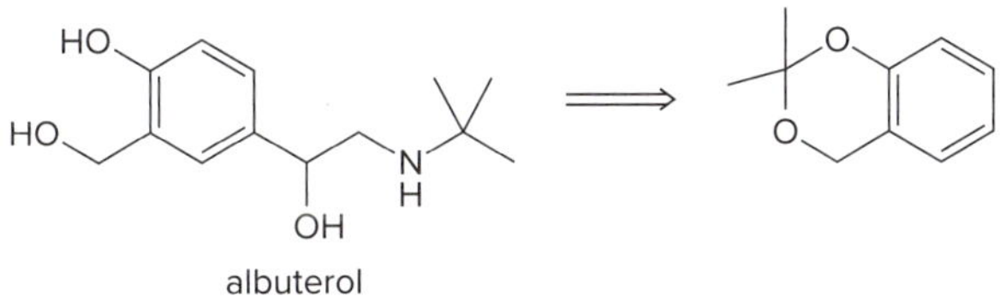

b. proparacaine ⟹ + +

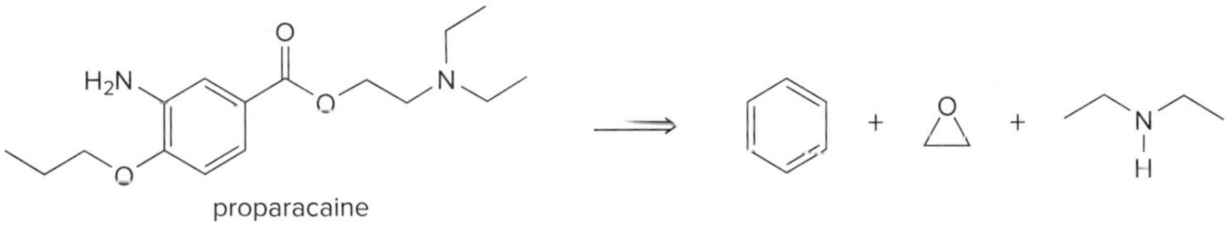

23.78 Heating compound **X** with aqueous formaldehyde forms **Y** ($C_{17}H_{23}NO$), which has been converted to a mixture of lupinine and epilupinine, alkaloids isolated from lupin. Identify **Y** and explain how it is formed.

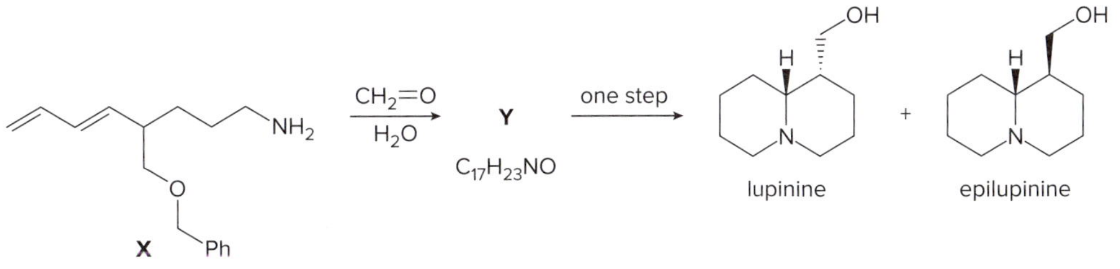

X —$CH_2{=}O$, H_2O→ **Y** $C_{17}H_{23}NO$ —one step→ lupinine + epilupinine

SELF-TEST ANSWERS

1. c 2. b 3. b 4. a 5. b 6. a 7. c 8. b 9. d 10. a

24 Carbon–Carbon Bond-Forming Reactions in Organic Synthesis

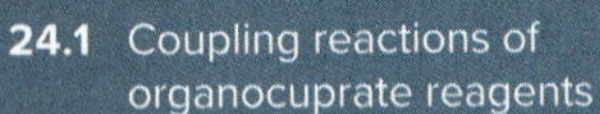

24.1 Coupling reactions of organocuprate reagents
24.2 Suzuki reaction
24.3 Heck reaction
24.4 Stille coupling
24.5 Carbenes and cyclopropane synthesis
24.6 Simmons–Smith reaction
24.7 Metathesis

DEA/M. GIOVANOLI/De Agostini Picture Library/Getty Images

Ingenol mebutate is an ester derived from ingenol, a natural product obtained from the sap of *Euphorbia peplus,* a type of milkweed native to Europe, northern Africa, and western Asia. Because ingenol derivatives exhibited useful biological activity and isolation from the natural source did not provide easy access to the material, scientists developed an efficient laboratory synthesis. A gel formulation of ingenol mebutate (trade name Picato) has been approved for the treatment of actinic keratosis, a skin condition resulting from over-exposure to the sun that may result in squamous cell carcinoma, a form of skin cancer. In Chapter 24, we learn about carbon–carbon bond-forming reactions that prepare complex compounds like ingenol.

Why Study . . .

Reactions That Form Carbon–Carbon Bonds?

To form the carbon skeletons of complex molecules, organic chemists need an extensive repertoire of carbon–carbon bond-forming reactions. In Chapter 17, for example, we learned about the reactions of organometallic reagents—organolithium reagents, Grignard reagents, and organocuprates—with carbonyl substrates. In Chapters 21 and 22, we studied the reactions of nucleophilic enolates that form new carbon–carbon bonds.

Chapter 24 presents more carbon–carbon bond-forming reactions that are especially useful tools in organic synthesis. Whereas previous chapters have concentrated on the reactions of one or two functional groups, the reactions in this chapter utilize a variety of starting materials and conceptually different reactions that form many types of products. All follow one central theme: they form new carbon–carbon bonds under mild conditions, making them versatile synthetic methods.

24.1 Coupling Reactions of Organocuprate Reagents

Several carbon–carbon bond-forming reactions involve the coupling of an organic halide (R'X) with an organometallic reagent or alkene. Four useful reactions are discussed in Sections 24.1–24.4:

[1] **Reaction of an organic halide with an organocuprate reagent (Section 24.1)**

$$R'-X + R_2CuLi \longrightarrow R'-R + RCu + LiX$$

organocuprate; new C–C bond

[2] **Suzuki reaction: Reaction of an organic halide with an organoboron reagent in the presence of a palladium catalyst (Section 24.2)**

$$R'-X + R-B< \xrightarrow[\text{NaOH}]{\text{Pd catalyst}} R'-R + HO-B< + NaX$$

organoboron reagent; new C–C bond

[3] **Heck reaction: Reaction of an organic halide with an alkene in the presence of a palladium catalyst (Section 24.3)**

$$R'-X + CH_2{=}CH-Z \xrightarrow[\text{Et}_3\text{N}]{\text{Pd catalyst}} R'-CH{=}CH-Z + Et_3\overset{+}{N}H \; X^-$$

new C–C bond

A complete list of reactions that form C–C bonds appears in Appendix F.

[4] **Stille coupling: Reaction of an organic halide with an organostannane in the presence of a palladium catalyst (Section 24.4)**

$$R'-X + R-SnBu_3 \xrightarrow{\text{Pd catalyst}} R'-R + X-SnBu_3$$

organostannane; new C–C bond

24.1A General Features of Organocuprate Coupling Reactions

In addition to their reactions with acid chlorides, epoxides, and α,β-unsaturated carbonyl compounds (Sections 17.13–17.15), **organocuprate reagents (R_2CuLi) also react with organic halides R'–X to form coupling products R–R' that contain a new C–C bond.** Only one R group of the organocuprate is transferred to form the product, while the other becomes part of RCu, a reaction by-product.

$$R'-X + R_2CuLi \longrightarrow R'-R + \underbrace{RCu + LiX}_{\text{by-products}}$$

organocuprate; new C–C bond

A variety of organic halides can be used, including methyl and 1° alkyl halides, as well as vinyl and aryl halides that contain X bonded to an sp^2 hybridized carbon. Some cyclic 2° alkyl halides give reasonable yields of product, but 3° alkyl halides are too sterically hindered. The halogen X in R'X may be Cl, Br, or I.

[1] 1-bromopentane + $(C_6H_5)_2CuLi$ ⟶ pentylbenzene

[2] (E)-1-bromohex-1-ene + $(CH_3)_2CuLi$ ⟶ (E)-hept-2-ene

new C–C bonds in red

Coupling reactions with vinyl halides are **stereospecific.** For example, reaction of (*E*)-1-bromohex-1-ene with $(CH_3)_2CuLi$ forms (*E*)-hept-2-ene as the only stereoisomer (Equation [2]).

Problem 24.1 Draw the product of each coupling reaction.

a. 1-chlorohexane $\xrightarrow{(CH_2{=}CH)_2CuLi}$

b. 3-bromocyclohexene $\xrightarrow{(CH_3)_2CuLi}$

c. 3,4-dimethoxyiodobenzene $\xrightarrow{(CH_3CH_2CH_2CH_2)_2CuLi}$

d. vinyl bromide (Br) $\xrightarrow{(\text{alkyl})_2CuLi}$

Problem 24.2 Identify reagents **A** and **B** in the following reaction scheme. This synthetic sequence was used to prepare the C_{18} juvenile hormone, a member of a group of compounds that regulate the complex life cycle of an insect.

vinyl iodide (OH) $\xrightarrow[\text{(excess)}]{\textbf{A}}$ (OH) ⟶ several steps ⟶ vinyl iodide (I, OH) $\xrightarrow[\text{(excess)}]{\textbf{B}}$ (OH) ⟶ several steps ⟶ C_{18} juvenile hormone

Juvenile hormones (Problem 24.2) maintain the juvenile stage of the cecropia moth and other insects until they are ready for adulthood. Synthetic juvenile hormones have been used to control insect populations by preventing maturation, so no sexually mature adults can propagate the next generation. *Matt Jeppson/Shutterstock*

24.1B Using Organocuprate Couplings to Synthesize Hydrocarbons

Because organocuprate reagents (R_2CuLi) are prepared in two steps from alkyl halides (RX), this method ultimately converts two organic halides (RX and R'X) to a hydrocarbon R–R' with a new carbon–carbon bond. A hydrocarbon can often be made by two different routes, as shown in Sample Problem 24.1.

$$R{-}X \xrightarrow[\text{(2 equiv)}]{Li} R{-}Li \xrightarrow[\text{(0.5 equiv)}]{CuI} R_2CuLi \xrightarrow{R'{-}X} R'{-}R$$

Two organic halides are needed as starting materials.

Sample Problem 24.1 Using an Organocuprate Coupling to Prepare a Hydrocarbon

Devise a synthesis of 1-methylcyclohexene from 1-bromocyclohexene and CH_3I.

1-methylcyclohexene ⟹ 1-bromocyclohexene + CH_3I

Solution

In this example, either halide can be used to form an organocuprate, which can then be coupled with the second halide.

Possibility [1] CH_3I —[1] Li (2 equiv); [2] CuI (0.5 equiv)→ $(CH_3)_2CuLi$ —1-bromocyclohexene→ 1-methylcyclohexene

Possibility [2] 1-bromocyclohexene —[1] Li (2 equiv); [2] CuI (0.5 equiv)→ (cyclohexenyl)$_2$CuLi —CH_3I→ 1-methylcyclohexene

Problem 24.3 Synthesize each product from the given starting materials using an organocuprate coupling reaction.

a. (3-methylbutyl)benzene ⟹ bromobenzene + 1-iodo-3-methylbutane

b. oct-2-ene ⟹ RX having 4 C's

c. 2,8-dimethyldecane ⟹ 1-chloro-3-methylbutane only

More Practice: Try Problems 24.26a, 24.27, 24.31.

The mechanism of this reaction may vary with the identity of R' in R'–X. Coupling occurs with organic halides having the halogen X on either an sp^3 or sp^2 hybridized carbon, so an S_N2 mechanism cannot explain all the observed results.

24.2 Suzuki Reaction

The **Suzuki reaction** is the first of three reactions that utilize a palladium catalyst and proceed by way of an intermediate organopalladium compound.

24.2A General Features of Reactions with Pd Catalysts

Reactions with palladium compounds share many common features with reactions involving other transition metals. During a reaction, **palladium is coordinated to a variety of groups**

called ligands, which donate electron density to (or sometimes withdraw electron density from) the metal. **A common electron-donating ligand is a phosphine,** such as triphenylphosphine, tri(*o*-tolyl)phosphine, or tricyclohexylphosphine.

PPh_3
triphenylphosphine

P(o-tolyl)$_3$
tri(o-tolyl)phosphine
abbreviated as **PAr_3**
Ar = an aryl group

PCy_3
tricyclohexylphosphine

A general ligand bonded to a metal is often designated as **L.** Pd bonded to four ligands is denoted as PdL_4.

Ac is the abbreviation for an acetyl group, **$CH_3C{=}O$,** so **OAc** (or **$^-$OAc**) is the abbreviation for acetate, **$CH_3CO_2^-$.**

Organopalladium compounds—compounds that contain a carbon–palladium bond—are generally prepared in situ during the course of a reaction, from another palladium reagent such as $Pd(OAc)_2$ or $Pd(PPh_3)_4$. In most useful reactions, only a catalytic amount of palladium reagent is utilized.

Two common processes, called **oxidative addition** and **reductive elimination,** dominate many reactions of palladium compounds.

- *Oxidative addition* is the addition of a reagent (such as RX) to a metal, often increasing the number of groups around the metal by two.

$$PdL_2 + R{-}X \xrightarrow{\text{oxidative addition}} L{-}Pd(R)(L){-}X$$

organopalladium compound

- *Reductive elimination* is the elimination of two groups that surround the metal, often forming new C–H or C–C bonds.

$$L{-}Pd(R)(L){-}H \xrightarrow{\text{reductive elimination}} PdL_2 + R{-}H$$

organopalladium compound

Reaction mechanisms with palladium compounds are often multistep. During the course of a reaction, the identity of some groups bonded to Pd will be known with certainty, while the identity of other ligands might not be known. Consequently, only the crucial reacting groups around a metal are usually drawn and the other ligands are not specified.

24.2B Details of the Suzuki Reaction

The Suzuki reaction is a palladium-catalyzed coupling of an organic halide (R'X) with an organoborane (RBY_2) to form a product (R–R') with a new C–C bond. $Pd(PPh_3)_4$ is the typical palladium catalyst, and the reaction is carried out in the presence of a base such as NaOH or $NaOCH_2CH_3$.

Suzuki reaction

R'—X + R—B(Y)(Y) —[$Pd(PPh_3)_4$, NaOH]→ R'—R + HO—B(Y)(Y) + NaX

X = Br, I; **organoborane**; new C–C bond

Vinyl halides and aryl halides, both of which contain a halogen X bonded directly to an sp^2 hybridized carbon, are most often used, and the halogen is usually Br or I. The Suzuki reaction is completely **stereospecific,** as shown in Example [2]; **a *Z* vinyl halide and an *E* vinylborane form a (*Z*,*E*)-1,3-diene.**

[1] Br + *E* vinylborane —[$Pd(PPh_3)_4$, NaOH]→ *E*

[2] *Z* vinyl bromide + *E* vinylborane —[$Pd(PPh_3)_4$, NaOEt]→ (*Z*,*E*)-1,3-diene

new C–C bonds in red

The organoboranes used in the Suzuki reaction are prepared from two sources.

- **Vinylboranes,** which have a boron atom bonded to a carbon–carbon double bond, are prepared by hydroboration of an alkyne using catecholborane, a commercially available reagent. **Hydroboration adds the elements of H and B in a syn fashion to form an *E* vinylborane.** With terminal alkynes, hydroboration always places the boron atom on the *less substituted* terminal carbon.

R—≡ + H—B (catecholborane) ⟶ *E* vinylborane

syn addition of H and B

- **Arylboranes,** which have a boron atom bonded to a benzene ring, are prepared from organolithium reagents by reaction with trimethyl borate [$B(OCH_3)_3$].

C_6H_5—Li + $B(OCH_3)_3$ (trimethyl borate) ⟶ C_6H_5—$B(OCH_3)_2$ (arylborane) + $LiOCH_3$

Problem 24.4 Draw the product of each reaction.

a. vinyl iodide (I) + vinylborane —[$Pd(PPh_3)_4$, NaOH]→

b. C_6H_5—$B(OCH_3)_2$ + vinyl bromide (Br) —[$Pd(PPh_3)_4$, NaOH]→

c. C_6H_5—Br —[[1] Li; [2] $B(OCH_3)_3$]→

d. cyclohexyl—≡ + H—B (catecholborane) ⟶

Problem 24.5 One step in the synthesis of the nonsteroidal anti-inflammatory drug rofecoxib (trade name Vioxx) involves Suzuki coupling of **A** and **B.** What product is formed in this reaction?

O, O, Br + CH_3S–(benzene ring)–$B(OH)_2$

A **B**

Problem 24.6 Draw the products formed in each reaction.

a. [1] (catecholborane: O, B–H, O) [2] Br (alkenyl bromide), $Pd(PPh_3)_4$, NaOH

b. Br, O (4-bromoanisole) [1] Li (2 equiv) [2] $B(OCH_3)_3$ → H, N, O, Br; $Pd(PPh_3)_4$, NaOH

The mechanism of the Suzuki reaction consists of oxidative addition of R'–X to the palladium catalyst, transfer of an alkyl group from the organoborane to palladium, and reductive elimination of R–R', forming a new carbon–carbon bond. A general halide R'–X and organoborane $R–BY_2$ are used to illustrate this process in Mechanism 24.1. The mechanism is often written

Mechanism 24.1 Suzuki Reaction

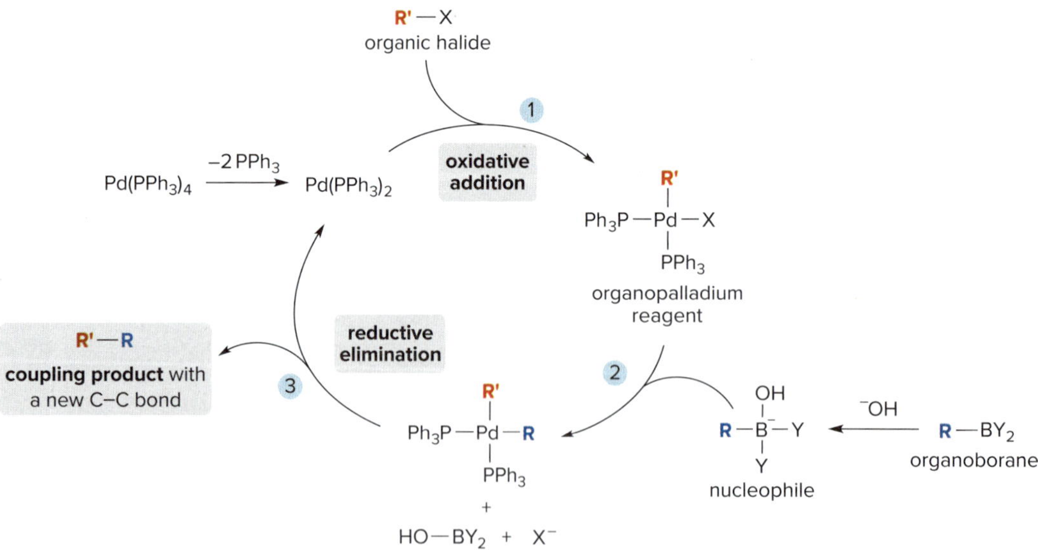

1 Loss of two triphenylphosphine ligands from $Pd(PPh_3)_4$ forms $Pd(PPh_3)_2$, which undergoes **oxidative addition of R'X to form an organopalladium reagent.**

2 Reaction of the organoborane RBY_2 with ^-OH forms a nucleophilic boron intermediate that **transfers an alkyl group from boron to palladium.**

3 **Reductive elimination of R'–R** forms a new carbon–carbon bond, and the palladium catalyst $Pd(PPh_3)_2$ is regenerated.

Figure 24.1 Synthesis of two natural products using the Suzuki reaction

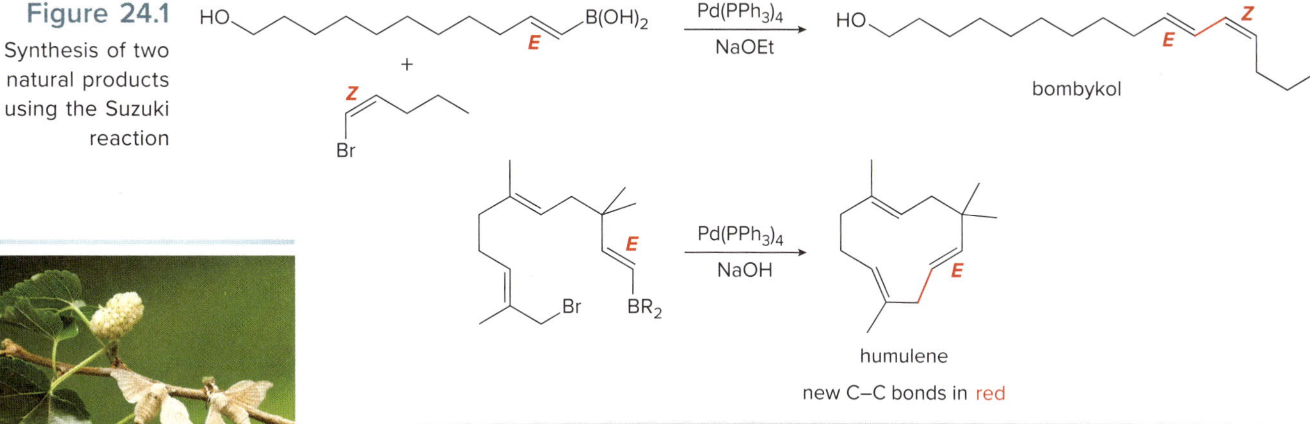

The structure of bombykol (Figure 24.1) the sex pheromone of the female silkworm moth *Bombyx mori,* was elucidated in 1959 using 6.4 mg of material obtained from 500,000 silkworm moths.
Alon Meir/Alamy Stock Photo

in a circle to emphasize that only a catalytic amount of palladium is needed, because the palladium reagent is regenerated during reductive elimination.

The Suzuki reaction was a key step in the synthesis of **bombykol,** the sex pheromone of the female silkworm moth, and **humulene,** a lipid isolated from hops, as shown in Figure 24.1. The synthesis of humulene illustrates that an intramolecular Suzuki reaction can form a ring. Sample Problem 24.2 shows how a conjugated diene can be prepared from an alkyne and vinyl halide using a Suzuki reaction.

Sample Problem 24.2 Devising a Synthesis with a Suzuki Coupling

Devise a synthesis of (1*Z*,3*E*)-1-phenylocta-1,3-diene from hex-1-yne and (*Z*)-2-bromostyrene using a Suzuki coupling.

(1*Z*,3*E*)-1-phenylocta-1,3-diene ⟹ hex-1-yne + (*Z*)-2-bromostyrene

Solution

This synthesis can be accomplished in two steps. Hydroboration of hex-1-yne with catecholborane forms a vinylborane. Coupling of this vinylborane with (*Z*)-2-bromostyrene gives the desired 1,3-diene. **The *E* configuration of the vinylborane and the *Z* configuration of the vinyl bromide are both *retained* in the product.**

[1] hydroboration

syn addition of H and B

[2] $Pd(PPh_3)_4$, NaOH — coupling

(1*Z*,3*E*)-1-phenylocta-1,3-diene

new C–C bond in red

Problem 24.7 Synthesize each compound from the given starting materials.

a.

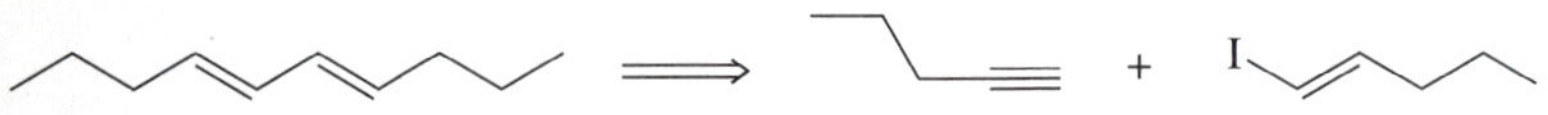

b.

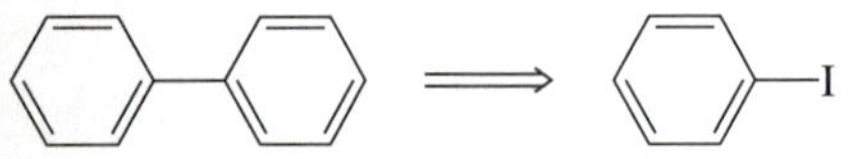

c. 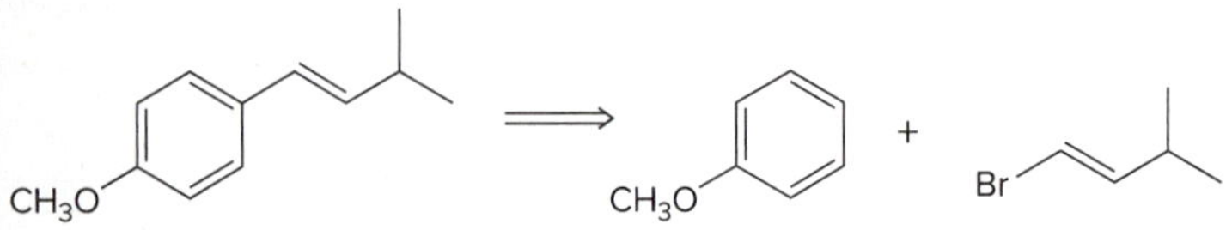 CH_3O CH_3O + Br

More Practice: Try Problems 24.28, 24.32a, 24.36a, 24.54.

24.3 Heck Reaction

Richard Heck, Akira Suzuki, and Ei-ichi Negishi won the 2010 Nobel Prize in Chemistry for their elegant work on palladium-catalyzed coupling reactions in organic synthesis.

The Heck reaction is a palladium-catalyzed coupling of a vinyl or aryl halide with an alkene to form a more highly substituted alkene with a new C–C bond. Palladium(II) acetate [$Pd(OAc)_2$] in the presence of a triarylphosphine [P(*o*-tolyl)$_3$] is the typical catalyst, and the reaction is carried out in the presence of a base such as triethylamine (Et_3N). The Heck reaction is a **substitution reaction** in which one H atom of the alkene starting material is replaced by the R' group of the vinyl or aryl halide.

Heck reaction

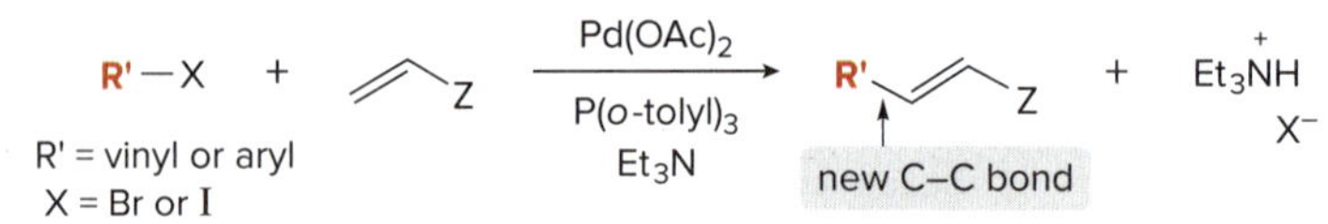

R'–X + Z → ($Pd(OAc)_2$; P(o-tolyl)$_3$; Et_3N) R' Z + $Et_3\overset{+}{N}H$ X^-

R' = vinyl or aryl
X = Br or I

new C–C bond

The alkene component is typically ethylene or a monosubstituted alkene (CH_2=CHZ), and the halogen X is usually Br or I. When Z = Ph, COOR, or CN in a monosubstituted alkene, **the new C–C bond is formed on the *less* substituted carbon to afford a trans alkene.** When a vinyl halide is used as the organic halide, the reaction is **stereospecific,** as shown in Example [2]; the *E* stereochemistry of the vinyl iodide is *retained* in the product.

[1] 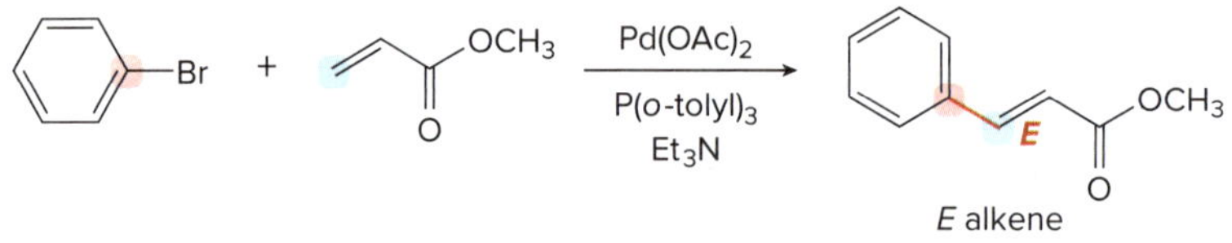 Br + OCH_3 O → ($Pd(OAc)_2$; P(o-tolyl)$_3$; Et_3N) *E* OCH_3 O

E alkene

[2] I + CN → ($Pd(OAc)_2$; P(o-tolyl)$_3$; Et_3N) *E* *E* CN

E vinyl iodide

(*E*,*E*)-1,3-diene
new C–C bonds in red

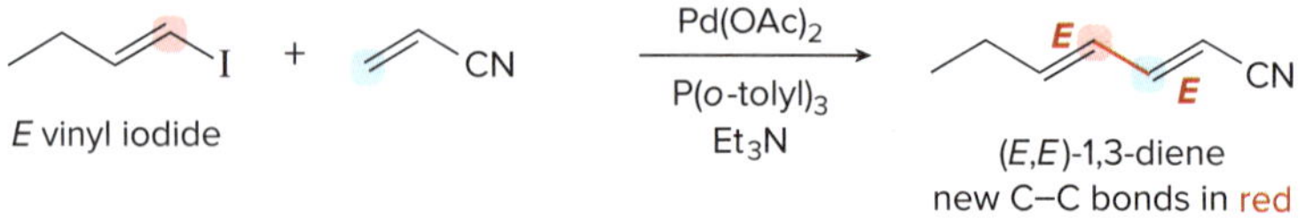

Problem 24.8 Draw the coupling product formed when each pair of compounds is treated with $Pd(OAc)_2$, P(*o*-tolyl)$_3$, and Et_3N.

a. 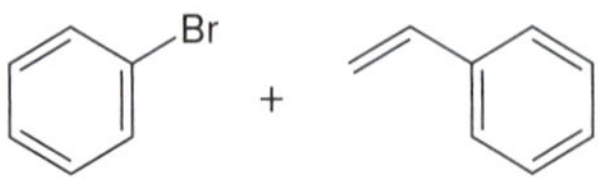 Br +

b. 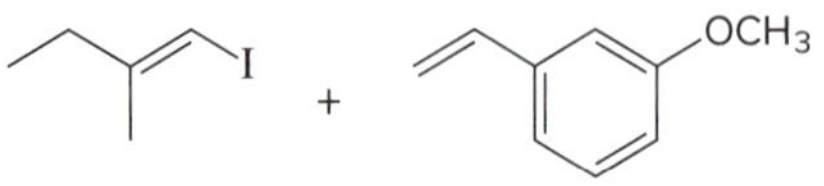 I + OCH_3

c. 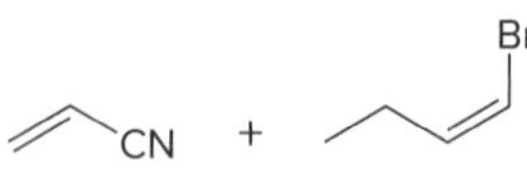 CN + Br

d. 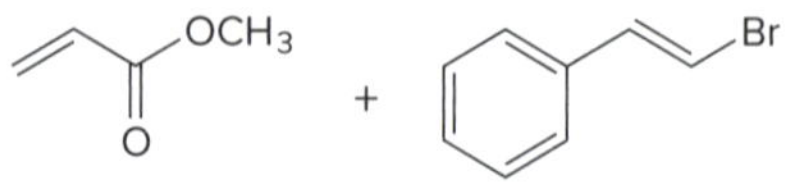 OCH_3 O + Br

To use the Heck reaction in synthesis, you must determine what alkene and what organic halide are needed to prepare a given compound. **To work backwards, locate the double bond with the aryl, COOR, or CN substituent, and break the molecule into two components at the end of the C=C *not* bonded to one of these substituents**. Sample Problem 24.3 illustrates this retrosynthetic analysis.

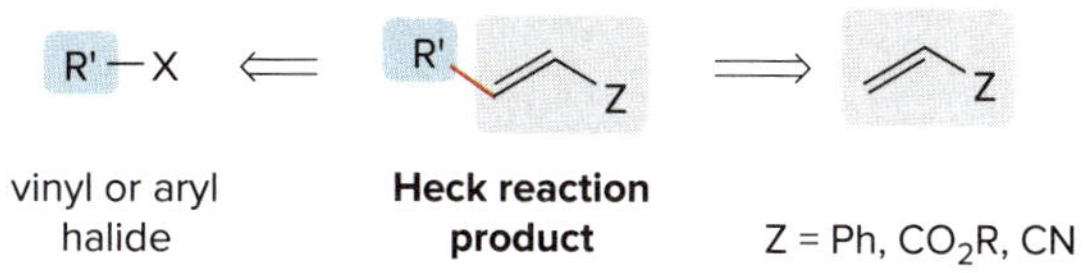

R'—X ⟸ R'CH=CHZ ⟹ CH_2=CHZ

vinyl or aryl halide | **Heck reaction product** | Z = Ph, CO_2R, CN

Sample Problem 24.3 Determining the Starting Materials Needed for a Heck Reaction

What starting materials are needed to prepare each alkene using a Heck reaction?

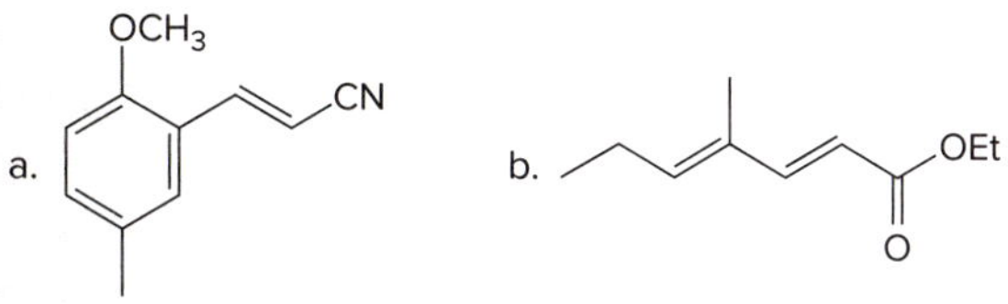

a. OCH_3, CN

b. OEt, O

Solution

To prepare an alkene of general formula R'CH=CHZ by the Heck reaction, two starting materials are needed—an alkene (CH_2=CHZ) and a vinyl or aryl halide (R'X).

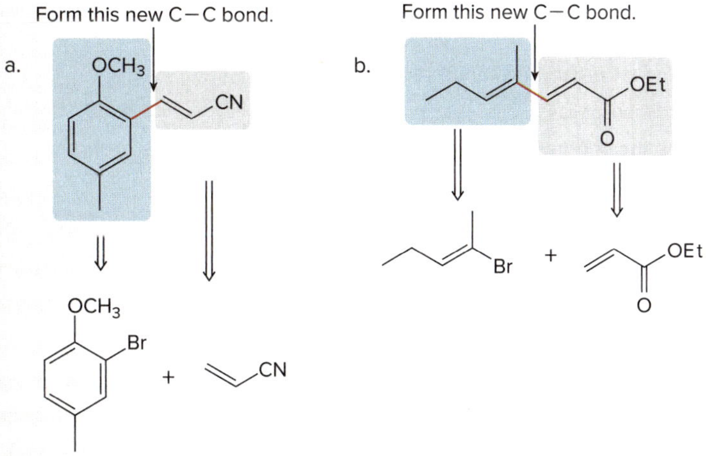

a. Form this new C–C bond. OCH_3, CN ⟹ OCH_3, Br + CN

b. Form this new C–C bond. OEt, O ⟹ Br + OEt, O

Problem 24.9 What starting materials are needed to prepare each compound using a Heck reaction?

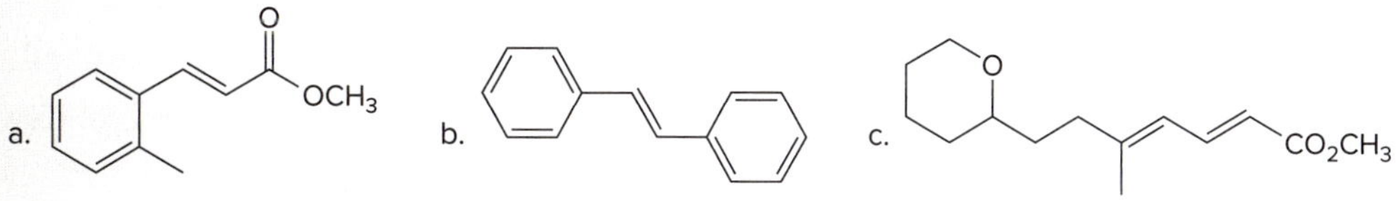

a. O, OCH_3 b. c. O, CO_2CH_3

More Practice: Try Problems 24.29, 24.36b, 24.56.

The actual palladium catalyst in the Heck reaction is thought to contain a palladium atom bonded to two tri(*o*-tolyl)phosphine ligands, abbreviated as $Pd(PAr_3)_2$. In this way it resembles the divalent palladium catalyst used in the Suzuki reaction. The mechanism of the Heck reaction consists of oxidative addition of the halide R'X to the palladium catalyst, **addition of the resulting organopalladium reagent to the alkene,** and **two successive eliminations.** A general organic halide R'X and alkene $CH_2{=}CHZ$ are used to illustrate the process in Mechanism 24.2, which is drawn in a circle to illustrate that the reaction is catalytic in palladium.

Mechanism 24.2 Heck Reaction

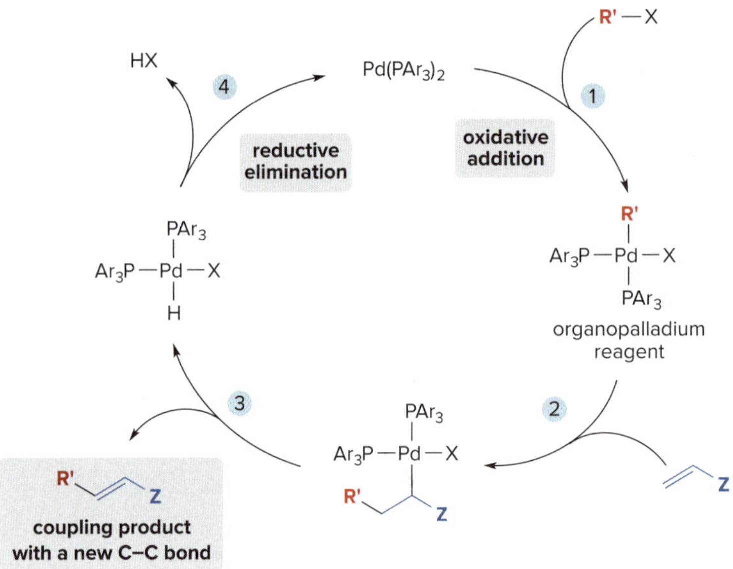

1 **Oxidative addition of R'X forms an organopalladium reagent.**

2 Addition of R' and Pd to the π bond of $CH_2{=}CHZ$ places the Pd on the carbon with the Z substituent.

3 **Elimination of H and Pd forms the π bond in the reaction product** and transfers a hydrogen to Pd.

4 **Reductive elimination of HX** regenerates the palladium catalyst $Pd(PAr_3)_2$.

24.4 Stille Coupling

The Stille coupling is a palladium-catalyzed reaction of an organic halide (R'X) with an organotin compound ($RSnBu_3$, also called an organostannane) to form a product (R–R') with a new C–C bond. This reaction can be carried out with a variety of palladium catalysts, including $Pd(PPh_3)_4$.

Stille coupling

$$R'{-}X \quad + \quad R{-}SnBu_3 \xrightarrow{Pd(PPh_3)_4} R'{-}R \quad + \quad X{-}SnBu_3$$

organostannane — new C–C bond

Most commonly, R and R' are vinyl or aryl groups and the halogen is usually Br or I. Other Stille couplings have been reported with acid chlorides as the organic halide, and organostannanes with R = allyl ($CH_2CH{=}CH_2$) and R = benzyl (CH_2Ph) can also be used. The reaction can occur in the presence of a variety of other functional groups (CO_2R, CN, OH, CHO), making it a widely used reaction in synthesis. The reaction is **stereospecific,** as shown in Example [2]; **an *E* vinyl stannane and an *E* vinyl halide form an (*E,E*)-1,3-diene.**

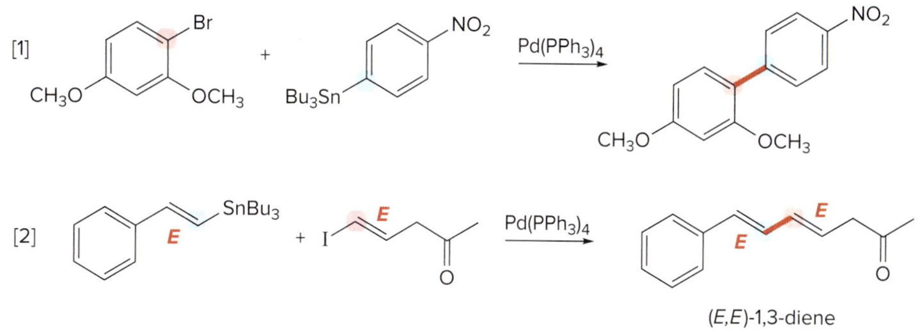

[1] Br, CH_3O, OCH_3 + Bu_3Sn, NO_2 → $Pd(PPh_3)_4$ → NO_2, CH_3O, OCH_3

[2] E, $SnBu_3$ + I, E, O → $Pd(PPh_3)_4$ → E, E, O

(*E,E*)-1,3-diene

The organostannanes used in the Stille coupling can be prepared from organolithium or Grignard reagents by reaction with tributyltin chloride.

Li + Cl—$SnBu_3$ (tributyltin chloride) → $SnBu_3$

MgBr + Cl—$SnBu_3$ (tributyltin chloride) → $SnBu_3$

Problem 24.10 Draw the products formed when each pair of compounds undergoes a Stille coupling in the presence of a palladium catalyst.

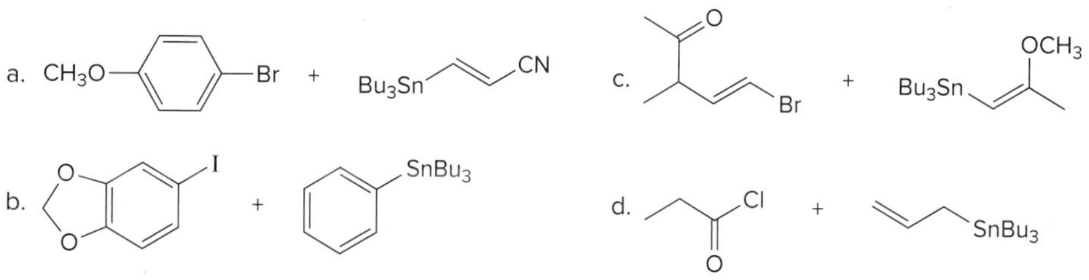

a. CH_3O—⟨⟩—Br + Bu_3Sn—CH=CH—CN

b. (benzodioxole)—I + (benzene)—$SnBu_3$

c. O, Br + Bu_3Sn, OCH_3

d. Cl, O + $SnBu_3$

Problem 24.11 Identify **A** and **B** in the following reaction sequence.

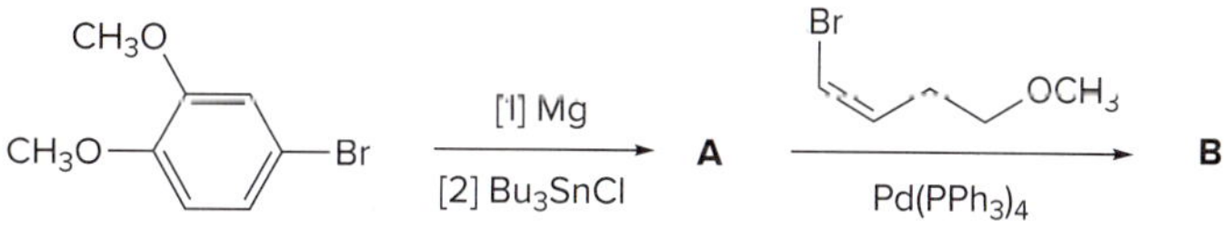

CH_3O, CH_3O—⟨⟩—Br → [1] Mg; [2] Bu_3SnCl → **A** → Br, OCH_3; $Pd(PPh_3)_4$ → **B**

Problem 24.12 An intramolecular Stille coupling can be used to form a ring. For example, compound **A** undergoes cyclization with $Pd(PPh_3)_4$ to form **X.** Treatment of **X** with acid results in conversion of the OR groups on the benzene ring to OH groups, forming the natural product zearalenone. Zearalenone is a toxin isolated from fungus of the *Gibberella* species that can infect cereal crops, posing a significant hazard to animals and humans. Identify the structures of **X** and zearalenone.

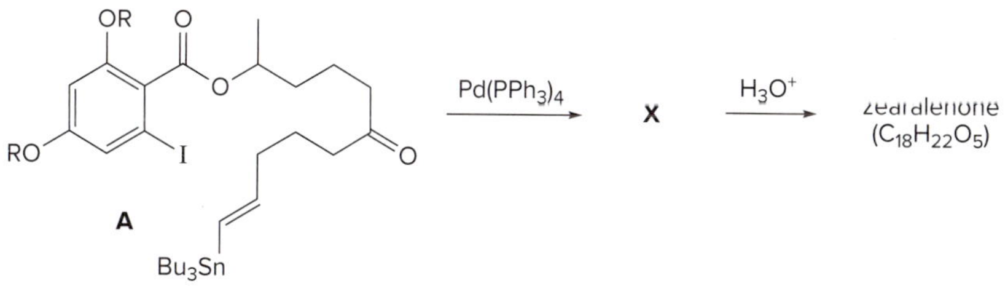

OR, O, O, RO, I, O, Bu_3Sn **A** → $Pd(PPh_3)_4$ → **X** → H_3O^+ → zearalenone ($C_{18}H_{22}O_5$)

The mechanism of the Stille coupling consists of oxidative addition of R'X to the palladium catalyst, transfer of an alkyl group from the organostannane to palladium (a **transmetallation**), and reductive elimination of R–R', forming a new carbon–carbon bond. Like the mechanisms for the Suzuki and Heck reactions, Mechanism 24.3 is written in a circle to emphasize the catalytic role of the palladium. Like the Heck reaction, the actual palladium catalyst is thought to be a divalent palladium species such as $Pd(PPh_3)_2$.

Mechanism 24.3 Stille Coupling

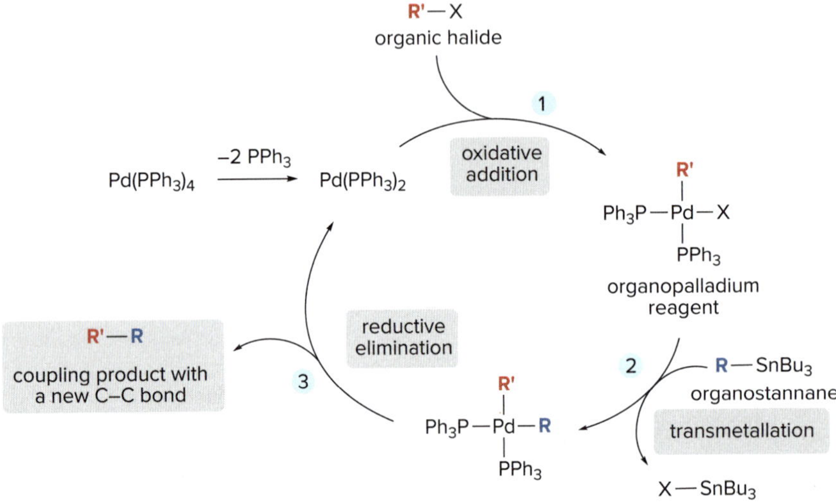

1 Loss of two triphenylphosphine ligands from $Pd(PPh_3)_4$ forms $Pd(PPh_3)_2$, which undergoes **oxidative addition with R'X to form an organopalladium compound.**

2 **Transmetallation** of the organostannane with the organopalladium reagent **transfers an alkyl group from tin to palladium.**

3 **Reductive elimination of R–R'** forms a new carbon–carbon bond and the palladium catalyst $Pd(PPh_3)_2$ is regenerated.

To use a Stille coupling in synthesis, you must determine what organic halide and organostannane can be used to prepare a given compound. Because the Stille coupling is generally carried out with reactants that contain aryl and vinyl groups, the new carbon–carbon bond usually joins two aryl groups, two double bonds, or an aryl group with a double bond. Often two pathways are possible, as shown in Sample Problem 24.4.

new C–C bond new C–C bond new C–C bond

Sample Problem 24.4 Determining the Starting Materials Needed for a Stille Coupling

Draw two different sets of starting materials that can be used to prepare valsartan (trade name Diovan), a drug used to treat high blood pressure.

valsartan

Solution

Locate the bond that joins two sp^2 hybridized C's together, and cleave the carbon–carbon bond into two components. One part becomes the organic halide and the other part becomes the organostannane. There are often two possible routes to carry out the Stille coupling.

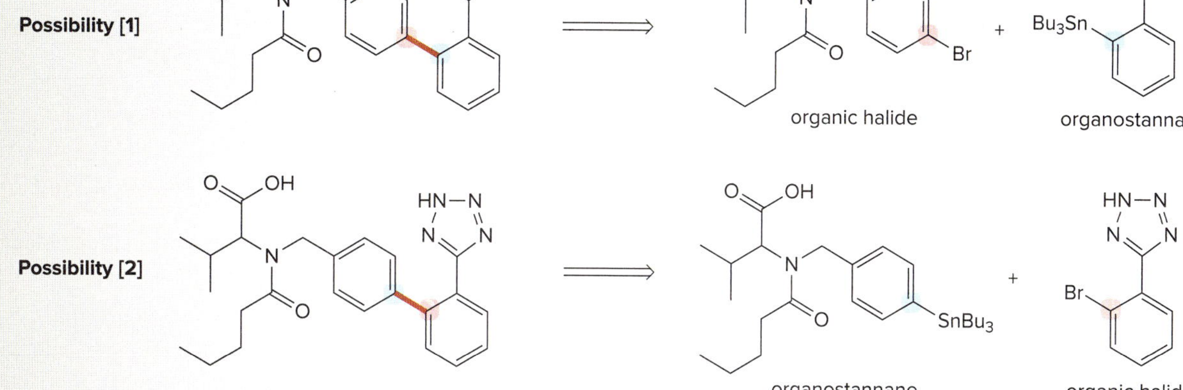

Both routes form valsartan from an aryl halide and an organostannane, so each method can be used.

Problem 24.13 Give two different sets of starting materials that can be used to prepare each compound by a Stille coupling.

a. (structure with OCH_3) b. (structure) c. (structure)

More Practice: Try Problems 24.30b, 24.36c.

24.5 Carbenes and Cyclopropane Synthesis

Another method of carbon–carbon bond formation involves the conversion of alkenes to cyclopropane rings using **carbene** intermediates.

$$\text{alkene} + :CR_2 \text{ (carbene)} \longrightarrow \text{cyclopropane}$$

new C–C bonds in red

John Thoeming/McGraw Hill

Pyrethrin I and **decamethrin** both contain cyclopropane rings. Pyrethrin I is a naturally occurring biodegradable insecticide obtained from chrysanthemums, whereas **decamethrin** is a more potent synthetic analogue that is widely used as an insecticide in agriculture.

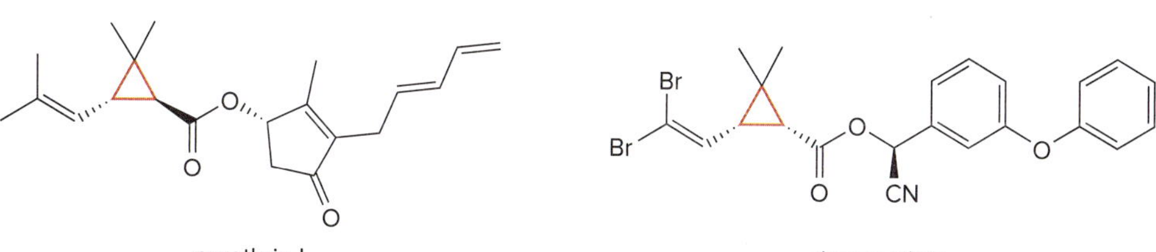

24.5A Carbenes

A *carbene*, R_2C:, is a neutral reactive intermediate that contains a divalent carbon surrounded by six electrons—the lone pair and two each from the two R groups. These three groups make the carbene carbon ***sp*^{2} hybridized,** with a vacant *p* orbital extending above and below the plane containing the C and the two R groups. The lone pair of electrons occupies an *sp*^{2} hybrid orbital.

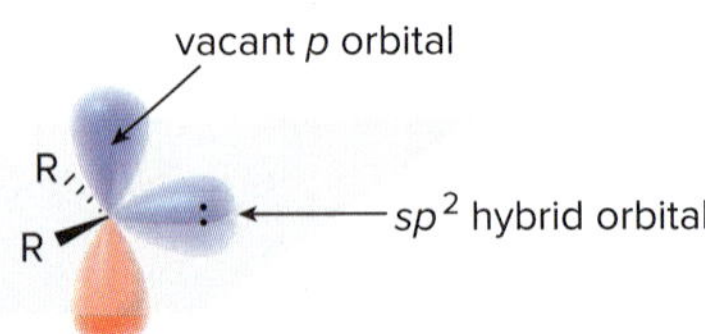

The carbene carbon is *sp*^{2} hybridized.

Carbenes share two features in common with carbocations and carbon radicals.

- **A carbene is highly reactive because carbon does not have an octet of electrons.**
- **A carbene is electron deficient, so it behaves as an electrophile.**

24.5B Preparation and Reactions of Dihalocarbenes

Dihalocarbenes, $:CX_2$, are especially useful reactive intermediates because they are readily prepared from trihalomethanes (CHX_3) by reaction with a strong base. Treatment of chloroform, $CHCl_3$, with $KOC(CH_3)_3$ forms dichlorocarbene, $:CCl_2$.

$$\underset{\text{chloroform}}{CHCl_3} \xrightarrow{KOC(CH_3)_3} \underset{\textbf{dichlorocarbene}}{\mathbf{:CCl_2}} + (CH_3)_3COH + KCl$$

Dichlorocarbene is formed by a two-step process that results in the elimination of the elements of H and Cl from the *same* carbon, as shown in Mechanism 24.4. Loss of two elements from the same carbon is called **α elimination,** to distinguish it from the β eliminations discussed in Chapter 8, in which two elements are lost from *adjacent* carbons.

Mechanism 24.4 Formation of Dichlorocarbene

$CHCl_3$ (Cl, Cl, Cl, H) + $^{-}:\ddot{O}C(CH_3)_3$ —[1]→ $^{-}:CCl_3$ + $H\ddot{O}C(CH_3)_3$ —[2]→ $:CCl_2$ (dichlorocarbene) + Cl^-

1. Three electronegative Cl atoms acidify the C–H of $CHCl_3$, so it can be removed by strong base to form a **carbanion.**
2. **Elimination of Cl^-** forms the carbene.

Dihalocarbenes are electrophiles, so they readily react with double bonds to afford cyclopropanes, forming two new carbon–carbon bonds.

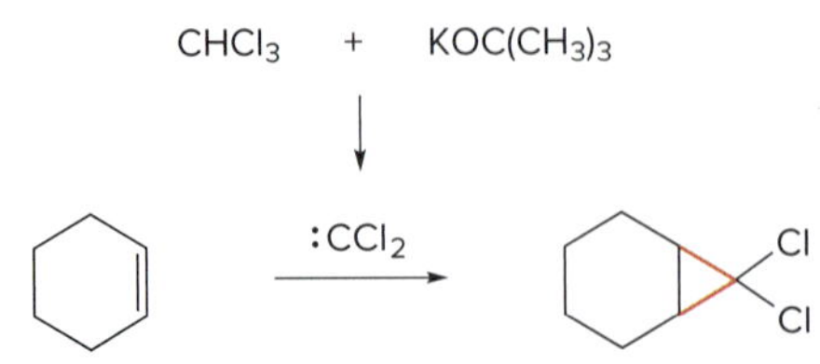

new C–C bonds in red

Cyclopropanation is a concerted reaction, so both C–C bonds are formed in a single step, as shown in Mechanism 24.5.

Mechanism 24.5 Addition of Dichlorocarbene to an Alkene

Carbene addition occurs in a **syn** fashion from either side of the planar double bond. The relative position of substituents in the alkene reactant is retained in the cyclopropane product. **Carbene addition is thus a stereospecific reaction,** because cis and trans alkenes yield different stereoisomers as products, as illustrated in Sample Problem 24.5.

Sample Problem 24.5 Drawing the Products of Carbene Addition

Draw the products formed when *cis*- and *trans*-but-2-ene are treated with $CHCl_3$ and $KOC(CH_3)_3$.

Solution

To draw each product, **add the carbene carbon from either side of the alkene, and keep all substituents in their original orientations.** The **cis** methyl groups in *cis*-but-2-ene become **cis** substituents in the cyclopropane. Addition from either side of the alkene yields the same compound—**an achiral meso compound that contains two stereogenic centers**—labeled in blue.

cis-but-2-ene → ($CHCl_3$, $KOC(CH_3)_3$)

cis CH_3 groups :CCl_2 added from **above** + **cis CH_3 groups** :CCl_2 added from **below**

Products are identical, an achiral meso compound.

The **trans** methyl groups in *trans*-but-2-ene become **trans** substituents in the cyclopropane. Addition from either side of the alkene yields an equal amount of two enantiomers—**a racemic mixture.**

trans-but-2-ene → ($CHCl_3$, $KOC(CH_3)_3$)

trans CH_3 groups :CCl_2 added from **above** + **trans CH_3 groups** :CCl_2 added from **below**

Products are **enantiomers.**

Problem 24.14 Draw all stereoisomers formed when each alkene is treated with $CHCl_3$ and $KOC(CH_3)_3$.

a. b. c.

More Practice: Try Problems 24.37c, d; 24.45a.

Finally, *dihalo* cyclopropanes can be converted to *dialkyl* cyclopropanes by reaction with organocuprates (Section 24.1). For example, cyclohexene can be converted to a bicyclic product having four new C–C bonds by the following two-step sequence: **cyclopropanation** with dibromocarbene (:CBr_2) and **reaction with lithium dimethylcuprate, $LiCu(CH_3)_2$.**

+ $CHBr_3$ —$KOC(CH_3)_3$ [1]→ Br, Br —$LiCu(CH_3)_2$ [2]→

two new C–C bonds in red two new C–C bonds in blue

Problem 24.15 What reagents are needed to convert 2-methylpropene [$(CH_3)_2C{=}CH_2$] to each compound? More than one step may be required.

a. Cl Cl b. Br Br c.

24.6 Simmons–Smith Reaction

Although the reaction of dihalocarbenes with alkenes gives good yields of halogenated cyclopropanes, this is not usually the case with **methylene, :CH_2,** the simplest carbene. Methylene is readily formed by heating diazomethane, CH_2N_2, which decomposes and loses N_2, but the reaction of :CH_2 with alkenes often affords a complex mixture of products. Thus, this reaction cannot be reliably used for cyclopropane synthesis.

$$:\bar{C}H_2-\overset{+}{N}{\equiv}N: \longrightarrow :CH_2 + :N{\equiv}N:$$

diazomethane **methylene**

Nonhalogenated cyclopropanes can be prepared by the reaction of an alkene with diiodomethane, CH_2I_2, in the presence of a copper-activated zinc reagent called zinc–copper couple [Zn(Cu)]. This process, the **Simmons–Smith reaction,** is named for H. E. Simmons and R. D. Smith, DuPont chemists who discovered the reaction in 1959.

R R R R + CH_2I_2 —Zn(Cu)→ R R R R + ZnI_2

Simmons–Smith reaction

+ CH_2I_2 —Zn(Cu)→ + ZnI_2

two new C–C bonds in red

The Simmons–Smith reaction does not involve a free carbene. Rather, the reaction of CH_2I_2 with Zn(Cu) forms (iodomethyl)zinc iodide, which transfers a CH_2 group to an alkene, as shown in Mechanism 24.6.

Mechanism 24.6 Simmons–Smith Reaction

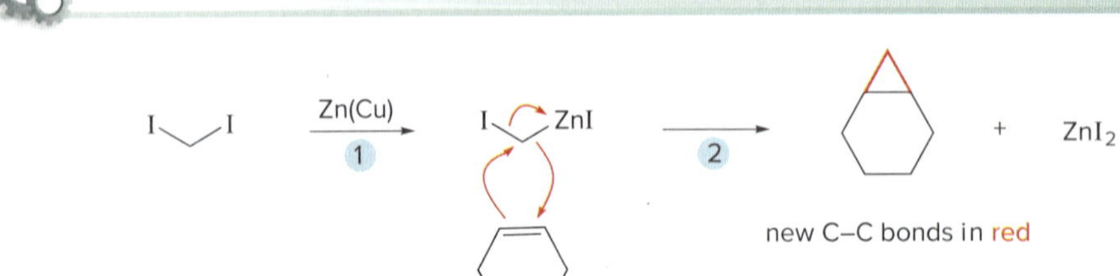

1 Reaction of CH_2I_2 with zinc–copper couple forms ICH_2ZnI [(iodomethyl)zinc iodide], the **Simmons–Smith reagent.** This intermediate is called a *carbenoid,* because the CH_2 does not exist as a free carbene.

2 **The Simmons–Smith reagent transfers a CH_2 to an alkene,** forming two new C–C bonds.

The Simmons–Smith reaction is stereospecific. The relative position of substituents in the alkene reactant is *retained* in the cyclopropane product, as shown for the conversion of *cis*-hex-3-ene to *cis*-1,2-diethylcyclopropane.

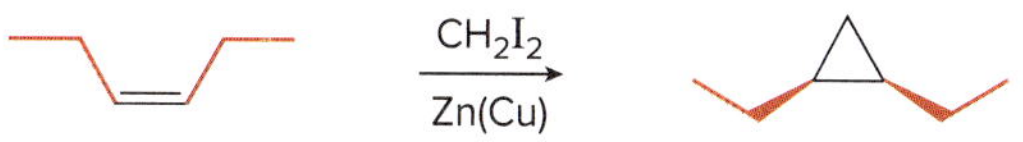

cis-hex-3-ene $\xrightarrow[\text{Zn(Cu)}]{CH_2I_2}$ *cis*-1,2-diethylcyclopropane

Problem 24.16 What product is formed when each alkene is treated with CH_2I_2 and Zn(Cu)?

a. 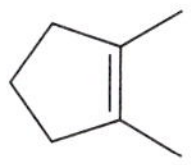 b. 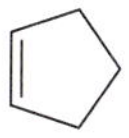 c.

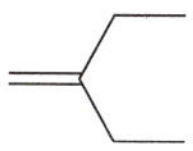

Problem 24.17 What stereoisomers are formed when *trans*-hex-3-ene is treated with CH_2I_2 and Zn(Cu)?

24.7 Metathesis

Recall from Section 10.1 that **olefin** is another name for an **alkene.**

Alkene metathesis, more commonly called **olefin metathesis,** is a reaction between two alkene molecules that results in the interchange of the carbons of their double bonds. Two σ and two π bonds are broken, and two new σ and two new π bonds are formed.

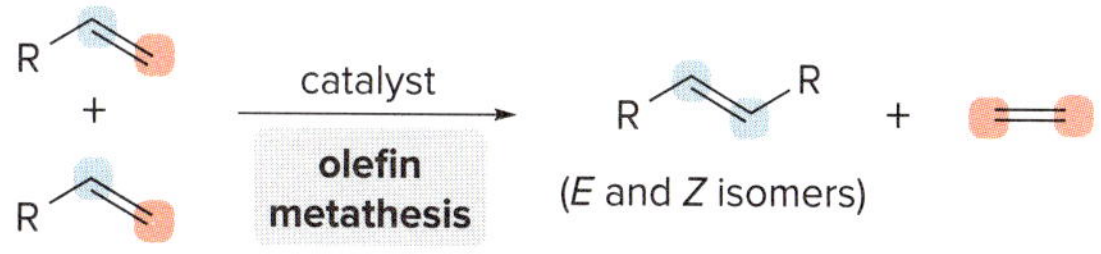

R + R $\xrightarrow{\text{catalyst}}$ R R (*E* and *Z* isomers) +

olefin metathesis

24.7A General Features of Metathesis

The word *metathesis* is derived from the Greek words *meta* (change) and *thesis* (position). The 2005 Nobel Prize in Chemistry was awarded to Robert Grubbs of the California Institute of Technology, Yves Chauvin of the Institut Français du Pétrole, and Richard Schrock of the Massachusetts Institute of Technology for their work on olefin metathesis.

Olefin metathesis occurs in the presence of a complex transition metal catalyst that contains a **carbon–metal double bond.** The metal is typically ruthenium (Ru), tungsten (W), or molybdenum (Mo). In a widely used catalyst, called **Grubbs catalyst,** the metal is Ru.

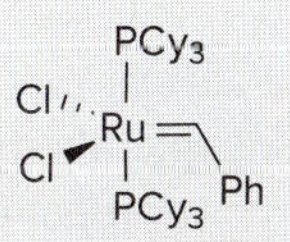

PCy_3 Cl Cl Ru Ph PCy_3

Grubbs catalyst

Olefin metathesis is an equilibrium process and, with many alkene substrates, a mixture of starting material and two or more alkene products is present at equilibrium, making the reaction useless for preparative purposes. With **terminal alkenes,** however, one metathesis product is $CH_2{=}CH_2$ (a gas), which escapes from the reaction mixture and drives the equilibrium to the right. As a result, **monosubstituted alkenes ($RCH{=}CH_2$) and 2,2-disubstituted alkenes ($R_2C{=}CH_2$) are excellent metathesis substrates** because high yields of a single alkene product are obtained, as shown in Equations [1] and [2].

[1] 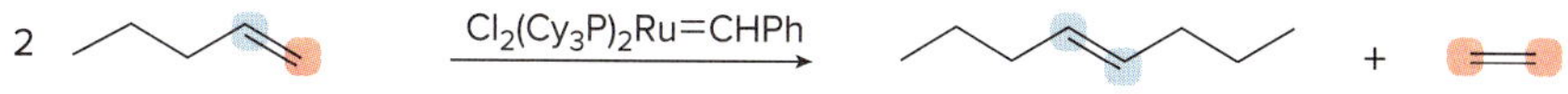 2 $\xrightarrow{Cl_2(Cy_3P)_2Ru{=}CHPh}$ (+ *Z* isomer) +

[2] 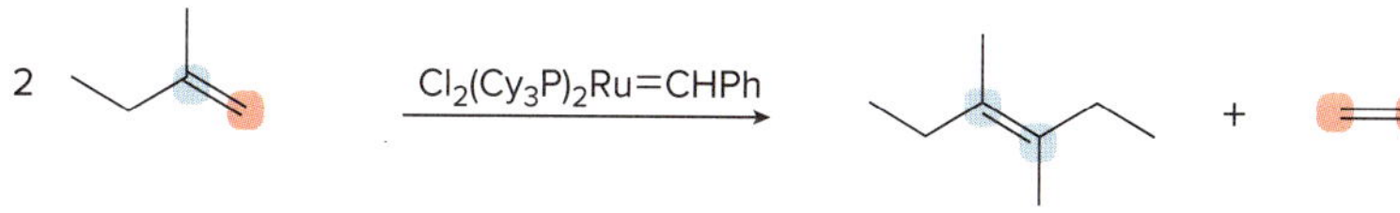 2 $\xrightarrow{Cl_2(Cy_3P)_2Ru{=}CHPh}$ (+ *Z* isomer) +

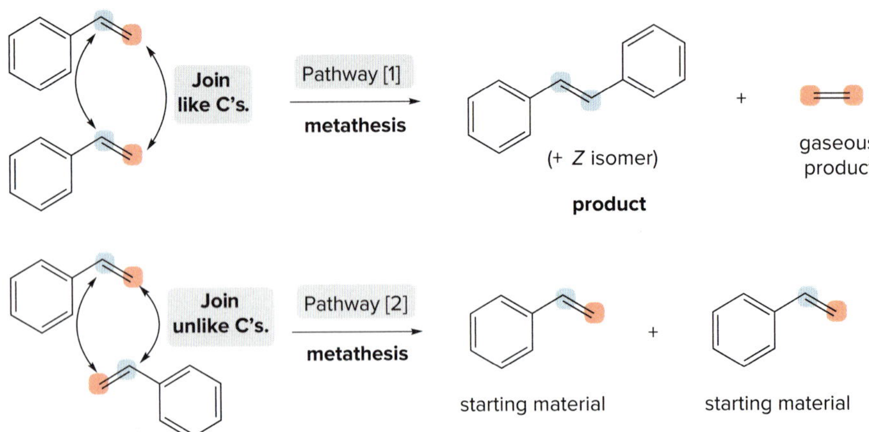

Figure 24.2 Drawing the products of olefin metathesis using styrene ($PhCH{=}CH_2$) as starting material

- Overall reaction: **2 $PhCH{=}CH_2$ → $PhCH{=}CHPh$ + $CH_2{=}CH_2$.**
- There are always two ways to join the C's of a single alkene to form metathesis products (Pathways [1] and [2]).
- When *like* C's of the alkene substrate are joined in the first reaction (Pathway [1]), $PhCH{=}CHPh$ (in a cis and trans mixture) and $CH_2{=}CH_2$ are formed. Because $CH_2{=}CH_2$ escapes as a gas from the reaction mixture, only $PhCH{=}CHPh$ is isolated as product.
- When *unlike* C's of $PhCH{=}CH_2$ are joined in the second reaction (Pathway [2]), starting material is formed, which can re-enter the catalytic cycle to form product by the first pathway.
- In this way, **a single constitutional isomer, $PhCH{=}CHPh$, is isolated.**

To draw the products of any metathesis reaction:

[1] Arrange two molecules of the starting alkene adjacent to each other as in Figure 24.2 where styrene ($PhCH{=}CH_2$) is used as the starting material.

[2] Then, break the double bonds in the starting material and form two new double bonds using carbon atoms that were *not* previously bonded to each other in the starting alkenes.

There are always two ways to arrange the starting alkenes (Pathways [1] and [2] in Figure 24.2). In this example, the two products of the reaction, $PhCH{=}CHPh$ and $CH_2{=}CH_2$, are formed in the first reaction pathway (Pathway [1]), whereas starting material is re-formed in the second pathway (Pathway [2]). Whenever the starting alkene is regenerated, it can go on to form product when the catalytic cycle is repeated.

Problem 24.18 Draw the products formed when each alkene is treated with Grubbs catalyst.

a. b. OCH_3 c.

Problem 24.19 What products are formed when *cis*-pent-2-ene undergoes metathesis? Use this reaction to explain why metathesis of a 1,2-disubstituted alkene ($RCH{=}CHR'$) is generally not a practical method for alkene synthesis.

The mechanism for olefin metathesis is complex and involves **metal–carbene intermediates—intermediates that contain a metal–carbon double bond.** The mechanism is drawn for the reaction of a terminal alkene ($RCH{=}CH_2$) with Grubbs catalyst, abbreviated as **Ru=CHPh,** to form RCH=CHR and $CH_2{=}CH_2$. To begin metathesis, Grubbs catalyst reacts with the alkene substrate to form two new metal–carbenes **A** and **B** by a two-step process: addition of Ru=CHPh to the alkene to yield two different metallocyclobutanes (Step [1]), followed by elimination to form **A** and **B** (Steps [2a] and [2b]). The alkene by-products formed in this process (RCH=CHPh and $PhCH{=}CH_2$) are present in only a small amount because Grubbs reagent is used catalytically.

Grubbs reagent
+
starting material
1
2a
A
(+ *Z* isomer)
by-product
2b
B
by-product
metallocyclobutanes
metal–carbene complexes

Each of these metal–carbene intermediates **A** and **B** then reacts with more starting alkene to form metathesis products, as shown in Mechanism 24.7. As was seen in Mechanisms 24.1–24.3, this mechanism is often written in a circle to emphasize the catalytic cycle. The mechanism demonstrates how two molecules of $RCH{=}CH_2$ are converted to RCH=CHR and $CH_2{=}CH_2$. The mechanism can be written beginning with reagent **A** or **B,** and all steps are equilibria.

Mechanism 24.7 Olefin Metathesis: $2\ RCH{=}CH_2 \rightarrow RCH{=}CHR + CH_2{=}CH_2$

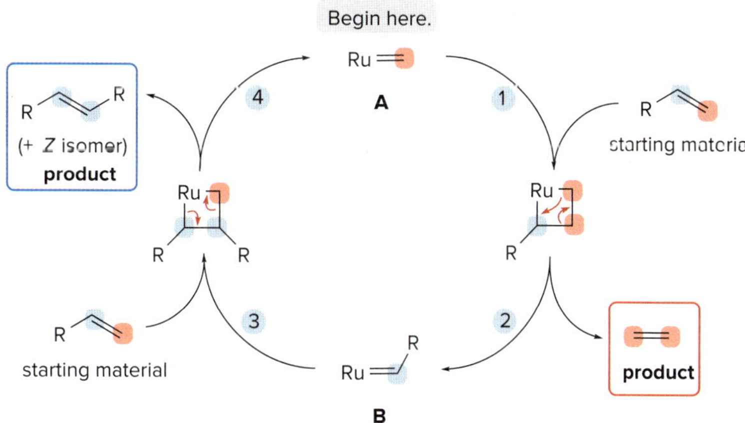

1. Reaction of $Ru{=}CH_2$ (**A**) with $RCH{=}CH_2$ forms a **metallocyclobutane.** Ru can bond to either the more or less substituted end of the alkene, but product is formed only when Ru bonds to the *more* substituted end, as shown.
2. **Elimination** forms one metathesis product, $CH_2{=}CH_2$, and metal–carbene complex **B.**
3. Reaction of **B** with $RCH{=}CH_2$ forms a **metallocyclobutane.** Ru can bond to either the more or less substituted end of the alkene, but product is formed only when Ru bonds to the *less* substituted end, as shown.
4. **Elimination** forms the other metathesis product, RCH=CHR, and metal–carbene complex **A.** The catalyst is regenerated and the cycle begins again.

24.7B Ring-Closing Metathesis

A metathesis reaction that forms a ring is called **ring-closing metathesis (RCM).**

When a diene is used as starting material, ring closure occurs.

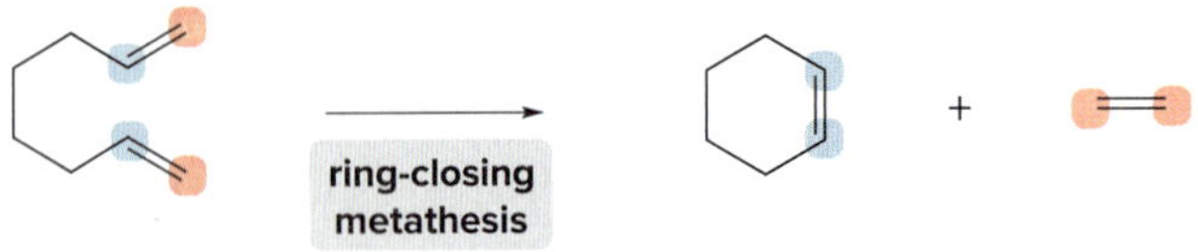

ring-closing metathesis

These reactions are typically run in very dilute solution, so that the two reactive ends of the *same* molecule have a higher probability of finding each other for reaction than two functional groups in *different* molecules. These high-dilution conditions thus favor ***intra*molecular** rather than *inter*molecular metathesis. Two examples are shown.

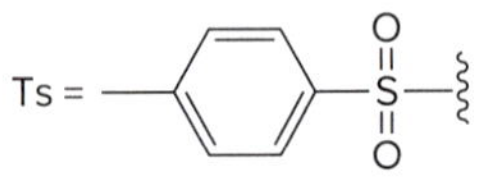

Ts = tosyl

Grubbs catalyst + $CH_2{=}CH_2$

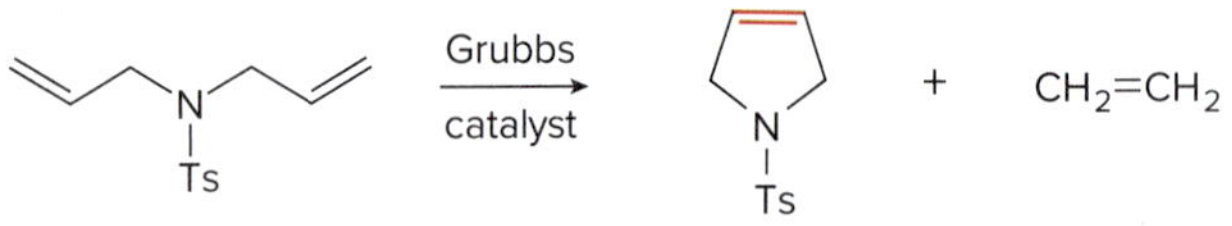

new C=C in red

Grubbs catalyst + $CH_2{=}CH_2$

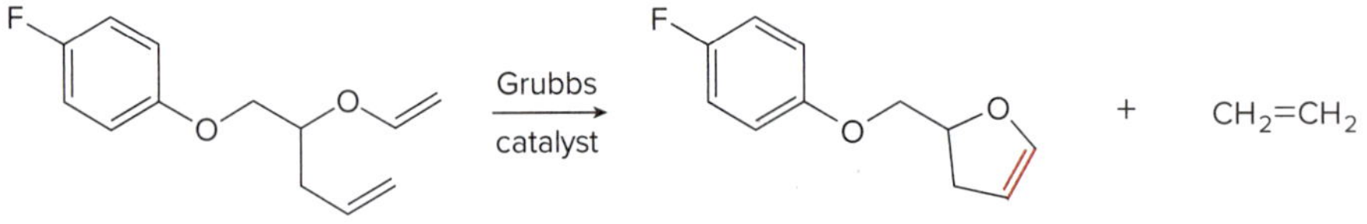

new C=C in red

Because metathesis catalysts are compatible with the presence of many functional groups (such as OH, OR, and C=O) and because virtually any ring size can be prepared, metathesis has been used to prepare many complex natural products such as epothilone A, shown in Figure 24.3.

Figure 24.3 Ring-closing metathesis in the synthesis of epothilone A

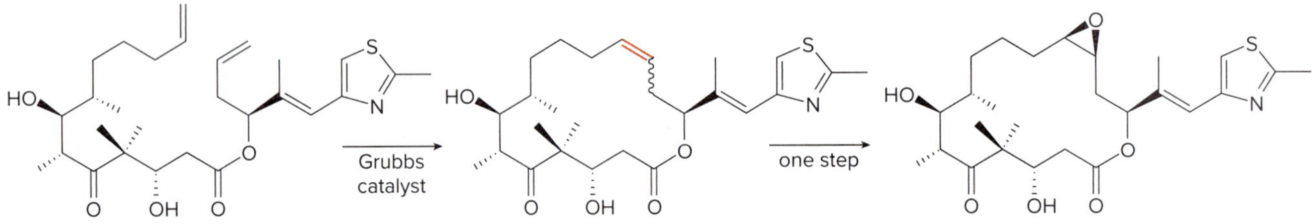

Grubbs catalyst

(*E* and *Z* isomers formed)

one step

epothilone A
anticancer drug

- **Epothilone A,** a promising anticancer agent, was first isolated from soil bacteria collected from the banks of the Zambezi River in South Africa.
- The new C–C bonds formed during metathesis are indicated in red. During metathesis, $CH_2{=}CH_2$ is also formed.

Problem 24.20 Draw the product formed from ring-closing metathesis of each compound.

a.

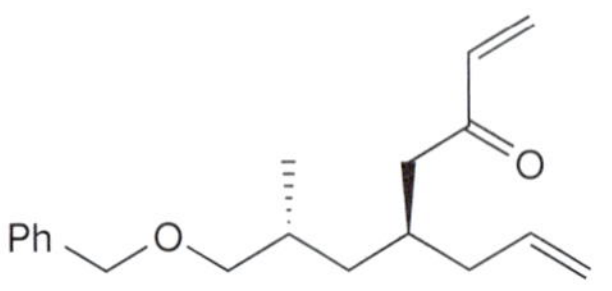

b.

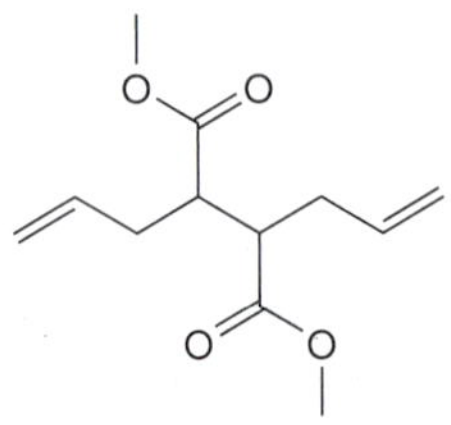

Problem 24.21 What product is formed when **B** is treated with Grubbs catalyst under high-dilution conditions? This reaction was used in the synthesis of oleocanthal, an antioxidant isolated from olive oil (Problem 21.68).

B

Ingenol (Problem 24.22) is isolated from the milky liquid obtained from *Euphorbia ingens,* a large cactus commonly called the candelabra tree, which is native to dry areas in southern Africa. *Papa Bravo/ Shutterstock*

Problem 24.22 What product is formed by ring-closing metathesis of compound **V,** a key intermediate in the synthesis of ingenol, a natural product mentioned in the chapter opener?

V ingenol

Sample Problem 24.6 Determining the Starting Material of a Ring-Closing Metathesis

What starting material is needed to synthesize each compound by a ring-closing metathesis reaction?

a. b.

Solution

To work in the retrosynthetic direction, cleave the C=C in the product, and **bond each carbon of the original alkene to a CH_2 group using a double bond.**

Break the C=C. $\Longrightarrow$ starting material

Add $=CH_2$ to both C's.

The resulting compound has a carbon chain with **two terminal alkenes.**

a. **Break the C=C.** $\Longrightarrow$ starting material

b. **Break the C=C.** $\Longrightarrow$ starting material

Problem 24.23 What starting material is needed to synthesize each compound by a ring-closing metathesis reaction?

a. b. CH_3O OH c. CO_2CH_3 CHO O

More Practice: Try Problems 24.41, 24.42.

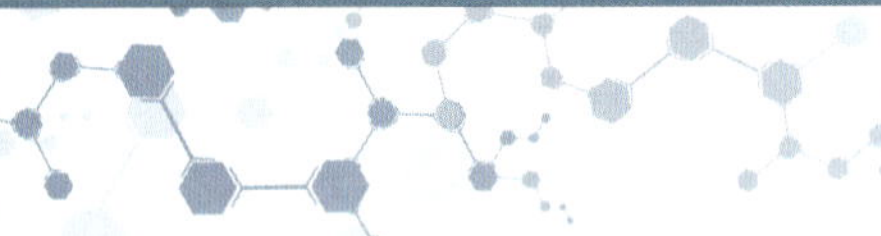

Chapter 24 REVIEW

KEY REACTIONS

Coupling Reactions

1 X + $(\diagup\!\!\diagdown)_2Cu^- Li^+$ → (24.1) + LiX + Cu
X = Cl, Br, I

2 X + B (catechol boronate) → Pd(PPh_3)$_4$, NaOH, Suzuki reaction (24.2) + NaX + HOB (catechol boronate)
X = Br, I

3 X + OCH_3 (methyl acrylate) → Pd(OAc)$_2$, P(o-tolyl)$_3$, Et_3N, Heck reaction (24.3) → OCH_3 + $Et_3\overset{+}{N}H$ X^-
X = Br, I

4 X + $SnBu_3$ → Pd(PPh_3)$_4$, Stille coupling (24.4) + Bu_3SnX
X = Br, I

Try Problems 24.24; 24.26; 24.29; 24.32a; 24.33–24.36; 24.45, b–d, g, h.

Cyclopropane Synthesis

1 → CHX_3, $KOC(CH_3)_3$, addition (24.5) → X X

2 → CH_2I_2, Zn(Cu), Simmons–Smith reaction (24.6) → + ZnI_2

Try Problems 24.37; 24.38; 24.45a, f.

Metathesis

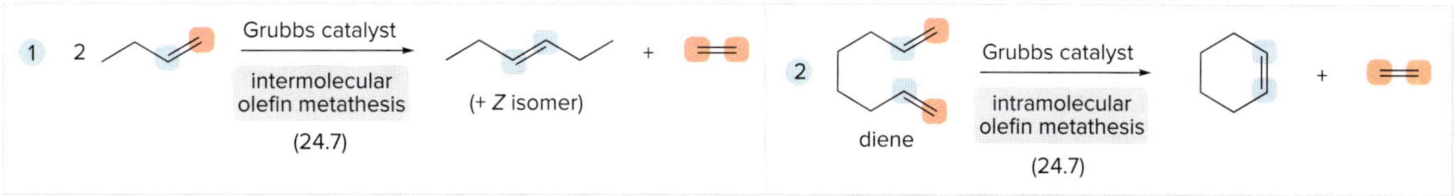

Try Problems 24.25, 24.39, 24.40, 24.44, 24.45e.

KEY SKILLS

[1] Drawing the product of a Suzuki reaction beginning with an alkyne and vinyl halide (Section 24.2); example: reaction of but-1-yne with (*E*)-2-bromobut-2-ene

1 Draw the product of hydroboration of the C≡C with catecholborane.

but-1-yne + H–B (catecholborane) ⟶ vinylborane (*E*)

- Hydroboration forms a vinylborane **by syn addition** of **H** and **B** to the C=C.

2 Couple the vinylborane with the vinyl bromide.

vinylborane (*E*) + Br (*E*) —$Pd(PPh_3)_4$, NaOEt, coupling⟶ *E*, *E* diene

- Coupling affords a conjugated diene with **retention of configuration.**

See Sample Problem 24.2. Try Problems 24.28, 24.36a, 24.54.

[2] Identifying the starting materials to synthesize an alkene using a Heck reaction (24.3); two possibilities

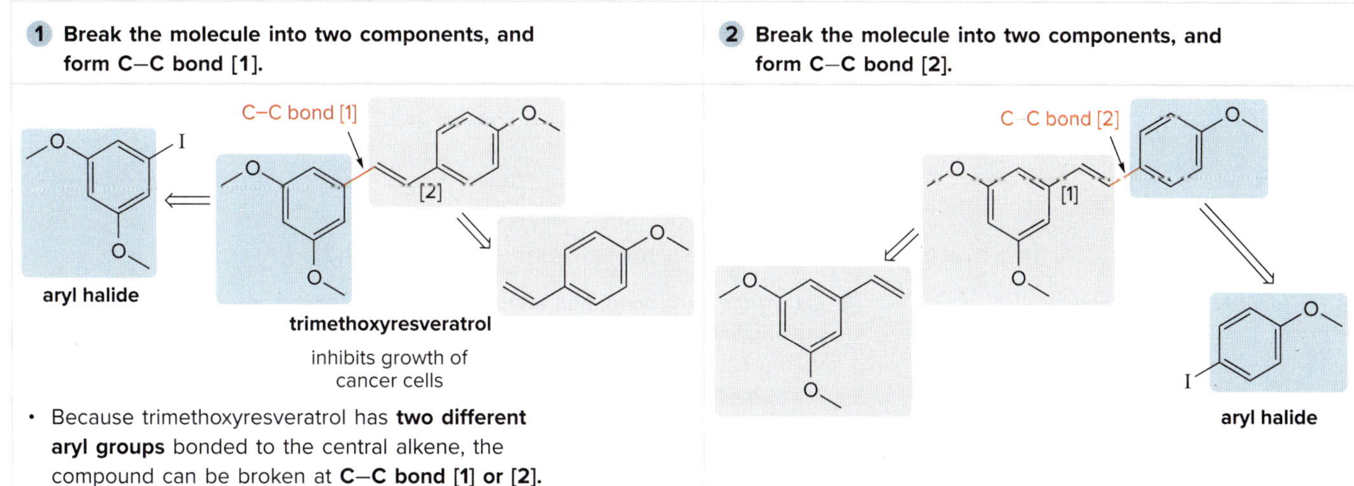

- Because trimethoxyresveratrol has **two different aryl groups** bonded to the central alkene, the compound can be broken at **C–C bond [1] or [2].**

See Sample Problem 24.3. Try Problems 24.29, 24.36b, 24.56.

[3] Identifying the starting materials to synthesize a 1,3-diene using a Stille coupling (Section 24.4); two possibilities

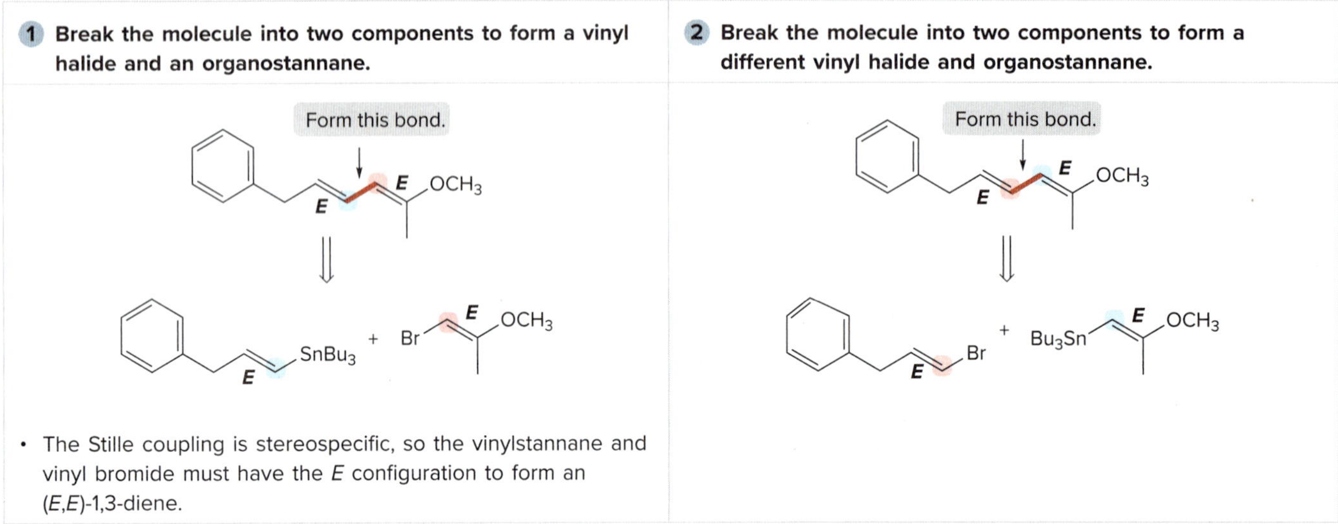

See Sample Problem 24.4. Try Problems 24.30, 24.32b, 24.36c.

[4] Drawing all stereoisomers that form in a cyclopropanation (24.5–24.6); example: cyclopropanation of 1-methylcyclohexene

1 **Use the reagents to identify the group added to the C=C.**	**2** **Add CH_2 from above the alkene.**	**3** **Add CH_2 from below the alkene.**	**4** **Determine the stereochemistry of the products.**
1-methylcyclohexene $\xrightarrow[\text{Zn(Cu)}\ (24.5)]{CH_2I_2}$			• Both stereogenic centers are opposite in configuration. • There is no plane of symmetry. • The compounds are **enantiomers.**

See Sample Problem 24.5. Try Problems 24.37; 24.38; 24.45a, f.

[5] Identifying the starting material in a ring-closing metathesis reaction (24.7)

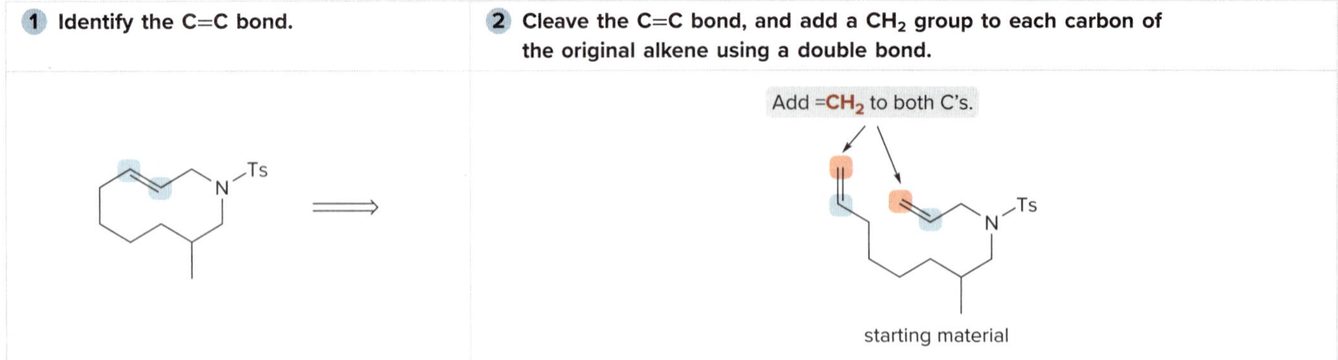

See Figure 24.3, Sample Problem 24.6. Try Problems 24.41, 24.42.

CHAPTER 24 MULTIPLE-CHOICE SELF-TEST

The Self-Test consists of multiple-choice questions similar to those found on the American Chemical Society organic chemistry exam. Answers are given at the end of the chapter.

1. What compound does *not* react with organocuprate reagents?

a. 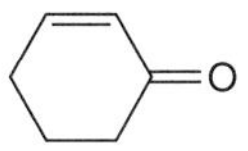 O b. 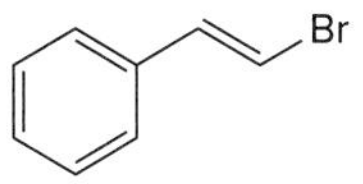 Br c. 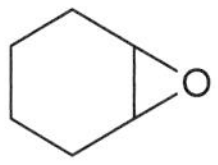 O d.  Br

2. Which statement about carbenes is *not* true?

a. The carbene carbon is sp^2 hybridized and the lone pair occupies a *p* orbital.
b. A carbene reacts as an electrophile.
c. The reaction of (*Z*)-hex-3-ene with dichlorocarbene forms a single achiral product.
d. A carbene is a neutral reactive intermediate.

3. What product is formed in the given reaction?

[1] O BH O

[2] Br

$Pd(PPh_3)_4$
NaOH

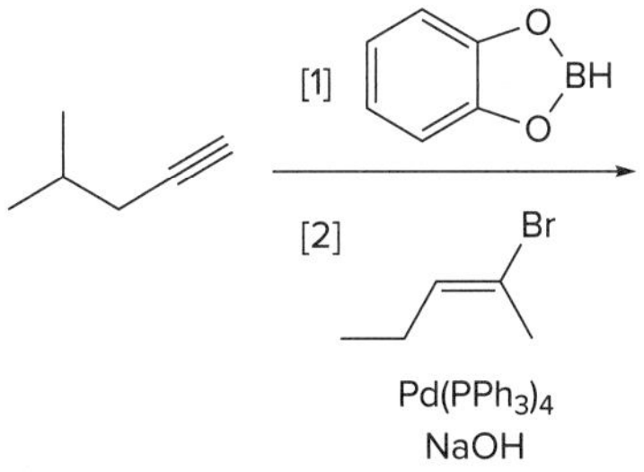

a. 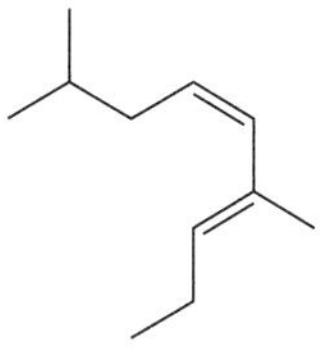 b. 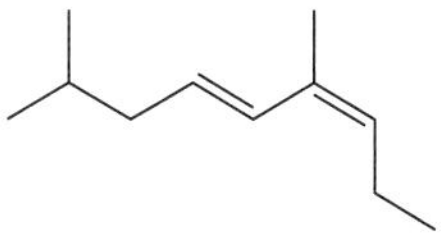 c. 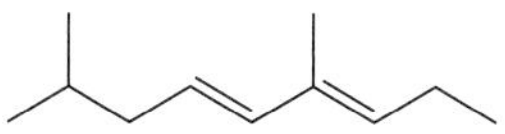 d.

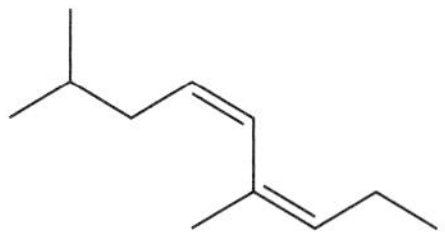

4. What product is formed in the given reaction?

$CHCl_3$ / $KOC(CH_3)_3$; $LiCu(CH{=}CH_2)_2$ (2 equiv)

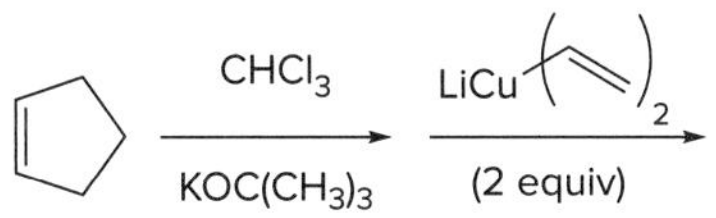

a. 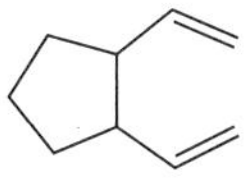 b. 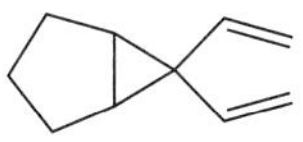 c. 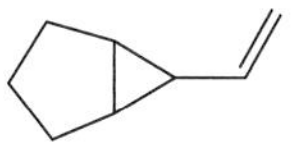 d.

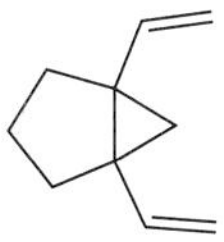

5. What product is formed when **X** is treated with Grubbs catalyst under high-dilution conditions?

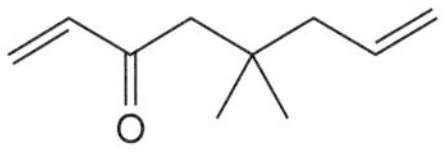

O
X

a. 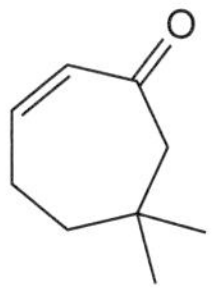 O b. 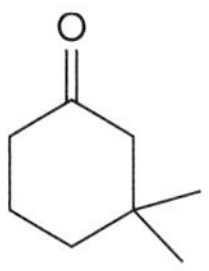 O c. 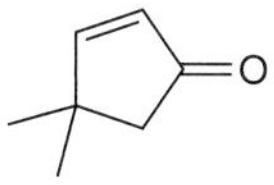 O d. 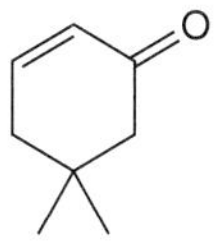 O

6. What starting material is needed to prepare **A** by a ring-closing metathesis?

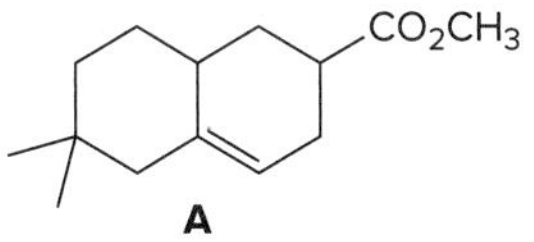

CO_2CH_3
A

a. 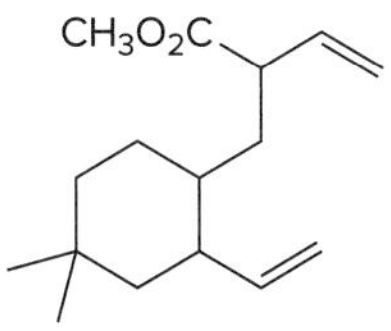 CH_3O_2C
b. 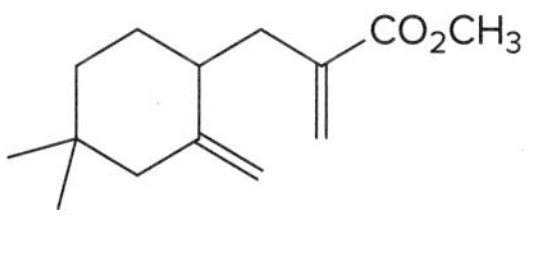 CO_2CH_3
c. 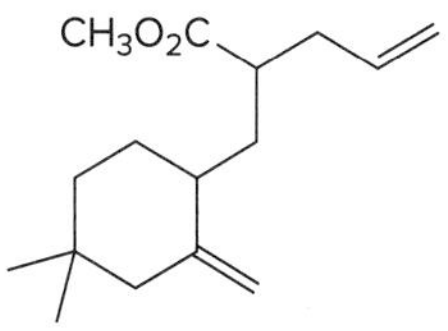 CH_3O_2C
d. 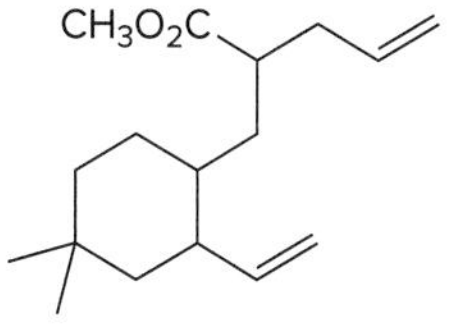 CH_3O_2C

7. What product is formed when **B** and **C** react in the presence of $Pd(OAc)_2$, P(*o*-tolyl)$_3$, and Et_3N?

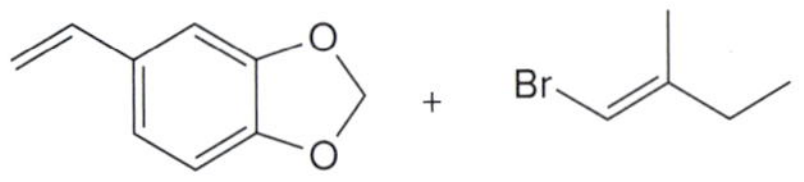

B + **C** (Br)

a. 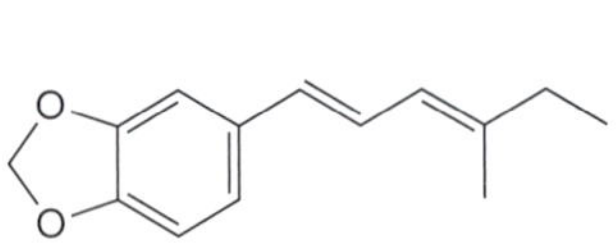 b. 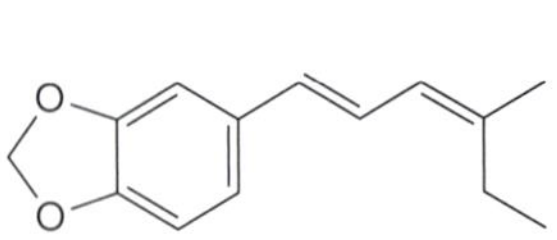 c. d.

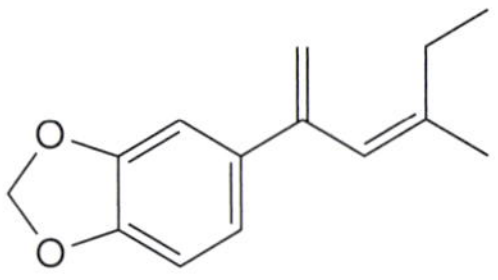

8. Which method does *not* form **D**?

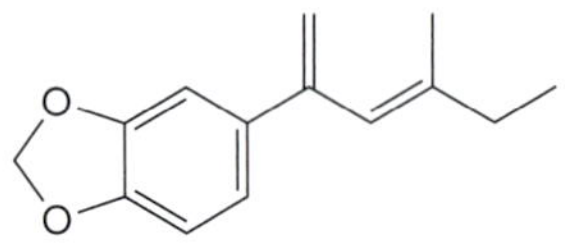

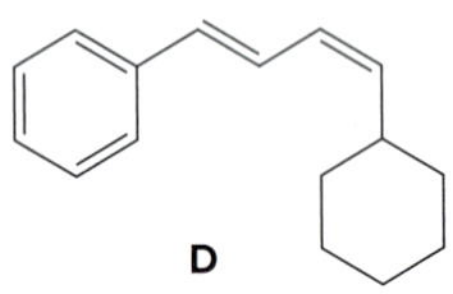

D

a. 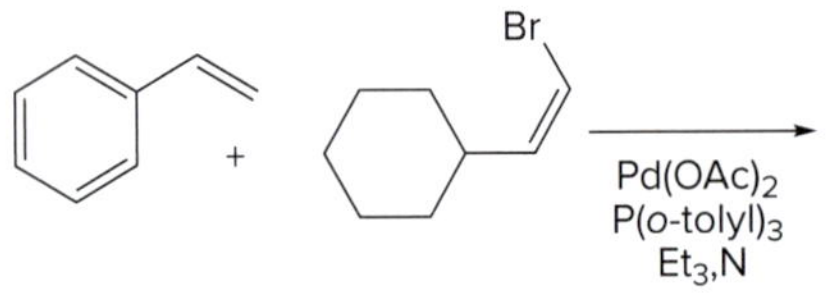 (Br) + ; $Pd(OAc)_2$, P(*o*-tolyl)$_3$, Et_3N

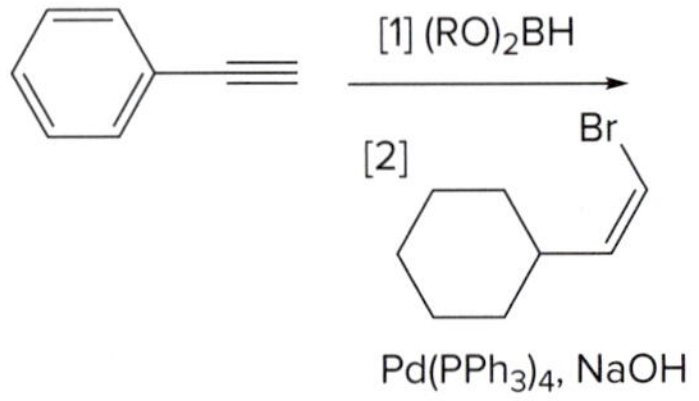

b. 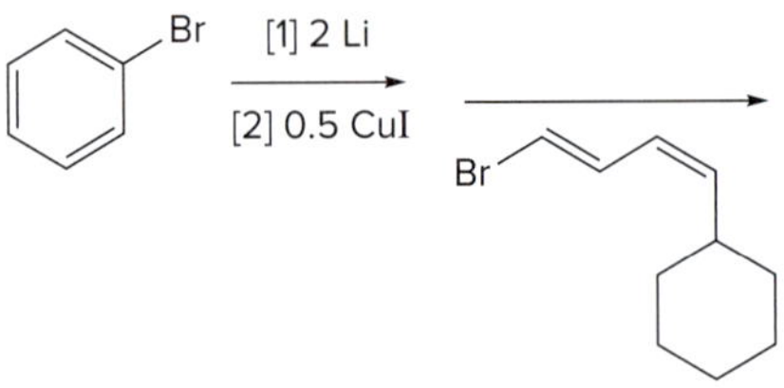 (Br) [1] 2 Li [2] 0.5 CuI ; (Br)

c. [1] $(RO)_2BH$ [2] (Br) $Pd(PPh_3)_4$, NaOH

d. 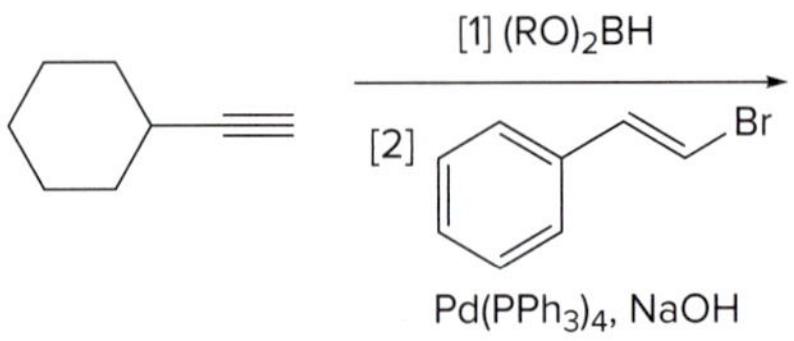 [1] $(RO)_2BH$ [2] (Br) $Pd(PPh_3)_4$, NaOH

9. What compound is *not* formed in the following reaction?

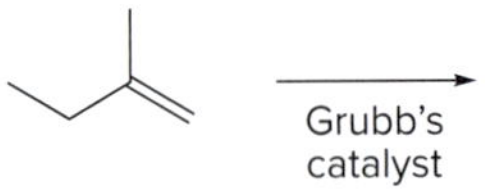

Grubb's catalyst

a. 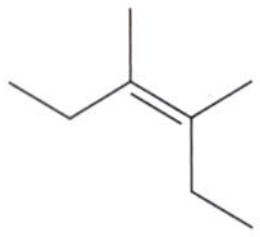 b. 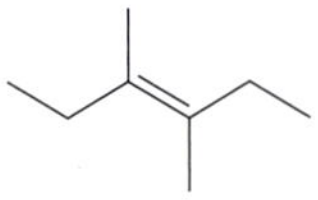 c. 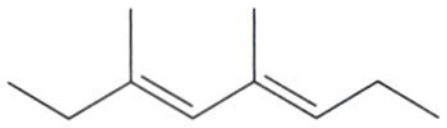 d. $CH_2{=}CH_2$

10. What reactant **E** is needed in the following Suzuki reaction?

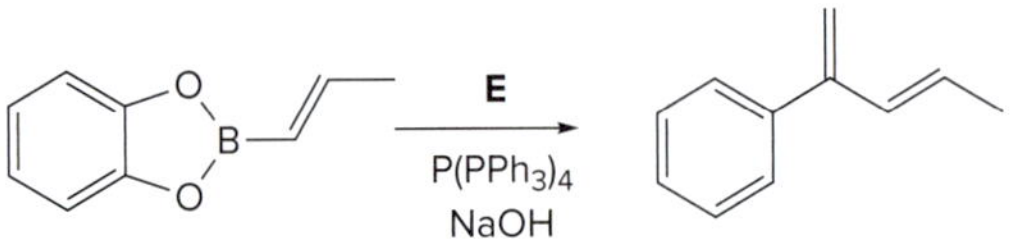

(O, B, O) **E** ; $P(PPh_3)_4$, NaOH

a. 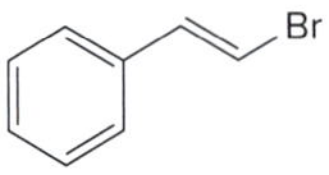 (Br) b. 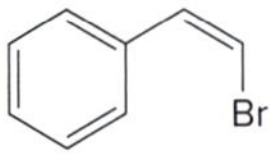 (Br) c. 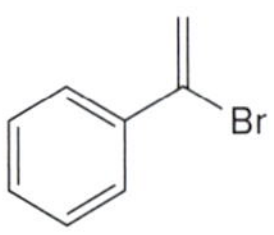 (Br) d. 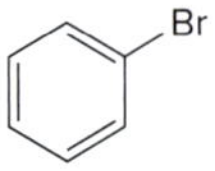 (Br)

PROBLEMS

Problems Using Three-Dimensional Models

24.24 In addition to organic halides, alkyl tosylates (R'OTs, Section 9.13) react with organocuprates (R_2CuLi) to form coupling products R–R'. When 2° alkyl tosylates are used as starting materials (R_2CHOTs), inversion of the configuration at a stereogenic center results. Keeping this in mind, draw the product formed when each compound is treated with $(CH_3)_2CuLi$.

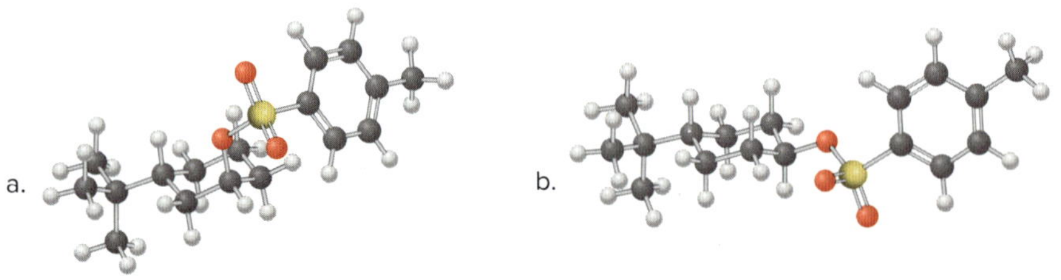

24.25 What product is formed by ring-closing metathesis of each compound?

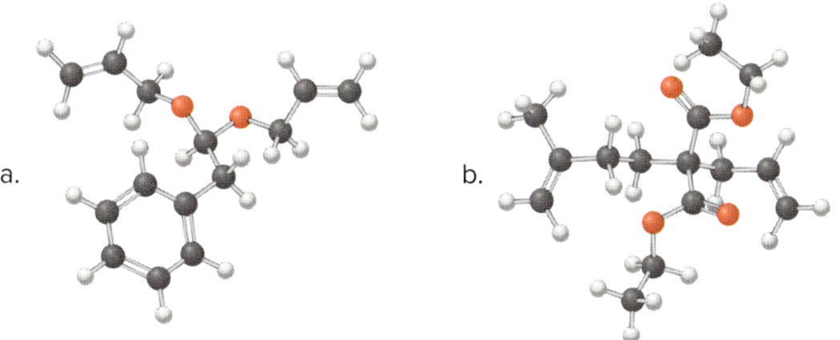

Coupling Reactions

24.26 Draw the products formed in each reaction.

a. 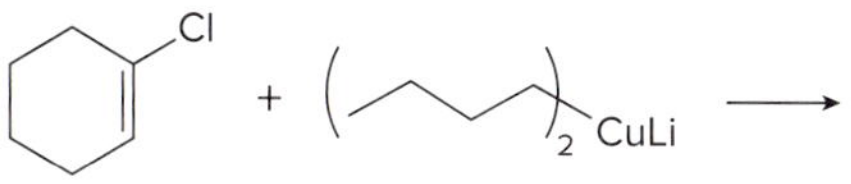 + $(\ldots)_2CuLi$ →

b. 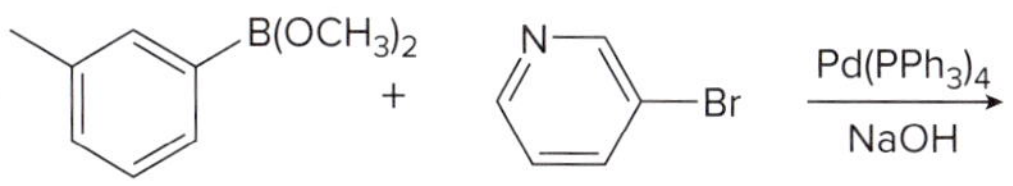 $B(OCH_3)_2$ + Br; $Pd(PPh_3)_4$, NaOH

c. 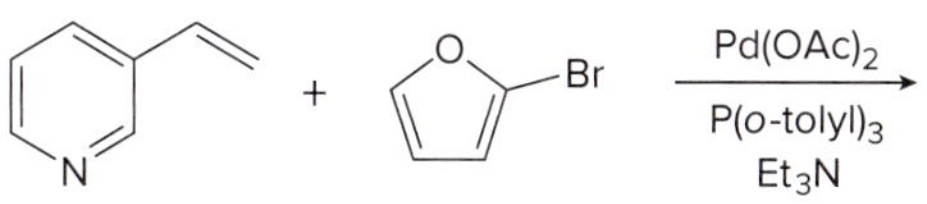 + Br; $Pd(OAc)_2$, P(o-tolyl)$_3$, Et_3N

d. 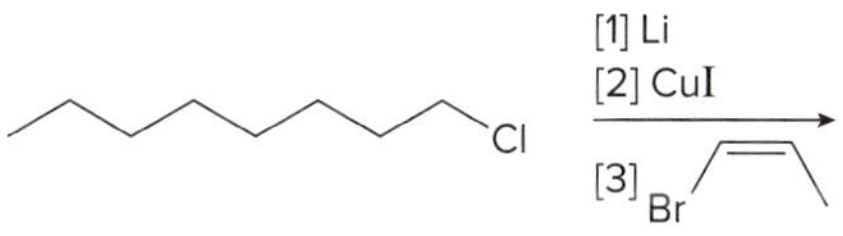 Cl; [1] Li [2] CuI [3] Br

e. 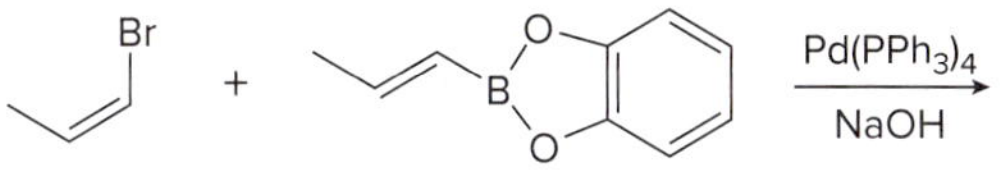 Br + B; $Pd(PPh_3)_4$, NaOH

f. 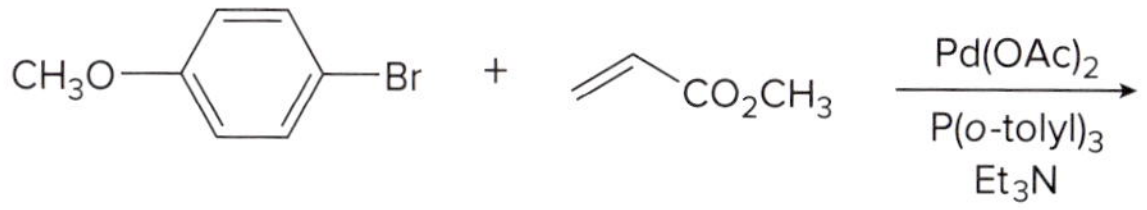 CH_3O–C$_6$H$_4$–Br + CO_2CH_3; $Pd(OAc)_2$, P(o-tolyl)$_3$, Et_3N

g. 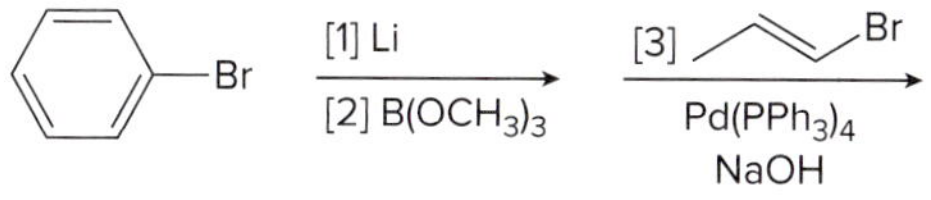 Br; [1] Li [2] $B(OCH_3)_3$; [3] Br, $Pd(PPh_3)_4$, NaOH

h. [1] H–B; [2] C_6H_5Br, $Pd(PPh_3)_4$, NaOH

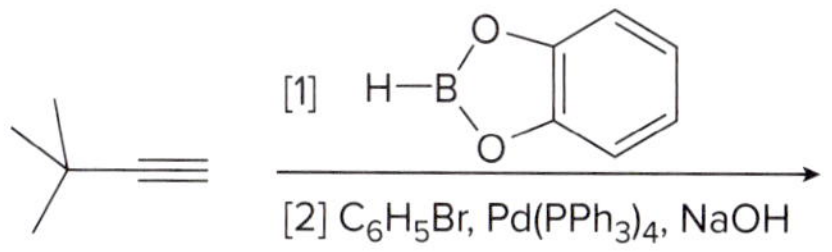

i. 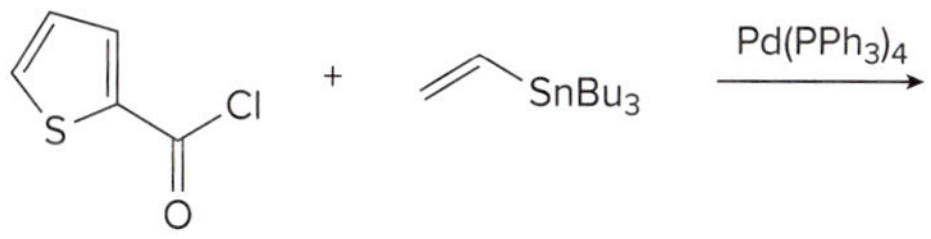 Cl + $SnBu_3$; $Pd(PPh_3)_4$

j. 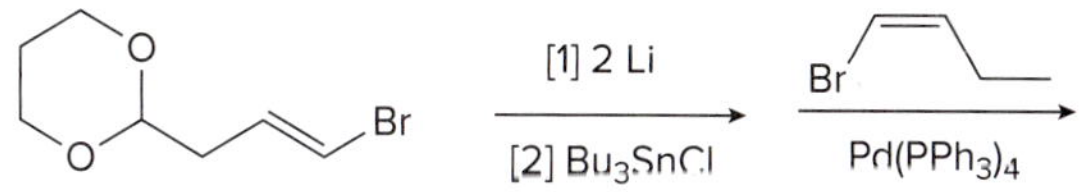 Br; [1] 2 Li [2] Bu_3SnCl; Br, $Pd(PPh_3)_4$

24.27 What organic halide is needed to convert lithium divinylcuprate [$(CH_2{=}CH)_2CuLi$] to each compound?

a. 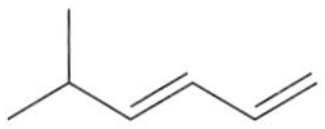 b. 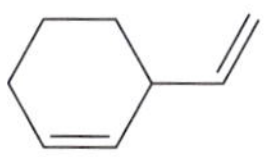 c. 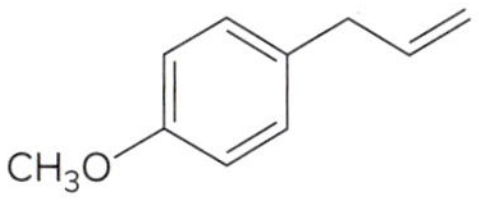 CH_3O

24.28 How can you convert ethynylcyclohexane to dienes **A–C** using a Suzuki reaction? You may use any other organic compounds and inorganic reagents. Is it possible to synthesize diene **D** using a Suzuki reaction? Explain why or why not.

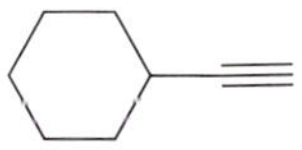

ethynylcyclohexane

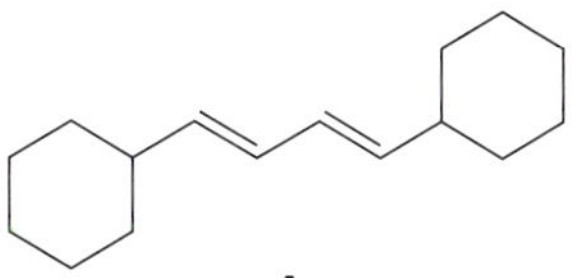

A 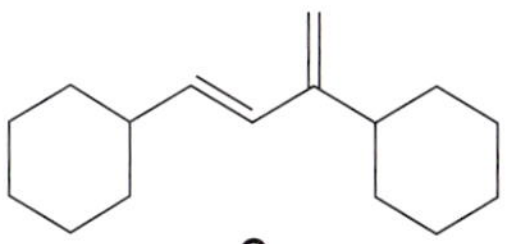 **C** 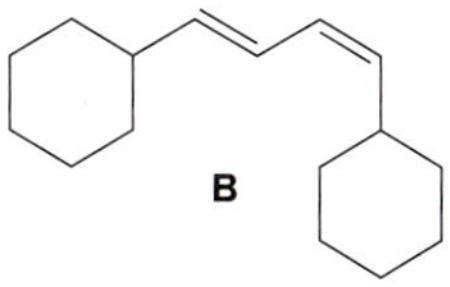 **B** 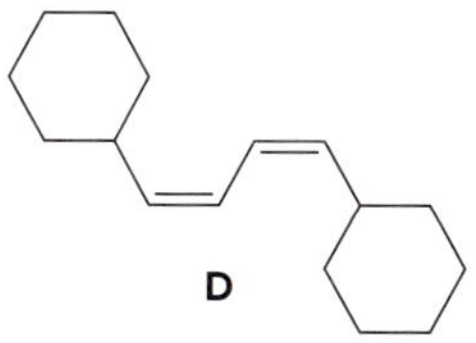 **D**

24.29 What compound is needed to convert styrene ($C_6H_5CH{=}CH_2$) to each product using a Heck reaction?

a. 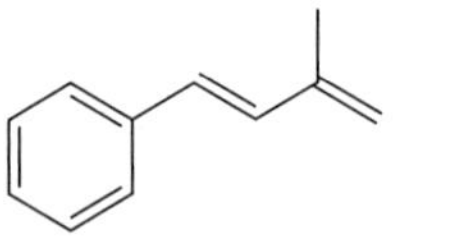 b. 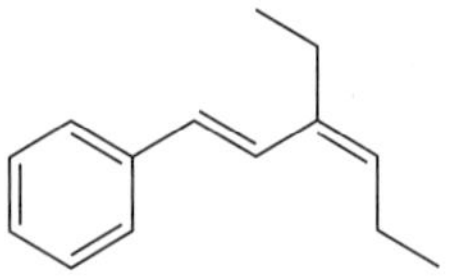 c.

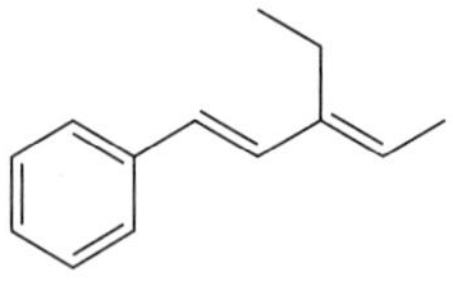

24.30 What starting materials are needed to synthesize each compound using a Stille coupling? Give one method only.

a. 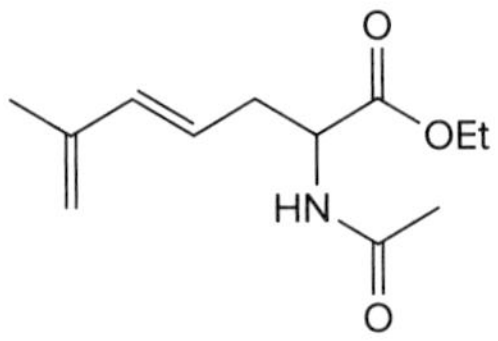 O, OEt, HN, O b. 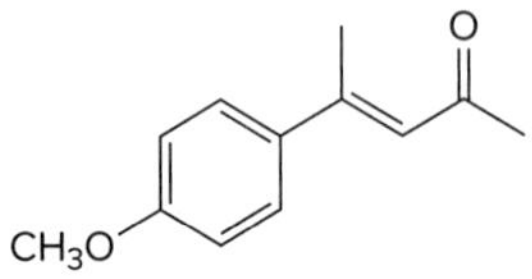 O, CH_3O c. 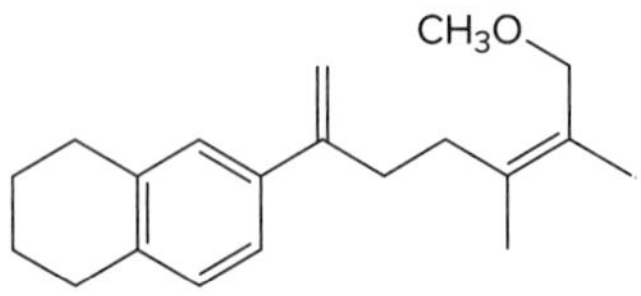 CH_3O

24.31 What steps are needed to convert but-1-ene ($CH_3CH_2CH{=}CH_2$) to octane [$CH_3(CH_2)_6CH_3$] using a coupling reaction with an organocuprate reagent? All carbon atoms in octane must come from but-1-ene.

24.32 Losartan is a drug used to treat hypertension. (a) What product is formed in the Suzuki coupling of **A** and **B?** (b) Give two different routes to losartan using a Stille coupling reaction.

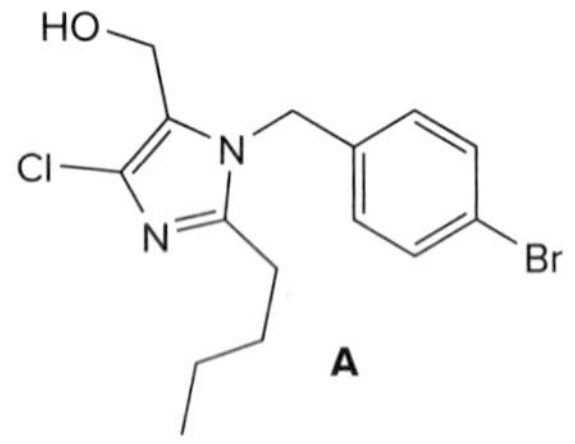

HO, Cl, N, N, Br — **A**

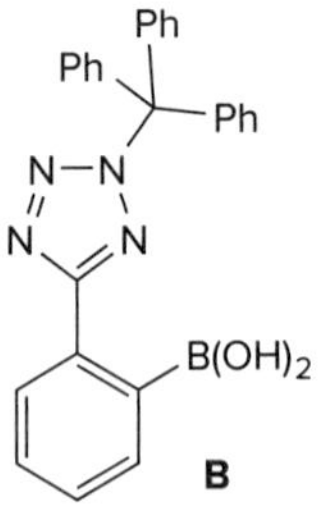

Ph, Ph, Ph, N–N, N, N, $B(OH)_2$ — **B**

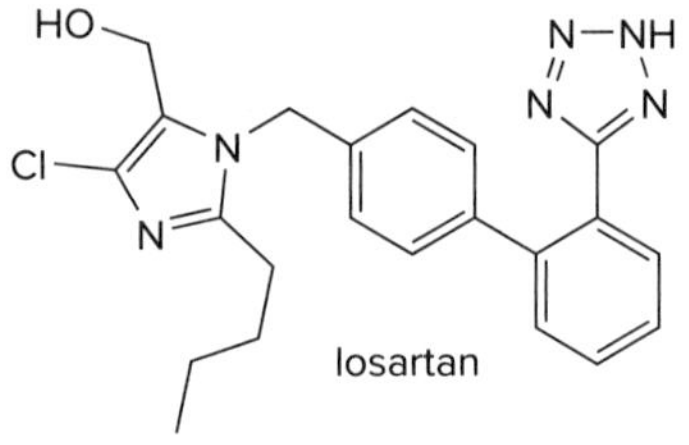

HO, Cl, N, N, N–NH, N, N — losartan

24.33 Draw the product formed when the following compound undergoes an intramolecular Heck reaction. Indicate the stereochemistry at all double bonds and tetrahedral stereogenic centers.

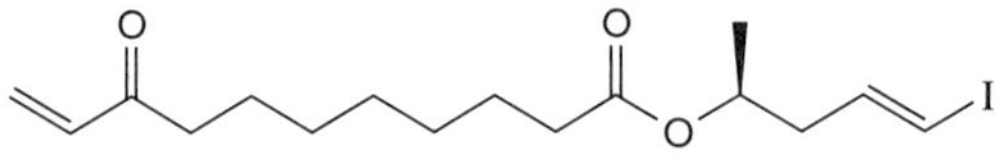

O, O, O, I

24.34 Identify **X,** an intermediate that was converted to eletriptan (trade name Relpax), a drug used to treat migraines.

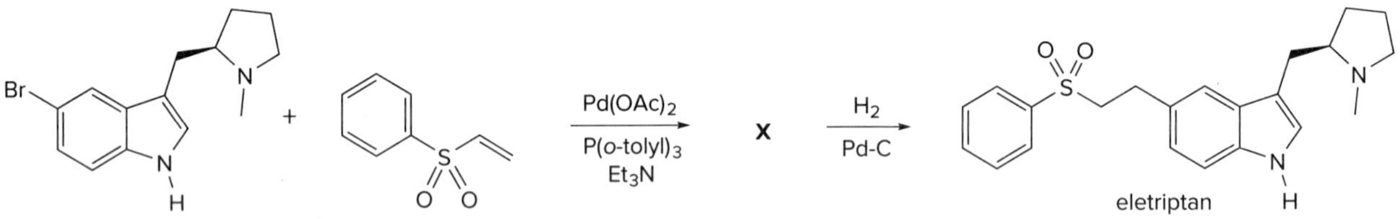

Br, N, H, N + S, O, O — $Pd(OAc)_2$, P(o-tolyl)$_3$, Et_3N → **X** — H_2, Pd-C → O, O, S, N, H, N — eletriptan

24.35 Identify the structure of the antibiotic indanomycin, which is formed in the following reaction.

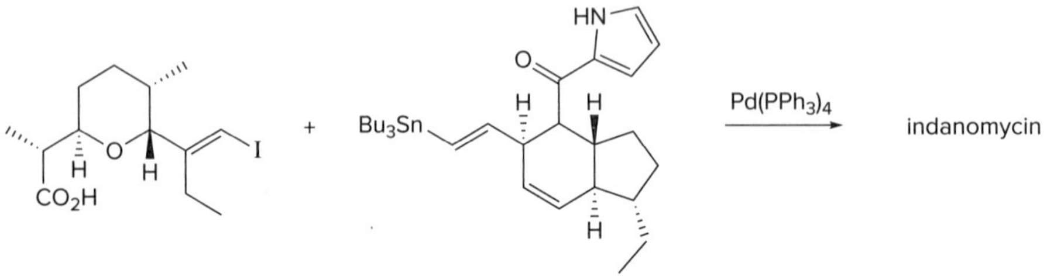

O, H, H, CO_2H, I + Bu_3Sn, HN, O, H, H, H — $Pd(PPh_3)_4$ → indanomycin

24.36 What starting materials are needed to synthesize **X** by each of the following coupling reactions: (a) Suzuki reaction; (b) Heck reaction; (c) Stille reaction? Give one method for each coupling reaction.

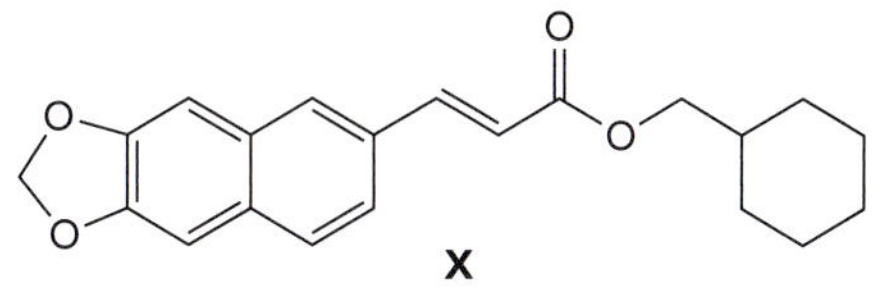

X

Cyclopropanes

24.37 Draw the products (including stereoisomers) formed in each reaction.

a. 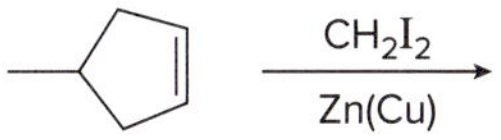 $\xrightarrow[Zn(Cu)]{CH_2I_2}$

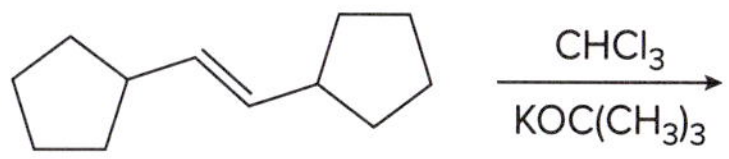

b. 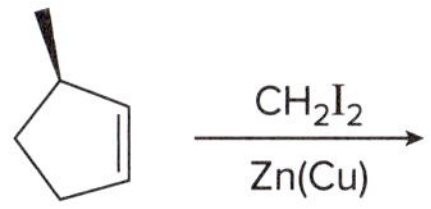 $\xrightarrow[Zn(Cu)]{CH_2I_2}$

c. $\xrightarrow[KOC(CH_3)_3]{CHCl_3}$

d. 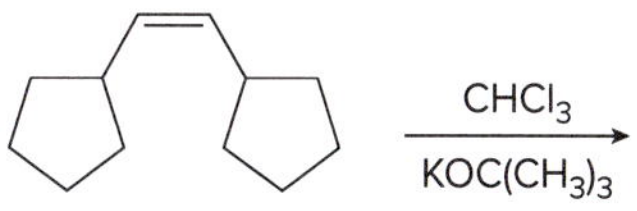 $\xrightarrow[KOC(CH_3)_3]{CHCl_3}$

24.38 Treatment of cyclohexene with $C_6H_5CHI_2$ and Zn(Cu) forms two stereoisomers of molecular formula $C_{13}H_{16}$. Draw their structures and explain why two compounds are formed.

Metathesis

24.39 What ring-closing metathesis product is formed when each substrate is treated with Grubbs catalyst under high-dilution conditions?

a. 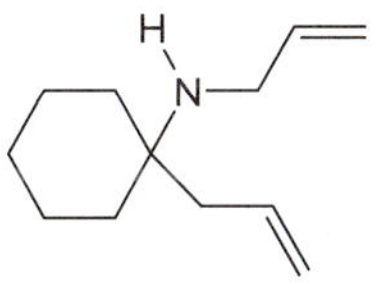 H N

b. 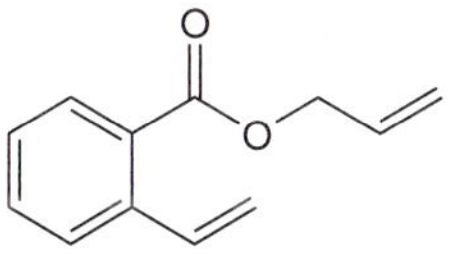 O O

c. 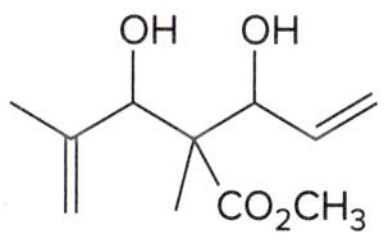 OH OH CO_2CH_3

d. 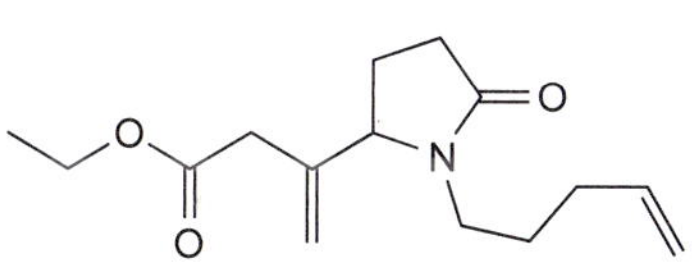 O O O N O

24.40 Draw the product formed when each compound is treated with Grubbs catalyst under high-dilution conditions. Indicate the stereochemistry at all stereogenic centers.

a. 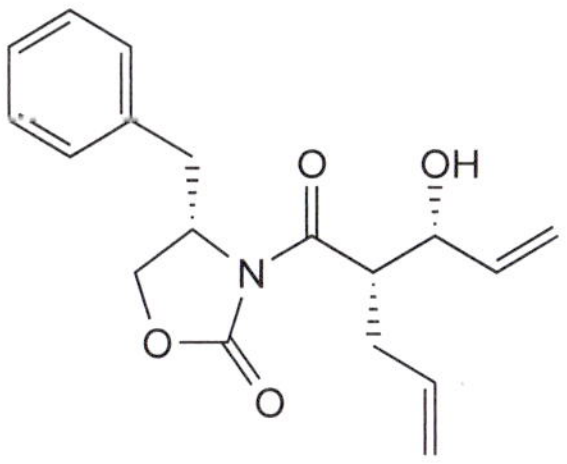 O OH N O O

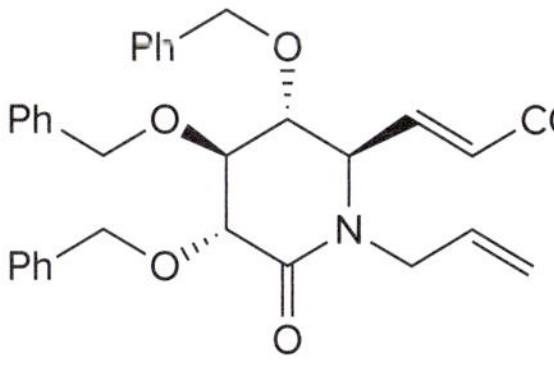

b. 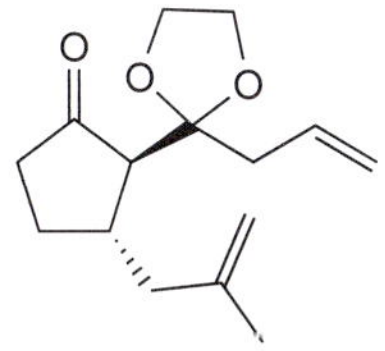 O O O

c. Ph O Ph O Ph O N O CO_2CH_3

d. 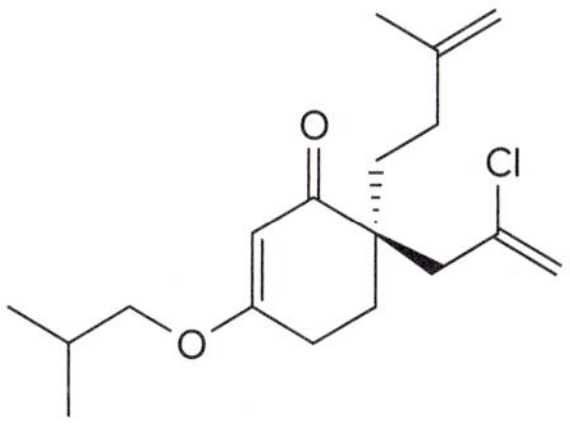 O Cl O

24.41 What starting material is needed to prepare each compound by a ring-closing metathesis reaction?

a. 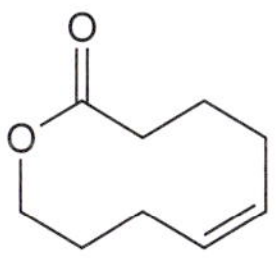 O O

b. 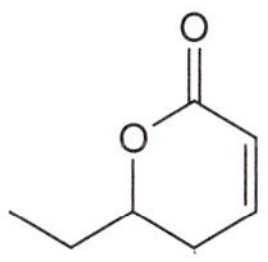 O O

c. 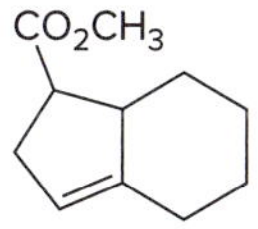 CO_2CH_3

24.42 A key step in the synthesis of migrastatin, a lactone obtained from a strain of *Streptomyces,* involved ring-closing metathesis to form alkene **B.**

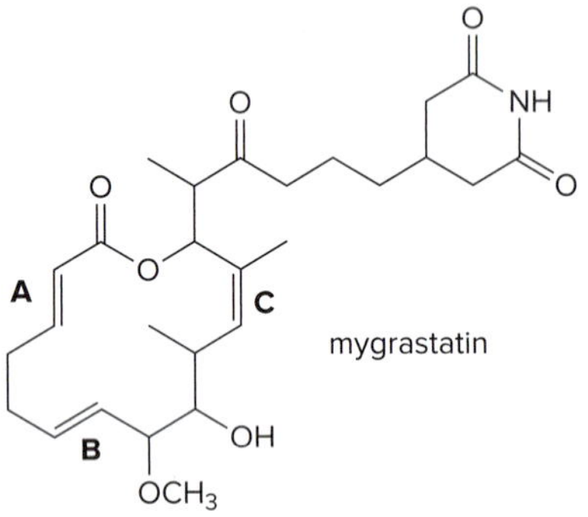

A C B OH OCH_3 O NH O O O mygrastatin

a. What starting material is needed to form alkene **B** by ring-closing metathesis?

b. What starting material would be needed to form alkene **A** by ring-closing metathesis?

c. What starting material would be needed to form alkene **C** by ring-closing metathesis?

24.43 Metathesis reactions can be carried out with two *different* alkene substrates in one reaction mixture. Depending on the substitution pattern around the C=C, the reaction may lead to one major product or a mixture of many products. For each pair of alkene substrates, draw all metathesis products formed. (Disregard any starting materials that may also be present at equilibrium.) With reference to the three examples, discuss when alkene metathesis with two different alkenes is a synthetically useful reaction.

a. 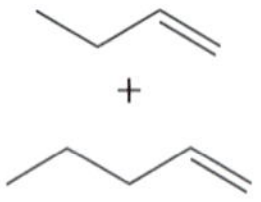 + b. 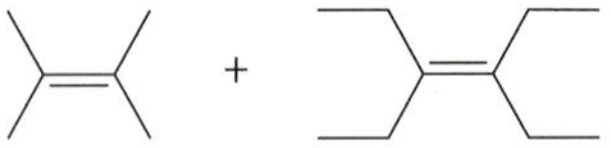 + c.  +

24.44 When certain cycloalkenes are used in metathesis reactions, **ring-opening metathesis polymerization (ROMP)** occurs to form a high-molecular-weight polymer, as shown with cyclopentene as the starting material. The reaction is driven to completion by relief of strain in the cycloalkene.

This C=C is cleaved.

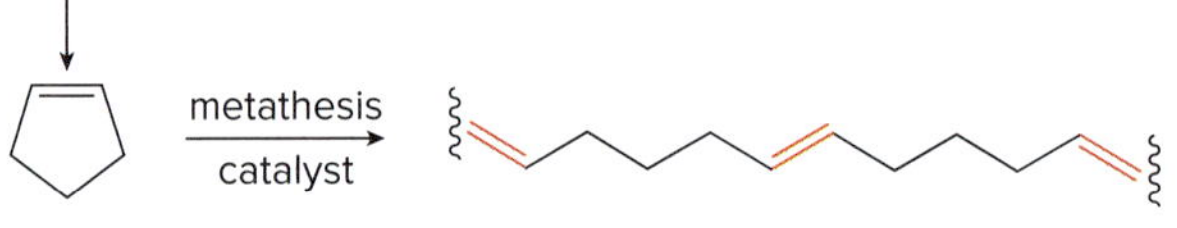

metathesis catalyst

new C=C's in red

What products are formed by ring-opening metathesis polymerization of each alkene?

a.  b. 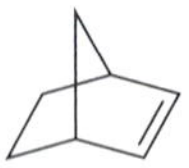 c.

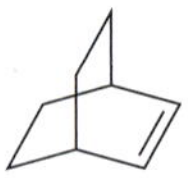

Problems That Combine Concepts

24.45 Draw the products formed in each reaction.

a. 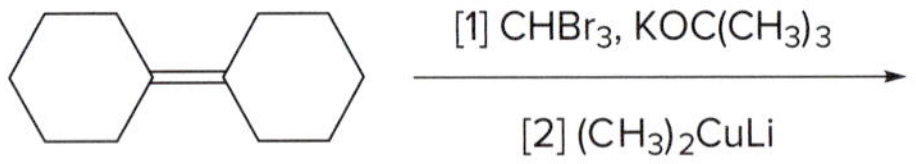 [1] $CHBr_3$, $KOC(CH_3)_3$ [2] $(CH_3)_2CuLi$

b. 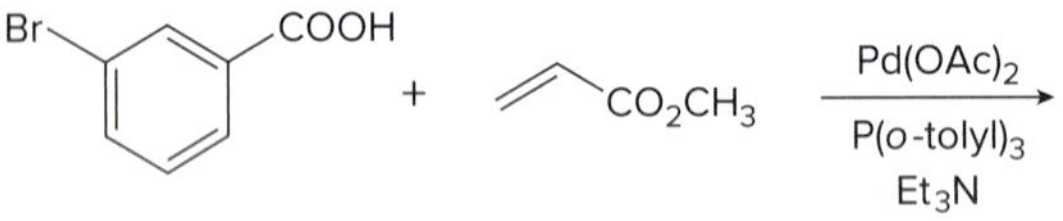 Br COOH + CO_2CH_3 $Pd(OAc)_2$ $P(o\text{-tolyl})_3$ Et_3N

c. 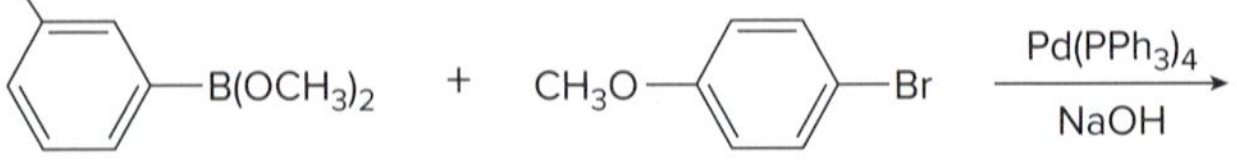 $B(OCH_3)_2$ + CH_3O Br $Pd(PPh_3)_4$ NaOH

d. 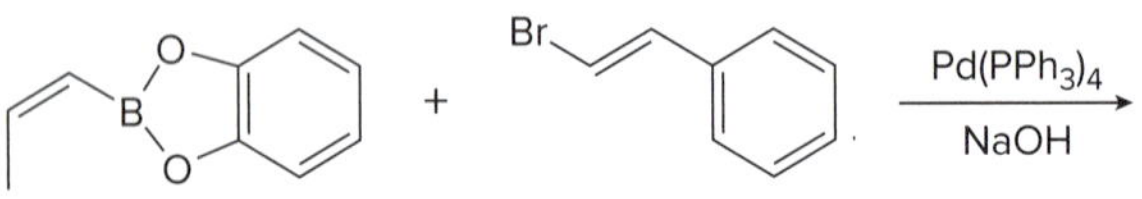 B O O + Br $Pd(PPh_3)_4$ NaOH

e. 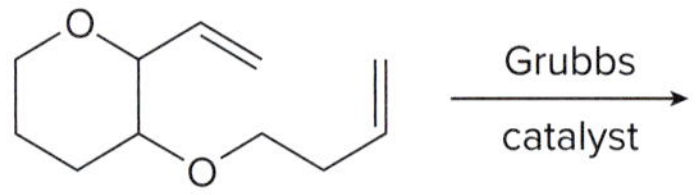 O O Grubbs catalyst

f. 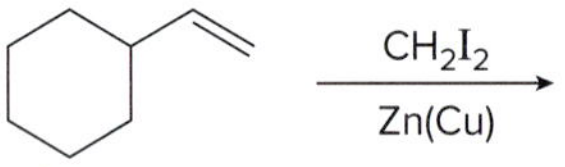 CH_2I_2 Zn(Cu)

g. 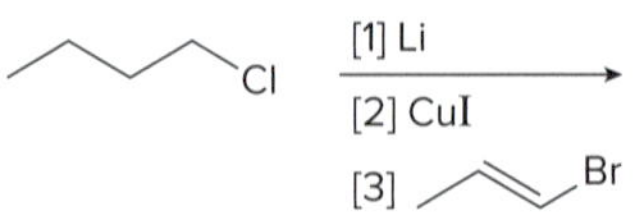 Cl [1] Li [2] CuI [3] Br

h. 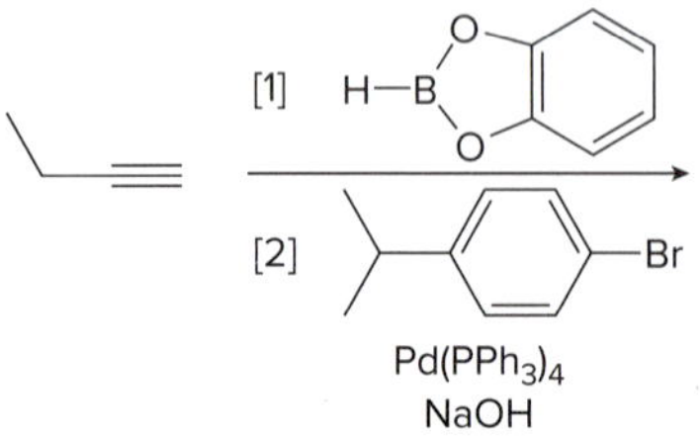 [1] H–B O O [2] Br $Pd(PPh_3)_4$ NaOH

24.46 Identify compounds **A–D**, intermediates in the synthesis of the antidepressant sertraline (Figure 15.5).

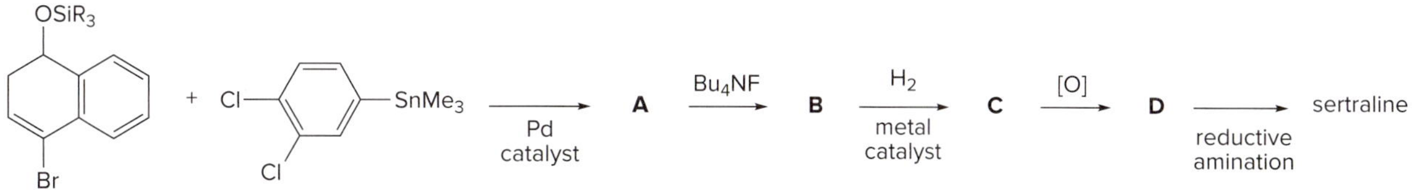

OSiR$_3$ Br + Cl Cl SnMe$_3$ $\xrightarrow{\text{Pd catalyst}}$ **A** $\xrightarrow{Bu_4NF}$ **B** $\xrightarrow[\text{metal catalyst}]{H_2}$ **C** $\xrightarrow{[O]}$ **D** $\xrightarrow{\text{reductive amination}}$ sertraline

24.47 Identify compounds **A–C** in the following reaction scheme.

H N O I + O B O $\xrightarrow[\text{NaOH}]{Pd(PPh_3)_4}$ **A** $\xrightarrow[\text{KOH}]{\text{Br}}$ **B** $\xrightarrow{\text{Grubbs catalyst}}$ **C** $C_{13}H_{15}NO$

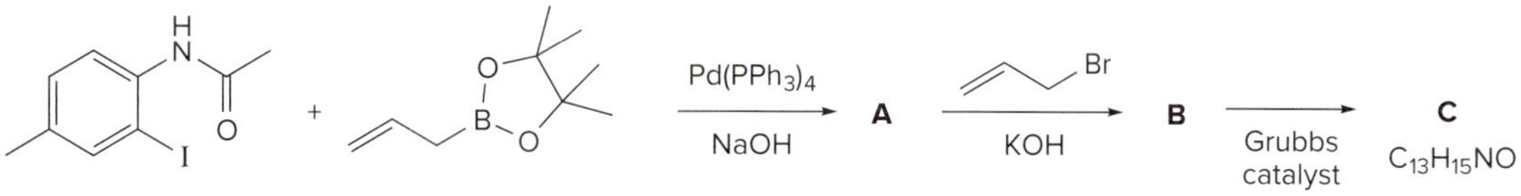

Mechanisms

24.48 In addition to using CHX_3 and base to synthesize dihalocarbenes (Section 24.5B), dichlorocarbene (:CCl_2) can be prepared by heating sodium trichloroacetate. Draw a stepwise mechanism for this reaction.

Cl Cl Cl O O^- Na^+ $\xrightarrow{\Delta}$ Cl Cl + CO_2 + NaCl

sodium trichloroacetate

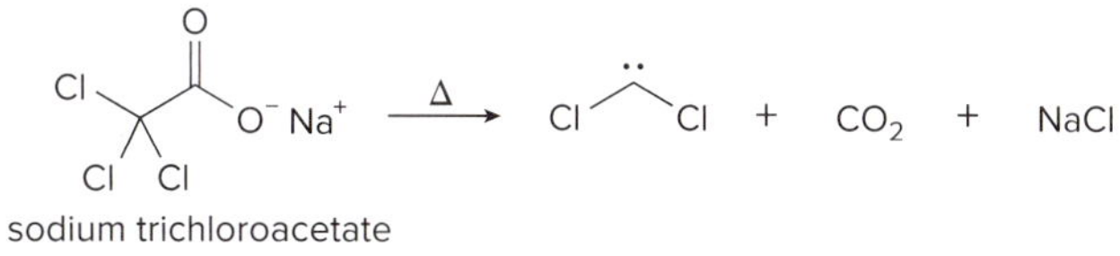

24.49 Draw a stepwise mechanism for the following reaction.

Br Br $\xrightarrow{Zn(Cu)}$ + $ZnBr_2$

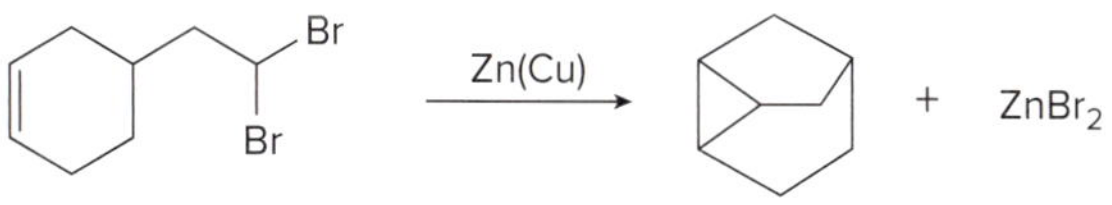

24.50 Identify **A** in the following reaction scheme, and draw a stepwise mechanism for the conversion of **A** to the furan **B**.

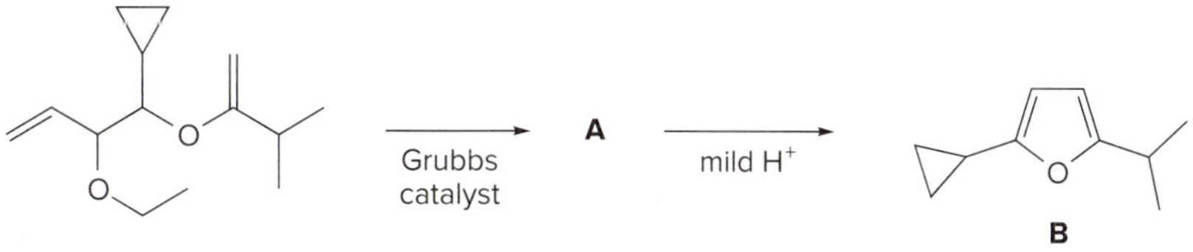

O O $\xrightarrow{\text{Grubbs catalyst}}$ **A** $\xrightarrow{\text{mild } H^+}$ O **B**

24.51 Sulfur ylides, like the phosphorus ylides of Chapter 18, are useful intermediates in organic synthesis. Methyl *trans*-chrysanthemate, an intermediate in the synthesis of the insecticide pyrethrin I (Section 24.5), can be prepared from diene **A** and a sulfur ylide. Draw a stepwise mechanism for this reaction.

O OCH_3 **A** + $Ph_2\overset{+}{S}$— a sulfur ylide $\longrightarrow$ OCH_3 O methyl *trans*-chrysanthemate $\longrightarrow$ O O O pyrethrin I

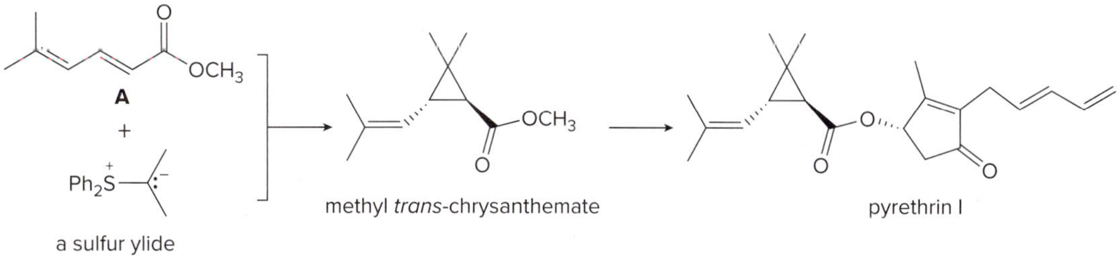

24.52 Although diazomethane (CH_2N_2) is often not a useful reagent for preparing cyclopropanes, other diazo compounds give good yields of more complex cyclopropanes. Draw a stepwise mechanism for the conversion of diazo compound **A** to **B,** an intermediate in the synthesis of sirenin, the sperm attractant produced by the female gametes of the water mold *Allomyces.*

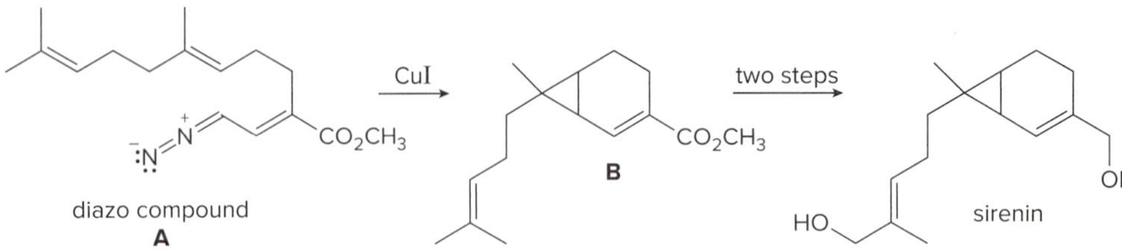

sperm attractant of the female water mold

24.53 The reaction of cyclohexene with iodobenzene under Heck conditions forms **E,** a coupling product with the new phenyl group on the allylic carbon, but none of the "expected" coupling product **F** with the phenyl group bonded directly to the carbon–carbon double bond.

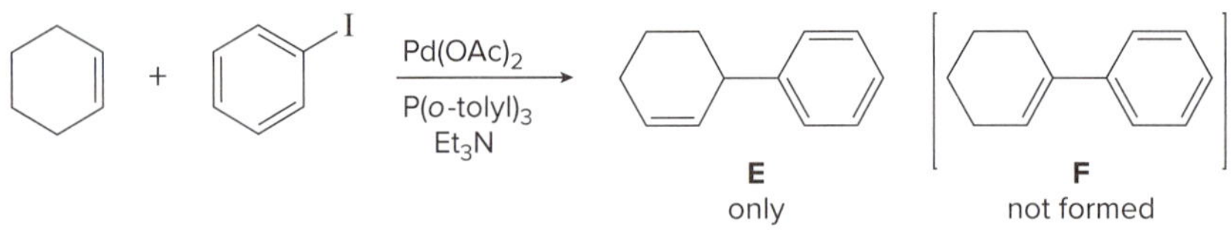

a. Draw a stepwise mechanism that illustrates how **E** is formed.

b. Step [2] in Mechanism 24.2 proceeds with syn addition of Pd and R' to the double bond. What does the formation of **E** suggest about the stereochemistry of the elimination reaction depicted in Step [3] of Mechanism 24.2?

Synthesis

24.54 Devise a synthesis of diene **A** from (*Z*)-2-bromostyrene as the only organic starting material. Use a Suzuki reaction in one step of the synthesis.

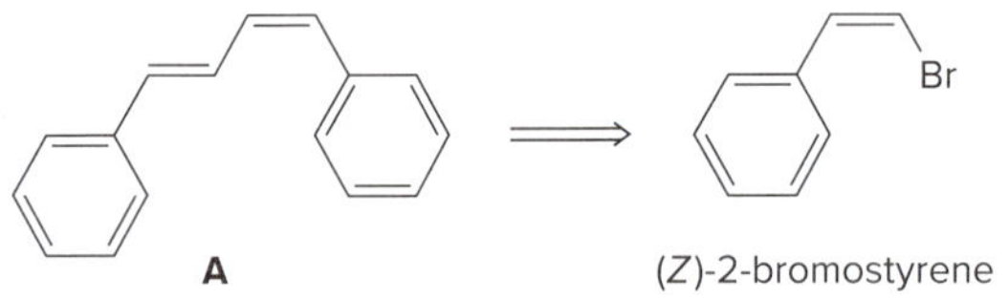

24.55 Devise a synthesis of the given trans vinylborane, which can be used for bombykol synthesis (Figure 24.1). All of the carbon atoms in the vinylborane must come from acetylene, nonane-1,9-diol, and catecholborane.

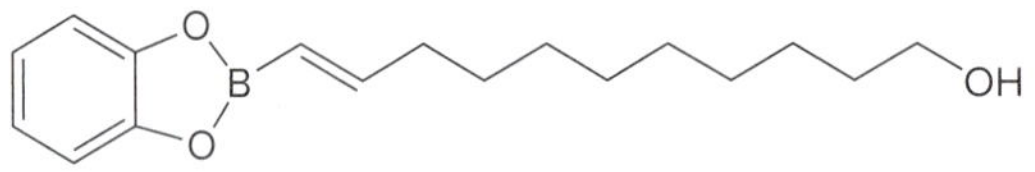

24.56 Devise a synthesis of each compound using a Heck reaction as one step. You may use benzene, $CH_2{=}CHCO_2Et$, organic alcohols having one or two carbons, and any required inorganic reagents.

a. 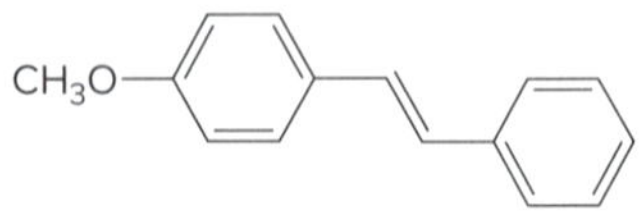

b. 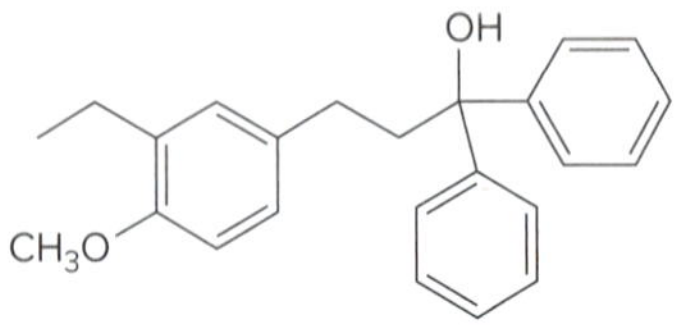

24.57 Devise a synthesis of each compound from cyclohexene and any required organic compounds or inorganic reagents.

a. 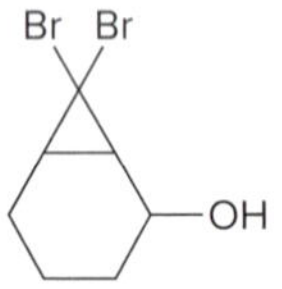

b. 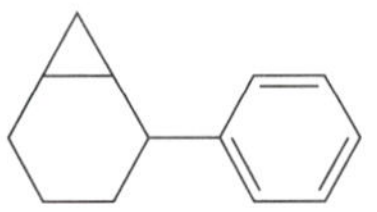

- **A sigmatropic rearrangement is a reaction in which a σ bond is broken in the reactant, the π bonds rearrange, and a σ bond is formed in the product.**

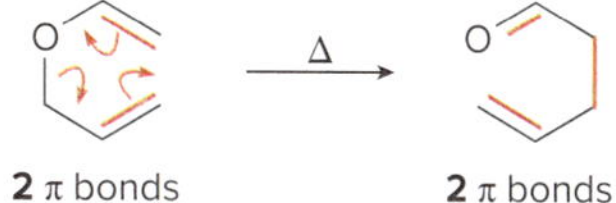

Hoffmann and Fukui received the 1981 Nobel Prize in Chemistry for developing theories that explain the course of pericyclic reactions.

Two features determine the course of the reactions: the **number of π bonds** involved and whether the reaction occurs in the presence of **heat** (thermal conditions) or **light** (photochemical conditions). These reactions follow a set of rules based on orbitals and symmetry first proposed by R. B. Woodward and Roald Hoffmann in 1965, and derived from theory described by Kenichi Fukui in 1954.

To understand pericyclic reactions, we must review and expand upon what we learned about the molecular orbitals of systems with π bonds in Chapter 15.

Problem 25.1 Classify each reaction as an electrocyclic reaction, a cycloaddition, or a sigmatropic rearrangement. Label the σ bonds that are broken or formed in each reaction.

a.

b. + $H_2C{=}CH_2$ ⟶

c.

d.

25.2 Molecular Orbitals

In Section 15.10, we learned that molecular orbital (MO) theory describes bonds as the mathematical combination of atomic orbitals that forms a new set of orbitals called **molecular orbitals (MOs). The number of atomic orbitals used *equals* the number of molecular orbitals formed.**

Because pericyclic reactions involve π bonds, let's examine the molecular orbitals that result from *p* orbital overlap in ethylene, buta-1,3-diene, and hexa-1,3,5-triene—molecules that contain one, two, and three π bonds, respectively. Keep in mind that the two lobes of a *p* orbital are opposite in phase, with a node of electron density at the nucleus.

25.2A Ethylene

The π bond in ethylene ($CH_2{=}CH_2$) is formed by side-by-side overlap of two *p* orbitals on adjacent carbons. Two *p* orbitals can combine in two different ways. As shown in Figure 25.1, when two *p* orbitals of similar phase overlap, a **π bonding molecular orbital** (designated as ψ_1) results.

Figure 25.1 The π and π* molecular orbitals of ethylene

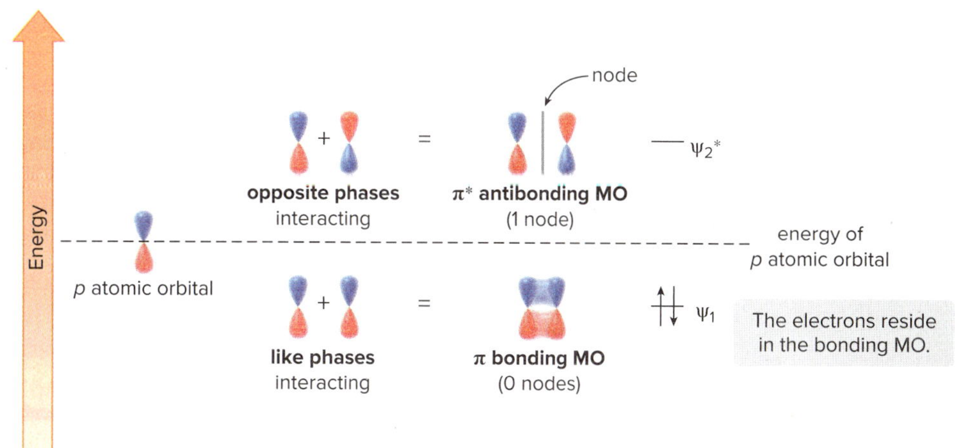

Two electrons occupy this lower-energy bonding molecular orbital. When two *p* orbitals of opposite phase combine, a **π* antibonding molecular orbital** (designated as $\psi_2{}^*$) results. A destabilizing node between the orbitals occurs when two orbitals of opposite phase combine.

25.2B Buta-1,3-diene

The two π bonds of buta-1,3-diene ($CH_2{=}CH{-}CH{=}CH_2$) are formed by overlap of four *p* orbitals on four adjacent carbons. As shown in Figure 25.2, four *p* orbitals can combine in four different ways to form four molecular orbitals designated as ψ_1–ψ_4. Two are bonding molecular orbitals (ψ_1 and ψ_2), and two are antibonding molecular orbitals ($\psi_3{}^*$ and $\psi_4{}^*$). The two bonding MOs are *lower* in energy than the *p* orbitals from which they are formed, whereas the two antibonding MOs are *higher* in energy than the *p* orbitals from which they are formed. **As the number of bonding interactions *decreases* and the number of nodes *increases*, the energy of the molecular orbital *increases*.**

Figure 25.2 The four π molecular orbitals of buta-1,3-diene

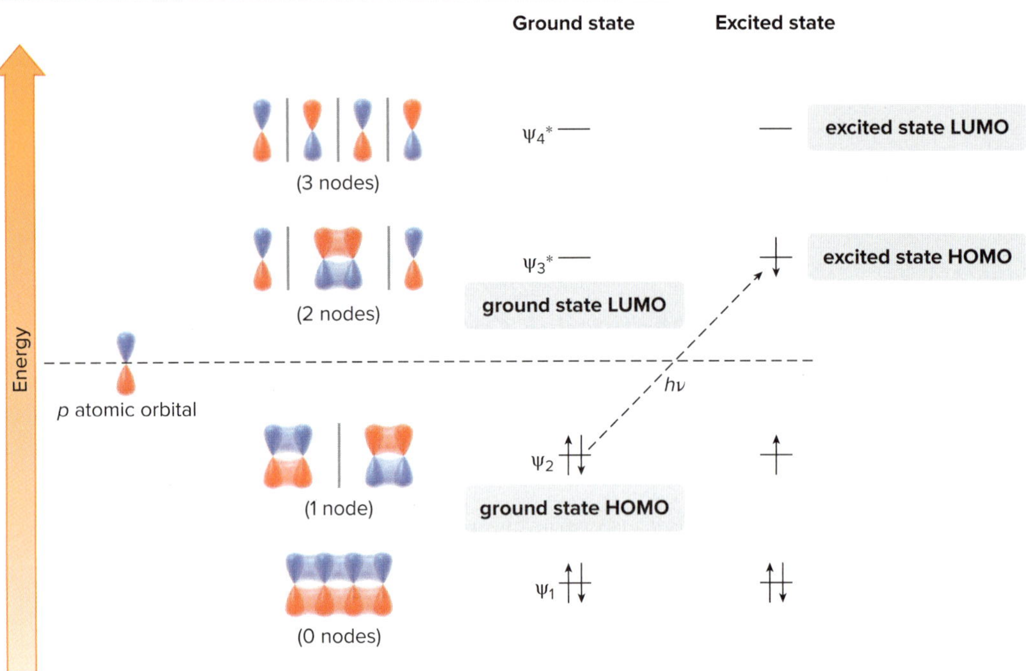

- The two lowest-energy molecular orbitals, ψ_1 and ψ_2, are **bonding MOs.**
- The two highest-energy molecular orbitals, $\psi_3{}^*$ and $\psi_4{}^*$, are **antibonding MOs.**

- **In the ground state electronic arrangement, the four π electrons occupy the two bonding molecular orbitals.**

Also recall from Section 15.10:

- The highest-energy orbital that contains electrons is called the highest occupied molecular orbital (HOMO). In the ground state of buta-1,3-diene, Ψ_2 is the HOMO.
- The lowest-energy orbital that contains no electrons is called the lowest unoccupied molecular orbital (LUMO). In the ground state of buta-1,3-diene, $\Psi_3{}^*$ is the LUMO.

The thermal reactions discussed in Section 25.3B utilize reactants in their ground state electronic configuration.

When buta-1,3-diene absorbs light of appropriate energy, an electron is promoted from ψ_2 (the HOMO) to $\psi_3{}^*$ (the LUMO) to form a higher-energy electronic configuration, the **excited state.**

In the excited state, the HOMO is now ψ_3*. **In the photochemical reactions in Section 25.3C, the reactant is in its excited state.** As a result, the HOMO is ψ_3* and the LUMO is ψ_4* for buta-1,3-diene.

All conjugated dienes can be described by a set of molecular orbitals that are similar to those drawn in Figure 25.2 for buta-1,3-diene.

Problem 25.2 For each molecular orbital in Figure 25.2, count the number of bonding interactions (interactions between adjacent orbitals of similar phase) and the number of nodes. (a) How do these two values compare for a bonding molecular orbital? (b) How do these two values compare for an antibonding molecular orbital?

25.2C Hexa-1,3,5-triene

The three π bonds of hexa-1,3,5-triene (CH_2=CH–CH=CH–CH=CH_2) are formed by overlap of six *p* orbitals on six adjacent carbons. As shown in Figure 25.3, six *p* orbitals can combine in six different ways to form six molecular orbitals designated as ψ_1–ψ_6. Three are bonding molecular orbitals (ψ_1–ψ_3), and three are antibonding molecular orbitals (ψ_4*–ψ_6*).

Figure 25.3 The six π molecular orbitals of hexa-1,3,5-triene

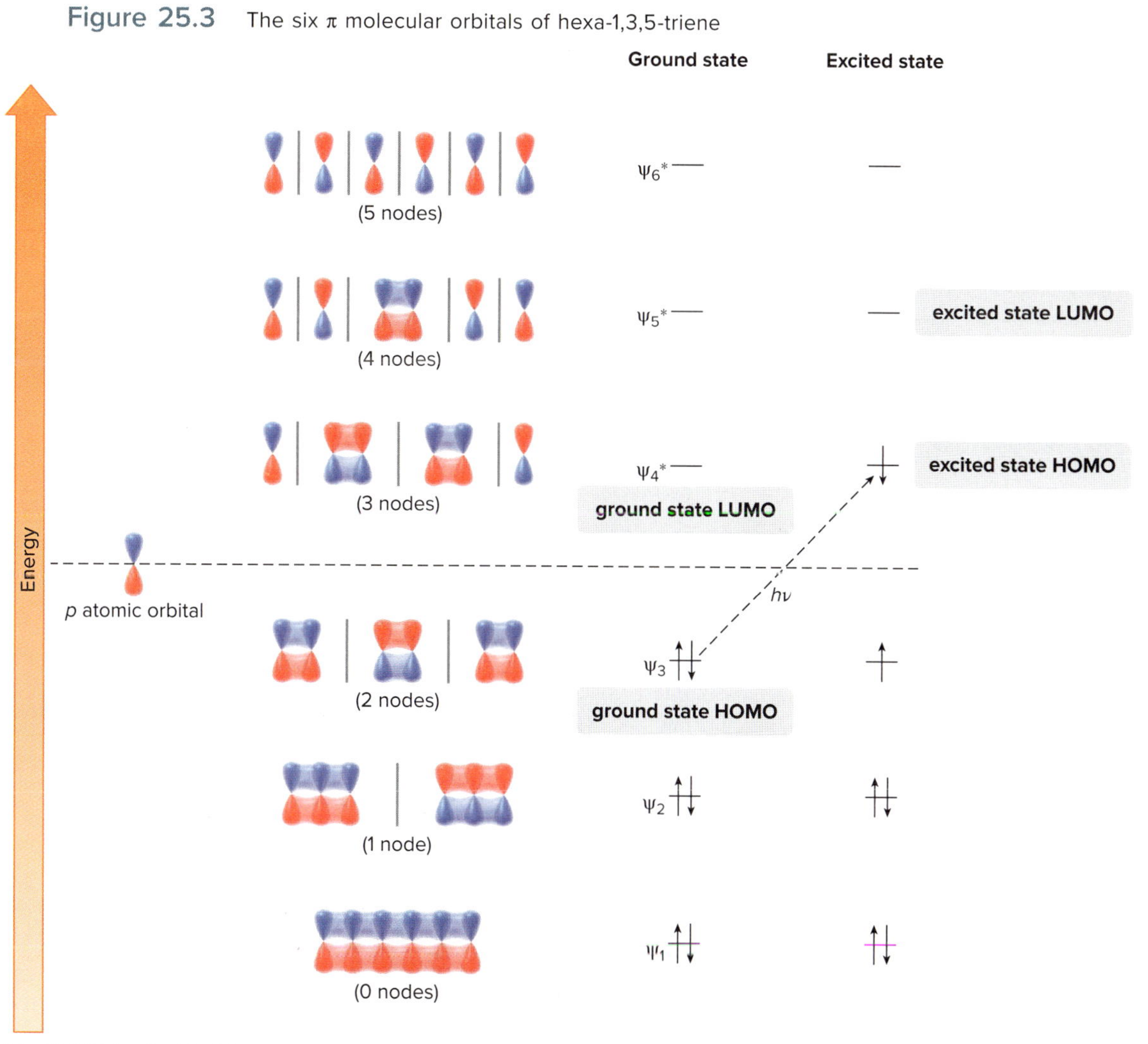

In the ground state electronic configuration, the six π electrons occupy the three bonding MOs, ψ_3 is the HOMO, and ψ_4* is the LUMO. In the excited state, which results from promotion of an electron from ψ_3 to ψ_4*, ψ_4* is the HOMO and ψ_5* is the LUMO.

Problem 25.3 (a) Using Figure 25.2 as a guide, draw the molecular orbitals for hexa-2,4-diene. (b) Label the HOMO and the LUMO in the ground state. (c) Label the HOMO and the LUMO in the excited state.

25.3 Electrocyclic Reactions

An electrocyclic reaction is a reversible reaction that involves ring closure of a conjugated polyene to a cycloalkene, or ring opening of a cycloalkene to a conjugated polyene. For example, ring closure of hexa-1,3,5-triene forms cyclohexa-1,3-diene, a product with one more σ bond and one fewer π bond than the reactant. Ring opening of cyclobutene forms buta-1,3-diene, a product with one fewer σ bond and one more π bond than the reactant.

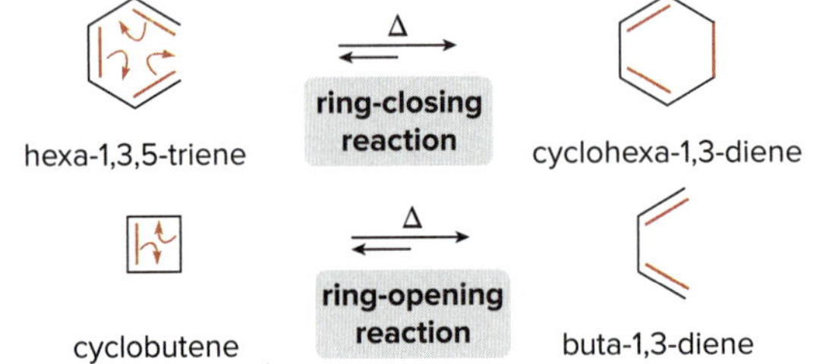

Arrows may be drawn in a clockwise or counterclockwise direction to show the flow of electrons.

- **To draw the product in each reaction, use curved arrows and *begin at a π bond*. Move the π electrons to an adjacent carbon–carbon bond and continue in a cyclic fashion.**

In a ring-closing reaction, this process forms a new σ bond that now joins the ends of the conjugated polyene. In a ring-opening reaction, this process breaks a σ bond to form a conjugated polyene with one more π bond.

Whether the reactant or product predominates at equilibrium depends on the ring size of the cyclic compound. Generally, a six-membered ring is favored over an acyclic triene at equilibrium. In contrast, an acyclic diene is favored over a strained four-membered ring.

Problem 25.4 Use curved arrows and draw the product of each electrocyclic reaction.

a. $\xrightarrow{\Delta}$ b. $\xrightarrow{\Delta}$ c. $\xrightarrow{\Delta}$

25.3A Stereochemistry and Orbital Symmetry

Electrocyclic reactions are completely stereospecific. For example, ring closure of (2*E*,4*Z*,6*E*)-octa-2,4,6-triene yields a single product with cis methyl groups on the ring. Ring opening of *cis*-3,4-dimethylcyclobutene forms a single conjugated diene with one *Z* alkene and one *E* alkene.

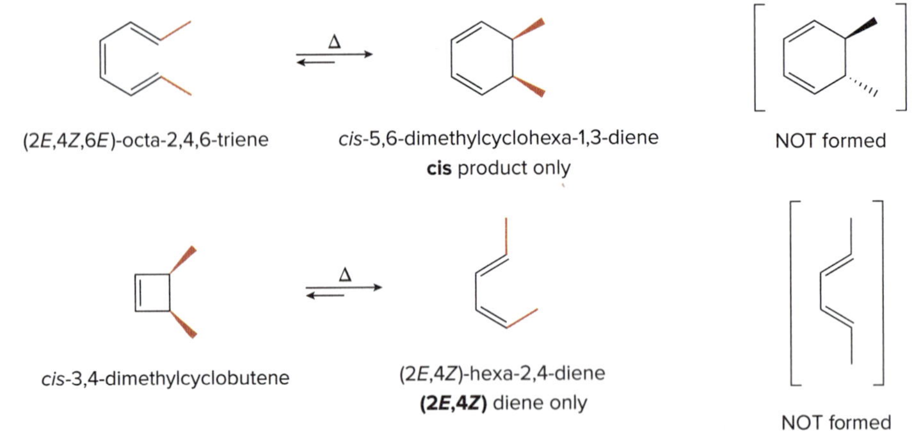

Moreover, the **stereochemistry of the product of an electrocyclic reaction depends on whether the reaction is carried out under thermal or photochemical reaction conditions**—that is, with heat or light, respectively. Cyclization of (2*E*,4*E*)-hexa-2,4-diene with heat forms a cyclobutene with trans methyl groups, whereas cyclization with light forms a cyclobutene with cis methyl groups.

Electrocyclic ring closure generally forms either an achiral meso compound or a mixture of chiral enantiomers. When enantiomers form, only one enantiomer is drawn in these reactions.

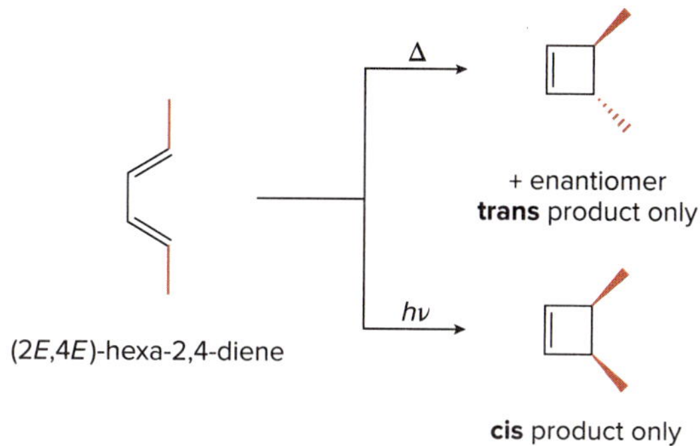

To understand these results, we must focus on the **HOMO of the acyclic conjugated polyene that is either the reactant or product** in an electrocyclic reaction. In particular, we must examine the *p* orbitals on the terminal carbons of the HOMO, and determine whether like phases of the orbitals are on the *same* side or on *opposite* sides of the molecule.

- An electrocyclic reaction occurs only when like phases of orbitals can overlap to form a bond. Such a reaction is *symmetry allowed.*
- An electrocyclic reaction can*not* occur between lobes of opposite phase. Such a reaction is *symmetry forbidden.*

To form a bond, the *p* orbitals on the terminal carbons must rotate so that like phases can interact to form the new σ bond. Two modes of rotation are possible.

- When like phases of the *p* orbitals are on the same side of the molecule, the two orbitals must rotate in *opposite* directions—one clockwise and one counterclockwise. Rotation in opposite directions is said to be *disrotatory.*

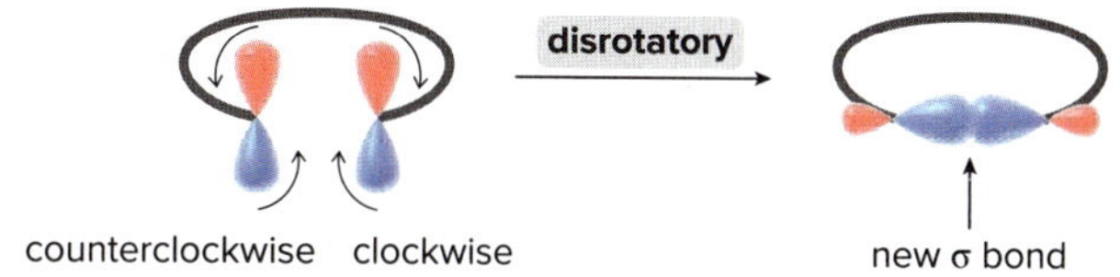

- When like phases of the *p* orbitals are on opposite sides of the molecule, the two orbitals must rotate in the *same* direction—both clockwise or both counterclockwise. Rotation in the same direction is said to be *conrotatory.*

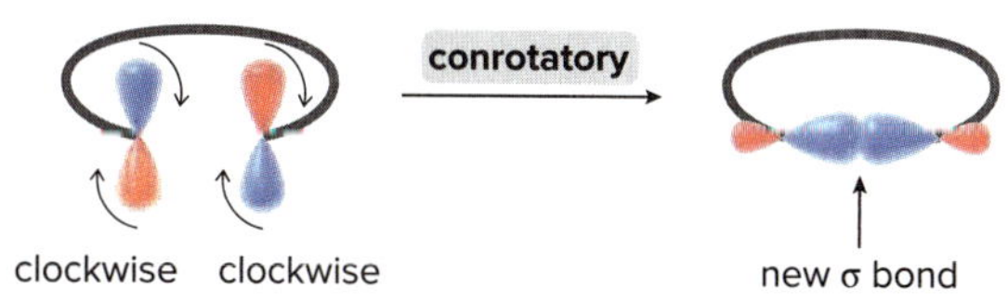

25.3B Thermal Electrocyclic Reactions

To explain the stereochemistry observed in electrocyclic reactions, we must examine the symmetry of the molecular orbital that contains the most loosely held π electrons. **In a thermal reaction, we consider the HOMO of the ground state electronic configuration.** Rotation occurs in a disrotatory or conrotatory fashion so that like phases of the *p* orbitals on the terminal carbons of this molecular orbital combine.

- **The number of double bonds in the conjugated polyene determines whether rotation is conrotatory or disrotatory.**

Two examples illustrate different outcomes.

Thermal electrocyclic ring closure of (2*E*,4*Z*,6*E*)-octa-2,4,6-triene yields a single product with cis methyl groups on the ring.

Only the *p* orbitals on the terminal carbons of the HOMO are drawn for clarity.

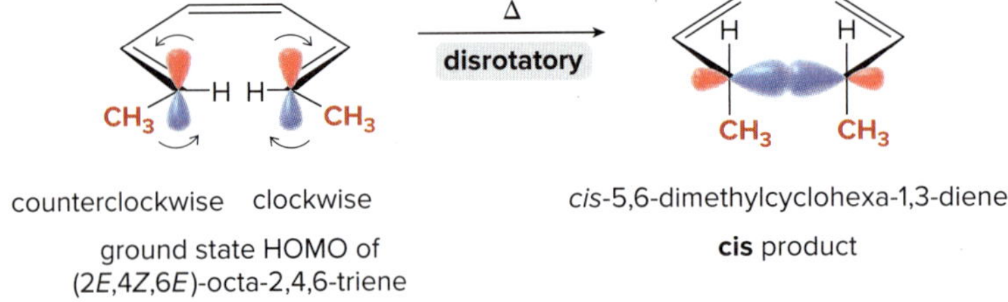

Cyclization occurs in a disrotatory fashion because the HOMO of a conjugated triene has like phases of the outermost *p* orbitals on the *same* side of the molecule (Figure 25.3). A disrotatory ring closure is symmetry allowed because like phases of the *p* orbitals overlap to form the new σ bond of the ring. In the disrotatory ring closure, both methyl groups are pushed *down* (or *up*), making them *cis* in the product.

This is a specific example of the general process observed for conjugated polyenes with an *odd* number of π bonds. The HOMO of a conjugated polyene with an odd number of π bonds has like phases of the outermost *p* orbitals on the *same* side of the molecule. As a result:

- **Thermal electrocyclic reactions occur in a *disrotatory* fashion for a conjugated polyene with an *odd* number of π bonds.**

In contrast, thermal electrocyclic ring closure of (2*E*,4*E*)-hexa-2,4-diene forms a cyclobutene with trans methyl groups.

The conrotatory ring closure of (2*E*,4*E*)-hexa-2,4-diene is drawn with two clockwise rotations. The conrotatory ring closure could also be drawn with two counterclockwise rotations, leading to the enantiomer of the trans product drawn. Both enantiomers are formed in equal amounts.

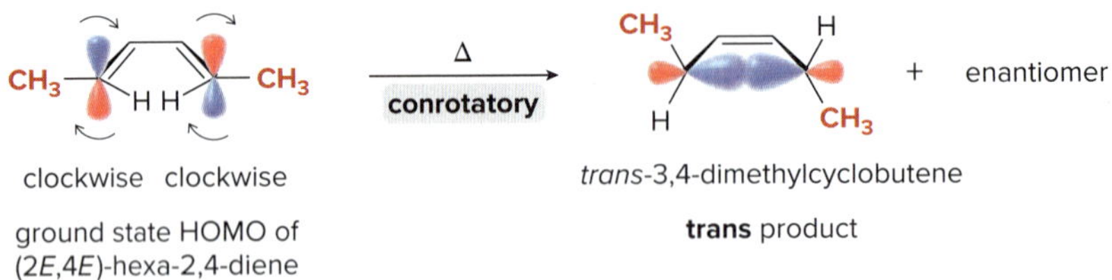

Cyclization occurs in a conrotatory fashion because the HOMO of a conjugated diene has like phases of the outermost *p* orbitals on *opposite* sides of the molecule (Figure 25.2). A conrotatory ring closure is symmetry allowed because like phases of the *p* orbitals overlap to form the new σ bond of the ring. In the conrotatory ring closure, one methyl group is pushed *down* and one methyl group is pushed *up*, making them *trans* in the product.

This is a specific example of the general process observed for conjugated polyenes with an *even* number of π bonds. The HOMO of a conjugated polyene with an even number of π bonds has like phases of the outermost *p* orbitals on *opposite* sides of the molecule. As a result:

- **Thermal electrocyclic reactions occur in a *conrotatory* fashion for a conjugated polyene with an *even* number of π bonds.**

Because electrocyclic reactions are reversible, **electrocyclic ring-opening reactions follow the same rules** as electrocyclic ring closures. Thus, thermal ring opening of *cis*-3,4-dimethylcyclobutene—which ring opens to a diene with an *even* number of π bonds—occurs in a *conrotatory* fashion to form (2*E*,4*Z*)-hexa-2,4-diene as the only product.

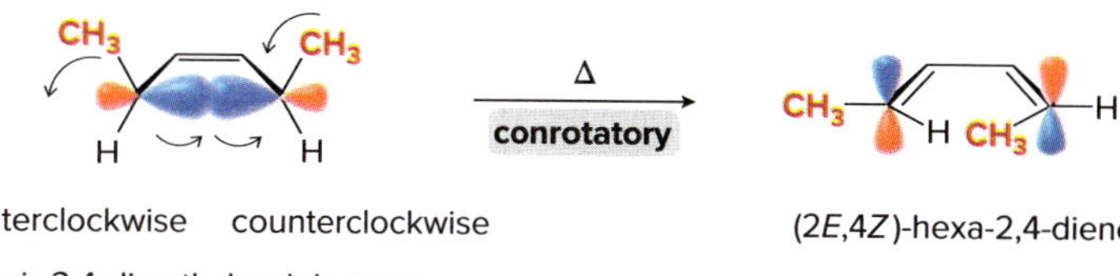

counterclockwise counterclockwise

cis-3,4-dimethylcyclobutene

(2*E*,4*Z*)-hexa-2,4-diene

Sample Problem 25.1 Drawing the Product of a Thermal Electrocyclic Ring Closure

Draw the product of each thermal electrocyclic ring closure.

a. (2*E*,4*Z*,6*Z*)-octa-2,4,6-triene $\xrightarrow{\Delta}$

b. CH_3O_2C–CH=CH–CH=CH–CO_2CH_3 (**B**) $\xrightarrow{\Delta}$

Solution

Count the number of π bonds in the conjugated polyene to determine the mode of ring closure in a thermal electrocyclic reaction.

- A conjugated polyene with an **odd** number of π bonds undergoes **disrotatory** cyclization.
- A conjugated polyene with an **even** number of π bonds undergoes **conrotatory** cyclization.

a. (2*E*,4*Z*,6*Z*)-Octa-2,4,6-triene contains three π bonds. The HOMO of a conjugated polyene with an *odd* number of π bonds has like phases of the outermost *p* orbitals on the *same* side of the molecule, and this results in *disrotatory* cyclization.

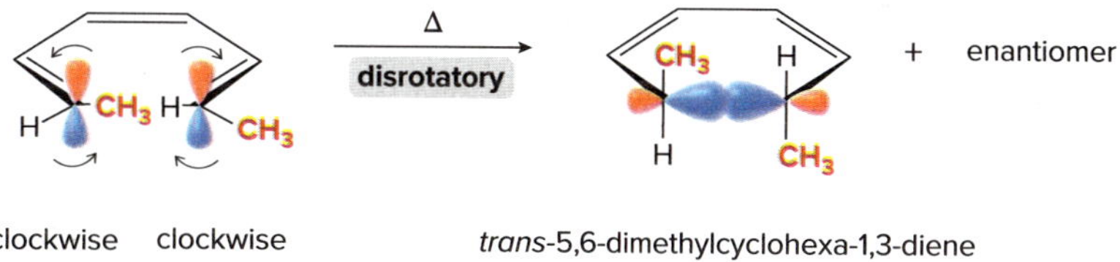

counterclockwise clockwise

(2*E*,4*Z*,6*Z*)-octa-2,4,6-triene

trans-5,6-dimethylcyclohexa-1,3-diene

trans product

b. Diene **B** contains two π bonds. The HOMO of a conjugated polyene with an *even* number of π bonds has like phases of the outermost *p* orbitals on *opposite* sides of the molecule, and this results in *conrotatory* cyclization.

CH_3O_2C H H CO_2CH_3 $\xrightarrow[\text{conrotatory}]{\Delta}$ CH_3O_2C H H CO_2CH_3 + enantiomer

clockwise clockwise

B

trans product

Problem 25.5 What product is formed when each compound undergoes thermal electrocyclic ring opening or ring closure? Label each process as conrotatory or disrotatory, and clearly indicate the stereochemistry around tetrahedral stereogenic centers and double bonds.

a. C_6H_5 C_6H_5

b.

More Practice: Try Problems 25.28a; 25.31; 25.34a, c; 25.37a; 25.54b.

Problem 25.6 What cyclic product is formed when each decatetraene undergoes thermal electrocyclic ring closure?

25.3C Photochemical Electrocyclic Reactions

Photochemical electrocyclic reactions follow similar principles as those detailed in thermal reactions with one important difference: **In photochemical reactions, we must consider the orbitals of the HOMO of the *excited* state to determine the course of the reaction.** As a photon is absorbed, an electron in the ground state HOMO is excited to the ground state LUMO. As a result, the excited state HOMO is one energy level higher than before (see Figures 25.2 and 25.3). The excited state HOMO has the *opposite* orientation of the outermost *p* orbitals compared to the HOMO of the ground state. As a result, **the method of ring closure of a photochemical electrocyclic reaction is *opposite* to that of a thermal electrocyclic reaction for the same number of π bonds.**

Photochemical electrocyclic ring closure of (2*E*,4*Z*,6*E*)-octa-2,4,6-triene yields a cyclic product with trans methyl groups on the ring.

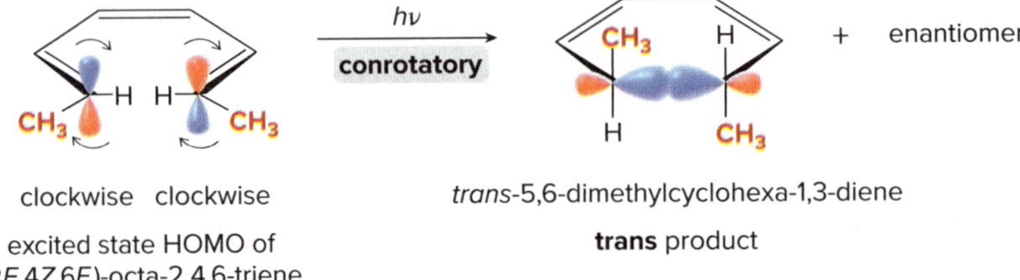

Cyclization occurs in a conrotatory fashion because the excited state HOMO of a conjugated triene has like phases of the outermost *p* orbitals on the *opposite* sides of the molecule (Figure 25.3). In the conrotatory ring closure, one methyl group is pushed *down* and one methyl group is pushed *up,* making them *trans* in the product. This is a specific example of the general process observed for conjugated polyenes with an *odd* number of π bonds.

- **Photochemical electrocyclic reactions occur in a *conrotatory* fashion for a conjugated polyene with an *odd* number of π bonds.**

Photochemical electrocyclic ring closure of (2*E*,4*E*)-hexa-2,4-diene forms a cyclobutene with cis methyl groups.

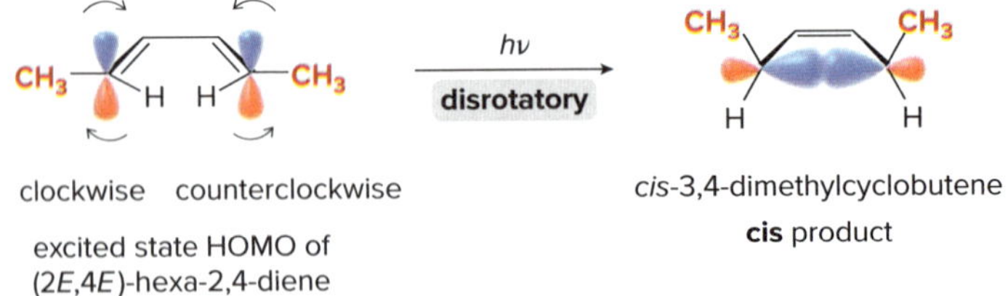

Cyclization occurs in a disrotatory fashion because the excited state HOMO of a conjugated diene has like phases of the outermost *p* orbitals on the *same* side of the molecule (Figure 25.3). In the disrotatory ring closure, both methyl groups are pushed *down* (or *up*), making them *cis* in the product. This is a specific example of the general process observed for conjugated polyenes with an *even* number of π bonds.

- **Photochemical electrocyclic reactions occur in a *disrotatory* fashion for a conjugated polyene with an *even* number of π bonds.**

Problem 25.7 What product is formed when each compound in Problem 25.5 undergoes photochemical electrocyclic ring opening or ring closure? Label each process as conrotatory or disrotatory and clearly indicate the stereochemistry around tetrahedral stereogenic centers and double bonds.

Milk sold in the United States is fortified with vitamin D (Problem 25.8), a compound that helps to regulate calcium and phosphorus metabolism.
Mary Reeg/McGraw Hill

Problem 25.8 Vitamin D_3, the most abundant of the D vitamins, is synthesized from 7-dehydrocholesterol, a compound found in milk and fatty fish such as salmon and mackerel. When the skin is exposed to sunlight, a photochemical electrocyclic ring opening forms provitamin D_3, which is then converted to vitamin D_3 by a sigmatropic rearrangement (Section 25.5). Draw the structure of provitamin D_3.

7-dehydrocholesterol $\xrightarrow{h\nu}$ provitamin D_3 $\longrightarrow$ vitamin D_3

25.3D Summary of Electrocyclic Reactions

Table 25.1 summarizes the rules, often called the **Woodward–Hoffmann rules,** for electrocyclic reactions under thermal or photochemical reaction conditions. The number of π bonds refers to the acyclic conjugated polyene that is either the reactant or product of an electrocyclic reaction.

Table 25.1 Woodward–Hoffmann Rules for Electrocyclic Reactions

Number of π bonds	Thermal reaction	Photochemical reaction
Even	Conrotatory	Disrotatory
Odd	Disrotatory	Conrotatory

Sample Problem 25.2 Determining the Product of an Electrocyclic Reaction

Identify **A** and **B** in the following reaction sequence. Label each process as conrotatory or disrotatory.

$\xrightarrow{\Delta}$ **A** $\xrightarrow{h\nu}$ **B**

Solution

Ring opening of a cyclohexadiene forms a hexatriene with **three** π bonds. A conjugated polyene with an odd number of π bonds undergoes a thermal electrocyclic reaction in a disrotatory fashion (Table 25.1). The resulting hexatriene (**A**) then undergoes a photochemical electrocyclic reaction in a conrotatory fashion to form a cyclohexadiene with cis methyl groups (**B**).

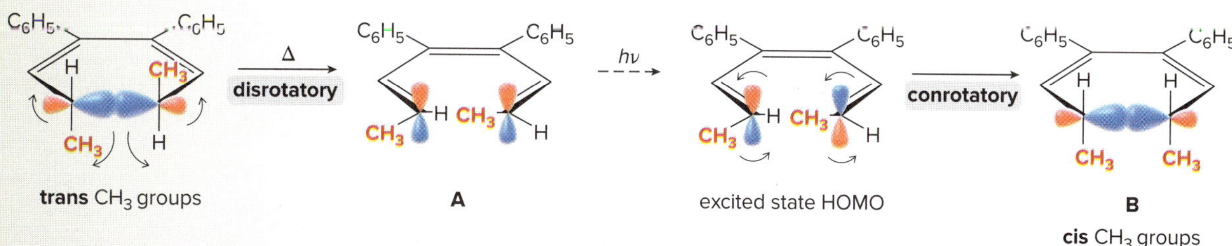

Problem 25.9 Draw the product formed when each triene undergoes electrocyclic reaction under [1] thermal conditions; [2] photochemical conditions.

a. b.

More Practice: Try Problems 25.28; 25.31–25.38; 25.54b, d.

Problem 25.10 What product would be formed by the disrotatory cyclization of the given triene? Would this reaction occur under photochemical or thermal conditions?

25.4 Cycloaddition Reactions

A cycloaddition is a reaction between two compounds with π bonds to form a cyclic product with two new σ bonds. Like electrocyclic reactions, cycloadditions are concerted, stereospecific reactions, and the course of the reaction is determined by the symmetry of the molecular orbitals of the reactants.

Cycloadditions can be initiated by heat (thermal conditions) or light (photochemical conditions). **Cycloadditions are identified by the number of π electrons in the two reactants.**

The Diels–Alder reaction is a thermal [4 + 2] cycloaddition that occurs between a diene with four π electrons and an alkene (dienophile) with two π electrons (Sections 14.12–14.14).

CO_2CH_3 CO_2CH_3 Δ [4 + 2] cycloaddition CO_2CH_3 CO_2CH_3

diene **4 π** electrons **dienophile** **2 π** electrons **Diels–Alder product** new σ bonds shown in red

A photochemical [2 + 2] cycloaddition occurs between two alkenes, each with two π electrons, to form a cyclobutane. Thermal [2 + 2] cycloadditions do *not* take place.

CH_2=CH_2 + maleic anhydride $\xrightarrow{h\nu}$ [2 + 2] cycloaddition

2 π electrons **2 π** electrons new σ bonds shown in red

Sample Problem 25.3 Classifying a Cycloaddition

What type of cycloaddition is shown in each equation?

a. cyclopentadiene + CH_2=CH_2 ⟶ b. cyclopentadiene + CH_2=CH_2 ⟶

Solution

Count the number of π electrons *involved in each reactant* to classify the cycloaddition.

a. **[2 + 2]** Cycloaddition b. **[4 + 2]** Cycloaddition

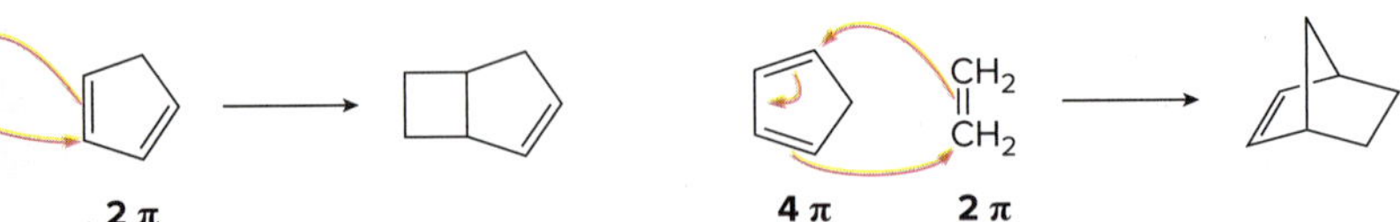

Problem 25.11 Consider cycloheptatrienone and ethylene, and draw a possible product formed from each type of cycloaddition: (a) [2 + 2]; (b) [4 + 2]; (c) [6 + 2].

=O + $CH_2{=}CH_2$

cycloheptatrienone

More Practice: Try Problem 25.39.

25.4A Orbital Symmetry and Cycloadditions

To understand cycloaddition reactions, we examine the *p* orbitals of the terminal carbons of both reactants. Bonding can take place only when like phases of both sets of *p* orbitals can combine. Two modes of reaction are possible.

- **A suprafacial cycloaddition occurs when like phases of the *p* orbitals of both reactants are on the *same* side of the π system, so that two bonding interactions result.**

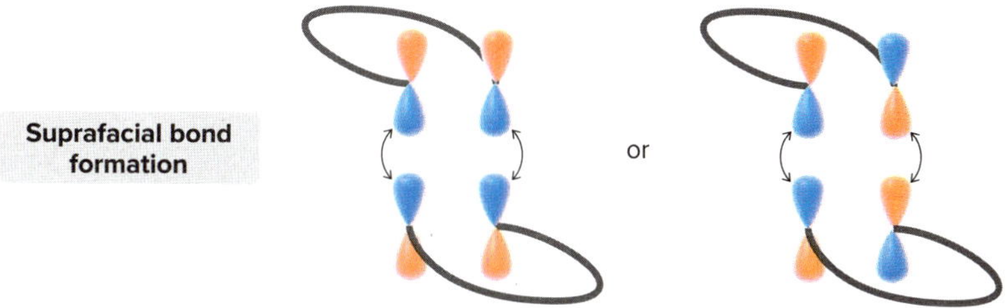

- **An antarafacial cycloaddition occurs when one π system must *twist* to align like phases of the *p* orbitals of the terminal carbons of the reactants.**

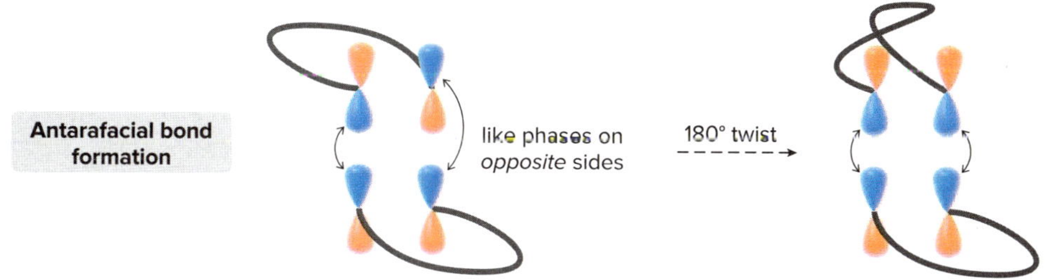

Because of the geometrical constraints of small rings, **cycloadditions that form four- or six-membered rings must take place by suprafacial pathways.**

Because cycloaddition involves the donation of electron density from one reactant to another, one reactant donates its most loosely held electrons—those occupying its **HOMO**—to a vacant orbital that can accept electrons—the **LUMO**—of the second reactant. The HOMO of either reactant can be used for analysis.

- **In a cycloaddition, we examine the bonding interactions of the HOMO of one component with the LUMO of the second component.**

25.4B [4 + 2] Cycloadditions

To examine the course of a [4 + 2] cycloaddition, let's arbitrarily choose the HOMO of the diene and the LUMO of the alkene, and **look at the symmetry of the *p* orbitals on the terminal carbons of both components.** Because two bonding interactions result from overlap of the like phases of both sets of p orbitals, **a [4 + 2] cycloaddition occurs readily by suprafacial reaction under thermal conditions.**

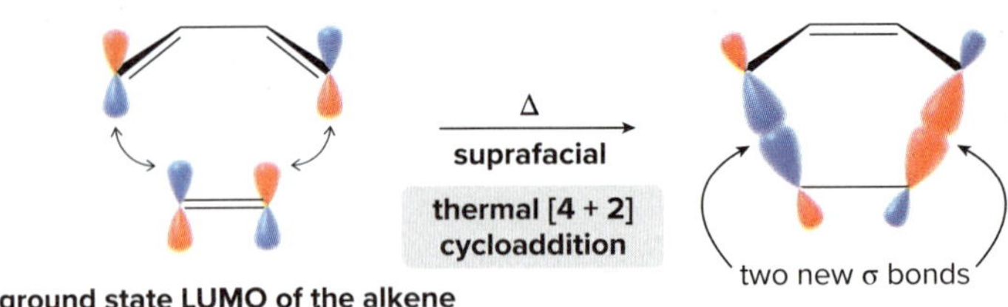

This is a specific example of a general cycloaddition involving an *odd* number of π bonds (three π bonds total, two from the diene and one from the alkene).

- **Thermal cycloadditions involving an *odd* number of π bonds proceed by a *suprafacial* pathway.**

In Section 14.13, we learned that the stereochemistry of the dienophile is retained in the Diels–Alder product.

Because a Diels–Alder reaction follows a concerted, suprafacial pathway, the **stereochemistry of the diene is retained in the Diels–Alder product.** As a result, reaction of (2*E*,4*E*)-hexa-2,4-diene with ethylene forms a cyclohexene with cis substituents (Reaction [1]), whereas reaction of (2*E*,4*Z*)-hexa-2,4-diene with ethylene forms a cyclohexene with trans substituents (Reaction [2]).

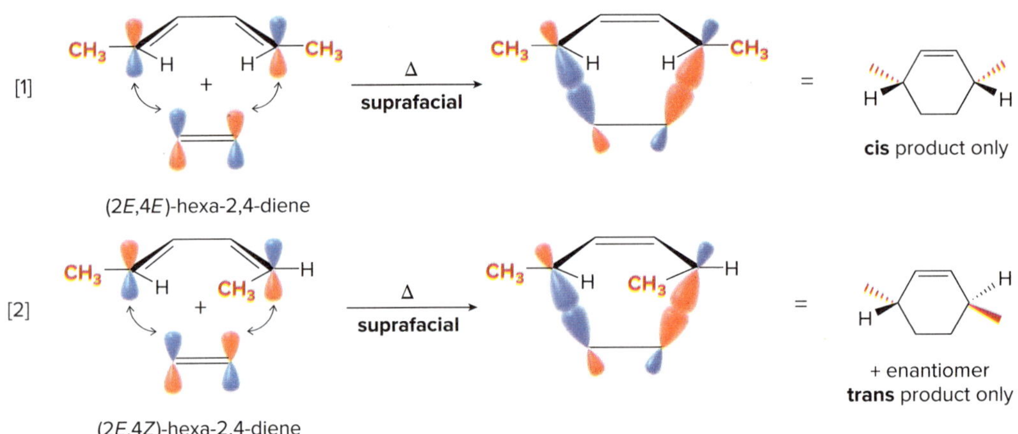

Problem 25.12 Show that a thermal suprafacial addition is symmetry allowed in a [4 + 2] cycloaddition by using the HOMO of the alkene and the LUMO of the diene.

Problem 25.13 Draw the product (including stereochemistry) formed from each pair of reactants in a thermal [4 + 2] cycloaddition reaction.

a. + $CH_2{=}CH_2$ b. + (CN)(CN)

25.4C [2 + 2] Cycloadditions

In contrast to a [4 + 2] cycloaddition, a [2 + 2] cycloaddition does *not* occur under thermal conditions, but *does* take place photochemically. This result is explained by examining the symmetry of the HOMO and LUMO of the alkene reactants.

In a thermal [2 + 2] cycloaddition, like phases of the p orbitals on only one set of terminal carbons can overlap. For like phases to overlap on the other terminal carbon, the molecule must twist to allow for an antarafacial pathway. This process *cannot* occur to form small rings.

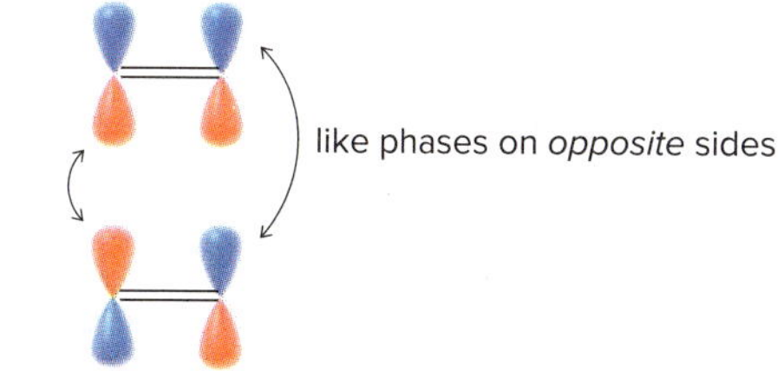

In a photochemical [2 + 2] cycloaddition, light energy promotes an electron from the ground state HOMO to form the excited state HOMO (designated as ψ_2^* in Figure 25.1). Interaction of this excited state HOMO with the LUMO of the second alkene then allows for overlap of the like phases of both sets of *p* orbitals. Two bonding interactions result and the reaction occurs by a suprafacial pathway.

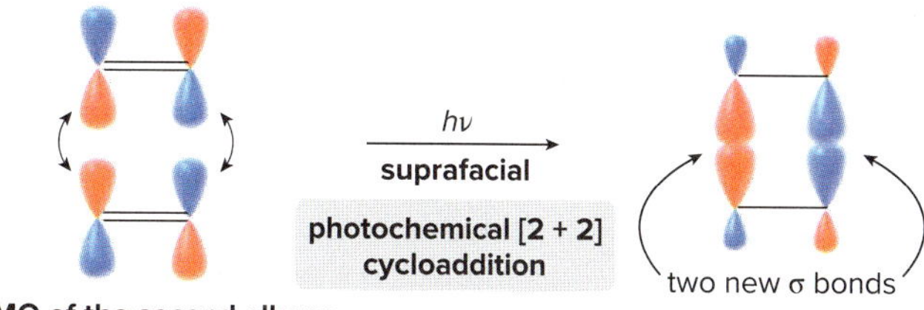

This is a specific example of a general cycloaddition involving an *even* number of π bonds (two π bonds total, one from each alkene).

- **Photochemical cycloadditions involving an *even* number of π bonds proceed by a *suprafacial* pathway.**

Problem 25.14 Draw the product formed in each cycloaddition.

a. $C_6H_5CH{=}CHC_6H_5$ + $(CH_3)_2C{=}C(CH_3)_2$ $\xrightarrow{h\nu}$ b. 3-cyano-2-cyclohexenone (O, CN) + $CH_2{=}CH_2$ $\xrightarrow{h\nu}$

25.4D Summary of Cycloaddition Reactions

Table 25.2 summarizes the Woodward–Hoffmann rules that govern cycloaddition reactions. The number of π bonds refers to the total number of π bonds from both components of the cycloaddition. For a given number of π bonds, the mode of cycloaddition is always *opposite* in thermal and photochemical reactions.

Table 25.2 Woodward–Hoffmann Rules for Cycloaddition Reactions

Number of π bonds	Thermal reaction	Photochemical reaction
Even	Antarafacial	Suprafacial
Odd	Suprafacial	Antarafacial

Problem 25.15 Using the Woodward–Hoffmann rules, predict the stereochemical pathway for each cycloaddition: (a) a [6 + 4] photochemical reaction; (b) an [8 + 2] thermal reaction.

Problem 25.16 Using orbital symmetry, explain why a Diels–Alder reaction does not take place under photochemical reaction conditions.

25.5 Sigmatropic Rearrangements

A sigmatropic rearrangement is an intramolecular pericyclic reaction in which a σ bond is broken in a reactant, the π bonds rearrange, and a new σ bond is formed in the product. In a sigmatropic rearrangement, the number of π bonds in the reactant and product is constant, and the σ bonds broken and formed are **allylic** C–H, C–C, or C–Z bonds (Z = N, O, or S). A sigmatropic rearrangement that results in cleavage and formation of a C–H bond is shown.

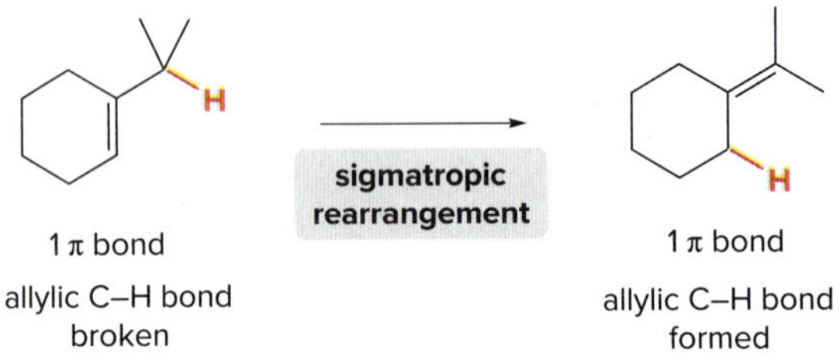

sigmatropic rearrangement

1 π bond
allylic C–H bond broken

1 π bond
allylic C–H bond formed

Sigmatropic rearrangements are characterized by a set of numbers in brackets, [*n*,*m*], to indicate the location of the new σ bond relative to the broken σ bond. To designate a sigmatropic rearrangement:

- **Locate the σ bond broken in the reactant and label both atoms in the bond with "1's."**
- **Locate the new σ bond in the product, and count the number of atoms from the broken σ bond to the new σ bond for each fragment.**
- **Place both numbers in brackets, with the lower number first. In a rearrangement involving a C–H bond, the first number is always "1."**

For example, a [3,3] sigmatropic rearrangement converts diene **A** to diene **B** when an allylic C–C bond in **A** is broken and a new allylic C–C bond is formed in **B.**

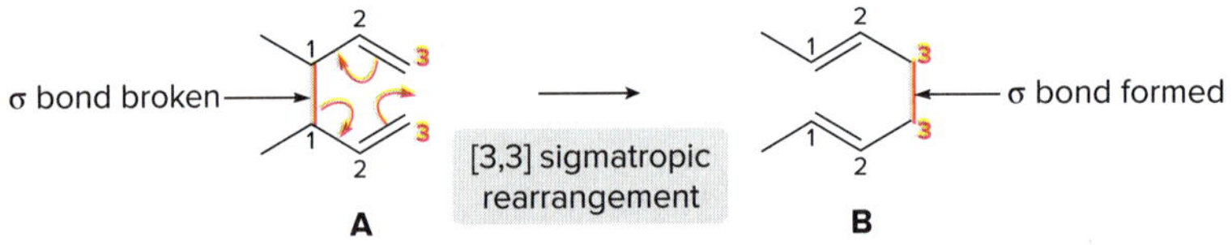

σ bond broken

[3,3] sigmatropic rearrangement

σ bond formed

A **B**

Sample Problem 25.4 Determining the Type of Sigmatropic Rearrangement

What type of sigmatropic rearrangement is illustrated in each equation?

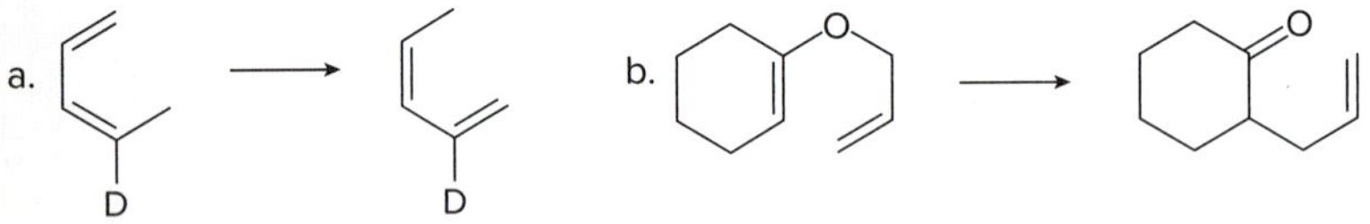

a.

b.

Solution

Locate the atoms in the broken σ bond and label them with 1's. Locate the atoms in the new σ bond, and count the number of atoms from the bond broken to the bond formed. When a C–H bond is broken, the first number in the [*n*,*m*] designation must be 1, because the H atom is bonded to no other atom.

a. A C–H bond is broken on the allylic C and a new C–H bond is formed on C5, so the reaction is a **[1,5] sigmatropic rearrangement.**

σ bond broken

new σ bond

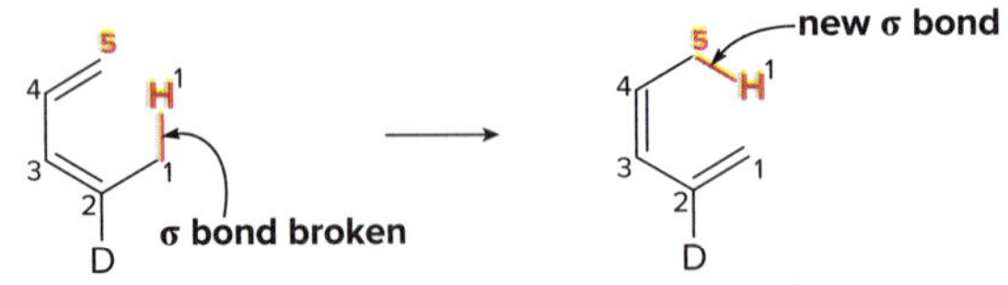

b. The reaction is a **[3,3] sigmatropic rearrangement,** because a C–O σ bond is broken and a new allylic C–C σ bond is formed between carbons that are three atoms removed from the broken bond.

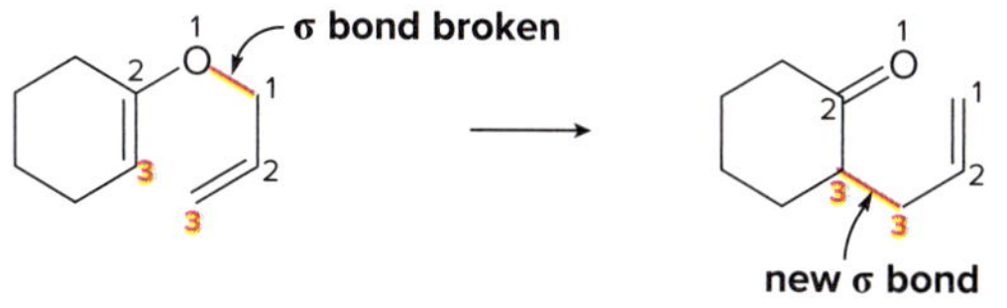

Problem 25.17 What type of sigmatropic rearrangement is illustrated in each equation?

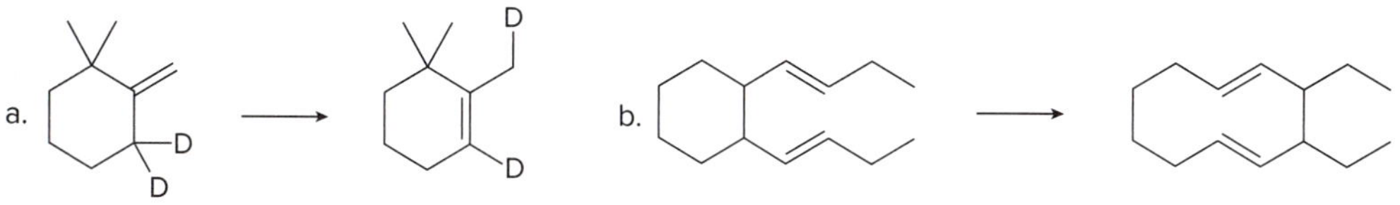

More Practice: Try Problems 25.45, 25.53[2].

25.5A Sigmatropic Rearrangements and Orbital Symmetry

The stereochemistry of a sigmatropic rearrangement, like that of other pericyclic reactions, is determined by the symmetry of the orbitals involved in the reaction. In sigmatropic rearrangements, we consider the orbitals of the σ bond that is broken and the terminal *p* orbital of the π bond at which the new σ bond forms. Two modes of rearrangement are possible: **suprafacial** and **antarafacial.**

- **In a suprafacial rearrangement, the new σ bond forms on the *same* side of the π system as the broken σ bond.**

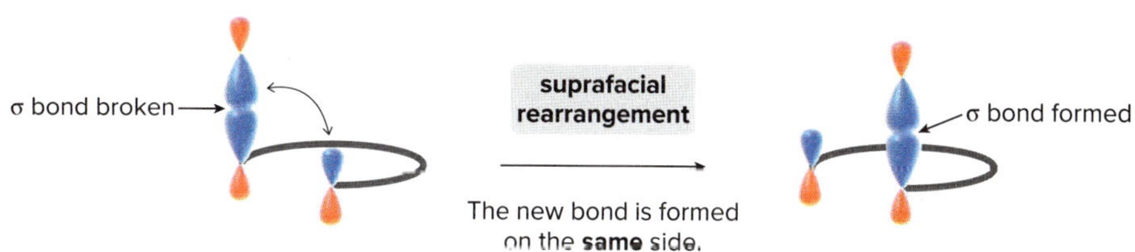

- **In an antarafacial rearrangement, the new σ bond forms on the *opposite* side of the π system as the broken σ bond.**

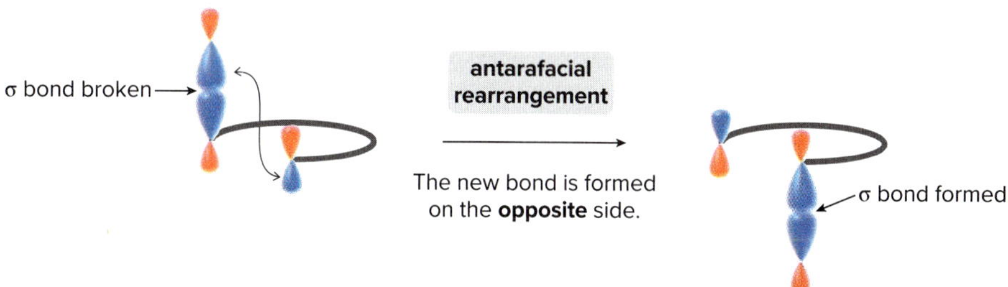

Sigmatropic rearrangements can occur under thermal or photochemical conditions, and follow the same rules observed in cycloaddition reactions. With sigmatropic rearrangements **we count the total number of electron pairs in the σ bond that is broken and the π bonds that rearrange** (Table 25.3). Because sigmatropic rearrangements involve cyclic transition states and small rings have geometrical constraints, **reactions involving six or fewer atoms must take place by suprafacial pathways.**

Table 25.3 Woodward–Hoffmann Rules for Sigmatropic Rearrangements

Number of electron pairs	Thermal reaction	Photochemical reaction
Even	Antarafacial	Suprafacial
Odd	Suprafacial	Antarafacial

For example, a [1,5] sigmatropic rearrangement of **X** to **Y** involves three electron pairs, one from the σ bond that is broken and two from the π bonds that rearrange.

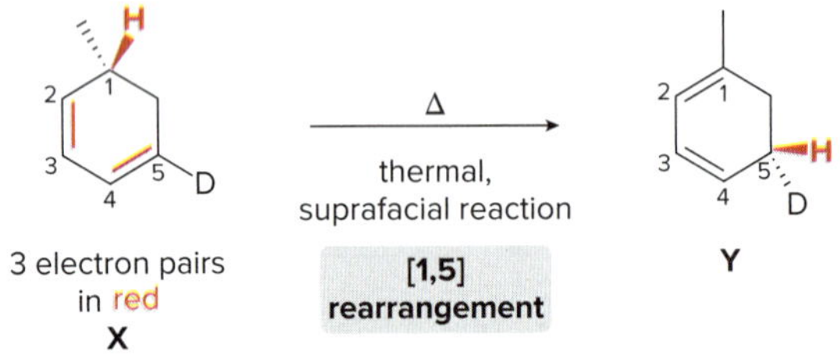

3 electron pairs in red
X

Δ
thermal, suprafacial reaction
[1,5] rearrangement

Y

According to Table 25.3, this reaction must occur in a suprafacial mode under thermal conditions and in an antarafacial mode under photochemical conditions. Because this reaction involves only six atoms (including the H atom that migrates), it must take place under thermal conditions in a suprafacial fashion.

Sample Problem 25.5 Determining Whether a Sigmatropic Rearrangement Occurs Under Thermal or Photochemical Conditions

Classify the following sigmatropic rearrangement and determine whether it takes place readily under thermal or photochemical reaction conditions.

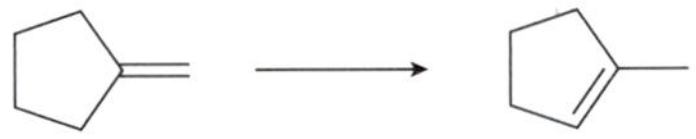

Solution

- Classify the rearrangement as in Sample Problem 25.4: Label the atoms in the broken σ bond with 1's, locate the new σ bond, and count the number of atoms from the bond broken to the bond formed.
- Count the number of electron pairs involved in the reaction, and use Table 25.3 to determine the stereochemical pathway of the reaction. Keep in mind that reactions involving six or fewer atoms must take place by suprafacial pathways.

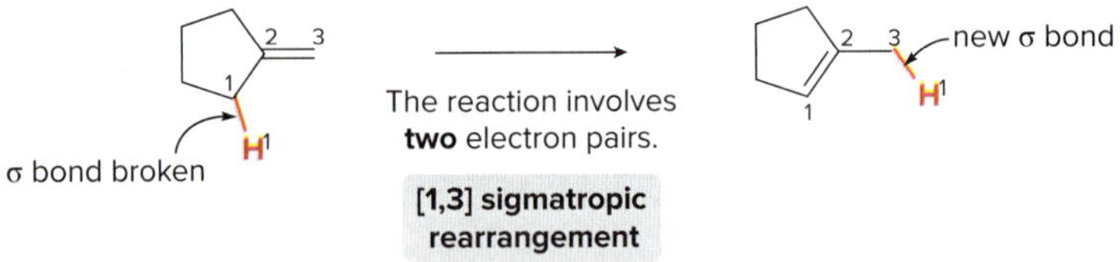

σ bond broken

The reaction involves **two** electron pairs.
[1,3] sigmatropic rearrangement

new σ bond

This reaction is a [1,3] sigmatropic rearrangement, involving **two** electron pairs: the C–H σ bond broken and one π bond. Because the reaction involves four atoms, it must take place via a **suprafacial pathway,** which occurs under **photochemical conditions.**

Problem 25.18 (a) What product is formed from the [1,7] sigmatropic rearrangement of a deuterium in the following triene? (b) Does this reaction proceed in a suprafacial or antarafacial manner under thermal conditions? (c) Does this reaction proceed in a suprafacial or antarafacial manner under photochemical conditions?

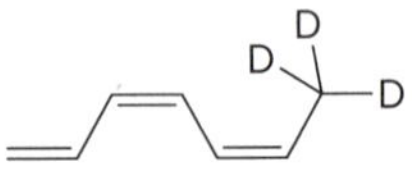

More Practice: Try Problems 25.49, 25.53[2].

25.5B [3,3] Sigmatropic Rearrangements

Two widely used [3,3] sigmatropic rearrangements in organic synthesis are the **Cope rearrangement** of a 1,5-diene to an isomeric 1,5-diene, and the **Claisen rearrangement** of an unsaturated ether to a γ,δ-unsaturated carbonyl compound.

σ bond broken → 1,5-diene ⇌ (Δ, **Cope rearrangement**) isomeric 1,5-diene ← σ bond formed

σ bond broken → unsaturated ether ⇌ (Δ, **Claisen rearrangement**) γ,δ-unsaturated carbonyl compound ← σ bond formed

Both reactions involve **three** electron pairs—two π bonds and one σ bond—and six atoms, and take place readily in a **suprafacial pathway under thermal conditions.**

Cope Rearrangement

Because a Cope rearrangement involves isomeric 1,5-dienes as reactant and product, the more stable diene is favored at equilibrium. Useful Cope rearrangements occur when the reactant 1,5-diene is considerably less stable than the product, as in the case of *cis*-1,2-divinylcyclobutane, which rearranges to cycloocta-1,5-diene with loss of strain from the cyclobutane ring.

cis-1,2-divinylcyclobutane → (Δ) cycloocta-1,5-diene ← σ bond formed

The **oxy-Cope rearrangement** is an especially powerful variation of a Cope rearrangement using an unsaturated alcohol. [3,3] Sigmatropic rearrangement forms an enol initially, which then tautomerizes to form a carbonyl group.

HO, 1, 2, 3, 3, 2, 1 — OH group at C3 → (Δ, [3,3], **oxy-Cope rearrangement**) [HO enol] ⇌ (**tautomerization**) **carbonyl compound**

3, 2, 1, 1, 2, 3, OH → (Δ, [3,3]) [OH] ⇌ (**tautomerization**) O

Moreover, ***anionic*** **oxy-Cope rearrangements** often give high yields of rearranged product under very mild reaction conditions. In an anionic oxy-Cope rearrangement, the unsaturated alcohol reactant is first treated with strong base, usually KH, to form an alkoxide. [3,3] Sigmatropic rearrangement then yields a **resonance-stabilized enolate,** which is protonated to form a carbonyl product.

HO → (KH) ⁻O alkoxide → (Δ, [3,3]) ⁻O ⟷ O resonance-stabilized enolate → (protonation) O

anionic oxy-Cope rearrangement

Sample Problem 25.6 Drawing the Product of a [3,3] Sigmatropic Rearrangement

Draw the product when the following compound undergoes a [3,3] sigmatropic rearrangement.

Solution

To draw the product of a [3,3] sigmatropic rearrangement:

- Locate the 1,5-diene unit, and draw the ends of the double bonds (C1 and C6) close to each other.
- Draw three arrows beginning with a π bond. Break two π bonds and one σ bond to draw the product.
- If an enol is formed, tautomerize the enol to a keto form.

1 Identify the 1,5-diene, and place C1 and C6 close to each other.

2 **Draw three arrows,** beginning at a π bond.

3 Draw the product.

4 Tautomerize.

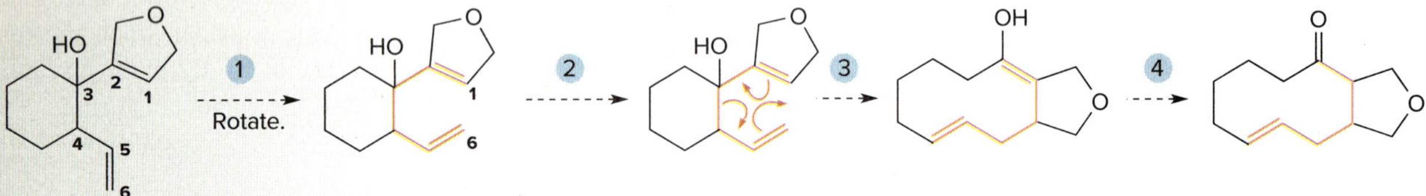

Problem 25.19 What product is formed from the Cope or oxy-Cope rearrangement of each starting material?

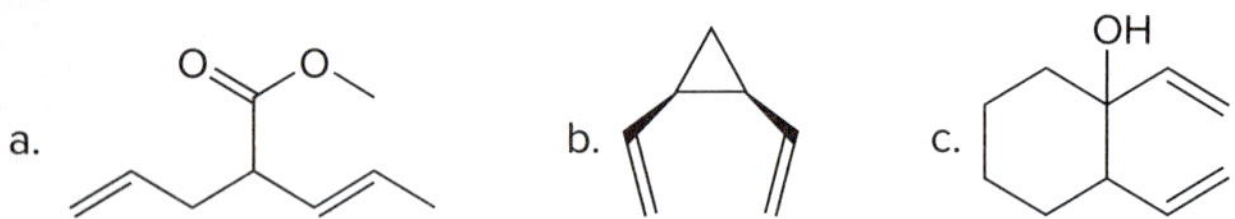

More Practice: Try Problems 25.29a; 25.46b, c; 25.47; 25.48a; 25.51; 25.55b; 25.56; 25.58b.

The structure of periplanone B (Problem 25.20), the sex pheromone of the female American cockroach, was determined in 1976 using 200 μg of material collected from 75,000 female cockroaches.
George Grall/National Geographic Creative/Alamy Stock Photo

Problem 25.20 One step in the synthesis of periplanone B (Problem 22.13), involves anionic oxy-Cope rearrangement of the following unsaturated alcohol. Draw the product that results after protonation of the intermediate enolate.

Problem 25.21 What compound forms geranial by a Cope rearrangement?

geranial

Claisen Rearrangement

A Claisen rearrangement is a [3,3] sigmatropic rearrangement of an unsaturated ether, either an allyl vinyl ether or an allyl aryl ether. With an allyl vinyl ether, a γ,δ-unsaturated carbonyl compound is formed directly by the concerted rearrangement. With an allyl aryl ether, Claisen rearrangement initially generates a cyclohexadienone intermediate, which tautomerizes to a phenol that contains an allyl group ortho to the OH group.

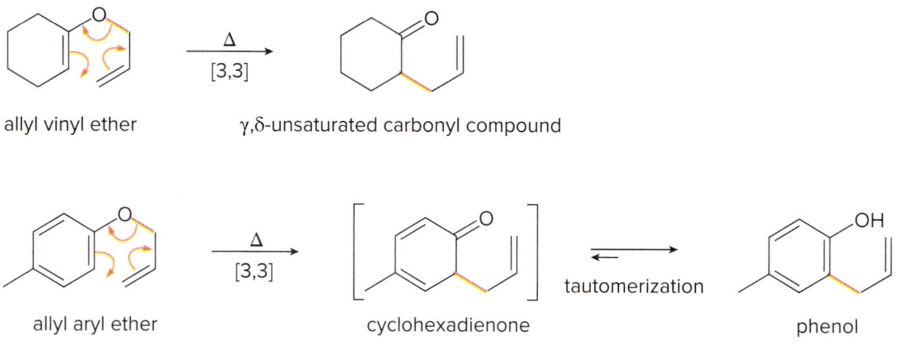

allyl vinyl ether —Δ, [3,3]→ γ,δ-unsaturated carbonyl compound

allyl aryl ether —Δ, [3,3]→ [cyclohexadienone] —tautomerization→ phenol

Problem 25.22 What product is formed from the Claisen rearrangement of each starting material?

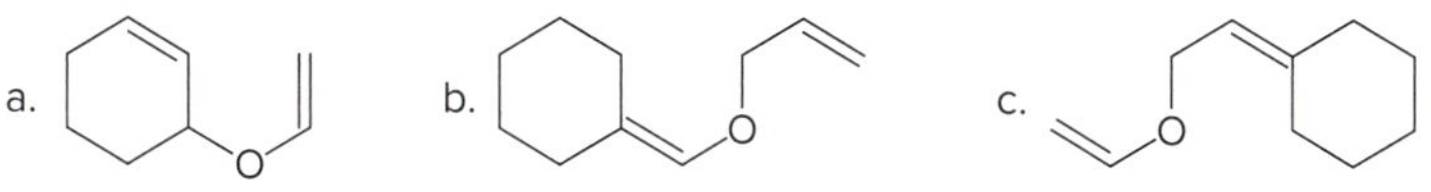

a. b. c.

Problem 25.23 Claisen rearrangements are known in biological systems. What product is formed from an enzyme-catalyzed rearrangement of chorismate?

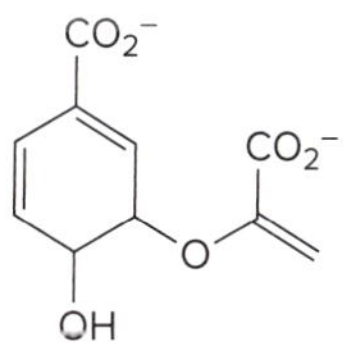

chorismate

Problem 25.24 (a) What product is formed by the Claisen rearrangement of compound **Z?** (b) Using what you have learned about ring-closing metathesis in Chapter 24, draw the product formed when the product in part (a) is treated with Grubbs catalyst. These two reactions are key steps in the synthesis of garsubellin A, a biologically active natural product that stimulates the synthesis of the neurotransmitter acetylcholine. Compounds of this sort may prove to be useful drugs for the treatment of neurodegenerative diseases such as Alzheimer's disease.

Garsubellin A (Problem 25.24) is isolated from the wood of *Garcinia subelliptica*, a tree grown in Okinawa, Japan.
Marina Khaytarova TopTropicals.com

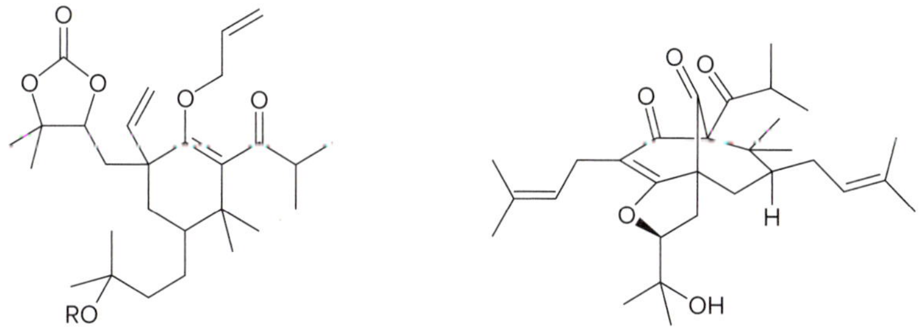

Z garsubellin A

Problem 25.25 Draw the product formed from the Claisen rearrangement of the given starting material. This reaction was a key step in the synthesis of the chapter-opening molecule, galantamine (Problem 23.54).

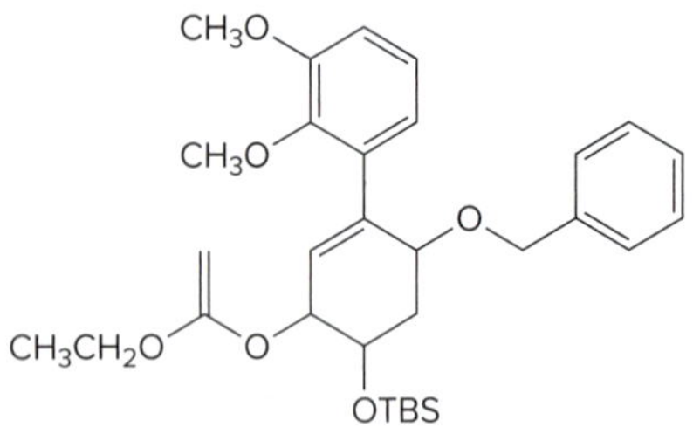

CH_3O CH_3O O CH_3CH_2O O OTBS

25.6 Summary of Rules for Pericyclic Reactions

Table 25.4 summarizes the rules that govern pericyclic reactions, and in truth, this table holds a great deal of information. To keep track of this information, it may be helpful to **learn one row in the table only,** and then note the result when one or more conditions change. For example:

- **A *thermal* reaction involving an *even* number of electron pairs is *conrotatory* or *antarafacial.***
- **If *one* of the reaction conditions changes—either from thermal to photochemical or from an even to an odd number of electron pairs—the stereochemistry of the reaction changes to disrotatory or suprafacial.**
- **If *both* reaction conditions change—that is, a photochemical reaction with an odd number of electron pairs—the stereochemistry does *not* change.**

Table 25.4 Summary of the Stereochemical Rules for Pericyclic Reactions

Reaction conditions	Number of electron pairs	Stereochemistry
Thermal	Even	Conrotatory or antarafacial
	Odd	Disrotatory or suprafacial
Photochemical	Even	Disrotatory or suprafacial
	Odd	Conrotatory or antarafacial

Problem 25.26 Using the Woodward–Hoffmann rules in Table 25.4, predict the stereochemistry of each reaction.

a. a [6 + 4] thermal cycloaddition
b. photochemical electrocyclic ring closure of deca-1,3,5,7,9-pentaene
c. a [4 + 4] photochemical cycloaddition
d. a thermal [5,5] sigmatropic rearrangement

Problem 25.27 Each of the following reactions was used in the synthesis of a complex natural product. What type of pericyclic reaction is depicted in each reaction?

a. O O → O O

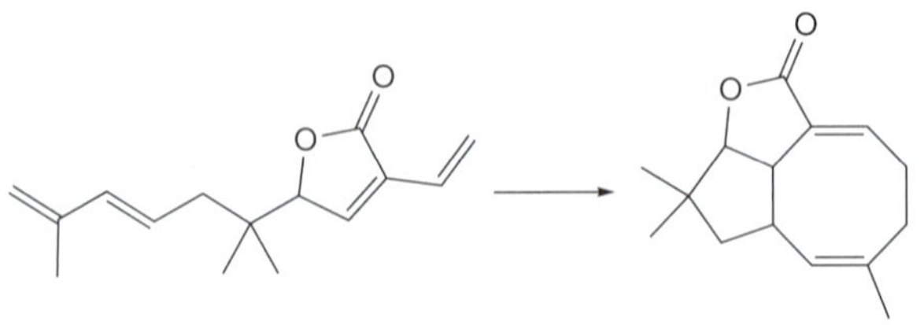

b. O O CH_3CO_2 → CH_3CO_2 O O

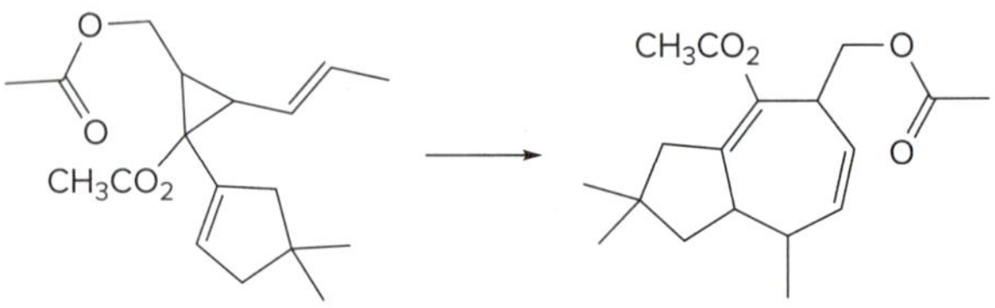

Chapter 25 REVIEW

KEY CONCEPTS

Woodward–Hoffmann rules for pericyclic reactions

1 Type of reaction	2 Number of electron pairs*	3 Thermal	4 Photochemical
Electrocyclic reactions (25.3)	Even Odd	Conrotatory Disrotatory	Disrotatory Conrotatory
Cycloaddition reactions (25.4)	Even Odd	Antarafacial Suprafacial	Suprafacial Antarafacial
Sigmatropic rearrangements (25.5)	Even Odd	Antarafacial Suprafacial	Suprafacial Antarafacial

*In electrocyclic reactions, count the number of π bonds in the acyclic conjugated polyene that is either the reactant or the product. In cycloaddition reactions, count the total number of π bonds from both components of the cycloaddition. In sigmatropic rearrangements, count the σ bond that is broken and the π bonds that rearrange.

See Tables 25.1–25.4.

KEY REACTIONS

[1] Electrocyclic reactions

Try Problems 25.28; 25.31–25.38; 25.54b, d.

[2] Cycloaddition reactions

Try Problems 25.39–25.44; 25.54a, c; 25.55d; 25.58a.

[3] Sigmatropic rearrangements

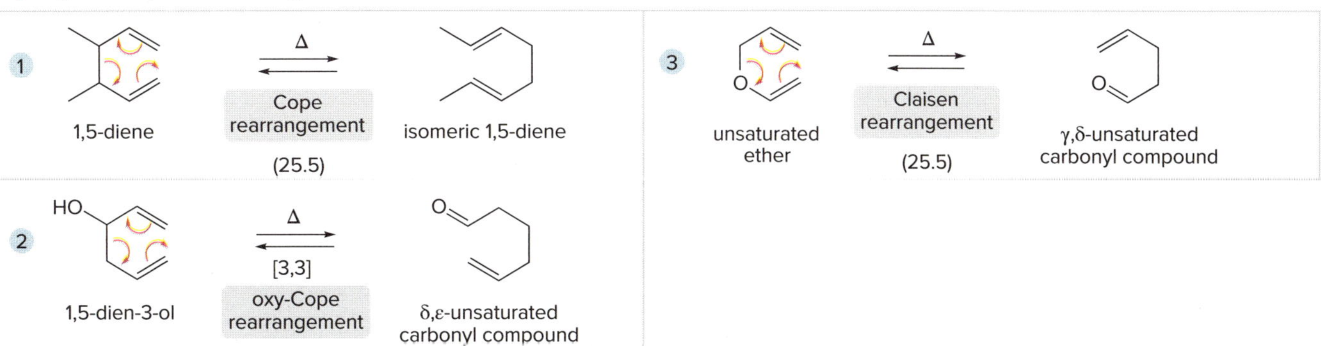

Try Problems 25.29, 25.46–25.52, 25.55a–c, 25.56, 25.58b.

KEY SKILLS

[1] Identifying the product of a thermal electrocyclic ring closure, and labeling a process as conrotatory or disrotatory (25.3A)

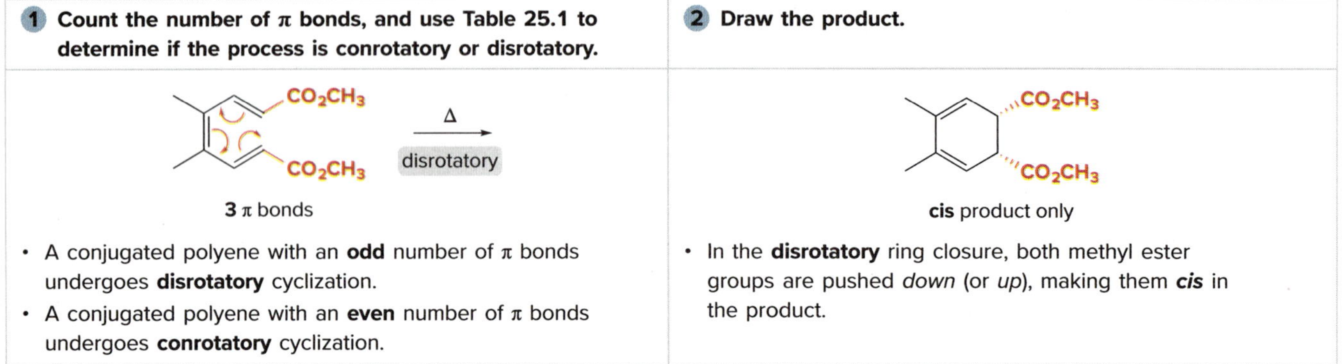

- A conjugated polyene with an **odd** number of π bonds undergoes **disrotatory** cyclization.
- A conjugated polyene with an **even** number of π bonds undergoes **conrotatory** cyclization.

- In the **disrotatory** ring closure, both methyl ester groups are pushed *down* (or *up*), making them ***cis*** in the product.

See Sample Problem 25.1 and Table 25.1. Try Problems 25.28a; 25.31; 25.34a, c; 25.37a; 25.54b.

[2] Identifying the product of a photochemical electrocyclic ring closure, and labeling a process as conrotatory or disrotatory (25.3C)

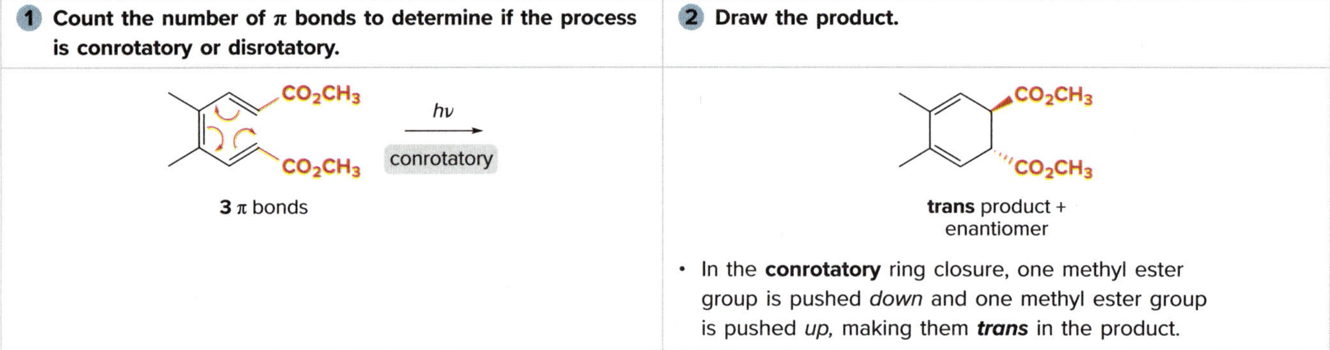

- In the **conrotatory** ring closure, one methyl ester group is pushed *down* and one methyl ester group is pushed *up*, making them ***trans*** in the product.

See Sample Problem 25.2 and Table 25.1. Try Problems 25.28b; 25.32; 25.33; 25.34b, d; 25.37b; 25.54d.

[3] Classifying a cycloaddition (25.4) and a sigmatropic rearrangement (25.5)

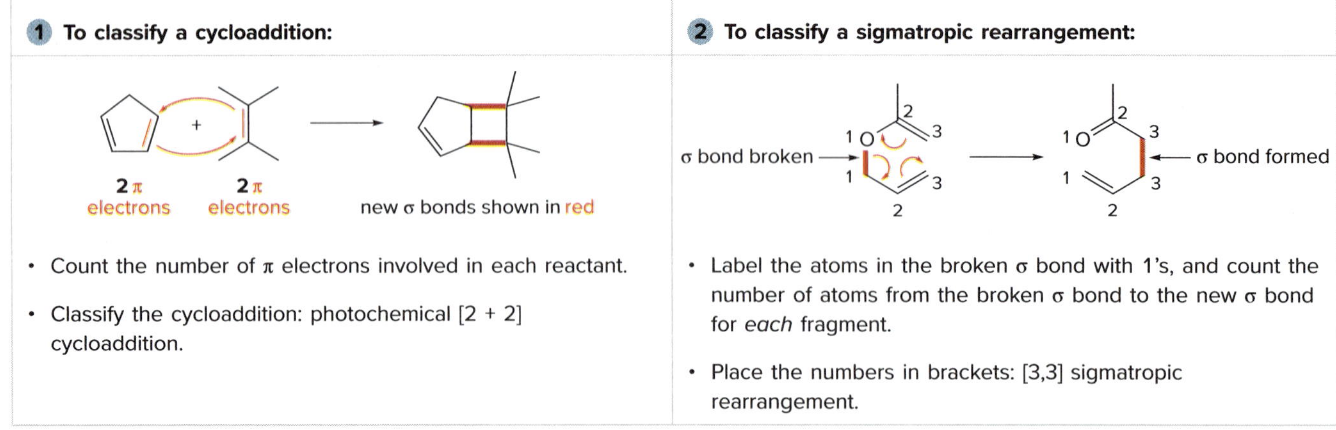

- Count the number of π electrons involved in each reactant.
- Classify the cycloaddition: photochemical [2 + 2] cycloaddition.

- Label the atoms in the broken σ bond with 1's, and count the number of atoms from the broken σ bond to the new σ bond for *each* fragment.
- Place the numbers in brackets: [3,3] sigmatropic rearrangement.

See Sample Problems 25.3 and 25.4. Try Problems 25.39, 25.45, 25.53[2].

[4] Determining the stereochemical pathway of a sigmatropic rearrangement (25.5)

1 Count the number of electron pairs involved in the reaction.

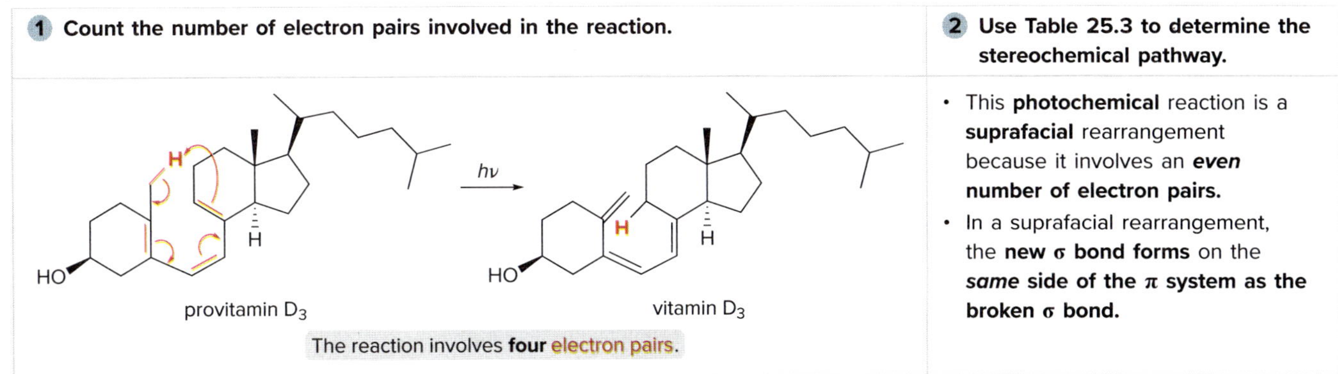

2 Use Table 25.3 to determine the stereochemical pathway.

- This **photochemical** reaction is a **suprafacial** rearrangement because it involves an ***even*** **number of electron pairs.**
- In a suprafacial rearrangement, the **new σ bond forms** on the ***same*** **side of the π system as the broken σ bond.**

See Sample Problem 25.5 and Table 25.3. Try Problem 25.49.

CHAPTER 25 MULTIPLE-CHOICE SELF-TEST

The Self-Test consists of multiple-choice questions similar to those found on the American Chemical Society organic chemistry exam. Answers are given at the end of the chapter.

1. How many π molecular orbitals are present in octa-1,3,5,7-tetraene and how many electrons are present in the HOMO of the ground state?

a. four orbitals and two electrons
b. four orbitals and one electron
c. eight orbitals and one electron
d. eight orbitals and two electrons

2. In the excited state of octa-1,3,5,7-tetraene, how many electrons are present in the HOMO and LUMO?

a. one electron in the HOMO and one electron in the LUMO
b. two electrons in the HOMO and zero electrons in the LUMO
c. zero electrons in the HOMO and two electrons in the LUMO
d. zero electrons in the HOMO and one electron in the LUMO

3. Which statement is *not* true about the conversion of **A** to **B?**

A → **B**

a. The reaction occurs under thermal conditions.
b. The reaction occurs in a disrotatory fashion.
c. The number of π bonds in the polyene product determines whether the rotation is conrotatory or disrotatory.
d. The reaction occurs in a conrotatory fashion.

4. What product is formed in the oxy-Cope rearrangement of **C?**

OH **C**

a. OH   b. CHO   c. CHO   d. O

5. Which statement about pericyclic reactions is *not* true?

a. Pericyclic reactions require light or heat.
b. In pericyclic reactions, bonds are broken and formed in more than one step.
c. Pericyclic reactions are stereospecific.
d. Pericyclic reactions are concerted.

6. What product is formed from the Claisen rearrangement of **E?**

O **E**

a. O   b. O   c. O   d. H O

7. What product is formed from conrotatory electrocyclic cyclization of **F?**

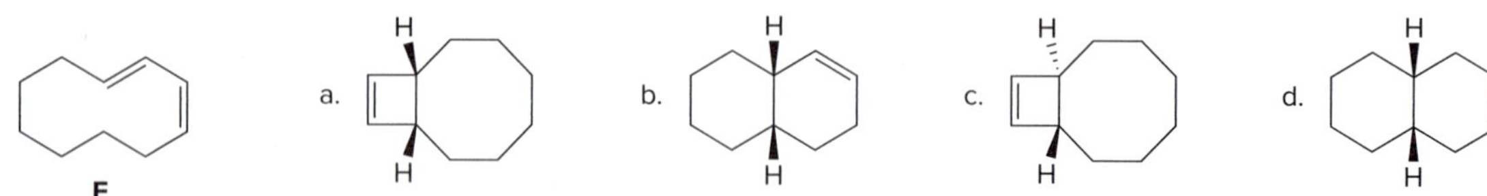

8. What type of cycloaddition is illustrated in the following reaction?

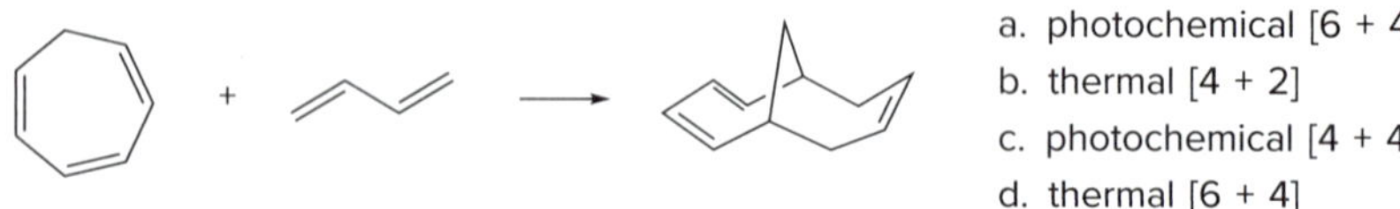

a. photochemical [6 + 4]
b. thermal [4 + 2]
c. photochemical [4 + 4]
d. thermal [6 + 4]

9. What type of sigmatropic rearrangement is shown in the following reaction?

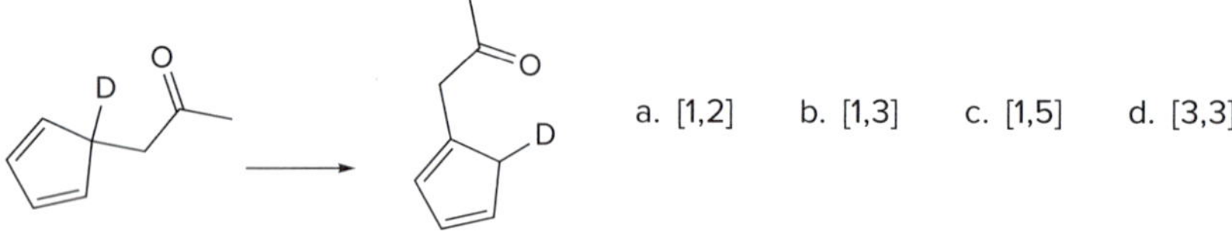

a. [1,2] b. [1,3] c. [1,5] d. [3,3]

10. Which statement about pericyclic reactions is *not* true?
a. The Claisen rearrangement involves three electron pairs—two in π bonds and one in a σ bond.
b. An electrocyclic ring closure forms a product with one more π bond and one fewer σ bond.
c. A thermal Cope rearrangement occurs in a suprafacial fashion.
d. Photochemical cycloadditions involving an even number of π bonds proceed by a suprafacial pathway.

PROBLEMS

Problems Using Three-Dimensional Models

25.28 (a) What product is formed when each compound undergoes a thermal electrocyclic ring opening? (b) What product is formed when each compound undergoes a photochemical electrocyclic ring opening?

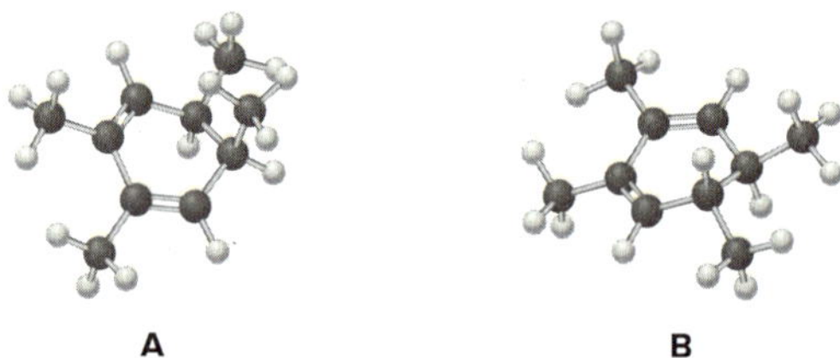

25.29 What product is formed by the [3,3] sigmatropic rearrangement of each compound?

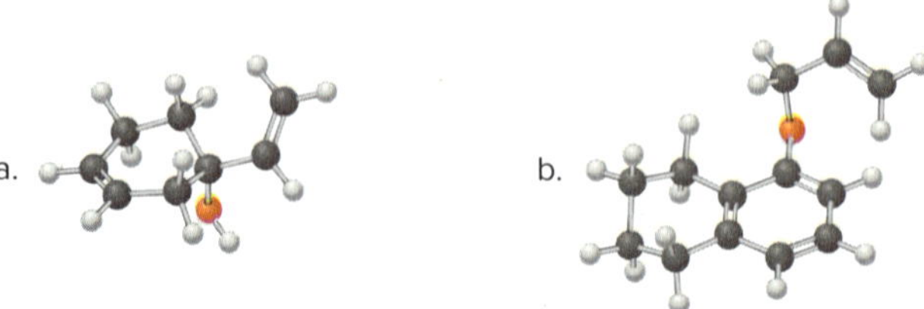

Types of Pericyclic Reactions

25.30 Classify each pericyclic reaction as an electrocyclic reaction, cycloaddition, or sigmatropic rearrangement. Indicate whether the stereochemistry is conrotatory, disrotatory, suprafacial, or antarafacial.

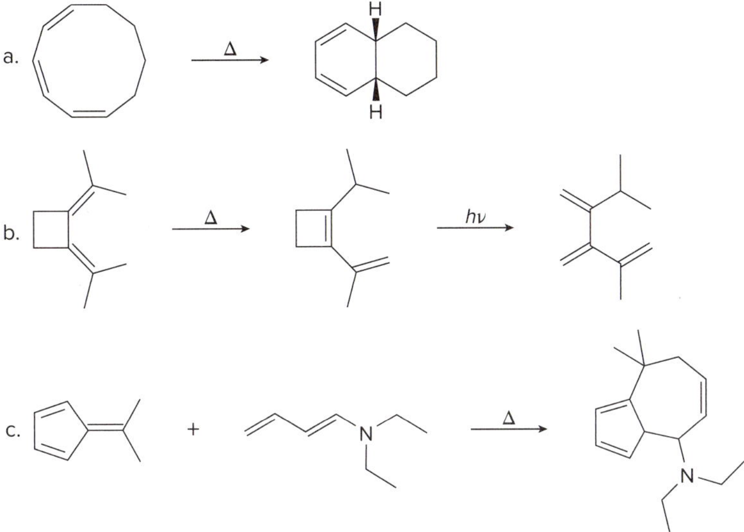

Electrocyclic Reactions

25.31 What product is formed when each compound undergoes thermal electrocyclic ring opening or ring closure? Label each process as conrotatory or disrotatory, and clearly indicate the stereochemistry around tetrahedral stereogenic centers and double bonds.

a. b.

25.32 What product is formed when each compound in Problem 25.31 undergoes photochemical electrocyclic reaction? Label each process as conrotatory or disrotatory, and clearly indicate the stereochemistry around tetrahedral stereogenic centers and double bonds.

25.33 What cyclic product is formed when each decatetraene undergoes photochemical electrocyclic ring closure?

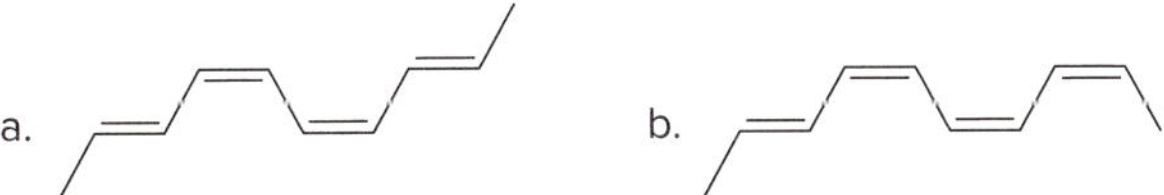

25.34 Draw the product of each electrocyclic reaction.

a. the thermal electrocyclic ring closure of (2*E*,4*Z*,6*Z*)-nona-2,4,6-triene
b. the photochemical electrocyclic ring closure of (2*E*,4*Z*,6*Z*)-nona-2,4,6-triene
c. the thermal electrocyclic ring opening of *cis*-5-ethyl-6-methylcyclohexa-1,3-diene
d. the photochemical electrocyclic ring opening of *trans*-5-ethyl-6-methylcyclohexa-1,3-diene

25.35 Consider the following electrocyclic ring closure. Does the product form by a conrotatory or disrotatory process? Would this reaction occur under photochemical or thermal conditions?

H
H

25.36 Draw the product formed when diene **M** undergoes disrotatory cyclization. Indicate the stereochemistry at new sp^3 hybridized carbons. Will the reaction occur under thermal or photochemical conditions?

M

25.37 (a) What product is formed when triene **N** undergoes thermal electrocyclic ring closure? (b) What product is formed when triene **N** undergoes photochemical ring closure? (c) Label each process as conrotatory or disrotatory.

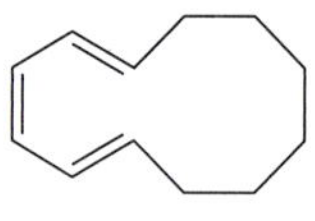

N

25.38 The bicyclic alkene **P** can be prepared by thermal electrocyclic ring closure from cyclodecadiene **Q** or by photochemical electrocyclic ring closure from cyclodecadiene **R.** Draw the structures of **Q** and **R,** and indicate the stereochemistry of the process by which each reaction occurs.

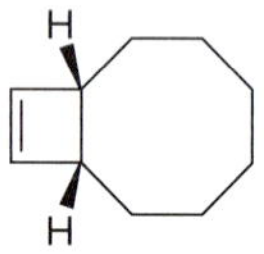

H

H

P

Cycloaddition Reactions

25.39 What type of cycloaddition occurs in Reaction [1]? Draw the product of a similar process in Reaction [2]. Would you predict that these reactions occur under thermal or photochemical conditions?

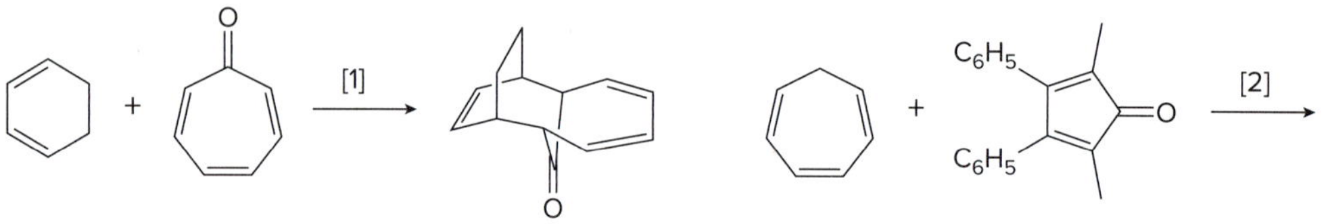

O + [1] → O

\+ C_6H_5 C_6H_5 O [2] →

25.40 Draw the product of each Diels–Alder reaction, and indicate the stereochemistry at all stereogenic centers.

a. 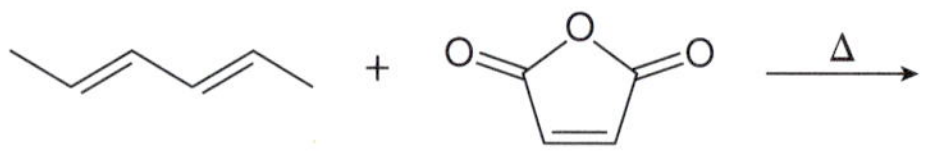 + O O O Δ →

b. 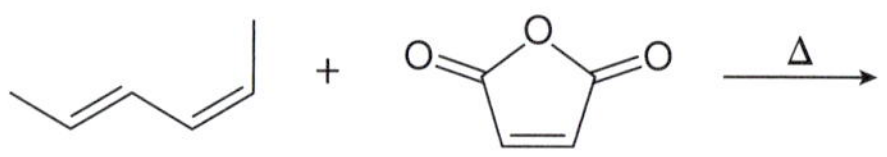 + O O O Δ →

25.41 Draw the product of each intramolecular cycloaddition.

a. 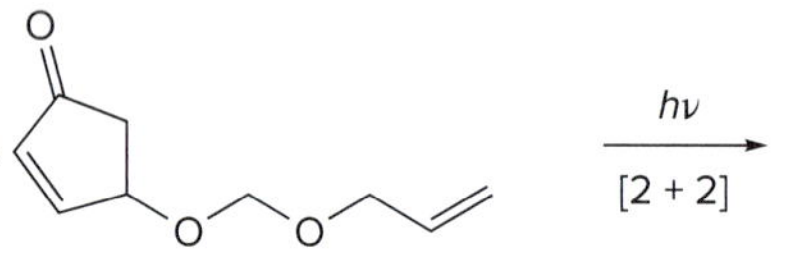 O O O $h\nu$ → [2 + 2]

c. 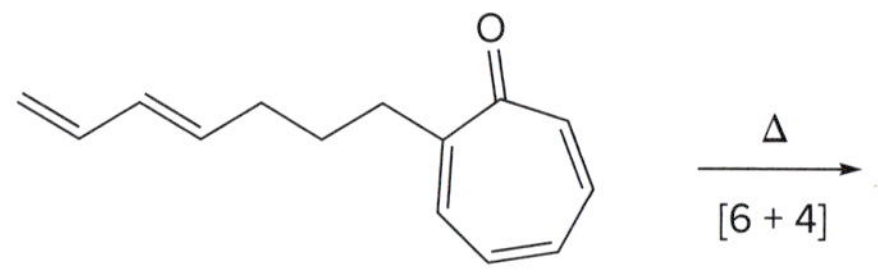 O Δ → [6 + 4]

b. 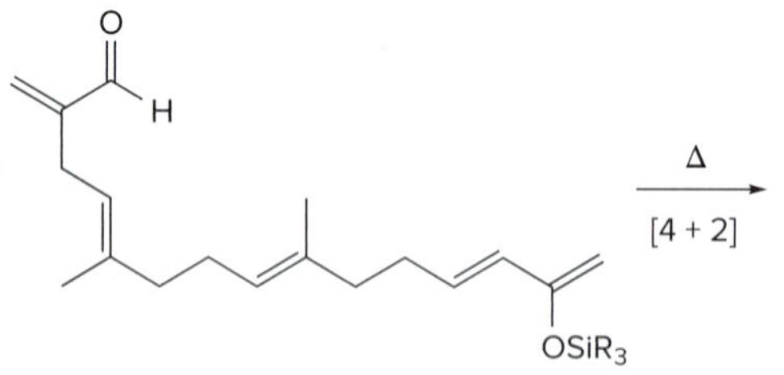 O H $OSiR_3$ Δ → [4 + 2]

d. 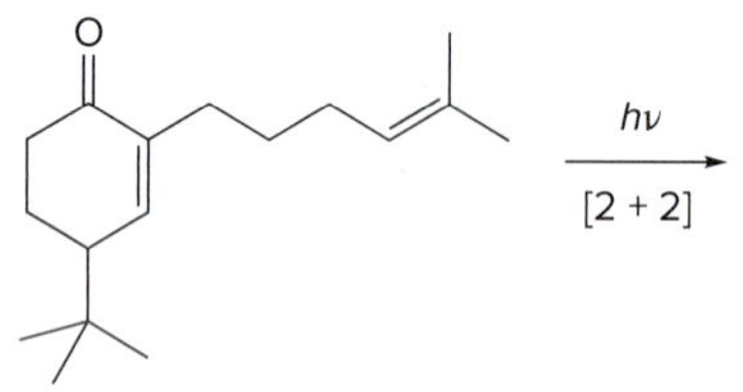 O $h\nu$ → [2 + 2]

25.42 What starting materials are needed to synthesize each compound by a thermal [4 + 2] cycloaddition?

a. 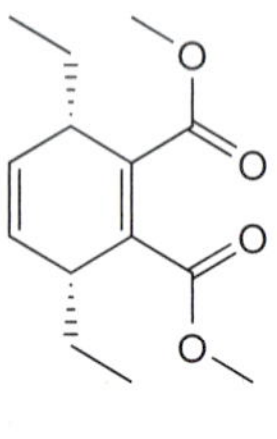 O O O O

b. 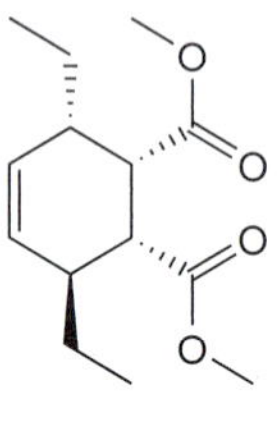 O O O O

c. 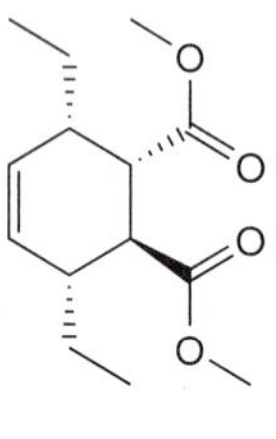 O O O O

25.43 Explain why heating buta-1,3-diene forms 4-vinylcyclohexene but not cycloocta-1,5-diene.

25.44 How can **X** be prepared from a constitutional isomer by a series of [2 + 2] cycloaddition reactions? Interest in molecules that contain several cyclobutane rings fused together has been fueled by the discovery of pentacycloanammoxic acid methyl ester, a lipid isolated from the membrane of organelles in the bacterium *Candidatus Brocadia anammoxidans*. The role of this unusual natural product is as yet unknown.

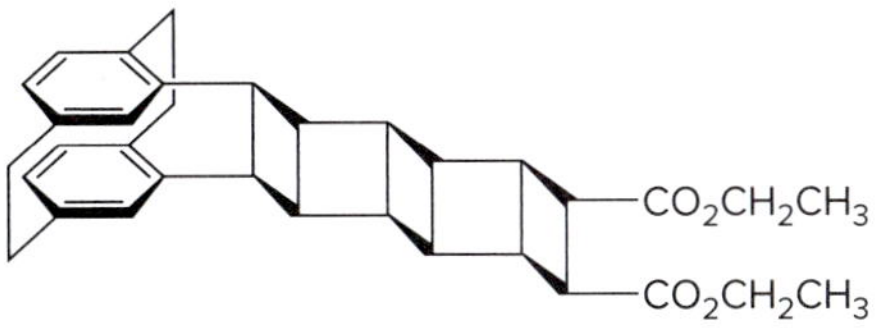

$CO_2CH_2CH_3$
$CO_2CH_2CH_3$
X

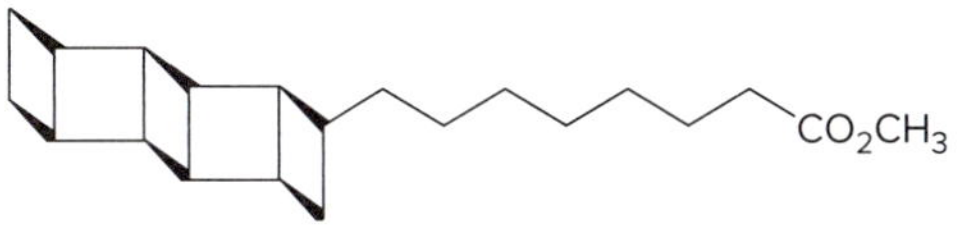

CO_2CH_3
pentacycloanammoxic acid methyl ester

Sigmatropic Rearrangements

25.45 What type of sigmatropic rearrangement is illustrated in each reaction?

a. 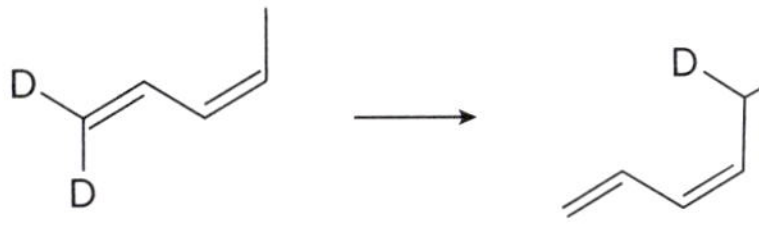 D D → D D

b.  O SC_6H_5 → O^- S C_6H_5

25.46 Draw the product of the [3,3] sigmatropic rearrangement of each compound.

a. 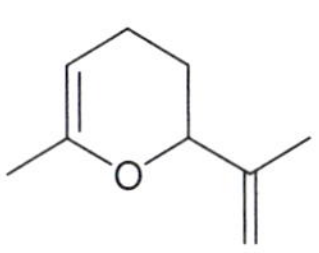 O b. 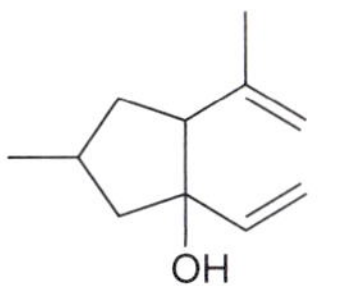 OH c.

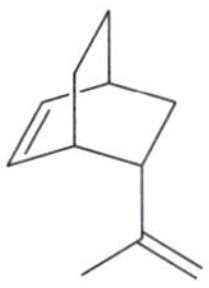

25.47 Draw the structure of **C** in the following reaction scheme, and show how **C** can be converted to **D** by a sigmatropic rearrangement.

O O → [1] MgBr [2] H_2O → **C** → [1] KH, Δ [2] H_2O → **D**

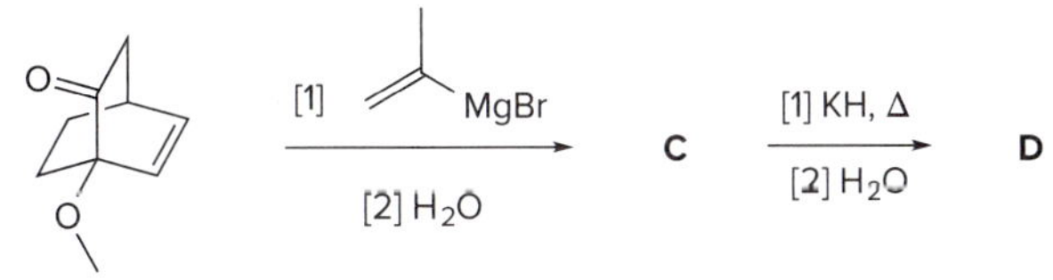

25.48 What product is formed from the [3,3] sigmatropic rearrangement of each compound?

a. 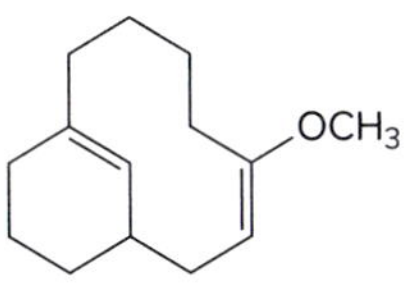 OCH_3 b. 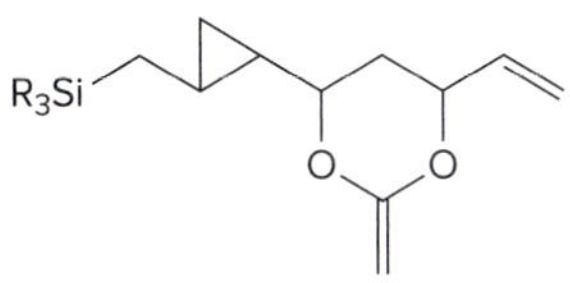 R_3Si O O c. 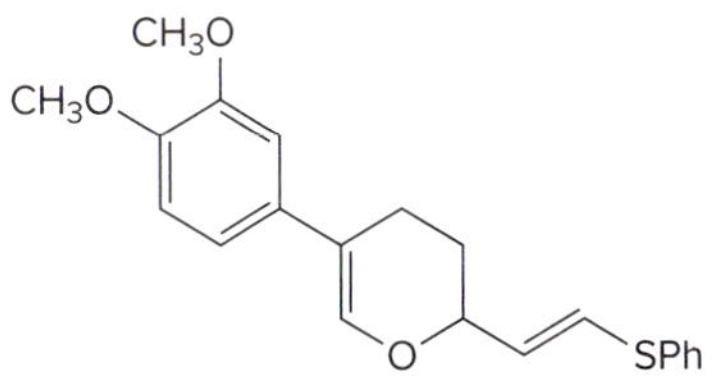 CH_3O CH_3O O SPh

25.49 A solution of 5-methylcyclopenta-1,3-diene rearranges at room temperature to a mixture containing 1-methyl-, 2-methyl-, and 5-methylcyclopenta-1,3-diene. (a) Show how both isomeric products are formed from the starting material by a sigmatropic rearrangement involving a C–H bond. (b) Explain why 2-methylcyclopenta-1,3-diene is not formed directly from 5-methylcyclopenta-1,3-diene by a [1,3] rearrangement.

25.50 What product is formed from the [5,5] sigmatropic rearrangement of the following unsaturated ether?

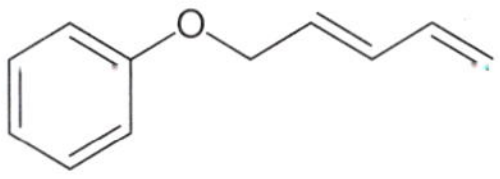

O

25.51 Identify the product of the following two-step reaction sequence. The initial intermediate formed from Step [1] undergoes a [3,3] sigmatropic rearrangement prior to reaction with CH_3I.

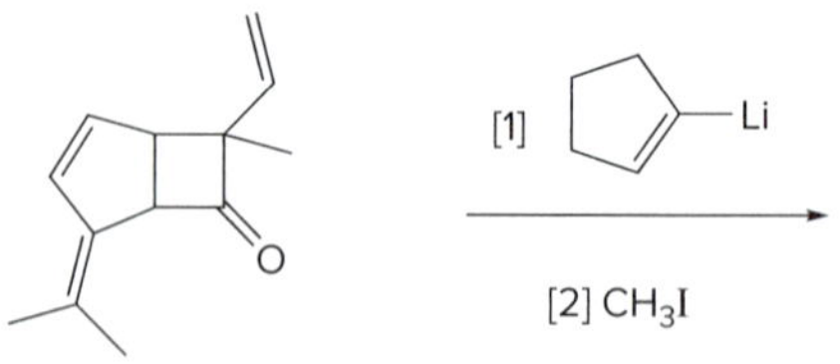

[1] Li
[2] CH_3I

25.52 Heating **A** results in two successive [3,3] sigmatropic rearrangements—Claisen reaction followed by Cope reaction—to afford β-sinensal, a component of mandarin orange oil. What is the structure of β-sinensal?

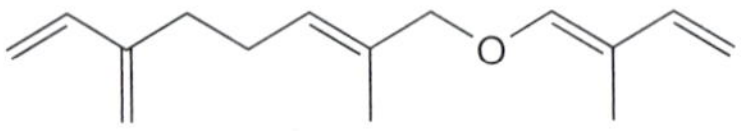

O

A

General Pericyclic Reactions

25.53 What type of pericyclic reaction is illustrated in each reaction?

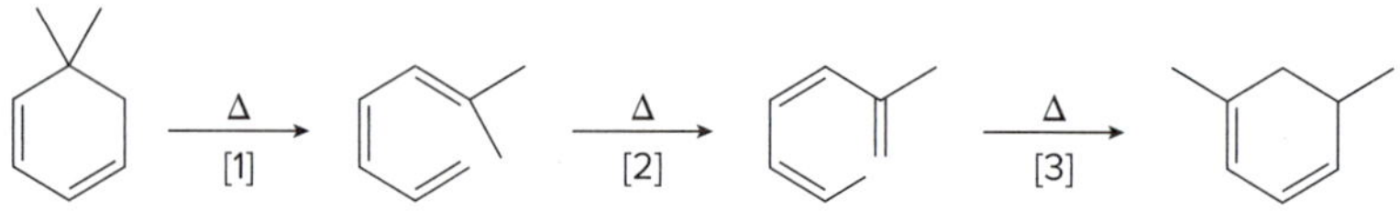

Δ [1] Δ [2] Δ [3]

25.54 Draw the product formed (including stereochemistry) in each pericyclic reaction.

a. 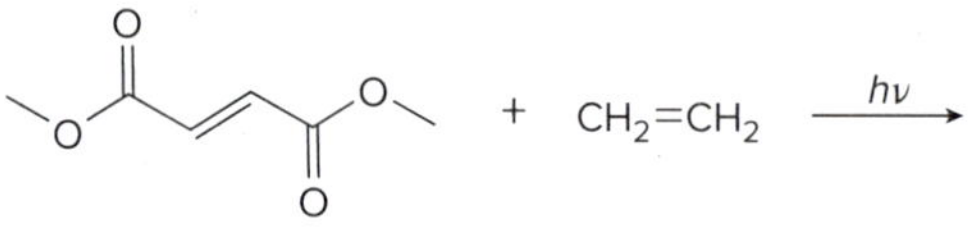 O O O O + $CH_2{=}CH_2$ $h\nu$

c. 2 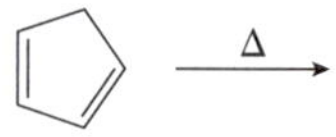 Δ

b. 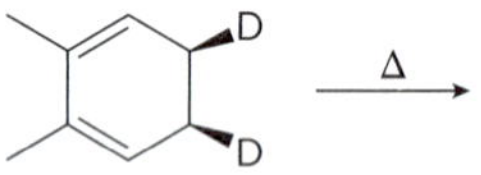 D D Δ

d. D D $h\nu$

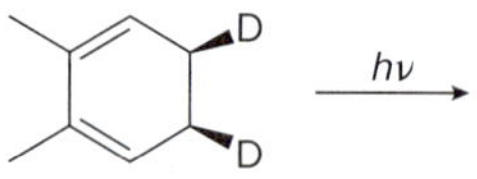

25.55 Draw the products of each reaction.

a. 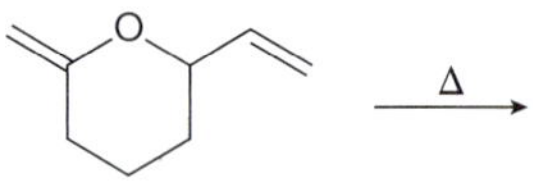 O Δ

c. 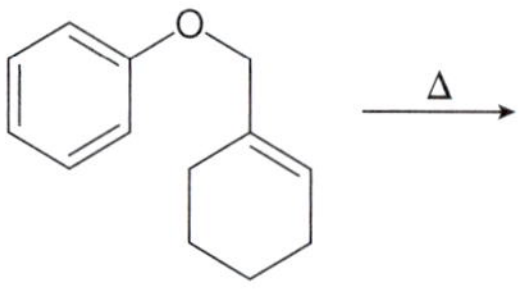 O Δ

b. 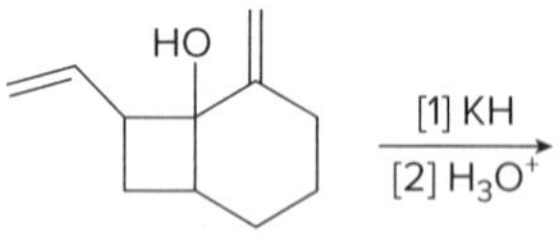 HO [1] KH [2] H_3O^+

d. 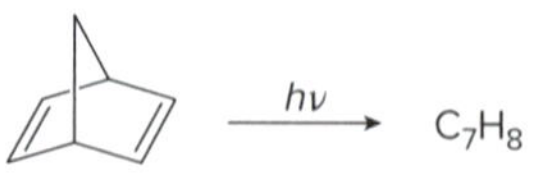 $h\nu$ C_7H_8

25.56 Identify **X** and **Y** in the following reaction scheme.

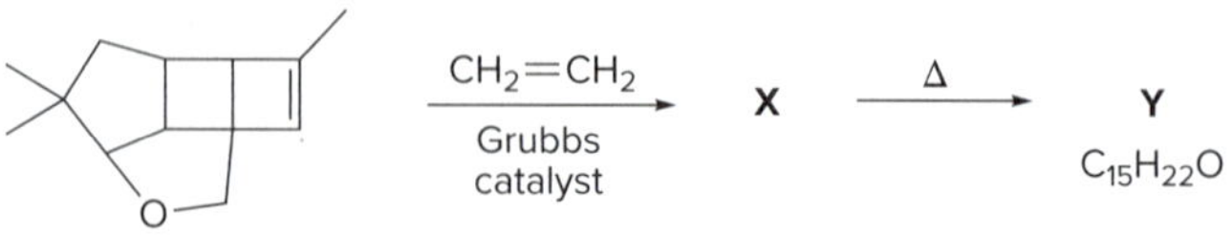

O $CH_2{=}CH_2$ Grubbs catalyst **X** Δ **Y** $C_{15}H_{22}O$

25.57 Identify compounds **A–D** in the following reaction sequence.

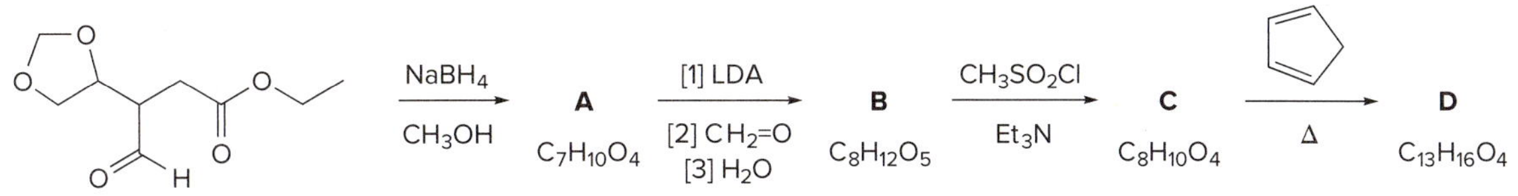

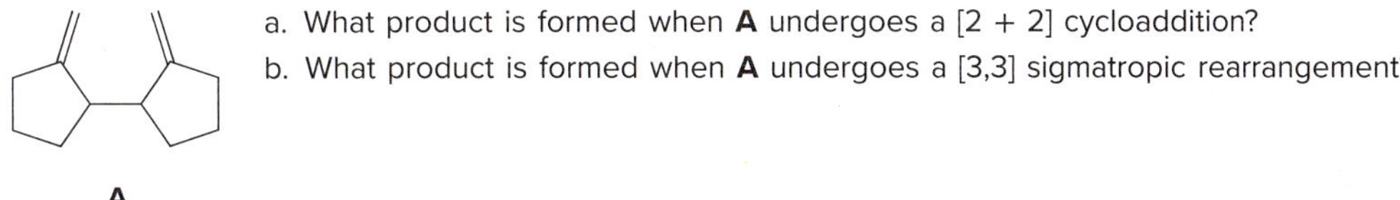

25.58 With reference to diene **A:**

a. What product is formed when **A** undergoes a [2 + 2] cycloaddition?

b. What product is formed when **A** undergoes a [3,3] sigmatropic rearrangement?

Mechanisms

25.59 When both carbons ortho to the aryl oxygen are not bonded to hydrogen, an allyl aryl ether rearranges to a para-substituted phenol. Draw a stepwise mechanism for the following reaction, which contains two [3,3] sigmatropic rearrangements.

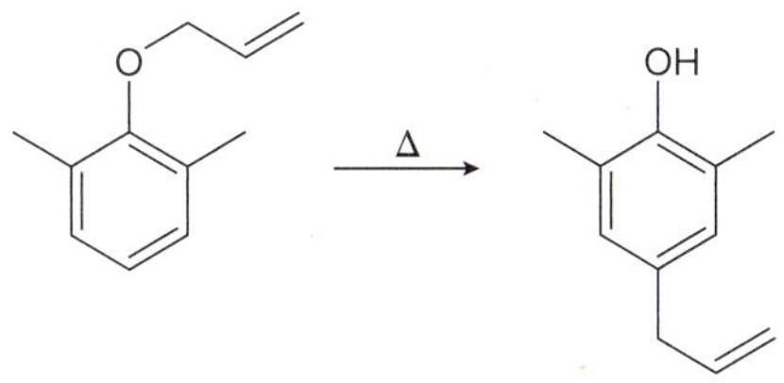

25.60 Draw a stepwise, detailed mechanism for the following reaction.

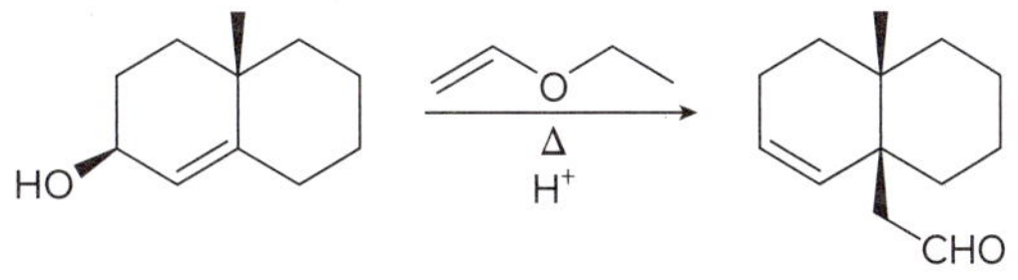

25.61 Show how the following starting material is converted to the given product by a series of two pericyclic reactions. Account for the observed stereochemistry.

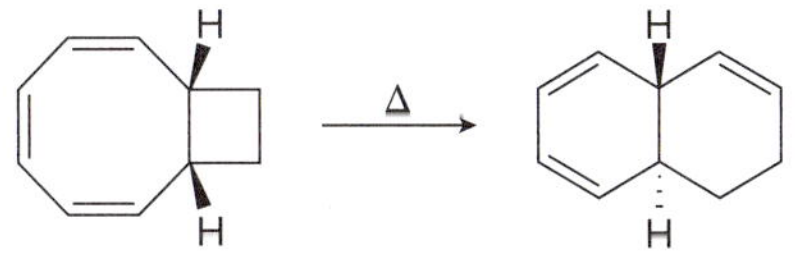

25.62 Use curved arrows to show how **E** is converted to **F** by a two-step reaction sequence consisting of a [1,5] sigmatropic rearrangement followed by a [4 + 2] cycloaddition.

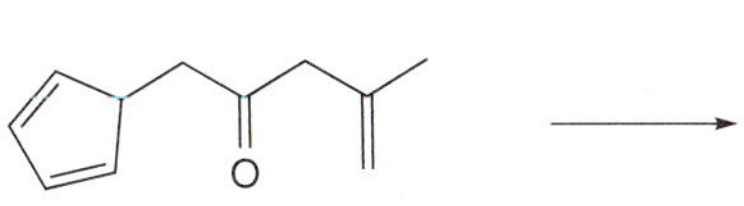

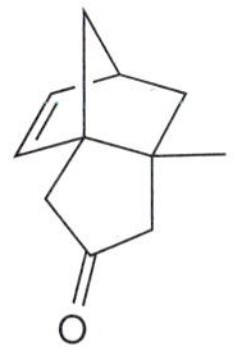

25.63 (a) Draw a stepwise mechanism for the conversion of **A** to **B.** (b) What product would be formed if **C** was exposed to similar reaction conditions?

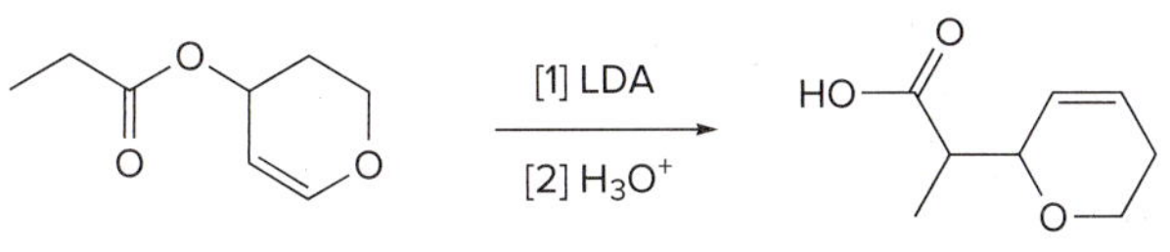

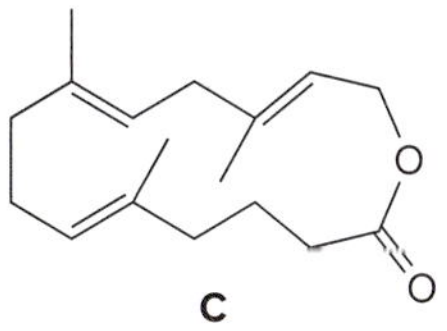

25.64 Show how the following starting materials are converted to the given product by a series of two pericyclic reactions. Account for the observed stereochemistry.

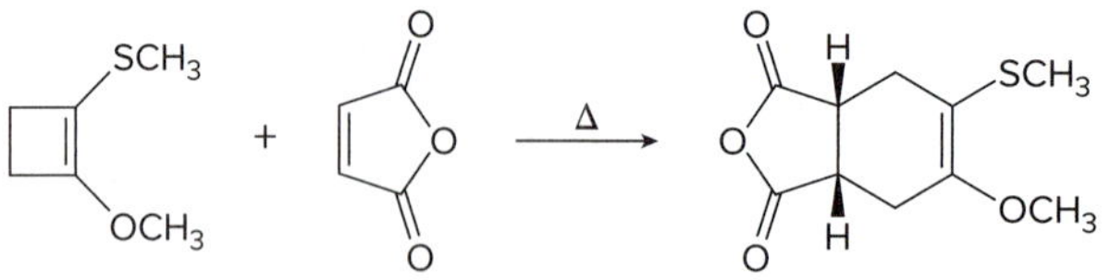

SCH₃ OCH₃ + O O O Δ → O O O H H SCH₃ OCH₃

25.65 Draw a stepwise, detailed mechanism for the following reaction.

O O + NH₂ Δ mild acid → N O H

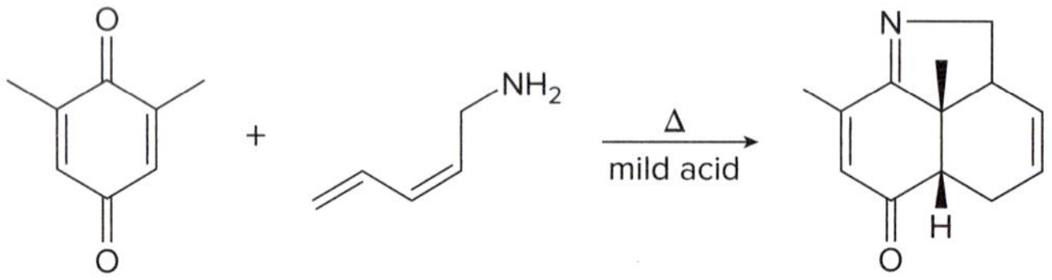

Challenge Problems

25.66 What product is formed by [3,3] sigmatropic rearrangement of the following compound? Clearly indicate the stereochemistry around all tetrahedral stereogenic centers.

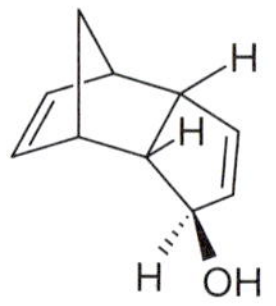

H H H OH

25.67 Draw a stepwise mechanism for the following reaction.

OH Cl Δ → O

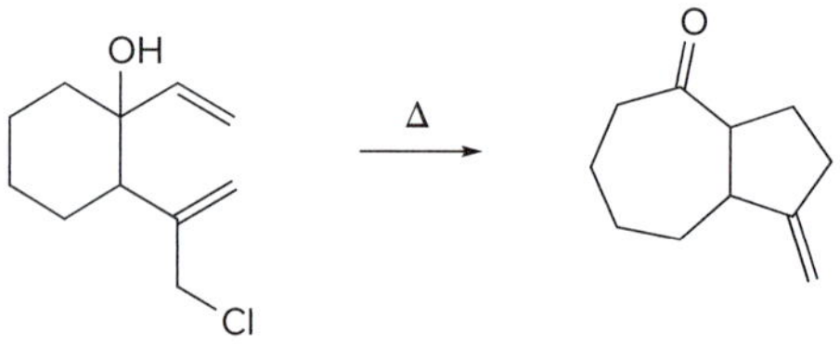

25.68 (a) What is the structure of **C,** which is formed by oxy-Cope rearrangement of **B** with NaOEt? (b) Draw a stepwise mechanism for the conversion of **C** to the bicyclic alcohol **D.**

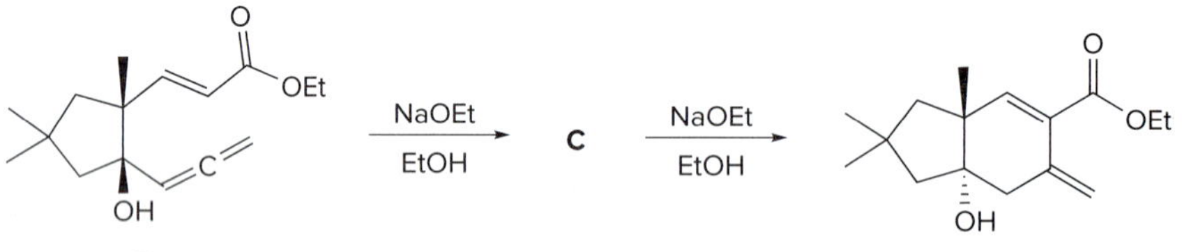

O OEt C OH **B** NaOEt EtOH **C** NaOEt EtOH O OEt OH **D**

25.69 Draw a stepwise mechanism for the Carroll rearrangement, a reaction that prepares a γ,δ-unsaturated carbonyl compound from a β-keto ester and allylic alcohol in the presence of base.

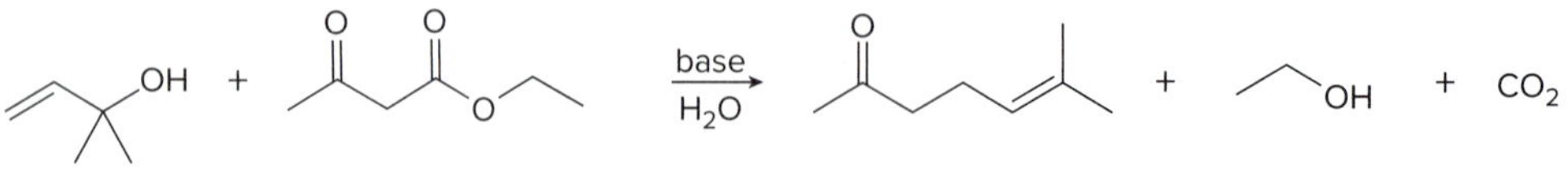

OH + O O O base H_2O → O + OH + CO_2

Why Study . . . Carbohydrates?

Chapters 26–29 discuss ***biomolecules,* organic compounds found in biological systems.** You have already learned many facts about these compounds in previous chapters while you studied other organic compounds having similar properties. In Chapter 10 (Alkenes), for example, you learned that the presence of double bonds determines whether a fatty acid is part of a fat or an oil. In Chapter 19 (Carboxylic Acids and Nitriles), you learned that amino acids are the building blocks of proteins.

Chapter 26 focuses on carbohydrates, the largest group of biomolecules in nature, comprising ~50% of the earth's biomass. Carbohydrates can be simple or complex, having as few as three or as many as thousands of carbon atoms. The glucose metabolized for energy in cells, the sucrose of table sugar, and the cellulose of plant stems and tree trunks are all examples of carbohydrates. Carbohydrates on cell surfaces determine blood type, and carbohydrates form the backbone of DNA, the carrier of all genetic information in the cell. Carbohydrates have many polar functional groups, whose structure and properties can be understood by applying the basic principles of organic chemistry.

26.1 Introduction

Carbohydrates were given their name because molecular formulas of simple carbohydrates could be written as $C_n(H_2O)_n$, making them **hydrates of carbon.**

Carbohydrates such as glucose and cellulose were discussed in Sections 5.1, 6.4, and 18.16.

Carbohydrates, commonly referred to as sugars and starches, are polyhydroxy aldehydes and ketones, or compounds that can be hydrolyzed to them. The cellulose in plant stems and tree trunks and the chitin in the exoskeletons of arthropods and mollusks are both complex carbohydrates. Four examples are shown in Figure 26.1. They include not only glucose and cellulose, but also doxorubicin (an anticancer drug) and 2'-deoxyadenosine 5'-monophosphate (a nucleotide base from DNA), both of which have a carbohydrate moiety as part of a larger molecule.

Carbohydrates are storehouses of chemical energy. They are synthesized in green plants and algae by **photosynthesis,** a process that uses the energy from the sun to convert carbon dioxide and water to glucose and oxygen. This energy is released when glucose is metabolized. The

Figure 26.1 Some examples of carbohydrates

β-D-glucose
most common simple carbohydrate

cellulose
main component of wood

doxorubicin
an anticancer drug

carbohydrate portion

carbohydrate portion

2'-deoxyadenosine 5'-monophosphate
a nucleotide component of DNA

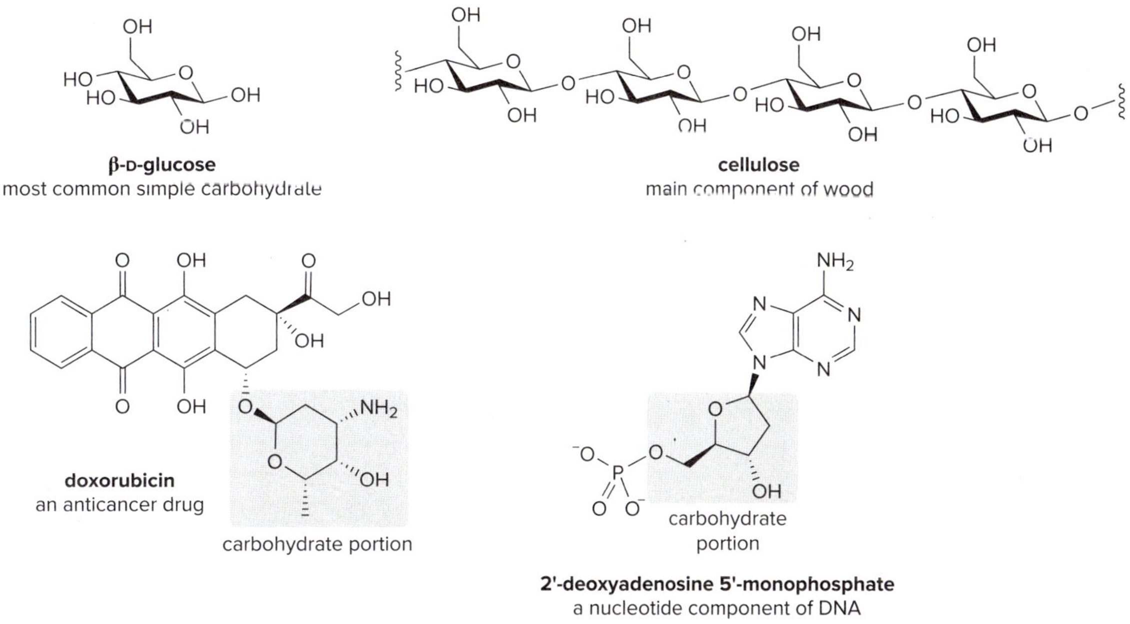

- These compounds illustrate the structural diversity of carbohydrates and their derivatives. **Glucose** is the most common simple sugar, whereas **cellulose,** which comprises wood, plant stems, and grass, is the most common carbohydrate in the plant world. **Doxorubicin,** an anticancer drug that has a carbohydrate ring as part of its structure, has been used in the treatment of leukemia, Hodgkin's disease, and cancers of the breast, bladder, and ovaries. **2'-Deoxyadenosine 5'-monophosphate** is one of the four nucleotides that form DNA.

Although the metabolism of lipids provides more energy per gram than the metabolism of carbohydrates, glucose is the preferred source when a burst of energy is needed during exercise. Glucose is water soluble, so it can be quickly and easily transported through the bloodstream to the tissues.

oxidation of glucose is a multistep process that forms carbon dioxide, water, and a great deal of energy (Section 6.4).

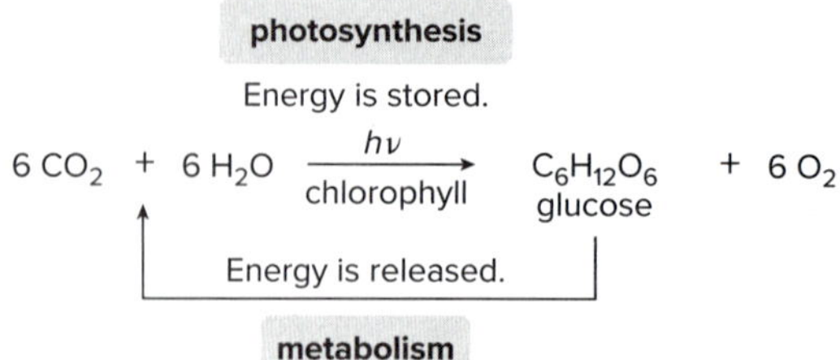

26.2 Monosaccharides

The simplest carbohydrates are called **monosaccharides** or **simple sugars. Monosaccharides have three to seven carbon atoms** in a chain, with a **carbonyl group** at either the terminal carbon (C1) or the carbon adjacent to it (C2). In most carbohydrates, each of the remaining carbon atoms has a **hydroxy group.** Monosaccharides are often drawn vertically, with the carbonyl group at the top. When this convention is used, monosaccharides look different from molecules encountered in prior chapters.

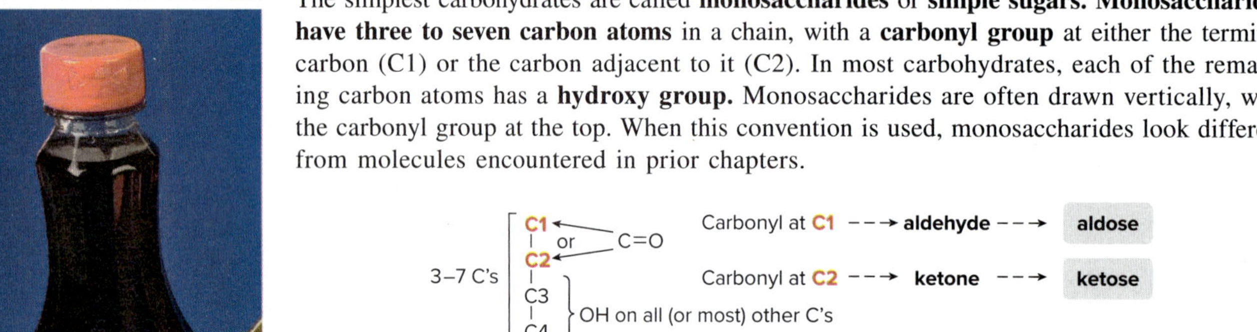

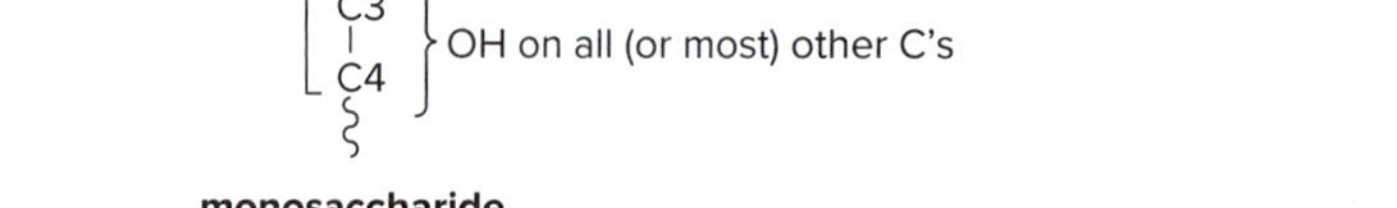

- **Monosaccharides with an aldehyde carbonyl group at C1 are called *aldoses*.**
- **Monosaccharides with a ketone carbonyl group at C2 are called *ketoses*.**

Several examples of simple carbohydrates are shown. D-Glyceraldehyde and dihydroxyacetone have the same molecular formula, so they are **constitutional isomers,** as are D-glucose and D-fructose.

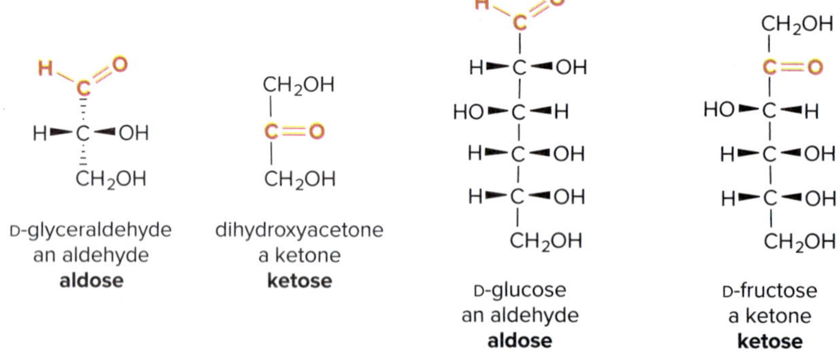

D-Fructose is almost twice as sweet as normal table sugar (sucrose) with about the same number of calories per gram. "Lite" food products use only half as much fructose as sucrose for the same level of sweetness, so they have fewer calories.
Matt Meadows/McGraw Hill

All carbohydrates have common names. The simplest aldehyde, glyceraldehyde, and the simplest ketone, dihydroxyacetone, are the only monosaccharides whose names do not end in the suffix ***-ose.*** (The prefix "D-" is explained in Section 26.2C.)

A monosaccharide is called

- **a triose if it has 3 C's;**
- **a tetrose if it has 4 C's;**
- **a pentose if it has 5 C's;**
- **a hexose if it has 6 C's, and so forth.**

The ketotriose dihydroxyacetone is the active ingredient in many artificial tanning agents.
Elite Images/McGraw Hill

These terms are then combined with the words *aldose* and *ketose* to indicate both the number of carbon atoms in the monosaccharide and whether it contains an aldehyde or ketone. Thus, glyceraldehyde is an **aldotriose** (three C atoms and an aldehyde), glucose is an **aldohexose** (six C atoms and an aldehyde), and fructose is a **ketohexose** (six C atoms and a ketone).

Monosaccharides have high melting points and are water soluble. Unlike most other organic compounds, monosaccharides are so polar that they are insoluble in organic solvents. Monosaccharides are all sweet tasting, but their relative sweetness varies a great deal.

Problem 26.1 Draw the structure of (a) a ketotetrose; (b) an aldopentose; (c) an aldotetrose.

26.2A Fischer Projection Formulas

A striking feature of carbohydrate structure is the presence of stereogenic centers. **All carbohydrates except for dihydroxyacetone contain one or more stereogenic centers.**

The simplest aldehyde, glyceraldehyde, has one stereogenic center, so there are two possible **enantiomers.** Only the enantiomer with the *R* configuration occurs naturally.

(*R*)-glyceraldehyde
naturally occurring enantiomer

(*S*)-glyceraldehyde

The stereogenic centers in sugars are often depicted following a different convention than is usually seen for other stereogenic centers. Instead of drawing a tetrahedron with two bonds in the plane, one in front of the plane and one behind it, the **tetrahedron is tipped so that horizontal bonds come forward (drawn on wedges) and vertical bonds go behind (on dashed wedges).** This structure is then abbreviated by a **cross formula,** also called a **Fischer projection formula.** In a Fischer projection formula:

- **A carbon atom is located at the intersection of the two lines of the cross.**
- **The horizontal bonds come forward, on wedges.**
- **The vertical bonds go back, on dashed wedges.**
- **In a carbohydrate, the aldehyde or ketone carbonyl is put at or near the top.**

Carbon atoms that are not stereogenic centers are generally drawn in. Using a Fischer projection formula, (*R*)-glyceraldehyde becomes:

Tip red bonds in the plane forward.

(*R*)-glyceraldehyde

- Horizontal bonds come *forward*.
- Vertical bonds go *back*.

Fischer projection formula
(*R*)-glyceraldehyde

Do *not* rotate a Fischer projection formula in the plane of the page, because you might inadvertently convert a compound to its enantiomer. When using Fischer projections, it is usually best to convert them to structures with wedges and dashed wedges, and then manipulate them. Although a Fischer projection formula can be used for the stereogenic center in any compound, it is most commonly used for monosaccharides.

Sample Problem 26.1 Drawing a Fischer Projection Formula

Convert each compound to a Fischer projection formula.

a.

b.

Solution

Rotate and re-draw each molecule to place the horizontal bonds in front of the plane and the vertical bonds behind the plane. Then use a cross to represent the stereogenic center.

a. 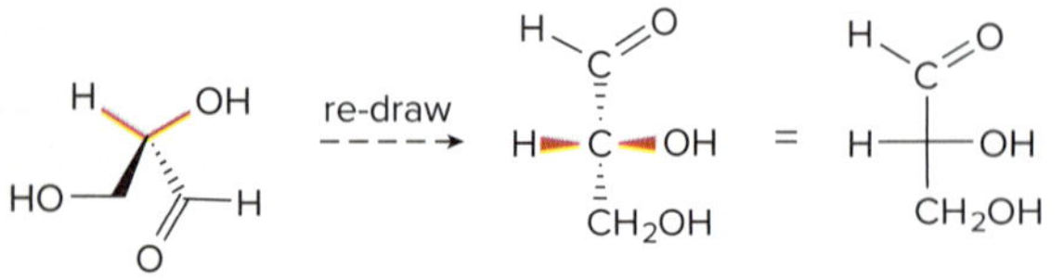 [structure] re-draw → H▸C◂OH (CHO top, CH_2OH bottom) = H—+—OH, CH_2OH

b. 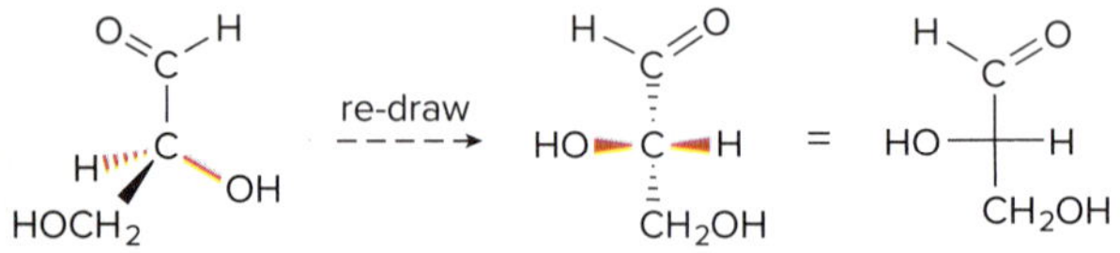 [structure] re-draw → HO▸C◂H (CHO top, CH_2OH bottom) = HO—+—H, CH_2OH

Problem 26.2 Draw each stereogenic center using a Fischer projection formula.

a. CH_3▸C◂OH (COOH top, CH_2CH_2OH bottom) b. [structure] c. [structure] d. [structure]

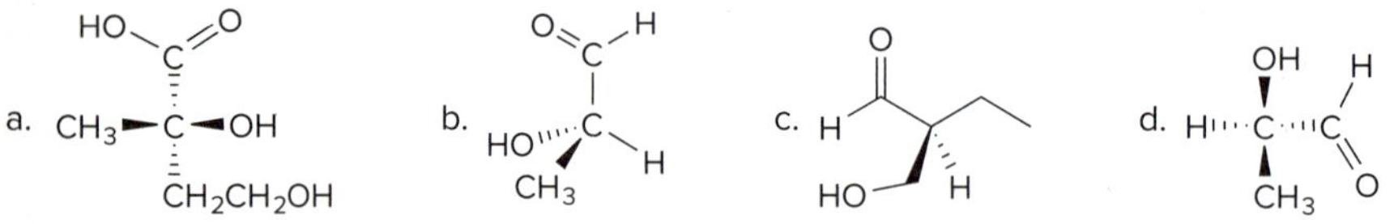

More Practice: Try Problems 26.38, 26.40.

R,S designations can be assigned to any stereogenic center drawn as a Fischer projection formula in the following manner:

[1] **Assign priorities (1 → 4)** to the four groups bonded to the stereogenic center using the rules detailed in Section 5.6.

[2] When the lowest-priority group occupies a **vertical bond**—that is, it projects *behind* the plane on a dashed wedge—tracing a circle in the **clockwise direction** (from priority group 1 → 2 → 3) gives the ***R* configuration.** Tracing a circle in the **counterclockwise direction** gives the ***S* configuration.**

[3] When the lowest-priority group occupies a **horizontal bond**—that is, it projects *in front of* the plane on a wedge—**reverse the answer** obtained in Step [2] to designate the configuration.

Sample Problem 26.2 Labeling a Fischer Projection as *R* or *S*

Re-draw each Fischer projection formula using wedges and dashed wedges for the stereogenic center, and label the center as *R* or *S*.

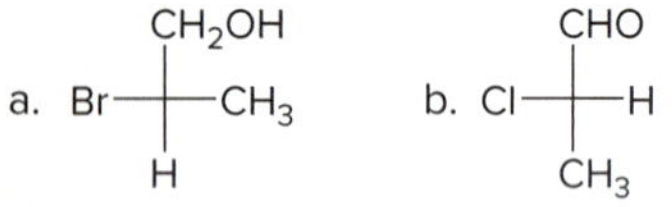

a. Br—+—CH_3 (CH_2OH top, H bottom) b. Cl—+—H (CHO top, CH_3 bottom)

Solution

For each molecule:

[1] Convert the Fischer projection formula to a representation with wedges and dashed wedges.

[2] Assign priorities (Section 5.6).

[3] Determine *R* or *S* in the usual manner. Reverse the answer if priority group [4] is oriented forward (on a wedge).

a. 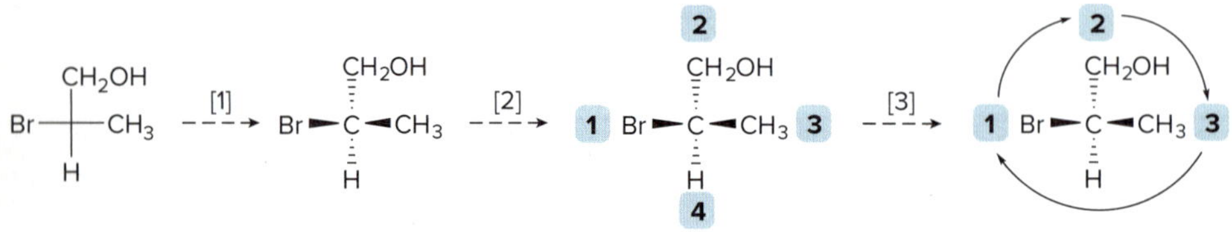 Br—+—CH_3 (CH_2OH, H) --[1]→ Br▸C◂CH_3 --[2]→ 1 Br▸C◂CH_3 3 (2 CH_2OH, 4 H) --[3]→ 1 Br▸C◂CH_3 3 (2 CH_2OH, H)

Clockwise circle and group [4] is oriented *behind:* ***R* configuration**

b. Cl–C(CHO)(CH$_3$)–H --[1]--> Cl▸C◂H (CHO, CH$_3$) --[2]--> 1 Cl▸C◂H 4 (2 CHO, 3 CH$_3$) --[3]--> 1 Cl▸C◂H (2 CHO, 3 CH$_3$)

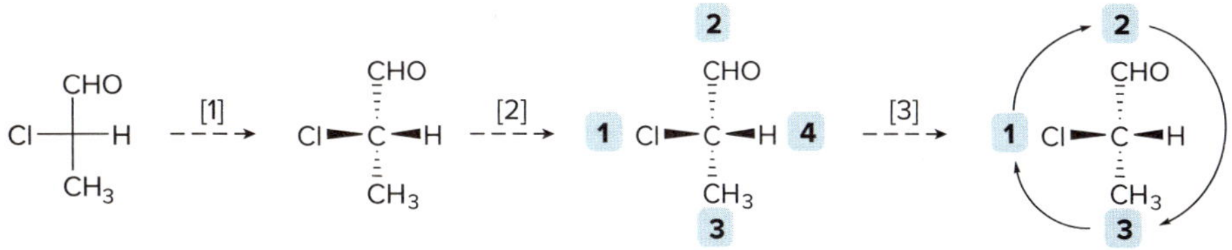

Clockwise circle and group [4] is oriented *forward:* ***S* configuration**

Problem 26.3 Label each stereogenic center as *R* or *S*.

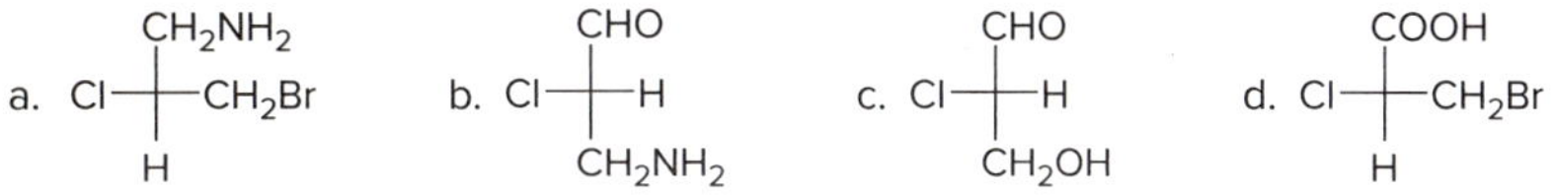

a. Cl–C(CH$_2$NH$_2$)(H)–CH$_2$Br b. Cl–C(CHO)(CH$_2$NH$_2$)–H c. Cl–C(CHO)(CH$_2$OH)–H d. Cl–C(COOH)(H)–CH$_2$Br

More Practice: Try Problem 26.41.

26.2B Monosaccharides with More Than One Stereogenic Center

The number of possible stereoisomers of a monosaccharide increases exponentially with the number of stereogenic centers present. **An aldohexose has four stereogenic centers, so it has $2^4 = 16$ possible stereoisomers,** or eight pairs of enantiomers.

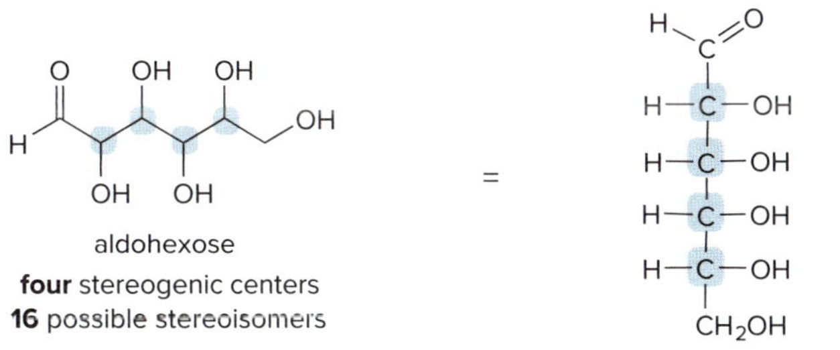

aldohexose
four stereogenic centers
16 possible stereoisomers

=

vertical representation

Fischer projection formulas are also used for compounds like aldohexoses that contain several stereogenic centers. In this case, the molecule is drawn with a vertical carbon skeleton and the stereogenic centers are stacked one above another. Using this convention, **all horizontal bonds project *forward* (on wedges).**

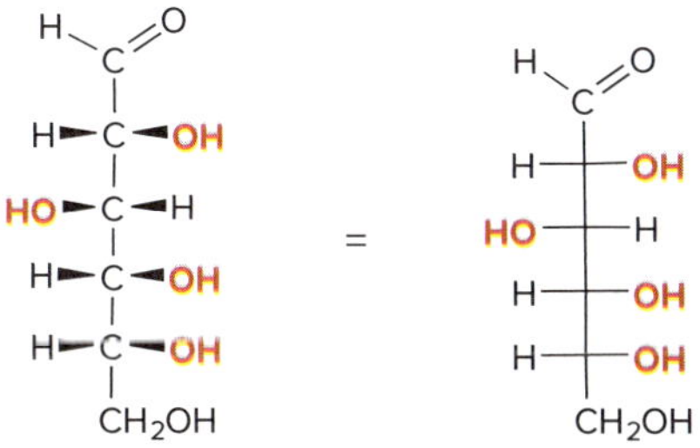

D-glucose
All horizontal bonds are drawn as **wedges.**

=

Fischer projection

Although Fischer projections are commonly used to depict monosaccharides with many stereogenic centers, care must be exercised in using them, because they do not give a true picture of the three-dimensional structures they represent. **Each stereogenic center is drawn in the**

Figure 26.2
A Fischer projection and the 3-D structure of glucose

All bonds are eclipsed in a Fischer projection.

D-glucose

Because all bonds are drawn **eclipsed,** the carbon backbone in a Fischer projection would curl around a cylinder.

less stable eclipsed conformation, so the Fischer projection of glucose really represents the molecule in a cylindrical conformation, as shown in Figure 26.2.

Sample Problem 26.3 Converting a Ball-and-Stick Model to a Fischer Projection

Convert the ball-and-stick model to a Fischer projection.

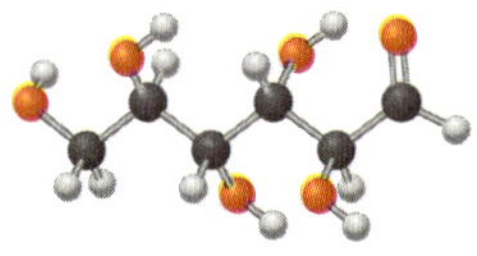

Solution

The ball-and-stick model is shown in the more stable staggered conformation, so it must be converted to the less stable eclipsed conformation used in a Fischer projection.

1. Re-draw the model as a skeletal structure (**A**), and rotate it to place the carbonyl group at the top (**B**).
2. To convert the all-staggered form to the all-eclipsed form, rotate around the bonds in **B** to swing two carbons (labeled in red) 180°, forming **C.**
3. Re-draw **C** so that all bonds to H and OH on the four stereogenic centers are drawn on wedges, forming **D.** Groups that are on wedges in **C** (in red) are on the left side of the carbon skeleton in **D,** and groups on dashed wedges in **C** (in blue) are on the right side of the carbon skeleton in **D.**
4. Replace the wedges with crosses to form the Fischer projection.

A skeletal structure of the ball-and-stick model → rotate (1) → **B** → Convert the staggered to the eclipsed conformation. (2) → **C** → re-draw (3) → **D** → (4) → **Fischer projection**

Problem 26.4 Convert the ball-and-stick model to a Fischer projection.

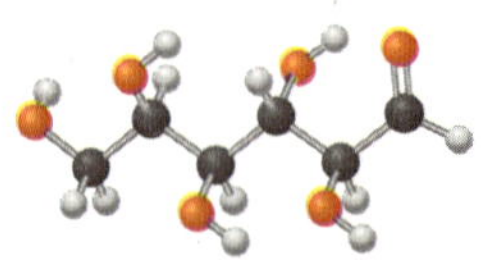

More Practice: Try Problem 26.38.

Problem 26.5 Assign *R,S* designations to each stereogenic center in glucose.

26.2C D and L Monosaccharides

Although the prefixes *R* and *S* can be used to designate the configuration of stereogenic centers in monosaccharides, an older system of nomenclature uses the prefixes D- and L- instead. **Naturally occurring glyceraldehyde with the *R* configuration is called the D-isomer. Its enantiomer, (*S*)-glyceraldehyde, is called the L-isomer.**

(*R*)-glyceraldehyde
D-glyceraldehyde

(*S*)-glyceraldehyde
L-glyceraldehyde

The letters D and L are used to label all monosaccharides, even those with multiple stereogenic centers. **The configuration of the stereogenic center *farthest* from the carbonyl group determines whether a monosaccharide is D- or L-.**

- **A D-sugar has the OH group on the stereogenic center farthest from the carbonyl on the *right* in a Fischer projection (like D-glyceraldehyde).**
- **An L-sugar has the OH group on the stereogenic center farthest from the carbonyl on the *left* in a Fischer projection (like L-glyceraldehyde).**

The two designations, D and *d,* refer to very different phenomena. The "D" designates the configuration around a stereogenic center in a monosaccharide. The "*d,*" on the other hand, is an abbreviation for "dextrorotatory"; that is, a *d*-compound rotates the plane of polarized light in the clockwise direction. A D-sugar may be dextrorotatory or it may be levorotatory. **There is no direct correlation between D and *d* or L and *l*.**

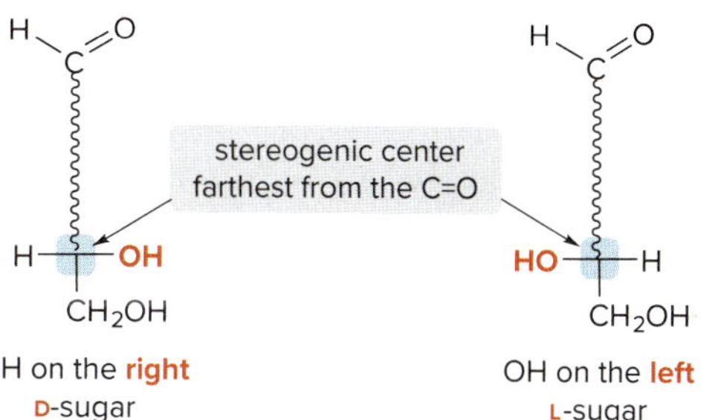

Glucose and all other naturally occurring sugars are D-sugars. L-Glucose, a compound that does not occur in nature, is the enantiomer of D-glucose. **L-Glucose has the opposite configuration at *every* stereogenic center.**

D-glucose
naturally occurring enantiomer

L-glucose

Problem 26.6 (a) Label compounds **A, B,** and **C** as D- or L-sugars. (b) How are compounds **A** and **B** related? **A** and **C**? **B** and **C**? Choose from enantiomers, diastereomers, or constitutional isomers.

A **B** **C**

26.3 The Family of D-Aldoses

The common name of each monosaccharide indicates both the number of atoms it contains and the configuration at each of the stereogenic centers. Because the common names are firmly entrenched in the chemical literature, no systematic method has ever been established to name these compounds.

Beginning with D-glyceraldehyde, one may formulate other D-aldoses having four, five, or six carbon atoms by adding carbon atoms (each bonded to H and OH), one at a time, between C1 and C2. **Two D-aldotetroses can be formed from D-glyceraldehyde,** one with the new OH group on the right and one with the new OH group on the left. Their names are D-erythrose and D-threose. They are two **diastereomers,** each with two stereogenic centers, labeled in blue.

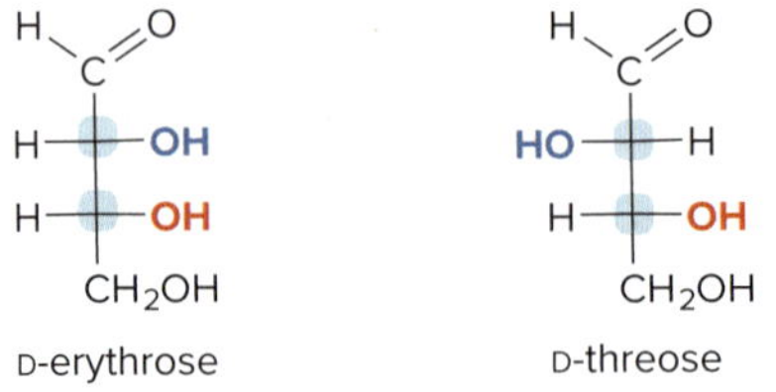

D-erythrose: H–C=O; H–C–OH; H–C–OH; CH_2OH

D-threose: H–C=O; HO–C–H; H–C–OH; CH_2OH

Because each aldotetrose has two stereogenic centers, there are 2^2 or four possible stereoisomers. D-Erythrose and D-threose are two of them. The other two are their enantiomers, called L-erythrose and L-threose, respectively. The configuration around each stereogenic center is exactly the opposite in its enantiomer. All four stereoisomers of the aldotetroses are shown in Figure 26.3.

Figure 26.3 The four stereoisomeric aldotetroses

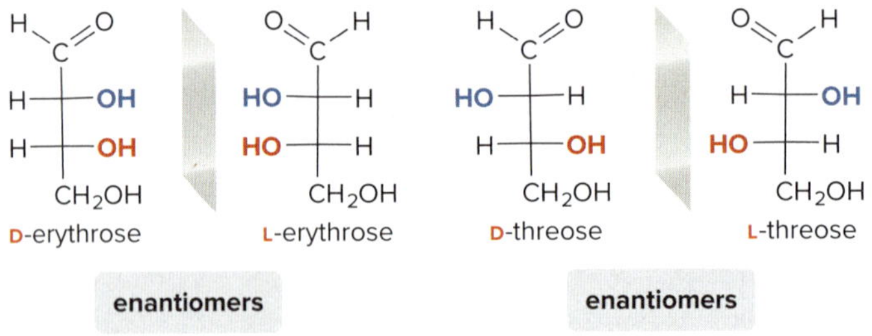

D-erythrose: H–C=O; H–C–OH; H–C–OH; CH_2OH
L-erythrose: O=C–H; HO–C–H; HO–C–H; CH_2OH
(enantiomers)

D-threose: H–C=O; HO–C–H; H–C–OH; CH_2OH
L-threose: O=C–H; H–C–OH; HO–C–H; CH_2OH
(enantiomers)

D-Ribose, D-arabinose, and D-xylose are all common aldopentoses in nature. D-Ribose is the carbohydrate component of RNA, the polymer that translates the genetic information of DNA for protein synthesis.

To continue forming the family of D-aldoses, we must add another carbon atom (bonded to H and OH) just below the carbonyl of either tetrose. Because there are *two* D-aldotetroses to begin with, and there are *two* ways to place the new OH (right or left), there are now *four* D-aldopentoses: D-ribose, D-arabinose, D-xylose, and D-lyxose. Each aldopentose now has *three* stereogenic centers, so there are 2^3 = **8** possible stereoisomers, or four pairs of enantiomers. The D-enantiomer of each pair is shown in Figure 26.4.

Finally, to form the D-aldohexoses, we must add another carbon atom (bonded to H and OH) just below the carbonyl of all the aldopentoses. Because there are *four* D-aldopentoses to begin with, and there are *two* ways to place the new OH (right or left), there are now *eight* D-aldohexoses. Each aldohexose now has *four* stereogenic centers, so there are 2^4 = **16** possible stereoisomers, or eight pairs of enantiomers. Only the D-enantiomer of each pair is shown in Figure 26.4.

The tree of D-aldoses (Figure 26.4) is arranged in pairs of compounds that are bracketed together. Each pair of compounds, such as D-glucose and D-mannose, has the same configuration around all of its stereogenic centers except for one.

Of the D-aldohexoses, only D-glucose and D-galactose are common in nature. **D-Glucose is by far the most abundant of all D-aldoses.** D-Glucose comes from the hydrolysis of starch and cellulose, and D-galactose comes from the hydrolysis of fruit pectins.

- **Two diastereomers that differ in the configuration around only one stereogenic center are called *epimers*.**

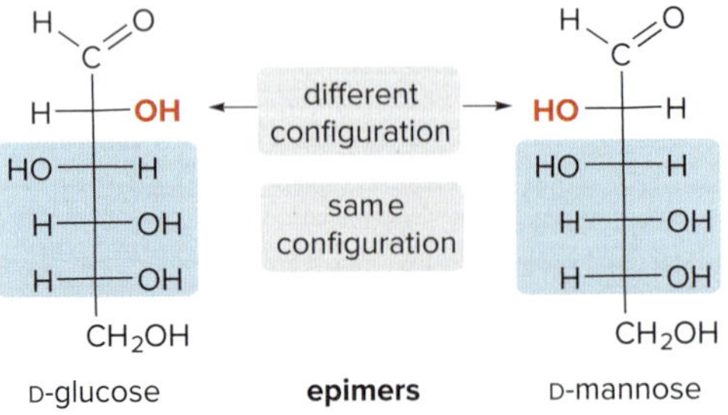

D-glucose: H–C=O; H–C–OH; HO–C–H; H–C–OH; H–C–OH; CH_2OH

D-mannose: H–C=O; HO–C–H; HO–C–H; H–C–OH; H–C–OH; CH_2OH

different configuration (C2); same configuration (C3–C5)

epimers

Figure 26.4
The family of D-aldoses having three to six carbon atoms

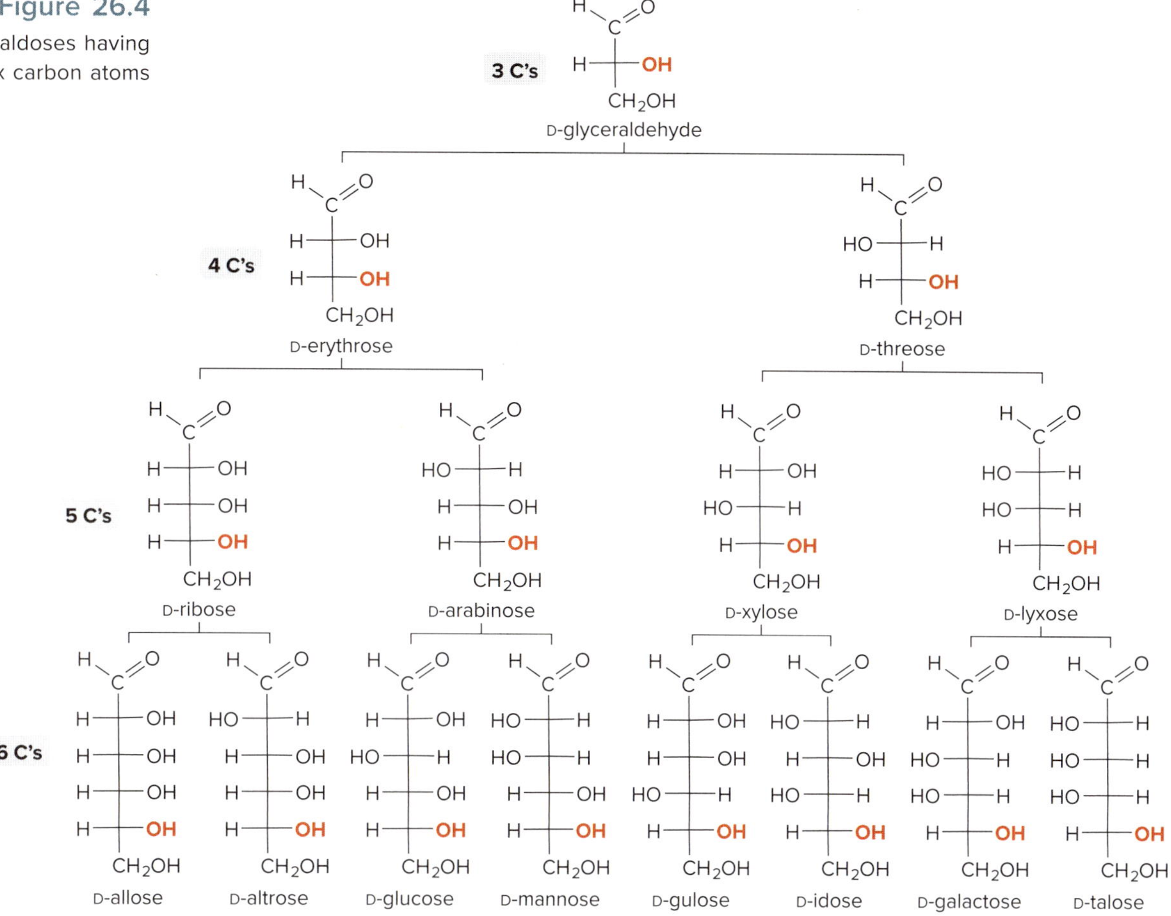

- All D-aldoses have the OH group on the stereogenic center farthest from the C=O (shown in red) on the **right**.

Problem 26.7 How many different aldoheptoses are there? How many are D-sugars? Draw all D-aldoheptoses having the *R* configuration at C2 and C3.

Problem 26.8 Draw two possible epimers of D-erythrose. Name each of these compounds using Figure 26.4.

26.4 The Family of D-Ketoses

The family of D-ketoses, shown in Figure 26.5, is formed from dihydroxyacetone by adding a new carbon (bonded to H and OH) between C2 and C3. Having a carbonyl group at C2 decreases the number of stereogenic centers in these monosaccharides, so that there are only four D-ketohexoses. The most common naturally occurring ketose is D-fructose.

Problem 26.9 Referring to the structures in Figures 26.4 and 26.5, classify each pair of compounds as enantiomers, epimers, diastereomers but not epimers, or constitutional isomers of each other.

a. D-allose and L-allose
b. D-altrose and D-gulose
c. D-galactose and D-talose
d. D-mannose and D-fructose
e. D-fructose and D-sorbose
f. L-sorbose and L-tagatose

Problem 26.10
a. Draw the enantiomer of D-fructose.
b. Draw an epimer of D-fructose at C4. What is the name of this compound?
c. Draw an epimer of D-fructose at C5. What is the name of this compound?

Figure 26.5
The family of D-ketoses having three to six carbon atoms

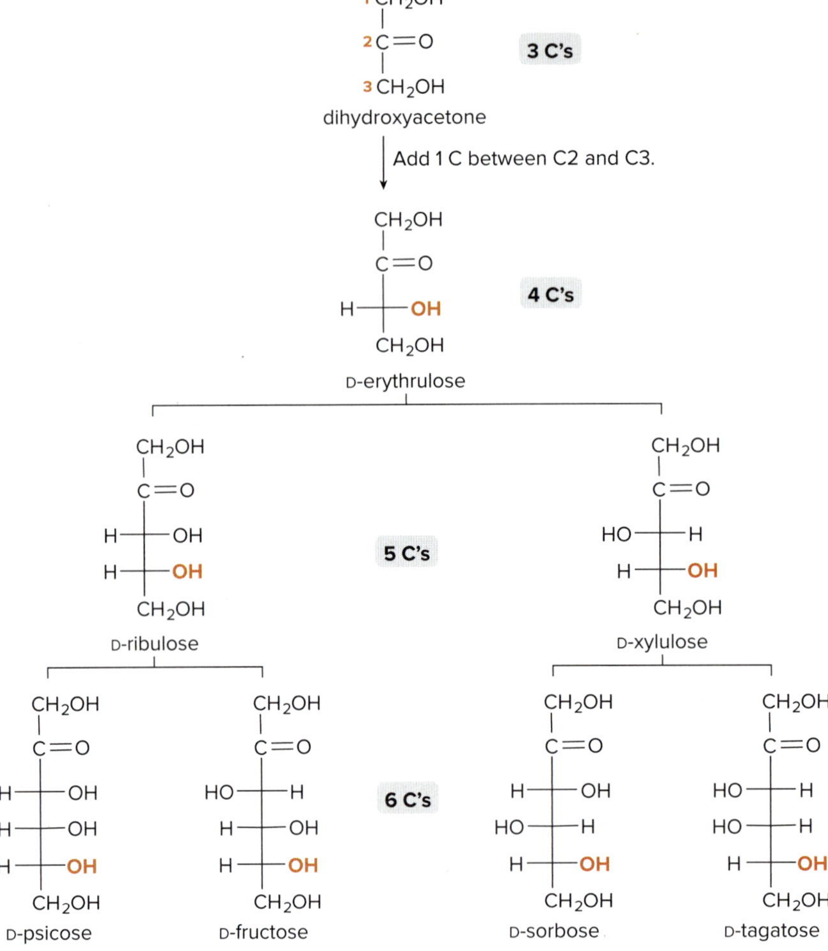

- All D-ketoses have the OH group on the stereogenic center farthest from the C=O (shown in red) on the **right.**

26.5 The Cyclic Forms of Monosaccharides

Although the monosaccharides in Figures 26.4 and 26.5 are drawn as acyclic carbonyl compounds containing several hydroxy groups, the hydroxy and carbonyl groups of monosaccharides can undergo intramolecular cyclization reactions to form **hemiacetals** having either five or six atoms in the ring. This process was first discussed in Section 18.15.

hemiacetal

pyranose ring
(a six-membered ring)

hemiacetal

furanose ring
(a five-membered ring)

- **A six-membered ring containing an O atom is called a *pyranose* ring.**
- **A five-membered ring containing an O atom is called a *furanose* ring.**

Cyclization of a hydroxy carbonyl compound always forms a stereogenic center at the hemiacetal carbon, called the **anomeric carbon.** The two hemiacetals are called **anomers.**

- **Anomers are stereoisomers of a cyclic monosaccharide that differ in the position of the OH group at the hemiacetal carbon.**

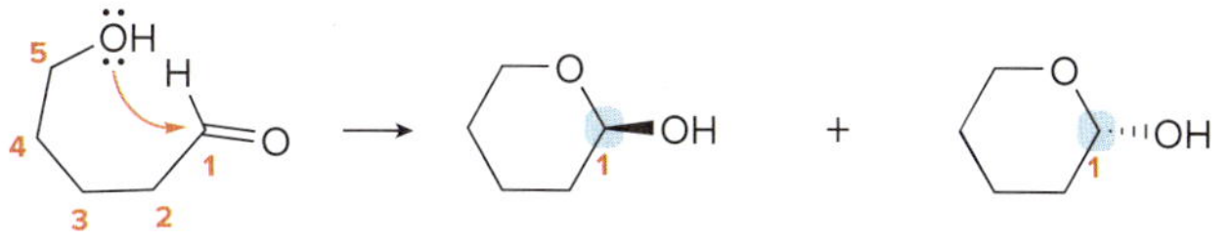

new stereogenic center at the anomeric carbon

Cyclization forms the more stable ring size in a given molecule. **The most common monosaccharides, the aldohexoses like glucose, typically form a pyranose ring,** so our discussion begins with forming a cyclic hemiacetal from D-glucose.

26.5A Drawing Glucose as a Cyclic Hemiacetal

Which of the five OH groups in glucose is at the right distance from the carbonyl group to form a six-membered ring? The **O atom on the stereogenic center farthest from the carbonyl** (C5) is six atoms from the carbonyl carbon, placing it in the proper position for cyclization to form a pyranose ring.

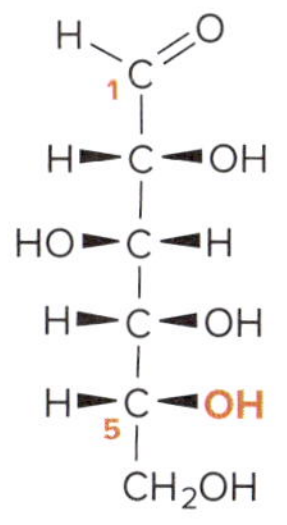

D-glucose

The OH at C5 forms the pyranose ring.

To translate the acyclic form of glucose into a cyclic hemiacetal, we must draw the hydroxy aldehyde in a way that suggests the position of the atoms in the new ring, and then draw the ring. **By convention the O atom in the new pyranose ring is drawn in the upper right corner of the six-membered ring.**

Rotating the groups on the bottom stereogenic center in **A** places all six atoms needed for the ring (including the OH) in a vertical line (**B**). Re-drawing this representation as a Fischer projection makes the structure appear less cluttered (**C**). Twisting this structure and rotating it 90° forms **D.** Structures **A–D** are four different ways of drawing the same acyclic structure of D-glucose.

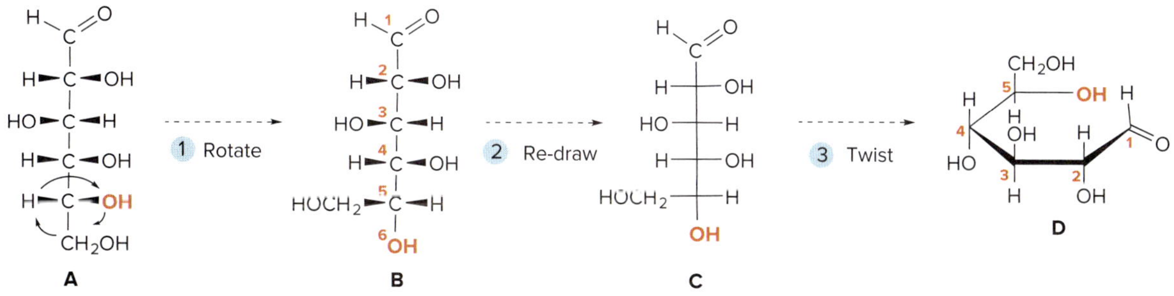

A → 1 Rotate → B → 2 Re-draw → C → 3 Twist → D

We are now set to draw the cyclic hemiacetal formed by nucleophilic attack of the OH group on C5 on the aldehyde carbonyl. Because cyclization creates a new stereogenic center, there are **two cyclic forms of D-glucose,** an **α anomer** and a **β anomer.** All the original stereogenic centers retain their configuration in both of the products formed.

- **The α anomer of a D monosaccharide has the OH group drawn *down*, trans to the CH_2OH group at C5. The α anomer of D-glucose is called α-D-glucose, or α-D-glucopyranose (to emphasize the six-membered ring).**
- **The β anomer of a D monosaccharide has the OH group drawn *up*, cis to the CH_2OH group at C5. The β anomer is called β-D-glucose, or β-D-glucopyranose (to emphasize the six-membered ring).**

acyclic D-glucose ⇌ **α anomer** α-D-glucose + **β anomer** β-D-glucose

new stereogenic center at the anomeric carbon (C1)

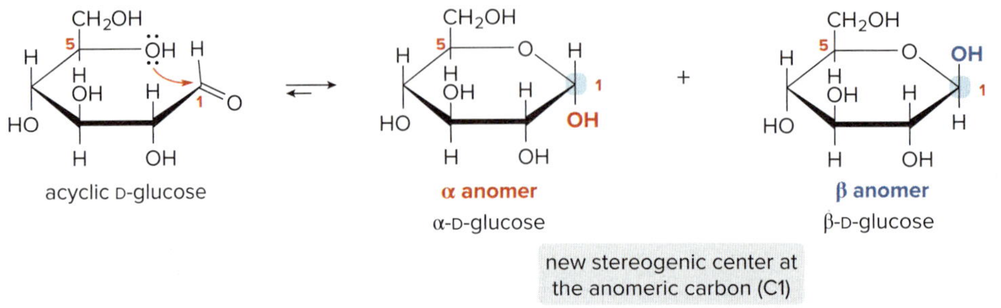

The **α anomer** in any monosaccharide has the anomeric OH group and the CH_2OH group **trans.** The **β anomer** has the anomeric OH group and the CH_2OH group **cis.**

These flat, six-membered rings used to represent the cyclic hemiacetals of glucose and other sugars are called **Haworth projections.** The cyclic forms of glucose now have **five stereogenic centers, the four from the starting hydroxy aldehyde and the new anomeric carbon.** α-D-Glucose and β-D-glucose are **diastereomers,** because only the anomeric carbon has a different configuration.

The mechanism for this transformation is exactly the same as the mechanism that converts a hydroxy aldehyde to a cyclic hemiacetal (Mechanism 18.12). **The acyclic aldehyde and two cyclic hemiacetals are all in equilibrium.** Each cyclic hemiacetal can be isolated and crystallized separately, but when any one compound is placed in solution, an equilibrium mixture of all three forms results. This process is called **mutarotation.** At equilibrium, the mixture has 37% of the α anomer, 63% of the β anomer, and only trace amounts of the acyclic hydroxy aldehyde (Figure 26.6). Also shown are representations of the three forms of glucose using wedges and dashed wedges.

Figure 26.6
The three forms of glucose

α anomer α-D-glucose ⇌ **acyclic aldehyde** ⇌ **β anomer** β-D-glucose

The CH_2OH and anomeric OH are **trans.**

The CH_2OH and anomeric OH are **cis.**

37% trace 63%

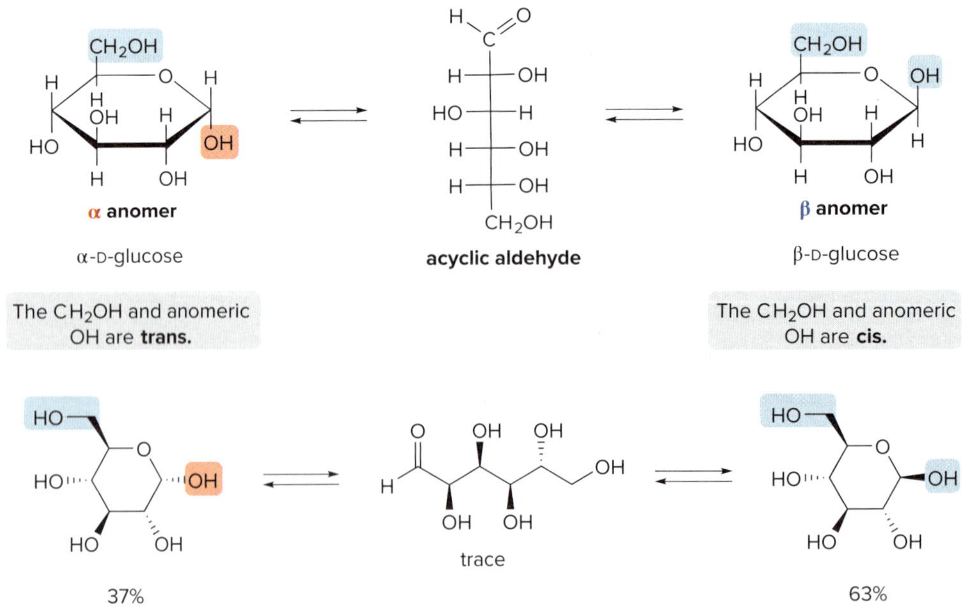

- Bonds above the ring in a Haworth projection are drawn as wedges.
- Bonds below the ring in a Haworth projection are drawn as dashed wedges.

Problem 26.11 Label each Haworth projection as an α or β anomer, and convert the Haworth projection to a six-membered ring with wedges and dashed wedges.

a. 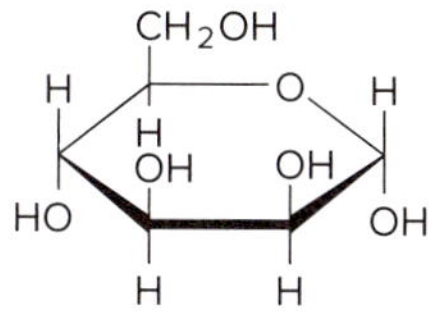 [Haworth projection: CH_2OH; ring O; substituents H, H, OH, OH, HO, OH, H, H, H]

b. 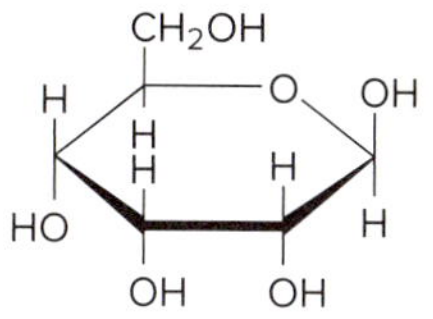 [Haworth projection: CH_2OH; ring O; substituents OH, H, H, H, H, HO, H, OH, OH]

26.5B Haworth Projections

To convert an acyclic monosaccharide to a Haworth projection, follow a stepwise procedure.

How To Draw a Haworth Projection from an Acyclic Aldohexose

Example Convert D-mannose to a Haworth projection.

[Fischer projection: H–C=O; HO–H; HO–H; H–OH; H–OH; CH_2OH]

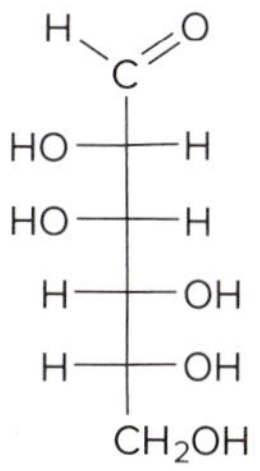

D-mannose

Step [1] **Place the O atom in the upper right corner of a hexagon, and add the CH_2OH group on the first carbon counterclockwise from the O atom.**

- For **D-sugars,** the CH_2OH group is drawn **up.** For **L-sugars,** the CH_2OH group is drawn **down.**

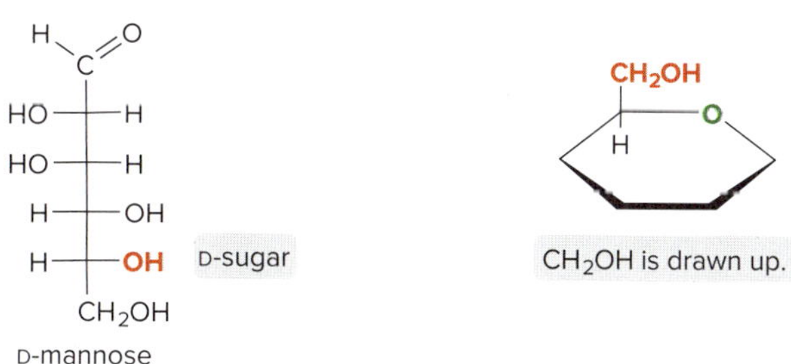

Step [2] **Place the anomeric carbon on the first carbon clockwise from the O atom.**

- For an **α anomer,** the **OH** is drawn **down** in a D-sugar.
- For a **β anomer,** the **OH** is drawn **up** in a D-sugar.

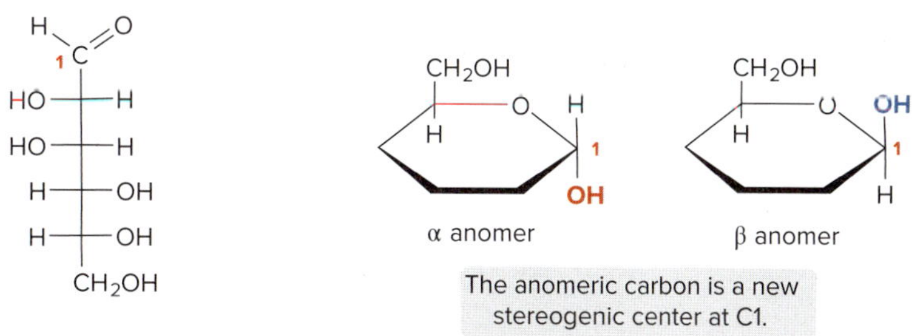

- Remember: **The carbonyl carbon becomes the anomeric carbon** (a new stereogenic center).

—*Continued*

How To, continued . . .

Step [3] **Add the substituents at the three remaining stereogenic centers clockwise around the ring.**

- The substituents on the **right side** of the Fischer projection are drawn **down.**
- The substituents on the **left** are drawn **up.**

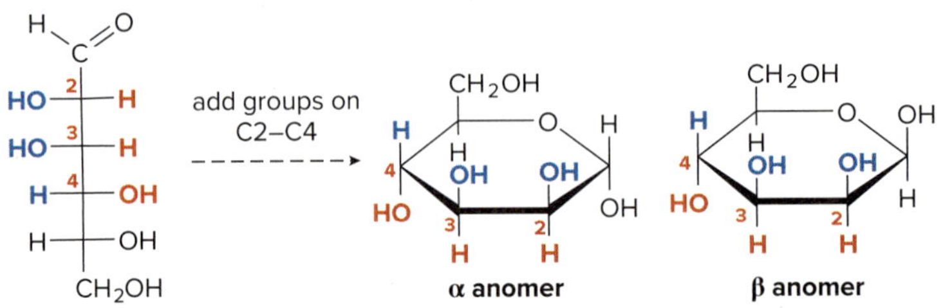

add groups on C2–C4

α anomer

β anomer

Problem 26.12 Convert each aldohexose to the indicated anomer using a Haworth projection.

a. Draw the α anomer of:

b. Draw the α anomer of:

c. Draw the β anomer of:

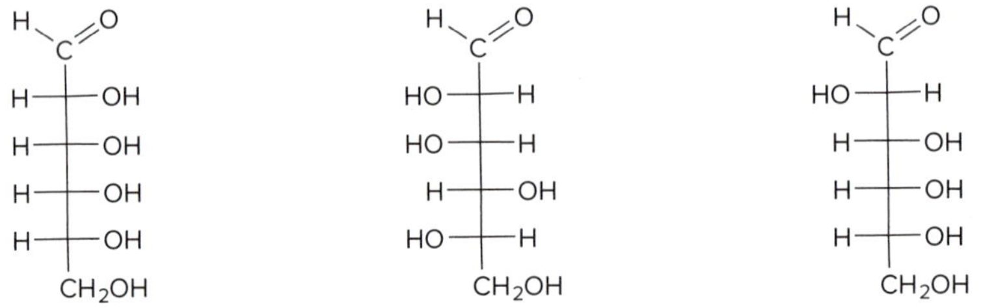

Sample Problem 26.4 shows how to convert a Haworth projection back to the acyclic form of a monosaccharide. It doesn't matter whether the hemiacetal is the α or β anomer, because **both anomers give the *same* hydroxy aldehyde.**

Sample Problem 26.4 Converting a Haworth Projection to a Fischer Projection

Convert the following Haworth projection to the acyclic form of the aldohexose.

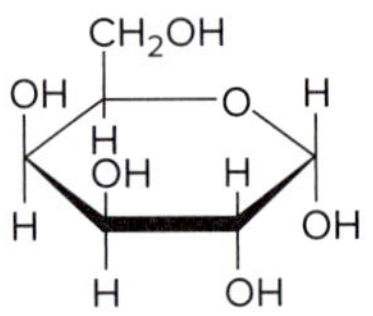

Solution

To convert the substituents to the acyclic form, **start at the pyranose O atom,** and work in a ***counterclockwise*** fashion around the ring, and from **bottom-to-top** along the chain.

[1] Draw the carbon skeleton, **placing the CHO on the top and the CH_2OH on the bottom.**

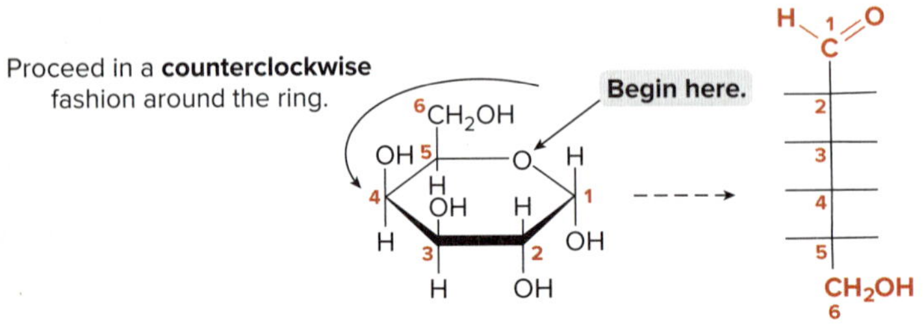

Proceed in a **counterclockwise** fashion around the ring.

Begin here.

[2] **Classify the sugar as D- or L-.**

- The CH_2OH is drawn **up,** so it is a **D-sugar.**
- A D-sugar has the OH group on the bottom stereogenic center on the **right.**

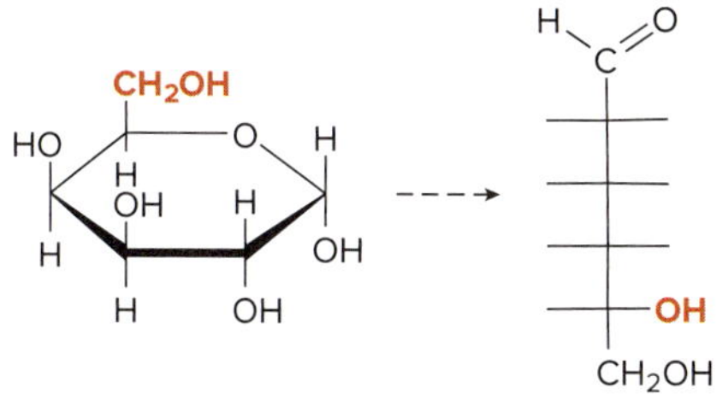

[3] Add the three other stereogenic centers.

- **Up** substituents go on the **left.**
- **Down** substituents go on the **right.**

Answer:

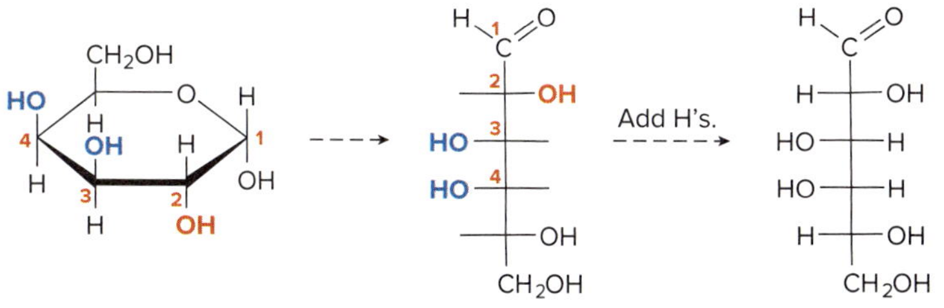

- **The anomeric carbon becomes the C=O at C1.**

Problem 26.13 Convert each Haworth projection to its acyclic form.

a.

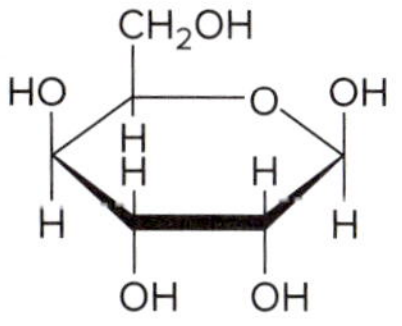

b.

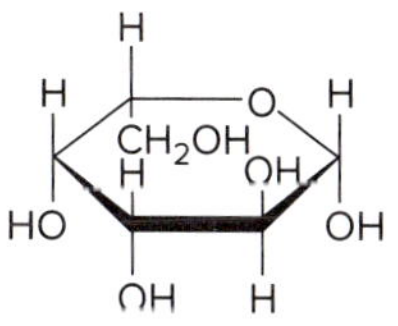

More Practice: Try Problems 26.39a, 26.47.

26.5C Three-Dimensional Representations for D-Glucose

Because the chair form of a six-membered ring gives the truest picture of its three-dimensional shape, we must learn to convert Haworth projections to chair forms.

To convert a Haworth projection to a chair form:

- Draw the pyranose ring with the O atom as an "up" atom.
- The "up" substituents in a Haworth projection become the "up" bonds (either axial or equatorial) on a given carbon atom on a puckered six-membered ring.
- The "down" substituents in a Haworth projection become the "down" bonds (either axial or equatorial) on a given carbon atom on a puckered six-membered ring.

As a result, the three-dimensional chair form of β-D-glucose is drawn in this manner:

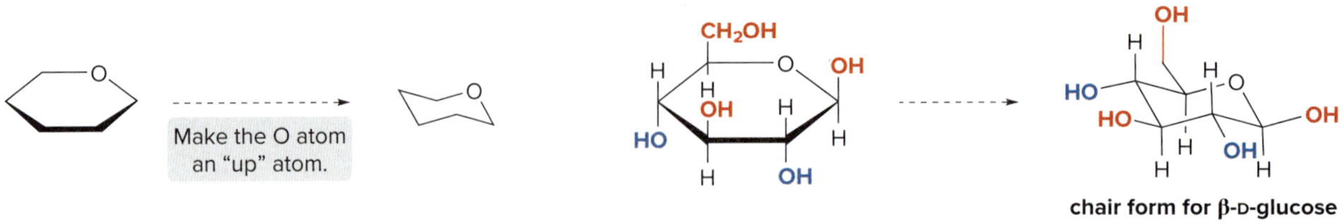

Make the O atom an "up" atom.

chair form for β-D-glucose

- "Up" substituents are labeled in red.
- "Down" substituents are labeled in blue.

Glucose has all substituents larger than a hydrogen atom in the more roomy equatorial positions, making it the most stable and thus most prevalent monosaccharide. The β anomer is the major isomer at equilibrium, moreover, because the hemiacetal OH group is in the equatorial position, too. Figure 26.7 shows both anomers of D-glucose drawn as chair conformations.

Problem 26.14 Convert each Haworth projection in Problem 26.13 to a three-dimensional representation using a chair pyranose ring.

Figure 26.7 Three-dimensional representations for both anomers of D-glucose

α anomer

β anomer

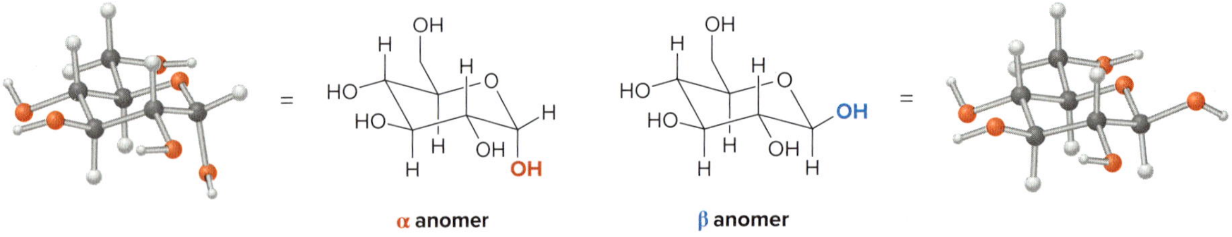

26.5D Furanoses

Certain monosaccharides—notably aldopentoses and ketohexoses—predominantly form furanose rings, rather than pyranose rings, in solution. The same principles apply to drawing these structures as for drawing pyranose rings, except the ring size is one atom smaller.

- Cyclization always forms a new stereogenic center at the anomeric carbon, so two different anomers are possible. For a D-sugar, the OH group is drawn *down* in the α anomer and *up* in the β anomer.
- Use the same drawing conventions for adding substituents to the five-membered ring. With D-sugars, the CH_2OH group is drawn *up.*

With D-ribose, the OH group used to form the five-membered furanose ring is located on C4. Cyclization yields two anomers at the new stereogenic center, which are called **α-D-ribofuranose** and **β-D-ribofuranose.**

Honey was the first and most popular sweetening agent until it was replaced by sugar (from sugarcane) in modern times. Honey is a mixture consisting largely of D-fructose and D-glucose. *Anastasy Yarmolovich/ iStockphoto/Getty Images*

D-ribose — The OH at C4 forms the furanose ring.

re-draw

α anomer: α-D-ribofuranose ⇌ ⇌ β anomer: β-D-ribofuranose

The same procedure can be used to draw the furanose form of D-fructose, the most common ketohexose. Because the carbonyl group is at C2 (instead of C1, as in the aldoses), the OH group at C5 reacts to form the hemiacetal in the five-membered ring. Two anomers are formed.

D-fructose — The OH at C5 forms the furanose ring.

re-draw

α anomer: α-D-fructofuranose ⇌ ⇌ β anomer: β-D-fructofuranose

The ketohexose D-allulose is present in small amounts in raisins and figs. Enzyme-catalyzed isomerization of fructose to allulose produces allulose in bulk quantities. *Daniel C. Smith*

Problem 26.15 Aldotetroses exist in the furanose form. Draw both anomers of D-erythrose.

Problem 26.16 D-Allulose, a ketohexose 70% as sweet as sucrose, is used as a low-calorie sweetener in food products. Because its caloric value is 10 times less than sucrose and it is readily prepared on an industrial scale from the monosaccharide fructose, there is widespread interest in using allulose as a sucrose substitute. (a) Convert allulose into a Fischer projection. Which monosaccharide in Figure 26.5 corresponds to allulose? (b) Allulose undergoes intramolecular cyclization to afford both furanose and pyranose forms. Label compounds [1]–[3] as allulose or a stereoisomer of allulose.

D-alluose [1] [2] [3]

26.6 Glycosides

Keep in mind the difference between a hemiacetal and an acetal:

hemiacetal
- **one OH group**
- **one OR group**

acetal
- **two OR groups**

Because monosaccharides exist in solution in an equilibrium between acyclic and cyclic forms, they undergo three types of reactions:

- **Reaction of the hemiacetal**
- **Reaction of the hydroxy groups**
- **Reaction of the carbonyl group**

Even though the acyclic form of a monosaccharide may be present in only trace amounts, the equilibrium can be tipped in its favor by Le Châtelier's principle (Section 9.8). Suppose, for example, that the carbonyl group of the acyclic form reacts with a reagent, thus depleting its equilibrium concentration. The equilibrium will then shift to compensate for the loss, thus producing more of the acyclic form, which can react further.

Note, too, that **monosaccharides have two different types of OH groups.** Most are "regular" alcohols and, as such, undergo reactions characteristic of alcohols. **The anomeric OH group, on the other hand, is part of a hemiacetal, giving it added reactivity.**

26.6A Glycoside Formation

Treatment of a monosaccharide with an alcohol and HCl converts the hemiacetal to an acetal called a glycoside. For example, treatment of α-D-glucose with CH_3OH and HCl forms two glycosides that are diastereomers at the acetal carbon. The α and β labels are assigned in the same way as anomers: with a D-sugar, **an α glycoside has the new OR group (OCH_3 group in this example) *down*, and a β glycoside has the new OR group *up*.**

CH_3OH, HCl

α-D-glucose → **α glycoside** + **β glycoside**

Only the hemiacetal OH reacts.

Mechanism 26.1 explains why **a single anomer forms two glycosides.** The reaction proceeds by way of a **planar carbocation,** which undergoes nucleophilic attack from two different directions to give a mixture of diastereomers. Because both α- and β-D-glucose form the same planar carbocation, each yields the same mixture of two glycosides.

The mechanism also explains why **only the hemiacetal OH group reacts.** Protonation of the hemiacetal OH, followed by loss of H_2O, forms a **resonance-stabilized carbocation** in Step [2]. A resonance-stabilized carbocation is *not* formed by loss of H_2O from any other OH group.

Unlike cyclic hemiacetals, **glycosides are acetals, so they do *not* undergo mutarotation.** When a single glycoside is dissolved in H_2O, it is *not* converted to an equilibrium mixture of α and β glycosides.

- Glycosides are acetals with an alkoxy group (OR) bonded to the anomeric carbon.

Problem 26.17 What glycosides are formed when each monosaccharide is treated with CH_3CH_2OH, HCl: (a) β-D-mannose; (b) α-D-gulose; (c) β-D-fructose?

Mechanism 26.1 Glycoside Formation

Part [1] Loss of H_2O from the hemiacetal

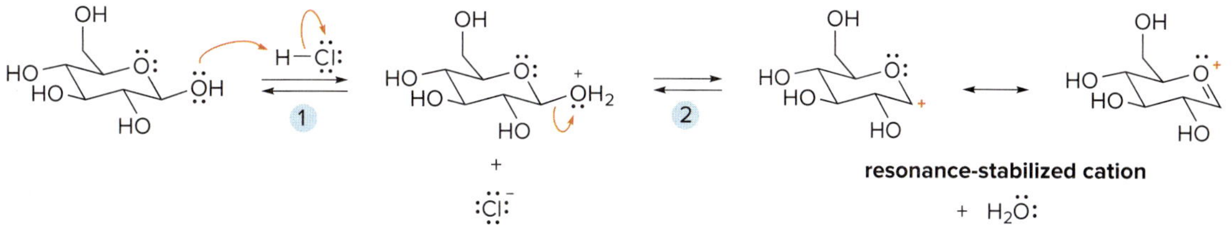

resonance-stabilized cation

1 – 2 Protonation of the hemiacetal OH followed by loss of H_2O forms a **resonance-stabilized carbocation.**

Part [2] Formation of the glycosides

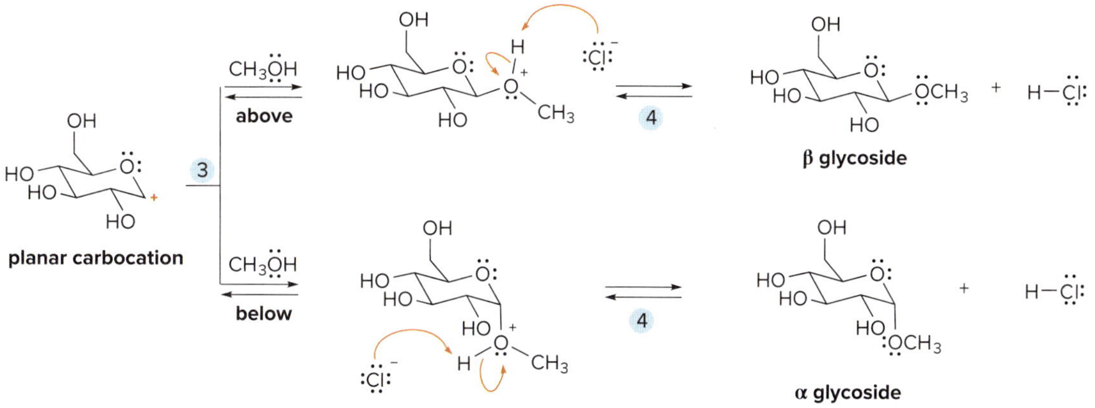

planar carbocation

above

below

β glycoside

α glycoside

3 – 4 **Nucleophilic attack by CH_3OH** occurs from both sides of the planar carbocation to yield α and β glycosides after loss of a proton.

26.6B Glycoside Hydrolysis

Because glycosides are acetals, **they are hydrolyzed with acid and water to cyclic hemiacetals and a molecule of alcohol.** A mixture of two anomers is formed from a single glycoside. For example, treatment of methyl α-D-glucopyranoside with aqueous acid forms a mixture of α- and β-D-glucose and methanol.

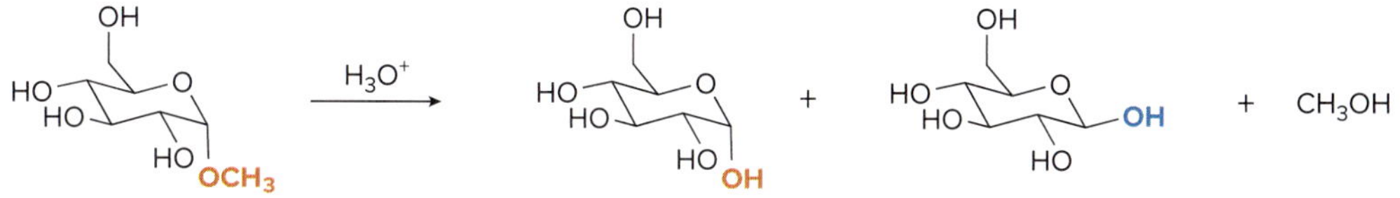

methyl α-D-glucopyranoside $\xrightarrow{H_3O^+}$ α-D-glucose + β-D-glucose + CH_3OH

The mechanism for glycoside hydrolysis is just the reverse of glycoside formation. It involves two parts: **formation of a planar carbocation,** followed by **nucleophilic attack of H_2O** to form anomeric hemiacetals, as shown in Mechanism 26.2.

Problem 26.18 Draw a stepwise mechanism for the following reaction.

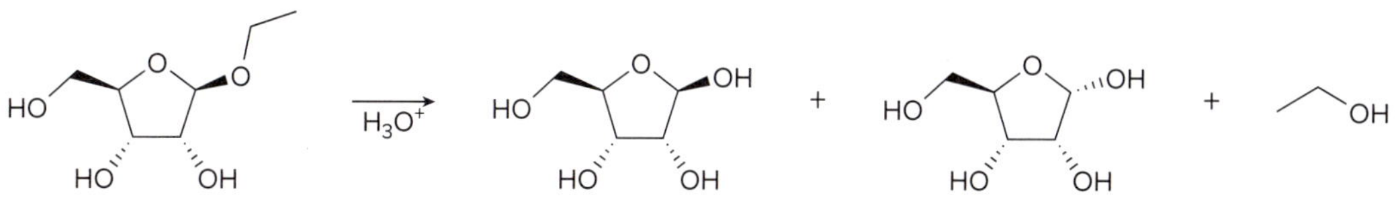

Mechanism 26.2 Glycoside Hydrolysis

Part [1] Loss of CH_3OH from the glycoside

1 – 2 Protonation of the acetal OCH_3 followed by loss of CH_3OH forms a **resonance-stabilized carbocation.**

Part [2] Formation of the hemiacetals

3 – 4 **Nucleophilic attack by H_2O** occurs from both sides of the planar carbocation to yield α and β anomers after loss of a proton.

26.6C Naturally Occurring Glycosides

The berries of the black nightshade plant (*Solanum nigrum*) are a source of the poisonous alkaloid solanine.
Westend61/Shutterstock

Salicin and **solanine** are two naturally occurring compounds that contain glycoside bonds as part of their structure. Salicin is an analgesic isolated from willow bark, and solanine is a poisonous compound produced in the leaves, stem, and green spots on the skin of potatoes. Solanine is also isolated from the berries of the deadly nightshade plant. It is believed that the role of the sugar rings in both salicin and solanine is to increase their water solubility.

Glycosides are common in nature. All disaccharides and polysaccharides are formed by joining monosaccharides together with glycosidic linkages. These compounds are discussed in detail beginning in Section 26.10.

Problem 26.19 Digoxin is a widely prescribed cardiac drug used to increase the force of heart contractions. (a) Label all the O atoms in digoxin that are part of a glycoside. (b) The alcohol or phenol formed from the hydrolysis of a glycoside is called an **aglycon.** What aglycon and monosaccharides are formed by the hydrolysis of digoxin?

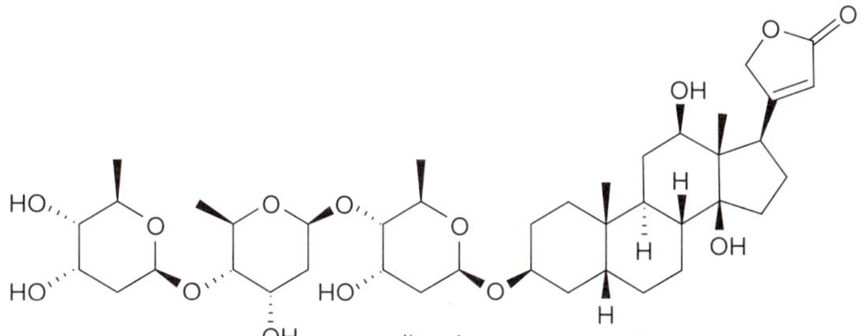

digoxin

26.7 Reactions of Monosaccharides at the OH Groups

Because monosaccharides contain OH groups, they undergo reactions typical of alcohols—that is, they are converted to **ethers** and **esters.** Because the cyclic hemiacetal form of a monosaccharide contains an OH group, this form of a monosaccharide must be drawn as the starting material for any reaction that occurs at an OH group.

All OH groups of a cyclic monosaccharide are converted to ethers by treatment with base and an alkyl halide. For example, α-D-glucose reacts with silver(I) oxide (Ag_2O, a base) and excess CH_3I to form a pentamethyl ether.

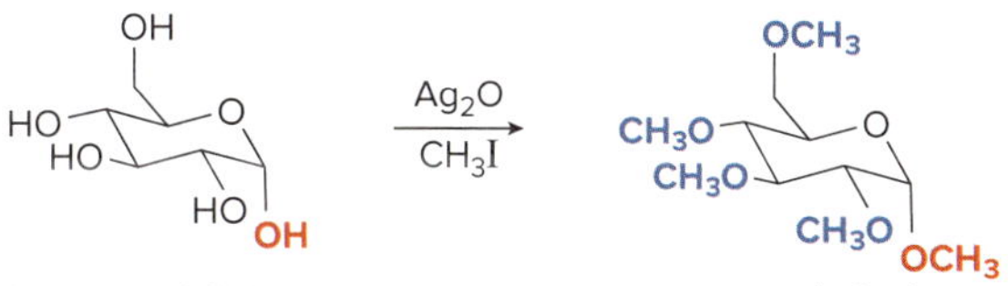

Ag_2O, CH_3I

hemiacetal OH in red

pentamethyl ether

- The acetal OCH_3 group is in red.
- Ether OCH_3 groups are in blue.

Ag_2O removes a proton from each alcohol, forming an alkoxide (RO^-), which then reacts with CH_3I in an S_N2 reaction. Because no C–O bonds are broken, the configuration of all substituents in the starting material is **retained,** forming a single product.

The product contains two different types of ether bonds. There are four "regular" ethers formed from the "regular" hydroxyls. The new ether from the hemiacetal is now part of an **acetal**—that is, a **glycoside.**

The four ether bonds that are *not* part of the acetal do not react with any reagents except strong acids like HBr and HI (Section 9.14). **The acetal ether, on the other hand, is hydrolyzed with aqueous acid** (Section 26.6B). Aqueous hydrolysis of a single glycoside (like the pentamethyl ether of α-D-glucose) yields both anomers of the product monosaccharide.

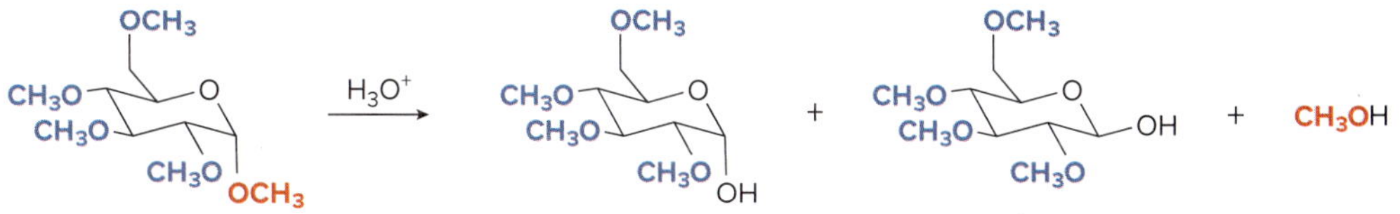

H_3O^+ → α anomer + β anomer + CH_3OH

The OH groups of monosaccharides can also be converted to esters. For example, treatment of β-D-glucose with either acetic anhydride or acetyl chloride in the presence of pyridine (a base) converts all OH groups to acetate esters.

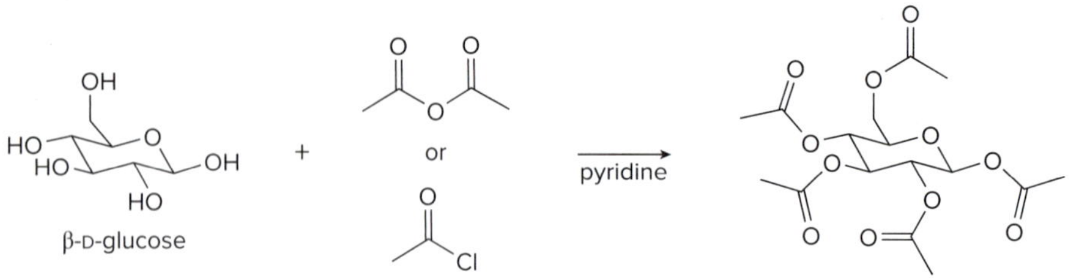

β-D-glucose + (acetic anhydride) or (acetyl chloride) —pyridine→ (pentaacetate ester)

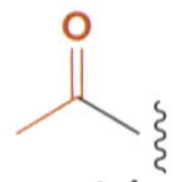

acetyl
Ac

Because it is cumbersome and tedious to draw in all the atoms of the esters, the abbreviation **Ac** is used for the acetyl group, **$CH_3C{=}O$.** The esterification of β-D-glucose can then be written as follows:

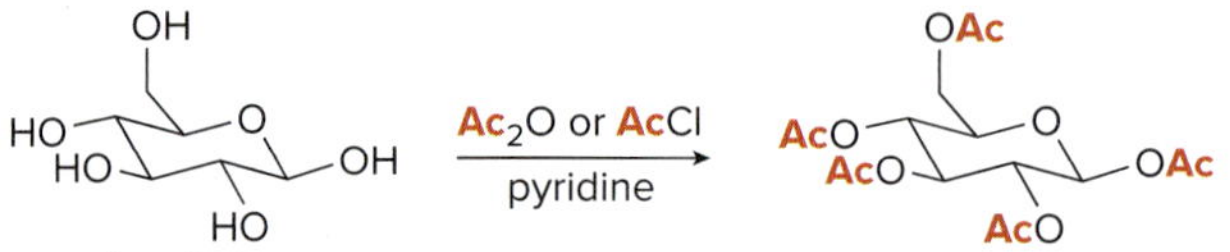

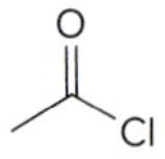

AcCl

β-D-glucose —Ac_2O or AcCl, pyridine→ (AcO)₅ product

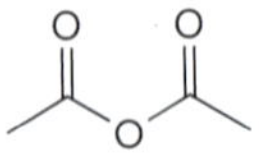

Ac_2O

Monosaccharides are so polar that they are insoluble in common organic solvents, making them difficult to isolate and use in organic reactions. **Monosaccharide derivatives** that have five ether or ester groups in place of the OH groups, however, **are readily soluble in organic solvents.**

Problem 26.20 Draw the products formed when β-D-galactose is treated with each reagent.

a. $Ag_2O + CH_3I$
b. $NaH + C_6H_5CH_2Cl$
c. The product in (b), then H_3O^+
d. Ac_2O + pyridine
e. C_6H_5COCl + pyridine
f. The product in (c), then C_6H_5COCl + pyridine

Problem 26.21 Draw the products formed when α-D-gulose is treated with each reagent.

a. CH_3I, Ag_2O
b. CH_3OH, HCl
c. Ac_2O, pyridine
d. The product in (a), then H_3O^+
e. The product in (b), then Ac_2O, pyridine
f. The product in (d), then $C_6H_5CH_2Cl$, Ag_2O

26.8 Reactions at the Carbonyl Group—Oxidation and Reduction

Oxidation and reduction reactions occur at the carbonyl group of monosaccharides, so they all begin with the monosaccharide drawn in the **acyclic form.** We will confine our discussion to aldoses as starting materials.

26.8A Reduction of the Carbonyl Group

Glucitol occurs naturally in some fruits and berries. It is sometimes used as a substitute for sucrose (table sugar). With six polar OH groups capable of hydrogen bonding, glucitol is readily hydrated. It is used as an additive to prevent certain foods from drying out.

Like other aldehydes, the **carbonyl group of an aldose is reduced to a 1° alcohol using $NaBH_4$.** This alcohol is called an **alditol.** For example, reduction of D-glucose with $NaBH_4$ in CH_3OH yields glucitol (also called sorbitol).

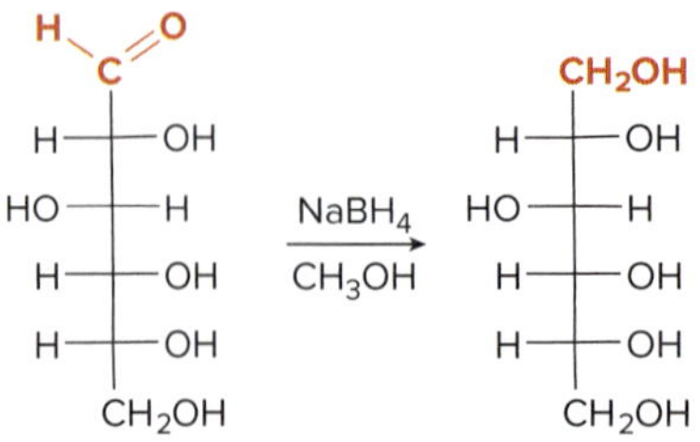

D-glucose (CHO; H–OH; HO–H; H–OH; H–OH; CH_2OH) —$NaBH_4$, CH_3OH→ glucitol (sorbitol) (CH_2OH; H–OH; HO–H; H–OH; H–OH; CH_2OH)

Problem 26.22 A 2-ketohexose is reduced with $NaBH_4$ in CH_3OH to form a mixture of D-galactitol and D-talitol. What is the structure of the 2-ketohexose?

Problem 26.23 Which D-aldopentoses are reduced to optically inactive alditols using $NaBH_4$, CH_3OH?

26.8B Oxidation of Aldoses

Aldoses contain 1° and 2° alcohols and an aldehyde, all of which are oxidizable functional groups. Two different types of oxidation reactions are particularly useful—**oxidation of the aldehyde to a carboxylic acid (an aldonic acid)** and **oxidation of both the aldehyde and the 1° alcohol to a diacid (an aldaric acid).**

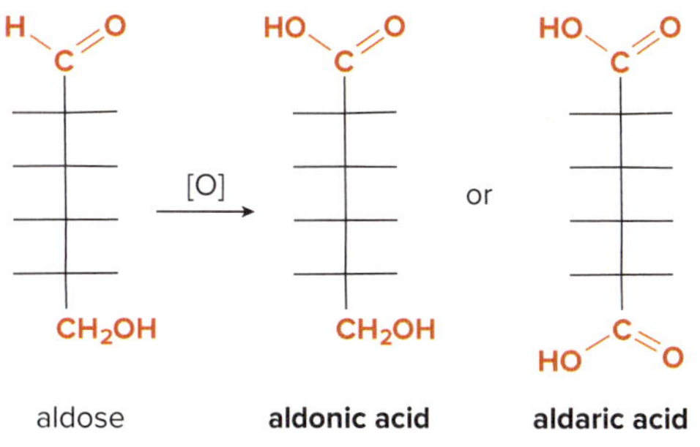

aldose —[O]→ aldonic acid or aldaric acid

[1] Oxidation of the aldehyde to a carboxylic acid

The aldehyde carbonyl is the most easily oxidized functional group in an aldose, so a variety of reagents oxidize it to a carboxy group, forming an **aldonic acid.**

Three reagents used for this process produce a characteristic color change because the oxidizing agent is reduced to a colored product that is easily visible. As described in Section 17.8, **Tollens reagent** oxidizes aldehydes to carboxylic acids using Ag_2O in NH_4OH, and forms a mirror of Ag as a by-product. **Benedict's** and **Fehling's reagents** use a blue Cu^{2+} salt as an oxidizing agent, which is reduced to Cu_2O, a brick-red solid. Unfortunately, none of these reagents gives a high yield of aldonic acid. When the aldonic acid is needed to carry on to other reactions, $\mathbf{Br_2 + H_2O}$ is used as the oxidizing agent.

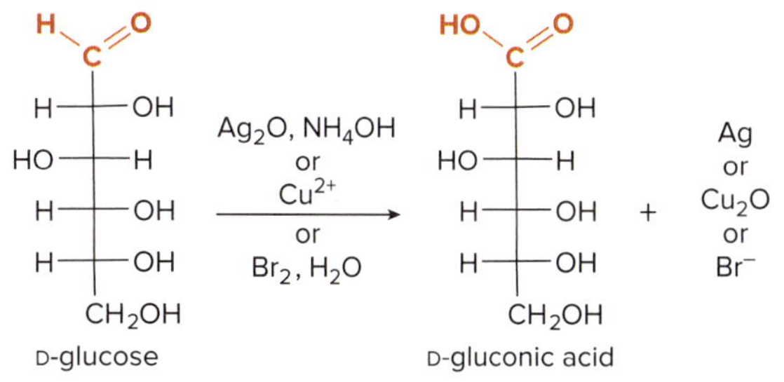

D-glucose —(Ag_2O, NH_4OH or Cu^{2+} or Br_2, H_2O)→ D-gluconic acid + Ag or Cu_2O or Br^-

- **Any carbohydrate that exists as a *hemiacetal* is in equilibrium with a small amount of acyclic aldehyde, so it is oxidized to an aldonic acid.**
- **Glycosides are acetals, not hemiacetals, so they are *not* oxidized to aldonic acids.**

Carbohydrates that can be oxidized with Tollens, Benedict's, or Fehling's reagent are called **reducing sugars.** Those that do not react with these reagents are called **nonreducing sugars.** Figure 26.8 shows examples of reducing and nonreducing sugars.

Figure 26.8 Examples of reducing and nonreducing sugars

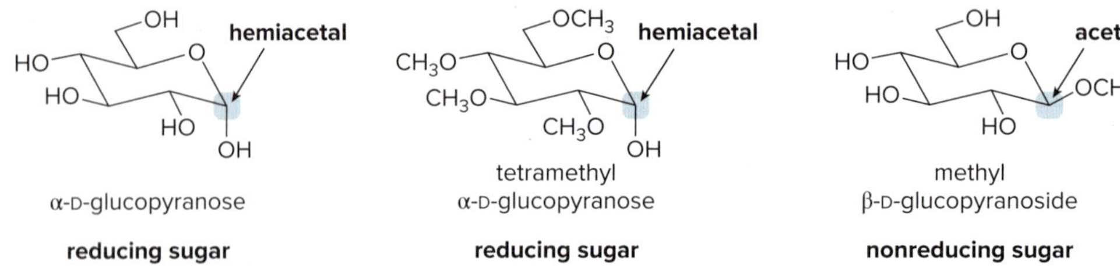

- Carbohydrates containing a hemiacetal are in equilibrium with an acyclic aldehyde, making them **reducing sugars.**
- Glycosides are acetals, so they are *not* in equilibrium with any acyclic aldehyde, making them **nonreducing sugars.**

Problem 26.24 Classify each compound as a reducing or nonreducing sugar.

a.

b.

c.

lactose

---

[2] Oxidation of both the aldehyde and 1° alcohol to a diacid

Both the aldehyde and 1° alcohol of an aldose are oxidized to carboxy groups by treatment with warm nitric acid, forming an aldaric acid. Under these conditions, D-glucose is converted to D-glucaric acid.

HNO_3, H_2O

D-glucose → D-glucaric acid **an aldaric acid**

Because aldaric acids have identical functional groups on both terminal carbons, some aldaric acids contain a plane of symmetry, making them achiral molecules. For example, oxidation of D-allose forms an achiral, optically inactive aldaric acid. This contrasts with D-glucaric acid formed from glucose, which has no plane of symmetry and is thus still optically active.

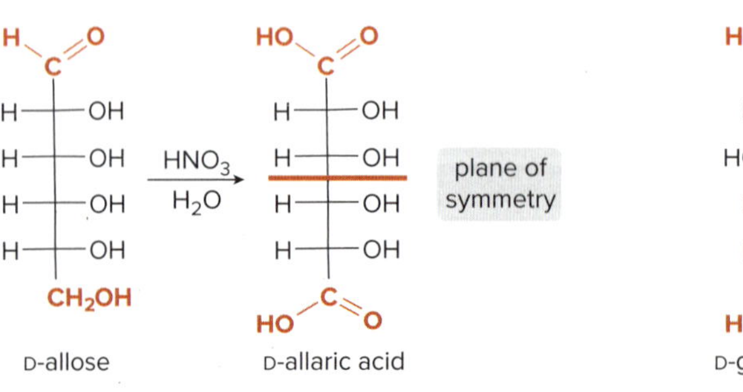

No plane of symmetry

D-allose

D-allaric acid **an achiral diacid**

D-glucaric acid **a chiral diacid**

Problem 26.25 Draw the products formed when D-arabinose is treated with each reagent: (a) Ag_2O, NH_4OH; (b) Br_2, H_2O; (c) HNO_3, H_2O.

Problem 26.26 Which aldoses are oxidized to optically inactive aldaric acids: (a) D-erythrose; (b) D-lyxose; (c) D-galactose?

26.9 Reactions at the Carbonyl Group—Adding or Removing One Carbon Atom

Two common procedures in carbohydrate chemistry result in adding or removing one carbon atom from the skeleton of an aldose. The **Wohl degradation** shortens an aldose chain by one carbon, whereas the **Kiliani–Fischer synthesis** lengthens it by one. Both reactions involve cyanohydrins as intermediates. Recall from Section 18.8 that cyanohydrins are formed from aldehydes by addition of the elements of HCN. Cyanohydrins can also be re-converted to carbonyl compounds by treatment with base.

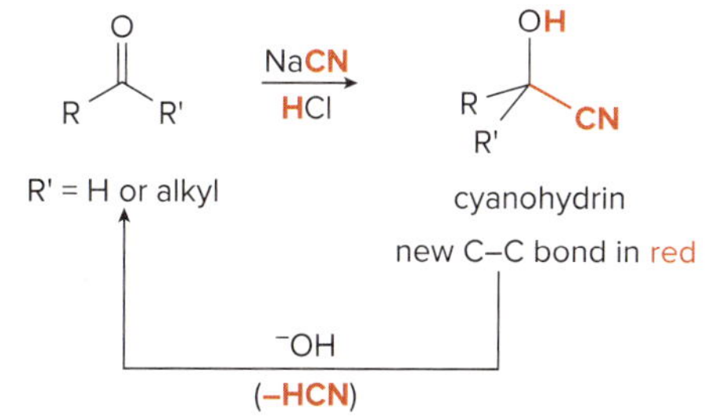

NaCN, HCl
R' = H or alkyl
cyanohydrin
new C–C bond in red
^-OH (–HCN)

- **Forming a cyanohydrin adds one carbon to a carbonyl group.**
- **Re-converting a cyanohydrin to a carbonyl compound removes one carbon.**

26.9A The Wohl Degradation

The Wohl degradation is a stepwise procedure that shortens the length of an aldose chain by cleavage of the C1–C2 bond. As a result, an aldohexose is converted to an aldopentose having the same configuration at its bottom three stereogenic centers (C3–C5). For example, the Wohl degradation converts D-glucose to D-arabinose.

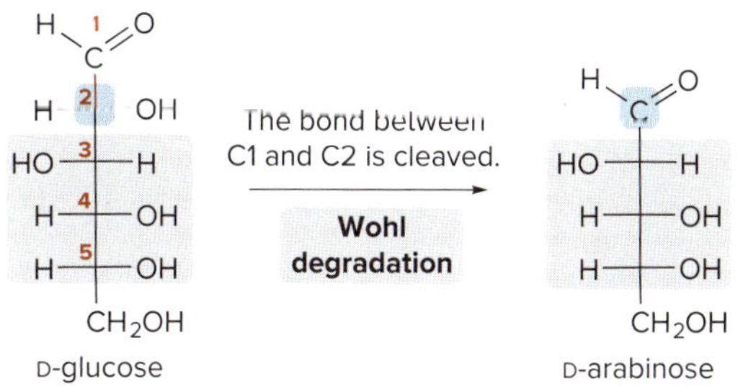

The bond between C1 and C2 is cleaved.
Wohl degradation
D-glucose
D-arabinose

The Wohl degradation consists of three steps, illustrated here beginning with D-glucose.

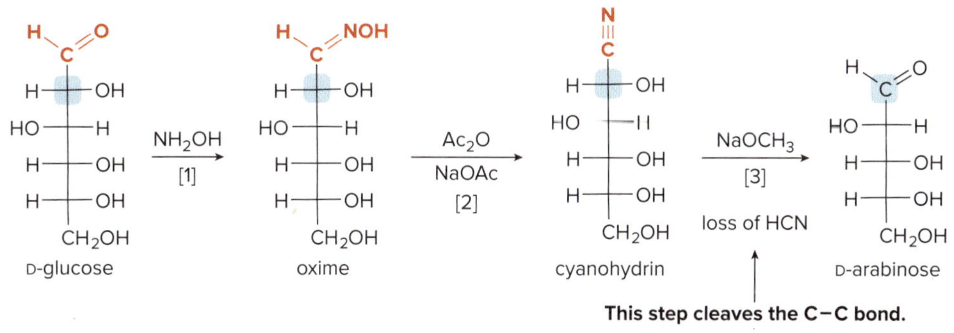

NH_2OH [1]
Ac_2O, NaOAc [2]
$NaOCH_3$ [3]
loss of HCN
D-glucose
oxime
cyanohydrin
D-arabinose
This step cleaves the C–C bond.

[1] Treatment of D-glucose with hydroxylamine (NH_2OH) forms an **oxime** by nucleophilic addition. This reaction is analogous to the formation of imines discussed in Section 18.10.

[2] Dehydration of the oxime to a nitrile occurs with acetic anhydride (Ac_2O) and sodium acetate (NaOAc). The nitrile product is a cyanohydrin.

[3] **Treatment of the cyanohydrin with base results in loss of the elements of HCN to form an aldehyde having one fewer carbon.**

The Wohl degradation converts a stereogenic center at C2 in the original aldose to an sp^2 hybridized C=O. As a result, a pair of aldoses that are epimeric at C2, such as D-galactose and D-talose, yield the *same* aldose (D-lyxose, in this case) upon Wohl degradation.

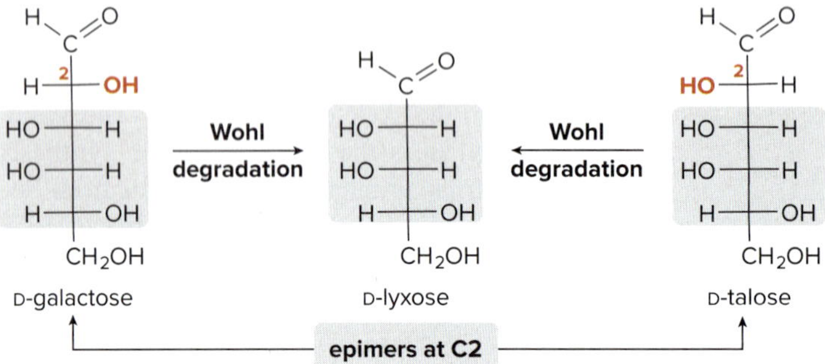

Problem 26.27 What two aldoses yield D-xylose on Wohl degradation?

26.9B The Kiliani–Fischer Synthesis

The Kiliani–Fischer synthesis lengthens a carbohydrate chain by adding one carbon to the aldehyde end of an aldose, thus forming a new stereogenic center at C2 of the product. The product consists of epimers that differ only in their configuration about the one new stereogenic center. For example, the Kiliani–Fischer synthesis converts D-arabinose to a mixture of D-glucose and D-mannose.

D-arabinose → Kiliani–Fischer synthesis → D-glucose + D-mannose

epimers

new C–C bond in red

The Kiliani–Fischer synthesis, shown here beginning with D-arabinose, consists of three steps. "Squiggly" lines are meant to indicate that two different stereoisomers are formed at the new stereogenic center. As with the Wohl degradation, **the key intermediate is a cyanohydrin.**

NaCN, HCl, "HCN" [1] → cyanohydrin → H_2, Pd-$BaSO_4$ [2] → imine → H_3O^+ [3]

[1] Treating an aldose with NaCN and HCl adds the elements of HCN to the carbonyl group, forming a **cyanohydrin** and a new carbon–carbon bond. Because the sp^2 hybridized carbonyl carbon is converted to an sp^3 hybridized carbon with four different groups, **a new stereogenic center is formed in this step.**

[2] Reduction of the nitrile with H_2 and Pd-$BaSO_4$, a poisoned Pd catalyst, forms an **imine.**

[3] **Hydrolysis of the imine with aqueous acid forms an aldehyde that has one more carbon than the aldose** that began the sequence.

Note that the **Wohl degradation and the Kiliani–Fischer synthesis are conceptually opposite transformations.**

- **The Wohl degradation *removes* a carbon atom from the aldehyde end of an aldose. Two aldoses that are epimers at C2 form the *same* product.**
- **The Kiliani–Fischer synthesis *adds* a carbon to the aldehyde end of an aldose, forming *two epimers* at C2.**

Problem 26.28 What aldoses are formed when the following aldoses are subjected to the Kiliani–Fischer synthesis: (a) D-threose; (b) D-ribose; (c) D-galactose?

26.9C Determining the Structure of an Unknown Monosaccharide

The reactions in Sections 26.8–26.9 can be used to determine the structure of an unknown monosaccharide, as shown in Sample Problem 26.5.

Sample Problem 26.5 Determining the Structure of an Unknown Aldose

A D-aldopentose **A** is oxidized to an optically inactive aldaric acid with HNO_3. **A** is formed by the Kiliani–Fischer synthesis of a D-aldotetrose **B,** which is also oxidized to an optically inactive aldaric acid with HNO_3. What are the structures of **A** and **B?**

Solution

Use each fact to determine the relative orientation of the OH groups in the D-aldopentose.

Fact [1] **A D-aldopentose A is oxidized to an optically *inactive* aldaric acid with HNO_3.**

An optically inactive aldaric acid must contain a **plane of symmetry.** Because the **OH group on C4 must be on the right for the D-sugar,** there are only two ways to arrange the OH groups in a five-carbon D-aldaric acid. Thus, only two structures are possible for **A,** labeled **A'** and **A''.**

Possible optically inactive D-aldaric acids:

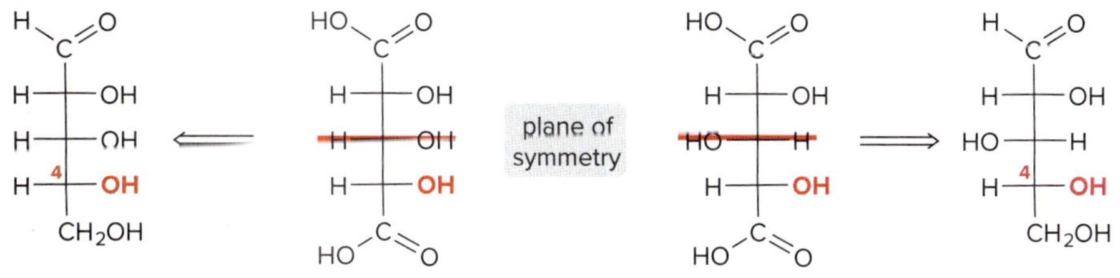

plane of symmetry

A' **A''**

Fact [2] **A is formed by the Kiliani–Fischer synthesis from a D-aldotetrose B.**

A' and **A''** are each prepared from a D-aldotetrose (**B'** and **B''**) that has the same configuration at the bottom two stereogenic centers.

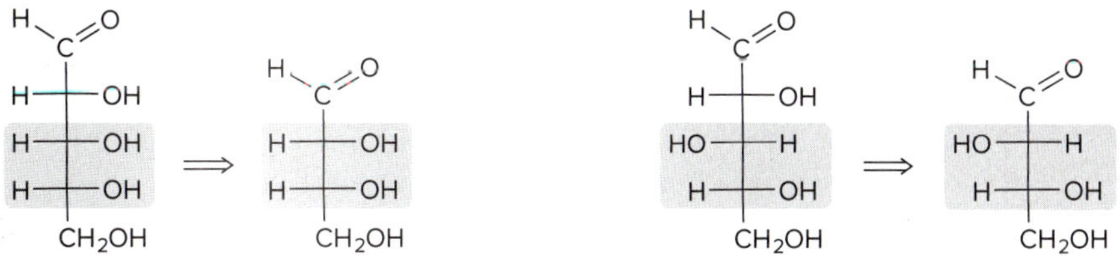

A' **B'** **A''** **B''**

Two possible structures for **B**

Fact [3] **The D-aldotetrose is oxidized to an optically *inactive* aldaric acid upon treatment with HNO_3.**

Only the aldaric acid from **B'** has a plane of symmetry, making it optically inactive. Thus, **B'** is the correct structure for the D-aldotetrose **B,** and therefore **A'** is the structure of the D-aldopentose **A.**

H–C=O; H–C–OH; H–C–OH; CH_2OH — HNO_3 → HO–C=O; H–C–OH; H–C–OH; HO–C=O (plane of symmetry)

B' — **optically inactive**

H–C=O; HO–C–H; H–C–OH; CH_2OH — HNO_3 → HO–C=O; HO–C–H; H–C–OH; HO–C=O

B" — **optically active**

Answer:

H–C=O; H–C–OH; H–C–OH; CH_2OH = **B**

H–C=O; H–C–OH; H–C–OH; H–C–OH; CH_2OH = **A**

Problem 26.29 A D-aldopentose **A** is oxidized to an optically inactive aldaric acid. On Wohl degradation, **A** forms an aldotetrose **B** that is oxidized to an optically active aldaric acid. What are the structures of **A** and **B?**

More Practice: Try Problems 26.64–26.66.

Problem 26.30 A D-aldohexose **A** is formed from an aldopentose **B** by the Kiliani–Fischer synthesis. Reduction of **A** with $NaBH_4$ forms an optically inactive alditol. Oxidation of **B** forms an optically active aldaric acid. What are the structures of **A** and **B?**

26.10 Disaccharides

Disaccharides contain two monosaccharides joined by a glycosidic linkage. The general features of a disaccharide include the following:

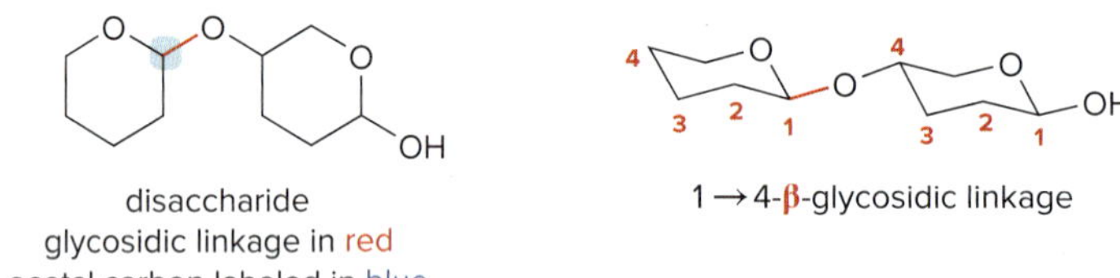

disaccharide
glycosidic linkage in red
acetal carbon labeled in blue

1→4-β-glycosidic linkage

[1] Two monosaccharide rings may be five- or six-membered, but six-membered rings are much more common. **The two rings are connected by an O atom that is part of an acetal, called a glycosidic linkage,** which may be oriented α or β.

[2] The **glycoside is formed from the anomeric carbon of one monosaccharide and any OH group on the other monosaccharide.** All disaccharides have **one acetal,** plus either a hemiacetal or another acetal.

[3] With pyranose rings, the carbon atoms in each ring are numbered beginning with the anomeric carbon. The most common disaccharides contain two monosaccharides in which the hemiacetal carbon of one ring (Cl) is joined to C4 of the other ring.

The three most abundant disaccharides are **maltose, lactose,** and **sucrose.**

26.10A Maltose

Maltose gets its name from malt, the liquid obtained from barley and other cereal grains. *Mir141/Shutterstock*

Maltose, a disaccharide formed by the hydrolysis of starch, is found in germinated grains such as barley. Maltose contains two glucose units joined by a 1→4-α-glycoside bond. Maltose contains one acetal carbon (in red) and one hemiacetal carbon (in blue).

maltose
1→4-α-glycosidic linkage
β anomer

Because one glucose ring of maltose still contains a hemiacetal, it exists as a mixture of α and β anomers. Only the β anomer is shown. Maltose exhibits two properties of all carbohydrates that contain a hemiacetal: it undergoes **mutarotation,** and it reacts with oxidizing agents, making it a **reducing sugar.**

Hydrolysis of maltose forms two molecules of glucose. The **C1–O** bond is cleaved in this process, and a mixture of glucose anomers forms. The mechanism for this hydrolysis is exactly the same as the mechanism for glycoside hydrolysis in Section 26.7B.

H_3O^+ → α-D-glucose + β-D-glucose

Problem 26.31 Draw the α anomer of maltose. What products are formed on hydrolysis of this form of maltose?

26.10B Lactose

Milk contains the disaccharide lactose. *Mitch Hrdlicka/Photodisc/Getty Images*

Lactose is the principal disaccharide found in milk from both humans and cows. Unlike many mono- and disaccharides, lactose is not appreciably sweet. Lactose consists of **one galactose** and **one glucose unit,** joined by a **1→4-β-glycoside bond** from the anomeric carbon of galactose to C4 of glucose.

lactose
β-glycosidic linkage
β anomer

Like maltose, lactose also contains a hemiacetal, so it exists as a mixture of α and β anomers. The β anomer is drawn. Lactose undergoes **mutarotation,** and it reacts with oxidizing agents, making it a **reducing sugar.**

Lactose is digested in the body by first cleaving the 1→4-β-glycoside bond using the enzyme *lactase.* Many individuals, mainly of Asian and African descent, lack adequate amounts of lactase, so they are unable to digest and absorb lactose. This condition, lactose intolerance, is associated with abdominal cramping and recurrent diarrhea when milk and dairy products are ingested.

Problem 26.32 Cellobiose, a disaccharide obtained by the hydrolysis of cellulose, is composed of two glucose units joined by a 1→4-β-glycoside bond. What is the structure of cellobiose?

26.10C Sucrose

Sucrose, the disaccharide mentioned in the chapter opener that is found in sugarcane and used as table sugar (Figure 26.9), is the most common disaccharide in nature. It contains **one glucose unit** and **one fructose unit.**

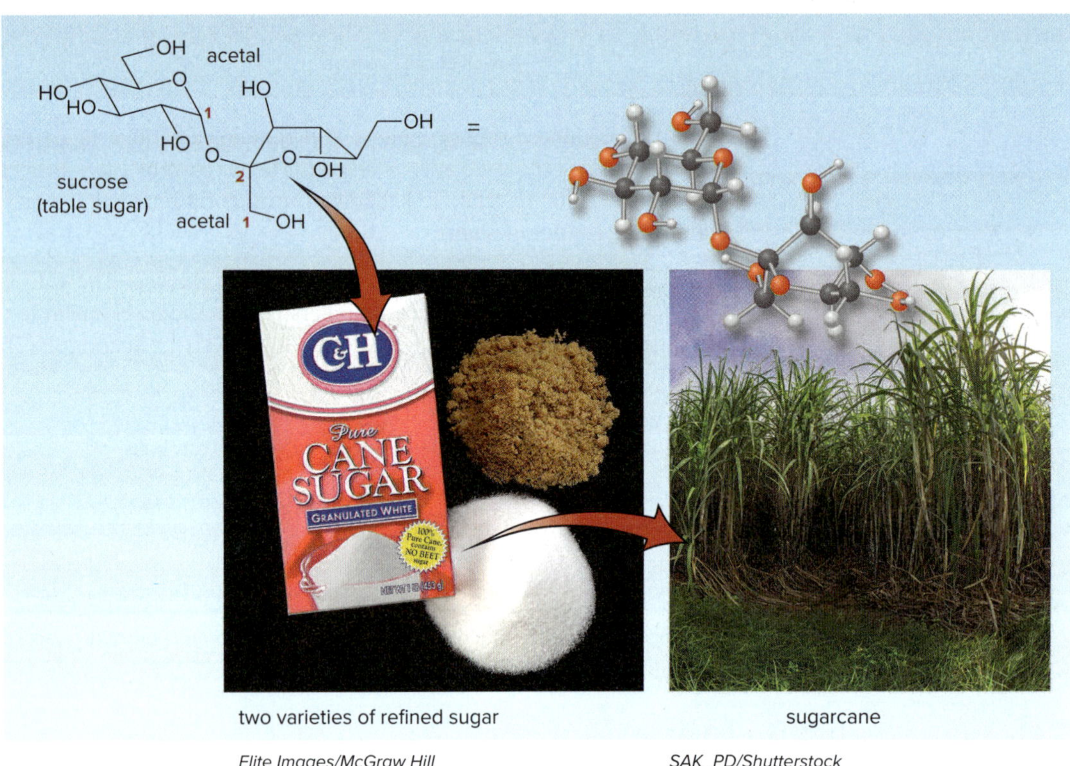

Elite Images/McGraw Hill

SAK_PD/Shutterstock

Figure 26.9 Sucrose

The structure of sucrose has several features that make it different from maltose and lactose. Sucrose contains one six-membered ring (glucose) and one five-membered ring (fructose), whereas both maltose and lactose contain two six-membered rings. In sucrose the six-membered glucose ring is joined by an α-glycosidic bond to C2 of a fructofuranose ring. The numbering in a fructofuranose is different from the numbering in a pyranose ring. The anomeric carbon is now designated as C2, so the anomeric carbons of the glucose and fructose rings are both used to form the glycosidic linkage.

As a result, **sucrose contains two acetals but no hemiacetal.** Sucrose, therefore, is a **nonreducing sugar** and **it does *not* undergo mutarotation.**

Sucrose's pleasant sweetness has made it a widely used ingredient in baked goods, cereals, bread, and many other products. It is estimated that the average American ingests 100 lb of sucrose annually. Like other carbohydrates, however, sucrose contains many calories. To reduce caloric intake while maintaining sweetness, a variety of artificial sweeteners have been developed. These include sucralose, aspartame, and saccharin (Figure 26.10). These compounds are much sweeter than sucrose, so only a small amount of each compound is needed to achieve the same level of perceived sweetness.

Figure 26.10
Artificial sweeteners

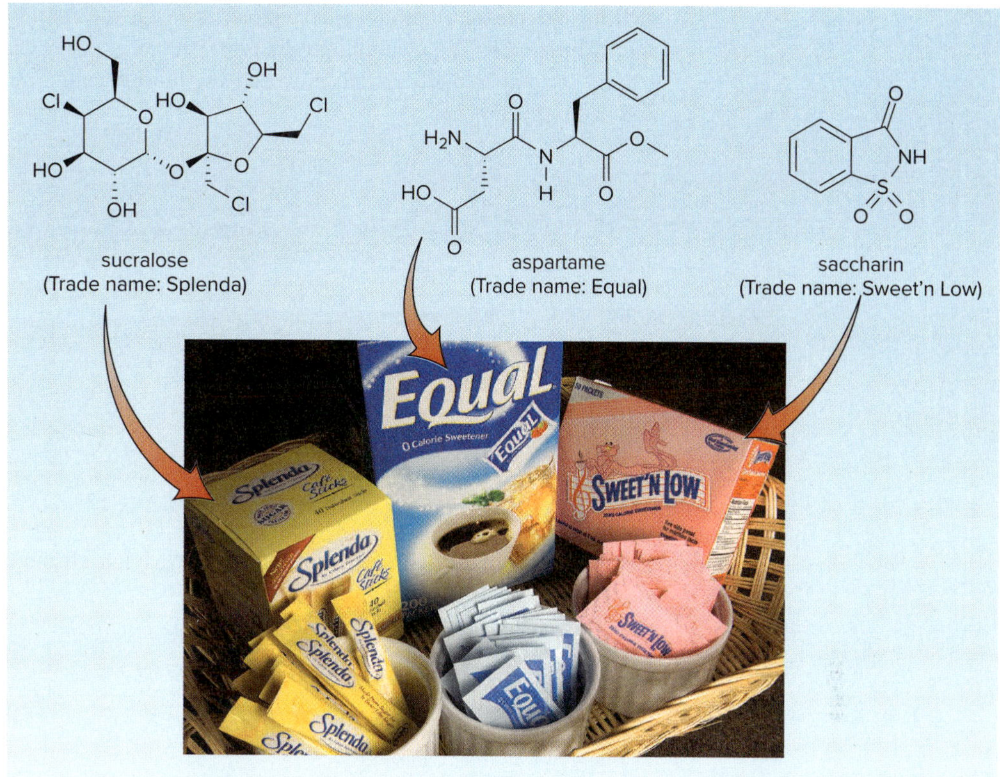

- The sweetness of these three artificial sweeteners was discovered accidentally. The sweetness of sucralose was discovered in 1976 when a chemist misunderstood his superior, and he *tasted* rather than *tested* his compound. Aspartame was discovered in 1965 when a chemist licked his dirty fingers in the lab and tasted its sweetness. Saccharin, the oldest-known artificial sweetener, was discovered in 1879 by a chemist who failed to wash his hands after working in the lab. Saccharin was not used extensively until sugar shortages occurred during World War I. Although there were concerns in the 1970s that saccharin causes cancer, there is no proven link between cancer occurrence and saccharin intake at normal levels. *Jill Braaten/McGraw Hill*

26.11 Polysaccharides

Polysaccharides contain three or more monosaccharides joined together. Three prevalent polysaccharides in nature are **cellulose, starch,** and **glycogen,** each of which consists of repeating glucose units joined by different glycosidic bonds.

26.11A Cellulose

The structure of cellulose was discussed in Section 5.1.

Cellulose is found in the cell walls of nearly all plants, where it gives support and rigidity to wood and plant stems. Cotton is essentially pure cellulose.

cellulose
1 → 4-β-glycosidic linkages in red

Ball-and-stick models showing the three-dimensional structures of cellulose and starch were given in Figure 5.2.

Cellulose is an unbranched polymer composed of repeating glucose units joined in a 1→4-β-glycosidic linkage. The β-glycosidic linkage forms long linear chains of cellulose molecules that stack in sheets, creating an extensive three-dimensional array. A network of intermolecular hydrogen bonds between the chains and sheets means that only the few OH groups on the surface are available to hydrogen bond to water, making this very polar compound water insoluble.

Cellulose acetate, a cellulose derivative, is made by treating cellulose with acetic anhydride and sulfuric acid. The resulting product has acetate esters in place of every OH group. Cellulose acetate is spun into fibers that are used for fabrics called *acetates,* which have a deep luster and satin appearance.

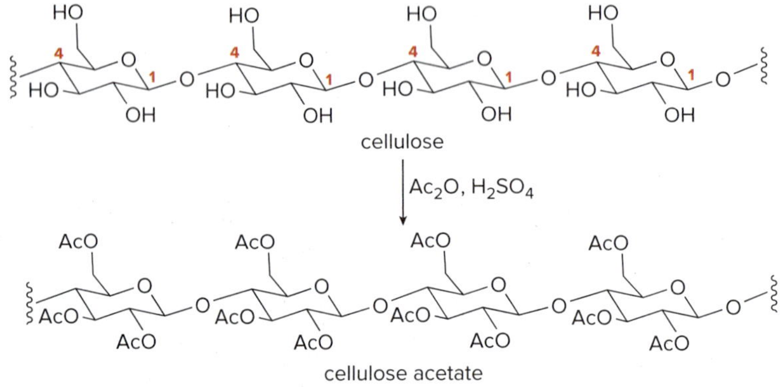

cellulose

Ac_2O, H_2SO_4

cellulose acetate

Cellulose can be hydrolyzed to glucose by cleaving all the β-glycosidic bonds, yielding both anomers of glucose.

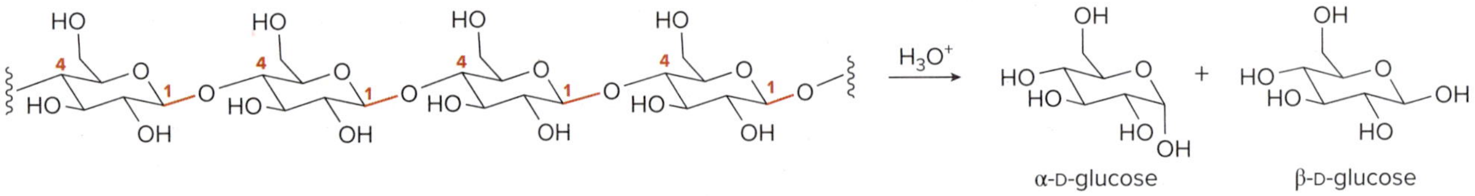

H_3O^+

α-D-glucose + β-D-glucose

A **β-glycosidase** is the general name of an enzyme that hydrolyzes a β-glycoside linkage.

In cells, the hydrolysis of cellulose is accomplished by an enzyme called a **β-glucosidase,** which cleaves all the β-glycoside bonds formed from glucose. Humans do not possess this enzyme and therefore cannot digest cellulose. Ruminant animals, on the other hand, such as cattle, deer, and camels, have bacteria containing a β-glucosidase in their digestive systems, so they can derive nutritional benefit from eating grass and leaves.

26.11B Starch

Starch is the main carbohydrate found in the seeds and roots of plants. Corn, rice, wheat, and potatoes are common foods that contain a great deal of starch.

Starch is a polymer composed of repeating glucose units joined in α-glycosidic linkages. Both starch and cellulose are polymers of glucose, but starch contains α glycoside bonds, whereas cellulose contains β glycoside bonds. The two common forms of starch are **amylose** and **amylopectin.**

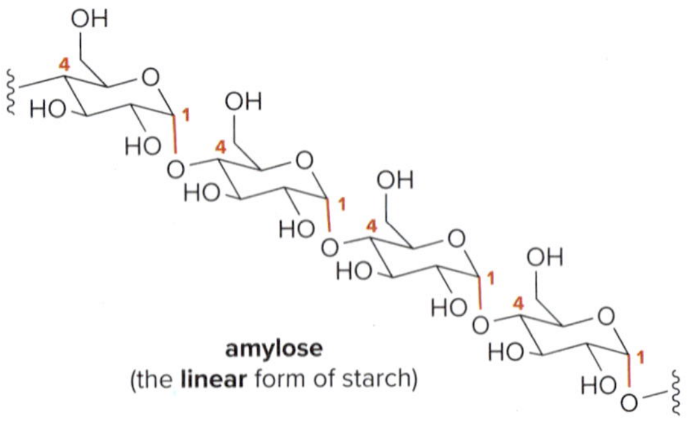

amylose
(the **linear** form of starch)

1→4-α-glycosidic linkages in red

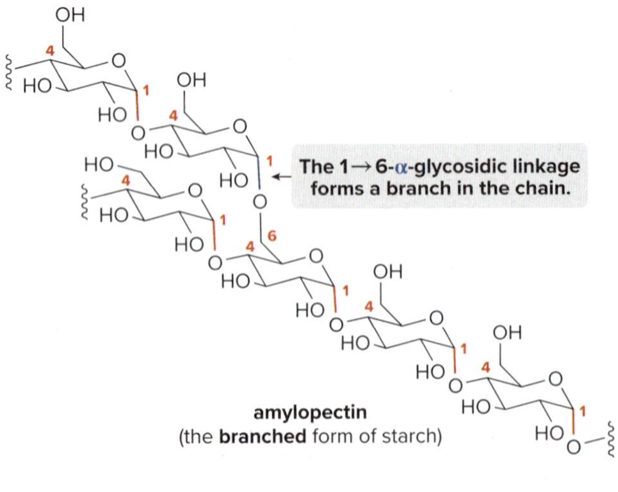

The 1→6-α-glycosidic linkage forms a branch in the chain.

amylopectin
(the **branched** form of starch)

1→4-α-glycosidic linkages in red
1→6-α-glycosidic linkage in blue

Amylose, which comprises about 20% of starch molecules, has an unbranched skeleton of glucose molecules with **1→4-α-glycoside bonds.** Because of this linkage, an amylose chain adopts a helical arrangement, giving it a very different three-dimensional shape from the linear chains of cellulose. Amylose was first described in Section 5.1.

Amylopectin, which comprises about 80% of starch molecules, likewise consists of a backbone of glucose units joined in **α-glycosidic bonds,** but it also contains considerable branching along the chain. The linear linkages of amylopectin are formed by **1→4-α-glycoside bonds,** similar to amylose. The branches are linked to the chain with **1→6-α-glycosidic linkages.**

Both forms of starch are water soluble. Because the OH groups in these starch molecules are not buried in a three-dimensional network, they are more available for hydrogen bonding with water molecules, leading to greater water solubility than cellulose has.

The ability of amylopectin to form branched polymers is a unique feature of carbohydrates. Other types of polymers in the cell, such as the proteins discussed in Chapter 27, occur in nature only as linear molecules.

Both amylose and amylopectin are hydrolyzed to glucose with cleavage of the glycosidic bonds. The human digestive system has the necessary **α-glucosidase** enzymes needed to catalyze this process. Bread and pasta made from wheat flour, rice, and corn tortillas are all sources of starch that are readily digested.

α-Glycosidase is the general name of an enzyme that hydrolyzes an α-glycoside linkage.

26.11C Glycogen

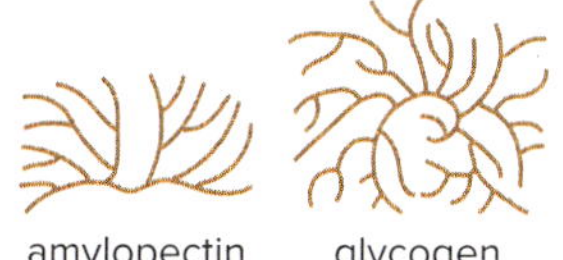
amylopectin glycogen

Glycogen is the major form in which polysaccharides are stored in animals. Glycogen, a polymer of glucose containing **α-glycosidic bonds,** has a branched structure similar to amylopectin, but the branching is much more extensive.

Glycogen is stored principally in the liver and muscle. When glucose is needed for energy in the cell, glucose units are hydrolyzed from the ends of the glycogen polymer, and then further metabolized with the release of energy. Because glycogen has a highly branched structure, there are many glucose units at the ends of the branches that can be cleaved whenever the body needs them.

Problem 26.33 Draw the structure of: (a) a polysaccharide formed by joining D-mannose units in 1→4-β-glycosidic linkages; (b) a polysaccharide formed by joining D-glucose units in 1→6-α-glycosidic linkages. The polysaccharide in (b) is dextran, a component of dental plaque.

26.11D Oligosaccharides

An **oligosaccharide** is a carbohydrate with a small number of monosaccharides—generally three to ten—joined together. **Human milk oligosaccharides (HMOs),** a group of carbohydrates found in breast milk, contain three or four monosaccharides joined together. 2'-Fucosyllactose is the most prevalent component, comprising about 30% of all HMOs. The two glycosidic linkages that join the three monosaccharides together are shown in red.

2'-fucosyllactose

The World Health Organization recommends that children are exclusively breast fed until six months of age, and then nursed along with other forms of nutrition until a child is two years old. *Daniel C. Smith*

HMOs, often called the fiber of breast milk, are not hydrolyzed by the gastric juices of the stomach, nor are they absorbed in the intestine. They nonetheless play a key role in the health of a newborn, by helping to establish the presence of beneficial bacteria in the infant's colon.

Moreover, harmful pathogens attach to the surface of HMOs and are eliminated in the feces of the nursing infant.

Ongoing research continues to study the hundreds of unique components of human breast milk in an effort to understand its benefits to both the mother and the child, even after infancy.

Problem 26.34 3-Fucosyllactose is another HMO found in breast milk. (a) Locate any acetal and hemiacetal. (b) What products are formed when 3-fucosyllactose is hydrolyzed in aqueous acid?

3-fucosyllactose

Problem 26.35 Raffinose is an oligosaccharide found in beans, cabbage, and broccoli. Because humans do not possess the enzyme needed to hydrolyze a glycosidic linkage in raffinose, it passes undigested to the large intestine, where it is fermented by bacteria, producing gas and causing flatulence.

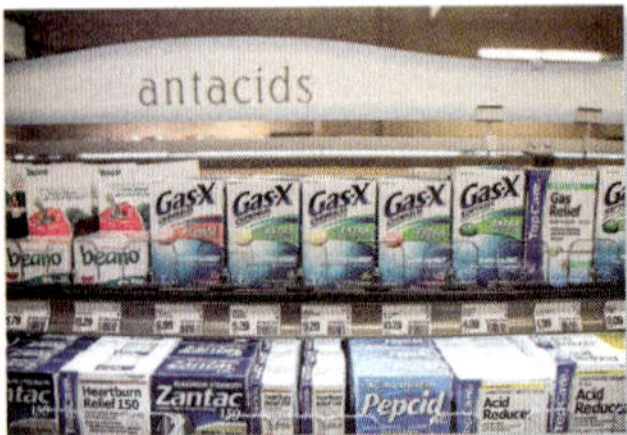

Beano and similar products have the enzyme needed to digest raffinose (Problem 26.35), thus helping to reduce bloating and ease discomfort caused when foods that contain raffinose are ingested. *Sara Stathas/Alamy Stock Photo*

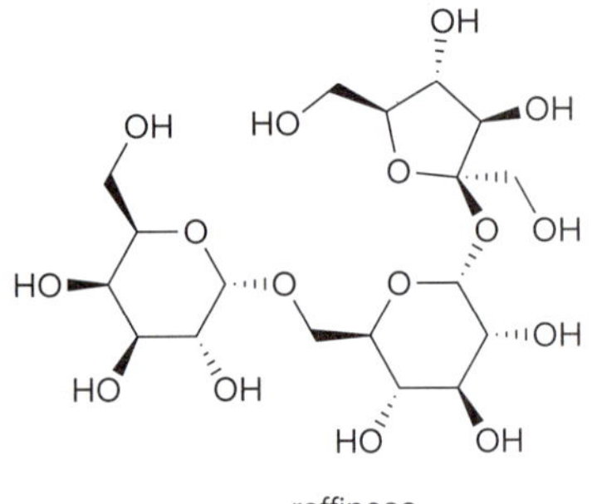

raffinose

a. Label any acetals and hemiacetals in raffinose.
b. What products are formed when raffinose is hydrolyzed with aqueous acid?

26.12 Other Important Sugars and Their Derivatives

Many other examples of simple and complex carbohydrates with useful properties exist in the biological world. In Section 26.12, we examine some carbohydrates that contain nitrogen atoms.

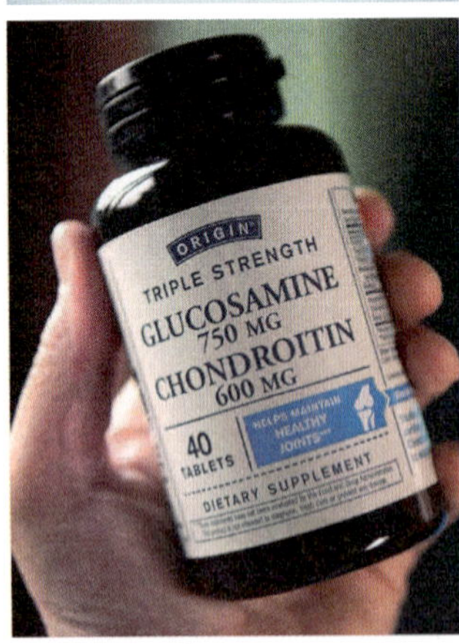

Dietary supplements containing glucosamine are used by individuals suffering from osteoarthritis. *Jill Braaten/McGraw Hill*

26.12A Amino Sugars and Related Compounds

Amino sugars contain an NH_2 group instead of an OH group at a non-anomeric carbon. The most common amino sugar in nature, **D-glucosamine,** is formally derived from D-glucose by replacing the OH at C2 with NH_2. Although it is not classified as a drug, and therefore not regulated by the U.S. Food and Drug Administration, glucosamine is available in many over-the-counter treatments for osteoarthritis.

D-glucosamine → (SCoA) → *N*-acetyl-D-glucosamine **NAG**

The rigidity of a crab shell is due to chitin, a high-molecular-weight carbohydrate molecule. Chitin-based coatings have found several commercial applications, such as extending the shelf life of fruits. Processing plants now convert the shells of crabs, lobsters, and shrimp to chitin and various derivatives for use in many consumer products. *Comstock/Getty Images*

Acetylation of glucosamine with acetyl CoA (Section 20.16) forms ***N*-acetyl-D-glucosamine,** abbreviated as **NAG. Chitin,** the second most abundant carbohydrate polymer, is a polysaccharide formed from NAG units joined together in **1→4-β-glycosidic linkages.** Chitin is identical in structure to cellulose, except that each OH group at C2 is now replaced by $NHCOCH_3$. The exoskeletons of lobsters, crabs, and shrimp are composed of chitin. Like those of cellulose, chitin chains are held together by an extensive network of hydrogen bonds, forming water-insoluble sheets.

1→4-β-glycosidic linkages in red

chitin

Several trisaccharides containing amino sugars are potent antibiotics used in the treatment of certain severe and recurrent bacterial infections. These compounds, such as tobramycin and amikacin, are called **aminoglycoside antibiotics.**

tobramycin

amikacin

Problem 26.36 Treating chitin with H_2O, ^{-}OH hydrolyzes its amide linkages, forming a compound called chitosan. What is the structure of chitosan? Chitosan has been used in shampoos, fibers for sutures, and wound dressings.

26.12B *N*-Glycosides

***N*-Glycosides are formed when a monosaccharide is reacted with an amine** in the presence of mild acid (Reactions [1] and [2]).

[1] β-D-glucopyranose → (mild H^+) α-*N*-glycoside + β-*N*-glycoside

[2] α-D-ribofuranose → (mild H^+) N-glycoside anomers

The mechanism of *N*-glycoside formation is analogous to the mechanism for glycoside formation, and both anomers of the *N*-glycoside are formed as products.

Problem 26.37 Draw the products of each reaction.

a. [pyranose sugar] $\xrightarrow[\text{mild H}^+]{CH_3NH_2}$ b. [pyranose sugar] $\xrightarrow[\text{mild H}^+]{C_6H_5NH_2}$

The *N*-glycosides of two sugars, **D-ribose** and **2-deoxy-D-ribose,** are especially noteworthy, because they form the building blocks of RNA and DNA, respectively, as you will learn in Chapter 28.

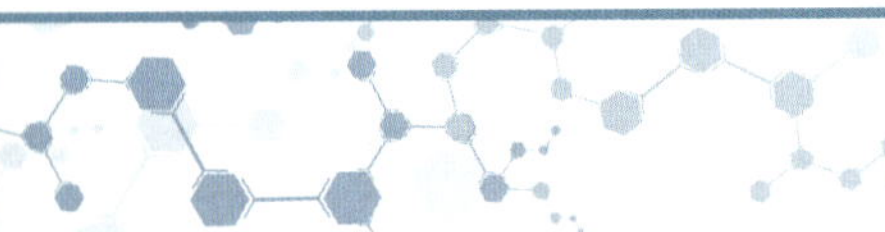

Chapter 26 REVIEW

KEY REACTIONS

[1] Reactions of monosaccharides involving the hemiacetal

1. α-D-glucose $\xrightarrow[\text{HCl}]{CH_3OH}$ α glycoside + β glycoside — glycoside formation (26.6A)

2. [methyl glycoside] $\xrightarrow{H_3O^+}$ α anomer + β anomer + CH_3OH — glycoside hydrolysis (26.6B)

Try Problems 26.50a, 26.51.

[2] Reactions of monosaccharides at the OH groups

1. [glucose] $\xrightarrow[CH_3X]{Ag_2O}$ [permethylated glucose, CH_3O groups, OCH_3] — ether formation (26.7)

2. [glucose] $\xrightarrow{Ac_2O \text{ or } AcCl, \text{ pyridine}}$ [peracetylated glucose, AcO groups, OAc] — ester formation (26.7)

Try Problem 26.50g.

[3] Reactions of monosaccharides at the carbonyl group

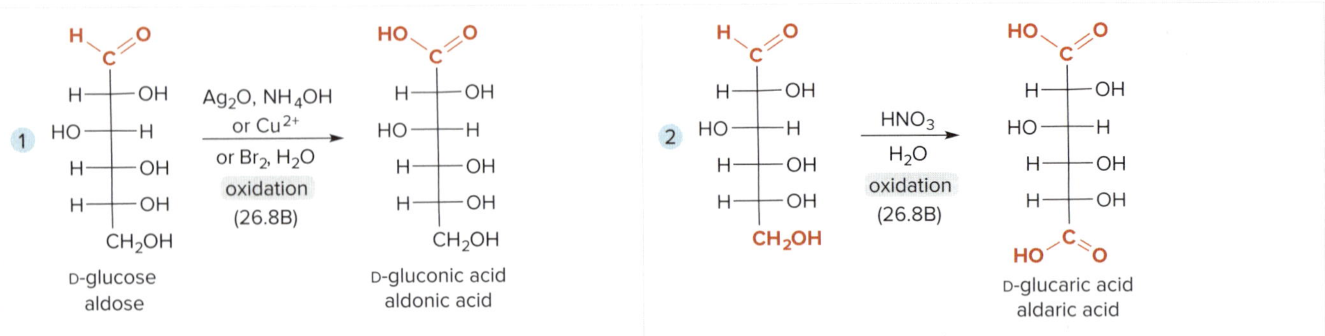

3
NaBH4
CH3OH
reduction
(26.8A)
glucitol (sorbitol)
alditol

5
[1] NaCN, HCl
[2] H2, Pd-BaSO4
[3] H3O+
Kiliani–Fischer synthesis
(26.9B)

4
[1] NH2OH
[2] Ac2O, NaOAc
[3] NaOCH3
Wohl degradation
(26.9A)

Try Problems 26.50b–f; 26.53c–e; 26.54c, d; 26.55–26.57.

[4] Other reactions

1
maltose
H3O+
hydrolysis
(26.10)
α-D-glucose
+
β-D-glucose

2
NH2
mild H+
N-glycoside formation
(26.12B)
α-N-glycoside
+
β-N-glycoside

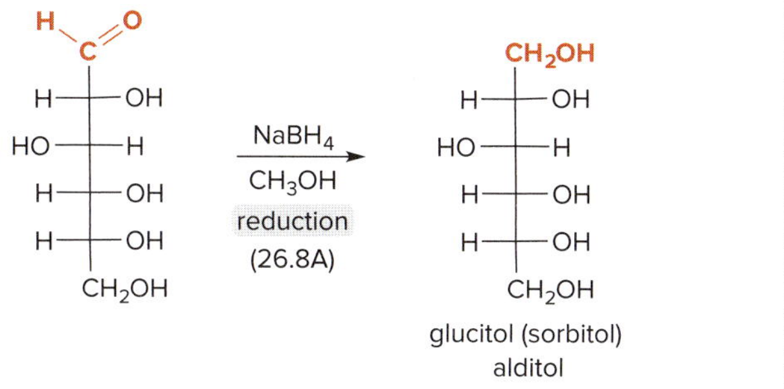

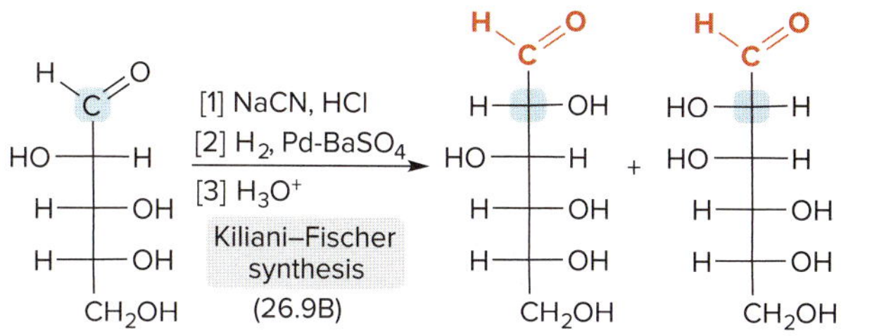

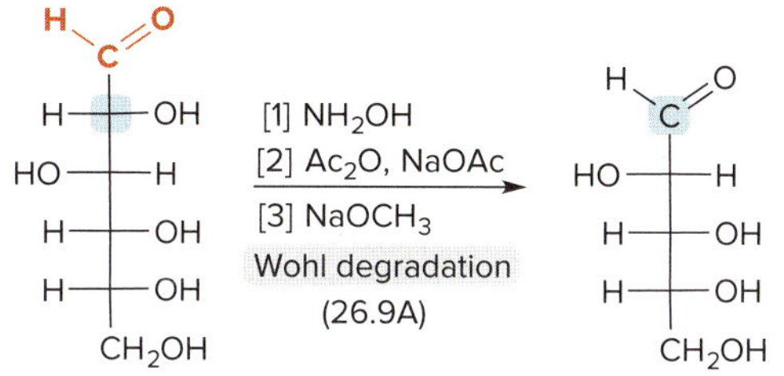

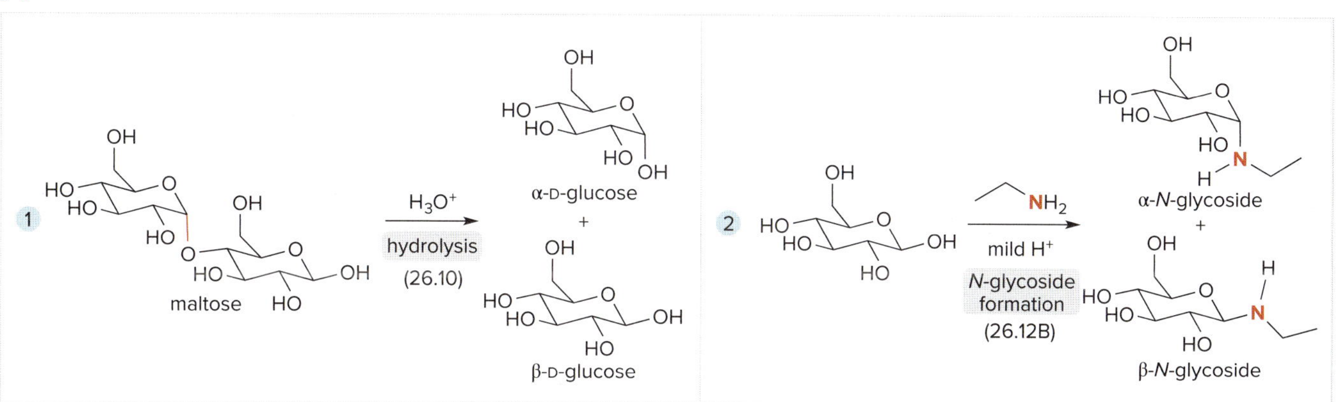

Try Problems 26.50h; 26.51; 26.59; 26.69b; 26.70c, e; 26.72b; 26.73c.

KEY SKILLS

[1] Converting a compound to a Fischer projection formula (26.2A); example: (S)-3-hydroxy-2-methylpropanal

1 **Rotate and re-draw the molecule to place the horizontal bonds in front of the plane and the vertical bonds behind the plane.**

re-draw
Tip red bonds in the plane forward.
• Horizontal bonds come forward.
• Vertical bonds go back.

2 **Use a cross to represent the stereogenic center.**

Fischer projection

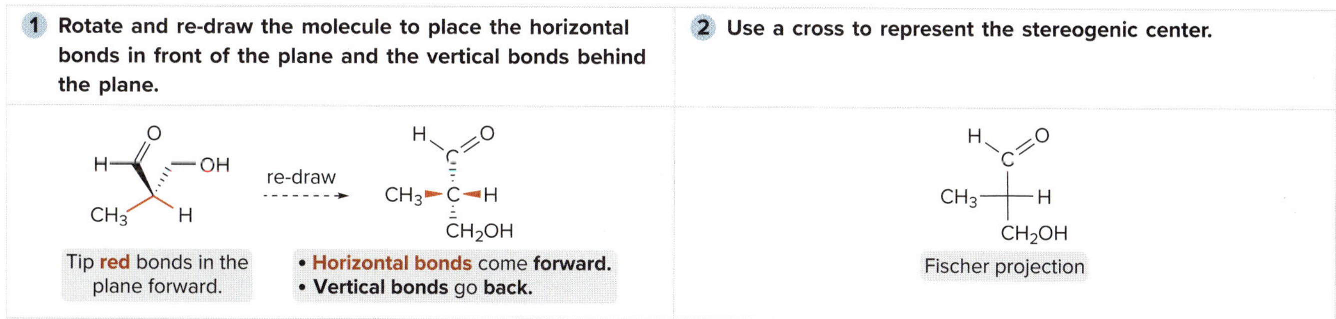

See Sample Problems 26.1 and 26.2. Try Problems 26.38, 26.40, 26.41.

[2] Converting a skeletal structure to a Fischer projection (26.2A); example: D-glucose

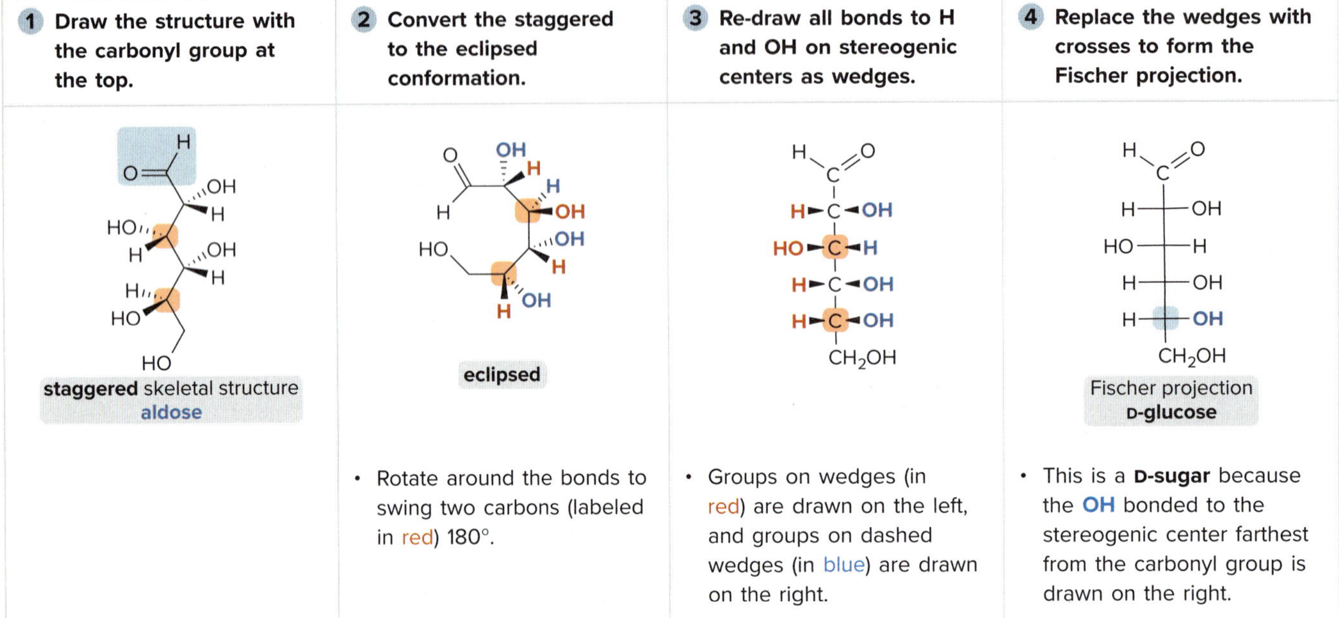

See Sample Problem 26.3. Try Problem 26.38.

[3] Drawing a Haworth projection from an acyclic aldohexose (26.5); example: D-glucose

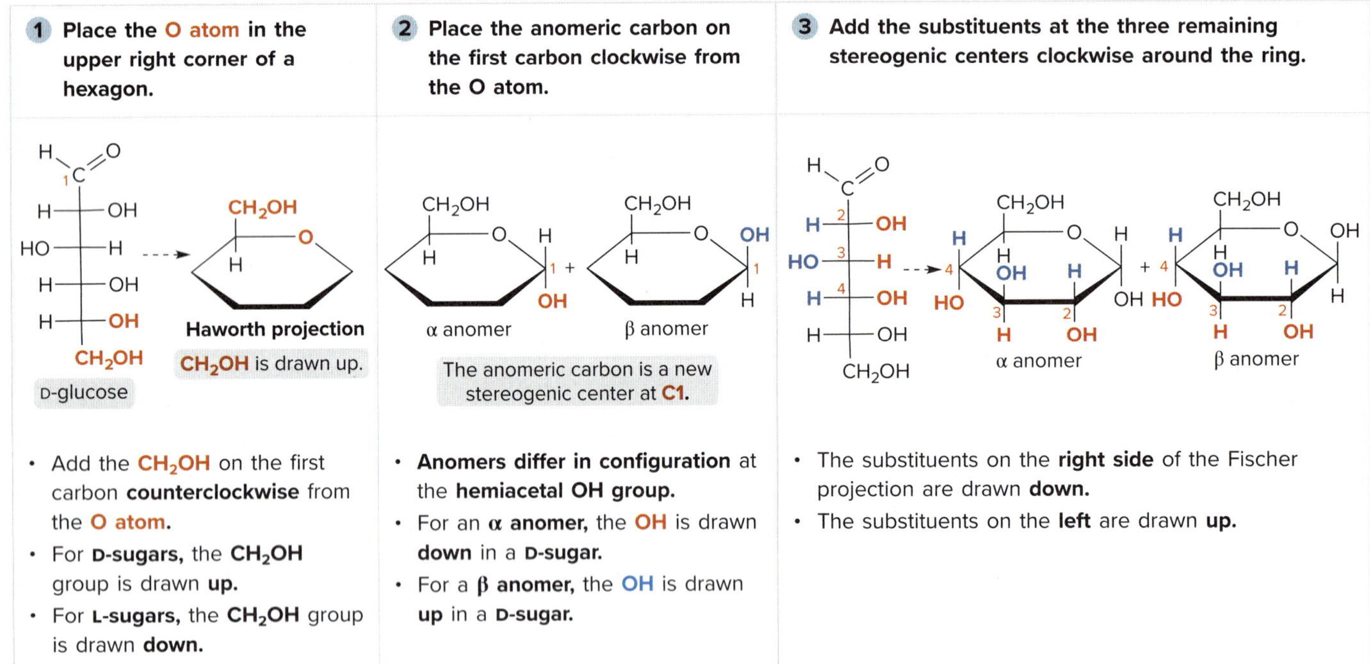

See *How To*, p. 1187. Try Problems 26.44, 26.45a, 26.53a, 26.54a.

[4] Converting a Haworth projection to its acyclic form (26.5B); example: D-glucose

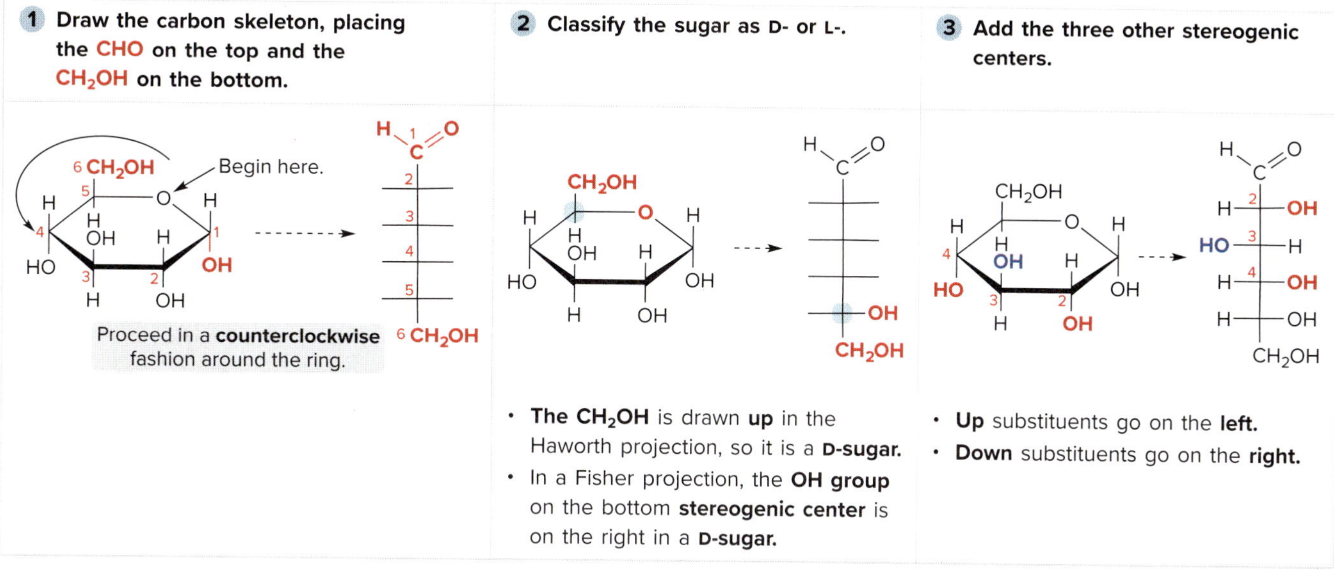

See Sample Problem 26.4. Try Problems 26.39a, 26.47.

[5] Converting a Haworth projection to a chair form (26.5C); example: D-glucose

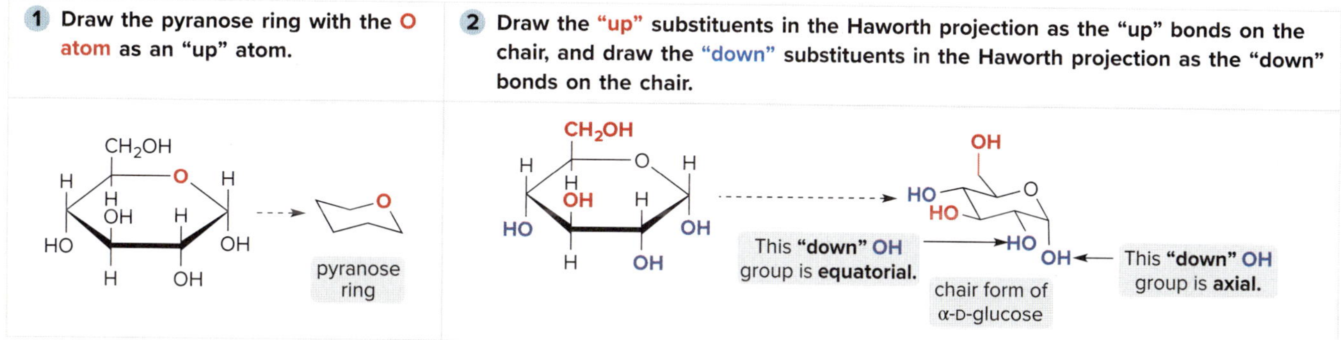

Try Problems 26.46, 26.53b, 26.54b.

[6] Determining the structure of an unknown D-aldopentose given a set of facts (26.8–26.9)

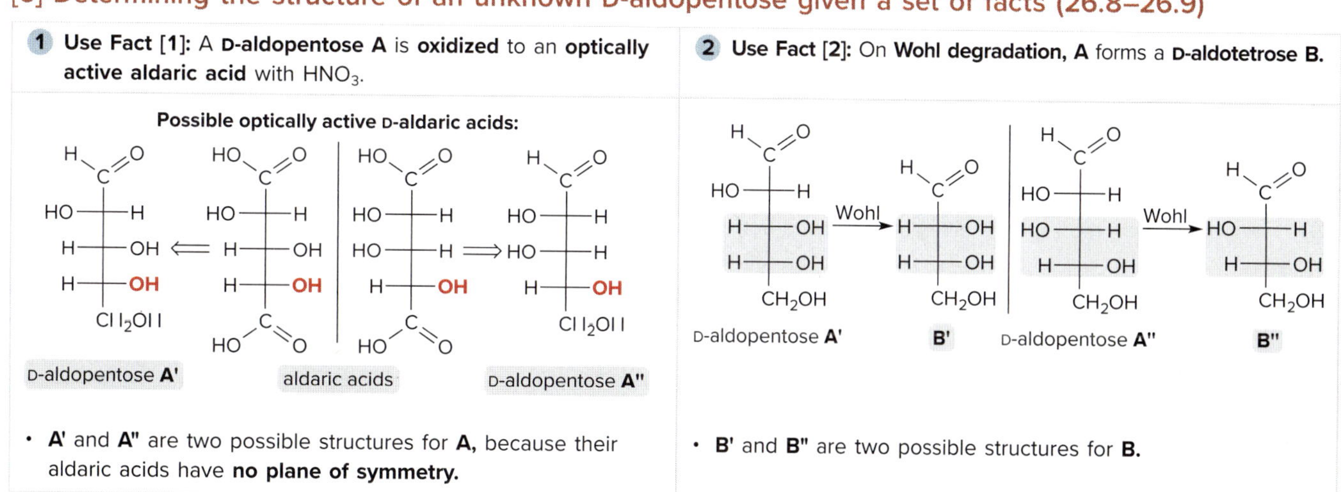

3 **Use Fact [3]:** The **D-aldotetrose B** is **oxidized** to an **optically inactive aldaric acid.**

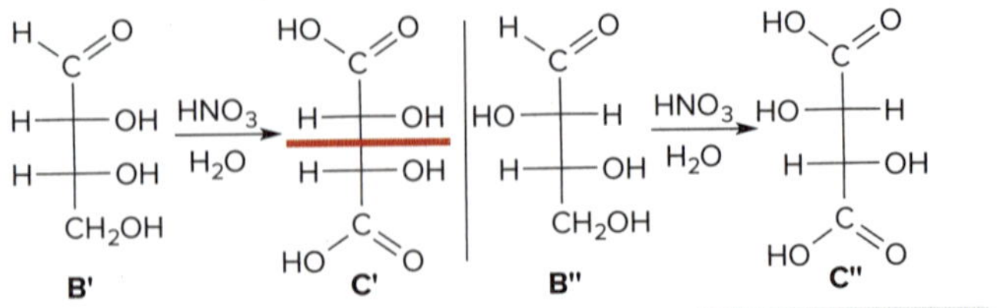

B' $\xrightarrow[H_2O]{HNO_3}$ C'

optically inactive aldaric acid plane of symmetry

B'' $\xrightarrow[H_2O]{HNO_3}$ C''

optically active aldaric acid no plane of symmetry

- The oxidation product of **B'** is **optically inactive,** because it has a **plane of symmetry.**

4 **Draw the unknown D-aldopentose A and D-aldotetrose B using Facts [1]–[3].**

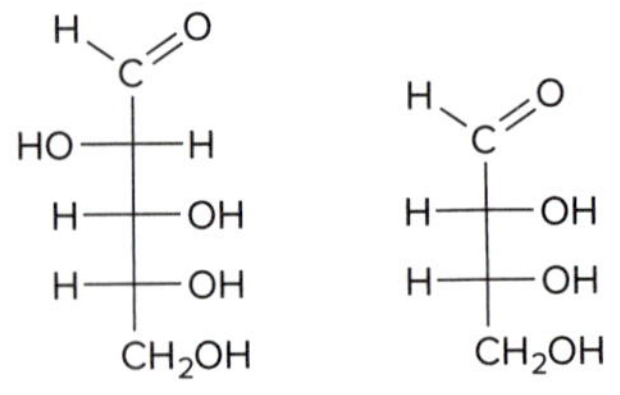

A' = A **B' = B**

- The precursor of **C'** is **B',** and thus **A' = A.**

See Sample Problem 26.5. Try Problems 26.64–26.66.

CHAPTER 26 MULTIPLE-CHOICE SELF-TEST

The Self-Test consists of multiple-choice questions similar to those found on the American Chemical Society organic chemistry exam. Answers are given at the end of the chapter.

1. Which term describes compound **A?**

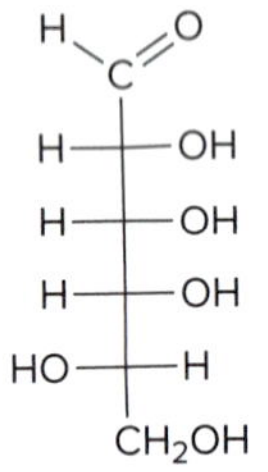

A

a. L-aldohexose b. aldonic acid c. aldaric acid d. D-aldohexose

2. Which compound is the α anomer of monosaccharide **B?**

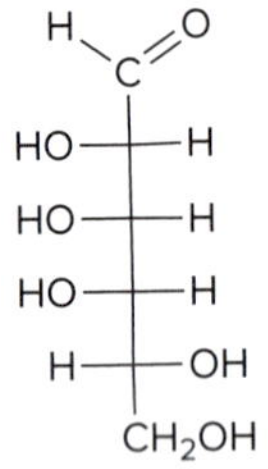

B

a. 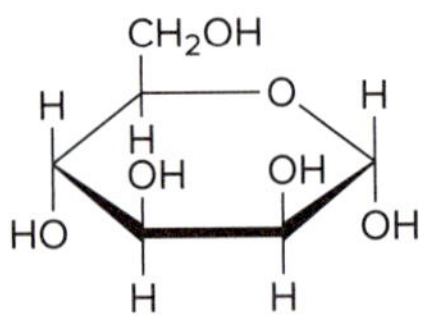 b. 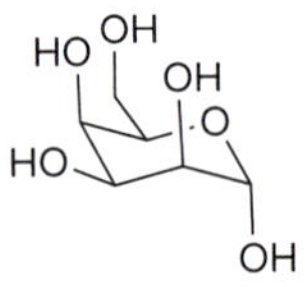 c. 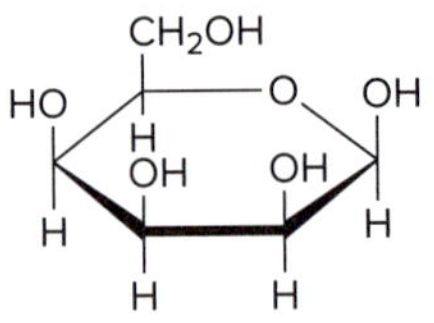 d.

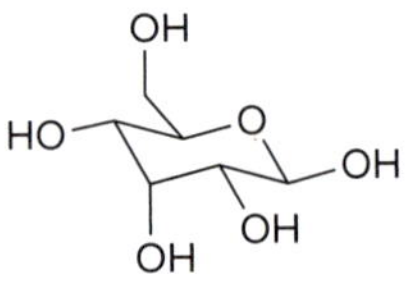

3. Which structure represents a different compound from the other three?

a. 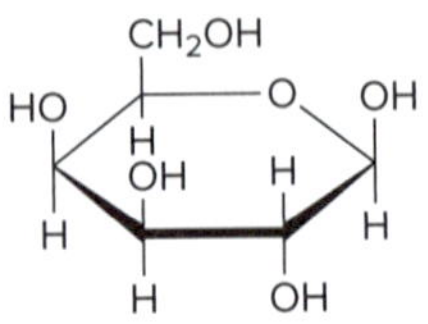 b. 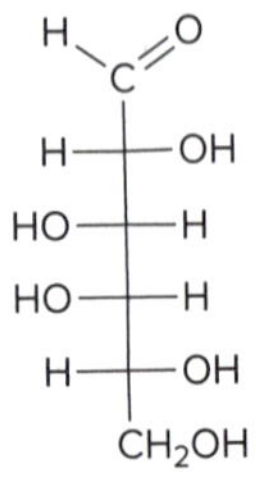 c. 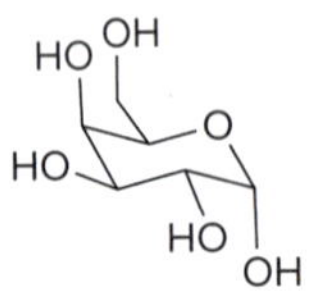 d.

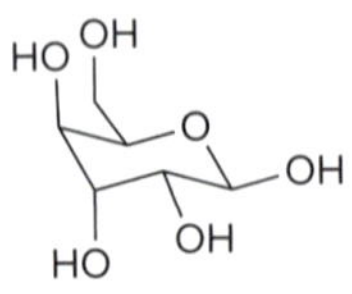

4. Which Fischer projection represents compound **C?**

C

a. CHO; H–C–OH; HO–C–H; CH_2OH

b. CHO; HO–C–H; H–C–OH; CH_2OH

c. CHO; H–C–OH; H–C–OH; CH_2OH

d. CHO; HO–C–H; HO–C–H; CH_2OH

5. Which compound(s) form an achiral diacid on treatment with HNO_3?

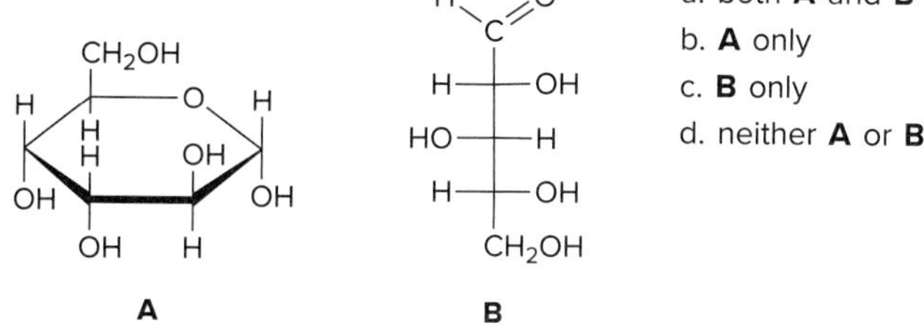

a. both **A** and **B**
b. **A** only
c. **B** only
d. neither **A** or **B**

6. Which statement about the physical properties of monosaccharides is *not* true?
a. Monosaccharides are soluble in diethyl ether.
b. Monosaccharides have high melting points.
c. Monosaccharides are generally sweet tasting.
d. Monosaccharides are soluble in water.

7. Which statement about monosaccharide **X** is *not* true?

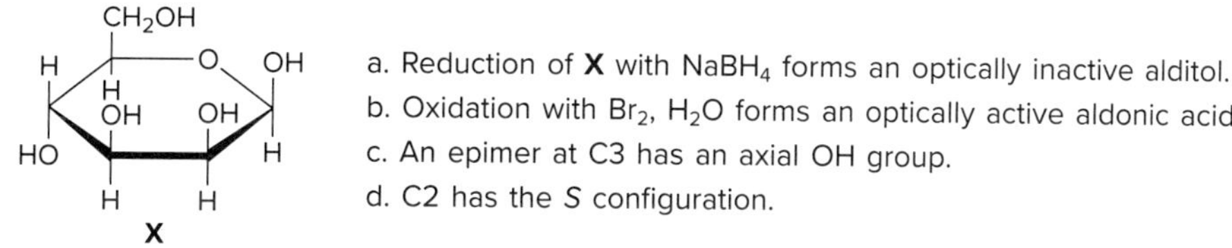

a. Reduction of **X** with $NaBH_4$ forms an optically inactive alditol.
b. Oxidation with Br_2, H_2O forms an optically active aldonic acid.
c. An epimer at C3 has an axial OH group.
d. C2 has the *S* configuration.

8. How many products are formed when **Y** is treated with excess CH_3I and Ag_2O followed by aqueous acid?

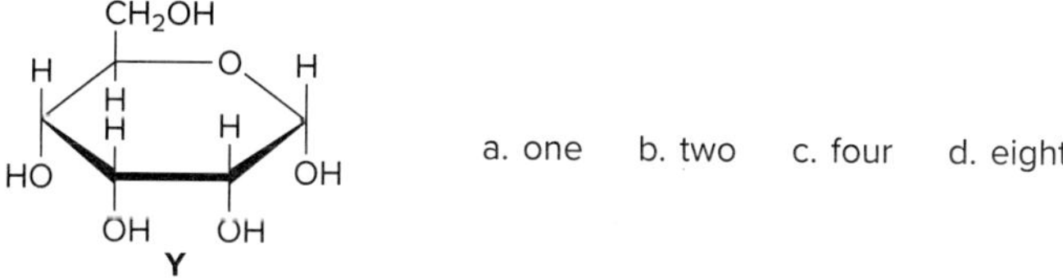

a. one b. two c. four d. eight

9. How many stereoisomers are possible for a ketohexose? How many are D monosaccharides?
a. four stereoisomers, two D monosaccharides
b. four stereoisomers, four D monosaccharides
c. eight stereoisomers, four D monosaccharides
d. 16 stereoisomers, eight D monosaccharides

10. How many glycosides does the given polysaccharide contain?

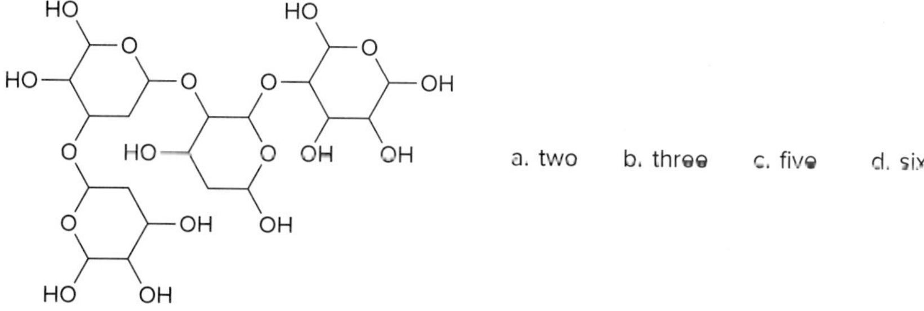

a. two b. three c. five d. six

PROBLEMS

Problems Using Three-Dimensional Models

26.38 Convert each ball-and-stick model to a Fischer projection.

a. 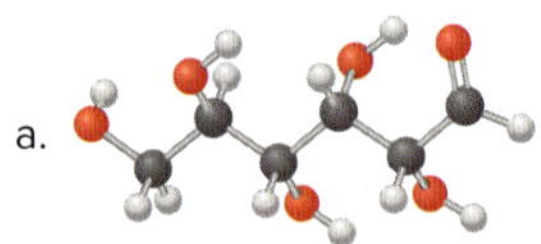

b. 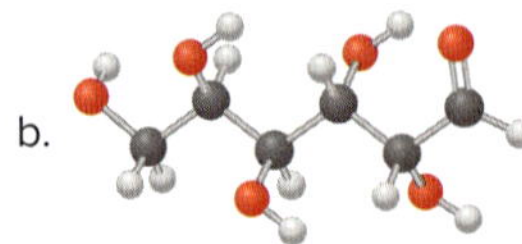

26.39 (a) Convert each cyclic monosaccharide to a Fischer projection of its acyclic form. (b) Name each monosaccharide. (c) Label the anomer as α or β.

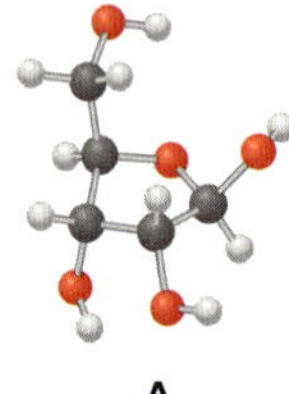

A

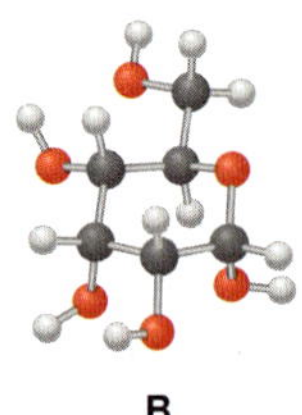

B

Fischer Projections

26.40 Classify each compound as identical to **A** or its enantiomer.

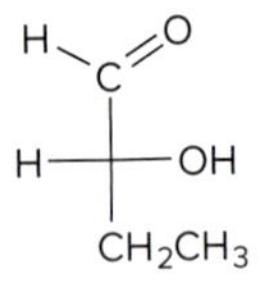

A

a. 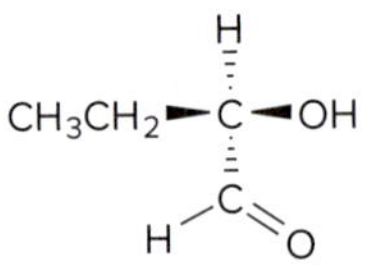 CH_3CH_2–C(H)(OH)–CHO

b. 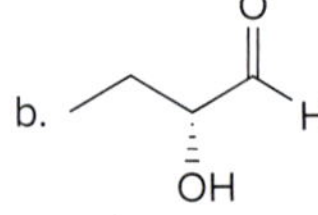

c. 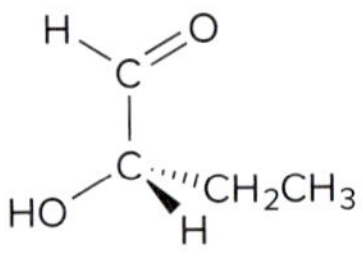 HO–C(H)(CH_2CH_3)–CHO

26.41 Convert each compound to a Fischer projection, and label each stereogenic center as *R* or *S*.

a. 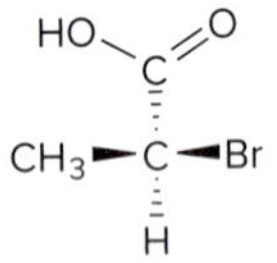 CH_3–C(Br)(H)–COOH

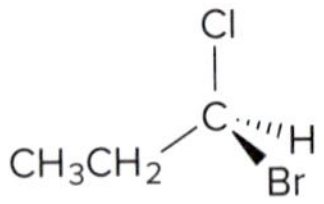

b. CH_3CH_2–C(Cl)(H)(Br)

c. CH_3(H)(Br)C–C(H)(Br)CH_3

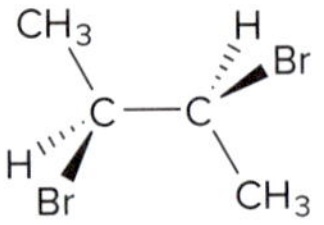

d. 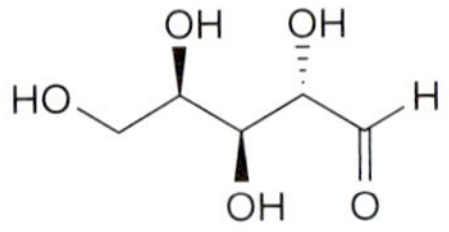 HO–CH₂–CH(OH)–CH(OH)–CHO

Monosaccharide Structure and Stereochemistry

26.42 For D-arabinose:

a. Draw its enantiomer.

b. Draw an epimer at C3.

c. Draw a diastereomer that is not an epimer.

d. Draw a constitutional isomer that still contains a carbonyl group.

26.43 Consider the following six compounds (**A–F**).

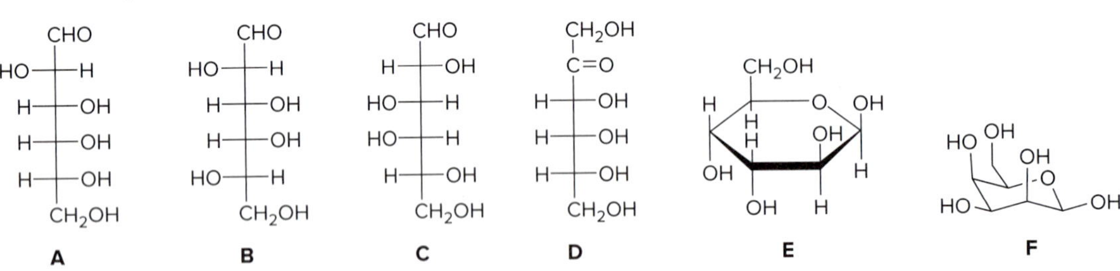

How are the two compounds in each pair related? Choose from enantiomers, epimers, diastereomers but not epimers, constitutional isomers, and identical compounds.

a. **A** and **B**
b. **A** and **C**
c. **B** and **C**
d. **A** and **D**
e. **E** and **F**

26.44 Draw a Haworth projection for each compound using the structures in Figures 26.4 and 26.5.

a. β-D-talopyranose
b. α-D-galactopyranose
c. α-D-tagatofuranose

26.45 Draw the structure of each compound and name it using the information in Figure 26.4.

a. the α anomer of a monosaccharide that is epimeric with D-glucose at C4 using a Haworth projection

b. the β anomer of a monosaccharide that is epimeric with D-gulose at C2 using a chair pyranose

26.46 Draw both pyranose anomers of each aldohexose using a three-dimensional representation with a chair pyranose. Label each anomer as α or β.

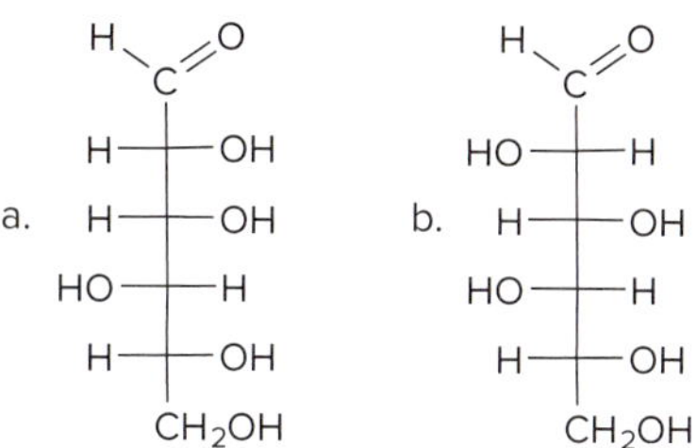

26.47 Convert each cyclic monosaccharide to its acyclic form.

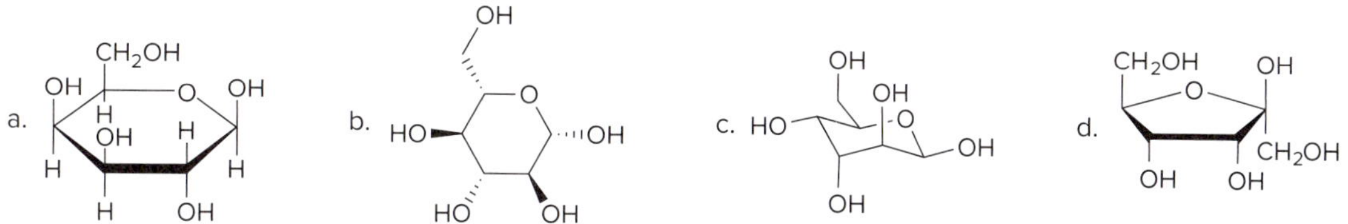

26.48 The most stable conformation of the pyranose ring of most D-aldohexoses places the largest group, CH_2OH, in the equatorial position. An exception to this is the aldohexose D-idose. Draw the two possible chair conformations of either the α or β anomer of D-idose. Explain why the more stable conformation has the CH_2OH group in the axial position.

Monosaccharide Reactions

26.49 Draw the structure (including stereochemistry) of the cyclic hemiacetal(s) formed when each hydroxy carbonyl compound is treated with aqueous acid.

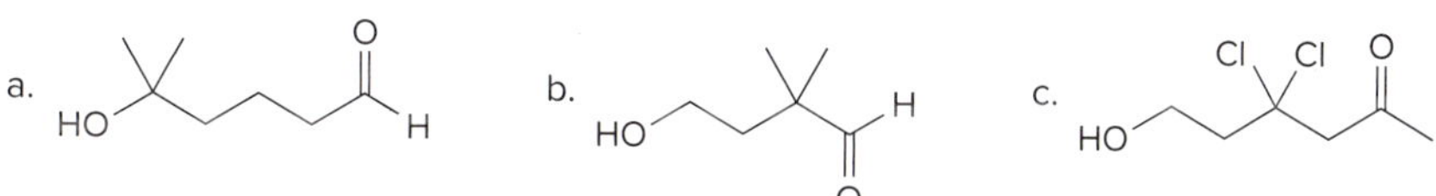

26.50 Draw the products formed when D-altrose is treated with each reagent.

a. $(CH_3)_2CHOH$, HCl

b. $NaBH_4$, CH_3OH

c. Br_2, H_2O

d. HNO_3, H_2O

e. [1] NH_2OH; [2] $(CH_3CO)_2O$, $NaOCOCH_3$; [3] $NaOCH_3$

f. [1] NaCN, HCl; [2] H_2, Pd-$BaSO_4$; [3] H_3O^+

g. CH_3I, Ag_2O

h. $C_6H_5CH_2NH_2$, mild H^+

26.51 What aglycon and monosaccharides are formed when salicin and solanine (Section 26.6C) are each hydrolyzed with aqueous acid?

26.52 Draw a Fischer projection of the monosaccharide from which each of the following glycosides was prepared.

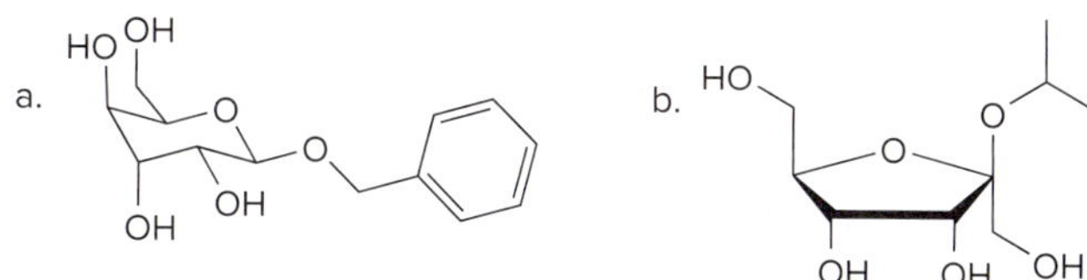

26.53 Answer the following questions about monosaccharide **A**.

a. Draw the α anomer of **A** in a Haworth projection.

b. Draw the β anomer of **A** in a three-dimensional representation using a chair conformation.

c. What two aldoses yield **A** in a Wohl degradation?

d. What product is formed when **A** undergoes a Wohl degradation?

e. What product is formed when **A** reacts with Ag_2O in NH_4OH?

A

26.54 Answer the following questions about monosaccharide **B**.

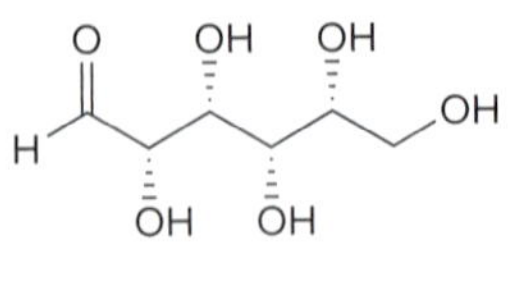

B

a. Draw the β anomer of **B** in a Haworth projection.
b. Draw the α anomer of **B** in a three-dimensional representation using a chair conformation.
c. What products are formed when **B** undergoes the Kiliani–Fischer synthesis?
d. What product is formed when **B** is treated with $NaBH_4$ in CH_3OH?
e. Draw the disaccharide formed when two molecules of **B** are joined by a 1→4-β-glycosidic linkage.

26.55 Draw the structure of two different aldohexoses that yield the following aldaric acid when oxidized with HNO_3. Use Figure 26.4 to name each aldohexose.

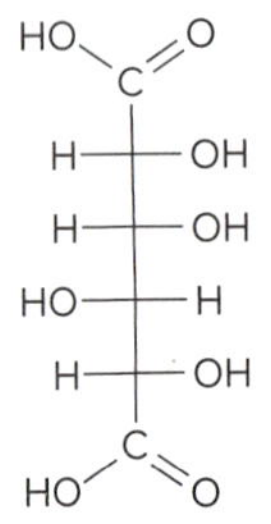

26.56 Treatment of D-glucose with $NaBH_4$ gives an alditol **A**. What L-aldohexose also yields **A** when treated with $NaBH_4$?

26.57 What products are formed when each compound undergoes a Kiliani–Fischer synthesis?

a.

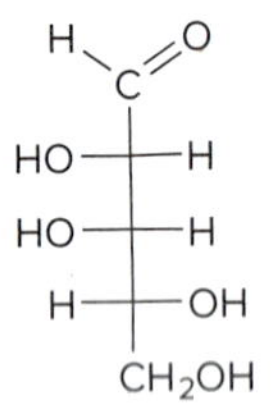

b.

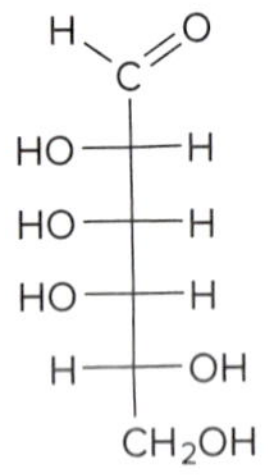

26.58 How would you convert D-glucose to each compound? More than one step is required.

a. 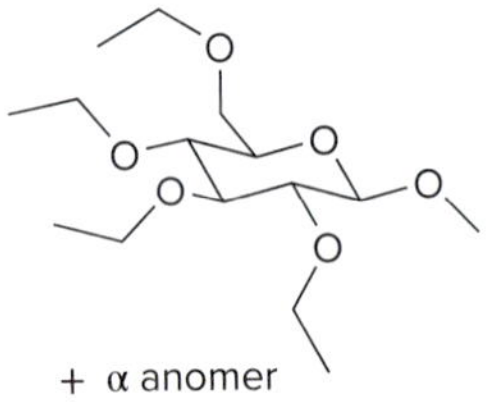 + α anomer

b.

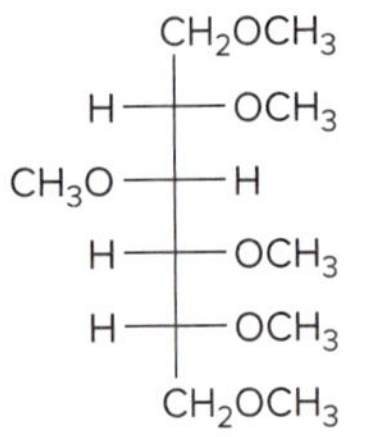

c.

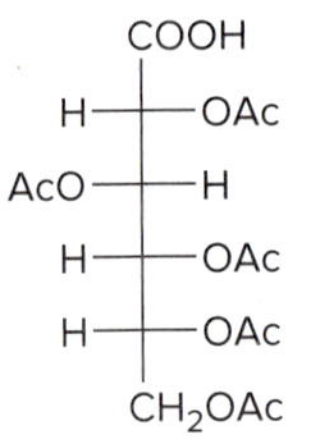

26.59 What products are formed when each compound is treated with aqueous acid?

a.

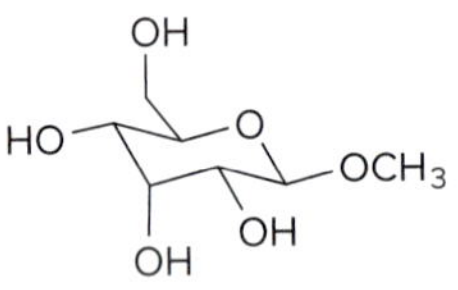

b.

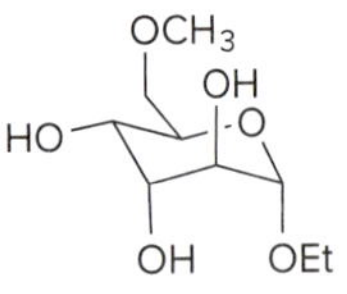

c.

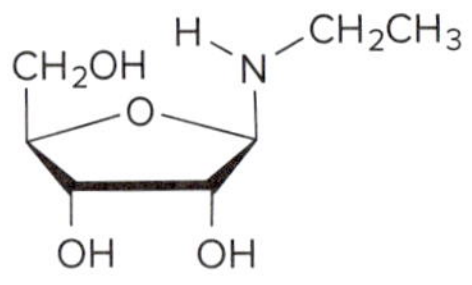

Mechanisms

26.60 Draw a stepwise mechanism for the following reaction.

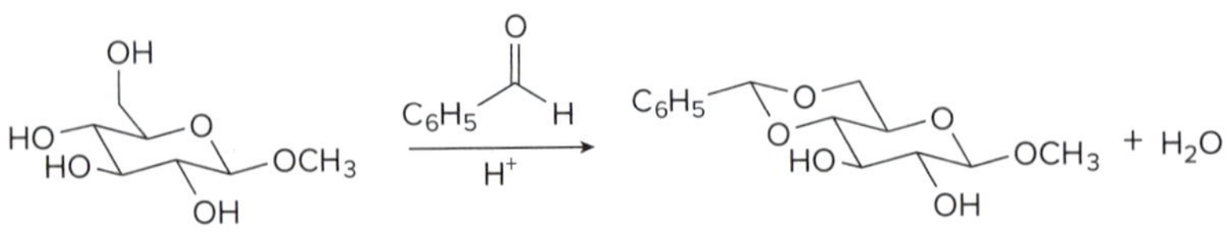

26.61 Draw a stepwise mechanism for the following hydrolysis.

H_3O^+ ; + CH_3OH

26.62 The following isomerization reaction, drawn using D-glucose as starting material, occurs with all aldohexoses in the presence of base. Draw a stepwise mechanism that illustrates how each compound is formed.

^-OH, H_2O

D-glucose

(recovered starting material)

26.63 The cyclic forms of some monosaccharides interconvert between furanose and pyranose rings. Draw a stepwise mechanism for the following conversion of a furanose to a pyranose in the presence of an acid (HA).

HA

furanose

pyranose

Identifying Monosaccharides

26.64 Which D-aldopentose is oxidized to an optically active aldaric acid and undergoes the Wohl degradation to yield a D-aldotetrose that is oxidized to an optically active aldaric acid?

26.65 Identify compounds **A–D.** A D-aldopentose **A** is oxidized with HNO_3 to an optically inactive aldaric acid **B. A** undergoes the Kiliani–Fischer synthesis to yield **C** and **D. C** is oxidized to an optically active aldaric acid. **D** is oxidized to an optically inactive aldaric acid.

26.66 A D-aldopentose **A** is reduced to an optically active alditol. Upon Kiliani–Fischer synthesis, **A** is converted to two D-aldohexoses, **B** and **C. B** is oxidized to an optically inactive aldaric acid. **C** is oxidized to an optically active aldaric acid. What are the structures of **A–C?**

Disaccharides and Polysaccharides

26.67 Draw the structure of a disaccharide formed from two mannose units joined by a 1→4-α-glycosidic linkage.

26.68 The average human body contains about 15 g of hyaluronic acid, largely in the lubricant around joints and the vitreous humor of the eye. It is used commercially in skin and joint care, wound healing, and eye surgery. Hyaluronic acid is a polysaccharide composed of alternating units of D-glucuronic acid and *N*-acetyl-D-glucosamine joined in 1→4-β-glycosidic linkages. Draw a short segment of hyaluronic acid in three dimensions using chair forms for the six-membered rings.

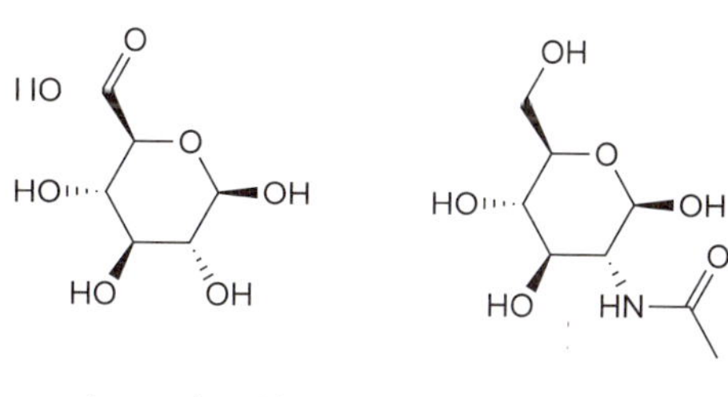

D-glucuronic acid

N-acetyl-D-glucosamine

26.69 a. Identify the glycosidic linkage in disaccharide **C**, classify the glycosidic bond as α or β, and use numbers to designate its location.

b. Identify the lettered compounds in the following reaction.

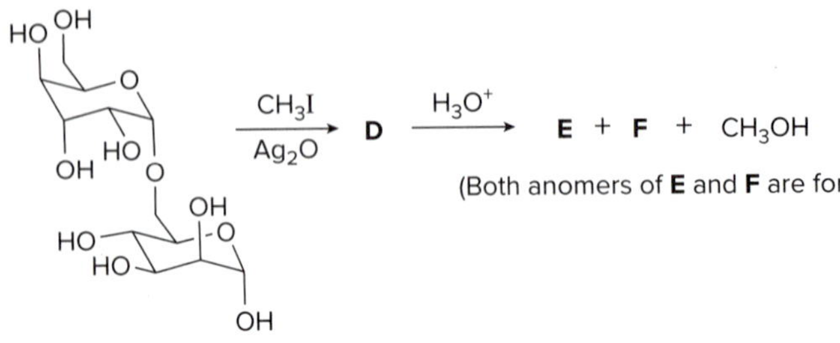

C $\xrightarrow[Ag_2O]{CH_3I}$ **D** $\xrightarrow{H_3O^+}$ **E** + **F** + CH_3OH

(Both anomers of **E** and **F** are formed.)

26.70 Lacto-*N*-neotetraose is another human milk oligosaccharide found in breast milk (Section 24.11D).

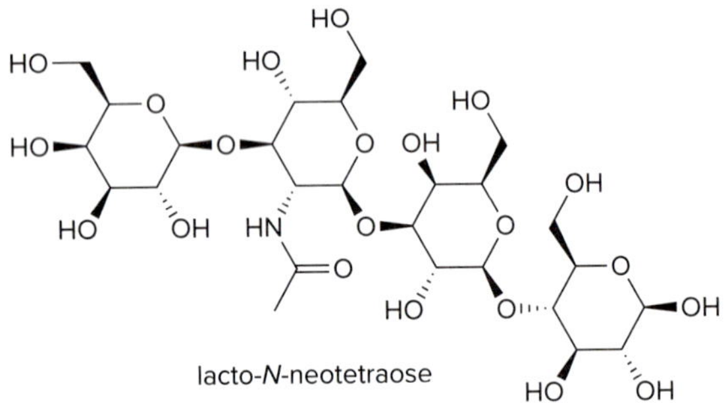

lacto-*N*-neotetraose

a. Label all glycosidic bonds.

b. Classify each glycosidic linkage as α or β and use numbers to designate the location between the two rings.

c. What products are formed when the glycosidic bonds in lacto-*N*-neotetraose are hydrolyzed with aqueous acid? Name each of the monosaccharides formed.

d. Is lacto-*N*-neotetraose a reducing sugar?

e. What products are formed when lacto-*N*-neotetraose is treated with excess CH_3I and Ag_2O?

26.71 Deduce the structure of the disaccharide isomaltose from the following data.

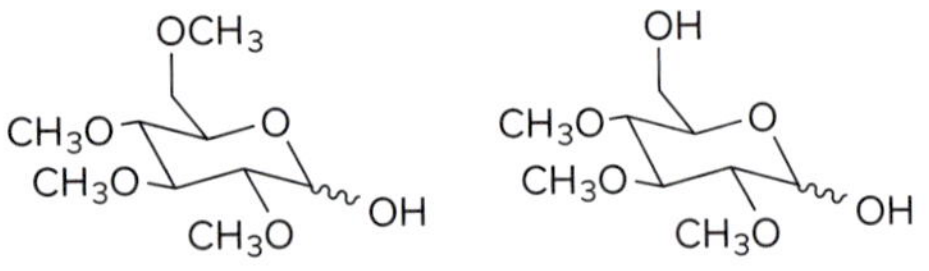

(Both anomers are present.)

[1] Hydrolysis yields D-glucose exclusively.

[2] Isomaltose is cleaved with α-glycosidase enzymes.

[3] Isomaltose is a reducing sugar.

[4] Methylation with excess CH_3I, Ag_2O and then hydrolysis with H_3O^+ forms the given products.

26.72 Draw the structure of each of the following compounds: (a) a polysaccharide formed by joining D-glucosamine in 1→6-α-glycosidic linkages; (b) an α-*N*-glycoside formed from D-arabinose and $C_6H_5CH_2NH_2$

26.73 Acarbose, a drug used to treat type 2 diabetes, acts by inhibiting the enzyme that cleaves glucose from starch and glycogen.

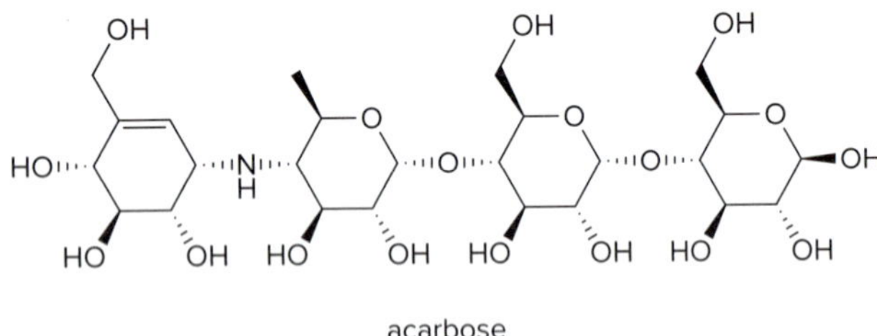

acarbose

a. Label all acetals and hemiacetals.

b. What products are formed when acarbose is treated with CH_3OH and HCl?

c. What products are formed when acarbose is treated with aqueous acid?

Challenge Problems

26.74 (a) Draw the more stable chair form of fucose, an essential monosaccharide needed in the diet and a component of carbohydrates on mammalian and plant cell surfaces. (b) Classify fucose as a D- or L-monosaccharide. (c) What two structural features are unusual in fucose?

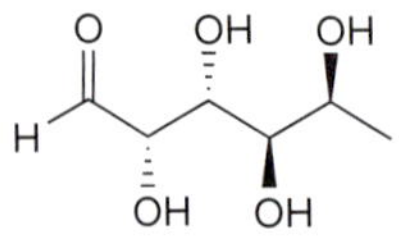

fucose

26.75 As we have seen in Chapter 26, monosaccharides can be drawn in a variety of ways, and in truth, often a mixture of cyclic compounds is present in a solution. Identify each monosaccharide, including its proper D,L designation, drawn in a less-than-typical fashion.

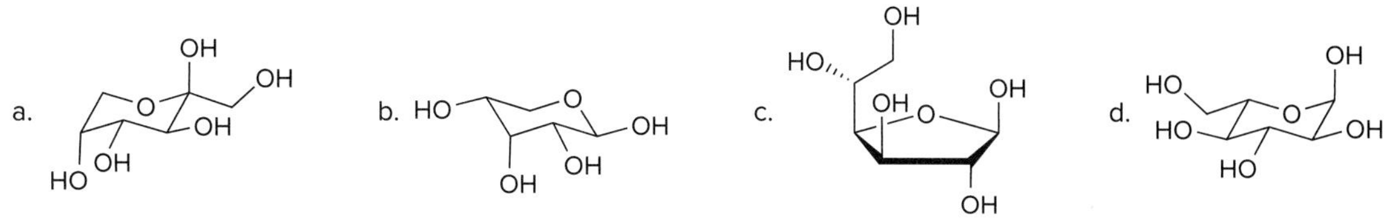

a. b. c. d.

26.76 Draw a stepwise mechanism for the following reaction.

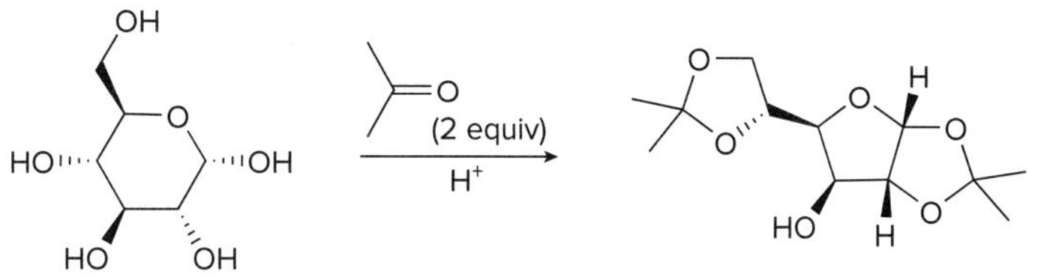

(2 equiv)
H^+

SELF-TEST ANSWERS

1. a 2. b 3. d 4. b 5. c 6. a 7. a 8. b 9. c 10. b

27 Amino Acids and Proteins

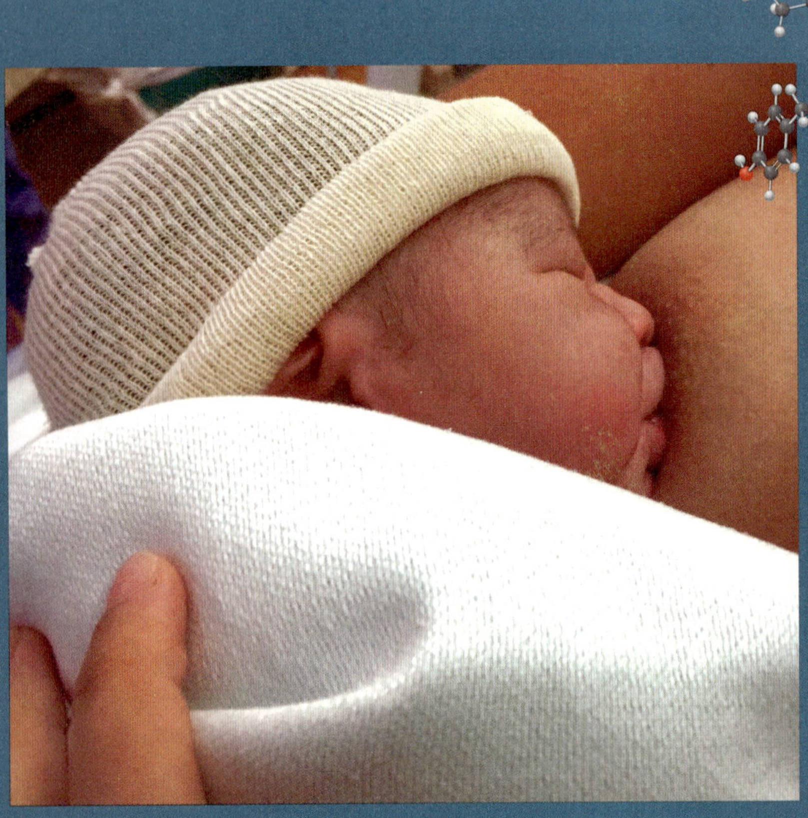

Daniel C. Smith

27.1 Amino acids
27.2 Separation of amino acids
27.3 Enantioselective synthesis of amino acids
27.4 Peptides
27.5 Peptide sequencing
27.6 Peptide synthesis
27.7 Automated peptide synthesis
27.8 Protein structure
27.9 Important proteins
27.10 Enzymes

Oxytocin, a peptide consisting of nine amino acids, is a hormone that causes cervical dilation in preparation for childbirth and uterine contractions during labor, and it also stimulates the flow of milk in nursing mothers. Oxytocin was the first peptide hormone synthesized, a feat for which Vincent du Vigneaud was awarded the 1955 Nobel Prize in Chemistry. Oxytocin, sold under the trade name Pitocin, is used to induce labor and to stop bleeding after a delivery. In Chapter 27, we learn about peptides like oxytocin and the amino acids that comprise them.

Why Study . . .

Amino Acids and Proteins?

Of the four major groups of biomolecules—lipids, carbohydrates, nucleic acids, and proteins—proteins have the widest array of functions. **Keratin** and **collagen,** for example, are part of a large group of structural proteins that form long insoluble fibers, giving strength and support to tissues. Hair, horns, hooves, and fingernails are all made up of keratin. **Collagen** is found in bone, connective tissue, tendons, and cartilage. **Enzymes** are proteins that catalyze and regulate all aspects of cellular function. **Membrane proteins** transport small organic molecules and ions across cell membranes. **Insulin,** the hormone that regulates blood glucose levels, **fibrinogen** and **thrombin,** which form blood clots, and **hemoglobin,** which transports oxygen from the lungs to tissues, are all proteins.

In Chapter 27 we discuss proteins and their primary components, the amino acids.

27.1 Amino Acids

Amino acids were first discussed in Section 19.11.

Naturally occurring amino acids have an amino group (NH_2) bonded to the α carbon of a carboxy group (COOH), so they are called **α-amino acids.**

- **All proteins are polyamides formed by joining amino acids together.**

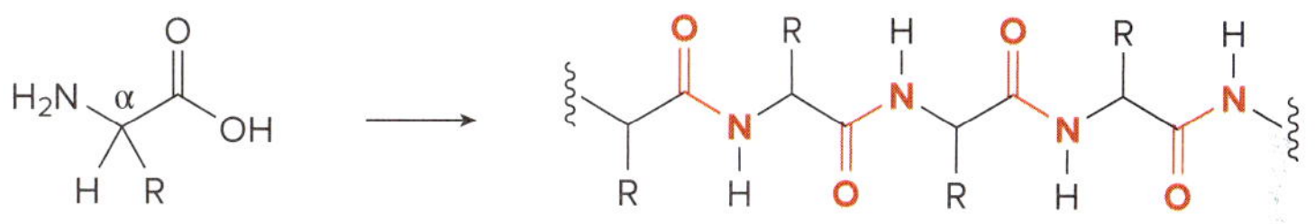

α-amino acid

portion of a protein molecule

27.1A General Features of α-Amino Acids

The 20 amino acids that occur naturally in proteins differ in the identity of the R group bonded to the α carbon. The R group is called the **side chain** of the amino acid.

The simplest amino acid, called glycine, has R = H. **All other amino acids (R ≠ H) have a stereogenic center on the α carbon.** As is true for monosaccharides, the prefixes D and L are used to designate the configuration at the stereogenic center of amino acids. Common, naturally occurring amino acids are called **L-amino acids.** Their enantiomers, D-amino acids, are rarely found in nature. These general structures are shown in Figure 27.1. According to *R,S* designations, all L-amino acids except cysteine have the ***S*** **configuration.**

Figure 27.1 The general features of an α-amino acid

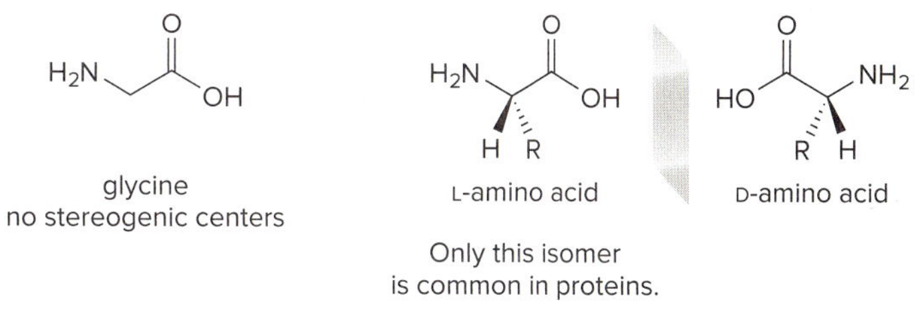

glycine
no stereogenic centers

L-amino acid

D-amino acid

Only this isomer
is common in proteins.

All amino acids have common names. These names can be represented by either a one-letter or a three-letter abbreviation. Figure 27.2 is a listing of the 20 naturally occurring amino acids, with their abbreviations. Note the variability in the R groups. A side chain can be a simple alkyl group, or it can have additional functional groups such as OH, SH, COOH, or NH_2.

- **Amino acids with an additional COOH group in the side chain are called *acidic* amino acids.**
- **Those with an additional basic N atom in the side chain are called *basic* amino acids.**
- **All others are neutral amino acids.**

Figure 27.2 The 20 naturally occurring amino acids

Neutral amino acids

Name	Structure	Abbreviations	Name	Structure	Abbreviations
Alanine	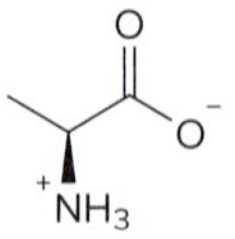	Ala A	Phenylalanine*	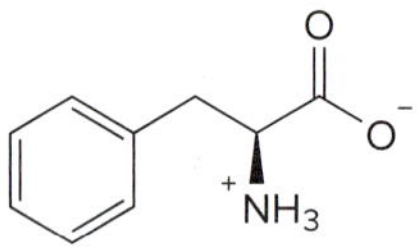	Phe F
Asparagine	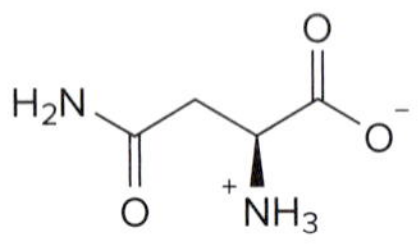	Asn N	Proline	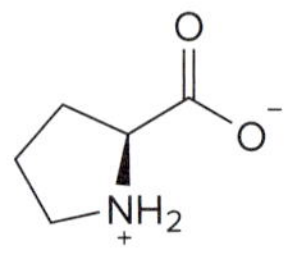	Pro P
Cysteine	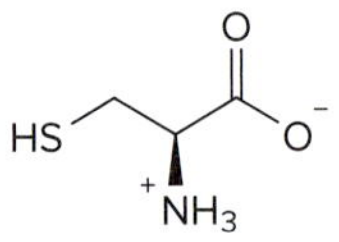	Cys C	Serine	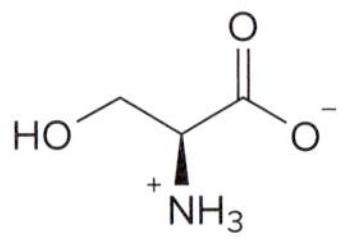	Ser S
Glutamine	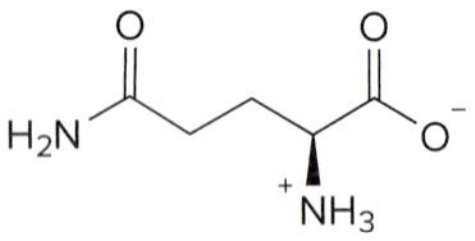	Gln Q	Threonine*	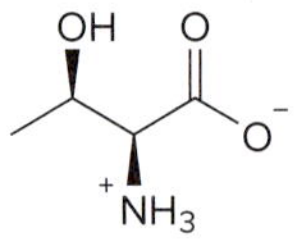	Thr T
Glycine	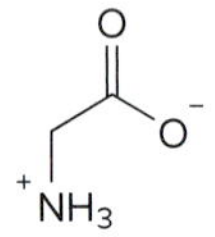	Gly G	Tryptophan*	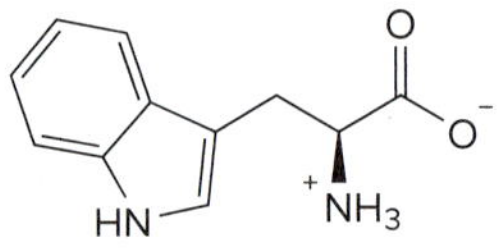	Trp W
Isoleucine*	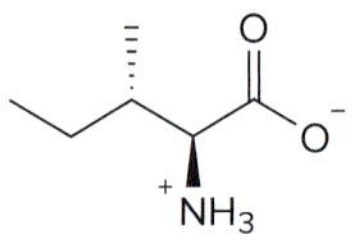	Ile I	Tyrosine	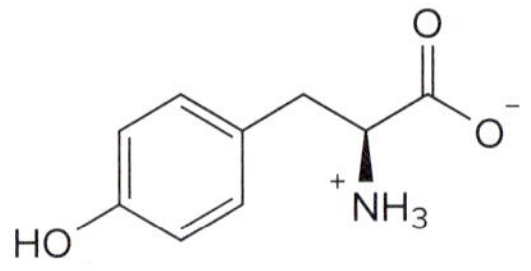	Tyr Y
Leucine*	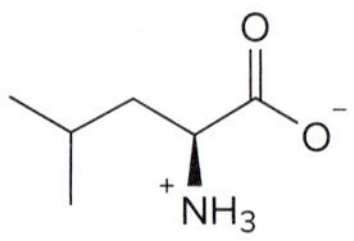	Leu L	Valine*	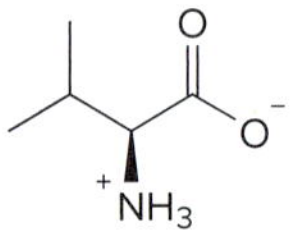	Val V
Methionine*	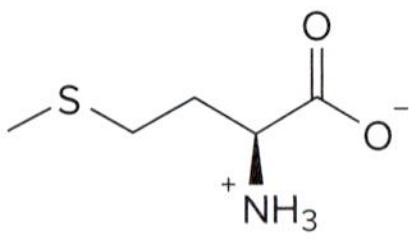	Met M			

Acidic amino acids

Name	Structure	Abbreviations
Aspartic acid	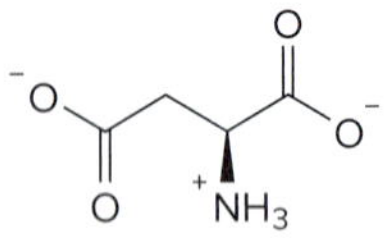	Asp D
Glutamic acid	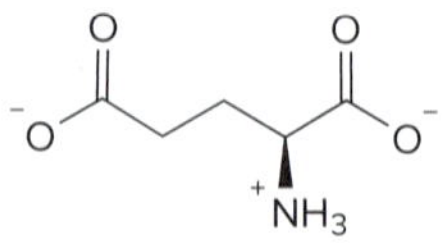	Glu E

Basic amino acids

Name	Structure	Abbreviations
Arginine*	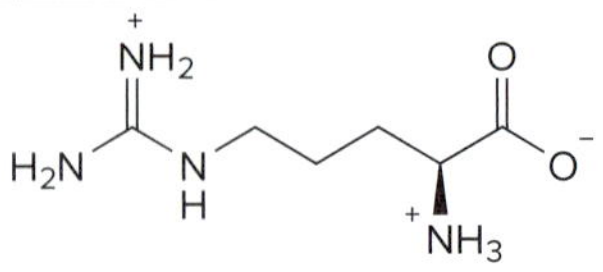	Arg R
Histidine*	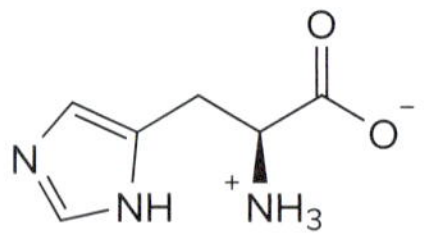	His H
Lysine*	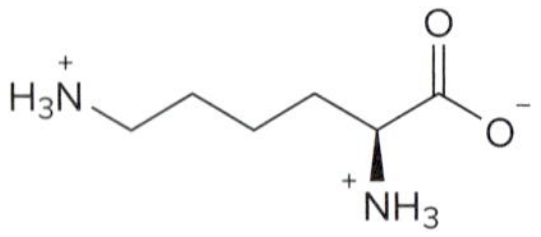	Lys K

Essential amino acids are labeled with an asterisk (*).

Although most grains are low in lysine, quinoa is relatively high in lysine content and a good source of essential amino acids for a vegetarian diet.
Sarka Babicka/Moment/Getty Images

Look closely at the structures of proline, isoleucine, and threonine.

- **All amino acids are 1° amines except for proline,** which has its N atom in a five-membered ring, making it a **2° amine.**
- **Isoleucine** and **threonine** contain an additional stereogenic center at the β carbon, so there are four possible stereoisomers, only one of which is naturally occurring.

L-proline
2° amine

L-isoleucine

L-threonine

Humans can synthesize only 10 of these 20 amino acids. The remaining 10 are called **essential amino acids** because they must be obtained from the diet. These are labeled with an asterisk in Figure 27.2.

Problem 27.1 Draw the other three stereoisomers of L-isoleucine, and label the stereogenic centers as *R* or *S*.

27.1B Acid–Base Behavior

Recall from Section 19.11B that an amino acid has both an acidic and a basic functional group, so proton transfer forms a salt called a **zwitterion.**

basic site — acidic proton — proton transfer → ammonium cation — carboxylate anion

This neutral form of an amino acid does *not* really exist.

The zwitterion is neutral.

This salt is the neutral form of an amino acid.

This form exists at **pH ≈ 6.**

The structures in Figure 27.2 show the charged form of the amino acids at the physiological pH of the blood.

- **Amino acids do not exist to any appreciable extent as uncharged neutral compounds. They exist as salts, giving them high melting points and making them water soluble.**

Amino acids exist in different charged forms, as shown in Figure 27.3, depending on the pH of the aqueous solution in which they are dissolved. For neutral amino acids, the overall charge is +1, 0, or –1. Only at pH ~6 does the zwitterionic form exist.

The –COOH and $-NH_3^+$ groups of an amino acid are ionizable, because they can lose a proton in aqueous solution. As a result, they have different pK_a values. The pK_a of the –COOH group is typically ~2, whereas that of the $-NH_3^+$ group is ~9, as shown in Table 27.1.

Figure 27.3 How the charge of a neutral amino acid depends on the pH

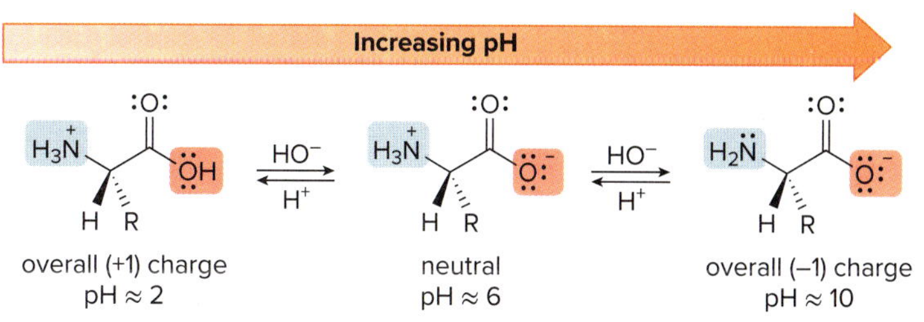

Table 27.1 pK_a Values for the Ionizable Functional Groups of an α-Amino Acid

Amino acid	α-COOH	$\alpha\text{-}NH_3^+$	Side chain	p*I*
Alanine	2.35	9.87	—	6.11
Arginine	2.01	9.04	12.48	10.76
Asparagine	2.02	8.80	—	5.41
Aspartic acid	2.10	9.82	3.86	2.98
Cysteine	2.05	10.25	8.00	5.02
Glutamic acid	2.10	9.47	4.07	3.08
Glutamine	2.17	9.13	—	5.65
Glycine	2.35	9.78	—	6.06
Histidine	1.77	9.18	6.10	7.64
Isoleucine	2.32	9.76	—	6.04
Leucine	2.33	9.74	—	6.04
Lysine	2.18	8.95	10.53	9.74
Methionine	2.28	9.21	—	5.74
Phenylalanine	2.58	9.24	—	5.91
Proline	2.00	10.60	—	6.30
Serine	2.21	9.15	—	5.68
Threonine	2.09	9.10	—	5.60
Tryptophan	2.38	9.39	—	5.88
Tyrosine	2.20	9.11	10.07	5.63
Valine	2.29	9.72	—	6.00

Some amino acids, such as aspartic acid and lysine, have acidic or basic side chains. These additional ionizable groups complicate somewhat the acid–base behavior of these amino acids. Table 27.1 lists the pK_a values for these acidic and basic side chains as well.

Table 27.1 also lists the isoelectric points (p*I*) for all of the amino acids. Recall from Section 19.11C that the **isoelectric point is the pH at which an amino acid exists primarily in its neutral form,** and that it can be calculated from the average of the pK_a values of the α-COOH and $\alpha\text{-}NH_3^+$ groups (for neutral amino acids only).

Problem 27.2 What form exists at the isoelectric point of each of the following amino acids: (a) valine; (b) leucine; (c) proline; (d) glutamic acid?

Problem 27.3 Explain why the pK_a of the $-NH_3^+$ group of an α-amino acid is lower than the pK_a of the ammonium ion derived from a 1° amine (RNH_3^+). For example, the pK_a of the $-NH_3^+$ group of alanine is 9.87 but the pK_a of $CH_3NH_3^+$ is 10.63.

Problem 27.4 L-Thyroxine, a thyroid hormone and oral medication used to treat thyroid hormone deficiency, is an amino acid that does not exist in proteins. Draw the zwitterionic form of L-thyroxine.

I I O O HO I I OH NH_2

L-thyroxine

27.2 Separation of Amino Acids

Common methods used to synthesize an amino acid yield a racemic mixture. Naturally occurring amino acids exist as a single enantiomer, however, so the two enantiomers obtained must be separated if they are to be used in biological applications. This is not an easy task. Two enantiomers have the same physical properties, so they cannot be separated by common physical methods, such as distillation or chromatography. Moreover, they react in the same way with achiral reagents, so they cannot be separated by chemical reactions either.

Nonetheless, strategies have been devised to separate two enantiomers using physical separation techniques and chemical reactions. We examine two different strategies in Section 27.2. Then, in Section 27.3, we will discuss a method that affords optically active amino acids without the need for separation.

- **The separation of a racemic mixture into its component enantiomers is called *resolution.* Thus, a racemic mixture is *resolved* into its component enantiomers.**

27.2A Resolution of Amino Acids

The oldest and perhaps still the most widely used method to separate enantiomers exploits the following fact: **enantiomers have the *same* physical properties, but diastereomers have *different* physical properties.** Thus, a racemic mixture can be resolved using the following general strategy.

[1] **Convert a pair of enantiomers to a pair of diastereomers,** which are now separable because they have different melting points and boiling points.

[2] **Separate the diastereomers.**

[3] **Re-convert each diastereomer to the original enantiomer,** now separated from the other.

This general three-step process is illustrated in Figure 27.4.

Figure 27.4 Resolution of a racemic mixture by converting it to a mixture of diastereomers

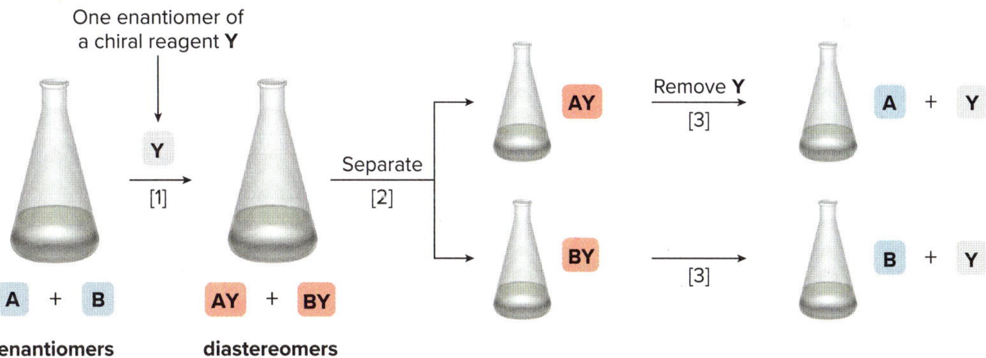

Enantiomers A and B can be separated by reaction with a single enantiomer of a chiral reagent, Y. The process of resolution requires three steps:

[1] Reaction of enantiomers **A** and **B** with **Y** forms two diastereomers, **AY** and **BY.**

[2] Diastereomers **AY** and **BY** have different physical properties, so they can be separated by physical methods such as fractional distillation or crystallization.

[3] **AY** and **BY** are then re-converted to **A** and **B** by a chemical reaction. The two enantiomers **A** and **B** are now separated from each other, and resolution is complete.

To resolve a racemic mixture of amino acids such as (*R*)- and (*S*)-alanine, the racemate is first treated with acetic anhydride to form ***N*-acetyl amino acids.** Each of these amides contains one stereogenic center and they are still enantiomers, so they are *still inseparable.*

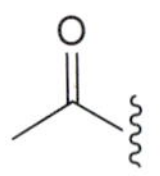

acetyl
Ac

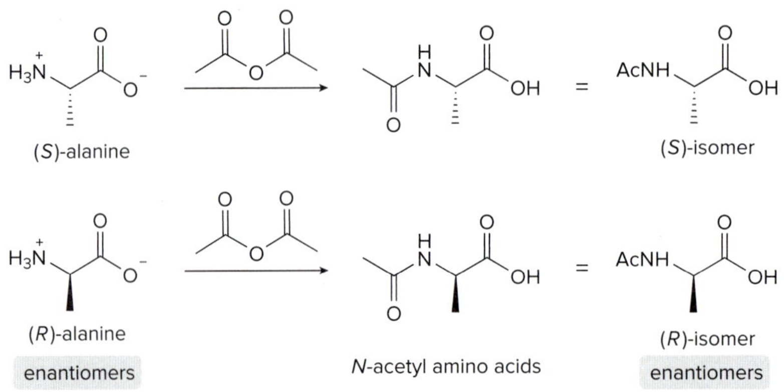

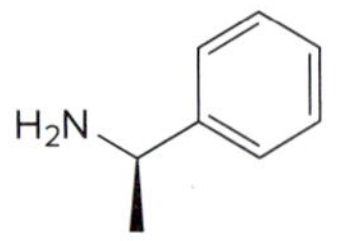

(*S*)-alanine → (*S*)-isomer

(*R*)-alanine → (*R*)-isomer

enantiomers | *N*-acetyl amino acids | enantiomers

(*R*)-α-methylbenzylamine
a resolving agent

Both enantiomers of *N*-acetyl alanine have a free carboxy group that can react with an amine in an acid–base reaction. **If a chiral amine is used, such as (*R*)-α-methylbenzylamine, the two salts formed are diastereomers, *not* enantiomers.** Diastereomers can be physically separated from each other, so the compound that converts enantiomers to diastereomers is called a **resolving agent.** Either enantiomer of the resolving agent can be used.

How To Use (*R*)-α-Methylbenzylamine to Resolve a Racemic Mixture of Amino Acids

Step [1] **React both enantiomers of an *N*-acetyl amino acid with the *R* isomer of the chiral amine.**

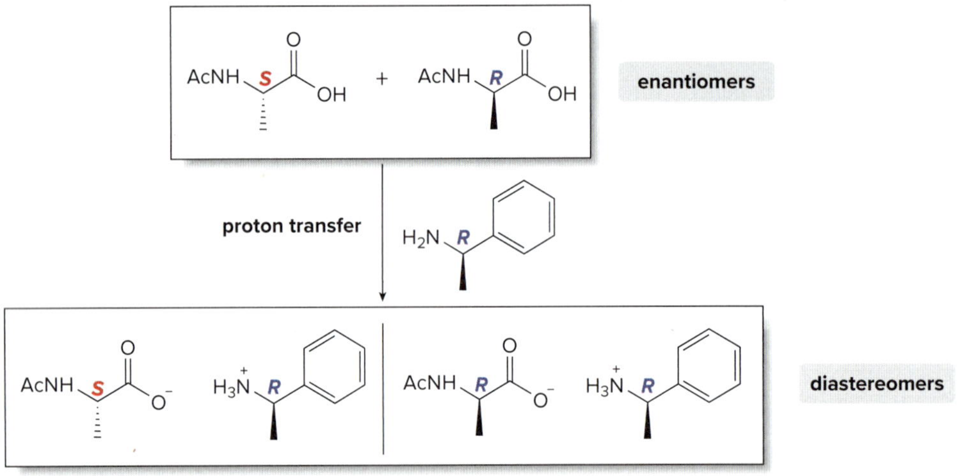

enantiomers

proton transfer

diastereomers

These salts have the *same* configuration around one stereogenic center, but the *opposite* configuration about the other stereogenic center.

Step [2] **Separate the diastereomers.**

separate

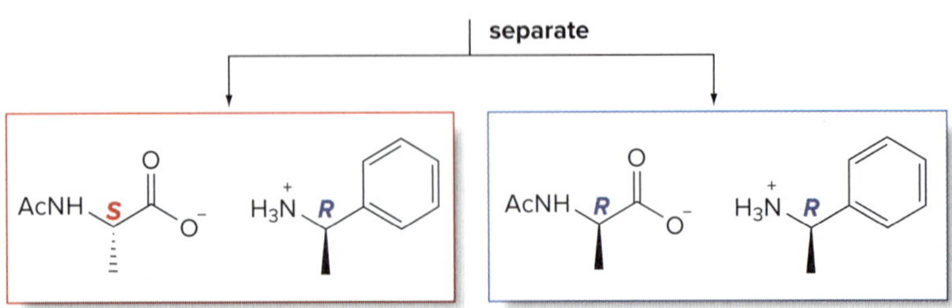

—*Continued*

How To, continued . . .

Step [3] **Regenerate the amino acid by hydrolysis of the amide.**

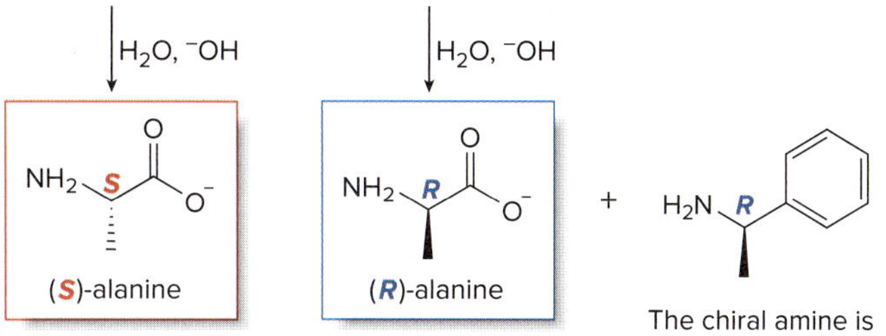

H_2O, ^-OH H_2O, ^-OH

(S)-alanine (R)-alanine + H_2N R

The chiral amine is also regenerated.

The amino acids are now separated.

Step [1] is just an acid–base reaction in which the racemic mixture of *N*-acetyl alanines reacts with the same enantiomer of the resolving agent, in this case (*R*)-α-methylbenzylamine. The salts that form are **diastereomers, *not* enantiomers,** because they have the same configuration about one stereogenic center, but the opposite configuration about the other stereogenic center.

In **Step [2],** the diastereomers are separated by some physical technique, such as crystallization or distillation.

In **Step [3],** the amides can be hydrolyzed with aqueous base to regenerate the amino acids. The amino acids are now separated from each other. The optical activity of the amino acids can be measured and compared to their known rotations to determine the purity of each enantiomer.

Problem 27.5 Which of the following amines can be used to resolve a racemic mixture of amino acids?

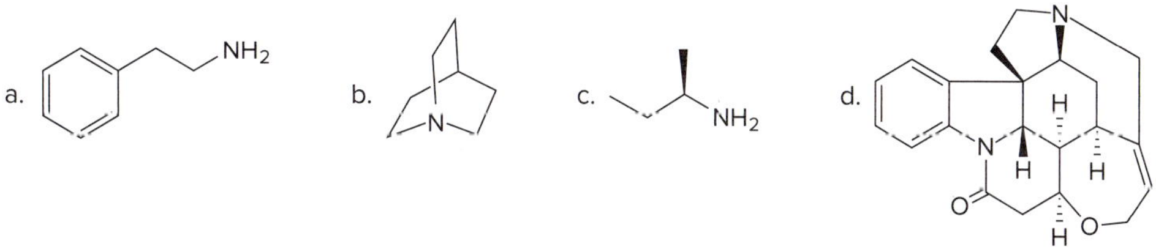

a. NH_2 b. c. NH_2 d.

strychnine
(a powerful poison)

Problem 27.6 Write out a stepwise sequence that shows how a racemic mixture of leucine enantiomers can be resolved into optically active amino acids using (*R*)-α-methylbenzylamine.

27.2B Kinetic Resolution of Amino Acids Using Enzymes

A second strategy used to separate amino acids is based on the fact that two enantiomers react differently with chiral reagents. An **enzyme** is typically used as the chiral reagent.

To illustrate this strategy, we begin again with the two enantiomers of *N*-acetyl alanine, which were prepared by treating a racemic mixture of (*R*)- and (*S*)-alanine with acetic anhydride (Section 27.2A). **Enzymes called acylases hydrolyze amide bonds, such as those found in *N*-acetyl alanine, but only for amides of L-amino acids.** Thus, when a racemic mixture of

N-acetyl alanines is treated with an acylase, only the amide of L-alanine (the *S* stereoisomer) is hydrolyzed to generate L-alanine, whereas the amide of D-alanine (the *R* stereoisomer) is untouched. The reaction mixture now consists of one amino acid and one *N*-acetyl amino acid. Because they have different functional groups with different physical properties, they can be physically separated.

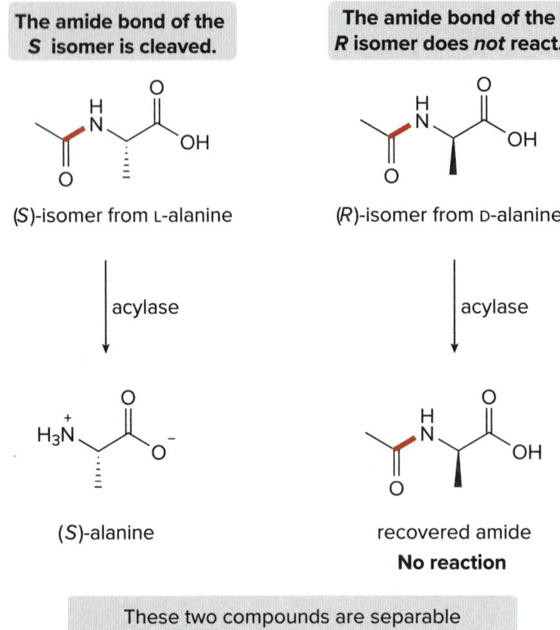

- **Separation of two enantiomers by a chemical reaction that selectively occurs for only one of the enantiomers is called *kinetic resolution*.**

Problem 27.7 Draw the organic products formed in the following reaction.

[1] Ac_2O
[2] acylase

(mixture of enantiomers)

27.3 Enantioselective Synthesis of Amino Acids

Although the two methods introduced in Section 27.2 for resolving racemic mixtures of amino acids make enantiomerically pure amino acids available for further research, half of the reaction product is useless because it has the undesired configuration. Moreover, each of these procedures is costly and time-consuming.

If we use a chiral reagent to synthesize an amino acid, however, it is possible to favor the formation of the desired enantiomer over the other, without having to resort to a resolution. For example, single enantiomers of amino acids have been prepared by using **enantioselective (or asymmetric) hydrogenation reactions.** The success of this approach depends on finding a chiral catalyst, in much the same way that a chiral catalyst is used for the Sharpless asymmetric epoxidation (Section 12.15).

The necessary starting material is an alkene. Addition of H_2 to the double bond forms an *N*-acetyl amino acid with a new stereogenic center on the α carbon to the carboxy group. With proper choice of a chiral catalyst, the naturally occurring *S* configuration can be obtained as product.

achiral alkene → (H_2, **chiral catalyst**) → new stereogenic center on the α carbon; With proper choice of catalyst, the naturally occurring *S* isomer is formed.

chiral hydrogenation catalyst (ClO_4^-, Ph = C_6H_5) = Rh*

Several chiral catalysts with complex structures have now been developed for this purpose. Many contain **rhodium** as the metal, complexed to a chiral molecule containing one or more phosphorus atoms. One example, abbreviated simply as **Rh*,** is shown.

This catalyst is synthesized from a rhodium salt and a phosphorus compound, 2,2'-bis(diphenylphosphino)-1,1'-binaphthyl **(BINAP).** It is the BINAP moiety (Figure 27.5) that makes the catalyst chiral.

Figure 27.5 The structure of BINAP

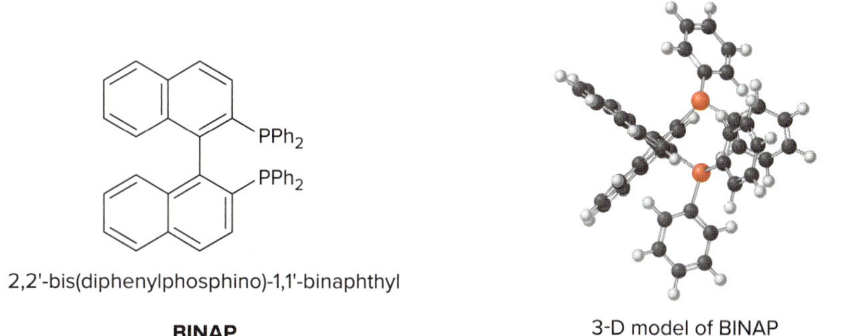

- The two naphthalene rings are oriented at right angles to each other, creating a rigid shape that makes the molecule chiral.

Ryoji Noyori shared the 2001 Nobel Prize in Chemistry for developing methods for asymmetric hydrogenation reactions using the chiral BINAP catalyst.

Twistoflex and helicene (Section 15.5) are two more aromatic compounds whose shape makes them chiral.

BINAP is one of a small number of molecules that is chiral even though it has no tetrahedral stereogenic centers. Its shape makes it a chiral molecule. The two naphthalene rings of the BINAP molecule are oriented at almost 90° to each other to minimize steric interactions between the hydrogen atoms on adjacent rings. This rigid three-dimensional shape makes BINAP nonsuperimposable on its mirror image, and thus it is a chiral compound.

Enantioselective hydrogenation can be used to synthesize a single stereoisomer of phenylalanine. Treating achiral alkene **A** with H_2 and the chiral rhodium catalyst Rh* forms the *S* isomer of *N*-acetyl phenylalanine in 100% *ee*. Hydrolysis of the acetyl group on nitrogen then yields a single enantiomer of phenylalanine.

A → (H_2, Rh*; **enantioselective hydrogenation**) → *S* enantiomer 100% *ee* → (H_2O, ^-OH; **hydrolysis**) → (*S*)-phenylalanine

Problem 27.8 What alkene is needed to synthesize each amino acid by an enantioselective hydrogenation reaction using H_2 and Rh*: (a) alanine; (b) leucine; (c) glutamine?

Problem 27.9 What starting material is needed to prepare sitagliptin (trade name Januvia, Problem 15.15), a drug used to treat type 2 diabetes, by asymmetric hydrogenation using a chiral metal catalyst? This reaction is used in the industrial synthesis of sitagliptin.

sitagliptin

27.4 Peptides

When amino acids are joined by amide bonds, they form larger molecules called **peptides** and **proteins.**

- A *dipeptide* has two amino acids joined together by *one* amide bond.
- A *tripeptide* has three amino acids joined together by *two* amide bonds.

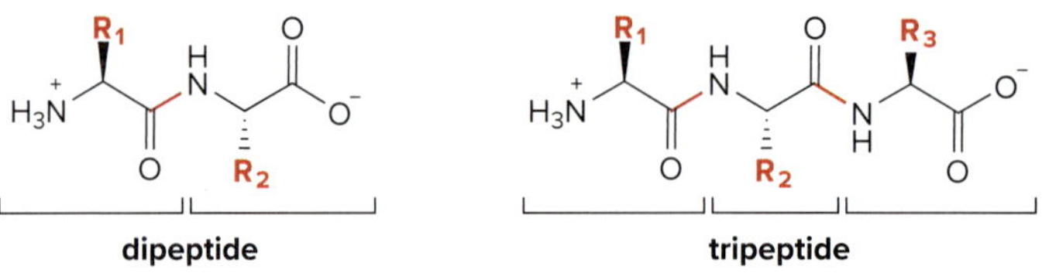

Two amino acids joined together. Three amino acids joined together.

[Amide bonds are drawn in red.]

Polypeptides and **proteins** both have many amino acids joined in long linear chains, but the term **protein** is usually reserved for polymers of more than 40 amino acids.

- The amide bonds in peptides and proteins are called *peptide bonds.*
- The individual amino acids are called *amino acid residues.*

27.4A Simple Peptides

To form a dipeptide, the amino group of one amino acid forms an amide bond with the carboxy group of another amino acid. Because each amino acid has both an amino group and a carboxy group, **two different dipeptides can be formed.** This is illustrated with alanine and cysteine.

[1] **The COO^- group of alanine can combine with the NH_3^+ group of cysteine.**

alanine + cysteine → Ala–Cys (peptide bond)

[2] **The COO^- group of cysteine can combine with the NH_3^+ group of alanine.**

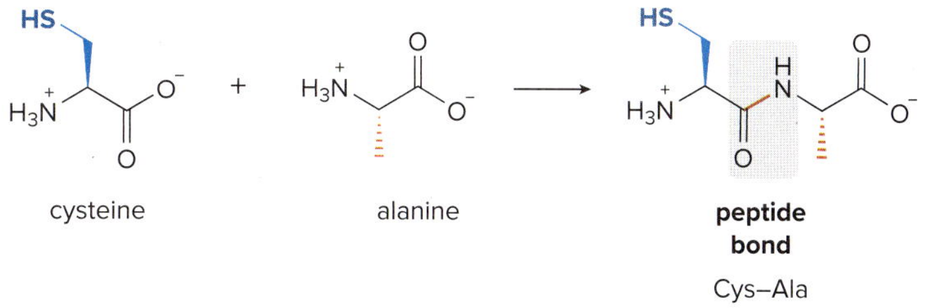

HS, $H_3\overset{+}{N}$, O, O^- + $H_3\overset{+}{N}$, O, O^- ⟶ HS, $H_3\overset{+}{N}$, H, N, O, O, O^-

cysteine alanine **peptide bond** Cys–Ala

These compounds are **constitutional isomers** of each other. Both have a free amino group (protonated as NH_3^+) at one end of their chains and a free carboxy group (deprotonated as a carboxylate anion, COO^-) at the other.

- **The amino acid with the free amino group is called the *N-terminal amino acid.***
- **The amino acid with the free carboxy group is called the *C-terminal amino acid.***

By convention, **the N-terminal amino acid is always written at the *left* end of the chain and the C-terminal amino acid at the *right*.** The peptide can be abbreviated by writing the one- or three-letter symbols for the amino acids in the chain from the N-terminal to the C-terminal end. Thus, Ala–Cys has alanine at the N-terminal end and cysteine at the C-terminal end, whereas Cys–Ala has cysteine at the N-terminal end and alanine at the C-terminal end. Sample Problem 27.1 shows how this convention applies to a tripeptide.

Sample Problem 27.1 Drawing the Structure of a Peptide from Three-Letter Symbols

Draw the structure of the following tripeptide, and label its N-terminal and C-terminal amino acids: Ala–Gly–Ser.

Solution

Draw the structures of the amino acids in order from **left to right, placing the COO^- of one amino acid *next* to the NH_3^+ group of the adjacent amino acid.** Always draw the **NH_3^+ group on the *left*** and the **COO^- group on the *right*.** Then, join adjacent COO^- and NH_3^+ groups together in amide bonds to form the tripeptide.

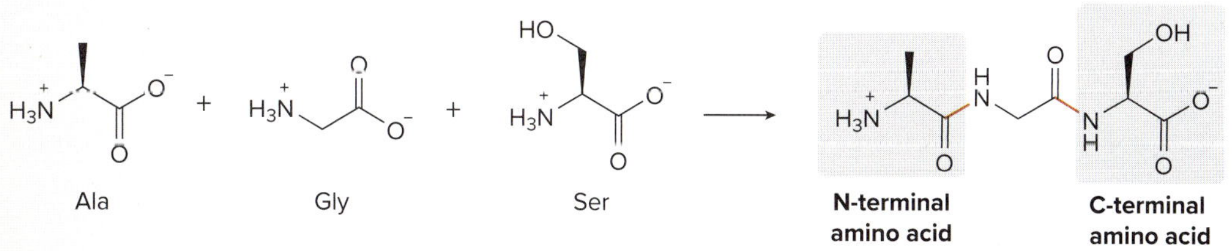

$H_3\overset{+}{N}$, O, O^- + $H_3\overset{+}{N}$, O, O^- + HO, $H_3\overset{+}{N}$, O, O^- ⟶ $H_3\overset{+}{N}$, O, H, N, O, OH, N, H, O, O^-

Ala Gly Ser **N-terminal amino acid** **C-terminal amino acid**

tripeptide **Ala–Gly–Ser**
[The new peptide bonds are drawn in red.]

The N-terminal amino acid is **alanine,** and the C-terminal amino acid is **serine.**

Problem 27.10 Draw the structure of each peptide. Label the N-terminal and C-terminal amino acids and all amide bonds.

a. Val–Glu b. Gly–His–Leu c. M–A–T–T

More Practice: Try Problems 27.29a; 27.44; 27.45b, d.

The tripeptide in Sample Problem 27.1 has one N-terminal amino acid, one C-terminal amino acid, and two peptide bonds.

- **No matter how many amino acid residues are present, there is only *one* N-terminal amino acid and *one* C-terminal amino acid.**
- **For *n* amino acids in the chain, the number of amide bonds is *n* – 1.**

Problem 27.11 Name each peptide using both the one-letter and the three-letter abbreviations for the names of the component amino acids.

a. b.

27.4B The Peptide Bond

Recall from Section 14.6 that buta-1,3-diene can also exist as *s*-cis and *s*-trans conformations. In buta-1,3-diene, the **s-cis conformation has the two double bonds on the same side of the single bond** (dihedral angle = 0°), whereas the **s-trans conformation has them on opposite sides** (dihedral angle = 180°).

The carbonyl carbon of an amide is ***sp*^{2} hybridized** and has **trigonal planar** geometry. A second resonance structure can be drawn that delocalizes the nonbonded electron pair on the N atom. Amides are more resonance stabilized than other acyl compounds, so the **resonance structure having the C=N makes a significant contribution to the hybrid.**

two resonance structures for the peptide bond

Resonance stabilization has important consequences. **Rotation about the C–N bond is restricted** because it has partial double-bond character. As a result, there are two possible conformations.

s-trans *s*-cis

- **The *s*-trans conformation has the two R groups oriented on *opposite* sides of the C–N bond.**
- **The *s*-cis conformation has the two R groups oriented on the *same* side of the C–N bond.**
- **The *s*-trans conformation of a peptide bond is typically more stable than the *s*-cis, because the *s*-trans has the two bulky R groups located farther from each other.**

The planar geometry of the peptide bond is analogous to the planar geometry of ethylene (or any other alkene), where the double bond between *sp*^{2} hybridized carbon atoms makes all of the bond angles ~120° and puts all six atoms in the same plane.

A second consequence of resonance stabilization is that **all six atoms involved in the peptide bond lie in the same plane.** All bond angles are ~120°, and the C=O and N–H bonds are oriented 180° from each other.

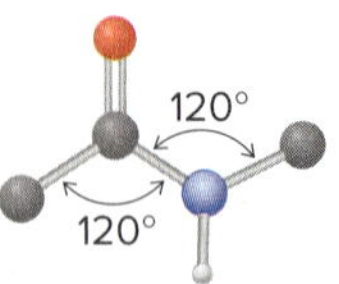

These six atoms lie in a plane.

The structure of a tetrapeptide illustrates the results of these effects in a long peptide chain.

- **The *s*-trans arrangement makes a long chain with a zigzag arrangement.**
- **In each peptide bond, the N–H and C=O bonds lie parallel and at 180° with respect to each other.**

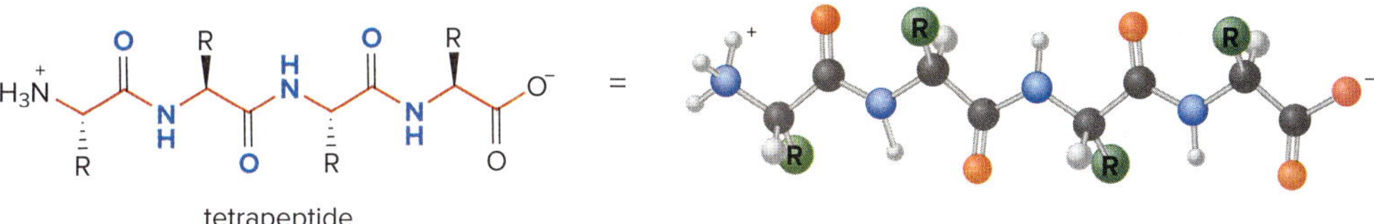

tetrapeptide

27.4C Interesting Peptides

Even relatively simple peptides can have important biological functions. **Bradykinin,** for example, is a peptide hormone composed of nine amino acids. It stimulates smooth muscle contraction, dilates blood vessels, and causes pain. Bradykinin is a component of bee venom.

Arg–Pro–Pro–Gly–Phe–Ser–Pro–Phe–Arg
bradykinin

Oxytocin and **vasopressin** are nonapeptide hormones, too. Their sequences are identical except for two amino acids, yet this is enough to give them very different biological activities. As mentioned in the chapter opener, oxytocin induces labor by stimulating the contraction of uterine muscles, and it stimulates the flow of milk in nursing mothers. Vasopressin, on the other hand, controls blood pressure by regulating smooth muscle contraction. The N-terminal amino acid in both hormones is a cysteine residue, and the C-terminal residue is glycine. Instead of a free carboxy group, both peptides have an NH_2 group in place of OH, so this is indicated with the additional NH_2 group drawn at the end of the chain.

Cys–Tyr–**Ile**–Gln–Asn–Cys–Pro–**Leu**–$GlyNH_2$ (S–S bond between the two Cys)
oxytocin

Cys–Tyr–**Phe**–Gln–Asn–Cys–Pro–**Arg**–$GlyNH_2$ (S–S bond between the two Cys)
vasopressin

- The N-terminal amino acid is labeled in red.
- The amino acids that differ are labeled in blue.

The oxidation of thiols to disulfides was discussed in Section 9.15.

The structure of both peptides includes a **disulfide bond,** a form of covalent bonding in which the –SH groups from two cysteine residues are oxidized to form a sulfur–sulfur bond. In oxytocin and vasopressin, the disulfide bonds make the peptides cyclic.

2 R–S–H (thiol) —[O]→ R–S–S–R (disulfide)

The artificial sweetener **aspartame** (Figure 26.10) is the methyl ester of the dipeptide Asp–Phe. This synthetic peptide is 180 times sweeter (on a gram-for-gram basis) than sucrose (common table sugar). Both of the amino acids in aspartame have the naturally occurring L-configuration. If the D-amino acid is substituted for either Asp or Phe, the resulting compound tastes bitter.

aspartame
the methyl ester of Asp–Phe
a synthetic artificial sweetener

Problem 27.12 Draw the structure of leu-enkephalin, a pentapeptide that acts as an analgesic and opiate, and has the following sequence: Tyr–Gly–Gly–Phe–Leu. (The structure of a related peptide, met-enkephalin, appeared in Section 20.5B.)

Problem 27.13 Glutathione, a powerful antioxidant that destroys harmful oxidizing agents in cells, is composed of glutamic acid, cysteine, and glycine, and has the following structure:

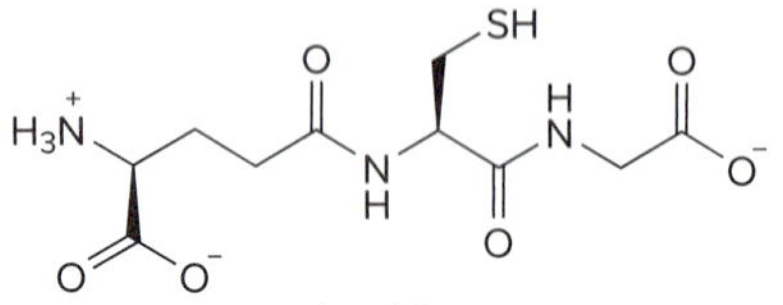

glutathione

a. What product is formed when glutathione reacts with an oxidizing agent?

b. What is unusual about the peptide bond between glutamic acid and cysteine?

27.5 Peptide Sequencing

To determine the structure of a peptide, we must know not only what amino acids comprise it, but also the sequence of the amino acids in the peptide chain. Although mass spectrometry has become an increasingly powerful method for the analysis of high-molecular-weight proteins (Section A.5C), chemical methods to determine peptide structure are still widely used and presented in this section.

27.5A Amino Acid Analysis

The structure determination of a peptide begins by analyzing the **total amino acid composition.** The amide bonds are first hydrolyzed by heating with hydrochloric acid for 24 h to form the individual amino acids. The resulting mixture is then separated using high-performance liquid chromatography (HPLC), a technique in which a solution of amino acids is placed on a column and individual amino acids move through the column at characteristic rates, often dependent on polarity.

This process determines both the identity of the individual amino acids and the amount of each present, but it tells nothing about the order of the amino acids in the peptide. For example, complete hydrolysis and HPLC analysis of the tetrapeptide Gly–Gly–Phe–Tyr would indicate the presence of three amino acids—glycine, phenylalanine, and tyrosine—and show that there are twice as many glycine residues as phenylalanine or tyrosine residues. The exact order of the amino acids in the peptide chain must then be determined by additional methods.

27.5B Identifying the N-Terminal Amino Acid—The Edman Degradation

To determine the sequence of amino acids in a peptide chain, a variety of procedures are often combined. One especially useful technique is to **identify the N-terminal amino acid using the Edman degradation.** In the Edman degradation, amino acids are cleaved one at a time from the N-terminal end, the identity of the amino acid determined, and the process repeated until the entire sequence is known. Automated sequencers using this methodology are now available to sequence peptides containing up to about 50 amino acids.

The Edman degradation is based on the reaction of the nucleophilic NH_2 group of the N-terminal amino acid with the electrophilic carbon of phenyl isothiocyanate, $\mathbf{C_6H_5N{=}C{=}S}$. When the N-terminal amino acid is removed from the peptide chain, two products are formed: **an *N*-phenylthiohydantoin (PTH) and a new peptide with one *fewer* amino acid.**

phenyl isothiocyanate + N-terminal amino acid → (Edman degradation) *N*-phenylthiohydantoin (PTH) + H_2N—PEPTIDE

This product characterizes the N-terminal amino acid.

This peptide contains a new N-terminal amino acid.

The *N*-phenylthiohydantoin derivative contains the atoms of the N-terminal amino acid. **This product identifies the N-terminal amino acid in the peptide** because the PTH derivatives of all 20 naturally occurring amino acids are known and characterized. The new peptide formed in the Edman degradation has one fewer amino acid than the original peptide. Moreover, it contains a new N-terminal amino acid, so the process can be repeated.

Mechanism 27.1 illustrates some of the key steps of the Edman degradation. The nucleophilic N-terminal NH_2 group adds to the electrophilic carbon of phenyl isothiocyanate to form an *N*-phenylthiourea, the product of nucleophilic addition (Part [1]). Intramolecular cyclization followed by elimination results in cleavage of the terminal amide bond in Part [2] to form **a new peptide with one fewer amino acid.** A sulfur heterocycle, called a thiazolinone, is also formed, which rearranges by a multistep pathway to form an *N*-phenylthiohydantoin. **The R group in this product identifies the amino acid located at the N-terminal end.**

Mechanism 27.1 Edman Degradation

Part [1] Formation of an *N*-phenylthiourea

phenyl isothiocyanate → ① → ② → ***N*-phenylthiourea**

① – ② Addition of the amino group of the N-terminal amino acid to phenyl isothiocyanate followed by proton transfer forms an ***N*-phenylthiourea.**

Part [2] Formation of the N-terminal amino acid and *N*-phenylthiohydantoin (PTH)

N-phenylthiourea → ③ → ④ ($-H^+$) → thiazolinone + H_2N—PEPTIDE → ⑤ (several steps) → ***N*-phenylthiohydantoin (PTH)**

③ Nucleophilic addition of the S atom to the amide carbonyl forms a five-membered ring.

④ Loss of the amino group forms two products—a thiazolinone ring and **a peptide chain that contains one fewer amino acid than the original peptide.**

⑤ The thiazolinone rearranges by a multistep pathway to form an ***N*-phenylthiohydantoin (PTH) that contains the original amino acid.**

In theory a protein of any length can be sequenced using the Edman degradation, but in practice, the accumulation of small quantities of unwanted by-products limits sequencing to proteins having fewer than approximately 50 amino acids.

Problem 27.14 Draw the structure of the *N*-phenylthiohydantoin formed by initial Edman degradation of each peptide: (a) Ala–Gly–Phe–Phe; (b) Val–Ile–Tyr.

27.5C Partial Hydrolysis of a Peptide

Additional structural information can be obtained by cleaving some, but not all, of the amide bonds in a peptide. Partial hydrolysis of a peptide with acid forms smaller fragments in a random fashion. Sequencing these peptides and **identifying sites of overlap** can be used to determine the sequence of the complete peptide, as shown in Sample Problem 27.2.

Sample Problem 27.2 Determining the Amino Acid Sequence of a Peptide Using Partial Hydrolysis

Give the amino acid sequence of a hexapeptide that contains the amino acids Ala, Val, Ser, Ile, Gly, Tyr, and forms the following fragments when partially hydrolyzed with HCl: Gly–Ile–Val, Ala–Ser–Gly, and Tyr–Ala.

Solution

Looking for points of overlap in the sequences of the smaller fragments shows how the fragments should be pieced together. In this example, the fragment Ala–Ser–Gly contains amino acids common to the two other fragments, thus showing how the three fragments can be joined together.

common amino acids

Tyr–Ala Gly–Ile–Val

Ala–Ser–Gly

→ **Answer:** Tyr–Ala–Ser–Gly–Ile–Val

hexapeptide

Problem 27.15 Give the amino acid sequence of an octapeptide that contains the amino acids Tyr, Ala, Leu (2 equiv), Cys, Gly, Glu, and Val, and forms the following fragments when partially hydrolyzed with HCl: Val–Cys–Gly–Glu, Ala–Leu–Tyr, and Tyr–Leu–Val–Cys.

More Practice: Try Problem 27.49.

Peptides can also be hydrolyzed at specific sites using enzymes. The enzyme carboxypeptidase catalyzes the hydrolysis of the amide bond nearest the C-terminal end, forming the C-terminal amino acid and a peptide with one fewer amino acid. In this way, **carboxypeptidase is used to identify the C-terminal amino acid.**

Other enzymes catalyze the hydrolysis of amide bonds formed with specific amino acids. For example:

- **Trypsin catalyzes the hydrolysis of amides with a carbonyl group that is part of the basic amino acids arginine and lysine.**
- **Chymotrypsin hydrolyzes amides with carbonyl groups that are part of the aromatic amino acids phenylalanine, tyrosine, and tryptophan.**

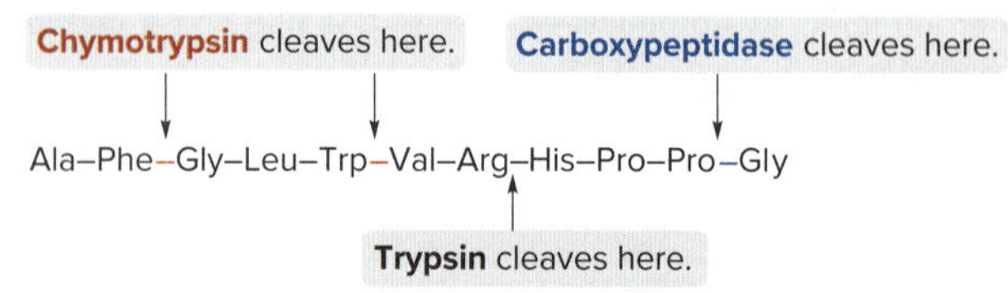

The structure of a tetrapeptide illustrates the results of these effects in a long peptide chain.

- **The *s*-trans arrangement makes a long chain with a zigzag arrangement.**
- **In each peptide bond, the N–H and C=O bonds lie parallel and at 180° with respect to each other.**

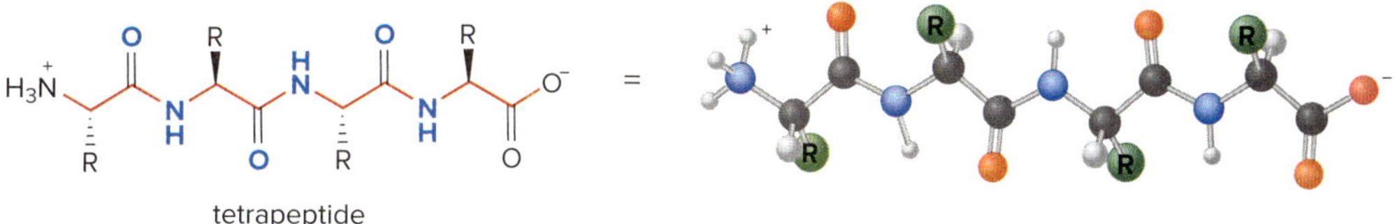

tetrapeptide

27.4C Interesting Peptides

Even relatively simple peptides can have important biological functions. **Bradykinin,** for example, is a peptide hormone composed of nine amino acids. It stimulates smooth muscle contraction, dilates blood vessels, and causes pain. Bradykinin is a component of bee venom.

Arg–Pro–Pro–Gly–Phe–Ser–Pro–Phe–Arg
bradykinin

Oxytocin and **vasopressin** are nonapeptide hormones, too. Their sequences are identical except for two amino acids, yet this is enough to give them very different biological activities. As mentioned in the chapter opener, oxytocin induces labor by stimulating the contraction of uterine muscles, and it stimulates the flow of milk in nursing mothers. Vasopressin, on the other hand, controls blood pressure by regulating smooth muscle contraction. The N-terminal amino acid in both hormones is a cysteine residue, and the C-terminal residue is glycine. Instead of a free carboxy group, both peptides have an NH_2 group in place of OH, so this is indicated with the additional NH_2 group drawn at the end of the chain.

Cys—Tyr—Ile—Gln—Asn—Cys—Pro—Leu—GlyNH2 (S—S bond between the two Cys)
oxytocin

Cys—Tyr—Phe—Gln—Asn—Cys—Pro—Arg—GlyNH2 (S—S bond between the two Cys)
vasopressin

- The N-terminal amino acid is labeled in red.
- The amino acids that differ are labeled in blue.

The oxidation of thiols to disulfides was discussed in Section 9.15.

The structure of both peptides includes a **disulfide bond,** a form of covalent bonding in which the –SH groups from two cysteine residues are oxidized to form a sulfur–sulfur bond. In oxytocin and vasopressin, the disulfide bonds make the peptides cyclic.

$$2\ R{-}S{-}H \xrightarrow{[O]} R{-}S{-}S{-}R$$

thiol → disulfide

The artificial sweetener **aspartame** (Figure 26.10) is the methyl ester of the dipeptide Asp–Phe. This synthetic peptide is 180 times sweeter (on a gram-for-gram basis) than sucrose (common table sugar). Both of the amino acids in aspartame have the naturally occurring L-configuration. If the D-amino acid is substituted for either Asp or Phe, the resulting compound tastes bitter.

aspartame
the methyl ester of Asp–Phe
a synthetic artificial sweetener

Problem 27.12 Draw the structure of leu-enkephalin, a pentapeptide that acts as an analgesic and opiate, and has the following sequence: Tyr–Gly–Gly–Phe–Leu. (The structure of a related peptide, met-enkephalin, appeared in Section 20.5B.)

Problem 27.13 Glutathione, a powerful antioxidant that destroys harmful oxidizing agents in cells, is composed of glutamic acid, cysteine, and glycine, and has the following structure:

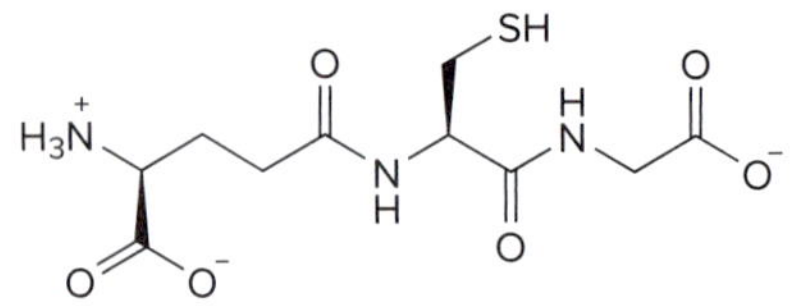

glutathione

a. What product is formed when glutathione reacts with an oxidizing agent?

b. What is unusual about the peptide bond between glutamic acid and cysteine?

27.5 Peptide Sequencing

To determine the structure of a peptide, we must know not only what amino acids comprise it, but also the sequence of the amino acids in the peptide chain. Although mass spectrometry has become an increasingly powerful method for the analysis of high-molecular-weight proteins (Section A.5C), chemical methods to determine peptide structure are still widely used and presented in this section.

27.5A Amino Acid Analysis

The structure determination of a peptide begins by analyzing the **total amino acid composition.** The amide bonds are first hydrolyzed by heating with hydrochloric acid for 24 h to form the individual amino acids. The resulting mixture is then separated using high-performance liquid chromatography (HPLC), a technique in which a solution of amino acids is placed on a column and individual amino acids move through the column at characteristic rates, often dependent on polarity.

This process determines both the identity of the individual amino acids and the amount of each present, but it tells nothing about the order of the amino acids in the peptide. For example, complete hydrolysis and HPLC analysis of the tetrapeptide Gly–Gly–Phe–Tyr would indicate the presence of three amino acids—glycine, phenylalanine, and tyrosine—and show that there are twice as many glycine residues as phenylalanine or tyrosine residues. The exact order of the amino acids in the peptide chain must then be determined by additional methods.

27.5B Identifying the N-Terminal Amino Acid—The Edman Degradation

To determine the sequence of amino acids in a peptide chain, a variety of procedures are often combined. One especially useful technique is to **identify the N-terminal amino acid using the Edman degradation.** In the Edman degradation, amino acids are cleaved one at a time from the N-terminal end, the identity of the amino acid determined, and the process repeated until the entire sequence is known. Automated sequencers using this methodology are now available to sequence peptides containing up to about 50 amino acids.

The Edman degradation is based on the reaction of the nucleophilic NH_2 group of the N-terminal amino acid with the electrophilic carbon of phenyl isothiocyanate, $\mathbf{C_6H_5N{=}C{=}S}$. When the N-terminal amino acid is removed from the peptide chain, two

How To, continued . . .

Step [2] **Protect the COOH group of glycine.**

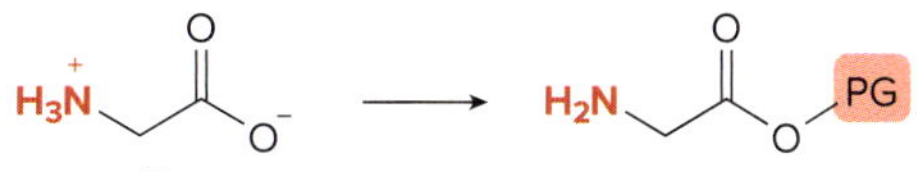

$H_3\overset{+}{N}CH_2COO^-$ (Gly) ⟶ $H_2NCH_2CO_2$–PG

- In the neutral amino acid, the COOH group exists largely as a carboxylate anion, $-COO^-$.

Step [3] **Form the amide bond with DCC.**

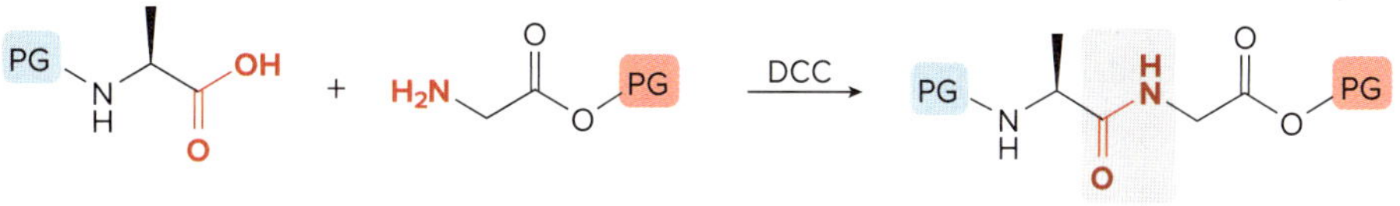

PG–NH–CH(CH_3)–COOH + $H_2NCH_2CO_2$–PG $\xrightarrow{\text{DCC}}$ PG–NH–CH(CH_3)–CO–NH–CH_2CO_2–PG

Dicyclohexylcarbodiimide (**DCC**) is a reagent commonly used to form amide bonds (see Section 20.9D). DCC makes the OH group of the carboxylic acid a better leaving group, thus **activating the carboxy group toward nucleophilic attack.**

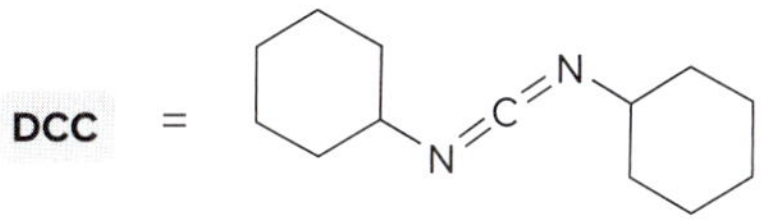

DCC = C_6H_{11}–N=C=N–C_6H_{11}

dicyclohexylcarbodiimide

Step [4] **Remove one or both protecting groups.**

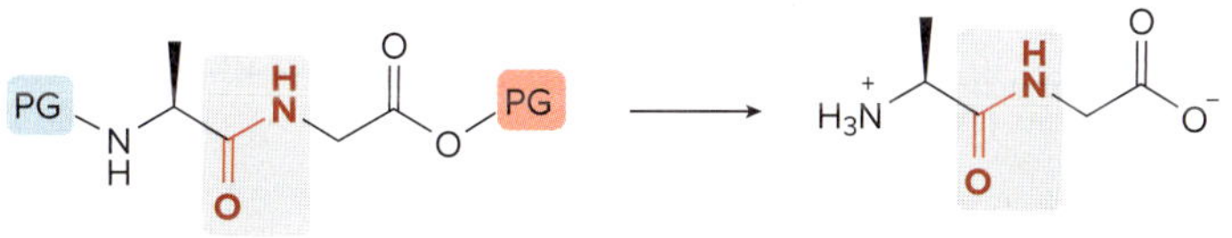

PG–NH–CH(CH_3)–CO–NH–CH_2CO_2–PG ⟶ $H_3\overset{+}{N}$–CH(CH_3)–CO–NH–CH_2COO^-

Ala–Gly

Two widely used amino protecting groups convert an amine to a **carbamate,** a functional group having a carbonyl bonded to both an oxygen and a nitrogen atom. Because the N atom of the carbamate is bonded to a carbonyl group, the protected amino group is no longer nucleophilic.

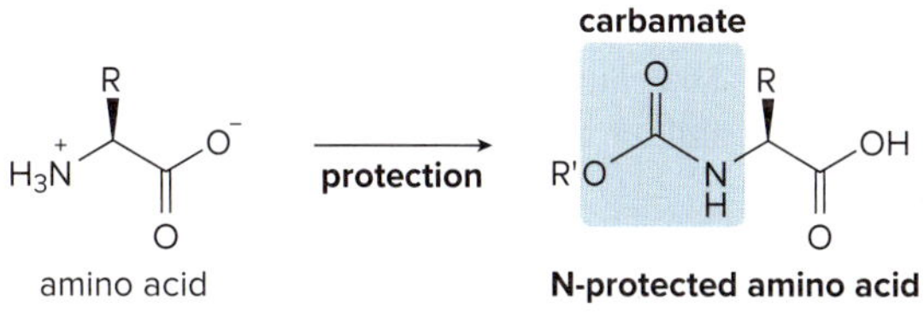

$H_3\overset{+}{N}$–CHR–COO^- (amino acid) $\xrightarrow{\text{protection}}$ R'O–CO–NH–CHR–COOH (**N-protected amino acid**; **carbamate**)

For example, the ***tert*-butoxycarbonyl protecting group,** abbreviated as **Boc,** is formed by reacting the amino acid with di-*tert*-butyl dicarbonate in a nucleophilic acyl substitution reaction.

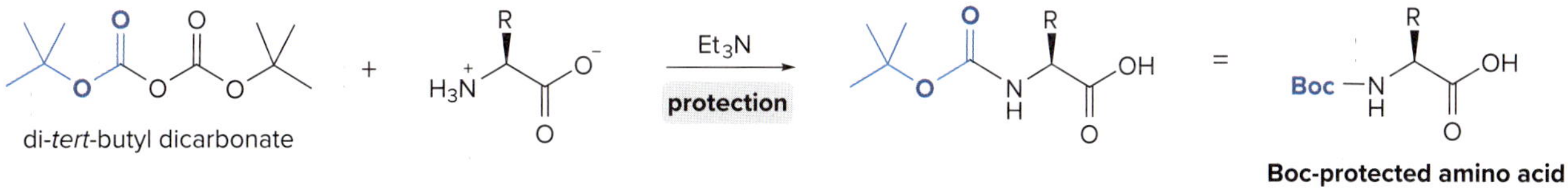

di-*tert*-butyl dicarbonate + $H_3\overset{+}{N}$–CHR–COO^- $\xrightarrow[\text{protection}]{Et_3N}$ $(CH_3)_3C$O–CO–NH–CHR–COOH = Boc–NH–CHR–COOH

Boc-protected amino acid

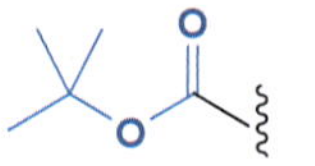

tert-butoxycarbonyl
Boc

(Boc)$_2$O

To be a useful protecting group, the Boc group must be removed under reaction conditions that do not affect other functional groups in the molecule. It can be removed with an acid such as **trifluoroacetic acid, HCl,** or **HBr.**

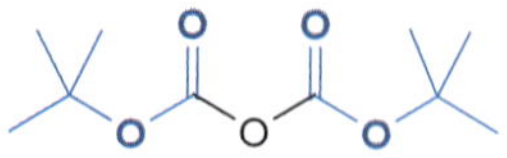

CF_3CO_2H or HCl or HBr — $H_3\overset{+}{N}$ … O^- + CO_2 +

deprotection

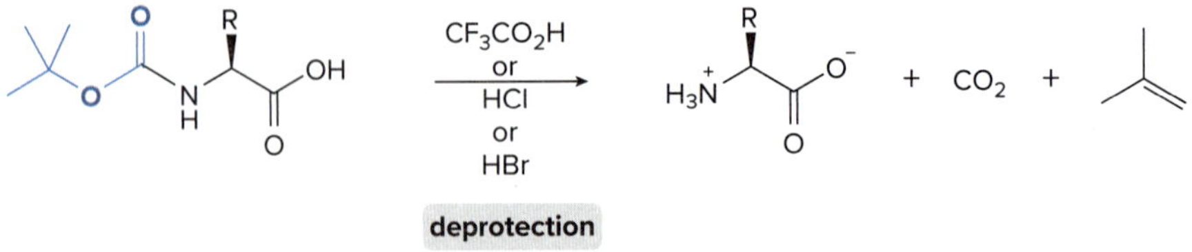

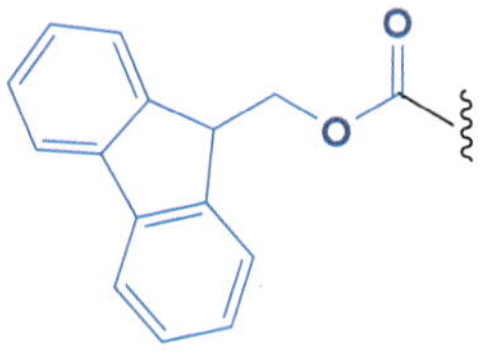

9-fluorenylmethoxycarbonyl
Fmoc

A second amino protecting group, the **9-fluorenylmethoxycarbonyl protecting group,** abbreviated as **Fmoc,** is formed by reacting the amino acid with 9-fluorenylmethyl chloroformate in a nucleophilic acyl substitution reaction.

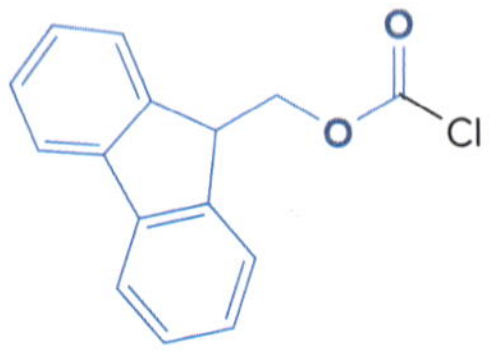

9-fluorenylmethyl chloroformate
Fmoc–Cl

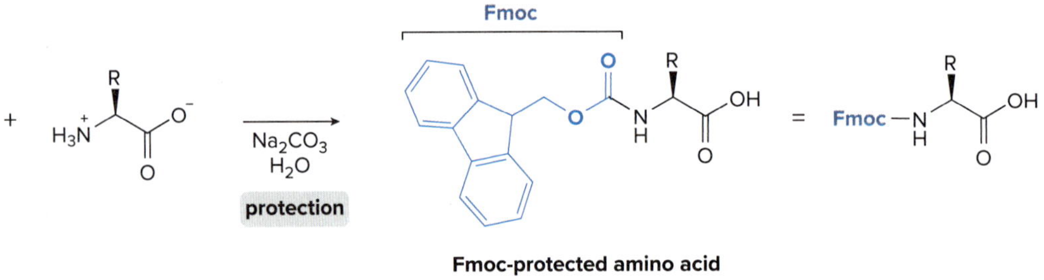

Fmoc–Cl + $H_3\overset{+}{N}$…O^- → (Na_2CO_3, H_2O) **protection** → Fmoc-protected amino acid = Fmoc–NH…OH

Fmoc-protected amino acid

Although the Fmoc protecting group is stable to most acids, it can be removed by treatment with base (NH_3 or an amine).

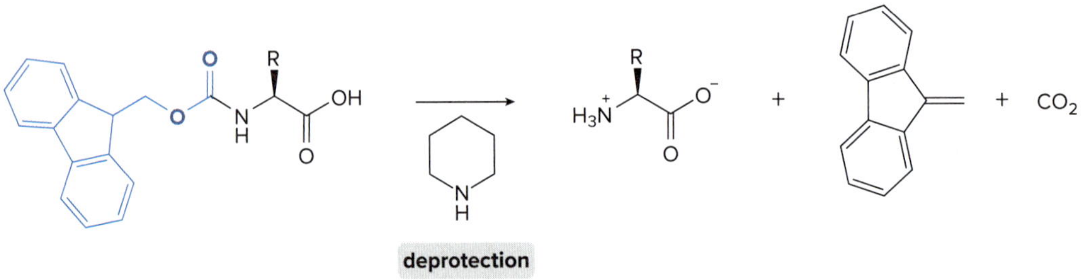

Fmoc-protected amino acid → (piperidine) **deprotection** → $H_3\overset{+}{N}$…O^- + dibenzofulvene + CO_2

The carboxy group is usually protected as a **methyl** or **benzyl ester** by reaction with an alcohol and an acid.

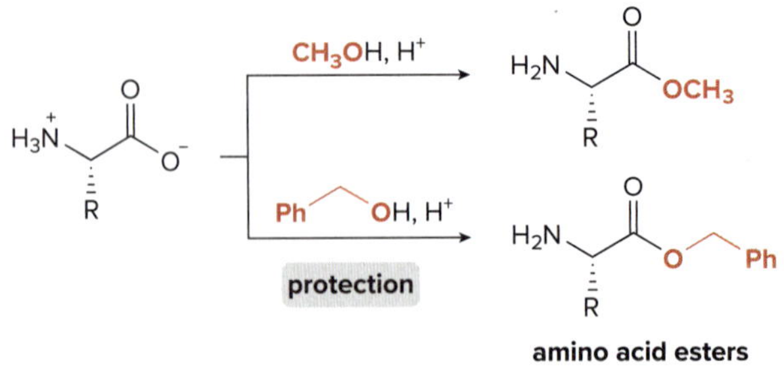

$H_3\overset{+}{N}$…O^- → CH_3OH, H^+ → H_2N…OCH_3
$H_3\overset{+}{N}$…O^- → $PhCH_2OH$, H^+ → H_2N…OCH_2Ph

protection

amino acid esters

These esters are usually removed by hydrolysis with aqueous base.

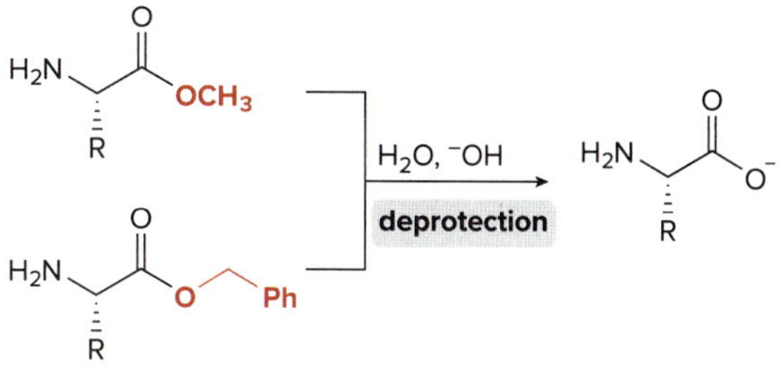

amino acid esters

H_2O, ^-OH

deprotection

One advantage of using a benzyl ester for protection is that it can also be removed with H_2 in the presence of a Pd catalyst. This process is called **hydrogenolysis.** These conditions are especially mild, because they avoid the use of either acid or base. Benzyl esters can also be removed with HBr in acetic acid.

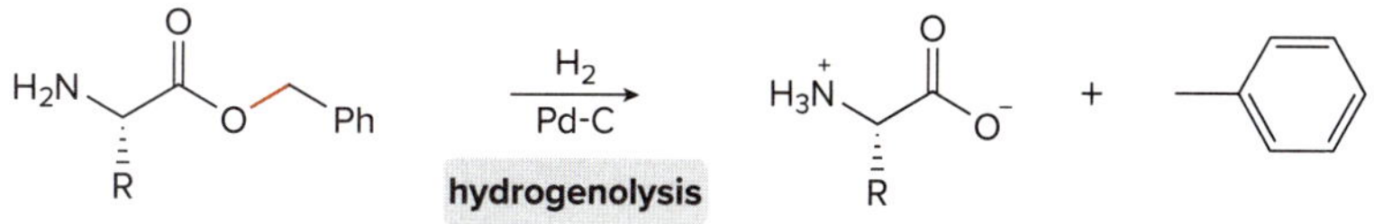

H_2, Pd-C

hydrogenolysis

The benzylic C–O bond (in red) is cleaved.

Problem 27.20 Draw the organic products formed in each reaction.

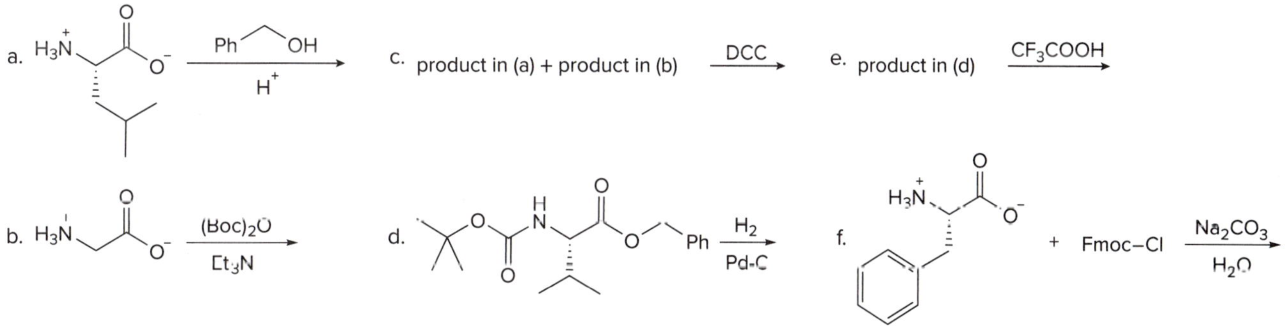

a. (leucine zwitterion) → Ph‒CH₂OH, H^+

b. (glycine zwitterion) → $(Boc)_2O$, Et_3N

c. product in (a) + product in (b) → DCC

d. (Boc-Val benzyl ester) → H_2, Pd-C

e. product in (d) → CF_3COOH

f. (phenylalanine zwitterion) + Fmoc–Cl → Na_2CO_3, H_2O

The specific reactions needed to synthesize the dipeptide Ala–Gly are illustrated in Sample Problem 27.4.

Sample Problem 27.4 Devising the Synthesis of a Dipeptide

Draw out the steps in the synthesis of the dipeptide Ala–Gly.

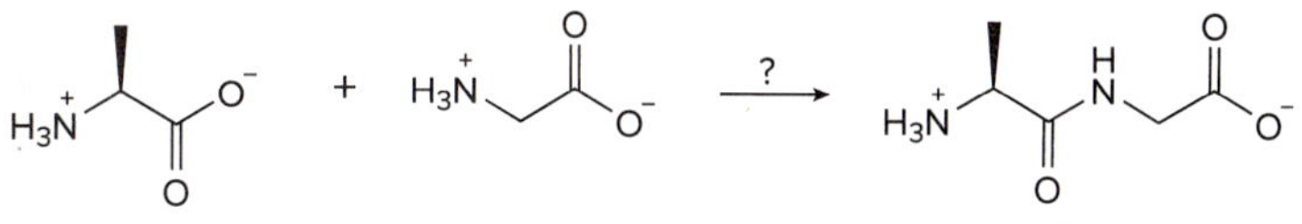

Ala + Gly → ? → Ala–Gly

Solution

Step [1] Protect the NH_2 group of alanine using a Boc group.

$$\text{Ala} \xrightarrow[\text{Et}_3\text{N}]{(\text{Boc})_2\text{O}} \text{Boc–Ala}$$

Step [2] Protect the COOH group of glycine as a benzyl ester.

$$\text{Gly} \xrightarrow{\text{PhCH}_2\text{OH, H}^+} \text{Gly–OCH}_2\text{Ph}$$

Step [3] Form the amide bond with DCC.

$$\text{Boc–Ala} + \text{Gly–OCH}_2\text{Ph} \xrightarrow{\text{DCC}} \text{Boc–Ala–Gly–OCH}_2\text{Ph}$$

Step [4] Remove one or both protecting groups.

The protecting groups can be removed in a stepwise fashion or in a single reaction.

$$\text{Boc–Ala–Gly–OCH}_2\text{Ph} \xrightarrow[\text{Pd-C}]{\text{H}_2} \text{Boc–Ala–Gly}$$

Remove the benzyl group.

$$\text{Boc–Ala–Gly} \xrightarrow{\text{CF}_3\text{COOH}} \text{Ala–Gly}$$

Remove the Boc group.

$$\text{Boc–Ala–Gly–OCH}_2\text{Ph} \xrightarrow[\text{CH}_3\text{COOH}]{\text{HBr}} \text{Ala–Gly}$$

Remove both protecting groups.

Problem 27.21 Devise a synthesis of each dipeptide from amino acid starting materials.

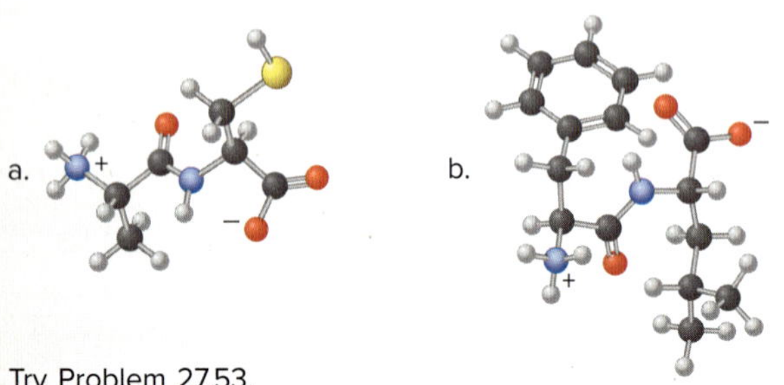

More Practice: Try Problem 27.53.

This method can be applied to the synthesis of tripeptides and even larger polypeptides. After the protected dipeptide is prepared in Step [3], only one of the protecting groups is removed, and this dipeptide is coupled to a third amino acid with one of its functional groups protected, as illustrated in the following equations.

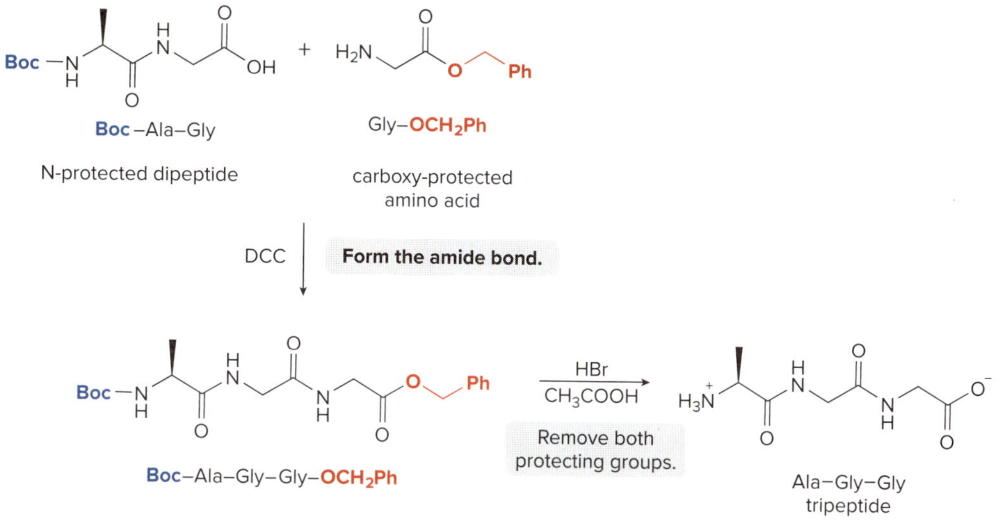

Boc–Ala–Gly

N-protected dipeptide

Gly–OCH_2Ph

carboxy-protected amino acid

DCC

Form the amide bond.

Boc–Ala–Gly–Gly–OCH_2Ph

HBr

CH_3COOH

Remove both protecting groups.

Ala–Gly–Gly tripeptide

Problem 27.22 Devise a synthesis of each peptide from amino acid starting materials: (a) Leu–Val; (b) Ala–Ile–Gly.

27.7 Automated Peptide Synthesis

Development of the solid phase technique earned Merrifield the 1984 Nobel Prize in Chemistry and has made possible the synthesis of many polypeptides and proteins.

The method described in Section 27.6 works well for the synthesis of small peptides. It is extremely time-consuming to synthesize larger proteins by this strategy, however, because each step requires isolation and purification of the product. The synthesis of larger polypeptides is usually accomplished by using the **solid phase technique** originally developed by R. Bruce Merrifield of Rockefeller University.

In the Merrifield method, an amino acid is attached to an insoluble polymer. Amino acids are sequentially added, one at a time, thereby forming successive peptide bonds. Because impurities and by-products are not attached to the polymer chain, they are removed simply by washing them away with a solvent at each stage of the synthesis.

A commonly used polymer is a **polystyrene derivative** that contains $–CH_2Cl$ groups bonded to some of the benzene rings in the polymer chain. The Cl atoms serve as handles that allow attachment of amino acids to the chain.

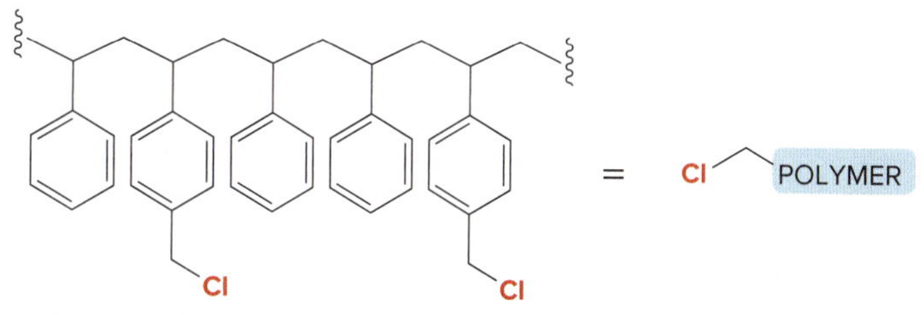

Cl POLYMER

polystyrene derivative with Cl leaving groups

An Fmoc-protected amino acid is attached to the polymer at its carboxy group by an S_N2 reaction.

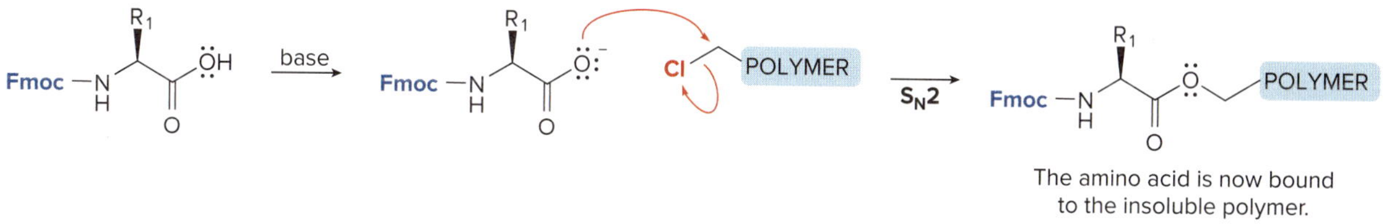

base

S_N2

The amino acid is now bound to the insoluble polymer.

Once the first amino acid is bound to the polymer, additional amino acids can be added sequentially. The steps of the solid phase peptide synthesis technique are illustrated in the accompanying scheme. In the last step, HF cleaves the polypeptide chain from the polymer.

How To Synthesize a Peptide Using the Merrifield Solid Phase Technique

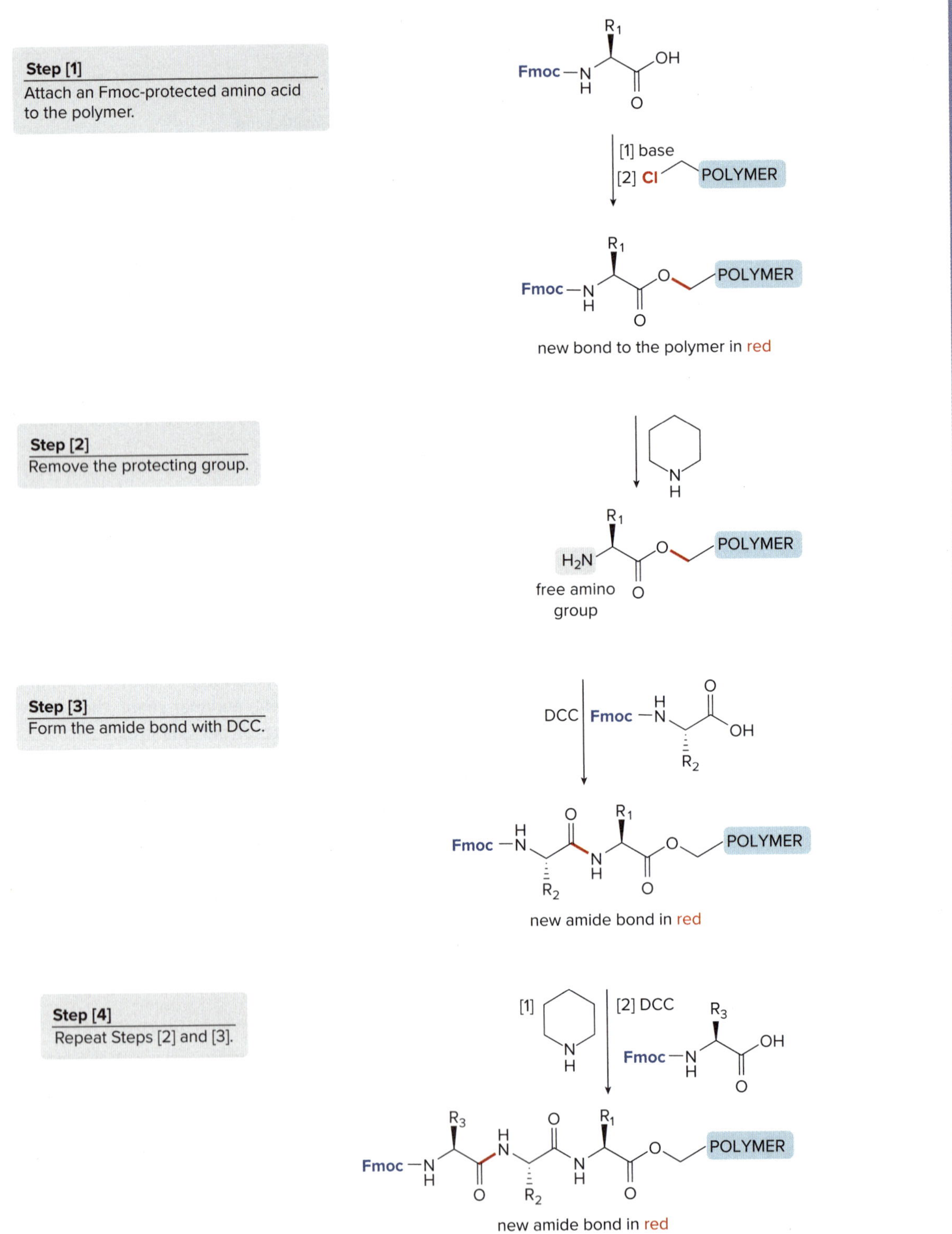

—Continued

How To, continued . . .

Step [5]
Remove the protecting group and detach the peptide from the polymer.

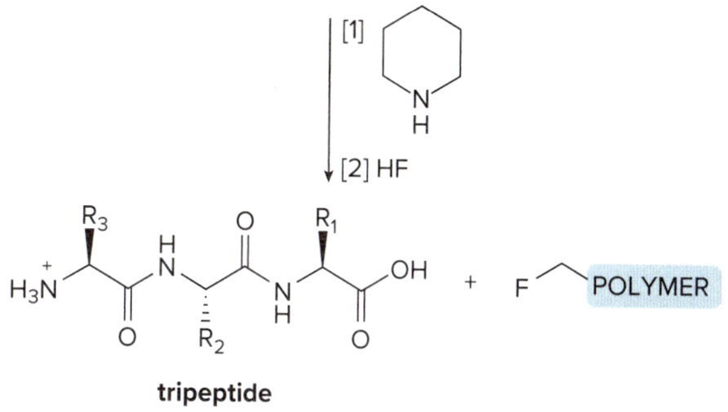

[1] piperidine
[2] HF

tripeptide + F–POLYMER

The Merrifield method has now been completely automated, so it is possible to purchase peptide synthesizers that automatically carry out all of the above operations and form polypeptides in high yield in a matter of hours, days, or weeks, depending on the length of the chain of the desired product. For example, the protein ribonuclease, which contains 128 amino acids, has been prepared by this technique in an overall yield of 17%. This remarkable synthesis involved 369 separate reactions, and thus the yield of each individual reaction was > 99%.

Problem 27.23 Outline the steps needed to synthesize the tetrapeptide Ala–Leu–Ile–Gly using the Merrifield technique.

27.8 Protein Structure

Now that you have learned some of the chemistry of amino acids, it's time to study proteins, the large polymers of amino acids that are responsible for so much of the structure and function of all living cells. We begin with a discussion of the **primary, secondary, tertiary,** and **quaternary structure** of proteins.

27.8A Primary Structure

The *primary structure* of proteins is the particular sequence of amino acids that is joined by peptide bonds. The most important element of this primary structure is the **amide bond.**

- Rotation around the amide C–N bond is *restricted* because of electron delocalization, and the *s*-trans conformation is the more stable arrangement.
- In each peptide bond, the N–H and C=O bonds are directed 180° from each other.

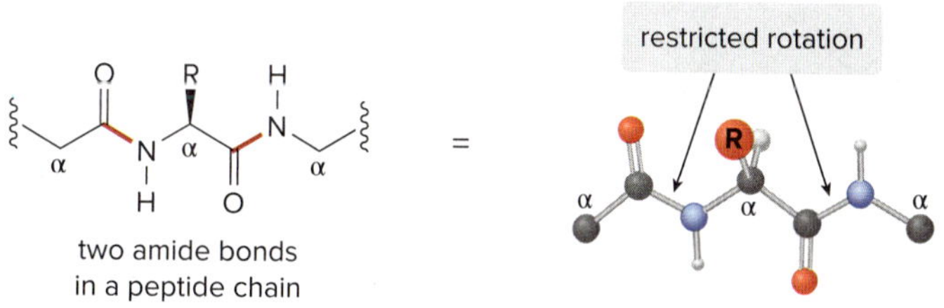

two amide bonds in a peptide chain

restricted rotation

Although rotation about the amide bonds is restricted, **rotation about the other σ bonds in the protein backbone is not.** As a result, the peptide chain can twist and bend into a variety of different arrangements that constitute the secondary structure of the protein.

27.8B Secondary Structure

The three-dimensional conformations of localized regions of a protein are called its *secondary structure*. These regions arise due to hydrogen bonding between the N–H proton of one amide and the C=O oxygen of another. Two arrangements that are particularly stable are called the **α-helix** and the **β-pleated sheet.**

hydrogen bond

α-Helix

The **α-helix** forms when a peptide chain twists into a right-handed or clockwise spiral, as shown in Figure 27.6. Four important features of the α-helix are as follows:

[1] **Each turn of the helix has 3.6 amino acids.**

[2] **The N–H and C=O bonds point along the axis of the helix.** All C=O bonds point in one direction, and all N–H bonds point in the opposite direction.

[3] **The C=O group of one amino acid is hydrogen bonded to an N–H group four amino acid residues farther along the chain.** Thus, hydrogen bonding occurs between two amino acids ***in the same chain.*** Note, too, that the hydrogen bonds are parallel to the axis of the helix.

[4] The **R groups of the amino acids extend outward** from the core of the helix.

Figure 27.6 Two different illustrations of the α-helix

a. The right-handed α-helix

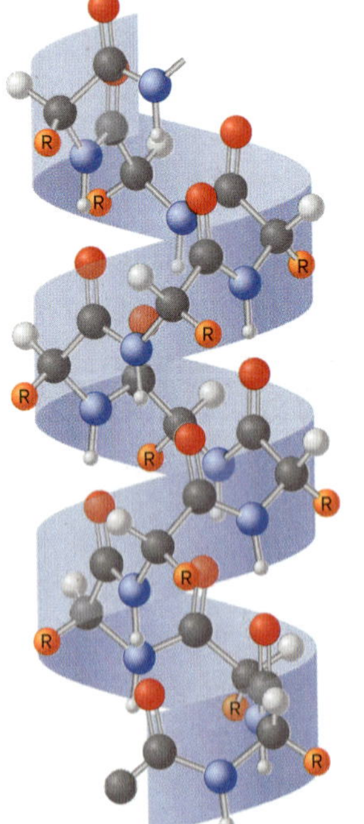

- All atoms of the α-helix are drawn in this representation. All C=O bonds are pointing up and all N–H bonds are pointing down.

b. The backbone of the α-helix

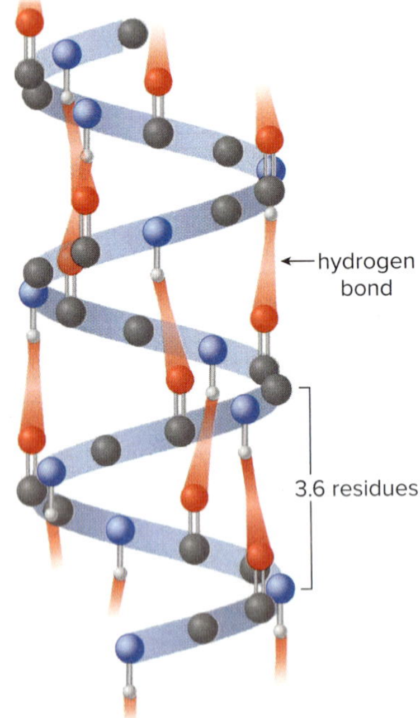

- Only the peptide backbone is drawn in this representation. The hydrogen bonds between the C=O and N–H of amino acids four residues away from each other are shown.

An α-helix can form only if there is rotation about the bonds at the α carbon of the amide carbonyl group, and not all amino acids can do this. For example, proline, the amino acid whose nitrogen atom forms part of a five-membered ring, is more rigid than other amino acids, and its C_α–N bond cannot rotate the necessary amount. Additionally, it has no N–H proton with which to form an intramolecular hydrogen bond to stabilize the helix. Thus, **proline cannot be part of an α-helix.**

Both the myosin in muscle and α-keratin in hair are proteins composed almost entirely of α-helices.

β-Pleated Sheet

The **β-pleated sheet** secondary structure forms when two or more peptide chains, called **strands,** line up side-by-side, as shown in Figure 27.7. All β-pleated sheets have the following characteristics:

[1] **The C=O and N–H bonds lie in the plane of the sheet.**

[2] **Hydrogen bonding often occurs between the N–H and C=O groups of nearby amino acid residues.**

[3] The **R groups are oriented above and below the plane** of the sheet, and alternate from one side to the other along a given strand.

Figure 27.7

Three-dimensional structure of the β-pleated sheet

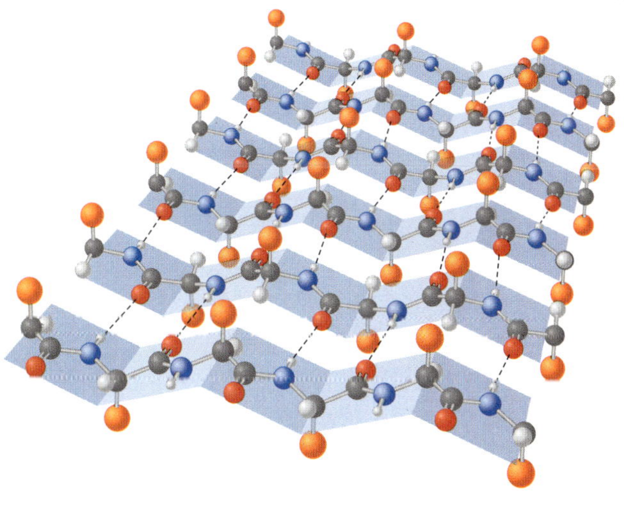

- The β-pleated sheet consists of extended strands of the peptide chains held together by hydrogen bonding. The C–O and N–H bonds lie in the plane of the sheet, and the R groups (shown as orange balls) alternate above and below the plane.

The β-pleated sheet arrangement most commonly occurs with amino acids with small R groups, like alanine and glycine. With larger R groups, steric interactions prevent the chains from getting close together, so the sheet cannot be stabilized by hydrogen bonding.

α-helix shorthand

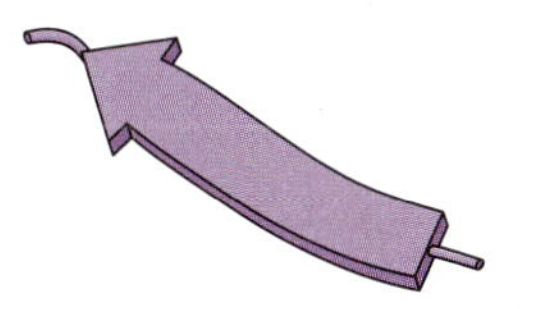

β-pleated sheet shorthand

Most proteins have regions of α-helix and β-pleated sheet, in addition to other regions that cannot be characterized by either of these arrangements. Shorthand symbols are often used to indicate regions of a protein that have α-helix or β-pleated sheet. A **flat helical ribbon** is used for the α-helix, and a **flat wide arrow** is used for the β-pleated sheet. These representations are often used in **ribbon diagrams** to illustrate protein structure.

Proteins are drawn in a variety of ways to illustrate different aspects of their structure. Figure 27.8 illustrates three different representations of the protein lysozyme, an enzyme found in both plants and animals. Lysozyme catalyzes the hydrolysis of bonds in bacterial cell walls, weakening them, often causing the bacteria to burst.

Figure 27.8 Lysozyme

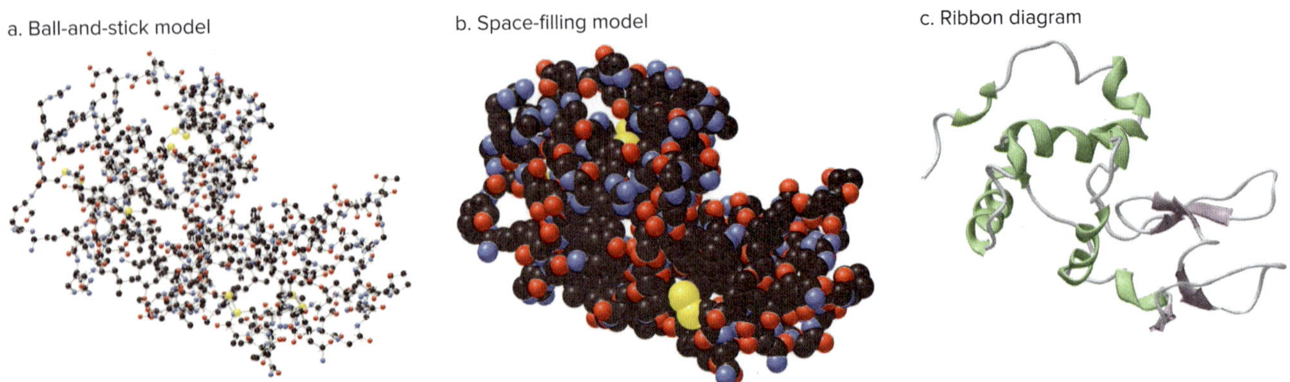

(a) The ball-and-stick model of lysozyme shows the protein backbone with color-coded C, N, O, and S atoms. Individual amino acids are most clearly located using this representation. (b) The space-filling model uses color-coded balls for each atom in the backbone of the enzyme and illustrates how the atoms fill the space they occupy. (c) The ribbon diagram shows regions of α-helix and β-pleated sheet that are not clearly in evidence in the other two representations.

Spider dragline silk is a strong yet elastic protein because it has regions of β-pleated sheet and regions of α-helix (Figure 27.9). α-Helical regions impart elasticity to the silk because the peptide chain is twisted (not fully extended), so it can stretch. β-Pleated sheet regions are almost fully extended, so they can't be stretched further, but their highly ordered three-dimensional structure imparts strength to the silk. Thus, spider silk suits the spider by comprising both types of secondary structure with beneficial properties.

Figure 27.9 Different regions of secondary structure in spider silk

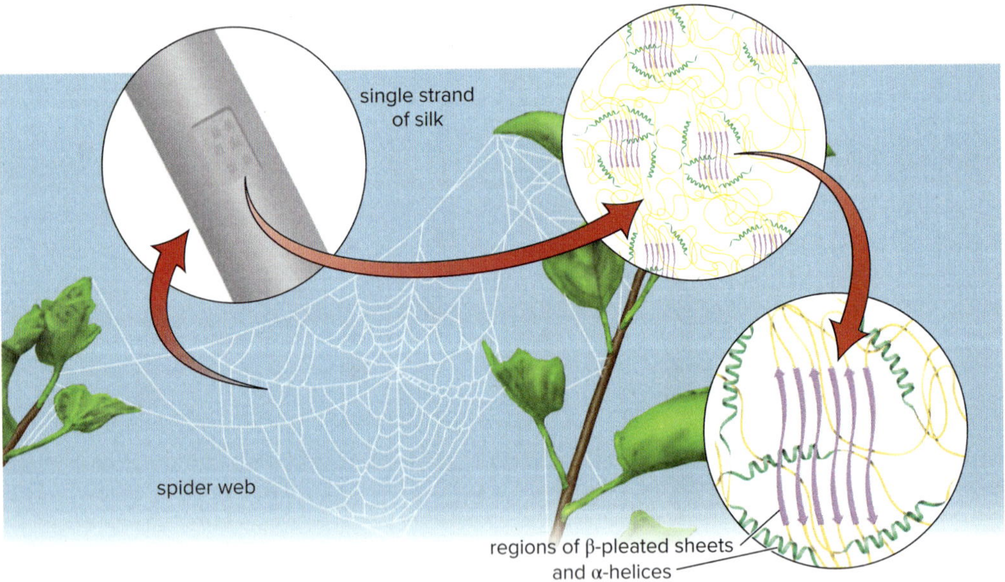

- Spider silk has regions of α-helix and β-pleated sheet that make it both strong and elastic. The green coils represent the α-helical regions, and the purple arrows represent the β-pleated sheet regions. The yellow lines represent other areas of the protein that are neither α-helix nor β-pleated sheet.

27.8C Tertiary and Quaternary Structure

The three-dimensional shape adopted by the entire peptide chain is called its *tertiary structure.* A peptide generally folds into a conformation that maximizes its stability. In the aqueous environment of the cell, proteins often fold in such a way as to place a large number of polar and charged groups on their outer surface, to maximize the dipole–dipole and hydrogen bonding interactions with water. This generally places most of the nonpolar side chains in the interior of the protein, where van der Waals interactions between these hydrophobic groups help stabilize the molecule, too.

In addition, polar functional groups hydrogen bond with each other (not just water), and amino acids with charged side chains like $-COO^-$ and $-NH_3^+$ can stabilize tertiary structure by electrostatic interactions.

Finally, **disulfide bonds are the only covalent bonds that stabilize tertiary structure.** As previously mentioned, these strong bonds form by oxidation of two cysteine residues on either the same polypeptide chain or another polypeptide chain of the same protein.

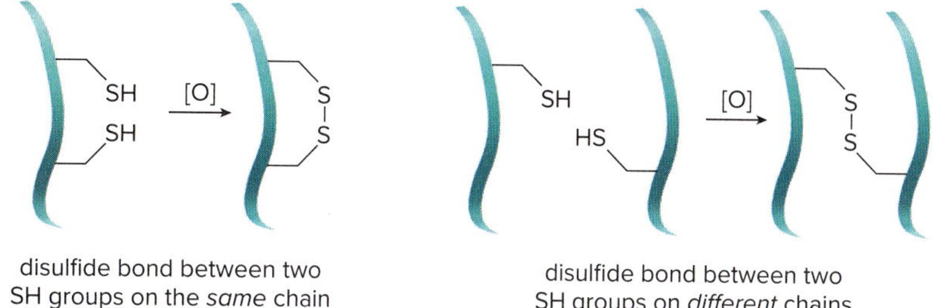

The nonapeptides **oxytocin** and **vasopressin** (Section 27.5C) contain intramolecular disulfide bonds. **Insulin,** on the other hand, consists of two separate polypeptide chains (**A** and **B**) that are covalently linked by two intermolecular disulfide bonds, as shown in Figure 27.10. The **A** chain, which also has an intramolecular disulfide bond, has 21 amino acid residues, whereas the **B** chain has 30.

Figure 27.10
Insulin

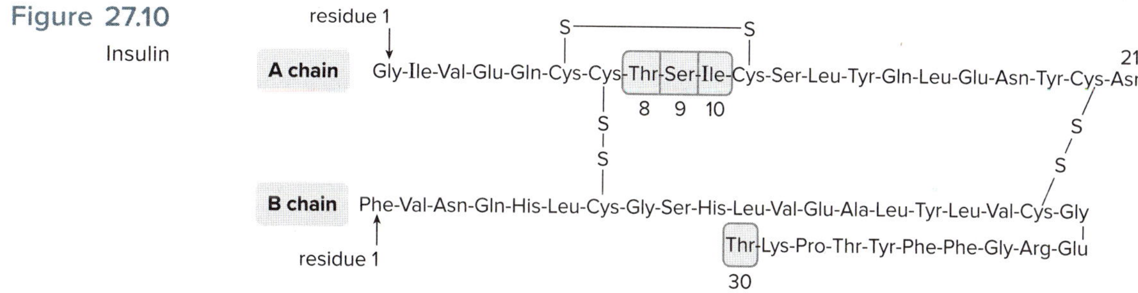

Insulin is a small protein consisting of two polypeptide chains (designated as the **A** and **B** chains) held together by two disulfide bonds. An additional disulfide bond joins two cysteine residues within the **A** chain.

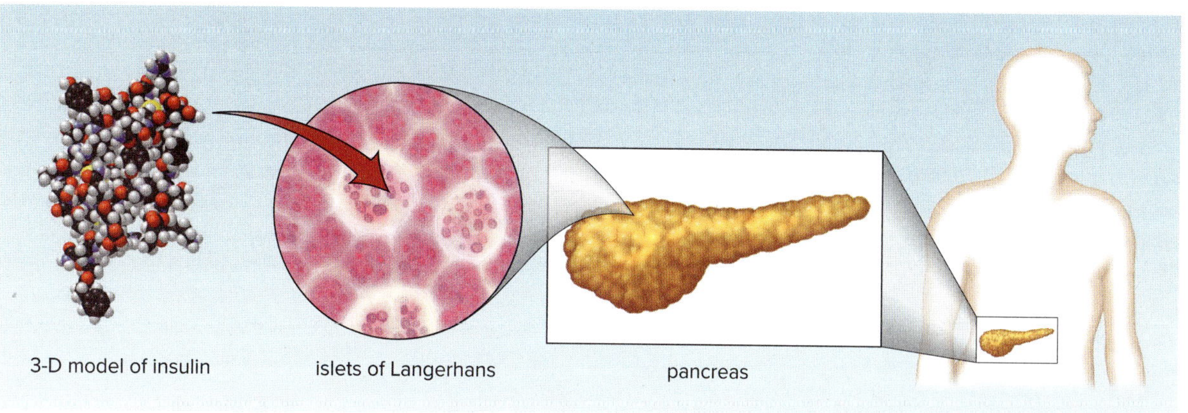

Synthesized by groups of cells in the pancreas called the islets of Langerhans, insulin is the protein that regulates the levels of glucose in the blood. Insufficiency of insulin results in diabetes. Many of the abnormalities associated with this disease can be controlled by the injection of insulin. Until the availability of human insulin through genetic engineering techniques, all insulin used by diabetics was obtained from pigs and cattle. The amino acid sequences of these insulin proteins is slightly different from that of human insulin. Pig insulin differs in one amino acid only, whereas bovine insulin has three different amino acids. This is shown in the accompanying table.

	Chain A			Chain B
Position of residue →	8	9	10	30
Human insulin	Thr	Ser	Ile	Thr
Pig insulin	Thr	Ser	Ile	Ala
Bovine insulin	Ala	Ser	Val	Ala

Figure 27.11 The stabilizing interactions in secondary and tertiary protein structure

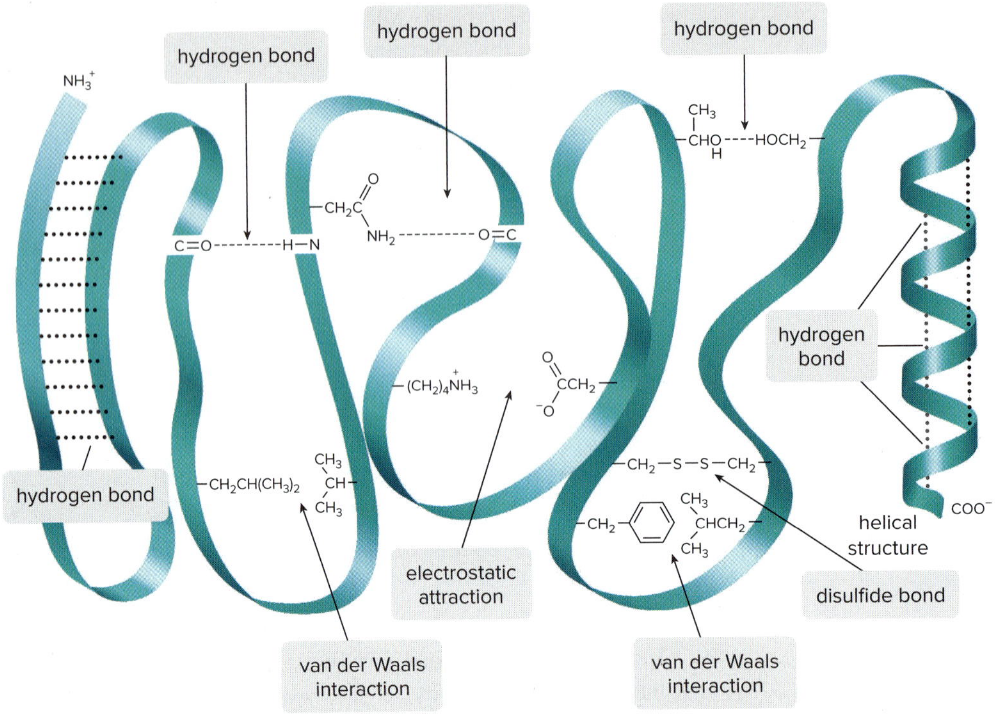

Figure 27.11 schematically illustrates the many different kinds of intramolecular forces that stabilize the secondary and tertiary structures of polypeptide chains.

The shape adopted when two or more folded polypeptide chains aggregate into one protein complex is called the *quaternary structure* of the protein. Each individual polypeptide chain is called a **subunit** of the overall protein. **Hemoglobin,** for example, consists of two α and two β subunits held together by intermolecular forces in a compact three-dimensional shape. The unique function of hemoglobin is possible only when all four subunits are together.

The four levels of protein structure are summarized in Figure 27.12.

Figure 27.12 The primary, secondary, tertiary, and quaternary structure of proteins

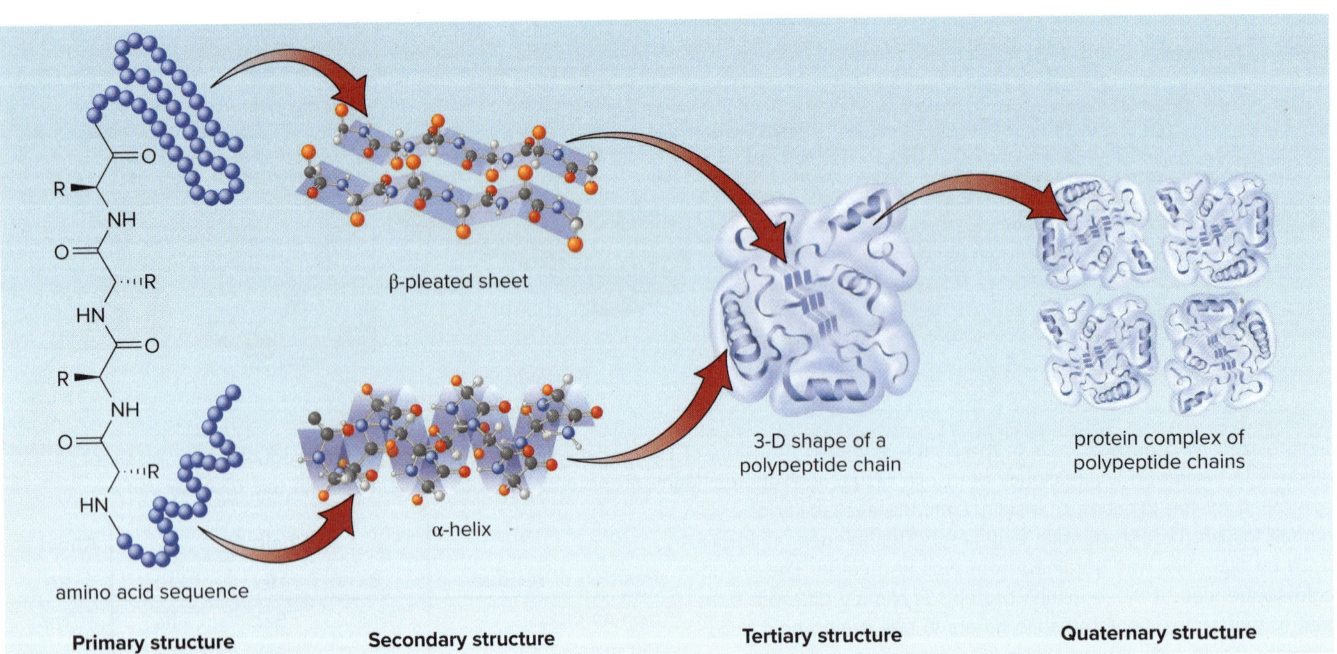

Problem 27.24 Which peptide in each pair has side chains that exhibit predominantly van der Waals forces?

a. Met–Gly–Leu–Phe–Gln–Ala or Lys–Gly–Arg–Tyr–Trp–Glu

b. Tyr–Asp–Leu–Lys–His or Phe–Asn–Leu–Leu–Met

Problem 27.25 The fibroin proteins found in silk fibers consist of large regions of β-pleated sheets stacked one on top of another. (a) Explain why having a glycine at every other residue allows the β-pleated sheets to stack on top of each other. (b) Why are silk fibers insoluble in water?

Problem 27.26 Hydrogen bonding stabilizes both the secondary and tertiary structures of a protein. (a) What functional groups hydrogen bond to stabilize secondary structure? (b) What functional groups hydrogen bond to stabilize tertiary structure?

27.9 Important Proteins

Proteins are generally classified according to their three-dimensional shapes.

- **Fibrous proteins** are composed of long linear polypeptide chains that are bundled together to form rods or sheets. These proteins are insoluble in water and serve structural roles, giving strength and protection to tissues and cells.
- **Globular proteins** are coiled into compact shapes with hydrophilic outer surfaces that make them water soluble. Enzymes and transport proteins are globular to make them soluble in the blood and other aqueous environments in cells.

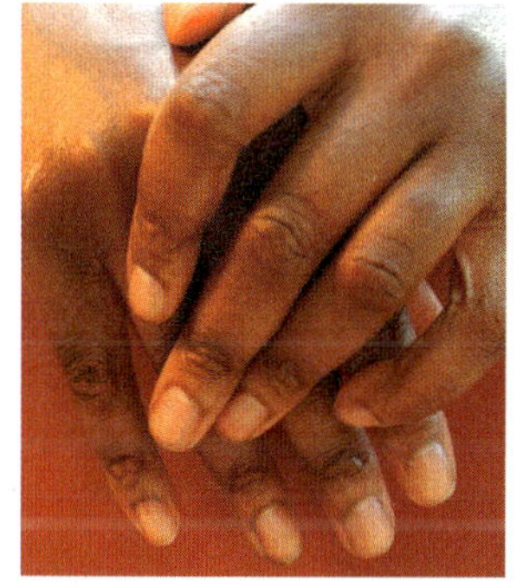

The many disulfide bonds in the proteins that compose fingernails make nails strong and hard. *Diffused Productions/Alamy Stock Photo*

27.9A α-Keratins

α-Keratins are the proteins found in hair, hooves, nails, skin, and wool. They are composed almost exclusively of long sections of α-helix units, having large numbers of alanine and leucine residues. Because these nonpolar amino acids extend outward from the α-helix, these proteins are very water insoluble. Two α-keratin helices coil around each other, forming a structure called a **supercoil** or **superhelix.** These, in turn, form larger and larger bundles of fibers, ultimately forming a strand of hair, as shown schematically in Figure 27.13.

α-Keratins also have a number of cysteine residues, and because of this, disulfide bonds are formed between adjacent helices. The number of disulfide bridges determines the strength of the material. Claws, horns, and fingernails have extensive networks of disulfide bonds, making them extremely hard.

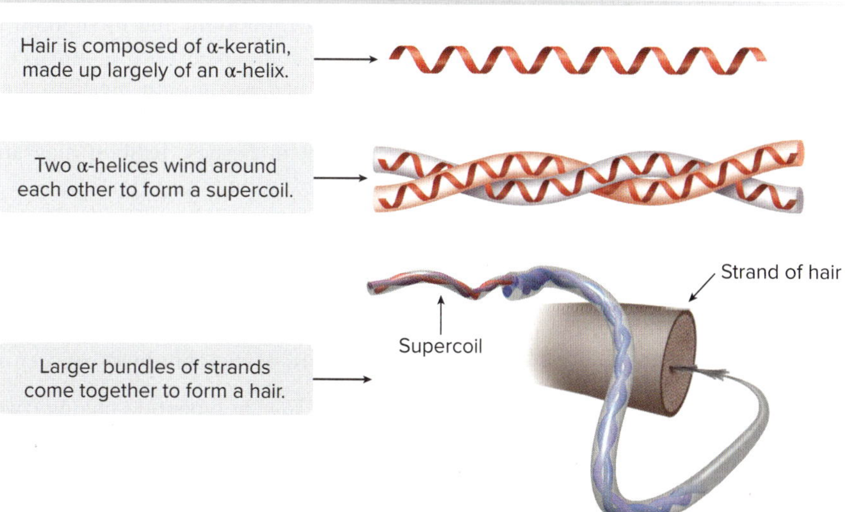

Figure 27.13 Anatomy of a hair—It begins with α-keratin.

Figure 27.14 The chemistry of a "permanent"—Making straight hair curly

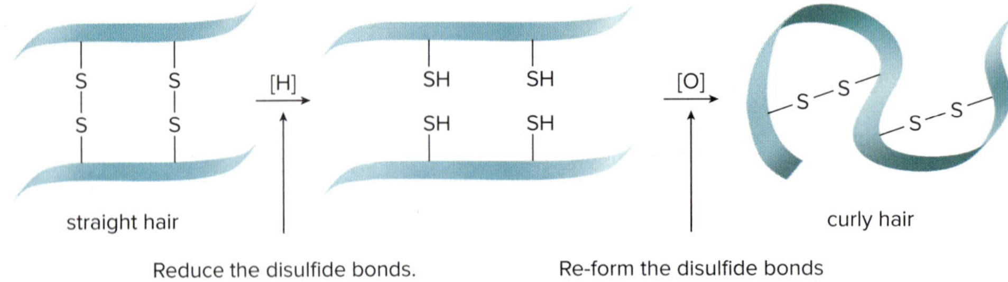

- To make straight hair curly, the disulfide bonds holding the α-helical chains together are cleaved by reduction. This forms free thiol groups (–SH). The hair is turned around curlers and then an oxidizing agent is applied. This re-forms the disulfide bonds in the hair, but between different thiol groups, now giving it a curly appearance.

Straight hair can be made curly by cleaving the disulfide bonds in α-keratin, and then rearranging and re-forming them, as shown schematically in Figure 27.14. First, the disulfide bonds in the straight hair are reduced to thiol groups, so the bundles of α-keratin chains are no longer held in their specific "straight" orientation. Then, the hair is wrapped around curlers and treated with an oxidizing agent that converts the thiol groups back to disulfide bonds, now with twists and turns in the keratin backbone. This makes the hair look curly and is the chemical basis for a "permanent."

Figure 27.15 The triple helix of collagen

- In collagen, three polypeptide chains having an unusual left-handed helix wind around each other in a right-handed triple helix.

27.9B Collagen

Collagen, the most abundant protein in vertebrates, is found in connective tissues such as bone, cartilage, tendons, teeth, and blood vessels. Glycine and proline account for a large fraction of its amino acid residues, whereas cysteine accounts for very little. Because of the high proline content, it cannot form a right-handed α-helix. Instead, it forms an elongated left-handed helix, and then three of these helices wind around each other to form a right-handed **superhelix** or **triple helix** (Figure 27.15). The side chain of glycine is only a hydrogen atom, so the high glycine content allows the collagen superhelices to lie compactly next to each other, thus stabilizing the superhelices via hydrogen bonding.

27.9C Hemoglobin and Myoglobin

Hemoglobin and **myoglobin,** two globular proteins, are called **conjugated proteins** because they are composed of a protein unit and a nonprotein molecule called a **prosthetic group.** The prosthetic group in hemoglobin and myoglobin is **heme,** a complex organic compound containing the Fe^{2+} ion complexed with a nitrogen heterocycle called a **porphyrin.** The Fe^{2+} ion of hemoglobin and myoglobin binds oxygen in the blood. Hemoglobin, which is present in red blood cells, transports oxygen to wherever it is needed in the body, whereas myoglobin stores oxygen in tissues. Ribbon diagrams for myoglobin and hemoglobin are shown in Figure 27.16.

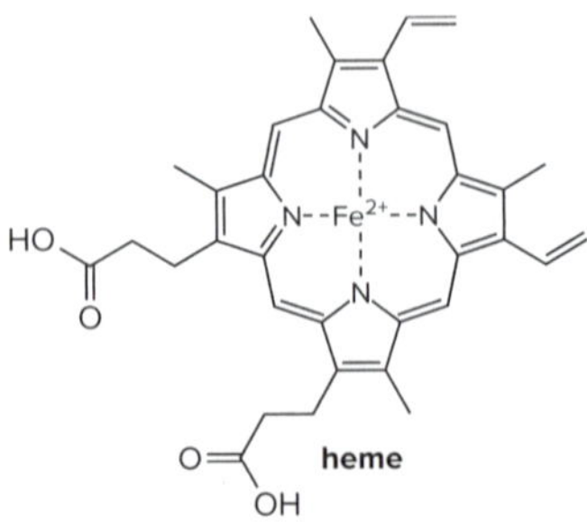

Myoglobin has 153 amino acid residues in a single polypeptide chain. It has eight separate α-helical sections that fold back on one another, with the prosthetic heme group held in a cavity inside the polypeptide. Most of the polar residues are found on the outside of the protein so that they can interact with the water solvent. Spaces in the interior of the protein are filled with nonpolar amino acids. Myoglobin binds oxygen in the blood and stores it in the tissues.

Hemoglobin consists of four polypeptide chains (two α subunits and two β subunits), each of which carries a heme unit. Hemoglobin has more nonpolar amino acids than myoglobin. When each subunit is folded, some of these remain on the surface. The van der Waals attraction between these hydrophobic groups is what stabilizes the quaternary structure of the four subunits.

Figure 27.16

Protein ribbon diagrams for myoglobin and hemoglobin

a. Myoglobin

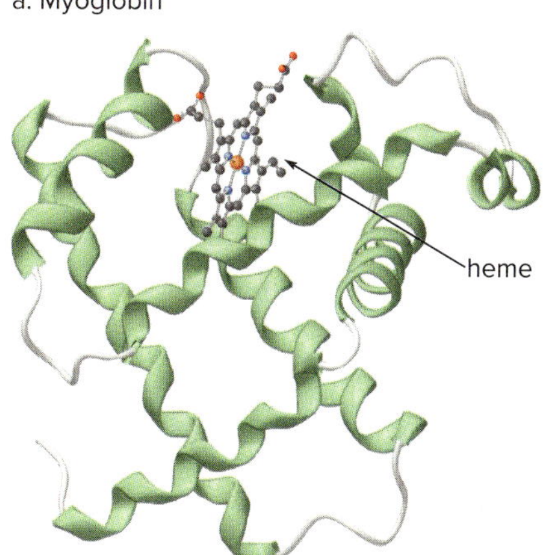

- Myoglobin consists of a single polypeptide chain with a heme unit shown in a ball-and-stick model.

b. Hemoglobin

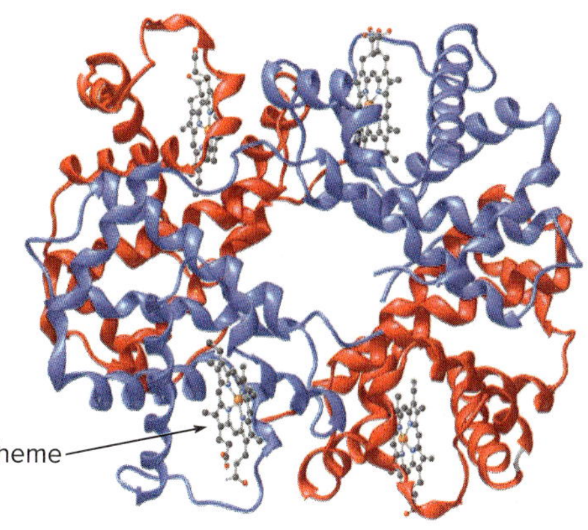

- Hemoglobin consists of two α and two β chains shown in red and blue, respectively, and four heme units shown in ball-and-stick models.

The high concentration of myoglobin in a whale's muscles allows it to remain underwater for long periods of time. *Daniel C. Smith*

Leghemoglobin, an oxygen-carrying protein (like hemoglobin) found in soybean roots, is used in plant-derived burgers. These meat alternatives are marketed to provide consumers with the nutritional benefits of meat without the negative environmental impact that occurs in large-scale livestock farming. *John D. Ivanko/ Alamy Stock Photo*

Carbon monoxide is poisonous because it binds to the Fe^{2+} of hemoglobin more strongly than does oxygen. Hemoglobin complexed with CO cannot carry O_2 from the lungs to the tissues. Without O_2 in the tissues for metabolism, cells cannot function, so they die.

The properties of all proteins depend on their three-dimensional shape, and their shape depends on their primary structure—that is, their amino acid sequence. This is particularly well exemplified by comparing normal hemoglobin with **sickle cell hemoglobin,** a mutant variation in which a single amino acid of both β subunits is changed from glutamic acid to valine. The replacement of one acidic amino acid (Glu) with one nonpolar amino acid (Val) changes the shape of hemoglobin, which has profound effects on its function. Deoxygenated red blood cells with sickle cell hemoglobin become elongated and crescent shaped, and they are unusually fragile. As a result, they do not flow easily through capillaries, causing pain and inflammation, and they break open easily, leading to severe anemia and organ damage. The end result is often a painful and premature death.

This disease, called **sickle cell anemia,** is found almost exclusively among people originating from central and western Africa, where malaria is an enormous health problem. Sickle cell hemoglobin results from a genetic mutation in the DNA sequence that is responsible for the synthesis of hemoglobin. Individuals who inherit this mutation from both parents develop sickle cell anemia, whereas those who inherit it from only one parent are said to have the sickle cell trait. They do not develop sickle cell anemia, and they are more resistant to malaria than individuals without the mutation. This apparently accounts for this detrimental gene being passed on from generation to generation.

27.10 Enzymes

Enzymes are water-soluble globular proteins that serve as biological catalysts for reactions in all living organisms. As we learned in Section 6.11, an enzyme contains an active site that binds a substrate, often in a small cavity that contains amino acids that are attracted to the substrate with various types of intermolecular forces. An enzyme-catalyzed reaction can be 10^6 to 10^{12} times faster than a similar uncatalyzed reaction.

Enzymes are specific. Some enzymes catalyze a single reaction of a single compound. Other enzymes, like trypsin and chymotrypsin (Section 27.5), catalyze a specific type of reaction, the cleavage of peptide bonds involving only certain amino acids.

Enzymes are crucial to the biological reactions that occur in the body, which would otherwise proceed too slowly. In humans, enzymes must catalyze reactions under specific physiological conditions, usually a pH around 7.4 and a temperature of 37 °C.

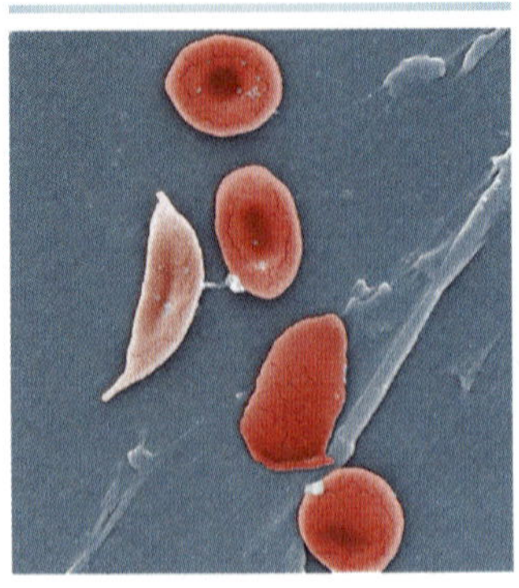

When red blood cells take on a "sickled" shape in persons with sickle cell disease, they occlude capillaries (causing organ injury) and they break easily (leading to profound anemia). This devastating illness results from the change of a single amino acid in hemoglobin (Section 27.9C). Note the single sickled cell surrounded by red cells with normal morphology.
Janice Haney Carr/CDC

27.10A Classification of Enzymes

Enzymes are classified into six categories by the type of reaction they catalyze.

- ***Oxidoreductases*** **catalyze oxidation–reduction reactions.**
- ***Transferases*** **catalyze the transfer of a group from one molecule to another.**
- ***Hydrolases*** **catalyze hydrolysis of esters, amides, and other functional groups that are cleaved when they react with water.**
- ***Isomerases*** **catalyze the conversion of one isomer to another.**
- ***Lyases*** **catalyze the addition of a molecule to a double bond or the elimination of a molecule to give a double bond.**
- ***Ligases*** **catalyze bond formation accompanied by energy release from a hydrolysis reaction.**

Table 27.3 summarizes the types of enzymes. Some enzyme classes are further subclassified by the functional group in the substrate or the type of molecule added or removed. For example, a transaminase is a transferase that catalyzes the transfer of an NH_2 group, whereas a kinase is a transferase that catalyzes transfer of a phosphate group.

Table 27.3 Classification of Enzymes

Enzyme Class or Subclass	Reaction Catalyzed
Oxidoreductases	**Oxidation–reduction**
• Oxidases	• Oxidation
• Reductases	• Reduction
• Dehydrogenases	• Addition or removal of 2 H's
Transferases	**Transfer of a group**
• Transaminases	• Transfer of an NH_2 group
• Kinases	• Transfer of a phosphate
Hydrolases	**Hydrolysis**
• Lipases	• Hydrolysis of lipid esters
• Proteases	• Hydrolysis of amide bonds in proteins
• Nucleases	• Hydrolysis of nucleic acids
Isomerases	**Isomerization**
Lyases	**Addition to a double bond or elimination to give a double bond**
• Dehydrases	• Removal of H_2O
• Decarboxylases	• Removal of CO_2
• Synthases	• Addition of a small molecule to a double bond
Ligases	**Bond formation accompanied by ATP hydrolysis**
• Carboxylases	• Bond formation between a substrate and CO_2

Problem 27.27 Classify the enzyme used in each of the following reactions.

a. [structure] ⟶ [structure] + CO_2

b. [structure] ⟶ [structure]

c. pyruvate ⟶ lactate

d. fumarate ⟶ malate

27.10B Using Enzymes to Diagnose and Treat Diseases

Measuring enzyme levels in the blood has aided greatly in diagnosing diseases. The concentration of some enzymes is higher within a cell than in the aqueous fluid outside the cell. When cells are damaged, the cells rupture and die, releasing the enzymes into the bloodstream. Measuring the activity of enzymes in the blood then becomes a powerful tool to diagnose the presence of disease or injury in some organs. For example, a higher-than-normal concentration of creatine phosphokinase (CPK) indicates whether a patient that has chest pain has had a heart attack.

Molecules that inhibit an enzyme can be useful drugs. An effective treatment of the human immunodeficiency virus (HIV), the virus that causes AIDS, uses protease inhibitors. These drugs inhibit the action of the HIV protease enzyme, an essential enzyme needed by HIV to make copies of itself that go on to infect other cells. Deactivating the HIV protease enzyme decreases the virus population, bringing the disease under control. Several protease inhibitors are currently available, and often an individual takes a combination of several drugs to keep the disease in check.

Another strategy for treating HIV is described in Section 28.9.

Fosamprenavir (trade name Lexiva) is a drug used to treat HIV infections. The body metabolizes fosamprenavir to amprenavir, the active drug that inhibits the HIV protease enzyme, so the virus cannot replicate. A ribbon diagram of the HIV-1 protease enzyme with amprenavir at the active site is shown in Figure 27.17.

fosamprenavir → amprenavir

Figure 27.17 Amprenavir at the active site of an HIV protease enzyme

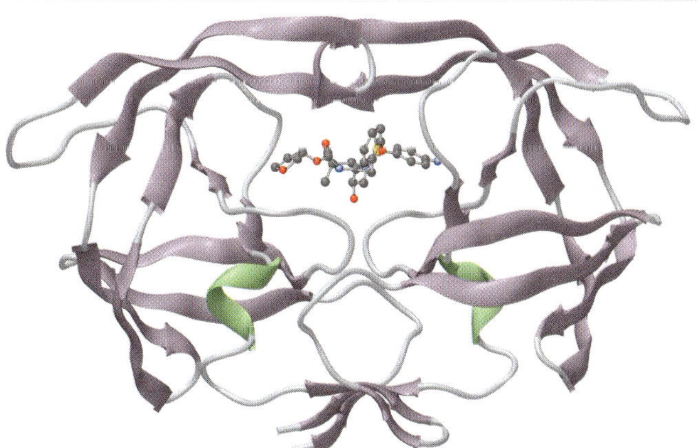

- Amprenavir, the active form of the drug fosamprenavir, inhibits the action of an HIV protease enzyme by binding to the active site.

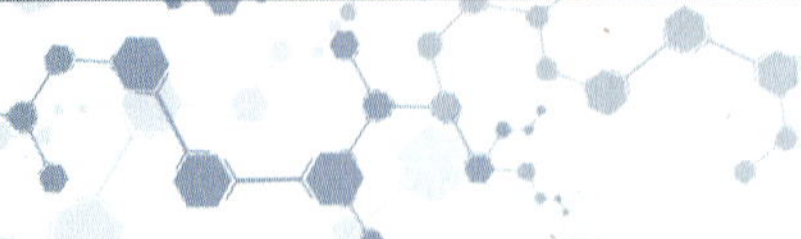

Chapter 27 REVIEW

KEY REACTIONS

[1] Enantioselective hydrogenation (27.3)

AcNH, H₂N — H_2 / Rh* → *S* enantiomer — ^-OH, H_2O → *S* amino acid

Rh* = chiral Rh hydrogenation catalyst

Try Problem 27.42b.

[2] Adding and removing protecting groups for amino acids (27.6)

1. L-phenylalanine — $(Boc)_2O$, Et_3N → Boc–NH product; protection (27.6)
2. Boc–NH phenylalanine — CF_3CO_2H or HCl or HBr → L-phenylalanine; deprotection (27.6)
3. L-leucine + Fmoc–Cl — Na_2CO_3, H_2O → Fmoc–NH product; protection (27.6)
4. Fmoc–NH leucine — piperidine → L-leucine; deprotection (27.6)
5. L-alanine — CH_3OH, H^+ → methyl ester (OCH_3); protection (27.6)
6. Methyl ester (OCH_3) — ^-OH, H_2O → L-alanine; deprotection (27.6)
7. L-leucine — Ph⁀OH, H^+ → benzyl ester (O⁀Ph); protection (27.6)
8. Benzyl ester (O⁀Ph) — ^-OH, H_2O or H_2, Pd-C → L-leucine; deprotection (27.6)

Try Problem 27.39a, d.

[3] Amide formation with DCC

Boc–NH (amino acid)–OH + H_2N (amino acid)–O⁀Ph — DCC (27.6) → Boc–NH dipeptide–O⁀Ph

KEY SKILLS

[1] Using (*R*)-α-methylbenzylamine to resolve a racemic mixture of amino acids (27.2A); example: separation of L- and D-leucine

1 React both enantiomers of an *N*-acetyl amino acid with the *R* isomer of the chiral amine.

2 Separate the diastereomers.

3 Regenerate the amino acid by hydrolysis of the amide.

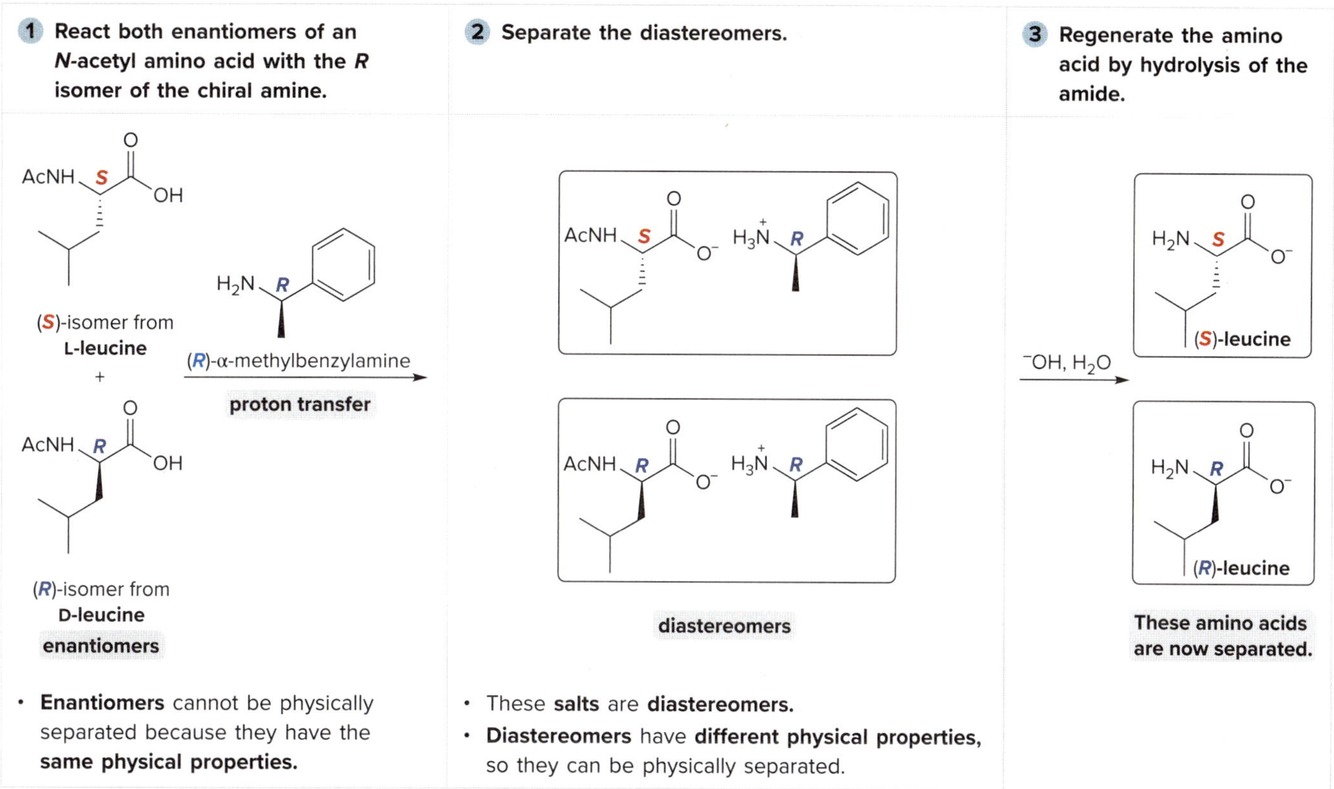

- **Enantiomers** cannot be physically separated because they have the **same physical properties.**

- These **salts** are **diastereomers.**
- **Diastereomers** have **different physical properties,** so they can be physically separated.

See *How To,* p. 1228. Try Problems 27.40, 27.41.

[2] Drawing the structure of a tripeptide, and labeling its N-terminal and C-terminal amino acids (27.4); example: Met–Ala–Arg

1 Draw the structures of the amino acids in order from left to right, placing the COO^- of one amino acid *next* to the NH_3^+ group of the adjacent amino acid.

2 Join adjacent COO^- and NH_3^+ groups together in amide bonds to form the tripeptide.

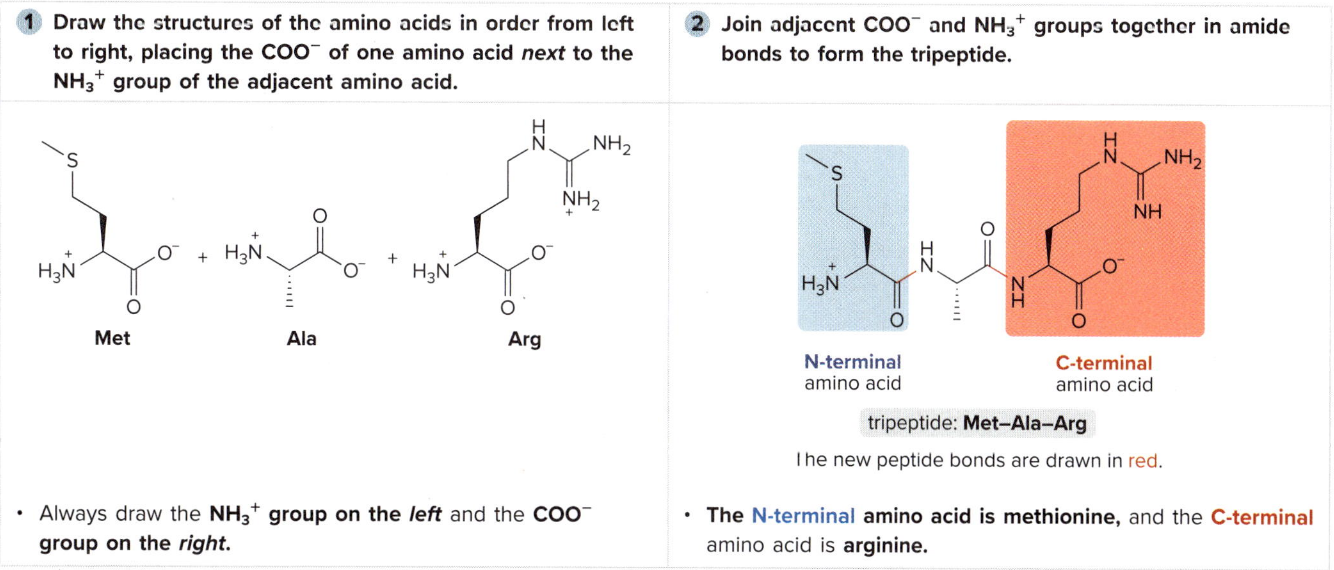

The new peptide bonds are drawn in red.

- Always draw the **NH_3^+ group on the *left*** and the **COO^- group on the *right.***

- **The N-terminal amino acid is methionine,** and the **C-terminal** amino acid is **arginine.**

See Sample Problem 27.1. Try Problems 27.44; 27.45b, d.

[3] Giving the amino acid sequence of a hexapeptide that contains the amino acids Gly, Pro, Val, Ser, Leu, His, and forms the following fragments when partially hydrolyzed with HCl: Ser–Val, Pro–His–Gly, Val–Leu–Pro (27.5)

① Look for points of overlap.	② Piece the fragments together.
common amino acids Ser–Val Pro–His–Gly Val–Leu–Pro	**Answer:** Ser–Val–Leu–Pro–His–Gly hexapeptide

See Sample Problem 27.2. Try Problem 27.49.

[4] Deducing the sequence of a pentapeptide that contains the amino acids Phe, Ile, Ala, Lys, Gly (27.5C)

① **Identify the N-terminal amino acid by Edman degradation.**	② **Identify the C-terminal amino acid by carboxypeptidase cleavage.**	③ **Identify the possible location of Lys or Arg, if applicable, from trypsin cleavage.**	④ **Complete the sequence, given the products from partial hydrolysis.**
Ala–__–__–__–__	Ala–__–__–__–Ile	• If a tripeptide and a dipeptide are obtained: Ala–Lys–__–__–Ile or Ala–__–Lys–__–Ile	• If Ile, Lys, Ala, and Phe–Gly are obtained: Ala–Lys–Phe–Gly–Ile

See Sample Problem 27.3, Table 27.2. Try Problems 27.49–27.52.

[5] Synthesizing a dipeptide from two amino acids (27.6): example: Leu–Ala

① Protect the NH_2 group of the N-terminal amino acid.

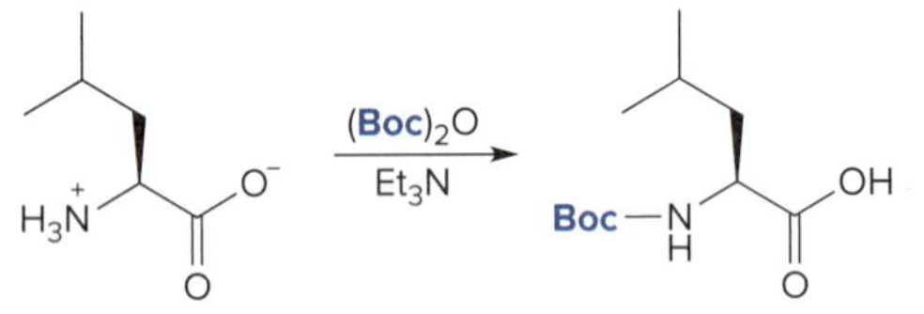

Leu $\xrightarrow[Et_3N]{(Boc)_2O}$ Boc–Leu

② Protect the COOH group of the C-terminal amino acid.

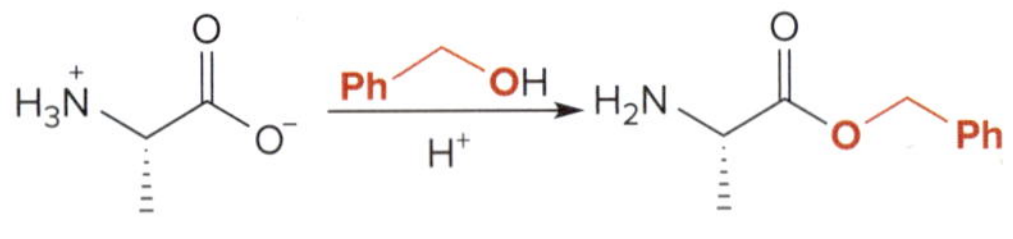

Ala $\xrightarrow[H^+]{PhCH_2OH}$ Ala–OCH_2Ph

③ Form the amide bond with DCC.

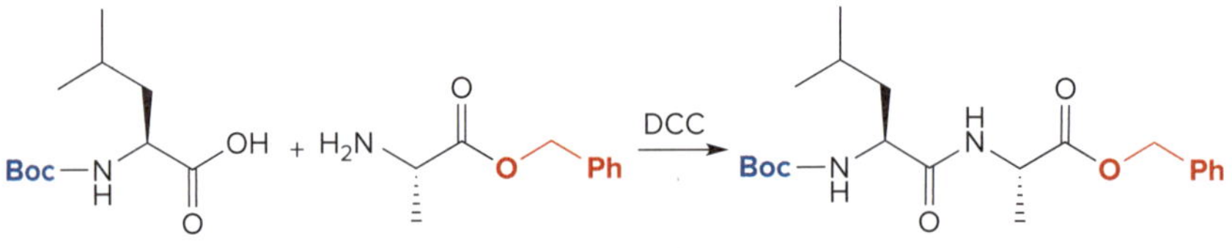

Boc–Leu + Ala–OCH_2Ph $\xrightarrow{DCC}$ Boc–Leu–Ala–OCH_2Ph

- **DCC** makes the OH group of the carboxylic acid a better leaving group, thus **activating the carboxy group toward nucleophillic attack.**

④ Remove one or both protecting groups.

Boc–Leu–Ala–OCH_2Ph $\xrightarrow[Pd\text{-}C]{H_2}$ Boc–Leu–Ala (Remove the benzyl group.)

Boc–Leu–Ala $\xrightarrow{CF_3COOH}$ Leu–Ala (Remove the Boc group.)

Boc–Leu–Ala–OCH_2Ph $\xrightarrow[CH_3COOH]{HBr}$ Leu–Ala (Remove both protecting groups.)

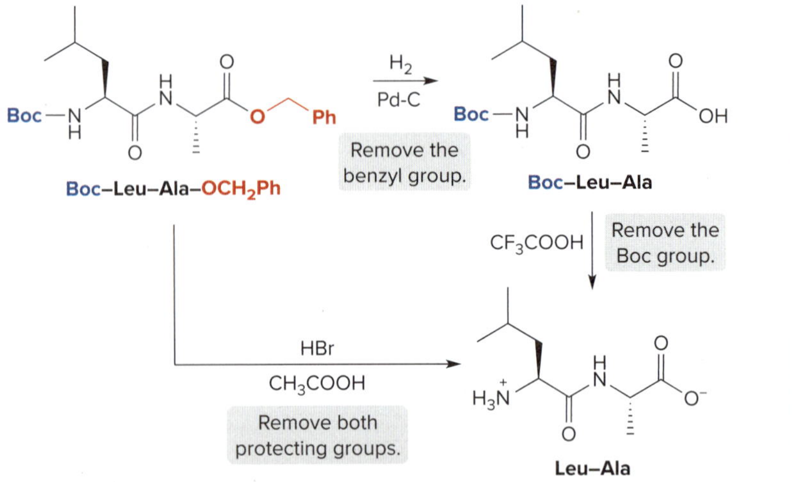

See *How To*, p. 1240, and Sample Problem 27.4. Try Problem 27.53.

[6] Synthesizing a tripeptide using the Merrifield solid phase technique (27.7)

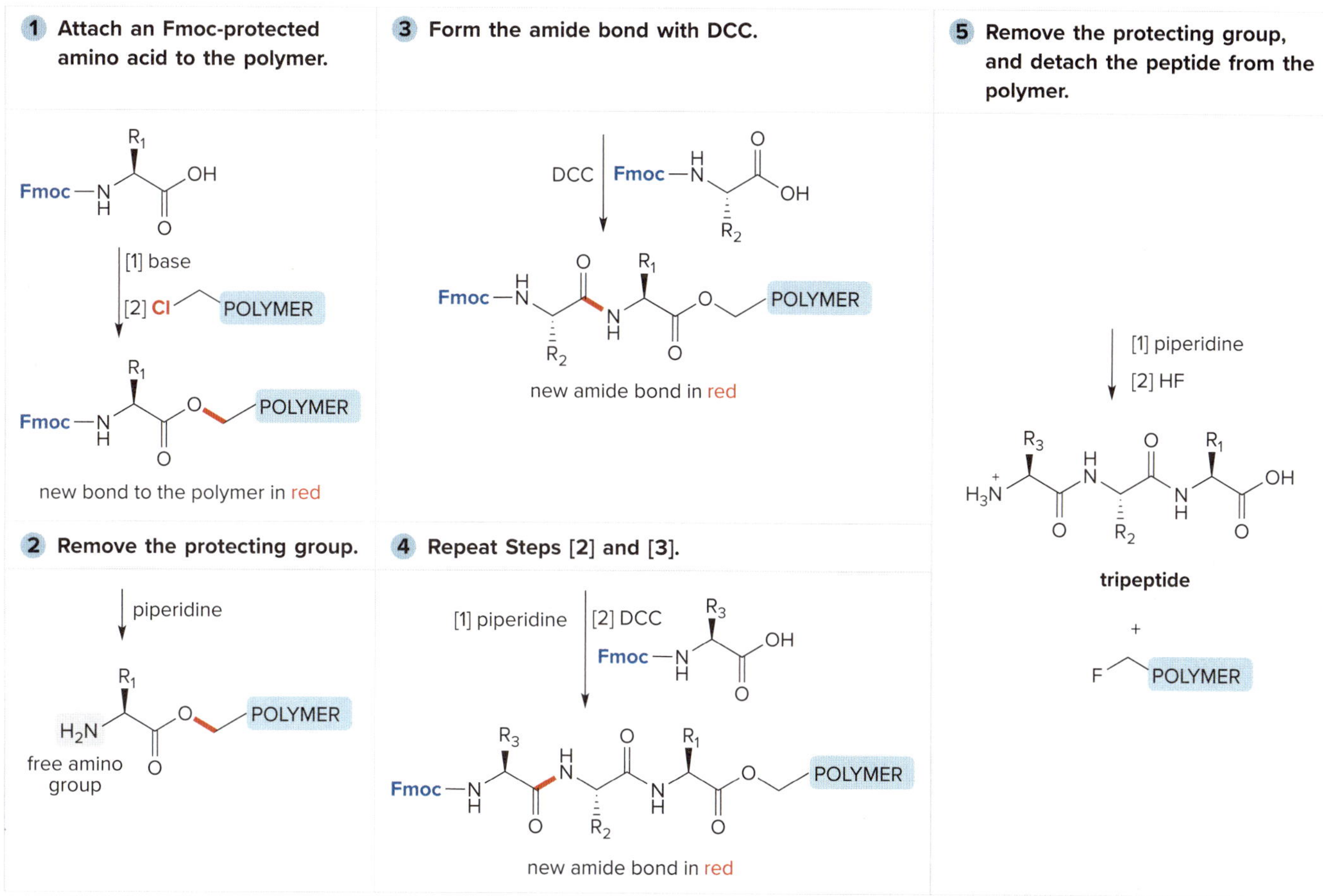

See *How To*, p. 1246. Try Problems 27.30, 27.54.

CHAPTER 27 MULTIPLE-CHOICE SELF-TEST

The Self-Test consists of multiple-choice questions similar to those found on the American Chemical Society organic chemistry exam. Answers are given at the end of the chapter.

1. Which statement about the levels of protein structure is *not* true?

a. Quaternary structure is present only when a protein is composed of more than one subunit.
b. Hydrogen bonding between amino acid residues in an α-helix is a component of a protein's tertiary structure.
c. The sequence of amino acids in a peptide chain constitutes its primary structure.
d. Van der Waals interactions between nonpolar amino acids are a component of tertiary protein structure.

2. What statement about the amino acid phenylalanine is *not* true?

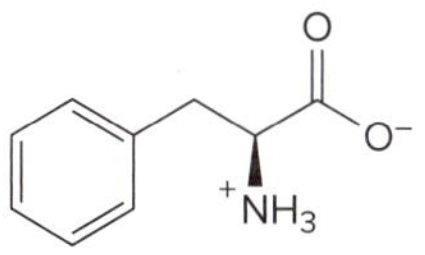

phenylalanine
pK_a (CO_2H) = 2.58
pK_a (α-NH_3^+) = 9.24

a. Naturally occurring phenylalanine has the *S* configuration at its stereogenic center.
b. The isoelectric point of phenylalanine is 5.91.
c. At pH = 10.5, phenylalanine exists in aqueous solution with a –1 charge.
d. Because of its carboxy group, phenylalanine is classified as an acidic amino acid.

3. Which of the following peptides is hydrolyzed by trypsin?

a. Lys–Trp–Gly–Gln–Met
b. Met–Phe–Ala–Leu–Arg
c. Ile–Val–Glu–Asn–Tyr
d. Val–Met–Asp–Gly–Phe

4. Which step is *not* used to synthesize the dipeptide Ala–Phe?

a. Phenylalanine is converted to its benzyl ester.

b. Alanine is converted to Boc–Ala.

c. An amide bond is formed between alanine and phenylalanine with DCC.

d. Two protecting groups are removed with HBr and CH_3CO_2H after the peptide bond is formed.

5. Which statement is *not* true about tripeptide **A**?

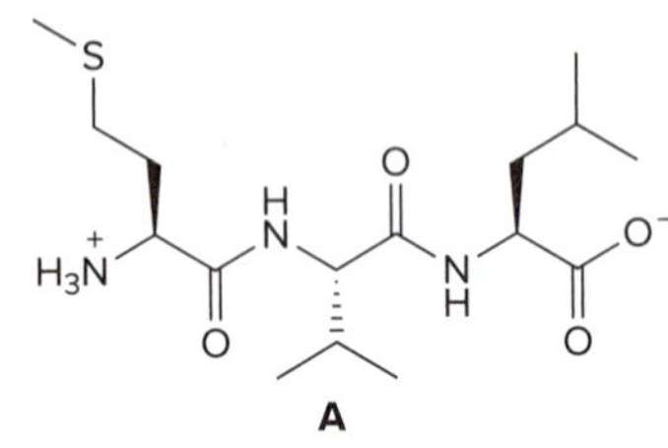

a. The peptide contains three peptide bonds.

b. The N-terminal amino acid is Met.

c. The C-terminal amino acid is L.

d. Edman degradation first removes M from the peptide chain.

6. What type of enzyme catalyzes each reaction?

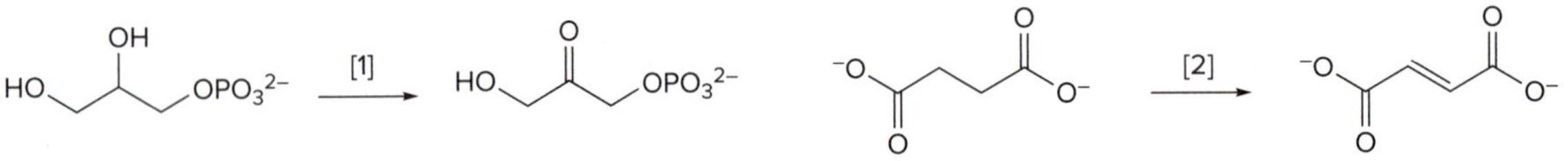

[1] [2]

a. Both Reactions [1] and [2] are catalyzed by dehydrogenase enzymes.

b. Reaction [1] is catalyzed by a dehydrogenase enzyme, and Reaction [2] is catalyzed by a reductase.

c. Reaction [1] is catalyzed by a reductase enzyme, and Reaction [2] is catalyzed by a dehydrogenase.

d. Reaction [1] is catalyzed by a kinase enzyme, and Reaction [2] is catalyzed by a reductase.

7. How many different tripeptides can be formed from two phenylalanines and one alanine?

a. one b. two c. three d. six

8. Determine the sequence of a pentapeptide from the following information. Partial hydrolysis forms Gly–Phe, Ala, Arg, and Leu. Edman degradation first cleaves Ala from the peptide. Carboxypeptidase cleaves Leu from the peptide. Treatment with trypsin forms Ala–Arg and one other fragment.

a. Gly–Phe–Ala–Arg–Leu

b. Ala–Arg–Gly–Phe–Leu

c. Leu–Gly–Phe–Ala–Arg

d. Gly–Phe–Ala–Arg–Leu

9. Which statement about hemoglobin is *not* true?

a. Hemoglobin is a water-soluble globular protein.

b. Normal hemoglobin differs from sickle cell hemoglobin by the presence of one amino acid in each of the β subunits.

c. Hemoglobin binds to both O_2 and CO.

d. Hemoglobin binds to O_2, transports it in the blood, and stores it in tissues.

10. Which amine can serve as a resolving agent for a racemic mixture of amino acids?

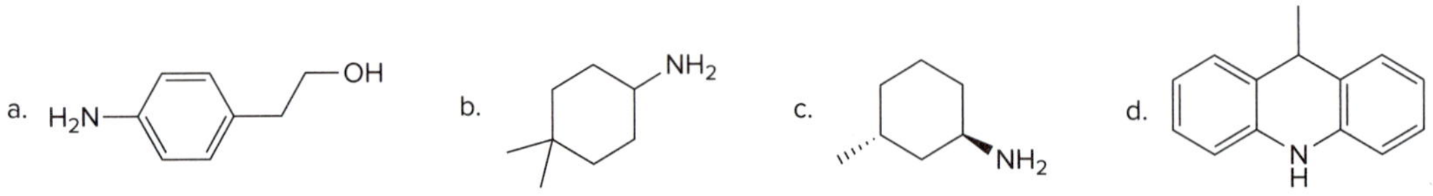

a. b. c. d.

PROBLEMS

Problems Using Three-Dimensional Models

27.28 Draw the product formed when the following amino acid is treated with each reagent: (a) CH_3OH, H^+; (b) CH_3COCl, pyridine; (c) HCl (1 equiv); (d) NaOH (1 equiv); (e) $C_6H_5N{=}C{=}S$.

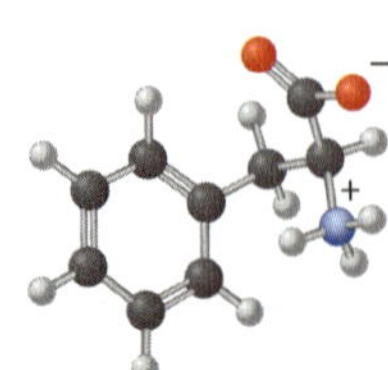

27.29 With reference to the following peptide: (a) Identify the N-terminal and C-terminal amino acids. (b) Name the peptide using one-letter abbreviations. (c) Label all the amide bonds in the peptide backbone.

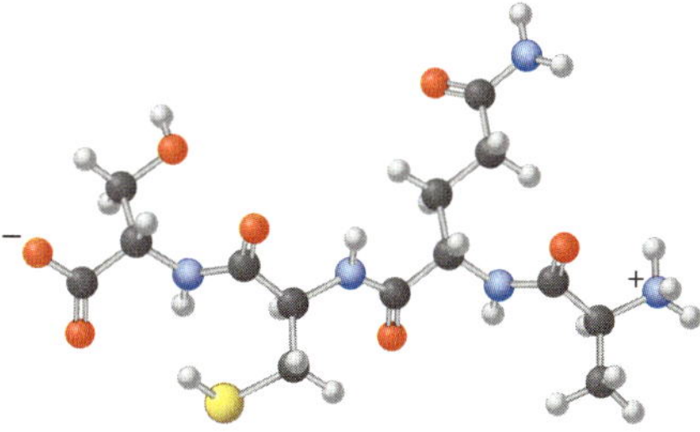

27.30 Write out the steps needed to synthesize the following peptide using the Merrifield method.

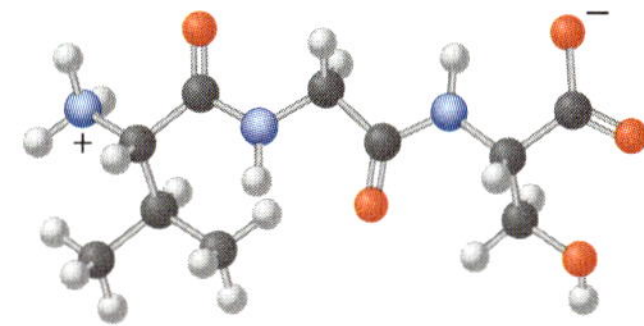

Amino Acids

27.31 penicillamine

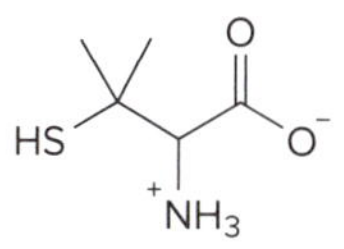

a. (*S*)-Penicillamine, an amino acid that does not occur in proteins, is used as a copper chelating agent to treat Wilson's disease, an inherited defect in copper metabolism. (*R*)-Penicillamine is toxic, sometimes causing blindness. Draw the structures of (*R*)- and (*S*)-penicillamine.

b. What disulfide is formed from oxidation of (*S*)-penicillamine?

27.32 Histidine is classified as a basic amino acid because one of the N atoms in its five-membered ring is readily protonated by acid. Which N atom in histidine is protonated and why?

27.33 Tryptophan is not classified as a basic amino acid even though it has a heterocycle containing a nitrogen atom. Why is the N atom in the five-membered ring of tryptophan not readily protonated by acid?

27.34 What is the predominant form of each of the following amino acids at pH = 1? What is the overall charge on the amino acid at this pH? (a) threonine; (b) methionine; (c) aspartic acid; (d) arginine

27.35 What is the predominant form of each of the following amino acids at pH = 11? What is the overall charge on the amino acid? (a) valine; (b) proline; (c) glutamic acid; (d) lysine

27.36 a. Draw the structure of the tripeptide A–A–A, and label the two ionizable functional groups.

b. What is the predominant form of A–A–A at pH = 1?

c. The pK_a values for the two ionizable functional groups (3.39 and 8.03) differ considerably from the pK_a values of alanine (2.35 and 9.87; see Table 27.1). Account for the observed pK_a differences.

Reactions Involving Amino Acids and Peptides

27.37 Identify **A–E** in the following reaction sequence.

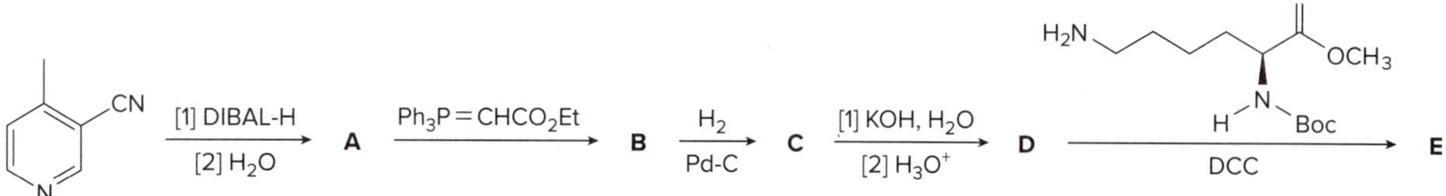

27.38 Neotame is an artificial sweetener prepared from aspartame by reductive amination with 3,3-dimethylbutanal.

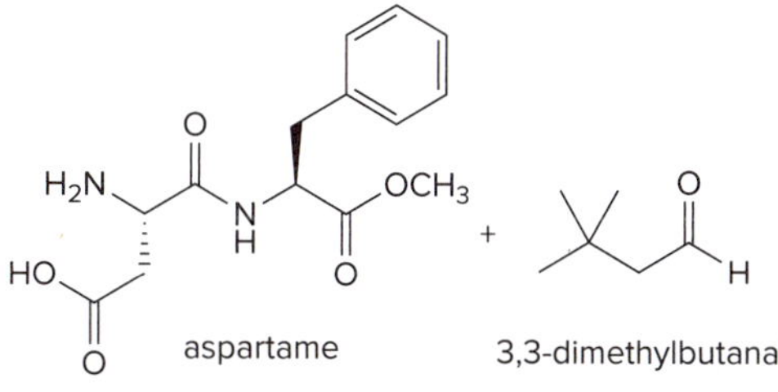

a. What is the structure of neotame?

b. The main metabolic pathway for the degradation of neotame is ester, but not amide, hydrolysis. What products are formed in this reaction? Unlike aspartame, which is metabolized to phenylalanine (Problem 20.48), neotame is not considered harmful for individuals with phenylketonuria.

27.39 Draw the product when the following tripeptide is treated with each reagent.

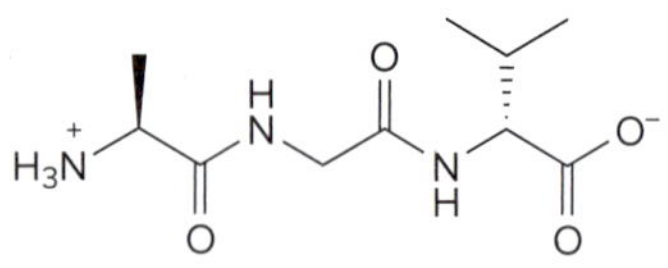

a. 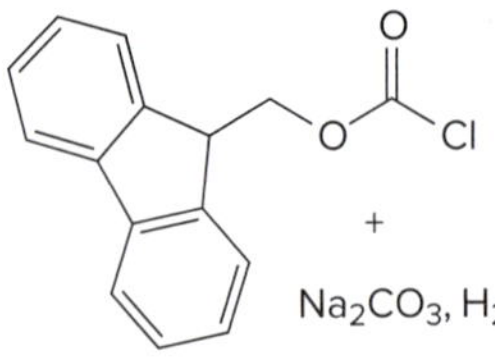 + Na_2CO_3, H_2O

b. H_3O^+

c. C_6H_5NCS

d. CH_3OH, H^+

Resolution; The Synthesis of Chiral Amino Acids

27.40 Another strategy used to resolve amino acids involves converting the carboxy group to an ester and then using a *chiral carboxylic acid* to carry out an acid–base reaction at the free amino group. Using a racemic mixture of alanine enantiomers and (*R*)-mandelic acid as resolving agent, write out the steps showing how a resolution process would occur.

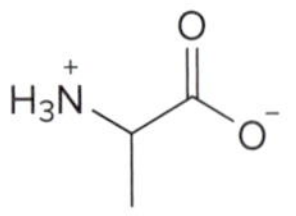

alanine (two enantiomers)

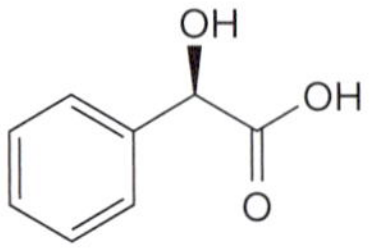

(*R*)-mandelic acid

27.41 Brucine is a poisonous alkaloid obtained from *Strychnos nux vomica,* a tree that grows in India, Sri Lanka, and northern Australia. Write out a resolution scheme similar to the one given in Section 27.2A, which shows how a racemic mixture of phenylalanine can be resolved using brucine.

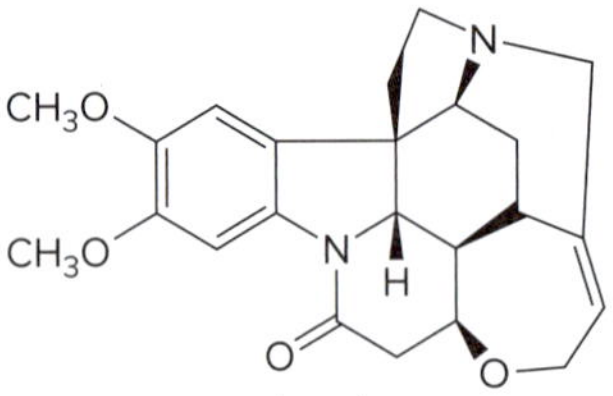

brucine

27.42 Draw the organic products formed in each reaction.

a. 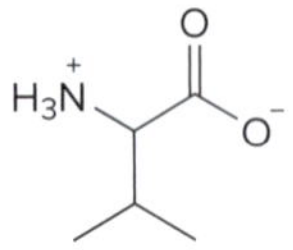 racemic mixture $\xrightarrow{Ac_2O}$ $\xrightarrow{\text{acylase}}$

b. 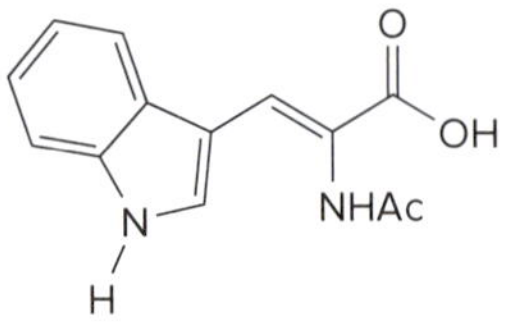 $\xrightarrow[\text{chiral Rh catalyst}]{H_2}$ $\xrightarrow[H_2O]{^-OH}$

27.43 What steps are needed to convert **A** to L-dopa, an uncommon amino acid that is effective in treating Parkinson's disease? These steps are the key reactions in the first commercial asymmetric synthesis using a chiral transition metal catalyst. This process was developed at Monsanto in 1974.

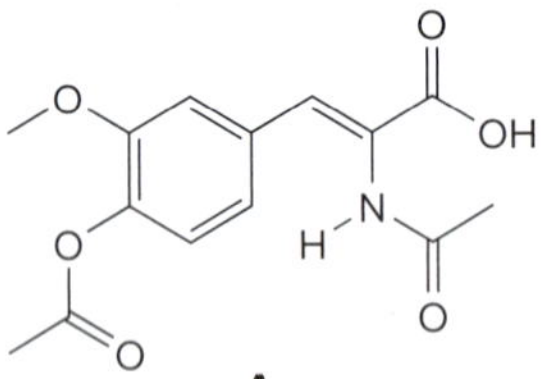

A 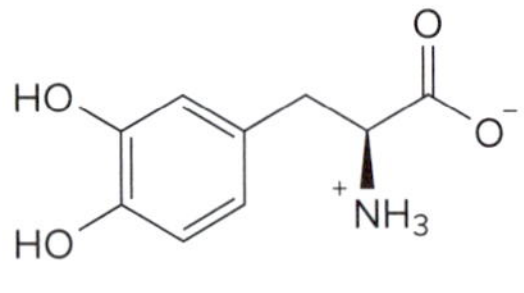 ⟶ L-dopa

Peptide Structure and Sequencing

27.44 Draw the structure for each peptide: (a) Phe–Ala; (b) Gly–Gln; (c) Lys–Gly; (d) R–H.

27.45 For the tetrapeptide Asp–Arg–Val–Tyr:

a. Name the peptide using one-letter abbreviations.

b. Draw the structure.

c. Label all amide bonds.

d. Label the N-terminal and C-terminal amino acids.

27.46 Name each peptide using both the three-letter and one-letter abbreviations of the component amino acids.

a. 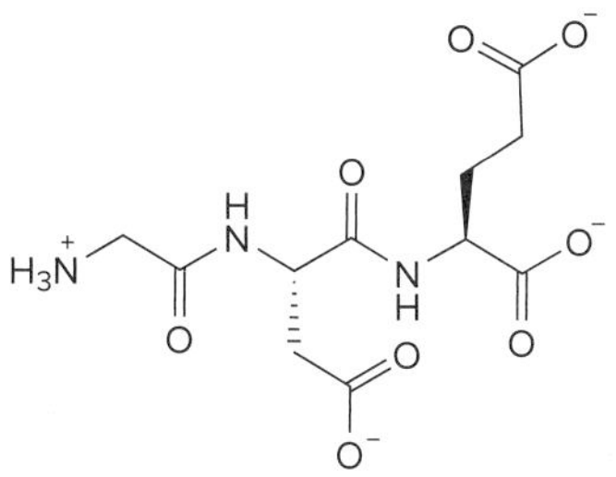 b.

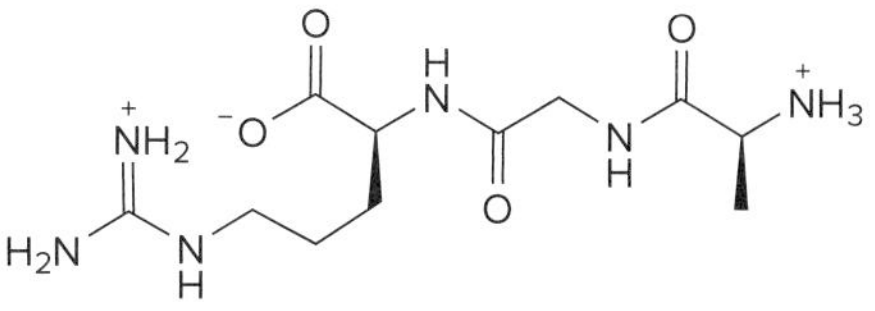

27.47 Somatostatin is a peptide hormone that regulates insulin and glucagon levels in the body.

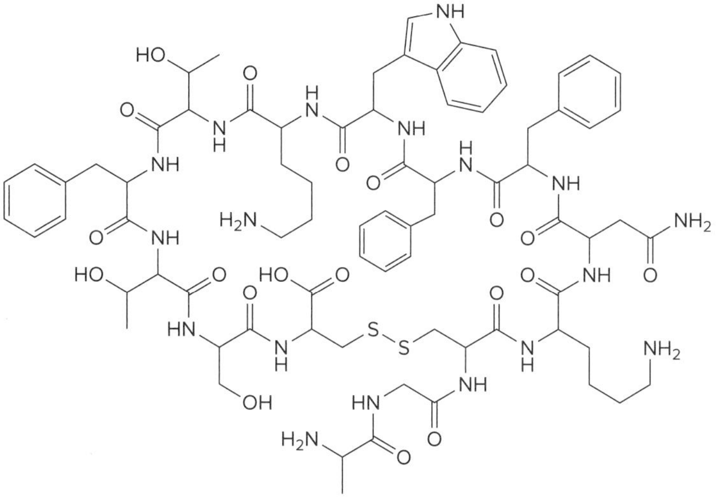

somatostatin

a. Identify the amino acids in somatostatin.
b. Assuming that somatostatin is composed of L-amino acids, add wedges and dashed wedges to all stereogenic centers to indicate the proper stereochemistry.
c. Label the N-terminal and C-terminal amino acids.
d. Using the three-letter abbreviations for the amino acids, write out the structure of the product formed when the disulfide bond in somatostatin is reduced.

27.48 Consider the decapeptide angiotensin I.

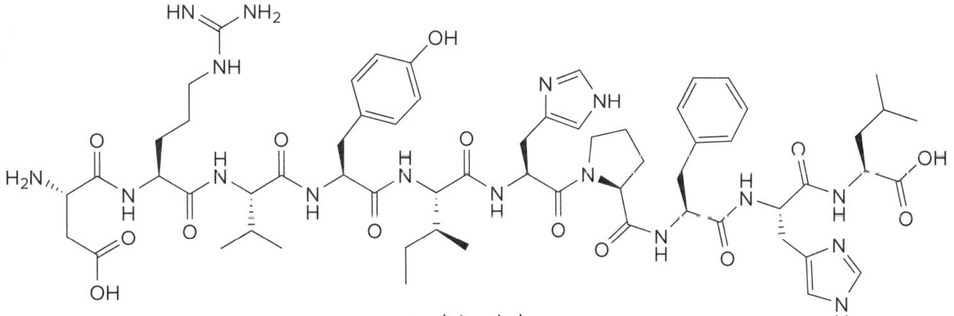

angiotensin I

a. What products are formed when angiotensin I is treated with trypsin?
b. What products are formed when angiotensin I is treated with chymotrypsin?
c. Treatment of angiotensin I with ACE (the angiotensin-converting enzyme) cleaves only the amide bond with the carbonyl group derived from phenylalanine to afford two products. The larger polypeptide is angiotensin II, a hormone that narrows blood vessels and increases blood pressure. Give the amino acid sequence of angiotensin II using three-letter abbreviations. ACE inhibitors are drugs that lower blood pressure by inhibiting the ACE enzyme.

27.49 Give the amino acid sequence of each peptide using the fragments obtained by partial hydrolysis of the peptide with acid.
a. a tetrapeptide that contains Ala, Gly, His, and Tyr, which is hydrolyzed to the dipeptides His–Tyr, Gly–Ala, and Ala–His
b. a pentapeptide that contains Glu, Gly, His, Lys, and Phe, which is hydrolyzed to His–Gly–Glu, Gly–Glu–Phe, and Lys–His

27.50 Glucagon, a hormone with 29 amino acids, is secreted by the pancreas. When the concentration of glucose in the bloodstream is too low, glucagon stimulates the liver to convert glycogen to glucose, thus increasing the blood glucose concentration. Deduce the amino acid sequence of glucagon from the following data. Treatment of glucagon with chymotrypsin forms: Thr–Ser–Asp–Tyr, Leu–Met–Asn–Thr, His–Ser–Gln–Gly–Thr–Phe, Ser–Lys–Tyr, Val–Gln–Trp, Leu–Asp–Ser–Arg–Arg–Ala–Gln–Asp–Phe. Treatment of glucagon with trypsin forms: Arg, Tyr–Leu–Asp–Ser–Arg, Ala–Gln–Asp–Phe–Val–Gln–Trp–Leu–Met–Asn–Thr, His–Ser–Gln–Gly–Thr–Phe–Thr–Ser–Asp–Tyr–Ser–Lys.

27.51 Use the given experimental data to deduce the sequence of an octapeptide that contains the following amino acids: Ala, Gly (2 equiv), His (2 equiv), Ile, Leu, and Phe. Edman degradation cleaves Gly from the octapeptide, and carboxypeptidase forms Leu and a heptapeptide. Partial hydrolysis forms the following fragments: Ile–His–Leu, Gly, Gly–Ala–Phe–His, and Phe–His–Ile.

27.52 An octapeptide contains the following amino acids: Arg, Glu, His, Ile, Leu, Phe, Tyr, and Val. Carboxypeptidase treatment of the octapeptide forms Phe and a heptapeptide. Treatment of the octapeptide with chymotrypsin forms two tetrapeptides, **A** and **B.** Treatment of **A** with trypsin yields two dipeptides, **C** and **D.** Edman degradation cleaves the following amino acids from each peptide: Glu (octapeptide), Glu (**A**), Ile (**B**), Glu (**C**), and Val (**D**). Partial hydrolysis of tetrapeptide **B** forms Ile–Leu in addition to other products. Deduce the structure of the octapeptide and fragments **A–D.**

Peptide Synthesis

27.53 Draw all the steps in the synthesis of each peptide from individual amino acids: (a) Gly–Ala; (b) Ile–Ala–Phe.

27.54 Write out the steps for the synthesis of each peptide using the Merrifield method: (a) Ala–Leu–Phe–Phe; (b) Phe–Gly–Ala–Ile.

27.55 Another method to form a peptide bond involves a two-step process:

[1] Conversion of a Boc-protected amino acid to a *p*-nitrophenyl ester

[2] Reaction of the *p*-nitrophenyl ester with an amino acid ester

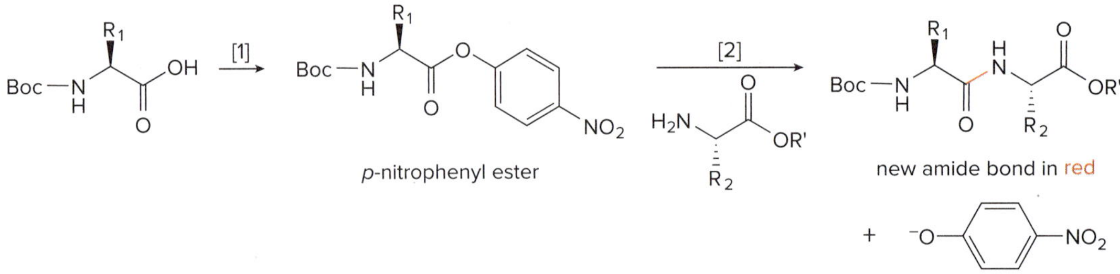

R_1 Boc–N(H) OH O [1] → Boc–N(H) R_1 O O NO_2

p-nitrophenyl ester

[2] → (with H_2N, O, OR', R_2) Boc–N(H) R_1 O H N O OR' R_2

new amide bond in red

+ ^-O–C_6H_4–NO_2

a. Why does a *p*-nitrophenyl ester "activate" the carboxy group of the first amino acid to amide formation?

b. Would a *p*-methoxyphenyl ester perform the same function? Why or why not?

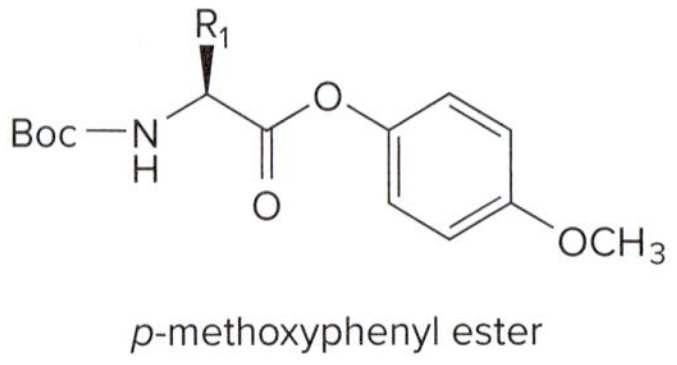

Boc–N(H) R_1 O O OCH_3

p-methoxyphenyl ester

27.56 Draw the mechanism for the reaction that removes an Fmoc group from an amino acid under these conditions:

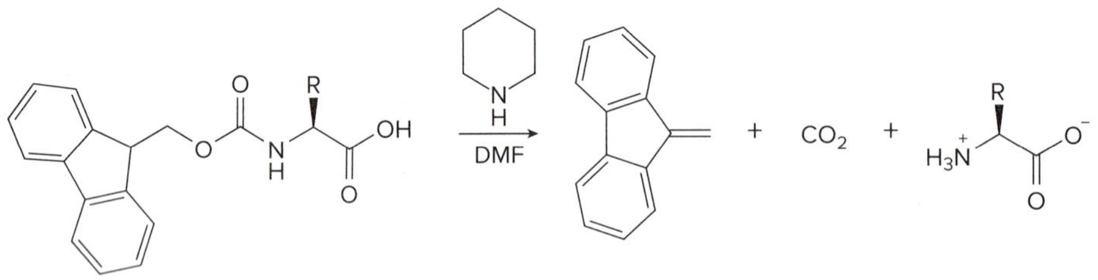

O R O N(H) OH O → (piperidine, N–H; DMF) + CO_2 + $H_3\overset{+}{N}$ R O^- O

Protein Structure and Enzymes

27.57 Which of the following amino acids are typically found in the interior of a globular protein, and which are typically found on the surface: (a) phenylalanine; (b) aspartic acid; (c) lysine; (d) isoleucine; (e) arginine; (f) glutamic acid?

27.58 Decide if the side chains of the following peptides are nonpolar or polar, and label the hydrophobic and hydrophilic end of each peptide.

a. VLLFGEDEK b. RKYSFLGAA

27.59 After the peptide chain of collagen has been formed, many of the proline residues are hydroxylated on one of the ring carbon atoms. Why is this process important for the triple helix of collagen?

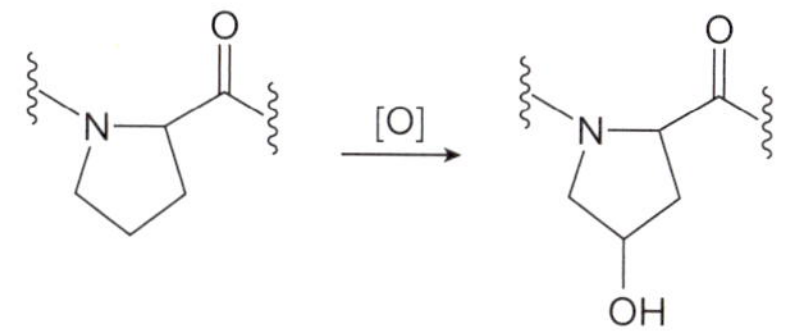

[O]

OH

27.60 What class of enzyme catalyzes each reaction?

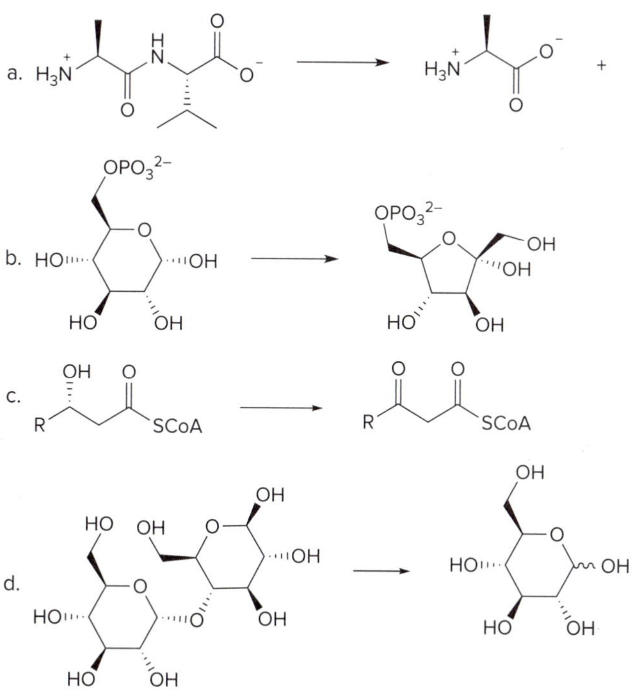

a.

b.

c.

d.

Challenge Problems

27.61 The anti-obesity drug orlistat works by irreversibly inhibiting pancreatic lipase, an enzyme responsible for the hydrolysis of triacylglycerols in the intestine, so that they are excreted without metabolism. Inhibition occurs by reaction of orlistat with a serine residue of the enzyme, forming a covalently bound, inactive enzyme product. Draw the structure of the product formed during inhibition.

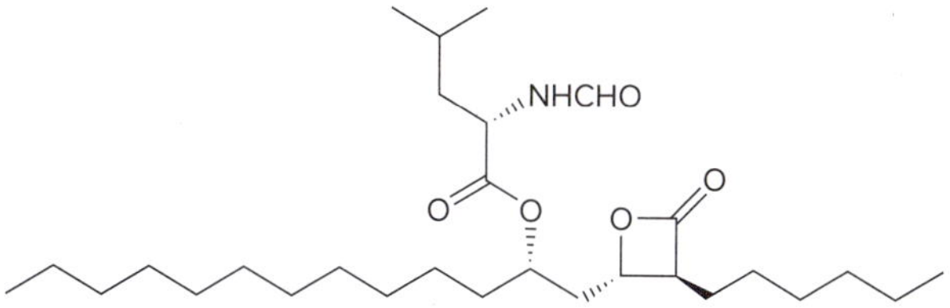

NHCHO

27.62 As shown in Mechanism 27.1, the final steps in the Edman degradation result in rearrangement of a thiazolinone to an *N*-phenylthiohydantoin. Draw a stepwise mechanism for this acid-catalyzed reaction.

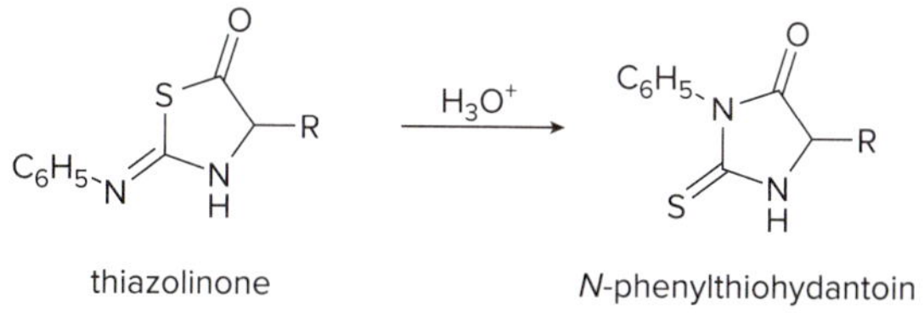

H_3O^+

thiazolinone

N-phenylthiohydantoin

SELF-TEST ANSWERS

1. b 2. d 3. a 4. c 5. a 6. a 7. c 8. b 9. d 10. c

28 Nucleic Acids and Protein Synthesis

Daniel C. Smith

28.1 Nucleosides and nucleotides
28.2 Nucleic acids
28.3 The DNA double helix
28.4 Replication
28.5 Ribonucleic acids and transcription
28.6 The genetic code, translation, and protein synthesis
28.7 DNA sequencing
28.8 The polymerase chain reaction
28.9 Viruses

Marine sponges are a rich source of natural products with promising pharmaceutical potential. Novel biologically active agents isolated from the shallow-water Caribbean sponge *Tectitethya crypta* by Bergmann in 1950 led to the development of **cytarabine,** a drug used to treat various forms of leukemia and non-Hodgkin's lymphoma. Related synthetic nucleosides interfere with the ability of a virus to synthesize nucleic acids, so they are used to treat viral infections. In Chapter 28, we learn about nucleosides and nucleotides, as well as the nucleic acids DNA and RNA, the polymers derived from them, which store and transmit the genetic information of an organism.

Why Study . . . Nucleic Acids?

Whether you are tall or short, fair-skinned or dark-complexioned, blue-eyed or brown-eyed, the nucleic acid polymers that reside in your cells determine your unique characteristics. The nucleic acid **DNA** stores the genetic information of a particular organism, whereas the nucleic acid **RNA** translates this genetic information into the synthesis of the proteins needed by cells for proper function and development. Minor alterations in the nucleic acid sequence can have significant effects on an organism, sometimes resulting in devastating diseases like sickle cell anemia and cystic fibrosis.

In Chapter 28, we learn about nucleic acids and the nucleotides from which they are formed.

28.1 Nucleosides and Nucleotides

***Nucleic acids* are unbranched polymers composed of repeating monomers called nucleotides.** DNA and RNA are two types of nucleic acids.

- **DNA, deoxyribonucleic acid, stores the genetic information of an organism and transmits that information from one generation to another.**
- **RNA, ribonucleic acid, translates the genetic information contained in DNA into proteins needed for all cellular functions.**

28.1A Identifying and Naming Bases, Nucleosides, and Nucleotides

The prefix *deoxy* means *without oxygen.*

The nucleic acids are composed of three components: a monosaccharide, a heterocyclic aromatic base, and a phosphate group. The monosaccharide component of RNA is **D-ribose,** whereas that of DNA is **2'-deoxy-D-ribose,** a monosaccharide that lacks a hydroxy group at C2'. Primes (') are used in numbering the carbons of the monosaccharide components.

D-ribose
(present in RNA)

2'-deoxy-D-ribose
(present in DNA)

The heterocyclic bases in DNA were first discussed in Section 15.9B.

Five common heterocyclic bases are present in nucleic acids. Three bases with one ring (**cytosine, uracil,** and **thymine**) are derived from the parent compound **pyrimidine.** Two bicyclic bases (**adenine** and **guanine**) are derived from the parent compound **purine.** Each base is designated by a one-letter abbreviation.

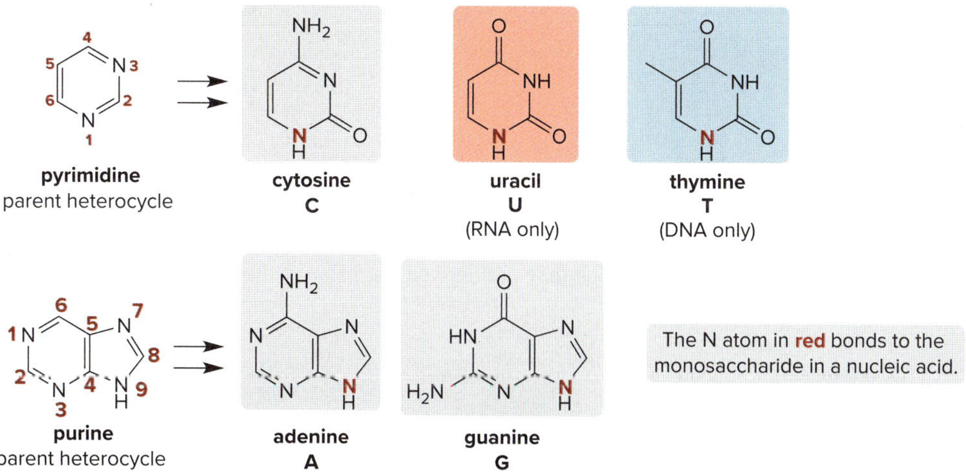

Uracil (U) occurs only in RNA, whereas thymine (T) occurs only in DNA, As a result,

- DNA contains the bases A, G, C, and T.
- RNA contains the bases A, G, C, and U.

A *nucleoside* is an *N*-glycoside, formed by joining the anomeric carbon (C1') of the monosaccharide with N1 of a pyrimidine base or N9 of a purine base in a β-glycosidic linkage.

- **Joining D-ribose with a base forms a *ribonucleoside*.**
- **Joining 2'-deoxy-D-ribose with a base forms a *deoxyribonucleoside*.**

For example, the ribonucleoside cytidine is formed from D-ribose and cytosine. The deoxyribonucleoside 2'-deoxyadenosine is formed from 2'-deoxy-D-ribose and adenine.

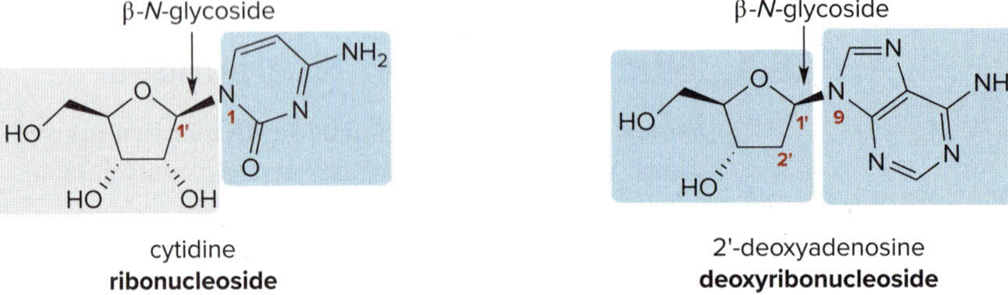

cytidine
ribonucleoside

2'-deoxyadenosine
deoxyribonucleoside

Nucleosides are named as derivatives of the bases from which they are formed.

- **To name a nucleoside derived from a pyrimidine base, use the suffix *–idine* (cytosine → cyt*idine*).**
- **To name a nucleoside derived from a purine base, use the suffix *–osine* (adenine → aden*osine*).**
- **Add the prefix *deoxy-* for deoxyribonucleosides, as in *deoxy*adenosine.**

Problem 28.1 Identify the base and monosaccharide in each nucleoside and then assign a name.

a.

b.

Problem 28.2 Novel antiviral agents isolated from Caribbean sponges led to the development of vidarabine, the first nucleoside drug used to treat herpes infections. From what you learned about monosaccharides in Chapter 26, determine what base and monosaccharide are present in vidarabine.

vidarabine

A *nucleotide* is formed by adding a phosphate group to the 5'-OH group of a nucleoside. Nucleotides are named by adding the term *5'-monophosphate* to the name of the nucleoside from which they are derived. At pH = 7 the phosphate is ionized, so the nucleotide bears a –2 charge.

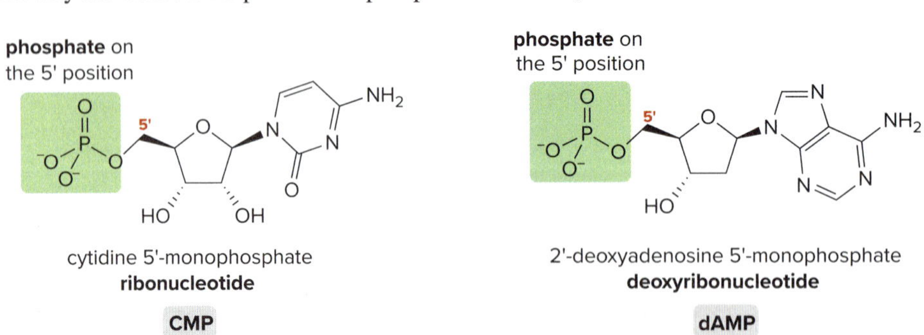

cytidine 5'-monophosphate
ribonucleotide
CMP

2'-deoxyadenosine 5'-monophosphate
deoxyribonucleotide
dAMP

Because of the lengthy names of nucleotides, three- or four-letter abbreviations are commonly used instead. Cytidine 5'-monophosphate is **CMP** and 2'-deoxyadenosine 5'-monophosphate is **dAMP.** Table 28.1 summarizes the names and abbreviations used for the bases, nucleosides, and nucleotides in nucleic acid chemistry.

Table 28.1 Names of Bases, Nucleosides, and Nucleotides in Nucleic Acids

Base	Abbreviation	Nucleoside	Nucleotide	Abbreviation
DNA				
Adenine	**A**	2'-deoxyadenosine	2'-deoxyadenosine 5'-monophosphate	**dAMP**
Guanine	**G**	2'-deoxyguanosine	2'-deoxyguanosine 5'-monophosphate	**dGMP**
Cytosine	**C**	2'-deoxycytidine	2'-deoxycytidine 5'-monophosphate	**dCMP**
Thymine	**T**	2'-deoxythymidine	2'-deoxythymidine 5'-monophosphate	**dTMP**
RNA				
Adenine	**A**	adenosine	adenosine 5'-monophosphate	**AMP**
Guanine	**G**	guanosine	guanosine 5'-monophosphate	**GMP**
Cytosine	**C**	cytidine	cytidine 5'-monophosphate	**CMP**
Uracil	**U**	uridine	uridine 5'-monophosphate	**UMP**

Sample Problem 28.1 Drawing the Structure of a Nucleotide

Draw the structure of each compound: (a) 2'-deoxyguanosine 5'-monophosphate; (b) UMP. Classify the nucleotide as a ribonucleotide or a deoxyribonucleotide.

Solution

Convert an abbreviation to the name of the nucleotide and then use these steps:

1. Draw the monosaccharide. If the name does not contain the prefix *deoxy,* the monosaccharide is D-ribose, making the compound a ribonucleotide. If the name contains the prefix *deoxy,* the monosaccharide is 2'-deoxy-D-ribose, making the compound a deoxyribonucleotide.
2. Add the base bonded to C1' of the monosaccharide ring, forming an *N*-glycoside with the β configuration.
3. Add the phosphate to the 5'-OH group.

a. For 2'-deoxyguanosine 5'-monophosphate:

2 Form an *N*-glycoside using N9 of guanine.

3 Add phosphate at the 5'-OH.

guanine

1 No OH group at C2': 2'-deoxy-D-ribose

2'-deoxyguanosine 5'-monophosphate
deoxyribonucleotide

b. UMP is the abbreviation for uridine 5'-monophosphate.

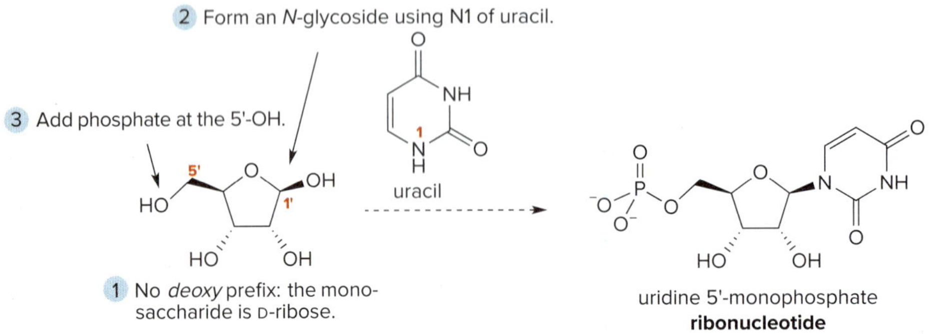

uridine 5'-monophosphate
ribonucleotide

Problem 28.3 Draw the structure of each nucleotide: (a) dTMP; (b) AMP.

More Practice: Try Problems 28.21, 28.22.

Problem 28.4 Two useful nucleoside drugs are cladribine and regadenoson. Cladribine [(part (a)], sold under the trade name Leustatin, was approved by the FDA in 2019 to treat multiple sclerosis, and regadenoson [part (b)], sold under the trade name Lexiscan, is used in imaging procedures to evaluate coronary artery disease. Use the given bases and monosaccharides to draw the structure of each drug.

a. 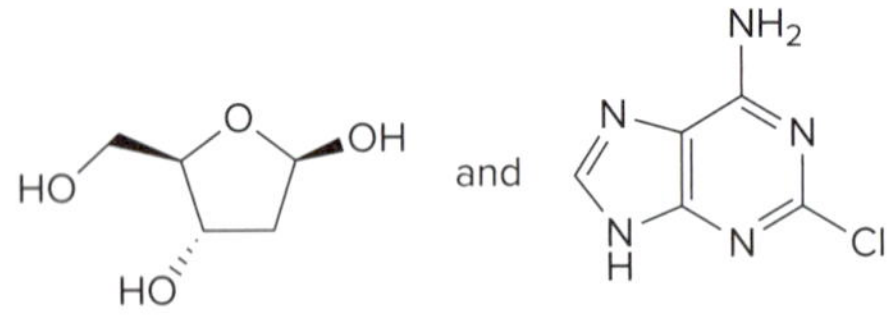

components of cladribine

b. 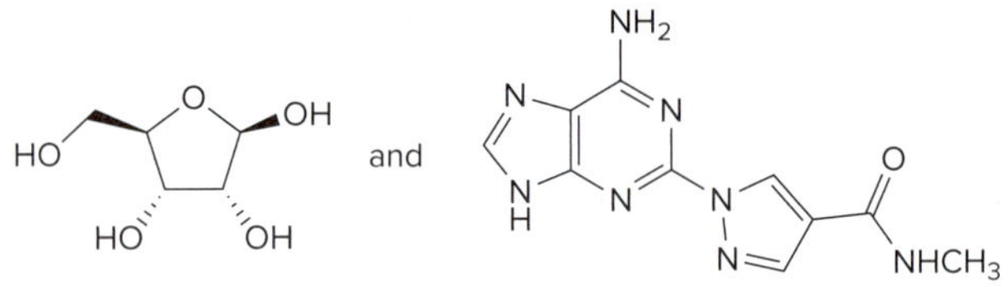

components of regadenoson

DNA molecules contain several million deoxyribonucleotides, whereas RNA molecules are smaller, containing perhaps a few thousand ribonucleotides. DNA is contained in the chromosomes of the nucleus, and the number of chromosomes differs from species to species. Humans have 46 chromosomes (23 pairs). An individual chromosome is composed of many genes. A **gene** is a portion of the DNA molecule responsible for the synthesis of a specific protein.

Problem 28.5 Which nucleic acid (DNA or RNA) contains each of the following components:
(a) ribose; (b) 2'-deoxyribose; (c) the base T; (d) the base U; (e) the nucleotide GMP;
(f) the nucleotide dCMP?

28.1B Nucleotide Diphosphates and Nucleotide Triphosphates

ATP was first introduced in Section 7.15A.

Di- and triphosphates can also be prepared from nucleosides by adding two and three phosphate groups, respectively, to the 5'-OH. For example adenosine can be converted to adenosine 5'-diphosphate (abbreviated as **ADP**) and adenosine 5'-triphosphate (abbreviated as **ATP**). The central role of these phosphates in energy production is discussed in Chapter 30.

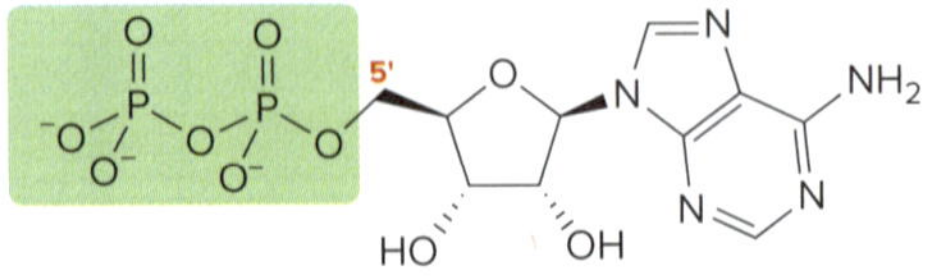

adenosine 5'-**di**phosphate
ADP

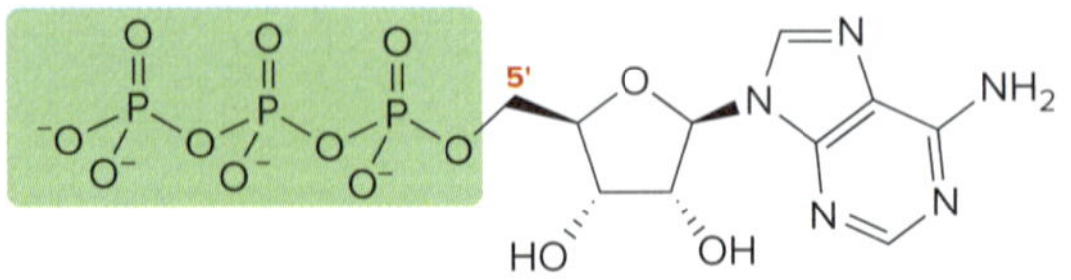

adenosine 5'-**tri**phosphate
ATP

Problem 28.6 Give the name that corresponds to each abbreviation: (a) GTP; (b) dCDP; (c) dTTP; (d) UDP.

28.2 Nucleic Acids

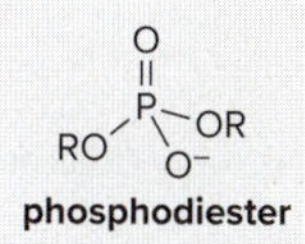

Nucleic acids—both DNA and RNA—are polymers of nucleotides, formed by joining the 3'-OH group of one nucleotide with the 5'-phosphate of a second nucleotide in a **phosphodiester** linkage.

For example, joining the 3'-OH group of dCMP and the 5'-phosphate of dAMP forms a **dinucleotide** that contains a 5'-phosphate on one end—the **5' end**—and a 3'-OH group on the other end—the **3' end.**

As additional nucleotides are added, the nucleic acid grows, each time forming a new phosphodiester linkage that holds the nucleotides together. **The primary structure of a polynucleotide is the sequence of nucleotides that it contains.** All polynucleotides contain a backbone of alternating sugar and phosphate groups. The identity and order of the bases distinguish one polynucleotide from another.

- **A polynucleotide has one free phosphate at the 5' end and a free OH group at the 3' end.**
- **A polynucleotide is named by the sequence of bases it contains, beginning with the 5' end and using the one-letter abbreviations for the bases.**

Figure 28.1 illustrates a polynucleotide that contains three phosphodiesters joining four different nucleotides.

Figure 28.1 Primary structure of a polynucleotide

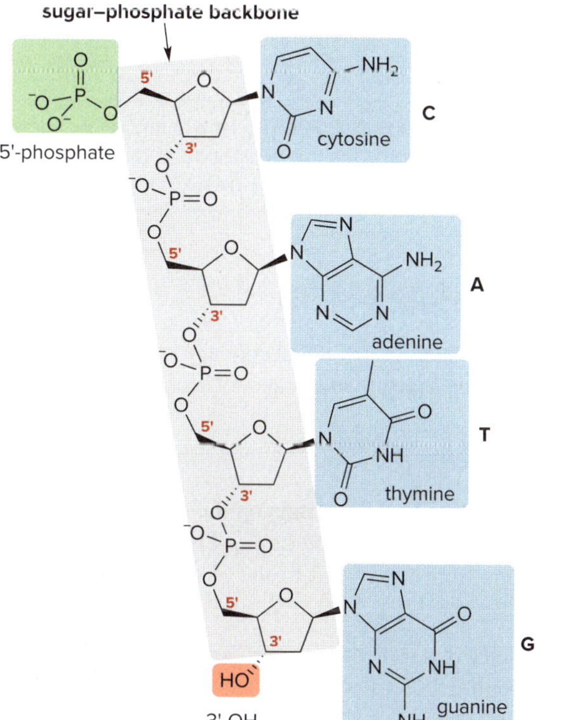

- A polynucleotide contains a backbone consisting of alternating sugar and phosphate groups, highlighted in gray.
- Phosphodiester bonds join the 3'-carbon of one nucleotide to the 5'-carbon of another.
- **A polynucleotide is named from the 5' end (in green) to the 3' end (in red),** using the one-letter abbreviations for the bases it contains.
- The structure of the polynucleotide CATG is drawn.

Sample Problem 28.2 Drawing the Structure of a Trinucleotide

Draw the structure of the trinucleotide AUG.

Solution

- Because the base U occurs only in **RNA,** draw the structure of the individual nucleotides using **D-ribose** (with an OH group at C2') as the monosaccharide.
- Abbreviations identify the bases of the trinucleotide in order from the 5' end to the 3' end: the nucleotide with the 5'-phosphate contains adenine (A) and the nucleotide with the 3'-OH group contains guanine (G).
- Join the nucleotides together with phosphodiester bonds between the 3'-OH groups and 5'-phosphates to form two phosphodiesters.

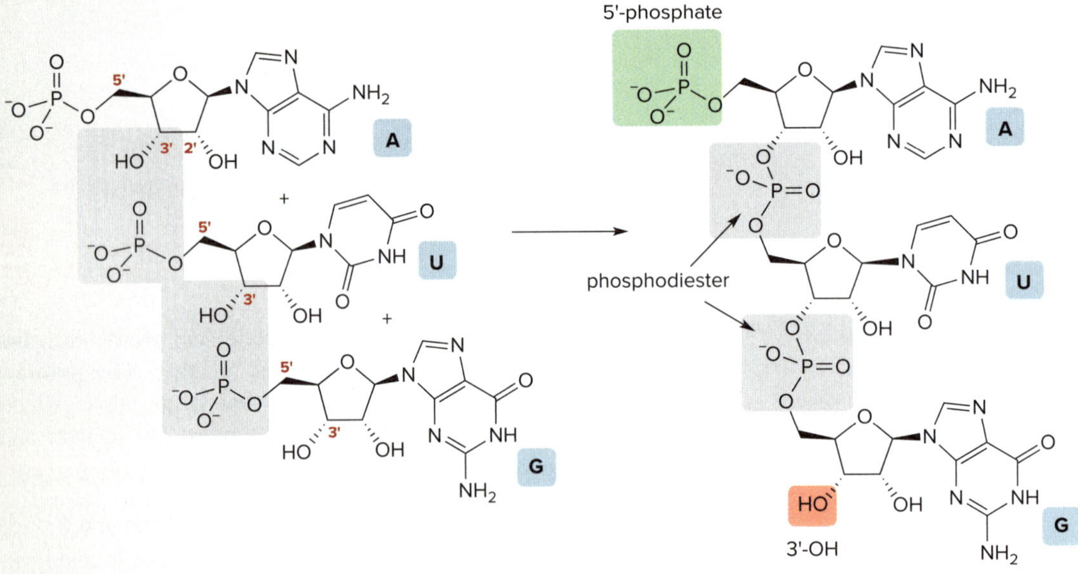

5'-phosphate

phosphodiester

3'-OH

trinucleotide AUG

Problem 28.7 Draw the structure of each polynucleotide: (a) CU; (b) TAG.

More Practice: Try Problems 28.30, 28.31.

Problem 28.8 Consider the polynucleotide ATGGCG. (a) How many phosphodiester linkages does the polynucleotide contain? (b) Does the nucleotide at the 5' end contain a purine or pyrimidine base? (c) Could the polynucleotide be part of a DNA or an RNA molecule?

28.3 The DNA Double Helix

Our current understanding of the secondary structure of DNA is based on the model initially proposed by James Watson and Francis Crick in 1953 (Figure 28.2).

- **DNA consists of two polynucleotide strands that wind into a right-handed double helix.**

The sugar–phosphate backbone runs on the outside of the helix and the bases lie on the inside, perpendicular to the axis of the helix. **The two strands of DNA are antiparallel;** that is, one strand runs from the 5' end to the 3' end, while the other strand runs from the 3' end to the 5' end.

The double helix is stabilized by hydrogen bonding between the bases of the two DNA strands as shown in Figure 28.3. A purine base on one strand always hydrogen bonds with a pyrimidine

Figure 28.2 The double helix of DNA

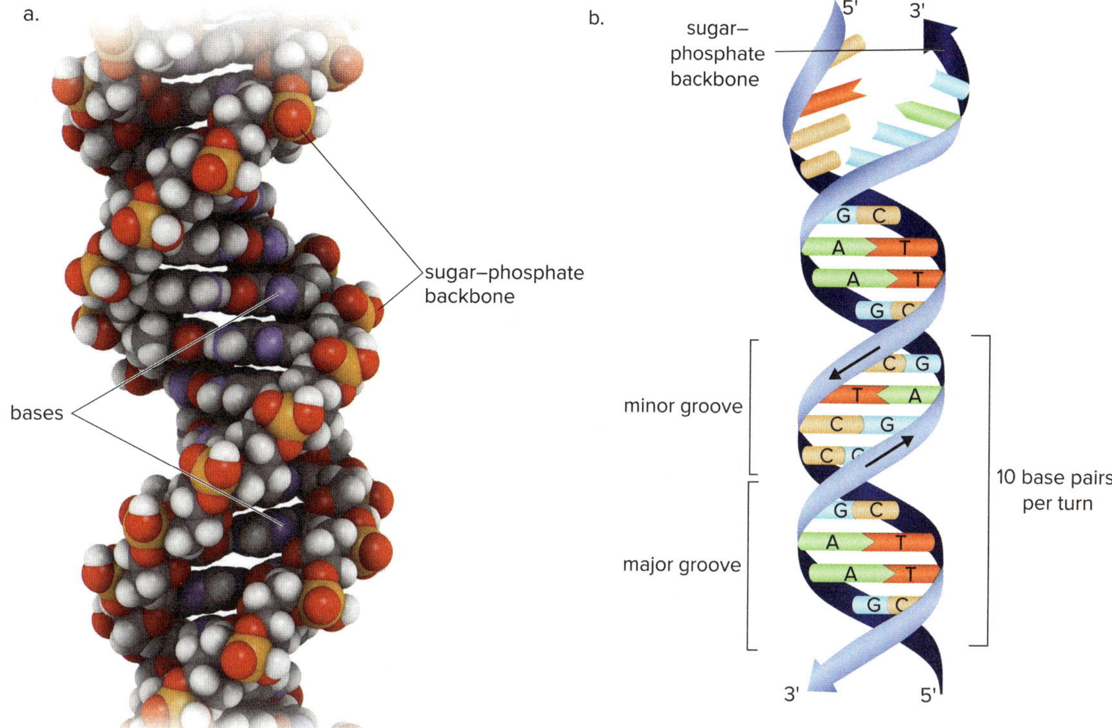

DNA consists of a double helix of polynucleotide chains. In view (a), the three-dimensional model shows the sugar–phosphate backbone visible on the outside of the helix. In the ribbon diagram in view (b), the bases in the interior are labeled, as are the major and minor grooves of the double helix.

Figure 28.3 Hydrogen bonding in the DNA double helix

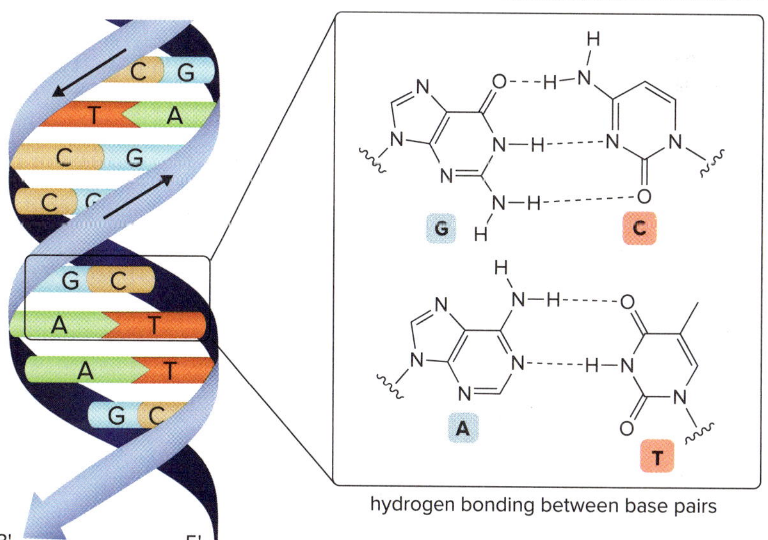

- Hydrogen bonding of base pairs (**A–T** and **G–C**) holds the two strands of DNA together.

base on the other strand. Two bases hydrogen bond in a predictable manner, forming **complementary base pairs.**

- **Adenine pairs with thymine using two hydrogen bonds, forming an A–T base pair.**
- **Cytosine pairs with guanine using three hydrogen bonds, forming a C–G base pair.**

The base pairs are stacked one on top of the other, with one complete turn of the helix containing 10 base pairs. The DNA double helix contains two grooves of different sizes, called the **major groove** and the **minor groove,** which run along the length of its cylindrical column (Figure 28.2b). Certain polycyclic aromatic compounds (Section 15.5) bind to the grooves in the DNA double helix.

Problem 28.9 Suggest reasons why the DNA double helix is arranged with the sugar–phosphate backbone on the outside of the double helix, and the base pairs on the inside.

Because of the consistent pairing of bases, we can write the sequence of the complementary strand of DNA when the sequence of one strand is known, as shown in Sample Problem 28.3.

Sample Problem 28.3 Predicting the Sequence of a Complementary Strand of a DNA Molecule

Write the sequence for the complementary strand of the following portion of a DNA molecule: 5'–TAGGCTA–3'.

Solution

The complementary strand runs in the opposite direction, from the 3' end to the 5' end. Use base pairing to determine the corresponding sequence on the complementary strand: A pairs with T and C pairs with G.

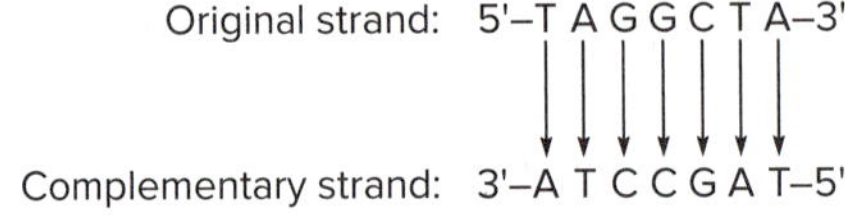

Problem 28.10 Write the complementary strand for each of the following strands of DNA.

a. 5'–AAACGTCC–3'
b. 5'–TATACGCC–3'
c. 5'–ATTGCACCCGC–3'
d. 5'–CACTTGATCGG–3'

More Practice: Try Problem 28.32.

Identical twins have the same genetic makeup, so that characteristics determined by DNA—such as hair color, eye color, or complexion—are also identical. *Daniel C. Smith*

The enormously large DNA molecules that compose the **human genome**—the total DNA content of an individual—pack tightly into the nucleus of the cell. **The genetic information of an organism is stored in the sequence of nucleotides in these DNA molecules.** How is this information transmitted from one generation to another? How, too, is the information in DNA molecules used to direct the synthesis of proteins?

To answer these questions, we must understand three key processes.

- ***Replication*—the process by which DNA makes a copy of itself when a cell divides.**
- ***Transcription*—the ordered synthesis of RNA from DNA. In this process, the genetic information stored in DNA is passed onto RNA.**
- ***Translation*—the synthesis of proteins from RNA. In this process, the genetic message contained in RNA determines the specific amino acid sequence of a protein.**

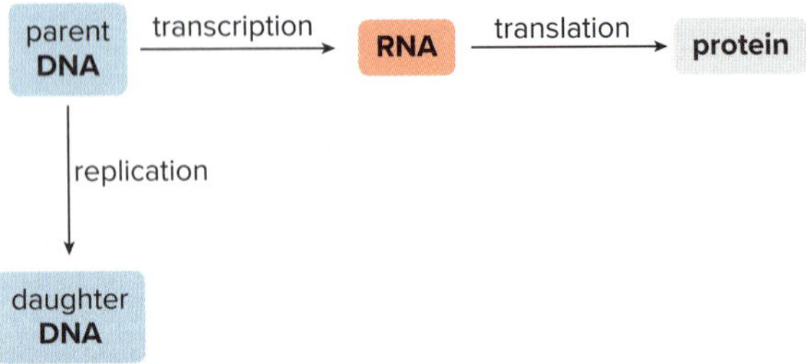

Each chromosome contains many **genes,** those portions of the DNA molecules that result in the synthesis of specific proteins. We say that the genetic message of the DNA molecule is *expressed* in the protein. Only a small fraction (1–2%) of the DNA in a chromosome contains genes that result in protein synthesis.

28.4 Replication

The genetic information in the DNA of a parent cell is passed onto daughter cells by the process of **semiconservative replication.** The strands of DNA separate and each serves as a template for a new strand. The original DNA molecule forms two DNA molecules, each of which contains one strand from the parent DNA and one new strand. The sequence of both strands of the daughter DNA molecules exactly matches the sequence of the parent DNA.

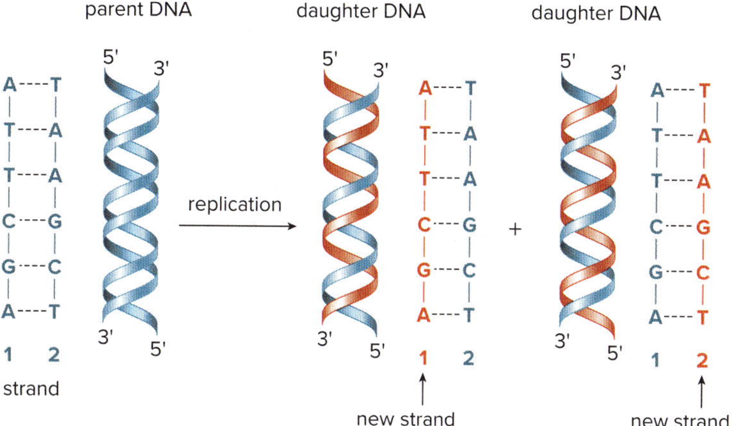

The first step in replication is the unwinding of the DNA helix to expose bases on each strand. Unwinding occurs at many places simultaneously along the helix, creating "bubbles" where replication can occur. Unwinding breaks the hydrogen bonds that hold the two strands of the double helix together (Figure 28.4).

Figure 28.4
DNA replication

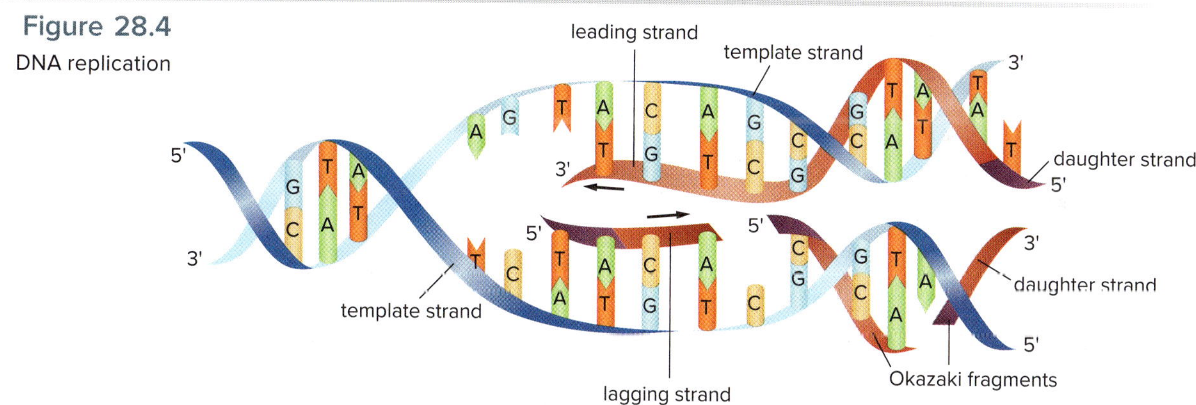

- Replication proceeds along both strands of the unwound DNA. Replication always occurs in the same direction, from the 3' end to the 5' end of the template. The leading strand grows continuously, whereas the lagging strand is synthesized in pieces that are then joined.

Once bases have been exposed in the unwound strands of DNA, the enzyme **DNA polymerase** catalyzes the replication process using the four nucleotide triphosphates (derived from the bases A, T, G, and C). A new phosphodiester bond is formed between the 5'-phosphate of the nucleoside triphosphate and the 3'-OH group of the new DNA strand. Two new strands of DNA grow from the ends of bubbles, called the replication forks.

- **The identity of the bases on the template strand determines the order of bases in the new strand: A must pair with T, and G must pair with C.**
- **Replication occurs in only *one* direction on the template strand, from the 3' end to the 5' end, so the newly synthesized DNA grows from its 5' end to its 3' end.**

Because replication proceeds in only one direction, the two new strands of DNA must be synthesized by different techniques. One strand, the **leading strand,** grows continuously from the 5' end to the 3' end, adding bases that are complementary to the template strand. The other strand, the **lagging strand,** is synthesized in small pieces called **Okazaki fragments,** which are joined together by a **DNA ligase** enzyme. The end result is two new strands of DNA, one in each of the daughter DNA molecules, both with complementary base pairs joining the two DNA strands together.

Problem 28.11 What is the sequence of a newly synthesized DNA segment if the template strand has each of the following sequences?

a. 3'–AGAGTCTC–5'
b. 3'–ATTGCTC–5'
c. 3'–ATCCTGTAC–5'
d. 3'–GGCCATACTC–5'

28.5 Ribonucleic Acids and Transcription

28.5A RNA

Ribonucleic acids (RNAs) are composed of nucleotides, but there are significant differences between DNA and RNA. In RNA,

- **The sugar is D-ribose.**
- **Uracil (U) replaces thymine (T) as one of the bases.**
- **RNA is single stranded.**

Although RNA molecules are much smaller than DNA molecules, a single strand of RNA can fold back on itself, forming loops and helical regions that are stabilized by intramolecular hydrogen bonding.

Three different types of RNA are involved in protein synthesis: **ribosomal RNA (rRNA), messenger RNA (mRNA),** and **transfer RNA (tRNA).**

Ribosomal RNA, the most abundant type of RNA, is found in the ribosomes in the cytoplasm of the cell. rRNA provides the site where polypeptides are assembled during protein synthesis.

Messenger RNA is the carrier of information from DNA in the nucleus to the ribosomes in the cytoplasm. Each gene of a DNA molecule corresponds to a specific mRNA molecule. **The sequence of nucleotides in the mRNA molecule determines the amino acid sequence in a particular protein.**

Transfer RNA interprets the genetic information in mRNA and brings specific amino acids to the site of protein synthesis in the ribosome. Each tRNA contains a sequence of three nucleotides called an **anticodon,** which is complementary to three bases in an mRNA molecule, and identifies what amino acid must be added to a growing polypeptide chain. tRNA molecules are often drawn in a cloverleaf fashion (Figure 28.5a). A model that depicts the three-dimensional structure of a tRNA is shown in Figure 28.5b. A particular amino acid may be recognized by one or more tRNA molecules.

Each tRNA also has an acceptor stem at the 3' end that always contains the nucleotides ACC (also shown in Figure 28.5a). The free 3'-OH group at this end is esterified with the α-carboxy group of a specific amino acid.

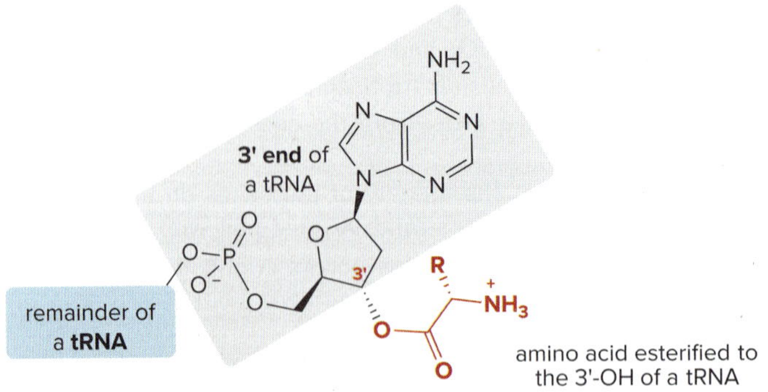

tRNA bonded to an amino acid

Figure 28.5 Transfer RNA

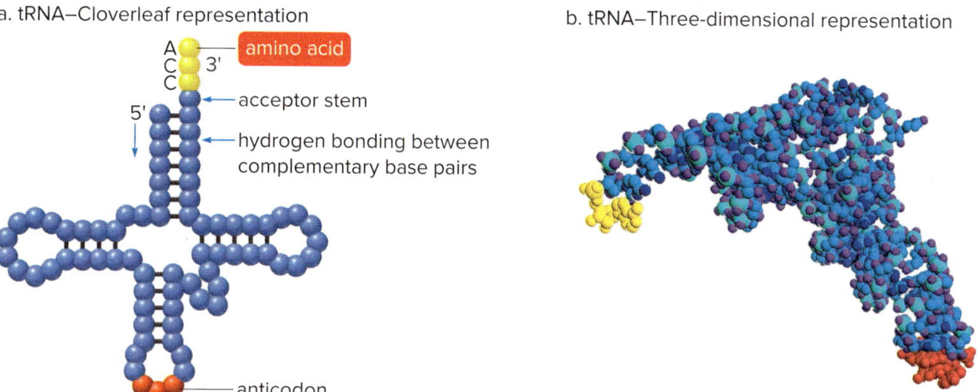

- tRNAs contain 70–90 nucleotides. Folding of the tRNA molecule creates regions in which complementary base pairs hydrogen bond to each other. Each tRNA binds a specific amino acid to its 3' end and contains an anticodon that identifies that amino acid for protein synthesis.

- In the three-dimensional model of a tRNA, the binding site for the amino acid is shown in yellow and the anticodon is shown in red. *Kenneth Edward/Science Source*

28.5B Transcription

The conversion of the information in DNA to the synthesis of proteins begins with ***transcription*—the synthesis of mRNA from DNA.**

RNA synthesis begins when the double helix of DNA unwinds, and a complementary strand of mRNA is synthesized from one strand of DNA, called the **template strand.** The strand of DNA not used for mRNA synthesis is called the **coding strand.**

Transcription proceeds from the 3' end to the 5' end of the template strand using an RNA polymerase enzyme (Figure 28.6). Complementary base pairing determines the order of RNA

Figure 28.6 Transcription of DNA to RNA

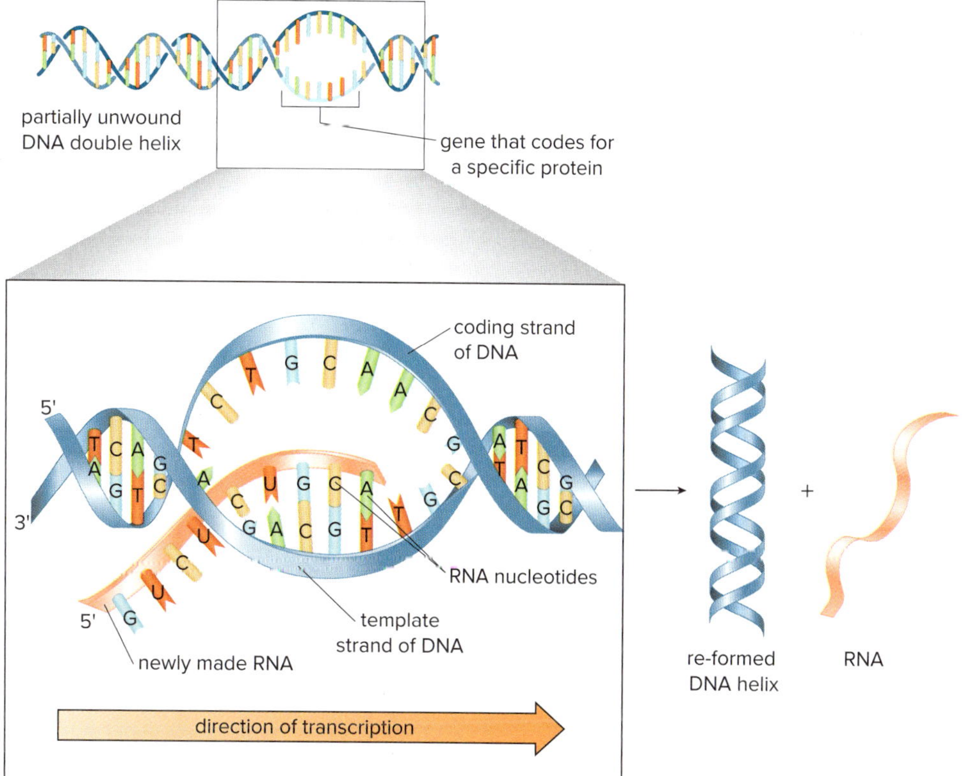

- Transcription proceeds from the 3' end to the 5' end of the template strand, so the mRNA bases are complementary to those in the DNA template.

nucleotides added to the growing RNA chain: **C pairs with G, T pairs with A, and A pairs with U.** Transcription is completed when a particular sequence of bases on the DNA template is reached. The new mRNA molecule is released and the double helix of the DNA molecule is re-formed.

In bacteria, the new mRNA molecule is ready for protein synthesis after it is prepared. In humans, the mRNA molecule first formed is modified before it is ready for protein synthesis by removing and splicing together pieces of mRNA by mechanisms that are not presented here.

- **mRNA has a sequence *complementary* to the DNA template strand from which it is prepared.**
- **mRNA is an *exact copy* of the coding strand of DNA, except that the base U replaces T on the mRNA strand.**

Sample Problem 28.4 Determining the Sequence of mRNA from DNA

Write the sequence of mRNA formed from the following template strand of DNA: 3'–CTAGGATAC–5'. Write the sequence of the coding strand of this segment of DNA.

Solution

mRNA has a base sequence that is *complementary* to the template from which it is prepared. On the other hand, mRNA has a base sequence that is *identical* to the coding strand of DNA, except that it contains the base U instead of T.

Template strand of DNA: 3'–C T A G G A T A C–5'

mRNA sequence: 5'–G A U C C U A U G–3'

Coding strand of DNA: 5'–G A T C C T A T G–3'

(complementary; complementary)

Problem 28.12 For each DNA segment: [1] What is the sequence of the mRNA molecule synthesized from each DNA template? [2] What is the sequence of the coding strand of the DNA molecule?

a. 3'–TGCCTAACG–5'
b. 3'–GACTCC–5'
c. 3'–TTAACGCGA–5'
d. 3'–CAGTGACCGTAC–5'

More Practice: Try Problems 28.37.

Problem 28.13 What is the sequence of the DNA template strand from which each of the following mRNA strands was synthesized?

a. 5'–UGGGGCAUU–3'
b. 5'–GUACCU–3'
c. 5'–CCGACGAUG–3'
d. 5'–GUAGUCACG–3'

28.6 The Genetic Code, Translation, and Protein Synthesis

28.6A The Genetic Code

How can the four different nucleotides in mRNA direct the synthesis of proteins that are formed from 20 amino acids? The answer lies in the **genetic code.**

- **The genetic code is the set of three-nucleotide units in mRNA called *codons* that correspond to particular amino acids. As a result, a series of codons in mRNA determines the amino acid sequence in a protein.**

For example, the codon UCA in mRNA codes for the amino acid serine, whereas the codon UGC codes for cysteine. The same genetic code occurs in almost all organisms, from bacteria to whales to humans.

Given four different nucleotides (A, C, G, and U), there are 64 different ways to combine them into groups of three, so there are 64 different codons. Sixty-one codons code for specific amino acids, so many amino acids correspond to more than one codon, as shown in Table 28.2. For example, GGU, GGC, GGA, and GGG all code for the amino acid glycine. Three codons—UAA, UAG, and UGA—do not correspond to any amino acids; they are called **stop codons** because they signal the stop of protein synthesis.

Table 28.2 The Genetic Code—Triplets in Messenger RNA

First Base (5' end)	Second Base								Third Base (3' end)
	U		C		A		G		
U	UUU	Phe	UCU	Ser	UAU	Tyr	UGU	Cys	U
	UUC	Phe	UCC	Ser	UAC	Tyr	UGC	Cys	C
	UUA	Leu	UCA	Ser	UAA	Stop	UGA	Stop	A
	UUG	Leu	UCG	Ser	UAG	Stop	UGG	Trp	G
C	CUU	Leu	CCU	Pro	CAU	His	CGU	Arg	U
	CUC	Leu	CCC	Pro	CAC	His	CGC	Arg	C
	CUA	Leu	CCA	Pro	CAA	Gln	CGA	Arg	A
	CUG	Leu	CCG	Pro	CAG	Gln	CGG	Arg	G
A	AUU	Ile	ACU	Thr	AAU	Asn	AGU	Ser	U
	AUC	Ile	ACC	Thr	AAC	Asn	AGC	Ser	C
	AUA	Ile	ACA	Thr	AAA	Lys	AGA	Arg	A
	AUG	Met	ACG	Thr	AAG	Lys	AGG	Arg	G
G	GUU	Val	GCU	Ala	GAU	Asp	GGU	Gly	U
	GUC	Val	GCC	Ala	GAC	Asp	GGC	Gly	C
	GUA	Val	GCA	Ala	GAA	Glu	GGA	Gly	A
	GUG	Val	GCG	Ala	GAG	Glu	GGG	Gly	G

A codon is written from the 5' to 3' end of mRNA. **The 5' end of the mRNA molecule codes for the N-terminal amino acid** in a protein, whereas **the 3' end of the mRNA codes for the C-terminal amino acid.**

Problem 28.14 Consider the following mRNA sequence: 5'–CAUAAAACGGAG–3'. (a) What is the N-terminal amino acid coded for by this sequence? (b) What is the C-terminal amino acid?

Problem 28.15 Sometimes codons for amino acids with similar types of side chains (i.e., acidic, basic, hydrophobic, or aromatic) have similarities. Compare the codons for the amino acids aspartic acid, glutamic acid, leucine, phenylalanine, and valine. Comment on the relationship between amino acid structure and codon identity.

28.6B Translation

The translation of the information in mRNA to protein synthesis occurs in the ribosomes at binding sites on rRNA.

mRNA contains the sequence of codons that determines the order of amino acids in the protein. Individual tRNAs bring specific amino acids to add to the peptide chain. Each tRNA contains an **anticodon** of three nucleotides that is complementary to the codon in mRNA and identifies individual amino acids. For example, a codon of UCA in mRNA corresponds to an anticodon of AGU in a tRNA molecule, which identifies serine as the amino acid.

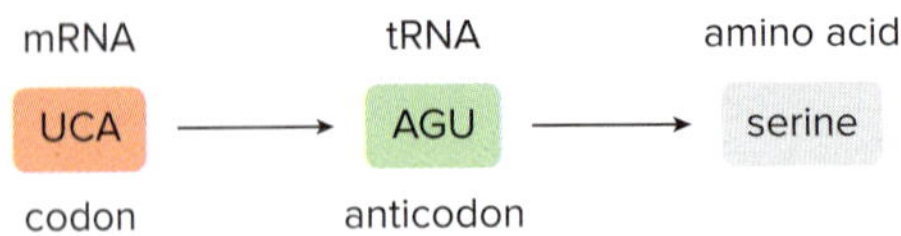

Translation begins when the first codon of an mRNA molecule binds to a ribosome, and a tRNA molecule, which contains the anticodon of the codon, carries the first amino acid of the peptide chain to the binding site. As mentioned in Section 28.5A, each tRNA is esterified to an individual amino acid. The new peptide bond is formed by **nucleophilic acyl substitution** of the amino group of one tRNA-bonded amino acid with the ester carbonyl of another, as shown in Figure 28.7.

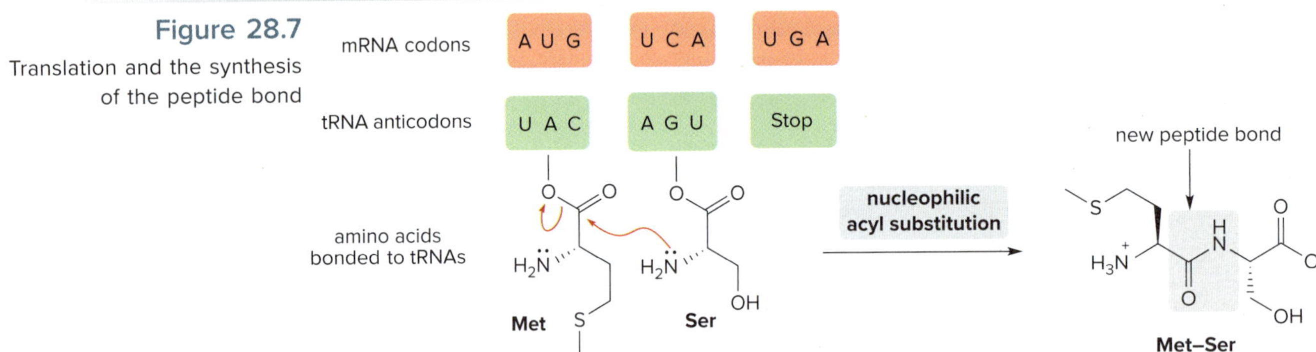

Figure 28.7 Translation and the synthesis of the peptide bond

- mRNA contains the codons that determine the sequence of amino acids of a peptide. Each tRNA is bonded to a specific amino acid and contains an anticodon that binds to the mRNA. A peptide bond forms between two amino acids by nucleophilic acyl substitution, and the peptide chain grows until a stop codon is reached. Depicted is the synthesis of a methionine–serine dipeptide.

As each successive codon on the mRNA is read, new tRNAs deliver the next amino acids, peptide bonds are formed, and the protein chain grows until a stop codon signals that synthesis is complete.

Figure 28.8 shows a representative segment of DNA, and the mRNA, tRNA, and amino acid sequences that correspond to it.

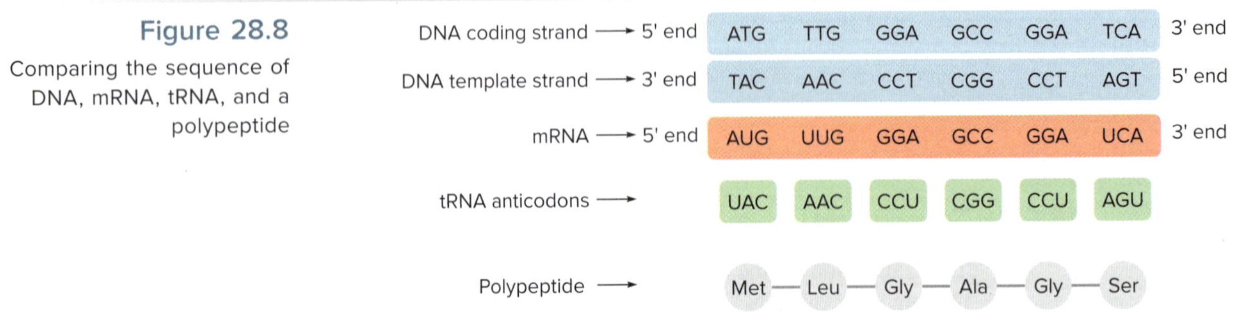

Figure 28.8 Comparing the sequence of DNA, mRNA, tRNA, and a polypeptide

Sample Problem 28.5 Deriving an Amino Acid Sequence from DNA

What polypeptide would be synthesized from the following template strand of DNA:

3'–CGGTGTCTTTTA–5'?

Solution

To determine what polypeptide is synthesized from a DNA template, two steps are needed.

- Use the DNA sequence to determine the transcribed mRNA sequence: C pairs with G, T pairs with A, and A (on DNA) pairs with U (on mRNA).
- Use the codons in Table 28.2 to determine what amino acids are coded for by a given codon in mRNA.

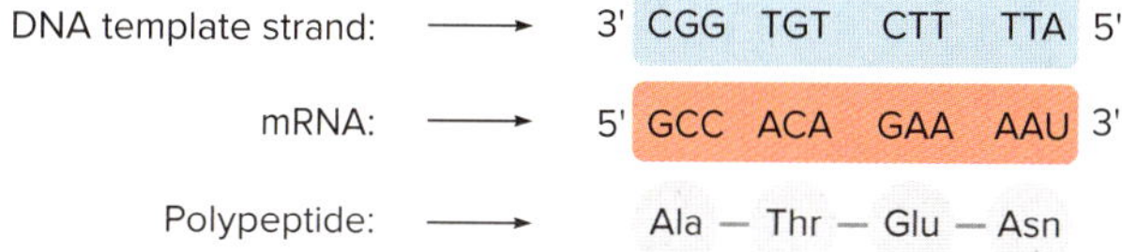

Problem 28.16 What polypeptide would be synthesized from each of the following template strands of DNA?

a. 3'–TCTCATCGTAATGATTCG–5' b. 3'–GCTCCTAAATAACACTTA–5'

More Practice: Try Problems 28.40, 28.44.

Problem 28.17 What sequence of amino acids would be formed from each mRNA sequence? List the anticodons contained in each of the needed tRNA molecules.

a. 5'–CCACCGGCAAACGAAGCA–3'
b. 5'–GCACCACUAAGAGAC–3'

Problem 28.18 Consider a template strand of DNA with the following sequence: 3'–ATGAAAGCCTTCTGT–5'. (a) What is the coding strand of DNA that corresponds to this template? (b) What mRNA is prepared from this template? (c) What polypeptide is prepared from the mRNA?

Problem 28.19 Fill in the base, codon, anticodon, or amino acid needed to complete the following table that relates the sequences of DNA, mRNA, tRNA, and the resulting polypeptide.

DNA coding strand:	5' end	AAC						3' end
DNA template strand:	3' end		CAT					5' end
mRNA codons:	5' end			UCA			AUG	3' end
tRNA anticodons:						GUG		
Polypeptide:					Thr			

28.7 DNA Sequencing

DNA sequencing has proven to be valuable methodology for determining the sequence of specific genes, individual chromosomes, and even the full genome of an organism. Determining the structure of genes that are associated with specific diseases has allowed scientists to understand how to prevent or cure them.

Because of the large size of DNA molecules, DNA is first cleaved into smaller units and the smaller fragments of DNA are then sequenced individually. Cleavage is carried out with **restriction endonucleases,** enzymes that cleave DNA at specific sequences of bases. Each restriction endonuclease recognizes a particular sequence of bases and cuts *both* strands of DNA in an identical manner.

For example, the enzyme EcoRI recognizes the sequence GAATTC and cuts the DNA molecule between G and A on both strands. The enzyme SmaI, on the other hand, recognizes the sequence CCCGGG and cuts the molecule between C and G.

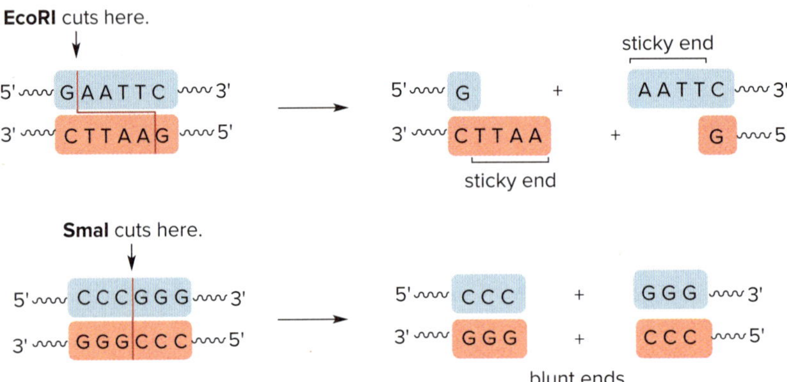

Cleavage with EcoRI affords strands of DNA of different length, with **sticky ends** that have unpaired bases, whereas cleavage with SmaI affords DNA fragments with **blunt ends.** Thousands of restriction enzymes are known and hundreds are commercially available. By cleaving DNA with a variety of restriction endonucleases, sequencing the fragments, and determining overlapping sequences, the DNA sequence of long strands of DNA has been determined.

Problem 28.20 Two other restriction endonucleases are HindIII, which cuts DNA between A and A in the sequence AAGCTT, and HaeIII, which cuts DNA between G and C in the sequence GGCC. Label the cleavage sites in the following segment of DNA. Only one strand of double-stranded DNA is provided.

5'–CGCGAATTGGCCGTAAGCTTACGTCCTAGGGCTACTCCTCGGCCCAATAAAGCTT–3'

In 1980 Frederick Sanger and Walter Gilbert shared the Nobel Prize in Chemistry for the development of methods to sequence DNA.

Early methods of DNA sequencing were developed in the 1970s by Frederick Sanger in Cambridge, England, and Walter Gilbert of Harvard University. Sanger sequencing was the most common method of DNA sequencing for 20 years, and early automated DNA sequencers were based on this technology. DNA sequencing methods have been used to sequence the entire human genome, which consists of 3.1 billion base pairs. It was first reported in preliminary form in 2001 and completed in 2003.

Next-generation DNA synthesizers have been developed since 2000, which have increased the speed and decreased the cost of DNA sequencing. As a comparison, the U.S. government spent \$2.7 billion on the Human Genome Project to sequence the human genome from 1990 to 2003. This figure includes the total cost of all activities related to the Human Genome Project, including technology development, ethics research, and program management, as well as determining the framework for organizing the data obtained from sequencing individual segments of DNA. It is estimated that sequencing itself cost somewhere between \$500 million and \$1 billion.

Using the sophisticated technology available today, as well as the competitive pricing offered by several commercial enterprises, the National Human Genome Research Institute estimated that in 2016, the DNA sequence of an organism could be obtained for under \$1000.

28.8 The Polymerase Chain Reaction

In order to study a specific gene, millions of copies of pure gene are needed. In fact, virtually an unlimited number of copies of any gene can be synthesized in just a few hours using a technique called the **polymerase chain reaction (PCR).** PCR *clones* a segment of DNA; that is, PCR produces exact copies of a fragment of DNA.

- **PCR *amplifies* a specific portion of a DNA molecule, producing millions of copies of a single molecule.**

PCR was developed by biochemist Kary Mullis of Cetus Corporation, who shared the 1993 Nobel Prize in Chemistry for its discovery.

Four elements are needed to amplify DNA by PCR:

- **The segment of DNA** that must be copied
- **Two primers**—short polynucleotides that are complementary to the two ends of the segment to be amplified
- **A DNA polymerase enzyme** that will catalyze the synthesis of a complementary strand of DNA from a template strand
- **Nucleoside triphosphates** that serve as the source of the nucleotides A, T, C, and G needed in the synthesis of the new strands of DNA

Each cycle of the polymerase chain reaction involves three steps, illustrated in Figure 28.9.

Figure 28.9 Using the polymerase chain reaction to amplify a sample of DNA

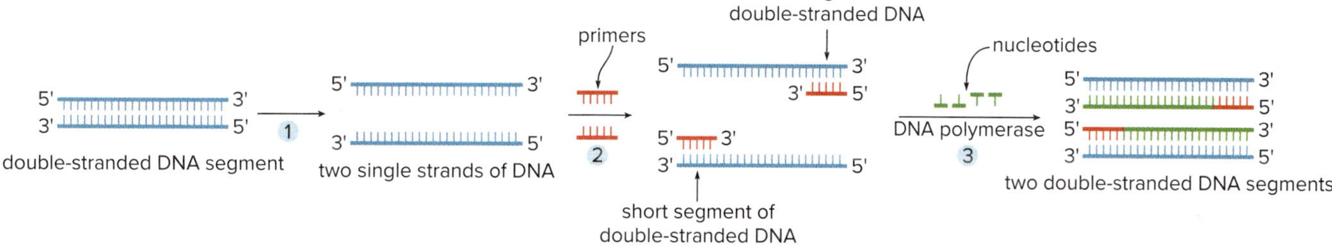

1. The DNA sample is heated to unwind the double helix into two single strands.
2. Primers are added to form a short segment of double-stranded DNA on each strand, to which a DNA polymerase enzyme can add new nucleotides to the 3' end.
3. DNA polymerase is used to add nucleotides to lengthen the DNA segment.
   - DNA polymerase catalyzes the synthesis of a new strand of DNA complementary to the existing strand using nucleoside triphosphates (dATP, dCTP, dGTP, and dTTP) available in the reaction mixture.
   - After one three-step cycle, *one* molecule of double-stranded DNA forms *two* molecules of double-stranded DNA.

Each double-stranded DNA molecule synthesized by this method contains one original strand and one newly synthesized strand. After each cycle, the amount of DNA doubles. After 20 cycles, about one million copies have been made.

Each step of a PCR cycle is carried out at a different temperature. PCR is now a completely automated process using a thermal cycler, an apparatus that controls the heating and cooling needed for each step. A heat-tolerant DNA polymerase called **Taq polymerase** is also typically used, so that new enzyme need not be added as each new cycle begins.

PCR is an indispensable method in clinical chemistry and forensic analysis. PCR is used in diagnosing genetic diseases and in determining paternity. Forensic scientists compare DNA collected from a crime scene with those of a suspect by cleaving DNA from both sources with restriction endonucleases, which are then amplified using the polymerase chain reaction.

28.9 Viruses

A *virus* is an infectious agent consisting of a DNA or RNA molecule that is contained within a protein coating. Because a virus has no enzymes or free nucleotides of its own, it is incapable of replicating until it invades a host organism and takes over the biochemical machinery of the host.

A virus that contains DNA uses the materials in the host organism to replicate DNA, transcribe DNA to RNA, and synthesize a protein coating, thus forming new virus particles that can infect new host cells. The common cold, influenza, and herpes are viral in origin.

Widely used vaccines contain an inactive form of a virus that causes an individual's immune system to produce antibodies to the virus to ward off infection. Many childhood diseases that were once very common, including mumps, measles, and chickenpox, are now prevented by vaccination. Polio has been almost completely eradicated, even in remote areas worldwide, by vaccination.

28.9A Coronaviruses

Coronaviruses are a group of viruses that cause a variety of diseases that include the common cold, SARS (severe acute respiratory syndrome), and MERS (Middle East respiratory syndrome). At the end of 2019, a novel coronavirus, **SARS-CoV-2,** was first reported to cause an unusual viral pneumonia known as **COVID-19,** coronavirus disease 2019. Because the disease is highly transmissible and can be spread by an individual before symptoms are evident, SARS-CoV-2 has led to a global pandemic, resulting in the infection and death of millions worldwide.

SARS-CoV-2 is a **retrovirus,** a virus that contains a core of RNA along with proteins that enable it to replicate. Surrounding the inner core is a viral envelope composed of a lipid bilayer, and projecting from the viral core are club-like spike proteins that allow the virus to attach to and thus infect cells of the host (Figure 28.10).

Figure 28.10 Schematic of SARS-CoV-2, the retrovirus that causes COVID-19

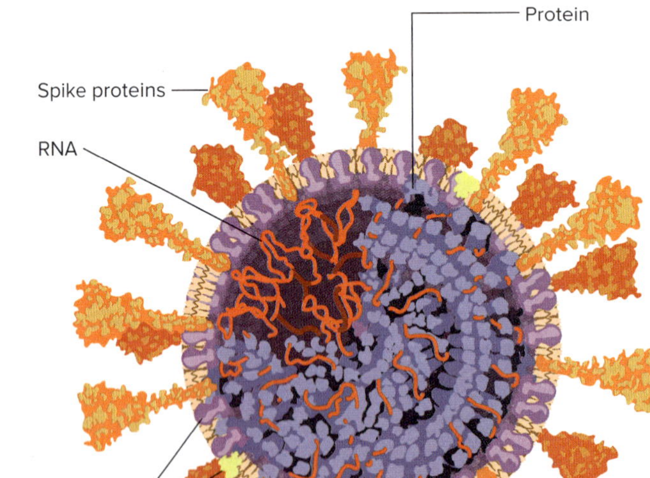

- SARS-CoV-2 consists of an RNA and protein core, surrounded by a lipid bilayer. The spike proteins that project from the surface latch onto receptors in human lung cells and move inside the cell.

Once a retrovirus invades a host organism, it must first make DNA by the process of **reverse transcription.** After viral DNA has been synthesized, the DNA can transcribe RNA, which can direct protein synthesis. New retrovirus particles are thus prepared and released to infect other cells.

viral RNA —(1) reverse transcription→ viral DNA —(2) transcription→ viral RNA —(3) translation→ viral proteins

The first vaccines approved to fight COVID-19 are unlike vaccines used to fight many childhood diseases, which use an inactive form of a virus. COVID-19 vaccines contain synthetic mRNA encapsulated in lipid nanoparticles. These mRNA vaccines instruct cells to make unusual proteins that the immune system does not recognize, resulting in an immune response that protects the body against infection.

28.9B Human Immunodeficiency Virus (HIV)

HEALTH NOTE

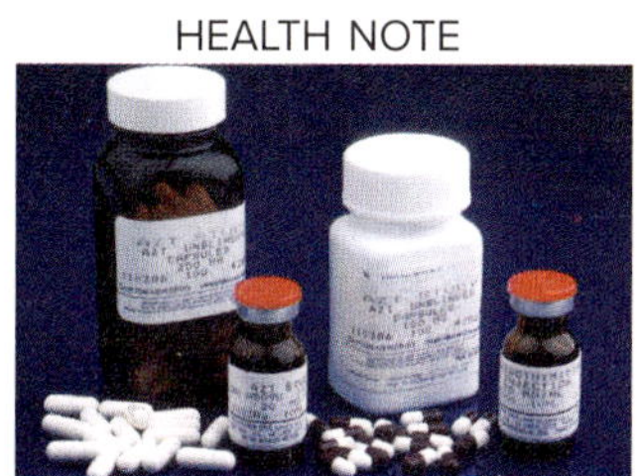

AZT, also called zidovudine and originally sold under the trade name Retrovir, has been available since the 1990s for the treatment of HIV. *Bill Branson/Jim Minor, NIH Pharmacy/National Cancer Institute (NCI)*

AIDS (acquired immune deficiency syndrome) is caused by HIV (human immunodeficiency virus), a retrovirus that attacks lymphocytes central to the body's immune response against invading organisms. As a result, an individual infected with HIV becomes susceptible to life-threatening bacterial infections. HIV is spread by direct contact with the blood or other body fluids of an infected individual.

HIV is often treated with a "cocktail" of drugs designed to destroy the virus at different stages of its reproductive cycle. One group of drugs, the protease inhibitors such as amprenavir (Section 23.10B), acts as enzyme inhibitors that prevent viral RNA from synthesizing needed proteins.

Other drugs are designed to interfere with reverse transcription, an essential biochemical process unique to the virus. Two drugs in this category are **AZT** (azidodeoxythymidine) and **ddI** (dideoxyinosine). The structure of each drug closely resembles the nucleotides that must be incorporated in viral DNA, and thus they are inserted into a growing DNA strand. Each drug lacks a hydroxy group at the 3' position, however, so no additional nucleotide can be added to the DNA chain, thus halting DNA synthesis.

HO, O, N, NH, O, O, 3', N_3

azidodeoxythymidine
AZT

HO, O, N, N, O, 3', N, NH

dideoxyinosine
ddI

Problem 28.21 Lamivudine is an antiviral drug formed from heterocycle **A** and cytosine. Draw the structure of lamivudine and explain why it is an effective antiviral agent.

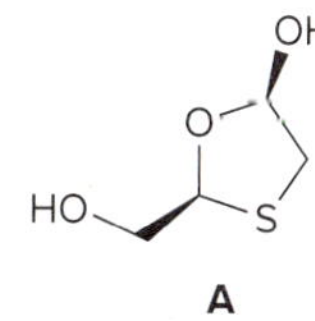

A

Chapter 28 REVIEW

KEY CONCEPTS

A comparison of DNA and RNA (28.1, 28.3, 28.5)

Nucleic acid	DNA	RNA
1 Structure	• DNA is composed of a right-handed **double helix with two strands** of deoxyribonucleotides winding in an antiparallel fashion.	• RNA contains a **single strand** of ribonucleotides.
2 Monosaccharide	• The monosaccharide component of DNA is **2'-deoxy-D-ribose.** 2'-deoxy-D-ribose	• The monosaccharide component of RNA is **D-ribose**. D-ribose
3 Bases	• DNA contains the bases A, G, C, and **T.** **thymine T** **cytosine, C** **adenine, A** **guanine, G**	• RNA contains the bases A, G, C, and **U.** **uracil U** **cytosine, C** **adenine, A** **guanine, G**

KEY SKILLS

[1] Drawing the structure of a dinucleotide (28.1, 28.2); example: TC

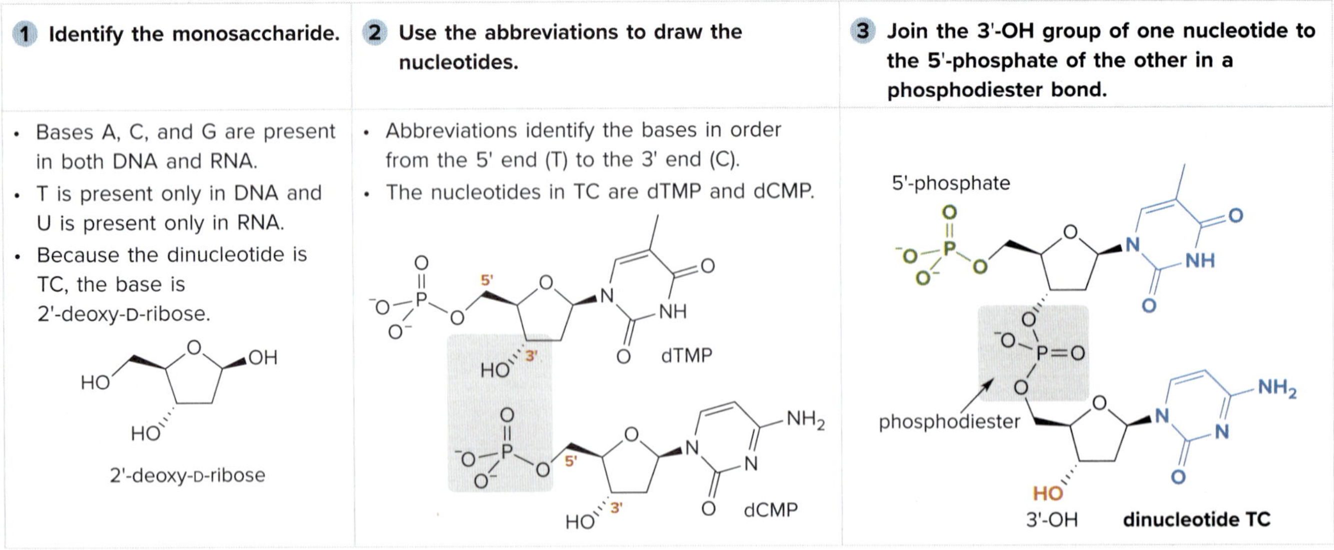

1 Identify the monosaccharide.	2 Use the abbreviations to draw the nucleotides.	3 Join the 3'-OH group of one nucleotide to the 5'-phosphate of the other in a phosphodiester bond.
• Bases A, C, and G are present in both DNA and RNA. • T is present only in DNA and U is present only in RNA. • Because the dinucleotide is TC, the base is 2'-deoxy-D-ribose.	• Abbreviations identify the bases in order from the 5' end (T) to the 3' end (C). • The nucleotides in TC are dTMP and dCMP.	

See Sample Problem 28.2. Try Problems 28.30, 28.31.

[2] Drawing the sequence of DNA bases from an original strand of DNA (28.3, 28.4)

1 Predicting the sequence of a complementary strand of DNA	**2 Drawing the new strand of DNA formed during replication**
• Write the original segment of DNA from the 5' end to the 3' end. • Use base pairing to write the complementary strand from the 3' end to the 5' end: **A** pairs with **T** and **G** pairs with **C.** Original strand: 5'–A T C C G T G T A–3' ↓↓↓↓↓↓↓↓↓ Complementary strand: 3'–T A G G C A C A T–5'	• Write the original segment of DNA from the 5' end to the 3' end. • Use base pairing to write the sequence of the segment formed after replication: **A** pairs with **T** and **G** pairs with **C.** Original strand: 5'–G C G A T T C C G T–3' ↓↓↓↓↓↓↓↓↓↓ Complementary strand: 3'–C G C T A A G G C A–5'

See Sample Problem 28.3. Try Problems 28.32, 28.36.

[3] Using a DNA template strand to determine an mRNA sequence (after transcription) and the sequence of the coding DNA strand (28.5); example: a DNA template strand with the sequence 3'–AGTATGACG–5'

1 Use base pairing to write the sequence of the mRNA segment formed after transcription.	**2 Use base pairing to write the sequence of the coding strand of DNA.**
• Write the complementary strand from the 5' end to the 3' end. • **G** pairs with **C, T** pairs with **A,** and **A** (on DNA) pairs with **U** (on RNA). DNA template strand: 3'–A G T A T G A C G–5' ↓↓↓↓↓↓↓↓↓ mRNA sequence: 5'–U C A U A C U G C–3'	• Write the coding strand from the 5' end to the 3' end. • **G** pairs with **C,** and **T** pairs with **A.** • The coding strand is identical to the mRNA strand except that **T** is present instead of **U.** DNA template strand: 3'–A G T A T G A C G–5' ↓↓↓↓↓↓↓↓↓ DNA coding strand: 5'–T C A T A C T G C–3'

See Sample Problem 28.4. Try Problem 28.37.

[4] Deriving an amino acid sequence from DNA (28.6); example: 3'–CCGTATCTT–5'

1 Use the DNA sequence to determine the transcribed mRNA sequence.	**2 Use the codons in Table 28.2 to determine what amino acids are coded for by a given codon in mRNA.**
• **G** pairs with **C, T** pairs with **A,** and **A** (on DNA) pairs with **U** (on RNA). DNA template strand: 3'– CCG TAT CTT –5' mRNA: 5'– GGC AUA GAA –3' Triplets correspond to RNA codons.	mRNA: 5'– GGC AUA GAA –3' ↓ ↓ ↓ **Polypeptide:** Gly – Ile – Glu

See Sample Problem 28.5. Try Problems 28.40–28.44.

CHAPTER 28 MULTIPLE-CHOICE SELF-TEST

The Self-Test consists of multiple-choice questions similar to those found on the American Chemical Society organic chemistry exam. Answers are given at the end of the chapter.

1. What is the correct structure for CMP?

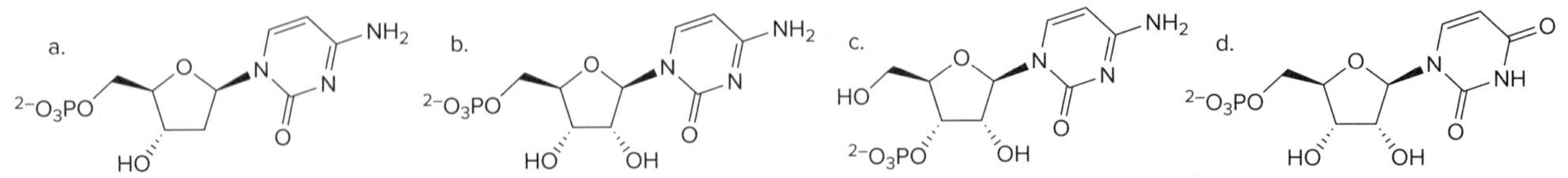

a. b. c. d.

2. What is the sequence of the complementary strand of the DNA fragment 3'–GGACTATA–5'?

a. 3'–CCTGATAT–5'
b. 5'–TATAGAGC–3'
c. 3'–TATAGTCC–5'
d. 5'–CCTGATAT–3'

3. Which abbreviation for a nucleotide in DNA or RNA is *not* valid: (a) TMP; (b) dAMP; (c) GMP; (d) dGMP?

4. What is the sequence of mRNA derived from the DNA fragment 3'–ATGACA–5'?

a. 5'–TACTGT–3'
b. 5'–UACUGU–3'
c. 3'–UGUCAU–5'
d. 3'–UACUGU–5'

5. Which statement about DNA is *not* true?

a. DNA consists of a template strand and a coding strand that have complementary base pairs.
b. In replication, DNA unwinds and daughter DNA is synthesized in one direction on only the template strand.
c. DNA has a sugar–phosphate backbone that winds on the outside of the double helix.
d. The order of bases in DNA distinguishes one DNA molecule from another.

6. What is the sequence of the coding strand of a DNA fragment if an mRNA prepared from the template strand has the sequence 5'–ACUGGA–3'?

a. 3'–ACTGGA–5'
b. 3'–TGACCT–5'
c. 5'–ACTGGA–3'
d. 5'–TGACCT–3'

7. What are the N-terminal and C-terminal amino acids present in the peptide synthesized from the mRNA fragment 5'–UUGGAAGUACGA–3'?

a. N-terminal Leu, C terminal Arg
b. N-terminal Trp, C terminal Leu
c. N-terminal Gln, C terminal Arg
d. N-terminal Arg, C terminal Leu

8. What amino acids would be present in a peptide if tRNAs containing the anticodons AGU and AAU are used to synthesize it?

a. Ser and Asn
b. Leu and Asn
c. Asn and Arg
d. Ser and Leu

9. What statement about polynucleotide AGTGCAC is *not* true?

a. The polynucleotide contains a free phosphate at the 5' end.
b. The polynucleotide contains a purine base at the 3' end.
c. The polynucleotide contains a purine base at the 5' end.
d. The nucleotide that contains cytosine has the free 3'-OH group.

10. Which statement about viruses is *not* true?

a. A retrovirus has an RNA core.
b. A retrovirus uses reverse transcription to synthesize viral proteins from RNA.
c. A virus is an infectious agent that must invade a host cell in order to replicate.
d. AZT is an effective antiviral medication because it lacks a functional group needed to synthesize viral DNA.

PROBLEMS

Problems Using Three-Dimensional Models

28.22 (a) Give the name of each compound shown as a ball-and-stick model. (b) Would the compound be a component of DNA, RNA, or both?

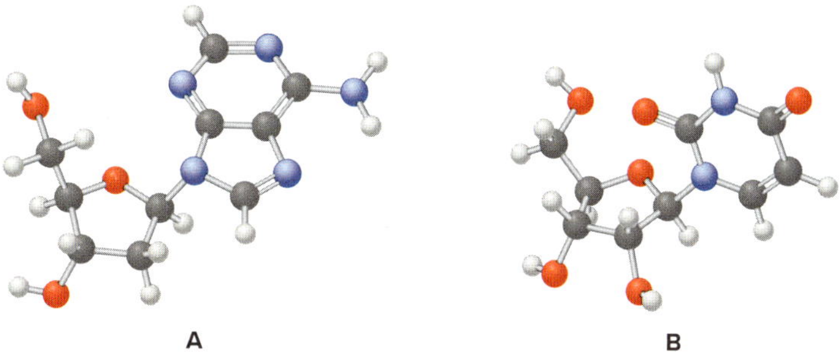

28.23 Give the name and the three- or four-letter abbreviation for each nucleotide.

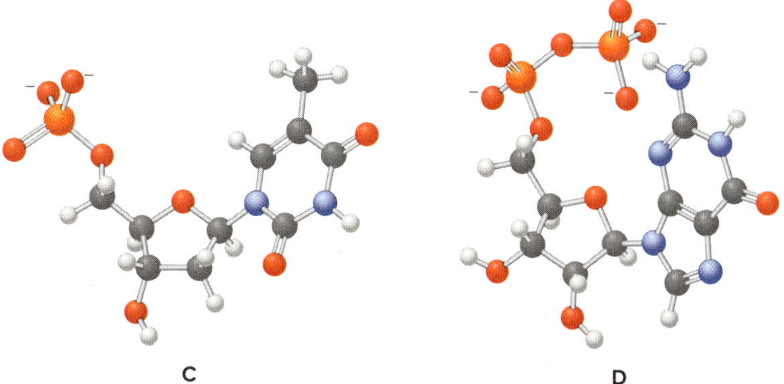

Bases, Nucleosides, Nucleotides, and Nucleic Acid Structure

28.24 Although the pyrimidine bases could exist as enol tautomers, making them hydroxy pyrimidines that contain a six-membered ring with 6 π electrons, these compounds are more stable as their amide tautomers. (a) Draw three different mono enol tautomers for thymine. (b) Draw a dienol tautomer for uracil. (c) How many enol tautomers can be drawn for caffeine, a natural product that contains a purine ring system? (d) Is caffeine an aromatic compound?

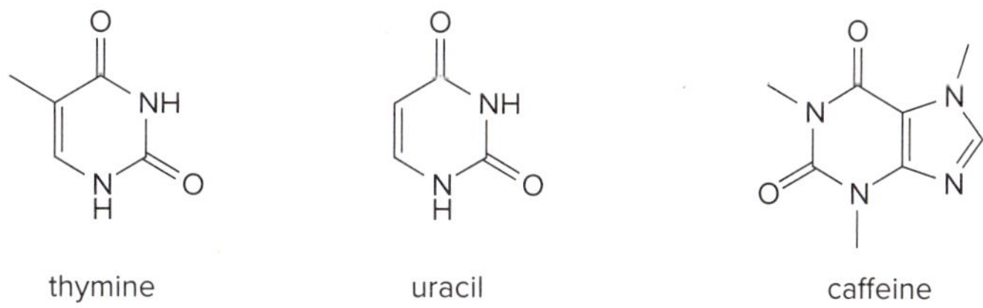

28.25 (a) Identify the most acidic proton in thymine and explain your choice. (b) If thymine is treated with two equivalents of very strong base, what dianion is formed?

28.26 Suppose 2,6-diaminopurine replaced adenine as one of the four bases in a nucleic acid. Draw the structure of 2,6-diaminopurine and the hydrogen-bonding interactions that would occur between 2,6-diaminopurine and thymine.

28.27 Idoxuridine is a nucleoside analogue used in ophthalmic solutions or topical ointments to treat herpes infections. (a) Why is the heterocyclic ring system in idoxuridine aromatic? (b) Draw two different enol tautomers of idoxuridine.

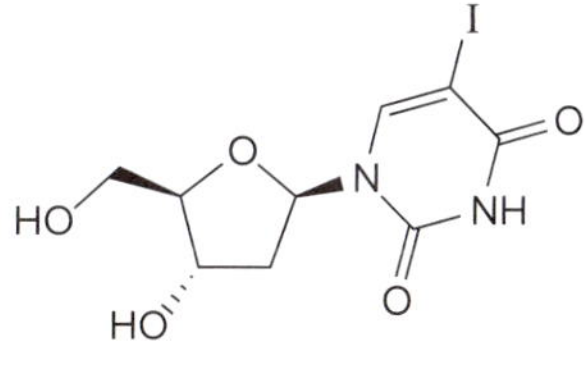

idoxuridine

28.28 DNA can be damaged by reaction of its bases with alkylating agents. For example, reaction of guanine with dimethyl sulfate [$(CH_3)_2SO_4$] forms O^6-methylguanine. Explain why alkylation occurs on O. What effect does this alkylation have on the ability to hydrogen bond with cytosine?

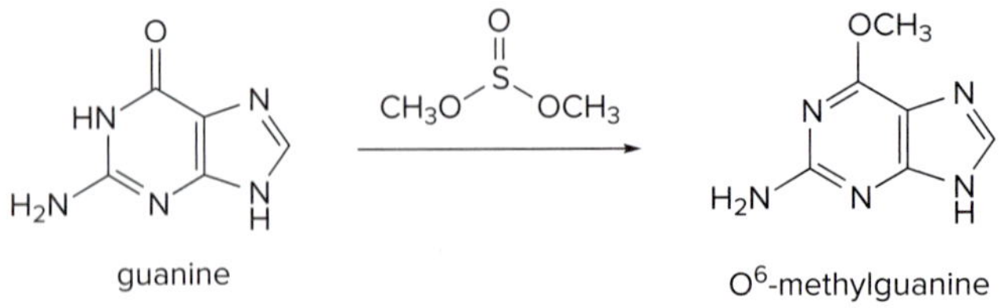

guanine

O^6-methylguanine

28.29 Spongothymidine is an *N*-glycoside isolated from *Tectitethya crypta,* a shallow-water Caribbean sponge. Identify the base and monosaccharide that compose spongothymidine, and draw the structure of the monosaccharide using a Fischer projection formula.

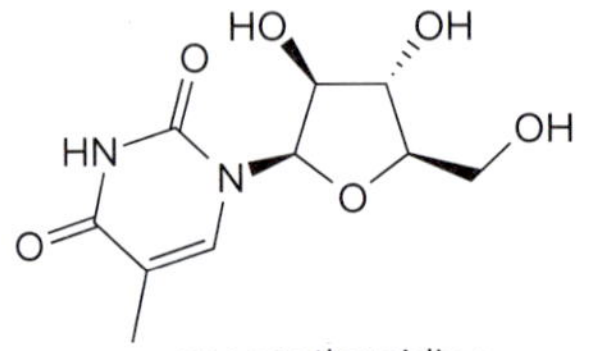

spongothymidine

28.30 Draw the structure of the two possible dinucleotides formed from each pair of nucleotides: (a) dTMP and dAMP; (b) uridine 5'-monophosphate and guanosine 5'-monophosphate. Name each dinucleotide.

28.31 Draw the structure of each polynucleotide: (a) GTA; (b) CGU.

28.32 Write the sequence of the complementary strand of each segment of a DNA molecule.

a. 5'–AAATAAC–3'
b. 5'–ACTGGACT–3'
c. 5'–CGATATCCCG–3'
d. 5'–TTCCCGGGATA–3'

28.33 If 27% of the nucleotides in a sample of DNA contain the base adenine (A), what are the percentages of bases T, G, and C?

28.34 DNA becomes denatured and unwinds when it is heated. Explain why the temperature required for unwinding increases as the G–C content of the double helix increases.

Replication, Transcription, Translation, and Protein Synthesis

28.35 Draw a complete structure of the ribonucleotide codon GCU.

28.36 What is the sequence of a newly synthesized DNA segment if the template strand has the sequence 3'–ATGGCCTATGCGAT–5'?

28.37 For each DNA segment: [1] What is the sequence of the mRNA molecule synthesized from each DNA template? [2] What is the sequence of the coding strand of the DNA molecule?

a. 3'–ATGGCTTA–5'
b. 3'–CGGCGCTTA–5'
c. 3'–GGTATACCG–5'
d. 3'–TAGGCCGTA–5'

28.38 How is the identity of the second base—whether it is a purine or pyrimidine—in a codon related to the polarity of the side chain of the amino acid it codes for?

28.39 If each of the 61 codons for amino acids occurs with equal frequency in mRNA, which amino acids are least commonly found in proteins?

28.40 Derive the amino acid sequence that is coded for by each mRNA sequence.

a. 5'–CCAACCUGGGUAGAA–3'
b. 5'–AUGUUUUUAUGGUGG–3'
c. 5'–GUCGACGAACCGCAA–3'

28.41 Write a possible mRNA sequence that codes for each peptide.

a. Ile–Met–Lys–Ser–Tyr
b. Pro–Gln–Glu–Asp–Phe
c. Thr–Ser–Asn–Arg

28.42 Considering each nucleotide sequence in an mRNA molecule: [1] write the sequence of the DNA template strand from which the mRNA was synthesized; [2] give the peptide synthesized by the mRNA.

a. 5'–UAUUCAAUAAAAAAC–3'
b. 5'–GAUGUAAACAAGCCG–3'

28.43 Using the given DNA template strand, determine the transcribed mRNA sequence and the polypeptide that would be synthesized from the template: 3'–AACGTCCTCACGATT–5'.

28.44 Met-enkephalin (Tyr–Gly–Gly–Phe–Met) is a painkiller and sedative (Section 20.5B). What is a possible nucleotide sequence in the template strand of the gene that codes for met-enkephalin, assuming that every base of the gene is transcribed and then translated?

28.45 Give a possible nucleotide sequence in the template strand of the gene that codes for each peptide.

a.

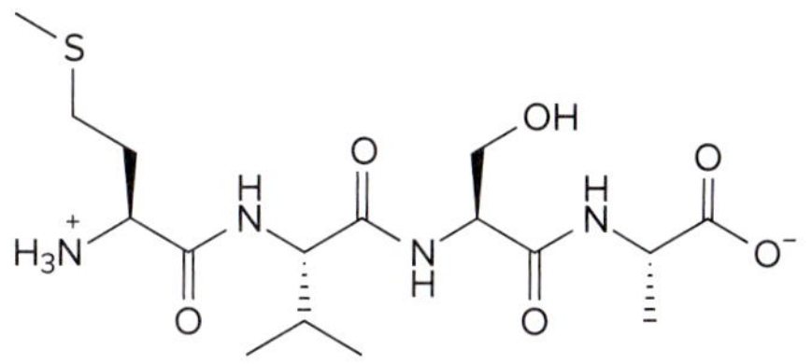

b.

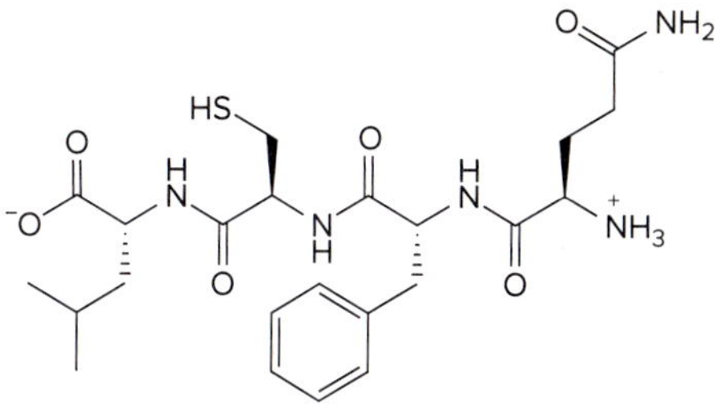

28.46 Draw a complete structure of the template strand of DNA responsible for the synthesis of the dipeptide Met–Trp.

28.47 Give a possible nucleotide sequence in the template strand of the gene that codes for the peptide angiotensin II. As we learned in Problem 27.48, ACE inhibitors are drugs that prevent the formation of angiotensin II, thus decreasing blood pressure.

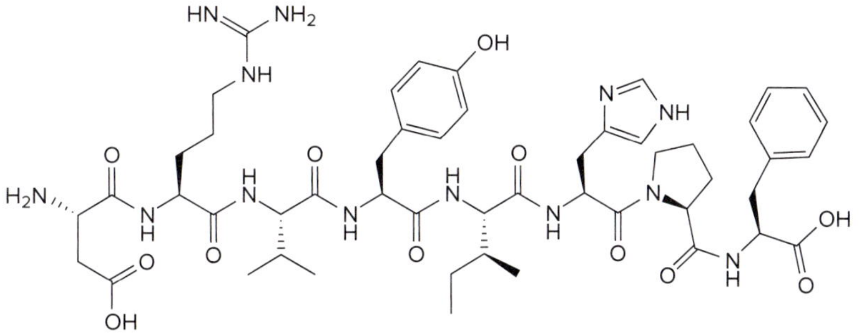

angiotensin II

Mechanism and Synthesis

28.48 Draw a stepwise mechanism for the acid-catalyzed hydrolysis of thymidine to 2'-deoxy-D-ribose and thymine.

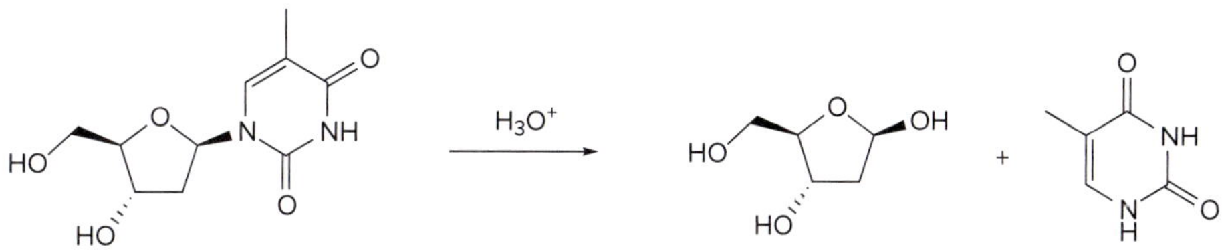

thymidine $\xrightarrow{H_3O^+}$ 2'-deoxy-D-ribose + α anomer + thymine

28.49 One way to synthesize uridine involves reaction of **A** with **B** to form **C,** followed by treatment with base. Draw a stepwise mechanism for the formation of **C.**

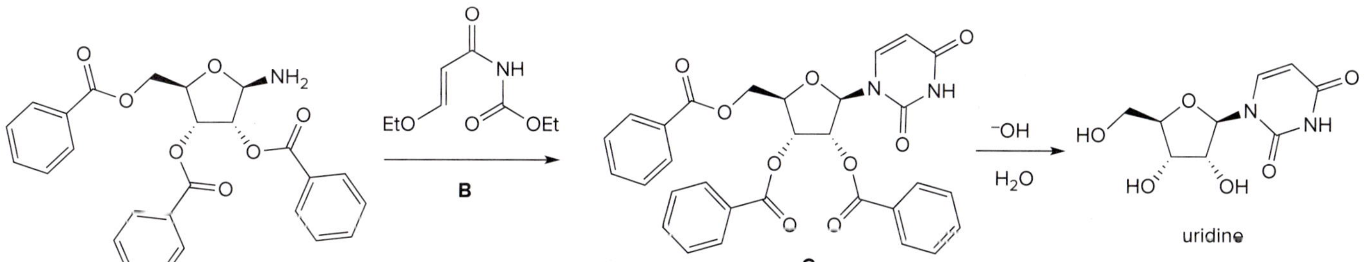

A $\xrightarrow{\mathbf{B}}$ **C** $\xrightarrow[H_2O]{^-OH}$ uridine

28.50 Acyclovir is an antiviral drug prepared by the following reaction sequence. Identify the structures of **A–C** and acyclovir.

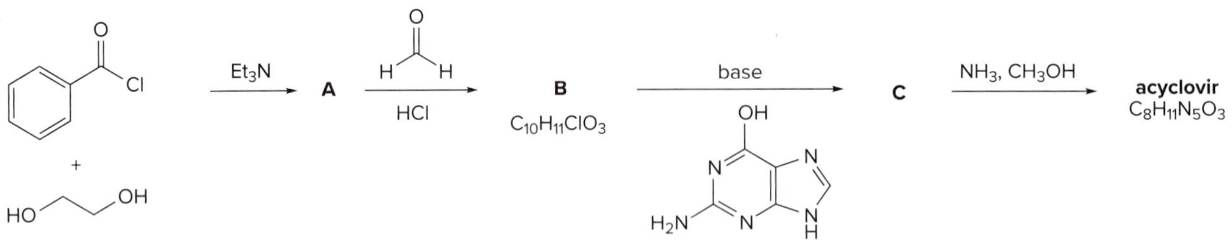

Et_3N → **A**; HCHO, HCl → **B** ($C_{10}H_{11}ClO_3$); base (guanine) → **C**; NH_3, CH_3OH → **acyclovir** ($C_8H_{11}N_5O_3$)

Challenge Problems

28.51 Nucleoside **C** can be synthesized by heating a nucleophilic base (**A**) and an electrophilic monosaccharide derivative (**B**). Suggest a mechanism that explains the stereochemistry of the observed nucleoside. (Hint: The acetate ester at C2' plays a role in the mechanism.)

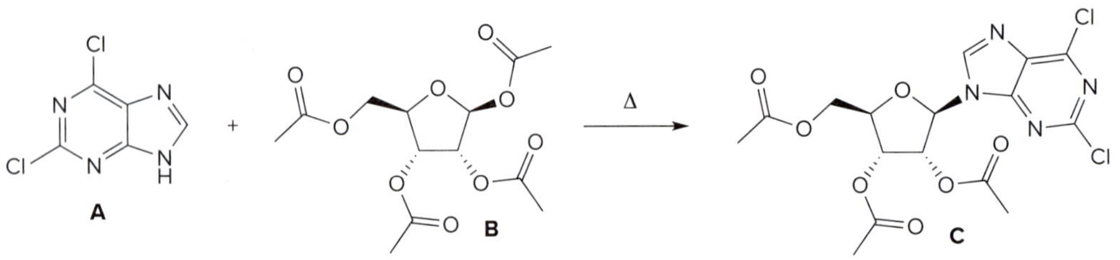

A + **B** —Δ→ **C**

28.52 Identify the nucleoside **X** (including stereochemistry) that is prepared by the following two-step reaction sequence.

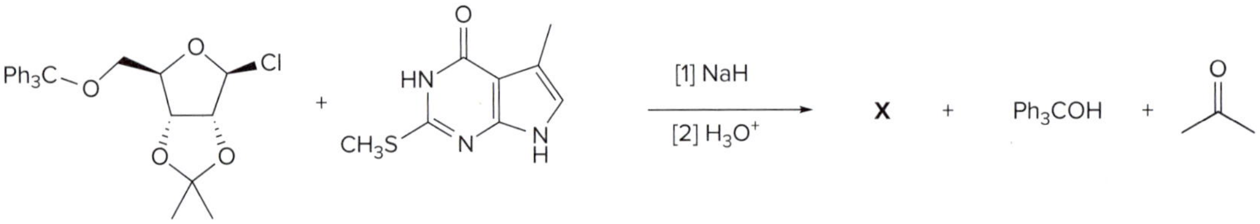

[1] NaH; [2] H_3O^+ → **X** + Ph_3COH + acetone

SELF-TEST ANSWERS

1. b 2. d 3. a 4. b 5. b 6. c 7. a 8. d 9. b 10. b

APPENDIX

Periodic Table of the Elements

	1A 1	2A 2	3B 3	4B 4	5B 5	6B 6	7B 7	← 8	8B 9	→ 10	1B 11	2B 12	3A 13	4A 14	5A 15	6A 16	7A 17	8A 18	
1	1 **H** 1.008																	2 **He** 4.003	1
2	3 **Li** 6.938	4 **Be** 9.012											5 **B** 10.81	6 **C** 12.01	7 **N** 14.01	8 **O** 16.00	9 **F** 19.00	10 **Ne** 20.18	2
3	11 **Na** 22.99	12 **Mg** 24.30											13 **Al** 26.98	14 **Si** 28.08	15 **P** 30.97	16 **S** 32.06	17 **Cl** 35.45	18 **Ar** 39.79	3
4	19 **K** 39.10	20 **Ca** 40.08	21 **Sc** 44.96	22 **Ti** 47.87	23 **V** 50.94	24 **Cr** 52.00	25 **Mn** 54.94	26 **Fe** 55.85	27 **Co** 58.93	28 **Ni** 58.69	29 **Cu** 63.55	30 **Zn** 65.38	31 **Ga** 69.72	32 **Ge** 72.63	33 **As** 74.92	34 **Se** 78.97	35 **Br** 79.90	36 **Kr** 83.80	4
5	37 **Rb** 85.47	38 **Sr** 87.62	39 **Y** 88.91	40 **Zr** 91.22	41 **Nb** 92.91	42 **Mo** 95.95	43 **Tc** (98)	44 **Ru** 101.07	45 **Rh** 102.91	46 **Pd** 106.42	47 **Ag** 107.87	48 **Cd** 112.41	49 **In** 114.82	50 **Sn** 118.71	51 **Sb** 121.76	52 **Te** 127.6	53 **I** 126.90	54 **Xe** 131.29	5
6	55 **Cs** 132.91	56 **Ba** 137.33	57–71 **Lanthanides**	72 **Hf** 178.49	73 **Ta** 180.95	74 **W** 183.84	75 **Re** 186.21	76 **Os** 190.23	77 **Ir** 192.22	78 **Pt** 195.08	79 **Au** 196.97	80 **Hg** 200.59	81 **Tl** 204.38	82 **Pb** 207.2	83 **Bi** 208.98	84 **Po** (210)	85 **At** (210)	86 **Rn** (222)	6
7	87 **Fr** (223)	88 **Ra** (226)	89–103 **Actinides**	104 **Rf** (261)	105 **Db** (262)	106 **Sg** (266)	107 **Bh** (272)	108 **Hs** (277)	109 **Mt** (276)	110 **Ds** (281)	111 **Rg** (280)	112 **Cn** (285)	113 **Nh** (285)	114 **Fl** (287)	115 **Mc** (289)	116 **Lv** (291)	117 **Ts** (293)	118 **Og** (294)	7

6	57 **La** 138.91	58 **Ce** 140.12	59 **Pr** 140.91	60 **Nd** 144.24	61 **Pm** (145)	62 **Sm** 150.36	63 **Eu** 151.96	64 **Gd** 157.25	65 **Tb** 158.93	66 **Dy** 162.5	67 **Ho** 164.93	68 **Er** 167.26	69 **Tm** 168.93	70 **Yb** 173.05	71 **Lu** 174.97	6
7	89 **Ac** (227)	90 **Th** 232.04	91 **Pa** 231.04	92 **U** 238.03	93 **Np** (237)	94 **Pu** (239)	95 **Am** (243)	96 **Cm** (247)	97 **Bk** (249)	98 **Cf** (252)	99 **Es** (252)	100 **Fm** (257)	101 **Md** (258)	102 **No** (259)	103 **Lr** (262)	7

metal | metalloid | nonmetal

APPENDIX

Common Abbreviations, Arrows, and Symbols

Abbreviations

Ac	acetyl, CH_3CO-
BBN	9-borabicyclo[3.3.1]nonane
BINAP	2,2'-bis(diphenylphosphino)-1,1'-binaphthyl
Boc	*tert*-butoxycarbonyl, $(CH_3)_3COCO-$
bp	boiling point
Bu	butyl, $CH_3CH_2CH_2CH_2-$
CBS reagent	Corey–Bakshi–Shibata reagent
DBN	1,5-diazabicyclo[4.3.0]non-5-ene
DBU	1,8-diazabicyclo[5.4.0]undec-7-ene
DCC	dicyclohexylcarbodiimide
DET	diethyl tartrate
DIBAL-H	diisobutylaluminum hydride, $[(CH_3)_2CHCH_2]_2AlH$
DMF	dimethylformamide, $HCON(CH_3)_2$
DMSO	dimethyl sulfoxide, $(CH_3)_2S{=}O$
ee	enantiomeric excess
Et	ethyl, CH_3CH_2-
Fmoc	9-fluorenylmethoxycarbonyl
HMPA	hexamethylphosphoramide, $[(CH_3)_2N]_3P{=}O$
HOMO	highest occupied molecular orbital
IR	infrared
LDA	lithium diisopropylamide, $LiN[CH(CH_3)_2]_2$
LUMO	lowest unoccupied molecular orbital
m-	meta
mCPBA	*m*-chloroperoxybenzoic acid
Me	methyl, CH_3-
MO	molecular orbital
mp	melting point
MS	mass spectrometry
MW	molecular weight
NBS	*N*-bromosuccinimide
NMO	*N*-methylmorpholine *N*-oxide
NMR	nuclear magnetic resonance
o-	ortho
p-	para
PCC	pyridinium chlorochromate
Ph	phenyl, C_6H_5-
ppm	parts per million
Pr	propyl, $CH_3CH_2CH_2-$
RCM	ring-closing metathesis
ROMP	ring-opening metathesis polymerization

TBS	*tert*-butyldimethylsilyl
THF	tetrahydrofuran
TMS	tetramethylsilane, $(CH_3)_4Si$
Ts	tosyl, *p*-toluenesulfonyl, $CH_3C_6H_4SO_2-$
TsOH	*p*-toluenesulfonic acid, $CH_3C_6H_4SO_3H$
UV	ultraviolet

Arrows

$\longrightarrow$	reaction arrow
$\rightleftarrows$	equilibrium arrows
$\longleftrightarrow$	double-headed arrow, used between resonance structures
$\curvearrowright$ (biological)	biological reaction arrow
$\curvearrowright$	full-headed curved arrow, showing the movement of an electron pair
$\frown$	half-headed curved arrow (fishhook), showing the movement of an electron
$\Longrightarrow$	retrosynthetic arrow
$\not\longrightarrow$	no reaction

Symbols

$\mapsto$	dipole
$h\nu$	light
Δ	heat
$\delta+$	partial positive charge
$\delta-$	partial negative charge
λ	wavelength
ν	frequency
$\tilde{\nu}$	wavenumber
HA	Brønsted–Lowry acid
B:	Brønsted–Lowry base
$:Nu^-$	nucleophile
E^+	electrophile
X	halogen
◀ (wedge)	bond oriented forward
·‧‧ǀǀ (hashed)	bond oriented behind
- - -	partial bond
$[\ \ ]^{\ddagger}$	transition state
[O]	oxidation
[H]	reduction

Common Element Colors Used in Molecular Art

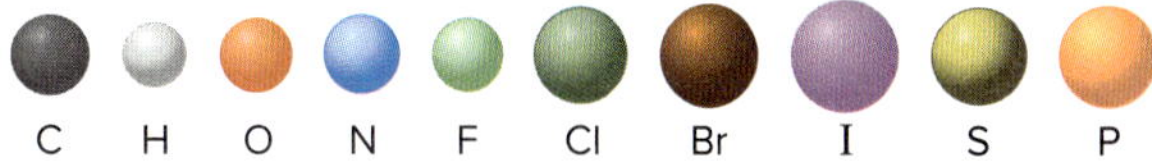

APPENDIX C pK$_a$ Values for Selected Compounds

Compound	pK$_a$
HI	−10
HBr	−9
H_2SO_4	−9
$(CH_3)_2C=\overset{+}{O}H$	−7.3
HCl	−6
$[(CH_3)_2OH]^+$	−3.8
$(CH_3OH_2)^+$	−2.2
$CH_3-C_6H_4-SO_3H$	−1.3
CH_3SO_3H	−1.2
H_3O^+	0
$CH_3C(=\overset{+}{O}H)NH_2$	0.0
CF_3CO_2H	0.5
CCl_3CO_2H	0.7
$O_2N-C_6H_4-\overset{+}{N}H_3$	1.0
Cl_2CHCO_2H	1.3
H_3PO_4	2.1
FCH_2CO_2H	2.7
$ClCH_2CO_2H$	2.8
$BrCH_2CO_2H$	2.9
ICH_2CO_2H	3.2
HF	3.2
$O_2N-C_6H_4-CO_2H$	3.4
HCO_2H	3.8
$Br-C_6H_4-\overset{+}{N}H_3$	3.9
$Br-C_6H_4-CO_2H$	4.0
$C_6H_5-CO_2H$	4.2
$CH_3-C_6H_4-CO_2H$	4.3
$CH_3O-C_6H_4-CO_2H$	4.5
CH_3CO_2H	4.8
$C_6H_5-\overset{+}{N}H_3$	4.9
$(CH_3)_3CCO_2H$	5.0
$CH_3-C_6H_4-\overset{+}{N}H_3$	5.1
$C_5H_5\overset{+}{N}-H$	5.3
$CH_3O-C_6H_4-\overset{+}{N}H_3$	5.3
H_2CO_3	6.4
H_2S	7.1
$O_2N-C_6H_4-OH$	7.1
C_6H_5-SH	7.8

Compound	pK_a
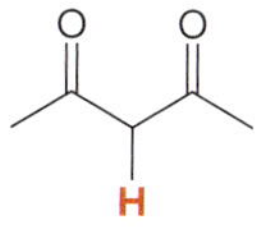 O O H	8.9
HC≡N	9.2
NH_4^+	9.4
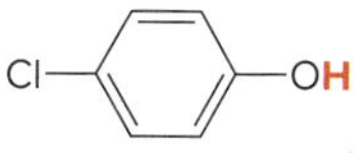 Cl– –OH	9.4
$H_3\overset{+}{N}CH_2CO_2^-$	9.8
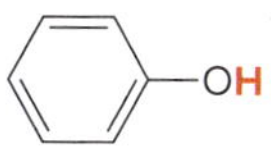 –OH	10.0
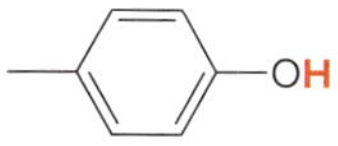 –OH	10.2
HCO_3^-	10.2
CH_3NO_2	10.2
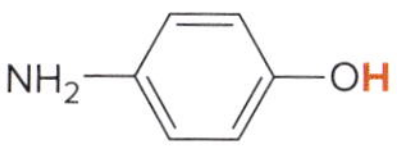 NH_2– –OH	10.3
CH_3CH_2SH	10.5
$[(CH_3)_3NH]^+$	10.6
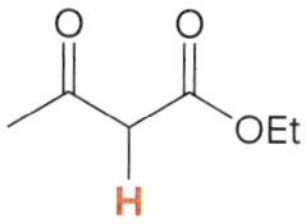 O O OEt H	10.7
$(CH_3NH_3)^+$	10.7
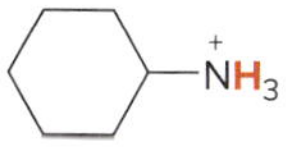 –$\overset{+}{N}H_3$	10.7
$[(CH_3)_2NH_2]^+$	10.7
CF_3CH_2OH	12.4
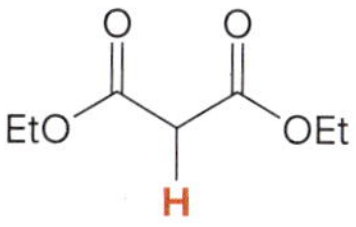 O O EtO OEt H	13.3

Compound	pK_a
H_2O	14
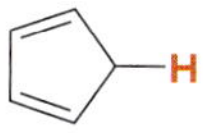 –H	15
CH_3OH	15.5
CH_3CH_2OH	16
CH_3CONH_2	16
CH_3CHO	17
$(CH_3)_3COH$	18
$(CH_3)_2C=O$	19.2
$CH_3CO_2CH_2CH_3$	24.5
HC≡CH	25
$CH_3C≡N$	25
$CHCl_3$	25
$CH_3CON(CH_3)_2$	30
H_2	35
NH_3	38
CH_3NH_2	40
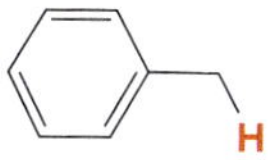 H	41
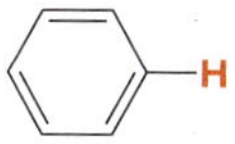 –H	43
$CH_2=CHCH_3$	43
$CH_2=CH_2$	44
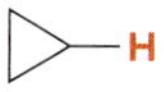 –H	46
CH_4	50
CH_3CH_3	50

APPENDIX D Nomenclature

Although the basic principles of nomenclature are presented in the body of this text, additional information is often needed to name many complex organic compounds. Appendix D concentrates on three topics:

- **Naming alkyl substituents that contain branching**
- **Naming polyfunctional compounds**
- **Naming bicyclic compounds**

Naming Alkyl Substituents That Contain Branching

Alkyl groups that contain any number of carbons and no branches are named as described in Section 4.4A: change the *-ane* ending of the parent alkane to the suffix *-yl.* Thus the seven-carbon alkyl group $CH_3CH_2CH_2CH_2CH_2CH_2CH_2-$ is called *heptyl.*

When an alkyl substituent also contains branching, follow a stepwise procedure:

[1] Identify the longest carbon chain of the alkyl group that begins at the point of attachment to the parent. Begin numbering at the point of attachment and use the suffix *-yl* to indicate an alkyl group.

4 C's in the chain ---→ butyl group

5 C's in the chain ---→ pentyl group

[2] Name all branches off the main alkyl chain and use the numbers from Step [1] to designate their location.

3-methylbutyl

1,3-dimethylpentyl

[3] Set the entire name of the substituent in parentheses, and alphabetize this substituent name by the first letter of the complete name.

(3-methylbutyl)cyclohexane

1-(1,3-dimethylpentyl)-2-methylcyclohexane

- Alphabetize the **d** of **d**imethylpentyl before the **m** of **m**ethyl.
- Number the ring to give the lower number to the first substituent alphabetically: place the dimethylpentyl group at C1.

Naming Polyfunctional Compounds

Many organic compounds contain more than one functional group. When one of those functional groups is halo (X–) or alkoxy (RO–), these groups are named as substituents as described in Sections 7.2 and 9.3B. To name other polyfunctional compounds, we must learn which functional group is assigned a higher priority in the rules of nomenclature. Two steps are usually needed:

[1] **Name a compound using the suffix of the highest-priority group,** and name other functional groups as *substituents.* Table D.1 lists the common functional groups in order of *decreasing* priority, as well as the prefixes needed when a functional group must be named as a substituent.

[2] Number the carbon chain to give the lower number to the highest-priority functional group that can be named as a suffix, and then follow all other rules of nomenclature. Examples are shown in Figure D.1.

Table D.1 Summary of Functional Group Nomenclature

Functional group (increasing priority ↑)	Suffix	Substituent name (prefix)
Carboxylic acid	-oic acid	carboxy
Ester	-oate	alkoxycarbonyl
Amide	-amide	amido
Nitrile	-nitrile	cyano
Aldehyde	-al	oxo (=O) or formyl (–CHO)
Ketone	-one	oxo
Alcohol	-ol	hydroxy
Amine	-amine	amino
Alkene	-ene	alkenyl
Alkyne	-yne	alkynyl
Alkane	-ane	alkyl
Ether	—	alkoxy
Halide	—	halo

Figure D.1
Examples of nomenclature of polyfunctional compounds

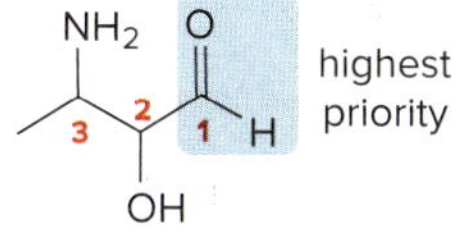

3-amino-2-hydroxybutanal

Name as a derivative of an **aldehyde,** because CHO is the highest-priority functional group.

o-cyanobenzoic acid (higher priority)

Name as a derivative of **benzoic acid,** because COOH is the higher-priority functional group.

methyl 4-oxohexanoate (higher priority)

Name as a derivative of an **ester,** because COOR is the higher-priority functional group.

4-formyl-3-methoxycyclohexanecarboxamide (highest priority)

Name as a derivative of an **amide,** because $CONH_2$ is the highest-priority functional group.

Polyfunctional compounds that contain C–C double and triple bonds have characteristic suffixes to identify them, as shown in Table D.2. The higher-priority functional group is assigned the lower number.

Table D.2 Naming Polyfunctional Compounds with C–C Double and Triple Bonds

Functional groups	Suffix	Example
C=C and OH	enol	5-methylhex-4-en-1-ol
C=C + C=O (ketone)	enone	(*E*)-hept-4-en-3-one
C=C + C≡C	enyne	hex-1-en-5-yne

Naming Bicyclic Compounds

Bicyclic ring systems—compounds that contain two rings that share one or two carbon atoms—can be bridged, fused, or spiro.

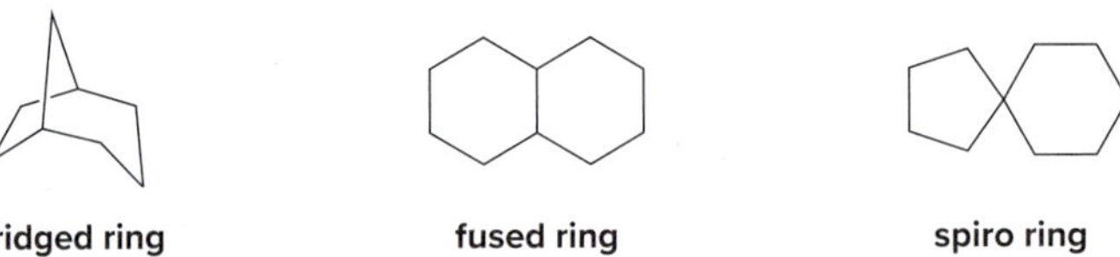

- A bridged ring system contains two rings that share two *non-adjacent* carbons.
- A fused ring system contains two rings that share a *common* carbon–carbon bond.
- A spiro ring system contains two rings that share *one carbon atom*.

Fused and bridged ring systems are named as bicyclo[*x.y.z*]alkanes, where the parent alkane corresponds to the total number of carbons in both rings. The numbers *x, y,* and *z* refer to the number of carbons that join the shared carbons together, written in order of *decreasing* size. For a fused ring system, *z* always equals zero, because the two shared carbons are directly joined together. The shared carbons in a bridged ring system are called the **bridgehead carbons.**

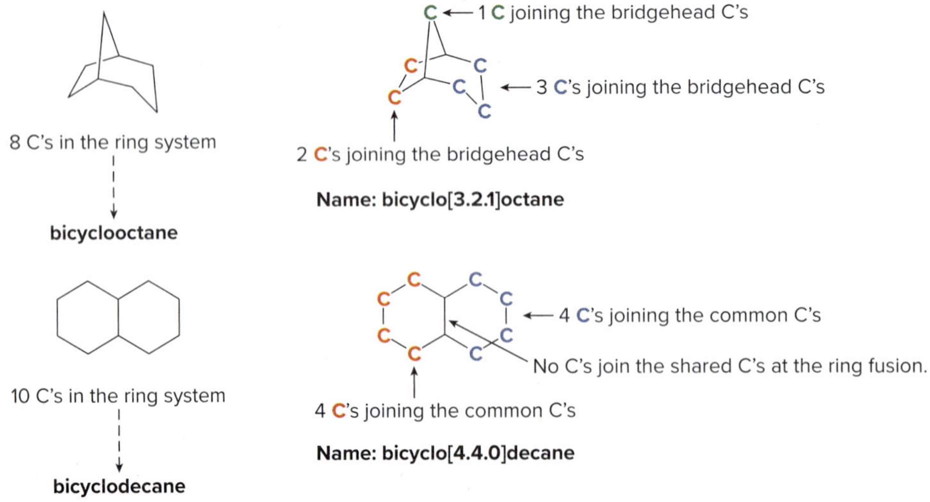

Rings are numbered beginning at a *shared* carbon, and continuing around the *longest* bridge first, then the next longest, and so forth.

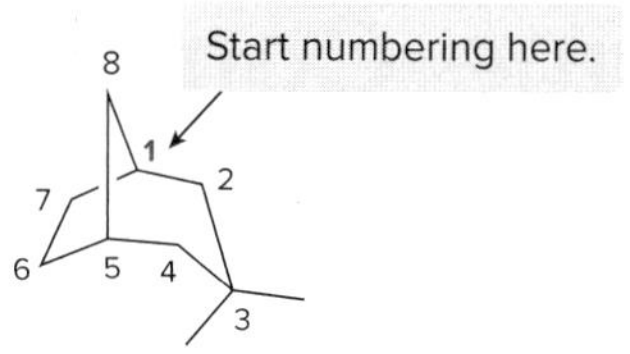

Start numbering here.

3,3-dimethylbicyclo[3.2.1]octane

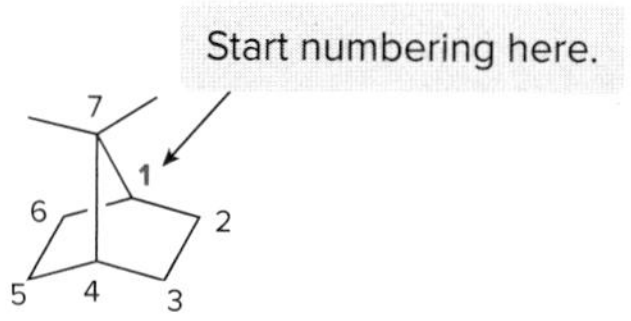

Start numbering here.

7,7-dimethylbicyclo[2.2.1]heptane

Spiro ring systems are named as spiro[*x*.*y*]alkanes where the parent alkane corresponds to the total number of carbons in both rings, and x and y refer to the number of carbons that join the shared carbon (the spiro carbon), written in order of *increasing* size. When substituents are present, the rings are numbered beginning with a carbon *adjacent* to the spiro carbon in the *smaller* ring.

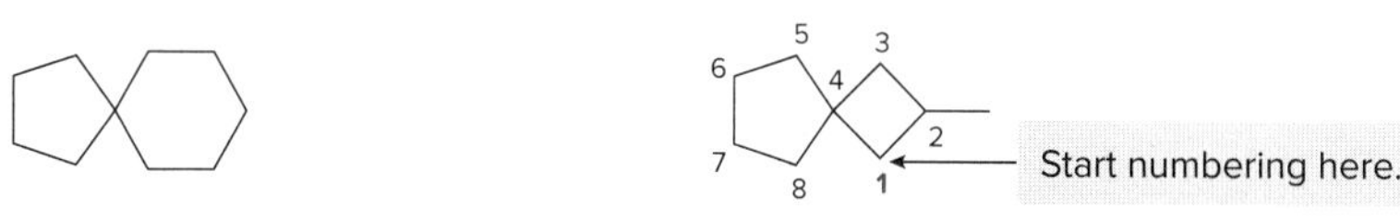

10 C's in the ring system

Name: spiro[4.5]decane

8 C's in the ring system

Name: 2-methylspiro[3.4]octane

APPENDIX

Bond Dissociation Energies for Some Common Bonds [A—B → A• + •B]

Bond	$\Delta H°$ kJ/mol	(kcal/mol)
H—Z bonds		
H—F	569	(136)
H—Cl	431	(103)
H—Br	368	(88)
H—I	297	(71)
H—OH	498	(119)
Z—Z bonds		
H—H	435	(104)
F—F	159	(38)
Cl—Cl	242	(58)
Br—Br	192	(46)
I—I	151	(36)
HO—OH	213	(51)
R—H bonds		
CH_3—H	435	(104)
CH_3CH_2—H	410	(98)
$CH_3CH_2CH_2$—H	410	(98)
$(CH_3)_2CH$—H	397	(95)
$(CH_3)_3C$—H	381	(91)
CH_2=CH—H	435	(104)
HC≡C—H	523	(125)
CH_2=$CHCH_2$—H	364	(87)
C_6H_5—H	460	(110)
$C_6H_5CH_2$—H	356	(85)
R—R bonds		
CH_3—CH_3	368	(88)
CH_3—CH_2CH_3	356	(85)
CH_3—CH=CH_2	385	(92)
CH_3—C≡CH	489	(117)

Bond	$\Delta H°$ kJ/mol	(kcal/mol)
R—X bonds		
CH_3—F	456	(109)
CH_3—Cl	351	(84)
CH_3—Br	293	(70)
CH_3—I	234	(56)
CH_3CH_2—F	448	(107)
CH_3CH_2—Cl	339	(81)
CH_3CH_2—Br	285	(68)
CH_3CH_2—I	222	(53)
$(CH_3)_2CH$—F	444	(106)
$(CH_3)_2CH$—Cl	335	(80)
$(CH_3)_2CH$—Br	285	(68)
$(CH_3)_2CH$—I	222	(53)
$(CH_3)_3C$—F	444	(106)
$(CH_3)_3C$—Cl	331	(79)
$(CH_3)_3C$—Br	272	(65)
$(CH_3)_3C$—I	209	(50)
R—OH bonds		
CH_3—OH	389	(93)
CH_3CH_2—OH	393	(94)
$CH_3CH_2CH_2$—OH	385	(92)
$(CH_3)_2CH$—OH	401	(96)
$(CH_3)_3C$—OH	401	(96)
Other bonds		
CH_2=CH_2	635	(152)
HC≡CH	837	(200)
O=C=O	535	(128)
O_2	497	(119)

APPENDIX F

Reactions That Form Carbon–Carbon Bonds

Section	Reaction
11.11A	S_N2 reaction of an alkyl halide with an acetylide anion, $^-C\equiv CR$
11.11B	Opening of an epoxide ring with an acetylide anion, $^-C\equiv CR$
13.14	Radical polymerization of an alkene
14.12	Diels–Alder reaction
16.5	Friedel–Crafts alkylation
16.5	Friedel–Crafts acylation
17.10	Reaction of an aldehyde or ketone with a Grignard or organolithium reagent
17.13A	Reaction of an acid chloride with a Grignard or organolithium reagent
17.13A	Reaction of an ester with a Grignard or organolithium reagent
17.13B	Reaction of an acid chloride with an organocuprate reagent
17.14A	Reaction of a Grignard reagent with CO_2
17.14B	Reaction of an epoxide with an organometallic reagent
17.15	Reaction of an α,β-unsaturated carbonyl compound with an organocuprate reagent
18.8	Cyanohydrin formation
18.9	Wittig reaction to form an alkene
19.12	S_N2 reaction of an alkyl halide with NaCN
19.12C	Reaction of a nitrile with a Grignard or organolithium reagent
21.8	Direct enolate alkylation using LDA and an alkyl halide
21.9	Malonic ester synthesis to form a carboxylic acid
21.10	Acetoacetic ester synthesis to form a ketone
22.1	Aldol reaction to form a β-hydroxy carbonyl compound or an α,β-unsaturated carbonyl compound
22.2	Crossed aldol reaction
22.3	Directed aldol reaction
22.5	Claisen reaction to form a β-keto ester
22.6	Crossed Claisen reaction to form a β-dicarbonyl compound
22.7	Dieckmann reaction to form a five- or six-membered ring
22.8	Michael reaction to form a 1,5-dicarbonyl compound
22.9	Robinson annulation to form a cyclohex-2-enone
23.13	Reaction of a diazonium salt with CuCN
24.1	Coupling of an organocuprate reagent (R_2CuLi) with an organic halide (R'X)
24.2	The palladium-catalyzed Suzuki reaction of an organic halide with an organoborane
24.3	The palladium-catalyzed Heck reaction of a vinyl or aryl halide with an alkene
24.4	The palladium-catalyzed Stille coupling of an organostannane with an organic halide
24.5	Addition of a dihalocarbene to an alkene to form a cyclopropane
24.6	Simmons–Smith reaction of an alkene with CH_2I_2 and Zn(Cu) to form a cyclopropane
24.7	Olefin metathesis
25.3	Electrocyclic reactions
25.4	Cycloaddition reactions
25.5	Sigmatropic rearrangements
26.9B	Kiliani–Fischer synthesis of an aldose
31.2	Chain-growth polymerization
31.4	Polymerization using Ziegler–Natta catalysts

APPENDIX

G Characteristic IR Absorption Frequencies

Bond	Functional group	Wavenumber (cm^{-1})	Comment
O–H			
	• ROH	3600–3200	broad, strong
	• RCO_2H	3500–2500	very broad, strong
N–H			
	• RNH_2	3500–3300	two peaks
	• R_2NH	3500–3300	one peak
	• $RCONH_2$, RCONHR	3400–3200	one or two peaks; N–H bending also observed at 1640 cm^{-1}
C–H			
	• C_{sp}–H	3300	sharp, often strong
	• C_{sp^2}–H	3150–3000	medium
	• C_{sp^3}–H	3000–2850	strong
	• C_{sp^2}–H of RCHO	2830–2700	one or two peaks
C≡C		2250	medium
C≡N		2250	medium
C=O			strong
	• RCOCl	1800	
	• $(RCO)_2O$	1800, 1760	two peaks
	• RCO_2R	1745–1735	increasing $\tilde{\nu}$ with decreasing ring size
	• RCHO	1730	
	• R_2CO	1715	increasing $\tilde{\nu}$ with decreasing ring size
	• R_2CO, conjugated	1680	
	• RCO_2H	1710	
	• $RCONH_2$, RCONHR, $RCONR_2$	1680–1630	increasing $\tilde{\nu}$ with decreasing ring size
C=C			
	• Alkene	1650	medium
	• Arene	1600, 1500	medium
C=N		1650	medium

APPENDIX

Characteristic NMR Absorptions

^{1}H NMR Absorptions

Compound type	Chemical shift (ppm)
Alcohol	
R–O–**H**	1–5
(OH)C–**H**	3.4–4.0
Aldehyde	
RC(=O)**H**	9–10
Alkane	0.9–2.0
RC**H$_3$**	~0.9
R_2C**H$_2$**	~1.3
R_3C**H**	~1.7
Alkene	
C=C–**H** sp^2 C–H	4.5 6.0
C=C–C–**H** allylic sp^3 C–H	1.5–2.5
Alkyl halide	
(F)C–**H**	4.0–4.5
(Cl)C–**H**	3.0–4.0
(Br)C–**H**	2.7–4.0
(I)C–**H**	2.2–4.0
Alkyne	~2.5
C≡C–**H**	
Amide	
RC(=O)N–**H**	7.5–8.5
Amine	
RN–**H**	0.5–5.0
(NR_2)C–**H**	2.3–3.0
Aromatic compound	
Ar–**H** sp^2 C–H	6.5–8
ArC–**H** benzylic sp^3 C–H	1.5 2.5
Carbonyl compound	
RC(=O)C–**H** sp^3 C–H on the α carbon	2.0–2.5
Carboxylic acid	
RC(=O)O**H**	10–12
Ether	
(OR)C–**H**	3.4–4.0

^{13}C NMR Absorptions

Carbon type	Structure	Chemical shift (ppm)
Alkyl, sp^3 hybridized C	—C— (tetrahedral C)	5–45
Alkyl, sp^3 hybridized C bonded to N, O, or X	Z—C (tetrahedral C) Z = N, O, X	30–80
Alkynyl, sp hybridized C	—C≡C—	65–100
Alkenyl, sp^2 hybridized C	C=C	100–140
Aryl, sp^2 hybridized C	(benzene ring)C—	120–150
Carbonyl C	C=O	160–210

APPENDIX

General Types of Organic Reactions

Substitution Reactions

[1] Nucleophilic substitution at an sp^3 hybridized carbon atom

a. Alkyl halides (Chapter 7) $\quad R{-}X + :Nu^- \longrightarrow R{-}Nu + X:^-$ (nucleophile)

b. Alcohols (Section 9.11) $\quad R{-}OH + HX \longrightarrow R{-}X + H_2O$

c. Ethers (Section 9.14) $\quad R{-}OR' + HX \longrightarrow R{-}X + R'{-}X + H_2O$

X = Br or I

d. Epoxides (Section 9.16) $\quad$ epoxide $\xrightarrow{[1]\ :Nu^-\ \ [2]\ H_2O \text{ or } HZ}$ product with Nu (Z) and OH

Nu or Z = nucleophile

[2] Nucleophilic acyl substitution at an sp^2 hybridized carbon atom

Carboxylic acids and their derivatives (Chapter 20) $\quad RCOZ + :Nu^- \longrightarrow RCONu + :Z^-$ (nucleophile)

Z = OH, Cl, OCOR, OR', NR'_2

[3] Radical substitution at an sp^3 hybridized C–H bond

Alkanes (Section 13.3) $\quad R{-}H + X_2 \xrightarrow{h\nu \text{ or } \Delta} R{-}X + HX$

[4] Electrophilic aromatic substitution

Aromatic compounds (Chapter 16) $\quad C_6H_5{-}H + E^+ \longrightarrow C_6H_5{-}E + H^+$ (electrophile)

[5] Nucleophilic aromatic substitution

Aromatic compounds (Chapter 16) $\quad A{-}C_6H_4{-}X + :Nu^- \longrightarrow A{-}C_6H_4{-}Nu + X^-$ (nucleophile)

X = F, Cl, Br, I
A = H or electron-withdrawing group

Elimination Reactions

β Elimination at an sp^3 hybridized carbon atom

a. Alkyl halides
(Chapter 8)

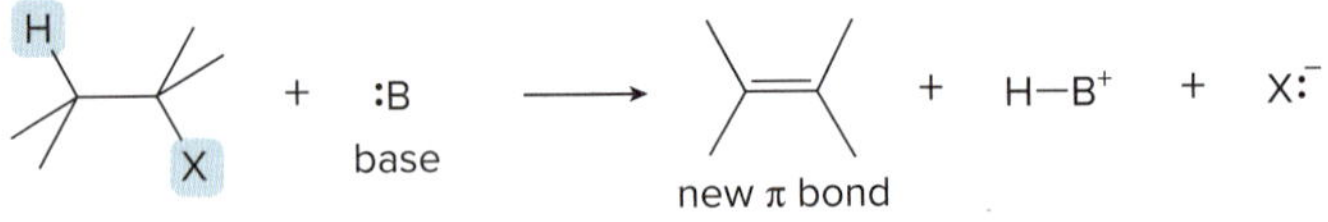

H, X + :B (base) ⟶ new π bond + H—B⁺ + X:⁻

b. Alcohols
(Section 9.8)

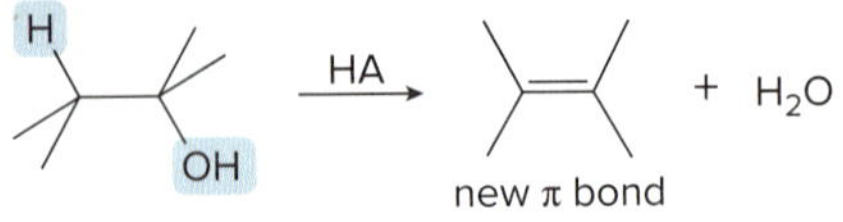

H, OH —HA⟶ new π bond + H_2O

Addition Reactions

[1] Electrophilic addition to carbon–carbon multiple bonds

a. Alkenes
(Chapter 10)

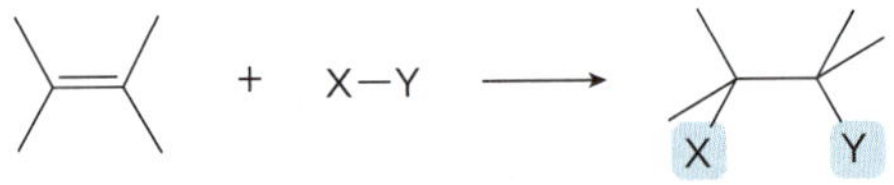

+ X—Y ⟶ X, Y

b. Alkynes
(Section 11.6)

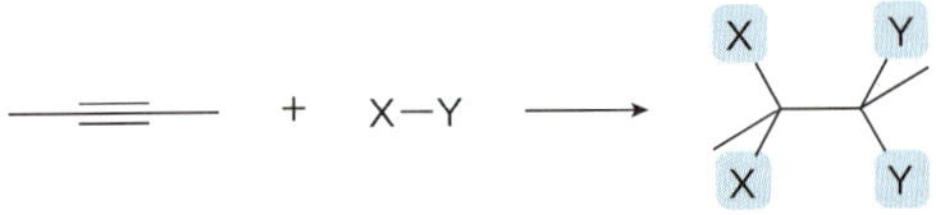

+ X—Y ⟶ X, Y, X, Y

[2] Nucleophilic addition to carbon–oxygen multiple bonds

Aldehydes and ketones
(Chapter 18)

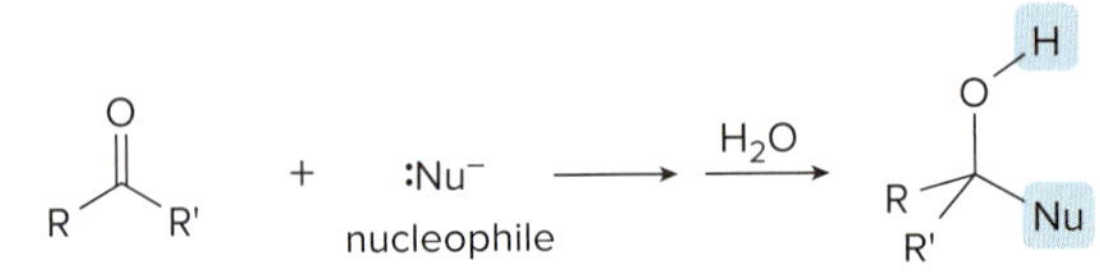

O, R, R' + :Nu⁻ (nucleophile) ⟶ —H_2O⟶ O, H, R, R', Nu

R' = H or alkyl

APPENDIX

How to Synthesize Particular Functional Groups

Acetals

- Reaction of an aldehyde or ketone with two equivalents of an alcohol (18.13)

Acid chlorides

- Reaction of a carboxylic acid with thionyl chloride (20.9)

Alcohols

- Nucleophilic substitution of an alkyl halide with ^{-}OH or H_2O (9.6)
- Hydration of an alkene (10.12)
- Hydroboration–oxidation of an alkene (10.16)
- Reduction of an epoxide with $LiAlH_4$ (12.6)
- Reduction of an aldehyde or ketone (17.4)
- Hydrogenation of an α,β-unsaturated carbonyl compound with H_2 + Pd-C (17.4C)
- Enantioselective reduction of an aldehyde or ketone with the chiral CBS reagent (17.6)
- Reduction of an acid chloride with $LiAlH_4$ (17.7)
- Reduction of an ester with $LiAlH_4$ (17.7)
- Reduction of a carboxylic acid with $LiAlH_4$ (17.7)
- Reaction of an aldehyde or ketone with a Grignard or organolithium reagent (17.10)
- Reaction of an acid chloride with a Grignard or organolithium reagent (17.13)
- Reaction of an ester with a Grignard or organolithium reagent (17.13)
- Reaction of an organometallic reagent with an epoxide (17.14B)

Aldehydes

- Hydroboration–oxidation of a terminal alkyne (11.10)
- Oxidative cleavage of an alkene with O_3 followed by Zn or $(CH_3)_2S$ (12.10)
- Oxidation of a 1° alcohol with PCC (12.12)
- Oxidation of a 1° alcohol with $HCrO_4^-$, Amberlyst A-26 resin (12.13)
- Reduction of an acid chloride with $LiAlH[OC(CH_3)_3]_3$ (17.7)
- Reduction of an ester with DIBAL-H (17.7)
- Hydrolysis of an acetal (18.13B)
- Hydrolysis of an imine or enamine (18.11B)
- Reduction of a nitrile (19.12B)

Alkanes

- Catalytic hydrogenation of an alkene with H_2 + Pd-C (12.3)
- Catalytic hydrogenation of an alkyne with two equivalents of H_2 + Pd-C (12.5A)
- Reduction of an alkyl halide with $LiAlH_4$ (12.6)

- Reduction of a ketone to a methylene group (CH_2)—the Wolff–Kishner or Clemmensen reaction (16.14C)
- Protonation of an organometallic reagent with H_2O, ROH, or acid (17.9)
- Coupling of an organocuprate reagent (R_2CuLi) with an alkyl halide, R'X (24.1)
- Simmons–Smith reaction of an alkene with CH_2I_2 and Zn(Cu) to form a cyclopropane (24.5)

Alkenes

- Dehydrohalogenation of an alkyl halide with base (8.3)
- Dehydration of an alcohol with acid (9.8)
- Dehydration of an alcohol using $POCl_3$ and pyridine (9.10)
- β Elimination of an alkyl tosylate with base (9.13)
- Catalytic hydrogenation of an alkyne with H_2 + Lindlar catalyst to form a cis alkene (12.5B)
- Dissolving metal reduction of an alkyne with Na, NH_3 to form a trans alkene (12.5C)
- Wittig reaction (18.9)
- β Elimination of an α-halo carbonyl compound with Li_2CO_3, LiBr, and DMF (21.7C)
- Hofmann elimination of an amine (23.11)
- Coupling of an organocuprate reagent (R_2CuLi) with an organic halide, R'X (24.1)
- The palladium-catalyzed Suzuki reaction of a vinyl or aryl halide with a vinyl- or arylborane (24.2)
- The palladium-catalyzed Heck reaction of a vinyl or aryl halide with an alkene (24.3)
- The palladium-catalyzed Stille coupling of an organostannane with an organic halide (24.4)
- Olefin metathesis (24.7)

Alkyl halides

- Reaction of an alcohol with HX (9.11)
- Reaction of an alcohol with $SOCl_2$ or PBr_3 (9.12)
- Cleavage of an ether with HBr or HI (9.14)
- Hydrohalogenation of an alkene with HX (10.9)
- Halogenation of an alkene with X_2 (10.13)
- Hydrohalogenation of an alkyne with two equivalents of HX (11.7)
- Halogenation of an alkyne with two equivalents of X_2 (11.8)
- Radical halogenation of an alkane (13.3)
- Radical halogenation at an allylic carbon (13.10)
- Radical addition of HBr to an alkene (13.13)
- Electrophilic addition of HX to a 1,3-diene (14.10)
- Radical halogenation of an alkyl benzene (16.14A)
- Halogenation α to a carbonyl group (21.7)
- Addition of a dihalocarbene to an alkene to form a dihalocyclopropane (24.5)

Alkynes

- Dehydrohalogenation of an alkyl dihalide with base (11.5)
- S_N2 reaction of an alkyl halide with an acetylide anion, $^-C\equiv CR$ (11.11)

Amides

- Reaction of an acid chloride with NH_3 or an amine (20.7)
- Reaction of an anhydride with NH_3 or an amine (20.8)
- Reaction of a carboxylic acid with NH_3 or an amine and DCC (20.9)
- Reaction of an ester with NH_3 or an amine (20.10)

Amines

- Nucleophilic aromatic substitution (16.13)
- Reduction of a nitro group (16.14D)
- Reduction of an amide with $LiAlH_4$ (17.7B)
- Reduction of a nitrile (19.12B)
- S_N2 reaction using NH_3 or an amine (23.6A)
- Gabriel synthesis (23.6A)
- Reductive amination of an aldehyde or ketone (23.6C)

Amino acids

- Enantioselective hydrogenation using a chiral catalyst (27.3)

Anhydrides

- Reaction of an acid chloride with a carboxylate anion (20.7)
- Dehydration of a dicarboxylic acid (20.9)

Aryl halides

- Halogenation of benzene with X_2 + FeX_3 (16.3)
- Reaction of a diazonium salt with CuCl, CuBr, HBF_4, NaI, or KI (23.13A)

Carboxylic acids

- Oxidative cleavage of an alkyne with ozone (12.11)
- Oxidation of a 1° alcohol with CrO_3 (or a similar Cr^{6+} reagent), H_2O, H_2SO_4 (12.12B)
- Oxidation of an alkyl benzene with $KMnO_4$ (16.14B)
- Oxidation of an aldehyde (17.8)
- Reaction of a Grignard reagent with CO_2 (17.14A)
- Hydrolysis of a cyanohydrin (18.8)
- Hydrolysis of a nitrile (19.12A)
- Hydrolysis of an acid chloride (20.7)
- Hydrolysis of an anhydride (20.8)
- Hydrolysis of an ester (20.10)
- Hydrolysis of an amide (20.12)
- Malonic ester synthesis (21.9)

Cyanohydrins

- Addition of HCN to an aldehyde or ketone (18.8)

1,2-Diols

- Anti dihydroxylation of an alkene with a peroxyacid, followed by ring opening with ^{-}OH or H_2O (12.9A)
- Syn dihydroxylation of an alkene with $KMnO_4$ or OsO_4 (12.9B)

Enamines

- Reaction of an aldehyde or ketone with a 2° amine (18.11)

Epoxides

- Intramolecular S_N2 reaction of a halohydrin using base (9.6)
- Epoxidation of an alkene with mCPBA (12.8)
- Enantioselective epoxidation of an allylic alcohol with the Sharpless reagent (12.15)

Esters

- S_N2 reaction of an alkyl halide with a carboxylate anion, RCO_2^- (7.18)
- Reaction of an acid chloride with an alcohol (20.7)
- Reaction of an anhydride with an alcohol (20.8)
- Fischer esterification of a carboxylic acid with an alcohol (20.9)

Ethers

- Williamson ether synthesis—S_N2 reaction of an alkyl halide with an alkoxide, ^-OR (9.6)
- Reaction of an alkyl tosylate with an alkoxide, ^-OR (9.13)
- Addition of an alcohol to an alkene in the presence of acid (10.12)
- Anionic polymerization of epoxides to form polyethers (31.3)

Halohydrins

- Reaction of an epoxide with HX (9.16)
- Addition of X and OH to an alkene (10.15)

Imine

- Reaction of an aldehyde or ketone with a 1° amine (18.10)

Ketones

- Hydration of an alkyne with H_2O, H_2SO_4, and $HgSO_4$ (11.9)
- Oxidative cleavage of an alkene with O_3 followed by Zn or $(CH_3)_2S$ (12.10)
- Oxidation of a 2° alcohol with any Cr^{6+} reagent (12.12, 12.13)
- Friedel–Crafts acylation (16.5)
- Reaction of an acid chloride with an organocuprate reagent (17.13)
- Hydrolysis of an imine or enamine (18.11B)
- Hydrolysis of an acetal (18.13B)
- Reaction of a nitrile with a Grignard or organolithium reagent (19.12C)
- Acetoacetic ester synthesis (21.10)

Nitriles

- S_N2 reaction of an alkyl halide with NaCN (7.18, 19.12)
- Reaction of an aryl diazonium salt with CuCN (23.13A)

Phenols

- Reaction of an aryl diazonium salt with H_2O (23.13A)
- Nucleophilic aromatic substitution (16.13)

Sulfides

- Reaction of an alkyl halide with ^-SR (9.15)

Thiols

- Reaction of an alkyl halide with ^-SH (9.15)

Glossary

A

Acetal (Section 18.13): A compound having the general structure $R_2C(OR')_2$, where R = H, alkyl, or aryl. Acetals are used as protecting groups for aldehydes and ketones.

Acetoacetic ester synthesis (Section 21.10): A stepwise method that converts ethyl acetoacetate to a ketone having one or two carbons bonded to the α carbon.

Acetylation (Section 20.8): A reaction that transfers an acetyl group (CH_3CO-) from one atom to another.

Acetyl coenzyme A (Section 20.16): A biochemical thioester that acts as an acetylating reagent. Acetyl coenzyme A is often referred to as acetyl CoA.

Acetyl group (Section 18.2E): A substituent having the structure $-COCH_3$.

Acetylide anion (Sections 11.11, 17.9B): An anion formed by treating a terminal alkyne with a strong base. Acetylide anions have the general structure $R-C\equiv C^-$.

Achiral molecule (Section 5.3): A molecule that is superimposable upon its mirror image. An achiral molecule is not chiral.

Acid chloride (Sections 17.1, 20.1): A compound having the general structure RCOCl.

Acidity constant (Section 2.3): A value symbolized by K_a that represents the strength of an acid (HA). The larger the K_a, the stronger the acid.

$$K_a = \frac{[H_3O^+][A:^-]}{[H-A]}$$

Active site (Section 6.11): The region of an enzyme that binds the substrate.

Acyclic alkane (Section 4.1): A compound with the general formula C_nH_{2n+2}. Acyclic alkanes are also called saturated hydrocarbons because they contain the maximum number of hydrogen atoms per carbon.

Acylation (Sections 16.5A, 20.16): A reaction that transfers an acyl group from one atom to another.

Acyl chloride (Section 16.5A): A compound having the general structure RCOCl. Acyl chlorides are also called acid chlorides.

Acyl group (Section 16.5A): A substituent having the general structure RCO–.

Acylium ion (Section 16.5B): A positively charged electrophile having the general structure $(R-C\equiv O)^+$, formed when the Lewis acid $AlCl_3$ ionizes the carbon–halogen bond of an acid chloride.

Acyl transfer reaction (Section 20.16): A reaction that transfers an acyl group from one atom to another.

Adenosine 5'-diphosphate (ADP, Section 30.1): A nucleoside diphosphate formed from ATP in phosphorylation reactions.

Adenosine 5'-triphosphate (ATP, Sections 7.15A, 30.1): A nucleoside triphosphate formed by adding three phosphates to the 5'-OH of adenosine. ATP is the most prominent member of a group of "high-energy" molecules that release energy by cleavage of a P–O bond during hydrolysis.

1,2-Addition (Sections 14.10, 17.15): An addition reaction to a conjugated system that adds groups across two adjacent atoms.

1,4-Addition (Sections 14.10, 17.15): An addition reaction that adds groups to the atoms in the 1 and 4 positions of a conjugated system. 1,4-Addition is also called conjugate addition.

Addition polymer (Section 31.1): A polymer prepared by a chain reaction that adds a monomer to the growing end of a polymer chain. Addition polymers are also called chain-growth polymers.

Addition reaction (Sections 6.2C, 10.8): A reaction in which elements are added to a starting material. In an addition reaction, a π bond is broken and two σ bonds are formed.

Aglycon (Section 26.6C): The alcohol formed from hydrolysis of a glycoside.

Alcohol (Section 9.1): A compound having the general structure ROH. An alcohol contains a hydroxy group (OH group) bonded to an sp^3 hybridized carbon atom.

Aldaric acid (Section 26.8B): The dicarboxylic acid formed by the oxidation of the aldehyde and the primary alcohol of an aldose.

Aldehyde (Section 11.10): A compound having the general structure RCHO, where R = H, alkyl, or aryl.

Alditol (Section 26.8A): A compound formed by the reduction of the aldehyde of an aldose to a primary alcohol.

Aldol condensation (Section 22.1C): An aldol reaction in which the initially formed β-hydroxy carbonyl compound loses water by dehydration.

Aldol reaction (Section 22.1A): A reaction in which two molecules of an aldehyde or ketone react with each other in the presence of base to form a β-hydroxy carbonyl compound.

Aldonic acid (Section 26.8B): A compound formed by the oxidation of the aldehyde of an aldose to a carboxylic acid.

Aldose (Section 26.2): A monosaccharide comprised of a polyhydroxy aldehyde.

Aliphatic (Section 3.2A): A compound or portion of a compound made up of C–C σ and π bonds but not aromatic bonds.

Alkaloid (Section 23.5A): A basic, nitrogen-containing compound isolated from a plant source.

Alkane (Section 4.1): An aliphatic hydrocarbon having only C–C and C–H σ bonds.

Alkene (Section 8.2A): An aliphatic hydrocarbon that contains a carbon–carbon double bond.

Alkoxide (Sections 8.1, 9.6): An anion having the general structure RO^-, formed by deprotonating an alcohol with a base.

Alkoxy group (Section 9.3B): A substituent containing an alkyl group bonded to an oxygen (RO group).

Alkylation (Section 21.8): A reaction that transfers an alkyl group from one atom to another.

Alkyl group (Section 4.4A): A group formed by removing one hydrogen from an alkane. Alkyl groups are named by replacing the suffix *-ane* of the parent alkane with *-yl.*

Alkyl halide (Section 7.1): A compound containing a halogen atom bonded to an sp^3 hybridized carbon atom. Alkyl halides have the general molecular formula $C_nH_{2n+1}X$.

1,2-Alkyl shift (Section 9.9): The rearrangement of a less stable carbocation to a more stable carbocation by the shift of an alkyl group from one carbon atom to an adjacent carbon atom.

Alkyl tosylate (Section 9.13): A compound having the general structure $ROSO_2C_6H_4CH_3$. Alkyl tosylates are also called tosylates and are abbreviated as ROTs.

Alkyne (Section 8.10): An aliphatic hydrocarbon that contains a carbon–carbon triple bond.

Allyl carbocation (Section 14.1B): A carbocation that has a positive charge on the atom adjacent to a carbon–carbon double bond. An allyl carbocation is resonance stabilized.

Allyl group (Section 10.3C): A substituent having the structure $-CH_2-CH=CH_2$.

Allylic bromination (Section 13.10A): A radical substitution reaction in which bromine replaces a hydrogen atom on the carbon adjacent to a carbon–carbon double bond.

Allylic carbon (Section 13.10): A carbon atom bonded to a carbon–carbon double bond.

Allylic halide (Section 7.1): A molecule containing a halogen atom bonded to the carbon atom adjacent to a carbon–carbon double bond.

Allyl radical (Section 13.10): A radical that has an unpaired electron on the carbon adjacent to a carbon–carbon double bond. An allyl radical is resonance stabilized.

Alpha (α) carbon (Sections 8.1, 18.2B): In an elimination reaction, the carbon that is bonded to the leaving group. In a carbonyl compound, the carbon that is bonded to the carbonyl carbon.

Ambident nucleophile (Section 21.3C): A nucleophile that has two reactive sites.

Amide (Sections 17.1, 20.1): A compound having the general structure $RCONR'_2$, where R' = H or alkyl.

Amide base (Sections 8.10, 21.3B): A nitrogen-containing base formed by deprotonating an amine or ammonia.

Amine (Sections 18.10, 23.1): A basic organic nitrogen compound having the general structure RNH_2, R_2NH, or R_3N. An amine has a nonbonded pair of electrons on the nitrogen atom.

α-Amino acid (Sections 19.11A, 27.1): A compound having the general structure $RCH(NH_2)COOH$. α-Amino acids are the building blocks of proteins.

Amino acid residue (Section 27.4): The individual amino acids in peptides and proteins.

Amino group (Section 23.3D): A substituent having the structure $-NH_2$.

Amino sugar (Section 26.12A): A carbohydrate that contains an NH_2 group instead of a hydroxy group at a non-anomeric carbon.

Ammonium salt (Section 23.1): A compound containing a positively charged nitrogen with four σ bonds; for example, $R_4N^+X^-$.

Anabolism (Section 30.2): In metabolism, the synthesis of large molecules from smaller ones, often absorbing energy.

Angle strain (Section 4.10): An increase in the energy of a molecule resulting when the bond angles of the sp^3 hybridized atoms deviate from the optimum tetrahedral angle of 109.5°.

Angular methyl group (Section 29.8A): A methyl group located at the ring junction of two fused rings of the steroid skeleton.

Anhydride (Section 20.1): A compound having the general structure $(RCO)_2O$.

Aniline (Section 23.3C): A compound having the structure $C_6H_5NH_2$.

Anion (Section 1.2): A negatively charged ion that results from a neutral atom gaining one or more electrons.

Anionic polymerization (Section 31.2C): Chain-growth polymerization of alkenes substituted by electron-withdrawing groups that stabilize intermediate anions.

Annulation (Section 22.9): A reaction that forms a new ring.

Annulene (Section 15.8A): A hydrocarbon containing a single ring with alternating double and single bonds.

α Anomer (Section 26.5): The stereoisomer of a cyclic monosaccharide in which the anomeric OH and the CH_2OH groups are trans. In a D monosaccharide, the hydroxy group on the anomeric carbon is drawn down.

β Anomer (Section 26.5): The stereoisomer of a cyclic monosaccharide in which the anomeric OH and the CH_2OH groups are cis. In a D monosaccharide, the hydroxy group on the anomeric carbon is drawn up.

Anomeric carbon (Section 26.5): The stereogenic center at the hemiacetal carbon of a cyclic monosaccharide.

Antarafacial reaction (Section 25.4): A pericyclic reaction that occurs on opposite sides of the two ends of the π electron system.

Anti addition (Section 10.8): An addition reaction in which the two parts of a reagent are added from opposite sides of a double bond.

Antiaromatic compound (Section 15.7): An organic compound that is cyclic, planar, completely conjugated, and has $4n$ π electrons.

Antibonding molecular orbital (Section 15.10A): A high-energy molecular orbital formed when two atomic orbitals of opposite phase overlap.

Anticodon (Section 28.5): A sequence of three nucleotides in a tRNA molecule, which is complementary to three bases in an mRNA molecule and identifies what amino acid must be added to a growing polypeptide chain.

Anti conformation (Section 4.9): A staggered conformation in which the two larger groups on adjacent carbon atoms have a dihedral angle of 180°.

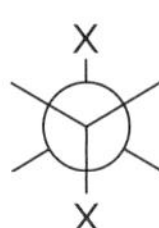

Anti dihydroxylation (Section 12.9A): The addition of two hydroxy groups to opposite faces of a double bond.

Antioxidant (Section 13.12): A compound that stops an oxidation from occurring.

Anti periplanar (Section 8.6A): In an elimination reaction, a geometry where the β hydrogen and the leaving group are on opposite sides of the molecule.

Aromatic compound (Section 15.1): A planar, cyclic organic compound that has p orbitals on all ring atoms and a total of $4n + 2$ π electrons in the orbitals.

Aryl group (Section 15.3D): A substituent formed by removing one hydrogen atom from an aromatic ring.

Aryl halide (Sections 7.1, 16.3): A molecule such as C_6H_5X, containing a halogen atom X bonded to an aromatic ring.

Asymmetric carbon (Section 5.3): A carbon atom that is bonded to four different groups. An asymmetric carbon is also called a stereogenic center, a chiral center, or a chirality center.

Asymmetric reaction (Sections 12.15, 17.6A, 22.2C, 27.3): A reaction that converts an achiral starting material to predominantly one enantiomer.

Atactic polymer (Section 31.4): A polymer having the substituents randomly oriented along the carbon backbone of an elongated polymer chain.

Atomic number (Section 1.1): The number of protons in the nucleus of an element.

Atomic weight (Section 1.1): The weighted average of the mass of all isotopes of a particular element. The atomic weight is reported in atomic mass units (amu).

Axial bonds (Section 4.11A): Bonds located above or below and perpendicular to the plane of the chair conformation of cyclohexane. Three axial bonds point upwards (on the up carbons) and three axial bonds point downwards (on the down carbons).

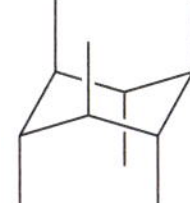

Azo compound (Section 23.14): A compound having the general structure RN=NR'.

B

Backside attack (Section 7.10C): Approach of a nucleophile from the side opposite the leaving group.

Barrier to rotation (Section 4.9): The energy difference between the lowest- and highest-energy conformations of a molecule.

Base peak (Section A.1): The peak in the mass spectrum having the greatest abundance value.

Basicity (Section 7.8): A measure of how readily an atom donates its electron pair to a proton.

Benedict's reagent (Section 26.8B): A reagent for oxidizing aldehydes to carboxylic acids using a Cu^{2+} salt, forming brick-red Cu_2O as a side product.

Benzoyl group (Section 18.2E): A substituent having the structure $-COC_6H_5$.

Benzyl group (Section 15.3D): A substituent having the structure $C_6H_5CH_2-$.

Benzylic halide (Sections 7.1, 16.14): A compound such as $C_6H_5CH_2X$, containing a halogen atom X bonded to a carbon that is bonded to a benzene ring.

Benzyne (Section 16.13B): A reactive intermediate formed by elimination of HX from an aryl halide.

Beta (β) carbon (Sections 8.1, 18.2B): In an elimination reaction, the carbon adjacent to the carbon with the leaving group. In a carbonyl compound, the carbon located two carbons from the carbonyl carbon.

Bimolecular reaction (Sections 6.9B, 7.9, 7.10A): A reaction in which the concentration of both reactants affects the reaction rate and both terms appear in the rate equation. In a bimolecular reaction, two reactants are involved in the only step or the rate-determining step.

Biodegradable polymer (Section 31.9B): A polymer that can be degraded by microorganisms naturally present in the environment.

Biomolecule (Section 3.9): An organic compound found in a biological system.

Boat conformation of cyclohexane (Section 4.11B): An unstable conformation adopted by cyclohexane that resembles a boat. The instability of the boat conformation results from torsional strain and steric strain. The boat conformation of cyclohexane is 30 kJ/mol less stable than the chair conformation.

Boiling point (Section 3.4A): The temperature at which molecules in the liquid phase are converted to the gas phase. Molecules with stronger intermolecular forces have higher boiling points. Boiling point is abbreviated as bp.

Bond dissociation energy (Section 6.4): The amount of energy needed to homolytically cleave a covalent bond.

Bonding (Section 1.2): The joining of two atoms in a stable arrangement. Bonding is a favorable process that leads to lowered energy and increased stability.

Bonding molecular orbital (Section 15.10A): A low-energy molecular orbital formed when two atomic orbitals of similar phase overlap.

Bond length (Section 1.7A): The average distance between the centers of two bonded nuclei. Bond lengths are reported in picometers (pm).

Branched-chain alkane (Section 4.1A): An acyclic alkane that has alkyl substituents bonded to the parent carbon chain.

Bridged ring system (Section 14.13D): A bicyclic ring system in which the two rings share non-adjacent carbon atoms.

Bromination (Sections 10.13, 13.6, 16.3): The reaction of a compound with bromine.

Bromohydrin (Section 10.15): A compound having a bromine and a hydroxy group on adjacent carbon atoms.

Brønsted–Lowry acid (Section 2.1): A proton donor, symbolized by HA. A Brønsted–Lowry acid must contain a hydrogen atom.

Brønsted–Lowry base (Section 2.1): A proton acceptor, symbolized by :B. A Brønsted–Lowry base must be able to form a bond to a proton by donating an available electron pair.

C

^{13}C NMR spectroscopy (Section C.1): A form of nuclear magnetic resonance spectroscopy used to determine the type of carbon atoms in a molecule.

Cahn–Ingold–Prelog system of nomenclature (Section 5.6): The system of designating a stereogenic center as either *R* or *S* according to the arrangement of the four groups attached to the center.

Carbamate (Sections 27.6, 31.6): A functional group containing a carbonyl group bonded to both an oxygen and a nitrogen atom. A carbamate is also called a urethane.

Carbanion (Section 2.5D): An ion with a negative charge on a carbon atom.

Carbene (Section 24.5): A neutral reactive intermediate having the general structure $:CR_2$. A carbene contains a divalent carbon surrounded by six electrons, making it a highly reactive electrophile that adds to C=C double bonds.

Carbinolamine (Section 18.6B): An unstable intermediate having a hydroxy group and an amine group on the same carbon. A carbinolamine is formed during the addition of an amine to a carbonyl group.

Carbocation (Section 7.12C): A positively charged carbon atom. A carbocation is sp^2 hybridized and trigonal planar, and contains a vacant *p* orbital.

Carbohydrate (Sections 18.16, 26.1): A polyhydroxy aldehyde or ketone or a compound that can be hydrolyzed to a polyhydroxy aldehyde or ketone.

Carbonate (Section 31.6D): A compound having the general structure $(RO)_2C{=}O$.

Carbon backbone (Section 3.1): The C–C and C–H σ bond framework that makes up the skeleton of an organic molecule.

Carbon NMR spectroscopy (Section C.1): A form of nuclear magnetic resonance spectroscopy used to determine the type of carbon atoms in a molecule.

Carbonyl group (Sections 3.2C, 11.9, 17.1): A functional group that contains a carbon–oxygen double bond (C=O). The polar carbon–oxygen bond makes the carbonyl carbon electrophilic.

Carboxy group (Section 19.1): A functional group having the structure COOH.

Carboxylate anion (Section 19.2C): An anion having the general structure RCO_2^-, formed by deprotonating a carboxylic acid with a Brønsted–Lowry base.

Carboxylation (Section 17.14): The reaction of an organometallic reagent with CO_2 to form a carboxylic acid after protonation.

Carboxylic acid (Section 19.1): A compound having the general structure RCO_2H.

Carboxylic acid derivatives (Section 17.1): Compounds having the general structure RCOZ, which can be synthesized from carboxylic acids. Common carboxylic acid derivatives include acid chlorides, anhydrides, esters, and amides.

Catabolism (Section 30.2): In metabolism, the breakdown of large molecules into smaller ones, often releasing energy.

Catalyst (Section 6.10): A substance that speeds up the rate of a reaction, but is recovered unchanged at the end of the reaction and does not appear in the product.

Catalytic hydrogenation (Section 12.3): A reduction reaction involving the addition of H_2 to a π bond in the presence of a metal catalyst.

Cation (Section 1.2): A positively charged ion that results from a neutral atom losing one or more electrons.

Cationic polymerization (Section 31.2C): Chain-growth polymerization of alkene monomers involving carbocation intermediates.

CBS reagent (Section 17.6A): A chiral reducing agent formed by reacting an oxazaborolidine with BH_3. CBS reagents predictably give one enantiomer as the major product of ketone reduction.

Cephalin (Section 29.4A): A phosphoacylglycerol in which the phosphodiester alkyl group is $-CH_2CH_2NH_3^+$. Cephalins are also called phosphatidylethanolamines.

Chain-growth polymer (Section 31.1): A polymer prepared by a chain reaction that adds a monomer to the growing end of a polymer chain. Chain-growth polymers are also called addition polymers.

Chain mechanism (Section 13.4A): A reaction mechanism that involves repeating steps.

Chair conformation of cyclohexane (Section 4.11A): A stable conformation adopted by cyclohexane that resembles a chair. The stability of the chair conformation results from the elimination of angle strain (all C–C–C bond angles are 109.5°) and torsional strain (all groups on adjacent carbon atoms are staggered).

Chemical shift (Section C.1B): The position of an absorption signal on the *x* axis in an NMR spectrum relative to the reference signal of tetramethylsilane.

Chirality center (Section 5.3): A carbon atom bonded to four different groups. A chirality center is also called a chiral center, a stereogenic center, and an asymmetric center.

Chiral molecule (Section 5.3): A molecule that is not superimposable upon its mirror image.

Chlorination (Sections 10.14, 13.5, 16.3): The reaction of a compound with chlorine.

Chlorofluorocarbons (Sections 7.4, 13.9): Synthetic alkyl halides having the general molecular formula CF_xCl_{4-x}. Chlorofluorocarbons, abbreviated as CFCs, were used as refrigerants and aerosol propellants and contribute to the destruction of the ozone layer.

Chlorohydrin (Section 10.15): A compound having a chlorine and a hydroxy group on adjacent carbon atoms.

Chromate ester (Section 12.12A): An intermediate in the chromium-mediated oxidation of an alcohol having the general structure $R-O-CrO_3H$.

***s*-Cis** (Sections 14.6, 27.4B): The conformation of a 1,3-diene that has the two double bonds on the same side of the single bond that joins them.

Cis isomer (Sections 4.12B, 8.2B): An isomer of a ring or double bond that has two groups on the same side of the ring or double bond.

Citric acid cycle (Section 30.7): A cyclic, eight-step metabolic pathway that begins with the addition of acetyl CoA to oxaloacetate. Overall the citric acid cycle forms two molecules of CO_2, four molecules of reduced coenzymes (NADH and $FADH_2$), and one molecule of GTP.

Claisen reaction (Section 22.5): A reaction between two molecules of an ester in the presence of base to form a β-keto ester.

Claisen rearrangement (Section 25.5): A [3,3] sigmatropic rearrangement of an unsaturated ether to a γ,δ-unsaturated carbonyl compound.

α Cleavage (Section A.3B): A fragmentation in mass spectrometry that results in cleavage of a carbon–carbon bond. With aldehydes and ketones, α cleavage results in breaking the bond between the carbonyl carbon and the carbon adjacent to it. With alcohols, α cleavage occurs by breaking a bond between an alkyl group and the carbon that bears the OH group.

Clemmensen reduction (Section 16.14C): A method to reduce aryl ketones to alkyl benzenes using Zn(Hg) in the presence of a strong acid.

Codon (Section 28.6): A set of three nucleotides in mRNA that corresponds to a particular amino acid. The order of codons in an mRNA molecule determines the amino acid sequence of a protein.

Coenzyme (Section 12.13): A compound that acts with an enzyme to carry out a biochemical process.

Combustion (Section 4.13B): An oxidation–reduction reaction, in which an alkane or other organic compound reacts with oxygen to form CO_2 and H_2O, releasing energy.

Common name (Section 4.4C): The name of a molecule that was adopted prior to and therefore does not follow the IUPAC system of nomenclature.

Compound (Section 1.2): The structure that results when two or more elements are joined together in a stable arrangement.

Concerted reaction (Sections 6.3, 7.10B): A reaction in which all bond forming and bond breaking occurs in one step.

Condensation polymer (Sections 20.15A, 31.1): A polymer formed when monomers containing two functional groups come together with loss of a small molecule such as water or HCl. Condensation polymers are also called step-growth polymers.

Condensation reaction (Section 22.1C): A reaction in which a small molecule, often water, is eliminated during the reaction process.

Condensed structure (Section 1.8A): A shorthand representation of the structure of a compound in which all atoms are drawn in but bonds and lone pairs are usually omitted. Parentheses are used to denote similar groups bonded to the same atom.

Configuration (Section 5.2): A particular three-dimensional arrangement of atoms.

Conformations (Section 4.8): The different arrangements of atoms that are interconverted by rotation about single bonds.

Conjugate acid (Section 2.2): The compound that results when a base gains a proton in a proton transfer reaction.

Conjugate addition (Sections 14.10, 17.15): An addition reaction that adds groups to the atoms in the 1 and 4 positions of a conjugated system. Conjugate addition is also called 1,4-addition.

Conjugate base (Section 2.2): The compound that results when an acid loses a proton in a proton transfer reaction.

Conjugated diene (Section 14.1A): A compound that contains two carbon–carbon double bonds joined by a single σ bond. Pi (π) electrons are delocalized over both double bonds. Conjugated dienes are also called 1,3-dienes.

Conjugated protein (Section 27.9C): A structure composed of a protein unit and a non-protein molecule.

Conjugation (Section 14.1): The overlap of *p* orbitals on three or more adjacent atoms.

Conrotatory rotation (Section 25.3): Rotation of *p* orbitals in the same direction during electrocyclic ring closure or ring opening.

Constitutional isomers (Sections 1.4, 4.1A, 5.2): Two compounds that have the same molecular formula, but differ in the way the atoms are connected to each other. Constitutional isomers are also called structural isomers.

Coordination polymerization (Section 31.4): A polymerization reaction that uses a homogeneous catalyst that is soluble in the reaction solvents typically used.

Cope rearrangement (Section 25.5): A [3,3] sigmatropic rearrangement of a 1,5-diene to an isomeric 1,5-diene.

Copolymer (Section 31.2D): A polymer prepared by joining two or more different monomers together.

Counterion (Section 2.1): An ion that does not take part in a reaction and is opposite in charge to the ion that does take part in the reaction. A counterion is also called a spectator ion.

Coupled reactions (Section 30.1): Two reactions paired together to drive an unfavorable process. The energy released by one reaction is used to drive the other reaction.

Coupling constant (Section C.6A): The frequency difference, measured in Hz, between the peaks in a split NMR signal.

Coupling reaction (Section 23.15): A reaction that forms a bond between two discrete molecules.

Covalent bond (Section 1.2): A bond that results from the sharing of electrons between two nuclei. A covalent bond is a two-electron bond.

Crossed aldol reaction (Section 22.2): An aldol reaction in which the two reacting carbonyl compounds are different. A crossed aldol reaction is also called a mixed aldol reaction.

Crossed Claisen reaction (Section 22.6): A Claisen reaction in which the two reacting esters are different.

Crown ether (Section 3.7B): A cyclic ether containing multiple oxygen atoms. Crown ethers bind specific cations depending on the size of their central cavity.

Curved arrow notation (Section 1.6A): A convention that shows the movement of an electron pair. The tail of the arrow begins at the electron pair and the head points to where the electron pair moves.

Cyanide anion (Section 18.8A): An anion having the structure $^{-}C\equiv N$.

Cyano group (Section 19.1): A functional group consisting of a carbon–nitrogen triple bond ($C\equiv N$).

Cyanohydrin (Section 18.8): A compound having the general structure $RCH(OH)C\equiv N$. A cyanohydrin results from the addition of HCN across the carbonyl of an aldehyde or a ketone.

Cycloaddition (Section 25.1): A pericyclic reaction between two compounds with π bonds to form a cyclic product with two new σ bonds.

Cycloalkane (Sections 4.1, 4.2): A compound that contains carbons joined in one or more rings. Cycloalkanes with one ring have the general formula C_nH_{2n}.

Cyclopropanation (Section 24.5): An addition reaction to a carbon–carbon double bond that forms a cyclopropane.

D

D-Sugar (Section 26.2C): A sugar with the hydroxy group on the stereogenic center farthest from the carbonyl on the right side in the Fischer projection formula.

Decalin (Section 29.8A): Two fused six-membered rings. *cis*-Decalin has the hydrogen atoms at the ring fusion on the same side of the rings, whereas *trans*-decalin has the hydrogen atoms at the ring fusion on opposite sides of the rings.

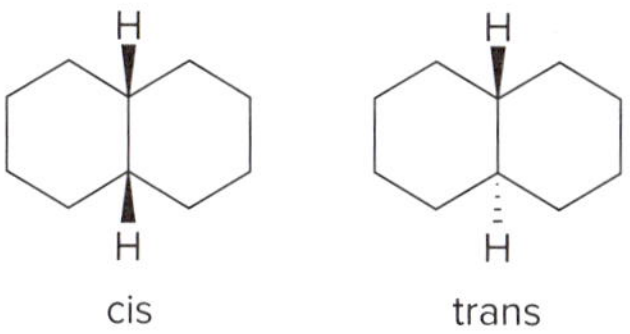

Decarboxylation (Section 21.9A): Loss of CO_2 through cleavage of a carbon–carbon bond.

Degenerate orbitals (Section 15.10B): Orbitals (either atomic or molecular) having the same energy.

Degree of unsaturation (Section 10.2): A ring or a π bond in a molecule. The number of degrees of unsaturation compares the number of hydrogens in a compound to that of a saturated hydrocarbon containing the same number of carbons.

Dehydration (Sections 9.8, 20.9B): A reaction that results in the loss of the elements of water from the reaction components.

Dehydrogenase (Section 27.10): An enzyme that catalyzes the addition or removal of two hydrogen atoms from a substrate.

Dehydrohalogenation (Section 8.1): An elimination reaction in which the elements of hydrogen and halogen are lost from a starting material.

Delta (δ) scale (Section C.1B): A common scale of chemical shifts used in NMR spectroscopy in which the absorption due to tetramethylsilane (TMS) occurs at zero parts per million.

Deoxy (Section 28.1): A prefix that means without oxygen.

Deoxyribonucleic acid (DNA, Section 28.1): The nucleic acid that stores the genetic information of an organism and transmits that information from one generation to another.

Deoxyribonucleoside (Section 28.1): An *N*-glycoside formed by the reaction of D-2-deoxyribose with certain amine heterocycles.

Deoxyribonucleotide (Section 28.1): A DNA building block having a deoxyribose and either a purine or pyrimidine base joined together by an *N*-glycosidic linkage, and a phosphate bonded to a hydroxy group of the sugar nucleus.

Deprotection (Section 17.12): A reaction that removes a protecting group, regenerating a functional group.

Deshielding effects (Section C.3A): An effect in NMR caused by a decrease in electron density, thus increasing the strength of the magnetic field felt by the nucleus. Deshielding shifts an absorption downfield.

Dextrorotatory (Section 5.12A): Rotating plane-polarized light in the clockwise direction. The rotation is labeled *d* or (+).

1,3-Diacid (Section 21.9A): A compound containing two carboxylic acids separated by a single carbon atom. 1,3-Diacids are also called β-diacids.

Dialkylamide (Section 21.3B): An amide base having the general structure R_2N^{-}.

Diastereomers (Section 5.7): Stereoisomers that are not mirror images of each other. Diastereomers have the same *R,S* designation for at least one stereogenic center and the opposite *R,S* designation for at least one of the other stereogenic centers.

Diastereotopic protons (Section C.2C): Two hydrogen atoms on the same carbon such that substitution of either hydrogen with a group Z forms diastereomers. The two hydrogen atoms are not equivalent and give two NMR signals.

1,3-Diaxial interaction (Section 4.12A): A steric interaction between two axial substituents of the chair form of cyclohexane. Larger axial substituents create unfavorable 1,3-diaxial interactions, destabilizing a cyclohexane conformation.

Diazonium salt (Section 23.12A): An ionic salt having the general structure $(R{-}N{\equiv}N)^+Cl^-$.

Diazotization reaction (Section 23.12A): A reaction that converts 1° alkylamines and arylamines to diazonium salts.

1,3-Dicarbonyl compound (Section 21.2): A compound containing two carbonyl groups separated by a single carbon atom.

1,4-Dicarbonyl compound (Section 22.4): A dicarbonyl compound in which the carbonyl groups are separated by three single bonds. 1,4-Dicarbonyl compounds can undergo intramolecular reactions to form five-membered rings.

1,5-Dicarbonyl compound (Section 22.4): A dicarbonyl compound in which the carbonyl groups are separated by four single bonds. 1,5-Dicarbonyl compounds can undergo intramolecular reactions to form six-membered rings.

Dieckmann reaction (Section 22.7): An intramolecular Claisen reaction of a diester to form a ring, typically a five- or six-membered ring.

Diels–Alder reaction (Section 14.12): An addition reaction between a 1,3-diene and a dienophile to form a cyclohexene ring.

1,3-Diene (Section 14.1A): A compound containing two carbon–carbon double bonds joined by a single σ bond. Pi (π) electrons are delocalized over both double bonds. 1,3-Dienes are also called conjugated dienes.

Dienophile (Section 14.12): The alkene component in a Diels–Alder reaction that reacts with a 1,3-diene.

Dihedral angle (Section 4.8): The angle that separates a bond on one atom from a bond on an adjacent atom.

Dihydroxylation (Section 12.9): Addition of two hydroxy groups to a double bond to form a 1,2-diol.

Diol (Section 9.3A): A compound possessing two hydroxy groups. Diols are also called glycols.

Dipeptide (Section 27.4): Two amino acids joined by one amide bond.

Diphosphate (Section 7.15): A good leaving group that is often used in biological systems. Diphosphate $(P_2O_7^{4-})$ is abbreviated as PP_i.

Dipole (Section 1.12): A partial separation of electronic charge.

Dipole–dipole interaction (Section 3.3B): An attractive intermolecular interaction between the permanent dipoles of polar molecules. The dipoles of adjacent molecules align so that the partial positive and partial negative charges are in close proximity.

Directed aldol reaction (Section 22.3): A crossed aldol reaction in which the enolate of one carbonyl compound is formed, followed by addition of the second carbonyl compound.

Disaccharide (Section 26.10): A carbohydrate containing two monosaccharide units joined by a glycosidic linkage.

Disproportionation (Section 31.2): A method of chain termination in radical polymerization involving the transfer of a hydrogen atom from one polymer radical to another, forming a new C—H bond on one polymer chain and a new double bond on the other.

Disrotatory rotation (Section 25.3): Rotation of *p* orbitals in opposite directions during electrocyclic ring closure or ring opening.

Dissolving metal reduction (Section 12.2): A reduction reaction using alkali metals as a source of electrons and liquid ammonia as a source of protons.

Disubstituted alkene (Section 8.2A): An alkene that has two alkyl groups and two hydrogens bonded to the carbons of the double bond ($R_2C{=}CH_2$ or $RCH{=}CHR$).

Disulfide (Sections 9.15A, 27.4C): A compound having the general structure RSSR', often formed between the side chain of two cysteine residues.

Diterpene (Section 29.7A): A terpene that contains 20 carbons and four isoprene units. A diterpenoid contains at least one oxygen atom as well.

Doublet (Section C.6): An NMR signal that is split into two peaks of equal area, caused by one nearby nonequivalent proton.

Doublet of doublets (Section C.8): A splitting pattern of four peaks observed when a signal is split by two different nonequivalent protons.

Downfield shift (Section C.1B): In an NMR spectrum, a term used to describe the relative location of an absorption signal. A downfield shift means the signal is shifted to the left in the spectrum to higher chemical shift on the δ scale.

E

E1 mechanism (Sections 8.3, 8.7): An elimination mechanism that goes by a two-step process involving a carbocation intermediate. E1 is an abbreviation for "Elimination Unimolecular."

E1cB mechanism (Section 22.1C): A two-step elimination mechanism that goes by a carbanion intermediate. E1cB stands for "Elimination Unimolecular, Conjugate Base."

E2 mechanism (Sections 8.3, 8.4): An elimination mechanism that goes by a one-step concerted process, in which both reactants are involved in the transition state. E2 is an abbreviation for "Elimination Bimolecular."

Eclipsed conformation (Section 4.8): A conformation of a molecule where the bonds on one carbon are directly aligned with the bonds on the adjacent carbon.

Edman degradation (Section 27.5B): A procedure used in peptide sequencing in which amino acids are cleaved one at a time from the N-terminal end, the identity of the amino acid determined, and the process repeated until the entire sequence is known.

Eicosanoids (Section 29.6): A group of biologically active compounds containing 20 carbon atoms derived from arachidonic acid.

Elastomer (Section 31.5): A polymer that stretches when stressed but then returns to its original shape.

Electrocyclic ring closure (Section 25.1): An intramolecular pericyclic reaction that forms a cyclic product containing one more σ bond and one fewer π bond than the reactant.

Electrocyclic ring-opening reaction (Section 25.1): A pericyclic reaction in which a σ bond of a cyclic reactant is cleaved to form a conjugated product with one more π bond.

Electromagnetic radiation (Section B.1): Radiant energy having dual properties of both waves and particles. The electromagnetic spectrum contains the complete range of electromagnetic radiation, arbitrarily divided into different regions.

Electron-donating inductive effect (Section 7.12A): An inductive effect in which an electropositive atom or polarizable group donates electron density through σ bonds to another atom.

Electronegativity (Section 1.12): A measure of an atom's attraction for electrons in a bond. Electronegativity indicates how much a particular atom "wants" electrons.

Electron-withdrawing inductive effect (Sections 2.5, 7.12A): An inductive effect in which a nearby electronegative atom pulls electron density toward itself through σ bonds.

Electrophile (Section 2.8): An electron-deficient compound, often symbolized by E^+, which can accept a pair of electrons from an electron-rich compound, forming a covalent bond. Lewis acids are electrophiles.

Electrophilic addition reaction (Section 10.9): An addition reaction in which the first step of the mechanism involves addition of the electrophilic end of a reagent to a π bond.

Electrophilic aromatic substitution (Section 16.1): A characteristic reaction of benzene in which a hydrogen atom on the ring is replaced by an electrophile.

Electrospray ionization (Section A.4C): A method for ionizing large biomolecules in a mass spectrometer. Electrospray ionization is abbreviated as ESI.

Electrostatic potential map (Section 1.12): A color-coded map that illustrates the distribution of electron density in a molecule. Electron-rich regions are indicated in red, and electron-deficient regions are indicated in blue. Regions of intermediate electron density are shown in orange, yellow, and green.

α Elimination (Section 24.5): An elimination reaction involving the loss of two elements from the same atom.

β Elimination (Section 8.1): An elimination reaction involving the loss of elements from two adjacent atoms.

Elimination reaction (Sections 6.2B, 8.1): A chemical reaction in which elements of the starting material are "lost" and a π bond is formed.

Enamine (Section 18.11): A compound having an amine nitrogen atom bonded to a carbon–carbon double bond [$R_2C{=}CH(NR'_2)$].

Enantiomeric excess (Section 5.12D): A measurement of how much one enantiomer is present in excess of the racemic mixture. Enantiomeric excess (*ee*) is also called optical purity; *ee* = % of one enantiomer − % of the other enantiomer.

Enantiomers (Section 5.3): Stereoisomers that are mirror images but are not superimposable upon each other. Enantiomers have the exact opposite *R,S* designation at every stereogenic center.

Enantioselective reaction (Sections 12.15, 17.6A, 22.2C, 27.3): A reaction that affords predominantly or exclusively one enantiomer. Enantioselective reactions are also called asymmetric reactions.

Enantiotopic protons (Section C.2C): Two hydrogen atoms on the same carbon such that substitution of either hydrogen with a group Z forms enantiomers. The two hydrogen atoms are equivalent and give a single NMR signal.

Endo position (Section 14.13D): A position of a substituent on a bridged bicyclic compound in which the substituent is closer to the longer bridge that joins the two carbons common to both rings.

Endothermic reaction (Section 6.4): A reaction in which the energy of the products is higher than the energy of the reactants. In an endothermic reaction, energy is absorbed and the $\Delta H°$ is a positive value.

Energy of activation (Section 6.7): The energy difference between the transition state and the starting material. The energy of activation, symbolized by E_a, is the minimum amount of energy needed to break bonds in the reactants.

Energy diagram (Section 6.7): A schematic representation of the energy changes that take place as reactants are converted to products. An energy diagram indicates how readily a reaction proceeds, how many steps are involved, and how the energies of the reactants, products, and intermediates compare.

Enolate (Sections 17.15, 21.3): A resonance-stabilized anion formed when a base removes an α hydrogen from the α carbon to a carbonyl group.

Enol tautomer (Sections 9.1, 11.9, 17.15): A compound having a hydroxy group bonded to a carbon–carbon double bond. An enol tautomer [such as $CH_2{=}C(OH)CH_3$] is in equilibrium with its keto tautomer [$(CH_3)_2C{=}O$].

Enthalpy change (Section 6.4): The energy absorbed or released in a reaction. Enthalpy change is symbolized by $\Delta H°$ and is also called the heat of reaction.

Entropy (Section 6.6): A measure of the randomness in a system. The more freedom of motion or the more disorder present, the higher the entropy. Entropy is denoted by the symbol $S°$.

Entropy change (Section 6.6): The change in the amount of disorder between reactants and products in a reaction. The entropy change is denoted by the symbol $\Delta S°$. $\Delta S° = S°_{products} - S°_{reactants}$.

Enzyme (Section 6.11): A biochemical catalyst composed of at least one chain of amino acids held together in a very specific three-dimensional shape.

Enzyme–substrate complex (Section 6.11): A structure having a substrate bonded to the active site of an enzyme.

Epoxidation (Section 12.8): Addition of a single oxygen atom to an alkene to form an epoxide.

Epoxide (Section 9.1): A cyclic ether having the oxygen atom as part of a three-membered ring. Epoxides are also called oxiranes.

Epoxy resin (Section 31.6F): A step-growth polymer formed from a fluid prepolymer and a hardener that cross-links polymer chains together.

Equatorial bonds (Section 4.11A): Bonds located in the plane of the chair conformation of cyclohexane (around the equator). Three equatorial bonds point slightly upwards (on the down carbons) and three equatorial bonds point slightly downwards (on the up carbons).

Equilibrium constant (Section 6.5A): A mathematical expression, denoted by the symbol K_{eq}, which relates the amount of starting material and product at equilibrium. K_{eq} = [products]/[starting materials].

Essential oil (Section 29.7): A class of terpenes isolated from plant sources by distillation.

Ester (Sections 17.1, 20.1): A compound having the general structure RCOOR'.

Esterification (Section 20.9C): A reaction that converts a carboxylic acid or a derivative of a carboxylic acid to an ester.

Ether (Section 9.1): A functional group having the general structure ROR'.

Ethynyl group (Section 11.2): An alkynyl substituent having the structure $-C{\equiv}C-H$.

Excited state (Sections 1.9B, 14.15A): A high-energy electronic state in which one or more electrons have been promoted to a higher-energy orbital by absorption of energy.

Exo position (Section 14.13D): A position of a substituent on a bridged bicyclic compound in which the substituent is closer to the shorter bridge that joins the two carbons common to both rings.

Exothermic reaction (Section 6.4): A reaction in which the energy of the products is lower than the energy of the reactants. In an exothermic reaction, energy is released and the $\Delta H°$ is a negative value.

Extraction (Section 19.10): A laboratory method to separate and purify a mixture of compounds using solubility differences and acid–base principles.

***E,Z* System of nomenclature** (Section 10.3B): A system for unambiguously naming alkene stereoisomers by assigning priorities to the two groups on each carbon of the double bond. The *E* isomer has the two higher-priority groups on opposite sides of the double bond, and the *Z* isomer has them on the same side.

F

FAD (flavin adenine dinucleotide, Section 30.3B): A biological oxidizing agent synthesized in cells from vitamin B_2, riboflavin. FAD is reduced by adding two hydrogens, forming $FADH_2$.

$FADH_2$ (Section 30.3B): A biological reducing agent formed when FAD is reduced.

Fat (Sections 10.6B, 29.3): A triacylglycerol that is solid at room temperature and composed of fatty acid side chains with a high degree of saturation.

Fatty acid (Sections 10.6A, 19.5): A long-chain carboxylic acid having between 12 and 20 carbon atoms.

Fehling's reagent (Section 26.8B): A reagent for oxidizing aldehydes to carboxylic acids using a Cu^{2+} salt as an oxidizing agent, forming brick-red Cu_2O as a by-product.

Fermentation (Section 30.6C): The anaerobic conversion of glucose to ethanol and CO_2 that occurs in yeast and other microorganisms.

Fibrous proteins (Section 27.9): Long linear polypeptide chains that are bundled together to form rods or sheets.

Fingerprint region (Section B.2B): The region in an IR spectrum at $< 1500\ cm^{-1}$. The region often contains a complex set of peaks and is unique for every compound.

First-order rate equation (Sections 6.9B, 7.9): A rate equation in which the reaction rate depends on the concentration of only one reactant.

Fischer esterification (Section 20.9C): An acid-catalyzed esterification reaction between a carboxylic acid and an alcohol to form an ester.

Fischer projection formula (Section 26.2A): A method for representing stereogenic centers with the stereogenic carbon at the intersection of vertical and horizontal lines. Fischer projections are also called cross formulas.

Fishhook (Section 6.3B): A half-headed curved arrow used in a reaction mechanism to denote the movement of a single electron.

Flagpole hydrogens (Section 4.11B): Hydrogens in the boat conformation of cyclohexane that are on either end of the "boat" and are forced into close proximity to each other.

Formal charge (Section 1.3C): The electronic charge assigned to individual atoms in a Lewis structure. The formal charge is calculated by subtracting an atom's unshared electrons and half of its shared electrons from the number of valence electrons that a neutral atom would possess.

Formyl group (Section 18.2E): A substituent having the structure –CHO.

Four-centered transition state (Section 10.16): A transition state that involves four atoms.

Fragment (Section A.1): Radicals and cations formed by the decomposition of the molecular ion in a mass spectrometer.

Freons (Sections 7.4, 13.9): Chlorofluorocarbons consisting of simple halogen-containing organic compounds that were once commonly used as refrigerants.

Frequency (Section B.1): The number of waves passing a point per unit time. Frequency is reported in cycles per second (s^{-1}), which is also called hertz (Hz). Frequency is abbreviated with the Greek letter nu (ν).

Friedel–Crafts acylation (Section 16.5A): An electrophilic aromatic substitution reaction in which benzene reacts with an acid chloride in the presence of a Lewis acid to give a ketone.

Friedel–Crafts alkylation (Section 16.5A): An electrophilic aromatic substitution reaction in which benzene reacts with an alkyl halide in the presence of a Lewis acid to give an alkyl benzene.

Frontside attack (Section 7.10C): Approach of a nucleophile from the same side as the leaving group.

Full-headed curved arrow (Section 6.3B): An arrow used in a reaction mechanism to denote the movement of a pair of electrons.

Functional group (Section 3.1): An atom or group of atoms with characteristic chemical and physical properties. The functional group is the reactive part of the molecule.

Functional group interconversion (Section 11.12): A reaction that converts one functional group to another.

Functional group region (Section B.2): The region in an IR spectrum at $\geq 1500\ cm^{-1}$. Common functional groups show one or two peaks in this region, at a characteristic frequency.

Furanose (Section 26.5): A cyclic five-membered ring of a monosaccharide containing an oxygen atom.

Fused ring system (Section 14.13D): A bicyclic ring system in which the two rings share one bond and two adjacent atoms.

G

Gabriel synthesis (Section 23.6A): A two-step method that converts an alkyl halide to a primary amine using a nucleophile derived from phthalimide.

Gas chromatography (Section A.4B): An analytical technique that separates the components of a mixture based on their boiling points and the rate at which their vapors travel through a column.

Gauche conformation (Section 4.9): A staggered conformation in which the two larger groups on adjacent carbon atoms have a dihedral angle of 60°.

GC–MS (Section A.4B): An analytical instrument that combines a gas chromatograph (GC) and a mass spectrometer (MS) in sequence.

***gem*-Diol** (Section 18.12): A compound having the general structure $R_2C(OH)_2$. *gem*-Diols are also called hydrates.

Geminal dihalide (Section 8.10): A compound that has two halogen atoms on the same carbon atom.

Gene (Section 28.1): A portion of a DNA molecule responsible for the synthesis of a specific protein.

Genetic code (Section 28.6): The set of three-nucleotide units in mRNA called codons that correspond to particular amino acids. The order of codons in mRNA determines the amino acid sequence of a protein.

Gibbs free energy (Section 6.5A): The free energy of a molecule. Gibbs free energy is denoted by the symbol $G°$.

Gibbs free energy change (Section 6.5A): The overall energy difference between reactants and products. The Gibbs free energy change is denoted by the symbol $\Delta G°$. $\Delta G° = G°_{products} - G°_{reactants}$.

Globular proteins (Section 27.9): Polypeptide chains that are coiled into compact shapes with hydrophilic outer surfaces that make them water soluble.

Glycol (Section 9.3A): A compound possessing two hydroxy groups. Glycols are also called diols.

Glycolysis (Section 30.5): An anaerobic 10-step metabolic pathway that converts glucose to two molecules of pyruvate.

Glycosidase (Section 26.11B): An enzyme that hydrolyzes glycosidic linkages. An α-glycosidase hydrolyzes only α-glycosidic linkages.

Glycoside (Section 26.6A): A monosaccharide with an alkoxy group bonded to the anomeric carbon.

***N*-Glycoside** (Section 26.12B): A monosaccharide containing a nitrogen bonded to the anomeric carbon.

Glycosidic linkage (Section 26.10): An acetal linkage formed between an OH group on one monosaccharide and the anomeric carbon on a second monosaccharide.

Green chemistry (Sections 12.13, 31.8): The use of environmentally benign methods to synthesize compounds.

Grignard reagent (Section 17.9): An organometallic reagent having the general structure RMgX.

Ground state (Sections 1.9B, 14.15A): The lowest-energy arrangement of electrons for an atom.

Group number (Section 1.1): The number above a particular column in the periodic table. Group numbers are represented by either an Arabic (1 to 8) or Roman (I to VIII) numeral followed by the letter A or B. The group number of a second-row element is equal to the number of valence electrons in that element.

Grubbs catalyst (Section 24.7): A widely used ruthenium catalyst for olefin metathesis that has the structure $Cl_2(Cy_3P)_2Ru{=}CHPh$.

Guest molecule (Section 9.5B): A small molecule that can bind to a larger host molecule.

H

^{1}H NMR spectroscopy (Section C.1): A form of nuclear magnetic resonance spectroscopy used to determine the number and type of hydrogen atoms in a molecule. ^{1}H NMR is also called proton NMR spectroscopy.

Half-headed curved arrow (Section 6.3B): An arrow used in a reaction mechanism to denote the movement of a single electron. A half-headed curved arrow is also called a fishhook.

α-Halo aldehyde or ketone (Section 21.7): An aldehyde or ketone with a halogen atom bonded to the α carbon.

Haloform reaction (Section 21.7B): A halogenation reaction of a methyl ketone ($RCOCH_3$) with excess halogen, which results in formation of RCO_2^- and CHX_3 (haloform).

Halogenation (Sections 10.13, 13.3, 16.3): The reaction of a compound with a halogen.

Halohydrin (Sections 9.6, 10.15): A compound that has a hydroxy group and a halogen atom on adjacent carbon atoms.

Halonium ion (Section 10.13): A positively charged halogen atom. A bridged halonium ion contains a three-membered ring and is formed in the addition of a halogen (X_2) to an alkene.

Hammond postulate (Section 7.13): A postulate that states that the transition state of a reaction resembles the structure of the species (reactant or product) to which it is closer in energy.

Haworth projection (Section 26.5A): A representation of the cyclic form of a monosaccharide in which the ring is drawn flat.

Head-to-tail polymerization (Section 13.14B): A mechanism of radical polymerization in which the more substituted radical of the growing polymer chain always adds to the less substituted end of the new monomer.

Heat of hydrogenation (Section 12.3A): The $\Delta H°$ of a catalytic hydrogenation reaction equal to the amount of energy released by hydrogenating a π bond.

Heat of reaction (Section 6.4): The energy absorbed or released in a reaction. Heat of reaction is symbolized by $\Delta H°$ and is also called the change in enthalpy.

Heck reaction (Section 24.3): The palladium-catalyzed coupling of a vinyl or aryl halide with an alkene to form a more highly substituted alkene with a new carbon–carbon bond.

α-Helix (Section 27.8B): A secondary structure of a protein formed when a peptide chain twists into a right-handed or clockwise spiral.

Heme (Section 27.9C): A complex organic compound containing an Fe^{2+} ion coordinated with a porphyrin.

Hemiacetal (Section 18.13A): A compound that contains an alkoxy group and a hydroxy group bonded to the same carbon atom.

Hertz (Section B.1): A unit of frequency measuring the number of waves passing a point per second.

Heteroatom (Sections 1.6, 3.1): An atom other than carbon or hydrogen. Common heteroatoms in organic chemistry are nitrogen, oxygen, sulfur, phosphorus, and the halogens.

Heterocycle (Section 9.3B): A cyclic compound containing a heteroatom as part of the ring.

Heterolysis (Section 6.3A): The breaking of a covalent bond by unequally dividing the electrons between the two atoms in the bond. Heterolysis generates charged intermediates. Heterolysis is also called heterolytic cleavage.

Hexose (Section 26.2): A monosaccharide containing six carbons.

Highest occupied molecular orbital (Section 15.10B): The molecular orbital with the highest energy that also contains electrons. The highest occupied molecular orbital is abbreviated as HOMO.

High-resolution mass spectrometer (Section A.4A): A mass spectrometer that can measure mass-to-charge ratios to four or more decimal places. High-resolution mass spectra are used to determine the molecular formula of a compound.

Hofmann elimination (Section 23.11): An E2 elimination reaction that converts an amine to a quaternary ammonium salt as the leaving group. The Hofmann elimination gives the less substituted alkene as the major product.

Homologous series (Section 4.1B): A group of compounds that differ by only a CH_2 group in the chain.

Homolysis (Section 6.3A): The breaking of a covalent bond by equally dividing the electrons between the two atoms in the bond. Homolysis generates uncharged radical intermediates. Homolysis is also called homolytic cleavage.

Homopolymer (Section 31.2D): A polymer prepared from a single monomer.

Homotopic protons (Section C.2C): Two equivalent hydrogen atoms such that substitution of either hydrogen with a group Z forms the same product. The two hydrogen atoms give a single NMR signal.

Hooke's law (Section B.3): A physical law that can be used to calculate the frequency of a bond vibration from the strength of the bond and the masses of the atoms attached to it.

Hückel's rule (Section 15.7): A principle that states for a compound to be aromatic, it must be cyclic, planar, completely conjugated, and have $4n + 2$ π electrons.

Human genome (Section 28.3): The total DNA content of an individual.

Hybridization (Section 1.9B): The mathematical combination of two or more atomic orbitals (having different shapes) to form the same number of hybrid orbitals (all having the same shape).

Hybrid orbital (Section 1.9B): A new orbital that results from the mathematical combination of two or more atomic orbitals. The hybrid orbital is intermediate in energy compared to the atomic orbitals that were combined to form it.

Hydrate (Sections 12.12B, 18.12): A compound having the general structure $R_2C(OH)_2$. Hydrates are also called *gem*-diols.

Hydration (Sections 10.12, 18.8A): Addition of the elements of water to a molecule.

Hydride (Section 12.2): A negatively charged hydrogen ion ($H:^-$).

1,2-Hydride shift (Section 9.9): Rearrangement of a less stable carbocation to a more stable carbocation by the shift of a hydrogen atom from one carbon atom to an adjacent carbon atom.

Hydroboration (Section 10.16): The addition of the elements of borane (BH_3) to an alkene or alkyne.

Hydrocarbon (Sections 3.2A, 4.1): A compound made up of only the elements of carbon and hydrogen.

Hydrogen bonding (Section 3.3B): An attractive intermolecular interaction that occurs when a hydrogen atom bonded to an O, N, or F atom is electrostatically attracted to a lone pair of electrons on an O, N, or F atom in another molecule.

Hydrogenolysis (Section 27.6): A reaction that cleaves a σ bond using H_2 in the presence of a metal catalyst.

α Hydrogens (Section 21.1): The hydrogen atoms on the carbon bonded to the carbonyl carbon atom (the α carbon).

Hydrohalogenation (Section 10.9): An electrophilic addition of hydrogen halide (HX) to an alkene or alkyne.

Hydrolase (Section 27.10): An enzyme that catalyzes the hydrolysis of an ester, amide, or another functional group.

Hydrolysis (Section 18.8A): A cleavage reaction with water.

Hydroperoxide (Section 13.11): An organic compound having the general structure ROOH.

Hydrophilic (Section 3.4C): Attracted to water. The polar portion of a molecule that interacts with polar water molecules is hydrophilic.

Hydrophobic (Section 3.4C): Not attracted to water. The nonpolar portion of a molecule that is not attracted to polar water molecules is hydrophobic.

β-Hydroxy carbonyl compound (Section 22.1A): An organic compound having a hydroxy group on the carbon β to the carbonyl group.

Hydroxy group (Section 9.1): The OH functional group.

Hyperconjugation (Section 7.12B): The overlap of an empty *p* orbital with an adjacent σ bond.

I

Imide (Section 23.6A): A compound having a nitrogen atom between two carbonyl groups.

Imine (Sections 18.6B, 18.10A): A compound with the general structure $R_2C{=}NR'$. Imines are also called Schiff bases.

Iminium ion (Section 18.10A): A resonance-stabilized cation having the general structure $(R_2C{=}NR'_2)^+$, where R' = H or alkyl.

Inductive effect (Sections 2.5B, 7.12A): The pull of electron density through σ bonds caused by electronegativity differences of atoms.

Infrared (IR) spectroscopy (Section B.2): An analytical technique used to identify the functional groups in a molecule based on their absorption of electromagnetic radiation in the infrared region.

Initiation (Section 13.4A): The initial step in a chain mechanism that forms a reactive intermediate by cleavage of a bond.

Inscribed polygon method (Section 15.11): A method to predict the relative energies of cyclic, completely conjugated compounds to determine which molecular orbitals are filled or empty. The inscribed polygon is also called a Frost circle.

Integration (Section C.5): The area under an NMR signal that is proportional to the number of absorbing nuclei that give rise to the signal.

Intermolecular forces (Section 3.3): The types of interactions that exist between molecules. Functional groups determine the type and strength of these forces. Intermolecular forces are also called noncovalent interactions or nonbonded interactions.

Internal alkene (Section 10.1): An alkene that has at least one carbon atom bonded to each end of the double bond.

Internal alkyne (Section 11.1): An alkyne that has one carbon atom bonded to each end of the triple bond.

Inversion of configuration (Section 7.10C): The opposite relative stereochemistry of a stereogenic center in the starting material and product of a chemical reaction. In a nucleophilic substitution reaction, inversion results when the nucleophile and leaving group are in the opposite position relative to the three other groups on carbon.

Iodoform test (Section 21.7B): A test for the presence of methyl ketones, indicated by the formation of the yellow precipitate, CHI_3, via the haloform reaction.

Ionic bond (Section 1.2): A bond that results from the transfer of electrons from one element to another. Ionic bonds result from strong electrostatic interactions between ions with opposite charges. The transfer of electrons forms stable salts composed of cations and anions.

Ionophore (Section 3.7B): An organic molecule that can form a complex with cations so they may be transported across a cell membrane. Ionophores have a hydrophobic exterior and a hydrophilic central cavity that complexes the cation.

Isocyanate (Section 31.6C): A compound having the general structure $RN{=}C{=}O$.

Isoelectric point (Sections 19.11C, 27.1A): The pH at which an amino acid exists primarily in its neutral zwitterionic form. Isoelectric point is abbreviated as p*I*.

Isolated diene (Section 14.1A): A compound containing two carbon–carbon double bonds joined by more than one σ bond.

Isomers (Sections 1.4A, 4.1A, 5.1): Two different compounds that have the same molecular formula.

Isoprene unit (Section 29.7): A five-carbon unit with four carbons in a row and a one-carbon branch on one of the middle carbons.

Isotactic polymer (Section 31.4): A polymer having all the substituents on the same side of the carbon backbone of an elongated polymer chain.

Isotope (Section 1.1): Two or more atoms of the same element having the same number of protons in the nucleus but a different number of neutrons. Isotopes have the same atomic number but different mass numbers.

IUPAC system of nomenclature (Section 4.3): A systematic method for naming compounds developed by the International Union of Pure and Applied Chemistry.

K

K_a (Section 2.3): The symbol that represents the acidity constant of an acid HA. The larger the K_a, the stronger the acid.

$$K_a = \frac{[H_3O^+][A:^-]}{[H{-}A]}$$

K_{eq} (Section 2.3): The equilibrium constant. K_{eq} = [products]/[starting materials].

Kekulé structures (Section 15.1): Two equilibrating structures for benzene. Each structure contains a six-membered ring and three π bonds alternating with σ bonds around the ring.

Ketal (Section 18.13): A compound having the general structure $R_2C(OR')_2$, where R = alkyl or aryl. Ketals are derived from ketones and constitute a subclass of acetals.

β-Keto ester (Section 21.10): A compound containing a ketone carbonyl on the carbon β to the ester carbonyl group.

Ketone (Section 11.9): A compound with two alkyl groups bonded to the C=O carbon atom, having the general structures $R_2C{=}O$ or RCOR'.

Ketose (Section 26.2): A monosaccharide comprised of a polyhydroxy ketone.

Keto tautomer (Section 11.9): A tautomer of a ketone that has a C=O and a hydrogen bonded to the α carbon. The keto tautomer is in equilibrium with the enol tautomer.

Kiliani–Fischer synthesis (Section 26.9B): A reaction that lengthens the carbon chain of an aldose by adding one carbon to the carbonyl end.

Kinase (Section 27.10): An enzyme that catalyzes the transfer of a phosphate from one compound to another.

Kinetic enolate (Section 21.4): The enolate that is formed the fastest—generally the less substituted enolate.

Kinetic product (Section 14.11): In a reaction that can give more than one product, the product that is formed the fastest.

Kinetic resolution (Section 27.2B): The separation of two enantiomers by a chemical reaction that selectively occurs for only one of the enantiomers.

Kinetics (Section 6.5): The study of chemical reaction rates.

L

L-Sugar (Section 26.2C): A sugar with the hydroxy group on the stereogenic center farthest from the carbonyl on the left side in the Fischer projection formula.

Lactam (Section 20.1): A cyclic amide in which the carbonyl carbon–nitrogen σ bond is part of a ring. A β-lactam contains the carbon–nitrogen σ bond in a four membered ring.

Lactol (Section 18.15): A cyclic hemiacetal.

Lactone (Section 20.1): A cyclic ester in which the carbonyl carbon–oxygen σ bond is part of a ring.

Leaving group (Section 7.6): An atom or group of atoms (Z) that is able to accept the electron density of the C–Z bond during a substitution or elimination reaction.

Leaving group ability (Section 7.7): A measure of how readily a leaving group (Z) can accept the electron density of the C–Z bond during a substitution or elimination reaction.

Le Châtelier's principle (Section 9.8D): The principle that a system at equilibrium will react to counteract any disturbance to the equilibrium.

Lecithin (Section 29.4A): A phosphoacylglycerol in which the phosphodiester alkyl group is $-CH_2CH_2N(CH_3)_3^+$. Lecithins are also called phosphatidylcholines.

Leukotriene (Section 9.17): An unstable and potent biomolecule synthesized in cells by the oxidation of arachidonic acid. Leukotrienes are responsible for biological conditions such as asthma.

Levorotatory (Section 5.12A): Rotating plane-polarized light in the counterclockwise direction. The rotation is labeled *l* or (−).

Lewis acid (Section 2.8): An electron pair acceptor.

Lewis acid–base reaction (Section 2.8): A reaction that results when a Lewis base donates an electron pair to a Lewis acid.

Lewis base (Section 2.8): An electron pair donor.

Lewis structure (Section 1.3): A representation of a molecule that shows the position of covalent bonds and nonbonding electrons. In Lewis structures, unshared electrons are represented by dots and a two-electron covalent bond is represented by a solid line. Lewis structures are also called electron dot structures.

Ligand (Section 24.2A): A group coordinated to a metal, which donates electron density to or sometimes withdraws electron density from the metal.

Ligase (Section 27.10): An enzyme that catalyzes bond formation accompanied by energy release from a hydrolysis reaction.

"Like dissolves like" (Section 3.4C): The principle that compounds dissolve in solvents having similar kinds of intermolecular forces; that is, polar compounds dissolve in polar solvents and nonpolar compounds dissolve in nonpolar solvents.

Lindlar catalyst (Section 12.5B): A catalyst for the hydrogenation of an alkyne to a cis alkene. The Lindlar catalyst is Pd adsorbed onto $CaCO_3$ with lead(II) acetate and quinoline.

Lipid (Sections 4.14, 29.1): A biomolecule with a large number of C–C and C–H σ bonds that is soluble in organic solvents and insoluble in water.

Lone pair of electrons (Section 1.2): A pair of valence electrons that is not shared with another atom in a covalent bond. Lone pairs are also called unshared or nonbonded pairs of electrons.

Lowest unoccupied molecular orbital (Section 15.10B): The molecular orbital with the lowest energy that does not contain electrons. The lowest unoccupied molecular orbital is abbreviated as the LUMO.

Lyase (Section 27.10): An enzyme that catalyzes the addition of a molecule to a double bond or the elimination of a molecule to give a double bond.

M

M peak (Section A.1): The peak in the mass spectrum that corresponds to the mass of the molecular ion. The M peak is also called the molecular ion peak or the parent peak.

M + 1 peak (Section A.1): The peak in the mass spectrum that corresponds to the mass of the molecular ion plus one. The M + 1 peak is caused by the presence of isotopes that increase the mass of the molecular ion.

M + 2 peak (Section A.2): The peak in the mass spectrum that corresponds to the mass of the molecular ion plus two. The M + 2 peak is caused by the presence of isotopes, typically of a chlorine or a bromine atom.

Magnetic resonance imaging (MRI) (Section C.12): A form of NMR spectroscopy used in medicine.

Malonic ester synthesis (Section 21.9A): A stepwise method that converts diethyl malonate to a carboxylic acid having one or two carbons bonded to the α carbon.

Markovnikov's rule (Section 10.10): The rule that states in the addition of HX to an unsymmetrical alkene, the H atom bonds to the less substituted carbon atom.

Mass number (Section 1.1): The total number of protons and neutrons in the nucleus of a particular atom.

Mass spectrometry (Section A.1): An analytical technique used for measuring the molecular weight and determining the molecular formula of an organic molecule.

Mass-to-charge ratio (Section A.1): A ratio of the mass to the charge of a molecular ion or fragment. Mass-to-charge ratio is abbreviated as *m*/*z*.

Megahertz (Section C.1A): A unit used for the frequency of the RF radiation in NMR spectroscopy. Megahertz is abbreviated as MHz; 1 MHz = 10^6 Hz.

Melting point (Section 3.4B): The temperature at which molecules in the solid phase are converted to the liquid phase. Molecules with stronger intermolecular forces and higher symmetry have higher melting points. Melting point is abbreviated as mp.

Merrifield method (Section 27.7): A method for synthesizing polypeptides using insoluble polymer supports.

Meso compound (Section 5.8): An achiral compound that contains two or more tetrahedral stereogenic centers.

Metabolism (Section 30.2): The sum of all the chemical reactions that take place in an organism.

Meta director (Section 16.7): A substituent on a benzene ring that directs a new group to the meta position during electrophilic aromatic substitution.

Meta isomer (Section 15.3B): A 1,3-disubstituted benzene ring. Meta substitution is abbreviated as *m*-.

Metal hydride reagent (Section 12.2): A reagent containing a polar metal–hydrogen bond that places a partial negative charge on the hydrogen and acts as a source of hydride ions ($H:^-$).

Metathesis (Section 24.7): A reaction between two alkene molecules that results in the interchange of the carbons of their double bonds.

Methylation (Section 7.15): A reaction in which a CH_3 group is transferred from one compound to another.

Methylene group (Sections 4.1B, 10.3C): A CH_2 group bonded to a carbon chain ($-CH_2-$) or part of a double bond ($CH_2=$).

1,2-Methyl shift (Section 9.9): Rearrangement of a less stable carbocation to a more stable carbocation by the shift of a methyl group from one carbon atom to an adjacent carbon atom.

Micelles (Section 3.6): Spherical droplets formed by soap molecules having the ionic heads on the surface and the nonpolar tails packed together in the interior. Grease and oil dissolve in the interior nonpolar region.

Michael acceptor (Section 22.8): The α,β-unsaturated carbonyl compound in a Michael reaction.

Michael reaction (Section 22.8): A reaction in which a resonance-stabilized carbanion (usually an enolate) adds to the β carbon of an α,β-unsaturated carbonyl compound.

Mixed aldol reaction (Section 22.2): An aldol reaction between two different carbonyl compounds. A mixed aldol reaction is also called a crossed aldol reaction.

Mixed anhydride (Section 20.1): An anhydride with two different alkyl groups bonded to the carbonyl carbon atoms.

Molecular ion (Section A.1): The radical cation having the general structure $M^{+\cdot}$, formed by the removal of an electron from an organic molecule. The molecular ion is also called the parent ion.

Molecular orbital theory (Section 15.10A): A theory that describes bonds as the mathematical combination of atomic orbitals to form a new set of orbitals called molecular orbitals. Molecular orbital theory is also called MO theory.

Molecule (Section 1.2): A compound containing two or more atoms bonded together with covalent bonds.

Monomers (Sections 5.1, 13.14): Small organic compounds that can be covalently bonded to each other (polymerized) in a repeating pattern.

Monosaccharide (Section 26.2): A simple sugar having three to seven carbon atoms.

Monosubstituted alkene (Section 8.2A): An alkene that has one alkyl group and three hydrogens bonded to the carbons of the double bond ($RCH=CH_2$).

Monoterpene (Section 29.7A): A terpene that contains 10 carbons and two isoprene units. A monoterpenoid also contains at least one oxygen atom.

Multiplet (Section C.6C): An NMR signal that is split into more than seven peaks.

Mutarotation (Section 26.5A): The process by which a pure anomer of a monosaccharide equilibrates to a mixture of both anomers when placed in solution.

N

***n* + 1 rule** (Section C.6C): The rule that an NMR signal for a proton with *n* nearby nonequivalent protons will be split into $n + 1$ peaks.

NAD^+ (nicotinamide adenine dinucleotide, Sections 17.6B, 30.3A): A biological oxidizing agent and coenzyme synthesized from the vitamin niacin. NAD^+ and NADH are interconverted by oxidation and reduction reactions.

NADH (reduced form of nicotinamide adenine dinucleotide, Sections 17.6B, 30.3A): A biological reducing agent and coenzyme formed when NAD^+ is reduced.

Natural product (Section 7.17): A compound isolated from a natural source.

Newman projection (Section 4.8): An end-on representation of the conformation of a molecule. The Newman projection shows the three groups bonded to each carbon atom in a particular C–C bond, as well as the dihedral angle that separates the groups on each carbon.

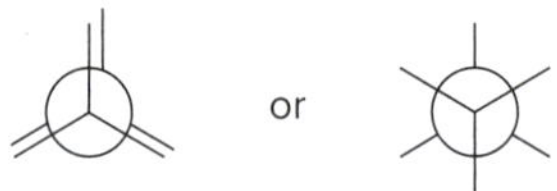

Nitration (Section 16.4): An electrophilic aromatic substitution reaction in which benzene reacts with $^+NO_2$ to give nitrobenzene, $C_6H_5NO_2$.

Nitrile (Sections 19.1, 19.12): A compound having the general structure $RC{\equiv}N$.

Nitronium ion (Section 16.4): An electrophile having the structure $^+NO_2$.

***N*-Nitrosamine** (Section, 23.12B): A compound having the general structure $R_2N-N=O$. Nitrosamines are formed by the reaction of a secondary amine with ^+NO.

Nitrosonium ion (Section 23.12): An electrophile having the structure ^+NO.

NMR peak (Section C.6A): The individual absorptions in a split NMR signal due to nonequivalent nearby protons.

NMR signal (Section C.6A): The entire absorption due to a particular kind of proton in an NMR spectrum.

NMR spectrometer (Section C.1A): An analytical instrument that measures the absorption of RF radiation by certain atomic nuclei when placed in a strong magnetic field.

Nonbonded pair of electrons (Section 1.2): A pair of valence electrons that is not shared with another atom in a covalent bond. Nonbonded electrons are also called unshared or lone pairs of electrons.

Nonbonding molecular orbital (Section 15.11): A molecular orbital having the same energy as the atomic orbitals that formed it.

Nonnucleophilic base (Section 7.8B): A base that is a poor nucleophile due to steric hindrance resulting from the presence of bulky groups.

Nonpolar bond (Section 1.12): A covalent bond in which the electrons are equally shared between the two atoms.

Nonpolar molecule (Section 1.13): A molecule that has no net dipole. A nonpolar molecule has either no polar bonds or multiple polar bonds whose dipoles cancel.

Nonreducing sugar (Section 26.8B): A carbohydrate that cannot be oxidized by Tollens, Benedict's, or Fehling's reagent.

Normal alkane (Section 4.1A): An acyclic alkane that has all of its carbons in a row. A normal alkane is an "*n*-alkane" or a straight-chain alkane.

Nuclear magnetic resonance spectroscopy (Section C.1): A powerful analytical tool that can help identify the carbon and hydrogen framework of an organic molecule.

Nucleic acid (Section 28.2): A polymer of nucleotides, formed by joining the 3'-OH group of one nucleotide with the 5'-phosphate of a second nucleotide in a phosphodiester linkage.

Nucleophile (Sections 2.8, 7.6): An electron-rich compound, symbolized by $:Nu^-$, which donates a pair of electrons to an electron-deficient compound, forming a covalent bond. Lewis bases are nucleophiles.

Nucleophilic acyl substitution (Sections 17.2B, 20.1): Substitution of a leaving group by a nucleophile at a carbonyl carbon.

Nucleophilic addition (Section 17.2A): Addition of a nucleophile to the electrophilic carbon of a carbonyl group followed by protonation of the oxygen.

Nucleophilic aromatic substitution (Section 16.13): A substitution reaction of an aryl halide with a strong nucleophile.

Nucleophilicity (Section 7.8A): A measure of how readily an atom donates an electron pair to other atoms.

Nucleophilic substitution (Section 7.6): A reaction in which a nucleophile replaces the leaving group in a molecule.

Nucleoside (Section 28.1A): A biomolecule having a sugar and either a purine or pyrimidine base joined by an *N*-glycosidic linkage.

Nucleotide (Section 28.1A): A biomolecule having a sugar and either a purine or pyrimidine base joined by an *N*-glycosidic linkage, and a phosphate bonded to a hydroxy group of the sugar nucleus.

O

Observed rotation (Section 5.12A): The angle that a sample of an optically active compound rotates plane-polarized light. The angle is denoted by the symbol α and is measured in degrees (°).

Octet rule (Section 1.2): The general rule governing the bonding process for second-row elements. Through bonding, second-row elements attain a complete outer shell of eight valence electrons.

Oil (Sections 10.6B, 29.3): A triacylglycerol that is liquid at room temperature and composed of fatty acid side chains with a high degree of unsaturation.

Olefin (Section 10.1): An alkene; a compound possessing a carbon–carbon double bond.

Oligosaccharide (26.11D): A carbohydrate with a small number of monosaccharides—generally three to ten—joined together.

Optically active (Section 5.12A): Able to rotate the plane of plane-polarized light as it passes through a solution of a compound.

Optically inactive (Section 5.12A): Not able to rotate the plane of plane-polarized light as it passes through a solution of a compound.

Optical purity (Section 5.12D): A measurement of how much one enantiomer is present in excess of the racemic mixture. Optical purity is also called enantiomeric excess (*ee*); *ee* = % of one enantiomer − % of the other enantiomer.

Orbital (Section 1.1): A region of space around the nucleus of an atom that is high in electron density. There are four different kinds of orbitals, called *s, p, d,* and *f.*

Order of a rate equation (Section 6.9B): The sum of the exponents of the concentration terms in the rate equation of a reaction.

Organoborane (Section 10.16): A compound that contains a carbon–boron bond. Organoboranes have the general structure RBH_2, R_2BH, or R_3B.

Organocopper reagent (Section 17.9): An organometallic reagent having the general structure R_2CuLi. Organocopper reagents are also called organocuprates.

Organolithium reagent (Section 17.9): An organometallic reagent having the general structure RLi.

Organomagnesium reagent (Section 17.9): An organometallic reagent having the general structure RMgX. Organomagnesium reagents are also called Grignard reagents.

Organometallic reagent (Section 17.9): A reagent that contains a carbon atom bonded to a metal.

Organopalladium compound (Section 24.2): An organometallic compound that contains a carbon–palladium bond.

Organophosphorus reagent (Section 18.9A): A reagent that contains a carbon–phosphorus bond.

Organostannane (Section 24.4): An organometallic reagent that contains a carbon–tin bond.

Ortho isomer (Section 15.3B): A 1,2-disubstituted benzene ring. Ortho substitution is abbreviated as *o*-.

Ortho, para director (Section 16.7): A substituent on a benzene ring that directs a new group to the ortho and para positions during electrophilic aromatic substitution.

Oxaphosphetane (Section 18.9B): An intermediate in the Wittig reaction consisting of a four-membered ring containing a phosphorus–oxygen bond.

Oxazaborolidine (Section 17.6A): A heterocycle possessing a boron, a nitrogen, and an oxygen. An oxazaborolidine can be used to form a chiral reducing agent.

Oxidation (Sections 4.13A, 12.1): A process that results in a loss of electrons. For organic compounds, oxidation results in an increase in the number of C–Z bonds or a decrease in the number of C–H bonds; Z = an element more electronegative than carbon.

β-Oxidation (Section 30.4): A catabolic process in which two-carbon units are sequentially cleaved from a fatty acid until all carbons of the fatty acid are degraded to acetyl CoA.

Oxidative addition (Section 24.2A): The addition of a reagent to a metal, often increasing the number of groups around the metal by two.

Oxidative cleavage (Section 12.10): An oxidation reaction that breaks both the σ and π bonds of a multiple bond to form two oxidized products.

Oxidoreductase (Section 27.10): An enzyme that catalyzes an oxidation–reduction reaction.

Oxime (Section 26.9A): A compound having the general structure $R_2C{=}NOH$.

Oxirane (Section 9.1): A cyclic ether having the oxygen atom as part of a three-membered ring. Oxiranes are also called epoxides.

Oxy-Cope rearrangement (Section 25.5): A [3,3] sigmatropic rearrangement of a 1,5-dien-3-ol to a δ,ε-unsaturated carbonyl compound.

Ozonolysis (Section 12.10): An oxidative cleavage reaction in which a multiple bond reacts with ozone (O_3) as the oxidant.

P

Para isomer (Section 15.3B): A 1,4-disubstituted benzene ring. Para substitution is abbreviated as *p*-.

Parent ion (Section A.1): The radical cation having the general structure $M^{+\cdot}$, formed by the removal of an electron from an organic molecule. The parent ion is also called the molecular ion.

Parent name (Section 4.4): The portion of the IUPAC name of an organic compound that indicates the number of carbons in the longest continuous chain in the molecule.

Pentose (Section 26.2): A monosaccharide containing five carbons.

Peptide bond (Section 27.4): The amide bond in peptides and proteins.

Peptides (Sections 20.5B, 27.4): Low-molecular-weight polymers of less than 40 amino acids joined together by amide linkages.

Percent *s*-character (Section 1.11B): The fraction of a hybrid orbital due to the *s* orbital used to form it. As the percent *s*-character increases, a bond becomes shorter and stronger.

Percent transmittance (Section B.2): A measure of how much electromagnetic radiation passes through a sample of a compound and how much is absorbed.

Pericyclic reaction (Section 25.1): A concerted reaction that proceeds through a cyclic transition state.

Peroxide (Section 13.2): A reactive organic compound with the general structure ROOR. Peroxides are used as radical initiators by homolysis of the weak O–O bond.

Peroxyacid (Section 12.7): An oxidizing agent having the general structure RCO_3H.

Peroxy radical (Section 13.11): A radical having the general structure ROO·.

Petroleum (Section 4.6): A fossil fuel containing a complex mixture of compounds, primarily hydrocarbons with 1 to 40 carbon atoms.

Phenol (Sections 9.1, 13.12): A compound such as C_6H_5OH, which contains a hydroxy group bonded to a benzene ring.

Phenyl group (Section 15.3D): A group formed by removal of one hydrogen from benzene, abbreviated as C_6H_5- or Ph–.

Pheromone (Section 4.1): A chemical substance used for communication in an animal or insect species.

Phosphate (Section 7.15): A PO_4^{3-} anion.

Phosphatidylcholine (Section 29.4A): A phosphoacylglycerol in which the phosphodiester alkyl group is $-CH_2CH_2N(CH_3)_3^+$. Phosphatidylcholines are also called lecithins.

Phosphatidylethanolamine (Section 29.4A): A phosphoacylglycerol in which the phosphodiester alkyl group is $-CH_2CH_2NH_3^+$. Phosphatidylethanolamines are also called cephalins.

Phosphoacylglycerols (Section 29.4A): A lipid having a glycerol backbone with two of the hydroxy groups esterified with fatty acids and the third hydroxy group as part of a phosphodiester.

Phosphodiester (Section 29.4): A functional group having the general formula $(RO)_2PO(OH)$ formed by replacing two of the H atoms in phosphoric acid (H_3PO_4) with alkyl groups.

Phospholipid (Sections 3.7A, 29.4): A hydrolyzable lipid that contains a phosphorus atom.

Phosphonium salt (Section 18.9A): An organophosphorus reagent with a positively charged phosphorus and a suitable counterion; for example, $R_4P^+X^-$. Phosphonium salts are converted to ylides upon treatment with a strong base.

Phosphorane (Section 18.9A): A phosphorus ylide; for example, $Ph_3P{=}CR_2$.

Photon (Section B.1): A particle of electromagnetic radiation.

Pi (π) bond (Section 1.10B): A bond formed by side-by-side overlap of two *p* orbitals where electron density is not concentrated on the axis joining the two nuclei. Pi (π) bonds are generally weaker than σ bonds.

pK_a (Section 2.3): A logarithmic scale of acid strength. $pK_a = -\log K_a$. The smaller the pK_a, the stronger the acid.

Plane-polarized light (Section 5.12A): Light that has an electric vector that oscillates in a single plane. Plane-polarized light, also called polarized light, arises from passing ordinary light through a polarizer.

Plane of symmetry (Section 5.3): A mirror plane that cuts a molecule in half, so that one half of the molecule is the mirror reflection of the other half.

Plasticizer (Section 31.7): A low-molecular-weight compound added to a polymer to give it flexibility.

β-Pleated sheet (Section 27.8B): A secondary structure of a protein formed when two or more peptide chains line up side by side.

Poisoned catalyst (Section 12.5B): A hydrogenation catalyst with reduced activity that allows selective reactions to occur. The Lindlar catalyst is a poisoned Pd catalyst that converts alkynes to cis alkenes.

Polar aprotic solvent (Section 7.8C): A polar solvent that is incapable of intermolecular hydrogen bonding because it does not contain an O–H or N–H bond.

Polar bond (Section 1.12): A covalent bond in which the electrons are unequally shared between the two atoms. Unequal sharing of electrons results from bonding between atoms of different electronegativity values, usually with a difference of ≥ 0.5 units.

Polarimeter (Section 5.12A): An instrument that measures the degree that a compound rotates plane-polarized light.

Polarity (Section 1.12): A characteristic that results from a dipole. The polarity of a bond is indicated by an arrow with the head of the arrow pointing toward the negative end of the dipole and the tail with a perpendicular line through it at the positive end of the dipole. The polarity of a bond can also be indicated by the symbols δ+ and δ–.

Polarizability (Section 3.3B): A measure of how the electron cloud around an atom responds to changes in its electronic environment.

Polar molecule (Section 1.13): A molecule that has a net dipole. A polar molecule has either one polar bond or multiple polar bonds whose dipoles reinforce.

Polar protic solvent (Section 7.8C): A polar solvent that is capable of intermolecular hydrogen bonding because it contains an O–H or N–H bond.

Polyanhydride (Section 31.6E): A step-growth polymer that contains many anhydride bonds in its backbone, formed from dicarboxylic acids.

Polyamide (Sections 20.15A, 31.6A): A step-growth polymer that contains many amide bonds. Nylon 6,6 and nylon 6 are polyamides.

Polycarbonate (Section 31.6D): A step-growth polymer that contains many $-OC(=O)O-$ bonds in its backbone, often formed by reaction of $Cl_2C{=}O$ with a diol.

Polycyclic aromatic hydrocarbon (Sections 9.18, 15.5): An aromatic hydrocarbon containing two or more benzene rings that share carbon–carbon bonds. Polycyclic aromatic hydrocarbons are abbreviated as PAHs.

Polyene (Section 14.7): A compound that contains three or more double bonds.

Polyester (Sections 20.15B, 31.6B): A step-growth polymer consisting of many ester bonds between diols and dicarboxylic acids.

Polyether (Sections 9.5B, 31.3): A compound that contains two or more ether linkages.

Polymer (Sections 5.1, 13.14): A large molecule composed of smaller monomer units covalently bonded to each other in a repeating pattern.

Polymerase chain reaction (PCR, Section 28.8): A technique that amplifies a specific portion of DNA, producing millions of copies of a single molecule.

Polymerization (Section 13.14A): The chemical process that joins together monomers to make polymers.

Polysaccharide (Section 26.11): A carbohydrate containing three or more monosaccharide units joined together by glycosidic linkages.

Polyurethane (Section 31.6C): A step-growth polymer that contains many –NHC(=O)O– bonds in its backbone, formed by reaction of a diisocyanate and a diol.

Porphyrin (Section 27.9C): A nitrogen-containing heterocycle that can complex metal ions.

Primary (1°) alcohol (Section 3.2): An alcohol having the general structure RCH_2OH.

Primary (1°) alkyl halide (Section 3.2): An alkyl halide having the general structure RCH_2X.

Primary (1°) amide (Section 3.2): An amide having the general structure $RCONH_2$.

Primary (1°) amine (Section 3.2): An amine having the general structure RNH_2.

Primary (1°) carbocation (Section 7.12): A carbocation having the general structure RCH_2^+.

Primary (1°) carbon (Section 3.2): A carbon atom that is bonded to one other carbon atom.

Primary (1°) hydrogen (Section 3.2): A hydrogen that is bonded to a 1° carbon.

Primary protein structure (Section 27.8A): The particular sequence of amino acids joined together by peptide bonds.

Primary (1°) radical (Section 13.1): A radical having the general structure $RCH_2\cdot$.

Propagation (Section 13.4A): The middle part of a chain mechanism in which one reactive particle is consumed and another is generated. Propagation repeats until a termination step occurs.

Prostaglandin (Section 19.5): A class of lipids containing 20 carbons, a five-membered ring, and a CO_2H group. Prostaglandins possess a wide range of biological activities.

Prosthetic group (Section 27.9C): The non-protein unit of a conjugated protein.

Protecting group (Section 17.12): A blocking group that renders a reactive functional group unreactive, so that it does not interfere with another reaction.

Protection (Section 17.12): The reaction that blocks a reactive functional group with a protecting group.

Proteins (Sections 20.5B, 27.4): High-molecular-weight polymers of 40 or more amino acids joined together by amide linkages.

Proton (Section 2.1): A positively charged hydrogen ion (H^+).

Proton NMR spectroscopy (Section C.1): A form of nuclear magnetic resonance spectroscopy used to determine the number and type of hydrogen atoms in a molecule.

Proton transfer reaction (Section 2.2): A Brønsted–Lowry acid–base reaction; a reaction that results in the transfer of a proton from an acid to a base.

Purine (Sections 15.9B, 28.1): A bicyclic aromatic heterocycle having two nitrogens in each of the rings.

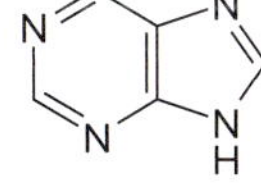

Pyranose (Section 26.5): A cyclic six-membered ring of a monosaccharide containing an oxygen atom.

Pyrimidine (Sections 15.9B, 28.1): A six-membered aromatic heterocycle having two nitrogens in the ring.

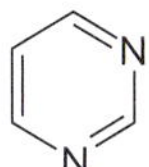

Q

Quantum (Section B.1): The discrete amount of energy associated with a particle of electromagnetic radiation (i.e., a photon).

Quartet (Section C.6C): An NMR signal that is split into four peaks having a relative area of 1:3:3:1, caused by three nearby nonequivalent protons.

Quaternary (4°) carbon (Section 3.2): A carbon atom that is bonded to four other carbon atoms.

Quaternary protein structure (Section 27.8C): The shape adopted when two or more folded polypeptide chains aggregate into one protein complex.

Quintet (Section C.6C): An NMR signal that is split into five peaks caused by four nearby nonequivalent protons.

R

Racemic mixture (Section 5.12B): An equal mixture of two enantiomers. A racemic mixture, also called a racemate, is optically inactive.

Racemization (Section 7.11C): The formation of equal amounts of two enantiomers from an enantiomerically pure starting material.

Radical (Sections 6.3B, 13.1): A reactive intermediate with a single unpaired electron, formed by homolysis of a covalent bond.

Radical anion (Section 12.5C): A reactive intermediate containing both a negative charge and an unpaired electron.

Radical cation (Section A.1): A species with an unpaired electron and a positive charge, formed in a mass spectrometer by the bombardment of a molecule with an electron beam.

Radical inhibitor (Section 13.2): A compound that prevents radical reactions from occurring. Radical inhibitors are also called radical scavengers.

Radical initiator (Section 13.2): A compound that contains an especially weak bond that serves as a source of radicals.

Radical polymerization (Section 13.14B): A radical chain reaction involving the polymerization of alkene monomers by adding a radical to a π bond.

Radical scavenger (Section 13.2): A compound that prevents radical reactions from occurring. Radical scavengers are also called radical inhibitors.

Rate constant (Section 6.9B): A constant that is a fundamental characteristic of a reaction. The rate constant, symbolized by k, is a complex mathematical term that takes into account the dependence of a reaction rate on temperature and the energy of activation.

Rate-determining step (Section 6.8): In a multistep reaction mechanism, the step with the highest-energy transition state.

Rate equation (Section 6.9B): An equation that shows the relationship between the rate of a reaction and the concentration of the reactants. The rate equation depends on the mechanism of the reaction and is also called the rate law.

Reaction coordinate (Section 6.7): The x axis in an energy diagram that represents the progress of a reaction as it proceeds from reactant to product.

Reaction mechanism (Section 6.3): A detailed description of how bonds are broken and formed as a starting material is converted to a product.

Reactive intermediate (Sections 6.3, 10.18): A high-energy unstable intermediate formed during the conversion of a stable starting material to a stable product.

Reciprocal centimeter (Section B.2): The unit for wavenumber, which is used to report frequency in IR spectroscopy.

Reducing sugar (Section 26.8B): A carbohydrate that can be oxidized by Tollens, Benedict's, or Fehling's reagent.

Reduction (Sections 4.13A, 12.1): A process that results in the gain of electrons. For organic compounds, reduction results in a decrease in the number of C—Z bonds or an increase in the number of C—H bonds; Z = an element more electronegative than carbon.

Reductive amination (Section 23.6C): A two-step method that converts aldehydes and ketones into amines.

Reductive elimination (Section 24.2A): The elimination of two groups that surround a metal, often forming new carbon–hydrogen or carbon–carbon bonds.

Regioselective reaction (Section 8.5): A reaction that yields predominantly or exclusively one constitutional isomer when more than one constitutional isomer is possible.

Replication (Section 28.3): The process by which DNA makes a copy of itself when a cell divides. The original DNA molecule forms two DNA molecules, each of which contains one strand of DNA from the parent DNA and one new strand.

Resolution (Section 27.2): The separation of a racemic mixture into its component enantiomers.

Resonance (Section C.1A): In NMR spectroscopy, when an atomic nucleus absorbs RF radiation and spin flips to a higher-energy state.

Resonance hybrid (Sections 1.6C, 14.4): A structure that is a weighted composite of all possible resonance structures. The resonance hybrid shows the delocalization of electron density due to the different locations of electrons in individual resonance structures.

Resonance structures (Sections 1.6, 14.2): Two or more structures of a molecule that differ in the placement of π bonds and nonbonded electrons. The placement of atoms and σ bonds stays the same.

Restriction endonuclease (Section 28.7): An enzyme that cleaves DNA at a specific sequence of bases.

Retention of configuration (Section 7.10C): The same relative stereochemistry of a stereogenic center in the reactant and the product of a chemical reaction.

Retention time (Section A.4B): The length of time required for a component of a mixture to travel through a chromatography column.

Retro-aldol reaction (Section 22.1B): The reverse of an aldol in which a β-hydroxy carbonyl compound is re-converted to carbonyl starting materials with base.

Retro Diels–Alder reaction (Section 14.14B): The reverse of a Diels–Alder reaction in which a cyclohexene is cleaved to give a 1,3-diene and an alkene.

Retrosynthetic analysis (Section 10.18): Working backwards from a product to determine the starting material from which it is made.

Retrovirus (Section 28.9): A virus that contains RNA.

RF radiation (Section C.1A): Radiation in the radiofrequency region of the electromagnetic spectrum, characterized by long wavelength and low frequency and energy.

Ribonucleic acid (Sections 28.1, 28.5): The nucleic acid that translates the genetic information contained in DNA into proteins needed for all cellular functions. Three types of RNA are involved in protein synthesis: ribosomal RNA (rRNA), messenger RNA (mRNA), and transfer RNA (tRNA).

Ribonucleoside (Section 28.1): An *N*-glycoside formed by the reaction of D-ribose with certain amine heterocycles.

Ribonucleotide (Section 28.1): An RNA building block having a ribose and either a purine or pyrimidine base joined by an *N*-glycosidic linkage, and a phosphate bonded to a hydroxy group of the sugar nucleus.

Ring-closing metathesis (Section 24.7): An intramolecular olefin metathesis reaction using a diene starting material, which results in ring closure.

Ring current (Section C.4): A circulation of π electrons in an aromatic ring caused by the presence of an external magnetic field.

Ring-flipping (Section 4.11B): A stepwise process in which one chair conformation of cyclohexane interconverts with a second chair conformation.

Ring-opening metathesis polymerization (Problem 24.44): An olefin metathesis reaction that forms a high-molecular-weight polymer from certain cyclic alkenes.

Robinson annulation (Section 22.9): A ring-forming reaction that combines a Michael reaction with an intramolecular aldol reaction to form a cyclohex-2-enone.

***R,S* System of nomenclature** (Section 5.6): A system of nomenclature that distinguishes the stereochemistry at a tetrahedral stereogenic center by assigning a priority to each group connected to the stereogenic center. *R* indicates a clockwise orientation of the three highest-priority groups and *S* indicates a counterclockwise orientation of the three highest groups. The system is also called the Cahn–Ingold–Prelog system.

Rule of endo addition (Section 14.13D): The rule that the endo product is preferred in a Diels–Alder reaction.

S

Sandmeyer reaction (Section 23.13A): A reaction between an aryl diazonium salt and a copper(I) halide to form an aryl halide (C_6H_5Cl or C_6H_5Br).

Saponification (Section 20.10B): Basic hydrolysis of an ester to form an alcohol and a carboxylate anion.

Saturated fatty acid (Section 10.6A): A fatty acid having no carbon–carbon double bonds in its long hydrocarbon chain.

Saturated hydrocarbon (Section 4.1): A compound that contains only C—C and C—H σ bonds and no rings, thus having the maximum number of hydrogen atoms per carbon.

Schiff base (Section 20.10A): A compound having the general structure $R_2C{=}NR'$. A Schiff base is also called an imine.

Secondary (2°) alcohol (Section 3.2): An alcohol having the general structure R_2CHOH.

Secondary (2°) alkyl halide (Section 3.2): An alkyl halide having the general structure R_2CHX.

Secondary (2°) amide (Section 3.2): An amide having the general structure RCONHR'.

Secondary (2°) amine (Section 3.2): An amine having the general structure R_2NH.

Secondary (2°) carbocation (Section 7.12): A carbocation having the general structure R_2CH^+.

Secondary (2°) carbon (Section 3.2): A carbon atom that is bonded to two other carbon atoms.

Secondary (2°) hydrogen (Section 3.2): A hydrogen that is attached to a 2° carbon.

Secondary protein structure (Section 27.8B): The three-dimensional conformations of localized regions of a protein.

Secondary (2°) radical (Section 13.1): A radical having the general structure $R_2CH\cdot$.

Second-order rate equation (Sections 6.9B, 7.9): A rate equation in which the reaction rate depends on the concentration of two reactants.

Separatory funnel (Section 19.10): An item of laboratory glassware used for extractions.

Septet (Section C.6C): An NMR signal that is split into seven peaks caused by six nearby nonequivalent protons.

Sesquiterpene (Section 29.7A): A terpene that contains 15 carbons and three isoprene units. A sesquiterpenoid also contains at least one oxygen atom.

Sesterterpene (Section 29.7A): A terpene that contains 25 carbons and five isoprene units. A sesterterpenoid also contains at least one oxygen atom.

Sextet (Section C.6C): An NMR signal that is split into six peaks caused by five nearby nonequivalent protons.

Sharpless asymmetric epoxidation (Section 12.15): An enantioselective oxidation reaction that converts the double bond of an allylic alcohol to a predictable enantiomerically enriched epoxide.

Sharpless reagent (Section 12.15): The reagent used in the Sharpless asymmetric epoxidation. The Sharpless reagent consists of *tert*-butyl hydroperoxide, a titanium catalyst, and one enantiomer of diethyl tartrate.

Shielding effects (Section C.3A): An effect in NMR caused by small induced magnetic fields of electrons in the opposite direction to the applied magnetic field. Shielding decreases the strength of the magnetic field felt by the nucleus and shifts an absorption upfield.

1,2-Shift (Section 9.9): Rearrangement of a less stable carbocation to a more stable carbocation by the shift of a hydrogen atom or an alkyl group from one carbon atom to an adjacent carbon atom.

Sigma (σ) bond (Section 1.9A): A cylindrically symmetrical bond that concentrates the electron density on the axis that joins two nuclei. All single bonds are σ bonds.

Sigmatropic rearrangement (Section 25.1): A pericyclic reaction in which a σ bond is broken in the reactant, the π bonds rearrange, and a σ bond is formed in the product.

Silyl ether (Section 17.12): A common protecting group for an alcohol in which the O–H bond is replaced by an O–Si bond.

Simmons–Smith reaction (Section 24.6): Reaction of an alkene with CH_2I_2 and Zn(Cu) to form a cyclopropane.

Singlet (Section C.6A): An NMR signal that occurs as a single peak.

Skeletal structure (Section 1.8B): A shorthand representation of the structure of an organic compound in which carbon atoms and the hydrogen atoms bonded to them are omitted. All heteroatoms and the hydrogens bonded to them are drawn in. Carbon atoms are assumed to be at the junction of any two lines or at the end of a line.

S_N1 mechanism (Sections 7.9, 7.11): A nucleophilic substitution mechanism that goes by a two-step process involving a carbocation intermediate. S_N1 is an abbreviation for "Substitution Nucleophilic Unimolecular."

S_N2 mechanism (Sections 7.9, 7.10): A nucleophilic substitution mechanism that goes by a one-step concerted process, where both reactants are involved in the transition state. S_N2 is an abbreviation for "Substitution Nucleophilic Bimolecular."

Soap (Sections 3.6, 20.11B): The carboxylate salts of long-chain fatty acids prepared by the basic hydrolysis or saponification of a triacylglycerol.

Solubility (Section 3.4C): A measure of the extent to which a compound dissolves in a liquid.

Solute (Section 3.4C): The compound that is dissolved in a liquid solvent.

Solvent (Section 3.4C): The liquid component into which the solute is dissolved.

Specific rotation (Section 5.12C): A standardized physical constant for the amount that a chiral compound rotates plane-polarized light. Specific rotation is denoted by the symbol [α] and defined using a specific sample tube length (*l* in dm), concentration (*c* in g/mL), temperature (25 °C), and wavelength (589 nm). $[\alpha] = \alpha/(l \times c)$

Spectator ion (Section 2.1): An ion that does not take part in a reaction and is opposite in charge to the ion that does take part in a reaction. A spectator ion is also called a counterion.

Spectroscopy (Section A.1): An analytical method using the interaction of electromagnetic radiation with molecules to determine molecular structure.

Sphingomyelin (Section 29.4B): A hydrolyzable phospholipid derived from sphingosine.

Spin flip (Section C.1A): In NMR spectroscopy, when an atomic nucleus absorbs RF radiation and its magnetic field flips relative to the external magnetic field.

Spin–spin splitting (Section C.6): Splitting of an NMR signal into peaks caused by nonequivalent protons on the same carbon or adjacent carbons.

Spiro ring system (Appendix D): A compound having two rings that share a single carbon atom.

Staggered conformation (Section 4.8): A conformation of a molecule in which the bonds on one carbon bisect the R–C–R bond angle on the adjacent carbon.

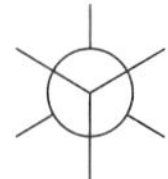

Step-growth polymer (Sections 20.15A, 31.1): A polymer formed when monomers containing two functional groups come together with loss of a small molecule such as water or HCl. Step-growth polymers are also called condensation polymers.

Stereochemistry (Sections 4.8, 5.1): The three-dimensional structure of molecules.

Stereogenic center (Section 5.3): A site in a molecule at which the interchange of two groups forms a stereoisomer. A carbon bonded to four different groups is a tetrahedral stereogenic center. A tetrahedral stereogenic center is also called a chirality center, a chiral center, and an asymmetric center.

Stereoisomers (Sections 4.12B, 5.1): Two isomers that differ only in the way the atoms are oriented in space.

Stereoselective reaction (Section 8.5): A reaction that yields predominantly or exclusively one stereoisomer when two or more stereoisomers are possible.

Stereospecific reaction (Section 10.14): A reaction in which each of two stereoisomers of a starting material yields a particular stereoisomer of a product.

Steric hindrance (Section 7.8B): A decrease in reactivity resulting from the presence of bulky groups at the site of a reaction.

Steric strain (Section 4.9): An increase in energy resulting when atoms in a molecule are forced too close to one another.

Steroid (Sections 14.14C, 29.8): A tetracyclic lipid composed of three six-membered rings and one five-membered ring.

Stille coupling (Section 24.4): A palladium-catalyzed coupling of an organic halide (R'X) with an organostannane ($RSnBu_3$) to form a product R–R'.

Straight-chain alkane (Section 4.1A): An acyclic alkane that has all of its carbons in a row. Straight-chain alkanes are also called normal alkanes.

Structural isomers (Sections 4.1A, 5.2): Two compounds that have the same molecular formula but differ in the way the atoms are connected to each other. Structural isomers are also called constitutional isomers.

Substituent (Section 4.4): A group or branch attached to the longest continuous chain of carbons in an organic molecule.

Substitution reaction (Section 6.2A): A reaction in which an atom or a group of atoms is replaced by another atom or group of atoms. Substitution reactions involve σ bonds: one σ bond breaks and another is formed at the same atom.

Substrate (Section 6.11): An organic molecule that is transformed by the action of an enzyme.

Sulfide (Section 9.15): A compound having the general structure RSR'.

Sulfonation (Section 16.4): An electrophilic aromatic substitution reaction in which benzene reacts with $^{+}SO_3H$ to give a benzenesulfonic acid, $C_6H_5SO_3H$.

Suprafacial reaction (Section 25.4): A pericyclic reaction that occurs on the same side of the two ends of the π electron system.

Suzuki reaction (Section 24.2): The palladium-catalyzed coupling of an organic halide (R'X) with an organoborane (RBY_2) to form a product R—R'.

Symmetrical anhydride (Section 20.1): An anhydride that has two identical alkyl groups bonded to the carbonyl carbon atoms.

Symmetrical ether (Section 9.1): An ether with two identical alkyl groups bonded to the oxygen.

Syn addition (Section 10.8): An addition reaction in which two parts of a reagent are added from the same side of a double bond.

Syn dihydroxylation (Section 12.9B): The addition of two hydroxy groups to the same face of a double bond.

Syndiotactic polymer (Section 31.4): A polymer having the substituents alternating from one side of the backbone of an elongated polymer chain to the other.

Syn periplanar (Section 8.6): In an elimination reaction, a geometry in which the β hydrogen and the leaving group are on the same side of the molecule.

Systematic name (Section 4.3): The name of a molecule indicating the compound's chemical structure. The systematic name is also called the IUPAC name.

T

Target compound (Section 11.12): The final product of a synthetic scheme.

Tautomerization (Sections 11.9, 21.2A): The process of converting one tautomer to another.

Tautomers (Section 11.9): Constitutional isomers that are in equilibrium and differ in the location of a double bond and a hydrogen atom.

Terminal alkene (Section 10.1): An alkene that has the double bond at the end of the carbon chain.

Terminal alkyne (Section 11.1): An alkyne that has the triple bond at the end of the carbon chain.

C-Terminal amino acid (Section 27.4A): The amino acid at the end of a peptide chain with a free carboxy group.

N-Terminal amino acid (Section 27.4A): The amino acid at the end of a peptide chain with a free amino group.

Termination (Section 13.4A): The final step of a chain reaction. In a radical chain mechanism, two radicals combine to form a stable bond.

Terpene (Section 29.7): A hydrocarbon composed of repeating five-carbon isoprene units.

Terpenoid (Section 29.7): A lipid that contains isoprene units as well as at least one oxygen heteroatom.

Tertiary (3°) alcohol (Section 3.2): An alcohol having the general structure R_3COH.

Tertiary (3°) alkyl halide (Section 3.2): An alkyl halide having the general structure R_3CX.

Tertiary (3°) amide (Section 3.2): An amide having the general structure $RCONR'_2$.

Tertiary (3°) amine (Section 3.2): An amine having the general structure R_3N.

Tertiary (3°) carbocation (Section 7.12): A carbocation having the general structure R_3C^+.

Tertiary (3°) carbon (Section 3.2): A carbon atom that is bonded to three other carbon atoms.

Tertiary (3°) hydrogen (Section 3.2): A hydrogen that is attached to a 3° carbon.

Tertiary protein structure (Section 27.8C): The three-dimensional shape adopted by an entire peptide chain.

Tertiary (3°) radical (Section 13.1): A radical having the general structure $R_3C\cdot$.

Tesla (Section C.1A): A unit used to measure the strength of a magnetic field. Tesla is denoted with the symbol "T."

Tetramethylsilane (Section C.1B): An internal standard used as a reference in NMR spectroscopy. The tetramethylsilane (TMS) reference peak occurs at 0 ppm on the δ scale.

Tetrasubstituted alkene (Section 8.2A): An alkene that has four alkyl groups and no hydrogens bonded to the carbons of the double bond ($R_2C{=}CR_2$).

Tetraterpene (Section 29.7A): A terpene that contains 40 carbons and eight isoprene units. A tetraterpenoid contains at least one oxygen atom as well.

Tetrose (Section 26.2): A monosaccharide containing four carbons.

Thermodynamic enolate (Section 21.4): The enolate that is lower in energy—generally the more substituted enolate.

Thermodynamic product (Section 14.11): In a reaction that can give more than one product, the product that predominates at equilibrium.

Thermodynamics (Section 6.5): A study of the energy and equilibrium of a chemical reaction.

Thermoplastics (Section 31.7): Polymers that can be melted and then molded into shapes that are retained when the polymer is cooled.

Thermosetting polymer (Section 31.7): A complex network of cross-linked polymer chains that cannot be re-melted to form a liquid phase.

Thioester (Section 20.16): A compound with the general structure RCOSR'.

Thiol (Section 9.15): A compound having the general structure RSH.

Tollens reagent (Sections 17.8, 26.8B): A reagent that oxidizes aldehydes, and consists of silver(I) oxide in aqueous ammonium hydroxide. A Tollens test is used to detect the presence of an aldehyde.

***p*-Toluenesulfonate** (Section 9.13): A very good leaving group having the general structure $CH_3C_6H_4SO_3^-$ and abbreviated as TsO^-. Compounds containing a *p*-toluenesulfonate leaving group are called alkyl tosylates and are abbreviated ROTs.

Torsional energy (Section 4.8): The energy difference between the staggered and eclipsed conformations of a molecule.

Torsional strain (Section 4.8): An increase in the energy of a molecule caused by eclipsing interactions between groups attached to adjacent carbon atoms.

Tosylate (Section 9.13): A very good leaving group having the general structure $CH_3C_6H_4SO_3^-$, and abbreviated as TsO^-.

***s*-Trans** (Sections 14.6, 27.4B): The conformation of a 1,3-diene that has the two double bonds on opposite sides of the single bond that joins them.

Transcription (28.3): The ordered synthesis of RNA from DNA in which the genetic information stored in DNA is passed onto RNA.

Trans diaxial (Section 8.6B): In an elimination reaction of a cyclohexane, a geometry in which the β hydrogen and the leaving group are trans with both in the axial position.

Transferase (Section 27.10): An enzyme that catalyzes the transfer of a group from one molecule to another.

Trans isomer (Sections 4.12B, 8.3B): An isomer of a ring or double bond that has two groups on opposite sides of the ring or double bond.

Transition state (Section 6.7): An unstable energy maximum as a chemical reaction proceeds from reactants to products. The transition state is at the top of an energy "hill" and can never be isolated.

Triacylglycerol (Sections 10.6, 20.11A, 29.3): A lipid consisting of the triester of glycerol with three long-chain fatty acids. Triacylglycerols are the lipids that comprise animal fats and vegetable oils. Triacylglycerols are also called triglycerides.

Triose (Section 26.2): A monosaccharide containing three carbons.

Triphosphate (Section 7.15): A good leaving group used in biological systems. Triphosphate ($P_3O_{10}^{5-}$) is abbreviated as PPP_i.

Triplet (Section C.6): An NMR signal that is split into three peaks having a relative area of 1:2:1, caused by two nearby nonequivalent protons.

Trisubstituted alkene (Section 8.2A): An alkene that has three alkyl groups and one hydrogen bonded to the carbons of the double bond ($R_2C{=}CHR$).

Triterpene (Section 29.7A): A terpene that contains 30 carbons and six isoprene units. A triterpenoid contains at least one oxygen atom as well.

U

Ultraviolet (UV) light (Section 14.15): Electromagnetic radiation with a wavelength from 200 to 400 nm.

Ultraviolet (UV) spectrum (Section 14.15): A plot of the absorbance of ultraviolet light versus wavelength, often recorded for conjugated systems.

Unimolecular reaction (Sections 6.9B, 7.9, 7.11A): A reaction that has only one reactant involved in the rate-determining step, so the concentration of only one reactant appears in the rate equation.

α,β-Unsaturated carbonyl compound (Section 17.15): A conjugated compound containing a carbonyl group and a carbon–carbon double bond separated by a single σ bond.

Unsaturated fatty acid (Section 10.6A): A fatty acid having one or more carbon–carbon double bonds in its hydrocarbon chain. In natural fatty acids, the double bonds generally have the *Z* configuration.

Unsaturated hydrocarbon (Section 10.2): A hydrocarbon that has fewer than the maximum number of hydrogen atoms per carbon atom. Hydrocarbons with π bonds or rings are unsaturated.

Unsymmetrical ether (Section 9.1): An ether in which the two alkyl groups bonded to the oxygen are different.

Upfield shift (Section C.1B): In an NMR spectrum, a term used to describe the relative location of an absorption signal. An upfield shift means a signal is shifted to the right in the spectrum to lower chemical shift.

Urethane (Section 31.6C): A compound that contains a carbonyl group bonded to both an OR group and an NHR (or NR_2) group. A urethane is also called a carbamate.

Valence bond theory (Section 15.10A): A theory that describes covalent bonding as the overlap of two atomic orbitals with the electron pair in the resulting bond being shared by both atoms.

Valence electrons (Section 1.1): The electrons in the outermost shell of orbitals. Valence electrons determine the properties of a given element. Valence electrons are loosely held and participate in chemical reactions.

Van der Waals forces (Section 3.3B): Very weak intermolecular interactions caused by momentary changes in electron density in molecules. The changes in electron density cause temporary dipoles, which are attracted to temporary dipoles in adjacent molecules. Van der Waals forces are also called London forces.

Vicinal dihalide (Section 8.10): A compound that has two halogen atoms on adjacent carbon atoms.

Vinyl group (Section 10.3C): An alkene substituent having the structure $-CH{=}CH_2$.

Vinyl halide (Section 7.1): A molecule containing a halogen atom bonded to the sp^2 hybridized carbon of a carbon–carbon double bond.

Virus (Section 28.9): An infectious agent consisting of a DNA or RNA molecule that is contained within a protein coating. A virus replicates when it invades a host organism and takes over the biochemical machinery of the host.

Vitamins (Sections 3.5, 29.5): Organic compounds needed in small amounts by biological systems for normal cell function.

VSEPR theory (Section 1.7B): Valence shell electron pair repulsion theory. A theory that determines the three-dimensional shape of a molecule by the number of groups surrounding a central atom. The most stable arrangement keeps the groups as far away from each other as possible.

Walden inversion (Section 7.10C): The inversion of a stereogenic center involved in an S_N2 reaction.

Wavelength (Section B.1): The distance from one point of a wave to the same point on the adjacent wave. Wavelength is abbreviated with the Greek letter lambda (λ).

Wavenumber (Section B.2): A unit for the frequency of electromagnetic radiation that is inversely proportional to wavelength. Wavenumber, reported in reciprocal centimeters (cm^{-1}), is used for frequency in IR spectroscopy.

Wax (Sections 4.14, 29.2): A hydrolyzable lipid consisting of an ester formed from a high-molecular-weight alcohol and a fatty acid.

Williamson ether synthesis (Section 9.6): A method for preparing ethers by reacting an alkoxide (RO^-) with a methyl or primary alkyl halide.

Wittig reaction (Section 18.9): A reaction of a carbonyl group and an organophosphorus reagent that forms an alkene.

Wittig reagent (Section 18.9A): An organophosphorus reagent having the general structure $Ph_3P{=}CR_2$.

Wohl degradation (Section 26.9): A reaction that shortens the carbon chain of an aldose by removing one carbon from the aldehyde end.

Wolff–Kishner reduction (Section 16.14C): A method to reduce aryl ketones to alkyl benzenes using hydrazine (NH_2NH_2) and strong base (KOH).

Woodward–Hoffmann rules (Section 25.3): A set of rules based on orbital symmetry used to explain the stereochemical course of pericyclic reactions.

Ylide (Section 18.9A): A chemical species that contains two oppositely charged atoms bonded to each other, and both atoms have octets of electrons.

Z

Zaitsev rule (Section 8.5): In a β elimination reaction, a rule that states that the major product is the alkene with the most substituted double bond.

Ziegler–Natta catalysts (Section 31.4): Polymerization catalysts prepared from an organoaluminum compound and a Lewis acid such as $TiCl_4$, which afford polymer chains without significant branching and with controlled stereochemistry.

Zwitterion (Sections 19.11B, 27.1B): A neutral compound that contains both a positive and negative charge.

Index

Page numbers followed by *f* indicate figures; those followed by *t* indicate tables. An A before page numbers indicates appendix pages.

A

Abbreviations, A–2-A-3
Abilify, 1099
Abyssomicin C, 1052
Acalabrutinib, 41
Acarbose, 1220
Accupril, 302
ACE (angiotensin-converting enzyme) inhibitors, 5, 46, 1265, 1293
Acebutolol, 127
Acetaldehyde
- common name used, 835
- dopamine reaction with, 883
- ethanol conversion to, 493
- 3-hydroxybutanal conversion to, 1018

Acetals
- in carbohydrates, 872, 1195, 1197, 1202
- cyclic hemiacetal formation and conversion, 868–871
- formation reactions, 862–867, 875
- hemiacetals versus, 1192
- as protecting groups, 867–868, 875

Acetamide, 55, 90, 934
Acetaminophen
- biological functions, 946
- mass-to-charge (*m/z*) ratio, 523
- product storage, 957
- synthesis from acetic anhydride, 946

Acetate
- ions, 493
- reaction with bromomethane, 267
- resonance structures, 71, 899, 900

Acetate fabrics, 1206
Acetic acid
- acidity versus ethanol, 71–72, 899–900
- basic features, 894
- as common name, 891
- deprotonation, 898
- ethanol conversion to, 163
- melting point, 106
- NMR spectrum, 588
- as polar protic solvent, 264*f*
- reaction with ethanol, 242
- relative strength, 76

Acetic anhydride, 51, 933, 946
Acetic benzoic anhydride, 933
Acetoacetic ester synthesis, 1001–1004, 1006
Acetone
- acidity, 72
- boiling point, 104*f*
- butan-2-one synthesis from, 1003–1004
- as common name, 837
- common uses, 840
- dipole–dipole interactions, 102
- enolate hybridization and geometry, 987*f*
- as polar aprotic solvent, 265*f*
- radical reactions forming, 639
- solubility, 108

Acetonitrile, 55, 265*f*
Acetophenone, 787, 837
N-Acetyl alanine, 1228–1230
Acetylation reactions, 946
Acetyl chloride, 250, 932
Acetylcholine, 126, 962, 1161
Acetyl coenzyme A
- biological functions, 962
- in citric acid cycle, 1047
- glucosamine reaction with, 1209
- molecular structure, 55, 961

1-Acetylcyclopentene, 1048
Acetylene
- acetylide anion reactions, 452
- acidity, 72–73
- amide reaction with, 65
- bond angles, 24
- bond dissociation energy, 437
- bond length and strength in, 41*t*
- common uses, 439
- hybridization in, 39
- Lewis structure, 13

N-Acetyl-D-glucosamine (NAG), 1209
Acetyl groups, 251, 837, 1110
Acetylide anions
- as organometallic reagents, 797
- retrosynthetic analysis of reactions, 455–456
- from terminal alkynes, 442–443, 450–454, 457, 459

6-Acetylmorphine, 119
Acetylsalicylic acid. *See also* Aspirin
- biological functions, 888, 896
- hydrogen bonding in, 108
- organic synthesis, 292, 292*f*
- product storage, 957
- properties, 895–896
- reaction with bases, 77–78
- synthesis from acetic anhydride, 946

Achiral molecules
- chiral versus, 181–184, 209
- epoxides, 483
- halogenation, 618–619
- hydrohalogenation reactions, 410–411
- interactions with polarized light, 202–203
- as meso compounds, 198, 200

Acid–base reactions
- acidity and, 62–64
- with alcohols, 353, 354–359
- with aspirin, 77–78
- Brønsted–Lowry, 59–62, 62*f*, 81, 260
- Lewis model, 79–80, 81, 259, 729–730
- in nucleophilic substitution, 259–261
- of organometallic reagents, 797, 798
- predicting results, 65–66, 898–899
- terminal alkynes, 442–443, 450–454

Acid chlorides
- amide formation from, 1077
- basic structure, 778, 929
- IR absorption, 937*t*
- ketone formation from, 810–811
- naming, 932, 935, 935*t*
- nucleophilic acyl substitution reactions of, 943–944, 947–948, 963
- reactivity, 781
- reduction, 790–792, 818
- stability, 932
- structures and functional groups, 99*t*, 556

Acidic amino acids, 1223, 1224*f*
Acidity
- amino acid charge and, 1225, 1225*f*
- basicity versus, 63
- carboxylic acids, 898–905, 918
- determinants, 66–75, 73*f*, 902, 918
- in periodic table, 67–69, 81, 83
- proton donation and, 62–63
- stability and, 932

Acidity constant (K_a), 63. *See also* pK_a values
Acid rain, 76
Acids
- alcohol treatment with, 353, 354–359, 362
- Brønsted–Lowry criteria for, 58–59, 58*f*
- Brønsted–Lowry reactions, 59–62
- as catalysts, 242

Acids (*continued*)
commonly occurring, 76
defined, 58, 78
epoxide reactions with, 376–378
ether reactions with, 370–371
halogenation in, 989
Lewis model, 78–80, 81
strength determinants, 66–75, 73*f*
strength measurement, 62–64, 63*t*, 82
Acquired immune deficiency syndrome, 1287. *See also* HIV treatment
Acrolein, 666*f*, 676
Acrylonitrile, 599
Actinic keratosis, 1106
Activated benzenes, 748
Activation energy. *See* Energy of activation (E_a)
Activators, 743–744
Active methylene compounds, 1023
Active sites of enzymes, 243, 243*f*
Acyclic aldohexoses, 1212
Acyclic alkanes
molecular structures, 131–133, 167
naming, 136–140
sample conformations, 145–151, 167
Acyclovir, 1009, 1294
Acylases, 1229–1230
Acylation
defined, 729
Friedel–Crafts, 724*f*, 729, 730–731, 734–735, 734*f*
Acyl compounds, 931*t*
Acyl groups
in amides, 934
basic structures, 929
naming, 837–838, 933
Acylium ion, 518, 730, 731
Acyl transfer reactions, 961
1,2-Addition, 813, 814
1,2-Addition product, 659–663
1,4-Addition, 813, 814, 822
1,4-Addition product, 659–663
Addition–elimination reactions, 753–755
Addition reactions. *See also* Nucleophilic addition reactions
alkene epoxidation, 479–482
alkene halogenation, 406*f*, 413–416, 428
alkene halohydrin formation, 406*f*, 416–418, 428
alkene hydration, 406*f*, 412–413, 427, 428, 429
alkene hydroboration–oxidation, 406*f*, 419–423, 428
alkene hydrohalogenation, 406*f*, 407–412, 428
alkyne halogenation, 442*f*, 445–446, 457
alkyne hydration, 442*f*, 446–448, 457
alkyne hydroboration–oxidation, 442*f*, 448–450, 457
alkyne hydrohalogenation, 442*f*, 443–445, 457
basic types listed, A–16
benzene resistance to, 687, 695
distinctive features in conjugated dienes, 658–660
kinetic versus thermodynamic products in conjugated dienes, 660–663
overview, 222–223, 244, 458
with radicals, 627–633
typical for alkenes, 405–406, 406*f*, 425, 427, 428
Adenine
base pairing with thymine, 1275, 1277
nucleosides and nucleotides, 1271*t*
π electrons, 705
as purine base, 1269
Adenosine, 210
Adenosine diphosphate (ADP), 1272
Adenosine triphosphate (ATP), 288, 374, 1272
S-Adenosylmethionine (SAM), 289, 374
-adiene (suffix), 398
Adrenaline
molecular structure, 98
physiological effects, 253
as physiologically active amine, 1060
production in the human body, 253, 289, 290*f*
Advil. *See* Ibuprofen
Afema, 915
Aflatoxin B1, 523
Agent Orange, 727*f*
Aglycons, 1195
AIDS (acquired immune deficiency syndrome), 1287. *See also* HIV treatment
-al (suffix), 834
L-Alanine, 1230
Alanine
abbreviations and R group, 909*t*
biological functions, 118
dipeptide synthesis from, 1240–1241
drawing enantiomer, 187
effects of pH on, 910, 910*f*
ionizable functional group values, 1226*t*
physical properties of enantiomer, 204
in simple peptides, 1232–1233
structure and abbreviation, 1224*f*
Albuterol
locating stereogenic center, 185
medical applications, 342
synthesis, 378, 378*f*, 1105
Alcohols
acetal formation reactions, 862–867, 875
from acetylide anion reactions with epoxides, 454
aldehyde and ketone reduction to, 783, 784–786, 798, 799–800, 1196
alkene production from, 405
biological oxidation, 492–493
carboxylic acid reduction to, 790–792, 793
carboxylic acid synthesis from, 897*t*, 918, 948
classifying, 97, 98, 119, 343
conversion to alkyl halides, 361–366, 366*t*
conversion to alkyl tosylates, 367–369
dehydration reactions, 354–360
determinants of reaction type, 424
ethanol as example, 94, 349–350
fermentation, 2, 349, 349*f*
formation by hydration, 412–413
formation by hydroboration–oxidation, 419–423
fragmentation patterns, 519
IR absorption, 542, 542*f*, 546*t*
monosaccharide treatment with, 1192
naming, 344–346, 382
oxidation, 488–490, 498
production, 350–351
properties, 348–349
reactivity, 353, 381, 902
structures and properties, 96, 97*t*, 343, 344
Aldaric acids, 1198, 1202, 1210
-aldehyde (suffix), 835
Aldehydes
addition reactions with alcohols, 862–867
alcohol oxidation to, 488, 490, 491
in aldol reactions, 1016, 1017
from alkyne hydroboration, 449
carbanion addition to, 846
carboxylic acid reduction to, 790–792
carboxylic acid synthesis from, 897*t*
in crossed aldol reactions, 1022
cyanohydrin formation from, 846–848
fragmentation patterns, 518
general reactions of, 842–844
hydration reactions, 860–862
hydride reactions with, 845–846
interesting examples, 840
IR absorption, 542, 543*f*, 546*t*
naming, 834–836, 836*f*
naming derivatives, A–7*f*
nitrile reduction reactions forming, 913–914, 917
nucleophilic addition reaction overview, 779
nucleophilic addition reactions with amines, 854–859
oxidation to carboxylic acids, 783, 795–796, 819, 1197, 1210

physical and spectroscopic properties, 838, 839*t*
pK_a values, 983*t*
protecting groups, 867–868
reactivity to nucleophilic attack, 780
reduction to alcohols, 783, 784–786, 798, 799–800
reductive amination, 1065–1068
structures and functional groups, 99*t*, 778, 833
synthesis, 840–841, 841*t*
Alder, Kurt, 663
Alditols, 1196
Aldohexoses, 1177, 1179, 1182
Aldol condensation, 1019
Aldol reactions
crossed, 1021–1025
directed, 1025–1027
general features, 1016–1018, 1017*f*
intramolecular, 1027–1029
product dehydration, 1019–1020
retrosynthetic analysis, 1020–1021
reversing, 1018–1019
Aldonic acids, 1197, 1210
D-Aldopentose, 1201, 1213–1214
Aldopentoses, 1182, 1189–1190
D-Aldoses, 1182, 1183*f*
Aldoses
carbon removal reactions, 1199–1201
defined, 1176
oxidation–reduction reactions, 1196–1199, 1210–1211
D-Aldotetrose, 1201–1202
Aldotetroses, 1182, 1182*f*
Aldotrioses, 1177
Alendronic acid, 16
Aliphatic hydrocarbons, 94, 901–902
Aliskiren, 186, 974
Alizarin, 1090, 1091
Alkaloids
brucine, 1264
from coca leaves, 1081
common examples, 1059
defined, 1059
galantamine, 1141
matrine, 1067, 1077
solanine, 1194
n-Alkanes, 132
Alkanes. *See also* Cycloalkanes
acyclic versus cyclic, 131–134, 167
alkene reduction to, 469
alkyl halides from, 254, 255
alkyne reduction to, 474
basic features, 94, 95
conversion from alkyl halides, 798
ethane as, 94
in ethers, 347
halogenation, 610–617
IR absorption, 541, 541*f*, 546*t*
lipid resemblance, 165–166
naming, 135–140, 169
natural occurrence, 142–143
oxidation, 162–164
properties, 144–145, 144*t*
radical reactions, 609
sample conformations, 145–151
Alkene oxides, 348
Alkenes
addition reaction overview, 405–406, 406*f*, 428
alcohol dehydration to, 354–360, 405
alkyne reduction to, 475–477
basic features, 94, 307–311, 394–395, 401–402
common uses, 402
conversion to alkynes, 441, 457
cyclopropane synthesis using carbenes, 1119–1122
determinants of reaction type, 424
dihydroxylation, 482–484, 499, 506
halogenation, 406*f*, 413–416, 428
halohydrin formation, 406*f*, 416–418, 428
Heck reaction producing, 1114–1116
Hofmann elimination producing, 1079–1080
hydration, 406*f*, 412–413, 427, 428, 429
hydroboration–oxidation, 406*f*, 419–423, 428
hydrohalogenation, 406*f*, 407–412, 428
IR absorption, 541, 541*f*, 546*t*
metathesis, 1123–1128, 1124*f*, 1129
naming, 397–400, 401*f*, 428
in organic synthesis, 425–427
oxidative cleavage, 479*f*, 485–487, 497
producing, 306–307, 405
proton equivalency in, 563–564
radical addition reactions with HBr, 627–629
reactivity, 117
reduction, 469–472, 497
spin–spin splitting, 580–582
stability, 310, 331
synthesis, 306, 849–853
unsaturation calculations, 395–397
Alkenols, 398
Alkoxides
common to elimination reactions, 306, 405
enolate formation from, 984
ethers produced from, 351
Alkoxy groups, 347
Alkylamines, 1072–1074, 1093
Alkylation
defined, 729
enolates, 988, 993–1004, 1005
Friedel–Crafts, 724*f*, 729–730, 731–733
Alkyl benzenes
aryl ketone reduction to, 759–760
carboxylic acid synthesis from, 897*t*
halogenation, 757–758
oxidation, 758
Alkylboranes, 419–420, 421*f*
Alkyl bromides, 365, 514, 627–629
Alkyl chlorides, 364–365, 514, 729–730
Alkyl diazonium salts, 1082–1083
Alkyl groups
in amides, 934
as electron donors, 278
in ethers, 343, 346–347
inductive effects, 737
naming, 135–136, 138, 167, 400, 401*f*, 933, A-6
in nitriles, 889
relation to radical stability, 608
Alkyl halides. *See also* Elimination reactions; Nucleophilic substitution reactions
acetylide anion reactions, 450–452, 453*f*, 459
alcohol and ether production from, 350–351
alcohol conversion to, 361–366, 366*t*
alcohols compared, 353
alkane production from, 798
alkene production from, 306–307, 405
basic features, 254
benzene treatment with, 749
classifying, 97, 98, 119
as common substrate in nucleophilic substitution, 261
common uses, 256–258, 256*f*
determinants of reaction type, 326–330, 327*t*, 332, 424
as determinants of substitution reaction mechanisms, 282–287
formation via hydrohalogenation, 407–412
identity as factor in E2 elimination reactions, 313–314
M + 2 peak and, 513–515
nomenclature, 255–256, 255*f*
nucleophilic substitution examples, 267–268
properties, 256
radical mechanisms forming, 610–617
reactivity, 117, 258–259, 271, 272*f*, 277, 293
reduction to alkanes, 477, 497
reduction to amines, 1064
structures and properties, 96, 97*t*
1,2-Alkyl shifts, 357
Alkyl shifts, 380
Alkylthio substituents, 373

Alkyl tosylates
alcohol conversion to, 367–369
alkene production from, 405
coupling reactions with organocuprates, 1132
reactivity, 381
Alkynes
acetylide anion reactions, 442–443, 450–454
addition reactions with, 442, 442*f*, 443–450, 458
basic features, 94, 437
carboxylic acid synthesis from, 897*t*
common uses, 439–440
elimination reactions producing, 324–326, 441, 457
IR absorption, 541, 541*f*, 546*t*
naming, 438–439, 439*f*
oxidative cleavage, 479*f*, 487–488, 497
physical properties, 439
reduction, 474–477, 497
synthesis, 306
Allenes, 218
Allicin, 134
Allomyces, 1138
D-Allulose, 1191
Allyl aryl ethers, 1161
Allyl groups, 400, 401*f*
Allylic alcohols, 391, 494, 498
Allylic carbocations, 647, 648
Allylic carbon atoms, 473, 622–627
Allylic diphosphates, 735
Allylic halides, 254, 254*f*
Allyl radicals, 622
Allyl type resonance, 649–650
Allyl vinyl ethers, 1161
Alpha (α) carbons (carbonyl groups)
aldol reactions overview, 1016–1021
amino acids, 908–909
Claisen reaction, 1030–1034
crossed aldol reactions, 1021–1025
defined, 835
Dieckmann reaction, 1034–1035
directed aldol reactions, 1025–1027
enolate alkylation, 988, 993–1004, 1005
enolate halogenation, 988, 989–993, 1004
general reactions of, 842, 843, 977, 988
identifying, 331
intramolecular aldol reactions, 1027–1029
Michael reaction, 1035–1037
Robinson annulation, 1037–1041
Alpha (α) cleavage, 518, 519
Alpha (α) elimination, 1120
Alprazolam, 523
Aluminum chloride ($AlCl_3$), 78
Alzheimer's disease, 1027
Amberlyst A-26 resin, 491
Ambident nucleophiles, 985
Ambien, 397, 719, 1099
-amide (suffix), 934
Amide bonds, 1232–1233, 1247. *See also* Peptide bonds
Amides
alkylamine basicity versus, 1073–1074
in alkyne synthesis, 325
anhydride conversion to, 945
basic structure, 778, 929, 1234
carbonyl absorption, 537–538
carboxylic acid conversion to, 950–951, 1241, 1258
classifying, 99, 119, 930
interesting examples, 938–939
IR absorption, 544, 544*f*, 546*t*, 937*t*
naming, 934, 935*t*
naming derivatives, A–7*f*
nucleophilic acyl substitution reactions of, 956–958, 965
physical properties, 936
pK_a values, 983*t*
reactivity, 781
reduction to amines, 793–794, 1065
stability, 932
as strong bases, 76, 76*f*
structures and functional groups, 99*t*
Amide tautomers, 912
-amine (suffix), 1055
Amine oxides, 484
Amines
acid chloride reactions with, 944
amide reduction to, 793–794
classifying, 98, 119, 854, 1054
common uses, 1090–1092
coupling reactions of aryl diazonium salts, 1088–1090, 1095
fragmentation patterns, 520
general reactions of, 1068–1069
Hofmann elimination, 1079–1081, 1094
interesting examples, 1059–1062
IR absorption, 545, 545*f*, 546*t*
nitrous acid reactions with, 1082–1083, 1094
nomenclature, 1055–1057
as nucleophiles, 1077–1078, 1094
nucleophilic addition reactions of, 854–859
physical and spectroscopic properties, 1058, 1059*f*
preparation, 1062–1068, 1093
reactions as bases, 1069–1077, 1094
relative basicity, 1068–1069, 1071–1077, 1076*t*, 1095–1096
structures and properties, 96, 97*t*, 1054–1055
substitution reactions of aryl diazonium salts, 1083–1088, 1094–1095
Amino acid residues, 1232
Amino acids
acid–base behavior, 1225–1226
basic features, 118
enantioselective synthesis, 1230–1232
enzymes, 1255–1257
general features, 1223–1225, 1223*f*
genetic information for, 1276, 1278, 1279*f*, 1281–1283, 1289
important proteins, 1253–1255
naturally occurring, 1224*t*
peptide and protein formation, 1232–1236
peptide sequencing, 1236–1240
peptide synthesis, 1240–1247
protein structure, 1247–1253
proton transfer reactions, 908–911
separating, 1227–1230, 1227*f*, 1264
α–Amino acids, 908–909
L-Amino acids, 1223*f*
Aminobenzene, 690
p-Aminobenzoic acid, 1092
1-Aminoethan-1-ol, 211
Aminoglycoside antibiotics, 1209
Amino groups
in amino acids, 908
defined, 1057
directing effects in electrophilic aromatic substitution, 746
lack of Friedel–Crafts reactions, 749
nitro group reduction to, 760–761
Amino sugars, 1208–1209
Ammonia
acid chloride reaction with, 944
amine basicity versus, 1071
hybridization in, 35
molecular shape, 25
Ammonium cations, 910
Ammonium hydroxide, 795
Ammonium salts, 1054, 1062–1063, 1069–1071
Amoxicillin, 3, 399, 939
Amphetamine, 78, 86, 1066
Amphotericin B, 218, 656
Amprenavir, 1257, 1257*f*, 1287
Amygdalin, 848
Amylopectin, 1206, 1207
Amylose, 1206–1207
Anabolic steroids, 720
Anacin, 54
Analgesics
acetaminophen, 946
aspirin, 77, 292, 888, 896, 946
codeine, 921
Darvon, 124, 185
fentanyl, 129
ibuprofen, 124, 1012, 1034
ketoprofen, 214
meperidine, 1011

morphine, 671, 921, 946
naproxen, 994
opium, 671
oxycodone, 129
phenacetin, 973
salicylic acid, 896
sodium salicylate, 896
Anastrozole, 895, 1013
-ane (suffix), 135
Anesthetics
benzocaine, 760
chloroethane, 54, 96
diethyl ether, 14, 97, 350
halothane, 256*f*
lidocaine, 600
proparacaine, 774, 1105
propofol, 692
sevoflurane, 115, 350
Angiotensin, 5, 969, 1265
Angiotensin-converting enzyme (ACE) inhibitors, 5, 46
Angle strain
defined, 152, 168
in epoxides, 344
relieving in cyclohexane, 153
Angstroms, 23
Anhydrides
amide formation from, 1077
basic structure, 929
IR absorption, 937*t*
naming, 933, 935*t*
nucleophilic acyl substitution reactions of, 945–946, 964
Aniline
aromatic amines named as derivatives, 1057
as common name of aminobenzene, 690
cyclohexanamine versus, 684
halogenation of, 748
nitration producing, 727
relative basicity, 1072–1073
resonance effects in electrophilic aromatic substitution, 736, 738, 738*f*
synthesis, 1078
Anionic oxy-Cope rearrangements, 1159, 1160
Anions
aromatic, 699–700
defined, 6
ionic bonding, 9
naked, 265
Anisole, 555
18-Annulene, 605
Annulenes, 698
Anomeric carbon, 1184
α Anomers
glucose, 1185–1186, 1195
maltose and lactose, 1203
mannose, 1187–1188
β Anomers, 1185–1188, 1195, 1203
Anomers, 1184–1185, 1186
Antabuse, 493
Antarafacial cycloadditions, 1153
Antarafacial rearrangements, 1157, 1158
Anthracene, 699
Anti addition, 406, 411, 414, 427
Antiaromatic compounds, 697, 704, 711, 712
Antibiotics
abyssomicin C, 1052
aminoglycoside, 1209
amoxicillin, 3, 399, 939
cephalosporins, 939
chloramphenicol, 185
cilastatin, 431
imipenem, 547
indanomycin, 1134
mechanisms, 957–958
monensin, 434, 502
mycomycin, 218
nigericine, 863
nonactin, 115, 116
penicillin, 529, 547, 939
polyene, 656
pretomanid, 195
sulfa drug discovery, 1091–1092
Antibonding orbitals, 706–707
Anticancer drugs
anastrozole, 895, 1013
cabazitaxel, 350
cytarabine, 1268
doxorubicin, 1175
epothilone A, 1126*f*
eribulin mesylate, 350
erlotinib, 461
etoposide, 880
exemestane, 190, 502
fadrozole, 915
iejimalide B, 432
imatinib, 1099
niraparib, 1099
paclitaxel, 178, 189, 782
raloxifene, 273
tamoxifen, 431, 995–996
trabectedin, 218, 792
vismodegib, 1077
Anticoagulants, 1049, 1099
Anticodons, 1278, 1279*f*, 1282, 1282*f*
Anti conformations
butane, 149*f*, 150
defined, 149, 167
ethylene glycol, 174
Antidepressants
bupropion, 727*f*, 1013
duloxetine, 826
fluoxetine, 3, 207, 502
paroxetine, 89, 351
(*S,S*)-reboxetine, 504
selective serotonin reuptake inhibitors, 1061
sertraline, 502, 735
trazodone, 1099
venlafaxine, 805
Anti dihydroxylation, 482, 483
Antifungal drugs, 461, 881
Antihistamines
brompheniramine, 525
cetirizine, 301, 515, 801
chlorpheniramine, 514, 727*f*
diphenhydramine, 302, 525, 689, 1071
fexofenadine, 463, 878, 1060
loratadine, 525
molecular ion characteristics, 525
physiological effects, 703, 1060
Anti-inflammatory drugs
aspirin, 888, 896, 971. *See also* Aspirin
ibuprofen, 207, 1034
naproxen, 207, 994
rofecoxib, 1112
Vioxx, 1049
Anti-obesity drugs, 1267
Antioxidants, 252, 626–627, 1236
Anti periplanar geometry, 317–319, 317*f*, 331
Antiplatelet agents, 976
Antipsychotics, 771
Antipyretics, 305
Antiseptics, 260
Antiulcer drugs, 1060
Antiviral drugs. *See also* HIV treatment
acyclovir, 1009, 1294
idoxuridine, 1291
lamivudine, 1287
molnupiravir, 938, 954
oseltamivir, 663, 970
PF-07321332, 128, 895
remdesivir, 3, 55
rimantadine, 62, 1068
Apixaban, 1099
Aprepitant, 300, 1102
Apricot pits, 848
Aprotic solvents, polar, 265, 285
D-Arabinose, 1199, 1200
Arachidonic acid
aspirin's effects, 896
hydrogen bonding in, 109
ibuprofen's effects, 1034
leukotriene synthesis from, 379, 641
$PGF_{2\alpha}$ synthesis from, 644
Arene oxides, 176
Arenes, 758
Arginine, 50, 1224*f*, 1226*t*
Aricept, 1027
Arimidex, 895

Aripiprazole, 1099
Aromatase inhibitors, 895
Aromatic amines, 1057
Aromatic compounds
  benzene properties, 125, 687
  benzene stability, 694–696, 723
  benzene structure, 94, 688–689
  criteria for, 696–698
  defined, 687
  electrophilic substitution, 723–725. *See also* Electrophilic aromatic substitution
  examples, 693–694, 698–701, 711
  heterocycles, 701–705
  identifying, 704, 712
  inscribed polygon method for predicting, 708–711
  IR absorption, 541, 541*f*, 546*t*
  naming, 690–692, 711
  spectroscopic properties, 692–693
Aromatic hydrocarbons, 94. *See also* Aromatic compounds
Aromaticity, effects on amine basicity, 1076*t*
Aromatic rings, 691
Arrows
  curved, 18–20, 56, 59–61
  for Diels–Alder reaction, 663
  for differing types of bond cleavage and formation, 224–228, 245–246
  double-headed, 16
  double reaction, 59, 60
  listed, A–3
  organic reaction conventions, 221, 226*t*
  for radical reactions, 607
  in retrosynthetic analysis, 454
Artemisinin, 422, 423*f*, 1008
Arylamines, 1072–1073, 1095
Arylboranes, 1111
Aryl bromides, 1084
Aryl chlorides, 727*f*, 754, 1084
Aryl diazonium salts, 1082–1090
Aryl ethers, 1161
Aryl fluorides, 1084
Aryl groups, 691
Aryl halides
  from benzene halogenation, 726
  configuration, 254, 254*f*
  in Heck reaction, 1114, 1129
  nucleophilic substitution and, 291, 753–756
  in Suzuki reaction, 1111
  unreactive in Friedel–Crafts alkylation, 731
Aryl iodides, 1085
Aryl ketones, 759–760
Ascorbate, 641
Ascorbic acid, 91, 641, 938
Asparagine, 1224*f*, 1226*t*
Aspartame
  hydrolysis, 970
  molecular structure, 118, 1205*f*, 1235
  perceived sweetness, 1204, 1235
  reductive amination, 1263
Aspartic acid, 1224*f*, 1226*t*
Aspirin
  biological functions, 888, 896
  hydrogen bonding in, 108
  organic synthesis, 292, 292*f*
  in over-the-counter drugs, 54, 888
  product storage, 957
  properties, 895–896
  reaction rate, 239
  reaction with bases, 77–78
  synthesis from acetic anhydride, 946
Asthma drugs
  albuterol, 342, 378, 378*f*
  fluticasone, 257
  mechanisms of action, 379
  theophylline, 1099
  zileuton, 379
Asthma symptoms, 379
Asymmetric epoxidation, 494, 498
Asymmetric organocatalysis, 1024
-ate (suffix), 933
Atenolol, 88, 100
Atherosclerosis, 963
Atmosphere (Earth)
  carbon dioxide from combustion, 164, 165*f*
  methane in, 130, 142
  ozone destruction, 257–258, 620–622, 621*f*
Atomic number, 6, 190
Atomic weight, 6
Atomoxetine, 553, 1101
Atoms, 6. *See also* Bonds and bonding
Atorvastatin, 1102
ATP (adenosine 5'-triphosphate), 288, 374, 1272
Atropine, 1059, 1059*f*
Attenol A, 879
Attention deficit hyperactivity disorder, 553, 1101
Automated peptide synthesis, 1245–1247
Avandia, 771
Avobenzone, 31, 126, 1033
Avocados, 92
Avodart, 970
Axial bonds, 153, 156–157, 157*f*, 180
Azaspiracid-1, 802, 802*f*
Azo compounds, 1084, 1088–1090, 1091, 1095, 1096
AZT (azidodeoxythymidine), 1287
Azulene, 715

B

Backside attack
  acetylide anion reactions with epoxides, 453
  of epoxides, 376–377, 483
  in halogenation of alkenes, 415
  S_N1 substitution reactions, 275
  S_N2 substitution reactions, 269–271, 368, 375
Bacterial cell membranes, 958. *See also* Antibiotics
Bactrim, 1092
Baeyer strain theory, 152
Ball-and-stick representations, 24, 1180, 1250*f*
Barbituric acid, 134
Barley, 1203
Barriers to rotation, 151
Base pairs, 1275–1276, 1279–1280, 1289
Base peak, mass spectrometry, 509
Bases
  for acetylide anions, 442–443, 443*t*
  in alkyne synthesis, 325
  amines' reactions as, 1069–1071, 1076*t*, 1093
  amino acids as, 909–910
  Brønsted–Lowry criteria for, 58–59, 58*f*
  Brønsted–Lowry reactions, 59–62
  commonly occurring, 76–77
  defined, 58, 78
  in E2 elimination reactions, 312–313
  in enolate formation, 983–984
  halogenation in, 990–991
  Lewis model, 78–80, 81
  organometallic reagents as, 798
  reaction mechanisms with alkyl halides, 330
  strength as determinant of elimination reaction type, 324, 327, 328
Basic amino acids, 1223, 1224*f*
Basicity
  acidity versus, 63
  amines versus other compounds, 1071–1077, 1093, 1095–1096
  leaving group ability and, 261, 262–263
  nucleophilic acyl substitution likelihood and, 942, 965
  nucleophilicity versus, 263–266
Beano, 1208
Bedaquiline, 1068
Beeswax, 165*f*
Benadryl, 302, 689, 1071
Benedict's reagent, 1197
Bent molecules, 26
Benzaldehyde
  aldol reactions, 1022, 1024*f*
  common name used, 835

organometallic reagent reactions with, 800
reactivity, 833
resonance effects in electrophilic aromatic substitution, 737, 738, 738*f*
treatment with zinc amalgam, 555
Benzamide, 934
Benzene. *See also* Aromatic compounds; Substituted benzenes
basic properties, 687
bromination, 687, 695, 726
chlorination, 726
derivative nomenclature, 690–692
formation from diazonium salt reactions, 1085
Friedel–Crafts alkylation and acylation, 729–735, 734*f*
as fuel additive, 94, 522, 693
halogenation, 726
heat of hydrogenation, 694, 695, 695*f*
IR absorption, 546*t*
melting point, 125
metabolism, 176
molecular orbitals, 707, 708*f*
molecular structure, 94, 688–689, 723
nitration and sulfonation, 727–728
NMR signals, 570, 583, 584*f*, 721
resonance structures, 650
unusual stability, 694–696, 723
Benzenecarboxylic acid, 891
Benzenesulfonic acid, 727, 728
Benzhydrocodone, 975
Benzocaine, 760, 972
Benzoic acid
arene oxidation to, 758
benzoyl chloride from, 947
as common name, 891
deprotonation, 898, 906
extraction, 906–907, 907*f*
naming derivatives, A–7*f*
substituted, 903–905, 918
Benzonitrile, 555
Benzonitriles, 1085
Benzophenone, 518, 837
Benzo[*a*]pyrene, 380, 693, 693*f*
Benzoquinone, 666*f*
Benzoyl group, 837
Benzyl bromide, 773
Benzyl esters, 1242–1243
Benzyl group, 691
Benzylic halides, 254, 254*f*, 757
Benzyl methyl ether, 773
Bergstrom, Sune, 896
Beryllium, 16
Beta (β) carbons, 331, 813, 835, 1035–1036
Beta (β) elimination
with alcohols, 353, 354, 369*f*
of alkyl tosylates, 368
dehydrohalogenation as, 306, 314–315
reaction mechanisms, 332, 340
BHA (butylated hydroxy anisole), 637
BHT (butylated hydroxy toluene), 626–627, 775
Biaryls, 1139
Bicyclic compounds, naming, A–8-A–9
Bicyclic products, 666, 667
Bicyclo[4.4.0]decane, 177
Biformene, 487
Bile acids, 176
Bile salts, 176
Bilobalide, 96
Bimatoprost, 896
Bimolecular elimination. *See* E2 elimination reactions
Bimolecular reactions, 240, 267, 268
Biodiesel, 220
Biofuels, 177, 220
Biological oxidation, 492–493
Biological reactions
allylic carbocations in, 649
enantioselective carbonyl reductions, 789–790
Friedel–Crafts, 735
nucleophilic substitution in, 288–290, 290*f*
Biological systems
alcohol dehydration, 356
allylic carbocations in, 649
amino acids' role, 908, 909
aspirin's effects, 888, 896
benzo[*a*]pyrene oxidation, 380
biomolecules, 118
bradykinin's functions, 1235
cell membrane, 113–116
cyclic compounds in, 188–190
E1cB mechanism, 1020
enantiomer activity differences, 189, 207, 342, 481, 482*f*
enantioselective reductions, 788*f*, 789–790
enols in, 980–981
enzyme functions, 243–244, 243*f*, 1255
ester activities in, 938
ethynylestradiol in, 436
Friedel–Crafts reactions in, 722, 735
imines' role, 855–856
β-lactam functions, 340
leukotriene synthesis in, 379
nicotine's effects, 1053
nucleophilic acyl substitution reactions in, 961–963
nucleophilic substitution reactions, 288–289, 290*f*
oxidation reactions, 492–493
penicillin's activity, 957–958
polymers in, 630
radical reactions in, 607
RNA's role, 88
steroid properties in, 671
sulfa drug actions, 1092
sunscreens, 674
vitamins' role, 92, 722
Biomolecules, 118, 1175
Birch reduction, 506
Bisphenol A (BPA), 773
Bisphenol F, 773
Biyouyanagin A, 1048
Black nightshade, 1194
Blattellaquinone, 944
Bleaching, coral reefs, 4
β Blockers, 88, 100
Blood
glucose regulation, 1251*f*
hemoglobin, 1223, 1252, 1254–1255, 1255*f*
measuring enzyme levels, 1257
Blood–brain barrier, 115
Boat form, 155, 155*f*
Boc group, 1241–1242
Boiling points
of alcohols, ethers, and epoxides, 348
aldehydes and ketones, 838
alkanes, 144*t*
alkenes, 401
alkyl halides, 256
alkynes, 439
amides, 936
amines, 1058
carboxylic acids, 893
melting points compared, 174
organic compound principles, 104–105, 104*f*, 120
predicting, 121
thiols, 371
Boll weevil, 1011
Bombykol, 1113
Bond angles, 23–27, 344
Bond cleavage, 224–225, 228–231, 229*t*
Bond dissociation energy
in alkynes, 437
calculating bonding energy from, 233, 246
common values, 229*t*, A–10
overview, 228–231, 244
as radical stability measure, 607
Bonding electrons, 10
Bonding orbitals, 706–707
Bond length, 23, 41–42
Bond polarity, 42–44
Bonds and bonding
basic principles, 8–10, 226
in condensed and skeletal structures, 27–33
determining molecular shapes, 23–27

Bonds and bonding (*continued*)
dissociation energy and strength, 228, 244
drawing Lewis structures, 10–15, 15*t*
electronegativity, 42–44
energy requirements, 228, 230–231
hybridization patterns, 33–41
infrared absorption, 534–540
isomers, 15, 17
molecular orbital theory, 706–708
polarity, 42–45
relative length and strength, 41–42, 534, 535, 536, 550
resonance structures, 16–22
strength in ethane, ethylene, and acetylene, 41–42
9-Borabicyclo[3.3.1]nonane (9-BBN), 420
Borane, 419–423, 787
Boron, 10, 16
Boron trifluoride (BF_3), 78, 79
Bosutinib, 90
Bovine insulin, 1251*f*
Bradykinin, 1235
Brain, passage of substances into, 115
Branched-chain alkanes, 132
Brassicadiene, 431
Breast milk, 1207–1208, 1222
Brevenal, 344
Brevibloc, 90
Bridged ring systems, 667, 667*f*, A–8-A-9
Bridgehead carbons, A–8
Bromination
of benzene, 687, 695, 726, 757–758
chlorination versus, 614–617
methylcyclopentane, 634
at selected allylic C–H bonds, 623–624
Bromine
alkane halogenation, 610
benzene reactions with, 687, 695, 726
isotopes, 514
3-Bromoacetophenone, 837*f*
m-Bromoacetophenone, 837*f*
m-Bromoaniline, 760
p-Bromobenzaldehyde, 1087–1088
Bromobenzene, 726
1-Bromobutane, 313
2-Bromobutane, 307
(*R*)-2-Bromobutane, 619
1-Bromobut-2-ene, 661
3-Bromobut-1-ene, 661
1-Bromo-1-chloroethylene, 563
Bromochlorofluoromethane, 183, 203
Bromochloromethane, 182, 203
Bromocyclodecane, 338
Bromocyclohexane, 410
2-Bromocyclohexanone, 991–992
1-Bromocyclohexene, 1109
1-Bromo-3,3-dimethylbutane, 640
2-Bromo-2,3-dimethylbutane, 640
2-Bromo-3,3-dimethylbutane, 434
(*R*)-6-Bromo-2,6-dimethylnonane, 303
Bromoethane, 351
Bromoetherification, 434
3-Bromohexane, 184, 186*f*
Bromohydrins, 417
Bromomethane, 267
1-Bromo-3-methylbut-2-ene, 624
3-Bromo-3-methylbut-1-ene, 624
cis-1-Bromo-4-methylcyclohexane, 283–284
1-Bromo-1-methylcyclopentane, 315
2-Bromo-2-methylpropane, 307, 313, 451
p-Bromonitrobenzene, 752
1-Bromopentane, 555
1-Bromo-1-phenylpropane, 304
1-Bromopropane, 578, 578*f*, 628
2-Bromopropane, 351, 514, 515*f*, 577, 577*f*, 628
(*Z*)-2-Bromostyrene, 1113
N-Bromosuccinimide (NBS), 417, 623–624, 757
Brompheniramine, 525
Bronchodilators. *See also* Asthma drugs
albuterol, 185, 342, 378, 378*f*, 1105
salmeterol, 300, 788
Brønsted–Lowry acids
basic features, 58–59
commonly occurring, 76
measuring strength, 62–64, 63*t*
reactions, 59–62, 65–66, 81
strength determinants, 66–75, 73*f*
Brønsted–Lowry bases
basic features, 58–59
carboxylic acid reactions with, 898–901, 899*t*
commonly occurring, 76–77
in nucleophilic substitution, 259, 260
reactions, 59–62, 65–66, 81
reactions with alkyl halides, 306
Brucine, 1264
BTX mixture, 94, 522, 693
Bufotenin, 1061
Bufo toads, 1061
Bupropion, 727*f*, 1013
Buta-1,3-diene
carbon–carbon sigma bond length, 656–657
conformations, 1234
molecular orbitals, 1144–1145, 1144*f*
penta-1,4-diene versus, 646–647
as simplest conjugated diene, 654
treatment with hydrogen bromide, 434, 661, 662*f*
UV light absorption, 672
Butanal, 103, 106
Butane
alkyl groups, 136
conformations, 148–151, 149*f*, 308, 309*f*
halogenation, 618–619
molecular structure, 132
solubility, 108
Butanedioic acid, 891
Butanenitrile, 366
Butanoic acid, 544*f*, 999
tert-Butanol, 264*f*
Butan-1-ol, 103, 106
Butan-2-ol, 186, 190, 192, 542*f*, 786
(*S*)-Butan-2-ol, 367
Butan-2-one, 543*f*, 575, 786, 1003–1004
Butene, 469–470, 470*f*
But-1-ene, 311, 336
But-2-ene, 308, 309*f*, 311
cis-But-2-ene, 336, 408, 415, 416*f*, 470*f*, 1121
trans-But-2-ene, 415, 470*f*, 1121
tert-Butoxide, 264
tert-Butoxycarbonyl protecting group, 1241
Butter yellow, 1088
(*R*)-*sec*-Butylamine, 219
tert-Butylbenzene, 690
Butyl α-cyanoacrylate, 642
Butylcycloheptane, 504
tert-Butyl cyclohexanecarboxylate, 933
tert-Butylcyclopentane, 141
1-*tert*-Butylcyclopentene, 391
tert-Butyldimethylsilyl ether, 806–807
sec-Butyl ethyl sulfide, 373
sec-Butyl group, 136
tert-Butyl group, 136, 157, 157*f*
tert-Butyl hydroperoxide, 494
Butyllithium, 76, 850
1-*sec*-Butyl-3-methylcyclohexane, 142
cis-1-(*tert*-Butyl)-2-methylcyclohexane, 170
tert-Butyl methyl ether, 560, 562
5-*tert*-Butyl-3-methylnonane, 139
tert-Butyl pentyl ether, 555
But-1-yne, 442*f*
Butyraldehyde, 835*t*

C

Cabazitaxel, 350
Cadaverine, 1059
Cade juniper, 502
Cadinenes, 502
Caffeic acid, 90
Caffeine
as aromatic compound, 716
in over-the-counter drugs, 54
passage through cell membranes, 115
solubility, 126
Caftaric acid, 921
Cahn–Ingold–Prelog system, 190
Calories, 41
Calquence, 41
Camphor, 205
Candidatus Brocadia anammoxidans, 1169

Candlenuts, 403
Cannabis, 522
Capnellene, 452, 453*f*
Caproaldehyde, 835*t*
Caproic acid, 895
Capsaicin, 53, 223
Captopril, 218
-carbaldehyde (suffix), 834
Carbamates, 1241
Carbanions
  defined, 72
  generation by heterolysis, 225–226, 244
  nucleophilic addition reactions, 845–846
  as strong bases, 76
Carbenes, 1119–1122
Carbenoids, 1122
Carbinolamines, 844, 854, 855, 1077
Carbocations
  in alcohol dehydration, 355, 357–359
  alkyne hydration, 447
  alkyne hydrohalogenation, 443–445
  allylic, 647, 648–651
  defined, 19, 80
  in electrophilic aromatic substitution, 730, 732–733
  formation in elimination reactions, 323–324
  formation in nucleophilic substitution, 267, 274, 275, 276, 279–281, 282*f*
  generation by heterolysis, 225–226, 244
  halonium ions versus, 414
  propene, 409
  rearrangement reactions, 252, 357–359, 363–364, 380, 383
  stability, 278–281, 282*f*, 294
  vinyl, 291
Carbohydrates
  D-aldoses, 1182–1183
  basic features, 871–872, 1175
  as common biomolecules, 1175
  common examples, 1175*f*
  common uses, 1203, 1204
  cyclic monosaccharides, 1184–1191
  disaccharides, 1202–1204, 1205*f*
  glycosides, 1192–1195
  importance to health, 1208–1210
  D-ketoses, 1183, 1184*f*
  monosaccharide reactions at carbonyl groups, 1196–1202, 1210–1211
  monosaccharide reactions at OH groups, 1195–1196
  monosaccharide variations, 1176–1181
  in plant fibers, 959, 959*f*
  polysaccharides, 1205–1208
  starch and cellulose, 179, 179*f*
Carbon
  in alkane isomers, 132–133
  allylic, 473
  atomic number, 6
  in carboxy groups, 889
  as central to organic chemistry, 1
  charged atoms, 32–33
  classifying in hydrocarbons, 95, 96, 119
  in common nucleophiles, 266*t*
  depicting in condensed structures, 27–28
  depicting in skeletal structures, 29–30
  formal charge, 15*t*
  isotopes, 6
  in periodic table, 6, 7*f*
  removal from aldoses, 1199–1201
  valence electrons, 8, 33–34
Carbon backbone, 93
Carbon–carbon bonds
  in acetylene, 39
  in alkene conversion to cyclopropane, 1119–1122
  in alkene metathesis, 1123–1128
  in alkenes, 307–309, 309*f*, 394, 395*t*
  in alkynes, 437
  in benzene, 687, 688
  in cell membranes, 114
  characteristic of organic compounds, 93
  conformation principles, 145–151
  conjugated dienes, 656–657
  depicting in condensed structures, 28
  in 1,3-dienes, 646
  in ethylene, 12, 37–38
  formation by metallic reagents, 798, 799, 815, 846
  formation by Wittig reagents, 849
  in Heck reaction, 1114–1116
  in hydrocarbons, 94–96
  length and strength, 41, 41*t*
  naming polyfunctional compounds with, A–8*t*
  in naphthalene, 716
  NMR signals, 570–572, 580–582
  as nonpolar bonds, 43
  from nucleophilic substitution with acetylide anions, 450–452
  organic halides to organocuprates, 1107–1109
  reactions forming, A–11
  role in organic synthesis, 455
  rotation of, 37, 145. *See also* Conformations
  as sigma bonds in ethane, 37, 93, 145
  in sigmatropic rearrangements, 1156, 1157
  in Simmons–Smith reaction, 1122–1123
  in Stille coupling, 1116–1119
  in Suzuki reaction, 1109–1113, 1113*f*
  in triacylglycerols, 403
Carbon chains, 135–140
Carbon compounds, covalent bond types, 40*t*
Carbon dioxide
  as combustion by-product, 164, 165*f*
  as nonpolar molecule, 44, 45*f*
  organometallic reagent reactions with, 811
Carbon–halogen bonds, 254, 256, 258–259
Carbon–hydrogen bonds
  in acetylene, 39
  alkyne reactions and, 441
  bond dissociation energy, 607, 614
  in cell membranes, 114
  characteristic of organic compounds, 93
  chemical shifts, 569
  conformation principles, 145–151
  length and strength, 42, 536
  in lipids, 166
  NMR spectroscopy and, 558
  as nonpolar bonds, 43
  in oxidation–reduction reactions, 163, 163*f*, 467
  as sigma bonds in ethane, 37, 93
  as sigma bonds in ethylene, 37–38
  as sigma bonds in methane, 34
  in sigmatropic rearrangements, 1156
Carbon monoxide, 1255
Carbon–nitrogen bonds
  classifying amines by, 98, 1054
  formation in amine synthesis, 1062
  in nitriles, 889
Carbon NMR spectroscopy
  aldehyde and ketone absorptions, 839*t*
  amine absorptions, 1058*t*
  basic features, 588–591
  benzene derivative absorptions, 692*t*
  carboxylic acid and nitrile absorptions, 894*t*
  carboxylic acid derivative absorptions, 937*t*
  characteristic absorptions, A–14
  defined, 558
  propanoic acid spectrum, 894*f*
  representative spectra, 591*f*
Carbon–oxygen bonds
  in alcohols, ethers, and epoxides, 343
  in carboxy groups, 889
  classifying functional groups by, 98–101
  depicting in condensed structures, 28, 28*f*
  in lipids, 166
  in oxidation–reduction reactions, 163, 163*f*, 467*f*
  as polar bonds, 43, 44
Carbon–palladium bonds, 1110
Carbon radicals, 607–608, 608*f*, 633
Carbon tetrachloride, 107, 108, 256
Carbonyl carbons
  1,2-addition, 813
  electrophilic properties, 833
  hybridization, 931, 1234

Carbonyl carbons (*continued*)
  in ketones, 795, 800
  leaving group presence or absence, 778–779
  reactions at, 842, 843
  stereochemistry, 786
Carbonyl compounds. *See also* Aldehydes; Ketones
  addition reactions with alcohols, 862–867
  aldehyde and ketone reduction, 784–786, 786*f*
  aldehyde oxidation, 795–796
  enantioselective reductions, 787–790
  general reactions of, 779–782, 977
  hydration reactions, 860–862
  as nucleophiles, 977
  organometallic reagents in acid–base reactions, 805–810
  organometallic reagents in nucleophilic addition, 796–804
  oxidation and reduction overview, 782–783
  reduction of carboxylic acids and derivatives, 790–795, 795*t*
  reduction stereochemistry, 786–787
  α,β-unsaturated, 812–814
Carbonyl condensation reactions
  aldol reactions overview, 1016–1021, 1041
  Claisen reaction, 1030–1034, 1041
  crossed aldol reactions, 1021–1025
  defined, 1016
  Dieckmann reaction, 1034–1035
  directed aldol reactions, 1025–1027
  intramolecular aldol reactions, 1027–1029
  Michael reaction, 1035–1037, 1041
  Robinson annulation, 1037–1041, 1041, 1043
Carbonyl groups
  in alkyne hydration, 446, 447
  dienophiles containing, 665, 666*f*
  functional groups with, 98–101
  IR absorption, 536–538
  major compounds containing, 778–779
  in monosaccharides, 1176, 1181, 1183, 1184, 1199–1202, 1210
  in oxidative cleavage, 485–486, 499
  structures and bonding, 931–932
Carboxy groups
  in amino acids, 908
  in carboxylic acids, 889, 890
  in simple peptides, 1232–1233
Carboxylate anions
  amino acid charge and, 910
  formation, 898, 899, 908
Carboxylation, 811, 822
Carboxylic acid derivatives
  acid chloride reactions, 943–944, 963
  amide reactions, 956–958, 965
  anhydride reactions, 945–946, 964
  basic structures, 778, 929–932
  characteristic reaction, 939–940
  ester reactions, 952–956, 964
  interesting examples, 938–939
  naming, 932–936, 935*t*
  physical and spectroscopic properties, 936–937
  reactivity, 781, 941–942, 958, 965
Carboxylic acids
  alcohol oxidation to, 488, 490
  aldehyde oxidation to, 783, 819, 1197, 1210
  from alkyl benzenes, 758
  alkynes oxidized to, 487–488
  amino acid reactions, 908–911
  aspirin's properties, 888, 895–896
  basic structure, 778, 929
  carbon dioxide reactions producing, 811
  diethyl malonate conversion to, 998–1000, 1006
  ester hydrolysis to, 952–954
  extraction, 905–908, 918
  functional group, 99*t*
  inductive effects, 901–902, 916
  interesting examples, 894–895
  IR absorption, 544, 544*f*, 546*t*
  naming, 890–892
  nucleophilic acyl substitution reactions of, 946–951, 947*f*, 964
  physical and spectroscopic properties, 893, 894*f*, 894*t*
  preparation, 897, 897*t*
  reactions with Brønsted–Lowry bases, 898–901
  reactivity, 941–942
  structure and bonding, 889
  substituted benzoic acids, 903–905, 904*f*
  in wine grapes, 494
Carboxypeptidase, 1238
Cardiac drugs, 1195
Carotatoxin, 126
β-Carotene, 111, 402*f*, 645
Carroll rearrangement, 1172
Carvacrol, 720
(*R*)-Carvone, 720
Carvone, 208
Cassava, 848
Catalysts
  in alkene reduction reactions, 469, 470–471
  in biological reactions, 243–244, 243*f*, 1255
  overview, 242–243, 242*f*
Catalytic hydrogenation, 469, 470–471, 785
Catecholborane, 1111, 1129
Cationic polymerization, 642
Cations, 6, 9, 700–701
CBS reagents, 787–788
Cecropia moth, 1108
Celebrex, 883
Celecoxib, 883
Cell membranes
  of bacteria, 958
  cholesterol's role, 166, 166*f*
  rigidity, 128
  structures and properties, 114–115, 114*f*
  transport across, 115–116, 116*f*
Cellobiose, 1204
Cellulose
  molecular structure, 1175*f*, 1205–1206
  sources, 1175
  starch structure versus, 179–180, 179*f*, 180*f*
Cellulose acetate, 1206
Cembrene A, 487
Cepacol, 260
Cephalexin, 189
Cephalosporins, 939
Cetirizine, 301, 515, 801, 802*f*
Cetylpyridinium chloride (CPC), 260
Chain mechanisms, 611
Chain propagation, 613, 627, 629, 632
Chair form
  cyclohexane, 153–154, 154*f*, 155, 156–162, 168
  glucose, 1189–1190, 1213
Chantix, 1098
s-Character, 42, 536, 550
Charged aromatic compounds, 699–701
Charged atoms, 32–33
Charged ions, 6
Chauvin, Yves, 1123
Chemical reactions
  bond dissociation energy, 228–231, 229*t*
  enthalpy and entropy, 233–234
  kinetics, 239–244
  mechanisms, 224–228
  organic reaction types, 222–223
  thermodynamics overview, 231–233
  writing equations, 221, 221*f*
Chemical shifts
  carbon NMR, 590, 590*t*
  proton NMR, 560, 561, 567–572, 567*f*, 569*t*, 593
Chiral drugs, 207
Chiral molecules
  achiral versus, 181–184, 209
  ammonium ions, 1055
  in drugs, 207
  halogenation, 619–620
  interactions with polarized light, 202–203
Chitin, 1175, 1209

Chloral hydrate, 861
Chloramphenicol, 185
Chlorination
  alkanes, 610, 611–615, 616–617
  of benzene, 726
  bromination versus, 614–617
Chlorine
  alkane halogenation, 610, 611–615, 616–617
  isotopes, 513
Chlorine–chlorine bonds, 228, 229*t*
Chlorobenzene, 690, 726
1-Chlorobutane, 618, 619
2-Chlorobutane, 618, 619
Chlorocyclohexane, 317–318, 318*f*
Chlorocyclopentane, 612
Chlorocyclopropane, 563, 639
Chloroethane, 54, 96, 589
Chloroethylene, 563
4-Chloro-1-ethyl-2-propylbenzene, 691
Chlorofluorocarbons, 2, 257–258, 620–622, 621*f*
Chloroform, 256
4-Chlorohexan-2-ol, 382
2-Chloro-4-isopropyl-6-methylheptan-3-one, 836
Chloromethane, 44*f*
cis-1-Chloro-2-methylcyclohexane, 320
trans-1-Chloro-2-methylcyclohexane, 318–319
o-Chloronitrobenzene, 754
p-Chloronitrobenzene, 753
(*R*)-2-Chloropentane, 639
meta-Chloroperoxybenzoic acid, 478
m-Chlorophenol, 1086–1087
2-Chloropropanal, 836
1-Chloropropane, 617, 648
2-Chloropropane, 514, 514*f*, 617
3-Chloroprop-1-ene, 648
3-Chloropropenoic acid, 581, 581*f*
Chloroquine, 75
N-Chlorosuccinimide (NCS), 465
p-Chlorotoluene, 756
Chlorpheniramine, 514, 727*f*
Cholesterol
  basic features, 166
  health effects, 404
  locating stereogenic centers, 189
  molecular structure, 165*f*
  reducers, 788, 1045, 1051, 1102
  solubility, 109
  as steroid, 671
  storage and activity in human body, 962–963
  trans fats and, 473
Cholesteryl esters, 962, 962*f*
Cholic acid, 176
Chorismate, 340
Chromate esters, 489
Chromium, 478
Chronic myeloid leukemia, 440
Chronic obstructive pulmonary disease, 257
Chrysanthemic acid esters, 590
Chymotrypsin, 1238
Cialis, 971
Cicutoxin, 27
Cilastatin, 431, 547
Cimetidine, 1060
Cinnamaldehyde, 110, 840
Cinnamoylcocaine, 969
Cipralex, 895
Cis alkenes, 475
s-Cis conformation, 664–665
Cis dienophiles, 666
Cis-fused systems, 666
Cis isomers
  of alkenes, 309, 310, 311, 394–395, 398–399, 401
  of cycloalkanes, 158–162, 168, 199–200
  cyclohexanes, 318
  hydrogenation, 473
Cis protons, 580
Citric acid, 51, 110
Citric acid cycle, 223, 1047
Citronellol, 556
Cladribine, 1272
Claisen reaction, 1030–1034, 1041, 1042
Claisen rearrangement, 1159, 1161–1162
Clavulanic acid, 399
Claws, 1253
Clemmensen reduction, 759
Climate change, 164
Clomid, 431
Clopidogrel, 694*f*, 976
CMP (cytidine 5'-monophosphate), 1271
^{13}C NMR. *See* Carbon NMR spectroscopy
Cocaine
  basic features, 928
  most basic electron pair, 88
  solubility in different forms, 127
  synthesis, 969
Cocaine hydrochloride, 127
Coca leaves, 928, 1081
Cockroaches, 131, 944, 1026
Coconut oil, 404
Codeine, 921, 1062
Coding strands (DNA), 1279
Codons, 1280–1283, 1281*t*
Coenzyme A, 92
Coenzymes, 492–493
Coffee, 57, 256*f*
Collagen, 1223, 1254, 1254*f*
Color coding in ball-and-stick models, 24, A–3
Combustion, 162–164
Common names
  alcohols, 345
  aldehydes, 835–836, 836*f*
  alkenes, 400, 401*f*
  alkyl halides, 255
  amino acids, 1223, 1224*f*
  carbohydrates, 1176–1177
  carboxylic acids, 891
  ethers, 346
  ketones, 837
  overview, 140
Complementary base pairs, 1275–1276, 1279–1280, 1289
Compounds defined, 8
Concentration, 239, 240
Concerted reactions, 224
Condensation polymers, 960–961
Condensed structures, 27–29
Configurations, isomer
  defined, 181
  inversion, 269–271, 270*f*, 362, 368, 369
  S_N1 substitution reactions, 275
  S_N2 substitution reactions, 269–271, 270*f*
Conformations
  buta-1,3-diene, 1234
  butane, 148–151, 149*f*, 308, 309*f*
  conjugated dienes, 654
  cycloalkanes, 152–162, 152*f*, 168
  cyclohexanes, 317–319, 318*f*
  defined, 145
  ethane, 145–146, 146*f*, 147–148
  peptide bonds, 1234
  propane, 147*f*
  stereoisomers versus, 654–655, 675
Conivaptan, 887
Conjugate acids
  basicity determination using, 1071–1075, 1076*t*
  defined, 59
  of good leaving groups, 261–262
  identifying and drawing, 60–61
Conjugate addition, 659, 813
Conjugate bases
  of aspirin, 77–78
  of carboxylic acids, 899–900, 903–905, 916
  defined, 59
  identifying and drawing, 60–61
  methane versus water, 67
  selected carbonyl compounds, 982*t*
  strength measurement, 63–64, 63*t*
Conjugated allyl carbocations, 648–651
Conjugated dienes
  basic features, 646–647, 654
  Diels–Alder reaction, 663–671, 675, 677
  distinctive features of addition reactions, 658–660

Conjugated dienes (*continued*)
- isolated dienes versus, 646–647, 656–658
- kinetic versus thermodynamic products in addition reactions, 660–663
- stability, 657–658, 675
- UV light absorption, 672–674, 675

Conjugated double bonds, 650
Conjugated molecules
- aromaticity criteria, 697
- benzene as, 688, 689
- defined, 646
- natural and synthetic dyes, 1091

Conjugated proteins, 1254
Conjugated systems
- carbonyl absorption, 536–538, 537*f*
- common types, 646–647
- determining hybridization, 653

Conjugation, 646, 978
Conrotatory movement, 1147, 1148, 1149, 1150, 1164
Constitutional isomers
- alkanes, 132–133, 169
- radical reactions forming, 634
- in simple carbohydrates, 1176
- stereoisomers versus, 15, 181, 181*f*, 201, 201*f*, 209

Contact lenses, 642
Contraceptives, 436, 439, 440*f*
Cope rearrangement, 1159–1160
Copper(I) bromide, 1084
Copper(I) chloride, 1084
Coral reefs, 4, 674
Corey, E. J., 777, 971
Coronaviruses, 1286–1287, 1286*f*
Corticatic acid A, 438
Cortisol, 119, 343
Cortisone, 671, 840
Coumarin, 1049
Counterions, 58
Coupled protons, 575
Coupling constant, 574, 580–581, 582
Coupling reactions
- of aryl diazonium salts, 1084, 1088–1090
- of organocuprate reagents, 1107–1109, 1128

Covalent bonds, 9, 40*t*, 532
Covalent compounds, 9, 101–103, 104, 110*t*
COVID-19, 3, 128, 938, 1286–1287, 1286*f*
Crab shells, 1209
Crack cocaine, 127
Cracking, 402
Crafts, James, 724
Creatine, 21
Creatine phosphokinase, 1257
Crestor, 1045
Crick, Francis, 1274
Crossed aldol reactions, 1021–1025
Crossed Claisen reactions, 1032–1034
Crown ethers, 115–116
Crutzen, Paul, 620
CS, 1024
C-Terminal amino acids, 1233, 1259, 1281
Cubane, 152*f*
Cucumber aldehyde, 838
Curcumin, 719
Curling hair, 1254, 1254*f*
Curry, 1015
Curved arrow notation
- applying, 56
- Brønsted–Lowry reactions, 59–61
- for Diels–Alder reaction, 663
- for differing types of bond cleavage and formation, 224–228, 245–246
- nucleophile–electrophile reactions, 122
- for radical reactions, 607
- in resonance structures, 18–20

Cyanide anions, 847
2-Cyanocyclohexanecarboxylic acid, 892
Cyano groups, 889, 892, 1064
Cyanohydrins, 846–848, 1199–1200
Cyclic anhydrides, 929
Cyclic compounds, 188–190
Cyclic dienes, 665
Cyclic dienophiles, 666
Cyclic ethers, 347
Cyclic hemiacetals, 868–871, 1184–1191
Cyclic molecules, 697
Cyclic monosaccharides, 1184–1191
Cyclization, 233, 872
Cycloaddition reactions, 1142, 1152–1155, 1163, 1164
[2 + 2] Cycloadditions, 1154–1155
[4 + 2] Cycloadditions, 1153–1154
Cycloalkanes. *See also* Alkanes
- acyclic versus, 131–134, 167
- conformations, 152–162, 152*f*
- disubstituted, 199–200
- molecular structures, 133–134
- naming, 141–142
- proton equivalency in, 563–564, 593
- unsaturation calculations, 395

Cycloalkenes, 394–395
Cyclobutadiene, 697, 706, 710–711
Cyclobutane, 134, 152*f*
Cyclodecane, 152*f*
Cycloheptane, 152*f*
Cyclohepta-1,3,5-triene, 697, 700
Cyclohexa-1,3-diene
- electrocyclic reaction producing, 1146
- heat of hydrogenation, 694, 695
- not aromatic, 697
- UV light absorption, 672

Cyclohexadienone, 1161
Cyclohexanamine, 684, 1069–1070, 1070*f*
Cyclohexane
- basic conformations, 153–155, 154*f*, 155*f*, 168
- conversion to cyclohexene, 163, 617
- formation, 242
- molecular structure, 134, 141
- as natural alkane, 131
- NMR signals, 583
- radical reactions with, 640
- skeletal structure, 29
- substituted conformations, 156–162, 249

Cyclohexanecarbaldehyde, 833
Cyclohexanecarbonyl chloride, 932
cis-Cyclohexane-1,2-diol, 483
Cyclohexanol
- conversion to MOM ether, 882
- from cyclohexene and water, 412
- dehydration, 519
- 1,2-dibromocyclohexane synthesis from, 425–426
- extraction, 906–907, 907*f*
- mass spectrum, 519*f*
- separation from cyclohexanamine, 1070, 1070*f*
- treatment with phosphorus oxychloride, 359

Cyclohexanone
- conversion to 2-ethylcyclohexanone, 994
- conversion to alkene, 853
- determining hybridization patterns in, 40
- halogenation, 992
- radical reactions forming, 639
- tertiary (3°) alcohol formation from, 800

Cyclohexene
- addition reactions, 406*f*. *See also* Addition reactions
- anti dihydroxylation reactions, 483
- conversion to cyclohexane, 163
- enantiomers, 414
- formation from cyclohexanol, 359, 412, 426
- heat of hydrogenation, 694, 695
- reaction with hydrochloric acid, 80
- reaction with hydrogen, 242
- reaction with hydrogen bromide, 410

Cyclohex-2-enone, 605, 785, 884, 991–992, 1040
Cyclohexyl acetate, 968
Cyclonona-1,3,5,7-tetraene, 717
Cyclooctane, 207, 208*f*
Cyclooctatetraene, 696, 712, 717,1105
trans-Cyclooctene, 394–395
Cyclooxygenase, 896
Cyclopentadiene, 667, 700, 717
Cyclopentadienide, 679
Cyclopentadienyl anion, 699–700, 701
Cyclopentadienyl cation, 700, 712
Cyclopentadienyl radical, 700

Cyclopentane
chlorination, 611–612
molecular structure, 134, 152*f*
radical reactions with, 640
Cyclopentanecarboxylic acid, 999–1000
cis-Cyclopentane-1,2-diol, 483
trans-Cyclopentane-1,2-diol, 345
Cyclopentene, 414–415
Cyclopropanation, 1121, 1122
Cyclopropane
molecular structure, 134
proton equivalency in, 563
Simmons–Smith reaction, 1122–1123, 1128
synthesis using carbenes, 1119–1122, 1128
Cyclopropenone, 884
Cymbalta, 826
Cysteine
abbreviations and R group, 909*t*
ionizable functional group values, 1226*t*
in keratin, 1253
in simple peptides, 1232–1233
structure and abbreviation, 1224*f*
Cystic fibrosis, 218, 389
Cytarabine, 1268
Cytosine
base pairing with guanine, 1275, 1277
nucleosides and nucleotides, 1271*t*
pi electrons, 705
as pyrimidine base, 1269

D

d (prefix), 203, 210
Dacron, 960
dAMP (2'-deoxyadenosine 5'-monophosphate), 118, 1175, 1175*f*, 1270, 1271
D and L Monosaccharides, 1181
Darunavir, 886
Darvon, 124
Daughter cells, 1277
DBN, 312
DBU, 312
DDE, 307
ddI (dideoxyinosine), 1287
DDT, 239, 257, 258, 307
Deactivators, 743–744, 749
Decaffeination, 256*f*
Decalins, 177
Decamethrin, 1119
Decane, 140
Decarboxylation, 997–998
Decongestants, 1068
DEET, 944
Degenerate orbitals, 708, 709
Dehydration
alcohols, 354–360, 383
of aldol reaction products, 1019–1020, 1022, 1023, 1027, 1038
alkene production from, 405
dicarboxylic acids, 948
7-Dehydrocholesterol, 1151
Dehydrohalogenation
alkyne synthesis from, 325, 325*f*
bromocyclodecane, 338
tert-butyl iodide, 320
common methods, 306, 311, 405
cyclohexanes, 318–319
defined, 306
drawing products, 307
Delocalization
in benzene, 689
in conjugate bases, 71–72, 72*f*
as conjugation advantage, 646–647, 658
hybridization and, 21, 648, 653
Delta (δ) carbons, 835
Delta (δ) scale, 560
Dementia, 1101, 1141
Demerol, 108, 1011
Dendrobates histrionicus, 443
Deoxy- (prefix), 1269, 1270
2'-Deoxyadenosine, 51
2'-Deoxyadenosine 5'-monophosphate, 118, 1175, 1175*f*, 1270, 1271
2-Deoxy-D-ribose, 1210
2'-Deoxy-D-ribose, 1269
2'-Deoxyguanosine, 32
2'-Deoxyguanosine 5'-monophosphate, 1271–1272
Deoxyribonucleic acid (DNA)
amino acid sequences, 1282*f*, 1283, 1289
aromatic bases, 705, 1269
biological functions, 118, 1269
double helix structure, 1274–1276, 1275*f*
nucleosides and nucleotides, 1269–1274, 1271*t*
PCR amplification, 1285, 1285*f*
replication, 1276, 1277–1278, 1277*f*
RNA versus, 1278, 1288
sequencing, 1283–1284
transcription, 1279–1280, 1279*f*
in viruses, 1286–1287
Deoxyribonucleosides, 1270
Deoxyribonucleotides, 1270, 1271
Depakote, 892
Deprotection, 805
Deprotonation
in acetoacetic ester synthesis, 1002
amino acids, 908, 910, 916
amphetamine, 78
base strength and, 65–66, 898–899
carboxylic acids, 898–901, 899*t*, 923
in malonic ester synthesis, 999
tautomerization in acid, 446, 447*f*
terminal alkynes, 442–443, 443*t*, 450–454
Deshielding effects in NMR, 567–572, 567*f*, 568*f*, 571*f*, 594, 839*f*
Desogestrel, 440, 772
Deuterium, 6
Dexamethasone, 98
Dextrorotatory compounds, 203, 210
Di- (prefix), 1056
Diabetes
bloodstream acetone levels, 840
causes, 1251*f*
drug treatments, 703, 770, 771, 880, 1220
β–Diacids, 997
Diacids, 891
Dialkylamides, 984
Dialkylborane, 449
Dialkyl cyclopropanes, 1122
Diastereomers
but-2-ene, 309
in disubstituted cycloalkanes, 200
glucose anomers, 1186
naming for alkenes, 398–399
overview, 195–197, 199, 201, 209
physical properties, 205–206
use in resolution, 1227–1229, 1227*f*
Diastereotopic protons, 565, 566, 592
Diaxial conformations, 159
Diazomethane, 1138
Diazonium salts, 1082–1090, 1094–1095
Diazotization, 1082
DIBAL-H, 914
Diborane, 419
Dibromobenzenes, 690, 692*f*
2,3-Dibromobutane, 197–198, 198*f*, 338
Dibromocarbene, 1122
1,2-Dibromocyclohexane, 425–426
1,3-Dibromocyclopentane, 199–200
2,3-Dibromopentane, 195–196, 197*f*, 199
1,4-Dicarbonyl compounds, 1027
1,5-Dicarbonyl compounds, 1028, 1036
β–Dicarbonyl compounds, 978, 1023, 1032, 1033
Dicarbonyl compounds, 486, 1032, 1033
Dicarboxylic acids, 758
2,5-Dichloroaniline, 691
4,4'-Dichlorobiphenyl, 109–110
1,4-Dichlorobutane, 593
Dichlorocarbene, 1120–1121, 1137
1,2-Dichloroethane, 151
1,1-Dichloroethylene, 563
Dichloromethane, 256*f*
2,4-Dichlorophenoxyacetic acid (2,4-D), 727*f*
(*E*)-1,3-Dichloropropene, 578–579, 579*f*
Diclofenac, 600
Dictyopterene D', 504
Dicyclohexylcarbodiimide, 950–951, 1241, 1258
Dicyclopentadiene, 670, 682
Dieckmann reaction, 1034–1035, 1041
Diels, Otto, 663

Diels–Alder reaction
drawing products, 664, 677
overview, 251, 663–664, 669–671
as pericyclic reaction, 1142, 1152, 1154
relative reactivity in, 675
rules governing, 664–668
1,3-Dienes, 646–647. *See also* Conjugated dienes
Dienes. *See also* Conjugated dienes
conjugated versus isolated, 646–647, 647*f*, 656–658
Cope rearrangement, 1159–1160
identifying for Diels–Alder reaction, 669–670, 669*f*, 677
naming, 398
ozonolysis, 486
Dienophiles
common examples with carbonyl groups, 665, 666*f*
defined, 663
identifying for Diels–Alder reaction, 669–670, 669*f*, 677
stereochemistry, 666
Diequatorial conformations, 159
Diesel fuel, 143
1,6-Diesters, 1034
1,7-Diesters, 1034
β–Diesters, 1033
Diethylbenzene, 718
Diethyl ether
flammability, 256*f*
IR absorption, 542*f*, 556
Lewis structure, 14
molecular structure, 97
solubility, 129
useful properties, 350
as weakly polar solvent, 107
Diethyl malonate, 997, 998–1000, 1024*f*
Diethyl sulfide, 373
Diethyl tartrate, 494
Digestion, 243, 243*f*, 1207
Digoxin, 1195
Dihalides, 325, 325*f*
Dihalocarbenes, 1120–1122
Dihalo cyclopropanes, 1122
Dihedral angles, 145, 148*f*, 150, 150*f*
Dihydroxyacetone
common uses, 51, 1177
as constitutional isomer, 1176
D-ketoses from, 1183, 1184*f*
Dihydroxyacetone phosphate, 980
Dihydroxylation, 482–484, 499, 506
1,4-Dihydroxynaphthoic acid, 735
Diisobutylaluminum hydride, 790
Diisopropylamine, 984
Dimers, 893
1,2-Dimethoxyethane, 561
5,5-Dimethoxypentan-2-one, 884
N,*N*-Dimethylacetamide, 537
Dimethylacetylene dicarboxylate, 685
Dimethylallyl diphosphate, 649
N,*N*-Dimethylbenzamide, 934
2,2-Dimethylbutane, 517–518
3,3-Dimethylbutan-2-ol, 357, 434
2,2-Dimethylbut-3-enal, 838
2,3-Dimethylbut-1-ene, 434, 640
3,3-Dimethylbut-1-ene, 640
2,3-Dimethylbut-2-ene, 434
cis-3,4-Dimethylcyclobutene, 1146, 1149
1,2-Dimethylcyclohexane, 161, 639
1,3-Dimethylcyclohexane, 141
trans-1,3-Dimethylcyclohexane, 160–161
1,4-Dimethylcyclohexane, 158–159
cis-1,4-Dimethylcyclohexane, 160*f*
1,6-Dimethylcyclohexene, 398*f*
1,2-Dimethylcyclopentane, 158
Dimethyl cyclopropanes, 217, 1140
Dimethyl ether, 15, 88, 546, 589
Dimethylformamide, 265*f*
N,*N*-Dimethylformamide, 105, 605
Dimethyl fumarate, 515
2,5-Dimethylhept-3-yne, 439*f*
2,2-Dimethylhexane, 526
4,5-Dimethylhexanoic acid, 890
2,4-Dimethylhexan-3-one, 816–817
2,6-Dimethyloctane, 504
6,6-Dimethyloct-3-yne, 438
2,2-Dimethyloxirane, 377–378, 377*f*
2,3-Dimethyloxirane, 376
2,3-Dimethylpentane, 138, 516–517
2,4-Dimethylpentanoic acid, 1006
3,3-Dimethylpentan-2-ol, 526
2,2-Dimethylpropane, 102, 106, 132
4,5-Dimethyl-2-propylhexanenitrile, 892
Dimethyl sulfide, 16, 485
Dimethyl sulfoxide (DMSO), 265*f*, 417
Dinucleotides, 1273, 1288
-dioic acid (suffix), 891
gem-Diol, 860, 861, 862
-diol (suffix), 345
Diols, 345
Diovan, 1118
Dioxybenzone, 126
Dipeptides, 1232–1233, 1240–1245, 1260
Diphenhydramine, 302, 525, 689, 1071
Diphosphates, 288–289
Dipole–dipole interactions
in alcohols, ethers, and epoxides, 348
in alkyl halides, 256
in amines, 1058
carboxylic acid derivatives, 936
in carboxylic acids and nitriles, 893
overview, 119
strength, 102, 103*t*
Dipoles, 43, 44–45, 101–102, 119
Directed aldol reactions, 1025–1027, 1041, 1042
Direct enolate alkylation, 993–996
Directing effects, 744–747, 748*f*, 752–753, 764
Disaccharides, 1202–1205
Disparlure, 481, 482*f*, 493
Disrotatory movement, 1147, 1148, 1149, 1150, 1164
Dissolving metal reductions, 468, 475–476, 506
Distillation, 143*f*
2,2-Disubstituted alkenes, 1123
Disubstituted alkenes, 308
Disubstituted benzene derivatives
electrophilic aromatic substitution with, 750–751, 767
naming, 690, 711
spectroscopic properties, 692, 692*f*
synthesis, 752–753
trans-1,2-Disubstituted cycloalkanes, 376
Disubstituted cycloalkanes, 199–200
Disubstituted cyclohexanes, 158–161, 168, 170
Disulfide bonds, 1235, 1251, 1251*f*, 1253–1254
Disulfides, 372
Diynes, 438
DNA. *See* Deoxyribonucleic acid (DNA)
DNA ligase, 1278
DNA polymerase, 1277, 1285, 1285*f*
Dodecahedrane, 140, 152*f*, 685
Dolutegravir, 200
Domagk, Gerhard, 1092
Donepezil, 100, 1027
L-Dopa, 1264
Dopamine
acetaldehyde reaction with, 883
molecular formula, 526
molecular structure, 98
as physiologically active amine, 1060
Double bonds
in alkenes, 307–309, 394, 395*t*
carbon–carbon, naming compounds with, A–8*t*
carbon–oxygen, 98–101
classifying functional groups by, 98–101
conjugated, 650
depicting in condensed structures, 28, 28*f*
in 1,3-dienes, 646
hybridization and, 38
length and strength, 41
in Lewis structures, 12
NMR signals, 570, 580–582
one atom more electronegative, 650
radical additions to, 609, 627–633
in resonance structures, 17, 18, 20, 21, 22
in triacylglycerols, 403, 403*t*

Double-headed arrows, 16
Double helix in DNA, 1274–1276, 1275*f*
Double reaction arrows, 59
Doublet of doublets, 581
Doublets, 573–574
Down carbons, 153, 154, 155, 156–157
Doxorubicin, 1175*f*
Duloxetine, 826
Dutasteride, 970
du Vigneaud, Vincent, 1222
Dyes, 1088, 1090–1091

E

E (prefix), 399–400
E1cB mechanism, 1020
E1 elimination reactions
  alcohol dehydration, 355–356
  overview, 320–322
  S_N1 substitution reactions compared, 323–324
  when favored versus E2 reactions, 324, 324*t*, 327*t*, 333
E2 elimination reactions
  alcohol dehydration, 355, 356
  alkyl halides versus amines, 1081*f*
  alkyne synthesis and, 324–326, 441
  kinetics, 312, 314*t*
  mechanism, 312–313, 314*t*
  R group effects, 313–314
  stereochemistry, 316–320
  when favored versus E1 reactions, 324, 324*t*, 327*t*, 333
  Zaitsev rule, 314–316
E-85 fuel, 349
Eclipsed conformations
  butane, 148–151, 149*f*
  defined, 145, 167
  ethane, 146, 146*f*, 147–148
  to find stereoisomers, 195
  propane, 147*f*
EcoRI enzyme, 1284
Edman degradation, 1236–1238, 1267
Efavirenz, 440
Effexor, 805
Ehrlich, Paul, 1091
Eicosapentaenoic acid, 110
Electric charge of amino acids, 1225
Electrocyclic reactions, 1142, 1146–1152, 1163
Electrocyclic ring openings and closures, 1142
Electromagnetic radiation, 530–531, 549, 558
Electromagnetic spectrum, 530, 530*f*
Electron cloud, 6
Electron donation
  in substituted anilines, 1072–1073
  in substituted benzenes, 736–739, 743, 744*f*, 766
  in substituted benzoic acids, 903–905, 904*f*, 918
Electron dot representation. *See* Lewis structures
Electronegativity
  acidity and, 67, 69–71, 902
  alkyl halide characteristics, 254, 258–259
  basic principles, 42–44
  effects on heterolysis, 224
  in functional groups, 96–98, 117, 779
  inductive effects in substitution reactions, 736, 737, 754
Electron pairs
  Brønsted–Lowry acid–base reactions, 59–61
  Brønsted–Lowry bases, 58
  determining relative basicity, 84
  diagramming movement, 225, 227, 245
  Lewis acids and bases, 58, 78–80, 81
Electrons
  in atomic structure, 6
  delocalizing density, 646, 647, 648, 653. *See also* Delocalization
  energy levels, 672
  magnetic field generation, 560, 567–572
  valence, 8
Electron shells, 7–9
Electron-withdrawing groups
  Diels–Alder reaction, 665, 675
  in electrophilic aromatic substitution reactions, 736–739, 743, 744*f*, 766
  in nucleophilic aromatic substitution reactions, 753–756
  stability and, 873
  in substituted anilines, 1072–1073
  in substituted benzoic acids, 903–905, 904*f*, 918
Electrophiles
  alkyne reactions with, 442
  benzene reactions with, 723
  carbenes as, 1120
  carbonyl carbon as, 98
  defined, 79
  in Diels–Alder reaction, 665
  structural features, 93, 116–118, 121
Electrophilic addition reactions
  alkyne hydrohalogenation, 443–445
  defined, 407
  distinctive features in conjugated dienes, 658–660
  Hammond postulate and, 409*f*
  hydration, 412–413
  mechanism, 408
  product quantities in conjugated dienes, 660–663
  stereochemistry, 410–412, 412*t*
Electrophilic aromatic substitution
  coupling reactions of aryl diazonium salts, 1088–1089
  disubstituted benzenes, 750–753
  energy diagram, 725*f*
  Friedel–Crafts acylation, 724*f*, 729, 730–731, 734–735, 734*f*, 765
  Friedel–Crafts alkylation, 724*f*, 729–730, 731–733, 765
  halogenation, 724*f*, 725–726, 727*f*, 765
  inductive effects in substituted benzenes, 736, 737–739
  nitration, 724*f*, 727, 765
  overview, 723–725, 724*f*
  resonance effects in substituted benzenes, 736–739
  substituted benzene activation or deactivation, 743–744
  substituted benzene limitations, 748–750
  substituted benzene orientation effects, 744–747, 748*f*
  substituted benzene variations, 739–743
  sulfonation, 724*f*, 727–728, 765
Electrospray ionization, 522
Electrostatic potential maps, 44, 44*f*, 45, 45*f*
Element effects in acidity, 67–69, 73*f*
Elements, 6–8
Elemicin, 123
Elenolic acid, 91
Eleostearic acid, 432
Eletriptan, 1134
Eleutherobin, 684
Elimination–addition reactions, 755–756
Elimination reactions
  in alcohols, 353, 355–356
  alkene products, 307–311, 425
  in alkyl halides, 259
  alkyne synthesis, 324–326, 441, 457
  basic types listed, A-16
  determinants, 326–330, 332
  E1 mechanism, 320–324
  E2 mechanism, 311–319, 324–325
  overview, 222, 244, 306–307
  predicting products, 314–316, 331
  substitution reactions with, 329–330
  when favored, 326, 330
Eliquis, 1099
Emend, 1102
Enals, 838
Enamines, 856–859, 858*f*, 1077
Enantiomeric excess, 204–205, 212
Enantiomers
  amino acids, 908–909
  chemical properties, 206–208
  defined, 183, 201, 209

Enantiomers (*continued*)
differing biological effects, 189, 207, 342, 481, 482*f*
in disubstituted cycloalkanes, 200
drawing, 186–187
1,2-epoxycyclohexane, 375
glyceraldehyde, 1177
in meso compounds, 197–198
3-methylcyclohexene, 188
with multiple stereogenic centers, 1179
physical properties, 202–205, 203*t*
pretomanid, 195
resolution, 1227–1230, 1227*f*, 1259
in S_N1 substitution reactions, 275
in S_N2 substitution reactions, 269
with two stereoisomers, 195
Enantioselective reactions
aldol reactions, 1024–1025, 1041
carbonyl reduction, 787–790, 788*f*
defined, 493
hydrogenation, 1230–1232, 1258
Sharpless epoxidation, 493–495, 496*f*
Enantiotopic protons, 565, 566, 592
Enclomiphene, 431
Endergonic reactions, 234
Endiandric acids, 1173
Endo addition rule, 667–668
Endothermic processes
bromination, 616
defined, 228
transition states, 280, 281*f*, 615
-ene (suffix), 397
Enediols, 980, 1009
Energy
benzene molecular orbitals, 709
bond dissociation, 228–231, 229*t*, 244, 437, 607, A–10
conformation and, 147–148, 159–160
conjugated versus isolated dienes, 657–658, 658*f*
diagrams, 235–239
of electromagnetic radiation, 530, 530*f*, 531, 532, 534*f*, 549
in electrophilic aromatic substitution reactions, 725*f*
from ethanol combustion, 349
of glucose metabolism, 1175–1176
hybridization and, 648, 652
in lipids, 166*f*
of molecular orbitals, 1144
of nuclear spin states, 558–560
released in bond formation, 226
of resonance structures, 651–652
storage in carbohydrates, 1175
thermodynamics concepts, 231–234
ultraviolet light, 672–674
Energy diagrams, 235–239, 236*f*, 245
Energy of activation (E_a)
defined, 235
effect of catalysts, 242, 242*f*
in energy diagrams, 236–237, 236*f*, 238, 238*f*
role in reaction rates, 239–240, 244
Enolates
aldehyde and ketone reactions with, 843
aldol reactions overview, 1016–1021
alkylation, 988, 993–1004
Claisen reaction, 1030–1034
crossed aldol reactions, 1021–1025
defined, 19
Dieckmann reaction, 1034–1035
directed aldol reactions, 1025–1027
formation and general reactions, 981–985
halogenation, 988, 990–991
intramolecular aldol reactions, 1027–1029
Michael reaction, 1035–1037
racemization, 987–988
reactivity, 988
Robinson annulation, 1037–1041
of unsymmetrical carbonyl compounds, 985–987
Enols
in alkyne hydration, 446–448, 458
biological functions, 980
configuration, 343
halogenation, 989
reactivity, 980–981, 988
synthesis and tautomerization, 977–980
tautomers, 446–447, 458, 978, 1005
Enones, 838
Entacapone, 50
Enthalpy changes ($\Delta H°$)
calculating with bond dissociation energy, 230–231, 246
defined, 228
in energy diagrams, 236–237, 236*f*, 238, 238*f*
reaction rates versus, 240
Entropy change ($\Delta S°$), 233–234
Enynes, 438
Enzymes
basic principles, 243–244, 243*f*, 1223, 1255
classifying, 1256, 1256*t*
in disease diagnosis and treatment, 1257
as oxidizing agents, 492–493
triacylglycerol hydrolysis, 955
use in amino acid hydrolysis, 1238–1239, 1239*t*
use in kinetic resolution of amino acids, 1229–1230
Enzyme–substrate complex, 243–244, 243*f*
Ephedrine, 185, 216
Epileptic seizure treatment, 1010
Epimers, 1182
Epinephrine, 253, 289, 290*f*
Eplerenone, 350
Epothilone A, 1126, 1126*f*
Epoxidation, 479–482, 483, 493–496, 498, 499
Epoxides
acetylide anion reactions with, 453–454
alcohol formation from, 477, 497
medical applications, 378–380
naming, 347–348
organometallic reagent reactions with, 812
oxidation reactions forming, 479–482
production, 352
properties, 348–349
reactivity, 353–354, 374–378, 382
structures and properties, 343, 344
useful properties, 350
Epoxyalkanes, 347
1,2-Epoxycyclohexane, 347, 375, 391
1,2-Epoxy-2-methylpropane, 347
cis-2,3-Epoxypentane, 347
Equations, 221, A–3
Equatorial bonds, 153, 156–157, 157*f*, 180
Equilibrium
direction in acid–base reactions, 65–66, 82, 898–899
in nucleophilic substitution, 262–263
reactants versus products in organic reactions, 231–233, 232*f*
Equilibrium arrows, 59, 60
Equilibrium constant (K_{eq})
in acid–base reactions, 62–64
defined, 231
determinants of size, 231–233, 232*f*, 233*t*
reaction rates versus, 240
Equivalent protons. *See* Proton equivalency
Erectile dysfunction drugs, 971
Ergosterol, 656
Ergot, 734
Eribulin mesylate, 350
Erivedge, 1077
Erlotinib, 461
Ertugliflozin, 880
β-Erythroidine, 682
D-Erythrose, 1182
Erythroxylon coca, 928
Escitalopram, 895
Esmolol, 90
Esomeprazole, 694*f*
Essential amino acids, 909, 1224*f*, 1225
Esters
acetoacetic ester synthesis, 1001–1004, 1006
basic structure, 778, 929
carbonyl absorption, 537–538
Claisen reaction in, 1030–1034
interesting examples, 938
IR absorption, 544, 544*f*, 546*t*, 937*t*
malonic ester synthesis, 996–1001, 1006

monosaccharide conversion to, 1195–1196, 1210
naming, 933, 935, 935*t*
naming derivatives, A–7*f*
nucleophilic acyl substitution reactions of, 948–949, 952–956, 964
pK_a values, 983*t*
in protecting group reactions, 1242–1243
reactivity, 781
reduction, 790–792, 819
structures and functional groups, 99*t*
in triacylglycerols, 403
Estradiol, 439
Estrogens, 440, 440*f*
Estrone
IR absorption, 552
as steroid, 671
synthesis, 418, 1036, 1037*f*
Eszopiclone, 715
Eth- (prefix), 135
Ethambutol, 215, 272
Ethane
acidity, 72–73
bond length and strength in, 41*t*
bond types in, 37, 93, 145
chlorination, 613, 613*f*
conformations, 145–146, 146*f*, 147–148
hybridization, 656
lack of functional group, 93
molecular structure, 131
Ethanedioic acid, 891
Ethanoic acid, 891
Ethanol
acidity, 69–72, 899–900
biological oxidation, 493
conversion to acetic acid, 163
detection in body, 127
fermentation, 2
functional group, 93–94
IR absorption, 546
isomers, 15, 88
NMR signals, 563, 582–583, 583*f*
as polar protic solvent, 264*f*
reaction with acetic acid, 242
starting material, 402*f*
useful properties, 349–350
water solubility, 109
Ethene, 400
Ethers
acetals versus, 862, 863
α cleavage, 527
Claisen rearrangement, 1161–1162
IR absorption, 542, 542*f*
monosaccharide conversion to, 1195, 1210
naming, 346–347
production, 350–351, 352
properties, 348–349
reactions with acids, 370–371
reactivity, 353, 382
structures and properties, 96, 97*t*, 343, 344
useful properties, 350
Ethoxide, 71, 264, 900
1-Ethoxy-2-methylcyclohexane, 426–427
4-Ethoxyoctane, 347
Ethyl acetate
carbonyl absorption, 537
common uses, 948
formation, 242
naming, 933
Ethyl acetoacetate, 1001–1004
Ethyl *p*-aminobenzoate, 760
Ethylbenzene, 690, 757–758
Ethyl benzoate, 1032
Ethyl butanoate, 938
Ethylcyclobutane, 142
N-Ethylcyclohexanamine, 1067–1068
2-Ethylcyclohexanecarbaldehyde, 834
1-Ethylcyclohexene, 629
Ethylcyclopentane, 526
4-Ethyl-3,4-dimethyloctane, 138
Ethylene
acidity, 72–73
bond angles, 24
bond dissociation energy, 437
bond length and strength in, 41*t*
as common name, 400
common uses, 402, 402*f*
drawing Lewis structures, 12
functions, 308
hydrogen bromide addition to, 407*f*
molecular orbitals, 1143–1144, 1143*f*
polymerization, 630–631
Ethylene glycol
in acetal formation reactions, 862
as common name, 345
conformations, 174
starting material, 402*f*
N-Ethylformamide, 934
Ethyl formate, 1032
5-Ethyl-2-hydroxy-3,6-dimethyloctanal, 834
Ethyl isopropyl ether, 351
Ethyl isopropyl ketone, 837*f*
1-Ethyl-3-methylcyclohexane, 141
N-Ethyl-*N*-methylcyclopentanamine, 1056
Ethyl methyl ether, 575
Ethyl methyl ketone, 837
4-Ethyl-5-methyloctane, 138
3-Ethyl-2-methylpentane, 526
Ethyl *p*-nitrobenzoate, 760
3-Ethylpentane, 615
Ethyl propanoate, 533–534, 533*f*
3-Ethylthio-5-methyloctane, 373
Ethyne, 439. *See also* Acetylene
Ethynylestradiol, 436, 439, 440*f*
Ethynyl groups, 439
1-Ethynyl-2-isopropylcyclohexane, 439*f*
Etonogestrel, 552
Etopophos, 880
Etoposide, 880
Euphorbia ingens, 1127
Euphorbia peplus, 1106
Excited state, 672, 1144–1145, 1150
Exemestane, 190, 502
Exergonic reactions, 234
Exo products, 667–668
Exothermic processes
chlorination, 616–617
defined, 228
transition states, 280, 281*f*, 615, 617*f*
Extraction
amine separation by, 1069–1070
carboxylic acids, 905–908, 918
defined, 905
Ezetimibe, 788

F

Fadrozole, 915
Faraday, Michael, 687
Farnesyl diphosphate, 649
Fat cells, 111
Fats, 404. *See also* Lipids
Fat-soluble vitamins, 111, 656
Fatty acids
basic features, 403–404, 403*t*
IR absorption, 547
in soybean oil, 466
Fehling's reagent, 1197
Fenn, John, 522
Fentanyl, 129, 304
Fermentation, 2, 349, 349*f*
Ferulic acid, 684
Fexofenadine, 53, 463, 770, 878, 1060
Fibrinogen, 1223
Fibrous proteins, 1253
Fingernails, 1253
Fingerprint region, 533, 533*f*
First-order kinetics, 241, 267, 274. *See also* S_N1 substitution reactions
Fischer esterification, 948–949, 961, 965
Fischer projections, 186, 1177–1180, 1180*f*, 1188–1189, 1211–1212
Fishhook, 225, 226*t*
Flagpole hydrogens, 155, 155*f*
Fleming, Alexander, 529, 939
Flosal, 1023
9-Fluorenylmethoxycarbonyl protecting group, 1242
Fluorine, 102, 257
Fluoroboric acid, 1084
Fluoroethane, 129
Fluoromethane, 104*f*

Fluoxetine
acid–base reactions, 86
enantiomer, 207
as important complex organic molecule, 3
nucleophilic substitution reactions in synthesis, 272, 272*f*, 755
as SSRI example, 3, 1061
synthesis step, 502
Fluticasone, 257
Fmoc protecting group, 1242
Folic acid, 1092
Formal charge, 13–15, 15*t*
Formaldehyde
aldol reactions, 1022
common name used, 835
common uses, 840
conversion to alcohol, 800
hydration reactions, 860
Formalin, 840
Formic acid, 134, 891, 894
Formyl group, 837
Fortified milk, 1151
Fosamax, 16
Fosamprenavir, 1257
Four-centered transitions, 420
Fragmentation patterns, 510, 515–520
Free energy change ($\Delta G°$), 232–234, 232*f*, 233*t*, 240
Freons, 257, 621
Frequency (electromagnetic)
IR absorption and, 532–533
as NMR variable, 559–560
overview, 530–531, 549
relationship to mass and bond strength, 534, 534*f*
relation to wavenumber, 532
Friedel, Charles, 724
Friedel–Crafts acylation, 724*f*, 729, 730–731, 734–735, 734*f*, 749–750, 759–760
Friedel–Crafts alkylation, 724*f*, 729–730, 731–733, 749, 759–760
(–)-Frontalin, 495, 496*f*
Frontside attacks, 269, 275
Fructofuranose rings, 1204
Fructose
D-fructose, 1176, 1183, 1191
furanose form, 1191
as naturally occurring ketose, 1176, 1183
nomenclature, 1177
stereogenic centers, 185
Fructose 1,6-bisphosphate, 1019
Fructose 6-phosphate, 980
Fucose, 1220
2'-Fucosyllactose, 1207
3-Fucosyllactose, 1208
Fukui, Kenichi, 1143
Full-headed curved arrows, 225, 226*t*, 245
Fumarate, 223
Fumaric acid, 129, 666
Functional groups
in amino acids, 908
characteristic of organic compounds, 93–94
common fragmentation patterns, 518–520
defined, 93
in different organic reaction types, 223
effects on intermolecular forces, 101–103
effects on physical properties of organic compounds, 104–111
effects on reactivity of organic compounds, 116–118
guide to synthesis reactions by group, A–17-A-20
identifying, 100–101, 532–534, 547, 550. *See also* Infrared spectroscopy
interconversions, 455
IR spectra of various classes, 540–547, 546*t*
learning reaction types for, 424
major types in organic compounds, 94–100
naming, A–7-A-8. *See also* Nomenclature
from nucleophilic substitution, 291–293
Furan, 712, 719
Furanose rings, 1184, 1190–1191, 1204
Fused ring systems, 666, 667*f*, A–8-A-9

G

Gabapentin enacarbil, 189
Gabriel synthesis, 1063–1064
Galactose, 243, 243*f*
Galantamine, 1101, 1141, 1162
Gamma (γ) carbons, 835
Gantrisin, 1092
Garcinia subelliptica, 1161
Garlic, 134
Garsubellin A, 1161
Gas chromatography–mass spectrometry (GC–MS), 521–522, 521*f*
Gasohol, 349
Gasoline, 143, 144
Gauche conformations
butane, 149*f*, 150
defined, 149, 167
ethylene glycol, 174
Geckos, 102
Geminal dichloride, 441
Geminal dihalides, 325, 441, 443
Geminal protons, 580
Generic names, 134
Genes, 1272, 1276
Genetic code, 1280–1281, 1281*t*
Genetic information, 1268, 1269, 1276, 1277, 1278
Geometry, molecular, 24–27
Geraniol, 289
Geranyl acetate, 110
Geranyl diphosphate, 289, 304, 649
German cockroaches, 944
Gibberella fungi, 1117
Gibberellic acid, 1048
Gibbs free energy, 232–234
Gilbert, Walter, 1284
Ginger root, 665
Ginkgo biloba, 96, 777, 895
Ginkgolide B, 777, 788
Glaucoma, 257, 686, 896
Gleevec, 1099
Global climate change, 164
Globular proteins, 1253
Glucagon, 1265
D-Glucaric acid, 1198, 1210
Glucitol, 1196
Glucocaffeic acid, 901
β-D-Glucopyranose, 1209
Glucosamine, 963, 1208–1209
D-Glucosamine, 1208
Glucose
acetals in, 872
biological functions, 118
from cellulose hydrolysis, 1206
conformations, 175
drawing, 1185–1186, 1186*f*, 1189–1190
entropy in metabolism, 234
Fischer projections, 1180*f*
formation from lactose, 243, 243*f*
from maltose hydrolysis, 1203
metabolism, 980, 1019
molecular structure, 1175*f*
nomenclature, 1177
oxidation, 231
regulation in blood, 1251*f*
in starch, 1206–1207
as D-sugar, 1181
starch hydrolysis to, 179
D-Glucose, 1176, 1212–1213
α-D-Glucose, 872
β-D-Glucose, 872, 885, 1175*f*
Glucose 6-phosphate, 980
α–Glucosidase, 1207
β–Glucosidase, 1206
Glutamic acid
abbreviations and R group, 909*t*, 1224*f*
ionizable functional group values, 1226*t*
molecular structure, 924, 1224*f*
Glutamine, 1224*f*, 1226*t*
Glutathione, 197, 1236
Glyceraldehyde, 1176, 1177, 1181
Glyceraldehyde 3-phosphate, 980
Glycerol, 345
Glycine
abbreviations, 909*t*
dipeptide synthesis from, 1240–1241
ionizable functional group values, 1226*t*
R group, 908, 909*t*

structure, 1223
structure and abbreviation, 1224*f*
Glycogen, 1207
Glycolic acid, 71
Glycols, 345
Glycopeptide transpeptidase, 958
α Glycosides, 1192, 1193, 1211
β Glycosides, 1192, 1193, 1211
Glycosides
in disaccharides, 1202, 1203–1204
formation and hydrolysis, 1192–1194, 1195
naturally occurring, 1194
N-form, 1209–1210
as nonreducing sugars, 1198
not converted to aldonic acids, 1197
in starch, 1206–1207
α Glycosidic bonds, 1206, 1207
β Glycosidic bonds, 1205, 1206, 1209
Glycosidic linkages, 1202
Grandisol, 1011
Green chemistry, 491–492
Greenhouse gases, 142
Grignard, Victor, 796
Grignard reagents
carbon dioxide reactions with, 811, 897*t*
in cetirizine synthesis, 801, 802*f*
nitrile reactions with, 915
origin of name, 796
preparation, 797
retrosynthetic analysis of reactions, 803–805
stereoisomers formed by, 801, 821
α,β-unsaturated carbonyl compound reaction with, 814
Ground state, 33, 672
Group numbers (periodic table), 6
Grubbs, Robert, 1123
Grubbs catalyst, 1123–1127, 1129
Guanine
base pairing with cytosine, 1275, 1277
nucleosides and nucleotides, 1271*t*
π electrons, 705
as purine base, 1269
Guanosine, 110
Gypsy moth, 481

H

HaeIII enzyme, 1284
Hair, 1253, 1253*f*
Half-headed curved arrows, 225, 226*t*, 246, 607
Halides
coupling reactions with organocuprates, 1107–1109, 1128
in Heck reaction, 1114–1116
as leaving groups in nucleophilic substitution, 259, 261–263
negative charge, 76
in Stille coupling, 1116–1119
in Suzuki reaction, 1109–1113
Hallucinogens, 1061, 1062
α–Halo carbonyl compounds, 989, 991–992, 1004
Halocyclohexanes, 317–319, 318*f*, 331
Haloethanes, 177
Haloform reaction, 990–991
Halogenation
of activated benzenes, 748
alkanes, 610–617
alkenes, 413–416
alkyl benzenes, 757–758
alkynes, 442*f*, 445–446, 457
at allylic carbons, 622–627
as electrophilic aromatic substitution, 724*f*, 725–726, 727*f*, 765
enolates, 988, 990–991
enols, 989
stereochemistry, 618–620, 618*t*
Halogens
in common nucleophiles, 266*t*
degrees of unsaturation in compounds containing, 396
effects on alkyl halide polarity, 258–259
in general alkyl halide structure, 254
in Heck reaction, 1114
naming in alkyl halides, 255
polarizability of, 413
Halohydrins, 352, 416–418
Halolactonization, 435
Halomon, 254
Halonium ions, bridged, 414, 416–417, 418*t*, 429
Halothane, 256*f*
Hammond postulate, 279–281, 409, 409*f*, 615, 616
Hand sanitizers, 393
Hansen's disease, 189
Hardening, 472
Haworth projections, 1186–1189, 1212–1213
HDL particles, 962–963
HDPE and LDPE, 630
Head-to-tail polymerization, 633
Heat
effect on electrocyclic reactions, 1147, 1148–1150
indicating in chemical equations, 221
role in pericyclic reactions, 1143
role in radical formation, 608
Heat of hydrogenation, 469, 657–658, 694–696
Heat of reaction. *See* Enthalpy changes (ΔH°)
Heck, Richard, 1114
Heck reaction, 1114–1116, 1128, 1129
Helicene, 693, 694*f*
Helium, 8
α-Helix, 1248–1250, 1248*f*
Heme, 1254
Hemiacetals
acetals versus, 1192
in carbohydrates, 872, 1184–1191, 1210
cyclic forms, 868–871, 1184–1191
drawing, 1185–1191
formation, 864–865, 866, 1184–1185
Hemibrevetoxin B, 97
Hemoglobin, 1223, 1252, 1254–1255, 1255*f*
Heptalene, 715
Heptane, 255
Heptan-2-ol, 526
Heptan-2-one, 1003
Herbicides, 727*f*, 979
Herniated disc, 592*f*
Heroin, 115, 552, 946
Herpes simplex virus, 1009, 1270, 1291
Heteroatoms
aromatic, 701–705
characteristic of organic compounds, 93
defined, 3, 20
electronegative, 96–98
role in reactivity, 93, 116–118
in skeletal molecular structures, 32
Heterocycles, 701–705
Heterocyclic aromatic amines, 1074
Heterocyclic aromatic bases, 1269
Heterolysis, 224, 225–226, 225*f*
Hexachloroethane, 207
Hexa-2,4-diene, 654, 1147, 1148, 1150, 1154
Hexa-1,3-diyne, 439*f*
Hexafluoroethane, 129
Hexamethylphosphoramide, 265*f*
Hexan-2-amine, 527
Hexan-3-amine, 520
Hexane
fragmentation patterns, 515–516, 516*f*
IR absorption, 541*f*
mass spectrum, 510*f*, 516, 516*f*
as nonpolar solvent, 107
skeletal structure, 29
Hexane-2,3-diol, 640
Hexane-2,5-dione, 1027–1028
Hexanoic acid, 895
Hexan-2-ol, 387, 526, 816, 816*f*
Hexan-2-one, 1006
Hexapeptides, 1260
Hexaphenylethane, 643
Hexa-1,3,5-triene, 696, 1145, 1145*f*
Hex-1-ene, 541*f*
cis-Hex-3-ene, 398
trans-Hex-3-ene, 398
Hexoses, 1176
Hexylcyclobutane, 142
Hex-1-yne, 541*f*

Highest occupied molecular orbital (HOMO)
 in cycloaddition reactions, 1153–1155
 defined, 708, 1144
 in electrocyclic reactions, 1147, 1148, 1150
 in excited state, 1144–1145, 1150
 in ground state, 1145
High-resolution mass spectrometry, 520–521
α-Himachalene, 525
HindIII enzyme, 1284
Hirsutene, 644
Histamine, 702–703, 1060, 1099
Histidine
 bacterial conversion to histamine, 702
 biosynthesis, 356
 classification, 1263
 ionizable functional group values, 1226*t*
 structure and abbreviation, 1224*f*
Histrionicotoxin, 124, 443
HIV treatment
 action of protease inhibitors, 1257, 1287
 AZT and ddI, 1287
 darunavir, 886
 dolutegravir, 200
 efavirenz, 440
 fosamprenavir, 1257
 ritonavir, 195
 saquinavir, 218
^{1}H NMR. *See* Proton NMR spectroscopy
Hodgkin's disease, 1175*f*
Hoffmann, Felix, 896
Hoffmann, Roald, 1143
Hofmann, Albert, 734
Hofmann elimination, 1079–1081, 1094, 1096
Homologous series, 133
Homolysis, 224, 225–226, 225*f*, 228
Homotopic protons, 565, 566, 592
Honey, 1191
Hooke's law, 534*f*
Hormones
 bradykinin, 1235
 glucagon, 1265
 insulin, 1223
 oxytocin, 1222, 1235, 1251
 somatostatin, 1265
 synthetic analogues, 439–440
 L-thyroxine, 1226
 vasopressin, 1235, 1251
Horns, 1253
5-HPETE, 379, 641
Hückel, Erich, 697
Hückel's rule
 basic criteria, 697, 698
 basis for, 706–708
 with charged aromatic compounds, 699, 700, 701
 inscribed polygon method and, 709
 with multiple-ring compounds, 699
 pyridine conformity, 701
Human genome, 1276
Human Genome Project, 1284
Human immunodeficiency virus (HIV), 1287. *See also* HIV treatment
Human milk oligosaccharides, 1207–1208
Humulene, 309, 1113
Hyaluronic acid, 1219
Hybridization
 acetate, 71
 acidity and, 72–73, 73*f*, 83
 in benzene, 688–689
 defined, 34
 delocalization and, 653
 effects on amine basicity, 1075, 1076*t*
 in ethane, ethylene, and acetylene, 36–41, 72–73
 heteroatom in aromatic heterocycle, 703
 impact on carbon–carbon bonds in conjugated dienes, 656–657
 patterns in organic compounds, 33–36
 using resonance structures to determine, 676
Hydration. *See also* Water
 aldehydes and ketones, 860–862
 alkenes, 412–413, 423, 427, 428, 429
 alkynes, 442*f*, 446–448, 457, 458
Hydride
 enolate formation from, 984
 nucleophilic addition reactions, 845–846
 as strong base, 76, 76*f*
Hydride reducing agents, 468, 784–786, 786*f*
Hydride shifts, 357, 358–359, 380, 732, 733
Hydroboration, 419–420, 421*f*, 1111
Hydroboration–oxidation
 alkenes, 419–423
 alkynes, 442*f*, 448–450, 457
Hydrocarbons
 basic features, 2, 95*t*
 classifying functional groups by, 94–96
 conversion to metabolites, 176
 IR absorption, 540–541, 541*f*
 oxidation and reduction, 467*f*
 in petroleum, 143
 saturated, 131
 synthesis by coupling reactions with organocuprates, 1108–1109
Hydrochloric acid
 alcohol treatment with, 362, 1192
 in amine extraction, 1070*f*
 reaction with cyclohexene, 80
 strength, 76
Hydrocodone, 975
Hydrofluoroalkanes, 622
Hydrofluorocarbons, 621–622
Hydrofluoroolefins, 621–622
Hydrogen. *See also* Carbon–hydrogen bonds
 acidity in bonds, 68–69
 axial and equatorial, 153, 154, 154*f*
 classifying in hydrocarbons, 95, 96, 119
 as compound, 8, 9
 depicting in skeletal structures, 29–30
 determining types in a molecule, 562
 isotopes, 6
 in Lewis structures, 10
 octet rule exception, 16
 orbital, 8
 reaction with cyclohexene, 242
 as reducing agent, 468
 treating oils with, 472–473
Hydrogenation reactions
 alkene stability and, 469–470, 657
 applications, 472–474
 catalysts, 469, 470–471
 degrees of unsaturation and, 471–472
 enantioselective, 1230–1232, 1258
Hydrogen bonding
 in alcohols, 348
 in amines, 1058
 in carboxylic acids and nitriles, 893
 determining sites, 108–109, 121
 in DNA, 1275*f*
 enols, 978
 overview, 102–103, 103*t*, 119
 in polar protic solvents, 264
Hydrogen bromide
 addition to buta-1,3-diene, 434, 661, 662*f*
 addition to conjugated dienes, 659
 addition to cyclohexene, 410
 addition to ethylene, 407*f*
 ether reactions with, 370
 radical addition reactions, 627–629, 630*f*
Hydrogen halides, 362–363
Hydrogen–hydrogen bond, 228, 229*t*
Hydrogen iodide, 370–371
Hydrogen ions, 58–59
Hydrogenolysis, 1243
Hydrohalogenation
 alkenes, 407–412, 427
 alkynes, 442*f*, 443–445, 457
Hydrolases, 1256, 1256*t*
Hydrolysis
 acetals, 865–867
 amides, 956–957
 cellulose, 1206
 esters, 952–956
 glycosides, 1193–1194
 imines and enamines, 858–859
 maltose, 1203, 1211
 nitriles, 912–913, 917
 peptides, 1238–1240
 starch, 179, 180
Hydroperoxides, 625, 641

Hydrophilic structures, 109, 112, 114, 114*f*
Hydrophobic structures, 109, 112, 114, 114*f*
Hydroxide
in alcohol synthesis, 351
as Brønsted–Lowry or Lewis base, 90
common to elimination reactions, 306, 405
of cyclic monosaccharides, 1195, 1210
in E2 elimination reactions, 312
enolate formation from, 984
nucleophilicity, 263
as strong base, 76, 76*f*
α-Hydroxy acids, 71
Hydroxybenzene, 690
p-Hydroxybenzoic acid, 927
3-Hydroxybutanal, 1018
2-Hydroxybutanedioic acid, 927
β–Hydroxy carbonyl compounds, 1016–1017, 1019
Hydroxychloroquine, 1076
Hydroxyethyl methacrylate, 642
Hydroxy groups
in alcohols, 343, 345
in ethanol, 93–94
glycol formation from, 482–484
Hydroxylamine, 1199
Hydroxyl groups, 1176
(*S*)-3-Hydroxy-2-methylpropanal, 1211
5-Hydroxypentan-2-one, 805, 806*f*, 807
Hygrine, 1081
Hymeniacidon agminata, 557, 792
Hyperconjugation, 278–279
Hypertension drugs
aliskiren, 974
captopril, 218
lisinopril, 5, 1100
losartan, 1134
valsartan, 1118–1119
Hypoglycin A, 199, 924
Hypophosphorus acid, 1085

I

Ibuprofen
enantiomer, 207, 1012
functional groups in, 124
most acidic proton, 88
names for, 135
synthesis, 1034
-ic acid (suffix), 891
Identical twins, 182, 1276
-idine (suffix), 1270
D-Idos, 1217
Idoxuridine, 1291
Iejimalide B, 432
Illicium jiadifengpi, 1049
Illudin-S, 1003
Imatinib, 1099
Imidazolides, 968
Imides, 1063
Imidic acid tautomers, 912, 913
Imines
from amine reactions with carbonyl compounds, 854–856, 858*f*, 1077
nitrile reduction forming, 914, 915, 1201
Iminium ions, 855
Imipenem, 547
Indanomycin, 1134
Indene, 717
Indigo, 1090, 1091
Indinavir, 124
Inductive effects
on acidity, 69–71, 73*f*, 83
on amine basicity, 1076*t*
on carbocations, 278
in carboxylic acids, 901–902, 916
in substituted benzenes, 736–739, 765
-ine (suffix), 255
Infrared radiation, 532
Infrared spectroscopy
aldehyde and ketone absorptions, 839*t*
amine absorptions, 1058, 1058*t*, 1059*f*
benzene derivative absorptions, 692*t*
bond absorptions, 534–540, 534*f*, 535*t*, A–12
carboxylic acid and nitrile absorptions, 894*t*
carboxylic acid derivative absorptions, 937*t*
general features, 530, 532–534
molecular structure determination, 547–549
spectra of common functional groups, 540–547, 546*t*
use with mass spectrometry, 548–549
Ingenol, 1106, 1127
Ingenol mebutate, 1106
Initiation
alkane halogenation, 611, 612
allylic bromination with NBS, 623
atmospheric ozone destruction, 621*f*
radical addition of HBr to alkenes, 628
radical polymerization of CH_2=CHZ, 632
Inorganic compounds, 1
Inscribed polygon method, 708–711, 710*f*
Insecticides, 507, 590, 1119
Insect repellants, 944
Insomnia, 715, 719, 1099
Insulin, 1223, 1251, 1251*f*
Integrals (NMR), 572
Integration ratio, 594
Intermediates, 426
Intermolecular forces
impact on boiling and melting points, 104, 105
impact on solubility, 107
main types, 101–103, 103*t*
Intermolecular hydrogen bonding, 1058
Internal alkenes, 394
Internal alkynes, 437, 449
International Union of Pure and Applied Chemistry, 134
Intramolecular aldol reactions, 1027–1029, 1037–1038, 1041, 1042
Intramolecular cyclization, 869, 875
Intramolecular Friedel–Crafts reactions, 734–735
Inversion of configuration, 269–271, 270*f*, 362, 368, 369
Iodine, 102
Iodoform test, 990
Iodomethane, 104*f*
1-Iodo-1-methylcyclohexane, 315
(Iodomethyl)zinc iodide, 1122
Ion–dipole interactions, 107, 107*f*
Ionic bonds, 9, 119
Ionic compounds
boiling points, 104
melting points, 105–106
principles of formation, 9
solubility, 107, 110*t*
strength of intermolecular forces, 101, 103*t*
Ionophores, 115–116, 116*f*
Ions
aromatic, 699–701
transport across cell membranes, 115–116, 116*f*
types, 6
IR inactive vibration, 535
Iron ions, 1254
Islets of Langerhans, 1251*f*
Iso- (prefix), 136
Isoamyl acetate, 938
Isobutyl group, 136
Isocomene, 435
Isoelectric points, 911, 1226, 1226*t*
Isolated dienes
conjugated dienes versus, 646–647, 647*f*, 656–658
electrophilic addition, 658
Isoleucine, 1224*f*, 1225, 1226*t*
Isomenthol, 175
Isomerases, 1256, 1256*t*
Isomeric 1,5-dienes, 1159
Isomers
of alkanes, 132–133, 136
of alkenes, 469–470
benzene, 690, 711
of cycloalkanes, 158–162, 168
defined, 15, 132, 180, 181, 209
ibuprofen, 1012
major classes, 181, 201, 201*f*, 209
physical properties of stereoisomers, 202–206

Isomers (*continued*)
resonance structures versus, 17
simple carbohydrates, 1176
starch and cellulose, 180
stereoisomers defined, 158
Isopentane, 140
Isopentenyl diphosphate, 649
Isophorone, 1051
Isophthalic acid, 921
Isoprene
functions, 655–656
from 3-methylbut-2-en-1-ol dehydration, 683
polymerization, 642
UV light absorption, 672, 673*f*
Isopropyl alcohol, 345, 393
Isopropylbenzene, 541, 541*f*, 760
Isopropylcyclopentane, 817–818
Isopropyl groups, 136
3-Isopropyl-4-methylcyclohexanone, 837
4-Isopropyl-3,7,8-trimethyldec-3-ene, 399
(*Z*)-4-Isopropyl-3,7,8-trimethyldec-3-ene, 399
Isopulegone, 556
Isotopes
bromine, 514
chlorine, 513
defined, 6
masses of common examples, 520*t*
nitrogen, 8
Itraconazole, 881
IUPAC nomenclature
alcohols, 344–345, 346
aldehydes, 834–835
alkanes, 135–140, 169
alkenes, 397–400
alkyl halides, 255–256, 255*f*
alkynes, 438–439, 439*f*
basic features, 134–135
carboxylic acids, 890–891
cycloalkanes, 141–142, 169
esters, 933
ethers, 347
ketones, 836–837
nitriles, 892–893
sulfides, 373
thiols, 372

J

Jaborandi, 686
Jack-o'-lantern mushroom, 1003
Januvia, 703
Japanese puffer fish, 670
Jasmine flowers, 475
cis-Jasmone, 475
Jatropha curcas, 220
Jiadifenin, 1049
Johnson, W. S., 1028
Joules, 41
Juvenile hormones, 1108

K

Kainic acid, 924
Karenia brevis, 344
Kavain, 430
Kekulé, August, 687
Kekulé structures, 687, 688
Kelsey, Frances Oldham, 189
Kempene-2, 682
Keratin, 1223, 1249, 1253–1254
α–Keratins, 1253–1254
Kerosene, 143
Ketene, 54
Keto–enol tautomers, 912
β–Keto esters, 1030–1031, 1033, 1036
Ketohexoses, 1177
Ketones
addition reactions with alcohols, 862–867
alcohol oxidation to, 488, 489, 491
in aldol reactions, 1016, 1017
alkylation, 994–995
carbanion addition to, 846
in crossed Claisen reactions, 1032
cyanohydrin formation from, 846–848
defined, 446
formation from carboxylic acid derivatives, 810
formation from nitrile reactions, 912, 915, 917
formation in alkyne hydration, 446, 447, 448, 449
formation in alkyne hydroboration, 449
fragmentation patterns, 518
general reactions of, 842–844
hydration reactions, 860–862
hydride reactions with, 845–846
interesting examples, 840
IR absorption, 543, 543*f*, 546*t*
naming, 836–837
nucleophilic addition reaction overview, 779
nucleophilic addition reactions with amines, 854–859
nucleophilic substitution reactions, 991
physical and spectroscopic properties, 838, 839*t*
pK_a values, 983*t*
as products of acetoacetic ester synthesis, 1001–1004
protecting groups, 867–868, 875
reactivity to nucleophilic attack, 780
reduction to alcohols, 783, 784–786, 798, 799–800
reductive amination, 1065–1068
structures and functional groups, 99*t*, 778, 833
synthesis, 840–841, 841*t*
Ketoprofen, 75
D-Ketoses, 1183, 1184*f*
Keto tautomers
basic features, 446
converting to enols, 458, 978–980, 1005
formation, 977–978
Kiliani–Fischer synthesis, 1199, 1200–1201
Kinetic enolates, 985–986, 987
Kinetic products of addition reactions in conjugated dienes, 660–663, 677
Kinetic resolution of amino acids, 1229–1230
Kinetics
activation energy and, 239–240
catalyst and enzyme role, 242–244
defined, 231, 239
rate equations, 240–241
substituent effects in aromatic substitution, 739, 741
Kodel, 961

L

l (prefix), 203, 210
Lactams, 930
β–Lactams, 340, 547, 930, 939, 957–958
Lactase, 243*f*, 1203
Lactate dehydrogenase, 789
Lactic acid, 29, 789, 895
Lacto-*N*-neotetraose, 1220
Lactones, 435, 930
γ–Lactones, 930
Lactose, 243, 243*f*, 872, 1203–1204
Laetrile, 848
Lagevrio, 938
Lagging strands (replication), 1277*f*, 1278
Lamisil, 461
Lamivudine, 1287
Latanoprost, 896
Laurencia genus, 258, 996
Laurencin, 996
Lauterbur, Paul C., 592*f*
Lavandulol, 804
Lavender oil, 804
LDA (lithium diisopropylamide), 984, 1025
LDL particles, 962–963
LDPE and HDPE, 630
Leading strands (replication), 1277–1278, 1277*f*
Leaving groups
alcohols, 353
basic features, 261–263, 262*t*
carbonyl condensation reactions, 1016
defined, 259
as determinants of substitution reaction mechanisms, 285, 780–782
in E2 elimination reactions, 312, 313
ethers and epoxides, 353–354

loss in nucleophilic acyl substitution, 940
phosphorus derived, 288–289
Le Châtelier's principle, 356
Leghemoglobin, 1255
Leptospermone, 979
Leucine, 89, 1224*f*, 1226*t*
Leu-enkephalin, 1236
Leukemia treatment
cytarabine, 1268
doxorubicin, 1175*f*
imatinib, 1099
ponatinib, 440
Leukotrienes
C_4, 101, 379
role in asthma, 379
synthesis, 379, 641
Leustatin, 1272
Levofloxacin, 88
Levonorgestrel, 440
Levorotatory compounds, 203, 210
Lewis acids and bases
in electrophilic substitution, 729–730
in nucleophilic substitution, 259, 408
overview, 78–80, 81
Lewis structures
basic rules for drawing, 10–15, 15*t*
benzene, 688
converting condensed structures to, 28–29
converting skeletal structures to, 31
determining hybridization patterns from, 36
drawing problems, 51–52
isomers, 15
molecular shape in, 23–27
octet rule exceptions, 16
resonance in, 16–22
simplified organic structures in, 27–29
Lexapro, 895
Lexiscan, 1272
Lexiva, 1257
Lidocaine, 600
Ligands, 1110
Ligases, 1256, 1256*t*
Light
effect on electrocyclic reactions, 1147, 1150–1151
indicating in chemical equations, 221
polarized, 202–204
role in pericyclic reactions, 1143
role in radical formation, 608
Lightning, 485
Lilac flowers, 310
Limonene, 486, 683
Limu lipoa, 504
Linalool, 804
Linalyl diphosphate, 304, 683
Linamarin, 848
Lindlar catalyst, 475, 477, 482
Linear shapes, 24
Lineatin, 827
Linoleic acid
in jatropha oil, 220
melting point, 403*t*
oxidation, 231
radical reactions, 626
from soybeans, 466
structure, 404*f*
Linolenic acid, 403*t*, 404, 404*f*
Lipases, 955
Lipid bilayers, 114, 114*f*, 1286, 1286*f*
Lipids
basic features, 118, 165–166
fatty acids, 403–404
hydrolysis, 954–956
metabolism, 1176
oxidation, 473, 625, 626*f*
physical properties, 404
Lipitor, 1102
Lipoxygenases, 379
LiquiBand, 642
Lisinopril, 5, 46, 1100
Lithium aluminum hydride, 468, 784–785, 790–795, 795*t*
Lithium diethylamide, 391
Lithium diisopropylamide (LDA), 984, 1025
Lithium dimethylcuprate, 555, 1122
Lithium tri-*tert*-butoxyaluminum hydride, 790
Livalo, 1051
Lone pairs
adjacent to positive charge, 650
in Brønsted–Lowry bases, 58, 58*f*
in charged atoms, 32
defined, 10
in Lewis structures, 11, 12–13
in nucleophiles, 259
role in reactivity, 117
Loratadine, 525
Losartan, 1134
Lowest unoccupied molecular orbital (LUMO), 708, 1144–1145, 1153–1155
LSD, 734, 1062
Lumigan, 896
Lunesta, 715
Lung cancer, 461
Lupin, 1067, 1077
Lyases, 1256, 1256*t*
Lychee, 199, 924
Lycopene, 655–656, 673
Lysergic acid, 685
Lysergic acid diethylamide (LSD), 734, 1062
Lysine
abbreviations and R group, 909*t*
classification, 924
ionizable functional group values, 1226*t*
structure and abbreviation, 1224*f*
Lysozyme, 1249, 1250*f*

M

M + 2 peak, 513–515
Madder root, 1090, 1091
Magnetic field effects on nuclei, 558–560
Magnetic resonance imaging, 591–592. *See also* Nuclear magnetic resonance spectroscopy
Ma huang, 216
Ma'ilione, 258
Major contributors to hybrids, 22, 651
Major groove, 1276
Malaria
drug treatments, 397, 422, 423, 1076, 1077
sickle cell trait resistance, 1255
Malate, 223
Maleic acid, 129, 666
Maleic anhydride, 666*f*
Malic acid, 29
Malonic acid, 90, 891
Malonic ester synthesis, 996–1001, 1006
Maltose, 886, 1203, 1211
Manganese, 478
D-Mannose, 1187–1188, 1200
Mansfield, Peter, 592*f*
Manzamine, 1096
Maple syrup, 1174
Marine sponges, 1268
Markovnikov's rule, 408–410, 412*t*, 413, 427, 444–445
Mass number, 6, 191
Mass spectrometer schematic, 508*f*
Mass spectrometry
analyzing unknowns with molecular ion, 510–513
fragmentation pattern identification in, 515–520
general features, 508–510
M + 2 peak, 513–515
types, 520–522
use with infrared spectroscopy, 548–549
Mass-to-charge (*m/z*) ratio, 508*f*, 509–510, 520
Matrine, 1067, 1077
Mauveine, 1091
Maxalt, 719
McLafferty rearrangement, 528
MDMA, 1103
Meadowsweet plant, 895, 896
Mefloquine, 397
Melanin, 674

Melting points
of alcohols, ethers, and epoxides, 348
alkanes, 144*t*
alkenes, 401
alkyl halides, 256
alkynes, 439
amides, 936
amino acids, 909, 1225
boiling points compared, 174
carboxylic acids, 893
fatty acids, 403*t*
monosaccharides, 1177
organic compound principles, 105–106, 120
thiols, 371
Membrane proteins, 1223
Menthol, 125, 175, 212
Menthone, 125
Meperidine, 108, 1011
Mercaptans. *See* Thiols
Mercapto groups, 371
Merrifield method, 1245–1247, 1261
Merrifield, R. Bruce, 1245
Mescaline, 1062
Meso compounds, 197–199, 200, 209, 666
Messenger RNA, 1278, 1279–1283, 1282*f*, 1289
Mestranol, 126, 536
meta- (prefix), 690, 711, 739
Metabolism, glucose, 1175–1176
Metabolites, hydrocarbon conversion to, 176
Meta directors, 740, 741, 744, 747, 752, 764
Metal–carbene intermediates, 1125
Metal catalysts, 469, 470
Metal hydride reagents, 784, 790–795, 795*t*, 913–914
Metallocyclobutanes, 1125
Metals as catalysts, 242
Metathesis, 1123–1128, 1124*f*, 1129, 1136
Met-enkephalin, 939, 1293
Meth- (prefix), 135
Methamphetamine, 69, 1060, 1066*f*
Methane
acidity versus water, 67
in atmosphere, 130, 142
bond angles, 24–25
as compound, 8
hybridization patterns, 33–34
molecular structure, 131
as natural gas component, 2, 142
oxidation, 164
oxidation–reduction reactions, 163*f*
Methanoic acid, 891
Methanol
as acid or base, 58–59
drawing Lewis structures, 11
human toxicity, 493
hybridization in, 36
as polar protic solvent, 264*f*
Methionine
abbreviations and R group, 909*t*
ionizable functional group values, 1226*t*
SAM synthesis from, 374
structure and abbreviation, 1224*f*
Methionine–serine dipeptide, 1282*f*
p-Methoxybenzoic acid, 904
Methyl acetate
acetyl chloride conversion to, 250
carbon NMR spectrum, 589
IR absorption, 556
NMR spectrum, 591*f*
reactions producing, 267
Methyl acrylate, 663, 666*f*
m-Methylaniline, 756
p-Methylaniline, 756, 1073
Methylation, 289
Methylbenzene, 690
Methyl benzoate, 556
(*R*)-α-Methylbenzylamine, 1228–1229, 1259
2-Methylbutane, 106, 132, 140
2-Methylbutanenitrile, 892
2-Methylbutanoyl chloride, 932
2-Methylbut-1-ene, 636
3-Methylbut-1-ene, 624
3-Methylbut-2-en-1-ol, 683
Methyl *trans*-chrysanthemate, 1137
3-Methylcycloheptene, 398*f*
Methylcyclohexane, 141, 156–158, 157*f*
2-Methylcyclohexanethiol, 372
2-Methylcyclohexanol, 426–427
cis-3-Methylcyclohexanol, 369
2-Methylcyclohexanone, 985, 1025
1-Methylcyclohexene, 426–427, 1109
3-Methylcyclohexene, 188
2-Methylcyclopentanamine, 1080
Methylcyclopentane, 188, 634
2-Methylcyclopentanecarboxamide, 934
1-Methylcyclopentene, 398*f*
Methyl dimethylphosphonate, 605
Methylene, 1122
α–Methylene-γ-butyrolactone, 996
Methylene chloride, 256*f*
Methylenecyclopentane, 635
Methylene groups, 133, 400, 401*f*
Methyl esters, 1242–1243
Methyl α-D-glucopyranoside, 1193
Methyl groups, 745
Methyl–halogen bonds, 229
4-Methylhexane-2-thiol, 372
2-Methylhexanoic acid, 1001
5-Methylhexan-2-one, 533–534, 533*f*
5-Methylhex-4-en-1-yne, 439*f*
Methyl ketone, 446, 449, 990
Methylmagnesium chloride, 805
Methylmagnesium iodide, 884
Methyl methacrylate, 642
Methyl 2-methylbutanoate, 938
N-Methylmorpholine *N*-oxide (NMO), 484
3-Methylpentanal, 836
4-Methylpentan-1-amine, 1056
2-Methylpentane, 517–518
4-Methylpentane-1,4-diol, 805, 806*f*, 807
2-Methylpentan-3-one, 837*f*
(*Z*)-3-Methylpent-2-ene, 498
4-Methylpent-3-en-2-one, 838
Methylphenidate, 390
2-Methylpropane, 132
2-Methylpropanenitrile, 600
Methyl propanoate, 544*f*
2-Methylpropanoic acid, 555
Methyl propyl ether, 368
Methyl salicylate, 556
1,2-Methyl shifts, 357–358
3-Methyl-3-sulfanylhexan-1-ol, 208
Methylthiocyclohexane, 373
Methyl vinyl ketone, 666*f*
Metoprolol, 89, 122
Micelles, 112, 113*f*
Michael acceptors, 1036
Michael reaction, 1035–1037, 1041
Microorganisms, 142
Migraines, 273, 719, 1134
Migrastatin, 1136
Milk, 1151, 1203
Mineral oil, 176
Minor contributors to hybrids, 22, 651
Minor groove, 1276
Miscibility, 108
Mixed anhydrides, 929, 933
Molecular formulas, 15, 512–513
Molecular ions
analyzing unknowns with, 510–511, 523
defined, 509
determining for alkyl chlorides, 514
proposing molecular formulas for unknowns with, 512–513
Molecular orbitals
buta-1,3-diene, 1144–1145, 1144*f*
ethylene, 1143–1144, 1143*f*
hexa-1,3,5-triene, 1145, 1145*f*
in inscribed polygon method, 708–711
theory of, 706–708, 1143
Molecules
defined, 9
determining polarity, 44–45
determining shapes, 23–27
drawing condensed and skeletal structures, 27–33
drawing Lewis structures, 10–15, 15*t*
isomers, 15
Molnupiravir, 938, 954
Molozonides, 485
Molybdenum, 1123

Mometasone, 283
Monensin, 434, 502
Monochlorination, 619–620, 634, 635
Monohalogenation, 610
Monomers, 179, 630
Monosaccharides
  D-aldoses, 1182, 1183*f*
  basic features, 118, 1176–1177
  cyclic forms, 1184–1191
  D and L designations, 1181
  depicting with two or more stereogenic centers, 1179–1180
  Fischer projections, 1177–1180
  D-ketoses, 1183, 1184*f*
  in nucleic acids, 1269
Monosubstituted alkenes, 308, 1123
Monosubstituted benzene derivatives, 583, 584*f*, 690, 766
Monosubstituted cyclohexanes, 156–158, 168, 249
Morphine
  as both acid and base, 59
  codeine synthesis from, 921
  Diels–Alder synthesis, 671
  heroin production from, 946
  IR absorption, 552
  passage through cell membranes, 115
  synthesis, 1051
Motrin. *See* Ibuprofen
Motuporamine B, 119
M peak, 513–515
MTBE (*tert*-butyl methyl ether), 109, 413
Mullis, Kary, 1285
Multiple bonds, 12–13, 17
Multiple sclerosis, 515, 1272
Multiplets, 575
Multistep synthesis, 762–764
(+)-α-Multistriatin, 495, 496*f*
Muscalure, 303
Muscone, 786*f*
Musk deer, 786*f*
Mutarotation, 1186, 1192, 1203
Myambutol, 272
Mycomycin, 218
Mycostatin, 656
Mylar, 960
Myoglobin, 1254, 1255*f*
Myosin, 1249
Myrcene, 504
Mytilus edulis, 802

N

n + 1 rule, 577, 594
(*n* + 1)(*m* + 1) rule, 579, 595
Nabumetone, 1004
NADH, 492, 789
NAG (*N*-acetylglucosamine), 963
Naked anions, 265
Naloxone, 128
Naphthalene, 693, 699, 716
Naphthea corals, 487
Naproxen, 86, 207, 994
Narcissus plant, 1141
Natural fibers, 959
Natural gas, 142
Natural products
  from acetylide anion reactions, 452, 453*f*
  alkaloids, 1059
  antioxidants, 626, 627
  biomolecules, 118
  carboxylic acids, 494
  β-carotene, 402
  cyanohydrin derivatives, 848
  defined, 292
  dehydration synthesis, 360, 360*f*
  dyes, 1090–1091
  ethanol from, 349–350
  fatty acids in, 403
  ferulic acid, 684
  garsubellin A, 1161
  from ginger root, 665
  glycosides, 1194
  isocomene, 435
  isoprene, 655, 656
  cis-jasmone, 475
  kavain, 430
  lycopene, 656
  from marine sponges, 1268
  muscone, 786
  niacin, 790
  occidentalol, 685
  oxytocin, 1222
  from petroleum, 292, 402, 402*f*
  pilocarpine, 686
  pygmaeocin C, 776
  quinine, 305, 314
  synthesis by Sharpless epoxidation, 495, 496*f*
  synthesis using aldehyde and ketone reduction, 786, 788
  tetrodotoxin, 670
  zingiberene, 402
Negatively charged bases, 76, 76*f*
Negishi, Ei-ichi, 1114
Neotame, 1263
Neral, 123, 838
Nerium oleander, 867
Nerolidol, 309
Net dipoles, 44–45
Neuroprotectin D1 (NPD1), 655
Neurotoxins, 97, 126. *See also* Toxins
Neurotransmitters, 1060, 1061*f*
Neutral amino acids, 1223, 1224*f*, 1225, 1225*f*, 1226
Neutrons, 6
Newman projections, 146, 170
Nexium, 694*f*
Niacin, 112, 790
Nicotinamide adenine dinucleotide (NAD^+), 492–493, 789–790
Nicotine, 115, 290, 1053
Nigericine, 863
Niphatoxin B, 452, 453*f*
Niraparib, 1099
Nirmatrelvir, 128
Nitration, 724*f*, 727, 765
Nitric oxide, 607
Nitriles
  interesting examples, 895
  IR absorption, 545, 545*f*, 546*t*
  naming, 892–893
  nucleophilic addition reactions, 911–915, 917
  reduction to amines, 1064
  selected pK_a values, 983*t*
  structure and bonding, 889
  synthesis, 911, 917
p-Nitroaniline, 1073
Nitrobenzene, 727, 739–740
p-Nitrobenzoic acid, 762, 905
Nitrogen
  in amines, 98, 1054–1055
  in common nucleophiles, 266*t*
  degrees of unsaturation in compounds containing, 396
  formal charge, 15*t*
  isotopes, 8
  as leaving group in nucleophilic substitution, 259
  relative basicity, 1071, 1096
  valence electrons, 10
Nitrogen-13, 8
Nitrogen-14, 8
Nitrogen heterocycles, 705, 1057, 1057*f*
Nitro groups
  directing effects in electrophilic aromatic substitution, 747
  lack of Friedel–Crafts reactions, 749
  reduction reactions, 760–761
  reduction to amines, 1064
Nitronium ions, 727
p-Nitrophenol, 753
N-Nitrosamines, 1082, 1083
Nitrosonium ions, 1082, 1083
p-Nitrotoluene, 750
Nitrous acid, 1082–1083, 1094
NMR spectrometers, 559*f*. *See also* Nuclear magnetic resonance spectroscopy
Noble gases, 9
Nodes, electron density, 7

Nomenclature
  alcohols, ethers, and epoxides, 344–348, 382
  aldehydes, 834–836
  alkanes, 135–140, 169
  alkenes, 397–400, 401*f*, 428
  alkyl halides, 255–256
  alkyl substituents with branching, A–6
  alkynes, 438–439, 439*f*
  amines, 1055–1057
  amino acids, 909, 909*t*, 1223, 1224*f*
  benzene derivatives, 690–692, 711
  carbohydrates, 1176–1177
  carboxylic acid derivatives, 932–936, 935*t*
  carboxylic acids and nitriles, 890–893
  cycloalkanes, 141–142, 169
  ketones, 836–837
  organic compounds, 134–135
  polyfunctional compounds, A–7-A-8
  thiols, 372
Nonactin, 115, 116
Nonane, 139
Nonaromatic compounds, 696–697, 704, 711, 712
Nonbonding electrons, 10
Nonequivalent protons
  complex examples, 577–580, 595
  spin–spin splitting, 573–577, 580–582
Nonnucleophilic bases, 264
Nonpolar bonds, 43, 144, 535
Nonpolar compounds, 107
Nonpolar molecules, 44, 101
Nonpolar tails, 112, 113*f*, 114–115, 114*f*
Nonreducing sugars, 1197, 1198*f*, 1204
Nonrenewable resources, 143
Nonsteroidal anti-inflammatory agents. *See* Anti-inflammatory drugs
Nootkatone, 507, 513
Noradrenaline, 289, 290*f*, 1060
Norethindrone, 109, 439, 440*f*
NSAIDs, 600
N-Terminal amino acids, 1233, 1236–1238, 1259, 1281
Nuclear magnetic resonance spectroscopy
  benzene derivative absorptions, 692*t*
  benzene rings and proton NMR signals, 570, 583, 584*f*
  characteristic absorptions, A–13-A-14
  common types, 558
  complex splitting examples, 577–580
  cyclohexane effects on proton NMR signals, 583
  identifying unknowns with, 585–588
  OH protons and proton NMR signals, 582–583
  overview, 558–561
  signal intensities, 572–573
  signal numbers, 561–566, 589–590
  signal positions, 566–570, 590–591
  spectra, 560–561, 588, 591*f*
  spin–spin splitting, 573–577, 580–582
Nuclei (atomic), 6, 558–560
Nucleic acids
  basic features, 1278, 1279*f*
  components, 1269–1274
  defined, 1269
  DNA double helix, 1274–1276
  DNA replication, 1277–1278
  DNA sequencing, 1283–1284
  genetic code and protein synthesis, 1280–1283
  PCR amplification, 1285
  RNA transcription, 1279–1280
  in viruses, 1286–1287
Nucleophiles
  acetylide anions as, 443
  in alcohol synthesis, 351
  ambident, 985
  amines as, 1069, 1077–1078
  carbonyl compounds as, 977
  carbonyl oxygen as, 98
  defined, 79, 259
  as determinants of substitution reaction mechanisms, 283–284
  in Diels–Alder reaction, 665
  frontside and backside attacks, 269–271, 275
  organometallic reagents as, 798–799
  reaction mechanisms with alkyl halides, 327–328, 330
  strength, 263–266
  structural features, 93, 116–118, 121
  susceptible to addition reactions, 844
Nucleophilic acyl substitution
  acid chloride reactions, 943–944, 947–948, 963
  amides, 956–958, 965
  anhydrides, 945–946, 964
  in biological systems, 961–963, 1282, 1282*f*
  carboxylic acid reactions, 946–951, 947*f*, 964, 977
  esters, 952–956, 964
  overview, 781, 930, 939–943
  summarized, 958, 959*t*
Nucleophilic addition reactions
  acetal formation, 862–867
  aldehyde and ketone susceptibility to, 843, 1017
  amines, 854–859, 1077, 1094
  carbonyl compound hydration, 860–862
  of carbonyl compounds in general, 779, 780, 822, 977
  cyanohydrin formation, 846–848
  cyclic hemiacetal formation and conversion, 868–871
  general mechanism, 843–844, 845*f*, 873
  hydride, 845–846
  in Michael reaction, 1036
  nitriles, 911–915
  Wittig reaction, 848–853, 873, 874
Nucleophilic attack
  in nucleophilic addition, 780, 784, 799, 843–844
  in nucleophilic substitution, 780–781, 940
Nucleophilic substitution reactions. *See also* Nucleophilic acyl substitution
  with acetylide anions, 450–454
  adrenaline from, 253
  alcohol and ether production, 350–351, 353
  of alcohols, 361–366, 369*f*
  of alkyl tosylates, 368
  in amine synthesis, 1062–1064, 1093, 1094
  of aromatic compounds, 291, 753–756, 765
  in biological reactions, 288–290, 290*f*
  carbocation stability, 278–279
  of carbonyl compounds in general, 780–782,822
  Claisen reaction as, 1031
  epoxides, 374–380
  of ethers, 370–371
  general features, 259–261, 294
  leaving groups overview, 259, 261–263
  mechanisms, 266–277, 282–287
  nucleophile strength, 263–266
  organic compounds from, 291–293
  in vinyl halides and aryl halides, 291, 753–756
Nucleosides, 1270, 1271*t*
Nucleoside triphosphates, 1285
Nucleotides
  basic features, 118
  defined, 1269
  drawing, 1271–1272
  naming, 1270–1271, 1271*t*
  in replication, 1277
  in RNA synthesis, 1280–1283, 1281*t*
Nuts, 627
Nylon, 960
Nystatin, 656

O

-o (suffix), 255
Observed rotation, 202–203
Occidentalol, 685
(*E*)-Ocimene, 310
Octane, 94, 693
Octanenitrile, 545*f*
(2*E*,4*Z*,6*E*)-Octa-2,4,6-triene, 1146, 1148, 1150

Octet rule, 9, 16
Octinoxate, 31, 674, 1050
Octylamine, 545*f*
Odors, 372, 1059
-oic acid (suffix), 890
Oil refineries, 143*f*
Oils
  defined, 404
  hydrogenation, 472–474
  IR absorption, 547
  oxidation, 473, 625, 626*f*
  from soybeans, 466
Oil spills, 144
Okazaki fragments, 1278
-ol (suffix), 344
Oleandrin, 867
Olefin metathesis, 1123–1128, 1124*f*, 1129
Olefins. *See* Alkenes
Oleic acid
  biological functions, 118
  IR absorption, 547
  melting point, 403*t*
  from soybeans, 466
  structure, 404*f*
Oleocanthal, 1013, 1127
Oligosaccharides, 1207–1208
Olives, 132
Omega (Ω) carbons, 835
Omega-3 fatty acids, 404
Omeprazole, 300
-one (suffix), 836–837
Open arrows, 454
Opium, 671
Opsin, 855, 856*f*
Optical activity, 202–206, 209, 376
Optical purity, 204–205
Orajel, 972
Oral contraceptives
  desogestrel, 440, 772
  ethynylestradiol in, 436, 439
  mechanism, 440*f*
Orbitals
  basic configuration, 7–8
  bonding and antibonding, 706–707
  hybrid, 33–41
1*s* Orbitals, 7–8, 8, 33
2*p* Orbitals, 8, 33–34
2*s* Orbitals, 8
Orchids, 505
Orders, rate equation, 240–241
Organic chemistry, elements of, 1
Organic compounds
  biomolecule overview, 118
  characteristic functional groups, 93–94
  common bonds in, 10, 10*f*
  common names, 140
  common uses, 1, 2*f*
  determining molecular shapes, 23–27
  determining polarity, 44–45
  determining relative acidity, 74
  drawing condensed and skeletal structures, 27–33
  drawing Lewis structures, 10–15, 15*t*
  examples of major properties, 111–116
  intermolecular forces, 101–103
  isomers, 15
  IUPAC nomenclature system, 134–135
  major functional group types, 94–101
  from nucleophilic substitution, 291–293
  octet rule, 9, 16
  oxidation–reduction reactions, 163, 164
  oxybenzone example, 45–46
  in periodic table, 7
  physical properties, 104–111
  reactivity, 116–118
  resonance structures, 16–22
Organic halides, 254, 254*f*
Organic molecules
  determining polarity, 44–45
  determining shapes, 23–27
  drawing condensed and skeletal structures, 27–33
  drawing Lewis structures, 10–15, 15*t*
  oxybenzone example, 45–46
  representative examples, 2–3
  simplified drawings, 27–33
Organic phosphates, 288
Organic reactions
  basic types, 222–223, 423, A-15-A-16
  bond dissociation energy, 228–231, 229*t*
  Brønsted–Lowry acids and bases, 59–62
  energy diagrams, 235–239
  enthalpy and entropy, 233–234
  kinetics, 239–244
  learning, 423–425
  mechanisms, 224–228
  nucleophilic substitution overview, 259–266. *See also* Nucleophilic substitution reactions
  role of functional groups, 116–118
  thermodynamics overview, 231–233
  writing equations, 221, 221*f*
Organic solvents, 107–108
Organic synthesis
  alkenes in, 425–427
  challenges, 815–816
  defined, 291
  elimination reaction's importance, 306
  halogenation in, 617–618
  multistep examples, 455–456
  from nucleophilic substitution, 291–293
  terminology and conventions, 454–455
Organoboranes, 420*f*, 1110–1113, 1129
Organocatalysts, 1024
Organocuprates
  coupling reactions with organic halides, 1107–1109, 1128
  producing, 797
  producing ketones with, 810
  reactivity, 796
  α,β-unsaturated carbonyl compound reaction with, 814, 822
Organolithium reagents
  in alcohol synthesis, 821
  general structure, 796
  nitrile reactions with, 915
  preparation, 797
  α,β-unsaturated carbonyl compound reaction with, 814
Organomagnesium reagents, 796
Organometallic reagents
  for acid–base reactions in carbonyl compounds, 805–807
  defined, 796
  general features of coupling reactions with organic halides, 1107–1109, 1128
  in Heck reaction, 1114–1116, 1128
  nitrile reactions with, 915
  for nucleophilic addition in carbonyl compounds, 796–804
  preparation, 797, 819
  reaction overview, 820
  reactions with carbon dioxide, 811
  reactions with carboxylic acid derivatives, 807–810
  reactions with epoxides, 812
  in Stille coupling, 1116–1119, 1128
  in Suzuki reaction, 1109–1113, 1113*f*, 1128
  α,β-unsaturated carbonyl compound reaction with, 814, 822
  in Wittig reactions, 849–853
Organopalladium compounds, 1110
Organophosphorus reagents, 849
Organosodium reagents, 797
Organostananes, 1116–1119, 1130
Orientation of substituents in electrophilic aromatic substitution, 739–741, 744–747, 748*f*, 752–753
Orlistat, 1267
Orlon, 632
ortho- (prefix), 690, 711, 739
Ortho, para deactivators, 740
Ortho, para directors, 739, 740, 741, 744, 745, 746, 752, 764
-ose (suffix), 1176
Oseltamivir, 100, 663, 970
-osine (suffix), 1270
Osmium tetroxide, 483–484
Oxalic acid, 891, 895
Oxaloacetate, 223, 1047
Oxaphosphetanes, 851
Oxazaborolidine, 787

Oxidation reactions
  alcohols, 488–490
  aldehydes, 783, 795–796, 819, 1197, 1210
  aldoses, 1197–1199, 1210–1211
  alkanes, 162–164
  alkenes, 479–487
  alkyl benzenes, 758
  alkynes, 487–488
  antioxidants and, 626–627
  carboxylic acid synthesis, 897, 897*t*
  defined, 163, 467, 782
  epoxides, 380
  green approaches, 491–492
  in hydroboration–oxidation reactions, 419, 421–422
  impact on oils, 473, 625, 626*f*
Oxidation–reduction reactions
  aldoses, 1196–1199, 1210–1211
  alkanes, 162–164
  defined, 163
Oxidative addition, 1110
Oxidative cleavage, 479*f*, 485–488, 497
Oxidizing agents, 478–479
Oxidoreductases, 1256, 1256*t*
Oximene, 504
Oximes, 1199–1200
Oxiranes, 344, 347–348
6-Oxoheptanal, 1048
Oxone, 491
-oxy (suffix), 347
Oxybenzone, 4, 45–46, 674
Oxycodone, 129, 552
Oxy-Cope rearrangement, 1159, 1160
Oxygen
  in alcohols, ethers, and epoxides, 344
  in common nucleophiles, 266*t*
  degrees of unsaturation in compounds containing, 396
  electronegativity, 67
  formal charge, 15*t*
  molecule as radical, 607
  storage in hemoglobin, 1254
  valence electrons, 8
Oxygen-containing compounds, IR absorption, 542–544
Oxygen–hydrogen bonds, 44, 898
Oxytocin, 930, 1222, 1235, 1251
-oyl (suffix), 837
Ozone, 485, 499
Ozone layer, 257, 620–622, 621*f*, 674
Ozonides, 485
Ozonolysis, 485, 486, 499

P

Paclitaxel, 178, 189
Pain relievers, 975, 1004, 1293
Palau, 557
Palau'amine, 557, 792
Palladium catalysts
  general features, 242
  in Heck reaction, 1114–1116
  Lindlar catalyst, 475, 477, 482
  in Stille coupling, 1116–1119
  in Suzuki reaction, 1109–1113, 1113*f*
Palythine, 684
Pancreas, 1251*f*, 1265
Pancreatic cancer, 461
Pancreatic lipase, 1267
Pantothenic acid, 92, 112
para- (prefix), 690, 711, 739
Parent ions, 509
Parent names, 135
Parkinson's disease, 515, 828, 1264
Paroxetine, 89, 351
Partial hydrogenation, 472–473, 473*f*
Partial hydrolysis, 1238–1240
Patchouli alcohol, 360*f*
PCBs (polychlorinated biphenyls), 109–110
Peach pits, 848
(*S*)-Penicillamine, 1263
Penicillin
  biological functions, 957–958
  discovery, 529, 547, 939
  physical properties, 939
Penicillin G
  amide ring, 529
  functional groups in, 122, 124, 177, 547
  structure, 939
Pentacycloanammoxic acid methyl ester, 1169
(*E*)-Penta-1,3-diene, 657
Penta-1,4-diene, 646–647, 657
Pentalene, 715
Pentane, 102, 103, 106, 132–133
Pentane-2,4-dione, 1009
Pentane-3-thiol, 372
Pentan-3-ol, 803
Pentan-3-one, 104*f*
Pentanoyl chloride, 555
Pentapeptides, 1260
Pentoses, 1176
2-Pentylcinnamaldehyde, 1023
Peptide bonds
  defined, 1232
  principles, 1234–1235
  protein primary structure importance, 1247
  translation and synthesis, 1282, 1282*f*
Peptide hormones, 1222, 1265
Peptides
  basic features, 1232—1233
  bonding principles, 1234–1235
  cellular functions, 938–939
  interesting examples, 1235–1236
  sequencing, 1236–1240, 1239*t*
  synthesis, 1240–1247
Percent *s*-character, 42, 536, 550
Percent transmittance, 533
Pericylic reactions
  cycloaddition reactions, 1152–1155, 1163
  defined, 1142
  electrocyclic reactions, 1146–1152, 1163
  rules summarized, 1162, 1163*t*
  sigmatropic rearrangements, 1156–1161, 1163
  types summarized, 1142–1143
Periodic table
  acidity trends in, 67–69, 81, 82
  bond dissociation energy and position in, 228–229
  bond lengths and position in, 23
  electronegativity in relation to, 42–43, 43*f*
  leaving group ability in, 261
  main features, 6–8, A–1
  nucleophilicity in, 263, 265, 294
  octet rule exceptions, 16
  valence electrons, 8, 9–10, 16
Periplanone B, 548, 1026, 1160
Perkin, William Henry, 1091
Peroxyacetic acid, 478
Peroxyacids, 478, 479–480
Pesticides, 258, 307
Petroformymic acid, 536
Petroleum, 143, 144, 221
PET scans, 8
PF-07321332, 128, 895
$PGF_{2\alpha}$, 248, 644, 896, 897
pH. *See* Acidity
Phenacetin, 973
Phenanthrene, 699, 720
Phenols
  acidity, 899–900
  benzene metabolism to, 176
  Claisen rearrangement yielding, 1161
  configuration, 343
  deprotonation, 923
  formation from water and diazonium salts, 1084
  halogenation of, 748
  nomenclature, 690
  radical reactions with, 627
Phenoxide, 900, 900*f*
Phentermine, 1068
Phenylacetaldehyde, 981
Phenyl acetate, 968
Phenylacetic acid, 968
Phenylacetonitrile, 1011
Phenylalanine
  abbreviations and R group, 909*t*
  aspartame metabolism to, 1263
  forms of, 916
  human inability to metabolize, 970
  ionizable functional group values, 1226*t*
  structure and abbreviation, 1224*f*
Phenylcyclohexane, 94

Phenylethanal, 836
2-Phenylethanamine, 1060, 1062
Phenyl groups, 94, 691
Phenyl isothiocyanate, 1236–1237
Phenylketonuria, 970
(1*Z*,3*E*)-1-Phenylocta-1,3-diene, 1113
2-Phenylpropanoic acid, 219
N-Phenylthiohydantoin, 1237, 1267
N-Phenylthiourea, 1237
Pheromones
  blattellaquinone, 944
  bombykol, 1113
  defined, 131
  disparlure, 481
  kempene-2, 682
  lineatin, 827
  muscalure, 303
  periplanone B, 548, 1026, 1160
  synthesis, 495, 496*f*
  (*E*)-tetradec-11-enal, 829
Phosphates, 288, 340, 1269
Phosphines, 1110
Phosphodiesters, 1273
Phospholipids, 114–115, 114*f*, 128
Phosphonium salts, 850, 874
Phosphorus, 16, 288–289, 849–850
Phosphorus oxychloride, 354, 359–360
Phosphorus tribromide, 365
Photochemical cycloadditions, 1152, 1154–1155, 1155*t*
Photochemical electrocyclic reactions, 1150–1151, 1164
Photochemical sigmatropic rearrangements, 1158, 1158*t*
Photons, 530, 531, 549
Photosynthesis, 1175, 1176
Phthalic acid, 921
Phthalimide, 1063
Phylloquinone, 112
Physical properties of organic compounds, 104–111. *See also* Boiling points; Melting points; Solubility
Phytyl diphosphate, 735
Pi (π) bonding molecular orbitals, 1143
Pi (π) bonds
  in acetylene, 39, 39*f*
  in addition reactions, 405–406
  in alkenes and alkynes, 306, 308, 394
  alkyne reactions and, 441, 442
  basic features, 38
  in benzene, 756
  in Brønsted–Lowry bases, 58*f*
  calculating numbers, 471–472
  in carbonyl groups, 98, 779
  characteristic of organic compounds, 93
  in conjugated molecules, 646–647
  in consecutive elimination reactions, 324–325
  in cycloaddition reactions, 1152
  in Diels–Alder reaction, 663
  in different organic reaction types, 222–223
  in electrocyclic reactions, 1146
  in ethylene, 38*f*, 1143–1144
  formation in dehydration, 354, 358
  in lone pairs, 58
  in molecular orbital theory, 706–708, 707*f*
  in nucleophiles, 259
  in organic halides, 254, 254*f*
  in pericyclic reactions, 1142–1143
  radical reactions with, 609, 632
  role in reactivity, 93, 116–118
  in sigmatropic rearrangements, 1156
  strength, 41
Picato, 1106
Picometers, 23
Pi (π) electrons
  aromaticity criteria, 696, 701
  in benzene, 688, 689, 723
  in inscribed polygon method, 708–711
  magnetic field effects on, 570
Pig insulin, 1251*f*
Pilocarpine, 686, 703
Pilocarpus microphyllus, 686
β-Pinene, 1011
Pinnatoxin A, 879
Pioglitazone, 770
Piperidine, 1057*f*, 1075
Pitavastatin, 1051
Pitocin, 930, 1222
pK_a values
  acidity and, 63–64, 63*t*
  alpha hydrogens in aldehydes and ketones, 981
  amino acids, 911, 1225, 1226*t*
  basicity determination using, 1071–1075, 1076*t*
  of common acids, 76
  defined, 63
  determinants, 66–75, 73*f*
  equilibrium direction and, 65–66, 82, 898–899
  of leaving group conjugate acids, 261–262
  selected compounds, 982*t*, 983*t*, A-4-A-5
Planar carbocations, 275, 276
Planar geometry of peptide bonds, 1234
Planar molecules, 697, 712
Planck's constant, 531
Plane-polarized light, 202–204
Planes of symmetry, 183, 184, 198
Plaque, in arteries, 963
Plavix, 694*f*, 976
β-Pleated sheet, 1249–1250, 1249*f*, 1250*f*
Plocamenol A, 301
Poison dart frogs, 443, 1100
Polar bonds, 42–45, 477, 478*f*, 889, 893
Polar C–X sigma (σ) bonds, 477, 478*f*
Polar heads, 112, 113*f*, 114–115, 114*f*
Polarimeters, 202
Polarity principles, 42–45
Polarizability
  defined, 102
  of halogens, 413
  impact on boiling point, 104, 104*f*
  inductive effects in substitution reactions, 736
Polarized light, 202–204
Polar molecules
  alkyl halides as, 256, 258–259
  defined, 44
  dipole–dipole interactions, 102
  monosaccharides as, 1196
  passage through cell membranes, 115–116
  solubility, 107
Polar reactions, 226
Polar solvents, 264–265, 285–286, 313
Polio, 1286
Poly(acrylic acid), 631
Polyacrylonitrile, 632
Polyalkylation, 749
Polyamides, 960
Polycyclic aromatic hydrocarbons, 380, 693, 694*f*, 699
Polyene antibiotics, 656
Polyesters, 960–961
Poly(ethyl acrylate), 641
Poly(ethyl α-cyanoacrylate), 632
Polyethylene
  common types, 630
  formation, 630–631
  molecular structure, 95
  oxidation resistance, 642
  starting material, 402, 402*f*
Poly(ethylene glycol), 126
Polyethylene terephthalate (PET), 960
Polyfunctional compounds, naming, A-7-A-8
Polyhalogenation, 748
Poly(hydroxyethyl methacrylate), 642
Polyisoprene, 642
Poly(lactic acid) (PLA), 961
Polymerase chain reaction, 1285, 1285*f*
Polymerization, 630–633
Polymers
  as benzene and toluene derivatives, 693
  cellulose, 1205
  common uses, 630, 631*t*
  defined, 179, 630
  natural and synthetic fibers, 959–961
  nucleic acids as, 1269
  polystyrene derivatives, 1245–1247
  proteins as, 938
  as radical reaction products, 630–633
  starch, 1206–1207

Poly(methyl methacrylate), 642
Polynucleotides, 1273, 1273*f*, 1274
Polypeptides, 1232, 1245–1247
Polypropylene, 631*t*
Polysaccharides, 1205–1208
Polystyrene
  common uses, 632
  derivative use in automated peptide synthesis, 1245–1247
  oxidation, 642
  starting material, 402*f*
  structure, 632
Polysubstituted benzenes, 691
Polysubstitution, 750
Polytetrafluoroethylene, 631*t*
Poly(vinyl acetate), 402*f*, 581, 632
Poly(vinyl chloride), 126, 257, 402*f*, 606
Ponatinib, 440
Poppy seed tea, 671
p Orbitals
  aromaticity criteria, 697
  in benzene, 689, 697
  in conjugation, 646
  defined, 7
  in 1,3-dienes, 646
  hybridization, 34
  in molecular orbital theory, 706–708, 707*f*, 1143–1145
  rotation in electrocyclic reactions, 1147
Porphyin, 1254
Porphyra umbilicalis, 674
Porphyrin, 1254
Positron emission tomography, 8
Potassium *tert*-butoxide, 306
Potassium iodide, 9
Potassium permanganate, 479, 483–484
Potassium peroxymonosulfate, 491
Potassium propanoate, 891*f*
Potassium sorbate, 895
Potential energy
  butane, 149, 150*f*
  cyclohexane, 159–160
  ethane, 147–148, 148*f*
  in lipids, 166*f*
Prednisone, 840
Prefixes
  cis-, 158
  d, 203, 210
  deoxy-, 1269, 1270
  di-, 1056
  E, 399–400
  eth-, 135
  iso-, 136
  l, 203, 210
  meta-, 690, 711, 739
  meth-, 135
  organic molecule names, 135
  ortho-, 690, 711, 739
  para-, 690, 711, 739
  R, 190–195, 199, 210
  S, 190–195, 199, 210
  sec-, 136
  tert-, 136
  trans-, 158
  tri-, 1056
  Z, 399–400
Pregabalin, 124
Pretomanid, 195
Prezista, 886
Prilosec, 300
Primary (1°) alcohols
  addition reactions producing, 800
  aldehyde reduction to, 784, 785
  dehydration, 355
  oxidation, 488, 490, 491
Primary (1°) alkyl chlorides, 733
Primary (1°) alkyl halides
  defined, 254
  E2 elimination reactions, 313
  in Friedel–Crafts alkylation, 732
  reaction mechanisms, 327, 332
  S_N1 substitution reactions, 277
  S_N2 substitution reactions, 271, 283
Primary (1°) amides, 930, 936
Primary (1°) amines
  amino acids as, 1225
  diazonium salt formation from, 1082–1083
  formation by reductive amination, 1065–1068
  Gabriel synthesis, 1063–1064
  IR absorption, 1058, 1059*f*
  naming, 1055
  nitrile reduction reactions forming, 913–914, 917
  nucleophilic addition reactions of, 854–856
  structure, 98, 1054
Primary (1°) carbocations, 278
Primary (1°) carbon radicals, 607
Primary (1°) carbons, 95, 488–490
Primary (1°) hydrogens, 95, 135–136
Primary structures of proteins, 1247, 1252*f*
Primers (PCR), 1285
Priority of enantiomers, 190–195, 191*f*, 211
Procyclidine, 828
Progesterone, 439, 440, 440*f*, 1028
Proline, 1224*f*, 1225, 1226*t*, 1249
Prontosil, 1092
Propagation
  alkane halogenation, 611, 612
  allylic bromination with NBS, 623
  atmospheric ozone destruction, 621*f*
  radical addition of HBr to alkenes, 628, 629, 630*f*
  radical polymerization of CH_2=CHZ, 632
Propanal
  aldol reactions, 1017
  IR absorption, 543*f*
  NMR spectra, 839*f*
  Wittig reaction, 851
Propanamide, 105, 544*f*
Propane
  alkyl groups, 135–136
  bromination, 615–616
  chlorination, 614, 615, 616–617
  conformations, 147*f*
  hexane-2,3-diol synthesis from, 640
  molecular structure, 132
  production from propyl tosylate, 368
Propanedioic acid, 891
Propan-1-ol, 366, 532–533, 591*f*
2-Propanol, 393
Proparacaine, 774, 1105
Propene
  borane added to, 420
  carbocations, 409
  hydrogen bromide addition to, 628–629
  polymerization, 631*t*
  radical formation, 248
Propionaldehyde, 835*t*
Propiophenone, 788, 990
Propofol, 692
Propoxyphene, 185
Propranolol, 85, 211, 390
Propylbenzene, 639, 759–760
Propyl groups, 136
Propyl tosylate, 368
Prostaglandins, 239, 644, 896
Prostate enlargement, 970
Prosthetic groups, 1254
Protease inhibitors, 1257, 1287
Protecting groups
  acetals as, 867–868, 875
  for acid–base reactions in carbonyl compounds, 805–807, 806*f*, 820, 821
  amides as, 1078, 1078*f*
  in peptide synthesis, 1240–1245, 1258
Protection defined, 805
Proteins
  in animal fibers, 959, 959*f*
  biological functions, 939
  defined, 1232
  folding, 125
  importance to health, 1253–1255
  lack of amide bond reactivity, 956
  physical properties, 938, 1247–1253, 1252*f*
  RNA synthesis, 1276, 1280–1283
Protic solvents, polar, 264–265, 266, 285
Protonation
  in alcohol dehydration, 355, 356, 358
  of amines, 1069
  amino acids, 910–911, 916

amphetamine, 78
in nucleophilic addition, 784, 799, 843–844
tautomerization in acid, 446, 447*f*
Proton equivalency
absorption splitting patterns and, 573, 577
in alkenes and cycloalkanes, 563–564, 593
general principles, 561, 562
homotopic and enantiotopic protons, 564–566
Proton NMR spectroscopy
aldehyde and ketone absorptions, 839*t*
amine absorptions, 1058*t*
benzene derivative absorptions, 692*t*
benzene rings and, 583, 584*f*
carboxylic acid and nitrile absorptions, 894*t*
carboxylic acid derivative absorptions, 937*t*
characteristic absorptions, A–13
complex splitting examples, 577–580
cyclohexane conformations and, 583
defined, 558
identifying unknowns with, 585–588
OH protons and, 582–583
propanoic acid spectrum, 894*f*
signal intensities, 572–573
signal numbers, 561–566
signal positions, 566–570
spectra, 560–561
spin–spin splitting, 573–577, 580–582
Protons
in acidity measures, 62–63
in atomic structure, 6
in Brønsted–Lowry criteria, 58–59, 58*f*
determining relative acidity, 74, 75, 84
magnetic field effects, 558–560
Proton transfer reactions
in aspirin, 77
Brønsted–Lowry acids and bases, 59–62, 62*f*, 81
predicting results, 65–66, 82, 898–899
Proximity effects, 662
Prozac, 3, 755, 1061. *See also* Fluoxetine
Pseudoephedrine, 69, 216, 1068
Psilocin, 1061
Puffer fish, 670
Pumiliotoxin C, 1100
Purine, 705, 1057*f*, 1269, 1274–1275
Putrescine, 1059
Pygmaeocin C, 776
Pygmaeopremia herbacea, 776
Pyranose rings, 1184, 1185–1186, 1198*f*, 1202, 1204, 1213
Pyrethrin I, 124, 1119, 1137
Pyridine
as aromatic compound, 701–702
as common nitrogen heterocycle, 1057*f*
dehydration using, 359–360
relative basicity, 1074, 1075
use in alcohol conversion to alkyl halides, 364–365
use in alcohol conversion to alkyl tosylate, 367, 369
as weak base, 77
Pyridinium chlorochromate, 478
Pyrimidines, 705, 1057*f*, 1269, 1274–1275
α-Pyrone, 716
Pyroxidine, 126
Pyrrole, 702, 717, 1057*f*, 1074
Pyrrolidine, 1057*f*
Pyruvic acid, 789

Q

Qing-hao, 422
Quantums, 530, 532
Quaternary (4°) carbons, 95
Quaternary ammonium salts, 1062–1063
Quaternary structures of proteins, 1252, 1252*f*
Quinapril, 128, 302
Quinic acid, 57, 75
Quinine, 217, 305, 314, 1077
Quinoa, 1225
Quinuclidine, 303

R

R (prefix), 190–195, 199, 210
Racemic mixtures
from achiral reactants, 375–376
albuterol sold as, 342
carbonyl reduction, 786
chiral drugs as, 207
defined, 203, 275
enantiomeric excess and, 204–205
resolution, 1227, 1227*f*, 1259, 1264
sample properties, 203*t*
in S_N1 substitution reactions, 275, 276*f*, 277
Racemization
defined, 275
enolates, 987–988
in S_N1 substitution reactions, 275, 276*f*, 277, 362
Radical cations, 509
Radical inhibitors, 609, 627
Radical intermediates, 607, 610–617
Radical reactions
addition to double bonds, 627–633
alkane halogenation, 610–617
at allylic carbons, 622–627
halogenation stereochemistry, 618–620, 618*t*
intermediates, 225
overview, 607–609
in ozone layer, 620–622, 621*f*
Radicals
bond formation, 226, 244
cyclopentadienyl, 700
defined, 225, 607
Radio waves, 558
Raffinose, 1208
Raloxifene, 273
Rate constant (*k*), 240
Rate-determining steps
defined, 238, 238*f*
in electrophilic aromatic substitution reactions, 725*f*
halogenation, 613, 615, 616
in nucleophilic substitution, 267, 274, 277, 279, 281
in rate equations, 240, 241
Rate equations, 240–241
Rate laws, 240
Razadyne, 1141
Reaction coordinates, 235
Reaction mechanisms, 224, 224–228
Reaction rates. *See* Kinetics
Reactions. *See* Organic reactions
Reactive intermediates
defined, 224, 237, 426
in stepwise reactions, 237, 238, 244
Reactivity
alcohols, ethers, and epoxides, 353–354
alkyl halides, 117, 258–259, 271, 272*f*, 277, 293, 313–314
carbonyl compounds, 779–782, 833
carboxylic acid derivatives, 781, 941–942, 958
of dienes in Diels–Alder reactions, 664–665
in electrophilic aromatic substitution, 742–743
epoxides, 374–378
hydrogen halides, 362–363
role of functional groups, 116–118, 121
Reagents
aldehyde and ketone reduction, 784
carboxylic acid reduction, 790–795, 795*t*
CBS, 787–788
dicyclohexylcarbodiimide, 950–951, 1241, 1258
indicating in chemical equations, 221
learning, 424
organometallic, 796–804
oxidation, 467, 478–479, 483–484, 1197
reduction, 467, 468, 477
Sharpless, 494
Rearrangement, in Friedel–Crafts alkylation, 732–733
Rebaudioside A, 832
(*S*,*S*)-Reboxetine, 504

Reciprocal centimeters, 532
Red algae, 258
Red tides, 97, 344
Reducing agents, 468
Reducing sugars, 1197, 1198*f*, 1203
Reduction reactions
  aldehydes and ketones, 783, 784–786, 818
  alkenes, 469–472, 497
  alkynes, 474–477, 497
  in amine synthesis, 1062, 1064–1068, 1093
  aryl ketones to alkyl benzenes, 759–760
  carbonyl groups of monosaccharides, 1196
  of carboxylic acids, 783, 790–795, 819
  defined, 163, 467, 782
  enantioselective, 787–790
  nitriles, 912, 913–914, 917
  nitro groups, 760–761
  polar C–X sigma (σ) bonds, 477, 478*f*
  types, 468
Reductive amination reactions, 1065–1068, 1093, 1095
Reductive elimination, 1110
Refining, 143, 143*f*
Regadenoson, 1272
Regioselective reactions
  defined, 315
  dehydration, 354, 405
  electrophilic addition of HX to alkenes, 408, 412*t*
  on epoxide rings, 377–378
  halohydrin formation, 418*t*
  Hofmann elimination, 1080–1081
  hydroboration–oxidation, 420, 422*t*
  ketone alkylation, 994–995
  overview, 496
Relafen, 1004
Relpax, 1134
Remdesivir, 3, 55
Replication forks, 1277
Replication of DNA, 1276, 1277–1278, 1277*f*
Resiniferatoxin, 827
Resolution of enantiomers, 1227–1230, 1227*f*, 1259, 1264
Resolving agents, 1228, 1259
Resonance, nuclear, 559
Resonance effects
  on acidity, 71–72, 73*f*, 83
  on amine basicity, 1076*t*
  in substituted benzenes, 736–739, 765
Resonance hybrids, 16, 21–22, 651–652, 688
Resonance-stabilized carbocations, 659
Resonance-stabilized molecules, 17, 899–900, 981, 1234
Resonance-stabilized radicals, 624, 625
Resonance structures
  acetate, 899
  acidity and, 71–72, 899–900
  alkyne hydration, 447
  alkyne hydrohalogenation, 445
  allylic carbocations, 648–651
  allyl radicals, 622, 624
  aniline, 736
  basic principles, 16–17
  benzene, 688
  of carbonyl groups, 779
  common examples, 649–651, 674
  cyclopentadienyl anion, 699–700
  drawing, 17–21, 648, 651, 676
  effect on IR absorptions, 536–538
  hybrids, 16, 21–22, 651–652
  peptide bonds, 1234
  stability and, 931–932
  in substituted benzenes, 744–747
Restricted rotation, 308–309
Restriction endonucleases, 1284
Resveratrol, 641
Retention of configuration, 269, 369
Retention time, 521
11-*cis*-Retinal, 855, 856*f*
Retinol, 111
Retro-aldol reactions, 1018–1019, 1041
Retro Diels–Alder reactions, 670
Retronecine, 98
Retrosynthetic analysis
  of acetoacetic ester synthesis, 1002–1003
  of aldol reaction, 1020–1021
  defined, 425
  Diels–Alder products, 669–670, 669*f*
  Grignard products, 803–805
  of malonic ester synthesis, 1000–1001
  of organic synthesis, 454–456, 459
  reductive amination, 1067, 1095
  Wittig reaction, 852, 874
Retroviruses, 1286–1287
Reverse transcription, 1287
RF radiation, 558, 591. *See also* Nuclear magnetic resonance spectroscopy
R groups
  alkene classification by, 308, 308*f*
  alkene stability and, 310, 331
  amino acids, 908–909, 909*t*, 1223
  carbon radical classification by, 607
  carbonyl group reactivity and, 833
  effects on S_N1 substitution reactions, 277, 277*t*, 278–281, 282*f*
  effects on S_N2 substitution reactions, 271–272, 272*f*
  identity as factor in E2 elimination reactions, 313–314
  relation to functional groups, 93
Rheumatoid arthritis, 715
Rhodopsin, 855–856, 856*f*
Ribbon diagrams, 1250*f*, 1255*f*
Riboflavin, 32
α-D-Ribofuranose, 1190–1191, 1209
β-D-Ribofuranose, 1190–1191
Ribonuclease, 1247
Ribonucleic acid (RNA)
  biological functions, 1269
  differences from DNA, 1278, 1288
  nucleosides and nucleotides, 1269–1274, 1271*t*
  transcription, 1276, 1279–1280
  translation, 1281–1283
  types and structure, 1278
  uracil structure, 88
  in viruses, 1286–1287
Ribonucleosides, 1270
Ribonucleotides, 1270, 1272
D-Ribose, 1210, 1269
Ribosomal RNA, 1278
Rimantadine, 62, 1068
Ring carbons, locating stereogenic centers, 188–190
Ring-closing metathesis, 1126–1128, 1140
Ring-flipping, 155, 156
Ring-opening metathesis polymerization, 1136
Rings. *See also* Aromatic hydrocarbons; Benzene; Pericylic reactions
  calculating numbers, 471–472
  carbonyl absorption, 543
  effect of formation on entropy, 233–234
Risperidone, 771
Ritonavir, 195
Rizatriptan, 273, 719
R labels, stereogenic centers, 1177–1179
RNA. *See* Ribonucleic acid (RNA)
Robinson annulation, 1037–1041, 1041, 1043
Rofecoxib, 1049, 1112
Roflumilast, 257
Rosemarinic acid, 627
Rosemary extracts, 627
Rose oxide, 525
Rosiglitazone, 771
Rosuvastatin, 1045
Rotation
  in alkenes, 308–309
  observed, 202–203
  in proteins, 1247
  specific, 204, 205
Rowland, F. Sherwood, 620
Ruthenium, 1123

S

S (prefix), 190–195, 199, 210
Saccharin, 1204, 1205*f*
Safinamide, 54, 515

Safrole, 867, 1103
Salicin, 77*f*, 895, 1194
Salicylates, 77
Salicylic acid, 895–896
Salinosporamide A, 124
Salmeterol, 300, 788
Salsolinol, 883
Salts
 in alcohol and ether production, 351
 amino acids as, 909, 1225
 Brønsted–Lowry bases and, 58
 nucleophiles as, 259
 sulfonium, 289
SAM (*S*-adenosylmethionine), 289, 374
Samuelsson, Bengt, 896
Sandmeyer reaction, 1084
Sanger, Frederick, 1284
Saponification, 952–954, 955
Saponins, 112
Saquinavir, 218
Sarcophytol B, 553
SARS-CoV-2 virus, 393, 938, 1286
Sassafras, 867
Saturated fats, 404, 625
Saturated fatty acids, 403
Schiff bases, 854
Schrock, Richard, 1123
Scombroid fish poisoning, 702
Scopolamine, 1059, 1059*f*
Sea sponge, 557
Sec- (prefix), 136
Secnidazole, 56
Secondary (2°) alcohols
 addition reactions producing, 800
 dehydration, 355–356
 ketone reduction to, 784, 785
 oxidation, 488, 489, 491
Secondary (2°) alkyl chlorides, 730
Secondary (2°) alkyl halides
 defined, 254
 in Friedel–Crafts alkylation, 732
 reaction mechanisms, 328, 329, 332
 S_N1 substitution reactions, 277, 283
 S_N2 substitution reactions, 271
Secondary (2°) amides, 930, 936
Secondary (2°) amines
 formation by reductive amination, 1065–1068
 IR absorption, 1058, 1059*f*
 naming, 1056
 nitrosonium ion reaction with, 1083
 nucleophilic addition reactions of, 856–859
 structure, 98, 1054
Secondary (2°) carbocations, 278, 730
Secondary (2°) carbon radicals, 607
Secondary (2°) carbons, 95
Secondary (2°) hydrogens, 95, 135–136
Secondary structures of proteins, 1247, 1248–1250, 1248*f*, 1249*f*, 1250*f*, 1252*f*
Second-order kinetics, 240, 267, 268, 312. *See also* S_N2 substitution reactions
Sedatives, 861, 1099, 1293
Selective serotonin reuptake inhibitors, 1061. *See also* Antidepressants
Semiconservative replication, 1277
Separatory funnels, 906
Septets, 577
Serevent, 300
Serine, 909*t*, 1224*f*, 1226*t*
Serotonin, 735, 1060–1061
Sertraline, 502, 694*f*, 735
β-Sesquiphellandrene, 665
Sevoflurane, 115, 350
Sharpless, K. Barry, 493
Sharpless epoxidation, 493–496, 499
Shellfish, toxins from, 802
Shells, electron, 7–9
Shielding effects in NMR, 567–572, 567*f*, 568*f*, 571*f*, 594
1,2-Shifts, 357–359
Shikimic acid, 100, 663
Shinorine, 674
Sickle cell anemia, 1255
Sickle cell hemoglobin, 1255, 1256
Side chains, 1223, 1226
Sigma (σ) Bonds
 in acetylene, 39*f*
 in addition reactions, 405–406
 in alcohols, ethers, and epoxides, 343
 in alkanes, 131
 in alkenes, 308
 basic features, 33
 breaking and forming in nucleophilic substitution, 266–267
 in butan-2-one, 575
 characteristic of organic compounds, 93
 classifying functional groups by, 96–98
 conjugated dienes, 656–657
 in cycloaddition reactions, 1152
 in Diels–Alder reaction, 663
 in 1,3-dienes, 646
 in different organic reaction types, 222–223
 in electrocyclic reactions, 1146
 in ethane, 37, 93, 145
 in ethylene, 37–38, 38*f*
 in ethyl methyl ether, 575
 functions in organic compounds, 93
 of hydrogen molecule, 33
 inductive effects, 278
 in methane, 34
 in pericyclic reactions, 1142–1143
 π bonds compared, 394
 radical reactions with, 608–609
 in sigmatropic rearrangements, 1156
 strength, 41
Sigmatropic rearrangements, 1143, 1156–1161, 1163, 1164–1165
[1,5] Sigmatropic rearrangements, 1156, 1158, 1159
[3,3] Sigmatropic rearrangements, 1156, 1157, 1159–1161
Silkworm moth, 1113
Silver oxide, 795
Silyl ethers, 806–807
Simmons, H. E., 1122
Simmons–Smith reaction, 1122–1123, 1128
Simple sugars, 118, 1176. *See also* Monosaccharides
β-Sinensal, 1170
Single bonds, 41
Single reaction arrows, 221
Single-ring aromatic compounds, 698
Singlets, 573
Sirenin, 1138
Sitagliptin, 703
Skeletal structures, 29–33, 53
S labels, stereogenic centers, 1177–1179
Sleep aids, 715, 719
SmaI enzyme, 1284
Smell sense, 207–208
Smith, R. D., 1122
Smoking cessation aids, 1098
S_N1 substitution reactions
 of alcohols, 362, 363–364, 383
 carbocation stability, 278–279, 282*f*
 conditions favoring occurrence, 282–287
 E1 elimination reactions compared, 323–324
 of ethers, 370–371
 general features, 274–277, 275*f*, 277*t*
 rate, 274, 277, 279–281
 when favored for alkyl halides, 327*t*
S_N2 substitution reactions
 acetylene, 452
 alcohol and ether production from, 350–351
 of alcohols, 361–362, 363, 366
 amine synthesis, 1062–1063
 conditions favoring occurrence, 282–287
 E2 elimination reactions compared, 312, 313
 of epoxides, 375
 of ethers, 370–371
 general features, 268–273, 270*f*, 272*t*
 of thiols, 372
 when favored for alkyl halides, 327*t*, 477
α-Snyderol, 523
Soaps, 112–113, 955–956
Sodium acetate, 106, 891*f*
Sodium amide, 325
Sodium benzoate, 895, 906
Sodium bisulfite, 484
Sodium borohydride, 468, 784–786, 786*f*

Sodium chloride, 9, 104, 105
Sodium cyanoborohydride, 1065–1066
Sodium ethoxide, 351
Sodium hypochlorite, 492
Sodium laurel sulfate, 113
Sodium methoxide, 351
Sodium salicylate, 895, 896
Sodium trichloroacetate, 1137
Solanine, 1194
Solanum nigrum, 1194
Solid phase technique, 1245–1247
Solubility
  of alcohols, ethers, and epoxides, 348
  aldehydes and ketones, 838
  alkanes, 144*t*
  alkenes, 401
  alkyl halides, 256
  alkynes, 439
  amino acids, 909, 1225
  ammonium salts, 1069–1071
  as extraction factor, 905–908
  monosaccharides, 1177
  organic compound applications, 111–116
  organic compound principles, 107–111, 120
Solutes defined, 107
Solvents
  alkyl halides as, 256, 256*f*
  defined, 107
  as determinants of substitution reaction mechanisms, 285–286
  in E2 elimination reactions, 313
  indicating in chemical equations, 221
  polar protic versus polar aprotic, 264–266
Somatostatin, 1265
s Orbitals, 7, 34
Sorbitol, 1196
Soybean oil, 466
Soybeans, 790
sp Hybrids
  in acetylene, 39
  chemical shifts, 570–572
  formation, 35–36, 35*f*, 36*t*
  length and strength, 42
sp^2 Hybrids
  of alkenes, 401
  benzene as, 688–689
  of carbenes, 1120
  carbon radicals as, 607
  carbonyl carbon atom, 778–779
  chemical shifts, 570–572
  in 1,3-dienes, 646
  electron density, 310–311
  in enols and phenols, 343
  in ethylene, 37–38
  formation, 35–36, 35*f*, 36*t*
  length and strength, 42
  in pilocarpine, 703
sp^3 Hybrids
  in alcohols, 343
  in alkanes, 131
  of alkenes, 401
  in alkyl halides, 254
  in amines, 1054
  in bilobalide, 96
  electron density, 310–311
  formation, 34–35
  length and strength, 42
  nucleophilic substitution, 259, 266–268
Space-filling representations, 1250*f*
Specific rotation, 204, 205, 217
Spectator ions, 58
Spectroscopy, 509. *See also* Infrared spectroscopy
SPF ratings, 674
Spider dragline silk, 1250, 1250*f*
Spike proteins, 1286, 1286*f*
Spin–spin splitting, 573–577, 580–582
Spirastrellolide A, 863
Spiro compounds, 219, A-9
Splitting patterns
  alkenes, 580–582
  carbon NMR versus proton NMR, 588
  causes, 573–575
  common examples, 576*t*
  complex examples, 577–580, 595
  rules, 575–577, 595
Sponges, 1268, 1270, 1292
Spongothymidine, 1292
Sporanox, 881
Spruce budworm, 829
Stability
  acidity and, 932
  aldehydes versus ketones, 860
  of alkenes, 310–311, 395*t*
  allyl radicals, 622, 633
  benzene, 694–696, 723
  of carbocations in electrophilic aromatic substitution, 743–744
  of carbocations in nucleophilic substitution, 278–281, 282*f*, 294
  of carbon radicals, 607–608, 608*f*, 633
  conjugated dienes, 657–658
  delocalization advantages, 646, 647, 648, 653, 658
  hydrate formation and, 860–861, 873
  relation to energy, 232, 244
Staggered conformations
  butane, 148–151, 149*f*
  cyclohexane, 153
  defined, 145, 167
  ethane, 146, 146*f*, 147–148
  propane, 147*f*
Stanozolol, 720
Starch, 179–180, 179*f*, 180*f*, 1206–1207
Stearic acid, 403*t*, 404*f*
Stearidonic acid, 404
Stegaltro, 880
Stepwise reactions, 224
Stereochemistry
  alkanes, 145–151
  carbonyl reduction, 786–787
  chemical properties of enantiomers, 206–208
  chiral versus achiral molecules, 181–184, 209
  defined, 145, 179
  diastereomers, 195–197, 199
  E2 elimination reactions, 316–320
  electrophilic addition of HX to alkenes, 410–412, 412*t*
  enolate alkylation, 994
  epoxidation, 480–481
  halogenation, 414–415, 416*f*
  halohydrin formation, 417–418, 418*t*
  hydroboration–oxidation reactions, 422*t*
  isomer classes, 181
  locating and drawing stereogenic centers, 184–190
  meso compounds, 197–199
  organometallic reagent addition, 801
  physical properties of stereoisomers, 202–206
  S_N1 substitution reactions, 275–277, 276*f*, 363
  S_N2 substitution reactions, 269–271, 270*f*, 363
  starch versus cellulose, 179–180, 179*f*, 180*f*
Stereogenic centers
  amino acids, 908–909, 1223
  D and L designations, 1181
  defined, 183, 184
  depicting in sugars, 1177–1180
  labeling as *R* or *S*, 190–195
  locating and drawing, 184–190, 210–211
  meso compounds, 197–198
  multiple *R* and *S* assignments, 199
  multiple stereoisomers, 195–197
  tetrahedral, 183, 184
Stereoisomerism, 395*t*
Stereoisomers
  alkene epoxidation, 480–481
  conformations versus, 654–655, 675
  conjugated dienes, 654
  constitutional isomers versus, 15, 181, 181*f*, 201, 201*f*, 209
  cyclohexane, 158–162
  defined, 158, 169
  disubstituted cycloalkanes, 199–200
  meso compounds, 197–198
  with multiple stereogenic centers, 197*f*, 1179–1180

naming for alkenes, 398–399
physical properties, 202–206
starch and cellulose, 180
with two stereogenic centers, 195–197
Stereoselective reactions, 315, 475, 496
Stereospecific reactions
defined, 415
in halogenation of alkenes, 415
pericyclic reactions as, 1142, 1146
vinyl halide coupling, 1108
Steric hindrance
in alcohols, 348
nucleophilicity and, 264, 266
in nucleophilic substitution with acetylide anions, 451
in S_N2 substitution reactions, 271–272, 272*f*
Steric strain, 149–150, 151*t*, 168
Steroids, 283, 671, 720
Steviol, 832
Stille coupling, 1116–1119, 1128, 1130
Stink bugs, 431
Stone, Edmund, 77*f*
Stop codons, 1281
Straight-chain alkanes, 132, 133*t*
Strands in β-pleated sheets, 1249, 1249*f*
Stress, 289, 290*f*
Strong nucleophiles, 283–284
Structural isomers. *See* Constitutional isomers
Strychnos nux vomica, 1264
Styrene, 632, 757–758
Styrofoam, 632
Substituents
alkylthio, 373
benzene activation and deactivation, 743–744
cyclohexane, 156–162
factors in acidity, 902
inductive and resonance effects in substituted benzenes, 736–739, 738*f*
naming, 135–136, 138, 344–348, 373, 382, 690–691
position in alkene reactions, 480–481, 631, 667–668, 669
reaction rate effects in substituted benzenes, 739–741
Substituted alcohols, 354
Substituted alkenes, 354–355
Substituted benzenes
activation or deactivation in substitution reactions, 743–744
disubstituted benzenes in substitution reactions, 750–753
formation from diazonium salt reactions, 1086–1088
inductive effects in substitution reactions, 736, 737–739
limitations in substitution reactions, 748–750
multistep synthesis, 762–764
orientation effects in substitution reactions, 744–747, 748*f*
resonance effects in substitution reactions, 736–739
variation in substitution reactions, 739–743
Substituted benzoic acids, 903–905, 904*f*, 916, 918
Substituted cycloalkanes, 156–162
Substituted phenols, 905
N-Substituted primary amines, 1056
Substitution reactions. *See also* Nucleophilic substitution reactions; S_N1 substitution reactions; S_N2 substitution reactions
alkylation at α carbons, 988, 993–1004
basic types listed, A–15
bromine with benzene, 687, 695
defined, 222
determinants, 326–330, 332
electrophilic aromatic, 723–725. *See also* Electrophilic aromatic substitution
elimination reactions with, 329–330
general features in alkyl halides, 258, 259–261
halogenation at α carbons, 988, 989–993, 1004
Heck reaction as, 1114
overview, 222, 244
radical intermediates in, 610–617, 624
when favored, 326, 330
Substrates, 243
Subunits of proteins, 1252
Succinate, 223
Succinic acid, 891
Sucralose, 97, 1204, 1205*f*
Sucrose, 189, 1174, 1204, 1204*f*
Sudafed PE, 69
Suffixes
-adiene, 398
-al, 834
-aldehyde, 835
-amide, 934
-amine, 1055
-ane, 135
-ate, 933
-carbaldehyde, 834
-dioic acid, 891
-diol, 345
-ene, 397
-ic acid, 891
-idine, 1270
-ine, 255
-o, 255
-oic acid, 890
-ol, 344
-one, 836–837
organic molecule names, 135
-ose, 1176
-osine, 1270
-oxy, 347
-oyl, 837
-thiol, 372
-yl, 135, 347, 837, A–6
-yne, 438
Sugarcane, 1191, 1204
Sugar–phosphate backbone, 1274, 1275*f*
Sugars. *See* Carbohydrates; Monosaccharides; Sweeteners
Sulfa drugs, 1091–1092
Sulfamethoxazole, 1092
Sulfanilamide, 1092
Sulfides
defined, 343, 371
reactivity, 382
structures and properties, 96, 97*t*, 373–374
Sulfisoxazole, 1092
Sulfonation, 724*f*, 727–728, 765
Sulfonium ions, 373–374
Sulfonium salts, 289
Sulfur, 16, 266*t*
Sulfuric acid, 16, 76, 242
Sulfur trioxide, 727–728
Sulfur ylides, 1137
Sunscreens
avobenzone in, 1033
environmental effects, 4, 674
functions of, 674
octinoxate in, 1050
palythine potential, 684
solubility, 126
Supercoils, 1253
Super Glue, 632
Superhelices, 1253, 1254
Suprafacial cycloadditions, 1153, 1154, 1155
Suprafacial rearrangements, 1157, 1158, 1159
Surface area, impact on boiling point, 104, 104*f*
Suzuki, Akira, 1114
Suzuki reaction, 1109–1113, 1113*f*, 1128, 1129
Sweating, 208
Sweeteners
aspartame, 970, 1204, 1205*f*, 1235, 1263
honey, 1191
maple syrup, 1174
neotame, 1263
rebaudioside A (Truvia), 832
saccharin, 1204, 1205*f*
sucralose, 1204, 1205*f*
sucrose, 1174, 1204

Symbols, A–3
Symmetrical anhydrides, 929, 933
Symmetry (molecular)
in ethers, 343
impact on melting point, 106
IR absorption and, 535
in meso compounds, 198
planes of, 183, 184, 198
Symmetry (orbital)
allowed and forbidden reactions, 1147, 1148
in cycloaddition reactions, 1153–1154
sigmatropic rearrangements and, 1157–1158
Syn addition, 406, 411, 427, 480
Syn dihydroxylation, 482, 483–484, 499
Syn periplanar geometry, 317, 317*f*
Synthesis. *See* Organic synthesis
Synthetically useful crossed aldol reactions, 1022–1024
Synthetic dyes, 1091
Synthetic intermediates, 426
Synthetic polymers, 630–631. *See also* Polymerization; Polymers
Systematic method (nomenclature), 134

T

Tadalafil, 971
Tagamet, 1060
Tamiflu, 663, 970
Tamoxifen, 431, 995–996
Tandem ring-opening–ring-closing metathesis, 1140
Taq polymerase, 1285
Tarceva, 461
Target compounds, 454
Tartaric acid, 205–206
Tautomerization
in alkyne hydration, 446–448
in alkyne hydroboration–oxidation, 449
defined, 446
mechanism, 979–980, 1005
in nitrile hydrolysis reactions, 912–913
Taxol, 178, 189, 782
TB Alliance, 195
Tecfidera, 515
Tectitethya crypta, 1268, 1292
Teflon, 257, 631*t*
Telfairine, 254
Temperature, 221, 239, 660–663
Template strands (DNA), 1277, 1277*f*, 1279, 1289
Temporary dipoles, 101–102, 119
Terbinafine, 461
Terminal alkenes, 394
Terminal alkynes
acetylide anion reactions, 442–443, 450–454, 457
defined, 437
hydroboration reactions, 449
reduction, 477
Termination
alkane halogenation, 611, 612
radical addition of HBr to alkenes, 628
radical polymerization of CH_2=CHZ, 632
Tert- (prefix), 136
Tertiary (3°) alcohols
acid–base reactions producing, 809
addition reactions producing, 800
dehydration, 355–356
oxidation resistance, 489
Tertiary (3°) alkyl chlorides, 730
Tertiary (3°) alkyl halides
defined, 254
E2 elimination reactions, 313
reaction mechanisms, 327
S_N1 substitution reactions, 277, 283
S_N2 substitution reactions, 271, 283
Tertiary (3°) amides, 930
Tertiary (3°) amines
formation by reductive amination, 1065–1068
IR absorption, 1058, 1059*f*
naming, 1056
structure, 98, 1054
Tertiary (3°) carbocations, 278, 730, 732
Tertiary (3°) carbon radicals, 607
Tertiary (3°) carbons, 95
Tertiary (3°) hydrogens, 95
Tertiary structures of proteins, 1250–1252, 1252*f*
Terylene, 960
Tetrabutylammonium fluoride, 806
Tetrachloromethane, 256
(*E*)-Tetradec-11-enal, 829
1,1,1,2-Tetrafluoroethane, 622
Tetrafluoroethylene, 631*t*
Tetrahalides, 445
Tetrahedral shapes
in alcohols, ethers, and epoxides, 344
in alkanes, 131
basic features, 24–25
drawing for carbohydrates, 1177–1178
in ethane, 37
hybrid orbitals in, 34
Tetrahedral sterogenic centers, 183, 184, 198
Tetrahedrane, 152*f*
Tetrahydrocannabinol (THC), 75, 127, 522, 522*f*
Tetrahydrofuran
boiling point versus furan, 719
as common cyclic ether, 347
complexed with borane, 419
as polar aprotic solvent, 265*f*
solubility, 129
α Tetralone formation, 734
Tetramethylsilane, 560
Tetrapeptides, 1235
Tetrasubstituted alkenes, 308
Tetravalent atoms, 3
Tetrodotoxin, 670
Tetroses, 1176
Tezacaftor, 389
Thalidomide, 188–189
Theophylline, 1099
Thermal cyclers, 1285
Thermal cycloadditions, 1152, 1154, 1155*t*
Thermal electrocyclic reactions, 1148–1150, 1164
Thermal sigmatropic rearrangements, 1158, 1158*t*, 1159
Thermodynamic enolates, 985–986, 987
Thermodynamic products of addition reactions, 660–663, 677
Thermodynamics
energy diagrams, 235–239
entropy and enthalpy, 233–234
overview, 231–233
Thiazolinone, 1237, 1267
Thioesters, 961
-thiol (suffix), 372
Thiols
defined, 343, 371
in hair, 1254, 1254*f*
reactivity, 382
structures and properties, 96, 97*t*, 371–372
Thionyl chloride, 364–365, 947–948
Threonine, 924, 1224*f*, 1225, 1226*t*
D-Threose, 1182
Thrombin, 1223
Thymine
base pairing with adenine, 1275, 1277
in DNA, 1269
nucleoside and nucleotide, 1271*t*
π electrons, 705
as pyrimidine base, 1269
Thymol, 718
Thyroid hormones, 1226
Thyrotropin-releasing hormone, 99
L-Thyroxine, 1226
Ticlopidine, 261, 883
Titanium(IV) isopropoxide, 494
Tofacitinib, 715
Tollens reagent, 795, 1197
Toluene
as common name, 690
as fuel additive, 522, 693
melting point, 125
substitution reaction kinetics, 739
synthetic polymers from, 693
p-Toluenesulfonic acid (TsOH), 76, 354, 367, 862
Torsional energy, 147, 148

Torsional strain, 147, 151*t*, 153, 168
Tosylates, 367–369
Toxins
  amine examples, 1059, 1059*f*
  azaspiracid-1, 802
  brucine, 1264
  carbon monoxide, 1255
  carotatoxin, 126
  cicutoxin, 27
  cyanohydrin derivatives, 848
  hemibrevetoxin B, 97
  histrionicotoxin, 124, 443
  hypoglycin A, 199, 924
  methanol, 493
  oleandrin, 867
  solanine, 1194
  tetrodotoxin, 670
  zearalenone, 1117
Trabectedin, 218, 792
Trade names, 134
Transacetalization, 883
Trans alkenes, 475–476
Transannular Diels–Alder reactions, 682
Transcription, 1276, 1279–1280, 1279*f*
Trans dienophiles, 666
Trans dihalides, 445
Transesterification, 971
Trans fats, 473
Transferases, 1256, 1256*t*
Transfer RNA, 1278, 1279*f*, 1282*f*
Trans isomers
  of alkenes, 309, 310, 311, 394–395, 398–399, 401
  of cycloalkanes, 158–162, 168, 199–200
  cyclohexanes, 318–319
  hydrogenation, 473
Transition state
  defined, 235
  in energy diagrams, 236, 237
  Hammond postulate, 280–281, 281*f*, 615
  in stepwise reactions, 238, 238*f*
Translation, 1276, 1281–1283
Transmetallations, 1117, 1118
Transmittance, 533
Trans protons, 580
Travoprost, 257
Trazodone, 1099
Tri- (prefix), 1056
Triacylglycerols
  hydrolysis, 954–956
  overview, 403–404, 472–473
  structure, 954*f*
Trialkylboranes, 420, 420*f*
Triarylphosphine, 1114
1,3,5-Tribromobenzene, 1085–1086, 1086*f*
1,1,1-Tribromopropane, 636
Tributyltin chloride, 1117
Tributyltin hydride, 643, 644
Trichlorofluoromethane, 2, 257
Trichloromethane, 256
2,4,5-Trichlorophenoxyacetic acid, 727*f*
(*Z*)-Tricosene, 505
Tricyclohexylphosphine, 1110
Trienes, 398
Triethylamine, 77, 303, 520, 1114
1,2,4-Triethylcyclopentane, 142
Trifluoroacetic acid, 1242
2,2,2-Trifluoroethanol, 69–70
Trigonal planar shapes
  of carbocations, 275, 276
  carbon radicals, 607
  in ethylene, 37
  of second-row elements, 24
Trigonal pyramidal shapes, 25
Trihalomethanes, 1120
Trimethoxyresveratrol, 1129
Trimethylamine, 1059
Trimethyl borate, 1111
1-chloro-1,3,3-Trimethylcyclohexane, 411
2,5,5-Trimethylcyclohexanecarboxylic acid, 890
1,2,2-Trimethylcyclohexanol, 391
2,5,5-Trimethylcyclohexanol, 346
1,3,3-Trimethylcyclohexene, 411
2,3,5-Trimethylhex-2-ene, 397
2,2,4-Trimethylpentane, 164
2,3,5-Trimethyl-4-propylheptane, 138
Trinucleotides, 1274
Triols, 345
Trioses, 1176
Tripeptides, 1232, 1245, 1259, 1261
Triphenylene, 720
Triphenylmethyl radical, 643
Triphenylphosphine, 850, 874, 1110
Triphenylphosphine oxide, 851
Triphosphates, 288
Triple bonds
  in acetylene, 13, 39
  in alkynes, 437
  carbon–carbon, naming compounds with, A–8*t*
  hybridization and, 39
  length and strength, 41
  NMR signals, 571–572
  reducing, 474–477, 476*f*
Triple helices, 1254, 1254*f*
Triplets, 573, 574–575
Trisubstituted alkenes, 308
Trisubstituted benzene derivatives, 711
Trisubstituted cyclohexane, 161–162
Tri(*o*-tolyl)phosphine, 1110
Triynes, 438
Tropolone, 927
Tropylium cation, 700–701
Tropylium radical, 712
Truvia, 832
Trypsin, 1238
Tryptophan
  classification, 924, 1263
  ionizable functional group values, 1226*t*
  solubility, 110
  structure and abbreviation, 1224*f*
Tung oil tree, 432
Tungsten, 1123
Turmeric, 719, 1015
ar-Turmerone, 1015, 1026
Twins, 182, 1276
Twistoflex, 693, 694*f*
Two-step reactions, diagramming, 237–239, 238*f*, 245
Tylenol, 946
Tyrian purple, 1090, 1091
Tyrosine, 1224*f*, 1226*t*

U

Ubrogepant, 215
Ultraviolet radiation, 656, 672–674
Undecane, 131
Unimolecular elimination. *See* E1 elimination reactions
Unimolecular reactions, 241, 268, 274
α,β-Unsaturated carbonyl compounds
  basic features, 812–814
  from dehydration of aldol reaction products, 1019, 1022, 1042
  in Michael reaction, 1035–1036, 1037*f*
  in Robinson annulation, 1037–1038, 1039
γ,δ-Unsaturated carbonyl compounds, 1159, 1161
Unsaturation
  in alkenes, 395–397, 428
  in alkynes, 437
  in benzene, 687
  calculating, 395–397, 471–472
  fats and oils, 404, 472–473, 625, 626*f*
  fatty acids, 403
  proposing molecular formulas for unknowns from, 513
Unsymmetrical epoxides, 375
Unsymmetrical ethers, 343
Unsymmetrical ketones, 994–995
Up carbons, 153, 154, 155, 156–157
Uracil
  molecular structure, 88
  nucleoside and nucleotide, 1271*t*
  as pyrimidine base, 1269
  in RNA, 1269, 1278
Uridine monophosphate, 302
Uridine 5′-monophosphate, 1272
Urine testing, 522

V

Vaccines, 1286, 1287
Valence bond theory, 706
Valence electrons
 carbon, 8, 33–34
 defined, 8
 formal charge and, 13–15, 15*t*
 in Lewis structures, 10–15, 15*t*
 role in bonding, 9–10
Valence shell electron pair repulsion theory, 23, 27
Valeraldehyde, 835*t*
Valine, 1224*f*, 1226*t*
Valproic acid, 89, 1010
Valsartan, 1118–1119
Van der Waals forces, 101–102, 103*t*, 119, 144
Vane, John, 896
Vanillin, 31, 840
Vaprisol, 887
Varenicline, 1098
Vasodilators, 1060
Vasopressin, 1235, 1251
Veklury, 3, 55
Venlafaxine, 805
Vetiver, 1014, 1046
β-Vetivone, 1014, 1046
Vicinal dibromide, 441
Vicinal dihalides, 325, 413, 441
Vidarabine, 1270
Vinyl acetate, 581–582, 581*f*, 582*f*
Vinylboranes, 1111, 1113, 1129
Vinyl chloride, 54, 606
Vinyl ethers, 1161
Vinyl groups, 400, 401*f*
Vinyl halides
 configuration, 254, 254*f*
 coupling reactions with organocuprates, 1107, 1108
 in Heck reaction, 1114
 hydrohalogenation forming, 443
 nucleophilic substitution and, 291
 in Suzuki reaction, 1111
 unreactive in Friedel–Crafts alkylation, 731
Vinylpyrrolidone, 642
Vioxx, 1049, 1112
Viruses, 128, 393, 938, 1286–1287. *See also* Antiviral drugs
Visible light, 530*f*
Vision, chemistry of, 855–856, 856*f*
Vismodegib, 1077
Vitamin A, 111, 645, 656
Vitamin B_3, 112
Vitamin B_5, 92, 112
Vitamin B_6, 126
Vitamin C
 biological functions, 112, 938
 radical reactions with, 641
 solubility, 112
 as stable enediol, 1009
Vitamin D, 656
Vitamin D_3, 1151
Vitamin E, 126, 626–627
Vitamin K_1, 112, 722
Vitamins defined, 111
Volatility, 104
Voltaren, 600
Von Baeyer, Adolf, 134

W

Walden, Paul, 269
Walden inversions, 269
Water. *See also* Hydration
 acid chloride reaction with, 943
 acidity versus methane, 67
 as acid or base, 58–59
 alcohols and ethers as derivatives, 343
 alkyne reactions with, 442*f*, 446–448, 457, 458
 in body's chemical environment, 111
 chirality, 182
 electrophilic addition to alkenes, 412–413
 formation in E2 elimination reactions, 312
 in halohydrin formation, 416–417
 hybridization in, 35
 in hydroboration–oxidation reactions, 419
 hydrogen bonding in, 102–103
 as leaving group in nucleophilic substitution, 259, 261
 as Lewis base, 79
 molecular shape, 23, 25, 26
 as polar molecule, 44, 45*f*
 predicting solubility in, 110–111
 reaction with boron trifluoride, 79
 reaction with Brønsted–Lowry acids, 62–63
 removing from alcohols, 354–360
 solvent properties, 107, 108, 109, 264
Water hemlock, 27
Water-soluble vitamins, 111, 112
Watson, James, 1274
Wavelength, 530–531, 532, 533, 549, 672
Wavenumber, 532
Waxes, 166
Weak bases, 261
Weakly polar compounds, 107, 256, 258–259
Weak nucleophiles, 283–284
Wedges in molecular models, 25
Williamson ether synthesis, 351
Willow bark, 77*f*
Willstatter, Richard, 1105
Wilson's disease, 1263
Wittig, Georg, 849
Wittig reaction, 848–853, 874
Wohl degradation, 1199–1200
Wolff–Kishner reduction, 759
Woodward, R. B., 1143
Woodward–Hoffmann rules, 1151, 1151*t*, 1158*t*, 1163*t*

X

Xadago, 515
Xalatan, 896
Xylene, 693
m-Xylene, 751
p-Xylene, 94, 522, 638, 693

Y

-yl (suffix), 135, 347, 837, A-6
Ylides, 849
-yne (suffix), 438
Youyou, Tu, 423

Z

Z (prefix), 399–400
Zaitsev rule, 314–316, 354, 1081
Zanamivir, 190
Z double bonds, 404
Zearalenone, 1117
Zejula, 1099
Zetia, 788
Zidovudine, 1287
Zileuton, 379
Zinc, 485
Zinc–copper couple, 1122
Zingiberene, 402*f*, 665
Zoloft, 694*f*
Zolpidem, 397, 719, 1099
Zwitterions, 909, 1225
Zyban, 1013
Zyrtec, 301, 515, 801